AF677471

PHYTOPHTHORA DISEASES WORLDWIDE

Phytophthora
Diseases Worldwide

Donald C. Erwin
Professor Emeritus, Department of Plant Pathology
University of California, Riverside

and

Olaf K. Ribeiro
Ribeiro Plant Labs
Bainbridge Island, Washington

APS PRESS
The American Phytopathological Society
St. Paul, Minnesota

Financial Contributors

The Academic Senate, University of California, Riverside
CIBA-GEIGY Corporation
Foundation for Microbiology
Petoseed Co., Inc.
Rhône-Poulenc Ag Company/CHIPCO

Cover: *Phytophthora cinnamomi* sporangia emitting zoospores.
Photograph by L. Harman. Reprinted, by permission, from
Bryan Shearer et al., 1991, Wildflower killers, Landscope 7:28-34.

Reference in this publication to a trademark, proprietary product, or company name is intended for explicit description only and does not imply approval or recommendation to the exclusion of others that may be suitable.

Library of Congress Catalog Card Number: 95-83805
International Standard Book Number: 0-89054-212-0

© 1996 by The American Phytopathological Society
Second printing 2005

All rights reserved.
No portion of this book may be reproduced in any form, including photocopy, microfilm, information storage and retrieval system, computer database, or software, or by any means, including electronic or mechanical, without written permission from the publisher.

Copyright is not claimed in any portion of this work contributed by U.S. government employees as a part of their official duties.

Printed in the United States of America on acid-free paper

The American Phytopathological Society
3340 Pilot Knob Road
St. Paul, Minnesota 55121-2097, U.S.A.

Contents

Foreword

One hundred fifty years after the Irish potato famine, it seems almost uncanny that the genus *Phytophthora* should be making headlines again. That famine resulted from disease caused by the type species of *Phytophthora, P. infestans.* The severe late blight epidemic in Ireland occurred prior to the development of the germ theory of disease, and the science of plant pathology was birthed in response to the Irish potato famine. Migrations of *P. infestans* led to the severe epidemic and resulting famine in 1845, and migrations are once again focusing popular and scientific attention on the genus.

We have learned much about *Phytophthora* during the intervening 150 years, and this book provides an excellent summary and guide to that literature. Chapters 1–7 describe generalities about the genus and provide useful hints for those of us who work with members of the genus. I found these chapters to be intriguing because of the new information I learned and because of the many challenges remaining. Chapters 8–66 are compilations of information about individual species. Each chapter provides specific information and the citations necessary to pursue specific questions. Although these chapters collate and summarize a vast body of information, it is still clear that there are many gaps in our knowledge about the genus. Thus, the book not only provides background and answers, it also provides challenges.

This volume amply indicates that this single genus contains species that illustrate the diversity of plant pathogens. Most of the *Phytophthora* species are soilborne pathogens and survive long periods without their hosts. However, with access to a susceptible plant in the presence of adequate moisture, they resume their remarkable pathogenic phase. In contrast, the type species, *P. infestans,* illustrates many of the characteristics associated with biotrophic pathogens. As an asexual organism, *P. infestans* is essentially an obligate parasite and can not survive for long periods in the absence of a living host. It shares many characteristics with powdery mildews, downy mildews, and many rusts.

The evolutionary relationships of the genus have become increasingly clear during the past two decades, and these are well described in the first portion of the book. During the past two decades, it became clear that the vegetative state of this genus was diploid (or higher ploidy) and that meiosis was gametic. It is only within the last 10 years that there has been general acceptance of the phylogenetic distance between the oomycetes and the "higher" fungi (members of Ascomycetes and Basidiomycetes). The clear presentation of these facts and events in this volume is of considerable benefit to the older generation of mycologists and plant pathologists.

The genus *Phytophthora* provides an endless supply of challenges to scientists. Although some species within the genus may be "models" for the genus or for the class, there is no member of the genus that has been domesticated sufficiently to enable basic studies in the same way that can be done with *Neurospora crassa* or *Saccharomyces cerevisiae.* The difficulties of working with this genus are illustrated by the relatively long time required to accurately understand the general life history.

The genus *Phytophthora* is important in a variety of respects, and so is this book. The book provides useful information for experienced and inexperienced students of the genus. Certainly, I learned quite a lot about the genus, but in this book I also now have a useful source for many facts and a tremendously useful guide to the literature of the genus. The book will serve as a valuable resource in my office and in my lab. It contains background information so that the beginning student can gain an overview of the genus. It provides very practical information for persons working with *Phytophthora,* and it challenges the specialist with unanswered questions.

It is certainly a tribute to the dedication of "Phytophthorologists" that they continue their investigations with members of the genus. Despite the many recent advances in concept and technology, working with this genus still requires a special dedication. Even the recent availability of molecular biology techniques does not entirely remove the barriers to successful investigation. Nonetheless, molecular biology has contributed to our understanding, and the many citations in this book clearly demonstrate that contribution and enhance the value of this volume. For example, the advent of molecular markers has enabled scientists to determine phylogenetic relationships, to decipher the life history of individuals, to detect migrations, to distinguish recombinant individuals from selfed individuals, and to learn about the importance of different sources of inoculum. Despite this impact, the genus remains recalcitrant to routine genetic studies possible with some members of Ascomycetes and Basidiomycetes.

My personal opinion is that individuals devoting their lives to *Phytophthora* are selfless scientists who sacrifice their personal careers to the very difficult task of learning more about this genus and the plethora of diseases for which it is responsible. Certainly, I commend Don Erwin and Olaf Ribeiro not only for their courage in devoting their careers to this genus, but also for their very significant effort in writing this book. They have performed a valuable service for the rest of us.

WILLIAM E. FRY

Preface

The late blight disease of potato first occurred in the northeastern United States in 1843 and in Europe in 1845 (Bourke 1991; Peterson et al. 1992). The genus *Phytophthora* became widely known after this important disease was determined to be caused by a previously unknown fungus, *Phytophthora infestans* (Mont.) de Bary, in 1876. Before 1876, there was great controversy over the cause of late blight and the naming of the pathogen. The conventional wisdom then attributed the disease to the rainy weather and even industrial pollution. The fascinating story of professional dispute about the cause of the disease and naming of the pathogen is told by Large (1940) and Bourke (1991).

P. infestans totally destroyed Ireland's staple potato crop during 1845 and 1846. The most immediate effects were poverty and mass starvation, which led to profound sociological and economic changes in that country (Large 1940; Gregory 1983; Bourke 1991) and to the emigration of large numbers of people to the United States and elsewhere. It is estimated that Ireland lost one-fourth of its 8 million inhabitants to starvation and emigration. Perhaps no other single plant disease has resulted in such widespread human suffering and sociological impact.

Although the host range of *P. infestans,* the type species, is limited to potato and tomato, the number of other *Phytophthora* species that subsequently have been described has expanded to more than 50 (excluding the marine species that were attributed to *Halophytophthora* by Ho and Jong [1990]) (Stamps et al. 1990) from the time the genus was first described by de Bary in 1876. Although *P. infestans* has a narrow host range, *P. cinnamomi* parasitizes more than 900 species of plants (Zentmyer 1980). *P. parasitica* has a comparably wide host range. Other *Phytophthora* species attack numerous economically important crop plants. *P. cinnamomi* caused one of the most severe epidemics in history in the jarrah tree forest in Western Australia (Podger et al. 1965; Podger 1972; Shea et al. 1983). The increasing concern regarding the impact of *Phytophthora* on food and fiber production is reflected by the large number of publications from all over the world that deal with virtually every aspect of the genus *Phytophthora.* Indeed, the genus *Phytophthora* now ranks high on the list of fungal pathogens most often cited in the literature (Steinhausen 1979).

The economic damage to crops caused by *Phytophthora* species in the United States alone runs into billions of dollars annually, and worldwide the cost is many times this amount. In Ohio (United States) alone, Phytophthora root rot of soybean causes a $50 million annual loss to growers (McBlain et al. 1991a, b). Crop losses from black pod disease of cocoa (*Theobroma cacao*), caused by *P. palmivora,* amount to 1,500 million pounds sterling per year on the basis of an estimated 10–30% world crop loss (Evans and Prior 1987).

Heavy losses are often the result of delays in recognition of *Phytophthora* as the causal agent of the disease under investigation. The symptoms of many *Phytophthora*-caused diseases can often be mistaken for damage from other pathogens or abiotic agents (Tsao 1990). Isolation and identification of the pathogen offers the only accurate method of early detection. Sensitivity of plants to "wet feet" can often be attributed to susceptibility to one or more *Phytophthora* species in the soil.

Unfortunately, information on isolation techniques for early identification of *Phytophthora* and detailed descriptions of diseases of various crops are scattered throughout a variety of journals written in many languages. This makes information retrieval a time-consuming task. We have brought together specific isolation techniques and detailed descriptions of symptoms on as many of the economically important host plants worldwide as possible. Furthermore, we have listed the geographical distribution for each *Phytophthora* species together with the host plant on which it was first isolated so that workers have but a single source to which they need resort. Photographs have been included to further aid in the diagnoses of diseases. It must be recognized that these lists cannot be considered absolutely complete, because we may have missed some citations in our search and because information continues to be published on the identity of new hosts.

Identification of *Phytophthora* species has always been difficult because of the general morphological similarities of various species (Waterhouse et al. 1983). An updated description of the characteristics of each *Phytophthora* species together with detailed illustrations of its morphology will aid in its identification. Tables and charts are included to facilitate identification of closely related species and are supplementary to the key of Stamps et al. (1990). Although much work has been conducted on taxonomy, the nomenclature has changed for some *Phytophthora* species, and as new knowledge is gained from modern molecular biology methods such as protein and isoenzyme patterns and DNA (deoxyribonucleic acid) restriction fragment length polymorphisms (Förster et al. 1987–1989; Förster and Coffey 1991), further changes are likely to be made.

There are many differences of opinion as to what constitutes a species. How best to key out species has always been and still is controversial. The information presented and documented here should provide a source of information for further studies on the various species and their interrelationships.

The chapter on isolation is aimed at describing the improvements in isolation techniques, which have changed from conventional tissue culture on water agar to the more specific isolation of *Phytophthora* on media containing antibiotics differentially toxic to contaminant fungi and bacteria. Although fraught with the difficulties inherent in isolation of any fungus from the soil, the methodology of quantitation of *Phytophthora* propagules in soil has advanced greatly within the last 20 years (Tsao 1983).

The chapters on control contain the leading principles, concepts, and methods. These are subdivided into control by cultural and biological means (Chapter 5), host resistance (Chapter 6), and application of fungicides (Chapter 7). We present specific control methods and descriptive information in the chapters devoted to each species.

We have written this book from the generalist's viewpoint. Our aim is to present an update of both current and historical information that will be of use not only to professional plant pathologists and graduate students, but also to those in agricultural businesses who must understand, diagnose, and control *Phytophthora*-caused diseases on a wide variety of host plants.

It seems pertinent to indicate other sources of information on *Phytophthora. Source Book of the Genus Phytophthora* by Ribeiro (1978) contains a wide variety of useful information on the genus *Phytophthora* and is widely quoted; however, this book is no longer in print. Some of the material has been incorporated into this book. The *Proceedings of the Workshop on Phytophthora Diseases of Tropical Cultivated Plants* (K. K. N. Nambiar, ed. 1982) resulted from an international workshop held in Kasaragod and Calicut, India, in 1980. The proceedings, which can be obtained from Central Plantation Crops Research Institute, Kasaragod, 670 124, Kerala, India, contains a large number of review and original research papers on *Phytophthora.* Novotel'nova (1974) published a book on the genus *Phytophthora* in the Soviet Union (now the Commonwealth of Independent States).

The book *Phytophthora: Its Biology, Taxonomy, Ecology, and Pathology* (D. C. Erwin et al.) contains state-of-the-art information on principles as seen by leaders in each subject area at that time. That book continues to be a valuable source, and since it approaches the subject from the viewpoint of principles, it supplements this book. We have drawn information from the 1983 book, but much of the information it presents has not been duplicated here.

The book *Phytophthora* (J. A. Lucas et al.) emanated from a symposium held in Dublin, Ireland, in September 1989. This book, with chapters by many authors, contains new information on the current molecular and biological trends that have a direct impact on taxonomy as well as on biology and control of *Phytophthora.* Another book, edited by Ingram and Williams (1991), *Phytophthora infestans, the Cause of Late Blight of Potato,* Academic Press, London, contains current information by several authorities on cytology, nutrition, cell biology, genetics, population biology, host resistance, and chemical control of *P. infestans.*

The book *Phytophthora infestans 150* (Dowley et al. 1995) contains research papers by many scientists from various areas of the world. The papers were presented at an international conference at Trinity College, Dublin, Ireland, that marked the 150th anniversary of the first record of potato blight in Ireland and the subsequent famine.

The *Phytophthora Newsletter* was initiated in the United Kingdom by C. M. Brasier with the assistance of F. J. Schwinn to facilitate interaction of research workers worldwide on an informal basis. The responsibility for this newsletter passed successively to several other laboratories throughout the world. Although some university libraries catalogue the *Phytophthora Newsletter,* it is not considered to be an organ of "published" information by general agreement among those who edit and write for it. Therefore, wherever we have used information from this source, we have considered the author the sole copyright owner and obtained permission for any quotations.

Botanical names and names of diseases used in the text are based on the following sources:

Index Kewensis, Plantarum Phanerogamarum and Supplements (1883–1970). Clarendon Press, Oxford, England.

Hortus Third. Liberty Hyde Bailey Hortorium, Macmillan Publishing Co., New York. 1,290 pp. (1976).

Lexicon of Plant Pests and Diseases, H. Merino-Rodriquez. Elsevier Publication Co., New York. 351 pp. (1966).

Multilingual Compendium of Plant Diseases, P. R. Miller and H. L. Pollard. American Phytopathological Society, St. Paul, Minnesota. 457 pp. (1976).

Common names of plant diseases follow the list compiled in *Plant Disease* (J. D. Hansen 1985) except for those diseases that occur in nations other than United States.

We greatly appreciate the generosity of many colleagues who gave of their time and talents to help us compile the necessary data. Ahmad Vaziri generously prepared drawings of the various species and several of the figures on methodology. R. A. Khan and John Kitasako gave effective technical assistance for photographic reproductions of *Phytophthora* spore structures. We received invaluable help in translating foreign-language papers from Helga Förster (German), M. Yoshikawa (Japanese), and Ta-Li Kuan (Chinese) and were aided by plant pathologists worldwide who readily shared photographs and offered their full cooperation in providing information from their respective countries. We are especially grateful to M. J. Griffin and T. M. O'Neill, ADAS, Cambridge, United Kingdom, who generously shared photomicrographs of spore structures of many species that occur in the British Isles. Many other investigators who have contributed photographs are individually acknowledged throughout the text. If any have been omitted, we extend our apologies.

We greatly appreciate the cooperation of several colleagues who read and critiqued parts of the text. These include Michael Dick for his thoughtful and incisive input in Chapter 1 and for sharing unpublished manuscripts; Peter H. Tsao for significant assistance in Chapter 2; G. A. Zentmyer for reading Chapter 2 and for sharing photomicrographs of sporangia of certain species of *Phytophthora;* Noel Keen for comments on Chapter 6; David Shaw, who read the parts on genetics in Chapter 3; David Nes for advice and discussion on sterols in Chapter 3; Helga Förster for sharing information and commenting on evaluation of species differences and similarities by use of DNA restriction fragment length polymorphism data in Chapter 4; and Bruce Grant, A. F. Schmitthenner, and M. D. Coffey, who generously supplied prepublication drafts of many manuscripts. Donald Barr was most cooperative in sharing unpublished manuscripts, photographs, and advice on the ultrastructure of the zoospore. Hon Ho and Geoffrey Hall made valuable suggestions on the tabular keys (Tables 4.2 and 4.3). However grateful we are to those who assisted us, we accept the responsibility for any errors in presentation of the material.

The use of the unpublished card index of literature citations, compiled by the late C. M. Tompkins, University of California, Berkeley, and left in the custody of G. A. Zentmyer, was an invaluable source for many early citations.

We thank Sylvia Fernandez for assistance in checking the literature and helping with library work; Pam Ferrin, BettyAnn Merrill, Doreen Alewine, Judy Ehlers, and Gale Matweeff for preparing drafts; but most especially Celena Turney for her indispensable cooperation in bringing the final copies together.

Last, but certainly not least, we thank Thor Kommedahl for creating the index and the editorial staff at APS Press for their advice and counsel.

Don Erwin expresses his appreciation and thanks to his wife, Veora, for her unwavering support and idealism.

CHAPTER 1

Introduction to the Genus *Phytophthora*

This chapter briefly presents an overview of pertinent information about the genus *Phytophthora* and its taxonomic place in the classification of living organisms as well as some of its unique features and the diseases it causes. Some of the material presented here is treated in more detail in Chapter 2 on methods of isolation of *Phytophthora,* in Chapter 3 on the physiology and genetics of *Phytophthora,* in Chapter 4 on morphology and identification, and in Chapters 5, 6, and 7 on control. The subsequent chapters contain particular information on individual species of *Phytophthora.*

Taxonomic Position of *Phytophthora* Among Living Organisms

The genus *Phytophthora* is closely related to the genus *Pythium,* and both genera are classified in the family Pythiaceae, so named because the genus *Pythium* was described first. Both genera are water molds, and, as the name implies, both are ecologically favored by free water in soil and on foliage. The genus *Pythium* includes many species that are pathogens of plants as well as some that are parasitic on some *Phytophthora* species. Many *Pythium* species are saprophytic (i.e., they subsist on dead organic material) (see the key by Dick [1990b]).

Both *Pythium* and *Phytophthora* have coenocytic mycelium with no, or few, septa and in water produce zoosporangia that bear biflagellate zoospores. Sexual oospores form singly within the oogonium after fertilization by a nucleus from the antheridium (only occasionally, two to six oospores form in the oogonium of certain species of *Pythium*). Morphological differences between the two genera are described and illustrated in Chapter 4.

The most important morphological feature of the oomycetes is sexual reproduction by production of oospores after union of two gametangia in which meiosis occurs prior to fertilization (Sansome 1965). The entire thallus is diploid. Zoospores typically form from zoosporangia (= sporangia). Several other features based on ultrastructural and biochemical criteria include mitochondria with tubular cristae and cell wall material composed of β-1,3-glucans, β-1,6-glucans, and β-1,4-glucans (cellulose) (Bartnicki-Garcia 1969, 1970; Bartnicki-Garcia and Wang 1983).

The taxonomic relationship of *Phytophthora* to other fungi is discussed here. The genera known as the downy mildews, *Basidiophora, Bremia, Bremiella, Sclerospora, Peronospora, Plasmopara, Paraplasmopara,* and *Pseudoperonospora,* all of which are obligate parasites (i.e., they grow only on their plant hosts), are closely related to *Phytophthora.* The downy mildews are considered to be more highly evolved than *Phytophthora* and *Pythium.* See the book on zoosporic fungi (Buczaki 1983) and Dick (1990a, 1995a, b) for taxonomic information on these groups. Although *Phytophthora* and other oomycetous microorganisms have long been included within the fungi, they are no longer classified in the kingdom Myceteae (the fungi) because, on the basis of their evolutionary phylogeny, they belong in another recently described kingdom, Chromista (Cavalier-Smith 1986, 1987; Barr 1992; Dick 1995a, b).

In older classifications, living organisms were simplistically divided into two kingdoms, animals and plants, and the fungi were included in the plant kingdom. In the 1970s, living organisms were divided into eukaryotes (those with true nuclei) and prokaryotes (those without membrane-bound nuclei). The fungi were classified as eukaryotes, but as a separate kingdom from plants, and designated Myceteae. Traditionally, the fungi have been classified as Ascomycetes, Basidiomycetes, Deuteromycetes (Fungi Imperfecti, i.e., fungi without a sexual stage), and Phycomycetes. Phycomycetes included two subclasses: Zygomycetes and Oomycetes. The genus *Phytophthora* was included in Oomycetes. Inclusion of both Zygomycetes (in which the sexual stage results from two equal plus and minus structures that fuse to produce a zygote) and Oomycetes (in which the sexual stage involves union of two different organs to produce a zygote) within the class Phycomycetes implies that the groups are related; however, the only feature that they have in common is their lack of septa in the mycelia. Since the oomycetes and zygomycetes are not related, the class Phycomycetes is no longer accepted or used in formal taxonomy.

This new concept may come as a shock to those who took formal courses in mycology and plant pathology years ago and always considered *Phytophthora* to be a typical fungus. However, the bottom line is that *Phytophthora* is indeed "fungus-like" and still commonly known as a fungus. Plant pathologists have always recognized that *Phytophthora* is unique and has special features that must be considered when conducting research on it or when trying to control it. The following is an attempt to condense the rationale for classification of *Phytophthora* and the oomycetes. In doing so, we hope to convince the reader that the newer classification actually supports and explains many of their unique features.

The evolution of the phylum Oomycota from an ancestral stock that possessed an anisokont and a heterokont (unequal or different lengths of the two flagella on the zoospore) (Figures 4.18–4.20) flagellate stage in the life cycle is not disputed. Although Dick (1990a) contributed a chapter on this phylum in *Handbook of Protoctista* (kingdom Protoctista), he has long held the concepts that characterize the kingdom Chromista (Dick 1969). Dick's current classification includes the oomycetes in the kingdom Chromista (Dick 1995a, b). This coherent evolutionary group, which has been discussed by many workers since Vlk (1939) first revealed heterokont flagellation on zoospores, was formally named the kingdom Chromista by Cavalier-Smith (1986), who characterized it as being comprised of all protists having tripartite, tubular hairs (mastigonemes) (retronemes; lateral, hairlike structures on flagella) arranged in two rows along the flagellar shaft (thereby reversing the thrust of the flagellum), a complete rough endoplasmic reticulum (RER) envelope around the chloroplast, or both. Currently, according to Dick (1995a), the term mastigoneme is obsolete (Andersen et al. 1991) for flagella with tripartite, tubular hairs.

Dick (1990a) divides Oomycota into three subclasses: Saprolegniomycetidae, Rhipidiomycetidae, and Peronosporomycetidae. Saprolegniomycetidae includes three orders, Saprolegniales (families Saprolegniaceae and Leptolegniellaceae), Leptomitales (families Leptomitaceae and Apodachlyellaceae), and Sclerosporales (families Verrucalvaceae and Sclerosporaceae); Rhipidiomycetidae contains a single order and family, Rhipidiales (Rhipidiaceae); and Peronosporomycetidae has two orders, Peronosporales (families Peronosporaceae and Albuginaceae) and Pythiales (families Pythiaceae and Lagenidiaceae). Other taxa were left *incertae sedis* (in an uncertain position), and some of these are being placed in new orders (Dick 1995a, b).

Pythiaceae includes the following genera listed in order of the dates of their descriptions: *Pythium* Pringsheim (1858), *Phytophthora* de Bary (1876), *Pythiogeton* Minden (1916), *Trachysphaera* Tabor and Bunting (1923), *Diasporangium* Höhnk (1938), and *Peronophythora* Chen ex Ko, Chang, Su, and Chen (1978). *Zoophagus* has been removed from the flagellate fungi (Dick 1990b). In a current paper on sexual reproduction in the phylum Peronosporomycetes, Dick (1995a) presents a somewhat different classification (Dick 1995b), which includes three subclasses, Peronosporomycetidae, Rhipidiomycetidae, and Saprolegniomycetidae. Within the subclass Peronosporomycetidae, he includes the orders Peronosporales and Pythiales. Within Pythiales, the single family Pythiaceae contains the genera *Pythium, Myzoctium, Lagenidium, Cystosiphon,* and *Phytophthora.*

Further morphological and biochemical studies are required before classification of the marine fungi that have been transferred from *Phytophthora* to *Halophytophthora* (Ho and Jong 1990) is certain; this genus may eventually be included either in Pythiales or with the marine fungi in Saprolegniomycetidae (M. W. Dick, *personal communication*). The marine species are classified as a new group VII in the *Phytophthora* key of Stamps et al. (1990) but are not included in our book.

Both Dick (1982, 1995a, b) and Barr (1983, 1992) reiterate the long-held concept that the oomycetes evolved with and are related phylogenetically to the heterokont algae. The concept that oomycetes descended with the heterokont algae is also supported by ultrastructural similarities (Beakes 1987). Research on small-subunit ribosomal RNA gene sequences by Förster et al. (1990a) is in agreement that the oomycetes differ at the molecular level from the true fungi. When gene sequence data from *Phytophthora megasperma* f. sp. *glycinea* (now known as *P. sojae,* an oomycete), *Lagenidium giganteum* (oomycete), and *Blastocladiella emersonii* (chytridiomycete) were compared with published data from *Achlya bisexualis* (oomycete), *Saccharomyces cerevisiae* (ascomycete), and *Neurospora crassa* (ascomycete), the evidence supported the concept that members of Oomycota (*sensu* Dick 1990a) are monophyletic and appear to have evolved with the heterokont photosynthetic algae. The chytridiomycetes and ascomycetes probably evolved on different phylogenetic branches (Förster et al. 1990a).

The kingdom Chromista includes the brown algae and all protists that have either tubular ciliary (Cavalier-Smith prefers the term "cilia" to "flagella") mastigonemes (fine, hairlike rodlets attached laterally in longitudinal rows on a flagellum) or chloroplast endoplasmic reticulum or both. These characteristics are considered to have evolved from a common ancestor. One of three phyla is Heterokonta, under which the class Pseudofungi (Heteromycotina) is included. This class contains Oomycetes (within which the genus *Phytophthora* is classified), Hyphochytrea, and Labyrinthulea. The oomycetes are characterized by the production of zoospores with two cilia (flagella) with anterior rootlets having a ribbed triplet and doublet, with posterior rootlets having an octet and doublet, and with cytoplasmic and nuclear-associated microtubules. See Figures 4.21 and 4.22, which contain illustrations from Barr (1992).

The recent paper by Barr (1992) discusses the evolution and kingdoms of organisms as they relate to mycology. Aptly, his lecture (presidential address to the Mycological Society of America in 1991) begins with the question, "Are oomycetes and chytrids fungi?" Barr traces the history of the five-kingdom concept of Whittaker (1969) and the eight-kingdom concept of Cavalier-Smith (1988) and the evidence for changing the classification of the oomycetous organisms such as *Phytophthora* from the kingdom Myceteae to the kingdom Chromista. In the following, we briefly present some of the history and evidence for this change.

In Sparrow's (1973) classification, the zoosporic fungi that include *Phytophthora* were included in Mastigomycotina on the basis of the concept "that the nonsexual propagative spore by the production of one or two flagella is adapted for locomotion in a liquid, usually water." In the classification by Alexopoulos and Mims (1979), the division Mastigomycota was subdivided into Haplomastigomycota and Diplomastigomycota. As implied by these names, fungi in the former division are haploid (some have only one flagellum) and those in the latter are diploid (these are all biflagellate). Diplomastigomycotina included only one class, Oomycetes, in which *Phytophthora* was classified within the order Peronosporales and the family Pythiaceae. In the taxonomic scheme of Sparrow (1973) and Waterhouse (1973) and in Barr's reports (1982b, 1983), however, Mastigomycotina is not divided into Haplomastigomycotina and Diplomastigomycotina.

In Parker's (1982) *Synopsis and Classification of Living* Organisms, the fungi were classified in the kingdom Eumycota (true fungi) by Barr (1982a), within which the subdivision Mastigomycotina is described (Barr 1982b). Three classes are distinguished by the types of flagellation. Zoospores of members of Chytridiomycetes are characterized by a single whiplash flagellum, Hyphochytriomycetes by a single tinsel flagellum, and Oomycetes by two heterokont flagella. Barr (1983) defines heterokont as the possession of two flagella, the whiplash flagellum being posteriorly directed and the other tinsel flagellum being anteriorly directed with two rows of mastigonemes (delicate fibrillar appendages; see Figure 4.20 for flagella on zoospores of *Phytophthora palmivora*). Although members of all three classes produce zoospores with two flagella, with the exception of the hyphochytriomycetes for which there is some evidence of a relationship, they are not phylogenetically related (Barr 1982a; Dick 1982). Only the oomycetes are related phylogenetically to the heterokont algae (Barr 1992).

Margulis et al. (1990) described the five-kingdom concept for living organisms that consisted of Monera, Protoctista, Myceteae (fungi), Animaliae, and Plantae. Myceteae was subdivided into Ascomycotina, Basidiomycotina, Zygomycotina, and Mastigomycotina; however, the oomycetes were removed from the kingdom Myceteae and transferred to the kingdom Protoctista, within which the phylum Oomycota was included by Dick (1990a). He states, "Physiologically and morphologically the Oomycota are 'fungi' in that they are heterotrophic, uninucleate or coenocytic protoplasts which are bounded by cell walls in their vegetative (assimilatory) states; however, they have no phylogenetic relationship to the true fungi (i.e., Zygomycotina, Ascomycotina, Basidiomycotina)." Dick (1990a) presents an evolutionary scheme for the various orders, families, and genera in Oomycota and traces the history of the various classification concepts. The history of the finding that meiosis occurs in the gametangia of *Phytophthora* and the closely related genus *Pythium* is related by Dick and Win-Tin (1973). The system, exemplified by *Phytophthora,* in which meiosis occurs in the antheridium and oogonium (Sansome 1965), is called haplomitotic B (mitosis confined to the diploid nuclei) life cycle (Dick 1987). This feature is unique and not found in any of the true fungi.

The evidence as stated above leads to the conclusion that *Phytophthora* in the phylogenetic sense can no longer be classified with the true fungi (Mycetae). Nonetheless, information about *Phytophthora* or other oomycetes (called pseudofungi by Cavalier-Smith [1989]) will probably still be taught in mycology courses to reach students who probably would not take courses devoted to the Chromista per se. Barr suggests including Oomycetes in a "Union of Fungi" under the phylum Heterokonta (Cavalier-Smith 1989). This Union of Fungi would include organisms that belong in three kingdoms: the true fungi in the kingdom Eumycota; oomycetes in Chromista; and myxomycetes (slime molds), protostelids, and plasmodiophormycota in Protozoa. Christensen (1990) suggests that the common term "fungi" be used in a biological rather than in a taxonomic sense. Practically speaking, Hawksworth (1991) states that microorganisms such as *Phytophthora* that are placed phylogenetically in kingdoms other than Myceteae should be included with "organisms studied by mycologists." Although, as stated above by Dick (1990a), the genus *Phytophthora* cannot be considered a true fungus, it will undoubtedly continue to be referred to as a fungus because of its many physiological and morphological similarities to the true fungi (Dick 1995a). In this sense, *Phytophthora* species are referred to in the common sense as fungi.

We, the authors, acknowledge the counsel and assistance of Michael Dick in preparation of this section on the place of *Phytophthora* and the oomycetes in the world of living organisms.

Evolutionary concepts are often considered to be inconsequential to practical plant pathologists because there is little or no fossil record for *Phytophthora* or *Pythium.* The important concept to be gained from what appears to be a complex change in our understanding of the taxonomy of the oomycetes, however, is that there are valid and now better understood reasons that *Phytophthora* and other oomycetes are fundamentally different from the traditional "true fungi." Most of the plant pathologists who work with *Phytophthora* have long realized these differences. For instance, it is known that seed-treatment fungicides must include a chemical that is active against *Pythium* as well as a chemical active against *Rhizoctonia,* a basidiomycetous fungus. Tsao (1990) states that many general diagnostic plant pathologists do not fully realize that methods for isolation of *Phytophthora* must be different from those used to detect members of the true fungi. Polyene antibiotics suppress growth of most true fungi but not that of *Phytophthora.* Many of these subjects are discussed at greater length in Chapter 2.

A chapter by Griffith et al. (1992) on the mode of action of fungicides and the target sites of fungicides to control oomycetous pathogens cogently takes the stand that failure of plant pathologists (especially those conducting research on new approaches to fungicidal control) to recognize that the oomycetes differ from the true fungi has retarded development of fungicidal control. They indicate several possibilities for exploiting some of the unique features of oomycetous microorganisms by development of fungicides that affect biological systems peculiar to oomycetes. Some of their ideas are discussed more fully in Chapter 7. They reiterate, as stated above, that oomycetous organisms differ from the true fungi because of their diploid nature; the morphology of mitochondrial cristae; the flagellar heterokont apparatus of the zoospores; the biochemistry of the cell walls, which contain only a cellulose microfibril skeleton and β-1,3-glucans amorphous material instead of chitin; their lack of epoxidation of squalene to sterols; their different metabolic pathways; and their unique molecular biology systems. Thus, the development of fungicides should be aimed at particular enzymes and other biochemical processes that are unique to oomycetous pathogens.

Life History of *Phytophthora*

The name of the genus *Phytophthora* is derived from the Greek that literally means *phyto* (plant) and *phthora* (destroyer). The salient features of the species of *Phytophthora* are described here briefly and in more detail in Chapter 4, where morphology and identification of species are discussed. The *life history,* a term recommended by Dick (1987) rather than *life cycle,* is illustrated in Figure 1.1, and a list of species and some of the synonyms that are currently used is given in Table 4.1.

We have drawn from the precise and detailed treatment of morphology and terminology as given in the classic paper by Blackwell (1949). See also the glossary presented at the end of this book. The thallus of the fungus is called *mycelium,* which consists of a large number of branched, tubular structures, each of which is individually called a *hypha* (plural: *hyphae*). These tubular filaments vary in diameter from 5 to 8 µm and can readily be seen with the low power of a compound microscope. When the mycelium is cultured on suitable media or when it grows out of infected tissue under moist conditions, it is unpigmented. When viewed with a microscope (100×), the young mycelium is hyaline (nearly transparent) and coenocytic; however, in cultures several days old, occasional cross walls can be seen. Hyphae may be smooth, swollen, nodose, or tuberculate. Growth is initiated at the tip of each hypha. This phenomenon is discussed by Bartnicki-Garcia (1990) in a review of the role of vesicles in apical growth. Side branches of hyphae are often slightly constricted at the base. The variation in branching patterns is associated with environmental effects (Ho 1978). *Phytophthora* species produce asexual spores under suitable environmental conditions, which usually include optimum temperature and moisture, and exhibit a transition from rapid vegetative growth in a relatively rich medium to slow growth under a limited nutrient status (see Chapter 3).

The most common and characteristic asexual spore is called a *sporangium* (plural: *sporangia*) or, to be more precise, a zoosporangium, which means a vessel containing zoospores. Sporangia are borne on sporangiophores (*phore* means "bearer of"), which are often similar in diameter to the hyphae. In some species, new sporangiophores emerge through the base of the old sporangium from which uninucleate zoospores (the number of zoospores produced within a sporangium varies among species) have been released (proliferation) (see Figure 4.7). In other species, new sporangiophores arise just beneath the bases of the old sporangia (i.e., they are *sympodial,* meaning "sharing the same foot") and produce more sporangia successively (Blackwell 1949) (see Figure 4.5).

Sporangia vary in shape and size (Figures 4.7–4.17). Shapes are somewhat distinctive for a particular species but are often variable and range from spherical, subspherical, ovoid, obovoid, ellipsoid, limoniform (lemon shape), pyriform (like a pear), obpyriform (upside down pear shape), turbinate (like a top), to obturbinate (like an upside down top). Other descriptive terms for shape appear in the literature (Figure 4.10). Sporangia appear hyaline to light yellow by transmitted light microscopy.

Some species produce sporangia, each with a conspicuous bud or nipple called a *papilla* (plural: *papillae*) at the tip. The papilla is defined as the plug that is composed of a hydrated material with a refractive index different from the wall material of hyphae and not the protuberance per se (Blackwell 1949). The material in the plug dissolves prior to emergence of zoospores. Species have been classified into groups on the basis of the presence or absence of conspicuous or inconspicuous papillae (Blackwell 1949; Waterhouse 1963; Newhook et al. 1978; Stamps et al. 1990). The sporangium germinates in aqueous solutions or on agar media by the production of a germ tube that usually, but not always, emerges from the tip of the sporangium (direct germination) or, in an aqueous medium after the temperature is reduced, by the production of uninucleate, biflagellate zoospores within the sporangium (indirect germination) (Figure 3.1). The zoospores emerge from the sporangium in a membranous vesicle that soon breaks, allowing zoospores to swim away (Gisi et al. 1979a, b; Gisi 1983). In contrast, zoospores of *Pythium* species form in the vesicle after the protoplasm has flowed from the sporangium (Figure 4.34). The vesicles are not the same, however. Those of *Pythium* are fibrillar, of extra-plasma membrane origin, and attached to the exit papilla; those of *Phytophthora* are of a persistent plasma membrane origin (M. W. Dick, *personal communication*).

The *zoospores* are reniform (like a kidney) in shape (Figure 4.19) with two heterokont *flagella* (singular: *flagellum*) (hair or lash) emerging from the concave side. The slightly flattened side with the groove is called the *ventral side* and the rounded side the *dorsal side* (Desjardins et al. 1969; Barr and Désaulniers 1989). The term *anisokont* implies different lengths of the flagella. More precisely, the term *heterokont* means that each flagellum is different in its external morphology (Barr 1983). In *Phytophthora,* one of the flagella is characterized by a long whiplash and the other by a shorter tinsel (ornamented with two lateral rows of hairs at the tip) (Desjardins et al. 1969; Figures 4.18–4.20).

Zoospores swim for hours (Bimpong and Clerk 1970) and eventually cease to swim, round up, and within minutes develop a cell wall (Bartnicki-Garcia 1973). At this stage, the spore is called a *cyst.* Encystment is induced by agitation produced artificially by shaking zoospores in a flask or naturally by their bumping against solid surfaces. Eventually the flagella are shed, as shown in *Phytophthora parasitica* by Reichle (1969a), in *P. megasperma* var. *sojae* by Ho and Hickman (1967a), and in *P. palmivora* by Bimpong and Hickman (1975). However, some flagella in encysting zoospores of *P. parasitica* retract (Reichle 1969a). Cysts germinate by producing germ tubes and mycelia. Occasionally, another zoospore may form within the cyst and be released. This is termed *repeated emergence* (Drechsler 1930; Blackwell 1949). Zoospores are considered to be major infectious propagules.

The *chlamydospore* (a seed within a cloak) is spherical to oval. It is either hyaline or deep brown and has either a thin or thick (about 0.5–1.5 μm) wall (Figures 4.23A and B and 4.25). The chlamydospore wall is usually not as thick as the oospore wall (up to 3 μm) (Hemmes 1983). Chlamydospores may form terminally at the tips of hyphae or may be intercalary (between the tip and base of the hypha). These structures can be distinguished from hyphal swellings because they are delimited from the mycelium by septa (Blackwell 1949; Hemmes and Wong 1975; Hemmes 1983).

Hyphal swellings are globose to irregular in shape, usually hyaline, terminal or intercalary, and often in clusters, and they are not delimited by septa (Blackwell 1949; Figure 4.4A and B). Conversely, in *Pythium,* hyphal swellings are delimited by septa. Hyphal swellings are common in *Phytophthora cinna-*

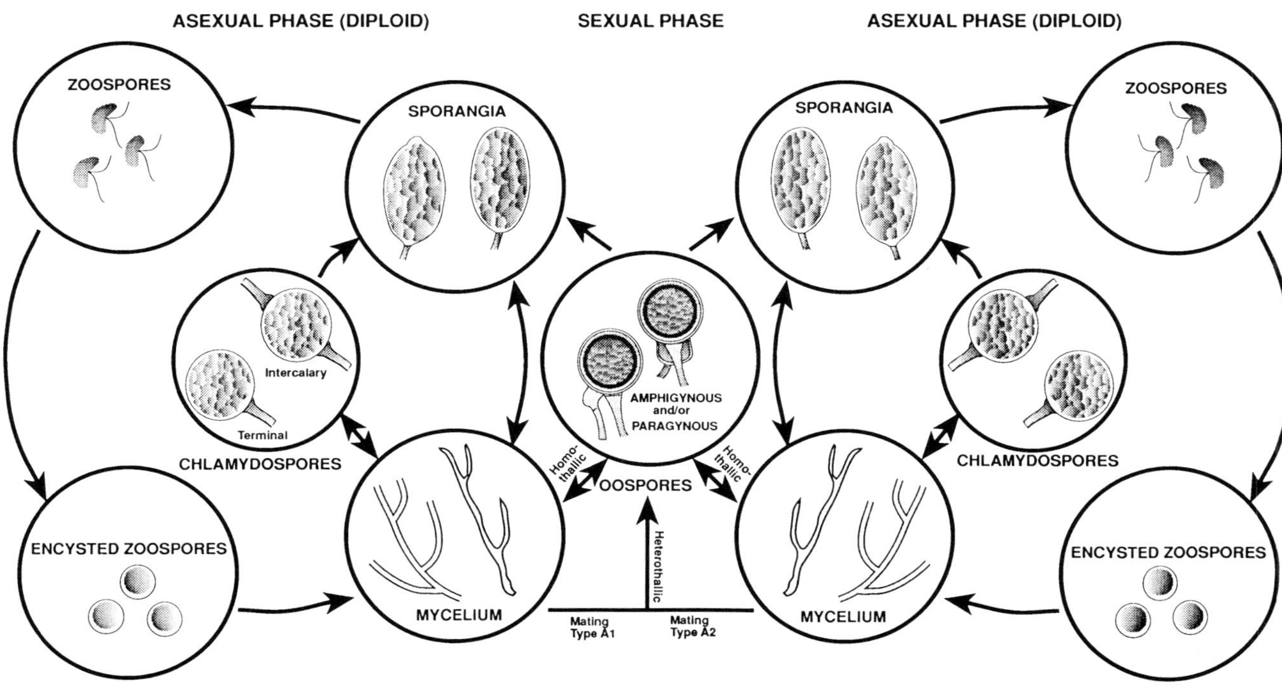

Figure 1.1. Life history of *Phytophthora.*

momi, P. cryptogea, P. drechsleri, P. megasperma, and *P. parasitica.* Wall thickness, in most cases, is less than 0.5 μm and similar to the thickness of mycelial walls.

The sexual structures of *Phytophthora* are composed of an *antheridium* (male component) and an *oogonium* (egg-containing female component). The oogonia are usually globose or subglobose but are occasionally pyriform. They are mostly hyaline, but in some cases the pigmented wall of the oogonium is yellowish brown. The oogonium is delimited from the hypha by a septum. The antheridium becomes delimited by a septum and attaches to the oogonial incept. The oogonial incept grows through the antheridial incept. This type of antheridium is classified as *amphigynous* (surrounding the female; Figure 4.26B). In some species, the antheridium attaches to the oogonium by contact to the lower hemispheric side of the oogonium and is called *paragynous* (beside the female; Figure 4.26C). Reduction division of the chromosomes from diploid to haploid (meiosis) in nuclei that occurs in the coenocytic antheridium and oogonium and the failure to form uninucleate gametes constitute unique differences separating *Phytophthora* and other oomycetes from the true fungi.

The supernumerary nuclei of the oogonium migrate to the periphery of the oogonium and are sequestered in the periplasm (aborted). A fertilization tube from the antheridium ruptures the oogonial wall and deposits the antheridial nucleus. A nucleus presumed to be the fusion nucleus remains alone in the oogonial cytoplasm (Figures 3.8 and 4.26A). The single oospore that forms within the oogonium is globose and characteristically develops a thick (0.5–6.0 μm) inner wall (Waterhouse 1963) composed largely of β-1,3-glucans (Figures 3.4 and 3.5) (Lippman et al. 1974; Bartnicki-Garcia and Wang 1983; see also Hemmes [1983] for photographs at the ultrastructure level). When viewed by electron microscopy, the outer oospore wall is electron dense and about 20 nm thick (Hemmes and Bartnicki-Garcia 1975; Hemmes 1983). The thicker (0.7–1.0 μm) endospore wall that forms within the oospore is not electron dense. Within the oospore a birefringent body of variable size, called an *ooplast,* and one or more nuclei, which have been called *pellucid* (translucent) *spots* (Dick 1969) or *pellucid bodies* (Erwin and McCormick 1971), can be seen by light microscopy (450×) (Figures 3.6–3.8). Prior to germination of the oospore, the haploid nuclei from the antheridium and the oogonium fuse to form the diploid nucleus. The fertilization process is discussed and illustrated by Hemmes (1983) via ultrathin electron microscopy. The process is described by Blackwell (1943) via light microscopy. See also Brasier and Sansome (1975) and Figure 3.10.

The diploid oospore germinates under suitable conditions by production of single or multiple germ tubes at the tips of which sporangia may or may not form. Sporangia of *Phytophthora medicaginis* form on germinating oospores in a medium containing a low level of nutrients, for instance, soil or root extracts (Förster et al. 1983; El-Hamalawi and Erwin 1986a); however, *P. capsici* sporangia form on oospores that germinate in distilled water (Hord and Ristaino 1991).

Some species of *Phytophthora* are *homothallic* (self-fertile), whereas others are *heterothallic* (self-sterile). Oospores form in heterothallic species at the juncture where A1 and A2 mating types grow together on a suitable medium. Oospores function in the homothallic species as persistent propagules in diseased plant material (Slusher and Sinclair 1973; Stack and Millar 1985c). The role in the field of the oospore in heterothallic types is not as well understood, but evidence strongly indicates that the crossing of A1 and A2 mating type isolates can be a source of new races or biotypes (see Chapter 3). When A1 and A2 mating types coexist in nature, for example, *Phytophthora infestans* on potato in the Toluca region of Mexico (Smoot et al. 1958; Niederhauser and Cobb 1959; Galindo and Gallegly 1960), recombination of genetic factors in the progeny can result in production of different races or more virulent biotypes on potato (Romero and Erwin 1969; Tooley et al. 1986; Spielman et al. 1989, 1990). The recent introduction of the A2 mating type into Europe has not induced perceptible changes in the development of new pathological races of *P. infestans* in some areas (Fry et al. 1991), but in Poland, new biotypes (races) have been found (Sujkowski et al. 1994). Crossing isolates of *P. infestans* that are marked by different isozyme patterns results in recombination of the genetic factors that control the production of certain enzymes (Shattock et al. 1986a, b). (See Chapters 3 and 6.)

Oospores can also result from selfing caused by production of a pheromone or hormonelike substance by the opposite mating type (Ko 1978). The selfing phenomenon was proved by placing A1 and A2 mating types on opposite sides of a polycarbonate membrane to prevent contact of the mycelia of the A1 and A2 types and noting that oospores formed on both sides of the membrane. Stimulation of oospore formation was attributed to an unidentified chemical substance that was capable of passing through the membrane and that stimulated the opposite mating type to produce oospores (Ko 1980). In such cases, the progeny are selfs (genetic material emanates from only one of the parent cultures), and genetic recombination of factors from both parental types does not occur.

A2 isolates can be stimulated to produce oospores by wounding a culture with a scalpel (Reeves and Jackson 1974) and by exposure to volatile components produced by the common soilborne fungus *Trichoderma viride* (Brasier 1971a; Reeves and Jackson 1972). Lipid substances from root extracts of avocado stimulate oospore formation in A2 but not in A1 isolates of *P. cinnamomi* (Zentmyer 1952, 1979; Zaki et al. 1983). If oospores in heterothallic species are incited by an external factor or by a hormonelike material produced by the opposite mating type, this phenomenon might favor survival and perpetuation of *Phytophthora* as oospores in natural sites. Although oospores are sensitive to drying, they can withstand dry conditions much better than asexual spores, with the possible exception of chlamydospores. (See Chapter 3 for more detail.)

Unique Features of *Phytophthora*

Phytophthora, Pythium, and other oomycetes have a number of biological characteristics that are relatively uncommon in other fungi (Zentmyer 1983; Griffith et al. 1992). The major part of the life history is primarily diploid whereas the higher fungi are haploid. The cell walls of *Phytophthora* are composed of cellulose and β-glucans and not chitin, which is the common cell wall component of fungi outside Oomycetes (Bartnicki-Garcia 1969; Bartnicki-Garcia and Wang 1983). Mycolaminarin, a β-1,3-glucan, is the characteristic storage carbohydrate (M. C. Wang and Bartnicki-Garcia 1974). The zoospores are biflagellate; one is a whiplash and the other a tinsel flagellum (Desjardins et al. 1969; Hemmes 1983). The flagellar rootlet morphology is similar to that of the heterokont algae (Barr 1981). *Phytophthora* species do not synthesize sterols but require an exogenous source of β-hydroxy sterols for sporulation (reviewed by Hendrix [1970]; Elliot 1983). *Phytophthora* and other members of Pythiaceae such as *Pythium* are uniquely resistant to polyene antibiotics such as pimaricin (Eckert and Tsao 1962), a characteristic that is correlated with a requirement for an exogenous source of β-hydroxy sterols (Ocana 1967; Hendrix 1970). Nonoomycetous fungi, which synthesize sterols, are sensitive to polyene antibiotics.

Epidemiology of *Phytophthora*, A Cause of Multicyclic Diseases

The epidemiology of root diseases caused by many soilborne fungi is the single-cycle type because the inoculum that initiates disease carries over from the previous year but does not increase during the crop year (Vanderplank 1963) (Figure 1.2); however, the inocula of *Phytophthora* species, which cause foliar as well as root diseases, increase from low, often undetectable levels to high levels within a few days or weeks (MacKenzie et al. 1983; Weste 1983b) (Figure 1.3). Other members of Oomycetes, particularly *Pythium* species, share this quality. Dick (1992) states that analyses of phenological (relationship of climate to periodic biological phenomena) changes indicate that the flux of the population density of the more ephemeral life form (e.g., sporangia and zoospores) rather than the survival propagule is more likely to determine the success of the organism. The increase of inoculum of *Phytophthora* is caused by the rapid production of sporangia and zoospores from infected plant tissues when environmental conditions, the most important of which is the presence of free water, are favorable. All of the oomycetous microorganisms have a short generation time and great reproductive capacity (Dick 1992). For this reason, *Phytophthora*-caused diseases of fruit, foliage, crowns, and roots of plants are generally considered to be multicyclic (Fry 1982; MacKenzie et al. 1983) (Figure 1.3).

The analogy of increase of money at continuous compound interest explains the dynamics of multicyclic diseases (Vanderplank 1963). New inoculum is produced continuously (like compound interest) within the crop year after the initial infection. Thus, with late blight of potato caused by *P. infestans*, a lesion on the stem (X_0 in Figure 1.3) frequently results from infection by zoospore cysts or water-splashed deciduous sporangia that form on a single diseased tuber. This lesion, in turn, is the source of several hundred additional sporangia, each of which can be splashed by rain or carried by wind during a rain storm to leaves of adjacent healthy plants. Those lesions become the source of more sporangia. Thus, inoculum "begets" inoculum within a relatively short time and continues to do so as long as environmental conditions favor rapid reproduction and dissemination of inoculum and subsequent infection of healthy plants (Fry 1982; MacKenzie et al. 1983).

Several studies in the literature on epidemiology of late blight of potato caused by *P. infestans*, black pod of cacao caused by *P. palmivora*, black shank of tobacco caused by *P. parasitica* var. *nicotianae*, and root rot of avocado caused by *P. cinnamomi* were analyzed by MacKenzie et al. (1983). They concluded that late blight of potato and black pod of cacao are best described as multicyclic or compound interest diseases. Avocado root rot caused by *P. cinnamomi* was difficult to classify as either a single cyclic or a multicyclic disease with the data available in the literature; however, since *P. cinnamomi* spreads by root grafts and irrigation water, the epidemic should most likely be classified as multicyclic (MacKenzie et al. 1983). The disease progress in 16 of 19 tobacco fields infested with *P. parasitica* var. *nicotianae*, the cause of black shank, was most adequately described by the multicyclic model (Campbell and Powell 1980).

Summary

Phytophthora is a genus that is mainly, if not entirely, parasitic on various plant hosts. Some species are host specific and some have broad host ranges. It occupies a small but pathologically significant niche in Oomycota, a phylum that had been previously classified in Myceteae (fungi) but cur-

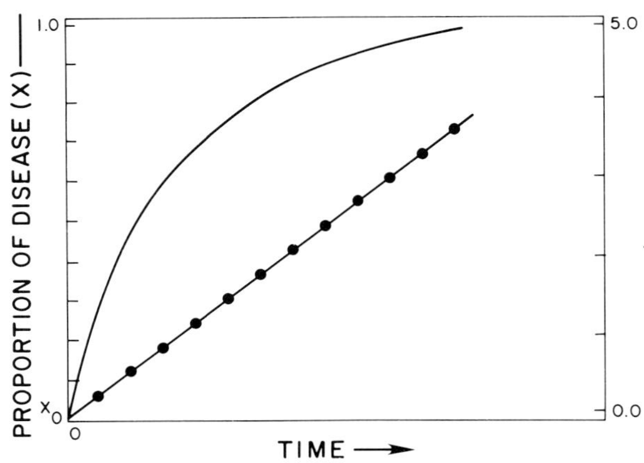

Figure 1.2. Disease progress curve for a simple interest type of epidemic (single cycle) plotted on an arithmetic axis (—) and a simple interest-transformed axis (•—•) for proportion of disease over time. The expected result of the transformation is a straight line described as $\ln[1/(1 - X)] = \ln(1/1 - X_0) + rt$, where X_0 is the initial proportion of disease and X is the proportion of disease after some time (t) at some simple rate of increase (r). (Reprinted, by permission, from MacKenzie et al. 1983)

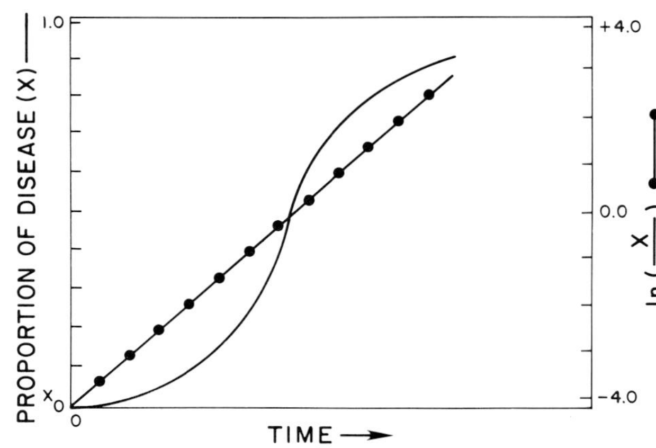

Figure 1.3. Disease progress curve for a compound continuous interest type of epidemic (multicycle) plotted on an arithmetic axis (—) and a logit-transformed axis (•—•) for proportion of disease over time. The expected result of the logit transformation is a straight line described as $\ln[X/(1 - X)] = [X_0/(1 - X_0)] + rt$, where X_0 is the initial proportion of disease and X is the proportion of disease after some time (t) at some compound continuous rate of increase (r), more commonly called the *apparent infection rate.* It is not possible to compare simple interest and compound continuous interest rates of increase directly. (Reprinted, by permission, from MacKenzie et al. 1983)

rently is placed in the kingdom Chromista (Cavalier-Smith 1986–1989; Barr 1992; Dick 1995a, b). *Phytophthora* species are diploid and produce biflagellate heterokont zoospores. Most produce oospores that are not only capable of surviving for long periods but also are a source of genetic variability in the progeny. *Phytophthora* species have a short regeneration time because of the rapid production of flagellate zoospores; thus, the inoculum from an initial infection can be amplified rapidly. The late blight of potato caused by *P. infestans* is a prime example. Multicyclic diseases can result in severe epidemics when environmental factors favor the rapid increase of *Phytophthora* propagules.

CHAPTER 2

Isolation and Detection of *Phytophthora*

In this chapter we discuss the various methods used to detect *Phytophthora* species, not only in diseased plants, but also in soil before disease outbreaks occur. *Phytophthora* species that are found to be associated with a diseased plant are likely to be the causal agent of the disease. Many other fungi, such as the closely related genus *Pythium* or the ubiquitous genus *Fusarium,* are often associated with diseased plants but are not necessarily the cause of the disease. *Phytophthora undulata* (Dick 1989) and *P. gonapodyides* may be exceptions in that they are isolated from aqueous systems and are either nonpathogenic or mild pathogens (Hoitink and Schmitthenner 1974a; Pittis and Calhoun 1984). Many species of the marine genus *Halophytophthora* (Ho and Jong 1990) are still classified in the genus *Phytophthora* (Stamps et al. 1990) but are not yet known to cause disease.

Most *Phytophthora* species cannot be isolated from diseased tissues as readily as many other fungal pathogens and often require somewhat different methods. As pointed out in Chapter 1, *Phytophthora, Pythium,* and other oomycetes are metabolically different from the "true fungi." This factor has been exploited in the development of a selective antibiotic medium, which will be covered later in this chapter. The relatively slow growth rates of isolates of most species of *Phytophthora* on culture media compared with those of many saprophytic fungi and bacteria often present a problem. Bacteria increase much more rapidly by binary fission (one bacterial cell splits to become two cells and on *ad infinitum*), and although bacteria do not grow radially, they suppress the growth of *Phytophthora* by competition, production of antibiotic substances, or direct parasitism of *Phytophthora.* Since *Phytophthora* species are often more difficult to isolate than those of the closely related genus *Pythium* and other unrelated types of fungi, much research has been devoted to the development of supplemental strategies to increase the frequency of selective isolation of *Phytophthora* (Tsao 1970b, 1983). Bait methods that exploit the pathogenicity of *Phytophthora* species on host tissue and that were developed before the advent of selective media (Eckert and Tsao 1960, 1962) are still useful today. Over the last three decades, however, progress has been made in the development of techniques utilizing selective antibiotics and chemicals in the culture medium to facilitate the isolation of *Phytophthora* from diseased tissues and soil. The comprehensive discussion of techniques and strategies for isolation of *Phytophthora* species, particularly by Tsao (1970b, 1983) and Weste (1983b) on population dynamics of *Phytophthora,* are excellent sources of information. We have drawn much material from Tsao and have included literature published since his work in 1983. The source book by Ribeiro (1978) contains a wide range of methods and media for isolation, some of which are still used. The methods manual of Dhingra and Sinclair (1985) briefly summarizes bait and selective media techniques for the isolation of *Phytophthora* and for many other pathogens. Singleton et al. (1992) describe methods for research on oomycetes as well as other fungal pathogens. The most useful principles, techniques, and media for the isolation of *Phytophthora* species are described and discussed below. Detailed descriptions of several media and techniques are presented in the appendix to Chapter 2.

Some *Phytophthora* species, such as *P. cinnamomi,* are more easily isolated than others. Ordinarily, *P. cinnamomi* is more easily isolated from avocado than from *Eucalyptus* and *Pinus* species; however, some *Phytophthora* species cannot be consistently isolated from diseased tissue. Initial growth of *P. medicaginis* (*P. megasperma* f. sp. *medicaginis*) from alfalfa is slow and tortuous and often cannot be detected from more than 10% of the roots plated when isolations are made from decaying plant tissue. After *P. medicaginis* was initially isolated from alfalfa roots and proved to be pathogenic in California (Erwin 1954), this slow-growing fungus was identified in other areas of the United States and the world mostly by research workers who understood how to isolate *Phytophthora*! At that time, before selective media were developed, isolation required frequent observation of plates and careful search for this slowly emerging fungus on water agar or weak media. This was often the only means of isolating the pathogen. Now, *P. medicaginis* is recognized in nearly every alfalfa-producing area of the world.

Phytophthora fragariae, the cause of red stele of strawberry (Montgomerie and Kennedy 1983), and *P. erythroseptica* from raspberries (O. Ribeiro, *personal communication*) are often not isolated consistently throughout the year. Such examples are common and suggest that it is likely that many diseases caused by *Phytophthora* species have not yet been detected.

It is safe to say that a large number of well-known root and crown diseases that are caused by various species of *Phytophthora* are not being properly diagnosed in many areas of the world. Also, many known diseases caused by *Phytophthora* are not reported in some countries. This often leads to an incorrect concept of the distribution of these diseases. For many of us who have been involved in the etiology of new *Phytophthora*-caused diseases, it is evident that there are likely to be many hosts on which the parasitism of *Phytophthora* has not yet been discovered. These gaps in knowledge are mainly the result of the lack of understanding of how to isolate *Phytophthora* from plant tissue. Our colleague Peter Tsao, who has traveled widely, was commonly told that *Phytophthora* diseases do not occur in certain countries. Yet, when he used proper methodology, several different species of *Phytophthora* were detected (Tsao 1990). In commercial diagnostic work, Olaf Ribeiro finds that it is often difficult to convince growers and householders that plant death is caused, not by "wet feet," but by *Phytophthora* species, which ecologically are favored by wet soil.

Tsao's article (1990) is an apt introduction to this chapter on the value of proper methods to detect *Phytophthora* species in plants and especially on roots. Tsao questions why *Phytophthora* species in many parts of the world are not recognized as causal agents of many diseases. In French Polynesia, root rot of vanilla was long erroneously attributed to *Fusar-*

ium, which is easily isolated without special techniques, but now this disease is known to be caused by *Phytophthora* (Mu and Tsao 1987; Tsao and Mu 1987; Tsao 1990). In China, Phytophthora root rot of various shrubs and trees in the Beijing and Shanghai Botanical Gardens was not recognized until proper methodology was used for isolation a few years ago. Even the widely known Phytophthora root rot diseases of avocado (*P. cinnamomi*) and citrus (*P. citrophthora* and *P. parasitica*) were discovered only recently in China, the Philippines, and Thailand.

A first report in 1989 of *Phytophthora* as a cause of peach tree decline and plant death in the Great Lakes region indicates that this disease may have been previously attributed to winter root injury or asphyxiation. In this case, use of the P_5ARPH medium (Appendix 2.4) facilitated isolation of *P. megasperma, P. cryptogea,* and one isolate of *P. cactorum.* Subsequent pathogenicity tests showed that most isolates of *P. cryptogea* and *P. cactorum* were highly virulent and isolates of *P. megasperma* were moderately virulent on peach plants in glasshouse tests (Wilcox and Ellis 1989). Similar examples from other parts of the United States and the world can be given.

Tsao (1990) cites two general ideas for why, in this relatively advanced age of agriculture, *Phytophthora* species have not been recognized more frequently. Primarily, most educated agricultural scientists receive relatively short and inadequate exposure to an understanding of *Phytophthora* in plant pathology and mycology classes. Since there are a myriad of other fungi to be considered, the genus *Phytophthora* is either neglected or lumped with the fungi in Ascomycetes, Basidiomycetes, or Fungi Imperfecti, all of which are physiologically much different from *Phytophthora* (see Chapter 1 and Griffith et al. [1992]), especially in response to detection methodology. Second, many diagnosticians are unaware that species of *Phytophthora* play such an important role in causing root and crown diseases.

Factors that are not considered seriously enough include the following:

1. *Phytophthora* species have limited saprophytic ability and do not grow and compete in soil with other microorganisms.
2. Most *Phytophthora* species attack only healthy, intact plant tissue or freshly made wounds and do not invade plant tissue previously invaded by other microorganisms; that is, they are primary, not secondary, invaders.
3. *Phytophthora* can infect the roots of woody trees months to years before foliage symptoms are detected. Often a tree can lose 50% of its lateral roots without loss of top growth. Thus, the disease may not even be suspected unless a proper isolation assay is conducted. Newhook (1959, 1970, 1980) tells the interesting story of 35-year-old *Pinus radiata* trees in windbreaks planted in New Zealand. These trees were infected but appeared to be healthy for years. After periods of unusually high rainfall during 1955 and 1956, however, the trees succumbed to root rot caused by *P. cinnamomi* and other *Phytophthora* species. Mircetich and Browne (1987) point out that stone fruit trees that are infected with root-rotting *Phytophthora* species do not usually show foliar symptoms until root and crown rot is well advanced, and at that time secondary fungi and bacteria make *Phytophthora* difficult to detect except by use of special isolation techniques. Mircetich and Browne (1987) give a wealth of detail on root disease technology.
4. *Phytophthora* species are responsible for most of the crown rots of woody plants. Tsao (1990) argues that if 90% of the crown diseases of woody plants is attributable to *Phytophthora,* how can one pass up such odds by not using proper techniques to detect it?

Bar-Joseph et al. (1981) tell an interesting story about *P. citrophthora* and *P. parasitica,* both of which indirectly became responsible for the devastating impact of the tristeza virus disease epidemic that during the early 1930s killed most of the citrus trees in Argentina and Brazil because of the extreme susceptibility of trees grafted on the *Phytophthora*-resistant sour orange rootstock. Fawcett (1936) reviews and describes the lethal disease, gummosis of the crowns of citrus trees, caused by *P. parasitica* and/or *P. citrophthora.* Although *P. parasitica* was not described until 1913 (Dastur) and *P. citrophthora* until 1925 (Leonian), the gummosis disease had been known in the Azores since 1836 (Bonavia 1888).

During the early 1800s, citrus trees were shipped around the world with no restrictions concerning plant health or disease, and undoubtedly these species of *Phytophthora* were carried on the roots of most of the trees that were planted in new areas of the world. The civilized world at that time developed a thirst for exotic plants. Several generations of "plant hunters" traveled as far as sailing ships would carry them to supply botanical gardens in Great Britain and other advanced nations. The book by Tyler-Whittle (1970) narrates the progress in that field, which was aided by the invention in 1827 of the simple glass terrarium by Ward (Wardian case) that allowed plants to be shipped long distances on sailing ships. Salt water and lack of attention had previously caused a loss of a large portion of the collected plants.

An early observation (prior to 1860) (Bar-Joseph et al. 1981) indicated that Phytophthora gummosis of citrus trees could be controlled by grafting sweet orange (*Citrus sinensis* (L.) Osbeck) trees on the resistant sour orange rootstock (*C. aurantium* L.). This information spread rapidly around the world, and consequently sweet orange trees grown in nearly all citrus-growing areas were grafted on sour orange rootstocks. Although this universal practice controlled gummosis caused by these *Phytophthora* species, little did the growers or scientists know that the unknown tristeza virus that was spread from tree to tree by the efficient aphid vector *Aphis citricida* would quickly cause the death of all sweet orange trees that had been grafted on sour orange rootstocks. Tristeza, first noted in 1930 in Argentina, was so named because the word means "sadness" in Spanish, an appropriate term to describe the almost total loss of large, fruit-bearing trees within a year after infection by the tristeza virus. In South America (including Brazil and Argentina), more than 30 million trees were killed. It was later learned that the tristeza virus caused the trees to die because of the severe hypersensitive necrotic reaction of the phloem tissue of the sour orange rootstock at the graft union. Although sweet orange trees on their own rootstock were infected by the tristeza virus, trees were killed only when grafted onto sour orange rootstocks. This problem has now been solved by replacement of the sour orange rootstock with Troyer and others that are not hypersensitive to the tristeza virus but are moderately resistant to *P. citrophthora* and *P. parasitica.*

Ironically, in retrospect, we can reflect that had these *Phytophthora* species not been transferred with the early import of citrus plants to almost every region of the world, sour orange would not have been used almost universally as a rootstock and the tristeza disease would not have had such a devastating effect on citrus. Perhaps the moral of the story is that genetic uniformity in an entire crop is frequently epidemiologically dangerous because of the probability that the crop may be more vulnerable to a specific, unknown pathogen.

More attention should be paid to the production and sale of healthy nursery stock, which often is contaminated by latent infection caused by *Phytophthora* and other pathogens. The plant nursery is the heart of the ornamental, woody tree, fruit, and forest industries. If plants become contaminated with *Phytophthora* in the seedbed, the chances are that the contaminating spores or mycelium of *Phytophthora* will be carried to the home, the orchard, or the forest. These infected plants are a source for further spread of this multicyclic type of disease to other healthy trees (E. M. Hansen et al. 1979, 1980; Matheron et al. 1988). To produce healthy trees, the manager of a nursery must know not only how to prevent the disease but also how to recognize different *Phytophthora* species in order to take measures to eradicate it and clean up the nursery. We discuss prevention and cleanup at greater length in Chapter 5, Cultural and Biological Control. We emphasize here how important it is to know how to recognize diseases caused by *Phytophthora* species and to isolate *Phytophthora* from the diseased tissue. Though there are occasional exceptions, one can almost always assume that if *Phytophthora* species are isolated, they are involved at least partially as the cause of the disease, and they will spread rapidly.

We do not imply that all nurseries are unaware of appropriate sanitation technology. Many have excellent programs and sell only clean nursery stock. There is also an economic side of the story that partially explains the lack of attention to production of clean stock. Far too often, the buying public fails to appreciate and is not willing to pay for the extra cost of sanitation. Too many customers look for the bargains and do not appreciate the value of disease-free plant material.

Isolation and proper identification of the causal agent of any disease is important, and we reiterate our previous premise that too little attention is given to this factor in many areas of the world—even in those where advanced technology is used. In the following pages we present the salient principles as well as state-of-the-art practices by which many *Phytophthora* diseases are diagnosed. We highly recommend the chapter by Tsao (1983) on isolation of *Phytophthora* species, which reflects his experience in this area.

The "art" of isolating *Phytophthora* species involves the interaction of carefully selecting freshly diseased plant tissue, placing diseased tissue on the proper selective agar medium, and looking carefully for slowly emerging, nonseptate hyphae on the agar medium. First, we explain the use of selective antibiotics and chemicals that can be incorporated into the isolation medium and follow with soil isolation and isolation methods that employ baiting with susceptible host tissue.

Selective Chemicals

One of the principles of selective isolation involves use of a chemical or chemicals in the medium that inhibit the associative bacteria and fungi but have little or no effect on the target *Phytophthora*. Before 1960, there was very little choice in selection of a differential chemical in the medium. Acidifying a medium such as potato-dextrose agar (PDA) to pH 4 with 25% lactic acid (about three drops in a petri plate) to suppress bacterial growth is a satisfactory method for isolation of many fungi but is generally not useful enough by itself to isolate selectively most *Phytophthora* species because spore germination is often inhibited at low pH values. Of the media without selective antibiotics, water agar, which does not support rapid bacterial growth, is often the most efficient medium

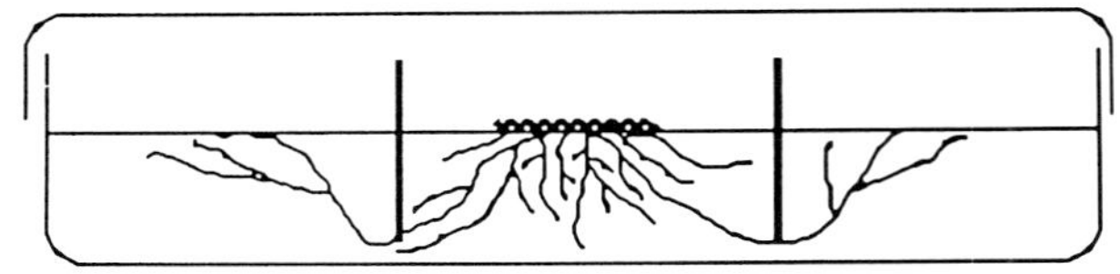

Figure 2.1. Van Tiegham cell method of purifying fungi from bacterial contamination. The bacteria-contaminated culture is placed inside the glass Van Tiegham cell, which forces the laterally growing mycelium down into water agar or a selective agar medium. When the mycelium emerges, it is usually free of bacteria because bacteria multiply slowly under the agar.

to use. Use of a Van Tiegham cell, which forces *Phytophthora* mycelium to grow under the surface of the agar, tends to free the mycelial strands of bacterial contaminants because bacteria do not increase as rapidly when submerged. The Van Tiegham cell method is illustrated in Figure 2.1.

Selective chemicals and their properties are listed in Table 2.1. For properties of antibiotics, the book by Reiner (1982) is useful. Also, see the appendix of this chapter for more detailed information on preparation and use of many selective media.

Before 1960, the methods for isolating *Phytophthora* were limited to baiting diseased plants or soil with living, susceptible plant material or plating infected plant tissue on water agar or weak agar media. The breakthrough came when J. W. Eckert ran a routine screen of many chemical compounds against a spectrum of several pathogenic fungi; he noted that *Phytophthora* and *Pythium* species grew well in the presence of polyene antibiotics, but none of several other soil fungi grew. This information (Eckert and Tsao 1960, 1962) demonstrates the characteristic of *Phytophthora* and the closely related genus *Pythium* that differentiates them from most other fungi: their resistance to the antifungal polyene antibiotics, e.g., pimaricin and nystatin. This highly significant discovery changed the approach to isolation of *Phytophthora* and stimulated research throughout the world on development of selective media for isolation of *Phytophthora* and/or *Pythium* free of other contaminating bacteria and fungi. Eckert and Tsao's medium, based on cornmeal agar and relatively weak in nutrients, was originally amended with pimaricin (antifungal) (100 μg/ml) and penicillin (50 μg/ml) to suppress gram-positive bacteria and polymyxin (50 μg/ml) to suppress gram-negative bacteria. Commonly called the 3P medium (Appendix 2.1), it was useful for isolation from freshly infected plant tissue but, as discovered later, not from soil or badly decayed plant tissue, in which *Phytophthora* persists mainly as spores. Although there are still occasional references to the use of the 3P medium, it is not the medium of choice because the high pimaricin concentration of 100 μg/ml (ppm) inhibits the germination of spores. The first discovery of the value of a lower dosage of pimaricin for isolation from soil was from a brief report that *P. sojae* (*P. megasperma* var. *sojae*) could be isolated from soil on a medium containing pimaricin at only 2 μg/ml (Haas 1964). Subsequent reports from Tsao's laboratory (reviewed by Tsao 1983) showed that pimaricin at 10 μg/ml allowed germination of spores of several other *Phytophthora* species from soil, whereas high concentrations (e.g., 100 μg/ml) inhibited spore germination (Ocana 1967). Since *Phytophthora* exists in soil as spores, this explains why the low dosage of pimaricin allowed isolation of *Phytophthora* from soil. This work led to the P_{10}VP medium (Tsao and Ocana 1969), which contains pimaricin (10 μg/ml), vancomycin (200 μg/ml), and PCNB (pentachloronitrobenzene) (100 μg/ml) (Appendix 2.2). Ocana (1967) showed that the percentage of germination of chlamydospores of *P. parasitica* and sporangia of *P. palmivora* was as high on cornmeal agar amended with pimaricin at 1.5, 3, 6, and 12 μg/ml as on unamended cornmeal agar; but at dosages at 25, 50, and 600 μg/ml, pimaricin reduced germination to 20–30% of the control. Other data with *P. cactorum, P. citrophthora, P. parasitica* var. *nicotianae* (from tobacco), as well as *Pythium irregulare* and *Pythium ultimum* showed that germination of propagules occurred at 10 but not at 100 μg of pimaricin per milliliter (Ocana 1967). Pimaricin at 100 μg/ml inhibited *P. cactorum* oospore germination by 80%, but con-

Table 2.1. Properties and sources of selective chemicals

Chemical	Properties	Source[a]
Ampicillin (Polycillin-N)	Antibiotic; active against gram-positive bacteria	Sigma Chemical Co. St. Louis, MO
Benomyl (Benlate, Tersan)	Systemic fungicide; methyl-1-(butylcarbamoyl)-2-benzimidazole carbamate; active against most fungi except Zygomycetes, Porosporae, and Oomycetes (Edgington et al. 1971); not active against bacteria; heat stable	E. I. duPont de Nemours, Inc. Wilmington, DE
Chloramphenicol	Antibiotic; active against gram-positive and gram-negative bacteria; reduces growth of some *Phytophthora* spp. more than that of others	Sigma Chemical Co. St. Louis, MO
Gallic acid	Reduces the colony size of fungi, allowing differentiation of various genera and species of fungi from each other	United States Biochemical Co. Cleveland, OH
Hymexazol (HMI) (Tachigaren)	Fungicide; 3-hydroxy-5-methyl isoxyazole; toxic to many but not all *Pythium* spp. and to many *Mortierella* sp.; not toxic (25–50 μg/ml) to many *Phytophthora* spp.	Sankyo Co., Ltd. Agrochemicals Planning Dept. Na 7-12, Ginza 2-chome Chuo-ku Tokyo 104, Japan
Iprodione (Rovral)	Fungicide; selectively inhibits *Pythium proliferum* but not *Phytophthora cinnamomi* (Solel and Pinkas 1984)	Rhone-Poulenc Chemical Co. Agrochemical Div., P.O. Box 125 Monmouth Junction, NJ
Neomycin sulfate	Antibiotic; active against both gram-positive and gram-negative bacteria	Sigma Chemical Co. St. Louis, MO
Nystatin (Mycostatin)	Polyene antibiotic; active against most fungi except members of the order Peronosporales in the Oomycetes; not active against some *Mortierella* spp.; similar to pimaricin but less active; light-sensitive	Sigma Chemical Co. St. Louis, MO
Quintozene (PCNB, Terraclor, Brassicol)	Fungicide; pentachloronitrobenzene; in soil it is active against many fungi but not Oomycetes	Uniroyal Chemical Co. Naugatuck, CT
Penicillin G	Antibiotic; active against gram-positive and gram-negative cocci, gram-positive bacilli, spirochaetes, and actinomycetes but not against gram-negative bacilli	Sigma Chemical Co. St. Louis, MO
Pimaricin (Natamycin, Pimafucin, Myprozine)	Polyene antibiotic; active against most fungi except the Pythiaceae in the Oomycetes; not active against some *Mortierella* spp.	Mycofarm-Delft Royal Netherlands Fermentation Ind. P.O. Box 1 Delft, Holland
Delvocid-Instant (50% a.i.)		Nelson-Jameson, Inc. Box 647, 2400 E. Fifth St. Marshfield, WI 54449
Polymyxin B sulfate	Antibiotic; active against gram-negative bacteria	Sigma Chemical Co. St. Louis, MO
Rifampicin (Rifamycin)	Macrocyclic antibiotic; predominately active against mycobacteria and gram-negative bacteria	Sigma Chemical Co. St. Louis, MO
Rose bengal	Toxic (50 ppm) to most bacteria but not to many fungi; however, it causes reduction in lateral growth; media amended with rose bengal should be stored in darkness because toxicity is increased by light (Pady et al. 1960); enhances the toxicity of streptomycin (Flowers and Hendrix 1969); somewhat toxic to *Phytophthora*	Aldrich Chemical Co. Milwaukee, WI
Streptomycin sulfate	Antibiotic; active against many gram-negative and some gram-positive bacteria; toxic to some *Phytophthora* spp.	Sigma Chemical Co. St. Louis, MO
Tetracycline (Aureomycin)	Antibiotic; active against both gram-negative and gram-positive bacteria; toxic to some *Phytophthora* spp.	Sigma Chemical Co. St. Louis, MO
Vancomycin hydrochloride	Antibiotic; active against most gram-positive and gram-negative bacteria	Eli Lilly Co. Research Laboratory Indianapolis, IN Sigma Chemical Co. St. Louis, Missouri
Solvents	Since many chemicals are not soluble in water and are difficult to distribute as wettable powders, some solvents can be used at concentrations of less than 0.5% of the total volume of the medium. These include ether, ethanol, dimethyl sulfoxide, diformamide, and methylene dichloride. Whenever a solvent is used, a solvent control should be tested alone.	

[a] Designation of commercial sources of chemicals is for the convenience of the reader and does not imply a recommendation of that company as the sole source.

centrations of less than 10 µg/ml were not inhibitory (Banihashemi and Mitchell 1976). Since mycelial growth is not inhibited by concentrations higher than 10 µg/ml, *Phytophthora* can be isolated from newly infected plant tissue in which there is viable mycelium, but the dosage must be limited to 10 µg/ml or less for use with decayed tissues or soil in which propagules exist almost entirely as encysted zoospores, sporangia, chlamydospores, or oospores. Some report that a dosage of 5 µg/ml is more effective than 10 µg/ml (Mircetich and Matheron 1976; Jeffers and Martin 1986; Larkin et al. 1995) (Appendix 2.4).

Although pimaricin is the antibiotic of choice, nystatin, also a polyene antibiotic, has activity similar to pimaricin's but is less potent and has a narrower spectrum of antifungal activity (Tsao 1983). Nystatin is utilized as a substitute for pimaricin in some media, especially in areas of the world where pimaricin is not as readily available (McCain et al. 1967; Schmitthenner 1973; Masago et al. 1977; Ribeiro 1978; Kliejunas and Nagata 1980; Ploetz and Parrado 1988). Since both pimaricin and nystatin are inactivated by light, plates must be stored, as well as incubated, in the dark.

PCNB (100 µg/ml), a soil fungicide with low water solubility, is added to Tsao's revised medium (Tsao and Ocana 1969) because it is toxic to a large number of fungi but not to *Phytophthora* and *Pythium.* PCNB is particularly active as a seed treatment against *Rhizoctonia* (Tsao 1970b) and is used routinely in a selective medium for isolation of *Fusarium* (Kerr 1963). PCNB suppresses radial growth of many fungi but does not suppress *Phytophthora* or *Pythium.* For example, PCNB is a highly effective amendment for isolation of *P. parasitica* var. *nicotianae* and *P. cinnamomi* from soil (Jeffers and Martin 1986). The majority of the selective media reported to date contain PCNB as a supplement to a polyene antibiotic.

In the selective medium (P_{10}VP) of Tsao and Ocana (1969), vancomycin (200 µg/ml) is substituted for polymyxin and penicillin because it possesses a broader spectrum of activity against both gram-positive and gram-negative bacteria (Tsao and Menyonga 1966) (Appendix 2.2). The chemicals are mixed as solutions or suspensions and added to sterilized, molten (45–48°C) cornmeal agar just before plates are poured. This is necessary because many antibiotics are degraded by autoclaving at 121°C.

Benomyl, a benzimadazole fungicide (sold as Benlate) that has systemic activity and moves upward in plants, is often used at 10–25 µg/ml as a replacement for or supplement to pimaricin (Follin 1971; Schmitthenner 1973; Papavizas et al. 1981; George and Milholland 1986b). Although it is active against many soil fungi (but not against *Phytophthora*), its spectrum is narrower than that of pimaricin. Growth of *P. fragariae* is only slightly sensitive to benomyl at 10 µg/ml but is highly sensitive at 100 µg/ml. *P. cactorum* can be isolated from soil on a dilute V8 juice agar containing benomyl plus other microbial inhibitors (Harris and Bielenin 1986). Benomyl is selectively antifungal to many competitive fungi but not to zygomycetes (e.g., *Mortierella, Mucor,* and *Rhizopus*) or dark-spored members of the Porosporae within the Fungi Imperfecti such as *Alternaria, Bipolaris, Drechslera, Stemphylium,* or others (Bollen and Fuchs 1970; Edgington et al. 1971). It is not antibacterial. Although benomyl is relatively insoluble in water, it has the advantage of being heat stable so that it can be autoclaved in the medium as a wettable powder (Masago et al. 1977) or in a stock solution of the wettable powder without loss of activity. Since it is insoluble in water, the medium should be agitated while it is poured into petri plates (Appendix 2.5).

A stock solution of benomyl can be made by suspending 10 g (active ingredient) in acetone (100 ml) to which 0.34 N HCl (320 ml) is added. When heated in a water bath at 70°C, benomyl dissolves and breaks down to methyl benzimidazole carbamate (carbendazim), which is equally toxic to most target fungi (Buchenauer and Erwin 1971).

Iprodione, an ergosterol-inhibiting fungicide, at 100 and 400 µg/ml selectively inhibits radial growth of *Pythium proliferum,* which is insensitive to hymexazol, a fungicide that is relatively more toxic to *Pythium* species than to *Phytophthora* species. For example, a greater percentage of isolations of *P. cinnamomi* from avocado seedlings was detected when P_{10}VPH or P_{10}ARPH (H stands for hymexazol at 50 µg/ml, R for rifampicin at 10 µg/ml, and A for ampicillin at 250 µg/ml) (Appendix 2.4) media were amended with iprodione (Solel and Pinkas 1984). Iprodione has not been used in enough situations to conclude that it would be equally effective for isolation of other *Phytophthora* species, however.

Gallic acid (425 µg/ml) reduces the size of colonies of some fast-growing nonpythiaceous fungi and has been used for isolation of *Phytophthora* (Flowers and Hendrix 1969). Gallic acid does not inhibit zoospore germination of *P. fragariae* (George and Milholland 1986b). With the availability of the P_{10}VP (Tsao and Ocana 1969) and P_{10}ARP media (see below) (Kannwischer and Mitchell 1978), however, gallic acid is no longer used widely (Tsao 1983).

The selection of the proper antibiotic and dosage is important for selective isolation. Tsao (1983) states that the most suitable antibacterial antibiotics for isolation of *Phytophthora* species are ampicillin, penicillin, rifampicin (active against gram-negative bacteria), and vancomycin (active against gram-positive and gram-negative bacteria). Masago et al. (1977) were the first to report the use of rifampicin (10 µg/ml) combined with ampicillin (500 µg/ml). Kannwischer and Mitchell (1978) substituted rifampicin (10 µg/ml) and ampicillin (250 µg/ml) for vancomycin in the medium of Tsao and Ocana (1969) for quantitative isolation of *P. parasitica* var. *nicotianae* from soil. This medium (P_{10}ARP), or modifications of it, is highly effective (Jeffers and Martin 1986). Jeffers and Martin (1986) and Pittis and Colhoun (1984) cited the superiority of rifampicin and ampicillin over vancomycin. Whether or not replacement of vancomycin with rifampicin and ampicillin is best for isolation from all soils is still a moot question. In some soils vancomycin has been more efficacious (P. H. Tsao, *unpublished*). Rifampicin and ampicillin are less expensive than vancomycin (Jeffers and Martin 1986) (Appendix 2.4.)

Jeffers and Martin's (1986) comprehensive study of the effect of amendment of cornmeal agar with different antibiotics on growth and isolation of several soilborne *Phytophthora* species from field soil showed that reduction of the dosage of pimaricin from 10 to 5 µg/ml, addition of hymexazol at 50 µg/ml, and substitution of ampicillin (250 µg/ml) and rifampicin (10 µg/ml) for vancomycin (Appendix 2.3) increased the number of colony-forming units (cfu) of *P. cinnamomi* 8 times and that of *P. parasitica* var. *nicotianae* 2.9 times. Combining ampicillin and rifampicin as a substitute for vancomycin (200 µg/ml) reduced the number of contaminant bacterial colonies by dilution isolation from orchard soil from 43 to 0 cfu/g.

Vancomycin and penicillin (up to 1,000 µg/ml) did not inhibit zoospore germination of *P. fragariae,* but neither antibiotic provided sufficient control of contaminant bacteria to isolate *P. fragariae.* Rifampicin (10 µg/ml) reduced zoospore germination and germ tube growth by about half but was the only compound that completely suppressed bacterial contamination (George and Milholland 1986b) (Appendix 2.5).

P. infestans has always been difficult to isolate from plant tissue and especially from decayed plant tissue. Hohl (1991a) reported that *P. infestans* can be readily isolated from plant material, even that heavily contaminated with bacteria and fungi, on rye agar (pH 5.5) (see Chapter 3) containing (in μg/ml) griseofulvin (20), nystatin (19), Benlate (10), methoxypurine (5), penicillin G (48), nalidixic acid (5), neomycin (30), and 8-azoguanine (40) (Hohl 1991a). This medium allowed 42–56% growth (compared with rye agar without antibiotics) of six isolates of *P. infestans* but no growth of *Mucor mucedo, Trichoderma viridans, T. alba, Fusarium* species, *Acremonium* species, *Byssochlamys nivea,* and *Pythium* species. The sensitivity of *P. infestans* and *P. palmivora* to a large range of antibiotics and other chemicals is presented in the chapter by Hohl (1991a) on nutrition of *P. infestans.* Drenth et al. (1995) bioassayed soil in which oospores of *P. infestans* had been incorporated. The soil was mixed with water (1 kg/liter). Leaves of a susceptible potato cultivar were incubated at 15°C on the surface of the water as bait. When lesions appeared on the leaves, sporangia from the lesions were transferred to potato tuber slices. After infection of the tuber slices, sporangia were purified from contaminants by transferring them to rye agar containing (in μg/ml) ampicillin (200), fenpliclonil (10), pimaricin (10), and rifampicin (30).

Because rose bengal (60 μg/ml) is toxic to bacteria, it has been used in isolation media (Hendrix and Kuhlman 1965). Although this compound is antibacterial, it is toxic to spore germination of *Phytophthora* (Tsao 1983 [citing Flowers and Hendrix 1969 and Mircetich 1970]). Rose bengal is especially toxic to *Phytophthora* spore germination when used in conjunction with streptomycin (Mircetich 1970). According to Papavizas et al. (1981), toxicity of rose bengal is greater when it is used in combination with benomyl or hymexazol. With the availability of antibiotics such as ampicillin, rifampicin, and vancomycin, there is little advantage in using rose bengal.

Although the antibiotics streptomycin, chlortetracycline, oxytetracycline, tetracycline, neomycin, and chloramphenicol are commonly used in many selective media for isolation of many fungi (Dhingra and Sinclair 1985), they are usually much more toxic to growth of many *Phytophthora* species than those mentioned above as most suitable.

Streptomycin is a broad-spectrum antibiotic active against both gram-negative and gram-positive bacteria, but it is toxic to most *Phytophthora* species (Tsao 1983) and thus is not a good choice for a selective medium. Streptomycin is inhibitory to growth of *P. citrophthora* and *P. heveae* but only partially inhibitory to that of *P. capsici, P. cinnamomi,* and *P. parasitica* (Eckert and Tsao 1962). Streptomycin is completely inhibitory to *P. infestans* (Érsek 1975) and several other species of *Phytophthora* (Cohen and Perl 1973). Streptomycin at 10 and 50 μg/ml, added to liquid wheat-seed medium, completely suppressed growth of *P. infestans;* however, when added 4 days later, growth was reduced only by half. When streptomycin was added to sporangium or zoospore inoculum, it inhibited lesion formation on potato leaves but was ineffective after infection had occurred.

Streptomycin should be used with due caution in isolation media (Cohen and Perl 1973). Streptomycin is more rapidly taken up by mycelium of *Phytophthora* and many other members of the Oomycetes than by other classes of fungi. Uptake was two to nine times greater in *Phytophthora* and *Pythium* species than it was in an insensitive member of the Mucorales, *Cuninghamella echinulata.* This was considered to be the result of the active absorption associated with nonchitinous cell walls in *Phytophthora* (Voros 1965).

Although reported to be useful in some media (Appendix 2.6), chloramphenicol is toxic to mycelial growth of many species of *Phytophthora* (Érsek 1975; Carnes 1976; Leary et al. 1982; Tsao 1983) and is thus not an optimal choice for a selective medium. Mitochondria were found to be the sites of activity of a chloramphenicol-resistant mutant of *P. palmivora* (Carnes 1976). Chloramphenicol at a dosage of 1 μg/ml in a minimal synthetic medium reduced the radial growth of 28 isolates of *P. cinnamomi* 11–75% compared with the control (no chloramphenicol). Radial growth of four isolates was completely inhibited at 25 μg/ml, and growth was inhibited by 70–80% at 5 μg/ml (Leary et al. 1982). Certain reports indicate, however, that chloramphenicol at 10 μg/ml (Schmitthenner 1973; Shew and Benson 1982) (Appendix 2.6) or at 30 μg/ml (Sneh 1972a; Sneh and Katz 1988) functions as an effective component of a selective medium. Hollomon (1965) utilized chloroamphenicol (50 μg/ml) for the isolation of *P. infestans* from potato leaves; Rahimian and Mitchell (1988) utilized chloramphenicol for isolation of *P. cactorum* from soil (Appendix 2.6). Érsek (1975) reported that chloramphenicol at 100, 200, and 400 μg/ml was increasingly toxic to mycelial growth but not to germination of sporangia or zoospores of *P. infestans* on both media and potato slices. A chloramphenicol-containing (10 μg/ml) medium yielded a higher number of *P. cinnamomi* colonies from two field soils than P_{10}ARP (Kannwischer and Mitchell 1978) and was effective when used with the wet-sieving technique for isolation of *P. cinnamomi* (Shew and Benson 1982) (Appendix 2.6).

Neomycin (50 and 100 μg/ml) totally inhibits zoospore and sporangium germination but not mycelial growth of *P. infestans,* whereas oxytetracycline is toxic to mycelial growth but not to sporangium and zoospore germination (Érsek 1975). Neomycin should not be considered an optimal choice for a selective medium.

Hymexazol inhibits many species of *Pythium* that are soil inhabitants. Members of the genus *Pythium* are active components of soils and the flora surrounding the roots of most plants. (See Chapter 4 for features of *Pythium* that differentiate it from *Phytophthora*). These fungi are often involved in preemergence and postemergence damping-off of seedlings, and since some are causal agents of root rot of mature plants, their presence in isolations should not be overlooked. In both quantitative and qualitative isolation of *Phytophthora* species from soil in media containing either pimaricin or benomyl (both noninhibitory to *Pythium*) as the antifungal component, however, *Pythium* colonies can be a hindrance because they grow much more rapidly than those of *Phytophthora* species. Consequently, the *Phytophthora* colonies may be overgrown and obscured by *Pythium.* As recently as 1977, the seed and soil fungicide hymexazol, also known commercially as Tachigaren, was found to be relatively more toxic to many *Pythium* species than to some *Phytophthora* species (Masago et al. 1977) (Appendix 2.3).

The uptake of hymexazol by *Pythium aphanidermatum,* which is sensitive, is reported to be relatively small and is completed within 2 h. Nakanishi and Sisler (1983) suggested that the small amount of uptake of hymexazol by *P. aphanidermatum,* even after toxicity was evident, indicates that the compound may exert toxicity at the surface of the plasma membrane. Although not yet well understood, the mode of action did not involve any biochemical process that governs energy production. The discovery of the greater activity of hymexazol against *Pythium* than against many *Phytophthora* species has stimulated much research on its use in selective media for soil isolation. In addition to some *Pythium* species, certain isolates of *Mortierella* species (which produce white petaloid, rosette, or stellate colonies that resemble those of *Phytophthora*) were inhibited on media containing hymexazol.

These unidentified *Mortierella* isolates from soil are not inhibited by pimaricin or nystatin. However, species such as *M. moisabellina* Oud. and Koning, *M. ramanniana* var. *angulispora* (Naum.) Linnemann, and *M. ramanniana* var. *ramanniana* (Moell.) Linnemann were inhibited by pimaricin in the P_{10}VP medium. Several other unidentified *Mortierella* isolates were tolerant to pimaricin and were suppressed on plates containing the antifungal antibiotic hymexazol (Tsao and Guy 1977). *Mortierella,* a zygomycete, has nonseptate mycelium like that of *Phytophthora* but is not related. After *Mortierella* sporulates by forming globular aerial sporangia, it can readily be distinguished from *Phytophthora* by its production of small, nonmotile, nonseptate spores in the sporangium. These spores are windborne and do not have flagella.

Although hymexazol is a useful selective chemical for isolation of some *Phytophthora* species, it is highly toxic to certain *Phytophthora* species and conversely is not inhibitory to all *Pythium* species (Tables 2.2 and 2.3). For instance, hymexazol does not inhibit *Pythium proliferum* (Solel and Pinkas 1984) or *Pythium vexans* (isolate Py-7) (Table 2.2) (Tsao and Guy 1977). Some report *P. vexans* to be sensitive (Masago et al. 1977; Kato et al. 1990), but the sensitivity may vary with different strains. Only six of 12 *Pythium* species isolated from forest soils were inhibited by hymexazol (E. M. Hansen et al. 1979). E. M. Hansen et al. (1979) discontinued the use of hymexazol for general isolation from forest soils because it is too inhibitory to *Phytophthora lateralis* and *P. cactorum* and to another pathogenic but unidentified *Phytophthora* species. Despite its inhibitory effect on growth of *P. lateralis,* hymexazol (25 µg/ml) added to the water in which Port Orford cedar branchlet baits were floated improved isolation of this fungus from soil (Hamm and Hansen 1984). The dosage of hymexazol was reduced to 10 µg/ml for isolation of *P. cactorum* from apple soils (Harris and Bielenin 1986).

Hymexazol severely inhibits the growth of *P. infestans, P. phaseoli, P. porri,* and *P. syringae,* retards the growth of *P. hibernalis* and *P. lateralis,* but slightly enhances the growth of *P. fragariae* and *P. ilicis; P. cactorum* and *P. pseudotsugae* are moderately inhibited (Ho 1987). Hymexazol at 25 or 50 µg/ml could not be used in a medium selective for isolation of *P. infestans* because it greatly inhibits both mycelial growth and sporangium germination (Shen and Tsao 1983). The sensitivity or tolerance to hymexazol recognized by several reports is summarized in Table 2.2, and the reactions of nearly all of the recognized species of *Phytophthora* are listed in Table 2.3 (Luo et al. 1988).

Many species of *Phytophthora* are resistant to hymexazol. Mycelial growth of several races of *P. fragariae* on pea agar amended with hymexazol (up to 100 µg/ml), nystatin (up to 250 µg/ml), and PCNB (up to 500 µg/ml) was good to excellent, but growth was inhibited by streptomycin sulfate (25 µg/ml) and rose bengal (30 µg/ml). Zoospore germination and germ tube growth of *P. fragariae* was not inhibited by hymexazol (George and Milholland 1986b) (Appendix 2.5).

Hymexazol (50 µg/ml) is useful for isolation of *P. cinnamomi* and *P. parasitica* var. *nicotianae* but is inhibitory to *P. cactorum* and highly inhibitory to *P. megasperma* f. sp. *glycinea* and *P. medicaginis* (*P. megasperma* f. sp. *medicaginis*) (Jeffers and Martin 1986). D. C. Erwin (*unpublished data*) has observed that the latter fungus is too highly sensitive to hymexazol for it to be useful for isolation from soil. Despite this limitation, Stack and Millar (1985a) cite the use of a medium (Masago et al. 1977) that contains hymexazol (50 µg/ml) for isolation of *P. medicaginis* from alfalfa seedlings used as bait. In some experiments, *Pythium* species overran the plates. This indicates, as others find, that not all *Pythium* species are inhibited by hymexazol; however, many sensitive *Phytophthora* species in *Pythium*-contaminated plugs of agar will grow out readily when transferred to a hymexazol medium (O. K. Ribeiro, *unpublished*).

Kato et al. (1990) measured growth rates of a wide range of genera and species within Oomycetes, including *Phytophthora* and *Pythium,* on cornmeal amended with hymexazol. Taxa within the order Leptomitales were most sensitive, followed by species within the Saprolegniales. The sensitivity of species of *Pythium* was variable: at 40 ppm hymexazol, *P. aquatile, P. aphanidermatum,* and *P. intermedium* were highly sensitive (>95% or more inhibition), and *P. paddicum* was less sensitive (about 83% inhibition); inhibition of *Pythium* species identified as the HS group was slightly more than 60%, whereas inhibition of *P. vexans* was only about 38%. In general, most species of *Phytophthora* were the least sensitive. Inhibition of radial growth of *P. palmivora* at 40 ppm was nearly 60%, *P. gonapodyides* about 30%, *P. cactorum* about 18%, and *P. undulata* about 10%. Other data indicated that *P. cactorum* was highly sensitive to hymexazol, but Luo et al. (1988) list *P. cactorum* as insensitive to hymexazol (Table 2.2).

Phytophthora undulata was recently reclassified from *Pythium undulatum* (Dick 1989) because of its ability to form zoospores within the sporangium. It is also interesting that *P. undulata* is relatively insensitive to hymexazol (Kato et al. 1990). *P. vexans* is also insensitive to hymexazol (see Table 2.2, in which two reports list *P. vexans* as resistant and one as sensitive). Since the ribosomal DNA of *P. vexans* is not typical of other *Pythium* species, Belkhiri and Dick (1988) suggest that its generic status should be restudied. Whether or not sensitivity to hymexazol implies a relationship to either *Phytophthora* or *Pythium* is still a moot point.

Because zoospores were considered to be the main dispersal and infective propagules of *P. cinnamomi* in the Western Australia epidemic in the jarrah forest (Shea et al. 1980), the effect of hymexazol on zoospore germination of *P. cinnamomi* and seven other *Phytophthora* species was determined (Tay et al. 1983) (Table 2.2). Some *P. cinnamomi* isolates behaved inconsistently on hymexazol-amended PDA. Growth of some isolates was slightly enhanced by hymexazol, but in others, both mycelial growth and germination of zoospores were reduced. Tay et al. (1983) predicted that if the P_{10}VP agar (Appendix 2.2) amended with hymexazol (Appendix 2.3) were used for quantitative evaluation of *P. cinnamomi* zoospores in soil, the numbers of viable propagules would be underestimated. Germination and germ tube growth of zoospores of *P. nicotianae* were markedly repressed by hymexazol in PDA. Motile zoospores were more sensitive than cysts (Tay et al. 1983).

Although hymexazol is a valuable chemical for selective isolation, it must be used discretely because it is not inhibitory to some *Phytophthora* species but is inhibitory to others and it is not inhibitory to all *Pythium* species.

Oospores persist in soil and diseased tissue much longer than most other propagules (Stack and Millar 1985c). Despite their importance, however, most antibiotics used in selective media have been tested only on germination of sporangia, zoospores, and chlamydospores and on radial mycelial growth rather than on oospore germination. Since oospores germinate slowly (see Chapter 3) and often erratically, they may not germinate soon enough to produce colonies on dilution plates.

There are few experimental data on the effects of antibiotics on isolation of oospores from soil by dilution plating. Germination of oospores of *P. cactorum* was not sensitive to vancomycin (200 µg/ml). Pimaricin (100 µg/ml) reduced oospore germination 80%, but at dosages of less than 10 µg/ml,

Table 2.2. Species of *Phytophthora* and *Pythium* that have been reported to be sensitive or resistant to hymexazol (50 μg/ml). (See also Table 2.3 for data by Luo et al. [1988].)

Genus Species	Sensitive	Moderately resistant	Reference
Phytophthora			
cactorum	+		Harris and Bielenin (1986)
	+		Jeffers and Martin (1986)
		+	Masago et al. (1977)
		+	Kato et al. (1990)
	+[a]	+	Tay et al. (1983)
cambivora		+	Jeffers and Martin (1986)
		+	Tay et al. (1983)
capsici	+	+	Jeffers and Martin (1986)
		+	Masago et al. (1977)
	+	+	Papavizas et al. (1981)
cinnamomi		+	Jeffers and Martin (1986)
		+	Masago et al. (1977)
		+	Shew et al. (1979b)
		+	Shew and Benson (1982)
	+[b]	+[b]	Tay et al. (1983)
		+	Tsao and Guy (1977)
citricola		+	Jeffers and Martin (1986)
		+	Masago et al. (1977)
		+	Tay et al. (1983)
citrophthora		+	Jeffers and Martin (1986)
		+	Tsao and Guy (1977)
cryptogea		+	Jeffers and Martin (1986)
		+	Masago et al. (1977)
		+	Tay et al. (1983)
		+	Serra et al. (1988a)
drechsleri		+	Masago et al. (1977)
		+	Jeffers and Martin (1986)
erythroseptica		+	Masago et al. (1977)
		+	Jeffers and Martin (1986)
fragariae		+	George and Milholland (1986b)
gonapodyides		+	Kato et al. (1990)
hibernalis	+		Ho (1987)
infestans	+		Ho (1987)
	+		Shen and Tsao (1983)
katsurae (= *P. castanae*)		+	Masago et al. (1977)
lateralis	+		Ho (1987)
	+		Hansen et al. (1979)
		+	Hamm and Hansen (1984)
medicaginis	+		Jeffers and Martin (1986)
(*P. megasperma* f. sp. *medicaginis*)	+		D. C. Erwin and R. A. Khan (*unpublished*)
		+	Stack and Millar (1985a)
megasperma		+	Jeffers and Martin (1986)
megasperma var. *sojae* (not from soybean)		+	Tay et al. (1983)
		+	Masago et al. (1977)
nicotianae var. *nicotianae*		+	Tay et al. (1983)
nicotianae var. *parasitica*		+	Tay et al. (1983)
palmivora		+	Masago et al. (1977)
		+	Ann and Ko (1980)
		+	Papavizas et al. (1981)
	+		Jeffers and Martin (1986)
		+[c]	Tay et al. (1983)
		+	Tsao and Guy (1977)
	+		Kato et al. (1990)
Phytophthora (continued)			
parasitica		+	Shen and Tsao (1983)
		+	Masago et al. (1977)
		+	Tsao and Guy (1977)
phaseoli		+	Masago et al. (1977)
	+		Ho (1987)
porri	+		Ho (1987)
sojae (*P. megasperma* f. sp. *glycinea*)	+		Jeffers and Martin (1986)
syringae		+	Jeffers and Martin (1986)
	+		Ho (1987)
undulata		+	Kato et al. (1990)
Pythium			
aphanidermatum	+		Masago et al. (1977)
	+		Tsao and Guy (1977)
betae	+		Masago et al. (1977)
debaryanum	+		Masago et al. (1977)
irregulare	+		Masago et al. (1977)
	+		Tsao and Guy (1977)
mamillatum	+		Masago et al. (1977)
	+		Tsao and Guy (1977)
monospermum	+		Masago et al. (1977)
nelumbium	+		Masago et al. (1977)
oligandrum	+		Tsao and Guy (1977)
periplocum	+		Tsao and Guy (1977)
proliferum		+	Solel and Pinkas (1984)
ostracodes	+		Masago et al. (1977)
spinosum	+		Masago et al. (1977)
splendens	+		Ann and Ko (1980)
sylvaticum	+		Tsao and Guy (1977)
ultimum	+		Masago et al. (1977)
	+		Tsao and Guy (1977)
vexans		+	Kato et al. (1990)
	+		Masago et al. (1977)
vexans (Py-7)[d]		+	Tsao and Guy (1977)
Unidentified isolates	+[e]	+[e]	Hansen et al. (1979)
Mortierella			
alpina Peyronel	+		Tsao and Guy (1977)
hyalina (Harz) W. Gams	+		Tsao and Guy (1977)
isabellina Oud and Koning	+[f]		Tsao and Guy (1977)
ramanniana var. *angulispora* (Haum.) Linnemann	+[f]		Tsao and Guy (1977)
ramanniana var. *ramanniana* (Moell.) Linnemann	+[f]		Tsao and Guy (1977)
spinosa Linnemann	+		Tsao and Guy (1977)
Mortierella			
species no. 1 (M4)	+		Tsao and Guy (1977)
species no. 2 (M6)	+		Tsao and Guy (1977)
species no. 3 (M-B551)	+		Tsao and Guy (1977)
species no. 4 (M-2C1)	+		Tsao and Guy (1977)
species no. 5 (M-14-4)	+		Tsao and Guy (1977)

[a] One-quarter of the isolates.
[b] Hymexazol was inhibitory to some isolates of *P. cinnamomi.*
[c] One-half the isolates.
[d] Py-7, although not identified in the reference, is *P. vexans* (P. H. Tsao, *personal communication*).
[e] Six of 12.
[f] Sensitive to pimaricin in the P_{10}VP selective medium (Tsao and Ocana 1969) as well as to hymexazol.

Table 2.3. Growth of *Phytophthora* isolates on various culture agar media and apple fruit[a]

	ATCC	Daily growth rate[b]				Apple inocu-	Relative growth rate[d]		Starch utilized[e]	Cardinal temperatures[f]		Pigment produc-	Colony patterns[h]		
Species	**no.**	**CM**	**V8**	**Malt**	**NO_3**	**lation[c]**	**HMI**	**Mala.**	**(S.H.I.)**	**Min.**	**Max.**	**tion[g]**	**Malt**	**V8**	**PDA**
P. aquatica	60288	7.6	9.0	2.8	2.2	3	101	76	83	1	5	2	3	3	3
P. arecae	46716	9.1	13.0	9.7	9.4	2	66	75	39	2	5	0	1	3	2
P. bahamensis[i]	28296	1.4	1.8	0.9	3.0	0	0	0	0	2	5	0	3	2	0
	28297	1.6	2.0	1.5	0.9	0	0	0	0	2	5	0	3	2	3
P. batemanensis[i]	56965	8.3	8.0	2.9	2.4	2	0	64	44	2	5	1	3	3	3
P. boehmeriae	52836	11.8	10.0	6.8	6.1	2	53	71	98	2	5	2	5	1	5
	46717	8.3	11.0	6.3	4.7	2	72	84	38	2	5	2	2	2	2
	60238	14.0	14.0	5.0	5.5	2	39	30	40	2	5	1	2	2	5
P. botryosa	58105	5.8	6.0	4.8	3.6	2	30	106	53	2	5	0	5	4	5
	58221	5.4	6.0	6.4	4.8	2	45	116	60	2	5	0	4	4	4
	26481	7.5	8.5	8.5	8.5	2	80	40	75	3	5	0	1	4	1
	62533	6.8	9.0	9.0	6.2	3	53	50	33	2	5	0	2	4	4
P. cactorum	52859	7.9	2.3	7.4	4.3	3	94	97	89	2	5	1	5	1	5
	60289	7.9	7.0	6.9	4.0	3	66	50	70	1	5	2	4	3	4
P. cambivora	38811	9.0	8.0	7.0	7.5	3	112	71	0	1	5	1	5	4	5
	46719	9.4	5.6	5.8	5.3	2	91	78	0	1	5	1	5	4	3
P. capsici	56600	6.2	11.0	4.4	5.3	3	127	143	56	2	5	2	4	4	3
	15399	12.5	9.0	5.7	2.2	3	120	117	40	2	5	0	4	3	4
	46012	11.7	11.0	7.3	1.7	2	112	130	0	2	5	0	5	3	5
	52776	13.6	12.0	12.0	7.4	3	410	57	68	1	5	0	4	3	1
P. cinnamomi	32992	14.1	14.0	8.2	5.5	3	72	36	0	2	6	0	3	4	4
	32993	14.0	15.0	4.8	5.5	2	72	36	0	2	6	0	3	3	5
	46671	9.8	10.0	54.0	7.2	2	50	70	0	1	5	0	3	4	4
P. citricola	52855	12.4	12.0	5.0	4.3	2	77	41	88	1	4	0	4	3	1
	28192	6.2	9.6	2.5	5.8	3	114	167	80	1	5	0	2	4	2
	52230	13.3	12.0	12.0	9.9	3	52	45	85	1	4	0	3	2	3
	46298	13.0	8.0	7.7	9.9	3	70	60	97	1	5	0	2	2	4
P. citrophthora	56195	16.0	14.0	19.9	5.9	3	53	36	86	2	5	2	2	3	4
	13613	13.7	12.0	11.0	6.5	3	85	21	93	1	5	1	2	3	4
	56599	8.4	8.0	8.9	5.6	3	102	58	94	1	5	1	2	3	4
P. clandestina	60438	2.3	3.0	0.0	1.0	3	0	0	90	2	4	0	0	4	0
P. colocasiae	56193	4.5	8.0	6.0	4.4	3	91	66	92	2	5	2	1	1	1
	52233	8.3	5.0	4.7	4.2	3	84	85	87	2	5	2	1	3	1
P. cryptogea	56962	13.2	11.0	11.0	5.4	3	120	28	64	1	5	1	1	3	1
	46721	12.7	12.0	11.0	10.0	3	141	38	98	1	5	1	4	3	4
	52404	15.0	15.0	13.0	9.7	2	152	44	83	1	5	2	3	3	3
P. drechsleri	45724	9.3	9.0	9.1	8.3	3	104	59	69	1	6	1	2	3	2
	26756	15.0	14.0	12.0	7.3	3	111	52	94	1	5	2	3	3	3
	15428	9.3	11.0	6.1	4.7	2	82	66	79	1	5	3	5	3	5
P. epistomium[i]	28293	0.5	5.3	0.7	0.0	1	97	0	0	2	5	1	2	2	2
P. erythroseptica	58104	13.0	14.0	9.0	4.4	2	129	81	80	1	5	2	5	3	5
	16699	11.4	13.0	12.0	3.8	0	107	112	76	2	5	1	5	3	5
P. fragariae	58229	0.8	1.0	0.0	0.0	1	0	0	0	1	4	1	0	1	0
	11374	0.7	1.5	1.0	0.0	0	0	0	0	1	4	1	0	4	0
	13974	0.8	2.0	0.0	0.0	1	0	0	0	1	4	1	0	4	0
	13977	0.4	1.8	0.0	0.0	1	0	0	0	1	4	1	0	3	0
P. gonapodyides	46726	8.0	8.0	4.7	2.0	3	116	68	96	1	5	1	2	3	3
	60351	5.5	6.0	3.9	2.8	3	89	55	92	1	5	1	2	3	3
P. heveae	16700	8.2	9.3	7.9	5.8	1	57	11	0	3	5	1	2	2	2
	58515	8.9	7.0	7.4	6.5	2	83	67	0	3	5	1	5	2	1
P. hibernalis	56353	7.2	7.5	1.9	4.5	2	45	0	0	1	4	0	4	3	4
	32995	4.6	9.2	1.8	2.2	2	47	0	0	1	4	0	1	4	1
	60352	4.8	9.0	2.1	1.6	2	43	63	0	1	4	0	5	4	5
P. humicola	52179	8.9	10.0	4.0	5.5	2	89	34	0	1	4	1	3	2	3
P. ilicis	56615	3.3	3.4	0.0	3.3	2	0	0	0	1	4	0	0	4	0
P. infestans	32835	7.2	12.0	2.3	0.4	1	50	117	0	1	5	1	5	2	5
	16980	7.5	4.0	0.0	0.9	0	0	0	0	2	5	2	5	3	0
	48718	6.6	3.5	2.2	1.5	2	46	94	0	2	5	1	5	2	5
	60180	7.1	11.0	3.0	2.3	2	65	80	0	2	5	2	5	3	5
P. insolita	38789	7.7	6.0	2.1	4.2	3	93	75	0	2	6	3	5	2	3
P. iranica	60237	5.9	6.4	3.1	4.5	1	71	103	93	3	5	2	5	3	5

(*continued on next page*)

Table 2.3. (*continued*)

Species	ATCC no.	Daily growth rate[b]				Apple inocu-lation[c]	Relative growth rate[d]		Starch utilized[e] (S.H.I.)	Cardinal temperatures[f]		Pigment produc-tion[g]	Colony patterns[h]		
		CM	V8	Malt	NO_3		HMI	Mala.		Min.	Max.		Malt	V8	PDA
P. katsurae	58598	4.2	8.8	8.8	4.8	3	135	188	0	2	5	3	5	1	4
	36818	2.4	12.0	9.5	8.4	3	89	36	0	1	5	3	5	1	5
P. lateralis	58399	4.2	5.0	5.1	3.1	2	94	15	0	1	4	2	1	2	1
	58816	4.4	5.0	2.7	3.0	2	72	0	0	1	4	[illegible]	1	2	1
P. meadii	58103	9.3	11.0	46.0	8.4	1	66	126	74	3	5	0	2	3	1
	52236	8.1	8.0	7.3	5.8	1	89	70	91	3	5	0	1	3	1
P. megakarya	42100	9.6	4.4	7.2	5.6	2	12	0	0	2	5	0	5	1	5
	46728	9.5	6.3	5.5	12.0	2	58	63	0	2	5	2	5	2	5
P. megasperma	38828	10.9	10.0	11.0	4.4	3	104	47	38	1	5	2	3	3	1
	58969	9.8	6.6	6.1	5.0	0	95	50	62	1	5	2	5	3	5
	48609	6.9	7.0	3.7	3.3	0	46	38	53	1	5	0	5	3	2
	58817	7.4	7.0	7.4	6.6	1	38	96	47	1	5	3	5	1	5
	16706	5.5	7.8	2.0	3.0	0	46	60	54	2	5	2	5	1	5
	44390	12.8	10.0	10.0	8.4[j]	0	69	0	77	1	5	2	5	3	5
P. melonis	58427	4.0	2.0	13.0	7.3	3	80	91	0	2	6	0	5	2	5
	52854	8.4	12.0	10.0	8.0	3	125	38	0	1	6	0	3	2	3
P. mexicana	46731	7.2	9.0	7.2	8.3	2	89	108	29	2	5	0	5	4	2
P. nicotianae	38606	7.3	8.0	4.1	4.0	3	66	55	49	2	6	2	2	2	2
	38607	7.5	8.0	1.8	3.7	3	60	61	65	2	6	1	4	3	2
	60439	3.8	7.6	3.4	5.8	3	66	43	80	2	5	1	3	3	3
P. operculata[i]	44952	5.0	3.0	2.0	3.0	3	20	41	0	2	5	1	0	2	5
P. palmivora	26200	9.0	7.0	3.9	5.5	2	69	67	78	2	5	2	4	1	4
	26201	7.7	9.0	4.4	5.4	2	68	64	0	2	5	2	3	4	2
P. phaseoli	60171	4.3	2.0	1.0	0.0	3	0	0	0	2	4	0	0	0	0
P. polymorphica[i]	56966	8.2	9.0	6.7	4.7	3	95	69	97	1	5	3	3	2	3
P. porri	56616	6.3	8.0	4.2	2.2	0	0	14	97	1	4	3	5	3	5
	52729	5.8	7.0	4.7	4.1	0	15	0	0	1	4	3	3	4	4
	38830	5.3	5.0	3.6	2.4	0	8	7	89	1	4	3	3	4	4
P. primulae	56963	4.0	5.0	3.5	1.8	2	100	128	0	1	5	1	3	3	2
P. pseudotsugae	52937	8.0	10.0	7.8	6.4	3	72	13	72	1	4	0	3	2	3
	52938	8.0	8.0	5.6	6.0	3	37	11	77	1	4	0	3	2	1
P. quininea	56964	6.0	6.0	5.3	5.7	2	94	101	0	2	5	0	4	2	1
	46733	8.5	7.4	6.3	5.0	3	96	88	0	2	5	0	1	2	4
P. richardiae	46734	10.1	10.0	10.0	7.3	1	97	22	0	1	5	0	5	3	5
	60353	6.7	4.0	5.3	6.1	3	73	83	0	2	5	0	5	2	5
P. sinensis	46538	6.8	9.0	9.1	4.8	2	38	43	42	2	5	1	5	2	5
P. spinosa															
var. *lobata*[i]	28291	3.8	2.2	3.2	0.8	1	32	32	0	1	5	1	2	3	2
var. *spinosa*[i]	28294	2.3	4.5	4.2	3.8	1	133	138	0	1	5	0	1	3	1
P. syringae	16709	8.5	9.0	7.1	5.7	2	97	13	0	1	4	2	2	3	2
	14837	3.7	6.0	5.0	6.4	1	24	65	86	1	4	2	5	1	1
P. vesicula[i]	48641	4.3	5.0	4.1	3.1	2	64	62	64	2	5	0	2	3	3
P. vignae	46735	4.7	7.7	2.0	3.4	0	106	0	82	2	5	0	5	2	5
	42466	6.7	7.0	2.9	4.8	0	107	0	72	2	5	0	5	4	5

[a] By permission of Mycotaxon Ltd., Ithaca, New York, courtesy Luo et al. (1988).

[b] Daily growth rate (mm/day) on cornmeal agar (CM), V8 juice agar (V8), malt agar (malt), and a synthetic medium with an NO_3 source of nitrogen was calculated from the following formula: d2 – d1/number of days between d1 and d2; d1 = the colony diameter 3 days after inoculation, and d2 = the diameter 2–3 days later.

[c] Growth on wounded McIntosh apple fruit after 2 weeks: 0, no infection; 1, lesion diameter less than 20 mm; 2, lesion diameter 20–50 mm; and 3, lesion diameter greater than 50 mm.

[d] HMI = hymexazol, 50 µg/ml; and Mala. = malachite green, 125 µg/ml. Relative growth rate = (daily growth rate on either hymexazol or malachite green medium/daily growth rate on cornmeal agar medium) × 100.

[e] Starch hydrolysis index (S.H.I.) = (diameter of clear zone after flooding with iodine solution/diameter of fungal colony) × 100.

[f] Minimal temperatures: 1 = less than 5°C; 2 = 5–10°C; and 3 = 10–15°C. Maximum temperatures: 4 = 25–30°C; 5 = 30–35°C; and 6 = above 35°C.

[g] The degree of brown pigmentation produced on tyrosine-casein hydrolysate agar medium (Timmer et al. 1970b) after 3–4 weeks was ranked as 0, none; 1, slight; 2, moderate; or 3, abundant.

[h] Colony patterns on malt and potato-dextrose agars (PDA) were classified as 0, no growth; 1, radiate (uniformly radiate hyphae); 2, chrysanthemum (narrow petaloid sectors); 3, rose or camellia (broad, rounded petaloid sectors); 4, stellate (intermediate between radiate and chrysanthemum); or 5, uniform (without any obvious pattern). Growth pattern on V8 agar medium was classified as 1, not fluffy; 2, slightly fluffy; 3, moderately fluffy; or 4, highly fluffy.

[i] Reclassified from *Phytophthora* to the genus *Halophytophthora* (Ho and Jong 1990).

[j] Corrected from 84 in original Luo et al. (1988); courtesy H. H. Ho.

germination was not inhibited. Polymyxin B (200 units/ml) reduced germination by 40% (Banihashemi and Mitchell 1976). Soft PDA (1%) containing (in µg/ml) benomyl (10), nystatin (25), PCNB (25), rifampicin (25), ampicillin (500), and hymexazol (50) was not inhibitory to oospore germination of *P. castaneae* (Masago et al. 1977).

When oospores of *P. medicaginis* (*P. megasperma* f. sp. *medicaginis*) were tested for germination in aqueous solutions, compounds (in µg/ml) that permitted germination included ampicillin (1, 10, 100), chloramphenicol (1, 10, 100), oligomycin (1, 10), chlortetracycline (1), vancomycin (1, 10), hymexazol (1), PCNB (1, 10, 100), and benomyl (1, 10, 50, 100). Compounds and dosages at which oospore germination did not occur included (in µg/ml) polymixin B (1, 10, 50, 100), chlortetracycline (10, 50, 100), vancomycin (100, 200), and hymexazol (50). Penicillin G (10, 50, 100) reduced germination about 50% (H. Förster, D. C. Erwin, and O. K. Ribeiro, *unpublished*). Although germination was evaluated, no attempt was made to determine whether these germinated oospores would establish colonies. Since germination of oospores is usually much slower than germination of sporangia and chlamydospores, more research should be conducted to determine the sensitivity of oospores to antibiotics used in selective media. Oospores are the most important survival propagules produced by homothallic species such as *P. cactorum, P. citricola, P. medicaginis,* and *P. sojae.*

Isolation by Baiting with Selective Hosts

Since most *Phytophthora* species are often difficult to isolate from decayed tissue or from soil, the bait method has been used for nearly half a century to aid in isolation. Before the advent of selective media in the early 1960s, this was the only selective technique that was known. The bait, sometimes termed the "trap," consists of a highly susceptible host that is readily infected by a *Phytophthora* species (Figure 2.2). A summary of baiting techniques and plant material used for baits is given in Table 2.4 (Tsao 1983). Tsao (1983) thoroughly discusses this subject. Much of the material here has been drawn from his definitive chapter. In principle, the baiting method exploits the selective pathogenicity of a *Phytophthora* species to living host tissue on which the lesion caused by the target *Phytophthora* can be recognized. A representational but incomplete list of references includes work with apple fruit (Tucker 1931; W. A. Campbell 1949; Luo et al. 1988), unripe cacao pods (P. Turner 1965; Chee and Foong 1968), pear fruit (McIntosh 1964; Mircetich and Matheron 1976; Greenhalgh 1978), avocado fruit (Zentmyer et al. 1960; Zentmyer 1980), lemon fruit (Klotz and Fawcett 1939; Klotz and DeWolfe 1958; Tsao 1960), citrus leaf disks (Grimm and Alexander 1973), carnation petals (Ponchet et al. 1972), pine needles (Dance et al. 1975), soybean leaf disks (Canaday and Schmitthenner 1982), asparagus seedlings (Falloon 1982), *Persea indica* seedlings (Zentmyer et al. 1960; Zentmyer 1980), lupine seedlings (Chee and Newhook 1965a), alfalfa seedlings (Marks and Mitchell 1970), soybean seedlings (Eye et al. 1976), and apple seedlings (McIntosh 1966) or cotyledons (Harris and Bielenin 1986; Jeffers and Aldwinckle 1987). See also Table 2.4 and Ribeiro (1978), Tsao (1983), and Dhingra and Sinclair (1985). No single bait is optimally effective for isolation of all species.

Slightly green and firm fruits such as tart apples or pears are inoculated with suspect soil or plant tissue by inserting the soil or plant tissue into a core made in the fruit. The soil is then wetted with sterile distilled water, and the hole is covered with tape or petroleum jelly (Vaseline) to prevent loss of moisture. After the fruit is invaded by *Phytophthora*, the necrotic tissue characteristically remains firm; however, *P. gonapodyides* appears to be an exception, since it induces a soft rot (Luo et al. 1988) (Table 2.3). Tissues of fruits that are infected by *Pythium* or *Rhizopus* species are usually soft. Once the bait fruit is infected, it is likely that the target *Phytophthora* species will be the only fungus occupying the margin of the lesion. Transferring a small portion of this infected tissue to an agar medium usually results in a pure culture of *Phytophthora.* Thus, the highly susceptible fruit or other plant tissue acts as a selective medium.

Isolation from roots of Douglas fir trees by inserting whole roots into apples for a few days and subsequently transferring the rotted roots to a selective agar medium is more successful for isolating *P. cactorum, P. cryptogea, P. drechsleri,* and *P. megasperma* than plating root pieces directly on a selective agar medium (E. M. Hansen et al. 1980).

Although apple fruits are susceptible to a large number of *Phytophthora* species (Tucker 1931; W. A. Campbell 1949), Luo et al. (1988) note that *P. fragariae, P. infestans, P. iranica, P. meadii,* some isolates of *P. megasperma, P. porri, P. vignae,* and the closely related *Halophytophthora bahmensis, H. epistomium,* and *H. spinosa* var. *lobata* are not pathogenic to apple fruits (Table 2.3). *P. medicaginis* (*P. megasperma* f. sp. *medicaginis*) is also not pathogenic to apple fruit (D. C. Erwin, *unpublished*).

Although the bait techniques exploit the pathogenicity of *Phytophthora* species to a particular host or host tissue, methods that induce or favor the production of sporangia and zoospores in a soil or diseased tissue sample give an added advan-

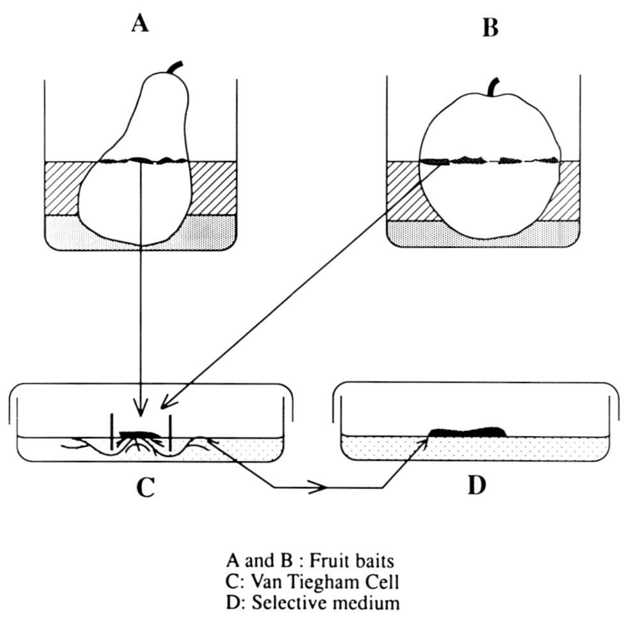

Figure 2.2. Bioassay for *Phytophthora* using pear (**A**) or apple (**B**) fruits as bait. Soil in the bottom of the beaker is flooded with water. *Phytophthora* usually causes lesions at the water line where zoospores accumulate. The infected tissue can be transferred to a glass Van Tiegham cell (**C**) placed in a plate of either plain water agar or a selective medium. To obtain a pure culture, transfer is made directly to a selective agar medium (**D**) or to water agar. The bioassay method is applicable for use with avocado fruit, *Persea indica* seedlings, leaf disks of various plants, alfalfa seedlings, lupine seedlings, and others (Table 2.4).

Table 2.4. Baiting techniques for the detection and isolation of *Phytophthora* species[a]

Species References	Bait material	Conditions of baiting and remarks
P. cactorum		
Schwinn (1961a)	Apple fruit	Buried partly in wet soil in large flat; 24°C, 90–100% RH, 5–6 days.
McIntosh (1964)	Pear fruit	Immersed partly in 50 cc of soil plus 100 cc of water; 6–7 days; pear better than apple, peach, and apricot fruits as bait.
Van der Scheer (1971)	Apple fruit, pear fruit	Buried partly in wet soil at 20°C until rot occurs; isolation from apple and strawberry soils; pear better than apple as bait.
Sewell et al. (1974)	Apple fruit	Immersed partly in 20-mm soil layer with 5 mm of water above soil; optimum 20°C, 14 days; *P. citricola* and *P. syringae* also recovered from apple soils at lower incubation temperatures.
Banihashemi and Mitchell (1975)	Safflower seedlings	Planted in steamed soil before adding 15-mm layer of test soil in contact with hypocotyl or directly to test soil; optimum 28°C, 3–4 weeks; better than apple fruit; used also in quantitative assays of apple soils.
Molot and Beyries (1976)	Strawberry fruit	Immature fruit inoculated with soil in small hole; 22°C, 3–6 days; used for isolation from strawberry soils; also susceptible to five other *Phytophthora* species.
Harris and Bielenin (1986)	Apple seedlings, cotyledons	Soil suspensions were diluted in up to 72-fold dilutions; seedlings and cotyledons incubated in 15–20 ml of suspension in glass vials; 14 days at 20°C in 16-h photoperiod.
Jeffers and Aldwinckle (1987)	Apple seedlings, cotyledons	Extended baiting (soil air dried, moistened for 3 days, and then flooded). See Stack and Millar [1985a] under *P. megasperma.*
P. cambivora		See Dance et al. (1975) under *P. cinnamomi* and Mircetich and Matheron (1976) under *P. megasperma.*
P. capsici		See Lee and Varghese (1974) and Ponchet et al. (1972) under *P. parasitica.*
Anandaraj and Sarma (1990)	*Albezia falcataria* leaflets	Leaflets were floated (adaxial side down) on soil suspension (25 g of soil/100 ml of glass-distilled water) at 26°C for 72 h. Infected leaves turn black.
Larkin et al. (1995)	Pepper (*Capsicum*) leaf disks	Leaf disks floated on saturation water of soil for 24 h and plated on Masago medium.
P. cinnamomi		
W. A. Campbell (1949)	Apple fruit	Holes made in fruit are filled with soil; incubated at 15–27°C, 5–10 days; suitable also for many other but not all *Phytophthora* species; however, *Pythium* and other soil fungi also colonize fruit.
Anderson (1951b)	Pineapple leaves and rooted crowns	Immersed in 400–800 cc of water added to about 10–20 cc of soil; 21–24°C, 4 days.
Zentmyer et al. (1960); Zentmyer (1980)	Avocado fruit, *Persea indica* seedlings	Fruit partly embedded in flooded soil; 27°C, 2–4 days; seedlings planted in wet soil, 2–3 days.
Klemmer and Nakano (1962)	Pineapple leaves and rooted crowns	Leaf base or root tips immersed in water-soil mixture at least 5 cm above the container bottom; 18–27°C.
Chee and Newhook (1965a)	Lupine radicles, excised	Root tips immersed in large volume of water over a 20-mm soil layer; 2 days; isolation from pine soils; better than apple fruit; also susceptible to four other *Phytophthora* species.
Marks et al. (1972)	Lupine radicles, excised	Pieces 1–2 cm long floated on water 2–3 mm over 30 cc of soil in petri plate; 2–3 days; used for quantitative assays of eucalyptus soils.
Pratt and Heather (1972)	Lupine radicles	Immersed in 150 cc of water added to 26 g of soil; 17–24°C, 5 days; lesions distinguishable from those caused by eight other *Phytophthora* species except *P. drechsleri.*
Gerrettson-Cornell (1974)	Apple slices, lupine radicles	Immersed in 200 cc of water added to 25 g of soil; 4–10 days; both baits were satisfactory.
Marks and Kassaby (1974)	Eucalyptus cotyledons	Detached cotyledons floated on water 5–10 mm over 50–60 cc of soil; 22–24°C, within 60 h; two to four times more sensitive than lupine radicles; used for quantitative assays of eucalyptus soils.
Dance et al. (1975)	Cedar needles, pine needles, lupine radicles	Floated on or immersed partly in water added to soil; both 16 and 22°C, 3 days; conifer needles also good for *P. cambivora, P. citricola, P. cryptogea, P. drechsleri, P. hibernalis,* and a few other *Phytophthora* species.
Linderman and Zeitoun (1977)	Eucalyptus leaf disks	Floated on water added to soil or roots; 20°C, 24 h; *Pythium* species interfere with results.
Pegg (1977a)	Avocado leaf pieces	Floated on water added to soil; 4 days; recovery up to 89%.
Greenhalgh (1978)	Eucalyptus cotyledons, lupine seedlings, pear fruit	Floated on water added to soil at various depths; 25°C under continuous light, 3–7 days; all baits successful but more frequently with pear; used in quantitative assays.
Zentmyer and Ohr (1978)	*Persea indica* seedlings	Immersed partly in large volume of water added to 1.2- to 2.5-cm soil layer; 5–7 days; isolation from avocado soils.
Shew et al. (1979a)	Azalea leaf disks, cedar needles, fir seedlings, lupine radicles	Floated on or immersed in 100 ml of water added to 5–50 g of soil; 20°C, 1–2 days; recovery was best with azalea leaf disks and least with lupine radicles.
P. citricola		See Dance et al. (1975) under *P. cinnamomi* and Sewell et al. (1974) under *P. cactorum.*
P. citrophthora		See Klotz and DeWolfe (1958) and Tsao (1960) under *P. parasitica.*
P. cryptogea		See Dance et al. (1975) under *P. cinnamomi.*
P. drechsleri		
Alavi and Strange (1979)	Cantaloupe seed and seedlings	Planted in wet soil mixed with diseased tissues of *Cucumis* species; 20–35°C, 7 days.
P. fragariae		
Duncan (1976, 1979, 1980a)	Strawberry rooted runners and seedlings of Baron Solemacher and *Fragaria vesca* clone VSI	Planted to wet soil in pots or in the field; 5 weeks or longer; used in quantitative assay of strawberry soils.

(*continued on next page*)

Table 2.4. *(continued)*

Species References	Bait material	Conditions of baiting and remarks
P. gonapodyides		
Robertson (1975)		Apples and pear baits were suspended in reservoirs for 9 days; *Begonia* petioles, citrus leaf pieces, lupine seedlings were suspended in water from reservoirs for 1–4 days and plated on P_{10}ARP.
P. heveae		See Lee and Varghese (1974) under *P. parasitica.*
P. hibernalis		See Dance et al. (1975) under *P. cinnamomi.*
P. infestans		
Lacey (1965)	Potato tuber slices	Soil placed on wet surface of tuber slice, 20°C, for 6 days.
P. lateralis		
Ostrofsky et al. (1977)	Port Orford cedar tissue	Branchlet or root sections floated on water added to soil or organic matter in petri plate or inserted deeper in soil; 18°C, 6 days; better than seedlings.
Hamm and Hansen (1984)	Port Orford cedar tissue	Baits floated over soil sample in water amended with hymexazol (25 mg/liter).
P. meadii		See Chee and Foong (1968) under *P. palmivora.*
P. medicaginis (= *P. megasperma* f. sp. *medicaginis*)		
Marks and Mitchell (1970)	Alfalfa seedlings, injured	Floated on water 2–3 mm over 30 cc of soil in petri plate; 20–22°C, 6 days; used for quantitative assays of alfalfa soils.
Stovold and Carratt (1978)	Alfalfa seedlings, injured	Same as in Marks and Mitchell (1970), except that nylon mesh is placed below seedlings to prevent roots from becoming enmeshed in the soil.
Stack and Millar (1985a)	Alfalfa seedlings	Same as Marks and Mitchell (1970), except that seedlings are not injured. Soil was dried to 5% moisture, wetted to field capacity for 3 days, and then flooded with water before the alfalfa seedlings were floated on the water (extended seedling bait assay).
Dale and Irwin (1990)		Modified from Marks and Mitchell (1970). Excised shoots of 6-day-old alfalfa seedlings were floated on 60 ml of soil in 150-ml cups with 2- and 10-fold dilutions in water. Field soil was diluted 2- and 10-fold. Sporangia on alfalfa shoots (6-day incubation) data were analyzed by Maximum Likelihood Program (most probable number of infective units per gram of soil).
P. megasperma		
Mircetich and Matheron (1976)	Pear fruit	Partly immersed in water 1 cm deep over 500 cc of soil; 20°C for 3 days; for isolation from cherry soils; also good for *P. cambivora.*
Hargreaves and Duncan (1978)	Pine needles, lupine seedlings, apple fruit, pear fruit	Needles or seedlings placed in water 5 mm deep over 100 g of soil; 16 and 22°C, 3 days; fruit buried in larger amounts of soil; 22°C, 10 days; for isolation from strawberry and raspberry soils.
P. megasperma f. sp. *glycinea*		See *P. sojae.*
P. megasperma f. sp. *medicaginis*		See *P. medicaginis.*
P. melonis		
Katsura (1971)	Cucumber fruit	Immersed partly in water 1 cm deep over 20–30 g of soil; 20 and 28°C, within 5 days.
P. palmivora (all morphological forms)		
Orellana (1954)	Cocoa pods	Placed on or in soil or inoculated with a loopful of soil suspension on the pod surface and incubated in a moist box; 1–4 days; for isolation from cocoa soils.
Holliday and Mowat (1963)	Apple fruit, black pepper leaves	Soil inserted into apple the same as in W. A. Campbell's (1949) method. Black pepper leaves immersed partly in water-soil mixture; for isolation (of MF4) from black pepper soils. Black pepper leaf disks used successfully by P. H. Tsao (*unpublished*).
P. D. Turner (1965)	Cocoa pods	Soil inserted beneath flaps of endocarp tissue of unripe pods; for isolation from cocoa soils.
Chee and Foong (1968)	Cocoa pods	Soil or diseased *Hevea* tissues inserted into unripe green pod as in W. A. Campbell's (1949) apple method; 26–30°C, 4–5 days; also good for *P. meadii* and *P. parasitica.*
Okaisabor (1971a, b)	Cocoa pod tissues	Small blocks of tissue placed in wet soil; 25°C, 4 days; premoistened for 5 days if soil samples were dry.
Newhook and Jackson (1977)	Cocoa pods and tissues	Four methods: pod on soil, soil in pod, pod tissue in flooded soil (the best), and pod tissue on soil; used in quantitative assays of cocoa soils.
Satyaprasad and Ramarao (1980)	*Colocasia esculenta* roots about 2.5 cm long	Roots autoclaved and incubated in moistened soil (50 g/petri plate) at 15°C for 1 week. Roots washed and plated on oatmeal agar.
P. parasitica (*P. nicotianae*)		
Klotz and DeWolfe (1958)	Apple fruit, lemon fruit, orange fruit	Soil or citrus tissues inserted into apple the same as W. A. Campbell's (1949) method. Lemon or orange placed on surface of saturated soil; 20°C, 4 or more days; also good for *P. citrophthora* and *P. syringae* from citrus soils.
Tsao (1960)	Lemon fruit	Immersed partly in 150 cc of water added to 25 cc of soil; 25°C, 6 days; also good for *P. citrophthora;* used for quantitative assays of citrus soils.
Jenkins (1962)	Tobacco leaves	Petiole end immersed in water-soil mixture as in Tsao (1960); used for quantitative assays of black shank fungus in tobacco soils.
Narasimhan and Ramakrishnan (1969)	Castor bean seed	Buried in soil; 63 h; for isolation from betel vine soils or irrigation water.
Ponchet et al. (1972), Ricci (1972)	Carnation petals	Immature petals from buds floated on soil suspension in petri plate, 24–35°C, 2–4 days; used for quantitative assays of carnation soils; also detects *P. capsici.*
Grimm and Alexander (1973)	Citrus leaf pieces	Small leaf pieces from various *Citrus* species, floated on water 1–2 cm above 100 cc of soil; 22–28°C, 3–4 days; calamondin fruit equally effective as bait.
Lee and Varghese (1974)	Apple fruit, eggplant fruit	Methods not given. Both fruits also good for *P. capsici.* Cocoa pod also allows isolation of *P. palmivora* and *P. heveae.*
Ioannou and Grogan (1977)	Tomato fruit	Green fruit placed in flooded soil from tomato fields after premoistening for 1 week.

(continued on next page)

Table 2.4. (*continued*)

Species References	Bait material	Conditions of baiting and remarks
P. sojae (= *P. megasperma* f. sp. *glycinea*)		
Eye et al. (1976)	Soybean seedlings	Placed in 30–40 cc of water added to 20–40 g of soil in petri plate; 3 days; for isolation from soybean soils.
Paxton (1976)	Soybean seed	Buried in wet soil; 20–30°C, 7 days; for isolation from soybean soil.
Canaday and Schmitthenner (1982)	Soybean leaf disks	Soil from soybean soils in the field moistened to −100 to −200 mb matric potential and stored for 1–2 weeks after which leaf disks were floated on 50 cc of water added to 12 g of soil for 3 h. Leaf disks blotted dry and plated on selective medium or observed directly for sporangia.
P. syringae		See Klotz and DeWolfe (1958) under *P. parasitica* and Sewell et al. (1974) under *P. cactorum.*
Unspecified *Phytophthora* species		
Waterhouse and Stamps (1969)	19 different plants	Seeds, seedlings, leaves, or fruits floated on or immersed in water-soil mixture; for isolation of various species of *Phytophthora* and *Pythium.*
Banihashemi (1970)	Cucumber fruit	Immersed partly in water 1 cm deep over 40–60 g of soil; 20°C, 4–7 days; for isolation of *Phytophthora* species causing cucurbit root rot. Cantaloupe, muskmelon, and watermelon fruits also are satisfactory.
Fatemi (1974)	Hemp and safflower seeds	Boiled seeds placed on nylon screen over 10–15 mm layer of saturated soil; 20°C, 2–3 days; good for various species of *Phytophthora* and *Pythium.*

[a] Adapted from Tsao (1983).

tage in the isolation technique because the original inoculum is amplified by the production of zoospores. Swimming zoospores are usually attracted chemotactically to a nutrient source (the bait) such as a root or plant tissue (Zentmyer 1970). Zoospores that move to an attractive site probably stick to or agglutinate at a particular site before they encyst. The ability to stick to a surface is a unique characteristic (Sing and Bartnicki 1975a, b).

Soil can be assayed for *Phytophthora* by placing unwounded fruit such as apples, pears, lemons, or avocados in a vessel such as a plastic or paper drink container of suitable size and inundating it with water (Appendix 2.7; Figures 2.2–2.4). The ratio of water to soil should be at least four or greater (Tsao 1983) to dilute any substances in the soil such as fertilizers or biocides that might be toxic to *Phytophthora.* Approximately one-third to one-half of the fruit bait should remain above the water line because zoospores characteristically migrate to the surface of water (Carlile 1983). After incubation of the fruits for several days, firm, brown lesions usually occur at the water line (Figure 2.2). The surface of the fruit is then disinfested with 0.5% sodium hypochlorite bleach or 70% ethanol, and infected tissue from the margin of the lesion is transferred to a selective medium or, if unavailable, to water agar (Figures 2.2 and 2.3). Although the fruit trap method has been used widely, a recent report indicated that apple seedlings or cotyledons are much more successful as baits for *P. cactorum* than apple or pear fruits (Jeffers and Aldwinckle 1987).

Dripping water continuously for 24–48 h on diseased citrus root or bark samples has been used to induce sporangium formation of *P. citrophthora* and *P. parasitica.* Lemons in the "silver stage" (between the green and ripe stages) are placed near the root samples to receive the splash from the diseased tissue; brown rot formation on the lemons indicates zoospore infection (Klotz and Fawcett 1939; L. J. Klotz, *personal communication* in the early 1950s). The diagnosis is confirmed by

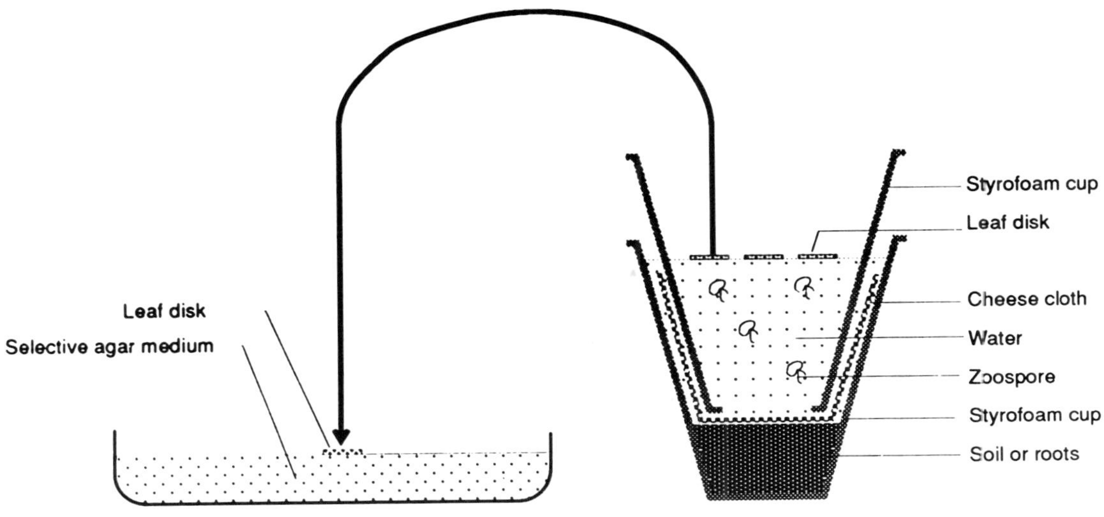

Figure 2.3. Double cup soil bait test in which *Eucalyptus* leaf disks are used as bait to isolate *Phytophthora.* After incubation, the leaf disks are transferred to a selective agar medium (Linderman and Zeitoun 1977).

microscopic examination of the fungus on the lesion and/or by transferring necrotic tissue to either water or water agar. Because the flesh of lemon fruit is acidic, pure cultures of *Phytophthora* can readily be obtained by culturing seeds from infected lemons. Since selective media were not available at that time, these methods were efficacious for obtaining pure cultures (see also Klotz and Fawcett 1939; Klotz and DeWolfe 1958; Tsao 1960).

Other baits such as soybean leaf disks (Canaday and Schmitthenner 1982), seeds, or seedlings (Marks and Mitchell 1970; Eye et al. 1976; Paxton 1976) are more specific to only one species (e.g., *P. sojae*). Despite this host specificity, these baits also attract *Pythium* species.

Prewetting of soil for various periods of time (e.g., Canaday and Schmitthenner 1982; Stack and Millar 1985c; Jeffers and Aldwinkle 1987) before submersion in water favors the production of sporangia. Larkin et al. (1995) report that prewetting soil before dilution plating or baiting with pepper leaf disks favors detection of oospore inoculum of *P. capsici*. *P. drechsleri* can form sporangia in soil as dry as –300 mb matric potential and also release zoospores. Although free water favors formation of sporangia, lack of aeration in water (submersion of a mycelium-agar disk to a depth of 5 mm in saturated sterile or nonsterile soil) suppresses formation of sporangia (Duniway 1975a). Duniway suggested that for bait isolation of *Phytophthora* from soil in soil-water suspensions, zoospores are probably the propagules that infect the plant tissue bait. Thus, it is possible that if sporangium and zoospore formation is suppressed by submersion in water, more moderate wetting of the soil sample prior to submersion in water might enhance the isolation of *Phytophthora* by the baiting method. Since Duniway conducted the experiments comparing sporangium formation on mycelial disks incubated on the surface of saturated soil with submersion to a 5-mm depth in both sterile and nonsterile soil, he concluded that the lack of aeration was caused by the low solubility of oxygen in water rather than by the soil microflora.

P. sojae (*P. megasperma* f. sp. *glycinea*) produces sporangia and zoospores after air-dried soil has been wetted to field capacity, incubated at room temperature for 1–2 weeks, and flooded with water for as little as 1 h. This method exploits the production of sporangia in the moist soil and subsequent isolation of the infective units (zoospores) on floating soybean leaf disks. *Phytophthora* colonies are detected after transfer to a selective medium. When leaf disks are floated for a short time (about 1 h), *P. sojae* is more frequently detected than *Pythium* species, but at longer incubation times (about 24 h), *Pythium* species are also detected on the leaf disks (Canaday and Schmitthenner 1982) (Appendix 2.8). This result supports Schmitthenner's (1973) earlier observation that *Phytophthora* species produce zoospores more rapidly and abundantly than *Pythium* species.

Most other reports recommend incubating leaf disks in water for much longer periods (e.g., 24–48 h) before they are transferred to a selective medium. Eucalyptus leaf disks are incubated in the test soil suspension for a 24-h period before

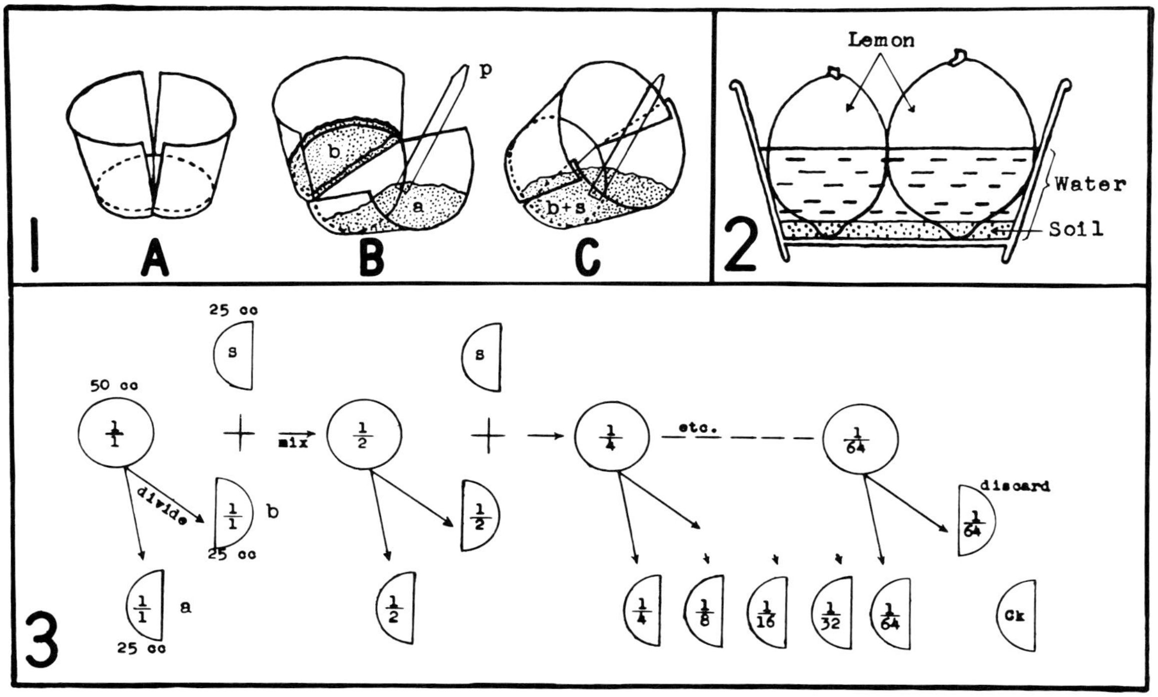

Figure 2.4. Semiquantitative serial dilution end-point bioassay for *Phytophthora*. **1A,** The "split-cup" is made from a 16-ounce paper cup by cutting the sides to form two connected, movable, equal halves. A different cup is used for each soil sample. **1B,** The split-cup shown in positions as a soil divider (left) and a soil mixer (right). The 50 cc of soil, after having been leveled at the bottom of the cup, is divided into two equal halves, as shown. One-half of the soil (a) is removed with the aid of a pot label stick (p) into a given dilution cup for the lemon trap test. To the remaining half (b), 25 cc of sterile soil (s) is then added and mixed thoroughly (**1C**) to provide 50 cc of soil (b+s) of the next higher dilution. The mixing is done in the split-cup with a pot label stick by a circular motion while the cup, with portions of the movable halves overlapping, is held at an angle. Efforts are made by scraping with the pot label to avoid adherence of wet soil particles on the cup walls. **2,** The lemon trap test with dilution cup, showing the position of two lemon fruits, 150 cc of water, and 25 cc of soil in the cup. **3,** Procedures involved in dividing and mixing the soil in the serial dilution end-point method. The eight semicircles in the bottom row, each representing 25 cc of soil, constitute a single replicate dilution series. (Reproduced by permission of the American Phytopathological Society, St. Paul, Minnesota, courtesy Tsao 1960)

they are transferred to a selective medium to isolate *P. cinnamomi* (Linderman and Zeitoun 1977) (Figure 2.3).

If the suspected *Phytophthora* species is not known, more than one type of bait should be used, since no single bait isolates all *Phytophthora* species present. Hargreaves and Duncan (1978) used a combination of pine needles, apple fruit, pear fruit, and lupine seedlings to isolate *Phytophthora* species from soil in infected red raspberry (*Rubus idaeus* L.) fields.

Eucalyptus (*Eucalyptus sieberi* L. A. S. Johnson) seedling bait is used by many laboratories. Six-week-old eucalyptus cotyledons are floated over samples of flooded soil to bait *P. cinnamomi,* which is subsequently identified by direct microscopic observation or plating of the infected cotyledons on a selective medium (Marks and Kassaby 1974). Leaf disks of *E. gunnii* Hook., *E. globulus* Labill., or *E. cinerea* F. Meull. ex Benth. are efficient baits. The soil or root sample is placed in the bottom of a polystyrene cup and covered with a layer of cheesecloth, which extends over the soil and sides of the cup. A second cup with the bottom removed is inserted to hold the cloth in place, after which the cups are flooded with distilled water (Figure 2.3). When three to 10 leaf disks (0.5 cm in diameter) were floated on the surface of the water and removed 24 h later, blotted dry, and plated on a selective medium, *P. cinnamomi* was the predominant species recovered (Linderman and Zeitoun 1977).

The serial dilution end-point technique semiquantitatively estimates the population of *P. parasitica* from citrus soils (Tsao 1960) (Appendix 2.7). Infested soil is diluted with sterilized soil in a series, for instance, 1/2, 1/4, 1/16, . . . , 1/64. The soil samples are then incubated in water and baited with a lemon fruit in the silver (slightly green) stage of ripeness. The disease potential index (DPI) is defined as the reciprocal of the highest dilution that yields brown rot lesions on lemon fruits. For instance, if brown rot on citrus fruit is noted on fraction 1/16 but not on higher dilutions, the DPI is 16. Although the method has merit because it is semiquantitative and is still being utilized (Magnano di San Lio and Perrotta 1983), the large numbers of replications, the amount of laboratory space, and the time required render it more time consuming than dilution plating of soil on a selective medium; it is also less quantitative (Tsao 1983; Larkin et al. 1995). See Appendix 2.7 and Figure 2.4 for more specific information.

The most probable number method, reviewed by Menzies (1963) for use in quantitating soilborne fungi, is discussed by Tsao (1983). He states that because the method is time consuming and the soil dilution plate method with selective media is more efficient, the most probable number method has been infrequently used for *Phytophthora.* The method has been used for quantitating *P. nicotianae* (Ricci 1972), *P. capsici* (Papavizas et al. 1981), and *P. medicaginis* (Dale and Irwin 1990).

Weste (1983b) defined DPI as the smallest fraction of undiluted soil that yields *P. cinnamomi* when baited with a susceptible host. Her methods were modified from the serial dilution end-point method of Tsao (1960). The soil sample is fractionated, but not diluted with sterile soil, in a series ranging from the original unfractionated sample (50 g) to 1/256 of the original 50-g sample. The fractions are incubated in 100 ml of deionized distilled water in plastic cups (three replicates). Three susceptible, germinated seedlings (radicles about 1 cm in length) of *Lupinus angustifolius* L. (Chee and Newhook 1965a) or *E. sieberi* (Marks and Kassaby 1974) are placed in each cup as bait (Weste and Ruppin 1977) and incubated at 20–25°C. After a 48- to 72-h incubation period, the seedlings are blotted dry and plated on a selective agar medium. The DPI value is the reciprocal of the highest fraction that yields *P. cinnamomi.* For instance, if *P. cinnamomi* is isolated from the 1/256 fraction, the DPI is 256 (see Appendix 2.7). Other methods and other uses of quantitative and semiquantitative methods of isolation are discussed by Weste (1983b) and Tsao (1983).

Alfalfa seedlings are successfully used for baiting *P. medicaginis* (*P. megasperma* f. sp. *medicaginis*) (Marks and Mitchell 1970; Pratt and Mitchell 1973; Wilkinson and Millar 1982; Stack and Millar 1985a, c) (see abbreviated assay in Appendix 2.8). Alfalfa seeds, surface sterilized in a solution composed of ethanol and sodium hypochlorite (0.5%) (1:1) for 15–20 min, are germinated on moistened filter paper in petri plates. The stems of the germinated seedlings are pinched below the cotyledonary leaves with a forceps to slow the growth (Marks and Mitchell 1970). Seedlings are then placed in petri plates containing the soil (30 ml) in 40 ml of water for 2–3 days. Sporangia are detected by microscopic examination of the seedling roots, or mycelial growth is detected by plating seedlings on a selective medium. This technique is essentially the same as the abbreviated seedling bait assay of Stack and Millar (1985a, c) (Appendix 2.8), except that the alfalfa seedlings were not pinched before being used as a bait.

Tsao (1983) suggests that the alfalfa seedling baiting method might be improved by increasing the water-soil ratio. Also, shorter incubation times (Canaday and Schmitthenner 1982) might facilitate more selective isolation of *Phytophthora* from soil that is also infested with *Pythium* species.

Both *P. medicaginis* and *P. sojae* (*P. megasperma* f. sp. *glycinea*) were isolated by a host-enrichment baiting technique from soil collected in a field in Wisconsin that had been under an alfalfa-soybean-corn rotation (Nygaard et al. 1989). The soil (100 ml) was placed on top of a 50-ml portion of sterilized vermiculite in a 250-ml beaker and then flooded with water. The excess was poured off. Alfalfa seed (susceptible cultivar Saranac) was planted in another layer of vermiculite and incubated at room temperature (about 22°C). When seedlings emerged, the soil was flooded, and the seedlings then damped off. These seedlings were surface disinfested with 0.5% sodium hypochlorite, air dried, and plated on P_5ARP (Appendix 2.4). Although sporangia could be seen on the isolated alfalfa seedlings, certain *Pythium* species were dominant. After eight cycles of recropping the soil with alfalfa, however, *P. medicaginis* became the dominant species isolated on diseased alfalfa seedlings. When a comparable subsample of the same soil was baited for *P. sojae* via soybean (susceptible cultivar McCall) seedlings, the pathogen was isolated infrequently; however, when cotyledons from 3-day-old soybean seedlings were wounded and incubated on the flooded soil for 2–3 days, *P. sojae* was isolated from red brown lesions at the wound sites on cotyledons (Nygaard et al. 1989). This example was the first report of isolation of both *Phytophthora* species from the same soil sample and shows how different *Phytophthora* species can be selectively isolated.

Dale and Irwin (1990), using a most probable number assay with alfalfa seedling bait similar to the method used by Marks and Mitchell (1970) (Table 2.4), report inoculum levels of *P. medicaginis* ranging from 0.01 to 10.95 infective units per gram of soil in fields cropped to susceptible chickpea (*Cicer arietinum* L.). Use of the extended assay (Appendix 2.8) did not augment the production of inoculum in the soil (Dale 1991).

The extended seedling bait bioassay amplifies the inoculum in soil by favoring the germination of dormant propagules prior to isolation. Isolation of *P. medicaginis,* an abundant oospore producer, involves drying soil samples (30 ml) in deep glass petri plates for 7 days to about 5% moisture (Stack and

Millar 1985c). The soil is rewetted to about field capacity with 10 ml of distilled water and incubated for 3 days at room temperature; 40 ml of water is then added. Six 3- to 5-day-old unwounded alfalfa seedlings are floated on the water for 3 days. The seedlings are examined for sporangia, blotted dry, and plated on a selective medium (Appendix 2.3) modified from that of Masago et al. (1977). Although drying the soil could reduce the viability of some *Phytophthora* propagules, it possibly stimulates oospore germination and apparently reduces the competitive effect of other soil microflora such as bacteria that might be killed by drying (Stack and Millar 1985c).

The extended seedling bait assay also yielded more *P. cactorum* propagules and in a higher percentage of samples than the abbreviated baiting method for isolation from soil. *P. cactorum* is also an abundant oospore producer. When apple seedlings or cotyledons were used as baits, the extended baiting method detected *P. cactorum* in 85% of 204 samples compared with only 6% for the abbreviated or direct baiting method (Jeffers and Aldwinckle 1987). Zoospore release from sporangia of *P. cactorum* on remoistened, air-dried soil was greatest at 8°C (Rahimian and Mitchell 1988). The low temperature allowed *P. cactorum* to be detected preferentially over *Pythium,* which does not produce zoospores at such low temperatures (Appendix 2.8).

If quantification of the inoculum that exists per se in soil is the primary purpose of isolation, the extended baiting method may overestimate the inoculum density because the primary inoculum, which probably exists as oospores or in some species as chlamydospores, will produce sporangia and subsequently many zoospores, each of which will produce a colony-forming unit. The method thus amplifies the primary inoculum and exemplifies the multiple cycle effect or rapid regeneration capacity so common to the epidemiology of *Phytophthora*-caused diseases. On the other hand, since drying the soil would probably kill mycelium, zoospores, and sporangia, the extended method would be unlikely to detect these propagules (see the discussion by Tsao [1983]). The results of Jeffers and Aldwinkle (1987), above, in identifying *P. cactorum,* however, suggest that this method is advantageous for diagnostic purposes.

P. clandestina, a pathogen of subterranean clover (*Trifolium subterraneum* L.), is not easily isolated in pure culture by conventional means. It was initially detected by observing sporangia on roots incubated in distilled water (Taylor 1984; 1985a). Propagules of *P. clandestina* were assayed in soil by use of the most probable number method, in which the air-dried soil was screened (5-mm sieve) and diluted with a yellow subsoil sand in twofold dilutions of 0, 1/2, . . . , 1/64 (Wong et al. 1986d). Clover seeds were planted in pots of soil that were saturated with water and maintained at 18°C. After 2 weeks, the seedlings were removed, roots were wounded, and 2-mm-long pieces were incubated in distilled water to detect formation of sporangia (Taylor et al. 1985a). Roots were examined 24, 48, and 72 h later. The most probable number was significantly highest in the samples of soil collected in May, which, in the Southern Hemisphere, is the autumn season. Surprisingly, the root disease index of the seedlings remained high, while the most probable number of propagules declined sharply.

Since the soil was air dried before saturation, the method resembles, to some extent, the extended soil bioassay (Appendix 2.8). When soil is air dried, the effective population detected is probably the result of the sporangia and zoospores that have emanated from dormant oospores and/or mycelium in infected root tissue.

Port Orford cedar branchlet bait is used to isolate *P. lateralis* from soil. Recovery is much improved by isolation from the organic matter separated from soil by wet-sieving (Ostrofsky et al. 1977) (Appendix 2.9). The washing process in the wet-sieving method may remove many of the bacteria and actinomycetes from the soil-organic matter suspension.

Diseased roots are often used to infest steamed soil in which seeds of a susceptible plant are planted. *Phytophthora* can then be isolated from newly infected tissue more readily than from the original roots. Alavi and Strange (1979) used this method to isolate *P. drechsleri* from diseased cantaloupe roots.

Healthy strawberry roots of a susceptible cultivar are used to bait and detect *P. fragariae* in soil (Montgomerie and Kennedy 1972; Duncan 1976) or in the field. Detection of *P. fragariae* is based on production of the red stele symptom and occurrence of oospores in the roots of the test plant. A modification of the "root tip bait" test, developed for plant certification, is capable of detecting one infected root in 99 healthy root systems (Duncan 1979).

Floating current-season needles of *Pinus radiata* and *Cedrus deodora* on flooded soil samples (Dance et al. 1975) allowed detection of 12 *Phytophthora* species, including *P. cactorum.* A novel technique to detect *Phytophthora* species in the field was based on *P. radiata* needles as bait (McCracken 1985). A hole was cored in the soil (25 cm deep) near apple trees. A vessel, 7 mm in diameter and 10 cm in height, containing *P. radiata* needle bait in distilled water was placed in the hole. A trough about 50 cm long was placed at an angle in the soil so that after a rainfall, the soil water drained into the vessel. The bait needles in the vessels were collected every 7 days for 14 months and plated on P_{10}VPH agar, after which the *Phytophthora* colonies were counted and identified. *P. syringae, P. cactorum, P. citricola,* and an unidentified *Phytophthora* species were consistently isolated. The frequency of isolation was greatest when the soil temperature was low (6–14°C) and least when it was above 18°C. This may be caused by production of zoospores at low temperatures. Surprisingly, frequency of isolation was not correlated with increased rainfall.

Isolation from Plant Tissue

When *Phytophthora* is suspected, the affected tissue selected for isolation should ideally be in an active stage of infection, since *Phytophthora* is especially difficult to isolate from necrotic tissue. The reasons for this difficulty are generally thought to be the suppressive effects of antagonistic bacteria and the absence of active viable mycelium. T. A. DeWolfe (*personal communication*) likened a *Phytophthora* infection to a prairie fire in which the activity is mainly at the margin. Thus, infected material for isolation should be selected from margins of plant tissue with actively progressing lesions, and tissue that has dried or has been corked off from healthy tissue should be avoided.

Successful isolation of *Phytophthora* from diseased plant tissue can be aided by careful preparation; however, if selective media are used, surface decontamination may not be necessary. Presumptive identification or detection of *Phytophthora* can be made by incubating well-washed (not surface-sterilized) roots, stems, or leaves in a shallow layer of pond water (Waterhouse and Stamps 1969), mineral salts solution (Chen and Zentmyer 1970), soil extract (10 g/liter of water), or sterile distilled water. If sporangia are present after 24–48 h, the sporangia-forming sector of the root can be transferred to a selective medium.

Although symptoms are not always dependable, almost all lesions on fruit, bark, or leaves are brown, purplish brown, or almost black, depending to a large extent on the pigments in the host tissue; the infected tissue is usually firm and not soft. Soft rots are more often caused by various bacteria, *Rhizopus* species, or *Pythium* species (Tsao 1983). In infected citrus roots, bark tissue can be readily stripped off, leaving a thread of the woody xylem tissue. *Phytophthora* does not usually invade the xylem tissue of stems or trunks of trees such as citrus; however, Shea et al. (1982) showed that *P. cinnamomi* extensively invades wood in vertical roots of Jarrah trees (*Eucalyptus marginata*) in Western Australia. *P. fragariae,* cause of the red stele disease, infects xylem tissue of strawberry. On diseased avocado trees, the fibrous roots become black and, because of their brittle texture, are readily broken (Zentmyer 1980).

Surface Preparation of Diseased Tissue

Diseased roots should be washed thoroughly in running water; often washing for several hours is helpful. This not only removes bacteria but also stimulates formation of sporangia.

If roots are well washed, chemical surface decontamination may not be necessary and could be undesirable because the target *Phytophthora* species could be killed (M. Stanghellini *personal communication;* Waterhouse and Stamps 1969; Tsao 1983; Weste 1983b). It is extremely important after washing or surface disinfestation to blot diseased tissue to dryness on clean paper towels before transferring it to an agar medium, because bacteria multiply rapidly in free water and tend to suppress growth of *Phytophthora* (Weste 1983b). Since selective antibiotics suppress but do not necessarily eliminate all contaminants, a selective medium should not be expected to substitute for other sanitation.

To isolate *Phytophthora* species in the presence of *Pythium* species, diseased roots can be washed in a sieve placed over a beaker with running water after adding 1 ml of Tween 20, a wetting agent to loosen the soil. Washing is continued until the water no longer foams (about 10 min). Roots are then incubated in a 1:2 dilution of pond water and sterile distilled water or in Schmitthenner's salt solution (Chapter 3, Appendix 3.5) in a petri plate to stimulate production of sporangia. These methods or variations can be used to induce production of sporangia on diseased roots (Schmitthenner 1973). If the infection extends deeply into the root, stem, leaf, or fruit, the surface of diseased tissue can be disinfested with a 1:10 dilution of commercial bleach (sodium hypochlorite 0.5%) and water. Addition of a wetting agent such as Tween 20 (0.05%) or ethanol (50%) increases the wetting capacity of the bleach but might also increase the chance of killing the target *Phytophthora.* Time of exposure to surface decontaminating agents should be relatively short (approximately 30 s) for small roots but can be extended to 2–5 min for larger pieces of plant tissue. The bleach solution should be washed off with sterile water. Chlorine (2 ppm for 2 min) in water is lethal to zoospores of *P. parasitica* from citrus. At 4 ppm, an exposure as short as 10 s killed zoospores that were bioassayed by inoculation of apple fruits with the treated zoospore suspension (Klotz and DeWolfe 1959).

Dipping root tissue in 70% ethanol for a few seconds followed by blotting on paper towels is effective for isolation of *P. cinnamomi* from avocado roots (Zentmyer 1980). Ethanol evaporates rapidly and does not leave a residue.

Surface disinfestation with 1.0 or 0.5× Teepol L (27% primary alkyd sodium sulfate), an anionic detergent (Shell Chemical Co., Ltd.), for 1 min increased the percentage (up to 93%) of isolation of *P. fragariae* from strawberry roots (Montgomerie and Kennedy 1983). The antifungal nature of Teepol is described by Mukherjee (1974, 1976). Teepol L is not sold in the United States; however, Teepol HB6 (34% active) (Sigma Chemical Co., St. Louis, Missouri, 63178) can be substituted (D. Kennedy, *personal communication*). George and Milholland (1986b) reported that Teepol L was deleterious for isolation of *P. fragariae* from strawberry roots, however, and that dipping roots or infected leaves in ethanol for about 10 s was more successful.

Small pieces of tissue, cut up with a sharp scalpel that has been dipped frequently between uses in fresh ethanol (dirty ethanol can harbor spore-forming *Bacillus* types of bacteria) and burned off can be placed on the surface of a petri plate containing a suitable selective medium or, if a selective medium is not available, on water agar or a weak medium such as cornmeal agar (Appendix 3.8) or a diluted V8 juice agar at about 1/5 or 1/10 the recommended concentration (Appendixes 3.1 and 3.2). Saprophytic bacteria and fungi grow rapidly in media that contain high concentrations of sugar and amino acids. Colonies of many *Phytophthora* species require some nutrients to grow out of infected tissue. Germination of most spore propagules produced by *Phytophthora* are stimulated by nutrients (see Chapter 3).

Direct Isolation by Tissue Plating

$P_{10}VP$ (Appendix 2.2) and $P_{10}ARP$ (Appendix 2.3) are two of the most effective media for direct isolation by tissue plating. Plates containing these media should be stored and incubated in the dark because pimaricin decomposes in light.

Incubation of isolation plates at 15–20°C, temperatures lower than optimum for growth of most *Phytophthora* species, favors recovery of many *Phytophthora* species from soil (McCain et al. 1967; Mircetich and Matheron 1976; Tsao 1983; Mircetich and Browne 1987). This effect is attributed to the reduction of the rate of multiplication of many bacteria and of fast-growing, high-temperature *Pythium* species. Tsao (1983) notes that the population of *Pythium splendans* and *P. vexans* from tropical soils is greatly reduced by incubation of plates at 18°C.

If a *Phytophthora* species is present in plant tissue, sporangia usually can be induced to form on diseased tissue in a shallow layer of copper-free distilled water or in various mineral salts solutions (see Chapter 3) within 2 days. Water or the salts solution should be changed frequently if sporangia do not form quickly. Zoospores can be induced to form by placing the petri dish at low temperatures (about 15°C) or at a temperature about 10°C lower than ambient for 30 min. Zoospores can be collected with a pipet and spread on a selective medium (Schmitthenner 1973). Kellam and Zentmyer (1986a) reported isolation of germinating oospores from an aqueous solution by transferring them with a fine capillary pipet to $P_{10}ARP$.

Since many *Phytophthora* species, such as *P. fragariae* (George and Millholland 1986b), grow more slowly than contaminating fungi, plates should be examined frequently. It is best in an unknown situation to plate many small (5 mm or less in diameter) pieces of diseased tissue and carefully look for slowly emerging *Phytophthora* hyphae for up to 7–10 days before discarding the plates.

The time of year is often a critical factor in the isolation of some *Phytophthora* species; for instance, *P. citricola* can be isolated from cankers on sugar maple trees in late summer or fall, but the same cankers do not yield the fungus in winter, spring, or early summer. Recovery from soil, however, is successful from early May to mid-November (Drilias et al. 1984). Similarly, in the state of Washington, isolations of *P. fragariae* from strawberry roots and *P. erythroseptica* (probably *P. fragariae sensu* Duncan et al. 1991) from raspberry roots are most successful from late fall to early spring (O. K. Ribeiro, *unpublished*). *P. parasitica,* a high-temperature pathogen (maximum growth temperature 35°C or higher), is isolated frequently in summer but is seldom isolated from citrus in California in winter (Lutz and Menge 1991), and *P. citrophthora,* a medium-temperature pathogen (maximum 30°C), is seldom isolated in the summer (Klotz 1978).

Phytophthora and most fungi can be conveniently isolated or purified free from bacterial contamination by placing a small piece of the contaminated culture or plant tissue on the inside of a sterile glass cell (Van Tiegham cell method; Ark and Dickey 1950), which is embedded in either water agar with a dry surface or an agar medium containing antibiotics (Figure 2.1). The glass cell is pushed firmly into the agar medium so that the fungus is forced to grow under the agar before growing to the surface. Since bacteria on the hyphae reproduce much more slowly deep in the agar than on the surface, the hyphal tips that emerge from the cell are usually free of bacteria. Glass Van Tiegham cells can be made by sawing Pyrex glass tubing about 1.5–2.0 cm in diameter into sections about 1.5–2.0 cm in length. Cells can also be made of plastic tubing if it is capable of withstanding autoclaving.

Pettit and Pegg (1991) comminuted strawberry crown tissue infected with *P. cactorum* with a Sorvall-Omni mixer for 30 s and mixed the fragments in a selective medium. They used the medium of Masago et al. (1977) but excluded hymexazol and substituted pimaricin for nystatin in cornmeal agar instead of PDA. More colonies of *P. fragariae* were detected when the strawberry plant tissue was submerged in cooled molten agar than when spread on the surface of the agar. Crown tissue of a partially resistant cultivar (Cambridge Favourite) yielded significantly fewer colonies of *P. fragariae* than a susceptible cultivar (Tamella). Although the yield of colonies might be of use to assess the degree of colonization and possibly resistance, Pettit and Pegg (1991) recognized that chlamydospores or oospores could give rise to sporangia and zoospores and that this might contribute to an overestimation of the population in host tissue.

To isolate *P. infestans* from plant material, Griffith et al. (1995) plated small pieces of diseased tissue on rye A agar amended (2 ml per liter) with an antibiotic cocktail (RAN) consisting of (µg/ml) nystatin (50), rifampicin (50), and ampicillin (25) dissolved in dimethyl sulfoxide (stored at –20°C). The medium is sufficiently "nutrient poor" to prevent rapid growth of contaminants. If *P. infestans* is sporulating, the infected leaf can be lightly brushed on the agar to dislodge caducous sporangia.

Isolation from Soil

Propagules in Soil

Although in many cases the propagule that is the persisting source of inoculum is well known (see Chapter 5, Table 5.3), detecting it is often difficult. Oospores are noted for their dormancy in soil and plant tissue. In most laboratory experiments, in which oospores germinate under ideal conditions, the time required for a high percentage to germinate varies from 2 to 3 days to 30 days (see Chapter 3). Thus, in most soil-dilution assays, in which the soil is washed off the surface of the agar after 1–2 days and results recorded a few days later, germinating oospores would not be detected.

Stack and Millar (1985a) selected a mutant of *P. medicaginis* (*P. megasperma* f. sp. *medicaginis*) that was naturally resistant to metalaxyl. Experimentally, this isolate could be readily isolated from soil on a selective medium amended with metalaxyl (20 mg/liter), a fungicide specifically toxic to *Phytophthora* and *Pythium* species in soil. They found that this *Phytophthora* species survived in soil mainly as oospores for at least 144 days. In New York soils, mycelium lysed almost completely in 9 days at 24°C and in 15 days at 15°C, but it persisted longer than 35 days at 4°C. Mycelium, sporangia, and oospores were capable of producing oospores in soil even in the absence of a host (Stack and Millar 1985c) (see Table 5.3).

Lutz and Menge (1986a) reported that soilborne propagules of *P. parasitica,* consisting of chlamydospores with only an occasional oospore, are dormant in rhizosphere soil obtained from fibrous roots of citrus during the cold winter months and cannot be detected by soil dilution methods except when soil samples are preincubated for 48 h at 28 or 34°C before isolation on dilution plates. With this method, the level of detectability increases from about two to 10–12 propagules per gram of soil. Although Lutz and Menge (1986a) first referred to this phenomenon as "heat shock," they later attributed this response to a degree-day heat regime necessary to overcome dormancy of chlamydospores in soil (Lutz 1987; Lutz and Menge 1991). Heat units were expressed in degree-days above a threshold temperature of 12°C. Calculation of degree-days was determined by the formula (incubation temperature – 12°C) × number of days of incubation. Soil collected from the field in the winter was stored at 9°C. To test the degree-day effect, soil in plastic bags was incubated at 32°C for different periods, after which isolation by dilution plating on P_{10}VPH (Appendix 2.2) was conducted. Soil was incorporated into molten P_{10}VPH agar cooled to 42°C (Tsao and Ocana 1969). The maximum number of colonies on the plates was recorded after the accumulation of 150 degree-day units by the soil sample (Lutz and Menge 1991; Lutz et al. 1992).

The maximal yield of *P. parasitica* propagules in Florida is obtained when citrus soil samples (from cores 18.5 cm deep and 8.0 cm in diameter) are collected directly from the field, moistened to a matric potential of about –3 centibars, and stored for 2–8 days at 21–24°C prior to isolation on a selective medium by dilution plating (Timmer et al. 1988). This procedure probably favors the production of sporangia and zoospores that form on chlamydospores. It differs in principle from the extended bait assay used by Stack and Millar (1985c) and Jeffers and Aldwinckle (1987) (see Appendix 2.8 and previous section on isolation from soil by baiting with selective hosts) because drying soil to air temperature reduced the population of *P. parasitica.* The data of Timmer and colleagues (1988) showed that mean propagule densities randomly selected from each of 4-ha blocks of citrus trees varied greatly; for instance, averages of 11.2, 0.8, 2.0, and 4.7 ppg of soil were recorded from each block. To obtain a statistically significant propagule value of 11.2 propagules per cubic centimeter of soil, analysis of 19 soil core samples per tree were required; however, to obtain a significant value of 2.0 propagules per cubic centimeter, it was necessary to assay 149 soil core samples. The yield of *P. parasitica* propagules was most consistent when core samples were collected beneath the tree canopy and within the drip line (L. V. Timmer, *personal communication*). These data support the concept that the *Phytophthora* population in soil is clustered and probably not random (Weste 1983b).

Weste (1983b) states that during cool months, *P. cinnamomi,* which persists mainly as chlamydospores, is undetectable in soil but that populations increase rapidly after rain.

Understanding which propagule or spore structure of *Phytophthora* is the target of isolation may help to establish a rationale for methods of isolation from soil. For instance, it is well known that *P. cinnamomi* probably survives in soil as chlamydospores, and if the population is high enough, each of these spores will germinate on a selective agar medium and produce a colony. Many *Phytophthora* species, however, either do not form chlamydospores or do not form them in numbers large enough to be a source of inoculum in soil. Examples include *P. sojae* (*P. megasperma* f. sp. *glycinea*) and *P. medicaginis* (*P. megasperma* f. sp. *medicaginis*). The latter species produces chlamydospores sparsely (Basu 1980; but could not be confirmed by Stack and Millar [1985c]) but forms oospores abundantly in diseased tissue. *P. cactorum* (Jeffers and Martin 1986; Rahimian and Mitchell 1988) and *P. fragariae* produce numerous oospores in infected root tissue. *P. fragariae* survives as oospores up to 3 years (Duncan and Cowan 1980). When the plant tissue rots, oospores are later released and persist in soil (Canaday and Schmitthenner 1982; Stack and Millar 1985c).

Successful means for isolation from soil in which oospores are the main source of inoculum seem to be associated with treatments that involve drying the soil followed by incubation after rewetting (Stack and Millar 1985c) (see extended baiting method, above, in this chapter and Appendix 2.8). Oospores germinate and produce sporangia and zoospores. Thus, for every oospore that germinates, the number of propagules could increase by the number of zoospores produced from the sporangium that forms on each germinating oospore. Since the effect of air drying followed by wetting probably increases the germination of oospores and subsequent production of sporangia and zoospores (Stack and Millar 1985c), it is not surprising that air drying soil from citrus groves in Florida reduces the recovery of *P. parasitica* drastically (Timmer et al. 1988). *P. parasitica* propagules are most likely composed primarily of chlamydospores and possibly some sporangia or mycelium growing from germinated chlamydospores. Since *P. parasitica* is heterothallic, oospores are comparatively rare in citrus soils (Lutz 1987; Lutz and Menge 1991).

Germination of oospores in laboratory experiments is enhanced by incubation in soil extract (Banihashemi and Mitchell 1975; Sneh et al. 1981; Jimenez and Lockwood 1982; Förster et al. 1983; Stack and Millar 1985a–c) and in root exudates (Förster et al. 1983; El-Hamalawi and Erwin 1986b). Although the precise reason for this stimulation is not known, it is probably associated with the presence of a low level of nutrients in soil or root extracts. Amino acids (e.g., aspartic acid) stimulate germination of *P. megasperma* f. sp. *medicaginis* and *P. cactorum* oospores (El-Hamalawi and Erwin 1986b; Jiang and Erwin 1990b; Jiang 1991) (see Chapter 3).

The concentration of sporangia per se in soil is difficult to determine because zoospores that may emanate from sporangia grow into individual colonies on dilution plates. These are then counted as colony-forming units (cfu) per gram of soil. Thus, if sporangia were the primary inoculum being assessed, germination by zoospore production could unduly increase the cfu values. Manipulation of the incubation temperature might correct this error. Sneh (1974) shows that the deciduous sporangia of *P. cactorum* can be washed off 7-day-old plates of hemp seed extract agar, washed by centrifugation to separate the sporangia from hyphal strands, and added in known concentrations to a test soil. When the soil-sporangium suspension was added to a selective synthetic agar medium, more than 80% of the sporangia germinated directly by germ tube at 24 and 30°C; but at 15 and 21°C, about 98% of the sporangia germinated indirectly by formation of zoospores.

Differences in methods of collecting soil samples to assay the inoculum level might be important. For instance, Weste (1985b) and Lutz and Menge (1986a) collected soil from the rhizosphere. Others collected samples randomly (Serra et al. 1988b; Timmer et al. 1988, 1989). The method chosen should depend on the rationale of the experimental goals. In most soil isolation reports, soil is screened through 2- to 5-mm sieves. The purpose of screening is to make the soil mass more uniform. Although removing the rocks and debris from soil is desirable, removing the fine roots may not be advantageous because *Phytophthora* propagules form in roots. According to Weste (1983b), the soil collected in the field should not be screened. Screening also will have a drying effect, which might render some *Phytophthora* propagules nonviable. Although not stated in the definitive chapter of Tsao on isolation (1983), Tsao agrees with Weste (1983b) and does not recommend screening the soil sample before isolation by bioassay or dilution plating, especially when accurate quantitation of inoculum per se is the goal. One has to keep in mind that many *Phytophthora* propagules are not as tolerant of heat, cold, and drying as are propagules of certain members of Fungi Imperfecti, Ascomycetes, and Basidiomycetes. Tsao (1983) presents the following considerations on selection and storage of soil samples. (Reproduced by permission of the American Phytopathological Society, St. Paul, Minnesota, courtesy of Tsao 1983):

Selection, Collection, Transport, Storage, and Preisolation Treatments of Samples. Negative recovery of *Phytophthora* from tissues or soils might not always be due to improper choice of isolation techniques or faulty isolation procedures in the laboratory. It could be simply due to the absence of *Phytophthora* in the selected sample at the time of collection or the loss of viable propagules during transport or storage if the samples are improperly handled.

Aside from certain ecological studies, e.g. on population distribution, in which a rigid or random scheme, depending on the experiment, must be followed, soil samples in a diagnostic study or disease survey can be discriminately selected to maximize *Phytophthora* recovery. Soils near the root zone or next to other underground plant parts are expected to harbor higher populations. Because *Phytophthora* succumbs easily to dryness and because soil near the ground surface is more subject to high temperature from solar radiation and to drying, soil samples for *Phytophthora* isolation should be chosen from at least 3–5 cm below the soil surface. For similar reasons, soil at the north or northeast corner (in the Northern Hemisphere) of a tree under foliage canopy would be the most desirable.

Although outside the realm of the topic of soil isolation, a brief discussion of selection and collection of infected plant tissues is pertinent here. *Phytophthora* infections on stems and roots result in discoloration and are usually restricted to the bark or cortical regions, with hyphae extending internally to the cambium but seldom (at most a few cells deep) into the wood. Dark lesions in the wood are largely from secondary invaders. *Phytophthora* lesions on fruits, leaves, and petioles are almost always brown, purplish brown, or almost black and are firm in texture. Soft, water-soaked lesions are probably not those produced by *Phytophthora* or may already have been invaded by secondary microorganisms. Proper recognition of *Phytophthora* symptoms of other kinds also improves the chances of isolation.

As with any fungus, *Phytophthora* would lose its viability if the collected tissue or soil samples were exposed to drying or high temperature during transit before reaching the laboratory. This often happens, especially in an enclosed vehicle in hot summer or in the tropics. Samples should be collected in plastic bags or similar closed containers. Loss of viability during sample transport can be avoided by the use of a portable, insulated "ice chest" or similar container. Most *Phytophthora* spp. are sensitive to extreme cold as well as heat, and excessively low temperatures during sample transport are also to be avoided. Therefore, samples should not be in direct contact with the ice and should be placed on a raised rack to prevent contact with the icy water from melted ice accumulated at the bottom. Volumes of ice in the container need not be large, because the purpose of the insulated container is only to maintain a cool, nonlethal temperature during transit.

For sample storage, before isolation experiments, both excessively high and low temperatures should be avoided. The general practice of storing samples in a "refrigerator" (4–5°C) to reduce growth of secondary bacteria and fungi also reduces the chance of *Phytophthora* recovery because of its sensitivity to extreme cold. A cool incubator at 10–15°C is generally more desirable than a cold refrigerator. Needless to say, lengthy storage at any temperature is not desirable, and chance of recovery is best if isolation is performed immediately after samples are collected.

We have reviewed some studies in which soil was assayed after premoistening (extended isolation method) (Stack and Millar 1985a; Jeffers and Aldwinckle 1987). This practice usually results in isolation of more propagules from the soil because the premoistening technique most likely induces germination of dormant chlamydospores and oospores, which in turn produce sporangia and zoospores. Strictly speaking, however, such data are not a quantitative measure of the inoculum per se that existed on the soil at the time of collection (Tsao 1983). Rather, they represent an amplification of the inoculum level. Unfortunately, the data are insufficient to determine by what factor the original inoculum is amplified. Even if known, it would probably be variable for different soils.

This observation does not negate the value of the extended isolation method, because there are so many instances in which *Phytophthora* species cannot be isolated by conventional direct plating techniques but at a later time the inoculum level increases dramatically (Weste 1983b). Jeffers and Aldwinkle (1987) could detect *P. cactorum* in some sites only when the extended isolation method was used. Since we no longer believe in the theory of spontaneous generation, the inoculum must have been in the soil but was nondetectable!

Direct Isolation from Soil

Isolation of *Phytophthora* directly from soil was seldom successful prior to the development of selective media. While the chemicals in selective media suppress many associative microorganisms, such as bacteria and fungi, the selective chemicals cannot be considered to be completely inhibitory to all microorganisms. Qualitatively, *Phytophthora* can be isolated on the $P_{10}VP$ medium (Appendix 2.2) by simply sprinkling soil crumbs on the surface of the agar plate (Zentmyer 1980; Tsao 1983). This method is considered to be a highly sensitive qualitative method to isolate *P. cinnamomi* (Zentmyer 1980). Transferring plugs of an agar-soil mixture to a selective medium is effective for isolation of *P. medicaginis* (*P. megasperma* f. sp. *medicaginis*) (Stack and Millar 1985a).

Since some *Phytophthora* species produce zoospores more rapidly than *Pythium* species in flooded soil (Schmitthenner 1973), *P. sojae* (*P. megasperma* f. sp. *glycinea*) can be differentially isolated from soil in the presence of *Pythium* by incubating field soil remoistened to 100–200 mb matric potential (slightly wetter than field capacity) for 1–2 weeks in polyethylene bags before isolation. As previously discussed, this regime probably favors sporangium formation from germinated oospores in the soil. Soil samples (5–20 g) are then placed in beakers and flooded with water to a depth of about 1–2 cm. When soybean leaf disks are incubated for as little as 1.5 h on the water, numerous *P. sojae* colonies, but very few *Pythium* colonies, are detected after the disks have been transferred to a selective medium. When leaf disks are incubated for longer times, however, *Pythium* species are also isolated (Canaday and Schmitthenner 1982) (Appendix 2.8).

P. cactorum was isolated from soil following the extended baiting assay by dilution plate assay (Appendix 2.7). The competitive effects of *Pythium* species were avoided by flooding moist soil with cold water (8°C) for 2 h to induce zoospore formation by *P. cactorum. Pythium* species usually do not form zoospores at that temperature (Rahimian and Mitchell 1988).

Chlamydospores of *P. cinnamomi* have been isolated by wet-sieving infested soil through a series of sieves and retaining the residue on the 38-µm-pore-size sieve (McCain et al. 1967) (Appendix 2.9). The method has also been used since then by several other laboratories (Broderick et al. 1976a; Greenhalgh 1978; Shew et al. 1979b; Kliejunas and Nagata 1980; Shea et al. 1980; Shew and Benson 1982). When Greenhalgh (1978) washed soil through sieves of 252-, 152-, and 41-µm mesh sizes and plated the residue from a 41-µm mesh sieve on a selective medium, he concluded that sieving was less sensitive than baiting with pear fruits. Although physical separation of *Phytophthora* propagules by screening allows a large volume of soil to be sampled and the screened propagules to be concentrated, the loss of propagules such as some chlamydospores, sporangia, and zoospores that are smaller than 38 or 41 µm in diameter is a distinct disadvantage. The recovery of *P. cinnamomi* by wet-sieving was less than 20% of that from soil not wet-sieved (Kliejunas and Nagata 1980). For these reasons, Tsao (1983) does not recommend the wet-sieving technique for quantitative assays.

Soil Dilution Plate Method

The soil dilution plate method for quantitative isolation of bacterial and fungal propagules from soil has been commonly used for many years (Johnson and Curl 1972). Quantitative assays for the population of *Phytophthora* species can be made from soil at different depths, different locations, and different times by use of the soil dilution plate method (e.g., Shea et al. 1980; Tsao 1983; Lutz and Menge 1986, 1991; Shearer and Shea 1987; Timmer et al. 1988). This approach has been facilitated by the great advancement in the development of agar media that are selective for *Phytophthora* species.

The percentage of moisture in the soil is determined so that the number of colony-forming units of *Phytophthora* that can be identified on a petri plate can be calculated per gram of oven-dry soil (approximately 105°C for 24 h, or small samples can be heated on a hot plate for a few minutes). The determination of soil moisture can be made on a parallel subsample of soil. One-milliliter portions of the soil-water slurry are pipetted into replicated, preweighed dishes or aluminum cups, which are then heated on a hot plate or in an oven to drive off the water and reweighed to determine the weight of the water lost. The percentage of moisture in soil is calculated by dividing the difference between weight of the wet soil and the dry soil (the weight of the water) by weight of the dry soil × 100.

The amount of soil to be suspended should be about 10 g/100 ml of water unless the population of *Phytophthora* is low, in which case the dilution can be increased to as much as 50 g/100 ml (Ferrin and Rhode 1992a, b). After vigorous shaking, the soil slurry is kept agitated (e.g., with a magnetic stirrer) while 0.5- or 1-ml portions are transferred with a widemouth pipet or specially made dipper to the surface of the selective agar medium. The water should be glass distilled or known to be free of heavy metal contaminants (e.g., copper) that are toxic to *Phytophthora* (Kennedy and Erwin 1961; Gerrettson-Cornell and Townsend 1976). If glass-distilled water is not available, tap water should be tested in comparison with a sample of glass-distilled water. For instance, a known concentration of zoospores could be suspended in both tap water and double-distilled water, and the yield of colonies from equal portions of each on V8 juice agar plates could be compared. Chlorine in tap water (toxic to zoospores at 4 ppm for 10 s) (Klotz and DeWolfe 1959) can be removed by boiling. Agar at 0.1–0.3% (w/v) can be added to the soil slurry to increase the viscosity of the suspension, thus reducing the rate of settling of propagules. The dilute water agar-soil suspension can be amended with the appropriate antibiotics (μg/ml): rifampicin (10), pimaricin (10), ampicillin (250), and if the target *Phytophthora* species will tolerate it, hymexazol (50) (Ferrin and Mitchell 1986a). Ferrin and Rhode (1992a, b) omitted pimaricin and hymexazol in subsequent reports. The soil slurry is spread with a glass "hockey stick" or similar device capable of spreading the soil slurry and distributing the propagules on the surface of a solid agar medium in standard 9-cm-diameter petri plates. Shea and colleagues (1982) used 15-cm-diameter petri plates to accommodate larger samples of the soil suspension. Each colony is assumed to grow from a single propagule (chlamydospore, sporangium, encysted zoospore, oospore, or hyphal fragment); however, there is no way to avoid some clumping.

After colonies have formed, usually in 1 to 2 days, the soil is washed off the plate with a gentle stream of running water or scraped off with a rubber spatula. *Phytophthora* and/or *Pythium* colonies, which have grown into the agar, do not wash off. The presumption is that each germinable propagule will have established a small colony on the agar in 1 to 2 days. The advantage of washing the plates is that the colonies are more easily observed and counted on the plates that have been washed free of soil; however, slowly germinating propagules, such as oospores, could be discarded by this practice. Probably the greatest advantage of this method is that large numbers of plates can be poured prior to application of the soil slurry to the plates.

Alternatively, the molten selective medium, maintained at a carefully controlled temperature of 43°C (slightly above the congealing point of agar) in a water bath or oven, can be poured over a 1-ml portion of the soil slurry in the petri plate. The soil slurry and the cooled molten medium are thoroughly mixed by gentle swirling and shaking of the petri plate before the agar solidifies (Tsao 1983; Lutz and Menge 1986). With this method, slowly germinating propagules have a better opportunity to form colonies, and the unwanted bacteria and fungal propagules in soil are surrounded by the antibiotics in the agar medium. The number of colonies of *P. parasitica* isolated from a soil sample was doubled by pouring the cooled antibiotic medium over the soil slurry compared with spreading the soil suspension sample on top of the agars (Tsao 1983).

Temperature of the molten medium should be precisely controlled at 43°C to prevent killing *Phytophthora* propagules (Tsao 1983). Although this method yields the largest number of colonies, it requires more time and more hands to pour the medium rapidly and to agitate each plate before the agar congeals.

Masago and colleagues (1977) introduced 1 ml of each antimicrobial chemical, formulated in 80% ethanol, directly into the petri plate and in parallel introduced 1 ml of the soil suspension with care not to mix the chemicals and soil suspension prior to addition of 8 ml of soft PDA (1% agar) at 45°C (Appendix 2.3).

Isolation from Bodies of Water

In irrigated agricultural systems, water can be taken from underground wells, which are not likely to be contaminated, or from reservoirs or canals that emanate from dams, which in turn are fed by runoff from streams or in many cases by runoff from infested fields. Since *Phytophthora* reproduces so readily by production of zoospores, which are capable of swimming for long periods, contamination of irrigation water would be expected. Indeed various species of *Phytophthora* have been isolated from bodies of water that are used for irrigation (Klotz et al. 1959), and, what may be of greater importance, even mountain streams in relatively pristine environments may be contaminated (Von Broembsden 1984a, b). Table 2.5 shows references that cite irrigation water as sources of *Phytophthora* inoculum.

Waste water from fields is often delivered back to the canal or recycled to a reservoir. Nurseries in California and other areas are being required to limit runoff partly to conserve water but also to limit the possibility that drainage water might increase the pollution level of ground water with pesticides or soluble fertilizers such as nitrates. To implement these requirements, some nurseries store runoff water in holding tanks and reuse it later (Ali-Shtayeh et al. 1991). While conservation of water by these means serves a useful purpose, conversely this technology can serve to redistribute *Phytophthora* propagules

Table 2.5. References that cite irrigation water as a source of *Phytophthora* inoculum

Species	Disease	Crop	Citation
P. cactorum	Crown rot	Apple	Braun and Schwinn (1963)
	Crown and fruit rot	Strawberry	Grove and Boal (1991)
	Crown rot	Apple	McIntosh (1964, 1966)
		Apple	Helton et al. (1984)
P. capsici	Boll rot	Cotton	Garber et al. (1986)
P. cinnamomi	Root rot	Ohia	Kliejunas and Ko (1976)
	Root rot	Heath	Oyler and Bewley (1937)
P. citricola, P. cactorum	Stem rot	Eucalyptus	Wardlaw and Palzer (1985)
P. citrophthora	Gummosis	Citrus	Klotz et al. (1959)
P. parasitica	Root rot and brown rot	Citrus	Klotz et al. (1959)
	Brown rot	Citrus	Whiteside and Oswalt (1973)
	Root rot	Citrus	Thomson and Allen (1974)
	Root rot and buckeye rot of fruit	Tomato	Neher and Duniway (1992)
Several species	Root rots	Ornamentals	Ribeiro and Linderman (1991); Taylor (1977)
Several species	Root rot	Pome and stone fruits	Mircetich et al. (1985)

in the form of zoospores to uninfected plantings. There are reports of various methods to alleviate this danger by use of chlorination and filtration (Daughtry 1984).

For experimental detection of *Phytophthora,* water samples can be passed through filters of low enough porosity that the *Phytophthora* propagules are caught on the filter. Fine nylon mesh materials such as Nitex cloth (Tobler, Ernst & Traber Inc., Elmsford, New York, 10523), which can be obtained in precise pore sizes of 10 µm and up, are useful as filters. Some use 0.45-µm filter pads (Ali-Shtayeh et al. 1991). The limiting factor with filtration is the amount of water that can be filtered before the filter becomes clogged by colloidal clay particles. The filter containing the residue can be incubated on one of the selective media to detect *Phytophthora.*

Low numbers of *Phytophthora* propagules in water-storage ponds in two California nurseries (Ali-Shtayeh et al. 1991) were detected with commercial ELISA (enzyme-linked immunosorbent assay) kits (Agri-Diagnostic Associates, Cinnaminson, New Jersey). Since the ELISA method correlated well with conventional isolation methods, this technique may be useful in establishing standards for risk or for guidelines indicating the necessity for application of chemicals to suppress *Phytophthora* or *Pythium* in runoff water in nurseries. See the following section, Detection by Serological Methods.

Klotz et al. (1959) detected *Phytophthora* in water in numerous irrigation canals and reservoirs in southern California by use of the lemon fruit bait method. Lemon fruits were placed in perforated plastic bags and incubated in the canal or reservoir for 2 weeks. Fruits with brown rot symptoms were then cultured to isolate and detect *Phytophthora.* Klotz et al. (1959) detected *P. parasitica, P. citrophthora,* and *P. syringae* and an unidentified *Phytophthora* species that grew at a temperature as high as 45°C. The frequency of isolation of these species was correlated with the time of year and temperature of the water. For instance, *P. syringae* was isolated most frequently at water temperatures of 6–17°C and *P. parasitica* during the summer and fall when water temperatures rose to 17–31°C. *P. citrophthora* was isolated throughout the year. Twelve of the 15 water bodies that supplied water to 40,000 acres of citrus yielded *Phytophthora* species pathogenic to citrus. Klotz et al. (1959) concluded that the source of *Phytophthora* in orchards was mainly drainage from infested orchards.

P. gonapodyides was isolated from irrigation water reservoirs by use of several leaf and seedling bait methods (Pittis and Colhoun 1984). The baits, which included *Begonia* petioles, citrus leaf pieces, cress seedlings, grass leaves, hemp seeds, and wheat coleoptiles, were incubated in 20-cc samples of water in glass petri plates. Lupine seedlings and cuttings of *Hebe pinguifolia* cv. Pageii were incubated in 100-cc samples of water for 24, 48, 72, and 96 h. The baits were washed, blotted dry, and plated on P_{10}ARP (Appendix 2.3). In some cases, filter paper disk baits (Robertson 1975) used for isolation of *Pythium* species were suspended in a reservoir for 9 days and plated on P_{10}ARP. *P. gonapodyides* was detected on all the baits but most frequently on the paper baits and on 8- and 20-µm disks through which the water had been filtered. Grass leaves and hemp seed were much more efficient baits for *Pythium* than for *Phytophthora* (see also Schmitthenner [1973]).

P. parasitica, P. citrophthora, and an unidentified *Phytophthora* species were found in reservoirs used for recycling irrigation water by baiting with whole leaves of rough lemon (*Citrus jambhiri*) (Thomson and Allen 1974). Whole leaves, feathered by removing 2-mm-wide strips from the margins and making several short cuts about 10 mm apart on the margins, were floated in irrigation water for 5 h and then transferred to sterile distilled water for observation, or small pieces (2 × 2 mm^2) of the water-exposed leaves were inserted into P_{10}VP agar (Appendix 2.2) to detect *Phytophthora.* Citrus leaves were more efficient than lemon fruits for detection. Water was passed through a series of filters (48-, 31-, and 10-µm porosity) of Nitex cloth and an 8-µm Millipore filter fitted into Millipore filter holders (Millipore Corp., Bedford, Massachusetts). Filters were then incubated in P_{10}VP agar to detect *Phytophthora.* Since all the propagules retained were on the 8-µm filters, zoospores were considered to be the main type of propagule (Thomson and Allen 1974).

Phytophthora species (mainly *P. cactorum*) were recovered from 27 of 31 water sources by suspending green pear fruits for 1 week in irrigation water sources in British Columbia, Canada (McIntosh 1966). The pear fruit baiting method is considered to be an optimal choice for isolation of *Phytophthora* species from stone and pome fruit orchards and from bodies of water supplying irrigation water to orchards (McIntosh 1964; Mircetich and Matheron 1976; Mircetich and Browne 1987). More than 16 *Phytophthora* species were recovered from cherry, walnut, apricot, and peach orchard soils and water sources (Mircetich and Keil 1970; Mircetich and Matheron 1976, 1983; Mircetich 1982; Mircetich et al. 1985, reviewed by Mircetich and Browne [1987]).

In the southwestern Cape Province of South Africa, *P. cinnamomi* was detected (up to 60 cfu/dm^3 of water were detected, but one to two per 4–5 dm^3 were more usual) in all the major rivers and even in the headwaters of the mountains (Von Broembsen 1984a, b). Water, which was collected from rivers and other bodies of water in volumes ranging from 5 to 50 cm^3 for mountain headwaters, was filtered through Nucleopore filters (45 mm in diameter). The filters were then transferred to petri plates and covered with cooled (45°C), molten P_{10}VPH agar (Appendix 2.2). Plates were incubated for 3 days at 24°C. In addition to *P. cinnamomi, P. cryptogea* (*sensu* Bumbieris 1974, who merges *P. cryptogea* with *P. drechsleri*) and other unidentified species of *Phytophthora* were found.

Hallett and Dick (1981) reported assay of zoospores of several oomycete fungi from lake water by use of a Sorvall RC2-B refrigerated centrifuge fitted with an SS-34 head and a Szent-Gyorgyi and Blum continuous flow system. Zoospore recovery was optimum at centrifuge speeds of 15,500 to 16,000 rpm with flow rates less than 600 cm per minute. Optimum counts of colonies were obtained from centrifugate that had been plated on cornmeal agar or glucose-yeast-soluble starch-tellurite (bacteriostatic). This latter medium was most favorable for *Phytophthora* isolates, most of which were considered to be *P. gonapodyides* with an occasional *P. cryptogea* isolate.

Detection by Serological Methods

There is a need to supplement conventional methods of diagnosis that utilize symptoms and direct isolation with newer technologies that might be less time consuming and possibly more accurate. Both serology and DNA technologies are rapidly becoming available for research and commercial purposes. A recent article by Fox (1990) strongly supports the concept that serological ELISA and DNA hybridization methods are becoming valuable tools. The use of ELISA methods should be thoroughly correlated with direct isolation (MacDonald et al. 1990) so that there is little or no chance that the ELISA method is detecting another fungus or bacterium. It is a useful tool but should not be considered a method that will totally replace direct isolation.

The principle for serological detection of *Phytophthora* is the well-known capacity of mammals to develop antibodies in the blood that react specifically to certain foreign proteins or carbohydrates (antigens). The antibodies recognize these unique compounds by binding with them. The antigen is the substance foreign to the animal, for example, a virus, a whole microorganism such as a bacterium or a fungus, or a chemical fraction of the target organism. Some components of the microorganism, such as cell wall material, are often more spescific as antigens.

The early work by Burrell et al. (1966) describes serological detection of different *Phytophthora* species. To produce the antibodies, a suitable animal, usually a rabbit, is injected subcutaneously, intramuscularly, or intravenously at intervals of several days with the antigen (e.g., comminuted mycelium of the *Phytophthora* species to be detected that has been fractionated into different components such as cell wall material). Three-day-old mycelial cultures grown on an ammonium sulfate-glucose synthetic medium (see Appendix 3.21) produce the maximum amount of soluble protein (which acts as the antigen) when rabbits are inoculated subcutaneously four times at weekly intervals. The use of the soluble protein fraction of young mycelium is critical for the induction of antibodies in the rabbit, since the most active antigenic material is found in mycelial tips. After the antibodies that are specific to the antigens from *Phytophthora* have increased in the rabbit (about 3 weeks), a sample of blood is drawn from the animal and allowed to clot in a test tube refrigerated at about 5–10°C, after which the serum is separated from the clot. The serum is centrifuged to remove precipitates and stored at –20°C. This serum will react specifically in a suitable system, such as the agar diffusion test described below, with the antigen (*Phytophthora*) (Burrell et al. 1966).

In the report of Burrell et al. (1966), the soluble protein fraction of mycelium was mixed in phosphate-buffered saline with the antiserum of the target *Phytophthora* for 24 h at 4°C. The precipitate was removed by two or three centrifugations at 1,200 × *g* at 24-h intervals. To determine by the immunodiffusion technique whether or not there is a specific reaction, the antiserum is usually placed in a well in the center of a petri plate containing an agar gel (0.1% Ion Agar No. 2 containing 0.05% glycerin is often used). The antigen (in comminuted mycelium of a *Phytophthora* species) is placed in another well in the petri dish and incubated for 10 days. The substances in the wells diffuse outward, and the zone of precipitation that occurs when the antiserum and antigen react after a 10-day period is evidence of a homologous reaction of the antigen and the antibody. The antigen from other species may produce one or more precipitation zones, some of which may or may not be similar to those produced by the species being differentiated. If other *Phytophthora* species produce a similar zone of precipitation, the antigens are common to more than the target species. These antigens can be cross-adsorbed by incubating the serum with nontarget *Phytophthora* species and removing the precipitate by centrifugation. Morton and Dukes (1967) used the immunodiffusion method to differentiate *Pythium aphanidermatum* from *Phytophthora parasitica* and *P. parasitica* var. *nicotianae,* both of which were serologically indistinguishable.

There are a number of variations to the method of producing an antiserum to *Phytophthora.* Halsall (1976a) used antigens composed of either soluble protein or cell wall fractions to induce antibodies in the rabbit. *P. cinnamomi, P. cambivora,* and the *P. cryptogea-P. drechsleri* group could be differentiated.

When antiserum is made against a whole organism, the antibodies are called polyclonal. The use of monoclonal antibody technology is much more specific (Hardham et al. 1986, 1991; Wycoff et al. 1987). The monoclonal antibody method implies that the specific antibody is derived from a single, cloned mammalian cell obtained by tissue culture technology (Trione 1988). Thus, monoclonal antibodies are more specific than polyclonal antibodies. Hardham et al. (1986, 1991) detected monoclonal antibodies that bound specifically to zoospores or cysts of *P. cinnamomi,* but not to both, and monoclonal antibodies that bound only to restricted parts of the surface of the zoospore. They cited unpublished data that showed evidence of taxonomic discrimination between species because the monoclonal antibodies for *P. cinnamomi* did not bind to those of *P. parasitica* or *P. vignae.* The method, along with ultrastructural electron microscopy, was used to label specific binding sites on zoospores and other organs (Hardham et al. 1991). Gabor et al. (1993) utilized the immunological techniques of Hardham in their report of two rapid diagnostic assays for *P. cinnamomi.*

In soil and root tissue, *Phytophthora* as well as most other fungi can be stained with cotton blue (0.01%) in lactophenol; however, cotton blue also stains other fungi and, to some extent, plant tissue. The use of fluorescent antibodies allows specific staining of the target *Phytophthora* propagule because of the specificity of the antigen-antibody relationship. Use of fluorescent antibodies, which can be viewed with a microscope equipped with ultraviolet light, has been developed for detection of *P. cinnamomi* (Malajczuk et al. 1975; MacDonald and Duniway 1979) and *P. megasperma* (MacDonald and Duniway 1979) in soil. To view the antibody-antigen reaction more readily, antirabbit globulin, which can be purchased from scientific supply houses, is conjugated with a fluorescent dye such as fluorscein isothiocyanate. The mycelium of the target *Phytophthora* species in a disk of agar is bathed in the *Phytophthora* antiserum, which fixes to the natural antigens in *Phytophthora.* The mycelium is then washed with phosphate-buffered saline to remove the excess, followed by addition of an antirabbit antiserum that has been conjugated to the fluorescein isothiocyanate dye. The antirabbit antibody will react only with the rabbit antiserum that has become fixed to its homologous antigen in the *Phytophthora* species being observed. The intensity of the dye, which fluoresces in ultraviolet light, is proportional to the concentration of the *Phytophthora* antibodies.

To prepare antibodies of *P. megasperma* (*P. medicaginis*) and *P. cinnamomi,* MacDonald and Duniway (1979) used washed cell wall fractions of mycelial extracts (the soluble portion was discarded) as the antigen to produce antibodies in rabbits. Details of the inoculation series are given by Malajczuk et al. (1975) and MacDonald and Duniway (1979). A fluorescence-labeled goat antirabbit gamma globulin (Antibodies Inc. P.O. Box 442, Davis, California, 95616) can be used to visualize microscopically under ultraviolet light the germinated zoospore cysts incubated in soil. The fluorescence serological technique is elegant and highly specific but requires laboratory personnel with expertise in serological methodology. Briefly, zoospore cysts of *P. megasperma* and *P. cinnamomi* are transferred from a soil suspension to petri plates containing a thin layer of a selective agar medium on nylon cloth. To obtain extremely thin layers of the agar medium suitable for observation microscopically, the molten selective agar medium is painted on pieces (15 × 25 mm) of Nitex nylon cloth with 10-µm mesh openings (Tetco Inc., Elmsford, New York, 10523) lying on the surface of agar plates. With this technique, a thin layer of agar is held together by the nylon cloth. Zoospore cysts from soil suspensions are pipetted onto the agar film and allowed to germinate. When the soil is gently washed off the agar film, the germinated cysts adhere to the film. The nylon cloth and agar films are prepared for visualization by immersion in a 1/16 dilution of the rabbit serum containing *Phytophthora*-specific antibodies. They are then immersed twice in 0.1 M phosphate-buffered saline for 15 min and once in a 1/64 dilution of fluorescence-labeled goat antirabbit gamma globulin for 45 min at 22–24°C and then washed twice in the phosphate-buffered saline for 15 min. The films are laid (with the surface down) on clean, 22- × 40-mm glass coverslips and dried for 2 h. The nylon cloth is then removed with forceps. The agar films, which adhere to the coverslips, are mounted on glass slides. The zoospores are visualized microscopically under ultraviolet light. The germinated zoospores in each 50-µl spot are then counted. Using this sensitive technique, MacDonald and Duniway (1979) concluded that zoospores of *P. megasperma* (*P. medicaginis*) from alfalfa were much more sensitive to desiccation in soil (killed in 2 days or less) than those of *P. cinnamomi* (viable up to 6 days).

The use of both polyclonal and monoclonal antibody assays to detect pathogens is now a well-advanced field, and commercially produced kits are available. ELISA utilizes a color reaction by which the presence of the antigen can be detected. These assays are sensitive, even when the target pathogen is present at very low levels, and can also be used to give a quantitative assessment of the pathogen in a soil or plant tissue sample.

The method requires experience and specialized laboratory equipment; however, the procedures are common technology in university and government diagnostic laboratories in which bacteria and viruses are the subjects of research. Recently, many reports of the use of ELISA technology for detecting fungal pathogens have been made. Since the technology has many variations in the published literature, the account here is limited mainly to the principles underlying the methods.

The ELISA method (Engvall and Perlman 1971; Voller et al. 1976, 1979; Clark and Adams 1977; Clark 1981) utilizes the affinity of antibodies to bind with proteins or carbohydrates or to adhere to certain solid surfaces such as those made of polystyrene. For sufficient detail to conduct the ELISA, see Clark and Adams (1977), Voller et al. (1979), and the reviews by Clark (1981) and Miller and Martin (1988). The method was first developed for assay of viruses, but recently the method has been applied for assay and detection of fungi (Miller and Martin 1988). ELISA plates, which are commercially made of polystyrene, are standard items in immunological research supply houses (e.g., Dynatech Co., Chantilly, Virginia). The plate consists of 96 small wells, each of which holds 250 µl of fluid. When either solutions of the antigen or the antibody, depending on the method used, are allowed to incubate in the wells, the antigens and/or antibodies adsorb to the solid polystyrene surface of the ELISA plate and remain adsorbed after washing. Voller et al. (1979) is an excellent source of information on the rationale and methodology.

Many variations of the method can be used (Voller et al. 1979). For instance, in the indirect method (Figure 2.5), the antigen (an aqueous suspension of the target pathogen either purified or in a crude sample) is added to the polystyrene wells. After a suitable period of incubation, the antigen suspension is irrigated with a wash bottle containing phosphate-buffered saline solution; however, the antigen continues to be bound to the polystyrene. The *Phytophthora* antiserum, appropriately diluted, is added and incubated for 1–4 h, during which time the *Phytophthora* antibodies will have made contact and become bound to the antigens adhering to the polystyrene wall of the vessel. The material in the wells is again washed with phosphate-buffered saline. This leaves the *Phytophthora* antibodies adhering to the antigens, which are held by surface forces to the polystyrene wall of the vessel. At this time an enzyme, which has been conjugated with another antiserum made from a goat to detect rabbit antigens, is added. The goat antirabbit antibodies adhere to the rabbit antibodies that have been adsorbed to the *Phytophthora* antigens. This mixture is allowed to incubate for an hour, and the wells are washed again with phosphate-buffered saline. Next, a procedure is used to detect an enzyme, for instance, alkaline phosphatase that has been conjugated to a goat antirabbit antiserum. This enzyme will react with another compound, paranitrophenyl phosphate in 10% diethanolamine, to produce a yellow color, the intensity of which is proportional to the amount of the original *Phytophthora* antigens that had become attached to the polystyrene ELISA plate. A commercial colorimetric apparatus will quantitatively read the color intensity and produce a printout of results. Controls, which consist of the antigen alone, the normal antiserum alone, and a positive control (a pure culture of the target pathogen), are tested in parallel with the unknowns.

Another variation of the ELISA is called the double antibody sandwich method (DAS-ELISA) (Figure 2.5). In this case, the *Phytophthora* antiserum is allowed to adsorb to the polystyrene vessel surface. After a suitable incubation time, the excess antiserum is removed by washing with phosphate-buffered saline. The antigen—for instance, the plant or soil specimen, mycelium, or spores of the target *Phytophthora* (control)—is added and incubated. The antigens bind to the *Phytophthora* antibodies and remain fixed after again being washed with phosphate-buffered saline. An antirabbit antiserum, which has been conjugated to the alkaline phosphatase enzyme, is added. These antibodies will adhere to the *Phytophthora* antigens after the vessels are washed with phosphate-buffered saline. When the reagent, paranitrophenyl phosphate in 10% diethanolomine, is added, a yellow color is produced, which can be read colorimetrically. As in the indirect ELISA, the intensity of the color is proportional to the amount of *Phytophthora* antigen in the sample.

The potential advantages of ELISA are 1) rapid identification, 2) detection at low levels, 3) detection *before* symptoms are manifested on the host plant, and 4) accuracy in the detection of *Phytophthora* despite the presence of plant material and other microorganisms. That ELISA will detect dead as well as living organism tissue might be a disadvantage (Timmer et al. 1993). Also, care must be taken because ELISA might detect a similar organism that contains a common antigen.

ELISA kits are now commercially available for the detection of *Phytophthora* (Agri-Diagnostic Associates, 2611 Branch Pike, Cinnaminson, New Jersey, 08077). These kits are easy to use, even for those with limited laboratory experience. ELISA kits are now being developed for *P. fragariae* and its many races and also for some other species that have many forms or strains. Detection of *P. sojae* (*P. megasperma* f. sp. *glycinea*) in soil samples is more rapid than the baiting technique (S. A. Miller et al. 1989; S. M. Miller et al. 1989). Timmer et al. (1993) reported that commercial kits quantitatively detected *P. parasitica* and *P. citrophthora* in citrus soils in Florida and California.

Pscheidt et al. (1992) tested immunoassay Kit E (Agri-Diagnostics) against *P. cinnamomi, P. sojae, P. cactorum,* and *P. megasperma* (from cherry) and against a large number of plant disease clinic samples in Oregon. The known *Phytophthora* species and plant disease specimens with typical *Phytophthora* disease symptoms tested positive. However, there was cross-reactivity with certain species of *Pythium* and *Pero-*

nospora. The variability of activity of the kit among several *P. cinnamomi* isolates was much greater than among *P. cactorum* isolates. They concluded that the immunoassay kit is a useful tool for clinical diagnosis.

The ELISA method has been utilized for detection of *P. fragariae,* the cause of red stele in strawberry roots (Werres and Casper 1987; Amouzou-Alladaye et al. 1988; Mohan 1988, 1989a, b). Mohan (1988) pooled mycelia of five isolates of *P. fragariae* for preparation of an antiserum to detect the fungus in diseased strawberry plants. The antiserum was developed by inoculating rabbits with mycelial suspensions. The serum collected from the blood of the rabbit contained poly-

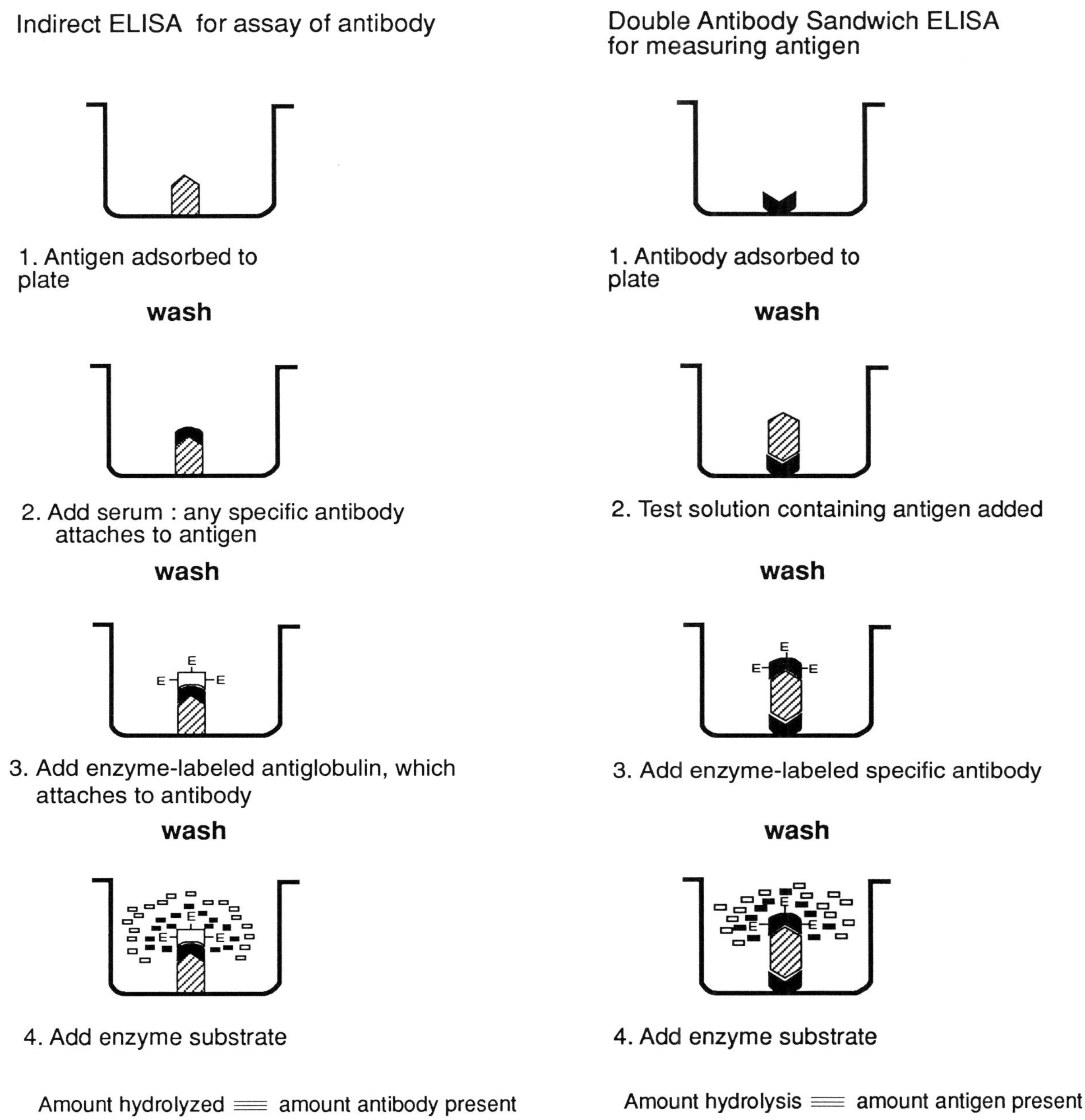

Figure 2.5. Enzyme-linked immunosorbent assay (ELISA) to detect *Phytophthora* by serological techniques using the indirect method and the double antibody sandwich method.

The indirect method can be used for the detection and measurement of antibodies. It is carried out as follows: 1) The relevant antigen is attached to the solid phase, which is then washed. 2) The diluted test serum is added, incubated, and washed. 3) An enzyme-labeled anti-species-globulin is added and allowed to react, and then washing is repeated. Unlabeled anti-species-globulin followed by labeled antigen to this second antibody can also be used. 4) The substrate is added. Degradation of substrate results in a color change. The amount and rate of color change is related to the amount of antibody in the test serum in step 2.

The double antibody sandwich method detects and measures antigen and is carried out as follows: 1) Immunoglobulin containing specific antibody or purified specific antibody (to the antigen to be measured) is attached to the solid phase, which is then washed. 2) The test solution thought to contain antigen is then incubated with the sensitized solid phase, which is then washed. 3) Enzyme-labeled antibody specific to the antigen is then incubated; the solid phase is followed by washing. 4) Substrate is added. The color change is proportional to the amount of antigen in the test solution in step 2. (Redrawn by Ahmad Vaziri by permission of Nufield Laboratories of Comparative Medicine, London, courtesy Voller et al. 1979) clonal antibodies sensitive to antigens specific to both *P. fragariae* and *P. cactorum,* the cause of leather rot of strawberry. Mohan (1988) suggested that the ELISA method would be quicker and more reliable than visual symptomology for certification of disease-free rootstocks, since positive ELISA values could be obtained from infected but symptomless plants. A subsequent study indicated that the antiserum from *P. fragariae* detected not only *P. fragariae* and *P. cactorum* but also *P. erythroseptica* (Mohan 1989a).

P. fragariae was detectable by DAS-ELISA and also by the indirect method (Figure 2.5) (Amouzou-Alladaye et al. 1988); however, DAS-ELISA was more specific than the indirect ELISA method. In strawberry plants inoculated with *P. fragariae,* positive reactions were detected about 15 days later. In addition to *P. fragariae,* isolates of *P. cactorum* from strawberry but not from other hosts could be detected by ELISA. *P. infestans, P. cinnamomi, P. parasitica, P. citricola,* and *P. megasperma* were not detectable. Only one of three isolates of *P. palmivora,* only one of three isolates of *P. syringae,* and only one of three isolates of *P. citrophthora* were detectable, but only at low ELISA values. *Pythium middletonii* was detectable by the DAS-ELISA method.

For detection of *P. cryptogea* in infected roots of large numbers of ornamental crops, commercial kits show approximately the same reliability as isolation on selective media (MacDonald et al. 1990).

Commercial kits were utilized to detect *P. cinnamomi* on azalea roots (Benson 1991). A rapid assay kit and a multiwell kit were compared with isolation on 3-P medium (Appendix 2.1). Agreement between recovery of *P. cinnamomi* on 3-P medium and the positive assay on the kits was consistent 3–5 weeks after inoculation when root symptoms were present. The rapid detection kit shows the most promise, but more experience and research are required before it can entirely supplant isolation.

Ali-Shtayeh et al. (1991) have shown ELISA kits to be useful for detection of *Phytophthora* in bodies of water. We quote (reprinted by permission of the American Phytopathological Society, St. Paul, Minnesota, courtesy of Ali-Shtayeh et al. 1991):

> Commercially available ELISA tests for detecting *Phytophthora* and *Pythium* species were used to detect propagules of these genera in water samples. Samples were filtered through 0.45-µm filters to concentrate propagules before extraction. Heating filter residues to 100°C for 5 min yielded extracts that were equal or superior in reactivity to those obtained by liquid nitrogen disruption. As few as 30–40 zoospore cysts were detected in filter pad extracts. Detection sensitivity (propagules per liter) depends on the volume of water that can be passed through the filters. A water sample collected in late winter from a recycling pond in a northern California nursery did not contain any viable propagules of pythiaceous fungi or detectable antigen. A similar water sample collected in early spring from a southern California nursery contained 442 viable propagules per liter of water. The species recovered on agar media were *Phytophthora parasitica, P. citrophthora, P. cryptogea,* and an unidentified *Phytophthora* species, in addition to *Pythium coloratum, P. rostratum, P. middletonii, P. ultimum* var. *sporangiferum,* and *Pythium* 'L' group. Reactivity of filter extracts in *Phytophthora* ELISA tests was proportional to the quantity of filter residue extracted. All of the *Phytophthora* species recovered reacted positively with *Phytophthora* kits and negatively with *Pythium* kits. All of the *Pythium* species recovered also reacted positively with *Phytophthora* kits and negatively with *Pythium* kits. With a relatively simple filter-extraction procedure, the ELISA kits effectively detected *Phytophthora* and *Pythium* in water samples. Although the tests are fallible at the genus level, they still could be used in commercial nurseries, where there is no known acceptable tolerance for these genera in irrigation water.

A study employing commercially available ELISA kits (Neogen Corporation, Lansing, Michigan, 48912) for assay of *P. parasitica* in Florida citrus orchards and *P. citrophthora* and *P. parasitica* in California orchards was reported by Timmer et al. (1993). The ELISA data on detection of the two *Phytophthora* species were compared with findings from dilution plate assay on selective media and with percentage root tip infection. In Florida, root ELISA values with commercial kits D and E were correlated with percentage root infection and propagule densities of *P. parasitica* on an orchard basis. In California, root ELISA values with kit D were correlated with propagule densities of *P. citrophthora* in soil samples from roots (rhizosphere soil) on an orchard basis. Detection of *Phytophthora* in rhizosphere soil from roots in kit D was more efficient than in kit E. Soil ELISA values obtained with kit E were not related to other measures of *Phytophthora* activity. Kit D was less sensitive to *P. citrophthora* hyphae than to hyphae of *P. parasitica.* Data from ELISA values with kit D were as useful a measure of *Phytophthora* activity as measurements of propagule densities obtained by dilution plate assay. Timmer et al. (1993) state, "Soil propagule determinations and root ELISA both provide useful information on *Phytophthora* activity in citrus orchards. The choice of method should depend on materials and personnel available, the ease of sample collection and processing, and the cost of the procedures."

For various studies, including evaluation of host resistance, the intensity at which leaves are infected is useful information. The lesion size can be measured (Umaerus and Lihnell 1976), but this does not indicate the amount of mycelium in the host tissue. The amount of *P. infestans* mycelium in infected potato leaf tissue can be estimated by ELISA (Harrison et al. 1990). ELISA detected *P. infestans* more strongly than *P. erythroseptica* but not other potato pathogens. ELISA results of infected potato tissue correlated well with known amounts of mycelium.

Monoclonal antibodies have been demonstrated to be useful for study of the composition and structure of cell walls of *P. sojae* (*P. megasperma* f. sp. *glycinea*). These antibodies were selected for specificity to extracellular glycoproteins and purified cell walls of this fungus. Although competitive binding assays showed dramatic differences between races and between different isolates of the same race, none of the 11 immunodominant epitopes that were identified was correlated with avirulence genes. A monoclonal antibody VIII was specific for β-1,3-linked terminal glucosyl residues and a monoclonal antibody X recognized nonterminal β-1,3-linked glucosyl residues (Wycoff and Ayers 1990). Although these interesting results will require much more research, there is a great possibility that this type of technology will eventually become important not only for pathogen detection but also for a basic understanding of the nature of pathogenesis of *Phytophthora* to plants with different degrees of resistance.

Use of monoclonal antibodies against surface components of *P. cinnamomi* zoospores by Hardham and her colleagues (1985, 1986, 1991a, b; Gabor et al. 1993; Cahill and Hardham 1994) has allowed practical as well as fundamental studies of the immunocytochemistry of organelles in zoospores. Although this subject is beyond the scope of this chapter, their recent review will be cited here. As an example, the use of immunolabeling allowed characterization of the peripheral vesicles in zoospores. Contents of these vesicles, which condition the adhesion of zoospores to surfaces, were labeled and clearly shown by this method (Hardham et al. 1985, 1991a; Hardham 1987). Adhesion of zoospores to surfaces is discussed in Chapter 3. Detection of β-1,3-glucans in hyphal wall tissue of *P. sojae* (*P. megasperma* f. sp. *glycinea*) was aided by use of monoclonal antibodies specific for certain isotopes combined with immunofluorescence and immunological labeling techniques (Wycoff and Ayers 1991). Such studies are useful to determine the nature of host-specific resistance to soybean.

Twenty-four monoclonal antibodies (Mabs) obtained from aldahyde-fixed zoospores of *P. cinnamomi* (Hardham et al. 1985, 1986, 1991a, b) were utilized as a test for several species of unrelated fungi. When these Mabs were used in an immunofluorescence assay (IFA), 11 of the 24 were specific for zoospores and cysts of *P. cinnamomi* but not for any other related or unrelated species. One Mab reacted with zoospores and cysts of all species of *Phytophthora* tested and was considered to be genus specific. By use of ELISA, 10 Mabs detected only *P. cinnamomi.* Of these 10 Mabs, two were *P. cinnamomi* specific (Gabor et al. 1993). Cahill and Hardham (1994) report a novel method in which the taxis of zoospores is exploited to identify *P. cinnamomi* and other *Phytophthora* species by use of a dipstick immunoassay with species-specific Mabs.

A summary of diagnostic assays for plant-pathogenic fungi in which methods involving hybridoma and DNA techniques are described is provided by a group of selected authors from Europe (Schots et al. 1994).

The report by Converse et al. (1988) of laboratory detection of viruses and mycoplasmalike organisms by ELISA in the United States, the United Kingdom, Canada, Germany, Chile, Israel, and Japan indicates the considerable world cooperation in detection of these pathogens. On the basis of several reports (Amouzou-Alladaye et al. 1988; Mohan 1988, 1989a, b), there is every reason to believe that the ELISA and similar techniques will become useful for detection of *Phytophthora* species in strawberry and other intensively grown crops in which detection of plant disease is highly important.

Detection in Soil by Artificially Prelabeling Propagules with Fluorescent Compounds

Continuous flow centrifugation is a method advantageous for collecting zoospores of aquatic fungi (Fuller and Poyton 1964). Water is collected in sterile bottles, refrigerated, and centrifuged in a Sorvall SS-3 automatic superspeed centrifuge in a 5°C room or in a Sorvall RC-2 automatic superspeed centrifuge run at 5°C. A Sorvall KSB-3 Szent-Gyorghi and Blum continuous flow system is necessary to concentrate the particulate matter. The centrifugate is resuspended and plated by streaking it onto a suitable selective medium. A *Phytophthora* species is reported to have been isolated from sea water, but no details were reported. Although a sophisticated centrifuge system is required, this should be an efficacious way to sample for *Phytophthora* zoospores in bodies of water. It is surprising that this method has not been more widely utilized.

In the study of activity of fungi in soil, the opaque nature of soil and the interference of soil particles prevent visualization of the fungi by conventional microscopic techniques. Above, we described the immunofluorescence methods of Malajczuk et al. (1975) and MacDonald and Duniway (1979), in which an antibody conjugated to a stain that fluoresces in ultraviolet light was used. These methods require specialized equipment and expertise. However, the use of "brightener" compounds that bind to the cell wall of the fungus propagules as reported by Darken (1961) is a much simpler method. This method was utilized to detect *Phytophthora* by Tsao (1970a) with Calcofluor and by Gisi (1975a, b) with "diethanol." A later report compared the use of diethanol and Calcofluor (Gisi and Schwinn 1976). The advantages and disadvantages of each compound are discussed below.

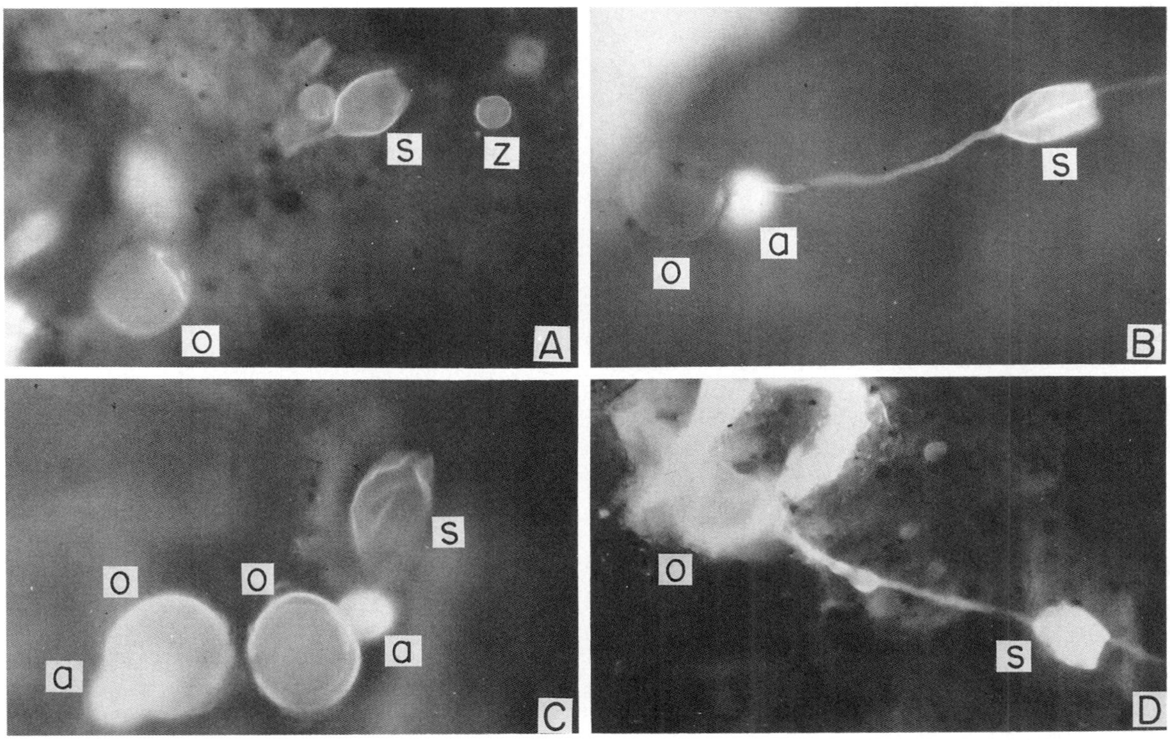

Figure 2.6. Use of fluorescent microscopy to visualize diethanol-labeled oospores of *Phytophthora medicaginis* (*P. megasperma* f. sp. *medicaginis*) germinating in saturated soil by formation of sporangia and release of zoospores. o = oogonium; a = antheridium; s = sporangium; and z = zoospore. (By permission and courtesy of T. L. Kuan. See Kuan and Erwin 1980b)

Calcofluor White M2R New (American Cyanamid Co., Princeton, New York, 08540), a glucan-labeling compound, is composed of the disodium salt 4,4′-*bis*(4-anilino-6-diethylamino-*s*-triazin-2-ylamino)-2,2′stilbene-disulfonic acid (70% active ingredient). *P. parasitica* can be grown in a V8 juice-$CaCO_3$ broth (Appendix 3.2) containing the M2R Calcofluor compound at 280 or 300 ppm. After the 5- to 7-day-old mycelial mat is washed three times, the propagules fluoresce a bright bluish white when viewed with ultraviolet light. Since propagules can be detected even in opaque soil smears on glass slides, Calcofluor is a valuable research tool for detection of the fungus (Tsao 1970).

In some cases, propagules (oospores, chlamydospores, and cysts) are labeled with Calcofluor (100–280 ppm) for 2–5 h in darkness. To prevent spurious fungal germination during the labeling period, the Calcofluor can be made up in 0.1 M phosphate buffer (pH 7.0), which is fungistatic to germination (Tsao 1970). After labeling, the spores are washed by centrifugation. Propagules are observed in soil by suspending them in a layer (2 mm) of soil in 60-mm-diameter petri plates wetted to 24–32% moisture. Soil samples (about 100 mg in a 0.05-ml drop of water) are spread on microscope slides. The large particles and sand grains are removed with a spear, and the slightly dried (about 4–5 min) sample is covered with a low fluorescence mineral oil on which a 22- × 40-mm glass coverslip is mounted (Tsao 1970).

An alternative method involves placement of Calcofluor-labeled propagules on fiberglass squares (2 × 2 cm) on the surface of an agar medium. The germinated propagules are then washed in water and incorporated either horizontally or vertically into soil at a desired depth. After the desired incubation time, the fiberglass squares are washed in water and observed in mineral oil on glass slides. This method is less quantitative than the soil-smear method (Tsao 1970).

Another brightener compound, diethanol, 4,4′-bis[2-diethanol-amino-4-(3-sulfophenylamino)-5-triazinyl-(6)-amino]-stilbene-2-2′disulfonic acid sodium salt, can be used at a concentration of 1,000 ppm as a vital stain (Gisi 1975a). Diethanol is advantageous over Calcofluor in that it has higher water solubility and a pH of 5.5, compared with the lower water solubility and higher pH (10.0) for Calcofluor. Both compounds have affinities for glucans, which are the major carbohydrate components of *Phytophthora* cell walls. Diethanol stimulates growth at 100–500 ppm and stimulates sporangia production by *P. cactorum* at 1,000 ppm (76–105% over the control). Neither compound stains the papillae of sporangia (Gisi and Schwinn 1976).

Oospore germination of *P. medicaginis* (*P. megasperma* f. sp. *medicaginis*) in soil can be demonstrated with diethanol-stained propagules (500 ppm for 12 h) (Kuan and Erwin 1980b). Mycelial mats (48 h old), washed twice with water, were incubated in the diethanol solution (250 ppm) for 14 h. The mats were washed by centrifugation three times to remove the excess diethanol and blended for 30 s in a Waring blender at low speed to break up the mycelium. Mycelial fragments were applied to autoclaved Nitex cloth (100-µm pore size) and incubated in soil at different matric potentials. When observed under ultraviolet light, mycelium at matric potentials of less than –50 mb lysed, but at matric potentials at 0 mb (saturated soil), mycelium grew and penetrated alfalfa roots. Oospores germinated and formed sporangia and zoospores in saturated soil (Figure 2.6).

Fluorescein diacetate stains mycelium and oospores of *P. sojae* (*P. megasperma* f. sp. *glycinea*) in soil. Fluorescein diacetate is dissolved in acetone (2 mg/ml), stored at 5°C, and diluted to 10 µg/ml for use in 60 mM phosphate buffer at pH 7.5. The stain is added to the soil sample (9 ml to 5 g of soil) to which mycelial fragments have been added. The fungus is examined by making a soil smear on a microscopic slide and using a Zeiss epi-fluorescence microscope (Cohen 1984). Oospores can be freed of mycelium and placed on Gelman filter disks (13 mm in diameter) or black Millipore (0.45-µm, 13-mm diameter) disks and incubated in soil between pieces of nylon mesh. The filter disks can be retrieved from soil and treated with a stain, blotted dry, and observed. Fluorescein diacetate detected oospores of *P. sojae* readily; however, the vital stain tetrazolium bromide was more effective and could be viewed by ordinary light microscopy (Cohen 1984).

Detection of a Metalaxyl-Resistant Isolate in Soil

P. medicaginis (*P. megasperma* f. sp. *medicaginis*), the cause of root rot of alfalfa, produces oospores, hyphal swellings, and mycelium in roots, all of which are eventually released into the soil. To study the fate of these propagules, Stack and Millar (1985a) isolated a wild-type, metalaxyl-resistant colony from a large population of zoospores by selection on a metalaxyl-containing medium. The isolate was used to artificially infest soil so that it could be detected among many naturally occurring isolates and species of *Phytophthora* and *Pythium* by isolation on Masago's medium (Appendix 2.3) amended with metalaxyl (20 mg/liter).

Detection of *Phytophthora* by DNA Probes

The use of DNA (deoxyribonucleic acid) probes to identify pathogens provides a highly specific tool. With DNA technology, both *P. parasitica* (P. H. Goodwin et al. 1989) and *P. citrophthora* (P. H. Goodwin et al. 1990b) can be specifically detected among a wide range of other species of *Phytophthora* and *Pythium.* This exciting new technology is rapidly becoming a new frontier for research on detection and identification of *Phytophthora* species and biotypes and may help not only to differentiate species with similar morphology but also to detect species in roots and soil with more precision.

Random cloned DNA fragments that are specific to *P. parasitica* were utilized to detect *P. parasitica* in soil and infected tomato tissue (P. H. Goodwin et al. 1990b). We quote from the first characterization of a DNA probe, as reported by P. H. Goodwin and colleagues (1989). (Reprinted by permission of the American Phytopathological Society, St. Paul, Minnesota, courtesy of P. H. Goodwin et al. 1989):

> Chromosomal DNA of *Phytophthora parasitica,* digested with *Hin*dIII and *Eco*RI, was ligated to pUC8 and used to transform *Escherichia coli.* Three recombinant plasmids were identified that hybridized to *P. parasitica* DNA but not to DNA of other *Phytophthora* species, healthy tomato roots, or *Pythium* species. DNA from all isolates of *P. parasitica* tested, including *P. parasitica* var. *nicotianae,* hybridized strongly with the probes. Southern blot hybridizations revealed that the probes hybridized to multiple restriction fragments, suggesting that the cloned sequences recognize repetitive DNA. There was no homology between the three cloned DNA fragments of *P. parasitica.* A non-radioactively labelled DNA probe, chemically modified by sulfonation, was as sensitive as a ^{32}P-labelled DNA probe for the detection of purified *P. parasitica* DNA. These species-specific DNA probes provide a sensitive new tool for the identification of *P. parasitica.*

In a subsequent report, P. H. Goodwin et al. (1990a) stated that DNA from *P. parasitica* had been detected by dilution plating of infested soil, in crushed infected leaf disks, and in infected tomato roots. Two different DNA fragments of *Phytophthora* were cloned, neither of which was homologous with DNA from plants or other fungi. The DNA hybridized with isolates from citrus but not from cacao. *P. parasitica* was not homologous with *P. cactorum, P. cambivora, P. capsici, P. cinnamomi, P. citricola, P. cryptogea, P. erythroseptica,* or *P. hibernalis.*

Summary

Phytophthora can be most successfully isolated from recently infected plant tissue by direct transfer of washed and blotted-dry infected tissue to a selective medium. Surface disinfestation with a sodium hypochlorite bleach solution at 0.525% concentration can be used if the infected part of the plant is large; however, on fine roots, chemical treatment may suppress the target *Phytophthora,* and it may be best to rely only on washing with sterile water. If a selective medium is not available, either water agar or dilute (1/10) V8 juice, cornmeal, or other suitable weak media should be used. *Phytophthora* is easily suppressed by contaminating fungi and bacteria.

An agar medium, such as cornmeal agar, can be made selective by addition of antibacterial and antifungal chemicals, thus allowing isolation and detection of *Phytophthora* species both qualitatively and quantitatively in soil. The most efficacious medium for isolation of *Phytophthora* species from infected plant material and soil by the dilution plate method is the P_{10} ARP medium in cornmeal agar. P_{10} indicates pimaricin at 10 ppm, which is a concentration toxic to most fungi except *Pythium* and *Phytophthora* species. However, *Mortierella* species, a member of the Mucorales, which occur in some soils, are also not inhibited (Tsao and Guy 1977). Pimaricin was originally reported to be used at 100 ppm (3-P medium), but later studies showed that concentrations of 5 or 10 ppm were more efficacious because the higher dosage was suppressive to spore germination of *Phytophthora* species. Vancomycin (V) at 200 ppm suppresses both gram-positive and gram-negative bacteria. Pentachloronitrobenzene (P) is antifungal to many species but not to *Phytophthora* or *Pythium.* Ampicillin (A) at 250 ppm is toxic to gram-positive bacteria, and rifampicin (R) is toxic to gram-negative bacteria. These last two antibiotics are less expensive than vancomycin and, according to some reports, are more efficient antibacterial components than vancomycin. Hymexazol at 25 and 50 μg/ml is added to some media because it has fungistatic activity against many *Pythium* species that grow more rapidly than *Phytophthora* on media containing pimaricin. Hymexazol also suppresses certain *Mortierella* species, which produce petallate colonies that superficially resemble *Phytophthora.* Hymexazol is too toxic to some species of *Phytophthora* to be of use, however (Tables 2.2 and 2.3). All antibiotics are added to cooled (42–45°C) agar media after autoclaving because they may be destroyed by heat. Since pimaricin is light sensitive, the selective agar plates should be stored and incubated in darkness.

Benomyl, the active ingredient of the systemic fungicide Benlate, is active at 10–25 μg/ml against a wide range of fungal species except dark-spored members of Dematiaceae in the Fungi Imperfecti and Mucorales. It is also ineffective against bacteria. Benomyl can be used as a supplement to pimaricin and as a substitute for pimaricin if it is not available; however, benomyl has a narrower range of activity against nonpythiaceous fungi than pimaricin. Since it is relatively heat stable, benomyl can be autoclaved with the medium.

Baiting infested soil or infected plant material by use of susceptible parts of plant hosts such as leaf disks, cotyledons, various fruits such as apple, lemons, and avocado, and seedlings of various plants is useful to detect and isolate *Phytophthora* species. The soil is inundated with copper-free water at a water-soil ratio of about 4 or greater in paper or plastic drink containers, and the bait is placed in the water. Leaf disks and seedlings are usually floated on the water, but fruit is usually submerged so that one-half to one-third of the fruit is above the water. If *Phytophthora* species are active, typical lesions are caused by zoospore infection near the water line on fruits such as avocado and lemons. *Phytophthora* can be detected microscopically on the lesion and, if necessary, isolated by tissue plating on a selective medium. Population of *Phytophthora* in a natural situation in the field fluctuates from nondetectable to a high inoculum density, often within a short time. Therefore, absence of *Phytophthora* on soil dilution plates, even with a selective medium, must be interpreted with caution. *Phytophthora* is generally accepted to be a multicyclic fungus with the capacity to increase rapidly in the soil within a short period of time. This is caused by the formation of sporangia on germinating survival propagules (e.g., oospores or chlamydospores) and subsequent formation of zoospores.

Quantitative estimates of the population of *Phytophthora* in soil can be made by the soil dilution plate method. Soil, collected in the field, is usually prepared for isolation as soon as possible after collection (cold storage suppresses *Phytophthora*). Soil is diluted with water at 1:10 or, if populations are high, at even greater dilutions. The soil-water suspension should be stirred before it is transferred to petri plates containing a selective medium. The preferred method is to transfer 1-ml portions to each petri plate, add the fresh selective molten agar medium (42–45°C), and then shake the plate horizontally to distribute the soil suspension uniformly in the medium before it congeals. An alternative method is to add the soil suspension to the top of a solidified selective agar medium in a petri plate and distribute it over the surface with a glass hockey stick. After incubation for 48 h, the soil is washed off, and the *Phytophthora* colonies that have become established are counted. The advantage of the latter method is that a large number of plates can be poured prior to the isolation procedure. There is a possibility that propagules that have not germinated within the 48-h period will be washed away. Distribution of the soil suspension in the cooled but still molten medium usually allows isolation of more colonies per gram of soil, perhaps because of the more intimate effect of the antibiotics on propagules of nontarget fungi or because none of the nonestablished *Phytophthora* propagules were washed away.

When colonies are counted, the population is expressed as propagules per gram of dry soil. The percentage of water in soil can be calculated by weighing appropriate samples of the soil that is being assayed before and after drying at 105°C to drive off the water.

A method called the extended assay involves incubating either dried or undried soil remoistened to about field capacity for several days followed by flooding the soil in water for a time prior to isolation by dilution plating or baiting. This method amplifies the inoculum concentration in the soil, probably because of the germination of propagules such as chlamydospores or oospores with subsequent production of sporangia and zoospores. Although this method of amplifying the inoculum may detect a higher count of *Phytophthora* propagules at a qualitative level, the measure does not strictly measure the active quantity of propagules in the field soil because the proportionate increase in inoculum caused by wetting cannot be determined.

Phytophthora can be detected by serological methods. Serology is based on the ability of antibodies from the blood of a mammal (e.g., rabbits or mice) to recognize specifically a foreign protein or carbohydrate. There is a range of specificities of different antibody systems. Some antibodies (polyclonal) recognize both *Pythium* and *Phytophthora;* some recognize only *Phytophthora;* and others recognize only single species. Monoclonal antibodies are derived from the response of a single mammalian cell in tissue culture to an antigen in *Phytophthora.* Monoclonal antibodies are much more specific but require more highly technical treatments. The most practical use of serology for detection is the well-known enzyme-linked immunosorbent assay (ELISA), which is now prepared at commercial levels to detect *Phytophthora* species in soil and plant samples. The ELISA method is also being used in research to quantitate the amount of *Phytophthora* in plant tissue as affected by the selective levels of host plant resistance.

Rivers, streams, and irrigation canals and reservoirs can be sources of *Phytophthora* inoculum. The inoculum can be detected by use of suitable baits, which are incubated in the body of water for a period varying from days to weeks. The baits are subsequently treated for isolation of *Phytophthora* on a selective medium. A more direct method involves passage of large volumes of water through appropriate filter disks, such as Millipore or Nitex nylon screens, with pore sizes large enough to retain *Phytophthora* propagules (e.g., 0.45–10 µm). Most *Phytophthora* propagules are retained on a 10-µm screen. The amount of water that can be passed through the filter prior to its becoming clogged with organic matter is the major limitation.

The filter pads or screens can be plated on a selective medium and the *Phytophthora* colonies counted and/or isolated. The data can be based on the number of *Phytophthora* colonies isolated per volume of water passed through the screen. The ELISA method has also been used to detect different *Phytophthora* species in bodies of water.

The fate of *Phytophthora* propagules in soil can be studied by use of biochemical techniques that label the propagules prior to placing them in soil. Compounds that become attached to glucan moieties in the cell walls of *Phytophthora* can be detected by ultraviolet light microscopy. Calcoflor and diethanol, known as brighteners, can be applied to propagules prior to their placement in soil. The propagules can be visualized microscopically in opaque soil smears because only the treated propagules fluoresce in ultraviolet light.

Fluorescent dyes, which are conjugated to an antibody specific to *Phytophthora,* can also be used for pretreatment of propagules before placement in soil. Hybridization of DNA specific to different *Phytophthora* species has also been used to detect *Phytophthora* in soil.

Since most of the natural population of *Phytophthora* in soil is sensitive to metalaxyl, resistance to metalaxyl by a selected single zoospore isolate in *P. medicaginis* (*P. megasperma* f. sp. *medicaginis*) has been exploited in field soil preinoculated with the metalaxyl-resistant isolate. Subsequently, the metalaxyl-resistant isolate can be detected by dilution plating among naturally recurring *Phytophthora* and *Pythium* species by conventional methods on an agar medium containing metalaxyl.

APPENDIX TO CHAPTER 2

Isolation Media and Methods

2.1. 3-P Medium (Eckert and Tsao 1960, 1962)

Difco cornmeal agar (17 g/liter) is autoclaved at 121°C (cooled to 45°C) and amended (μg/ml = mg/liter = ppm) with pimaricin (100), penicillin (50), and polymyxin (50). Plates should be stored under refrigeration in the dark because pimaricin is light sensitive. This is the first selective medium to utilize the antifungal pimaricin and is effective for isolation from freshly diseased plant material in which *Phytophthora* species exist as mycelium but not from freshly diseased soil in which propagules exist as spores. Since pimaricin at the high dosage of 100 mg/liter inhibits spore germination (Tsao and Ocana 1969; Tsao 1983), $P_{10}VP$ and $P_{10}ARP$ or P_5ARP (see Appendixes 2.2 and 2.4) are preferable to the 3-P medium.

2.2. $P_{10}VP$ Medium (Tsao and Ocana 1969)

Difco cornmeal agar (17 g/liter) is autoclaved (cooled to 45°C) and amended (μg/ml) with low dosages of pimaricin (10), vancomycin (200), and pentachloronitrobenzene (100). Plates should be stored in darkness because of sensitivity of pimaricin to light. This medium is effective for isolation of propagules from soil and plant tissue because the reduced dosage of pimaricin does not inhibit spore germination. Hymexazol (25–50 mg/liter) can be added after autoclaving to suppress some species of *Pythium* (Masago et al. 1977; Tsao and Guy 1977). See Tables 2.2 and 2.3 for relative sensitivities of different *Phytophthora* and *Pythium* species.

2.3. Selective Isolation of *Phytophthora* from *Pythium* spp. by Use of Hymexazol (Masago et al. 1977)

Potato-dextrose agar (1% agar) is amended (μg/ml) with benomyl (10), pentachloronitrobenzene (25), nystatin (25), ampicillin (500), rifampicin (10), and hymexazol (25–50). Hymexazol suppresses many but not all *Pythium* species; many *Phytophthora* species are tolerant, but others are too sensitive to be isolated on a medium containing hymexazol (see Tables 2.2 and 2.3). Cornmeal or dilute V8 agar are better choices than potato-dextrose agar (see Appendix 2.4 and Tables 2.3 and 2.4).

2.4. $P_{10}ARP$ (Kannwischer and Mitchell 1978) and P_5ARP Media (Papavizas et al. 1981; Jeffers and Martin 1986)

Cornmeal agar is amended (μg/ml) after autoclaving (cooled to 45°C) with pimaricin (10), ampicillin (250), rifampicin (10), and pentachloronitrobenzene (100). $P_{10}ARP$ is modified by Jeffers and Martin (1986) by reduction of pimaricin from 10 to 5 μg/ml for isolation of *P. cactorum* (P_5ARP). The addition of rifampicin to control gram-positive bacteria has made $P_{10}ARP$ and P_5ARP the media of choice by many for isolation of most *Phytophthora* species. Hymexazol can be added at 25–50 μg/ml after autoclaving to suppress some species of *Pythium* (Masago et al. 1977; Tsao and Guy 1977) if the target *Phytophthora* species will tolerate it (see Tables 2.2 and 2.3).

2.5. Media Utilizing Benomyl (Sneh and Katz 1988)

Benomyl suppresses many fungi (see text). Amendments to a selective synthetic medium (Sneh 1972a; McIntosh 1975) include (μg/ml) benomyl (25), pentachloronitrobenzene (30), penicillin (60), chloramphenicol (30), penicillin (60), polymyxin (50), and hymexazol (25) for isolation of *P. nicotianae* and *P. citrophthora* from citrus soil. Chloramphenicol and polymyxin are not the optimal antibiotics (see text).

George and Milholland (1986b). Amendments to frozen pea medium (μg/ml) include benomyl (10), pimaricin (10), rifampicin (10), and hymexazol (50) for isolation of *P. fragariae* from plant tissue.

Papavizas et al. (1981). Amendments to cornmeal agar (adjusted to pH 4.0) (μg/ml) include benomyl (25), pimaricin (5), vancomycin (200), penicillin (100), pentachloronitrobenzene (100), and hymexazol (20) for isolation of *P. capsici* from soil.

Ponchet et al. (1972); Ricci (1972). Amendments to commercial malt agar (μg/ml) include benomyl (15), penicillin G (250), polymyxin B (250), and pentachloronitrobenzene (100) for isolation of *P. nicotianae*. See PBNC medium (Schmitthenner 1973; Rahimian and Mitchell 1988) in Appendix 2.6.

Bischt and Nene (1988). BHM PPVR containing (μg/ml) Benlate (20), hymexazol (20), mycostatin (10), pentachloronitrobenzene (20), pimaricin (5), vancomycin (200), and rifamycin (10) was selective for *P. drechsleri* f. sp. *cajani* from soil and pigeon pea debris.

2.6. Media Utilizing Chloramphenicol

PBNC (Schmitthenner 1973; Rahimian and Mitchell 1988). A weak V8 juice (40 ml/liter) medium buffered with $CaCO_3$ (0.6 g/liter) and amended with (g/liter) yeast extract (0.2), sucrose (1.0), cholesterol (0.01 in 2 ml) *N, N*-dimethylformamide, and agar (20) includes (μg/ml) Benlate (50% benomyl) (20), pentachloronitrobenzene (27), neomycin sulfate (100), and chloramphenicol (10) and facilitates isolation of several *Phytophthora* species. Also, *P. cactorum* was isolated from pear and apple fruit (Grove and Boal 1991). A subsequent report by Canaday and Schmitthenner (1982) substituted rifampicin (90 μg/ml) for chloramphenicol and added hymexazol (29 μg/ml) for a soybean leaf bait assay of *P. sojae* (*P. megasperma* f. sp. *glycinea*). Chloramphenicol and neomycin are probably too toxic to *Phytophthora* growth to be useful for isolation of most *Phytophthora* species (see text).

PCH medium (Shew and Benson 1982). PCH is a semisynthetic medium that includes (g/liter) KH_2PO_4 (1.0), $MgSO_4 \cdot 7H_2O$ (0.5), KCl (0.5), $CaCl_2 \cdot H_2O$ (0.01), $FeSO_4 \cdot H_2O$ (0.02), thiamine·HCl (0.001), yeast extract (0.3), $NaNO_3$ (1.0), glucose (15.0), pentachloronitrobenzene (35.0), chloramphenicol (0.01 in 2 ml of ethanol), and hymexazol (50.0) and is used for isolation of *P. cinnamomi.*

Soil samples (up to 50 mg) are blended in a Waring blender at low speed for 5–30 s and passed through nested 20-cm-diameter sieves of 25- and 38-μm pore sizes. The residue on the 25-μm sieve is washed with running tap water and discarded. The residue on the 38-μm sieve is rinsed with tap water, suspended in 20–30 ml of deionized water, and distributed over the surface of five to 15 PCH selective agar plates and incubated in the dark for 48–72 h at 20°C. The soil on the surface of the agar is washed off, and the colonies of *P. cinnamomi* are counted. The yield of colony-forming units (87% from chlamydospores) of *P. cinnamomi* was 3.5 per gram from an azalea soil and 12.6 from a Fraser fir soil. The recovery values were significantly higher on the PCH medium than on P_{10}ARP (Appendix 2.4). PCH medium was also effective for isolation of *P. cinnamomi* by direct plating of unsieved soils. Addition (μg/ml) of chloramphenicol (10), penicillin (5), and pimaricin (5) to the soil suspension prior to dispensing on PCH agar aided maximum recovery. PCH was effective for assaying *P. cinnamomi* in soil suspensions obtained by mechanical elutriation, a method originally designed to assay nematodes in soil (Byrd et al. 1976). The medium was less effective for isolation of *P. parasitica* var. *nicotianae* (H. D. Shew, *personal communication*).

Chloramphenicol may be effective for isolation of *P. cinnamomi,* but for certain other species it should be used with caution (see Tsao [1983] and text). *P. nicotianae* spores, however, germinated in media containing chloramphenicol at 60 μg/ml (Sneh and Katz 1988).

2.7. Serial Dilution End-Point Method (Tsao 1960, 1983; Weste 1983b)

According to the report by Tsao (1960), soil from the field is pulverized to a particle size of 2 mm or less and mixed. Fifty cubic centimeters of the soil (measured in a 50-ml or larger beaker) is placed in a wax paper cup (about 500-ml capacity) and divided into two equal portions by splitting the cup vertically (Figure 2.4). Half of the soil sample is transferred to another cup for use in the lemon fruit bait bioassay. The other half is diluted with an equal volume of an autoclaved soil and labeled 1/2. This sample is further diluted with sterile soil to provide a 1/4 dilution. This process is repeated until diluted to 1/64.

Duplicate cups containing 25 cc of the undiluted soil, the soil in the dilution series, and an autoclaved soil as control are flooded with 150 ml of deionized water. Lemon fruit, used as baits, should be hand picked (to avoid wounding), preferably from trees that have not been sprayed with a pesticide, and need not be surface sterilized. Unwounded lemon fruits are relatively resistant to most soilborne fungi except *Phytophthora.* Two lemon fruits (Eureka or Lisbon) are incubated (peduncle ends up to prevent fruit rot caused by residual latent fungi in the "button") in the soil-water suspension for 6 days at room temperature. Brown rot lesions on each fruit are recorded. A disease potential index (DPI) is defined as the reciprocal of the highest dilution at which brown rot lesions on the fruit occur.

The method detects both *P. citrophthora* and *P. parasitica* and is semiquantitative. The evidence indicates that zoospores forming in flooded soil infect the fruits and cause brown rot lesions at or near the water line. This is consistent with the negative geotropism habit of zoospores (Carlile 1983; see Chapter 3).

The dilution end-point method was modified by Weste (1983b) to isolate *P. cinnamomi* from forest soils in Australia. Four 0.5-kg soil samples from diseased, but not dead, plants containing fine roots are collected and mixed well but not screened. Stored samples should not be refrigerated, and the assay should be conducted within a few hours of sample collection. The mixed soil sample is spread on a table in a circular pattern, and subsamples from alternate quadrants are taken until a 50-g portion is obtained. This procedure is necessary because the original sample contains unscreened, unevenly sized particles (Leeper 1967). The smallest fraction is obtained, not by dilution, but by collecting smaller fractions in a series of dilutions, for instance, 1/1, 1/2, . . . , 1/256, of the original 50-g sample. These soil samples (three replications) are flooded with 100 ml of deionized water in plastic cups and baited by floating germinated seedlings of lupines (*Lupinus angustifolius*) (Chee and Newhook 1965a) or *Eucalyptus sieberi* (Marks and Kassaby 1974) on the soil suspension (Weste and Ruppin 1977). Weste (1983b) states that collection of soil should be from the root systems of diseased plants because of the clustered, uneven distribution of *P. cinnamomi* in the field.

After incubation, the baits should be dried on a paper towel before plating on the selective medium because the population of bacteria increases rapidly in free water. After cotyledons of eucalyptus seedlings become infected, a change in color is characteristic of infection; however, the length of time (4 weeks) to grow seedlings is a disadvantage. Lupine seedlings are nearly as satisfactory and can be used in 3 days. Seedlings can be observed microscopically to detect sporangia and in addition are plated on P_{10}VP selective medium (Appendix 2.2). Positive detection of *P. cinnamomi* gives semiquantitative evidence of disease potential by recording the fraction of the original 50-g sample in which *P. cinnamomi* is recovered; for instance, a 256 reading indicates that *P. cinnamomi* is recorded in the 1/256 portion of a 50-g soil sample.

2.8. Direct and Extended Bioassay Methods (Canaday and Schmitthenner 1982; Stack and Millar 1985c; Jeffers and Aldwinckle 1987; Rahimian and Mitchell 1988)

Direct or Abbreviated Assay. Soil (about 25% H_2O) from a field infested with *P. medicaginis* (*P. megasperma* f. sp. *medicaginis*) is screened (6-mm openings), added to 90-mm petri plates (30 ml/plate), and flooded with 40 ml of distilled water. Six 3- to 5-day-old alfalfa seedlings are floated in the water above the soil for 3 days. The seedlings are removed, blotted dry, and plated on a modified Masago et al. (1977) medium (Appendix 2.3). This method is similar to that used by Marks and Mitchell (1970) and Stovold and Carratt (1978), except that alfalfa seedlings are not injured by pinching with forceps (Stack and Millar 1985c). For use with a metalaxyl-resistant isolate, metalaxyl (20 μg/ml) is added (Stack and Millar 1985a).

Extended Assay. Soil naturally infested with *P. medicaginis* is allowed to dry in petri plates to about 5% moisture (7 days), remoistened with 10 ml of water, and incubated at room temperature for 3 days. The soil is then flooded with 40 ml of water on which six 3- to 5-day-old alfalfa seedlings are floated for 3 days. The seedlings are removed, blotted dry, and plated on a modified Masago et al. (1977) medium (Stack and Millar 1985c).

Isolation by the extended soil dilution method yields an incidence of recovery of *P. medicaginis* higher than that from nonmoistened and incubated soil, probably because sporangia that were formed from germinated oospores during the incubation in moist soil during the flooding period subsequently released zoospores.

In a secondary assay, the water is drained off the soil after removal of the seedlings. The soil is then mixed in 15 ml of water agar containing the antibiotics in the Masago et al. (1977) medium. Soil-agar plugs are then transferred to the modified metalaxyl-containing selective medium and incubated at room temperature (Stack and Millar 1985a, c).

P. sojae (*P. megasperma* f. sp. *glycinea*) is isolated from soil by baiting with soybean (cultivar Amsoy) leaf disks (Canaday and Schmitthenner 1982). Isolation of *Pythium* contaminants is minimized by incubating leaf disks for very short periods (1.5 h) via a method somewhat similar to the extended baiting method described above. Soil from the field is moistened to a matric potential of –100 to –200 mb and stored for 1–2 weeks at 21–25°C before baiting. Presumably, this allows sporangia to form on germinated oospores in the soil. A sample of the soil (5–20 g moist weight) is placed in a small beaker and flooded with distilled water to a depth of 1–2 cm above the soil. After the organic debris is swept from the surface with tissue paper, about 20–30 soybean leaf disks from unifoliate leaves are floated on the water. After 1.5 h, leaf disks are blotted dry on paper toweling and plated on the selective PBNRH medium, which is the same as PBNC medium (Appendix 2.6) except that rifampicin (0.09 µg/ml) is substituted for chloramphenicol and hymexazol (29 µg/ml) is added to suppress *Pythium.* When leaf disks are incubated on the soil suspension for longer periods of time, certain *Pythium* species are more frequently isolated. Plates containing the soybean leaf baits are incubated in the dark for 3 days and observed microscopically for *P. sojae* sporangia. *P. sojae* is successfully isolated from soybean and noncropped fields at all times of the year.

P. cactorum is isolated from soil by a modification of the extended method (Rahimian and Mitchell 1988). Soil samples are stored in polyethylene bags at 5°C until used and then sieved through a 2-mm-mesh screen, thoroughly mixed, and air dried for 2–3 days at room temperature (23–27°C). Subsamples (10 g) of dried soil in each of three 9-cm-diameter glass petri plates are remoistened by adding 4–7 ml of distilled water and incubating for 1–3 days at 24°C under light. The soil is then flooded with 10 ml of distilled water for 2 days, after which the excess water is removed and replaced with 10 ml of sterilized distilled water cooled to 8°C. The plates are incubated at 8°C for 2 h. The soil suspension in test tubes is shaken on a Vortex mixer for 2–3 min. Three 0.5-ml samples are then spread on each of three PBNC (Appendix 2.6) medium plates and incubated at room temperature for 5 days. The plates are examined for *P. cactorum* colonies, each of which has emanated from a single zoospore. The method detects as few as four zoospore propagules per gram of soil. The interference by fast-growing *Pythium* species is reduced because zoospores of *P. cactorum* are released at the low temperature of 8°C (optimum temperature for zoospore release) more readily than those of *Pythium* species such as *P. aphanidermatum,* which releases zoospores optimally at 20°C (Rahimian and Banihashemi 1979).

2.9. Wet Sieving for Isolation of Chlamydospores of *Phytophthora cinnamomi* from Soil (McCain et al. 1967; see also Shew and Benson 1982 in Appendix 2.6)

The purpose of wet sieving is to isolate selectively all *Phytophthora* propagules greater than 38 µm in diameter. Although many chlamydospores of *P. cinnamomi* are greater than 35 µm in diameter, many others range from less than 25 µm up to 35 µm (Stamps et al. 1990). Therefore, some of the chlamydospores pass through the sieve. Soil (50 g) is sieved (6.35-mm mesh) and comminuted with a Waring blender for 1 min at slow speed (about 3,000 rpm). The supernatant is poured through a set of nested sieves with mesh sizes of 149, 61, 44, and 38 µm. The supernatant that passes through the 38-µm sieve is collected and restirred with the Waring blender twice in 200 ml of water at slow speeds for 20 s and again poured through the nested sieves. The residue that collects on the 149-µm sieve and on each of the remaining sieves is washed into separate beakers by holding the screen at an 80° angle and carefully backwashing on the bottom side of the screen with a moderate stream of water from a small rubber tube. The suspension that passes through the 38-µm sieve is rescreened three times through the 38-µm sieve (held at a 60–70° angle), and the residue is collected as above. The residue containing the chlamydospores is plated on a medium containing Campbell V8 juice (1%) and agar (4%) amended (µg/ml) with nystatin (50), vancomycin (100), and pentachloronitrobenzene (10). (On the basis of current knowledge [Tsao 1983], the dosage of nystatin probably should be reduced to 10 µg/ml.) The colonies are counted 24 h later. Sieved chlamydospores germinate readily and grow faster than chlamydospores from unsieved natural soil, probably because bacteria and small-spored fungi are removed by sieving and dilution. This method was adapted from Hendrix and Kuhlman (1965).

CHAPTER 3

Culture, Physiology, and Genetics of *Phytophthora* Species

The genus *Phytophthora* clearly possesses some of the most sophisticated weaponry known among plant pathogens (Gregory 1983).

In the preceding chapter, we describe and discuss the efficacy of different methods of isolation of *Phytophthora* from plant tissue and soil. Now we discuss the different methods and rationale for the culture of *Phytophthora* species and examine the nutritional requirements for growth and production of the asexual chlamydospores, sporangia, and zoospores and the sexual oospore stages. Why is such information necessary? First, the various spore stages must be produced to determine the identity of *Phytophthora* species and to study their responses to various ecological factors. Second, it is necessary to develop inoculum for use in pathogenicity experiments on different hosts and to establish controlled epidemics to identify host plant resistance. Third, knowledge of the nutritional requirements of the fungus is important because such an understanding might lead to the development of a control mechanism. *Phytophthora* species have an absolute growth requirement for thiamine (vitamin B_1). If a regime could be set up in which thiamine was eliminated from the environment, *Phytophthora* simply could not grow. This is probably not possible in nature; however, an in vitro study showed that certain soil bacteria that synthesize thiamine augment growth of *Phytophthora* in a thiamine-deficient medium and other bacteria that utilize thiamine reduce growth of *Phytophthora* (Erwin and Katznelson 1961b). Since thiamine is such a common constituent of the plant flora and is even found in soil, there is probably little chance that this concept could be utilized for control. Nevertheless, this requirement is a vulnerable site in the life cycle of *Phytophthora.*

In addition to other features, the unique nature of sterol metabolism by *Phytophthora* and *Pythium* is important in evolutionary considerations (W. R. Nes and W. D. Nes 1980; W. D. Nes 1987). Griffith et al. (1992) made a strong plea that information on the unique physiology of oomycetes like *Phytophthora* should be more thoroughly exploited in the search for fungicides that act on vulnerable sites. Their essay and review is discussed in Chapter 7.

The physiology of *Phytophthora* needs to be understood because such information can be fundamental to development of better control practices and in understanding how control methods work. An example is the role of water. Many fundamental studies indicate the significant effect of soil water status on production of sporangia in vitro and in soil (Duniway 1983; Gisi 1983). These studies show that sporangia are produced when the matric potential (soil suction) value in soil is slightly more negative (drier) than zero (free water) and that zoospores form and disseminate more readily when the matric potential of soil is zero (saturated soil). This information gives a quantitative explanation for the control of root diseases by the proper management of soil water. Irrigation technology, involving the use of soil moisture tensiometers and neutron probes to measure the water status in soils, is well advanced; however, it probably is not yet as well utilized as it should be (Cook and Papendick 1971; Cook and Duniway 1981). This information has become valuable for the development of an integrated pest management approach to control of *Phytophthora*-caused root disease of tomato (Ristaino et al. 1988, 1989). Indeed, these concepts are the foundation of soil water management for the control of *Phytophthora* in soil on many crops (Duniway 1983). The management of soil water is probably the first line of defense in any control strategy. Often, when a resistant cultivar is not available or fungicidal treatment is not feasible or economical, it is the only strategy.

Phytophthora species, as well as most other oomycetes, are epidemiologically unique because of their ability to increase rapidly in soil and on foliage. They have a short regeneration time, which results from their ability to produce sporangia rapidly and subsequently to produce numerous zoospores under suitable conditions. Although we know much about the production of sporangia and zoospores from in vitro research, many aspects of this process in the real world are still not understood well enough.

Although *Phytophthora* species usually grow best on natural media, it is desirable for research purposes, in which all the variables should be controlled, to grow *Phytophthora* on a synthetic medium, which, by definition, contains only substances that are defined chemically. The results from use of synthetic media should be more reproducible than those from variable organic substances. Probably of more importance, however, is the use of synthetic media to define the roles of different components in the physiology of the fungus. For instance, we know that thiamine at 10–100 mg/liter is absolutely required for growth. A β-hydroxysterol (e.g., sitosterol) at 10–30 mg/liter is absolutely necessary for sporulation but not for growth of mycelium. Sterols are functional as constituents of cell membranes (W. R. Nes 1974; Elliott 1983) but are not synthesized by *Phytophthora* (W. R. Nes 1974; W. R. Nes and W. D. Nes 1980; W. D. Nes 1987) and must be obtained from exogenous sources. This explains to a large extent why the sterol-inhibiting fungicides such as triadimefon and imazalil are ineffective against *Phytophthora* species.

Phytophthora species grow on relatively simple sugars such as glucose and sucrose, and most of the species utilize simple inorganic nitrogen compounds such as ammonium sulfate and potassium nitrate. The growth processes, however, are probably less complex than the production of sporangia. Some *Phytophthora* species, especially foliar pathogens such as *P. palmivora* and *P. infestans,* produce sporangia on many organic substrates; however, the soilborne species have more subtle requirements. *P. cinnamomi* produces sporangia abundantly on nonsterile soil extract but produces much fewer on soil extract that has been sterilized by filtration or heat (Zentmyer 1980; Ribeiro 1983). What is the cause? Is this phenomenon the result of what the bacteria in nonsterile soil extract contribute or to their depletion of components such as sugars? These questions are discussed below.

To determine the answers to many of these questions, the study of growth and sporulation on synthetic media is necessary. In general, growth requires an adequate balance of a carbohydrate and a nitrogen source; sporangial formation occurs when the carbohydrate is depleted, whereas oogonium initiation leading to oospore formation occurs when the nitrogen in the medium is nearly exhausted (Elliot 1989). Although these generalizations are nearly always true, there is some variability between and among the responses of different species of *Phytophthora.*

In an extensive and comprehensive review of *Phytophthora*-caused diseases in Australia, Irwin et al. (1995) concluded that a greater research effort should be made on *Phytophthora* per se, especially on the isolation and characterization of genes that are involved in "development, reproduction, and pathogenesis."

In this chapter, we review the essential information about maintaining pure cultures, nutrition, and the various physiological processes involved in the development of sporangia, zoospores, chlamydospores, and oospores and in germination of these spores.

The explanation of why *Phytophthora* species are variable in so many ways and especially why they are capable of developing new races that are pathogenic to resistant plant cultivars has eluded researchers since this genus was first described in 1876 (de Bary). New information and new techniques are now at hand to explain these phenomena on a genetic basis. Why is *P. infestans* capable of overcoming the most efficient fungicide (metalaxyl) that has ever been produced by developing fungicide-resistant biotypes in the field? These questions are addressed later in this chapter and in Chapter 7, Chemical Control.

Purification from Contaminants

Although fungi that can contaminate cultures are usually obvious because of different growth patterns, bacteria are the unseen plague of *Phytophthora* cultures. Some people are convinced that mycelium can be infected internally by bacteria, but there is little evidence for this. Bacteria can usually be eliminated from a culture by placing a small piece of mycelium on the inside of a glass Van Tiegham cell (Figure 2.1) that has been placed on water agar (Ark and Dickey 1950) or on a selective medium (Appendixes 2.2–2.4) that contains appropriate antibiotics such as ampicillin and rifampicin or vancomycin. The glass cell forces the fungus to grow below the surface of the agar rather than on top of it. Even when the medium contains antibiotics, some bacteria will multiply on the surface of hyphae growing on the top of the agar; however, bacteria usually reproduce sparingly within the agar matrix. Placing a piece of the culture under a flap of agar or inverting a slab of agar over the top of a contaminated culture in a petri plate accomplishes the same thing. *Phytophthora* spores can be transferred to the surface of the agar in a petri plate, after which the entire agar culture is inverted. *Phytophthora* can grow up through the agar, but bacteria and actinomycetes cannot (Schmitthenner 1964). Since many fungi are inhibited by pimaricin or benomyl but *Phytophthora* species are not, either or both of these compounds in an agar medium are selective for *Phytophthora* (see Chapter 2 for selective media useful for isolation of *Phytophthora*).

Phytophthora isolates can be checked for bacterial contamination by incubating a mycelial plug transferred from a test tube culture to a clear broth such as Difco Eugon broth. If the culture is contaminated with bacteria, the broth becomes cloudy after 24–48 h of incubation, but if bacteria-free, the broth remains clear. This simple test saves experimental time that would be lost if the original culture were contaminated.

Single-Spore Isolation

To limit variability, an isolate should be obtained preferably from a single germinated cyst (a zoospore that has lost the flagella) or hyphal tip to make sure that only one biotype is isolated. Hyphal tips are usually multinucleate, but single zoospores are, in the main, uninucleate (Ho and Hickman 1967a; Long et al. 1975). In a study by Converse and Shiroishi (1962), more than 98% of *P. fragariae* zoospores were uninucleate. Thus, a single germinated zoospore should produce a genetically pure culture. However, since *Phytophthora* is diploid in its vegetative state (Shaw 1983; Sansome 1987), genetic homogeneity of subsequent generations cannot be assumed, as it might be with haploid fungi that produce uninucleate spores. Despite the relatively rare occurrence of binucleate zoospores in most studies, the report of Castronovo (1955), which illustrates some binucleate zoospores of *P. infestans,* suggests that a parasexual mechanism for variation (Pontecorvo 1956) could be operative. Mortimer et al. (1977) showed by use of Feulgen stain that 96 and 97% of the zoospores from two sectors of a colony of *P. infestans* were uninucleate. Although this information indicates that the majority of zoospores are uninucleate, it seems possible that the 3 or 4% that were shown to be binucleate could be functional in a parasexual manner. In rare instances, accidental cleavage of zoospores results in fusion of two zoospores. Such a zoospore will be larger and possess four flagella (Desjardins et al. 1970). A *P. palmivora* zoospore with six flagella was assumed to possess three nuclei (Desjardins and Zentmyer 1977). Cleavage of zoospores from aged sporangia of *P. palmivora* is often aberrant, resulting in production of binucleate and multinucleate zoospores. Isolation of abnormally large zoospores should be avoided, since they could have two or more nuclei.

Critopoulos (1955) reported that an isolate of *P. capsici* from Greece produced atypical zoospores that were largely nonmotile. Many of these zoospores settled rapidly to the bottom of the petri plate and many of them fused. The protoplast from the fused zoospores was larger than that of the nonfused zoospores; however, other isolates of *P. capsici* from the United States produced motile zoospores that did not fuse.

Single-spore isolation is readily accomplished by isolation of germinated cysts after they have been spread thinly on the surface of hard, 3–4% water agar. Although hard agar is not absolutely necessary, transfer of the agar containing a single germinated spore is somewhat easier than that of a soft agar. Spores must be distributed far enough apart so that they can be cut out of the agar with a sharp needle and transferred singly to a suitable medium. After the cysts have germinated (about 2–4 h), the germinating spores can be detected on the surface of the agar by use of a dissecting microscope equipped with 25 or 50× objectives and 10× eyepieces if the mirror and light source are properly adjusted. Oblique top lighting with some dissecting microscopes and bottom light with adjustment of the mirror to create a semidark field with others facilitates observation. The germ tubes of spores are much easier to detect on the agar surface than ungerminated spores because they reflect light. The beginner will probably have more trouble finding the upper surface of the agar than actually picking off germinated spores. A handy technique is to make a slight scratch on the surface of the agar to make that plane more visible. The surface can also be marked by deposition of a large number of spores in one spot to help locate the plane of the agar on which spores are germinating. Marking that spot on the bottom of the petri dish with a wick pen makes a convenient reference point. After the surface of the agar is located, the light can be adjusted in intensity and direction to detect the germinated spores by reflected light. If the light intensity is too high, the germinating spores will be invisible.

Tools to isolate single spores vary with the ingenuity of the operator. Sharp needles or small "biscuit cutter" devices (about 1 mm in diameter) can be prepared with flattened nichrome transfer wire to make a thin, flat strip. A small hole (0.5 mm or less) drilled through the flattened tip of a nichrome wire often helps to pick up small pieces of agar containing single germinated spores (Tuite 1969). Some researchers use a sewing needle with a small eye to pick up single spores. Single-spore isolation is an art that can be learned quickly with practice.

The method described above is essentially a modification of the technique described by H. N. Hansen and Smith (1932). Many other single-spore isolation techniques have been described (Hildebrand 1950; Ainsworth 1968). Constantinescu (1988) describes a biscuit cutter mechanism constructed to attach to the objective of a compound microscope. This allows the operator to find the spore and rotate the objective downward to cut out the spore. See Tuite (1969) for a description of the Darley isolator and several other methods.

These techniques would also suffice for isolation of single sporangia, chlamydospores, and oospores; however, since oospores germinate more slowly, they would have to be incubated up to 6 or 7 days for germ tubes to form (see the discussion of oospore germination below).

Organic Media

Several common natural media that are useful for growth of *Phytophthora* are presented in Appendixes 3.1–3.10. The amount of the agar medium (usually 15 or 20 ml) added to standard petri plates (90 mm in diameter) made of glass or plastic should be uniform because radial growth varies with the depth of the agar medium. Usually the diameter of the radial growth is greater on a thin plate of agar.

Use of a standard medium for identification of *Phytophthora* species is important; however, there has been little or no consensus about which medium is best to use. Sporangia often vary considerably in size and shape when produced on different substrates. Some investigators compared isolates of *P. palmivora* and related species on carrot agar (Brasier and Griffin 1979) (Appendix 3.7). Cornmeal agar (Appendix 3.8) was the standard medium utilized by Tucker (1931) and Waterhouse (1963). V8 juice agar (V8A) (Appendixes 3.1 and 3.2) has become one of the most frequently used agar media for growth of *Phytophthora* and other fungi. This medium was first described by Diener (1955) and later modified by Miller (1955). Diener adjusted the pH of Campbell's V8 juice (4.3–4.5) to pH 6.0 with NaOH; but Miller used 3 g of $CaCO_3$ per liter to adjust to pH 7.5 and 2 g/liter to adjust to pH 7.0. Various modifications of V8A and V8 juice broth are given in Appendixes 3.1, 3.2, 3.49, and 3.50. Despite the widespread use of V8A made up with different concentrations of V8 juice and $CaCO_3$, there is little evidence that one recipe is better than another. V8 juice media are now used worldwide, but there is some danger in the universal use of such a medium because the company (Campbell) that markets this product could change the composition. In fact, at one time ascorbic acid (vitamin C), which was found to be repressive to oospore production, was added (O. K. Ribeiro, *unpublished*).

Cornmeal agar and 2% V8A (pH 4.5–4.6) are utilized by Gerrettson-Cornell (1983) in a description of the morphology of several *P. cinnamomi* isolates in Australia. A 2% V8A medium is advantageous for use with the lupine cotyledon bait isolation method (Gerrettson-Cornell 1975, 1980).

Although V8A is favorable for growth of most *Phytophthora* species, some isolates of *P. infestans* do not grow or sporulate as well on V8A as on rye agar (Caten and Jinks 1968; Hohl 1983, 1991a; Galindo and Hohl 1985) (Appendix 3.9). Potato-dextrose agar (PDA) (Appendix 3.4), a rich medium, is often used to demonstrate colony morphology; however, *P. sojae* does not grow on PDA (Kaufmann and Gerdemann 1958; Hildebrand 1959), except when diluted 1:10 (Erwin et al. 1968). Because of the high sugar content of PDA, it is not as useful as weaker media for sporulation or isolation.

Synthetic Media

The nutritional requirements of different *Phytophthora* species for growth in synthetic media (in which all components are chemically defined) are thoroughly reviewed by Hohl (1975, 1983, 1991a). Hohl's latter chapter treats *P. infestans* in greater detail than other species. See Hohl's (1983) Table 1 for a listing of numerous studies of nutrition on synthetic media. See Appendixes 3.11–3.23 for several synthetic media that have been used to culture various *Phytophthora* species.

By strict definition, media containing agar are not synthetic because agar is derived from seaweed and contains low levels of minor elements, sterols, and other compounds (W. D. Nes 1987). Mycelial growth on liquid media that is measured in dry weight (dried at 55°C for 24 h) after suitable periods of incubation is considered to reflect the effect of different nutritional components more accurately than radial growth on agar. Nonetheless, data from radial growth on agar gives useful information. For instance, W. D. Nes and Stafford (1984) reported that a synthetic medium containing cholesterol supported oospore production of *P. cactorum* when solidified with agar but not as a liquid medium.

The appropriate length of the incubation period varies with isolate and species. Since the nutrients in a medium are eventually depleted and the metabolic byproducts of growth (with the exception of CO_2) remain in the medium during the growth period, growth reaches a peak when one or more of the major components have been utilized. After the maximum growth has occurred, it is common for mycelium to break down or autolyze, and consequently the dry weight decreases. For this reason, a study of the effects of varying nutrients or other factors on growth is best conducted as a rate experiment in which replicated cultures are harvested at appropriate intervals, depending on when the maximum growth is reached. Usually, 2-day intervals are satisfactory. The time at which maximum growth is reached can be used as a standard to measure the effects of different factors.

Efficiency of a medium to support mycelial growth is important because lack of the proper balance can be a limiting factor. For instance, the yield in 10 ml of a synthetic medium (containing 15 g of sucrose and 2 g of asparagine per liter) in a 50-ml Erlenmeyer flask should be about 80–100 mg of dry mycelium (Erwin and Katznelson 1961a). An efficient medium should yield dry weight of mycelium that is about 55% of the weight of the carbohydrate in the medium (Steinberg 1948). See also J. W. Hendrix et al. (1969) for a discussion of efficiency of utilization of carbohydrate.

Growth of *Phytophthora* is experimentally examined by preparing a liquid medium, dispensing 10- or 20-ml portions into 125- or 250-ml Erlenmeyer flasks, and autoclaving for a minimal period (20–30 min at 121°C). In some cases, when glucose is used as a carbohydrate (sugar) source, it should be autoclaved separately and combined with the other components when cool to avoid formation of a toxic product (McKeen 1955). In our experience, sucrose does not contribute to this problem; however, autoclaving the carbohydrate separately in a preliminary experiment is good experimental procedure. The mineral content of the medium should be balanced so that precipitation does not occur. Precipitation removes certain components that contribute to the nutrition of the fungus. To avoid precipitation during autoclaving, the phosphate component is often autoclaved separately from the calcium component, and the two are combined when cool. The calcium salt can be autoclaved with the carbohydrate source, after which the nutrients are combined when cool (Davies 1959). The concentration of phosphate is critical for some species of *Phytophthora;* an excess often leads to retardation of growth (D. C. Erwin, *unpublished*).

Aeration of cultures increases the rate of growth (Mitchell and Zentmyer 1971a). Incubating the culture on a laboratory shaker to induce mild reciprocal or gyrating agitation of the media is commonly used. If shaking is to be used for incubation, the original plug of inoculum should be allowed to incubate for 24–48 h in still culture before the shaking regime is initiated. This allows the plug to produce a mycelial colony or mat that will grow three dimensionally when the shaking regime is started. If shaking is started before establishment of the colony, several small colonies break off the original, usually resulting in erratic growth rates of *P. medicaginis* and much replicate variation (Erwin and Katznelson 1961a). J W. Hendrix et al. (1966) also noted improved growth of *P. parasitica* var. *nicotianae* in shake culture when shaking was delayed 3 days after inoculation of the liquid medium.

The initial inoculum most often used is a small agar plug (3–4 mm in diameter) cut from the margin of a culture on a minimal medium to prevent transfer of large amounts of the components into the medium with the inoculum. Use of axenic inoculum composed of zoospores is an ideal method of inoculation of the medium (Byrt and Grant 1979), but since zoospore production usually must be induced by a procedure involving one or more washings of the culture and each step of the procedure must be conducted under total sterility to prevent contamination, this practice is not often used. Methods to induce synchronous production of sporangia are given in Appendix 3.39.

A large number of inoculum units can be introduced into the medium by mincing a washed mycelium culture from a liquid medium for 15–20 s with a Sorvall Omnimixer or Waring blender. Comminution time of 30 s at 15,000 rpm in an Omnimixer cup (50 ml) produced more viable colonies of *P. cactorum* than longer periods (Pettit and Pegg 1991). An aliquot of the minced mycelium can be added to each flask containing the medium. If quantitation is necessary, dilutions of the mycelial inoculum can be distributed on petri plates containing a suitable agar medium and incubated for 2–3 days. The concentration is expressed as viable colony-forming units per milliliter of inoculum. Actual dry weight of mycelium can be used to quantitate inoculum (Davies 1959). Use of minced mycelium rather than a single inoculum plug often results in more synchronous production of mycelial growth, sporangia, or oospores because the lag period of growth is shortened.

Most *Phytophthora* species grow well on relatively simple synthetic media containing thiamine (vitamin B_1); a suitable carbohydrate, such as glucose or sucrose (see Table 2.3 by Luo et al. 1988); nitrogen as the L form of an amino acid (e.g., L-asparagine) for some species or an ammonium or nitrate salt for other species (Hohl 1983) (Appendixes 3.11–3.23); a balance of inorganic salts containing magnesium, iron, potassium, sulfur (oxidized or unoxidized forms), phosphate, and calcium; and a minor element solution containing zinc, molybdenum, manganese, and copper (Appendix 3.15). Sterols such as β-sitosterol (subsequently referred to simply as sitosterol) stimulate growth of some species and are required for sporulation of all species. Lecithin augments growth of some species (Hohl 1975, 1983, 1991a). See the subsequent section on sterols and the review by W. D. Nes (1987) on sterol metabolism for discussion of the interaction between sterols and components in lecithin. Some of the synthetic media that have been used for growth of different *Phytophthora* species are given in Appendixes 3.11–3.23.

Thiamine Requirement

Phytophthora species uniquely require an exogenous source of the total molecule thiamine but cannot utilize the pyrimidine and thiazole components of thiamine individually or when added together (Leonian and Lilly 1938; Payette and Perrault 1944). In contrast, members of the closely related genus *Pythium* can synthesize thiamine and do not require an exogenous source in a synthetic medium (Robbins 1938; Robbins and Kavanagh 1942; Ridings et al. 1969). The amount of thiamine required in a synthetic medium is about 0.2–1.0 mg/liter (Payette and Perrault 1944; Davies 1959; Erwin and Katznelson 1961a). A dosage of 0.8 mg/liter is considered to be optimum for *P. parasitica* (Lopatecki and Newton 1956). If the dosage of thiamine is critically close to the minimal threshold concentration, the thiamine solution should be filter sterilized (0.22-μm filter) rather than autoclaved because thiamine is heat sensitive; however, when dosages approach 1.0 mg/liter, enough thiamine survives heating at 121°C to promote growth. Stock solutions of thiamine that are heat stable can be prepared in 25% ethanol acidified to pH 3.5 with HCl (Erwin and Katznelson 1961a). The chapter by Hohl (1983) contains an extensive set of references on use of thiamine for growth of *Phytophthora* species. No other vitamins are known to be required for growth (Fothergill and Child 1964).

Carbohydrates

Phytophthora species grow on a wide range of carbohydrates (Table 2.3 in Luo et al. 1988). Glucose and sucrose support growth of all tested species. However, sucrose was not utilized by *Halophytophthora* (*Phytophthora*) *bahamensis* (Hohl 1983). Maltose is an acceptable source of carbon for many species but was not utilized by *H. bahamensis* and *H. spinosa* var. *spinosa* (Bahnweg 1978). *Halophytophthora* species, originally described as *Phytophthora,* were relegated to the new genus *Halophytophthora* by Ho and Jong (1990). Glucose, maltose, sucrose, and dextrin are good sources for *P. fragariae* (Davies 1959). Glucose and sucrose are favorable sources of carbohydrate for *P. parasitica, P. cactorum, P. erythroseptica,* and *P. megasperma;* the optimum concentration is 4% (Lopatecki and Newton 1956) (Appendix 3.15). *P. medicaginis* (misnamed *P. cryptogea*) grows well on sucrose and glucose (Erwin and Katznelson 1961a). Glucose is the most favorable source for *P. infestans* (Hohl 1983).

The reducing sugar, glucose, but not the nonreducing sugar, sucrose, can cause a toxic reaction when added to organic media before autoclaving. Sucrose is less reactive than glucose and can usually be autoclaved with other components. Autoclaving glucose with glycine as a nitrogen source produced a toxin in the medium; however, when glucose was

than when added with lecithin. For instance, the Tween 80 wetting agent contains 31 μg of sitosterol per 100 g, and soybean lecithin contains 52 μg of sitosterol per 100 g. These small amounts of sterol can act in a synergistic or, to use W. D. Nes's term, "sparking" capacity.

Although a β-hydroxysterol has been proved to be necessary for sexual reproduction (Elliott et al. 1964; Haskins et al. 1964; J. W. Hendrix 1964; Leal et al. 1964), Ko and Ho (1983) and Ko (1985) concluded that sterols should not be considered strictly essential, because oospore production could also be enhanced by the addition of a source of lecithin, which they found to be free of sterols, to a synthetic medium. Later reports indicated that soybean lecithins (99% pure) were much more effective than sitosterol or cholesterol for stimulation of oospore production in *P. cactorum.* The lecithin dioleoyl-L-α-phosphatidylcholine is the only one of four synthetic lecithins that was active (Ko and Ho 1983; Ko 1985). Conversely, the sources of lecithin tested by Hohl (1975) did not support oospore production by *P. cactorum.* Paired A1 and A2 cultures of *P. capsici* in Brazil did not form viable oospores in media to which either lecithin or cholesterol had been added; however, oospores were formed abundantly in an unclarified V8 juice medium (Ansani and Matsuoka 1987).

The induction of oosporogenesis by lecithins (containing phospholipids) in the absence of sterols (Ko and Ho 1983; Ko 1985) was reexamined by Kerwin and Duddles (1989). They used *P. cactorum, P. citricola, P. megasperma* (from asparagus), and *Pythium ultimum* to test the effects of phospholipids and cholesterol on oosporogenesis. Kerwin and Duddles (1989) question Ko's (1985) conclusion that sterols are not necessary for oosporogenesis in a medium containing certain lecithins because trace levels of sterols in the inoculum and even in the medium were at a level high enough to induce some oosporogenesis, especially when the cultures were supplemented with dioleoyl phosphatidylcholine. When inoculum that had been maintained by seven passages on a sterol- or phospholipid-free synthetic medium was used, no viable oospores formed in the presence of the synthetic lecithin dioleoyl phosphatidylcholine (1 g/liter). When cholesterol (10 mg/liter) was tested alone on *P. cactorum,* the number of viable oospores was 3.2×10^4/ml, but with both compounds present, the oospore number increased to 5.8×10^4/ml. The increase in numbers of oospores per milliliter was similar for *P. citricola, P. megasperma,* and *Pythium ultimum.* More detailed studies with lower dosages of both cholesterol and dioleoyl phosphatidylcholine with *Pythium ultimum* proved that the two compounds exerted a synergistic effect on the production of viable oospores (Kerwin and Duddles 1989).

In a study of lipids in *P. cactorum* that elucidated the nature of certain compounds, called "phytophthorols" by W. D. Nes (1988), phospholipids (lecithin) alone did not turn on oosporogenesis except when small amounts of a dietary sterol were added. Phytophthorols are lipid compounds that were identified in mycelium of *P. cactorum.* They are composed of "a series of saturated and unsaturated odd chain fatty alcohols (C_{17}–C_{21}) esterified to myristic acid (14:0) with each possessing two hydroxyl functions in a 1,2-diol relationship centered toward the ester terminus." Their function appears to be involved with the molecular architecture of the membranes. We quote another interpretation of the possible lecithin effect on oosporogenesis as noted by Ko (1985) (reproduced by permission and courtesy of W. D. Nes 1988):

> The ability of Ko to induce reproduction with commercial soybean and egg lecithins in *Phytophthora* is relevant to the proposed essential fatty acid requirement by the Oomycetes (Ko 1985; Ko and Ho 1983). In order to stimulate oospore formation, Ko added the commercial phospholipids at 1 g/liter; a two orders of magnitude lower level, 10 mg/l (10 ppm), failed to induce oosporogenesis. It should be remembered that 10 ppm sterol induces maximal sexual sporulation (W. R. Nes and W. D. Nes 1980). However, as has been shown, *Phytophthora* has the biosynthetic capacity to synthesize the head groups and dominant fatty acids that are present in the dietary phospholipid preparations. Clearly, the commercial phospholipids are not essential dietary requirements for reproduction, since these same lipids are produced by *Phytophthora.* It may be that a trace contaminant of an EPA (eicosapentaenoic acid) (or other prostaglandin) precursor, e.g., α-linoleic acid, together with a contaminant sterol is present in the commercial samples which include agar. The results of this study imply that oosporogenesis may require a dual lipid nutritional supplement: one for a select sterol and another for a select unsaturated fatty acid. If either of the two lipids are provided alone, growth, not reproduction, is affected. If the two lipids are provided together, reproduction is turned on; the number of spores produced will depend on their respective levels in the media. Presumably, the phytophthorols and VLFA (very-long-chain-fatty alcohols) replace sterols in mycelial membranes during vegetative growth. VLFA, but not phytophthorols, are present in the sterol-producing Oomycetes, *Saprolegnia ferax* (Nes, W. D., unpublished). Addition of sterols fails to change the total fatty acid composition of *P. cactorum,* although following cholesterol-induced sparking of growth the amount of fatty acid increases, and individual fatty acids associated with key ester linkages change dramatically (Nes, W. D., unpublished). This coupled with our observation that VLFA biosynthesis is compartmentalized from the fatty acid pathway leading to neutral and polar acyl lipids indicates that regulation of the fatty acid biosynthetic pathway, e.g., by use of desaturase and cyclooxygenase inhibitors, may be an approach to control pathogenesis. There are preliminary data to support this view (Herman and Herman 1985; Kerwin et al. 1986). Interestingly, AA (arachidonic acid) and EPA have been implicated in disease resistance as sesquiterpene phytoalexin elicitors (Kuć et al. 1984). In conclusion, fatty acids and alcohols play multiple physiological and ecological roles, some of which are interchangeable, in *P. cactorum.* A determination of their relative importance with sterols as developmental triggers (cell-cycle mediated process [Nes, W. D. 1987]) vs fluidity modulators of mycelial membranes to initiate and then sustain growth and govern reproduction is under active investigation.

Authors' note: We acknowledge with appreciation the assistance of W. D. Nes in the preparation of this section.

Lecithins

Lecithins, which are phospholipids, increase the growth rate of some but not all *Phytophthora* species, and the effect is not always consistent within species. Likewise, different sources give different responses (Hohl 1975, 1983). The growth stimulatory effect could be replaced by use of linoleic acid. Linoleic acid is a predominant part of the lipid component of *P. infestans* mycelium (Creamer and Bostock 1986). Lecithin increased radial growth of *P. palmivora, P. megakarya,* and two of three isolates of *P. palmivora* MF4 (probably *P. capsici;* Table 9 in Hohl 1983). A synthetic lecithin containing the unsaturated fatty acid dilineoyl was active in promotion of growth, but another lecithin containing a saturated fatty acid was not. Since concentrations of lecithin and/or linoleic acid as low as 10 mg/liter markedly increased radial growth on agar and dry weight 10-fold, it is thought that these compounds do not function as carbon sources but probably affect membrane stability or enzyme action (Hohl 1975, 1983, 1991a). See also the discussion above on interaction of sterols and lecithins.

Minor Elements

Most synthetic media contain the minor elements manganese, copper, and molybdenum at about 0.05 μg/ml and zinc and iron at 1.0 μg/ml (Appendix 3.15). Iron depresses growth at a concentration of 2 μg/ml or higher. The essential nature of these elements is shown by Lopatecki and Newton (1956) for all except molybdenum; however, this element as well as others may occur as contaminants in other nutrients. Cuppett and Lilly (1973) reported that the ferrous ion is more favorably taken up than the ferric ion. Inclusion of ascorbic acid, a reducing agent, maintained iron in the ferrous state and was effective in improving growth of *P. infestans.* Sakai (1961) reported that molybdenum increased growth of *P. infestans.* The discussion by Hohl (1991a) contains more references that apply to *P. infestans.*

Calcium

A number of reports indicate a necessary role of calcium ion (Ca^{2+}) in synthetic media (Lopatecki and Newton 1956; Davies 1959; Erwin and Katznelson 1961a; Hwang et al. 1965; Lin and Liang 1965; Erwin 1968; Elliot 1972, 1986, 1988; Lin et al. 1982; see also Hohl 1983, 1991a). Growth of *P. parasitica* var. *nicotianae* has a nearly complete requirement for Ca^{2+} when nitrate is the nitrogen source and a sterol is absent. When asparagine is used as the nitrogen source, Ca^{2+} increases growth markedly and cholesterol added to media with Ca^{2+} has an additive effect (J. W. Hendrix et al. 1969; J. W. Hendrix and Guttman 1970).

Omission of $CaSO_4 \cdot 2H_2O$ (0.1 g/liter) from a synthetic medium reduced mycelial growth of *P. erythroseptica* by 70%, *P. infestans* by 50%, and *P. parasitica* by only 7%. Variation in the balance of K_2HPO_4, $MgSO_4 \cdot 7H_2O$, and $CaSO_4 \cdot 2H_2O$ concentrations greatly affected growth of *P. parasitica* but not of *P. erythroseptica* (Fothergill and Child 1964). Maximum growth of *P. fragariae* occurred at a Ca^{2+} concentration of 70 mg/liter (Fleetwood Walker 1954). Lin et al. (1982) concluded that Ca^{2+} plays a role in the buffering of the medium and/or as an essential metabolite. When the organic acid-salt of malate and Ca^{2+} were available, ammonium ion was a better source of nitrogen than nitrate. Lin et al. (1982) concluded that Ca^{2+} counteracted the deleterious effects of organic acids, but this is probably not the only role (Hohl 1991a).

Calcium stimulates both asexual and sexual reproduction (Elliott 1972, 1986, 1988; Halsall and Forrester 1977). Use of the compound trifluoroperazine, which inhibits or interacts with calmodulin, a calcium-binding protein, inhibited both sporangium production by *P. palmivora* and sex organ formation by *P. cactorum* (Elliott 1986, 1988), thus supporting the conclusion that Ca^{2+} promotes oogonium initiation. In *Lagenidium giganteum,* an oomycete in the Lagenidiales, extracellular calcium was necessary for oosporogenesis (Kerwin and Washino 1986). Kerwin and Washino (1986) stated, "Oosporogenesis can be disrupted using compounds which antagonize calcium metabolism at concentrations which do not subsequently affect asexual zoosporogenesis, suggesting that oosporogenesis is highly regulated by calcium fluxes and (or) sequestration."

In a study of synchronous differentiation of zoospores (swimming to cyst formation) of *P. palmivora,* Grant et al. (1985) found that addition of Tris-pectate rapidly increased levels of phosphatidic acid (250% within 20 s). Phosphatidic acid (3 μM) applied exogenously induced zoospore differentiation but was dependent on an optimum concentration of Ca^{2+} in the suspension medium. Both phosphatidic acid and zoospore differentiation were calcium dependent (Zhang et al. 1992).

Although the data indicate that calcium augments mycelial growth, the mode of action is not understood well enough.

Maintenance of *Phytophthora* Collections

Storage of *Phytophthora* species in culture collections has always presented problems because frequent transferring is necessary, and in many cases mutations or adaptive changes occur after prolonged culturing. *P. infestans* often loses virulence rapidly in culture (Jeffrey et al. 1962; Pietkiewicz 1978; Goth 1981). *P. parasitica* var. *nicotianae* loses virulence with continuous culturing (Apple 1957). Continuous transfer could also lead to undesirable variability.

V8 juice agar (V8A) (Appendix 3.1) is an almost universally used medium for maintenance of cultures, but oatmeal agar (Appendix 3.5) is also used. Some isolates of *P. infestans* do not grow readily on V8A but grow better on rye seed agar (Caten and Jinks 1968) (Appendix 3.9). Wernham and Miller (1948) maintained *P. infestans* cultures on lima bean agar for 1 year. Cultures maintained 2 years were less virulent when tested on potato plants in the greenhouse.

For many years, Zentmyer and Erwin maintained a culture collection at 15°C on V8A slants to which sterile mineral oil had been added (see Ainsworth [1968]). When the colony had grown over almost all of the slant surface, it was inundated with a sterilized light mineral oil, which can be purchased in drug stores. Mineral oil can be autoclaved at 121°C and heated in an oven at 150°C to remove the water. Cultures incubated at 15°C (lower temperatures are less favorable) can be kept up to 2 years and often longer; however, reculturing after 2 years is usually necessary. The main disadvantage is that recovery from the oil is a messy process, and occasionally some cultures die.

Cultures can be stored successfully in distilled water (Boesewinkel 1976; Ann and Ko 1990b). Small blocks of agar media on which the fungus is growing are transferred to autoclaved, screw-capped bottles containing distilled water. The cap is tightened for storage. The agar blocks are removed when necessary and transferred to a suitable medium. *P. cactorum, P. cinnamomi,* and *P. megasperma* var. *sojae* survived from 1969 to 1975 (Boesewinkel 1976). Chlamydospores of *P. parasitica* were germinable in sterile distilled water after storage for 50 days at 9°C (Holdaway and Tsao 1971).

P. infestans can be maintained as sporangia on infected potato leaf tissue at –30°C for at least 12 months; however, this method is subject to contamination by other fungi and bacteria. Several races of *P. infestans* from Denmark and Sweden infected and formed sporangia on leaflets of differential cultivars (Black et al. 1953) or on the cultivar Bintje in petri plates lined with moist filter paper. The petri plates were stored at –30°C. After 2 and 12 months of storage, sporangia were washed off the leaves and incubated at 8°C to stimulate zoospore formation. The zoospores were then inoculated onto cultivar Bintje leaves from which a pure culture could be obtained (Gürtler 1984). *P. infestans* was maintained for more than a year on a broth made from sweet corn cultivar Golden Bantam seeds that had been immersed in glass-distilled water for 16 h. Ten seeds were placed in 50 ml of glass-distilled water in 125-ml flasks and autoclaved for 30 min at 2-day intervals. Each flask was inoculated with *P. infestans,* incubated, and stored at 15°C. Fresh transfers were made by transferring a mycelium-colonized seed either to another corn seed broth or to frozen lima bean agar in petri plates. *P. infestans* sporulated profusely on the corn seed when incubated in the dark. Isolates maintained virulence after 5 years in storage (Goth 1981) (Appendix 3.28).

Corn seed was previously reported for maintenance of *P.*

phaseoli (Goth and Wester 1963). Oospores formed abundantly in the embryo of corn seed. Screw-capped tubes containing corn seeds can also be used to maintain cultures at 15°C. After growth has occurred, the screw caps can be tightened to prevent loss of moisture. The method is applicable to a wide range of *Phytophthora* species including *P. capsici, P. cryptogea, P. drechsleri, P. parasitica,* and *P. phaseoli* (R. W. Goth, *personal communication*).

During recent years, the technique of cryopreservation (storage of cultures in a frozen condition) has been developed (San Antonio and Blount [1973] for *P. phaseoli*). Now most large culture collections (Alexander et al. 1980; M. D. Coffey, University of California, Riverside, *personal communication*; Dahmen et al. 1983; American Type Culture Collection, Jong and Atkins 1985; D. Smith et al. 1986; Hohl and Iselin 1987; Tooley 1988) store cultures under liquid nitrogen (–196°C). Cryopreservation has a distinct advantage in that it is not only labor saving and dependable but, perhaps most important, also preserves the fungus in its original genetic state; however, cryopreservation is relatively expensive in terms of cost of equipment and liquid nitrogen. The strategy of cryopreservation is described by D. Smith (1982, 1983) and D. Smith and Onions (1983).

Preparation of *Phytophthora* cultures involves use of cryoprotective solutions such as 8.5% sterile skim milk, 10% glycerol, or 15% dimethyl sulfoxide (Dahmen et al. 1983) and controlled freezing at a rate of 1°C per minute to –40°C, after which the cryogenic polypropylene vials or screw-capped tubes containing the fungus in disks of agar are stored in either liquid or gaseous nitrogen. The report by D. Smith et al. (1986) shows microscopically that glycerol-protected mycelium of *P. nicotianae* survives cooling at rates of 10°C per minute or less, but at faster cooling rates, a glycerol-induced osmotic shrinkage rather than ice nucleation causes loss of viability. Viable cultures are recovered by either rapid or slow thawing.

Controlled rates of freezing require specialized equipment. Tooley (1988) reported a method in which the rate of freezing was not controlled but viable cultures were maintained in liquid nitrogen for long periods. *P. infestans* and several other species were grown on 20% V8A and stored as small mycelium-agar plugs in polypropylene vials in either DMSO (5 or 10%) or glycerol (10%). Cultures were first cooled in freezers at –20°C and then at –40°C before storage in liquid nitrogen. Isolates of *P. infestans, P. boehmeriae,* and *P. sojae* (*P. megasperma* f. sp. *glycinea*) had survived 9 months by 1988 (Tooley) and were still viable in 1991 (P. Tooley, *personal communication*).

Sporulation of *Phytophthora*

The physiology of sporangium production is reviewed and discussed in detail by Waterhouse (1931, 1962), Ribeiro (1978, 1983), and Gisi (1983). The physiology of oospore production and germination is reviewed by Eliott (1983), Ribeiro (1983), Shaw (1983), Shaw and Shattock (1991), and Hohl (1991a). The paper by Lilly (1966), one of the pioneer physiological plant pathologists, aptly summarizes the role of light and sterols on production and germination of spores. The effect of various environmental factors on production of spores of *Phytophthora* are shown by Ribeiro (1983, his Figure 3).

The developmental sequence of nuclei in sporangia of *P. infestans* consists of migration of elongate nuclei from the sporangiophore followed by division of nuclei in the sporangium. After movement of nuclei into the sporangium, the septum forms at the base followed by degeneration of some of the nuclei. At this time, the papilla forms. The entire process is completed in about 18 h (Evola Maltese et al. 1995).

Physiology of Sporangial Production

Zoospore production by sporangia, which is completed in a few minutes, is one of the most exciting events to watch in the study of *Phytophthora* (Figure 3.1). See also Plate 3.1 on zoospore formation of *P. cinnamomi* (Shearer et al. 1991).

Production of sporangia and zoospores is the principal means by which the numbers of infective units are increased and accounts for the rapid regeneration time of *Phytophthora* species. Several reports of the time required for sporangium production by different species are summarized in Table 3.1. The physiology of this process is complex. The induction of sporangium production, at least in soilborne species, requires a series of precise environmental manipulations. Every teacher of mycology has a "pet" isolate that will readily sporulate and demonstrate the fascinating process of zoospore emission from sporangia for students to see. One of the most useful is P113 of *P. palmivora* (M. Coffey, *Phytophthora* Collection, University of California, Riverside), originally isolated from papaya in Hawaii (Tokunaga and Bartnicki-Garcia 1971). Although the protocol may seem to be foolproof, occasionally the *Phytophthora* isolate remains vegetative and does not sporulate in time for the class! This lack of sporulation probably stems from a failure to meet some precise requirement of water and temperature management for this process to occur. Grant et al. (1984) described an elegant method for synchronous production of zoospores (2×10^6/ml) from the P113 isolate of *P. palmivora* (Appendix 3.39); however, another Australian isolate of *P. palmivora* from papaya failed to produce zoospores under the same regime! We have also heard unpublished reports that some isolates of *P. cinnamomi,* which is reported to form sporangia readily only when treated with precise methodologies involving the washing of young mycelium in a salt solution and reducing temperatures (Chen and Zentmyer 1970; Byrt and Grant 1979) (Appendixes 3.38 and 3.39), fail to produce sporangia under these protocols. We believe that these exceptions do not reflect on the validity of the published methods but that *Phytophthora* is telling us that we do not really understand all the factors we need to know about its physiology and variability!

In laboratory studies, water free of contaminant metals is absolutely essential. *Phytophthora* species are extremely sensitive to metal ions such as copper and perhaps others. Copper pipes are a source of contamination. For optimal results, only glass-distilled water or double glass-distilled water should be used. Glass-distilled water was superior to deionized water for bait isolation of *P. cinnamomi* on lupine seedlings in water to which a block of mycelium and agar had been transferred (Gerrettson-Cornell 1976b). *P. cinnamomi* could not be recovered from water obtained from a metal still. The primary inoculum in this case would most likely have been zoospores. Sporangium formation by *P. medicaginis* (*P. megasperma* f. sp. *medicaginis*) is sensitive to as little as 0.02 ppm of copper ion (Kennedy and Erwin 1961) and by *P. cinnamomi* to copper at 1.0 ppm or less (Zentmyer and Marshall 1959).

Copper ion (Cu^{2+}) at 5×10^{-5} M prevented sporangium formation of *P. cinnamomi* and infection of *Pinus radiata* D. Don needles, *Eucalyptus sieberi* E. Labill cotyledons, and safflower (*Carthamus tinctorius* L.) seedlings in a salt solution. Cu^{2+} at 1 and 5×10^{-7} M concentrations inhibited *P. drechsleri* sporangium formation and at 10^{-6} M prevented infection of eucalyptus cotyledons by zoospores. The chelating agent ethylenediaminetetraacetic acid (10^{-4} M) reversed the inhibitory effect (Halsall 1977). Halsall conjectured that soil might have a similar chelating effect.

Soilborne *Phytophthora* species usually produce sporangia optimally on new mycelial growth only after the culture is changed from rapid growth on a relatively rich medium to a starvation regime by frequent washing with water or a salt solution. For instance, exudates and extracts from soybean roots that are rich in nutrients stimulated mycelial growth of *P. sojae* (*P. megasperma* var. *sojae*) but not formation of sporangia (Ho 1969a). Sporangia of *P. cactorum* formed in the dark when the sugar in a synthetic medium was exhausted but not when the medium was amended with additional glucose or sucrose (Elliott 1989). Sporangium production appears to require exhaustion of the sugar in the medium, whereas oogonial and subsequent oospore production occurs when the nitrogen content of the medium has been nearly exhausted but when there is adequate carbohydrate (Elliot 1989). These phenomena, as noted by Klebs for *Saprolegnia* (1899, 1900), are commonly referred to as Klebs's principles, which, simply stated, say that the fungus sporulates when partial starvation conditions prevail. Sporangia production appears to be associated with catabolite repression (Yoshikawa and Masago 1977). Elliott's (1989) paper precisely defines the role of sugars and nitrogen sources on the production of sporangia and oogonia of *P. cactorum*. More research in areas such as that reported by Elliott (1989) might further clarify the physiology of sporangial formation.

The foliage-infecting species produce sporangia at high relative humidity (approaching or at 100%), but most of the root-infecting species produce sporangia only in free water or in aqueous salt solutions and soil extract. A pH level of 6.0 or 7.0 is most favorable. Temperature effects vary with species. For all species, the nutrients in the medium must include a sterol of the β-hydroxysterol type (e.g., sitosterol) and be adequate to ensure that the fungus grows actively and rapidly. Most plant extracts contain natural sterols (W. D. Nes et al. 1982). Adequate atmospheric oxygen is essential. In some saturated soils, the oxygen supply is a limiting factor. Where CO_2 levels increase above that in air, sporangium production is suppressed (Mitchell and Zentmyer 1971b).

Each of the above factors can be simply stated, but overall it is the balance of such environmental factors that appears to govern whether or not high numbers of sporangia form. The studies in which synchronous development of sporangia is well defined demonstrate these principles (Chen and Zentmyer 1970; Zaki and Zentmyer 1976; Yoshikawa et al. 1977; Byrt and Grant 1979; Grant et al. 1984) (Appendixes 3.38 and 3.39).

Sporangia production by most *Phytophthora* species can be induced by washing and transferring marginal portions of the culture on a vegetable decoction medium (Appendixes 3.24–3.30) to water (Appendixes 3.31–3.34), certain salt solutions (Appendixes 3.35–3.43), or soil extract (Appendix 3.44). Although potato-dextrose agar (Appendix 3.4) is occasionally reported to support sporangium production, the high sugar content (20 g of glucose per liter) tends to promote mycelial growth rather than sporangia production. Weak nutrient solu-

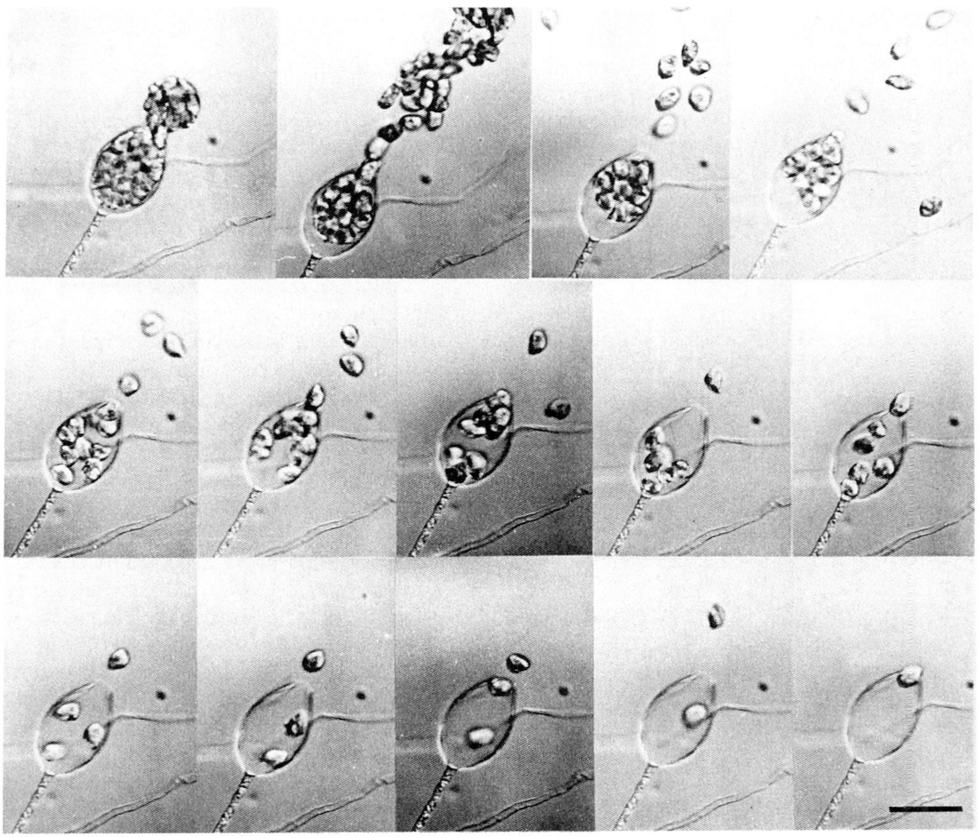

Figure 3.1. Sequence of zoospore release from a sporangium of *Phytophthora medicaginis* (*P. megasperma* f. sp. *medicaginis*). Note in frame 1 that the emitted zoospores are retained in the membranous sac. In frame 2, the membrane has broken and the zoospores are released. Bar = 30 µm. (Reproduced by permission of the American Phytopathological Society, St. Paul, Minnesota, courtesy Ta-Li Kuan, Saratoga, California)

tions such as V8 juice broth (Appendixes 3.1, 3.2, and 3.31–3.33) or carrot broth (Appendix 3.7) diluted at a ratio of one part broth to four to 10 parts of distilled water can be inoculated with mycelium as either plugs of agar or mycelium minced in a Waring blender or colonized disks of cloth (Halsall and Forrester 1977) and incubated for 24 h. The medium is then decanted off, and the colonies are washed two or three times with double distilled water, overlaid with a shallow layer of sterile double distilled water, one of several inorganic salt solutions (Appendixes 3.35–3.43), or soil extract (Appendix 3.44), and incubated under continuous light from 40-watt fluorescent lamps (e.g., General Electric Cool White or Vitalite bulbs are effective). Growth of *Phytophthora* on V8 juice agar that is overlaid with cloth such as organdy (Appendix 3.41) (Halsall and Forrester 1977), nylon, Tygan mesh, or cellophane (Appendix 3.40) allows residual nutrients to be more rapidly depleted from the culture during the washing process.

Since sporangium formation is sensitive to an adequate supply of atmospheric oxygen, sporulation occurs most abundantly at the surface of an aqueous culture. The fact that changing the water or frequent rinsing enhances sporangia formation is probably the result of increased aeration as well as depletion of nutrients. *P. cryptogea* forms sporangia maximally in soil at –25 to –300 mb matric potential, but markedly less production occurs in saturated soil. This is likely the result of the lack of adequate oxygen in flooded soil (0.0 mb matric potential). Mycelial plugs on the surface of soil, where aeration is adequate, sporulate best in free water. A water-saturated flow of nitrogen over the surface of soil at –300 mb matric potential inhibited sporangium formation (Duniway 1975b, 1983). Sporangium formation was reduced by oxygen concentrations less than that of air and by CO_2 concentrations greater than that of air (Mitchell and Zentmyer 1971b). Consequently, sporulation occurred less frequently on submerged pieces of agar and mycelium than on those covered by a thin layer of water (Ayers and Zentmyer 1971). Mycelium mats of *P. parasitica* var. *nicotianae* wetted with 0.01 M KNO_3 (Appendix 3.43) produced 600 sporangia per square millimeter but at a depth of 1 mm, only 325 sporangia per square millimeter and at 14 mm, only 40 (Gooding and Lucas 1959a). Copious numbers of sporangia of *P. parasitica* were produced on well-washed mycelium mats incubated in a shallow layer of sterile water (1–2 mm deep) to allow adequate aeration; however, in water at 50-mm depth at a lower temperature (15–18°C), large numbers of chlamydospores but no sporangia formed (Menyonga and Tsao 1966; Tsao 1969b) (Appendix 3.31).

Light in the near-ultraviolet (UV; 320–400 nm on the spectrum) or blue region is most favorable for sporangium production (Ribeiro et al. 1976a); thus, the fluorescent light bulbs selected for incubation of cultures should have a radiation peak in the near-UV region of the spectrum. Ribeiro (1978) shows the spectral distribution of wavelength peaks for several commercial sources of light. Some of those emitting a peak within the near-UV include General Electric Daylight, General Electric Cool White, Verd-A-Ray Indorsun, Vitalite, and Westinghouse Plant Gro. Sylvania Blacklight emits only between 312 and 400 nm (near-UV). Such information can also be obtained from the manufacturer. Measurement of intensity should be in terms of energy and not lux (which implies the quality of light to the human eye). Since some commercial fluorescent or incandescent bulbs do not emit light in the near-UV or blue regions, they would not be satisfactory light sources for the production of sporangia.

The effect of light on *Phytophthora* is most likely caused by the presence of a biochemical receptor as yet to be determined. Light in the blue and near-UV regions of the spectrum enhances sporangium production of many airborne *Phytophthora* species such as *P. capsici* and *P. palmivora* (Ribeiro et al. 1976a), but sporangium production of *P. infestans* was reported to be inhibited by light in the blue region of the spectrum (Cohen et al. 1975). *P. hibernalis* and *P. syringae* sporulate better in the dark (Harnish 1965). Light is usually not necessary for soilborne *Phytophthora* species such as *P. parasitica* var. *nicotianae* (Gooding and Lucas 1959a), *P. cinnamomi*, or *P. sojae* (*P. megasperma* var. *sojae*) (Ribeiro et al. 1976a). Sporangium production by *P. erythroseptica* is stimulated by light (Vujičić and Colhoun 1966). *P. cinnamomi* is stimulated by light only in the near-UV region and at intensities less than 100 μWm^{-2} (Zentmyer and Ribeiro 1977).

Light affects morphology of some species. In *P. capsici*, a sporangium formed under continuous light is mainly elongate with a tapered base, but one formed in the dark has a more rounded base. In certain species, more sporangia are caducous when formed under light than in the dark (Tsao 1991; see also Ribeiro [1983]). Sporangia of *P. cinnamomi* formed under near-UV were longer than those formed under laboratory light (Zentmyer and Ribeiro 1977).

Sporangium production of several *Phytophthora* species can be induced by flooding cultures on various media with double distilled water followed by several replacements of the water at various intervals. These methods are described in Appendix 3.31 for *P. parasitica*, Appendix 3.32 for *P. cactorum*, Appendix 3.33 for *P. medicaginis*, and Appendix 3.34 for *P. sojae*. Alizadeh and Tsao (1985b) report standardized methods for producing sporangia of *P. capsici* and *P. palmivora* by use of an agar disk in water technique (Figures 4.16 and 4.17). See also Appendixes 3.24–3.30 for use of different media that support sporangia production by several species of *Phytophthora*. Grass leaf cultures were more stimulatory for sporangia production by *Pythium* species (Dick 1990a), whereas hemp seed was better for *Phytophthora* species (Appendix 3.26).

Petri's solution (Appendix 3.35) stimulated sporangia of various *Phytophthora* species (Tucker 1931). Mycelium of *P. parasitica* var. *nicotianae*, incubated in 0.01 M KNO_3, pro-

Table 3.1. Representative times required for sporangial production by some *Phytophthora* species after transfer to different induction media[a]

Species	Sporangial production	Reference
P. cactorum	3.5 days	Waterhouse (1931)
	12–48 h	Buddenhagen (1958)
P. capsici	36–48 h	Katsura et al. (1968)
P. cinnamomi	8–36 h	Chen and Zentmyer (1970)
	48 h	Chee and Newhook (1966)
P. citrophthora	12–15 h	Fawcett and Klotz (1934)
P. colocasiae	24 h	Waterhouse (1931)
P. cryptogea	5 days	Waterhouse (1931)
P. fragariae	24–48 h	Converse and Scott (1962)
P. infestans	1 week	Hodgson and Grainger (1964)
P. meadii	5–7 days	Peries and Fernando (1966)
P. megasperma (some isolates)	12–24 h	O. K. Ribeiro (*unpublished*)
P. melonis	36–48 h	Katsura et al. (1968)
P. nicotianae var. *parasitica*	24–48 h	Waterhouse (1931)
P. palmivora	36–48 h	Katsura et al. (1968)
P. phaseoli	24 h	Goth and Wester (1963)
P. sojae	72 h	Hilty and Schmitthenner (1962)

[a] Modified from Ribeiro (1978) by permission of A. R. Gantner Verlag, J. Cramer, Vaduz.

duced sporangia (Gooding and Lucas 1959a) (Appendix 3.43). Hemp seeds (Appendix 3.26) or hot-water-sterilized (100°C) leaves of bent grass (*Agrostis alba* L.) or blue grass (*Poa pratensis* L.) can be colonized by incubation of the host tissue on an agar culture of a *Phytophthora* species for 24 h and then transfer to a petri plate containing 10–15 ml of water, a salt solution (Schmitthenner 1973) (Appendix 3.37), or soil extract (Appendix 3.44). Incubation of V8A (neutralized with NaOH) plugs of *P. medicaginis* (*P. megasperma* f. sp. *medicaginis*) in a plant-growth nutrient solution (Hoagland and Arnon 1950) stimulated sporangia (Kennedy and Erwin 1961) (Appendix 3.42).

Incubation of mycelial disks of *P. cinnamomi* in nonsterile soil extract induces sporangia (Mehrlich 1935; Zentmyer and Marshall 1959; Ayers and Zentmyer 1971; King 1980; Schoulties et al. 1980) (Appendix 3.44). Nonsterilized soil extract was more effective than sterilized for production of sporangia of *P. cinnamomi.* The amount of sporulation was much greater on soil extract that had been incubated for 4 days at 25°C prior to use (King 1980). The stimulatory factors in soil extract were removed when it was passed through a 0.2-μm pore size filter. Thus it appears that bacteria in soil extract are involved.

Zentmyer (1965) first showed that a bacterium, *Chromobacterium violaceum,* isolated from soil extract stimulated sporangium formation of *P. cinnamomi.* Ayers and Zentmyer (1971) found that two isolates of *Pseudomonas* bacteria from soil stimulate sporangium formation. In the study by King (1980), the total bacterial population in nonsterile soil extract increased up to day 4 but declined after that. The increase in bacterial population in soil extract was correlated with sporangial production. This evidence indicates that heat-stable metabolite(s) from the bacteria in soil extract are likely to be responsible for inducing sporulation. Ayers (1971) showed that a substance from several bacteria and soil extract stimulated sporangium production by *P. cinnamomi;* and in a related study, a heat-stable substance from nonsterilized soil extract stimulated axenic production of sporangia and zoospores. The stimulatory substance supported production of up to 236,000 zoospores per mycelium mat compared with only 34,200 zoospores per mat in distilled water. The stimulatory substance counteracted the need for light and to some extent suppressed the negative effects of nutrients on sporangium formation (Schoulties et al. 1980). Several other papers report the effect of soilborne bacteria on sporangium formation by *P. cinnamomi* (Marx and Haasis 1965; Manning and Crossan 1966).

Although sporangium production by *P. cinnamomi* usually requires incubation of mycelial disks in nonsterilized soil extract (Appendix 3.44), a method to produce zoospores in axenic culture was developed by Chen and Zentmyer (1970). The fungus is grown on dilute pea or V8 juice broth and washed four times with a salt solution (Appendix 3.38).

Halsall and Forrester (1977) modified the Chen and Zentmyer (1970) method by growing *P. cinnamomi, P. cambivora,* and *P. drechsleri* on sterile, 8-mm disks of organdy cloth that were incubated for 24 h at 25°C for *P. cambivora* and at 27.5°C for *P. cinnamomi* and *P. drechsleri* in 15 ml of sterile dilute V8 juice broth (30 ml/liter) containing 1 ml of a comminuted suspension of mycelium (Appendix 3.41). When the mycelial mats were washed four successive times at 1-h intervals in a sterile mineral salts solution, sporangia formed. The optimum concentrations of salts for each of four *Phytophthora* species are given in Table 3.2. In a subsequent report, the chelating agent was omitted because of the possibility that it might bind other cations (Halsall 1977). These and other data (Kennedy and Erwin 1961; Chen and Zentmyer 1970; Allen and Nandra 1975; Halsall 1977) show that combinations of Ca^{2+}, Mg^{2+}, Fe^{3+}, and K^+ stimulate sporangium formation in many species.

Use of the Chen and Zentmyer (1970) mineral salts solution induced sporangia of *P. cinnamomi* to be produced synchronously with yields of 10^7–10^8 zoospores per liter from axenic cultures via precise, stepwise washing procedures (Zaki and Zentmyer 1976; Byrt and Grant 1979) (Appendix 3.39). A similar method was used for *P. palmivora* (Grant et al. 1984).

Physiology of Sporangial Germination

The production of zoospores from sporangia is the most important part of the life cycle of *Phytophthora,* especially because it allows the population to increase rapidly and disperse widely when free water is available. Classical information for *P. infestans* indicates that production of zoospores is most frequent at low temperatures (about 12°C) and direct germination of sporangia by germ tubes is most common at an elevated temperature of 24°C (Crosier 1934). In general, this concept applies to all species of *Phytophthora,* although many other species produce zoospores at higher temperatures because they grow better at higher temperatures than *P. infestans.* Cooling an aqueous sporangial culture of most species to about 10°C below the optimum temperature for growth stimulates zoospore formation (Figure 3.1).

Although sporangia can germinate by production of germ tubes or by production of zoospores, the sporangium is always programmed to germinate by the production of zoospores. When sporangia of *P. palmivora* were viewed in ultrathin sections by electron microscopy, flagella were always detectable within the sporangium protoplasm (Hemmes and Hohl 1969, 1972). Whether or not sporangia germinate by production of zoospores is governed by environmental factors such as reduction of temperature and the aqueous nature of the medium (e.g., in soil, on a leaf, or in culture). In the early stages of development, two highly organized flagella form. The axoneme (microtubular structure of the flagellum) consists in cross section of a pair of axonemal microtubules in the center and nine pairs of axonemal microtubules surrounding the central pair (Hemmes and Hohl 1971; Barr and Désaulniers 1990;

Table 3.2. Concentrations of cations in washing solutions that induce maximum formation of zoosporangia by *Phytophthora cinnamomi, P. drechsleri, P. cambivora,* and *P. cryptogea*[a,b]

Cation	*P. cinnamomi*	*P. drechsleri*	*P. cambivora*	*P. cryptogea*
Ca^{2+}	0.71 mM	0.12 mM	0.23 mM	0.68 mM
Mg^{2+}	0.16 mM	4×10^{-9} mM[c]	0.16 mM	0.05 mM
K^+	0.62 mM	12.0 mM	0.82 mM	0.73 mM
Fe^{3+}	0.11 mM	0.02 mM	0.05 mM	0.06 mM

[a] Reproduced by permission of the National Research Council, Ottawa, Canada, courtesy of Halsall and Forrester (1977).
[b] All the stationary points are maxima and hence represent the optimum concentration of each cation for zoosporangial production.
[c] This value lies outside the concentration range tested and is the result of a linear Mg^{2+} effect without any quadratic effect (Halsall and Forrester 1977).

Barr 1992) (Figures 4.21 and 4.22). See Chapter 4 for a more complete discussion of zoospore morphology (Figures 4.18–4.20). If conditions are unfavorable for zoospore formation, the sporangia contain a mechanism by which the flagella are absorbed. The breakdown of the flagella within the cytoplasm of the sporangium involves destruction of the membrane around the flagellum associated with the production of small, enzyme-containing vesicles within the surrounding vacuole (Hemmes 1983).

When environmental conditions favor zoospore production, the cytoplasm in the sporangium cleaves around each nucleus, and subsequently biflagellate zoospores form. The production of zoospores increases the inoculum rapidly. After a few hours, zoospores cease to swim and form a cell wall during the encystment process. The ultrastructural features in a zoospore include a Golgi apparatus, elongated dictyosomes, mitochondria, rough endoplasmic reticulum, microtubules, microbodies, membrane-free ribosomes, and a large number of vesicles. Both *P. sojae* (*P. megasperma* var. *sojae;* Ho et al. 1968a, b) and *P. parasitica* (Reichle 1969a, b; Hemmes and Hohl 1971) have been studied at the ultrastructural level. The cytology of zoospores is described and illustrated in detail by Hemmes (1983). See also detailed reviews of zoospore physiology and function by Hickman (1970) and Hickman and Ho (1966). Barr (1981) discusses flagellar rootlet morphology (Figures 4.21 and 4.22A and B) and in subsequent reports describes the ultrastructure of the flagellar apparatus for *P. sojae* and *P. megasperma* (Barr and Désaulniers 1989) and *P. infestans* (Barr and Désaulniers 1990). The flagellar apparatus of *P. parasitica* is compared with that of *Saprolegnia diclina* Humphrey (Saprolegniales), *Thraustochytrium aureum* Goldstein (Thraustochytriales), and *Rhizidiomyces apophysatus* Zopf (Hyphochytriales) (Barr and Allan 1989).

How are zoospores expelled from the sporangium? The sporangium content has a higher osmotic potential (about −6 bars) than the external water (0.0 bar) or soil solution (−0.1 to −1.0 bar). Water moves from a region of low osmotic potential through a membrane to a region of high osmotic potential. Thus, when water enters the sporangium, the turgor pressure within the sporangium wall increases. The plug or apical thickening at the tip of the sporangium, which is probably a gellike substance (MacDonald and Duniway 1978a) and is distensible (Gisi et al. 1979a), dissolves prior to emission of the zoospores. The increase in osmotic pressure within the sporangium causes the contents to be pushed out through the apical opening (Gisi 1983). The outer wall at the apex ruptures, and the membrane distends outward rapidly (about 200 μm/s) to form a large, spherical vesicle into which many of the swimming zoospores flow (Figure 3.1, frame 1). Eventually, the thin membrane surrounding the zoospores bursts, and zoospores are released into the surrounding medium. In some cases, the membrane breaks after the emergence of the first few zoospores. The process is elegantly described and discussed by Gisi et al. (1979a, b), Gisi and Zentmyer (1980), and Gisi (1983).

Zoospore discharge by *P. cryptogea* and *P. medicaginis* (*P. megasperma*) (alfalfa isolate) is prevented at a soil matric potential of −25 mb and is restricted at −5 and −10 mb, but at 0 and −0.1 mb discharge occurs freely. The negative matric potential effect of soil appears to be a much greater factor than the osmotic effect since sporangia can release zoospores in solutions of KCl and $MgSO_4$ at solute potentials at or greater than −450 mb and of NaCl and sucrose at solute potentials as low as −600 mb. These data suggest that release of zoospores might be caused by swelling of a gel matrix and not by osmotic effects (MacDonald and Duniway 1978a). See Chapter 5 (Appendix 5.1), Cook and Papendick (1972), Duniway (1983), and Cook and Duniway (1981) for discussions of the matric potential concept in soil.

Zoospore formation is strongly inhibited in some species by glucose and other catabolites, such as citric acid cycle acids and amino acids. This inhibition can be partially reversed by cyclic adenosine monophosphate (AMP), which promotes glycolosis at concentrations of 10^{-6} to 10^{-5} M. Reversal of inhibition by cyclic AMP, however, is not possible at glucose concentrations greater than 5 mM. Repression of sporangium production caused by noncatabolites such as antimycin A or cycloheximide (inhibitors of protein synthesis) are not reversed by cyclic AMP (Yoshikawa and Masago 1977). This effect suggests that cyclic AMP acts directly on the regulation processes that control the change from mycelial growth to zoosporangium formation. Cyclic AMP may be a key sporogenic factor mediating a process that changes the fungus from mycelial growth to sporulation. These metabolic alterations are controlled at the transcriptional and translational levels (Yoshikawa and Masago 1977).

Physiology of Zoospore Locomotion and Encystment

Zoospores of *Phytophthora,* which are normally uninucleate and form within the sporangium (Figure 3.1), swim by means of two flagella of unequal length, one whiplash (naked) and the other tinsel (see Chapter 4; Figures 4.18–4.20). Zoospores with these features are considered to be heterokont (having two types of flagella) and indicate a relationship to the algae, for instance, the Chrysophyceae, Phaeophyceae, and Xanthophyceae (Barr 1983). Zoospores and their function are discussed in greater detail by Hickman and Ho (1966) and Lange and Olson (1983). These reports review all of the zoosporic fungi including other members of Oomycetes, Hyphochytriomycetes, Chytridiomycetes, and Plasmidiophoromycetes.

Does the swimming-zoospore phase of *Phytophthora* give it a competitive advantage? Cameron and Carlile (1977) and Carlile (1983), who comprehensively describe the motility, taxis, and tropism of *Phytophthora* zoospores, suggest that the negative geotaxis of zoospores (i.e., their tendency to move toward the surface of water) may be a significant factor in migration of zoospores upward in flooded soil. The attraction of zoospores to plant roots (chemotropism) appears to be advantageous to their parasitism. The ability of zoospores to stick to surfaces during the encystment process (Sing and Bartnicki-Garcia 1975a, b; Bartnicki-Garcia and Wang 1983) also might be ecologically advantageous for a plant pathogen. At least four types of vesicles, some of which excrete a sticky substance, exist in the periphery of a zoospore. These vesicles have been identified by immunolabeling techniques with monoclonal antibodies specific for the substrate associated with each vesicle. Three of these vesicles excrete during encystment. One type of vesicle occurs on the ventral surface and contains a protein with a molecular weight greater than 200 kDa (kilodaltons), and the other forms on the dorsal surface and contains glycoproteins rich in *N*-acetyl-D-galactoseamine with a molecular weight greater than 300 kDa. Preliminary assumptions indicate that some of these materials serve to stick the zoospore to the root on which it encysts prior to penetration (Hardham et al. 1985, 1986). Review and discussion of these and other phenomena associated with the surface properties of *Phytophthora* zoospores are elegantly presented with full literature citations and illustrations by Hardham et al. (1991a). Methodology is reviewed by Hardham (1991b). Such studies at the molecular level may eventually explain how a pathogen recognizes a host.

The review by Carlile (1983) discusses zoospore motility, taxis, and tropism in greater detail. The review by Lange and Olson (1953) discusses the fungal zoospore in a holistic way, contrasting and comparing zoospores produced by fungi such as the chytrids, which have only one flagellum, with *Phytophthora* and *Pythium,* which have two flagella. Their review has a wealth of information on zoospore ultrastructure and function of various zoospore types that should be read by anyone interested in this biological structure. Studies on the ultrastructural level of the encystment and germination of zoospores of *P. palmivora* (Cho and Fuller 1989b) and on the mechanism of development of the water expulsion vacuole (Cho and Fuller 1989a) are of fundamental biological interest. Since zoospores do not have a cell wall and their outer surface is the plasma membrane, the cells are not able to osmoregulate by a buildup of turgor pressure. Water expulsion vesicles located at the anterior end of the zoospore function to pump water out of the cell (Hardham et al. 1991a). The water expulsion vesicle is illustrated by Hemmes (1983). See also Figure 4.21 in Chapter 4.

Zoospores swim by means of two flagella in a helical path for several hours (Ho and Hickman 1967a; Hickman 1970; R. N. Allen and Newhook 1973) (Figure 3.2) at speeds of about 121–182 μm/s (R. N. Allen and Newhook 1973). The velocity of *P. sojae* zoospores is 157 μm/s at 26°C (Ho and Hickman 1967a). Higher and lower rates of travel for other species have been recorded. The flagella undulate from base to tip, thereby producing a thrust. The anterior flagellum exerts about 10 times the thrust of the posterior flagellum; however, the posterior flagellum seems to control the direction of movement. The helical path (amplitude of 26–70 μm) (Figure 3.2) of zoospores has been shown by time lapse photomicrography in dark fields (Ho and Hickman 1967a; R. N. Allen and Newhook 1973, 1974).

Movement of zoospores against surfaces causes a disorienting effect and markedly restricts their activity, generally resulting in cessation of swimming and formation of a cyst. Zoospores of *P. sojae* lose motility within 30 min when suspended in water only 0.3 mm in depth, but in water 0.5 mm in depth, the zoospores continue to swim for hours. Glass beads glued to the bottom of a petri dish hasten the encystment process, a result of the zoospores bumping into solid objects (Hickman and Ho 1966). Encystment of zoospores can be readily induced by rapidly shaking them in a flask or by adding pectin to the aqueous solution in which they are suspended (Irving and Grant 1984b; Grant et al. 1985). A series of reports by Grant et al. (1984–1986) discusses differentiation of swimming zoospores to cysts.

Prior to encystment, the zoospores become sluggish and move in a less coordinated way, finally stop swimming, and round up. Just prior to their being shed, a swelling forms at the base of each flagella and moves toward the flagellar tip (Ho and Hickman 1967a; Reichle 1969a). The flagella then become shorter as the axoneme coils up within the swelling. Each flagellum usually becomes detached after two-thirds to three-fourths of it has disappeared. Hemmes (1983) illustrates detached flagella floating among encysted zoospores. Most reports indicate that the flagella are shed during encystment (Hickman and Ho 1966; Ho and Hickman 1967a); however, one report (Reichle 1969a) indicates that in 10–20% of the zoospores of *P. parasitica* var. *nicotianae,* the flagella are retracted. When zoospores of *P. parasitica* var. *nicotianae* were viewed with phase contrast optics, it was noted that when they encysted, the flagella became lumpy and then detached. The anterior flagellum had a helical appearance (Rich 1978).

Since the impacting of zoospores with solid surfaces causes premature encystment (Ho and Hickman 1967a; Bimpong and Clerk 1970), it might be asked whether the possession of flagella confers an ecological advantage. Do zoospores actually move actively in soil? In soil, the matric potential must be near 0 (saturation) for zoospores to migrate. Zoospores of *P. cryptogea* swim 25–35 mm in the surface water of flooded soils or in coarse-textured UC mix (fine sand and peat moss, 50:50) (Baker 1957) at a matric potential of about –1 mb. At the more negative matric potential of –10 mb (drier), movement in soil is limited to 5 mm, and at –100 mb, zoospores do not move. In soils with smaller pores, such as a fine sandy loam or clay, movement of zoospores is limited to 5 mm at –1 mb, and there is no movement at –10 mb (Duniway 1976). Zoospores of *P. cinnamomi* failed to reach root pieces in petri dishes when placed 7.6 cm (3 inches) away (Kuhlman 1964).

Duniway (1976) points out that a zoospore with its flagella extended is less than 60 μm in diameter, but because of its helical swimming habit (Figure 3.2), much larger pores are required to allow uninhibited locomotion. Unobstructed locomotion of *P. cinnamomi* zoospores in soil requires that the pores in soil be approximately 50–140 μm in diameter (Allen and Newhook 1974). Zoospores of *P. medicaginis* (*P. megasperma*) migrated 65 mm upward through a layer of a sandy loam soil (average pore size >190 μm) but only 24 mm in a silt loam soil (average pore size <120 μm) (Pfender et al. 1977). *P. cinnamomi* zoospores traveled 76 mm in soil leachate and 24 mm in wet soil (Kuhlman 1964). In soil, the tortuous pattern of pores increases the likelihood that a swimming zoospore will contact a solid surface and encyst before moving very far. Research by Newhook et al. (1981) indicates that autonomous movement of zoospores does occur, although passive dispersal is probably a more important mode of transport. By use of "ideal soils" consisting of glass microbeads of different sizes and different sand-soil mixtures, they show that *P. cinnamomi* zoospores can move autonomously, provided the size of the pore necks is greater than the wavelength of the helical swimming pathway (Figure 3.2). Their evidence indicates that many zoospores retain their swimming mode within the soil mass and do not encyst immediately.

The answer to the question of whether zoospores move actively in soil is that zoospores are capable of locomotion in certain soils but overall their movement in soil is limited.

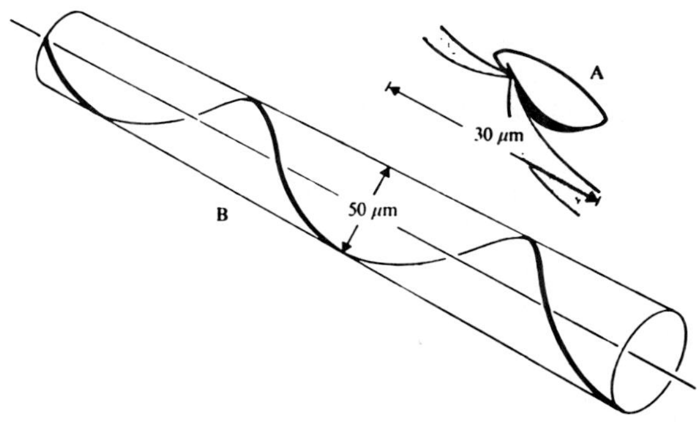

Figure 3.2. Diagrammatic representation of a swimming zoospore of *Phytophthora cinnamomi* showing (**A**) the effective diameter as defined by the distance between the tips of the fore and aft flagella and (**B**) the typical helical path of movement when a zoospore swims in water. (Reproduced by permission of C.A.B. International, Wallingford, Oxon, United Kingdom, courtesy Allen and Newhook 1974)

Under flooded conditions caused by rainfall or running irrigation water, however, passive movement could be great. Inoculum of *P. parasitica* that was detected in irrigation water passed through a 10-µm sieve, a result that indicated that zoospores were the active unit (Thomson and Allen 1974). Detection of *Phytophthora* inoculum in canals and reservoirs is discussed in Chapter 2. See Table 2.5, in which references are cited on detection of *Phytophthora* in irrigation water.

The duration of motility of zoospores in aqueous solutions varies under different conditions. Increasing temperature above the optimum for growth usually decreases the length of the motility period; lowering temperature below optimum increases the motility period (Hickman and Ho 1966; Ribeiro 1983). The temperature for optimum motility varies with the species. Examples of different lengths of the period of zoospore motility of some *Phytophthora* species are as follows: *P. citrophthora,* 24 h at 12.5°C, 1 h at 24°C, and a few seconds at 7°C (Klotz 1940); *P. fragariae,* 5 h at 13–14°C, 1.5 h at 20°C, and 15–30 min above 20°C (Goode 1956); *P. infestans,* 22 h at 3°C and 30 min at 24°C (Crosier 1933); *P. sojae* (*P. megasperma* var. *sojae*), 24–48 h at 24–25°C (O. K. Ribeiro, *unpublished*); and *P. phaseoli,* 24 h at 10°C, 12 h at 15°C, 6 h at 20°C, and <3 h at 25°C (Hyre and Cox 1953). *P. cinnamomi* zoospores are known to swim for up to 84 h (Bimpong and Clerk 1970). The motility of zoospores is reduced when pH values are on the acidic side. Motility of *P. fragariae* zoospores was reduced from 4 h at pH 6.8 to 5 min at pH 4.4 (Goode 1956).

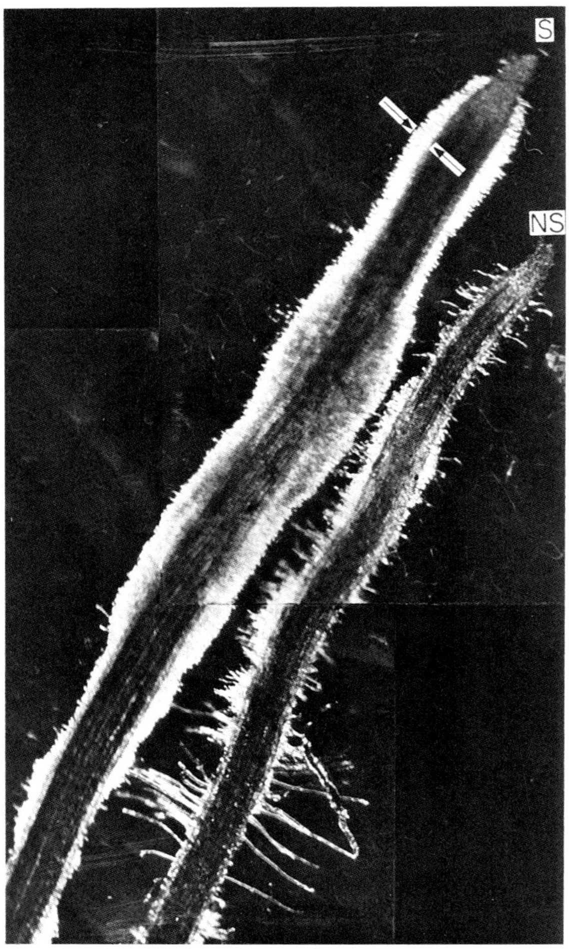

Figure 3.3. Attraction of zoospores of *Phytophthora megasperma* f. sp. *medicaginis* (*P. medicaginis*) to alfalfa roots grown in saturated soil (S) and nonsaturated soil (NS). Area between arrows is the zone of encysted zoospores. (Reproduced by permission of the American Phytopathological Society, St. Paul, Minnesota, courtesy Kuan and Erwin 1980b)

Zoospores remain motile longer if the concentration is high. For instance, at 10^6 zoospores per milliliter, more than 50% of *P. parasitica* var. *nicotianae* zoospores remained motile after 4 h, but at a concentration of 10^3 zoospores per milliliter, motility ceased after 5 min (Gooding and Lucas 1959a). Zoospores maintain activity while swimming by utilization of endogenous lipids, which seem to be the main source of energy (Bimpong 1975). Bimpong suggested that fatty acids were utilized via the glyoxalate cycle. Motile and germinating zoospores of *P. drechsleri* catabolize glucose, acetate, serine, aspartate, glutamate, and asparagine (Barash et al. 1965).

Érsek et al. (1994c) report that "approximately 1/5 of the cell respiration, measured in oxygen consumption, is used to keep zoospores motile," according to Hölker et al. (1993). They suggest that the great demand for oxygen might in part explain the high toxicity of a new antibiotic, mucidin (active against zoospore release at 0.25 µg/ml and motility of zoospores at 0.08 µg/ml). Érsek et al. (1994c) noted that the effects of mucidin on retardation of zoosporogenesis and the formation of "large undifferentiated zoospores are indicative of an affected respiratory function (Oertel and Jelke 1986)."

Zoospores migrate toward stimuli such as amino acids and encyst to form a wall within minutes. Grant and coworkers studied synchronous production of zoospores in detail with the view toward explaining the physiology by biochemical criteria. See Byrt and Grant (1979) for information on *P. cinnamomi* and Grant et al. (1984) and Paktitis et al. (1986) for *P. palmivora.*

The zoospore is a cell that undergoes a stimulus-mediated secretion that allows it to recognize certain substrates in nature (Grant 1984). In various aqueous solutions, zoospores swim for hours; however, when they come in contact with a root, they apparently recognize the substances in root slime because they differentiate from the swimming to the cyst stage and germinate quickly (within 90 min) (Byrt et al. 1982a, b; Irving and Grant 1984a). Germination and penetration of the host root is complete within an hour (Ho and Zentmyer 1977b).

Swimming zoospores chemotactically respond to certain chemical stimuli exuded by roots in aqueous solutions (Figure 3.3) (Kuan and Erwin 1980b; see also Goode [1955] and Hickman and Ho [1966]). After germination, the germ tube follows a chemotactic and also a directional or tropic response toward roots (see Zentmyer [1966, 1970] for information on *P. cinnamomi* and Ho and Hickman [1967b] for *P. sojae* [*P. megasperma* var. *sojae*]). A recent report (Carlile and Tew 1988) indicates that germinating cysts of *P. citricola* were negatively aerotrophic (grew away from a source of air), whereas germ tubes of *Geotrichum candidum* were positively aerotrophic. Orientation of germ tubes toward roots could be caused by growth down an oxygen gradient or up a carbon dioxide gradient. Added nutrients (e.g., casein hydrolysate) did not affect the direction of germ tube growth.

Zoospores in aqueous solutions migrate to the zone of elongation (the region of maximum exudation) on growing roots (Goode 1956; Dukes and Apple 1961; Zentmyer 1961; Hickman and Ho 1966; Ho and Hickman 1967b; Bimpong and Clerk 1970; Kuan and Erwin 1980b). The facts that the zone of elongation is an area in which maximum exudation of plant materials occurs and that zoospores are positively attracted to cationic fractions (containing amino acids) lead to the concept that chemotaxis is ecologically and pathologically significant. The response is apparently nonspecific, since zoospores swim to the zone of elongation on nonhost as well as host roots (Ho and Hickman 1967b; Mehrotra 1970; Hinch and Weste 1979); however, *P. cinnamomi* zoospores respond more effectively to host roots such as avocado than to nonhost roots. Avocado roots killed by heat or by propylene oxide were not attractive to zoospores (Zentmyer 1961, 1966, 1970).

The report by Kliejunas and Ko (1974) illustrated the accumulation of zoospores on a papaya root in water. They also showed that the percentage of seedlings killed was markedly higher (about 93%) when motile zoospores (9×10^5 per pot of soil) were used as inoculum than when encysted zoospores were used (about 21%). It is highly likely that the accumulation of a large number of zoospores on a host root may overcome the natural resistance more rapidly than a lower number.

The chemotactic effect of roots on zoospores can be demonstrated in soil. When safflower roots were separated from *P. drechsleri* zoospores by use of a cellulose acetate membrane that allowed diffusion of root exudates, the accumulation of zoospores near the root surface was visualized by staining the membrane with cotton blue in lactophenol. Another method involved applying a zoospore suspension to soil and transferring roots from soil to glass slides at 2- to 22-h intervals after inoculation. When the roots and zoospores were stained with the fluorescent vital stain Calcoflor White RW (0.01% in water), the accumulation of zoospores at the zone of elongation on roots was visualized by microscopic observation under ultraviolet light (Mehrotra 1970). See Chapter 2 for use of Calcoflor and diethanol as ultraviolet light fluorescing stains.

Experimentally, zoospores of *P. cactorum, P. capsici, P. cinnamomi, P. citrophthora,* and *P. palmivora* move chemotactically in water toward capillary tubes containing various compounds. Vitamins, phenolic compounds, nitrogenous bases of nucleic acids, nucleotides, growth regulators, sugars, organic acids, and amino acids are indicated to be chemotactically active because they induce zoospores to accumulate at the mouths of the capillary tubes containing these compounds. Positively charged molecules are more active than those negatively charged. The amino acids arginine, aspartic acid, glutamic acid, and methionine (10^{-1} to 10^{-5} M), which have a positive charge, exert a greater chemotactic index on *P. cactorum, P. capsici,* and *P. palmivora* than on *P. cinnamomi* and *P. citrophthora* zoospores. A chemotactic index can be calculated by dividing the number of zoospores that encyst near the end of the capillary tube containing the compound by the number of zoospores randomly encysted in a similar field beyond the influence of the solution in the capillary tube. A chemotactic index of 1.0 indicates no specific activity of the test solution, and values greater than 1.0 indicate a positive chemotaxic response (Khew and Zentmyer 1973). Although all compounds shown to be chemotactically active occur in soil near roots and thus could be important in nature, they conclude that the mechanism of chemotaxis is still obscure.

When capillary tubes containing nutrients were placed in water, zoospores of *P. drechsleri, P. cryptogea, P. cinnamomi, P. nicotianae* var. *parasitica,* and *P. citricola* responded chemotactically most readily to amino acids with the structural components of an α-amino acid group with a short carbon chain that terminated with an amide group (Halsall 1976b). Molecules with a neutral charge were more attractive than negatively charged molecules. Glutamine and asparagine were the most attractive amino acids. Slight attractiveness of some sugars and ethanol were also noted. Zoospores carried a negative surface charge, which was confirmed by their affinity for cationic dyes such as basic fuchsin, orange G, and eosin (1:1,000). Anionic dyes stained zoospores weakly. Ethanol and sucrose in combination with some amino acids acted synergistically. Halsall (1976b) states that these interactions may explain why effects of any single compound seldom equal that of root exudates and extracts.

Increasing concentrations of calcium ion (Ca^{2+}) stimulate differentiation of *P. palmivora* zoospores to the cyst stage (Byrt et al. 1982a; Irving and Grant 1984b; Grant et al. 1986; Zhang et al. 1992); pectin materials also appeared to mimic the effect of root slime on differentiation (Grant et al. 1985). Both citrus pectin and other uronic acids effectively induced differentiation of zoospores to cysts. The rate of establishment of colonies from *P. infestans* cysts was increased from 0 to about 17% by adding pectic substances (e.g., apple pectin) to a balanced medium (D. D. Clarke 1966).

An isolate of *P. parasitica* from tomato, obtained from a saline soil, was more tolerant to increasing concentrations of Na:Ca (1:5) (up to 50 milliequivalents [meq] per liter) in a nutrient solution, as evidenced by the decrease in motility and suppression of germination, than an isolate from nonsaline soil; however, after 2 months in culture, the tolerance was lost. The percentages of motility and germination both decreased with increasing concentration of the salts. Conversely, the percentage of lysed and encysted zoospores increased with increasing concentration of salts (Bouchibi et al. 1990).

Addition of pectin or strontium chloride ($SrCl_2$) (3 mM) to an aqueous medium caused surface changes in *P. palmivora* zoospores that had been synchronously produced and induced to differentiate to cysts (Paktitis et al. 1986). Within 10 s after addition of pectin, "blebs" (protuberances) and "microvilli" appeared on the cell and flagellum surfaces, and within 30–40 s, pits appeared on the surfaces. Zoospores rounded up and formed a sphere with a wrinkled surface after 4 min. Germ tubes emerged in 40 min, and as germ tubes enlarged, the cell wall surfaces became smooth but with a fibrous texture. They noted that when zoospores differentiated to cysts in the presence of pectin (500 μg/ml), the flagella remained attached to the cyst up to the germling stage; but in the presence of $SrCl_2$ (3 mM), the flagella dropped off. These observations were similar to those of Reichle (1969a) and Tokunaga and Bartnicki-Garcia (1971). In isoosmotic media (i.e., osmotic pressure in the bathing solution is the same as inside the cell) (sorbitol at 200 mM concentration), the flagella remained attached during the encystment stage, and the germ tubes always emerged on the side on which the flagella were attached. Paktitis et al. (1986) cited the reports on chemotropism by Ho et al. (1968), Mehrotra (1970), Zentmyer (1970), and Hinch et al. (1985) (illustration of cyst germination) that showed that germ tubes always emerged on the side of the cyst closest to the host surface (chemotropism).

Zoospores of *P. sojae* were attracted to capillary tubes containing the isoflavones daidzein (1.0 μM) and genistein (0.6 μM), both of which are components of root exudate from soybean (Morris et al. 1991). Both compounds also induced rapid encystment and germination of zoospores. Zoospores of two isolates of *P. megasperma* from alfalfa and an isolate each of *P. megasperma* from Douglas fir, *P. parasitica* from tomato, *P. capsici* from pepper, *P. cryptogea* from alfalfa, and *Pythium irregulare* from pea were attracted only at concentrations of 50.0–60.0 μM. It appears that the flavones from soybean are active signals that allow *P. sojae* to recognize its soybean host (Morris and Ward 1992).

Electrotaxis (the attraction of zoospores to low levels of electrical current) has been demonstrated in roots and artificial systems. Although electrotaxis can be demonstrated, little is understood about the function of this phenomenon in nature.

We quote from Carlile (1983) (by permission of the American Phytopathological Society, St. Paul, Minnesota, courtesy of Carlile 1983):

> The effect of electric currents on zoospore motility was studied by Troutman and Wills (1964), Ho and Hickman (1967b), Katsura and Miyata (1971), and Katsura et al. (1966). Discrepancies among the results obtained by these different authors led Khew and Zentmyer (1974) to study the response of zoospores of seven *Phytophthora* species to a range of current intensities. Differences among species were slight. At low current intensities, usually those below 0.5 μÅ, zoospores swam toward the anode and ultimately encysted around the electrode. At higher intensities, usually those above 0.5 μÅ, zoospores swam toward the anode but were repulsed near the anode. Hence a zoospore-free zone, its margin corresponding to an equipotential line and its width depending on the current intensity, surrounded the anode. Zoospores ultimately encysted just outside the zøne of repulsion. At these high current intensities, zoospores swimming near the cathode were immobilized and thus accumulated. These trapped zoospores often burst and rarely germinated.
>
> Khew and Zentmyer (1974) published dark-field photomicrographs of the tracks of zoospores showing electrotaxis. Precise orientation of zoospores does not seem to occur, but attraction seems to be due to the suppression of turning in zoospores swimming toward the anode. At high intensities, sharp turning occurred at the boundary of the zone of repulsion. Electric currents probably suppress or cause turning by altering transmembrane potential. The role of electrotaxis in nature is unclear, but weak electric currents do occur in the vicinity of roots, so electrotaxis may perhaps contribute to zoospore accumulation at the root surface.

A. L. Miller et al. (1988) showed that a positive current of about 10 mA^{-2} enters meristematic and elongating cells of primary roots of *Nicotiana tabaccum* var. *havana,* about 20 mA^{-2} at the growing tips of emerging lateral roots, and about 70 mA^{-2} at induced wound sites. These values correlated well with those of areas on roots on which zoospores of *P. parasitica* var. *nicotianae* accumulated (Troutman and Wills 1964).

In weak electrical fields (50–100 mV cm^{-1}) comparable to those generated in the region of plant roots, zoospores of *P. palmivora* moved toward the anode (positive pole) while zoospores of *Pythium aphanidermatum* moved toward the cathode (negative pole). The surface charge of the posterior flagellum of *P. palmivora* was positive while the surface charge of the posterior flagellum of *Pythium aphanidermatum* was negative. Although not well understood, B. M. Morris and Gow (1993) postulate that knowledge about electrotaxis might be useful to identify and locate regions on roots that are most susceptible to infection.

When *Phytophthora* zoospores are mechanically disturbed by contacting solid surfaces or by vigorous shaking or centrifugation, they rapidly cease swimming and encyst (Hickman 1970; Sing and Bartnicki-Garcia 1975a, b). When the zoospore stops swimming, the flagella in most cases are shed, but withdrawal of flagella into the zoospore body was reported for *P. parasitica* (Reichle 1969a). After an amorphous cell wall forms rapidly, the zoospore becomes a cyst. Ultrastructural studies show that the peripheral vesicles on the membrane of the zoospore discharge their contents and initiate formation of a cell wall within 30 s after encystment (Tokunaga and Bartnicki-Garcia 1971; Sing and Bartnicki-Garcia 1975a, b; Bartnicki-Garcia 1973; Bartnicki-Garcia and Wang 1983). The cell wall becomes adhesive during the encystment process, but after encystment the adhesiveness is lost. The alkali-resistant cell wall of the cyst, which consists of a "thin fabric of long interwoven, randomly oriented microfibrils" (Tokunaga and Bartnicki-Garcia 1971), forms rapidly within minutes (Bartnicki-Garcia 1973). Cysts begin to germinate within 30 min, and germ tubes form in 2–3 h. Germination can also occur by production of a secondary zoospore by repeated emergence (also called diplanetism) (Drechsler 1930; Blackwell 1949; Singh and Chauhan 1988).

See Hemmes (1983) and Paktitis et al. (1986) for illustrations and a general discussion of zoospore development. The reports by Tokunaga and Bartnicki-Garcia (1971) and Sing and Bartknicki (1975a) show excellent electron microscopic photography of the fine structure of the zoospore encystment process.

Appressorium-like bodies produced by encysted zoospores of *P. parasitica* in water that had been amended with plant materials such as orange peels or citrus leaves survived for at least 60 days in irrigation water (Thomson and Allen 1976). Removal of the nutrient source from appressorium-like bodies of *P. parasitica* gave rise to microsporangia, some of which survived in irrigation water for more than 60 days (Thomson and Allen 1976). Appressoria are also produced on potato leaves by *P. infestans* (Pristou and Gallegly 1954).

A study of hyphal tip growth from germinated cysts of *P. palmivora* (Meyer et al. 1976) showed that the turgor pressure of germinating cysts was 1.8 atmospheres; the results favored the concept that weakening of the apical wall of the hypha by the removal of β-1,3-glucan (the major component of *Phytophthora* cell walls) (Bartnicki-Garcia 1966) with β-1,3-glucanase allowed uncontrolled growth of hyphae that was characterized by production of hyphal swellings and apical branching and that exo-β-1,3-glucanases are involved in normal hyphal tip growth. Enzymes may be transferred to the wall by two kinds of vesicles, one containing the enzyme β-1,3-glucanase and the other containing uridine diphosphate glucose (UDPG)-transferase and possibly acid phosphatase.

Root exudates and nutrients increase the percentage of germination and also increase the percentage of germination by germ tubes (Ho and Hickman 1967a, b). Amino acids are required for germination of cysts in pure water (Tsao 1969b).

For some studies it is important to quantify the concentration of zoospores. The most common method utilizes a hemacytometer (blood cell counter). An electronic Coulter counter can be used, but the equipment is specialized and expensive. Both methods are described by the manufacturers and are relatively simple to use. Another simple, but adequate, method utilizes a microsyringe (Ko et al. 1973). The volume-control sleeve attached to the plunger guide of a 10-μl microsyringe is preadjusted to 2 μl. A 2-μl aliquot of a diluted zoospore suspension is divided into two 1-μl drops placed separately on a glass slide. The zoospores in each drop are counted. The total area of one drop of suspension is approximately equal to two microscopic fields utilizing a 10× objective on a compound microscope. The number of zoospores per milliliter is calculated from the average number of zoospores per microliter. Dilution of the spore suspension to five motile zoospores per microliter facilitates counting and produces counts comparable to those obtained with the hemacytometer (Ko et al 1973). The viability of either sporangia or zoospores can be tested by staining subsamples with rose bengal (300 ppm); dead protoplasm is indicated by a red coloration, but live cells remain colorless (Gisi and Schwinn 1976).

Production and Germination of Chlamydospores

Chlamydospores are usually globose and are formed by some but not all species of *Phytophthora.* Chlamydospores have walls somewhat thicker those of sporangia and are capable of persisting in soil and plant tissue for long periods (Mircetich and Zentmyer 1966; Zentmyer and Erwin 1970; Holdaway and Tsao 1972; Hwang and Ko 1978; Weste 1983b).

Chlamydospores are of particular importance as survival propagules, especially to *P. parasitica* and *P. cinnamomi.* See Ribeiro (1983) for a general review of production and germination of chlamydospores.

Chlamydospores are produced in infected host tissue and in most vegetable decoction media. Like other spores produced by *Phytophthora,* an adequate supply of a β-hydroxysterol (e.g., sitosterol) added to the medium often increases the quantity produced (Huguenin 1974; Englander and Turbitt 1979; Englander and Roth 1980). Although V8 juice agar and V8 juice broth contain β-hydroxysterols, Englander and Turbitt (1979) could not detect sterols in extracted, centrifuged, clarified V8 juice with anhydrous ether; however, they stated that esterified sterols could be present. Addition of sitosterol to clarified V8 juice media enhanced the production of chlamydospores by *P. lateralis* about 10-fold (Englander and Roth 1980); but addition of sitosterol to V8 juice broth did not affect the production of chlamydospores by *P. cactorum* (Darmono and Parke 1990). Chlamydospores of *P. palmivora* formed abundantly in media rich in nutrients such as V8 juice or papaya juice supplemented with 0.1% $CaCO_3$ (Chee 1973a; Kadooka and Ko 1973). In a synthetic medium, carbon-nitrogen ratios of 15:6 and 28:9 were more favorable than ratios of 10:1 or 50:6 and 28:9 (Huguenin 1974). *P. parasitica* produced chlamydospores abundantly when cultures were grown for several days at 25°C in clarified V8 juice broth, washed in water, and incubated for several days at 50-mm depth in bottles of water at 15–18°C. Since sporangia were not produced by this method (Tsao 1971), chlamydospores can be selectively produced (without sporangia) for quantitative establishment of inoculum in soil (Graham 1990).

A carbon-nitrogen ratio of 30:1 in media was optimum for production of chlamydospores of *P. palmivora.* Casein hydrolysate was the best nitrogen source, and ammonium sulfate inhibited formation of chlamydospores (Huguenin 1974). The synthetic medium of Ribeiro et al. (1975a) (Appendix 3.54) supported chlamydospore formation of numerous species. Conflicting reports exist on the importance of pH for chlamydospore production of *P. palmivora.* Huguenin (1974) found no significant difference between pH 4.5 and 7.0, but Chee (1973a) reported an optimum pH of 5.8. *P. cactorum* produced chlamydospores optimally on V8 juice broth containing $CaCO_3$ but not when $CaCO_3$ was omitted. This may have been caused by a pH effect, since $CaCO_3$ buffers media at pH 5.5–7.0 (Darmono and Parke 1990). Chlamydospores withstood freezing at –23°C for 24 h.

Light has variable effects on the production of chlamydospores. *P. cactorum* produced chlamydospores at 4°C on V8 juice broth in dark and in light and also when buried in pasteurized potting mix soil at 4°C. *P. cinnamomi* produced chlamydospores most abundantly on V8 juice agar plus sitosterol (20 mg/liter) when incubated at 23–28°C under near-UV light (310- to 420-nm region of the light spectrum). Fluorescent Blacklight Blue lamps (near-UV) used at 100–200 μWcm^2 were a useful light source (Englander and Turbitt 1979). Stimulation by light was not observed in V8 juice agar or broth not amended with sitosterol. *P. lateralis* produced chlamydospores on V8 juice agar or broth optimally in the dark but not when colonies were exposed to light at several wavelengths, including Blacklight Blue and Cool White fluorescent lamps (Englander and Roth 1980). Light (Cool White fluorescent bulbs) did not affect formation of chlamydospores by *P. cactorum* (Darmono and Parke 1990).

Although not all species of *Phytophthora* produce chlamydospores (Stamps et al. 1990), the work by Darmono and Parke (1990) with *P. cactorum* may indicate that chlamydospores are produced, as stated by Blackwell (1943), only under certain environmental conditions, such as temperature or moisture stress. Chlamydospore formation has been considered by most to be either absent or sparse in *P. cactorum* (Meyer and Schönbeck 1975); however, when mycelial mats were incubated in V8 juice broth or water (moistened but not submerged) at 4°C for 7–20 days, chlamydospores formed optimally (Darmono and Parke 1990).

Although *P. citrophthora* does not ordinarily produce chlamydospores in culture, chlamydospores were formed in soil and colonized roots during warm summer months in California. During the summer, soil temperatures are usually too high (above 30°C) to be favorable for growth or infection. Isolation from soil by dilution plating (see Chapter 2) was augmented by incubation of soil at 16°C for 3–10 days prior to plating (Sjoerdsma et al. 1988).

Chlamydospores of *P. parasitica* (Tsao 1971) and *P. palmivora* MF4 (*P. capsici* according to Alizadeh and Tsao [1985b] and Tsao [1991]) are produced by growing mycelial mats in cleared V8 juice broth for 7 days in 450-ml glass prescription bottles (incubated horizontally). The broth is decanted and replaced with 10 ml of sterile deionized water. The submerged mycelial mats in the bottles are then incubated vertically at 18°C for 21 days in the dark. Chlamydospores form abundantly, but sporangia do not. By use of the submerged culture method (Tsao 1971), *P. capsici* (= *P. palmivora* MF4) cultures produced chlamydospores optimally at 15°C, but *P. capsici* cultures from *Capsicum annum* L. did not produce chlamydospores at any temperature (Alizadeh and Tsao 1985b). The isolate P113 of *P. palmivora* produced chlamydospores at 12–24°C. Data for *P. cinnamomi* were similar. Although lack of aeration (oxygen) has been considered to be the cause of preferential production of chlamydospores but not sporangia when mycelium is submerged deeply in water (Hugenin 1974), increased levels of CO_2 were shown to be responsible for increased chlamydospore production by *P. parasitica* from tomato (Ionnou and Grogan 1985). When a stream of CO_2-O_2 mixture was passed across the mycelial mats in petri plates containing 5 ml of sterile distilled water, the greatest number of chlamydospores and the lowest number of sporangia were produced when CO_2 was 6% and O_2 was 10–12%. When CO_2 was tested at 0.03% (normal content of air), abundant sporangia formed at 2–20% oxygen but only about one-fourth as many chlamydospores formed. The physiological explanation of the effect of CO_2 is not yet known. The concept appears to be ecologically logical. Since neither chlamydospores nor sporangia formed on infected tomato stem tissue buried in water-saturated soil, it is likely that the chlamydospore production process requires oxygen.

The effect of incubation temperature on chlamydospore formation varies with different species. Optimum temperatures reported for several species were *P. parasitica,* 15–18°C (Tsao 1971); *P. capsici* (= *P. palmivora* MF4), 15°C; *P. palmivora,* 31°C (Chee 1973a); another isolate of *P. palmivora,* 18°C (Huguenin 1974); *P. cinnamomi,* 25°C (Englander and Turbitt 1979); *P. lateralis,* 24–25°C (Englander and Roth 1980); and *P. cactorum,* 4°C (Darmano and Parke 1990). It is interesting but unexplainable that formation of chlamydospores by *P. lateralis* was optimum at a much higher temperature than that optimum for growth (Englander and Roth 1980). In the report by Chee (1973a), production of chlamydospores by *P. palmivora* occurred at 31°C, which is also at the upper limit for growth. Conversely, the abundant production by *P. cactorum* at 4°C, at which little growth occurs, indicates, as Blackwell (1943) postulated, that chlamydospore formation could be a stress-related phenomenon.

Figure 3.5. Sequence of cell wall differentiation during sexual morphogenesis in *Phytophthora* (events listed inside dash-line rectangles are shown in their suspected sequence). (Reproduced by permission of Springer-Verlag, Heidelberg, Germany, courtesy Hemmes and Bartnicki-Garcia 1975)

Unpublished observations by the late D. Chen (to D. C. Erwin), during his work with sporangium production by *P. cinnamomi* in washed mycelial mats, indicated that sporangia

Some species of *Phytophthora* produce oospores mainly when the A1 and A2 (opposite) mating types grow together. These species are considered to be self-sterile and are called

milliliter). This could be the result of improved aeration. A pH of 7.0 was more favorable than pH 4.0 for production of oospores. Oospores caused root rot when used as inoculum in soil at concentrations of >100 oospores per gram of soil (Kittle and Gray 1979b).

Synthetic media (Appendixes 3.51–3.54) with a high carbon-nitrogen ratio of 94:1 (Ribeiro et al. 1975a) (Appendix 3.54) are reported to support oospore formation optimally (Leal et al. 1964; Huguenin and Boccas 1971; Elliott 1972; Ko 1985). In synthetic media, amino acids such as aspartic acid, glutamic acid, and asparagine were best for oospore formation, but valine and leucine inhibited oosporogenesis (Ribeiro et al. 1975a). V8 juice agar amended with sitosterol and tryptophan supported oospore formation by pairing A1 and A2 mating types of *P. palmivora* (Chee et al. 1976).

Maximum numbers of oospores of *P. cactorum* were produced in a synthetic medium containing sucrose (10 g/liter) when the nitrogen as glutamate, asparagine, or glycine (2.5–20 μM) neared exhaustion (Elliott 1989). Elliott presented the hypothesis "that oosporogenesis [of *P. cactorum*] is specifically induced by exhaustion of the nitrogen source, and that the numbers of oospores subsequently produced depend on the amount of sugar remaining in the medium. On the contrary sporangium production is induced by sugar deficiency." Elliott's paper is rich in elegant data and discussion that draw on many sources, including the classical research by Klebs (1889, 1900).

The effects of β-hydroxysterols, such as cholesterol (animal origin), stigmasterol, sitosterol (plant origin), and many others, on oogonium and oospore formation are reviewed by J. W. Hendrix (1970), Elliott (1983), and W. D. Nes (1987) and discussed in more detail in the previous section on sterols. In general, sitosterol is the most effective sterol for oospore production. Minimum dosages for activity are about 10^{-6} M (0.415 mg/liter) (J. W. Hendrix 1970). Sitosterol was utilized at a concentration of about 30 mg/liter for production of oospores of A1 × A2 isolates of *P. infestans* (Romero and Erwin 1969) and at 9 mg/liter for *P. cactorum* (W. D. Nes et al. 1979). Corn, linseed, and wheat germ oils, all of which contain sterols, stimulate oospore production by *P. syringae* (Doster and Bostock 1988a).

Oospore production was greater at O_2 concentrations of 1 and 5% than in air but decreased with increasing CO_2 concentrations at O_2 levels of 1, 5, or 20% (Mitchell and Zentmyer 1971b).

The temperature optimum for oospore formation of *P. palmivora* is usually lower than the optimum for growth (P. D. Turner 1960b; Satchuthananthavale 1963; Brasier 1969b); however, in *P. heveae* isolates, the temperature that was optimal for production of oospores was the same as that for growth (Brasier 1969b). *P. syringae* produced oospores optimally at 5–9°C but not at temperatures of 12°C or greater (Doster and Bostock 1988a). The optimal temperatures for production of oospores of *P. cinnamomi* by pairing A1 and A2 cultures was 15–21°C and by incubating A2 cultures in avocado root extract was 18–24°C. Maximum numbers of oospores formed after 2 weeks of incubation on carrot agar (Zentmyer et al. 1979; Zentmyer 1980).

For production of oospores by a homothallic *Phytophthora* species, a minced mycelial culture can be blended aseptically in an Omnimixer cup for 60 s at 10-s intervals to prevent heating and then spread over the surface of carrot or V8 juice agar in a petri plate or mixed with a comparable liquid medium in petri plates (Förster et al. 1983). Use of a mycelium-seeded plate rather than a single plug of mycelium placed in the center of the plate allows more synchronous production of oospores than when the plate is inoculated in the center with a single mycelial disk.

The minced mycelium of A1 and A2 types of heterothallic *Phytophthora* species can be mixed and the cultures incubated for at least 30 days in darkness at a suitable temperature (Honour and Tsao 1974) (Appendix 3.49).

Heterothallism

The heterothallic species contain two mating types designated A1 and A2. Although oospores form after contact of the mycelia of A1 and A2 mating types, isolates of each type are bisexual and self-incompatible. Relative degrees of maleness and femaleness occur within *P. infestans* (Gallegly and Galindo 1958). Thus, some isolates that are strongly male will form more antheridia than oogonia; some isolates are intermediate in sexual strength. See also the review by Elliott (1977). The A2 characteristic is not always stable, and occasionally self-fertile isolates arise from the A2 type. Some examples of this phenomenon are cited below. Some isolates of certain species are neuter and do not produce oospores alone or when paired with A1 or A2 types (Ko 1981).

As early as 1911, Clinton (1911) noted oospores by pairing *P. infestans* and *P. phaseoli* (see Shaw [1983, 1987, 1990]). The designation of A1 or A2 for mating types in all heterothallic species is derived from the original reports of mating types in *P. infestans* (Gallegly and Galindo 1958; Smoot et al. 1958; Galindo and Gallegly 1960). Nearly all heterothallic species form oospores when paired with the opposite mating type of most other species (Savage et al. 1968). See Appendix 3.55 for methods of determination of A1 and A2 types. Some isolates of *P. parasitica* are neuter, and others induce both A1 and A2 cultures to produce oospores (Ko 1981). Oospores of *P. cinnamomi* (which is heterothallic) were produced in single A2 culture when incubated in avocado root extract (Zentmyer 1979) and were larger than those from A1 and A2 crosses (Zentmyer et al. 1979). See Shaw (1983, 1988, 1990, 1991) for a more detailed review of the nature of heterothallism in *Phytophthora.*

According to Shaw (1988),

> There is now convincing evidence that both mating types of all heterothallic species so far examined are reciprocal translocation heterozygotes, as multiple associations of four chromosomes are always present. No multiple associations are found in meiosis of homothallic species (Sansome, 1980). Because self-fertility in some progeny of *P. drechsleri* was associated with tertiary trisomy (the additional chromosome being associated with the translocation complex), it has been argued that the complex carries the gene(s) determining mating type and is thus a 'mating type complex' (Mortimer et al. 1977; Sansome, 1980). Field isolates of one heterothallic species, *P. nicotianae* var. *parasitica,* which showed tendencies to self, have now been shown to be tertiary trisomics (Sansome, 1985).

Shaw (1987), in a discussion of the "breeding" system of A1 and A2 isolates of *P. infestans,* illustrated hyphal interactions in mating of a recently discovered Egyptian A2 isolate (Shaw et al. 1985) with an A1 from Wales. When the A1 and A2 isolates were inoculated into adjacent halves of the same nutrient agar plate, the advancing edge of the A2 colony nearest to the A1 colony responded to diffusates from the latter by an increase in growth rate and a derepression of lateral hyphal branching. Soon after contact of the rapidly growing A2 hyphae with the A1 colony, a small proportion of A2 hyphal tips coiled around A1 hyphae to produce ropelike structures that cated that the antheridial initial encircled the oogonial initial but did not penetrate. Photographic evidence appeared to be convincing, but later work utilizing ultrathin sections observed by electron microscopy indicated conclusively that the antheridial initial was penetrated by the oogonial incept (Hemmes and Bartnicki-Garcia 1975), as concluded by Pethybridge (1913) (Figure 3.4).

The sexual process was visualized by time-lapse cinemicrophotography utilizing A1 and A2 mating types of *P. drechsleri.* This film, which is available for purchase or rent from the photographic unit of Media Resources, University of California, Riverside 92521, depicts not only the mating behavior, but also the production of zoospores (Erwin and Kraft 1967).

In some species, such as *P. medicaginis* (*P. megasperma* f. sp. *medicaginis*), both amphigynous and paragynous antheridia occur in the same single-zoospore culture (Erwin 1965). *P. clandestina* also produces both types of oospores in single culture (Taylor et al. 1985a). Since amphigyny and paragyny are used as characters for classification of species (see Chapter 4), production of both characters in the same culture makes this feature of questionable value for classification, at least for species in which both types occur.

After meiosis occurs in the antheridia and oogonia (see Shaw's reviews [1983, 1988]), the antheridial nucleus passes into the oogonium. One or more haploid antheridial nuclei pass into the oogonium, and the oogonium expands to a globose structure (Figure 3.4). The oogonium increases in size, and within it a single oospore forms. The oospore becomes delimited from the periplasm, and in 1 or 2 days, the oospore wall (1–3 μm in thickness) forms from peripheral vesicles. This wall is composed of an outer wall (five discernible layers by electron microscopy) and a thicker inner wall (electron transparent endospore wall) (Hemmes 1983). The endospore wall is composed almost entirely of glucans (Lippman et al. 1974; Bartnicki-Garcia and Wang 1983). The fine structure (shown via electron microscopy) of the oospore wall is described for an amphigynous species, *P. capsici,* by Hemmes and Bartnicki-Garcia (1975) and for *P. megasperma* (Ho 1979b) by scanning electron microscopy. Fine structure of a paragynous species, *P. sojae* (*P. megasperma* var. *sojae*), is shown by Hemmes and Ribeiro (1977). The morphology of sexual structures as well as the cytology of sporangia, zoospores, and chlamydospores is reviewed by Hemmes (1983).

Two or more nuclei and a well-defined ooplast (often referred to as the refringent body, reserve body, or reserve

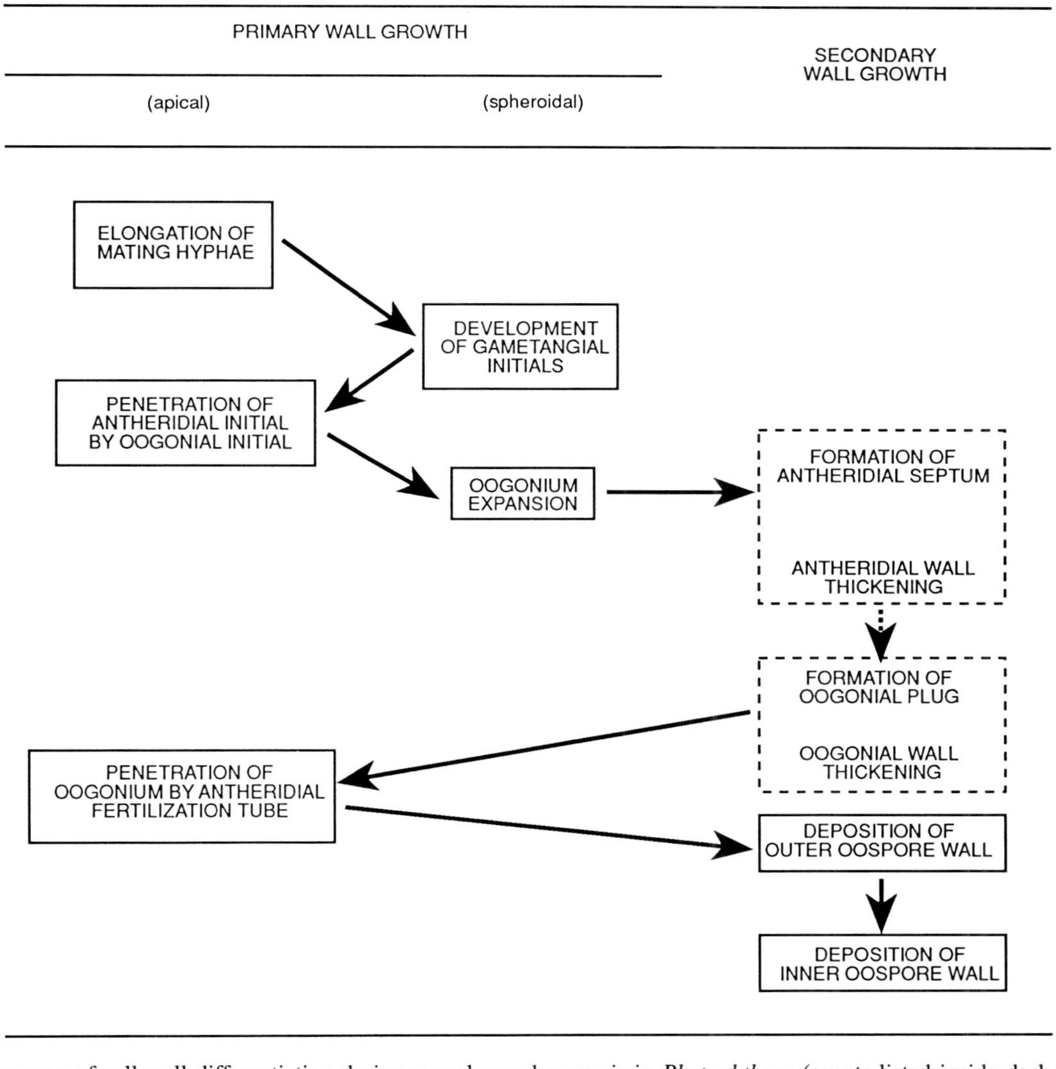

Figure 3.5. Sequence of cell wall differentiation during sexual morphogenesis in *Phytophthora* (events listed inside dash-line rectangles are shown in their suspected sequence). (Reproduced by permission of Springer-Verlag, Heidelberg, Germany, courtesy Hemmes and Bartnicki-Garcia 1975)

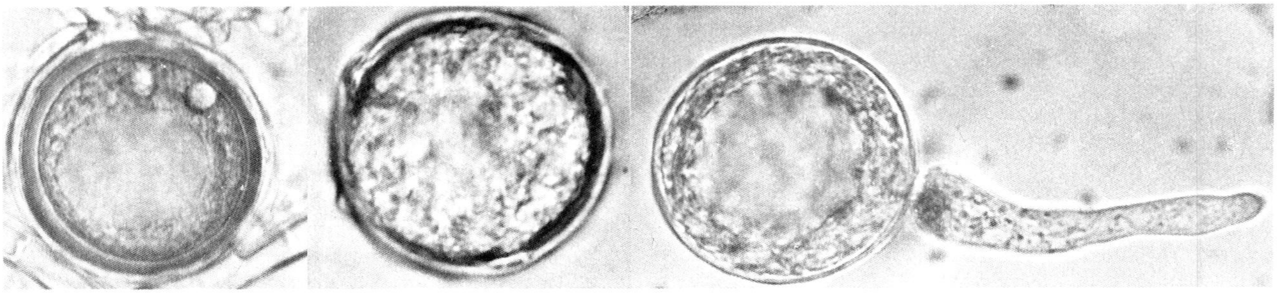

Figure 3.6. Representative stages in the germination of three oospores of *Phytophthora sojae* (*P. megasperma* var. *sojae*). Left, the oospore has a thick wall, a central body called an ooplast, and two bodies at the periphery that are called pellucid bodies, each of which contains a nucleus. These nuclei fuse before germination. Center, the oospore is activated. The thick inner cell wall has eroded (activation), apparently because of enzymatic digestion of the glucans that make up the inner wall (Lippman et al. 1974). Right, the oospore has germinated. Eventually the protoplasm flows out of the oospore, leaving an empty shell. (Reproduced by permission of the American Phytopathological Society, St. Paul, Minnesota, courtesy Erwin and McCormick 1971)

globule) can be readily seen in an oospore by phase or light microscopy (Figures 3.6–3.8); the ooplast is a globular body that does not contain organelles when examined by electron microscopy (Hemmes 1983). During formation of the thick endospore wall in some species of *Phytophthora,* the diameter of the oospore shrinks, often leaving a space between the inner oogonial wall and the outer oospore wall (a condition described as *aplerotic*). Oospores go through a maturation phase, which usually requires about 30 days in many species but is longer in others, prior to germination. In some cases, oospores of *P. medicaginis* germinate without a maturation phase in carrot broth culture (J. Jiang and D. C. Erwin, *unpublished*; Jiang 1991). This would seem to indicate that under some conditions, oospores may not be dormant. Oospores of *P. cactorum, P. citricola,* and *P. medicaginis* were mature in 30 days, but oospores of an isolate of *P. sojae* (*P. megasperma* f. sp. *glycinea*) did not become mature until much later (Jiang et al. 1989) (Figure 3.8). One of the first indications of maturity is fusion of the two nuclei (Dick 1969; Sansome 1976; Jiang et al. 1989) (Figures 3.6 and 3.8B).

Gametic nuclei that can be readily seen in oospores have

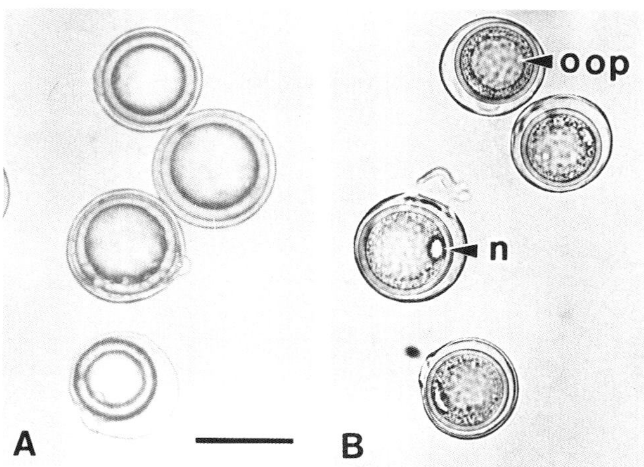

Figure 3.7. Nonviable and viable oospores (30 days old) of *Phytophthora sojae.* **A,** Nonviable, abnormally thick-walled oospores without nuclei or ooplasts. **B,** Viable oospores with thinner walls and one or more nuclei (n) (pellucid bodies) and an ooplast (oop). Bar = 35 μm. (Reproduced by permission from Mycologia 82: 107–113, J. Jiang and D. C. Erwin, copyright 1990, The New York Botanical Garden, Bronx, New York)

been referred to as pellucid (translucent) bodies (Erwin and McCormick 1971; Jiang et al. 1989) or pellucid spots (Blackwell 1943, 1949; Dick 1969; Brasier and Sansome 1975; Brasier and Brasier 1978). Although fusion of nuclei per se is not always readily observed, one can assume that if more than one nucleus can be seen in oospores, they are probably not yet mature.

Brasier and Brasier (1978) visualized both gametic and zygotic nuclei of *P. capsici* (MF4 form of *P. palmivora*) by use of lactophenol and dilute cotton blue stain. Oospores from an A1 × A2 pairing on carrot agar were covered with the stain, warmed gently to soften the agar, and gently squashed under a number 1 cover slip. They photographed presumptive gametic nuclei and also a single, larger nucleus (zygotic) that was approximately twice the size of each gametic nucleus.

After fusion of the nuclei, the first evidence of activation, just prior to germination, is the erosion of the inner oospore wall (Figures 3.6 and 3.8D and E), which is composed mainly of glucans (Beakes and Bartnicki-Garcia 1989; Lippman et al. 1974). The erosion of the inner oospore wall requires less than a day, after which the activated or developing oospore enlarges to completely fill the oogonium (Figures 3.6 and 3.8E). If conditions are favorable, one or more germ tubes emerge from the oospore, break through the oogonial wall, and grow outward. The germ tube can emerge from any region of the oospore, including the area to which the antheridium is attached (see Hemmes [1983], his Figure 40).

The erosion of the endospore wall in oospores of *P. cactorum* is associated with an increase of β-1,3-glucanase enzyme(s) and a corresponding increase in the production of proteins and reducing sugars. Enzyme activity was not detected in dormant oospores, but in germinated oospores, crude extracts of the enzyme obtained in the precipitate from treatment with 35–70% ammonium sulfate yielded β-1,3-glucanase with an optimal pH of 7.0 in a citrate-phosphate buffer (20.0 mM) (Jiang 1991; Jiang et al. 1991).

After the germ tube emerges, it either develops into mycelium or produces a germ sporangium, which in turn is capable of producing zoospores. Germinated oospores of *P. medicaginis* (*P. megasperma* f. sp. *medicaginis*) formed sporangia in dilute alfalfa root exudate but in distilled water formed only germ tubes (El-Hamalawi and Erwin 1986a, b). Strictly speaking, continued germ tube growth or germ sporangium formation are events separate from the germination process per se.

Oospores of homothallic species usually form in diseased plant tissue, for instance, *P. syringae* in apple leaves (D. C. Harris 1979a; D. C. Harris and Cole 1982), *P. sojae* in soybean roots (Slusher and Sinclair 1973; Slusher et al. 1974), *P. fragariae* in strawberry roots (Duncan 1977), and *P. cactorum* in apple crown tissue (Braun and Nienhaus 1959). Oospores are thick walled and have great resistance to external factors that destroy mycelium, sporangia, zoospores, or chlamydospores and are long lived in soil. Oospores that form in plant tissue are considered to be the most important means for survival of *P. medicaginis* (*P. megasperma* f. sp. *medicaginis*), the cause of root rot of alfalfa (Stack and Millar 1985a–c). Since rotation of soybean to nonhosts of *P. sojae* for periods of 4 years does not eliminate the pathogen, Schmitthenner (1985) postulates that there may be an internal timing mechanism that prevents germination of all oospores at one time. Germination of *P. cactorum* oospores in soil began at a temperature of 7°C and increased to 87% within 22 days as the temperature increased to 13°C (Braun and Nienhaus 1959).

Several factors influence oospore production. Oospore production is optimal in darkness and reduced in light (Harnish 1965; Savage et al. 1968; Brasier 1969a; Ribeiro et al. 1976a; Kittle and Gray 1979b; Schecter and Gray 1987; see also the reviews of Lilly [1966] and Elliott [1983]). Continuous light (700 footcandles) inhibits oospore production, but a low intensity of light (3 footcandles) does not. Light in the blue region but not in the red region of the spectrum inhibits oospore formation. Conversely, blue light increases sporangial formation of many species (Harnish 1965). A similar relationship was reported by Brasier (1969a). Wavelengths in the red and far red regions of the spectrum stimulated oospore formation (Huguenin and Jaques 1973; Ribeiro et al. 1976a). Elliott (1983) comments that the deleterious effect of a 24-h period of light on oospore formation, which is double the effect of a 12-h period of light (Huguenin and Boccas 1971), is presumptive evidence that the inhibition is caused by a simple photochemical reaction; however, there is no chemical evidence for the hypothesized photochemical receptor.

Oospores are produced on vegetable decoction media such as bean meal (Appendix 3.46), rape seed extract (Appendix 3.47), V8 juice supplemented with tryptophan (Appendix 3.48), V8 juice broth or agar (Appendix 3.49), or carrot broth (Appendix 3.50), all of which apparently contain naturally occurring β-hydroxysterols. Carrot agar was most favorable for oospore production by *P. cinnamomi* (A1 × A2). By the third week of incubation at 21°C, the yield of oospores per milliliter of carrot agar (196,616) was about double that on V8 juice agar (101,000). Production on Ribeiro's synthetic medium (Appendix 3.54) and cornmeal agar was low (2,088 and 1,064, respectively). More oospores formed in agar than in broth media (Zentmyer et al. 1979a).

P. sojae produced oospores more abundantly in V8 juice broth when flasks were shaken for 6 days followed by 6 days in still culture (2.0×10^5 oospores per milliliter) than when incubated in still culture for 14 days (2.4×10^3 oospores per

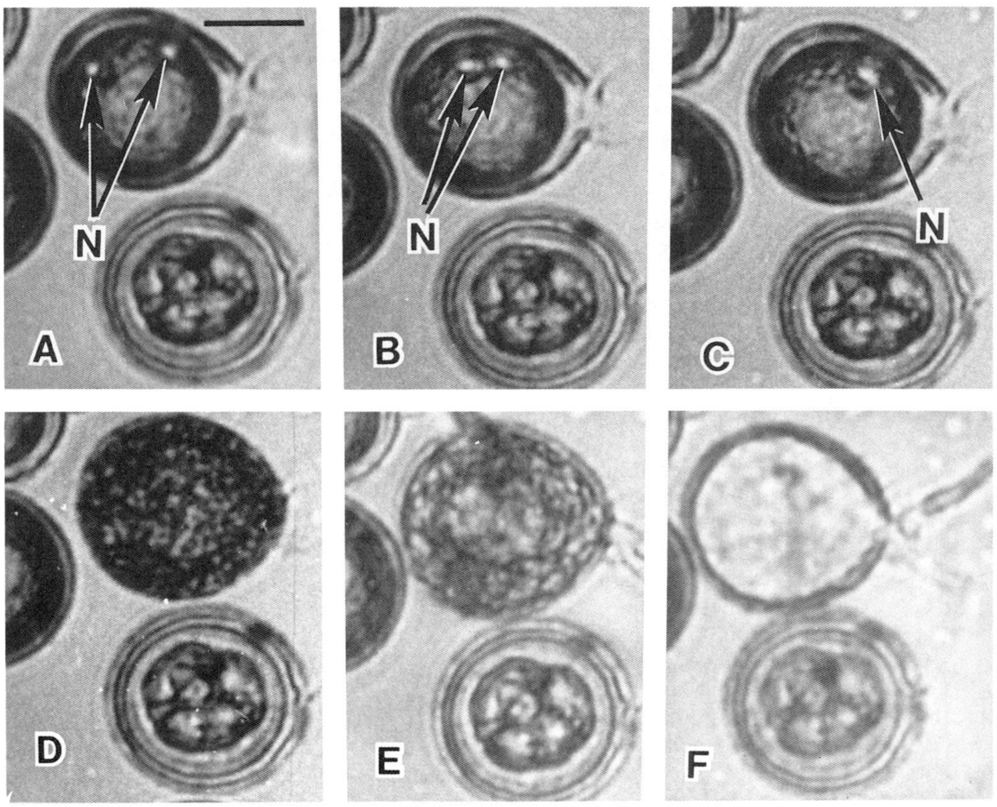

Figure 3.8. Nuclear fusion and germination stages of a 40-day-old oospore of *Phytophthora sojae* (*P. megasperma* f. sp. *glycinea*) recorded at a fixed position by time-lapse photomicrography over a period of 22 days. The oospore at the top of each picture was one of a population of about 400 incubated in alfalfa seedling exudate. The aborted oospore (shown below in each picture) is included as a marker to show that the same oospore was photographed throughout the series. **A,** After 8 days, two nuclei (N) are readily seen. **B,** After 10 days, the two nuclei are in contact. **C,** After 12 days, the two nuclei have fused. **D and E,** After 14 and 15 days, the inner wall has been digested, the oospore has swelled, and germination has initiated. **F,** After 22 days, the protoplasm in the oospore has flowed into the germ tube, leaving only the outer oospore wall. All figures share the same bar, which equals 20 μm. (Reproduced by permission of the British Mycological Society, courtesy Jiang et al. 1989)

milliliter). This could be the result of improved aeration. A pH of 7.0 was more favorable than pH 4.0 for production of oospores. Oospores caused root rot when used as inoculum in soil at concentrations of >100 oospores per gram of soil (Kittle and Gray 1979b).

Synthetic media (Appendixes 3.51–3.54) with a high carbon-nitrogen ratio of 94:1 (Ribeiro et al. 1975a) (Appendix 3.54) are reported to support oospore formation optimally (Leal et al. 1964; Huguenin and Boccas 1971; Elliott 1972; Ko 1985). In synthetic media, amino acids such as aspartic acid, glutamic acid, and asparagine were best for oospore formation, but valine and leucine inhibited oosporogenesis (Ribeiro et al. 1975a). V8 juice agar amended with sitosterol and tryptophan supported oospore formation by pairing A1 and A2 mating types of *P. palmivora* (Chee et al. 1976).

Maximum numbers of oospores of *P. cactorum* were produced in a synthetic medium containing sucrose (10 g/liter) when the nitrogen as glutamate, asparagine, or glycine (2.5–20 μM) neared exhaustion (Elliott 1989). Elliott presented the hypothesis "that oosporogenesis [of *P. cactorum*] is specifically induced by exhaustion of the nitrogen source, and that the numbers of oospores subsequently produced depend on the amount of sugar remaining in the medium. On the contrary sporangium production is induced by sugar deficiency." Elliott's paper is rich in elegant data and discussion that draw on many sources, including the classical research by Klebs (1889, 1900).

The effects of β-hydroxysterols, such as cholesterol (animal origin), stigmasterol, sitosterol (plant origin), and many others, on oogonium and oospore formation are reviewed by J. W. Hendrix (1970), Elliott (1983), and W. D. Nes (1987) and discussed in more detail in the previous section on sterols. In general, sitosterol is the most effective sterol for oospore production. Minimum dosages for activity are about 10^{-6} M (0.415 mg/liter) (J. W. Hendrix 1970). Sitosterol was utilized at a concentration of about 30 mg/liter for production of oospores of A1 × A2 isolates of *P. infestans* (Romero and Erwin 1969) and at 9 mg/liter for *P. cactorum* (W. D. Nes et al. 1979). Corn, linseed, and wheat germ oils, all of which contain sterols, stimulate oospore production by *P. syringae* (Doster and Bostock 1988a).

Oospore production was greater at O_2 concentrations of 1 and 5% than in air but decreased with increasing CO_2 concentrations at O_2 levels of 1, 5, or 20% (Mitchell and Zentmyer 1971b).

The temperature optimum for oospore formation of *P. palmivora* is usually lower than the optimum for growth (P. D. Turner 1960b; Satchuthananthavale 1963; Brasier 1969b); however, in *P. heveae* isolates, the temperature that was optimal for production of oospores was the same as that for growth (Brasier 1969b). *P. syringae* produced oospores optimally at 5–9°C but not at temperatures of 12°C or greater (Doster and Bostock 1988a). The optimal temperatures for production of oospores of *P. cinnamomi* by pairing A1 and A2 cultures was 15–21°C and by incubating A2 cultures in avocado root extract was 18–24°C. Maximum numbers of oospores formed after 2 weeks of incubation on carrot agar (Zentmyer et al. 1979; Zentmyer 1980).

For production of oospores by a homothallic *Phytophthora* species, a minced mycelial culture can be blended aseptically in an Omnimixer cup for 60 s at 10-s intervals to prevent heating and then spread over the surface of carrot or V8 juice agar in a petri plate or mixed with a comparable liquid medium in petri plates (Förster et al. 1983). Use of a mycelium-seeded plate rather than a single plug of mycelium placed in the center of the plate allows more synchronous production of oospores than when the plate is inoculated in the center with a single mycelial disk.

The minced mycelium of A1 and A2 types of heterothallic *Phytophthora* species can be mixed and the cultures incubated for at least 30 days in darkness at a suitable temperature (Honour and Tsao 1974) (Appendix 3.49).

Heterothallism

The heterothallic species contain two mating types designated A1 and A2. Although oospores form after contact of the mycelia of A1 and A2 mating types, isolates of each type are bisexual and self-incompatible. Relative degrees of maleness and femaleness occur within *P. infestans* (Gallegly and Galindo 1958). Thus, some isolates that are strongly male will form more antheridia than oogonia; some isolates are intermediate in sexual strength. See also the review by Elliott (1977). The A2 characteristic is not always stable, and occasionally self-fertile isolates arise from the A2 type. Some examples of this phenomenon are cited below. Some isolates of certain species are neuter and do not produce oospores alone or when paired with A1 or A2 types (Ko 1981).

As early as 1911, Clinton (1911) noted oospores by pairing *P. infestans* and *P. phaseoli* (see Shaw [1983, 1987, 1990]). The designation of A1 or A2 for mating types in all heterothallic species is derived from the original reports of mating types in *P. infestans* (Gallegly and Galindo 1958; Smoot et al. 1958; Galindo and Gallegly 1960). Nearly all heterothallic species form oospores when paired with the opposite mating type of most other species (Savage et al. 1968). See Appendix 3.55 for methods of determination of A1 and A2 types. Some isolates of *P. parasitica* are neuter, and others induce both A1 and A2 cultures to produce oospores (Ko 1981). Oospores of *P. cinnamomi* (which is heterothallic) were produced in single A2 culture when incubated in avocado root extract (Zentmyer 1979) and were larger than those from A1 and A2 crosses (Zentmyer et al. 1979). See Shaw (1983, 1988, 1990, 1991) for a more detailed review of the nature of heterothallism in *Phytophthora*.

According to Shaw (1988),

> There is now convincing evidence that both mating types of all heterothallic species so far examined are reciprocal translocation heterozygotes, as multiple associations of four chromosomes are always present. No multiple associations are found in meiosis of homothallic species (Sansome, 1980). Because self-fertility in some progeny of *P. drechsleri* was associated with tertiary trisomy (the additional chromosome being associated with the translocation complex), it has been argued that the complex carries the gene(s) determining mating type and is thus a 'mating type complex' (Mortimer et al. 1977; Sansome, 1980). Field isolates of one heterothallic species, *P. nicotianae* var. *parasitica,* which showed tendencies to self, have now been shown to be tertiary trisomics (Sansome, 1985).

Shaw (1987), in a discussion of the "breeding" system of A1 and A2 isolates of *P. infestans,* illustrated hyphal interactions in mating of a recently discovered Egyptian A2 isolate (Shaw et al. 1985) with an A1 from Wales. When the A1 and A2 isolates were inoculated into adjacent halves of the same nutrient agar plate, the advancing edge of the A2 colony nearest to the A1 colony responded to diffusates from the latter by an increase in growth rate and a derepression of lateral hyphal branching. Soon after contact of the rapidly growing A2 hyphae with the A1 colony, a small proportion of A2 hyphal tips coiled around A1 hyphae to produce ropelike structures that later bore gametangia and oospores. Microcultures in which A1 and A2 were grown separately in droplets (0.15 ml) of liquid medium allowed observation of hyphae as they grew together on the plastic surface between adjacent droplets. This revealed that A1 hyphal tips stopped growing on contact with A2 and that A2 hyphae tracked along the A1 hyphae as the A1 colony was "invaded." Later, oospores formed within the A1 culture but not within the A2, reflecting the colonization of A1 by A2 but not vice versa.

The function of mating of the A1 and A2 types has been considered to be a true hybridization phenomenon in which factors that condition parasitism and many other characters would recombine in the progeny. This phenomenon is discussed in more detail later in this chapter. In brief, it can be categorically stated that true hybridization does occur when A1 and A2 isolates pair and form oospores. Results of tests in which nuclear gene markers are used show that recombination of factors in the oospore progeny occurs. This is discussed in the reports of Shattock et al. (1986a, b, 1987), Shaw (1988, 1990, 1991), and Spielman et al. (1989, 1990), who report their own work and that of others.

Oospore progeny from the pairing of A1 and A2 types also can be the result of selfing caused by hormonelike substances that are produced by the opposite mating type. Brasier (1972) first suggested that a diffusible chemical from the opposite mating type might be responsible for induction of oospores. When oospores from the pairing of A1 and A2 cultures were separated with cellophane, oospores formed; but since the cellophane barrier was not entirely impervious to hyphal growth, the selfing phenomenon could not be proved conclusively. Physical separation of A1 and A2 cultures was achieved in another laboratory by using a polycarbonate membrane that was impervious to mycelial growth but allowed free passage of water-soluble substances (Ko 1978) (Appendix 3.56). This method resulted in the production of oospores on both sides of the polycarbonate membrane caused by reciprocal stimulation. These experiments gave unequivocal proof of the production of water-soluble substances, considered to be hormonal or pheromone-like, that stimulated selfed-oospore formation without hybridization (Ko, 1978; Uchida and Aragaki 1980; Ko and Kunimoto 1981; Ko 1988). Ko postulated that α-1 and α-2 hormones produced by the A1 and A2 mating types, respectively, stimulate the opposite mating type to form oospores; however, an active compound has not been identified.

Selfing can also occur by stimulation between different species (Shen et al. 1983; Skidmore et al. 1984; Shattock et al. 1986b). By use of the polycarbonate membrane method, the A1 mating type of *P. infestans* produced oospores when paired opposite A2 isolates of *P. parasitica, P. capsici,* and *P. palmivora,* as well as *P. infestans* (Shen et al. 1983). Some homothallic species are capable of stimulating production of oospores by heterothallic species. *P. heveae* (self-fertile) induced oospore formation in both A1 and A2 types of *P. parasitica; P. sojae* (*P. megasperma* f. sp. *glycinea*) induced oospores in A2 but not A1 isolates of *P. parasitica;* and *P. cactorum* and *P. katsurae* stimulated oospore production in A1 but not A2 cultures of *P. parasitica* (Ko 1980). When an A2 *P. colocasiae* isolate was separated from an A1 *P. parasitica* isolate by a polycarbonate membrane, oospores were formed (Yu et al. 1981).

Sansome (1970) suggested that selfing may be a cause of abnormal segregation in the F_1 of crosses of A1 × A2 mating types. Selfing is described in more detail by Elliott (1983) and Ko (1988) and is discussed by Shaw (1987, 1988, 1990).

A2 isolates occasionally become self-fertile spontaneously, especially in older cultures (Timmer et al. 1970; Brasier 1972; Tsao et al. 1980), but this characteristic is not usually sustained after several transfers (Brasier 1972). Occasionally, reversals of A1 and A2 types occur (Ko 1981). Brasier (1972) reported that 12 of 23 isolates of the self-sterile *P. palmivora* formed oospores when transferred from cultures stored under sterile paraffin oil.

Cultures of *P. parasitica,* which is ordinarily heterothallic, were grown on V8 juice agar and other media and aged for 12 months. Transfers to fresh media resulted in formation of oospores in single culture. Oospores continued to form in single culture after four to five transfers, but colonies from frequently transferred young mycelium did not produce oospores (Tsao et al. 1980). Tsao et al. hypothesized that an oospore-inductive chemical or chemicals are formed after long periods of incubation. Since washings or broth from oosporogenic cultures did not induce young, frequently transferred colonies to produce oospores, the inductive substance may be in the mycelium. Old cultures of *P. palmivora* and *P. cinnamomi* did not produce oospores by this method; however, aged cultures of the black pepper strain of *P. capsici* (*P. palmivora* MF4) (Tsao and Alizadeh 1988) produced oospores similarly to *P. parasitica.*

Oospore production in A2 mating types is also stimulated by volatile products of *Trichoderma viride* (Brasier 1971a, b, 1975a, b, 1978). Haskins and Gardner (1978) showed that a *Trichoderma* sp. stimulated production of sex organs in both *Phytophthora* and *Pythium.* The intensity of the response varied among different heterothallic *Phytophthora* species (Brasier 1971a), and the activity of the effect on reproduction was correlated with antibiotic activity (Brasier 1975a). A volatile substance from *Trichoderma koningii* induced oospores in *P. cinnamomi* (B. H. Pratt et al. 1972a). Two isocyanide substances, homothallin I and homothallin II, were associated with the *Trichoderma* effect (Sakata and Rickards 1980; see also Elliott [1983]). The ecological significance of the interactive effect of *T. viride* and some other species of *Trichoderma* with *Phytophthora* (B. H. Pratt et al. 1972a; Brasier 1975a) is discussed by Brasier (1978).

Cutting mycelium in an agar culture with a scalpel induced oosporogenesis in A2 but not A1 cultures of *P. cinnamomi* and *P. cryptogea* (Reeves and Jackson 1974). The fungicide chloroneb induced oospore formation in A2 but not in A1 cultures of *P. capsici* (Noon and Hickman 1974). Some single-zoospore cultures from chloroneb-treated A2 cultures were A1 and occasionally A1A2 (reacted both ways) (Ko 1981). Oospores were sometimes induced in A1 mating types of *P. infestans* when a thick mycelial mat was placed in contact with a polycarbonate membrane overlaying water agar. Some of these oospores were later germinated on water agar (A. M. Campbell et al. 1985). When A1 and A2 isolates of *P. parasitica* and *P. cinnamomi* were grown in a medium containing chloroneb or ethazole (etradiazole), oospores formed in sectors. Cultures derived from single zoospores from the sectors were of both A1 and A2 mating types, but cultures from the untreated colonies retained the original A1 or A2 mating type. The water-soluble hormone (Ko 1978) associated with A1 or A2 also changed (Ann and Ko 1989).

Oospores of the A2 mating type of *P. cinnamomi* have been found occasionally in avocado roots, and extracts from roots stimulated the production of oospores. Oogonial walls were more pigmented and thicker in homothallic oospores produced on avocado root extract than in oospores formed by pairing A1 and A2 mating types (Mircetich and Zentmyer 1966; Zentmyer 1979; Zentmyer et al. 1979a). The stimulatory substance extracted from roots was thermostable and withstood autoclaving and was extractable in water, soluble in ether and chloroform, and dialyzable but has not been identified. Extracts from chestnut, pine, and heather (all hosts) did not stimulate oospore production (Zentmyer et al. 1979a). An oleic acid component of avocado root extract stimulated oospore formation in culture (Zaki et al. 1983). Thus, it is possible that mechanisms exist in nature that allow oospore formation by ordinarily self-sterile *Phytophthora* species. This could be important epidemiologically because oospores are capable of survival for relatively longer periods than other propagules (see Table 5.3). Several A1 isolates of *P. cinnamomi* produced oospores on sterile oat grains (Royle and Hickman 1964).

Mortimer et al. (1977) noted that up to 14.5% of isolates of F_1 single-oospore progeny from A1 and A2 crosses of *P. drechsleri* acted as A1A2 (produced oospores when paired with either A1 or A2 tester cultures). When grown alone, the A1A2 isolates were self-fertile. Mortimer et al. (1977) "... suggested that the self-fertile condition was due to the presence of an extra chromosome containing the mating type locus, the A^1A^2 type being *Aaa*."

The different mechanisms that are operative for production of heterothallic oospores include direct hybridization, stimulation by aging and exotic substances, and stimulation of the opposite mating type, presumably by the water-soluble α-1 or α-2 hormonelike substances described by Ko (1978). The review by Ko (1988) strongly implied that production of selfed oospores rather than hybrid oospores was the most common; however, considerable genetic evidence indicates that many oospores from A1 × A2 pairings are true hybrids (Shaw 1988, 1990, 1991). Work by Shattock et al. (1986a, b, 1987) with *P. infestans* shows that factors for virulence on potato leaves and expression of isoenzyme markers recombined when A1 and A2 strains were crossed. See the following section, Genetics and Cytology of *Phytophthora*.

Separation of Oospores from Mycelium

Experimental germination of oospores is fraught with difficulties because oospores usually germinate much more slowly than chlamydospores, sporangia, zoospores, or mycelium. Hence, for genetic analysis, it is necessary to separate germinated oospores from other fungal propagules that may overgrow the oospores that germinate more slowly and to be sure that a rescued colony is from a single germinated oospore and not from an asexual propagule. See Appendix 3.57 for methods to separate oospores from mycelium.

Freezing cultures at temperatures as low as –20°C for several hours has been considered a method capable of selectively separating oospores from asexual structures because oospores are less sensitive to freezing than mycelium or asexual spores (Appendix 3.57A). This method has been used by several investigators (Timmer et al. 1970; Erwin and McCormick 1971; Banihashemi and Mitchell 1976; Kittle and Gray 1979b; Plourde and Green 1982; D. C. Harris and Cole 1982). Although the method is useful, occasionally hyphal fragments survive freezing (O. K. Ribeiro, *unpublished*), and the germination percentage of oospores of *P. medicaginis* (*P. megasperma* f. sp. *medicaginis*) was decreased by freezing (El-Hamalawi and Erwin 1986a) (Appendix 3.57A).

Oospores can be freed from mycelium by filtering an agar-mycelium-oospore mat through a fine-mesh nylon screen (approximately 50-µm pores) or by mincing the culture in a Sorvall Omnimixer cup for 1–2 min (on ice to prevent heating) and then sonicating the minced mycelium on ice for about 4–5 min at 100 W to break up the mycelium (Sauve and Mitchell 1977; Förster et al. 1983; El-Hamalawi and Erwin 1986a) (Appendix 3.57D). Time of the treatment and intensity of sonication should be experimentally tested before use; however, oospores in cultures of *P. medicaginis* (*P. megasperma* f. sp. *medicaginis*) will withstand 5 min of sonication at medium speed with a Braun-Sonic 1510 sonicator (Braun Instruments, South San Francisco, California) (El-Hamalawi and Erwin 1986a). Although mature oospores are thick walled, such severe treatment can cause injury, especially to young oospores (Duncan 1985a; Jiang and Erwin 1990a). Oospores of *P. cactorum* were collected on 26- and 36-µm sieves (D. C. Harris and Cole 1982). The oospore-mycelium suspension can be allowed to settle for 1–2 h in a graduated cylinder or in a separatory funnel (Romero and Erwin 1969) to allow the heavier oospores to settle to the bottom. The bottom layer containing oospores and some fragmented mycelium can be collected and resuspended and washed by centrifugation at 1,200 × *g* several times. The lighter mycelium can be removed from the heavier oospore pellet by gently aspirating and discarding the mycelial fragments. Usually it is necessary to centrifuge and subsequently discard the lighter mycelium in the supernatant several times to obtain a clean oospore preparation (Appendix 3.57E).

An oospore-mycelium suspension of *P. infestans* treated with $KMnO_4$ (potassium permanganate; 0.25% for 15 min) can be screened through 53- and 20-µm sieves (Appendix 3.57E). Oospores that collect on the 20-µm sieve are washed free of $KMnO_4$ (Chang and Ko 1991). The $KMnO_4$ has a killing effect on mycelial fragments, sporangia, and zoospores.

To eliminate much of the mycelium, the oospore-mycelium suspension can be treated with certain enzymes (Appendix 3.57B) such as 1% Glusulase (crude snail enzyme consisting of a mixture of β-glucuronidase and arylsulfatase; Calbiochem Corp., La Jolla, California). Oospores usually survive this treatment, but most of the mycelium is digested (Salvatore et al. 1973; Banihashemi and Mitchell 1976; Sauve and Mitchell 1977; Förster et al. 1983; El-Hamalawi and Erwin 1986a). A similar method has been used for *P. cactorum* and *P. citricola* (Jiang and Erwin 1990a). NovoZym 234 digests mycelium of *P. infestans* (Spielman et al. 1989; 1990) (Appendix 3.57B). Sneh (1972b) washed mycelial mats of *P. cactorum* in distilled water and incubated them in a solution of 0.5–2.0% cellulase and hemicellulase in 0.1 M phosphate buffer at pH 6.0 at 30°C for 2–7 days (Appendix 3.57B).

Oospores of *Pythium aphanidermatum, Pythium ultimum,* and *Pythium myriotylum* were separated from mycelial mats grown in V8 juice broth supplemented with cholesterol (30 mg/liter) by flooding the mats with nonsterile soil extract (10 g/liter) for 1–2 weeks. Mycelium and asexual spores had been lysed by bacterial action. Unlysed oospores were separated by rinsing in tap water and comminuting in a Tekmar Tissue Mizer for 30 s. Oospores were filtered off by passing the suspension through two layers of cheesecloth and one layer of lens paper (W. A. Ayers and Lumsden 1975). Oospores can be separated from mycelium by feeding the agar containing oospores and mycelium to garden snails, *Helix aspersa* (Gregg 1957), or water snails, *Planorbarius carneus,* after which the oospores are collected from the excrement (Shaw 1967; Shattock et al. 1986a) (Appendix 3.57C). Density gradient centrifugation in sucrose solutions can be used to separate oospores from mycelium (Appendix 3.57F).

Viability Tests for Oospores

Although germination of oospores is the ultimate viability test, it does not distinguish between oospores that are dormant and oospores that are not viable. In many cases, oospores are produced but may not germinate. While this failure to germinate may be the result of dormancy, many oospores produced have characteristic thick walls, and although they superficially appear normal, they are not. Figure 3.7A shows oospores that have thick inner walls but are devoid of nuclei or ooplasts. Figure 3.7B shows germinable oospores with typical ooplasts and pellucid bodies that contain nuclei. The nucleus in the center photograph is larger than those in other oospores and is assumed to be a fusion (antheridial and oogonial) nucleus.

Staining of oospores with tetrazolium bromide, a vital stain, is an indicator of viability (El-Hamalawi and Erwin 1986a; Jiang and Erwin 1990) (Appendix 3.58) and also can be used to detect oospores in root tissue (Ribeiro 1978; Sutherland and Cohen 1983; Cohen 1984). According to Altman (1976), the tetrazolium staining reaction is the result of reduction of the salt by cellular dehydrogenase to form its colored formazan product (Gahan and Kalinda 1968). Staining of most fungal species is more effective at 37 ± 1°C than at 23 ± 2°C. Dormant oospores stain rose, and germinated oospores stain blue, but nonviable oospores either stain black or remain unstained. Oospores that are killed by autoclaving do not stain (Sutherland and Cohen 1983). Results of staining oospores of *P. sojae* (*P. megasperma* f. sp. *glycinea*) with tetrazolium bromide are similar at concentrations of 0.05, 0.5, and 5%. Although the number of oospores staining with tetrazolium bromide in phosphate buffer was reported to be similar to the number that stained in water (Sutherland and Cohen 1983), Jiang and Erwin (1990) observed that staining was more consistent when a phosphate buffer was used. The use of vital stains allows easy distinction among the dormant and activated stages of oospores and nonviable oospores (El-Hamalawi and Erwin 1986a) (Appendix 3.58).

The ability of normal oospores to plasmolyze in 4 M NaCl and recover when transferred to distilled water as a viability test is comparable to but more rapid than the tetrazolium bromide test (Jiang and Erwin 1990) (Appendix 3.59). Viability percentages (tested by either tetrazolium bromide or plasmolysis) of immature 15-day-old and mature 30-day-old oospores of *P. medicaginis* (*P. megasperma* f. sp. *medicaginis*) and *P. cactorum* that had been chilled at 2°C to induce dormancy were highly correlated with the percentages of germination (Jiang and Erwin 1990).

Germination of Oospores

Oospore germination has been the subject of much research based on the potential importance of oospores as the most persistent of the propagules produced by *Phytophthora* and on the hypothesis that the sexual oospores might be a source of genetic variation. Of special interest are heterothallic species that are self-sterile and produce oospores only when an A1 and A2 mating type are grown together (e.g., *P. infestans*). Various methods used to germinate oospores are presented in Appendix 3.60. See Ribeiro (1978, p. 167) for a historic listing of methods and Ribeiro (1983) and Elliott (1983) for reviews.

Although there are many reports of high percentages of germination of oospores (Table 3.3), many oospores in some cultures abort (fail to produce a normal oospore in the oogonium) (Figure 3.8) (Förster et al. 1983). Rutherford and Ward (1985b) observed that, in certain cultures of *P. sojae,* the percentage of oospores that aborted occurred at a 3:1 (normal to aborted) ratio in five pathologic races and at a 9:7 ratio in two other races. They interpreted the 3:1 ratio to indicate that oospore abortion is caused by a single recessive gene and the 9:7 ratio by two recessive genes. We have noted also that after many mass mycelial transfers of a single-zoospore culture of *P. sojae,* the culture produced mostly aborted oospores. Oospore production by the culture was recovered by propagation of single-oospore cultures (D. C. Erwin, *unpublished*).

The germination of oospores obtained from artificial pairing of A1 and A2 mating types is often inconsistent and seldom approaches 100% (Romero and Erwin 1969; Klisiewicz 1970; Timmer et al. 1970; Kaosiri et al. 1980; Förster and Coffey 1990) (Table 3.3). Low percentages of germination of oospores from heterothallic isolates may be caused by incompatibility or lethal factors in the parental lines resulting in aborted or malformed oospores. In a recent report, it was noted that some oospores that might be superficially considered to be normal were, under closer observation, abnormally thick walled and contained no evidence of nuclei (pellucid bodies) or ooplasts. When these oospores were stained with the vital stain (tetrazolium bromide) or tested for viability with a solution of 4 M NaCl, they were found to be nonviable (Jiang and Erwin 1990; see preceding section on viability). A recent report showing a high percentage of germination of *P. parasitica* oospores from an A1 × A2 cross (Ann and Ko 1988b; Chang and Ko 1990) and of *P. infestans* (Chang and Ko 1991) indicated that selection of the appropriate A1 and A2 parent cultures was important (see also Förster and Coffey [1990]).

In most cases, oospores must be about 30 days old or older to become mature, although this time varies depending on species and isolates. Usually the percentage of germination increases with age (Meyer 1975; see also Ribeiro [1983] for numerous citations). Germination of oospores of an A1 × A2 cross of *P. infestans* was up to 40% when oospores were 4 days old and up to 70% when they were 20 days old (Chang and Ko 1991). Oospores of *P. katsurae* (homothallic) germinated when 4–8 weeks old but did not germinate at 1 week of age (Ko and Arakawa 1980).

Oospores of *P. cactorum* and *P. medicaginis* (*P. megasperma* f. sp. *medicaginis*) germinated only after the two gametic nuclei (pellucid bodies) had fused (Jiang et al. 1989). Germination percentages correlated well with the increase in percentage of nuclear fusion. Although nuclear fusion and germination of these two species was optimal at about 30 days, nuclei in many oospores, such as those in a *P. sojae* (*P. megasperma* f. sp. *glycinea*) isolate, did not fuse and did not germinate within the 30-day period. Time-lapse photography showed that this isolate required an additional 20-day period in soil extract for nuclei in the oospore to fuse and for the oospores to germinate and that nuclear fusion may be an indication of maturity (Figure 3.8) (Jiang et al. 1989).

The time course of germination of *Phytophthora* oospores is illustrated by Blackwell (1943) and Meyer (1975) for *P. cactorum,* by Erwin and McCormick (1971) and Jiang et al. (1989) for *P. megasperma* var. *sojae* (*P. sojae*) from soybean (Figures 3.6 and 3.8), by Zentmyer and Erwin (1970) for *P. drechsleri,* and by Harris and Cole (1982) for *P. syringae.*

The first indications of germination are the swelling of the oospore within the oogonium and dissolution or erosion of the inner oospore wall. This phase has been called *activation.* Activation is illustrated in frame two of Figure 3.6 and in Figure 3.8D. Erosion of inner oospore walls is described by Beakes and Bartnicki-Garcia (1989), who treated broken oospore walls with enzymes and showed that the thick inner oospore wall, which is electron dense (no structure was evident), can be eroded by treatment with an exo-β-1,3-glucanase. This would be expected, since the carbohydrate β-1,3-glucans accounts for 35–50% of the weight of the oogonium-oospore wall complex (Lippman et al. 1974). The origin of the β-1,3-glucanase that digests the inner oospore wall is not known. Beakes and Bartnicki-Garcia (1989) state (excerpted by per-

Table 3.3. Methods of inducing oospore germination of homothallic and heterothallic species[a]

Species	Culture medium	Treatment after oospore harvest	Germination (%)	Age of culture (days)	Reference
Homothallic species					
P. cactorum	Not given	Chill oospores 1–2 wk, and then water soak (pH 7.6–8.0) for 1–3 wk	Not given	30+	Blackwell (1943)
	GAAML[b]	Dark 30 days; fluorescent light (600–800 ft-c) for 7 days	90+	30+	Berg (1966)
	Pea	Continuous light for 14 days at 25°C on 1% water agar	74	21+	Elliott and MacIntyre (1973)
	Synthetic agar + ether extract of oats and pea infusion	Ingestion by water snail, *Planorbarius corneus,* followed by 4–8 days at 18–22°C in diffuse light	67	21+	Shaw (1967)
	Bean meal broth	Natural daylight	50–60	30–210	Meyer 1975
	V8 juice broth	Freeze (–20°C 24 h); treatment with 2% glusulase; continuous cool white fluorescent light (2,165 lux) 7 days	76	120	Banihashemi and Mitchell (1976)
	V8 juice broth	Fluorescent white light (2,594 lux) for 104 h	91	120	Banihashemi and Mitchell (1976)
	Carrot broth	Blue light; amino acids, soil extract	Up to 90	30	Jiang and Erwin (1993)
		Incubated under blue or green light	Up to 90	30	Chang and Shu (1988)
P. citricola	Lima bean agar	Fluorescent light and moderate sunlight, unregulated	50	Not given	Henry and Stelfox (1968)
	V8 juice broth	Freezing (–3°C); blue and ultraviolet light	70-91	60	Plourde and Green (1982)
P. erythroseptica	Oat agar	Ingestion by garden snail, *Helix aspersa*	Not given	30–60	Gregg (1957)
	GAAML	Dark 30 days; fluorescent light (600–800 ft-c) 7 days	50+	30+	Berg (1966)
P. fragariae	Roots of strawberry	Surface sterilize 2 min in 0.001% mercuric chloride; wash 10–15 times in distilled water	5–60	29–30	Duncan (1977)
	French bean agar, roots of strawberry	Treat oospores and roots with 0.001% mercuric chloride; vary temperature and osmotic potential	Up to 70	30–180	Duncan (1985a)
P. heveae	Synthetic agar	Incubation dark 25 days; light 12–15 days	65–70	39+	Leal and Gomez-Miranda (1965)
		Incubation under blue or green light	Up to 90	30	Chang and Shu 1988
	GAAML	Dark 30 days; fluorescent light (600–800 ft-c) for 7 days	75	30+	Berg (1966)
P. katsurae		Incubation under blue or green light	Up to 90	30	Chang and Shu (1988)
P. medicaginis (*P. megasperma* f. sp. *medicaginis*)	V8 juice agar	Ingestion by snails	93	35+	Salvatore et al. (1973)
	V8 juice broth	β-Glucuronidase	52	35+	Salvatore et al. (1973)
	Carrot broth	Glusulase; sucrose density gradient; germinated under blue light (20°C) in root exudate	Up to 87	11–50	Förster et al. (1983)
	Carrot broth	$KMnO_4$ (0.05%); pregermination treatment at 33°C; germination under blue light (21°C) in root exudate and dilute root exudate plus amino acids	Up to 84	30	El-Hamalawi and Erwin (1986a, b)
P. sojae	Lima bean agar	Growth under cool white fluorescent light 9 h per day at 24°C; germinated on water agar plus cholesterol (10 mg/ml) and rifampicin (10 mg/ml)	Not given	30	Bhat et al. (1993a)
	Lima bean agar	Growth in dark; incubated in 12 h light at 27°C	30	80	Schecter and Gray (1987)
	V8 juice agar plus cholesterol	Mycelium killed at –10°C; fluorescent light; soil smear; soybean seedling	Up to 66	30–45	Jimenez and Lockwood (1982)
	V8 juice broth	Dark 10 days; blue light 10 days; dark 10 days; varying treatments	23	30	Ribeiro et al. (1975b)
	V8 juice agar plus sitosterol	Frozen 24 h; fluorescent light 2 h/day (200 ft-c) 7 days; incubated in water 4 days	40–70	10–30+	Erwin and McCormick (1971)
	V8 juice agar	Water soak 48 h at 36°C; overlay on water agar	45–67	Up to 49	Long et al. (1975)

(continued on next page)

Table 3.3. *(continued)*

Species	Culture medium	Treatment after oospore harvest	Germination (%)	Age of culture (days)	Reference
Heterothallic species					
P. botryosa	V8 juice agar	Growth in dark	Not given	30–84	Chee (1973b)
P. cambivora	Pea	Inter- and intraspecific pairing	1–5	15	Boccas (1981)
P. capsici	Naturally infected cacao pods	Distilled water under fluorescent light (21–22°C)	46	Unknown	Kellam and Zentmyer (1986a)
	Pea	Inter- and intraspecific pairing	1–5	15	Boccas (1981)
A1 × A2	V8 juice agar, rape seed A	Mixture of daylight and fluorescent light (160 ft-c)	15–20	45–60	Satour and Butler (1968)
	V8 juice agar	Laboratory light and temperature	6–8	60+	Polach and Webster (1972)
	V8 juice broth	Dark 10 days; blue light 10 days; dark 10 days; varying treatments	8	30	Ribeiro et al. (1975a)
× *P. parasitica*	Synthetic agar[c]	Laboratory light 10 days	1	32–34	Ribeiro et al. (1975a)
× *P. palmivora*	Synthetic agar[c]	Laboratory light 10 days	7	32–34	Ribeiro et al. (1975a)
	Pea	Growth in dark plus under light for 10 days	4	30	Boccas (1981)
P. cinnamomi	Pea	Inter- and intraspecific pairing	1–5	15	Boccas (1981)
A1 × A2	V8 juice broth	Dark 10 days; blue light 10 days; dark 10 days; varying treatments	23	30	Ribeiro et al. (1975b)
× *P. cambivora*	Synthetic agar[c]	Laboratory light 10 days	2	32–34	Ribeiro et al. (1975a)
× *P. drechsleri*	Synthetic agar[c]	Laboratory light 10 days	2	32–34	Ribeiro et al. (1975a)
P. drechsleri					
A1 × A2	Cornmeal agar, safflower agar	Sterile tap water (pH 7.5) at 24°C; exposed to white fluorescent light (45 ft-c, 10 h; dark 14 h); pregermination heat to 33° C for 2–8 days	40–53	35–56	Klisiewicz (1970)
	Synthetic agar[c]	Laboratory light 10 days	42	32–34	Ribeiro et al. (1975a)
	Bean agar	Diffuse light in laboratory; water snail ingestion of media	20	30–37	Hooley and Shaw (1985)
P. infestans	Soybean	Incubate in horse dung infusion	3–4	No data	Smoot et al. (1958)
	Lima bean agar	Dark 30 days; fluorescent light (600–800 ft-c) for 7 days	10+	30+	Berg and Gallegly (1966); Berg (1966); Gallegly (1968)
	V8 juice broth, rye A	Pregermination $KMnO_4$ (0.25%); white fluorescent light	40–70	5–25	Chang and Ko (1991)
A1 × A2	Clarified V8 juice agar plus sitosterol	Fluorescent light; chilled to 4°C for 45 days	20–48	20+	Romero and Erwin (1969)
	Clarified V8 juice agar plus sitosterol	Maceration of mycelium with NovoZym, 24 h	~11		Spielman et al. (1989)
P. megakarya	Pea	Inter- and intraspecific pairing	1–5	15	Boccas (1981)
P. nicotianae var. *parasitica*	Pea	Inter- and intraspecific pairing	1–5	15	Boccas (1981)
P. palmivora	V8 juice agar	Growth in dark	Not given	28–56	Chee (1973b)
		Sonication of oospores; treatment with hemicellulase and cellulase at 40°C followed by maturation for 30 days	15	14	Sharma (1990)
		Incubation of oospores in snail gut extract	52%, 12 h; 20%, 2 h	No data	Kumari et al. (1988)
	Carrot agar	Ingestion of oospores on a cabbage leaf by snails	0%, 0 h; 87%, 96 h	30 34	Dutta et al. (1984)
A1 × A2	Carrot agar	Aging, pregermination treatment at 12 and 30°C	Up to 40	15–180	Kaosiri et al. (1980)
P. parasitica					
A1 × A2	V8 juice agar	$KMnO_4$ (0.25%) for 20 min; germinated on lecithin-glucose-mineral antibiotic medium; light and dark varied	Up to 95	10–80	Ann and Ko (1988a, 1990b); Chang and Ko (1990)
	V8 juice agar	$KMnO_4$ (0.25%) for 20 min; incubated in filter-sterilized soil extract 2 days under blue light; plated out on a salts-lecithin agar (Ruben et al. 1980)	Up to 70, depending on parents	50–60	Förster and Coffey (1990)

[a] Modified from Ribeiro (1978), by permission A. R. Gantner Verlag, J. Cramer, Vaduz.
[b] Glucose asparagine agar plus methyl linoleate (2.5 ml/liter).
[c] Appendix 3.51.

mission of the British Mycological Society, United Kingdom, courtesy of Beakes and Bartnicki-Garcia 1989):

> The endogenous glucanases may be synthesized *de novo* in a reactivated cytoplasmic endomembrane system and then transported to the oospore wall in cytoplasmic vesicles. Alternatively the glucanases might be present in an inactive, repressed, or latent form either within the cytoplasm or bound to the plasma membrane or to the cell wall as shown in hyphae of *Saprolegnia* (Fevre 1979).

Their hypothesis is presented in Figure 3.9.

When nongerminated oospores of *P. cactorum* were assayed for β-1,3-glucanase (with mycolaminarin as a substrate), activity could not be detected; however, the activity of β-1,3-glucanase gradually increased over a period of 7 days with increasing percentage of germination of oospores in soil extract. Proteins and sugars also increased as the percentage of germination increased (Jiang 1991; Jiang et al. 1991).

Oospores of many species that do not germinate readily are considered to be dormant. Two types of dormancy are recognized in fungi: constitutive, which implies that there is a metabolic requirement, and exogenous, which implies that there is a requirement for a stimulus from the environment such as a nutrient (Sussman 1966). A case could be made for both types of dormancy in oospores. Since fusion of nuclei occurs before germination (Jiang et al. 1989), a delay in this phenomenon might be considered constitutive. Germination of oospores of many species is stimulated by root exudates, amino acids, and soil extract (Sneh et al. 1981; Jimenez and Lockwood 1982; Förster et al. 1983; El-Hamalawi and Erwin 1986a, b); this evidence could indicate an exogenous effect.

Oospores (30 days old) from a cross of A1 and A2 *P. capsici* isolates contained cytochrome c but not cytochromes a and b. In mycelium, however, cytochromes a, b, and c were detected. This suggested that the electron transport chain might

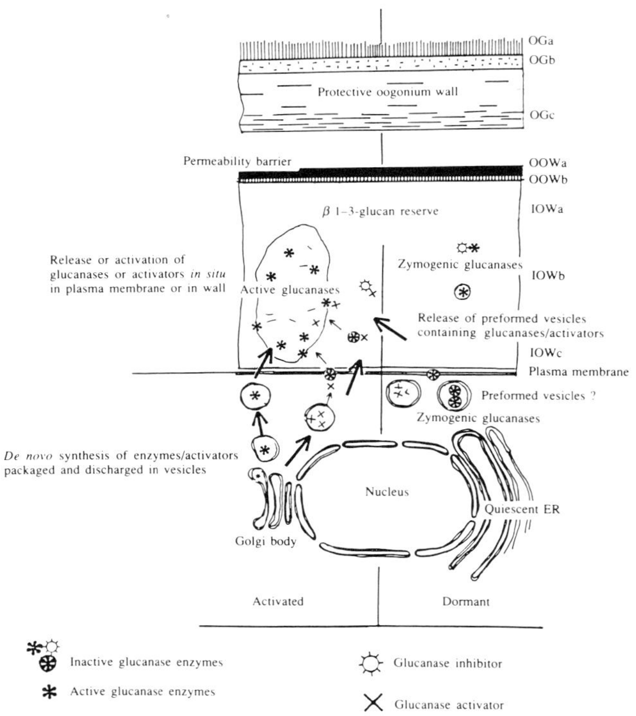

Figure 3.9. Summary of possible cytological events in the mobilization of β-glucan reserves from the inner oospore wall during oospore germination. (Reproduced by permission of the British Mycological Society, courtesy Beakes and Bartnicki-Garcia 1989)

be incomplete and that some change might be necessary to initiate the germination process (Ribeiro et al. 1976b). Ribeiro et al. (1976b) state (excerpted by permission of the American Phytopathological Society, St. Paul, Minnesota, courtesy of Ribeiro et al. 1976b):

> Recently Hemmes and Bartnicki-Garcia (1975) indicated that mitochondria disintegrate or decrease in number during oosporogenesis. Their electron micrographs of dormant oospores also suggest that much cell differentiation must occur during the germination process. Also, Leary et al. (1974) reported the lack of intact functional 80S ribosomes, a basic component of protein synthesis, in dormant oospores of *Phytophthora.* However, subribosomal ribonucleoprotein particles, ribosomal RNA, and ribosomal protein were isolated from oospores, suggesting that intact ribosomes may not be required for the initial germination process.

After oospores are produced in the dark, they require a period of exposure to light from the blue region of the spectrum to germinate (Berg and Gallegly 1966; Huguenin and Boccas 1971; Banihashemi and Mitchell 1976; Ribeiro et al. 1976a; Plourde and Green 1982). Exposure to far red as well as blue light is also effective (Ribeiro et al. 1976a). The requirement for light appears to be a general phenomenon (see reviews by Lilly [1966], Elliott [1983], and Ribeiro [1983]).

Oospores of *P. infestans* were incubated on soft water agar (0.6%) under continuous blue light (430–490 nm) at 18°C (Shattock et al. 1986a, b) prior to germination. Oospores of *P. infestans* germinated well (up to 70%) when incubated under Cool White fluorescent light (Chang and Ko 1991). We have used a Sylvania FI 5TB-B blue light fluorescent bulb routinely in research on oospore germination (Förster et al. 1983; El-Hamalawi and Erwin 1986a, b; Jiang and Erwin 1990, 1993).

Oospores germinate best at temperatures somewhat lower than those optimum for mycelial growth; germination of *P. syringae* oospores was optimal at 12–15°C (Harris and Cole 1982). The percentage of germination of oospores of *P. cactorum* at 16°C was 41%; at 20°C, 88%; at 24°C, 62%; and at 28°C, 0% (Banihashemi and Mitchell 1976).

Heating oospores at different temperatures prior to germination sometimes increases germination percentages. Exposure to a temperature of 33°C for 2–8 days increased the subsequent percentage of germination of A1 × A2 *P. drechsleri* oospores (Klisiewicz 1970). Treatment of *P. palmivora* oospores at 12 or 30°C for 1 week enhanced the germination but at 33°C suppressed germination (Kaosiri et al. 1980). Germination percentage of oospores of *P. sojae* was higher when oospores were presoaked at 36°C for 48 h prior to germination compared with treatment at 25°C (Long et al. 1975). Pregermination treatment of 30-day-old *P. medicaginis* oospores with double distilled water at 33°C for 5 h increased the percentage of germination from 41 to 89%. Activation of oospores, as measured by use of tetrazolium bromide vital stain (Appendix 3.58), was increased from 1% for the 24°C pretreatment (control) to 75% for the 33°C pretreatment (El-Hamalawi and Erwin 1986a). Oospores of *P. fragariae,* which has a low temperature optimum (about 21°C), were stimulated by exposure at 25°C for 24–48 h (Duncan 1977). The mechanism of these effects has not been explained physiologically. It is possible that higher than normal temperature affects membranes in some way and allows leakage of substances from oospores. Exudates from germinating oospores or mycelium of *P. cactorum* experimentally stimulated oospore germination (Jiang 1991; Jiang and Erwin 1993).

Chilling oospores presumably to break dormancy (Blackwell 1943) stimulated germination only for some species, for instance, *P. sojae* (Erwin and McCormick 1971) and *P. infes-*

tans (Romero and Erwin 1969). Oospores of *P. drechsleri* (Klisiewicz 1970) and *P. palmivora* (Kaosiri et al. 1980) were not stimulated by reduction of temperatures to 3 and 5–9°C, respectively.

Development of oospores of *P. sojae* at optimal temperatures for growth (24 and 27°C) results in much better germination than development at 15 or 30°C (Erwin and McCormick 1971). A similar response is shown for *P. cactorum* (Banihashemi and Mitchell 1976).

Oospore germination occurs at a wide range of pH values, but in most cases a pH of 6.5–7.0 is considered to be optimal. Glucose in the germination medium retarded germination of *P. medicaginis* (Förster et al. 1983). Several studies show that the percentage of germination of oospores is increased by incubation in soil extract or root exudates (Sneh et al. 1981; Jimenez and Lockwood 1982; Förster et al. 1983; El-Hamalawi and Erwin 1986a, b; Jiang and Erwin 1990, 1993; Jiang 1991) (Appendix 3.60) and amino acids at low concentrations (10 µM) (Jiang and Erwin 1993). A study of various physical and chemical factors that affected the germination of the homothallic *P. medicaginis* showed that 30-day-old oospores germinated optimally (60–80%) in root exudate from alfalfa (host) and from pea and tomato (nonhosts) but that only 20% germinated in water. The cationic fraction of root exudate (which contains amino acids) was more active than anionic and neutral fractions. Aspartic and glutamic acids (which are present in root exudate) supported germination but not at as high a percentage as in root exudate (El-Hamalawi and Erwin 1986b).

Alanine, aspartic acid, and several other amino acids at 10 µM stimulated germination of *P. cactorum* oospores when buffered at a pH near 6.0 (Jiang and Erwin 1993). Since the aborted oospores that occur in almost all populations could be a source of amino acids, this could explain why low percentages of oospores of *P. medicaginis* occurred in water (El-Hamalawi and Erwin 1986a, b). Exudation products of either germinated oospores or aborted oospores were capable of stimulating germination when added to water. When freshly isolated oospores were incubated in water, germination often reached about 20%, but if oospores were stored at 2°C for 14 days, they became dormant and no germination occurred except in soil extract (1 kg of soil per liter of water; Appendixes 3.44 and 3.60) (Jiang and Erwin 1993). Soil extracts stimulated germination of *P. sojae* (Sneh et al. 1981; Jiminez and Lockwood 1982) (Appendix 3.60). Although the mechanism of the stimulating effect of soil extract on oospore germination is not explainable (Jiang and Erwin 1993), soil extract is a useful medium for experimental germination of oospores, because oospores do not grow so rapidly that the mycelium from the early germinating oospores obscures those that germinate later. Also, soil extract does not support rapid growth of airborne contaminant bacteria that infest the germination medium while germination is being observed over a 6- to 10-day period.

Treatment of *P. medicaginis* oospores with dilute potassium permanganate ($KMnO_4$) solution (0.1%) for 10 min stimulated oospores to germinate in water to the maximum percentage of those in root exudate (El-Hamalawi and Erwin 1986a, b) (Appendix 3.60). The reason for stimulation by $KMnO_4$ is not clear, since $KMnO_4$ did not induce a perceptible change in the ultrastructure of the oospore wall (Beakes et al. 1986). We noted a release of amino acids into the germination solution following oospore treatment with $KMnO_4$, which suggested that amino acid release could account for the stimulation (J. Jiang and D. C. Erwin, *unpublished*).

$KMnO_4$ (0.25%) increased the percentage of germination of oospores from an A1 × A2 pairing of *P. parasitica* to about 90% (Ann and Ko 1988a, 1990b; Chang and Ko 1990). A similar effect was noted for oospores of *P. infestans* from an A1 × A2 pairing. After treatment with $KMnO_4$ (0.25% for 15 min on a shaker), oospores were collected on a 20-µm sieve and washed free of $KMnO_4$ with tap water. After incubation for 20 days at 19°C on the S + L medium of Ruben et al. (1980), supplemented with 100 µg of ampicillin, 50 µg of nystatin, and 10 µg of pentachloronitrobenzene per milliliter to prevent the growth of contaminants, about 68% of the oospores germinated (Chang and Ko 1991).

$KMnO_4$ kills mycelial fragments and chlamydospores and additionally stimulates oospore germination. When *P. parasitica* oospores were incubated 12 min in 0.025% $KMnO_4$, mycelial fragments were killed prior to germination (Förster and Coffey 1990). $KMnO_4$ might be an aid for isolation of oospore-producing *Phytophthora* species from soil suspensions because of its ability to kill other microbes but not thick-walled oospores.

Gregg (1957) was the first to report that the oospores of *P. erythroseptica* could be induced to germinate after being fed to snails and collected from the excrement. Some reports (Salvatore et al. [1973] for *P. megasperma* f. sp. *medicaginis* and Banihashemi and Mitchell [1976] for *P. cactorum*) show that germination was markedly stimulated by pregermination treatment with snail enzymes such as Glusulase (Calbiochem, La Jolla, California) (Appendix 3.57B); however, Long et al. (1975) showed that Glusulase was suppressive to germination of *P. sojae*. El-Hamalawi and Erwin (1986a) found that Glusulase did not stimulate oospore germination of *P. medicaginis,* and J. Jiang and D. C. Erwin (*unpublished*) confirmed this. Förster et al. (1983) treated an oospore suspension with Glusulase because it digested mycelium but not oospores, but an increase in oospore germination was not reported. Since oospores are stimulated to germinate by one or more amino acids (El-Hamalawi and Erwin 1986b; Jiang and Erwin 1993), crude snail enzymes or snail ingestion may provide amino acids by breaking down components in the material that is digested. Also, feeding oospores to snails (Appendix 3.57C) may screen out nongerminable oospores by digestion of aborted and nonliving oospores (Sauve and Mitchell 1977), leaving only intact germinable oospores. We think that the stimulatory effect of snails and snail enzymes on germination is likely the result of a nutritive effect.

Single, germinated oospores can be transferred from a plate of a suitable agar medium by single-spore methods presented above and in Appendix 3.60. The oospores of the slow-growing *P. infestans* can be spread on soft water agar (0.6%) overlaid after germination is detected (3–5 days) with a thin layer of particle-free rye A broth (Appendix 3.9). Colonies can then be transferred to a V8 juice-lima bean agar amended with vancomycin and rifampicin for control of bacterial contamination (Shattock et al. 1986a, b).

Oospores of *P. parasitica* germinated well when spread on the S + L medium of Ruben et al. (1980), which was amended with asparagine (0.01%) and agar (2.0%). After autoclaving, (in mg/liter) ampicillin (100), nystatin (50), and pentachloronitrobenzene (10) were added to prevent growth of contaminants (Chang and Ko 1990). The dosage of nystatin probably could be lowered to 5 or 10 mg/liter because of the possibility that a polyene antibiotic such as nystatin might inhibit spore germination (discussed in Chapter 2).

Germination of *P. medicaginis* oospores was stimulated by rifampicin at 1 and 20 ppm. Vancomycin at 100 ppm completely suppressed germination and at 50 ppm caused germ tubes to be abnormal and reduced germination to about 33% of the control. At 10 ppm, germination was 65% of the control (H. Förster and D. C. Erwin, *unpublished*). Banihashemi and

Mitchell (1976) reported that pimaracin at 100 μg/ml reduced *P. cactorum* oospore germination by 80% but had little effect at 50 μg/ml. A dosage of 10 μg/ml allowed the highest percentage of germination. Vancomycin (200 μg/ml) allowed maximum germination, but polymyxin B (200 units/ml) reduced germination by 50%. Relatively little other work has been conducted on the effect of antibiotics on oospore germination.

Genetics and Cytology of *Phytophthora*

General Discussion

The life history of members of Oomycetes has been a subject of considerable controversy since the late 1800s. Dick (1969, 1972) and Dick and Win-Tin (1973) present meticulous and scholarly historical accounts of the research and opinions on Oomycetes, which varied from the concept of *Phytophthora* as a haploid fungus, typical of the majority of the fungal world (Ascomycetes, Basidiomycetes, and Fungi Imperfecti), to the presently accepted concept that all of the oomycetes are diploid in the vegetative state, with meiosis (reduction division of chromosomes) occurring in the gametangial antheridia and oogonia of *Pythium debaryanum* (Sansome 1961, 1963) and *Phytophthora* (Sansome 1965, 1976; reviewed by Boccas [1976]). It is now accepted that all *Phytophthora* species are uniquely diploid in the major part of their life cycle (Figure 3.10) (Brasier and Sansome 1975).

The cytology of the sexual stage and genetics of *Phytophthora* are reviewed by Win-Tin and Dick (1975) and Shaw (1983, 1987, 1988, 1990, 1991). Shaw and Shattock (1991) and Shaw (1991) specifically treat *P. infestans,* the cause of late blight of potato. Shaw (1983, 1988) gives an interesting historical account of the controversy on the diploid hypothesis.

Brasier (1992) reviewed the evolutionary biology of *Phytophthora* with emphasis on the genetic system, sexuality, and generation of variation. He comments on the possible role of heterothallism in the development of variant types in nature. A second review by Brasier and Hansen (1992) develops information on phylogeny, speciation, and population structure. They discuss several homothallic and heterothallic species in which variants are now recognized and the role of molecular methods for recognition of genetic similarities and differences.

With remarkable foresight, E. Sansome was able to show in cytological work during 1961–1965 that meiosis occurs in the oogonium and antheridium rather than in the germinating oospore. Even more remarkable, Sansome conducted most of this research after she had retired from university work. Since the chromosomes in the nuclei of *Phytophthora* are extremely small, observation by light microscopy and preparation of convincing photographs is difficult. Also, the thick wall that develops on the oogonium and antheridium and even thicker wall on the internal oospore are barriers to nuclear stains and to the passage of light.

The evidence for the concept of diploidy is now beyond question (Maia et al. 1976; Erselius and Boccas 1978; Eriksson and Motzfeldt-Laane 1982; Shaw 1983, 1990). The report by Erselius and Boccas (1978), who used a giemsa stain for chromosomes, presented exceptionally clear configurations. Detailed techniques for staining chromosomes by use of aceto-orcein and aceto-carmine are given by Sansome (1976), for giemsa stain by Erselius and Boccas (1978), and for Feulgen fluorescence microscopy and aceto-orcein by Eriksson and Motzfeldt-Laane (1982).

Sansome's contributions changed the direction of research on *Phytophthora.* Her pioneering efforts were recognized by the presentation of an engraved plaque at the first International Symposium on the Biology, Ecology, and Pathology of *Phytophthora* held at the University of California, Riverside, in April 1981.

Sansome's conclusions that *Phytophthora, Pythium,* and other members of Oomycetes are diploid at first met with skepticism. In our laboratory, two theses presented data based on genetics (Timmer et al. 1970a) and cytology (Stephenson et al. 1974b) that meiosis in *P. capsici* did not occur in the gametangia. The latter study was repeated by Sansome (1976),

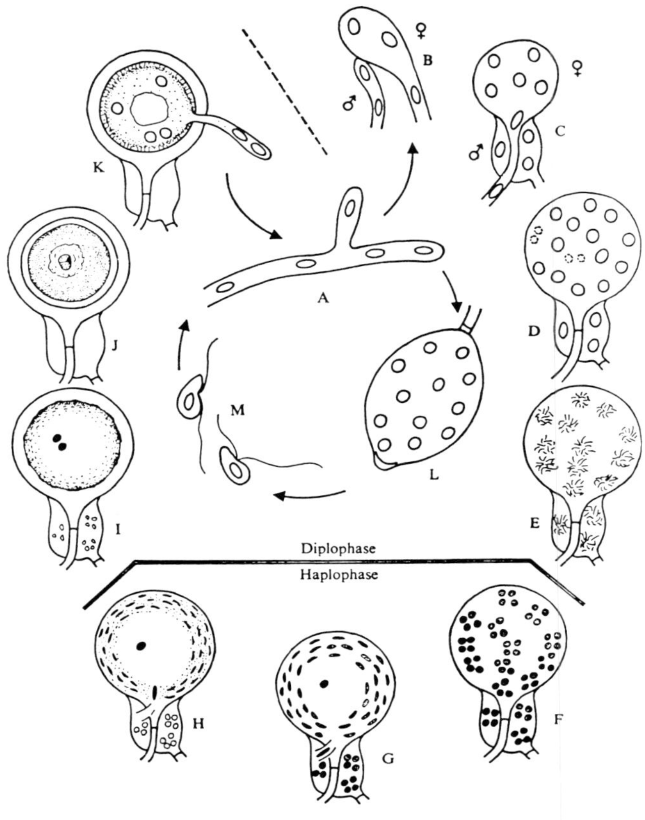

Figure 3.10. The nuclear cycle in *Phytophthora,* as seen in *P. infestans* (not to scale). **A,** Somatic mycelium with diploid nuclei (A1 or A2 mating type in a heterothallic species). **B,** Formation and association of antheridial (male) and oogonial (female) incepts. **C,** Emergence of the young oogonium (following penetration of the amphigynous antheridium by the oogonial incept); migration of nuclei into the developing oogonium. **D,** Abortion of nuclei in pairs in the expanded oogonium (about 38 μm in diameter); increase in size of remaining nuclei. **E and F,** Meiosis in the oogonium and antheridium; near-synchronous first division with a long prophase followed by a very short second division with no increase in nuclear size between the two divisions, thus resulting in formation of tetrads of haploid nuclei. **G,** Beginning of oosphere delimitation; migration of the male gametic nucleus into the oogonium from the antheridium. **H,** Penetration of the oosphere by the male nucleus. **I,** Association of the male and female gametic nuclei in the oosphere. **J,** Formation of the thick-walled oospore (about 30 μm in diameter) with a diploid zygotic nucleus in the "pellucid spot." **K,** Germination of the oospore to give the diploid vegetative stage. **L,** Diploid multinucleate sporangium (about 29 × 19 μm). **M,** Diploid uninucleate zoospores. (Reproduced by permission of the British Mycological Society, courtesy Brasier and Sansome 1975)

and the conclusions were shown to be in error. We realize now in retrospect with new information that the data could be interpreted differently.

A recent study (Eriksson and Motzfeldt-Laane 1982), in which light, fluorescence, and electron microscopy were used with *P. palmivora* A1 and A2 isolates of the "L" chromosome type, previously reported by Sansome et al. (1975) and Sansome (1977), presented excellent illustrations of gametangial meiosis. After the antheridium and oogonium become delimited, many of the nuclei degenerate. The remaining nuclei go through two meiotic divisions in both oogonium and antheridium, and three of the four products from each nucleus degenerate. When the oogonium is penetrated by the antheridial fertilization tube, nuclear fusion occurs, after which the oosphere forms and the periplasm degenerates. Formation of the thick oospore wall follows.

The chromosome number in metaphase 1 chromosomes ranges from $n = 5$ in *P. megakarya* (Sansome 1980) to 22–27 in *P. megasperma* var. *megasperma* (Sansome and Brasier 1974). Hansen et al. (1986) reports $n = >30$ in certain isolates of *P. megasperma.* See Brasier (1983; Table 1) and Waterhouse et al. (1983; Table 3) for lists of chromosome numbers reported for various species. There is evidence of polyploidy in *P. infestans;* isolates from Wales had nuclei with twice as many chromosomes as certain isolates from Mexico (Sansome 1977).

One of the convincing proofs that *Phytophthora* species are diploid comes from the finding by microfluorometry that the DNA content of gametangial nuclei (haploid phase) is about half that of somatic nuclei (diploid phase) (Mortimer and Shaw 1975; see reviews by Shaw [1983, 1988]). As would be expected, the DNA content of zoospore nuclei for all races of *P. sojae* is similar (Rutherford and Ward 1985a).

In most cases, the DNA content of nuclei correlates with the chromosome numbers in nuclei (Shaw 1988). Shaw (1988) states (excerpted by permission of Academic Press Ltd., London, United Kingdom, courtesy of Shaw 1988):

> A more detailed study of isolates of *P. megasperma* from a variety of hosts, using nuclei isolated from zoospores, showed a correlation between DNA content and chromosome number (Hansen et al. 1986). For example, isolates from alfalfa and soyabean with the smallest chromosome counts had the lowest DNA contents, broad host range isolates with the largest oospores had the highest chromosome number and the highest DNA contents. The standards used for comparison here were *P. infestans* and *P. palmivora,* with contents larger and smaller respectively than any isolate of *P. megasperma.* Relative to the 0.2 pg estimated for soyabean isolates (Rutherford and Ward, 1985a), the *P. infestans* isolate (from Scotland) would have a content of 0.95 pg, that of the *P. palmivora,* 0.15–0.17 pg. Hooley et al. (1985) estimated the DNA content of nuclei of *P. drechsleri* strain 6503 to be five times that of his 2n *A. nidulans* standard, namely 0.45 pg. Volumes of cytoplasm associated with one nucleus in zoospore germlings was constant (1000 μm^3) and was approximately five times that of *A. nidulans* (166 μm^3). If this relationship holds in general, cytoplasmic volumes may be useful in estimating DNA contents.

Both diploid and polyploid isolates exist in populations of *P. infestans.* Nuclei of diploid isolates can be treated with Feulgen's reagent and the DNA measured by the concentration of the stain (Tooley and Therrien 1987). Whittaker et al. (1991a) utilized fluorescence cytometry and staining with DAPI (4,6-diamidino-2-phenylindole) and showed that the DNA content of the isolates with the lowest DNA (arbitrarily labeled 2C) had eight to 12 bivalent chromosomes while others with a higher DNA content (4C) had 14–20 bivalent chromosomes.

Since *Phytophthora* is diploid in its vegetative state, genetic factors that govern its phenotypic expression can be heterozygous; that is, homologous chromosomes may carry dominant and recessive alleles of a single gene. A brief explanation here as to why certain mutations that are expressed as drug resistance are necessary for research on genetics of *Phytophthora* may be in order. Proof for the type of inheritance of certain traits has come mainly from the study of "marked" genes that are derived from selected mutations induced by use of artificial mutagens; however, in many cases, naturally occurring mutations have been found by selection.

The methodology for selecting mutants may be simple—for instance, Stack and Millar (1985a) selected a natural metalaxyl-resistant mutant from a population of about 12 million zoospores by plating the zoospores on a medium containing metalaxyl—or it may be complex, involving the use of a chemical mutagen such as *N*-methyl-*N*′-nitroso-*N*-nitrosoguanadine (MNNG) (Shattock and Shaw 1975; Davidse 1981; Layton and Kuhn 1988a; Lucas et al. 1990b), after which isolates that grow on a metalaxyl-containing medium are selected.

In a method described by Layton and Kuhn (1988a) and adapted from Davidse (1981), mutations were induced in *P. sojae* by incubating zoospores (5×10^5 per petri plate) in 10 ml of Petri's solution (Appendix 3.35) containing 30 μg of MNNG per milliliter for 30 min at 25°C. About 10% of the zoospores will stick to the bottom of the petri plate. After decanting off the MNNG, 5 ml of V8 juice broth is added to each plate and decanted off after 10 min. To select mutants, the drug-resistant plates are overlaid with V8 juice agar supplemented with (μg/ml) metalaxyl (20), fluorophenylalanine (150), or fluorotryptophan (10) and incubated for 1–2 weeks at 25°C. Drug-resistant colonies are transferred to fresh plates amended with one of the drugs. Each mutant culture is tested for resistance to the indicated drug after about five generations of single-spore isolations to indicate whether or not the resistance is stable.

Drug-resistance mutations can be, but are not necessarily always, inherited from a single gene. If a mutant isolate with resistance to a drug or fungicide that is conditioned by a nuclear gene is otherwise comparable to wild types, such an isolate is "marked" with that drug resistance and can be identified by its ability to grow in a medium containing that drug on which wild types will not grow. For instance, resistance to the fungicide metalaxyl in *P. infestans* (Shattock 1988) and *P. parasitica* (Chang and Ko 1990) is conditioned by a single gene with incomplete dominance. Resistance to chloroneb is also inherited in *P. parasitica* as a dominant nuclear gene (Chang and Ko 1990). Drug-resistance markers make it possible to determine whether or not the recombination of genetic factors, which may result from either a parasexual or a sexual process, occurs.

Many other characteristics such as growth rate, colony morphology, and oospore production (Erwin and Li 1986), pigment production (Galindo and Zentmyer 1967a, b; Timmer et al. 1970b; Shepherd 1976), and even pathogenicity to some hosts are probably inherited as multiple genes that are complexly inherited, thus making phenotypes of progeny from genetic crosses difficult to interpret. Variability in colony morphology in progeny of numerous A1 × A2 crosses is common; however, the nature of its heritability may be under either polygenic or cytoplasmic (maternal) control. Although such variability is frequently reported (Galindo and Zentmyer 1967a, b; Timmer et al. 1970b), it has not been a useful criterion for genetic studies.

Pigment production by *P. drechsleri* that occurs in a medium containing peptone and beef extract (Galindo and Zentmyer 1967b) can be used as a genetic marker. A synthetic medium containing casein hydrolysate and tyrosine supported the greatest amount of pigment, but even small amounts of glucose suppressed pigment formation (Timmer et al. 1970b). The pigment-producing ability of single-oospore isolates from a *P. drechsleri* A1 × A2 cross of two pigment producers varied from none to abundant. Timmer et al. (1970b), however, concluded that pigment production was probably under polygenic control and would not be a useful marker for genetic research. Since Shepherd (1976) showed that pigment production differentiated northern from southern Australia ecotypes of *P. drechsleri,* this marker might be useful to characterize biotypes. Pigment production by A1 *P. parasitica* isolates was double that of A2 mating types. *P. cinnamomi* isolates did not produce any pigment (Shepherd 1976).

Although auxotrophic markers (mutations conferring a specific nutrient requirement, e.g., for an amino acid such as methionine or arginine) (Timmer et al. 1970a) can be used for genetic studies, genes controlling auxotrophy are usually recessive and thus are difficult to detect in a diploid fungus (Shaw and Elliott 1968). Streptomycin, a common antibiotic, is not toxic to most of the higher fungi, but it is toxic to most species of *Phytophthora* (see Chapter 2). Drugs at low concentration (10 µg/ml) that inhibit growth of *P. infestans* include acriflavine, actidione, *p*-fluorophenylalanine, iodoacetic acid, streptomycin, and tetracycline. Chloramphenicol at 200 and 400 µg/ml is inhibitory to growth (Shattock and Shaw 1975). If resistance to a drug such as streptomycin were dominant, resistance should be useful for genetic study of a diploid organism. Shaw and Elliott (1968) reported finding a few isolates of *P. cactorum* that were resistant to streptomycin (100 µg/ml) by mass selection methods. One isolate was streptomycin dependent. Streptomycin resistance in *P. capsici* was also used in genetic studies of *P. capsici* (Timmer et al. 1970a). In *P. parasitica,* however, streptomycin resistance was found to be cytoplasmically and not nuclearly inherited (Chang and Ko 1990) (Figure 3.11). Shaw and Khaki (1971) indicate that resistance to *p*-fluorophenylalanine in *P. drechsleri* is dominant.

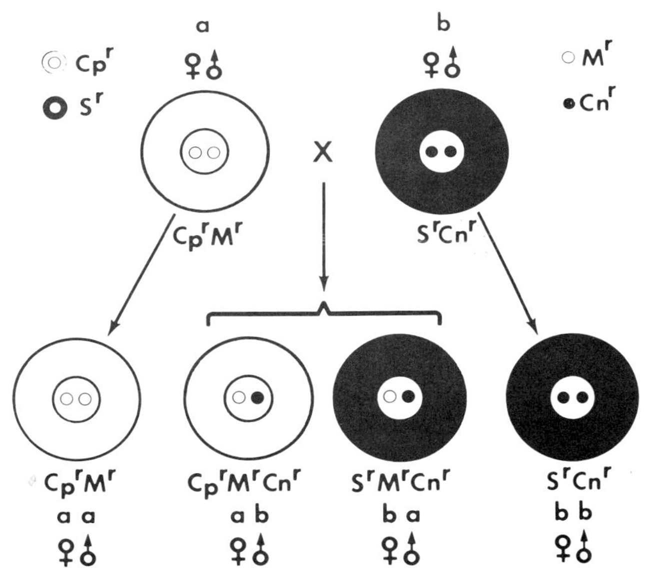

Figure 3.11. Results of the pairing of A1 and A2 isolates of *Phytophthora parasitica.* Isolate a carries homozygous dominant metalaxyl-resistant (M^r) alleles in the nucleus and a chloramphenicol-resistant (Cp^r) gene in the cytoplasm, and isolate b carries homozygous dominant chloroneb-resistant (Cn^r) alleles in the nucleus and a streptomycin-resistant (S^r) gene in the cytoplasm. Both parents produce oogonia (female) and antheridia (male). Sex symbols of the progeny represent the origins of gametangia for the formation of each type of offspring. (Reproduced by permission of the American Phytopathological Society, St. Paul, Minnesota, courtesy Chang and Ko 1990)

A problem with many of the genetic analyses that have been made is that certain factors governing resistance to some antibiotics such as streptomycin and chloramphenicol may not be under nuclear chromosomal control but are inherited via extrachromosomal cytoplasmic (DNA within mitochondria) control and would thus be inherited only through the maternal parent with no normal recombination of genes between parents. For example, in a zoospore of *P. infestans* that had been treated with DAPI (which stains specifically for DNA), DNA was detectable in both the mitochondria and the nucleus (Figure 4 in Shaw 1983). A recent report showed that chloramphenicol and streptomycin resistance in *P. parasitica* is controlled by cytoplasmic genes whereas resistance to metalaxyl and chloroneb is controlled by single dominant nuclear genes (Chang and Ko 1990). This work corrected an earlier inference that oospores from A1 × A2 crosses were mainly, if not altogether, the result of selfing by pheromone stimulation by the opposite mating type (Ko 1988).

As reviewed by Shaw (1988), the fluorochrome stain DAPI visualizes nuclei by ultraviolet light microscopy (Tooley and Therrien 1987; Therrien et al. 1989; Tooley et al. 1989). The DNA contents of nuclei of *P. infestans* isolates from Mexico, the United States, United Kingdom, Egypt, Brazil, Ethiopia, and Canada were similar to those of nuclei in mycelium; however, the DNA values for isolates of *P. infestans* (both A1 and A2) from different countries varied from 2C (arbitrary number indicating relative DNA content), typical of a diploid nucleus, to 4C, presumed to be tetraploid. One isolate contained nuclei of both 4C and 2C DNA content, indicating that such an isolate could be heterokaryotic. The number (n) of chromosomes in the 2C nuclei = 8–12, and in 4C isolates n = 14–20 (Whittaker et al. 1991a). Thus, both diploid and polyploid isolates coexist in the field. This probably indicates an increased potential for variability.

A report from the Commonwealth of Independent States (C.I.S., formerly U.S.S.R.) (Poedinok et al. 1982) discusses a parasexual cycle for *P. infestans.* In a conventional parasexual system, diploid nuclei arise within heterokaryons; crossing over between homologous chromosomes may occur; and gradual haploidization occurs by nondisjunction, yielding haploid recombinants. Perhaps similar events occur in *Phytophthora* following somatic fusion of diploid nuclei of different strains. By use of mutants resistant to acriflavine, oxytetracycline, and streptomycin, Poedinok et al. obtained evidence of mitotic crossing over in the parasexual cycle of *P. infestans.* Haploidization of nuclei in mycelium was measured by use of cytospectrophotometry after staining with 2% aceto-orcein. The frequency of haploidization of nuclei was increased by use of para-fluorophenylalanine (dosages of 1, 3, and 5 µg/ml in the medium). Heterokaryons and heterozygous diploids of *P. infestans* have been reported from the C.I.S. (D'yakov and Kuzovnikova 1974; Kulish and D'yakov 1979).

The A2 mating type of *P. infestans* from potato occurs at about a 50:50 ratio with the A1 type in the Toluca and Chapingo regions of Mexico, but until 1994 (Hohl and Iselin), only the A1 mating type had been found in other regions of the world. Potato pathologists have long been concerned that oospore progeny from A1 × A2 mating types, occurring naturally in the field in Mexico, could give rise to new and more highly virulent races (Niederhauser and Cobb 1959; Niederhauser 1961, 1991). Indeed, the most virulent races of *P. infestans* in the world are found consistently in the Toluca region of Mexico. No one knows why the A2 mating type had not been transported from Mexico, the area of origin of *P.* infestans (Reddick 1943; reviewed by Fry and Spielman [1991] and Zentmyer [1988]), to new areas of the world until recently. The A2 mating type was first found in Switzerland (Hohl and Iselin 1984) and later in Great Britain (Malcomson 1985; Shattock et al. 1990), Scotland (Malcomson 1985), Egypt (Shaw et al. 1985), Germany (Schöber and Rullich 1986; Götz 1991), Wales (Tantius et al. 1986), Sweden (Kadir and Umaerus 1987), the United States and Canada (Deahl et al. 1991), Ireland (O'Sullivan and Dowley 1991), and in the Middle East, Asia, and South America (reviewed by Spielman [1991]). In Great Britain, isolates of the A2 mating type were found to be rather widely distributed, but the percentage of A2 types varied between 6.7 and 15.1 over a 4-year period, with no indication as yet of its increasing to 50% as it is in Mexico (Shattock et al. 1990). A2 frequencies of up to 50% have been reported in parts of the C.I.S. (Gorborova et al. 1989). The migration of new biotypes of *P. infestans* to Europe, Africa, and the Americas was reviewed by Fry et al. (1993), Gillis (1993), S. B. Goodwin et al. (1994a, b), and Fry and Goodwin (1995).

How long the A2 mating type has existed in various countries other than Mexico has been addressed by Fry and Goodwin (1995), who discussed the migration of the A2 mating type of *P. infestans* to various countries and state that "prior to the 1980s, all populations of *P. infestans* outside of Mexico were very simple (Tooley et al. 1985; Spielman et al. 1991), and most of the world (including Europe) was dominated by a single clonal lineage (S. B. Goodwin et al. 1994b). However, in the United States, there were several clonal lineages (S. B. Goodwin et al. 1994b)." Fry and Goodwin (1995) further state that the migration of the A2 mating type, which probably occurred during the 1970s, was detected in the 1980s. Work in their laboratory with allozymes of *P. infestans* isolates collected widely in the world showed that the "new" population in Europe after 1982 had much different allozyme characteristics than isolates collected before then. See excerpt below for a discussion of isozymes by Spielman (1991). Niederhauser (1991) postulated that the migration of A2 mating types probably occurred in 1976 when many tons of potatoes were exported to Europe from Mexico, where both mating types have coexisted for many years.

The introduction of the A2 mating type of *P. infestans* to potato-growing regions outside Mexico may lead to production of new pathologic races in the field. Fry et al. (1991, 1992), who are quoted below, had not yet (in 1993) noted any changes in the race situation in Europe. S. B. Goodwin et al. (1991) show by allozyme patterns and DNA probes that a greater amount of genetic diversity occurs in central Mexico, where A1 and A2 mating types have coexisted for a long period, than in northeastern or northwestern Mexico, where only the A1 mating type was noted before 1970. Several studies indicate that recombination of genetic factors for virulence can occur in oospore progeny from A1 × A2 crosses (Romero and Erwin 1969; Laviola and Gallegly 1983; Tooley et al. 1986; Spielman et al. 1989, 1990; Shaw 1991; see also Table 4.5). Research is in progress by Fry's group to determine whether introduction of the A2 type to areas of the world other than Mexico will lead to an increase in new, more virulent pathologic races. Fry et al. (1993) indicated that there was no convincing evidence then that the sexual process (A1 × A2) influences the life cycle outside central Mexico; however, the reports by Sujkowski et al. (1994) and S. B. Goodwin et al. (1995) now support the concept that the sexual process could increase the production of new races and lead to greater genetic diversity.

The oospore has greater persistence than other propagules and thus could become a source of inoculum. Andrivon (1995) in a review of the biology, ecology, and epidemiology of *P. infestans* in the soil states that there is much experimental information to indicate that oospores could be an important source of inoculum, but there is a paucity of data from the field. Drenth et al. (1995) demonstrated that oospores of *P. infestans* survived the winter in the soil in the Netherlands and caused late blight on potato plants. Pittis and Shattock (1994) reported that oospores from several A1 × A2 crosses of *P. infestans* isolates survived for up to 8 months in soil. Oospores were produced in potato stems that had been inoculated with zoospores of A1 and A2 isolates. Oospores produced in vitro were capable of infecting potato shoots.

Wilde (1961) showed a camera lucida drawing of hyphal anastomosis in *P. infestans* A1 mating type and considered that this may be a mechanism for exchange of genetic material because at that time, the A2 mating type was not known to exist in Europe. Stephenson et al. (1974a) showed photomicrographic evidence for hyphal anastomosis in isolates of *P. capsici;* however, hyphal anastomosis has been infrequently detected in subsequent observations (D. C. Erwin, *unpublished*).

Attempts have been made over the past several years to determine how virulence and the ability to develop new races that would overcome resistant cultivars of the host plant are inherited. Several of the early studies with oospore progeny of A1 × A2 mating types of *P. infestans* show that recombination of factors for virulence occur in the F_1 progeny. For instance, a cross of an A1 race 0 with A2 race 1.2.3.4 (multiple genes for virulence) yielded single-oospore progeny that were race 1, race 2, race 1.3, and race 2.3.4, which suggests that recombination had occurred (Romero and Erwin 1969). Similar phenomena are reported by Gallegly (1968, 1970) and Laviola and Gallegly (1983).

Shaw (1991) reported that matings of A1 × A2 isolates of *P. infestans,* both of which were virulent on potato plants, yielded all virulent oospore progeny. Mating avirulent × virulent isolates yielded both avirulent and virulent progeny, usually in equal numbers. Avirulent × avirulent isolates yielded oospore progeny that segregated to approximately one-quarter virulent progeny. It appears that a gene-for-gene (Flor 1956) system may be operative and that avirulent parents may be heterozygous and recessive for virulence. Spielman et al. (1989) indicates that avirulence may sometimes be recessive. According to Vorb'eva and Gridnev (1983), single-oospore variants from a cross of *P. capsici* × *P. infestans* were pathogenic to both pepper and potato, whereas the parent cultures were pathogenic only to their respective hosts.

New expressions of virulence in *P. infestans* were noted in isolates that emanated from sexual reproduction as well as in those that had been treated with the mutagen parafluorophenylalanine. New expressions of virulence arising from avirulent isolates were not the result of spontaneous mutation but of the rearrangement or isolation of recessive virulence alleles previously present within a heterozygous genome (LeGrand-Pernot 1988). This report also supports the early hypothesis that the sexual stage of *P. infestans* could be a source of pathogenic variability.

According to Spielman (1989), results from research with single-oospore progeny from crosses of A1 × A2 *P. infestans* cultures have been interpreted to indicate that there are several kinds of genetic control of virulence to potato or tomato leaves from plants with *R* (resistance) genes:

- dominant virulence (*P. infestans* gene *St2* was virulent on *Solanum tuberosum* with the *R2* gene, and the *P. infestans* gene *St4* was virulent on *S. tuberosum* with the *R4* gene);
- dominant avirulence (*P. infestans* gene *Le1* was virulent on tomato, *Lycopersicon esculentum*, containing an *R1* gene with an epistatic inhibitor locus);
- dominant avirulence (*P. infestans* gene *St3* with a complementary locus).

Al-Kherb et al. (1995) mated A1 and A2 types of *P. infestans* and obtained cultures from germinated oospores to study the inheritance of virulence to potato. Their data supported the hypothesis that avirulence alleles are dominant and virulence alleles are recessive; however, unexpected segregations were genetically difficult to explain.

Analysis of the literature on experimental variability of oospore progeny from A1 × A2 crosses of *P. infestans* supports the hypothesis that these progeny vary in virulence and can be a source of new races.

Although variation in virulence (or avirulence) and other qualities occurs in progenies of A1 and A2 crosses, there has always been a question as to whether or not crossing of the A1 and A2 types was actually important in nature. A1 and A2 mating types of *P. infestans* coexist in central Mexico and have been thought to contribute to the apparently greater array of highly virulent pathogenic races that occur in that area (Niederhauser 1991).

Although the data from many studies indicate that genes conditioning race specificity in A1 × A2 *P. infestans* crosses can recombine after experimental production and germination of single-oospore progeny, much of the field evidence had not indicated that sexual reproduction had an impact on the development of new races in Europe following recent introductions of the A2 mating type (Fry et al. 1991; Fry et al. 1992).

We quote from Fry et al. (1991; excerpted by permission of the American Pathological Society, St. Paul, Minnesota, courtesy of Fry et al. 1991):

> An important question in the Netherlands and in other locations where both A1 and A2 mating types occur is whether or not sexual reproduction is contributing to the epidemiology of late blight. Although data from the present study do not conclusively resolve this issue, the infrequent occurrence of A2 isolates, and the discoveries that the A2 mating type is associated with certain allozyme genotypes and that A1 mating types are associated with other allozyme genotypes, is more consistent with the notion of asexual reproduction than with significant sexual reproduction. Associations between mating type and allozyme genotype have not been detected in Central Mexico (Tooley et al. 1985). If sexual reproduction were common and contributing to the diversity of the Netherlands population, one would expect A1 and A2 individuals to have similar allozyme genotypes and both to be represented in all regions of the country. Conformation or lack of conformation of allele frequencies to those exposed at Hardy-Weinberg equilibrium is not at all conclusive in this issue because of the limited diversity and many generations of asexual reproduction during epidemics.

In contrast to other areas of Europe, the genetic evidence following introduction of the A2 mating type of *P. infestans* to Poland (Spielman et al. 1991) indicates that sexual reproduction may complicate disease-control measures (Sujkowski et al. 1994). Sujkowski et al. (1994) cite four lines of evidence based on allozyme patterns of isolates collected over 5 years that support the possibility that sexual reproduction increases variability in the field. 1) Genotypes at most loci were in Hardy-Weinberg equilibrium; 2) there was no genetic differentiation between A1 and A2 isolates; 3) there was a high proportion of genetically unique individuals that would not be expected in populations restricted to asexual reproduction; and 4) a new genotype (90/90) was noted at the allozyme locus glucose-G-phosphate isomerase that could have been produced from sexual recombination. They also state that the cultivar Bronka was highly resistant (20% blighted) during the early 1980s, but by the late 1980s, it had become more susceptible (90% blighted).

As discussed in Chapter 7 on fungicidal control, resistance to the systemic fungicide metalaxyl (a fungicide specifically toxic to *Phytophthora*) has occurred frequently in natural populations of *P. infestans* causing late blight of potatoes in Europe. Resistance of *P. infestans* to metalaxyl is based on its inhibition of ribosomal ribonucleic acid (RNA) (Davidse et al. 1983a). It is chromosomally inherited by a single nuclear locus exhibiting incomplete dominance (Shattock 1988). Laboratory crosses of A1 and A2 mating types, resistant and susceptible to metalaxyl, yielded F_1 progeny that in the majority were intermediately sensitive to metalaxyl, with a few being either totally sensitive or resistant; however, two back-cross generations to a metalaxyl-resistant parent gave unexpected phenotypic frequency distributions that were unexplainable. Intermediately resistant isolates were considered to be heterozygous for the dominant metalaxyl resistance gene (Shattock 1988).

Crosses of A1 × A2 mating types of *P. infestans* produced single-oospore progeny that varied quantitatively in growth rate and zoospore production at 16°C. The data suggest that parental strains are heterozygous for these characters (LeGrand-Pernot 1986b).

The use of isozyme loci as markers (described in more detail in Chapter 4 for use in distinguishing between closely related species) has allowed the identification of different populations of *P. infestans* in different areas of the world. An excerpt from Spielman (1991) that explains the rationale for use of isozyme technology is given here (excerpted by permission of Cambridge University Press, Cambridge, United Kingdom, courtesy of L. Spielman 1991):

> Isozymes are ideal tools for this kind of investigation, because they are easily assayed, stable, codominant (which is important in a diploid organism like *P. infestans*), and generally under simple genetic control. Isozymes are defined as variant forms of an enzyme which share the same catalytic function, and allozymes are isozymes whose variant forms are coded at the same locus. Allozymes are therefore allelic to each other, and are ideal for population investigations, because the best analytical methods depend on knowledge of allelic frequencies and distributions.
>
> Isozymes are assayed by means of starch gel electrophoresis, the methods for which have been standardized over the past 30 years (Shaw and Prasad, 1970; Siciliano and Shaw, 1976; Micales, Bonde and Peterson, 1986). Over 50 different enzyme systems have been tested in *P. infestans* using various extraction methods, buffers, running conditions, and staining recipes to optimize the resolution of each system (Spielman et al., 1990). So far, 11 monomorphic enzymes and 18 polymorphic enzymes have been detected, but most of the latter were either so poorly resolved or genetically complex that it was not possible to deduce genotypes from the gel patterns observed. The two systems for which simple genetic control has been demonstrated are glucosephosphate isomerase (GPI, E.C.5.3.1.9) and peptidase (PEP, E.C.3.4.3.1) (Shattock et al., 1986a; Spielman et al., 1990).
>
> The enzymes commonly used in population studies are either monomeric or dimeric. For a monomeric enzyme, the gene product is

the active enzyme unit, so homozygotes are one-banded and heterozygotes are two-banded. One of our polymorphic systems, mannose-phosphate isomerase (MPI), is such an enzyme, and segregation of four electromorphs (93, 100, 105, and 111) is observed in crosses. However, the data suggest that there is more than one locus coding for MPI, and it is not possible to determine which alleles belong to which locus (Spielman et al., 1990).

In dimeric enzymes, the initial gene products are inactive. Enzyme activity occurs only when two products, or subunits from a single gene, are joined. These subunits combine randomly, so if there are two different kinds present, as would be the case in a heterozygote, three kinds of active enzyme molecule are formed: two different homodimers and a heterodimer. Thus, homozygotes are single-banded, but heterozygotes are three-banded. Both GPI and PEP produce banding patterns on gels which are typical of dimeric enzymes and if two parents with different, single bands (homozygotes) are crossed, progeny which are all three-banded (heterozygotes) are obtained. If two identical heterozygotes are crossed, the three expected classes of progeny (1 homozygote: 2 heterozygote: 1 homozygote) are observed. Finally, if a heterozygote is crossed with a homozygote, the same two genotypes occur in the progeny in equal numbers. Unexpected classes do sometimes occur in low frequencies, probably as a result of selfing (Spielman et al., 1990).

Unexpected ratios are sometimes observed in crosses between related parents (Spielman et al., 1990). For instance, crosses between two heterozygous F_1 sibs have yielded excesses or deficiencies of expected homozygous classes, and in one backcross an expected class was absent. A likely explanation for these abnormalities is that inbreeding exposes deleterious recessive alleles linked to marker alleles. Nevertheless, the overall result of the crossing studies is that the electromorphs of GPI and PEP behave like Mendelian alleles.

It should be obvious from the foregoing discussion that the development of isozymes as genetic markers can be rather involved. Studies of *P. infestans* have so far yielded only two suitable polymorphic loci: *Gpi* with 7 alleles and *Pep* with 4 alleles. A few of the alleles we have detected are cryptic variants, isozymes which are distinguishable only under certain running conditions. For example, on gels buffered with histidine at pH 7, three alleles of *Gpi* (90, 98, and 100) are indistinguishable, but the same buffer at pH 6 reveals the differing mobilities of these alleles.

In recent years, evidence for segregation and recombination of factors for virulence and avirulence in *P. infestans* (discussed above) and of the genes specifying polymorphic enzymes, which are detected by electrophoretic methods, indicates strongly that matings of A1 × A2 mating types yield hybrid progeny. Matings with A1 and A2 types of *P. infestans* from Mexico, characterized for their electrophoretic patterns at two enzyme loci, glucosephosphate isomerase (GPI-1) and peptidase (PEP), via starch gel electrophoresis (previously characterized by Tooley et al. [1985]) yielded evidence that the parents were sexually compatible and that hybridization had occurred. The progeny from the cross yielded genetic data indicating that the majority of the single-oospore progeny were hybrids and only a few of the progeny were selfs. Single-oospore cultures from parents that were heterozygous for alleles at the GPI-1 and PEP loci yielded the expected ratios of 1:2:1 for homozygous fast, heterozygous, and homozygous slow electromorphs, respectively ("fast" and "slow" refer to rates of movement of the protein on the electrophoretic field of the starch gel). These ratios were consistent with that expected for a diploid organism. Only 6 of 685 progeny were selfs (Shattock et al. 1986a).

When *P. infestans* A1 and A2 cultures with different isozyme characteristics were crossed, these loci controlling the isozyme characteristics segregated normally in the F_1 progeny, but when sibling cultures were backcrossed to parental isolates, segregations were abnormal. This abnormality was not explainable with the data at hand but may be caused by the presence of linked deleterious recessive alleles. Since *P. infestans* is a diploid and obligately outcrossed, this is a likely cause of abnormal segregations with backcrosses or sibling crosses (Spielman et al. 1990).

Spielman et al. (1990) utilized genotypes at two polymorphic allozyme loci, *Gpi-1* (glucose phosphate isomerase, E.C.5.3.1.9) and *Pep-1* (peptidase, E.C.3.4.3.1), to characterize isolates from different regions. The alleles *Gpi83, Gpi98,* and *Pep83* occur in isolates found only in central Mexico in low to moderate frequencies, and *Gpi90, Gpi111,* and *Pep96* are found in isolates outside central Mexico (Spielman 1991). Spielman et al. (1990) found that the isolates of *P. infestans* could be characterized by allozyme type as the "old" population (before the A2 mating type was found outside Mexico) and the "new" population (after the A2 population was found). Each population was unique for allozyme alleles and genotypes. They concluded that the change in populations of *P. infestans* in Europe was probably the result of migration of new biotypes and not of mutation. The new population in Europe appears to be more "fit" than the old population; however, the mechanism by which it became more fit is not understood. Migration of *P. infestans* populations, marked by isozyme characteristics, from Mexico to various regions of the world is reported by Fry et al. (1991) and reviewed by Fry et al. (1992). See also S. B. Goodwin et al. (1994a, b) and Fry and Goodwin (1995).

S. B. Goodwin et al. (1995) showed that a single clonal line of *P. infestans* moved from Mexico to the northwest United States, from there to the British Isles, and thence to other parts of the world. See Chapter 33 on *P. infestans.*

Both A1 and A2 mating types of *P. infestans* detected in 1989 (Deahl et al. 1991) in British Columbia, Canada, contributed to the production of unique genotypes that most likely resulted from sexual recombination. This was the first report of the "probable occurrence of sexual reproduction of *P. infestans* in North America north of Mexico" (S. B. Goodwin et al. 1995).

In the Netherlands, oospores that were produced by pairing A1 and A2 isolates of *P. infestans* on leaves of potato plants in the field survived in the soil during the winter of 1992–1993. Oospores survived in soil held for 48 h at temperatures ranging from 33 to −80°C but not at 40°C or higher. Eighteen isolates were shown to be recombinants by use of DNA fingerprinting with an R57 probe (Drenth et al. 1995).

Double-stranded ribonucleic acid (dsRNA) has been detected in Mexican isolates of *P. infestans* but not in isolates from the United States or Europe. Although the presence of dsRNA sometimes indicates a virus infection, no viruses have been found. The presence of dsRNA was not associated with low virulence except for one isolate. The occurrence of dsRNA was considered to be a useful cytoplasmic marker, especially since dsRNA was not found in non-Mexican isolates (Tooley et al. 1989). A subsequent report from the same laboratory (Newhouse et al. 1992) analyzed more isolates of *P. infestans* and confirmed that dsRNA was found most abundantly in Mexican isolates (83.3%), followed by the Netherlands (20%) and Peru (2.9%). Isolates from the United States and Israel did not contain dsRNA. Sizes of dsRNA ranged from 1.35 to 11.10 kilobase pairs (kbp). Newhouse et al. (1992) suggest that the A2 isolates recently detected in the Netherlands came from Mexico after the emergence of metalaxyl resistance in 1980 (Davidse et al. 1981) because all of the dsRNA isolates from the Netherlands were metalaxyl sensitive but 80% of the dsRNA-free isolates were resistant.

When isolates of *P. infestans* with 2C × 4C and 4C × 4C (arbitrary DNA values) were crossed, the germination percentage of oospores was lower in progeny than from 2C × 2C crosses. Isozyme analysis of progeny from the 2C × 2C crosses showed that most germinated oospore progeny were hybrids. The proportion of progeny that were selfs was higher in the 2C × 4C crosses, however. Hybrid progeny from the 2C × 2C crosses were all 2C, but those from 2C × 4C crosses ranged from 2C to 3C, and progenies from the 4C × 4C crosses included both 2C and 4C characteristics. Although 2C and 4C isolates can cross and produce germinating progeny, it is still not understood what impact this has on the establishment of such progeny in the field (Whittaker et al. 1991b).

The duplication cycle of nuclei in germinating zoospores of *P. infestans* required about 7–8 h for a presumed diploid isolate but 10–12 h for a presumed tetraploid isolate. Whittaker et al. (1991a) concluded that zoosporic nuclei at the germinating stage were at the G1 stage of the cell cycle. This indicated that the variation in zoosporic DNA among isolates of *P. infestans* (Whittaker et al. 1991a) was the result of differences in ploidy (Whittaker et al. 1992). These studies add to the evidence that both diploid and tetraploid isolates exist in nature, a finding that further explains why the genetic system in *Phytophthora* is so complex.

Numerous reports cite that cultural and pathogenic variability of progeny of sexual crosses of A1 and A2 isolates of *P. palmivora* are more variable than asexual progeny (Chee 1973b; Kaosiri et al. 1980). Although single-zoospore progeny of *P. palmivora* and *P. botryosa* were not variable, progeny of intraspecific matings of A1 and A2 types of each species differed in morphological appearance, mating behavior, temperature-growth relations, rate of production, shape and size of sporangia, abundance of chlamydospores, and pathogenicity to *Hevea* (rubber tree) leaf petioles and cacao pods. Oospores from a *P. palmivora* × *P. botryosa* interspecific cross did not germinate, however (Chee 1973b).

The possibility that *P. meadii* may be derived from a species hybrid was addressed by Sansome et al. (1991). They found that some isolates from Sri Lanka, where the predominant mating type was A1 (Dantanarayana et al. 1984), were predominantly polyploid and others predominantly diploid. Oogonia of polyploids were larger (25–30 µm in diameter) than those of the diploids (20–24 µm). Cytological examination by use of the aceto-orcein squash method (Sansome and Brasier 1974) revealed that the large oogonial types had 18–24 chromosome configurations, and they were considered to be tetraploid. Some smaller oogonial cultures were self-sterile but when paired with *P. megakarya,* produced small oogonia (20–24 µm) or a mixture of small and large oogonia (up to 30 µm); however, oosphere formation frequently failed. Nuclei of the small oogonia had nine to 12 chromosome configurations. There were no differences in the protein patterns of the polyploid and diploid cultures.

Although the above hypothesis is interesting, the finding that the isozyme pattern of *P. palmivora* is distinctly different from that of *P. meadii* indicates that it would be an unlikely parent of *P. meadii* (Oudemans and Coffey 1991c) if it were possibly an allopolyploid species, as suggested by Brasier (1987) and Sansome et al. (1991).

There is convincing evidence based on DNA restriction fragment length polymorphism (RFLP) markers that factors controlling RFLP markers recombine in progeny of oospores from matings between A1 and A2 types of *P. parasitica.* When A1 and A2 mating types of *P. parasitica* were marked by unique RFLPs, the majority of the single-oospore progeny carried both parental markers, indicating that sexual recombination had occurred. However, mitochondrial DNA (mtDNA) was, as expected, inherited only from the parent from which the oospore was obtained (Förster and Coffey 1990).

Since mtDNA is uniparentally inherited (cytoplasmic inheritance) and does not recombine in sexual progeny, Förster and Coffey (1990) suggested that resistance to streptomycin or chloramphenicol might be conditioned by mtDNA and not by nuclear genes. Ann and Ko (1988b) had concluded that resistance to these drugs was not inherited independently in oospore progeny and made the generalization that recombination probably did not occur in progeny of most A1 and A2 crosses. A subsequent report by Chang and Ko (1990), however, showed that resistance to streptomycin and chloramphenicol in *P. parasitica* was indeed cytoplasmically and not nuclearly inherited. In the report by Chang and Ko (1990) that recognized this anomaly, a unique genetic method for separation of nuclear and cytoplasmic inheritance of factors was given. Progeny of crosses of A1 × A2 cultures and of selfed, A1, or A2 cultures of *P. parasitica* by the polycarbonate membrane method were analyzed. See Figure 3.11 by Chang and Ho (1990) for a graphic method useful for separation.

Chang and Ko (1990) produced oospores by the polycarbonate membrane method (Appendix 3.56), which allows hormonelike materials from A1 or A2 mating types to diffuse through the membrane to the opposite culture but does not allow sexual union with the opposite mating type. The single-oospore progeny from four metalaxyl-resistant (M^r) mutants segregated in a ratio of three resistant to metalaxyl and one sensitive. Chloroneb-resistant (Cn^r) single-oospore progeny also segregated in a ratio of three resistant to one sensitive. These results indicate that resistance to both fungicides is inherited by means of a single dominant gene that is in a heterozygous condition. Their work showed that resistance to chloramphenicol and streptomycin is uniparentally inherited with no recombination and thus is under cytoplasmic control.

When A1 and A2 isolates of *P. parasitica* were paired, colony types and growth rates of single-oospore progeny varied, but colony types and growth rates of single-zoospore cultures did not (Ann and Ko 1990b). Ann and Ko (1990b) suggested that nuclear polygenes in a heterozygous condition might be involved in variation through the sexual oospore stage. Ability to produce the enzymes, amylase, or pectin depolymerase is variable among different isolates of *P. parasitica.* Although the response of single-zoospore progeny was similar to that of the parent culture, single-oospore progeny from pairing A1 or A2 isolates by use of the polycarbonate membrane method (progeny would be selfs) were found by Ann and Ko (1990a) to be variable in their ability to produce amylase. They concluded that amylase production was controlled by heterozygous multiple genes. These results were similar to those of Boccas (1972), who showed variability in growth rate from single-oospore isolates of the homothallic *P. syringae.*

Although A1 and A2 heterothallic isolates can produce sexual oospores by pairing of unlike mycelial tips as described by Galindo and Gallegly (1960) and Shaw (1987) for *P. infestans,* it is not known how crossing takes place between homothallic isolates.

Long and Keen (1977) paired genetically marked isolates of *P. sojae* in an inoculum drop and showed evidence that crossing had occurred because some progeny were heterokaryotic. Recombination for marked characteristics occurred in single zoospores and oospore progeny. Heterokaryons were forced by mixing inoculum of different isolates (Layton and Kuhn 1988a). Presumably, the antheridium incept from one marked isolate pairs with the oogonial incept of a different isolate. In members of Ascomycetes, hyphal anastomosis is a common method by which genetic material is exchanged between unlike isolates, but hyphal anastomosis in oomycetous fungi is uncommon.

P. sojae is homothallic and not ordinarily considered to be subject to hybrid variability because it has been assumed that all oospore progeny are selfs; however, recent reports (Bhat et al. 1992; Bhat and Schmitthenner 1993a) present new and interesting developments. Hybrid progeny of *P. sojae* were obtained by pairing genetically marked isolates of *P. sojae*. Wild-type isolates were treated with the mutagen MNNG, after which mutants were selected. A race 1, marked by resistance to *p*-fluorophenylalanine (FPA^r), an amino acid analogue, and a race 3, marked by resistance to metalaxyl (MEX^r), were paired by mixing the cultures. The analysis of F_2 single-oospore progeny cultures showed that inheritance of MEX^r was incompletely dominant and inheritance of FPA^r was dominant. Hybrids were obtained in the F_2 generation of progeny that grew on a medium containing both metalaxyl and FPA. Bhat et al. (1992) detected progeny on metalaxyl-containing media that were resistant (MEXMEX), intermediately resistant (MEXmex), and sensitive (mexmex) (capital letters indicate dominance and small letters indicate recessiveness of the genes) at a ratio of 1:2:1. Resistance to FPA was dominant. They propose genotypes for FPA^r as FPAFPA and FPAfpa. The ratio of FPA^r (resistant) to fpa^s (sensitive) was 3:1. This study is fundamental to understanding how new races of *P. sojae* might arise in nature (Bhat et al. 1992; Bhat and Schmitthenner 1993a, b). These results are also discussed in Chapter 6 on control by host resistance and in Chapter 59 on *P. sojae*. They show that pathologic race characteristics can recombine in progeny from paired cultures of different races.

Outcrossing by the homothallic species *P. sojae* has also been reported by Long and Keen (1977) (discussed above), Bhat (1991), Bhat et al. (1990, 1992, 1993a, b), Bhat and Schmitthenner (1992, 1993a, b) (discussed in Chapter 6), Förster et al. (1994), and Whisson et al. (1994). It now appears to be an accepted fact that the evolution of new races of *P. sojae* can be attributed in part to variability resulting from the sexual cycle.

Förster et al. (1994) determined DNA RFLP patterns for several pathologic races of *P. sojae* and found that some isolates were nearly identical to race 1 (the first to be described), but several other isolates showed a distribution of "RFLP alleles" that suggested that most of the genetic variation in the species was found to exist in isolates P1658 (race 1), P7064 (race 7), P7074 (race 14), and P7076 (race 19). All other isolates tested "appear to have been produced by rare outcrosses between representatives of these four genotypes." They proposed that occasional outcrosses were probably a major contributor to the origin of new races of *P. sojae*.

Whisson et al. (1994) found that 10 of 354 hybrid oospores were identified from a cross by "co-culturing" of two isolates of *P. sojae* that had been characterized for random amplified polymorphic DNA (RAPD) markers. Genetic analysis indicated that avirulence genes were dominant to virulence. They conclude that "sexual recombination under field conditions may play an important role in generating and maintaining genetic diversity in populations of *P. sojae*."

P. medicaginis, which is homothallic, shows considerable variation among single-oospore isolates. When oospores from a single-zoospore isolate were germinated by methods described by El-Hamalawi and Erwin (1986a, b), the progeny were variable in growth rate and colony formation. The colony of the original single-zoospore isolate was appressed, but oospore progeny contained fluffy as well as appressed colony-forming isolates. The appressed isolates produced oospores much more abundantly than fluffy colonies. Single-oospore isolates from both fluffy and appressed colony isolates produced both colony types. When isolates were tested on a highly resistant alfalfa germplasm, A77-10B (Erwin et al. 1982), the oospore isolates varied from high virulence to low virulence, but the disease incidence caused by single-zoospore isolates was uniformly mild on the resistant alfalfa germplasm (Erwin and Li 1986). The results were interpreted to indicate that recombination of factors for virulence and growth (probably multigenic) occurred in the germinating oospores.

Protoplast Fusion: A Genetic Method

Since germination of oospores from crosses of A1 and A2 types has not been consistently high in many studies (Table 3.3), the question arises as to whether a low percentage of germination adequately samples the population. This unpredictability of oospore germination has stimulated an interest in protoplast fusion for genetic research. A parasexual genetic system (Roper 1966; Tinline and MacNeil 1969) consisting of forcing a heterokaryon (unlike nuclei occupying the same cytoplasm), karyogamy (fusion of nuclei), and genetic recombination can be utilized to study the mode of inheritance of certain traits. Observations of fusion of hyphae of *Phytophthora* are rare; however, hyphal anastomosis has been noted in *P. capsici* (Stephenson et al. 1974a) and *P. infestans* (Wilde 1961).

Hybrids were made between *P. capsici* protoplast cells with resistance to metalaxyl, phosphorus acid, and fluorotryptophan (Lucas et al. 1990a). The use of protoplast fusion is also discussed by Pesti and Ferenczy (1979) and Campbell et al. (1989) for *P. infestans*, Jahnke et al. (1987) for *P. parasitica*, in recent chapters on parasexuality in *P. sojae* by Kuhn (1991), and on gene structure and expression in *P. infestans* by Kinghorn et al. (1991).

The use of protoplasts is attractive because of the greater possibility that individual cells that do not have cell walls will fuse, allowing nuclei from one cell to flow into another (heterokaryosis). This is essentially what happens in the more complex phenomenon of fusion of the male antheridium and female oogonium. As stated above, however, germination problems with oospores have often been a barrier. With the protoplast method, matings of cells with different genetic characteristics can be made in a petri plate, and the mode of inheritance of the factor being studied can be determined. Many cultivars of soybean possess single gene resistance to different races (Athow and Laviolette 1983; Schmitthenner 1985; Athow 1987; Ward 1990) (Chapter 6). Prior to the recent reports by Layton and Kuhn (1988a, b), Bhat et al. (1992), Bhat and Schmitthenner (1993a), Förster et al. (1994), and Whisson et al. (1994), little was known about the mode of inheritance of virulence or avirulence of *P. sojae* to resistant cultivars of soybean.

The preparation of protoplasts of *Phytophthora* was first reported by Bartnicki-Garcia and Lippman (1966), who used the technique to study cell wall morphogenesis of *P. cinnamomi*. Layton and Kuhn (1988a) developed a method for producing and maintaining protoplasts of *P. sojae* and *P. medicaginis* obtained from zoospores and/or minced mycelium. Protoplasts are derived from encysted zoospores or mycelial fragments from which the cell walls have been removed by enzymatic digestion.

Protoplasts tend to form after the protoplasm squeezes out through holes in the cell wall and pinches off to form spheres. The nuclear content of individual protoplasts can vary from none to several per protoplast; however, those without nuclei do not form cell walls or subsequent colonies (Kuhn 1991).

To produce protoplasts, zoospores or mycelial fragments must be treated with suitable enzymes that will remove the cell wall without damaging the cell membrane. Also, since removal of the cell wall deprives the membrane of rigidity, the fluid in which the naked *Phytophthora* protoplasts are incubated must be isotonic (i.e., the osmotic pressure of the bathing medium is the same as the osmotic pressure of the internal protoplasm) so that the cell will not absorb water and burst. Since the methodology is complex and the technology is relatively new for *Phytophthora,* the reports by Layton and Kuhn (1988a) for information on protoplast fusion of *P. sojae,* Wagner and Wilkinson (1993) for *P. sojae* and *P. medicaginis,* and Lucas et al. (1990a) for *P. capsici* should be consulted.

In principle, either rapidly growing mycelium or germinated zoospore cysts can be treated with an enzyme such as Drieselase (a crude preparation of laminarase, xylanase, and cellulase) in an isotonic solution of 1 M mannitol and 7 mM $MgSO_4$ (Layton and Kuhn 1988a). In the report of Wagner and Wilkinson (1993), protoplasts from cysts of *P. sojae* were prepared after treatment with Meicelase (Meijii Seika Kaisha, Tokyo, Japan), Rhozyme (Genecor, South San Francisco, California), and NovoZym 234 (Novo BioLabs, Wilton, Connecticut) followed by treatment with an isotonic solution of $MgSO_4$ (0.1 M). Another method utilized incubation of protoplasts in a dithiothreitol solution prior to treatment with enzymes and a solution of sorbitol as an isotonic stabilizer. Viable protoplasts of *P. sojae* were most efficiently produced by digesting the cell walls with NovaZym 234 and cellulase from mycelium and from germinated cysts grown in broth media (Judelson et al. 1993).

P. sojae and the single-gene resistance of soybean is a model system that has received much attention in both biochemical research and practical agriculture (Chapters 6 and 59) (Athow 1987; Ward 1990). Understanding more about the genetics and physiology of the interaction of race-specific virulence of the pathogen and the evolution of new pathologic races is a major research goal (Keen and Yoshikawa 1983a). How new races that are pathogenic on currently used, resistant cultivars evolve in nature is not known. Is there a recombination of genetic factors within the fungus that results in these new races? Are new races the results of mutations, and, if so, how are they inherited? These questions are addressed in more detail in Chapter 6. Protoplast fusion as used by Layton and Kuhn (1988a) and Kuhn (1991) is becoming a new and profitable approach to the study of genetics of *Phytophthora.*

When protoplasts of a fluorotryptophan-resistant mutant of *P. sojae* were allowed to fuse with a metalaxyl-resistant mutant in a polyethylene glycol-$CaCl_2$ medium, colonies that were resistant to both compounds were rescued on agar plates amended with both fluorotryptophan and metalaxyl (Layton and Kuhn 1988a). Layton and Kuhn (1988a) stated that heterokaryons (cells containing genetically unlike nuclei) were present because from 2,554 colonies studied, they recovered 498 single-zoospore (zoospores are single-celled and uninucleate) colonies that were resistant to both metalaxyl and fluorotryptophan (Kuhn 1991). Although Layton and Kuhn (1988a) stated that heterokaryons had been isolated, according to Shaw and Shattock (1991, p. 221), the phenomenon should be called *somatic hybridization,* since a heterokaryon is defined as a cell containing unlike nuclei and single zoospores are uninucleate. Whatever the interpretation, it is significant that the marked genes for drug resistance can recombine in progeny of a protoplast cross. This work is significant in that it establishes that this phenomenon can be important in nature for the study of resistance of *Phytophthora* isolates to metalaxyl and in the development of new pathologic races.

The protoplast-fusion technique has become useful for study of the mode of inheritance of virulence and avirulence factors in homothallic species in which A1 and A2 types do not exist. Protoplast fusion provides a method of generating hybrids. By use of protoplast technology, Layton and Kuhn (1988b) crossed a race 1 heterokaryon with avirulence to the soybean cultivar Harosoy 63 with a race 3 heterokaryon that was virulent to Harosoy 63. They obtained progeny (22 of 26) that were of the race 1 phenotype (avirulent on Harosoy 63). This proved that the gene for avirulence was dominant. They interpreted the data to support Flor's (1956) gene-for-gene hypothesis.

The protoplast technique can be useful to show genetic relationship or nonrelationship of *Phytophthora* species. Protoplasts of *P. sojae* and *P. medicaginis,* both of which had at one time been classified as formae speciales of *P. megasperma,* did not fuse (Layton and Kuhn 1988a).

Zoospore Fusion

Swimming zoospores are natural protoplasts because they do not have cell walls. To slow the activity of zoospores and prevent encystment, a lithium salt was added to an incubation solution containing polyethylene glycol (Érsek et al. 1991). This allows fusion of zoospores to occur naturally (Figure 3.12). Replacement of lithium with calcium and potassium ions allows the zoospores to regenerate (Figure 3.13). Érsek et al. (1993) showed that interspecific hybridization had occurred among *P. parasitica, P. capsici,* and *P. citrophthora.* This might indicate a genetic relationship. *P. parasitica* was resistant to fluorophenylalanine, *P. capsici* to metalaxyl, and *P. citrophthora* to geneticin. Since double drug resistance from

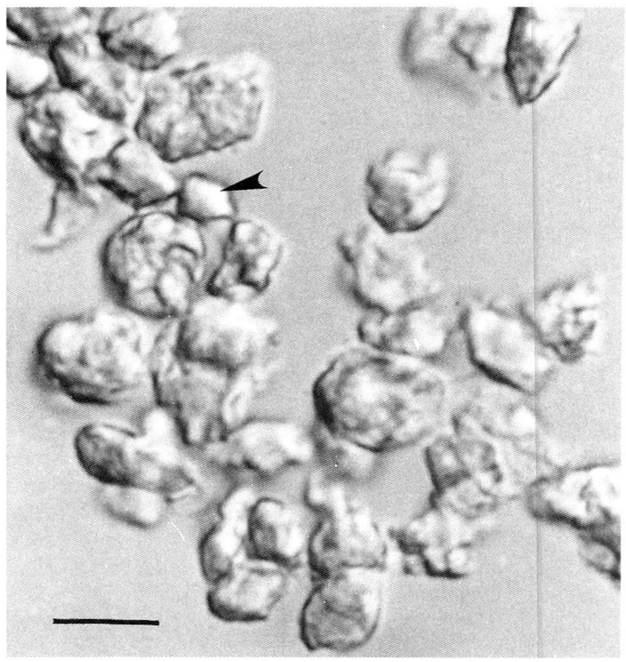

Figure 3.12. Fusion of zoospores of *Phytophthora.* Zoospores of *P. parasitica, P. capsici, P. citrophthora,* and *P. infestans* are capable of fusing when incubated in a polyethylene glycol-lithium-chloride (PEG-Li^+) solution, which markedly slows the activity of zoospores. The arrow indicates an unfused single zoospore. Bar = 15 µm. (Reproduced courtesy T. Érsek, The Plant Protection Institute, Budapest, Hungary, and J. T. English and J. S. Schoelz 1993, Department of Plant Pathology, University of Missouri, Columbia)

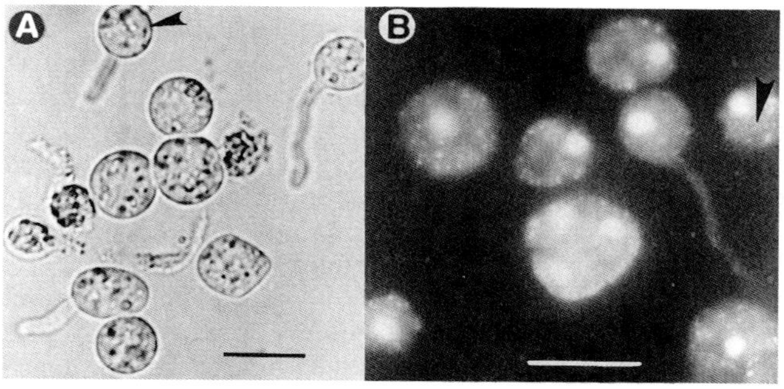

Figure 3.13. A, Zoospores of *Phytophthora parasitica* regenerate after fusion when PEG-Li$^+$ is replaced by Ca^{2+} and K$^+$ ions in the incubation solution. Arrow indicates an unfused single zoospore; regenerated fused zoospores are larger. **B,** Zoospores are stained with DAPI (4-6-diamidino-2-phenylindole, a stain specific for DNA) and photographed in ultraviolet light to show the nuclei. Arrow indicates a zoospore that presumably has not fused; regenerated, fused zoospores are larger and may include a larger nucleus or more than one nucleus. Bar = 15 µm. (Reproduced courtesy T. Érsek, The Plant Protection Institute, Budapest, Hungary, and J. T. English and J. S. Schoelz 1993, Department of Plant Pathology, University of Missouri, Columbia)

inter- and intraspecific crosses was identified on media containing the drugs, it was assumed that heterokaryosis (different nuclei in the same cell or cytoplasm) had occurred. Regenerated zoospores of *P. parastica* that had fused were larger than unfused zoospores (Figure 3.13A), and staining with DAPI showed that fused zoospores had more than one nucleus or larger nuclei (Figure 3.13B) (Érsek et al. 1993).

Hybrid cultures from zoospore fusion between isolates of *P. nicotianae* (*P. parasitica*) and *P. capsici* had modified host ranges compared with the parent cultures (Érsek et al. 1995). Hybridity was proved by detection of species-specific DNA sequences (Érsek et al. 1994a) of each parent culture in the hybrid zoospore cultures. While fusion of zoospores in nature is highly unlikely, this novel method will be useful for studies involving genetic analysis of pathogenesis in *Phytophthora* species (Érsek et al. 1995).

Genetic Transformation of *Phytophthora*

Development of efficient DNA-transfer systems in *Phytophthora* is important for study of the nature and inheritance of genes that govern host specificity and virulence or avirulence that may aid in constructing new strategies in the development of host resistance. Transformation in fungi was reviewed by Fincham (1989). Kinghorn et al. (1991) reviewed the current research on attempts to transform *Phytophthora* species and predicted that the development of gene transfer systems would require 1) a protoplasm and regeneration system; 2) a selectable marker; and 3) a functional promoter and terminator. Since their review, considerable progress has been made. Transformation of DNA in *Phytophthora* has only recently been accomplished (Bailey et al. 1991; Judelson and Michelmore 1991; Judelson et al. 1992).

The first report of a successful transformation system utilizing isolates of *P. capsici* (P1314) and *P. parasitica* (P991) (both from the M. D. Coffey collection at the University of California, Riverside) was by Bailey et al. (1991). They developed protoplasts by the method of Layton and Kuhn (1988a). Both *P. parasitica* and *P. capsici* were transformed to hygromycin B resistance by use of plasmids pCM54 and pHL1, which carried the bacterial hygromycin B phosphotransferase gene (*Hpt*) fused to promoter elements of the *Ustilago maydis* heat shock *Hsp70* gene that serves as a marker for isolation of transformed colonies.

Judelson et al. (1991) were the first to show that *P. infestans* could be transformed by utilization of vectors that contained oomycete promoters (*Ham34* and *Hsp70* promoters of *Bremia lactucae*) (Judelson and Michelmore 1989, 1990) and terminators fused to a hygromycin B marker gene. The pCM54 vector from *U. maydis,* reported to be useful for *P. capsici* and *P. parasitica* by Bailey et al. (1991), was not successful for *P. infestans* or *P. sojae* (*P. megasperma* f. sp. *glycinea*) (Judelson et al. 1991, 1993a).

Promoters from the heat shock gene *Hsp70* and *Ham34* that had been cloned for transformation experiments with *B. lactucae* (Judelson and Michelmore 1989, 1990) were tested as promoters for introduction of DNA in *P. infestans* by Judelson and Michelmore (1991) and Judelson et al. (1991, 1992). They found that the *Ham34* and *Hsp70* promoters functioned well in *P. infestans,* but promoters from higher fungi (including those from *U. maydis*), plants, and animals were inactive (Judelson et al. 1991, 1992). A new procedure involving the use of DNA-cationic liposome complexes proved to be optimal for transformation (Judelson and Michelmore 1991).

Stable transformants of *P. infestans* were obtained from fusions between *B. lactucae* promoters and genes for resistance to the antibiotics G418 (geneticin), hygromycin, or streptomycin (Judelson et al. 1991). Judelson et al. (1992) report, "Oomycete promoters isolated from the *Ham34* and *Hsp70* genes of *Bremia lactucae* and the actin gene of *P. sojae* resulted in high levels of the β-glucuronidase gene in *P. infestans* and in *Achlya ambisexualis.*" Because promoters from higher fungi were inactive, they conclude that the transcriptional machinery of oomycetous fungi differs significantly from that of higher fungi. Highly efficient cotransformation was obtained by use of selected and nonselected vectors (reporter β-glucuronidase and *LacZ* genes) cleaved with the same restriction enzyme. Fragments were efficiently ligated in vivo by *P. infestans* and integrated into chromosomal DNA (Judelson 1993). Judelson et al. (1993b), using antisense RNA techniques, identified several additional promoters that are active in *P. infestans* for expressing genes in vivo and in planta and for inhibiting gene expression.

H. S. Judelson (*personal communication*) studied the A1 and A2 mating system of *P. infestans.* DNA markers that flanked closely to the A1 and A2 loci were selected. The nature of progeny from A1 × A2 crosses showed that the inheritance of mating type was not Mendelian, as had previously been proposed, and suggested that lethal genes were involved.

Genetic transformation of race 2 of *P. sojae* was accomplished by use of a vector containing a bacterial marker (*Escherichia coli*) gene for resistance to the antibiotic hygromycin B. The bacterial gene was fused to a promoter and terminator from the *Ham34* gene from *B. lactucae.* By hybridization to electrophoretically separated chromosomes, it was shown that vector DNA had integrated into only one chromosome in four transformants and into multiple chromosomes in one transformant (Judelson et al. 1993a).

Érsek et al. (1994b) reported on transformation of *P. capsici.* "Insertional mutagenesis of *P. capsici* zoospores by transformation with available plasmid vectors containing a gene for resistance to hygromycin-B (Judelson et al. 1991) resulted in mutants resistant to the corresponding antibiotic at a low (ca. 10^{-6}) frequency. The vector element could only be detected in a few hygromycin-resistant isolates with labeled plastic probes. Transformants exhibited almost no variation in colony morphology but their sporulation was retarded with no release of zoospores."

Concerning the potential of transformation in *Phytophthora,* Judelson et al. (1991) stated in their report on transformation of *P. infestans,* "Additional experiments indicated that the gene transfer procedure will be useful for studying *P. infestans* interactions in late blight. Transformants were obtained at similar frequencies using 3 different strains of *P. infestans,* suggesting that the technology will be applicable to most strains. Also, 1 randomly selected hygromycin-resistant transformant was tested for pathogenicity and found to be vigorously pathogenic on a tomato host, despite having experienced extended in vitro culture (>10 subcultures) and transformation procedures involving regeneration from protoplasts."

Summary

Some *Phytophthora* species grow on inorganic nitrogen sources, such as nitrate or ammonium ions; some grow only on one of the two; and others require an amino acid such as asparagine. Most species will grow on several carbohydrates, but glucose and sucrose are preferred sources. In addition to the usual inorganic chemicals commonly required for growth by most fungi, such as phosphate, sulfur, magnesium, zinc, potassium, and iron and minute amounts of manganese, molybdenum, and copper, calcium augments growth, but its role is not well understood. The intact thiamine (vitamin B_1) molecule is an absolute growth requirement. A source of the vitamin thiamine and a β-hydroxysterol such as sitosterol is required for production of sporangia, zoospores, chlamydospores, and oospores. Some sources of lecithin, especially soybean, enhance growth and, in combination with minute quantities of a β-hydroxysterol, will stimulate oospore formation in a synthetic medium. Synthetic media are described in Appendixes 3.51–3.54; however, most routine culturing of *Phytophthora* species utilizes organic media such as V8 juice, carrot, bean, oatmeal, or grain agars. These are described in Appendixes 3.46–3.50.

Phytophthora species can be stored on a suitable medium for long periods of time after being frozen (–196°C) in liquid nitrogen. Storage under mineral oil at 15°C is successful, but specimens must be transferred every 1 or 2 years. Disks of agar and mycelium can be stored in sterile distilled water or on sterile corn seeds in distilled water for well over 1–2 years.

Sporangia, particularly of foliar *Phytophthora* pathogens, form directly on suitable substrates, but most soilborne *Phytophthora* species form sporangia only when the substrate is placed in distilled water, soil extract, or certain dilute salt solutions. In general, *Phytophthora* species will sporulate when sugars are depleted. Most sporangia-inducing procedures allow soluble sugars to be washed away or diluted from the substrate on which the culture is growing. Solutions that stimulate sporangium formation are described in Appendixes 3.24–3.46. Light stimulates sporangium formation of most foliar and some soilborne *Phytophthora* species but is inhibitory to *P. infestans.*

Sporangia germinate by production of germ tubes (usually at optimal temperatures for growth) or by production of many zoospores that emanate from each sporangium at reduced temperatures. After the plug at the tip of the sporangium dissolves, zoospores are expelled into the thin membrane of the sporangium, which expands into the aqueous medium. When the membranous sac breaks, the biflagellate zoospores swim away. In nature, the zoospores are chemotactically attracted to roots, especially at the zone of elongation or to any region at which the maximum amount of root exudation occurs. Zoospores lack cell walls and thus are sensitive to osmotic effects in aqueous solutions. Zoospores are negatively geotactic (swim toward the surface) and are attracted to positively charged substances such as certain cationic amino acids. Electrophoretically, they move toward the anode in a weak electrical field. Zoospores possess two flagella of unequal length, one, a long whiplash and the other, a shorter, tinsel type. Zoospores swim for extended lengths of time (1–84 h) depending on the environmental conditions. Artificial agitation or natural movement of a zoospore against a solid barrier triggers the encystment process, which is characterized by loss of the two flagella and rapid formation of a cell wall. After the zoospore encysts, it germinates by production of a germ tube or occasionally by production of a second zoospore (diplanetism).

Chlamydospores are produced in cultures of many but not all species. These globose spores have walls thicker than those of sporangia but thinner than those of oospores and tend to persist longer than sporangia. Chlamydospore production by some species, for instance, *P. parasitica,* is enhanced by incubating well-nourished mycelial mats in water several centimeters in depth. The effect could be caused by development of high CO_2 levels in deep cultures. The germination of chlamydospores is dependent on exogenous sources of amino acids. Chlamydospores of *P. parasitica* become dormant in cold soils, and germination is enhanced by exposure to high temperatures.

Oospores are thick-walled sexual structures that form after the antheridium (male) and oogonium (female) fuse, nuclei from the antheridium flows into the oogonium, and antheridial and oogonial nuclei fuse in the oogonium. The oospores of homothallic (self-fertile) species form in agar or liquid media and in infected plant issue. Species that are heterothallic (self-sterile) form oospores after A1 and A2 mating types have grown together, either in a suitable medium or in an infected plant. Oospore progeny from pairing A1 × A2 cultures can be the result of hybridization or selfing from the production of hormone- or pheromone-like substances that stimulate the spore to form without hybridization.

Oospore production is suppressed by light and favored by culturing in the dark. Prior to germination, the oospore requires a short period of light, preferably in the blue region of the spectrum. Oospores usually mature within a 30-day incubation period. Oospores can become dormant. When an oospore germinates, it produces either a germ tube or a sporangium, which in turn produces either mycelium or zoospores.

Meiosis occurs in nuclei in the antheridium and oogonium, and the nuclei that come together in the oogonium to form the oospore are haploid. This is the only haploid stage in the life cycle. After fusion of the oogonial and antheridial nuclei, the oospore and the germ tube that emanates from it are diploid. Chromosomes in gametangia are extremely small but can be observed within squashed oogonia stained with aceto-orcein or other nuclear stains. In most species, the chromosomes equal about 10, but because of their small size, estimates vary from eight to 10 to nine to 12. Some isolates of *P. infestans* are polyploid. There is evidence that both mating types of the heterothallic species are reciprocal translocation heterozygotes because multiple associations of four chromosomes are always present, whereas such associations are not found in homothallic species.

Oospores can be freed from mycelium by mincing cultures in a blender and separating the heavier oospores from the lighter mycelium by several centrifugations. Oospores are sometimes fed to snails and collected from the excrement, washed by centrifugation, and surface-decontaminated by use of mercuric chloride (30 ppm) to separate them from mycelium, which is readily digested by snail gut enzymes. Certain crude enzymes, such as the snail gut enzymes Glusulase, Helicase, and NovoZym, are capable of digesting sporangia and mycelium and oogonium wall but not oospores.

The viability of oospores can be tested by use of tetrazolium bromide, which produces a red color in response to succinic dehydrogenase activity in the viable oospore, or by the oospore's ability to plasmolyze in 4 M NaCl (Appendixes 3.58 and 3.59). The ooplast in viable, but not in nonviable, oospores will contract (plasmolyze) and rehydrate when the NaCl is replaced with water.

Oospores germinate on agar or in an aqueous solution. In some cases, a shallow layer of an aqueous solution promotes better germination than agar. In some species, the percentage of germination in water is especially low in A1 × A2 crosses mainly because of the genetic incompatibility of the parents. In most cases, the percentage of germination of well-washed oospores is enhanced by incubating them in soil extract, root extract, root exudates, soil extract, or amino acid solutions (10^{-1} to 10^{-5} mM). Treatment of oospores with the oxidant $KMnO_4$ for short periods of time (about 10 min) is stimulatory to germination in water. $KMnO_4$ also kills mycelial fragments, sporangia, zoospores, and chlamydospores. Incubation of oospores at 2°C induces dormancy, which can be overcome by incubation in soil extract, root exudate, or amino acid solutions. Some progeny from oospores (paired A1 and A2 isolates) are selfed because of cross stimulation by hormonelike substances, but true hybridization also occurs in progeny of paired A1 × A2 cultures. In *P. infestans,* the majority of the progeny from paired A1 and A2 isolates are hybrids.

Development of protoplasts by dissolution of the cell walls of propagules of *Phytophthora* in isotonic solutions has allowed genetic crosses of isolates without recourse to the sexual process. Isolates were obtained that incorporated resistance to two toxic drugs in a uninucleate, single-zoospore isolate. This demonstrates that heterokaryosis or somatic hybridization can occur and might be a means for natural development of new degrees of virulence. When protoplasts of *P. sojae* with a gene for avirulence were crossed with protoplasts with a gene for virulence, avirulence was found to be dominant.

Transmission of DNA in a plasmid with resistance to hygromycin into protoplasts from one isolate to another (transformation) was shown for the first time in *Phytophthora* with *P. parasitica* and *P. capsici.* Transformants with resistance to hygromycin retained their pathogenic fitness. Transformation of genes for antibiotic resistance has been accomplished for *P. infestans* and *P. sojae* by using promoters from *B. lactucae.* Transformation technology will become useful for studies of the genetics of parasitism on host plants.

Spore production and germination by *Phytophthora* are processes that are intimately responsive to many ecological factors such as temperature, moisture, and nutrients. In a chapter on major epidemics caused by *Phytophthora,* the late Philip Gregory (1983), a famous pioneer epidemiologist from Rothamsted Experimental Station in England, characterized *Phytophthora* by use of an analogy to ammunition in time of war (excerpted by permission of the American Phytopathological Society, St. Paul, Minnesota, courtesy of Gregory 1983):

> The genus *Phytophthora* clearly possesses some of the most sophisticated weaponry known among plant pathogens. The mycelium is an active undercover communications network. Oospores and perhaps chlamydospores may be likened to time bombs, programmed for activation over widely different periods. The sporangium is a multipurpose weapon capable of emitting a flotilla of 20 or so minisubmarines, each oriented by gravity and able to home in on a chemical target against which it is equipped for biological warfare.

To extend Gregory's analogy, *Phytophthora,* like a powerful enemy, has an informational code that has puzzled the most formidable resources known. It produces new pathologic races in rapid response to plant breeders' incorporation of genes for resistance and also can become resistant to fungicides.

The recent breakthroughs in understanding how to analyze DNA patterns and to use isozymes as unique markers and DNA probes that will identify known genotypes have given the code breakers in this war tools that are beginning to explain how *Phytophthora* can adapt to adverse circumstances. DNA technology has also been shown to be of prime importance in settling identification problems (see also Chapter 4).

As in war, plant pathologists have long attempted to suppress this enemy, *Phytophthora,* but like most generals, they greatly respect the prowess of their enemy! A fuller understanding of how *Phytophthora* produces its ammunition and the code by which it gives instruction to its progeny will allow development of more efficient control measures for diseases that have for too long cost the world too much (billions of dollars) in the value of its food and fiber.

APPENDIX TO CHAPTER 3

Growth Media and Methods

Growth of *Phytophthora*

Natural Media

3.1. V8 Juice Agar or Broth (P. M. Miller 1955)

V8 juice (Campbell Soup Co.) (200 ml) is diluted to 1 liter with deionized distilled water and mixed with 2 or 3 g of $CaCO_3$ for several minutes with a magnetic stirrer or similar device. Agar (15 g) is added, and the medium is autoclaved at 121°C for 20 min. The flask is agitated while the medium is dispensed to ensure uniform dispersion of $CaCO_3$. A broth can be made by omitting the agar. The pH is buffered at 7.0 with 2 g and at 7.5 with 3 g of $CaCO_3$ per liter. There are several variations of this recipe in the literature.

3.2. V8 Juice Agar or Broth (Clarified)

V8 juice (354 ml) is buffered with 5 g of $CaCO_3$, mixed for several minutes with a magnetic stirrer, and centrifuged for 20 min at 4,000 rpm; the clear solution is decanted and diluted (1:4) with deionized distilled water, autoclaved at 121°C, and dispensed. For solid media, 15 g of agar per liter is added to the broth before autoclaving. In some cases, the centrifuged V8 juice is further clarified by filtration through filter paper. Although improved clarity allows better microscopic observation of *Phytophthora,* removal of particulate material may also decrease nutrients such as sterols (Englander and Turbitt 1979).

3.3. V8 Juice-Vermiculite Medium for Soil Infestation (Varney 1961; Ristaino et al. 1988; Ristaino and Duniway 1989)

Vermiculite (Terralite) can be amended with a liquid broth medium, such as V8 juice, lima bean, or potato-dextrose (3 parts Vermiculite to 1 part liquid), sterilized at 121°C, inoculated with *Phytophthora,* and incubated 1–2 weeks prior to use as inoculum in soil. To infest soil in the field with *P. parasitica,* the medium is made with 500 ml of Vermiculite to 250 ml of V8 broth medium and incubated for 4 weeks. The inoculum is incorporated into soil at a rate of 500 cm^3 per 1.2 m of row placed laterally about 20 cm on both sides of the row of tomato plants. To augment sporulation, the Vermiculite-V8 broth medium may be washed prior to use to remove nutrients (Browne and Mircetich 1988).

3.4. Potato-Dextrose Agar

Potatoes (200 g) are washed and sliced (with peels), autoclaved or steamed until well cooked in 500 ml of distilled water, and filtered through a layer of muslin cloth. Glucose (dextrose) (20 g) and agar (15 g) are added prior to autoclaving at 121°C for 20 min. This medium is probably the most commonly used in plant pathology; however, because of its high carbohydrate concentration, it is not recommended as the optimal medium to support asexual or sexual reproduction or for isolation of *Phytophthora* from septic plant tissue. Addition of wheat germ oil (0.5 ml/liter) augments sporulation (Ainsworth 1968). Patterns of growth on potato-dextrose agar are often quite characteristic of certain species such as *P. citrophthora,* which produces a petaloid colony, and *P. parasitica,* which produces an arachnoid (spider web) type of colony. *P. sojae* will not grow on potato-dextrose agar (Kaufmann and Gerdemann 1958; Hildebrand 1959; Hendrix 1964).

3.5. Oatmeal Agar

Oatmeal (50–60 g), amended with 17 g of agar, is autoclaved in 1 liter of distilled water for 15 min at 121°C. If a less viscous medium is needed, the oatmeal can be strained in a muslin bag. Addition of 0.5 ml of wheat germ oil augments growth of *Phytophthora* and *Pythium* (Ainsworth 1968). Oatmeal agar is an excellent substrate for all *Phytophthora* species; however, it is viscous and inconvenient to dispense (Savage et al. 1968).

3.6. Oat Grain Medium for Soil Infestation

Nonhulled grains are soaked in distilled water 2–3 h or longer. The free water is poured off, and the grain is autoclaved 30 min at 121°C. To eliminate spore-forming bacteria, autoclaving should be repeated 1 day later at 121°C. The medium in jars or wide mouth flasks is inoculated with a mycelial plug or an aseptic mycelial suspension and incubated for 10–30 days. Shaking the cultures periodically serves to redistribute the inoculum and aerate the medium. Milo and millet grain can also be used in a similar manner to produce inoculum for soil infestation (Royle and Hickman 1964).

3.7. Carrot Broth or Agar

Sliced, fresh carrots (200 g) are comminuted in a blender in 500 ml of distilled water for 40 s at high speed and filtered through four layers of cheesecloth. The juice is then squeezed from the residue. If necessary, the medium can be clarified by centrifugation. The filtrate and 15 g of agar are brought to 1 liter with distilled water and autoclaved at 121°C for 15 min. This medium is useful for oospore production of many *Phytophthora* species, particularly *P. palmivora.* A clearer medium is obtained when carrots are sliced but not comminuted prior to autoclaving (Brasier 1967; Kaosiri et al. 1978, 1980; Brasier and Griffin 1979). Carrot broth supports production of oospores of *P. medicaginis* (Förster et al. 1983; El-Hamalawi and Erwin 1986a, b).

3.8. Cornmeal Agar

Cornmeal (yellow) (60 g/liter) is autoclaved at 121°C for 15 min and filtered. The filtrate is dispensed as a broth or solidified with agar (17 g/liter) (Savage et al. 1968). It is not a rich medium and is used routinely for isolation media (Eckert and Tsao 1960, 1962; Tsao 1983). This medium was used by Waterhouse (1963) for growth of isolates included in a taxonomic key.

3.9. Rye Seed Agar (Including Rye A and Rye B) for *Phytophthora infestans*

Rye seed agar (Snieszko et al. 1947) is superior to other media for growth and sporulation of *P. infestans.* Some isolates will not grow on V8 juice agar but will grow on rye agar. See Hohl (1991a) for a review of media favorable for *P. infestans.* Rye seeds (100 g) are steamed in 500 ml of water for 30 min and filtered through cheesecloth and filter paper pulp. The filtrate is diluted with water and amended with 5 g of glucose. Hodgson and Grainger (1964) modified the medium by soaking 50 g of rye seeds in 1 liter of water, comminuting the seeds with a Waring blender for 3 min, adding 10 g of glucose and 15 g of agar, steaming to melt the agar, and autoclaving. *P. infestans* sporulates in 1 week at 20°C. In a modification by Caten and Jinks (1968) (rye A), 60 g of rye grains are soaked in distilled water at room temperature for 36 h, after which the supernatant is retained. The soaked rye grain is comminuted, heated for 3 h at 50°C in distilled water, and filtered through several sheets of muslin cloth. The filtrate is added to the original filtrate, amended with 20 g of sucrose and 15 g of agar, brought to 1 liter with distilled water, and autoclaved. Rye B agar, which is a clearer medium than rye A, is similar except that the soaked (36 h) rye seeds are boiled gently for 1 h in distilled water but are not comminuted. The filtrate is amended with glucose and agar as above.

3.10. Charcoal Amendment in *Phytophthora* Growth Media (Parmentier 1970)

Activated charcoal (1–10%) often improves growth of *Phytophthora* because it absorbs impurities. It does not prevent the production of sporangia and oospores.

Synthetic Media

3.11. Medium of Lin and Liang (1965) and Lin et al. (1982) for *Phytophthora infestans*

The basal medium contains (g/liter): glucose (20), $(NH_4)_2SO_4$ (1.3), KH_2PO_4 (0.5), K_2HPO_4 (0.5), $MgSO_4{\cdot}7H_2O$ (0.5), $Fe_2(SO_4)_3$ (0.004), $ZnSO_4$ (4×10^{-4}), $MnSO_4$ (4×10^{-4}), malic acid (2.7), $CaCl_2$ (1.0), thiamine (0.001), and distilled water to 1 liter. The stock solutions of glucose, phosphates, and thiamine are sterilized separately by filtration. The medium is adjusted to pH 5.6–5.8 before autoclaving. The report by Lin et al. (1982) states that this medium supports growth of *P. boehmeriae, P. colocasiae, P. palmivora, P. parasitica,* and *P. parasitica* var. *nicotianae.* Both malic acid and Ca^{2+} are critical ingredients.

3.12. Media of Hohl (1975, 1983, 1991a) and Henniger (1959) for *Phytophthora infestans* and Other Species

Medium P-1 of Hohl. The medium contains (g/liter) sucrose (30.0), asparagine·H_2O (0.130), $NaNO_3$ (3.0), KH_2PO_4 (1.0), $Na_2HPO_4{\cdot}2H_2O$ (0.35), $MgSO_4{\cdot}7H_2O$ (0.5), KCl (0.5), $CaCl_2{\cdot}2H_2O$ (0.1), $FeSO_4{\cdot}7H_2O$ (0.01), $FeCl_3$ (0.001), $ZnCl_2$ (0.001), ethylenediaminetetraacetic acid (0.01), thiamine·HCl (0.0005), guanine (0.05), histidine·HCl (0.190), serine (0.05), leucine (0.065), cysteine (0.06), agar (15.0), cholesterol or stigmasterol in ether (0.05), and double distilled water to 1 liter. Purine and pyrimidines are ground with a mortar and pestle before they are added to the medium. Lecithin (L) (0.2) can be added to this medium (P-1L) and any of the following media if necessary and is homogenized with 5–10 ml of the medium with a Virtis high-speed homogenizer or in a Braun-Sonic ultrasonic homogenizer. The medium is useful for culture of *P. infestans* (Hohl 1983, 1991a).

Medium P-3 of Hohl. The medium contains (g/liter) glucose (20.0), L-asparagine (1.0), K_2HPO_4 (0.5), KH_2PO_4 (0.5), $MgSO_4{\cdot}7H_2O$ (0.5), $CaCl_2{\cdot}2H_2O$ (0.02), thiamine (0.001), agar (15.0), and distilled water to 1 liter.

Medium P-4 of Hohl. The medium is the same as P-3 except that $NaNO_3$ (1.5 g/liter) is substituted for asparagine.

Medium P-5 of Hohl. Medium P-5 is the same as P-3 except that NH_4Cl (1.0) is substituted for asparagine.

Medium of Henniger. The medium contains (g/liter) KH_2PO_4 (0.2), $NaNO_3$ (0.5), $CaCl_2$ (0.2), $MgCO_3$ (0.15), $(NH_4)_2SO_4$ (0.05), $FeSO_4{\cdot}7H_2O$ (0.01), tartrate (0.2), succinate (0.2), α-ketoglutarate (0.4), L-glutamic acid (0.3), DL-alanine (0.3), L-aspartic acid (0.2), L-leucine (0.2), π-aminobutyric acid (0.2), tryptophan (0.08), cystein·H_2O (0.2), arginine (0.1), dextrose (glucose) (10.0), and thiamine·HCl (0.0002).

According to Hohl (1991a), medium P-1 and Henniger's medium (1959) have been successfully used to culture fastidious *P. infestans* isolates. The Henniger medium utilizes organic acids, and the Hohl P-1 medium emphasizes use of lipids as nutrients. The amino acid cystein in the Henniger medium may act as a reducing agent.

3.13. Complex Amino Acid Medium of Davies (1959) for *Phytophthora fragariae*

The medium is composed of (g/liter) glucose (5.0), KH_2PO_4 (0.5), $MgSO_4{\cdot}7H_20$ (0.25), thiamine (0.001), Fe (2×10^{-4}), Zn (1.6×10^{-4}), Cu (4×10^{-5}), Mn (2×10^{-5}), and Mo (1.4×10^{-5}). The nitrogen source consists of a complex of amino acids based on analyses of casein hydrolyzate: (mg/liter) glycine (60), L-lysine (180), DL-serine (140), L-cystine (8), DL-valine (160), DL-leucine (200), L-proline (250), DL-alanine (70), DL-phenylalanine (110), DL-methionine (60), L-glutamic acid (500), DL-aspartic acid (160), DL-threonine (1.0), DL-tyrosine (140), L-histidine (70), L-arginine (90), and DL-iso-leucine (140). Ca^{2+} is added as a chloride at 40 mg/liter. The carbohydrate and Ca^{2+} are autoclaved separately from the other components and combined when cool to prevent precipitation of Ca^{2+} salts. The medium is adjusted to pH 6.7 with NaOH or HCl. The medium (20 ml in 100-ml conical flasks) is inoculated with homogenized mycelium (10 s in a blender) equilibrated to 1 mg (dry weight) per milliliter of medium. No growth occurs in the absence of Ca^{2+}. Growth on single amino acids such as L-asparagine, DL-aspartic acid, DL-alanine, and L-proline is nearly as good as on the complex amino acids in casein hydrolyzate. No growth occurs on inorganic nitrogen (NH_4^+ or NO_3^-).

A semisynthetic medium (HC) is made by substituting vitamin-free casein hydrolyzate (2.5 g/liter) for amino acids.

Tariq (1990) utilizes a modification of Davies's (1959) semisynthetic medium HC for production of sporangia. Maximum numbers of sporangia of *P. fragariae* are produced by incubating 11- or 15-mm-diameter disks from the margin of 15- to 20-day-old colonies on HC medium in 6 ml of sterile glass-distilled water in a 4.5-cm-diameter petri dish at 13°C or 22°C. Light does not affect sporangia formation. Addition of dilute nutrient solution increases sporangia formation on one isolate but decreases sporangia formation on another.

The medium contains (g/liter) glucose (5.0), casein hydrolyzate (2.5), KH_2PO_4 (0.5), $MgSO_4{\cdot}7H_2O$ (0.25), $CaCl_2{\cdot}H_2O$ (0.053), thiamine·HCl (0.001), $ZnSO_4{\cdot}7H_2O$ (1.46×10^{-4}), $CuSO_4{\cdot}5H_2O$ (7.9×10^{-4}), $MnSO_4{\cdot}4H_2O$ (4.1×10^{-4}), H_3BO_3 (2.9×10^{-4}), molybdic acid (1.8×10^{-4}), ferric citrate (4.39 ×

10^{-3}), sitosterol (0.020), and Oxoid agar no. 1 (15). Thiamine and trace elements are added as stock solutions. Sitosterol is dissolved in chloroform (1 ml) before it is added to the medium. The chloroform boils off during autoclaving at 121°C for 15 min.

3.14. Medium of Payette and Perrault (1944) for *Phytophthora infestans*

The medium contains (g/liter) glucose (5), glycine (1), K_3PO_4 (1), Na_2HPO_4 (1), $MgSO_4$ (1), NH_4NO_3 (1), ammonium tartrate (1), asparagine (0.5), fumaric acid (0.25), succinic acid (0.25), and $CaCl_2$ (0.02). Boric acid, ferrous chloride, and manganese sulfate are added at a concentration of 1 ppm. The pH is 6.2. Thiamine, added at various concentrations from 0.0001 to 4.0 mg/liter, optimally supports growth at 0.2–0.4 mg/liter.

3.15. Medium of Lopatecki and Newton (1956)

The medium contains (g/liter) glucose (50), asparagine (4), KH_2PO_4 (1), $MgSO_4$ (0.1), $CaSO_4$ (0.1), $FeSO_4$ (0.001), thiamine (0.02), and a stock solution of trace minerals that provides (μg/ml in the prepared medium) zinc (1), molybdenum, copper, and manganese (0.02). A table showing dry weights of mycelium resulting from variations of this medium for *P. parasitica, P. cactorum, P. megasperma,* and *P. erythroseptica* is reproduced in Ribeiro (1978). DL-Alanine (4 g/liter) supports growth nearly as well as asparagine.

Modifications of the Lopatecki and Newton (1956) medium are reported by Fothergill and Child (1964) for *P. infestans, P. erythroseptica,* and *P. parasitica.* Medium A contains (g/liter) glucose (25), *dl*-asparagine (4.0), K_2HPO_4 (1.0), $MgSO_4{\cdot}7H_2O$ (0.1), $CaSO_4{\cdot}2H_2O$ (0.1), $FeSO_4{\cdot}7H_2O$ (0.001), and thiamine (0.8). The trace elements are amended (μg/ml) per Lopatecki and Newton (1956). The pH is 4.7. Omission of $CaSO_4{\cdot}2H_2O$ decreases growth of *P. erythroseptica* by 70%, *P. infestans* by 50%, and *P. parasitica* by only 7%. Physiological balance of concentrations of the phosphate, $MgSO_4$, and $CaSO_4$ components is important. Optimal concentrations for *P. erythroseptica* (g/liter) are K_2HPO_4 (0.018), $MgSO_4{\cdot}H_2O$ (8×10^{-4}), and $CaSO_4{\cdot}2H_2O$ (4.8×10^{-3}) (yield 237 mg [dry weight] per 20 ml of medium); and for *P. parasitica* K_2HPO_4 (0.018), $MgSO_4{\cdot}7H_2O$ (0.0016), and $CaSO_4{\cdot}2H_2O$ (0.0024) are optimal for growth (191 mg of mycelium [dry weight] per 20 ml of medium). This is one of the few reports on optimizing the nutritional balance of compounds.

3.16. Medium of Erwin and Katznelson (1961a), Modified from Lopatecki and Newton (1956) for *P. medicaginis*

The medium contains (g/liter) sucrose (10–15), L-asparagine (2.0), $FeSO_4{\cdot}7H_2O$ (0.001), KH_2PO_4 (0.43), K_2HPO_4, (0.30), $MgSO_4{\cdot}7H_2O$ (0.1), $CaCl_2{\cdot}2H_2O$ (0.010), thiamine·HCl (0.25–1.0), minor element solution (1 ml), which provides 1 ppm of Zn^{2+}($ZnSO_4{\cdot}7H_2O$) and 0.02 ppm of Cu^{2+}($CuSO_4{\cdot}5H_2O$), Mo^{6+}($NaMoO_4{\cdot}2H_2O$), and Mn^{2+}($MnCl_2{\cdot}4H_2O$), and distilled water to 1 liter. If precipitation occurs, phosphates should be autoclaved separately and combined with other components when cooled.

For growth of *P. capsici, P. citrophthora,* and *P. palmivora,* the above medium is modified by the addition of 0.03 g of β-sitosterol (in 50 ml of dichloromethane) and 5.3 g of 2-(*n*-morpholino)-ethanesulfonic acid (MES), a nonmetabolized buffer (Mitchell and Zentmyer 1971a).

3.17. Medium of Bartnicki-Garcia (1966)

The medium contains (g/liter) glucose (20), L-asparagine (2.0), KH_2PO_4 (3.0), $MgSO_4{\cdot}7H_2O$ (0.5), $CaCl_2$ (3.4×10^{-3}), $FeSO_4{\cdot}7H_2O$ (0.001), citric acid (1.4×10^{-3}), $ZnSO_4{\cdot}7H_2O$ (1.8×10^{-3}), $MnSO_4{\cdot}H_2O$ (3×10^{-4}), $CuSO_4{\cdot}5H_2O$ (4×10^{-4}), $(NH_4)_6Mo_7O_{24}{\cdot}4H_2O$ (3×10^{-4}), thiamine·HCl (0.001), and distilled water to 1 liter. The pH is adjusted to 6.0 with KOH. Glucose is autoclaved separately from other ingredients at 121°C for 20 min.

3.18. Medium of Cuppett and Lilly (1973) for Growth of *Phytophthora infestans*

The medium contains (g/liter) D-glucose (5), L-asparagine (anhydrous) (0.5), KH_2PO_4 (1), $MgSO_4{\cdot}7H_2O$ (0.5), Zn^{2+} (2×10^{-4}), Ca^{2+} (0.01), Cu^{2+} (2×10^{-4}), Mn^{2+} (5×10^{-4}), Mo^{6+} (1×10^{-5}), β-sitosterol (0.01), thiamine (1.0), ascorbic acid (0.2), ferric iron (1.0 ppm), and deionized distilled water to 1 liter. Ascorbic acid maintains iron in a reduced (ferrous) state. The *P. infestans* isolates tested are sensitive to the lack of iron in the ferrous state. Isolates vary in their ability to reduce ferric to ferrous iron, and long lag periods of growth can result.

3.19. Medium of Hendrix et al. (1969) for Growth of *Phytophthora parasitica* var. *nicotianae*

The medium contains (g/liter) glucose (21.6), L-asparagine (5.8), KH_2PO_4 (0.5), $CaCl_2{\cdot}2H_2O$ (0.100), ferric iron as chloride or sequestrene (sodium ferric diethylenetriamine pentaacetate) (0.001), Zn^{2+} (0.001), and (μg/liter) Cu^{2+} (10), Mn^{2+} (10), Mo^{6+} as $(NH_4)_6Mo_7O_{24}{\cdot}4H_2O$ (10), EDTA (ethylenediaminetetraacetic acid disodium salt) (0.0108) and distilled water to 1 liter. Utilization of carbohydrate (economic coefficient) exceeds 45% (up to 63%), indicating that the medium is highly efficient. The report shows that sterols, Ca^{2+}, and shaking are important factors affecting growth. The greatest amount of mycelium growth (562 mg per 50 ml of medium) occurred in the presence of EDTA plus Ca^{2+} and minor elements compared with the control of Ca^{2+} plus minor elements (148 mg).

3.20. Medium of Chee and Newhook (1965c) for Growth of *Phytophthora cinnamomi*

The medium contains (g/liter) sucrose (20), L-glutamic acid (2.5), KH_2PO_4 (1.0), $Ca(NO_3)_2$ (0.5), KCl (0.5), and $MgSO_4{\cdot}7H_2O$ (0.5) (dissolved separately and added last to minimize precipitation with Ca^{2+}), Fe-EDTA (ethylenediaminetetraacetic acid; iron chelate with a final concentration of 6 ppm Fe^{2+}) (0.6 ml), thiamine (0.005), cholesterol dissolved in ether (0.050), agar (12), a microelement solution (1 ml) consisting of (ppm in final solution) B^+ (as H_3BO_3) (0.5), Mn^{2+} (as $MnSO_4{\cdot}4H_2O$) (0.05), Zn^{2+} (as $ZnSO_4{\cdot}7H_2O$) (0.02), Cu^{2+} (as $CuSO_4{\cdot}5H_2O$) (0.02), Mo^{6+} (as $(NH_4)_6Mo_7O_{24}{\cdot}4H_2O$) (0.02), and distilled water to 1 liter. The pH is adjusted to 6.3–6.5 with 1 N NaOH.

3.21. Medium of Burrell et al. (1966) and Merz et al. (1969) for Growth of Several Species

The medium contains (g/liter) D-glucose (10), $(NH_4)_2SO_4$ (2), fumaric acid (2), $MgSO_4{\cdot}7H_2O$ (0.5), KH_2PO_4 (1), $FeNH_4(SO_4)_2{\cdot}12H_2O$ (0.002), $ZnSO_4{\cdot}7H_2O$ (0.002), $MnSO_4{\cdot}H_2O$ (0.01), thiamine·HCl (0.100), β-sitosterol (0.002), and distilled water to 1 liter. The pH is adjusted to 6.0.

The medium is also used to culture *P. cinnamomi, P. erythroseptica,* and *P. cactorum* for use as antigens to produce antibodies in rabbits. Flasks containing 25 ml of medium are inoculated with 10 mg (equivalent dry weight) of minced mycelium and incubated 3 days. Mycelium grown for 3 days has the optimal protein nitrogen content (10 mg/ml) to inject rabbits for production of antibodies. Concentration of protein nitrogen of 14-day-old cultures decreases to a range of 0.01 to 0.1 mg/ml (Burrell et al. 1966).

3.22. Casein Hydrolyzate Medium for *Phytophthora fragariae* (Maas 1972)

The medium contains (g/liter) sucrose (15), casein hydrolyzate, enzymatic (13% total N) (3.39), $MgSO_4 \cdot 7H_2O$ (0.1), KH_2PO_4 (0.43), K_2HPO_4 (0.3), $FeSO_4 \cdot 7H_2O$ (0.001), $CaCl_2 \cdot 2H_2O$ (0.01), $ZnSO_4 \cdot H_2O$ (0.001), $CuSO_4 \cdot 5H_2O$ (2×10^{-6}), and distilled water to 1 liter. Because of the addition of casein hydrolyzate, this medium is not strictly a synthetic medium.

3.23. Sucrose-Asparagine-EDTA Medium for *Phytophthora cactorum* (Elliott 1972)

The medium contains (g/liter) sucrose (10.0), L-asparagine (1.0), KH_2PO_4 (0.5), $MgSO_4 \cdot 7H_2O$ (0.25), thiamine·HCl (0.001), anhydrous $CaCl_2$ (0.1), cholesterol in ether (0.200) or sitosterol (0.010) in 10 ml of acetone (Elliott 1989), ethylenediaminetetraacetic acid disodium salt (5), and microelement solution (1 ml) consisting of (mg/liter of distilled water) $Na_2B_4O_7 \cdot 10H_2O$ (88), $CuSO_4 \cdot 5H_2O$ (393), $Fe_2(SO_4)_3 \cdot 9H_2O$ (910), $MnCl_2 \cdot 4H_2O$ (72), $Na_2MoO_4 \cdot 2H_2O$ (50), and $ZnSO_4 \cdot 7H_2O$ (4,403).

Sporangia and Zoospore Production

Media

3.24. Media Prepared from Bean Seed

Frozen lima beans (285 g) are autoclaved at 121°C for 15 min in 1 liter of distilled water (not blended). The beans are removed by straining through a cotton cloth, and 17 g of agar is added to the filtrate. The volume is adjusted to 1 liter before reautoclaving (Savage et al. 1968). *P. meadii* produces sporangia when incubated at 28–30°C for 5–7 days (Peries and Fernando 1966). Frozen lima bean agar should be stored in darkness, since light turns the medium from green to yellow, which makes it unsuitable for growth (M. E. Gallegly, *personal communication,* cited in Calvert et al. 1960; Goth 1981).

Frozen lima beans (141 g) are steamed in 0.5 liter of distilled H_2O for 20–30 min, filtered through a cotton towel or cheesecloth, solidified with 20 g of agar, and adjusted with distilled water to 1 liter. The medium is autoclaved, dispensed in petri plates, and inoculated with minced mycelium from liquid lima bean medium. After culture for 5–6 days under continuous fluorescent light (500–700 ft-c) at 20–25°C, sporangia are produced by *P. cactorum, P. heveae, P. capsici, P. hibernalis,* and *P. syringae. P. erythroseptica* forms sporangia on lima bean agar in light after transfer to distilled water, pond water, or compost percolate (Bywater and Hickman 1959). When 2- to 3-week-old cultures of *P. fragariae* on autoclaved lima beans are transferred to nonsterile stream water at 13°C, abundant sporangia develop after 24–48 h (Converse and Scott 1962). *P. infestans* produces sporangia on lima bean agar when incubated in the dark at 18°C (Harnish 1965) or at 16°C (Goth 1981).

As many as 100,000 zoospores of races 2 and 4 of *P. fragariae* are produced per milliliter of red kidney bean agar and 250,000 per milliliter of frozen lima bean agar. The petri plates of agar media are inoculated by flooding with them zoospores (10,000–15,000 per plate). After 5–9 days of growth, the colonies are cut into quarters, and each quarter is incubated at 20°C in 50 ml of sterile distilled water in a 100-cm-diameter petri plate. Changing the water after 24 h increases zoospore production. Frozen lima bean agar is better than red kidney bean agar. Zoospore production is slightly less on Davies's (1959) modified synthetic medium (Appendix 3.13). Wheat germ agar, Difco lima bean agar, and Difco potato-dextrose broth do not support zoospore formation by *P. fragariae* (Mussel and Fay 1973).

P. parasitica from tomato and *P. parasitica* var. *nicotianae* from tobacco produce large numbers of sporangia and zoospores on semisolid Difco lima bean agar (1/10 strength) (20 ml/125-ml Erlenmeyer flask). After incubation at 20–14°C for 15–18 days followed by addition of 40 ml of sterile distilled water, colonies produce 20,000–180,000 zoospores per flask, depending on the isolate. *P. cactorum* produces about 120,000 zoospores per flask in 9 days. Zoospore production for instructional purposes can be demonstrated by transferring a drop of the culture grown on semisolid lima bean agar and adding a drop of distilled water on a glass microscope slide. The sporangia and zoospores are produced under a coverslip a few minutes later (Dale and Jones 1969).

Dried lima beans (1 g) are autoclaved in 1 liter of distilled water for 30 min. After *P. megasperma* f. sp. *glycinea* is incubated at 25°C in the broth for 72 h, sporangia form. Zoospore production is induced by reducing temperature to 14–15°C for 8–10 h (Hilty and Schmitthenner 1962).

Sporangia of *P. primulae* form from mycelial plugs on bean meal agar when incubated in aerated, sterile, filtered soil extract (100 g of soil/200 ml of distilled water) at 15–20°C (Tomlinson 1952).

3.25. Frozen Pea Medium

Frozen peas (160 g) are comminuted with a blender, amended with sucrose (5 g) and agar (15 g) in 1 liter of distilled water, and autoclaved at 121°C. Isolates of *Phytophthora* are incubated on pea agar for 3–7 days. Sporangia form on mycelial plugs incubated in pond water (filtered through Whatman no. 1 filter paper) for 2 days at 20°C. Zoospore release is stimulated by chilling for 30 min at 4°C prior to incubation at room temperature (Dance et al. 1975).

3.26. Hemp Seed Broth or Agar

Hemp seed (50 g/liter) (Carolina Biological Supply Co., Burlington, North Carolina 21275) is soaked 12–24 h, after which the autoclaved filtrate is used as a broth or an agar (17 g/liter). Sporangia of many species form on hemp seed medium (Savage et al. 1968). About five to 10 nonsterilized hemp seeds in 20 ml of distilled water in a 250-ml flask or a 20-ml portion of the water extract of the hemp seeds after soaking for 24 h are inoculated with mycelium of *P. cinnamomi* stripped from potato-dextrose agar plates. After incubation at 27°C for 5 days, sporangia form in nonautoclaved but not in autoclaved hemp seed decoction. Zoospores form after mycelial mats containing sporangia are rinsed in distilled water and incubated at 8°C for 15 min. When the mats are returned to room temperature, zoospores are released (Manning and Crossan 1966; Schmitthenner 1973).

3.27. Prune Juice Broth

P. citrophthora, P. parasitica, and *P. cactorum* are cultured for 10–14 days in prune juice broth (60 g of prune pulp autoclaved at 121°C in 1 liter of tap water) at 2–4°C below the optimum temperature for growth. Mycelial mats are transferred to a petri dish containing sterile tap water so that about half of the mycelial mat is submerged and incubated in light. Sporangia of *P. citrophthora* will form in 12–15 h. For zoospore release, the petri plate is floated for 3–5 min in a water bath at 28–30°C and transferred to a water bath at 15–18°C for 1 h (Fawcett and Klotz 1934).

3.28. Seeds for Production of Sporangia

Seeds of sunflower, squash, pea, corn, or soybean are autoclaved in tap water, inoculated with cultures of *P. fragariae,* and incubated for 2–3 weeks at 20°C. When the colonized seeds are transferred to petri plates containing a shallow layer of tap water and rinsed daily, abundant sporangia of *P. fragariae* are produced in 4–5 days (Felix 1962).

Golden Bantam sweet corn seeds (2 g) are autoclaved in a flask containing 100 ml of deionized distilled water. After *P. phaseoli* is incubated for 14 days at 20°C in the broth, sporangia develop within 24 h on mycelium-covered seeds when placed at 90–100% relative humidity at 18–22°C (Goth and Wester 1963). Several species of *Phytophthora* were successfully stored on colonized corn seeds (R. W. Goth, *personal communication*). *P. infestans* remained viable for 5 years (Goth and Wester 1963).

3.29. Alfalfa Stem Method for *Phytophthora parasitica* and *P. citrophthora* (Klotz and DeWolfe 1960)

P. citrophthora or *P. parasitica* from citrus are grown on autoclaved alfalfa stems (about 8 cm in length) in test tubes containing 0.5 g of $CaCO_3$ (buffer) and deionized water to a depth of 5 cm. The stems are inoculated with mycelium and incubated at 26°C for 4–5 days. The stems are placed on a cloth spread on the top of a container of water and irrigated with a stream of water (for aeration). In 4–5 h at 20–25°C, sporangia of *P. citrophthora* are initiated and in 48 h become abundant. Sporangia of *P. parasitica* form best at a water temperature of 25–27°C. Incubation at 23–24°C of the sporangia-laden alfalfa stems in water results in zoospore release for *P. parasitica* and at 16–18°C for *P. citrophthora.*

3.30. *Eucalyptus sieberi* Seedlings for *Phytophthora drechsleri* (Gerrettson-Cornell 1975)

One-month-old seedlings of *Eucalyptus sieberi,* surface sterilized for a few seconds in 70% ethanol, are floated in water in petri plates containing mycelial disks of *P. drechsleri* that have been incubated in darkness at 25°C for 7 days. The cultures are then incubated under laboratory light at 23–28°C. Abundant sporangia form in 2–5 days on the cotyledons, hypocotyls, roots, and agar close to the seedlings. This technique also induces sporangia production by *P. cactorum* and *P. cryptogea* but by only a few cultures of *P. cinnamomi* and none by *P. megasperma.*

Washing Techniques for Production of Sporangia

3.31. Sporangial Formation by *Phytophthora parasitica* (Menyonga and Tsao 1966)

Clarified V8 juice broth containing Campbell V8 juice (20%), $CaCO_3$ (0.2%), and distilled H_2O (to 1 liter) is strained through a metal strainer, twice through several layers of cheesecloth, and finally through a layer of absorbent cotton, a process that reduces the volume to about 700–800 ml. The medium is filtered through Whatman no. 1 filter paper in a Buchner funnel until clear (pH should be 6.4–6.6). The clear broth, diluted (1:1) with an equal volume of distilled water, is dispensed (15 ml) in 6-oz glass prescription bottles and autoclaved for 15 min at 121°C. The medium in each bottle is inoculated with a 7-mm mycelial plug from the edge of a colony on V8 juice agar. The bottles are incubated for 24 h at 25°C and shaken vigorously by hand (20–30 strokes) to fragment the mycelium. The bottles are incubated in a horizontal position for 5 days at 25°C, by which time the mycelium will have covered the surface. The contents of each bottle are poured aseptically into a sterile, 35-ml, screw-capped centrifuge tube and centrifuged for 10 min at 1,610 × *g* (3,200 rpm). The mycelial mat is washed in sterile, triple distilled water (20 ml) and recentrifuged. After the third centrifugation, the mycelial mat is resuspended in 10 ml of distilled water and poured into a deep, flat-bottomed glass petri plate. The original inoculum plug is removed, and the mycelial mat is spread out. The plates are incubated 5 days to produce sporangia. Zoospores are induced to form by washing three times with distilled water and incubating for 15 min or longer at 20°C prior to incubation at room temperature (25°C). After 30 min, the zoospores (up to $2–5 \times 10^5$ zoospores per milliliter) are pooled by transferring them to another vessel with a pipette. Zoospore subsamples are counted by a hemacytometer after immobilization with iodine (1:50,000) or after agitation on a reciprocal shaker for 10 min. This method is also applicable to *P. palmivora.*

3.32. Sporangial and Zoospore Production by *Phytophthora cactorum* (D. C. Harris 1986)

Abundant sporangia and zoospores form in a medium made of V8 juice (20%) that is filtered and adjusted to pH 7.0 with 1 M KOH and supplemented with sitosterol (0.2 g/liter in 10 ml of chloroform). A circular T980 Tygan mesh about the same diameter as the bottom of a 150-ml conical flask is placed in the flask as a solid, nondegradable substrate to which mycelium adheres during the growth phase. Thirty milliliters of the broth is autoclaved in each of the flasks. The V8 broth is inoculated with one or two disks of mycelium and incubated in the dark for 2 days at 20°C. The cultures are shaken by hand to break up the mycelium and incubated for an additional 3 days, after which the broth is poured off. The mycelium, which has become attached to the Tygan mesh, is gently washed with distilled water, suspended in 3.0 ml of sterile distilled water, and incubated on a rotary shaker (100 cycles per minute) under warm white fluorescent light (16 h/day) for 3–4 days, at which time sporangia form abundantly. To obtain zoospores, the cultures are chilled (5–6°C) for 30 min and then placed at 15°C for 30 min. Yield is usually 5×10^6 to 3×10^7 zoospores per flask.

To obtain motile zoospores free of sporangia and mycelial fragments, the suspension of zoospores is filtered through Whatman 113V filter paper (retention size >30 μm), which allows free passage of zoospores. Whatman nos. 1 and 41 (retention sizes 11 and 20–25 μm, respectively) remove one-third to one-half of the zoospores.

A reliable method of counting zoospores in a suspension involves mixing equal amounts of the zoospore suspension with 0.1- or 0.5-ml quantities of 0.1% Victoria blue stain (C.I. 44045) in stoppered glass vials (5 × 1 cm). The concentration of zoospores in the suspension is estimated by using a hemacytometer. The samples can be refrigerated if time does not permit immediate examination. Motile spores are distinguishable from cysts. For dilutions of zoospores (kept at 15°C), wide mouth pipettes or graduate cylinders are used to avoid mechanical induction of encystment.

Among the many salts solutions and buffers that have been described, soil extract (1 g of soil per 100 ml of water) is consistently the best diluent to maintain maximum motility of zoospores (D. C. Harris 1986). Harris (1986) reported that autoclaving soil extract did not affect its ability to maintain motility of zoospores. The East Malling soil extract has a pH of 6.5–6.8, with the following range of inorganic components (µg/ml): calcium (7–20); magnesium (2); potassium (7–11); sodium (2–7); nitrate-N (0.2–2); and chloride (5–10). There was an indication that a mixture of suitable ions at the right concentration was important to maintain viability, but none were as suitable as soil extract. Calcium and magnesium have each been found to support mobility of *P. palmivora* zoospores (Cameron and Carlile 1977). Dilution with distilled water gives inconsistent results. Harris (1986) cites reports that distilled water induces lysis and/or encystment (Gooding and Lucas 1959a; McKee 1964; Ho and Hickman 1967a, b; Cameron and Carlile 1977).

3.33. Flooding Cultures of *Phytophthora medicaginis*

The fungus is grown in petri dishes containing 10 ml of either clarified V8 juice agar or Ribeiro synthetic agar (Appendix 3.54) until the colony reaches the edge of the plate. The plates are flooded with sterile distilled water and incubated on a shaker at about 60 excursions per minute for 48 h. The water is changed, and the cultures are incubated overnight at 12°C, after which sporangia and zoospores form.

Alternatively, the fungus is grown in petri dishes containing 15 ml of either V8 juice broth or the Ribeiro synthetic medium (Appendix 3.54) until colonies reach the edges of the plates. The medium is decanted, and the colonies are rinsed twice with sterile distilled water, after which they are incubated at 12°C or at room temperature overnight for production of sporangia (O. K. Ribeiro and J. Baumer, *unpublished*).

3.34. Flooding Cultures of *Phytophthora sojae*

When 6-day-old cultures of *P. sojae* on Difco lima bean agar (10 ml per plate) are flooded and decanted five times at 30-min intervals with double distilled water at 20°C and then incubated in darkness for at least 8 h, 10^4 to 10^5 zoospores per milliliter are formed. Zoospores are counted by use of a hemacytometer after immobilization of a 1-ml subsample with a drop of 0.1% aniline blue in lactophenol. Culture age, flooding temperature (20°C), flooding interval (30 min), and incubation after flooding in darkness are critical factors (Eye et al. 1978a). Zoospores were used to determine pathologic races and race-specific resistance in soybean lines (Eye et al. 1978b). Injection of soybean hypocotyls with zoospores (10^5/ml) requires one-third to one-quarter less time than the conventional inoculation of hypocotyls with mycelium (Schwenk et al. 1979).

Modifications of this method were utilized by Layton and Kuhn (1988a) and by Judelson et al. (1992) to produce a large number of zoospores (10^4 to 10^5/ml). Layton and Kuhn (1988a) washed plates of seeded (with homogenized mycelium) lima bean agar with Petri's solution (Appendix 3.35). Judelson et al. (1992) grew *P. sojae* on V8 juice agar for 1 week, after which the culture was cut into 5- × 5-mm blocks and overlaid with lima bean broth. After 5 days, the broth was removed with an aspirator and cultures were flooded with Petri's solution for 30 min followed by a final wash with sterile soil extract (21 g of soil in 2 liters of water, mixed for 30 min, filtered through Whatman no. 42 filter paper, and incubated overnight at 14°C under fluorescent light).

Inorganic Salts Solutions for Sporangial Production

Molarity of salts was calculated from recipes in publications.

3.35. Petri's Solution (Petri 1917; Tucker 1931; Hildebrand 1959)

The solution contains (g/liter) $Ca(NO_3)_2$ (0.4) (0.0024 M), $MgSO_4$ (0.15) (0.001 M), KH_2PO_4 (0.15) (0.001 M), KCl (0.06) (8×10^{-4} M), and distilled water to 1 liter. Small disks of inoculum are transferred from a suitable organic medium (e.g., V8 juice agar) to 10 ml of solution per petri plate at 12–20°C. The salts solution is replaced with distilled water after 3–4 days. See also Appendix 3.34 for use of Petri solution to stimulate sporangia of *P. sojae.*

3.36. Wills' Mineral Salts Solution (Wills 1954b)

The solution contains (g/liter) KCl (0.05) (0.001 M), KH_2PO_4 (0.10) (7×10^{-4} M), $MgSO_4 \cdot 7H_2O$ (0.05) (2×10^{-4} M), $K_2CO_3 \cdot 3H_2O$ (0.30) (9×10^{-4} M), and distilled water to 1 liter. Inoculum of *P. parasitica* var. *nicotianae* is transferred from potato-dextrose agar or other media to the solution (sterile or nonsterile) and observed after 1–3 days for sporangia. Sporangia form optimally when mycelial mats are incubated in the salts solution or in soil extract, tap water, or charcoal-amended solution.

3.37. Schmitthenner (1973) Salts Solution

Solution A contains (g/0.95 liter of distilled water) KH_2PO_4 (0.075) (5×10^{-4} M), K_2HPO_4 (0.075) (4×10^{-4} M), $MgSO_4 \cdot 7H_2O$ (0.05) (2×10^{-4} M), $ZnSO_4 \cdot H_2O$ (0.0022) (7.7×10^{-6} M), $FeSO_4 \cdot 7H_2O$ (0.0005) (1.8×10^{-6} M), and $MnCl_2 \cdot 4H_2O$ (3.5×10^{-5}) (2×10^{-7} M). Solution B contains (g/50 ml of distilled water) $Ca(NO_3)_2 \cdot 2H_2O$ (0.1) (5.0×10^{-4} M). Solutions A and B are autoclaved separately and mixed before use to prevent precipitation of calcium salts. Disks of mycelium are incubated in petri plates containing the solution.

3.38. Salts Solutions of Chen and Zentmyer (1970)

A *P. cinnamomi* culture is comminuted in a blender, and 200–400 small pieces of mycelium are transferred to either pea or V8 juice broth. Pea broth consists of 200 g of frozen green peas in 500 ml of deionized water blended for 5 min, after which the mixture is clarified by centrifuging for 10 min at 4,080 × *g*; the supernatant is brought to 1 liter with double distilled water. V8 juice broth consists of 200 ml of V8 juice thoroughly mixed with 2 g of calcium carbonate, after which the mixture is clarified by centrifuging as for pea broth. The supernatant is brought to 1 liter with double distilled water. The media are diluted with double distilled water to 1:8 for pea broth and 1:10 for V8 juice. After 16–18 h, the mycelial mats in petri dishes are washed four times at 1-h intervals with 15–20 ml of an autoclaved salts solution consisting of (g/liter) $Ca(NO_3)_2$ (1.64) (0.01 M), KNO_3 (0.05) (0.005 M), $MgSO_4$ (0.48) (0.004 M), deionized H_2O to 1 liter, and a solution of chelated iron (1 ml) consisting of (g/liter) ethylenediaminetetraacetic acid (EDTA) (13.05), KOH (7.5), $FeSO_4$ (24.9) (0.16 M), and deionized water to 1 liter.

Cultures are incubated at 24°C in petri plates under fluorescent cool-white or day-light lamps suspended 40 cm above the colonies. Sporulation begins within 8 h from the time of the first washing and reaches a maximum in 24–36 h. Zoospores are induced by incubation at about 5°C for 15–20 min and thereafter at 24°C. Zoospores are released approximately 1 h after chilling (Chen and Zentmyer 1970).

3.39. Synchronous Production of Sporangia and Zoospores

Twenty mycelial plugs (5 mm in diameter) of *P. cinnamomi* on V8 juice agar are transferred to four petri plates of clear, sterilized V8 juice broth (25 ml diluted 1:50) and incubated at 25–27°C for 24 h. The V8 broth is replaced with the Chen-Zentmyer salts solution (Appendix 3.38) and incubated for 30 min, after which the salts solution is changed three more times at 30-min intervals. After the final wash, the mycelial plugs are incubated under fluorescent light (200 $\mu W/cm^{-2}$) at 25–27°C for 24 h. Sporangia form particularly well at the periphery of each mycelial plug. Mycelial plugs from all four plates are combined, and the salts solution is replaced with 25 ml of sterile distilled water. Zoospores are induced by chilling plates at 4°C for 30 min and incubating at 24°C for 2 h. The zoospore suspension (in 1-ml aliquots) is transferred to each of 20 small petri plates (50 mm in diameter) containing 5 ml of dilute V8 broth (1:50) and incubated for 30 h. The nutrient solution is decanted, and the thin mycelial mat that forms in each plate is covered with 7 ml of the Chen-Zentmyer salts solution, which is changed again after 10 min; the plate is then incubated under light for 24 h. About 150–200 sporangia per 40× microscopic field form uniformly in each plate (Zaki and Zentmyer 1976).

Sporangia of *P. cinnamomi* are induced to form from axenic cultures that yield a high number of zoospores (10^7 to 10^8/liter). A siliconized cellulose fiber cloth (Miracloth, Chicopee Mills, New York) is overlaid on a V8 juice agar culture. The plates are incubated at 24°C for 4–6 days in darkness (critical because light inhibits growth), after which the Miracloth, containing the mycelium, is transferred from the agar to a 250-ml Erlenmeyer flask containing 100 ml of cleared V8 juice broth, which consists of V8 juice (100 ml), $CaCO_3$ (4 g), and glass-distilled water (100 ml), and centrifuged at 5,000 × *g* for 20 min. The supernatant is passed through Whatman no. 1 filter paper, diluted 1:10, adjusted to pH 6.0–6.5, and autoclaved at 121°C for 20 min. The flasks are incubated on a shaker (160 orbits per minute) for 19 h at 23°C. The mycelial mats are then washed twice with the mineral salts solution of Chen and Zentmyer (Appendix 3.38), suspended in 40 ml of the solution, and shaken for an additional 24-h period, during which time the sporangia form. To obtain zoospores, mats are washed with the Chen-Zentmyer solution (18°C) and incubated for an additional 50–90 min at 18°C. Zoospores are collected by passage of the suspension through Whatman no. 541 filter paper to remove chlamydospores and mycelial fragments. Zoospores are induced to encyst by shaking with a wrist-action shaker (Byrt and Grant 1979).

A similar method was reported for *P. palmivora* (Grant et al. 1984). Sorbitol (200 μM) can be used as an isotonic solution for study of zoospore differentiation (Paktitis et al. 1986).

3.40. Cellophane Overlayer Technique for *Phytophthora cinnamomi* (S. C. Hwang et al. 1975)

An overlayer of cellophane on solid medium allows *Phytophthora* to take up nutrients that pass through the cellophane, but when the cellophane and mycelium are stripped off, the residual nutrients can be easily washed away.

Several mycelial plugs are transferred from a 2- to 3-day-old culture of *P. cinnamomi* from V8 juice agar (Appendix 3.1) to a sterilized sheet of washed, uncoated cellophane (90 mm in diameter) placed on the surface of a V8 juice agar plate and incubated for 24 h (24°C). The sheet of cellophane and mycelium is transferred to a petri dish containing 25 ml of 5% clarified V8 juice broth consisting of V8 juice (50 ml) and $CaCO_3$ (2 g) in 1 liter of distilled water and centrifuged at 1,500 rpm for 5 min. Mycelium that is produced on the cellophane after 24 h is rinsed twice with 20 ml of the Chen-Zentmyer salts solution (Appendix 3.38) and incubated in 20 ml of the solution at 24°C under continuous cool white-fluorescent light (200 ft-c). Sporangia are produced after 9 h. The salts solution is then decanted to prevent premature release of zoospores. To induce release of zoospores, cultures are washed three times with sterile distilled water, chilled at 16°C for 30 min, and returned to 24°C.

3.41. Salts Solutions of Halsall (1977) and Halsall and Forrester (1977)

Organdy cloth disks (8 mm in diameter) are floated on the surface of a sterile broth consisting of 30 ml of V8 juice per liter of water to which 1 ml of a comminuted culture of *P. cinnamomi* (incubated at 27.5°C) or *P. cambivora* (incubated at 25°C) is added. After 24 h of incubation, the disks are transferred to the sterile salts solution and washed four times at 1-h intervals. The disks are incubated in the final wash at 25°C for 24 h under daylight-fluorescent lights. Halsall and Forrester (1977) present data on the interactive effects of varying concentrations of the salts that affect the numbers of sporangia produced (Table 3.2). Halsall (1977) reported that a salts solution containing (g/liter) $Ca(NO_3)_2$ (0.16) (0.001 M), $Mg(NO_3)_2$ (0.26) (0.001 M), KNO_3 (0.10) (0.001 M), and $FeSO_4$ (0.03) (1×10^{-4} M) is optimum for *P. cinnamomi* but that the concentration of $Ca(NO_3)_2$ should be reduced to 1×10^{-4} M for *P. drechsleri*. Copper was toxic to sporangium formation at 10^{-6} M. Molybdenum at 10^{-6} M had only a minor suppressive effect, but zinc at 10^{-6} and 10^{-7} M increased the numbers of sporangia. Halsall cited the report of Hendrix et al. (1969), which noted a zinc requirement for growth of *P. parasitica* var. *nicotianae* from tobacco.

3.42. Hoagland and Arnon (1950) Plant Nutrient Solution (Kennedy and Erwin 1961)

The solution contains (g/liter) KH_2PO_4 (0.136) (0.001 M), KNO_3 (0.505) (0.005 M), $Ca(NO_3)_2{\cdot}4H_2O$ (1.180) (0.005 M), $MgSO_4{\cdot}7H_2O$ (0.493) (0.002 M), and iron chelate (1.0 ml), containing (g/liter) ethylenediaminetetraacetic acid (EDTA) (26.1 g) (0.089 M), KOH (15 g) (0.268 M), $FeSO_4{\cdot}H_2O$ (24.9 g) (0.146 M), and distilled water to 1 liter (Jacobson 1951). Disks of *P. medicaginis* (*P. megasperma* f. sp. *medicaginis*) from V8 juice agar are incubated in the solution on a reciprocal shaker for 24 h at room temperature. Sporulation was enhanced by the four-salts solution (minus the chelated iron) when the fungus was grown on V8 juice agar neutralized to pH 7.0 with NaOH. On disks of V8-$CaCO_3$ agar, the iron chelate plus four-salts solution stimulated sporangia. Copper (as low as 0.2 ppm) suppressed sporulation; however, the EDTA negated the effect of copper up to 1.0 ppm.

3.43. KNO_3 Solution (Leonian and Geer 1929; Schiffman-Nadel and Cohen 1968)

P. citrophthora is grown for 10 days at 20°C on 2% potato-dextrose agar or 2% oatmeal agar medium, after which the mycelium is flooded with either tap water or 0.01 M KNO_3

solution and then incubated at 25°C. Sporangium growth is initiated within 4–5 days and is abundant within 7–9 days. Cultures of the fungus are grown for 4–6 days at 25°C in potato-dextrose broth or 5% oatmeal broth. Mycelial fragments are washed in tap water and transferred to a petri dish containing tap water or 0.01 M KNO_3 at 28°C. Sporangia form on the mycelium after 1–2 days and are abundant in 3 days. Sporangia can be produced by these two methods at a lower temperature with longer periods of incubation.

The mycelium from 6- to 20-day-old cultures of *P. parasitica* var. *nicotianae* on oatmeal agar is transferred to petri plates and wetted but not flooded with 0.01 M KNO_3 (Gooding and Lucas (1959b). The cultures are incubated 8–12 days at 26°C under continuous light (250 ft-c) until sporangia form. To obtain zoospores, double distilled water (10 ml) is added to mycelial mats bearing sporangia, and the plates are chilled for 25 min at 8°C. Most of the sporangia discharge zoospores about 15 min after the plates are returned to room temperature.

3.44. Soil Extract

Soil extract is prepared by mixing 10 g of soil with 1 liter of distilled water, after which the particulate material is removed by centrifugation and/or filtration. Centrifugation prior to filtration is usually necessary to prevent clogging of the filter, especially if the filter is fine enough to remove bacteria (0.2-µm pore size). Chee and Newhook (1965a) allowed the soil particles to settle for 2 days prior to filtering through Whatman no. 1 paper. Although autoclaved soil extract is effective for stimulation of sporangia of most *Phytophthora* species, sterilization by filtration or autoclaving markedly reduces the stimulatory effect on *P. cinnamomi* (Zentmyer and Marshall 1959).

Disks of mycelium and agar from the margins of a rapidly growing culture of *P. cinnamomi* on V8 juice agar are placed in a shallow layer of nonsterile soil extract in petri plates. Sporangia usually form in 3–4 days. See also Mehrlich (1935), Chee and Newhook (1965a), W. A. Ayers and Zentmyer (1971), King (1980), and Schoulties et al. (1980). W. A. Ayers and Zentmyer (1971) report that although some sporangia from young, 1- to 2-day-old colonies were produced on sterile soil extract, many more formed on nonsterile extract.

To extract a sporangium-stimulatory substance from nonsterile soil, 10 g of garden soil per liter of water was allowed to incubate at room temperature for 3 days. The suspension was filtered under vacuum through Whatman no. 20 or 42 paper, and 800 ml of the soil extract was centrifuged at 13,200 × *g* for 1 h. The small amount of the centrifugate (precipitate) was extracted with 20 ml of acetone and filtered through Whatman no. 1 filter paper; the filtrate was reduced in volume to 6 ml in a rotary evaporator at 30°C. Extracts of whole soil were made by mixing 200 g of soil in 200 ml of acetone in a stoppered flask and shaking for 4 h at 25°C. This mixture was filtered through Whatman no. 1 paper and the filtrate reduced to 10 ml in vacuo. Extracts from a broth culture of a pseudomonad bacterium from soil were made similarly. Acetone extracts diluted up to 10^{-9} in water increased production of sporangia by *P. cinnamomi* markedly (W. A. Ayers 1971). Similar results were reported by Schoulties et al. (1980).

3.45. Stimulation of *Phytophthora cinnamomi* Sporangium Formation by Bacteria

The bacterium *Chromobacterium violaceum* and several other bacteria and combinations of bacteria isolated from soil (Appendix 3.44) stimulated sporangium production when added to sterile soil extract (Zentmyer 1965). Certain pseudomonads from nonsterile soil extract also stimulated sporangial formation. Glucose and glutamine suppressed sporulation, but when the bacteria were added to sterile soil extract, abundant sporulation occurred (W. A. Ayers and Zentmyer 1971). See also Marx and Haasis (1965) and Manning and Crossan (1966).

Oospore Production and Germination

3.46. Bean Meal Agar Media for *Phytophthora fragariae*

P. fragariae forms oospores on a medium consisting of red kidney bean meal (28 g) mixed with 13.6 g of agar in 800 ml of water in a 2-liter flask. One drop of Antifoam, FG-10 (food grade) emulsion (Dow Corning Corp., Midland, Michigan), is added to 100 ml of medium to prevent boiling over during autoclaving. Some, but not all, races of *P. fragariae* produce oospores on this medium (Maas 1972).

French bean (dwarf *Phaseolus vulgaris* cv. Canadian Wonder) agar, which is used to isolate *P. fragariae* (Montgomerie and Kennedy 1983), supports production of oospores. An aqueous extract (120 g of dry French beans in 190 ml of distilled water steamed 0.5–1.0 h at 100°C) is mixed with clarified V8 juice agar (Appendix 3.2) or Elliott's synthetic medium (Appendix 3.23) at a concentration of 10 ml/100 ml of medium and autoclaved 15 min at 121°C. The extract imparts a deep wine color to the media. A deep orange brown color develops in areas where oospores of *P. cactorum, P. erythroseptica* var. *erythroseptica, P. nicotianae* var. *nicotianae,* and *P. palmivora* form. The color change does not occur on an extract of white-seeded beans. The color is considered to be caused by anthocyanins from red beans (Duncan 1988).

3.47. Rape Seed Extract Agar for *Phytophthora capsici*

Rape seed (100 g) is thoroughly washed and boiled in 1 liter of distilled water for about 30 min. The extract is filtered through cotton and adjusted to 1 liter with distilled water. The pH is adjusted to 7.0. Agar (20 g) is added to the extract and autoclaved at 121°C for 15 min. The zone in which oospores form becomes dark in color (Satour and Butler 1968).

3.48. Tryptophan-Supplemented V8 Juice Agar

Clarified V8 juice broth is supplemented (mg/liter) with β-sitosterol (30), tryptophan (20), $CaCl_2{\cdot}2H_2O$ (100), and thiamine (1) and incubated in the dark at 20°C. *Phytophthora* isolates that did not normally produce oospores in such standard media as V8 juice, oatmeal, and cornmeal agar formed oospores (Chee et al. 1976).

3.49. V8 Juice Broth for *Phytophthora parasitica* (Honour and Tsao 1974)

V8-$CaCO_3$ broth is prepared by clarifying V8 juice by centrifugation at 6,000 rpm (5,860 × *g*) for 10 min and filtering the supernatant under vacuum three times through one layer and two times through two layers of Whatman no. 42 filter paper. V8 juice (2%) is filtered through one layer of filter paper to remove suspended material. The cleared medium consists of 100 ml of filtered V8 juice and 100 ml of filtered $CaCO_3$ (2%) in 1 liter of distilled water. The final pH after autoclaving is 4.3–4.5. Inoculum (7-mm plugs) from the periphery of 7-day-old cultures growing on cornmeal agar plates is transferred to glass prescription bottles (~350 ml) containing 25 ml of Bartnicki-Garcia's synthetic medium (Appendix 3.17) amended with 30 mg of ergosterol in 30 ml of dichloromethane per liter (sitosterol would be a better choice than ergosterol [W. D. Nes et al. 1982b]). The bottles are incubated for 22–24 h at 25°C in darkness in a vertical position, after which each bottle is shaken 30 times to fragment the mycelium. To avoid formation of sporangia, hyphae should not be allowed to adhere to the walls of the bottle. The bottles are then incubated in a horizontal position for 6 days at 25°C in darkness. Three cultures of an isolate are comminuted in 100 ml of water for 10 s in a Sorvall Omnimixer, after which a 0.5-ml aliquot of the mycelial suspension of each mating type (A1 and A2) is transferred to a bottle and incubated at 25°C in darkness. Oospores are initiated in 3 days and continue to form for 10 days (about 1.5–3.0 × 10^6 oospores per culture). This method is also useful for oospore production of *P. palmivora, P. drechsleri, P. cinnamomi, P. megasperma* var. *sojae* (*P. sojae*), and *P. cactorum.* (About 4 × 10^4 to 9 × 10^5 oospores per culture are formed.) *P. parasitica* oospores form at a range of 1 8–30°C; the optimum is 21–24°C (Honour and Tsao 1974).

3.50. V8 Juice and Carrot Broth

Oospores of *P. medicaginis* (*P. megasperma* f. sp. *medicaginis*) (El-Hamalawi and Erwin 1986a, b; Förster et al. 1983) and *P. palmivora* (Kaosiri et al. 1978) were produced on carrot broth (first reported by Brasier 1967).

One-half of a 5-day-old culture (on V8 juice agar) is comminuted in a Waring blender cup containing 100 ml of sterile distilled water (60 s at 10-s intervals to prevent heating). For self-sterile species, one-half of a plate each of the A1 and A2 mating types is mixed (Kaosiri et al. 1978). Fifteen-milliliter portions of clarified V8 juice broth (Appendix 3.2) or carrot broth (Appendix 3.7) are dispensed in petri plates, inoculated with 1 ml of the minced mycelium, and incubated in darkness for 30 days at 25°C. In most cases, oospores are formed in 10 days but are not usually mature in less than 30 days. Plates should be wrapped in aluminum foil to maintain darkness. Shake culture of *P. sojae* (*P. megasperma* f. sp. *sojae*) in V8 broth for 8 days followed by still culture for 8 days in flasks covered with aluminum foil produced up to 1.7 × 10^5 oospores per milliliter of medium (Kittle and Gray 1979b). For some heterothallic species, better oospore production results if each mating type is comminuted separately and 1 ml of each type is added to the V8 juice broth plates (O. K. Ribeiro, *unpublished;* Förster et al. 1983). In some studies, β-sitosterol (30 mg/liter) is added to V8 juice media to ensure that a deficiency does not exist (Romero and Erwin 1969; Erwin and McCormick 1971). Addition of cholesterol (20 mg/liter) to V8 juice broth did not increase production of oospores of *P. sojae* (*P. megasperma* var. *sojae*), however (Kittle and Gray 1979b).

To obtain mycelium-free oospores, broth cultures are preferable to agar cultures because small particles of comminuted agar tend to settle out with the oospores during the oospore separation process.

3.51. Sucrose-Asparagine-Tyrosine Medium

The medium contains (g/liter) sucrose (30), $Ca(NO_3)_2$ (0.05), L-asparagine (2.0), L-tyrosine (0.5), KH_2PO_4 (0.6), $MgSO_4{\cdot}7H_2O$ (0.3), thiamine (0.001), agar (15), and a microelement solution (1 ml) consisting of (mg/100 ml of distilled water) $FeSO_4{\cdot}7H_2O$ (50), $ZnSO_4{\cdot}7H_2O$ (50), $CuSO_4{\cdot}5H_2O$ (2), and $MnCl_2{\cdot}4H_2O$ (2) in distilled water to 1 liter. Sucrose and $CaCO_3$ are autoclaved separately from the other constituents. *P. cactorum* and *P. heveae* produce oospores on this medium (Leal et al. 1964).

3.52. Glucose-Asparagine Media (High C:N Ratio)

The medium contains (g/liter) glucose (5.0), asparagine (0.1), KH_2PO_4 (0.5), $MgSO_4{\cdot}7H_2O$ (5 × 10^{-4}), $Fe(NO_3)_3{\cdot}9H_2O$ (1 × 10^{-4}), $ZnCO_3{\cdot}7H_2O$ (2 × 10^{-4}), $CaCl_2{\cdot}2H_2O$ (0.01), $CuSO_4{\cdot}5H_2O$ (2 × 10^{-5}), $MnSO_4{\cdot}H_2O$ (5 × 10^{-5}), thiamine·HCl (0.001), sitosterol (in ether) (10), agar (15), and distilled water to 1 liter. The pH is adjusted to 6.0 with 6 N KOH or 6 N H_2SO_4 before adding agar (15 g) and distilled water to 1 liter. *P. cactorum, P. cambivorum, P. hevae, P. megasperma,* and *P. palmivora* formed oospores at carbon and nitrogen ratios of 94:1 (glucose, 5 g/liter and asparagine, 0.1 g/liter). *P. cactorum, P. cambivora, P. capsici, P. cinnamomi, P. citricola, P. heveae, P. ilicis, P. megasperma,* and *P. palmivora* formed oospores when asparagine was replaced with an equivalent amount of KNO_3 (Leal et al. 1967a).

3.53. Glucose-KNO_3 Medium for *Phytophthora palmivora*

The medium contains (g/liter) D-glucose (10.0), K_2HPO_4 (0.1), KH_2PO_4 (0.8), KNO_3 (1.0), K_2SO_4 (0.5), $MgSO_4{\cdot}7H_2O$ (0.5), $CaCl_2{\cdot}2H_2O$ (0.2), thiamine·HCl (0.02), and microelement solution (2 ml), which contains (mg/liter) $FeSO_4{\cdot}7H_2O$ (0.5), $ZnSO_4{\cdot}7H_2O$ (0.5), $CuSO_4{\cdot}5H_2O$ (0.02), $MnCl_2{\cdot}7H_2O$ (0.02), $TiSO_4{\cdot}5H_2O$ (0.02), $Mo_7(NH_4)_6{\cdot}4H_2O$ (0.07), sitosterol agar (20), and deionized water to 1 liter. The pH of the medium is adjusted to 5.5 (Huguenin and Boccas 1971).

3.54. Glucose-Asparagine-Nitrate (High C:N Ratio) Medium

The medium contains (g/liter) glucose (4.5), L-asparagine (0.1), KNO_3 (0.15), KH_2PO_4 (1.0) $MgSO_4{\cdot}7H_2O$ (0.5) $CaCl_2$ (0.1), thiamine·HCl (Millipore filter sterilized and added after autoclaving) (0.001), β-sitosterol (dissolved in 30 ml of CH_2Cl_2) (0.03), chelated iron solution (1 ml), a microelement solution (1 ml), and distilled water to 1 liter. The pH is adjusted to 6.2. The chelated iron solution contains (g/100 ml) $FeCl_3{\cdot}6H_2O$ (0.05) dissolved in 100 ml of distilled H_2O containing ethylenediaminetetraacetic acid (EDTA) (2.6) and KOH (1.5). The microelement solution contains (mg/100 ml of distilled water) $Na_2MoO_4{\cdot}2H_2O$ (41.1), $CuSO_4{\cdot}5H_2O$ (7.85), $Na_2B_4O_7$ (0.5), $ZnSO_4{\cdot}7H_2O$ (87.8), and $MnSO_4{\cdot}H_2O$ (15.4) (Ribeiro et al. 1975a).

This medium successfully induced formation of oospores by 20 *Phytophthora* species. In liquid form, the medium readily induces oospore formation of homothallic species, but heterothallic species produce more oospores in solid than in liquid medium (Ribeiro et al. 1975a). This medium also supports production of sporangia when inoculum plugs of *P. palmivora* are incubated at room temperature in water 30 cm below a 15-watt black-light fluorescent lamp. Synchronous production of sporangia by *P. capsici* is reported (Yoshikawa and Masago 1977).

Determination of Mating Types

3.55. Determination of A1 and A2 Mating Types by Pairing with Known Tester Cultures

Determination of A1 and A2 mating types involves placing a 4- to 5-mm agar and mycelium plug of the unknown isolate on one side of a petri dish containing a suitable medium such as V8 juice agar and placing an agar plug from the known A1 or A2 isolate about 20–25 mm away from the unknown isolate. The culture should be incubated in the dark (light suppresses oospore formation) at a suitable temperature and then observed for about 7 days for oospores at the junction where the two colonies meet.

The sandwich method of Kellam and Zentmyer (1986a) allows several pairings between unknown isolates and A1 or A2 tester types to be made in a single petri dish. Carrot agar plugs (8 mm in diameter) from the margins of growing cultures are made into a sandwich by placing an 8-mm plug of uninoculated carrot agar between two plugs: one of the A1 or A2 culture and the other of the unknown culture to be tested. Ten stacks could be incubated (darkness) in each petri plate, which is sealed with Parafilm. After 6 days, plugs are separated and the center plug is observed microscopically for oospores. Known A1 and A2 testers should be included as controls.

3.56. Determination of Ability of A1 and A2 Mating Types to Self

For determination of a diffusible hormonelike substance by pairing cultures of A1 and A2 mating types separated by a polycarbonate membrane (Ko 1978), A1 and A2 isolates are grown on V8 juice agar (10%, v/v) neutralized with $CaCO_3$ (0.02%, w/v) and solidified with agarose (Sea Kem HGT-P Agarose, Maine Colloids, Rockland, Massachusetts). Agarose does not bind or inactivate chemicals as much as ordinary agar. An A2 mating type culture (15 × 10 × 3 mm) in the center of a petri plate is covered with a polycarbonate membrane (CPR, 0.2 µm, 90 mm in diameter; Nucleopore Corporation, Pleasanton, California). An A1 culture is placed on the Millipore membrane directly over the A2 culture. After 7 days at 24°C in darkness, the oospores in each culture are counted. Since mycelium does not penetrate the polycarbonate membrane or grow across it, a diffusible substance that passes through the filter to the opposite mating type is the inducer of the oospore production (selfing). The hormonelike substance produced by the A1 type is designated $\alpha 1$ and the hormone from the A2 type $\alpha 2$. A1 isolates produce more oospores than the A2 isolates (Ko 1978). See also Uchida and Aragaki (1980) and Ko and Kunimoto (1981).

3.57. Separation of Oospores from Mycelium

A. Freezing. In some reports, the oospore-containing agar culture is frozen (–5 to –20°C) for 24 h to kill mycelial fragments (Timmer et al. 1970; Erwin and McCormick 1971; Banihashemi and Mitchell 1976; Kittle and Gray 1979b). The method should be used with caution because some mycelial fragments will survive freezing (El-Hamalawi and Erwin 1986a; O. K. Ribeiro, *unpublished*). Chlamydospores of *P. cactorum* germinated (68%) after freezing at –20°C for 24 h (Darmono and Parke 1990). Freezing at –20°C also causes some loss in viability of oospores of *P. megasperma* f. sp. *medicaginis* (El-Hamalawi and Erwin 1986a).

B. Enzymes. Crude snail gut enzymes, such as Helicase (Endo Labs Inc., Garden City, New York) (0.1–0.5%) or Glusulase (Cal Biochem Corp., San Diego, California) (1–5%), are incubated with a 1:1 (v/v) mycelium-oospore suspension on a reciprocal shaker (Förster et al. 1983) for 24 h at 33°C. Snail gut enzymes destroy mycelium and asexual spores, but because of the thickness of the oospore wall, oospores are not injured unless either the concentration or time of treatment is too high (El-Hamalawi and Erwin 1986a). Mycelium mats can be comminuted in a Waring blender for 2 min at 10-s intervals to avoid heating. The minced mycelium can be filtered through a sterile, 53-µm, heavy-duty nylon mesh (Tobler, Ernst & Traber, Inc., 420 Saw Mill River Rd., Elmsford, New York), which can be autoclaved. The nylon can be supported on a Buchner funnel mounted on a sterile, 250-ml flask.

A mixture of cellulase and hemicellulase at concentrations of 0.5% completely lyses the mycelium and sporangia after 7 days while freeing the oospores from mycelium of *P. cactorum.* The oospores are filtered by passage of the lysed mycelium suspension through a 0.22-µm nylon screen (Nitex brand) by use of a Millipore apparatus (Swinnex-25) attached to a plastic syringe. Oospores are washed from the Nitex screen by reversing the screen and applying pressure with the syringe. The cellulase-hemicellulase treatment is claimed to be more efficient than use of Glusulase. Oospores are decontaminated by incubating them in $HgCl_2$ (30 ppm) for 5 min, washing them in sterile water by centrifugation, and then adding sterile water. About 40% of the oospores germinate on 0.8% water agar when incubated under fluorescent light (Sneh 1972b).

NovoZym (Novo Industries, Bagsvard, Denmark, or Wilton, Connecticut, United States) is used to macerate mycelium and other propagules in suspensions of oospores. An oospore-mycelium culture of *P. infestans* (A1 × A2) is blended in 10–20 ml of sterile water for 2 min, NovoZym (0.4%, w/v) is added, and the culture is incubated at 18°C in the dark for 24 h or until no intact mycelium or sporangia remain. Oospores are filtered with a nylon cloth (pore size of 20 µm) and rinsed with sterile water (Spielman et al. 1990; method ascribed to A. M. Fyfe, *personal communication*).

C. Snail Ingestion. Snails have enzymes in the gut that digest mycelium but not oospores. Various workers separate oospores from mycelium and other propagules of *Phytophthora* by feeding mycelial mats, either in water or agar, to the garden snail (*Helix aspersa*) (Gregg 1957) or the water snail (*Planorbarius corneus*) (Shaw 1967). The snails are starved for a day or so and sometimes fed water agar to clean out the gut. After the snails feed on the mycelium-oospore mat, the feces is collected, suspended in water, surface sterilized with $HgCl_2$ (60 ppm) for 5 min, and washed by centrifugation.

D. Sonication. To free oospores from mycelium by fragmentation, mycelial mats are sonicated in a beaker placed in an ice bath to prevent heating. Mycelial mats of *P. medicaginis* (*P. megasperma* f. sp. *medicaginis*) are sonicated (Braun-Sonic 1510, S. Braun Instruments, South San Francisco, California 94680) for 30 s at 50 W (Förster et al. 1983) or for 4 min at 100 W. Sonication for limited periods is not harmful to germination (El-Hamalawi and Erwin 1986a).

E. Sieving and Differential Centrifugation. Oospore-mycelial mats can be comminuted with an Omnimixer for 1–5 min (on ice to prevent heating). This breaks up most of the mycelium. The suspension can be filtered through nylon screens of different mesh sizes. The first screen should be large enough to allow the oospores to pass through the mesh but small enough (usually about 50 µm, depending on the size of the oospores, which vary from about 20 µm to more than 50 µm on different species) to catch much of the mycelium. Since the density of oospores is greater than that of mycelium, oospores can be separated by sedimentation with a column or a graduate cylinder (Romero and Erwin 1969; Timmer et al. 1970). Most of the oospores of *P. cactorum* or *P. medicaginis* settle out in a separatory funnel in 1–3 h (J. Jiang, *personal communication*). The lower fraction can be drawn off, resuspended in water, and centrifuged several times at 1,200 × *g* for 2–3 min until no mycelial fragments remain.

This process should be repeated until the oospore suspension is free of viable mycelial fragments. The presence of viable mycelial fragments in the oospore suspension can be detected by plating a sample of the suspension on V8 juice agar, because the mycelial fragments will germinate and form colonies more rapidly than germinating oospores.

Oospores can be obtained from a block of agar and mycelium by triturating the agar with an Omnimixer at 4,500 rpm for 1 min and filtering it successively through 53- and 20-μm sieves. The oospores retained on the 20-μm sieve can be washed with tap water and resuspended in sterile distilled water (Ann and Ko 1988b, 1989b; Chang and Ko 1990). Chee (1973b) separated comminuted mycelium from oospores by filtration through two layers of Kleenex tissues.

F. Sucrose Density Centrifugation. Oospores of *P. sojae* can be produced in 1-liter Roux flasks containing 100 ml of clarified V8 juice-$CaCO_3$ broth (Appendix 3.2) at 22–24°C. Mycelial mats are blended for 8 min in 0.1 M sucrose (3.42%) in a Sorvall Omnimixer at 6,000 rpm and shaken on a reciprocal shaker at 130 excursions per minute for 2 h. The suspension is centrifuged, suspended in 1 M sucrose, and frozen for 12 h at –5°C to kill mycelial fragments. The suspension is rewashed, resuspended in deionized distilled water, layered over 1 M (3.42%) sucrose, and centrifuged for 1 min at 1,500 × *g*. Because of their greater density, oospores will differentially collect at the bottom of the centrifuge tube, and the mycelium will collect in the less dense fraction of sucrose. The oospores are collected and rewashed by centrifugation. About 500 mg of oospores can be collected from 20 flasks (Partridge and Erwin 1969).

Oospores of *P. medicaginis* can be produced in autoclaved carrot broth (Appendix 3.7). After digestion in Glusulase (5%, w/v) (27°C for 12 h), the mycelium-oospore suspension is sonicated for 10 s at 30 W and layered over sucrose gradients (0, 10, and 20%), with the most dilute layer on top. Oospores are collected in the highest sucrose concentration. Sucrose is removed by repeated centrifugation and resuspension in double distilled water (Förster et al. 1983).

Viability and Germination of Oospores

3.58. Tetrazolium Test for Viability

A suspension of oospores is immersed in 0.1% tetrazolium bromide (Aldrich Chemical Co., Milwaukee, Wisconsin 53223) in 0.1 M phosphate buffer (pH 5.8); use of the buffer is critical (Jiang and Erwin 1990a). The suspension is incubated in a water bath at 35°C for 2–2.5 days (note: recommendation of 2–2.5 h of exposure to tetrazolium bromide by Ribeiro [p. 162, 1978] is an error). When examined microscopically, the color of the viable oospores ranges from rose to dark blue, depending on the species used. Dormant oospores stain rose, and activated (ready to germinate) oospores stain blue. Nonviable oospores either do not stain or stain black (Ribeiro 1978; Sutherland and Cohen 1983; S. D. Cohen 1984; El-Hamalawi and Erwin 1986a, b; Jiang and Erwin 1990a).

Pittis and Shattock (1994) reported that the tetrazolium test gave false positives in heat-killed oospores of *P. infestans*.

3.59. Plasmolysis for Viability

When oospores are incubated in 4 M NaCl for 5–60 min, only viable oospores plasmolyze, as noted by microscopic examination. The percentage of plasmolysis is directly proportional to the percentage of germination. When the salts solution is diluted with water, viable oospores recover and germinate in soil extract (Jiang and Erwin 1990a). Plasmolysis (2 M NaCl) of oospores of *P. infestans* was a good measure of viability (Pittis and Shattock 1994).

3.60. Germination of Oospores

As explained in Appendix 3.57, oospores must be freed of mycelium, sporangia, chlamydospores, and any other propagules that germinate more rapidly than oospores. Oospores are distributed on plates of water agar (about 50–100 oospores per plate) widely enough that single, germinated oospores can be rescued and transferred. Although there is no conclusive evidence, results from some reports (Romero and Erwin 1969; Klisiewicz 1970; Förster et al. 1983; El-Hamalawi and Erwin 1986a, b; Kellam and Zentmyer 1986a) suggest that incubation of oospores in an aqueous solution is favorable to that on water agar. Conceivably, water agar, over the period of 7–10 days required for germination, could become dry on the surface. The oospore suspension should be applied to the agar in a thin film of a suitable germination solution such as those described below. Water agar is more convenient than aqueous solutions because germinated oospores can be readily picked off the solid agar. Some workers transfer oospores singly before germination from a water agar plate to another plate marked on the bottom with a grid (25 oospores per plate) so that germinated oospores can be readily counted, recorded, and transferred to a suitable growth medium such as V8 juice agar (Long et al. 1975; J. M. Duncan, *personal communication*).

Oospore germination of all species studied thus far is stimulated by incubation under a constant source of light that emits radiation in the blue (400- to 480-nm) region of the spectrum (Ribeiro 1983). Low levels of radiation (2 μWcm^{-2}) on the far red region enhanced germination of some species (Ribeiro et al. 1975b). We have routinely used a blue Sylvania FI 5TB-B fluorescent bulb at 800-μWcm^{-2} intensity for incubation of oospore cultures (Förster et al. 1983; El-Hamalawi and Erwin 1986a, b; Jiang and Erwin 1990a, b).

Oospores can be diluted to a concentration of one oospore per 5 μl, after which 5-μl drops can be pipetted separately onto the water agar in 1-cm^2 grids (Schecter and Gray 1987). Some investigators use a soft agar (0.8%) (Shaw 1967; Sneh 1972b). Oospores of *P. infestans* are germinated on 0.6% water agar, and, after germination is detected (3–5 days), the soft agar is overlaid with a thin layer of particle-free rye A broth (Appendix 3.9) (Shattock et al. 1986a, b). *P. parasitica* oospores can be germinated on the basal salts-lecithin-glucose agar medium of Ruben et al. (1980) to which is added (mg/liter) ampicillin (100), nystatin (50), and pentachloronitrobenzene (10) to suppress contaminants (Ann and Ko 1988b, 1990b; Chang and Ko 1990).

Several studies indicate that nutrients enhance the germination percentage. To supply these, root exudates and extracts (Förster et al. 1983; El-Hamalawi and Erwin 1986a, b) and soil extracts (Sneh et al. 1981; Jiminez and Lockwood 1982; Förster et al. 1983; Jiang and Erwin 1993) are used.

Root exudates can be obtained by germinating seedlings (from seeds surface sterilized in 0.5% sodium hypochlorite bleach and then washed with sterile distilled water) in sterile distilled water. After removal from the seedlings (3–4 days of incubation), the exudate is sterilized by filtration (0.2-μm filter). About 100–300 oospores in about 3 ml of fluid are incubated in a 35-cm-diameter plastic petri dish. Because oospores are heavy, most of them sink and stick to the bottom of the dish (Förster et al. 1983; El-Hamalawi and Erwin 1986a). Since the petri dishes must be opened to read the percentage of germination microscopically, some contamination is inevitable.

Soil extract can be prepared by mixing 1 kg of soil with 1 liter of water (see Appendix 3.44). The reason for the favorable effect of soil extract on oospore germination is not well understood (Förster et al. 1983; Jiang and Erwin 1990b, 1991, 1993).

Although it is easier to transfer single germinated oospores from a solid agar plate than from an aqueous fluid, single oospore transfers can be made from aqueous cultures by use of fine glass capillary pipettes. For example, a glass Pasteur pipette can be heated in a Bunsen burner flame and drawn out to a narrow diameter (Kellam and Zentmyer 1986a). The pipette can be attached to thin-walled soft rubber tubing, one end of which is placed in the mouth, and gentle suction is applied when the operator, using a dissecting microscope, locates a germinated oospore in the dish (Laviola 1968; Romero and Erwin 1969; Erwin and McCormick 1971; Chee 1973b; Laviola et al. 1978). The germinated oospore is transferred to a suitable medium to determine whether a colony can be established. In most cases, nearly all of the germinated oospores produce colonies; however, in one case, most of the F_2 progeny of an F_1 pairing of A1 × A2 *P. infestans* cultures did not grow (Castro and Zentmyer 1969). Hohl (1991b) states that establishment of *P. infestans* spores from germinated sporangia or cysts is often a problem. Clarke (1966) reported that addition of pectin to the agar medium increased the percentage of spores established.

When oospore suspensions *P. medicaginis* are treated with potassium permanganate (0.05%) for 10 min and then washed in distilled water by centrifugation, the percentage of germination increases from about 23 to 73% (El-Hamalawi and Erwin 1986a). Treating *P. parasitica* oospores with 0.25% potassium permanganate for 20 min and then washing them in distilled water kills sporangia and chlamydospores but allows a high percentage of oospores to germinate (Ann and Ko 1988b, 1990b). Potassium permanganate treatment of oospores of *Pythium aphanidermatum* that have been desiccated enhances their germination (Ruben et al. 1980).

Oospores of several races of *P. sojae* were produced in lima bean broth after incubation for 30 days at 24°C (Bhat et al. 1993a). The mycelium-oospore culture was mechanically sheared (homogenized) in a 5-ml vial at 1,000 rpm for 60 s with a Tissue Tearor and incubated in β-glucaronidase (2,000 units/ml) at 37°C for 14 h. The digested mycelium-oospore mixture was centrifuged at 1,900 × *g* for 5 min and washed five times with distilled water. The oospore suspension was frozen at –20°C overnight and then thawed in water at 45°C for 5 min. The floating mycelial fragments were removed aseptically with a Pasteur pipette, and the oospores were pelleted by centrifugation. Oospores were spread on a 90-mm-diameter petri plate containing 1.5% water agar, which contained 10 μg of cholesterol (in dimethyl sulfoxide, 1:100), and rifampicin (in methanol, 1:100) to suppress bacteria. Oospores germinated in 4 days and formed colonies by 7 days. Residual oospores were stored at –20°C.

To eliminate bacterial or fungal contaminants, the colonies were grown on a selective medium consisting (units per liter) of dilute V8 juice agar (40 ml of clarified V8 juice and 0.6 g of calcium chloride that had been filtered through diatomaceous earth and solidified with 15 g of agar), pentachloronitrobenzene (0.04 g), benomyl (0.02 g), iprodione (0.02 g), neomycin sulfate (0.1 g), and chloramphenicol (0.01 g) (Bhat et al. 1993a).

CHAPTER 4

Morphology and Identification of *Phytophthora* Species

Some of the earlier literature on the development of nomenclature in the genus *Phytophthora* includes a discussion by Wilson (1914) on *P. parasitica, P. colocasiae, P. arecae, P. phaseoli, P. erythroseptica, P. infestans, P. thalictri, P. fagi, P. cactorum, P. syringae, P. nicotianae, P. faberi,* and *P. omnivora* as well as *P. agaves* and *P. jatrophae.* Wilson stated that these last two species were of doubtful standing. Currently, these two as well as *P. thalictri, P. fagi, P. faberi,* and *P. omnivora* are not recognized (Waterhouse 1963; Stamps et al. 1990). On the basis of the wide host range of some of these species, Wilson predicted that because the parasitism of *Phytophthora* was of such a low order, host-specific races, such as those found in the rust fungi, were not likely to occur! This prediction has proved to be wrong, because the development of new pathologic races of *P. infestans, P. sojae,* and *P. vignae* have overcome the resistance in many cultivars (see Chapter 6).

Identification of many *Phytophthora* species is relatively simple; however, the morphological differences among other species are so small and some characteristics are so variable that the genus is considered, even by experts, to be difficult. "*Phytophthora* has always been recognized as a notoriously difficult genus—especially for those who wished only to identify an isolate and were not concerned with taxonomic problems and/or minute details" (Waterhouse et al. 1983). The history of the development of taxonomy within *Phytophthora,* as given by Waterhouse et al. (1983), begins with the type species, *P. infestans* de Bary (1886), which was characterized mainly by the monopodial branching habit of the sporangiophore and papillate sporangia that release zoospores in free water. The great controversy that preceded the description of *P. infestans* is presented in fascinating historical accounts by Buczaki (1991), Bourke (1991), and Peterson et al. (1992). After the discovery of *P. infestans,* additional species were described, and a comparative study of 11 of these species was reported by Rosenbaum (1917), who began his study at Cornell University and finished while a member of the Office of the Cotton, Truck, and Forage Crop Disease Investigations in the United States Department of Agriculture. His identification key utilized the type of antheridium, the prominence of papillae on sporangia, the sporangial size and length-breadth ratio, presence or absence and size of chlamydospores, and size and abundance of oospores to differentiate species. He paid particular attention to biometry and apparently was one of the first to utilize the length-breadth ratio of sporangia as an aid to characterization of a species.

The subsequent key of Tucker (1931) was based not only on morphology but also on physiology and pathology. Tucker did his initial work on coconut bud rot caused by *P. palmivora* at a research station in Puerto Rico. He recognized 21 species on the basis of their the ability to grow on certain media, type of antheridium, the presence or absence of sex organs, the nature of the papillae on sporangia, presence or absence of chlamydospores, pathogenicity, oospore size, presence of hyphal swellings, cardinal temperatures for growth (minimum, optimum, and maximum), and host specificity.

Shortly after publication of Tucker's key, Leonian (1934), at West Virginia University, utilized most of Tucker's criteria but added growth on malachite green and production of sporangia by transfer of mycelium in pea broth to water for differentiation of species. Leonian and Geer (1929) questioned the value of sporangia size to differentiate species because of the great variation in size of sporangia grown under different conditions. Although Leonian's (1934) key included 22 species, his discussion contained the often quoted statement, "In the light of the foregoing discussion the writer is inclined to think that sometime in the future so many of the present species will be discarded that there will be no more than two or perhaps three firmly established species *viz. Phytophthora infestans, P. cactorum,* and *P. palmivora.*" Although time has not validated his predictions (58 species, excluding the marine species, and *Sclerophthora macrospora* are listed in Table 4.1), he expressed the frustration experienced by plant pathologists and mycologists who try to separate species that differ only slightly in certain morphological characteristics. Leonian was an outspoken pioneer who strongly objected to the restraint of adhering to the type species concept. He had a keen realization of the variability that is now recognized to be a problem in speciation (Erwin et al. 1963; Erwin 1983).

Leonian's statements are quoted frequently in Brasier's (1991) philosophical chapter on questions about *Phytophthora* systematics, primarily because Leonian saw the variability within species and exposed the need for physiological parameters to separate species. As we show below, DNA restriction fragment length polymorphism (RFLP) pattern data have supported the continued separation of certain species such as *P. cryptogea* and *P. drechsleri,* even though they are morphologically similar (Mills et al. 1991). The use of DNA and isozyme data is discussed later in the chapter. A discussion of isozyme technology by Spielman (1991) is excerpted in Chapter 3.

Several other taxonomic keys and studies based on *Phytophthora* species in local areas were subsequently published by Wager (1940) in South Africa, Frezzi (1950) in Argentina, Waterhouse and Blackwell (1954) in Great Britain, and Schwinn (1959) in Germany.

The taxonomic key produced by Waterhouse (1963), a mycologist at The Commonwealth Mycological Institute at Kew, United Kingdom, was the first to categorize species into morphological groups. Because of the breadth of fungal specimens received for identification from throughout the British Commonwealth and the slides prepared by S. F. Ashby during the 1940s and 1950s, her key represented the first attempt to look at *Phytophthora* from a global standpoint (Waterhouse 1963). Waterhouse's separation of *Phytophthora* species into six groups based on a series of morphological and physiological parameters was a marked advance in the study of taxonomy and is still useful and accepted. These parameters include branching patterns of the sporangiophores; the sporangium apex; abundance of sporangia on solid media; the noncaducous or caducous nature of the sporangia; internal proliferation of sporangia; production of oogonia and oospores in single culture (homothallic); production of oospores and oogonia only when opposite mating types are paired in a culture (heterothallic); nature of the antheridium; abundance or absence of oospores on host tissue or in culture; and for certain species, sporangium shape and dimensions, e.g., length-breadth ratios, sporangiophores, hyphal swellings, presence or absence of chlamydospores, oogonium size, ornamentation of the oogonium wall, host specificity, and cardinal temperatures. In the Waterhouse (1963) taxonomic key, *P. arecae, P. citricola, P. hibernalis,* and *P. meadii* were reintroduced following their relegation to synonymy by Tucker (1931). *P. parasitica* was rejected because the name *P. nicotianae* had been used previously for the same species. (This change, which has not been universally accepted, is discussed in a later section, Molecular Technology for Differentiation of Species, and in Chapter 50 on *P. nicotianae.*) A revision of the Waterhouse (1963) key included six new species and was based on the same parameters as the 1963 key but was presented in a tabular key format (Newhook et al. 1978). The latest revision of this tabular key contains 67 species, 19 of which have been described since 1978 (Stamps et al. 1990). It incorporates the original groups I–VI of the previous key with the addition of group VII for the marine species. The marine species were not included in the Newhook et al. (1978) key; they were judged to be unacceptable because of their unusual characters compared with other *Phytophthora* species (Waterhouse et al. 1983). Because of the recent relegation of the marine species to the genus *Halophytophthora* (Ho and Jong 1990; Pan et al. 1994), we are not including them here; however, for the convenience of the reader, the marine species are listed in Table 4.1. Brasier and Hansen (1992) suggest, "If these taxa prove to be close to *Phytophthora* on molecular grounds, there may be a case for a redefinition of the genus [*Phytophthora*] to accommodate them."

Kröber (1985; in German) classified 17 species of *Phytophthora* commonly found in Germany on the basis of the occurrence of paragynous or amphygynous antheridia, papillation of sporangia, and several other characters. His publication contains the best photomicrograph illustrations of the various spore stages of any of the published keys. The key of Frezzi (1950; in Spanish) includes species from Argentina and presents excellent photomicrographs of spore structures for each species.

A taxonomic key to the identification of 41 *Phytophthora* species from Australia by Gerrettson-Cornell (1989) was based on four general characteristics: 1) homothallic oospore production and sporangia formed on agar media that had not been flooded with water or aqueous solutions; 2) homothallic oospore production and sporangia formed on agar media only when flooded with water or aqueous solutions; 3) heterothallic oospore production and sporangia produced on agar media without flooding the culture with water or aqueous solutions; and 4) heterothallic oospore production and sporangia formed on agar media only if flooded with water or aqueous solutions. Various morphological criteria were used to differentiate species within these broad categories. This approach was a development of a previous key in which Gerrettson-Cornell (1985) categorized 39 species.

The synoptic key by Ho (1981) included 38 culturable species, including five species known only from hosts. Although Ho's (1981) synoptic key should be workable, we have not utilized it enough to present a judgment on its efficacy. The literature cited by Ho (1981) indicates that the synoptic key concept is well accepted for identification of certain other fungal genera.

Ho (1992) and Ho et al. (1995) presented dichotomous and synoptic keys for the 23 *Phytophthora* species that occur in Taiwan. The latter publication in particular is illustrated with high-quality light photomicrographs of morphological structures produced by each species. A complete description of each species in included.

Despite the rather large number of available keys, the identification of species is still considered to be difficult, principally because there are relatively few morphological features by which species are distinguished and great variability and overlapping features exist within species (Leonian and Geer 1929; Leonian 1934; Erwin et al. 1963; Erwin 1983; Ho and Jong 1987; Gerrettson-Cornell 1989; Förster et al. 1990; Mills et al. 1991; Oudemans and Coffey 1991). Many *Phytophthora* isolates have not been identified to species (e.g., E. M. Hansen et al. 1979).

Because of the considerable variation within the genus *Phytophthora,* Brasier (1983, 1991) questions whether morphology alone should be used for identification and whether the "type" species concept as espoused by the International Botanical Code of Nomenclature should be religiously followed. The type species concept does not reflect the broad variability within populations. As an example of the recognition of broad variability in *Phytophthora, P. capsici* has been redescribed to include isolates previously designated *P. palmivora* MF4 from cacao (Brasier and Griffin 1979) and black pepper (*Piper nigrum*) as well as isolates from pepper (*Capsicum annuum*) (Tsao and Alizadeh 1988; Tsao 1991; Mchau and Coffey 1994). In opposition to the broad variability concept, a new species, tentatively named *P. tropicalis,* has been proposed but not formally described for isolates of this complex that attack Macadamia nut and black pepper (Uchida and Aragaki 1989; Aragaki and Uchida 1992).

Brasier's (1987, 1991, 1992) thoughtful articles on the dynamics of fungal speciation treat species as emerging subpopulations that may differ genetically yet not be morphologically divergent. Thus, the fact that different species might be almost morphologically identical should not be surprising. Brasier (1992) discusses *P. meadii,* an important pathogen of rubber trees in Sri Lanka and southern India. Since isolates of *P. meadii* exist as both diploid and tetraploid forms and since the diploid forms are sterile, Sansome et al. (1991) suggest that *P. meadii* is an alloploid or a hybrid species and that it may be a product of hybridization between *P. palmivora,* which is also a pathogen of rubber trees, and another *Phytophthora* species, possibly *P. botryosa.* Although *P. meadii* and *P. botryosa* can be differentiated morphologically, an isozymic comparison of them shows that these two species share 10 of 18 isozyme loci, indicating their close relationship (Oudemans and Coffey 1991c). This example is discussed in more detail in Genetics and Cytology in Chapter 3. In this chapter, see the section Molecular Technology for Differentiation of Species for an explanation of isozyme technology.

Brasier (1992) and Brasier and Hansen (1992) discuss the evolutionary biology of *Phytophthora.* E. M. Hansen (1987, 1991) addresses the evolution of species within the *P. megasperma* complex. Results from the use of isozymes and RFLP patterns of mitochondrial DNA show that *P. megasperma* f. sp. *medicaginis* and *P. megasperma* f. sp. *glycinea* are distinct species, and they are now known as *P. medicaginis* and *P. sojae,* respectively (E. M. Hansen and Maxwell 1991). Other components of this heterogeneous species appear to be distinct by molecular analysis but have not yet been resolved.

Brasier (1983) points out the difficulties of using the type species concept as the norm for classifying unknown isolates. More recently, he has supported the continued study of the relationship of morphological criteria to the modern molecular

Table 4.1. Species of *Phytophthora,* including certain synonyms that are still in use[a]

Classification[b]	Taxonomic group	Chapter	Species
S	II	8	*P. arecae* (Coleman) Pethybridge (1913)
M S	VII		*P. avicennae*[b] Gerrettson-Cornell and Simpson (1984)
M S	VII		*P. bahamensis*[b] Fell and Master (1975)
M S	VII		*P. batemanensis*[b] Gerrettson-Cornell and Simpson (1984)
S	II	9	*P. boehmeriae* Sawada (1927)
S	II	10	*P. botryosa* Chee (1969a)
S	I	11	*P. cactorum* (Lebert and Cohn) Schröter (1886)
S	VI	12	*P. cajani* Amin, Baldev and Williams (1978) is probably synonymous with *P. drechsleri* var. *cajani* Pal, Grewal and Sarbhoy (1970) and *P. drechsleri* f. sp. *cajani* Kannaiyan, Ribeiro, Erwin and Nene (1980) according to Mills et al. (1991)
S	VI	13	*P. cambivora* (Petri) Buisman (1927)
S	II	14	*P. capsici* Leonian (1922)
S	VI	15A	*P. cinnamomi* Rands (1922)
	VI	15B	*P. cinnamomi* Rands var. *parvispora* Kröber and Marwitz (1993)
S	III	16	*P. citricola* Sawada (1927)
S	II	17	*P. citrophthora* (R. E. Smith and E. H. Smith) Leonian (1925)
S	I	18	*P. clandestina* Taylor, Pascoe and Greenhalgh (1985a)
S	IV	19	*P. colocasiae* Raciborski (1900)
S	VI	20	*P. cryptogea* Pethybridge and Lafferty (1919)
S	III	21	*P. cyperi* (Ideta) S. Ito (1935)
S	III	22	*P. cyperi-bulbosi* Seethalakshmi and Ramakrishnan (1953)
S	VI	23	*P. drechsleri* Tucker (1931) var. *drechsleri*
			P. drechsleri var. *cajani* Pal, Grewal and Sarbhoy(1970) = *P. cajani*
			P. drechsleri f. sp. *cajani* Kannaiyan, Ribeiro, Erwin, and Nene (1980) = *P. cajani*
M S	VII		*P. epistomium* Fell and Master (1975)
S	II	24	*P. eriugena* Clancy and Kavananagh (1977)
S	VI	25A	*P. erythroseptica* Pethybridge var. *erythroseptica* (1913)
S	VI	25B	*P. erythroseptica* var. *pisi* Bywater and Hickman (1959)
S	V	26A	*P. fragariae* Hickman (1940) var. *fragariae*
	V	26B	*P. fragariae* var. *oryzobladis* Wang and Lin (1979)
	V	26C	*P. fragariae* var. *rubi* Wilcox and Duncan (1993) (in Wilcox et al. 1993)
S	VI	27	*P. gonapodyides* (Petersen) Buisman (1927)
S	II	28	*P. heveae* Thompson (1929)
S	IV	29	*P. hibernalis* Carne (1925)
S	V	30	*P. humicola* Ko and Ann (1985)
	I	31	*P. idaei* Kennedy and Duncan (1995)
S	IV	32	*P. ilicis* Buddenhagen and Young (1957)
S	IV	33	*P. infestans* (Montagne) deBary (1876)
S	III	34	*P. inflata* Caroselli and Tucker (1949)
S	V	35	*P. insolita* Ann and Ko (1980)
S	I	36	*P. iranica* Ershad (1971)
		37	*P. italica* Cacciola et al. (1996)
S	VI	38	*P. japonica* Waterhouse (1974a)
S	VI	39	*P. katsurae* Ko and Chang (1979) (= *P. castaneae* Katsura and Uchida [Katsura 1976])
S	V	40	*P. lateralis* Tucker and Milbrath (1942)
	V		*P. leersiae*[c] Sawada (1941) ex Ho and Chang (1992)
S	III	41	*P. lepironiae* Sawada (1919)
	IV	42	*P. macrochlamydospora* Irwin (1991)
		66	*P. macrospora* (Saccardo) S. Ito and Tanaka (1940) species is now considered to be *Sclerophthora macrospora* (Stamps et al. 1990)
S	II	43	*P. meadii* McRae (1918)
	V	44	*P. medicaginis* E. M. Hansen and Maxwell (1991) (= *P. megasperma* f. sp. *medicaginis*)
S	II	45	*P. megakarya* Brasier and Griffin (1979)
S	V	46	*P. megasperma* Drechsler (1931)
S			*P. megasperma* var. *sojae* Hildebrand (1959) and *P. megasperma* var. *megasperma* were rejected by Faris et al. (1989) and E. M. Hansen and Maxwell (1991). Soybean isolates are now classified as *P. sojae* (E. M. Hansen and Maxwell 1991)
			P. megasperma f. sp. *glycinea* Kuan and Erwin (1980a) = *P. sojae* (E. M. Hansen and Maxwell 1991)
			P. megasperma f. sp. *medicaginis* Kuan and Erwin (1980a) = *P. medicaginis* (E. M. Hansen and Maxwell 1991)
			P. megasperma f. sp. *trifolii* Pratt (1981) = *P. trifolii* (E. M. Hansen and Maxwell 1991)
S	VI	47	*P. melonis* Katsura (1976) (considered to be synonymous with *P. drechsleri* by Ho [1986b] but according to Mills et al. [1991] should be retained)
S	II	48	*P. mexicana* Hotson and Hartge (1923)

(continued on next page)

Table 4.1. *(continued)*

Classification[b]	Taxonomic group	Chapter	Species
S	IV	49	*P. mirabilis* Galindo-A. and Hohl (1985)
M S	VII		*P. mycoparasitica* Fell and Master (1975)
S	II	50	*P. nicotianae* Breda de Haan (1896) = *P. parasitica* Dastur (1913)
M S	VII		*P. operculata*[b] Pegg and Alcorn (1982)
S	II	51A	*P. palmivora* (E. Butler) E. Butler (1919) var. *palmivora*
	II	51B	*P. palmivora* var. *heterocystica* Babacauh (1983a)
			P. parasitica Dastur (1913). This species is considered to be synonymous with *P. nicotianae* Waterhouse (1963); however, the name is still used widely. See discussion in Chapter 50 on *P. nicotianae*
			P. parasitica var. *nicotianae* was described to be host-specific to tobacco (Tucker 1931); however, the name cannot be used interchangeably for *P. nicotianae* var. *nicotianae, sensu* Stamps et al. (1990). See Chapter 50.
S	IV	52	*P. phaseoli* Thaxter (1889a)
M S	VII		*P. polymorphica*[b] Gerrettson-Cornell and Simpson (1984)
S	III	53	*P. porri* Foister (1931)
S	III	54	*P. primulae* Tomlinson (1952)
S	I	55	*P. pseudotsugae* Hamm and Hansen (1983)
S	V	56	*P. quininea* Crandall (1947)
S	VI	57	*P. richardiae* Buisman (1927)
			P. robinicola[d] Xu, Lu, and Fang 1990
S	V	58	*P. sinensis* Yu and Zhuang (1982) (considered to be synonymous with *P. drechsleri* by Ho [1986b] but according to Mills et al. [1991] should be included with *P. melonis*)
	V	59	*P. sojae* Kaufmann and Gerdemann (1958)
			P. sojae f. sp. *glycines* Faris, Sabo, Barr, and Lin (1989) = *P. sojae* Kaufmann and Gerdemann 1958 (Hansen and Maxwell 1991); *P. sojae* f. sp. *medicaginis* Faris, Sabo, Barr, and Lin (1989) = *P. medicaginis* (E. M. Hansen and Maxwell 1991)
M S	VII		*P. spinosa* Fell and Master var. *spinosa*[b] (1975)
M S	VII		*P. spinosa* var. *lobata*[b] Fell and Master (1975)
S	III	60	*P. syringae* (Klebahn) Klebahn (1909)
	I	61	*P. tentaculata* Kröber and Marwitz (1993)
	V	62	*P. trifolii* E. M. Hansen and Maxwell (1991) = *P. megasperma* f. sp. *trifolii*)
S	VI	63	*P. undulata* (Petersen) Dick (1989)
S	V	64	*P. verrucosa* Alcock and Foister (1940). Published by Foister (1940)
M S	VII		*P. vesicula*[b] Anastasiou and Churchland (1969)
S	VI	65	*P. vignae* Purss (1957)
		66	*Sclerophthora macrospora*[e] (Saccardo) Thirumalachar, Shaw, and Narasimhan (1953)

[a] Some species are included for information purposes, since either the names are still widely used, e.g., *Phytophthora parasitica,* or they are relatively new. The revival of *P. sojae* by Faris et al. (1989) and E. M. Hansen and Maxwell (1991) and the description of *P. medicaginis* and *P. trifolii* by E. M. Hansen and Maxwell (1991) have made use of the term formae speciales for *P. megasperma* f. sp. *glycines* (*P. sojae*), *P. megasperma* f. sp. *medicaginis* (*P. medicaginis*), and *P. megasperma* f. sp. *trifolii* (*P. trifolii*) redundant. The designations should be discontinued. The names are retained here for reference purposes.

[b] The marine species (M) are considered by some to be distinct from *Phytophthora* and recently were reclassified in the new genus *Halophytophthora* by Ho and Jong (1990). They were excluded in the *Phytophthora* key of Newhook et al. (1978) but included in the key of Stamps et al. (1990) (S) as a new group VII. Their taxonomic position is still not certain (Brasier and Hansen 1992). Although references are made to these species in some chapters, they are not covered here in detail.

[c] A Latin description of *P. leersiae* by Ho and Chang (1992) based on Sawada's original description (1941) fulfills the requirement for its acceptance as a valid species. *P. leersiae* has not been cultured, however, and is known only from Sawada's herbarium collection on the host *Leersia hexandra* Swartz. Large, thick-walled, globose oogonia (often compressed with an average diameter of 53.9 × 48.9 μm) with tapered bases and an undulating surface are formed in host tissue. The oospores are aplerotic and thick walled. Antheridia are amphigynous with dimensions of 20–28 × 20–28 μm, average 23.9–23.5 μm. Sporangia have not been seen (Ho and Chang 1992).

[d] *Phytophthora robinicola* was described by Xu et al. (1990) in Chinese, but a Latin description was not made.

[e] *Sclerophthora macrospora* is included because there is considerable literature that cites the name of this fungus as *Phytophthora macrospora* (Saccardo) Ito and Tanaka (1940). The keys of Waterhouse (1963) and Newhook et al. (1978) listed *P. macrospora* and did not accept *S. macrospora* as a valid name; however, the key of Stamps et al. (1990) excluded the epithet *P. macrospora* but accepted *S. macrospora* as the valid name.

techniques used to study such things as protein patterns, isozyme patterns, and mitochondrial and nuclear DNA RFLPs (Brasier 1991). The RFLP technology has revealed new insights into genetic relatedness and is becoming a valuable adjunct to the determination of a species (Förster and Coffey 1991). (See the discussion below.) For instance, DNA RFLP data of Förster and Coffey (1991) support the merging of the two varieties of *P. nicotianae, nicotianae* and *parasitica* (Waterhouse 1963), as recommended by Tsao and Sisemore (1978), who showed that the morphological features used to separate these varieties in the Waterhouse (1963) and Newhook et al. (1978) keys overlapped.

However, many isolates that had been classified into an accepted species, for example, *P. citrophthora,* contain different subgroups that can be distinguished at the DNA level (Förster and Coffey 1991; Oudemans and Coffey 1993; Mchau and Coffey 1994b). The articles by Brasier (1992) and Brasier and Hansen (1992) on evolutionary biology discuss many of these examples in more detail.

Identification of some *Phytophthora* species is relatively

Table 4.2. Asexual characteristics of *Phytophthora* species[a]

SPECIES NAME	Chapter in text	Taxonomic group	Hyphal swellings	CHLAMYDOSPORES Production	CHLAMYDOSPORES Average diameter (µm)	SPORANGIA Sporangiophore branching	Caducity	Pedicel length (µm)	Shape	Length (µm)	Breadth (µm)	Average (µm)	L:B ratios	Papillate[b]	Proliferation
cactorum	11	I	-	+/-	25-40	Simple sympodial	+	>4	Ovoid-obpyriform	19-55	16-46	38 x 26	1.3	P	-
clandestina	18	I	+	-		Simple sympodial	-/+[c]	1-6	Ovoid-ellipsoid-globose	16-66	15-44	41 x 32	1.3	P	-
idaei	31	I	-	-		Lax sympodial	-	-	Spherical to ovoid	31-71	28-40	49 x 36	1.4	P	+
iranica	36	I	-	+/-	28.7	Simple sympodial	-	-	Ovoid-ellipsoid irregular	30-72	22-51	48 x 37	1.3	P	-
italica	37	I	-	-		Simple sympodial	-	-	Ovoid-obpyriform	14-56	11-38	39 x 29	1.3	P	-
pseudotsugae	55	I	+	-		Unbranched	-	-	Ovoid-obturbinate	35-45	28-35	39 x 22	1.2	P	-
tentaculata	61	I	+	+	10-45	No data	-	-	Spherical, ovoid, obpyriform	10-81	13-52	36 x 27	1.3	P	
arecae = palmivora	8	II	-	+/-	35-40	Simple sympodial	+	1-6	Ellipsoid-spherical	30-71	20-48	28 x 30	<1.4	P	-
boehmeriae	9	II	-	+	26-51	Simple sympodial	+	<5	Ellipsoid-spherical	20-70	15-51	44 x 32	1.4	P	-
botryosa	10	II	-	+/-	14-30	Simple to compound sympodial	+	5-20	Ovoid-ellipsoid variable	26-31	14-16	28 x 15	1.9	P	-
capsici	14	II	+	+/-	28-29	Simple sympodial; some umbellate	+	10-200	Ellipsoid to pyriform	30-100	25-90	variable	1.3-2.1	P	-
citrophthora	17	II	-	+ /-	28	Irregularly branched; loose sympodial	-	-	Variable	30-90	27-60	41 x 30	1.4	P	-
heveae	28	II	-	-		Irregularly branched	+	<10	Variable ellipsoid	27-66	20-49	45 x 30	1.5	P	-
katsurae	39	II	-	+/-	12-19	Irregularly branched	-	-	Ovoid liminoform	10-42	10-38	28 x 22	<1.6	P	-
meadii	43	II	-	rare	16-30	Simple sympodial	+	10-20	Obpyriform	36-67	14-38	48 x 21	<2.0	P	-
megakarya	45	II	-	+	30	Simple sympodial	+	10-30	Ovoid-obpyriform	20-60	13-41	36 x 26	1.4	P	-
mexicana	48	II	+/-	+/-	28-44	Simple sympodial	-	-	Ovoid-fusiform	16-77	16-33	46 x 24	<1.8	P	-
nicotianae =[d] *parasitica*	50	II	+	+	33	Loose sympodial or unbranched	-	-	Spherical,ovoid-ellipsoid, obturbinate	14-74	12-60	43 x 36	1.3	P	-
palmivora var *palmivora*	51 A	II	-	+	30-45	Simple sympodial	+	<5	Ovoid-ellipsoid obpyriform	35-60	20-40	55 x 33	1.7-1.9	P	-
palmivora var. *heterocystica*	51B	II	-	+	<25	Simple sympodial	+	<5	Ovoid-ellipsoid obpyriform	35-75	17-43	No data	1.2-1.7	P	-

Table 4.2. Asexual characteristics of *Phytophthora* species (continued)

SPECIES NAME	Chapter in text	Taxonomic group	Hyphal swellings	CHLAMYDOSPORES Production	CHLAMYDOSPORES Average diameter (μm)	SPORANGIA Sporangiophore branching	Caducity	Pedicel length (μm)	Shape	Length (μm)	Breadth (μm)	Average (μm)	L:B ratios	Papillate[b]	Proliferation
citricola	16	III	-	-		Simple sympodial	-	5	Obpyriform to irregular	30-75	21-44	47 x 34	1.4	SP	-
cyperi	21	III	-	-		Sympodial	+	4	Ovoid-pyriform	32-50	19-30	39 x 23	1.7	SP	-
cyperi-bulbosi	22	III	-	-		Sympodial	+	<5	Obpyriform-ovoid	23-50	20-34	40 x 25	>1.6	SP	-
inflata	34	III	-	-		Unbranched	-		Limoniform-ovoid elongated	20-67	15-32	38 x 23	>1.3	SP	+
lepironiae	41	III	+	-		Unbranched	+	5-20	Ovoid	40-64	36-52	49 x 38	<1.6	SP	-
S. macrospora	66	III	-	-		Sympodial	+	14	Ovoid-ellipsoid	65-112	32-55	87 x 47	1.6	SP	-
porri	53	III	+	-		Sympodial	-		Obpyriform-ellipsoid	35-82	23-52	46 x 37	1.4	SP	-
primulae	54	III	-	-		Irregularly branched	-	>50	Ovoid-obpyriform	33-165	23-52	77 x 44	2.1	SP	-
syringae	60	III	+	-		Sympodial	-		Ovoid-obpyriform	27-75	11-42	47 x 30	1.8	SP	-
colocasiae	19	IV	-	+	27	Irregularly branched	+	3-10	Ellipsoid-fusiform	40-70	17-28	59 x 23	1.6	SP	-
eriugena	24	IV	+/-	-		Sympodial	-		Ovoid-ellipsoid	23-104	18-58	59 x 36	1.8	SP	-
hibernalis	29	IV	-	-		Irregulary branched to sympodial	+	23-73	Ovoid-elongated	29-53	14-22	40 x 19	2.1	SP	-
ilicis	32	IV	-	-		Sympodial	+	5-15	Ovoid-obpyriform	30-50	18-30	39 x 24	1.5	SP	-
infestans	33	IV	-	-		Compound sympodial	+	<3	Ovoid-limoniform	21-38	12-23	29 x 19	1.6	SP	-
macrochlamydospora	42	IV	+	+	55	Simple sympodial	-		Ellipsoid-obpyriform	20-50	18-36	No data		SP	+
mirabilis	49	IV	-	-		Compound sympodial	+	2-4	Ovoid-ellipsoid	26-38	No data		1.9	SP	-
phaseoli	52	IV	-	-		Simple sympodial	+	5-20	Ovoid-limoniform	35-50	20-24	26 x 22	1.4	SP	-

Table 4.2. Asexual characteristics of *Phytophthora* species (continued)

SPECIES NAME				CHLAMYDOSPORES		SPORANGIA									
	Chapter in text	Taxonomic group	Hyphal swellings	Production	Average diameter (µm)	Sporangiophore branching	Caducity	Pedicel length (µm)	Shape	Length (µm)	Breadth (µm)	Average (µm)	L:B ratios	Papillate[b]	Proliferation
fragariae var. *fragariae*	26 A	V	+	-		Simple sympodial	-		Ellipsoid-obpyriform	32-90	22-52	57 x 36	1.6	NP	+
fragariae var. *oryzobladis*	26B	V	-	-		Simple sympodial	-		Ellipsoid-obpyriform	43-93	28-57	67 x 43	1.6	NP	+
fragariae var. *rubi*	26C	V	-	-		Simple sympodial	-		Ellipsoid-obpyriform	53-68	34-44	59 x 40	1.4	NP	+
humicola	30	V	+	-		Unbranched/sympodial	-		Ovoid	38-60	50-72	44 x 58	1.3	NP	+
insolita	35	V	+	+	24-48	Unbranched	-		Ovoid	36-68	29-59	51 x 35	1.6	NP	+
lateralis	40	V	+	+	40	Simple sympodial	-		Ovoid-obpyriform	20-60	12-20	26 x 15	1.7	NP	+
medicaginis	44	V	+	+/-	6-16	Unbranched	-		Ovoid-ellipsoid	40-60	24-30	39 x 27	1.3	NP	+
megasperma [e]	46	V	+	-		Unbranched	-		Ovoid-obpyriform	35-60	25-45	41 x 28	1.5	NP	+
quininea	56	V	+	+	45-90	Unbranched	-		Ovoid-obpyriform	46-68	26-38	54 x 31	1.7	NP	+
sojae	59	V	+	+/- rare		Unbranched	-		Ovoid-obpyriform	23-89	17-52	58 x 38	1.5	NP	+
trifolii	62	V	+	-		Unbranched	-		Ovoid-ellipsoid	44-50	30-35	44 x 31	1.4	NP	+
verrucosa	64	V	-	-		Unbranched	-		Ovoid-obpyriform	31-56	24-36	42 x 30	1.4	NP	+

Table 4.2. Asexual characteristics of *Phytophthora* species (continued)

SPECIES NAME	Chapter in text	Taxonomic group	Hyphal swellings	CHLAMYDOSPORES Production	CHLAMYDOSPORES Average diameter (µm)	SPORANGIA Sporangiophore branching	Caducity	Pedicel length (µm)	Shape	Length (µm)	Breadth (µm)	Average (µm)	L:B ratios	Papillate[b]	Proliferation
cajani[f]	12	VI	+	-		Simple sympodial	-		Ovoid-obpyriform	49-82	19-44	60 X 32	1.7	NP	+
cambivora	13	VI	+	-		Simple sympodial	-		Ovoid-ellipsoid	22-95	17-66	41 X 31	1.6	NP	+
cinnamomi var. *cinnamomi*	15A	VI	+	+	41	Unbranched to simple sympodial	-		Ovoid-ellipsoid	27-114	20-63	75 x 40	>1.5	NP	+
cinnamomi var. *parvispora*	15B	VI	+	+	27.1	Simple sympodial	-		Ovoid,ellipsoid, obpyriform	14-83	13-43	37 X 26	1.4	NP	+
cryptogea	20	VI	+	-		Simple sympodial	-		Ovoid-irregular	35-63	24-35	52 x 30	1.7	NP	+
drechsleri[f]	23	VI	+	+	7.9	Simple sympodial	-		Ovoid-obpyriform	24-38	15-24	31 x 21	1.6	NP	+
erythroseptica var. *erythroseptica*	25 A	VI	+	-		Simple sympodial	-		Ellipsoid-obpyriform	43-69	26-47	44 x 27	1.6	NP	+
erythroseptica var. *pisi*	25B	VI	?	-		Simple sympodial	-		Ellipsoid-obpyriform	54-171	25-40	86 x 35	2.5	NP	+
gonapodyides	27	VI	+	-		Unbranched to simple sympodial	-		Obpyriform	42-75	20-32	44 x 22	1.6	NP	+
japonica (*oryzae*)	38	VI	+	+		Sympodial	-		Ovoid-ellipsoid	41-84	26-48	60 x 36	1.9	NP	+
melonis = drechsleri ?	47	VI	+	+	31.2	Simple sympodial	-		Ovoid-obpyriform	33-75	23-46	49 x 34	1.5	NP	+
richardiae	57	VI	+	-		Unbranched	-		Variable	40-60	20-40	52 x 33	1.8	NP	+
sinensis = drechsleri ?	58	VI	+	-		Simple sympodial	-		Ovoid-elongate	24-80	20-40	52 x 31	1.7	NP	+
undulata	63	VI	-	+	21-61	Unbranched	-		Ellipsoid	27-156	12-50	77 x 33	>2.0	NP	+
vignae	65	VI	+	+	17	Sympodial	-		Ovoid-obpyriform	24-72	15-54	48 x 27	<1.8	NP	+

a. Dimensions of spore structures are compiled from data of several reports or from the original description. Please consult the appropriate chapter for additional data.

b. Papillate, P; semipapillate, SP ; nonpapillate, NP.

c. ◯ Predominantly of this type.

d. The *P. nicotianae* varieties *nicotianae* and *parasitica* (Waterhouse 1963) are no longer recognized (Förster et al. 1990b; Förster and Coffey 1991; Hall 1993)

e. Data are from Drechsler (1931). The varieties *megasperma* and *sojae* are now redundant; *P. sojae*, *P. medicaginis*, and *P. trifolii* are transferred to species classification (Hansen and Maxwell 1991).

f. See chapters 12 (*P. cajani*) and 23 (*P. drechsleri*) for discussion of *P. drechsleri* f. sp. *cajani*.

Table 4.3. Sexual characteristics and temperature relations of *Phytophthora* species[a]

SPECIES NAME	Chapter in text	Taxonomic group	Homothallic	Heterothallic	ANTHERIDIA			OOGONIA		OOSPORES				TEMPERATURE (°C)		
					Amphigynous	Paragynous	Size (µm) (L X B)	Ornamented wall	Size (µm)	Size (µm)	Average size (µm)	Wall thickness (µm)	Formed in host tissue	Minimum	Optimum	Maximum
cactorum	11	I	+	-	-	+	9-18 x 9-20	-	18-40	15-36	25	2	+	2	25	31
clandestina	18	I	+	-	+	⊕[b]	9-45 x 4-17	-	21-36	18-31	25	1.4	+	<5	25	31
idaei	31	I	+	-	+	⊕	7.8-15.8 x 12.6-27.5	-	26-38	22-34	27	2.1	+	5	20-22.5	27
iranica	36	I	+	-	+	⊕	11-21 x 8-19	-	21-45	15-37	29	3	-	10	27.5	35
italica	37	I	+	-		+	6-12 x 4-10	-	15-29	11-23	19			10	26	35
pseudotsugae	53	I	+	-	+	⊕	6-21 x 10-24	-	29-45	27-33	29		-	2	20-25	30
tentaculata	61	I	+	-	+	⊕	16.3 x 11.9	-	29-49	14-38	28.1			7	15-25	32
arecae = *palmivora*	8	II	+	⊕	+		14 x 15 (avg.)	-	23-44	23-36	28	3	-	10-12	27-30	35
boehmeriae	9	II	+	-	+		14-17 x 14-16	-	19-41	16-35	23-30	<2	+	5-6	25	32.5
botryosa	10	II	-	+	+		14 x 13	-	22-31	18-24	22	1.5	-	9	26	32
capsici	14	II	-	+	+		14-19 x 13-15		18-50	15-40	26	2-6	-	6-9	25-32	>35
citrophthora	17	II	Sex organs not produced in nature											<5	24-28	>32
heveae	28	II	+	-	+		10-11 x 8-9	-	17-32	15-28	21	3	+	11.5	25	31-33
katsurae	39	II	+	-	+		18-14 x 8-13	+/-	19-31	15-28	20	1.6	+	9	26-28	32
meadii	43	II	⊕	+	+		10-16	-	20-48	16-33	24	4	+/-	>5	25-30	>33
megakarya	45	II	-	+	+			-	19-37	18-34	26	2.4	-	10-11	24-26	29.5
mexicana	48	II	-	+	+		15 x 14 (avg.)	-	30-38	24-37	25	4.4	-	>10	27.5	<35
nicotianae = *parasitica*[c]	50	II	-	+	+		10-12 x 9-10	-	16-36	12-34	24	2	+	5-7	27-32	37
palmivora var. *palmivora*	51A	II	-	+	+		15 x 14 (avg.)	-	21-40	16-30	23	2	-	11	28-30	<35
palmivora var. *heterocystica*	51 B	II	-	+	+		No data									

Table 4.3. Sexual characteristics and temperature relations of *Phytophthora* species (continued)

SPECIES NAME	Chapter in text	Taxonomic group	Homothallic	Heterothallic	ANTHERIDIA			OOGONIA		OOSPORES				TEMPERATURE (°C)		
					Amphigynous	Paragynous	Size (µm) (L X B)	Ornamented wall	Size (µm)	Size (µm)	Average size (µm)	Wall thickness (µm)	Formed in host tissue	Minimum	Optimum	Maximum
citricola	16	III	+	-	+	⊕	9-13 x 7-8	-	18-35	16-30	22	1.5	+	1-3	25-28	31
cyperi	21	III	+	-	-	+	20 x 12 (avg.)	-	30-45	25-38	30	3	+	Not culturable		
cyperi-bulbosi	22	III	+	-	-	+	16 x 12 (avg.)	+	31-53	25-42	33	1.5	+	Not culturable		
inflata	34	III	+	-	-	+	50 x 15 (avg.)	-	30-43	26-39	31	3-4	+		25-30	<35
lepironiae	41	III	+	-	-	+	23 x 15 (avg.)	-	40-52	30-38	34	3.5	+	Not culturable		
macrospora	66	III	+	-	-	+	27 x 15 (avg.)	-	50-100	45-94	58	4	+	Not culturable		
porri	53	III	+	-	+	⊕	13 x 12 (avg.)	-	29-46	22-39	32	4.5	+	8	20	27
primulae	54	III	+	-	-	+	15 x 15 (avg.)	-	23-43	17-33	30	3	+	1-2	15-20	<27
syringae	60	III	+	-	+	⊕	12 x 8 (avg.)	-	33-41	30-35	33	2	+	<5	15-20	23
colocasiae	19	IV	-	+	+	-	11-13 x 7-11	-	20-35	18-30	23	2.5	-	>10	27-30	>35
eriugena	24	IV	+	-	+	-	12 x 18 (avg.)	-	23-65	20-60	35	3.1	+	0	12	22
hibernalis	29	IV	+	-	⊕	+	10 x 15 (avg.)	-	33-35	29-32	31	1-3	-	<5	15	25-26
ilicis	32	IV	+	-	+	-	13 x 11 (avg.)	-	25-32	18-26	21	2	+	5	20	25
infestans	33	IV	-	+	+	-	17 x 16 (avg.)	-	31-50	24-46	30	3.5	-	3-4	20	26
macrochlamydospora	42	IV					Sex organs unkown							6	23-30	34
mirabilis	49	IV	-	+	+	-	14 x 12	-	26	18-25	21	2.5	-	7	21	27
phaseoli	52	IV	+	-	+	-	15 x 10 (avg.)	-	16-36	14-27	21	3	+	5	15-20	25-30

Table 4.3. Sexual characteristics and temperature relations of *Phytophthora* species (continued)

SPECIES NAME	Chapter in text	Taxonomic group	Homothallic	Heterothallic	ANTHERIDIA			OOGONIA		OOSPORES				TEMPERATURE (°C)		
					Amphigynous	Paragynous	Size (µm) (L X B)	Ornamented wall	Size (µm)	Size (µm)	Average size (µm)	Wall thickness (µm)	Formed in host tissue	Minimum	Optimum	Maximum
fragariae var. *fragariae*	26A	V	+	-	(+)	+	22 x 16 (avg.)	-	28-44	24-60	34	3	+	3	18-22	30
fragariae var. *oryzobladis*	26 B	V	+	-	+	-	No data	-	40-58	27-48	36		+		No data	
fragariae var. *rubi*	26 C	V	+	-	(+)	+	13-16	-	30-45	28-37	35	3	+	4	19-22	25-28
humicola	30	V	+	-	+	(+)	14 x 18 (avg.)	-	25-52	20-40	34	3.8	-	9	28-30	32
insolita	35	V	+	-	-	-	Antheridia unknown	-	29-36	27-31	30	2.4	-	9	32	<40
lateralis	40	V	+	-	-	+	12-18 x 13-16	-	33-50	28-46	40	2	-	<2	20	<26
medicaginis	44	V	+	-	+	+	14 x 14 (avg.)	-	31-38	23-34	31	2-4	+	8	25	33
megasperma[d]	46	V	(+)	+[d]	+	(+)	10-18 x 14-20	-	42-52	37-47	41	3.6	+	5	20-25	<33
quininea	56	V	+	-	-	+	Variable	-	68-82	49-64	52		+	10-15	25	30
sojae	59	V	+	-	+	(+)	15 x 13 (avg.)	-	27-54	24-46	31	2.4	+	5	25	32-35
trifolii	62	V	+	-	+	(+)	No data	-	37-46	33-37	35	2		5	22	30
verrucosa	64	V	+	-	+	+	16 x 13 (avg.)	+/-	23-47	17-31	24	3	+		Not culturable	

Table 4.3. Sexual characteristics and temperature relations of *Phytophthora* species (continued)

SPECIES NAME					ANTHERIDIA			OOGONIA		OOSPORES				TEMPERATURE (°C)		
	Chapter in text	Taxonomic group	Homothallic	Heterothallic	Amphigynous	Paragynous	Size (µm) (L X B)	Ornamented wall	Size (µm)	Size (µm)	Average size (µm)	Wall thickness (µm)	Formed in host tissue	Minimum	Optimum	Maximum
cajani[e]	12	VI	+	(+)	+	-	15-16	-	22-40	20-37	32		-		No data	
cambivora	13	VI	-	+	+	-	25-35 x 19	+	23-62	32-38	34	2	+/-	2	22-24	32
cinnamomi var. *cinnamomi*	15 A	VI	-	+	+	-	19-21 x 16-17	-	21-58	19-54	33	2	rare	5-15	20-32.5	30-36
cinnamomi var. *parvispora*	15 B	VI	-	+	+	-	large	-	23-41	22-36	28			10-11	26-32	36-37
cryptogea	20	VI	-	+	+	-	13-19 x 15-17	-	28-37	24-32	27	3.5	-	<1	20-25	30-35
drechsleri[e]	23	VI	-	+	+	-	14 x 13 (avg.)	-	22-53	17-45	26	3	-	10	30-32	37.5
erythroseptica var. *erythroseptica*	25 A	VI	+	-	+	-	14-19 x 13-18	-	26-46	28-35	30	2.5	+	<2.5	27.5	34
erythroseptica var. *pisi*	25 B	VI	+	-	+	-	19 x 18 (avg.)	-	37	14-34	26	2.9	+	12	25	30
gonapodyides	27	VI	-	+	+	-	variable	+/-	27-48	18-32	21		-	5	25	30-35
japonica (*oryzae*)	38	VI					Sex organs unknown							9-11	26-28	35
melonis = *drechsleri* ?	47	VI	+	+	+	-	15 x 20	-	28-35	25-32	29	2.8	-	9	28-32	37
richardiae	57	VI	+	-	(+)	+	14-18 x 14-18	-	34-48	19-41	31	4-5	+	3.4	24	30
sinensis = *drechsleri* ?	58	VI	-	+	+	-	19x18 (avg.)	-	19-31	16-28	24	2-3	-	9	29-32	36.5
undulata	63	VI					Sex organs unknown							5	20-25	35
vignae	65	VI	+	-	+	-	16x15 (avg.)	-	33-31	18-32	26	3	+	10-12	28-30	<35

a. Dimensions of spore structures are compiled from data of several reports or from the original description. Please consult the appropriate chapter for additional data.

b. ◯ Predominantly of this type.

c. See discussion of *P. nicotianae* (Chapter 50). The *P. nicotianae* varieties and *parasitica* (Waterhouse 1963) are no longer recognized (Förster et al. 1990b; Förster and Coffey 1991; Hall 1993).

d. Data are from Drechsler (1931). The varieties *megasperma* and *sojae* are now redundant (Faris et al 1989; Hansen and Maxwell 1991). The only heterothallic cultures are large oogonial isolates from alfalfa (Barr 1980, Ho 1986a).

e. See chapters 12 *(P. cajani)* and 23 *(P. drechsleri)* for discussion of *P. drechsleri* f. sp. *cajani*.

easy even for those with little background. For instance, *P. cinnamomi* uniquely produces tuberculate (knobby) mycelium, large chlamydospores (Figure 4.1), and nonpapillate sporangia (Figure 4.12N) when incubated in nonsterile soil extract. Also, data from molecular studies show that the protein patterns (Hall et al. 1969; Gill and Zentmyer 1978; Zentmyer 1980) and isozyme patterns of many isolates collected worldwide are relatively uniform (Old et al. 1984, 1988).

Many other species, such as *P. cryptogea* and *P. drechsleri,* are not readily distinguished with certainty (Bumbieris 1974; Erwin 1983; Ho and Jong 1986; Mills et al. 1991). The two species are similar in morphology, and both produce nonpapillate sporangia that according to some are similar in shape (Bumbieris 1974; Halsall 1976a; Gerrettson-Cornell 1985, 1989; Ho and Jong 1986) but according to others are different (Tucker 1931; Waterhouse 1963; Newhook et al. 1978) (Chapters 20 and 23). Tucker (1931) differentiated *P. drechsleri* by growth at the maximum temperature of 35°C, a temperature at which *P. cryptogea* does not grow. Some of those who study isolates of the *P. cryptogea-P. drechsleri* complex state that the criterion for growth at 34–36°C is too variable to be of use taxonomically (Gerrettson-Cornell 1985) and judge that the two taxa should be merged under the earlier species name, *P. cryptogea.* When a large number of isolates of this group collected worldwide were tested for isozyme and mitochondrial DNA RFLP patterns, however, it was found that the *P. cryptogea-P. drechsleri* complex contained several discrete groups (Mills et al. 1991). Thus, the concept that the two species should be merged because of the similarity in morphology (Bumbieris 1974; Ho and Jong 1986) appears to be questionable.

Data from molecular DNA technology apparently indicates that there may be morphological differences among species not readily discernible with the methods at hand. Brasier and Hansen (1992) suggest "that several morphologically similar taxa may have arisen via geographical isolation and climatic and host specialization, and that their distributions are now overlapping through introductions."

The discussion of the taxonomy of *Phytophthora* by Ho and Jong (1987) documents many critical problems. The current controversies over identifying species of *Phytophthora* indicate that some names used now may or may not be valid in the future. The new techniques now available to study these fungi at the DNA level (Förster et al. 1987, 1988, 1990) will undoubtedly become more available to many more laboratories. These methods are discussed below.

In the following pages, we explain the known morphological and physiological features that are unique to each species. The species of *Phytophthora* are listed alphabetically in Table 4.1 and designated by morphologic groups (Stamps et al. 1990) (Appendix 4.1). The criteria that are useful for identifying a *Phytophthora* species are given in Appendix 4.2. In Appendix 4.3, a tally sheet useful for characterizing *Phytophthora* species (Kröber 1985) is given. Appendix 4.4 shows the association of certain *Phytophthora* species with sizes and shapes of sporangia and types of sporangiophores, production of chlamydospores, size and shape of oogonia, types of antheridia, and host range. Illustrations of useful characteristics include structure of sporangiophores (Figures 4.5–4.8), the shapes of sporangia (Figures 4.10–4.12), the papillate, semipapillate, and nonpapillate nature of sporangia (Appendix 4.5; Figures 4.7 and 4.9), the caducous nature of sporangia (Appendix 4.6; Figures 4.15–4.17), chlamydospore morphology (Figures 4.23–4.25), antheridium configuration (Appendix 4.7; Figure 4.26), oogonium shape and ornamentation (Figures 4.26 and 4.27), sporangium size, chlamydospore production, oogonium-oospore size and host range (Appendix 4.4), and maximum temperatures for growth (Appendix 4.8). Some species listed in Tables 4.1–4.3 have been described since the key of Stamps et al. (1990). Asexual characteristics are summarized in Table 4.2, and sexual characteristics and maximum, optimum, and minimum temperatures for growth are summarized in Table 4.3.

Morphological Characteristics

The morphological characteristics by which most *Phytophthora* species are classically identified include the following, based on the groups I–VI of Stamps et al. (1990), originally proposed by Waterhouse (1963):

1. Whether the antheridium is amphigynous (around the female) (Figure 4.26B; groups II, IV, and VI) or paragynous (beside the female) on the oogonium (Figure 4.26A and C) (groups I, III, and V of Stamps et al. [1990]); ornamentation on the oogonium wall (typical of *P. cambivora*) (Figures 4.27 and 4.28); whether oospores are formed in single culture (self-fertile or homothallic) or only if paired with an opposite compatibility type (self-sterile or heterothallic) (Appendix 4.7).
2. Sporangium morphology (shape, size, and length-breadth ratio) (Figures 4.10–4.12); presence of an apical thickening on the sporangia called a papilla (nipple) (groups I, II, III of Waterhouse [1963]) (Appendixes 4.5 and 4.6; Figures 4.5–4.9).
3. Presence or absence of caducity (shedding of the sporangium at maturity) and length of the pedicel on the shed sporangia (short, intermediate, or long) (Appendix 4.6; Figures 4.12 and 4.15–4.17).
4. Proliferation of sporangia (production of new sporangia within a sporangium that has germinated directly (Figures 4.5, 4.7D and E, and 4.8).
5. Branching of sporangiophores (unbranched, monochasial, i.e., single branch; sympodial, i.e., beneath the foot or base of the sporangium) (Figures 4.5, 4.6, and 4.8).
6. Presence or absence of chlamydospores (Figures 4.23–4.25).
7. Presence or absence of hyphal swellings (globose to irregular protuberances on mycelium by a septum) (Figures 4.4 and 4.23).
8. Maximum temperature for growth (Appendix 4.8).

Some of these criteria that are useful to identify species of *Phytophthora* are summarized in Tables 4.2 and 4.3. All characteristics do not have equal weight; some are primary and others secondary in importance. See Appendix 4.3, in which a tally sheet is included for recording information.

Mycelium

The thallus of *Phytophthora* is called mycelium (Greek *mykes,* a fungus; *-lium,* after epithelium). It is composed of hyaline, branched, coenocytic (nonseptate) filaments, except in old cultures where septa can sometimes be seen. In young cultures, the cytoplasm flows freely within the mycelium. The diameter of the mycelium (5–8 μm) is variable and dependent on the physical and chemical nature of the medium and on whether the mycelium is on the surface, aerial, submerged, or within host cells (mycelia of six species are shown in Figure

4.1). Mycelium is sometimes swollen, nodose, or tuberculate (typical of *P. cinnamomi;* Figure 4.1) and seldom grows symmetrically. Hyphae (single branches of the mycelium) branch at nearly right angles and are sometimes constricted at the base (Ho 1978). Although certain species show characteristic mycelial patterns (Figure 4.1) (Frezzi 1950), characteristics of mycelium are not useful enough to differentiate species. *P. porri,* however, characteristically produces coiled hyphae .

Hyphal branching systems of *Phytophthora* have not thus far been used for taxonomic purposes. Ho's (1978) investigation shows that certain species such as *P. cinnamomi, P. cactorum,* and *P. sojae* (*P. megasperma* f. sp. *glycinea*) differ. The apical segment ratio (the ratio between the length of the apical segment and the mean length of internodes) of growth at the margin of a colony growing on one-tenth strength clear V8 juice agar (Appendix 3.2) was 1.7 for *P. cinnamomi,* 1.3–1.5 for *P. cactorum,* and 4.3 for *P. megasperma.* This result simply quantitates the fact that leading hyphae of *P. megasperma* grow longer before they branch than do those of *P. cinnamomi* or *P. cactorum.* Hyphal diameters were greater for *P. megasperma* (6.5 µm) than for *P. cinnamomi* (5.8–6.0 µm) or *P. cactorum* (5.0–5.1 µm). As might be expected, the total branch ratio (total number of branches borne by leading hyphae to the number of hyphae observed) was 20–26 for *P. cinnamomi* and only 11.2 for *P. megasperma* and 8.8–10.6 for *P. cactorum.* Again, this result simply quantitates what is immediately observed when colonies of these species are examined microscopically (Ho 1978).

The hyphal branching angle (between a leading hypha and its primary branch) do not differ greatly among the three species (70–80° for *P. cinnamomi,* 80° for *P. megasperma,* and 65–70° for *P. cactorum*); however, the branching angle for the closely related *Pythium aphanidermatum* was about 29° (Ho 1978). Although branching patterns might be useful enough to separate *Pythium* from *Phytophthora,* there are many other more obvious differential morphological characteristics covered below in this chapter. Much more data, on many more species and growth conditions, are necessary to determine the significance of Ho's (1978) study for the genus as a whole.

The most distinct cultural patterns occur on potato-dextrose agar. Patterns can also be detected on cornmeal agar or carrot agar.

Although Waterhouse (1970b) suggests that colony types and hyphal branching should be examined, these parameters have not been useful for identification. Note the variability in colony types produced by the *P. drechsleri-P. cryptogea-P. melonis* complex (Figure 4.3). According to Waterhouse (1970b), cultural form (colony type) depends on the frequency,

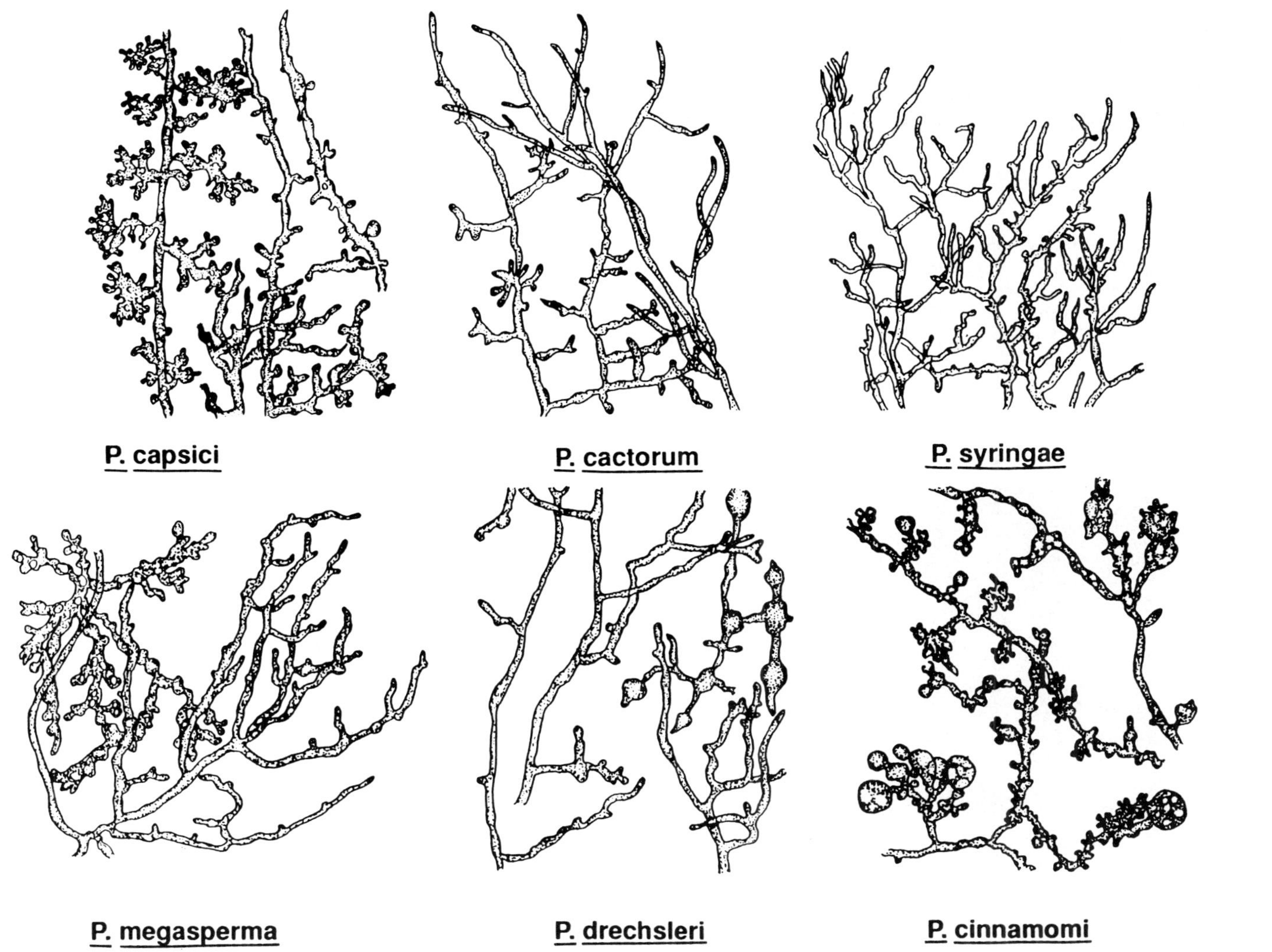

Figure 4.1. Mycelia typical of several *Phytophthora* species. Although these mycelial types are of presumptive value, they are not dependable for positive identification. (Reproduced from Frezzi 1950 by permission of E. H. Hijano, Estacion Experimental Manfredi, Argentina)

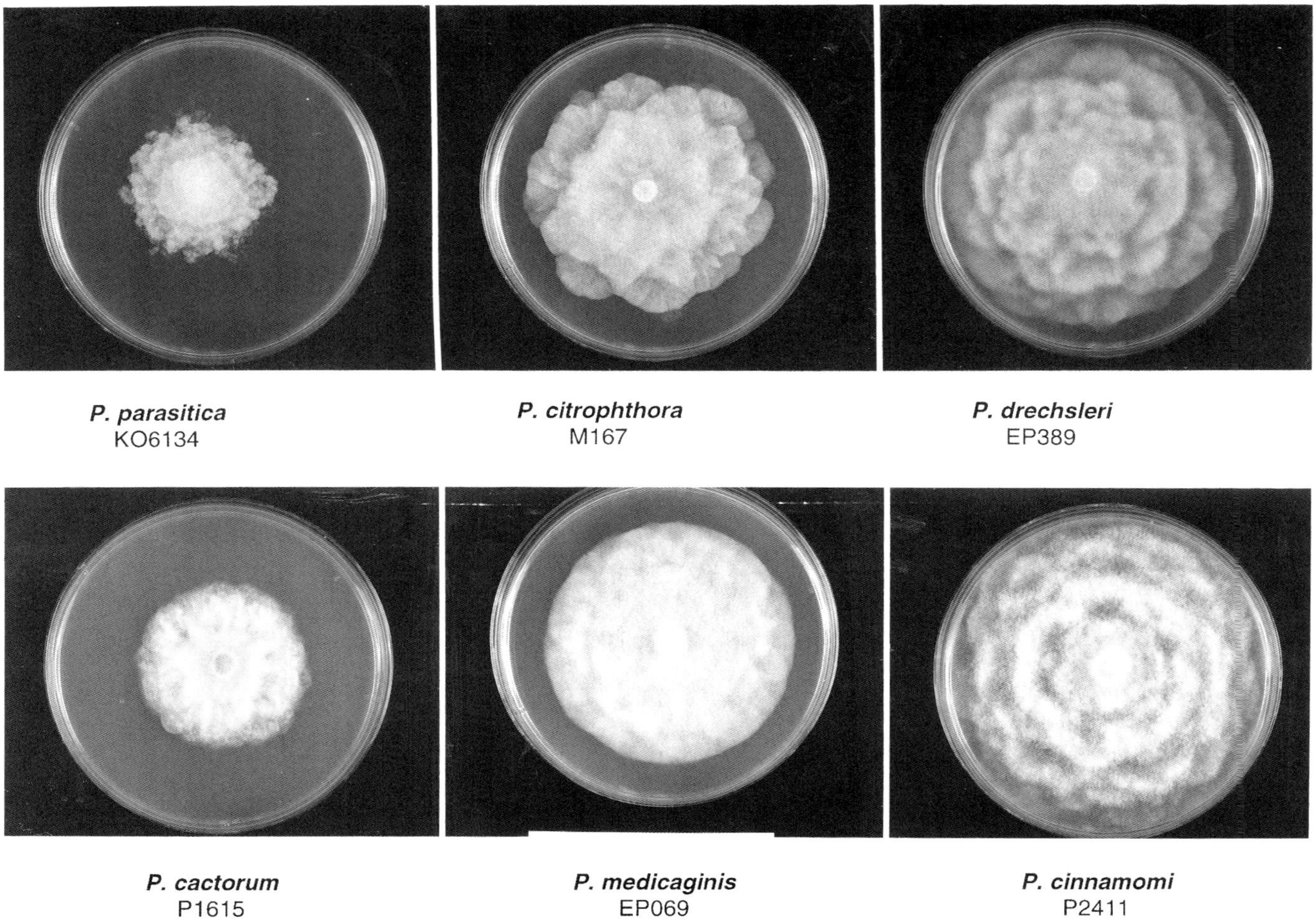

Figure 4.2. Colony types of several *Phytophthora* species grown on potato-dextrose agar. *P. parasitica* (K06134 from W. Ko) produces a stoloniferous colony; *P. citrophthora* (M167 from J. A. Menge) produces a petaloid colony; *P. drechsleri* (EP389 from D. C. Erwin) produces a rosaceous colony; *P. cactorum* (P1615 from S. N. Jeffers) produces a less defined petaloid colony; *P. medicaginis* (EP069 single-oospore isolate from D. C. Erwin) produces only a slight petaloid pattern; and *P. cinnamomi* (P2411 from G. A. Zentmyer) produces a rosaceous colony. Although colony types are somewhat typical, they are too variable to be definitive.

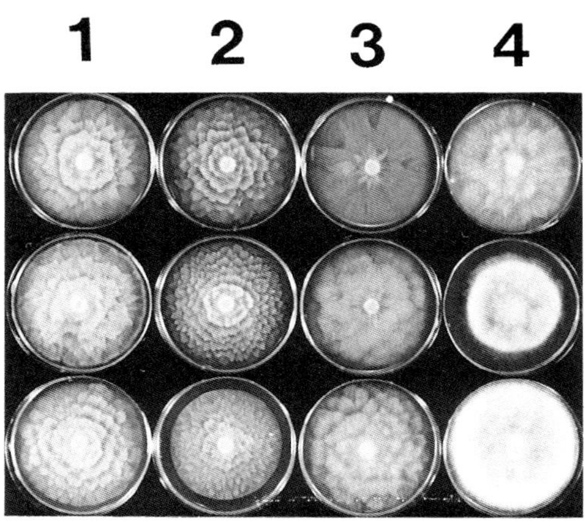

Figure 4.3. Variability of colony types of isolates from the *Phytophthora cryptogea-P. drechsleri* complex showing rosaceous, stellate, and cottony patterns. Isolates 1–4 are from the isozyme groups J, K, H, and C, respectively. Colonies of number 1 are stellate, number 2 are rosaceous, number 3 are stellate to rosaceous, and number 4 are petallate (top) to cottony. (Reproduced by permission of the British Mycological Society, courtesy Mills et al. 1991)

angle, and extent of branching of the hyphae. She cites Caten and Jinks (1968), who imply that cultural type is variable and may be under cytoplasmic control. Despite this possibility, she notes that certain cultural types are typical. In some species, the hyphae radiate uniformly. In the chrysanthemum type, the sectors are narrowly petaloid (*P. citricola*). In the rose or camellia types, the sectors are broadly rounded and petal-like (e.g., *P. cinnamomi, P. syringae,* and *P. drechsleri*); in the stellate type, the colony radiates like a star (e.g., *P. palmivora*); and in some types, the mycelium grows out without an obvious pattern (e.g., *P. infestans*) (Waterhouse et al. 1983).

Some of the colony types are illustrated in Figure 4.2. *P. nicotianae* (*P. parasitica*) has a distinctive arachnoid (spider web) or stoloniferous growth on potato-dextrose agar. The colony type rapidly differentiates *P. parasitica* from *P. citrophthora,* which produces petaloid colonies. Both are common pathogens of citrus trees. *P. medicaginis* (*P. megasperma* f. sp. *medicaginis*), commonly isolated from alfalfa, produces cottony mycelium with only a slight pattern to the colony (Figure 4.2). Colonies of some isolates are "densely fluffy" or "scanty-fluffy," but some colonies are appressed (Erwin 1965; R. G. Pratt and Mitchell 1973; Stack and Millar 1985a; Erwin and Li 1986). Erwin and Li (1986) reported that appressed colonies produced more oospores and more rapid radial growth than fluffy colonies. *P. drechsleri* produces a petaloid colony on potato-dextrose agar (Figure 4.2); however, when more isolates are observed, different colonies vary from no pattern through stellate to petaloid types (Figure 4.3) (Mills et al. 1991). Despite the ease of examining colony types, they are not usually consistently distinctive enough to be more than supplementally useful for identification of species. Leonian (1934) showed that single-spore isolates of *P. parasitica* varied widely in colony patterns.

Hyphal swellings (Figure 4.4) form in the mycelium in chains or clusters. These structures are noted most readily in water cultures of many species, such as *P. cinnamomi, P. cambivora, P. drechsleri, P. cryptogea, P. megasperma,* and others (see Chitzanidis and Kouyeas [1970]). The function of these bodies is not known, but it is supposed that they might serve to perpetuate the fungus, like chlamydospores. They can be differentiated from chlamydospores by the lack of delineation from the mycelium by septation. Hyphal swellings are produced by *P. sojae* (*P. megasperma* f. sp. *glycinea*), *P. medicaginis* (*P. megasperma* f. sp. *medicaginis*), *P. megasperma* var. *megasperma, P. megasperma* var. *sojae* (*sensu* Newhook et al. 1978), *P. cryptogea, P. drechsleri* var. *drechsleri, P. drechsleri* var. *cajani, P. nicotianae* (*P. parasitica*), *P. syringae, P. porri, P. quininea, P. cambivora, P. cinnamomi, P. erythroseptica* var. *erythroseptica, P. erythroseptica* var. *pisi, P. richardiae, P. vignae,* and *P. japonica.* In *P. cinnamomi,* they are large and distinctive in shape and numbers, as are chlamydospores (Figure 4.1).

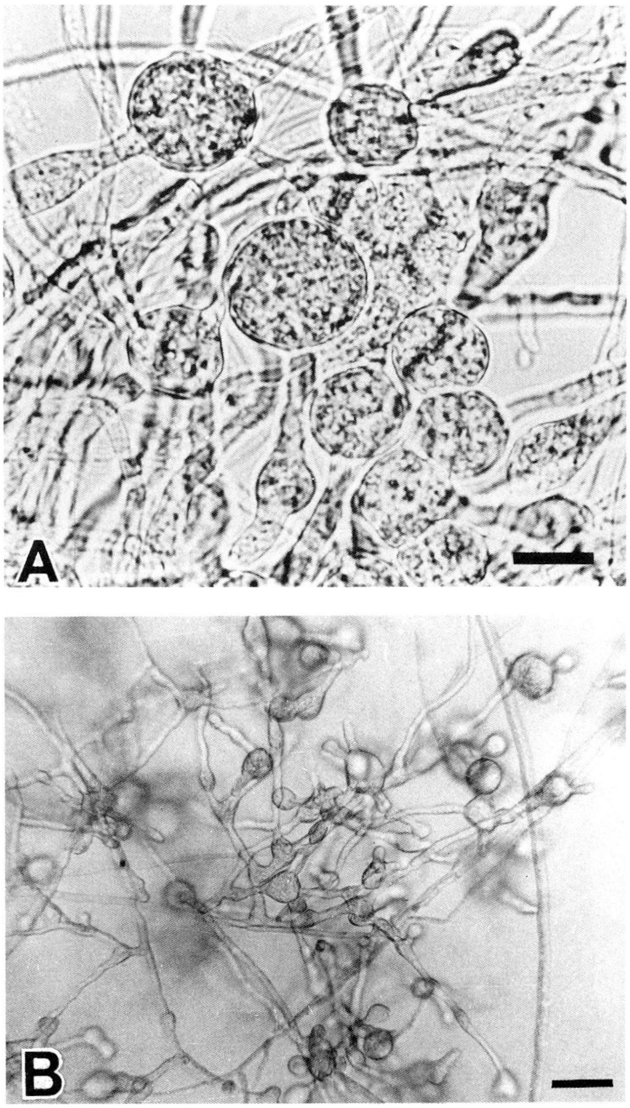

Figure 4.4. Hyphal swellings of *Phytophthora* that form in an aqueous culture. **A,** *P. drechsleri* (P389) from sugar beet. Bar = about 26 µm. **B,** *P. medicaginis* (P1057) from alfalfa. Bar = about 42 µm. (Courtesy R. A. Khan and D. C. Erwin)

Hyphal swellings are not distinctive enough in most species to be of diagnostic use. Waterhouse et al. (1983) point out that in group III, their absence in *P. citricola* differentiates this species from *P. syringae, P. porri,* and *P. primulae.* Failure to produce hyphal swellings might be useful as one criterion to differentiate species with similar characteristics (Newhook et al. 1978).

Sporangia and Zoospores

Morphology of Sporangia

Asexual sporangium development is the means of most rapid reproduction (Figure 3.1). Some species produce sporangia on natural agar media, but others require subsequent incubation in aqueous solutions (see Chapter 3 on nutrition and physiology). Sporangia (Greek *spora,* a seed; *angeion,* a vessel) are asexual spores that are produced on stalks called sporangiophores (Greek *phoros,* a vessel), which differ slightly or not at all from vegetative hyphae. In some species, a new sporangiophore emerges externally from the base of the previous sporangium (*sympodial,* at the foot of) and produces a new sporangium (simple sympodium) (Figures 4.5A and 4.8A). When the sporangiophore is branched, it is called a compound sympodium, which is typical of *P. infestans* (Figures 4.5B and C and 4.6). *P. infestans* characteristically produces swellings on the sporangiophore above the place from which the sporangia have been shed. *P. capsici* isolates from cacao, formerly called *P. palmivora* MF4, form sporangia in an umbel pattern (inverse umbrella) (Figures 4.5D and 4.16B–F). In some species, the sporangiophore resumes growth through the base of the evacuated sporangium; the new sporangium may proliferate within the walls of the empty one (internal proliferation) (Figures 4.7C and D and 4.8C) or the sporangiophore may

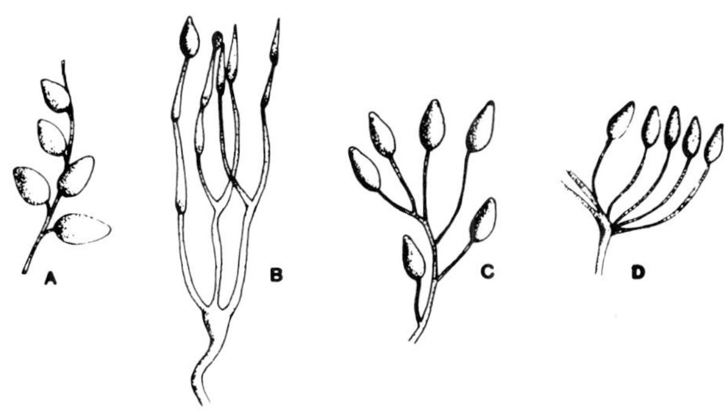

Figure 4.5. Morphology of sporangiophores. **A,** Simple sympodium (typical of *Phytophthora cactorum*). **B and C,** Compound sympodia. **B** is representative of a *Phytophthora* species such as *P. infestans,* in which the sporangiophores have a swollen node at the place at which the caducous sporangia on four of the five sporangiophores have shed (see Figure 4.6). **D,** Umbel (umbrella) type, characteristic of isolates of *P. capsici* from black pepper or cacao (see also Figure 4.16). (Drawn by A. Vaziri)

grow out through the exit pore of the previous sporangium and form the next sporangium some distance from the last (extended proliferation) (Figures 4.7E and 4.8B). External or internal proliferation appears to be much more typical of taxonomic groups V and VI than of groups I, II, III, and IV (Table 4.2). An exception is *P. inflata,* classified in group III, which produces sporangia that proliferate (Hall et al. 1992).

The apical thickening on the sporangium is called the papilla (plural, papillae) (Figures 4.7A and B and 4.9B and C). Sporangia are classified on the basis of the apical thickening as papillate (conspicuous thickening) (Figure 4.9C), semipapillate (Figure 4.9B), or nonpapillate (or inconspicuous thickening) (Figure 4.9A; Appendix 4.5) (Waterhouse et al. 1983).

Since the term papilla is derived from the Latin word for nipple and a large number of species produce a protuberance on the apical end of the sporangium that is nipplelike, this term can be confusing because in *Phytophthora* terminology, the papilla refers to the plug or thickening at the apical end of the sporangium through which the zoospores exit (Blackwell

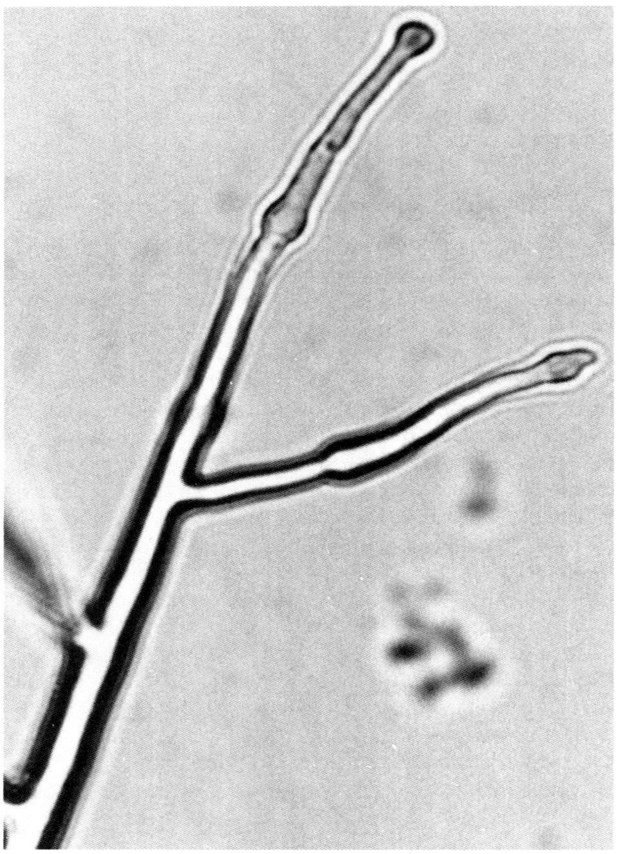

Figure 4.6. Sporangiophore of *Phytophthora infestans* showing characteristic swellings above the point from which the caducous sporangia have shed. (Courtesy M. J. Griffin)

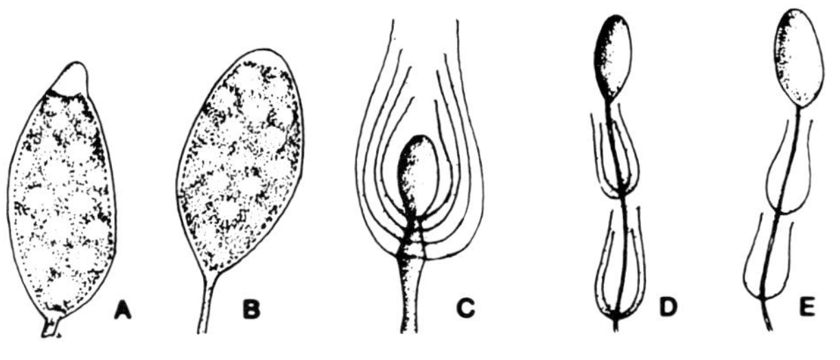

Figure 4.7. Morphology of sporangia. **A,** Papillate (note the conspicuous apical plug). **B,** Semipapillate (note that although there is a protuberance at the tip of the sporangium, the protoplasm extends nearly to the sporangium wall). **C,** Nested proliferation of a sporangium. **D,** Nested and extended proliferation of a nonpapillate sporangium. **E,** Extended proliferation of a nonpapillate sporangium. (Drawn by A. Vaziri)

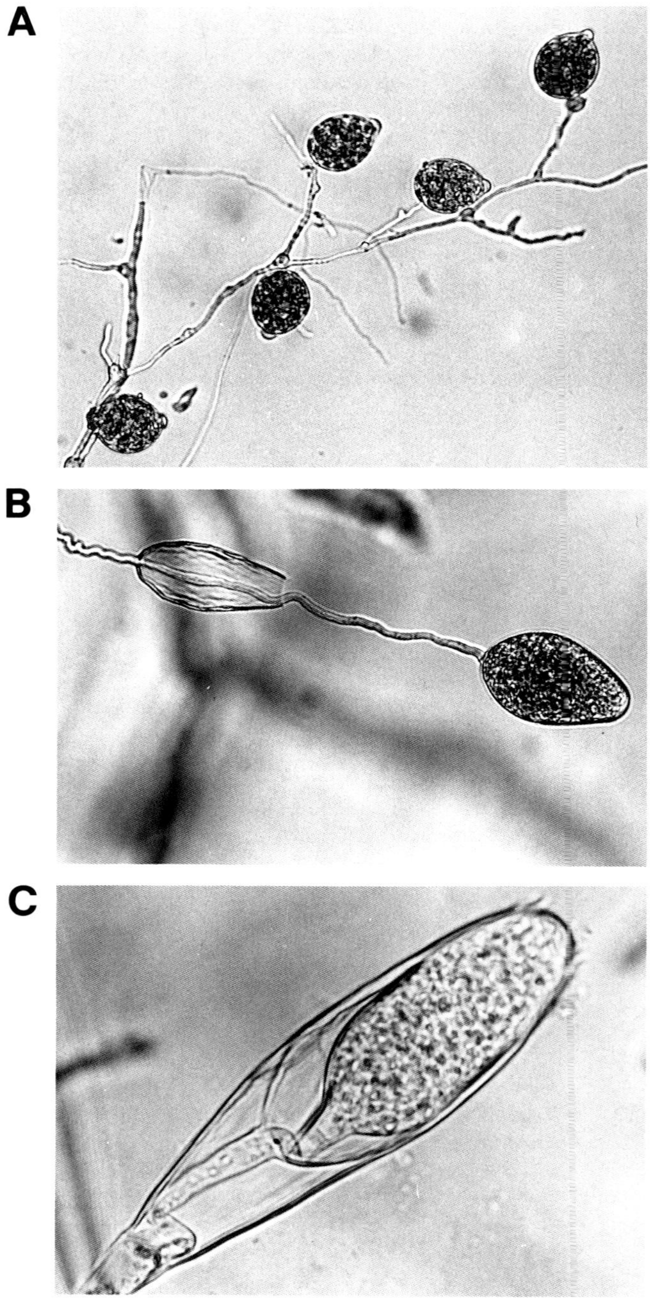

Figure 4.8. Morphological types of sporangiophore structures. **A,** Sympodium with semipapillate sporangia of *Phytophthora citricola* (P566). **B,** External proliferation from a nonpapillate sporangium of *P. cambivora.* **C,** Internal proliferation (nesting of a nonpapillate sporangium of *P. megasperma*). (A, Courtesy L. J. Klure and G. A. Zentmyer; B, courtesy P. Oudemans; C, courtesy M. J. Griffin)

1949) and not to the shape of the sporangium. The thickness of this apical structure varies among species from nonpapillate, in which there is only a slightly clear area between the outer sporangium wall and the surface of the protoplasm, as in *P. cambivora* (Figure 4.12P), *P. drechsleri* (Figure 4.12M), and *P. erythroseptica* (Figure 4.12O); through semipapillate, which implies a shallow thickening, as in *P. infestans* (Figure 4.12G); to papillate, in which there is an apical thickening that is hemispherical and 3–5 μm deep, as in *P. cactorum* (Figure 4.12A), *P. nicotianae* (Figure 4.12C–E), and *P. palmivora* (Figure 4.12F). The exit pore is usually narrow (5.7 μm) for papillate and semipapillate sporangia but wide (about 12 μm) for nonpapillate sporangia (Waterhouse et al. 1983) (Appendix 4.5).

Species that are definitely papillate include *P. cactorum* (Figure 4.12A), *P. parasitica* (= *P. nicotianae*) (Figure 4.12C–F), and *P. palmivora* (Figures 4.9C and 4.12F). Among those with semipapillate sporangia are *P. infestans, P. primulae* (Figure 4.12G and Q), and *P. hibernalis* (Figure 4.12K). Species with nonpapillate sporangia include *P. cinnamomi* (Figure 4.12N), *P. cambivora* (Figure 4.12P), *P. drechsleri* (Figure 4.12M), *P. erythroseptica* (Figure 4.12O), and *P. medicaginis* (Figure 4.12L). See Appendix 4.5 for a listing of species that are nonpapillate, semipapillate, or papillate.

The differentiation of nonpapillate from semipapillate sporangia is the most difficult to interpret. Staining sporangia with cotton blue (0.01%) in lactophenol (Waterhouse 1963) will differentiate the apical thickening from the protoplasm (Figure 4.9C). Rose bengal (50 ppm in water) also differentiates the apical plug from the protoplasm. Lactophenol-based stain is used partly because slides can be preserved in a herbarium for future study, but lactophenol often causes the protoplasm in sporangia to shrink. Shrinkage can be avoided by using trypan blue or cotton blue stain in 0.85% sodium chloride, which will stain the protoplasm but not cause shrinkage; however, these stains will not preserve the specimen.

Although the morphological characteristics of papillate, semipapillate, and nonpapillate species differ, the basic sequence of events leading to zoospore formation within the sporangium and dissolution of the apical plug before emission of zoospores from the sporangium is the same for all species (Gisi et al. 1979a, b; Gisi 1983; Waterhouse et al. 1983) (see Chapter 3).

Caducity of sporangia and pedicel length are important features that are unique to certain species (Waterhouse 1974c). None of the nonpapillate species is caducous (Waterhouse et al. 1983) (Appendix 4.6).

Waterhouse (1974c) reviewed the literature on caducity and listed *P. infestans* and *P. phaseoli* as the only species with caducous sporangia, which, when shed, are both airborne and waterborne. All other species listed as caducous are waterborne, but there is little evidence that they are capable of being airborne. In dry air, the sporangiophores of *P. infestans* twist and bend, causing the sporangia to be discharged into the air in much the same manner as sporangia of *Peronospora,* a genus of the downy mildews. Waterhouse (1974c) points out that the sporangia of *Phytophthora* differ from those of *Peronospora* in that each has a stalk, the length of which varies with species and ranges from short to medium (Appendix 4.6). A cross wall that delimits the sporangium forms in the sporangiophore, and above the cross wall a plug of wall material can be seen. In determination of caducity as a character, it is extremely important to note the plug at the breakage point because sporangia can be broken off by harsh shaking methods and considered to be caducous when they are not (Al-Hedaithy and Tsao 1979a,b).

Undoubtedly, caducity gives these species an epidemio-

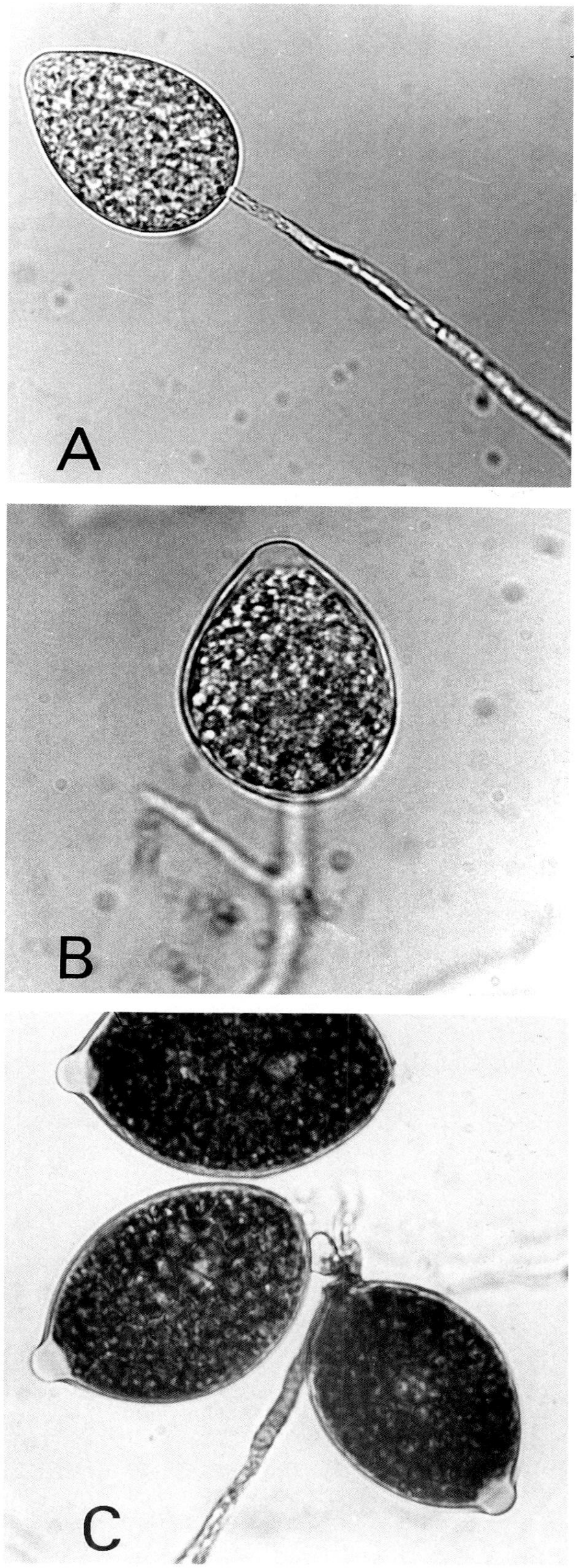

Figure 4.9. A, Nonpapillate sporangium of *Phytophthora drechsleri.* **B,** Semipapillate sporangium of *P. citricola* (P704). **C,** Papillate, caducous sporangia of *P. palmivora.* The apical plug is differentiated from the protoplasm by use of cotton blue-lactophenol stain. (A, courtesy R. A. Khan and D. C. Erwin; B and C, courtesy L. J. Klure and G. A. Zentmyer)

logical advantage in dissemination. The caducous types are categorized as follows:

1. short (<5 μm), as in *P. palmivora* (Figures 4.9C, 4.12F, 4.15G–I, and 4.17C and D), *P. infestans* (Figure 4.12E), and *P. cactorum* (Figure 4.12A);
2. intermediate (5–20 μm), as in *P. colocasiae, P. botryosa* (Figure 4.15A–C), and *P. meadii* (Figure 4.15D–F);
3. long (>20 μm), as in *P. hibernalis* (Figure 4.12K) and *P. capsici* (Figures 4.9C; 4.12A, E, F, and K; and 4.17A and B) (Al-Hedaithy and Tsao 1979a, b). The pedicel length for *P. capsici* ranges from 43 to 148 μm (Tsao 1991).

A method for quantitative determination of the percentage of caducous sporangia in a culture is given by Al-Hedaithy and Tsao (1979b). To dislodge the detachable sporangia, a mycelial colony containing mature (3- to 5-day-old) sporangia is washed in a test tube with a small amount of water by shaking on a Burrell wrist action shaker at setting 2 for 1 min (330 excursions per minute). The suspension of sporangia is poured into a 100-ml glass bottle, and the mycelial mat in the tube is rinsed again four times with 5 ml of sterilized, distilled water. The rinse water is added to the bottle, and the volume is adjusted to 25 ml. The residual mat in the test tube is then blended in a Sorvall Omnimixer 50-ml cup in 10 ml of water at setting 8 for 1 min. The blended mycelium and sporangia

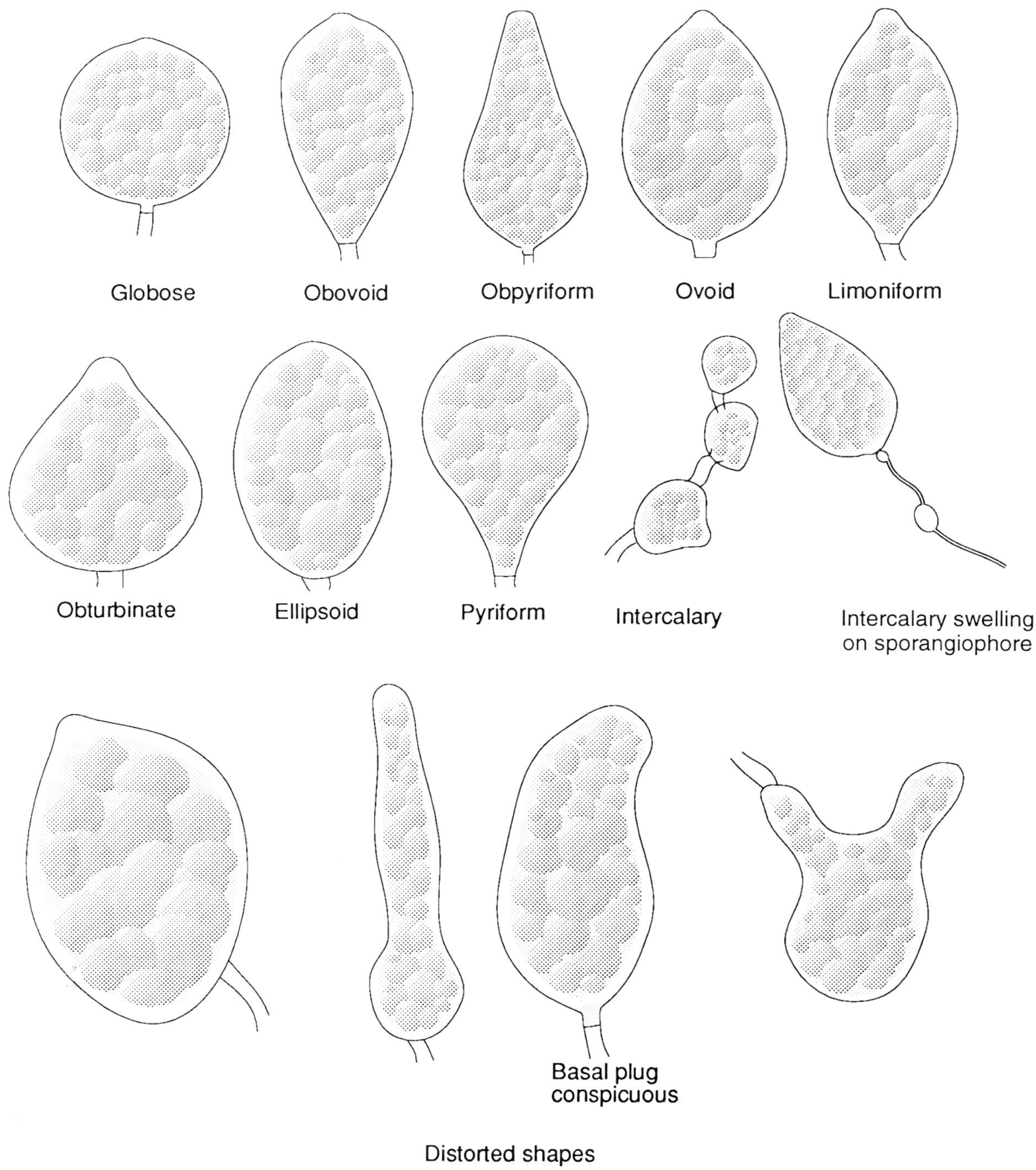

Figure 4.10. Some typical shapes of the sporangia produced by different species of *Phytophthora.* (Drawn by A. Vaziri)

are poured into another 100-ml bottle. The cup is rinsed again three times with 5-ml aliquots of water, which are added to the suspension with enough water to make up a 25-ml quantity. The percentage of sporangia that are deciduous in each bottle (shaken and blended) is evaluated separately by examination of the suspensions in a Hawksley nematode counting cell by use of a Leitz microscope with a long working distance (32× objective). The percentages of sporangia that are caducous and have pedicel lengths that are short (<5 µm), intermediate (5–20 µm), and long (>20 µm) are recorded. Al-Hedaithy and Tsao (1979b) were the first to report a quantitative comparison of a number of *Phytophthora* species for different degrees of caducity.

A qualitative observation of the length of pedicels can be more quickly conducted by dipping a disk of colonized, sporangia-bearing agar about 10 times into a drop of clear lactophenol (see Ainsworth [1968] for formula) to dislodge the sporangia. Also, sporangia-bearing hyphae on the edge of a disk of colonized carrot agar that has been incubated in water under light can be cut with a scalpel, mounted in lactophenol, and observed (Alizadeh and Tsao 1985a). Oudemans and Coffey (1991c) and Mchau and Coffey (1994c) dipped a sporulating disk of agar repeatedly into water to determine pedicel length of several *Phytophthora* species.

Uniformity of pedicel length is more diagnostic of caducity than the percentage of caducous sporangia. Sporangia must be mature before they will shed. For instance, when mycelium bearing sporangia of a noncaducous species such as *P. cinnamomi* is agitated vigorously, some of the sporangia will accidentally break off; however, the length of the pedicels will not be uniform (Al-Hedaithy and Tsao 1979a, b). Although the percentage of sporangia that are caducous may be affected by light, age, temperature, and the culture medium (Al-Hedaithy and Tsao 1979a, b), the relative lengths of the pedicels on sporangia of a truly caducous species are similar under all environmental conditions.

In *P. capsici,* most of the sporangia have elongated bases with long pedicels (43–148 µm) when formed under continuous fluorescent light but are elongated and noncaducous when formed in the dark (Hendrix 1967; Brasier 1969; Alizadeh and Tsao 1985a; Tsao 1991) (see Figures 4.16 and 4.17 for variations in caducity and sporangial shapes for *P. capsici*). A recent comparative study of *P. capsici* isolated from pepper (17 isolates) and cucurbits (seven isolates) showed a wide range of variability in the pedicel length of isolates within and between the host groups; however, since the total range in length varied between 31 and 99 µm (Ristaino 1990), these isolates would be classified within the "long" pedicel group (Stamps et al. 1990).

About 7–9% of the sporangia of *P. parasitica* (*P. nicotianae*) produced on clear V8 juice agar were considered to be caducous with varying lengths of pedicels; however, on blighted ti plant (*Cordyline terminalis* (L.) Kunth) leaves, about 21–27% were caducous with uniformly short pedicels (<5 µm). Trichillo and Aragaki (1982) concluded that the fungus should be considered noncaducous, despite the variation on ti leaves, and recommend that to call a species caducous, more than 50% of the sporangia with pedicels of uniform length should be dislodged by agitation (Vortex mixer at 250 rpm for 1 min). Since they show that on naturally infected ti leaves about one-fourth of the sporangia are caducous, in nature some sporangia of these *P. nicotianae* isolates may be capable of becoming detached. A callose plug was detected in the pedicel above the place where the pedicel had broken (Trichillo and Aragaki 1982).

Caducity of sporangia has been applied as a differential characteristic for *P. nicotianae* var. *parasitica* to separate it from *P. nicotianae* var. *nicotianae* by Waterhouse (1974c) and Newhook et al. (1978). In the key of Stamps et al. (1990), however, the variety *nicotianae* is considered to be occasionally caducous. The concept of caducity in either variety has not been supported by studies of Ho and Jong (1989). Extensive research data by Al-Hedaithy and Tsao (1979a, b) indicate conclusively that sporangia of *P. parasitica* (*P. nicotianae*) are not caducous. Although a small percentage of sporangia became detached, the lengths of the pedicels were randomly distributed, indicating that they had broken accidentally. Ho and Jong (1989) also state that sporangia of both varieties are primarily noncaducous, although a few sporangia with short pedicels or of uneven length are found in some isolates. Hall (1993) reported a small percentage of sporangia that

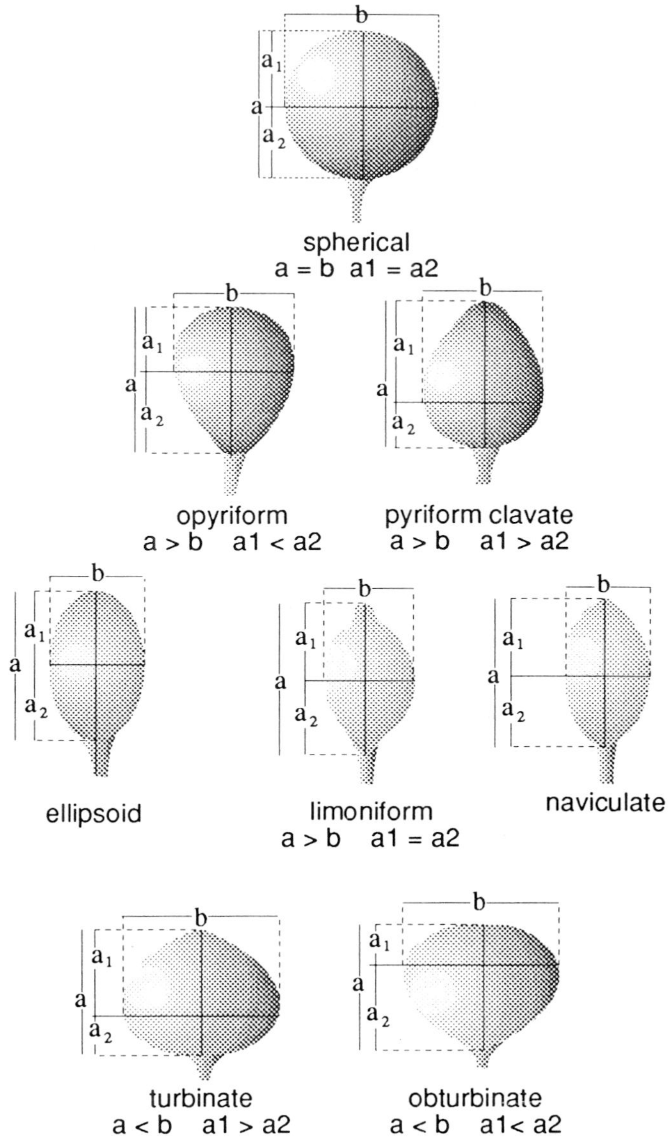

Figure 4.11. Examples of how dimensions of different sporangium shapes can be quantified to define different characteristics. Shapes depend on the a:b ratio and deviations from ellipsoid curvature. This concept was suggested by M. Dick, who utilized the biometry of oogonium, oospore, and ooplast diameters and oospore wall thickness as an aid for differentiation of *Pythium* species (Dick 1990b; Shahzad et al. 1992).

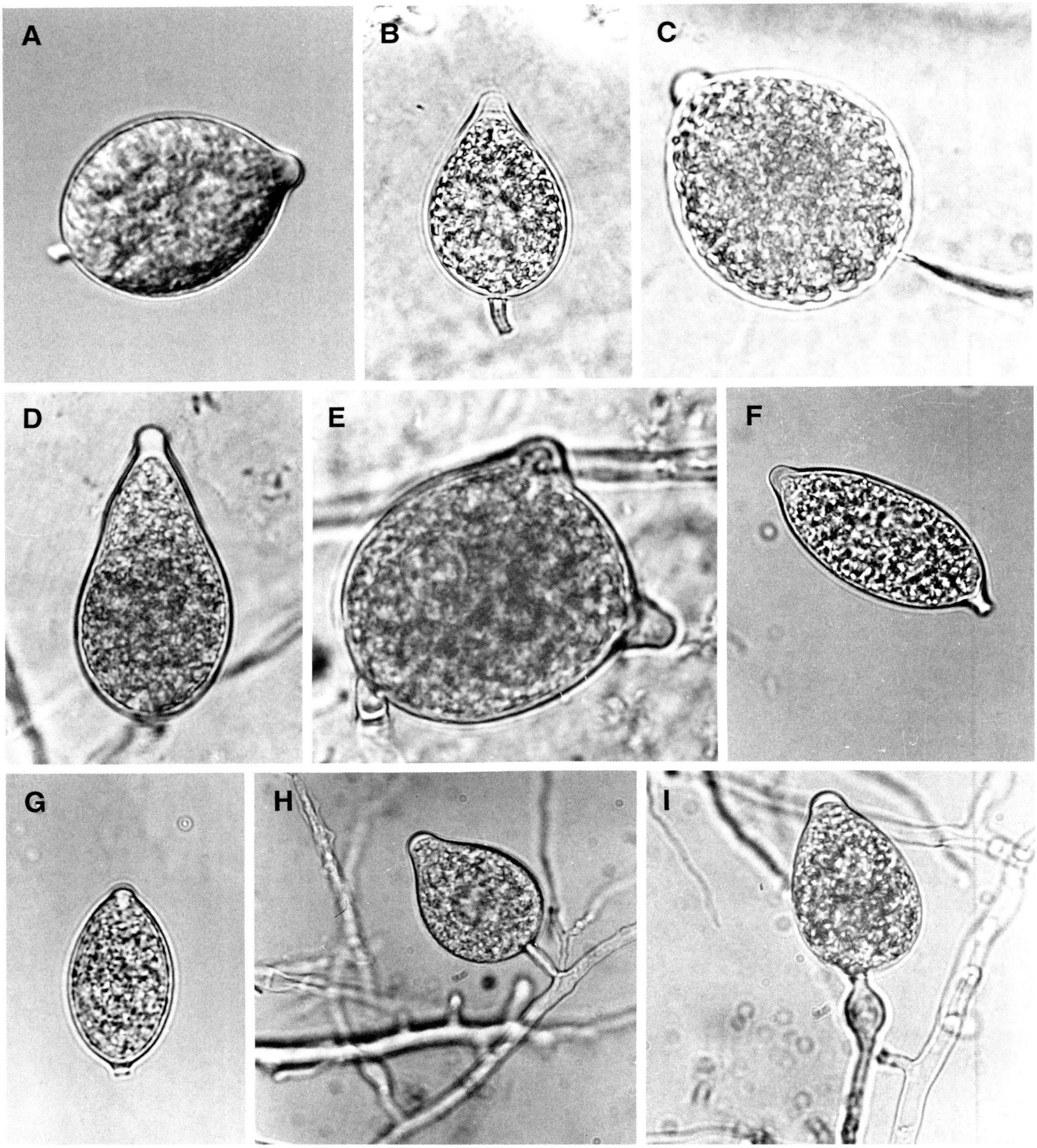

Figure 4.12. Various shapes, caducity, and papillate nature of sporangia produced by different species of *Phytophthora*. **A,** Ovoid, papillate, caducous sporangium of *P. cactorum.* **B,** Ovoid-obpyriform semipapillate sporangium of *P. citricola* (P586). **C,** Subspherical, papillate sporangium of *P. nicotianae.* **D,** Obpyriform, papillate sporangium of *P. nicotianae.* **E,** Spherical, bipapillate sporangium of *P. nicotianae.* **F,** Ellipsoid, papillate, caducous (short stalk) sporangium of *P. palmivora* (P3108). **G,** Ellipsoid, semipapillate, caducous (short stalk) sporangium of *P. infestans.* **H,** Obpyriform, semipapillate sporangium of *P. citricola* (P586) **I,** Ovoid-obpyriform, semipapillate sporangium with intercalary swelling on the sporangiophore of *P. citricola.* **J,** Bilobed (two apices), semipapillate sporangium of *P. citricola* (P475). **K,** Ellipsoid, semipapillate, caducous (long stalk) sporangium of *P. hibernalis.* **L,** Ovoid, nonpapillate, noncaducous sporangium of *P. medicaginis* (EP1057). **M,** Ovoid-ovate, nonpapillate, noncaducous sporangium of *P. drechsleri* (EP389). **N,** Ovoid, nonpapillate sporangia of *P. cinnamomi* (Pc428). **O,** Ovoid, nonpapillate sporangium of *P. erythroseptica.* **P,** Ovoid, nonpapillate sporangium of *P. cambivora.* **Q,** Obpyriform, semipapillate sporangium of *P. primulae.* (A, Nomarski optics, courtesy P. Oudemans; B, F, and H–J, courtesy L. J. Klure and G. A. Zentmyer; C–E, G, K, and O–Q, courtesy M. J. Griffin; L–N, courtesy R. A. Khan and D. C. Erwin)

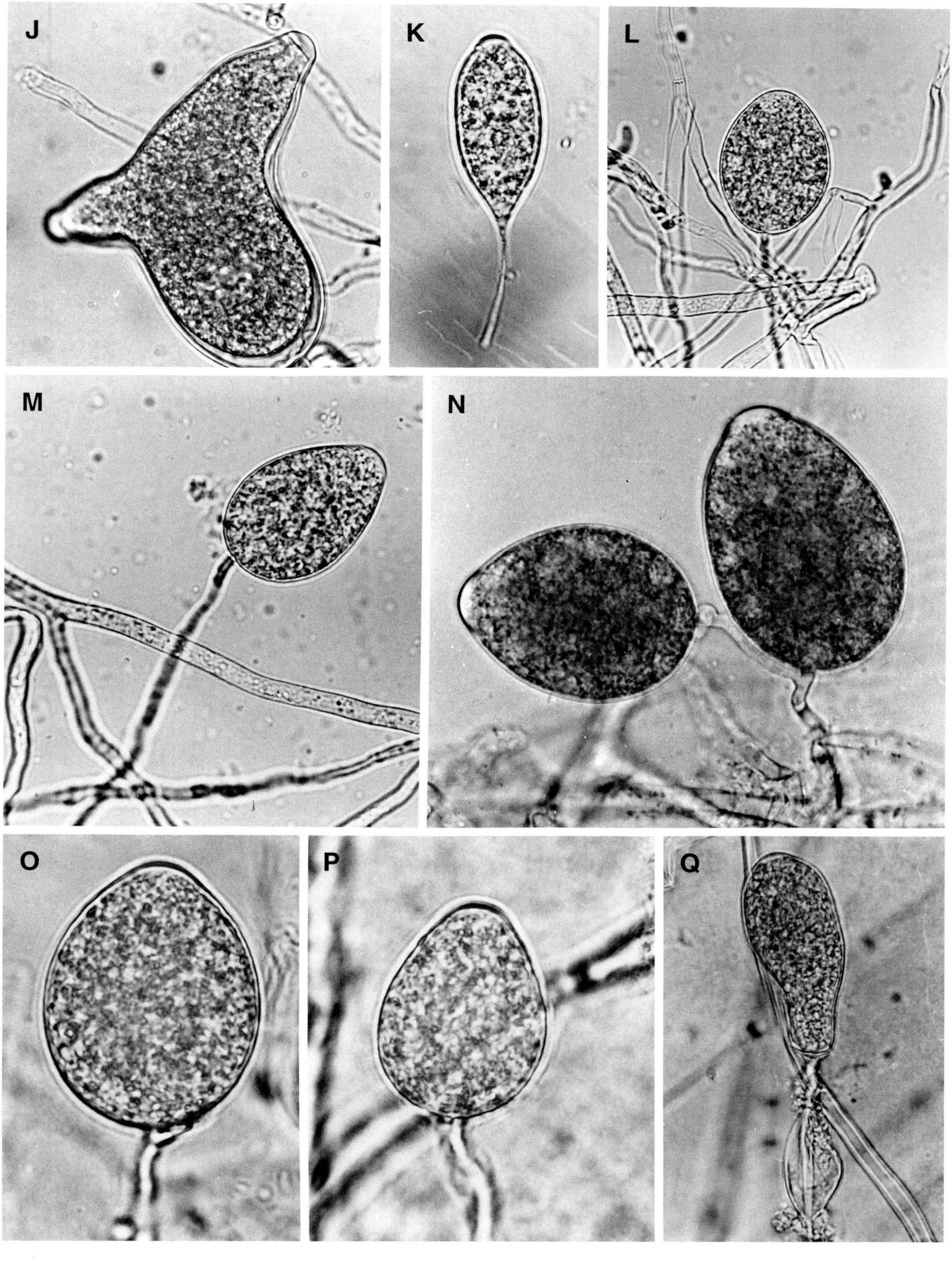

were caducous after shaking a water culture on a wrist-action shaker for 1 min (330 excursions per minute). It would appear that *P. nicotianae* cannot be considered caducous in the sense that *P. cactorum* or *P. palmivora* are caducous.

Lateral attachment, distorted shapes, elongated necks in water, hyphal projections, and intercalary formation of sporangia are also considered to be useful features for identification (Appendix 4.4) (Waterhouse et al. 1983). Some species produce small sporangia (<45 µm in length; e.g., *P. cactorum, P. iranica, P. botryosa,* and *P. katsurae*), and some produce large sporangia (>75 µm in length; e.g., *P. capsici, P. citrophthora, P. mexicana, P. porri, P. primulae, P. cyperi, P. fragariae, P. cambivora, P. cinnamomi,* and *P. japonica*) (Waterhouse et al. 1983). Except for these extremes, size differences for diagnostic purposes should be treated with caution because size can be influenced by many environmental factors such as light, sterols (Brasier 1967), and other nutrients in media (Ribeiro 1983). The length-breadth ratio of sporangia is considered a useful diagnostic characteristic, but data are not absolutely definitive; however, these data quantitate the terms such as ovoid-pyriform (high length-breadth ratio) or ovoid or round (low length-breadth ratio) (Figure 4.11).

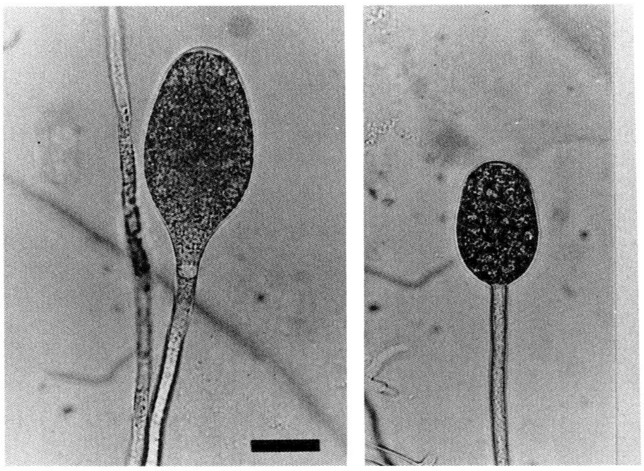

Figure 4.13. Variation in sporangium morphology of an isolate of *Phytophthora drechsleri* f. sp. *cajani* (P1795). Sporangium on left has a tapered base; sporangium on right has a rounded base. Bar = 25 µm. (Reproduced by permission of the British Mycological Society, courtesy Mills et al. 1991)

Sporangia are variable in shape (Figures 4.10–4.12) and usually can be classified as subspherical, ovoid, obovoid, ellipsoid, limoniform (lemon shaped), pyriform (pear shaped), obpyriform, or obturbinate. Within *P. cambivora, P. cinnamomi,* and *P. cryptogea,* shapes of sporangia were categorized into 14 classes (Gerrettson-Cornell and Lind 1981). These shapes correspond to the categories of shape as defined by Hawksworth et al. (1983) (Figure 4.10):

Category A: Elliptical or nearly so; symmetrical or slightly asymmetrical; the apex sometimes slightly protruding.

Category B: Egg shaped or limoniform; symmetrical or slightly asymmetrical; the apex protruding; the membrane at times forming a structure reminiscent of a "cap" above the apex.

Category C: Ovate to obpyriform; the apex somewhat flat to normal or round.

Category D: Between pyriform and oval.

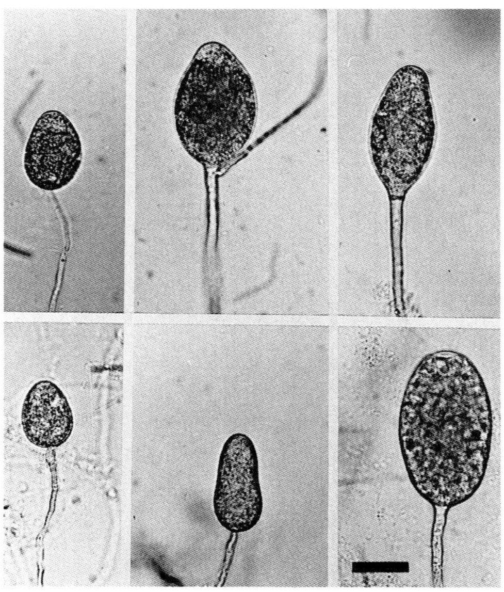

Figure 4.14. Range in sporangium morphology for isolates of *Phytophthora cryptogea* and *P. drechsleri.* Upper row (left to right): Isolates P3104 (group D), P3850 (group E), and P3279 (group K) of *P. cryptogea.* Lower row (left to right): Isolates P1087 (group A), P1899 (group A), and P3239 (group F) of *P. drechsleri.* Bar = 25 µm. (Reproduced by permission of the British Mycological Society, courtesy Mills et al. 1991)

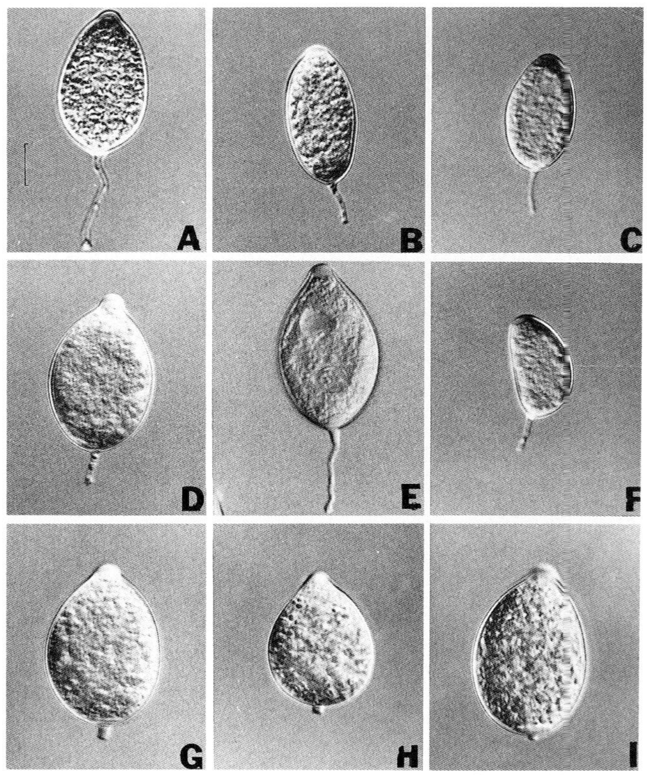

Figure 4.15. Caducous sporangia with different pedicel lengths produced by several *Phytophthora* species. **A–C,** Medium pedicel length of *P. botryosa;* **D–F,** medium pedicel length of *P. meadii;* and **G–I,** short pedicel length of *P. palmivora.* Bar (in A) = 10 µm. Sporangia with long stalks (*P. capsici*) are shown in Figure 4.16B–F. (Reproduced by permission of the British Mycological Society, courtesy Oudemans and Coffey 1991b)

Category E: Obovate to oblong; at times constricted at or near the center; the apex slightly to markedly protruding; with or without "cap."
Category F: Conical; tapering.
Category G: As F, but with round apex.
Category H: Pyriform.
Category I: One side bent immediately below the apex.
Category J: "Ventral type," i.e., one side flat or concave.
Category K: Slender; cylindrical.
Category L: Strongly asymmetrical without being "ventral."
Category M: Oddly shaped.
Category N: Sporangia with two apices.

Gerrettson-Cornell and Lind (1981) found that sporangia of *P. cinnamomi* occurred more often as shapes B and E, those of *P. cryptogea* as shapes I, J, and L, and those of *P. cambivora* as shapes D, F, and G; however, shape A (elliptical) occurred with nearly identical frequency in all three species. Gerrettson-Cornell and Lind (1981) were dubious that any shape could be primarily diagnostic of a species because one isolate of *P. cinnamomi* did not produce any sporangia of type B. Frequency of type A was 72% for one isolate of *P. cryptogea* but only 7% for another isolate. They concluded that sporangium shape is "not a character of primary importance" for identification of these three species but that shape may be an important secondary criterion.

In the dichotomous key of Tucker (1931), *P. cryptogea* was described as producing nonpapillate sporangia with rounded bases while nonpapillate sporangia of *P. drechsleri* were illustrated as having sloping bases. Klisiewicz (1977) (Figure 23.2) also illustrates and differentiates the two species on the basis of these characteristics. Mircetich and Matheron (1976; their Figure 2H) illustrated the sloping base on a sporangium of *P. drechsleri* from cherry. The sloping-base characteristic is considered to be of doubtful value by others because both types of sporangia can often be seen in the same culture (Ho and Jong 1986, 1991; Mills et al. 1991). Both sloping and rounded bases occur in *P. drechsleri* f. sp. *cajani* (Figure 4.13). A range of sporangium shapes for the *P. cryptogea-P. drechsleri* complex studied by Mills et al. (1991) is shown in Figure 4.14.

Morphology of Zoospores

Biflagellate zoospores (Figure 4.18) are reniform to pyriform in shape; the anterior end is tapered, and there is a deep longitudinal groove on one side (Figure 4.19) (Desjardins et al. 1973a). Two flagella of unequal length that emerge from the groove (Figure 4.18) can be seen by light microscopy after special treatment to immobilize the zoospores. In Figure 4.18, zoospores of *P. infestans* have been immobilized but not killed

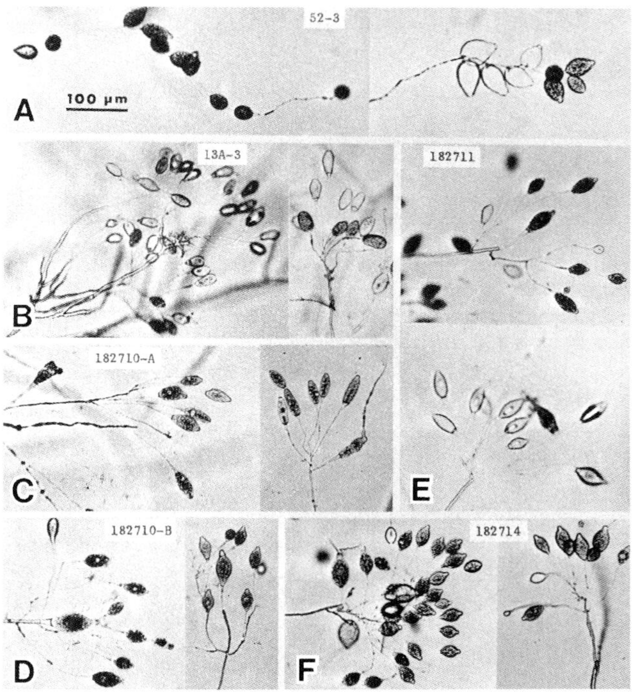

Figure 4.16. Branching patterns of sporangiophores. **A,** *Phytophthora palmivora sensu stricto* (Brasier and Griffin 1979). Note the characteristic sympodial (lax or close) formation of successive sporangia in *P. palmivora.* **B–F,** *P. capsici sensu lato* (Tsao 1991). Note the unique umbellate or fan-shaped ontogeny of sporangium formation in *P. capsici,* a process that involves little or no elongation of the sporogenous hypha. The sporangia of both species were produced with the "agar-disk-in-water" technique under continuous light. *P. palmivora* isolate 52-3 was from rubber in Thailand; *P. capsici* isolates 13A-3, IMI 182710-A, IMI 182710-B, IMI 182711, and IMI 182714 were from black pepper in Thailand or Malaysia. (Reproduced by permission of the National Research Centre for Spices, Calicut, Kerala, India, courtesy Tsao 1991)

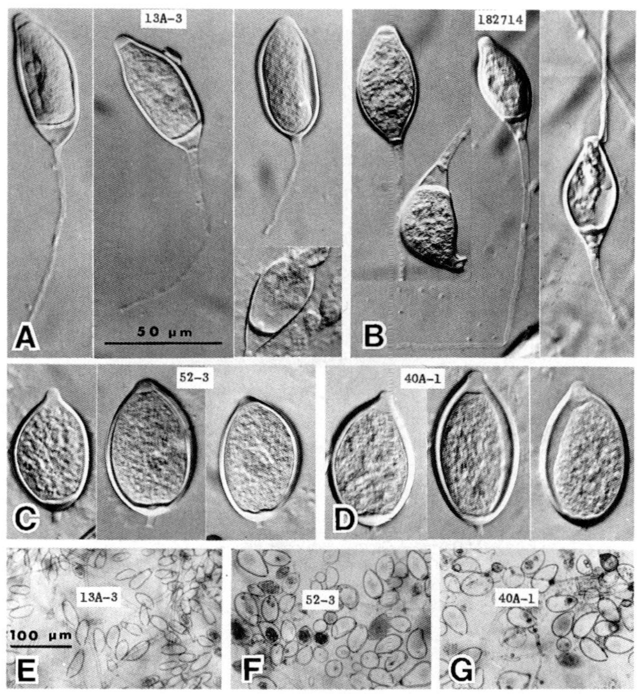

Figure 4.17. Shapes of sporangia produced by *Phytophthora capsici sensu lato* (**A, B, and E**) and *P. palmivora sensu stricto* (**C, D, F, and G**). Note the long pedicels, tapered bases, and higher length-breadth ratios in the sporangia of *P. capsici* and the short pedicels, more rounded bases, and generally lower length-breadth ratios in the sporangia of *P. palmivora. P. capsici* isolates 13A-3 (**E**) and IMI 182714 (**B**) were from black pepper in Thailand and Malaysia, respectively, and *P. palmivora* isolates 52-3 (**F**) and 40A-1 (**G**) were from rubber and durian, respectively, in Thailand. Scale bars represent 50 µm in **A–D** and 100 µm in **E–G.** (Reproduced by permission of the National Research Centre for Spices Calicut, Kerala, India, courtesy Tsao 1991)

by addition of Li^+ as the chloride salt (100 µM), which induces leakage of Ca^{2+}. Encystment was induced by replacement of Li^+ with Ca^{2+} and K^+ (Érsek et al. 1991). The longer, anteriorly directed flagellum (whiplash) is nearly smooth except for fine lateral hairs that can be seen only by electron microscopy (Figure 4.20) (Desjardins et al. 1969, 1973b). The shorter flagellum is posteriorly directed, and the tinsel is ornamented with a row of lateral hairs (mastigonemes or "tripartite tubular hairs" according to Andersen et al. [1991]) on each side. Each mastigoneme is about 13 nm wide (Desjardins et al. 1969). Morphologically, there are no species-oriented differences in shape of zoospores or the cysts into which they form.

During the 1980s and early 1990s, the flagellar apparatus in several *Phytophthora* species has received much more attention because of the development of high resolution electron microscopy. (See Barr [1981] for a comparison of the ultrastructural morphology of the flagellar apparatus of the members of Oomycetes, excluding Thaustochytriales, with that of members of other related classes.) The rootlet morphology in zoospore flagella indicates that the oomycetes evolved with or from the golden algae (Chrysophycophyta). Barr and Désaulniers (1990) summarize previous research on zoospores of *P. parasitica.* See also the reports of Barr and Allan (1985). The flagellar apparatus in *P. sojae* f. sp. *glycinea* (*P. sojae*) and *P. megasperma* is contrasted with that of *P. cinnamomi* and *P. parasitica* (Barr and Désaulniers 1989). The flagellar apparatus of *P. cinnamomi* was studied by Hardham (1987).

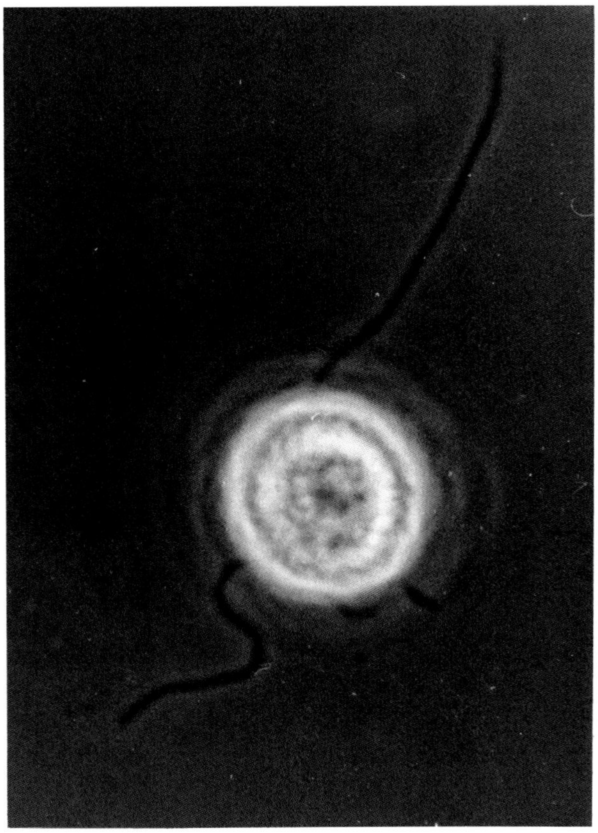

Figure 4.18. Zoospore of *Phytophthora infestans* with a long whiplash flagellum and a short tinsel flagellum. The zoospore (about 10 µm in diameter) has been immobilized by addition of PEG-Li^+ (100 µg/ml) to distilled water. (Reproduced by permission of the British Mycological Society, courtesy Érsek et al. 1991)

The typical flagellar apparatus in the zoospore is composed of the two flagella, the kinetosomes (basal bodies), a transition zone that lies between the flagellum and the kinetosome, and the microtubule rootlet system that anchors the flagella to the zoospore. Figure 4.21 presents Barr's (1992) concept of the flagellar apparatus of members of Pythiaceae. The

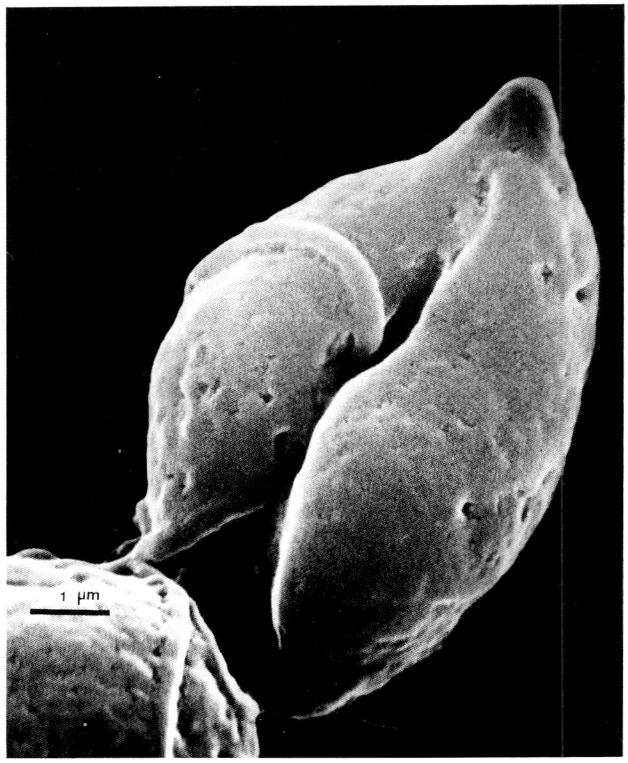

Figure 4.19. Scanning electron micrograph of a zoospore of *Phytophthora palmivora* showing the longitudinal groove. (Reproduced by permission of Springer Verlag, Heidelberg, Germany, courtesy Desjardins et al. 1973a)

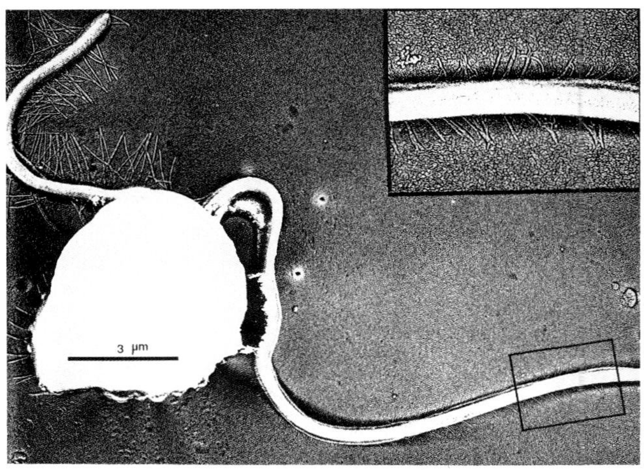

Figure 4.20. Zoospore of *Phytophthora palmivora* fixed with glutaraldehyde and rotary shadow cast showing tinsel (shorter) and whiplash (longer) flagella. The inset is an enlargement of the whiplash flagellum. (Reproduced by permission of the National Research Council of Canada, Ottawa, Canada, courtesy Desjardins et al. 1969)

rootlet systems are genetically highly conserved; therefore, any deviation from this pattern might be significant to classification. Barr and Désaulniers (1990) stated that it was not surprising that *P. sojae, P. megasperma, P. cinnamomi,* and *P. parasitica* were similar; however, the absence of the multistranded rootlet (Figure 4.21 of Barr and Désaulniers [1990]) and the presence of structurally more developed posterior and anterior ribbed rootlets in the *P. infestans* zoospore (Figure 4.22A) were unexpected. The rootlets of *P. infestans* flagella are significantly different from those of *P. parasitica, P. cinnamomi, P. sojae,* and *P. megasperma* (Figure 4.22B). In the microtubule rootlet system of the *P. infestans* flagella, there are only five rootlets compared with six in the other species (Figure 4.22A). Also, one of the rootlets in the *P. infestans* zoospore contains six microtubules compared with only four in the other species. Thus, *P. infestans,* the type species of *Phytophthora,* differs significantly in this respect from the other species studied. This characteristic may be unique at the genus level and may become useful for classification purposes (Barr and Désaulniers 1990).

Chlamydospores

Normally, chlamydospores are globose; however, irregular to oblong shapes have been described (Cother and Griffin 1973b; Hemmes and Lerma 1985) (see Appendix 4.4). The literature on the cytology of chlamydospores is summarized and described by Hemmes (1983). Chlamydospores, which can be intercalary or terminal (Figure 4.23), are delimited by septation from the mycelium and can be confused with hyphal swellings (Figure 4.4), which by definition are not delimited by septation from mycelium (Blackwell 1949). Chlamydospore walls are thicker than sporangium walls. An increase in wall thickness has been found to occur when *P. palmivora* cultures are grown in a papaya juice medium (Kadooka and Ko 1973). *P. parasitica* and *P. palmivora* have thicker walls than *P. cinnamomi* (Figure 4.24). Chlamydospore morphology of several isolates of *P. capsici* (formerly classified as *P. palmivora* MF4; Brasier and Griffin 1979) is shown in Figure 4.25 (Alizadeh and Tsao 1985b). Since the morphology of chlamydospores does not vary appreciably among species, its

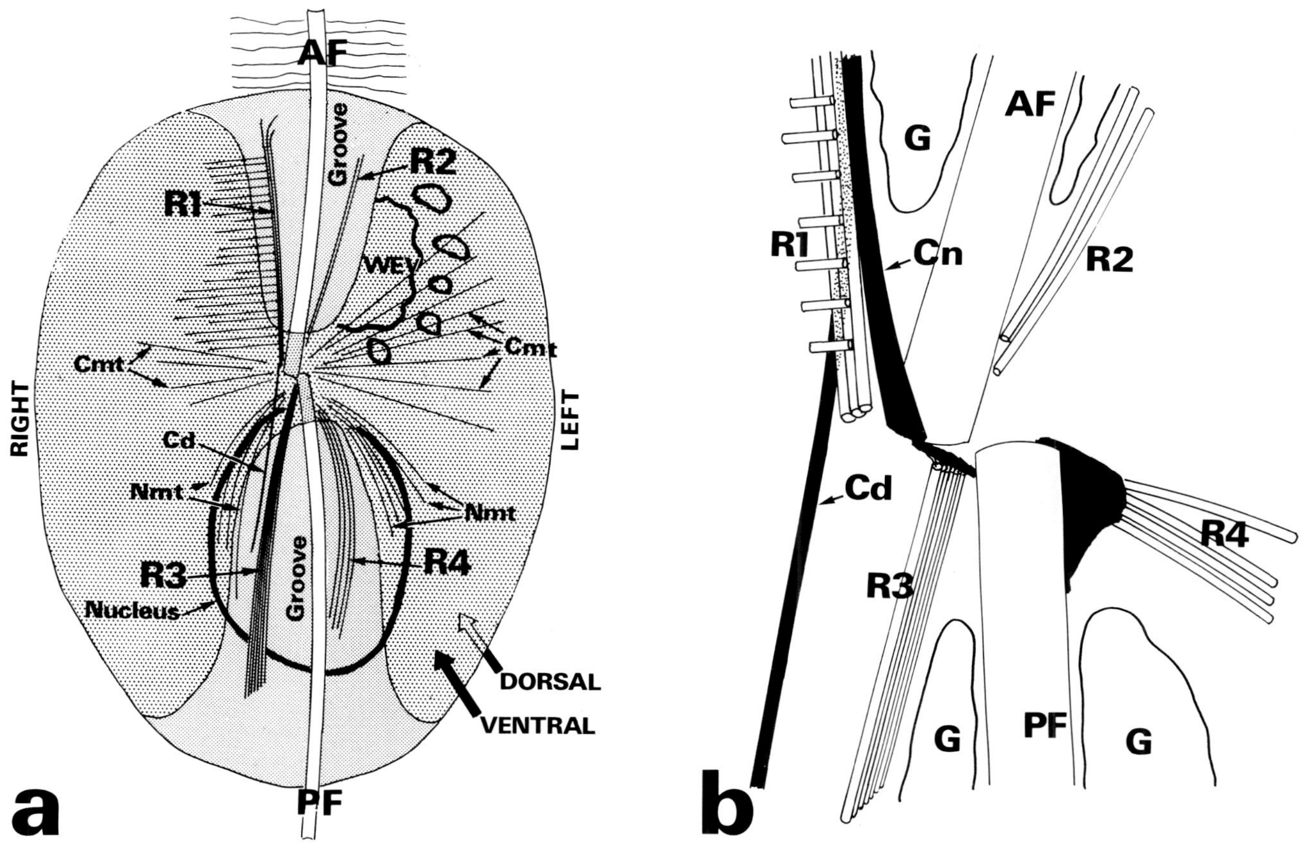

Figure 4.21. Flagellar apparatus of zoospores produced by members of Pythiaceae. **a,** The whole zoospore and **b,** enlarged drawing of the root systems viewed from the ventral side. The anterior (AF) and posterior (PF) flagella lie in grooves (G) on the ventral surface and are attached to the zoospore body near the center. The flagella are anchored in the zoospore by six roots (R1–4, Cmt, and Nmt). The ribbed triplet root (R1) extends from the basal body of the anterior flagellum and runs close to the cell membrane under the anterior groove on the right side of the zoospore (left in drawing, which is viewed from ventral side). There are three or four microtubules running at right angles into the zoospore body. The R1 root is attached to the side of the anterior basal body by electron opaque material (Cn). There is also an electron-opaque, cordlike structure (Cd) that runs parallel to the root and extends into the posterior of the zoospore. The anterior doublet root (R2) lies under the groove on the left side and comprises two microtubules. The multistranded, band-shaped root (R3) extends from electron-opaque material that joins the ends of the basal bodies. This root comprises five to nine microtubules that run close to the cell membrane under the right side of the posterior groove. The posterior root (R4) comprises two to six microtubules that run close to the cell membrane under the left side of the posterior groove. The nucleus lies dorsal to (behind) the R3 and R4 roots and extends from the basal bodies into the posterior of the zoospore body. Nuclear-associated microtubules (Nmt) ensheath the nucleus. Cytoplasmic microtubules (Cmt) extend from near the basal bodies into the zoospore body (i.e., away from viewer). A water expulsion vesicle (WEV) lies in the anterior of the zoospore body. (Reproduced by permission of the National Research Council Canada, Ottawa, Canada, courtesy Barr 1992)

significance for identification is limited. Nonetheless, *P. lateralis* is characterized by production of chlamydospores sessile (without a stalk) on the mycelium. Since many species do not produce chlamydospores, the abundant chlamydospore production and numerous hyphal swellings in addition to nonpapillate sporangia are good presumptive evidence for *P. cinnamomi* (Figures 4.1, 4.23, and 4.24E). *P. macrochlamydospora* produces large (28–84 μm in diameter) chlamydospores (Irwin 1991). *P. palmivora* produces chlamydospores abundantly, but only some isolates of *P. megakarya* or *P. meadii* produce them. In this case, presence or absence may be a useful presumptive indicator (Waterhouse et al. 1983).

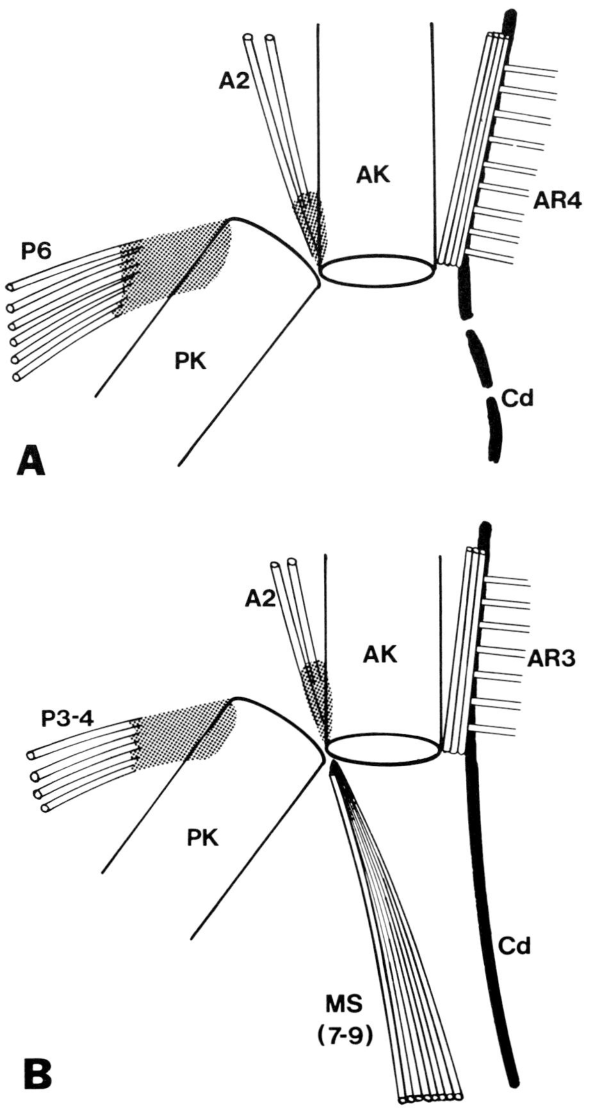

Figure 4.22. Numbers of rootlet microtubules in zoospores of *Phytophthora* spp. **A,** Zoospore of *P. infestans* that does not have the MS rootlet. **B,** Zoospore that is representative of *P. parasitica, P. cinnamomi* (this species does not have the Cd), *P. sojae,* and *P. megasperma.* The cytoplasm and nucleus-associated arrays of microtubules, common in all species, have been omitted. The anterior ribbed rootlet (AR3 and AR4) consists of three or four main microtubules, secondary microtubules at right angles, and associated cord (Cd). There is a second anterior rootlet consisting of two microtubules (A2). The posterior rootlet (P3, P4, and P6) consists of three, four, or six microtubules in different species. A band-shaped, multistranded rootlet (MS) consists of seven to nine microtubules that run parallel. AK = anterior kinetosome; PK = posterior kinesotome. (Reproduced by permission of the National Research Council Canada, Ottawa, Canada, courtesy Barr and Désaulniers 1990)

Sex Organs

Morphology of Antheridia, Oogonia, and Oospores

The sexual process in the *Phytophthora* life cycle (Hemmes 1983) (Figures 3.4 and 3.5) described in Chapter 3 involves the production of the female oogonium (the sac in which the oospore is formed) and the antheridium (the male structure), both of which emanate from mycelial tips that make contact. The oogonial incept either grows through the antheridium (amphigynous) (Figure 4.26A and B) or appresses the oogonial incept laterally (paragynous) (Figure 4.26C). If paragynous, the antheridium of *Phytophthora* becomes attached to the lower hemisphere near the stalk of the oogonium. In *Pythium* (depending on the species), paragynous antheridia are produced by most species but are not limited to the lower part of the oogonium (Table 4.7 and Figure 4.35).

The position of the antheridium, whether amphigynous or paragynous, is a major diagnostic character. Paragynous antheridia are characteristic of groups I, III, and V (Newhook et al. 1978; Stamps et al. 1990). Species that have paragynous or amphigynous antheridia are shown in Appendix 4.7.

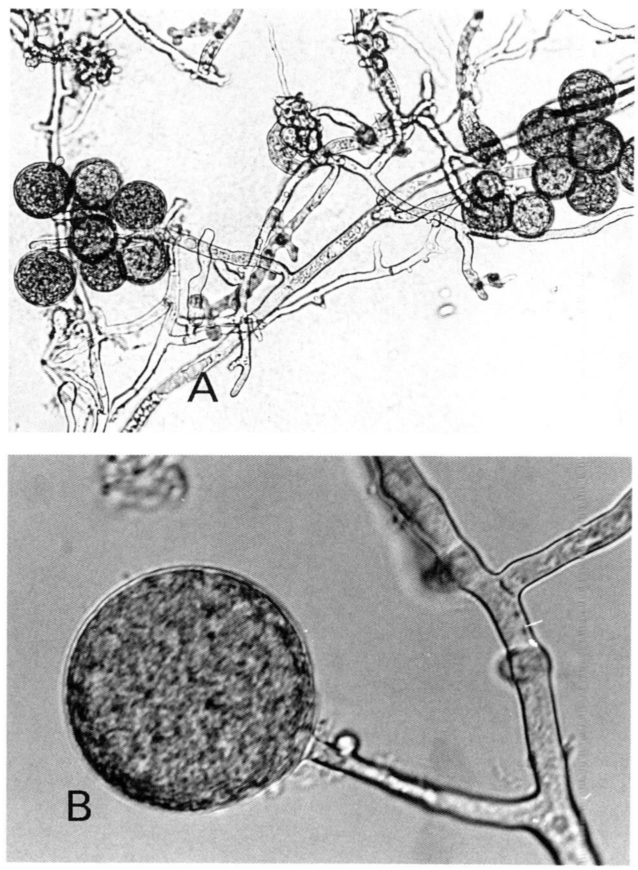

Figure 4.23. Chlamydospores of *Phytophthora cinnamomi.* **A,** Formation of chlamydospores, hyphal swellings, and coralloid mycelium. **B,** Note the relatively thin wall. (A, Courtesy M. J. Griffin; B, courtesy P. Oudemans)

In the mature oospore shown in Figure 4.26A, the thick inner wall can be seen within the oogonium. In the center of the oospore is an ooplast (also called a refringent body), and at the upper left is a pellucid body that contains a nucleus. In this photomicrograph, the oospore fills the oogonium and thus is described as *plerotic.* When the oospore does not fill the oogonium, it is described as *aplerotic.*

Usually, the morphology of oogonia and oospores per se does not differ enough among most species to be of diagnostic use; however, ornamentation in the form of bullate (bubble- or blisterlike) protuberances on the oogonial wall is characteristic of *P. cambivora* (Figure 4.27) (Ho et al. 1977). Verrucose, small, rounded, wartlike protuberances are also common for *P. katsurae* (Figure 4.28) (Gerrettson-Cornell 1985), *P. verrucosa,* and *P. cyperi-bulbosa* (Stamps et al. 1990). Note in Figures 4.27 and 4.28 that the bullate condition is not always consistently formed on all oogonia. Funnel-shaped oogonium stalks are characteristic of *P. katsurae* and *P. heveae.* Note in Figure 4.28 that this characteristic is not always consistent (P. Oudemans, *personal communication*).

When viewed by scanning electron microscopy, oogonial walls of *P. cactorum, P. citricola, P. heveae,* and *P. palmivora* are smooth, whereas those of *P. megasperma* (*P. medicaginis* from alfalfa) and *P. megasperma* var. *sojae* (*P. sojae* from soybean) are finely granular and those of *P. parasitica* are coarsely granular (Ho 1979b).

Species that have paragynous or amphigynous antheridia are shown in Appendix 4.7. Amphigynous antheridia are characteristic of groups II, IV, and VI (Newhook et al. 1978). Amphigyny (groups II, IV, and VI) and paragyny (groups I, III, and V) are unique characters that differentiate these groups (Waterhouse 1963; Newhook et al. 1978; Stamps et al. 1990) (Appendix 4.1). Waterhouse (1983) notes that amphigyny is a consistent character in species of group II, but *P. hibernalis* and *P. richardiae* (in groups IV and VI) occasionally produce some paragynous antheridia. In groups III and IV, which contain most of the species with paragynous antheridia, a few (about 1%) are amphigynous. In some single-zoospore isolates of *P. medicaginis* (*P. megasperma* f. sp. *medicaginis;* group V), however, about half of the antheridia are amphigynous and half paragynous (Erwin 1966). The reason for variability within single-spore isolates of such important characters as amphigyny and paragyny is not known.

The large size of oogonia (averaging 47 μm in diameter) was used as a character to distinguish the new species *P. megasperma* isolated from hollyhock (*Althea rosea*) (Drechsler 1931) from *P. erythroseptica* and similar species; however, the Tompkins et al. (1936b) amendment to Tucker's (1931) key broadened this character to include *P. megasperma* isolates from other hosts such as cauliflower with oogonia greater than 30 μm in diameter. To accommodate two size ranges for oogonia, Waterhouse (1963) created two varieties: *P. megasperma* var. *sojae* (not necessarily from soybean) with oogonia <40 μm in diameter and *P. megasperma* var. *megasperma* with

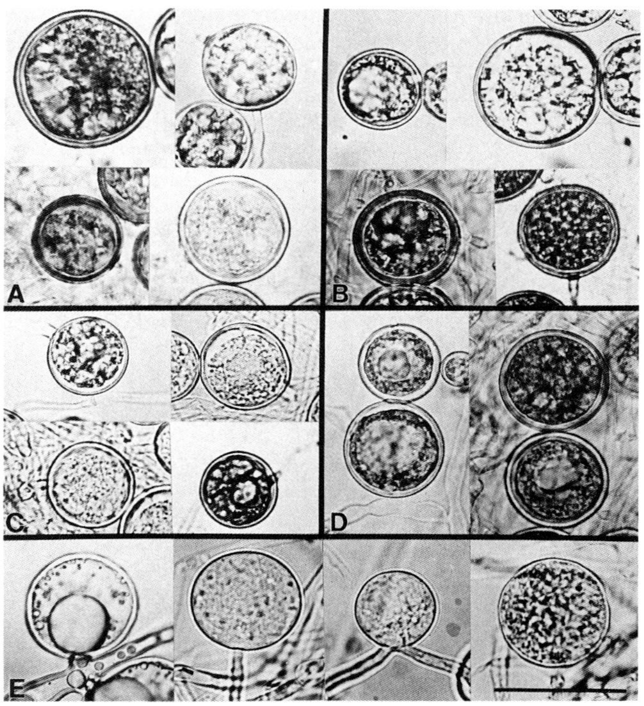

Figure 4.24. Chlamydospore morphology of various *Phytophthora* isolates and species. **A and B,** *P. parasitica* (T131 and P731); **C and D,** *P. palmivora* (P113 and 52-3); **E,** *P. cinnamomi* (T139). All figures share the same scale bar, 50 μm. Note that the wall of *P. cinnamomi* chlamydospores is thinner than walls of other species. See also Figure 4.23. (Reproduced by permission of the British Mycological Society, courtesy Alizadeh and Tsao 1985b)

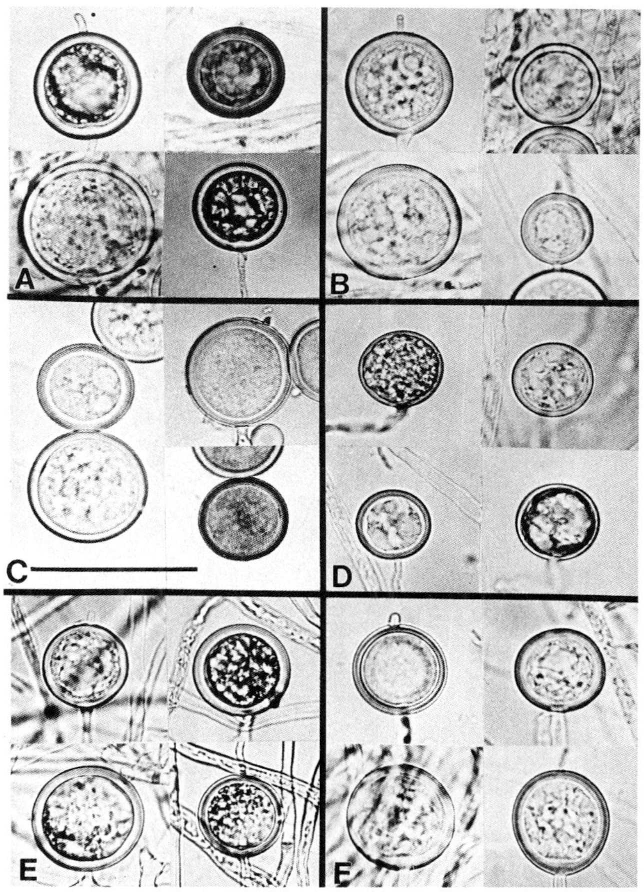

Figure 4.25. Chlamydospore morphology of *Phytophthora capsici sensu lato* (*P. palmivora* MF4 isolates) (Tsao 1991). **A,** P1072 from *Macadamia ternifolia;* **B–D,** P1027, S-1, and P558, respectively, from *Piper nigrum;* **E,** P1246 from *Sechium edule;* and **F,** P1249 from *Spondias purpurea.* All figures share the same scale bar, 50 μm. (Reproduced by permission of the British Mycological Society, courtesy Alizadeh and Tsao 1985b)

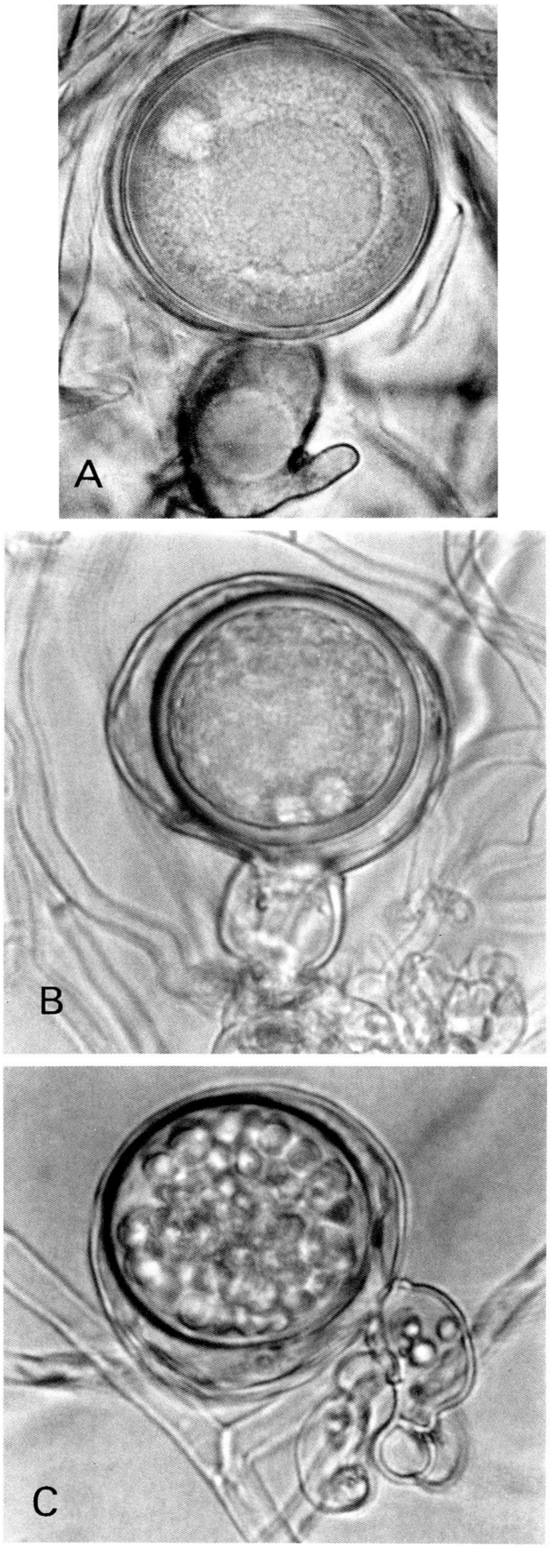

Figure 4.26. Morphology of oogonia, antheridia, and oospores. **A,** Mature oospore of *Phytophthora erythroseptica* with an amphigynous antheridium. Note the single pellucid body (nucleus) at the margin of the ooplasm and the distinct reserve globule (ooplast) in the center of the oospore. The oospore has a thick inner wall and entirely fills the oogonium (plerotic). **B,** Amphigynous antheridium on a mature *P. medicaginis* oospore. **C,** Paragynous antheridium on an immature *P. medicaginis* oogonium within which a thick-walled oospore will form. (R. A. Khan and D. C. Erwin)

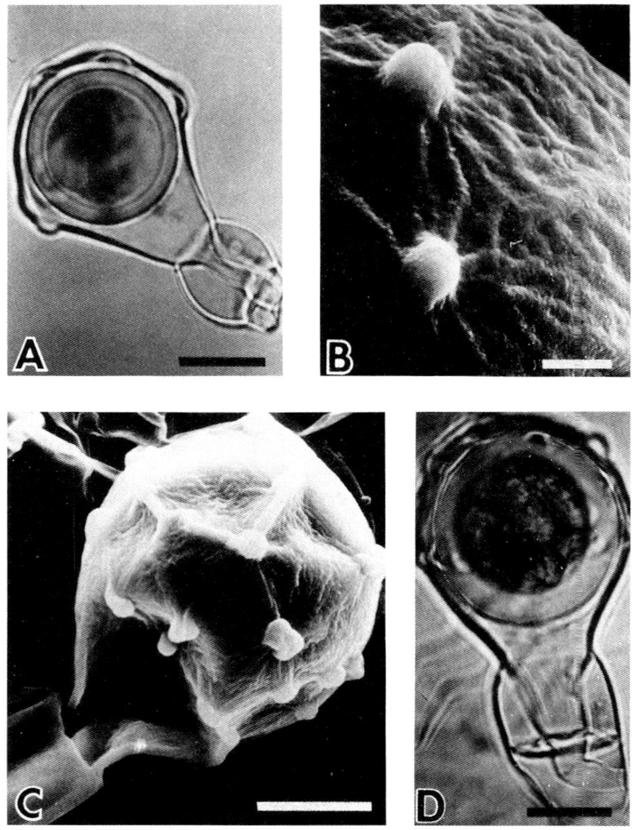

Figure 4.27. Knobby or bullate oogonia and bicellular antheridia of *Phytophthora cambivora.* **A and D,** Bullate oogonia with sloping bases and bicellular antheridia. Bar = about 15 µm. **B and C,** Papillae on the oogonial wall shown by scanning electron microscopy. Bar in **B** = about 1.7 µm and in **C,** about 20 µm. (Reproduced by permission from Mycologia 69:641-646, Ho et al., copyright 1977, The New York Botanical Garden, Bronx, New York)

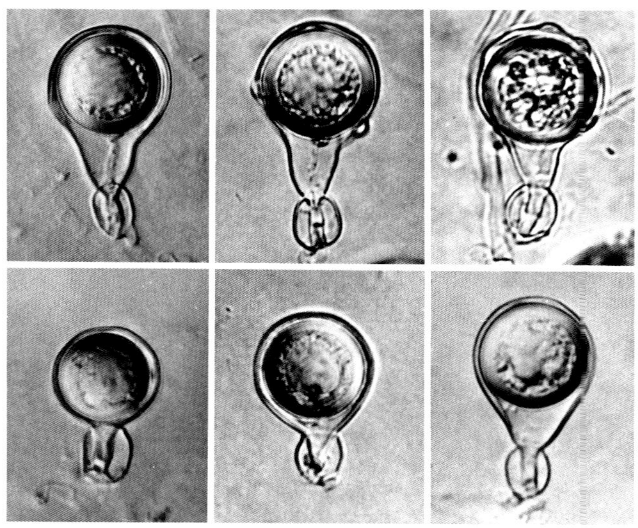

Figure 4.28. Upper row: oogonia of *Phytophthora katsurae* with funnel-shaped stalks and bullate protuberances on the walls of two of the oogonia; lower row: funnel-shaped stalks on only two oogonia produced by *P. heveae.* Feature is not absolutely consistent. (Reproduced by permission of the British Mycological Society, courtesy Oudemans and Coffey 1991b)

oogonia >45 μm in diameter. Since many isolates from different hosts produced oogonia with diameters that overlapped the 40- to 45-μm threshold, the validity of this designation has been questioned (Kuan and Erwin 1980a; Hamm and Hansen 1982a). Faris et al. (1989) proposed to retain *P. megasperma,* characterized by large oogonia as it was originally described by Drechsler (1931), to reject the broadened oogonia size concept of Tompkins et al. (1936b), and to retain *P. sojae* as a species as first described by Kaufmann and Gerdemann (1958). The name *P. sojae,* the narrow-host-range pathogen on soybean, had been rejected by Hildebrand (1959). Subsequently, Hansen and Maxwell (1991) also proposed to retain *P. sojae* but created two new species, *P. medicaginis* and *P. trifolii,* both of which had been previously classified as formae speciales of *P. megasperma* (Kuan and Erwin 1980a; see also E. M. Hansen [1991] for further discussion). When consulting the key of Stamps et al. (1990), one should realize that *P. megasperma* var. *sojae* does not necessarily pertain to the soybean root rot pathogen (Tables 4.2 and 4.3).

Growth at High Temperatures

Growth at certain high temperatures is a physiological character that has been used as an aid to classify species that grow well at temperatures as high as 35°C (Appendix 4.8). Tucker (1931) differentiated *P. parasitica* partially by its ability to grow at temperatures of 35°C or above. *P. drechsleri* grows well at 35°C and was differentiated from *P. cryptogea* on that basis. Species that have a maximum temperature for growth of less than 30°C, between 30 and 35°C, and 35°C or above are listed in Appendix 4.8. Although this type of data has some presumptive use and was used by Tucker (1931), temperature alone is not absolutely dependable, especially for separation of *P. cryptogea* from *P. drechsleri* (Mills et al. 1991). However, the dichotomous key of Ho (1992) (Appendix 4.9) separates *P. drechsleri* from *P. cryptogea* on the basis of growth at 35°C. Isolates within *P. capsici* varied in growth at 35°C (see Chapter 14).

The Identification Process

The known species and the morphological groups into which they have been placed in the tabular key of Stamps et al. (1990) are shown in Appendix 4.1. Tabular keys that contain pertinent information on each species are given in Tables 4.2 and 4.3. A dichotomous key for 22 species of *Phytophthora* that occur in Taiwan is presented in Appendix 4.9 (Ho 1992; Ho et al. 1995).

The only major differences in nomenclature from the key of Stamps et al. (1990) are in the treatment of the varieties of *P. megasperma* and *P. nicotianae.* Considerable research now shows that *P. megasperma* var. *megasperma* and *P. megasperma* var. *sojae* are not useful epithets (E. M. Hansen and Maxwell 1991). Likewise, *P. nicotianae* var. *parasitica* and *P. nicotianae* var. *nicotianae* are not separable on either morphological or genetic bases (see G. Hall [1993] and other citations given in the text and in Chapter 50 on *P. nicotianae*).

To facilitate the identification process, the culture should be obtained preferably from a hyphal tip or single, germinated cyst, sporangium, or oospore. The culture then should be incubated at the optimum temperature on a natural medium such as V8 juice agar or carrot broth agar. As pointed out in Chapter 3, *Phytophthora* species require a medium that contains a β-hydroxy sterol to sporulate. Potato-dextrose agar contains a low concentration of this component and is more favorable for mycelial growth than for sporulation, so this medium should be avoided.

Some *Phytophthora* species form sporangia readily on the surface of an agar medium, but many require subsequent culturing in water or mineral solutions or in dilute soil extracts under a suitable light. Methodologies for these techniques are given in Appendixes 3.24–3.45 of Chapter 3. Species of *Phytophthora* that sporulate well on agar and those that usually sporulate only in aqueous solutions or soil extract are presented below. Gerrettson-Cornell (1989) utilizes such criteria as a presumptive step in the identification process; however, such information is not definitive. Many of these species, e.g., *P. nicotianae* and *P. capsici,* that will produce sporangia on agar produce them more abundantly when transferred to dilute aqueous solutions.

Species that usually produce sporangia on agar:

arecae
boehmeriae
botryosa
cactorum
capsici
clandestina
iranica
heveae
hibernalis
ilicis
infestans
inflata
megakarya
mirabilis
nicotianae (= *parasitica*)
palmivora
parasitica (= *nicotianae*)
phaseoli
porri
primulae
syringae

Species that usually produce sporangia only in soil extract or in aqueous salts solutions:

cajani
cambivora
cinnamomi
clandestina
citricola
cryptogea
drechsleri
erythroseptica
fragariae
gonapodyides
lateralis
medicaginis (= *megasperma* f. sp. *medicaginis*)
megasperma
quininea
sojae (= *megasperma* f. sp. *glycinea*)
trifolii (= *megasperma* f. sp. *trifolii*)

The formation of sporangia must be induced in the species listed above. Incubation in a shallow layer of distilled water, soil extract, or various salts solutions usually gives the best results. Commonly used culture media include V8 agar or carrot agar or broth. See the appendixes to Chapter 3 for various media and treatments that induce formation of sporangia. Sporangia can be induced by incubating mycelium and agar disks (Appendixes 3.24–3.30), cut from the margin of a colony, in double distilled water or in one of several salts solutions. We recommend sterile or nonsterile soil extract (Appendix 3.44) or the Chen-Zentmyer solution (Appendixes 3.38 and 3.39); however, others described in Appendixes 3.31–3.45 should also be tried. A suitable light regime is necessary for sporangium formation for most species and especially for such foliage-infecting species as *P. palmivora* (see Chapter 3).

The section Production and Germination of Chlamydospores in Chapter 3 should be consulted for methods. Methods for production of oogonia, antheridia, and oospores are presented in Appendixes 3.46–3.54. For species that are heterothallic, A1 and A2 mating type isolates will be required for use as testers (Appendix 3.55). The polycarbonate membrane method, in which a known A1 or A2 mating type is separated from the unknown culture by the membrane, can be used to stimulate oospore production by an unknown isolate (Appendix 3.56).

After sporangia, chlamydospores (if present), oogonia, antheridia, and oospores are induced, the next task is a process of elimination to determine the relationship of descriptions to the isolate at hand. The appendixes noted above help in this process of elimination. Tables 4.2 and 4.3 contain characteristics in tabular form. Each chapter on individual species contains a brief description and illustrations of spores. The tabular key of Stamps et al. (1990) is a standard; however, the synoptic and dichotomous keys of Ho (1981, 1992) and the key of Gerrettson-Cornell (1989) may also be useful. The dichotomous key of Tucker (1931), although old, may be useful for the older, well-established species. The synoptic key for Taiwan species of *Phytophthora* is reproduced in Appendix 4.9 (Ho 1992; Ho et al. 1995). The Frezzi (1950) key (in Spanish), Kröber's (1985) key (in German), and that of Ho et al. (1995) have excellent photomicrographs of spore structures. The compilation of all the known original descriptions of species to 1970 by Waterhouse (1970a) is a valuable tool for identification purposes.

The known *Phytophthora* species are listed alphabetically in Table 4.1 and designated by the taxonomic group number according to the scheme of Waterhouse (1963) and Stamps et al. (1990) (also used in Tables 4.2 and 4.3). Appendix 4.1 presents the criteria used to group species and the species that are listed in each group. Appendix 4.2 presents a list of important criteria to consider when an isolate is observed. Appendix 4.3 is a tally sheet (Kröber 1985), which can be reproduced to make notes on the different criteria utilized for identification. The various appendixes and tables noted above can be used to narrow the search for the proper name of the species for the unknown isolate. The use of molecular technology as discussed later in the chapter is becoming necessary to properly define new species and to differentiate some that cannot be distinguished morphologically.

The identification of *Phytophthora* species is fraught with problems because distinct morphologic characters are few, variable, and overlapping. In some cases, certain isolates do not produce spore structures on any media or if they do, do not conform to any of the described species (Klotz et al. 1959; Mircetich and Matheron 1976; Jeffers et al. 1982; Wilcox et al. 1985; Wilcox and Mircetich 1985a; Mircetich and Browne 1987).

Phytophthora cultures in plastic dishes should be incubated under a light bank with fluorescent bulbs that emit light with a peak in the 300- to 400-nm wavelength region (near ultraviolet) of the spectrum (Ribeiro et al. 1976a). General Electric Cool White F40D 40-watt bulbs are usually adequate. Continuous light at an intensity of 1,700 μWcm^{-2} placed 60 cm above the cultures supported sporangium formation of *P. capsici* and other species (Al-Hedaithy and Tsao 1979a). Incandescent bulbs do not have the proper light spectrum and are less effective. When young mycelium was grown on clarified V8 juice broth, washed, and transferred to double distilled water at 25°C for 3 days, sporangia of *P. hibernalis, P. infestans, P. palmivora sensu stricto, P. palmivora* MF4 (now known as *P. capsici* [Tsao 1991]), *P. cactorum, P. botryosa, P. colocasiae, P. parasitica, P. cinnamomi,* and *P. megakarya* (cited then as *P. palmivora* MF3) were produced abundantly. *P. hibernalis* and *P. infestans,* which are low-temperature species, were incubated under a light bank at 1,300 μWcm^{-2} at 18°C. *P. cinnamomi* was induced to produce sporangia via the washing-solution method of Chen and Zentmyer (1970) (Appendix 3.38; Al-Hedaithy and Tsao 1979a, b).

P. heveae, P. arecae, P. meadii, P. botryosa, P. megakarya, P. palmivora, and *P. capsici* (*P. palmivora* MF4) form sporangia on lima bean agar; however, sporangium production on young, washed mycelium incubated in double distilled water occurs more consistently and rapidly (within 24 h) than on unwashed cultures (Dantanarayana et al. 1984). Formation of *P. capsici* (*sensu* Tsao 1991) sporangia by isolates from black pepper and other hosts was affected by light. Sporangiophores formed in an umbel (Figure 4.16) under light, but formation was irregular in the dark. Also, the bases of sporangia formed in the dark were more round and the length-breadth ratio was lower than that of sporangia formed in light (Alizadeh and Tsao 1985a).

Once sporangia are present, they can be characterized as papillate, semipapillate (inconspicuously papillate), or nonpapillate (Figure 4.7 and Appendix 4.5). The most important feature of sporangia is the nature of the apical thickening (papillae). In most cases, the papillate sporangia are readily recognized, but deciding whether they are semipapillate or nonpapillate requires close observation of several hundred sporangia from different petri plates and different isolates. Staining the sporangia with lactophenol-cotton blue (0.01%), trypan blue (0.05%) in 0.085% sodium chloride, or rose bengal (50 ppm) in water helps to differentiate the apical thickening on the sporangium from the protoplasm of the sporangium. Applying heat to the slide on which the sporangia are mounted increases uptake of the stain; however, heat can cause the protoplasm to shrink. See Figure 4.9C, in which the sporangium is stained without shrinkage of the protoplasm.

Length and breadth of 50–100 sporangia should be recorded and the length-breadth ratio calculated. Figure 4.11 shows a suggested method of obtaining biometrical characteristics of sporangia (Dick 1990b; Shazad et al. 1992). Spore sizes can vary, depending on isolates and conditions; thus, it is best to state the range of variation as well as the mean.

The branching habit of sporangiophores should be observed directly without mounting them in an aqueous solution. The branching habit can be unbranched, a simple sympodium (Figures 4.5A and 4.8A), a compound sympodium (Figure 4.5B and C), or an umbel (umbrella) pattern (Figures 4.5D and 4.16). A series of shapes by which sporangia might be classified is presented in Figures 4.10–4.12.

The caducous nature of sporangia is a critical feature that characterizes some species (Appendix 4.6; Figures 4.15 and 4.17). Caducity is expressed more readily in light than in the dark (Alizadeh and Tsao 1985a). The stalks on sporangia that are caducous can be classified as short, intermediate, and long (more than 50 μm) pedicels. The lengths are characteristic of certain species (Appendix 4.6). Methods to dislodge sporangia and to determine caducity are given above and in Chapter 3 (Al-Hedaithy and Tsao 1979a, b). Caducity should not be attributed to a species in an aqueous culture if the length of pedicels is variable. *P. cinnamomi* does not produce caducous sporangia, but if the culture is agitated in water long enough, a few sporangia will become dislodged.

Chlamydospores are produced by some but not all species. (See Chapter 3 for protocols reported to induce chlamydospores.) These structures are produced abundantly in culture media by some species such as *P. cinnamomi* and *P. parasitica,* but they are produced sparsely by many species. The effect of light varies among species. *P. cinnamomi* produced the greatest number (almost fourfold) of chlamydospores when sitosterol-amended cultures were incubated under light at 310–420 nm (near ultraviolet). The numbers produced at 470–610 and 550–725 nm were similar to that produced in the dark (Englander and Turbitt 1979). Chlamydospores of *P. lateralis* were suppressed by light but formed in the dark (Englander and Roth 1980). The submerged-culture method of Tsao (1971) supported chlamydospore production in the dark with *P. parasitica* (cacao and black pepper isolates) and *P. capsici* (Alizadeh and Tsao 1985b). See Figures 4.23–4.25 for illustrations.

Three species, *P. cambivora, P. katsurae,* and *P. verrucosa,* produce ornamented oogonia (Appendix 4.4; Figures 4.27 and 4.28). Oogonia with tapered bases are characteristic of *P. heveae.* Several other features, including pathogenicity, that are unique to certain species are shown in Appendix 4.4.

The relationship of antheridia to the oogonia is a major differential character. Some species produce mainly paragynous antheridia and others amphigynous antheridia (Appendix 4.7; Figures 4.26A–C).

Most of the features explained above are listed for each taxonomic group in Tables 4.2 and 4.3 according to criteria used in the key of Stamps et al. (1990). Our Tables 4.2 (asexual characteristics) and 4.3 (sexual and growth-temperature characteristics) should be used as supplements to the more detailed information in the Stamps et al. (1990) key; however, there are some differences between the two keys based on information subsequent to 1990.

The two varieties of *P. nicotianae, nicotianae* and *parasitica,* according to Stamps et al. (1990) have been omitted because several reports indicate that they cannot be consistently differentiated by morphology (Tsao and Sisemore 1978; Ho and Jong 1989), DNA RFLP patterns (Panabieres et al. 1989; Förster et al. 1990b), isozyme patterns (Oudemans and Coffey (1991c), or a combination of morphological, physiological, and biochemical characters (Hall 1993). The two varieties of *P. megasperma, sojae* and *megasperma,* have been omitted on the basis of the report of E. M. Hansen and Maxwell (1991).

Molecular Technology for Differentiation of Species

Identification or characterization of species, especially those with similar morphological features, is often difficult. Brasier (1991) makes a case for adoption of physiological parameters as well as morphology and indicates that since *Phytophthora* is a successful parasite and pathogen on many crop hosts, new species could evolve relatively rapidly and changes in morphology might lag behind pathological capability. See discussions by Brasier and Hansen (1992) and Brasier (1992) on the evolution of *Phytophthora* species.

The technology adapted from molecular biology studies has contributed several important tools for confirmation of the validity of species. These approaches are rapidly evolving as aids to help answer questions as to whether certain morphological features are valid for differentiating species and whether certain species that are similar morphologically actually are similar genetically.

Some interesting results that would not have been anticipated have arisen after the use of molecular techniques. For instance, in the key of Stamps et al. (1990), *P. nicotianae* var. *parasitica* and *P. nicotianae* var. *nicotianae* are differentiated by morphological features; however, DNA and isozyme technology (Panabieres et al. 1989; Förster et al. 1990b; Oudemans and Coffey 1991c; G. Hall 1993) have confirmed the premise of Ho and Jong (1989), based on morphology, that they are not different (see Chapter 50). Oudemans and Coffey (1991c) recommend that use of these varieties should be discontinued (Figure 4.31). Brasier and Hansen (1992) agreed with this concept.

The possible merger of *P. cryptogea* and *P. drechsleri* has been controversial. Many claim that these morphologically similar species should be designated *P. cryptogea* because of priority (Bumbieris 1974; Ho and Jong 1986). On the basis of results from isozyme and DNA technology, as many as seven genetic groups can be delineated within the *P. cryptogea*-*P. drechsleri* complex (Mills et al. 1991) (Figure 4.33). These examples were previously reviewed as taxonomic problems (Erwin 1983). Brasier and Hansen (1992), who discuss this group in light of the results of Mills et al. (1991), link *P. gonapodyides* with this complex and speculate that molecular groups J and K of Mills et al. (1991) may represent *P. gonapodyoides.*

See Chapter 20 on *P. cryptogea* and Chapter 23 on *P. drechsleri.* These examples are discussed further in the following sections and in the chapters covering the *Phytophthora* species under discussion. The use of protein patterns obtained by gel electrophoresis, isozyme patterns, and DNA RFLP to resolve taxonomic problems are also reviewed.

Protein Electrophoresis Patterns

The rationale for the use of protein patterns as a distinguishing character in *Phytophthora* taxonomy is that proteins extracted from mycelium move on a gel matrix (starch gel or polyacrylamide gel) at different speeds in an electrophoretic field. (See Table 4.4 for references to several reports on use of protein patterns to resolve or supplement taxonomic problems.) The measure of differences and similarities of protein patterns among *Phytophthora* species is discussed by Brasier (1983, 1991) and Brasier and Hansen (1992) and is aptly described by Gallegly (1983a) (excerpted by permission of the American Phytopathological Society, St. Paul, Minnesota, courtesy of Gallegly 1983a):

> Perhaps the most promising of the physiologic methods for use in classification and identification of species of *Phytophthora* is the determination of protein patterns by gel electrophoresis. The rationale used is that valid species should differ in their protein makeup and that such differences can be detected by the rate at which the proteins move in a gel aided by an electric current. Electrophoretic measurement of protein differences can be much less difficult and time-consuming than measurement by serologic methods. Although the reliability of protein patterns as a taxonomic tool for other fungi is controversial, the data thus far obtained with species of *Phytophthora* appear to offer reliability. Starch and acrylamide gels have been used in slab and tubular electrophoresis to produce these patterns.
>
> The first evidence that protein patterns may be useful in identification of pythiaceous fungi was provided by Clare (1963), who used starch gel electrophoresis to produce patterns for six species of *Pythium.* Each species produced a distinctly different pattern, with one band common to all species. Three isolates of one species had identical patterns as did two isolates of another species. Using the same method, Clare and Zentmyer (1966) found that the protein patterns of *P. cinnamomi* and *P. palmivora* were distinctly different and that the patterns of different isolates of the same species were identical or similar, regardless of geographic or host source. Using polyacrylamide gel electrophoresis, Hall et al. (1969) obtained similar results with *P. cactorum, P. cinnamomi,* and *P. palmivora,* but the *P. palmivora* isolates showed two bands, which were different quantitatively and identical qualitatively. Esterase patterns were also examined, but 13 different patterns were found among the isolates of the three species. In retrospect, one wonders whether the two different bands among the isolates of *P. palmivora* are associated with any of the different morphological forms of the species (Brasier and Griffin, 1979).

Gill and Powell (1968b), using polyacrylamide gel (disc) electrophoresis, found distinct protein patterns for *P. cactorum, P. fragariae,* and *P. sojae*. However, this technique did not differentiate the eight physiologic races of *P. fragariae* (Gill and Powell 1968a); all isolates of *P. fragariae* had nearly identical protein patterns regardless of race. Later, Gill and Zentmyer (1978) found that the protein pattern from polyacrylamide gel (disc) electrophoresis of 30 isolates of *P. cinnamomi* were identical except for a minor variation by one isolate. Likewise, five isolates of *P. cactorum* had identical patterns. The *P. cinnamomi* isolates were from 17 different hosts and from widely separated geographic areas. Further, neither the length of time an isolate had been in culture nor the compatibility types A1 and A2 affected the protein patterns. Boccas and Zentmyer (1976) and Boccas (1981) determined the protein patterns of the parents used in their genetic studies of interspecific hybrids and compared the protein patterns of cultures established from single germinating oospores. They concluded that all but one of the progeny cultures were selfed oospores. Previously, Zentmyer et al. (1974) used protein patterns to help identify *P. citricola* obtained from avocado roots.

For protein pattern analysis, the isolates are grown in a suitable liquid medium for about 7 days, after which the mycelium is washed three or more times with double distilled water and in some cases with a suitable buffer (e.g., 0.1 M Na_2HPO_4 and 0.1 M KH_2PO_4) (Gill and Zentmyer 1978; Kaosiri and Zentmyer 1980). Extraction of proteins involves freezing the mycelium, grinding it with a mortar and pestle with acid-washed sand in a phosphate buffer (pH 7.0; 0.5 ml/g of mycelium), and centrifuging it for 30 min at 47,800 × *g*. The buffer-soluble proteins in the supernatant can be dispensed into small polyethylene tubes and frozen at –20°C. The proteins in the mycelium suspension are extracted by variations of several methods. See Davis (1964) and Ornstein (1964) for basic methods and theory. The general subject was reviewed by R. Hall (1969), and techniques on starch gel electrophoresis are given by O'Malley et al. (1980).

The buffer-soluble protein can be absorbed onto small, 5- × 8-mm pieces of chromatography paper and placed in slots at the base of a horizontal polyacrylamide gel slab. The gel in the apparatus is set up at 4°C and contains *N,N,N,N′*-tetramethylethylenediamine, 0.15 ml; ammonium persulfate, 0.07 g; and 0.076 M Tris-HCl buffer (pH 9.0), 100 ml. Borate buffer (0.1 M, pH 9.0) is placed in the electrode vessels, and then a constant current of 30 mA is applied until the front has moved 7.5 cm past the sample slot in the gel. The gel can be split and stained with a 1% solution of amido black in 7% acetic acid to detect protein bands and, if necessary, destained over several days by rinsing with 7% acetic acid. The other half of the gel can be stained for esterase activity with a solution of α-naphthyl acetate (20 mg) and fast blue RR (25 mg) in 50 ml of 0.076 M Tris buffer at pH 7.0 (R. Hall et al. 1969).

Bielenin et al. (1988) used a "discontinuous system of electrophoresis" (Laemmli 1970), in which the soluble protein extracts were mixed with a "disruption buffer" consisting of 0.5 M Tris-HCl at pH 6.8, glycerol (amount not given), 4% sodium dodecyl sulfate (SDS), and 10% 2-mercaptoethanol and brought to 100°C for 2–3 min in a water bath. The dissociated proteins (10–20 μg per well in the polyacrylamide gel) were subjected to electrophoresis at 60–70 V at room temperature. Gels were stained by immersion in Coomassie brilliant blue R250 in acetic acid-methanol-water (1:5:4) and destained with several changes of water-methanol-acetic acid (7:2:1) to detect protein bands (Blakesley and Boezi 1977).

Bielenin et al. (1988) state, "Based on our procedures and the isolates analyzed, it appears that electrophoretic banding patterns of SDS-dissociated proteins provide a more conservative taxonomic criterion for indicating differences among these three *Phytophthora* spp. (*P. cactorum, P. syringae,* and *P. cambivora*) than do patterns of native proteins." Electrophoresis of dissociated proteins provided distinct patterns that could be grouped and readily differentiated. The protein patterns for *P. cryptogea* and *P. drechsleri* each contained two subgroups, one of which was common to isolates of both species. Isolates of *P. megasperma* from deciduous fruit crops

Table 4.4. References on use of protein patterns to differentiate species of *Phytophthora*

Species differentiated or characterized	Reference
P. cactorum, P. cambivora, P. cryptogea, P. drechsleri, P. syringae	Bielenin et al. (1988)
P. cactorum, P. cinnamomi, P. palmivora	R. Hall et al. (1969)
P. cactorum, P. citrophthora, P. citricola, P. nicotianae	Magnano di San Lio et al. (1987a)
P. cactorum, P. fragariae	Gill and Powell (1968a, b)
P. cinnamomi isolates; A1 mating type differentiated from A2 by a minor band	Gill and Zentmyer (1978)
P. cinnamomi, P. citrophthora	Clare and Zentmyer (1966)
P. cinnamomi, P. parasitica	Boccas and Zentmyer (1976); Boccas (1981)
P. citricola	Zentmyer et al. (1974)
P. cryptogea	Matsumoto and Sato (1979); Magnano di San Lio (1987b; 1992)
P. cryptogea-P. drechsleri not differentiated[a]	Masago et al. (1989)
P. cryptogea-P. drechsleri	Bielenin et al. (1988)
P. cryptogea, P. drechsleri, P. erythroseptica	Cacciola et al. (1990)
P. drechsleri	Thompson and Phillips (1988); E. M. Hansen et al. (1988)
P. heveae. First report in Taiwan. Isolates from forest and citrus soils were similar.	Chang and Shu (1988)
P. meadii	Sansome et al. (1991)
P. megasperma complex	E. M. Hansen et al. (1986); E. M. Hansen (1987)
P. megasperma f. sp. *medicaginis* compared with *P. megasperma* f. sp. *glycinea, P. vignae,* and *P. nicotianae*	Irwin and Dale (1982)
P. megasperma f. sp. *medicaginis* and large oogonium producing isolates of *P. megasperma*	Faris et al. (1986)
P. nicotianae, P. cactorum, P. citrophthora, P. citricola, P. capsici, P. cryptogea	Erselius and de Vallaveille (1984)
P. nicotianae	Cacciola and Magnano di San Lio (1988)
P. palmivora	Erselius and Shaw (1982)
P. palmivora	Magnano di San Lio et al. (1986a)
P. palmivora (typical and atypical)	Idosu and Zentmyer (1978)
P. palmivora, P. capsici, P. cinnamomi	Kaosiri and Zentmyer (1980)
P. parasitica var. *parasitica* differentiated from *P. parasitica* var. *nicotianae*[a]	Masago et al. (1989)
P. pseudotsugae	Hamm and Hansen (1983)

[a] Used two-dimensional electrophoresis.

exhibited protein patterns similar to those of the broad host range group defined by E. M. Hansen et al. (1986), who established six protein pattern subgroups within the *P. megasperma* complex. The other isolates belonged to the apple, cherry, and apricot group. As in other studies (Förster et al. 1987–1989), *P. megasperma* f. sp. *glycinea* (*P. sojae*) and *P. megasperma* f. sp. *medicaginis* (*P. medicaginis*) were widely separated.

In summary, protein patterns provide confirmatory evidence in the identification process and supplement morphology data that might not be definitive. As far as we know, protein patterns have been used more as confirmatory than as primary evidence for the establishment of species. Brasier (1991) suggested that protein electrophoresis shows excellent discrimination at the morpho-species level and has been underused for taxonomy purposes. Although the analysis of protein patterns reveals differences among species, the large number of bands that can be discerned on the gels is often difficult to quantify. Also, many minor bands, which may be as important as major or darker-staining bands, could be obscured. Biosynthesis of some proteins might be under environmental or developmental control and thus would not necessarily be constant (Förster and Coffey 1991). These authors favored detection of isozymes and analysis of mitohondrial DNA (mtDNA). In light of the rapid progress in development of diagnostic isozyme and DNA methodology (covered in subsequent sections), current research emphasizes these areas to differentiate species.

Isozyme Patterns

Development of isozyme patterns is a means of identifying genetic diversity within populations of organisms (Micales et al. 1986). The technology and rationale for the use of isozymes by Spielman (1991) to study population movement and genetics of *P. infestans* is presented in Chapter 3 under Genetics and Cytology of *Phytophthora.* Nygaard et al. (1989) pointed out that isozyme banding patterns are less complex than total protein patterns and can be differentiated and interpreted more readily. By use of isozyme techniques, Old et al. (1984, 1988) noted that the genetic variability of many isolates of *P. cinnamomi* from various regions of Australia was relatively low; however, the isozyme patterns of the A1 and A2 mating types differed. Some references that have utilized isozyme technology are listed in Table 4.5.

The methodology for isozyme analysis is well worked out and straightforward. The excerpt from Spielman (1991) in Chapter 3 under Genetics and Cytology of *Phytophthora* gives the principles of the methods. See Spielman et al. (1989, 1990, 1992), Nygaard et al. (1989), Oudemans and Coffey (1991a), and other references in Table 4.5 for specific methodology.

Oudemans and Coffey (1991a) showed distinct differentiation of *P. cactorum, P. cambivora,* and *P. cinnamomi. P. megasperma* has long been known as a heterogeneous species (E. M. Hansen and Hamm 1983; E. M. Hansen 1987, 1991; E. M. Hansen et al. 1986). Several biotypes were readily differentiated by use of isozymes (Nygaard et al. 1989). Six isozyme pattern groups included each of the three formae speciales, *P. megasperma* f. sp. *glycinea, P. megasperma* f. sp. *medicaginis* (Kuan and Erwin 1980a), and *P. megasperma* f. sp. *trifolii* (R. G. Pratt 1981), and the apple-apricot-cherry, Douglas fir, and broad host range groups. These groups corresponded to the protein pattern groups reported by E. M. Hansen et al. (1986). Data from isozyme patterns agreed closely with analyses of mtDNA RFLPs (Förster et al. 1987–1989). Although *P. megasperma* f. sp. *medicaginis* and *P. megasperma* f. sp. *glycinea* have morphologic similarity, Nygaard et al. (1989) point out that the small but consistent differences of in vitro colony habit (E. M. Hansen et al. 1986), growth-temperature response (E. M. Hansen and Hamm 1983), nuclear DNA content and karyotypes (E. M. Hansen et al. 1986), mtDNA size and arrangement (Förster et al. 1987–1989; Förster et al. 1991), sensitivity to metalaxyl and phosphorus acid (Coffey and Bower 1984a, b), and metabolism of nitrate (Hohl 1983) readily differentiate these two formae speciales. These results strongly support the subsequent separation of these formae speciales into separate species, *P. sojae* and *P. medicaginis* (Hansen and Maxwell 1991). See Chapters 44, 46, and 59 on *P. medicaginis, P. megasperma,* and *P. sojae,* respectively.

The possibility had been suggested that *P. cambivora* and *P. cinnamomi* might be related because both produce nonpapillate, noncaducous, ovoid sporangia and some antheridia with two cells (Ashby 1929c; Royle and Hickman 1964; Ho et al. 1977) (Figure 4.27). Oudemans and Coffey (1991a) delineated eight electrophoretic types (ETs) within isolates of *P. cambivora,* five of which originated in Australia. Genetic distances between *P. cambivora* and *P. cinnamomi* showed clearly that they were not related, despite the production of bicellular antheridia in each species. Within *P. cinnamomi,* however, eight ETs were noted. The A1 mating type was separable from the A2 mating type, thus confirming the findings of Old et al. (1980, 1988). A new biotype from citrus reported as *P. cinnamomi* by Ann and Ko (1985) was clearly distinct from the other ETs and did not support the inclusion of this isolate in *P. cinnamomi.* Although *P. cactorum, P. cambivora,* and *P. cinnamomi* were readily differentiated by isozyme patterns, separate electrophoretic subtypes were differentiated within each species. *P. cactorum* contained the lowest level of intraspecies diversity. Similar DNA RFLP data were reported by Förster and Coffey (1991) (Figure 4.32).

Isozyme analysis of 12 papillate species in group II (Oudemans and Coffey 1991b, c) confirmed the validity of *P. botryosa, P. heveae, P. katsurae, P. meadii, P. palmivora,* and *P. parasitica* (= *P. nicotianae*) as distinct species.

Since the tobacco isolates of *P. parasitica,* designated *P. parasitica* var. *nicotianae sensu* Tucker (1931), could not be distinguished from isolates from other hosts, Oudemans and Coffey (1991b) concluded that maintaining both *P. nicotianae* var. *nicotianae* and *P. nicotianae* var. *parasitica* (*sensu* Waterhouse 1963) should be discontinued (Figure 4.31). This was also noted by Brasier and Hansen (1992). Previous reports based on morphology (Tsao and Sisemore 1978; Ho and Jong 1989) and DNA RFLPs (Panabieres et al. 1989; see Oudemans and Coffey 1991c) (Figure 4.31) support deletion of the two varieties of *P. nicotianae.* Following the recent report by Hall (1993), the two varieties of *P. nicotianae* can be deleted from future keys. See Chapter 50 on *P. nicotianae.*

The summary of the Oudemans and Coffey (1991b) report on isozyme evaluation of papillate sporangium species (groups I and II) is as follows (reproduced by permission of the British Mycological Society, courtesy of Oudemans and Coffey 1991b):

> Isozyme analysis was used to examine intraspecific diversity and interspecific relatedness of *P. cactorum* (group I) and twelve papillate species of *Phytophthora* within group II of Waterhouse's (1963) taxonomic scheme. Isolates of six species, specifically *P. botryosa, P. heveae, P. katsurae, P. meadii, P. palmivora* and *P. parasitica* (*P. nicotianae*), contained low levels of genetic diversity and were judged to be valid species. Based upon isozyme analysis, all isolates of *P. arecae* were found to be identical with some isolates of *P. palmivora.* Thus, it is proposed that these two species are synonymous. For *P. parasitica,* differences in oospore size or host plant

were contrasted with the isozyme data [by] using single locus coefficients, and no evidence was found to support the existence of distinct varieties. In contrast, several other species, particularly *P. capsici, P. citrophthora* and *P. megakarya,* demonstrated much higher levels of variation in isozyme pattern, and subgroups were readily identified. Isolates of *P. capsici,* including those previously identified as *P. palmivora* 'MF4', separated into three subgroups CAP1, CAP2 and CAP3. Those in CAP1 were from widely distributed geographical locations on a range of hosts including *Capsicum* sp., tomato, cucurbits, as well as cacao and black pepper and it contained the greatest amount of intraspecific diversity. A single representative of *P. mexicana* also clustered with this subgroup and as such is considered

Table 4.5. Use of isozyme technology to differentiate *Phytophthora* species or as markers for genetic research

***Phytophthora* species**	**Results**	**Reference**
P. arecae	Conspecific with *P. palmivora.*	Oudemans and Coffey (1991c); Mchau and Coffey (1994a)
P. capsici, P. citrophthora, P. palmivora, and *P. megakarya*	All species were differentiated.	Blaha (1990)
P. cryptogea, P. drechsleri, and *P. erythroseptica*	All species were differentiated.	Cacciola et al. (1990)
P. cambivora, P. cinnamomi, P. erythroseptica, P. lateralis, and *P. richardiae*	All species were differentiated.	Mills et al. (1991)
P. drechsleri, P. cryptogea, and *P. melonis*	Isolates differed from those above but separated into 10 ET[a] subgroups.	
P. nicotianae (*P. parasitica*)	The two varieties, *nicotianae* and *parasitica,* were not separable.	
P. capsici	Isolates separated into three ET subgroups.	
P. citrophthora	Isolates separated into two ET subgroups.	
P. megakarya	Isolates separated into two ET subgroups.	
P. boehmeriae	Isolates separated into four ET subgroups.	
P. capsici	Isolates were related to *P. citrophthora.*	
P. botryosa and *P. meadii*	Both species were clustered together. Data were supported by DNA RFLP technology.	
P. megasperma	Differentiated into six ET subgroups in which *P. megasperma* f. sp. *glycinea* and *P. megasperma* f. sp. *medicaginis* were unrelated to each other or to other *P. megasperma* subgroups.	Nygaard et al. (1989)
P. cactorum, P. cryptogea, and *P. parasitica* var. *nicotianae*	All species were differentiated.	
P. cinnamomi	Isolates were homogenous, but A1 mating types were differentiated from A2 mating types.	Old et al. (1984, 1986)
P. cactorum, P. cambivora, and *P. cinnamomi*	*P. cactorum* contained two ET subgroups, *P. cambivora* eight ET subgroups, and *P. cinnamomi* eight ET subgroups; A2 isolates were differentiated from A1 isolates.	Oudemans and Coffey (1991a, c)
Phytophthora species in Waterhouse (1963), taxonomic groups I and II	*P. cactorum* is differentiated from other species in groups I and II.	Oudemans and Coffey (1991b, c)
	Group II: *P. botryosa, P. heveae, P. katsurae, P. meadii, P. palmivora,* and *P. parasitica* (*P. nicotianae* var. *nicotianae* and *P. nicotianae* var. *parasitica* were indistinguishable). Each species contained low levels of genetic diversity. *P. arecae* isolates were identical to *P. palmivora.*	
	Some species were highly diverse. *P. boemeriae,* four ET subgroups; *P. botryosa,* clustered with *P. meadiae; P. capsici,* ET CAP1 isolates from *Capsicum annuum* grouped with *P. mexicana,* ET CAP2 included isolates from *Piper* and *Macadamia,* ET CAP3 included isolates from cacao in Brazil; *P. citrophthora,* ET CTR1 included isolates from citrus and kiwi, CTR2 included isolates from cacao in Brazil; *P. heveae* isolates clustered with *P. katsurae; P. meadii* isolates clustered with *P. botryosa; P. megakarya* isolates included ET MGK1 from Nigeria and Guinea and ET MGK2 from Cameroon.	
P. capsici	Distinct subgroups ET CAPA and ET CAPB were differentiated. Majority of CAPA isolates had rounded sporangia and irregular sporangiophores. Most CAPB isolates had ellipsoid-lanceolate sporangia, but about 50% had umbellate sporangiophore branching. See Chapter 14.	Mchau and Coffey (1995)
P. citricola	10 ET types clustered in five subgroups; subgroup CIT1 included isolates from citrus; CIT2 and CIT3, from diverse locations and hosts; CIT4, from native plant hosts in South Africa; CIT5, from avocado may be a biological species. See also Table 4.6.	Oudemans et al. (1994)
P. citrophthora	Delineated ET subgroups CTR1, CTR2, and CTR3 from 91 isolates obtained worldwide.	Mchau and Coffey (1994b)
P. infestans	Isozyme characteristics were inherited in germinated oospore progeny of A1 × A2 crosses.	Shattock et al. (1986a, b)
	Isozyme patterns used as markers to assay geographical distribution of types. Review and excellent explanation of methodology and principles (see Chapter 3).	Spielman et al. (1989, 1991)
	Isozyme patterns used as markers to characterize germinated oospore progeny of A1 × A2 crosses.	Tooley et al. (1985)
	Geographical distribution of populations traced by isozyme characteristics; evidence supports the concept of Mexico (not Peru) as the origin of *P. infestans.*	Tooley et al. (1989)
	Population is genetically more diverse in central Mexico (Toluca) where A1 and A2 isolates have coexisted for a long time than in northwestern and northeastern regions, where only the A1 mating type existed before 1970.	Goodwin et al. (1992)
P. palmivora and *P. arecae*	Assay of 93 isolates. *P. arecae* clustered with *P. palmivora* supporting conspecificity.	Mchau and Coffey (1994a)

[a] Electrophoretic type.

conspecific to *P. capsici*. CAP2 were found primarily on black pepper and from Hawaii on *Macadamia integrifolia* as well as other hosts. In some instances, specifically with isolates from India and Indonesia on black pepper, both CAP1 and CAP2 were found. However, the evidence from isozymes suggests that the two groups are genetically isolated from one another. All representatives of CAP3 were derived from diseased cacao in Brazil and this subgroup was monomorphic for all loci examined. *P. citrophthora* was divided into two subgroups. Isolates of CTR1 were derived from a large range of hosts including citrus and kiwi fruit. CTR2 was composed of only Brazilian isolates from cacao. *P. megakarya*, which has only been reported on cacao from West Africa, was also divided into two subgroups which were separated geographically. Isolates of MGK1 originated from either Nigeria or Guinea and isolates of MGK2 were from Cameroon. *P. boehmeriae* was the most genetically diverse species examined with four isozymically diverse ETs representing only 11 isolates. Interspecific comparisons revealed relationships which would not have been predicted based on morphological comparisons alone. For example, *P. meadii* and *P. botryosa* clustered together indicating a very close genetic relatedness. *P. katsurae* and *P. heveae* also formed a single cluster. However, comparison of oospores of these two species showed that the oogonial stalk length of *P. katsurae* was consistently much longer. Unexpectedly, *P. capsici* and *P. citrophthora* also formed a cluster indicating that they were genetically related. Finally, the results of this isozyme analysis suggest that the papillate species of Group II of Waterhouse's (1963) scheme may form at least six distinct evolutionary lines.

Mchau and Coffey (1994c) extended the study of *P. capsici* with more isolates and found that isolates clustered in two subgroups, CAPA and CAPB (see Chapter 14 and Table 4.5).

P. drechsleri and *P. cryptogea* produce nonpapillate sporangia and amphigynous antheridia and are placed in Waterhouse (1963) group VI. Tucker (1931) separated these species by growth at 35°C for *P. drechsleri* and no growth for *P. cryptogea* and also noted that sporangia of *P. drechsleri* were more elongated than those of *P. cryptogea*. Variability in temperature response and their similarity in morphology prompted the proposal that the two species should be merged to *P. cryptogea*, which has priority (Bumbieris 1974; Halsall 1976a; Shepherd 1978; Ho and Jong 1986).

Mills et al. (1991) utilized isozyme analysis (Oudemans and Coffey 1991a, b) to study the relationship of 123 isolates from various regions of the world. Surprisingly, isozyme analysis delineated 10 distinct groups on the basis of numerical analysis of 24 putative enzyme loci. The electrophoretic patterns were not associated with previously identified species on the basis of morphology or temperature tolerance. A parallel study of mtDNA RFLPs yielded data that led to similar conclusions (Figure 4.33). Also, morphological shapes of sporangia were variable among isolates. The sloping bases, characteristic of sporangia attributed to *P. drechsleri* by Tucker (1931), were too inconsistent to reliably differentiate this species from *P. cryptogea* (Figures 4.13 and 4.14). Mills et al. (1991) conclude, on the basis of the isozyme and mtDNA RFLP patterns, that at least seven molecular species exist within the *P. cryptogea*-*P. drechsleri* group.

P. drechsleri f. sp. *cajani* formed a discrete population, indicating that the described species, *P. cajani*, might be correct. The *P. melonis* and *P. sinensis* species formed different groups (Figure 4.33). The variability between each of the groups was as great as that between *P. cinnamomi*, *P. cambivora*, *P. lateralis*, *P. erythroseptica*, and *P. richardiae*.

An interesting but unexpected sidelight of the report by Mills et al. (1991) was the finding that the culture considered by several laboratories to be the type species of *P. erythroseptica* fell into the group that contained the type species of *P. cryptogea*. Other isolates of *P. erythroseptica* were uniform and completely different from the other species. Mills et al. (1991) conclude that inadvertent mislabeling of the culture was the most plausible explanation. This example points out the powerful potentialities of this molecular technique.

Oudemans et al. (1994) studied 125 isolates of *P. citricola* by isozyme and mtDNA RFLP analysis and temperature-growth responses. Subgroups CIT1, CIT2, CIT3, CIT4, and CIT5 were delineated. An avocado pathogen subgroup, CIT5, did not grow at 30°C, but isolates CIT1, CIT2, and CIT3 grew well at 30°C. The CIT5 group, which causes trunk canker on avocado from Mexico, California, and Guatemala (Zentmyer et al. 1974), appeared to be a distinct group and may constitute another taxon. Since CIT5 also clustered closely to *P. capsici* and *P. citrophthora*, Oudemans et al. (1994) suggested that taxonomic revision should await further study. The relatedness of certain subgroups of *P. citricola*, *P. citrophthora*, and *P. capsici* was unexpected; however, isozyme patterns (Oudemans and Coffey 1991b) and ribosomal DNA analysis (Lee and Taylor 1992) had previously indicated a close relationship between *P. capsici* and *P. citrophthora*.

Oudemans et al. (1994) state that these data suggest that papillation of sporangia, sporangium caducity, heterothallism and homothallism, and paragyny and amphigyny of the antheridia might not be as reliable taxonomic parameters as had been considered in the past. *P. citricola* is inconspicuously papillate, nondeciduous, and homothallic and has paragynous antheridia, whereas *P. capsici* has papillate and deciduous sporangia and is heterothallic with amphigynous antheridia and *P. citrophthora* has nonpapillate, nondeciduous sporangia, is most often sterile, and antheridia, when produced on crosses with other species, are amphigynous. Because of these anomalies, Oudemans et al. (1994) questioned whether sex organ and sporangium morphology will be useful for construction of a phylogeny model as suggested by Brasier (1983). Mchau and Coffey (1994b) delineated three subgroups of *P. citrophthora* isolates by isozyme technology (see Chapter 17).

The different expressions of isozymes have been used as genetic markers in crosses of *P. infestans* isolates. The oospore progeny of these crosses segregated for isozyme types in the F_1 generation (Shattock et al. 1986a). Isozyme patterns have also been used to study the geographic distribution of different populations of *P. infestans* (Spielman 1991; Fry and Goodwin 1995; S. B. Goodwin et al. 1994a, b, 1995) (Table 4.5). These results are discussed in the genetics and cytology section of Chapter 3.

DNA RFLPs

Analysis of DNA RFLPs has become a powerful tool to evaluate the relatedness of certain biotypes and species in the taxonomy of *Phytophthora* (Förster et al. 1987–1989, 1990b; Förster and Coffey 1991). Klimczak and Prell (1984) characterized mtDNA in *P. infestans*. A restriction fragment map is reproduced from Klimczak and Prell (1984) by Carter et al. (1991). In principle, the DNA from either mtDNA or chromosomes in nuclei (chrDNA) is extracted by suitable biochemical methods (Förster et al. 1987, 1989, 1990b; Panabieres et al. 1989). The DNA is then treated with certain endonuclease enzymes that cut the DNA molecule at specific sites. The resultant fragments of DNA are then categorized for size by electrophoresis on gels that can be photographed and evaluated by methods that characterize the DNA. Two gels with DNA RFLP patterns are shown in Figure 4.29 (Förster et al. 1988). A phylogenic network compiled by use of a computerized program for evaluating differences in DNA RFLPs between species is shown in Figure 4.30 (Förster et al. 1988). Figures 4.31 and 4.32 show phenograms that quantitatively

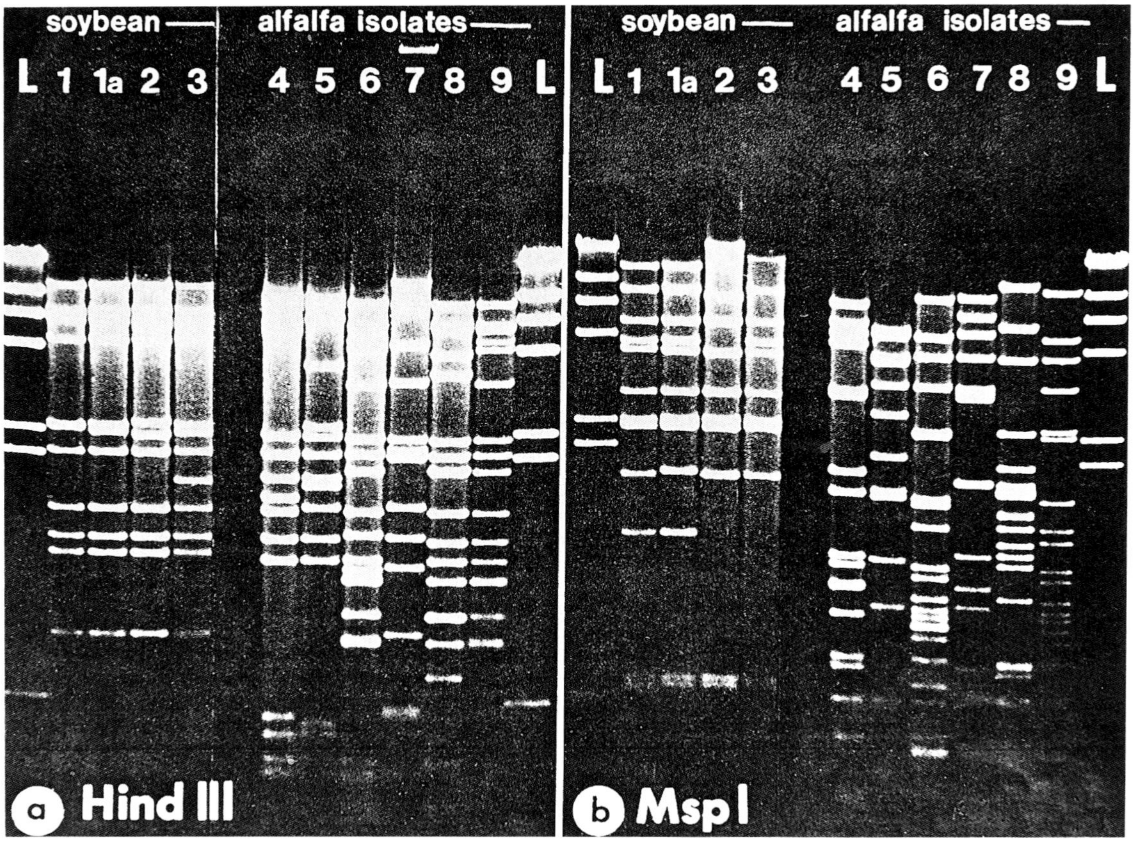

Figure 4.29. Restriction fragment length polymorphism (RFLP) patterns of mitochondrial DNA (mtDNA) obtained by gel electrophoresis of *Phytophthora megasperma* f. sp. *glycinea* (*P. sojae*) from soybean and *P. megasperma* f. sp. *medicaginis* (*P. medicaginis*) from alfalfa. mtDNAs were digested with **a,** *Hin*dIII and **b,** *Msp*I. *Hin*dIII-digested λ DNA (L) was used as a size marker. Note that the RFLP patterns of the soybean and alfalfa isolates differ markedly, thus showing that there is little genetic relationship between them. The soybean isolates have since been reclassified as *P. sojae* and the alfalfa isolates as *P. medicaginis* by E. M. Hansen and Maxwell (1991). (Reproduced by permission of the National Research Council Canada, courtesy Förster et al. 1989)

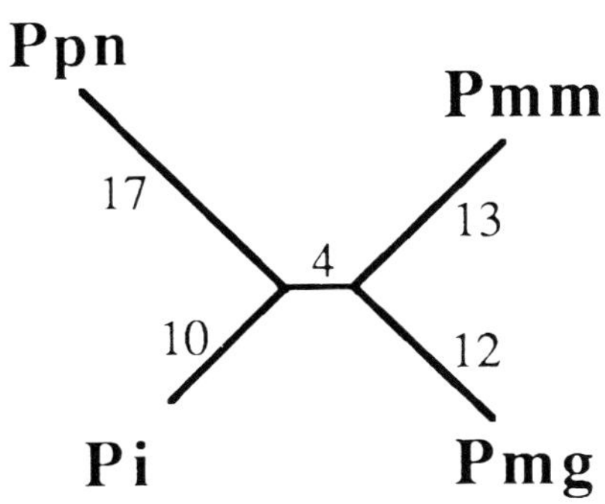

Figure 4.30. Phylogenetic network for DNA restriction fragment length polymorphism (RFLP) patterns of *Phytophthora megasperma* f. sp. *glycinea* (Pmg), *P. megasperma* f. sp. *medicaginis* (Pmm), *P. parasitica* var. *nicotianae* (Ppn), and *P. infestans* (Pi) based on restriction cleavage site mutations, insertions or deletions, and inversions derived from a composite mtDNA restriction site map. This phylogenetic tree was derived by PAUP computer analysis with Wagner parsimony. The numbers on the branches represent the number of character changes. (Reproduced by permission from Mycologia 80:466-478, Förster et al. 1988, copyright The New York Botanical Garden, Bronx, New York)

differentiate the DNA RFLPs of different species (Förster and Coffey 1991).

The series of papers by Förster and colleagues, which was initiated in the laboratory of D. P. Maxwell at the University of Wisconsin, Madison, and continued in the laboratory of M. D. Coffey at the University of California, Riverside, has laid the groundwork and presented interesting and convincing new evidence concerning problems in the speciation of *Phytophthora.* The following were just a few of the sources of concern and confusion: whether the morphologically similar species *P. drechsleri* and *P. cryptogea* should be merged (Bumbieris 1974; Halsall 1976a; Ho and Jong 1986); whether *P. nicotianae* var. *parasitica* and *P. nicotianae* var. *nicotianae* were really different (Tsao and Sisemore 1978; Ho and Jong 1989); and whether all the entities included within *P. megasperma* were one species or several (reviewed by Erwin 1983; E. M. Hansen et al. 1986; E. M. Hansen 1987, 1991; E. M. Hansen and Maxwell 1991), as discussed above. Many of these problems and others have been addressed by Förster and her colleagues, and from these studies, answers to many of these questions have emerged. See Table 4.6 for references on the use of DNA RFLP patterns to study relationships of *Phytophthora* species.

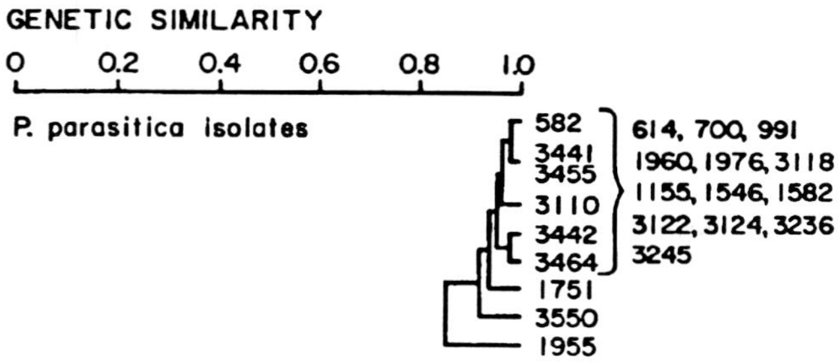

Figure 4.31. Phenogram for *Phytophthora parasitica* isolates based on analysis of restriction fragment length polymorphisms (RFLP) of mtDNA after digestion of DNA with *Hind*III, *Hac*III, *Msp*I, and *Eco*RV restriction endonucleases. Isolates P582, P3110, P1751, P1955, and P1960 were from tobacco and would be named *P. parasitica* var. *nicotianae sensu* Tucker (1931). Other isolates were from various hosts. Isolates P582, P3441, P3110, P3550, and P1955 had been identified *sensu* Waterhouse (1963) as *P. nicotianae* var. *nicotianae* and P3442 and P3464 as *P. nicotianae* var. *parasitica.* The variety status of some isolates was unknown (see Table 11.2 in Förster and Coffey 1991). Graph shows that RFLPs of *P. parasitica* isolates (*P. nicotianae sensu* Waterhouse 1963) do not differ significantly and that retaining the two varieties is unnecessary. (Reproduced by permission, courtesy Förster et al. 1990b and Cambridge University Press, courtesy Förster and Coffey 1991)

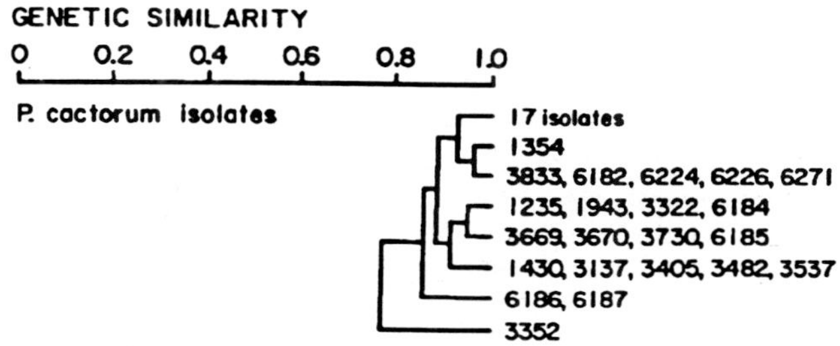

Figure 4.32. Phenogram for *Phytophthora cactorum* isolates based on analysis of RFLPs of mtDNA after digestion with restriction endonucleases. The graph shows the similarity of these isolates. The 17 identical isolates were P472, P714, P1103, P1598, P1721, P1722, P1723, P1724, P3072, P3074, P3075, P3083, P3084, P3128, P3406, P3468, and P3634 obtained from a wide variety of hosts (listed in Förster and Coffey 1991). (Reproduced by permission, Cambridge University Press, courtesy Förster and Coffey 1991)

Phytophthora species can now be evaluated at the DNA level, which is the basis of life of the organism. DNA patterns are effective in typing species because DNA molecules genetically determine the nature of the organism. We quote by permission of the National Research Council of Canada, courtesy of Förster et al. (1989):

> While morphological, physiological, and biochemical differences may be based on minor genetic variation, investigations of DNA sequences can provide information about the genetic background of organisms and thereby allow conclusions about relatedness. Moreover, since minor nucleotide variation may not be expressed at the protein level and since parts of the DNA consist of noncoding regions, DNA sequence studies are more informative than morphological or biochemical studies.

The analysis of DNA requires special laboratory skills and experience and a laboratory equipped for such studies. Those who wish to follow such a line of research should not only read the rationale and methodology (Maniatis et al. 1982; Sambrook et al. 1989), but also obtain hands-on experience. The chapter by Carter et al. (1991) contains an informative discussion of both mtDNA and chrDNA analyses by use of polyacrylamide electrophoresis for mtDNA and radiolabeled (^{32}P) clone fragments for chrDNA. The DNA map shown by Carter et al. (1991) for *P. infestans* used the restriction endonuclease enzymes *Cla*I, *Eco*RI, *Bgl*II, and *Bcl*I. The use of DNA probes to identify *P. parasitica* in soil and diseased tomato tissue has been discussed in Chapter 2 (S. B. Goodwin et al. 1989, 1990a).

The mycelium for DNA extraction should be grown in clear V8 juice broth or other suitable liquid media that does not contain extraneous residues (see Chapter 3) and should be in an actively growing stage. The mycelium can be filtered through Miracloth and washed two or three times with distilled water, frozen at –80°C, and lyophilized. The mtDNA is extracted from the mycelium by biochemical techniques, treated with one or several endonuclease restriction enzymes, which are commercially available, and assayed by electrophoresis on an acrylamide gel. The DNA fragments are identified by the rates at which they move on the gel. chrDNA fragments from the gel are assayed on nylon membranes by detection of hybridization with known DNA fragments radiolabeled with ^{32}P (see Maniatis et al. [1982] and Carter et al. [1991] for methods used).

The DNA RFLP pattern analysis used to characterize several species by Panabieres et al. (1989) was claimed to be rapid. A purified sample of DNA, which they concluded was mostly nuclear DNA, was obtained in 2 h. Several *Phytophthora* species were grown on a glucose-asparagine synthetic medium (R. Hall et al. 1969), harvested by vacuum filtration on filter paper, rinsed with distilled water, and stored frozen. *P. megakarya, P. cactorum,* and *P. infestans* were grown on cellophane film spread over green pea agar plates. The DNA purification protocol is given by Panabieres et al. (1989).

The estimated size of the circular mtDNA molecules of *P. sojae* is about 45.3 kilobases (kb), of *P. medicaginis,* 41 kb, and of *P. parasitica* var. *nicotianae,* 39.5 kb. Physical maps constructed for restriction endonuclease enzyme sites indicated that four genes (*l-rRNA, s-rRNA, oxi-2,* and *cob*) had the same order in the three DNAs (Förster et al. 1987). This report laid the groundwork for subsequent research comparing and contrasting the mtDNA RFLP patterns of different *Phytophthora* species.

Restriction enzymes that recognize specific base sequences cut the circular mtDNA molecule at different regions, resulting in different fragment sizes. The sample is analyzed by horizontal agarose gel electrophoresis. The DNA fragments migrate at different rates on the gel, depending on their size. The analysis of the DNA fragments, called RFLPs, can indicate distinct DNA characteristics that vary among species. Photographs of gels with mtDNA RFLP patterns comparing the soybean isolate *P. sojae* (*P. megasperma* f. sp. *glycinea*) and the alfalfa isolate *P. medicaginis* (*P. megasperma* f. sp. *medicaginis*) are shown in Figure 4.29. The RFLP patterns after use of several different restriction enzymes can be readily seen (Förster et al. 1989).

Although differences between *P. sojae* and *P. medicaginis* can be readily seen in Figure 4.29, DNA RFLP patterns can also be scored numerically and the data expressed as degrees of genetic similarity of relatedness by several mathematical techniques referenced by Förster et al. (1989).

The genetic similarities or differences between isolates can be shown in a phylogenetic network or phenogram in which the similarity is rated on a numerical scale (Figure 4.30). This phenogram differentiates *P. parasitica* var. *nicotianae, P. infestans, P. megasperma* f. sp. *medicaginis* (*P. medicaginis*), and *P. megasperma* f. sp. *glycinea* (*P. sojae*). Figure 4.31 shows the genetic similarity of several *P. parasitica* (*P. nicotianae*) isolates, which previously had been divided into two varieties (Stamps et al. 1990).

Previous to the report by Mills et al. (1991), Pananbieris et al. (1989) used isozyme and DNA RFLP analysis to characterize several species, including *P. cryptogea* (two isolates) and *P. drechsleri* (two isolates). They likewise found that the isolates representing the two species were different. The mtDNA RFLP and isozyme pattern data of Mills et al. (1991) led to the conclusion that there were seven distinct molecular species within 123 isolates of *P. drechsleri* and *P. cryptogea*. Therefore, merging these two species, as advocated by Bumbieris (1974), Halsall (1976a), Shepherd (1978), Matsumoto and Sato (1979), and Ho and Jong (1986), would not be warranted.

RFLP patterns distinguished isolates of well-established species such as *P. cactorum, P. megakarya, P. parasitica* (both tobacco and *Citrus* sp. isolates), *P. capsici, P. citrophthora, P. infestans, P. cambivora,* and *P. cinnamomi*. Notably, *P. parasitica* var. *nicotianae* (tobacco isolates) *sensu* Tucker (1931) could not be distinguished from other *P. parasitica* isolates (Panabieres et al. 1989). This result is in agreement with the data of Förster et al. (1990b) and Förster and Coffey (1991) (Figure 4.31). These data also further confirm the conclusion by Ho and Jong (1989) that *P. nicotianae* var. *nicotianae* could not be differentiated from *P. nicotianae* var. *parasitica* on morphological grounds.

P. palmivora, P. parasitica, P. citricola, and *P. citrophthora* from citrus showed distinct DNA RFLP patterns. *P. citrophthora* from *Arbutus* sp., however, had an index of similarity of only 44/100 compared with an isolate from citrus (Ricci et al. 1990). Likewise, the cacao pathogens *P. palmivora sensu stricto, P. megakarya, P. citrophthora,* and *P. capsici* were differentiated by DNA RFLP profiles. The restriction enzyme *Bam*HI was most useful (Blaha 1990). RFLPs for mtDNA among isolates of *P. capsici* from pepper (*Capsicum annuum*) separated into four homogeneous groups. Isolates from New Mexico (United States) were closely related to those from Korea (B. K. Hwang et al. 1991).

mtDNA RFLP patterns of *P. infestans* of isolates from several countries that showed distinctive patterns were useful for identifying the sources of various isolates and their degree of relatedness. One DNA insertion of about 2 kbp occurred in the mtDNA of isolates from Brazil, Egypt, United Kingdom, and

Table 4.6. References on the delineation of species of *Phytophthora* by DNA restriction fragment length polymorphisms (RFLP) or other DNA technology

Species	Result	Reference
P. cactorum, P. cambivora, P. capsici, P. cinnamomi, P. citrophthora, P. infestans, P. megakarya	Differentiated each species.	Panabieres et al. (1989)
P. cajani, P. cryptogea, P. drechsleri, P. melonis, P. sinensis	Differentiated seven molecular species within 123 isolates.	Förster and Coffey (1991); Mills et al. (1991)
P. capsici	Populations from the United States, Korea, and various countries in Europe were closely related.	Hwang et al. (1991)
P. capsici, P. citrophthora, P. megakarya, P. palmivora	Differentiated each species.	Blaha (1990)
P. capsici, P. citricola, P. citrophthora, P. megakarya, P. palmivora, P. parasitica	Diversity of mtDNA and chrDNA patterns was shown for each species.	Förster and Coffey (1990b, 1991)
P. citrophthora	*P. citrophthora* specifically detected by use of DNA probe; however, isolates from cacao did not hybridize, indicating a lack of genetic relationship.	P. H. Goodwin et al. (1990b)
P. citrophthora	Delineated three subgroups, CTR1, 2, and 3, among 77 isolates.	Mchau and Coffey (1994b)
P. citrophthora, P. palmivora, P. megakarya, P. capsici (cacao isolates); *P. cinnamomi* (polymerase chain reaction amplification and direct sequencing)	*P. capsici* and *P. citrophthora* were closely related; and *P. megasperma* had a common lineage for *P. palmivora. P. cinnamomi* was distinctly different.	Lee and Taylor (1992)
P. fragariae, P. erythroseptica, P. megasperma	Root rot isolates from red raspberry (now *P. fragariae* var. *rubi*) were closely related to *P. fragariae* from strawberry.	Förster and Coffey (1992)
P. infestans	A1 and A2 isolates and other genetic types in Mexico were differentiated.	S. B. Goodwin (1990, 1992)
P. megasperma (broad host range group)	Physical and genetic map of the mtDNA (43.5-kb circular genome) indicated a short (0.5- to 0.9-kb) inverted repeat.	Shumard-Hudspeth and Hudspeth (1990)
P. megasperma complex	*P. medicaginis* and *P. sojae* were differentiated from the *P. megasperma* complex.	Förster et al. (1988, 1989); Förster and Coffey (1991)
	Distinct subgroups (nine), included *P. sojae, P. medicaginis, P. trifolii,* an asparagus group, and a Douglas fir group. The broad host range group separated into four groups. mtDNA data were supported by chrDNA data. A group A was a distinct genetic entity.	Förster and Coffey (1994)
P. megasperma f. sp. *glycinea* (*P. sojae*)	Relationship of U.S. and Australian isolates is close enough to suggest that both sets are derived from a common ancestor.	Whisson et al. (1992)
P. nicotianae var. *nicotianae, P. nicotianae* var. *parasitica*	Showed similarity of the two varieties.	Panabieres et al. (1989); Förster and Coffey (1990b, 1991)
P. parasitica	Cloned DNA probe differentiates *P. parasitica* from *Pythium* and 10 other *Phytophthora* spp. The tobacco pathogen *P. parasitica* var. *nicotianae* was not different from *P. parasitica* from tomato.	P. H. Goodwin et al. (1989)
	DNA species-specific probe detects isolates from soil and plant tissue.	P. H. Goodwin et al. (1990a)
P. sojae (*P. megasperma* f. sp. *glycinea*)	Evaluation and differentiation from other zoosporic orders.	Förster et al. (1990a); Förster and Coffey (1990b, 1991)
	Showed genetic diversity among a world collection of races and that outcrosses can be sources of new races.	Förster et al. (1994)
P. vignae	*P. vignae* population from Australia is related to that from Japan.	Liew et al. (1991)

the United States, thus suggesting that this biotype had a global distribution (S. B. Goodwin 1991; S. B. Goodwin et al 1994; Fry and Goodwin 1995).

DNA RFLP data show that the Australian isolates of *P. vignae,* a pathogen of cowpea (*Vigna unguiculata* L. (Walp.)) (Purss 1972), were genetically similar to Japanese isolates from adzuki bean (*Vigna angularis* (Wild.) Ohwi. and Ohashi) (Kitazawa et al 1978, 1979; Liew et al. 1991). Although the Japanese and Australian isolates were found to be similar with RFLP and high copy probes, dendrograms indicated that they were in different clusters. This finding indicates that the groups probably were derived from a single but unknown source (Liew et al. 1991).

DNA RFLP analyses will soon become available for routine studies, even though characterization of mtDNA and nuclear DNA is time consuming and requires specialized expertise. DNA technology is a definitive tool for the study of the validity of *Phytophthora* species.

DNA RFLP data give credibility to small differences that may or may not be important. The old controversies about "splitting" or "lumping" in the categorization of isolates into species may be resolved. For instance, isolates within *P. parasitica* (= *P. nicotianae*) (Figure 4.31) and *P. cactorum* (Figure 4.32) are genetically similar. Already, RFLP data show that differentiating *P. nicotianae* var. *nicotianae* from *P. nicotianae* var. *parasitica* by small differences in morphology is untenable (Panabieres et al. 1989; Förster et al. 1990b; Förster and Coffey 1991) (Figure 4.31). However, the proposals to merge *P. drechsleri* and *P. cryptogea* on the basis of similar morphology have been shown to be unreliable, since at least seven subpopulations were delineated from 123 isolates by RFLP and isozyme data (Mills et al. 1991) (Figure 4.33). In this case, DNA RFLP data indicate that more research is necessary to clarify the taxonomy of this group of *Phytophthora* species. Phenograms constructed from DNA RFLP analyses for *P. capsici, P. citricola, P. citrophthora, P. megakarya,* and *P. palmivora* (Förster et al. 1990b) are presented in chapters on some of these species.

Using the variable internal transcribed spacer I (ITSI) system, Förster et al. (1995) compared a wide range of *Phytophthora* species. In addition, they reviewed information derived from isozyme and DNA RFLP studies from previous reports.

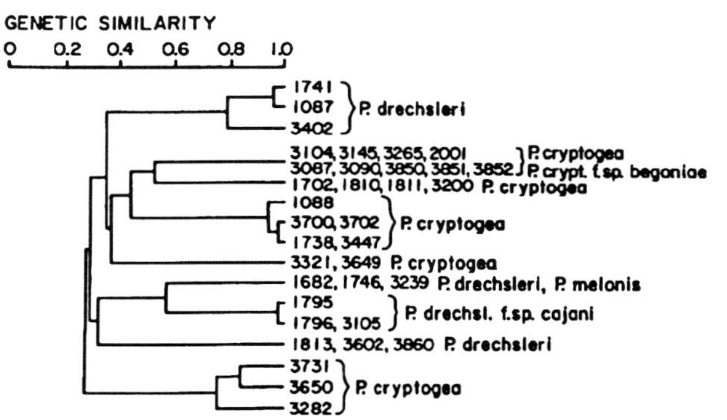

Figure 4.33. Phenogram for *Phytophthora cryptogea, P. drechsleri,* and *P. melonis* isolates based on analysis of mtDNA RFLP after digestion of DNA with *Msp*I, *Hae*III, *Eco*RV, and *Hin*dIII restriction endonucleases. Similar data were obtained by isozyme patterns (from Mills et al. 1991). Graphs show the lack of genetic similarity among this group of isolates considered to be morphologically similar (Bumberis 1974, Halsall 1976; Ho 1980; Ho and Jong 1986). (Reproduced by permission of the British Mycological Society, courtesy Mills et al. 1991 and Cambridge University Press, courtesy Förster and Coffey 1991)

They concluded that the genetic variability within a world collection of *P. cactorum, P. cambivora, P. cinnamomi, P. infestans, P. palmivora,* and *P. parasitica* was limited. However, *P. botryosa, P. heveae, P. katsurae,* and *P. meadii* were moderate in genetic variability. Other species with more extensive genetic variability included *P. capsici, P. citricola, P. citrophthora,* and *P. megakarya.* The most genetically divergent species in the world were the *P. cryptogea-P. drechsleri* complex and *P. megasperma.* Their results show that the evolution of the ITSI region of rDNA does not correlate well with many of the morphological criteria used currently to establish species. Thus, they warn that the new ITSI system "will not immediately benefit the practical plant pathologist."

The contributions of the molecular biological technology in which isozyme and DNA RFLP patterns and DNA probes are used firmly support the conclusion that creation of new species and varieties of *Phytophthora* should, and we might be so bold as to say must, include evaluation by these technologies before new species are described.

Comparison of the Morphology of *Pythium* and *Phytophthora*

Phytophthora and *Pythium* are classified in the family Pythiaceae of the order Peronosporales within Oomycetes. See Chapter 1 for a discussion of the taxonomic position of these

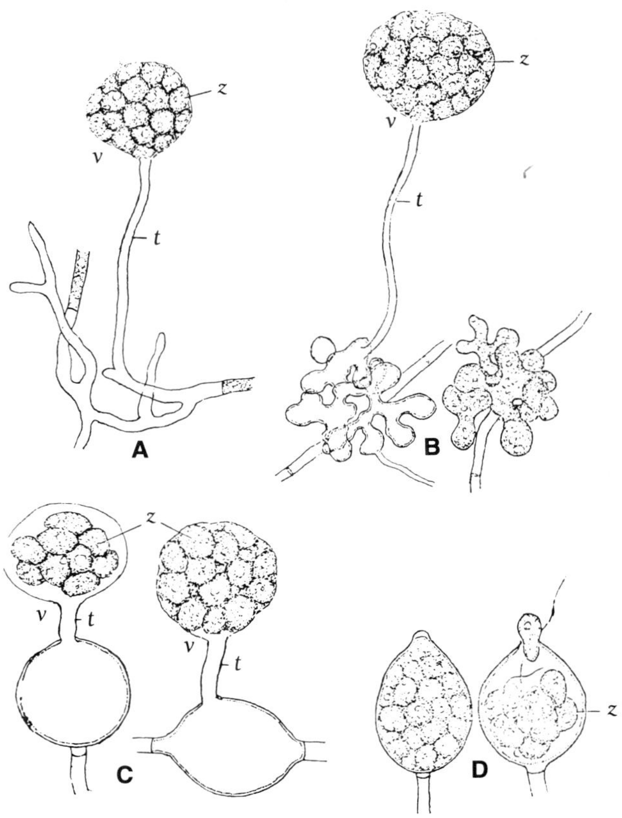

Figure 4.34. Comparison of zoospore formation by *Pythium* and *Phytophthora* species. Sporangia of *Pythium* species: **A,** filamentous; **B,** lobate; and **C,** globose. Zoospores (z) become delimited in the vesicle (v) after the protoplasm has flowed through a tube (t) from the sporangium into the vesicle. **D,** In *Phytophthora,* zoospores are delimited within the sporangia. (Reproduced by permission, courtesy Frezzi 1950)

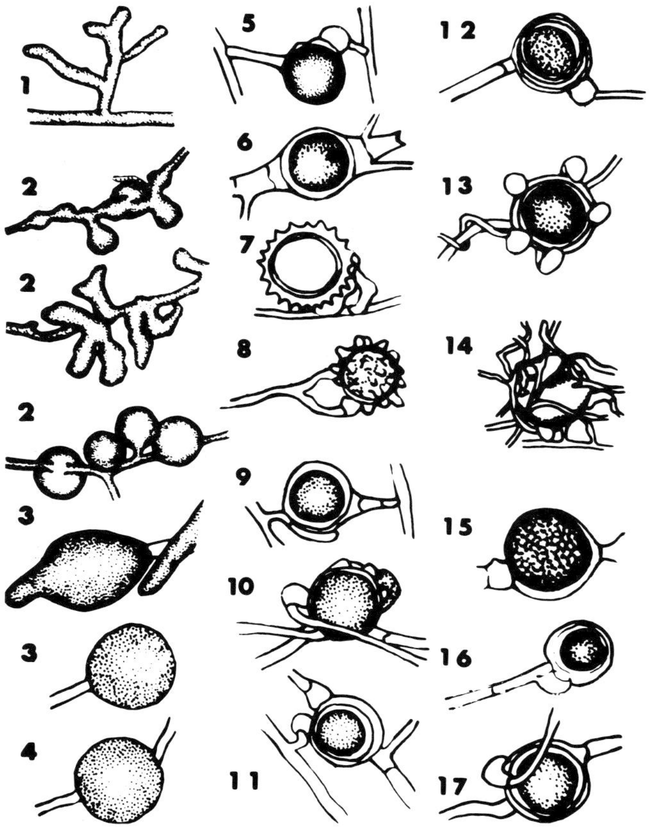

Figure 4.35. Taxonomically important morphological characteristics of the genus *Pythium*. Sporangia: 1, filamentous; 2, lobate; 3 and 4, spherical. Oogonia: 5, terminal; 6, intercalary; 7, pointed spines; 8, rounded spines; and 9, smooth. Oospores: 10, plerotic; and 11, aplerotic. Antheridia: 12, one to two; 13, one to five (usually two to four); 14, usually more than six; 15, hypogynous; 16, monoclinous; and 17, diclinous. (Reproduced by permission of the American Phytopathological Society, St. Paul, Minnesota, courtesy F. F. Hendrix and Papa 1974)

species in the kingdom Chromista. The two genera are similar in that both produce nonseptate, coenocytic mycelium and both produce oospores. Both are diploid in their vegetative life history, and meiosis occurs in the oogonia and antheridia prior to oospore formation (Sansome 1963, 1965). Taxonomic keys for the genus *Pythium* have been published by Middleton (1943a), Frezzi (1956), Hendrix and Papa (1974), Plaats-Niterink (1981), Kröber (1985), and Dick (1990b). The physiology of growth and reproduction was reviewed by Hendrix (1974). Stanghellini's (1974) review of spore germination and growth in soil shows that *Pythium* has many similarities to *Phytophthora.* Hendrix and Campbell (1973) presented a general review of plant pathogenic information on *Pythium* as a plant pathogen; however, *Pythium* species in general tend to be more saprophytic and weaker parasites than *Phytophthora* species.

Brasier and Hansen (1992) cite references that show that *Pythium* and *Phytophthora* can be differentiated by use of monoclonal antibodies in a serological procedure (Hardham et al. 1986) and that *Phytophthora* species lack an inverted repeat sequence of mtDNA that occurs in *Pythium* (Hudspeth et al. 1983; McNabb and Klassen 1988).

Differences between *Phytophthora* and *Pythium* include the fact that all species of *Phytophthora* require thiamine for growth in synthetic media but all *Pythium* species, with the exception of *P. vexans,* grow in the absence of thiamine or its moieties, pyrimadine and thiazole (Ridings et al. 1969).

Pythium and *Phytophthora* can be readily differentiated by the morphology of sporangia, chlamydospores, and oospores (Figures 4.34 and 4.35 and Table 4.7). The formation of zoospores within the sporangium of *Phytophthora* (Figures 3.1 and 4.34) is the most distinctive diagnostic difference compared with *Pythium,* in which zoospores form after the protoplasm has flowed from the sporangium into a vesicle (Figure 4.34). The mycelium of *Phytophthora* is usually thicker (6–14 μm in width) than that of *Pythium* (5–10 μm) and usually grows in a more tortuous pattern than *Pythium;* however, there are many exceptions to these characteristics. The sporangia of *Phytophthora* are always terminal and are mostly variations of

Table 4.7. Differences between *Phytophthora* and *Pythium*[a]

Characteristic	*Phytophthora*	*Pythium*
Sporangium shape	Usually ovoid or obpyriform; seldom globose.	Filamentous, inflated hyphae, spherical, occasionally ovoid or pyriform.
Sporangium position	Always terminal.	Terminal or intercalary.
Sporangium apical thickening (papilla)	Yes	No
Sporangium caducity	Caducous (deciduous) or persistent.	Always persistent.
Zoospore differentiation	Protoplasm differentiates into zoospores within the sporangium.	Protoplasm (undifferentiated) flows from the sporangium into a vesicle in which zoospores differentiate.
Hyphae	Coarse and stiffly branched (6–14 μm wide).	Fine and flexuous (5–10 μm wide).
Antheridia	Amphigynous or paragynous; if paragynous attached only to the lower hemisphere.	Paragynous or hypogynous. Never amphigynous. Can be one or many and attached to any part of the oogonium.
Oogonia	Smooth or in a few species, ornamented. Pigmented walls.	Smooth or spiny. Colorless walls.
Oospore	Plerotic or aplerotic.	Plerotic or aplerotic.
Chlamydospores	Yes	Yes
Hyphal swellings	Yes	Rare
Thiamine requirement	Yes	No (except for *P. vexans,* which requires thiamine for growth).
Habitat	Soil or aquatic; aerial parts of plants; rarely saprophytic.	Soil or aquatic; saprophytic or facultatively parasitic.
Addition of hymexazol (3-hydroxy-5-methylisoxazole) to isolation medium	Some species are not inhibited.	Many species are inhibited.
Host specificity	Specific or with a wide host range, depending on species.	Numerous hosts, rarely host specific.

[a] Adapted from Ribeiro (1978) by permission of AR Gantner Verlag, Vaduz.

ovoid or obpyriform shapes, whereas sporangia of *Pythium,* depending on the species, can be globose, lobulate (many lobes), or even filamentous and frequently intercalary (not terminal on hyphae) (Figures 4.34 and 4.35). The antheridia of *Phytophthora* can be amphigynous or paragynous; however, if paragynous, the antheridium makes contact only on the lower hemisphere of the oogonium (Figure 4.26C), but in *Pythium,* the antheridia are always paragynous but may become attached to the oogonium at any location. Also, depending on the species, many antheridia may be attached to a single oogonium (see Figure 4.35).

Summary

The history of taxonomy within the genus *Phytophthora* began in 1876 with the description of *P. infestans* by de Bary. The discovery of new species since then has resulted in several approaches to its classification. Because the number and nature of morphological differences among species is not large, identification of *Phytophthora* species in some cases is difficult. The grouping of *Phytophthora* species based on different morphological features that was initiated by Waterhouse (1963) is described in the current tabular key by Stamps et al. (1990) and continued with modifications by us (Tables 4.2 and 4.3). In addition to nonseptate mycelium, *Phytophthora* species are characterized by hyphal swellings, sporangia, biflagellate zoospores, chlamydospores, and the sex organs, antheridia, oogonia, and oospores. The oospores form within the oogonia.

Species are classified on the basis of the papillate, inconspicuously papillate, and nonpapillate nature of sporangia; whether the sporangia or sporangiophores are persistent or caducous (deciduous), and, if caducous, whether they have short, medium, or long stalks; whether the sexual stage is homothallic (isolates self-fertile) or heterothallic (isolates form oospores only when paired with a complementary A1 or A2 mating type); and whether the type of antheridium is paragynous or amphigynous on the oogonium. Secondary factors to separate species are sporangium shape and size, nature of sporangiophores (unbranched, formed in a simple, compound sympodium, or umbel), whether sporangia proliferate (a new sporangium grows out from within the old sporangium), whether antheridia are unicellular or bicellular, the cardinal temperatures for growth, and production or nonproduction of chlamydospores. These features are described and illustrated by drawings and photomicrographs in the text. Tabular keys are presented in Tables 4.2 and 4.3.

Several relatively new molecular techniques (protein patterns, isozymes, and DNA RFLPs) are useful to determine genetic differences or similarities among species and in combination with morphological data may resolve questions concerning separation and/or merger of species. For example, DNA RFLP patterns show that *P. nicotianae* var. *nicotianae* and *P. nicotianae* var. *parasitica* cannot be differentiated. RFLP and isozyme data have determined that the proposal to merge *P. cryptogea* with *P. drechsleri* is not tenable because, within 123 isolates collected worldwide, at least seven electrophoretic types were delineated. More study will be required to resolve this nomenclature problem. *P. drechsleri* var. *cajani* or *P. drechsleri* f. sp. *cajani* should probably revert to *P. cajani.* The *P. megasperma* var. *sojae* and *P. megasperma* var. *megasperma* complex has been restudied by DNA RFLP and isozyme technology, and from this complex the epithet *P. sojae* was reestablished and the new species *P. medicaginis* and *P. trifolii* were described. Several other molecular species within the *P. megasperma* complex remain to be delineated.

APPENDIX
TO CHAPTER 4

Methods of Identifying *Phytophthora* Species

4.1. Morphological Criteria Used to Group *Phytophthora* Species

This information is taken from Waterhouse (1963), Newhook et al. (1978), and Stamps et al. (1990). For convenience of identification, the taxa are segregated into six groups; however, these groups are not necessarily intended to imply a natural classification.

Group I

Sporangia are conspicuously papillate; the hyaline apical thickening is hemispherical or at least 4 μm deep; the exit pore is narrow (5–7 μm). Sporangia are usually abundant on solid substrata and usually caducous (deciduous) and do not proliferate internally. Oospores are always formed in the host and in culture. Antheridia are all or nearly all paragynous.

Species: *P. cactorum, P. clandestina, P. idaei, P. iranica, P. italica, P. pseudotsugae, P. tentaculata.*

Group II

Sporangia are conspicuously papillate; apical thickening of sporangium is similar to that of species in group I. Oospores are not often formed in single-strain culture but form when A1 and A2 mating types are paired. Antheridia are amphigynous.

Species: *P. arecae* (considered to be conspecific with *P. palmivora*), *P. boehmeriae, P. botryosa, P. capsici, P. citrophthora, P. heveae, P. katsurae, P. meadii, P. megakarya, P. mexicana, P. nicotianae* (= *P. parasitica*), *P. palmivora* var. *palmivora, P. palmivora* MF4 (= *P. capsici*), *P. palmivora* var. *heterocystica.*

Group III

Sporangia are semipapillate; the apical thickening is shallower than that in groups I and II; sometimes scarcely papillate; hyaline thickening is evident but less hemispherical; exit pore is narrow (5–7 μm). Sporangia can be caducous (deciduous) or not. Oospores are always present in the host or in single-isolate culture. Antheridia are predominantly or all paragynous.

Species: *P. citricola, P. cyperi* (not culturable), *P. cyperibulbosi* (not culturable), *P. inflata, P. lepironiae* (not culturable), *P. macrospora* (considered to be *Sclerophthora* by Stamps et al. 1990), *P. porri, P. primulae, P. syringae.*

Group IV

Sporangia are semipapillate; the apical thickening is shallower than that in groups I and II. Sporangia are usually deciduous. New sporangia do not proliferate internally. Sporangiophores are sympodial. Oospores seldom form in single-isolate culture but form when A1 and A2 mating types are paired. Antheridia are predominantly amphigynous.

Species: *P. colocasiae, P. eriugena, P. hibernalis, P. ilicis, P. infestans, P. macrochlamydospora, P. mirabilis, P. phaseoli.*

Group V

Sporangia are nonpapillate; apex of the sporangium does not protrude beyond the general contour of the sporangium (i.e., nonpapillate); very thin apical thickening, scarcely detectable without staining; the exit pore is broad, 12 μm or more. Sporangia are nondeciduous, rarely or never produced on agar media, but form in water or aqueous salts solutions and proliferate internally. Oospores are always produced in the host and/or in agar culture. All or a majority of the antheridia are paragynous.

Species: *P. fragariae* var. *fragariae, P. fragariae* var. *oryzobladis, P. fragariae* var. *rubi, P. humicola, P. insolita, P. lateralis, P. leersiae* (sporangia have not been seen [Ho and Chang 1992]; not listed in Tables 4.2 and 4.3), *P. medicaginis* (formerly *P. megasperma* f. sp. *medicaginis*), *P. megasperma, P. quininea, P. sojae* (formerly *P. megasperma* f. sp. *glycinea*), *P. trifolii* (formerly *P. megasperma* f. sp. *trifolii*), *P. verrucosa.*

Note: *P. megasperma* var. *sojae sensu* Waterhouse (1963) and Stamps et al. (1990) should not necessarily be equated with *P. sojae* or *P. megasperma* f. sp. *glycinea* because they separate it from *P. megasperma* var. *megasperma* on the basis of the smaller size of the oogonia and not on pathogenicity to soybean (Waterhouse 1963). E. M. Hansen and Maxwell (1991) revised the description of *P. megasperma* and did not accept the variety *sojae* (oogonia <45 μm in diameter) or the variety *megasperma* (oogonia >45 μm in diameter).

Group VI

Sporangia are nonpapillate as in group V species. Sporangia proliferate externally and internally. Oospores are not always produced in single-strain culture. In some species, oospores form when A1 and A2 mating types are paired. Antheridia are all or mostly amphigynous.

Species: *P. cajani,* P. cambivora, P. cinnamomi* var. *cinnamomi, P. cinnamomi* var. *parvispora, P. cryptogea, P. drechsleri, P. drechsleri* var. *cajani** (not listed in Tables 4.2 and 4.3), *P. drechsleri* f. sp. *cajani** (= *P. cajani*), *P. erythroseptica* var. *erythroseptica, P. erythroseptica* var. *pisi, P. gonapodyides, P. japonica, P. melonis* (= *P. drechsleri*), *P. richardiae, P. sinensis* (= *P. drechsleri*), *P. undulata, P. vignae* (*P. vignae* f. sp. *vignae* and *P. vignae* f. sp. *adzukicola* [Tsuchiya et al. 1986]).

Note: *P. sinensis* and *P. melonis* are similar to and considered to be identical to *P. drechsleri* (Ho 1981; Ho and Jong 1986, 1991). These species names are listed here mainly for informational purposes and because the isozyme pattern data of Mills et al. (1991) indicates that the isolates associated with *P. melonis* and *P. sinensis* are distinct from *P. drechsleri* and *P. cryptogea*. See Chapter 20 (*P. cryptogea*) and Chapter 23 (*P. drechsleri*) for discussion.

*These species names are apparently synonymous. The designation of *P. drechsleri* var. *cajani* may be incorrect since host-specific pathogenicity to pigeon pea and not morphology was used to justify the epithet (Pal et al. 1970). The designation *P. drechsleri* f. sp. *cajani* implies and is based on host specificity to pigeon pea (Kannaiyan et al. 1980); however, the isozyme data of Mills et al. (1991) indicates that the pigeon pea isolates may be unique and different from *P. drechsleri*. If so, *P. cajani* might be the most correct epithet.

4.2. Criteria Useful for Describing and/or Identifying a *Phytophthora* Species

1. Sporangia: nonpapillate, semipapillate, or papillate.
2. Width of exit pore on sporangia.
3. Persistence or caducity of sporangia and pedicel length.
4. Sporangium shape and size. Presence or absence of a basal plug.
5. Sporangium length-breadth (L:B) ratio.
6. Sporangiophore: unbranched, irregularly branched, in a simple sympodium, in a compound sympodium, or umbellate; do new sporangia form internally (proliferate)?
7. Oogonia: size and shape.
8. Oogonia: wall ornamentation.
9. Oospore: diameter and wall thickness.
10. Oospores: plerotic or aplerotic.
11. Antheridia: amphigynous or paragynous or both.
12. Antheridia: unicellular and/or bicellular.
13. Antheridia: size and shape.
14. Homothallic (produces oospores in single culture) or heterothallic (produces oospores only or mostly when paired with an opposite A1 or A2 mating type).
15. Cardinal temperatures for growth: minimum, optimum, and maximum.
16. Hyphal swellings: single, catenulate, or clustered.
17. Chlamydospores: terminal, lateral, intercalary, or sessile.
18. Cultural pattern in agar media: petalloid, stellate, rosette, irregular, appressed, or fluffy.
19. Hyphae: smooth, sinuous, coralloid, or coiling.
20. Hyphal branching systems (Ho 1978).
21. Pigment production: casein hydrolyzate-tyrosine medium (Shepherd 1976).
22. Pathogenicity: single host or many hosts.
23. Protein profiles (Gill and Zentmyer 1978; Erselius and Shaw 1982; Erselius and de Vallavielle 1984) (Table 4.4).
24. Isozyme patterns (Mills et al. 1991; Oudemans and Coffey 1991a–c, 1992; Spielman 1991) (Table 4.5).
25. DNA restriction fragment length polymorphism (RFLP) patterns (Förster et al. 1987; 1988; Mills et al. 1991) (Table 4.6; Figures 4.29–4.33).

Note: Numerical data should include not only the average but also the range (Gerrettson-Cornell 1989).

4.3. Tally Sheet for Compiling Information on Identification of *Phytophthora* Species

Isolate No.: ______________

Culture patterns (Figures 4.2 and 4.3)
- Smooth ____
- Appressed ____
- Zonate ____
- Fluffy ____
- Cottony ____
- Irregular ____
- Radiate ____
- Rosette ____
- Chrysanthemum ____
- Stellate ____
- Zonate ____

Hyphae (Figure 4.1)
- Smooth ____
- Corraloid ____
- Sinuous ____
- Coiled ____

Sporangia
- Length
  - Minimum ____
  - Maximum ____
  - Average ____
- Width
  - Minimum ____
  - Maximum ____
  - Average ____
- L:B ratio
  - >1.6:1.0 ____
  - <1.6:1.0 ____
- Sporangium base tapered ____
- Basal plug present ____
- Appendage at base of sporangium ____
- Extended or nested proliferation (Figure 4.7E) ____
- Nested (Figure 4.7C and D) ____
- Exit pore
  - Narrow, <7 µm ____
  - Broad, 7–12 µm ____
  - Very wide, >12 µm ____
- Formed in agar (see text) ____
- Formed in liquid only ____
- Formed in agar and liquid ____
- Papilla apex
  - 1 apex (Figure 4.12A–D) ____
  - 2 apices (Figure 4.12E) ____
  - 3+ apices ____
- Apical thickening (Appendix 4.5; Figure 4.9C)
  - Hemispherical (conspicuously papillate; Figure 4.9C) ____
  - Protruding slightly beyond contour of the sporangium (semipapillate) (Figure 4.9B) ____
  - Almost in line with contour of sporangium (nonpapillate; Figure 4.9A) ____
  - Not protruding beyond contour of sporangium (nonpapillate; Figure 4.12L) ____

Sporangia (*continued*)
- Persistent (Figure 4.12L and M) ____
- Caducous (Figures 4.12A, 4.15, and 4.17; Appendix 4.6) ____
- Pedicel length
  - <5 µm (short; Figures 4.15G–I and 4.18C–G) ____
  - 6–15 µm (intermediate; Figure 4.15A–F) ____
  - >16 µm (long; Figures 4.16B–F and 4.17A and B) ____
  - Irregular ____
- Shapes (Figures 4.10–4.14)
  - Ovoid-ovate ____
  - Ovoid-pyriform ____
  - Ovoid-obpyriform ____
  - Ovoid-ellipsoid ____
  - Ovoid-fusiform ____
  - Ovoid-limoniform ____
  - Ovoid-obclavate ____
  - Obpyriform ____
  - Broadly turbinate ____
  - Spherical (globose) ____
  - Ellipsoid-spherical ____
  - Pyriform ____
  - Distorted ____
  - Variable ____

Sporangiophores (Figures 4.5, 4.6, and 4.8A)
- Unbranched ____
- Irregularly branched ____
- Branched sympodium (Figure 4.8A) ____
- Lax sympodium ____
- Closed sympodium ____
- Compound sympodium (Figure 4.5B) ____
- Umbel (Figures 4.5D and 4.16F) ____
- Proliferating through base of empty sporangium (Figure 4.8B and C) ____

Chlamydospores (Figures 4.23–4.25)
- Absent ____
- Few ____
- Many ____
- In agar media ____
- Terminal ____
- Intercalary ____
- Thick or thin wall (Figure 4.24A–E) ____
- Diameter
  - <35 µm ____
  - >35 µm ____

Hyphal swellings (Figure 4.4A and B)
- Absent ____
- Relatively few ____
- Many ____
- Most chainlike ____
- Most corraloid ____
- Most single ____
- With radiating hyphae ____
- Angular ____

Oogonia (Figures 4.26A–C, 4.27, and 4.28; Appendix 4.7)
- Many ____
- Few ____
- Occasional (rare) ____
- Formed in single culture (homothallic) ____
- Formed in dual A1 × A2 culture only (heterothallic) ____
- Present in plant tissue ____
- Smooth walled ____
- Ornamented wall (Figures 4.27A–D and 4.28) ____
- Diameter
  - <29 µm ____
  - 30–45 µm ____
  - >45 µm ____
- Round ____
- Tapered base (Figures 4.27 and 4.28) ____

Antheridia (Figures 4.26A–C, 4.27, and 4.28; Appendix 4.7)
- Present ____
- Absent ____
- Paragynous (Figure 4.26C) ____
- Paragynous and amphigynous in same culture ____
- Amyphigynous (Figure 4.26A and B) ____
- One per oogonium ____
- More than one per oogonium ____
- Some two-celled (Figure 4.27A and D) ____
- Side attachment ____
- Subterminal attachment ____
- Square ____
- Rounded ____
- Short (length < 10 µm) ____
- Elongated (length > 20 µm) ____

Oospores (Figure 4.26A–C)
- Plerotic (Figure 4.26A) ____
- Aplerotic ____
- Wall thickness
  - <1.5 µm ____
  - 1.5–2.5 µm ____
  - >4.0 µm ____
- Diameter
  - <25 µm ____
  - 26–30 µm ____
  - 31–40 µm ____
  - >40 µm ____

Growth temperatures (Appendix 4.8)
- Minimum
  - <5°C ____
  - 5–10°C ____
- Optimum
  - <25°C ____
  - 25–30°C ____
  - >30°C ____
- Maximum
  - <28°C ____
  - 28–35°C ____
  - >35°C ____

Adapted from Table 2, courtesy of Kröber (1985) by permission of the Biologische Bundesanstalt für Land- und Forstwirtschaft, Berlin, Germany.

4.4. Some *Phytophthora* Species Characterized by Unique Sporangium Size, Chlamydospore Production, Oogonia Size and Shape, or Host Range (Waterhouse et al. 1983; Stamps et al. 1990)

1. Sporangium size (range) and shapes

A. Small (<45 μm in length): *boehmeriae; botryosa; cactorum; iranica; katsurae.*
B. Large (>75 μm in length): *cambivora; capsici; cinnamomi; citrophthora; cyperi; fragariae; japonica; mexicana; palmivora; porri; primulae; vesicula.*
C. Length-breadth ratio. Large: *capsici; cyperi; hibernalis; mexicana; palmivora* MF4 (*P. capsici sensu lato* Tsao and Alizadeh 1988).
D. Sporangium base tapered: *capsici; colocasiae; hibernalis; palmivora.*
E. Sporangia with more than one apex: *capsici; citricola; citrophthora; nicotianae* (= *parasitica*); *mexicana; tentaculata.*
F. Distorted shapes: *citricola; citrophthora; primulae.*
G. Compound sympodial sporangiophores: *infestans; phaseoli; mirabilis.*
H. Closed sympodial sporangiophores: *cactorum; palmivora; syringae.*
I. Umbellate sporangiophores: *capsici* (particularly cacao and black pepper isolates; e.g., *P. palmivora* MF4).

2. Chlamydospores

A. Production of numerous chlamydospores and tuberculate mycelium: *cinnamomi.*
B. Many chlamydospores formed sessile (no stalk) and lateral to the mycelium: *lateralis.*
C. Chlamydospores large (~55 μm in diameter): *macrochlamydospora.*

3. Oogonia

A. Small (diameter < 40 μm): *colocasiae; medicaginis* (*megasperma* f. sp. *medicaginis*); *megasperma* var. *sojae* (*sensu* Stamps et al. 1990); *heveae; sojae* (*megasperma* f. sp. *glycinea*); *nicotianae* (= *parasitica*).
B. Large (diameter > 49 μm): *cambivora; leersiae* (not listed; see Table 4.1 footnote c); *megasperma* var. *megasperma* (*sensu* Stamps et al. 1990); *quininea.*
C. Thick-walled: *capsici; cryptogea; erythroseptica* var. *pisi; leersiae; lepironiae; megakarya; megasperma; palmivora; verrucosa.*
D. Oogonia with tapered bases: marked in *heveae* and *megakarya* but occurs in several other species such as *katsurae; ilicis; infestans; fragariae; cambivora; cinnamomi; cryptogea; drechsleri; sinensis* (see Stamps et al. 1990).
E. Oogonial wall ornamented: *cambivora; cyperi-bulbosi; katsurae; verrucosa.*
F. Absence of oogonia: *P. japonica; P. macrochlamydospora; P. undulata; P. citrophthora.*

4. Antheridia

A. Amphigynous (groups II, IV, VI; Table 4.3).
B. Paragynous (groups I, III, V; Table 4.3).

5. Oospores

A. Oospore markedly aplerotic: *cactorum; clandestina; colocasiae; erythroseptica* var. *pisi; fragariae; heveae; humicola; infestans; katsurae; meadii; megasperma; nicotianae; phaseoli; porri; pseudotsugae; quininea; verrucosa; vignae.*

6. Pathogenicity

A. Narrow host range (one or two hosts): *cajani* (= *drechsleri* var. *cajani; drechsleri* f. sp. *cajani*); *cyperi; cyperi-bulbosi; clandestina; eriugena; fragariae* var. *fragariae; fragariae* var. *oryzobladis; fragariae* var. *rubi; ilicis; infestans; inflata; japonica; katsurae; lepironiae; meadii; medicaginis* (*megasperma* f. sp. *medicaginis*); *melonis; mexicana; mirabilis; porri; quininea; richardiae; sinensis; sojae* (*megasperma* f. sp. *glycinea*); *trifolii* (*P. megasperma* f. sp. *trifolii*); *verrucosa; vignae.*
B. Wide host range: *cactorum; cambivora; cinnamomi; citricola; cryptogea; drechsleri; megasperma; nicotianae* (= *parasitica*); *palmivora.*
C. Weak pathogens: *gonapodyides; iranica; undulata.*

4.5. Species of *Phytophthora* with Papillate, Semipapillate, and Nonpapillate Sporangia

Papillate

P. arecae
P. boehmeriae
P. botryosa
P. cactorum
P. capsici
P. citrophthora
P. clandestina
P. heveae
P. idaei
P. iranica
P. italica
P. katsurae
P. meadii
P. megakarya
P. mexicana
P. nicotianae
P. palmivora
P. palmivora var. *heterocystica*
P. pseudotsugae
P. tentaculata
P. vesicula

Semipapillate

P. citricola
P. colocasiae
P. cyperi
P. cyperi-bulbosi
P. eriugena
P. hibernalis
P. ilicis
P. infestans
P. inflata
P. insolita
P. leprioniae
P. macrochlamydospora
P. macrospora
P. miribilis
P. phaseoli
P. porri
P. primulae
P. syringae

Nonpapillate

P. cajani
P. cambivora
P. cinnamomi
  var. *cinnamomi*
  var. *parvispora*
P. cryptogea
P. drechsleri
P. erythroseptica
  var. *erythroseptica*
  var. *pisi*
P. fragariae
  var. *fragariae*
  var. *oryzobladis*
  var. *rubi*
P. gonapodyides
P. humicola
P. insolita
P. japonica
P. lateralis
P. medicaginis
P. megasperma
  var. *megasperma*
  var. *sojae*
P. melonis
P. quininea
P. richardiae
P. sinensis
P. sojae
P. trifolii
P. undulata
P. verrucosa
P. vignae

4.6. Species of *Phytophthora* that Produce Caducous or Noncaducous (Persistent) Sporangia

Caducous

Short pedicel (< 5 µm)	Intermediate pedicel (5–20 µm)	Long pedicel (> 20 µm)
P. arecae	*P. botryosa*	*P. capsici*
P. boehmeriae	*P. colocasiae*	*P. hibernalis*
P. botryosa	*P. heveae*	
P. cactorum	*P. ilicis*	
P. clandestina	*P. meadii*	
P. cyperi	*P. megakarya*	
P. cyperi-bulbosi		
P. infestans		
P. lepironiae		
P. mirabilis		
P. palmivora		
P. palmivora var. *heterocystica*		
P. phaseoli		

Noncaducous (persistent)

P. cajani
P. cambivora
P. cinnamomi
  var. *cinnamomi*
  var. *parvispora*
P. citricola
P. citrophthora
P. cryptogea
P. drechsleri
P. eriugena
P. erythroseptica
  var. *erythroseptica*
  var. *pisi*
P. fragariae
  var. *fragariae*
  var. *oryzobladis*
  var. *rubi*
P. gonapodyides
P. humicola
P. idaei
P. inflata
P. insolita
P. iranica
P. italica
P. japonica
P. katsurae
P. lateralis
P. macrochlamydospora
P. medicaginis
P. megasperma
P. melonis
P. mexicana
P. nicotianae (= *P. parasitica*)
P. porri
P. primulae
P. pseudotsugae
P. quininea
P. richardiae
P. sinensis
P. sojae
P. syringae
P. tentaculata
P. trifolii
P. undulata
P. verrucosa
P. vignae

4.7. *Phytophthora* Species Classified According to Sexual Type and Antheridial Configuration (amended from Ribeiro 1978)

Homothallic species

Paragynous

P. cactorum
P. citricola
P. clandestina[a]
P. cyperi
P. cyperi-bulbosi
P. eriugena[a]
P. fragariae
  var. *fragariae*[a]
  var. *rubi*[a]
P. hibernalis[a]
P. humicola[a]
P. idaei[a]
P. inflata
P. iranica
P. italica
P. lateralis
P. lepironiae
P. medicaginis[a]
P. megasperma[a]
P. porri[a]
P. primulae
P. pseudotsugae
P. quininea
P. richardiae[a]
P. sojae[a]
P. syringae[a]
P. tentaculata[a]
P. trifolii[a]
P. verrucosa[a]

Amphigynous

P. boehmeriae
P. cajani
P. clandestina[a]
P. eriugena[a]
P. erythroseptica
  var. *erythroseptica*
  var. *pisi*
P. fragariae
  var. *fragariae*[a]
  var. *oryzobladis*
  var. *rubi*[a]
P. heveae
P. hibernalis[a]
P. humicola[a]
P. idaei[a]
P. ilicis
P. katsurae
P. meadii
P. medicaginis[a]
P. megasperma[a]
P. phaseoli
P. porri[a]
P. richardiae[a]
P. sojae[a]
P. syringae[a]
P. tentaculata[a]
P. trifolii[a]
P. verrucosa[a]
P. vignae

Heterothallic species (amphigynous)

P. arecae
P. botryosa
P. cambivora
P. capsici
P. cinnamomi
  var. *cinnamomi*
  var. *parvispora*
P. colocasiae
P. cryptogea
P. drechsleri
P. gonapodyides[c]
P. infestans
P. meadii
P. megakarya
P. megasperma[b]
P. melonis
P. mexicana
P. mirabilis
P. nicotianae (= *P. parasitica*)
P. parasitica (= *P. nicotianae*)
P. palmivora
  var. *palmivora*
  var. *heterocystica*
P. sinensis

[a] Species produce varying proportions of both paragynous and amphigynous antheridia. Sexual bodies are unknown in *P. japonica, P. macrochlamydospora, P. undulata,* and *P. citrophthora.*
[b] Large oogonial isolate from alfalfa (Barr 1980).
[c] When paired with opposite mating type of another species.

4.8. Grouping of *Phytophthora* Species by Maximum Temperature for Growth

Below 30°C

P. eriugena
P. fragariae
  var. *fragariae*
  var. *rubi*
P. hibernalis
P. ilicis
P. infestans
P. lateralis
P. megakarya
*P. megasperma**
P. mirabilis
P. phaseoli
P. primulae
P. pseudotsugae
P. quininea
P. syringae
P. vignae

30–35°C

P. boehmeriae
P. botryosa
P. cactorum
P. cambivora
P. cinnamomi
  var. *cinnamomi*
  var. *parvispora*
P. citricola
P. citrophthora
P. clandestina
P. cryptogea
P. erythroseptica
  var. *erythroseptica*
  var. *pisi*
P. gonapodyides
P. heveae
P. humicola
P. inflata
P. japonica
P. katsurae
P. macrochlamydospora
P. medicaginis
P. megasperma
P. mexicana
P. palmivora
  var. *heterocystica*
  var. *palmivora*
P. porri
P. richardiae
P. sojae
P. tentaculata
P. trifolii
P. vignae

35°C or above

P. arecae
P. cajani
P. capsici
P. colocasiae
P. drechsleri
P. insolita
P. iranica
P. japonica
P. meadii
P. melonis
P. nicotianae (= *P. parasitica*)
P. palmivora
  var. *palmivora*
  var. *heterocystica*
P. parasitica (= *P. nicotianae*)
P. sinensis
P. undulata

Note: Unknown, *P. cyperi, P. cyperi-bulbosi, P. lepironiae, P. verrucosa.*

*Some isolates of *P. megasperma* have temperature maxima below or above 30°C.

4.9. A Dichotomous Key to Taiwan Species of *Phytophthora* (Ho 1992; Ho et al. 1995)[a,b]

1. Fungus cannot be cultured on agar media.......................... 2
1. Fungus can be cultured on agar media.............................. 5

2. Oogonia greater than 50 μm in diameter.......................... 3
2. Oogonia less than 50 μm in diameter.............................. 4

3. Antheridia paragynous.................................. *macrospora*[c]
3. Antheridia amphigynous...................................... *leersiae*

4. Sporangia oblong-ovoid (length-breadth ratio 1.6–1.7) .... *cyperi*
4. Sporangia spherical-ovoid (length-breadth ratio 1.1–1.2) .... *lepironiae*

5. Sex organs produced readily in single culture 6
5. Sex organs not or scarcely produced in single culture 13

6. Antheridia absent .... *insolita*
6. Antheridia present.. 7

7. Antheridia predominantly paragynous............................ 8
7. Antheridia amphigynous... 9

8. Sporangia nonpapillate .... *humicola*
8. Sporangia semipapillate...................................... *citricola*
8. Sporangia papillate .... *cactorum*

9. Sporangia nonpapillate 10
9. Sporangia papillate 11

10. Good growth on agar at 35°C .... *drechsleri*
10. No or slight growth on agar at 35°C .... *vignae*

11. Oogonia spherical .... *boehmeriae*
11. Oogonia pyriform with tapered bases 12

12. Oogonial wall smooth.. *heveae*
12. Oogonial wall verrucose .... *katsurae*

13. Sporangia nonpapillate 14
13. Sporangia papillate to semipapillate 15

14. Hyphae coralloid with large, spherical swellings in clusters .... *cinnamomi*
14. Hyphae uniform to uneven, with small swellings in network; sporangia regularly ovoid to obpyriform in short, close sympodia; mostly no or limited growth at 35°C .... *cryptogea*
14. Hyphae uniform to uneven, sometimes with small swellings; sporangia ovoid or ellipsoidal, usually single, terminal on unbranched or lax sympodially branched sporangiophore; good growth at 35°C.. *drechsleri*

15. Sporangiophore branching compound sympodial, nodose .... *infestans*
15. Sporangiophore branching simple sympodial, undifferentiated... 16

16. Sporangia nondeciduous... 17
16. Sporangia deciduous... 18

17. Good growth on agar medium at 35°C; mycelium tufted.............. *nicotianae* (*parasitica*)
17. No or slight growth on agar medium at 35°C; mycelium finely radiate .... *citrophthora*

18. Sporangial pedicel greater than 20 μm long............ *capsici*
18. Sporangial pedicel 5–20 μm long 19
18. Sporangial pedicel less than 5 μm long.......................... 20

19. Sporangia regularly elongate, ellipsoidal to fusiform .... *colocasiae*
19. Sporangia ellipsoidal to variable, often asymmetrical, bifurcate.. *meadii*

20. Sporangia broadly ellipsoidal to obturbinate (length-breadth ratio less than 1.4) in short irregular sympodia .... *arecae*[d]
20. Sporangia ovoid to ellipsoidal (length-breadth ratio greater than 1.4) in long regular sympodia .... *palmivora*

[a] Reproduced by permission of *The Plant Pathology Bulletin*, courtesy H. H. Ho (1992).

[b] Species name and first report (reference) in Taiwan:

P. arecae (Coleman) Pethybridge (T. H. Huang, *unpublished*)
P. boehmeriae Sawada (Sawada 1927b)
P. cactorum (Lebert and Cohn) Schroeder (Sawada 1927b)
P. capsici Leonian (Kao and Leu 1977)
P. cinnamomi Rands (Anon. 1965b)
P. citricola Sawada (Sawada 1927b)
P. citrophthora (R. E. Smith and E. H. Smith) Leonian (Sawada 1915a)
P. colocasiae Raciborski (Sawada 1911)
P. cryptogea Pethybridge and Lafferty (Kao 1978)
P. cyperi (Ideta) Ito (Kawakami and Suzuki 1908)
P. drechsleri Tucker (*melonis* Katsura) (Kao and Leu 1977)
P. heveae Thompson (Ann 1984)
P. humicola Ko and Ann (Ko and Ann 1985)
P. infestans (Montagne) de Bary (Kawakami and Suzuki 1908)
P. insolita Ann and Ko (Ann and Ko 1980)
P. katsurae Ko and Chang (Ko and Chang 1979)
P. leersiae Sawada (Sawada 1941)
P. lepironae Sawada (Sawada 1919b)
P. macrospora (Saccardo) Ito and Tanaka (Sawada 1927b)
P. meadii McRae (Ann 1992)
P. nicotianae Breda de Haan (*P. parasitica* Dastur) (Sawada 1942b)
P. palmivora (E. Butler) E. Butler (Huang and Chen 1974)
P. vignae Purss (Kao and Leu 1977)

[c] Treated as *Sclerophthora macrospora* (Stamps et al. 1990).

[d] Treated as conspecific with *P. palmivora* (Oudemans and Coffey 1991b; Mchau and Coffey 1994a).

CHAPTER 5

Cultural and Biological Control

All the options must be investigated to develop efficient strategies to achieve as much control of the pathogen as possible and at the least cost. To do this, we must take advantage of the pathogen at the weakest part of its life cycle. In this chapter, we discuss some areas in which disease is highly dependent on environment and how this dependency will affect control.

The principles of how control measures affect disease can be simply illustrated by use of three circles depicting the host, the pathogen, and the environment (Figure 5.1). When these factors (circles) overlap or interact, disease ensues; conversely, practices that move the circles apart will control disease. Measures that affect interaction of the host, the pathogen, and the environment, none of which acts independently, are effective for control of disease. Sometimes the triangle model is used to explain disease. The pathogen, the host, and the environment each represent one side of the triangle. Any factor that affects one of the three sides of the triangle controls the disease. Although models are theoretical, the soundness of the concepts is substantiated by numerous data. As shown in Figure 5.1, there are three basic areas on which to base control methods: 1) the pathogen, 2) the host, and 3) the environment.

The environmental factor to which plants are most vulnerable and on which *Phytophthora* has a critical dependency is moisture. The science and the art is to set up conditions in which the plant will thrive but *Phytophthora* will not. Later in the chapter, we discuss the epidemiological principles that are useful in establishing the rationale for successful control practices. Several environmental situations in which the severity of *Phytophthora* diseases is increased are shown in Table 5.1.

Phytophthora, the pathogen, can be prevented from becoming a factor by eradication methods or total sanitation in some agricultural situations. Planting stock should be free of contamination at the nursery. If planting soil in the nursery is likely to be infested, the pathogen should be destroyed by steam heat or methyl bromide. In the field, a suitable fungicide such as methyl bromide, applied by soil fumigation, or metham (Vapam), applied by soil drenching, will suppress *Phytophthora* for at least a year if applied before the crop is planted. Total sanitation is expensive, however, and in the real world is seldom obtainable.

If cultural control is insufficient or resistance is unavailable, fungicides effectively control many *Phytophthora* diseases. There is a difference in the approaches taken by entomologists to control insects and plant pathologists to control diseases with chemicals. Because crops will tolerate a certain threshold number of insects before damage is sufficient to cause a significant yield loss, a chemical application can often be withheld; however, this is seldom the case with foliar diseases such as late blight caused by *P. infestans,* because the population increases rapidly in and on infected tissue when environmental conditions are favorable. Once *P. infestans* infects a few leaves on potato plants, more inoculum is readily produced during the next few days if rain and temperature favor the pathogen. Penetration of the host by spores of an airborne *Phytophthora* species can be prevented only by applying the appropriate fungicides, and in most cases, even with systemic fungicides, it is important that they be applied *before* disease occurs.

Certain systemic compounds, called the phenylamide fungicides, which include metalaxyl, furalaxyl, benalaxyl, and oxadixyl, are capable of moving upward (acropetally) from soil or foliage. Because phenylamide fungicides have a spe-

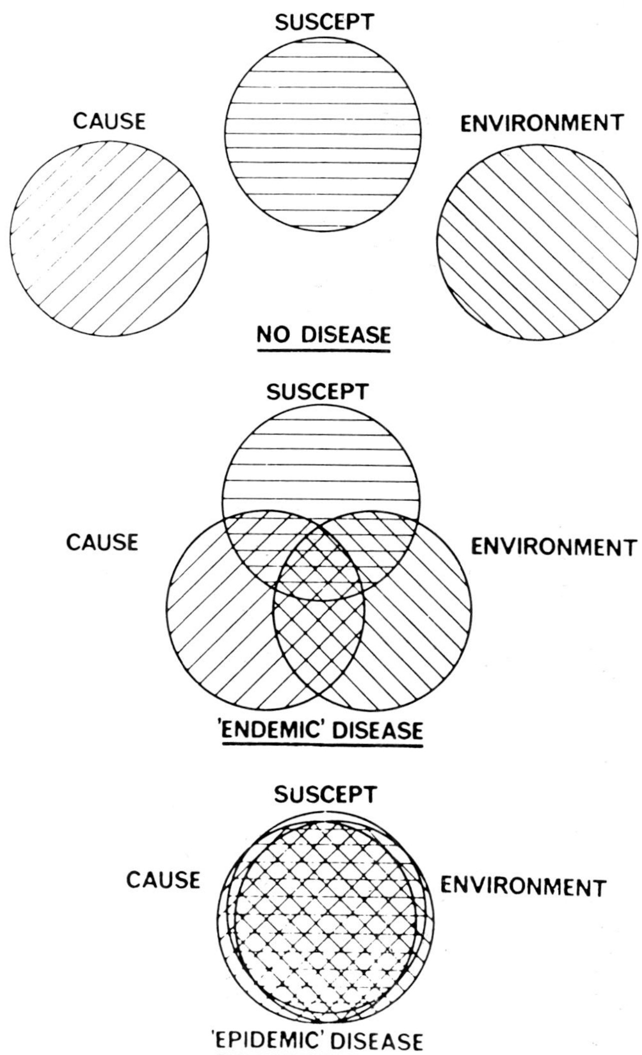

Figure 5.1. Interaction of the host (suscept), pathogen, and environment. This theoretical diagram shows that the absence of the causal agent, the total resistance of the host (suscept), or the unfavorable nature of the environment will result in no disease. As these factors change, the severity of disease increases or decreases. (Courtesy O. T. Maloy 1971)

cific mode of action, metalaxyl-resistant mutants of *P. infestans* often quickly displace the sensitive wild-type isolates in field populations. Management of resistance has become a problem for control of late blight of potato and is covered in more depth in Chapter 7.

Fosetyl-Al is another type of systemic fungicide. It is taken up into the plants and translocated both upward and downward (basipetally). This is the first and only known fungicide that moves both upward and downward in the plant. Fosetyl-Al has been effective for control of many *Phytophthora*-caused diseases, but it is not consistently effective against late blight of potato caused by *P. infestans*. Fosetyl-Al and the phenylamide fungicides as well as others are discussed in Chapter 7 on fungicidal control.

To turn back to the plant in our model of disease and the means to control disease, host resistance has been exploited for control of the late blight disease of potato by use of race-specific R-genes from a wild species of potato (Gallegly 1968). This type of resistance, called vertical resistance, actually acts as a sanitary factor because sporulation by *Phytophthora* is prevented, and hence the inoculum is reduced to nearly zero (Vanderplank 1963). Although the control of late blight of potato caused by *P. infestans* with host-specific resistance was initially highly successful (Black et al. 1953) and plant breeders and plant pathologists had high hopes, the rapid and uncontrollable evolution of new pathologic races of *P. infestans* rapidly overcame this vertical resistance. The breeding approach in potatoes has now changed to the search for and use of general resistance, referred to by Vanderplank as horizontal resistance and by others as field resistance, rate-limiting resistance (Mackenzie et al. 1983; Umaerus et al. 1983), or tolerance (Schmitthenner and Walker 1979). Overall, on cultivars with general resistance, lesions form less frequently and expand less rapidly than on susceptible plants; however, on plants with host-specific resistance, only a hypersensitive fleck follows infection and, furthermore, *P. infestans* does not sporulate. In the long run, although general resistance is less complete, it is more durable because it is not as vulnerable to the development of populations of new virulent biotypes (pathologic or physiologic races) of *P. infestans*. A similar situation exists with soybean root rot, caused by *P. megasperma* f. sp. *glycinea* (*P. sojae*), and a few other diseases. Nonetheless, host-specific resistance in soybean is still useful (Schmitthenner 1985). These and other systems are discussed more fully in Chapter 6 on resistance.

Moisture, which includes rainfall, dew deposition, and irrigation, is the main environmental factor that favors the pathogen. For control of soilborne diseases, any practice that reduces the time that soil remains flooded will help to reduce the probability that severe root disease will ensue. Planting on beds allows for better drainage. In potted plants, porous soils are better drained. These factors are discussed in more detail in the section Management of Soil Water. Although irrigation can be controlled by the grower, rainfall cannot. As discussed in the section The Hardpan Effect in Agricultural Soils, root rot of jarrah trees in Western Australia was shown to occur mainly in areas in which soil water collected on a concreted lateritic layer of soil (Shea et al. 1982, 1983).

Temperature extremes, low and high, are also modifying factors. For instance, *P. infestans* will not sporulate at temperatures greater than 27–30°C. In this case, weather prediction helps immensely to determine when to apply fungicides. If it is predicted that rainfall will occur for several days and the air temperature will be about 20°C, the growers of potatoes should plan to apply a suitable protectant or systemic fungicide against foliar disease such as late blight, even if disease is not seen in the field. If no rain is forecast and temperatures are predicted to remain high, fungicides may not be necessary. Disease forecasting is covered in more detail in Chapter 7.

In some cases, environmental factors that predispose a host to infection, such as excess nitrogen fertilizer, soil salinity, drought stress, and excess soil moisture, can be prevented by appropriate cultural management. Optimal fertilization practices and irrigation management have a direct bearing on control. Predisposition to disease and its effect on susceptibility and host resistance is discussed at greater length in the section Predisposition.

The recent essay "Biological Control—The Need for a New Scientific Framework" by Gabriel and Cook (1990) makes a bold suggestion that all control measures should be conceptualized as physical, biological, or chemical rather than as the traditional chemical, resistance, cultural, and biological categories. They point out that cultural methods can be a means to either biological or physical control. According to their concept, host resistance should be discussed as a means of biological control. Although we recognize the validity of their argument, our sections on control here are organized under the traditional categories of cultural and biological control (Chapter 5), control by host resistance (Chapter 6), and control with fungicides (Chapter 7) because the information is more readily

Table 5.1. Some factors that are reported to increase severity and spread of *Phytophthora* diseases

Treatment *Phytophthora* **species**	**Reference**
Excess flooding of soil	
cinnamomi	Wager (1942a, b)
	Zentmyer and Richards (1952)
capsici	Bowers and Mitchell (1990)
	Bowers et al. (1990)
cryptogea	Bumbeiris (1976)
	Duniway (1979, 1983)
	Duniway and Gordon (1986)
medicaginis	Kuan and Erwin (1980b, 1982)
cambivora, drechsleri, megasperma	Duncan and Kennedy (1989)
fragariae	Duncan et al. (1991)
parasitica	Neher and Duniway (1991, 1992)
Impervious layers of soil (hardpans)	
cinnamomi	Burns et al. (1960)
	Shea et al. (1984)
	Zentmyer (1980)
Drought stress	
cryptogea	Duniway (1977b, 1983)
drechsleri	Zimmer and Urie (1967)
cinnamomi	Blaker and MacDonald (1981)
parasitica	Ristaino and Duniway (1989)
High soil salinity	
cryptogea	Blaker and MacDonald (1986)
	Bouchibi et al. (1990)
	MacDonald (1982, 1984)
	MacDonald et al. (1984)
Increase in soil pH from 5 to 7	
parasitica	Kincaid et al. (1970)
Increase in duration of foliar wetness	
cactorum	Grove et al. (1985a)
parasitica	Kuske and Benson (1983a)
Tent-building ants	
palmivora	Newhook and Jackson (1977)
	Okaisabor (1971a)
megakarya	McGregor and Moxon (1985)
	B. Taylor and Adedoyin (1981)
	B. Taylor and Griffin (1981)
Excess nitrogen-containing fertilizers	
cactorum	Hoitink et al. (1986)
	Utkhede (1984a, b)

presented. Nonetheless, the ideas of Gabriel and Cook should be kept in mind because obviously there are no sharp demarcations between some of the cultural controls covered in the first part of this chapter and some of the biological controls discussed in the latter part. For example, the Ashburner system used in Queensland, Australia, in which amendments and cover-cropping are used to control *P. cinnamomi* on avocado, is a cultural system that leads to biological control.

In this chapter, we discuss initially factors that increase disease and then give some classical examples of how the pathogen, the host, and the environment interact and discuss the importance of this interaction. The rapid regeneration of *Phytophthora* inoculum, as described in Chapter 3, increases its effectiveness as a pathogen. Agricultural methods can often be manipulated in ways that affect the environment so that infection is prevented or reduced. The degree of resistance of the host plant can either reduce production of symptoms and inoculum (host-specific resistance) or slow the rate of infection and symptom production (rate-limiting resistance) (Chapter 6). Biological control exploits the effects of other microbes that are antagonistic or antibiotic to *Phytophthora* but are not harmful to the plant (R. Baker 1968). Information on this type of control is presented in the latter part of this chapter. As an adjunct to these types of control, fungicides are being used extensively, especially on crops that are particularly vulnerable to epidemics (Chapter 7).

The Susceptible Host, the Virulent Pathogen, and the Favorable Environment

Consider the three-circle model for the host, pathogen, and environment. A historically important, classic example of a susceptible host, favorable environment, and a newly introduced pathogen (Figure 5.1) is the late blight epidemic of potato, which first became severe during the years after 1846 in the United States, Ireland, and other areas of Europe. Late blight resulted from the unexpected appearance of a newly introduced fungus, *P. infestans,* which apparently was carried to Ireland in potato tubers imported from Mexico (Large 1940; Zentmyer 1988; Niederhauser 1991). At this time in Ireland, potato crops produced large yields and were relied upon as the most stable food crop that was available. Potatoes, grown in large acreages, were killed by the blight, and the tubers rotted. After many years of controversy, the causal agent was named *P. infestans* in 1876 by de Bary. However, the late blight epidemic was the result not only of the presence of a new pathogen on a susceptible crop, but equally important, of the cool temperatures and frequent rains during 1846 and the years after, conditions that were optimal for the spread of the pathogen and colonization of potato leaf tissue. When these conditions do not prevail, late blight is usually not a problem. Late blight of potato is favored by leaf wetness (caused by rain or dew deposition) that lasts at least 8 h at temperatures near 21°C for several days (Cox and Large 1960). During the years following 1846, the uncontrolled late blight disease killed entire fields of potatoes and led to mass starvation of the people of Ireland. This is probably the most dramatic and tragic example in recorded history of the necessity for plant disease control (Large 1940). During the mid-1800s, there were no fungicides nor resistant cultivars; however, the epidemic eventually abated because environmental conditions favorable to it did not occur every year. It was not until many years later that suitable fungicides became available, a few resistant varieties were discovered, and breeding programs for development of resistant cultivars were initiated in many nations. During the years immediately following 1846, however, the loss of the main food crop wreaked havoc on the entire nation of Ireland (Large 1940; Carefoot and Sprott 1967).

Regional variations in severity of disease exemplifies the importance of environment. For instance, 96% of the potato acreage in the humid, high-rainfall area of the northeastern United States is sprayed with fungicides to prevent late blight caused by *P. infestans* and early bight caused by *Alternaria solani,* while in the arid, low-rainfall region of the western United States, fungicides are applied to only 8% of the acreage (Andrilenas 1975, cited by Fry [1977b]). Nonetheless, late blight of potato and tomato in the western United States, an area of essentially low summer rainfall, can occur sporadically when weather conditions lead to increased dew deposition and rainfall (Vartanian and Endo 1985a, b; M. D. Coffey, *personal communication*). Even in the eastern United States and in Europe, environmental conditions are not always "good blight years." Thus, the strategy to control such a disease by use of appropriate fungicides is dependent on the frequency of rainfall and optimal temperatures for sporulation of *P. infestans.* Even when fungicides are necessary, general resistance in some potato cultivars, which works in concert with fungicides to augment their effects, reduces the amount and frequency of fungicide applications needed to achieve control (Fry 1977a).

The sensitivity of *Phytophthora* inoculum to temperature varies among species (Zentmyer 1980). In Germany, *P. capsici* and *P. cinnamomi* are unable to survive in soil during the cold winters (Kröber 1980). *P. cinnamomi* becomes inactivated in soils only when the temperature drops below 0°C, regardless of whether the pathogen is in root segments, colonized oat grains, or naturally infested soil (Benson 1982). In another study, *P. cinnamomi* became inactivated in a naturally infested soil in 2 days at 6–7°C, whereas at 14°C, 16 days were required for inactivation. *P. citrophthora,* which survives freezing to –21°C in nutrient agar media (Gerlach et al. 1976b), survives low temperatures only if the cultures are first conditioned at 4°C for 4 days prior to freezing. Since oospores of many *Phytophthora* species will withstand freezing at –20°C in the laboratory (Chapter 3), they are probably also capable of overwintering in cold climates.

Although temperatures above 18–22°C and approaching 30°C are restrictive for epidemics of a potato disease such as late blight caused by *P. infestans,* in contrast, the root disease of alfalfa, caused by *P. medicaginis* (*P. megasperma* f. sp. *medicaginis*), occurs readily at a wide range of soil temperatures (17–27°C) (Wilkinson and Millar 1982) and is not suppressed in soils by freezing temperatures (Stack and Millar 1985b, c).

Examples of some practices that increase the severity and/or spread of *Phytophthora* diseases are shown in Table 5.1. Some of these predisposing factors, such as soil salinity in potted plants (MacDonald 1984; MacDonald et al. 1984) and drought stress of safflower prior to flooding, are covered in greater detail under the section on predisposition (see also Duniway [1983] and Shea and Broadbent [1983]). Increase in disease when douglas fir sawdust was used as a mulch on strawberry beds was apparently caused by the retention of moisture in the bed, because when sawdust was mixed into soil, the disease severity was not increased (Vaughn et al. 1954). Severity of red raspberry root rot caused by *P. cambivora* was greatest after a 4-day period of waterlogging (Duncan and Kennedy 1989). Ponding of water on the upper surface of lateritic hard pans in the jarrah forest of Western Australia where dieback disease was caused by *P. cinnamomi* is discussed in the following section, The Hardpan Effect in Agricultural Soils, based on Shea et al. (1984) and Shea (1988).

Thus, it is the balance of the three major factors—the pathogen, the degree of susceptibility of the host, and the specific set of environmental conditions for pathogenesis—that is important. Circumventing the effects of the various components of these interactions is what integrated control is all about. Several examples of methods of control other than host resistance and fungicides are given in Table 5.2.

Emeritus Professor Frank Newhook (1988), from the University of Aukland in New Zealand, published an interesting, popular article that shows how the host, pathogen, and environment interact on root disease of pine trees. Newhook (1988) points out that propagules such as oospores or chlamydospores persist in soil and, after the soil is wetted for 24 h or more, may germinate and produce sporangia and zoospores. Thus, the inoculum increases rapidly, and roots or crowns of trees can become infected. Crown or collar rots are the most noticeable symptoms of infection of trees; however, fibrous root infection is probably more frequent but not as often detected. An example of collar rot on a bottle tree caused by *P. drechsleri* is shown in Plate 5.1. On apple trees (see Chapter 11), *P. cactorum* causes a serious collar and crown rot that, if allowed to progress, will kill the tree. Citrus trees are often subject to the gummosis crown rot disease caused by *P. citrophthora* and/or *P. parasitica.* These diseases can be prevented to some extent, especially in areas where rainfall is sparse, by preventing irrigation water from touching the trunk of the tree. In some cases, the diseased bark tissue can be cut away and the area treated with an acceptable fungicide (see Chapter 7).

Newhook (1988) states that *Phytophthora* disease on fibrous roots is much less readily detected but is probably more common. In many crops, the tips of a large number of the newly emerging roots are killed, but if the percentage of such killed roots is low enough that the uptake of water or nutrients is not materially affected, the foliar part of the plant may not show any symptoms. If the soil is well drained and does not remain saturated for long, a new crop of roots will regenerate and replace the killed roots. In fact, a plant can often survive, even when a large portion of the fibrous roots is infected. For example, Newhook (1959, 1988) shows a series of photographs that exemplify the progress of fibrous root rot caused by *Phytophthora* on *Pinus radiata* seedlings grown in hydroponic culture (Plates 5.2–5.4). In Plate 5.2, although many lateral roots are killed, new adventitious roots are replacing the

Table 5.2. Methods other than host resistance and fungicides used to control *Phytophthora* species in various crops

Control method	*Phytophthora* species	Host	Reference
Cultural			
Control of soil moisture	Many	Tomato Safflower Asparagus	Duniway (1983) (review)
	parasitica		Duniway and Gordon (1985) Falloon et al. (1991) Ristaino (1991) Ristaino et al. (1988) Ristaino and Duniway (1989)
Sanitation	Many	Many	Shea and Broadbent (1983) (review)
Suppressive soils	*cinnamomi*	Many	Broadbent and Baker (1974a, b, 1975); Broadbent et al. (1971); Baker (1978); Cook and Baker (1983)
Virgin soil in planting hole	*palmivora*	Papaya	Ko (1971, 1982)
Site selection	*cinnamomi*	Several	Goodall (1955); Burns et al. (1960); Zentmyer and Ohr (1978)
Planting on raised beds	*cinnamomi*	Avocado	Coffey (1987)
	fragariae	Strawberry	Newhook (1988)
Organic amendments			
Alfalfa straw and/or meal	*cinnamomi*	Avocado	Zentmyer (1963, 1980); Gilpatrick (1969a, b); Tsao (1977a)
Barley straw, fowl manure	*cinnamomi*	Avocado	Broadbent and Baker (1974a, b, 1975)
Corn straw and ammonium nitrate	*parasitica*	*Piper betle*	Mehrotra and Tiwari (1976)
Hardwood bark composts	Many	Ornamentals	Hoitink et al. (1975, 1977); Hoitink (1980); Hoitink and Fahey (1986)
	cactorum	Apple	Spring et al. (1980)
	cinnamomi	Several	Sivasithamparam (1981); Sivasithamparam et al. (1981); Hardy and Sivasithamparam (1989, 1991a–d)
Fowl manure	Several	Several	Broadbent and Baker (1975)
Ammoniated douglas fir bark	*fragariae*	Strawberry	Vaughn et al. (1954)
Other amendments			
Superphosphate	Many	Eucalyptus	Newhook (1970)
Elemental sulfur	*cinnamomi*	Pineapple	Pegg (1977a)
Calcium carbonate	*cinnamomi*	Avocado	Broadbent and Baker (1974a)
	cinnamomi and others	*Eucalyptus marginata* and others	Boughton et al. (1978)
Calcium nitrate	*cinnamomi*	Avocado	Lee and Zentmyer (1982)
Ammonium sulfate	*cinnamomi*	Avocado	Lee and Zentmyer (1982)
Ammonia	*cinnamomi, parasitica*	Avocado, citrus	Tsao and Oster (1981)
Nitrate and ammoniac fertilizers	*parasitica*	Citrus	Klotz et al. (1958)
	cinnamomi	Several	Broadbent (1977a); Tsao and Oster (1981)
	Several	Several	Shea and Broadbent (1983)
Nitrapyrin	*cinnamomi*	Avocado	Lee and Zentmyer (1982)
Urea	*cinnamomi*	Avocado	Tsao and Zentmyer (1979)
Rhizobium spp.	Several	Several	Tu (1978, 1980)

dead roots. However, when favorable conditions for disease continue, the second crop of roots will also be killed by zoospores that emanate from the previously infected roots (Plate 5.3). Eventually, when the rate of production of zoospores by *Phytophthora* and subsequent infection of new roots exceeds the rate of production of new roots, the plant is killed. Plate 5.4 shows the root of a pine seedling infected by *P. cinnamomi* in a nursery after heavy rainfall. Although such plants can survive by producing new adventitious roots, they often die if soil moisture becomes excessive. The initial inoculum from the contaminated nursery probably remains in a viable state for the life of the plant.

Newhook (1988) relates the spectacular death of large *P. radiata* trees planted in New Zealand to provide windbreaks. These trees were apparently infected but grew well until 1956, a year of abnormally heavy rainfall, after which many trees died suddenly as a result of root rot caused by *P. cinnamomi* (Newhook 1958, 1959; Plate 5.5).

Newhook (1958, 1960) concludes that these pine trees died suddenly because of the death of most of their lateral roots. Since rainfall was unusually heavy during 1956, regenerating roots became infected, and thus the necessary water for growth and transpiration of the tall trees could not be supplied. The trees first showed their foliar symptoms of stress by September and died during October and November (spring and early summer in the southern hemisphere). These examples "illustrate the principle that the margin between outward health and disease can depend on the state of balance between rootlet death and rootlet replacement . . . [which] in turn [is] governed by environmental factors such as moisture and soil temperature. Air temperature, sunshine and wind interact to affect transpiration and hence water demand by the plant. It is this which is the critical factor when rootlet loss is extensive" (Newhook 1988).

Newhook's (1988) account of the deleterious effect of *P. cinnamomi* on mature stands of *P. radiata* trees (Plate 5.5) following heavy rainfall is a good example of the importance of latent infection and excess soil moisture. Diagnosis of the cause of such a situation can be controversial if one does not understand the classic principle that disease is the result of the interaction of the host, pathogen, and environment.

Since the mature pine trees died so suddenly, some forest scientists in New Zealand attributed the deaths to waterlogging of soil; however, Newhook and colleagues showed that *P. cinnamomi* was widely distributed in New Zealand. It also appeared likely that the initial inoculum had originated from the planting stock. Separate studies showed that flooding soil for 1 month did not kill *P. radiata* trees unless the flooded soil had been previously infested with *P. cinnamomi.* A 5-month period of flooding was required to kill the trees in sterilized soil (Newhook 1959, 1960, 1970; Sutherland et al. 1959).

So, what strategies can be employed to combat *Phytophthora*? What are its weak points, and what are its strong features? *Phytophthora* requires free water or very high relative humidity to produce sporangia and zoospores (Chapter 3). These factors are discussed in the following section.

Cultural Factors that Increase Disease

Soil and Atmospheric Water

Excessive irrigation by flooding and sprinkling has a serious impact on disease (Rotem and Palti 1969). Excess irrigation and rainfall are probably the most important factors that increase the severity and spread of *Phytophthora* diseases (Table 5.1). As Newhook (1988) states, the most important environmental factor is the duration of free water, either in soil (Duniway 1983) or on foliage or fruit (Grove et al. 1985a), because this sets up a favorable situation for an increase in the population of *Phytophthora* species. Obviously, the reproductive capacity of *Phytophthora* allows it to increase rapidly by formation of sporangia and zoospores.

The most important ways that zoospores or cysts are moved in soil is by flowing irrigation water, rainfall runoff, and movement of soil by any means. When gravel and soil were inadvertently moved from *P. cinnamomi*-infested areas to make new roads in the jarrah forests in Australia, a few years later a path of diseased and dying trees became apparent because inoculum of *P. cinnamomi* (probably chlamydospores) had been moved with the soil (Shea et al. 1983; Weste 1983b). *P. cinnamomi* inoculum moves downhill (probably as zoospores) with water runoff under culverts, resulting in fan-shaped patterns of disease in forests. Disease progress on drainage slopes reached as much as 400 m per year (Weste 1983b). Kinal et al. (1993) showed that zoospores of *P. cinnamomi* in subsurface water moved laterally in a hardpan layer up to 120 m down slopes in the Western Australia jarrah forest.

Although zoospores are motile, they do not move very far by themselves in soil. In column experiments, zoospores moved behind a wetting front (free water) 35 cm in sand, 44 cm in sandy clay, and 48 cm in loam but failed to move at all in silt soil (finest texture). Swimming zoospores moved about twice the distance of cysts, indicating that motility has some advantage but not enough to account for more than a local effect (Wilkinson et al. 1981; see also Chapter 3). In flowing water, however, movement of zoospores is one of the most important means of spread (see Chapter 2, Isolation from Bodies of Water).

In Table 2.5 and in the section Isolation from Bodies of Water in Chapter 2, many references indicate that *Phytophthora* inoculum can be readily detected in bodies of water such as canals, reservoirs, lakes, and streams. Klotz et al. (1959) showed that both *P. citrophthora* and *P. parasitica* were present in canals used for irrigation of citrus trees in southern California. In a most unusual instance, a severe outbreak of boll rot of cotton occurred in a field near Bakersfield, California, where rain does not fall during the growing season. Inoculum of *P. capsici* came from infested water in a reservoir used to sprinkle-irrigate the cotton field (Garber et al. 1986). Von Brombsen (1984b) detected *P. cinnamomi* in head waters of rivers of the Southwestern Cape Province of South Africa.

Even though potting containers used in glasshouse industries have drainage holes in the bottoms, overwatering is still a common problem because drainage of soil in pots is much less efficient than in the field, and if plants are growing slowly and transpiration is limited, the soil actually may be saturated much of the time. Drainage of soil mixes can be improved by adding peat moss, perlite, vermiculite, sand, or redwood sawdust, which increase the porosity of soil (K. F. Baker 1957), or a decomposed composted hardwood bark (Hoitink and Fahy 1986). Composted hardwood bark, in addition to increasing porosity, has a suppressive effect on *P. cactorum* when applied around apple seedlings (Spring et al. 1980).

Increase in the porosity of the soil mixture in container-grown plants is highly important in reducing the severity of root disease of Toyon (*Heteromeles arbutifolia*), caused by *P. cinnamomi.* When air-filled porosity values in soil were less than 10%, severe root rot followed inoculation with zoospores, but at porosity values of 10–20%, there was little root rot (Filmer et al. 1986). Even ideal soil mixes will not control the predisposing effects of overwatering, however. An excellent discussion of control of root disease and growth of potted ornamental plants is found in the book on propagation of container-grown plants edited by K. F. Baker (1957). See also the discussion of Hoitink and Powell (1990) on the effect of hardwood bark compost on the production of rhododendron plants in commercial nurseries, discussed later in this chapter.

In the ornamental plant industry, plants are rooted from cuttings or started from seeds in small containers before they are transplanted to homes, landscapes, or forests. In such cases, natural infection often occurs. Several reports indicate that the severity of disease, e.g., on *Rhododendron,* is greatest in heavy potting mixes (high bulk density) that hold a large amount of water. Peat and pine bark mixes that decrease the bulk density and are better drained (hold less water) reduce the severity of disease (Ownley et al. 1990; Ownley and Benson 1991). Other examples are given in the section Potting Mixes for Nursery Use.

Management of Soil Water

Free water in soil will cause the numbers of infective units (sporangia and/or zoospores) to increase, which will probably increase the severity of disease (Mitchell et al. 1978; Mitchell and Kannwischer-Mitchell 1983). In turn, the increase in inoculum exacerbates the severity (proportion of roots affected and size and depth of root lesions) of Phytophthora root rot.

Irrigation management is the first line of defense in low-rainfall areas. Diseases of roots occur only after the soil has been saturated with water for a certain period. Therefore, either rainfall, irrigation, or any condition that prevents optimal vertical drainage of water in soil leads to soil saturation, which favors the onset of Phytophthora root rot. In land irrigated by flooding or via furrows, Phytophthora root rot usually occurs first in low areas or at the ends of poorly drained fields. Citrus growers who irrigate direct the water away from the trunks of the trees to prevent gummosis. In housing areas in California that are developed in old walnut groves, trees that are saved for home landscapes often die because *P. cactorum* causes crown rot when lawn sprinklers keep the trunks wet.

The most decisive factor that governs the severity of root disease is the length of time that the soil remains saturated or near saturation (Pfender et al. 1977; Duniway 1983). Unfortunately, there are few, if any, specific recommendations on a field basis about how long the soil must be in a saturated condition to constitute a significant hazard (Schmitthenner 1985). Since soil texture and structure are highly variable, such data, if available, would be difficult to extrapolate to diseases of different crops.

Water and temperature are the major factors governing the development of sporangia and zoospores (Chapter 3). Environmentally, soil water is more amenable to management than temperature. Both laboratory and field research indicates that sporangia are the predominant propagules that have the capacity to increase inoculum rapidly in a multicyclic manner, and sporangia are formed only in the presence of free water if aeration is adequate. Zoospore formation is induced by a shift downward in temperature by about 10°C below ambient. Optimal temperatures also favor production of sporangia and zoospores, but water is the primary factor. Thus, to prevent buildup of *Phytophthora* propagules, both factors should be considered in developing a strategy in a management system. In many cases, soil water can be manipulated or managed culturally in a cropping system.

In the life cycle of *Phytophthora,* free water is necessary for release of zoospores. Zoospores have the ability to stick to leaf and other surfaces because they rapidly produce an adhesive substance during the encystment process (Sing and Bartnicki 1975a). Although the adhesive material has not been identified, data from use of a monoclonal antibody that is capable of detecting the material presented evidence that adhesion is associated with lectin-ligand-type interactions (Hardham et al. 1991a; Hohl 1991b). Apparently, the adhesive material is not specific to the host.

In many field situations, such as established citrus, avocado, and stone fruit orchards, the inoculum cannot be eradicated economically. In such cases, the grower should attempt to set up conditions that will be most favorable for the plant and least favorable for the pathogen. Since presence of free water is critical for production of inoculum, preventing the contact of irrigation water with the trunks of trees is one of the most important methods of preventing the crown rot type of disease previously referred to by Newhook (1988) (Plate 5.1). Certain irrigation practices, such as alternate tree row watering, reduce the population of *Phytophthora* in citrus orchards (Lutz et al. 1989). Other examples are discussed below in the section on soil inoculum.

Duniway's (1983) review of the relationship between water and *Phytophthora*-caused diseases aptly covers the effect of water on the infection stage that is usually caused by zoospores. Tippett and Hill (1983) and Tippett et al. (1987) introduced the concept of phloem-moisture relations as they affect the progress of lesions in woody plants. Tippett et al. (1985) noted an increase in lesion development caused by *P. cinnamomi* after summer rains. In 1987, they presented data to show that the susceptibility of jarrah trees in Western Australia increased directly with the increase in phloem (inner bark) relative water content (RWC). Growth of *P. cinnamomi* in stem pieces at 85% RWC was one-half that at RWC of 95% or greater. When RWC was reduced to 70%, growth of *P. cinnamomi* ceased. Tippett et al. pointed out that *P. cinnamomi* was much more sensitive to low water potentials in the bark than it was to those in the soil and spoke of a possible "balance of rates" between fungal invasion and host resistance. The jarrah dieback problem is discussed at more length in a later section, The Hardpan Effect in Agricultural Soils, and in Chapter 15A on *P. cinnamomi.*

Water should not be sprayed on trunks of trees in orchard irrigation systems; however, this practice is often seen. The phloem-water concept developed by Tippett et al. (1985) shows that a high internal water content of plant tissue favors infection. To control canker production caused by *P. citricola, P. cactorum,* and other *Phytophthora* species on woody orchard trees, this concept should be kept in mind. Gummosis of citrus trees caused by *P. citrophthora* and *P. parasitica* is much more severe when soil moisture is excessive. Klotz (1978) advocates preventing contact of irrigation water with the trunks of citrus trees because the trees become more susceptible and the inoculum of *Phytophthora* increases.

Duniway's (1983) explanation of the role of soil water and the measurement of water status is excerpted in Appendix 5.1. For a more detailed description of measurement of soil water, see also Cook and Papendick (1972).

Much irrigation management is based on subjective judgments on when the amount of water administered is sufficient for plant growth. When subjective judgment is tempered by experience, this method may be adequate; however, in many agricultural systems, the soil tensiometer has become a management tool that is extremely valuable for regulation of soil water where irrigation is practiced. This instrument (Figure 5.2A) (Marsh 1978) measures the tension (or suction) of the water caused by capillary and absorption forces in the soil from zero (free water is arbitrarily specified as zero) to about −800 mb (millibars) matric potential (dry). Practically, when the matric potential is about −300 mb, soil is at field capacity (loosely defined as the moisture status in natural field soil at which water no longer percolates downward with the force of gravity); at 0 mb, the soil is saturated with water (all pores in the soil profile are filled). As water in the soil percolates downward or is used by plants, the matric potential gradually becomes more negative until rain falls or the soil is irrigated. At this time, the matric potential increases again to zero. The tensiometer measures the availability of water in soils of all textures, whether they contain clay, silt, or sand (Figure 5.2 and Appendix 5.1) (Cook and Papendick 1972; Duniway 1983).

The proper use of tensiometers not only prevents wasting expensive irrigation water but also aids in maintaining soil at a moisture status at which Phytophthora root rot can be controlled. Tensiometers have been recommended for many years for use in avocado orchards in California where Phytophthora root rot occurs (Zentmyer and Ohr 1978). Use of tensiometer data for the integrated control strategy of Phytophthora root rot of citrus caused by *P. parasitica* and *P. citrophthora* has particular potential on land underlaid by hard pans (Lutz et al. 1986).

Role of Soil Moisture

Laboratory Research

The precise effect of soil moisture on the severity of disease and the development of *Phytophthora* sporangia in soil can be determined with soil moisture tension plates, which are based on the same principle as the tensiometer. Small samples

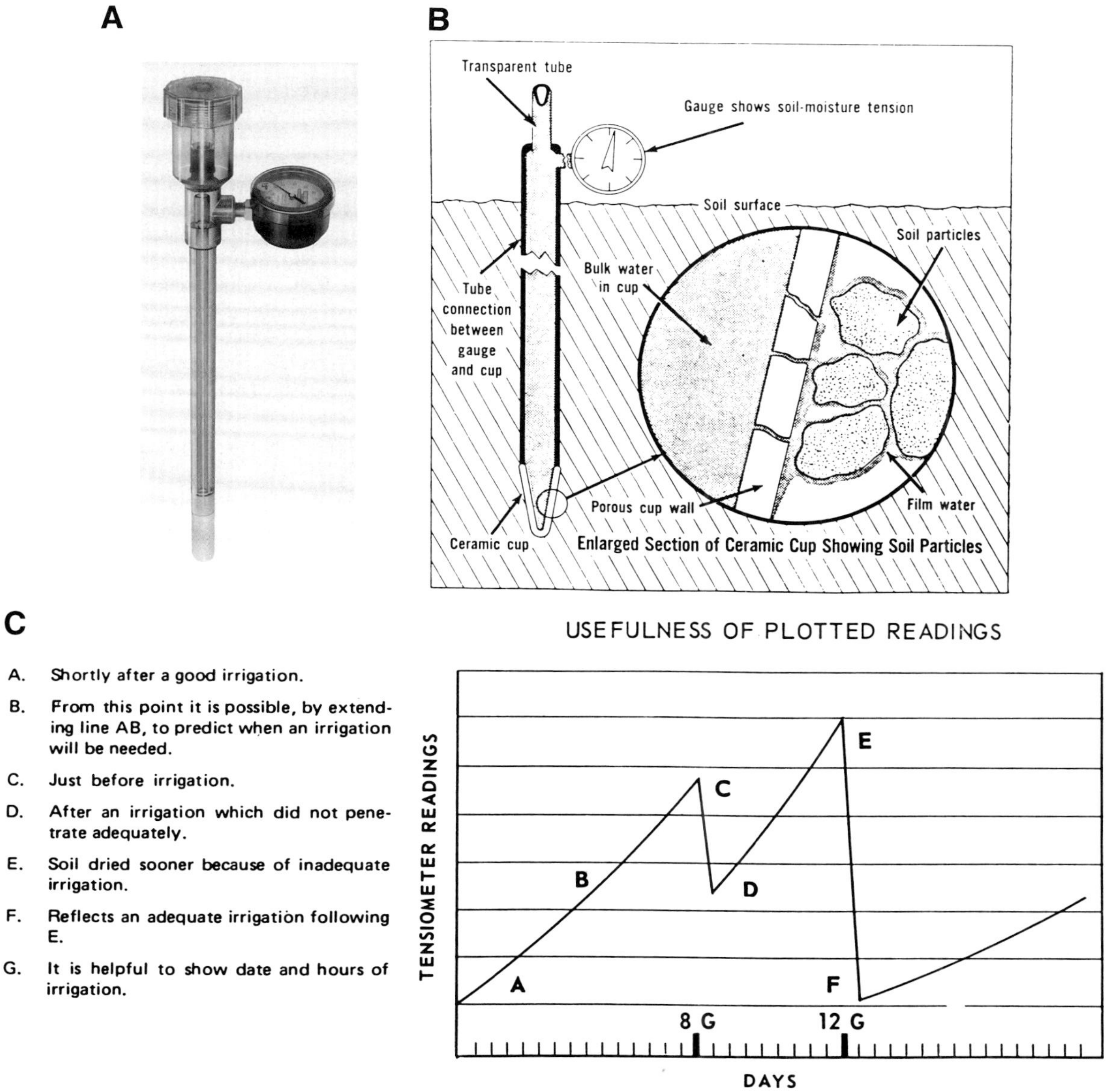

Figure 5.2. Use of the soil moisture tensiometer to record status of soil water in soil. **A,** A tensiometer equipped with a vacuum gauge, which measures negative matric potential (soil suction) in millibars (mb). **B,** Cross section of a tensiometer showing its essential parts: a porous ceramic cup, a connecting tube, and a vacuum gauge. The ceramic cup is porous to water, but its pores are so small that they are sealed to air by the water film when the cup is wet. **C,** Increase in negative matric potential (increase in negative tensiometer readings) with time between irrigations. (A, Courtesy William Pogue, Irrometer Co., Inc., Riverside, California; B, courtesy Richards and Hagen 1958; C, adapted from Marsh 1978)

of soil are maintained at precise matric potential values by adjustment of the height of the source of water (see Duniway [1975a, b, 1976, 1983] for rationale and technology). The tension plates are constructed from 9-cm-diameter Büchner funnels with fine-porosity fritted glass plates (Kimble 28400-90F) over which a layer of filter paper is placed (Figure 5.3) to prevent clogging the fritted glass. The soil is screened (1.4-mm pores), if necessary, and autoclaved (or steamed) to eliminate other pathogens before it is placed in the Büchner funnel. The desired matric potential can be adjusted by regulating the distance between the porous plate and the surface of a water reservoir. The soil suction increases (becomes drier) –1 mb with each 1.002-cm increase in height above the water level. When the height of a growth chamber does not permit raising the soil layer to a sufficient distance above the water source to maintain maximum negative matric potential values, a split column described by Duniway (1976) can be used (Figure 5.4). Matric potential values, given in negative millibars (mb), are equivalent to soil-water suction or tension values obtained with the commercial tensiometer (Figure 5.2A and B). One millibar is equivalent to the pressure exerted by a 1.022-cm-high column of water, and 1,000 mb = 1 bar = 0.99 atmosphere = 100 kPa (kilopascals). Thus, to maintain the soil in a tension plate at –50 mb, the water reservoir should be placed 51.1 cm (50 × 1.022) below the soil level. If a 0-mb matric potential (saturated soil) were desired, the water source would be placed at the same level as the soil. If the soil is to remain flooded, the water source is raised slightly above the soil. After the tension plates are set to maintain a matric potential of –50 mb, soil is added to a depth of 1.5 cm and then wetted with 200 ml of water. The tension plates are adjusted to –250 mb for 12 h before being set at the matric potential desired. After seeds are planted in the soil, the tension plates are loosely covered with polyethylene bags during the test period to reduce evaporation.

This technique and modifications of it, in which soil moisture is maintained either constant or adjusted by periodic flooding, have been used by many investigators to study model systems of the relationship of soil moisture to the fate of zoospores, sporangia, chlamydospores, oospores, and mycelium in soil. Production of sporangia at various soil matric potentials is well documented (P. D. Turner 1965a; Sneh and McIntosh 1974; Duniway 1975a, b, 1976, 1979, 1983; Pfender et al. 1977; MacDonald and Duniway 1978a–c; Weste and Vithanage 1979; Gisi et al. 1980; Kuan and Erwin 1982; Gisi et al. 1983; Weste 1983b; Hardy and Sivasithamparam 1991a, b). See also Figure 5.5, in which data from several investigators on production of sporangia in soil are compared.

Pfender et al. (1977) found that sporangia were produced on alfalfa seedling radicles infected with *P. medicaginis* (*P. megasperma*) in unsterilized soil at matric potential values of 0

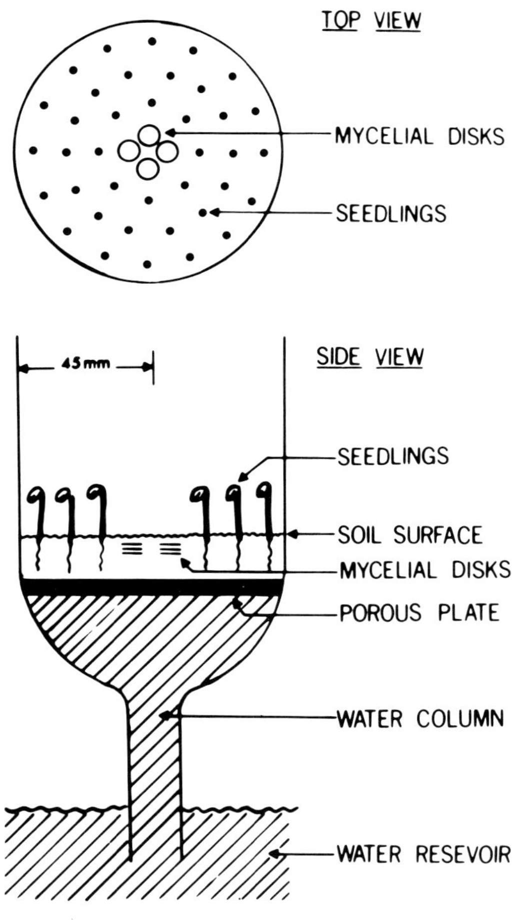

Figure 5.3. Spatial arrangement of safflower seedlings and mycelial disks of *Phytophthora cryptogea* on tension plates. At matric potential = 0, the surface of the water reservoir should be at the level of the porous plate. The water surface should be 5 mm above the soil surface in flooded treatments. For each 1.022 cm that the water level is lowered, the matric potential decreases by 1 mb. This system, with modifications, has been utilized to study the effect of matric potential on root disease caused by several *Phytophthora* species (Duniway 1970, 1983; Stern et al. 1977a, b; MacDonald and Duniway 1978; Bernhardt and Grogan 1982; Kuan and Erwin 1982; Bowers and Mitchell 1990; Bowers et al. 1990; Hardy and Sivasithamparam 1991b). (Reprinted by permission of the American Phytopathological Society, St. Paul, Minnesota, courtesy Duniway 1976)

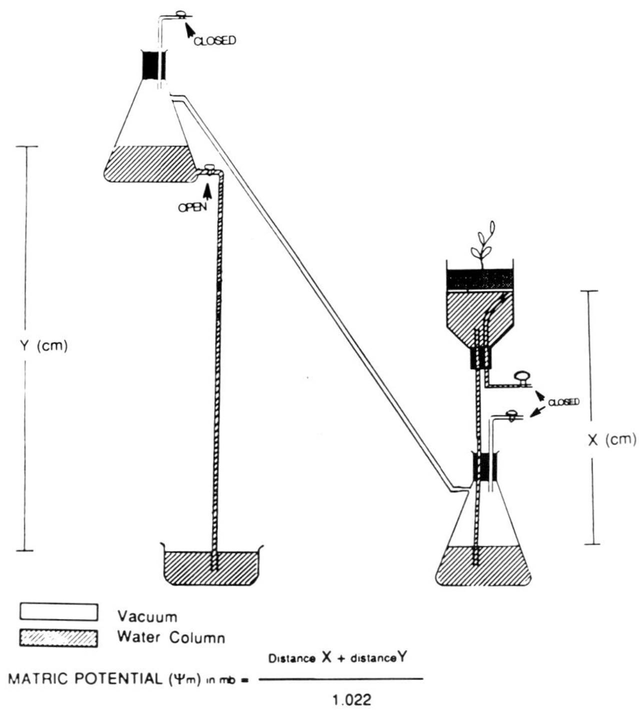

Figure 5.4. Split column method of controlling matric potential of soil in a Büchner funnel. Matric potential is measured in negative millibars (mb). In soil, a matric potential of zero indicates saturation. (Drawn by Ahmad Vaziri from information by Duniway 1976 and Kuan and Erwin 1982)

to –100 mb at 16°C, but very few formed in drier soil at –600 mb. Flooded soil was the most favorable, where the greatest amount of sporangial production occurred at 16°C and none at 28°C or above. Pfender et al. (1977) proposed that perhaps temperature and soil aeration interact to affect sporulation in flooded soil because very little sporangial formation occurred at 28°C, a temperature at which *P. medicaginis* grows well. At higher temperatures, oxygen could become limiting in flooded soil because of increased oxygen consumption by microorganisms and decreased solubility of oxygen in water.

The apparent infection rate (Vanderplank 1963) of Phytophthora root rot of alfalfa seedlings caused by *P. medicaginis* (*P. megasperma* f. sp. *medicaginis*) was much lower at a soil matric potential of –350 mb compared with that in wetter soils at –10 and 0 mb (Kuan and Erwin 1982). The reduction achieved in incidence of seedling disease by maintaining the soil at matric potentials more negative than –50 mb was significant. When oospores were used as inoculum under these drier conditions, the reduced incidence of seedling disease was correlated with a decrease in the percentage of oospores that germinated in soil. More oospores germinated in wet than in drier soil. Both oospores and infected root pieces served as effective inoculum. At 0 mb soil matric potential (flooded), oospores produced sporangia and zoospores, and the sporangia that formed on the infected root inoculum increased the percentage of diseased seedlings (Kuan and Erwin 1982). This

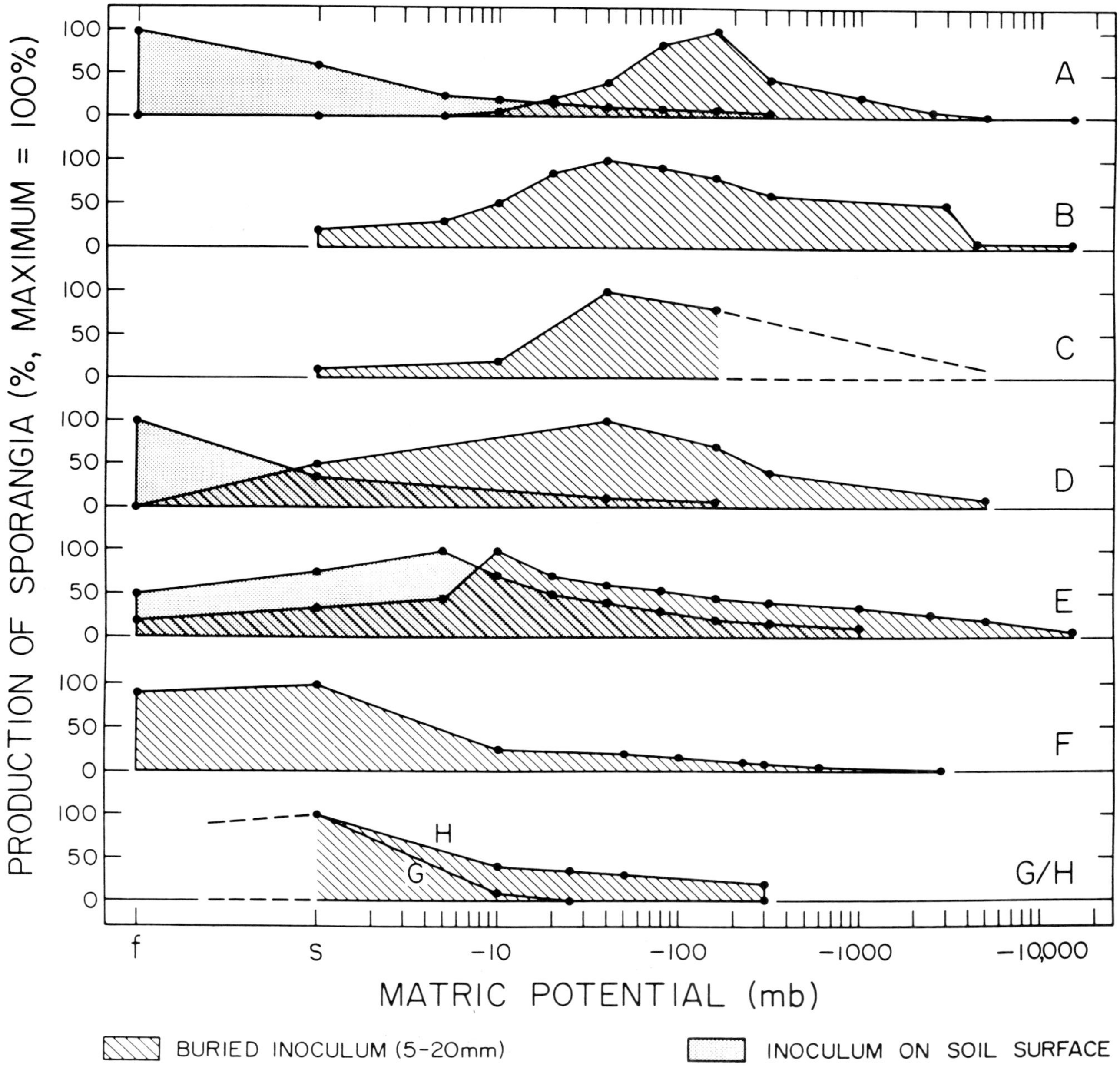

Figure 5.5. Effect of soil matric potential on production of sporangia by *Phytophthora* species, expressed as a percentage of the maximum number (100%) observed after 2–4 days (f = flooded soil; s = saturated soil). **A,** *P. cinnamomi* (Gisi et al. 1980). **B,** *P. cryptogea* (Duniway 1975a, 1975b). **C,** *P. megasperma* (MacDonald 1977; Kuan and Erwin 1982). **D,** *P. cactorum* (Sneh and McIntosh 1974 [inoculum is buried]; Gisi 1975b [inoculum on the surface]; Meyer and Schönbeck 1975). **E,** *P. palmivora* (Gisi et al. 1980). **F,** *P. medicaginis* (*P. megasperma*) (Pfender et al. 1977; Kuan and Erwin 1982). **G/H,** *P. cambivora* (**G**) and *P. drechsleri* (**H**) (Sugar 1977; data from Gisi et al. 1980. Reprinted by permission of the American Phytopathological Society, St. Paul, Minnesota, courtesy Gisi 1983)

increase in production of infective inoculum plus a predispositional effect on alfalfa roots from flooding soil for 3 days before inoculation (Kuan and Erwin 1980b) and the better growth of hyphal tips in wet soil help to explain why diseases caused by *Phytophthora* are more severe in soil with free moisture. When the inoculum was standardized, the apparent infection rate (r), in which disease incidence (x) was transformed to the multiple cycle disease equation $r = \log_e (x/1 - x)$, showed a straight line relationship of increase in disease incidence with wetter soil over time. The apparent infection rate was reduced from 0.24 unit per day at zero matric potential to only 0.06 unit per day in drier soil at –10 mb matric potential (Kuan and Erwin 1982).

P. cinnamomi caused very little root disease on *Persea indica* plants when the soil matric potential was maintained in a sandy loam soil above –250 mb, but the percentage of diseased roots approached 100% at 0, –50, and –100 mb (Sterne et al. 1977a, b). Since the addition of a small amount of asparagine to soil at –250 mb increased the percentage of diseased roots to 86%, it appeared that the low disease occurrence at –250 mb was probably caused by reduced availability of root exudates that stimulate germination of chlamydospores. (See also Mircetich et al. [1968], who showed that asparagine stimulates chlamydospore germination.) In contrast to results with the sandy soil, decreasing the matric potential of clay soil from 0 to –250 mb did not reduce the amount of disease on *P. indica* roots (Sterne et al. 1977a, b).

The responses of apple seedlings to different species of *Phytophthora* at several soil moisture regimes are not always the same. In a glasshouse experiment by Browne and Mircetich (1988) in which soil in large crocks was maintained at –20 mb matric potential, *P. cryptogea* did not cause root rot of apple seedlings, *P. cactorum* caused only mild root rot, and *P. cambivora* caused severe root rot. When soil that had been maintained at –50 mb was flooded (0 mb) for different periods (40–48 h), however, *P. cryptogea* and *P. cambivora* caused increasingly severe root disease while *P. cactorum* caused only moderate disease. Browne and Mircetich (1988) showed that the increase of the flooding period was positively correlated with the increasing severity of disease for *P. cryptogea* and *P. cambivora* but not for *P. cactorum*. They concluded that careful management of soil water might be successful for control of root rot of apple trees caused by *P. cryptogea* but less successful for control of *P. cactorum* and *P. cambivora*.

Data from several reports on the effect of matric potential on sporangial production as summarized by Gisi (1983) (Figure 5.5) show that the production of sporangia in soil at different matric potentials varies with the *Phytophthora* species and experimentally depends on whether the inoculum (agar disks or infected plant parts) is buried or placed on the surface. Sporangia usually form better in buried inoculum when the matric potential is more negative than zero, probably because oxygen, which has a low solubility coefficient in water, is more limiting in free water than at more negative matric potentials near –300 mb, which are favorable for root growth of most plants. These graphs (Figure 5.5) do not explain the total epidemiological picture, however. In order for sporangia to produce zoospores, free water is necessary. This situation occurs when dry soil is reflooded by rain or irrigation.

The matric potential values that favor sporangial formation are variable among species of *Phytophthora*, and results from studies of some species vary among investigators. The latter variability probably results from differences in techniques. For instance, maximum numbers of sporangia were produced by *P. cinnamomi* at –160 mb (Gisi et al. 1980), but in another study maximum numbers of sporangia formed at –15 to –75 mb (Benson 1984) and in another at –100 mb (Hardy and Sivasithamparam 1991b). Because sporangia produce zoospores on roots of container-grown plants only when the matric potential returns to zero (free water) for at least 4 h (MacDonald and Duniway 1978a), Hardy and Sivasithamparam (1991b) suggested that the "stop-start" effect on zoosporangiogenesis could be a useful control strategy. The shorter the time that the soil remains at a matric potential of zero, the less likely it is that sporangia will produce zoospores, and even if zoospores germinate directly by germ tube, the inoculum level will not have increased in the soil. They also suggested that during this stop-start interval, the lytic activity of a high population of microorganisms might have the opportunity to affect and suppress or kill *Phytophthora* propagules.

In testing the pathogenicity of *Phytophthora* isolates or resistance to root rot in a glasshouse, methods that favor pathogenesis must be used. Although some researchers use a continuous flooding period of several days or weeks (Frosheiser and Barnes 1973, 1984; Wilcox and Mircetich 1985a, b), we prefer a method developed by Tsao and Garber (1960) for research on citrus plants in which the period of soil flooding is alternated with a nonflooded treatment that allows drainage of the soil. After the potted plants are inoculated, saucers are placed beneath the pots, and the pots are kept flooded for 3 days. During the next 4 days, the saucers are turned over and the soil is allowed to drain but kept moist as needed. Such a regime allows zoospore release and activity during the flooding period and, during the 4-day drainage period, reaeration of the soil and root mass as the water in the soil pores is replaced by air and sporangia form on infected roots. In the following repeat 3-day flooding cycle, the zoospores release and infect the roots. We have used this method successfully for testing the resistance of alfalfa cultivars to *P. medicaginis* (Erwin et al. 1982).

One of the dangers of continuous flooding to induce Phytophthora root rot is the development of an anaerobic condition. Some root diseases that are caused by anaerobic conditions (anoxia) could be confused with Phytophthora root rot. A serious root disease of alfalfa occurred when land was flooded during periods of high temperatures (38–40°C) in the Imperial Valley of southern California (Erwin et al. 1959). Flooding injury to alfalfa resulting from anaerobic conditions was also shown by Zook et al. (1986).

The important principle learned from the many laboratory studies on the effect of matric potential on sporangium formation is that maintenance of soil at or near zero matric potential is likely to allow a rapid increase in the soil inoculum. Thus, on the basis of laboratory results it can be stated that if soil is at the saturation status long enough, *Phytophthora* inoculum is more likely to increase sufficiently to cause a serious amount of disease on roots.

Field Research

Many examples show that high water tables or excessive irrigation provide the ideal conditions under which various *Phytophthora* species cause root rot of many crops. Excess water provided by a high water table allowed severe infection of citrus tree rootlets by *P. citrophthora* and *P. parasitica* (Klotz et al. 1958). Wager (1942a) and Zentmyer and Richards (1952) concluded that high water tables and excess irrigation in avocado orchards resulted in conditions suitable for zoospore infection of avocado roots by *P. cinnamomi*. In New Zealand, excessively high rainfall during 1956 led to the deaths of established shelter belt trees (*Pinus radiata*) by *P. cinnamomi* (Newhook 1958, 1959, 1988; see the section above, The Susceptible Host, the Virulent Pathogen, and the Favorable Environment). Newhook (1982) states that well-fertilized trees with rapidly growing foliage are killed more often than trees with sparse foliage growing in infertile soils because the more sparse the foliage, the less the demand for water.

Results from much of the field research parallel those obtained from the laboratory closely enough to substantiate the principle that inoculum of *Phytophthora* species increases rapidly in wet soils. Excessive soil water increases the incidence and severity of Phytophthora root rot of alfalfa caused by *P. medicaginis* (*P. megasperma* f. sp. *medicaginis*) (Pulli and Tesar 1975; Wilkinson and Millar 1982; Alva et al. 1985, 1986; Barta and Schmitthenner 1986; Mueller and Fick 1989). Although the effect of high soil moisture on increased severity of root rot caused by many species of *Phytophthora* and on many crops is well known, the critical length of time that field soil must be saturated for severe expression of root rot to occur is still quantitatively unknown (Schmitthenner 1985). Qualitatively, where irrigation is used, soil should not be allowed to remain flooded for more than a few hours, but the precise timing varies from field to field because of different soil types and cropping systems. Where irrigation is practiced for alfalfa production, fields should be leveled so that water does not accumulate in ponds. Ponding of water at the lower ends of alfalfa fields is common. The length of the irrigation run on flooded fields is also important. In the Imperial Valley of California, where many fields are one-half mile in length, the soil at the upper end of the field often remains saturated for a longer period than that at the lower end, especially when the grower stops the irrigation as soon as water reaches the lower end. There is a trend in irrigated agriculture to attempt zero leveling of land to allow as little runoff as possible. Although this practice conserves water, horizontal drainage is impeded, and consequently the soil remains wet for a longer time. This increases the danger that the expression of Phytophthora root rot will be more severe.

Flowing soil water applied for irrigation spread inoculum of *P. parasitica* on tomato from a point source. The incidence of buckeye fruit rot increased linearly with logarithmic increases in distance of 0–68 m downstream from original points of inoculum. Dispersal was less marked in the causation of root rot upstream from inoculum, however (Neher and Duniway 1992).

In a field experiment with tomato and pepper plants at Davis, California, both *P. parasitica* and *P. capsici* were dispersed by furrow irrigation (14-day cycles) up to 70 m from the source of inoculum (V8 juice-vermiculite), which had been placed in the furrow early in the season. Fruit (tomato and pepper) infection increased with increasing distance downstream, but disease severity on roots was highest near the source of inoculum and gradually decreased up to 32 m downstream from the inoculum source (Cafe Filho and Duniway 1995).

Citrus trees should be irrigated in furrows between the rows of trees, so that free water does not contact the trunk (Klotz et al. 1967; Klotz 1978). The lower trunk rot, called gummosis, caused by *P. parasitica* or *P. citrophthora,* can be lethal to a large tree within a single season. Although fibrous root rot is also important, the effect is usually chronic and seldom kills a citrus tree.

Planting row crops on raised beds prevents free water from contacting the plants. Nearly all row crops in California are grown on beds. This allows better drainage after irrigation and prevents free water from contacting the plant directly. Coffey (1987, 1991) and Pegg and Whiley (1987) advocate planting perennial crops on mounds or ridges to increase the drainage quality of the soil.

Newhook (1988) advocates planting strawberries on mounds to reduce the severity of red stele caused by *P. fragariae* in New Zealand and states that root rot of pineapple caused by *P. cinnamomi* in Australia can be prevented by planting on raised beds.

Phytophthora root rot of avocado, caused by *P. cinnamomi,* is a major problem in nearly all areas of the world in which avocados are grown (Zentmyer 1980). *P. cinnamomi* is capable of killing large, bearing avocado trees by destruction of the lateral roots, which occur mainly in the upper level of soil. Attempts to control this disease by irrigation management (lengthening the interval between irrigations) has been fraught with difficulty, partly because avocados are often grown on steep hillsides or on decomposed granite soils and irrigation water tends to collect in the planting holes. The planting of trees on mounds about 1–1.5 m in diameter and 0.5–1 m in height is experimentally somewhat successful but not widely practiced in California. This practice not only places the root system above the surrounding area, thus allowing better drainage, but also assures that the roots are in friable and noncompacted soil, thus allowing better root growth (Coffey 1987, 1991; Pegg and Whiley 1987). This practice also allows incorporation of organic matter such as poultry manure, which has been shown to be beneficial in some areas of Australia (Broadbent and Baker 1975; Shea and Broadbent 1983).

A comparative physiological study of water relations of healthy (no Phytophthora root rot) and diseased (roots infected with *P. cinnamomi*) avocado trees showed that the xylem pressure potential, measured by the pressure chamber method (Elfving et al. 1972), was reduced markedly in trees infected by *P. cinnamomi.* Trees stressed by low soil moisture recovered during the night when transpiration was reduced, but trees infected by *P. cinnamomi* did not recover (Sterne et al. 1978).

Root infection of safflower by *P. drechsleri* also greatly increased the resistance to water flow in the xylem of stems. Duniway (1975, 1977) noted that the loss of root surface caused by *P. cryptogea* infection was not significant enough to explain the total resistance to uptake of water. The physiological mechanism of these effects is not well understood. The evidence for wilt-inducing properties of certain glucans polysaccharides produced by *P. cinnamomi* and other species is discussed in reports by Keen et al. (1975), Halsall (1978), and Woodward et al. (1980). Woodward et al. (1980) reviewed the effects of β-glucans and β-glucans hydrolases as wilt-inducing toxins in plants.

Furrow irrigation of citrus trees where soil had dried to −500 to −700 mb matric potential favored root and tree growth and reduced the amount of root damage by *P. parasitica* when compared with irrigation when the matric potential readings were lower (more frequent irrigation) (Feld et al. 1990). Drip and furrow irrigation also had different effects on the population of *P. parasitica* in the field. We quote from the report of Feld et al. (1990) (reproduced by permission of the American Phytopathological Society, St. Paul, Minnesota, courtesy of Feld et al. 1990):

> During a 2-year field study in a 70-year-old citrus orchard, the soil population of *P. parasitica* was significantly lower under drip irrigation than under furrow irrigation; however, this difference was evident only during the summer months when the fungus was most active. The abundance of citrus roots under furrow irrigation increased with distance from the furrow center. Citrus roots under drip irrigation were uniformly distributed in the area wetted by the emitter. Distribution of *P. parasitica* propagules correlated directly with the distribution of feeder roots.
>
> In both greenhouse and lathhouse, Phytophthora root rot of citrus seedlings was most severe where seedlings were watered by drip

irrigation or kept constantly moist by furrow irrigation. Seedlings that received water by furrow irrigation that allowed the soil to dry sufficiently between irrigations were able to produce more feeder roots than seedlings grown under drip irrigation. Data obtained from greenhouse and lathhouse studies indicate that the commonly recommended method of furrow irrigation that allows the soil to dry to –50 to –70 cb [–500 to –700 mb] matric potential between irrigations favors the host plant and reduces damage by *P. parasitica* more than the other irrigation methods investigated.

Experimental evidence by Lutz et al. (1989) also indicates that modification of irrigation treatments decreases the population of propagules of *P. parasitica* and increases the health of roots.

According to Lutz et al. (excerpted by permission of *The California Grower,* courtesy of Lutz et al. 1989):

> Citrus plants in the greenhouse irrigated with longer time intervals between irrigations supported lower populations of *P. parasitica* and had healthier root systems compared to plants irrigated with the same volume of water, but with shorter time intervals between irrigations.
>
> Reduction in the time interval between irrigations caused the soil to be saturated more frequently. This may have increased the number of opportunities suitable for production of sporangia prior to saturation and release of zoospores after saturation, which would increase the inoculum level. It appeared that the number of times soil reached saturation rather than the length of time it remained saturated was the most important in determining disease severity in the pot experiments.

Although reduction of the soil water status or frequency of irrigation often induces a measure of control of Phytophthora root rot of citrus trees, an integrated pest management experiment indicated that reduction of soil water does not always produce maximum yields. A large, complex experiment by Menge et al. (1990) on 35-year-old Washington navel orange trees in the San Joaquin Valley in California in which soil moisture, fertilizer, Phytophthora fibrous root rot (caused by *P. parasitica* in summer and *P. citrophthora* in the winter), and insect control were monitored and assayed showed that the highest production level (12% yield increase of orange fruit) occurred when soil water was 120% of the evapotranspiration demand of trees. Menge et al. (1990) had anticipated from previous data (see Lutz et al. [1989], summarized above; Feld et al. 1990) that this treatment would exacerbate fibrous root rot and decrease yield. Treatments with a soluble systemic fungicide (metalaxyl) and a soluble nematicide injected into the irrigation drip system line increased yield by 10%, and spray treatment to control citrus red mites increased the yield 9%.

Phytophthora root rot of soybean caused by *P. sojae* (*P. megasperma* f. sp. *glycinea*) is often a problem in soils in which there is a perched water table or impedance of vertical drainage and in low areas with inadequate horizontal drainage. A brief report from Ontario, Canada, showed that artificially packing field soil from a bulk density of 1.2 to 1.49 g/cc increased the percentage of soybean plants affected by Phytophthora root rot to 45% compared with only 2.5% at the bulk density of 1.2 g/cc. The effect was similar when the soil was maintained at a low, unspecified moisture tension (J. M. Fulton et al. 1961). In Illinois, compaction of soil with a tractor increased the number of soybean plants killed by Phytophthora root rot and also decreased the yield of the susceptible cultivar Corsoy (Gray and Pope 1986; Moots et al. 1988). Thus, soils of high bulk density, which are usually of fine particle clay content and hold more water, are more conducive to Phytophthora root rot (J. M. Fulton et al. 1961; Kittle and Gray 1979a; Schmitthenner 1985; Gray and Pope 1986). Even the practice of minimal tillage (i.e., the land is not plowed before planting), which has become widely used in the middle west part of the United States as a soil conservation measure, increases the bulk density of soil and correspondingly increases the severity of Phytophthora root rot (Schmitthenner 1985; Schmitthenner and Van Doren 1985).

Thus, it appears that compressing the soil could set up an anaerobic condition. An effect of anaerobic conditions was shown by a report of Murch and Paxton (1979), who infused nitrogen gas throughout the soil. This treatment suppressed the resistance of a soybean cultivar by repressing glyceollin, a phytoalexin in plants.

A recent report from Queensland, Australia, indicated that the severity of Phytophthora root rot of soybean was essentially the same with continuous irrigation as with conventional irrigation in which irrigation in the furrows was made after 60 mm of water had evaporated from a pan (pan evaporation method; adjusted for rainfall). The continual irrigation consisted of maintaining a trickle of water in the furrows between two beds (1–2 m wide). This comparison included the cultivars Nessen, resistant to race 1, which was predominant; Davis, tolerant; Bragg and Forest, moderately tolerant; Wills, Semstar, and Fitzroy, susceptible; and Ross, highly susceptible. The percentages of diseased plants of each cultivar, related to the two irrigation regimes, were not significantly different; however, the percentages of diseased plants reflected the levels of resistance and tolerance of the particular cultivar (Troedson et al. 1991). In light of our previous discussion, it is difficult to explain why Phytophthora root rot is not more severe with continuous irrigation than with intermittent irrigation. However, the matric potential would likely be more negative than zero (free water) at the plant level because the trickle irrigation water in the furrows would have to move upward by capillarity to plants growing in the bed (see equation for relationship of height above the water table on matric potential in Figure 5.4). Agronomically, continuous irrigation was found in previous work to be advantageous because it enhanced nodulation by *Rhizobium* spp., nitrogen fixation, and stimulation of photosynthesis through enhanced leaf water status. We can conclude that continuous irrigation did not make the plants more susceptible. Troedson et al. (1991) pointed out that because the plants were on beds and the moving irrigation trickle was in the furrow, the soil did not become saturated and thus anaerobic. This indicates that both irrigation methods were equally conducive to the maximum amount of disease development commensurate with the genetic resistance of the cultivar. Although the experiments did not include a treatment utilizing even fewer irrigations, it is possible that such a treatment might have resulted in lower disease severity.

Increased frequency and duration of irrigation accelerated the development of an epidemic caused by *P. parasitica* on tomatoes (Ristaino et al. 1988).

As stated above, the extensive, integrated pest management field experiment on orange trees reported by Menge et al. (1990) showed that irrigation, equated with 120% of water and based on the pan evaporation demand method, produced the highest yield of fruit when compared with less frequent irrigations. These field studies indicate the necessity in agriculture of balancing the effects of different treatments against the expected yield of the crop. What is important is to prevent a total loss, for instance, death of a large percentage of plants. When root rot affects fibrous roots, new roots continually replace decayed roots if the main roots are not infected. If the new roots replace decaying roots rapidly enough, the crop often yields well despite disease. As Newhook stated (1988; see previous section), if the collar rot phase occurs, the plant is usually killed; however, if only fibrous root rot occurs, the disease may not be debilitating if new roots regenerate rapidly enough.

In summary, root diseases caused by *Phytophthora* can be alleviated to a large extent by proper irrigation practices, but on a field basis, control is often only partial at best. When the main source of water is rainfall, attention to optimal horizontal as well as vertical drainage in soil and prevention of ponding add to the degree of control.

In container-grown crops, such as ornamentals, soil often remains saturated for a long time. To alleviate this problem, most efficient growers have replaced field soil with a mixture of sand, redwood shavings, vermiculite, perlite, or peat moss as well as other components that increase the porosity of soil and allow better drainage of water. The development of this approach, which utilizes fine sand-peat moss mixtures, is thoroughly discussed by Baker and his colleagues (1957).

Hardwood bark composts not only have excellent drainage characteristics but also have a biologically suppressive effect on *Phytophthora* (Hoitink et al. 1977; Daft et al. 1979; Hoitink 1980; Hoitink and Fahey 1986). Hardwood bark composts are commonly used in the eastern United States. According to H. A. Hoitink (*personal communication*), they have replaced peat moss in many of the soil mixes used in glasshouse and nursery operations. The interaction of soil porosity and irrigation management with the use of hardwood bark composts is discussed below in Potting Mixtures for Nursery Use.

The Hardpan Effect in Agricultural Soils

A hardpan or, for that matter, any change in the texture of soil in layers below the surface creates an interface that either temporarily (in the case of sandy layers) or permanently (clay pans) impedes the downward percolation of soil water. Even the bottom surface of a pot with drainage holes and filled with a permeable soil mix creates an interface that temporarily impedes the downward percolation of soil water. This impedance extends the time that the soil remains saturated and provides an environment that favors the production of sporangia and the release of zoospores into the root zone. When the soil moisture eventually reaches the "optimal" field capacity (matric potential of about –300 mb), plants can extract water, but *Phytophthora,* in most cases, produces fewer sporangia and zoospores (Figure 5.5).

Phytophthora root rot of avocado is severe in southern California on soils underlaid by a hardpan. Burns et al. (1960) showed that Phytophthora root rot of avocado was least severe in Vista soil, which had adequate internal drainage, but was more severe in Merriam soils, which had been developed on old alluvial deposits and were underlaid by a compact clay pan horizon. Consequently, this soil is considered to be hazardous for production of avocado. Zentmyer and Ohr (1978) list several such soil types that are hazardous for planting avocado in California.

In California, fibrous root rot of citrus, caused by *P. parasitica* during the summer and *P. citrophthora* during the winter, occurs on land near the foothills of the Sierra Nevada mountains in the San Joaquin Valley, which is underlaid by hardpans 18–24 inches below the surface. These hardpan layers in soil limit the availability of water, oxygen, and soil nutrients. From the standpoint of root disease, these relatively impervious layers prevent vertical drainage of water. Lutz et al. (1986) discuss the various types of hardpans and conclude that healthy citrus trees can be grown on land underlaid by a hardpan with careful management utilizing modern drip or minisprinkler irrigation systems and monitoring of soil moisture with tensiometers. The population of *P. parasitica* in rhizosphere soil was more than twice as high when soil was irrigated when tensiometer values reached –350 mb (slightly drier than field capacity) than when irrigation was limited to matric potential values reaching –700 mb (Lutz et al. 1989).

The soil hardpan effect that occurs in some areas of the native jarrah (*Eucalyptus marginata* Sm.) forest of Western Australia is a classic example and is described in fascinating popular articles by Shea et al. (1984) and Shea (1988).

Shea et al. (1984) and Shea (1988) give a history of the decline of the jarrah forest from its outset in 1965 to 1988 (see Chapter 15A). Although the death of large jarrah trees was attributed to *P. cinnamomi* as early as 1965 (Podger et al. 1965), this explanation was somewhat controversial because it was difficult to explain physiologically how the death of fibrous roots that occurred only in the upper layer of soil and only during a certain period of the year, when rain was plentiful, could cause the death of large trees with roots deep in the subsoil.

Shea et al. (1984) stated, "The surprising conclusion from these studies was that despite the fact that the fungus caused mass destruction of the jarrah forest on some upland sites, all measurements of the soil environment and the activity of the fungus in the surface layers on these sites indicated that the trees should not be subject to severe disease. Two of the most perplexing questions which confronted jarrah dieback research workers were: how did *Phytophthora cinnamomi* operate in a soil environment which was only marginally favorable, and how could it cause so much damage at apparently low soil population levels?"

When Shea and his colleagues found that, even in a wet year, they could not detect *P. cinnamomi* consistently in fibrous roots, even those who strongly supported the concept that *P. cinnamomi* was the cause of the decline began to waiver (Shea 1988)! Shea then proposed that maybe they had been looking for *P. cinnamomi* in the wrong place. Immediately they excavated a recently killed tree with a backhoe and washed the soil off the entire root system. Unexpectedly, they found large lesions, typical of those caused by *P. cinnamomi*, on the large vertical roots that had penetrated the hardpan layer some 10–70 cm below the forest floor. When isolation from these roots yielded *P. cinnamomi* cultures within 48 h, Shea employed forest crews equipped with backhoes and fire-control tankers with high-pressure water hoses to excavate many more large trees. In repeated experiments, *P. cinnamomi* was isolated from lesions on large vertical roots that had penetrated the hardpan layer. Giant "Rocsaws" from a mining company cut cross sections of the soil profile to expose the vertical holes in the hardpan layer where the roots of jarrah trees had penetrated (illustrated in Shea et al. [1984]; Shea 1988). They later showed that soil water that collected in the holes allowed production of sporangia and zoospores and infection of the vertical roots (Shea et al. 1982, 1983) (Figure 5.6). Shea et al. (1984) and Shea (1988) describe the immense, expensive effort to determine the effect of *P. cinnamomi* on the large vertical roots that maintained the tall trees during the dry summer months. Numerous excavations and isolation of *P. cinnamomi* from lesions on large vertical roots finally proved that the increase in free water in the hardpan layer was the environmental condition necessary for *P. cinnamomi* to kill the large jarrah trees.

Although recovery of *P. cinnamomi* from soil at these depths had not been reported previously (Shea et al. 1983, 1984), vertical roots were found to be infected only where they had penetrated the hardpan. Shea and his colleagues (1983) found that *P. cinnamomi* produced sporangia at high populations only near the hardpan layer because of the retention of soil water. They determined that this layer was favorable for reproduction of *P. cinnamomi* by placing autoclaved and inoculated leaf disks of *Banksia grandis* (highly susceptible) in gauze envelopes. When these leaf disks were incubated at different depths ranging from just below the surface of the soil down to the layer immediately above the hardpan (from 3 to 75 cm deep), sporangia formed only in the zone just above the hardpan. This finding proved that only the hardpan layer was wet enough to support the development of *P. cinnamomi* at sufficient concentrations to rot and kill the large vertical roots that supported the trees (Shea et al. 1983).

The relationship of the hardpan to the killing of vertical roots that have penetrated the concreted lateritic layer is shown in Figure 5.6. Shea et al. (1985) state, "Whereas fungal invasion of a horizontal root (unless it takes place near the trunk) appears to have little impact on the tree's capacity to obtain water, the fungus has only to invade a small distance into a vertical root to cut off the water supply. Once the tree's vertical roots are destroyed, its unsated demand for water results in desiccation and death." We refer the reader to the popular article "A New Perspective on Jarrah Dieback" by Shea and colleagues (1984) and to a more recent article by Shea (1988) for a fuller treatment of this fascinating story.

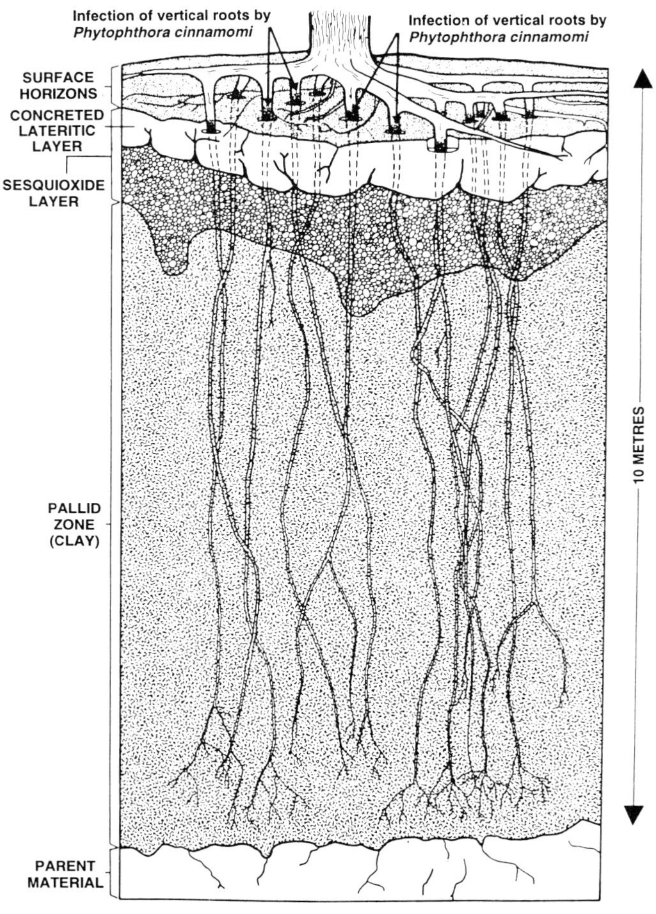

Figure 5.6. Vertical root system of a jarrah tree (*Eucalyptus marginata*) penetrating a soil hardpan in Western Australia. Jarrah's ability to tap the groundwater at great depths permits it to use water at a high rate during summer months. But on specific sites, water collects just above a concreted lateritic layer (hardpan), allowing *Phytophthora cinnamomi* to produce sporangia and zoospores that infect and kill the vertical roots. This in turn, causes the trees to die during the dry summer. (Reprinted by permission of the Department of Land Management of Western Australia, Crawley, Australia, courtesy Shea et al. 1983; Shea 1984)

What use can be made of such information? The understanding that only certain sites underlaid by a concreted lateritic layer were particularly conducive to Phytophthora root rot of jarrah trees encouraged the continued use of sanitation to prevent movement of soil from infested to clean areas of the forest on trucks or any wheeled vehicle. Wash racks were set up to remove soil before traffic proceeded into clean forest areas. All of these procedures are expensive and had to be justified politically. Although control of severe diseases in forests is difficult, forest managers in Western Australia have set up programs to eliminate the most susceptible understory species, such as *B. grandis,* on which inoculum of *P. cinnamomi* builds up rapidly, and to plant resistant leguminous trees, such as *Acacia* species, in infested sites. This may have a retarding effect on the buildup and persistence of *P. cinnamomi.*

The discovery that *P. cinnamomi* had invaded large roots as well as fine feeder roots of jarrah trees (Dell and Wallace 1981; Marks et al. 1981; Shearer et al. 1981; Shea et al. 1982, 1983) caused a shift in research emphasis on the jarrah dieback problem from studies of fine root susceptibility to studies of root and collar invasion. Tippett et al. (1983) reported that spread of *P. cinnamomi* hyphae was most rapid in the outer phloem tissue, in which the expanded parenchyma cells are loosely packed. Necrophylactic periderm tissue that formed during winter and spring inhibited progress of *P. cinnamomi;* however, the fungus broke out during the summer to invade new areas of phloem tissue. Phloem tissue was most susceptible when the amount of moisture in the bark was highest. Such a condition occurs when the roots have a maximal supply of water or the trunk is kept wet by rain or irrigation (Tippett et al. 1987).

In subsequent studies in Western Australia, Kinal et al. (1993) found that in certain soils in the jarrah forest, *P. cinnamomi* inoculum (shown to emanate chiefly from zoospores) moved downward in the subsurface gravelly layers at depths greater than 2 m. The propagule density in water collected at the lower end of the slope averaged 60 propagules per liter (Kinal et al. 1993).

Several interesting, popular, and well-illustrated summaries of the *Phytophthora* epidemic in the jarrah forest are given by Shea (1975b, 1988), Shearer and Bailey (1989), and Shearer (1990).

Predisposition

Predisposition to disease is caused by environmental factors (nongenetic) that affect plants prior to infection in such a way that susceptibility to disease is increased (Schoenweiss 1975). Understanding this concept is important because plants considered to be resistant under ordinary circumstances can become susceptible under predisposing conditions. The concept should be distinguished from the effect of environmental factors that tend to increase the population of the pathogen. An experiment to determine whether an environmental effect is predisposing should include controls (nontreated plants and noninoculated plants) and treatments with inoculation alone, the predisposing factor alone, and the predisposing factor plus inoculation.

Most research on predisposition caused by environmental factors has been conducted on the effects of drought stress, preinoculation flooding, and soil salinity (Duniway 1983). The sensitivity of safflower to drought stress prior to irrigation was noted shortly after the development of safflower as an irrigated oil seed crop in California during the early 1950s. The senior author personally remembers in 1951 observing a field of safflower (which is relatively drought resistant) in the upper San Joaquin Valley of California that had just been flood-irrigated after a long dry period. A few days later, nearly all of the plants in the entire field were killed by root rot caused by *P. drechsleri*. At that time, Paul Knowles, an agronomist at the University of California, Davis, observed that such a problem could be avoided by maintaining the field in a moist condition but not allowing it to become dry. This advice is still useful, and subsequent research has validated this observation (Knowles et al. 1965).

Root rot of safflower caused by *P. drechsleri* was shown experimentally to be much more severe when flood irrigation was applied to the crop after a period of drought stress (Zimmer and Urie 1967). As stated above, entire fields were killed within a week after flood irrigation. This phenomenon of predisposition by drought stress was later demonstrated quantitatively under controlled soil moisture conditions in glasshouse experiments. Water stress was induced in safflower plants of the susceptible cultivar N-10, growing in crocks (clay pots) of soil, by withholding water. When leaf water potential was reduced from a range of –4 to –6 bars to –13 to –17 bars (drier) before inoculation, the disease severity was much greater on stressed plants than on plants that had been watered daily. Resistance was also overcome by drought stress. When the resistant cultivar Biggs was stressed similarly, the severity of root rot increased and root weight decreased (Duniway 1977b). The increase in severity of disease was associated with an increase in resistance to water transport in the plant (Duniway 1977a).

Drought stress in 3- to 4-week-old tomato plants (leaf water potential was reduced to –9 bars) before or after inoculation of roots with zoospores of *P. parasitica* increased the severity of root rot symptoms compared with that of unstressed plants. The lengths of diseased roots and the numbers of colonies of *P. parasitica* that could be detected on the roots were also increased markedly by preinoculation drought stress. In plants that were nearing maturity, preinoculation water stress increased the subsequent disease severity, but postinoculation water stress did not. Water stress without inoculation did not materially affect root length or fresh weight of plants (Ristaino and Duniway 1989).

The physiological reason that drought stress makes plants more susceptible is not known with certainty. Ristaino and Duniway (1989) cite references that show that physiological processes such as expansive growth and net photosynthesis and increased root-shoot ratios (Hsiao 1973; Sharp and Davies 1979) could be involved. The mechanisms are too complex to be fully explained. There are reports that desiccation of roots increases the release of amino acids into the soil (Katznelson et al. 1954, 1955). This increase could favor the germination of oospores or chlamydospores or increase the attractiveness of roots to zoospores, but this hypothesis has not been proved.

Ornamental rhododendron plants, grown in cans for resale, were predisposed to root rot (*P. cinnamomi*) by both water stress and salinity (Baker and MacDonald 1981). In this case, severity of disease on the susceptible cultivar Purple Splendor was not increased by water stress before inoculation, but the normally resistant cultivar Caroline became highly susceptible after preinoculation water stress (leaf water matric potential was reduced from a range of –2 to –6 bars to –16 to –20 bars).

Salinity stress increased the susceptibility of chrysanthemum (*Chrysanthemum morifolium* Ramat.) to root rot caused by *P. cryptogea*. Plants were subjected to salinity by adding NaCl (0.1 M and 0.2 M concentrations) to the soil prior to inoculation with zoospores. On the stressed plants, 70–88% of the roots had lesions compared with only 20% on unstressed plants (MacDonald 1982, 1984). When chrysanthemum roots were given a 24-h pulse treatment with 200 milliequivalents of sodium chloride per liter in Hoagland and Arnon (1950) nutrient solution and subsequently inoculated with zoospores, roots of stressed plants were penetrated deeply by *P. cryptogea*, but on nonstressed plants, *P. cryptogea* penetrated only to a depth of three to four cell layers within 6–12 h after inoculation (Swiecki and MacDonald 1988). Excellent photomicrographs of the infection process indicated that salinity inhibits active root defense systems. Phytoalexins have not been implicated in resistance of chrysanthemum plants; however, in soybean roots resistant to *P. sojae* (*P. megasperma* var. *sojae*), salinity repressed the production of the phytoalexin glyceollin (Murch and Paxton 1980). Although soil salinity enhanced the severity of root rot of tomato, the asexual reproduction of *P. parasitica* was hindered by soil salinity (Swiecki and MacDonald 1991). Infection of safflower roots by *P. cryptogea* increased the amount of sodium uptake even though the degree of infection was moderate. Weicht and MacDonald (1992) concluded that in breeding plants for salt tolerance, the possibilities for the effect of root infection should also be taken into consideration. High soil salinity significantly increased the severity of root rot of tomato caused by *P. parasitica* (Snapp et al. 1991).

High soil salinity increased the susceptibility of citrus rootstocks Troyer citrange, sour orange, and rough lemon to stem rot caused by *P. citrophthora*. Salinity also reduced the accumulation of the phytoalexin 6,7-dimethoxycoumarin (Sulistyowati and Keane 1992).

Excess water in soil, applied for 3 days before inoculation with zoospores, predisposed alfalfa plants to root rot caused by *P. medicaginis* (*P. megasperma* f. sp. *medicaginis*) (Kuan and Erwin 1980b). The mechanism of predisposition caused by flooding appears to some extent to be associated with oxygen depletion in saturated soil. Since the solubility of oxygen in free water is low, filling the soil pore spaces with water will cause much less oxygen to be available to roots than in soil in which the pore spaces are not filled with water. The alfalfa roots from soil flooded for at least 3 days were much more attractive to *P. medicaginis* zoospores (chemotaxis) than roots from unflooded soil (Kuan and Erwin 1980b; Chapter 3, Figure 3.3). Increased electrical conductivity and concentration of sugars and amino acids in root exudate indicated that roots from flooded soils were more leaky than roots from non-flooded soil.

The presence of aerenchyma tissue in flood-stressed plants of many species is a common phenomenon (Drew and Lynch 1980). Ethylene was produced in soil in which alfalfa roots were kept flooded for 3 days; this caused the production of aerenchyma (tissue that contains large air spaces) in the xylem tissue of flooded plants. Ethylene caused these roots to become more susceptible to *P. medicaginis* (Zook and Erwin 1986).

Barta and Schmitthenner (1986) compared a 6-day flooding period before inoculation with zoospores to a 3-day flooding period after inoculation on a series of alfalfa cultivars ranging from susceptible to resistant. The most severe disease expression occurred with postinoculation flooding. They concluded that either postinoculation flooding caused an increase in inoculum or it negatively affected root metabolism. The severity of Phytophthora root rot (*P. medicaginis*) was greater under postinoculation flooding conditions among 10-week-old plants clipped (foliage removed) before inoculation than among unclipped plants. Theoretically, in clipped plants, less oxygen would be available to translocate downward to the roots.

To mimic the effects of waterlogged soil, Fraedrich and Tainer (1989) maintained low oxygen levels (0–0.25 mg/liter for 6–24 h) in nutrient culture of loblolly pine (*Pinus taeda* L.) and shortleaf pine (*P. echinata* Mill.) by artificial dilution of oxygen with nitrogen. Reduction of oxygen to <0.25 mg/liter for 6–24 h increased the susceptibility of roots to *P. cinnamomi.* When concentrations of oxygen were maintained at 0.25–1.0 mg/liter in nutrient solution for 12- and 24-h periods, plants were not predisposed to infection. Low oxygen concentration appears to affect root susceptibility more than it affects the pathogen (Fraedrich and Tainer 1989).

Although flooding before inoculation predisposed cherry trees to Phytophthora root rot (Wilcox and Mircetich 1985a), Fraser fir seedlings were not predisposed by preinoculation flooding (Kenerley et al. 1984b). The lack of a predisposing effect may mean that the susceptibility of the host was so great and/or the virulence and inoculum level of the *Phytophthora* isolate was so high that the effect of the predisposing factor was not measurable.

Stems of young *Eucalyptus sieberi* L. A. S. Johnson trees were more susceptible (larger lesions were produced by *P. cinnamomi*) when maintained at a high plant water potential than when drought stressed (I. W. Smith and Marks 1986). Tippett and Hill (1983) showed that increased water in phloem tissue allowed *P. cinnamomi* to invade *E. marginata* trees more readily. The physiological reason for this is not well understood. Newhook (1988) notes that aging has a predisposing effect on susceptibility of *Pinus radiata* trees to *P. cinnamomi.*

In pots containing chrysanthemums, soil temperature often rises to 45–47°C when pots are exposed to sunlight, whereas the soil temperature of pots in shade does not exceed 34–36°C. The severity of root rot caused by *P. cryptogea* increased significantly in pots that had been subjected to the higher temperatures (heat stress). As short a time as 30 min of exposure to a soil temperature of 40–45°C increased susceptibility. This response was similar in controlled hydroponic culture and in containers in the nursery (MacDonald 1991).

Reduction of vigor of loblolly pine resulting from heavy, poorly drained soils and decreased aeration rendered trees more susceptible to *P. cinnamomi,* the cause of little leaf disease on pine (Zak 1961).

Strawberry transplants that had been injured by frost were more susceptible than nonfrozen plants to crown rot caused by *P. cactorum.* Uninjured plants became susceptible only when wounded (Lederer and Seemüller 1992).

Experimentally, incubation of zoospore-inoculated soybean hypocotyls at temperatures above 27.5°C increased the susceptibility of a cultivar with race-specific resistance to *P. sojae.* In this case, the production of glyceollin (the phytoalexin associated with resistance) remained high (1,559 μg per gram, fresh weight, of hypocotyl tissue) (Ward and Lazarovits 1982). Incubation of the inoculated hypocotyls at 25°C resulted in a resistant reaction. Although glyceollin production in the resistant cultivar was about twice as high as it was at 27.5°C, the difference was not considered sufficient to account for the loss of resistance, especially since the glyceollin concentration in hypocotyls incubated at 20°C (which retained resistance) was 1,059 μg per gram of hypocotyl tissue (even at 27.5°C). Ward and Lazarovits (1982) postulated that this result provided circumstantial evidence that glyceollin may not be the only essential factor that conditions resistance.

Welty et al. (1980) showed that the incidence and severity of Phytophthora root rot increased when alfalfa plants were coinoculated with *Meloidogyne hapla* or *M. incognita* (Kofoid and White) Chitwood. Infection of alfalfa roots by the root-knot nematode *M. hapla* prior to inoculation with *P. medicaginis* increased the amount of infected root tissue of alfalfa on the cultivar Deseret from 25.3 to 61.3%. In correlation, infection by the nitrogen-fixing bacterium *Rhizobium meliloti* was also increased by nematode infection. The physiological alteration of root tissue by *M. hapla* infection and increased root exudation may have caused the increased susceptibility. The nodules of the *P. medicaginis*-resistant cultivar Apollo II were more susceptible in the presence of *M. hapla* (Gray et al. 1990).

Tobacco plants that are resistant to *P. parasitica* var. *nicotianae* (*sensu* Tucker 1931) became susceptible in the presence of infection by *M. incognita,* the cause of root knot disease (Powell and Nusbaum 1960; Powell 1979).

The age of leaves can be a predisposing factor to infection of potato by *P. infestans.* The degree of infection of leaves (15–46 days old) and plants (18–53 days old) decreased with leaf and plant age; however, the ability of the fungus to sporulate increased slightly with increase in leaf and plant age (Rotem and Sari 1983). The findings of Rotem and Sari (1983) on the predisposing effect of age on infection of leaves are similar to those of Lowings and Acha (1959) and Lapwood (1961a, b).

As Duniway (1983) points out, the "inoculum of those few species [of *Phytophthora*] that have been examined under stress can, nevertheless, effectively persist through the stress cycles that cause predisposition." Of course, the flooded soil or saturated soil conditions that usually relieve a drought or salinity stress, and which may also predispose a host, are also suitable for zoospore release and dispersal (Chapter 3; Figure 5.5).

Research has tended to emphasize the effect of physical factors such as water, drying, and temperature on the pathogen (see Chapter 3 for the numerous examples of such effects on formation of sporangia) but not on the host. The assumption has been that what happens to the pathogen will be sufficient to reduce the effect of the disease. The results of research on predisposition to disease caused by physical factors show that this assumption may be in part fallacious, however, and that conditions that affect the susceptibility of the plant should also be considered in greater depth (Duniway 1983). This conclusion is similar to that of Fraedrich and Tainer (1989).

Predisposition by physical factors such as salinity, drying, or excess water needs to be more thoroughly researched at the physiological level in more *Phytophthora*-crop systems. There is a danger in accepting the assumption that root diseases caused by *Phytophthora* can always be controlled in the field simply by a drastic reduction in irrigation water.

Effects Caused by the Pathogen

Inoculum: Its Survival and Epidemiological Impact in Soil

In the absence of host plants, most species of *Phytophthora* do not survive as long as fungi that are more saprophytic (i.e., have the ability to colonize nonliving organic material) such as *Fusarium oxysporum* or the more closely related *Pythium* species, both of which are competitive soil inhabitants. Evidence that certain species of *Phytophthora* can persist in soil indicates, however, that the absence of a host plant in a crop rotation system will not suppress them all. Table 5.3 gives several examples of persistence of certain spore forms in soil for various periods.

Although the data in Table 5.3 on survival of different

Table 5.3. Representative survival times for propagules of *Phytophthora* species in field soil, except where otherwise noted[a]

Species	Sexual status	Mycelium	Sporangia/ zoospores	Chlamydospores	Oospores	Reference
P. cactorum	Hom			3.5 mo		Gisi and Meyer (1973)
					9 mo (in fruit)	Grove et al. (1985b)
					54 wk	Jeffers and Aldwinkle (1987, 1988)
				6 yr (A); 2 yr (B)		Kröber (1980)
					2 yr	Madden et al. (1991)
			35 days (21°C)			McIntosh (1972)
		3 days (29°C)				Sneh and McIntosh (1974)
		45 days (4°C)				Sneh and McIntosh (1974)
P. capsici	Het	<120 days (infected tissue)	< 75 days			Ansani and Matsuoka (1983a)
Summer			4–8 wk		12 wk	Bowers et al. (1990)
Winter					8 wk	Bowers et al. (1990)
P. cinnamomi[b]	Het	1–60 days	2–8 wk	1–6 yr		Hwang and Ko (1978)
					84–365 days	Kassaby et al. (1977)
			2 yr (A); >1 yr (B)			Kröber (1980)
			>2 mo			MacDonald and Duniway (1979)
				100 days		MacKay et al. (1985)
		>15 days				Reeves (1975)
		>2 yr				Shea et al. (1980)
				5 yr		Weste (1975)
				30 days		Weste (1983a, b)
				21–42 days		Weste and Vithanage (1979)
				6 yr (20°C) (in roots)		Zentmyer and Mircetich (1966)
P. citricola	Hom				6 yr (A); 3 yr (B)	Kröber (1980)
P. citrophthora		40 wk (leaves)				Gerlach et al. (1976b)
				4 yr (A); 3 yr (B)		Kröber (1980)
		5 mo				Sneh and Katz (1988)
P. colocasiae	Het		2 wk			Gollifer et al. (1980)
P. cryptogea	Het		14 days			Bumbieris (1979)
P. drechsleri	Het	4 yr (A); 2 yr (B)				Kröber (1980)
		10–12 days; >30 days				Mehrotra (1972)
P. erythroseptica	Hom		6–8 mo			Weste (1983b) (review)
P. fragariae	Hom	63 days			<3 yr[d]	Bain and Demaree (1945)
					8 mo[d]	J. M. Duncan (1980b)
					3 yr[d]	Duncan and Cowan (1980)
					1.5 yr[d]	Hickman (1958); Montgomerie (1977)[c]
P. infestans	Het	6–8 mo				Weste (1983b) (review)
			63–77 days			Zan (1962)
			15–77 days			Andrivon (1995) (review)
					35 wk (winter)	Drenth et al. (1995)
					5–8 mo	Pittis and Shattock (1994)
P. meadii	Het	3 wk	3 wk	22 wk		Lyanage and Wheeler (1991)
P. medicaginis	Hom		3–4 wk	7 mo (5–30°C)		Basu (1980)
		7–8 days	7–14 days			MacDonald and Duniway (1979)
					2.5–3.5 yr	Pratt and Mitchell (1975)
		9 days (24°C) 15 days (15°C) >35 days (4°C)			4.5 mo	Stack and Millar (1985c)
P. palmivora	Het	6 mo–2 yr	10–22 days			P. Turner (1965)
		14–28 days				P. Turner (1974)
					6 yr (A); 4 yr (B)	Kröber (1980)
P. parasitica	Het			180 days		Holdaway and Tsao (1971)
				6 yr (A); 4 yr (B)		Kröber (1980)
				30 days (A); 60 days (B)		Ramarao and Umabala (1982)
		<7 days; >14 days				Tsao (1969b)
		7 mo				Kuske and Benson (1983b)
P. phaseoli	Hom				8 mo	Wester et al. (1966)
P. sojae	Hom			>5 days (A); <5 days (B)		Ho (1969b)
P. syringae	Hom				2.5 yr	D. C. Harris (1985c)

[a] Hom = homothallic; het = heterothallic; (A) = sterile soil; and (B) = nonsterile soil. The cited reports should be consulted for more information on effects of temperature and soil moisture.
[b] See also Zentmyer (1980).
[c] Hickman (1958) thought *P. fragariae* survived as mycelium in bean agar disks, but Montgomerie (1977) suggested that it survived as oospores.
[d] Naturally infested soil in which oospores were present in roots.

propagules indicate that survival times are relatively short for many species, the difficulty in detecting *Phytophthora* in soils may preclude and obscure long-term survival. Also, as discussed below, the ability of a few propagules to increase rapidly when soil is flooded may override the significance of an apparent short survival time.

The effectiveness of control strategies depends on the ability of each *Phytophthora* species to survive in the absence of the host plants. *Phytophthora* species differ in their ability to survive, either saprophytically or as dormant spores. In general, mycelium and zoospores produced by *Phytophthora* survive only for a few weeks; however, chlamydospores may survive for as long as 6 years and oospores up to 13 years (Table 5.3). *P. cinnamomi* appears to have the greatest ability to survive for a long period. Survival for up to 6 years in soil maintained at 20°C (Zentmyer and Mircetich 1966) indicates that *P. cinnamomi* would be unlikely to be controlled by crop rotation. Kröber (1980) reported survival for 3 years in nonsterile soil and 6 years in sterile soil. *P. cinnamomi* also has the unique capability to invade and colonize wheat, straw, and dead avocado roots in nonsterile as well as sterile soil, especially in soils nearing saturation (16% water). It also has the ability to grow up to 3 cm from potato-dextrose agar disks in sterile and nonsterile soil (Zentmyer and Mircetich 1966; Zentmyer 1980). Such data led Zentmyer (1980) to conclude that *P. cinnamomi* may be a competitive saprophytic colonizer. His results were similar to those of Reeves (1975).

Other *Phytophthora* species grow only short distances and lyse in soil; thus, they are considered to be poor saprophytes (Tsao 1969b; Sneh and McIntosh 1974; Reeves 1975). This characteristic explains why *P. palmivora* could be controlled on papaya seedling roots by filling the transplant holes in infested fields with uninfested virgin soil through which the pathogen was unable to grow (Ko 1971). However, *P. medicaginis,* the alfalfa pathogen, colonizes organic material and produces oospores in soil at low temperatures (Stack and Millar 1985c).

Although *P. infestans* is not considered to be soilborne, there is evidence that it can in some undetermined way move through soil (Gregory 1983). Gregory quotes Zan's (1962) finding that soil infectivity persists up to 11 weeks—not long enough to explain the persistence of *P. infestans* in soil for the 6-month period between potato crops. Although planting infected tubers often initiates epidemics, only about 0.7% of the infection of stems could be traced to blighted tubers (Hirst and Stedman 1960). Hirst and Stedman (1960) experimentally planted "eyeless" but infected potato seed tubers (would not produce a new plant) in artificially sprinkle-irrigated field plots at Rothamsted Experimental Station in England. Because infection of some plants occurred near the eyeless infected seed tubers but not near healthy tubers, the most likely explanation was that *P. infestans* produced sporangia and zoospores on the infected tubers in soil and that these propagules were splashed by rain to susceptible foliage of other plants nearby.

Quantitative methods for isolation of *Phytophthora* species from soil have been a focus of research for more than 30 years based principally on the hypothesis that the incidence and/or severity of disease is likely to be greatest where and when the incidence of inoculum is highest (see Chapter 2 for isolation methods). Of probably more importance is the use of quantitative assays for the comparison of effects of various control measures such as fungicides or biological or cultural methods on populations of *Phytophthora* in soil. Inoculum levels of fewer than 1–30 propagules per gram (ppg) of soil from perennial hosts are common. Prior to a severe disease onset, the initial inoculum concentrations of *Phytophthora* in many annual crops may be even lower or undetectable. See Mitchell and Kannwischer-Mitchell's review (1983) of the relationship of inoculum density to disease incidence. After disease onset, the population often increases to several hundred propagules per gram of soil if conditions are favorable. This rapid increase in populations of *Phytophthora* species in soil is a unique feature of the epidemiology of *Phytophthora*-caused diseases. We discuss determination of inoculum density in soil in Chapter 2. References on the impact of increase of inoculum density in soil on disease are listed in Table 5.4 (Weste 1983b).

The rapid reproductive capacity and short regeneration time of *Phytophthora* under favorable environmental conditions is shared by other oomycetous fungi with a zoosporic stage. Phenology is the study of the relationship between climate and periodic biological phenomena. Dick (1992) deals with phenologic patterns as they affect populations of zoosporic fungi and indicates that water and temperature are the prime climatic factors. Although Dick (1992) uses examples drawn mainly from the life history of *Pythium* and members of the zoosporic fungi other than *Phytophthora,* his treatment of phenology is also applicable to *Phytophthora.*

In a survey of epidemics of tobacco black shank, caused by *P. parasitica* var. *nicotianae,* the initial inoculum density, as determined by the wet sieving method of Shew et al. (1979a, b), ranged from 0.04 to 0.33 ppg at one site and 0.09 to 0.16 ppg of soil at another site. Despite these relatively low levels of initial inoculum, significant epidemics of black shank resulted in disease incidences of more than 90% by the end of the crop year (C. L. Campbell et al. 1984). As indicated for other soilborne *Phytophthora* diseases (Weste 1983b), the populations of *P. parasitica* var. *nicotianae* increase rapidly in the rhizosphere (Kannwischer and Mitchell 1978). C. L. Camp-

Table 5.4. Examples of increase in disease associated with increase in pathogen populations of several *Phytophthora* species[a]

Species Initial infection or inoculum density	Final disease or pathogen population	Dispersal unit	Reference
P. infestans			
One infection in 10^5 potatoes	Epidemic disease	Sporangia, zoospores	Newhook (1982)
P. palmivora			
0.5 chlamydospores per gram of soil	50% infection	Chlamydospores	Ramirez and Mitchell (1975)
10.0 chlamydospores per gram of soil	95% infection	Chlamydospores	Ramirez and Mitchell (1975)
10^5 zoospores per plant	75% infection	Zoospores	Ramirez and Mitchell (1975)
One infection per cocoa pod	$5–10 \times 10^6$ sporangia per pod	Sporangia	P. Turner (1974)
P. parasitica var. *nicotianae*			
10^5 spores per tobacco plant	100% infection of plant	Sporangia	Gooding and Lucas (1959b)
P. cinnamomi			
One sporangium per seedling	100 sporangia per seedling	Sporangia	Byrt and Holland (1978)

[a] Reproduced by permission of the American Phytopathological Society, St. Paul, Minnesota, courtesy Weste (1983b).

bell and Powell (1980) and C. L. Campbell et al. (1984) conclude that black shank of tobacco is potentially a multicyclic disease (see discussion in Chapter 1) during the growing season, but how the inoculum increased rapidly in the field is unclear because the field sites studied were well drained and water did not stand for more than short periods. C. L. Campbell et al. (1984) suggested that increased growth of roots through soil presents opportunities for infection because growing roots contact more inoculum. In 28 cases, a logistic model of disease progress and in 14 cases, the Gompertz model (Berger 1981) best fit the disease progress data for black shank of tobacco. With the susceptible cultivar NC2326, an asymmetrical growth function (Gompertz model) best described disease progress. See C. L. Campbell et al. (1984) for a more complete description of the statistical models used.

Failure to detect *Phytophthora* in soil does not necessarily indicate its absence. For this reason, a minimal inoculum status or even the absence of inoculum has to be interpreted with caution because inoculum can change from undetectable to a high concentration in a short period. In Western Australia soils, *P. cinnamomi* is rarely detected by baiting during much of the year, particularly during the dry summer months (Old 1979; Weste 1983b). Shea et al. (1980) report that the population of *P. cinnamomi* in the jarrah forest of Western Australia is high only during September and October, the months with the highest rainfall. Artificial irrigation of plots in the forest during dry periods, when recovery of *P. cinnamomi* is low or nil, increases the number of propagules that can be recovered by dilution plate isolation. This finding supports the general principle that a high soil moisture regime favors an increase in population of *Phytophthora* in soil. The demonstration of the presence of *Phytophthora* can be highly useful; however, reports of its absence must be interpreted skeptically.

As our discussion of the importance of inoculum in soil demonstrates, data from quantitative assays of soil may or may not be an absolutely positive indicator of the prospects for plant health during the subsequent crop season. In most case histories that have been analyzed, both soilborne and foliar diseases caused by *Phytophthora* are multicyclic disease types in which inoculum increases continuously if environmental conditions are favorable (Fry 1982; MacKenzie et al. 1983; see also Chapter 1). Following this scenario, the population of *P. parasitica* on citrus changes with time and in response to irrigation and temperature changes (Lutz and Menge 1991). Conversely, Tooley and Grau (1984a) characterized root rot of soybean (caused by *P. sojae*) as a monocyclic disease because there was little evidence of plant-to-plant spread and they did not feel that evidence of secondary increase in inoculum precluded the use of the monocyclic (simple interest) model. In their plots, the initial inoculum was high because of previous cropping to soybean.

Vanderplank (1963) states that the initial inoculum is a prime factor that should be studied as thoroughly as the subsequent disease progress in analyzing epidemics. However, few reports deal with the quantitative status of the initial inoculum in soil for *Phytophthora*-caused disease (Mitchell and Kannwischer-Mitchell 1983). Furthermore, in areas in which *Phytophthora* species are known to cause disease, the initial inoculum is often not detectable, despite the recent advances in technology for detection and isolation. The failure to isolate propagules could result from the inadequacy of techniques to detect low numbers of active or dormant propagules (e.g., oospores) at certain times of the year.

Mitchell and Kannwischer-Mitchell (1983) cite reports of initial *P. cactorum* inoculum levels of one to 19 ppg of apple soil (McIntosh 1977) and of *P. parasitica* levels of 0.2–32 ppg of citrus soil (M. E. Kannwisher-Mitchell, *unpublished*). The population of *P. capsici* increased from a nondetectable level to 24 ppg of pepper field soil within 3 months (Papavizas et al. 1981). Several research reports indicate that the initial inoculum, which initiates both soilborne and foliar epidemics, is usually low (Table 5.4).

Pfender (1982), in an apt review in which monocyclic and multicyclic root diseases were evaluated by the nature of the disease cycle and the shape of the disease progress curve, states that determining the occurrence of secondary infection cycles gives more accurate information than data on disease progress over time because environmental factors and host susceptibility also affect the rate of infection over a season.

Final disease loss of bell pepper (*Capsicum annuum* L.) caused by *P. capsici* in artificially infested field plots in North Carolina was independent of the concentration of initial inoculum applied. Initial inoculum densities were established at 0, 3.9, 5.9, and 9.4 colony-forming units (cfu) per gram of soil. Although the rate of disease incidence at the lower inoculum density was slightly less than at higher inoculum densities, by the end of the season, yield was reduced by approximately the same amount at all inoculum levels (Ristaino 1991). The data adequately confirmed that Phytophthora root rot of pepper is a multicycle disease as stated by C. L. Campbell (1986).

Gregory (1983) reviewed several epidemics: late blight on potato (*P. infestans*), leaf blight on taro (*P. colocasiae*), fruit rot of apples (*P. syringae*), black pod of cacao (*P. megakarya* and *P. palmivora*), and root rot of Australian forest trees (*P. cinnamomi*). In each case, the dormant-season propagule that presumably initiates an epidemic is different. *P. infestans* survives as mycelium in infected potato tubers and for a limited time in soil, but the propagule involved is unknown. *P. colocasiae* produces both chlamydospores and oospores in diseased rhizomes of the taro plant, *P. syringae* produces oospores in apple leaves on the orchard floor, and *P. palmivora* produces chlamydospores in the black pod-affected fruit of cacao. The initial inoculum of *P. megakarya,* which produces only a few chlamydospores in culture and apparently none in the field, is more difficult to explain; however, both *P. palmivora* and *P. megakarya* survive in infected flower cushions. It is suspected that both species can survive in infected but apparently symptomless roots (Gregory 1983). *P. cinnamomi* survives as chlamydospores both in infected roots of jarrah trees and in soil (Weste 1983b).

In conclusion, the most convincing hypothesis is that almost all soilborne and foliar diseases caused by *Phytophthora* are multicyclic, since they can be initiated by low, often undetectable levels of inoculum, and that the population in soil or on foliage can increase exponentially if environmental conditions are favorable.

Obviously, disease will not occur without inoculum, so in new plantings or glasshouse operations, prevention of disease by use of sanitation is one of the most important lines of defense (Ribeiro and Linderman 1991). *Phytophthora* can be killed in soil used in glasshouses and nurseries by heating it to 82°C (180°F) with flowing steam for 30–40 min (Baker and Roistacher 1957; see also the following section on control by heat) and by ensuring that transplant stocks are pathogen free. In many intensive agricultural industries, such as strawberry culture in Great Britain, much attention is given to production of clean plants free of the red stele fungus, *P. fragariae* (Duncan 1976, 1979, 1980a; Duncan et al. 1986). Health of strawberry nursery stock can be monitored by use of the root tip detection method before it is transplanted in the field (see Chapter 2). Because oospores of both *P. fragariae* (Duncan 1977; Milholland et al. 1989) and *P. cactorum* are produced in the stele of infected strawberry plants and because oospores cannot be eradicated by fungicides (Duncan 1985b), any plant with infected roots is a source of contamination.

Research data on long-term survival of *Phytophthora* species in soil and plant debris are very limited for the following reasons: 1) the difficulty of separating the various chemical, physical, and biological factors in the environment that affect *Phytophthora* propagules; 2) the difficulty in ascertaining the survival value of each propagule type produced by *Phytophthora* species, since most species produce at least five structures (mycelium, sporangia, zoospores, chlamydospores, and oospores) that are capable of survival for short or long periods, depending on the environmental conditions (Table 5.3); and 3) the lack of methods sensitive enough to detect the presence of the various *Phytophthora* propagules at low concentrations in the soil. Some species are more sensitive to certain selective media or baits than other species and thus are more difficult to detect (Jeffers and Martin 1986; see also Chapter 2).

Although data obtained with different species and by different investigators cannot be directly compared, overall the oospore has the longest record of survival. Oospores of *P. phaseoli* formed in inoculated lima bean pods in the field. Pods containing oospores can overwinter on the plant in Maryland and when used as inoculum will cause disease (Wester et al. 1966).

The next longest survival period is that of chlamydospores, and the shortest survival period is that of sporangia, zoospores, and mycelium. Although the times given for survival are not as long as would generally be expected for long-lived fungi such as *Fusarium oxysporum* or *Verticillium dahliae* (10–15 years), they are long enough to assume that the fungus could probably persist in soil or in infected plant material for at least 2 years, although in the absence of oospores, survival rates could be quite low. Such projections are necessary for predicting whether or not the *Phytophthora* species in question will survive through a crop-rotation plan.

Longevity of propagules depends on their ability to colonize plant material and persist in gravel, sand, or soil, independent of a host (Weste 1983b). The major factors that favor the survival of *Phytophthora* in the soil environment are temperatures lower than those optimal for growth, soil moisture at field capacity (matric potential value about –300 mb), and low soil microbial activity, a quality that is typical of gravelly silt. Increase in the populations of both fungi and bacteria often results in suppression of *Phytophthora* propagules (Weste 1983a, b).

Drying suppresses sporangium, zoospore, and mycelium production in plant residue (Schmitthenner 1970). Nearly all spore stages of *Phytophthora* species except oospores (Tsao 1969b) and chlamydospores (P. D. Turner 1965a; Mircetich and Zentmyer 1966; Schmitthenner 1970) are highly susceptible to drying. Drying naturally infected (*P. parasitica*) leaf disks of rhododendron to 20–30% moisture suppressed the sporangium-producing capability to nearly zero. After 18–22 days, *P. parasitica* could not be recovered (Kuske and Benson 1983a). *P. parasitica* survived during two winters in nondried leaf disks that were buried in soil (Kuske and Benson 1983b).

Vartanian and Endo (1985a) were able to isolate *P. infestans* from infected tomato fruit; however, when seeds, 96% of which were infected, were dried at 29.5–37.5°C, *P. infestans* could not be recovered. Vartanian and Endo (1985a) concluded that seed was probably not an important source of inoculum, because commercial seed is fermented and dried prior to sale. In the field, however, seedborne inoculum from volunteer plants is a source of primary inoculum.

Because the thick-walled oospores, chlamydospores, or mycelium in plant tissue that survive can quickly germinate and produce sporangia and zoospores, inoculum increases rapidly under favorable conditions (Weste 1983b). Oospores, chlamydospores, or mycelium (called residual inoculum by Mitchell and Kannwischer-Mitchell [1983]) give rise to sporangia when water is available. In turn, each sporangium produces about 20–40 swimming zoospores (transitory inoculum). The zoospores stop swimming, round up, germinate, and penetrate the host plant. This entire process requires less than 48–96 h under optimal temperature and moisture conditions. (See Chapter 3 and Figure 1.1 for the life history of *Phytophthora*.) Thus, *Phytophthora* species have the unique ability to "explode" by rapid production of inoculum in soil and on foliage during the crop season (MacKenzie et al. 1983). An epidemic can occur quickly if the environmental conditions are conducive.

Despite the relatively ephemeral nature of mycelium in soil (Tsao 1969b; Sneh and MacIntosh 1974; Reeves 1975), *P. cinnamomi* not only can be disseminated by zoospores in flood water, but it can also move uphill to a limited extent, an ability that implies its spread by infection of roots as well as by passive movement of zoospores (Weste 1983b). Zoospores of *P. medicaginis* (*P. megasperma* f. sp. *medicaginis*) can move as much as 2 cm in a silt loam soil (Stack and Millar 1985b); however, in a moving front of water, zoospores can move much farther (Wilkinson et al. 1981). Mycelium moved small distances (at least 2.0 cm) in soil and was capable of forming sporangia and oospores at matric potentials of about –100 mb (Stack and Millar 1985b). Although 24–27°C was most favorable for growth, colonization of host roots occurred more frequently at 4°C, a temperature at which the fungus grows much more slowly. Competition and parasitism by associative microorganisms in soil, which likely contribute to hyphal lysis, decrease greatly at low temperatures (4°C) (Stack and Millar 1985c). The scarcity of reports on this aspect reflects the difficulty of observing and recording such data in normal soil. In the study of Stack and Millar (1985a), the use of a unique metalaxyl-resistant isolate facilitated its isolation and detection by use of metalaxyl in the isolation medium.

Sneh and Katz (1988) reported that mycelia of *P. citrophthora* and *P. nicotianae* var. *parasitica* survived for >150 days in both air-dried and moist soils, but approximately twice as many propagules survived at 20% soil moisture than in air-dried soil. Propagules (most likely chlamydospores) of *P. cinnamomi* survived in high numbers up to 3 years in naturally infested, moist, sandy loam soil maintained in jars at 20°C. Thereafter, survival decreased rapidly; however, a small percentage of propagules (2%) could still be isolated 6 years later. The pathogen survived only 3 months in soil dried to 2–3% moisture (Zentmyer 1980) (Table 5.3).

Although *P. citrophthora* from citrus has not usually been known to produce chlamydospores, chlamydospores were detected in soil and in dead citrus tree roots during summer months in California orchards (Sjoerdsma et al. 1988). During the summer months, the activity of *P. citrophthora* is more limited by high soil temperatures than that of *P. parasitica*, which is active during the summer at temperatures approaching or above 30°C. The chlamydospores of *P. citrophthora* may be dormant at high temperatures, since soil populations were shown to increase when field soil samples were incubated at 16°C for 3–10 days before isolation by dilution plating (Sjoerdsma et al. 1988). Mchau and Coffey (1994b) reported that chlamydospores were formed only by Brazilian isolates of *P. citrophthora* from cacao.

Oospores may exist in an inactive state in dry soil and,

when wetted, become rapidly activated. Stack and Millar (1985a–c) and Jeffers and Aldwinckle (1987) report that drying and rewetting soil before isolation increases the recovery of *P. medicaginis* and *P. cactorum*. This regeneration is probably caused by stimulation of oospore germination (see Chapters 2 and 3). Oospores of *P. medicaginis* persist in dry or moist soil up to 140 days (Stack and Millar 1985c). El-Hamalawi and Erwin (1986a), however, reported that germination of oospores produced in carrot broth was reduced from 70 to 18% by air drying.

Oospores develop within 9 days in fallen apple leaves infected by zoospores of *P. syringae*. These oospores survive for more than 2 years and contribute substantially to inoculum in the soil (Harris 1985a, c). A low but significant percentage of oospores of *P. syringae* in apple leaves survived burial for 32 months in soil and in mesh bags on the surface of an apple orchard floor. Although the internal parasite *Microdochium fusariodes* destroyed a large percentage of oospores that were buried in soil, a detectable number survived (Harris 1985b). Oospores of *P. cactorum* overwinter and germinate in apple leaves (Sneh and McIntosh 1974) and in infected, mummified strawberry fruit (Grove et al. 1985c).

Oospores of the homothallic *P. fragariae*, which are produced in naturally infected strawberry root tissue (Duncan 1976, 1977, 1985b), germinate rapidly (see Chapter 3). In inoculated roots, oospores began to form in 4–6 days and reached a maximum in 12 days (Law and Milholland 1991). Since oospores can withstand disinfestation with the potent antimicrobial agent mercuric chloride (5 min at 10–60 μg/ml) (Shaw 1967; Sneh 1972a) without reduction of the germination percentage, it is not surprising that fungicidal (metalaxyl) treatment of infected strawberry roots does not eradicate the oospore inoculum of *P. fragariae* (Duncan 1985b). The thick wall prevents rapid uptake of toxic materials. Most homothallic species, such as *P. cactorum*, *P. fragariae*, *P. sojae*, and *P. medicaginis*, produce oospores in host tissue.

P. capsici occurs on pepper in some soils as either A1 or A2 mating type, or both can occur together. Since oospores from pairing A1 × A2 mating types on V8 juice agar survived in field soil during the summer for more than 12 weeks and during the winter for at least 16 weeks (much longer than sporangia or zoospores), it appeared that oospores could be a major overwintering propagule (Bowers et al. 1990) (Table 5.3). A subsequent report shows that the LD_{50} (infective dose of oospores that causes 50% loss of plants) was 41 oospores per gram of soil (Bowers and Mitchell 1991).

P. infestans rarely if ever produces oospores in host tissues unless the opposite mating type is also present (Galindo and Gallegly 1960). Fry and Spielman (1991) propose that oospores could be produced by *P. infestans* in areas such as Toluca, Mexico, where A1 and A2 mating types coexist, and in areas of Europe and Egypt, where A2 as well as A1 types have recently been found. If oospores are formed, they could be important sources of inoculum and potentially sources of new and more virulent isolates. Because the A2 mating type has recently been introduced into the Netherlands, Poland, the British Isles, and the United States, major genetic changes in virulence levels of *P. infestans* resulting from the production of oospores might be expected (Drenth et al. 1993, 1995; Fry et al. 1992; Spielman et al. 1991; Sujkowsky et al. 1994). See also the review of *P. infestans* in the soil by Andrivon (1995).

In the field, the inoculum in soil is most often aggregated or clumped and seldom randomly dispersed (Weste 1983b; Magnano di San Lio et al. 1990). Statistical equations to treat aggregated inoculum values in soil are discussed and explained by Ferrin and Mitchell (1986a). Their study of the epidemiology of the tobacco black shank pathogen, *P. parasitica* var. *nicotianae*, in a naturally infested field indicated that plant mortality in the susceptible cultivar Hicks was not related to the initial mean inoculum density, which ranged during 2 years of study from 0.01 to 1.31 ppg; however, increased summer plant mortality was related to the aggregation of inoculum in the field. Plant mortality in the resistant cultivar Speight G-28 was related throughout the epidemic to the initial mean inoculum density but not to the aggregation of inoculum. Thus, as the concentration of initial inoculum increased, the plant mortality of the resistant cultivar Speight G-28 became greater and the pattern of disease was essentially monocyclic, whereas in the susceptible cultivar Hicks, the increase in disease was multicyclic. This appears to indicate that the rate of reproduction of *P. parasitica* on the cultivar Speight G-28 was slower than that on Hicks.

The main factors that increased inoculum of *P. cinnamomi* in the Brisbane Ranges and Wilson's Promontory forests in Australia were increases in soil moisture and temperature (Weste and Rupin 1977). Populations of *P. cinnamomi* varied from undetectable to a high population density index of 256, defined as the reciprocal of the lowest fraction (1/256) of soil from which *P. cinnamomi* could be detected by baiting with lupine seedlings (see method described by Chee and Newhook [1965a] and in Chapter 2).

The severity of disease often increases in proportion to the increase of inoculum in soil. Watercress (*Nasturtium officinale* R. Br.) plants suffer from a severe stem and root rot caused by *P. cryptogea* (Mitchell et al. 1978). When 2-week-old plants growing in vermiculite were inoculated with dosages of 10^3, 10^4, 10^5, and 10^6 motile zoospores per plant, infection after 7 days was 83% at the 10^3 dosage and 100% at all higher dosages. On plants 21 days of age, however, plant mortality increased directly with higher inoculum dosages: 30% at 10^3, 51% at 10^4, 67% at 10^5, and 81% at 10^6 zoospores per plant. These data show a direct relationship between inoculum levels and incidence of disease. Mitchell et al. (1978) estimated that inoculum containing 276 zoospores per plant was required to infect 50% of the plants. They concluded that the severity of disease expression is related to the number of infections on each root system.

Populations of *P. citrophthora* in a citrus grove in Sicily increased within 24 h after trickle irrigation in lines under the trees. Populations 24 h after irrigation increased from about 2–3 to 12 ppg of soil in two other orchards and from about 14 to 55 ppg in another (Magnano de San Lio et al. 1990). These results indicate that the inoculum dynamics in a citrus orchard have a multicyclic pattern. Magnano di San Lio et al. consider a population of about 15–20 ppg to be a threshold value for a significant risk of root infection (Magnano di San Lio et al. 1987b, 1988; Menge et al. 1988; Timmer et al. 1988a). Inoculum density in soil at two locations was similar on the rootstocks sour orange, Cleopatra mandarin, trifoliate orange, Troyer, and Carizzo citrange used on sweet orange trees. In other studies on different orchards in Sicily, the population of *P. citrophthora* in nonirrigated root zones was similar to that in irrigated zones (Magnano di San Lio et al. 1988).

Several field studies with *P. parasitica*, the cause of root rot and fruit rot of canning tomatoes, conducted in Davis, California, included both inoculum level and frequency of irrigation treatments (Neher and Duniway 1991). Disease severity increased with an increase in initial inoculum density; average inoculum levels were 1–2, 3–4, 7–28, and 29–65 cfu per gram of soil. Results were similar at irrigation durations of 4–24 h. Although the percentage of plants affected increased in direct proportion to the inoculum density, yield was reduced significantly (20%) only in 1 year and at the highest initial inoculum level, contrasting with the results of a previous study (Ristaino et al. 1988) in which increased length of furrow irrigation (from 4 to 24 h) and frequency accelerated the disease epidemic.

When alfalfa cultivars are screened for general resistance to *P. medicaginis,* the amount of inoculum should be limited to a minimal threshold value to differentiate seedling plants with resistance from those that are susceptible, because an excessive amount of inoculum overcomes plant resistance (F. A. Gray et al. 1973; R. G. Pratt et al. 1975; Irwin et al. 1979). This principle applies to several other host-pathogen systems (see Chapter 6).

When plants are produced in starter fields or containers for later resale, in forestry nurseries, or in intensive agriculture such as strawberry culture or ornamental agriculture, root diseases that occur at this stage are unprofitable not only because the nursery growers lose many infected plants, but more importantly because inoculum on infected plants is introduced into clean areas. Dissemination of Phytophthora root rot diseases to surrounding plants by sale of infected nursery stock to home gardens, fields, or forest plantings is one of the major ways by which epidemics are initiated (E. M. Hansen et al. 1979). Epidemics of crown rot have been initiated in new plantings of apple trees because of nursery contamination (Jeffers and Aldwinckle 1988; Matheron et al. 1988).

As stated above, an epidemic caused by *Phytophthora* can emanate from a small amount of inoculum. An epidemic of root rot of tobacco caused by *P. parasitica* var. *nicotianae* that resulted in the death of 60% of the resistant and 100% of the susceptible cultivar plants by the end of the season emanated from the low inoculum density value of 0.20–0.75 ppg of soil (Mitchell and Kannwischer-Mitchell 1983).

Phytophthora propagules in soil can cycle from undetectable to as high as 100 ppg within a given season. Inoculum density associated with roots of perennial hosts, such as apple, ranged from 1 to 19 ppg for *P. cactorum* (McIntosh 1977) and from 0.2 to 32 ppg for *P. parasitica* on *Citrus* species (M. E. Kannwischer-Mitchell, *unpublished,* and cited by Mitchell and Kannwischer-Mitchell [1983]). In glasshouse studies, the rapidity of root disease development on tobacco was directly proportional to the number of zoospores of *P. parasitica* var. *nicotianae* applied per plant. The disease index (0–100) after a 20-day inoculation period ranged from 3.8 to 17.3 at an inoculum density of 10^3 zoospores per plant, from 13.0 to 45.0 at 10^4, from 82.5 to 87.5 at 10^5, and from 89.3 to 92.7 at 10^6 on the susceptible tobacco cultivar 402. On the resistant cultivar Dixie Bright 101, no disease occurred at inoculum levels of 10^3 and 10^4 zoospores per plant, but at 10^5 the disease index ranged from 0.2 to 2.0 and at 10^6 the disease index ranged from 6.3 to 20.3 (Kannwischer and Mitchell 1981). In this case, resistance was overcome by an increase in inoculum levels in soil. This characteristic is common to nearly all cultivars that have general (horizontal) resistance.

In citrus, *P. citrophthora* causes fibrous root rot in climates with moderate temperatures, e.g., in California during the fall, winter, and spring. *P. parasitica* causes root rot in climates with summer temperatures that reach up to 35°C. The cycle is described in a popular article by Lutz and Menge (1986b). Wherever citrus trees are grown throughout the world, these diseases are present, probably because the causal agents were transported on rootstocks exchanged by research and commercial personnel when the industry was becoming established during the 1800s. Fibrous root rot of citrus seldom kills trees but does reduce the vigor and yield. Thus, the effect of the size of the population of *Phytophthora* species in soil has been of considerable practical as well as fundamental interest. Such questions as whether or not to apply an expensive systemic fungicide such as metalaxyl in irrigation water (Menge 1986) might be answered if the population of *Phytophthora* species could be monitored. Of even more importance, can citrus trees tolerate a low population of *Phytophthora* and not suffer damage?

A recent report by Timmer et al. (1993) comparing commercial ELISA serological kits (see Chapter 2) with soil dilution plate methods to assay soil populations of *Phytophthora* quantitatively indicated that in California, fungicidal treatment is recommended when populations reach 15–20 ppg of soil based on plating out rhizosphere soil on a selective medium (Lutz and Menge 1986). In Florida, where different assay methods are used, a population of 0–5 propagules per cubic centimeter of soil is considered low, 5–15 is moderate, and more than 15 is considered high (Timmer et al. 1988a). In Florida, population levels are based on assay of random soil cores (Timmer et al. 1988a) and in California on rhizosphere soil (Lutz and Menge 1991) (see Chapter 2). Despite the different numbers attained by different methodologies, it is becoming accepted that minimal thresholds that might be critical for management can be established.

In Sicily, Italy, soil populations of *P. citrophthora* in citrus groves ranged from about 1 to 100 ppg over a 3-year period (1984–1986). The populations were highest during the cooler part of the year and lowest during late spring and midsummer. Populations were higher at matric potentials nearing zero (free water in soil), except during summer months when high temperatures prevailed. Expression of disease on foliage (caused by root rot) was not always correlated with population, but epidemics of fruit rot on the trees (inoculum was splashed from soil to fruit by rain) were associated with soil populations of more than 20 ppg (Magnano di San Lio et al. 1988). A later study shows significant increases in the *Phytophthora* populations after 24-h periods of trickle irrigation from nearly zero (undetectable) to 13 in one orchard, from 14 to 55 in another, but from 2 to only 3 ppg in another orchard. The latter set of data indicates that factors other than irrigation in some orchards suppress populations in soil (Magnano di San Lio et al. 1990). Levels of inoculum of *P. citrophthora* in some citrus soils in Italy are higher in coarse sand or gravel; however, more rapid regeneration of roots offsets the effects of increased inoculum. Calcium carbonate applied to soil in orchards tended to depress inoculum levels (Ippolito et al. 1990).

Populations of *P. parasitica* in rhizosphere soil of 70-year-old Washington Navel trees on sweet orange rootstocks (susceptible) increased from 17 ppg shortly before furrow irrigation to 70 ppg 2 days after furrow irrigation (Lutz 1987; Lutz and Menge 1991). These data support the multicyclic concept that the inoculum of *Phytophthora* in soil increases within the crop year. The increase in propagules is most likely the result of formation of sporangia and zoospores from germinated chlamydospores that persist in soil. As more roots become infected, more sporangia are formed, which infect new roots. During November and from May through August, sporangia and zoospore cysts accounted for the majority of the population, but from December through April, the most frequently isolated propagules were chlamydospores (74%). Only a small number of oospores were detected throughout the year. Since *P. parasitica* is heterothallic, oospores are unlikely to be a source of inoculum (Lutz 1987).

On citrus, *Fusarium solani* (Mart.) Sacc. is the predominant fungus isolated from soil or roots. Although it colonizes roots and can reduce root growth, it is considered to be a relatively weak pathogen that is probably most active during quiescent periods of root growth. *F. solani* reduces rhizosphere and soil populations of *P. parasitica* but not of *P. citrophthora.* Although the mechanism is not well understood, it is proposed that "*F. solani* suppresses populations of *P. parasitica* by a combination of 1) reducing the availability of infection sites as a result of root dieback and 2) decreasing the number of microsites available for encystment [of zoospores] and infection." These factors apparently are less important for the buildup of *P. citrophthora* than of *P. parasitica* (Danderand and Menge 1993).

Dispersal of Inoculum: Sporangia and Caducity

Long-range dispersal of *Phytophthora* occurs mainly when infected plant tissue or soil is transported to uninfected areas. Since potato tubers can become infected in the field and are used to vegetatively propagate new potato plants, it is obvious how *P. infestans* was spread around the world. The migration of the "new" genotype and A2 mating type throughout the world may have been associated with large shipments of potatoes to Europe from Mexico during 1976 (Niederhauser 1991). Infection spreads from the tubers to the stems of emerging potato plants on which lesions then form. Sporangia from these lesions are splashed from the infected stems to the leaves and from plant to plant by windborne rain. Rapid production of sporangia of *P. infestans* is a characteristic that contributes to its success as a cause of epidemics. McAlpine (1910) stated that its life cycle, consisting of sporangium to sporangium, occurs in less than 7 h.

"Caducity" is the morphological capability of the sporangia produced by some species to readily detach from the sporangiophores, thus allowing air currents, wind, or splashing water to disseminate them from several meters to several kilometers away. Although different in ontogeny, caducity is somewhat analogous to the detaching of a leaf caused by the formation of the abscission layer on the petiole.

Caducous sporangia and zoospores of all species are termed "xenospores" by Gregory (1966, 1983). Their most important role is in the spread of inoculum from the initial foci of infection in the fields. Caducity is undoubtedly an evolutionary trait that gives certain species a selective advantage on foliage and contributes to their success as multicyclic pathogens. Species that produce caducous sporangia are listed in Appendix 4.6.

Species that produce caducous sporangia, for instance, *P. botryosa, P. capsici, P. colocasiae, P. heveae, P. hibernalis, P. infestans, P. palmivora,* and *P. phaseoli,* are also efficient foliar pathogens. According to Waterhouse (1974b), *P. infestans* and *P. phaseoli* are spread by both wind and water splash, but other caducous species are spread only by water splash. One of the reasons *P. infestans* is thought to be dispersed so widely is that the numerous sporangia produced in a field are caducous and thus can be both waterborne and airborne.

Fry and Spielman (1991) state that "little is known about long-range dispersal of *Phytophthora infestans.*" Although the early literature (Crosier 1934) showed that drying sporangia for various periods prior to incubating them in water killed them, it has been assumed that all sporangia are killed by drying; however, Minogue and Fry (1981) showed that dried sporangia survived much better when rehydrated slowly by increasing relative humidity to 100% than by the usual experimental method of plunging sporangia into free water. This finding indicates that sporangia that are moved long distances in the air and later rehydrated gradually could survive and initiate an epidemic of late blight in potato fields.

Robinson (1976) notes that "sporangia [of *Phytophthora infestans*] are windborne and have a high dissemination efficiency" and proposes without any data that a new vertical pathotype (race) will spread for hundreds of miles. Bonde and Schultz (1943) noted that late blight occurred in a potato field about 200 feet (60 m) away from a cull pile of infected tubers on which sporangia formed during rainy weather. They noted that 1% of the plants became blighted when piles of infected tubers were experimentally placed as far as 180 m away. Even higher percentages of infection were noted when infected tubers were placed closer to the field. Because spread was always in the direction of the prevailing wind, they concluded that the sporangia were windborne. Hirst (1953), using petroleum jelly-covered slides in a spore trap, reported that the number of sporangia of *P. infestans* in the air, collected 1 m above infected potato plants, increased to a peak by 11 A.M. and decreased during the afternoon.

Hirst and Stedman (1960) showed indirect evidence for "distant spread" of caducous sporangia of *P. infestans* in addition to local spread from initial sources, such as infected tubers in the potato field. Although it is well known that *P. infestans* spreads from plants emanating from infected tubers, a yearly examination over several years of a 30-acre field continuously planted to potatoes at the Rothamsted Experimental Station in England showed that dissemination from individually infected plants occurred only during 1951 and 1959. In other years, all the plants in the field became infected uniformly. They interpreted this to mean that inoculum of *P. infestans* originated at a distance, "so that the spore cloud had become uniform in concentration."

Although late blight of potato is favored by wet and cool weather, late blight epidemics can develop on potato and tomato in semiarid regions of the world, such as Israel, depending on the dispersal and survivability of *P. infestans* sporangia as affected by hot and dry conditions. Although low relative humidity deters sporangia production, it actually increases the rate of dispersal of the caducous sporangia (Rotem and Cohen 1974). Rotem and Cohen showed by use of the spore trap method of Kramer and Pady (1966) that during a few hours on a hot, dry day (maximum 42°C and minimum 15% relative humidity), 2,000 sporangia were detected on the trap compared with about 400 on a cool, humid day (maximum 22°C and minimum 78% relative humidity). Increased dispersal of sporangia is overridden, however, by the suppressive effects of high temperature and low relative humidity on survival. However, if infection occurs, the mycelium in leaves can survive and renew the epidemic when environmental conditions become more favorable (Rotem and Cohen 1974). See Rotem et al. (1978) for a general review of host and environmental influences on sporulation.

Sporangia of *P. cactorum,* which are caducous with short pedicels, are dispersed in strawberry fields by rain splash (Grove et al. 1985a, c; Reynolds et al. 1987–1989; Madden and Ellis 1990; Madden et al. 1991) and cause epidemics of leather rot of fruit (Ellis and Grove 1983). Studies with rain simulations indicate that sporangia are carried in droplets of water, and since much of commercial strawberry culture utilizes polyethylene plastic sheets on beds to control weeds, plastic was compared with straw and sand to determine whether a solid surface would increase splash dispersal. The percentage of strawberry fruit affected by leather rot was significantly higher where plastic was used than where beds were mulched with straw. Apparently the straw absorbs more of the water droplets and is less reflective than plastic surfaces. Straw mulch is recommended rather than plastic for control of leather rot of strawberries (Ellis and Grove 1983).

P. palmivora, the cause of black pod of cocoa, persists in plantations in the tropical, high-rainfall Solomon Islands (up to 3,000 mm of rainfall per year) in plant material such as flower cushions, immature foliage on seedlings, chupons (shoots at the base of a cocoa tree), and leaves of flush shoots but rarely on mature leaves (Gollifer et al. 1980). *P. palmivora* is caducous and readily dispersed by rain splash. Inoculum can also be introduced or transported by ants, rats (Newhook and Jackson 1977), or pruning tools. Water-transferable propagules can be detected on the surface of healthy canopy leaves. Up to 200 sporangia per leaf were found on fallen leaves of the non-host *Leucaena leucocephala*. The inoculum sources are so numerous that heavy roguing of infected pods and cherelles (immature pods) did not prevent the progress of disease in the Solomon Islands (Jackson and Newhook 1978).

Caducous sporangia of *P. palmivora* that sporulated on papaya fruits were not recovered on Hirst spore trap slides until windy, rainy conditions prevailed, after which the sporangia were readily released from diseased lesions. Detached sporangia dehydrated rapidly (2–3 min) at relative humidities of less than 100%; however, sporangia that remained attached to the lesions survived drying (Hunter and Kunimoto 1974).

Caducity of sporangia affects control practices. For example, spraying the lower "skirts" of citrus trees (about 1 m above the ground) with a fungicide controls brown rot of fruit caused by *P. parasitica, P. citrophthora,* or *P. syringae* because the sporangia from these species are noncaducous and can be transported to fruit on the lower part of the tree only by splashing of soil particles. The caducous sporangia of *P. hibernalis,* however, may be carried in water droplets by wind to the tops of trees during rainy weather; thus, when *P. hibernalis* is present in an orchard, a protective fungicide must be applied to the entire tree (Klotz 1978).

Spread of noncaducous sporangia, chlamydospores, and oospores depends on movement of irrigation water, rain splash of soil particles, or other mechanical means.

Although sporangia of *P. parasitica* are noncaducous, rainfall increased the splash dispersal of infested soil particles among rhododendron plants grown in containers (Kuske and Benson 1983a). Use of gravel (average diameter, 1.9 cm) beds 5–7 cm deep reduced the splash dispersal markedly compared with pine bark. On gravel, only 5% of the plants were killed, but on bark, 70% of the plants died. When propagules of *P. parasitica* in soil are splashed to leaves of *Rhododendron* sp., lesions occur when conditions are warm and moist (Kuske et al. 1983a). Because *P. parasitica* produces only persistent sporangia, the propagules probably consisted of chlamydospores in the soil.

Blight of pigeon pea (*Cajanus cajan* (L.) Millsp.), caused by *P. drechsleri* f. sp. *cajani,* causes stem lesions but little or no root rot (Kannaiyan et al. 1980). When plots in the field were maintained weed-free by hand weeding, about 35% of the plants in the weeded plots were killed compared with only 18% in unweeded plots (Chauhan and Singh 1991). Since the sporangia are persistent and noncaducous, the primary inoculum in infested soil is more likely to splash onto the stems when the soil is free of weeds. *P. syringae,* also noncaducous, is splashed from soil onto apple fruits (Upstone and Gunn 1978). Other examples of infection by splash dispersal are given by Gregory (1983).

Cultural Control

Although cultural control includes practices other than the fungicides or resistant cultivars that growers use in agricultural systems to reduce or eliminate disease, these categories are not clear-cut, and in most cases in soil an interaction of microbial antagonism and antibiosis results in biological control. The microbial antagonism aspect, which suppresses or eliminates inoculum in soil, is discussed in the section on biological control in this chapter. Shea and Broadbent (1983) review both cultural and biological control. Schmitthenner and Canaday (1983) present an interesting overview of the role of various chemical amendments to soil in the degree of severity of *Phytophthora*-caused diseases.

In the introductory portion of this chapter, we discussed some of the principles for control and various characteristics of *Phytophthora* that facilitate cultural control measures. Here we present an overview of cultural methods and how they function.

Prevention of disease is the primary defense in some situations. If there is no inoculum, there will be no disease; however, elimination of inoculum is difficult and in many cases unfeasible. Prevention can be accomplished in some unique cases by quarantine. Quarantines are relatively rare, especially now that plants are moved so frequently on a national, regional, and worldwide basis. In Western Australia, however, quarantines have been established within the jarrah forest to prevent the spread of *P. cinnamomi* into unaffected areas by road-building equipment or on the tires of trucks. Trucks must be washed before proceeding from infested to clean areas of the forest. This aspect of control is discussed by Shea (1977, 1988) and Shea et al. (1984). Many avocado growers in southern California enforce their own rules to prevent the introduction of *P. cinnamomi* into their orchards by means of mud on tires or on boots worn by people. Many glasshouse nursery operations conduct strict sanitary measures. Many other examples could be given of successful prevention strategies.

Soil in glasshouse beds or containers should be composed of either known noninfested soil or, better, a noninfested, well-drained potting mixture of fine sand and peat moss (Baker 1957) or hardwood bark compost that suppresses pathogens such as *P. cinnamomi* (Hoitinck and Fahey 1986). Every effort should be made in nurseries or areas where plants are grown for resale to prevent introduction of disease-causing pathogens, and if they are introduced into a nursery, they should be eradicated by heat (see following section) or fumigation of soil with methyl bromide-chloropicrin (see Chapter 7). The use of copper fungicides in gravel walkways is an excellent sanitation method to suppress the production of sporangia in the gravel and subsequent spread to other areas. Wooden benches in nurseries or research greenhouses should be painted routinely with copper naphthenate to suppress fungi (Baker and Roistacher 1957). Expanded metal benches are easily decontaminated without chemicals.

In the absence of a registered or an economical fungicide or resistance of the crop plant, cultural control is the only method that can be used; and in many cases, cultural control methods are adequate. Certainly, even if an adequate fungicidal control method or host resistance were available, cultural control methods should be integrated into the system so that the use of fungicides can be reduced to a minimum. Some fertilizer amendments to soil enhance control, but others do not (Schmitthenner and Canaday 1983). Control of Phytophthora root rot of avocado involves management of soil water, resistant rootstocks, and fungicides (Zentmyer 1980, 1984; Coffey 1987). Integrated approaches to control root rot of avocado, caused by *P. cinnamomi;* root rot of soybean, caused by *P. sojae;* foot rot of black pepper (*Piper nigrum*), caused by *P. capsici;* and black pod and stem canker of cocoa, caused by *P. megakarya* and *P. palmivora* are discussed by Coffey (1991). The different strategies, which include cultural practices, resistance, and an appropriate fungicide if necessary, used before infection occurs are intermeshed and should be considered as a system.

Sanitation—Its Use and Limitations

Although sanitation (prevention of the deposition of inoculum and/or removal of diseased plant material from the field) is a highly effective method for control of epidemics of single-cycle diseases such as wilt caused by *Fusarium oxysporum,* which produces only one cycle of inoculum in the fall of the crop year (Vanderplank 1963), sanitation is less effective for epidemics caused by *Phytophthora* species that cause foliar disease. For instance, removal or destruction of residual inoculum of *P. infestans* by planting disease-free seed potatoes and destroying refuse piles of contaminated tubers near fields delayed early onset of late blight but did not prevent epidemics later in the season in Maine (Bonde and Schultz 1943, 1944; Vanderplank 1963). Farmers who had noted the delay of early-season blight after these sanitary measures were first practiced (infected potato cull piles were destroyed) and discontinued spraying with fungicides suffered losses late in the season. The ensuing epidemics showed that sanitary practices alone are insufficient to control this disease because inoculum of *P. infestans* is capable of multiplying rapidly from a small initial amount. The apparent paradox is explained when the multicyclic nature of *P. infestans* is taken into consideration. Vanderplank (1963) shows mathematically that the degree of sanitation would have to exceed 99.9% to be fail-safe and that this degree of sanitation is simply impossible in a field situation. Although this explanation should not be construed to argue against sanitation as a desirable primary and supplementary control measure (MacKenzie et al. 1983), it indicates that sanitation in a field or orchard crop has a limited effect on control of a multicyclic disease because inoculum can build up rapidly.

Sanitation at the nursery level before transplants are moved to the field is of primary importance (Pegg 1978; E. M. Hansen 1979). Forsberg (1985) reported a wide range of ornamental plants from which *P. cinnamomi, P. nicotianae* var. *parasitica, P. palmivora, P. megasperma* var. *megasperma, P. drechsleri,* and unidentified *Phytophthora* species were isolated. An accreditation scheme that assured high standards of hygiene in avocado (Pegg 1978) (Appendix 5.2) was successful, and Forsberg (1985) recommended that it should be adopted by ornamental plant nurseries.

As a means to control the red stele disease of strawberry in Scotland, strawberry transplants must be shown to be free of *P. fragariae* before they are planted in fields (Duncan 1976, 1979, 1980a; Duncan et al. 1986). To determine whether strawberry planting stock is free of *P. fragariae,* root tips (25–50 mm long) are cut from strawberry runners sampled in a field from which runners are to be sold. Cross-contamination is prevented by using disposable overshoes and washing implements in sodium hypochlorite solution. The root tips are collected in polyethylene bags and later mixed with a soilless compost in plastic pots. Plants of the highly susceptible cultivar Baron Solemacher (*Fragaria semperflorens* var. *alpina*) are planted in the pots in which the root tip samples have been mixed with the compost. In 3–4 weeks, the presence of red stele symptoms and oospores in the roots indicates that plants from the field in which runners were selected are contaminated by *P. fragariae* (Duncan et al. 1986). The test includes control pots inoculated with known infected roots and uninoculated pots of bait plants. During the period following the introduction of the root tip bait test (1980–1984), the number of commercial fields in Scotland showing red stele disease decreased dramatically, and from 1989 to 1991, no infected planting stocks were found (J. M. Duncan, *personal communication*). In this type of intensive agriculture, sanitation is extremely important.

The record of some commercial nurseries in producing disease-free apple rootstocks is not good. Sampling of 13 nurseries in North Carolina yielded *P. cactorum* or *P. cambivora* in 16.5% of the seedling rootstock (Julis et al. 1978). Tidball and Linderman (1990) isolated *P. cactorum* from stool beds in Oregon. Both *P. cactorum* and *P. syringae,* isolated previously, were pathogenic on apple rootstocks. Although root-dip treatments with the systemic fungicides metalaxyl and fosetyl-Al, applied singly or together, reduced the percentage of infected plants, they did not eradicate these species of *Phytophthora* in British Columbia, Canada (McIntosh 1964, 1966).

Sanitation at the nursery level is important; however, studies in British Columbia show that *P. cactorum* or *P. cambivora* can be introduced via irrigation water (McIntosh 1964, 1966). See the section in Chapter 2, Isolation from Bodies of Water. Several other references cite irrigation water as a source of inoculum (Table 2.5). If irrigation water is contaminated with *Phytophthora,* the beneficial effect of clean planting stock is much reduced.

Methods for monitoring *Phytophthora* in nursery stock by selective media, baits, and ELISA probes are presented in Chapter 2. We have discussed the danger that *Phytophthora* poses in an established nursery because it can spread rapidly by means of drainage water and runoff. Nurseries are sometimes required to recirculate stored runoff water because percolation of nitrates into the groundwater is an environmental hazard. Because *Phytophthora* has been detected frequently in bodies of water (see Table 2.5; Chapter 2), some nurseries chlorinate water. One nurseryman (Daughtry 1983) found that injecting chlorine gas into water with a V500 remote vacuum chlorinator (Wallace and Tiernan, Belleville, New Jersey 07019) to a concentration of 0.3 ppm was more economical than use of calcium hypochlorite. McIntosh (1964) stated that *P. cactorum* and other species were detected in many fruit orchards in British Columbia but could not be detected in orchards irrigated with chlorinated water. See Chapter 7 on chemical control for further discussion of chlorine as a disinfectant.

Areas in southeastern Arizona were in either native grassland or crops not susceptible to *P. cactorum* or *P. cambivora* before apple orchards were established during 1979–1985. Yet in many orchards that were planted to apple trees, decline and death of trees associated with crown and root rot became a problem. Isolations from infected plants yielded *P. cactorum, P. cambivora, P. drechsleri, P. parasitica,* and an unidentified species. Isolations of *Phytophthora* from nursery trees that had come originally from both Maryland and Oregon yielded *P. cactorum* and *P. cambivora,* thus supporting the conclusion that the field infection was the result of contamination of rootstocks at the nursery level (Matheron et al. 1988).

These are but a few examples of how infected nursery stock can and does initiate epidemics. New tree crop industries could well afford to pay more for clean nursery stock if nurseries could deliver disease-free plants. After diseases become established, runoff water can and does become a secondary source of inoculum.

In certain farming systems, especially when the unit value per hectare is low, sanitation is the only method of control. The devastating jarrah dieback disease (Podger et al. 1965) was spread by the introduction of *P. cinnamomi* from infested sites by means of earth-moving equipment used to make roads. In this forest situation, exclusion of inoculum from healthy tree sites has been a primary strategy for control (Shea et al. 1984). Shea (1975b, 1977) concludes that the entire jarrah forest in Western Australia is not uniformly susceptible to *P. cinnamomi* and that in many areas, principally upland areas, the soil-moisture environment is only marginally suitable for pathogenicity and survival because it is too dry during most of the year. Most of the infested areas lie close to "water catchment" areas, where the soil moisture is more plentiful. The presence of a concreted laterite hardpan layer in the soil profile of some sites causes accumulation of enough water to allow sporangial formation by *P. cinnamomi* and subsequent infection of large vertical roots (Shea et al. 1984; see The Hardpan Effect in Agricultural Soils in this chapter). Subsoil water (1 m deep) in the jarrah forest carried zoospores up to 120 m down a hill slope after a heavy rain (Kinal et al. 1993).

Suppression by Heat

In any intensive operation, whether in research, nursery, and glasshouse operations or even a home environment, control of diseases is much simplified if *Phytophthora* or any pathogen can be eradicated. Steam heat is the cleanest form of control, but it is expensive. If sterility of soil is required for research purposes, the growing medium or soil can be sterilized by steam autoclaving under pressure at 120°C for 1 or 2 h,

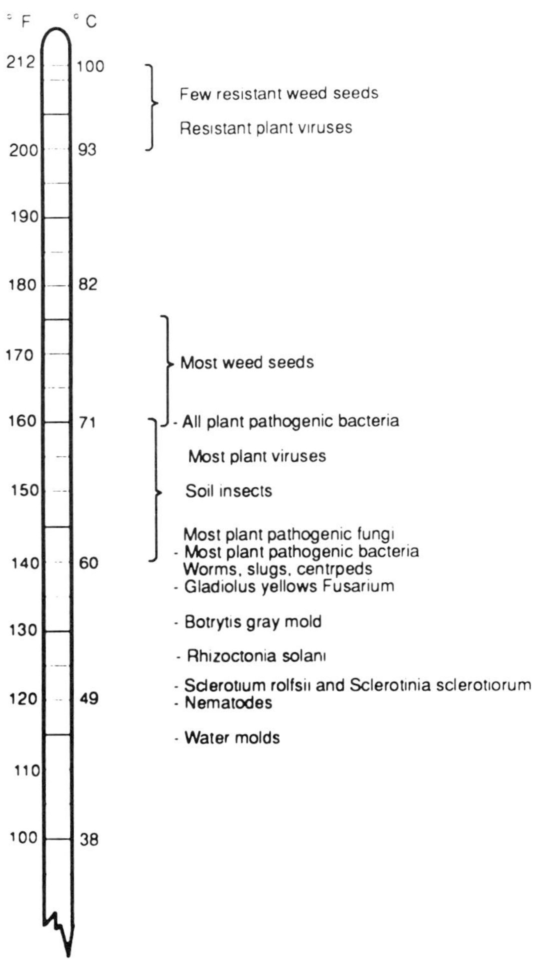

Figure 5.7. Temperatures necessary to kill pathogens and other microorganisms harmful to plants. Most of the temperatures indicated are for 30-min exposures under moist conditions. (Redrawn by permission from Baker and Roistacher 1957 by A. Vaziri)

depending on the size of the load. To eliminate spore-forming bacteria, the soil should be autoclaved a second time after a 2-day period of incubation at room temperature to allow heat-resistant bacterial spores to germinate, a stage at which they are more vulnerable to heat. However, because to our knowledge there are no important bacterial pathogens of plants that produce heat-resistant spores and most plant pathogens are vulnerable to temperatures much lower than 100°C (Figure 5.7), in agricultural practice or in most research projects it is not necessary to achieve complete sterility of soil by autoclaving. The soil needs be steamed only long enough to increase soil temperature at the coolest spot to at least 82°C for 30 min; with flowing steam, the temperature in the entire mass of soil will eventually reach 100°C. When the desired temperature is reached, the flow of steam should be reduced to a trickle so that cold corners or pockets in the soil reach 100°C (Baker and Roistacher 1957). We quote by permission of Baker and Roistacher (1957):

> Because steam thus moves through soil as an advancing front, temperature rise at a given point is usually quite sudden under efficient operating conditions. Steam condenses on the nearest cold soil particles and does not penetrate farther until it has heated them to 212°F. For this reason, several thermometers may be placed at varying distances from the steam inputs to plot the advancing heat front.

Solarization of soil in the field is a relatively new and unique method that consists of trapping radiant heat from the sun with clear polyethylene tarps that have been laid on the surface of the soil. The method is particularly effective where the summers are hot and a large proportion of the days are cloudless. Solarization can increase soil temperatures up to 44–55°C at 5-cm depth and to 36–45°C at 20-cm depth (Katan 1981), temperatures at which fungal spores, bacteria, nematodes, and weed seeds are killed. Control of soilborne microorganisms by solarization was first reported by Katan et al. (1976; see the review by Katan [1981]).

Solarization involves laying a clear polyethylene tarp (about 25–100 µm thick) on moist soil. The plastic must be clear, and it is important that the soil be moist (Pullman et al. 1979). The tarps are weighted with soil to prevent removal by wind and maintained on the land for varying periods, usually 1 month or more, during which the heat from the rays of the sun is trapped under the tarp. The method is expensive because of the cost of tarps and layoff of the land, but it is less expensive than soil fumigation by methyl bromide-chloropicrin and is environmentally sound with no known undesirable side effects. In addition to control of nematodes, bacteria, and many soil fungi, solarization suppresses seed germination of most weed species and increases growth response and yields of the subsequent crop (Stapleton and DeVay 1982, 1986). A recent book edited by Katan and DeVay (1991) contains state-of-the-art information from a wide spectrum of world authorities on the practical application of soil solarization.

Heat developed by solarization effectively suppressed propagules of three *Phytophthora* species in soil. In an experiment at Davis, California, a region in which almost every day in summer is clear, the soil temperature under a polyethylene tarp rose to 45°C at 15-cm depth compared with 32°C in soil not covered. At 45-cm depth, the maximum temperature rose to 33°C compared with 28°C in soil not tarped. Soil was assayed for *P. cinnamomi, P. cactorum,* and *P. megasperma* by flooding 25-cc soil samples in petri plates with tap water followed by floating peach (cultivar Lovell) leaf disk baits on the water for 48 h at 18°C. The survival of each species was based on the percentage of leaf disks that yielded *Phytophthora* when plated on the PARP (Chapter 2, Appendix 2.4) selective medium (Juarez-Palacios et al. 1991). A summary of this report follows (excerpted by permission of the American Phytopathological Society, St. Paul, Minnesota, courtesy of Juarez-Palacios et al. 1991):

> Heat sensitivity and the effect of soil solarization on survival of hyphae and spores of *Phytophthora cinnamomi, P. cactorum,* and *P. megasperma* were determined in inoculated walnut twigs and artificially infested Reiff silty clay loam soil. Viable chlamydospores of *P. cinnamomi* were not detected in infested soil which was exposed to 45°C for 20 min. In contrast, oospores of a high temperature isolate of *P. megasperma* survived exposure for 30 min at 45°C, whereas *P. cactorum* was killed within 30 min at 45°C. In field studies, solarized soil reached a maximum temperature of 45°C at the 15-cm depth and 33°C at the 45-cm depth compared with nonsolarized soil where maximum temperatures of 31° and 28°C were recorded at 15- and 45-cm, respectively. No activity of *P. cinnamomi* was detected in infested soil after 2 wk at the 30-cm depth or after 4 wk at 45-cm depth in solarized soil, whereas, there was little or no reduction in survival in nonsolarized soil. *P. cactorum* withstood the effects of soil solarization at the 30- and 45-cm depths, but was killed within 2 wk at the 15-cm depth, with little or no change in the percentages of infection of peach leaf disks used as bait in nonsolarized soil. Some propagules of the high temperature isolate of *P. megasperma* survived the treatment, although percentage leaf-disk infection was greatly reduced after soil solarization at the 15-cm depth during 4 wk. The heat sensitivity of the isolates of *Phytophthora* in laboratory experiments closely reflected their inactivation in solarized soil and indicated the possible use of soil solarization for management of these pathogens.

P. cambivora, a root pathogen of cherry and almond in South Australia, was controlled by solarization for more than 12 months beginning in December or January (summer in the southern hemisphere) in an irrigated cherry orchard, but solarization was not effective in a nonirrigated orchard. Although the inoculum of *P. cambivora* (in the form of artificially infected beet seed) was suppressed by solarization, established infections on trees were not controlled (Wicks 1988a). This result is not surprising, because the pathogen in roots is protected by living or necrotic tissue.

P. cinnamomi was suppressed by soil solarization in South Africa at depths of 10 and 30 cm (Barbercheck and Von Broemsden 1986).

Solarization of naturally infested soil in Victoria, Australia, with clear plastic (50 μm thick) for 2 months during the summer increased soil temperatures to 45–52°C for 2–4 h each day, an increase of about 8°C compared with uncovered soil. Use of black plastic, which absorbs radiant heat, resulted in a maximum temperature of only 35°C. *P. cinnamomi, Pythium* sp., and *Fusarium oxysporum* could not be recovered after solarization with clear plastic. *P. cinnamomi* was undetectable up to 16 months later, but low levels of a *Pythium* sp. could be detected. The percentage of damping-off was significantly reduced in *Pinus radiata* seedlings grown in soil solarized with clear plastic but not in soil solarized with black plastic (Kassaby 1985).

In addition to the inhibiting effect of heat on fungal pathogens, control by solarization may be associated at least partially with a shift to a higher population of heat-tolerant antagonistic microflora consisting of gram-positive bacteria and thermotolerant fungi or thermophilic bacteria (Stapleton and DeVay 1982, 1986). If so, solarization is an example of a cultural control that is enhanced by a biological control mechanism.

Thermal inactivation of pathogens in infected plants (thermotherapy) has been an attractive method because some plants can withstand higher temperatures than *Phytophthora.* Although a certain amount of success with the method has been shown, the margin between the temperature necessary to kill the pathogen and the temperature that is toxic to the plant is often narrow. Yarwood (1975) discusses the rationale and presents a quantitative approach. Thermotherapy is most useful for operations at the nursery level, where eradication of *Phytophthora* from infected plants might be important. For example, *P. citrophthora* can be eliminated from infected lemon fruits by a 2-min soak at 48.9°C (Barrett and Fawcett 1922; Klotz et al. 1960; Klotz and DeWolfe 1961a–c). *P. richardiae* was suppressed in calla lily rhizomes (*Zandeschia* sp.) by a 60-min soak at 50°C (Dimock and Baker 1944).

Two isolates of *P. cinnamomi* from Fraser fir (*Abies fraseri* (Pursh) Poir) showed different levels of tolerance to high temperatures (Benson 1978). When V8 juice agar disks of inoculum were immersed in water, the ED_{50} (time required for 50% inactivation) at 39°C was 26.3 min for one isolate and 51.7 for the other. At 43°C, the ED_{50} was 2.7 min for one isolate and 3.3 for the other. Hot water treatment (45°C for 15 min) applied to fir seedlings that had been inoculated 14 days previously suppressed *P. cinnamomi;* however, this regime was unsuccessful for naturally infected seedlings. Benson's (1978) explanation indicated that in naturally infected seedlings, the infection had spread to larger roots in which the lethal temperature could not be attained within a 15- to 120-min time frame.

Rotation of Crops

Rotation of land, especially to crops that are not susceptible, has long been a principle of control based on the rationale that inoculum buildup in soil planted continuously to the same crop will eventually reach a level high enough to become a limiting factor (Glynne 1965). However, Glynne (1965) points out that with some disease-crop situations, this principle is not always true. For instance, the take-all disease of wheat appears to become less severe over a period of several years. However, the red stele disease of strawberry, caused by *P. fragariae,* is listed under "pathogens with irreversible disease patterns" following monoculture (Shipton 1977). In Scotland, the amount of land planted to strawberry declined by two-thirds within 10 years after the first occurrence of red stele (Alcock et al. 1930). With monoculture of strawberry on limestone soils in England, the effect of the disease was of lesser consequence (Hickman and English 1951a). The cause of this phenomenon is thought to be associated with a higher population of antagonistic microorganisms in calcareous soil (Duncan 1975b).

Most orchard crops such as avocado, citrus, and apple are essentially monocultures for many years. Many orchards of citrus in California are more than 70 years old and probably have coexisted with *P. parasitica* and *P. citrophthora* for most or all of that time. The root disease caused by *P. cinnamomi* on avocado is a severe, limiting factor and eventually causes death of large trees, especially when soil water drains poorly or rainfall is frequent (Zentmyer 1980). On the other hand, Shipton (1977) notes that in some areas of Queensland, Australia, avocado has existed in the presence of *P. cinnamomi* in suppressive soils for more than 30 years. See also Broadbent and Baker (1974a, b, 1975) and the excerpt from Baker (1978) in the subsection The Ashburner System under the section Biological Control.

For annual crops such as tomato, pepper, and safflower, rotation is undoubtedly an important strategy that should be followed. The inoculum produced by *Phytophthora* is subject to microbial antagonism. Although there are instances of long life in the soil as oospores and chlamydospores (Table 5.3), *Phytophthora* is usually not a competitive saprophyte (Tsao 1969b) and would not be expected to persist in high populations in the absence of a susceptible host.

Fertilizers

Maintaining the proper nutritional status of plants is of course necessary to obtain maximum yield of all crops. Books have been written on the subject, and the state of the art is well advanced for most crops, albeit there are differences in opinion and strategies on how to use fertilizers optimally. Many people have the erroneous concept that adequately fertilized plants will not be as susceptible to disease. This may be true in some cases, but actually many examples show that a well-nourished plant is more susceptible to many diseases than a poorly nourished one!

Nitrogen is applied to most crops because its deficiency is certainly one of the most limiting factors in agriculture, but Schmitthenner and Canaday (1983) point out that nitrogen has both beneficial and detrimental effects on *Phytophthora* diseases. Their review of the effects of nutrients on *Phytophthora*-caused diseases surprisingly shows that some forms of nitrogen favor an increase in disease and other forms of nitrogen suppress disease (Schmitthenner and Canaday 1983) (Table 5.5). Too much nitrogen can increase succulence and susceptibility of foliage. Rhododendron plants stimulated to rapid growth by the increased use of nitrogen fertilizers became more succulent and consequently more susceptible to leaf blight caused by *P. cactorum* than less-fertilized plants (Hoitink et al. 1986). A similar dieback disease in North Carolina is caused by four species of *Phytophthora* (Benson and Jones 1980).

Few fertilizers and nutrients or chemical soil amendments by themselves are effective for control (Schmitthenner and Canaday 1983). The following is excerpted by permission of the American Phytopathological Society, St. Paul, Minnesota, courtesy of Schmitthenner and Canaday (1983):

> Nonetheless, five chemical factors may have direct application in control programs: (1) incorporation of high rates of organic N before planting, to use the toxicity to *Phytophthora* propagules in soil of NH_3 [ammonia] and HNO_2 [nitrite] where economically and culturally feasible; (2) reduction of pH to below 4.0 for acid-tolerant plants; (3) reduction of pH to below 5.0 in high-aluminum soil for aluminum-tolerant plants; (4) judicious application of deficient minerals to optimize uptake by root systems reduced by infection by *Phytophthora* species; and (5) avoidance of indiscriminate over-fertilization in attempts to maximize productivity.

Table 5.5. Effects of nitrogen (N) on *Phytophthora* diseases[a]

Pathogen Host and disease	Factor[b]	Effect[c]	References
P. infestans			
Potato late blight	N	+	Awan and Struchtemeyer (1957)
		+	Gallegly and Niederhauser (1959)
		+	Weindlmayr (1965)
	Nitrate	+	Walerych et al. (1970)
	Ammonium, urea	0	Walerych et al. (1970)
	N	–	Langbein and Pehl (1962)
		–	Lowings and Acha (1959)
		–	Schaffnit and Volk (1927)
Potato tuber rot	N	+	Herlihy (1970)
		+	Herlihy and Carroll (1969)
P. capsici			
Pepper blight	N	+	Elenkov and Bakharieva (1975)
P. parasitica			
Tobacco black shank	N	+	Apple (1961)
		+	McCarter (1965)
Roselle stem rot	N	+	Olunloyo and Adeniji (1976)
Tomato stem lesions	N	–	Weststeijn (1973)
Betelvine wilt	N	–	Thyagarajan et al. (1972)
Philodendron leaf spot	N	–	Harkness and Reynolds (1964)
Citrus root rot	Nitrate	–	Klotz et al. (1958)
	Ammonium, urea	+	Klotz et al. (1958)
P. sojae (P. megasperma f. sp. *glycinea)*			
Soybean rot	N	+	Canaday and Schmitthenner (1979)
P. drechsleri			
Pigeon pea blight	N	+	Pal and Grewal (1976)
P. cinnamomi			
Avocado root rot	Ammonium	–	Broadbent and Baker (1974)
	Nitrite	–	Zentmyer and Bingham (1956)
	Nitrate in solution culture	–	Bingham et al. (1958)
	Ammonia	–	Gilpatrick (1969a)
Pineapple root rot	N	0	Anderson (1951a)
Pine little leaf	N	+	Newhook and Podger (1972)
		–	Campbell and Copeland (1954)
		–	Hepting et al. (1945)
		–	Roth et al. (1948)
P. cinnamomi and *P. parasitica*			
In vitro test	Ammonia, pH 8	–	Tsao and Oster (1981)
	Nitrate, pH 6	–	Tsao and Zentmyer (1979)

[a] Reproduced by permission of the American Phytopathological Society, St. Paul, Minnesota, courtesy Schmitthenner and Canaday (1983).
[b] N = any form of nitrogen. Where nitrogen forms were compared, the forms are indicated.
[c] Disease increased (+), decreased (–), or did not change (0) with high levels of N.

Of all factors—nitrogen, phosphorus, potassium, and minor elements—that have been considered, only the addition of sulfur to soil, which reduces the pH to about 3.8, markedly suppresses the severity of pineapple root rot, caused by *P. cinnamomi* (Pegg 1977b). This practice suppresses formation of sporangia in soil (Pegg 1977b). Although reduction of the pH of soil to about 4.0 by addition of sulfur or aluminum sulfate is suppressive to *Phytophthora* (Pegg 1977a; Schmitthenner and Canaday 1983), this measure must be limited to crops that are favored by low soil pH such as rhododendron, azalea, and pineapple.

Treatment of an acidic soil (pH 4.8) with calcium carbonate and calcium hydroxide in greenhouse research in Brazil to increase the pH increased the severity of a seedling blight caused by *P. capsici* not because of change in pH, but because free aluminum was immobilized in the soil. Addition of calcium as calcium sulfate or calcium chloride did not produce the same effect (Muchovej et al. 1980).

Aluminum sulfate, applied as a drench to a peat moss-vermiculite growing medium (pH 4.1) that had been previously infested with a high concentration of *P. parasitica* inoculum (3,000 cfu per gram of medium) controlled seedling disease of *Catharanthus roseus* (L.) G. Don at a minimal concentration of 17 milliequivalents (meq) of aluminum ion (Al^{3+}) per 100 cm^3 of medium (Benson 1993). Calcium carbonate (dolomitic limestone) is often used to buffer media toward a neutral pH that is more favorable for plant growth. When lime was added (3–6 g of calcium carbonate per 1,000 cm^3 of medium), the concentration of applied Al^{3+} necessary to control seedling disease increased. At 3 g of lime per 1,000 cm^3 of medium, 41 meq of Al^{3+}/100 cm^3 was required, and at 6 g of lime per 1,000 cm^3, 100 meq of Al^{3+}/100 cm^3 was required to control seedling disease. However, as the pH of the medium increased as a result of the addition of lime at 3 and 6 g per 1,000 cm^3, the effective concentration of exchangeable Al^{3+} necessary to control *P. parasitica* actually decreased from 5.7 meq/100 g in unlimed medium to 1.2 meq at 3 g of lime per 100 g of medium and to 0.5 meq at 6 g of lime per 100 g of medium. The population of *P. parasitica* in the plant growth medium was reduced from more than 100 cfu per gram of medium to fewer than 40 9 days after the aluminum sulfate treatment (Benson 1993). Benson (1993) suggested that the increased fungistatic effectiveness of aluminum might be caused by the presence of a large part of the soluble aluminum as the $Al(OH)_2^+$ ion. The results suggest that addition of lime to aluminum sulfate could render the treatment less phytotoxic but still fungistatic to *P. parasitica.* In vitro studies showed that dosages of 1.25 meq of aluminum per 100 ml of water at pH 5.0 inhibited sporangia production of *P. parasitica* on mycelial mats.

The suppression of germination of chlamydospores, sporangia, and zoospores by many nitrogenous amendments encourages further research (Tsao 1977a). Many nitrogenous amendments are phytotoxic when used at high concentrations. Preplant amendment of *Phytophthora*-infested soil with ammonia-generating fertilizers suppressed *P. cinnamomi* and *P. parasitica* (Tsao and Oster 1981); however, the phytotoxicity of ammonia is a major problem (Gilpatrick 1969a, b). Boughton et al. (1978) observed that infection of *Eucalyptus marginata* seedlings by *P. cinnamomi* was suppressed from 75 to 1% by application of calcium carbonate to the soil. Lee and Zentmyer (1982) noted that fertilizing naturally infested soil with calcium nitrate (100–300 µg/g of soil) and ammonium sulfate (300 µg/g of soil) reduced Phytophthora root rot of avocado.

The role of fertilizers or nutrients in controlling or suppressing *Phytophthora* is unclear. Chemical factors that have been studied in relation to reducing or controlling *Phytophthora* diseases include calcium, magnesium, sulfur, nitrogen, potassium, phosphorus, micronutrients, pH, salinity, and various combinations of fertilizers. Reports on several chemicals have resulted in contradictory findings. Out of 30 studies conducted by various investigators on the effects of different nitrogen sources on *Phytophthora* diseases, 16 found nitrogen to be beneficial in decreasing the severity of *Phytophthora* diseases, but 14 found that nitrogen increased the severity of the disease (Schmitthenner and Canaday 1983) (Table 5.5). Schmitthenner and Canaday (1983) concluded that the form of nitrogen applied, the soil type, and the host-pathogen combination all influenced the severity of the disease.

Kincaid et al. (1970) used multiple regression statistics to determine the overall effects in the field of several soil factors on black shank of tobacco, caused by *P. parasitica.* They showed that as pH increased from 5 to 7, the black shank index increased linearly. Although low soil pH would be an effective control, they pointed out that the prevailing practice of adding lime to soil to increase leaf quality precluded use of this type of cultural control. Despite this limitation, they indicated that rotation and use of resistant cultivars achieved reasonably good control of black shank.

In a study of the relationships of inorganic nutrients on potato growth and the production of lesions caused by *P. infestans* on leaves, an increase in amount of potassium applied in a Hoagland's nutrient solution (Hoagland and Arnon 1950) was correlated with an increase in lesion size. Excess magnesium competed with potassium and had the same effect (Umaerus 1969). The size of lesions on the cultivar President (which has general but not race-specific resistance) was not greatly different from that on the susceptible cultivar Bintje when low or medium levels of potassium were maintained in the nutrient system. Umaerus (1969) confirmed the conclusions of Main and Gallegly (1964) that control of host nutrition is necessary for accurate assessment of general resistance. Thus, this study has more application to methods of testing for resistance than for field application.

The review by Schmitthenner and Canaday (1983) illustrates the many contradictory effects of agricultural chemicals on control of *Phytophthora.* This does not indicate that proper management of agricultural amendments is not important, but that there are many interactions caused by several opposing factors. If these effects can be quantified by modern computerized research, probably these factors could be placed in better perspective. Much research (particularly on a field scale) needs to be conducted before the role of chemical factors in *Phytophthora* disease complexes can be fully elucidated.

Potting Mixes for Nursery Use

Baker's (1957) comprehensive publication on container-grown crops showed the value of using well-drained mixes containing peat moss, commonly called UC mixes. Even though peat moss is acidic (pH 4.0), it does not usually lower the pH of the soil mix enough to suppress soilborne pathogens. As an alternative to peat moss, composted hardwood bark has become a highly useful component in greenhouse soil mixes (Hoitink et al. 1975, 1977; Daft et al. 1979; Hoitink 1980; Hoitink and Fahy 1986). Composted hardwood bark used in container-grown nursery stock in a 4:1 (v/v) mixture of bark to Canadian or German peat moss controls rhododendron root rot caused by *P. cinnamomi.* The suppressive nature of the bark was lost if the volume of peat equaled or exceeded that of the bark. Also, if the hardwood bark compost contained more than 60% wood chips, the suppressive effect was lost (Hoitink 1980). Because of the widespread success in nurseries in the eastern United States, composted hardwood bark has become a much-used component of soil mixes and has replaced or supplemented peat moss for greenhouse propagation and growth of many container-grown ornamental crops (H. A. J. Hoitink, *personal communication*). Composted pine bark has also been used successfully for the reduction of *P. cinnamomi* in soil (Gerrettson-Cornell et al. 1976).

Composted hardwood bark is a convenient method of controlling *Phytophthora* species on container-grown stock, since it reduces the need for soil sterilization and soil fungicides. However, the techniques for preparing composted hardwood bark may vary. According to Hoitink (1980), "Success or failure of any compost treatment for disease control depends on the nature of the raw product from which it was prepared, the maturity of the compost, as well as the process of composting used." Control of root rot of rhododendron and other ornamental plants by use of composts and other means is covered by Hoitink and Powell (1989, 1990).

The nature and mechanisms of control of *Phytophthora* and other soilborne pathogens by hardwood bark composts were reviewed by Hoitink and Fahy (1986). Hardwood bark decomposes slowly and is much more effective than pine bark, which resists decomposition. Composting of hardwood bark is an aerobic process in which the internal temperature in the compost windrow rises to 60–70°C. According to Hoitink et al. (1976) and Hoitink (1986), pathogens are suppressed by the high temperature in the compost pile, the toxic products from the compost, and the microbial antagonists that recolonize the compost. Most pathogens within the compost pile are suppressed by temperatures of 55°C for 30 min, but recolonization of the cooled compost material by antagonistic fungi such as *Trichoderma* spp. and *Gliocladium virens* contributes to continued suppressiveness. This biological process associated with composting causes a release of chemical inhibitors of sporangia and zoospores (Hoitink et al. 1976; Hoitink and Powell 1990). Although hardwood bark composts are biologically suppressive to *Phytophthora* species, these effects do not usually last longer than a year (Hoitink and Powell 1990).

Although organic additives in soil mixes improve drainage by increasing the amount of air-filled pore spaces in the mix, thus also improving the control of Phytophthora root rot (Baker 1957; Filmer et al. 1986), the suppressiveness of composted hardwood bark was shown to be the result of biological factors other than better drainage per se (Hoitink et al. 1977). A planting medium composed of composted hardwood bark-peat-perlite-sand (5:2:2:1, v/v) was compared with a peat-perlite-sand (2:1:1, v/v) medium for growth of apple seedlings. When apple trees were inoculated with zoospores or oospores of *P. cactorum,* the percentage of trees killed in the composted hardwood bark medium was much lower than that in the peat-perlite-sand medium. Sporangium production and zoospore viability were reduced by leachates from the composted bark but not by leachates from the medium not containing composted bark. Suppression of disease on apple seedlings is caused by the production of inhibitory substances in the composted bark leachate (Spring et al. 1980).

Composted bark from marri (*Eucalyptus calophylla* R. Br. ex Lindl) and karri (*E. diversicolor* F. J. Muell) trees was suppressive to *P. cinnamomi* in a soil mix used in a container-plant industry in Australia (Sivasithamparam 1981; Sivasithamparam et al. 1981; Hardy and Sivasithamparam 1991c). During the composting period, the numbers of soilborne *Bacillus* spp. and pseudomonads increased (Hardy and Sivasithamparam 1989). Extracts from the composted eucalyptus bark medium (a mix of composted *E. diversicolor* and *E. calophylla* bark and sand [3:1] suppressive to Phytophthora root rot) were compared with a nurseryman's mix (pine bark, peat, sand, and jarrah [*E. marginata*] sawdust [2:3:1.5:2] conducive to Phytophthora root rot). Leachates were prepared by soaking 2 liters of the composted eucalyptus bark medium and the nurseryman's mix for 24 h in 1 liter of deionized water. Filtrates were passed through 10 layers of cheesecloth and centrifuged, after which the pellet (particulate matter) was discarded. When necessary, the filtrate was sterilized by passage through a 0.22-µm pore size filter. Mycelia of *P. cinnamomi, P. cryptogea, P. citricola, P. drechsleri,* and *P. nicotianae* on colonized Miracloth disks were then incubated in the extracts to determine the effects of leachate from bark samples on sporulation. Production of sporangia of all species was suppressed by the extract of nonsterile, composted eucalyptus bark medium, but sporangia production in the sterile medium was stimulated more than in the nurseryman's mix extract. Chlamydospore production by *P. cinnamomi* was greatest in extracts from the nurseryman's mix (sterile and nonsterile), but about one-tenth as many sporangia were formed in the sterile, composted eucalyptus extract. Nonsterile extract suppressed chlamydospore formation, and those formed in nonsterile eucalyptus bark extract were rapidly lysed by bacteria and actinomycetes (Hardy and Sivasithamparam 1991a).

In a study of the effect of matric potential in the composted eucalyptus bark medium and the nurseryman's mix on sporulation and germination of different *Phytophthora* species, the container medium and matric potential of the medium exerted different effects on each of the *Phytophthora* species tested. In the suppressive eucalyptus bark medium and in the nurseryman's mix, sporangium formation (on disks of mycelium-colonized Miracloth) was similar, but little or no zoospore formation by *P. cinnamomi* or *P. drechsleri* occurred in the eucalyptus bark mix. *P. citricola* sporangia formed only in the nurseryman's mix and after a 10-day incubation time. Oospore and chlamydospore formation was greatly reduced in the eucalyptus medium (Hardy and Sivasithamparam 1991b). Since the eucalyptus bark medium contained much higher numbers of fungi, bacteria, yeasts, and actinomycetes than the nurseryman's mix, Hardy and Sivasithamparam (1989) concluded that this larger population could possibly account for the suppressiveness of the composted eucalyptus mix. In saturated soil (matric potential = 0), *P. citricola* did not produce any oospores on either composted eucalyptus bark or nurseryman's mix media. However, as the matric potential declined (soil became drier) to values below –2.5 kPa, oospores formed; but they were rapidly lysed by associative microorganisms. Bacteria and actinomycete filaments were closely associated with all propagules in the composted eucalyptus bark mix, particularly in drier soils (–2.5 kPa).

Isolates of actinomycetes and fungi from the eucalyptus bark compost were tested for their effects on sporangium production and germination. Production of abnormal sporangia, direct and indirect germination of sporangia, and lack of viability were associated with exposure to isolates from soil mixes in varying degrees, but none affected survival or infectivity of *P. drechsleri* when tested on roots of the susceptible plant waratah (*Telopea speciosissima* L.), although there was a consistent reduction in the percentage of disease incidence resulting from the presence of actinomycete or fungal isolates from composted eucalyptus bark (Hardy and Sivasithamparam 1991d).

Nonsteamed eucalyptus bark compost was suppressive (in descending order) to *P. cryptogea, P. nicotianae, P. citricola, P. drechsleri,* and *P. cinnamomi* infection of the highly susceptible waratah and *Banksia accidentalis* R. Br. plants. These results indicate that the control of *Phytophthora* species in the nonsterile, composted eucalyptus bark is biological. Also, since the drainage characteristics of eucalyptus bark medium and the nurseryman's mix (conducive) conformed to accepted standards, the degree of control on the composted eucalyptus bark is not likely to be the result of improved drainage (Hardy and Sivasithamparam 1991d).

The results of a study of sporangial responses by *P. drechsleri* by Hardy and Sivasithamparam (1991d) suggested that the isolates from microflora in composted eucalyptus bark were capable of suppressing root disease but not *P. drechsleri.* They concluded that suppression was the result of biological factors because steaming the eucalyptus bark mix caused it to become conducive.

The total porosity of soil mixes is an important component. In general, the severity of root rot decreases with an increase in the total porosity of the soil mix after incorporation of organic components (Ownley et al. 1990). The severity of root disease of *Rhododendron* sp. caused by *P. cinnamomi* was reduced by use of noncomposted, hammer-milled pine bark in soil mixes. This additive to soil mixes affected the matric potential and in correlation reduced the degree of Phytophthora root rot. We quote from the summary of the report by Ownley and Benson (1991) (reprinted by permission of the American Phytopathological Society, St. Paul, Minnesota, courtesy of Ownley and Benson 1991):

> At low and high inoculum densities (nine and 30 infested oat grains per container) of *Phytophthora cinnamomi,* severity of Phytophthora root rot of *Rhododendron* sp. was significantly less in pine bark media than in peat/sand/soil. Differences in root rot severity between the pine bark media and the peat/sand/soil medium were best explained by differences in air-filled porosity and matric water potential among media. Matric water potential generally fluctuated from –1.0 to –3.0 kPa between irrigations in the pine bark media and from 0 to –1.0 kPa between irrigations in the peat/sand/soil medium. Pine bark media had volumetric air-filled porosities that fluctuated from about 40% just after irrigation to about 45% 48 h later, whereas peat/sand/soilhad an air-filled porosity near 10% after irrigation and about 20% 48 h later. Irrigation water ponded on the surface of containers with peat/sand/soil immediately after trickle irrigation. This was not observed in containers with pine bark media. Relatively large pores that provide high air-filled porosity in pine bark media account for the rapidity with which pine bark media drain after irrigation, which resulted in matric potentials unfavorable for zoospore discharge and dispersal.

When *Aucuba japonica* (gold dust plants) were grown in hardwood bark compost, hammer-milled pine bark, or a soil-sand medium and inoculated with *P. citricola* and *P. cinnamomi,* root rot was suppressed in both hardwood bark compost and pine bark, but severe root rot occurred in the soil-sand mix. The suppressive qualities of hardwood bark compost were lost if the medium was saturated, but the pine bark remained suppressive even when saturated. When the pH of pine bark (4.5) was neutralized to 6.5 with lime, the suppressiveness was lost (Spencer and Benson 1981, 1982).

Although composted hardwood bark mulched around apple trees was found to be effective for control of collar rot in an experimental test (Spring et al. 1980) and in orchards in Ohio (H. A. J. Hoitink, *personal communication*), in British Columbia this treatment was unsuccessful in the field (Utkhede 1984a, b). Softwood and hardwood bark composts (1 kg per tree per site) were reapplied on apple trees inoculated with *P. cactorum;* however, neither component used as a mulch decreased Phytophthora root rot. Application of several fertilizers (g per tree)—calcium nitrate (540), urea (180), sewage sludge (1,000), ammonium sulfate (400), and ammonium nitrate (240)—increased infected trees from 14% for the untreated control to 50–80%.

Negative interactions resulting indirectly from the use of a successful pathogen-suppressive soil mix are possible. For instance, growers, particularly of *Rhododendron* sp., who have come to depend on the *Phytophthora*-suppressive nature of composted hardwood bark, tried to shorten the production cycle by increasing the rate of nitrogen fertilization. However, epidemics of dieback caused by *P. cactorum,* which previously had been a relatively minor foliar disease, became more prevalent. Hoitink et al. (1986) showed that increase in susceptibility of *Rhododendron* sp. to *P. cactorum* was directly correlated with an increase in nitrogen uptake in host tissue. This is an example of a strategy of control that emphasized suppression of one target pathogen but increased the susceptibility to another. Despite this problem, the economics of producing container-grown plants requires that they grow rapidly, so this secondary problem is approached by sanitation, judicious use of irrigation water, and preventative fungicides (Hoitink and Powell 1990).

In some cases, waste products that are readily available have been mixed into soil with the intention of changing the microorganism complex with the introduced substrate and thereby either antagonizing and/or competing with the pathogen or releasing products that might be toxic to soil pathogens (R. J. Cook and Baker 1983).

Organic Amendments to Soil

In a review of principles of use of organic amendments in soil to control root diseases, Linderman (1989) states that certain amendments to soil can either enhance disease, as they do for *Sclerotium rolfsii* because germination of dormant sclerotia is stimulated, or they can result in reduced populations of certain pathogens. For example, the population of *P. cinnamomi* is reduced in the deep humus layer of the Australian rain forest (R. J. Cook and Baker 1983). In Australia, the natural rain forest ecosystem was reconstructed in a naturally infested avocado grove by the Ashburner system (see section on biological control). Despite several empirical examples, the mechanisms of this type of control are still obscure. Linderman (1989) states (excerpted by permission of the National Research Council of Canada, courtesy of Linderman 1989):

> It is difficult to draw other than conceptual conclusions from this discussion and from our research. It seems clear that organic amendments to soil can have a range of effects on soil-borne plant diseases, and these effects are strongly related to the kind of material added and its state of decomposition. At intermediate stages of substrate degradation, decomposition products can enhance or decrease disease, and the response is largely dependent on the pathogen involved and whether the amendment supports specific stages of its life cycle, or favors other microbes which can effectively out-compete and antagonize the pathogen. At the end of the decomposition process are humic compounds that induce a variety of responses from plants and soil microbes. The challenge to harness these humics for specific purposes will certainly involve improved fractionation and purification procedures that will allow specific humic fractions to be associated with specific responses. Without these refinements, we will continue to juggle the wide range of kinds and sizes of humic compounds in relation to a wide range of plant and microbial responses without any justifiable expectation of a predictable response when humics are added to agricultural and forest soils.

Several amendments that have a deleterious effect on *Phytophthora* include alfalfa meal (2–4%), cotton waste, soybean meal, wheat straw, chicken manure (2–4%), hydrolyzed feather meal, and urea (0.1–0.2%) (Zentmyer 1963; Gilpatrick 1969a, b; Tsao 1977a; Tsao and Zentmyer 1979). Although the history of use of organic amendments indicates some positive results in glasshouse tests, success has not been obtained consistently in the field.

Saponins present in alfalfa meal could be a factor in disease control. Saponin fractions, at concentrations present in alfalfa, prevented or greatly reduced the formation of sporangia, decreased zoospore germination to a low level, and retarded mycelial growth (Zentmyer and Thompson 1967).

The reduction of Phytophthora root rot of avocado by alfalfa meal is associated with production of ammonia and an increase in the number of competitive microorganisms in the soil. Gilpatrick (1969a) found that ammonia levels rose to 60–197 ppm in potted soil 1 week after amendment with 5% (w/w) alfalfa meal but decreased in 4–8 weeks to 2–8 ppm. He concluded that ammonia produced by the alfalfa meal may not only be fungitoxic to *P. cinnamomi* but also phytotoxic to avocado roots, thus making them less attractive to the pathogen. Further evidence of the role of organic amendments in control of Phytophthora root rot from ammonia-killed avocado roots indicates that organic amendments such as alfalfa meal reduce the amount of inoculum of *P. cinnamomi* in the soil by preventing growth of the pathogen (Gilpatrick 1969b). No sporangia were produced on the surface of diseased avocado roots from alfalfa meal-amended soil, but sporangia formed abundantly on roots from nonamended soil. Also, in some cases, *P. cinnamomi* could not be isolated from naturally infected avocado roots that had been incubated for 2–4 weeks in soils amended with alfalfa meal. Organic amendments with high C:N ratios (e.g., grain straw) were less effective in reducing root rot than those with low C:N ratios (e.g., alfalfa meal).

After the successful results with alfalfa meal in the glasshouse, a large number of field trials were initiated in southern California on mature avocado trees in early stages of disease development. Avocado trees are generally shallow rooted, and the majority of the feeder roots are in the top layer of soil. Alfalfa meal, 0.23–0.69 kg per 1.1 m^2 of soil surface (0.5–1.5 lb/ft^2), was applied under each tree and worked lightly into the soil. The results were variable, showing good response in some cases but no general trend for improvement (Zentmyer 1963). The pot tests may have achieved greater success because the alfalfa meal was mixed with all of the infested soil rather than applied as a mulch. See also Zentmyer (1980), who discusses the lack of success in controlling Phytophthora root rot in the field with alfalfa meal.

Despite these conclusions that alfalfa meal was unsuccessful in California, a recent report from Mexico indicated that avocado trees treated with alfalfa straw (25 kg per tree) or bovine fresh manure (360 kg per tree) suppressed the population of *P. cinnamomi* in the soil and induced an increase in root growth and yield of fruit (Juarez et al. 1989).

Biological Control

Concepts

The concept of biological control has gained much attention during the past several years and is defined by Cook and Baker (1983):

> Biological control is the reduction of the amount of the inoculum or disease-producing activity of a pathogen accomplished through one or more organisms other than man.

Within this definition, Cook and Baker include manipulation of genetically controlled resistance in the host and cultural practices as well as microorganismal antagonists of the pathogen. The United States National Academy of Science (NAS) (1987) defined biological control as follows:

> The use of natural or modified organisms, genes, or gene products to reduce the effects of undesirable organisms (pests) and to favor desirable organisms such as crops, trees, animals, and beneficial insects and microorganisms.

A more specific concept was originally defined by DeBach (1964), a well-known entomologist, who states that biological control consists of the use of parasites, predators, or pathogens in maintaining another organism's population density at a lower average than would occur in their absence. Gabriel and Cook (1990) defend the broader NAS (1987) definition and propose that conceptually, control of pests and diseases should be divided into physical, biological, and chemical methods. Physical methods include quarantine, meristem culture (freeing a plant of naturally occurring pathogens), light reflection, tillage, heat, and fire. Chemical control includes inorganic, organic, synthetic, and natural chemicals. Biological control would include natural or modified (e.g., genetic recombinant) organisms (hyperparasites and microbial antagonists) and genes or gene products as well as host resistance. Their idea is thought provoking, and it points out how different factors interact. Gabriel and Cook (1990) properly state that cultural methods can be a means to biological control, just as plant breeding is a form of biological control.

Microbial antagonism and hyperparasitism, phenomena that exploit the effects of one or more organisms in the environment against the pathogen, comprise a viable control alternative in plant pathology that is being vigorously researched in many areas of the world for control of various diseases because a large number of chemicals used for control are no longer available or are likely to be removed from the market as a result of environmental concerns. Widespread commercial use of this type of control is still in the experimental stage.

Early research in which certain bacterial or fungal antagonists or antibiotic producers were added to sterilized soil before infesting it with the pathogen showed excellent control. As is true in many of the contemporary studies, however, most of these microbial candidates did not fare well when placed in the field (K. F. Baker 1978; Malajczuk 1983). Many bacteria, actinomycetes, and fungi are antagonistic to *Phytophthora,* but their activity in the field is limited. The use of microbial antagonism against *Phytophthora* is an important component of control, and undoubtedly such systems are operative in nature (Malajczuk 1983; Shea and Broadbent 1983), but the prospect of implementing control by use of biological antagonism is still uncertain, especially on a practical, short-term basis (Shea and Broadbent 1983).

The idea of using one microorganism against another, the predator versus the prey, offers an environmentally compatible and intellectually appealing approach for the control of plant diseases. Biological control was used with considerable success at the turn of this century to save the citrus industry in California from the ravages of the cottony cushion scale insect. The predator used in this particular case was the lady beetle imported from Australia. Although many other insect problems were successfully controlled by natural means, the advent of new and powerful pesticides that could eradicate pests by either a single or a few well-timed spray applications caused biological control to be less appealing. Only in the last decade or two have we realized some of the problems that have developed with the indiscriminate use of pesticides.

The growing interest by the public in safer pest control that will protect the environment and the problem of pests developing resistance to ever-increasing dosages of pesticides make it obvious that more effort is needed in developing biological control methods to be incorporated into integrated pest management programs for the control of plant pathogens. The many successes and failures of biological control imply that much more research is needed before it will become practical.

The development of effective biological control of *Phytophthora* species has been fraught with difficulties because of their ability to produce several forms of inoculum (zoospores, sporangia, chlamydospores, oospores, and mycelium) rapidly and repeatedly, their ability to penetrate and infect a host plant within a few hours, their propensity to exist in soil to depths that allow them to escape most antagonists, and in some cases their wide host range.

The concept of suppressiveness of certain soils to pathogens is well documented (Broadbent et al. 1971; Broadbent and Baker 1974a, b; K. F. Baker 1978; Halsall 1982a, b; Cook and Baker 1983; Linderman et al. 1983). The cited workers and others too numerous to mention here have contributed greatly to the concept that biological control is feasible, but it is not yet well enough understood to deliver a system to the field.

The Ashburner System

A much-quoted example of a cultural system for control that apparently works in the field is known as the Ashburner system. Soil amendments and cover crops are used in certain avocado groves in New South Wales and Queensland to control root rot caused by *P. cinnamomi.* The system is described by K. F. Baker (1978) (see also Pegg 1977b; Cook and Baker 1983; Shea and Broadbent 1983).

One of the most intriguing examples of biological control is given by K. F. Baker (1978), one of the most productive synthesizers of information on biological control of our time (K. F. Baker and Cook 1974; Cook and Baker 1983). He and Patricia Broadbent (Broadbent and Baker 1974a, b) researched the now famous Ashburner system of farming in Queensland, Australia, which allows growth and development of healthy avocado trees despite the presence of *P. cinnamomi* in the soil. A portion of a paper by K. F. Baker (1978), "The Ashburner System to Control *Phytophthora cinnamomi* in Australia," follows (excerpted by permission of H. T. Hartman, editor of *Proceedings, International Plant Propagators Society,* Davis, California, courtesy of K. F. Baker 1978):

Most of the avocado growers in Queensland and New South Wales now used this so-called "Ashburner system" and the disease losses have been greatly reduced (Pegg 1977b). The sequence of the system: New Zealand blue lupine is planted in the fall (March-April). This is disced down in spring (October-November) when in flower, and chicken manure (2 tons/acre) broadcast, plus an NPK fertilizer (1 lb. per tree). A mixture of *Lablab purpureus* and corn or sorghum is immediately thickly planted. This is disced in fall, and chicken manure (2 tons/acre) and NPK (1 lb. per tree) applied. Dolomite limestone is added whenever the pH falls below 6.0. Blue lupine is then planted, and so on *ad infinitum.* This procedure supplies a great deal of organic matter on the surface of the soil. Surface roots are never disturbed by cultivation or plowing. These cover crops are grown in place two years before a new planting of avocados is made. The organic matter is piled around the base of young trees, but not against the trunk, for the first five years. After that, fallen leaves maintain the organic matter under the trees, but organic matter (barley straw, sorghum, or Rhodes grass hay) may be added. Use of container-grown trees free of *Phytophthora* is emphasized in planting (Broadbent and Baker 1974a).

Old, diseased groves may be pulled and started anew with the above procedure. When an old orchard (Ware) was replanted after following the procedure outlined by K. F. Baker (1978), the population of *P. cinnamomi* was reduced below the detectable limit, and the trees made excellent growth (Pegg 1977b).

K. F. Baker (1978) continues:

There are at least two general means by which suppressive soils operate to decrease activity of *P. cinnamomi.* The fungus requires a stimulatory compound produced by soil bacteria to form abundant zoosporangia in soil. These bacteria occur in all soils, but may be repressed by inhibitory microorganisms, or the compound may be destroyed by them. This apparently is the dominant effect in the Ashburner suppressive soil. The inhibitors seem to be inactivated in waterlogged suppressive soil, and infection then occurs. Other antagonists may operate more directly by attacking the mycelium, chlamydospores, vesicles, or zoosporangia of the pathogen (Broadbent and Baker 1974a, b). This is clearly shown by the Ware soil, in which the fungus produces copious zoosporangia. Why then is it suppressive? Probably the germ tubes of the zoospores are attacked, preventing infection. [This is shown in Table 5.6.]

It is of interest how Ashburner first devised his system 40 years ago. He felled rain forest for his planting area. Reading that avocado was a rain forest tree in Central America, he tried to maintain rain forest conditions in his grove. He was told that this meant high organic matter on the surface, fairly high calcium, magnesium, phosphate, and nitrogen (mainly in the ammonium form), and a pH near neutrality. The trees he planted were from a nursery that supplied trees to many others who subsequently sustained heavy root-rot losses. Probably the trees were infected, or at least infested, when planted. The rest is history! Ashburner might well be called a precocious organic gardener, and the pertinence of his reasoning was exceptional.

In rain forests, nutrients are brought up from deep soils by roots and are recycled in the surface by fallen leaves. Calcium is locked up in the organic cycle, with almost none lost, but is one of the first

Table 5.6. Mechanisms of biocontrol of *Phytophthora cinnamomi* in avocado soils in Queensland, Australia[a,b]

		Microbial quality of soils			
Soil type	**Pathogen**	**Sporangial stimulators**	**Sporangial inhibitors**	**Lytic microorganisms**	**Root rot**
Suppressive					
Moist	Sporangia sparse (Ashburner)[c]	+	++++	+++	Slight
	Mycelium lysed (Ware)[c]	++++	+	++++	Moderate
Waterlogged	Sporangia and mycelium moderate	+++	+	+++	Moderate to severe
Conducive	Sporangia and mycelium abundant	++++	+	+	Severe

[a] Reprinted by permission of H. T. Hartman, editor, Proceedings of the International Plant Propagator's Society, Davis, California, courtesy K. F. Baker (1978).
[b] Table utilizes data from Broadbent and Baker (1974a, b). Plus signs indicate relative increase in effect.
[c] Refers to name of grower's field.

cations to be leached from mineral soil. Thus, Queensland rain forests have 3,200–10,400 ppm exchangeable Ca, but the organic matter is lost and Ca is quickly reduced to 180–270 ppm in cultivated pineapple fields. Suppressive avocado soils range from 3,000 to 6,000 ppm (Broadbent and Baker 1974a; Pegg 1977b). The highly conducive western Australia Gosnell sand has only about 160 ppm. Cultivation has, in this sense, been a largely exploitive process. Nitrogen (as NH_4) and Mg are also involved in the organic cycle. With loss of organic matter and Ca, the soil becomes too acid for bacteria. Microorganisms also decline because of loss of Ca and N, and the soil rapidly becomes biologically impoverished (K. F. Baker and Cook 1974; Broadbent and Baker 1974a).

See Table 5.6 for K. F. Baker's (1978) concept of mechanisms of biocontrol by the Ashburner System.

According to Baker (1978), the rain forest ecology may suppress *P. cinnamomi* because 1) high organic matter, calcium, and nitrogen levels stimulate antagonistic microorganisms; the soil is biologically very active; 2) soil pH of 6.0–7.0 is favorable for bacteria; 3) high organic matter and calcium levels improve soil structure and drainage; 4) high calcium levels may affect host resistance; and 5) healthy plants remove much water from soil and decrease waterlogging.

The principles in K. F. Baker's (1978) paper are still valid today and exude optimism for the future success of biological control. Unfortunately, the application of effective biological control practices is still elusive in much of the world. Several review articles espouse the development of biological control of many diseases (Tsao 1977a; R. Baker 1968, 1987, 1990; Linderman et al. 1983; Linderman 1985, 1989; Malajczuk 1983, 1985; R. Baker 1990; Linderman and Ribeiro 1991). Biological control has become a high-priority goal in agricultural research programs in the United States and many other countries. We review these concepts and the progress in the field of biological control below.

Recent Examples of Experimental Biological Control

References on the antagonistic effect of fungi other than basidiomycetes (Table 5.7) and basidiomycetes (Table 5.8) indicate the potential for control of many *Phytophthora* species. These antagonistic fungi exert lytic effects on mycelium, chlamydospores, sporangia, and oospores. Although most of these effects are demonstrable, very few have been utilized successfully in field situations.

Coffey (1987, 1991), who has worked with *P. cinnamomi*, the major fungal pathogen on roots of avocado, indicates that "biological control is still only a good idea, with practically no practical applications in California." At present there are few

Table 5.7. Antagonism of fungi to *Phytophthora* species[a]

Antagonistic fungi	*Phytophthora* species	Reaction[b]	Reference
Chytridiales			
Canteriomyces stigaoclonii	*P. megasperma*	OP	Sneh et al. (1977); Hsu and Lockwood (1984)
Hypochytrium catenoides	*P. cactorum*	OP	Drechsler (1938)
	P. erythroseptica	OP	Wynn and Epton (1979)
	P. megasperma	OP	Sneh et al. (1977); Hsu and Lockwood (1984)
Pleolpidium inflatum	*P. erythroseptica*	OP	Waterhouse (1940)
	P. megasperma	Z/SP	Waterhouse (1940)
Rhizidiomycopsis japonicus	*P. erthroseptica*	OP	Waterhouse (1940)
	P. megasperma	OP	Sneh et al. (1977); Hsu and Lockwood (1984)
Oomycetales			
Leptolegnia sp.	*P. megasperma*	OP	Sneh et al. (1977); Hsu and Lockwood (1984)
Pythium sp.	*P. megasperma*	OP	Sneh et al. (1977); Hsu and Lockwood (1984)
Mucorales			
Mucor spinosus	*P. infestans*	HL	Lacey (1965)
Moniliales (= Hyphomycetes)			
Alternaria alternata	*P. erythroseptica*	OP	Waterhouse (1940)
	P. megasperma	OP	Sneh et al. (1977); Hsu and Lockwood (1984)
Cephalosporium sp.	*P. megasperma*	OP	Sneh et al. (1977); Hsu and Lockwood (1984)
Cladosporium herbarum	*P. infestans*	HL	Lacey (1965)
Dactylella spermatophaga	*P. cactorum*	HP, OP	Sneh et al. (1977); Hsu and Lockwood (1984)
	P. cinnamomi	HP, OP	Sneh et al. (1977); Hsu and Lockwood (1984)
	P. megasperma	HP	Drechsler (1938)
		OP	Sneh et al. (1977); Hsu and Lockwood (1984)
Diheterospora chlamydosporia	*P. megasperma*	OP	Sneh et al. (1977); Hsu and Lockwood (1984)
Fusarium oxysporum	*P. erythroseptica*	OP	Waterhouse (1940)
	P. megasperma	OP	Sneh et al. (1977); Hsu and Lockwood (1984)
Gliocladium roseum	*P. erythroseptica*	OP	Waterhouse (1940)
	P. megasperma	HL	Vaartaja et al. (1979)
Humicola fuscoatra	*P. erythroseptica*	OP	Waterhouse (1940)
	P. megasperma	OP	Sneh et al. (1977); Hsu and Lockwood (1984)
Monosporium sylvaticum	*P. infestans*	HL	Lacey (1965)
Penicillium patulum	*P. cryptogea*	HL	El-Goorani et al. (1976)
Rhizoctonia	*P. infestans*	HL	Lacey (1965)
Trichoderma harzianum	*P. cinnamomi*	HL	Lacey (1965)
T. lignorum	*P. megasperma*	HP	Durrell (1968)
T. polysporum	*P. cinnamomi*	STIM	Kelley (1977)
T. viride	*P. cactorum*	HP	Dennis and Webster (1971)
	P. cinnamomi	HL, HP	Brasier (1975a); Reeves (1975)
	P. erythroseptica	HP	Dennis and Webster (1971)

[a] Reprinted by permission of the American Phytopathological Society, St. Paul, Minnesota, courtesy Malajczuk (1983).
[b] HL, hyphal lysis from antibiotic production; HP, hyphal parasitism; Z/SP, zoospore/sporangia parasitism; OP, oospore parasitism; and STIM, stimulation of the fungus.

choices of biocontrol agents or techniques to deliver them to the field at an economical cost. Although there are difficulties, Coffey cited the possibility of applying a suitable microorganism to the planting mixes in nurseries for control of *P. cinnamomi* in avocados. If such an agent could successfully colonize the rhizosphere, it might persist. Currently, the optimal control practices in California include an integrated system of optimal irrigation management in the crop system, resistance of rootstocks, and chemical control when necessary.

Myrothecium verrucaria and a *Streptomyces* sp. inhibited growth of *P. cinnamomi* on potato-dextrose agar. Munnecke (1984) used the strategy of infesting field soils with these microorganisms after methyl bromide fumigation of land naturally infested with *P. cinnamomi* in an avocado orchard. The fumigation reduced the population of *P. cinnamomi* to undetectable levels. The introduction of the antagonistic microorganisms by incorporation into the soil was followed by addition of cotton gin waste, a high-carbon organic amendment, and ammonium phosphate. It was found that although methyl bromide had suppressed *P. cinnamomi* initially in the field, the fungus became reestablished after 9 months. The cotton gin waste amendment supported an increase in population of the antagonists, but it was insufficient to control root rot on *Persea indica,* a highly susceptible relative of *P. americana,* the common avocado. Both the amended and nonamended methyl bromide-fumigated plots became reinfested with *P. cinnamomi,* as measured by baiting with highly susceptible *P. indica* plants. This research further demonstrates that biological control is complex and not likely to be effected by addition of a single antagonist to the soil, especially in land to be cropped with trees.

Table 5.8. Antagonism to *Phytophthora* species by members of Basidiomycetes[a]

Basidiomycete	*Phytophthora* species	Reference
Agaricus langei	*P. cinnamomi*	Marx (1969)
Suillus luteus	*P. cactorum*	Marx (1969)
	P. cambivora	Marx (1969)
	P. cinnamomi	B. H. Pratt (1971); Marx (1969)
	P. cryptogea	Marx (1969)
	P. palmivora	Marx (1969)
	P. parasitica	Marx (1969)
Bovista brunnea	*P. cinnamomi*	B. H. Pratt (1971)
Clavulina amethystina	*P. cinnamomi*	B. H. Pratt (1971)
Clitocybe eucalyptorum	*P. cinnamomi*	B. H. Pratt (1971)
Clitocybe infundibuliformis	*P. cinnamomi*	B. H. Pratt (1971)
Collybia abutyracea	*P. cinnamomi*	B. H. Pratt (1971)
Collybia sp.	*P. cinnamomi*	B. H. Pratt (1971)
Cortinarius austro-venetus	*P. cinnamomi*	B. H. Pratt (1971)
Cortinarius walkeri	*P. cinnamomi*	B. H. Pratt (1971)
Geastrum	*P. cinnamomi*	B. H. Pratt (1971)
Gymnopilus	*P. cinnamomi*	B. H. Pratt (1971)
Hyprophoropsis aurantiaca	*P. cinnamomi*	B. H. Pratt (1971)
Hygrophorus niveus	*P. cinnamomi*	B. H. Pratt (1971)
Laccaria laccata	*P. heveae*	Marx (1969)
Lactarius deliciosus	*P. cactorum*	Marx (1969)
	P. cambivora	Marx (1969)
	P. cinnamomi	B. H. Pratt (1971); Marx (1969)
	P. palmivora	Marx (1969)
	P. parasitica	Marx (1969)
Leucopaxillus cerealis var. *piceina*	*P. cactorum*	Marx (1969)
	P. cambivora	Marx (1969)
	P. cinnamomi	Marx (1969)
	P. citrophthora	Marx (1969)
	P. cryptogea	Marx (1969)
	P. drechsleri	Marx (1969)
	P. heveae	Marx (1969)
	P. palmivora	Marx (1969)
	P. parasitica	Marx (1969)
Naematoloma fasciculare	*P. cinnamomi*	B. H. Pratt (1971)
Omphalina sp.	*P. cinnamomi*	B. H. Pratt (1971)
Rhizopogon vinicolor	*P. cinnamomi*	Zak (1971)
Scleroderma bovista	*P. cactorum*	Marx and Bryan (1969)
	P. cinnamomi	Marx and Bryan (1969)
	P. citricola	Marx and Bryan (1969)
	P. drechsleri	Marx (1969)
	P. heveae	Marx and Bryan (1969)

[a] Reprinted by permission of the American Phytopathological Society, St. Paul, Minnesota, courtesy Malajczuk (1983).

The first successful application of *Trichoderma* to control *P. cactorum* on apple was by Valdebenito-Sanhueza (1987) in Brazil. *Trichoderma* significantly reduced mortality after natural or artificial inoculations of young apple trees.

Root rot of soybean, caused by *P. sojae,* was partially controlled in a series of field experiments by row application of spore formulations of *Bacillus cereus* strain UW85, previously reported as an antagonist of *P. medicaginis* by Handelsman et al. (1990) (Osburn et al 1995). *B. cereus* in clay granules, clay dust, or peat moss formulations placed under the seed row was associated with yield increases, particularly on susceptible cultivars of soybean. Zwittermycin A, a product of *B. cereus,* caused inhibition of growth of *P. sojae* at 100 μg/ml.

Suppressive Soils

The concept of certain soils being "suppressive" or "conducive" to disease was noted by K. F. Baker (1978). This concept has intrigued investigators for years, and much research has been conducted on this subject (Broadbent and Baker 1974a, b; K. F. Baker and Cook 1982). Soils that favor the expression of disease are conducive, but soils that are inhospitable to some plant pathogens to the extent that either the pathogen cannot become established or, if established, cannot initiate disease are suppressive. Conversely, in some cases with continued culture of the crop, the intensity of disease lessens. This unusual phenomenon occurs with the take-all disease of wheat (caused by *Gaeumannomyces tritici*) in Washington State, United States (Cook and Baker 1983). Malajczuk (1983) in Australia suggests that the principal cause of suppressiveness is microbial because sterilization by autoclaving, steam pasteurization, and irradiation renders these soils conducive. The percentage of microbes that were antagonistic to *P. cinnamomi* was higher in suppressive than in conducive soil (Tables 5.6 and 5.9).

Some of the first reports on natural control of *Phytophthora* were by Broadbent et al. (1971) and Broadbent and

Table 5.9. Estimation of total populations of bacteria and actinomycetes in suppressive loam and conducive lateritic soils and in the rhizosphere of *Eucalyptus calophylla* (resistant) and *E. marginata* (susceptible) seedlings growing in these soils[a]

Soil type / Microorganisms	Soil	*E. calophylla*	*E. marginata*
Loam (suppressive)			
Bacteria	32.8×10^5	18.8×10^7	14.5×10^6
Actinomycetes	11.2×10^6	22.2×10^6	69.8×10^6
Antagonists (%)	4.2	10.68	15.92
Laterite (conducive)			
Bacteria	52.4×10^5	72.5×10^6	36.6×10^6
Actinomycetes	13.6×10^5	32.3×10^6	16.3×10^6
Antagonists (%)	0.75	11.68	2.00

[a] Reprinted by permission of the American Phytopathological Society, St. Paul, Minnesota, courtesy Malajczuk (1983).

Baker (1974a, b), who showed that *P. cinnamomi* had been suppressed in certain soils by the agricultural system practiced by the Australian grower Ashburner, described by K. F. Baker (1978) above. The soil microorganisms that thrive in the Ashburner ecosystem interfere with zoospore production by inactivating sporangia production and/or lysing hyphae and sporangia. Broadbent and Baker (1974a, b, 1975) established three criteria to characterize suppressiveness of soil in the Mt. Tamborine region: 1) disease was absent in the field but *P. cinnamomi* could be isolated; 2) there was reduced sporangial production and release of zoospores; and 3) healthy jacaranda and eucalyptus seedlings would grow as well in untreated soil as in soil that had been steamed for 30 min at 60°C after inoculation with *P. cinnamomi.* This much cited example of suppressiveness of Phytophthora root rot in soil in Australia is also discussed by Shea and Broadbent (1983).

There are also several documented instances in Australia in which *P. cinnamomi* is established in natural forest ecosystems but has not caused a serious amount of disease (Weste and Vithange 1977, 1978a, b; Malajczuk and McComb 1977; Malajczuk 1979a, b). In these cases, suppressiveness was associated with an increase in the percentage of antagonistic bacteria and actinomycetes.

The soil of the Tallaganda State Forest in southern New South Wales, Australia, is suppressive to *P. cinnamomi* (Halsall 1982a). This soil differs from the suppressive Ashburner soils in Queensland (Broadbent and Baker 1974) or the loam soil in Queensland (Malajczuk 1979a) in that the New South Wales soil is relatively infertile and relatively low in organic matter. Hyphae were more rapidly lysed, the numbers of germinated chlamydospores were reduced by 50%, and survival of chlamydospores was reduced from 50% for the conducive soil to 20% for the suppressive soil over a period of 3 months. Germination of chlamydospores by sporangia was also reduced in the suppressive soil. Since the treatment of the suppressive soil at 60°C for 30 min did not change the suppressiveness, it was hypothesized that suppressiveness was caused by a "heat stable condition or an organism such as spore-forming bacteria or actinomycetes" (Halsall 1982a).

Interesting but difficult to explain is the fact that the New South Wales soil was suppressive to *P. cinnamomi* but not to *P. cryptogea.* The New South Wales soil contained a greater number of actinomycetes that were antagonistic to both aerobic bacteria and *P. cinnamomi.* In soil extract, the soluble component inhibited the growth of *P. cinnamomi* more than that of *P. cryptogea.* The hypothesis was presented that *P. cinnamomi,* which apparently sporulates more efficiently in nonsterile soil extract (Mehrlich 1935) and in association with certain aerobic bacteria (Zentmyer 1965), may be at a disadvantage in a soil in which *Streptomyces* spp. inhibit the multiplication of aerobic bacteria (Halsall 1982b). *P. cryptogea* produces sporangia in sterile soil extract or in water and perhaps was not inhibited by the suppressive soil. In addition, certain soluble components of the soil extract were more toxic to *P. cinnamomi* than to *P. cryptogea.*

P. cactorum is one of the agents that cause root rot of ginseng (*Panax ginseng*) in Korea. Certain soils are conducive, and others are suppressive. Growth of *P. cactorum* as well as *Fusarium solani* and *Sclerotinia sclerotiorum* was inhibited more on suppressive than on conducive soil extract agar. The numbers of total fungi and total bacteria were highest in suppressive soil. Magnesium and sodium levels were lower and the proportion of clay was higher in suppressive soils (Chung et al. 1984).

There is much evidence to demonstrate that most *Phytophthora* spp. are relatively nonsaprophytic and do not grow well in competition with other microorganisms in soil. The seedling disease of papaya caused by *P. palmivora* was controlled by simply filling in the transplanting holes with noninfested soil. This practice exploited the inability of *P. palmivora* to compete with naturally occurring microorganisms in soil (Ko 1971, 1982a).

The relative fragility of *Phytophthora* propagules in soil and the many biotic factors that affect them were lucidly presented and summarized by Malajczuk (1983), who has extensively conducted research in that area. These data indicate that *Phytophthora* is vulnerable to a large number of microorganismal species. Some highlights from Malajczuk's review are included here.

Microbial Activity

Several genera of bacteria, actinomycetes, fungi, amoebas, nematodes, and mites have been shown to parasitize and lyse *Phytophthora* propagules in soil (Sneh et al. 1978; Weste and Vithanage 1978a; Wyn and Epton 1979; Malajczuk 1983, 1988; Daft and Tsao 1984; Sutherland and Lockwood 1984; Sutherland et al. 1984; Harris 1985a, c; Nesbitt et al. 1985). A wide range of fungi are antagonistic to several species of *Phytophthora* (Malajczuk 1983) (Tables 5.7–5.9). Weste and Vithanage (1978a) in Australia demonstrated that survival of *P. cinnamomi* in infested soil varied inversely with the number of microorganisms present. They concluded that the lower microbial populations in gravels probably contributed to the longer survival of *P. cinnamomi* compared with its ability to survive in heavier Krasnozem soils (Weste 1983b). Adding soils containing large quantities of microorganisms to microbially deficient soils resulted in a virtual elimination of *P. cinnamomi* propagules in plant roots (Tippett 1978).

Several fungi show promise for biological control in greenhouse research. For instance, *Trichoderma* and *Gliocladium* species suppressed *P. cactorum* on apple seedlings in glasshouse tests (V. L. Smith et al. 1990). A strain of *Myrothecium roridum* was active in suppression of *P. cinnamomi* on *Persea indica,* a highly susceptible relative of avocado, in potting mixes and also in three naturally infested soils. That *M. roridum* could be isolated from actively growing root tips indicates that the fungus might be adapted to the rhizosphere (Gees and Coffey 1989). These examples and others are listed in Table 5.10. Coffey (1991) noted that the lack of sensitivity of these fungi to metalaxyl might allow fungicidal and biological control to work in unison.

A commercial preparation (Binab T) containing about 50×10^6 spores of *Trichoderma viride* per gram inhibited formation of sporangia and lysed mycelium of *P. cactorum* in vitro. Binab T reduced (54–74%) the size of lesions on inoculated apple twigs. Lesion size was also reduced on apple trees that were sprayed with Binab T (Orlikowski and Schmidle 1985).

Although oospores are long lived in soil, they are parasitized by various microorganisms. More than a dozen microorganisms are capable of infecting oospores (Sneh et al. 1977), including certain oomycetes, hyphomycetes, chytrids, actinomycetes, and *Pseudomonas* spp. Harris and Cole (1982) reported that most of the oospores of *P. syringae* that had been stored in soil became abnormal (vesiculose) and were infected with a fungus later identified as *Microdochium fusariodes* (Harris 1985b). More than 70% of the oospores were nonfunctional within 8 months of burying the leaves, but the remaining oospores that escaped remained functional for more than 2 years. Oospores in surface-stored leaves declined less rapidly; approximately 50% were still functional a year later. *Catenaria anguillulae,* a chytrid fungus, parasitized oospores

Table 5.10. Biological control agents that either suppress a *Phytophthora* species or show potential for control of a disease caused by a *Phytophthora* species

***Phytophthora* species** Host	**Disease or cultural study**	**Fungal or bacterial agent**	**Reference**
P. cactorum			
Apple	Crown rot	*Enterobacter aerogenes; Bacillus subtilis*	Utkhede (1987b); Utkhede and Smith (1991)
	Crown rot	*B. subtilis*	Utkhede (1984b, d, e)
	Crown rot	*B. subtilis; E. aerogenes*	Gupta and Utkhede (1986)
	In vitro	*B. subtilis; E. aerogenes*	Gupta and Utkhede (1987)
	In vitro	Bacterium B8	Utkhede and Gaunce (1983); Utkhede (1983)
	Crown rot	*Trichoderma* sp., *Gliocladium* spp.	V. L. Smith et al. (1990)
	Crown rot	*E. aerogenes*	Utkhede and Smith (1993)
	Cultural study with calcium and ammonia	*Bacillus cereus* UW85	Gilbert et al. (1990)
	Seedling disease	*Trichoderma* spp.	Roiger et al. (1991)
	Seedling disease	*Trichoderma koningii; Trichoderma* spp.	Roiger and Jeffers (1991)
	Stem and crown canker	*Trichoderma viride* (Binab T)	Orlikowski and Schmidle (1985)
Gerbera	Root rot	*T. koningii; T. piluliferum; T. viride*	Orlikowski (1980/1981)
P. cinnamomi			
Avocado	Root rot	*Myrothecium roridum*	Gees and Coffey (1989)
Pine	Root rot	*Trichoderma harzianum*	Kelley (1976)
Casuarina littoralis	Root rot	Actinomycetes	Gerrettson-Cornell (1983–1984)
Eucalyptus marginata	Root rot	Ectomycorrhizae	Malajczuk (1988)
Pine	Root rot	Ectomycorrhizae	Marx (1969, 1970, 1972, 1973a, b)
(also *P. cryptogea, P. citricola, P. cactorum*)	In vitro	*Epicoccum purpuroscens*	Brown et al. (1987)
Eucalyptus marginata	Parasitism	Several Basidiomycetes	B. H. Pratt (1971)
	Parasitism	*Catenaria anguillulae*	Daft and Tsao (1984)
(also *P. parasitica*)	Parasitism	*Humicola fuscoatra; Anguillospora pseudolongissima; Hypochytrium catonoides*	Daft and Tsao (1983)
Eucalyptus sieberi; E. obliqua	Suppressive soil		Marks and Smith (1981)
Avocado (*Persea indica*)	Root rot	*Myrothecium verrucaria* *Streptomyces* sp.	Munnecke (1984)
	In vitro	Bacteria	Nesbit et al. (1985)
Acacia pulchella, Banksia grandis, Eucalyptus marginata	In vitro	Rhizosphere fungi and bacteria	Murray (1987)
Ceonothus velutinus		*Streptomycetes griseolaoalbus*	Rose et al. (1980)
Eucalyptus spp.	Suppressive soil	*Streptomyces* spp.	Halsall (1982a, b)
P. cryptogea			
Gerbera jamesonii	Root rot	*T. koningii; T. piluliferum; T. viride*	Orlikowski (1980)
P. erythroseptica	Parasitism of oospores	Hyphomycetes; Chytridiomycetes; bacteria	Wynn and Epton (1979)
P. medicaginis (*P. megasperma* f. sp. *medicaginis*)			
Alfalfa	Root rot	*Bacillus cereus*	Handelsman et al. (1990)
P. palmivora			
Cacao	Black pod	*Botryodiplodia theobromae; Gliocladium roseum; Penicillium* spp.; *Bacillus cereus; B. subtilis*	Odigie and Ikotun (1982)
P. parasitica (*P. nicotianae*)			
Citrus	Root rot	*Glomus intraradices*	J. H. Graham and Egel (1988)
Tobacco	Root rot	*Bacillus cereus*	Handelsman et al. (1991)
P. sojae (*P. megasperma* f. sp. *glycinea*)	Parasitism of oospores; ultrastructure	*Actinoplanes missouriensis, Humicola fuscoatra*	Sutherland et al. (1984)
Soybean	Root rot	*Hyphochytrium catenoides; Actinomyetes* species	Filonow and Lockwood (1985)
	Root rot	*H. catenoides*	Hsu and Lockwood (1984)
		Humicola fuscoatra; A. missouriensis; Pythium monospermum; Verticillium chlamydosporium	Sutherland and Lockwood (1984)
	In vitro	*Gliocladium roseum; Trichoderma harzianum; Trichothecium roseum*	Al-Heeti and Sinclair (1988)
	Parasitism	Oomycetes, Chytridiomycetes, Hyphomycetes, Actinomycetes, bacteria	Sneh et al. (1977)
		Streptomyces verzuelae; S. lavendulae	Knauss (1976)
	Root rot	*Bacillus cereus*	Osburn et al. (1995)
P. syringae			
Apple	Parasitism of oospores	*Microdochium fusarioides*	Harris (1985)

of *P. cinnamomi* and *P. parasitica* and chlamydospores of *P. parasitica*. *C. anguillulae* was detected by baiting in six of seven citrus soils (Daft and Tsao 1984). Oospores of *P. cinnamomi* were parasitized by *Humicola fascoatra, Anguillospora pseudolongissima,* and *Hypochytrium catenoides* (Daft and Tsao (1983). *Phlyctochytrium* sp. infected resting spores of *P. nicotianae* and *P. cactorum* (Kenneth 1982).

Many fungi, bacteria, and actinomycetes are capable of producing antibiotics that will suppress the growth and/or sporulation of *Phytophthora* in a petri dish; however, a petri dish contains an ecosystem much simpler than that of soil. Many antibiotics or toxic byproducts of antagonistic microorganisms, either volatile or not volatile, may cause lysis (disintegration of the propagule or mycelium; amensalism) (Malajczuk 1983). Many microorganisms from soil can be shown to produce antibiotics, but there is disagreement as to whether these antibiotics are active per se in soil. This is because antibiotics are difficult to extract from soil and may be adsorbed on clay or humus particles (Brian 1957).

Detection and Role of Microbial Antagonists

There is now strong evidence that the mode of action of many bacterial and fungal antagonists is the production of antibiotics or other extracellular compounds, which in turn initiate lysis of sporangia as well as oospores (Malajczuk 1983). Microorganisms thus far reported to show antagonism to *Phytophthora* are listed in Tables 5.7–5.9 (Malajczuk 1983). Table 5.9 shows that the percentage of antagonistic microorganisms is higher in suppressive soils than in conducive soils. Table 5.10 includes references on control or suppression of several species of *Phytophthora* by antagonistic bacteria or fungi.

Malajczuk (1983) summarized the effect of associated soil microorganisms with *Phytophthora* in soil (excerpted by permission of the American Phytopathological Society, St. Paul, Minnesota, courtesy of Malajczuk 1983):

> Microbial antagonism to *Phytophthora* species may be expected in many situations involving a wide range of microflora and fauna. The ability of the pathogens to counter these antagonistic interactions by producing propagules that enable them to survive under these adverse conditions has evolved over long periods of time. The challenge for pathologists is to anticipate any possible changes in the life stage of *Phytophthora* in which control might be brought about. Control by use of antagonistic microorganisms is considered the ultimate goal and in many cases may be the only means available. However, to date the state of the art is severely deficient not only in the number of examples of biological control achieved by these means, but also in our understanding of it. We need to develop an understanding of the antagonists as well as of the behavior of the pathogen in soil, in the rhizosphere, and within plant tissue. Bowen's approach (1980) of developing theoretical models to interpret the changes of the pathogen in relationship to the antagonist under different environmental conditions will help to establish principles for predicting epidemics or control of the pathogen. Using guides provided in situations where *Phytophthora* species are naturally suppressed, e.g. in suppressive soil, will help not only to formulate model systems but also to test models. Selective stains, fluorescent brighteners (Tsao 1969a), and immunofluorescent techniques (Malajczuk et al. 1975, 1978) would be powerful tools to help visualize the activity of both *Phytophthora* and antagonist(s) *in situ*. Scanning and transmission electron microscopy may then indicate the type and nature of interaction. Only with more of this type of knowledge will we gain an understanding of and subsequent realization of biocontrol of *Phytophthora*.

Although *Trichoderma* species antagonize *Phytophthora* species (Tables 5.7 and 5.9), isolates of this fungus also have been shown to stimulate the formation of oospores in normally self-sterile isolates of *Phytophthora* (Brasier 1971a, 1972, 1975a, b; P. H. Pratt et al. 1972; Reeves and Jackson 1972; Reeves 1975). It is thus not sufficient to merely identify an organism as an antagonist of *Phytophthora* spp. One also needs to ascertain what response is elicited in *Phytophthora* as a result of the antagonism. It would defeat the purpose of biological control if *Phytophthora* produced resting spores as a response to antagonism by other microorganisms. Reeves (1975) reported both lysis of hyphae and induction of oospores of *P. cinnamomi* by *T. viride*.

The potential for biological control by *Trichoderma* and *Gliocladium* species was reviewed by Papavisas (1985). A formulation of the beneficial fungus *G. virens* is now commercially available from W. R. Grace and Co., Columbia, Maryland, United States, under the trade name GL-21. This product has efficacy against *Pythium* and *Rhizoctonia* in greenhouse crops. Tests are also underway to register this product for use against *Phytophthora, Thielaviopsis,* and *Fusarium.*

Some *Phytophthora* species, including *P. cinnamomi,* are inhibited on pine tree roots by certain beneficial ectomycorrhizal fungi (Marx 1969, 1972, 1973a, b; Marais and Kotze 1976; Malajczuk 1979a, b, 1983, 1985, 1988).

Marx (1972, 1973) reviewed the literature on control of diseases of feeder roots of plants by ectomycorrhizae, which consist of certain fungal species that "penetrate intercellulary and particularly replace the middle lamellae between cortical cells of the feeder roots. This hyphal arrangement around the cortical cells is called the Hartig net. Ectomycorrhizal fungi also form a dense, usually continuous, hyphal network called the *fungal mantle* over the feeder root surface" (Marx 1972).

Certain mycorrhizal fungi, such as the ectomycorrhizal fungus *Lactarius cerealis* var. *picaina,* produce antibiotic diatretyne compounds. These compounds were functional in the resistance of feeder roots of shortleaf pine seedlings to *P. cinnamomi* (Marx 1970, 1972, 1973). Host plants infected by ectomycorrhizae also produce mono-terpene compounds that are inhibitory to *P. cinnamomi.* Mycorrhizal roots do not attract zoospores (chemotaxis) as do nonmycorrhizal roots (Marx 1972). Mycorrhizae constitute an important form of biological control on pine.

An extensive discussion of the effect of ectomycorrhizae on the reduction of infection by *Phytophthora* is given by Malajczuk (1983) and Shea and Broadbent (1983).

On the other hand, endomycorrhizae, which are normal and necessary components of the soil microflora, reduced the resistance of avocado seedlings to infection by *Phytophthora* but increased the resistance of citrus seedlings (Davis et al. 1978; Davis and Menge 1980). However, Graham and Egel (1988) reported little difference in resistance or tolerance of sweet orange trees that were grown under mycorrhizal or nonmycorrhizal conditions.

Tu (1978, 1979) showed that increasing the population of the nitrogen-fixing symbiont *Rhizobium japonicum* in soil reduced the severity of root rot of soybean caused by *P. sojae* (*P. megasperma* var. *sojae*). Tu (1980) obtained similar results when *R. meliloti* was incorporated into soil before inoculation of alfalfa with *P. medicaginis* (*P. megasperma* f. sp. *medicaginis*). Drapeau et al. (1973) showed that some *Rhizobium* spp. produce antibiotics active against *P. cactorum.* Chou and Schmitthenner (1974) found, however, that reinoculation of soybeans with *R. japonicum* had no effect on disease severity. Gray and Hine (1976) reported that nodules on alfalfa roots, induced by *R. meliloti,* were more readily infected by *P. medicaginis* than were roots. Thus, it would not appear that *Rhizobium* spp. consistently protect plant roots.

Prospects for Biological Control

This brief summation of biological control is intended to give a broad overview of the various types of natural control agents available. Table 5.10 lists a series of reports in which biological control agents showed potential for control of diseases. The choices are many. Research on biological control of *Phytophthora* is active in many laboratories and experimental stations throughout the world. Despite much information, the field is still exploratory and descriptive. The many contradictory results, some of which are detailed above, reflect more a lack of understanding of the complex interaction that occurs in soil than lack of prospects for success. Refinements in technique and standardization of experimental procedures may eventually resolve these contradictions. Although the phenomenon of biological control is the most complex of all the control measures attempted, it has great potential as an environmentally safe method. Because of its complexity, much more research is required.

Biological control by its very nature requires a high degree of precision and timing. It also requires a basic knowledge of the epidemiology of the pathogen. These considerations have discouraged many growers from adopting biological control. The development of techniques that will make biological control a more feasible reality is a worthwhile research challenge.

To make a biological control method effective, new systems and research methods will have to be implemented. In most cases, scientists use the scientific method, which reduces all variables to one, but with biological control there are many factors that work either in unison or in opposition and may vary from location to location. The Ashburner system in Australia (Baker 1978; Shea and Broadbent 1983) apparently evolved from grower practice and is probably adapted mainly to that particular area in Australia. Thurston (1990) discusses many systems of control that have evolved from grower practice with little knowledge of the causal agent. A systems approach on a long-term basis with industry as well as government support might synthesize new practices or combinations of practices for control, especially in crops such as those from fruit-bearing trees.

The reader should also refer to the reports in *Biological and Cultural Tests for Control of Plant Disease* published annually by the American Phytopathological Society, St. Paul, Minnesota.

Summary

It is a well-known concept that disease of plants is the result of the interaction of the pathogen (*Phytophthora*), the susceptibility of the host plant, and the environment. *Phytophthora* species, most of which are plant pathogens, cause problems only under certain environmental conditions. Different species of *Phytophthora* are active under various ranges of temperature; however, all species require free water in the environment to become active. A large part of this chapter is devoted to the role of soil and atmospheric water and its management. Since water is of prime importance, regulation of water is one of the most important cultural ways to control diseases. Where irrigation is practiced, the administration of water to plants can be managed. The concept of matric potential in soil, an important component of water potential, is explained by Duniway (1983) (Appendix 5.1). Through methods to regulate the matric potential of soil, laboratory experiments showed that *Phytophthora* is not active when the matric potential of soil measures at or below –300 mb, which is loosely considered to be the "field capacity" of soil (water does not move downward with the force of gravity). Several examples of regulation of frequency of irrigation in the field show that such methods are to some extent successful for control of root diseases. Impermeable soil layers (hardpans) that impede the vertical drainage of water in soil often allow free water to accumulate above the hardpan. This phenomenon sets up a favorable environment for infection by *Phytophthora*. A dramatic example of the hardpan phenomenon occurs in Western Australia, where *P. cinnamomi* caused death of a large number of jarrah (*Eucalyptus sieberi*) forest trees despite the relatively dry summers and moderate resistance of the jarrah host. Certain sites in the jarrah forest are underlaid by a concreted lateritic layer that jarrah tree roots penetrate. By excavation of large trees, Shea and his colleagues showed that water had accumulated in the hardpan area and caused the population of *P. cinnamomi* in this wet area of the soil to increase to a level high enough to kill the large roots that had penetrated the hardpan to the subsoil.

Drought predisposes safflower plants to be more severely affected by *P. drechsleri* when the soil is subsequently flooded. Plants growing at excess soil moisture for long periods or under saline conditions are also predisposed to be more susceptible to *Phytophthora* than is normal.

Inoculum increase in the soil often leads to a corresponding increase in incidence and severity of disease expression. Increase in *Phytophthora* inoculum is dependent on temperature and the availability of free water for extended periods of time. The ability of propagules of *Phytophthora* to increase rapidly in the soil and on foliage by the production of sporangia and zoospores explains why soilborne inoculum often increases from imperceptible quantities (as dormant oospores or chlamydospores) to large quantities in only a few days. For this reason, epidemiologically, most *Phytophthora*-caused diseases are considered to be multicyclic or, as analogized by Vanderplank (1963), compound-interest diseases. *Phytophthora* propagules can be dispersed in either splashing rain water or irrigation water. Most foliage-infecting *Phytophthora* species produce deciduous (caducous) sporangia, thus implying that the mature sporangia break off from the mycelium and become freely transmissible by rain splash or wind-driven rain.

Manipulation of factors that cause an increase or suppression of inoculum can be used for cultural control. These methods include lengthening the time between irrigations, decreasing the duration of irrigation, planting on beds or mounds, planting in permeable, well-drained soils, and preventing flowing irrigation water from contacting the trunks of trees. Sanitation by transplanting uninfected plants to clean, noninfested land is the ideal, but because of the wide dispersal of many *Phytophthora* species, it is not always practical. Some intensive industries, such as those involved with strawberry culture in Scotland, carefully monitor the health of plants and land before planting.

Use of steam heat to kill *Phytophthora* in soil prior to use in glasshouses is described. Solar heating in the field by laying out clear polyethylene tarps increases soil temperature above 45°C. This method is experimentally most efficacious in areas with a large proportion of cloudless days (e.g., Israel, California, and Arizona). Thermotherapy of infected plant tissue can be used to free plants from infections if the seeds or plant parts are more capable than the *Phytophthora* species of withstanding high temperatures.

Rotation, especially for annual crops, is an important control measure if the alternative crop is not susceptible to the pathogen. *Phytophthora* is not usually a successful competitive saprophyte; however, exceptions exist.

Some fertilizing regimes have been suppressive to *Phytophthora,* but the literature is replete with inconsistencies. Methods to control *Phytophthora* diseases culturally include use of high levels of organic nitrogenous materials that release ammonia and nitrous acid, use of fertilizers and amendments that reduce the pH to <4.0 with sulfur for acid-tolerant plants, reduction of pH to <5.0 in high-aluminum soils (for plants with a tolerance for aluminum), application of minerals to make up for loss of uptake by rotting fibrous roots, and avoidance of excessive nitrogen fertilization, which makes the resulting succulent foliage more susceptible to *Phytophthora.* The use of sulfur to reduce pH of soil to about 3.8 has given marked control of root rot of pineapple caused by *P. cinnamomi.* Addition of composted hardwood bark in potting mixes has increased the degree of control of *Phytophthora* markedly. Composted hardwood bark not only increases the "air-filled porosity" of soil but also allows the buildup of antagonistic soil fungi such as *Trichoderma* species.

An agricultural system that in practice controls *P. cinnamomi* is known as the Ashburner system for avocado production in Queensland, Australia. The system involves interplanting New Zealand blue lupine between the trees during the fall and winter and disking the residue into the soil, using fowl manure in the spring, and interplanting with corn, sorghum, or *Lablab purpureus* in the summer and at the beginning of fall and disking the residue into the soil. The total practice results in control of *P. cinnamomi.* The Ashburner system involves cultural practices that favor the development and activity of soilborne antagonistic fungi and bacteria. Several examples of the discovery of naturally suppressive soils have encouraged research on the use of antagonistic fungi and bacteria that are rhizosphere competent and inhibit *Phytophthora* on roots.

Many citations show that propagules of *Phytophthora* species are sensitive to many soil organisms such as bacteria, fungi, and amoebas. There are some commercial approaches to the implementation of antagonistic microorganisms in soil to supplant or supplement fungicidal control, but most of these are still experimental.

APPENDIX

TO CHAPTER 5

Soil-Water Terminology and Hygienic Criteria for Nurseries

5.1. Concepts and Terminology in Plant-Soil-Water Relationships (Duniway 1983)

The status of water in fungi and the substrates on which they grow can best be described by using the concept of water potential. Water potential is the chemical potential of water per unit volume and has the dimensions of pressure. The units used here are bars and millibars (mb). One bar equals 100 kPa and nearly equals 1 atm. The water potential of pure water at atmospheric pressure is defined arbitrarily as zero. At any given temperature, the potential of liquid water is a logarithmic function of its activity (a_w) and therefore its equilibrium relative humidity (RH). For example, at a water potential of −15 bars, a value at which crop plants frequently wilt, the equilibrium RH of air is 98.9%; at a water potential of −100 bars, which is much too dry for *Phytophthora* spp. to be active (Duniway, 1979), the RH is 92.8%. Useful tables relating water potential to humidity and showing the conversion of units can be found in Slavík (1974).

The major forces acting on water can be expressed as components of water potential in the equation

$$\Psi = \psi_s + \psi_p + \psi_m,$$

where Ψ is the total water potential and ψ_s, ψ_p, and ψ_m are the solute, pressure, and matric components, respectively. Solute potential, ψ_s, represents the influence of dissolved substances on water and is equivalent to, but opposite in sign from, osmotic pressure. As it is used here, the pressure potential, ψ_p, represents a hydrostatic pressure, such as the positive ψ_p identified with turgor or the negative ψ_p in the xylem elements of transpiring plants. Matric potential, ψ_m, represents a reduction in Ψ resulting from absorption and capillary forces that are of primary importance in soil. Older expressions of soil moisture suction, sometimes given as centimeters of water, as pF values (Griffin 1972), and as negative pressure terms used in soil hydrology, can usually be equated to the negative ψ_m component of soil Ψ. The soil contains a system of interconnecting capillary pore spaces; although soil pores are heterogeneous in shape, there are fairly definite relationships between the size of soil pores that can retain liquid water and ψ_m (Griffin 1972; Slavík 1974). Therefore, as a fully saturated soil begins to dry or drain, ψ_m is the first component of Ψ to change appreciably, and as it decreases, the water in progressively smaller pore spaces is replaced by air. By the time a soil dries to $\psi_m = -0.3$ bar, a ψ_m value at which soil water content is frequently considered to represent field capacity, most pore spaces and pore necks larger than 10 μm in diameter will have drained. As the soil continues to dry, both ψ_m and ψ_s change, and partitioning soil Ψ into its component potentials becomes increasingly difficult. Because the configuration of water in very wet soils is clearly linked to ψ_m, zoospore dispersal, solute diffusion, and aeration in soil all interact strongly with ψ_m, and it is one of the more useful parameters in the epidemiology of diseases caused by soilborne species of *Phytophthora*.

5.2. Hygienic Criteria Recommended to Produce *Phytophthora*-Free Plants for Sale as Transplants to the Field

Adapted by permission of *Queensland Agricultural Journal,* courtesy K. G. Pegg (1978).

Install concrete or bitumen floors in the nursery. If too expensive, the concrete or bitumen can be limited to walkways and the remaining area covered with 5 cm of clean gravel.

Benches with steel mesh tops are more readily sanitized than wooden benches. Wooden benches should be treated with copper naphthenate once or twice a year. Clean containers or propagating material should never be placed on the ground.

Use new or sterilized containers, and store them where there is no opportunity for contamination.

Disinfect tools with steam, or dip them in formalin (1 part of commercial formalin to 18 parts of water) or sodium hypochlorite solution containing 0.2% available chlorine.

Keep the nozzle of the hose off the ground, because it may become contaminated.

Hold contaminated material in a bin away from the growing area prior to fumigation with methyl bromide or steam treatment.

Treat floors with formalin before dumping treated potting mixtures on them (formalin may not be registered for use in some countries).

Segregate avocado trees or plants introduced from other nurseries, because they may be diseased.

Restrict entry of people to propagating areas. Insist on use of foot baths containing formalin (1 part of commercial formalin to 18 parts of water), if registered for use, or copper oxychloride powder at the entrance to growing areas. Disposable plastic shoes are a suitable alternative.

Control dust in and around the nursery by spraying dusty paths with sump oil or covering them with gravel.

CHAPTER 6

Control by Host Resistance

Host resistance is probably the most valuable control measure in agriculture. Its success is the result of extensive research in many state experiment stations, the U.S. Department of Agriculture, commercial companies, and comparable organizations throughout the world. Host resistance in many crops is most successful if sanitation measures and regulation of irrigation and soil drainage are also practiced. In this chapter, the types of resistance in certain crop types and the problems that emanate from the development of pathologic (physiologic) races that are virulent on cultivars with race-specific resistance are emphasized. Variability in *Phytophthora* was reviewed by Erwin et al. (1963) and Erwin (1983). The foliar disease of potato called late blight, caused by *P. infestans,* and root and stem rot of soybean, caused by *P. sojae* (*sensu* Kaufmann and Gerdemann 1958; Hansen and Maxwell 1991), are covered in the greatest detail because of the relatively greater amount of research that has been conducted on resistance to these diseases. A section on control of major diseases is also included in each chapter devoted to an individual species. A selected group of references on the use of resistance to control important diseases caused by several species of *Phytophthora* is presented in Table 6.1.

Host-specific resistance in potato to late blight, caused by *P. infestans,* has been identified in wild potato and incorporated into commercial cultivars. Host-specific resistance in soybean to root rot, caused by *P. sojae,* has been found within soybean (*Glycine max* L. Merr.), and many genes for resistance have been incorporated into agriculturally adapted cultivars (Schmitthenner 1985). In strawberry, host-specific resistance to root rot caused by *P. fragariae* has been identified in several cultivars since the work of Hickman and English (1951b), and subsequently many new resistant cultivars have been released in Europe and the United States.

Dutt's (1982) review of physiologic specialization in *Phytophthora* species covers *P. infestans* on potato and *P. parasitica* on tobacco and several other hosts. He cites a few examples of *P. parasitica* that are labeled in some reports as varieties. *P. parasitica* var. *macrospora* causes fruit rot of *Annona squamosa* L. (Rao et al. 1963), *P. parasitica* var. *sesame* is specific to *Sesamum orientale* (Kale and Prasad 1957), and *P. parasitica* var. *piperina* affects *Piper betle* and some other hosts (Saxena 1977).

Resistance is defined as the ability of the host to hinder development of the pathogen (R. A. Robinson 1969, 1976); however, as discussed below, there are many differences among plant species in the extent of this ability. Host resistance is the most important of the various means of control available because the grower need not expend funds for extraneous control measures such as fungicides. Even where fungicides must be used, resistance is a useful complementary control measure. The goal of plant breeders is to obtain a high level of resistance but retain the agronomic or horticultural qualities of the crop that are necessary for public consumption. The discussion of problems and progress in breeding plants for resistance to *Phytophthora* by Umaerus et al. (1983) emphasizes control of the late blight disease of potato, caused by *P. infestans,* by development of general resistance. Keen and Yoshikawa (1983) discuss the physiologic nature of host-specific resistance with emphasis on the root rot disease of soybean, caused by *P. sojae.* The remainder of this chapter is devoted to examples of resistance, problems in maintaining resistance to *Phytophthora,* mechanisms of resistance, and its integration into control programs.

Types of Resistance and Terminology

The terminology for different types of resistance in the literature lacks uniformity, and although definitions have been standardized by G. A. Robinson (1969) and Parlevliet (1979), usage of terms is still variable. Umaerus et al. (1983) developed their discussion of resistance under two categories, *specific resistance* and *general resistance* (Thurston 1971; Browning et al. 1977); however, G. A. Robinson (1969) says the term general resistance is too vague and should be avoided. Plants with host-specific resistance are nearly immune, and hence this type of resistance was referred to as *true resistance* by Müller (1953) and by the epidemiological term coined by Vanderplank (1963), *vertical resistance.*

A *physiologic race* is defined as one of a group of forms that are alike in morphology but differ by certain cultural, physiological, biochemical, or other characteristics (Hawksworth et al. 1983). Physiologic race is considered to imply a biotype that is pathogenic on certain cultivars of plants but not on others. We share G. A. Robinson's (1969) objection to the term because the definition is too broad. We instead use *pathologic race.* In most cases, resistance in the host to pathologic races is inherited by a single gene and is dominant.

General resistance, as Umaerus et al. (1983) and Thurston (1971) use the term, is synonymous with *horizontal resistance.* General resistance has also been called *field resistance,* as in the recent chapter on breeding for resistance in potato to late blight (Wastie 1991). *Rate-limiting resistance* (Parlevliet 1979; MacKenzie et al. 1983) is an apt term if rate-disease data are available. Schmitthenner and Walker (1979) argue convincingly that the term *tolerance* is more descriptive and better understood by growers than rate-limiting or horizontal resistance.

Vertical and horizontal resistance are epidemiological terms based on a bar graph by Vanderplank (1963), who showed that the race-specific potato cultivars Kennebec and Maritta had R-gene resistance to race 1 of *P. infestans,* but Maritta had a higher level of resistance (but not immunity) than Kennebec to other races. When the cultivars are grown side by side, Kennebec will be blighted completely and Marritta will remain green (Vanderplank 1984). Vanderplank called this horizontal resistance; it is not race specific, and there is no interaction between host and pathogen genotypes. Vertical resistance, however, is race specific and is characterized by single-gene interactions between host and pathogen genotypes. Disease does not occur on a plant that possesses vertical resistance.

Although plants with horizontal resistance can be infected, the rate of disease progress (exemplified by the rate of increase in size of the lesion on leaf or root) is usually slower than on the susceptible plant. This reduction in rate may be sufficient to prevent the death of a root or leaf during relatively short periods when enough free moisture is present to allow spread or multiplication of the pathogen. To express this, the term *rate-reducing resistance* (Parlevliet 1979) was used by MacKenzie et al. (1983) and Tooley and Grau (1982, 1984a, b) to characterize resistance that is not race specific. The definition of rate-reducing resistance conceptually implies that there are different components of resistance that govern the rate of disease development. Resistance of this type can be subdivided into three components: 1) infection, 2) colonization, and 3) reproduction, because each of these factors can be measured (Parlevliet 1979).

Schmitthenner and Walker (1979) used the terms tolerance, to define the ability of a susceptible soybean plant to endure infection by *P. sojae* without showing severe symptoms, and resistance, to characterize the ability of the plant to prevent infection (Schmitthenner and Walker 1979). They consider the term tolerance to be apt because growers understand it as it is meant, whereas the term resistance, even if qualified, is often misunderstood, especially by farmers, to mean absolute. Schmitthenner and Walker (1979) defend the use of the term tolerance, which was defined and established as they use it by the American Phytopathological Society (1940), the British Mycological Society (1950), and the National Academy of Sciences (1968). Schmitthenner and Walker (1979) recognize that tolerance has been defined differently by Browning et al. (1977), who state that tolerance implies the state of a plant on which disease occurs, for instance, stem rust on wheat, but despite the disease, yields more than a plant of a susceptible cultivar with the same amount of disease. It is impossible, however, to quantitate how much disease occurs on roots of a plant without destroying it. The amount of disease on a wheat plant infected with rust (*Puccinia graminis* f. sp. *tritici*) can be quantitated by the number of pustules that have formed on the leaves. Mussell (1980) has the same concept of the term tolerance as Schmitthenner and Walker (1979) and states that tolerant plants can allocate minerals, energy, and food even though they are infected. He recommends the use of the term tolerance for non-race-specific resistance.

Stating an opposite viewpoint, Clarke (1986) would not use the term tolerance for the soybean horizontal or rate-limiting resistance phenomenon and disagrees with the use of tolerance as a form of resistance. He accepts the definition that tolerance involves putting up with or bearing with the disease but states that the term should not be used to indicate any capability of the host plant to prevent, reduce, or delay the development of the pathogen. He states that these qualities are forms of resistance.

G. A. Robinson (1969) states that tolerance should be used to describe a host-parasite relationship but not a suscept-pathogen relationship, and thus this term would be alien to plant pathology. He states that the term is appropriate to describe a symptomless carrier or a host in which a virus multiplies but does not cause symptoms. He accepts that *disease tolerance* can be ascribed to a cultivar that develops disease symptoms as severe as those of another cultivar but that suffers less loss of yield. Robinson states categorically that tolerance should not be used as a synonym for horizontal resistance. Despite his opinion, the term tolerance is used frequently to characterize this type of resistance.

Thurston (1971), in a review of general resistance of potato to *P. infestans,* also objected to the use of the term tolerance. Thurston's (1971) review on the relationship of general resistance to late blight of potato recognized that many terms in the literature described essentially the same phenomenon. He utilized the terms specific resistance and general resistance as explained by Hooker (1967) for rust resistance: "specific resistance (race-specific) which functions against certain races or biotypes but not against others; and generalized resistance (race non-specific) which functions against all biotypes."

Since most of the reviews during the last 30 years (Wilson and Coffey 1980; Coffey and Wilson 1983a, b; Umaerus et al. 1983; Wastie 1991) have used general resistance to describe non-race-specific resistance, we continue its use but also use other, apparently synonymous terms as appropriate for clarity. We tend to agree with Schmitthenner and Walker (1979) that tolerance is a useful term to describe the moderate resistance situation in soybean, alfalfa, avocado, and citrus; however, it is beyond our scope to settle the various cogent arguments that have been made for the different terms. Identification of the different degrees of resistance in cultivars that are useful, with or without race-specific resistance, is the most important concept to keep in mind.

Race-Specific Resistance

Phytophthora infestans—Late Blight of Potato

Variability in *P. infestans* is discussed in Chapter 3, and the late blight disease caused by *P. infestans* is covered in Chapter 33. Resistance to the late blight disease of potato is covered briefly for the United States by Thurston and Schultz (Hooker 1981) and for Europe by I. M. Smith et al. (1988). Both indicate that race-specific resistance has contributed little to control because of the continuous development of new races of *P. infestans.* When potato cultivars with single-gene-controlled, race-specific resistance to corresponding races of *P. infestans* are infected, the lesions on leaves are limited to small flecks with no other growth of the lesion over time. The infection process is reviewed by Coffey and Wilson (1983) and Coffey and Gees (1991). Continued infection and subsequent spread of the pathogen do not occur. Vanderplank (1963, 1984) makes the point that vertical (race-specific) resistance reduces the initial inoculum to near zero, thus preventing an epidemic because the infected lesion does not extend beyond a few cells and *P. infestans* cannot produce secondary sporangia. Because vertical resistance prevents increase of the initial inoculum, the use of vertical resistance is essentially the same as sanitation (Vanderplank 1984). Since vertical resistance is usually controlled by a single gene, however, mutation allows development of a new race of *P. infestans* capable of infecting leaves of the host with single-gene resistance. This phenomenon rapidly negates the value of the race-specific cultivar, because the resistant *P. infestans* can increase rapidly and spread over the entire field or area. Alternatively, horizontal resistance is most often, but not necessarily always, conditioned by two or more genes. Therefore, horizontal resistance is considered to be stable, whereas race-specific (vertical) resistance is not stable and historically has become ineffective because of the development of new pathologic races (Vanderplank 1971). Stability of horizontal resistance has been explained by the improbability that chance mutations of two or more genes in the pathogen will occur simultaneously. For a more documented discussion of resistance to late blight, see the chapter by Wastie (1991).

Dowley et al. (1991) credited Salaman (1911) as being one of the first to discover that potato progeny of a cross of *Solanum tuberosum* L. and *S. demissum* Lindl., a wild species

Table 6.1. Diseases caused by *Phytophthora* species for which host resistance is non-host-specific (general, G) or host-specific (S) and mechanisms of resistance

Species Crop	Subject	Type of resistance	Reference
P. cactorum			
Apple	Histology of infection	G	Alt (1980)
	Resistance of 13 rootstocks	G	Browne and Mircetich (1993)
	Excised shoot assay for resistance	G	Barritt et al. (1990)
	Resistant clone, Novole	G	Cummins et al. (1983)
	Effect of time of inoculation on resistance	G	Dakawa and Sewell (1981)
	Review	G	D. C. Harris (1991b)
	Screening methods	G	Harris and Tobutt (1986)
	Excised shoot assay	G	Jeffers et al. (1981)
	Resistant cv. Freedom	G	Lamb et al. (1985)
	Seedling inoculation	G	McIntosh (1968)
	Resistance in wild species	G	Ram (1983)
	Review	G	Sansavini and Rosati (1986)
	Rootstocks	G	Sewell and Wilson (1959, 1964)
	Review	G	Utkhede (1986)
	Excised shoot assay	G	Utkhede and Quamme (1988)
Cherry	Cankers on certain scions	G	Bielenin and Jones (1988b)
	Rootstocks	G	Mircetich and Matheron (1976, 1981)
	Rootstocks	G	Wilcox and Mircetich (1985a)
Strawberry	Resistance sources	G	Govorova (1983)
Walnut	Variation in susceptibility	G	C. O. Smith and Barrett (1931)
	Resistance of *Juglans* species rootstocks	G	Mircetich and Matheron (1983)
P. cambivora			
Almond	Rootstocks	G	Wicks and Lee (1985)
Apple, pear, and stone fruits	Review	G	Mircetich and Browne (1987)
	Survey	G	Mircetich and Matheron (1976)
P. capsici			
Black pepper (*Piper nigrum*)	Review	G	Waard, de (1980)
	Review; *P. columbrinum* is immune	G	Sarma et al. (1991)
Pepper (*Capsicum annuum*)	Resistance to blight and crown rot	G	Barksdale et al. (1984)
	Recurrent selection breeding program	G,S	Bartual et al. (1991)
	Nature of age-related resistance	G	Jeun and Hwang (1991)
	Genetics of resistance	G	Gil Ortega et al. (1990, 1991)
	Virulence of isolates from different areas on resistant cultivars	G	Kim and Hwang (1992)
	Effect of age of plant on resistance	G	Kim et al. (1989)
	Sources of resistance	?	Matsuoka et al. (1984b)
	Methods of inoculation and techniques in breeding for disease resistance	S?	Palloix et al. (1988a, b, 1990)
	Time of infection, pathogen strain, recurrent selection	G	Palloix et al. (1988a, 1989, 1990)
	Phospholipid from pathogen induces resistance	G	Pivot et al. (1991)
	Greenhouse evaluation of strains	G	Polach and Webster (1972)
	Physiology of resistant cultivars	S	Redondo et al. (1989)
	Enzyme activity in resistant plants	S	Saimmaime et al. (1991)
	Inheritance of resistance	G	P. G. Smith et al. (1967)
	Pathogen variability	G	Yang et al. (1989)
Tomato	Greenhouse evaluation of resistance	G	Bolkan (1985)
P. cinnamomi			
Acacia pulchella	Cytology of resistant host	G	Tippett and Malajczuk (1979)
Avocado	Resistance evaluated by zoospore encystment on roots	G	Aveling and Rijkenberg (1989, 1991)
	Rootstock resistance	G	Botha et al. (1990)
	Review	G	Coffey (1987, 1991, 1992)
	Cross protection against *P. cinnamomi* by *P. parasitica*	G	Dolan et al. (1986)
	Rootstock evaluation for resistance	G	Dolan and Coffey (1986); Gabor and Coffey (1990, 1991)
	Field resistance of rootstocks	G	Gabor et al. (1990)
	Mechanism of resistance	G	Kellam and Coffey (1985)
	Resistance in tissue culture	G	Phillips et al. (1991)
	Review	G	Zentmyer (1980)
	Rootstock evaluation for resistance in nutrient solution	G	Zentmyer and Mircetich (1965)
	Resistant rootstocks found in Latin America	G	Zentmyer and Schieber (1987)
Banksia spp.	Field stem and root inoculation	G	McCredie et al. (1985)
Eucalyptus	Evaluation of seedling roots with zoospore inoculum	G	Byrt and Holland (1978)
	Compares physiology of resistance of several *Eucalyptus* spp.	G	Cahill et al. (1985, 1986, 1989, 1992)
	Resistant host list in Australia	G	R. W. Edmiston in Shea (1975b)
	Effect of temperature on colonization of *Eucalyptus marginata* (susceptible) and *E. calophylla* (resistant)	G	Grant and Byrt (1984)
	Root inoculation of *E. sieberi* (susceptible) and *E. maculata* (resistant)	G	Halsall (1978)

(*continued on next page*)

Table 6.1. (*continued*)

Species Crop	Subject	Type of resistance	Reference
P. cinnamomi (*continued*)	Variation in mortality of families of *Eucalyptus regnans*	G	J. A. Harris et al. (1985)
	Selection of resistant *E. marginata* (jarrah) clones	G	McComb et al. (1991)
	Phloem water content effects on resistance	G	Tippett et al. (1985)
Pine	Genetic variation in resistance	G	Butcher et al. (1984)
	Callus tissue of loblolly and loblolly × short leaf pine hybrids were more resistant than short leaf or Virginia pine	G	Jang and Tainter (1990a, b)
	Mycorrhizal effects on resistance (review)	G	Marx (1972)
	Mycorrhizal effects on resistance	G	Marx (1973a, b)
Rhododendron	Evaluation by root inoculation; lists tolerant clones; cv. English Roseum is resistant to 8 *Phytophthora* spp.	G	Hoitink and Schmitthenner (1970, 1974b)
P. citrophthora			
Citrus	Phytoalexin scoparone	G	Afek et al. (1986)
	Resistance expressed in excised shoots	G	Afek et al. (1990
	Phytoalexin, scoparone	G	Afek and Sztejnberg (1986, 1988, 1990, 1993)
	Mode of infection	G	Broadbent (1969)
	Review	G	Broadbent (1977b)
	Review	G	H. S. Fawcett (1936)
	Root inoculation of young trees with chlamydospores	G	Graham (1990)
	Root inoculation of young trees	G	Grimm and Hutchinson (1973)
	Rootstocks inoculated in the field	G	Klotz et al. (1958, 1965a, 1968)
	Review	G	Magnano di San Lio et al. (1986b)
	Seasonal effects on resistance on excised shoots	G	Matheron and Matejka (1989)
	Soil infestation with chlamydospores	G	M. A. L. Smith et al. (1991)
	Salinity effects on reduced phytoalexin production	G	Sulistyowati et al. (1990, 1992)
	Review	G	Timmer and Menge (1988)
	Failure of culture filtrate to select resistance in callus tissue	G	Vardi et al. (1986)
P. cryptogea (*P. drechsleri*)			
Safflower	Sources of resistance	S?	Harrigan et al. (1982)
	Environmental factors on resistance	G	Heritage and Harrigan (1984)
P. drechsleri (*P. cryptogea*)			
Safflower	Accumulation of phytoalexin, safynol, over time	G	Allen and Thomas (1971)
	Effect of temperature and light on resistance	S?	Johnson and Klisiewicz (1969)
	Inheritance of resistance of cv. VFR 1	G	C. A. Thomas (1976)
	Screening for resistance	G	C. A. Thomas et al. (1960)
	Assay of resistance on cotyledons	G	C. A. Thomas and Hill (1977)
	Resistance of cv. Biggs	G	C. A. Thomas and Zimmer (1970)
	Effect of irrigation on resistance		Zimmer and Urie (1967)
P. drechsleri and other species			
Pome and stone fruits	Resistance in rootstocks (review)	G	Mircetich and Browne (1987)
P. drechsleri f. sp. *cajani*			
Pigeon pea	Screening for resistance	S	Kannaiyan et al. (1981)
P. fragariae			
Strawberry	Physiologic races described	S	Converse (1967)
	Physiologic race A10 in California	S	Converse (1970)
	Susceptibility of *Potentillla* and *Geum* to races	S	Converse and Moore (1966)
	Evaluation of resistance by root inoculation	S	George and Milholland (1986a)
	Relative susceptibility of cultivars	S	Hickman and English (1951b)
	Evaluation of races by symptoms and oospore production in roots	G,S	Law and Milholland (1992)
	Criteria for race designation	S	Milholland et al. (1989)
	Description of race situation and control in Scotland	S	Montgomerie (1977)
	Races in Scotland	S	Montgomerie and Kennedy (1975)
	Phytoalexin-like compounds associated with resistance	S	Mussell and Staples (1971)
	Comparison of indicator cultivars to determine race resistance	S	Pepin and Daubeny (1969)
	Physiologic races by differential variety response	S	Scott et al. (1950)
	Screening for resistance	G,S	Scott et al. (1976)
	Review on breeding for resistance	G	Scott et al. (1984)
	Describes resistance (1 gene) and partial resistance	G,S	Van de Weg (1989)
	Assay of resistance to races	G,S	Wynn (1968)
P. heveae			
Coconut palm	Screening for resistance	G	de Franqueville et al. (1989)
P. infestans			
Potato	Assay of general resistance on leaves	G	Berggren et al. (1988)
	General resistance	G	Bjor and Mulelid (1991)
	International race designation	S	Black et al. (1953)
	Elicitors of resistance response	S	Bostock (1989)
	Seedling blight test for general resistance	G	Caligari et al. (1985)
	Field resistance	G	Cassells et al. (1991)
	Adaptation of pathogen to general resistance	G	Caten (1974)

(*continued on next page*)

Table 6.1. *(continued)*

Species Crop	Subject	Type of resistance	Reference
P. infestans (*continued*)	Cytology of pathogen development (review)	G	Coffey and Gees (1991)
	Histology of infection of a resistant and susceptible cultivar	G	Coffey and Wilson (1983a, b)
	General resistance in wild *Solanum* species	G	Colon and Budding (1988)
	Epidemics in the world	G,S	Cox and Large (1960)
	Immunohistochemical study of resistant and susceptible cultivars	G	Cuypers and Hahlbrock (1988)
	Variability in different biotypes of the pathogen	G	Denward (1967, 1970)
	Induced resistance with hyphal wall components	S	Doke et al. (1987)
	Integration of general resistance and metalaxyl for control of late blight	G	Doster et al. (1990)
	Complex races from wild hosts	G,S	Fernández and Galindo (1989)
	Integration of polygenic resistance and protective fungicides	G	Fry (1975, 1977a, b)
	Plant disease management	G	Fry (1982)
	Late blight forecasting, effect of host resistance, and fungicide weathering	G	Fry et al. (1983)
	Review; genetics of specific resistance	S	Gallegly (1968)
	Review; host-pathogen interaction	G,S	Gallegly and Niederhauser (1959)
	Assay of resistance in vitro	S	Gotz (1985)
	Physiologic races in Denmark	S	Gurtler (1984)
	Rate of penetration, time for sporulation, and quantities of sporangia related to general resistance	G	Guzmán-N (1964)
	Assessment of specific resistance by ELISA	S	Harrison et al. (1991)
	Dominant gene for resistance is expressed in tissue culture	S	Helgeson (1972, 1976)
	Resistance gene in hybrid of *Solanum brevidens* × *S. solanum*	S	Helgeson et al. (1986)
	Formation of haustorial encasements in resistant potato plants	S	Hohl and Stössel (1976); Hächler and Hohl (1984)
	Ultrastructural changes in *Capsicum frutescens* caused by incompatibility of *P. infestans* and production of the phytoalexin capsidiol	G	D. R. Jones et al. (1975a, b)
	Superoxide dismutase-dependent cytochrome c and resistance	G,S	Ivanova et al. (1991a, b)
	Races and field resistance	G	Jeffrey et al. (1962)
	Field methods to evaluate resistance	G	Lapwood (1982)
	Apparent infection rates of late blight	G	Latin et al. (1981)
	Multigenic resistance disease cycle	G	Main and Gallegly (1964)
	Assay of races of *P. infestans*	S	Malcomson (1969a, b, 1979)
	Assessment of field resistance	G	Malcomson (1976)
	A2 mating type found in Great Britain	G,S	Malcomson (1985)
	R genes from *Solanum demissum*	S	Malcomson and Black (1966)
	Nature of race-specific resistance	S	Müller (1953); Müller and Burger (1940)
	Multigenic field resistance in Mexico	G	Niederhauser (1962)
	Review of races in Mexico and development of field resistance	G,S	Niederhauser (1991)
	Review of races in Mexico and development of field resistance	G	Niederhauser and Cobb (1959)
	Resistant cultivars	G,S	O'Brien and Rich (1976)
	Field response of cultivars to late blight	G	Platt and McRae (1990)
	Late blight resistance in commercial and wild potato	S	Reddick (1928, 1932)
	Role of rishitin inhibition of *P. infestans* on cut potato tubers	S	Sato et al. (1971)
	Resistant clones from protoplasts	G,S	Shepard et al. (1980)
	Comparison of glasshouse and field tests	S	H. E. Stewart et al. (1983)
	Detached leaf assay of resistance; effect of inoculum concentration and plant age	S	H. E. Stewart (1990)
	Induced resistance to *P. infestans* by prior inoculation with *P. cryptogea*	G	Strömberg and Brishammar (1991)
	Field observation	G,S	Stuart (1906)
	Detection of virulence genes in *P. infestans*	S	Sujkowski (1990)
	Resistance related to maturity	G,S	Sweizynski (1990)
	Resistance from wild species	G,S	Sweizynski et al. (1991)
	Review of general resistance	G	Thurston (1971)
	Elba, late maturing resistant cultivar	G	Thurston et al. (1985)
	Resistance in *Solanum verrucosum*	G	Tooley (1990)
	Field assessment of fitness of isolates	S	Tooley and Fry (1985)
	Fitness in asexual and sexual populations	S	Tooley et al. (1986)
	Race-specific resistance and origin of new races	S	Toxopeus (1956, 1960)
	Effect of mineral nutrition and mechanisms of general resistance	G	Umaerus (1969, 1970)
	Laboratory method to test for general resistance	G	Umaerus and Lihnell (1976)
	Review of resistance—general and race-specific	G	Umaerus et al. (1983)
	Stability of horizontal resistance	G	Vanderplank (1971)

(*continued on next page*)

Table 6.1. (*continued*)

Species Crop	Subject	Type of resistance	Reference
P. infestans (*continued*)	Specificity of the phytoalexin rishitin	S	Varns et al. (1971)
	Production of a phytoalexin capsidiol in pepper leaves	G	Ward (1976)
	Review, breeding for resistance	G	Wastie (1991)
	Cytology of plants with general resistance	G	V. E. Wilson and Coffey (1980)
	Nature of resistance	G,S	Yamamoto (1982)
P. lateralis			
Port Orford cedar	Testing for resistance	G	E. M. Hansen et al. (1989)
P. medicaginis (= *P. megasperma* f. sp. *medicaginis*)			
Alfalfa	Development of resistance in central Mexico	G	Aguirre et al. (1983)
	Interaction of resistance and excess soil moisture stress	G	Alva et al. (1985, 1986)
	Interaction of resistance and excess soil moisture	G	Barta and Schmitthenner (1986)
	Correlation of disease severity index and percentage of resistant plants	G	Basu (1988)
	Screening for resistance	G	Bray and Irwin (1978)
	Screening for resistance	G	Damirgi et al. (1983)
	First report of resistance in cv. Lahontan	G	Erwin (1966)
	Variability of virulence in isolates	G	Faris (1985)
	Yield and survival of resistant cultivars	G	Faris and Sabo (1981)
	Variability of virulence in isolates	G	Faris et al. (1983)
	Review	G	Frosheiser (1980)
	Testing by inoculation in glasshouse and field	G	Frosheiser and Barnes (1973, 1984)
	Assessment of resistance	G	Gramshaw et al. (1985)
	Screening for resistance; resistant cultivars increase stand and yield	G	F. A. Gray et al. (1973, 1988)
	Interaction of *Meloidogyne hapla* and resistance	G	F. A. Gray et al. (1990)
	Resistance and decline in yield	G	Havey and Grau (1985)
	Inheritance of resistance in diploid and tetraploid alfalfa	G	Havey and Maxwell (1987, 1988); Havey et al. (1987)
	Recurrent selection for resistance in *Medicago cerrulea* and *M. falcata*	G	Heisey and Murphy (1985)
	Resistance found in cv. Hayden alfalfa by selection of resistant seedlings after standardization of inoculum concentration	G	Hine et al. (1975)
	Flooding technique for selection for resistance in greenhouse	G	Hohrein et al. (1983)
	Assay of resistance is optimal at 16–28°C; cv. Apollo II resistance compared with resistance to *Aphanomyces euteiches*	G	Holub and Grau (1990a, b)
	Resistance of cultivars in Australia; mode of infection	G	Irwin (1974, 1976)
	Inheritance of resistance	G	Irwin and Maxwell (1980)
	Seedling resistance and inheritance of factors for resistance in diploid and tetraploid alfalfa	G	Irwin et al. (1979, 1981a, b)
	Review		Leath et al. (1988)
	Selection for resistance in nondormant cultivars	G	Lehman et al. (1967)
	Yield and stand persistence related to resistance in Australia	G	Lowe et al. (1985)
	Inheritance of factors for resistance	G	Lu et al. (1973)
	Effect of resistance on yield	G	Lueschen et al. (1976)
	Infection process in resistant and susceptible cultivars	G	Marks and Mitchell (1971a, b)
	Evaluation of disease resistance in tissue culture	G	Miller and Maxwell (1983)
	Infection of resistant and susceptible germplasm by light microscopy and at the ultrastructural level	G	Miller and Maxwell (1984a, b)
	Comparison of resistant and susceptible cultivars at low flooding stress	G	Mueller and Fick (1987)
	Interrelationship of inoculum, soil moisture, temperature, and host and pathogen genotype	G	Pratt and Mitchell (1976)
	Resistance and susceptibility is shown on inoculated cotyledons	G	Pratt et al. (1975)
	Development of resistance in poorly aerated soils	G	Rogers et al. (1978)
	Review and description of symptoms, disease development, and control	G	Stuteville and Erwin (1990)
	Interaction of *Meloidogyne hapla* and *M. incognita* exacerbates symptoms on both resistant and susceptible cultivars	G	Welty et al. (1980)
	Anatomical, physiological, and morphological effects due to flooding soil	G	Zook et al. (1986)
Chickpea	Screening for resistance	G	Brinsmead et al. (1985)
	Development of resistant cultivars	G	Dale and Irwin (1990, 1991a, b)
P. nicotianae (= *P. parasitica*)			
African violet	Evaluation of two cultivars	G	Strider (1978, 1979)

(*continued on next page*)

Table 6.1. (*continued*)

Species Crop	Subject	Type of resistance	Reference
P. nicotianae (= *P. parasitica*) (*continued*)			
Tomato	Inheritance of resistance	G	Boukema (1983)
	Methods of detection of resistance	G	Skadow (1955)
P. palmivora			
Cocoa	Sources of resistance	G	Soria (1974)
	Combining ability and additive inheritance of resistance to pod rot	G	G. Y. Tan and Tan (1990)
	Review	G	Toxopeus (1974)
P. parasitica (= *P. nicotianae*)			
Citrus	Rootstock effect on soil population	G	Agostini et al. (1991)
	Tolerant root stocks predisposed by salinity	G	Blaker and MacDonald (1986)
	Mode of infection of rootstocks and review of control with rootstocks	G	Broadbent (1969, 1977b)
	Root inoculation of young trees	G	Graham (1990)
	Root inoculation of young trees	G	Grimm and Hutchinson (1973)
	Review of resistance in rootstocks	G	Klotz (1973, 1978)
	Review of resistance in rootstocks	G	Klotz and Calavan (1965, 1978)
	Resistant rootstocks	G	Klotz et al. (1958, 1965a, 1968)
	Excised shoot inoculation vs. season	G	Matheron and Matejka (1989, 1990)
	Review of acceptable rootstocks	G	Naidu et al. (1982)
	Shoot inoculation with chlamydospores	G	M. A. L. Smith et al. (1991)
	Review	G	Timmer and Menge (1988)
Peperomia	Formation of a phytoalexin in inoculated stems	G	Sirodhana (1968)
Tomato	Seedling and field resistance	G	Blaker and Hewitt (1987a, b)
Tomato, pepper	Techniques to evaluate resistance	G	Bolkan (1985)
Pineapple	Techniques to evaluate resistance	G	Louvel (1975)
P. parasitica var. *nicotianae*			
Tobacco	Physiologic specialization of pathogen on tobacco	S	Apple (1962)
	Cytology of defense in a susceptible cultivar	S	Benhamou and Côté (1992)
	Resistance heritable by 1 gene	S	Boukema (1983)
	Phytoalexins produced in resistant callus tissue	S	Budde and Helgeson (1981)
	Inheritance of resistance factors	S	Collins et al. (1971)
	Inoculum density in soil affected by resistant and susceptible cultivars	G	Ferrin and Mitchell (1986a, b)
	Inoculum concentration on roots	G	Gooding and Lucas (1959b)
	Resistance in callus and roots	S	Helgeson et al. (1972, 1976, 1986)
	Review	G,S	Lucas (1975)
	Resistance expressed in callus	G,S	Maronek and Hendrix (1978)
	Kestose not produced by race 3	S	McIntyre and Hanken (1977)
	Designation of races by seedling inoculation	S	McIntyre and Taylor (1976, 1978)
	Production of capsidiol in tobacco callus in response to cryptogein	S	Milat et al. (1991a, b)
	Free sterols in stems of colonized tobacco	G	Moore and Orcutt (1982)
	Parasiticein not produced by *P. parasitica* var. *nicotianae*	S	Ricci et al. (1992)
	Pathogen movement in field from inoculated resistant and susceptible cultivars	G	Shew (1987)
	Inoculum concentration on seedlings	G	Taylor and Waggoner (1979)
	Detached leaf test to detect resistance	G	Tedford et al. (1990)
	Resistance test by inoculation of leaves	G	Wills and Crews (1964)
P. phaseoli			
Lima bean	Resistance to race B	S	Wester (1968)
	Multiple resistance	S	Wester et al. (1972)
P. sojae (= *P. megasperma* f. sp. *glycinea;* *P. megasperma* var. *sojae*)			
Soybean	Efficacy of metalaxyl use on tolerant cultivars; defeated *Rps*1 resistance gene; linkage of *Rps*7 gene	G	T. R. Anderson and Buzzell (1982, 1988, 1992)
	Inheritance of race-specific resistance	S	T. R. Anderson and Buzzell (1982)
	Review	S	Athow (1987)
	*Rps*6 gene in cv. Altona	S	Athow and Laviolette (1982)
	Review	G,S	Athow and Laviolette (1983)
	Genetics of resistance to races	S	Athow et al. (1979)
	Nature of elicitors of glyceollin activity	S	A. R. Ayers et al. (1976a–c)
	Histological study of infection in resistant and susceptible cultivars	S	Beagle-Ristaino and Rissler (1983b)
	Designation of gene in cv. Harasoy 63 as *Rps*1	S	Bernard et al. (1957)
	*Rps*1-k gene for resistance in cv. Kingwa	S	Bernard and Cremeens (1981)
	Plant loss and resistance	S	Buzzell and Anderson (1982)
	Effects of race, temperature, and cultivar on reaction of soybean to zoospore inoculation	S	Buzzell and Anderson (1982)

(*continued on next page*)

Table 6.1. *(continued)*

Species Crop	Subject	Type of resistance	Reference
P. sojae (*continued*)	Cross protection and heat-induced susceptibility	S	Chamberlain (1972); Chamberlain and Gerdemann (1966)
	New races in Alabama	S	Collins et al. (1985)
	Determination of field resistance	G	Irwin and Langdon (1982)
	Determination of field tolerance	G	Jimenez and Lockwood (1980)
	Methods of testing for resistance; description of races 10–20 and single genes for resistance in soybean	S	Keeling (1976, 1980, 1982, 1984)
	Linkage of *Rps*1 and *Hm* (resistance to metribuzin)	S	Kilen (1986) (also Kilen and Barrentine 1983)
	Resistance gene *Rps* 2 in cv. CNS	S	Kilen et al. (1974)
	Screening for resistance to races; gene frequency changes in bulk populations of soybean	S	Kilen and Keeling (1977, 1990)
	New physiologic races 7, 8, 9, and 21, 22	S	Laviolette and Athow (1977, 1981, 1983)
	Inheritance of resistance to races 5 to 9	S	Laviolette et al. (1979)
	Inheritance of resistance and a new physiologic race	S	Layton et al. (1984, 1986); Lazarovits et al. (1980)
	Virulence and drug-resistance of protoplasts and inheritance of resistance; heterokaryosis in *P. sojae*	S	Layton and Kuhn (1988a, b, 1990)
	Determination of races in flooded soil	S	Lockwood and Cohen (1978)
	Tolerance in the cross of cv. Ripley × cv. Harper	G	McBlain et al. (1991a)
	Determination of tolerance by three methods	G	McBlain et al. (1991b)
	First report of physiologic races	S	Morgan and Hartwig (1965)
	Determination of races by inoculation of cuttings	S	Nguyen et al. (1985)
	Glyceollin does not accumulate in tolerant plants	G	Olah and Schmitthenner (1985)
	Determination of tolerance in a growth chamber test	G	Olah et al. (1985)
	Review on role of phytoalexins in resistance	S	Paxton (1983, 1988)
	Phytoalexins and host resistance	S	Paxton and Chamberlain (1967, 1969)
	Races in Australia	G,S	Rose et al. (1982)
	Race-specific resistance in soybean	S	Ryley and Obst (1992)
	Structure of elicitor-active and inactive compounds from walls of *P. sojae*	S	J. K. Sharp et al. (1984)
	New race	S	Schmitthenner (1972)
	Review of root rot control	G,S	Schmitthenner (1985)
	Review of integrated control strategies	G,S	Schmitthenner and Van Doren (1985)
	Tolerance and resistance in soybean	G,S	Schmitthenner and Walker (1979)
	Race 4 described	S	Schwenk and Sim (1974)
	Determination of race by inoculation of soybean with zoospores	S	Schwenk et al. (1979)
	Gene *Rps*1 in soybean	S	Singh and Lambert (1985)
	In vitro assay of soybean root rot in a microculture	G,S	M. A. L. Smith et al. (1990, 1991)
	Pathogen specificity in tolerant lines	G	Thomison et al. (1988, 1991)
	Quantification of rate-reducing resistance	G	Tooley and Grau (1982, 1984a, b)
	Races of *P. sojae*	S	Tooley et al. (1982, 1984)
	Determination of field resistance by continuous row irrigation in the field	G	Troedson et al. (1991)
	Determination of resistance by aeroponic culture of plants	S	Wagner and Wilkinson (1992b)
	New race	S	Wagner and Wilkinson (1992a)
	Modeling of rate-reducing resistance	G	Wagner et al. (1993)
	Development of tolerant varieties	G	Walker and Schmitthenner (1982)
	Review of host-pathogen relations	G,S	Ward (1990)
	Elicitors of phytoalexins	S	Ward and Stössell (1976)
	Cytological differences between resistant and susceptible cultivars	S	Ward et al. (1989)
	Influence of light, temperature, and wounding on resistance	S	Ward and Lazarovits (1982); Ward and Buzzell (1983)
	Age-related resistance is related to race specific resistance	S	Ward et al. (1981)
	Relationship of metalaxyl to glyceollin production	S	Ward et al. (1980)
	Races in Nebraska	S	White et al. (1983)
	Review, breeding for resistance	G,S	Wilcox (1976)
P. syringae Apple	Review	G	D. C. Harris (1991b)
P. vignae Adzuki bean	Isolates not pathogenic to cowpea	S	Kitazawa et al. (1978, 1979, 1986)
	Proposed f. sp. *adzukicola* and *vignae* because of differential pathogenicity	S	Tsuchiya et al. (1986)
Cowpea	Kievitone and phenylalanine ammonia lyase produced	S	Munn and Drysdale (1975)
	Phytoalexin kievitone produced	S	Partridge and Keen (1976)
	Cultivars resistant to races	S	Purss (1958, 1972, 1975)
	Production of phenylalanine ammonia lyase as a resistance response	S	Ralton et al. (1988)

that grows in Mexico, were resistant to late blight. Single genes from *S. demissum* were transferred to potato (*S. tuberosum*) by Schick (1932) and Black (1952). A wide range of wild *Solanum* species is distributed on the western slopes of the volcano Nevada de Toluca in Mexico (Rivera-Peña and Molina-Galan 1989). The incidence of late blight of wild *Solanum* species was reported by Rivera-Peña (1990a, b). *P. infestans* is probably native to Mexico, where it has existed for centuries on wild solanaceous plants (Niederhauser 1956, 1991; Zentmyer 1984; Fry and Spielman 1991). *S. demissum* plants survived epidemics of late blight in natural native settings because of the evolution of resistant genes naturally selected by centuries of exposure and subjection to selection pressure by *P. infestans.* When these genes are incorporated by transfer of pollen into commercial potato cultivars, plants with a high degree of resistance (near immunity) can be selected and propagated into germplasm lines. Subsequent reselection for desirable agronomic characters and asexual propagation by use of the eyes in tubers as "seed" result in a resistant potato cultivar that is genetically uniform. At one time, potato breeders were optimistic that single-gene-controlled, race-specific resistance would successfully control late blight of potato (Black et al. 1952, 1953; Schick 1932; Niederhauser and Cobb 1959; Niederhauser 1961, 1991). In practice, however, single-gene resistance has been rapidly overcome by the development of biotypes, called pathologic or physiologic races, of *P. infestans,* which are indistinguishable but capable of causing late blight on cultivars previously considered resistant (Niederhauser 1962, 1991; Vanderplank 1984; Wastie 1991).

Vanderplank (1984) cites the example of the cultivar Kennebec, which was released in 1948 in the New England region of the United States and has R1 gene resistance to late blight, to show that vertical resistance has become ineffective for control of late blight of potato. Although initially the resistance of Kennebec was highly successful, by 1954 and 1955 new races that caused severe blight had become dominant in the population of *P. infestans,* and a late blight epidemic destroyed the potato crop. Vanderplank states categorically that farmers now must rely on fungicides to control late blight and that plant breeders have turned to the development of horizontal resistance (i.e., general resistance). Although vertical resistance (i.e., race-specific resistance) may still be useful in some areas, the consensus is that overall this means of control is doomed to failure (Wastie 1991). Some of the many references on race-specific as well as general resistance to late blight are listed in Table 6.1

Although resistance in potato to *P. infestans* can be assessed by inoculation of intact plants with a suspension of zoospores, the detached leaf test (in which leaves are incubated in moist chambers) has the advantages of being more readily controlled and using less space. When Stewart (1990), using Malcomson's (1976) method, compared inoculations with different concentrations of zoospores on leaves of different ages, the immune reaction was most easily detected when a drop of inoculum containing a low number of zoospores (2.5 and 5.0 × 10^3/ml) was placed on detached leaves from flowering plants 8–10 weeks old.

Pathologic races of *P. infestans* are designated by the numbers that correspond to the resistance conferred by R-genes from *S. demissum* according to the international rating method organized by Black et al. (1953). To designate and differentiate races of *P. infestans,* a series of differential potato cultivars (each with resistance or susceptibility to different races with different R-genes) can be inoculated with different isolates. An isolate that causes disease on a potato cultivar with R-gene 1 is labeled race 1 (R1), and an isolate that causes disease on potato with R-gene 2 is designated race 2 (R2). The numbers increase, corresponding to the numbers of R-genes that have been identified on potato. An isolate that causes disease only on a potato cultivar without R-genes is labeled race 0 (Table 6.2).

Eventually, complex races (those resistant to more than one R-gene) of *P. infestans* capable of causing disease on cultivars with more than one gene for resistance were identified. These are labeled, appropriately, R1.2 (capable of attacking R-genes 1 and 2) or R1.2.3 (capable of attacking potato plants with R-genes 1, 2, and 3). Although in many publications, the race numbers designating multiple R-genes in *P. infestans* are set off by commas (e.g., R1,2,3,4), the period is used here (e.g., R1.2.3.4) because R1,2,3,4 could be mistaken to mean a succession of different genotypes. Complex races are reported in numerous areas (e.g., Malcomson and Black 1966; Leach and Rich 1969; Fernández and Galindo 1989; Swiezynski et al. 1991).

After decades of active research on single-gene resistance, it became evident that the genome of *P. infestans* contained an

Table 6.2. Interrelationship of dominant genes for resistance to late blight of potato and pathologic races of *Phytophthora infestans* (the gene-for-gene relationship)[a]

Host resistance genotype	Pathologic race of *P. infestans*[b]															
	0	1	2	3	4	1.2	1.3	1.4	2.3	2.4	3.4	1.2.3	1.2.4	1.3.4	2.3.4	1.2.3.4
R	S	S	S	S	S	S	S	S	S	S	S	S	S	S	S	S
R1	—	S	—	—	—	S	S	S	—	—	—	S	S	S	—	S
R2	—	—	S	—	—	S	—	—	S	S	—	S	S	—	S	S
R3	—	—	—	S	—	—	S	—	S	—	S	S	—	S	S	S
R4	—	—	—	—	S	—	—	S	—	S	S	—	S	S	S	S
R1R2	—	—	—	—	—	S	—	—	—	—	—	S	S	—	—	S
R1R3	—	—	—	—	—	—	S	—	—	—	—	S	—	S	—	S
R1R4	—	—	—	—	—	—	—	S	—	—	—	—	S	S	—	S
R2R3	—	—	—	—	—	—	—	—	S	—	—	S	—	—	S	S
R2R4	—	—	—	—	—	—	—	—	—	S	—	—	S	—	S	S
R3R4	—	—	—	—	—	—	—	—	—	—	S	—	—	S	S	S
R1R2R3	—	—	—	—	—	—	—	—	—	—	—	S	—	—	—	S
R1R2R4	—	—	—	—	—	—	—	—	—	—	—	—	S	—	—	S
R1R3R4	—	—	—	—	—	—	—	—	—	—	—	—	—	S	—	S
R2R3R4	—	—	—	—	—	—	—	—	—	—	—	—	—	—	S	S
R1R2R3R4	—	—	—	—	—	—	—	—	—	—	—	—	—	—	—	S

[a] Reproduced by permission of the American Phytopathological Society, St. Paul, Minnesota, courtesy M. E. Gallegly (1968).
[b] S = susceptibility, and — = resistance.

arsenal of genes capable of causing late blight on every potato cultivar derived from plant genes of wild *S. demissum.* Eventually, races were identified that were capable of attacking plants with several R-genes (Table 6.2) (Gallegly 1968; Dowley et al. 1991; Niederhauser 1991; Wastie 1991).

How new pathologic races arise within populations of *P. infestans* is not known with certainty. Denward (1920) produced evidence for somatic recombination in the mycelium. Leach and Rich (1969) found three genotypes from one sporangium, indicating somatic recombination. Since the increase in the number of propagules (sporangia and zoospores) produced in a potato field under optimum blight conditions is rapid—for instance, McAlpine (1910) reported a sporangium-to-sporangium cycle within a 6¾-h period—mutation or somatic recombination with survival of the most fit individuals that are capable of causing disease on plants with new R-genes is a likely explanation for the increase in the number of new races.

Because *P. infestans* and all other *Phytophthora* species are diploid and in some cases polyploid in the vegetative state (Chapter 3), crossing over of chromosomes and recombination of genes in the sexual oospore stage is likely to be involved in the development of new biotypes. Until recently, the Toluca region of Mexico was the only area in the world where both the A1 and A2 mating types existed together (in other regions of the world, only A1 was found). It was thought, therefore, that the sexual stage was at least partly responsible for the large numbers of new races of *P. infestans* in the Toluca region of Mexico (Niederhauser 1956, 1961, 1991). Since A1 and A2 mating types occur at about a 50:50 ratio in field plots in the Toluca area and weather is ideal for expression of late blight, plants with resistance to *P. infestans* from breeding programs throughout the world, including Scotland, Germany, the Netherlands, and many other regions, were susceptible when tested in Toluca (Niederhauser 1991). Because the number of different races of *P. infestans* was greater there than in any other region of the world, it was assumed, but not proved, that new races could have emanated from sexual hybridization. Recent data from crosses of *P. infestans* A1 and A2 isolates with markers for both virulence and biochemical characters now lend credibility to this hypothesis (Le Grand-Pernot 1986a, b; Shattock et al. 1986b; Shattock et al. 1987; Spielman et al. 1989; Spielman et al. 1990). See Chapter 3 for more information on results of hybridization experiments with *P. infestans.* Although the A2 and A1 mating types now occur in the United Kingdom, the Netherlands, other countries in Europe, and the United States (Chapters 3 and 33), several reports contain evidence that there is not a perceptible increase in new biotypes of *P. infestans* that could be attributed to sexual reproduction (Fry et al. 1991, 1992).

However, Sujkowski et al. (1994) recently reported that after introduction of the A2 mating type into Poland (Spielman et al. 1991), there was evidence from isozyme pattern analysis that sexual reproduction had occurred from 1987 to 1991. Most genotypes were at Hardy-Weinberg equilibrium; A1 and A2 mating types were not genetically differentiated; a high proportion of individuals were found that would not be expected from asexual reproduction; and a new genotype occurred that could have been produced from sexual recombination. They also suggested that the rapid breakdown in resistance of the cultivar Bronka, which was resistant during the mid-1980s but which by the late 1980s had become susceptible to severe blight, was the result of sexual reproduction in the field population.

Although the incorporation of single, race-specific genes from *S. demissum* into the commercial potato, *S. tuberosum,* was a breakthrough in plant breeding for control of a plant disease, race-specific resistance has not proved to be a solution to the control of late blight of potato because of the rapid evolution of new races. General resistance, which is less effective but not vulnerable to new races, has proved to be more durable. See Table 6.1 for a listing of references on both specific and general resistance to several *Phytophthora* species.

Vanderplank (1988) states that the emphasis on obtaining vertical resistance to *P. infestans* in a crop such as potato can result in a loss in horizontal resistance, which he terms the *vertifolia* effect (Vanderplank 1963, 1984). High-yielding cultivars with vertical resistance often lose horizontal resistance because plant breeders inadvertently tend to select only plants that show race-specific resistance. He points out that resistance and high yield potential often oppose each other; for instance, an increase in yield potential can result in a decrease in horizontal resistance. This phenomenon is explained on the basis of the source-sink hypothesis, in which the high yield component is a sink that competes with the horizontal resistance sink. This concept apparently may hold true for other diseases as well as late blight. A slight yield loss resulting from incorporation of single-gene resistance to *P. sojae* has been noted in soybean (Schmitthenner 1985). When seven resistant, near-isogenic (*Rps, Rps*) lines and seven susceptible (*rps, rps*) lines were compared in nondiseased fields in Minnesota, however, yield and other agronomic characteristics were similar in all lines with and without resistance (Singh and Lambert 1985).

ELISA (enzyme-linked immunosorbent assay; see Chapter 2) was used to detect the degree of infection by *P. infestans* in inoculated, detached potato leaves (Harrison et al. 1990). This method was also used to determine the expression of major genes for race-specific resistance. Although in most cases visual symptoms and ELISA results were correlated, in a few cases, moderate growth of *P. infestans* was noted in leaves without symptoms. Also, the absence of *P. infestans* in a spreading lesion appeared to indicate that a toxin rather than the fungus per se was involved (Harrison et al. 1991).

Phytophthora sojae (*P. megasperma* f. sp. *glycinea*)—Soybean Root and Stem Rot

Race-specific resistance to the root rot disease of soybean (*G. max*) caused by *P. sojae* (see discussion in Chapters 3 and 59) is currently one of the major control methods (see the reviews by Schmitthenner 1985; Athow 1987; Ward 1990). A list of references on resistance to *P. sojae* is given in Table 6.1. Single-gene, race-specific resistance to Phytophthora root rot was first reported by Bernard et al. (1957) and Kaufmann and Gerdemann (1958). Although resistance has been incorporated into numerous cultivars in the United States and Australia, new pathologic races of this pathogen were noted a few years after the first resistant cultivar was introduced. A current report now indicates that there are seven genes for race-specific resistance and 37 races (*Phytopathology News*, December 1993, page 171). According to Ward (1990), theoretically 2,048 new races would be possible from 11 genes known for resistance in soybean.

The current designation of races based on pathogenicity to a series of cultivars with different genes for resistance is shown in Table 6.3 (Ward 1990). The cultivar Harosoy was originally considered to be susceptible to all races; however, it was later reported to be resistant to races 12 and 19 by Athow (1987) and to 12, 15, 18, and 19 by Ward (1990) (Table 6.3). The soybean plant introduction germplasm PI 157,409 possesses the resistance gene combination *Rps*1-b and *Rps*4, which confers resistance to races 1 through 22 (Layton et al. 1984).

Despite the problem with new races, single-gene resistance is still one of the major methods being utilized for control of root rot of soybean (Buzzell and Anderson 1982; Schmitthenner 1985; Athow 1987; Ward 1990). Since Phytophthora root rot of soybean is a soilborne disease and *P. sojae* does not produce windborne sporangia, dissemination of new races over wide distances is probably more restricted than that of *P. infestans* on potato, which produces rain- and windborne deciduous sporangia.

Screening for race-specific resistance is relatively easy in soybean by use of the stem-wound inoculation method because susceptible plants die within a few days after inoculation (Keeling 1976, 1980; Schmitthenner 1985). A comparison of several methods showed that hypocotyl (stem) inoculation was consistent and more efficient than root inoculation by infestation of sand or a hydroponic solution with inoculum (Keeling 1976). Because of its 99% effectiveness (freedom from disease escape), stem inoculation is routinely used by both commercial and public agency plant breeders to screen plants for race-specific resistance. The method has several variations, but essentially it involves placement of mycelium from a soft agar culture or zoospore inoculum (10^5/ml) (Schwenk et al. 1979) into a needle- or scalpel-puncture wound in the hypocotyl just below the cotyledons of 7- to 10-day-old plants. After inoculation, plants are incubated in a moist chamber or the inoculated wound on the stem is covered with petrolatum. The susceptible plants die within 2–4 days, but the plants with race-specific resistance continue to develop normally and the wound heals. Although wounds are always disruptive of normal physiological development and may not be optimal for cytological and physiological studies (Coffey and Wilson 1983), the method is a boon to plant breeders because thousands of plants can be screened rapidly and very few susceptible plants escape infection.

P. sojae can be grown on oatmeal agar (or several other media such as V8 juice agar or lima bean agar; see Chapter 3). A mycelial mat can be cut into 2- × 2-mm pieces, each of which is inserted with a spear into a longitudinal slit in the hypocotyl of 10-day-old seedlings (Athow and Laviolette 1983). Tweezers with fine points (the type commonly used for transfer of electron microscope grids) can be used to pick up and insert aerial mycelium into the wound (J. S. Baumer, *personal communication*).

Incubation of soybean seedlings about 10 cm in height in a 25% Hoagland's nutrient system (Hoagland and Arnon 1950) was used to screen for resistance to race 2. Plants were supported individually in holes in thick polystyrene sheets that floated on the nutrient solution and inoculated with *P. sojae* 3 days after they were transplanted by adding inoculum to the nutrient system. Ten days later, plants were determined to be either susceptible (dead or with lesions on the hypocotyl) or resistant (Kilen et al. 1974).

A laboratory method utilized 10-day-old soybean seedlings of cultivars with race-specific resistance that were cut off at the base of the stem, bundled, and inoculated by placing mycelium on the base of each cut stem. The bundles of stems were supported in holes made in a Styrofoam sheet. The seedlings in the Styrofoam sheet were floated in a shallow tray containing tap water and incubated for 7 days under artificial light. The plants were rated as resistant (roots had initiated from the healthy hypocotyl) or susceptible (roots had not formed, and hypocotyl was decayed). The method gave results comparable to those of the stem-puncture inoculation method and was considered to be quicker and to use less space than the stem-puncture method (Nguyen et al. 1985).

Since the stem-inoculation technique does not consistently evaluate resistance conferred by the *Rps*2 gene, Wagner and Wilkinson (1992) developed a method of testing the effects of *P. sojae* on roots of soybean plants grown in an aeroponic system. A nutrient system was applied to roots in a closed chamber (method of Bohm [1979]). The root lesions were measured several times 4 days after inoculation without sacrificing or harming the plants. Cultivars with genes *Rps*7 or *Rps*1 were susceptible, and cultivars with genes *Rps*1-b, *Rps*1-c, *Rps*1-d, *Rps*2, *Rps*3, or *Rps*6 were resistant to race 3 of *P. sojae*.

The history of breeding for resistance to *P. sojae* is summarized by Athow and Laviolette (1983), Schmitthenner (1985), Athow (1987), and Ward (1990).

The reason for the rapid increase in the number of races of *P. sojae* is unknown, but it could be partly the result of mutation and/or genetic recombination at the sexual oospore stage (Hobe and Schmitthenner 1981; Bhat et al. 1990, 1992; Schmitthenner et al. 1994). Hobe and Schmitthenner (1981) reported that 58 single-zoospore cultures, obtained by baiting soil samples with leaf disks of a susceptible soybean cultivar, yielded 12 new races based on a standard set of differential cultivars. When two additional differential cultivars of soybeans with different genes for resistance were used to assay the race characteristics of the isolates, 10 more races were identified that were more diverse than the predominant isolates from diseased plants growing in field soils. This concept was further developed in the report by Schmitthenner et al. (1994).

Table 6.3. Interaction between physiologic (pathologic) races of *Phytophthora sojae* (*P. megasperma* f. sp. *glycinea*) and *Rps* alleles for resistance in soybeans[a]

Source	Allele	Race[b]																								
		1	2	3	4	5	6	7	8	9	10	11	12	13	14	15	16	17	18	19	20	21	22	23	24	25
Mukden	*Rps*1-a	R	R	S	S	S	S	S	S	S	R	R	S	R	S	R	R	R	R	S	S	S	S	S	R	S
Sanga	*Rps*1-b	R	S	R	R	R	R	R	R	R	S	S	S	R	R	R	S	S	R	S	S	R	R	S	S	S
Mack	*Rps*1-b	R	R	R	S	S	R	R	R	R	R	R	S	R	S	R	S	R	S	S	S	R	S	R	R	S
Kingwa	*Rps*1-k	R	R	R	R	R	R	R	R	R	R	R	S	R	R	R	S	R	R	S	S	R	R	—	R	S
CNS	*Rps*2	R	R	R	R	R	S	S	S	R	R	R	R	R	R	R	R	R	R	R	R	—	—	—	—	—
PI 171,442	*Rps*3	R	R	R	R	R	S	S	R	R	S	R	S	R	R	S	R	S	R	S	S	S	S	R	S	R
PI 172,901	*Rps*3-b	R	R	R	R	R	R	R	R	R	R	R	R	R	R	R	R	S	—	—	—	R	R	—	R	R
PI 300,046	*Rps*3-c	R	R	R	R	S	—	S	—	S	—	—	R	R	—	—	—	—	—	—	—	—	—	—	—	—
PI 86,050	*Rps*4	R	R	R	R	S	S	S	S	S	R	S	R	R	R	R	R	S	R	R	R	R	S	—	S	R
T240	*Rps*5	R	R	R	R	R	S	S	R	R	S	R	R	R	R	S	S	S	R	S	R	S	S	—	S	R
Altona	*Rps*6	R	R	R	R	S	S	S	S	S	R	S	R	S	R	R	R	S	R	R	R	R	S	S	S	R
Harosoy	*Rps*?	S	S	S	S	S	S	S	S	S	S	S	R	S	S	S	R	S	R	R	S	S	S	S	S	S
PI 103,091	?	R	R	R	R	R	S	R	S	R	R	R	S	R	R	R	R	S	R	S	R	R	R	R	R	R

[a] Reproduced by permission of C. A. B. International, Wallingford, Oxon, U.K., courtesy Ward (1990).
[b] Table complete up to December 1988, compiled from a number of sources by R. I. Buzzell (*personal communication,* 1988). R = resistant host response; S = susceptible host response; and — = not reported. The report of Förster et al. (1994) lists 37 races. Races 36 and 37 from Japan were unique.

Since new, undescribed races of *P. sojae* could be isolated from propagules persisting in field soil (Hobe and Schmitthenner 1981; Schmitthenner et al. 1994), a logical presumption could be made that such variability could exist in the *Phytophthora* genome and be expressed by progeny from individual oospores that germinate. Over a certain period, a new race may build up enough inoculum to become a field problem; however, the fitness of such new races may be limited by unknown environmental or biological parameters and thus may take years to become limiting to crop production.

Ward (1990) reviewed the literature on the competitive ability of races. In many cases, isolates from single zoospores or even single oospores are similar in virulence to the parental culture. Rutherford et al. (1985) reported, however, that new races were found in progeny of single-zoospore cultures taken from mass-mycelial cultures. Wagner and Wilkinson (1992a) also reported this. Rutherford et al. (1985) reported that a mass-mycelial isolate of *P. sojae* race 6 yielded single-zoospore cultures in the first generation that were similar to that of the parent race 6; but in a subsequent generation, several single-zoospore progenies differed from the parent in that they were virulent on a soybean cultivar with the *Rps*1-k resistance gene. In a second generation, several single-zoospore progeny were avirulent on soybean cultivars with *Rps*1-a, *Rps*2, *Rps*3, *Rps*4, and *Rps*6 resistance genes but virulent on a cultivar with the *Rps*1-b gene. One isolate was avirulent on all cultivars. Ward (1990) gives some theoretical explanations such as mutation, mitotic crossing over, and possibly the effect of extra chromosomal genes as discussed by Shaw (1983). Whatever the explanation, the diploid genome of *P. sojae* is a potential source of new races.

Bhat and colleagues presented new information about the role of oospores produced abundantly in diseased plant tissue and about the variability and the evolution of new races of *P. sojae*. Bhat et al. (1990, 1992a) and Bhat and Schmitthenner (1992, 1993a, b) reported that selfing of isolates of races 1 and 4 by reisolation of single oospores yielded progeny from race 1 that segregated into race 1 and a putative race 4. Progeny from race 4 segregated into other undesignated races as well as race 4. Germinated oospore progeny of a race 3 isolate resulted in not only race 3, but also races 7, 8, and 21 virulence types. Thus, field isolates of *P. sojae* are heterogenic in nature (Bhat and Schmitthenner 1992).

When single oospores of race 1 were germinated, the progeny segregated at a 3:1 ratio of race 1 to avirulence but were identical in cultural characteristics. The avirulent strains produced only a hypersensitive response when inoculated into the universally susceptible cultivar Williams by Bhat et al. (1992) and Bhat and Schmitthenner (1993a). They concluded that the avirulent strains either lost their race specificity for soybeans or became universally avirulent strains that had lost all race-specific virulence genes. The segregation of single-oospore progeny of race 1 from virulence to avirulence at a 3:1 ratio indicates that the allele for avirulence at the race 1 locus was probably homozygous recessive (Bhat 1991).

Since single-oospore progeny of some races were found to be heterozygous (did not breed true), successive single-oospore isolations through several generations were necessary to fix the genetic characteristics (Bhat et al. 1993a). Previous work by Long and Keen (1977a, b) showed that variability in growth rate was eliminated by five generations of selfing, indicating the heterozygosity of *P. sojae* for different traits.

To prove that isolates of *P. sojae* could be crossed, even though mating types have not been described, Bhat and Schmitthenner (1993a, b) studied resistance to certain chemical inhibitors. Strains of *P. sojae* resistant to the fungicide metalaxyl (MEX^r) and the amino acid analog fluorophenylalanine (FPA^r) were selected after mutagenesis treatment with *N*-methyl-*N′*-nitro-*N′*-nitrosoguanidine (MNNG) (Bhat and Schmitthenner 1993a). They proved that by mixing cultures with resistance to the markers MEX^r and FPA^r that a cross could be made and that some progeny of the crosses were indeed sexual hybrids. A single-oospore culture from a cross of FPA^r race 1 × MEX^r race 3 yielded cultures with resistance to both inhibitors. They concluded that when cultures were grown on lima bean medium, resistance to MEX (10 µg/ml) and FPA (50 µm/ml) was controlled by major genes. "Three phenotypes were distinguished in F_2 progeny: resistant ($MEX^{r\text{–}n}$), intermediate ($MEX^{r\text{–}i}$), and sensitive (MEX^s) to MEX, which fit a monohybrid genotypic segregation ratio of 1:2:1" (incomplete dominance of MEX) (Bhat and Schmitthenner 1993a).

On the FPA medium, the only recognized phenotypes were resistant; single-oospore progeny were either resistant (FPA^r) or sensitive (FPA^s). When progeny of crosses were assessed, the F_2 progeny yielded three FPA^r to one FPA^s, thus indicating that resistance to FPA was dominant (Bhat and Schmitthenner 1993b).

The reports by Bhat (1991), Bhat et al. (1990, 1992, 1993a, b), Bhat and Schmitthenner (1992, 1993a, b), Förster et al. (1994), and Whisson et al. (1994) strongly support the concept that the great variability among pathologic races and the continuous production of new races is in a large part the result of the sexual recombination within the oospore stage of *P. sojae*. See also the discussion in Chapter 3 under Genetics and Cytology of *Phytophthora*.

Phytophthora vignae—Cowpea Stem and Root Rot

Race-specific resistance to root and stem rot of cowpea (*Vigna unguiculata* L. (Wolp.)), caused by *P. vignae* (Purss 1957, 1958, 1972) in Australia, is an important control measure (Table 6.1; see also Chapter 65).

The cultivars Blackeye 5 and Iron were resistant. Resistance in Blackeye 5 was considered to be near immunity, whereas cultivars Malabar and Giant were field resistant (Purss 1958). The resistant cultivar Caloona emanated from a cross of cultivars Poona × Blackeye 5 with three successive backcrosses to Poona (Purss 1963). Two new races were later found that caused disease on both Blackeye 5 and Caloona. Resistance in the cultivar Chinese Red, resistant to all four races, was transferred to Caloona, from which a new cultivar, Red Caloona, with resistance to four races was produced (Purss 1975). This type of single-gene, race-specific resistance remains useful in Australia.

Phenylalanine ammonium lyase activity increases rapidly in infected tissue in the resistance response to race 2 (Ralton et al. 1988). The phytoalexin kievetone is associated with resistance (Munn and Drysdale 1975; Partridge and Keen 1976).

In Japan, *P. vignae* was isolated and proved to be pathogenic to adzuki bean (*Vigna angularis* (Willd.) Ohwi and Ohashi) (Kitazawa et al. 1978). Adzuki bean isolates were more virulent on adzuki bean than on cowpea. Cowpea isolates (races 1, 2, 3, and 4) were pathogenic to some adzuki bean cultivars, but adzuki bean isolates were virulent only on the cowpea cultivars Poona and Kurotone-Sonjaku. There was some evidence of host specificity (Kitazawa et al. 1978, 1979). These isolates were designated *P. vignae* f. sp. *adzuki* (Tsuchiya et al. 1986).

Phytophthora fragariae—Strawberry Red Stele

Red stele (called red core in Great Britain) on strawberry (*Fragaria* × *ananasa* L.) is caused by *P. fragariae,* which was originally described by Hickman (1940) and redescribed by

Ho and Jong (1988). See Table 6.1 for several references on control by host resistance and Chapter 26A.

Pathologic races were first reported by Scott et al. in the United States in 1950 and later by Hickman and English (1951b) and Hickman (1962) in Great Britain. The Hickman and English report distinguished races 1, 2, and 3 by differential pathogenicity to the cultivars Huxley, Perle de Prague, and Climax. Since then, a series of reports from Europe and the United States have extended the number to at least 10 races on the basis of the red stele symptom. In inoculated plants, the strawberry cultivars Blakemore, MD-683, Aberdeen, Del Norte, and Stelemaster differentiated six pathogenic races designated A1–A6 (Converse and Scott 1962). At the time of this last report, Montgomerie (1961) had distinguished 13 races in Scotland. McKeen (1958) listed six races in British Columbia (Canada).

Milholland et al. (1989) distinguished several races from the United States, not only on the basis of pathogenicity but also on the ability of *P. fragariae* to produce oospores in the roots of the differential host cultivars. They found that the technique of inoculation by spraying young (10- to 21-day-old) roots with encysted zoospores (10–30 × 10^3/ml) followed by incubation in a growth chamber at 15°C consistently differentiated six races. They inoculated runner plants that had been grown in a 1:1:1 mix of Metromix 220 (W. R. Grace Co., Cambridge, Massachusetts), sand, and soil under intermittent mist at 25–30°C. After 10–21 days, roots on paper towels were washed and sprayed with encysted zoospores (1.5 ml of inoculum per plant). Plants on the paper towels were incubated in clear plastic bags in the dark for 48 h at 15°C. Plants were then transplanted to plastic containers filled with Metromix that had been saturated with water and allowed to drain. The containers were each sealed in a plastic bag and incubated at 15°C under a 12-h light regime. After 2 weeks, roots were examined to determine the number of oospores that had developed. Ten root tip segments (6–8 mm long) were washed, and each was mounted on a glass slide and sealed under a cover slip (60 × 22 mm) with Permount (Fisher Scientific Co.). The number of oospores per root segment as well as the percentage of root segments containing oospores were recorded (George and Milholland 1986a). A disease severity index (DSI) consisted of the number of oospores per root segment × percentage of root segments infected × 100. A strawberry cultivar with a DSI greater than 1.0 was considered to be susceptible, and one with a DSI less than 1.0 was considered to be resistant (Milholland et al. 1989). By use of the strawberry differential genotypes, cultivars Tennessee Beauty (susceptible to all races), Del Norte, Aberdeen, Climax, MD-683, Stelemaster (or Siletz), and Surecrop, the races Pf-1, Pf-2, Pf-3, Pf-4, Pf-5, and Pf-6 were differentiated. With this system, Milholland et al. (1989) found that races A2 and A7 (Converse 1970) and the North Carolina isolates would be classified as race Pf-2, which was the only race found to occur naturally in North Carolina (Law and Milholland 1992).

Scott et al. (1984) reviewed various methods that have been developed for screening and breeding cultivars for resistance to *P. fragariae*. Development of resistance is complicated by the necessity for cultivars to have acceptable pomological characteristics and to be adaptable to different regions of the world and by the presence of pathologic races that vary by region. Strawberry is also heterozygous, and most characters are inherited quantitatively (many genes involved). Although resistance to red stele is partially dominant, incorporation of resistance in a horticulturally acceptable cultivar requires up to 10–12 years after a cross is made; however, when the source of resistance is from another species such as *F. chiloensis* (clones Yaquina and Del Norte), the time required is even longer than 10–12 years (Scott et al. 1984).

Resistance to *P. fragariae* appears to be race specific or vertical. Nonetheless, general resistance is also found The presence of different pathologic races of *P. fragariae* in different parts of the world has made exchange of breeding material on a worldwide basis very risky, except under controlled quarantine conditions (Scott et al. 1984).

General Resistance

General resistance is usually equated with the terms horizontal resistance, field resistance, and rate-limiting resistance (see above). The history of race-specific resistance to several species, especially of *P. infestans,* indicates that new races can evolve rapidly enough to render the race-specific resistance ineffective. New races may evolve more slowly in the soilborne species such as *P. sojae* because inoculum is not disseminated rapidly by airborne sporangia. The development of new races in *P. sojae* is important but not as limiting as it is in *P. infestans,* which produces airborne sporangia. Nonetheless, the principle is the same. Because the potential exists for new races to arise that will overcome race-specific resistance, the trend in breeding for resistance is to search for and develop general resistance that is not as vulnerable to attack by new races. In most cases, general resistance is inherited by multigenes, and the actual resistance is rate limiting and not immunity. Therefore, the resistant host cultivar is still vulnerable to the pathogen but less so than susceptible cultivars.

Many reports in addition to those covered here indicate that general resistance in potato to late blight is a successful control measure but not always adequate without fungicides (see Chapter 7). After perusal of the many reports, the conclusion can be drawn that the amount of inoculum and the environmental conditions (optimal temperatures and association of free water in the soil or on the foliage) are critical in assessing the degree of general resistance. Since inhibition of the rate of lesion expansion, rather than immunity, is the basis of general resistance, it is obvious that even the plant with general resistance can be overcome by *Phytophthora,* and thus general resistance could not be detected if the inoculum level or environmental conditions unduly favored the pathogen.

Environmental conditions, especially soil moisture, can play different roles. The most obvious is the evidence that free soil water is associated with increases in inoculum concentration (see section in Chapter 5 on soil moisture). Another is that excess soil water can lead to anoxia (lack of oxygen) of roots, which in itself increases host susceptibility (reviewed by Duniway [1983]; Fraedrich and Tainter 1989).

In the following sections, we discuss examples of the use of general resistance under headings of different host-pathogen systems. See Table 6.1 for references on use of both general and race-specific resistance. Also, we discuss resistance to many other species of *Phytophthora* in chapters devoted to individual species.

Phytophthora infestans—Late Blight of Potato

Thurston's (1971) well-documented discussion of resistance to late blight states that after discovery of R-genes, almost all potato breeders discontinued research on general resistance and concentrated on host-specific resistance with R-genes (seven R-genes had been identified by 1971). As discussed above, however, new races appeared within 2–3 years after release of each host-specific cultivar. Failure of cultivars with R-genes to maintain a stable level of immunity to late blight revived interest in the search for general resistance.

Wastie (1991) attributes the interest in the development of general resistance to the pioneering thinking of Niederhauser (1961). Niederhauser (1991) related some of his early experience in a chapter on *P. infestans* entitled "*Phytophthora infestans*—The Mexican Connection." He was awarded the World Food Prize for his research in Mexico on resistance to late blight, which was begun during the late 1940s and conducted under the Rockefeller program (McCoy 1991; Van der Zwet 1991). Early in his career, Niederhauser noted that blight-immune cultivars from the United States and Europe were killed year after year by *P. infestans* races in the Toluca Valley of Mexico, where both A1 and A2 mating types were found in equal proportions and frequent rainfall and moderate temperatures provided optimal environmental conditions for infection. Adequate control was not obtained until general (field) resistance was developed from selections within *Solanum tuberosum* and from crosses with *S. demissum,* a wild potato that had coexisted for centuries with *P. infestans* in Mexico, the center of origin for *P. infestans* (Reddick 1932, 1943; Zentmyer 1988; Niederhauser 1991).

During more than 40 years of field testing in central Mexico, none of the cultivars of *S. tuberosum* or other tuber-forming species, which had been developed in Mexico and in other regions of the world, were immune to all races of *P. infestans* causing late blight, despite the incorporation of R-genes obtained from the wild potato, *S. demissum,* into many tuber-bearing potato cultivars. Yet among many of these cultivars, different levels of resistance were expressed, not as immunity, but by the presence of fewer and smaller lesions on the leaves and by reduced sporulation (Niederhauser 1991). Niederhauser noted that much information has been developed on the use of race-specific R-genes but comparatively little is known about the nature of the more durable horizontal resistance (general resistance or field resistance). Despite the greater difficulty in evaluating and developing cultivars with general resistance, during the period of 40 years that Niederhauser was involved with the potato program, some 20 blight-resistant potato cultivars were released for use in central Mexico, and many of these blight-resistant cultivars have been successfully introduced into other countries, such as Costa Rica, Rwanda, and the Philippines.

Niederhauser (1991) recommended several research approaches that are necessary at an international level to further the course of development of general resistance to late blight in potato:

- What basic chemical and physical factors contribute to horizontal resistance?
- How are such factors genetically controlled?
- How can screening techniques for the physiological and chemical mechanisms involved in horizontal resistance be improved?
- What levels of general (horizontal) resistance are needed to minimize dependence on fungicides in integrated disease control programs?

The chapter by Umaerus et al. (1983) reflects the authors' wide experience in breeding for late blight resistance in Sweden and other countries in Europe. Umaerus (1970) recognized three components of general resistance to *P. infestans:* 1) resistance to penetration, 2) restriction of growth of the fungus in the host, and 3) reduced sporulation capacity of the fungus on the host. In a cytological study comparing the general resistance of the potato cultivars Shamrock and Pimpernel with the susceptible Majestic cultivar, fungal growth was restricted in leaf tissue of cultivars with general resistance. The frequency of penetration in cultivar Pimpernel was low and could be considered to be a major factor in general resistance (Coffey and Wilson 1983b; Wilson and Coffey 1980). Reduced frequency of penetration and localization of the fungus in the mesophyll cells of the leaves may be useful indicators of resistance.

Many wild *Solanum* species contribute an abundant supply of sources of general resistance (Wastie 1991). Wastie cites van Soest et al. (1984), who show that general resistance to late blight (which ranked from 1 to 3 on a scale of 1 to 9, with 1 being the lowest severity of disease), can be found in *S. demissum* and 12 other species from accessions maintained in the Germany-Netherlands collection.

Vanderplank (1971) states that horizontal resistance is stable, despite claims by Toxopeus (1956) that new biotypes of *P. infestans* could become "trained" to adapt to certain resistant cultivars without R-genes. He cites disease evaluations in the Netherlands on the potato cultivar Voran during 1937–1970 on which disease resistance ratings of about 7 on tubers and foliage (on a scale of 1 to 10) remained the same over that period. Caten (1974) showed experimentally that strains of *P. infestans* within a race may become differentially adapted to the cultivars with general resistance from which they have been isolated. This adaptation could result in some erosion of horizontal resistance; however, he concluded that complete breakdown of horizontal resistance in the field was unlikely. Although disease severity changes from year to year in some cases, factors other than late blight also are involved. Thurston (1971) states that the purported loss in the general resistance of the cultivar Champion, which was introduced in 1877 and grown in Ireland for more than 50 years, resulted from the gradual buildup of viruses in seed tubers perpetuated by vegetative propagation (Davidson 1928).

The cultivars Surprise and Pimpernel have maintained the same level of general resistance for the past 30 years (Colon and Budding 1988). Colon and Budding (1988) analyzed the general resistance of several species in a field in the Netherlands over a 2-year period. Plants were inoculated with zoospores of the complex race isolate R1.2.3.4.5.7.10.11, and foliage disease was measured over time by the area under the disease progress curve. Evidence of clones with high resistance was obtained in *S. tuberosum* cv. Pimpernel and also in *S. berthaultii, S. chacoense, S. microdontum, S. sucrense, S. venturii,* and *S. verrucorsum* (Colon and Budding 1988).

Although general resistance in potato is accessible, practical, and heritable, it is usually conditioned by multiple genes (Umaerus et al. 1983). Therefore, the assay of much larger numbers of progeny is needed from crosses of an agronomically susceptible cultivar with a source of general resistance than from crosses with plants with resistance conditioned by a single R-gene. Also, the detection of general resistance is much more difficult than that of race-specific resistance. Many factors affect the expression or lack of suppression. We previously mentioned inoculum concentration and environmental conditions as key factors. General resistance can be detected in foliage only with the appropriate concentration of zoospore inoculum (Hodgson 1962).

Dowley et al. (1991) reviewed methods of testing for resistance, and although several laboratory methods were cited, they concluded that field testing with disease assays based on the rate of defoliation was necessary (Cox and Large 1960). They described a breeding program in Ireland, where conditions are favorable for late blight nearly every year. The program requires at least 12 years to evaluate a cultivar thoroughly.

Phytophthora sojae (*P. megasperma* f. sp. *glycinea*)—Soybean Root and Stem Rot

The discovery of many new races of *P. sojae*—up to 25 according to Ward (1990) (Table 6.3), 27 according to Wagner and Wilkinson (1992a), even higher according to Schmitthenner et al. (1994), and 37 according to Förster et al. (1994)—has stimulated research to find a type of resistance more durable than single-gene resistance. Many lines of soybean germplasm with rate-reducing resistance (Tooley and Grau 1982, 1984a, b)—or, as termed by Schmitthenner and Walker (1979) and Walker and Schmitthenner(1984a, b), tolerance to the disease—have been developed. Schmitthenner (1985) reviewed the control of Phytophthora root rot of soybean and stated that tolerance is quantitatively heritable as shown by the data of Walker and Schmitthenner (1984b). A large number of cultivars in the Midwest region of the United States have a significant level of tolerance. References on methods of determination of tolerance in cultivars are listed in Table 6.1.

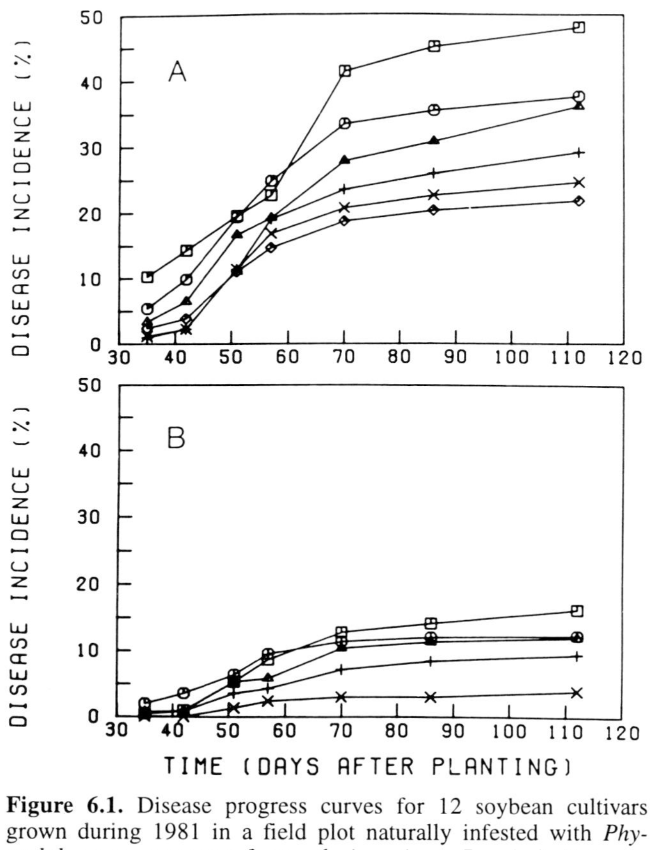

Figure 6.1. Disease progress curves for 12 soybean cultivars grown during 1981 in a field plot naturally infested with *Phytophthora megasperma* f. sp. *glycinea* (now *P. sojae*). Data are means of 10 observations on plants not treated with metalaxyl. **A,** Squares = the cultivar Steele, circles = Harosoy 63, triangles = Marshall, plus signs = Amsoy 71, multiplication signs = Wells, and diamonds = Corsoy. **B,** Cultivars show increased rate-reducing resistance. Squares = cultivar Asgrow A2656, circles = Beeson, triangles = Asgrow A2575, plus signs = Century, multiplication signs = Wayne, and diamonds = Northrup King S1492. (Reproduced by permission of the American Phytopathological Society, St. Paul, Minnesota, courtesy Tooley and Grau 1984a)

Field tolerance in many soybean cultivars is useful in both the Midwest region of the United States (Schmitthenner 1985) and Australia (Irwin and Langdon 1982). In many areas, race-specific resistance against prevalent races is incorporated into cultivars with field tolerance. Wilcox (1976) indicates that yield of the tolerant cultivar Wayne is reduced by only 10% in fields in which that of more susceptible cultivars is reduced by 50%. In Indiana, progenies of advanced breeding lines have the cultivar Wayne in their pedigree. Cultivars Tracy, Davis, Hill, and CNS were tolerant in Australia (Irwin and Langdon 1982).

Disease progress data developed by recording the disease incidence caused by *P. sojae* for different cultivars against time after planting indicate that resistance results from the reduction in the rate of disease onset (rate-reducing resistance) (Tooley et al. 1984a, b). When the data are plotted graphically, the area under the disease progress curve significantly and quantitatively shows that rate-reducing resistance is characteristic of some cultivars of soybean in Wisconsin fields (Figure 6.1). Disease incidence at progressive growth stages R3, R5, and R7 (see Fehr et al. [1971] for growth stages of soybean) and disease severity at the R5 growth stage were also correlated with the area under the disease progress curve.

Tooley and Grau (1982, 1984a, b) concluded that the primary component of rate-reducing resistance is the ability of the soybean plant to restrict or slow the colonization of root and lower stem tissue by *P. sojae.* During the years 1981 and 1982, the high amount of inoculum in the field resulted in a "simple-interest" (single-cycle) type of epidemic (see Chapter 1 for a brief discussion of single- and multicycle diseases). Tooley and Grau (1982, 1984a, b) prefer the single-cycle model to describe the epidemic on soybean because critical experiments have not been conducted to prove that secondary infection cycles occur in the soil. For instance, the number of diseased plants did not increase in rows in which dead plants were not removed compared with rows in which dead plants were removed soon after disease symptoms were seen. They quote Vanderplank (1963), who states that if the inoculum in the soil is the main source at the beginning of the season, the simple-interest model is appropriate. Since the inoculum level of *P. sojae* was high in their experiments, secondary production of inoculum (multicycle epidemic) probably would not have increased the disease severity or incidence. The effect of *P. sojae* on yield of cultivars without race-specific resistance during 1980 and 1981 in Wisconsin was caused mainly by the reduction in the number of yielding plants per row (Tooley and Grau 1984b).

Tooley and Grau (1984a) evaluated fungicide equivalency. This value is the equivalence of resistance with the effect of the systemic fungicide metalaxyl (granular, 1.12 kg of active ingredient per hectare) applied as a band over the seeds at planting time to reduce the final incidence of disease to 15%. Metalaxyl is an efficient systemic fungicide (see Chapter 7 on fungicides) that controls *Phytophthora,* but it may be too expensive to be cost effective on a field crop. Fry (1975) showed that potato cultivars with general resistance required less metalaxyl to achieve control of late blight than susceptible cultivars.

It is evident from the data of Tooley and Grau (1984b) (Figure 6.1) that reduction in the rate of disease onset is an important factor, and hence the use of the term rate-reducing resistance (MacKenzie et al. 1983) to characterize general or horizontal resistance is well substantiated. The different rates of disease onset are likely caused by factors in the plant that slow the rate of colonization. When the soil is not wet, the rate of colonization might be reduced enough to allow for development of new roots, which tend to compensate for the infected roots that have died. This type of resistance allows the plant to live with the pathogen, and unless there are long periods of waterlogging of soil from rainfall or excessive irrigation, the plant may not suffer a loss of yield.

Although selection of cultivars with tolerance or general resistance has been made in the field, a more controlled system would be advantageous to speed up the selection process. Field screening has some disadvantages, such as nonuniform inoculum density in the soil and lack of soil conditions that favor the disease. Distribution of races in the field can be nonuniform, and the test field may not be representative of a wide agricultural area. In the field, whether a soybean cultivar has multirace resistance or is highly tolerant is difficult to distinguish, and only one screening experiment can be conducted during 1 year (McBlain et al. 1991b). Evaluation of tolerance is not as easy as that of race-specific resistance, in which the plant is either resistant or it is killed.

Tooley and Grau (1982) developed a cotyledon-wound inoculation method in which a 10-µl drop of inoculum was placed on the cotyledon. The expression of resistance in different cultivars was affected by inoculum density, photoperiod, and plant age. The method correlated well with the performance of the same cultivars in naturally infested soil in the field. Although cotyledons are not the natural infection court for *P. sojae,* the correlation with field performance indicated that the method may be of use for plant-breeding evaluations.

Among the many methods used to evaluate tolerance in the laboratory and glasshouse, most evaluate the relative growth of soybean plants in artificially or naturally infested soil. Zoospores (4,000 in 16–20 ml of water) are applied to trays (17 × 25 × 6 cm) containing seedlings of different field-tolerant cultivars grown in a greenhouse mix. The degree of field tolerance is measured by the length of stems and dry weights compared with uninoculated control plants (Jimenez and Lockwood 1980). This method correlates well with results of a field test by Wilcox (1976); however, when the inoculum concentration is increased to 20,000 zoospores per tray, the moderately field-tolerant cultivar Williams becomes fully susceptible. Thus, it appears that tolerance can be overcome by an increase in inoculum concentration.

In another method, the inoculum layer test, a lima bean agar petri plate culture of *P. sojae* was placed 2.5 cm below the seeds in a pot containing a soil mix in which candidate soybean cultivars were growing (Walker and Schmitthenner 1982; Thomison et al. 1991). The growth of plants relative to that of plants of an uninoculated control cultivar indicates the degree of tolerance. Thomison et al. (1991), however, stated that it was difficult to distinguish between race-specific root resistance and tolerance by use of the inoculum layer test.

Irwin and Langdon (1982) mixed comminuted mycelium (0.036–0.1442 g, dry weight basis) with 1 kg of a soil mix. Three days after the seeds were planted, the soil mix was flooded and maintained in a flooded condition until the seventh day. At that time, data on the numbers of seedlings that emerged, numbers killed after emergence, numbers with nonlethal hypocotyl lesions, and the hypocotyl length (percentage of control) were recorded. These data for the tolerant cultivars Davis and Tracy compared favorably with the response in the field.

Although race-specific resistance of soybean cultivars can be assayed by inoculation of wounded hypocotyls, tolerance or rate-limiting resistance can be assayed most reliably on roots. In normal greenhouse or soil assays, however, evaluating root lesion progress without sacrificing the roots is difficult. M. A. L. Smith et al. (1991) reported a sophisticated method in which the expansion of lesions on intact roots was measured. Roots were inoculated by dipping them into aliquots with 1,000 zoospores per 500 ml and then grown in a clear Gelrite solidified medium (Gellan Green, Scott Laboratories, Inc., Carson, California) containing a growth culture medium (Pieper and Smith 1988). The lesions on the roots were visually imaged by use of a video camera (Pieper and Smith 1988; M. A. L. Smith et al. 1989, 1990). The lesion length was greatest on the susceptible cultivar Corsoy and least on the cultivars Williams, Asgrow 2575, and Williams 82, all known to be field tolerant.

A slant-board method was used to evaluate the extent of development of root lesions on an intact root (Olah and Schmitthenner 1985). Six-day-old plants grown in vermiculite were arranged on slant boards (Kendall and Leath 1974) and overlaid with polyester greenhouse wicking material. The boards (or plastic trays that supported the wicking material) were set at a slant with the lower end in a weak (25% concentration) nutrient solution. When the plants were covered with the wicking material, they adhered to the board. After 7 days of growth, the wicking was removed briefly in order to wound and inoculate the taproots. Each wound, a longitudinal cut 2 cm below the hypocotyl junction to the depth of the stele made with a single-edge razor blade, was inoculated with two drops of a soft agar-mycelium suspension. The roots were then covered again with the wicking material for an additional 6 days. Evaluations of the lesion lengths showed marked differences between the tolerant cultivar Voris 295 (7 mm) and the low-tolerance cultivar 0X 20-8 (30 mm). On some plants within cultivars, the lesion expansion was slight. These plants were selected, and the upper segment was allowed to root in nutrient solution for further testing (Olah and Schmitthenner 1985). A parallel study in which both tolerant and race-specific resistant cultivars were tested showed that the phytoalexin glyceollin was produced by cultivars with race-specific resistance but not by cultivars with tolerance (Olah et al. 1985).

In Canada, selection made in the field for tolerance is based on the lowest numbers of plants killed. These selections are then backcrossed to cultivars with specific resistance to races that are predominant in certain localities (Buzzell and Anderson 1982). Athow (1987) states that race-specific and rate-limiting types of resistance should complement one another and provide long-term disease control.

All of the research results reviewed previously on tolerance as a means of obtaining durable resistance have assumed that new races of *P. sojae* will not develop and overcome tolerance. Perhaps new races will not develop as quickly as in the case of race-specific resistance. Nonetheless, in Queensland, Australia, the tolerant (field-resistant) cultivar Davis (Irwin and Langdon 1982) became susceptible after years of cultivation (Ryley and Obst 1992). Ryley and Obst concluded that field resistance in the cultivar Davis was race specific.

Phytophthora medicaginis (*P. megasperma* f. sp. *medicaginis*)—Alfalfa Root Rot

After the initial discovery of Phytophthora root rot of alfalfa, *Medicago sativa* L. (Erwin 1954), the disease was found in most major alfalfa-producing regions of the world. The disease is a cause of severe stand loss when flooding caused by irrigation or rainfall is prevalent.

Within most alfalfa cultivars, a small percentage of tolerant plants can be found. Since alfalfa is obligately outcrossed and naturally heterozygous, resistance to many disease and insect problems has been readily found. If the inoculum level is high enough or if soil is flooded frequently, resistant plants will be killed. Resistance to Phytophthora root rot in alfalfa acts like rate-reducing resistance in that both resistant and susceptible plants become infected, but the rate of colonization of roots is slower in the resistant line (Marks and Mitchell 1971a; Frosheiser 1980). After inoculation, root tissue of the resistant clone (single plant) M193 (Irwin et al. 1981) and susceptible Vernal cultivars of alfalfa were examined by light microscopy (Miller and Maxwell 1984a) and by electron microscopy of ultra-thin root sections (Miller and Maxwell 1984b); both became infected. Resistance was associated with a slower rate of colonization of root tissue. Infection courts on alfalfa roots that are highly susceptible to penetration and infection include the zone of elongation on lateral roots (Irwin 1976; Marks and Mitchell 1971a), the place where lateral roots emerge from the taproot (Irwin 1976), and wounds (Chi and Sabo 1978).

Many resistant alfalfa cultivars have been developed by selection (Frosheiser and Barnes 1973) (see Table 6.1 for several references). Because selfing alfalfa plants results in loss of vigor, obtaining a cultivar or germplasm with 100% resistant plants is nearly impossible. Variability is evolutionarily favorable for alfalfa as a crop, since resistance in alfalfa to numerous diseases can be readily obtained by selection within agronomically adapted cultivars (Elgin et al. 1988). To date, new pathologic races of *P. medicaginis* have not been detected; however, the obligately heterozygous genetic makeup of alfalfa is such that it would be difficult to detect a new race.

The cultivar Lahontan was one of the first to be recognized for its general resistance to *P. medicaginis* (Erwin 1966). Plants that varied from 1 to 12 months of age were inoculated with minced mycelium (four mats blended in 1 liter of distilled water) or with a suspension of zoospores. Inoculum was placed in a trench beside rows of 4- to 6-week-old plants growing in pots of UC mix (50% sand and 50% peat moss). The incidence of plants with a high root rot score (based on a scale of 1–5) was about 15% for Lahontan and 75–97% for the susceptible cultivar Africa.

Several reports have given attention to the role of inoculum concentration in tests to screen for resistance. Selection for resistance in 3- to 4-week-old alfalfa plants has been conducted in naturally infested field soils and under greenhouse conditions in which the soil is infested with minced agar or colonized grain inoculum. To select from a seedling population, several reports indicate that standardizing a threshold level of inoculum that is low enough to allow the resistant seedlings to survive but high enough to kill the susceptible seedlings is important. F. A. Gray et al. (1973) and Hine et al. (1975) mixed cornmeal-sand inoculum in soil and selected a small population of survivors that were resistant. Irwin et al. (1979) incorporated about 0.1 g (dry weight basis) of mycelium into 1 kg of a soil mix. Hohrein et al. (1983) poured 65 ml of a minced suspension of inoculum (two plates per liter of water blended 10 s in a Waring blender) between every other row (60 cm in length) of 12-day-old plants, after which plants were flooded for 3 days. Pratt et al. (1975) showed that inoculation of turgid cotyledonary leaves with a drop (0.01 ml) containing about 79 zoospores differentiated resistant from susceptible plants.

Frosheiser (1980) and Frosheiser and Barnes (1973, 1983) developed a program in Minnesota to evaluate the resistance of alfalfa cultivars and selections to *P. medicaginis*. The primary screen consists of a 3-cm-deep layer of gravel overlaid with sand in watertight, galvanized sheet metal tanks (85 × 60 × 23 cm) equipped with drain holes. The sand is infested with mycelium of *P. medicaginis* (mycelium and agar from one 9-cm petri plate comminuted in 200 ml of distilled water). The mycelial suspension is poured over the sand at a rate of 1,600 ml per tank. The water level is raised to the surface by stoppering the drain holes in the tank. Alfalfa seedlings are started after draining the sand bed and watered sparingly to allow establishment. When plants in rows (about 3.5 cm apart) are 3–4 weeks of age, the drainage holes are again plugged, and the sand bed is maintained in a flooded condition for 3–4 weeks, at which time the plants are removed and the symptoms evaluated on a five-class scale.

Although this flooding method in the greenhouse has been useful in the Minnesota system, there is a danger that lack of aeration could cause secondary root damage. At low temperatures, an oxygen deficiency might not be a problem, but at high temperatures (30–40°C), alfalfa roots do not tolerate long periods of flooding (Erwin et al. 1959; Zook et al 1986). We prefer a 3-day flood and a 4-day drainage period (Tsao and Garber 1960), which favors disease expression but does not result in anoxia (oxygen deficiency). The method described by Hohrein et al. (1983) in which plants are kept flooded for several days has in our work with sugar beet inoculated with *P. drechsleri* produced artifact results caused by prolonged flooding (D. C. Erwin, *unpublished*).

A method for field evaluation of Phytophthora root rot was developed by Frosheiser and Barnes (1973, 1984). Level plots (15 × 40 m) of land without horizontal surface drainage are infested with roots and soil from infected plants in the greenhouse. Plots are kept moist to near saturation by sprinklers during most of the test period. After the seedlings become established (4 weeks), plots are sprinkled twice a day for about 2 weeks, after which sprinkling is withheld for 2 weeks. The sprinkling and dry periods are repeated until September, when the plants are lifted and evaluated for symptoms of Phytophthora root rot on taproots. Plants are rated on a scale of 1–5, where 1 = no symptoms on the taproot or on the fine lateral roots; 2 = no root lesions, but most fine roots are destroyed; 3 = distinct localized lesions on the taproot or on one or two secondary roots with tip rot; 4 = part or all of the taproot rotted; and 5 = plant is dead (Frosheiser and Barnes 1973). After evaluation, about 150 plants in the least disease class (1 or 2) are transplanted to pots in the greenhouse and intercrossed by hand. The seed produced is reevaluated for resistance the following year. Two or three cycles of selection have increased the population of resistant plants in several cultivars sufficiently to nearly double the yield when tested on infested land. The cultivar Agate is considered to be a standard resistant cultivar (Frosheiser and Barnes 1973).

The survival level of a Mexican cultivar, INIA-76, was increased from 5.6 to 27.8% by two cycles of selection and intercrossing plants from artificially infested soil. A suspension of comminuted mycelium and oospores was adjusted by dilution so that the light transmittance at 550 nm on a spectronic spectrophotometer was 20%. The inoculum was diluted with water (1:4) and mixed in soil (50% field Gila sandy loam and 50% sand) at a rate of 5 liters to 35 kg of soil. Lower levels of inoculum allowed too high a percentage of survivors for selection of resistant plants (Aguirre et al. 1983).

The percentage of resistant alfalfa plants in a population is positively correlated with the disease severity index. Also, values obtained in greenhouse assays for resistance are strongly correlated with the disease response in the field (Basu 1988).

Breeding for resistance to *P. medicaginis* has been a major contribution to the increase in stand and yield of alfalfa wherever alfalfa is grown. A large number of resistant cultivars have been developed by commercial companies, the U.S. Department of Agriculture, state agricultural experiment stations in the United States, and comparable organizations in other regions of the world. Resistance is a major control measure when coupled with cultural management of irrigation and proper drainage. The improved field performances of *Phytophthora*-resistant cultivars is documented by Frosheiser and Barnes (1973) and Lueschen et al. (1976).

The high level of disease resistance of germplasm line A77-10B (Erwin et al. 1982) was correlated with the production of the highest amount of the phytoalexin medicarpin (Vaziri et al. 1981). A77-10B also ranked as highly resistant in inoculation tests reported by Barta and Schmitthenner (1986).

Two mechanisms have been proposed to explain the inheritance of resistance in alfalfa (tetraploid) to *P. medicaginis*. Lu et al. (1973) concluded that susceptibility was conditioned by one tetrasomically inherited gene, *Pm*. The data of Irwin et al. (1981a, b) indicated that resistance was controlled by two incompletely dominant, independently segregating, complementary genes, Pm_1 and Pm_2. The simplex at one locus and the duplex at the second locus are required for the expression of resistance. Resistance in the related diploid species *M. falcata* is conditioned by two independently segregating loci, Pm_3 and Pm_4 (Havey and Maxwell 1987, 1988).

Although no evidence indicates the occurrence of new races of *P. medicaginis*, there are references to variability in virulence of different isolates (Erwin 1966; Aguirre et al. 1983). An isolate from Mexico was more virulent than isolates from Arizona that had been used to develop the moderately resistant cultivar Hayden PX-III. Hayden PX-III was susceptible to the Mexican isolates but resistant to the Arizona isolates (Aguirre et al. 1983). Faris et al. (1983) reported in Canada that 26 isolates of *P. megasperma* (now called *P. medicaginis*) (Hansen and Maxwell 1991) from alfalfa, which produced small oogonia, were uniformly pathogenic in glasshouse tests to the field-tolerant cultivars Apollo, DeKalb 120, and Trident. Three other isolate groups, which would be included in the broad-host-range *P. megasperma* var. *megasperma* that produces large oogonia (*sensu* Waterhouse 1963), were milder in their virulence on alfalfa than *P. medicaginis*. Considering all the heterogeneous groups of isolates together, they showed (see also Faris [1985]) a statistically high *Phytophthora* × cultivar interaction; however, since these large-oospore groups within *P. megasperma* are now considered to be from a species different from *P. medicaginis*, as has been clearly shown by DNA-RFLP and isozyme data (Förster et al. 1988, 1989; Hansen and Maxwell 1991; Chapter 4), the *Phytophthora* × alfalfa cultivar interaction in Faris et al. (1983) is probably the result of interspecific differences among the *Phytophthora* isolates and not of variability within *P. medicaginis*. Virulence of single-oospore cultures obtained from a single-oospore isolate of *P. medicaginis* varied from slight to severe (Erwin and Li 1986).

Dale and Irwin (1991a) noted that chickpea (*Cicer arietinum* L.), an additional host of *P. medicaginis*, was not only infected via roots but was also penetrated and infected on the stem just below the cotyledons. Inoculation of chickpea seedlings by pouring a suspension of zoospores at different concentrations on the surface of flooded soil was unsuccessful in differentiating resistant lines, but a standardized concentration of oospores incorporated into the soil was successful (Dale and Irwin 1991a). Dale and Irwin (1991a) suggested that resistance in the hypocotyl (stomatal infection) might be different from that expressed in the roots.

Phytophthora cinnamomi—Root Rot of Avocado

General resistance can be found in most crops by selection of plants within a wide range of cultivars and wild types after inoculation of a large population; however, in general, it is more difficult to find host resistance to pathogens with wide host ranges than with narrow host ranges. Plants infected by *P. cinnamomi* apparently fit into this generalization. Because of the long life of tree crops in orchards and forests, selection for resistance is more time consuming than for annual crops. Zentmyer's (1980) monograph on *P. cinnamomi* contains an extensive host list and a full discussion of the development of resistance to *P. cinnamomi* in avocado and some other woody hosts.

After many years of collecting avocado plants from Guatemala and Mexico, Zentmyer found moderate resistance in *Persea americana* in Mexico and in *P. schiediana* and other sources from Guatemala, where avocado is native (Zentmyer 1980; Zentmyer and Schieber 1987). There are now many reports of general resistance in avocado (Zentmyer and Mircetich 1965; Zentmyer 1980; Kellam and Coffey 1985; Dolan et al. 1986; Aveling and Rijkenberg 1989, 1991; Gabor et al. 1990; and Gabor and Coffey 1990) (Table 6.1). The cultivars Martin Grande, Thomas, and Barr Duke are more resistant than Topa Topa and UCR 2023. Earlier work on resistance is documented by Zentmyer (1980) and subsequently by Coffey (1990), who discusses the history of selection for tolerance or moderate resistance.

The moderate resistance of avocado rootstocks to *P. cinnamomi* is not great enough to give maximum control in all situations, especially in high-hazard sites (Zentmyer and Ohr 1978), for example, where vertical drainage of irrigation water is impeded by hardpans or in tropical regions where heavy rainfall might cause infested soil to be waterlogged for long periods. The cultivars Duke 7 and Thomas regenerate new roots more rapidly than susceptible rootstock cultivars such as Topa Topa (Kellam and Coffey 1985; Gabor and Coffey 1990). The Martin Grande rootstock seems to be capable of restricting development of *P. cinnamomi* in infected roots (Dolan and Coffey 1986; Gabor and Coffey 1991).

When avocado rootstocks were grown in the greenhouse in soil naturally infested with *P. cinnamomi*, 55% of the lateral roots were infected on the susceptible cultivar Topa Topa but only 2 and 11% on the cultivars Duke 7 and G6, respectively. A greater increase in the *P. cinnamomi* propagule count in soil occurred on the susceptible cultivar Topa Topa than on G6. The greater capacity of cultivars Duke 7 and G6 to regenerate roots may be at least one characteristic that gives them resistance (Kellam and Coffey 1985). Field experiments in California indicated that the rootstocks of cultivars Thomas, Martin Grande, Barr Duke, and D9 grafted on Haas were more resistant than those of cultivars Topa Topa, Borchard, and G6.

The number of zoospores that encysted on the root tips was significantly reduced on the tolerant cultivar Martin Grande rootstock (15.6 encysted zoospores per root tip) compared with the susceptible Edranol cultivar (50.3). Zoospore encystment on cultivars G6 (37.1) and Duke 7 (41.3) indicated an intermediate level of tolerance (Botha et al. 1990). The attraction of *P. cinnamomi* zoospores to exudates from excised avocado roots that diffused through a dialysis membrane was shown by Botha et al. (1990) to reflect known tolerance or susceptibility of avocado rootstocks in South Africa. They compared seedling roots of the susceptible rootstock cultivar Edranol with those of Duke 7 and G7 (moderately tolerant) and G755 (*P. schiedeana* Haas) (tolerant). Development of lesions on detached roots and leaves and field performance correlated well with the root-exudate method for evaluation of tolerance.

Resistance expressed in callus tissue from the susceptible cultivar Topa Topa and the moderately resistant cultivars Duke 7 and Martin Grande was similar to that in the field. Callus tissue was grown in the laboratory by aseptically sampling tissue from these clones aseptically grown on the Murishige-Skoog nutrient medium. Phillips et al. (1991) found that rates of extension of *P. cinnamomi* mycelium on callus tissue of the resistant cultivars Duke 7 and Martin Grande clones (1–3 mm per day) was much less than on the susceptible cultivar Topa Topa (4–7 mm per day), thus indicating that resistance was the result of a biochemical rather than an anatomical effect. They suggested that this technique might be useful to screen for resistance. A general evaluation of host resistance in callus tissue was reviewed by Miller and Maxwell (1983).

Phytophthora cinnamomi— Root Rot of Jarrah Trees

Tippett et al. (1985) reviewed a number of reports that indicate that a wide range in degree of resistance to *P. cinnamomi* occurs in *Eucalyptus* species in Western Australia. They found by use of wound inoculation with *P. cinnamomi* that several *Eucalyptus* species were resistant, but that the jarrah tree, *E. marginata* Donn ex. Sm., was susceptible. *E. wandoo* was the most resistant of several species. Remarkably, *P. cinnamomi* did not become established in *E. wandoo* as it did in some moderately resistant species such as *E. calophylla* and *E. patens.* Lesions formed on inoculated stems of the *Monocalyptus* and *Corymbia* subgenetic groups but not on stems of species of the subgenus *Symphyomyrtus* (Tippett et al. 1985).

Shea et al. (1984) stated in a popular article that jarrah trees were tolerant to *P. cinnamomi* even though large trees were killed at some sites (see Chapter 5). "A number of ideas were formulated and tested in field trials. In one trial, some 3000 root and stem inoculations were carried out, over a period of 18 months, and the progress of the fungus was monitored by microscopic examination of the tree's tissue. The conclusion from this study was that jarrah trees have a range of defense mechanisms which curtail the spread of the fungus, even when it is introduced directly into a wound. Seasonal variation in susceptibility was observed, but it was proved that the trees have the capacity to form a barrier to the invading fungus" (Shea et al. 1984). They later found that large jarrah trees were killed because zoospore inoculum, which increased near a hardpan layer where water collected, infected and killed large vertical roots that had penetrated the layer (Shea et al. 1982, 1983). See the section on the hardpan effect in Chapter 5.

Conceptually, Tippett et al. (1985) refer to the different degrees of lesion extension as *aggressive* (unchecked by the host) or *confined* (seldom exceeding 6 cm in length), terms used by Sewell and Wilson (1973) to describe effects of *P. cactorum* on apple trees. In the susceptible jarrah trees, *E. marginata,* the lesions can be confined, but *P. cinnamomi* can survive, possibly as chlamydospores. As noted in other crops, susceptibility in jarrah and other *Eucalyptus* species varies with the season. The inner phloem (bark) tissue is most vulnerable during early summer and least vulnerable during the fall of the year (Tippett et al. 1985). An increasing amount of bark moisture directly causes an increase in the lesion size (Tippett and Hill 1983; Tippett et al. 1987).

Resistance in roots of selected plants from the jarrah forest (*E. marginata*) of Western Australia indicates that selection for resistance is a useful goal for future control measures (Cahill et al. 1992). In the resistant roots of selected plants, the lesions develop much more slowly from a point of inoculation with zoospores than in roots of an unselected plant. On the field-resistant marri tree (*E. calophylla* R. Br.), lesions at the point of inoculation are restricted and do not progress (Cahill et al. 1992). See Table 6.1 for other citations on development of resistance in jarrah trees.

Phytophthora citrophthora and *P. parasitica*— Citrus Gummosis and Root Rot

The review on *Phytophthora*-caused citrus diseases by Broadbent (1977b) aptly covers losses, causal agents, variability, epidemiology, and control and contains numerous citations on the development of host resistance (Klotz et al. 1965a; Klotz and Calavan 1969, 1978; see also Klotz [1978]) (Tables 6.1 and 17.4). General resistance to *P. citrophthora* and *P. parasitica* is now available in many rootstocks that can be grafted at the nursery level to scions of commercially acceptable citrus fruit trees such as lemon, lime, orange, and grapefruit. Probably the first to be discovered was sour orange, which was utilized in many regions of the world by the middle or late 1800s (Fawcett 1936). The sour orange rootstock is no longer recommended because sweet orange trees that are grafted on sour orange rootstocks are killed by the tristeza virus (Bar-Joseph et al. 1981). The widespread use of sour orange as a rootstock resulted in 1931 in a major epidemic that caused the death of large acreages of citrus trees grafted on sour orange rootstocks in Brazil and Argentina (Bar-Joseph et al. 1981). As discussed previously in Chapter 2, the death of sweet orange trees grafted on sour orange rootstocks resulted from the hypersensitivity of the phloem tissue of sour orange trees just below the graft union caused by the tristeza virus. The tristeza virus had been introduced via budwood imported from China and was later shown to be transmitted from tree to tree by an efficient aphid vector (*Toxoptera citricida* Kirk). Fortunately, sweet orange on other rootstocks, such as Troyer citrange, was tolerant to the tristeza virus. Although these trees were susceptible to infection by the tristeza virus, they did not suffer the lethal effects that the sweet-on-sour orange combination did. The search for rootstocks other than sour orange that are resistant to *Phytophthora* has yielded many useful cultivars. Since the devastating epidemics in Brazil and Argentina, many other *Phytophthora*-resistant rootstocks that are not subject to the lethal effect of the tristeza virus have been found (see reviews by Broadbent [1977b] and Klotz [1978]). Klotz et al. (1968) in California reported that rootstocks of the tristeza-tolerant trifoliate orange and hybrids of certain cultivars, Nanshodaidai (*Citrus taiwanica*) and Siamelo (*C. reticulata* × *C. sinensis* × *C. paradisi*), did not develop gummosis when grown in soil infested with *P. citrophthora* and *P. parasitica.*

In several different experiments by Klotz et al. (1968), citrus rootstock cultivars and species were placed in a field that was naturally infested with both *P. citrophthora* and *P. parasitica.* The effect of inoculation was measured by the diameters of lesions at points of inoculation on the trunks above and below the bud union. The most effective rootstock for use with the cultivar Valencia orange scions were Nanshodaidai, King Mandarin, Citremon 1449, Battangas mandarin, and Pon kan mandarin; the best rootstocks for the cultivar Lisbon lemon were Battangas mandarin, Citremon 1449, Nanshodaidai, Cleopatra mandarin, and Ichang lemon (Klotz et al. 1965). Six years later, the trees were irrigated to saturation and pulled from the ground with a forklift, and the washed roots were examined. The order of best rootstocks based on symptoms was different from the evaluation based on growth. The best rootstocks on the cultivar Lisbon lemon in one experiment were Texas, USDA, and Barnes trifoliates, Ichang lemon 1219, Pomeroy trifoliate, Nanshodaidai, Alemow, Siamelo, King Mandarin, and Carrizo citrange. The best rootstocks to be used on the cultivar Valencia orange were Nanshodaidai, USDA, Texas, Barnes, and Pomeroy trifoliates, King Mandarin, Carrizo citrange, Cleopatra mandarin, Citremon 1449, and Siamelo (Klotz et al. 1968).

To test rootstocks for resistance to gummosis on the trunk, inoculations were made with disks of mycelium actively growing on agar placed on the cambium after removal of the bark with a cork borer about 3/16 inch (4.8 cm) in diameter. Although trifoliate orange cultivars and their hybrids (Citrange and Citremon), *C. taiwanica,* and Siamelo were resistant to gummosis on the trunk in a natural situation, artificial inoculation of the cambium resulted in infection. This indicated that avoidance of wounding and prevention of excessive irrigation near the crown of the tree are important (Klotz et al. 1968).

Citrus trees 10–12 months old were tested for resistance to *P. parasitica* by submerging the roots in an aerated rainwater-mycelium suspension in which sporangia and zoospores formed readily (Grimm and Hutchison 1973). The young trees were transplanted to beds of a fine sand and vermiculite mix and flooded at the time of transplanting and every 7–10 days thereafter by the alternate flooding and draining method of Tsao and Garber (1960). In a secondary test, mycelium inoculum was placed in a V-shaped flap under the bark of each small tree. The resulting lesion was evaluated on a scale of 1 to 5 (1 = no symptoms, and 5 = severe symptoms). Grimm and Hutchison (1973) emphasized the well-known principle that rootstocks with known resistance or susceptibility should be used as markers because of the variability in severity between experiments. Sweet orange was highly susceptible, rough lemon was susceptible, and the cultivar Carrizo citrange was moderately resistant. Plants were rated relative to these controls.

Graham (1990) found that the cultivars Trifoliate orange and Swingle citrumelo were tolerant when planted in soil infested with one to 10 propagules of *P. parasitica* per gram. In addition to a reduction in the rate of infection progress, tolerance was also expressed by the ability of plants to regenerate new roots after infection. Although sour orange and Carrizo citrange were not tolerant in the greenhouse test, field studies showed that these rootstocks are tolerant (Klotz et al. 1968; Timmer and Menge 1988).

Resistance to colonization of citrus rootstocks and scions by *P. parasitica* and *P. citrophthora* varies with the season. When the citrus trees stop growing in the winter, their rates of starch consumption and water transport are lowest. Matheron and Matejka (1989) found that during this period, the extent of colonization of stem tissues was at a minimum. The maximum degree of invasion (size of lesions) occurred from March through October. They also noted that the disease severity on excised twig tissue did not always correspond to relative susceptibility in the field. For example, lesions were larger on *C. auriantium* trees than on *C. jambhiri;* but in the field, *C. auriantium* trees are resistant and *C. jambhiri* trees are susceptible.

Afek et al. (1990) reported a method of evaluating resistance to *P. citrophthora* by inoculating branches of citrus rootstock tree sources and comparing the relative sizes of the lesions. Lesions were smallest on inoculated branches of *C. macrophylla* and *Poncirus trifoliata* trees, intermediate on *C. aurantium* and *P. trifoliata* × *C. sinensis,* and largest on *C. jambhiri* and *C. sinensis.*

A known quantity of chlamydospores of *P. parasitica* in soil-cornmeal agar blocks was used as inoculum to test citrus trees for resistance. Agar-soil blocks of different sizes (0.5 × 0.5 or 1.5 × 0.5 cm) were placed under the bark of small citrus trees, and the lesions were measured 8 weeks later. Inoculation with the quantified chlamydospore inoculum concentrations was superior to use of carrot agar-mycelium blocks, because the threshold minimal amount of inoculum was sufficient to cause disease but not high enough to overwhelm the inoculated host. The lesions on the rootstock cultivar Carrizo citrange were smaller than those on the well-known, field-tolerant sour orange rootstock. The chlamydospore-soil-agar inoculum was advantageous in that it could be stored for long periods and produced results similar to those observed on rootstocks in the field (M. A. L. Smith et al. 1991).

Resistance of trifoliate orange cultivars and hybrids (e.g., Troyer citrange) is expressed in mature, suberized root tissue as well as in the crown or trunk (Broadbent 1968; Klotz et al. 1968). A compound, 6,7-dimethoxycoumarin, also known as scoparone, a phytoalexin, is associated with resistance to *P. citrophthora* in bark tissues of sour orange trees (*C. aurantium*) (Musumeci et al. 1975; Afek et al. 1986; Afek and Sztejnberg 1988; Sulistyowati et al. 1990). The compound is produced in quantities sufficient to inhibit growth of *P. citrophthora.* Sulistyowati et al. (1990) showed that the concentration of scoparone in infected roots and bark of seedlings was positively correlated with resistance. Scoparone production was more rapid in the cultivar Troyer citrange (*P. trifoliata* × *C. sinensis*) than in the susceptible rough lemon (*C. aurantium*).

Resistance of Other Host Species

Resistance of many other plants is covered in the following chapters on individual *Phytophthora* species. See Table 6.1 for selected references on general and specific resistance of hosts to several different species of *Phytophthora.*

When the susceptibility of shortleaf pine (*Pinus echinata* Mill.) and loblolly pine (*P. taeda* L.) was tested by inoculation of roots of 6- to 7-week-old seedlings with a suspension of *P. cinnamomi* zoospores, differences between the responses of the two pine species (loblolly pine was less susceptible) were evident only at inoculum concentrations between 80 and 280 zoospores per milliliter (Fraedrick et al. 1989). In a related study, symptoms progressed more rapidly in soil maintained at low oxygen concentrations (Fraedrich and Tainter 1989). Higher inoculum levels induced similar disease responses on both species. The inoculum level was also critical for detection of resistance to *P. parasitica* var. *nicotianae* on tobacco (Gooding and Lucas 1959b). The concentration of inoculum is one of the most important principles in an evaluation of general resistance (see also M. A. L. Smith et al. [1991] in the previous section).

Mechanisms of Resistance

The large volume of research reports and reviews that treat the nature of host resistance to *Phytophthora* permits only a brief review here. The reviews of Keen and Yoshikawa (1983a, 1990) are pertinent to the expression of resistance in soybean for *P. sojae.* Other useful reviews include those of Coffey and Wilson (1983a), Paxton (1983, 1988), Yoshikawa (1983), Ebel (1986), Ebel and Griesbach (1988), Stössel (1988), and Ward (1990). For *P. infestans,* the chapters by Keenan et al. (1986), Preisig and Kuć (1987), Friend (1991), Hohl (1991b), and Parker et al. (1991) are pertinent. The comparative cytology of plants resistant and susceptible to *P. infestans* and *P. sojae* is reviewed by Coffey and Wilson (1983a).

Race-specific resistance to both *P. infestans* on potato and *P. sojae* on soybean has been developed and utilized in the field; however, it is now well recognized that pathologic races have the capability to overcome newly introduced genes for race-specific resistance in the host. The race-specific resistance of potato to *P. infestans* and of soybean to *P. sojae* (*P. megasperma* f. sp. *glycinea*) has been studied since the late

1950s to determine the biochemical mode of resistance. In addition to the importance of root and stem rot as a disease of soybean, a major crop on the world market, the biochemical and physiological mechanisms of resistance of soybean to *P. sojae* has been of scientific interest throughout the world, even where soybeans are not grown. Resistance to both *P. infestans* and *P. sojae* has been assumed to be of the "gene-for-gene" model proposed by Flor (1956). Flor's concept, in brief, is that for every resistance gene in the host, there is a specific complementary gene in the pathogen that conditions virulence or avirulence. This concept is discussed by Keen (1982) and Keen and Yoshikawa (1983a). Keen and Yoshikawa (1983b) state that the *P. sojae*-soybean system is assumed to be a gene-for-gene interaction. Although the finding of several race-specific genes in soybean and corresponding pathologic races of *P. sojae* strongly indicates that the gene-for-gene system is operative, formal genetic proof that virulence genes occurred in *P. sojae* had not been obtained at the time of their reports.

The assumption that the *P. sojae*-soybean system is a gene-for-gene interaction was strengthened by the report of Layton and Kuhn (1988b). When *P. sojae* heterokaryon cultures, made by protoplast fusion (Layton and Kuhn 1988a), containing race 1 and race 3 nuclei were tested on the susceptible soybean cultivar Harosoy and the resistant cultivar Harosoy 63 (the latter is an isoline with one gene for resistance to race 1), 22 of 26 heterokaryon cultures had a race 1 phenotype, thus implying that race 1 was dominant to race 3 (virulent on Harosoy 63). In addition, two heterokaryons with race 1 phenotype yielded single-zoospore cultures with race 3 phenotypes. The data of Layton and Kuhn (1988b) might also imply that new races result from somatic recombination of factors in the diploid genome.

Much of the discussion on genetics of race-specific resistance to *Phytophthora* utilizes the currently held concept of

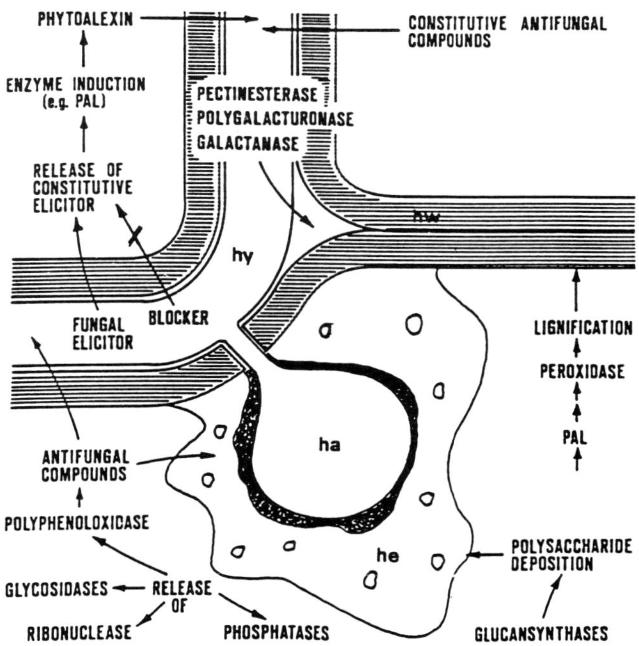

Figure 6.2. Interplay between host and pathogen. Factors (probably) involved in host tissue colonization and host defense. Haustorium, ha; haustorial encasement, he; host cell wall, hw; hypha, hy. (By permission, copyright Gordon and Breach Science Publishers S.A., Montreaux, Switzerland, courtesy Stössel 1988)

avirulence genes. The allele in plants for resistance to a pathogen is usually dominant over the corresponding allele for susceptibility. The allele for avirulence in the pathogen is usually recessive to the allele for virulence. Keen (1990) discusses avirulence genes and gene products produced by *Pseudomonas syringae* var. *glycinea,* a bacterial pathogen of soybean. Goodell et al. (1985) present a model showing that an avirulence gene can produce a product, glycocyl transferase, which induces production of a race-specific glycoprotein determinant that is recognized on a membrane receptor of the host. Subsequently, the defense mechanisms of the host are turned on to produce phytoalexins, which have antibiotic properties against the pathogen. In this situation, the pathogen without the avirulence gene would not initiate this set of events, and disease would ensue. Although the avirulence gene concept is accepted widely, Vanderplank (1986, 1990) disputes it.

Coffey and Wilson (1983a) review the *P. sojae*-soybean interaction that was reported by Stössel et al. (1981) and the *P. infestans*-potato interaction as reported by Hohl and Suter (1976), Shimony and Friend (1975), and Wilson and Coffey (1980). Because *P. sojae* causes death of many individual infected host cells at an early stage in pathogenesis (Stössel et al. 1981), a high degree of compatibility (biotrophy) does not seem to be required to facilitate successful colonization of soybean tissue.

Coffey and Wilson (1983a) indicate that information on the cytology of the *P. infestans*-potato interaction is more comprehensive than that of the *P. sojae*-soybean system. The relationship of *P. infestans* with potato appears to be highly compatible, as indicated by the relative sparsity of necrotic cells after foliage infection (Hohl and Suter 1976; Wilson and Coffey 1980) and tuber infection (Hohl and Stössel 1976; Shimony and Friend 1976, 1977). This compatibility probably reflects an active biotrophic phase of fungal development that is critical for rapid colonization of the host. Farrell et al. (1969) showed that in early phases of foliar infection, the stomata remain open, and carbon assimilation is enhanced at the infection sites. The currently accepted physiologic concept of the susceptible *P. infestans*-potato interaction indicates that soluble glucans produced by *P. infestans* suppress the resistance response of the host (Doke and Tomiyama 1980b) and can serve as specific suppressors unique to certain pathologic races. In contrast, a nonspecific reaction in host tissues is elicited by insoluble glucans of *P. infestans* (Kuć et al. 1976).

The data on the *P. infestans*-potato system support the physiologic nature of susceptibility; however, Coffey and Wilson (1983a) question whether there are sufficient data to confirm physiologic concepts of resistance because physiologic data indicate that resistance in potato tubers is a nonspecific event that operates where the specific situation, susceptibility, is absent (Doke and Tomiyama 1980a, b).

Coffey and Wilson (1983a) cite reports on the nature of resistance in tubers of potato to *P. infestans* that suggest that resistance mechanisms conditioned by the R1-gene and the R3-gene are different. "With the R1-gene in the cultivar Orion, host cell death occurs after 48 hours and is the principle feature characterizing the resistance response (Shimony and Friend 1976). With the R3-gene in the cultivar Elba, host cell necrosis was observed much earlier, at 20 hours, but the main feature thought to relate to resistance was the rapid development of encasements around haustoria (Hohl and Stössel 1976). A single, detailed cytological study of R-gene resistance, using several different cultivars for each resistance genotype, is urgently required" (Coffey and Wilson 1983a).

Stössel states (by permission, copyright Gordon and Breach

Science Publishers S. A., Montreaux, Switzerland, courtesy of Stössel 1988):

> In young soybean seedlings, race-specific resistance and age-related (general) resistance to *Phytophthora megasperma* f. sp. *glycinea* are very similar (Paxton and Chamberlain 1969; Lazarovits et al. 1981; Stössel et al. 1981; Ward et al. 1981). The potential for resistance remains, while the ability for a susceptible response is lost very rapidly with aging. Since there appears to be no unique mechanism in race-specific resistance that does not occur in general resistance, it is questionable whether resistance mechanisms are specifically induced by the pathogen or whether susceptibility is induced by specific blockage of non-specifically elicited general defense mechanisms. A blocker that suppresses phytoalexin accumulation has been found in compatible races of *P. infestans* (Doke and Tomiyama 1977, 1980b; Doke et al. 1979, 1980; Garas et al. 1979). Therefore, instead of 'race-specific resistance' the term 'race-specific susceptibility' may be more appropriate. This latter view is supported by several authors (Ward and Stössel 1976; Kuć and Preisig 1984).

The intensive and critical review by Stössel (1988) poses many questions about the biochemical explanation of resistance. Stössel's (1988) concept of resistance is shown in Figure 6.2.

Phytoalexins

A wide body of evidence indicates that phytoalexins, defined as antifungal compounds produced by the plant in response to the invasion of a pathogen, are involved in the physiology of resistance. The phytoalexins that have been studied most extensively are rishitin in potato and glyceollin in soybean. These and other phytoalexins that have been implicated in resistance to *Phytophthora* are listed in Table 6.4.

The variety of review articles on phytoalexins associated with resistance to numerous diseases indicates the great interest in this subject. These reviews include those by Kuć (1972, 1982), Kuć et al. (1976), Keen and Bruegger (1977), Gross (1979), Keen (1982), Dixon et al. (1983), who reviews the enzymology and molecular biology of a wide range of phytoalexins, Keen and Yoshikawa (1983a, 1990), Paxton (1983, 1988), Ebel (1986), Ebel and Griesbach (1988), Stössel (1988), and Keen (1990), who emphasizes compounds that elicit the phytoalexin response in plants.

Phytoalexins are antibiotic products of secondary plant biosynthesis. They are not generally present in significant amounts in normal plant tissue but are inducibly formed in response to inoculation with incompatible pathogens or after treatment of plant tissue with certain chemicals called *elicitors,* which emanate from pathogens (Keen and Yoshikawa 1983a; Keen 1990). Phytoalexins are not specific in their inhibitory action and can be induced (elicited) by nonpathogens, chemical irritants such as heavy metal salts, or physical injury such as that caused by ultraviolet light or freezing. Some compounds produced by pathogens that elicit the phytoalexin response include complex carbohydrates from fungal cell walls, lipids, enzymes, and polypeptides.

The phytoalexins from legumes, for instance, glyceollin from soybean and medicarpin from alfalfa, are derived from isoflavins, and those from solanaceous plants, for instance, rishitin from potato and capsidiol from pepper, are derived from sesquiterpenoids (reviewed by Dixon et al. [1983]) (Table 6.4).

The triggering of the hypersensitive response by compounds called elicitors or elicitins and the stimulation of the plant to produce phytoalexins, especially plants resistant to certain *Phytophthora* species, has received much attention. See Stössel (1988), Keen (1990), and Keen and Yoshikawa (1990) for a discussion of this concept. Two recent reports have implicated a protein from mycelial extracts of *P. cryptogea,* named cryptogein, that triggers the defense mechanisms in tobacco plants (Milat et al. 1991a, b). Cryptogein also elicits the production of the phytoalexin capsidiol in tobacco leaves, which protects against invasion by *P. parasitica* var. *nicotianae* (*sensu* Tucker 1931), the black shank pathogen (Bonnet

Table 6.4. Host-parasite systems involving *Phytophthora* spp. in which the production of phytoalexins is associated with resistance expression[a]

Host-parasite system	Type of resistance	Phytoalexin	Reference
Pepper-*P. infestans*	General	Capsidiol	Jones et al. (1975b); Ward (1976)
Pepper-*P. capsici*	General	Capsidiol	Stössel et al. (1972); Ward and Stössel (1974); Ward (1976; Turelli et al. (1984, 1986); Saimmaime et al. (1991)
Lupine and *Neonotonia wightii-P. sojae*	General	Wighteone	Ingham et al. (1977)
Peperomia-*P. parasitica* var. *nicotianae*	General	Unidentified	Siradhana et al. (1969)
Soybean-*Phytophthora* spp.	General	Glyceollin	Yoshikawa et al. (1978b)
Strawberry-*P. fragariae*	Specific?	Unidentified	Mussell and Staples (1971)
Carnation-*P. parasitica*	Specific?	Unidentified	Bouillant et al. (1983)
Cucumber-*P. drechsleri*	Specific?	Unidentified	A. Alavi and R. N. Strange (*unpublished*)
Tobacco-*P. parasitica* var. *nicotianae*	Specific	Rishitin and two others	Budde and Helgeson (1981)
	Specific	Capsidiol	Milat et al. (1991a, b)
Alfalfa-*P. medicaginis*	Specific?	Medicarpin	Vaziri et al. (1981)
Soybean-*P. sojae*	Specific	Glyceollin	Keen (1971); Yoshikawa et al. (1978a)
Potato-*P. infestans*	Specific	Rishitin and related compounds	Sato et al. (1971); Varns et al. (1971)
	Specific	Elicitation of phytoalexins by arachidonic and eicosapentaenoic acids	Bostock et al. (1982); Bloch et al. (1984)
Cowpea-*P. vignae*	Specific	Kievitone	Partridge and Keen (1976)
	Specific	Kievitone	Munn and Drysdale (1975)
Citrus-*P. citrophthora*	General	Scoparone	Afek and Sztejnberg (1988, 1989); Afek et al. (1986)
Safflower-*P. drechsleri*	Specific	Safynol	Allen and Thomas (1971)

[a] Excerpted and modified by permission of the American Phytopathological Society, St. Paul, Minnesota, courtesy Keen and Yoshikawa (1983a).

1988). Eicosapentaenoic and arachidonic acids from *P. infestans* incite the production of the phytoalexin rishitin and the stress metabolite lubimim in potato tuber tissue (Bostock et al. 1981, 1982; Bostock 1989). Csinos and Hendrix (1977) noted that a nonpathogenic isolate of *P. cryptogea* released a toxin into culture media that caused necrosis on tobacco plants.

Another protein produced in mycelium of *P. parasitica* isolates (from hosts other than tobacco plants), called parasiticein (Billard et al. 1988), induces a resistance reaction in tobacco plants to *P. parasitica* var. *nicotianae*. The lack of parasiticein production appears to be involved in the specific pathogenicity of *P. parasitica* var. *nicotianae* to tobacco (Ricci et al. 1992; see Chapter 50).

Elicitins from various *Phytophthora* species induce a hypersensitive defense response (HR) both locally and at some distance from the point of inoculation in some plants of the families Cruciferae and Solanaceae. In Solanaceae, the response to elicitins is restricted to species of *Nicotiana,* but in Cruciferae, the HR is limited to certain radish and turnip cultivars (Kamoun et al. 1993a). The HR response induces resistance in tobacco to infection by *P. parasitica* var. *nicotianae* (Ricci et al. 1988, 1992).

The gene encoding the host-specific elicitor protein from *P. parasitica* was identified by Kamoun et al. (1993b). Kamoun et al. (1994) examined 85 isolates of 14 *Phytophthora* species from a number of hosts for their ability to produce elicitins and induce resistance in tobacco plants to *P. parasitica* var. *nicotianae*. The degree of virulence of various isolates on tobacco plants was measured by the progression rate of the lesion on the stem. Elicitins were produced only by isolates of *P. parasitica* that had no virulence or only weak virulence on tobacco. Isolates of several other species that were not virulent on tobacco produced elicitins. These included *P. cactorum, P. capsici, P. cinnamomi, P. citricola, P. citrophthora, P. cryptogea, P. drechsleri, P. infestans, P. megasperma, P. melonis, P. palmivora,* and *P. parasitica* (isolates from tobacco and other hosts) (Kamoun et al. 1994).

A protein from *P. cinnamomi,* which is nonpathogenic on tobacco, was characterized and designated cinnamomin (Billard et al. 1988). This protein is a basic, low molecular weight (98 amino acid residues) compound that has 80% homology with the elicitors cyptogein, parasiticein, and capsicein.

Zanetti et al. (1992) demonstrated that the elicitors cryptogein and capsicein are capable of moving from a wound in a stem of a nonhost tobacco plant to the leaves and that these compounds reach the target in the leaf via the sap stream in the xylem. Although these elicitors caused drastic damage to the leaf lamina, they did not decrease the flow of sap in the xylem.

Although viewpoints differ on how phytoalexins function, (e.g., Stössel 1988 and Keen 1990), the concept that phytoalexins have a causal role in disease resistance in soybean has become highly convincing. Glyceollin was identified after wound inoculation from hypocotyls of soybean plants resistant to *P. sojae* by Keen (1971), Sims et al. (1972), Burden and Bailey (1975), Lyne et al. (1976), and Yoshikawa et al. (1978a, b). In potato infected with *P. infestans,* rishitin has been consistently associated with race-specific resistance in tubers; however, it has not been found in leaves (Kuć 1972).

The *P. sojae*-soybean system has been of particular interest to scientists throughout the world who are interested in determining why a certain race of *P. sojae* can cause disease on one genotype of plant and not on another. These fundamental studies have practical implications because the rapid production of new races of *Phytophthora* erodes the arsenal of plant resistance genes that are available for production of commercial cultivars. Understanding why *P. sojae* will cause disease on certain cultivars but not on others is of basic interest and, if thoroughly understood, could be exploited for novel control measures.

Probably the earliest research on the production of a phytoalexin against *P. sojae* was by Klarman and Gerdemann (1963b), who demonstrated that a phytoalexin was produced by cultivar Harosoy soybean plants that were inoculated in the hypocotyl with mycelium of *P. cactorum,* a nonpathogen. The discovery of phytoalexins initially evolved from an ingenious experiment in which the phytoalexin compound was continuously removed by use of a string wick that was passed from a beaker of water through the inoculation wound. When the plant fluid was wicked away, the resistant plant behaved as though it were genetically susceptible. Inoculation of a soybean plant with an incompatible (avirulent) race before inoculation with a compatible (virulent) race prevented infection. Paxton and Chamberlain (1967) reported that a phytoalexin was produced by cultivar D60-9647 when it was inoculated with race 1, to which it was resistant, but not when inoculated with race 2, to which it was susceptible. Also, prior inoculation of cultivar D60-9647 with race 1 cross-protected it from secondary inoculation with race 2.

Glyceollin is produced in soybean cultivars resistant to *P. sojae* at a more rapid rate and at higher antifungal concentrations than in susceptible cultivars (Yoshikawa et al. 1978a, b). These data were confirmed by Hahn et al. (1985) by use of antibodies specific for glyceollin. Keen and Yoshikawa (1990) state, "Thus, these two studies have established the fact that glyceollin is made specifically in the incompatible reaction—it is present at the right time, at the right place and at fungitoxic concentrations that account for inhibition of the fungus." The sequence of biochemical events in resistant soybean plants leading up to glyceollin production is shown in Table 6.5 (Keen and Yoshikawa 1983a).

Glyceollin toxicity to *P. sojae* appears to affect plasma membranes (Kaplan et al. 1980). Yoshikawa et al. (1987a, b) reported that glyceollin also caused damage to *P. capsici* (nonpathogenic to soybean) that was visible at the ultrastructural level on membranes of mycelium in infected soybean hypocotyl tissue. Glyceollin at about 100 μg per milliliter causes inhibition of growth of different races of *P. sojae* on agar (Yoshikawa et al. 1978a) and Millipore filter pads (Stössel 1983). Although sensitivity of different races to glyceollin varies, it does not explain the function of different races. Sensitivity of different races is determined by certain race-specific elicitors produced by *P. sojae* (Keen and Legrand 1980). These elicitors have not yet been identified chemically

Table 6.5. Approximate sequence of events in soybean plants during an incompatible reaction to *Phytophthora sojae*[a]

Approximate time (hours)	Event
0	Inoculation or elicitor treatment
0–4	Host recognition
2	Increased poly-adenosine mRNA synthesis
4	DNA transcription required to here
5	mRNA translation required to here
6	Increased phenylalanine ammonia-lyase (PAL) activity
?	Inducible prenyl-transferase activity appears
6.5	Glyceollin accumulation starts
8	Glyceollin reaches an in vivo ED_{90} concentration against the fungus
9	Fungus growth stops

[a] Reproduced by permission of the American Phytopathological Society, St. Paul, Minnesota, courtesy Keen and Yoshikawa (1983a).

because of their apparent lability or sparsity of production in plant tissue (Keen 1990).

Goodell et al. (1985) discuss various explanations of disease resistance of different soybean plant cultivars to different races. First, race-specific elicitors mediate the accumulation of phytoalexins, a concept first proposed by Keen and Legrand (1980). Second, race-specific suppressors of phytoalexin accumulate in the plant (Ziegler and Pontzen 1982). Third, race-specific interactions precede phytoalexin accumulation. In all of these complementary hypotheses, certain glucans or mannans are implicated as elicitors. Ayers et al. (1976a–c) used monoclonal antibodies to identify glycoproteins that acted as elicitors at membrane sites.

In addition to soybean (*G. max*), glyceollin has been detected in 61 accessions of 10 species of the genus *Glycine* (Keen et al. 1986), thus indicating that glyceollin is genetically highly conserved among members of the genus (Keen 1990). The review by Ebel and Griesbach (1988) on analysis of the resistance mechanisms of host-specific cultivars of soybean indicates that glyceollin production is an important early defense mechanism. Even though considerable biochemical evidence supports the role of glyceollin in the resistance mechanism, they also cite other inducible biochemical responses such as callose (1,3-β-glucan) deposition at the infection site, which can be detected within 2–4 h after infection (Bonhoff et al. 1987).

Müller and Börger (1940) first reported that potato tubers attacked by *P. infestans* were cross-protected by prior inoculation with a nonpathogenic race of *P. infestans.* This experiment laid the groundwork for much more extensive research, which eventually led to the identification of rishitin in tubers of resistant plants (Tomiyama et al. 1968); however, it is not known why rishitin is not formed in resistant foliage (Kuć 1972). Coffey and Wilson (1983a) and Friend (1991) indicate a concern about accepting the phytoalexin explanation for resistance to *P. infestans,* since rishitin has not been found in leaves.

Friend (1991) questions whether antifungal compounds are produced before or after cell death and whether the biotrophic nature of *P. infestans,* especially at the early growth stage, might prevent its progress through hypersensitively killed cells. He cites indirect evidence that oxygen-free radicals such as superoxide and hydroxyl might be involved in the hypersensitive reaction of Kennebec tuber disks on which elicitors or sporangia of an avirulent race of *P. infestans* were placed. When inhibitors of oxygen-free radicals were used, phytoalexin (rishitin) accumulation and cell browning did not occur (Keenan et al. 1986). Ivanova et al. (1991a, b) in Russia present evidence that production of superoxide dismutase-dependent cytochrome c is associated with resistance of potato leaf tissue to *P. infestans.* Friend (1991) concludes that more evidence in needed before the mechanism of race specificity and resistance can be explained.

Although much progress has been made in the determination of the mechanisms of resistance, the biochemical mode of race-specific resistance has not yet been fully elucidated, and many questions remain unanswered (Stössel 1988).

Summary

Host resistance to *Phytophthora* species is the most important type of control measure; however, it is not attainable for all disease systems and against all species (see Table 6.1 for references on resistance of hosts to several *Phytophthora* species). In potato, control of late blight was developed by incorporating single genes for resistance (R-genes) from *Solanum demissum,* a wild species of potato. Although the level of host resistance that was developed was near immunity, biotypes, called pathologic or physiologic races, capable of blighting plants with R-genes developed within natural populations of *P. infestans.* Races are identified by their pathogenicity to a series of cultivars with different R-genes. Races can be either simple, e.g., R1 or R2, or complex, e.g., R1.2.3.4 or R1.4.6.10 (Table 6.2). Extensive research on a worldwide basis has shown that the appearance of new races has kept up with the development of race-specific resistant cultivars; however, in some areas, races have been found in field populations that have not been exposed to R-genes. The emphasis now on breeding potatoes for resistance rests mainly on the development of general resistance or non-race-specific resistance. This type of resistance does not induce immunity or prevent infection but acts against all pathologic races by slowing the rate of infection and reducing the size of lesions. Although this type of resistance can be overcome by environmental conditions favorable to the pathogen and/or an increase in inoculum, it has been durable and stable over the years. In most areas where potatoes are grown in high-rainfall areas, this type of resistance has to be supplemented with the use of fungicides (discussed in Chapter 7).

In soybeans, cultivars with both race-specific and general resistance are utilized in the field. Currently, more than 24 races of *P. sojae* can be differentiated by the disease reaction on cultivars carrying different dominant *Rps* genes (Table 6.3). In addition to race-specific resistance governed by *Rps* genes, a type of resistance referred to as general resistance (also known as field resistance, rate-limiting resistance, or tolerance) has been identified in numerous cultivars. This trait is heritable and has not been subject to attack by new races of *P. sojae;* however, the degree of resistance can be reduced by increased inoculum or extensive flooding of soil.

Cowpea plants grown in Australia are attacked by *P. vignae,* which is specific to cowpea and adzuki bean. Three races of *P. vignae* have been identified. Race-specific resistant cultivars have been developed, and cultivars with general resistance are recognized.

More than 10 races of *P. fragariae* are capable of causing the red stele (red core) disease on different resistant cultivars of strawberry. Both resistance and sanitation are used for control in the field. The fungus is homothallic and produces oospores in the infected stele of strawberry plants. A recent method of categorizing resistance in different cultivars includes not only the lack of red stele symptoms in roots but also the suppression of oospore formation in infected roots.

Resistance is a major control measure against root rot of alfalfa caused by *P. medicaginis.* Resistance is associated with two genes with incomplete dominance and is best characterized as general rate-limiting resistance (or tolerance), because there is no evidence of pathologic races and resistance can be overcome by an increase in inoculum or waterlogging of the soil. Waterlogging of soil can either lead to an increase in inoculum or predispose alfalfa roots to increased susceptibility.

Resistance of rootstocks of avocado trees to root rot caused by *P. cinnamomi* is characterized as a general rate-limiting type. Some rootstocks express general resistance by the rapid regeneration of roots, while in others the progress of infection in the root is reduced. The degree of resistance is not usually high enough by itself to exert satisfactory control under environmental conditions in which disease pressure is severe.

Resistance to *P. citrophthora* and *P. parasitica,* the causes of gummosis and root rot of citrus, is of the general or rate-limiting type, and there is no known immunity. Resistant rootstocks usually give adequate control when used in combination with cultural control practices.

The mechanisms of race-specific resistance are associated with the production of phytoalexins (chemicals produced by the plant in response to infection). The evidence for the function of phytoalexins in potato in response to *P. infestans* is associated with the production of rishitin (a sesquiterpenoid compound) in resistant tubers; however, rishitin has not been identified in foliage of resistant plants. The phytoalexin glyceollin (an isoflavonoid compound) is associated with race-specific resistance in soybean. Glyceollin is produced after infection more rapidly in resistant plants than in susceptible plants.

CHAPTER 7

Chemical Control

Fungicides are chemical compounds that either destroy a fungus or inhibit or suppress its growth. The term *fungistat* implies that the compound will suppress sporulation and growth but will not kill the fungus. Despite the preciseness of the definitions, the term *fungicide* is most commonly used in both the trade and scientific publications to include both fungicides and fungistats. This may be because of the difficulty in accurately defining the role of a compound that may act as a fungistat at some dosages and conditions but as a fungicide at others. Most of the modern fungicides are fungistats. It is important to realize this because one could expect that use of a fungicide will actually eradicate the pathogen, but this is seldom the case.

One of the earliest uses of a successful fungicide started with Millardet's discovery of Bordeaux mixture in France (see Millardet [1885] and Large [1940]). Schwinn and Margot (1991) cite reports of the first use of Bordeaux mixture in 1888. The story is told that Millardet, from the city of Bordeaux, walked by a vineyard every day and noted that the vines near the edge of the road did not have downy mildew symptoms. He associated this absence of downy mildew caused by *Plasmopara viticola,* a fungus that is closely related to *Phytophthora,* with a poison that the farmer had applied to keep people from stealing his grapes. The poison happened to be a mixture of copper sulfate and lime. Millardet followed up this observation by preparing the mixture for use as a mildew fungicide, and eventually it became known as the famous Bordeaux mixture for control of downy mildew of grapes (Millardet 1885). Bordeaux mixture has also been used successfully to control late blight of potato, caused by *Phytophthora infestans,* and many other diseases caused by other species of *Phytophthora.*

Bordeaux mixture has probably the greatest ability of any fungicide ever developed to stick tenaciously to foliage after rainfall; however, its disadvantage is that its active ingredient, copper, causes a significant phytotoxic effect if absorbed by some plants. After World War II, the dithiocarbamates, for example, maneb and zineb, became available, and to a large extent these and other fungicides displaced Bordeaux mixture; however, both Bordeaux mixture and other copper-containing fungicides are still used today.

The fungicidal control of oomycete fungi, including *Phytophthora,* has been the subject of a large number of investigations. A useful general treatment of fungicidal control of plant diseases is the book by R. Heitefuss of the George-August University of Gottingen-Weende in Germany (translated to English by J. Welch [1989]). Heitefuss puts fungicides in perspective with other methods of control and presents a brief commentary on the modes of action of different fungicides. The pesticide manual of Worthing and Walker (1987), which is updated and issued periodically, is a valuable source of the properties of fungicides. The use of pesticides as well as other methods for control of diseases and pathogens in greenhouse crops is presented by Jarvis (1992).

A more detailed and specific review of fungicidal control of oomycetous fungi can be obtained from the comparative coverage of fungicidal control of all the zoosporic plant pathogens (monoflagellate and biflagellate) by Bruin and Edgington (1983). They introduce the subject by pointing out some of the unique features of zoosporic plant pathogens that impact on chemical control. For instance, cell walls of zoosporic fungi contain cellulose but not chitin, which is common to most other fungi. Oomycetes are thus insensitive to polyoxin antibiotics.

Since oomycetous fungi do not require sterols for growth, they are insensitive to the sterol-synthesis-inhibiting fungicides such as triadimefon, triadimenol, bitertanol, and propiconazole (Heitefuss 1989; Griffith et al. 1992). Benomyl is a potent systemic fungicide that is active at low dosages on many fungi but not on *Phytophthora* or other members of Oomycetes. The biochemical mode of action of benomyl is the result of its ability to bind to tubulin, the protein involved in the mitotic phase of cell division. Since the structure of tubulin is different in *Phytophthora,* benomyl is not an effective fungicide and is used in several media to selectively isolate *Phytophthora* (see Chapter 2).

Bruin and Edgington (1983) point out that the unique feature of diploidy in *Phytophthora* makes this oomycetous fungus extremely versatile and adaptable. (See Chapter 3, in which genetics of *Phytophthora* is discussed.) This versatility has been demonstrated in the many reports on development of resistance in *Phytophthora* populations to fungicides with a specific mode of action, such as metalaxyl (Davidse et al. 1981, 1988; Davidse and DeWaard 1984), and on development of new races that can infect cultivars with host-specific resistance. The latter aspect is discussed in Chapter 6 on host resistance, and resistance to fungicides is covered later in this chapter.

Davidse and DeWaard (1984) compare the uses and modes of action of several fungicides and discuss the development of resistance in populations of many fungal pathogens, including that of *P. infestans* to metalaxyl, which has become common in Europe and Israel. Y. Cohen and Coffey (1986) more specifically treat the control of pathogens within the class Oomycetes. They, as well as Davidse and DeWaard (1984), discuss the direct effects of systemic fungicides (which actually function more as fungistats) and the indirect effects, for example, the possible interactive effect of a fungicide such as fosetyl-Al on the host as well as on *Phytophthora* (Griffith et al. 1992; Guest and Grant 1991).

Griffith et al. (1992), in a review entitled "Fungicides to Control Oomycetes," take a unique approach by documenting the many ways in which *Phytophthora* and other oomycetous organisms differ from the so-called higher fungi (e.g., members of Zygomycetes, Ascomycetes, and Basidiomycetes). Griffith et al. (1992) follow the concept that *Phytophthora* is a member of the kingdom Protista (also called Protoctista) (Corliss 1984; Margulis and Schwartz 1988). They relate the modes of action of fungicides, when warranted, to the unique biochemical features of *Phytophthora* and other oomycetes (see Chapter 1).

Table 7.1. Chemistry of new fungicides active against members of Oomycetes[a]

Classification/Common name	Trade name/Formulation(s)[b]	First report/Chemical name
1. Carbamates		
prothiocarb (Schering, SN 41703)	*Previcur 70; SCW*	*Bastiaansen et al. (1974)*
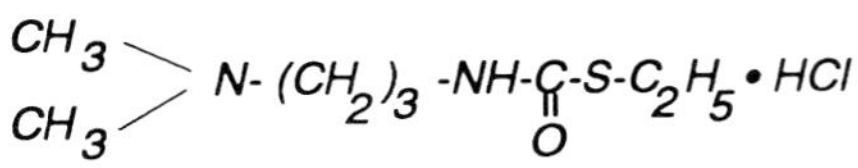		*S-ethyl (3-dimethylaminopropyl) thiocarbamate (DEP 1,567,169)*
propamocarb (Schering, SN 66752)	*Previcur N; SCW*	*Pieroh et al. (1978)*
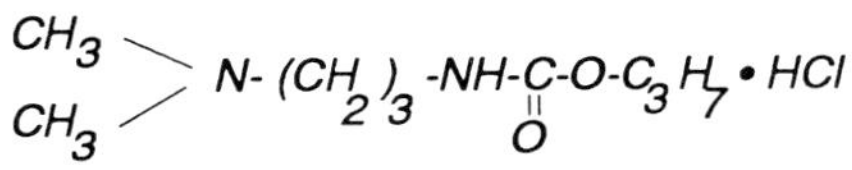		*propyl 3-(dimethylamino) propyl carbamate (DEP 1,567,169;1,643,040)*
2. Isoxazoles		
hymexazol (Sankyo, F319, SF-6505)	*Tachigaren; EC, SD, D*	*Takahi et al. (1974 a, b)*
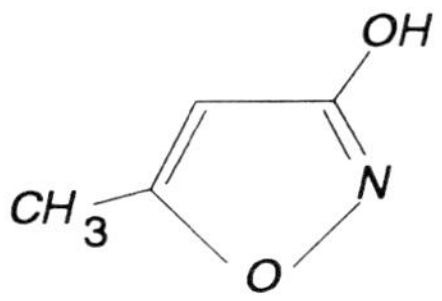		*5-methylisoxazol -3-ol (JPP 518,249;532,202)*
3. Cyanoacetamide-oximes		
cymoxanil (Du Pont, DPX 3217)	*Curzate; WP*	*Serres and Carraro (1976)*
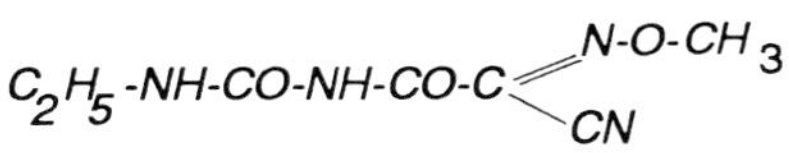		*1-(2-cyano-2-methoxyiminoacetyl)-3-ethylurea (USP 3,957,847)*
4. Ethyl phosphonates		
fosetyl (Rhône-Poulenc, LS 74-783)	*Alliette; WP*	*Bertrand et al. (1977), Williams et al. (1977)*
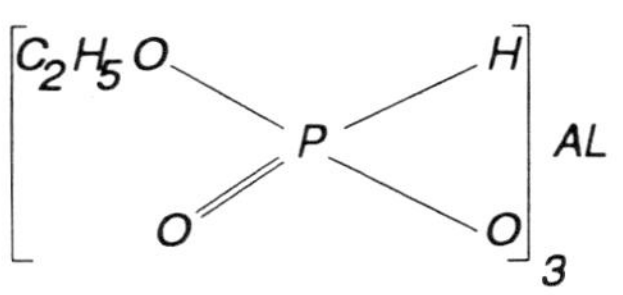		*ethylhydrogen phosphate, aluminium salt (FP 2,254,276)*
5. Phenylamides		
5.1. Acylalanines		
furalaxyl (Ciba-Geigy, CGA 38140)	*Fongrid; WP, G*	*Schwinn et al. (1977 a, b)*
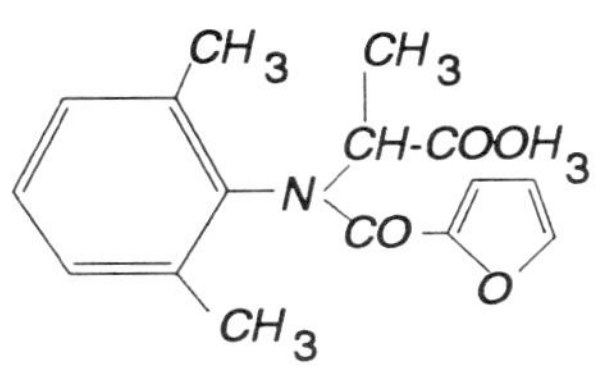		*methyl N-(2-furoyl)-N-(2,6-xylyl)-DL-alaninate (BEP 827,419; GBP 1,448,810)*

(continued on next page)

Table 7.1. (*continued*)

Classification/Common name	Trade name/Formulation(s)[b]	First report/Chemical name
metalaxyl (Ciba-Geigy, CGA 48988) 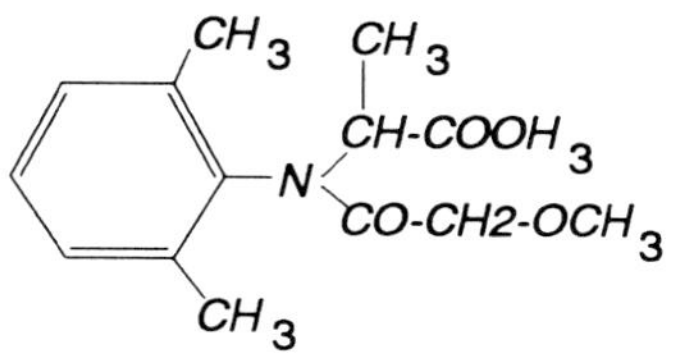	*Ridomil; WP, G* *Acylon; WP* *Apron; SD*	*Urech et al. (1977)* *methyl N-(2-methoxyacetyl)-N-(2,6-xylyl)-DL-alaninate (BEP 827,671; GBP 1,500,581)*
benalaxyl (Montedison, M9834) 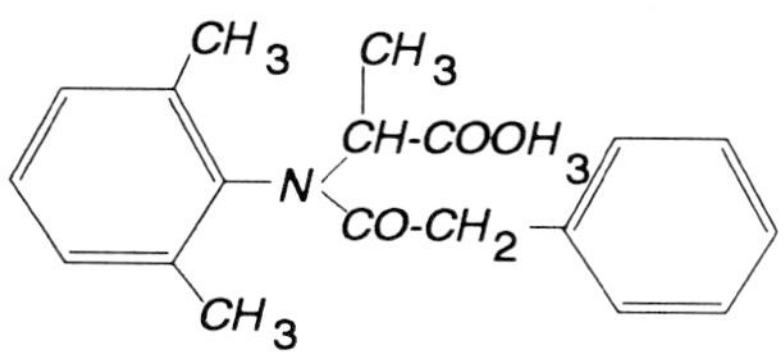	*Galben; WP, G*	*Bergamaschi et al. (1981)* *methyl N-phenylacetyl-N-2,6-xylyl-DL-alaninate (BEP 873,908; DEP 2,903,612; ITPA 19',896-1978)*
5. 2. Acylamino-butyrolactones		
ofurace (Chevron, RE 20615) 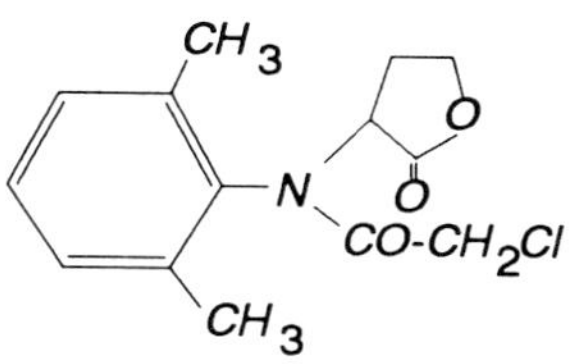	*Patafoll, Caltan; WP*	*Lukens et al. (1978)* *DL-3[N-chloroacetyl-N-(2,6-dimethylphenyl)-amino] γ-butyrolactone (US 3,933,860)*
cyprofuram (Schering, SN 78314) 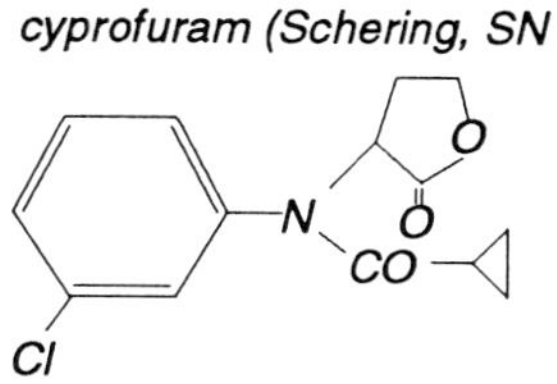	*Vinicur; WP*	*Baumert and Buschhaus (1982)* *(±)-d-[N-(3-chlorophenyl)cyclo-propane carboxamido] γ - butyrolactone (BP 1,603,730)*
5. 3. Acylamino-oxazolidionones		
oxadixyl (Sandoz, SAN 371 F) 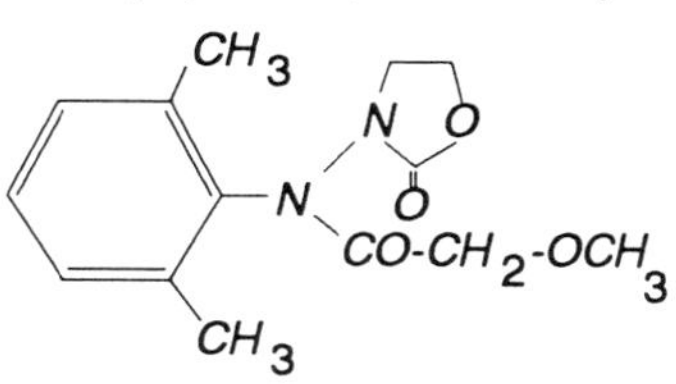	*Sandofan; WP*	*Gisi et al. (1983)* *2-methoxy-N-(2-oxo-1,3-oxazolidin-3yl) acet-2',6'-xylidide (BE 884,661; GBP 2058-059)*

[a] Common names, chemical names, and patent numbers after Worthing and Walker (1983), except ofurace and oxadixyl. Reprinted by permission of Gustav Fischer Verlag, Jena, Germany, courtesy Schwinn and Staub (1987).

[b] D = dust; EC = emulsifiable concentrate; G = granules; SCW = soluble concentrate in water basis; SD = seed treatment; and WP = wettable powder.

The excellent chapter on plant disease management by Fry (1977b) integrates the theory and strategy for use of chemicals with epidemiological models of disease expression. Fry (1977b) predicts that chemicals will probably always be needed for control of diseases such as late blight of potato but that their use should be integrated with other methods of control that include resistance and sanitation. A chapter on control of seed and soilborne diseases covers application methods as well as a wide range of chemical controls of diseases caused by soilborne fungi (Rodriguez-Kabana 1977).

Late blight of potato, caused by *P. infestans,* is an international problem that is controlled to a large extent by a program of appropriately spaced fungicide applications. On a worldwide basis, control of late blight of potato utilizes almost one-fourth of the annual money output for fungicides (Schwinn and Staub 1987). Schwinn and Margot (1991) present an apt and practical evaluation of the susceptibility to late blight of potato plants grown under different environmental conditions and control of the disease. Weather monitoring as an aid to deciding when to apply fungicides is discussed later in this chapter. Schwinn and Margot (1991) point out that despite the use of forecasting systems such as BLITECAST and Beaumont Rule (spraying only when the relative humidity exceeds 75% and temperatures are above 10°C for more than 48 h), all systems are subject to error because infection at suboptimal conditions, albeit minor, can be a source of inoculum to start an epidemic at a later, more favorable time. In Europe, a rule of thumb is that if blight-conducive weather has not occurred by the time the foliage of the plants meets between the rows, the first fungicidal spray should be applied, irrespective of weather conditions.

Also, the element of cost-benefit ratio has to be considered. For instance, MacKenzie et al. (1983) advocate the intelligent spacing of fungicidal sprays to save money (about $19 per hectare for each application of a fungicide at that time in Pennsylvania) because failure to spray at a critical time in the epidemic could cause loss of the crop. They posed the question that farmers who prefer to "spray weekly and sleep at night" ask: "Would you risk $144,000 (cost of production of potatoes on 50 ha) to save $2,850 (cost of three fungicidal sprays)?"

Despite the current popular trend toward totally organic, pesticide-free farming, diseases such as late blight of tomato and potato in intensive farming operations will probably never be totally controlled without pesticides. As discussed in Chapter 6 and in the chapter by Wastie (1991), the critical problem associated with control of late blight of potato by host resistance obtained through single-gene, host-specific resistance is the evolution of new pathologic races of *P. infestans.* Experience has shown that control by use of single-gene, race-specific resistance is doomed to failure. Although general resistance is more durable because it is usually controlled by multiple genes, it has been difficult to incorporate into commercial cultivars and does not always adequately control late blight without fungicides. For this reason, Fry (1977a) and Fry and Doster (1991) support the integrated use of general resistance and fungicides for control of late blight.

Schwinn (1983) and Schwinn and Staub (1987, 1995) review control of oomycetous fungi, which include the obligate downy mildew fungi in Peronosporaceae and Albuginaceae and the nonobligate members of Pythiaceae, which include *Pythium* and *Phytophthora.* They cover the use and chemistry of the carbamates (e.g., prothiocarb), the isoxazoles (e.g., hymexazol), the cyanoacetamide oximes (e.g., cyamoaxanil), the ethyl phosphonates (e.g., fosetyl-Al), and the phenylamides, which include the acylalines (e.g., metalaxyl and furalaxyl [Table 7.1]). Another review specifically covers chemical control of late blight of potato, caused by *P. infestans* (Schwinn and Margot 1991). Schwinn and Staub (1995) extended their 1987 review of oomycete fungicides. See also other chapters in the book edited by Lyr (1995).

Resistance to metalaxyl within populations of *P. infestans* has occurred in Northern Ireland (Cooke 1991), continental Europe (Davidse et al. 1981, 1991), Israel (Y. Cohen and Samoucha 1984), and the United States (Deahl et al. 1993). Schwinn and Staub (1987) and Staub (1991) discuss the prevention or suppression of buildup of resistant *Phytophthora* and downy mildew fungal biotypes by use of mixtures of broad-spectrum fungicides with the more highly specific metalaxyl.

The development of the systemic fungicide metalaxyl was a landmark in the history of control of fungal pathogens classified within Peronosporales, namely the downy mildews, *Pythium,* and *Phytophthora.* The activity of metalaxyl (as CGA-48944) was first reported by Urech et al. (1977) and Schwinn et al. (1977a, b) of the Ciba Geigy Company. Metalaxyl is translocated acropetically (upward) in plants, is toxic to *Phytophthora* at low concentrations (<1 ppm), and is water soluble.

Fosetyl-Al is another remarkable systemic fungicide and was developed about the same time as metalaxyl. This ethyl phosphonate fungicide was first reported in France (Bertrand et al. 1977; Williams et al. 1977) by the Rhone-Poulenc Agrochemie Company. Fosetyl-Al (Alliette) is chemically known as aluminum Tris-O-ethyl phosphonate. This product breaks down into ethanol and phosphorous acid (H_3PO_3) in plant tissue (Williams et al. 1977) and has been highly effective for control of many *Phytophthora*-caused diseases (Beach et al. 1979; Farih et al. 1981a, 1981b; Vo-Thi-Hai et al. 1979; see Fenn and Coffey [1984], Y. Cohen and Coffey [1986], and Coffey and Ouimette [1989] for more citations). Fosetyl-Al is widely used for control of many diseases, but unfortunately it is not effective against the foliar phase of late blight of potato or tomato. Although the reason is not well understood, fosetyl-Al gives good control of tuber blight but not of foliar blight (Schwinn and Margot 1991). The use of the modern systemic fungicides and the older protectant fungicides is discussed below. We also discuss the role of certain eradicative fungicides that are toxic to plants but useful for suppressing *Phytophthora* in soil and water and on glasshouse benches. Bear in mind that all fungicides must be used according to directions on the label; and for this reason, our discussion here is limited to descriptions of the fungicides. We do not recommend dosages or procedures.

Suppression of Inoculum by Eradicant Biocides

If the purpose of control is to kill the fungus at its source, for instance, on benches, in soil, or in water in an intensive glasshouse operation, the chemicals to be used are general biocides that kill all living organisms, including plants, so they must be used with due caution. These compounds must be applied to the source of *Phytophthora* inoculum and used in such a way that the plants will not be affected.

Treatment of Benches with Copper Naphthenate

To prevent carryover of *Phytophthora* and other pathogens in a glasshouse, benches can be painted once or twice a year with 2% copper naphthenate, a wood preservative that suppresses residual inoculum of most fungi that reside in cracks in the wood. This product can injure roots if they are allowed to contact the material directly; however, there is no danger when pots or wooden flats are set on painted benches (Baker et al. 1957).

Sodium Hypochlorite

Sodium hypochlorite, the active ingredient in commercial bleach (5.25%), releases free chlorine. When diluted for use to about 1:10, it can be used to disinfest benches and tools. This product will kill most pathogens if the exposure time is long enough; but unlike copper naphthenate, it does not have a lasting effect, and treatment must be repeated for maximum effectiveness. Since sodium hypochlorite is a strong oxidant, it is corrosive to metal tools if not washed off soon after use.

As discussed in Chapter 2, many *Phytophthora* species have been detected in bodies of water such as reservoirs, ponds, and streams. This residual inoculum is often a primary source when the water is used for irrigation (Table 2.5). P. M. Smith (1979) reports that chlorine released from sodium hypochlorite at 2 mg (active ingredient) per liter killed zoospores of *P. cinnamomi* after only 1 min of exposure at 18°C; however, 100–200 mg of chlorine per liter was required to kill mycelium, probably because *P. cinnamomi* produces many relatively thick-walled chlamydospores. Smith quotes Bewley and Buddin (1921), who reported that a dosage of only 20–50 mg of chlorine per liter killed mycelium of *P. cryptogea.* This species produces few, if any, chlamydospores. Control of inoculum in water by injection of chlorine to provide a concentration of 1 ppm has been an effective control measure in the northwestern United States (Ribeiro and Linderman 1991). Klotz (1978) recommends addition of a chlorine source to give a 1.0 ppm concentration in a reservoir.

When chlorine is used in irrigation water or reservoirs, the amount must be adjusted to accommodate its absorption by organic matter, which reduces the active concentration of the chlorine. Morrow and Martin (1977) discuss measurement of residual chlorine in water. The dosage of chlorine compounds must be kept low enough to prevent toxicity to plants that receive the treated water for irrigation. Pegg (1978) devised a system for production of disease-free avocado plants in nurseries and recommended use of chlorine at 2 ppm to kill zoospores of *Phytophthora.* He recommends treatment of contaminated water with a chlorine gas-injection unit or with a liquid sodium hypochlorite (5.25%) bleach or calcium hypochlorite, a dry white compound sold in granular or tablet form. Holding tanks of water can be treated. When the chlorine content has decreased from 2 to 0.6 ppm, it is suitable for irrigation of avocado trees, which are sensitive to chlorine (it causes a marginal burning of leaves). Pegg (1978) suggests that the chlorine content can be measured with the water-testing kits used for swimming pools.

Greenhouse studies by Grech and Rijkenberg (1992) indicated that chlorine at 200 and 500 μg/ml reduced propagules of *Phytophthora* in soil and in some cases eradicated the fungus; however, in field studies with citrus trees, chlorinating the water did not affect the *Phytophthora* population.

Miscellaneous Compounds

Copper-containing compounds (13–45 mg of free copper per liter) kill mycelium of *P. cinnamomi* at exposures of 24 h. High dosages (in milligrams of active ingredient per liter) of the fungicides prothiocarb (1,500), etridiazole (1,000), and furalaxyl (600) killed mycelium but only when it was immersed in solutions or suspensions for 3–6 days. Mycelium survived a 3-day immersion in fosetyl-Al at a concentration of 4,000 mg per liter (P. M. Smith 1979). Klotz (1978) suggests treatment of contaminated water from reservoirs or canals with copper sulfate sufficient to produce a concentration of 20 ppm. Although these fungicides are effective for use on plants to suppress the fungus, they would not suffice to clean up irrigation water.

Quaternary Ammonium Products

Noske and Shearer (1985) tested several chemicals to determine whether inoculum of *P. cinnamomi* could be killed. Because epidemics in virgin forested areas have been caused by this fungus, they saw the need for compounds that nurseries could use to eradicate *P. cinnamomi* and prevent further infection of transplants. They inoculated autoclaved leaf disks of the highly susceptible plant *Banksia grandis* with *P. cinnamomi* to test the efficacy of fungicides. Colonized leaf disks were dipped in solutions (10% active ingredient) of various chemicals for 30- and 120-s periods, after which the disks were plated out on a selective cornmeal antibiotic medium. Several quaternary ammonium products (Superquat, Applied 3-300, Liquitain, and Chemene) were more effective for suppression of *P. cinnamomi* than a phenolic product (Biogram). Sodium hypochlorite was even less effective, probably because of the short immersion times.

Sodium Tetrathiocarbonate

Sodium tetrathiocarbonate (Unocal Corporation, Los Angeles, California) releases carbon disulfide when added to water and is being developed as a prospective nematicide for soil application. A soil drench with sodium tetrathiocarbonate at 122, 245, and 490 μg/ml reduced sporangium production by *P. parasitica, P. capsici,* and *P. cactorum* by 23, 65, and 98%, respectively (Matheron and Matejka 1988a). A concentration of 2,450 μg/ml in agar media was required to inhibit completely growth of all species. Since citrus nematodes (*Tylenchulus semipenetrans*) and *P. parasitica* coexist on roots of citrus trees in many orchards, the material appears to have possibilities for control of both Phytophthora root rot and citrus nematode infection.

Methyl Bromide and Chloropicrin

For eradication of *Phytophthora* from soil in the field or in glasshouse operations, the fumigant fungicide methyl bromide is effective for suppression of *P. cinnamomi* chlamydospores in roots and from agar culture (Munnecke and Ferguson 1953; Munnecke et al. 1974). General reviews of control by use of methyl bromide and other soil fungicides are given by Rodriguez-Kabana et al. (1977) and Munnecke (1967, 1972). Methyl bromide mixed 50:50 with chloropicrin (tear gas) effectively eradicates *P. cinnamomi.* It is extremely important that the soil be dry enough to allow dispersion of the fumigant through the pore spaces. When pore spaces are saturated with water, the gases will not penetrate and microorganisms will not be killed (Munnecke et al. 1978). Both compounds are hazardous and must be used with caution by qualified personnel, and a legal permit must be obtained for their use. The cost of soil fumigation for strawberry culture exceeds $1,000 per acre (Yuen et al. 1991), so the practice is limited to intensively developed agriculture.

Land planted to strawberry in California is usually fumigated on a yearly basis with methyl bromide-chloropicrin prior to setting out disease-free strawberry plants (Wilhelm et al. 1974; Wilhelm and Paulus 1980). Wilhelm and Paulus (1980) tell an interesting history of the use of soil fumigation to control weeds; Verticillium wilt; red stele, caused by *P. fragariae;* root-knot nematodes; and other pathogens causing root diseases. Before methyl bromide and chloropicrin were used as soil fumigants, yields of strawberry in California averaged only 5–6 tons per acre (11.2–13.4 metric tons per hectare [mT/ha]), but after the introduction of soil fumigation with a methyl bromide-chloropicrin mixture in the early 1960s, the yield by 1978 had increased to 22 tons per acre (49.3 mT/ha) (Wilhelm and Paulus 1980). Yuen et al. (1991) report that soil fumigation with methyl bromide-chloropicrin markedly increases the percentage of healthy roots and yield of strawberry fruit at different locations in California. If methyl bromide and chloropicrin are removed from the market because of a potential environmental hazard, control of these diseases in California will have to be approached by other means.

Although eradicant fumigants such as methyl bromide and chloropicrin effectively control soil pathogens, an indirect deleterious effect can occur. The beneficial endomycorrhizal fungi that naturally colonize roots of citrus trees are also eradicated in fumigated nursery soils in which rootstocks are grown (Menge 1982; Menge et al. 1978b). During the 1950s and 1960s, growers noted that after citrus trees grown in fumigated soil in some nurseries were transplanted to the field, they remained stunted, especially in soils low in phosphorus, but the cause of the stunting was not known (Martin et al. 1963). Not until almost a decade later was the cause of the stunting found to be the destruction of the natural mycorrhizal fungi by fumigation of nursery soil (Kleinschmidt and Gerdemann 1972). When mycorrhizal fungi were added to the fumigated soil, citrus trees grew normally. The mycorrhizal fungus *Glomus mosseae* infects roots and functions symbiotically with the host by aiding in the uptake of phosphorous and some micronutrients by the roots (Menge 1983; Menge et al. 1977). When it became known that the stunting problem on citrus was caused by the lack of mycorrhizae, methods of inoculation of nursery stock by incorporating mycorrhizal inoculum on seed or by layering or banding it around intact plants were developed (Menge et al. 1977).

Isothiocyanates (Metham Sodium)

Metham sodium, sold in the United States as Vapam, is an alkaline, water-soluble carbamate that breaks down in soil to methyl isothiocyanate (Munnecke et al. 1962; Munnecke 1967). Metham sodium can be drenched on soil as an aqueous solution about 2–4 weeks before planting. Metham is a highly toxic chemical that should be utilized with appropriate caution and safeguards. Its activity as a general biocide follows the downward movement of the water in which it is dissolved. Its advantage over methyl bromide is that it can be applied in irrigation water, whereas methyl bromide must be applied to dry soil so that the gas will diffuse through the pore spaces in the soil. Metham sodium has also been applied by overhead sprinkler irrigation. This practice suppressed root diseases caused on peanut by *Pythium myriotylum* Drechs. and *Verticillium dahliae* Kleb. (Krikun and Frank 1982). Although control was obtained for 1 year, metham sodium did not kill oospores of *Pythium myriotylum.* Thus, it might not be effective in killing *Phytophthora* oospores. Since metham sodium does not kill mycorrhizae, its use in nursery soils is advantageous over that of methyl bromide, which is lethal to mycorrhizae (Linderman 1987; Ribeiro and Linderman 1991). Klotz (1978) and Klotz and Calavan (1978) give methods of treating infested soil with metham sodium prior to planting disease-free citrus trees.

Protectant Fungicides

Heitefuss (1989) states that fungicides are often classified as eradicative, protective, or curative; however, with the modern fungicides, many of which are systemic and are translocated in the plant, such distinctions are not sharply defined because they may act as either eradicants, protectants, or curatives at different dosages or under different conditions.

Copper Compounds

As stated above, the Bordeaux mixture discovered by Millardet in 1882 in France was one of the earliest fungicides (Heitefuss 1989) and was used as early as 1885 (Schwinn and Staub 1987). This fungicide consists of a mixture of copper sulfate and calcium hydroxide (slaked lime). The ratio of these components governs how much free Cu^{2+} is released into solution on the leaf or stem. Although this fungicide probably has the best tenacity (sticks to the foliage after rainfall) of any fungicide on the market, it is not used frequently, probably because of the inconvenience of preparation and the labor required for mixing and dispensing it. When used as a field application with overhead sprinklers, it was superior to copper oxychloride for control of brown rot caused by *P. citrophthora, P. parasitica,* and *P. hibernalis* on orange and grapefruit (Oren and Solel 1978).

Under heavy rainfall conditions in tropical areas of the world, such as India, where monsoons are a regular part of the weather pattern, Bordeaux mixture holds up well on foliage throughout the monsoon season if allowed to dry before ensuing rain. A single spray with 1% Bordeaux mixture plus 0.5% zinc sulfate decreased the incidence of brown rot of Coorg mandarin (*Citrus reticulata* Blanco) fruit, caused by *P. nicotianae* var. *parasitica,* from 97.5 to 100% in unsprayed fruits to 5.6 to 6.8%. Bordeaux mixture without zinc sulfate was less effective than captan (0.2%); difolatan (0.2%) was even less effective (Rao 1985).

Cuprous oxide was not introduced as a foliage treatment until 1932 (Schwinn and Staub 1987). Other copper salts that are used frequently as protectant foliar sprays include copper oxychloride, called green copper, and cuprous oxide often called red copper; however, the latter compound is not frequently used (Heitefuss 1989). Copper compounds and organic tin compounds are often recommended for use on potato against late blight during the latter part of the growing season because they are tolerated better by older plants. Copper compounds are also sometimes combined with metalaxyl to broaden the spectrum of target pathogens (for instance, bacteria) and, probably more important, to repress development of isolates of *P. infestans* resistant to the more specific fungicide metalaxyl (Schwinn and Margot 1991). Free copper (Cu^{2+}) is released on the surfaces of leaves and, if taken up by spores, poisons enzyme reactions. This occurs when copper displaces other heavy metals in natural compounds. Copper also blocks the activity of sulfhydryl groups. According to Heitefuss (1989), copper fungicides are particularly effective against oomycetous fungi. This may be because of the hydrophilic (waterloving) nature of copper salts. The ability of copper fungicides to be redistributed on the leaf by rainfall is an important advantage. After infection occurs, an application of a copper fungicide is no longer effective, however. On the other hand, copper, which penetrates leaves to some extent, can cause growth retardation and fruit russeting. This phytotoxic factor is one of the main disadvantages of using copper fungicides.

In Florida (United States), a fixed-copper fungicide applied by spraying the skirts (lower third) of young citrus trees provided the best control of brown rot of citrus fruit caused by *P. citrophthora* or *P. parasitica* (Timmer and Fucik 1975). In this case, the inoculum in the soil is splashed to the lower part of the tree during rainstorms. These species produce noncaducous sporangia; however, in certain areas of the world where *P. hibernalis,* a producer of caducous sporangia, is a cause of brown rot, the entire tree must be sprayed because the detached sporangia can be disseminated throughout the tree by wind-driven rain (Klotz 1978).

Paint containing fungicides was applied to young citrus trees under polyurethane foam wraps that are used to prevent freeze damage in Texas. Several copper-containing compounds (60 mg/ml) (copper ammonium carbamate, cupric hydroxide, basic copper sulfate, and copper salts of fatty rosin and acids) were effective as trunk paints against infection by *P. parasitica.* Captafol (60 mg/ml) was more effective, but captan was ineffective (Timmer 1977).

A series of reports from Nigeria discusses the use of copper fungicides to control pod rot of cacao (*Theobroma cacao* L.) caused by *P. palmivora* (Hislop and Park 1962a–c). The effective threshold deposit of copper for control is about 1 $\mu g/cm^2$ if coverage is uniform. Another coverage factor to be considered is that the surface area on pods increases during rapid growth. Hislop and Park (1962c) point out that little profit is to be gained by spraying during rainfall because a copper fungicide must first become dry on the pod to be maximally tenacious. Application of Bordeaux mixture and cuprous oxide fungicides at 3-week intervals controlled pod rot of cacao (Thorold 1959).

Organic Tin Compounds

Organic tin compounds, for example, fentin acetate or fentin hydroxide, inhibit oxidative phosphorylation in fungal cells, that is, the energy-generating reaction of the cells. These compounds are used as foliar sprays for control of *P. infestans* (Heitefuss 1989). Fentin acetate reduced infection of lemon fruits by *P. citrophthora* zoospores. Soil drench with fentin acetate applied by overhead sprinklers was partially effective in reducing gummosis symptoms in a lemon tree orchard (Çinar et al. 1990).

Organic tin compounds are particularly effective antisporulant fungicides. They translocate only translaminarly (from one side of the leaf to the other) and are particularly effective on potato when used late in the growing season because they reduce the number of sporangia formed on leaves that will later infect the tubers in the soil. The relatively high phytotoxic quality of organic tin fungicides has been alleviated by mixing them with maneb or mancozeb and by applying them late in the season (reviewed by Schwinn and Margot [1991]).

Dithiocarbamates (Zineb, Maneb, and Mancozeb)

The dithiocarbamates were developed between 1951 and 1962 and have been used extensively as broad-spectrum fungicides, many of which have been successful for control of *Phytophthora.* Dithiocarbamic acid is combined with different cations to make up fungicides that differ in certain properties. The simplest compound is metham sodium (Vapam), a sodium salt that is soluble in water. This compound is discussed above with other eradicant fungicides as a preplant soil drench to kill weed seed, nematodes, fungi, and other microorganisms a few weeks before planting. Because of its phytotoxicity, metham sodium is not used as a protective spray.

Ziram and ferbam are dialkyl dithiocarbamates, while nabam, zineb, maneb, and propineb are alkylbisdithiocarbamates. Mancozeb is a highly stable manganese-zinc salt and is more active than either maneb or zineb alone.

The dithiocarbamates are preferred over copper fungicides because they are much less phytotoxic. Some dithiocarbamates increase the green color of leaves, probably because of the added cation, such as zinc, which corrects a minor element deficiency (Heitefuss 1989). The dithiocarbamate fungicides act in a protective role but do not translocate in foliage. They are probably the most important general group of fungicides used for control of late blight of potato (33% of the world market) and tomato. They are used as admixtures with metalaxyl in many countries because they have a broader spectrum of target fungi and also tend to suppress the development of metalaxyl-resistant biotypes of *P. infestans* (Schwinn and Margot 1991). In Northern Ireland, however, metalaxyl was introduced in a mixture with dithiocarbamate fungicides in 1978, and the frequency of metalaxyl-resistant *P. infestans* strains became so great by 1981 that metalaxyl was withdrawn for commercial use (Cooke 1991).

Chlorothalonil

Chlorothalonil, sold as Bravo, has been available since the 1960s and is effective against *Phytophthora* and other members of Fungi Imperfecti, Ascomycetes, and Basidiomycetes. Its use to control late blight of potato, complemented by its effectiveness against early blight caused by *Alternaria solani,* makes it an effective fungicide. Chlorothalonil has greater persistence in rainy weather than other protectant fungicides, such as zineb (Fry 1982). Chlorothalonil has a world market share of 4% on potato and 17% on tomatoes for control of late blight (Schwinn and Margot 1991).

Phthalimides (Captan, Captafol, and Folpet)

The phthalimides have been used to a large extent for control of downy mildews, but only captafol and folpet have become important for control of late blight of potato. These fungicides exert a protective effect but are not translocated. Thus, application has to be frequent enough to cover new growth (Schwinn and Margot 1991). Captafol (60 mg/ml) was much superior to captan when used as a trunk paint to control foot rot of young citrus trees caused by *P. parasitica* (Timmer 1977).

Systemic Fungicides

A systemic fungicide is characterized as a compound that can be taken up passively or actively through roots, stems, leaves, or flowers and will translocate to another area in the plant. Translocation can be across the leaf (*translaminar*), upward to new growth (*apoplastic*), or downward (*symplastic*). Most systemic fungicides translocate upward with the transpiration stream; however, fosetyl-Al translocates both upward and downward (in the transpiration stream and in the phloem sap). Systemic fungicides are less subject to loss by rainfall than protectant fungicides and can suppress the pathogen after infection has occurred. Thus, the effectiveness of a systemic fungicide is greater than that of a protective fungicide over a longer period. To a limited extent, a systemic fungicide can exert a curative effect; however, under heavy disease pressure, metalaxyl failed to control late blight of potato (Platt 1985a, b). Because of the rapid increase in the populations of metalaxyl-resistant strains of *Phytophthora,* some regulations forbid the use of metalaxyl after late blight is established.

Most of the modern effective fungicides have been developed since the mid-1970s. The development of systemic fungicides for control of oomycetous fungi, which include the obligate parasitic downy mildews and the more necrotrophic *Phytophthora* and *Pythium* pathogens, began in 1976 with cymoxanil (Serres and Carraro 1976), followed by metalaxyl (Urech et al. 1977), furalaxyl (Schwinn et al. 1977a, b), ofurace (Lukens et al. 1978), oxadixyl (Gisi et al. 1983), and fosetyl-Al (Bertrand et al. 1977; Williams et al. 1977). These and other systemic fungicides are listed in Table 7.1 (Schwinn and Staub 1987). Schwinn and Staub (1995) extended the list to include morpholines, e.g., dimethomorph (Albert et al. 1988), 2,6-dinitroanilines, strobilurine analogues, benzamides, and anilino-pyrimidines. See also other chapters in the new book *Modern Selective Fungicides* edited by Lyr (1995).

The development of systemic fungicidal control of *Phytophthora* and other related fungi stimulated an immense amount of research, especially on the phenylamide metalaxyl and the ethyl phosphonate fosetyl-Al (the names phosethyl aluminum and fosetyl-Al are synonymous). Several excellent and up-to-date reviews discuss systemic fungicidal control of *Phytophthora* in great detail (Coffey and Ouimette 1989; Schwinn and Staub 1982, 1987; Bruin and Edgington 1983; Schwinn 1983; Davidse and De Waard 1984; Y. Cohen and Coffey 1986; Schwinn and Urech 1986; Davidse et al. 1991; Schwinn and Margot 1991). The chapter by Schwinn and Margot (1991) specifically discusses control of late blight of potato and tomato, caused by *P. infestans.*

The modern systemic compounds effective against oomycetes are classified into several chemical groups. These include the carbamates, isoxazoles, cyanoacetamide oximes, ethyl phosphonates (fosetyl-Al), and phenylamides (e.g., metalaxyl and several related compounds). The phenylamides, formerly called acylalanines, are subclassified into the acylalanines, which includes metalaxyl, the acylamino-butyrolactones, and the acylamino-oxazolidones (Schwinn and Staub 1987) (Table 7.1; see Table 7.2 for the spectrum of target fungi that are controlled).

Carbamates (Prothiocarb and Propamocarb)

Prothiocarb (Iwan et al. 1977; Kerkenaar and Kaars Sijpesteijn 1977) (Table 7.1) has a limited systemic capability and is not translocated away from the point of application (Table 7.2). A half-life of 59 days for one soil and 144 for another was reported (Iwan et al. 1977). In South Africa, application of prothiocarb on land heavily infested with *P. nicotianae* var. *nicotianae,* cause of black shank of tobacco, increased the percentage of survival from about 10% in the nontreated plots to nearly 70% in 1 year and to more than 70% in the next year at a dosage of 6.3 kg/ha. Data from a series of lower dosages showed that 50% survival was attained with a dosage of 3.8 kg/ha (Prinsloo et al. 1979). Prothiocarb is active against *Aphanomyces,* a member of Saprolegniales, and *Bremia,* a downy mildew in Peronosporales (Schwinn and Staub 1987). See Y. Cohen and Coffey (1986) and Schwinn and Staub (1995) for more references.

Propamocarb is an analogue of prothiocarb in which an oxygen molecule replaces sulfur (Table 7.1). When applied as a soil drench at high concentrations (1,000 µg/ml), it suppressed root rot of rhododendron caused by *P. cinnamomi* but was inferior to metalaxyl (75 µg/ml) (Englander et al. 1980). Root rot of tobacco caused by *P. parasitica* var. *nicotianae* was controlled by soil application (Reilly 1980). According to Schwinn and Staub (1987), propamocarb and prothiocarb are used on vegetable and ornamental crops. Propamocarb is translocated from roots to shoots. For instance, downy mildew of cucumber, caused by *Pseudoperonospora cubensis,* was controlled by soil drench (Cohen 1979); however, translocation via application to leaves was poor (Table 7.4). Propamocarb had poor activity against *P. citrophthora* on citrus fruits (Tuset and Portilla 1990). Despite the indications that both propamo-

Table 7.2. Spectrum of activity of new fungicides active against *Phytophthora* and other oomycetous fungi[a]

Common name	Pathogens on roots or stems	Foliar pathogens	Activity against pathogens other than oomycetes
Hymexazol	*Aphanomyces* *Pythium*	. . .	*Corticium sasaki* *Fusarium* spp.
Prothiocarb and propamocarb	*Aphanomyces* *Pythium* *Phytophthora*	*Bremia* *Peronospora* *Pseudoperonospora*	None
Cymoxanil	. . .	*Plasmopara* *Peronospora* *Phytophthora* *Pseudoperonospora*	
Fosetyl-Al	*Phytophthora* *Pythium*	*Bremia* *Plasmopara* *Pseudoperonospora*	*Phomopsis viticola* *Guignardia bidwelli* *Pseudopeziza tracheiphila*
Phenylamides	*Peronosclerospora* *Phytophthora* *Pythium* *Sclerospora* *Sclerophthora*	*Albugo* *Bremia* *Peronospora* *Peronosclerospora* *Phytophthora* *Plasmopara* *Pseudoperonospora* *Sclerophthora*	None

[a] Reprinted by permission of Gustav Fischer Verlag, Jena, Germany, courtesy Schwinn and Staub (1987).

carb and prothiocarb are systemic and active against *Phytophthora,* they are not yet widely used commercially. Y. Cohen and Coffey (1986) review both propamocarb and prothiocarb.

Isoxazoles (Hymexazol)

The only fungicide within the isoxazoles is hymexazol, which is sold as Tachigaren (Table 7.1). Takahi et al. (1974a, b) describe its use as a soil fungicide in Japan. Hymexazol is a useful component in a *Phytophthora*-selective medium because it is relatively more toxic to *Pythium* than to *Phytophthora* species (Chapter 2, Tables 2.2 and 2.3) (Masago et al. 1977). Although some species of *Phytophthora* are sensitive to hymexazol, it has not become a commercial fungicide for these pathogens. Commercially, hymexazol has been found to be useful as a seed treatment for sugar beets against *Aphanomyces* and *Pythium* species that cause seedling diseases. Hymexazol also has activity against *Corticium sasaki* and *Fusarium* species when used as a drench in rice seedling boxes (Schwinn and Staub 1987).

Cyanoacetamide Oximes (Cymoxanil)

The cyanoacetamide oxime cymoxanil was developed by DuPont and is sold as Curzate. It is systemic in an apoplastic manner (see Table 7.1 for chemical structure and Table 7.2 for spectrum of activity). Cymoxanil has a marked local penetration advantage over strictly protective fungicides, but it is only weakly transported upward in the plant (Klopping and Delp 1980). Cymoxanil has a greater capacity to penetrate foliage and has more curative properties than prothiocarb and propamicarb. The fungicide is rapidly metabolized in plant tissue and has a half-life of only a few days (Klopping and Delp 1980). It is effective as a foliar spray against late blight of potato, caused by *P. infestans,* and downy mildew of grapes, caused by *Plasmopara viticola* (Douchet et al. 1977; Schwinn and Staub 1987, 1995). The activity of cymoxanil in a leaf is shown to be shorter than that of metalaxyl or fosetyl-Al but longer than that of a protective fungicide (Schwinn and Staub 1987, 1995; Schwinn and Margot 1991).

Cymoxanil has a synergistic effect when used with mancozeb (Gisi et al. 1986; Gisi 1991). *P. infestans* was optimally controlled on tomato by use of mixtures of cymoxanil, oxadixyl, and mancozeb, even when the inoculum included strains of *P. infestans* resistant to oxadixyl (Samoucha et al. 1987; Y. Cohen and Reuveni 1983). Because of its relatively short half-life, cymoxanil is used chiefly in combination with a protectant fungicide such as mancozeb or folpet. It has been used successfully in combination with oxadixyl (Tables 7.8 and 7.9) (Schwinn and Staub 1987; see Y. Cohen and Coffey [1986] for more references). Synergism is discussed in more detail later in this chapter.

Phenylamide Fungicides

Among the phenylamide compounds, furalaxyl, metalaxyl, and benalaxyl are included in the acylalamines, ofurace and cyprofuram in the acylamino-butyrolactones, and oxadixyl in the acylamino-oxazolidionones (Table 7.1) (Schwinn and Staub 1987). Another phenylamide analogue of metalaxyl (CGA 8000) has activity against soilborne members of Peronosporales (Margot et al. 1988). Although all of these compounds are active against the oomycetous *Pythium, Phytophthora,* and downy mildew (obligate parasites) fungi (Table 7.2), metalaxyl is the most widely used. The chemical structures of the phenylamides are shown in Table 7.1 (Schwinn and Staub 1987). A conceptual comparison of the activities of phenylamides, cymoxanil, fosetyl-Al, and maneb (a protectant fungicide) are shown by Schwinn and Margot (1991). Schwinn and Staub (1995) compared the activity of local systemic fungicides, protectants, phenylamides, and dimethomorph on the different stages of the life cycle of *P. infestans* on potato (Figure 7.1). Metalaxyl is water soluble and effective against all *Phytophthora* species at relatively much lower in vitro dosages (<1 ppm) than are used with protective fungicides (Table 7.3); however, none of the phenylamide compounds have any effect on nonoomycetous fungi. To obtain control of these fungi, other compounds must be added to metalaxyl or other phenylamide compounds. Research on the growth response of several species of *Phytophthora* to different dosages of ofurace, oxadixyl, and metalaxyl in cornmeal agar showed that *P. cactorum* was more sensitive than *P. cryptogea, P. capsici, P. citrophthora, P. megasperma* var. *sojae, P. citricola,* and *P. cinnamomi* (Zaviezo et al. 1993). For *P. cactorum,* the ED_{50} value for metalaxyl (μg/ml) was 0.04, for ofurace 0.01, and for oxadixyl 0.17. For *P. cryptogea,* the ED_{50} for metalaxyl was (μg/ml) 0.22, for ofurace 1.1, and for oxadixyl 1.77. *P. citrophthora* was least sensitive to oxadixyl with an ED_{50} of 144.8 μg/ml. *P. citricola* was the least sensitive to metalaxyl with an ED_{50} of 1.3 μg/ml and to ofurace with an ED_{50} of 91.6 μg/ml.

The dosage responses of several species of *Phytophthora* to metalaxyl are shown in Table 7.3. *P. bohmeriae* stands out as the most sensitive of all the *Phytophthora* species (ED_{50} 2–30 μg/ml) (Bailey and Coffey 1984). *P. drechsleri* f. sp. *cajani* is the least sensitive (ED_{50} 19 μg/ml) (Chauhan and Singh 1987).

Metalaxyl is translocated apoplastically from seeds, roots, and leaves to new growth (Staub et al. 1978; Zaki et al. 1981; J. P. Gupta et al. 1985); hence, the systemic and curative effect that metalaxyl exerts in plants makes it advantageous over protectant fungicides, especially when infection has occurred before application. The activity and use of metalaxyl and the other related phenylamide compounds are compared with other systemic compounds in Table 7.4. The relative systemic activity is shown in Table 7.5. Uses of the various classes and members of classes of systemic compounds are shown in Table 7.6.

In some reports, downward translocation (basipetal) of metalaxyl has been reported (Staub et al. 1978; Zaki et al. 1981), but its actual importance is minor. Since metalaxyl is water soluble, it can be applied to the soil, where it is taken up readily by the roots and translocated upward (Schwinn 1979, 1983; Bruck et al. 1980; Schwinn and Staub 1982, 1987).

Metalaxyl, an example of the phenylamide fungicides, retains its activity within the plant (Figure 7.1). When pepper (*Capsicum annum* L.) plants in pots were drenched with either captafol (a protectant fungicide) or metalaxyl at rate of 3.0 mg per plant, extracts from the tops of metalaxyl-treated plants (3 days after treatment) mixed in an agar medium reduced the growth of *P. capsici* by 57% and oospore production by nearly 100%, but extracts from captafol-treated or nontreated plants had no effect. Even when it was collected as long as 21 days after treatment, the metalaxyl-treated plant extract caused *Phytophthora* mycelium to be malformed (Papavisas and Bowers 1981). Captafol is more toxic than metalaxyl to *P. capsici* on the surface of the plant, where zoospore release and direct germination of sporangia occur; however, metalaxyl is more toxic to *Phytophthora* growth and infection in the plant (Schwinn and Margot 1991; Schwinn and Staub 1987, 1995). Hence, metalaxyl has eradicative properties and is less susceptible to loss by rainfall.

Metalaxyl translocates upward in plant tissue when applied

to soil, seeds, and foliage. Bioassays of tomato plants inoculated with *P. infestans* showed this upward translocation (Staub et al. 1978). When metalaxyl is synthesized with a radioactive carbon called ^{14}C, the movement of metalaxyl in the plant can be monitored by measuring the amount of radioactivity in samples of plant tissue or by exposing the tissue to X-ray film. These tests showed that metalaxyl moved from root to stems and leaves above the point of application. In avocado, ^{14}C-metalaxyl translocated chiefly in an upward diection; less than 1% of foliarly applied ^{14}C-metalaxyl was recovered in roots of *Persea indica* (Zaki et al. 1981). Analysis of the radioactivity in avocado plant extracts with a scintillation counter indicated that only a relatively small proportion of the metalaxyl applied to foliage was taken up and translocated. More than 90% of

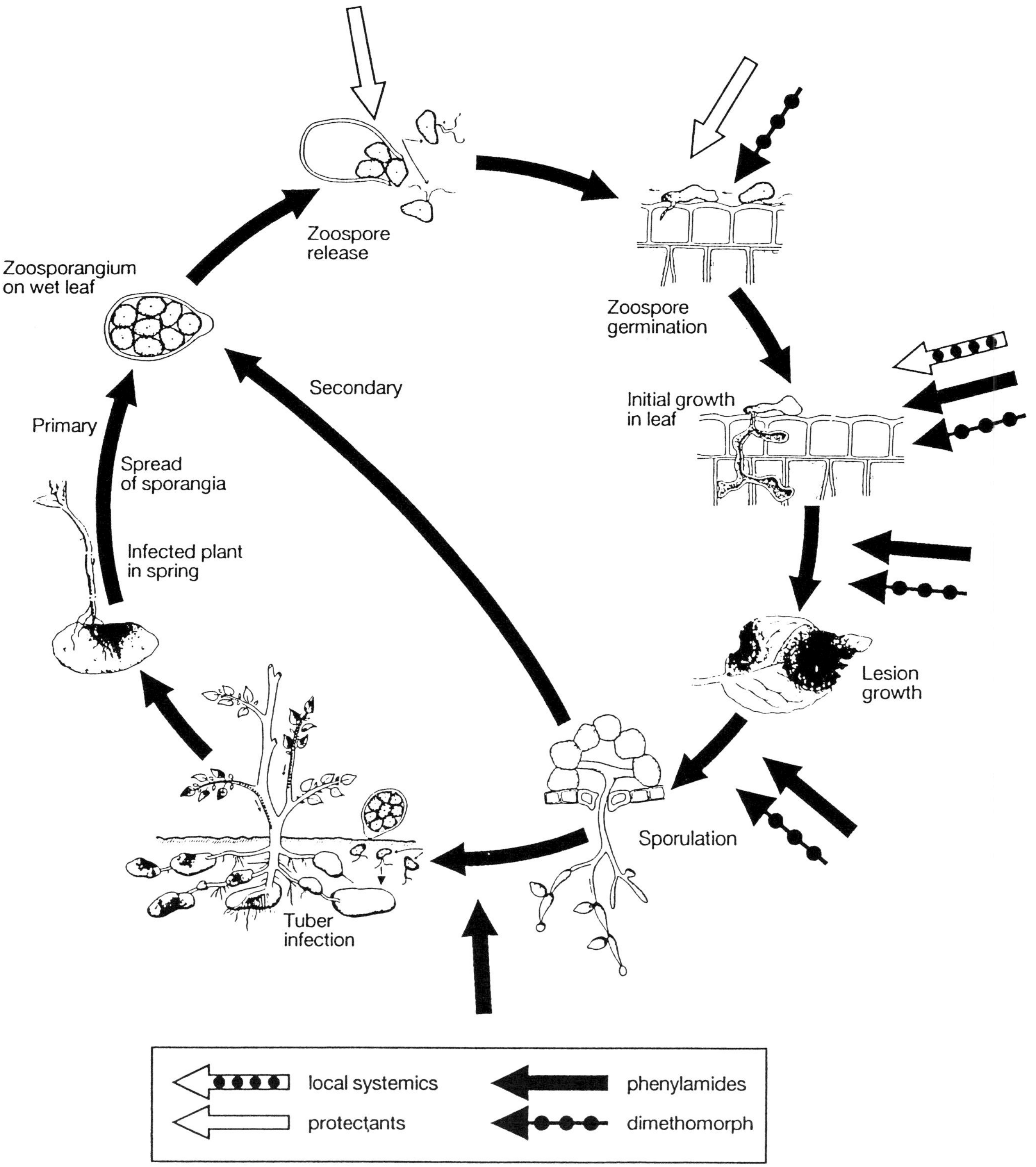

Figure 7.1. Target steps in the life cycle of *Phytophthora infestans* for the different groups of oomycete fungicides. (Reproduced by permission of Gustav Fischer Verlag, Jena, Germany, courtesy Schwinn and Staub 1995)

the metalaxyl applied to leaves remained on the surface (Zaki et al. 1981).

When soybean seeds were treated with ^{14}C-metalaxyl, 91.5% of the radioactivity remained in the cotyledons of the treated seeds, 5.8% was detected in the leaves and stems, and 2.5% in the roots. When ^{14}C-metalaxyl was applied to a leaf or cotyledon, 99% of the recovered radioactivity remained in the treated leaf or cotyledon. When ^{14}C-metalaxyl was applied as a drench to soil, 80.5% of the radioactivity taken up by the plant accumulated in the cotyledons, 15.1% in other aerial

Table 7.3. Growth of *Phytophthora* species in response to different dosages of metalaxyl

Species	Minimal dosage	Reference
P. boehmeriae	2–30 ng/ml	Bailey and Coffey (1984)
	0.48 μg/ml	Chauhan and Singh (1987)
	ED_{50} <0.025 μg/ml	Coffey and Bower (1984a)
P. cactorum	ED_{50} 2.0 μg/ml	Fuller and Gisi (1985)
	ED_{50} 0.4 μg/ml	Schwinn and Staub (1982)
	ED_{50} 0.04 μg/ml	Zaviezo et al. (1993)
P. capsici	Variable: 11–54% inhibition at 0.1 μg/ml	Coffey and Bower (1984a)
P. capsici (variability among 113 isolates)	100% to 25–42% inhibition at 0.5 μg/ml; some insensitive	Mchau and Coffey (1995)
P. cinnamomi (A1 and A2)	ED_{50} 0.07–0.29 μg/ml	Coffey et al. (1984a)
	ED_{50} 0.075 μg/ml	Coffey and Bower (1984a)
	ED_{50} 0.11 μg/ml	Benson (1979a)
	ED_{50} 0.9 μg/ml	Fuller and Gisi (1985)
	90% inhibition of zoospore release at 10 μg/ml (furalaxyl) and 25 μg/ml (metalaxyl)	Geissler and Katekar (1983)
	ED_{50} 0.2 μg/ml	Schwinn and Staub (1982)
P. citricola	ED_{50} 1.18–4.6 μg/ml	Coffey et al. (1984a)
	ED_{50} 1.17 μg/ml	Coffey and Bower (1984a)
	ED_{50} 6.45 μg/ml	Chauhan and Singh (1987)
	ED_{50} 1.3 μg/ml	Zaviezo et al. (1993)
P. citrophthora	ED_{50} 0.5 μg/ml	Coffey and Bower (1984a)
	ED_{50} 0.56 μg/ml	Farih et al. (1981a)
	ED_{50} 0.2 μg/ml	Schwinn and Staub (1982)
P. cryptogea	ED_{50} 0.22 μg/ml	Zaviezo et al. (1993)
P. drechsleri f. sp. *cajani*	ED_{50} 19 μg/ml	Chauhan and Singh (1987)
P. drechsleri	ED_{50} 0.5 μg/ml	Chauhan and Singh (1987)
P. heveae	ED_{50} 0.025 μg/ml	Coffey and Bower (1984a)
P. infestans	ED_{50} 1.0 μg/ml	Coffey and Young (1984)
	ED_{50} 0.3 μg/ml	Schwinn and Staub (1982)
P. medicaginis (*P. megasperma* f. sp. *medicaginis*)	ED_{50} 0.013 μg/ml	Coffey and Bower (1984a)
	ED_{50} 0.09 μg/ml	Chauhan and Singh (1987)
	100% at 1 μg/ml	Hansen and Hamm (1983)
	100% at 1 μg/ml	Hunger et al. (1982)
P. megasperma	ED_{50} >0.1 μg/ml	Schwinn and Staub (1982)
	ED_{50} 0.95 μg/ml	Coffey and Bower (1984a)
	Variable with isolate: 28–100% at 1 μg/ml	Hansen and Hamm (1983)
P. megasperma BHR[a]	47% at 1 μg/ml	Hunger et al. (1982)
P. megasperma DF[b]	76% at 1 μg/ml	Hunger et al. (1982)
P. nicotianae var. *parasitica*	ED_{50} 0.6 μg/ml	Schwinn and Staub (1982)
P. nicotianae	ED_{50} 0.2 μg/ml	Fuller and Gisi (1985)
P. palmivora	0.2 μg/ml	Fuller and Gisi (1985)
P. parasitica	ED_{50} 0.04 μg/ml	Farih et al. (1981a)
	ED_{50} 0.1 μg/ml	Coffey and Bower (1984a)
P. parasitica var. *nicotianae* (*sensu* Tucker 1931)	ED_{50} 0.2 μg/ml	Staub and Young (1980)
P. sojae (*P. megasperma* f. sp. *glycinea*)	ED_{50} 0.05 μg/ml	Coffey and Bower (1984)a
	ED_{50} 0.52 μg/ml	Chauhan and Singh (1987)
	100% at 1 μg/ml	Hunger et al. (1982)
P. trifolii (*P. megasperma* f. sp. *trifolii*)	100% at 1 μg/ml	Hansen and Hamm (1983)
	100% at 1 μg/ml	Hunger et al. (1982)

[a] Broad-host-range isolate.
[b] Douglas fir isolate.

Table 7.4. Main practical uses of fungicides active against *Phytophthora* and other oomycetous fungi[a]

Compound	Main use	Main crops	Application method
Prothiocarb and propamocarb	Diseases of roots and stems	Ornamentals, vegetables	Drench
Hymexazol	Diseases of roots and stems in seedling stage	Rice, sugar beets	Drench, seed dressing, dust
Furalaxyl	Diseases of roots and stems	Ornamentals	Drench
Cymoxanil	Foliar diseases	Grapes, potatoes	Spray
Fosetyl-Al	Foliar, stem, and root diseases	Grapes, avocado, pineapple, citrus, ornamentals	Spray, drench, dip, injection
Metalaxyl and related compounds	Foliar, stem, and root diseases	Grapes, potatoes, avocado, pineapple, citrus, tobacco, hops, maize, sorghum, millet	Spray, drench, dip, granules, seed dressing

[a] Reprinted by permission of Gustav Fischer Verlag, Jena, Germany, courtesy Schwinn and Staub (1987).

parts of the plant, and 4.4% remained in the roots (J. P. Gupta et al. 1985).

Metalaxyl is formulated as a wettable powder for use as a spray and as a granular material for use in soil at planting time. The compound is marketed under several names: Ridomil for foliar application, Apron for seed treatment, and Subdue for soil application. Characteristics of metalaxyl as a systemic fungicide are reviewed by Bruin and Edgington (1983), Schwinn and Staub (1982, 1987, 1995), Schwinn (1983), Cohen and Coffey (1986), Schwinn and Urech (1986), Davidse et al. (1991), and Schwinn and Margot (1991). Davidse and DeWaard (1984) present an overall picture of systemic chemicals and discuss different modes of action and a wide range of diseases controlled by different compounds, including the systemic antibiotics. Schwinn and Staub (1991) specifically address control of late blight of potato.

Metalaxyl evolved from research on herbicides at the Ciba-Geigy company in Basel, Switzerland. The history of this compound shows an interesting synthetic development utilizing chemistry and biological techniques (Figure 7.2). We quote from Schwinn and Urech (1986) (reprinted by permission of the American Chemical Society, Washington, DC, courtesy of Schwinn and Urech 1986):

> The history of the acylalanines began in herbicide research of Ciba-Geigy, Ltd., Basel, Switzerland. In the early seventies, a synthesis program was established aiming at new herbicides with a broader spectrum of biological activity in the class of anilides. The replacement of the alkoxyalkyl moiety of the rape seed herbicide dimethachlor (Teridox®) by the ethyl propionate substituent, as found in the wild oats herbicide benzoylprop-ethyl (Suffix®), led to a compound (CGA 22,574). This compound showed good herbicidal activity and, surprisingly, some systemic and curative fungicidal effect against late blight [Figure 7.2]. This novel, yet weak, fungicidal performance of CGA 22,574 could have been easily overlooked if the corresponding methyl ester (CGA 29,212) had not been prepared in the course of the herbicide synthesis program. This molecule exhibited much stronger fungicidal action. It outperformed all standards in our fungicide screening against the Peronosporales, both in respect to residual and systemic behavior, as well as to protective and curative activity. However, its phytotoxicity was the limiting factor. In further synthesis programs, the herbicidal effect could be eliminated by replacing the chloroacetyl group by two other acyl fragments, namely the 2-furoyl group, thus leading to furalaxyl in 1973, and the methoxyacetyl group, leading to metalaxyl in 1974.

Metalaxyl has a specific suppressive effect on ribosomal RNA (rRNA) (Davidse 1987, 1995; Davidse et al. 1983b). Davidse (1987) explains that the inhibition of rRNA synthesis ultimately leads to inhibition of fungal growth because reduction of rRNA deprives the cell of ribosomes that regulate protein synthesis. Thus, metalaxyl does not usually inhibit germination of sporangia or encysted zoospores as effectively as it does mycelial growth because intact spores apparently have sufficient ribosomes for germ tubes to form; however, when *Phytophthora* germ tubes penetrate the leaf, the metalaxyl that has been translocated into treated foliage is taken up by the mycelium. Metalaxyl then causes malformation and cessation of growth of the infecting mycelium. A protectant fungicide such as maneb, however, kills germinating spores in a few hours on a leaf but has no effect on the mycelium after it has penetrated the leaf (Figure 7.1). Once infection has occurred, the purely protective fungicides have little or no effect on disease progress because *Phytophthora* can progress unimpeded if it has infected the internal cells of the leaf before the fungicide was applied. The effect of maneb, a typical protectant fungicide, is conceptually contrasted with metalaxyl, a systemic phenylamide fungicide (Figure 7.1) (Schwinn and Staub 1987, 1995). Metalaxyl affects the development of mycelium in the leaf only after penetration (Grohmann and Hoffman 1989). Tuber rot of potatoes, caused by sporangia that wash off infected leaves and stems, is prevented by metalaxyl because *P. infestans* does not sporulate on treated leaf or stem tissue.

Since metalaxyl is water soluble and is taken up by roots, it has been successfully utilized as a soil drench or as granules on crops such as soybean (Schmitthenner 1985) and many ornamental plants growing in containers in nurseries (e.g., Benson 1979a, 1980, 1985, 1990). Because of the long half-life of metalaxyl in soils (15–30 days) and its high mobility, activity in soil is excellent (Y. Cohen and Coffey 1986). *P. capsici* on pepper can be controlled by metalaxyl as a furrow treatment but not as a seed treatment (Siqueira and Reifschneider 1981). Many of the downy mildew fungi infect seeds; and in these cases, seed treatment with metalaxyl is highly effective (Margot 1983).

When metalaxyl is used repeatedly in some avocado soils, it biodegrades rapidly because of the buildup of fungal and bacterial flora (Bailey 1985). Using *P. bohmeriae*, a highly sensitive species (Table 7.3) (Bailey and Coffey 1984), Bailey

Table 7.5. Characteristics of translocation of several oomycete fungicides[a]

Chemical	Local (penetration)	Apoplastic	Symplastic
Prothiocarb and propamocarb	++[b]	+	–
Hymexazol	++	+	–
Cymoxanil	+++	+	–
Fosetyl-Al	+++	++	+++
Phenylamides	+++	+++	+

[a] Reprinted by permission of Gustav Fischer Verlag, Jena, Germany, courtesy Schwinn and Staub (1987).

[b] +++ = Rapid, major factor for performance; ++ = intermediate, contributing factor for performance; + = weak, slow transport, effective only in special circumstances; and – = no transport in effective quantities.

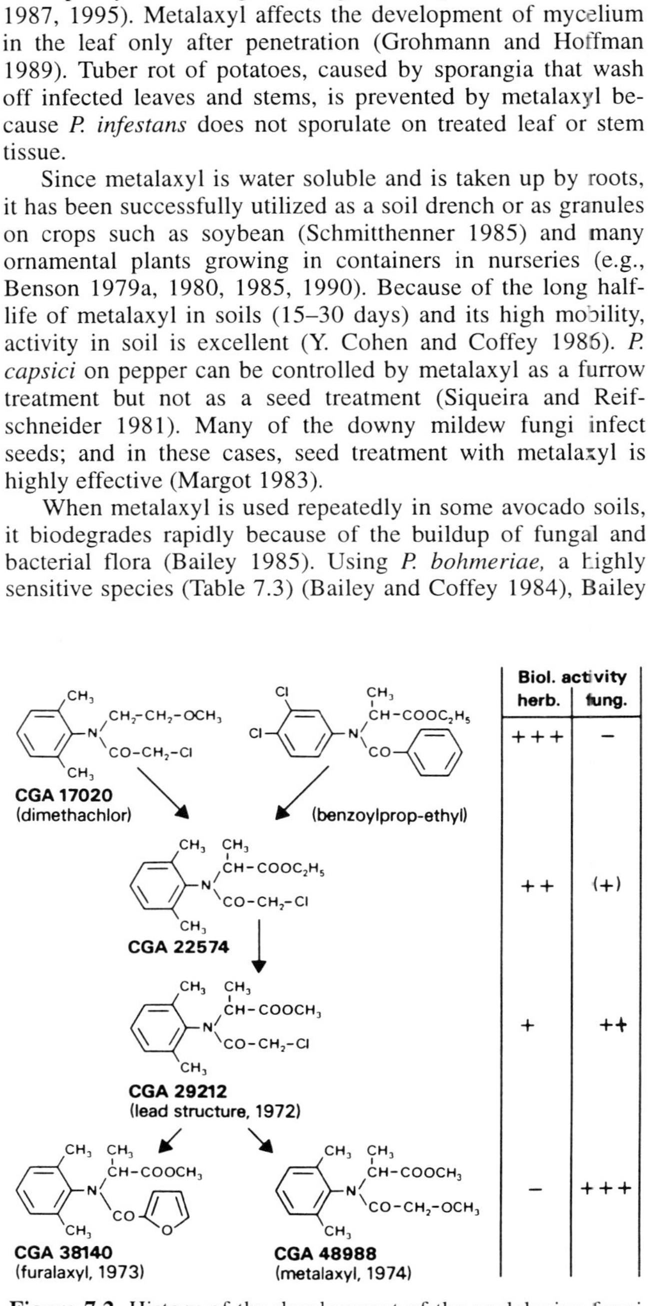

Figure 7.2. History of the development of the acylalanine fungicide metalaxyl. (Reproduced with permission from the American Chemical Society, Washington, DC., copyright 1986, courtesy Schwinn and Urech 1986)

and Coffey (1985) found that the half-life of metalaxyl was markedly reduced in metalaxyl-treated soils. Cometabolizing bacteria and fungi were associated with the biodegration of metalaxyl. Zaviezo et al. (1993) compared the mobility of metalaxyl, ofurace, and oxadixyl in sand and a clay soil. The greatest mobility was shown by metalaxyl, intermediate mobility by ofurace, and least mobility by oxadixyl. These data were correlated with the relative solubilities of these compounds. As would be expected, mobility was lower in clay than in sand (also shown by Benson [1979]). Increased mobility of metalaxyl in sand versus soil was also shown by Bailey and Coffey (1985).

Metalaxyl can be effective as a soil application for the control of root rot of soybean caused by *P. sojae* (*P. megasperma* f. sp. *glycinea*) but is not always economical (see Sinclair and Backman 1989). Schmitthenner (1985) states that metalaxyl is most effective as a seed treatment in Ohio when combined with use of a highly tolerant cultivar. The yield of a highly tolerant cultivar was increased 86% over that of a low-tolerance cultivar; when seed was treated with metalaxyl, the yield increased by 116%. When the highly tolerant cultivar was treated with metalaxyl in the furrow, the increase in yield was 13%. More data are given in Schmitthenner et al. (1985). Although in principle both seed and soil application increase yield, the cost may be too high. The in-furrow application was not considered to be practical on a field scale (Schmitthenner and Van Doren 1985).

Metalaxyl (0.5–1.5 kg/ha in the furrow) increased the yield and decreased the plant loss of soybean in an extremely susceptible cultivar but was not effective in a moderately susceptible or tolerant cultivar in Ontario, Canada (Anderson and Buzzell 1982). In Queensland, Australia, seed treatment of soybean with metalaxyl (0.7 g/kg of seed) improved emergence, reduced initial disease incidence, and increased yield (Troedson et al. 1991). Treatment of soybean seed with metalaxyl (1.65 g/kg) in the field in Canada increased plant survival and yield, but lower dosages were ineffective (Vaartaja et al. 1979).

When metalaxyl was used as a pot-soil drench (1.2 μg/ml) on 7- to 10-day old soybean plants that had been inoculated by hypocotyl stem puncture, the plant mortality and the extent of stem necrosis on soybean cultivars resistant and susceptible to *P. sojae* was markedly reduced. The reduction in disease severity was positively correlated with concentration of metalaxyl (0.2–1.2 μg/ml). The severity of disease was greater on plants grown at 32°C than at 24 or 28°C (Guy et al. 1989). See Table 7.6 for other examples of control of various species of *Phytophthora* by different modes of application of metalaxyl.

One of the earlier reports indicated that a dosage of 10 ppm applied to potato foliage inhibited lesion formation, expansion, and sporulation of *P. infestans* on leaves and reduced the germination of sporangia. Even when metalaxyl was applied 3 days after inoculation (by which time infection had occurred), lesion expansion was reduced but only at dosages above 100 ppm. Soil applications to potted plants (1,000 μg of metalaxyl applied to about 1,000 cc of soil) 3 days before inoculation of leaves suppressed lesion formation on leaves (Bruck et al. 1980). Since this report, numerous examples testify to the efficacy of metalaxyl for control of late blight of potato (Table 7.6).

Eckert and Ogawa (1985) reviewed postharvest control of brown rot on citrus fruit (caused by *P. citrophthora* and *P. parasitica*). Citrus fruits (lemons, oranges, and grapefruit) can be affected by brown rot. The fungicide biphenyl, incorporated in paper wraps, controls brown rot, but the fungicide is odoriferous and leaves objectionable residue on fruit. Metalaxyl is highly effective for eradication of incipient infection of fruit. *Phytophthora* species can penetrate and infect intact fruit if the surface of the fruit remains moist for a time after natural inoculation by soil splash during a rain storm. Metalaxyl (1–2 g/liter in an aqueous water-wax formulation) will delay or prevent brown rot and also suppress surface growth of *Phytophthora,* which often spreads from infected grapefruit to other fruits in storage for up to 3 months. Chitzanidis and Lascaris (1981) in Greece, E. Cohen (1981a, b, 1982) in Israel, and Eckert et al. (1981) in California also report control with metalaxyl.

Although there is no question that metalaxyl affects *Phytophthora* in plant tissue after infection, there is a limit to its effectiveness as an eradicant foliar treatment. This is shown in two reports by Platt (1985a, b), in which metalaxyl and metalaxyl combined with a protectant fungicide did not control late blight of potato on Prince Edward Island, Canada, when application was made after an epidemic had started (disease severity was about 15–25%). This area has high rainfall, and temperatures are nearly always favorable for late blight; however, when either a protectant fungicide or the systemic fungicide metalaxyl was applied as a preventive spray before disease was initiated, control was usually adequate.

Soil-drench treatments of pepper plants with metalaxyl retarded lesion formation on the stems of the plants that were inoculated with *P. capsici.* This result indicated that enough metalaxyl translocated to prevent infection of stems. In correlation with control of lesion formation, the phytoalexin capsidiol increased in the treated plants. The increase in concentration of capsidiol was even greater in metalaxyl-treated plants that were normally resistant to *P. capsici* (B. K. Hwang and Sung 1989). An ultrastructural study of pepper stems inoculated with *P. capsici* after a treatment with metalaxyl (5 μg/ml) showed that metalaxyl caused a shrinking of the *Phytophthora* cells in infected plant tissue, a separation of the plasma membrane from the hyphal wall, an invagination and breakdown of the plasma membrane, the presence of vesicles in the invaginated spaces and within the damaged fungal cells, and an indistinct structure of cell organelles. There was also electron-dense material in the plant tissue opposite the site of intimate contact of the *Phytophthora* cell wall with the host. B. K. Huang et al. (1990) suggested that deposition of this electron-dense material might be interpreted as a defense mechanism of the plant, since this material was not seen in untreated infected plant material.

Production of the phytoalexin glyceollin was increased when soybean plants were treated by soil drench with metalaxyl (Börner et al. 1983; Lazarovits et al. 1981; Ward et al. 1980). The importance of the stimulation of phytoalexins by metalaxyl is interesting but not yet understood, and although such an indirect phenomenon might increase the control potential, it may be coincidental. The ultrastructure studies of Grohmann and Hoffman (1989) showed no changes in potato leaf tissue after treatment with metalaxyl at 10, 100, and 200 μg/ml. They concluded that defense mechanisms were secondary. These reports and others cited in Table 7.6 indicate that living plant tissue is compatible with metalaxyl despite its fungitoxicity. Other related phenylamide fungicides also act this way.

Mode of Action of Metalaxyl and Related Phenylamide Fungicides

Interference with RNA synthesis in target *Phytophthora* species results from treatment with metalaxyl, cyprofuram, and ofurace, all related acylalanine fungicides (Davidse 1981,

Table 7.6. Control of diseases caused by *Phytophthora* species with metalaxyl and related phenylamide compounds

Species	Disease	Crop	Application	Reference
P. botryosa	Black stripe	Rubber	Foliar and trunk	A. M. Tan (1983)
P. cactorum (see also *P. cambivora, P. cryptogea, P. megasperma*)	Collar rot	Apple	Soil drench	Bielenin and Jones (1988b)
P. cactorum	Collar rot	Apple	Apply to lesion	Harris (1987)
	Collar and fruit rot	Apple	Trunk, soil, fruit	Harris (1991) (review)
	Collar rot	Apple	Foliar spray, soil drench, trunk paint	Orlikowski et al. (1986)
	Collar rot	Apple	Soil drench	Rana and Gupta (1984)
	Root rot	Apple	Root dip	Tidball and Linderman (1990)
	Collar rot	Apple	Soil drench, granular	Utkhede (1987a, b)
	Collar rot	Apple	Soil drench	Utkhede and Gupta (1988)
	Collar rot	Apple	Soil drench	Utkhede and Smith (1991a, b)
	Collar rot	Peach	Soil drench	Taylor and Washington (1984)
P. cambivora	Collar rot	Apple	Soil drench	Bielenin and Jones (1988b)
	Crown rot	Almond	Soil drench	Wicks (1988b)
	Root rot	Apple	Soil drench	Utkhede and Smith (1991a, b)
	Crown rot	Almond	Soil drench	Wicks and Lee (1985)
	Crown rot	Almond and cherry	Soil drench	Wicks and Hall (1988)
	Ink disease	Chestnut	Soil drench	Skoudridakis and Bourbos (1990)
P. capsici	Root rot	Pepper	Soil drench	P. K. Hwang and Sung (1989)
	Stem rot	Pepper	Soil drench	Palazón et al. (1990)
	Root rot	Pepper	Soil drench	Papavisas and Bowers (1981)
P. cinnamomi	Root rot	Azalea	Soil drench	Benson (1979)
	Root rot	Azalea	Soil drench	Benson (1985)
	Root rot	Azalea	Soil drench, foliar spray	Benson (1990)
	Root rot	Rhododendron	Soil drench	Benson and Blazich (1989)
	Root rot	Avocado	Soil drench	Coffey et al. (1984b)
	Root rot	Avocado	Several methods	Coffey (1987) (review)
	Root rot	Banksia	Soil drench	Cho (1981)
	Root rot	Rhododendron	Soil drench	Englander et al. (1980)
	Root and heart rot	Pineapple	Plant dip	Rohrbach and Schenck (1985)
	Root rot	*Chaemaecyparis lawsonianae*	Soil drench	P. M. Smith (1980)
	Root rot	Avocado	Soil drench	Whiley et al. (1986)
P. citricola	Root and crown rot	Walnut	Soil drench, stem application	Matheron and Mircetich (1985a)
P. citrophthora	Root rot and crown canker	Walnut	Soil drench and stem application	Matheron and Mircetich (1985a)
	Brown rot of fruit	Citrus	Fruit wash	Eckert and Brown (1986) (review)
	Gummosis, root rot	Citrus	Soil drench	Farih et al. (1981a, b)
	Gummosis, root rot	Citrus	Soil drench	Feichtenberger (1990)
	Root rot	Rough lemon	Soil drench	Matheron and Matejka (1991)
	Root and crown rot	Orange	Soil drench	Magnano di San Lio (1986b)
	Root rot, collar rot	Citrus	Soil drench	McKenzie (1985)
	Root rot	Citrus	Soil drench	Ohr et al. (1986)
	Brown rot of fruit	Citrus (orange)	Foliar spray	Tuset and Portilla (1990)
P. colocasiae	Leaf blight	Taro (*Colocasia esculenta*)	Foliar spray	Aggarwal and Mehrotra (1987)
P. cryptogea	Collar rot	Apple	Soil drench	Bielenin and Jones (1988b)
	Root rot	Cucumber	Hydroponic system	Price and Fox (1986)
P. cryptogea (also *P. megasperma* var. *sojae*)	Root rot	Asparagus	Soil drench	Falloon et al. (1991)
P. cryptogea	Root rot	Raspberry	Soil drench	Washington (1988)
	Root rot	Cineraria	Soil drench	Orlikowski et al. (1984)
	Root rot	Kiwi	Soil drench	Zaviezo et al. (1993)
P. drechsleri	Root rot	Douglas fir	Soil drench	Hamm et al. (1984)
P. drechsleri f. sp. *cajani*	Collar rot	Pigeon pea	Seed treatment	Kannaiyan and Nene (1984)
P. erythroseptica	Pink rot	Potato	Soil drench, granular	Torres et al. (1985)
P. fragariae	Red stele	Strawberry	Root drench	Duncan (1985b)
	Red stele	Strawberry	Soil drench	Jamart et al. (1985)
	Red stele	Strawberry	Soil drench	McIntyre and Walton (1981)
P. heveae	Foliar blight	Rhododendron	Foliar and soil drench	Benson (1980)
P. infestans	Late blight	Potato	Foliar	Bashan et al. (1989)
	Late blight	Potato	Foliar	Bruck et al. (1980, 1981)
	Late blight	Potato	Foliar	Bruin et al. (1982)
	Late blight	Potato	Foliar	Y. Cohen and Coffey (1986) (review)
	Late blight	Potato	Foliar	Y. Cohen et al. (1979)
	Late blight	Potato	Foliar	Doster and Fry (1991)
	Late blight	Potato	Foliar and soil	Easton and Nagel (1985)
	Late blight	Potato	Foliar	Fry et al. (1979, 1983)
	Late blight	Potato	Foliar	Gujer (1991)

(continued on next page)

Table 7.6. *(continued)*

Species	Disease	Crop	Application	Reference
P. infestans (*continued*)	Late blight	Potato	Foliar	Cooke et al. (1982)
	Late blight	Potato	Foliar	Hide and Cayley (1991)
	Late blight	Potato	Foliar	Kadish and Cohen (1988a–c, 1989)
	Late blight	Potato	Foliar	Milgroom and Fry (1988a, b)
	Late blight	Potato	Foliar	Milgroom et al. (1988)
	Late blight	Potato	Foliar	Olofsson (1991)
	Late blight	Potato	Foliar	O'Sullivan and Dowley (1983)
	Late blight	Potato	Foliar	Pappas (1985)
	Late blight	Potato	Foliar	Platt (1985a, b)
	Late blight	Potato	Foliar and soil	Rowe (1982)
	Late blight	Potato	Foliar	Samoucha and Cohen (1988, 1989)
	Late blight	Potato	Foliar	Schwinn and Margot (1991) (review)
	Late blight	Potato	Foliar	Schwinn and Staub (1982) (review)
	Late blight	Potato	Foliar	Shattock (1988)
	Late blight	Potato	Foliar	Shtienberg and Fry (1990)
	Late blight	Potato	Foliar	Shtienberg et al. (1989)
	Late blight	Potato	Foliar	Staub et al. (1980)
	Late blight	Potato	Foliar	Stewart and McClamont (1982)
P. meadii	Koleroga, fruit rot	Arecanut	Foliar and fruit spray	Sastry and Hegde (1988)
P. medicaginis (= *P. megasperma* f. sp. *medicaginis*	Damping off	Alfalfa	Seed treatment	Rhodes and Myers (1989)
P. megasperma	Collar rot	Apple	Soil drench	Bielenin and Jones (1988b)
	Root rot	Douglas fir in nurseries	Soil drench	Hamm et al. (1984)
P. megasperma var. *sojae* (not *P. sojae*)	Root rot	Asparagus	Soil spray, plant dip	Falloon et al. (1985, 1991)
Phytophthora sp. (probably *P. megasperma*)	Root rot	Asparagus	Plant dip, soil drench	Falloon and Fraser (1991)
P. nicotianae	Root rot	Tomato	Soil drench	Flett et al. (1991)
P. nicotianae and *P. citrophthora*	Gummosis, root rot	Citrus	Foliar, soil drench, granular	Feichtenberger (1990)
	Root rot	Citrus	Soil, granular/aldicarb	Le Roux et al. (1991)
		Citrus (orange)	Soil drench	Magnano di San Lio et al. (1986b)
	Root rot, collar rot	Citrus	Soil drench	McKenzie (1985)
P. palmivora	Black pod	Cocoa	Pod rot	McGregor (1984a, b)
	Black pod	Cocoa	Spray in combination with cuprous oxide	Prior and Smith (1982); Holderness (1992)
	Black pod, canker, seedling blight	Cocoa	Spray (ineffective under high rainfall)	R. D. Anderson and Guest (1990)
	Black stripe	Rubber	Foliar and trunk	A. M. Tan (1983)
P. palmivora and *P. botryosa*	Black stripe (tapping panel disease	Rubber	Spray	A. M. Tan (1990)
P. palmivora MF4 (probably *P. capsici*)	Foot rot	Black pepper	Spray	Ramachandran and Sarma (1985)
P. parasitica	Root rot, fruit rot	Citrus	Soil drench	E. Cohen (1981a, b, 1982)
	Root rot	Citrus	Soil drench	Timmer (1979)
		Citrus	Soil drench, foliar, trunk paint	Timmer and Castle (1985)
	Brown rot, fruit rot	Citrus	Fruit wash	Eckert and Brown (1986) (review)
	Root and fruit rot	Pineapple	Foliar spray	Rohrback and Schenck (1985)
	Gummosis	Citrus	Soil drench	Farih et al. (1981b)
	Root rot	Citrus	Soil drench	Ohr et al. (1986)
		Pineapple	Plant dip	Rohrback and Schenk (1985)
	Root rot	Tomato	Soil drench	Ioannou and Grogan (1984a)
P. parasitica var. *nicotianae*	Root rot	Tobacco	Soil drench	Staub and Young (1980)
	Root rot	Tobacco	Soil drench	Shew (1985)
P. pseudotsugae	Root rot	Douglas fir	Soil drench	Hamm et al. (1984)
P. syringae	Pruning wound canker	Almond	Aqueous paint on wound	Doster and Bostock (1988c)
	Fruit rot	Apple, pear	Soil drench	Harris (1979b)
	Collar and fruit rot	Apple	Soil trunk, fruit	Harris (1991b) (review)
	Collar and fruit rot	Apple	Soil and fruit dip	Edney and Chambers (1981a, b)
Phytophthora sp.	Root rot	Asparagus	Plant dip, soil drench	Falloon and Fraser (1991)
P. sojae (*P. megasperma* f. sp. *glycinea*)	Root rot	Soybean	Soil granular	T. R. Anderson and Buzzel (1982)
	Root rot	Soybean	Soil, seed, foliage	Gupta et al. (1985)
	Root rot	Soybean	Soil drench	Guy et al. (1989)
	Root rot	Soybean	Seed, soil drench	Lazarovits et al. (1980)
	Root rot	Soybean	Seed, soil drench	Papavizas et al. (1979)
	Root rot	Soybean	Seed, soil drench	Schmitthenner (1985) (review)
	Root rot	Soybean	Soil, drench, granular	Tooley and Grau (1984a)
	Root rot	Soybean	Seed, soil drench (field)	Ryley et al. (1989)
	Root rot	Soybean	Seed, soil drench, granular	Vaartaja et al. (1979)
	Root rot	Soybean	Soil drench	Ward et al. (1980)
	Root rot	Soybean	Seed	Willis (1983)

1995; Davidse et al. 1981, 1983b; Kerkenaar 1981; Fisher and Hayes 1982, 1984, 1985). There is a possibility of a second mode of action in cyprofuram because it inhibits germination of sporangia of the downy mildew *Plasmopara viticola* (Roos and Buchenauer 1985) and directly affects the plasma membrane of *P. medicaginis* (*P. megasperma* f. sp. *medicaginis*), but metalaxyl does not (Fisher and Hayes 1985). This second mode of action, however, is probably of "limited practical importance for counteracting phenylamide resistance" (Davidse et al. 1988).

Resistance to Phenylamide Fungicides

The development of resistance in target *Phytophthora* species to such an important group of fungicides as the phenylamides (metalaxyl and related compounds) is economically important. Methods of monitoring metalaxyl resistance in field populations of *P. infestans* include growth on rye agar amended with increasing concentrations of metalaxyl up to 100 µg/ml and treatment of whole plants, detached leaves, or leaf disks with increasing dosages of metalaxyl, after which the plant material is inoculated with sporangia (Sozzi and Staub 1987).

Resistance in strains of *P. infestans* to metalaxyl has become a serious problem on potato for control of late blight. Resistance to metalaxyl extends to all fungicides of similar phenylamide composition, e.g., benalaxyl, furalaxyl, ofurace, cyprofuram, and oxadixyl, but not to other major groups shown in Table 7.1 (Davidse 1987, 1995; Schwinn and Staub 1987, 1995; and Davidse et al. 1991).

Since inhibition of *Phytophthora* ribosomal RNA by metalaxyl and other phenylamide fungicides is a specific mode of action (Davidse et al. 1983b, 1991), it is not surprising that resistance to these compounds occurred in *P. infestans* and other *Phytophthora* species experimentally.

The possibility that resistance to metalaxyl by *Phytophthora* species could occur was shown by Bruin and Edgington (1982). Resistance to metalaxyl was induced in *P. capsici* and *Pythium ultimum* by irradiation of zoospores or mycelium with ultraviolet light (11 watts at 254 nm on the light spectrum) on a V8 agar medium amended with metalaxyl (250 µg/ml). Davidse (1981) detected resistance in both natural and induced mutants of *P. megasperma* f. sp. *medicaginis.*

Although initial greenhouse tests indicated that resistance did not occur in *P. infestans* after experimental usage of metalaxyl (Staub et al. 1979), resistance in *P. infestans* to metalaxyl appeared in the field shortly after its introduction into Europe (reviewed by Y. Cohen and Coffey [1986]; Davidse 1987; Schwinn and Staub 1987; Schwinn and Morton 1990; Cooke 1991; Davidse et al. 1991; Staub 1991). When metalaxyl was used widely during the first year on potato fields infested with *P. infestans* in the Netherlands, the population of resistant strains developed in the field to the extent that metalaxyl became ineffective for control of late blight (Davidse et al. 1983a, 1991). Since then, resistance has been reported in Israel (Y. Cohen and Reuveni 1983), Northern Ireland (Cooke et al. 1982; Cooke 1991), Ireland (Dowley and O'Sullivan 1981, 1991), Scotland (Holmes and Channon 1984), the northwestern United States, and British Columbia in Canada (Gillis 1993). See Table 7.7 for other citations. Deahl et al. (1993) reported that 81% of the isolates found in northwestern Washington after a late blight epidemic were resistant to metalaxyl (10 µg/ml).

Fry and Goodwin (1995) discussed the migration of *P. infestans* clonal lineages US-6, US-7, and US-8 (Goodwin et al. 1995). All of these lineages are resistant to metalaxyl and had been identified in Mexico by use of allozyme markers prior to their immigration to the United States. Sujkowski et al. (1995) reported that the majority of 75 isolates from Mexico showed a low sensitivity to metalaxyl (>100 µg/ml).

Some biotypes of *Phytophthora* appear to be naturally metalaxyl resistant. Although the pigeon pea (*Cajanus cajan* (L.) Millsp.) stem rot pathogen, *P. drechsleri* f. sp. *cajani,* was controlled in a greenhouse test in Hyderabad, India, by seed treatment with metalaxyl (Kannaiyan and Nene 1984), Chauhan and Singh (1987, 1988) reported that all isolates from Delhi, Kampur, and Varanasi, India, were naturally resistant to metalaxyl (ED_{90} was 199 µg/ml in a growth test compared with 0.9 µg/ml for a sugar beet isolate of *P. drechsleri* from the United States).

The inheritance of resistance to metalaxyl in *P. infestans* was studied by crossing a sensitive isolate with a resistant isolate. The reaction of progeny from single oospores resulting from the cross was consistent with the explanation that resistance is governed by a single nuclear locus exhibiting incomplete dominance (Shattock 1988). In *P. sojae,* resistance is governed by an incompletely dominant gene (Bhat et al. 1993a).

Shattock states that the mechanisms by which metalaxyl resistance arose are not known for certain, but since *Phytophthora* is diploid, there is a strong possibility that mutation of chromosomal genes could give rise to heterozygotes that are intermediately sensitive to metalaxyl. Further mutations or, alternatively, crossing over during the mitotic prophase in the heterozygous diploid nuclei would produce homozygous segregants with resistance to metalaxyl. Field spraying with metalaxyl would thus actually select a metalaxyl-resistant population that would rapidly become dominant because it would have a selective advantage over the sensitive wild types. Since both A1 and A2 mating types are now found in several areas of Europe, Africa, and North America as well as in Mexico, meiotic recombination and segregation could also lead to the production of homozygotes resistant to metalaxyl (Shattock 1988). (See Shattock et al. [1986a, b, 1987] and Chapter 3 for a discussion of genetics of *P. infestans* and other heterothallic *Phytophthora* species.) Since resistance to metalaxyl is considered to be controlled by a single gene, it would be expected that resistant biotypes would emerge rapidly and become dominant within a population of a rapidly sporulating species such as *P. infestans,* which has a short regeneration time. The situation is similar to that of introducing a new cultivar with single-gene-mediated resistance, for example, potato, which has resulted in the development of numerous new pathologic races that have rapidly become dominant in the field population (see Chapter 6).

In Israel, populations of *P. infestans* that were resistant to metalaxyl were actually more fit; that is, their ability to infect, sporulate, and colonize host tissue was greater than that of nonresistant populations (Kadish and Cohen 1989). A model system to compare the effect of mixtures of phenylamide fungicides and other fungicides with different modes of action was developed and validated in the field. Levy et al. (1991) reported that a three-way mixture consisting of oxadixyl, mancozeb, and cymoxanil was essential for preventing a buildup of resistance in *P. infestans.*

An increase in resistance of soilborne isolates of *Phytophthora* after repeated use of metalaxyl as a soil drench does not appear to pose as great a danger as the use of metalaxyl as a foliar spray against *P. infestans.* Nonetheless, the genetic evidence gained from research on *P. infestans* and other isolates strongly suggests that an increase in resistance of a soilborne population of *Phytophthora* could occur. Conjecturally, one might expect that the chances of this happening could be much less because a soil drench treatment with metalaxyl affects a much smaller portion of the total population than foliar spraying. This might result in a larger percentage of wild-type,

Table 7.7. Reports of resistance of *Phytophthora* species to metalaxyl and related phenylamide fungicides

Species	Laboratory (L) or field (F)/nation	Reference
Several species	L, F/several	Staub (1991) (review)
Several species	L, F/several	Schwinn and Morton (1990) (review)
P. cactorum	L/Canada	Utkhede and Gupta (1988)
P. capsici	L/Great Britain	Lucas et al. (1990a, b)
P. cinnamomi	L/United States	Bower and Coffey (1985a)
	L/United States	Coffey et al. (1984a)
P. citricola	L/United States	Bower and Coffey (1985)
	F/United States	Ferrin and Kabashima (1991)
	L/United States	Joseph and Coffey (1984)
	L/United States	Coffey et al. (1984a)
P. drechsleri f. sp. *cajani*	L, F/India	Chauhan and Singh (1987, 1988)
P. fragariae	F/Germany	Seemüller and Sun (1989)
P. infestans	F/Israel	Bashan et al. (1989, 1991)
	L, F/Ireland	Coffey and Young (1984)
	F/Israel	Y. Cohen and Reuveni (1983)
	F/Israel	Y. Cohen and Samoucha (1984)
	F/Israel	Y. Cohen and Samoucha (1990
	F/Northern Ireland	Cooke (1991) (also review)
	F/The Netherlands	Davidse et al. (1981, 1983a, 1991)
	F/Canada	Gillis (1993)
	F/Israel	Kadish and Cohen (1988a–c, 1989, 1992)
	F/Israel	Kadish et al. (1990)
	F/Ireland	O'Sullivan and Dowley (1983)
	F/Greece	Pappas (1985)
	F/Israel	Samoucha et al. (1987)
	F/Israel	Samoucha and Cohen (1988, 1989)
	F/Great Britain	Shattock (1988); Shattock et al. (1990); Shaw and Shattock (1991) (review)
	F/United States	Deahl et al. (1993)
P. medicaginis (*P. megasperma* f. sp. *medicaginis*)	L/Netherlands	Davidse (1981)
	L/United States	Hunger et al. (1982)
	L/United States	Stack and Millar (1985a)
P. palmivora	L/Great Britain	Lucas et al. (1990a, b)
P. parasitica	L, F/United States	Ferrin and Kabashima (1991)
	L/United States	Ferrin and Wadsworth (1992)
	L, F/United States	Ferrin and Rohde (1992b)
P. parasitica var. *nicotianae*	L, F/United States	Shew (1985)
P. sojae (*P. megasperma* f. sp. *glycinea*)	L/United States	Lamboy and Paxton (1992)
Pythium aphanidermatum	L/United States	Sanders et al. (1990)

metalaxyl-sensitive propagules remaining active in the ecological system.

Evidence from a tobacco-*P. parasitica* var. *nicotianae* agricultural system in which metalaxyl was used as a soil-drench treatment indicates that continuous use of metalaxyl in soil enhances the selection of resistance in *P. parasitica* var. *nicotianae* populations, making them less sensitive to metalaxyl than soilborne populations that have not been exposed to metalaxyl (Shew 1985). Shew (1985) proposed consideration of periodic rotation of fungicidal treatment of the tobacco crop.

The antiresistance strategies in current use and the approaches that must be made on an international basis for control of resistant populations of plant pathogens are discussed by Staub (1991) and Schwinn and Morton (1990). Thus far, much progress has been made (Delp 1980, 1984, 1988; Delp and Dekker 1985). After the development of resistance to metalaxyl, it is now accepted that resistance to a fungicide with a specific mode of action must be expected and that measures to cope with resistance must be developed and utilized consistently in the field.

As a measure to alleviate the problem, metalaxyl and other phenylamides are often marketed in combination with a protectant fungicide, such as maneb, which has a different mode of action and a broader fungal spectrum (Tables 7.8–7.10). This has become a standard practice since the discovery of metalaxyl-resistant biotypes within *P. infestans* populations after the widespread use of metalaxyl alone in the Netherlands (Davidse et al. 1983a). Isolates resistant to metalaxyl are also resistant to other phenylamide compounds such as furalaxyl and oxadixyl. The methods used to prevent the buildup of resistant populations are not based on a large experimental database, and more information is needed (Schwinn and Staub 1987, 1995).

The international Fungicidal Resistance Action Committee (FRAC), composed of research workers from academic and

Table 7.8. Advantageous use of mixtures of systemic and protectant fungicides[a]

New fungicide	Mixture partner	Rationale
Hymexazol	Metalaxyl	Enhances activity, broadens the spectrum
Cymoxanil	Protectants, oxadixyl	Improves performance, broadens the spectrum
Fosetyl-Al	Protectants	Stabilizes performance, broadens the spectrum
Phenylamides	Protectants, cymoxanil	Useful as an antiresistance strategy, broadens the spectrum, improves the end-of-season performance

[a] Adapted and reprinted by permission of Gustav Fischer Verlag, Jena, Germany, courtesy Schwinn and Staub (1987).

commercial institutions (Delp 1984), mutually agreed on certain principles based on available information and experience to alleviate the resistance problem. The FRAC recommendations for international use of systemic, site-specific fungicides include use of all cultural methods available to reduce disease pressure, prepackaged mixtures with nonphenylamide protectant fungicides, a partner fungicide at three-fourths to full rate, only two to four sprays with the systemic fungicide per crop season, restriction to 14-day intervals, no curative or eradicative use (i.e., on plants with well-developed infection), and no soil use against foliar pathogens (Staub 1991). Some countries recommend that metalaxyl not be used on fields grown for seed potatoes because resistant inoculum from seed potatoes could be a source of resistant biotypes during the following season. Because an epidemic of late blight can start with a minimum of one blighted plant per square kilometer of field (Van der Zaag 1956), these recommendations for suppression of resistant strains are based on the sound epidemiological principle that *P. infestans* is a multicyclic pathogen with a rapid regeneration time.

Developing a world strategy is important because the phenylamide fungicides are valuable tools for control. If the

Table 7.9. Mixtures of systemic and protectant fungicides in common use on potatoes and tomatoes[a]

Combination	Trade name	Countries	Crop
Phenylamide dual mixtures			
Metalaxyl + maneb or mancozeb	Ridomil MZ	Worldwide	Potato, tomato
	Fubol	UK	Potato
	Acylon P	France	Potato
	Ridomil M	Morocco	Potato
Metalaxyl + Cu-oxychloride	Ridomil plus	Eastern Europe	Potato, tomato
Metalaxyl + chlorothalonil	Ridomil CT	USA	Potato, tomato
Metalaxyl + polycarbazin	Arzerid	USSR	Potato
Oxadixyl + mancozeb	Sandofan M	Worldwide	Potato, tomato
	Recoil	UK, South Africa	Potato, tomato
Oxadixyl + propineb	Fruvit	Argentina, Malaysia	Tomato, potato
	Vitiril	Italy	Tomato
Oxadixyl + copper	Sandofan C	Eastern Europe	Potato
Benalaxyl + mancozeb	Galben M	Worldwide	Potato, tomato
Benalaxyl + copper	Galben copper	Egypt	Potato
Ofurace + mancozeb	Patafol plus	UK, Ireland	Potato
Phenylamide triple mixtures			
Oxadixyl + mancozeb + cymoxanil	Pulsan	France	Potato
	Ripost	Ireland	Potato
	Sandofan M	Switzerland	Potato
Metalaxyl + maneb + fentin acetate	Ridomil delta	Netherlands	Potato
Cymoxanil dual mixtures			
Cymoxanil + mancozeb	Fytospore	UK	Potato
	Ciluan	Germany, Austria	Potato
	Curzate M	Worldwide	Potato, tomato
	Fulvax	France	Potato
	Remiltine	France	Potato
	Galaxy	Switzerland	Potato
	Systol	UK	Potato
Cymoxanil + metiram	Aviso C	Belgium	Potato
	Polycur	Portugal	Potato
Cymoxanil + folpet	Vitipec	Portugal	Potato
Cymoxanil + propineb	Milraz	Portugal	Potato, tomato
	Fitoraz	Peru	Tomato
Cymoxanil + chlorothanil	Daconil combi	Switzerland	Potato
Cymoxanil triple mixtures			
Cymoxanil + folpet + Cu-oxychloride	Anteor C	Morocco	Tomato
Cymoxanil + mancozeb + fentin acetate	Curam	Netherlands	Potato

[a] Reprinted by permission of Academic Press, New York, courtesy Schwinn and Margot (1991).

Table 7.10. Fungicidal activities (EC_{90} mg l^{-1}) of oxadixyl (A), mancozeb (B), and cymoxanil (C), alone and in mixtures, and synergy ratio of mixtures against sensitive and phenylamide-resistant strains of *Phytophthora infestans* on tomato[a]

Fungicides and ratios of components	EC_{90} (mg l^{-1}) Sensitive	EC_{90} (mg l^{-1}) Resistant	Synergy ratio Sensitive	Synergy ratio Resistant
A	34	>5,000	...	...
B	776	495	...	...
C	48	23	...	...
A:B = 1:7	69 (9 + 60)[b]	406 (51 + 355)	3.0	1.4
B:C = 7:0.4	134 (127 + 7)	51 (48 + 3)	3.2	4.6
A:C = 1:0.4	22 (16 + 6)	41 (29 + 12)	1.7	2.0
A:B:C = 1:7:0.4	74 (9 + 62 + 4)	63 (8 + 53 + 3)	2.4	4.2

[a] Reprinted by permission of Butterworth-Heinemann Ltd., Salem, Massachusetts, courtesy of Grabski and Gisi (1987) and Cambridge University Press, Cambridge, courtesy Gisi (1991).

[b] Numbers in parentheses are concentrations of the individual components in the mixture.

total *Phytophthora* population becomes resistant, the phenylamide fungicides could be lost for use. Examples of mixtures of systemic and protectant fungicides that are in use are given by Schwinn and Margot (1991) (Tables 7.8 and 7.9).

After the introduction of metalaxyl in 1979 for commercial use in Ireland, where late blight epidemics are frequent because the temperatures are cool and rainfall is frequent, resistance to metalaxyl became so dominant in the population of *P. infestans* that metalaxyl had to be withdrawn from the market by 1981; however, by 1985 the level of resistant isolates had decreased to the extent that the Irish government allowed reintroduction of metalaxyl, but it can be used only when combined with a protectant fungicide and only twice per season. This approach has kept the resistance to metalaxyl in a manageable state (Dowley and O'Sullivan 1991; Staub 1991).

Metalaxyl was introduced into the Netherlands in 1979. Its use was suspended from 1981 to 1984 because of a buildup of resistant strains of *P. infestans,* but it was reintroduced in 1985 for restricted use with a protectant fungicide and only twice per season. Schwinn and Staub (1995) discuss the use of prepackaged mixtures of metalaxyl and protectant fungicides.

Synergism Among Different Fungicides

Fungicides with different modes of action are often used together to give a broader spectrum of target pathogens (Tables 7.8 and 7.9), for example, a protectant fungicide with metalaxyl, as discussed above. The rationale for combining fungicides is to increase their efficiency. If two pesticides used together induce a greater degree of control than could be explained as additive, they are said to be synergistic. Gisi et al. (1985) demonstrated that several systemic and protectant fungicides are synergistic.

Gisi (1991) reviewed and discussed synergism between and among different protective and systemic fungicides on the basis of theoretical and observed criteria as a means of enhancing the activity of different fungicides and reducing the impact and development of fungicide-resistant strains (Table 7.10). As a method of quantifying synergism, he explained the Abbott formula (Levy et al. 1986), which calculates the percentage of control (% C expected) as

$$\%\ C\ exp = A + B - (AB/100),$$

in which A and B are the control levels given by single fungicides. Synergism (SF, synergy factor) is the ratio between the observed experimental efficacy of the mixture (*Cobs*) and the expected efficacy of the mixture:

$$\mathrm{SF} = Cobs/C\ exp.$$

If SF has a value higher than 1.0, the mixture is synergistic. This method is not accurate when levels of control by single components are higher than 70%, however, because the SF values decrease toward the value of 1.0 (Levy et al. 1986).

An alternative method that is more exact but more labor intensive is the formula of Wadley (Gisi et al. 1985; Levy et al. 1986) with which dosage-response curves of A and B alone and $A + B$ are constructed (Figure 7.3). With a logit-log transformation, the dosage response curves are converted to linear regressions that can be used for calculating equally effective concentrations (EC values) of different levels of control, e.g., EC_{50} (EC that is 50% inhibitory) or EC_{90}.

We quote (by permission of Cambridge University Press, Cambridge, U. K., courtesy of Gisi 1991):

> When a and b are the absolute amounts of the components in the mixture, the expected effective concentration (EC exp) providing a given control level can be calculated. For example, for 90% control:
>
> $$EC_{90exp} = (a + b)/(a/EC_{90A} + b/EC_{90B}).$$

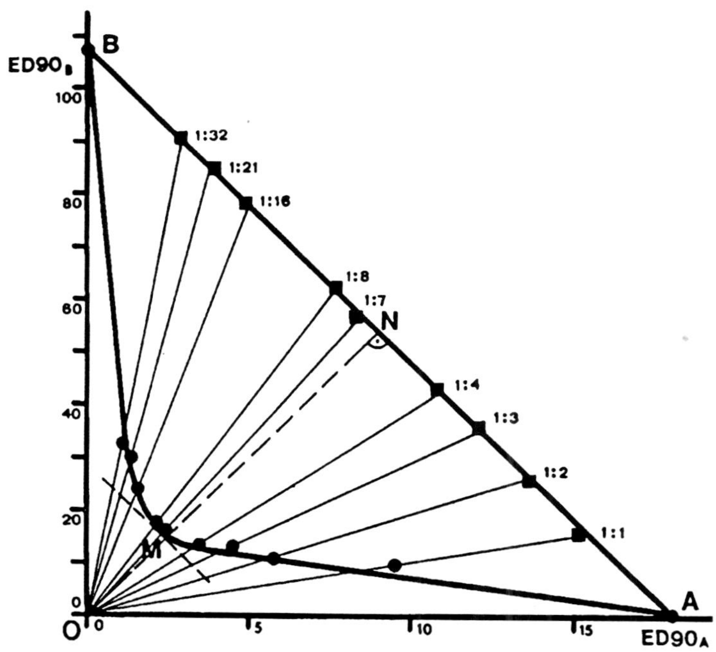

Figure 7.3. Isobologram for the two fungicides, oxadixyl (A) and mancozeb (B), active against *Phytophthora infestans* on potato plants, giving a theoretical (ANB) and an experimental (AMB) isobole for A + B mixtures with ratios between 1 + 1 and 1 + 32. The level of synergy can be measured by the ratio R = ON/OM, which is maximal for the mixture A + B = 1 + 7 (Levy et al. 1986). (Reproduced by permission of the European and Mediterranean Plant Protection Organization, Paris, France, courtesy Levy et al. 1986 and the Cambridge University Press, Cambridge, U. K., courtesy Gisi 1991)

By dividing the expected and the observed experimental EC_{90}, the interactions between the two fungicides A and B in the mixture can be calculated; if this value is 1, interactions are additive, if it is higher than 1, interactions are synergistic. Unlike the Abbott method, the Wadley approach allows estimation of the interactions at any fungicide concentration. In addition, the Wadley approach can be used to define the optimum ratio of the components in the mixture to achieve the highest control levels with the minimum amounts of the components [Figure 7.3] (Gisi 1991, Fig. 24.1); this ratio represents maximum synergy levels (Levy et al. 1986).

In Figure 7.3, which was constructed from greenhouse tests (Levy et al. 1986), the maximum (M) efficacy was with a mixture of 1 part of oxadixyl (a phenylamide fungicide) and 7 parts of mancozeb.

A mixture of three fungicides, oxadixyl (systemic), mancozeb, (protectant), and cymoxanil (systemic), was tested against late blight on tomato plants, and maximum levels of synergism were calculated for the response of each chemical and combination of chemicals. This experiment tested metalaxyl-resistant strains of *P. infestans* and phenylamide-sensitive strains (Table 7.10). Oxadixyl mixed with mancozeb and/or cymoxanil showed the highest EC_{90} ratio (synergism) in the in vitro tests. When tested against *P. infestans* on tomato in a glasshouse, the highest synergism was found with oxadixyl plus mancozeb, cymoxanil, or fosetyl-Al. Oxadixyl with fentin acetate, folpet, or thiram also gave a synergistic reaction (Gisi et al. 1985). Mancozeb and cymoxanil (4:1 ratio; Mancur 71.4) and mancozeb, cymoxanil, and oxadixyl (7:2:1 ratio; Sandocur-M69) controlled late blight of potato in Israel (Y. Cohen 1986).

Samoucha and Cohen (1988, 1989) showed that late blight, caused by both metalaxyl-resistant and metalaxyl-sensitive isolates, was better controlled in growth chambers by Sandocur-M69, Mancur 71.4, and Pulsan 67 (oxadixyl, cymoxanil, and mancozeb at a 1.0:0.4:7.0 ratio) than by each of the individual ingredients. They concluded that cymoxanil mixtures were suitable for control of resistant populations of *P. infestans.* Similar results were obtained for control of late blight of tomato with a mixture of oxadixyl, mancozeb, and cymoxanil (Samoucha et al. 1987).

When synergy ratios were calculated for single fungicides and combinations of three fungicides, phenylamide-resistant strains were better controlled with some combinations than with others. Oxadixyl-mancozeb-cymoxanil (1.0:7.0:0.4) induced the highest synergy ratio on both phenylamide-resistant strains and phenylamide-sensitive strains of *P. infestans* (Table 7.10).

Combinations of oxadixyl (phenylamide systemic) and mancozeb (protectant), which are most commonly used, were much less effective for control of phenylamide-resistant strains than the three-way mixture of oxadixyl, mancozeb, and cymoxanil. Gisi (1991) presents more data to support the concept that mixtures of a phenylamide fungicide and a protectant fungicide may be insufficient to delay the buildup of resistant strains. As discussed above, cymoxanil penetrates leaves rapidly and has a mode of action different from that of phenylamide fungicides such as metalaxyl or oxadixyl. The nature of synergism is not yet well understood, but the experimental and practical data presented by Gisi and others indicate that mixtures of fungicides are beneficial. The calculations and method of determining optimal ratios of concentrations quantify the efficacy of fungicides used in combination and support this concept. Gisi (1991) cites evidence for several possible mechanisms that might explain synergism (excerpted by permission of Cambridge University Press, Cambridge, courtesy of Gisi 1991):

increased uptake and binding of fungicides to the site of action may lead to higher concentrations in the target cell;
decreased biodegradation of fungicides in fungal and plant cells may prolong activity;
different sites of action of fungicides both in the fungal cell (biochemical mechanism) and in the life cycle of the fungus (pathogenic mechanism) may decrease the aggressiveness of the fungus.

Phosphonate Fungicides

The phosphonate fungicides include fosetyl-Al and its breakdown product, phosphorous acid. The development of fosetyl-Al is a major landmark in the fungicidal control of *Phytophthora*-caused diseases, despite its relatively poor activity against late blight of potato, caused by *P. infestans,* and root rot of soybean, caused by *P. sojae.* Fosetyl-Al and phosphorous acid, which is often referred to as a phosphite but is more correctly termed a phosphonate (Coffey and Ouimette 1989), are remarkably effective for control of *Phytophthora*-caused root diseases of pineapple (Rohrbach and Schenck 1985), avocado (Coffey et al. 1984b; Darvas et al. 1984; Pegg et al. 1985; Coffey 1987; De Boer et al. 1990), citrus (Laville 1979; De Boer et al. 1990; Le Roux et al. 1991), and walnut (Matheron and Mircetich 1985). See Table 7.10 for other citations on the efficacy of fosetyl-Al and phosphorous acid for control.

Fosetyl-Al (see Table 7.1 for the chemical structure; also cited in some reports as phosethyl-Al) is sold as Alliette. Fosetyl-Al has a broad spectrum of activity, ranging from control of downy mildew on grapes to root rot of avocado (Coffey et al. 1984) and in some cases to nonoomycete fungi (Table 7.2). Although highly active on some *Phytophthora* species, the activity of fosetyl-Al is unexplainably not consistent among *Phytophthora* species. For instance, it is not understood why fosetyl-Al has such weak activity on the foliage phase of late blight of potato, caused by *P. infestans,* but has activity for control of tuber rot (Bertrand et al. 1977). The fungicide can be applied by soil drench or trunk injections, by foliar spray to avocados for control of root rot caused by *P. cinnamomi,* or by a dip application on pineapple tops before planting to control *P. cinnamomi* and *P. parasitica* on roots (Rohrbach and Schenck 1985). Postharvest control of brown rot of citrus fruit (*P. citrophthora* and *P. parasitica*) was reported by Gutter (1983). See Table 7.1 for other references on control.

Fosetyl-Al rapidly converts to phosphorous acid or the phosphonate salt and other minor components (Fenn and Coffey 1984; Saindrenan et al. 1985; reviewed by Cohen and Coffey [1986]; Coffey and Ouimette 1989; Guest and Grant 1991; Griffiths et al. 1992). The in vitro toxicity of fosetyl-Al against different *Phytophthora* species on conventional media is relatively low (1,000 µg/ml or greater) (Clerjeau and Beyries 1977; Bompeix et al. 1980; Farih et al. 1981a); however, fungitoxicity to *P. cinnamomi* on a synthetic medium (Ribeiro et al. 1975a) on which the phosphate level is low (0.084 mM) occurs at concentrations as low as 10 µg/ml for phosphorous acid and 50 µg/ml for fosetyl-Al. The in vitro activity of fosetyl-Al was greatly increased if the phosphate concentration of the medium was minimal, but activity of phosphorous acid was not affected by the phosphate concentration in the medium (Fenn and Coffey 1984).

Differences in sensitivity of a resistant isolate (P113) and a sensitive isolate (P376) of *P. palmivora* (Dolan and Coffey 1988) were associated with the ability to take up the phosphonate anion. "The percentage of inhibition on a synthetic medium was proportional to the internal concentration of phosphonate, but the slope of the plot of inhibition against internal phosphonate differed between strains" (Griffith et al. 1993).

See Table 7.11 for numerous examples of control by fosetyl-Al. In plants, phosphonate was as effective as fosetyl-Al for control of root rot of *Persea indica* caused by *P. cinnamomi* (Fenn and Coffey 1984). Both fosetyl-Al and phosphonate translocated downward through the phloem tissue of plants as well as upward in the xylem. This is a unique feature that has not been shown by any other fungicide. Both compounds are water soluble (fosetyl-Al, 12%; phosphonate, >50%) and can be sprayed on foliage or mixed into soil. Translocation patterns in plants are similar to those of radioactive ^{14}C-sucrose. Uptake of fosetyl-Al by phloem tissue was inhibited by 2,4-dinitrophenol, an indication that uptake is an active process (Ouimette and Coffey 1989b, 1990). See Cohen and Coffey (1986) and Schwinn and Staub (1987, 1995) for additional information.

The discovery that the phosphonate anion was the active ingredient of fosetyl-Al (Bompeix and Saindrenan 1984; Bower and Coffey 1985; Fenn and Coffey 1984, 1985; Coffey and Joseph 1985) stimulated research on phosphonate as a fungistat and also as a means of controlling *Phytophthora*-caused diseases. The patent on phosphonate held by Rhone-Poulenc in Australia includes its use only with a detergent and not with water. As a result, Australia is the only nation in which phosphonate in water can be legally used as a fungicide. Guest and Grant (1991) relate that several studies show that trunk injections of avocado trees with phosphonate have successfully controlled root rot caused by *P. cinnamomi.* They cite the following reports on control of *Phytophthora*-caused diseases by phosphonate in Australia: Taylor and Washington (1984); Pegg et al. (1985); Piccone et al. (1987); and de Boer et al. (1990).

The efficiency of application of fosetyl-Al and phosphorous acid on 40-year-old lemon trees, applied to trunks as a paint (300 g/liter) or by trunk injection (1.5 g of active ingredient per square meter of canopy area), was determined by gas-liquid chromatography analysis of roots and leaves for phosphite (phosphonate). The distribution of phosphonate was more rapid after foliar application than after trunk injection, but the phosphonate levels in roots declined more rapidly with foliar spray than with trunk injection. Schutte et al. (1991) concluded that the treatment interval should not exceed 42 days to maintain an effective concentration.

Fosetyl-Al (2 g per square meter of leaf canopy) used as a trunk injection on 4-year-old navel orange trees on rough lemon rootstock every 3 months increased the yield 10 and 40 months after treatment. Trunk injection with fosetyl-Al did not affect the population of *P. nicotianae* in the soil, whereas metalaxyl applied as 5% granules reduced the population (Le Roux et al. 1991).

Fosetyl-Al is efficacious for control of diseases after *Phytophthora* has become established (Griffith et al. 1992). El-Hamalawi and Menge (1994a) found that trunk canker of avocado caused by *P. citricola* occurs mainly after the trunk or crown of a tree is wounded by sucker shoot removal or mechanical means. Treatment of wounds with a fosetyl-Al solution (0.5 g/ml) or as a 1:1 mixture with the commercial product Tree Seal before or 18 h after inoculation with *P. citricola* completely prevented canker formation. Development of *P. citricola* in plant tissue was also prevented, as measured by isolation on a selective medium. Metalaxyl or Bordeaux mixture were ineffective when applied after inoculation (El-Hamalawi and Menge 1994b).

Fosetyl-Al has a relatively low activity against growth of some *Phytophthora* species in vitro. In a defined medium containing a low concentration of phosphate (0.084 mM), the EC_{50} varied from 5.2 μg/ml for *P. citrophthora* to 91 μg/ml for *P. medicaginis* (*P. megasperma* f. sp. *medicaginis*) from alfalfa (Fenn and Coffey 1984). The EC_{50} for *P. infestans* on rye seed medium was 224 μg/ml (Coffey and Ouimette 1989). The EC_{50} at the relatively high dosage of 250–1,000 μg/ml for some species suggested that fosetyl-Al acted by indirectly affecting resistance mechanisms in the plant rather than by direct toxicity to *Phytophthora* (Bompeix et al. 1981; Durand and Safle 1981; Guest 1984b). See Y. Cohen and Coffey (1986), who support the direct-mode-of-action hypothesis, and Guest and Grant (1991) and Griffiths et al. (1992), who support the hypothesis that both direct fungistatic effects and host resistance are involved.

Many workers have concluded that phosphonate from the breakdown of fosetyl-Al inhibits *Phytophthora* directly. Bower and Coffey (1985) isolated phosphonate-resistant strains of *P. capsici* by mutagenic techniques in the laboratory. The EC_{50} value for a resistant mutant was 415 μg/ml compared with 21–37 μg/ml for wild-type isolates. The resistant mutant retained its pathogenicity to pepper plants that were either treated or not treated with phosphonate or fosetyl-Al (Bower and Coffey 1985; Cohen and Coffey 1986; Fenn and Coffey 1985). The induction of resistance supported the contention that fosetyl-Al exerts a direct fungitoxic effect on *Phytophthora.* To date, however, resistance to fosetyl-Al or phosphonate has been reported in the field only once (*P. cinnamomi*) (Vegh et al. 1985). Coffey and Ouimette (1989) comment that this finding would not be unexpected in light of laboratory results. Further information on experimental induction of resistance in *P. capsici* and *P. palmivora* was reported by Lucas et al. (1990a, b) and in *P. infestans* by Y. Cohen and Samoucha (1984).

Griffiths et al. (1992) state that the action of phosphonates is unlikely to be the result mainly of the effects on the plant because of evidence that *Phytophthora* mutants with resistance to phosphonates were developed in the laboratory (Bower and Coffey 1985; Dolen and Coffey 1988; Neema et al. 1988) and that these *Phytophthora* mutants were also less sensitive to phosphonate in plants. The complexity of the mode of action of phosphonates indicates, however, that phosphonate also has an effect on the plant (Griffiths et al. 1992). Griffiths et al. (1992) present a well-documented, more detailed discussion of the biochemical impact of phosphonate compounds on both oomycetous pathogens and plants. See Table 7.12 for a listing of reports that support either the direct or indirect mode of action of phosphonates.

Fosetyl-Al has filled a practical need for efficient control of chronic disease problems such as avocado root rot caused by *P. cinnamomi* (Darvas et al. 1984), but its development and use also has introduced some interesting new research concepts, which are discussed by Coffey and Ouimette (1989) (by permission of Cambridge University Press, Cambridge, U.K., courtesy of Coffey and Ouimette 1989):

> The active ingredient of fosetyl aluminum, a unique phloem mobile xenobiotic possessing remarkable antifungal activity against some important Oomycete plant pathogens, is phosphonate. This simple anion is believed to be an important component of the phosphorus cycle in soils (Alexander 1977). Many microorganisms have retained the enzyme systems necessary to oxidise it to phosphate. What is different about the biology or metabolism of some Oomycetes that causes phosphonate to be inhibitory to their growth? A simple inability by Oomycetes to utilize phosphonate as a phosphorus source (Fothergill & Child 1964) would appear unlikely to provide an explanation of this selective inhibition. Neither higher plants, nor mammals (Robertson & Boyer 1956) can utilize phos-

Table 7.11. Control of *Phytophthora* species on different crops with phosphonate compounds

***Phytophthora* species** / **Disease**	**Crop**	**Application method**	**References**
Many (review)			Coffey and Ouimette (1989); Y. Cohen and Coffey (1986); Guest and Grant (1991)
P. cactorum			
Trunk rot	Peach	Trunk injection	De Boer et al. (1990)
Crown rot	Peach	Foliar and trunk paint	Taylor and Washington (1984)
Trunk canker	Cherry	Foliar spray	Bielenin and Jones (1988b)
Trunk canker	Apple	Root soak	Jeffers (1992)
Collar rot	Apple	Trunk injection, foliar spray, paste on cankers	Harris (1987)
Collar and fruit rot	Apple	Trunk, soil, fruit	Harris (1991) (review)
Trunk rot	Apple	Foliar, soil drench, trunk paint	Orlikowski et al. (1986)
		Soil mix	Rana and Gupta (1984)
Crown rot	Apple	Soil drench, granular	Utkhede (1987a, b)
Crown rot	Apple	Soil drench, granular	Utkhede and Smith (1991a)
P. cambivora			
Trunk canker	Almond and cherry	Soil drench, foliar spray, trunk injection	Wicks and Hall (1988, 1990)
Trunk canker	Cherry	Foliar spray	Bielenin and Jones (1988b)
Ink disease	Chestnut	Foliar spray	Skoudridakis and Bourbos (1990)
P. capsici			
Stem canker	Pepper	Root drench	Ouimette and Coffey (1989a)
Leaf spot	Tomato	Floated detached leaves	Fenn and Coffey (1985)
P. cinnamomi			
Root rot	Azalea	Foliar spray	Benson (1990)
Root rot	Avocado	Root drench, foliar spray	Coffey (1987)
Root rot	Avocado	Trunk injection, preplant soil drench, root drench, foliar spray	Coffey et al. (1984b)
Root rot	Avocado	Root drench	De Boer et al. (1990)
Root rot	Avocado	Trunk injection	Darvas et al. (1983, 1984)
Root rot	*Persea indica*	Foliar spray, soil drench	Fenn and Coffey (1984)
Stem rot	*Rhododendron, Leucadendron, Eucalyptus*	Soil drench, foliar spray	Marks and Smith (1990)
	Avocado		
Root rot	Avocado	Soil drench, foliar spray	Ouimette and Coffey (1989b)
Root rot	Avocado	Soil drench, trunk injection	K. G. Pegg et al. (1985)
	Avocado and pineapple	Trunk injection	K. G. Pegg and Whiley (1987)
	Chamaecyparis lawsoniani cv. Elwoodii	Trunk injection, foliar spray	K. G. Pegg et al. (1990)
Root rot		Soil drench	P. M. Smith (1980)
	Pineapple		
Root rot	Avocado	Plant dip	Rohrbach and Schenk (1985)
Root rot	Walnut	Foliar spray	Whiley et al. (1986)
Root rot, crown canker		Root drench, stem application	Matheron and Mircetich (1989e)
P. citricola			
Trunk canker, root rot	*Persea indica*	Root drench	Ouimette and Coffey (1989a)
Root rot, crown canker	Walnut	Root drench, stem application	Matheron and Mircetich (1989e)
Canker	Avocado	Stem application	El-Hamalawi and Menge (1994b)
P. citrophthora			
Trunk canker	Citrus (Mandarin orange)	Foliar spray	De Boer et al. (1990)
Root rot, gummosis	Citrus	Soil drench, trunk paint	Farih et al. (1981a, b)
Root rot	Citrus (orange and lime)	Foliar spray	Feichtenberger (1990)
Brown rot	Citrus (orange)	Fruit dip	Gauliard and Pelossier (1983)
Root rot	Citrus (orange)	Foliar spray	Magnano di San Lio et al. (1986b)
Trunk canker	Citrus (tangelo)	Trunk paint	Matheron and Matejka (1988b)
Root rot	Citrus (rough lemon)	Soil drench	Matheron and Matejka (1991)
Crown and root rot	Citrus	Soil drench	Ohr et al. (1986)
Brown rot	Citrus (orange)	Fruit dip	Tuset and Portilla (1990)
P. clandestina			
Root rot	Subterranean clover	Foliar spray	De Boer et al. (1990)
P. colocasiae			
Leaf blight	Taro (*Colocasia esculenta*)	Foliar spray	Aggarwal and Mehrotra (1987)
P. cryptogea			
Crown rot	Cherry	Foliar spray	Bielenin and Jones (1988b)
Leaf lesion	Cowpea	Floated detached leaves	Saindrenan et al. (1988a, b)
Root rot	Raspberry	Soil drench	Washington (1988)
P. fragariae			
Red stele	Strawberry	In hydroponic solution	Jamart et al. (1985)
Red stele	Strawberry	Foliar spray	O'Neill and Griffin (1987)
P. heveae			
Dieback	Rhododendron	Foliar and soil drench	Benson (1980)
Meristem decay, nutfall	Coconut palm	Trunk injection	de Franqueville and Renard (1989)

(*continued on next page*)

Table 7.11. *(continued)*

Phytophthora species Disease	Crop	Application method	References
P. infestans			
Late blight	Potato	Foliar treatment of leaf disks	Bashan et al. (1990)
Late blight	Potato	Foliar	Schwinn and Margot (1991)
P. meadii			
Koleroga, fruit rot	Arecanut	Fruit and foliar spray	Sastry and Hegde (1988)
P. megasperma			
Crown rot	Cherry	Foliar spray	Bielenin and Jones (1988b)
P. nicotianae			
Root rot	Tomato	Foliar spray	De Boer et al. (1990)
Root rot	Citrus (Mandarin orange)	Foliar spray	De Boer et al. (1993)
Root rot	Tomato	Foliar spray	Flett et al. (1991)
Root rot	Citrus (navel orange on rough lemon rootstock)	Trunk injection	Le Roux et al. (1991)
Root rot	Citrus (sour orange)	Foliar spray and drench	Magnano di San Lio et al. (1986b)
Root rot	Citrus (Mandarin orange)	Foliar spray	K. G. Pegg et al. (1990)
Root rot	Tomato	Foliar spray	K. G. Pegg et al. (1990)
Root rot	Citrus	Foliar spray, trunk injection, trunk paint	Schutte et al. (1991)
Root rot	Citrus (Cleopatra mandarin)	Soil drench, foliar spray	Walker (1988)
P. palmivora			
Black pod	Cocoa	Trunk injection gave good control; foliar spray did not	Holderness (1992)
Black stripe (tapping panel disease)	Rubber	Foliar spray	A. M. Tan (1990)
Black pod, canker, seedling blight	Cocoa	Trunk injection (highly effective); foliar spray (ineffective)	R. D. Anderson et al. (1989); R. D. Anderson and Guest (1990)
P. palmivora MF4 (probably *P. capsici*)			
Foot rot	Black pepper	Foliar spray effective but less so than by metalaxyl	Ramachandran and Sarma (1985)
P. parasitica			
Root rot	Tomato	Foliar spray	R. M. Davis (1989)
Trunk canker, root rot	Citrus (orange and lemon)	Soil drench, stem paint	Farih et al. (1981a, b)
Root rot	Citrus (orange and lemon)	Soil drench	Feichtenberger (1990)
Root rot	*Catharanthus roseus*	Foliar	Ferrin and Rohde (1992a, b)
Brown rot	Citrus	Fruit dip	Gauliard and Pelossier (1983)
Trunk canker	Citrus (tangelo)	Trunk paint	Matheron and Matejka (1988b)
Crown and root rot	Citrus	Soil broadcast	Ohr et al. (1986)
Root and heart rot	Pineapple	Plant dip	Rohrbach and Schenck (1985)
Root rot	Citrus	Soil, foliar and trunk paint	Timmer and Castle (1985)
P. syringae			
Collar and fruit rot	Apple	Soil, trunk, fruit	Harris (1991) (review)
Pruning wound canker	Almond	Wound paint	Doster and Bostock (1988c)

phonate, and yet it exhibits no toxic effects. We believe that the answer to the ability of phosphonate to inhibit selectively certain Oomycetes, may lie in unique features of their phosphorus metabolism. Wang and Bartnicki-Garcia (1973) have demonstrated that novel phosphoglucans are abundant in the storage vacuoles of some *Phytophthora* species at certain stages in their life cycle, notable during sporangium development. Polyphosphates are common in filamentous fungi and yeasts, but have not been detected in Oomycetes (Dietrich 1976). Could it be that Oomycetes store their phosphate in a unique fashion, quite unlike that of other fungi, or higher organisms? Are specific metabolic steps in this storage process highly sensitive to phosphonate? Certainly, the ability of extremely low concentrations of phosphonate (1 to 2 μg ml^{-1}) to inhibit zoosporangium production by *Phytophthora* (Coffey & Joseph 1985; Dolan & Coffey 1988) may indicate selective interference with key biosynthetic events. It is our belief that phosphonate will prove not only to be a valuable tool in the control of important plant diseases, but may also be used experimentally to unmask new and important properties concerning both phloem translocation in plants and unique features of phosphorus metabolism in Oomycetes.

The mode of action of fosetyl-Al and phosphonate is still controversial. Authors cited above who discuss direct fungitoxicity to target *Phytophthora* maintain that the action is directly on *Phytophthora,* because isolates of *Phytophthora* can be induced to become resistant by use of mutagenic compounds (Coffey and Ouimette 1989). Guest and Grant (1991) agree with the direct-action hypothesis but cite evidence that phosphonate increases host resistance as well. The summary and conclusions from their review are as follows (excerpted by permission of Cambridge University Press, Cambridge, U.K., courtesy of Guest and Grant 1991):

1. Although phosphonates are not always more active against oomycetes *in vitro,* they are much more selective for this group *in vivo.*
2. Phosphonate activity *in vitro* and therefore presumably *in vivo* is very sensitive to phosphate concentration in the environment surrounding the pathogen.
3. Phosphonates cause disturbances at several metabolic sites in the mycelial phase of the life-cycle, and inhibit sporulation at low concentrations (<0.1 mM), without affecting mycelial growth. Even at high concentrations (1–10 mM), phosphonates are fungistatic rather than fungitoxic.
4. Diseases in plants are sometimes controlled by lower application rates of phosphonates than would be predicted from the ED_{50} values determined in phosphate-limiting media.
5. Phosphonate does not always control the same pathogen in different hosts, or in different varieties of the some host. In general, phosphonates are more effective in species or culti-

vars with some level of resistance to the pathogen than in those with none.

6. Phosphonates stimulate the defence responses of the host following infection. Inhibition of these responses reduces the effectiveness of phosphonates.
7. Although low levels of resistance to phosphonates *in vitro* have been obtained using mutagenic treatments, the biochemical basis of this resistance has not been described, nor has the resistance in the field yet been observed. The probability of high levels of resistance emerging in the field is reduced by the multi-site action of phosphonates on the metabolism of target organisms.

Conclusions. We suggest that all of the available evidence relating to the mode of action of phosphonates may be accommodated by the following hypothesis: Phosphonates stop pathogens growing in plants utilizing a complex mode of action, which works in the following ways:

1. They slow the growth of the pathogen, and inhibit its sporulation. This is accomplished by direct fungistatic action, the extent of which is dependent only upon the concentration of phosphonate which accumulates in the fungus. This in turn will be influenced by the concentration of phosphate present and the effectiveness of any phosphonate oxidation system.
2. The reduction in pathogen growth rate allows the host defence system extra time to develop and kill the invading organism.
3. Phosphonates alter the metabolism of the pathogen in such a way that a more rapid and more effective defence response can be mounted by the host. These alterations involve a reduction in the amount of suppressor molecules present at the pathogen surface, or an increase in the amount of elicitor molecules exposed to the receptors on the host cells, or both. These changes are not inhibitory to the fungus in axenic culture, but lethal in the presence of an active host defence system.

The relative importance of each of these components varies with each host-parasite interaction. This explains why conclusions about the mode of action have sometimes differed. This complex mode of action accounts for the variation in results using different plants and pathogen strains or species. It also suggests why resistance to phosphonates has not been found in practice. Resistance to fungicides is associated with a mutation at a single site in the pathogen, because fungicides usually inhibit a single enzyme or process in the pathogen. It appears likely that phosphonate may have multiple biochemical sites of action.

Table 7.12. Reports on the modes of action of phosphonates

Evidence for activity[a]	Reference
DP	Afek and Szteinberg (1986, 1989)
DP	Barchietto et al. (1988)
DP	Barchietto et al. (1989)
D	Coffey and Ouimette (1989) (review)
D	Y. Cohen and Coffey (1986) (review)
DP	Dunstan et al. (1990)
D	Fenn and Coffey (1984, 1985)
DP	Griffith et al. (1992)
DP	Guest (1984a, b)
DP	Guest et al. (1989)
DP	Guest and Grant (1991) (review)
DP	Niere et al. (1990)
D	Ouimette and Coffey (1989a, b)
DP	Saindrenan and Bompeix (1985)
DP	Saindrenan et al. (1988a, b)
DP	Saindrenan et al. (1990)
DP	Schwinn and Staub (1987) (review)
DP	Smillie et al. (1989, 1990)
DP	Soulie et al. (1989)
DP	Nemestothy and Guest (1990)

[a] D = control is claimed to result from the direct effect of the phosphonate ion per se; P = the phosphonate ion also induces a physiologic response in the host plant that increases its resistance.

Dimethomorph

Dimethomorph, produced by Shell Research, Ltd., in the United Kingdom, is a new curative fungicide that translocates upward (Albert et al. 1988). Its effectiveness against *Phytophthora* species but not *Pythium* species is a unique feature. It is active at low concentrations (0.25–0.75 µg/ml) (Kuhn et al. 1989a, b). Its activity is compared with that of the phenylamides, local systemics, and protectant fungicides (Figure 7.1) (Schwinn and Staub 1995). The mode of action may be novel in that a biochemical target has not yet been found. Dimethomorph is inhibitory to germination of sporangia and zoospores and in some cases can be lethal. In comparison, the phenylamide compounds such as metalaxyl are fungistats and do not inhibit germination. An ultrastructural study indicates that dimethomorph causes extensive proliferation and aberrant deposition of cell wall material in *Phytophthora* hyphae. False septa occur in hyphae, thus leading to a beaded appearance. Dimethomorph appears to act on the biogenesis of the *Phytophthora* cell wall (Kuhn et al. 1991).

Dimethomorph at soil-drench rates of 0.6 and 1.2 mg/liter inhibited stem lesion infections of *Rhododendron ponticum, Eucalyptus sieberi,* and *Leucodendron salignum* × *L. leureolum* caused by *P. cinnamomi* in glasshouse studies. The activity was about as effective as that of fosetyl-Al but somewhat less effective than that of metalaxyl. Dimethomorph was ineffective as a foliar spray (Marks and Smith 1990).

Y. Cohen et al. (1995) showed that dimethomorph was highly effective against *P. infestans* on potato when applied by spraying at dosages above 62 µg/ml. ED_{90} values varied among different isolates from 62 to 193 µg of active ingredient per milliliter. Dimethomorph had translaminar (across the leaf) activity, local systemic translocation, and strong resistance to washing. Although dimethomorph translocated in the xylem tissue, it did not translocate in phloem tissue. Dimethomorph was active against metalaxyl-resistant isolates of *P. infestans.* The activity of dimethomorph was mainly associated with its suppression of the cell wall-formation process of zoospores.

Zarilamide

A new systemic benzamide fungicide, zarilamide, is active against oomycetous fungi and is equally effective on metalaxyl-resistant and metalaxyl-sensitive strains of *P. infestans* (Heaney et al. 1988a, b). The chemical formula is 4-chloro-*N*-[cyano(ethoxy)-methyl]benzamide. The mode of action involves destruction of the microtubule cytoskeleton and inhibition of mitosis in *P. capsici* (Young 1991).

Disease Forecasting

Phytophthora species have a rapid regeneration time and consequently are nearly all multicyclic. The inoculum increases and reproduces several times within the crop year. The single and multicyclic disease concepts, first established by Vanderplank (1963), an astute observer with a unique ability to distill knowledge to its essential principles, have strongly influenced epidemiological thought in the control of plant disease. With a few exceptions, most diseases caused by *Phytophthora* are multicyclic because of the rapid increase in inoculum in a short time during the crop year (MacKenzie et al. 1983). Basically, the rapid production of zoospore inoculum from sporangia that emanate from colonized plant tissue, oospores, or chlamydospores is the unique feature that all species of *Phytophthora* as well as other oomycetous fungi have in common

(Griffith et al. 1992). Fry and Doster (1991) and several other authors cited below summarize the progress in forecasting epidemics of late blight on the basis of epidemiological principles and present quantitative information on application of fungicides based on forecasting procedures.

The late blight disease of potato, caused by *P. infestans,* is characterized by production of many cycles of inoculum within the crop year and is a classic example of a multiple-cycle disease. The primary inoculum occurs mainly on an occasional infected tuber. As few as one blighted plant from an infected tuber per square kilometer of land in the Netherlands is sufficient to initiate an epidemic (Van der Zaag 1956; Vanderplank 1963; Boyd 1980). When the infected potato sprout emerges, the fungus in the infected seed piece infects the stem and causes a lesion on which sporangia form if weather conditions are favorable. If there is rainfall and the temperature is optimal (Crosier 1934), caducous (deciduous) sporangia and the zoospores that emanate from them are splashed from this lesion to the foliage and subsequently to nearby plants. The pathogen spreads from this "focus" of infection by repeating this cycle many times throughout the field of potatoes.

The film by James and Fry (1982), obtainable from Cornell University, the Department of Plant Pathology, Ithaca, New York, shows by time-lapse photography how *P. infestans* spreads from a primary focus of infection to cause widespread devastation of potato plants in a field. This film is an excellent teaching tool to demonstrate this phenomenon.

In contrast to multiple-cycle diseases, single-cycle diseases depend almost entirely on the quantity of initial soilborne inoculum for a serious epidemic to occur. Most soilborne diseases, especially Fusarium wilt and Verticillium wilt, are not multicyclic and fit the single-cycle concept of Vanderplank (1963), because inoculum that is responsible for the relatively high incidence of disease increases only after the infected plant material from the previous crop decomposes and releases the inoculum into the soil; however, most root-rotting *Phytophthora* species produce propagules intermittently during the crop year, and these increase rapidly in number in the soil as well as on foliage when environmental conditions are favorable. Therefore, most soilborne diseases caused by *Phytophthora* appear to be exceptions and are generally considered to be multicyclic diseases (Mackenzie et al. 1983; Mitchell and Kannwischer-Mitchell 1983); however, it is possible for the previous crop to leave a large number of propagules from decaying, previously infected roots in soil. For instance, soybean roots infected with *P. sojae* usually contain many oospores that can germinate during the spring and infect plants at the seedling stage. Tooley and Grau (1984b) interpreted their field data to indicate that Phytophthora root rot of soybean was a single-cycle disease because of the initial inoculum in the soil and no evidence of spread down the row from a single focal point.

The multicyclic nature of a disease epidemic directly affects the type of control measure selected (Fry 1982) and explains why certain control measures fail. For instance, soil fumigation with methyl bromide usually kills almost all propagules produced by *Phytophthora* (Munnecke et al. 1974, 1978); however, fumigation in the field seldom eradicates 100% of propagules, so if a small concentration of residual inoculum should persist in the biological vacuum that is created by fumigation of soil, the residual inoculum could increase rapidly to a high level in a short time. This occurred in a field experiment with *P. cinnamomi* on avocado (Munnecke 1984).

The multicyclic nature of *Phytophthora* also explains the rapid increase in the population of metalaxyl-resistant *P. infestans* isolates in the Netherlands. Strains of *P. infestans* that were resistant to metalaxyl became dominant within 1 or 2 years after the first use of this highly specific fungicide as a control measure. Why should this happen? The widespread use of metalaxyl suppressed the residual population of *P. infestans* but allowed the resistant population to increase and eventually to become dominant (Davidse et al. 1983a). Since *P. infestans* is disseminated widely by deciduous sporangia (Hirst et al. 1960), the metalaxyl-resistant population spreads rapidly, whereas the metalaxyl-sensitive population does not produce sporangia.

Although rain cannot be controlled, predictions of impending rain and temperature can be obtained from weather forecasts. Since it is not wise economically to apply fungicides to protect plants from late blight unless rain and favorable temperatures are imminent, forecasting systems such as BLITECAST for potato culture are practical technologies utilized to some extent by many growers in the eastern United States (Krause et al. 1975; Fry et al. 1983; Mackenzie et al. 1983; Fry and Fohner 1985; Fry and Doster 1991). Similar systems of forecasting are used in Europe (Hyre 1954; Schrödter and Ullrich 1966; Bourke 1970) to determine when application of fungicides would be optimal for control of late blight.

Many forecasting systems based on weather parameters can result in more efficient management of agricultural resources. Forecasting can decrease the risk of large crop loss, since additional fungicidal applications can be made if the forecast indicates a probability that the disease will reach epidemic proportions. Also, reducing the amount of fungicides applied to a crop to a minimum is an environmentally responsible objective (Fry and Fohner 1985). Forecasting is highly useful with diseases that are sporadic, but in areas where the environment is continuously favorable for late blight of potato, such as Long Island, New York, forecasting is of little value. Forecasting for multicyclic diseases such as late blight is based on the rapidity of the secondary cycles or the amount of secondary inoculum and on weather factors, for example, relative humidity exceeding 90% and temperatures of approximately 20–26°C. We quote (by permission of Academic Press, Orlando, Florida, courtesy of Fry and Fohner 1985):

> In the United States computers have streamlined the use of several different forecasts. Probably the best known of these is BLITECAST, a computerized forecast for potato late blight developed by Krause and colleagues at Pennsylvania State University (Krause et al. 1975; Krause and Massie 1975). BLITECAST combined a potato late blight forecasting system developed by Hyre (1954) with a forecast system developed by Wallin (1962). Hyre's system forecasts late blight 7–14 days after 10 consecutive rain-favorable days. 'Severity values' are assigned to periods of high relative humidity. Severity values increase to a maximum of 4/day with increases in the duration of high relative humidity and with increases of temperature to 26.6°C. The initial appearance of late blight is predicted after the accumulation of 18–21 severity values from the time of plant emergence, and further occurrences are predicted after each subsequent accumulation of three severity values. BLITECAST was initially programmed on a large mainframe computer, but subsequently has become available on a microcomputer coupled with a weather data logger (Mackenzie 1981). The microcomputer-data logger is a self-contained unit and can be placed into a potato field (Fry and Fohner 1985).

BLITECAST and the system of Fry et al. (1983) in New York State determine when fungicides should be applied and appear to give control equivalent to that of regular weekly spray applications (Fohner et al. 1984). Fry and Spielman (1991) state that BLITECAST is effective if the primary inoculum is from infected tubers, but if primary inoculum comes from a neighboring field, the inoculum might arrive before the BLITECAST system can predict it. BLITECAST (Krause et al. 1975) is cited as the major forecasting system used in South America (Thurston and Schultz 1981).

Doster and Fry (1991) evaluated the timing of metalaxyl-chlorothalonil applications on potatoes for late blight control by using a computer simulation strategy to test different spray programs over a recorded 50-year weather pattern. They utilized a computer model that simulated development of *P. infestans* (Bruhn and Fry 1981) and the dynamics of chlorothalonil (a protectant fungicide) (Bruhn and Fry 1982a, 1982b) and metalaxyl (a systemic fungicide) (Milgroom and Fry 1988a, b; Milgroom et al. 1988). This simulator model was validated by current field data (Doster et al. 1990). We quote the summary of Doster and Fry (1991) (by permission of Butterworth Heinemann Ltd., Oxford, U.K., courtesy of Doster and Fry 1991):

> A complex and realistic computer simulation model was used to determine the type of spray scheme utilizing metalaxyl (in mixture with a protectant fungicide) (chlorothalonil), that would allow a reduction in the amount of fungicides applied but still maintain suppression of potato late blight caused by *Phytophthora infestans*. Timing spray applications according to some weather-based decision rules improved disease suppression relative to the standard treatment when the same or a slightly reduced total amount of fungicide was applied. However, weather-dependent spray schemes did not maintain disease control equivalent to the standard schedule when the total amount of fungicide applied was reduced by amounts >10%. Some programmes in which spray dosages and intervals were adjusted according to time in the season resulted in much better disease suppression than the standard treatment. Reduction of the dosage in each spray application resulted in better disease suppression than reduction in the number of sprays. The dosage of sprays applied late in the season can be reduced substantially (50%) without greatly affecting disease suppression. More frequent applications at a lower dosage resulted in better disease suppression than the standard treatment even though less total fungicide was applied.

MacHardy (1979) describes a noncomputerized program based on BLITECAST that is suitable for New Hampshire, a small state without a major potato-growing industry. It gives the daily maximum, minimum, and mean temperatures, the daily rainfall, and the consequent blight condition. A 5-day average temperature below 25.6°C and a 10-day total rainfall of 1.2 inches (3 cm) or more give a positive blight condition. Negative blight values are recorded at temperatures below 7.2°C, 5-day average temperature at 25.6°C or above, or 10-day total rainfall of less than 1.2 inches (3 cm). On Prince Edward Island, Canada, the BLITECAST principles, consisting of temperature, relative humidity, and rainfall, were broadcast by radio to alert growers to epidemic conditions (Bootsma 1979).

The Status of Fungicidal Control of *Phytophthora*

The introduction of metalaxyl was a breakthrough in the control of foliar diseases caused by *Phytophthora* because of its water solubility, activity at low dosages, and translocation upward in plant tissue. All of these factors allow control of *Phytophthora* after the plant has been penetrated by germ tubes and also allow for greater efficacy, even after some rainfall. Some of the other phenylamide compounds, such as ofurace and oxadixyl, share many of the beneficial qualities of metalaxyl, but the usage of these compounds has not been as extensive as that of metalaxyl. Although metalaxyl has curative properties (capable of controlling the disease after infection has taken place), this usage is not recommended because it places selection pressure on *Phytophthora* populations that could result in the development of resistant biotypes.

The efficacy of metalaxyl as a drench treatment of soil for control of root and collar rot diseases has also been a breakthrough, because no other fungicide has effected control of root diseases to the extent that metalaxyl has. Since soil adsorbs a high quantity of a suspension or solution of a fungicide used as a drench, the amount that must be used is often great and the cost can be prohibitive, except on high-income crops such those grown intensively in nurseries. A major disadvantage of metalaxyl and similar fungicides as drenches or soil treatments stems from their failure to destroy *Phytophthora* propagules, since they act as fungistats. Once metalaxyl is chemically broken down in the soil, the latent inoculum of *Phytophthora* can revive. Because of the rapid regeneration ability of *Phytophthora,* the inoculum can again increase to levels high enough to cause lethal disease when the soil moisture increases to near saturation. In such cases, infected nursery plants that are transported to the field can serve as a source of inoculum at a later date. Thus, use of a fungicide to cover up poor sanitation in a nursery is not good practice. Nonetheless, metalaxyl is probably frequently used in this way.

Although the advantages of metalaxyl are many, the development of resistance within populations of target pathogens such as *P. infestans* has become a problem. Since metalaxyl acts on RNA synthesis of *Phytophthora,* the mode of action is specific and probably governed by one gene (Shattock 1988). When and if this gene mutates, these resistant propagules can dominate the population because the production of the majority of metalaxyl-sensitive propagules has been suppressed. When metalaxyl was used frequently in the Netherlands and Ireland, resistant populations of *P. infestans* became dominant within a year, thus rendering metalaxyl useless as a fungicide. After use of metalaxyl was banned in these countries, the chemical industry initiated the practice of marketing metalaxyl with a partner fungicide, such as maneb or mancozeb, that has a broader mode of action (Tables 7.8 and 7.9) and is therefore less likely to be overcome by resistance in a *Phytophthora* population. An unlikely, simultaneous mutation of a large number of genes would be required to allow the *Phytophthora* population to become resistant. With usage restricted to two times per year and in combination with a partner protectant fungicide, metalaxyl has now been reintroduced for use in the Netherlands and Ireland for control of late blight.

In addition to crop rotation and use of resistant tobacco cultivars, root rot of tobacco caused by *P. parasitica* var *nicotianae* can be controlled with metalaxyl applied to soil. In North Carolina, Shew (1985) found that a 3- and 4-year continuous use resulted in the gradual buildup of a population of *Phytophthora* that was less sensitive to metalaxyl. Although total resistance of *P. parasitica* var. *nicotianae* did not occur, the range of isolate sensitivity was extended. For instance, after year 0, the average ED_{50} of a population was 0.4, but after treatment for 3 years, the ED_{50} increased to 1.2. Also, a higher percentage of isolates was inhibited to a lesser degree at a specific fungicide concentration. In year 0 about 45% of the isolates were inhibited by metalaxyl at 1 μg/ml, but in year 4 none were inhibited by the same dosage (Shew 1985); however, these isolates were sensitive to higher dosages of metalaxyl. Thus, resistance to metalaxyl can be expressed not only as total resistance, as shown above for *P. infestans,* but also by a gradual increase in tolerance by the target *Phytophthora* species.

Thus far, there has been little evidence that other soilborne *Phytophthora* species, such as *P. cinnamomi,* have developed resistance in the field. This may be because drench treatment of soil with metalaxyl reaches only a small proportion of the total *Phytophthora* population compared with foliar spraying, which affects the major portion of the population of the foliar pathogen. Although resistance has not become a problem in soilborne *Phytophthora* species in the field, there is no question that these species can become resistant if the population is subjected to enough selection pressure by metalaxyl and for a long enough time, as indicated above for *P. parasitica* on tobacco (Shew 1985). Stack and Millar (1985a–c) obtained a resistant isolate of *P. medicaginis* by selection from a population of 12×10^6 zoospores. Previously, Davidse (1981) isolated metalaxyl-resistant colonies of the same species from a population that had been treated with a mutagen. Isolation of metalaxyl-resistant isolates of *P. sojae* in the laboratory has been reported (Layton and Kuhn 1988a, 1990; Cahill and Ward 1989; Lamboy and Paxton 1992; Bhat et al. 1993b).

Although *P. infestans* is most active on the foliage of potato, tuber rot follows when the caducous sporangia are washed to the soil by rainfall (Cox and Large 1960; Lacey 1965). If planted, these infected tubers can become a source of a subsequent epidemic of late blight. Thus, the accepted practice of forbidding the use of metalaxyl late in the season by some nations is wise because if infected tubers have been treated with metalaxyl, the resistant isolates will thrive and become dominant and could give rise to a population of resistant propagules during the following crop year.

The status of fosetyl-Al and the phosphonates, the second outstanding breakthrough in fungicidal control of *Phytophthora,* is highly favorable. It is clearly the first fungicide with the capacity to be translocated upward in the xylem with water uptake and downward in the phloem with the sugars formed in the process of photosynthesis in the leaves. Trunk injection, soil application, and foliar spray have resulted in control and suppression of severe pathogens such as *P. cinnamomi,* which causes root rot of avocado (Coffey et al. 1984b) and root and heart rot of pineapple (Rohrbach and Schenck 1985). Unfortunately, late blight of potato, caused by *P. infestans,* is poorly controlled by fosetyl-Al.

According to Griffith et al. (1992), the full potential of control by phosphonates has not yet been realized. They suggest that subtle alterations in the phosphonate chemical structure might yield even more efficacious products. They cite data by Dunstan et al. (1990) that show that high molecular weight, water-soluble components (polysaccharides or glycoproteins) of *P. infestans* propagules and mycelium suppress the wound response in potato leaves. Phosphonate at noninhibitory dosages reduced the production of these surface molecules (Dunstan and Grant 1990; Smillie et al. 1990). Griffith et al. (1992) and Guest and Grant (1991) propose that the phosphonate ion exerts an indirect effect that increases resistance of the host as well as a fungistatic effect on *Phytophthora.* Coffey and Ouimette (1989) conclude that the effect of phosphonate is direct on *Phytophthora* per se.

Summary

Suppression of the initial inoculum of *Phytophthora* by chemicals can be a primary control measure. For treatment of benches in glasshouse operations, copper naphthenate is effective. Sodium hypochlorite is inexpensive and effective for application to physical items such as tools but is highly corrosive. It and other chlorine products can be added to infested irrigation water reservoirs to suppress *Phytophthora.* Sodium tetrathiocarbonate releases carbon disulfide into soil and shows promise for control of *Phytophthora.* Methyl bromide-chloropicrin fumigation of soil is effective but too expensive, except for high-value crops such as strawberry. Metham sodium is a carbamate that breaks down in soil to isothiocyanate and can be applied with irrigation water.

Fungicides are used commonly to protect plants against foliar infection by various *Phytophthora* species. Copper compounds, exemplified by Bordeaux mixture and the fixed-copper compounds, have been used since 1885 and are still highly effective, especially in the tropics where rainfall is heavy. The dithiocarbamate fungicides such as maneb, mancozeb, and zineb are used as protectant fungicides. Chlorothalonil effectively controls late blight of potato and also early blight caused by *Alternaria solani.* Organic tin compounds are effective but somewhat more phytotoxic. Phthalimides such as captan, difolatan, and folpet are often used as protectant foliar sprays but are less effective than the dithiocarbamates and chlorothalonil. All of these fungicides are nonsystemic and exert their effects only on the germinating spores before the plant becomes infected.

Several classes of improved compounds that act both as protectant and systemic fungicides have become available during the 1980s and 1990s. The degree of systemic activity varies from a translaminar effect (movement across the leaf) to apoplastic (upward from roots to shoots, e.g., metalaxyl) to symplastic (both upward and downward, e.g., fosetyl-Al and its breakdown product, phosphonate). The chemical structures of the cyanoacetamide oximes (cymoxanil), the carbamates (prothiocarb and propamocarb), isoxazoles (hymexazole), phenylamide compounds (e.g., metalaxyl, furalaxyl, and oxadixyl), and phosphonates (fosetyl-Al and phosphorous acid) are illustrated and their properties explained. Dimethomorph and zarilamide are promising new compounds about which there is as yet little experimental information.

Metalaxyl is the most effective of the phenylamide compounds and is active only against *Phytophthora, Pythium,* and the downy mildews. It has been effective as a foliar spray because *Phytophthora* can be suppressed even after infection has occurred. The suppressive effect of metalaxyl is longer than that of the other protectant but nonsystemic fungicides. Metalaxyl is also effective as a soil drench for some crops that are of high enough value for treatment to be cost effective. It has a relatively long half-life in soil.

Metalaxyl and similar fungicides are marketed mainly as mixtures with a protectant fungicide such as mancozeb because shortly after extensive use in the Netherlands and Ireland, metalaxyl-resistant strains of *P. infestans* became dominant. The efficacy of using computerized forecasting to determine when to apply fungicides is discussed. Several two- and three-component systemic fungicide mixtures have been tested, and some are considered to be synergistic in their effect.

Fosetyl-Al is a phosphonate compound that is taken up by the plant through roots, leaves, and stems; translocates upward and downward; and effectively controls root diseases such as Phytophthora root rot of avocado by trunk, soil, or foliar application. Fosetyl-Al breaks down to the phosphonate anion of phosphorous acid. Phosphorous acid, when buffered, can be injected into tree trunks to control root diseases but can be used legally only in Australia because of patent regulations. An anomaly is the relatively ineffective control of *P. infestans* on potato and *P. sojae* on soybean by fosetyl-Al. There is controversy as to whether fosetyl-Al controls *Phytophthora* species by direct fungitoxic activity or by a combination of fungitoxicity and enhancement of resistance by the host.

CHAPTER 8

Phytophthora arecae

(Coleman) Pethybridge (1913)

Synonyms of *Phytophthora arecae* include *P. omnivora* Sydow and Butler (1906), *P. omnivora* var. *arecae* (Coleman) (1910), and *P. cactorum* var. *arecae* (Coleman) Sacc. and Trotter (1912). See Stamps (1985c) for a summary. *P. arecae* was first isolated from the leaves, fruits, peduncles, and apexes of arecanut trees, *Areca catechu* L., in India by Butler (1906), who originally described the fungus as *P. omnivora,* a species that is no longer accepted. Coleman (1910) later changed the name to *P. omnivora* var. *arecae;* however, Pethybridge (1913) showed that it was quite different from *P. omnivora* and proposed that the fungus be named *P. arecae.* Butler (1918) eventually formalized the name *P. arecae* (Coleman) Pethybridge. Rao (1970) reviewed reports of *P. arecae* and other *Phytophthora* species that are important to Indian agriculture. Although listed here, *P. arecae* should now be classified as *P. palmivora* (Mchau and Coffey 1994a).

Tucker (1931) combined *P. arecae* and *P. meadii* with *P. palmivora,* which has priority. This combination was also supported by Leonian (1934) and Thomas et al. (1947). The recent reports of isozyme (Oudemans and Coffey 1991b, c) and mtDNA restriction fragment length polymorphism (RFLP) patterns (Förster and Coffey 1990b) indicate a genetic similarity between *P. arecae* and *P. palmivora.* Thomas et al. (1947) found that oospores formed when different isolates from arecanut (all of which they considered to be *P. palmivora*) were paired among themselves and also when paired with opposite mating types of certain other species. *P. arecae* was recognized as a valid species by Waterhouse (1963, 1974a), who states that *P. arecae* differs from *P. palmivora sensu* Butler because of the production of more abundant aerial mycelium on cornmeal and oatmeal agars, absence of haustoria in infected plant tissue, irregular sporangiophores with swellings, irregular sporangiophores with longer stalks, none or few chlamydospores that are small (maximum 36 µm, most are less that 30 µm), formation of sex organs in single culture, and large oogonia and oospores, usually about 33 µm in diameter compared with 30 µm for *P. palmivora.*

Sastry and Hedge (1987a–c) present evidence that supports reclassifying *P. arecae* as *P. meadii.* However, *P. meadii* differs markedly from *P. arecae* in isozyme patterns (Oudemans and Coffey 1991c). Although the keys of Newhook et al. (1978), Ho (1981), Gerrettson-Cornell (1985, 1989), and Stamps et al. (1990) list *P. arecae* as a valid species, there is evidence that isolates classified as *P. arecae* should be included in *P. palmivora. P. arecae* is serologically related to *P. palmivora* (Merz et al. 1969). Isozyme (Oudemans and Coffey 1991c; Mchau and Coffey 1994a) and mtDNA RFLP patterns of *P. palmivora* and *P. arecae* show that isolates of both species are genetically similar (Förster and Coffey 1990b) (Figure 8.1). Förster and Coffey (1990b) state that the minor differences in lack of or diminished chlamydospore formation and formation of irregularly shaped sporangia and sporangiophores for *P. arecae,* as described by Waterhouse (1963), are probably caused by isolate variability. Ho (1992) lists *P. arecae* as conspecific with *P. palmivora,* which has priority. We agree with Oudemans and Coffey (1991b, c) and Mchau and Coffey (1994a) that *P. arecae* should be merged with *P. palmivora.* See Chapter 4 for an excerpt from Oudemans and Coffey (1991b).

Characteristics of *Phytophthora arecae*

P. arecae is classified in group II (Stamps et al. 1990). Detailed descriptions of the fungus are given by Coleman (1910), Waterhouse (1974b), and Stamps (1985c). The morphology of spore structures shown in Figure 8.2 is redrawn from Rosenbaum (1917). See Tables 4.2 and 4.3 for tabular

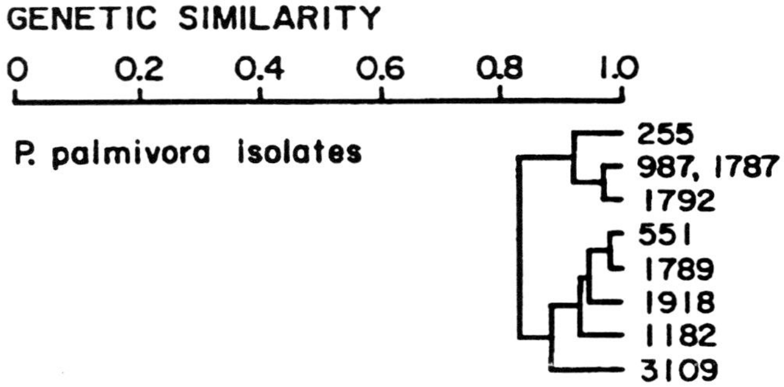

Figure 8.1. Phenogram showing the similarity of mtDNA RFLP patterns for *Phytophthora palmivora* isolates and *P. arecae* (3109 isolated from a *Chaemaedora* hybrid of Ploetz and Mitchell 1989). *P. palmivora* isolates include P255, P551, P987 from *Theobroma cacao,* P1182 from *Morrenia odorata,* P1787 from *Carica papaya,* P1789 from *Dendrobium* sp., P1792 from *Cattleya* sp., and P1918 from *Vanilla planifolia.* (Reproduced by permission of Academic Press, courtesy of Förster et al. 1990b)

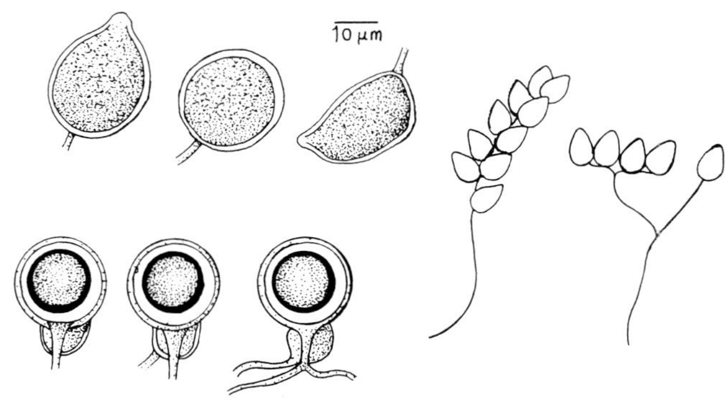

Figure 8.2. Morphology of *Phytophthora arecae.* Lower row, globose oogonia, each containing an oospore. Amphigynous antheridia are attached at the bases of the oogonia. Right, sporangia produced in close sympodia. Upper row, caducous sporangia and an asymmetric sporangium. Scale bar applies only to drawings on the left. (Redrawn from Rosenbaum 1917 by A. Vaziri)

keys and the dichotomous key of Ho (1992) (Appendix 4.9). See also Tables 51A.4 and 51A.5 for morphological comparisons with *P. palmivora* and related species.

Sporangia

Sporangia are papillate, with shapes varying from very broad ellipsoidal to obturbinate to almost spherical with a tapered base (Figure 8.2). The sporangia measure 20 to 72 µm long × 20 to 48 µm wide, averaging 47.92 × 30.05 µm (Tables 4.2 and 4.3) (Rosenbaum 1917); 33.93 × 24.78 µm (Tucker 1931); and 40 to 50 × 35 to 40 µm (Waterhouse 1963; Stamps 1985c). The sporangia are caducous with short pedicels measuring 1–6 µm. The length-breadth ratios of sporangia range from 1.1 to 1.6 with an average of 1.37. Mchau and Coffey (1994a) report a length-breadth ratio of 1.6 (45.4 × 29.5 µm), which is similar to that of *P. palmivora.* Sporangiophores are irregularly branched or in a simple sympodium. Each branch ends in a short sympodium of one to several sporangia (Figure 8.2).

Chlamydospores

Chlamydospores are not normally produced, but the diameter in those isolates that produce chlamydospores ranges from 18 to 40 µm, averaging 25.74 µm (Tucker 1931); 35 to 40 µm with a maximum diameter measurement of 60 µm (Waterhouse 1963); and 18 to 60 µm, averaging 25 to 40 µm (Gerrettson-Cornell 1989). The wall thickness is usually 1.0 µm or less. Two of five isolates produced chlamydospores (average diameter 32 µm), which were similar to those produced by *P. palmivora.*

Sex Organs

This species is normally heterothallic (Ashby 1929a; Uppal and Desai 1939; Marudarajan 1941; Waterhouse 1963; Gerrettson-Cornell 1989). Oogonia and oospores are produced abundantly when isolates of opposite compatibility types are paired in culture; however, homothallism has been reported by Ramakrishnan and Seethalakshmi (1956a). The antheridia are amphigynous and average 14 × 15 µm. Oogonia measure 23 to 44 µm with an average diameter of 28 to 30 µm (Figure 8.2). Oospores range from 22 to 48 µm, average 31.5 µm (Rosenbaum 1917). Waterhouse (1974b) gives oospore diameters ranging from 23 to 36 µm, average 28 µm (Table 4.3).

Growth Temperatures

The minimum temperature for growth is 10 to 12°C, the optimum between 27 and 30°C, and the maximum 35°C (Stamps 1983c).

Growth Characteristics in Culture Media

The mycelium is slightly radiate with abundant aerial mycelium often produced in culture media. No distinct culture pattern can be discerned in various culture media.

Some Diseases Caused by *Phytophthora arecae*

Koleroga (*kole,* rotting; *roga,* disease)

Koleroga is also called mahli, or nut and flower disease. It is important on *Arecae catechu* L. (betel nut palm or arecanut), which is a perennial crop that bears nuts used extensively for masticatory purposes and in religious and social functions in

Table 8.1. Distribution of hosts reported for *Phytophthora arecae*

Host	Common name	Disease	Geographical distribution
Agave sisalana Perrine	Sisal	Zebra disease	Tanzania (Peregrine 1964)
Areca catechu L.	Areca palm	Koleroga or mahli disease	India (Butler 1906, Coleman 1910; Reddy et al. 1978)
		Bud and fruit rot	Sri Lanka (Park 1930); Assam (Barooah 1955)
Artocarpus heterophyllus Lam. (*A. integrifolia*)	Jak, jackfruit	Leaf and flower blight	India (Narasimhan 1927)
Bryophyllum calycinum	Kalanchoe	Leaf blight	India (Narasimhan 1927)
Cereus formosus	Cactus	Stem rot	India (Coleman 1910)[a]
Chamaedorea seifrizzi Burret × *erumpens*	Bamboo palm	Root rot	United States (Ploetz and Mitchell 1989)
Citrus limon (L.) Burm.	Lemon	Leaf blight	India (Narasimhan 1927)
Citrus medica L.	Citron	Leaf spot	India (Narasimhan 1927)
Citrus sinensis (L.) Osbeck	Orange	Leaf blight	India (Narasimhan 1927)
Clarkia elegans Dougl.	Clarkia	Seedling blight	India (Coleman 1910)[a]
Cocos nucifera L.	Coconut palm	Nut-fall disease, mahli disease	India (Sundararaman and Ramakrishnan 1924); Andaman Islands (Mitra 1929b)
Clocasia antiquorium Schott.	Elephant's ear	Leaf spot	India (Narasimhan 1927)
Ficus benjamina L. (*F. nitida*)	Weeping fig, Java fig	Fruit rot	India (Narasimhan 1927)
Hevea brasiliensis Muell. Arg.	Rubber	Leaf rot	India (Narasimhan 1927)
Jatropha curcas L.	Barbados nut	Leaf blight	India (Venkata Rao 1930)
Jatropha glandulifera Roxb.		Fruit rot	India (Narasimhan 1927)
Kalanchoe pinnata (Lam.) Pers. (*Bryophyllum calycinum*)	Air plant, good luck leaf	Blight	India (Narasimhan 1927)
Loranthus longiflorus Desr.	Mistletoe, bandarike	Leaf blight	India (Narasimhan 1927)
Lycopersicon esculentum L.	Tomato	Stem dieback, foot rot	Netherlands (Verhoeff and Weber 1966)
Malus pumila Mill.	Apple	Brown rot	United States (Tucker 1931)[a]
Mangifera indica L.	Mango	Fruit rot	India (Narasimhan 1927)
Oenothera biennis L.	German rampion	Seedling blight	India (Coleman 1910)[a]
Salpiglossis sinuata Ruiz and Pav.	Painted tongue	Seedling blight	India (Coleman 1910)[a]
Santalum album L.	White sandalwood	Leaf, flower, and fruit rot	India (Narasimhan 1927)
Schizanthus wisetonensis Hort.	Butterfly flower	Seedling blight	India (Coleman 1910)[a]
Solanum melongena L.	Eggplant	Seedling blight	India (Coleman 1910)[a]
Solanum tuberosum L.	Potato	Pink tuber rot	United States (Rosenbaum 1914)[a]

[a] Artificially inoculated host.

India and the Far East. Koleroga was recorded first in India (Butler 1906), but Coleman (1910) was the first to show that the disease was caused by *Phytophthora.* See also Reddy et al. (1978), Reddy and Anandaraj (1982), and Holliday (1980) for descriptions of koleroga. The review by Reddy and Anandaraj (1982) comprehensively covers epidemiology and control.

The disease is usually first observed 2–3 weeks after the start of the monsoon season (June–September) when high humidity prevails. The pathogen causes dark green water-soaked lesions on the nut and the flowering shoots, after which the nuts fall before maturity. In infected plants, the stalks blacken. Infected nuts that remain on the tree are marked on the surface by dark brown radial strands, and large vacuoles are found inside the nut (Reddy et al. 1978). Infections late in the season result in shriveling of nuts clinging to the branches (Manudarajan 1950). Eventually the fungus destroys the entire center of the bud, and the leaves become yellow, droop, and drop off. Invasion by secondary organisms results in a slimy ooze that emits a characteristic fetid odor (Coleman 1910). Severe infection results in the death of the entire tree. Annual losses of 10 to 90% of the crop have been reported (Coleman 1910; Reddy and Anandaraj 1982).

The *Phytophthora* isolates from arecanut and the isolates from the flowering parasite *Loranthus* are of a different mating type from that found on *Sandal* and *Jatropa* (Narasimhan 1931). It was suggested that the close proximity of these hosts in *Areca* plantations could result in the production of oospores by the mating of the two strains and that oospores might allow the pathogen to survive through the hot summer; however, there is no proof that this occurs. The fungus also survives in numerous secondary hosts (Table 8.1), making control difficult, and in infected crowns and fallen nuts (Reddy and Anandaraj 1982).

Disease Development. The disease is most severe during wet seasons and results in heavy losses. In India, the disease is most severe during the southwest monsoon season (Reddy and Anandaraj 1982). The disease intensity is greatest wherever humid conditions exist, such as in valleys surrounded by dense forests. Heavy rain, wind, insects, and even small birds disseminate the pathogen (Coleman 1910).

Control. Spraying trees with a 1% Bordeaux mixture prior to the monsoon season followed by a second spray 6 weeks later has proved effective (Sundararaman and Ramakrishnan 1924; Reddy and Anandaraj 1980). In some reports the addition of a resin soap to Bordeaux mixture increases the adhesive qualities against torrential rains (Coleman 1910) and the degree of control (Reddy and Anandaraj 1982); however, adhesives did not increase the copper retention in a recent report (Anandaraj 1985). If registered for use, metalaxyl or fosetyl-Al should be effective. Disease incidence was lowest on trees sprayed with fosetyl-Al (Anandaraj and Saraswathy 1986).

The presence of oospores in tissues and soil make long-term control difficult. Removal and destruction of plant debris and nuts from the field reduces the inoculum potential. Removal of alternate hosts such as *Colocasia* and other plants and trees in the vicinity of arecanut plantations is also advocated (Reddy et al. 1978; Reddy and Anandaraj 1982).

Root Rot of *Chamaedorea seifrizii* Burret × *erumpens* (Bamboo Palm)

Root rot of bamboo palm caused by *P. arecae* was first reported in Florida, United States, by Ploetz and Mitchell (1989). This isolate is considered to be the same as *P. palmivora* on the basis of isozyme patterns (Oudemans and Coffey 1991c).

Blackened, water-soaked necrosis occurs on roots and in those portions of the stem below the soil surface. In experimental tests, the predominant area of infection was approximately 5 mm behind the tip of primary roots (Ploetz and Mitchell 1989).

Disease Development. Root rot severity was not significantly altered by different water regimes (Ploetz and Mitchell 1989), but it is favored by wet soil conditions.

Control. Both metalaxyl and fosetyl-Al provide good control of root rot.

Foot Rot of *Lycopersicon esculentum* L. (Tomato) (see also Chapter 57, *Phytophthora richardiae*)

Foot rot of tomato was first described in the Netherlands. Symptoms on the aerial parts of the plants are either very slight or not evident, but soon after extensive rotting of the root systems, the plants wilt rapidly. Losses of greater than 50% of the tomato crop 6 weeks after planting have been reported (Verhoeff and Weber 1966).

Disease Development. Infection is favored by high soil temperatures (25 to 28°C) and moist soil conditions.

Control. Tomatoes should be grown in well-drained soils. Excessive treatment with nitrogen fertilizers should be avoided because it increases the susceptibility of plants.

Zebra Disease of *Agave sisalana* Perrine (Sisal)

See the description in Chapter 50, *Phytophthora nicotianae.*

CHAPTER 9

Phytophthora boehmeriae

Sawada (1927)

Phytophthora boehmeriae was first isolated in 1927 from leaves of the Chinese silk tree, ramie, *Boehmeria nivea* (L.) Gaud-Beau., by Sawada in Taiwan. Isolation of the fungus from soil was reported by Frezzi (1941) and Gerrettson-Cornell (1976a). Fruit rot of citrus was reported by Frezzi (1941). *P. boehmeriae* has been recorded on only a few hosts (Table 9.1). Although described in 1927 by Sawada in Taiwan, it has not been reported there since (Ho 1990).

Characteristics of *Phytophthora boehmeriae*

P. boehmeriae is classified in group II (Stamps et al. 1990). See Stamps (1978a) for a description, Tables 4.2 and 4.3 for tabular keys, and Ho (1992; Appendix 4.9) for a dichotomous key. Morphology is shown in Figure 9.1.

Sporangia

Sporangia are papillate and spherical, ovoid, ellipsoidal, or obturbinate. The spores measure 28 to 69 × 20 to 51 µm, average 51.77 × 40.08 µm (Tucker 1931); 20 to 70 × 15 to 51 µm, average 44 × 32 µm (Frezzi 1950) (Table 4.2); 50 × 35 to 40 µm (Waterhouse 1963); 27 to 72 × 20 to 46 µm, average 49.9 × 34.5 µm (Katsura 1971); and 42 to 50 × 34 to 39 µm (Gerrettson-Cornell 1989). Length-breadth ratios reported are 1.25 to 1.4 (Waterhouse 1963) and 1.0 to 1.6 (Gerrettson-Cornell 1989). Sporangia are caducous with a pedicel length of 5.0 µm or less. Sporangiophores are sympodial.

Chlamydospores

Chlamydospores are produced infrequently. Diameters range from 26 to 51 µm, averaging 41.42 µm (Tucker 1931); 17 to 42 µm, averaging 29.5 µm (Frezzi 1950); 40.0 µm (Waterhouse 1963); and 30 to 41 µm (Gerrettson-Cornell 1989). Walls are 2 µm thick.

Sex Organs

P. boehmeriae is homothallic, and oospores form abundantly in host tissues. Antheridia are amphigynous and almost spherical in shape, often with a residual oil globule. Dimensions reported are 8 to 16 × 12.5 to 21 µm (Frezzi 1950); 14 to 21 × 13 µm (Waterhouse 1963); and 13 to 21 × 8 to 19 µm, average 14 to 17 × 14 to 16 µm (Gerrettson-Cornell 1989) (Table 4.3). Oogonia are smooth-walled and hyaline to yellow. Diameters are 21.7 to 31.7 µm, average 28 µm (Waterhouse 1963); and 19 to 41 µm, average 26 to 38 µm (Gerrettson-Cornell 1989). Oospores nearly fill the oogonium (plerotic). The wall is 2 µm or less in thickness. Oospore diameters are 20.9 to 27.6 µm, average 25.1 µm (Tucker 1931); 16 to 34 µm, average 27.5 µm (Frezzi 1950); average 24 µm (Waterhouse 1963); 9 to 12 × 9 to 14 µm (Katsura 1971); 16 to 35 µm, average 23 to 30 µm (Gerrettson-Cornell 1989) (Table 4.3).

Growth Temperatures

Minimum temperature for growth is 5 to 6°C, optimum is 25°C, and maximum is 32.5°C.

Growth Characteristics in Culture Media

Colonies are uniform and compact with well-defined margins and dense aerial mycelia. Oospores, chlamydospores, and sporangia form abundantly in an agar medium in 7 to 10 days.

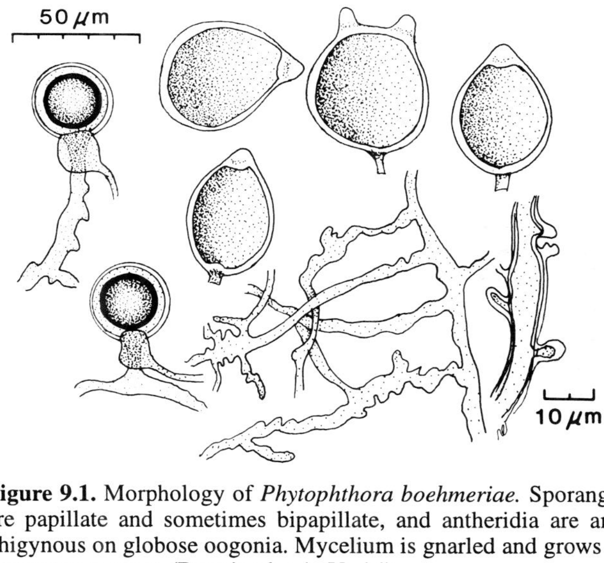

Figure 9.1. Morphology of *Phytophthora boehmeriae*. Sporangia are papillate and sometimes bipapillate, and antheridia are amphigynous on globose oogonia. Mycelium is gnarled and grows in a tortuous manner. (Drawing by A. Vaziri)

Table 9.1. Distribution of hosts reported for *Phytophthora boehmeriae*

Host	Common name	Disease	Geographical distribution
Boehmeria nivea (L.) Gaud-Beaup.	Ramie, Chinese silk plant	Leaf blight	Taiwan (Sawada 1927a, b); Japan (Ito and Tokunaga 1935); China (H. F. Chen 1985)
Broussenitia papyrifera L'Her	Paper-mulberry		China (Zheng and Lu 1989)
Cedrus deodara (G. Don) D. Don	Desdar cedar		China (Zheng and Lu 1989)
Citrus sinensis (L.) Osbeck	Sweet orange	Brown rot	Argentina (Frezzi 1941)
Eucalyptus piluloris Sm.	Eucalyptus, black butt	Root rot	Australia (Gerrettson-Cornell 1976a)
Gossypium spp.	Cotton	Boll blight	China (Liang 1964)
Gossypium hirsutum L.	Cotton	Boll rot	Greece (Paplomatas et al. 1995)
Pinus patula Schlectend. and Cham.	Mexican yellow pine	Root rot	Australia (Anon. 1962c)

P. boehmeriae is one of the most sensitive of all *Phytophthora* species to metalaxyl (linear growth on cornmeal agar was inhibited at 2 to 30 ng/ml). Because of its sensitivity, it has been used as a bioassay for metalaxyl in soil (A. M. Bailey and Coffey 1984–1986), plant tissue (Milgroom and Fry 1987), and in vitro (Chauhan and Singh 1987).

Distinguishing Characteristics

P. boehmeriae can be recognized by the production of abundant oogonia and oospores in single culture, amphigynous antheridia, and broadly ovoid to subspherical conspicuously papillate sporangia (Figure 9.1). *P. boehmeriae* differs from *P. parasitica* and *P. citrophthora* by production of oospores in single culture and from *P. megasperma* by production of papillate sporangia (Waterhouse 1963). Although 11 isolates of *P. boehmeriae* showed much genetic diversity by isozyme analysis, *P. boehmeriae* differed significantly from the morphologically similar species, *P. katsurae* and *P. heveae* Oudemans and Coffey (1991b).

Disease Caused by *Phytophthora boehmeriae* (see Holliday 1980)

Brown Rot of Citrus

Citrus fruits are infected when soil is splashed onto the fruits during rains (Frezzi 1941). The symptoms, disease development, and control measures are similar to those described for brown rot of citrus caused by *P. citrophthora* (Chapter 17). Fruit rot should be readily controlled by metalaxyl or fosetyl-Al if registered for use.

CHAPTER 10

Phytophthora botryosa

Chee (1969a)

Phytophthora botryosa was first isolated and described by Chee (1969a) from leaf petioles of the rubber tree (*Hevea brasiliensis*) in Malaysia and Thailand. *P. botryosa* is the causal agent of leaf fall and pod rot of rubber trees. It also causes pod rot of cocoa (Chee and Wastie 1970). It was reported from China on Taro, *Colocasia esculenta* (L.) Schott (Zheng and Lu 1989).

Characteristics of *Phytophthora botryosa*

P. botryosa is classified in group II (Stamps et al. 1990). Morphology is described by Waterhouse (1974c) and Stamps (1985e). See Tables 4.2 and 4.3 for tabular keys. The following description is from Chee (1969a). Morphology is shown in Figure 10.1. See also Tables 51A.4 and 51A.5 for morphological comparisons with *P. palmivora* and related species.

Sporangia

Sporangia form in yellowish clumps (average about 90 sporangia per clump) and are inconspicuously papillate and oval or occasionally ovoid (28 × 15 μm) (Stamps 1985e); the length-breadth ratio varies from 1.6 to 2.1. See Table 4.2 for measurements by Chee (1969a). Sporangia are caducous with pedicels 5 to 20 μm long and are produced on sympodially branched sporangiophores with nodal swellings. Sporangia form as yellowish clumps on lima bean agar when incubated under light.

Chlamydospores

Chlamydospores are rare but when formed are terminal; diameters range from 14 to 30 μm (average 18.7 μm) (Chee 1969a).

Sex Organs

P. botryosa is heterothallic and produces oospores when A1 and A2 mating types are paired. Chee (1969a) reported that one-third of the isolates produced a few oospores in single culture (homothallic). Oospores have been found in pods of the rubber tree. Antheridia are amphigynous (14 × 13 μm); spherical oogonia are 25 to 30 μm in diameter, averaging 25 μm, and contain plerotic oospores that are 22 to 24 μm in diameter (Chee 1969a; Stamps 1985e).

Growth Temperatures

The minimum temperature for growth is 9°C, optimum 26°C, and maximum 32°C (Chee 1969a; Stamps 1985e).

Distinguishing Characteristics

P. botryosa differs from *P. palmivora* by its slower growth with little aerial mycelium on lima bean agar and its infrequent production of chlamydospores, which are smaller (Chee 1969a; Waterhouse 1974c) (Figure 10.1). *P. botryosa* closely resembles *P. meadii,* but it can be distinguished from *P. meadii* by its production of smaller sporangia with medium-length pedicels produced in conspicuous clumps. Oudemans and Coffey (1991c) stated that distinction between these species by sporangial morphology was problematic, although sporangia of *P. botryosa* were consistently smaller; however, isozyme patterns of *P. meadii* and *P. botryosa* showed a high level of genetic similarity. Both species were distinct from *P. palmivora. P. botryosa* was compared with isolates of several other species of rubber by Liyanage and Wheeler (1989a). Rose bengal (3 mg/liter) and streptomycin hydrochloride (5 mg/liter) in malt extract agar inhibited growth of *P. palmivora* but not of *P. botryosa* (Chee 1969a).

Diseases Caused by *Phytophthora botryosa*

Green Pod Decay of Para-Rubber (*Hevea brasiliensis* Meull. Arg.)

Symptoms, disease development, and control measures are similar to those caused by *P. meadii* (Chapter 43) and *P. palmivora* (Chapter 51A). Secondary leaf fall caused by *Glomerella cingulata* and *Oidium heveae* occurs only on young leaves, but leaf fall caused by *P. botryosa* occurs on leaves of any age (Holliday 1980).

Disease Development and Control. Development and control of diseases caused by *P. botryosa* are similar to those covered in the chapters on *P. meadii* and *P. palmivora.*

Metalaxyl controlled black stripe better than Difolatan in Malaysia (Tan 1983). Tapping panel disease was controlled by spraying with an oxadixyl-mancozeb mixture (Sandofan), an oxadixyl-propineb mixture (Fruvit), an ofurace-folpet mixture (Caltan), a metalaxyl-folpet mixture (Ridomil Combi), di-

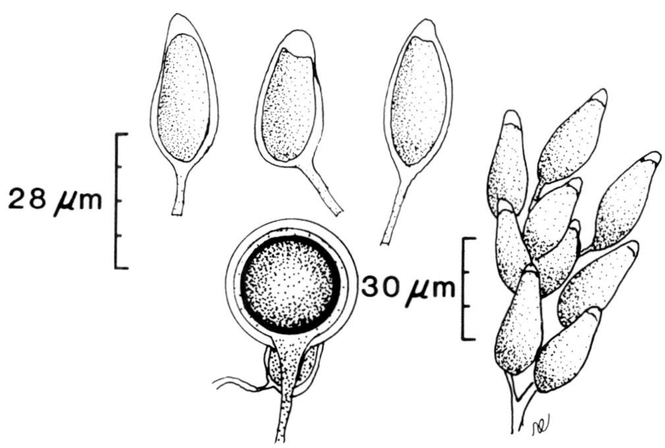

Figure 10.1. Morphology of *Phytophthora botryosa.* Sporangia are caducous with medium-length stalks and are produced in a sympodium. Oogonia are globose, and antheridia are amphigynous. (Drawing by A. Vaziri)

folatan (Captafol), and fentin hydroxide. Phosphorous acid (active ingredient of fosetyl-Al) was also effective. Formulation of fungicides with 3% Methocel F4M, a sticker, allowed reduction of sprays from twice weekly to once weekly (Tan 1990).

Pod Rot of Cocoa

Symptoms, disease development, and control are similar to those of *P. palmivora* (Chapter 51A) (Chee 1969a; Waterhouse 1974c; Holliday 1980).

CHAPTER 11

Phytophthora cactorum

(Leb. and Cohn) Schröeter (1886)

Synonyms of *Phytophthora cactorum* include *Peronospora cactorum* Leb. and Cohn (1870), *Peronospora sempervivi* Schenk (1875), *Phytophthora fagi* (Hartig) Hartig (1876, 1880), *Phytophthora omnivora* de Bary (1881), *Phloeophthora cactorum* (Leb. and Cohn) G. W. Wilson (1907), *Nozemia cactorum* (Leb. and Cohn) Pethybridge (1913), *Phytophthora pini* Leonian (1925), and *Phytophthora paeoniae* Cooper and Porter (1928).

P. cactorum, first reported from rotting cacti (*Cereus giganteus* Engelm and *Melocactus nigrotomentosus*) in Czechoslovakia by Lebert and Cohn (1870), occurs worldwide (see CMI Map 280 and Table 11.1) but is most commonly found in temperate regions. It causes root and collar rots, fruit rots, cankers, leaf blights, wilts, and seedling blights. It parasitizes more than 200 plant species in 150 genera representing 60 plant families (Tucker 1933; Nienhaus 1960). Isolates of *P. cactorum* are not generally host specific; for instance, an isolate of *P. cactorum* from apple infected a large number of weed species, and isolates from 23 host genera were pathogenic to pear trees (Nienhaus 1960). Marked differences occur in degree of pathogenicity to different hosts, however. Although specialization to a single host is rare (Braun and Schwinn 1963), forest soil isolates of *P. cactorum* did not infect ginseng (Darmano et al. 1991).

A summary of diseases caused by *P. cactorum* is given by Gjaerum et al. (1988).

Characteristics of *Phytophthora cactorum*

In 1963 (Waterhouse), *P. cactorum* was the only species listed in group I, but since then, *P. clandestina* (Taylor et al. 1985), *P. iranica* (Ershad 1971), *P. pseudotsugae* (Hamm and Hansen 1983), *P. idaei* (Kennedy and Duncan 1995), *P. italica* (Cacciola et al. 1996), and *P. tentaculata* (Kröber and Marwitz 1993) have been described. Morphology of spore structures is shown in Figures 11.1 and 11.2. See Waterhouse and Waterston (1966a) for a description of *P. cactorum,* Tables 4.2 and 4.3 for tabular keys, and Appendix 4.9 for a dichotomous key (Ho 1992; Ho et al. 1995).

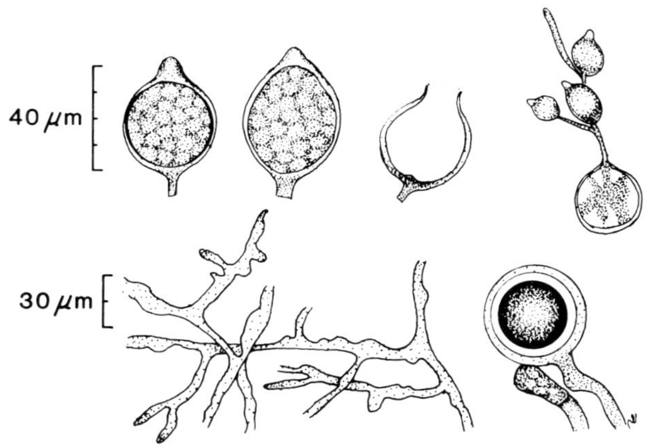

Figure 11.1. Morphology of *Phytophthora cactorum.* Lower row, gnarled mycelium and an oospore in an oogonium with a paragynous antheridium. Upper row, papillate sporangia, empty sporangium, and germinated oospore producing sporangia. (Drawing by A. Vaziri)

Sporangia

Sporangia are distinctively papillate, normally borne terminally, and occasionally intercalary. Shapes of sporangia range from broadly ellipsoidal, obpyriform, or ovoid to spherical. The sporangia are caducous with short pedicels (less than 4 μm in length). Sporangiophores are normally simple or in a close or lax sympodium with sporangia often clustered (Figure 11.2). Dimensions of sporangia are 31.4 ± 4.8 × 26.4 ± 4.0 with a length-breadth ratio of 1.2 ± 0.1 (Oudemans and Coffey 1991a). Data reported in several references are given in Table 11.2. Media for sporangia production are evaluated by Fortes and Pecknold (1981). Several methods are given in the Appendix of Chapter 2.

Chlamydospores

Chlamydospores are produced by some but not all isolates. The somewhat erratic production of chlamydospores by certain isolates in different reports (Waterhouse and Waterston

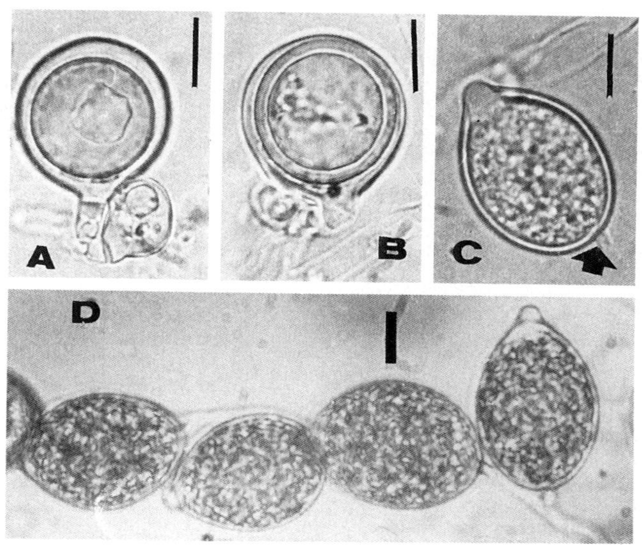

Figure 11.2. Morphology of *Phytophthora cactorum.* **A and B,** Oospores in oogonia with paragynous antheridia. **C,** Papillate, deciduous sporangium with a short pedicel (arrow). **D,** Sporangia produced on a sympodium. Bars = 10 μm. (Reproduced by permission of Edagricole, Bologna, Italy, courtesy Magnano di San Lio et al. 1985)

Table 11.1. Distribution of hosts of *Phytophthora cactorum*

Host	Common name	Disease	Geographical distribution
Abies Mill.	Fir	Seedling rot	Germany (Hartig 1882)
Abies amabilis Doug. ex. J. Forbes	Cascade fir, Pacific silver fir	Root rot, necrosis, chlorosis	United States (Hamm and Hansen 1982a)
Abies procera Rehd.	Fir	Crown rot	United States (Adams and Bielenin 1988)
Abies procera Rehd.	Noble fir	Root rot, necrosis, chlorosis	United States (Hamm and Hansen 1982a)
Acacia Mill.	Acacia	Seedling rot	Germany (Hartig 1883)
Acer macrophyllum Pursh.	Big leaf maple	Stem canker	United States (Stuntz and Séliskar 1943)
Acer palmatum Thunb.	Japan maple	?	Europe (Stevenson 1926)
Acer platanoides L.	Norway maple	Stem canker, seedling rot	Germany (Hartig 1880); France (Prilleux 1895); United States (Caroselli and Howard 1939)
Acer pseudoplatanus L.	Sycamore, mock plane	Bleeding canker	Germany (Hartig 1880); France (Prilleux 1895); United States (Crandall and Hartley 1938)
Acer rubrum L. (*A. sanguineum* Spach.)	Red, scarlet soft or swamp maple	Bleeding canker	United States (Caroselli and Howard 1939)
Acer saccharum Marsh. (*A. saccharophorum* C. Koch)	Sugar, rock or hard maple	Bleeding canker	United States (Caroselli and Howard 1939)
Acer saccharinum L. (*A. dasycarpum* J. F. Ehrh.)	Silver, soft, white, or river maple	Bleeding canker	United States (Caroselli and Howard 1939)
Acer spp.	Maple	Seedling rot	Germany (Hartig 1882); England (Collinge 1912); United States (Marshall and Waterman 1948)
Actinidia chinensis Planch.	Kiwi fruit	Root rot	New Zealand (Robertson 1953)
Aesculus hippocastanum L.	Common or European horse chestnut	Bleeding canker	United States (Caroselli 1953)
Aesculus sp.	Horse chestnut	Gummosis, stem canker	United Kingdom (Brasier 1971b; Brasier and Strouts 1976)
Agonis flexvosa Schav.		Root rot	Australia (Hardy and Sivasithamparam 1988)
Alnus oregona Nutt. (*A. rubra* Bong.)	Red alder, red acacia	Stem canker	United States (Stuntz and Seliskar 1943)[a]
Anemone coronaria L.	Poppy anemone	Root rot	Italy (Garibaldi and Gullino 1975)
Ananas comosus L.	Pineapple	Fruit rot	Holland (Vlasveld 1954)
Annona cherimola Mill.	Custard apple		United States (C. O. Smith 1937)[a]
Antirrhinum majus L.	Snapdragon	Stem girdling, root rot	India (Sundararaman and Ramakrishnan 1928); United States (Harris 1934); South Africa (Mes 1934); Germany (Nienhaus 1960)
Aquilegia L.	Columbine	Stem blight	Greece (Papaioannoy 1954)
Arbutus menziesii Pursh.	Madrone	Stem canker	United States (Stuntz and Seliskar 1943)
Aster L.	Michaelmas daisy, frost flower	Stem blight	Switzerland (Osterwalder 1906; Müller-Thurgau et al. 1922); Czechoslovakia (Smolak 1925)
Banksia dryandroides Bayt. and Sweet.	Banksia	Root rot	Australia (Hardy and Sivasithamparam 1988)
Banksia hookerana Meissn.	Banksia	Root rot	Australia (Hardy and Sivasithamparam 1988)
Banksia laricina Gardn.	Banksia	Root rot	Australia (Hardy and Sivasithamparam 1988)
Banksia occidentalis R. Br.	Banksia	Root rot	Australia (Hardy and Sivasithamparam 1988)
Banksia praemorsa Andrews.	Banksia	Root rot	Australia (Hardy and Sivasithamparam 1988)
Bayera sesiflora F. Muell.	Banksia	Root rot	Australia (Hardy and Sivasithamparam 1988)
Begonia L.	Begonia	Seedling rot	Germany (Pape 1933)
Beta vulgaris var. *crass* Alef.	Sugar beet	Seedling damping-off	United States (Stevens and Plunkett 1925)
Betula L.	Birch	Bleeding stem canker	United States (Howard 1941)
Boehmeriae nivea (L.) Gaud.-Beaup.	Ramie, Chinese silk plant	Leaf blight	Taiwan (Sawada 1927a)
Brachychiton populineus (Schott. and Endl.)	Kurrajong, bottle tree	Stem canker	Argentina (Frezzi 1950)
Brassica oleracea L. var. *bullata* D. C.	Cabbage	Leaf blight	United States (Young 1926)
Bryophyllum pinnatum (L. Am.) Kurz.	Kalanchoe		Bermuda (Leonian and Geer 1929)
Cactus sp.	Cactus	Root rot	Romania (Tudosescu-Banescu 1966)
Calceolaria integrifolia J. Murr. (*C. rugosa* Ruiz. and Pav.)	Slipperwort, pocketbook or pouch flower	Stem canker	Switzerland (Osterwalder 1909)
Callistephus chinensis (L.)	China aster		Czechoslovakia (Nicklova-Navratilova 1949)
Calytrix angulsta Lindl.		Root rot	Australia (Hardy and Sivasithamparam 1988)
Capsicum annum L.	Green, red, or chili pepper	Seedling rot	United States (Brown and Evans 1933)[a]; Italy (Trotter 1924)
Capsicum frutescens L.	Tabasco pepper	Fruit rot	El Salvador (Crandall et al. 1951)
Caragana arborescens Lam. (*C. sibirica* Medic.)	Siberian pea tree	Stem canker, seedling damping-off	United States (Crandall and Hartley 1938)
Carthamus tinctorius L.	Safflower	Root and stem rot	United States (Thomas 1951)
Carya illinoensis (Wangenheim) K. Koch.	Pecan	Shuck and kernel rot	United States (Reilly et al. 1989)
Castanea sativa Mill. (*C. vesca* Gaertn.)	Spanish or European chestnut	Stem canker	Italy (Curzi 1933)
Castanea spp.	Chestnut		United States (C. O. Smith 1937)[a]; Italy (Cristinzio and Verneau 1954)[a]
Cattleya Lindl. sp.	Orchid	Leaf blight	Philippines (Schwarz 1927)
Ceanothus thyrsiflorus Eschsch.	Blue blossom	Blight	France (Vegh and Berre 1985)

(*continued on next page*)

Table 11.1. *(continued)*

Host	Common name	Disease	Geographical distribution
Cedrus deodara (D. Don) G. Don	Deodar	Seedling rot	Germany (Hartig 1883); United States (C. O. Smith 1937); Soviet Union (Shishkina 1969)
Centaurea L.	Sweet sultan, cornflower, knapweed	Wilt	South Africa (Wijers 1937)
Ceratonia siliqua L.	Carob, St. John's bread, locust bean		United States (C. O. Smith 1937)[a]
Cereus giganteus	Cactus	Tissue rot	Germany (Lebert and Cohn 1870)
Cereus martianus Zucc.	Cactus	Tissue rot	Germany (Himmelbaur 1911)[a]
Cereus peruvianus (L.) Mill.	Hedge citrus	Tissue rot	Germany (de Bary 1881)[a]
Cereus speciusus (Cav.)	Cactus	Tissue rot	Germany (de Bary 1881)[a]
Chrysalidocarpus lutescens Wendl. (*Areca lutescens* Bary)	Areca, yellow, or golden-feather palm		India (Coleman 1910)
Chrysanthemum cinerarifolium (Trevir.) Vis.	Pyrethrum, Dalmation flower	Stem canker	Peru (Crandall and Dieguez 1953)
Citrus limon (L.) Burm. f.	Lemon	Fruit rot	Brazil and Argentina (Fawcett and Bitancourt 1940)
Citrus maxima (Burm.) Merrill. (*C. grandis* (L.) Osbeck)	Shaddock, pomelo	Stem canker	United States (Wagener and Cave 1944); Argentina (Frezzi 1950)
Citrus sinensis (L.) Osbeck	Sweet orange	Stem canker, fruit rot	Formosa (Sawada 1927); Brazil (Fawcett and Bitancourt 1940)
Citrus spp.	Citrus	Stem canker, fruit rot	United States (Fawcett 1920); Puerto Rico (Cook 1939); Argentina (Frezzi 1950); Italy (Verneau 1954)
Citrullus vulgaris Schrad.	Watermelon	Fruit rot	United States (Brown and Evans 1933)
Clarkia elegans Doug.	Clarkia	Stem and foot rot	Germany (de Bary 1881); Sweden (Lindfors and Holmberg 1941)
Clarkia pulchella Pursh.	Farewell-to-spring, godetia		Germany (Himmelbaur 1911)[a]
Cleome violacea L.	Spider flower	Stem and foot rot	Germany (de Bary 1881)
Cleome viscosa L.	Spider plant	Stem and foot rot	Germany (de Bary 1881)
Cornus L.	Dogwood, cornel	Stem canker	United States (Crandall and Hartley 1938)
Cornus florida L.	Flowering dogwood	Crown rot	United States (Creager 1937)
Cornus nuttallii Aud.	Mountain dogwood	Stem canker	United States (Stuntz and Seliskar 1943)
Corylus avellana L.	European hazel or filbert	Fruit rot	Netherlands (Meurs 1928)
Corylus cornuta var. *californica* (A. DC.) Sharp	Beaked hazel or filbert	Stem canker	United States (Stuntz and Seliskar 1943)[a]
Cotoneaster horizontalis Decne.	Rock cotoneaster	Fruit rot	Canada (Conners 1939)
Crataegus L.	Hawthorn, thornapple, red haw		Germany (Himmelbaur 1911)
Crowea saligna Andr.		Root rot	Australia (Hardy and Sivasithamparam 1988)
Cucumis melo L. var. *inodorus* Naud.	Honeydew melon, cantaloupe	Fruit rot	United States (Wiant 1939)
Cucumis sativus L.	Cucumber	Fruit rot	United States (Brown and Evans 1933)[a]
Cydonia oblonga Mill. (*C. vulgaris* Pers.)	Common quince		Formosa (Sawada 1927)[a]; United States (C. O. Smith 1937)[a]
Dahlia spp.	Dahlia	Damping-off	Czechoslovakia (Baudys 1935)
Daphne spp.	Daphne	Root rot	Germany (Nienhaus 1960)
Daucas carota L.	Carrot	Soft rot	United Kingdom (P. H. Williams et al. 1938)
Dianthus caryophyllus L.	Carnation	Stem blight	Switzerland (Müller-Thurgau and Osterwalder 1924)
Diospyros kaki L. f. (*D. chinensis* Blume)	Japanese persimmon, date plum		United States (C. O. Smith 1937)[a]
Echinopsis eyriesii Zucc.	Sea-urchin cactus	Tissue rot	Germany (de Bary 1881)[a]
Elaeagnus angustifolia L.	Russian olive	Canker	United States (Carter 1953)
Epilobium L.	Fireweed, willow herb		Germany (de Bary 1881)[a]
Erica hyemalis Nichols.	White winter or French heather	Stem canker	United Kingdom (Williams et al. 1937)
Eriobotrya japonica Lindl.	Loquat, Japanese plum	Crown rot	Japan (Dufrenoy 1927); United States (Miller 1940)
Escallonia Multis. ex L.f. sp.		Blight	France (Vegh and Berre 1985)
Eschscholtzia californica Cham.	California poppy	Foot and stem rot	Netherlands (Meurs 1928)
Eucalyptus L'Her.	Australian gum iron or stringy bark	Root rot	United States (C. O. Smith 1937)[a]
Euphorbia resinifera Berger.	Spurge		Formosa (Sawada 1927)
Fagopyrum esculentum Moench (*F. vulgare* Nees)	Buckwheat	Seedling rot	Germany (Himmelbaur 1911)
Fagopyrum fagopyri Takm.	Buckwheat	Root rot	Japan (Takimoto 1935)[a]
Fagopyrum tataricum Gaertn.	India or duck wheat	Foot and stem rot	Germany (de Bary 1881)
Fagus grandifolia Ehrh.	American beech	Bleeding canker	Europe (Stevenson 1926)
Fagus spp.	Beech	Seedling damping-off	Germany (Hartig 1876); Switzerland (Osterwalder 1906); England (Hiley 1919); Denmark (Jorgensen 1934); United States (Crandall and Hartley 1938); Poland (Kozlowska et al. 1961); Czechoslovakia (Jancarik 1961)

(continued on next page)

Table 11.1. (*continued*)

Host	Common name	Disease	Geographical distribution
Fatsia japonica (Thunb.) Decne and Planch.	Japanese fatsia rice tree, paper plant	Stem canker	Argentina (Frezzi 1950)
Feijoa sellowiana O. Berg	Feijoa	Brown rot of fruit	Italy (Magnano di San Lio and Tuttobene 1985)
Forsythia Vahl.	Golden-bells		Germany (Himmelbaur 1911)[a]
Fragaria vesca L. × *ananassa* Duch.	Cultivated strawberry	Leather rot of fruit; stem, root, and crown rot	Switzerland (Osterwalder 1912); United States (Rose 1924); France (Arnaud and Arnaud 1930); England (Beaumont and Hodson 1931); Ireland (Anon. 1932); Italy (Passavalli 1937; D'Ercole and Canova 1974; Messina et al. 1989); Germany (Deutschmann 1954); Iran (Ershad 1971); Taiwan (Kao and Leu 1979); Egypt (El-Din et al. 1981); Morocco (Farih and Vanderweyen 1983)
Fraxinus L.	Ash	Seedling rot	England (Collinge 1912)
Fremontodendron californicum (Torr.) Cov.	Flannel bush	Seedling rot, stem canker	United States (Hansen and Thomas 1945)
Fremontodendron mexicanum Davids.	Flannel bush	Seedling rot, stem canker	United States (Hansen and Thomas 1945)
Glycine max (L.) Merrill (*G. soja* Hort.)	Soybean	Seedling and root rot	United States (Skotland 1955)
Godelia willdenowiana Spach.		Foot and stem rot	Netherlands (Meurs 1928)
Gossypium herbaceum L.	Levant cotton	Pod rot	West Indies (Anon. 1929)
Grammatophyllum Blume.	Orchid, kalanchoe	Leaf blight	Philippines (Schwarz 1927)
Grevillea drummondii Meissn.	Grevillea	Root rot	Australia (Hardy and Sivasithamparam 1988)
Hakea francisiana F. Muell.	Hakea	Root rot	Australia (Hardy and Sivasithamparam 1988)
Hakea platysperma Hook.	Hakea	Root rot	Australia (Hardy and Sivasithamparam 1988)
Hebe speciosa Lock. and Al. 'Buxifolia'	Box-leaf	Root rot	Australia (Hardy and Sivasithamparam 1988)
Heteromeles arbutifolia (Ait.) Roem. (*Photinia arbutifolia* (Ait.) Lindl.)	Toyon, Christmas berry	Crown rot	United States (Keim et al. 1976a)
Hevea brasiliesis Muell. Arg.	Rubber	Leaf fall, stem canker, black stripe, root rot	Soviet Union (Novotel'Nova 1974); China (Ho et al. 1984a)
Hibiscus rosa-sinensis L.	Hibiscus	Foot rot	United States (James and Plakidas 1951)
Hibiscus schizopetalus L.	Hibiscus	Foot rot	United States (James and Plakidas 1951)
Humulus lupulus L.	Hop	Black root rot	New Zealand (Brien 1938; Christie 1956); England (Keyworth 1943); United States (Hoerner 1949); Canada (Conners 1955)
Hydrastis Ellis	Orange root, yellow puccoon	Root and stem rot	United States (Westcott 1950)
Hypocalyma angustifolium Endl.		Root rot	Australia (Hardy and Sivasithamparam 1988)
Incarvillea Juss.			Netherlands (Anon. 1963c)
Iris L.	Flag iris lily, Fleur-de-Lis		United States (Stevens and Plunkett 1925)[a]
Jasiminum L.	Jasmine		Germany (Himmelbaur 1911)[a]
Juglans L.	Walnut	Black disease	Italy (Cristinzio and Verneau 1954)
Juglans californica Wats.	Southern black walnut	Crown rot	United States (Barrett and Smith 1926)
Juglans hindsii Jep.	Northern black walnut	Crown rot	United States (R. E. Smith and Smith 1925)
Juglans nigra L.	Black walnut	Seedling rot	United States (R. E. Smith et al. 1912)
Juglans pyriformis Liebm.	Mexican walnut	Stem canker	United States (C. O. Smith and Barrett 1931; C. O. Smith 1937)[a]
Juglans regia L. (*J. regia* var. *sinensis* C. DC.)	English walnut, Madeira nut	Stem canker	United States (Barrett 1928)
Juniperus procera Hochst. ex. Endl.	African juniper	Stem canker	United States (C. O. Smith 1937)[a]
Kalanchoe Adans. (*Bryophyllum* Salisb.)	Palm-beach bells	Collar and stem rot	Italy (Gullino 1975)
Kalanchoe blossfeldiana Poelln.	Palm-beach bells	Stem rot	Germany (Kotthoff 1937)
Larix decidua Mill.	European larch	Seedling rot	Germany (Hartig 1882); England (Hiley 1919); Netherlands (van Poeteren 1930)
Lathyrus L. (*Orobus* L.)	Vetchling, wild pea	Root rot	United States (Clinton 1920)
Lepidium sativum L.	Garden cress	Seedling rot	Germany (de Bary 1881)
Ligustrum L. (*Parasyringa* W. W. Sm.)	Privet, hedge plant		Germany (Himmelbaur 1911)[a]
Lilium L.	Lily	Foot and stem rot	United States (Weiss and Brierley 1926, in Martin 1927); Japan (Asuyama 1935)
Lilium candidum L.	Madonna lily	Foot rot	United States (Drechsler 1926)
Lilium dauricum Ker-Gawl.	Candlestick lily		Japan (Tasugi 1931)
Lilium pyrenaicum Gouan	Yellow Turk's cap lily	Basal stem rot	United States (Drechsler 1926)
Linum usitatissimum L.	Flax	Seedling rot, root rot	United States (Stevens and Plunkett 1925)[a]; Italy (Petri 1941)
Liquidamber styraciflua L.	Sweet, red, or American gum	Bleeding stem canker	United States (Caroselli 1953)
Liriodendron L.	Tulip tree	Seedling rot	United States (Crandall and Hartley 1938)
Lycopersicon esculentum L.	Tomato	Fruit rot	Japan (Tasugi and Kumazawa 1932); United States (Brown and Evans 1933); South Africa (Wijers 1937); England (P. H. Williams et al. 1938)

(*continued on next page*)

Table 11.1. (*continued*)

Host	Common name	Disease	Geographical distribution
Malus pumila Mill.	Apple	Fruit rot, crown and collar rot, sour sap, Moutere dieback	Switzerland (Osterwalder 1906); United States (Barrett 1917; R. E. Smith and E. H. Smith 1925; Baines 1939); United Kingdom (Wormald 1919); New Zealand (Waters 1916); Netherlands (van Poeteren 1922); South Africa (Wijers 1937); Argentina (Lindquist 1938); Italy (Petri 1941); Canada (Welsh 1942); Norway (Ramsfjell 1951); India (Bose and Metha 1951); Poland (Bielenin and Borecki 1970);Spain (Tuset-Barachina 1977); China (Zhuang 1981); Pakistan (Shah and Hasan 1991)
Mammillaria Haw.	Pincushion or strawberry cactus	Tissue rot	Ireland (Lafferty and Pethybridge 1922)
Melocactus nigrotomentosus L.	Cactus	Tissue rot	Germany (Lebert and Cohn 1870)
Melilotus alba Medik.	Clover	Root rot	United States (Jones 1939); Canada Cormack 1940)
Melilotus officinalis (L.) Lam.	Yellow sweet clover	Root rot	Canada (Cormack 1940); United States (Gerdemann 1950)
Myristica fragrans Hout.	Nutmeg		Java (Zimmerman 1901)
Nicotiana tabacum L.	Tobacco	Root rot	England (P. H. Williams et al. 1938)[a]
Nyssa silvatica Marsh (*N. multiflora* Wangenh.)	Pepperridge sour gum	Seedling rot	United States (Crandall and Hartley 1938)
Oenothera biennis L. (*O. muricata* L.)	Evening primrose, German rampion	Seedling rot	Germany (de Bary 1881)
Olea europaea L.	Common olive	Canker	United States (C. O. Smith 1937)[a]
Ostyra Scop.	Hop hornbeam	Stem canker	United States (Crandall and Hartley 1938)
Paeonia albiflora Pallas.	Peony	Bud, leaf and stem blight, downy mildew blight	United States (Thurston and Orton 1921); Japan (Tasugi and Kumasawa 1932); Canada (Conners and Savile 1948)
Panax L.	Ginseng	Root rot	Soviet Union (Bunkina 1957)
Panax quinquefolium L.	Ginseng	Mildew, stem canker, root rot, leaf blight, Koshi-ore	United States (Van Hook 1906; Rahimian and Mitchell 1988; Darmano et al. 1991); Japan (Hori 1907); Korea (Nakata and Takimoto 1922)
Pelargonium radens Moore. (*P. radula* (Cav.))	Geranium, crowfoot	Stem rot	Germany (Schwinn 1960)
Pelargonium tomentosum Jacq.	Herb-scented geranium	Stem rot	Germany (Schwinn 1960)
Persea americana Mill.	Avocado, alligator pear	Bark canker	United States (Barrett 1917)
Photinia arbutifolia (see *Heteromeles* above)	Toyon	Crown rot	United States (Keim et al. 1976a)
Picea L.	Spruce	Seedling blight	United States (Van Hook 1906); United Kingdom (Collinge 1912)
Picea abies (L.) Karst. (*P. excelsa* Link)	Norway spruce	Seedling disease	Germany (Hartig 1883); Italy (Voglino 1916)
Pimilea ferruginea Labill.	Rosy rice flower	Root rot	Australia (Hardy and Sivasithamparam 1988)
Pinus coulteri D. Don.	Big cone pine	Seedling root rot	United States (C. O. Smith 1937)[a]
Pinus mugo Turra (*P. montana* Mill.)	Swiss montana pine		Europe (Stevenson 1926)
Pinus muricata D. Don. (*P. remorata* Mason)	Bishop pine	Seedling rot	United States (C. O. Smith 1937)[a]
Pinus nigra Arnold. (*P. labrico* Poir)	Austrian pine	Seedling root rot	United States (Crandall and Hartley 1938); Canada (Vaartaja 1955)
Pinus ponderosa Dougl. ex P. Laws and C. Laws.	Western yellow pine	Seedling root rot	Germany (Hartig 1883)[a]; United States (Rathbun-Gravatt 1925)
Pinus resinosa Ait.	Red or Norway pine	Seedling root rot	United States (Hartley 1921)
Pinus strobus L.	Eastern white pine	Seedling root rot	Germany (Sorauer 1928)
Pinus sylvestris L.	Scotch pine	Seedling disease	United Kingdom (Collinge 1912); Bulgaria (Dimitroff 1926); Sweden (Hammarlund 1933)
Pisum sativum L.	Garden or English pea	Root rot	United States (Clinton 1920); United Kingdom (P. H. Williams et al. 1938); India (V. B. Singh 1946)
Platanus occidentalis	Eastern sycamore, button-wood, plane tree	Seedling blight	Europe (Hartley 1921)
Primula japonica Gray	Primula	Stem canker	United Kingdom (Dennis and Foister 1942)
Primula obconica Hance	German or poison primula		United States (P. H. Williams et al. 1938)
Prunus americana Marsh	Wild, American, or goose plum	Stem canker	United States (Baines 1939)
Prunus armeniaca L.	Apricot	Bark canker, fruit rot	United States (Barrett 1917); Greece (Sarejanni 1935); India (V. B. Singh 1943); Switzerland (Bolay 1962)
Prunus avium (L.) L.	Sweet cherry, mazzard	Crown and trunk canker	Spain (Unamuno 1910); United States (R. E. Smith and E. H. Smith 1925); Canada (McIntosh 1953)

(*continued on next page*)

Table 11.1. *(continued)*

Host	Common name	Disease	Geographical distribution
Prunus cerasifera Ehrh. (*P. myrobalana* (L.) Loisel)	Cherry palm, myrobalan plum	Stem canker	Germany (Braun and Kröber 1958)[a]
Prunus domestica L.	Plum	Stem canker	United States (R. E. Smith and E. H. Smith 1925)
Prunus dulcis Mill. D. Webb (*Prunus amygdalus* Batsch.)	Almond	Bark canker, collar rot, foot rot	United States (E. H. Smith 1915); Chile (Pinto 1966)
Prunus ilicifolia (Nutt. ex Hook and Arnott) Walp.	Evergreen cherry, mountain holly	Stem canker	United States (C. O. Smith 1937)[a]
Prunus mahaleb L.	Mahaleb cherry, St. Lucie cherry	Stem canker	United States (Baines 1939)[a]
Prunus mume Siebold and Zucc.	Japanese apricot		United States (C. O. Smith 1937)[a]
Prunus persica (L.) Batsch	Peach	Bark canker, seedling blight	United States (R. E. Smith and E. H. Smith 1925); France (Arnaud and Arnaud 1931); Italy (Curzi 1933)
Prunus sp.	Prunus		Spain (Unamuno 1911)
Pseudotsuga menziesii (Mirb.) Franco	Douglas fir	Crown rot	United States (Pratt et al. 1976; Adams and Bielenin 1988)
Psidium guajava L.	Guava	Fruit rot, storage rot	United States (Clinton 1920); India (Singh 1943); Canada (Conners 1955)
Pyrus communis L.	Pear	Collar rot, stem canker, fruit rot	Switzerland (Osterwalder 1906); Belgium (Marchal 1908); Spain (Unamuno 1911); England (Wormald 1919); Canada (Güssow 1920); United States (Clinton 1920; Rose and Lindegren 1925); France (Arnaud and Barthelet 1929); Czechoslovakia (Baudys 1930); India (Singh 1943); New Zealand (H. C. Smith 1955)
Quercus agrifolia Nee	California live oak	Trunk canker	United States (Mircetich et al. 1977)
Quercus lobata Nee	Valley oak, California white oak	Seedling rot	United States (C. O. Smith 1937)[a]
Rheum rhaponticum L.	Rhubarb, pie plant	Crown rot	United States (Beach 1922); Germany (Nienhaus 1959)
Rhododendron L.	Azalea, rhododendron	Tuber, stem, and seedling rot	United States (White 1930); France (Barthelet 1934); Germany (Kaven 1934); Australia (Anon. 1950)
Ribes grossularia L.	European gooseberry	Twig and flower blight	Switzerland (Müller-Thurgau et al. 1922); Netherlands (van Poeteren 1922); New Zealand (H. C. Smith 1950); United Kingdom (Brooks 1953)
Ribes spp.	Currant, gooseberry	Leaf scorch wilt	Germany (Schmidle 1961)
Robinia pseudoacacia L.	False acacia, locust	Seedling blight	Germany (Hartig 1883); Europe (Stevenson 1926)
Robinia sp.	Locust	Seedling blight	England (Collinge 1912)
Rosa laevigata Michx.	Cherodee rose	Root rot	United States (Dennis and Foister 1942)
Saintpaulia ionantha Wendl. (*S. kewensis* Clarke)	African violet		Greece (Sarejanni 1935)
Salix scoulerana Barrett	Mountain willow	Stem canker	United States (Stuntz and Seliskar 1943)
Salix scoulerana Brtt. ex. Hook	Scouler willow	Seedling blight	United States (Stuntz and Seliskar 1943)
Salix sp.	Willow	Bleeding stem canker	United States (Caroselli 1953)
Salpiglossis sinuata Ruiz and Pav.	Painted-tongue	Stem canker	France (de Bary 1881)[a]
Schizanthus pinatus Ruiz	Butterfly flower, poor man's orchid	Stem canker	France (de Bary 1881)[a]
Schinus molle L.	Pepper tree, Peruvian mastic tree, molle, pirul	Stem canker	Argentina (Frezzi 1950)
Sempervivum glaucum Schmittsp. and Lehm.	Houseleek, hen and chickens	Stem, leaf, and flower rot	Germany (Schenk 1875)
Sempervivum spp.		Root rot	Ireland (Lafferty and Pethybridge 1922)
Sempervivum tectorum L.	Houseleek	Stem, leaf, and flower rot	Germany (Schenk 1875)
Sesamum indicum L. (*S. orientale* L.)	Sesame	Stem canker, fruit canker	Peru (Crandall and Dieguez 1948)
Sinningia speciosa (Lodd.) Hiern.	Gloxinia, violet slipper	Stem canker	Netherlands (van Poeteren 1926)
Solanum capiscastrum Link ex. Schauer.	False Jerusalem cherry		United Kingdom (P. H. Williams et al. 1938)[a]
Solanum melongena L.	Eggplant, Brinjal	Seedling blight	Italy (Trotter 1924); United States (Caroselli and Howard 1939)
Solanum tuberosum L.	White or Irish potato	Root rot	South Africa (Wijers 1937)[a]; United Kingdom (P. H. Williams et al. 1938)[a]; United States (Caroselli and Howard 1939)[a]; Germany (Braun and Kröber 1958)
Syringa vulgaris L.	Common lilac	Leaf and twig blight	United States (White 1930); France (Vegh and Berre 1985)
Theobroma cacao L.	Cacao	Pod rot	West Indies (Massee 1899)
Tsuga orientalis L.	Cypress	Stem canker	United States (C. O. Smith 1937)[a]

(continued on next page)

Table 11.1. (*continued*)

Host	Common name	Disease	Geographical distribution
Tilia sp.	Linden, basswood, lime tree	Bleeding stem canker	United States (Caroselli 1953)
Tulipa L.	Tulip	Blossom blight, stem canker	United States (Stevens and Plunkett 1925); Netherlands (Schenk 1926); England (Foister 1930); Canada (Conners 1939); Germany (Schönbeck 1967)
Ulmus sp.	Elm	Bleeding stem canker	United States (Howard 1941)
Vanda coerulea Griff. ex Lindl.	Blue orchid	Stem canker	India and Philippines (Schwarz 1927)
Verbena L. (Glandularia Gmel.)	Verain	Wilt	South Africa (Wijers 1937)
Vicia faba L.	Broad bean	Blight	United States (Cooper and Porter 1928)[a]
Vicia sp.	Vetch	Black rot	Canada (Conners 1955)
Vigna unguiculata L. Walp. (*V. sinensis* (L.) Savi ex Hassk.)	Cowpea	Red stem canker	United States (Weimer 1949)
Viola sp.	Pansy	Collar rot, stem canker	Germany (Stahl 1965); United Kingdom (Moore 1948)

[a] Artificially inoculated host.

Table 11.2. Sizes (length and breadth) of sporangia of *Phytophthora cactorum* recorded in different reports

Host	Range of length × breadth (μm)	Average (μm)	Length-breadth ratio	Reference
Panax	. . .	36.62 × 26.06	1.41 : 1	Rosenbaum (1917)
Lillium	. . .	35.30 × 27.60	1.27 : 1	Drechsler (1926)
Several	19–46 × 16–32	30.39 × 22.77	1.33 : 1	Tucker (1931)
Several	21–54 × 17–46	34.0 × 25.5	1.45 : 1	Frezzi (1950)
Malus	29–43 × 21–32	33.5 × 26.7	1.25 : 1	Braun and Kröber (1958)
?	55 × 40 (max.)	36–50 × 28–35	1.3–1.4 : 1	Waterhouse (1963)
Fragaria	24–50 × 19–33	36.5 × 27.0	1.33 : 1	Ershad (1971)
Several	24–55 × 19–40	35.7 × 26.8	1.33 : 1	Kröber (1985)
?	19–73 × 16–41	27–58 × 19–32	1.3–1.4 : 1	Gerrettson-Cornell (1989)
Several	. . .	31 ± 4.8 × 26 ± 4	1.2 ± 0.1 : 1	Oudemans and Coffey (1991a)

1966) may have been caused by unfavorable environmental conditions. Several isolates from different hosts, including ginseng (*Panax quinquefolium* L.), formed chlamydospores abundantly in V8 juice-calcium carbonate broth when incubated for 20 days at 4°C, but none were produced at 8, 12, 16, 20, or 32°C. A few developed at 24 and 28°C. Chlamydospores did not form in potato-dextrose broth or pea seed broth. Chlamydospores in mycelial mats formed when buried in potting mix soil at 4°C. Chlamydospores germinated well (60 to 80%) even when subjected to freezing at –23°C for 24 h (Darmano and Parke 1990). A majority of the chlamydospores produced are terminal, although occasionally some are intercalary. The two-layered, single-wall thickness is usually 1 to 1.5 μm. Diameters (25 to 39.7 μm) of chlamydospores recorded in different reports are given in Table 11.3.

Sex Organs

P. cactorum is homothallic. Most isolates produce oospores in culture media and in diseased plant tissue. Antheridia are paragynous and unicellular and nearly spherical to club-shaped and often occur in a knot of hyphae. Antheridia measure 8.5 to 21 × 12 to 21 μm (Frezzi 1950); 15 × 13 μm (Waterhouse 1963); 9 to 20 × 7 to 13 μm, average 13.1 × 10.1 μm (Ershad 1971); and 9 to 12 × 9 to 20 μm (Gerrettson-Cornell 1989). Oogonia are smooth-walled and usually hyaline. Oospores have been reported to be plerotic with an average wall thickness of 2 μm; however, Oudemans and Coffey (1991a) reported oospores were aplerotic. Dimensions of oogonia and oospores in different reports are given in Table 11.4. Germination of oospores by production of sporangia or mycelium has been reported many times. See Table 3.3 in Chapter 3. Germination in the field begins after a rest period during the winter at soil temperatures of about 7.5°C and then increases rapidly within 22 days to 82% as the soil temperature increases to 13°C (Braun and Nienhaus 1959).

Growth Temperatures

The minimum temperature for growth is 2°C, the optimum 25°C, and the maximum 31°C.

Growth Characteristics in Culture Media

Colonies are slightly radiate, compact without a definite border, and fluffy but not dense and have short aerial hyphae. See Chapter 4.

Distinguishing Characteristics

In group I, *P. cactorum* is distinguished from *P. iranica* and *P. pseudotsugae* by its production of caducous sporangia with short pedicels. *P. iranica* and *P. pseudotsugae* produce noncaducous sporangia. Some of the antheridia of *P. clandestina* are amphigynous and some paragynous, whereas those of *P. cactorum* are mostly paragynous. Oospores of *P. cactorum* are plerotic, but those of *P. clandestina* are aplerotic. Isozyme analysis and molecular characterization (mtDNA RFLP) of *P.*

Table 11.3. Diameters of chlamydospores recorded in different reports

Range (μm)	Average (μm)	Reference
23–43	31.6	Tucker (1931)
19–53.5	33.2	Frezzi (1950)
53 (max.)	33.0	Waterhouse (1963)
14–55	25–33	Gerrettson-Cornell (1989)
28–56	39.7	Darmano and Parke (1990)

cactorum (Förster and Coffey 1991; Oudemans and Coffey 1991a) showed it to be highly uniform and distinct from other species (Figure 4.32 in Chapter 4).

Diseases Caused by *Phytophthora cactorum*

Because of the importance of *P. cactorum* as the causal agent of so many important diseases, especially on woody plants (e.g., apple, *Malus pumila* Mill., or pear, *Pyrus communis* L.), we emphasize description and control of the root, collar, and crown rot diseases of these crops and especially of apple. The factors that favor disease development and control are similar for diseases of many crops. More detailed information on these diseases can be obtained in the American Phytopathological Society's compendiums on diseases of apple and pear (A. L. Jones and Aldwinckle 1990), rhododendron and azalea (Coyier and Roane 1986), and strawberry (Maas 1984).

General recommendations for control include planting clean nursery stock on well-drained soil, preventing irrigation water from contacting the trunks of trees, planting the most resistant cultivar that is recommended, and, if necessary, applying a fungicide such as metalaxyl (Table 7.7) or fosetyl-Al (Table 7.11) if registered for use. Many references pertinent to host resistance are listed in Chapter 6 (Table 6.1); cultural methods of control are discussed at greater length in Chapter 5. Fungicidal control is covered in Chapter 7, and specific control measures are discussed for each of the diseases cited here.

Crown, Collar, and Root Rot of *Malus pumila* Mill. (Apple) and *Pyrus communis* L. (Pear)

Baines (1935, 1939) was probably one of the first to describe trunk rot and collar rot of apple. A similar disease occurs on pear trees (*Pyrus communis* L.) (Rose and Lindegren 1925; Braun and Schwinn 1963). These diseases are of importance in all countries where apples, pears, and many other woody plants are grown.

The impact of *P. cactorum* and *P. syringae* as causal agents of crown, collar, and root rot of apple with special emphasis on the problem in the United Kingdom is reviewed by D. C. Harris (1991). Mircetich (1981) and Mircetich and Browne (1987) review and summarize the information on *Phytophthora*-caused diseases of deciduous fruit trees in California. The biology and control of crown rot of apple in southwestern Canada is reviewed by Utkhede (1986). Recognition and control of crown, collar, and root rot of apple and pear trees is reviewed in the American Phytopathological Society's compendium by Jones and Aldwinckle (1990).

Other species of *Phytophthora* that cause crown and root rot of apple include *P. megasperma* (Jeffers et al. 1982), *P. citricola, P. cambivora,* and *P. cryptogea* (Jeffers and Aldwinckle 1988). Utkhede and Smith (1991) report that *P. cactorum, P. cambivora,* and *Pythium ultimum* are pathogenic on apple seedlings. Mircetich and Browne (1987) also emphasize that several species of *Phytophthora* in addition to *P. cactorum* are responsible for crown and collar rot of deciduous fruit trees such as cherry, peach, and apricot.

Symptoms of root and crown rot are shown in Plates 11.1–11.7. Apple trees initially infected by *Phytophthora* are difficult to distinguish from healthy trees unless one looks carefully for symptoms on the crown (the rootstock; Plate 11.1), collar (the scion), or roots (Plate 11.2). By the time visible symptoms are observed on the foliage of the tree, considerable development of root, collar, and/or crown rot has already occurred, and at this stage it is difficult to save the tree by applying standard therapeutic control measures. When rainy conditions favor disease, even the large scaffold branches can become infected (Plate 11.3).

Crown and collar rot usually cause an abnormally light green color of the foliage. Unaffected areas of the tree retain their normal green color and may bear a crop for several years. Later, terminal growth is stunted and the small, chlorotic leaves droop. In some instances, only one or two segments of the foliage turn purple or bronze, but in more severe cases the entire tree is affected. As the disease progresses, the leaves drop and fruits are exposed to sunburn, which results in a loss of quality (Plates 11.4 and 11.5).

Other symptoms, depending on the cultivar that is infected, include a slow decline or a sudden collapse and death of the tree in a single growing season, delayed bud break in the spring, and twig dieback during the growing season. These symptoms are secondary and related to the initial infection that occurs at the collar or crown of the tree. Crown or collar rot eventually leads to girdling of the trunk and death of the infected trees.

Care must be taken to distinguish aboveground symptoms from those caused by the viral disease union net necrosis, which kills the tissue at the graft union; by *Sclerotium rolfsii, Rosellinia,* and Fusarium root rots; or by drought, rodent injury, freeze damage, and early natural coloration of leaves in the fall on trees grafted on rootstocks such as MM.26.

To diagnose collar or crown rot, the soil from the crown region of the plant should be removed so that characteristic lesions that occur on the crown, collar, or roots can be seen (Plates 11.1 and 11.2). Isolations can then be made on selective media (see Chapter 2) to detect *Phytophthora.*

Collar rot on apple trees (Plate 11.1) affects the scion cultivar if the trees were planted with the graft union close to or in contact with the soil. Crown rot, on the other hand, usually affects the rootstock. Scraping or cutting away the outer bark from infected crowns reveals dead bark (often with a reddish

Table 11.4. Diameters of oogonia and oospores of *Phytophthora cactorum* recorded in different reports

Oogonia		Oospores		
Range (µm)	Average (µm)	Range (µm)	Average (µm)	Reference
23.4–36.7	26.5–29.4	17.5–30.0	23.3–26.5	Tucker (1931)
23.0–34.0	28.2	20.0–29.5	25.0	Frezzi (1950)
...	25.0–32.0	...	20.0–26.0	Waterhouse (1963)
...	...	18.0–30.0	24.6	Ershad (1971)
23.0–35.0	27.4	18.0–27.0	24.0	Kröber (1985)
18.0–40.0	23.0–30.0	15.0–35.0	25.0–27.0	Gerrettson-Cornell (1989)
20.0–36.0	27.5	18.0–30.0	26.4	Darmano and Parke (1990)
...	32.9 ± 2.5	...	27.7 ± 2.5	Oudemans and Coffey (1991a)

tinge). A sharp demarcation line can be observed between diseased, reddish brown bark and healthy, white bark. Young cankers are difficult to detect unless the outer bark is peeled back so that the discolored (infected) areas can be seen. In some instances, a characteristic sour odor can be detected when the rotted bark is lifted (Robertson and Dance 1971). In some cultivars, a moist exudate covers the cankered areas. If crown and collar rots are not diagnosed before secondary foliar symptoms appear, the damage is usually too extensive to be treated successfully. Rapid extension of the cankered areas leads to girdling and death of the tree in a single season, although in the majority of cases, the cankers are confined to one side of the trunk during the first season of infection.

The root rot phase of the disease (Plate 11.2) can be detected only by removal of sufficient soil to expose a substantial part of the root system. Failure of some rootstocks to remain upright during windstorms is often the result of the rotting of feeder roots. Many trees, particularly those on MM.104 and MM.106 rootstocks that lean or blow over in strong winds, are affected by root rot. These trees rarely show aboveground symptoms and appear to be growing normally.

Fruit rot caused by *P. cactorum* is generally of minor economic importance; however, fruit rot caused by *P. syringae* can lead to economic losses (see Chapter 60; D. C. Harris 1991). Fruit infection usually follows contact with soil particles that splash upward during heavy rain. The infected tissues become spongy, mealy, or granular because of the dissolution of the middle lamellae between the cells (Plate 11.6). After the infected fruit becomes mummified (Plate 11.7), it can be a source of inoculum in the form of oospores that form in the decayed tissue.

On nursery rootstocks symptoms are different from those observed on older bearing trees. On young rootstocks, *Phytophthora* infection may be indicated by 1) broken and red brown discoloration of roots, 2) lesions where the roots emerge from the rootstocks, 3) reddish brown discoloration that is visible when clumps of roots are cut, and 4) dark brown, soft, mushy lesions on older roots (Julis et al. 1978). Since nursery trees may be infected without showing any visible symptoms, rating their health only by visual examination is often inadequate. If disease-free plants are necessary, inspection should be accompanied by isolation technology (Chapter 2).

Because early symptoms of *Phytophthora* are difficult to detect, *P. cactorum* can be detected by plating of soil or roots directly on a selective medium or by baiting soil with susceptible apple seedlings or pear fruit (see Chapter 2 and D. C. Harris and Bielenin [1986]). Baiting soil with 2- to 3-week-old apple seedlings (Jeffers and Aldwinckle 1987) is a sensitive bioassay for the detection of *P. cactorum* in apple orchard soils. Seedlings are floated on a slurry containing 30 ml of soil flooded with 60 ml of water. After incubation at 21°C for 4 to 7 days, *P. cactorum* can be detected microscopically by noting the formation of sporangia on infected seedlings (abbreviated bioassay). The modification of the method in which the soil is air dried at room temperature, remoistened with 10 to 15 ml of water, and incubated for a few days before flooding (extended bioassay) greatly increased the percentage of detection of *P. cactorum* in soils collected in several orchards (Jeffers and Aldwinckle 1987; Appendix 2.8). Drying the soil probably reduces the bacterial population in the soil and allows resistant oospores to germinate in the remoistened soil. Flooding favors production of sporangia and zoospores, which increases the amount of inoculum in the soil. When soil was diluted and spread on a selective medium, germinated oospores were the only propagules confirmed to be a source of *Phytophthora* colonies (D. C. Harris and Bielenin 1986). This shows the importance of oospores as a means of perpetuating *Phytophthora* in soil. Rahimian and Mitchell (1988) improved isolation of *P. cactorum* from soil by use of the extended method but modified it by flooding the soil at 8°C and plating the soil slurry on a selective medium. Zoospores of *P. cactorum* formed readily at 8°C, but those of *Pythium* did not.

Species of *Phytophthora* such as *P. cambivora* or *P. cryptogea* that do not produce oospores are more readily detected by omitting the drying of soil step (see Chapter 2, Appendix 2.8) and baiting with excised apple twigs (Jeffers et al. 1981).

Disease Development. Infection of apple or pear trees occurs in the late fall prior to dormancy or during damp, cool weather in the spring, usually from pre-bud through blossom formation. The period of the greatest susceptibility is during leaf and flower development (Braun and Nienhaus 1959). *P. cactorum* can penetrate unwounded bark; however, wounds caused by insects, rodents, or mechanical injury are commonly the most favorable infection courts.

P. cactorum is commonly brought into orchards on infected seedling trees or rootstocks. It spreads from diseased to healthy trees within an orchard by means of surface irrigation, run-off water, and flooding of the soil. Custom operators who move machinery and implements from orchard to orchard increase the danger of spread of *Phytophthora* (Mircetich and Browne 1987). Also, some sources of irrigation water can be contaminated by different *Phytophthora* species (McIntosh 1966).

The incidence and severity of Phytophthora crown and collar rot are usually higher in orchards in which the soil is subjected to prolonged periods of flooding. Severe symptoms can occur on sandy, loamy, or clay soils; however, heavier soils are most conducive to disease because water drains more slowly. The rate of infection and severity depends on the initial level of inoculum in soil, the soil temperature, and the amount of soil moisture present. In many plants, the trunk tissue is more susceptible and allows more rapid infection when the tree is heavily irrigated (Tippett and Hill 1983).

Control. Control is difficult, since *P. cactorum* survives for years in orchard soils either as mycelium or oospores (Braun and Schwinn 1963; Kröber 1980; D. C. Harris and Bielenin 1986; O. K. Ribeiro, *unpublished*; see Table 5.4 in Chapter 5).

The use of sanitation in production of disease-free nursery trees is important, but unfortunately disease-free trees are often difficult to find because of inadequate methods of growing seedlings free of pathogens. In a sampling of trees from 15 commercial nurseries in the United States and Canada, 88% of grafted planting stock and as much as 97% of unbudded rootstocks were infected with *P. cactorum* and/or *P. cambivora* (Jeffers and Aldwinckle 1988). Infected nursery rootstocks rarely show visible symptoms well enough to be eliminated by visual inspection. In Arizona, an epidemic of crown rot caused by *P. cactorum* on clean land was traced to contaminated nursery trees (Matheron et al. 1988). One of us (Ribeiro) has documented the loss of a large number of trees in fumigated soils within the first year because seedling trees were contaminated. Tidball and Linderman (1990) isolated both *P. cactorum* and *P. cambivora* from nursery trees in stool beds. The protocol recommended by Pegg (1978) for development of disease-free avocado plants in Australia (Appendix 5.2) is a worthy approach to nursery sanitation.

Phytophthora contamination of nursery tree stock can be effectively reduced or eliminated by soaking roots in metalaxyl, fosetyl-Al, or copper hydroxide for 1 h at a concentration of 1,000 mg (active ingredient) per liter (Jeffers 1992). Similar results are also reported by Tidball and Linderman (1990). Although this method may be an effective control measure, especially if metalaxyl and fosetyl-Al are registered for use as root dips, there is always the danger that latent inoculum in roots may survive after the fungicide has broken down in the field. The report by Duncan (1985b) that oospores of *P. fragariae* were resistant to fungicides (probably because of their thick, impenetrable walls) implies that fungicides to eradicate latent inoculum should be used with caution.

Therapeutic treatments with some chemicals are effective if applied to trees during the early stages of infection. Because metalaxyl, in particular, is systemic and highly effective, it is presently one of the chemicals of choice in many apple-growing regions. Metalaxyl plus mancozeb applied as a soil drench is also effective (Utkhede 1984c, 1986, 1987a). Soil drenches with metalaxyl plus captafol prevented infection of newly planted trees (Ellis et al. 1986). Shanking in metalaxyl close to the root zone by a specially modified plow has been successfully used in several orchards in the Pacific Northwest apple-growing region of the United States. The machinery for shanking was devised by R. Trumbull, Complete Fertilizer and Chemical Company, Kennewick, Washington 99337.

Fosetyl-Al is also effective and can be used if registered. Several reports indicate that fosetyl-Al, which is systemic and moves upward and downward, has curative properties when used as a foliar, trunk, or trunk injection application on infected trees. Trunk injection (De Boer et al. 1990) and foliar and trunk paint with fosetyl-Al (P. A. Taylor and Washington 1984) on peach, foliar spray on cherry (Bielenin and Jones 1988), rootsoak on apple (Jeffers 1992), and foliar spray trunk injection or fungicidal paste on cankers (D. C. Harris 1987) on apple were effective treatments.

Copper formulations painted on the trunk at the base of the tree are effective for prevention of infection from the splashing of contaminated soil but are not effective for control of internal infection that progresses from the roots into the crown and trunk.

See Chapter 7 for more information on chemical control, Table 7.7 for references on control with phenylamide fungicides such as metalaxyl, and Table 7.11 for references on control with the phosphonates such as fosetyl-Al. Once visible foliar symptoms appear, trees rarely respond to any form of treatment.

Biological control of *P. cactorum* by use of the bacterium *Enterobacter aerogenes* has been reported by Utkhede (1987a, b). When the fungus *Trichoderma* was added to the planting hole and watered in 2 weeks prior to planting the trees, colonization of apple trees was prevented (Valdebenito-Sanhuenza 1987). A 1:1 mixture of composted hardwood bark (Spring et al. 1980) and field soil in the planting hole effectively reduced the number of infected apple trees (Ellis et al. 1986). Ground covers of crown vetch and grasses such as *Lolium perenne* and *Festuca rubra* reduced the incidence of crown rot, whereas a straw mulch resulted in as much as 35% incidence of crown and root rot (Merwin et al. 1992).

Resistance of various rootstocks is one of the most important control measures. Resistance is relative, since all rootstocks can become infected if subjected to adverse growing conditions, such as poor soil drainage, excessive fertilizer application, or winter injury, or to a high inoculum level of the pathogen. Apple rootstocks M.104 and M.106 are very susceptible to *P. cactorum;* MM.111 and M.26 rootstocks are less susceptible; open pollinated seedling rootstock and M.9 are usually the most resistant (A. L. Jones and Aldwinckle 1990).

Since many species other than *P. cactorum* also cause crown rot (Mircetich and Browne 1987; Browne and Mircetich 1993), resistance in a rootstock to one or more of these species of *Phytophthora* does not necessarily imply that it will be resistant to others. Browne and Mircetich (1993) determined that rootstocks M.9EMLA, Mark, Bud.118, and Bud.9 were highly resistant.

The variation in susceptibility of plants during different seasons complicates the evaluation of cultivars for resistance. This subject is reviewed by Mircetich and Browne (1987), who cite examples by Bielenin (1977), Gates and Milliken (1972), Jeffers and Aldwinckle (1986), and Sewell and Wilson (1964). In Jeffers and Aldwinckle's report (1986), testing for resistance and susceptibility of rootstocks is often conducted by inoculation of twigs. When twigs from two rootstocks (MM.111 is field-resistant and MM.106 is susceptible) were inoculated and incubated in sealed jar moist chambers at 22°C, the susceptibility of the rootstocks could be measured by the size of the lesion formed 15 days after inoculation. The size of lesions differed significantly, depending on the season and on the species of *Phytophthora* tested. *P. cactorum* and *P. cambivora* were more active on twigs collected during the spring and summer, but *P. megasperma, P. cryptogea,* and an unidentified species were most active on twigs collected during the summer and winter. The maximum amount of necrosis caused by all species was greatest on collections made in early June.

After trees break dormancy in the spring, major physiological changes occur that increase susceptibility. D. C. Harris (1991) and Mircetich and Browne (1987) review the complexity of evaluating root rot resistance in pome fruit trees. Browne and Mircetich (1993) list rootstocks of apple that are resistant to *P. cactorum, P. cambivora,* and *P. cryptogea.*

Inarch grafting of crown rot-infected trees from the trunk to the roots with resistant rootstocks has been accomplished successfully in some orchards (Plate 11.8). This treatment is rarely effective if more than 50% of the crown of the tree is infected and is expensive.

Fruit Rot of Pear and Apple

P. cactorum also causes a rot of apple and pear fruit. Infection results in dark brown lesions that quickly cover the whole fruit. At this stage, the disease on fruit is often mistaken for the fire blight disease caused by *Erwinia amylovora. P. cactorum* infection is indicated by a white mycelial fuzz that develops on the surface of the fruit. Apple fruits normally become infected close to maturity, whereas pear fruits are susceptible from an early stage of development. If not controlled, the infection can move from fruit into the spurs and cause twig dieback. Disease development is most rapid in orchards where overhead sprinklers are used or when soil is splashed onto low-hanging fruit during heavy rain showers.

If irrigation water comes from a source that is contaminated with *Phytophthora,* addition of a low concentration of copper (1 to 2 ppm) to the water will kill the spores and reduce inoculum available for fruit infection (Ribeiro and Linderman 1991). Higher levels of copper can result in fruit russetting, so care has to be taken when applying copper. Precise computerized meters are available for this purpose but are expensive. Infected fruit from the trees as well as those that have fallen on the ground should be removed whenever feasible because oospores form and overwinter in infected fruit.

Bleeding Canker of *Acer* Species (Maple)

Two types of symptoms are recognized. 1) Chronic symptoms, which usually occur in older trees, include a reduction in the size and number of leaves and change in color to a yellow green with dieback of the branches. Infected trees are eventually killed (Caroselli and Howard 1939). 2) Acute symptoms, in which the cambium is killed, include a furrowed appearance on the trunk with exudation of deep red fluid from the fissures in the cankered bark. The appearance of dried blood is the source of the name, bleeding canker. Transverse sections show characteristic reddish brown necrotic areas with well-defined olive green margins. Mildly infected trees usually recover, but severely infected trees succumb to the disease (Caroselli and Howard 1939).

Collar rot and basal canker of sugar maple (*Acer saccharum* Marsh.), caused by *P. cactorum,* has been described in detail by Drileas et al. (1982). Several other tree crops are also susceptible (Table 11.4).

Disease Development. The disease occurs mainly in areas with excessive irrigation and poor soil drainage.

Control. Factors that contribute to succulent growth, e.g., large amounts of high-nitrogen fertilizer and excessive irrigation, should be corrected (Caroselli and Howard 1939). Fungicides, such as copper formulations, metalaxyl, or fosetyl-Al, can be used if economically feasible.

Fruit Rot of *Citrullus vulgaris* Schrad. (Watermelon)

Fruit rot of melons usually appears late in the picking season. Tan spots often with brown zonate margins occur on melons. The rot of the internal tissues is watery, which results in a total collapse of the fruit, particularly during storage in the warehouse. The disease is described by Brown and Evans (1933), Wiant (1939), and Wiant and Tucker (1940).

Disease Development. The disease progresses rapidly at temperatures between 18 and 30°C, especially if the fruit has been exposed to irrigation water or rain. Insects in the field are also known to spread inoculum from rotted to healthy fruit.

Control. Flooding of land should be prevented, and melons should not be replanted in infested fields for at least 4 years.

Crown Canker of *Cornus* spp. (Dogwood)

Crown canker was described by Creager (1937) and Stuntz and Seliskar (1943). Crandall and Hartley (1938) showed that infection by *P. cactorum* was an important seedling disease in forest nurseries. Cankers occur on the stem either at or near the soil line on the main root and are discernible only by close examination. As infection progresses, the first observable symptoms that appear in the foliage in the upper portions of the tree consist of a lighter green color, after which the leaves prematurely turn bronze. Branches and twigs often die back in patches, and by the end of the second year all branches become infected. These symptoms are caused by the girdling effects of the cankers on the stems. The bark of severely diseased trees sloughs off, exposing the wood.

Disease Development. The disease has been reported only on transplanted trees. The pathogen is believed to enter the tree through wounds caused by cultivating tools or through injuries caused by other means.

Control. Trees should be obtained from a disease-free source. If cankers are detected in the early stages, surgical removal of the diseased tissue is effective, followed by painting with a fungicidal copper formulation. Metalaxyl or fosetyl-Al should be effective if registered for use. Replanting in infested nurseries should be conducted only in soil that has been steamed or fumigated with methyl bromide or another suitable fumigant.

Crown Rot of *Juglans regia* L. (English Walnut)

The first signs of crown rot infection are the appearance of sparse, yellow green foliage, poor twig growth, and somewhat stunted growth of the tree. Infected trees often set a heavy crop of nuts because of the girdling effect of the disease on the crown tissue. If infection of the tree is in an advanced stage, the tree may wilt in mid-summer; this results in premature leaf fall and usually death of the tree.

Cankers form on the trunk below the soil line and are dark brown or black and irregularly shaped. The infected areas are soft or spongy and usually limited to the bark. Although the wood is not infected, discoloration often occurs below the infected bark. A black fluid often oozes out from the affected cambial region, and in advanced stages, the lateral roots and taproots are killed. See C. O. Smith and Barrett (1931) for a thorough description and illustrations of symptoms.

Several other species of *Phytophthora* that are involved in root and crown rot of walnuts include *P. cactorum, P. citricola, P. cinnamomi, P. citrophthora, P. cryptogea, P. megasperma,* and four unidentified *Phytophthora* species (Mircetich and Matheron 1983). This was the first report of *Phytophthora* species other than *P. cactorum* being involved with crown rot of walnut in California. *P. cryptogea, P. parasitica,* and two unidentified species of *Phytophthora* were shown later to be the cause of crown rot of English walnut for the first time in California (Matheron and Mircetich 1985b). Other *Phytophthora* species that also cause root rots of seedling walnut trees in nurseries include *P. cinnamomi* (Crandall 1936) and *P. citricola* (Green and Pratt 1970).

Disease Development. The root and crown canker diseases occur most frequently on heavy, poorly drained, or excessively wet soils that are improperly irrigated or subject to prolonged heavy rains.

Oak trees are often killed by *P. cactorum* (Mircetich et al. 1977). In urban areas where walnut or oak trees are allowed to remain in landscape plantings after houses are built in old groves, trees are commonly killed by *Phytophthora* because the trunks are sprinkled frequently.

Control. New plantings should be started with disease-free plants grafted on resistant rootstocks on land that is well-drained. The Paradox hybrid (*Juglans regia* × *J. hindsii* (Jeps.) Jeps. California black walnut) rootstock and *Pterocarya stenoptera* C. DC. rootstock that have been used in China were more resistant than *J. hindsii* or *J. regia* to several *Phytophthora* species (Mircetich and Matheron 1983; Matheron and Mircetich 1985c). See also Chapter 16 on *P. citricola.*

On trees with established infection, removal of the soil from the base of the cankered area allows the tissue to dry out. C. O. Smith and Barrett (1931) stated that in California (dry summers), the English walnut was resistant under ordinary orchard conditions, and prevention of irrigation water from running close and ponding near trees aided in control. Monthly soil drenches with metalaxyl or fosetyl-Al controlled root rot of English walnut in soil artificially infested with *P. citricola* or *P. cinnamomi.* The treatments also suppressed canker formation caused by *P. citricola* or *P. citrophthora* (Matheron and Mircetich 1985e).

Leather Rot of *Fragaria* spp. Fruit (Strawberry)

Leather rot of fruit caused by *P. cactorum* was described by Rose (1924) and is now of economic importance to strawberry growers worldwide (Ellis and Grove 1983; Maas 1984; Grove et al. 1985c; Madden et al. 1991). Under conditions favorable for disease development, more than 80% of the strawberry fruit crop may become infected in affected areas. Berries at all stages of development are susceptible. If excessive rainfall occurs during late spring and early summer, calyxes, peducles, and fruit stalks may also become infected. Dark brown lesions occur on infected green fruit as the fruit ripens. These lesions later turn light brown with purple edges. Infected, partially ripe fruit rarely fully ripen. Diseased berries become tough, rubbery, or leathery in texture, hence the name leather rot. A brown discoloration of the fruit occurs both internally and externally, and although the berry itself may become slightly soft, it is never mushy or watery. Discolored vascular bundles in green fruit are symptomatic. A characteristic bitter taste is often imparted to the fruit, and an unpleasant odor often emanates from diseased plantings. In addition to *P. cactorum, P. citrophthora, P. citricola,* and a high-temperature, unidentified *Phytophthora* species also cause fruit rot in Taiwan (Chang 1988).

Crown rot of strawberry plants caused by *P. cactorum* has been reported by Wilcox (1989). The typical symptoms are a wilt and dieback of the plants. Roots show necrosis and red brown lesions within the crown (compare with red stele of strawberry caused by *P. fragariae,* Chapter 26).

Disease Development. The severity of leather rot increases with a corresponding increase in the period of wetness from 0 to 5 h. At temperatures of 17 to 25°C, maximum infection (>80%) resulted when the wetness period on fruit exceeded 1 h (Grove et al. 1985a).

Control. Removal of infected fruit from the plants and disposal to an area far away from a strawberry planting is a good sanitation practice, since oospores of *P. cactorum* survive in mummified fruit in the field even when winter temperatures drop to –1.1°C (Grove et al. 1985b). Avoidance of close spacing allows the fruit to dry more rapidly. Nitrogenous fertilizers should be applied during the fall or summer rather than in the spring, since application in the spring increases the succulence and density of foliage. Dense foliage, which shades and prevents the berries from drying out after rains and dew formation, enhances berry rot. Mulching with plastic film or straw is helpful in preventing direct contact of berries with the soil; however, spread of inoculum by rain splash was reduced more by a straw mulch than by plastic film (Ellis and Grove 1983; Madden and Ellis 1990).

Fungicide sprays have been used but may be ineffective if conditions are conducive to disease development. Bordeaux mixture or other copper-based fungicides should be applied every 2 weeks until berries are one-third grown; after that, they should be applied only if rot is present or if cool rainy weather prevails. Metalaxyl and fosetyl-Al give effective control (Madden et al. 1991).

Cultivars vary in resistance to leather rot (El-Din et al. 1981).

Root Rot and Stem Canker of *Panax quinquefolium* L. (Ginseng)

Ginseng has been used in Asian countries for more than 200 years as a stimulant and health food additive. Its cultivation in recent years has been greatly threatened by a destructive root rot and blight caused by *P. cactorum.* Serious losses occur in Japan, the United States, and Russia. The disease was first described by Van Hook (1906) in the United States.

The earliest recognizable symptoms are drooping leaflets that hang limply from the tops of the stems. Dark green, water-soaked lesions, which later turn white in the center, occur on leaf blades. The size of the lesions varies from a few centimeters to coverage of the entire leaf surface. A watery discoloration is followed by a hollowing and collapse of infected stems. A semisoft rot often occurs on roots. Secondary microorganisms often invade the tissue, resulting in a characteristic decaying odor. Because these symptoms can often be confused with several other diseases of ginseng, the identification of the causal agent is necessary. A highly efficient method of isolation of *P. cactorum* from soil was described by Rahimian and Mitchell (1988) (see Chapter 2).

Disease Development. High moisture and cool temperatures favor infection of stems by *P. cactorum,* which moves rapidly down the stalk. Entire ginseng gardens can be destroyed in a few weeks under favorable wet conditions (Darmano et al. 1991). During periods of warm, sunny weather when the diseased tissues dry rapidly, the disease is temporarily arrested (Rosenbaum 1915).

Darmano et al. (1991) found that isolates of *P. cactorum* from forest soils near cultivated ginseng fields in the state of Wisconsin, United States, were not pathogenic to ginseng, thus indicating that in these areas, the primary inoculum for ginseng root rot was from sources other than the native soil.

Control. Destruction of diseased plants and deep plowing have helped to alleviate the spread of the disease. Preventative spraying with Bordeaux mixture (3:3:50) has been used with some success. A single application of metalaxyl in June reduced disease incidence (Putnam and Mitchell 1984). An application rate of 0.25 kg a.i./ha resulted in a 66% reduction in disease incidence and an increase of more than 50% in yield. In Korea, some soils in fields were reported to be suppressive to *P. cactorum* (Chang et al. 1984).

CHAPTER 12

Phytophthora cajani

Amin, Baldev, and Williams (1978)

Phytophthora cajani was described in 1978 after it was reported as a cause of a new stem rot disease of pigeon pea (*Cajanus cajan* (L.) Millsp.), one of the most important pulse crops in India (F. J. Williams et al. 1975). Previous to these reports, Pal et al. (1970) had reported a stem rot disease of pigeon pea and named the causal agent *P. drechsleri* var. *cajani;* however, Amin et al. (1978) and F. J. Williams et al. (1975) did not cite the report by Pal et al. (1970). Kannaiyan et al. (1980) reexamined several isolates of the pigeon pea pathogen from India and renamed it *P. drechsleri* f. sp. *cajani* on the basis of its morphological similarity to *P. drechsleri* and its host-specific nature on pigeon pea and *Atylosia* species. Although authentic *P. cajani* isolates were not available, they expressed the opinion that stem blight caused by *P. drechsleri* var. *cajani* and that caused by *P. cajani* are probably the same.

Since the taxon *P. cajani* is listed in the *Phytophthora* key of Stamps et al. (1990), we include it here; however, the name *P. cajani* has not appeared in the literature since the original publication, and it is known as *P. drechsleri* f. sp. *cajani* (Singh and Chauhan 1992). As explained in Chapter 23 on *P. drechsleri,* the isolates causing stem blight of pigeon pea that are currently identified as *P. drechsleri* f. sp. *cajani* (Kannaiyan et al. 1980) were found by isozyme and mtDNA RFLP technology to form a distinct group (G) that differed markedly from the typical *P. drechsleri* group (A) (Mills et al. 1991). Their data could be interpreted to support designation of the pigeon pea pathogen as a distinct species (e.g., *P. cajani*). *P. cajani* was considered by Amin et al. (1978) to be unique because of the large exit pore on sporangia (9.4 μm); however, whether this is a unique difference or is even consistent has not been determined. Homothallism was also considered to be an important difference from *P. drechsleri,* which is usually heterothallic; however, the literature indicates that both homothallism and heterothallism exist in isolates of *P. drechsleri,* and thus this is not a unique difference (Kannaiyan et al. 1980). Without more study it would be premature to conclude here that the name of *P. drechsleri* f. sp. *cajani* should be changed to *P. cajani.* Nonetheless, it should be investigated. A review of the stem blight on pigeon pea and characteristics of the pathogen named *P. drechsleri* f. sp. *cajani* is reported by Singh and Chauhan (1992).

Characteristics of *Phytophthora cajani*

P. cajani is classified in group VI by Stamps et al. (1990). See Tables 4.2 and 4.3 for tabular keys. Characteristics given here are from the description of *P. cajani* by Amin et al. (1978).

Sporangia

Sporangia form readily on infected pigeon pea stems incubated in distilled water for 36 to 48 h. Sporangia are ovoid, obpyriform, and persistent. The apical thickening is subhemispherical and 1.5 μm deep ("slightly conspicuous"). First-formed sporangia average 60 × 32 μm with a length-breadth ratio of 1.7:1.0. Sporangia formed later are smaller and more irregular in shape. The diameter at the exit pore averages 9.4 μm, a feature that Amin et al. (1978) considered to be unique.

Chlamydospores

Chlamydospores have not been described.

Hyphal Swellings

Hyphal swellings have not been described.

Sex Organs

Oogonia formed in 2 to 3 weeks when *P. cajani* was grown on lima bean, oatmeal, or cornmeal agar and on autoclaved leaves of *Cynodon dactylon* (L.) Pers. incubated in distilled water at 30°C for 6 to 7 days (homothallic). Oogonia are 22 to 40 μm in diameter (average is 33.8 μm on lima bean agar, 40.2 μm on oatmeal agar, and 35.5 μm on cornmeal agar). Antheridia are amphigynous and 15 to 16 μm in diameter. Oospores are spherical, smooth, plerotic, and 20 to 37 μm in diameter.

Growth Temperatures

Growth temperatures were not reported by Amin et al. (1978).

Distinguishing Characteristics

Amin et al. (1978) report that sporangium size is close to that of *P. cinnamomi, P. cambivora, P. erythroseptica* var. *pisi,* and *P. oryzae* (*P. japonica*), but *P. cajani* differs from *P. cinnamomi* by not producing botryose hyphal swellings or chlamydospores, from *P. cambivora* by production of narrower sporangia and no protuberances on the oogonial wall, from *P. erythroseptica* var. *pisi* by "lacking a medial sporangium construction," and from *P. oryzae* by production of single sporangiophores that bear a single terminal sporangium and by production of sex organs in single culture (homothallic). Amin et al. (1978) recognized that *P. cajani* resembled *P. drechsleri* but noted that *P. cajani* produced larger sporangia and undifferentiated sporangiophores. They also considered that the homothallic nature of *P. cajani* differentiated it from *P. drechsleri,* which is considered to be heterothallic according to Savage et al. (1968). As stated above, homothallism has been noted in several reports on *P. drechsleri* (Kannaiyan et al. 1980).

Our conclusion, based on Kannaiyan et al. (1980), is that *P. cajani* is probably the same as *P. drechsleri* var. *cajani* and *P. drechsleri* f. sp. *cajani.* The forma specialis designation is supported by its host specificity to pigeon pea and the closely related wild *Atylosia* sp. and the lack of pathogenicity to other hosts that are susceptible to *P. drechsleri.* The data of Mills et al. (1991) strongly suggest that the pigeon pea blight pathogen is significantly different from other *P. drechsleri* isolates and if redescribed could be renamed *P. cajani.*

Disease Caused by *Phytophthora cajani*

Description of the stem blight disease of pigeon pea, *C. cajan*, is given in Chapter 23, *Phytophthora drechsleri* f. sp. *cajani.*

CHAPTER 13

Phytophthora cambivora

(Petri) Buisman (1927)

The only known synonym of *Phytophthora cambivora* is *Blepharospora cambivora* by Petri (1917), who was first to describe it. It is best known as the causal agent of *maladie de l'encre* or *enfermedad de tinta* in Europe and of ink disease of chestnuts in Britain and the United States. *P. cambivora* also causes root rot on several woody plant species and has been shown to infect a number of succulent plants when they are artificially inoculated (Table 13.1). Distribution is indicated in Commonwealth Mycological Institute (CMI) Map 70 (1961).

Characteristics of *Phytophthora cambivora*

P. cambivora is classified in group VI (Stamps et al. 1990) because it produces nonpapillate sporangia and amphigynous antheridia. Morphology is shown in Figures 13.1 and 13.2. Tabular keys are shown in Tables 4.2 and 4.3. Waterhouse and Waterston (1966) summarize the morphology and diseases caused.

Sporangia

Shapes of sporangia range from broadly ovate, subglobose, and obpyriform to ellipsoidal with well-rounded bases but are variable in shape depending on the culture media utilized. Dimensions of sporangia have been recorded as 40 to 54 × 60 to 75 µm (Petri 1917); 55 to 65 × 40 to 45 µm with maximum diameters of 85 × 60 µm (Waterhouse and Waterston 1966); 48 to 83 × 32 to 61 µm with an average of 59 to 64 × 40 to 45 µm (Chitzanidis and Kouyeas 1970) for an isolate obtained from diseased chestnut trees; 44 to 69 × 33 to 48 µm with an average diameter of 38 × 54 µm (Mircetich and Matheron 1976);

Table 13.1. Distribution of hosts of *Phytophthora cambivora*

Host	Common name	Disease	Geographical distribution
Acer pennsylvanicum L.	Snake-bark maple, whistlewood	Root rot	United Kingdom (Brasier 1971b)
Aesculus hippocastanum L.	Horse chestnut	Root rot, dieback	United Kingdom (Brasier and Strouts 1976)
Castanea spp. Mill.	Chestnut	Ink disease	United Kingdom (Day 1932)
Castanea crenata Sieb. & Zucc.	Japanese chestnut	Root rot	United Kingdom (Day 1932)
Castanea dentata (Marsh.)	American chestnut	Black root rot	United States (Milburn and Gravatt 1932)
Castanea pumila (L.) Borkh. Mill.	Chinquapin	Black root rot	United States (Milburn and Gravatt 1932)
Castanea sativa Mill.	Eurasian chestnut	Ink disease	France (Delacroix 1897); Italy (Petri 1917); United Kingdom (Day 1932); now throughout Europe
Casuarina equisetifolia J. R. Forst and G. Forst	Filao, horsetail tree	Wilt	Mauritius (Orian 1951)
Chrysanthemum cinerariafolium (Trev.) Vis.	Pyrethrum	Wilt disease	India (Ramakrishnan and Soumini 1948)
Cineraria spp.	Cineraria	Wilt	United Kingdom (Munro 1945)
Erica sp.	Heather	Wilt	United States (White 1937); Spain (Landaluze 1941)
Fagus sylvatica L.	Beech	Root rot	United Kingdom (Day 1938)
Juglans L.	Walnut	Root rot	Italy (Petri 1935); Spain (Landaluze 1941)
Lupinus albus L.	Lupins	Blight	France (Allain 1935)
Malus spp.	Apple hybrids	Collar and root rot	Japan (Suzui and Hoshino 1979)
Malus pumila Mill.	Apple	Root rot and trunk cankers	Canada (McIntosh 1966); United States (Mircetich et al. 1976); Japan (Suzui and Hoshino 1979)
Nothofagus Blume.	Beech-like trees	Root rot	United Kingdom (Day 1938)[a]
Persea americana Mill.	Avocado	Root rot	South Africa (Wager 1931)
Pisum sativum L.	Pea	Seedling blight	Italy (Arnaudi 1928)[a]
Prunus amygdalus Batsch. (= *P. dulcis* (Mill.) D. A. Webb.)	Almond	Trunk canker	Australia (Wicks and Lee 1985)
Prunus armeniaca L.	Apricot	Root rot and trunk canker	United States (Mircetich et al. 1974)
Prunus avium (L.) L.	Mazzard cherry	Root rot and trunk canker	United States (Mircetich et al. 1974)
Prunus cerasus L.	Sour cherry	Root and crown rot	United States (Wilcox et al. 1985)
Prunus domestica L.	European plum	Root rot and trunk canker	United States (Mircetich et al. 1974)
Prunus dulcis (Mill.) D. Webb.	Almond	Trunk canker	Australia (Wicks and Lee 1986)
Prunus mahaleb L.	Mahaleb cherry	Root rot and trunk canker	United States (Mircetich et al. 1974)
Prunus persica (L.) Batsch.	Peach	Root rot and trunk canker	United States (Mircetich et al. 1974)
Rhododendron sp.	Rhododendron	Blight	United States (White 1937); Denmark (Koch 1971)
Rubus idaeus L.	Red raspberry	Root rot	Scotland (Duncan and Kennedy 1989)
Senecio sp.	Groundsel	Root rot	United Kingdom (Munro 1945)
Ulmus sp.	Elm	Root rot	United Kingdom (Day 1938)[a]

[a] Artificially inoculated host.

30.37 to 47.35 × 22.58 to 30.92 μm (Gerrettson-Cornell 1978); 25 to 65 × 17 to 31 μm with an average diameter of 46 × 25 μm (Suzui and Hoshino 1979); and average diameters of 50.4 ± 10.9 × 35.2 ± 6.7 μm for 12 isolates from several different countries (Oudemans and Coffey 1991a). Length-breadth ratios of sporangia have been recorded as 1.8 by Suzui and Hoshino (1979) and as 1.4 by Oudemans and Coffey (1991a). Sporangia proliferate internally and sometimes are nested within the original sporangium. Sporangiophores are simple, unbranched, or sympodial (Figures 13.1 and 13.2a and b).

Sex Organs

P. cambivora is normally heterothallic. Detailed morphology of the sex organs of *P. cambivora* shown by scanning electron microscopy and light microscopy is reported by Ho et al. (1977). Oogonia are spherical and produced terminally. The outer wall of more than 50% of the oogonia is distinctly bullate or verrucose (Figures 4.27, 13.1, and 13.2d). Oogonia are normally produced abundantly only when paired with a suitable isolate of the opposite mating type of the same or different species. Some A2 isolates produce oogonia in single culture (Gerrettson-Cornell 1977).

Diameters of oogonia vary from 43 μm, with a maximum of 62 μm (Waterhouse and Waterston 1966); 30 to 58 μm, average 48 μm (Chitzanidis and Kouyeas 1970); 38 to 48 μm, average 44 μm (Mircetich and Matheron 1976); average 46.3 μm (Gerrettson-Cornell 1977); average 37 μm (Suzui and Hoshino 1979); and average 40.5 ± 5.5 μm for 12 isolates from different countries (Oudemans and Coffey 1991a). The average oogonial wall thickness is 2 μm (Waterhouse and Waterston 1966).

Oospores nearly fill the oogonium (plerotic). The average reported diameters of oospores are 36 μm with a wall thickness of 3 μm (Waterhouse and Waterston 1966); 37 to 44 μm, average of 40 μm for isolates from cherry trees (Mircetich and Matheron 1976); and 35.14 to 39.42 μm for isolates obtained from eucalyptus trees (Gerrettson-Cornell 1978). The diameters of oospores of 12 isolates obtained from different countries averaged 33.8 ± 5.6 μm (Oudemans and Coffey 1991a).

The antheridia of *P. cambivora* are always amphigynous and are normally elongated and either unicellular or bicellular (Figures 13.1 and 13.2d); 60% of the antheridia that Ho et al. (1977) observed were bicellular. The average length of the antheridium is reported to be 25 μm with a maximum length of 35 μm (Waterhouse and Waterston 1966). Other reported diameters are 14 to 32 × 12 to 21 μm, average 19 × 16 μm (Chitzanidis and Kouyeas 1970); 19 to 38 × 15 to 21 μm (Mircetich and Matheron 1976); 16.6 to 29.4 μm, average length of 27.3 μm for the bicellular type; and 12.9 to 27.6 μm, average length 19.3 μm for the unicellular type (Ho et al. 1977).

Growth Temperatures

The minimum temperature for growth is 2°C, optimum is 22 to 24°C, and maximum (cessation of growth) is above 32°C.

Cultural Characteristics

The mycelium has coralloid hyphae often distorted into irregular shapes (Figure 13.1) with moderate to profuse growth of aerial mycelium. Mircetich and Matheron (1976) report that cherry isolates produce uniform and profusely fluffy cultures on cornmeal agar at 21°C. Older mycelium is often septate

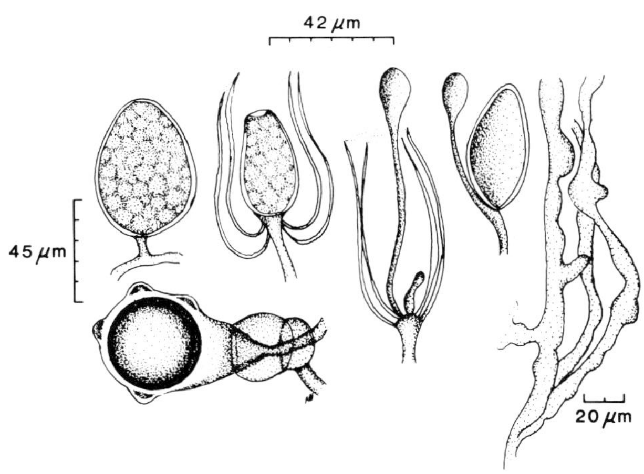

Figure 13.1. Morphology of *Phytophthora cambivora.* Lower row, bullate (verrucose) oogonium with a bicellular antheridium. Upper row, nonpapillate, ovoid sporangium; internal proliferation (nesting of a sporangium); internal and external proliferation; sympodial production of a new sporangium; and knobby, coralloid mycelium. (Drawing by A. Vaziri)

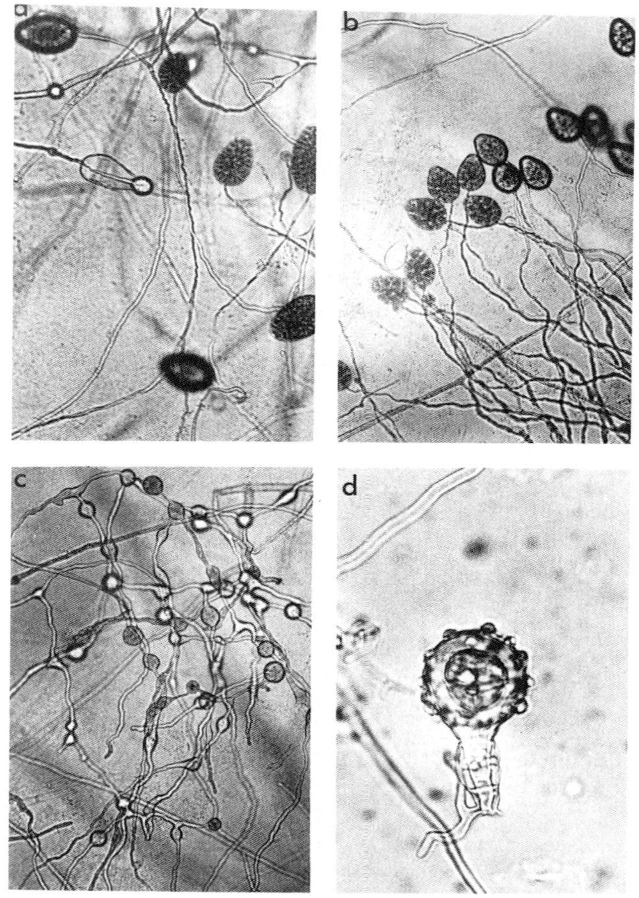

Figure 13.2. Morphological characteristics of *Phytophthora cambivora,* cause of crown rot of almonds in Australia. **a and b,** Ovoid sporangia; a sporangium in **a** is germinating by internal proliferation; **c,** hyphal swellings that develop on lima bean agar disks flooded with soil extract; and **d,** a bullate oogonium and a bicellular amphigynous antheridium. (Reproduced by permission of the Australian Journal of Agricultural Research, courtesy Wicks and Lee 1986)

(Mircetich and Matheron 1976; O. K. Ribeiro, *unpublished*). Luo et al. (1988) report that the colony patterns of isolates varied from stellate to uniform (no discernible pattern).

Distinguishing Characteristics

P. cambivora has been considered to be similar to *P. cinnamomi,* but it differs in that it seldom produces the lightly pigmented, botryose, thin-walled hyphal swellings and chlamydospores that are produced by *P. cinnamomi.* Contrast the hyphal swellings illustrated in Figure 13.2 with those illustrated for *P. cinnamomi* in Figure 4.1. Some of the oogonia of *P. cambivora* are verrucose (bullate) (Figures 4.27, 13.1, and 13.2); *P. cinnamomi* oogonia are always smooth walled.

P. cambivora is serologically distinguished from *P. cinnamomi* and other species (Halsall 1976). Polyacrylamide gel electrophoresis of native proteins in the mycelium of several isolates of *P. cambivora,* obtained from various sources in the United States, showed the presence of only one protein-banding pattern that distinguished them from *P. cactorum* and *P. syringae* (Bielenin et al. 1988). Extensive isozyme analyses by starch gel electrophoresis showed that *P. cambivora* is distinctly separated from *P. cinnamomi* and *P. cactorum.* Eight distinct electrophoretic types were delineated in isolates of *P. cambivora* from different parts of the world. Isolates from Europe were of a single electrophoretic type, but five of the types were characteristic of isolates from Australia (Oudemans and Coffey 1991a).

Diseases Caused by *Phytophthora cambivora*

Although *P. cambivora* is best known as the cause of ink disease of chestnut, it has also become increasingly important as the causal agent of crown and root rots of many fruit trees. Ink disease and root rot of *Prunus* species and other fruit trees are described below to give a general idea of the types of symptoms caused by *P. cambivora.*

Ink Disease of *Castanea dentata* (Marsh.) Borkh. (Chestnut)

P. cambivora usually infects the tree near the base of the trunk and on the larger roots. Trunk or root rot may develop either rapidly or slowly. If trees are girdled at the collar, leaves at the top of the tree wilt, and death of the tree usually follows within the first year of infection. When infection develops slowly, leaves and flowers become smaller during the first year, and the tree generally dies by the end of the second year. Occasionally, an inky fluid, from which the name of the disease is derived, exudes from dead and dying bark at the base of the trunk. Root lesions also exude a blue black substance that stains the soil close to the roots. A similar disease can also be caused by *P. cinnamomi* (Crandall et al. 1945; Crandall 1950). Both species appear to be equally pathogenic. A short discussion of the ink disease in Europe is given by I. M. Smith et al. (1988).

Disease Development. The pathogen is soilborne. Consequently, disease develops rapidly when the soil remains excessively wet for prolonged periods and temperatures do not exceed 30°C.

Control. Methods of control of ink disease in *Castanea* species include 1) sanitation, which includes planting disease-free plants, 2) establishment of plants on well-drained soil, 3) use of resistant rootstocks (high levels of resistance to *P. cambivora* has been reported in *C. crenata* and in some F_3 hybrids of *C. crenata* Sieb. and Zucc. × *C. sativa*) (Crandall et al. 1945), and 4) removal of the diseased bark followed by treatment of the wound and soil with copper fungicides or with metalaxyl or fosetyl-Al (if registered for use). Copper fungicides tend to slow the rate of infection when applied in the early stages of disease (Landaluze 1941). In Greece, metalaxyl, applied as a soil drench in the spring and autumn, controlled canker formation. Fosetyl-Al, applied as a foliar spray, controlled the disease but was ineffective when used as a soil drench (Skoudridakis and Bourbos 1990).

Root and Crown Rot of *Prunus* (Cherry, Peach, Plum, and Apricot) and *Malus* (Apple) Species

Fruit trees such as cherry, peach, plum, apricot, and apple are subject to infection by *P. cambivora* and by several other *Phytophthora* species (Mircetich 1982). At early stages of infection, trees with symptoms are difficult to distinguish from healthy trees. In later stages, leaves on infected trees are small and chlorotic, droop, and do not grow in terminal shoots, or the trees die suddenly in early summer without manifestation of observable symptoms. The aboveground symptoms of *Phytophthora*-infected trees often resemble those caused by other root-rotting pathogens. Positive diagnosis by isolation of the pathogen from the roots of diseased trees is often the only way to determine the cause (see Chapter 2 for methods). Crown and root rot may occur on the same trees. Crown rot can be recognized by the presence of decayed bark at the base of the tree, which is often initiated at several points around the base of the tree. These infections progress and eventually coalesce until the lower trunks are entirely girdled. Infected roots become brown, firm, and brittle in contrast to the soft, watery rot typical of some other root-rotting diseases (Mircetich and Matheron 1976; Wicks and Lee 1986).

Disease Development. Progress of the disease after infection by *P. cambivora* and other *Phytophthora* species depends on the type and species of rootstock, soil water status, age of infected trees, the *Phytophthora* species involved, and the prevalence of moderate temperatures less than 30°C. Trees up to 35 years of age are susceptible to infection; however, trees 3 to 8 years old are more frequently infected. Wicks and Lee (1986) reported that in Australia, only A1 but not A2 mating types of *P. cambivora* were pathogenic to almond trees.

Control. The most important consideration for control is planting clean nursery stock. Contaminated rootstocks are a source of primary inoculum. Jeffers and Aldwinckle (1988) showed that apple rootstocks from different nurseries in the United States and Europe were contaminated with *P. cactorum* and *P. cambivora* in varying amounts. The proper management of irrigation water and drainage to avoid excessively wet soil conditions for prolonged periods is the most efficacious cultural practice. Contamination of otherwise healthy areas of the orchards can be minimized by sanitary measures such as cleaning shoes and implements before entering noninfested areas.

The fungicides metalaxyl as a drench (Wicks 1988b; Wicks and Lee 1988; Wicks and Hall 1988) and fosetyl-Al as a foliar spray (Wicks and Lee 1985; Bielenin and Jones 1988; Wicks and Hall 1988) have good efficacy against the *Phytophthora* species that infect cherry and almond fruit trees. See Tables 7.6 and 7.11 in Chapter 7 for more citations on fungicidal control. Registration for use of fungicides on fruit trees varies in different countries.

Solarization of soil in the field (discussed in Chapter 5) by spreading clear plastic sheeting on moist soil increased soil temperature to about 40°C at 10 cm in depth. This treatment controlled *P. cambivora* in an irrigated cherry orchard, but in a nonirrigated orchard solarization was ineffective. Although it did not affect established infection of trees, solarization improved the growth of trees (Wicks 1988a).

Since the *Phytophthora* species that infect deciduous fruit and nut trees are not host specific (Wilcox and Mircetich 1985a), care must be exercised in replanting orchards known to have problems with Phytophthora root rot. Selection of rootstocks for replanting depends on the *Phytophthora* species present in each planting site. Trunk cankers on Mahaleb (cherry) rootstocks can extend 60 to 80 cm above the ground, whereas those on Mazzard are limited to the crown region. Although Mazzard rootstock was more resistant to infection by *P. cambivora,* this rootstock was as susceptible as Mahaleb rootstock to infection by *P. cinnamomi, P. citricola, P. cryptogea,* and *P. drechsleri* (Mircetich and Matheron 1976).

Selection of a rootstock with as high a degree of resistance as possible is important; however, none of the adapted rootstocks for either pome or stone fruit tree propagation are immune (Mircetich and Browne 1987). The success of different rootstocks varies from area to area and is dependent on the severity of the conditions that favor the disease and on the virulence of the *Phytophthora* species. Mircetich and Browne (1987), discussing the variability in severity of disease on different rootstocks, state that the cherry rootstock Mazzard was more resistant than Mahaleb to crown rot, but there was no difference between the rootstocks in their resistance to root rot. On the other hand, in Australia, Wicks (1989) found that Mazzard was more resistant than Mahaleb to *P. cambivora.* Wicks (1989) reported that Nemaguard peach rootstock and the almond rootstock Titan were resistant to *P. cambivora.* Both Mahaleb and Mazzard were equally susceptible to a cherry isolate of *P. cryptogea* (Mircetich and Browne 1987). Authorities in local areas should be consulted for specific recommendations. See Chapter 6, Table 6.1; Mircetich and Matheron (1976); Mircetich and Browne (1987); Wicks and Lee (1986); and Wicks (1989) for other reports on resistance of trees to *P. cambivora* and other *Phytophthora* species.

CHAPTER 14

Phytophthora capsici

Leonian (1922) emend. A. Alizadeh and P. H. Tsao (1988); Tsao (1991); Mchau and Coffey (1995)

Synonyms of *Phytophthora capsici* include *P. hydrophila* Curzi (1927), *P. parasitica* var. *capsici* (Leonian) Sarejanni (1936), and *P. palmivora* MF4 Griffin (1977). *P. capsici* was first described by Leonian (1922) as the causal agent of a blight of *Capsicum annuum* L. (chili pepper) in New Mexico, United States. Although originally it was considered to be host specific in the key of Tucker (1931), host specificity can no longer be used as a taxonomic criterion because *P. capsici* is now known to infect a number of other plant species, including eggplant fruit (Katsura and Tokura 1955), cotton bolls (Garber et al. 1986), black pepper (Tsao 1991), cacao (Zentmyer et al. 1977), tomato (J. B. Jones et al. 1991), and many others (Table 14.1). World distribution is shown in Commonwealth Mycological Institute Map 277, edition 4 (1977). The diseases caused include foliar blights, fruit rots, and stem and root rots. *P. capsici* has been redescribed to accommodate a wider range of biotypes and to include the *P. palmivora* MF4 isolates that cause black pod of cocoa and wilt of black pepper (Tsao and Alizadeh 1988; Tsao 1991) and a world collection from several hosts (Mchau and Coffey 1991c). Uchida and Aragaki (1989) and Aragaki and Uchida (1992) proposed that the isolates pathogenic to black pepper, cacao, macadamia, papaya, and others that produce chlamydospores and elongate sporangia and are avirulent to pepper (*C. annuum*) should be classified as *P. tropicalis;* however, Tsao and Alizadeh (1988) and Mchau and Coffey (1994c) include these isolates in *P. capsici.* The taxonomic situation involving the MF4 type from cacao is discussed further in Chapter 51 on *P. palmivora.*

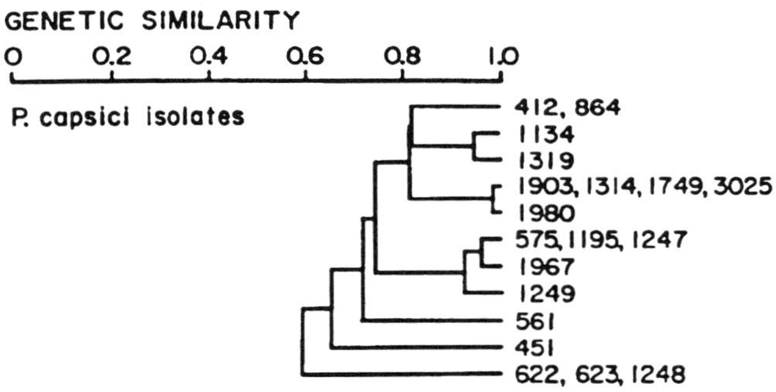

Figure 14.1. Phenogram showing relatedness of *Phytophthora capsici* isolates by mtDNA restriction fragment length polymorphism patterns. Sources of isolates of *P. capsici* are as follows:

Isolate	Host	Origin
P412 MF4	*Piper nigrum*	Guatemala
P451 MF4	*Theobroma cacao*	Brazil
P561 MF4	*Piper nigrum*	Brazil
P575 MF4	*Theobroma cacao*	Mexico
P622 MF4	*Theobroma cacao*	Brazil
P623 MF4	*Theobroma cacao*	Brazil
P864 MF4	*Theobroma cacao*	Cameroon
P1134	*Capsicum annuum*	New Mexico
P1195 MF4	*Theobroma cacao*	Mexico
P1247	*Sechium edule*	Costa Rica
P1248	Unknown	Unknown
P1249	*Spondias purpurea*	Costa Rica
P1314	*Capsicum annuum*	California
P1319	*Capsicum annuum*	California
P1749	*Cucumis sativus*	Taiwan
P1903	*Gossypium* sp.	California
P1967	*Macadamia integrifolia*	Hawaii
P1980	*Lycopersicon esculentum*	California
P3025	*Capsicum annuum*	Chile

(Reproduced by permission of the British Mycological Society, courtesy Förster et al. 1990b; Förster and Coffey 1991)

Since the species *P. capsici* has been broadened, there is considerable diversity in host-specific pathogenicity. Macadamia isolates are not pathogenic on pepper (Uchida and Aragaki 1989). Pepper and some cucurbit isolates are virulent on pepper, but some cucurbit isolates are less virulent on pepper (Ristaino 1990).

Characteristics of *Phytophthora capsici*

Isozyme analysis of 84 isolates from a world collection indicates that *P. capsici* is a genetically complex species containing three subgroups (Oudemans and Coffey 1991b). One subgroup, CAP1, contains a majority of the isolates from annual solanaceous and cucurbit hosts as well as some isolates from black pepper and cocoa that had been previously described as *P. palmivora* MF4 (Kaosiri and Zentmyer 1980). A second subgroup, CAP2, contains isolates mainly from tropical crops such as black pepper, cocoa, papaya, macadamia, and rubber and isolates from Hawaii tentatively proposed as *P. tropicalis* by Uchida and Aragaki (1989) and Aragaki and Uchida (1992). The third subgroup, CAP3, with the least genetic diversity, includes cocoa isolates from Brazil. Isolates previously designated *P. palmivora* MF4 occurred in all three subgroups. Förster and Coffey (1991), in a summary of data from Förster et al. (1990a) that involved mtDNA RFLP patterns, also noted that *P. palmivora* MF4 grouped with isolates of *P. capsici* from *C. annuum* (pepper) (Figure 14.1).

Mchau and Coffey (1995), utilizing 113 isolates of *P. capsici,* most of which were included in the report of Oudemans and Coffey (1991b), conducted further isozyme analyses. By use of 15 enzymes, the isolates were shown to be separable into two subgroups, CAPA and CAPB. Each subgroup contained isolates from a wide range of hosts and geographical locations. However, some members of each subgroup varied in morphology. This supports the redescription by Tsao and Alizadeh (1988) that consolidates these types under *P. capsici.* A subsequent redescription, which adds the genetic isozyme data to morphological data, is given by Mchau and Coffey (1995).

Blaha (1990) differentiated *P. palmivora, P. megakarya,* and *P. capsici* by use of two isozyme systems. Hwang et al. (1991) distinguished four groups from *P. capsici* pepper isolates on the basis of mtDNA RFLP patterns.

Table 14.1. Distribution of hosts of *Phytophthora capsici*

Host	Common name	Disease	Geographical distribution
Abelmoschus esculentus (L.) Moench.	Okra	Root rot	United States (Satour and Butler 1967)[a]
Allium cepa L.	Onion		Taiwan (Leu and Kao 1981; cited by Ho 1990)
Aloe sp.	Aloe	Root rot	Taiwan (Ho 1990)
Anthurium andreanum Linden	Flamingo lily; oilcloth flower	Spathe blight	Hawaii (Aragaki and Uchida 1980)
Brassica oleracea var. *botrytis* L.	Cauliflower	Root rot	United States (Satour and Butler 1967)[a]
Bryonia dioica Jacq.	Red bryony; wild hop	Root rot	Italy (Curzi 1927)[a]
Capsicum annuum L.	Red or sweet pepper	Fruit and root rot; stem blight; Moria; Tristeza O Seca	United States (Leonian 1922); Italy (Curzi 1927); Puerto Rico (Tucker 1928); Argentina (Marchionatto 1938); Venezuela (Malaguti and Pontis-Videla 1950a); Brazil (Do Amaral 1952); Japan (Katsura and Tokura 1955)a; Bolivia (Bell and Alandia 1957); Spain (Alfaro Moreno and Vegh 1971); Iran (Ershad 1971); Serbia (Aleksic and Miladinovic 1975); Taiwan (Chang 1978)
Capsicum annuum L.	Red or sweet pepper	Seedling damping-off	Argentina (Godoy 1940); Korea (Kim et al. 1975)[a]; China (Ho et al. 1983b)
Capsicum annuum var. *grossum*	Bell or green pepper; pimento	Root rot	United States (Tompkins and Tucker 1941b)
Carthamus tinctorius L.	Safflower	Root rot	United States (Satour and Butler 1967)[a]
Chenopodium amaranticolor Coste and Reyn.	Spinach	Root rot	United States (Satour and Butler 1967)[a]
Citrullus lanatus (Thunb.) Matsum. and Nakai (*C. vulgaris* Schrad.)	Watermelon	Brown fruit rot; vine and leaf wilt	Japan (Katsura and Tokura 1954); United States (Kreutzer et al. 1940; Wiant and Tucker 1940)
Citrullus Schrad. sp.	Melon	Collar rot	France (Clerjeau 1973); Spain (Tuset-Barrachina 1977)
Citrus spp.	Citrus	Root rot	United States (Wiant and Tucker 1940)[a]
Cosmos Cav. sp.	Cosmos	Not stated	Taiwan (Ho 1990)
Crataegus oxyacantha L.	Hawthorne	Fruit rot	Italy (Curzi 1927)[a]
Cucumis melo var. *inodorus* Naud.	Honeydew melon	Fruit rot; postharvest decay	United States (Tompkins and Tucker 1937)
Cucumis sativus L.	Cucumber; pepino	Fruit rot	United States (Kreutzer et al. 1940); Korea (Kim et al. 1975); Taiwan (Ho 1990)
Cucurbita maxima Dcne.	Squash; pumpkins	Wilt and basal stem rot	United States (Kreutzer et al. 1940); Argentina (Pontis 1945); Italy (Noviello et al. 1977); Japan (Kamjaipai and Ui 1978)
Cucurbita pepo L. var. *condensa* Bailey	Summer squash; bush pumpkin	Leaf and stem blight	United States (Tucker 1936)
Datura stramonium L.	Jimson weed	Wilt	Italy (Curzi 1927)[a]
Daucus carota L. var. *sativa* DC	Carrot	Root rot	United States (Tompkins and Tucker 1941)[a]
Dianthus barbathus L.	Carnation	Not stated	Taiwan (Leu and Kao 1981; cited by Ho 1990)
Diospyros kaki L.	Persimmon	Fruit rot	Italy (Curzi 1927)[a]
Enkianthus quinquefolius Lour.		Not stated	Taiwan (Leu and Kao 1981; cited by Ho 1990)
Ficus carica L.	Fig	Fruit rot	Japan (Katsura and Tokura 1955)[a]
Gossypium hirsutum L.	Cotton	Root rot; boll rot	United States (Satour and Butler 1967; Garber et al. 1986)[a]
Leucospermum R. Br.	Pincushion flower	Stem and root rot	United States (Hawaii) (Uchida and Aragaki 1979)
Linum sp.	Flax	Root rot	United States (Satour and Butler 1967)[a]
Lycopersicon esculentum Mill.	Tomato	Fruit rot; root and crown rot; seedling damping-off	United States (Kreutzer et al. 1940); U.S.S.R. (Osnitzkaya 1949); Venezuela (Malaguti and Pontis-Videla 1950a); Japan (Katsura and Tokura 1955)[a]; Korea (Kim et al. 1978); Taiwan (Ho 1990)
Macadamia integrifolia Maiden and Betche	Macadamia nut	Raceme blight	United States (Hawaii) (Hunter et al. 1971; Kunimoto et al. 1976)
Macadamia ternifolia Muell.	Macadamia nut	Raceme blight	United States (Hawaii) (Kunimoto et al. 1976)
Malus pumila Mill.	Apple	Fruit rot	United States (Tompkins and Tucker 1941)[a]
Medicago sativa L.	Alfalfa; lucerne	Root rot	United States (Satour and Butler 1967)[a]
Nicotiana glutinosa L.	Tobacco	Root rot	United States (Satour and Butler 1967)[a]
Opuntia ficus-indica Mill.	Indian fig	Cladode rot	Italy (Curzi 1927)[a]
Persea americana Mill.	Avocado	Root rot	United States (Tompkins and Tucker 1937a)
Phaseolus sp.			Taiwan (Ho 1990)
Phaseolus lunatus L.	Butter or civet bean	Root rot	Argentina (Frezzi 1950)
Piper betle L.	Betle	Foot rot	Thailand (Tsao and Tummakate 1977); Taiwan (Ho 1990)[b]
Piper nigrum L.	Black pepper	Foot rot	Thailand (Tsao and Tummakate 1977); Indonesia (Tsao et al. 1985); India (Sarma et al. 1991)[b]
Pisum sativum L.	Pea	Root rot	United States (Critopoulos 1955)
Prunus persica (L.) Batsch	Peach	Seedling wilt; stem canker	United States (Tompkins and Tucker 1937a)
Raphanus sativus L.	Radish	Root rot	United States (Satour and Butler 1967)[a]
Solanum marginatum L.		Fruit rot	Italy (Curzi 1927)[a]
Solanum melongena L. var. *esculentum* Nees	Eggplant	Brown rot; fruit rot	Italy (Curzi 1927; Tompkins and Tucker 1941b); Argentina (Frezzi 1950); Venezuela (Pontis and Rodriguez 1953); Japan (Katsura 1958); Korea (Kim et al. 1975); Taiwan (Ho 1990)
Theobroma cacao L.	Cocoa	Black pod	Brazil and Cameroun (Zentmyer et al. 1977); Cameroun (Griffin 1977); Brazil (Brasier and Medeiros 1978)
Vanilla planifolia Andr.	Vanilla	Root rot	French Polynesia (Mu and Tsao 1987; Tsao and Mu 1987)
Vicia faba L.	Broadbean	Root rot	United States (Satour and Butler 1967)[a]

[a] Artificially inoculated host.

[b] Most references to *P. palmivora* as the causal agent should be corrected to *P. capsici* (Tsao and Tummakate 1977; Tsao et al. 1985; Tsao and Alizadeh 1988).

Morphology is shown in Figure 14.2. *P. capsici* is classified in group II (Stamps et al. 1990). See Tables 4.2 and 4.3 for tabular keys, Appendix 4.9 for a dichotomous key (Ho 1992; Ho et al. 1995), and Stamps (1985f) for a summary. See also Tables 51A.4 and 51A.5 for morphological comparisons with *P. palmivora* and related species.

Sporangia

The morphology of sporangia is shown in Figures 4.16, 4.17A, B, and E, 14.2, and 14.3. Sporangia are mostly papillate, but in some cases they appear to be semipapillate (apical thickening ranges from 1.4 to 9.2 µm). Occasionally sporangia have two or three apices. Sporangial shapes are influenced by light and other cultural conditions (Tsao and Alizadeh 1988; Tsao 1991) and range from subspherical, ovoid, obovoid, ellipsoid, fusiform, pyriform, to distorted shapes. Sporangia are predominately tapered at the base and are caducous with long pedicels varying in length from 35 to 138 µm (Mchau and Coffey 1995). Length-breadth ratios in different reports are 1.70:1 (Frezzi 1950); 1.72:1 (Ershad 1971); 1.57 to 2.19, average 1.76:1 in agar and 1.52 to 2.10, average 1.73:1 in water (Kröber 1985); in light 1.73 and in dark 1.27 (Tsao and Alizadeh 1988); and 1.4 to 1.8 (range for 24 isolates incubated in water under light) (Ristaino 1990). Mchau and Coffey (1995) report that sporangia are extremely variable in shape; shapes include ellipsoid, spherical, subspherical, broadly ovoid, obturbinate, obovoid, fusiform, and pyriform. Dimensions (length × breadth) varied from 32.8 to 65.8 × 17.4 to 38.7 µm. Length-breadth ratio varied from 1.3:1 to 2.1:1. The larger length-breadth ratio was more typical of the CAPB subgroup and the smaller of the CAPA subgroup. See Table 14.2 for sporangial dimensions.

Sporangiophores that are formed under light are irregularly branched, and sympodia form only in water. Black pepper and cocoa (*P. palmivora* MF4) isolates produce sporangia on sporangiophores that are characteristically umbellate (Figures 4.5 and 4.16B–F). Caducous sporangia with long stalks are shown in Figures 4.17A and B and 14.3. Stalk lengths varied among isolates from pepper and cucurbit hosts from 37.5 to 98.6 µm for pepper isolates and 31.5 to 85.3 for cucurbit isolates (Ristaino 1990).

Most isolates of subgroup CAPA produced rounded ellipsoid sporangia, some of which had more than one papilla. The majority of the isolates of subgroup CAPB produced ellipsoid-lanceolate sporangia. Since ellipsoid-lanceolate sporangia were more characteristic of CAPB isolates than of CAPA isolates, the length-breadth ratio for CAPB was higher than that for CAPA. Sporangia in CAPB were produced on umbellate sporangiophores. However, these morphological traits were not characteristic of all isolates within the subgroup. Pedicel length varied from 34.7 to 138 µm, but the mean value was significantly higher for CAPB. Since both ovoid and ellipsoid-lanceolate sporangia occurred within each subgroup, Mchau and Coffey (1995) concluded that the elongated sporangial morphology that was suggested by Uchida and Aragaki (1989) and Aragaki and Uchida (1992) to be typical of a proposed new species, *P. tropicalis,* was not justified.

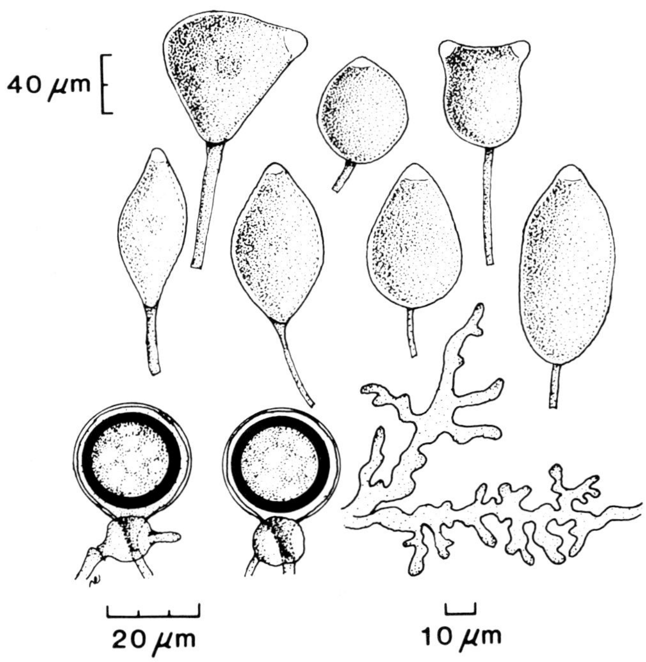

Figure 14.2. Morphology of spore types of *Phytophthora capsici.* Upper row, variation in shapes of papillate, caducous sporangia with long pedicels. Lower row, globose oogonia with paragynous antheridia (oospores are plerotic); torulose mycelium. (Drawing by A. Vaziri)

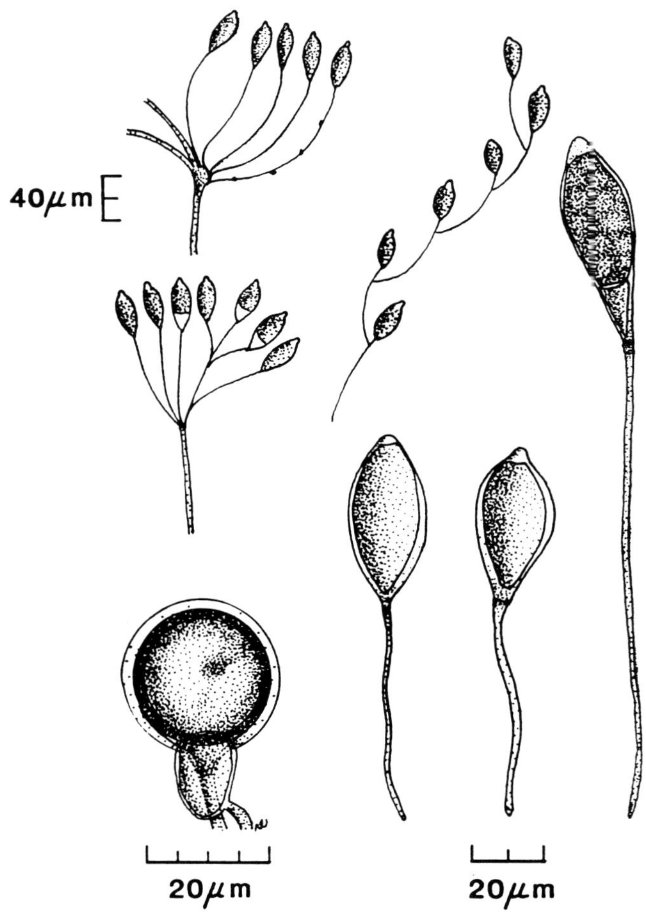

Figure 14.3. Morphology of *Phytophthora capsici* (*sensu* Tsao 1991) formerly known as the atypical *P. palmivora* MF4 (*sensu* Griffin 1977). Lower row, oogonium with an amphigynous antheridium containing a globose oospore; ellipsoid, caducous sporangia with long pedicels. Upper row, umbellate formation of sporangiophores; formation of sporangia in a sympodium. (Drawing by A. Vaziri)

Chlamydospores

Chlamydospores of *P. capsici* isolates from pepper (*C. annuum*) are rare in culture according to Tucker (1931). Pepper or cucurbit isolates did not form chlamydospores (Ristaino 1990); however, chlamydospores were reported on some pepper isolates from Iran (Ershad 1971). Chlamydospores are abundant in isolates from black pepper (*Piper niger*), cocoa, and macadamia (Alizadeh and Tsao 1985b; Uchida and Aragaki 1985; Tsao 1991). Diameters are 28 to 29 µm with wall thickness of 2.4 to 2.7 µm (Tsao 1991; Figure 4.25). Mchau and Coffey (1995) report chlamydospore production exclusively in a population of isolates from cacao, black pepper, and other hosts (but not from *Capsicum* pepper) that were differentiated isozymically into the subgroup CAPB. Chlamydospores were terminal or intercalary and varied from 22 to 52 µm in diameter.

Production of chlamydospores in some isolates is conditioned to some extent by the choice of cultural methods (Uchida and Aragaki 1985). When mycelial mats were grown for 5 days on clarified V8 juice neutralized with calcium carbonate and then submerged in sterile distilled water for 4 to 8 weeks and incubated in the dark, 20 of 29 isolates of *P. capsici* produced chlamydospores; however, isolates from solanaceous hosts, except for three isolates from eggplant, did not produce chlamydospores. A 50% papaya juice broth previously used to induce oospores of *P. palmivora* (Kadooka and Ko 1973) did not induce chlamydospores of any *P. capsici* isolate (Uchida and Aragaki 1985). Walls were 1.0 to 1.5 µm thick.

Hyphal Swellings

Hyphal swellings are occasionally produced by some isolates in aqueous cultures.

Sex Organs

P. capsici is predominately heterothallic; however, oospores have been noted to form in single culture in media containing the fungicide chloroneb (Noon and Hickman 1974) after wounding of the mycelium by exposing cultures to the fungus *Trichoderma* (Brasier 1975a, b, 1978) and in single A1 cultures (Kamjaipai and Ui 1978). Oospores formed in roots and stems of *C. annuum* inoculated with A1 and A2 mating types (Matsuoka and Ansani 1984). Both A1 and A2 mating types are isolated from pepper and cucurbit plants in the field, but in general A1 or A2 mating types occur singly and seldom together on a plant (Ristaino 1990).

Antheridia are amphigynous. Diameters given in several reports include 12 to 21 × 12 to 17 µm (Frezzi 1950); 17 × 15 µm (Stamps 1985); 9 to 20 × 8 to 17 µm, average 14.1 × 13.5 µm (Kröber 1985); 12 to 17 × 14 to 16 µm, average 14 × 15 µm (Tsao 1991).

Oogonia are spherical or subspherical and hyaline to brown. Dimensions of oogonia from different hosts vary from 23 to 50 µm (Table 14.2).

Oospores are predominately plerotic with a wall thickness of 2 to 6 µm (Tsao 1991). Mean diameters of oospores from several pepper isolates ranged from 23.7 to 34.9 µm; mean diameters of several cucurbit isolates ranged from 27.8 to 34.2 µm (Ristaino 1990). Dimensions of oospores from various hosts are given in Table 14.2.

Growth Temperatures

Minimum temperature for growth is 10°C, optimum is 28°C, and maximum is >35°C (Stamps 1985). Several reports show different data for minimum, optimum, and maximum temperatures: 7.5, 30, and 35°C (Kröber 1985); 6 to 9, 27 to 30, and 33 to 39°C (Tsao 1991); and 12, 28, and 32°C (Leu and Kao 1981). Isolates from CAP1 with the exception of three isolates grew well at 35°C. Four isolates of CAP2 grew poorly at 35°C, but the remaining isolates grew well (0.7–7.3 mm per day). Isolates from CAP3 grew poorly or not at all at 35°C. Optimum temperatures for growth among all isolates varied from 24 to 33°C (Mchau and Coffey 1995).

Distinguishing Characteristics

P. capsici is distinguished from other species in group II by its production of long pedicels on caducous sporangia with shapes varying from nearly spherical to elongated with a tapering base. *P. parasitica* produces more rounded sporangia that are noncaducous; *P. palmivora* produces ovoid sporangia that are caducous but have short pedicels.

Diseases Caused by *Phytophthora capsici*

Blight and Root and Crown Rot of *Capsicum annuum* L. (Pepper)

Phytophthora blight of peppers is an important disease worldwide. It is also known as *pourriture des fruits* or *mildou du piment* in French, *podredumbre del fruto* in Spanish, and *Phytophthora-Fäule* in German. The damping-off phase of this disease can result in the death of seedlings within 5 days of infection. The disease on the pod (Plate 14.1) first appears as a small, water-soaked, dull green spot that rapidly elongates

Table 14.2. Dimensions of reproductive structures of *Phytophthora capsici* from various hosts

Host	Substrate	Sporangia (µm)	Oogonia (µm)	Oospores (µm)	Reference
Pepper	Oatmeal	35–105 × 21–56 (60 × 36)[a]	NR[b]	25.0–35.0 (NR)	Leonian (1922)
Pepper	Oatmeal	16–69 × 12–31 (30 × 20.8)	23.4–33.4 (28.1)	20.9–29.2 (24.9)	Tucker (1931)
Pepper	Oatmeal	27–58 × 23–48 (46.9 × 35.9)	25.1–32.6 (29.1)	19.2–30.1 (24.4)	Tucker (1931)
Melon	Oatmeal	21.4 × 18–40.7 (36.5 × 27.0)	24.1–41.4 (29.9)	22.7–31.4 (26.5)	Wiant and Tucker (1940)
Pepper		28–123 × 21–50 (53.0 × 30.5)	29.5–46.0 (36.2)	25.0–42.0 (31.5)	Frezzi (1950)
Pepper	Cornmeal	30–60 × 25–35 (NR)	<39.0 (NR)	NR	Waterhouse (1963)
Pepper	Water	29–68 × 17–38 (44.7 × 28.1)	21.0–50.0 (31.8)	18.0–43.0 (28.8)	Ershad (1971)
Pumpkin	V8 juice	24–60 × 30–92.4 (39.0 × 65.3)	28.8–33.6 (30.3)	24.0–28.8 (26.3)	Kamjaipai and Ui (1978)
?	Water	32–92 × 25–46 (51.4 × 34.5)	21.0–50.0 (31.8)	18.0–43.0 (28.8)	Kröber (1985)
?	Malt	29–68 × 17–38 (NR)	<39.9 (30.0)	NR	Waterhouse (1963)
Various	Carrot	40–52 × 20–31 (47.0 × 27.0)	27.0–43.0 (33.0)	22.0–37.0 (28.0)	Tsao (1991)
Various	Carrot	32.8–65.8 ± 5.8 × 17.4–38.7 ± 3.8	NR	22.0–36.6 ± 2.9	Mchau and Coffey (1995)

[a] Average dimension is given in parentheses.
[b] Not recorded.

under favorable conditions to cover more than one-quarter of the pod. The progression of the lesions is usually more rapid along the stem end and the blossom tip of the pod than in the center. Infected tissues become dry, sunken, and parchment-like and turn to a straw color. In severely diseased fruit, the seeds become infected, turn brown, and shrivel. The pathogen spreads by water splash or growth from the fruit to the branches where it readily kills the young and tender tissues.

When large plants are infected, the main branches are girdled and killed. Infection of the main stem is indicated by dark purplish lesions close to the soil line. These lesions rapidly coalesce and entirely girdle the stem. It is not unusual for infected plants to produce an excessive amount of pods as a result of the girdling effect before succumbing (Plate 14.2).

Disease Development. Prolonged rain or irrigation that leads to excessively wet soil conditions favors the spread of the disease. Sporangia or zoospores in splashing water can infect the branches or fruit (Schlub 1983). When zoospore inoculum was poured on the soil, a concentration of 10^4 zoospores per plant was the minimum required to kill 100% of the inoculated plants (Ansani and Matsuoka 1983a). When inoculum was mixed with the soil, the minimum number was 1.6×10^4 zoospores per plant to kill 100% of transplanted seedlings. The severity of disease decreased with reduction in numbers of zoospores. Infectivity of zoospores was similar in plants varying in age from 17 to 59 days, although symptom expression was delayed with increasing age. Survival of transplants in soil was least (18%) when soil contained 25 ppg (propagules per gram), intermediate (55%) at 5 ppg, and greatest at 1 ppg (84%) (Schlub 1983).

Periodic flooding increases the number of pepper plants killed by *P. capsici*. Distribution in the field follows the irrigation pattern (Schlub 1983). In soil infested with 25 oospores per gram of soil, the plant mortality increased with the increase in number of 24-h flooding periods at 10-day intervals (Bowers and Mitchell 1990). Ristaino (1991) also showed that severity of disease on pepper was reduced by irrigating less frequently. A high level of disease from oospore inoculum in soil occurred after only 2 h of soil saturation (Hord and Ristaino 1992). A 2-h period of saturation had an effect similar to that of longer periods (up to 12 h) of flooding.

An epidemiological examination of three pepper fields by Ristaino et al. (1993) in North Carolina indicated that naturally formed inoculum of *P. capsici* dispersed down the row and across the rows. A cross-row spread in some areas was associated with ponding of water. Their data on a cross-row distribution agreed with the finding of Bowers et al. (1990), who showed that *P. capsici* spread from point sources of introduced inoculum. Infection of roots followed by spread of infection to the crowns was the predominant method of infection. Plants often wilted even before crown rot was evident (Ristaino et al. 1993).

P. capsici survived in infected plant tissue in moist soil but not in dry soil (Schlub 1983). Although Ansani and Matsuoka (1983c) found that mycelium, sporangia, and zoospores did not survive more than 75 days when incorporated into the soil at different moisture levels and soil depths, oospores obtained from pairing A1 and A2 isolates survived in the soil for 210 to 240 days (Ansani and Matsuoka 1983a). If oospores form from natural crosses in the field, rotation with crops known to be susceptible to *P. capsici* may not successfully control the disease.

Inoculation of pepper plants with A1 and A2 isolates resulted in production of oospores in the cortex of infected plants (Matsuoka and Ansani 1984). When concentrations as low as 2×10^1 oospores (germination was only 3.5%) per plant were used as inoculum, a large percentage of pepper plants were killed (Ansani and Matsuoka 1983b). This indicates, as explained in Chapter 5, that a rapid increase in inoculum density occurs when conditions are favorable.

Although the practice of using black plastic mulch enhances yields because of greater control of weeds, it has greatly increased the incidence of Phytophthora blight by increasing the spread to higher areas in the field (Springer and Johnston 1982).

Control. Selection of cultivars that are genetically resistant and planting in soil that is well drained are the most effective methods of control. Growing plants on a ridge or bed is also effective, provided that the ridge is high enough that roots and crowns of the plants are above any standing water in the field. Since oospores can develop in roots and stems of diseased plants, removal of the plant debris from infected fields should reduce the overwintering inoculum (Matsuoka and Ansani 1984).

A large number of cultivars with resistance have been developed (see Chapter 6 for listing and a discussion of resistance). Resistance in some cultivars is governed by two dominant genes (Smith et al. 1967). Resistance in line 92 is conditioned by three dominant genes (Gil Ortega et al. 1990). Pepper varieties and some germ plasm lines are resistant to certain races of *P. capsici* but not to others (Matsuoka et al. 1984). Mejia and Bustamente (1993) characterized races of *P. capsici* by differential resistance to several cultivars of red pepper. A comparison of *P. capsici* isolates from various parts of the world showed that the virulence of isolates varied on different cultivars and that the cultivar × isolate effect was significant (Kim and Hwang 1992). Both host-specific and horizontal resistance have been found to be useful (Bartual et al. 1991). The phytoalexin capsidiol is involved in host-specific resistance (Stössel et al. 1972; Turelli et al. 1984, 1986; see also Table 6.1 for references on host resistance). A discussion of phytoalexins, compounds produced by the plant in response to infection, is given in Chapter 6.

Since *P. capsici* from pepper also infects several other plant species, it may not be advisable to rotate *C. annuum* or *C. fructescens* with the following susceptible crops: *Carica papaya, Citrullus lunatus, Cucumis melo, C. melo* vars. *inodorus* and *reticulatus, Cucurbita maxima, C. cucurbita* var. *condensa, Lycopersicon esculentum, Solanum melongena,* or *S. melongena* var. *esculentum.* Other hosts are listed in Table 14.2; however, all hosts are not susceptible to all biotypes of *P. capsici.*

The collar rot phase of the disease can be controlled with repeated applications of metalaxyl (Ridomil 2E); the aerial blight phase can be controlled by Difolatan 4F, begun as soon as conditions become favorable for disease and repeated at 10-day intervals for a total of five applications (Springer and Johnston 1982). Metalaxyl was more effective than captafol for control of collar rot (Papavizas and Bowers 1991).

Spray application with triphenyl tin acetate at 20-day intervals was effective in Turkey (Çinar and Bicci 1977). Bimonthly foliar spray applications with Terrazole and Dexon controlled pepper rot in Italy (Mancini et al. 1977). In Brazil, maneb (0.02%) or a copper fungicide (0.2%), applied every 3 to 4 days after a rain, controlled the disease (Cruz and Da Silveira 1965). Aluminum and sodium ethyl phosphites (fosetyl-Al), prothiocarb, and pyroxychlor protected plants up to 4 weeks. Captan seed dressings prevented infection of seedlings and controlled seedborne inoculum (Clerjeau and Beyries 1977).

Fruit and Root Rot of Cucurbit Plants

Fruit rot of tomato and cucurbits (watermelon, cantaloupe, cucumber, squash) was reported by Kreutzer et al. (1940). Two types of lesions occur on cucurbit fruit: 1) initially, tiny, reddish brown discolorations on the fruit increase in size and become dark olive green with little evidence of water-soaking; later the lesions turn various shades of brown with irregular margins; 2) large, brown, slightly zonate, water-soaked lesions initially have irregular margins but later show well-defined margins (Plate 14.3). In some instances, lesions are mottled or have alternating light and dark concentric bands. In advanced stages, the white mycelium of *P. capsici* is visible on the affected areas. The internal tissues have a soft, watery, mushy consistency without the typical melon fruit odor. Fruits split easily when handled at harvest time.

A collar rot of melons caused by *P. capsici* and *P. nicotianae* that occurs in France (Clerjeau 1973) is characterized by a wilt of the leaves that progresses from the base to the extremities of the vines (Plate 14.4.). The collar of the plant eventually becomes constricted and the tissues corky and light brown.

In Japan, only the A1 mating type of *P. capsici* was isolated from rotted pumpkins (Kamjaipai and Ui 1978). Root rot of both pepper and cucurbit crops occurs in North Carolina. Some cucurbit isolates were virulent on pepper while others were not (Ristaino 1990).

Disease Development. Collar rot usually begins in areas of the field in which excessive water accumulates as a result of poor drainage and then progresses to other areas if conditions favorable for disease development continue to prevail. Melon fruit is more likely to become infected in wet areas of a field.

Control. Regulation of irrigation and host resistance (see Chapter 6) are important for control of root and collar rot (Ristaino 1991). Maintenance of temperature at 7 to 10°C during transit of fruit to market arrests development of fruit rot. Captafol applied by spray application in the field controls the root rot phase of the disease, and if registered for use, metalaxyl or fosetyl-Al should be effective.

Blight of *Macadamia integrifolia* (Macadamia)

Phytophthora blight of macadamia, a disease of the flowers and nuts, was first reported in Hawaii to be caused by *P. parasitica* and *P. palmivora* (Hunter et al. 1971); however, the isolates then identified as *P. palmivora* are now known to be *P. capsici* (Kunimoto et al. 1976). *P. capsici* usually infects the rachis first and a few days later rapidly invades and kills the buds. Affected racemes remain attached to the branches for several months. Racemes and nuts are susceptible to the pathogen at all stages of growth. A brownish black discoloration on the nuts is the first symptom. If the hard shell has not developed under the husk at the time of infection, *P. capsici* penetrates and destroys the kernel. Diseased nuts drop a few days after infection. Foliar blight is limited to the tender, expanding young leaves and the elongating green shoots on which extensive, irregularly shaped necrotic lesions occur. Mature leaves are seldom affected. Both *P. capsici* and *P. palmivora* are implicated in some areas in the development of the foliar phase. *P. cinnamomi* causes a trunk canker on macadamia trees (Zentmyer and Mitchell 1985/86).

Disease Development. Macadamia blight develops rapidly following prolonged periods of rain and at temperatures between 16 to 22°C. The spread of the disease is much reduced by dry weather. Continuance of heavy rain for several days increases the severity and spread of blight (Hunter et al. 1971). The foliar phase is believed to have little economic significance except as a means of carrying the pathogen over from one season to the next.

Control. No control measures have been recommended. If registered for use, metalaxyl or fosetyl-Al, used as a foliar spray, should be efficacious.

Buckeye Rot of *Lycopersicon esculentum* (Tomato)

Buckeye rot may also be caused by *P. parasitica* (*nicotianae*) or *P. drechsleri* (J. B. Jones et al. 1991). Green fruit of all ages that are close to or in contact with soil are most subject to infection (Kreutzer et al. 1940). The first symptoms are irregularly shaped, water-soaked lesions usually, but not always, at the blossom end of the fruits. Sometimes the lesions enlarge and the centers turn brown to blackish brown with complete or incomplete concentric rings of narrow, dark brown, and light brown bands. Diseased portions of the fruits are usually firm in texture, although internally the rotted tissue turns mushy.

Other references on fruit rot of tomato include *P. parasitica* (Tucker 1931), *P. mexicana* (Hotson and Hartge 1923), *P. drechsleri* (Tompkins and Tucker 1941a), and *P. palmivora* (Ramakrishnan and Soumini 1947). *P. capsici* and *P. parasitica* can also cause a root and crown rot of tomato plants (see below).

Disease Development. Abundant soil moisture, caused by excess rain or irrigation (Kreutzer and Bryant 1946), and soil temperatures between 18 and 30°C enhance the rapid progress of the disease.

Control. Poorly drained areas of the field should be avoided when planting tomatoes. Staking tomato plants prevents the developing fruit from contacting the soil surface. Growing plants on a ridge or bed when furrow irrigating reduces the chance that free water will stand near plants. Rotation with crops that are not susceptible will reduce the inoculum in soil. Cultivars with a high level of resistance are available. If registered, metalaxyl or fosetyl-Al should be useful as foliar sprays.

Root and Crown Rot of *Lycopersicon esculentum* (Tomato)

Plants of all ages are susceptible. Typical damping-off symptoms occur on seedlings. In young plants, infection of the roots and crowns results in rapid wilt followed by complete collapse of the plant. On older plants, cankers that are originally moist but later become soft, dry, and crumbly develop on the roots. Occasionally, the vascular tissues become discolored immediately above and below the cankers. Leaves become bronze and later necrotic. Necrosis begins at the tip and progresses toward the petiole, after which the entire leaf dies. Satour and Butler (1967) show excellent photographs of disease symptoms. The pathogen is not host specific. Root rot can also be caused by *P. parasitica.*

Disease Development. Depending on the type of irrigation applied, symptoms may vary from cankers limited to the crown and upper part of the root (sprinkler irrigation) to cankers forming on a majority of the roots (furrow irrigation).

The severity of symptoms was increased by excessive applications of nitrogen fertilizer, whereas applications of potassium helped to retard infection (Elenkov and Bakharieva 1975).

Control. Planting tomatoes in well-drained soil on raised beds helps to prevent severe expression of disease. Saturation of the soil for 5 h or longer favors infection (J. B. Jones et al.

1991). Resistance to *P. capsici* is available (Bolkan 1985; see also Chapter 6). Rotation to nonsusceptible crops will decrease the inoculum in soil.

Foot Rot of *Piper nigrum* L. (Black Pepper)

Probably the most recent general compilation on the epidemiology and control of black pepper wilt disease is by Sarma et al. (1991) in India, who describe and illustrate the collar rot, root rot, and leaf diseases caused by *P. capsici.* This chapter updates a previous account by Sarma and Nambiar (1982) in which the causal agent was attributed to *P. palmivora.* Foot rot was first reported from Indonesia in 1885 and has become limiting to the successful cultivation of black pepper in several areas of the world (Sarma and Nambiar 1982). Losses of 40 to 50% caused by foot rot have been reported in some black pepper plantations (de Waard 1980, 1986). Black pepper in India is grown on an area of 136,000 hectares and accounts for 54% of the world production of this spice crop. Foot rot, sometimes known as "quick wilt," and pepper yellows (associated with nematodes and soilborne fungi) are the most limiting agricultural problems in India. The disease is also important in Indonesia, Sarawok, and Thailand.

The causal agent of black pepper wilt has been restudied over the past 15 to 20 years. H. R. A. Muller (1936) identified the causal agent in Indonesia as *P. palmivora* var. *piperina;* however, this varietal name has not been accepted (Stamps et al. 1990). Much of the older literature attributes the disease to *P. palmivora* (Holliday and Mowatt 1963). G. J. Turner (1973) noted considerable variability in pathogenicity among isolates then known in Sarawok as *P. palmivora.* Holliday (1980) succinctly described foot rot of black pepper in his book on diseases of tropical crops as caused by *P. palmivora* but recognized that these causal isolates were not typical of *P. palmivora.* As discussed at greater length in Chapter 51 on *P. palmivora,* one of the atypical *P. palmivora* biotypes (MF4) was found to be *P. capsici.* The history of the nomenclatural problems is given by Tsao et al. (1985), who determined that isolates from black pepper and betle (*Piper betle* L.) plants in Indonesia fit into what was known then to be *P. palmivora* MF4 but was later redescribed as *P. capsici sensu lato* (Tsao and Alizadeh 1988; Tsao 1991; Mchau and Coffey 1995). In Thailand, isolates from black pepper and betle vine fit into the species now known as *P. capsici* (Tsao and Tummakate 1977).

Information from the mtDNA RFLP data confirms that the black pepper isolates, the *P. palmivora* MF4 isolates from cocoa, and the *P. capsici* isolates from pepper and other hosts such as macadamia are genetically related (Förster et al. 1990a). The literature contains many citations that have misidentified *P. palmivora* as the cause of the decline of *Piper nigrum.* It is now known to be caused by *P. capsici* (Tsao 1991).

Foot rot first appears as a slight wilt of the vine. The leaves become slightly pale green, and then the vines droop and leaves drop prematurely. Prior to dropping, leaves often curl inward along the edges and turn yellow. After defoliation, the fruit becomes dried and wrinkled. Seven to 14 days after the initial wilt symptoms are observed, approximately three-fourths of the plant defoliates (Plate 14.5). All parts of the plant—roots, collar, aerial stems, leaves, and spikes—are susceptible. Occasionally, only a single vine may be affected. The aboveground wilt symptoms result from the complete destruction of the main lateral roots and the girdling of the stem at the crown region. The infection of the roots is usually undetected until after the appearance of foliar yellowing. In advanced stages, the affected tissues emit a foul odor. All parts of the plant—roots, unsuberized stems, leaves, flowers, and fruit—are susceptible. Vines more than 3 years old are more susceptible.

Immature leaves of *Piper betle* are particularly susceptible (G. J. Turner 1969). On *P. nigrum,* deep brown lesions appear on leaves near the ground. The fimbriate edges of the lesions are a diagnostic symptom. In dry weather, lesions are not active, and the fimbriations become indistinct. With periods of recurrent rainfall, lesions may become concentrically zoned. Leaf infection results in defoliation. The leaf spot phase does not occur in some areas in which foot rot is prevalent.

Stems can be infected at any place on the vine. The infected portion of the vine then becomes water soaked, the foliage turns yellow, and leaves fall. The infection gradually progresses both up and down the vine, resulting in dieback of the affected stem. It is more difficult to isolate the pathogen from infected stems and roots than from leaf lesions; however, with the availability of improved selective media, isolation of *P. capsici* from roots can be made (see Chapter 2).

Disease Development. The primary source of inoculum is in infected plant debris. Infection by *P. capsici* usually occurs through one or more of the main roots, which arise from the underground nodes. Leaf infections initially occur from inoculum that is splashed by rain from the soil. The disease increases in severity during periods of heavy rainfall, characteristic of the monsoon season in the tropics, and when day and night temperatures vary between 19 and 23°C. Symptom expression is delayed at temperatures above 28°C (Nambiar and Sarma 1982). Disease incidence has been found to be greater in soils that are low in inorganic nutrients such as calcium, magnesium, and potassium and high in nitrogen (Nambiar et al. 1965). Termites and slugs can transmit inoculum to healthy trees. Black pepper wilt spreads in a centrifugal pattern and has been found to fit the compound interest model of Vanderplank (1963) (unpublished data of Ramachandran, cited by Nambiar et al. [1991]).

Control. Planting on well-drained sites not previously grown to black pepper for at least 1 year is the first step in control. After the disease is detected, removing the diseased vines and treating the soil around the diseased roots with a suitable copper fungicide is usually effective in preventing the spread to uninfected trees. Of several fungicides tested, Bordeaux mixture has given a measure of control. In general, fungicides are most effective in conjunction with other preventative measures. Efficacy of metalaxyl as a foliar and soil drench is reviewed by Nambiar et al. (1991). Metalaxyl and, to a lesser extent, fosetyl-Al as a foliar spray were effective (Ramachandran and Sarma 1985). If registered and economically feasible, these fungicides should be useful (see Chapter 7).

Resistance to Phytophthora foot rot has been reported in *Piper aduncum, P. colubrinum, P. guineese, P. obliquum* var. *eximum, P. scabrum,* and *P. treleasanum* (Ruppel and Almeyda 1965; Albuquerque 1968; G. J. Turner 1971). In commercially grown cultivars of *P. nigrum,* a high degree of resistance is lacking, but different degrees of tolerance are detectable by stem lesion and root rot rating systems (reviewed by Nambiar et al. [1991]). Although *P. colubrinum* is immune, grafting *P. niger* scions on *P. colubrinum* rootstocks has been "variable and less encouraging" (Nambiar et al. 1991).

CHAPTER 15A

Phytophthora cinnamomi

Rands (1922)

var. *cinnamomi*

The only known synonym of *Phytophthora cinnamomi* is an undescribed fungus from pineapple named *Pseudopythium phytophthoron* by Sideris and Paxton (1931) because it did not produce sporangia. After Mehrlich (1932) obtained sporangia that were typical of *P. cinnamomi,* Sideris withdrew his description. White (1930) and Mehrlich (1936) considered *P. cinnamomi* to be a synonym for *P. cambivora.* The uncertainty concerning these two species was resolved by Waterhouse (1963), who differentiated *P. cinnamomi* as a valid species distinct from *P. cambivora. P. cinnamomi* is described in Waterhouse and Waterston (1966c). See also the monograph on *P. cinnamomi* by Zentmyer (1980) and description of diseases by Holliday (1980).

P. cinnamomi was first described as the causal agent of stripe canker of *Cinnamomum burmannii* Blume in Sumatra by Rands (1922). Since then, this species has proved to be a serious root-rotting pathogen of numerous woody host plants. Tucker (1933) published a host list for all *Phytophthora* species then known. Subsequent lists were reported by Thorne and Zentmyer (1954), Crandall and Gravatt (1967a, b), Titze and Palzer (1969, 1970), and Weste (1977). A complete host range of *P. cinnamomi* up to 1980 covering approximately 950 plant species was published by Zentmyer (1980). Recent additions to this host list include plants infected by *P. cinnamomi* in Australia (Weste 1979) and China (Ho et al. 1983a, 1984a–c). Forsberg (1985) listed a large number of ornamental hosts in Queensland, Australia; an accessory list for Australia that includes species in 21 families was given by Weste (1977); and an annotated bibliography of little leaf and tree decline diseases was reported by Tainter et al. (1987). A list of ornamental plant hosts in Australia was reported by Hardy and Sivasithamparam (1988). The host list presently covers well over 1,000 plant species.

Some economically important plants affected by this pathogen include *Ananas comosus* (L.) Merrill (pineapple) (Mehrlich 1934, 1936); *Castanea dentata* Siebold and Zucc. (chestnut) (Crandall 1936; Crandall et al. 1945); *Chamaecyparis lawsonianae* (A. Murr.) Parl. (cypress) (Torgeson 1954; Torgenson et al. 1954; Newhook 1959); *Cinchona* spp. (quinine) (Crandall 1947, 1950, 1954); *Cinnamomum* spp. (cinnamon) (Rands 1922); species in the family *Ericaceae,* e.g., *Rhododendron* (R. P. White 1930, 1937; White et al. 1933; Dingley 1969; Benson 1980; Coyier and Roane 1986); *Eucalyptus* spp., especially *E. marginata,* the jarrah tree (Podger et al. 1965; Podger 1972; Weste 1994); *Juglans* spp. (walnut) (Crandall 1936); *Pinus* spp. (pine) (Campbell and Copeland 1954; Zentmyer and Munnecke 1952; Newhook 1958–1960); *Prunus* spp. (almond, cherry, peach, and plum) (Mircetich and Keil 1970; Mircetich and Browne 1987); and *Quercus* spp. (oak) (Crandall 1936; Mircetich et al. 1977; see also Table 1 in Zentmyer 1980). In the United States, *P. cinnamomi* is the most important pathogen of avocado (*Persea americana* Mills) (Tucker 1929; Wager 1942a; Zentmyer's 1980 review).

See the monograph by Zentmyer (1980) for comprehensive coverage and Waterhouse and Waterston (1967f) for a summary of characteristics. World distribution is shown in Commonwealth Mycological Institute Map No. 302 (ed. 4, 1978). Weste (1984, 1994) reviews effects on native vegetation in the Australasian area of the world.

A recent first report indicates that *P. cinnamomi* causes root rot of cranberry (*Vaccinium macrocarpon* Aiton) in the east coast region of Massachusetts (Caruso and Wilcox 1990). Flooding in December to prevent freezing and flooding in the spring and fall to provide insect control, facilitate harvest of cranberries, and remove organic debris are standard horticultural practices. The authors hypothesize correctly that resulting conditions are ideal for Phytophthora root rot. They isolated *P. cinnamomi* from 25 cranberry beds in which a common water system for flooding and drainage was used.

Methods for isolation of *P. cinnamomi* by baiting with susceptible plant parts and selective media are given by Zentmyer (1980). See also Chapter 2.

P. cinnamomi has become an important pathogen in several Australian forests. As a result, a large volume of information on all aspects of this pathogen has been published by researchers in Australia. The dieback epidemic on jarrah (*Eucalyptus marginata* Sm.) has caused a great loss of this important forest tree in Western Australia (Podger et al. 1965; Shea 1977). See Chapter 5 for excerpts by Shea covering this epidemic.

Newhook (1978) aptly summarizes the Phytophthora root rot disease situation in the forests in the Australia-New Zealand area. Most of the problems have occurred within the last 30 to 40 years—a relatively short time in the history of forest pathology. First, *P. cinnamomi* has been found to kill eucalyptus trees and other woody herbaceous understory species in several areas, notably on jarrah, *E. marginata,* in Western Australia (Podger et al. 1965; Podger 1972), in Victoria in the Brisbane Ranges (Weste and Taylor 1971), and in Wilson's Promontory National Park (Weste 1974). A second situation involves extensive killing of trees following abnormally heavy rainfall during the summer and fall, e.g., in east Gipland, Victoria (Marks et al. 1973; Weste and Marks 1974), and near Auckland, New Zealand, on Kauri trees, *Agathis australis* Salisb. (Podger and Newhook 1971). A third situation involves a flora of trees in some areas that tolerate *P. cinnamomi* such as forests in New South Wales (Pratt and Heather 1973; Pratt et al. 1973), Queensland (Pratt and Heather 1973), and New Zealand (Podger and Newhook 1971), which are comparable to forests in Western Australia and Victoria. Newhook (1978) cites literature and arguments to support why *P. cinnamomi* is tolerated by trees in certain of the above situations and not in others. Certainly the environment, e.g., lack of high soil moisture at critical times, is involved (Newhook and Podger 1972; Marks et al. 1973; Weste 1983), and high microbial populations of saprophytes in the soil could adversely affect the pathogenesis of *P. cinnamomi* (Broadbent and Baker 1974; Weste 1983b). Despite these reports, the answers to many of these questions are complex and not yet well enough understood.

Much speculation in recent years has centered on the origin of *P. cinnamomi* (Ko et al. 1978; Zentmyer 1988). Ann and Ko (1985) present evidence to support the hypothesis that *P. cinnamomi* has an Asian origin, i.e., Taiwan and other nearby countries, based in part on the isolation of both A1 and A2 mating types from native vegetation (Zentmyer 1983, 1988). *P. cinnamomi* has probably been introduced into other warm-season areas of the world by plant-moving activities of people. Newhook (1978) discounts the theory that *P. cinnamomi* is indigenous to Australia (Pratt et al. 1973).

Characteristics of *Phytophthora cinnamomi*

The morphology of *P. cinnamomi* has been described in detail by Royle and Hickman (1964), Ho and Zentmyer (1977a), Gerrettson-Cornell (1983), and Waterhouse and Waterston (1966c). *P. cinnamomi* is listed in group VI (Stamps et al. 1990). Morphology is illustrated in Figure 15A.1. See Tables 4.2 and 4.3 for tabular keys and Appendix 4.9 for a dichotomous key (Ho 1992; Ho et al. 1995).

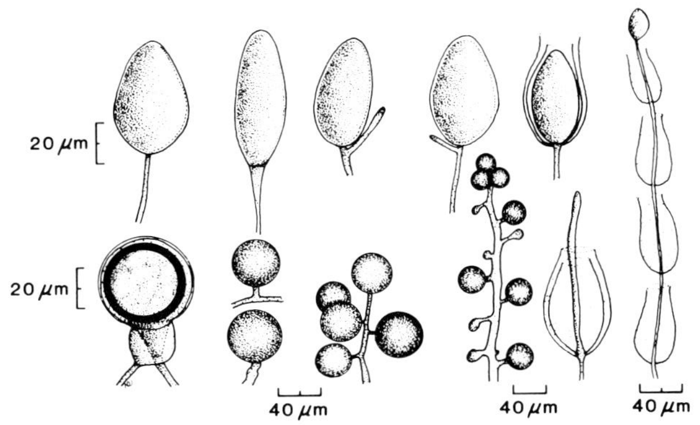

Figure 15A.1. Morphology of *Phytophthora cinnamomi*. Nonpapillate sporangia, proliferation of sporangia, globose oogonia and oospores, globose chlamydospores, and numerous hyphal swellings. (Drawing by A. Vaziri)

Sporangia

Sporangia are ovoid, obpyriform, or ellipsoid to elongate-ellipsoid with an inconspicuous apical thickening (nonpapillate). They are tapered or rounded at the base, noncaducous, and borne terminally. New sporangia are produced by internal or external proliferation or by sympodial development of the sporangiophore immediately below empty sporangia (Figure 15A.1).

Sporangia average 75 × 40 µm (length × breadth) with a length-breadth ratio of 1.54 (Ho and Zentmyer 1977); however, dimensions of sporangia vary in different reports (Table 15A.1). Royle and Hickman (1964) show an array of shapes ranging from elongate to ovoid. Sporangia are not produced readily in axenic culture, but incubation of mycelial disks in nonsterilized soil extract (10 g of soil per liter) (Mehrlich 1935) or frequent washing of mycelium-agar disks in a salt solution (Chen and Zentmyer 1970) usually induces sporangium formation (see Appendixes 3.38 and 3.39 in Chapter 3). Sporangiophores are simple or sympodially branched either in a lax or close arrangement and typically proliferate through an empty sporangium (Figure 4.7).

Chlamydospores

Chlamydospores form abundantly in culture and infected tissues and are borne on the parent hyphae or on new hyphal branches; sizes range from 31 to 50 µm in diameter (average 41 µm) (Rands 1922). Chlamydospores are globose and have thinner walls than most other *Phytophthora* species (Zentmyer 1980; Figures 4.23A, B and 4.24E). They are either terminal or intercalary in the mycelium and in culture and often occur in characteristic grapelike clusters of three to 10 chlamydospores (Figure 4.1). Hyphal swellings are also formed more profusely than by most other species.

Sex Organs

P. cinnamomi is heterothallic; oospores form when A1 and A2 mating types are paired (Haasis and Nelson 1963; Galindo and Zentmyer 1964; Zentmyer et al. 1979). The A2 mating type is more commonly found than A1 (Zentmyer 1980; Old et al. 1984). Selfed oospores sometimes form in cultures of a single A2 mating type when incubated in avocado root extracts (Zentmyer 1952; Zaki et al. 1983), in diseased avocado

Table 15A.1. Dimensions of sporangia of *Phytophthora cinnamomi* recorded by various investigators

Host	Range (µm)	Average (µm)	L-B[a]	Reference
Cinnamomum sp.	88–84 × 27–39	57.0 × 33.0	NR	Rands (1922)
Avocado	30–64 × 22–39	46.1 × 31.3	NR	Tucker (1931)
Pineapple	47.9–101.6 × 22.7–55.9	NR	NR	Mehrlich (1936)
Avocado	39–66 × 26–40	50.0 × 32.0	NR	Wager (1942)
Erica sp.	37.1–117.4 × 25.7–62.9	67.2 × 45.9	NR	Royle and Hickman (1964)
Avocado	31.4–122.9 × 24.3–61.4	65.0 × 42.0	NR	Royle and Hickman (1964)
Vaccinium	32.9–112.9 × 22.9–57.1	62.8 × 40.4	NR	Royle and Hickman (1964)
NR	100 × 40 (max.)	57.0 × 33.0	NR	Waterhouse and Waterston (1963)
Various	42.3–112.8 × 32.9–51.7	75.2 × 40.2	1.94	Ho and Zentmyer (1977)
Various	47.0–89.3 × 28.2–56.4	61.1 × 39.8	1.54	Ho and Zentmyer (1977)
Cedrus sp.	37.1–85.5 × 25.7–45.6	61.2 × 29.4	NR	Ho et al. (1984)
Citrus	51.0–85.0 × 25.0–50.0	69.0 × 40.0	1.70	Ann and Ko (1985)
Pineapple	55.0–100.0 × 30.0–56.0	76.0 × 44.0	1.80	Ann and Ko (1985)
Camellia	50.0–75.0 × 38.0–45.0	59.0 × 40.0	1.50	Ann and Ko (1985)
Various	11.0–103.0 × 11.0–53.0	50.6 × 34.2	1.48	Kröber (1985)
Various	22.0–113.0 × 17.0–62.0	48–75 × 31–46	1.3–1.9	Gerrettson-Cornell (1989)
Beaucarnea sp. (*P. cinnamomi* var. *parvispora*)	14–83 × 13–43	36.9 × 25.5	1.45	Kröber and Marwitz (1993)

[a] L-B = length-breadth ratio. NR = not recorded.

root tissue (Mircetich and Zentmyer 1966), or on oat grains (Royle and Hickman 1964). Oospore production can be stimulated by the presence of the fungus *Trichoderma viride* (Reeves and Jackson 1972; Brasier 1975a, b, 1978; see also Chapter 3).

Antheridia are amphigynous, averaging 19 × 17 µm. Some are bicellular (Royle and Hickman 1964; Ho and Zentmyer 1977a). Oogonia are round, often with a tapered base, smooth walled and hyaline to yellow brown and range in size from 21 to 58 µm in diameter, averaging 40 µm in diameter. Dimensions of sex organs from various plant species and reports are given in Table 15A.2. A fertilization tube from the antheridium that penetrates the oogonium is shown by Ho and Zentmyer (1977a).

Oospores are round, hyaline to yellow brown, and plerotic. Size ranges from 19 to 54 µm in diameter, averaging 20 µm in carrot agar medium, 33 µm in V8 juice medium, 34 µm in cornmeal agar, and 40 µm in Ribeiro's synthetic medium (Ribeiro 1975a; see Chapter 3). Dimensions of oogonia, oospores, and antheridia of isolates from various plant species are given in Table 15A.2.

Growth Temperatures

The minimum temperature for growth is 5 to 6°C, optimum between 24 and 28°C, and maximum 32 to 34°C. On the basis of a large number of reports, Zentmyer (1980) states, "The cardinal temperatures for this species are minimum 5–15°, optimum 20–32.5° and maximum 30–36°."

Growth Characteristics in Culture Media

Distinctive coralloid-type mycelium with abundant hyphal swellings, swollen vesicles, and sessile terminal or lateral protuberances produced singly or in clusters are characteristic (Figure 4.1). In most cases, the colony produced on potato-dextrose agar is a rose or camellia type (Figure 4.2); however, there are exceptions (see Zentmyer 1980, Figures 24 and 25).

Distinct differences in pathogenicity and morphology among isolates exist (Zentmyer 1980). Strain differences expressed as host preference have been reported on walnut (Crandall et al. 1945), Lawson cypress, Irish yew, Douglas fir (Torgeson 1954), and pineapple (Klemmer and Nakano 1964). Chee and Newhook (1965b) in New Zealand report that pathogenic variation is minor.

Distinguishing Characteristics

P. cinnamomi is one of the most easily identified of all the species because of its unique morphological features described above and in Chapter 4. Worldwide, the main mating type is A2. Although variability exists (Zentmyer 1980), the range as detected by isozyme analysis indicates that this species is quite homogeneous and genetically clearly different from *P. cambivora* and *P. cactorum* (Old et al. 1984, 1988; Oudemans and Coffey 1991); however, eight electrophoretic types can be identified within *P. cinnamomi* isolates. The isolates from citrus that were considered by Ann and Ko (1985) to be closely related to *P. cinnamomi* by protein pattern analysis were shown by Oudemans and Coffey (1991a) to be clearly distinct by isozyme analysis; consequently, they did not support identification of these isolates as *P. cinnamomi*.

A1 isolates can be differentiated from A2 isolates on the basis of the electrophoretic migration pattern on gels of isocitrate dehydrogenase (Oudemans and Coffey 1991). See Chapter 4 for further discussion of isozyme and mtDNA technology used for confirmation of identity and genetic relationships of *Phytophthora* species.

Diseases Caused by *Phytophthora cinnamomi*

Diagnostic features of root diseases caused by *P. cinnamomi* are given by Zentmyer (1980) and Coffey (1992). Usually after root rot occurs, the foliage of plants becomes chlorotic and wilted and rapidly dies back (depending on the severity of the root rot). Sometimes the infected plants suddenly collapse, but in other cases they survive for several years, particularly in cooler, damp climates. Symptomless plants are often a major source for spread of *P. cinnamomi* to previously clean areas, and this is a major problem for intensive growing systems in the nursery industry.

It is unusual for monocotyledonous plants to be susceptible to *Phytophthora*, but the grass tree *Xanthorrhoea australis* R. Br. is severely affected in native Victoria forests of Australia (Weste 1975, 1984) (Figure 15A.6). Even wheat plants become infected, and roots rot when they are artificially inoculated (Weste 1978).

Plants that are stressed because of excessively moist or dry

Table 15A.2. Dimensions of sex organs of *Phytophthora cinnamomi* from various hosts[a]

Host	Antheridia: Range or average (µm)	Oogonia: Range (µm)	Oogonia: Average (µm)	Oospores: Range (µm)	Oospores: Average (µm)	Reference
NR[b]	NR	21–42	32	19–36	27.2	Ashby (1929)
Cinnamomum	NR	22.5–35.9	28.1	20.9–33.4	25.7	Tucker (1931)
Avocado	NR	27–48	37.0	NR	NR	Wager (1942)
Avocado	NR	22.9–54.3	40.4	20.0–51.4	37.7	Royle and Hickman (1964)
Erica sp.	NR	24.3–51.4	39.6	20.0–47.1	37.1	Royle and Hickman (1964)
Juglans regia	NR	28.6–44.3	37.1	25.7–40.0	34.5	Royle and Hickman (1964)
Vaccinium corymbosum	NR	25.7–54.3	41.4	22.9–51.4	38.7	Royle and Hickman (1964)
Various	19.8	31.3–42.3	35.7	27.6–36.8	31.7	Ho and Zentmyer (1977)
NR	22 × 17	58 (max.)	40.0	NR	NR	Waterhouse and Waterston (1966)
Cedrus sp.	18.8	25.7–45.6	43.5	22.8–42.8	36.9	Ho et al. (1984)
Pineapple	18 × 16	30–50	40.0	23–48	36.0	Ann and Ko (1985)
Avocado	20 × 17	23–45	36.0	20–40	32.0	Ann and Ko (1985)
Camellia	18 × 16	26–58	37.0	22–43	29.0	Ann and Ko (1985)
Macadamia	20 × 17	27–48	37.0	22–43	33.0	Ann and Ko (1985)
Various	19–21 × 16–17	21–58	28.0–42	20–54	29.0–40	Gerrettson-Cornell (1989)
Beaucarnea sp. (*P. cinnamomi* var. *parvispora*)	. . .	23–41	31.9	22–36	28.1	Kröber and Marwitz (1993)

[a] See also Zentmyer (1980; Table 3)
[b] Not recorded by the reference cited.

soil conditions are sometimes more prone to infection. Excessive use of nitrogen-based fertilizers further increases the susceptibility of plants and the severity of the disease. See Chapter 5 for a discussion on the role of soil moisture and predisposition to disease.

P. cinnamomi has more capabilities for survival as a saprophyte than most other *Phytophthora* species and is known to survive for as long as 6 years in moist soil (Zentmyer and Mircetich 1966). Mycelium is readily lysed by soil microorganisms, however, particularly *Trichoderma* spp. and *Gliocladium* spp. (Reeves 1975; Malajczuk 1983). The main means of long-term survival is the chlamydospore (Weste 1983b). Zentmyer (1980) shows that *P. cinnamomi* is capable of colonizing dead organic matter and presents evidence, both supporting and conflicting, from the literature. Survival is discussed in Chapter 5.

Control of *P. cinnamomi* is difficult because of its wide host range and its ability to survive in symptomless plants as chlamydospores in roots at soil depths at which fumigation is not always effective (Munnecke 1972, 1984).

Planting disease-free stock is the first line of control (Pegg 1978; Zentmyer and Ohr 1978). A set of criteria for strict nursery hygiene set up by Pegg (1978) is discussed in Chapter 5 and reproduced in Appendix 5.2. Fumigation of soil with methyl bromide is effective for control in areas such as nursery seedbeds in which penetration of the gas is effective (Munnecke and Ferguson 1953; Munnecke et al. 1974).

In the field, control should include an integrated combination of cultural, biological, and chemical methods with the use of host resistance (Coffey 1987, 1991, 1992). Cultural and biological control methods are discussed in considerable detail in Chapter 5. Biological control of avocado root rot in some areas of Australia is reproduced from Baker (1979).

A list of woody shrubs and trees that are resistant to *P. cinnamomi* was reported by Edmiston (in Shea 1975). Zentmyer and Mitchell (1985/1986) discuss control of fruit tree diseases in the tropics, and the monograph by Zentmyer (1980) summarizes disease-control methods. In Chapter 6, Table 6.1 contains citations on examples of resistance to *P. cinnamomi.*

In Chapter 7, Table 7.6 contains citations on diseases reported to be controlled by metalaxyl or similar phenylamide systemic fungicides, and Table 7.11 contains examples of diseases controlled by fosetyl-Al.

Heart Rot of *Ananas comosus* (Pineapple)

Heart rot is also known as *pourriture de coeur de l'ananas; podredumbre de la raiz; seccume della frutescenza dell'nanas;* and *Phytophthora-Herzfäule. P. cinnamomi* infects young as well as older pineapple plants just prior to fruit set and causes root and heart rot (Mehrlich 1934). In addition to *P. cinnamomi, P. palmivora* (Mehrlich 1934, 1936) and *P. parasitica* (Hine et al. 1964) also cause heart rot of pineapple. In the early stages of infection after the crown has been invaded (Figure 15A.2), the terminal crown of leaves can be easily detached with a slight pull. A gradual change in leaf color from a normal green to yellowish green to red is the first recognizable field symptom (Plates 15A.1 and 15A.2). The leaves then curl back along the edges, eventually become smoky brown in appearance, disintegrate, and fall to the ground. Outer (older) leaves often remain normal until the disease is well advanced. The bases of the central leaves become yellowish white, bordered by a distinct and characteristic margin. The interior of the fruit becomes brown (Figure 15A.2), and a distinct foul odor often emanates from the center of the plant as secondary bacteria invade the diseased tissues. A disease with similar symptoms, caused by *P. parasitica,* occurs at warmer temperatures (30°C) (Hine et al. 1964). *P. cinnamomi* also causes a rot in green fruits that are in contact with the soil (Plate 15A.3).

Figure 15A.2. Fruit rot of a pineapple (cut lengthwise and crosswise) caused by *Phytophthora cinnamomi.* (Courtesy K. G. Pegg, Department of Primary Industries, Queensland, Australia)

Disease Development. This disease is most severe in areas of high rainfall or excessive irrigation and at temperatures between 20 and 27°C; however, the pathogen is often present but not active in soils in warmer and drier areas. Physical conditions of the soil that prevent vertical drainage of soil water favor the disease.

Control. Planting suckers or clones in well-drained soil and on raised beds helps to alleviate the severity of root rot. Dipping suckers in Bordeaux mixture before planting them in infested soil helps to prevent infection. Captafol has also been used extensively as a preplant dip and foliar spray. In Australia, preplant soil applications with elemental sulfur at 1,200 kg/ha, sufficient to lower the pH to 3.8, controlled root rot (Pegg 1977a, b; Pegg and Whiley 1987). According to Schmitthenner and Canady (1983), heart rot is one of the few diseases that can be controlled by use of a chemical fertilizer amendment—in this case, primarily because pineapple growth is not adversely affected by the increase in aluminum released by acidification of soil. Some cultivars of *Ananas comosus* (e.g., Pernambuco and Red Spanish) and some species, such as *A. ananassoides* (Bak.) L. B. Sm. and *A. bracteaus* (Lindl.) Schult., are resistant to heart and root rot.

Treatment of fruit with a preharvest spray of potassium phosphite (phosphonate), the active ingredient of fosetyl-Al, effectively controlled heart rot in the field in Australia (Allen et al. 1980; Pegg and Whiley 1987; De Boer et al. 1990) and in Hawaii, United States (Rohrbach and Schenck 1985). Phosphorous acid is registered for use in Australia but not in the United States, where only fosetyl-Al is registered. See Chapter 7 for more references and a discussion of control with fungicides.

Collar and Root Rot of *Actinidia chinensis* Planch. (Kiwi Fruit)

Severe collar and root rot of kiwi fruit trees occurs in many areas. After infection of roots, the bark sloughs away and exposes the characteristic brown to black stained vascular tissues (Figure 15A.3). The lesions are usually confined to the lower stem at or just below the soil level. Root systems are entirely rotted from the extremities back to the crowns, and plants defoliate and eventually die (Figure 15A.4) (G. I. Robertson 1982b).

In addition to *P. cinnamomi, P. citricola, P. cryptogea, P. gonopodyides, P. megasperma,* and two sterile *Phytophthora* species were isolated by the cedar needle baiting method of Dance et al. (1975) from rotted kiwi roots in New Zealand. All species except *P. megasperma* were highly pathogenic in glasshouse tests; however, previous records had indicated this species to be pathogenic. *P. cinnamomi* and *P. cryptogea* were the most virulent (Stewart and McCorrison 1991). Conn et al. (1991) in California reported that *P. citrophthora, P. cryptogea,* and *P. megasperma* were associated with root rot and were pathogenic. Although *P. cactorum* was isolated, it was avirulent on kiwi fruit. Waterlogging of soil in absence of *Phytophthora* will also kill kiwi fruit plants (Robertson 1982b).

Disease Development. Root rot develops rapidly when vines are grown in poorly drained, heavy soils.

Control. Adherence to the fundamental principles of planting disease-free stock and prevention of waterlogging of soil will control root rot. If necessary, however, and if it is registered, an application at 10 g/m^2 of a 25% solution of metalaxyl should provide control (Robertson 1982b).

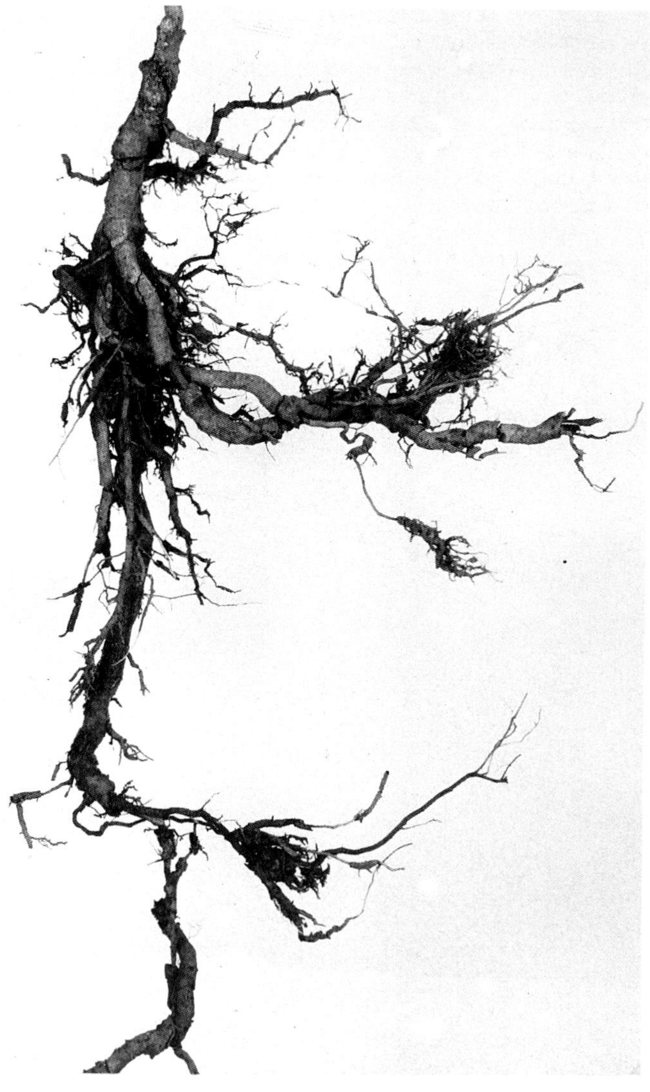

Figure 15A.3. Fibrous feeding roots from a 5-year-old vine of kiwi fruit infected by *Phytophthora cinnamomi.* The rotted outer tissue is sloughing off. (Courtesy G. I. Robertson and A. P. Underhill, D.S.I.R., Auckland, New Zealand; Robertson 1982b)

Root Rot of *Camellia japonica* L. and *C. sasanqua* Thunb. (Camellia)

Root rot is prevalent wherever camellia plants are grown, probably because *P. cinnamomi* can be carried on infected but symptomless plants in nurseries. To a large extent, information on root rot of *Camellia* spp. can be applied to a wide range of woody nursery stock planted in landscapes for ornamental purposes (Zentmyer and Munnecke 1952).

Depending on the severity of the infection, leaves become yellowish green and the plants become stunted, gradually wilt, and eventually die. Brown, necrotic areas characteristically occur initially on tips of the small feeder roots. New roots that form on larger roots above the zone of infection may delay the death of the plant, but if soil moisture remains high, these roots will also become infected.

The presence of chlorotic leaves and gradual wilting of the camellia plants can be confused with the dieback disease caused by *Glomerella cingulata.* Camellia dieback results in sudden death compared with the slow decline in vigor associated with Phytophthora root rot (Can and Baxter 1977).

Disease Development. The disease is most severe during the warmer periods of the year on plants growing under moist or waterlogged soil conditions.

Control. Extensive control measures as given by Can and Baxter (1977) include prevention of waterlogging of soil and use of resistant cultivars. Susceptible cultivars such as Pink Perfection, Herme, Imura, and Prince Eugene Napoleon should be avoided. Resistant camellia (*C. japonica*) cultivars include Debutante, R. L. Wheeler, and Professor Sargent. *C. sasanqua* (especially the cultivar Kanjiro) is more resistant than *C. japonica* and three other species and is recommended as a rootstock because of its vigorous growth (Marshall and Cole 1987). Cultivars of camellia should be grafted onto this rootstock whenever possible because it is reported to be adapted to a wide range of climatic conditions and soil types.

The pathogen can be eradicated from mildly infected nursery stock by submerging diseased roots in water at 45°C (113°F) for 15 min. Treatment of roots or soil with metalaxyl and foliar spray with fosetyl-Al are effective; however, neither fungicide completely eradicates *P. cinnamomi* in infected roots.

Root Rot of Conifer and Hardwood Trees under Forest Conditions

P. cinnamomi and tree diseases are also discussed in Chapter 5 under The Susceptible Host, the Virulent Pathogen, and the Favorable Environment and under The Hardpan Effect in Agricultural Soils.

The biology of *P. cinnamomi* and its invasion of Australian forests are reviewed by Weste and Marks (1987) and Weste (1984, 1994). *P. cinnamomi* is characterized as a "root pruner" of forest trees because it primarily infects the fine root system (Newhook 1959; Weste and Taylor 1971; Marks 1978). In general, root infection either leads to a slow decline that eventually causes the death of the plant or spreads rapidly and causes a sudden wilt and death (Figures 15A.5–15A.7). Plate 15A.4 shows a similar decline of ohia trees caused by nutrient deficiency and not by *P. cinnamomi.*

P. cinnamomi is involved in many of the disease-induced problems in Australian forests (Podger 1972; Podger et al. 1965; Weste 1974, 1984, 1994; Weste and Marks 1974, 1987).

As stated and explained in Chapter 5 in The Susceptible Host, the Virulent Pathogen, and the Favorable Environment, the presence of *P. cinnamomi* does not necessarily predict that severe disease will develop since soil temperature and particularly episodes of high soil moisture influence the severity of root rot on trees.

The status of *Phytophthora*-caused diseases of forest trees in Australia and New Zealand is comprehensively reviewed by Newhook and Podger (1972). Newhook (1982) documents the loss of 35-year-old stands of *Pinus radiata* that had been planted as wind-shelter belts in New Zealand. These trees had been infected years before during a year of abnormally high rainfall. Hepting and Newhook (1962) describe a disease on *P. radiata* (Plate 5.5) that resembles the little leaf disease that had been originally described by Campbell and Copeland (1954) on loblolly and short leaf pine in the southwestern United States.

A wide range of symptoms is associated with root rot in forest situations. Resin deposits and irregular, reddish brown, wedge-shaped streaks of infected tissue are often visible on trunks of infected hardwood trees (see Plate 15A.7, which shows a resinous canker on a plane tree in Argentina). In walnut (*Juglans regia*), the streaks in the bark are black rather than reddish brown; in oriental plane trees, a characteristic premature reddening of the leaves is associated with root rot; in American boxwood (*Buxus sempervirens* L.), leaves turn pale green and then bronze or full yellow and eventually are shed. Similar symptoms occur on the umbrella pine (*Sciadopitys verticillata* (Thunb.) Siebold and Zucc.) (Lambe and Wills 1983). In conifers, the gradual loss of color in the needles is an early symptom of infection. *Phytophthora*-infected conifers rarely wilt. In the little leaf disease of pines, the needles become smaller than normal and the foliage turns yellow. Shoot growth decreases, and the affected trees decline slowly and die prematurely (W. A. Campbell and Copeland 1954).

In Western Australia, the dieback disease of *Eucalyptus marginata* and many more of the understory species that is caused by *P. cinnamomi* has become a severe threat to these natural forests (Podger et al. 1965; Podger 1972; Weste and Taylor 1971). This devastating epidemic is considered to be one of the most serious in the history of plant pathology (Holiday 1980). In Queensland forests, *P. cinnamomi* is widespread but mainly causes damage only in the tropical rain forest areas in central Queensland. The episodes of severe tree loss are associated with wallow formation by wild feral pigs. In New South Wales, *P. cinnamomi* is widespread, but disease and loss of trees are rare. In Victoria, root rot is severe in the forest areas. The root disease of jarrah in Western Australia, as described by Shea et al. (1984), is covered in Chapter 5. The book edited by Old (1979) contains several summary papers on the role of *Phytophthora* diseases in Australia's forests. Symptoms of diseases caused by *P. cinnamomi* on many Australian native plants have been described in detail by Weste and Law (1973). Although root decay is a primary symptom on all hosts, secondary symptoms vary with the host involved. In *Banksia serrata,* the symptoms range from a chlorosis to a change in color of leaves to gray green with apical browning that eventually extends to the petioles (Plate 15A.6). In *Eucalyptus obliqua,* the leaves of infected trees become smaller than normal, veins are prominent, and leaves eventually turn brown; the entire tree dies with the brown leaves still attached. Other trees such as *Sprengelia incarnata* do not become

Figure 15A.4. Symptoms in a 5-year-old planting of kiwi fruit caused by *Phytophthora cinnamomi.* Unthrifty vine (foreground) and defoliated vine (middle distance) are showing aboveground symptoms of root infection compared with the healthy vines in the background at center and left. (Courtesy G. I. Robertson and B. Eykel, D.S.I.R., Auckland, New Zealand; Robertson 1982b)

chlorotic but wilt suddenly and die. Infection of *Xanthorrhoea australis* results in a chlorosis of the crown followed by a darkening of the leaves; this eventually leads to a change of the crown to a cinnamon color (Figure 15A.6). The plant then dies, and the crown collapses into an umbrella shape. Two months after the first symptoms appear, trees on waterlogged sites usually die, but on well-drained sites, infected trees may survive up to 12 months (Weste 1984).

Weste and Taylor (1971) characterized the symptom development of some Australian plants in the Victoria forest in three categories: 1) progressive chlorosis of all foliage and rapid death of infected plants; 2) simultaneous chlorosis of all foliage followed by death of the whole plant with leaves still attached; and 3) progressive chlorosis of leaves and dieback of branches with subsequent shedding of the affected leaves.

Disease Development. Disease caused by *P. cinnamomi* in the jarrah (*Eucalyptus marginata* Sm.) forests in Western Australia (Plate 15A.5) develops optimally at temperatures between 15 and 25°C and declines rapidly at temperatures above 30°C (Old 1978). Severity of disease increases during periods of heaviest rainfall or when plants are experimentally irrigated. Disease seldom occurs at temperatures lower than 12 to 15°C (Zentmyer 1980). When the environment favors the host, the pathogen may be present but not cause obvious symptoms. Other conditions that favor disease development include the following: 1) susceptible hosts; for instance, in Australia most members of the subgenus *Monocalyptus* of the genus *Eucalyptus* show varying degrees of susceptibility; *Banksia grandis* is highly susceptible; 2) seasonal water stress or waterlogging of soils; 3) shallow soil subtended by hardpan layers (Shea et al. 1983); 4) poor drainage resulting in excessive water accumulation around plant roots; and 5) lack of organic matter in soils, which is associated with low microbial populations and activity.

Although the effect on fine roots is well documented, Shea et al. (1983, 1984) state that fine-root necrosis alone does not explain the death of tall jarrah trees because jarrah trees have root systems that penetrate to great depths in soil. They found that a hardpan layer that exists at varying depths in the forest through which the vertical roots of jarrah trees penetrate allowed an accumulation of a layer of water just above the hardpan that supported the development of sporangia and zoospores. The increase in the population of *P. cinnamomi* in the region above the hardpan was sufficient to allow infection and subsequent girdling and death of the large vertical roots, thus explaining how large trees with deep root systems could be killed. For more detail see Chapter 5 under The Hardpan Effect in Agricultural Soils.

Control. The management guide for control of Phytophthora root rot diseases in conifer nurseries in the Pacific Northwest of the United States contains useful information on control of diseases at the nursery level. It covers the use of fumigants and fungicides and the chlorination of irrigation water and lists western red cedar as resistant, many hardwood trees, pine, spruce, larch, and incense cedar as tolerant, and western and mountain hemlock, Douglas fir, and true fir as highly susceptible to *P. cinnamomi* and other *Phytophthora* species that infect forest trees. *P. cinnamomi* is not as impor-

Figure 15A.5. *Eucalyptus macrorhyncha* and *E. baxteri* trees killed by *Phytophthora cinnamomi* in the Brisbane Ranges, Victoria, Australia. (Courtesy G. Weste, University of Melbourne, Parkville, Victoria, Australia)

tant in conifer nurseries in the Pacific Northwest of the United States as *P. cactorum, P. cryptogea, P. drechsleri, P. pseudotsugae,* and two biotypes of *P. megasperma* (Cooley et al. 1985).

Chemical control of root rot in forests is not economically feasible. Old (1978) summarized several basic management principles that minimize the spread of *P. cinnamomi* in forest areas: 1) minimum disturbance of the forest area; in particular, strict control of the movement of soil from diseased areas; 2) insurance of good drainage in forested areas; 3) maintenance of a wide diversity of both overstory and understory species to reduce the number of potentially susceptible species; and 4) strict hygiene and quarantine in areas likely to be affected by the introduction of a pathogen such as *P. cinnamomi.*

In Australia, the movement of gravel for road making from infested to noninfested areas of the forest has been regulated since it was found that an epidemic causing extensive tree death often followed road making. Weste (1977a) mapped areas in the Wilson's Promontory National Park in Victoria, Australia, that show how dieback of several woody plants followed the preparation of roads with infested gravel that had been moved from areas where root rot had occurred. In the jarrah forest in Western Australia, Shea et al. (1984; see Chapter 5) recommended reducing the understory vegetation, which is composed mainly of highly susceptible *Banksia* plants, promoting resistant flora such as *Acacia* trees, minimizing the disturbance of soil in infested sites, and preventing the spread of infectation by road building or by transportation vehicles.

Trees and shrubs that currently show some resistance to *P. cinnamomi* are listed by Zentmyer (1980) and Edmiston (in Shea 1975).

Root Rot of Species in the Family Ericaceae (*Erica, Gaulteria, Pieris, Rhododendron, Azalea*)

P. cinnamomi is known to be a major limiting factor in the production of many plants in the Ericaceae family that are used in ornamental landscape plantings. In general, the fibrous root system is infected, resulting in the eventual death of the diseased plants (Benson et al. 1982; Benson and Campbell 1985).

In *Rhododendron* species, *P. cinnamomi* causes a serious dieback disease that is prevalent wherever rhododendrons are grown (Figure 15A.8). In the early stages of infection, the foliage of infected plants becomes an off-green to a purplish color that resembles phosphate deficiency. A distinct darkening occurs on either side of the midrib in some rhododendron cultivars. Leaves eventually droop or wilt, and complete defoliation usually follows. In some species and/or under less favorable conditions for disease development, plants may remain unthrifty for one to several growing seasons before suddenly collapsing. On plants with incipient infections, leaves are usually smaller than normal and shoot growth is less vigorous than on healthy plants. A cut made lengthwise in the branches of infected rhododendrons often shows dark streaks in the woody tissues above the rotted roots or crowns. Infected fibrous roots have a distinct light-dark banding pattern. These roots then turn a dark reddish brown and decay. In severe cases, the infection extends into the lower part of the stem, causing a distinct reddish brown discoloration of the basal

Figure 15A.6. *Xanthorrhoea australis,* grass tree, killed at the edge of a disease front in the Brisbane Ranges, Victoria, Australia. (Courtesy G. Weste, University of Melbourne, Parkville, Victoria, Australia)

3.1. *Phytophthora cinnamomi* sporangia emitting swimming zoospores in an aqueous substrate. Since sporangia and zoospores are hyaline in nature, the photographer has infused color by use of a green filter. (Reproduced from a photomicrograph by L. Harman by permission of the Department of Conservation and Land Management, courtesy B. Shearer in Shearer et al. 1991)

5.1. Collar rot on a bottle tree (*Brachychiton populneus* Schott & Endler) caused by *Phytophthora drechsleri*. (Courtesy H. S. Gill, Riverside County Agricultural Commissioner's Office, Riverside, California)

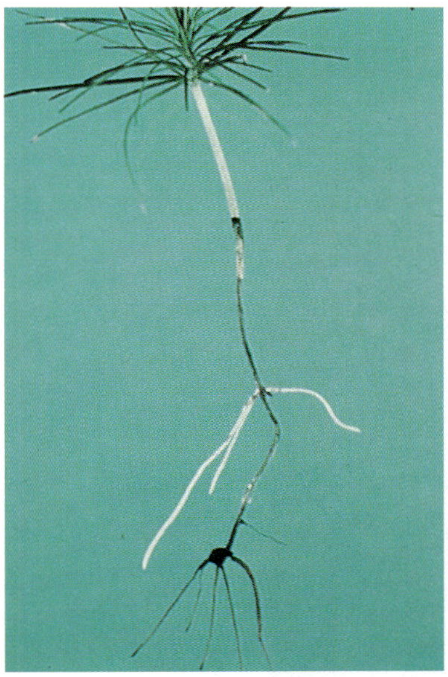

5.2. *Pinus radiata* seedling growing in a hydroponic solution into which *Phytophthora cactorum* zoospores have been introduced. Existing rootlets were killed, but the plant replaced these with new adventitious rootlets. (Reprinted by permission of New Zealand Tree Crops Association, Hamilton, courtesy Newhook 1988)

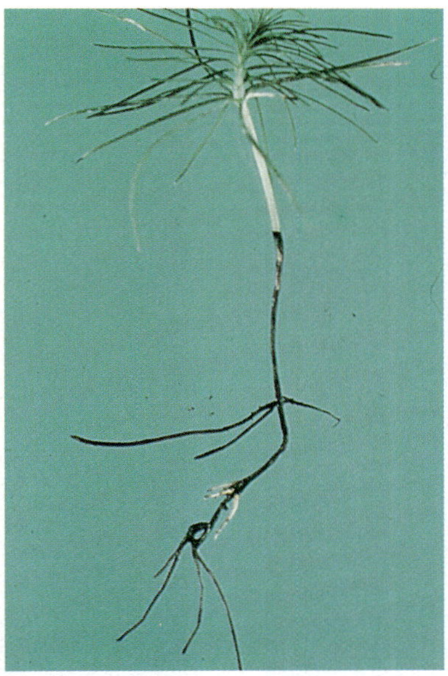

5.3. Same seedling as in Plate 5.2. Three weeks later, infection of the replacement rootlets has occurred as a result of release of *Phytophthora cactorum* zoospores from sporangia borne on the original killed rootlets. Yet another group of replacement rootlets can be seen. These in turn were invaded, and the seedling died. (Reprinted by permission of New Zealand Tree Crops Association, Hamilton, courtesy Newhook 1988)

5.4. Pine seedling from a commercial nursery. After heavy January rainfall, extensive infection of rootlets by *Phytophthora cinnamomi* occurred, killing many seedlings. After wrenching (loosening soil with a spade) had improved soil drainage, plants that were still alive produced adventitious roots. (Reprinted by permission of New Zealand Tree Crops Association, Hamilton, courtesy Newhook 1988)

5.5. Forty-year-old *Pinus radiata* shelter belt trees killed by *Phytophthora cinnamomi* during the epidemic that occurred throughout the Auckland Province of New Zealand in 1956. After massive destruction of feeding rootlets during an abnormally wet late summer and autumn, trees were unable to meet water requirements in the spring and died by early summer. (Courtesy Newhook 1988)

11.1. Collar rot of apple caused by *Phytophthora cactorum.* (Photograph by O. K. Ribeiro)

11.2. Root necrosis of apple caused by *Phytophthora cactorum.* (Photograph by O. K. Ribeiro)

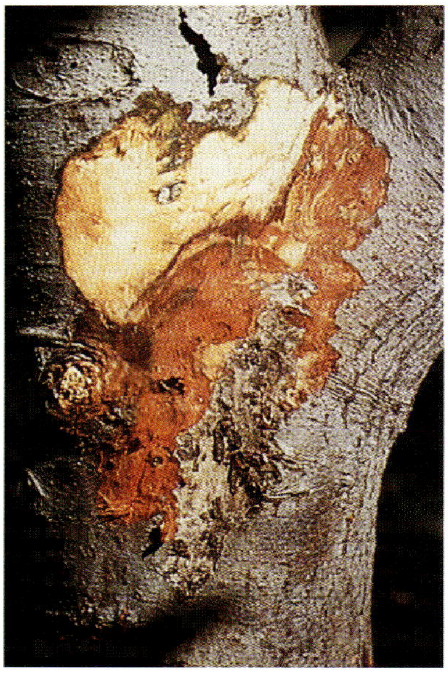

11.3. Canker on a limb of apple (Cox Orange Pippin) caused by *Phytophthora cactorum* in New Zealand. (Courtesy G. A. Zentmyer, University of California, Riverside)

11.4. Young apple tree with crown and root rot caused by *Phytophthora cactorum*. (Photograph by O. K. Ribeiro)

11.5. Cherry tree with crown rot caused by *Phytophthora cactorum*. (Photograph by O. K. Ribeiro)

11.6. Pear fruit infected at the blossom end by *Phytophthora cactorum*. (Photograph by O. K. Ribeiro)

11.7. Mummified pear fruit infected by *Phytophthora cactorum*. (Photograph by O. K. Ribeiro)

11.8. Control of crown or collar rot of an apple tree by inarching. Stems of a resistant clone are grafted from a root to the trunk to bypass the rotted crown. (Photograph by O. K. Ribeiro)

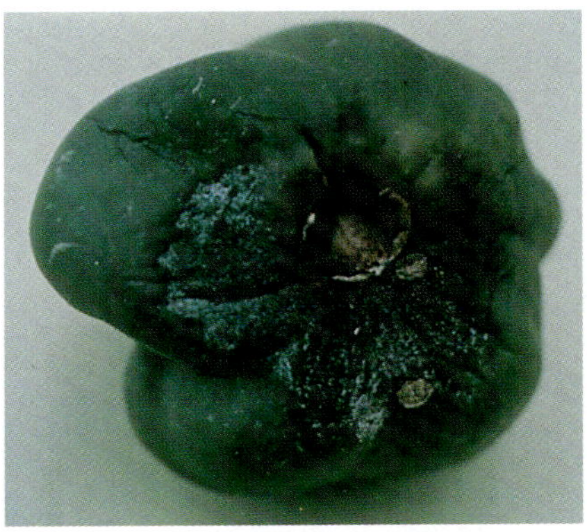

14.1. Fruit rot of bell pepper caused by *Phytophthora capsici*. (Courtesy G. C. Papavizas, USDA, Beltsville, Maryland)

14.2. Root rot of *Capsicum frutescens* caused by *Phytophthora capsici* in Taiwan. (Courtesy H. S. Chang, Institute of Botany, Academic Sciences, Taiwan)

14.3. Buckeye rot of tomato caused by *Phytophthora capsici* in Louisiana. (Courtesy L. W. Black, Louisiana State Cooperative Extension Service, Baton Rouge)

14.4. Crown rot of summer squash in the field caused by *Phytophthora capsici*. (Courtesy G. C. Papavizas, USDA, Beltsville, Maryland)

14.5. Foot rot of black pepper in Thailand. Left, healthy tree and right, tree affected by foot rot caused by *Phytophthora capsici sensu lato* Tsao (1991). (Courtesy T. Kaosiri, Thailand Department of Agriculture, Bangkok)

15A.1. Right, pineapple plants killed by *Phytophthora cinnamomi*. (Courtesy K. G. Pegg, Department of Primary Industries, Queensland, Australia)

15A.2. Reddening of pineapple leaves resulting from root rot caused by *Phytophthora cinnamomi*. (Courtesy K. G. Pegg, Department of Primary Industries, Queensland, Australia)

15A.3. Rot of green pineapple fruit caused by *Phytophthora cinnamomi*. (Courtesy K. G. Pegg, Department of Primary Industries, Queensland, Australia)

15A.4. Decline of ohia caused by nutrient deficiency. (Courtesy W. Ko, University of Hawaii, Hilo)

15A.5. Dieback of jarrah trees in Western Australia caused by *Phytophthora cinnamomi*. (Courtesy F. J. Newhook, University of Auckland, Auckland, New Zealand)

15A.6. Change in color of a *Banksia serrata* plant in Victoria, Australia, associated with root rot caused by *Phytophthora cinnamomi*. (Courtesy G. Weste, University of Melbourne, Parkeville, Victoria, Australia)

15A.7. Resinous canker on a plane tree caused by *Phytophthora cinnamomi* in Argentina. (Courtesy G. A. Zentmyer, University of California, Riverside)

15A.8. Death of a *Pieris japonica* plant caused by *Phytophthora cinnamomi* in a nursery. (Courtesy R. C. Lambe, Lambe Consulting, Port Ludlow, Washington)

15A.9. Dieback in the upper branches of an avocado tree infected with *Phytophthora cinnamomi*. (Courtesy G. A. Zentmyer, University of California, Riverside)

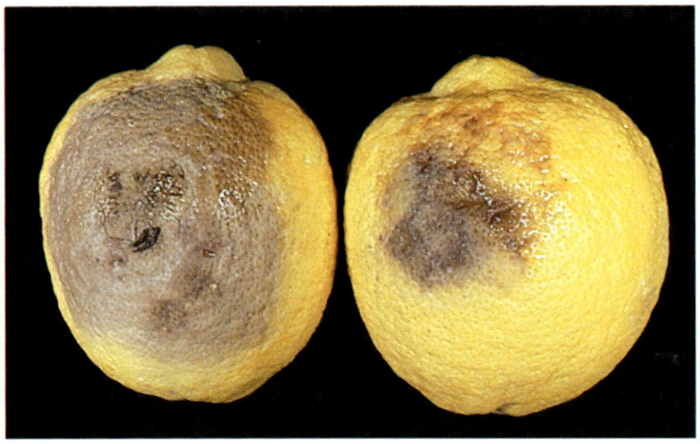

17.1. Brown rot of lemon caused by *Phytophthora citrophthora*. (Courtesy J. W. Eckert, University of California, Riverside)

17.2. Gummosis on an orange tree trunk infected by *Phytophthora citrophthora*. (Courtesy G. A. Zentmyer, University of California, Riverside)

17.3. Accelerated fruit set on a young Valencia orange tree resulting from girdling caused by *Phytophthora citrophthora*. (Photograph by D. C. Erwin)

17.4. Leaf blight of *Pieris japonica* caused by *Phytophthora citrophthora*. (Courtesy H. W. Hoitink, University of Ohio, Wooster)

17.5. Dieback of *Pieris japonica* caused by *Phytophthora citrophthora*. (Courtesy H. W. Hoitink, University of Ohio, Wooster)

19.1. Leaf blight of taro caused by *Phytophthora colocasiae*. (Courtesy G. V. H. Jackson)

20.1. Wilt of a gerbera plant resulting from root rot caused by *Phytophthora cryptogea*. (Courtesy P. Bonnet and S. Mercier, I.N.R.A., Antibes, France)

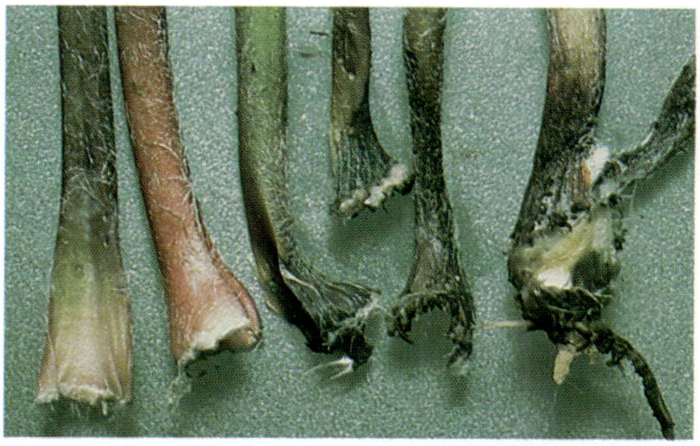

20.2. Left, two healthy gerbera stems and right, four stems rotted by *Phytophthora cryptogea*. (Courtesy P. Bonnet and S. Mercier, I.N.R.A., Antibes, France)

20.3. Crown rot of a gerbera plant caused by *Phytophthora cryptogea*. (Courtesy P. Bonnet and S. Mercier, I.N.R.A., Antibes, France)

20.4. Wilt and death of African daisy in a freeway landscape caused by *Phytophthora cryptogea*. (Courtesy H. S. Gill, Riverside County Agricultural Commissioner's Office, Riverside, California)

20.5. Healthy planting of African daisy in a freeway landscape. (Courtesy H. S. Gill, Riverside County Agricultural Commissioner's Office, Riverside, California)

23.1. Root rot of sugar beet caused by *Phytophthora drechsleri* in California. (Photograph by R. A. Khan and D. C. Erwin)

23.2. Stem rot of pigeon pea caused by *Phytophthora drechsleri* f. sp. *cajani*. Note that roots are not rotted. (Courtesy J. Kannaiyan, International Crops Research Institute for the Semi-Arid Tropics, Hyderabad, India)

23.3. Lesion on a pigeon pea leaf caused by *Phytophthora drechsleri* f. sp. *cajani*. (Courtesy J. Kannaiyan, International Crops Research Institute for the Semi-Arid Tropics, Hyderabad, India)

25A.1. Pink rot of potato caused by *Phytophthora erythroseptica*. (Courtesy R. Loria, Cornell University)

26A.1. Red stele of strawberry caused by *Phytophthora fragariae* in New Zealand. Plant on left is healthy. (Courtesy F. J. Newhook, University of Auckland, Auckland, New Zealand)

26A.2. Stunting of strawberry plants caused by *Phytophthora fragariae* in a low area of a field in New Zealand. (Courtesy F. J. Newhook, University of Auckland, Auckland, New Zealand)

32.1. Leaf spots caused by *Phytophthora ilicis* on holly. (Courtesy D. L. Coyier, USDA ARS, Corvallis, Oregon)

32.2. Spreading lesions on a holly leaf caused by *Phytophthora ilicis.* (Courtesy D. L. Coyier, USDA ARS, Corvallis, Oregon)

32.3. Blight of holly berries caused by *Phytophthora ilicis.* (Courtesy D. L. Coyier, USDA ARS, Corvallis, Oregon)

32.4. Blight and defoliation on the lower part of a holly tree caused by *Phytophthora ilicis*. (Courtesy D. L. Coyier, USDA ARS, Corvallis, Oregon)

32.5. Complete defoliation of a holly tree resulting from blight caused by *Phytophthora ilicis*. (Courtesy D. L. Coyier, USDA ARS, Corvallis, Oregon)

33.1. Late blight of potato, caused by *Phytophthora infestans*. (Courtesy Robert Goth, USDA ARS, Beltsville, Maryland)

33.2. Lesions on potato leaves caused by *Phytophthora infestans*. (Courtesy Robert Goth, USDA ARS, Beltsville, Maryland)

33.3. Left, potato cultivar resistant to late blight and right, the susceptible cultivar Alpha. (Photographed in 1965 by D. C. Erwin at experimental plots of John Niederhauser, Mexican Agricultural Program of the Rockefeller Foundation, Toluca, Mexico)

40.1. Bronzed foliage of Port Orford cedar resulting from root rot caused by *Phytophthora lateralis* in an Oregon forest. (Courtesy J. Kliejunas and D. H. Adams, USDA Forest Service, San Francisco, California, and Department of Forestry, Sacramento, California, respectively)

40.2. Discoloration of the inner bark at the collar region of a Port Orford cedar infected with *Phytophthora lateralis*. (Courtesy J. Kliejunas and D. H. Adams, USDA Forest Service, San Francisco, California, and Department of Forestry, Sacramento, California, respectively)

40.3. Sign posted at the entrance of a Port Orford cedar forest to prevent spread of *Phytophthora lateralis* into the forest. (Courtesy E. H. Hansen, Department of Plant Pathology, Oregon State University, Corvallis)

40.4. Trucks being washed with water to prevent infested soil or plant material from being carried into a healthy Port Orford cedar forest in Oregon. (Courtesy E. H. Hansen, Department of Plant Pathology, Oregon State University, Corvallis)

44.1. Root rot of alfalfa caused by *Phytophthora medicaginis*. Note the diffuse margins and light brown color of the lesions, symptoms that distinguish this disease from other root diseases of alfalfa. (Photograph by D. C. Erwin)

50.1. Quick wilt of tobacco plants caused by *Phytophthora nicotianae*. (Courtesy K. G. Pegg, Department of Primary Industries, Queensland, Australia)

50.2. Stem and root rot of tobacco caused by *Phytophthora nicotianae*. (Courtesy K. G. Pegg, Department of Primary Industries, Queensland, Australia)

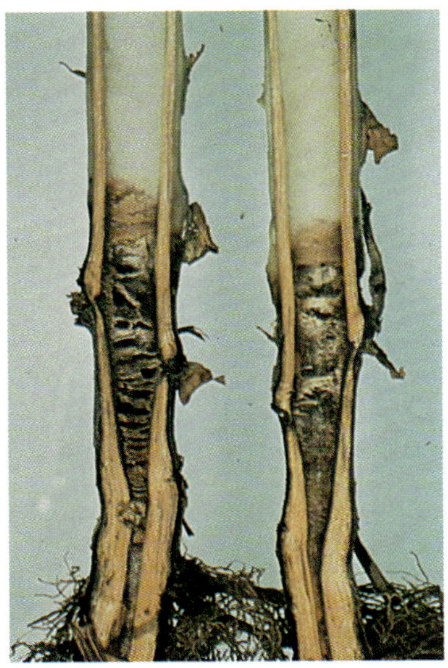

50.3. Internal checking symptom in the pith of tobacco stems infected with *Phytophthora nicotianae*. (Courtesy K. G. Pegg, Department of Primary Industries, Queensland, Australia)

50.4. Root and stem rot of a boxwood plant (right) caused by *Phytophthora nicotianae*. (Courtesy R. C. Lambe, Lambe Consulting, Port Ludlow, Washington)

50.5. Root rot of tomato caused by infestation of soil with *Phytophthora parasitica* inoculum. (Photographed by D. C. Erwin in field plots of J. Ristaino and J. Duniway, University of California, Davis)

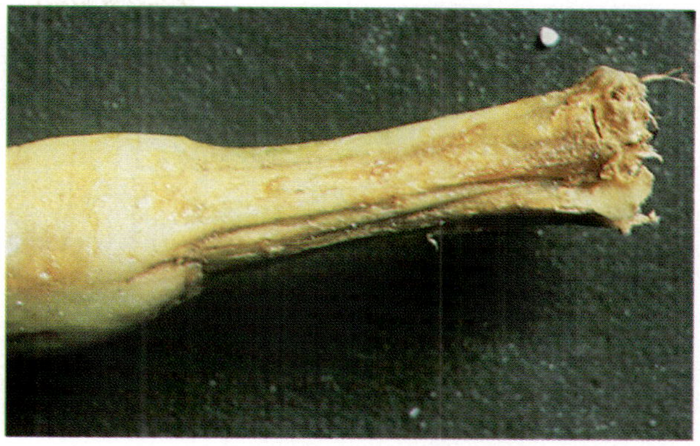

50.6. Collar rot of carnation caused by *Phytophthora parasitica*. (Courtesy P. Ricci, I.N.R.A., Antibes, Cedex, France)

50.7. Stem rot of orchid (*Vanda* sp. × *Rotchiloliana*) caused by *Phytophthora parasitica*. (Courtesy F. Kohler and B. Boccas, ORSTOM, Papeet, Tahiti, Polynesie Francais)

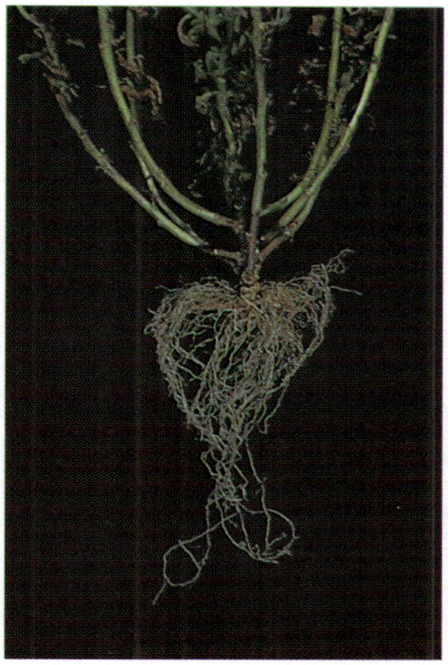

50.8. Stem and root rot of petunia caused by *Phytophthora parasitica*. (Courtesy H. S. Gill, Riverside County Agricultural Commissioner's Office, Riverside, California)

51A.1. Black pod of cocoa caused by *Phytophthora palmivora*. (Courtesy G. A. Zentmyer, University of California, Riverside)

51A.2. Devastation of a papaya planting caused by *Phytophthora palmivora*. (Courtesy W. H. Ko, University of Hawaii, Hilo)

51A.3. Damping-off of young papaya plants caused by *Phytophthora palmivora*. (Courtesy W. H. Ko, University of Hawaii, Hilo)

51A.4. Pod rot of papaya caused by *Phytophthora palmivora*. (Courtesy W. H. Ko, University of Hawaii, Hilo)

51A.5. Black stripe of rubber caused by *Phytophthora palmivora*. (Courtesy K. H. Chee, Rubber Research Institute of Malaya, Kuala Lumpur)

51A.6. Pod rot of rubber caused by *Phytophthora palmivora*. (Courtesy K. H. Chee, Rubber Research Institute of Malaya, Kuala Lumpur)

53.1. Stem canker of carnation caused by *Phytophthora porri.* (Courtesy H. Kouyeas, Benaki Phytopathological Institute, Athens, Greece)

59.1. Root rot of soybean seedlings caused by *Phytophthora sojae.* Plant on the left is healthy. Other plants have varying degrees of severity of root rot. (Courtesy G. C. Papavizas, USDA SEA, Beltsville, Maryland)

59.2. Preemergence root rot of soybean caused by *Phytophthora sojae.* (Reproduced by permission of the American Pathological Society, St. Paul, Minnesota, courtesy Schmitthenner 1985)

59.3. Foliage symptoms of root rot of soybean caused by *Phytophthora sojae.* (Courtesy A. F. Schmitthenner, Department of Plant Pathology, University of Ohio, Wooster)

59.4. Stem rot of soybean caused by *Phytophthora sojae*. (Courtesy A. F. Schmitthenner, Department of Plant Pathology, University of Ohio, Wooster)

65.1. Stem rot of cowpea caused by *Phytophthora vignae* occurring in spots in a field in Queensland, Australia. (Courtesy G. S. Purss, Queensland Department of Primary Industries, Indooroopilly, Queensland, Australia)

65.2. Foliar blight and stem rot of cowpea caused by *Phytophthora vignae*. (Courtesy G. S. Purss, Queensland Department of Primary Industries, Indooroopilly, Queensland, Australia)

65.3. Stem rot of cowpea in Queensland, Australia. The plants of the susceptible Poona cultivar on the right have all been killed by *Phytophthora vignae,* but plants of the resistant cultivar Caloona on the left are not affected. (Courtesy G. S. Purss, Queensland Department of Primary Industries, Indooroopilly, Queensland, Australia)

portion of the stem, which is known as crown rot. In some wilt-resistant cultivars, roots may rot but crowns are not affected. These plants rarely wilt because new fibrous roots regenerate from healthy crowns at the soil line.

P. cactorum, P. cambivora, P. citricola, P. citrophthora, P. cryptogea, P. gonapodyides, P. heveae, P. lateralis, P. megasperma, and *P. parasitica* have also been implicated in the dieback or wilt of rhododendrons. Nonetheless, *P. cinnamomi* is the most important species involved in the root rot complex (Hoitink and Schmitthenner 1974). *P. syringae* has recently been reported to cause a leaf spot and stem or branch canker of rhododendron (Coyier and Roane 1986). *P. cinnamomi* also causes a seedling wilt of rhododendron plants. Death of *Pieris japonica* is shown in Plate 15A.8.

Disease Development. The disease is favored by wet soils and soil temperatures of about 26°C and above. In nurseries, dieback frequently develops even when temperatures are high (30°C) and overhead irrigation is frequent (Benson and Jones 1980). The severity of disease is reduced on sandy soil even if the plants are infected, whereas in heavy soils infected plants wilt and die. Nitrogen fertilization of rhododendron plants increases the growth rate but also makes the plant more susceptible to infection (Hoitink et al. 1986).

Control. Planting noncontaminated plants obtained from a disease-free nursery is the most important control method. Soil with good drainage should be provided. Growing plants in raised beds decreases the possibility of standing water (Benson et al. 1982). Soil that is well aerated and drained is essential for container-grown plants (see Chapter 5). A composted hardwood bark medium suppresses *Phytophthora* without the use of fungicides (Hoitink et al. 1977; Hoitink and Fahy 1986). Since rhododendron and azalea plants are grown largely in containers by nurseries for resale to retailers, the medium in which rooted cuttings or seedlings are grown is important. In principle, the proper amount of pore space in the potting medium is important to achieve adequate drainage. Too little pore space is associated with a lower amount of aeration in soil and longer periods that the soil remains saturated (see Potting Mixtures for Nursery Use in Chapter 5). Ownley et al. (1990) found that root rot severity in nursery stock in pots is least in pine bark or in a pine bark-sand (3:1) mixture and greatest in field soil. Biological control using composted hardwood bark, which is successful in Ohio and other regions in the eastern United States, is discussed in more detail in Chapter 5.

Planting susceptible cultivars of the species *Rhododendron carolinianum* Rehd., *R. catawbiense* Michx., and *R. maximum* L. should be avoided. Rhododendron species resistant to *Phytophthora* include *R. davidsonianum* Rehd. and E. H. Wils., *R. delavayi* French, *R. glomerulatum* Hutch., *R. hyperythrum* Hayata, *R. lapponicum* (L.) Wahlenb., *R. occidentale* Torr. and Gray, *R. pseudochrysanthum* Balfif., *R. poukhanense* (= *R. yedoense* Maxim. ex Regel), *R. quinquefolium* Bisset and S. Moore, *R. sanctum, R. simsii* Planch., and *R. websterianum.* Extensive lists of *Rhododendron* species and cultivars that are

Figure 15A.7. *Eucalyptus regnans,* the tallest larchwood tree in the world, killed by *Phytophthora cinnamomi* in Sherbrooke Forest. (Courtesy G. Weste, University of Melbourne, Parkville, Victoria, Australia; Weste and Marks 1974)

Figure 15A.8. *Rhododendron* cultivar Nova Zembla infected by *Phytophthora cinnamomi.* (Courtesy R. C. Lambe, Port Ludlow, Washington)

resistant or susceptible to *P. cinnamomi* were reported by Hoitink and Schmitthenner (1974b) and Benson and Cochran (1980). For instance, some *Rhododendron* hybrids such as cultivars Caroline, Martha Isaacson, Pink Trumpet, Professor Hugo de Vries, and Red Head are resistant, but none is immune. It is important to note that normally resistant rhododendron plants that are drought stressed or flooded for 48 h or more can be predisposed to root and crown rot (Blaker and MacDonald 1981). Although a cultivar may be resistant to *P. cinnamomi,* it is not necessarily resistant to other species of *Phytophthora.* In addition to *P. cinnamomi, P. heveae, P. cactorum, P. citricola,* and *P. nicotianae* cause dieback of rhododendron (Benson and Jones 1980).

The fungicides metalaxyl (trade name Subdue) (Benson 1979, 1985, 1990; Englander et al. 1980; Table 7.7) and fosetyl-Al (trade name Aliette) (Benson 1990; Marks and Smith 1990; Table 7.12) are efficacious fungicides for the control of *Phytophthora* diseases of plants in the Ericaceae family if registered for use. See Coyier and Roane (1986) for a discussion of etiology and control.

Blight and Canker of *Macadamia integrifolia* Maiden and Betche (Macadamia)

Leaves of infected trees become chlorotic, and growth is stunted. The bark in infected areas of the tree is roughened and sunken, developing into deep, furrowed lesions (see Cook [1975] for excellent photographs of symptoms). Cankers that often extend several feet up the trunk eventually girdle the trunk and lateral branches. This disease is often confused with branch canker caused by *Dothiorella gregaria* Sacc., which can be distinguished from Phytophthora canker by the presence of cracked branches and cankers with definite margins between dead and healthy tissue that are easily visible when the bark is peeled back. Small black pustules (pycnidia) of *Dothiorella* are often visible on the infected branches and branches killed by the fungus.

A blight of immature racemes and nuts and a foliar blight caused by *P. capsici* and *P. palmivora* was reported from Hawaii by Hunter et al. (1971) and by Aragaki and Uchida (1980).

Disease Development. *P. cinnamomi* enters only through root and trunk wounds, usually caused by cultural practices. Uninjured plants are not as susceptible to infection. Prolonged heavy rains increase the severity of the disease.

Control. Trees from a noncontaminated source should be planted on well-drained sites that have not been previously planted with a susceptible crop such as avocado, and wounding should be prevented. Spraying or painting the trunks of trees with a copper fungicide before the rains begin has been effective in preventing infection. Defoliation prevented infection on racemes of macadamia (Hunter and Kunimoto 1972). *Macadamia tetraphylla* L. A. S. Johnson has been reported to be more resistant than *M. intergrifolia* (Zentmyer and Storey 1961).

Root Rot of *Persea americana* (Avocado)

Root rot is now recognized to be the most important disease wherever avocado trees are grown and is an economically limiting factor to production in many areas of California, Australia, South Africa, and many other countries. In California, between 60 and 70% of the orchards are affected by Phytophthora root rot, and it is estimated that the disease loss amounts to $30,000,000 annually (Coffey 1992). See the monograph by Zentmyer (1980) on *P. cinnamomi* for an extensive review of Phytophthora root rot on avocado and other diseases caused by *P. cinnamomi.* The cause, development, and control of root rot have been reviewed by Zentmyer and Ohr (1978) and Zentmyer (1984). A general description of root rot and its control by Coffey (1992) and on strategies for integrated control (Coffey 1987, 1991) are more recent sources of information.

Occasionally, *P. cinnamomi* infects large roots and sometimes causes a canker at the base of the tree that extends upward approximately 2 m; however, the more common cause of trunk canker of avocado is *P. citricola* (Zentmyer et al. 1974; Coffey 1992). *P. heveae* has also been shown to be a cause of trunk canker (Zentmyer et al. 1978). A majority of the feeder roots of infected trees are blackened, brittle, or necrotic. Roots of pencil-sized thickness or larger are rarely infected. After Phytophthora root rot occurs, the leaves on infected trees become smaller than normal and pale green to yellow. Wilting occurs as the disease progresses followed by heavy leaf fall and branch dieback. In older, more mature trees, dieback of branches and sparse foliage give the tree a very characteristic "stags head" appearance (Figure 15A.9 and Plate 15A.9). Diseased trees quite often set an abnormally heavy crop of small fruit, probably because of the girdling effect. Trees eventually are reduced to leafless skeletons and die.

Aboveground symptoms of container-grown plants consist of yellowing and stunting followed by defoliation and dieback of the branches from the top of the plant. Infected roots are blackened, necrotic, and brittle. The disease can be positively diagnosed by baiting soil with avocado fruit or susceptible *Persea indica* seedlings or by plating roots on a selective medium (Zentmyer 1980; Chapter 2).

Disease Development. The disease is more severe and develops most rapidly in soils with poor drainage. Soils underlaid with hardpans are listed as high-hazard soils (Burns et al. 1960; Zentmyer and Ohr 1978; Zentmyer 1980). Disease progress is optimum in wet soil with a pH of 6.5 and at a temperature of about 24°C; however, disease occurs within a range of 13 to 32°C. The pathogen is easily spread in moist soil by vehicles, implements, and shoes. *P. cinnamomi* can survive in small diseased or dead roots up to 6 years if the soil is moist. Chlamydospores can be disseminated by passage through the intestinal tracts of termites and wild birds (Keast and Walsh 1979). Trees downhill from an infested area are often infected by surface or subsurface drainage of irrigation water or by runoff of rain water. Zentmyer (1980) and Weste (1983b) illustrate graphically how the disease spreads over time in flood water drainage areas.

Control. In a discussion of strategies for integrated control of avocado root rot that utilize sanitation, irrigation, management, host resistance, and fungicides, Coffey (1987, 1991a, 1992) summarizes the approaches used in several regions of the world where Phytophthora root rot is a problem. The most effective control measure is to exclude *P. cinnamomi* by planting disease-free nursery stock on clean land by use of a strict standard of nursery and field hygiene. Unfortunately, this goal is not always attainable. The nursery practices necessary to develop disease-free plants are given in Appendix 5.2 in Chapter 5, which is excerpted from Pegg (1978). Seeds from fruits that have fallen to the ground might be infected and should not be used in nurseries. If necessary, *P. cinnamomi* in infected seed can be eradicated by treatment with hot water at 49 to 50°C for 30 min (Zentmyer 1980; Zentmyer and Ohr 1978).

Resistance to *P. cinnamomi* is discussed in Chapter 6. Several resistant avocado rootstocks are currently available, but none is immune. Resistance in rootstocks of cultivars and species of *Persea* from Guatemala and Mexico has been found in some Mexican race avocados and in *Persea schiedeana* Nees. See Zentmyer and Schieber (1987), Zentmyer (1980), and

Coffey (1987, 1991, 1992) for documented descriptions of the status of resistance in avocado. The cultivar Duke, a selection of Mexican origin, was found to be moderately resistant (Zentmyer and Mircetich 1965). Gabor et al. (1990) review previous greenhouse work in testing cultivars for resistance and give the results of field testing of several rootstock cultivars. A natural hybrid cultivar from a *Persea americana* × *P. schiedeana* cross called Martin Grande was most resistant. Martin Grande is named after a Mayan assistant to E. Schieber from Guatemala (Zentmyer and Schieber 1987). The resistant cultivar Thomas was selected in the field from a Mexican race of avocado (Coffey 1987). The resistance of the cultivars G6 and Duke 7 was associated with their ability to regenerate roots rapidly (Kellam and Coffey 1985). These cultivars also appear to have physiologic resistance, indicated by the production of smaller lesions on inoculated roots (Dolan and Coffey 1986; Gabor and Coffey 1990, 1991). The reports by Phillips et al. (1987) indicated a physiologic response by the cultivar Duke 7, which was also expressed in unorganized callus tissue (Phillips et al. 1991).

Certain soils in Queensland, Australia, have a suppressive effect on *P. cinnamomi* (Broadbent and Baker 1974; see Chapter 5 and the portion of a paper by K. Baker that is reprinted there). Suppressive soils have a higher total nitrogen and calcium content and an abundant microflora and are high in organic matter. A detailed discussion of this suppressive effect and other forms of biological control are given in Chapter 5. In greenhouse experiments, calcium nitrate or ammonium sulfate was found to significantly reduce root rot (Lee and Zentmyer 1982).

A thorough discussion of the role of fungicidal control of Phytophthora root rot of avocado is given by Coffey (1992). Although many protective fungicides that have been tested in the past are in general ineffective, the systemic fungicides in the acylalanine class (e.g., metalaxyl) and the phosphonates (e.g., fosetyl-Al or phosphorous acid) suppress root rot and in many cases are cost effective. These fungicides are discussed in more detail in Chapter 7.

Metalaxyl can be used as a soil drench to suppress the development of root rot on trees growing in infested field soil. Metalaxyl has been efficacious when soil treatment is integrated with establishment of the resistant cultivar Duke 7 rootstock (Coffey et al. 1984; Coffey 1987). However, on bearing trees, metalaxyl is not an economical control measure, and when used repeatedly, it degrades rapidly as a result of the activity of certain microorganisms that build up in soil (Bailey and Coffey 1985, 1986).

The phosphonates are highly effective because they are translocated both upward and downward, a unique quality for control of root rot (see Chapter 7). In a field experiment, fosetyl-Al, at a dosage of 1 liter per tree at 3 g (a.i). per liter, controlled root rot up to 40 weeks (Coffey 1987). Foliar spray gives excellent control in the field (Coffey et al. 1984). Control by these means is also reported in Australia (Pegg et al. 1980; Pegg and Whiley 1987) and in South Africa (de Boer 1990).

Figure 15A.9. A mature avocado tree, approximately 25 years old, with symptoms of root rot caused by *Phytophthora cinnamomi*. Dieback of the branches and sparse foliage give the tree a characteristic "stag's head" appearance. (Courtesy G. I. Robertson, D.S.I.R., Auckland, New Zealand)

Trunk canker can be avoided by prevention of excessive wetting of the trunk and wounding by implements and by removal of suckers. Early infection can be controlled by excising the infected tissue and painting the affected areas with a suitable fungicide paint such as fosetyl-Al, metalaxyl, or a copper-based compound.

In 1983 and 1984, Darvas et al. reported that trunk injection of avocado trees in South Africa with a 7 to 10% fosetyl-Al solution applied twice a year controlled root rot. This was confirmed by Coffey (1987). Phosphorous acid, the active ingredient of fosetyl-Al (Coffey and Ouimette 1989; see Chapter 7), which is registered for use only in Australia (Guest and Grant 1991), has been efficacious when a buffered (potassium) solution is used as a foliar spray or trunk injection (Pegg et al. 1985; de Boer et al. 1990). Trunk injection requires a much lower dosage than foliar or soil application; however, wound damage to the trunk of the tree may be a disadvantage (Coffey 1991).

Coffey (1987, 1991) recommends an integrated control strategy that includes healthy nursery stock; well-drained, low-hazard soil; and one of the moderately resistant rootstocks. In replanting trees where *P. cinnamomi* exists in the soil, planting on mounds assures better drainage and tilth of soil. With resistant rootstocks and judicious use of appropriate fungicides, this method prolongs the life of trees (Goodall et al. 1987; see also Coffey 1987). Fungicides should be an alternative used only when necessary, not only because of cost, but also because resistant biotypes of *P. cinnamomi* could develop.

CHAPTER 15B

Phytophthora cinnamomi var. *parvispora*

Kröber and Marwitz (1993)

In 1993, *Phytophthora cinnamomi* var. *parvispora* var. nov. (Kröber and Marwitz 1993) was described and proved to be pathogenic on the ornamental plant *Beaucarnea* sp. growing in a commercial greenhouse in Germany.

Characteristics of *Phytophthora cinnamomi* var. *parvispora* var. nov.

See Figure 15B.1 for morphology. The following description is reproduced by permission of the Journal of Phytopathology, courtesy Kröber and Marwitz (1993):

Mycelium on carrot agar hyaline; in strands; transparent; only a few downy airmycelia.

Hyphal swellings abundant; mostly side twigs coralloidly blown up, sometimes allantoid or tailshaped, in part several times constricted, or spherical.

Chlamydospores abundant; spherical; terminal; intercalary, rarely lateral, sometimes in clusters on short side stalks; thin-walled; 16–43 (27.1) µm.

Zoosporangia without papilla; ovoid, ellipsoidal, obpyriform or rarely irregularly shaped; single or several in succession with internal proliferation, sometimes nest-shaped; 14–83 × 13–43 (36.9 × 25.5) µm, l/b = 1.45/1.

Generative organs only when complementary strains grown together in intraspecific mating or with *P. cinnamomi* or *P. cryptogea* in interspecific mating.

Oogonia in intraspecific mating culture abundant; spherical; 23–41 (31.9) µm.

Antheridia 1 per oogonium; amphigynous; spherical, broad ellipsoidal; relatively large.

Oospores spherical; with a moderately thick wall; hyaline; generally almost plerotic; 22–36 (28.1) µm.

Minimum 10–11, optimum 26–32, maximum 36–37°C; growth rate 10–14 mm/24 hr at optimum temperatures.

Isolated from stem bases of *Beaucarnea* spec.; Germany (in greenhouses), 1990.

P. cinnamomi var. *parvispora* is in [sic] some extent closely connected with *P. cryptogea* Pethybridge et Lafferty and *P. drechsleri* Tucker but nearest with *P. cinnamomi* Rands. But it is clearly distinct even from this fungus by considerably smaller chlamydospores and zoosporangia and it has a higher demand of temperature.

Distinguishing Characteristics

Kröber and Marwitz (1993) state that *P. cinnamomi* var. *parvispora* differs from *P. cinnamomi* because it produces smaller chlamydospores (27.1 µm in diameter) than *P. cinnamomi* (41 µm) (Rands 1922). They also state that sporangia are smaller, a statement that is borne out by sporangial size data shown in Table 15A.1. Nonetheless, since sporangial size is influenced by so many factors (see Chapter 3), the size of *P. cinnamomi* var. *parvispora* sporangia might simply be within the range of normal variability.

A higher maximum temperature for growth of *P. cinnamomi* var. *parvispora* (36–37°C) is also cited as a difference; however, Table 4 in Zentmyer (1980) lists maximum temperatures from various reports on *P. cinnamomi* as 30–36°C.

Since DNA RFLP technology has become such a useful tool in determining genetic differences between species, it will be of interest to determine how *P. cinnamomi* var. *parvispora* compares with other isolates of *P. cinnamomi* and/or *P. cryptogea* that Kröber and Marwitz (1993) say it resembles. Oudemans and Coffey (1991a) showed that isozyme patterns of *P. cinnamomi* isolates from a worldwide collection differed distinctly from those of *P. cryptogea*. The various isolates of *P. cinnamomi* were relatively uniform; however, eight electrophoretic types could be distinguished.

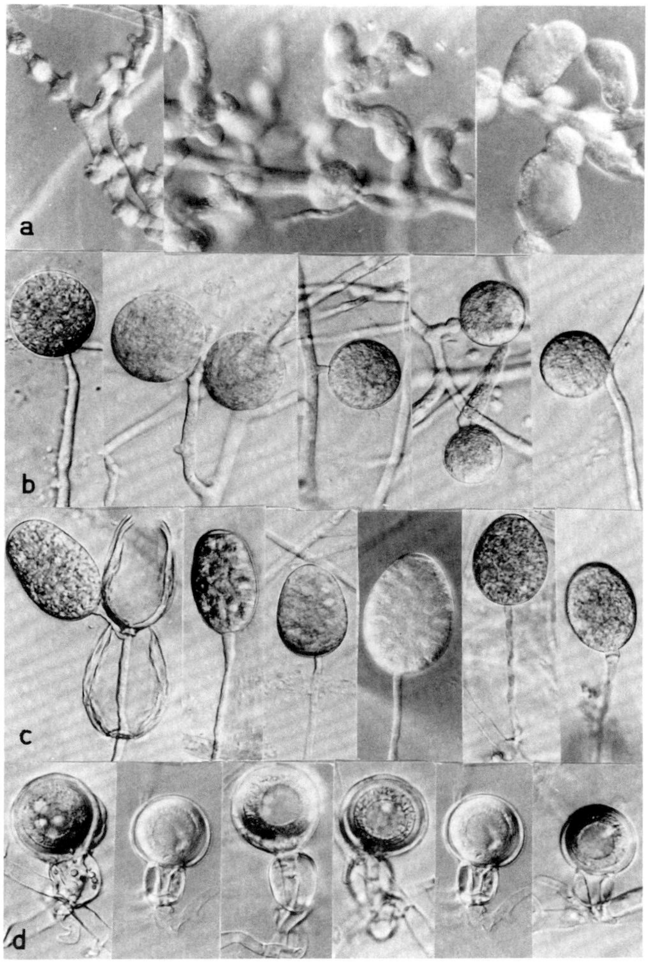

Figure 15B.1. Morphology of *Phytophthora cinnamomi* var. *parvispora* var. nov. **a,** Hyphal swellings; **b,** chlamydospores; **c,** zoosporangia; and **d,** oogonia, antheridia, and oospores (after intraspecific crossing). (500×) (Reproduced by permission of the Journal of Phytopathology, courtesy Kröber and Marwitz 1993)

CHAPTER 16

Phytophthora citricola

Sawada (1927)

Synonyms of *Phytophthora citricola* include *P. cactorum* var. *applanata* Chester (1932); *P. pini* Leonian (1934); and *P. pini* var. *antirrhini* Sundaraman and Ramakrishnan (1928).

P. citricola was first isolated from Sekkan orange fruit with brown rot symptoms in Formosa (Taiwan) by Sawada (1927). *P. citricola* is now known also to cause root rot and trunk canker on many economically important plants (Table 16.1). Shoot tip blight of lemon seedlings growing in plastic tunnels occurred when plants were sprayed frequently with water (Von Maltitz and Von Broembsen 1985). Root rot of kiwi fruit (Figure 16.1) (Robertson 1982b) and grape (Figure 16.2) (Robertson 1982a) occurs in New Zealand.

Characteristics of *Phytophthora citricola*

P. citricola is classified in group III of Stamps et al. (1990). Morphology is shown in Figures 16.3 and 16.4. See Tables 4.2 and 4.3 for tabular keys and Appendix 4.9 for a dichotomous key (Ho 1992). See Waterhouse (1957) and Zent-

Table 16.1. Distribution of hosts of *Phytophthora citricola*

Host	Common name	Disease	Geographical distribution
Abies balsamea (L.) Mill.	Balsam fir	Crown rot	United States (Adams and Bielenin 1988)
Abies concolor Gord. and Glend.) Lindl. ex Hildebr.	White fir	Twig blight; canker	United States (McCain and Scharpf 1986)
Abies fraseri (Pursh) Poir.	Fraser fir	Crown rot	United States (Adams and Bielenin 1988)
Abies magnifica A. Murr.		Twig blight; canker	United States (McCain and Scharpf 1986)
Acer saccharum Marsh	Sugar maple	Collar rot	United States (Drilias et al. 1982)
Actinidia chinensis Planch.	Kiwifruit	Collar and root rot	New Zealand (G. I. Robertson 1982b; Stewart and McCarrison 1991)
Aesculus hippocastanum L.	Horse chestnut	Bleeding canker; root rot; dieback	United Kingdom (Brasier and Strouts 1976)
Antirrhinum sp.	Snapdragon	Root rot	United Kingdom and New Zealand (Anon. 1956)
Aucuba japonica	Gold dust plant	Root rot	United States (Spencer and Benson 1981)
Azalea indica L.	'White wings'	Root rot	Australia (Hardy and Sivasithamparam 1988)
Beaufortia sparsa R. Br.		Root rot	Australia (Hardy and Sivasithamparam 1988)
Brachychiton populenus R. Br.	Bottle tree	Root rot	Argentina (Frezzi 1950)
Ceanothus griseus (Trel.) McMinn	Ceanothus	Trunk canker	See Waterhouse and Waterston (1966d)
Chamaecyparis lawsoniana (A. Murr.) Parl.	Lawson cypress; Port Orford cedar	Wilt	Netherlands (Rattink and Van Steekelenburg 1973)
Chamaelaucium uncinatum Schau.		Root rot	Australia (Hardy and Sivasithamparam 1988)
Citrus limon (L.) Burm.	Lemon	Shoot tip blight; fruit rot	United Kingdom (Smith 1953); South Africa (Von Maltitz and Von Broembsen 1985)
Citrus × *paradisi* Macfady	Grapefruit	Fruit rot	South Africa (Wager 1940)
Citrus sinensis (L.) Osbeck	Sweet orange	Fruit rot	Taiwan (Sawada 1927); Japan (Ito and Tokunaga 1935)
Citrus spp.	Citrus	Brown fruit rot	Taiwan (Sawada 1927); Antilles (Cook 1939); South Africa and Japan (Klotz 1943); Australia (Doepel 1966); Portugal (Telhada 1990)
Citrus tankan Hyata	Citrus	Fruit rot	Japan (Ito and Tokunaga 1935)
Corylus avellana L.	European hazel	Collar and trunk canker	Greece (Psarros 1978; Psarros and Tzavella-Khomari 1982)
Cyclamen L.	Persian violet	Leaf blight	Greece (Kouyeas 1973)
Cydonia oblonga Mill.	Quince	Root rot	Formosa (Sawada 1927)[a]
Cytisus praecox Bean.	Legume	Root rot	Netherlands (Rattink and Van Steekelenburg 1973)
Erica hiemalis Nichols.	White heather	Decline	Australia (Anon. 1965)
Euphorbia resinifera A. Berger	Spurge	Root rot	Formosa (Sawada 1927)[a]
Feijoa sellowiana O. Berg.	Feijoa	Brown rot of fruit	Italy (Magnano di San Lio and Tuttobene 1985)
Fragaria chilensis Ehrh.	Strawberry	Fruit rot	Bulgaria (Ilieva and Koleva 1984); Taiwan (Chang 1988)
Hakea francisiana F. Muell.		Root rot	Australia (Hardy and Sivasithamparam 1988)
Hakea laurina R. Br.		Root rot	Australia (Hardy and Sivasithamparam 1988)
Hebe 'Pagei' Hort.		Root and stem rot	United Kingdom (Jones and Griffin 1974)
Hevea brasiliensis Muell. Arg.	Rubber		Sri Lanka (Liyanage 1989)
Hibiscus rosa-sinensis L.	China Rose	Root rot; wilt	United States (James and Plakidas 1952)
Hibiscus schizopetalus Hook.	Japanese lantern	Root rot; wilt; blackroot	New Zealand (Brien 1938); United Kingdom (Keyworth 1943); United States (James and Plakidas 1952)
Humulus lupulus L.	Hop	Collar and trunk canker; crown and root rot	United Kingdom (Keyworth 1943); Greece (Kouyeas 1973)

(*continued on next page*)

myer et al. (1974) for microphotographic exposition of sporangium and oospore morphology and Waterhouse and Waterston (1964d) for a general description.

Sporangia

Shapes of sporangia are highly variable (Figures 16.3 and 16.4) and range from obovoid, obclavate, and obpyriform to slightly flattened on one side; occasionally deeply bifurcated with two apices; or irregularly shaped with three to four lobes. Zentmyer et al. (1974) show a wide range of variability in shapes. Sporangia are 30 to 75 μm long × 21 to 44 μm wide (average 47 × 34 μm), noncaducous, and persistent on the stalk. The septum is flush with the base of the sporangium. The papillae are wide and flat (semipapillate). Sporangiophores are simple or occasionally sympodial. A single terminal sporangium is sometimes borne on an unbranched sporangiophore.

Chlamydospores

Chlamydospores are rare (reported on oatmeal agar only).

Sex Organs

P. citricola is homothallic. Antheridia are paragynous (rarely amphigynous); oogonia are spherical and 18 to 35 μm in diameter, average 25.5 μm, sometimes tapering to a funnel-shaped base; oospores are spherical, almost plerotic, and 16 to 30 μm in diameter, average 22.0 μm.

Growth Temperatures

The minimum temperature for growth is 3°C; optimum is 25 to 28°C; and maximum is 31°C (Waterhouse 1963; Zentmyer et al. 1974). Avocado isolates produced radiate growth with pointed petallate zones that were more obvious on cornmeal agar than on V8 juice agar (Zentmyer et al. 1974).

Distinguishing Characteristics

Waterhouse's (1957) restudy of *P. citricola* in relation to reports by Sawada (1927), Chester (1932), and Ito and Tokunaga (1935) shows conclusively that *P. citricola* can be readily differentiated from *P. cactorum* var. *applanata* of Chester (1932). *P. cactorum* is conspicuously papillate (see Figure

Table 16.1. (*continued*)

Host	Common name	Disease	Geographical distribution
Juglans nigra L.	Black walnut	Seedling root rot	United States (Green and Pratt 1970)
Juglans regia L.	English walnut	Root and crown rot	United States (Mircetich and Matheron 1983; Matheron and Mircetich 1985a–e)
Lycopersicon esculentum L.	Tomato	Basal rot; seedling wilt	Italy (Salerno and Calabretta 1960)
Malus pumila Mill.	Apple	Root rot	Formosa (Sawada (1931); United Kingdom (Smith 1953)[a]; Netherlands (Buddenhagen 1955)[a]; Canada (McIntosh 1966); New Zealand (Robertson 1971)
Olea europea L.	Olive	Collar and trunk canker	Greece (Kouyeas and Chitzanidis 1978)
Persea americana Mill.	Avocado	Trunk canker; fruit rot	United States (Zentmyer et al. 1974; Kolke et al. 1987); Mexico (Fucikovsky and Luna 1987)
Phaseolus vulgaris L.	Kidney bean	Root rot	Taiwan (Sawada 1927)[a]
Pieris japonica (Thunb.) Don.	Japanese andromeda	Shoot blight; stem dieback	New Zealand (Newhook 1959); United States (Gerlach et al. 1974)
Pinus radiata D. Don.	Monterey pine	Root rot	United States (Sandlin et al. 1992)
Pistacia vera L.	Pistachio tree	Apoplexy; wilt	Greece (Kouyeas 1973)
Platanus orientalis L.	Oriental plane tree	Collar and trunk rot	Greece (Tzamos 1978)
Primula japonica A. Gray	Primrose	Root rot	United Kingdom (Dennis and Foister 1942)
Prunus amygdalus Batsch. (*Prunus dulcis* (Mill.) Webb)	Almond	Collar and trunk rot	Greece (Kouyeas 1973)
Prunus armeniaca L.	Apricot	Collar rot	Iran (Ershad 1979)
Prunus avium (L.) L.	Mazzard cherry	Root and crown rot	United States (Wilcox and Mircetich 1985)[a]
Prunus mahaleb L.	Mahaleb cherry	Root and crown rot	United States (Wilcox and Mircetich 1985)[a]
Prunus persica (L.) Batsch.	Peach	Root and crown rot	Rhodesia (Rothwell 1975)
Pseudotsuga menziesii (Mirb.) Franco.	Fir	Crown rot	United States (Adams and Bielenin 1988)
Psidium guajava L.	Guava	Fruit rot	Hawaii (Ko et al. 1982)
Quercus agrifolia Nee	California live oak	Trunk canker	United States (Mircetich et al. 1977)
Quercus alba L.	White oak	Seedling root rot	United States (Green and Pratt 1970)
Quercus rubra L.	Red oak	Seedling root rot	United States (Green and Pratt 1970)
Rhododendron catawbiense Michx.	Mt. Rosebay; purple laurel	Bud, leaf, and twig blight	Germany (Kröber 1959); United States (Hoitink and Schmitthenner 1969)
Rhododendron spp.	Rhododendron	Wilt	Germany (Kröber 1959); Australia (Anon. 1964); United States (Hoitink and Schmitthenner 1969); Netherlands (Rattink and Vansteekelenburg 1973); France (Vegh and Bourgeois 1975)
Rubus idaeus var. *idaeus* L.	Raspberry	Root rot	United Kingdom (Dennis and Foister 1942); United States (Wilcox 1989)
Sambucus sp.	Elder	Crown rot	Canada (Henry and Stelfox 1966)
Schinus molle	Australian or California pepper	Crown rot	Argentina (Frezzi 1950)
Syringa vulgaris L.	Lilac	Shoot blight of suckers	United States (Chester 1932); Canada (Henry and Stelfox 1966)
Tilia sp.	Lime tree	Bark canker	United Kingdom (Young and Strouts 1975)
Vitis vinifera L.	Grape	Root rot	New Zealand (G. I. Robertson 1982a)

[a] Artificially inoculated host.

11.1, Chapter 11), whereas *P. citricola* has only a shallow but distinguishable thickening on the apex of sporangia (Figures 16.3 and 16.4). Also, *P. cactorum* sporangia are shed in water (caducous with short pedicels; Figure 11.1), but those of *P. citricola* are not (Table 16.2). The oogonia of *P. citricola* and *P. cactorum* are very similar, but oogonia of *P. cactorum* are usually smaller and less pigmented than those of *P. citricola.* The radial growth of *P. citricola* is almost twice as rapid as that of *P. cactorum.* Other differences are given in Table 16.2.

Genetic Relationships

On the basis of isozyme and mtDNA RFLP patterns, *P. citricola* isolates are divided into five subgroups (CIT1–CIT5) (Oudemans and Coffey 1993). CIT1 is comprised of isolates from various *Citrus* species, including the Sawada (1927) type culture, an isolate previously known as *P. pini* (Leonian 1934), and an isolate identified as *P. cactorum* var. *applanta* (Chester 1932), now considered a synonym of *P. citricola.* Isolates in subgroups CIT1, CIT2, and CIT3 differ only at a few enzyme loci and are from diverse locations and hosts. Isolates in subgroup CIT4 are from South African plants, and CIT5 is composed of isolates pathogenic only to avocado (Zentmyer et al. 1974). The CIT5 isolates from avocado survive at lower temperatures (30°C) than isolates of the other groups (33°C), and oogonia are consistently slightly smaller than those of the other groups. When cluster analyses were made, CIT5 clustered closer to *P. capsici* (from cocoa) and *P. citrophthora* than to the other CIT groups. The CIT5 group was considered to be a host-specific, genetically distinct group that could be regarded as a separate biological species (Förster and Coffey 1991; Oudemans and Coffey 1993). All of the CIT groups differ substantially from *P. cactorum,* thus further supporting the general conclusion that the *P. cactorum* var. *applanta* erected by Chester (1932) is not valid (Waterhouse 1957).

Förster et al. (1990) also showed via mtDNA RFLP patterns that isolates of *P. citricola* are genetically diverse. Protein pattern analysis also showed diversity within *P. citricola* isolates (Erselius and de Vallavielle 1984).

Isolates of CIT3 were shown to be more sensitive to metalaxyl than isolates of CIT1 (Ferrin and Kabashima 1991).

Some Diseases Caused by *Phytophthora citricola*

Bleeding Canker of *Aesculus hippocastanum* L. (Horse Chestnut)

Both *P. citricola* and *P. cactorum* have been implicated as causal agents of bleeding canker of *Aesculus hippocastanum* L. Dead and dying scaffold branches are normally the first observable symptoms. A rusty red or occasionally yellow brown fluid oozes out of recently killed bark, hence the name bleeding canker. In older lesions, the dried fluid appears black and is crusty. Wood beneath the infected bark is stained blue black. The necrotic inner bark stains brown, but at the active margins of the lesion the affected tissue is pale and water soaked (Brasier and Strouts 1976).

Disease Development. Root and crown rot are favored by heavy rainfall. *P. citricola* was recovered from a bark canker in England. Inoculation of soil and roots did not reproduce disease, but inoculation of stems was successful (Brasier and Strouts 1976).

Control. Effective control measures have not been developed.

Figure 16.1. Young kiwi fruit vines killed by collar rot (left) and root and crown rot (right) caused by *Phytophthora citricola.* (Courtesy G. I. Robertson and D. M. McVicker, D.S.I.R., Auckland, New Zealand; Robertson 1982b)

Root Rot, Leaf Spots, Shoot Blight, and Stem Dieback of *Pieris japonica* (Thunb.) Don (Japanese Andromeda), *Rhododendron*, and Other Ericaceous Species

A blight occurs on young twigs or leaves of *Pieris japonica*. Olivaceous spots on the leaves are surrounded by a dark red margin. Infected leaves usually drop. The mature wood is rarely affected (Gerlach et al. 1974). Root and crown rot, which results in foliar chlorosis and stunting of the plant, can also be caused by *P. citrophthora* or *P. cinnamomi*. In advanced stages of disease, leaf epinasty and dieback occur (Gerlach et al. 1974). A dieback caused by *Botryosphaeria dothidea* does not occur on healthy, vigorously growing leaves and shoots. *B. dothidea* usually kills entire branches, whereas *Phytophthora* usually kills only terminal growth.

P. citrophthora is believed to be more important than *P. citricola* in the dieback complex of *Pieris japonica* (Gerlach et al. 1974).

In *Rhododendron* hybrids, *P. citricola* causes a stem dieback. Wilting of the leaves first occurs in the terminal portions of the plant. These leaves die as the disease progresses down the stem. In landscape areas, the lower stems and leaves are infected by sporangia or zoospores that are presumably splashed from the soil by rain or sprinklers. Bark necrosis also develops and extends up the stem until it envelopes the terminal node (Weiss 1943; Kröber 1959). In some cases, the entire plant wilts (Hoitinck and Schmitthenner 1969).

Disease Development. The disease occurs primarily during the spring and early summer and is favored by excess moisture and warm conditions. Since the fungus persists as dormant oospores in infested plant material, it may be more difficult to control than *P. cinnamomi* (Hoitink and Schmitthenner 1969).

Control. In container culture, noninfested, steamed or fumigated, well-aerated, and well-drained soil should be used, and resistant varieties, if available, should be planted. In nursery conditions, application of metalaxyl or fosetyl-Al should be efficacious, if registered for use.

Root Rot of *Juglans nigra* L. (Black Walnut) and *J. regia* L. (English Walnut)

On seedlings, the symptoms include chlorosis and wilting of the foliage followed by blackening of the entire plant. The roots become extensively covered with brown, water-soaked lesions. Similar symptoms are found on seedlings of *Quercus* spp. (oak) except that plants do not become black. Infected seedlings are either killed or infection continues to develop, and eventually the whole root system rots. This results in heavy losses during handling and shipping of nursery plants of *Juglans nigra* in the spring (Green and Pratt 1970).

In California, root rot of commercial walnut (*J. regia*) trees has become common on heavy soils that are irrigated (Mircetich and Matheron 1983). Both decay of roots and crown rot occur on both *J. hindsii* and the cultivar Paradox (*J. hindsii* × *J. regia*), which are standard rootstocks. A wide range of *Phytophthora* species has been isolated and proved to be patho-

Figure 16.2. Grape plants affected by root rot from which *Phytophthora citricola* was isolated. Left, dieback of grape (cultivar Palamino on its own root). Right, healthy Palamino plant on Mouvedre × Rupestris no. 1202 rootstock at Mt. Albert, New Zealand. (Photograph by B. Eykel, courtesy G. I. Robertson, D.S.I.R., Auckland, New Zealand; Robertson 1982a)

genic. In addition to *P. citricola,* these include *P. cactorum, P. cinnamomi, P. citrophthora, P. megasperma, P. cryptogea,* and four unidentified species (Mircetich and Matheron 1983). The disease is described in more detail in Chapter 11, *P. cactorum.*

Disease Development. Black walnut seedlings usually become infected in the seedbed. The disease is more severe after heavy rain or excessive irrigation causes prolonged waterlogging of the soil.

Control. In nurseries, seedlings should be stored at <5°C for only short periods before planting; long periods of cold storage should be avoided, since the pathogen can grow even at low temperatures. Soil at planting sites should be well drained.

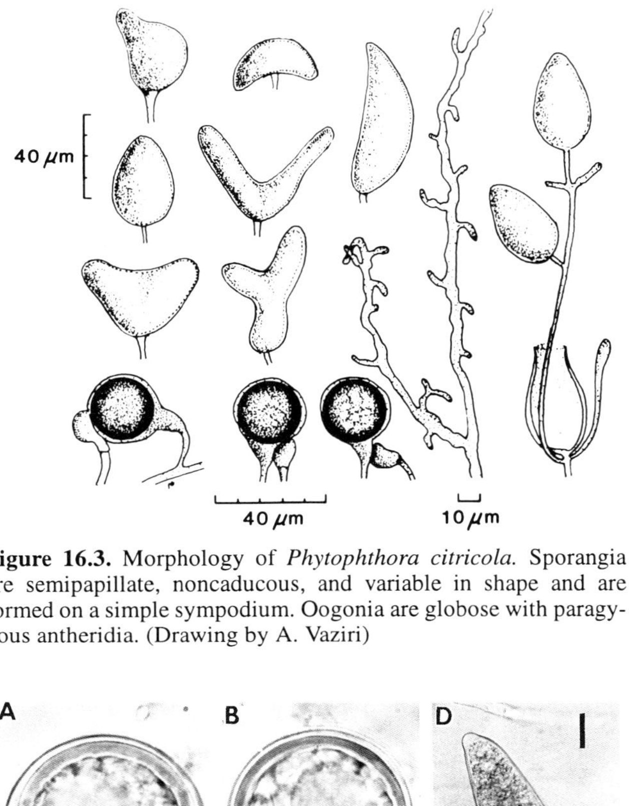

Figure 16.3. Morphology of *Phytophthora citricola.* Sporangia are semipapillate, noncaducous, and variable in shape and are formed on a simple sympodium. Oogonia are globose with paragynous antheridia. (Drawing by A. Vaziri)

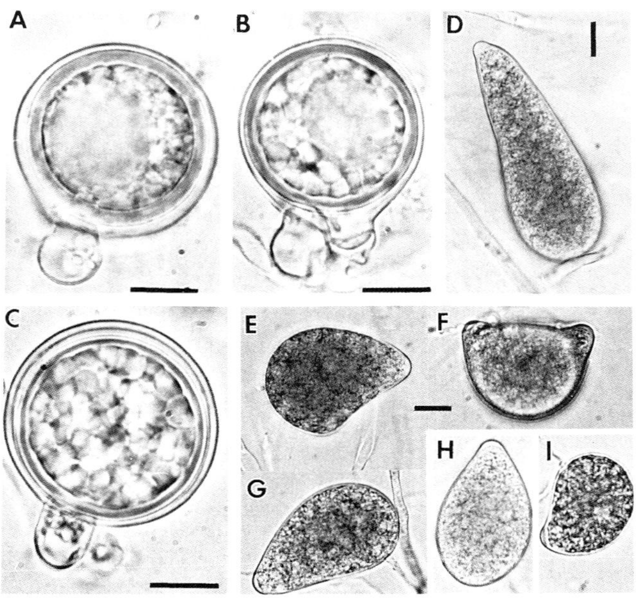

Figure 16.4. Morphology of *Phytophthora citricola* from feijoa. **A–C,** Oogonia with paragynous antheridia containing globose oospores. **D–H,** Semipapillate sporangia with variable shapes. **I,** Bipapillate sporangium. Bars = 10 μm. (By permission of Edagricole, Bologna, Italy, photographs courtesy Magnano di San Lio and Tuttobene 1985)

Rootstock plants of *J. hindsii* and the cultivar Paradox (*J. hindsii* × *J. regia*) were more resistant during October to November and January to May than during June. This was the result of the physiological status of the plants at the time of inoculation, since all pathogenicity tests were conducted in a growth chamber under standard optimum temperatures and 48-h soil flooding conditions (Matheron and Mircetich 1985a). The disease was more severe after flooding for 48 h (Matheron and Mircetich 1985b).

When soil was flooded for 48 h after inoculation with *P. citricola,* the rootstock *Pterocarya stenoptera* was highly resistant, but *J. hindsii* was highly susceptible to both *P. citricola* and *P. cinnamomi* under both flooded and nonflooded conditions. The cultivar Paradox and the species *J. ailantrifolia* and *J. nigra* (black walnut) were resistant to both *P. citricola* and *P. cryptogea* under nonflooded conditions, but resistance was overcome by flooding. *Pterocarya stenoptera* was resistant to *P. citricola, P. cinnamomi, P. cryptogea, P. citrophthora, P. cactorum,* and *P. megasperma* even under 48-h flooded conditions (Matheron and Mircetich 1985d).

Root and crown rot and trunk canker diseases induced by artificial inoculation of roots or stems were suppressed by soil drench and stem application of either metalaxyl or fosetyl-Al (Matheron and Mircetich 1985e).

Crown and Collar Rot of *Persea americana* Mill. (Avocado)

Although *P. cinnamomi* can cause a crown rot of avocado, *P. citricola* is the most frequently found causal agent in California (Coffey and Cohen 1984). The disease caused by *P. citricola* was described by Zentmyer et al. (1974). Although they did not consider *P. citricola* to be a serious threat to avocado production, the disease has since become more serious (Coffey and Cohen 1984).

Cankers form at the base of the tree and may expand so slowly that they are not noticed until the girdling effect causes yellowing of the foliage. In this respect, the effect is very much like that of *P. cactorum* on apple (Chapter 11).

The morphology of *P. citricola* from avocado was reported and illustrated by Zentmyer et al. (1974). Protein pattern analysis indicated a distinct resemblance to other isolates from other crops. When avocado isolates were characterized for isozyme and mtDNA RFLP patterns, the avocado isolates formed a discrete group (CIT5). Because it is separated from other *P. citricola* biotypes, the possibility exists that it may be a different species (Oudemans and Coffey 1993).

Disease Development. The development of cankers is favored by high soil moisture; however, the incidence and occurrence of the disease is not as frequent as root rot caused by *P. cinnamomi.*

El-Hamalawi and Menge (1994) reported that infection by *P. citricola* occurred only on avocado trees that had been wounded. *P. citricola* did not penetrate intact bark or lenticels. Although *P. citricola* survived in infected adventitious roots, it did not advance beyond the base of the root or into the woody portion of the root. Stressing the plant by pruning the roots predisposed plants to an increase in severity of disease expressed by greater lesion size. Increase in lesion size resulting from root-pruning stress was associated with a lower level of carbohydrates and a higher level of amino acids in the bark tissue. The rate of canker development on wounded but unstressed plants was increased sevenfold by spray misting (3 s each hour) inoculated sites with water.

Table 16.2. Differences between *Phytophthora citricola* and *P. cactorum*

Characteristic	*P. citricola*	*P. cactorum*
Apex of sporangium	Flat, wide, shallow thickening	Prominent papilla; deep thickening
Caducity of sporangia	Persistent, not caducous	Caducous with short pedicel
Chlamydospores	Rare	Numerous
Septum delimiting sporangium	Flush with the base	1–2 μm below the base
Septal plug	Absent	Present
Sporangiophores	Long and lax, with occasional swellings	Compact, monochaisal sympodium; no swellings
Oogonia	Sometimes <17 μm up to >40 μm in diameter	Rarely <19 μm or >37 μm in diameter
Colony growth	Radiate; petaloid	Uniformly fluffy

The susceptibility of wounded avocado trees to trunk canker caused by *P. citricola* varies with the physiological state of the tree, which is influenced in turn by the season of the year. Artificial wound inoculation with *P. citricola* resulted in formation of large cankers in May and June and November and December, but during other months the lesions remained small. These periods of increased susceptibility followed flushes of growth. The free amino acid content of the bark tissues was also increased during the period of high susceptibility. Correlation with lesion size was highly significant (El-Hamalawi and Menge 1995).

Control. Control methods should include prevention of wounds and the removal of suckers with subsequent application of a registered fungicide. Both metalaxyl and fosetyl-Al are effective (El-Hamalawi and Menge 1994b). After infection has become established, even systemic chemicals may not be effective.

Prevention is of prime importance. Coffey (1984) reported that when trees from one nursery were transplanted to the field, trunk canker occurred on nearly all the trees, but none of the trees from another nursery became infected.

Because fosetyl-Al is systemic and moves downward in plants, it may be effective if applied after wounding caused by removal of suckers from the base of trees (El-Hamalawi and Menge 1994b).

CHAPTER 17

Phytophthora citrophthora

(R. E. Smith and E. H. Smith) Leonian (1925)

Synonyms of *Phytophthora citrophthora* include *Pythiacystis citrophthora* Smith and Smith (1906) and *P. imperfecta* var. *citrophthora* (Smith and Smith) Sarejanni (1936). *P. citrophthora* was first isolated by Smith and Smith (1906) from rotted lemons; since then it has become recognized not only as a major pathogen of citrus trees, causing crown rot, gummosis, fibrous root rot, and brown rot of fruit (Table 17.1), but of several other crops as well (Table 17.2). Similar diseases on *Citrus* species are caused by *P. nicotianae* (= *P. parasitica*) (Klotz 1978).

A historical review and results of an investigation of the role of *Pythiacystis* (*Phytophthora*) *citrophthora* and *P. terrestria* (*parasitica*) as a cause of the gummosis disease on citrus in California are given by H. S. Fawcett (1923). A further review of the diagnosis and control of diseases of citrus caused by *P. citrophthora* and other species (*P. parasitica, P. hibernalis,* and *P. syringae*) by Klotz (1978) cites reports of the destructive gummosis disease occurring as early as the 10th century in Spain. During the late 1800s, nearly all of the citrus trees in Sicily were destroyed. Gummosis was noted in California as early as 1875. Among the complex of *Phytophthora* species attacking citrus, *P. citrophthora,* active at moderate temperatures (less than 30°C), and *P. parasitica* (*P. nicotianae*), active at high temperatures (above 30°C), are the most important. In Mediterranean climates such as that in California, *P. citrophthora* is active during the fall, winter, and spring but not during the summer; conversely, *P. parasitica* is most active during the summer. *P. hibernalis* and *P. syringae* are low-temperature pathogens and are active at temperatures near 15 to 20°C. Ricci et al. (1990) review *P. citrophthora* and other species of *Phytophthora* that are pathogenic to citrus in France.

In addition to those species cited above, *P. palmivora, P. hibernalis, P. syringae, P. cactorum, P. citricola, P. megasperma, P. boehmeriae, P. cinnamomi,* and *P. drechsleri* have also been reported to be pathogens on citrus trees (Table 17.1). Wager (1942) summarized information on the world distribution of pythiaceous species on citrus. *P. palmivora* attacks citrus mainly in the tropics and occasionally in the subtropics and Mediterranean regions during the hot and humid seasons of the year (Boccas and Laville 1976).

In addition to being a serious pathogen on citrus, *P. citrophthora* has become identified as important on several other crops, including ornamental plants such as *Pieris japonica* and Japanese andromeda (Gerlach et al. 1974; see below in this chapter; Table 17.2; Mchau and Coffey 1994b), and has been isolated from and proved to cause black pod on cocoa in Brazil (Campelo and Luz 1981; Kellam and Zentmyer 1981, 1986), Mexico (Lozano and Romero 1984), and Indonesia (Mchau and Coffey 1994b). *P. citrophthora* is reported on rubber trees in China (Yu et al. 1986) and in the Ivory Coast and Indonesia (Liyanage and Wheeler 1989). See Commonwealth Mycological Institute Map 35 and Table 17.2.

Isolates from cocoa pods proved to be pathogenic to cocoa pods, stems, and roots, but these isolates were not pathogenic to citrus; conversely, citrus isolates were not pathogenic to cocoa pods or seedlings (Kellam and Zentmyer 1986). A few oospores were found in infected cocoa pods that had been inoculated with single isolates.

The report that *P. citrophthora* was pathogenic on milk-

Table 17.1. Distribution of *Phytophthora* species isolated from diseased citrus trees (see also Wager 1942b)

Species	Geographical distribution	Reference	Part of plant infected
P. boehmeriae	Argentina	Frezzi (1941, 1950)	Fruit
P. cactorum	Argentina, Brazil, Formosa, Italy, Puerto Rico, United States	See Table 11.1	Roots and crowns; fruit
P. cinnamomi	Brazil, Formosa	Fawcett and Bitancourt (1940)	Fruit
P. citricola	Australia, Greece, Italy, Japan, South Africa	Boccas and Laville (1976)	Fruit
P. citrophthora	Australia, Belgian Congo, Brazil, Corsica, Cyprus, Egypt, Greece, India, Israel, Italy, Japan, Mozambique, New Zealand, Puerto Rico, Rhodesia, Sardinia, Sicily, Spain, South Africa, United States, Zambia	Smith and Smith (1906); Fawcett (1936); Boccas and Laville (1976); see also Table 17.2	Roots and crowns; leaf; fruit
P. drechsleri	United States	Fawcett (1936)	Roots
P. hibernalis	Australia, Greece, Israel, Italy, Portugal, South Africa, United States	Boccas and Laville (1976); Fawcett (1936)	Fruit; leaf
P. megasperma	Argentina, United States	Gondell (1946); Fawcett (1936)	Roots and crowns
P. palmivora	Congo, Guyana, India, Indonesia, Malaysia, Philippines, Puerto Rico, Tanzania	Boccas and Laville (1976); Fawcett (1936)	Roots and crowns; fruit
P. parasitica (*P. nicotianae*)	Argentina, Australia, Brazil, Cameroun, Congo, Corsica, Cuba, Formosa, Greece, Guatamala, India, Indonesia, Iran, Israel, Italy, Japan, Ivory Coast, Lesser Antilles, Malaysia, Mali, Philippines, Puerto Rico, Sicily, South Africa, Spain, Tunisia, Uruguay, United States, Venezuela	Boccas and Laville (1976); Fawcett (1936)	Roots and crowns; fruit; seedling damping-off

weed, *Morrenia odorata* Lindl., but not on citrus in Florida (Burnett et al. 1974) was the result of a misidentification. The causal agent was later found to be *P. palmivora* (Feichtenberger et al. 1984b).

Characteristics of *Phytophthora citrophthora*

P. citrophthora is classified in group II (Stamps et al. 1990). Morphology is shown in Figure 17.1. Tabular keys are presented in Tables 4.2 and 4.3 in Chapter 4 and a dichotomous key in Appendix 4.9 (Ho 1992). A summary on the morphology of *P. citrophthora* and diseases caused is reported by Waterhouse and Waterston (1964c). Mchau and Coffey (1994b) reported results of a study of 77 isolates from a world collection and redescribed the species. The redescription of *P. citrophthora* is reproduced by permission from Mchau and Coffey (1994b):

Mycelium. Hyphae smooth or coarse, 3–7 μm diameter. Colonies dense cottony like, or rosette, or stellate, or patternless on V8C agar at 24°. Radial growth rate on V8C agar at 24°, 7.6 ± 2.4 mm d^{-1} [per day]. Hyphal swellings present in some isolates, spherical to ovoid 1–6 ± 1.2 μm diameter, some irregularly shaped, forming within 3–14 days at 15° in sterilized 1% soil extract water. Stasis at 5°, optimum growth at 18–27°, lethal at 35° on V8C agar in the dark.

Sporangia. Produced in sterilized 1% soil extract, also on solid agar by a few isolates. Extremely variable in shape; ellipsoid, broadly ovoid, globose, limoniform, or extremely distorted with prominent papillae and some semipapillate in same isolates: 27 ± 3.4–65.3 ± 13.9 × 18.9 ± 1.8–40.4 ± 5.6 μm; length:breadth ratio 1.3:1 ± 0.2 to 1.8:1 ± 0.1; sporangia often with 2 or more papillae. In soil extract, sporangia borne singly or in very loose sympodia of 2–3 sporangia, often laterally attached to sporangiophore; sporangiophore often with globose swellings at branching points. Non-caducous.

Chlamydospores. Present in some isolates, terminal or intercalary, spherical, 10–43 ± 5.2 μm diameter some with irregular shapes, hyaline, forming within 7 days, walls 0.5–1.5 μm thick when produced in V8C agar in darkness at 25°. Sexuality. Sterile, though can sometimes induce oospore formation in other heterothallic species, e.g., A1 mating types of *P. capsici*. Isozyme phenotype. Monomorphic in MDH2, PGD1, PGD2, IDH2, LDH2, ME, FDP2, PGM, AK, and SOD. Polymorphic up to 25% in ACO, GPI, MDH1, IDH1, G6PDH, HEX2, and PEP. Paratypes: P318, P451, P1200, P1320, P1323, P1837, P3521, P3527, P3771, P3934, P6295, P6310, P7481.

Sporangia

Sporangia are mostly papillate, persistent, and not caducous. Although sporangia are described by Stamps et al. (1991) to be caducous in some isolates but not in others, Mchau and Coffey (1994b) reported that all of 77 isolates produced noncaducous sporangia. The shapes are variable, ranging from spherical, ovoid, obpyriform, obturbinate, to ellipsoidal; dimensions are 23 to 90 × 18 to 60 μm, average 45 × 30 μm. The length-breadth ratio is 1.2:1 to 2.0:1, average 1:4. Mchau and Coffey (1994b) report length-breadth ratios of 1.3:1 to 1.8:1 with a length range of 27.4–64.3 μm and a breadth of 18.9–40.4 μm. See Table 17.3 for a range of dimensions in other reports. Sporangia are occasionally laterally attached and some are intercalary. Sporangiophores are irregularly branched, some singly and some in a loose sympodium with a swelling at the point of branching.

Chlamydospores

Chlamydospores do not ordinarily occur on citrus isolates; however, chlamydospores formed in citrus roots from a California orchard during the warm summer months (Sjoerdsma et al. 1988). The diameter was 25 to 35 μm, average 28 μm. Mchau and Coffey (1994b) report chlamydospores only on isolates from cacao in Brazil. Chlamydospores were 10–43 μm in diameter and hyaline and were produced both singly and in clusters.

Sex Organs

In general, it has been assumed that the oospore stage is not a part of the life history of *P. citrophthora.* Sex organs do not occur in nature on citrus. Sex organs are not included in current keys (Stamps et al. 1990) and were not noted in the original description. We have illustrated oospore formation here because, experimentally, oospores have been shown to form when certain *P. citrophthora* isolates are paired on carrot agar. Four isolates from cocoa black pod in Brazil formed oospores when paired with an A2 isolate from citrus (P318 from Australia), and one cocoa isolate formed oospores when paired with P717 (A1), a *P. citrophthora* isolate (host unknown) from New Zealand. None of the cocoa isolates produced oospores when crossed among themselves, however, although some were A1 and another was A2 mating type. Kellam and Zentmyer (1986) concluded from these data that the compatibility or mating system was probably more complex than the usual A1/A2 system in *Phytophthora.*

Mchau and Coffey (1994b) report that less than 30% of the 77 *P. citrophthora* isolates studied produced oospores when paired with an A2 mating type of *P. capsici.* By use of the polycarbonate membrane method of Ko (1978; see Chapter 3), *P. citrophthora* isolates acted only as inducers of oospores in the *P. capsici* partners. Although oospores from pairing A1 and A2 isolates of *P. citrophthora* were reported by Savage et al. (1968), no evidence of oospores from pairing of *P. citrophthora* isolates was found. When isolates are paired with A1 or A2 mating types of other species, antheridia are amphigynous and oogonia are 26 to 36 μm in diameter and contain oospores 21 to 28 μm in diameter.

Growth Temperatures

The minimal temperature for growth is <5°C, the optimum 24 to 28°C, and the maximum 32 to 33°C.

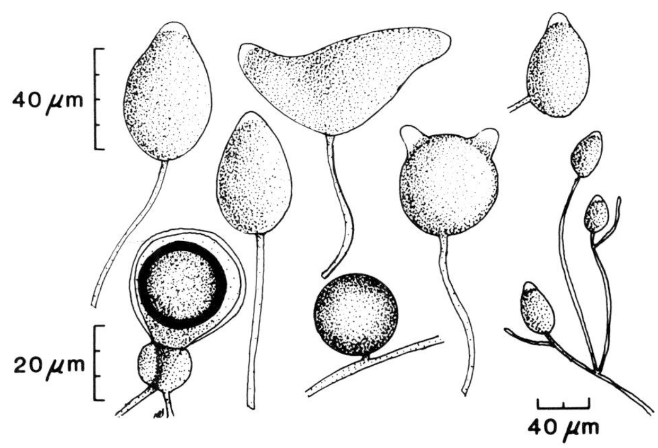

Figure 17.1. Morphology of *Phytophthora citrophthora.* Lower row, oospore within a globose oogonium with an amphigynous antheridium (oospores have not been seen in nature but have been induced in culture in a few cases [Kellam and Zentmyer 1986b]); a globose chlamydospore, which is also rarely found; and sporangia forming in a simple sympodium. Upper row, papillate, ovoid to ellipsoid sporangia, some of which have two papillae. According to Mchau and Coffey (1994b), the morphology is more diverse than illustrated here. (Drawing by A. Vaziri)

Table 17.2. Distribution of hosts of *Phytophthora citrophthora*

Host	Common name	Disease	Geographical distribution
Acer negundo L.	Box elder	Root rot; collar and trunk canker	Argentina (Frezzi 1950)
Actinidia chinensis Planch.	Kiwi fruit	Root rot	Chile (LaTorre et al. 1991)
Araucaria augustifolia (Bertol.) Kuntze	Parana or Brazilian pine	Apical and flower necrosis	Argentina (Frezzi 1950)
Averrhoa carambola L.	Carambola	Root rot	China (Ho et al. 1983b)
Bauhinia acerifolia Cum.	Bauhinia	Root rot and stem necrosis	Argentina (Frezzi 1950)
Bauhinia forticata Link (*B. candicans*)		Root rot and stem necrosis	Argentina (Frezzi 1950)
Bauhinia populneum R. Br.		Root rot and stem necrosis	Argentina (Frezzi 1950)
Brachychiton acerifolius Muell.	Flame tree	Root rot and stem necrosis	Argentina (Frezzi 1950)
Capsicum annuum L.	Pepper	Root rot and stem necrosis	Argentina (Frezzi 1950); Peru (Bazan de Segura 1958)
Capsicum pubescens Ruiz and Pav.	Red pepper	Root rot and stem necrosis	Argentina (Frezzi 1950)
Carica papaya L.	Papaya	Root rot	China (Ho et al. 1983b)
Castanea spp.	Chestnut	Root rot	United States (C. O. Smith 1937)
Catharanthus roseus L.	Rose periwinkle; old maid	Root rot and stem necrosis	Argentina (Frezzi 1950)
Cedrus deodora (D. Don.) G. Don.	Cedar	Stem canker	United States (Smith 1937)[a]
Celtis australis L.	European nettle tree; honeyberry	Root rot and stem necrosis	Argentina (Frezzi 1950)
Cicer arietinum L.	Chick pea; gram, or garbanzo bean	Root rot and stem necrosis	Argentina (Frezzi 1950)
Citrus sp.	Citrus	Gummosis; root rot	Portugal (Telhada 1990)
Citrus aurantifolia (Christm.) Swing	Lime	Gummosis; foot rot	United States (H. S. Fawcett 1923, 1936); Congo (Staner 1929); Argentina (Frezzi 1950)
Citrus aurantium L.	Sour orange	Brown rot of fruit	United States (H. S. Fawcett 1936); Greece (Kouyeas 1978); Italy (Magnano di San Lio et al. 1986b)
Citrus bergamia Risso	Bergamot	Blight	Italy (Magnano di San Lio et al. 1986b)
Citrus limon (L.) Burm.	Lemon	Brown rot of fruit; gummosis; foot rot	United States (R. E. Smith and Smith 1906; H. S. Fawcett 1936); New Zealand (Cockayne and Cunningham 1921); Argentina (Fawcett 1916); Australia (Darnell-Smith 1918); Italy (Traverso 1921); Rhodesia (Hopkins 1940)
Citrus medica L.	Citron	Collar rot	Corsicca (Dufrenoy 1925)
Citrus microcarpa Bunge	Shiki-Kitsu	Brown rot of fruit; dieback	Malaysia (Kueh 1980)
Citrus mitis	Calamondin	Root rot	Italy (Magnano di San Lio et al. 1986b)
Citrus reticulata Blanco	Mandarin tangerine	Gummosis; trunk canker	Congo (Staner 1929)
Citrus paradisi Macf.	Grapefruit; pomelo	Foot rot; trunk canker; gummosis	United States and Cuba (Fawcett 1915); Congo (Staner 1929); Puerto Rico (Guiscafre 1931); India (Sharangapani 1930); South Africa (Wager 1931); Australiia (Fraser 1942); New Zealand (Brien 1946); Israel (Schiffman-Nadel and Cohen 1968). Now worldwide.
Citrus sinensis (L.) Osbeck	Sweet orange	Brown rot of fruit; leaf fall; twig blight; gummosis	United States (Fawcett 1915); Taiwan (Sawada 1919); South Africa (Doidge 1925); Congo (Staner 1929); Philippines (Anon. 1938); Australia (Fraser 1942); Argentina (Frezzi 1950); New Zealand (Fletcher 1957); Czechoslovakia (Cejp and Jechova 1962). Now worldwide.
Citrus spp.	Citrus	Brown rot of fruit; leaf fall; twig blight; gummosis	Brazil (Averna-Sacca 1912); Sardinia (Petri 1925); Spain, Sicily, and Egypt (Fawcett 1930); Puerto Rico (Guiscafre 1932); Tunisia (Chabrolin 1932); Mozambique (Cardoso 1934); Australia (McLennan 1936); Argentina (Frezzi 1940); South Africa (Loest 1950); Mauritius (Orian 1951); Malaysia (Anon. 1960)
Clerodendron L.	Glory-blower; tube-flower	Seedling damping-off	Argentina (Frezzi 1950)
Cucumis melo L.	Melon	Fruit rot	United States (Tompkins and Tucker 1937a)[a]
Cucurbita L.	Squash	Foot rot; seedling blight	United States (Merrill 1929); Argentina (Frezzi 1950)
Cyclamen sp.	Persian violet	Collar, leaf, and petiole blight	Greece (Kouyeas and Chitzanidis 1978)
Dianthus caryophyllus L.	Carnation	Collar rot; foot rot	Greece (Anon. 1965)
Diospyros kaki L.	Japanese persimmon	Fruit rot	Iran (Ershad 1971)
Eucalyptus sp.	Eucalyptus	Stem canker	United States (Smith 1937)[a]
Euonymus japonicus Thunb.	Spindle tree	Blight	China (Zheng and Lu 1989)
Euonymus radicans var. *argenteo marginatus*	Spindle tree	Twig blight	United States (Gerlach et al. 1976b)
Fagara coco (Gil.) Engl. (*Zanthoxylum*)	Cocoa	Stem and apical necrosis	Argentina (Frezzi 1950)

(*continued on next page*)

Table 17.2. (*continued*)

Host	Common name	Disease	Geographical distribution
Ficus carica L.	Fig	Brown rot of fruit; leaf necrosis	Japan (Katsura and Yamamoto 1969); Argentina (Frezzi 1950
Ficus elastica Roxb. ex. Hornem.	India rubber tree	Leaf blight	Greece (Pappas 1978)
Fortunella japonica (Thunb.) Swing × *C. aurantifolia* (Christm.) Swing	Eustis limequat	Fruit rot	Italy (Magnano di San Lio et al. 1986b)
Fortunella margarita (Lour.) Swingle	Kumquat	Brown rot; gummosis	Japan (Takimoto 1931); United States (Fawcett 1936); Italy (Magnano di San Lio et al. 1986b)
Fragaria chilensis Ehrh.	Strawberry	Fruit rot	Taiwan (Kao and Leu 1979; Chang 1988)
Gleditschia triancanthos L.	Honey locust	Crown rot	United States (Miller 1940)
Gomphrena L.	Amaranth	Root rot	Argentina (Frezzi 1950)
Hedera helix L.	Ivy	Root rot; stem necrosis	Argentina (Frezzi 1950); United States (Gerlach et al. 1976)
Hevea brasiliensis Muell. Arg.	Rubber tree	Leaf fall; stem canker; black stripe	China (Ho et al. 1983b, 1984a; Tu et al. 1986); Ivory Coast (Liyanage and Wheeler 1989); Indonesia (Liyanage and Wheeler 1989)
Ilex crenata Thunb.	Japanese holly	Root rot	United States (Gerlach et al. 1976b)[a]
Juglans spp.	Walnut	Stem canker	United States (C. O. Smith 1937)[a]
Juglans californica Walt.	Southern black walnut	Stem canker	United States (C. O. Smith and Barrett 1930)
Juglans hindsii Jep.	Walnut	Crown rot	United States (Matheron and Mircetich 1985)
Juglans regia L.	English walnut	Mal de la tinta; trunk canker; gummosis	Argentina (Pontis 1941); United States (C. O. Smith 1937[a]; Mircetich and Matheron 1983)
Lycopersicon esculentum L.	Tomato	Fruit rot	Argentina (Frezzi 1950)
Magnolia grandifolia L.	Bull-bay	Root rot	Argentina (Frezzi 1950); Iran (Ershad 1971)
Malus pumila Mill.	Apple	Trunk and crown canker	United States (R. E. Smith and Smith 1925)
Melia azedarach L.	Chinaberry; paradise tree	Root rot; apical necrosis	Argentina (Frezzi 1950)
Morrenia ororata	Milkweed vine	Stem and root rot	United States (Burnett et al. 1974; reidentified as *P. palmivora* by Feichtenberger et al. 1984b)
Persea americana	Avocado	Collar rot and canker	United States (H. S. Fawcett 1916)
Picea engelmanni Parry ex. Engelm.	Engelmann spruce	Pre-emergence damping-off	United States (Rathbun-Gravatt 1931)
Pieris japonica (Thunb.) Don.	Japanese andromeda	Shoot blight and stem dieback	United States (Gerlach et al. 1974)
Pinus banksiana Lamb.	Gray pine; northern scrub pine	Pre-emergence damping-off	United States (Rathbun-Gravatt 1931)
Pinus halepensis Mill.	Aleppo pine	Root rot	Argentina (Frezzi 1950)
Pinus pinea L.	Italian stone pine	Root rot; apical necrosis	Argentina (Frezzi 1950)
Pinus radiata Don.	Monterey pine	Root rot; apical necrosis	Argentina (Frezzi 1950)
Pinus resinosa Ait.	Red pine; Norway pine	Pre-emergence damping-off	United States (Rathbun-Gravatt 1931)
Pistacia vera L.	Pistachio tree	Apoplexy; sudden wilt	Iran (Ershad 1971)
Poncirus trifoliata Raf. (L.) Raf. (*Citrus trifoliata* L.)	Trifoliate orange	Stem canker; root rot	Argentina (Frezzi 1950); Iran (Ershad 1971); New Zealand (Fletcher 1957)
Potentilla fructicosa L.	Cinquefoil	Foliage and twig blight	United States (Gerlach et al. 1976b)
Prunus amygdalus Batsch	Almond	Foot rot	Greece (Anon. 1965)
Prunus armeniaca L.	Apricot	Stem girdling cankers	United States (Smith and Smith 1925)
Prunus avium L.	Cherry	Stem girdling cankers	United States (Smith and Smith 1925)
Prunus doestica L.	Plum	Collar rot	Greece (Kouyeas and Chitzanidis 1978)
Prunus dulcis (Mill.) Webb	Almond	Bark canker	United States (Smith 1914); Greece (Kouyeas and Chitzanidis 1978); Australia (Wicks and Lee 1986)
Prunus persica (L.) Batsch.	Peach	Trunk canker; root rot	United States (Smith 1914); Iran (Ershad 1971)
Pyrus communis L.	Pear	Bark canker	United States (Smith 1915)
Robinia pseudo-acacia L.	Black locust; false acacia	Root rot	Argentina (Frezzi 1950)
Rosa L.	Rose	Collar rot	Greece (Kouyeas and Chitzanidis 1978)
Rosmarinus officinalis L.	Rosemary	Leaf blight; root rot	United States (Gerlach et al. 1976b)[a]
Salvia splendens Sellow ex. Roem and Schult.	Scarlet salvia	Root rot; flower blight	Argentina (Frezzi 1950)
Sequoiadendron giganteum (Lindl.) Buchholz	Giant redwood	Foliar blight; root rot	United States (Sandlin and Ferrin 1993)
Taxus spp.	Yew	Root and crown rot	United States (Gerlach et al. 1976b)
Theobroma cacao L.	Cocoa	Black pod; seedling rot	Brazil (Campelo and Luz 1981; Kellam and Zentmyer 1986b); Mexico (Lozano and Romero 1984); Indonesia (Mchau and Coffey 1994b)
Utica L.	Nettle	Collar rot	Greece (Kouyeas and Chitzanidis 1978)

[a] Artificially inoculated host.

Genetic Relationships

An isozyme pattern study of 32 isolates of *P. citrophthora* from various hosts, including citrus, cocoa, *Actinidia deliciosa,* rubber, walnut, strawberry, pistachio, *Ficus elastica, Prunus domestica, Amygdalus communis, Ribes sanguineum,* and *Coptus japonica,* indicated that *P. citrophthora* was genetically related to *P. capsici,* but within the *P. citrophthora* isolates studied, two subgroups, CITR1 and CITR2, could be defined. Isolates from citrus grouped into CITR1, and those from cocoa grouped into CITR2 (Oudemans and Coffey 1991a). In a subsequent report, a group CIT5 (*P. citricola* from avocado) showed a close relationship with CITR groups (Oudemans and Coffey 1993). By use of polymerase chain reaction amplification of rDNA, Lee and Taylor (1992) showed a close relationship between cocoa isolates of *P. capsici* (formerly *P. palmivora* MF4) and *P. citrophthora.*

Mchau and Coffey (1994b) found that three subgroups of *P. citrophthora* could be differentiated isozymically. Subgroup CITR1 contained isolates from a broad range of hosts, including citrus from various geographical areas. Subgroup CITR2 was composed exclusively of isolates from cacao grown in Brazil. Subgroup CITR3 included isolates from cacao grown in Indonesia.

Major Diseases of Citrus Caused by *Phytophthora citrophthora* and/or *P. parasitica* (*P. nicotianae*)

Several useful references on citrus diseases include Klotz (1978), Boccas and Laville (1978), Whiteside et al. (1988), and Ricci et al. (1990). The last citation is a review of conventional and molecular methods of identification of species and a map of the world distribution of *Phytophthora* species affecting citrus.

The discussions on diagnosis and control of citrus diseases by Klotz (1978) and Klotz and Calavan (1978) are still useful references and lack only information on control by the systemic fungicides known as phenylamides (e.g., metalaxyl) and phosphonate (e.g., fosetyl-Al), which had not been developed in 1978. The Klotz (1973) handbook on citrus diseases has excellent color renditions of symptoms of brown rot and gummosis caused by *P. citrophthora* as well as symptoms of a wide range of diseases. The *Compendium of Citrus Diseases* is an excellent source of information (Timmer and Menge in Whiteside et al. 1988).

Brown Rot of Citrus Fruit

Diagnosis, development, and control of fruit diseases, including brown rot, are discussed by Eckert and Brown (1986) and Whiteside et al. (1988). Feld et al. (1979) review the control of brown rot in California by use of copper fungicides.

In the field, infection often starts on the bottom end of the fruit because inoculum in infested soil is splashed by rainfall from the ground onto the fruit. Several infected areas, which may develop on a single fruit, coalesce and eventually cover the whole fruit. The diseased tissue is firm and leathery, even in advanced stages (Plate 17.1). After becoming mummified, fruits drop and *Phytophthora* can be isolated from the infected seeds. In wet weather, a white fungal growth often appears on the surface of the diseased fruit. The characteristic pungent, rancid odor that emanates from rotted fruit readily distinguishes this disease from other fruit rots (Eckert and Brown 1986). This is particularly obvious in storage areas, where picked fruit with incipient infections rot rapidly.

In areas of high rainfall, dark, water-soaked areas develop on leaves, particularly at the tips and along the margins, and leaf tips die back. Severely affected leaves become black at advanced stages of disease and drop prematurely. Affected orange and mandarin trees can be completely defoliated.

In Florida, United States, *P. nicotianae* (*P. parasitica*) is more commonly found in soil and on roots than *P. citrophthora;* however, *P. citrophthora* has a greater capability to cause brown rot on fruit than *P. nicotianae* (Whiteside 1970). Brown rot can also be caused by *P. hibernalis* and *P. syringae.*

Disease Development. Fruit loss caused by brown rot is directly related to the total amount and frequency of rainfall. Brown rot develops rapidly during rainy weather from zoospore infection of any part of the fruit, particularly on fruit low enough to be splashed with soil. The susceptibility of fruit increases with maturity.

Fruits that are infected prior to harvest may not show any external symptoms, but if the fruit is boxed for shipment, the disease can spread to healthy fruit. By 3 days after harvest, tan spots develop on the fruit stored at 25°C or after 10 days at

Table 17.3. Dimensions of *Phytophthora citrophthora* sporangia from various hosts

Host	Sporangia (μm)		Length-breadth ratio		Reference
	Range	Average	Range	Average	
Citrus	30–90 × 20–60	50 × 35	NM[a]	NM	Smith and Smith (1906)
Citrus	27–44 × 17–30	35.3 × 24.7	NM	NM	Tucker (1931)
Citrus	23–75 × 18–42	43 × 30.8	NM	1.40:1	Frezzi (1950)
Almond	23–54 × 18–35	41 × 31	1.15–1.6	1.37:1	Kouyeas and Chitzanidis (1968)
Citrus	29–56 × 22–42	40 × 29	1.10–1.54	1.35:1	Kouyeas and Chitzanidis (1968)
Pistachio	24–63 × 17–41	41 × 31	1.05–1.84	1.33:1	Kouyeas and Chitzanidis (1968)
Pistachio	26–72 × 20–41	37 × 28	1.1–2.04	1.32:1	Kouyeas and Chitzanidis (1968)
Pistachio	30–67 × 21–42	45.9 × 30.7	N/M	1.5:1	Ershad (1971)
Diospyros kaki	34–99 × 21–37	51.7 × 30.0	N/M	1.7:1	Ershad (1971)
Citrus	30–60 × 25–47	42.9 × 32.6	N/M	1.31:1	Ershad (1971)
Citrus	35–62 × 22–37	48.1 × 30.8	N/M	1.6:1	Ershad (1971)
Magnolia	21–41 × 17–27	30.3 × 21.5	N/M	1.42:1	Ershad (1971)
Citrus	32–110 × 21–54	63 ± 15.3 × 34.8 ± 5.9	1.1–3.6	1.8:1	Feichtenberger (1980)
Citrus	33–108 × 22–64	59.5 ± 17 × 35.3 ± 5.7	0.8–3.9	1.7:1	Feichtenberger (1980)
Citrus	29–98 × 23–56	47 ± 11.4 × 33.5 ± 6.3	0.6–3.1	1.4:1	Feichtenberger (1980)
Citrus	35–89 × 25–50	50 ± 8.3 × 35.1 ± 4.0	0.7–2.6	1.5:1	Feichtenberger (1980)
Various (30 species)	27–65 × 19–40	. . .	1.3:1–1.8:1	. . .	Mchau and Coffey (1994b)

[a] Not mentioned by author in publication cited.

10°C. Under favorable conditions, the entire fruit becomes brown in approximately 7 days (Klotz and Dewolfe 1961b). Fruit infected with *Phytophthora* are readily colonized by *Penicillium italicum, P. digitalis,* or *Geotrichum candidum,* all of which are wound pathogens (Eckert and Brown 1986).

When orange or lemon fruits were maintained in packing boxes at 5°C, spread of disease did not occur, but at 10°C, *P. citrophthora* spread rapidly to cause fruit rot after 12 days and at 14°C after 3 days. *P. syringae,* which causes brown rot in some areas of the world, spread to healthy fruits at 5°C after only 6 days of incubation (Chitzanidis and Argyri 1990).

Control. See Klotz et al. (1961a–c, 1965c), Klotz (1974), Whiteside (1970), or Feld et al. (1979) for detailed information on control.

Before the first rains, the skirts of trees and all fruit within 1 to 1.5 meters above the soil should be sprayed with Bordeaux mixture, another suitable copper formulation, or any other fungicide that is registered for use, such as the systemic fungicide metalaxyl, to protect fruit against contamination by rain-splashed soil. If *P. hibernalis,* which produces deciduous sporangia that are airborne during rain storms, is known to occur, trees should receive a full-coverage spray (Klotz 1978).

Although Bordeaux mixture is not as frequently used as it was in the past (see Chapter 7), it is inexpensive and has the best sticking quality of all fungicides if allowed to dry before rain falls. Its use in California is summarized by Feld et al. (1979). The fungicide contains 1.4 kg of copper sulfate pentahydrate (25% metallic copper) and 2.0 kg of hydrated lime in 400 liters (approximately 100 gallons) of water. If there is a danger of copper injury, a mixture containing zinc sulfate, copper sulfate pentahydrate, and hydrated lime (1.4:1.0:2.7:400) can be used (Klotz 1978). Several other formulations are given by Feld et al. (1979). Captan 50W can be applied (Klotz 1978), but it is much less efficacious than Bordeaux mixture (Feld et al. 1979). In a season during which severe brown rot is likely to occur, a second application should be applied approximately 1 month after the first application. One application is usually sufficient in a normal season unless rainfall is frequent. Helicopter application leads to poor control because the lower skirts of trees are not well covered (Feld et al. 1979).

Postharvest development of brown rot can be prevented by immersing the fruit for 2 to 3 min either in water or fungicidal solutions at 46 to 49°C (Klotz and Dewolfe 1961a). Fruit from the winter crop must first be conditioned by drying until slightly wilted before treatment to avoid the liberation of toxic rind oil. Lemons picked under dry weather conditions should be held for 36 to 48 h before hot water treatment. If fruit is picked during wet weather, treatment should be delayed for at least 4 days. Oranges should be held at least 24 h before hot water treatment. There is a danger that some fruits may be injured by hot water treatment because it increases the susceptibility of fruit to secondary decay diseases (Eckert and Brown 1986).

Metalaxyl (see Chapter 7), a systemic fungicide, has excellent efficacy for postharvest control of *Phytophthora.* Metalaxyl may be incorporated in the wax at a dosage of 1,000–2,000 mg/liter prior to packing for shipment (Cohen and Schiffman-Nadel 1978a, b; Cohen 1981a, b, 1982). Eckert and Brown (1956) reviewed control of fruit diseases of citrus caused by *P. citrophthora* and other pathogens.

Postharvest dips of orange and lemon fruits with metalaxyl (0.05 and 0.1%) or with fosetyl-Al (0.1, 0.2, and 0.3%) prevented the spread of *P. citrophthora* in packing boxes kept at 11°C for periods of 30, 40, and 60 days. The Penicillium rots were not affected by either fungicide, but when the fungicides were combined with thiabendazole (0.1%) or imazalil (0.5%), both diseases were controlled (Chitzanidis and Argyri 1990).

Gummosis, Foot Rot, and Fibrous Root Rot

Citrus gummosis (lesion is below the bud union) was first described in the Azores in 1834 (Fawcett 1925). Foot rot is similar, except the lesion is above the bud union (Timmer et al. 1993). Fibrous root rot involves the small feeder roots that are produced in the upper layer of soil. These diseases occur in every citrus-growing area worldwide and were most likely spread on infected rootstocks. A change to the *Phytophthora*-resistant sour orange rootstock was needed to ensure continued production. The relationship of the widespread use of sour orange rootstock to the subsequent development of the viral tristeza disease, which is lethal to sweet orange grafted on sour orange rootstock (Bar-Joseph et al. 1981), is discussed in Chapter 6.

P. citrophthora and *P. parasitica* (*P. nicotianae*) are both causal agents of gummosis and root rot, but in Mediterranean-type climates where rains come in the winter and summers are dry, *P. citrophthora* is the most important cause of gummosis and root rot because it is most active during the winter months. In warmer areas (e.g., Queensland, Australia, and Florida, United States), *P. parasitica* is considered to be the major cause of gummosis and root rot of citrus (Whiteside 1970; Klotz 1978). Since the diseases caused by *P. parasitica* and *P. citrophthora* on citrus are similar, both species are discussed in this chapter.

The types of lesions on the trunks of trees caused by *P. citrophthora* and *P. parasitica* vary in appearance among cultivars (see Klotz 1973). Environmental factors play an important part in lesion development. Aboveground symptoms include the following:

1. Dark, water-soaked areas on the trunk close to the soil line; small or large amounts of gum exude from these lesions, depending on the cultivar and the weather (Plate 17.2).
2. The diseased bark remains firm and does not slough off. In advanced stages of disease, patches of dead bark may break off, leaving bare areas.
3. A brown stain occurs in the underlying wood. The diseased wood is clearly defined from the healthy portions of the wood.
4. The cambium beyond the diseased areas becomes yellow and gummous.
5. Eventually, drying and longitudinal cracking of the bark occur.
6. A foul-smelling odor develops from the decaying inner bark caused by secondary invading bacteria.
7. Infected trees often bear a heavy crop of poor quality fruit because of the girdling effect on the phloem (Plate 17.3). The leaves turn yellow and drop prematurely, and depending on environmental conditions, the tree may die rapidly, especially on highly susceptible rootstocks, or some regrowth may occur.
8. Trees affected with crown rot show severe yellowing of foliage followed by the rapid death of the top branches of the tree.

Fibrous root rot symptoms caused by *P. citrophthora* or *P. parasitica* include a brown discoloration and dieback of feeder roots (Fraser 1942; Klotz et al. 1958). Distinct frog-eye lesions are obvious on highly susceptible rootstocks (Klotz 1978).

Although many rootstocks are resistant, resistance does not always extend to fibrous roots. Root rot on gummosis-resistant cultivars (Table 17.4) may occur when the population of *P. parasitica* exceeds 15–25 propagules per gram of soil (Lutz and Menge 1986b). Dead areas of the root remain firm, but the bark can easily be peeled or stripped off the stelar portion of affected feeder roots (Klotz 1978). Fibrous root rot leads to a slow decline of trees; in some cases trees survive and bear fruit despite the infection. Root rot can be distinguished from gummosis by the lack of gumming and the dusky to black discoloration of the wood below the bark (Klotz et al. 1967). Fibrous root rot can also be caused by anaerobic conditions in soil resulting from excess soil water or salinity (Lutz and Menge 1986b; Timmer et al. 1993).

Disease Development. The severity of the root rot phase varies with the soil conditions. Under unfavorable conditions for the growth of the fungus, the infection is limited to the fibrous roots. If the soil is waterlogged for a prolonged period, infection may extend to the cortical and phloem tissues of large branch roots. Smaller lateral roots are killed back, and lesions often form on larger roots.

Infection of crowns of trees is aided by wounds in the bark. Heaping soil against the trunk increases the likelihood of infection. Soil pH between 6.0 and 6.5 and excessively wet soils are favorable for disease development. Infection of the trunk occurs readily if water is in contact with the trunk for as few as 5 h. This condition often occurs when groves are flood-irrigated or when trees are grown in heavy, compacted soils with poor drainage (Klotz et al. 1967) because water in irrigation canals is frequently infested with either *P. citrophthora* or *P. parasitica* (Klotz et al. 1959).

Control. Extensive control measures are given by Klotz (1978), Klotz and Calavan (1969, 1978), and Whiteside et al. (1988).

Integrated management of root disease includes use of noninfested soil or soil fumigated with methyl bromide or steam in nursery beds to produce clean seedlings prior to planting, use of highly resistant rootstocks, selection of well-drained sites for planting trees, avoidance of excessive irrigation, removal of low branches touching the soil to avoid brown rot on fruit, avoidance of damage to the bark and placement of mulch or soil against the trunk, and control of weeds prior to use of fungicides if necessary.

Resistant rootstocks are listed in Table 17.4; also see the section in Chapter 6 on control by host resistance (Klotz et al. 1965a, 1968; Klotz 1978; Klotz and Calavan 1978). Sour orange is probably the most resistant, but it is highly susceptible to tristeza, a viral disease (see Chapter 6). *Poncirus trifoliata* and its hybrids, the cultivars Troyer and Carrizo citranges, have been widely and successfully used as resistant rootstocks for orange, grapefruit, some mandarins, and the cultivar Lisbon lemon (Broadbent et al. 1971; Çinar et al. 1976; Klotz and Calavan 1978); however, these hybrids are incompatible with the cultivar Eureka lemon. The cultivar Benton citrange (3697) has shown promise as a resistant rootstock for Eureka lemon (Long et al. 1977). Efficient use of resistant rootstocks depends on the availability of scion budwood that is free of the viroid that causes exocortis, scion compatibility with the rootstock, and a knowledge of the soil types to which the rootstocks are adapted. Scions of the susceptible but commercially desirable cultivars should be grafted 10 to 18 inches (25 to 45 cm) high on the rootstocks. After planting, the base of the rootstock should be painted or sprayed with a Bordeaux or other suitable copper-based paste to prevent infection. Wounding increases the susceptibility of the trunk.

If gummosis is detected early enough, infected trees can be treated by surgical removal of the diseased bark together with a strip (1 cm wide) of healthy bark beyond the margin of the diseased lesion. The wound should be disinfected with a 1% solution of potassium permanganate if registered for use. The crown and lower trunk can be sprayed or painted with a copper fungicide or captan (approximately 0.9 kg/400 liters of water) (Klotz 1978). Copper fungicides such as copper ammonium carbonate, cupric hydroxide, or captafol at 60 μg of active ingredient per milliliter provide protection against foot rot for 33 to 52 weeks (Timmer 1977).

Copper-containing compounds, such as Bordeaux mixture, captan, and captafol, and the dithiocarbomates, such as zineb and manzate, prevent infection when applied to the trunks or bases of trees; however, none of these fungicides are capable of eradicating *Phytophthora* after infection has taken place. The systemic fungicides—e.g., the phenylamides such as metalaxyl, furalaxyl, benalaxyl, ofurace, cyprofuram, and oxadixyl (see Chapter 7, Table 7.7) or the phosphonate, fosetyl-Al (Table 7.12)—have systemic action and are highly toxic to most *Phytophthora* species.

Farih et al. (1981a, b) reported that both metalaxyl and fosetyl-Al (efosite aluminum in their reports) significantly reduced or stopped the expansion of lesions on susceptible lemon trees in the field and also reduced the severity of root rot on experimentally inoculated trees.

In an extensive pest management experiment with established orange trees in Tulare County, California, use of a dilute solution of metalaxyl in the drip irrigation system increased yield of fruit by 10% (Menge et al. 1990).

Ohr et al. (1986) reported control of *Phytophthora* in container-grown citrus by soil drench with metalayxl. A problem with control in container-grown plants is that metalaxyl does not eradicate *Phytophthora*. When plants are transplanted to the field, latent *Phytophthora* inoculum can cause disease if conditions become favorable.

In Florida, Timmer et al. (1979) and Timmer and Castle

Table 17.4. Approximate integrated response to natural and artificial inoculation of citrus rootstocks with *Phytophthora parasitica* and *P. citrophthora*[a]

Susceptibility	Rootstock
Highly susceptible	Lemons (*Citrus limon*)
	Some citranges (*Poncirus trifoliata* × *C. sinensis*), except Troyer and Carrizo
Susceptible[b]	Sweet orange (*C. sinensis*)
	Limes (*C. aurantifolia*)
	Sampson tangelo (*C. reticulata* × *C. paradisi*)
	Rough lemon
	Grapefruit (*C. paradisi*)
	Mandarins (*C. reticulata*)
	Troyer and Carrizo citranges
	Citremon
	Nansho daidai (*C. taiwanica*)
	Siamelo
	Most selections of trifoliate orange (*P. trifoliata*), including U.S.D.A. and Pomeroy
Resistant	Ichang lemon 1219 (*C. ichangensis* × *C. grandis*)
	Alemow (*C. macrophylla*)
	Sour orange (*C. aurantium*)
	Texas, Barnes, English, Rubidoux, and Davis B trifoliates
	Kumquat (*Fortunella* species)
	Severinia (*Severinia buxifolia*)

[a] Reproduced from Klotz and Calavan (1978).
[b] In order of susceptibility from highest to lowest.

(1985) reported effective control of foot rot of young orange trees caused by *P. parasitica* with use of metalaxyl and fosetyl-Al. Control of foot and root rot of citrus was reported in South Africa by McKenzie (1985). Le Roux et al. (1991), also in South Africa, showed that over a 2-year period, treatment with metalaxyl as a soil drench and fosetyl-Al as a trunk injection of navel orange trees on a rough lemon rootstock infected with *P. nicotianae* increased the girth of trees; the yield increase was cost effective. Both treatments were combined with an aldicarb soil treatment to suppress the citrus nematode, *Tylenchulus semipenetrans* Cobb. As might be expected, metalaxyl reduced the population of *P. nicotianae* in the soil, but trunk injection with fosetyl-Al did not. Similar results were recorded by Magnano di San Lio et al. in Italy (1984). Fosetyl-Al and metalaxyl controlled root rot and gummosis in Brazil in both field and glasshouse trials. At least two foliar sprays or trunk paints with fosetyl-Al or soil drench with metalaxyl during the rainy season are necessary to control root rot and gummosis (Feichtenberger 1990).

Research on the relationship of inoculum density in soil to the severity of fibrous root rot of citrus is necessary to determine when metalaxyl and/or fosetyl-Al should be applied. Difficulties in obtaining accurate data are discussed in Chapter 5 on the effect of inoculum level in soil on disease expression. Timmer and Menge (1993) showed that use of a commercial ELISA (enzyme-linked immunosorbent assay) compares favorably with more conventional dilution plate methods to determine the inoculum level in soil (see Chapter 2; Timmer et al. 1980, 1988, 1989; Lutz and Menge 1991). Data on soil inoculum density (Menge 1986; Timmer et al. 1988b) may be helpful to predict when fungicides should be used. A precise threshold has not been established in either California or Florida, however. In Florida, 0.5 propagules per cubic centimeter of soil is considered low, 5 to 15 is moderate, and more than 15 is high. In California, where propagules are measured in the rhizosphere soil taken from roots, the propagule count is about 1.5 to 3.5 times higher (Timmer et al. 1993).

Fosetyl-Al is the only fungicide that is translocated both upward and downward in trees (Chapter 7). Phosphite concentration in citrus trees after treatment by foliar application, trunk injection, or trunk paint with fosetyl-Al or phosphorous acid was measured by gas-liquid chromatography (Schutte et al. 1991). Rapid transport to roots and foliage followed foliar application and trunk paint. Trunk injection was also effective. Distribution was greater with phosphorous acid, the active ingredient of fosetyl-Al, than with fosetyl-Al. Phosphite levels in roots, where activity is most needed to control root rot, was maintained at a higher level for a longer time when application was made by trunk injection than by foliar spray. Schutte et al. (1991) concluded that treatment intervals should not exceed 42 days.

Some success in curing infected citrus trees with gummosis symptoms by heat treatment has been reported by Hough et al. (1979). When margins of lesions were heat-treated with a kerosene blowtorch, *P. citrophthora* was killed. Treated areas heal rapidly, and trees fully recover. The flame is slowly moved over the infected bark until the temperature of the bark reaches or exceeds 60°C.

In Brazil, induction of dwarfing of trees by inoculation with the exocortis viroid is under experimental investigation. Lesions resulting from artificial inoculation with *P. citrophthora* on exocortis-inoculated trees were consistently smaller than those on more vigorously growing trees uninoculated with exocortis viroid (Rosetti et al. 1984). Similar results were shown on trees previously inoculated with a mild tristeza virus to control severe strains of tristeza virus by cross protection (Feichtenberger et al. 1984a).

Damping-Off of Citrus Seedlings

Disease Development. When temperatures are favorable and seedbeds are wet, *P. citrophthora, P. parasitica, Pythium* spp., and *Rhizoctonia solani* cause damping-off of citrus seedlings.

Control. If soil is not known to be free of pathogens, the seedbed should be fumigated with methyl bromide, drenched with vapam, or treated with another suitable fumigant (if registered for use) at recommended dosages before planting (see Klotz 1978). If feasible, soil can be steamed so that the temperature reaches 180°F for 30 min (Baker 1957; see Figure 5.7 in Chapter 5).

If seeds are infected, both *P. citrophthora* and *P. parasitica* can be eradicated by immersing them in water at 51.7°C for 10 min (Klotz et al. 1960).

Shoot Blight and Stem Dieback of *Pieris japonica* (Japanese Andromeda)

Dieback of *Pieris japonica,* an ornamental plant, can be caused by a number of factors, e.g., *Botryosphaeria dothidea,* cold injury, severe mite infestation, and *Phytophthora* spp. (Gerlach et al. 1974). In addition to *P. citrophthora, P. citricola, P. parasitica,* and *P. cinnamomi* are also recognized as important causal agents of the dieback complex. Of these, *P. citrophthora* is considered to be the most important.

The first symptoms are olivaceous spots on leaves, which turn necrotic and become surrounded by a dark red margin (Plate 17.4). These lesions rarely extend down into the mature wood. The root and crown rot phase is characterized by chlorosis, followed by leaf epinasty, dieback, stunting, and eventual death of the affected plants (Plate 17.5).

Phytophthora dieback differs from that caused by *B. dothidea* in that *Phytophthora* attacks only healthy, vigorous new growth whereas *B. dothidea* requires wounds for entry, and *Phytophthora* normally kills only the terminal growth while *B. dothidea* kills entire branches.

Disease Development. Young, immature leaves and twigs are most susceptible. Uninjured, mature leaves and stems are resistant (Gerlach et al. 1974). *P. citrophthora* survives as mycelium in infected plant tissue. In culture, *P. citrophthora* can survive freezing to –21°C when incubated at 4°C for 4 days prior to freezing. It survived 44 weeks in infected leaves that were incubated in or on the surface of nonsterilized soil (Gerlach et al. 1976b). Young leaf and the inflorescence tissues are the most susceptible parts of the plant. The optimum temperature for disease development is 24 to 30°C. Excessive fertilization with nitrogen increases the susceptibility of plants in nurseries.

Control. Plants in containers should be started in steamed or fumigated soil. Fungicidal sprays that have been reported to control the dieback disease successfully include captan 50WP (0.9 kg/400 liters). If registered, both metalaxyl and fosetyl-Al should be effective (see Chapter 7).

CHAPTER 18

Phytophthora clandestina

Taylor, Pascoe, and Greenhalgh (1985)

Phytophthora clandestina was first isolated in Victoria, Australia, from rotted taproots of subterranean clover, *Trifolium subterraneum* L., an important pasture legume. It was proved to be pathogenic by P. A. Taylor (1984) and was described by P. A. Taylor et al. (1985). As is common among *Phytophthora* species, *P. clandestina* remained obscure until recently because of inadequate isolation methods (P. A. Taylor et al. 1985). Reliance on testing the pathogenicity of associated fungi only in sterilized soil had led to the problematic conclusion that *Pythium irregulare* Buis. and *Fusarium avenaceum* (Corda ex. Fr.) Sacc., both of which are readily isolated from roots, were the most important pathogens (Greenhalgh and Taylor 1985; Taylor and Greenhalgh 1987b).

P. clandestina was first detected by the observation of sporangia on infected roots. To isolate the unidentified *Phytophthora* species, sterile clover roots were used as baits in a water culture containing infected roots. The sporangia were picked off with a finely drawn-out capillary pipette and transferred to a clean root; or zoospores produced by chilling the aqueous culture (5°C) for 5 min were transferred to sterile seedling roots. Zoospores were produced more rapidly by *P. clandestina* than by associated *Pythium* species. This was also noted by Canady and Schmitthenner for *P. sojae* (1982). *P. clandestina* was separated from bacteria by plating the infected bait root on water agar and inverting the agar. *P. clandestina* grows through the agar, whereas the bacteria cannot (see discussion of this method in Chapter 2).

According to Greenhalgh and Taylor (1985), *P. clandestina* is more important as a root rot pathogen of subterranean clover than *F. avenaceum* and *Pythium irregulare,* both of which had been implicated as the major causes of root rot in previous reports. *P. clandestina* caused typical orange brown to dark brown lesions on taproots when inoculum was incorporated into nonsterile field soil, but inoculum of the other pathogens did not induce this symptom, and root rot symptoms were milder. Rot of the taproot caused by *P. clandestina* was subsequently found in Western Australia (P. A. Taylor et al. 1985b).

Further evidence that *P. clandestina* was the main cause of root rot was obtained by use of the *Phytophthora*-selective fungicide metalaxyl. The excellent degree of control of clover root rot by soil drench with metalaxyl and foliar spray with fosetyl-Al, both of which are toxic mainly to oomycetes, and the lack of control by benomyl, which has no activity against *Phytophthora* but is active against *Fusarium* spp., supports *P. clandestina* as the major pathogen of subterranean clover (P. A. Taylor et al. 1985b). Metalaxyl as a soil drench controlled taproot rot caused by *P. clandestina,* but it is not considered to be economical (Greenhalgh et al. 1988).

Greenhalgh et al. (1988) compared *P. clandestina* and *Aphanomyces euteiches* Drechs. as pathogens of subterranean clover. In a pasture experiment, *A. euteiches* caused more disease when soil was flooded for 24 h each week, but *P. clandestina* caused the most severe degree of disease when soil was flooded for only 4 h.

Characteristics of *Phytophthora clandestina*

P. clandestina is classified in group I (Stamps et al. 1990) because it produces papillate sporangia; however, the production of both amphigynous and paragynous antheridia is an exception for this group. See Tables 4.2 and 4.3 in Chapter 4 for tabular keys. Morphology is described by P. A. Taylor et al. (1985) and is shown in Figures 18.1–18.3. A summary is provided by Hall (1991a).

Sporangia

Papillate sporangia (often two or three papillae) are produced most abundantly on infected roots incubated in water but are also produced on solid lima bean media. The apical thickening is about 4 to 6 µm deep with an exit pore about 6 to 7 µm wide. Sporangia are broadly ellipsoid to ovoid, sometimes subglobose, 16 to 66 µm (±11 µm) long, and 15 to 44 µm (mean 32 ± 11 µm) wide and have a length-breadth ratio of 1.35:1 (range 1.2 to 1.4:1). A conspicuous dome-shaped or

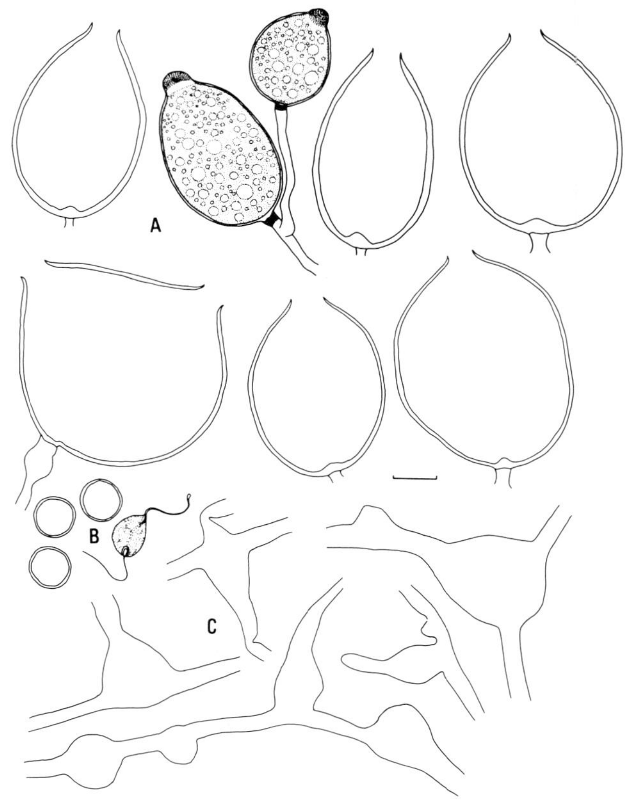

Figure 18.1. Morphology of *Phytophthora clandestina.* **A,** Sporangia. **B,** Zoospore and encysted zoospores. **C,** Hyphal swellings on V8 juice agar. Scale bar = 10 µm. (Reproduced by permission of Mycotaxon Ltd., Ithaca, New York, courtesy Taylor et al. 1985a)

broadly conical basal plug protrudes into the sporangium. Sporangia are caducous and do not readily separate from sporangiophores at first but become deciduous after zoospores have been emitted. The sporangial pedicel is 1 to 6 µm in length (Figure 18.1). Sporangia are formed sympodially on sporangiophores.

Chlamydospores

Chlamydospores have not been seen.

Sex Organs

Oogonia, which are usually terminal, have a diameter of 21 to 36 µm (mean 30 ± 3 µm). Both paragynous and amphigynous antheridia are formed in host tissue, in which they are predominately paragynous (70 to 90%), but in lima bean agar they are more frequently amphigynous (50 to 60%).

Antheridia are terminal or subterminal, occasionally intercalary, and variable in shape. The diameter ranges from 9 to 45 µm (mean 18 ± 5 µm) to 4 to 17 µm (mean 11 ± 3 µm). Antheridia are delimited by thick septa.

Oospores are markedly aplerotic with diameters of 18 to 31 µm (mean 25 ± 2 µm) and a wall thickness of 0.3 to 3.6 µm (mean 1.4 ± 1.0 µm).

Growth-Temperature Relations

P. clandestina grows slowly; e.g., on lima bean agar, growth is about 1.5 to 2.0 mm per day at 25°C. The optimum temperature for growth is 25°C, minimum is <5°C, and maximum is 31°C (P. A. Taylor et al. 1985a). Growth is maximum at a pH of 6.0 at 25°C. Survival in soil is best at 5 to 10°C (Wong et al. 1986).

Distinguishing Characteristics

P. clandestina differs from other members of group I because of its slow growth rate, characteristic production of hyphal swellings, and subterminal or digitate antheridia. A distinct basal plug in sporangia of *P. clandestina* is also unique. Unlike *P. cactorum, P. clandestina* produces a large proportion of amphigynous antheridia, and unlike *P. pseudotsugae* and *P. iranica,* it produces aplerotic oospores (P. A. Taylor et al. 1985).

Diseases Caused by *Phytophthora clandestina*

The most important known host is *Trifolium subterraneum;* however, Clarke and Greenhalgh (1986) reported that *Medicago truncatula* cv. Borung was susceptible; *M. rugosa* cv. Paraponto and *M. scutellata* L. Miller were slightly susceptible. Barbetti (1989) reported that *P. clandestina* was only weakly pathogenic on the above *Medicago* spp.

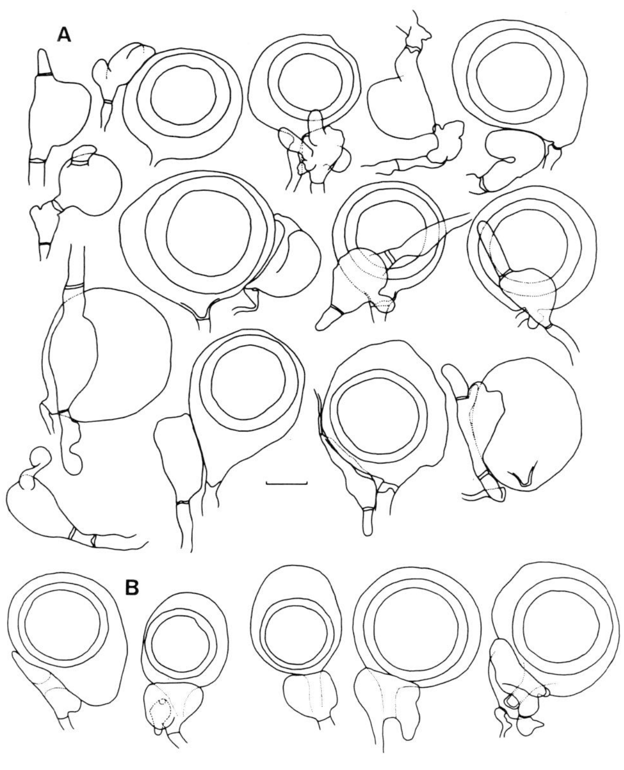

Figure 18.2. Morphology of *Phytophthora clandestina* gametangia in roots of subterranean clover seedlings. **A,** Oogonia and paragynous antheridia. **B,** Oogonia and amphigynous antheridia. Note the digitate protuberances of antheridia and aplerotic oospores. (Reproduced by permission of Mycotaxon Ltd., Ithaca, New York, courtesy Taylor et al. 1985a)

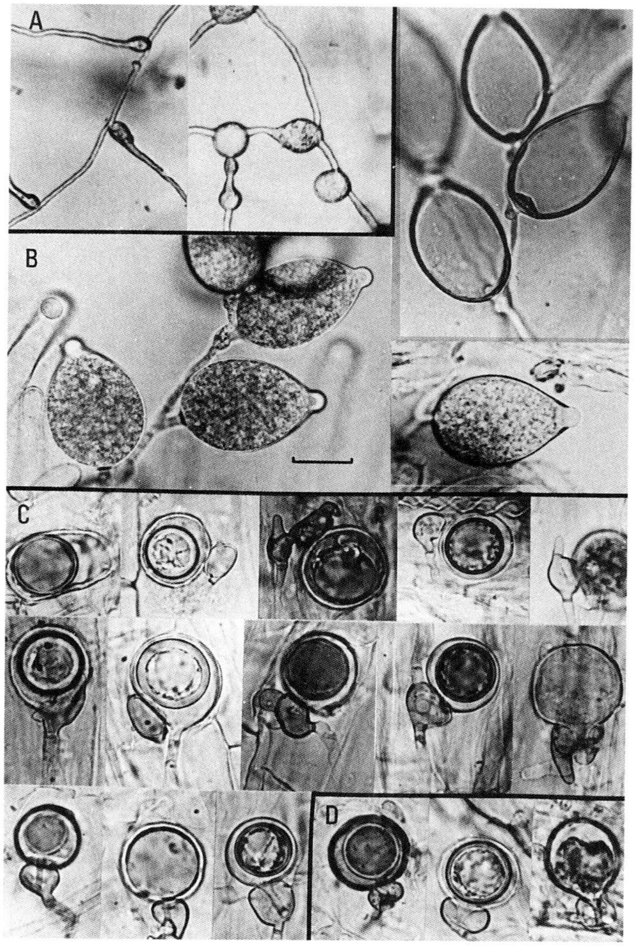

Figure 18.3. Morphology of *Phytophthora clandestina.* **A,** Hyphal swellings. **B,** Papillate sporangia. **C,** Gametangia in roots. **D,** Gametangia in lima bean agar. Scale bar = 20 µm. (Reproduced by permission of Mycotaxon Ltd., Ithaca, New York, courtesy Taylor et al. 1985a)

On subterranean clover, *P. clandestina* causes an orange brown to dark brown discoloration of the cortical and stellar tissue of the taproots. In most cases, the taproot is killed. Other pathogens, such as *F. avenaceum,* cause black discoloration of the taproot, and *Pythium irregulare* causes brown to black discoloration on the tips of lateral roots (Greenhalgh and Taylor 1985).

Disease Development. When *P. clandestina* was tested for pathogenicity at 10, 15, 20, and 25°C and plants were watered twice weekly by 6 or 16 h of drip irrigation, root rot severity was greatest at 15 and 20°C (P. A. Taylor and Greenhalgh 1987a). When seed was planted in artificially infested soil in another laboratory, the most severe disease occurred at 10°C; a significant degree of severity also occurred at 15, 20, and 25°C, but no disease occurred at 30°C. The greatest reduction in seedling survival occurred under flooded or saturated conditions (Wong et al. 1986).

Control. Although management of soil moisture should be one of the control measures because *P. clandestina* depends on high soil moisture, this is difficult under pasture conditions. Studies in Victoria, Australia, show that the cultivars Larisa, Trikkala, and Meteora are resistant; however, Trikkala and Larisa are not agronomically suitable for dryland pasture (P. A. Taylor and Greenhalgh 1987b).

CHAPTER 19

Phytophthora colocasiae

Raciborski (1900)

Synonyms include *Kawakamia colocasiae* (Racib.) Sawada (1911) and *Phytophthora parasitica* var. *colocasiae* (Racib.) Sarejanni (1936).

Phytophthora colocasiae was first described by Raciborski (1900) as the causal agent of a leaf spot of taro, *Colocasia esculenta* (L.) Schott., in Java (Indonesia). The pathogen is now recognized as the major limiting factor in the cultivation of taro in many locations in Southeast Asia and the Pacific regions. A reduction in yield of 30–50% of the crop may occur if conditions are favorable for disease development. Taro is a tropical crop used as a staple food source for people in many developing countries. Since most of the crop is consumed locally and not exported, the disease problem of taro has been more neglected than those of exportable tropical crops such as cocoa (Gollifer et al. 1980; Gregory 1983). *P. colocasiae* is considered to be primarily a foliar pathogen with a limited host range (Table 19.1). There is little information on the origin of *P. colocasiae*, but there are indications of an Asiatic origin. *P. colocasiae* has probably been distributed in vegetatively propagated material and by soil (Zentmyer 1988). Aggarwal et al. (1990) reviewed the taxonomy, physiology, and control.

Characteristics of *Phytophthora colocasiae*

P. colocasiae is classified in group IV (Stamps et al. 1990). See Tables 4.2 and 4.3 for tabular keys and Appendix 4.9 for a dichotomous key (Ho 1992). Morphology is shown in Figure 19.1.

Sporangia

Sporangia are semipapillate; caducous (pedicel length 3.5 to 10 µm); ovoid, ellipsoid, or sometimes fusiform; and 40 to 70 µm long × 17 to 28 µm wide. Sawada (1911) reported sporangia 114 µm in length, average 59 × 23 µm. They have a length-breadth ratio of 1.6:1 and a tapered base with an occasional lateral attachment and are sometimes intercalary. A conspicuous basal plug is present at the point of attachment of the sporangium to the sporangiophore. Sporangiophores in culture are irregularly branched with basal swellings, but on leaf surfaces they are usually short and unbranched.

Chlamydospores

Chlamydospores are abundant in some isolates and rare in others. They are 17 to 38 µm in diameter (average 27 µm), and

Table 19.1. Distribution of hosts of *Phytophthora colocasiae*

Host	Common name	Disease	Geographical distribution
Alocasia spp.	Yam	Leaf blight	Ceylon (Park 1939)
Alocasia macrorrhiza (L.) G. Don.	Pai, taro, ape	Leaf blight	Solomon Islands (Gollifer et al. 1980)
Amorphophallus campanulatus Blume	Elephant-foot yam	Foliage blight	India (Paharia and Mathur 1961)
Araceae Juss.	Arum, aroids	Foliage blight	Italian East Africa (Castellani 1939)
Catharanthus roseus L. Don. (*Vinca roseus* L.)	Periwinkle	Foliar blight	Hawaii (Anon. 1938)
Colocasia antiquorum Schott.	Elephant's ear gabi	Corm rot and leaf blight	India (Butler and Kulkarni 1913); Java (Raciborski 1913); Philippines (Mendiola and Espino 1916); China (Reinking 1919a)
Colocasia esculenta (L.) Schott.	Taro, dasheen, gabi dalo	Leaf spot, wet rot of corms, storage rot of corms	Java (Raciborski 1900); India (Butler and Kulkarni 1913); Taiwan (Sawada 1911); Philippines (Mendiola and Espino 1916); Ceylon (Petch 1918); Guam (Weston 1918); Hawaii (Carpenter 1920; Parris 1941); Burma (Butler and Bisby 1931); China (Teng 1932); Malaysia (Thompson 1939); Fernando Po (Gomez-Moreno 1942); Fiji (Parham 1947); British Solomon Islands (Johnston 1960); also reported from Hong Kong, Nepal, Thailand, Ethiopia, Papua, and New Guinea
Hevea brasiliensis Muell.-Arg.	Rubber	Leaf fall, stem canker, black stripe	China (Ho et al. 1984a)
Piper betle L.	Betel, pan	Wilt	Malaysia (Thompson 1928); India (Thompson 1919); Burma (Su 1933)
Vinca rosea L.	Periwinkle	Blight	Hawaii (Anon. 1938); Argentina (Frezzi 1950)
Xanthosoma mafaffa Schott.		Foliar blight	Fernando Po (Gomez-Moreno 1942)
Xanthosoma sagittifolium (L.) Schott.	Yautia, malanga, tannia, ocumo	Foliar blight	Dominican Republic (Ciferri 1954)
Xanthosoma violaceum Schott. (*Alocasia violacea* Hort.)	Blue taro, blue ape	Foliar blight	Fernando Po (Gomez-Moreno 1942)

the wall is 2 to 3 μm thick. Formation is either terminal or intercalary in the mycelium.

Sex Organs

Antheridia are amphigynous and subterminal; oogonia are 20 to 35 μm in diameter, average 29.0 μm; oospores are 18 to 30 μm in diameter, average 23.0 μm, and aplerotic. *P. colocasiae* is heterothallic, although some isolates produce oogonia in single culture. Ko (1979) reports that only the A1 type is found in Hawaii; but in Taiwan, only the A2 mating type was found in 799 isolates sampled (Ann et al. 1986).

Growth Temperatures

The minimum is >10°C, optimum 27 to 30°C, and maximum >35°C.

Disease Caused by *Phytophthora colocasiae*

Leaf Blight of *Colocasia esculenta* (Taro, Dasheen, or Gabi)

Holliday (1980) briefly summarizes the disease on taro. *P. colocasiae* produces sporangia that are readily shed in water and thus can be disseminated by rain splash but not by wind. Since *P. colocasiae* survives in the rhizomes and produces both oospores and chlamydospores, there is a continual source of inoculum (Gregory 1983). Comprehensive reviews were published by Ghosh et al. (1989) and Thankappan (1985).

The early stages of blight are characterized by water-soaked, purple to brown circular lesions on the leaf that vary from a few millimeters to approximately 2 cm in diameter. Infections commonly begin on the lobes and sides of the leaf where water often collects. The lesions gradually enlarge and become dark brown with characteristic yellow margins (Plate 19.1). A delicate white fuzz indicating the presence of *P. colocasiae* can often be seen around the lesion, particularly in the early morning. Another diagnostic feature is the yellow to red fluid that exudes from the centers of the spots. The exudate becomes dark brown and hard after drying. In advanced stages of the disease, irregularly shaped lesions of varying sizes partially or completely cover the leaf blade. Rotted leaves do not drop from the petioles. Petioles occasionally become infected if favorable temperature and moisture conditions prevail. Rhizomes of *Alocasia macrorrhiza* often rot while in storage (Gollifer et al. 1980).

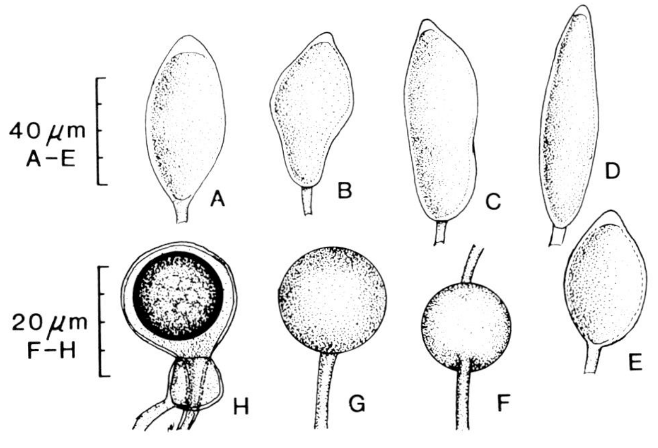

Figure 19.1. Morphology of *Phytophthora colocasiae.* **A–E,** Semipapillate ellipsoid sporangia. **F,** Intercalary chlamydospore. **G,** Terminal chlamydospore. **H,** Globose oogonium with an amphigynous antheridium. (Drawing by A. Vaziri)

Phytophthora blight of *Colocasia* may be confused in the early stages with a disease caused by *Phyllosticta colocasiae* (*P. coliocasiophila* Weedon). This pathogen causes brown spots with yellow borders similar to those caused by *P. colocasiae;* however, the spots do not enlarge or coalesce, and the centers usually fall out, causing a shot-hole type of symptom.

A postharvest rot of the corms can also occur. The diseased tissue is firm and leathery in contrast to the soft rot caused by *Pythium* and *Botryodiploidia* species. *Phytophthora*-infected corms may become completely rotted in 5 to 10 days.

Disease Development. The disease is disseminated by splash dispersal of caducous sporangia (Gregory 1983). The disease is enhanced at temperatures between 25 and 28°C and relative humidity at or above 65% during the day followed by temperatures of 20 to 22°C and 100% relative humidity during the night. An epidemic may occur when night temperatures and relative humidity are at an optimum for 6 to 8 h on 3 to 4 consecutive days with light rain or dew in the morning (Trujillo 1965). Spread of the disease is by water-splashed sporangia and zoospores onto nearby leaf surfaces. The pathogen overwinters in plant debris and rhizomes in the form of resistant oospores (Gollifer et al. 1980).

Control. Burning infected leaves and removing corms from the field after harvest are effective. Since the severity of disease is greatly increased by high relative humidity, close planting and planting in shady areas should be avoided. A copper-based fungicide amended with a sticker/spreader applied as soon as the disease is first observed and at 7- to 14-day intervals thereafter is effective. Copper oxychloride sprays (4.5 kg in 100 liters of water per hectare by motorized knapsack mist blower or 4.5 kg in 1,000 liters of water per hectare by hydraulic-type sprayer) have given good control in the Solomon Islands (Jackson 1977). Bordeaux mixture has also been successful. Although in one study control by copper fungicides was poor, yield was increased (Gollifer et al. 1974). Dithane M-45 sprayed at 5- to 7-day intervals at the rate of 2.24 or 4.48 kg per hectare controls the disease (Bergquist 1974). If registered for use, metalaxyl or fosetyl-Al should be effective but may be too expensive. Metalaxyl (Apron 35F), captafol (Difolatan), and chloroneb (Demosan 65W) used as sprays at 15-day intervals gave good control in India. See Aggarwal and Mehrotra (1987) for more citations.

Postharvest decay of corms can be much reduced if the leaf spot phase is controlled prior to harvest. Harvesting corms with petiole bases (about 30 cm long) attached helps to prevent rot in storage by keeping the corms more physiologically active. Corms from diseased plants should be treated with a 1% solution of sodium hypochlorite for 2 min and dried before storage in polyethylene bags. Varietal screening of cultivars in Pradesh, India, showed that five cultivars were immune and one was moderately resistant (Chaudhary et al. 1988).

Extensive control measures are given by Jackson (1977).

CHAPTER 20

Phytophthora cryptogea

Pethybridge and Lafferty (1919)

No synonyms are known for *Phytophthora cryptogea.*

P. cryptogea was first described as the causal agent of tomato foot rot in Ireland by Pethybridge and Lafferty in 1919. It is now recognized worldwide as an important pathogen of ornamentals, particularly those grown under glass or plastic, and many tree crops as well. The host range for *P. cryptogea* includes a large number of genera (many of which are in Compositae) distributed among more than 23 plant families (Table 20.1). See Commonwealth Mycological Institute (CMI) Map 99, ed. 4, 1976 and Stamps (1978b) for the CMI description. Kröber (1981) described *P. cryptogea* f. sp. *begoniae* because of host specificity to begonia. Some of these isolates were characterized by unique mitochondrial DNA restriction fragment length polymorphism (mtDNA RFLP) and isozyme patterns (Mills et al. 1991).

P. cryptogea has been recovered from irrigation water (Bewley and Buddin 1921; Taylor 1977), thus indicating that the fungus can survive in water in the absence of a suitable host. The saprophytic ability of *P. cryptogea* to colonize dead organic matter was established by Bumbieris (1978, 1979). The fungus has also been recovered directly from soil (McIntosh 1964; Davison and Bumbieris 1973; Flowers et al. 1973; Orlikowski 1978; Matsumoto and Sato 1979; Kröber 1980; O. K. Ribeiro, *unpublished data*). Kröber (1980) reported that *P. cryptogea* persisted in the soil for more than 4 years in the absence of a suitable host.

P. cryptogea is not a pathogen of tobacco but is capable of causing stunting of tobacco plants when soil is artificially infested with mycelium (J. W. Hendrix 1976). Since both living and killed mycelium incorporated into soil caused stunting of tobacco plants and plants were not killed by stem inoculation (Kanalong and Hendrix 1976), it appeared that the stunting effect was the result of a toxic reaction and not of parasitism by *P. cryptogea* (Csinos and Hendrix 1977).

Characteristics of *Phytophthora cryptogea*

P. cryptogea is morphologically very similar to *P. drechsleri.* Differentiation of *P. cryptogea* from *P. drechsleri* is discussed in Chapter 4. The major distinguishing feature that separates *P. cryptogea* from *P. drechsleri* is the growth of *P. drechsleri* isolates at 35°C (Tucker 1931). See Table 20.2 for a summary of classical differences between the two species. Bumbieris (1974) studied the morphology and physiology of both *P. cryptogea* and *P. drechsleri* and concluded that the species were identical; he suggested that *P. drechsleri* should be included within *P. cryptogea,* since this species was described earlier. Similar conclusions were reached by Hallsall (1976), who found isolates of both species to be serologically related; by Shepherd (1978) on the basis of mating behavior; by Matsumoto and Sato (1979), who noted that the two species had similar protein patterns; and by Ho and Jong (1986), who examined the morphology and physiology of a large number of species described as either *P. cryptogea* or *P. drechsleri.* Ho and Jong found that the sporangial shapes (tapered or rounded bases) and sizes of these two species varied considerably with much overlapping of characteristics, making it difficult categorically to separate *P. cryptogea* from *P. drechsleri.* An isolate from bean produced sporangia with both tapered and rounded bases (Flowers et al. 1973). In a later report, Ho and Jong (1991) decided that both species should be retained and differentiated by growth at 35°C for *P. drechsleri* and no growth for *P. cryptogea* isolates, a character used by Tucker (1931).

Protein profiles obtained by polyacrylamide gel electrophoresis (see Chapter 4) showed that a type culture of *P. cryptogea* and an authentic culture of *P. drechsleri* were clearly differentiated. Several other isolates with similar morphology had protein profiles intermediate between those of *P. cryptogea* and *P. drechsleri,* however, thus making it impossible to resolve the identities of these isolates with either of the above species (Cacciola et al. 1990). The protein profile was distinctly different from that of *P. erythroseptica.* An isolate from lentisk yielded a protein profile identical to that of the type culture of *P. cryptogea* and was considered to be *P. cryptogea* (Magnano di San Lio 1992).

Recently, Mills et al. (1991) studied the isozyme and mtDNA RFLP patterns of 123 isolates designated either *P. cryptogea* or *P. drechsleri* that had been isolated from various host plants from the United States and overseas. Their findings indicated that groups within these two species varied greatly, and consequently they concluded that merging *P. cryptogea* with *P. drechsleri* was not warranted. Instead, their data indicate that there are at least seven distinct genetic groups within these two species. Some of these genetic groups differ more in mtDNA patterns than do some accepted species.

A revised description of *P. cryptogea* and *P. drechsleri* recently presented by Ho and Jong (1991) broadened the description of *P. cryptogea* to include isolates that are intermediate between *P. cryptogea* and *P. drechsleri* and retained the criterion of growth of *P. drechsleri* at 35°C to separate it from *P. cryptogea.* In the report by Mills et al. (1991) on isozymic and mtDNA RFLP patterns, all isolates in group A grew at 35°C. However, another group, H, which differed genetically from group A, also grew at 35°C. Isolates from pigeon pea, named *P. cajani* or *P. drechsleri* f. sp. *cajani,* which are included in *P. drechsleri* because of growth at 35°C, also were genetically different from all other groups (Mills et al. 1991). So these two studies agree in some ways but not in others that growth at 35°C is a critical criterion. If the criterion of growth at 35°C is valid, the causal agents of certain diseases listed in Table 20.1 would have to be revised. For example, the causal agent of Phytophthora root rot of safflower grows at 35°C; however, the name of some of these isolates was changed from *P. drechsleri* to *P. cryptogea* by Klisiewicz (1977) on the basis of production of sporangia with rounded bases (Figure 23.1).

P. cryptogea is assigned to taxonomic group VI of the genus (Waterhouse 1963; Newhook et al. 1978; Stamps et al.

Table 20.1. Distribution of hosts of *Phytophthora cryptogea*

Host	Common name	Disease	Geographical distribution
Abies concolor (Gord.) Lindl. ex Hildebr.	White fir	Root rot; chlorosis; necrosis	United States (Hamm and Hansen 1982)
Abies magnifica A. Murr. var. *shastensis* Lemm.	Shasta red fir	Root rot; chlorosis; necrosis	United States (Hamm and Hansen 1982)
Abies procera Rehd.	Noble fir	Root rot; chlorosis; necrosis	United States (Hamm and Hansen 1982)
Achimenes cardinalis Dietr.	Pansy	Seedling wilt	United States (Middleton et al. 1944)[a]
Achimenes grandiflora (Schiede) DC (*A. robusta*)	Pansy	Seedling wilt	United States (Middleton et al. 1944)[a]
Achimenes longiflora DC	Pansy	Seedling wilt	United States (Middleton et al. 1944)[a]
Actinidia chinensis Planch.	Kiwi fruit	Root rot; wilt	New Zealand (Sale 1985); United States (Conn et al. 1988); France (Youssef and Bergamini 1989); Chile (Latorre et al. 1991)
Aeschynanthus radicans Jack.	Lipstick plant	Seedling wilt	United States (Middleton et al. 1944)[a]
Aeschynanthus speciosus Hook.	Basket plant	Seedling wilt	United States (Middleton et al. 1944)[a]
Alloplectus schlimii Planch. and Linden	Gesneriaceae	Seedling wilt	United States (Middleton et al. 1944)[a]
Antirrhinum majus L.	Snapdragon	Stem rot	Australia (Brittlebank and Fish. 1927); Tasmania (Anon. 1953c); United Kingdom (Foister 1961)
Apium graveolens L.	Celery	Seedling blight	United Kingdom (Cotton 1921)
Asparagus officinalis L.	Asparagus	Root rot	United States (Faloon and Grogan 1988)
Aster spp.	Aster	Black neck; foot rot; wilt	United Kingdom (Robinson 1915); Ireland (Pethybridge and Lafferty 1919)
Avena sativa L.	Oat	Root rot	Sweden (Larsson and Gerhardson 1990)[a]
Baccharis pilularis D.C.	Coyote brush; chaparral broom	Dieback	United States (Gill and Zentmyer 1982)
Banksia laricina Gardn.	Banksia	Root rot	Australia (Hardy and Sivasithamparam 1988)
Banksia occidentalis R. Br.	Banksia	Root rot	Australia (Hardy and Sivasithamparam 1988)
Banksia praemorsa Andrews	Banksia	Root rot	Australia (Hardy and Sivasithamparam 1988)
Bavera sessiflora F. Muell.		Root rot	Australia (Hardy and Sivasithamparam 1988)
Begonia elatior	Begonia	Root rot	Germany (Kröber 1981)
Beta vulgaris L.	Sugar beets	Collar and root rot	Japan (Matsumoto and Sato 1979)[a]; United States (Vincelli et al. 1990)
Brassica napus L.	Oil-seed rape	Root rot	Sweden (Larsson and Gerhardson 1990)[a]
Brassica oleracea L.	Cabbage	Root rot	Japan (Matsumoto and Sato 1979)[a]
Brassica oleracea var. *botrytis* L.	Cauliflower	Root rot	United Kingdom (Pethybridge et al. 1934)
Brassica rapa L.	Turnip	Root rot	Ireland (Pethybridge and Lafferty 1919)[a]
Cajanus cajan (L.) Millsp.	Pigeon pea	Root rot	Australia (Stirling and Irwin 1986)
Calceolaria crenatiflora Cav.	Slipperwort	Root rot	United States (Middleton et al. 1944)[a]
Calendula officinalis L.	Marigold	Wilt	United States (Pirone 1937)
Callistephus chinensis (L.) Nees.	China aster	Root and foot rot; blackleg; damping-off	Ireland (Pethybridge and Lafferty 1919); Australia (Brittlebank and Fish. 1927); United Kingdom (Bewley 1927; Beaumont 1950); United States (Thompson et al. 1935); Czechoslovakia (Nicklova-Navratilova 1949); Argentina (Pontis and Feldman 1959); Greece (Chitzanidis and Kouyeas 1970); Japan (Matsumoto and Sato 1979)[a]
Capsicum annuum L.	Green pepper	Collar rot	Bulgaria (Ileva 1979); Japan (Matsumoto and Sato 1979)[a]
Capsicum annuum var. *grossum* L.	Green or bell pepper	Fruit rot	United States (Tompkins and Tucker 1937)[a]
Carthamus tinctorius L.	Safflower	Root and stem rot	United States (Klisiewicz 1974); Australia (Stirling and Irwin 1986)
Castanea sp.	Chestnut	Wilt	Australia (Wicks and Volle 1976)
Ceanothus prostratus Benth. (Schiede) DC (*A. robusta*)	Mahalamat	Root rot	United States (Middleton et al. 1944)
Celosia argentea L. var. *cristata* (L.) Ktze.	Cockscomb	Root rot	United States (Middleton et al. 1944)
Centaurea moschata L.	Knapweed; sweet sultan	Foot rot	United Kingdom (Foister 1961)
Cheiranthus cheiri L.	Wallflower	Foot rot	Ireland (Pethybridge and Lafferty 1919); Australia (Brittlebank and Fish. 1927); United States (Tompkins and Tucker 1937)[a]
Chrysanthemum cinerariifolium (Trevir.) Vis.	Pyrethrum	Blight	Rhodesia (Whiteside 1966)
Chrysanthemum frutescens L.	Paris daisy; marguerite	Black neck disease	United Kingdom (Williams et al. 1937)
Chrysanthemum morifolium Paragon	Chrysanthemum	Root rot	United States (MacDonald 1984)
Cicer arietinum L.	Chick pea; gram; garbanzo	Seed decay; root rot; seedling blight	United States (Bhelwa 1962)[a]
Cichorium intybus L.	Chicory	Root rot	United States (Stanghellini and Kronland 1982)
Cichorium intybus L. var. *foliosum* Hegi	Chicory	Root and stem rot	France (Forlot et al. 1966)
Cineraria sp. (Senecio L.)	Groundsel; dusty miller	Foot rot	United States (Tompkins and Tucker 1937)[a]
Citrullus vulgaris Schrad.	Watermelon	Root and crown rot	United States (Brown and Evans 1933); Egypt (Satour and El-Shinnawy 1973)

(*continued on next page*)

Table 20.1. (*continued*)

Host	Common name	Disease	Geographical distribution
Cotoneaster lacteus Smith	Cotoneaster	Root rot	United States (Middleton et al 1944)
Crowea saligna Andr.		Root rot	Japan (Matsumoto and Sato 1979)[a]
Cucumis melo L. var. *inodorus* Naud.	Honeydew melon	Root and crown rot	Egypt (Satour and El-Shinnawy 1973)
Cucumis sativus L.	Cucumber	Root and crown rot	United Kingdom (Bewley 1930); United States (Tompkins and Tucker 1937)[a]; Egypt (Satour and El-Shinnawy 1973); Greece (Kouyeas 1978); Japan (Matsumoto and Sato 1979); Bulgaria (Ileva 1979)
Cucurbita L.	Pumpkin; squash	Fruit rot	United States (Tompkins and Tucker 1937)
Cucurbita maxima Duchesne	Pumpkin; winter squash	Storage fruit rot	New Zealand (Hawthorne 1988)
Cucurbita moschata (Duchesne) Poir.	Pumpkin; winter squash	Storage fruit rot	New Zealand (Hawthorne 1988)
Cupressus L.	Cypress	Root rot	United Kingdom (Channon 1972)
Cyamopsis tetragonoloba (L.) Taub.	Guar; cluster bean	Root rot	Australia (Stirling and Irwin 1986)
Cyphomandra betacea (Cav.) Sendt.	Tamarillo	Collar and root rot	New Zealand (Fletcher 1975)
Cytisus albus Hacq.		Root rot	Japan (Matsumoto and Sato 1979)[a]
Dahlia sp.	Dahlia	Root rot	United Kingdom (Pethybridge et al. 1934)
Dianthus barbatus L.	Sweet William	Root rot	United Kingdom (Foister 1961)
Dianthus caryophyllus L.	Carnation	Root rot	United States (Middleton et al. 1944)
Episcia cupreata (Hook.) Hanst.	Flame violet	Seedling blight	United States (Middleton et al. 1944)
Erica sp. Dawn		Root rot	Australia (Hardy and Sivasithamparam 1988)
Eucalyptus sp.	Eucalyptus, jarrah	Root rot	Tasmania (Halsall ATCC)?
Eucalyptus ficifolia F. Muell.		Root rot	Japan (Matsumoto and Sato 1979)[a]
Eutaxia oborata Gardn.		Root rot	Japan (Matsumoto and Sato 1979)[a]
Fragaria spp.	Strawberry	Root rot	Australia (Kerr 1957)
Gerbera jamesonii H. Bolus ex Hook.	Transvaal or Barberton daisy	Wilt	United States (Tompkins and Tucker 1937); New Zealand (Robinson 1957); Germany (Pag 1960); Poland (Orlikowski 1976/77); Australia (Kaewruang et al. 1988)
Gerbera sp.	Daisy	Crown rot; wilt	New Zealand (Say 1963); Netherlands (Wegman 1966); Iran (Ershad 1971); U.S.S.R. (Kulibaba 1972); Greece (Kouyeas and Chitzanidis 1978)
Gilia tricolor Benth.	Bird's eyes	Root rot	Ireland (Pethybridge and Lafferty 1919)
Gladiolus sp.	Gladiolus	Root rot	United Kingdom (Alcock 1929)
Gloxinia spp.	Gloxinia	Leaf and stem rot	United States (Middleton and Tucker 1939; Middleton et al. 1944)
Glycine max Merr.	Soybean	Collar rot	Japan (Matsumoto and Sato 1979)[a]
Godetia grandiflora Lindl.	Godetia	Root rot	United States (Middleton et al. 1944)
Grevillea dielsiana Gardn.		Root rot	Australia (Hardy and Sivasithamparam 1988)
Gypsophila paniculata L.	Baby's breath	Wilt	Germany (Kröber 1971)
Hakea francisiana F. Muell.		Root rot	Australia (Hardy and Sivasithamparam 1988)
Hebe Pagei Hort.		Root and stem rot	United Kingdom (Jones and Griffin 1974)
Helianthus annuus L.	Sunflower	Stem rot	United States (Klisiewicz and Beard 1976); Australia (Stirling and Irwin 1986)[a]
Iris sp.	Flag lily; fleur-de-lis	Root rot	United States (Middleton et al. 1944)[a]
Juglans regia L.	English walnut	Root and crown rot	United States (Mircetich and Matheron 1983; Matheron and Mircetich 1985))
Juniperus chinensis	Chinese juniper	Root rot	United States (MacDonald 1984)
Kohleria amabilis Fritsch. (*Isoloma amabile* Hort.)	Gloxinia tree	Seedling blight	United States (Middleton et al. 1944)[a]
Kohleria eriantha (Benth.) Hort. (*Isoloma hirsutum* Hort.)	Gloxinia tree	Seedling blight	United States (Middleton et al. 1944)[a]
Lactuca sativa L.	Lettuce	Root rot	United States (Linde et al. 1990)
Limonium sinuatum (L.) Mill.	Sea lavender; statice	Root rot	Israel (Sneh and Ruben 1978, *personal communication*)
Lycopersicon esculentum L.	Tomato	Foot rot; seedling damping-off	Ireland (Pethybridge and Lafferty 1919); United Kingdom (Bewley 1922); Denmark (Weber 1922); Australia (Brittlebank and Fish. 1927); New Zealand (Brien 1940); Tasmania (Wade 1954); Poland (Orlikowski 1976/77); Japan (Matsumoto and Sato 1979)[a]; Bulgaria (Ilieva 1979)
Lupinus sp.	Lupine	Leaf wilt; stem discoloration	Australia (Brittlebank and Fish. 1927); Greece (Kouyeas and Chitzanidis 1978)
Malus pumila Mill.	Apple	Root rot	United States (Jeffers and Aldwinkle 1986; Bielenin and Jones 1988b)
Mattiola incana R. Br. var. *annua* Voss.	Annual stock; gilliflower	Foot rot; blight	United States (Tompkins and Tucker 1937); United Kingdom (Foister 1961)
Medicago sativa L.[b]	Alfalfa; lucerne	Root rot; seedling blight	Australia (Purss 1959); United States (Johnson and Morgan 1965); Greece (Kouyeas and Chitzanidis 1978); Japan (Matsumoto and Sato 1979)

(*continued on next page*)

Table 20.1. *(continued)*

Host	Common name	Disease	Geographical distribution
Melilotus alba Desr.	Sweet clover	Collar rot	Japan (Matsumoto and Sato 1979)[a]
Nasturtium officinale R. Br.	Watercress	Dieback; root necrosis	United States (Mitchell et al. 1978)
Nemeria spp.		Root rot	United Kingdom (Foister 1961)
Nicotiana sp.	Tobacco	Stunting	United States (Hendrix 1976)
Osteospermum ecklonis (DC) Norl.	African daisy	Root and crown rot	United States (Gill et al. 1976)
Osteospermum hybrids	African daisy	Root and crown rot	United States (Gill et al. 1976)
Papaver nudicaule L.	Iceland poppy	Wilt; collar rot; bud blight	Australia (Brittlebank and Fish. 1927)
Parthenium argentatum	Guayule	Root rot	United States (Beaupré and Cheo 1983)
Petunia hybrida Hort. Vilm. Andr.	Petunia	Seedling blight	Ireland (Pethybridge and Lafferty 1919)
Phaseolus angularis Wight	Adzuki beans	Collar rot	Japan (Matsumoto and Sato 1979)[a]
Phaseolus aureus Roxb.	Mungi beans	Collar rot	Japan (Matsumoto and Sato 1979)[a]
Phaseolus vulgaris L.	String bean	Stem and root rot	United States (Flowers et al. 1973); Japan (Matsumoto and Sato 1979)[a]
Pieris floribunda (Pursh.) Benth. and Hook.	Fetterbush	Root rot	United States (Humphrey and Wood 1935)
Pimilea ferruginea Labill.		Root rot	Australia (Hardy and Sivasithamparam 1988)
Pinus lambertiana Dougl.	Sugar pine	Root rot	United States (Hamm and Hansen 1982a)
Pinus radiata D. Don	Monterey pine	Decline	Australia (Bumbieris 1976)
Pistacea lentiscus L.	Lentisk	Root rot	Italy (Magnano di San Lio et al. 1992)
Pisum sp.	Alaska pea	Seedling rot	United States (Johnson and Morgan 1965)
Pisum sativum L.	Pea	Root rot	Australia (Stirling and Irwin 1986)[a]
Populus simonii Carriere (*P. amygdalus* Batsch.)	Poplar; aspen; cottonwood	Stem and root rot	Argentina (Frezzi 1950)
Prunus armeniaca L.	Apricot	Root rot	Australia (Anon. 1958b)
Prunus avium (L.) L.	Mazzard cherry	Root and crown rot	United States (Wilcox et al. 1985)[a]
Prunus cerasus L.	Sour cherry	Root and crown rot	United States (Wilcox et al. 1985)
Prunus dulcis (Mill.) D. Webb.	Almond	Collar and trunk canker	United States (Middleton et al. 1944); Greece (Helena 1978); Iran (Fatemi 1980); Australia (Wicks and Lee 1986)
Prunus mahaleb L.	Malaheb cherry	Root and crown rot	United States (Wilcox et al. 1985)[a]
Prunus persica (L.) Batsch.	Peach	Root rot	Australia (Anon. 1958b)
Pseudotsuga menziesii (Mirb.) Franco	Douglas fir	Root rot	United States (Pratt et al. 1976); Canada (Hamm et al. 1985)
Rhododendron carolinianum Rehd.	Carolina rhododendron	Root rot	United States (White and Hamilton 1935)
Rhododendron catawbiense Michx.	Catawba purple laurel	Root rot	United States (White and Hamilton 1935)
Rhododrendron maximum L.	Great laurel	Root rot	United States (White and Hamilton 1935)
Rubus idaeus L.	Raspberry	Root rot	Australia (Washington 1988)
Saintpaulia ionantha Wendl.	African violet	Root rot	United States (Middleton et al. 1944)
Schizanthus spp.	Butterfly flower	Wilt	United Kingdom (Foister 1961)
Senecio cruentus (Mass.) DC	Cineraria	Foot rot; wilt	United States (Middleton et al. 1944); France (Moreau and Moreau 1957)
Simmondsia chinensis Nutt.	Jojoba	Seedling blight	United States (Gill and Zentmyer 1981)
Sinningia cardinalis (Lehm.) Moore	Cardinal or helmet flower	Root rot	United States (Middleton et al. 1944)
Sinningia speciosa Benth. and Hook.	Gloxinia	Root and crown rot	United States (Middleton and Tucker 1939); Austria (Vukovits 1956); Denmark (Jorgenson et al. 1964)
Smithiantha cinnabarina (Linden) Kuntze	Temple-bells	Seedling blight	United States (Middleton et al. 1944)[a]
Smithiantha multiflora (Martn. and Galeotti) Fritsch.	Naegelia	Seedling blight	United States (Middleton et al. 1944)
Smithiantha zebrina (Paxt.) (Linden) Kuntze	Temple-bells	Seedling blight	United States (Middleton et al. 1944)[a]
Solanum melongena L.	Eggplant	Collar and trunk rot	Greece (Chitzanidis and Kouyeas 1970); Iran (Ershad 1971)
Solanum tuberosum L.	Potato	Pink root rot	Australia (Anon. 1957); United States (Rowe and Schmitthenner 1977)
Spinacia oleracea L.	Spinach	Root rot; seedling damping-off	Germany (Kröber and Beckmann 1973); Sweden (Larsson and Gerhardson 1990)
Streptocarpus kewensis N.E. Br.	Cape primrose	Wilt	United States (Middleton et al. 1944)
Tagetes erecta L.	African marigold	Collar rot; wilt; seedling damping-off; seed decay	United States (Pirone 1937)
Tulipa sp.	Tulip	Root and bulb rot; blossom blight; shanking	United Kingdom (Alcock 1929)
Triticum aestivum L.	Wheat	Root and crown rot	Sweden (Larsson and Gerhardson 1990)[a]
Vicia dasycarpa Ten.	Woolly-pod vetch	Seedling rot	United States (Johnson and Morgan 1965)
Vicia sativa L.	Spring vetch	Seedling rot	United States (Johnson and Morgan 1965)
Vigna sinensis Savi, et Hassk.	Cowpea	Collar rot	Japan (Matsumoto and Sato 1979)[a]
Viola spp.	Violets	Root rot	United Kingdom (Foister 1961)
Vitis vinifera L.	Grape	Root rot	South Africa (Marais 1979)
Zantesdeschia aethiopica Spreng.	Calla lily	Root rot	United Kingdom (Pethybridge et al. 1934)
Zinnea elegans Jacq.	Zinnia	Foot rot	United Kingdom (Anon. 1934); Canada (Conners 1955); Japan (Matsumoto and Sato 1979)[a]

[a] Artificially inoculated host.
[b] Pathogen later was reclassified as *P. megasperma* (Erwin 1965) and as *P. medicaginis* (Hansen and Maxwell 1991).

1990). The morphology of the asexual and sexual structures of this species is illustrated in Figures 20.1 and 20.2. See Chapter 4, Tables 4.2 and 4.3 for tabular keys (data of Ho and Jong [1991]), Appendix 4.9 for a dichotomous key (Ho 1992), Table 20.2 for a summary of characteristics recorded for *P. cryptogea* and *P. drechsleri,* and Table 20.3 for dimensions of spores recorded in different references.

The amended description by Ho and Jong (1991) is as follows (reproduced by permission of Mycotaxon Ltd., Ithaca, New York, courtesy of Ho and Jong 1991):

Mycelium with no distinct or slightly floral growth pattern on clarified V-8 juice agar medium, slightly fluffy to fairly fluffy; aerial mycelium often profuse on unfiltered yellow corn meal agar medium; main hyphae uniform to uneven 5–7 μm wide, commonly forming in water; conspicuous network of small hyphal swellings (under 25 μm), spherical to irregular; sporangia rarely produced on agar but

Table 20.2. Morphological features and temperature relations of *Phytophthora cryptogea* and *P. drechsleri*

Characteristic	*P. cryptogea*	*P. drechsleri*
Temperature (minimum, optimum, maximum)	<1, 22–25, <35°C	>5, 28–31, 37.5°C
Mycelium	Coralloid	Regular
Hyphae	Uneven, broad	Uniform, narrow
Hyphal width	~8 μm	>8 μm
Sporangia in agar media	None	±
Sporangia in liquid media	Abundant	Abundant
Sporangial shapes	Oval, pyriform, variable	Oval, pyriform, variable
Sporangial base	Predominately rounded (Klisiewicz 1977)	Predominately tapered (Klisiewicz 1977)
Sporangial apex	Nonpapillate	Nonpapillate
Sporangial caducity	Persistent	Persistent
Sporangia dimensions	20–63 × 15–38 μm (av. 31 × 33 μm)	36–70 × 26–40 μm (av. 39 × 27 μm)
Length-breadth ratios	1.1–2.6 (av. 1.48)	1.34–1.91 (av. 1.58)
Sporangiophores	Proliferating	Proliferating
Chlamydospores	None	±
Oogonia	30–32 μm: tapered base 31 ± 3 (Ho and Jong 1986)	36 μm: tapered base 33 ± 3 (Ho and Jong 1986)
Oospores	22.8 μm: plerotic	25.6 μm: plerotic
Antheridia	12–19 μm (av. 14 ± 1 μm)	10–14 μm (av. 15 ± 1)
Mating types	A1, A2	A1, A2
Homothallic	Some isolates only	Some isolates only
Heterothallic	Yes	Yes
Ecotypes[a]	None	Yes
Chromosomes	9–12	9

[a] Southern type resembles *P. cryptogea* in temperature (maximum <36°C; optimum 25°C). A1 type. Northern type resembles *P. drechsleri.* Homothallic and sterile isolates are reported (Shepherd and Pratt 1973).

Table 20.3. Growth temperatures and dimensions of spores produced by *Phytophthora cryptogea* from various hosts

Host	Sporangia (μm)	L-b[a] ratio	Oogonia (μm)	Oospores (μm)	Temperature (°C) Min.	Opt.	Max.	Reference
Tomato	24–50 × 17–30 (av. 40 × 27)	1.48:1	NR[b]	av. 25	NR	NR	NR	Pethybridge and Lafferty (1919)
Tomato	26–49 × 16–29 (av. 36.7 × 21.9)	1.67:1	av. 25.8	av. 22.8	NR	25	30–32	Tucker (1933)
Daisy	23–45 × 11.6–21 (av. 32.2 × 17.2)	NR	NR	NR	4.5	25	<35	Tompkins and Tucker (1937)
Gloxinia	av. 35.0 × 20.0	1.7:1	av. 31	av. 25	8	25	30–33	Middleton et al. (1944)
Populus simoni	29–52 × 16–38 (av. 44 × 28)	1.57:1	22–30	av. 24	NR	NR	NR	Frezzi (1957)
Tomato	Max. 55 × 30 (av. 30–32 × 23)	1.7:1	Max. 38 (av. 30–32)	NR	1	22–25	31–33	Waterhouse (1963)
Gypsophila	32–50 × 21–37 (av. 39.4 × 28.4)	1.39:1	21–32 (av. 27.7)	20–32 (av. 26.4)	5	25	30–34	Kröber (1971)
Gerbera	20–63 × 15–34 (av. 39.1 × 24.0)	1.62:1	26–33 (av. 29.3)	19–33 (av. 26.3)	2.5	25–27	32.5	Ershad (1971)
Spinach	32–54 × 22–35 (av. 43.6 × 28.7)	1.5:1	NR	NR	2.0	25	33	Kröber and Beckman (1973)
African daisy	16–43 × 14–30 (av. 31 × 23)	NR	NR	NR	5.0	21–28	33	Gill et al. (1976)
Various	35–63 × 24–35 (av. 52 × 30)	1.7:1	28–37 (av. 31)	24–32 (av. 27)	0–5	20–25	30–35	Ho and Jong (1991)

[a] Length-breadth.
[b] Not recorded by investigator.

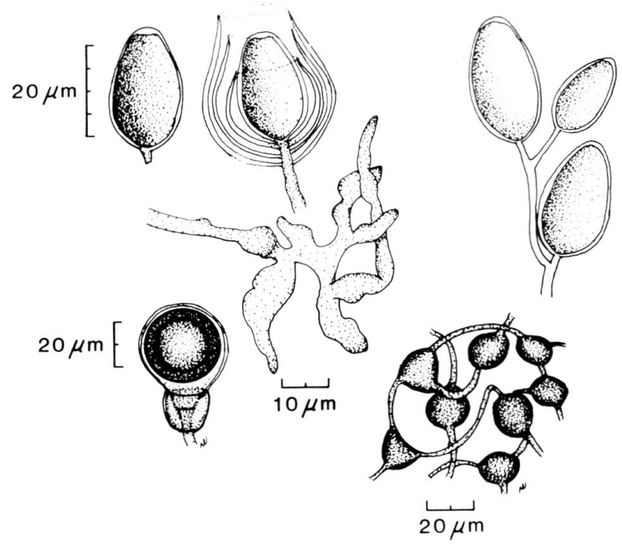

Figure 20.1. Morphology of *Phytophthora cryptogea.* Upper row, ellipsoid sporangium; internal proliferation of sporangia; and sympodial production of sporangia. Center, inflated hyphae. Lower row, globose oogonium with an amphigynous antheridium; and hyphal swellings. (Drawing by A. Vaziri)

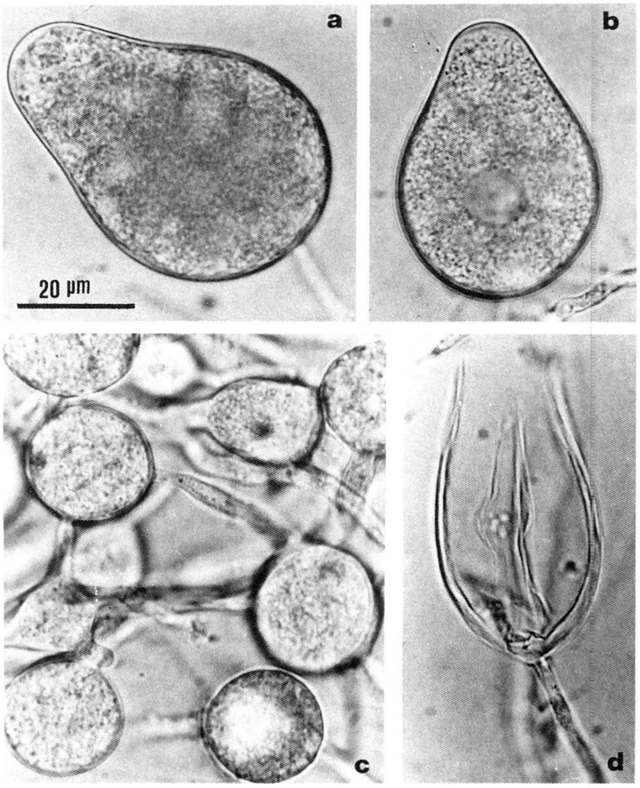

Figure 20.2. Morphology of *Phytophthora cryptogea* from lentisk (*Pistacea lentiscus* L.). **a and b,** Obpyriform, nonpapillate sporangia; **c,** hyphal swellings; and **d,** germination of a sporangium by internal proliferation. (Reproduced by permission of Edagricole, Bologna, Italy, courtesy Magnano di San Lio et al. 1992)

abundantly in water, nonpapillate, nondeciduous, internally proliferating, terminal on unbranched or often close sympodially branched sporangiophores (1–3 µm wide); first-formed sporangia regularly obpyriform to ovoid, mostly with rounded base but later-formed sporangia tend to be more elongated, asymmetrical or variable in shape; sporangia av. 52 × 30 µm (35–63 × 24–35 µm), L/B ratio 1.7 (1.4–2.3), apex flattening on mounting; sporangia collapse after zoospore release which often requires chilling; exit pore over 8 µm wide, pore/breadth ratio over 0.3; heterothallic though sex organs sometimes formed sparsely in aged culture; oogonia spherical to subspherical, smooth, av. 31 µm (28–37 µm) diam., wall becoming yellowish to brown; oospores plerotic, spherical, av. 27 µm (24–32 µm) diam., wall 2–4 µm thick; antheridia amphigynous, short cylindrical, 14 µm (12–17 µm) × 14 µm (13–17 µm), unicellular; no chlamydospores formed in culture.

Cardinal temperatures for growth: 0–5°C, 20–25°C, 30–35°C; growth rate on clarified V8 agar medium 5-10 mm per day at 20°C.

Until more research is conducted to identify the numerous genetically different biotypes, the description as amended by Ho and Jong (1991) should be used to characterize *P. cryptogea.* See Chapter 23, *P. drechsleri,* for a more complete discussion of the *P. cryptogea-P. drechsleri* complex. It would appear doubtful that isolates that grow at 35°C but otherwise have morphological features similar to those of *P. drechsleri* or *P. cryptogea* should be classified as *P. cryptogea,* e.g., *P. cryptogea* on safflower (*Carthamus tinctorius*) (Table 20.1).

Sporangia

Sporangia are oval to obpyriform with rounded bases, nonpapillate, and persistent on the stalk. Sporangia that develop later tend to be more elongated or variable in shape (Ho and Jong 1991). Dimensions of sporangia range from 20 to 93 µm in length and 15 to 51 µm in breadth, with average measurements of 37 to 40 × 23 µm (Stamps 1978b). Measurements of sporangia from various host plants are given in Table 20.2. As stated in Chapter 3, nutritional status has a great ef-

Table 20.4. Variability in sizes of sporangia produced by isolates of *Phytophthora cryptogea* under three nutrient regimes[a]

Isolate or source	Potato-dextrose agar (µm)	Pea broth (µm)	Nonsterile soil extract (µm)
Schwinn (1959)	24–53 × 18–38	42–74 × 32–51	. . .
Pethybridge and Lafferty (1919)	19–45 × 14–34	38–67 × 27–40	. . .
Callistephus chinensis	24–56 × 18–38	54–93 × 35–56	. . .
Solanum melongena	18–67 × 11–31	26–83 × 19–51	. . .
IMI 21278	. . .	. . .	32–60 × 25–30
Native Australian shrubs	. . .	. . .	35–55 × 28–35
Apricot	. . .	. . .	35–55 × 24–40
Apple	. . .	. . .	35–55 × 28–44
Pine	. . .	. . .	35–65 × 25–38
Cupressus arizonica	. . .	. . .	35–77 × 28–44
Peach	. . .	. . .	30–50 × 20–30

[a] From Ribeiro (1978). Information from Bumbieris (1974) and Chitzanidis and Kouyeas (1970).

fect on the size of sporangia. In Table 20.4, the ranges of sizes given in various reports show that sporangia grown initially on potato-dextrose agar are smaller than those grown on pea broth. Thus, comparison of sporangial sizes is of doubtful value for taxonomic work unless the cultural conditions (e.g., nutrients in the substrate), the type of washing solution used on mycelium (e.g., water, mineral salts, or soil extract), and incubation under light are the same. See Chapter 3 for a discussion of sporangial formation.

The length-breadth ratio is 1.45:1 to 1.57:1. Ho and Jong (1991) report ratios of 1.4 to 2.3 with an average of 1.7. Sporangia rarely form in solid culture media but form in abundance in a suitable aqueous substrate. See Chapter 3 for methods of inducing sporangia. Sporangiophores are 2 to 3.5 µm in width and proliferate through the empty sporangium (Figure 20.2d) or develop sympodially from below the sporangium (Figure 20.1).

Hyphal Swellings

Hyphal swellings are rarely observed in solid culture media but are abundant in aqueous culture. Hyphal swellings form in chains or clusters and average 11 µm in diameter with a maximum of 20 µm (Figures 20.1 and 20.2c).

Chlamydospores

Chlamydospores are not usually known to occur; however, chlamydospores formed on pine roots in soil (Bumbieris 1971).

Sex Organs

P. cryptogea is heterothallic and requires pairing with an opposite mating type to form oogonia and oospores. Antheridia are amphigynous and spherical and average 10 µm in diameter. Antheridia may sometimes be oval or cylindrical, depending on the species used for the mating. Oogonia rarely form in single culture. Oogonia have tapered bases and measure 28 to 40 µm in diameter with an average of 30 µm. They are smooth walled and become yellow/brown with age. Oospores are plerotic or nearly so, measure 20 to 32 µm in diameter with an average of 26 µm, and are thick-walled (3.5 µm) (Figure 20.1).

Growth Temperatures

The minimum temperature for growth is <1°C; optimum temperature for growth is 22 to 25°C; and the maximum temperature for growth is 31 to 33°C. Temperature relations of isolates from various hosts (Table 20.3) show that there are variations from these criteria, particularly at minimal temperatures.

Cultural Characteristics

Cultures are uniform, with some isolates displaying a slight to distinct floral pattern. Mycelium in V8 juice agar medium is fairly fluffy.

Distinguishing Characteristics

The hyphae are very uneven in width. In liquid culture, characteristic angular hyphal swellings form (Figures 20.1 and 20.2). No chlamydospores are observed in culture media. Although they are not common, chlamydospores have been reported for some *P. drechsleri* isolates (Cother and Griffin 1973b, 1974).

Kröber (1981a, b) described the forma specialis *begoniae* of *P. cryptogea* on the basis of host specificity to begonia. The sporangia of *P. cryptogea* f. sp. *begoniae* averaged 46.7 × 30.5 µm; oogonia averaged 30.8 µm; oospores averaged 28.3 µm in diameter; and hyphal swellings averaged 20.8 µm. This forma specialis was A2 mating type. Isolates from begonia formed a group D when assayed for isozyme patterns (Mills et al. 1991).

Some isolates differ in pathogenicity to different hosts. For example, the cultivar Transvaal daisy isolates are pathogenic to *Mattiola* sp. and *Callistephus chinensis* only, whereas isolates from *Callistephus* are pathogenic to both Transvaal daisy and *Mattiola* sp. as well as cucumber, *Cheiranthus,* eggplant, tomato, *Capsicum* sp., pumpkin, and watermelon (Tompkins and Tucker 1937). Isolates from *Gloxinia* sp. (Middleton et al. 1944) and chicory (Forlot et al. 1966) are not host specific.

Some Diseases Caused by *Phytophthora cryptogea*

Root and Collar Rot of *Cyphomandra betacea* (Cav.) Sendt (Tamarillo)

A collar rot develops at or just below ground level. Root infection results in a complete rotting of bark tissue, leaving a black-stained stele (Figures 20.3 and 20.4). The aboveground symptoms consist of a reduction of foliage, defoliation, and eventual dieback of branches (Fletcher 1975; Robertson, *unpublished*).

Disease Development. Root and collar rot poses a serious threat on poorly drained, heavy soils.

Control. Planting in poorly drained areas should be avoided. Fosetyl-Al or metalaxyl should be effective, if registered for use.

Figure 20.3. One-year-old tamarillo tree (*Cyphomandra betacea* Car. Sendt.) showing symptoms of collar rot caused by *Phytophthora cryptogea.* The bark tissue of the lower portion of the stem is extensively rotted. (Photograph by J. W. Endt, courtesy G. I. Robertson, D.S.I.R., Auckland, New Zealand)

Foot and Root Rot of *Lycopersicon esculentum* L. (Tomato)

P. cryptogea was first described to cause root rot of tomato (Pethybridge and Lafferty 1919) and occurs in several countries (Table 20.1), but it has not been reported in the United States. When plants are approximately 15 cm high, a brownish black discoloration of the stem occurs at or near the soil line. Secondary adventitious roots often emerge above the lesions on the stem. The lesions gradually advance to girdle the stem, and the plant topples over. The root system of affected plants becomes entirely rotted. The foliage does not normally show any visible symptoms, and even when the stem is girdled, the leaves may remain green and turgid for a short time. Eventually, the margins of the leaflets roll upward and inward and the leaves turn yellow.

An isolate from guar, *Cyamopsis tetragonoloba* (L.) Taub., was not pathogenic to tomato (Stirling and Irwin 1986) but was pathogenic to safflower, pigeon pea, sunflower, and pea.

Production of tomatoes in nutrient solution culture in glasshouses has become important in Great Britain. *P. cryptogea* can be a limiting factor in rockwool nutrient culture (Pegg and Jordan 1990). *P. nicotianae* and *Pythium* species are also involved in root rot (Evans 1979). Pegg and Jordan (1990) discuss a method of analysis of root infection, and Kennedy and Pegg (1990) report that the low temperature (15°C) of the rockwool substrate favors more severe infection than 25°C (optimum temperature for growth of *P. cryptogea*). Plants grown at 15°C were killed in 21 days; however, at 25°C, plants remained symptomless for 3 months, partly because of the rapid regeneration and growth of tomato roots. Kennedy and Pegg (1990) concluded that aerial symptoms are a function of the net amount of available healthy roots.

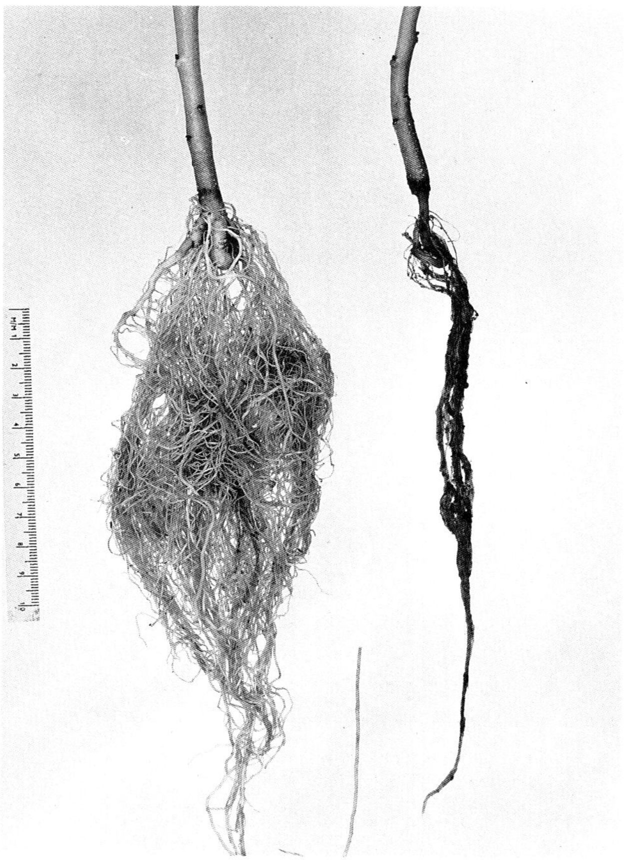

Figure 20.4. Collar rot of tamarillo. Right, seedling grown in infested soil and left, seedling grown in sterilized soil. (Photograph by J. W. Endt, courtesy G. I. Robertson, D.S.I.R., Auckland, New Zealand)

Other *Phytophthora* species infecting tomato plants include *P. arecae* (Chapter 8); *P. cactorum* (Chapter 11); *P. citrophthora* (Chapter 17); *P. drechsleri* (Chapter 23); *P. infestans* (Chapter 33); *P. nicotianae* (*P. parasitica*) (Chapter 50); *P. palmivora* (Chapter 51); and *P. richardiae* (Chapter 57).

Disease Development. The disease is most severe in wet or waterlogged soils.

Control. Fumigation of the seedbeds with methyl bromide or other suitable soil fumigant before planting is effective in reducing the incidence of the disease. Diseased plants should be removed and destroyed before replanting, and overwatering or planting in poorly drained areas should be avoided, since these practices greatly increase the incidence and severity of the disease. Whenever possible, pathogen-free nursery stock should be planted. Desirable but susceptible tomato cultivars can be grafted onto resistant rootstocks such as R.S.4, R.S.9, R.S.10, and R.S.11 if the agriculture is intensive enough to justify the cost (Glasshouse Crops Research Institute, Littlehampton, England). Tomato plants become resistant approximately 2 months after planting in infested soil (Upstone and Finney 1966).

Foot, Root, and Crown Rot of Ornamentals

P. cryptogea is an important pathogen of many ornamentals because of the speed with which it can destroy a crop under ideal conditions of high humidity and temperatures that are favorable for the pathogen. Root and crown rot is one of the limiting factors in the production of *Gerbera jamesoni* Bolus (Orlikowski 1978; Orlikowski et al. 1984). In general, such ornamentals infected with *P. cryptogea* appear chlorotic and are often stunted. Affected roots and stem tissue are soft and water soaked, dark brown, and odorless. Under a favorable environment consisting of warm conditions and wet soils, lesions extend up the stems. Cankers may also girdle the crown, resulting in a rapid, permanent wilting of the leaves. Plants then collapse and topple over. In some cases, adventitious roots may develop above the discolored portions of the crown. Plants are susceptible at all stages of development.

In *Gerbera,* the first observable symptom is a loss of turgidity in the leaves similar to that of plants under water stress (Plate 20.1). The leaves then darken, and the flower stems develop water-soaked lesions at the bases (Plate 20.2). Roots blacken and decay. When these plants are uprooted, most of the roots remain in the soil (Plate 20.3 and Figure 20.5). These symptoms are often misdiagnosed as being caused by *Rhizoctonia solani.* It is thus important to isolate the pathogen from diseased plants to obtain a positive identification (Bonnet et al. 1980).

In greenhouse culture, infection spreads rapidly from one or two infected plants, resulting in heavy losses. The pathogen is most active at temperatures between 10 and 25°C. Maintaining the planting substrate above 25°C is reported to reduce the mortality of gerbera plants (Stilton 1970). Sandofan M (8% oxadixyl + 56% mancozeb), used as a drench immediately after planting or 3 to 12 days after planting, protected gerbera plants for 16 weeks (Orlikowski and Wojdyla 1988). Biological control with *Trichoderma viride* Pers. ex S. F. Gray mixed in soil reduced the population of *P. cryptogea,* but disease on gerbera plants was not controlled (Orlikowski 1981). Soil solarization reduced the root disease index of gerbera from 52.0 to 28.6% (Kaewruang et al. 1989).

No resistant gerbera cultivars have been reported to date.

Figure 20.5. Root rot of African daisy (*Gerbera jamesonii* Bolus ex Hook.) caused by *Phytophthora cryptogea.* (Courtesy H. S. Gill, Riverside County Agricultural Commissioner's Office, Riverside, California)

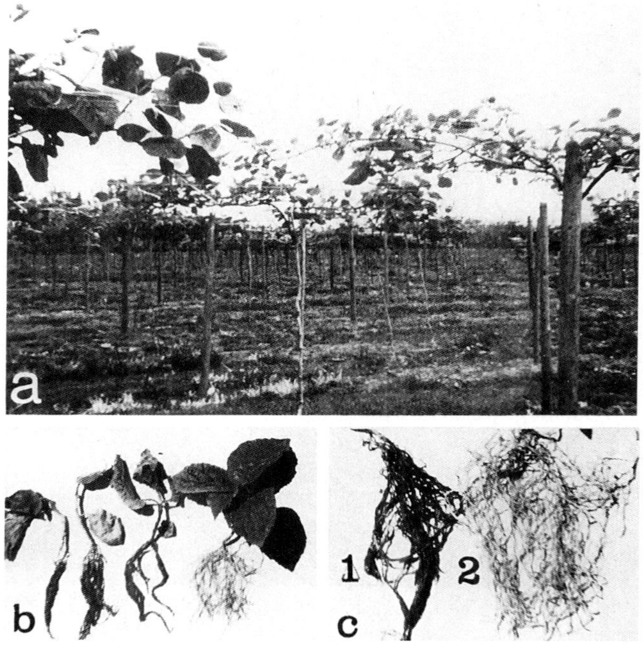

Figure 20.6. Kiwi fruit (*Actinidia chinensis*). **a,** Right, 3- to 4-year-old vines with symptoms typical of those caused by Phytophthora root rot and left, a symptomless vine. **b,** Small seedlings (from left to right) inoculated with *Phytophthora cryptogea* isolates K-13-4 and K-15-2 and *P. citrophthora* isolate K-14-1 and noninoculated control. **c,** Left, roots from plant inoculated with *P. cryptogea* isolate K-13-4 and maintained under continuous flooding for 20 days and right, noninoculated control flooded for the same time. (Reproduced by permission of the American Phytopathological Society, courtesy Latorre et al. 1991)

In China, the stems and roots of aster, *Callistephus chinensis* (L.) Hees, appear water soaked, and a black soft rot develops (see Table 20.1 for references).

Root Rot of Woody Plants

In woody plant species such as *Rhododendron* spp., a brown discoloration of the bark and wood occurs and infected branches wilt and die. Symptoms in *Pinus* spp. frequently resemble those described for little leaf disease of short-leaf and loblolly pines by Campbell and Copeland (1954) and Hamm and Hansen (1982). The affected trees show a general decline and die prematurely. Death of trees soon after the first appearance of symptoms has been reported in some cases. The management guide on control of root rot of forest trees in nurseries by Cooley et al. (1985) is useful.

Washington (1988), in apparently the first report of the disease, proved that *P. cryptogea* caused root rot of red raspberry in Australia. In other areas, *P. megasperma* and *P. erythroseptica* have been implicated—possibly erroneously, however, because Duncan et al. (1991) showed that the root rot pathogen in Scotland, Germany, and the United States should be identified as *P. fragariae* var. *rubi* (Chapter 26C), which is not pathogenic to strawberry.

Raspberry plants infected with *P. cryptogea* show wilting, dieback, stem lesions, and root rot. The cultivars Glen Clova and Canby are very susceptible, whereas the cultivars Chilcotin, Nootka, Haida, and Puyallup are resistant (Washington 1988). Root rot was controlled by soil treatment with metalaxyl, phosphorous acid, and fosetyl-Al.

In kiwi fruit (*Actinidia chinensis*), symptoms consist of leaf chlorosis, lack of shoot growth, leaf scorch, root rot, and eventual plant death (Figure 20.6) (Latorre et al. 1991). Occasionally, cankers may develop at the crown region of the plant. Other species of *Phytophthora* associated with root rot of kiwi fruit include *P. cactorum, P. cinnamomi, P. citricola,* and *P. lateralis* (Robertson 1982).

CHAPTER 21

Phytophthora cyperi

(Ideta) S. Ito (1935)

Peronospora cyperi Ideta (1903), *Phytophthora cyperi-rotundati* Sawada (1927), *P. cyperi-iriae* Sawada (1931), and *Kawakamia cyperi* (Miyabe and Ideta) Miyabe (1942) (see Waterhouse 1970a for citations) are synonyms of *Phytophthora cyperi.* It was first described by Ideta (1935) (see Waterhouse 1970a) as a pathogen of *Cyperus malaccensis* Lam. (shichito mat grass) in Japan. The disease was described as Akayamibyo (red disease) or Bekko (tortoise-shell). *P. cyperi* is now recognized as a specialized foliar and stem pathogen of the following sedges in Taiwan: *Cyperus iria* L., *C. monophyllus* Vahl., and *C. rotundus* L.

In India near Coimbatore, both *P. cyperi* (cited as *P. cyperi-rotundati*) and *P. cyperi-bulbosi* cause a leaf blight of the sedges *C. rotundus* and *C. bulbosus* Vahl. Both sedges are weeds, but the disease does not extend to the underground tubers. Although some plants are killed, the disease is not severe enough to limit sedge as a weed (Seethalakshmi 1953). Waterhouse (1963) listed *P. cyperi-rotundati* (Sawada 1927) but stated that it was not seen and probably was the same as *P. cyperi.*

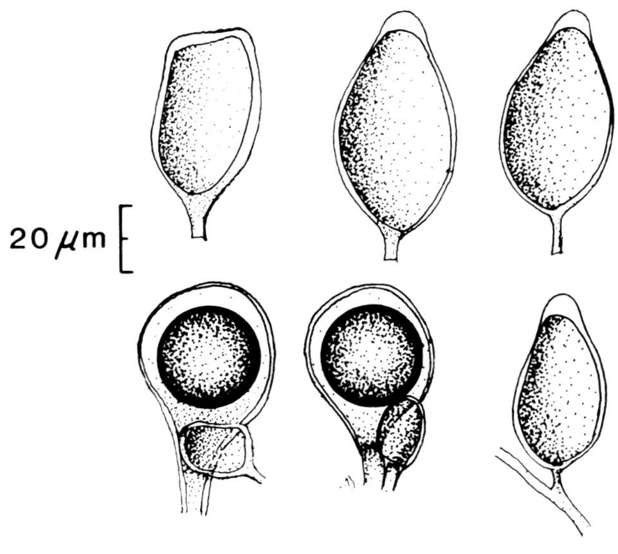

Figure 21.1. Morphology of *Phytophthora cyperi.* Sporangia are semipapillate. Antheridium can be either paragynous or amphigynous on globose oogonia. (Drawing by A. Vaziri)

Characteristics of *Phytophthora cyperi*

P. cyperi is classified in group III (Stamps et al. 1990). See Figure 21.1 for morphology, Tables 4.2 and 4.3 for tabular keys, and Appendix 4.9 for a dichotomous key (Ho 1992). Since the fungus has not been cultured, data are from observations on the plant. See Ho et al. (1995).

Sporangia

Sporangia are semipapillate (flattened apex); caducous (pedicel 4 μm in length); ellipsoid, ovoid, or obpyriform; and 32 to 50 μm long by 19 to 30 μm wide (average 39 × 23 μm). Similar dimensions are reported for *P. cyperi-rotundati* (considered to be the same as *P. cyperi* [Waterhouse 1963]). Sporangiophores are sympodial with intercalary swellings.

Chlamydospores

Chlamydospores are not produced.

Sex Organs

Antheridia are paragynous but occasionally are amphigynous (Stamps et al. 1990); oogonia are 30 to 45 μm in diameter, average 37 μm; oospores are plerotic and 25 to 38 μm in diameter, average 30 μm. The fungus is homothallic and forms oospores readily in host tissue (Ho et al. 1995).

Growth Temperatures

Temperatures are not reported because the fungus cannot be cultured.

Disease Caused by *Phytophthora cyperi*

Stem and Leaf Spot of *Cyperus*

The pathogen attacks stems and leaves of the plant and forms small, yellow, water-soaked lesions that gradually enlarge during the rainy season in India and other tropical regions (Seethalakshmi 1953). These lesions eventually turn reddish brown or amber. In advanced stages of the disease, the lesions girdle the stem.

Disease Development. The disease is more severe in alkaline soils. On wet days, whitish spore masses surrounding the lesions are visible.

Control. Recommendations for control of the disease have not been reported.

CHAPTER 22

Phytophthora cyperi-bulbosi

Seethalakshmi and Ramakrishnan (1953)

Phytophthora cyperi-bulbosi was first isolated from the leaves and leaf sheaths of *Cyperus bulbosus* Vahl in India near Coimbatore by Seethalakshmi and Ramakrishnan (1953). It has not been reported on any other host or from any other country and has not been cultured.

Characteristics of *Phytophthora cyperi-bulbosi*

P. cyperi-bulbosi is classified in group III (Stamps et al. 1990). See Tables 4.2 and 4.3 for tabular keys. The fungus has not been grown in axenic culture. Morphological descriptions (Figure 22.1) are from material taken directly from host tissue, on which sporangia and oospores are readily produced.

Sporangia

Sporangia are semipapillate; caducous with short pedicels approximately 4 to 5 µm in length; ellipsoid, ovoid, or obpyriform; and 23 to 50 µm long × 20 to 34 µm wide, average 40 × 25 µm. Sporangiophores are sympodially branched.

Chlamydospores

Chlamydospores are unknown.

Sex Organs

The fungus is homothallic. Antheridia are paragynous; oogonia are brown and tuberculate, 31 to 53 µm in diameter, average 40 µm; oospores are 25 to 42 µm in diameter, average 33 µm, and always plerotic.

Growth Temperatures

Temperature relationships are unknown.

Distinguishing Features

P. cyperi-bulbosi differs from *P. cyperi* in that it has distinctly tuberculate oogonia and host specificity to *C. bulbosus.*

Disease Caused by *Phytophthora cyperi-bulbosi*

Initially, infection is indicated by one or more isolated, water-soaked, brown lesions on individual leaves. Infected leaves rapidly wither, and the aerial shoots die back. Symptoms are similar to those caused by *P. cyperi* (Seethalakshmi 1953). The mycelium permeates the intercellular spaces. Sporangia emerge from the stomata. Numerous oospores are formed in the air spaces in mesophyll tissue of leaf sheaths.

Disease Development. The disease develops rapidly after periods of prolonged rainfall. During the dry season, the disease is arrested.

Control. Control measures have not been reported.

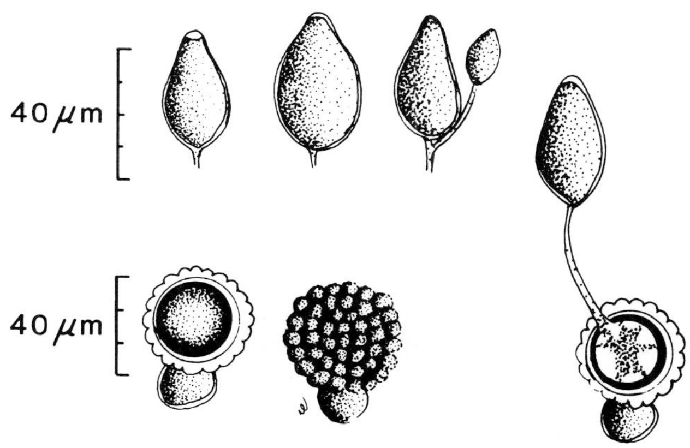

Figure 22.1. Morphology of *Phytophthora cyperi-bulbosi.* Sporangia are semipapillate. Oogonial walls are tuberculate on the surface. (Drawing by A. Vaziri)

CHAPTER 23

Phytophthora drechsleri

Tucker (1931)

Synonyms include *Pythium teratosporon* Sideris (1932) and *Phytophthora erythroseptica* var. *drechsleri* (Tucker) Sarejanni (1936). *Phytophthora drechsleri* was originally isolated from rotting potato tubers by Drechsler and described by Tucker (1931). It has a broad host range including approximately 40 plant families (Table 23.1). It is primarily a root pathogen (Plate 23.1 shows root rot of sugar beet) (Tompkins et al. 1936a), but some isolates also attack ripening fruits such as melons and cucumber. E. M. Hansen et al. (1988) speculated that *P. drechsleri* was indigenous in southeast Alaska because isolates that were recovered from undisturbed soil in declining forests of Alaska cedar, *Chaemaecyparis nootkatensis* (D. Don) Spach., were not capable of causing more than mild damage to Alaska cedar and were nonpathogenic to Douglas fir. Normally, these trees are susceptible to *P. drechsleri* isolates of agricultural origin. These isolates were later assigned to the "aquatic species *P. gonapodyides*" (Brasier et al. 1989; Brasier 1991). Previous tests showed that both *P. drechsleri* and *P. cryptogea* were associated with root rot and were pathogenic to Douglas fir in Oregon (Hamm et al. 1985). Holliday (1980) and Stamps (1985j) provide a brief description of *P. drechsleri* and some of the diseases it causes.

Characteristics of *Phytophthora drechsleri*

P. drechsleri is classified in group VI (Stamps 1985j; Stamps et al. 1990). Figure 23.1 shows its morphology. See Tables 4.2 and 4.3 in Chapter 4 for tabular keys and Appendix 4.9 for a dichotomous key (Ho 1992). Classical differences between *P. drechsleri* and *P. cryptogea* are given in Chapter 20, Table 20.2 (*P. cryptogea*). Morphologically, *P. drechsleri* and *P. cryptogea* are similar and differ mainly by the ability of *P. drechsleri* to grow at 35°C compared with *P. cryptogea*, which does not (Tucker 1931); however, differentiating these species has become one of the most vexing taxonomic problems in *Phytophthora* (Erwin 1983; Mills et al. 1991). Ho and Jong (1986) cite several reports that attest to the confusion and lack of agreement in the literature on the identification of isolates as either *P. cryptogea* or *P. drechsleri.* To indicate the uncertainty that exists, Keim et al. (1979) described a root rot of Washington palm as being caused by the *P. drechsleri-cryptogea* complex. We discuss *P. drechsleri* in the following paragraphs and then discuss current progress in resolving the taxonomic problem.

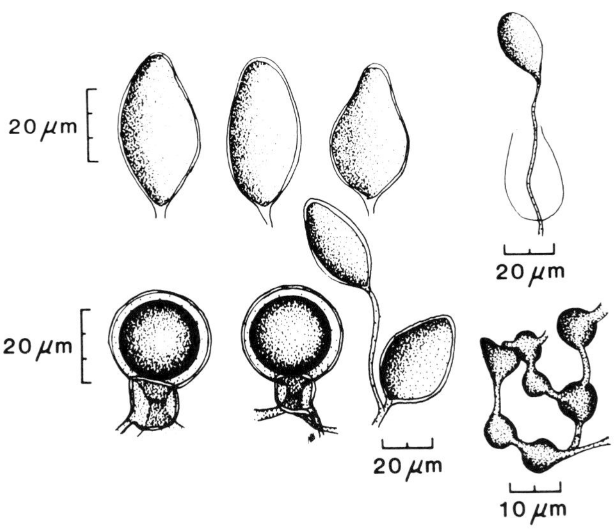

Figure 23.1. Morphology of *Phytophthora drechsleri.* Upper row, nonpapillate, ellipsoid to obpyriform sporangia; internal proliferation. Lower row, globose oogonia with amphigynous antheridia; sporangia forming in a sympodium; hyphal swellings formed in aqueous culture. (Drawing by A. Vaziri)

Sporangia

Sporangia vary in shape from broadly obpyriform to elongated obpyriform, ovoid to elongate (occasionally tapering at the base) (Klisiewicz 1977). They are nonpapillate and persistent on the stalk. Sizes are variable (Table 23.2), in general 36 to 70 μm long × 26 to 40 μm wide, average 39 × 24 μm, and they proliferate internally or externally. Data in Tables 4.2 and 4.3 are from Tucker (1931). (See Figures 4.9A, 4.13, and 4.14 in Chapter 4.) Sporangiophores are usually narrow but sometimes widen just below the sporangium and form in a sympodium (Figure 23.1).

Hyphal Swellings

Hyphal swellings are round or angular, form in chains or clusters, and are similar to those of *P. cryptogea* (see Figure 4.4A).

Chlamydospores

Chlamydospores are reported only in some isolates (Frezzi 1950; Katsura 1971; Cother and Griffin 1973b, 1974) and measure 4.2 to 11.0 μm in diameter, average 7.9 μm.

Sex Organs

P. drechsleri is usually heterothallic, forming oospores only when paired with A1 or A2 mating types; however, Tucker (1931), Tompkins et al. (1936a), and Klisiewicz (1977) report that some isolates produce oospores in single culture. Although *P. drechsleri* f. sp. *cajani* was found to be heterothallic when paired with A2 isolates of other species, oospores formed in single culture when incubated at 30°C (Kannaiyan et al. 1980). Singh and Chauhan (1988) reported that oospores formed in single culture when pigeon pea leaves were inoculated with zoospores. Antheridia are amphigynous; oogonia have tapered bases and are 22 to 53 μm in diameter, average 33 μm; and oospores are plerotic and 17 to 50 μm in diameter, average 26 μm. See Table 23.2 for references on oospore and sporangium sizes.

Growth Temperatures

Minimum temperature for growth is 5°C, optimum 28 to 31°C, and maximum 35 to 37°C (Tucker 1931). Many isolates produce optimal growth at 30°C with some growth at 35°C (Tucker 1931; Erwin 1952; Tompkins et al. 1936a; Mills et al. 1991) (Table 23.2). Colonies form in rosaceous, stellate, or cottony patterns; however, these patterns are not consistent for all isolates (Figures 4.2 and 4.3).

Distinguishing Characteristics

Phytophthora root rot of safflower was originally reported to be caused by *P. drechsleri,* as identified by Tucker (see Erwin 1950). The shape of the base of the sporangium was used as a criterion to differentiate *P. cryptogea* from *P. drechsleri* (Klisiewicz 1977) (Figure 23.2). Kliesewicz (1977) described cultures from safflower that grew at 35°C, some of which he designated *P. drechsleri* because the sporangia were elongated to ellipsoid and the bases were sloping and others *P. cryptogea* because sporangia had rounded bases (Figure 23.2). Although Klisiewicz (1977) did not present data for the population of sloping-base or rounded-base types, he stated that the majority of sporangia of *P. drechsleri* had sloping bases and sporangia of *P. cryptogea* had rounded bases. Mircetich and Matheron (1976) also illustrate sporangia with sloping bases to be typical of *P. drechsleri* from cherry trees (their Figure 2H). Isolates from bean produced sporangia with both tapered and rounded bases and were designated *P. cryptogea* (Flowers et al. 1973). Ho and Jong (1986) compared 15 isolates previously identified as *P. cryptogea* and 14 isolates previously identified as *P. drechsleri* from a wide ranges of sources, but morphological characters including rounded or tapered sporangial bases were inconsistent. Likewise, Mills et al. (1991) found that this character was too inconsistent to be reliable.

Ho and Jong (1986) considered merging the two species but refrained pending further study on more isolates. Their evidence for merging the two species was convincing until a study of a larger number of isolates of both species by isozyme and mitochondrial DNA restriction fragment length polymorphism (mtDNA RFLP) patterns showed that 10 groups could be differentiated within the *P. cryptogea-P. drechsleri* complex. Differences between some of the groups were as great as those among certain other valid *Phytophthora* species (Mills et al. 1991). Both reports, as well as a subsequent 1991 report by Ho and Jong in which *P. cryptogea* was redescribed without being merged with *P. drechsleri,* are reviewed below.

Ho and Jong (1986) noted that colony morphology of *P. drechsleri* and *P. cryptogea* isolates was similar on cornmeal agar; sporangia of most isolates were obpyriform, ovoid, or elongate, with either tapered or rounded bases. In seven of 19 cultures of *P. cryptogea* and in seven of 14 isolates of *P. drechsleri,* sporangia with both rounded and tapered bases formed; however, tapered bases were more common on *P. drechsleri* sporangia. Most isolates were heterothallic, and oogonia were spherical to subspherical without tapered bases. Smooth, globose oospores were mostly plerotic. Chlamydospores were not noted in culture. All isolates produced pigmentation on a casein hydrolysate medium (Timmer et al. 1970), tolerated malachite green (dilution of $1{:}18 \times 10^6$), and utilized inorganic nitrates in a synthetic medium. When cultures were incubated at 35°C, all but five of 14 isolates previously designated *P. drechsleri* grew. At 35°C, one of 19 isolates previously designated *P. cryptogea* grew well, three showed only slight growth, and the remainder did not grow.

Broad hyphae (>8 µm in diameter) were cited by Newhook et al. (1978) as characteristic of *P. cryptogea,* whereas narrower hyphae (<8 µm in diameter) are characteristic of *P. drechsleri,* but Ho and Jong (1986) did not note any consistent difference in hyphal breadth between the two species.

Size of sporangia for *P. cryptogea* (small, 37 to 40 µm long) and *P. drechsleri* (large, 36 to 50 µm long) was cited by Waterhouse (1963) to be a differential character; however, Ho and Jong (1986) found that the sizes of sporangia of *P. drechsleri* isolates was 52 ± 10 µm versus 49 ± 8 µm for *P. cryptogea.* The length-breadth ratio was 1.7 ± 0.2 for the *P. cryptogea* isolates versus 1.9 ± 0.3 µm for the *P. drechsleri* isolates (Ho and Jong 1986), hardly enough to differentiate species.

In 1991, Ho and Jong examined more isolates of *P. drechsleri* and *P. cryptogea,* concluded that these species should not be merged, redescribed *P. cryptogea* to include those isolates intermediate between *P. cryptogea* and *P. drechsleri* that did not grow at 35°C, but retained *P. drechsleri* to include those isolates that grew well at 35°C. *P. cajani* (Amin et al. 1978), *P. erythroseptica* var. *drechsleri* (Sarejanni 1936), *P. melonis* (Katsura 1976), and *P. sinensis* (Yu and Zhuang 1982) were included as conspecific with *P. drechsleri* (also by Ho in 1981). Their use of growth at 35°C as a differential character actually revives the criterion first established by Tucker (1931) to differentiate *P. cryptogea* from *P. drechsleri.*

Mills et al. (1991) concluded from their molecular studies, in which isozyme and mtDNA RFLP patterns (see Chapters 3 and 4 for a discussion of the methods and rationale [Spielman 1991]) were used to categorize isolates, that the *P. cryptogea-P. drechsleri* complex was too diverse to warrant a merger of

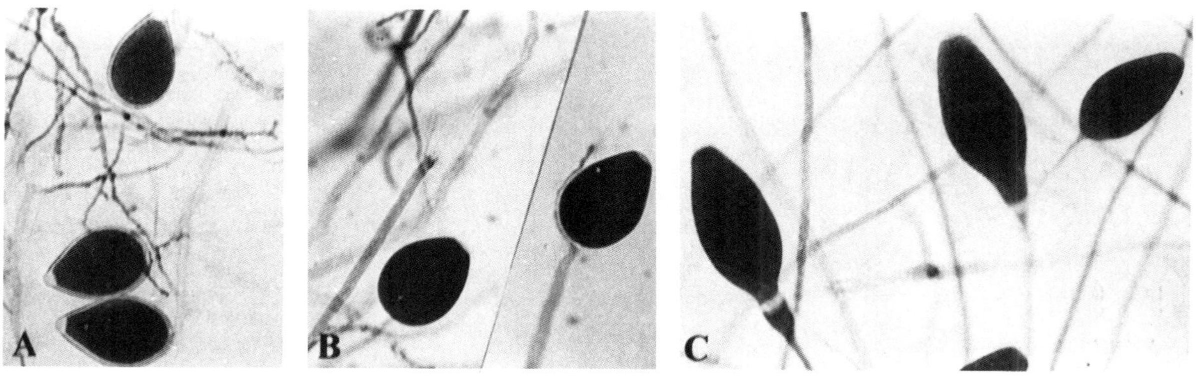

Figure 23.2. Comparison of sporangial shapes. **A,** *Phytophthora parasitica;* **B,** *P. cryptogea;* and **C,** *P. drechsleri* from safflower. All three isolates grow at 35°C. (Reproduced by permission of the American Phytopathological Society, St. Paul, Minnesota, courtesy J. Klisiewicz 1977)

Table 23.1 Distribution of hosts reported for *Phytophthora drechsleri*

Host	Common name	Disease	Geographical distribution
Actinidia deliciosa (Chevalier) Liang and Ferguson	Kiwi fruit	Root rot	United States (Conn et al. 1991)
Adenanthos sericeus Labill.		Root rot	Australia (Hardy and Sivasithamparam 1988)
Albizia stipulata (Roxb.) L. H. Boivin.	Mimosoideae	Bark canker; dieback	Malagassy Republic (Bouriquet et al. 1959)
Asimina triloba (L.) var. *crassa* Alef.	Pawpaw	Fruit, trunk, collar, and root rot	Sarawak (Kueh 1976)
Atylosia albicans Benth.		Stem rot	India (Kannaiyan et al. 1980)[a]
Atylosia cajanifolia Haines		Stem rot	India (Kannaiyan et al. 1980)[a]
Atylosia lineata Wight and Arn.		Stem rot	India (Kannaiyan et al. 1980)[a]
Atylosia scarabaeoides Benth.		Stem rot	India (Kannaiyan et al. 1980)[a]
Atylosia volubis Gamble		Stem rot	India (Kannaiyan et al. 1980)[a]
Azalea indica L.	Azalea	Root rot	Australia (Hardy and Sivasithamparam 1988)
Banksia dryandroides Baxt. and Sweet.	Banksia	Root rot	Australia (Hardy and Sivasithamparam 1988)
Banksia hookerana Meissn.	Banksia	Root rot	Australia (Hardy and Sivasithamparam 1988)
Banksia laricina Gardn.	Banksia	Root rot	Australia (Hardy and Sivasithamparam 1988)
Banksia occidentalis R. Br.	Banksia	Root rot	Australia (Hardy and Sivasithamparam 1988)
Banksia prionotes Lind.	Banksia	Root rot	Australia (Hardy and Sivasithamparam 1988)
Banksia sparsa R. Br.	Banksia	Root rot	Australia (Hardy and Sivasithamparam 1988)
Beta vulgaris (L.) var. *crassa* Alef.	Sugar beet	Wet rot; taproot rot	United States (Tompkins et al. 1936a); Iran (Ershad 1971)
Boronia megastigma Nees.		Root rot	Australia (Hardy and Sivasithamparam 1988)
Brassica oleracea var. *capitata* L.	Cabbage	Root rot	South Africa (Thompson and Phillips 1988)
Brassica rapa L.	Turnip	Root rot	United States (Tompkins and Tucker 1941a)
Cajanus cajan (L.) Millsp.	Pigeon pea	Stem rot	India (Pal et al. 1970; Kannaiyan et al. 1980)
Callistemon citrinus (Curtis) Stapf.	Bottlebrush	Root rot	Australia (Hardy and Sivasithamparam 1988)
Callistemon viminalis (Soland. ex Gaertn.)	Bottlebrush	Root rot	Australia (Hardy and Sivasithamparam 1988)
Callitris pressei Miq.	Cypress pine	Root rot	Australia (Hardy and Sivasithamparam 1988)
Calluna vulgaris Hull.	Heather	Root rot	Australia (Hardy and Sivasithamparam 1988)
Calytrix angulsta Lindl.		Root rot	Australia (Hardy and Sivasithamparam 1988)
Capsicum annuum L.	Chili pepper	Fruit rot	United States (Tompkins et al. 1936a)[a]
Carthamus lunatus L.	Saffron thistle	Root rot	Australia (Stovold 1973)
Carthamus tinctorius L.	Safflower; false saffron	Root rot; preemergence damping-off	United States (Erwin 1950); Iran (Ershad 1971); Australia (Stovold 1973)
Cedrus deodara (Don) G. Don	Deodar	Root and crown rot; chlorosis	United States (Gill 1979); China (Ho et al. 1984d)
Celosia plumosa	Woolflower	Root rot; stem necrosis	Argentina (Frezzi 1950)
Chamaecyparis nootkatensis (D. Don) Spach.	Alaska cedar	No disease caused	Alaska, United States (Hansen et al. 1988)
Chamaelaucium uncinatum Schauer.	Geraldton wax plant	Root rot	Australia (Hardy and Sivasithamparam 1988)
Chamaelaucium uncinatum	'Burgundy Blush'	Root rot	Australia (Hardy and Sivasithamparam 1988)
Chondrilla juncea L.	Skeleton weed	Necrosis; wilt	Australia (Cother and Griffin 1973a)[a]
Chrysanthemum cinerariifolium (Trevir.) Vis.	Pyrethrum	Root rot	Argentina (Frezzi 1950)
Cicer arietinum L.	Chickpea; garbanzo bean	Root rot	Iran (Ershad 1971)
Citrullus lunatus (Thunb.) Matsumura and Nakai	Camel melon	Necrosis; wilt	Australia (Cother and Griffin 1973a)[a]
Citrullus vulgaris Schrad.	Watermelon	Stem, leaf, root, and seedling blight; fruit rot	Japan (Katsura 1958); Egypt (El-Helaly et al. 1968)[a]; Iran (Ershad 1971)
Coleonema pluchum Hook. 'Rubrum'		Root rot	Australia (Hardy and Sivasithamparam 1988)
Coleus Lour.	Flame nettle; painted leaves	Root rot	Argentina (Frezzi 1950)
Coprosma repens Rich.		Root rot	Australia (Hardy and Sivasithamparam 1988)
Cordyline australis Hook.	Dracena	Root rot	Australia (Hardy and Sivasithamparam 1988)
Crowea saligna Andr.		Root rot	Australia (Hardy and Sivasithamparam 1988)
Cucumis melo L. var. *inodorus* Naud.	Honeydew melon	Fruit rot; stem rot; green death	United States (Tompkins and Tucker 1937a)[a]; Iran (Ershad and Mostowfipoor 1969)
Cucumis sativus L.	Cucumber	Root and foot rot	Japan (Katsura 1968, identified as *P. melonis*); Egypt (El-Helaly et al. 1968)[a]; Iran (Ershad and Mostowfipoor 1969); Greece (Kouyeas and Chitzanidis 1978); China (Ho et al. 1984c)
Cucurbita pepo L.	Vegetable-marrow; pumpkin	Fruit rot	Egypt (El-Helalay et al. 1968)
Daucus carota L. var. *sativa* DC	Carrot	Root rot	United States (Tompkins et al. 1936a)[a]; Japan (Katsura 1958)[a]
Dodonea viscosa Jacq.		Root rot	Australia (Hardy and Sivasithamparam 1988)
Dodonea proteiodes Lindl.		Root rot	Australia (Hardy and Sivasithamparam 1988)

(*continued on next page*)

Table 23.1 (*continued*)

Host	Common name	Disease	Geographical distribution
Dodonea speciosa Meissn.		Root rot	Australia (Hardy and Sivasithamparam 1988)
Eufaxia oborata Gardn.		Root rot	Australia (Hardy and Sivasithamparam 1988)
Euphorbia pulcherrima Willd. ex Klotzsch	Poinsettia	Foliage blight	United States (Yoshimura et al. 1985)
Gerbera jamesonii Bolus ex Hook.	Transvaal daisy; barberton daisy	Root rot	United States (Tompkins and Tucker 1937b)
Grevillea biternata Meissn.	Spider flower	Root rot	Australia (Hardy and Sivasithamparam 1988)
Grevillea dielsiana Gardn.	Spider flower	Root rot	Australia (Hardy and Sivasithamparam 1988)
Grevillea 'Honey Gem'	Spider flower	Root rot	Australia (Hardy and Sivasithamparam 1988)
Grevillea 'Robyn Gordon'	Spider flower	Root rot	Australia (Hardy and Sivasithamparam 1988)
Grevillea 'Sandra Gordon'	Spider flower	Root rot	Australia (Hardy and Sivasithamparam 1988)
Hakea francisiana F. Meull.	Pincushion tree	Root rot	Australia (Hardy and Sivasithamparam 1988)
Hakea laurina R. Br.	Pincushion tree	Root rot	Australia (Hardy and Sivasithamparam 1988)
Hakea platysperma Hook.		Root rot	Australia (Hardy and Sivasithamparam 1988)
Hebe speciosa Lock. and Al. 'Buxifolia'	Hebe	Root rot	Australia (Hardy and Sivasithamparam 1988)
Hebe tetrapterum Turez.	Hebe	Root rot	Australia (Hardy and Sivasithamparam 1988)
Helianthus annuus L.	Sunflower	Stem rot	Iran (Ershad 1971); United States (Klisiewicz and Beard 1976)
Indigofera australis Willd.	Indigo	Root rot	Australia (Hardy and Sivasithamparam 1988)
Isopogon cuneatus R. Br.		Root rot	Australia (Hardy and Sivasithamparam 1988)
Kunzea barteri Schau.		Root rot	Australia (Hardy and Sivasithamparam 1988)
Kunzea pulchella George		Root rot	Australia (Hardy and Sivasithamparam 1988)
Lagerstroemia indica L.	Crape myrtle	Root rot	Australia (Hardy and Sivasithamparam 1988)
Lycopersicon esculentum L.	Tomato	Green fruit rot	United States (Tompkins and Tucker 1941a); Argentina (Frezzi 1950)
Malus pumila Mill.	Apple	Root rot; trunk canker	United States (Mircetich et al. 1974; Matheron et al. 1988)
Manihot esculenta Crantz.	Cassava; manioc	Soft rot; root rot	Brazil (Figueiredo and Albuquerque 1970); Colombia (Oliveros et al. 1974)
Medicago sativa L.	Alfalfa; lucerne	Root rot	Australia (Cother 1975)[a]; South Africa (Thompson 1987)
Melaleuca armillaris Sm.	Honey myrtle; bottlebrush	Root rot	Australia (Hardy and Sivasithamparam 1988)
Melaleuca diosmaefolia Andr.	Honey myrtle; bottlebrush	Root rot	Australia (Hardy and Sivasithamparam 1988)
Melaleuca hypercifolia Sm.	Honey myrtle; bottlebrush	Root rot	Australia (Hardy and Sivasithamparam 1988)
Melaleuca lateriflora Benth.	Honey myrtle; bottlebrush	Root rot	Australia (Hardy and Sivasithamparam 1988)
Melaleuca nesophila F. Muell.	Honey myrtle; bottlebrush	Root rot	Australia (Hardy and Sivasithamparam 1988)
Mucuna deeringianum (Bort.) Merr.	Velvet bean	Root and stem rot; blight	Australia (Mungomery 1957); Rhodesia (Bates 1961)
Parthenium argentatum Gray.	Guayule	Root and crown rot	United States (Campbell and Presley 1946)
Pastinaca sativa L.	Parsnip	Root rot	United States (Tompkins et al. 1936a)[a]
Pimilea ferruginea Labill.		Root rot	Australia (Hardy and Sivasithamparam 1988)
Pelargonium zonale (L.) L'Her ex Ait.	Geranium	Root rot	Argentina (Frezzi 1950)
Phaseolus vulgaris L.	Bean	Seedling blight	Japan (Katsura 1958)
Pinus spp.	Pine	Pre- and postemergence damping-off	Australia (Heather et al. 1977)
Pinus radiata D. Don	Pine	Dieback	Australia (Heather and Pratt 1975)
Prunus amygdalus Batsch. (*P. dulcis* (Mill.) Webb)	Almond	Collar rot; trunk canker	Greece (Kouyeas and Chitzanidis 1978)
Prunus armeniaca L.	Apricot	Root rot; trunk canker	United States (Mircetich et al. 1974)
Prunus avium L.	Mazzard sweet cherry	Root rot; trunk canker	United States (Mircetich et al. 1974)
Prunus domestica L.	European plum	Root rot; trunk canker	United States (Mircetich et al. 1974); Japan (Katsura 1958)
Prunus mahaleb L.	Mahaleb cherry	Root rot; trunk canker	United States (Mircetich et al. 1974)
Prunus persica (L.) Batsch.	Peach	Root rot; trunk canker	United States (Mircetich et al. 1974)
Pseudotsuga menziesii (Mirb.) Franco	Douglas fir	Root rot	United States (Pratt et al. 1976)
Schinus molle L.	Pepper tree; pirul; mastic tree	Bud and flower necrosis	Argentina (Frezzi 1950)
Senecio cruentus (Masson) DC	Cineraria; groundsel	Collar and root rot	Argentina (Carranza 1964)
Sesamum indicum L.	Sesame	Root rot	Rhodesia (Bates 1961)
Solanum melongena L.	Eggplant	Seedling blight	United States (Tucker 1931)[a]
Solanum tuberosum L.	Potato	Tuber rot	United States (Drechsler 1929b)
Viburnum tinus L.	Viburnum	Root rot	Australia (Hardy and Sivasithamparam 1988)
Vigna unguiculata (L.) Walp.	Cowpea	Root rot	United States (Erwin et al. 1991)
Washingtonia robusta Wendl.	Thread palm	Collar rot	United States (Keim et al. 1979)
Westringia rosemariniformis Smith		Root rot	Australia (Hardy and Sivasithamparam 1988)
Xanthium spinosum L.	Bathurst burr	Leaf necrosis	Australia (Cother and Griffin 1973a)[a]

[a] Artificially inoculated host.

the two species and postulated that there were probably seven genetic species within this group.

After analysis via isozymes and mtDNA RFLP technology of 129 isolates previously identified as *P. drechsleri, P. cryptogea, P. melonis,* and *P. drechsleri* f. sp. *cajani* and selected from a wide range of geographical areas and crops, Mills et al. (1991) grouped the cultures according to their genetic relatedness. A phenogram demonstrating the diversity in mtDNA RFLP patterns is shown in Figure 4.33 (Förster and Coffey 1991; summarized from Mills et al. 1991). Group A was homogeneous and included the *P. drechsleri* culture from potato originally described by Tucker (1931), three isolates from rotted roots of safflower that had originally been named *P. drechsleri* by Tucker (Erwin 1952) but had been renamed *P. cryptogea* on the basis of the production of sporangia with tapered bases (Klisiewicz 1977), isolates causing root rot of sugar beet (described by Tompkins et al. [1936a]) (Plate 23.1), and an isolate causing root rot of sweet pepper. All of these isolates grew well at 35°C, and the nonpapillate, ovoid, ellipsoid, obpyriform sporangia were typical of the criteria for *P. drechsleri* formulated by Tucker (1931) and continued by Waterhouse (1963), Newhook et al. (1978) in the revisions of her key, and Stamps et al. (1990); however, the criterion of rounded or tapered sporangial bases was not consistent. Many isolates produced both rounded- and tapered-base sporangia. They noted a few isolates in groups G and H that had predominately tapered sporangial bases and some isolates in groups J and K that produced mainly sporangia with tapered bases (see Figures 4.14 and 23.2). Isolates of *P. drechsleri* f. sp. *cajani* (group G) produced sporangia with both rounded and tapered bases (Figure 4.13).

A molecularly different group B contained what Mills et al. (1991) considered to be authentic cultures of *P. cryptogea* that did not grow at 35°C but were similar morphologically to both *P. drechsleri* and *P. cryptogea.* The three cultures previously designated *P. erythroseptica* that occurred in group B were not explainable but could have resulted from labeling errors or to misidentification by those who donated the cultures.

Group F was distinct and included a number of isolates from melons (*Cucumis sativa*), some of which had been previously identified as *P. melonis.* These isolates were similar morphologically and by growth at 35°C to group A and were pathogenic to cucumber but not to tomato. Mills et al. (1991) recommended that these not be placed in *P. drechsleri* but that an amended description of *P. melonis* should be considered. To date, an amended description has not been made. This description is needed because the sporangia of *P. melonis* were originally described as semipapillate by Katsura (1976); however, Ho (1986) observed that *P. melonis* and *P. sinensis* produced nonpapillate sporangia. He considered them to be synonymous with *P. drechsleri.*

Group G of Mills et al. (1991) was comprised of isolates from pigeon pea (*Cajanus cajan*) originally designated *P. cajani* (Amin et al. 1978) and reassigned to *P. drechsleri* f. sp. *cajani* (Kannaiyan et al. 1980). Mills et al. (1991) recommended that these host-specific isolates, which grew at 35°C, be returned to *P. cajani* (see Chapter 12). We agree that this recommendation should be considered, but more definitive data are needed. Amin et al. (1978) described the pigeon pea pathogen as homothallic, but this was not confirmed by Kannaiyan et al. (1980). The cause of Phytophthora blight is described here under *P. drechsleri* f. sp. *cajani.*

Group C, which did not grow at 35°C, consisted of isolates from woody hosts that had been identified in different reports as *P. cryptogea.* They were unique in that tests with the isozyme G-6-pi (glucose-6-phosphate isomerase), known to be a dimeric enzyme (Tooley et al. 1985), showed five-banded patterns. These isolates with heterozygous phenotypes, in which three or four alleles were involved, appeared to be polyploids.

Group D, which did not grow at 35°C, included only the host-specific *P. cryptogea* f. sp. *begoniae* isolates from *Begonia elatior* in Germany (Kröber 1981b). This group should be restudied and possibly described as a new species, if warranted.

Group E isolates, which did not grow at 35°C, were from grape, pine, and apple and were isozymically similar to group D.

Group H contained seven isolates, all of which grew at 35°C, from deciduous fruit trees and vines (walnut, apple, cherry, and kiwi) in California and Arizona that had been previously identified as *P. drechsleri.*

Groups J and K, which did not grow at 35°C, were isozymically and genetically unrelated to each other and contained isolates previously identified as *P. cryptogea* from peach, cherry, apple, fir, and pine trees. Brasier and Hansen (1992) suggest that isolates in group J may represent *P. gonapodyides.*

Another group was isozymically too diverse to categorize

Table 23.2. Characteristics of various isolates of *Phytophthora drechsleri*

Host	Sporangia (µm)	Chlamydospores	Oogonia µm)	Oospores (µm)	Temperature (°C)			Reference
					Min.	Opt.	Max.	
Potato	24–38 × 15–24 (av. 31.4 × 21)	–	21.7–53.4 (av. 31.3)	16.7–45.1 (av. 25.6)	10	30–32	37.5	Tucker (1931)
Sugar beet	24–56 × 22.8–40 (av. 38.6 × 26.7)	–	27–44 (av. 35.8)	23.6–35.2 (av. 28.6)	8	28–31	35+	Tompkins et al. (1936a)
Chrysanthemum	34–67 × 20–34 (av. 50 × 26.5)	+	28.5–34	25.5–29.5	NR[a]	30	35	Frezzi (1950)
Pumpkin	36.4–75.6 × 22.4–39.4 (av. 50 × 30)	–	36.9	NR	7.5	28–32	38	Ershad and Mostowfipoor (1969)
Watermelon	23.9–76.9 × 15.9–45.1 (av. 42.4 × 29.2)	+	16.8–36.4 (av. 28)	11.2–28 (av. 21.24)	9	28–30	35–36	Katsura (1971)
Safflower and gooseberry	35–55 × 26–37 (av. 44.3 × 31.2)	NR	NR	NR	NR	25–30	35	Bumbieris (1974)
Several	40–71 × 22–34 (av. 52 × 28)	NR	33 (28–38)	28 (16–37)	5–10	25–30	35–37	Ho and Jong (1991)

[a] Not reported.

and consisted of seven isolates previously designated *P. drechsleri,* nine *P. cryptogea,* and two unidentified. Data on growth at 35°C was not given for these.

The morphology of sporangia or oogonia was insufficient to separate species as similar as *P. drechsleri* and *P. cryptogea,* but Mills et al. (1991) concluded that these two species should not be merged because many of the groups within them were unrelated in isozyme and mtDNA RFLP patterns. An interesting sidelight was the inclusion in group A of an isolate that had been received with the identity of *P. erythroseptica.* Upon reexamination of the morphology, it was found to be heterothallic and grew well at 35°C, traits that were characteristic of *P. drechsleri* but not *P. erythroseptica.*

Förster and Coffey (1991) referred to the seven groups as molecular species instead of biological species. Many of the groups differed genetically as much as the known species *P. cinnamomi, P. lateralis,* and *P. erythroseptica,* which were included as "out species" in the study.

After reviewing the careful studies by Ho and Jong (1986, 1991) and Mills et al. (1991), we are left with the practical dilemma that previous methods of separating *P. cryptogea* from *P. drechsleri* are inadequate and that more undescribed species probably exist within the complex. Brasier and Hansen (1992) postulate that the data of Mills et al. (1991) suggest that morphologically similar species may have evolved "via geographical isolation and climatic and host specialization, and that their distributions are now overlapping through introductions. This overlapping may also have led to hybridization and the emergence of new taxa. Sexual crosses between some of the groups might help test their biological species status and whether or not they need to be formally recognized." These ideas are speculative.

One problem in interpretation of the data to distinguish *P. cryptogea* from *P. drechsleri* lies in the lack of a firm establishment of what qualities characterize the species that Tucker (1931) established as *P. drechsleri.* Apparently, the species he described was represented by only one isolate. The grouping from a sample of 129 isolates tested by Mills et al. (1991) shows that group A includes what could be considered typical *P. drechsleri sensu* Tucker (1931), because all of these isolates grew well at 35°C. From a practical standpoint, growth at 35°C appears to be a consistent feature in the groups associated with *P. drechsleri.* Groups F, G, and H, however, all of which were previously considered to be *P. drechsleri* by other references, also grow at 35°C, and each of these groups is as genetically different from group A (probably typical *P. drechsleri*) and from each other as they are from many other species.

Ho and Jong (1986) showed that a few of the isolates previously designated *P. cryptogea* grew at 35°C and a few of those designated *P. drechsleri* did not grow at 35°C. The problem here is the unclear basis on which isolates were designated as one or the other species in the Mills et al. (1991) or Ho and Jong (1986) reports. For instance, within group A (Mills et al. 1991), three isolates listed as *P. cryptogea* from safflower had been identified as *P. drechsleri* for many years before being reassigned to *P. cryptogea* by Duniway (1976) and Klisiewicz (1977) on the basis of sporangial morphology.

Another example of the lack of a firm baseline for comparison of *P. cryptogea* and *P. drechsleri* is in the report by Bumbieris (1974), who compared mycelial width, sporangial shapes, growth in a medium amended with malachite green (0.1 ppm), and growth at 35°C for five isolates designated *P. drechsleri* and seven isolates designated *P. cryptogea;* however, only the isolate considered to be an authentic *P. drechsleri* isolate (I.M.I. 40499) obtained from the Commonwealth Mycological Institute (CMI) in the United Kingdom grew at 35°C. All other isolates of either species did not grow at 35°C and were indistinguishable by any of the other criteria measured. Although this study has long been quoted to justify merging the two species, if the isolate I.M.I. 40499 from CMI were the only valid *P. drechsleri* isolate and if growth at 35°C were a valid criterion, this study does not support merging the two species because all of the isolates with one exception may have been *P. cryptogea.* Previous identification of an isolate does not necessarily make it a reliable item for comparison.

Because it is now clear that sporangial shape or mycelial width morphology cannot separate the two species (Figures 4.13 and 4.14 show both rounded- and tapered-base sporangia produced in one culture of *P. drechsleri*), the only character left is growth at 35°C. Admittedly, growth at a high temperature is not a morphological character, but it is the only character until a better one is found. The mtDNA RFLP and isozyme data tell us that there may be differences that have not been seen and should be looked for. In the absence of more definitive data, the amended description by Ho and Jong (1991), which cites growth at 35°C, has to be accepted. The molecular-species concept of Mills et al. (1991) should be considered, but it is not yet well enough defined to use in a practical way. The amended description of *P. drechsleri* by Ho and Jong (1991) follows (reproduced by permission of Mycotaxon Ltd., Ithaca, New York, courtesy of Ho and Jong 1991):

Phytophthora drechsleri Tucker emend., 1931.
= *P. cajani* Amin, Baldev et Williams, *Mycologia* 70:174, 1978.
= *P. drechsleri* var. *cajani* Pal, Grewal et Sarboy, *Indian Phytopathol.* 23:585, 1970.
= *P. erythroseptica* var. *drechsleri* Sarejanni, *Ann. Inst. Phytopathol. Benaki* 2:48, 1936.
= *P. melonis* Katsura, *Trans. Mycol. Soc. Japan* 17:238, 1976.
= *P. sinensis* Yu et Zhuang, *Mycotaxon* 14:183, 1982.

Mycelium with no distinct or slightly floral growth pattern on clarified V-8 juice agar medium, slightly fluffy to fairly fluffy; aerial mycelium usually not profuse on unfiltered yellow corn meal agar medium; main hyphae uniform to arbuscular 5–7 μm wide, sometimes forming in water network of small hyphal swellings (under 25 μm), spherical to irregular; sporangia occasionally produced on agar but more abundantly in water, nonpapillate, nondeciduous, internally proliferating; terminal on unbranched or sometimes lax sympodially branched sporangiophores (1–3 μm wide); first formed sporangia often variable in shape, broadly obpyriform, ovoid, ellipsoidal or elongated with tapered base; sporangia av. 52 μm (40–71 μm) × 28 μm (22–34 μm), L/B ratio 1.9 (1.4–2.5), apex flattening on mounting, sporangia release zoospores promptly and then collapse; exit pore over 8 μm wide, pore/breadth ratio over 0.3; heterothallic though some isolates self-fertile especially when freshly isolated; oogonia spherical to subspherical, smooth, av. 33 μm (28–38 μm) diam., wall becoming yellowish to brown; oospores plerotic to aplerotic, spherical, av. 28 (16–37 μm) diam., wall 2–4 μm thick; antheridia amphigynous variable in size but commonly short cylindrical, 15 μm (13–17 μm) × 15 (13–19 μm), unicellular or occasionally bicellular in longer antheridia; no chlamydospores found in culture though chlamydospore-like vesicles may be produced.

Cardinal temperatures for growth: 5–10°C, 25–30°C, 35–37°C; good and steady growth at 35°C (over 3 mm per day); growth rate on clarified V8 agar medium 4–9 mm per day at 20°C.

P. drechsleri is readily distinguished from other species in group VI (Stamps et al. 1990) but is considered in some reports to be conspecific with *P. cryptogea* (the older name) and in others to be different.

Since *P. drechsleri* f. sp. *cajani* from pigeon pea formed a separate group, replacing the name *P. drechsleri* f. sp. *cajani* by the older species name, *P. cajani,* following the suggestion of Mills et al. (1991) should be considered (see Chapter 12); however, because the name *P. drechsleri* f. sp. *cajani* is still in use, we discuss the disease it causes on pigeon pea in this chapter.

Ho (1986) examined *P. melonis* (Katsura 1968) and *P. sinensis* (Zu and Zhuang 1982) and concluded that both species should be merged under *P. drechsleri* because of growth at 35°C and the similarity of sporangium and oogonium morphology. In the group F of Mills et al. (1991), an isolate of *P. sinensis* from melon (*Cucumis sativa*) differentiated into the same mtDNA RFLP group as *P. drechsleri* isolates from *C. sativa.* These isolates were mostly from Asia (Japan, Taiwan, and China), with one isolate from Iran. These data support in part the conclusion of Ho (1986) and Ho et al. (1984a, b) to combine *P. drechsleri, P. melonis,* and *P. sinensis;* however, if group A is considered typical of *P. drechsleri,* group F is genetically distinct, and Mills et al. (1991) suggest that an amended description of *P. melonis* (Katsura 1968) would be appropriate. Although we do not attempt to redescribe *P. melonis,* this species is listed in a separate chapter to emphasize its possible significance. We list *P. sinensis* and *P. melonis* separately in Tables 4.2 and 4.3 but do not have enough data formally to retain these names as species. Morphological and molecular studies should be continued to determine whether *P. sinensis* and *P. melonis* should be retained as separate species.

Diseases Caused by *Phytophthora drechsleri*

Root and Stem Rot of *Carthamus tinctorius* (Safflower)

Safflower is a source of oil used for food. Root and stem rot was first described in Nebraska (Erwin 1950, 1952) and is of major importance wherever safflower is grown. Although the name of the causal agent for some isolates was changed to *P. cryptogea* (Duniway 1976; Klisiewicz 1977) on the basis of sporangial morphology, the recent molecular data (isozyme patterns) of Mills et al. (1991) indicate that this change was not warranted if growth at 35°C is an acceptable criterion. See discussion of the data of Mills et al. (1991) above.

On young safflower plants, infection by *P. drechsleri* produces symptoms closely resembling those of damping-off; necrotic lesions form on roots and lower stems. Plants at all stages of growth from preemergence to near maturity are susceptible. The infected lower stem shrivels, and the plant eventually falls over and wilts. On older plants, black necrotic lesions form on the roots (Figure 23.3). These lesions, which gradually encompass the roots and extend up the stem (Figure 23.4), cause wilt and death of the entire plant within 10 days. Leaves of older plants often turn yellow prior to wilting.

P. cactorum was isolated from soil by use of safflower seedlings as a bait (Banihashemi and Mitchell 1975), and *P. parasitica* has also been cited as a root pathogen of safflower (Klisiewicz 1977).

Disease Development. Stem rot develops rapidly under irrigated agriculture conditions when stems are in contact with water or wet soil for an extended time. Plants close to maturity are particularly vulnerable, especially when soil temperatures remain as high as 20 to 30°C. Root disease is favored by high temperature (Erwin 1952). According to Klisiewicz (1977), root rot caused by certain homothallic isolates developed more slowly than that caused by heterothallic species. Both susceptible and resistant cultivars can be predisposed to the severe expression of root rot if the soil in the field is allowed to dry to near the wilting point before flood irrigation (Knowles et al. 1965; Zimmer and Urie 1967; Duniway 1977b). Predisposition of safflower by water stress is covered more fully in Chapter 5.

Control. Prevention of excessive irrigation and avoidance of drought stress prior to irrigation are necessary cultural practices (Knowles et al. 1965). See Chapter 5 for more documentation.

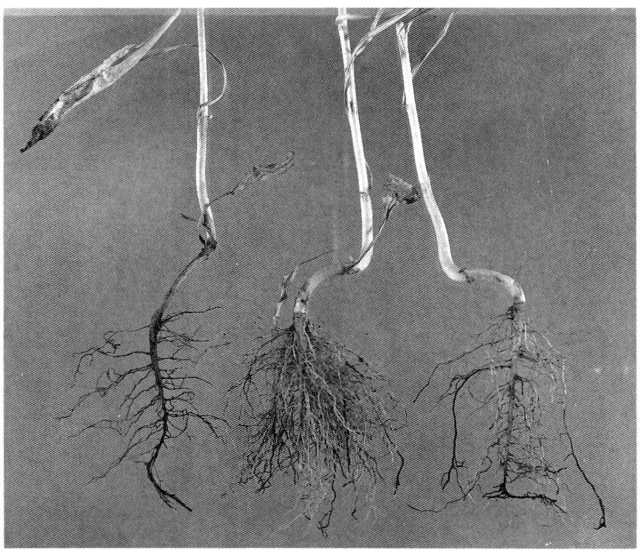

Figure 23.3. Root rot of safflower caused by *Phytophthora drechsleri.* (Courtesy J. Klisiewicz, USDA ARS, University of California, Davis)

Figure 23.4. Stem rot of sunflower caused by *Phytophthora drechsleri.* (Courtesy J. Klisiewicz, USDA ARS, University of California, Davis)

Many cultivars with adequate resistance to Phytophthora root rot have been developed (Thomas et al. 1960). See Chapter 6 for a discussion and Table 6.1 for references listing resistant cultivars. On cultivars Gila, U.S. 10, and VFR 1, roots are resistant but hypocotyls are susceptible to infection, whereas both roots and hypocotyls of cultivar U.S.B. are resistant. Root resistance of Gila and U.S. 10 appear to be governed by a single dominant gene. In U.S.B. and VFR 1, however, which have a higher level of resistance than Gila or U.S. 10, different dominant factors appear to be involved (Thomas and Hill 1977). The hypocotyl resistance of U.S.B. is recessive (Thomas and Zimmer 1970). By use of cultivars with differential resistance, Thomas and Klisiewicz (1963) noted pathologic races. The cultivar Nebraska 10 was susceptible to all races; a selection from the cultivar Biggs was resistant to all races tested.

The value of crop rotation may be reduced by the ability of the pathogen to survive in alternate hosts, including weeds (Cother and Griffin 1973a).

Root Rot of *Beta vulgaris* L. (Sugar Beet)

Root rot of sugar beet, which was first noted in 1927 in Utah, Idaho, and California, was proved to be caused by *P. drechsleri* by Tompkins et al. (1936a). Roots are susceptible from the seedling stage to the fully developed root stage. Under severe conditions, plants may be killed, but in plants that survive, the sugar content and dry weight are reduced. An important secondary effect is the contamination of the crude sugar products during the extraction process by dark material from the necrotic roots. A nearly mature root that has turned dark brown to black after infection is shown in Plate 23.1.

Tompkins et al. (1936a) described the symptoms and the isolation and pathogenicity of *P. drechsleri.* Plants affected by root rot may wilt but recover during the night. If infection is severe, the plants may die. The infection usually starts at the lower part of the enlarged taproot and progresses upward and eventually encompasses the entire root (Plate 23.1). The infected root tissue remains firm.

Disease Development. The disease occurs in heavy soils and especially in land that is irrigated excessively. Root rot often occurs near the end of irrigation runs where water ponds. Root rot occurs most frequently when summer temperatures are high. Although seedlings are susceptible, Tompkins et al. (1936a) attributed the finding that *P. drechsleri* did not occur on seedling roots to the usually lower temperatures when beets are planted in the spring.

Control. Regulating the frequency and duration of irrigation and planting on well-drained soil are the most frequently used control measures (Tompkins et al. 1936a). Resistant germ plasm has been identified, but resistance has not been incorporated into commercial cultivars (D. C. Erwin, *unpublished*). Planting on raised beds, which is a common practice in California, appears to be successful in most fields.

Dieback or Root Rot of Cucurbits

The causal agents of this disease were categorized into a molecular group F (by isozyme and mtDNA RFLP patterns), some of which had been identified as *P. melonis* and others as *P. drechsleri* (Mills et al. 1991); however, the morphological features and ability of all isolates to grow at 35°C were similar to those of *P. drechsleri.*

Cucurbit plants are susceptible at all stages from preemergence, seedling (Figure 23.5), to fruiting. In mature plants, extensive root rot results in a loss of turgidity of leaves and a general dieback of the plant. Sunken areas and fluffy white mycelial growth are often visible on infected fruit. In vegetable marrow, squash fruits (*Cucurbita pepo* L.), or cucumber (Figure 23.6), a brown discoloration starts from the blossom end and progresses toward the stem end. Formation of a cottony mat over the diseased areas is characteristic under moist conditions.

Disease Development. The disease is most destructive when soil temperature and moisture are high, conditions that are typical in irrigated fields during the summer.

Control. According to Ershad and Mostowfipoor in Iran (1969), the disease can be controlled by planting on raised beds, which offer better drainage and reduce water accumulation in soil near the plants, by irrigating judiciously to avoid excess wetting of the collar region of the plant, and by practicing long-term crop rotation or leaving fields fallow between crops. Evaluation of resistance of seedlings in Iran indicated that certain cultivars of cucumber, watermelon, long melon, and cantaloupe were more resistant than others. Cultivars of

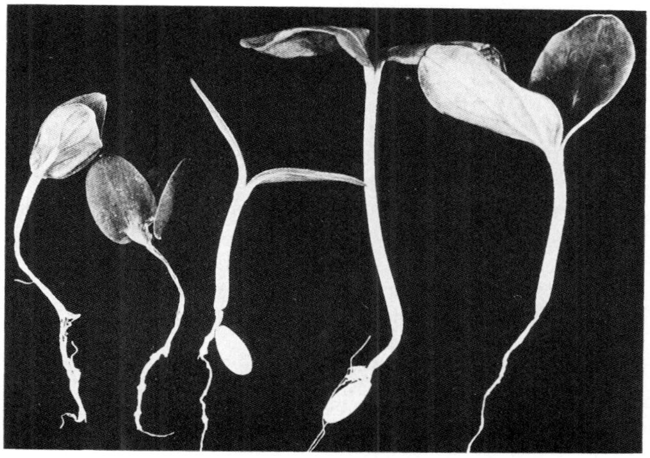

Figure 23.5. Seedling root rot of cucumber caused by *Phytophthora drechsleri.* (Courtesy D. Ershad, Institute of Pests and Diseases, Department of Agriculture, Tehran, Iran)

Figure 23.6. Fruit rot of cucumber caused by *Phytophthora drechsleri.* (Courtesy D. Ershad, Institute of Pests and Diseases, Department of Agriculture, Tehran, Iran)

cantaloupe were more susceptible than those of cucumber. Ohio MR17, a local cucumber cultivar from Iran, watermelon Klondike no. 7, and cantaloupe Gold & Silver were highly resistant (Mansoori and Banihashemi 1982).

Stem Rot of *Helianthus annuus* L. (Sunflower)

P. drechsleri causes a black rot that partially or totally girdles the stem and causes the plant to fall over and die. This disease has been reported in California (Klisiewicz and Beard 1976) and Iran (Ershad and Mostowfipoor 1969).

Disease Development. This disease is serious in irrigated sunflower if surface water is allowed to come into contact with the stem. Soil placed against stems, as a result of cultivation, favors infection by the pathogen during subsequent irrigation.

Control. Fields should be well drained, and heaping soil around stems should be avoided. In Iran, some sunflower hybrids are reported to have resistance (Ershad 1971).

Disease Caused by *Phytophthora drechsleri* f. sp. *cajani*

Stem Rot and Blight of *Cajanus cajan* (L.) Mill. (Pigeon Pea)

Stem rot of *Cajanus cajan* (L.) Mill. and the causal agent, *P. drechsleri* f. sp. *cajani* Kannaiyan et al. (1980), are described in Chapter 12, *P. cajani.*

P. drechsleri f. sp. *cajani* causes stem rot and does not affect roots. The first observable symptoms on foliage are desiccation and an upward rolling of the leaflets, usually without chlorosis, followed by a withering of the petioles and small stems. Dark brown to black lesions encircle the stem at or near the base and may extend 15 to 29 cm up the stem (Plate 23.2).

Table 23.3. Comparison of Fusarium wilt and Phytophthora blight of pigeon pea

Fusarium wilt	Phytophthora blight
Leaves turn yellow.	Leaves remain light green and roll upward.
Leaves eventually turn brown.	Leaves eventually turn brown.
Dark brown to purple streak develops up one side of stem.	Girdling, necrotic lesions develop and extend up the stem.
Wilt develops on side of streak only.	General wilt of whole plant occurs.
Roots discolor and decay.	Roots remain white and healthy.
In early stages, older xylem tissue is discolored.	In early stages, older xylem tissue is not affected.

Lesions with well-defined margins may also develop on the branches, petioles, or leaves (Plate 23.3). These lesions initially have a flat surface but later become slightly depressed. In newly formed lesions, the bark and cambium, but not the older xylem tissue, may become discolored. Roots are not attacked, even in advanced stages of disease development. Occasionally, large galls may develop on the stem at the soil line. Stems break easily at lesion sites on the main stems during strong winds or attempts to uproot plants. Plants are susceptible to infection from the seedling stage to maturity.

This disease may be confused with Fusarium wilt, caused by *Fusarium udum* Butl., another serious disease of pigeon pea (Table 23.3). The wilting symptom of this disease closely resembles that of Phytophthora blight. It is important to distinguish between these two diseases when breeding for resistance since there is apparently no correlation between resistance to Fusarium wilt and resistance to Phytophthora blight (Amin et al. 1976).

CHAPTER 24

Phytophthora eriugena

Clancy and Kavanagh (1977)

No synonyms are presently known for *Phytophthora eriugena,* which was first isolated by Clancy and Kavanagh in 1968 from diseased *Chamaecyparis lawsoniana* (Murr.) Parl. seedlings in Ireland. It has been reported only on that host growing in seedbeds (Clancy and Kavanagh 1977; Clancy 1979).

Characteristics of *Phytophthora eriugena*

P. eriugena has been isolated only once, and data in Clancy (1979) and Clancy and Kavanagh (1977) are based on only that isolate, which is no longer available. An English but not a Latin description is given by Clancy and Kavanagh (1977). Morphology of *P. eriugena* is shown in Figure 24.1. It is classified in group IV (Stamps et al 1990); however, Clancy (1979) had rejected group IV because other species in that group produced deciduous sporangia. In Tables 4.2 and 4.3, we list *P. eriugena* in group IV; however, it is an exception as noted by Clancy (1979). Though the species may be of doubtful validity because of the lack of a Latin description, the information of Clancy (1979) is presented in case the pathogen is isolated again.

Sporangia

Sporangia are papillate, persistent on the stalk, nonproliferating, hyaline, ovoid to ellipsoid, occasionally asymmetric, and 23 to 104 μm long × 18 to 58 μm wide, average 59 × 36 μm; an empty sporangium has a unique conical or fingerlike basal plug projecting inside the sporangial case. Sporangiophores are simple but occasionally occur as close or lax sympodial branches.

Hyphal Swellings

Hyphal swellings are abundant in water; globose, elongate, or irregular; and terminal, intercalary, or catenulate.

Chlamydospores

Chlamydospores are unknown.

Sex Organs

P. eriugena is homothallic. Antheridia are amphigynous but occasionally paragynous (12 to 18 μm); oogonia are globose, ovoid, to reniform (23 to 65 μm in diameter, average 41 μm) and abundant in host tissue and axenic culture. Oospores are mostly globose but may be ovoid to reniform. Oospores that are globose almost fill the oogonium but are aplerotic in nonglobose oogonia. The diameters of globose oospores on V8 juice agar ranged from 20 to 60 μm (average 35 μm), on Lawson cypress-dextrose agar from 38.4 to 62.4 μm (average 52.8 μm), and on naturally infected needles of Lawson cypress from 26.0 to 43.2 μm (average 31.2 μm). Data on agar media were obtained at 12°C (Clancy 1979).

Growth Temperatures

Minimum temperature for growth is <0°C, optimum 12°C, and maximum 22°C.

Disease Caused by *Phytophthora eriugena*

Stem and Collar Rot of *Chamaecyparis lawsoniana* (A. Murr.) Parl. (Lawson Cypress)

According to Clancy (1979), the disease has been observed in County Tipperary since 1968 in seedbeds where plants are grown for 2 years prior to commercial sale. The symptoms occur during the winter months as a progressive browning necrosis of the stem and older leaves. The disease does not spread during April to November. A well-defined line separates stem lesions from the root system, which remains disease free until the aerial part of the plant dies. Young seedlings (1 year old) are rapidly killed, and older plants are severely stunted. Heavy losses (more than 50%) of Lawson cypress seedlings occurred in nurseries infested with *P. eriugena* (Clancy and Kavanagh 1977).

The disease is distinct from other diseases of Lawson cy-

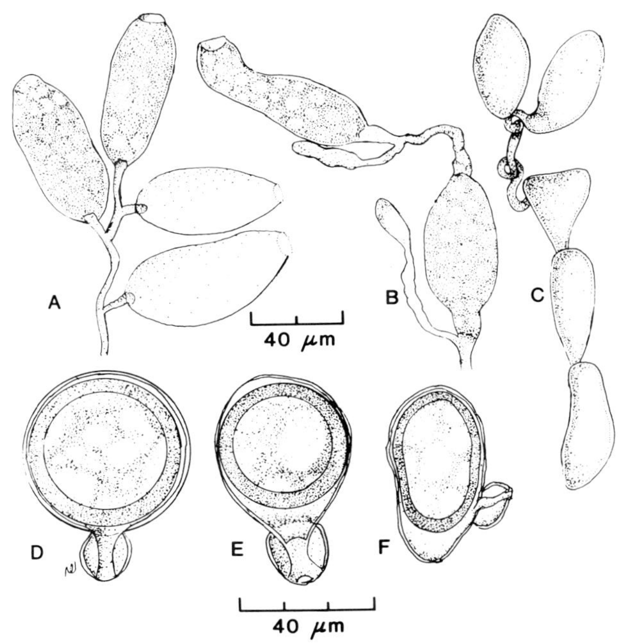

Figure 24.1. Morphology of *Phytophthora eriugena.* **A,** Papillate, elongate, ellipsoid sporangia. Note the prominent basal plugs in empty sporangia. **B,** Variation in sporangium morphology. **C,** Catenulate sporangium formation. **D and E,** Oogonia with amphigynous antheridia and aplerotic oospores. **F,** Reniform oogonium and oospore. (Drawing by A. Vaziri by permission, courtesy K. Clancy, University College, Dublin, 1979)

press caused by *P. cinnamomi* (Torgeson 1954; Torgeson et al. 1954) or *P. lateralis* (Trione 1959). Both of these species initially cause root rot. The sharp line of demarcation of the lesions caused by *P. eriugena* is between the root and upper stem, but in diseases caused by *P. cinnamomi* and *P. lateralis,* the line of demarcation is between the root and the stem.

Isolation of pure cultures is difficult because of its slow growth. *P. eriugena* can best be detected by washing infected needles in running tap water for 1 h followed by three washes in sterile distilled water, drying, and transferring advancing lesions to sterile distilled water in petri dishes. After incubation for 48 h at 8°C, sporangiophores grow out of the stomata.

Disease Development. Stem and collar rot occurs only in cool, damp weather and only during the winter months. The disease develops in young seedlings during the cold season of the year (December through March) and continues to develop until late spring. Plants affected in their second year of growth are usually stunted and lack vigor. The disease is arrested during the warm period of the year. Oospores form in diseased needles and are probably a source of inoculum in seedbeds.

Control. There is little information on control, but it could be safely presumed that principles of control would be similar to those for other woody nursery plants; i.e., seedbeds should be fumigated or steamed, or if that is not feasible, seedlings should be started on land not cropped previously by Lawson cypress.

CHAPTER 25A

Phytophthora erythroseptica
Pethybridge (1913)
var. *erythroseptica*

Phytophthora himalayensis Dastur 1948 was considered a synonym by Waterhouse (1963). *Phytophthora erythroseptica* Pethybridge (1913) is subclassified into the varieties *erythroseptica* Waterhouse (1963) and *pisi* (Bywater and Hickman 1959) in the key of Stamps et al. (1990). *P. erythroseptica* var. *pisi* is described separately in Chapter 25B.

P. erythroseptica was first described by Pethybridge (1913, 1914) as the causal agent of pink rot of potato tubers in Ireland. It is now present in virtually all potato-growing areas of the world (Table 25A.1). Although *P. erythroseptica* is primarily known as a pathogen of solanaceous hosts, it also has been reported to be a serious pathogen of other plants, e.g., cineraria (*Senecio cruentus* Masson ex L'Her DC) and calla lilies (*Zantedeschia* spp.). Recently, a serious disease of wild rice, *Leymus chinensis* (Trin.) Tzvelev, was reported to be caused by *P. erythroseptica* (Gunnell and Webster 1988). Ho and Jong (1989b) indicate that the rice isolate differs from other *P. erythroseptica* isolates in that it produces a fluffy colony and larger sporangia, grows poorly on malt extract agar, is unable to hydrolyze starch, is sensitive to hymexazol, and

Table 25A.1. Distribution of hosts of *Phytophthora erythroseptica*

Host	Common name	Disease	Geographical distribution
Atropa belladonna L.	Belladonna	Root rot; damping-off; seedling rot	Netherlands (Westerdijk and van Luijk 1920); United Kingdom (Alcock 1926)
Brassica oleracea var. *acephala* DC	Kale	Root rot	Ireland (Wheland and Loughnane 1969)[a]
Brassica oleracea var. *capitata* L.	Cabbage	Root rot	Ireland (Wheland and Loughnane 1969)[a]
Brassica oleracea var. *rapa* L.	Turnip	Root rot	Ireland (Whelan and Loughnane 1969)[a]
Carduus arvensis Robs.	Creeping thistle	Root rot	Ireland (Whelan and Loughnane 1969)[a]
Calceolaria L.	Slipperwort, slipper flower	Root rot	New Zealand (Anon. 1977)
Cymbidium Swartz	Orchid	Root rot; leaf spot	Mexico (Romero 1977); Australia and New Zealand (G. Hall 1989b)
Cynosurus cristatus L.	Crested dogtail	Root rot	Ireland (Whelan and Loughnane 1969)[a]
Hordeum vulgare L.	Barley	Root rot	Ireland (Whelan and Loughnane 1969)[a]
Legmus chinensis (Trin.) Tzvelev	Wild rice	Crown and root rot	United States (Gunnell and Webster 1988)
Lolium multiflorum Lam.	Ryegrass	Root rot	Ireland (Whelan and Loughnane 1969)[a]
Lycopersicon esculentum Mill.	Tomato	Fruit rot	Canada (W. Jones 1954)
Phleum pratense L.	Timothy	Root rot	Ireland (Whelan and Loughnane 1969)[a]
Plantago media L.	Plantain	Root rot	Ireland (Whelan and Loughnane 1969)[a]
Saccharum officinarum L.	Sugarcane	Seed piece rot	United States (Steib and Chilton 1950)[a]
Secale cereale L.	Rye	Root rot	Ireland (Whelan and Loughnane 1969)[a]
Senecio cruentus (Mass.) DC	Cineraria	Wilt	New Zealand (Lucas 1977)
Solanum tuberosum L.	Potato	Pink tuber rot	Ireland (Pethybridge 1912); Netherlands (Pethybridge 1915); United Kingdom (Cotton 1922); Indonesia (Van Hall 1923; Bulgaria (Atanasoff and Kovacevski 1929); Canada (W. Jones 1945, 1954); Australia (Chambers 1946); United States (Blodgett 1945); India (Dastur 1949); New Zealand (Anon. 1953); Yugoslavia (Heddergott 1953); Tasmania (Wade 1954); Egypt (Abo-el-Dahab and El-Goorani 1968); Peru (Torres et al. 1970); Iran (Ershad 1971)
Spinacia oleracea L.	Spinach	Seedling rot	Canada (W. Jones 1954)[a]
Trifolium incarnatum L.	Crimson clover	Root rot	United States (Pratt 1980)
Trifolium vesiculosum Savi.	Crimson clover	Root rot	United States (Pratt 1980)
Triticum vulgare	Wheat	Root rot	Ireland (Whelan and Lougnane 1969)[a]
Tulipa L.	Tulip	Root rot; shanking	United Kingdom (Buddin 1938): Greece (Kouyeas 1978)
Tussilago farfara L.	Coltsfoot	Root rot	Ireland (Whelan and Loughnane 1969)[a]
Vicia angustifolia L.	Tare; vetch	Root and crown rot	United States (Morgan and Johnson 1965)
Vicia dasycarpa Ten.	Wooly-pod; vetch	Root and crown rot	United States (Morgan and Johnson 1965)[a]
Vicia faba L.	Broad bean	Root rot	Ireland (Whelan and Loughnane 1969)[a]
Vicia sativa L.	Spring vetches	Root rot	Ireland (Whelan and Loughnane 1969)[a]
Zantedeschia aethiopica (L.) Spreng.	White calla; arum lily	Rhizome rot	United States (Tompkins and Tucker 1950)
Zantedeschia elliottiana (Engl.)	Yellow calla	Leaf and petiole blight	United States (Tompkins and Tucker 1947a)
Zantedeschia rehamanii (Engl.)	Pink calla	Leaf and petiole blight	United States (Tompkins and Tucker 1947a)

[a] Artificially inoculated host.

lacks pigment on casein hydrolysate agar medium. Geographical distribution is cited in Commonwealth Mycological Institute Map 83, ed. 4, 1973. References on distribution of hosts are listed in Table 25A.1.

Characteristics of *Phytophthora erythroseptica*

A summary of the morphology and the diseases caused by *P. erythroseptica* is given by Stamps (1978c). *P. erythroseptica* was redescribed by Ho and Jong (1989b). See Tables 4.2 and 4.3 for tabular keys. It is classified in group VI (Stamps et al. 1990). Morphology is shown in Figure 25A.1.

Sporangia

Sporangia are nonpapillate and ellipsoid, ovoid, obpyriform, or distorted in shape, often with a constriction distal to the middle. They have tapered bases, germinate by internal proliferation, and are 43 to 69 µm long × 26 to 47 µm wide (average 44.2 × 27.2 µm). The length-breadth ratio averages 1.65:1. Sporangiophores are sympodial, branching immediately below the sporangium. Zoospores are released from sporangia after chilling of the aqueous culture (Vujičić and Calhoun 1966).

Hyphal Swellings

Hyphal swellings (11–27 µm in diameter) are round or angular and form in chains or clusters in water.

Chlamydospores

Chlamydospores have not been found.

Sex Organs

Antheridia are amphigynous, elongated, and cylindrical; oogonia are 30 to 46 µm in diameter (average 34.9 µm); oospores are aplerotic, 28 to 35 µm in diameter (average 30.5 µm), and homothallic. Oogonia with two, three, or four antheridia were observed by Gerrettson-Cornell (1985) but were not noted by Ho and Jong (1989b).

Growth Temperatures

Minimum temperature for growth is 2.5°C, optimum 27.5°C, and maximum 34°C.

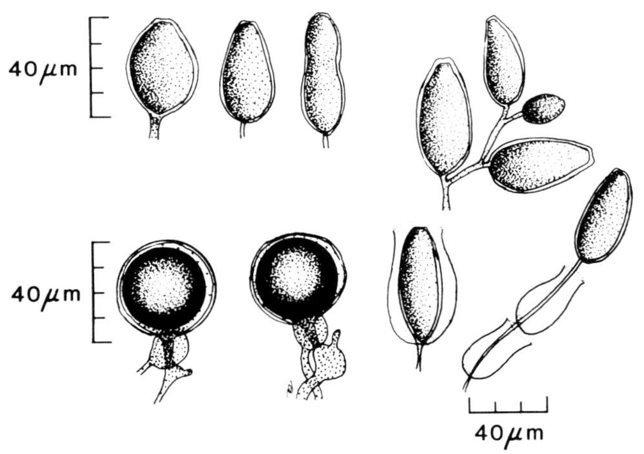

Figure 25A.1. Morphology of *Phytophthora erythroseptica.* Upper row, nonpapillate sporangia; sympodium. Lower row, homothallic oospores with amphigynous antheridia; internal and external proliferation. (Drawing by A. Vaziri)

Some Diseases Caused by *Phytophthora erythroseptica*

P. erythroseptica has been misidentified over the years as one of the *Phytophthora* species that causes root rot of red raspberry (*Rubus idaeus* L.) (Converse and Schwartz 1968; Montgomerie and Kennedy 1980; Müller et al. 1986); the cause of red raspberry root rot has since been resolved to be *P. fragariae* var. *rubi* (Duncan et al. 1991; Wilcox et al. 1993). (See Chapter 26C for a discussion of red raspberry root rot.)

Characterization of mitochondrial DNA restriction fragment length polymorphism (mtDNA RFLP) patterns has confirmed the similarity of red raspberry isolates, formerly attributed to *P. erythroseptica,* to *P. fragariae* (54 to 55% similarity) (Förster and Coffey 1992). This pathogen has been identified and described as *P. fragariae* var. *rubi* Wilcox and Duncan var. nov. (Wilcox et al. 1993). The raspberry isolates produce occasional paragynous antheridia, thus deviating from the accepted description of *P. erythroseptica* characterized by production of amphigynous antheridia (Pethybridge 1913). Ho and Jong (1989b) stated that isolates that produce any paragynous antheridia should not be classified as *P. erythroseptica.* This requirement excludes the isolates from red raspberry (Converse and Schwartz 1968; Duncan et al. 1990) and fast-growing isolates from arrowleaf clover (R. G. Pratt 1981) and vetch (Morgan and Johnson 1965). The isolates in the latter paper were reported to produce only amphigynous antheridia but subsequently were found by Savage et al. (1968) to produce some paragynous antheridia. The mtDNA RFLP patterns showed that the genetic distance between the clover isolates (R. G. Pratt 1981) and the potato isolate were large enough to raise questions regarding their identity as *P. erythroseptica* (Förster and Coffey 1992). Ho and Jong (1989b) recommended reclassifying these isolates to *P. megasperma.* Although this may be an acceptable solution, Förster and Coffey (1993) found that many of the biotypes that are lumped into *P. megasperma* by Hansen and Maxwell (1991) show greater genetic diversity in mtDNA RFLP patterns than exists between many valid species.

Pink Rot or Tuber Rot of *Solanum tuberosum* L. (Potato)

The disease and pathogen are described by Rowe and Nielson in the *Compendium of Potato Diseases* (Hooker 1981).

Pink rot occurs primarily in the tubers, although stem decay and wilt of foliage also occur. In general, affected tubers have a dull brown appearance, and the lenticels and eyes turn dark brown. The internal decay usually begins at the stem end, and the affected tissue becomes somewhat rubbery and exudes water readily under pressure. In cut tubers, a clear line is not always visible between the healthy and diseased tissue, although at times there is a definite blackish band near the periphery of the lesion. After exposure to air, the cut surface of the tuber becomes faint pink, and within 30 to 45 min, the entire surface of the diseased tuber becomes deep salmon pink. This characteristic feature gives the disease its name, pink rot (Plate 25A.1). The penetration of *P. erythroseptica* by mycelium and zoospores was shown on stems, stolons, and roots of growing potato plants (Lonsdale et al. 1980).

The tough, leathery, rubberlike texture of tubers infected by *P. erythroseptica* distinguishes it from bacterial rot diseases in which the diseased tissue becomes soft and pulpy and contains numerous cavities.

Disease Development. The disease is probably initiated by dormant oospore inoculum and is favored by high soil moisture and temperatures between 20 and 30°C. The pathogen is disseminated rapidly by means of zoospores. Lonsdale et al. (1980) found that zoospores or mycelial inoculum applied to stems, stolons, or roots of growing potato plants induced symptoms; however, infection occurred on growing tubers (oospores formed in surface tissue), but symptoms did not. They discuss the possibility that oospores in infected but symptomless tubers could be a source of inoculum on uninfested land. In addition to *P. erythroseptica, P. drechsleri* is capable of causing pink rot of tubers at higher temperatures. *P. cryptogea, P. megasperma,* and *P. parasitica* also cause pink rot (Hooker 1981). See also Rowe and Schmitthenner (1977), who reviewed the literature and showed that *P. erythroseptica* and *P. cryptogea* were endemic in Ohio potato fields and were capable of causing pink rot.

Control. Since *P. erythroseptica* is soilborne and can persist for many years as resistant oospores, rotation of infested land for at least 4 years to a nonsusceptible crop is advisable.

The use of disease-free seed and destruction of all infected plants and tubers before planting in fields are sanitary preventative measures. Fields with good drainage of water from irrigation or rain should be selected. Excessive irrigation should be avoided late in the growing season.

Seed-Piece Rot of *Saccharum officinarum* L. (Sugarcane)

Seed-piece rot was first described by Steib and Chilton (1950) on stem pieces of sugarcane used to start new plantings in Louisiana, United States. Infected tissues become water soaked. As the disease progresses, salmon pink to orange red streaks appear throughout the interior of the stalk. In yellow-stalked cultivars, the streaking can usually be observed throughout the intact cane. In advanced stages, the whole cane becomes water soaked and reddish brown and has a distinct etherlike odor (Steib and Chilton 1950). Seed pieces fail to develop roots, and buds do not germinate when planted in infested soil. The disease has not been reported from any other country. Ho and Jong (1989b) express reservations about classifying the sugarcane isolates in *P. erythroseptica* because the oospores and sporangia are much larger than those of *P. erythroseptica* and the hyphal swellings resemble those of *P. megasperma.*

P. megasperma has also been isolated from decaying seed pieces and appears to be more virulent than *P. erythroseptica* in causing seed-piece rot of sugarcane (Van der Zwet and Forbes 1961).

Disease Development. The disease is most destructive in poorly drained soils after a cold, wet winter and a cool spring (Steib and Chilton 1950).

Control. Improvement of soil drainage aids in control, since the disease is most prevalent in wet soils.

The sugar cane cultivar C.P. 36-13 is very susceptible, while C.P. 29-116, C.P. 44-101, C.P.47-193, C.P. 48-103, and N.Co. 310 are resistant. The level of resistance available in 24 sugarcane cultivars to Phytophthora seed-piece rot is listed by Van der Zwet et al. (1960).

CHAPTER 25B

Phytophthora erythroseptica Pethybridge var. *pisi* Bywater and Hickman (1959)

No synonyms are presently known for *Phytophthora erythroseptica* var. *pisi,* which was first isolated from diseased pea (*Pisum sativum* L.) roots in England by Bywater and Hickman (1959) and is specific to pea. Whelan and Loughnane (1969) reported that pea was a nonsolanaceous host of *P. erythroseptica.* Although *P. erythroseptica* var. *pisi* was originally rejected by Waterhouse (1963) as an invalidly published variety, it has now been reinstated in the key by Newhook et al. (1978) and Stamps et al. (1990). Ho and Jong (1989b) recommend that it be retained but that its "acceptance . . . is debatable."

Characteristics of *Phytophthora erythroseptica* var. *pisi*

Sporangia

Sporangia are nonpapillate and persistent on the stalk and proliferate internally. They are ellipsoid, ovoid, obpyriform, often with a median constriction, or often distorted in shape; 54 to 171 μm long × 25 to 40 μm wide (average 86 × 35 μm) on bean agar; and 43 to 95 μm long × 19 to 40 μm wide (average 63 × 29 μm) in water. Sporangiophores are sympodially branched.

Hyphal Swellings

Hyphal swellings form readily in water; they are round, angular, and in chains or clusters; the average diameter is 20 μm.

Chlamydospores

Chlamydospores are unknown.

Sex Organs

Antheridia are round, isodiametric, and amphigynous; oogonia are abundant in the cortex and the stele of the host and 28 to 44 μm in diameter (average 37 μm); oospores are aplerotic, 14 to 34 μm in diameter (average 26 μm), and homothallic.

Growth Temperatures

Minimum temperature for growth is <6°C, optimum 24°C, and maximum >30°C.

Distinguishing Features

P. erythroseptica var. *pisi* differs from *P. erythroseptica* var. *erythroseptica* by its lateral, tortuous hyphae (2 to 14 μm wide) and larger sporangia. It also differs in that it is specific to pea, whereas *P. erythroseptica* has a fairly wide host range.

Disease Caused by *Phytophthora erythroseptica* var. *pisi*

Root Rot of *Pisum sativum* L. (Pea)

Early symptoms of root infection by *P. erythroseptica* var. *pisi* resemble those caused by *Aphanomyces euteiches* and *Pythium debaryanum.* The infected roots are soft and have a water-soaked appearance. Roots are also faintly discolored. In severe cases, shoots may wilt.

Disease Development. The disease is most severe at low temperatures, i.e., around 15°C, particularly in glasshouses.

Control. In glasshouse culture, only noninfested soil should be used and cool, damp conditions should be avoided. In the field, peas should be planted only in well-drained soils; low-lying areas and heavy soils should be avoided.

CHAPTER 26A

Phytophthora fragariae

Hickman (1940)

var. *fragariae*

No synonyms are known for *Phytophthora fragariae,* which was first associated with strawberry red core (term used in Great Britain) or red stele (term used in the United States) by Alcock (1929) in Scotland; however, the causal fungus was not described as *P. fragariae* until 1940 by Hickman. It has since been reported to cause a root rot of considerable importance in all strawberry-growing countries (Table 26A.1). The pathogen appears to have a limited host range because it occurs in the field only on strawberry (*Fragaria* × *Ananassa*) derived from hybridization of *F. chiloensis* (L.) Duch. and *F. virginiana* Duch. and loganberry (*Rubus ursinus* Cham. and Schlechtend. var. *longanobaccus* (L. H. Bailey) L. H. Bailey) (Table 26A.1). *P. fragariae* var. *oryzobladis* was described on rice (*Oryza sativa* L.) by J. S. Wang and Lu (1978) and is discussed in Chapter 26B. *P. fragariae* var. *rubi,* the cause of root rot of red raspberry (*Rubus idaeus* L.), was described by Wilcox et al. (1993). This new variety, which was characterized by analysis of mtDNA (Förster and Coffey 1992; Stammler et

Table 26A.1. Distribution of hosts of *Phytophthora fragariae*

Host	Common name	Disease	Geographical distribution
Dryas drummondii Rich	Dryas	Red stele	Canada (Pepin 1967)[a]
Fragaria × *Ananassa*	Strawberry	Red stele; red core; brown-core root rot; black stele; lanarkshire disease	England (Wardlaw 1926; Wormald 1936; Hickman 1940); Scotland (Alcock 1929); United States (Bain and Demaree 1938; Anderson 1940); Canada (Conners and Saville 1946); New Zealand (Anon. 1949b); France (Trebuchet 1950); Austria (Schreier 1954); Northern Ireland (Anon. 1955a); Australia (Kerr 1957); Russia (Maklakova 1958); Czechoslovakia (Cejp 1961); Japan (Morita 1965); Netherlands (Vlasveld 1969); Italy (Canova and D'Ercole 1972); Bulgaria (Nakov 1976); Taiwan (Anon. 1978b); Sweden (Graberg 1979)
Fragaria chiloensis (L.) Duchesne	Beach strawberry	Root rot	United States (Bain and Demaree 1945)[a]
Fragaria moschata Duchesne	Strawberry	Root rot	United States (Bain and Demaree 1945)[a]
Fragaria vesca L.	Woodland strawberry	Root rot	United States (Bain and Demaree 1945)[a]
Fragaria virginiana Duchesne	Strawberry	Root rot	United States (Bain and Demaree 1945)[a]
Geum coccineum Sibth. and Sm.	Avens	Red stele	Canada (Pepin 1967)[a]
Geum trifolium Pursh.	Avens	Red stele	Canada (Pepin 1967)[a]
Geum unbanum L.	Avens	Root rot	United States (Converse and Moore 1966)[a]
Lycopersicum esculentum Mill.	Tomato	Root rot	Canada (Anon. 1955b)
Nicotiana tabacum L.	Tobacco	Root rot	Canada (Anon. 1955b)
Potentilla andicola Benth.	Cinquefoil; five-finger	Root rot	United States (Moore et al. 1964)[a]
Potentilla anserina L.	Cinquefoil; five-finger	Red stele	Canada (Pepin 1967)[a]
Potentilla argentea L.	Cinquefoil; five-finger	Red stele	Canada (Pepin 1967)[a]
Potentilla erecta (L.) Rauschel	Tormentil	Root rot	Canada (Pepin 1967)[a]
Potentilla glandulosa Lindl.	Cinquefoil; five-finger	Red stele	United States (Moore et al. 1964)[a]; Canada (Pepin 1967)[a]
Potentilla glandulosa var. *pseudorupestris* (Rydb.) Breit	Cinquefoil	Root rot	Canada (Pepin 1967)[a]
Potentilla gracilis var. *flabelliformis* (Lehm.) Nutt. ex T. and G.	Cinquefoil	Root rot	Canada (Pepin 1967)[a]
Potentilla leucopolitana Muell ex Schultz	Cinquefoil	Root rot	Canada (McKeen 1958)[a]; United States (Moore et al. 1964)[a]
Potentilla norvegica L.	Cinquefoil	Root rot	Canada (Pepin 1967)[a]
Potentilla recta L.	Cinquefoil; five-finger	Red stele	Canada (Pepin 1967)[a]
Potentilla rupestris var. *villosa* Lec. and Lamotte	Cinquefoil; five-finger	Root rot	Canada (McKeen 1958); United States (Moore et al. 1964)[a]
Rubus idaeus L. var. *idaeus* L.	Red raspberry	Red rot	Canada (Pepin 1967)[a]
Rubus loganobaccus Bailey	Loganberry	Root rot	Canada (McKeen 1958); United States (Wilcox 1989); United Kingdom (Duncan et al. 1991)
Rubus parviflorus Nutt.	Thimbleberry	Root rot	Canada (Pepin 1967)[a]
Spinacia oleracea L.	Spinach	Seedling rot	Canada (Anon. 1955b)[a]

[a] Artificially inoculated host.

al. 1993), is discussed in Chapter 26C. Neither of these varieties is pathogenic on strawberry, and the strawberry pathogen is not pathogenic on either of the other hosts.

Typical root rot symptoms on several members of the family Rosaceae have been caused by artificial inoculation with the strawberry pathogen, *P. fragariae* var. *fragariae* (hereafter referred to as *P. fragariae*) (Table 26A.1). Ho and Jong (1988) summarized the literature, restudied the characteristics of 20 isolates collected from strawberry in Europe and North America and one isolate from loganberry from Canada, and redefined the species. Montgomerie (1961, 1967, 1977) and Scott et al. (1950, 1976, 1984) reviewed the early literature on red stele, its control by use of host resistance, and the pathologic races of *P. fragariae.*

Characteristics of *Phytophthora fragariae*

Ho and Jong's (1988) redescription of *P. fragariae* includes a summary table of morphological data from 24 reports (1929 to 1986). *P. fragariae* is classified in group V (Stamps et al. 1990). See Tables 4.2 and 4.3 in Chapter 4 for tabular keys. Maas (1984) summarizes information on the disease and *P. fragariae.* Morphology is shown in Figure 26A.1.

Sporangia

Sporangia are produced by most isolates in 24 to 48 h when disks of a culture on V8 juice agar are incubated in water, preferably nonsterile stream water. Sporangia are nonpapillate and nondeciduous and proliferate internally. Sporangia are formed on undifferentiated sporangiophores, which often branch sympodially to bear sporangia at the tips. Sporangia are mostly large, 32 to 90 µm × 22 to 52 µm, and length-breadth ratios average 1.8 ± 0.2 (Ho and Jong 1988). The overall mean in the literature is 1.5 to 1.8. Sporangia are obpyriform or ovoid to elongate with rounded bases tapering to blunt beaks.

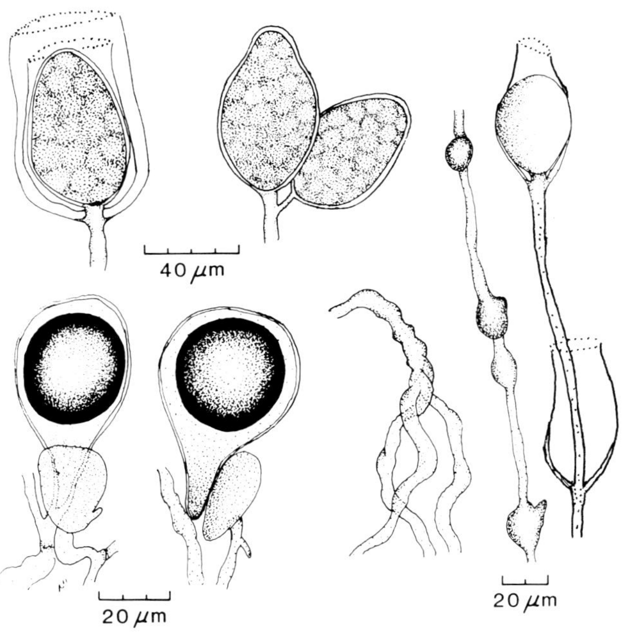

Figure 26A.1. Morphology of *Phytophthora fragariae.* Sporangia are nonpapillate, form sympodially, and proliferate internally and externally. Antheridia are mostly amphigynous but are sometimes paragynous. (Drawing by A. Vaziri)

Chlamydospores and Hyphal Swellings

Cultures incubated for a week in water often develop small hyphal swellings, but chlamydospores are not formed (Ho and Jong 1988).

Sex Organs

P. fragariae is homothallic and produces sex organs abundantly in host tissue but sparsely in culture (Ho and Jong 1988). Oogonia are globose to subglobose and often have a tapered, funnel-shaped base. The antheridia are predominantly amphigynous and single celled, but some paragynous antheridia are also formed. Ho and Jong (1988) question placement of *P. fragariae* in group V (Stamps et al. 1990), because that group is primarily characterized by production of paragynous antheridia. Oospores are globose and about 34 µm in diameter and do not fill the oogonium (aplerotic). Oogonia are about 28 to 44 µm (average 33 µm) in diameter. Oospores of *P. fragariae* are germinable (Duncan 1975a, 1977, 1985a). Since fungicides do not affect germination of oospores (Duncan 1985b), these structures are important for survival. See Chapter 3 for a discussion of oospore germination.

Growth Temperatures

Minimum temperature for growth is 3°C, optimum is 18 to 22°C, and the maximum is 30°C. Maas (1972) reports that some isolates grow at 30°C. Kröber (1985) in Germany cites 25°C as the maximum temperature.

Distinguishing Characteristics

P. fragariae is a low-temperature fungus with optimum growth recorded at temperatures ranging from 10 to 15°C and 20 to 24°C in different reports (Ho and Jong 1988). It grows slowly (about 2 to 4 mm per day) and is more nutritionally fastidious than most other *Phytophthora* spp. Davies (1959) concluded that a complex mixture of amino acids comparable to that in casein hydrolyzate was required for optimum growth in synthetic media. It will not grow on an inorganic nitrate source (Fleetwood and Walker 1955; Hohl 1983). Ho and Jong (1988) found that *P. fragariae* does not grow well on malt agar or on agar containing reducing sugars such as glucose. The raspberry isolates (*P. fragariae* var. *rubi*), however, grow better on malt agar than strawberry isolates. McKeen (1955) reported that a product toxic to *P. fragariae* formed in media containing glycine and glucose. Starch is not utilized well. Such data indicate that *P. fragariae* lacks certain enzymes necessary to utilize inorganic nitrogen or certain carbohydrate sources. Growth is good on cleared V8 juice agar supplemented with sitosterol (30 mg per liter) (Ho and Jong 1988).

Since most of the antheridia are amphigynous, *P. fragariae* resembles *P. cambivora,* but *P. cambivora* produces a bullate oogonial wall. Ho and Jong (1988) state that the presence of large sporangia (56 × 35 µm) does not necessarily separate *P. fragariae* from *P. megasperma* as stated by Waterhouse (1963), because some isolates of *P. fragariae* produce small sporangia. *P. fragariae* can be differentiated from *P. megasperma* by the following: its slower growth rate on cleared V8 juice agar; the production of thick and fluffy aerial mycelium with a light brown discoloration of the medium on the bottom of the plate; the meandering and coiling nature of the hyphae with sparse but elongated branches; the absence of chlamydospores or hyphal swellings; the slightly more elongated sporangia; production of oogonia with tapered bases; its inability to grow at 30°C; its sparse growth on potato-dextrose agar, malt agar, starch agar, or inorganic nitrate agar; and its insensitivity to malachite green. Differences from *P. fragariae* var. *oryzobladis* and *P. fragariae* var. *rubi* are discussed in Chapters 26B and 26C.

Disease Caused by *Phytophthora fragariae*

Red Stele or Red Core of *Fragaria* (Strawberry)

Red stele is now recognized as the most important fungal disease of strawberry worldwide. In New Zealand, red stele is a major factor in the production of strawberries as a perennial crop. A complete loss of the crop usually occurs after the first bearing year in afflicted fields. Montgomerie (1977) and Plakidas (1964) reviewed the characteristics of red stele and its control. Maas (1984) covers recognition and control in the United States.

Red stele is detected in the spring or early summer by the presence of severely stunted plants. On warm days, entire plants or only their outer leaves wilt suddenly. In general, the older leaves of infected plants senesce prematurely and take on various shades of yellow, red, and brown; the leaves then wilt and dry up. New leaves produced on infected plants are small, blue green, and borne on short petioles. Plants in advanced stages of infection produce either no fruit or fruit that often shrivels up before ripening. Slightly infected plants may be indistinguishable from healthy ones, but growth is progressively more restricted as disease severity increases.

The root systems of diseased plants are devoid of lateral roots, thus giving the larger roots with black tips a characteristic smooth, rat's tail appearance (Plate 26A.1). When the stele is cut, a distinctive red or reddish brown color, sharply defined from the surrounding white cortical tissue, is seen. The coloration may extend the whole length of the root up to the crown. This distinctive symptom gives the disease its name. The crown is not affected. The roots that develop during the warmer summer months do not become infected, and this accounts for the partial or complete recovery of plants during the warm season. Thus, plants should be diagnosed for red stele in the spring, since diseased roots decay during the summer, leaving little or no evidence of disease. Isolation of *P. fragariae* is extremely difficult during the summer.

Oospores of *P. fragariae* form abundantly in diseased steles and survive 3 years or longer in the field (Duncan and Cowan 1980). They cite field evidence of persistence up to 15 years in soil. See also Table 5.3. In some hosts, oospores have also been found in the leaves, petioles, and stolons (Duncan 1976); however, they develop sparsely in culture (Ho and Jong 1988). Oospores have been shown to germinate and undoubtedly account for persistence of *P. fragariae* in soil (Duncan 1975a, 1977, 1985a; see Chapter 3 for a discussion of oospore germination).

Disease Development. The amount of soil water is important, since infection does not occur until the water-holding capacity of the soil reaches 50% or more. Plate 26A.2 shows loss of plants in low spots in a field. The disease spreads during the fall, winter, and spring, because the pathogen is favored by soil temperatures below 18°C. Slightly infected plants may escape detection by inspectors. These plants are important sources for initiation of the disease at new sites (Duncan and Cowan 1980).

Control. Only planting stock that is free of the pathogen should be planted. See Chapter 5 for a discussion of control by the planting of clean stock (Duncan 1975b, 1976, 1979, 1980a; Duncan et al. 1986). Disease-free nursery stock is extremely important for control, because the disease can spread so rapidly when conditions are cool. Various methods of detection, ranging from baiting with susceptible roots to use of the serological enzyme-linked immunosorbent assay (ELISA), are given in Chapter 2. Duncan et al. (1986) reported that testing with strawberry root bait to determine the presence or absence of *P. fragariae* in planting stocks, with subsequent rejection of plants with infected roots, had resulted in a dramatic reduction in the red stele disease in fields in Scotland. During 1989–1990, no infected planting stocks in commercial fields were found (J. M. Duncan, *personal communication*).

Soil fumigation with a mixture of methyl bromide and chloropicrin has been routinely used to control red stele and other diseases in areas of intense cultivation such as in California (Wilhelm et al. 1974; Wilhelm and Paulus 1980). The land is fumigated each year before new plants are started (see Chapter 7).

Although routine fumigation of soil has been highly effective in California, a comparison of dazomet, chloropicrin, and methyl bromide in England used for control of red core of strawberry yielded varying results. Dazomet (570 kg/ha) and methyl bromide (450 kg/ha) applied in beds were ineffective for control of red stele; however, the incidence of crown rot caused by *P. cactorum* was reduced, and the yield of berries was increased by 40 to 60%. Chloropicrin (150 liters/ha) had no effect on either crown rot or red core. Methyl bromide (750 kg/ha) reduced but did not eliminate the incidence of crown rot and the severity of red core (D. C. Harris 1991b).

Because methyl bromide contributes halogen residues to the air and is considered to react with the ozone layer in the atmosphere, its use as a soil fumigant will probably be discontinued. If so, the red stele disease will become much more important in areas such as California, where fumigation successfully controls *P. fragariae.* More emphasis will then have to be placed on host resistance and sanitation measures.

A single postplant drench with captafol (0.2%) and prothiocarb (0.28%) or a preplant dip in LS1038 (sodium ethyl phosphite) has been reported to be successful for control of red stele (Upstone 1976; Montgomerie and Kennedy 1977).

Application of metalaxyl as a soil drench gave good control of red stele (McIntyre and Walton 1981; Duncan 1985; Jamart et al. 1985). In England, fosetyl-Al applied once as a spray after fruit picking in autumn reduced the incidence of red stele, increased the number of healthy roots, and increased the yield during the following year. A second treatment that included fosetyl-Al in the autumn and again the following spring further increased the numbers of healthy roots but did not increase the yield (O'Neill and Griffin 1987). Control of red core by fungicides is discussed further in Chapter 7. McIntyre and Walton (1981) reported that plants treated with metalaxyl produced an average of 161% more fruit by weight. They advocated use of metalaxyl as a transplant drench followed by spray applications in the spring after bud break and each fall prior to dormancy for plants to be grown in fields infested with *P. fragariae.* Since metalaxyl is a site-specific fungicide, resistance has developed in *P. infestans* (Chapter 7). Müller and Sun (1989) reported that resistance in field populations of *P. fragariae* occurred after regular applications of metalaxyl. An extensive review of chemical control is given by Montgomerie (1977).

Duncan (1985b) noted that because most fungicides do not kill oospores, the use of fungicides as root dips to control latent infection should be looked upon skeptically. See Chapter 7 for pertinent literature and discussion of chemical control.

Heat treatment of planting stock is ineffective, since the fungus can withstand temperatures that kill the plant. Because the fungus can persist for 8 to 10 years in the soil, crop rotations are of little value.

Houck (1962) obtained good control by incorporating ammoniated Douglas fir bark at 90–225 tons/ha into infested soil. After 2 years, however, the suppression effect was lost, and after the fourth year, the disease became severe. Douglas fir sawdust increases the incidence of red stele (Vaughn et al. 1954). This is probably because of the high wood content (Hoitink 1980).

The best approach is to plant disease-free, red stele-resistant cultivars in clean soil. However, the frequent occurrence of 12 to 15 pathologic races of *P. fragariae* complicates breeding new cultivars of strawberry for control of red stele (Montgomerie 1961, 1967, 1977; Maas 1972, 1984; Scott et al. 1950, 1976, 1984; Hickman 1951a, b, 1962; Converse and Scott 1962; Pepin and Daubeny 1964). See also Chapter 6 for more discussion. Red stele-resistant cultivars must be planted with the understanding that new, more virulent races of the fungus may develop in areas where resistant material is grown continuously. In the United States, the following cultivars have varying degrees of resistance to different races of *P. fragariae:* Columbia, Darrow, Earliglow, Guardian, Hood, Marshall, Midway, Milalla, Olympus, Rainier, Redchief, Redcrop, Shuksan, Siletz, Sparkle, Surecrop, Sunrise, Temple, Totem, and Vermillion. Some of these are better adapted to certain areas of the country than others. The cultivars Midway, Sparkle, and Surecrop are presently resistant to *P. fragariae* races in the eastern United States, while cultivars Hood, Molalla, Olympus, Rainier, Shuksan, Siletz, and Totem are resistant to several races known to occur in the western United States. In the United Kingdom, cultivars with resistance to one or more races include Aberdeen, Cambridge Vigour, Perle de Prague, Siletz, and Frith (Scott et al. 1984). See Chapter 6 for a more extensive discussion of the evaluation and use of cultivar resistance to red stele.

Status of a High-Temperature Variant from Taiwan

A "high-temperature" variant from strawberry in Taiwan that grows optimally at 30°C and maximally at 39°C was tentatively assigned to *P. fragariae* by H. S. Chang (1988); however, Ho and Jong (1988) examined the high-temperature isolate and reported that it differed from *P. fragariae* by its rapid production of sporangia (in 6 to 7 h), shorter and wider sporangia, resistance to malachite green, utilization of starch and inorganic nitrate, and lack of production of sex organs in culture. They objected to inclusion of the high-temperature Taiwan isolate in *P. fragariae* and noted its resemblance to the high-temperature isolate that grows at 39°C from alfalfa that had been tentatively assigned to *P. megasperma* (Ribeiro et al. 1978). On the basis of current knowledge, the latter fungus is probably not *P. megasperma* because the mtDNA RFLP pattern of the alfalfa high-temperature isolate differs markedly from that of *P. megasperma* f. sp. *medicaginis, P. megasperma* f. sp. *glycinea,* and large- and small-oospore isolates of *P. megasperma* from cherry and apple (Förster et al. 1989). Förster et al. (1989) concluded that each subgroup was probably representative of a biological species. The mtDNA of this high-temperature variant of *P. fragariae* (Chang 1988) has not been analyzed (Förster and Coffey 1992; Seemüller et al. 1993), and it probably should not be considered to be *P. fragariae* until this analysis has been conducted.

According to Ann and Ko (1994), the high-temperature variant from strawberry is similar in morphology, temperature relations, and electrophoretic protein patterns and in production of the α^1 hormone to the high-temperature isolate from alfalfa (Ribeiro et al. 1978) and to *P. insolita.* They conclude that this isolate from strawberry should be designated *P. insolita* (see Chapter 35).

CHAPTER 26B

Phytophthora fragariae var. *oryzobladis*

Wang and Lu (1978)

No synonyms are known for *Phytophthora fragariae* var. *oryzobladis,* which was first described by J. C. Wang and Lu (1978; in Chinese with an English summary) in the Jianzsu province of China as the causal agent of a leaf blight of rice (*Oryza sativa* L.) seedlings. The pathogen appears to be specific to rice plants, and the disease it causes is of minor importance.

Characteristics of *Phytophthora fragariae* var. *oryzobladis*

P. fragariae var. *oryzobladis* is classified in group V (Stamps et al. 1990). The following description is from J. C. Wang and Lu (1978).

Sporangia

Sporangia are obpyriform to elliptical and nonpapillate, proliferating internally, terminally, or laterally. They are 43.1 to 93.1 µm long × 28.8 to 57.5 µm wide (average 67 × 42.8 µm); length-breadth ratio is 1.6:1. Sporangiophores are simple or sympodially branched and 405 to 675 µm long. Sporangia germinate by proliferation.

Chlamydospores

Chlamydospores are not produced.

Sex Organs

The fungus is homothallic and produces amphigynous antheridia; subglobose or irregularly spherical, yellowish brown oogonia 39.9 to 58.1 µm in diameter (average 46.9 µm); and globose oospores 27.2 to 48.0 µm in diameter (average 35.5 µm).

Growth Temperatures

Growth temperatures have not been recorded.

Distinguishing Characteristics

J. C. Wang and Lu (1978) report that the sporangia and other characteristics are similar to those of *P. fragariae* from strawberry, but *P. fragariae* var. *oryzobladis* produces amphigynous antheridia, whereas *P. fragariae* produces both paragynous and amphigynous antheridia. They also note its specific pathogenicity to rice.

Ho and Jong (1988) noted that the type culture is no longer available for examination. They had reservations about its affinity to *P. fragariae* because of its rapid growth rate (4 to 6.5 cm per day) and its abundant production of chlamydospores.

Disease Caused by *Phytophthora fragariae* var. *oryzobladis*

The following is from J. C. Wang and Lu (1978).

Leaf Blight of *Oryza sativa* L. (Rice)

In the early stages of infection, light brown, circular spots occur on leaf blades. Under favorable conditions of temperature and humidity, the spots enlarge, turn gray green, and coalesce. The tips of the leaf blades curl downward. In advanced stages of the disease, the lesions appear as purple brown streaks along the margins of the leaf, gradually turning brown toward the center of the leaf blade. During periods of high relative humidity, a white, fluffy growth of mycelium can be seen on the leaf blades. Sporangia and oospores form abundantly on and in diseased tissues.

Disease Development. The disease is most destructive between 16 and 22°C, lessening in severity as temperatures increase to 25°C.

Control. Control measures have not been reported.

CHAPTER 26C

Phytophthora fragariae var. *rubi*

Wilcox, Scott, Hamm, Kennedy, Duncan, and Brasier (1993)

The finding that *Phytophthora fragariae* var. *rubi* is the cause of the root rot disease of red raspberry, *Rubus idaeus* L. (Duncan et al. 1991; Wilcox et al. 1993), follows much controversy and disagreement on the proper name for this pathogen (Wilcox et al. 1993).

The highly pathogenic isolates of *Phytophthora* from red raspberry grown in North America and Europe had been previously assigned to *P. erythroseptica* (Converse and Schwartze 1968; Seemüller et al. 1986), *P. megasperma* (Duncan et al. 1987), and *P. fragariae* (Nourrisseau and Baudry 1987; Wilcox 1989). Citations on isolation of *P. cactorum, P. cambivora, P. cinnamomi, P. citricola, P. drechsleri, P. erythroseptica, P. megasperma,* and *P. syringae* from *Rubus* spp. are listed by Duncan (1991). However, the major pathogen is now known as *P. fragariae* var. *rubi* (Duncan et al. 1991; Wilcox et al. 1993).

A comparative study of a large number of isolates from red raspberry and strawberry by Duncan et al. (1991) showed that isolates from strawberry (*P. fragariae*) and red raspberry were similar enough in morphology and protein patterns to be considered *P. fragariae.* Although the biotypes from strawberry and red raspberry were similar, neither was pathogenic to the other host. The red raspberry isolates are now named *P. fragariae* var. *rubi* (Wilcox et al. 1993).

Characteristics of *Phytophthora fragariae* var. *rubi*

P. fragariae var. *rubi* is classified in group V (Stamps et al. 1990). See Tables 4.2 and 43. for tabular keys.

Duncan et al. (1991) showed that *P. fragariae* var. *rubi* is similar to *P. fragariae* in production of sporangia and oospores and in temperature relations. Electrophoretic assay on polyacrylamide gels showed only slight differences in protein patterns. Although *P. fragariae* var. *rubi* was distinguishable from *P. fragariae* var. *fragariae* by DNA technology, most mtDNA and nuclear DNA-developed probes from both varieties hybridized, thus indicating that they are related. That some hybridization occurred with *P. cambivora* indicated that its relationship with *P. fragariae* should be investigated further (Stammler et al. 1993).

A comparative study of the molecular characterization by DNA restriction fragment length polymorphisms (RFLP) of the strawberry and red raspberry biotypes showed a 54 to 55% similarity between the two groups but did not show a close genetic relationship with *P. megasperma* or *P. erythroseptica,* which had previously been named as causal agents (Förster and Cof fey 1992).

In an analysis of a wide range of *P. fragariae* var. *fragariae* (from strawberry) and *P. fragariae* var. *rubi* isolates via Southern blot hybridization, DNA from *P. fragariae* var. *fragariae,* which had been cloned in *Escherichia coli* bacteria, hybridized with *P. fragariae* var. *rubi;* however, little or no hybridization occurred with DNA of other *Phytophthora* species, including *P. megasperma* and *P. erythroseptica* (Stammler et al. 1993). Since isolates of either *P. fragariae* var. *fragariae* or *P. fragariae* var. *rubi* were very homogenous (little variability), it could be assumed that both populations had been spread worldwide from a common source, probably by horticultural movement of plant material. The 54–55% similarity in mtDNA RFLP patterns of *P. fragariae* from strawberry and *P. fragariae* var. *rubi* from red raspberry supports the hypothesis by Duncan et al. (1991) that both groups might have evolved from a common ancestor pathogenic on wild members of the tribe Potentilleae in the Pacific Northwest of the United States (Förster and Coffey 1992). Wilhelm and Sagen (1974) suggested that the red stele disease of strawberry caused by *P. fragariae* probably had its origin in this area.

The description of the new variety, *P. fragariae* var. *rubi,* is given below (reproduced by permission of the British Mycological Society, Manchester, United Kingdom, courtesy of Wilcox et al. 1993):

> Mycelial colonies grow slowly but well on sucrose asparagine, french bean, lima bean, and V8-juice agars, but growth is poor or absent on media with nitrate as the sole nitrogen source. Colonies uniform with profuse aerial mycelium on most agars including sucrose asparagine and V8 agars. Minimum growth at or below 4°, optimum 19–22°, and maximum 25–28°. Hyphae hyaline, non-septate or sparingly so in older parts: width 5.3 µm to 7.4 µm, mean 6.5 µm. Sporangiophores simple or sparingly branched. Sporangia produced only when irrigated; non-deciduous, terminal, non-papillate, proliferating internally by nesting and extension, and less often in sympodial fashion on a short lateral branch immediately beneath the original sporangium; ovoid to obpyriform in shape; sometimes tapered at the base; usually 53–68 × 34–44 µm (59 × 40 µm); wall thin with slight apical thickening; germinating by zoospores, rarely by germ tube. Zoospores 13 × 10 µm. Sexual organs formed in abundance in the vascular tissue of raspberry roots; also formed on some agar media, sometimes more commonly in the area around the inoculum plug. Oogonial terminal, usually globose with a tapered base 30–45 µm in diam. (mean ca. 39 µm); becoming golden brown with age. Antheridia terminal, amphigynous or less commonly paragynous; sphaeroidal to ovoidal; averaging 13–16 µm. Oospores spherical and aplerotic; 28–37 µm diam. (mean ca. 35 µm) with a smooth hyaline wall 2.7–3.4 µm (mean 3.0).
>
> Habitat: In the roots, crown, and canes of cultivated raspberries in most countries of northern Europe and North America.
>
> The varietal epithet of this fungus is named for its host, *Rubus* spp. The primary cultural traits distinguishing *P. fragariae* var. *rubi* from *P. fragariae* var. *fragariae* are the former's more robust growth on agar media and far greater tendency to form oospores in single agar culture. The two varieties also differ in the electrophoretic patterns of their mycelial proteins, in the composition of their mitochondrial and nuclear DNA as indicated by RFLP analysis and hybridization studies, and in host range, with *Rubus* and *Fragaria* spp. as primary hosts, respectively.

Disease Caused by *Phytophthora fragariae* var. *rubi*

Root Rot of *Rubis idaeus* L. var. *idaeus* (Red Raspberry)

P. fragariae var. *rubi* Wilcox and Duncan (Wilcox et al. 1993) has during the last 10 years become the cause of one of the most important diseases affecting the production of red raspberries. Considerable evidence indicates that the disease spread by the movement of infected plants among and from various nurseries. The relatively recent discovery of the highly pathogenic raspberry isolates can be attributed in part to the use of antimicrobial chemicals in *Phytophthora*-selective media (see Chapter 2), particularly because both *P. fragariae* var. *fragariae* and *P. fragariae* var. *rubi* are slow growing and notoriously difficult to isolate, especially during the summer (Duncan et al. 1991).

A symptom observed in the early spring is the dull color of the leaves on the newly emerged primo canes. Later, there is a slight wilting of the terminal leaves. Infected raspberry plants may continue in an unthrifty condition for several seasons before the plants die. As the infection progresses, the leaves become small and yellowish and petioles shorten; older leaves often wither and appear scorched during hot, dry periods. Plants often appear to recover. These symptoms are often confused with those caused by *Verticillium dahliae;* however, in the case of Phytophthora root rot, the dead leaves remain attached to the primo canes. Roots of plants infected with *Phytophthora* become necrotic, and feeder roots are killed. Lesions often extend several millimeters above and below the point of infection. When infected roots are slit open lengthwise, a reddish brown discoloration of the stele can be seen. New roots may arise above the points of infection; however, these roots become infected during the cool months of the year. Distinct necrotic lesions often eventually girdle the whole crown. Grayish, water-soaked lesions often occur at the bases of diseased canes. Diseased fruiting canes do not normally survive the winter, and those that do often collapse before producing harvestable fruit. Berries remain small and often shrivel before ripening.

Disease Development. Raspberry root rot is favored by cool conditions and wet and flooded soil, much the same conditions that favor the red stele disease of strawberry (Duncan and Kennedy 1989). *P. fragariae* var. *rubi* is most active from November to March; thus, this is the best time of the year to isolate it. *P. fragariae* var. *rubi* probably overwinters in the soil and in plant roots as thick-walled oospores. Chlamydospores have not been observed (Wilcox et al. 1993).

Root rot in infested fields reduced the yield of the susceptible cultivar Willamette by 77% (compared with plots on which root rot was controlled by soil application of metalaxyl) and by 49% (compared with plots on which root rot was controlled by foliar sprays of fosetyl-Al) (Bristow and Hummel 1993).

Control. Since the disease has spread throughout the world on infected canes or roots (Duncan et al. 1987, 1991), only noncontaminated nursery stock grown under sanitary conditions should be used for planting in noninfested land. The existence of pathologic races will complicate the maintenance of resistance by different cultivars (Kennedy and Duncan 1993).

Cultivars known to have resistance include Latham, Newburgh, Durham, Chief, and WSU 458 whereas Lloyd George, Skeena, Glen Clova, Chilcotin, Canby, and Willamette are susceptible. The cultivars Meeker and Sumner are moderately resistant. An extensive list of cultivars resistant and susceptible to Phytophthora root rot is given by Barritt et al. (1979). Wilcox and Nedill (1992) list the cultivar Cherokee as resistant and Titan as susceptible. In greenhouse tests, seedlings of *R. glaucus* Benth. and *R. summatranus* Miq. survived inoculation with *P. fragariae* var. *rubi* (then known as *P. erythroseptica*). Selections from populations of *R. idaeus* and *R. spectabilis* show potential for incorporation into red raspberry, *R. idaeus* (Bristow et al. 1988). See Bristow and Moore (1992, 1994) for evaluation of resistance among red raspberry cultivars grown in the northwest United States.

Regulating irrigation to prevent inordinate flooding of land is a suitable control practice (Duncan and Kennedy 1989). In inoculation experiments, disease did not ensue if plants were not subjected to prolonged waterlogging of the soil (Duncan et al. 1991). In heavy soils, growing plants on raised beds or on ridges that are sufficiently high to keep roots and crowns of the plants above the areas where water accumulates is effective (Maloney et al. 1993). In fields where waterlogging is associated with impervious hardpan layers in soil, breaking layers with subsoil equipment is often helpful.

In planting sites in Washington State that are infested with *P. fragariae* var. *rubi,* the systemic fungicide metalaxyl can be applied as a soil drench shortly after planting in the spring and again in the fall. Fosetyl-Al as a foliar spray applied four times during the growing season was effective in recovery of affected plants (Bristow and Hummel 1993). Fosetyl-Al was 60% as effective as metalaxyl in field trials (Bristow and Windom 1992). In contrast, field studies in New York State showed that soil treatment with metalaxyl was ineffective (Maloney et al. 1993). This could be because of differences in soil types. In Washington State, the half-life of metalaxyl in soil was more than 150 days. Soil application of metalaxyl in the spring increases the development of new roots and foliage. Subsequent foliar sprays of fosetyl-Al (which translocates downward) are more effective if the foliage is well developed (P. Bristow, *personal communication*).

CHAPTER 27

Phytophthora gonapodyides

(Petersen) Buisman (1927)

Pythiomorpha gonapodyides Petersen (1909) is the only known synonym of *Phytophthora gonapodyides* and was first isolated and described by Petersen (1910) from decaying apples in a pond. The 1909 reference in Danish is not cited because it is essentially the same as the English version in 1910 (Waterhouse 1970a). Later, Buisman (1927) transferred *Pythiomorpha gonapodyides* to the genus *Phytophthora.* The history of the species *P. gonapodyides* was reviewed by Waterhouse (1958), who indicated that the genus *Pythiomorpha* was invalid (see also Blackwell et al. 1941) and formally transferred three species to *Phytophthora* (including *P. gonapodyides*) and one to *Pythium. P. gonapodyides* is a minor pathogen on only a few hosts (Table 27.1). Brasier et al. (1989) reassigned the *P. drechsleri*-like isolates from remote forested areas of Alaska and Oregon (Hansen et al. 1988) to *P. gonapodyides* partly on the basis of similarity of protein banding patterns to isolates of *P. gonapodyides* from Great Britain (Brasier 1991). Brasier and Hansen (1992) stated that although *P. gonapodyides* is sterile, it may serve as a stimulator of selfing in A2 mating types of other species. They also cited unpublished information and Brasier (1991) that protein banding patterns of North American and British isolates were similar (their Figure 8.6) and suggested that the sterile isolates in group J and the chlamydospore-producing isolates in group K of the Mills et al. (1991) study of the *P. cryptogea-P. drechsleri* complex (Chapters 20 and 23) may be *P. gonapodyides.* They also speculated that isolates in group K may have originated from hybridization with *P. cryptogea* isolates. Brasier et al. (1993) compared the "*P. drechsleri*" isolates from Douglas fir trees in the Pacific Northwest (Hansen et al. 1988) with *P. gonapodyides* isolates from Great Britain and concluded that the "*P. drechsleri*" isolates were by morphological and protein pattern characteristics *P. gonapodyides.* Most of the *P. gonapodyides* isolates exhibited protein bands distinct from those of *P. drechsleri* and *P. cryptogea.* They concluded that *P. gonapodyides* is a moderate pathogen on seedling Douglas fir and *Abies* spp. and a more aggressive pathogen on seedling mountain hemlock. See R. G. Pratt et al. (1976), Hamm et al. (1985), and S. J. Campbell and Hamm (1989) for discussions of root rot caused by isolates now known to be *P. gonapodyides.*

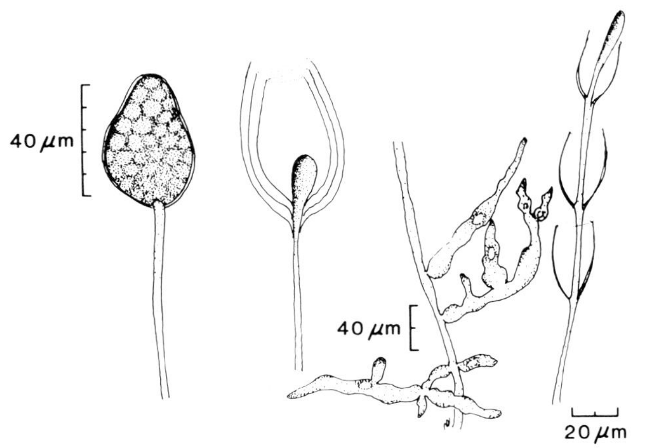

Figure 27.1. Morphology of *Phytophthora gonapodyides.* Left to right, obpyriform sporangium; nesting or internal proliferation; irregular shapes of hypae; and external proliferation of sporangia. (Drawing by A. Vaziri)

Characteristics of *Phytophthora gonapodyides*

P. gonapodyides is classified in group VI (Stamps et al. 1990). See Tables 4.2 and 4.3 for tabular keys. Morphology of spores is shown in Figures 27.1 and 27.2A–C. A typical colony is shown in Figure 27.2D. See Waterhouse (1958) for a comprehensive description.

Sporangia

Sporangia are nonpapillate, proliferating internally; nondeciduous; ellipsoid, ovoid, to obpyriform; and 42 to 75 μm long × 20 to 32 μm wide (average 44 × 22 μm) (Petersen 1910). Sporangia are broadly pyriform (Figure 27.2A), elongated pyriform, and occasionally ovate, 23 to 48 μm × 43 to 91 μm, mean 35.8 × 58.7 μm. The length-breadth ratio is 1.2:1.0 to 2.4:1.0, mean 1.6:1.0. Proliferation may be internal, nested (Figures 27.1 and 27.2B), or external. Sporangiophores are sympodial only in water (Pittis and Colhoun 1984).

Hyphal Swellings and Chlamydospores

Hyphal swellings and chlamydospores are not produced.

Table 27.1. Distribution of *Phytophthora gonapodyides* on different hosts

Host	Common name	Disease	Geographical distribution
Alnus B. Ehrh.	Alder	Twig blight	Denmark (Petersen 1910)
Fremontodendron Cov. (*Fremontia*)	Flannel bush	Root rot	United States (Hoitink and Schmitthenner 1974a)
Malus pumila (L.) Mill.	Apple	Fruit rot	Denmark (Petersen 1910)
Pyrus malus (= *Malus sylvestris*)	Pear	Fruit rot	Denmark (Petersen 1910)
Rhamnus sp. L.	Buckthorn	Root rot	United States (Hoitink and Schmitthenner 1974a)
Rhododendron spp.	Rhododendron	Root rot	United States (Hoitink and Schmitthenner 1974a)
Sciadopitys verticillata (Thunb.) Siebold and Zucc.	Umbrella pine	Root rot	United States (Hoitink and Schmitthenner 1974a)

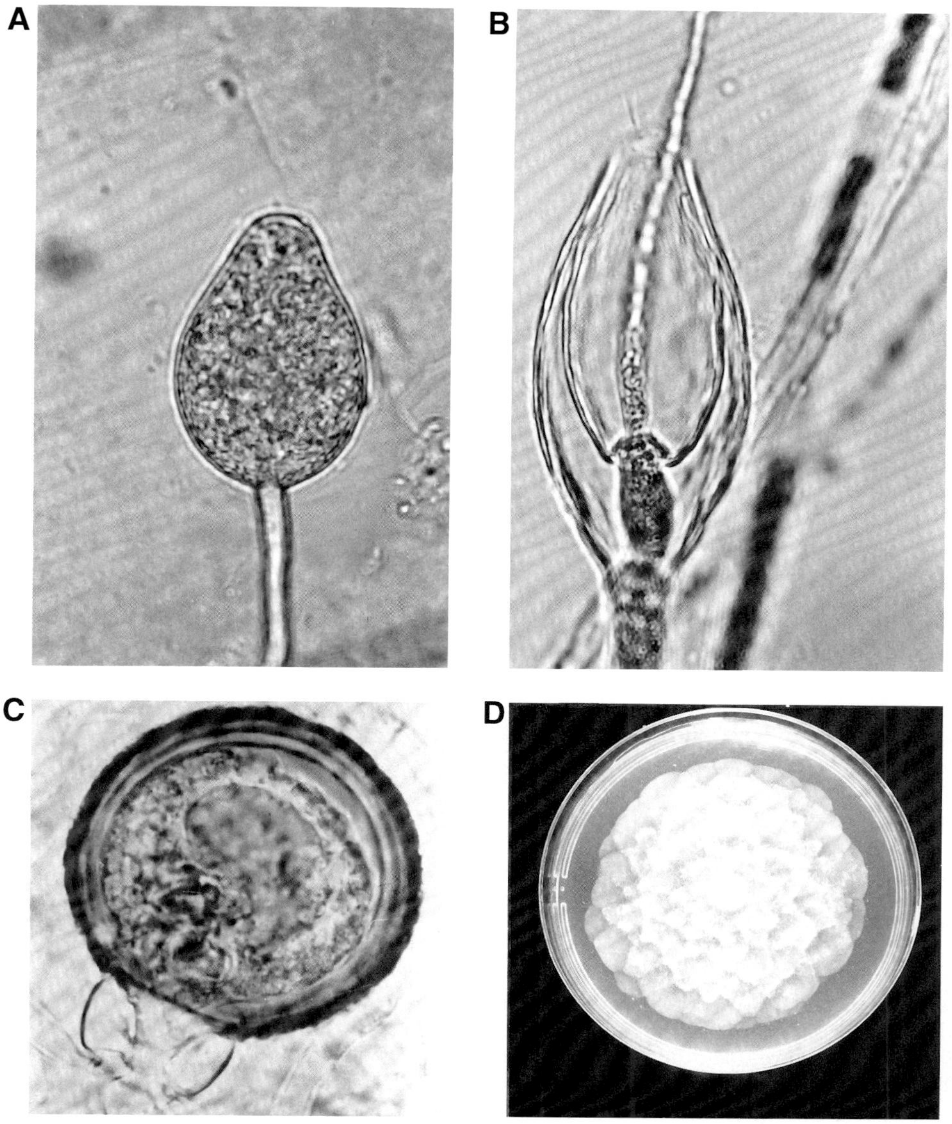

Figure 27.2. Morphology of *Phytophthora gonapodyides*. **A,** Obpyriform, nonpapillate sporangium. **B,** Nesting of sporangia; internal proliferation. **C,** Globose oogonium with a rough exterior wall; amphigynous antheridium. **D,** Rosette growth pattern. (Courtesy M. J. Griffin, ADAS, Cambridge, United Kingdom)

Sex Organs

P. gonapodyides is heterothallic; however, Kanouse (1925) reported that oogonia were produced in single culture, but this report has not been confirmed (Waterhouse 1958). Oogonia with amphigynous antheridia were produced by pairing an A1 culture with an A2 culture of *P. drechsleri*. Oogonia are globose (27 to 48 µm in diameter, mean 40.1 µm). Walls are frequently rough (Figure 27.2C). Antheridia are amphigynous and variable in shape. Oospores in culture are aplerotic, thick walled, 18 to 32 µm in diameter, mean 21.1 µm, and yellow (Pittis and Colhoun 1984). All isolates studied by Brasier et al. (1993) were self sterile but when paired with A2 isolates of *P. cambivora, P. megakarya,* or *P. meadii* produced gametangia that were characteristic of these species.

Growth Temperatures

Mycelial growth forms a rosette pattern (Figure 27.2D). The minimum temperature for growth is <5°C, optimum 25°C, and maximum 30 to 35°C. Brasier et al. (1993) reported slow growth at 20, 25, and 30°C but no growth at 35°C.

Disease Caused by *Phytophthora gonapodyides*

The symptoms produced by *P. gonapodyides* are much like those produced by other *Phytophthora* spp. that cause root rot. Since it is considered to be a minor pathogen, the damage done by this species is overshadowed by that of other *Phytophthora* species.

Root Rot of Ornamentals

The disease caused by this pathogen on flannel bush is mild but similar to that caused by *P. cinnamomi* (Chapter 15) (Hoitink and Schmitthenner 1974a). *P. gonapodyides* has been reported as a saprophyte on twigs (no host species mentioned) in Great Britain (Barnes and Melville 1932), on *Fraxinus* twigs in the United States (Sparrow 1933), and on grapes, prunes, oranges, and rose hips floating in pond water in Victoria, Australia (Crooks 1937). When inoculated on *Atirrhinum majus* L. and *Chamaeocyprus lawsoniana* (A. Murr.) Parl., it was nonpathogenic (Pittis and Colhoun 1984).

CHAPTER 28

Phytophthora heveae

Thompson (1929)

No synonyms are known for *Phytophthora heveae,* which was first described by Thompson (1929) as the causal agent of "black stripe" of *Hevea* rubber in Malaysia (Malaya). This was the first record of a *Phytophthora* species causing a disease of an economically important crop in Malaysia. For many years, *P. heveae* was believed to be confined to Malaysia, but it has since been reported from Brazil, Australia, Guatemala, New Zealand, and the United States on several different hosts (Table 28.1). *P. heveae* causes bud rot and nut fall diseases of coconut on the Ivory Coast of Africa. Quillec et al. (1984) discuss symptoms and the development of the diseases. The fungus has also been isolated from forest soils in the absence of diseased trees (W. A. Campbell and Gallegly 1965; Gadgil 1974a, b; Gerrettson-Cornell 1976a). Zhang et al. (1995) also noted that isolates from soil beneath a diseased Chinese eaglewood tree were not pathogenic. In the United States, dieback of rhododendron occurs in nurseries; the disease can be controlled by foliar spray with metalaxyl (Benson 1980). A brief description of *P. heveae* is given by Stamps (1978d) and Holliday (1980).

Characteristics of *Phytophthora heveae*

P. heveae is classified in group II (Stamps et al. 1990), and its characteristics are summarized by Waterhouse (1974c) and Stamps (1978d). Morphology is illustrated in Figure 28.1. See Tables 4.2 and 4.3 for tabular keys and Appendix 4.9 for a dichotomous key (Ho 1992). Its characteristics are compared with *P. palmivora* and several other species of *Phytophthora* in Tables 51A.4 and 51A.5. Zentmyer et al. (1978) discuss the characteristics of several isolates. Zhang et al. (1995) summarize the characteristics of *P. heveae* as reported by several workers.

Sporangia

Sporangia are papillate, irregular in shape, obpyriform to ellipsoidal, and often asymmetrical and form laterally on the stalk. They are 27 to 66 μm long × 20 to 49 μm wide (average 45 × 29.6 μm) with a length-breadth ratio of 1.1 to 2.9:1 (average 1.5) and caducous (pedicel <10 μm) (Albuquerque et al.

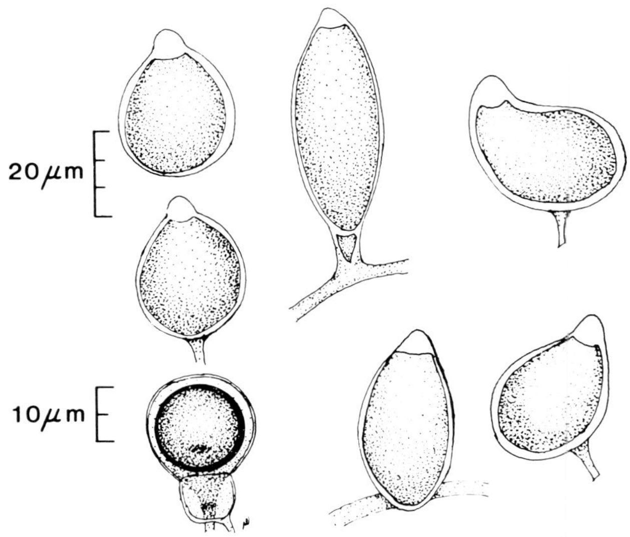

Figure 28.1. Morphology of *Phytophthora heveae.* Papillate sporangia with variable shapes often form laterally on the sporangiophores. Globose oogonium (lower left) with an amphigynous antheridium; oospore is globose and often aplerotic. See Figure 4.28 for a photomicrograph of a funnel-shaped oogonium stalk produced by *P. heveae.* (Drawing by A. Vaziri)

Table 28.1. Distribution of hosts of *Phytophthora heveae*

Host	Common name	Disease	Geographical distribution
Agathis australis Salisb.	Kauri	Root rot	New Zealand (Gadgil 1974a, b); New Guinea (Arentz 1980)
Bertholletia excelsa Humb. and Bonpl.	Brazil nut	Leaf blight	Brazil (Albuquerque et al. 1974)
Cocos nucifera L.	Coconut	Bud rot; nut fall	Ivory Coast (Quilec et al. 1984)
Eucalyptus pilularis		Trunk canker	Australia (Gerrettson-Cornell 1976a)
Hevea brasiliensis (H.B.K.) Mull. Arg.	Para rubber tree	Black stripe	Malaysia (Thompson 1929; Chee 1970, 1972, 1976; Chee and Foong 1968; Chee et al. 1967; Chee and Greenwood 1968; Chee and Wastie 1970, 1980)
Mangifera indica L.	Mango	Leaf blight	India (Thankamma 1983)
Persea americana Mill.	Avocado	Trunk canker	Guatemala (Zentmyer et al. 1978)
Pinus patula Schlechtend. and Cham.	Mexican yellow pine	Root rot	Australia (Oxenham and Winks 1963)
Psidium guajava L.	Apple guava	Leaf blight	India (Thankamma 1983)
Rhododendron spp.	Rhododendron hybrids	Dieback; wilt	United States[a] (Benson and Jones 1980)
Rhododendron maximum L.	Rhododendron	Dieback; wilt	United States (Benson and Jones 1980)
Theobroma cacao L.	Cocoa	Pod rot	Malaysia (P. D. Turner 1968)

[a] Previously reported only from soil in the United States (W. A. Campbell and Gallegly 1965; Hendrix and Campbell 1970).

1974). However, in isolates obtained from avocado as well as in several reference isolates, caducity was not noted, and sporangiophores that did detach from the stalk were of varying length (Zentmyer et al. 1978). Caducity was not noted by Thompson (1929). Sporangia develop profusely on agar media. Light increased sporangium production in vitro (Brasier 1969a). Branching of sporangiophores is irregular, and intercalary swellings are often present.

Chlamydospores

Chlamydospores are not produced.

Hyphal Swellings

Hyphal swellings are small to large, resembling abortive reproductive organs.

Sex Organs

P. heveae is homothallic. Antheridia are amphigynous, usually spherical, occasionally cylindrical (11 × 10 μm), and sometimes bicellular (reported first by Zentmyer et al. [1978] on an avocado isolate and on authentic reference isolates). Although Albuquerque et al. (1974) noted some paragynous antheridia, none were seen by Zentmyer et al. (1978) in eight cultures from rubber, cocoa, rhododendron, and avocado. Oogonia are globose and form readily in culture, often in close groups; they are broadly funnel shaped at the base and 17 to 32 μm in diameter (average 22.3 μm) (see Figure 4.28). Oospores are round, smooth, thick walled, 15 to 26.8 μm in diameter (average 21.5 μm), and markedly aplerotic (see Stamps 1978d and Holliday 1980). Zentmyer et al. (1978) noted that the average oospore diameter of isolate P640 from rubber on carrot agar was 23 μm, smaller than the 28-μm average diameter cited by Waterhouse (1963), and that this difference was probably the result of the use of carrot medium rather than cornmeal. Berg et al. (1966) and Leal et al. (1965) described oospore germination. Up to 90% of 30-day-old oospores germinated when they were incubated under blue and green light (Chang and Shea 1988).

Growth Temperatures

The minimum temperature for growth is 11.5°C, optimum 25°C, and maximum 31.0 to 32.5°C. Glutamine, urea, and ethanol reduced growth in synthetic media (Leal et al. 1968).

Distinguishing Characteristics

P. heveae is distinguished from *P. meadii* by lack of aerial mycelium, funnel-shaped oogonia, smaller and more numerous oospores, and more elongated sporangia (the length-breadth ratio of sporangia is greater in *P. heveae*). *P. heveae* is distinguished from *P. colocasiae, P. parasitica,* and *P. palmivora* by the lack of aerial mycelium and by the absence of chlamydospores (see Table 51A.4 for a comparison of characteristics with other species).

Some Diseases Caused by *Phytophthora heveae*

Pod Rot and Black Stripe of *Hevea brasiliensis* L. (Para Rubber)

The symptoms are similar to those caused by *P. palmivora* or *P. meadii* on *Hevea.* A detailed account of symptoms is discussed under *P. palmivora* (Chapter 51). *P. heveae* causes watery green patches covered with mycelium on pods. Oospores form abundantly in the infected tissue, a characteristic that contrasts with the absence of oospores in pods infected with *P. palmivora* or *P. meadii.*

Disease Development. See *P. palmivora* (Chapter 51A).

Control. See *P. palmivora* (Chapter 51A).

Trunk Canker of *Persea americana* Mill. (Avocado)

In Guatemala, cracked and bleeding cankers were found to occur on the lower trunk and upper portion of the rootstock. The lesions are normally confined to the outer bark but extend into the inner xylem under favorable conditions. Excision of the cankers reveals reddish brown margins encroaching into healthy tissue. Diseased trees are stunted, the foliage turns a light green, and a dieback of parts of the branches sometimes occurs. Diseased trees eventually die (Zentmyer et al. 1978).

Disease Development. See root rot of *Persea americana* (avocado) by *P. cinnamomi* (Chapter 15).

Control. See root rot of avocado by *P. cinnamomi* (Chapter 15).

Pod Rot of *Theobroma cacao* (Cocoa)

Pod rot of cacao is similar to that caused by *P. palmivora* and is discussed in detail in Chapter 51A. The paler brown color of lesions caused by *P. heveae* on cacao pods differentiates the disease from that caused by *P. palmivora.* Oospores are produced abundantly by *P. heveae* in the affected tissues, but few sporangia form.

Disease Development. Similar to that of *P. palmivora* (see Chapter 51A).

Control. Similar to that of *P. palmivora* (see Chapter 51A).

CHAPTER 29

Phytophthora hibernalis

Carne (1925)

No synonyms are known for *Phytophthora hibernalis*, which was first isolated from diseased citrus fruit in Western Australia by Carne (1925). *P. hibernalis* is a pathogen of citrus fruit and occasionally of leaves during the cooler seasons of the year in Argentina, Australia, France, Greece, Israel, Italy, New Zealand, the United Kingdom, the United States, and Venezuela (Holiday 1980). It has a limited host range predominantly on citrus species (Table 29.1). Since sporangia are caducous, fruit in the upper part of the tree are commonly infected (Fawcett 1936; Klotz 1978). There is little evidence of root and crown rot caused by *P. hibernalis*.

Characteristics of *Phytophthora hibernalis*

P. hibernalis was described briefly by Waterhouse and Waterston (1964a) and Holliday (1980) and is classified in group IV (Stamps et al 1990). Ho and Jong (1993) reported a comprehensive comparison of *P. hibernalis* and *P. syringae,* which had erroneously been considered to be conspecific (Tucker 1931). See Tables 4.2 and 4.3 for tabular keys of *Phytophthora* species. Morphology is shown in Figure 29.1.

Sporangia

Sporangia are elongate, ellipsoid, ovoid, or obovoid, often with the broadest point near the apex; 29 to 53 μm long × 14 to 22 μm wide (average 40 × 19 μm); caducous (pedicel 23 to 73 μm long); and semipapillate. Sporangiophores are unbranched or formed as long, lateral branches in an irregular sympodium.

Hyphal Swellings and Chlamydospores

Hyphal swellings and chlamydospores are not formed. According to Ho and Jong (1993), the chainlike "chlamydospores" observed by Nadel-Schiffman (1947) were most likely monilioid hyphal swellings.

Sex Organs

P. hibernalis is homothallic; antheridia are mostly amphigynous but occasionally are paragynous; oogonia are 22 to 56 μm in diameter (average 35 μm); oospores are 22 to 45.6 μm in diameter (average 30 μm) and plerotic.

Growth Temperatures

Minimum temperature for growth is <5°C, optimum 15°C, and maximum <25°C.

Distinguishing Characteristics

P. hibernalis is distinguished from *P. syringae* by production of elongate, caducous sporangia with long pedicels and of mostly amphigynous antheridia (Waterhouse and Waterston

Table 29.1. Distribution of hosts of *Phytophthora hibernalis*

Host	Common name	Disease	Geographical distribution
Agathis australis Salisb.	Kauri pine	Canker	New Zealand (Gadgil 1974b)
Carthamus tinctorius L.	Safflower	Root rot	Venezuela (Malaguti 1953)[a]
Chamaecyparis lawsoniana (A. Murr.) Parl.	Port Orford white cedar	Canker	United Kingdom (Young and Strouts 1976)
Citrus limon (L.) N. L. Burm.	Lemon	Leaf blight; brown rot of fruits	Australia (Carne 1925); Portugal (Moniz da Maia 1925); Israel (Nadel-Schiffmann 1947)
Citrus nobilis var. *deliciosa* (Ten.) Swingle	Mandarin orange, tangerine	Brown rot of fruits	Australia (Carne 1925); Portugal (Moniz da Maia 1925)
Citrus paradisi Macf.	Grapefruit	Brown rot of fruits	Australia (Carne 1925)
Citrus sinensis (L.)	Sweet orange	Defoliation; twig blight; fruit decay	Australia (Carne 1925); Portugal (Moniz da Maia 1925); Fiji Islands (Parham 1942); Israel (Nadel-Schiffmann 1947); Turkey (Salih 1974)
Citrus spp.	Citrus	Brown rot of fruits; leaf blight; twig dieback	Portugal (Moniz da Maia 1925); Australia (Carne 1925); South Africa (Doidge 1925); United States (Fawcett 1933, 1936); Fiji Islands (Parham 1942); Israel (Nadel-Schiffmann 1947); Italy (Verneau 1954); France (Waterhouse and Waterston 1964a)
Lycopersicon esculentum L.	Tomato	Root rot	Israel (Nadel-Schiffmann 1947)
Malus pumila Mill.	Apple	Root rot	Israel (Nadel-Schiffmann 1947)
Sesamum indicum L.	Sesame; gingelly	Root rot and stem canker	Venezuela (Malaguti 1953)
Solanum melongena L.	Eggplant; brinjal	Root rot	Israel (Nadel-Schiffmann 1947)

[a] Artificially inoculated host.

1964a). *P. syringae* (Chapter 60) produces noncaducous sporangia and mostly paragynous antheridia. See Ho and Jong (1993) for a comprehensive literature review and results of a comparative study of both species.

Disease Caused by *Phytophthora hibernalis*

Brown Rot of Citrus Fruits

Symptoms of brown rot caused by *P. citrophthora, P. parasitica,* and *P. syringae* are similar to those caused by *P. hibernalis.* A description of symptoms is given under *P. citrophthora* (Chapter 17).

Disease Development. Brown rot is restricted to the cool seasons of the year. During these periods, the disease develops rapidly under conditions of high humidity and soil moisture. Greatest crop losses occur after rainfall and wind during the autumn and winter (Waterhouse and Waterston 1964a). Since sporangia are caducous, they can be disseminated by wind and rain and infect fruit on the upper parts of trees (Klotz 1978). In warmer climates, *P. citrophthora* and *P. parasitica* are the dominant pathogens causing brown rot.

Control. Measures for control of brown rot are similar to those described for *P. citrophthora* (Chapter 17) and *P. nicotianae* (*P. parasitica*) (Chapter 50). If *P. hibernalis* is suspected of being active, the entire tree must be sprayed with a preventative fungicide because the deciduous sporangia can be carried to fruit on tops of trees by windborne rain; if *P. citrophthora* or *P. nicotianae* is suspected, only the skirts (lower 3 m) of trees need be sprayed.

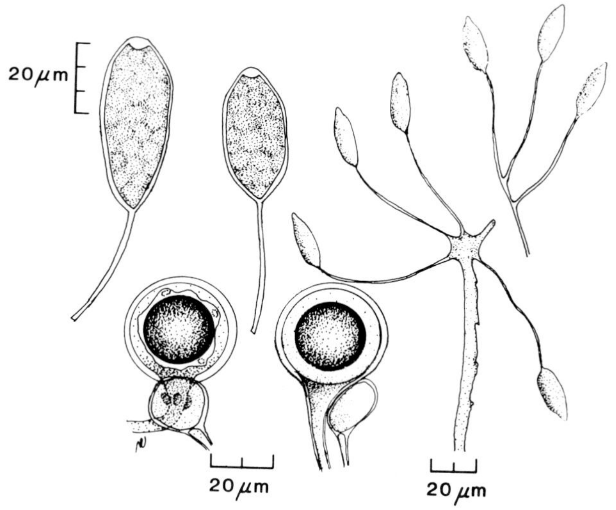

Figure 29.1. Morphology of *Phytophthora hibernalis.* Upper left, ellipsoid, caducous, papillate sporangia; right, sporangia borne on an irregular sympodium; and lower left, sex organs with amphigynous and paragynous antheridia. (Drawing by A. Vaziri)

CHAPTER 30

Phytophthora humicola

Ko and Ann (1985)

Phytophthora humicola was isolated from soil in a citrus orchard in Taiwan at the same time that *P. insolita* was isolated (Ann and Ko 1980) by baiting with citrus leaves (method of Grimm and Alexander [1973]). The species name *humicola* implies soil-inhabiting. It is pathogenic after wound-inoculation of apple and cucumber fruits but is not pathogenic to orange or eggplant fruits. An isolate obtained from a rotted kidney bean root at Peikang, Yunlin, Taiwan, was not pathogenic to kidney bean.

Description of *Phytophthora humicola*

P. humicola is classified in group V (Stamps et al. 1990). See Tables 4.2 and 4.3 for tabular keys and Appendix 4.9 for a dichotomous key (Ho 1992). Morphology is shown in Figure 30.1. Data presented below are from Ko and Ann (1985).

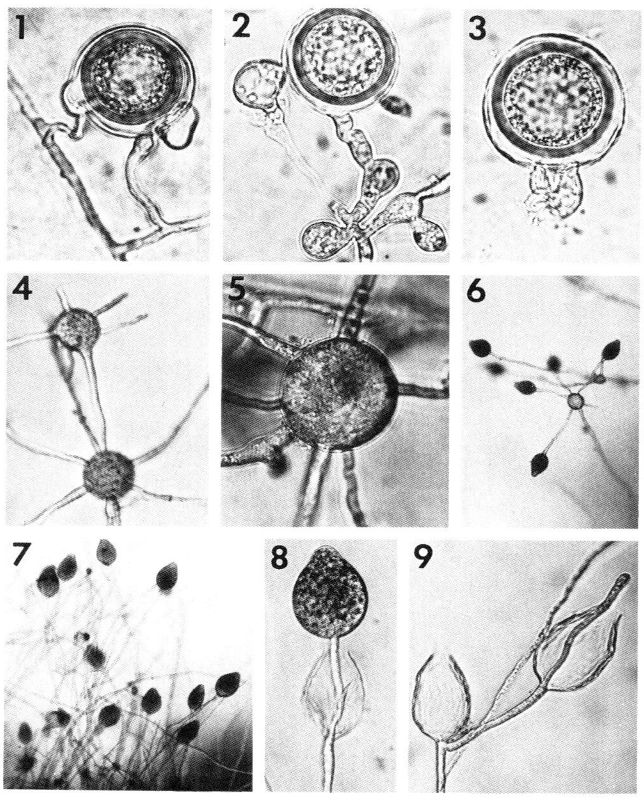

Figure 30.1. *Phytophthora humicola.* **1, 2, and 3,** Sex organs. (378×) **4,** Radiating hyphal swellings. (190×) **5,** Radiating hyphal swelling. (378×) **6,** Sporangia produced on the tips of radiating hyphae of a hyphal swelling. (55×) **7,** Sporangia. (55×) **8,** Sporangiophore proliferating internally. (190×) **9,** Sporangiophore proliferating externally. (190×) (Reproduced by permission from Mycologia 77:631–636, W. H. Ko and P. H. Ann, Copyright 1985, The New York Botanical Garden)

Sporangia

Sporangia are formed only when pieces of newly colonized V8 juice agar are incubated in nonsterile soil extract under diffuse light. Incubation in water or other conventional salt solutions (see Chapter 3) does not induce sporangial formation. Nonpapillate, ovoid sporangia, measuring 38 to 60 µm × 50 to 72 µm (average 44 × 58 µm), are noncaducous. Zoospores are released after the temperature of the sporangia in soil extract is reduced to 8°C for 30 min. Sporangia proliferate internally but rarely externally. They occasionally are produced on the tips of hyphae radiating from germinating hyphal swellings.

Chlamydospores

Chlamydospores have not been observed.

Hyphal Swellings

Large, spherical hyphal swellings measuring 22 to 56 µm, average 40 µm, are produced in sterile distilled water. Radiating hyphae emanate from the hyphal swellings. Sometimes the dominant hypha produces another hyphal swelling and forms a chain of hyphal swellings.

Sex Organs

P. humicola is homothallic. About 47,000 oospores per square centimeter were produced on the surface of a petri plate of 20% V8 juice agar incubated at 25 to 28°C in the dark. Oospores were not produced at 23°C or below or above 30°C. Continuous light prevented oospore formation. Oogonia are spherical and smooth walled, measure 25 to 52 µm in diameter, average 39 µm, and are borne on long stalks. Antheridia are mostly paragynous and ovate to elliptical or irregular in shape and are borne on stalks 8 to 89 µm, average 31 µm, in length. Amphigynous antheridia occur but are rare. Paragynous antheridia measure 10 to 16 × 11 to 26 µm, average 14 × 18 µm, and amphigynous antheridia measure 15 to 19 × 15 to 21 µm, average 16 × 17 µm.

Growth Temperatures

The minimum temperature for growth is 9°C, optimum 28 to 30°C, and maximum 32°C.

Distinguishing Characteristics

Ko and Ann (1985) state that *P. humicola* resembles *P. megasperma* because of the production of nonpapillate, noncaducous sporangia that proliferate internally and the relatively large oogonia and oospores with predominantly paragynous antheridia (see Chapter 46). They state that the production of large hyphal swellings and the long stalks on oogonia and antheridia distinguish *P. humicola* from *P. megasperma* and that the protein pattern (by disk gel electrophoresis) of *P.*

humicola is distinct from that of *P. megasperma* f. sp. *glycinea* (now known as *P. sojae* [Hansen and Maxwell 1991]). However, since *P. sojae* is now known not to be a member of the *P. megasperma* species complex (see Chapter 46), these data do not prove a difference from *P. megasperma.* Use of molecular restriction fragment length polymorphisms (RFLP) of mitochondrial DNA (mtDNA) pattern data have show[illegible] marked variability in the genetic makeup of various biotype[illegible] *P. megasperma* (Förster and Coffey 1993). A comparis[illegible] mtDNA RFLP pattern of *P. humicola* with the differen[illegible] of *P. megasperma* that have been described would p[illegible] more convincing evidence that *P. humicola* is distinct.

HAPTER 31

Phytophthora idaei Kennedy and Dunca (1995)

as isolated from rotting raspberry
Phytophthora i Britain and described as a new species
(*Rubus idaeus*) in 1995). Although 5–20% of the roots of
(Kennedy and nts were infected, aerial symptoms were
certain ei appeared to be limited in host range to *R.*
not evi rasd may have been carried along with apparently
idae berry cultivars used for breeding purposes.

acteristics of *Phytophthora idaei*

Kennedy and Duncan (1995) classified *P. idaei* in group I (Stamps et al. 1990). A comparison of the morphology of sporangia, sporangiophore branching, and sexual organs of *P. idaei, P. cactorum, P. citricola,* and *P. syringae* is shown in Figure 31.1 (Kennedy and Duncan 1995). See Tables 4.2 and 4.3 for tabular keys. Colonies of *P. idaei* have limited aerial mycelium. Chlamydospores and hyphal swellings are not formed.

Sporangia

Sporangia form optimally at 20°C and are papillate and predominantly spherical to ovoid (49 × 36 μm). Germination by production of zoospores is most common. Direct germination is rare. Sporangiophores are simple and in a lax sympodium. Sporangia are persistent (noncaducous).

Sex Organs

Oogonia (26 × 38 μm) form in single culture (homothallic) and in the cortex of roots of infected raspberry plants and are smooth and spherical. Oospores (22–34 μm) are aplerotic in the oogonia. Antheridia are predominantly paragynous.

Growth Temperatures

Growth, sporangial formation, and oogonial and oospore formation are optimal at 20–22°C.

Distinguishing Characteristics

P. idaei resembles *P. cactorum* because the sporangia are nearly spherical and are papillate. Also, antheridia of both species are paragynous. Sporangia of *P. idaei* are persistent while those of *P. cactorum* are caducous. Hyphae of *P. idaei* are consistently wider than those of *P. cactorum.* The optimal temperature for sporulation and growth is lower for *P. idaei* (about 20°C) than for *P. cactorum* (about 25–28°C). *P. idaei* isolates do not utilize nitrate as a nitrogen source, whereas *P. cactorum* isolates do.

P. idaei differs from *P. iranica* by its failure to produce chlamydospores and from *P. pseudotsugae* by its much lower production of sporangia in liquid culture. *P. idaei* differs from *P. clandestina* by its absence of hyphal swellings, persistent sporangia, subterminal or digitate antheridia, and much lower levels of amphigyny (see also Taylor et al. [1983] for a description of *P. clandestina*).

Description of *Phytophthora idaei* Kennedy, sp. nov.

The following is from Kennedy and Duncan (1995) courtesy of the British Mycological Society:

Mycelial colonies grow moderately slowly on a defined sucrose/asparagine/mineral salts agar and more slowly on corn meal, cleared V-8 juice and malt extract agars. Growth is poor or absent on media with nitrate as the sole nitrogen source. Colonies uniform with no

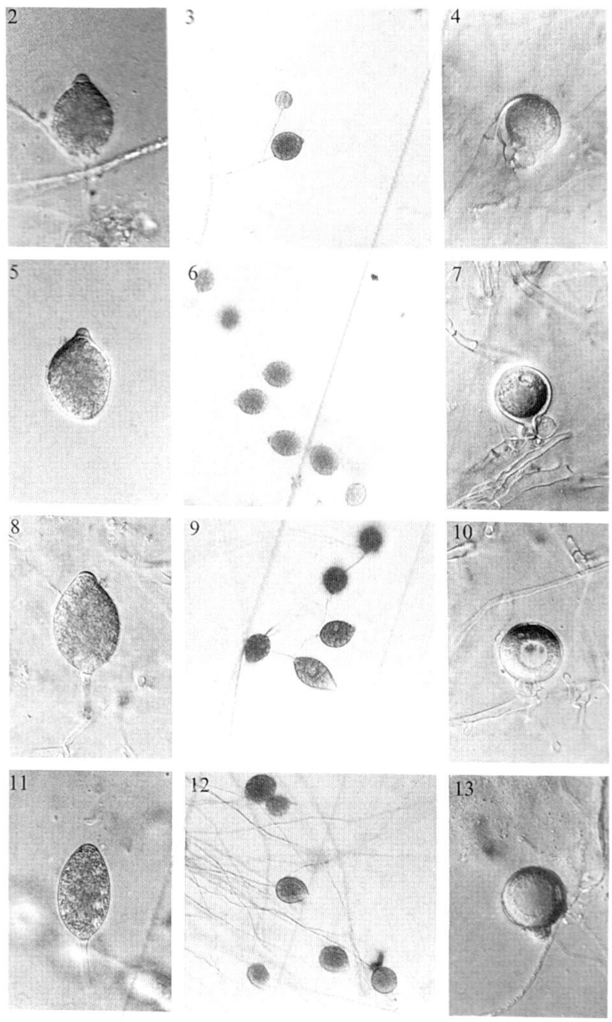

Figure 31.1. Comparison of sporangial and oogonial morphology of four *Phytophthora* species. Sporangium, sporangial proliferation, and an oospore of *P. idaei* (**2–4**), *P. cactorum* (**5–7**), *P. citricola* (**8–10**), and *P. syringae* (**11–13**). (Reproduced by permission of the British Mycological Society, courtesy Kennedy and Duncan 1995)

aerial mycelium on corn meal agar and limited aerial mycelium on the other test agars. Minimum growth at or below 5°, optimum 20–22.5°, and maximum 27°. Hyphae hyaline, non-septate; width 4.8–6 µm (mean 5.1 µm). Sporangiophores simple or sparingly branched with irregular lax sympodia; width 2.4–3.6 µm (mean 3 µm) length variable averaging 3.1–33 µm (mean 16 µm); attached at base of sporangia, but occasionally laterally. Sporangia produced in small numbers on solid media but more numerous when irrigated; persistent, terminal or occasionally intercalary, proliferating externally; spherical to ovoid, 41.4–55.2 × 29.9–43.7 µm (mean 48.9 × 34.4 µm); papilla conspicuous protruding beyond contour of sporangium; apical thickening hyaline, depth 2.4–6 µm (mean 4.1 µm). Zoospores 12–16.8 × 7.2–9.1 µm (mean 13.9 × 8.3 µm) discharged through a spore 6–7.4 µm wide (mean 7 µm); direct germination of sporangia rare. Sexual organs formed in abundance in the cortex of raspberry roots; also formed in the test media. Oogonia terminal on stalks 5.5–10.5 µm wide (mean 7.5 µm) and 6.1–20.9 µm long (mean 15 µm); spherical, smooth-walled, 27–38.4 µm diam. (mean 32.1 µm), sometimes pigmented when mature. Antheridia hyaline, simple, club-shaped, occasionally spherical, 9.1–15.7 µm wide (mean 11.9 µm) and 12.6–26.2 µm long (mean 16.7 µm), with stalks 4.1–6.6 µm wide (mean 5.4 µm) and 6.9–23.1 µm long (mean 14.1 µm). Oospores aplerotic 22.8–31.2 µm diam. (mean 27 µm) with a wall 1.9–2.7 µm (mean 2.2 µm).

Habitat: In the roots of cultivated raspberries in Britain. It is named for its original host *Rubus idaeus*. The culture derived from the holotype (isolate R77) on which the above description is based has been placed with the International Mycological Institute.

CHAPTER 32

Phytophthora ilicis

Buddenhagen and Young (1957)

No synonyms are known for *Phytophthora ilicis,* which was first described by Buddenhagen and Young (1957) as host specific on English holly in the western United States. Waterhouse (1970b) questioned the validity of this species because the type was not cited; however, in the keys of Newhook et al. (1978) and Stamps et al. (1990), *P. ilicis* is recorded as a valid species. It has been reported only on English holly in the United States and Canada (Punja and Ormrod 1979). See G. Hall (1991b) for a summary of the morphology and the disease caused.

Characteristics of *Phytophthora ilicis*

P. ilicis is classified in group IV (Stamps et al. 1990). The morphology of *P. ilicis* is shown in Figures 32.1 and 32.2. See Tables 4.2 and 4.3 for tabular keys.

Sporangia

Sporangia are obpyriform, semipapillate, caducous (pedicels 5 to 15 µm long), 30 to 50 × 18 to 30 µm (average 39 × 24) on holly leaves, 26 × 18 µm in distilled water, and 39 × 29 µm in Petri's mineral solution. Length-breadth ratio ranges from 1.3:1 to 1.6:1 (Figure 32.2). Sporangiophores are sympodial.

Hyphal Swellings and Chlamydospores

Hyphal swellings are not produced, and chlamydospores are rare.

Sex Organs

P. ilicis is homothallic; antheridia are amphigynous; oogonia have tapered bases, 25 to 32 µm in diameter (average 27 µm) in leaves, 17 to 28 µm in diameter (average 25 µm) in pea broth; oospores are plerotic, 18 to 26 µm in diameter (average 21 µm) in leaves, 15 to 23 µm (average 20 µm) in pea broth.

Growth Temperatures

The minimum temperature for growth is 5°C, optimum 20°C, and maximum 25°C.

Disease Caused by *Phytophthora ilicis*

Leaf and Twig Blight of *Ilex aquifolium* L. (English Holly)

Leaf and twig blight was originally attributed to *Boydia insculpta* and *Phomopsis crustosa,* but inoculation tests proved that these fungi were not responsible for the disease (Buddenhagen and Young 1957). The first observable symptom is the appearance of black spots on the periphery of leaves on the lower part of the tree during cool, rainy weather in the fall (Plates 32.1 and 32.2). Active cankers on twigs and large stems are black and rarely exceed 12 cm in length. The cankers become lighter in color when they become inactive during warm weather. Ripe berries are also susceptible to infection, particularly during late winter or early spring (Plate 32.3). The stem below infected berries is often girdled, resulting in the death of the terminal portion of the stem. Leaves on twigs beyond the girdled stem turn purple and then tan and remain attached. Defoliation commonly occurs after the appearance of the leaf spots (Plate 32.4). The disease gradually progresses upward, and in severe cases, large trees may be completely defoliated with the exception of the topmost branches (Plate 32.5). Sporangia are produced on the lower surfaces of infected leaves. Oospores form in infected leaf and twig tissue.

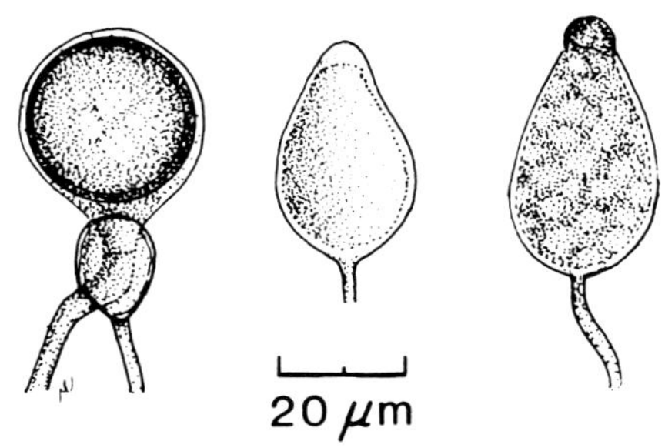

Figure 32.1. Morphology of *Phytophthora ilicis.* Left, globose, smooth-walled oogonium containing a plerotic oospore; antheridium is amphigynous. Center, ovoid, semipapillate sporangium. Right, ovoid sporangium in which zoospores have formed immediately before emission. (Drawing by A. Vaziri)

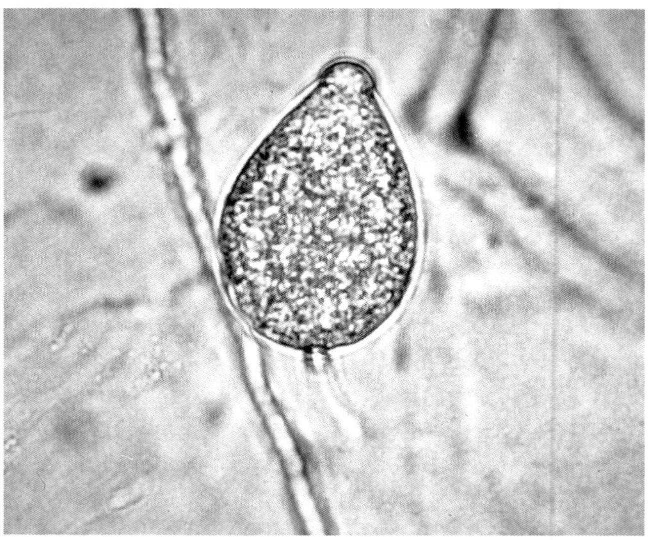

Figure 32.2. Obpyriform, semipapillate sporangium of *Phytophthora ilicis.* Sporangium is caducous with a short pedicel. (Courtesy D. L. Coyier, USDA ARS, Corvallis, Oregon)

In some areas, the disease causes postharvest defoliation.

Disease Development. The disease develops during the rainy, cool period of the year. It is arrested during dry periods in winter and is completely inactive during the summer.

Control. Soil drench or foliar spray with metalaxyl are reported to be effective control measures (Coyier 1981). After holly foliage is cut for sale, it should be dipped in a solution of copper sulfate (30 ppm) at pH 6 to 6.5 (Coyier 1981). Lower branches of trees should be pruned to prevent contact with rain-splashed, infested soil.

CHAPTER 33

Phytophthora infestans

(Mont.) de Bary (1876)

Synonyms of the causal agent of late blight of potato include *Gangraena tuberum solani* Martius (1842), *Botrytis vastatrix* Libert (1845) (see Waterhouse 1970a), *Botrytis infestans* Montagne (1845), *Botrytis fallax* Desmazieres (1845), *Botrytis solani* Harting (1846) (see Waterhouse 1970a), *Peronospora trifurcata* Unger (1847) (see Waterhouse 1970a), *Peronospora fintelmani* Caspary (1852), *Peronospora devastatrix* (Libert) Caspary (1855), *Peronospora infestans* (Montagne) de Bary (1863), *Phytophthora thalictri* Wilson and Davis (1907) (see Waterhouse 1970a), and *Phytophthora devastatrix* (Libert) Puttemans (1937).

Martius (1842) in Prussia was the first to claim that a fungus caused the malady that devastated the foliage and tubers of potato in nearly all of Europe and named the fungus *Gangraena tuberum solani.* According to Bourke (1991), who reviewed the scientific controversy, Martius's report was unorthodox at that time and "to some smacked of heresy." Berkeley, a country parson in England who was also interested in plant diseases, viewed Martius positively and published the translated paper in the *Gardener's Chronicle.* The volume *Phytophthora* (Lucas et al. 1991) was dedicated to Reverend Berkeley because of his active participation in the quest to determine the cause of late blight of potato (Buczacki 1991).

It must be remembered that in 1842 the germ theory of disease was not universally accepted, and many considered fungi and bacteria the results but not the causes of disease. Bourke (1991) relates an interesting saga about the many theories to explain this new disease of the potato that appeared in continental Europe, the British Isles, and Ireland. The majority of comments blamed the miserable weather, an idea we still hear frequently. I once told an Irish priest that my research involved problems such as the potato blight, and his reply was, "Ah sure, Don, and that is caused by the damp!"

In 1845, Dr. August Morren from Belgium wrote several letters that supported the thesis that a fungus was the cause of the potato blight. His theory was vigorously contested, as Bourke (1991) relates, by Dieudonné, who declared that Morren's concept would hang forever like Damocles's sword over his head and that this depressing thesis must be exposed so that the fungus did not "become the terror of the farmers." Morren's replies were equally sharp. After some doubts, Montagne eventually accepted Morren's thesis and with information that he had developed, described *Botrytis infestans,* the fungus that we now know as *Phytophthora infestans.* Little did Montagne know that his description would be so vigorously denounced that even he would back off for a time. Bourke (1991) graphically relates the many controversies that followed Montagne's report until Anton de Bary (1876) determined conclusively that the fungus, which he renamed *Phytophthora infestans,* was the cause of late blight.

P. infestans is perhaps the best known of all the species of *Phytophthora* because of the tremendous destruction it wreaked on the potato crops in Ireland during the 1840s. The economic chaos from the loss of this vital food crop directly led to so many deaths that a high proportion of the surviving Irish people emigrated overseas to Europe and America. A fascinating and detailed account of this tragic period in history is told by Large (1960). Although *P. infestans* has a limited host range (mainly solanaceous hosts) (Table 33.1; Commonwealth Mycological Institute Map 109 ed. 5, 1982), it is of considerable importance, since it severely affects potato, a food crop that is vital to the diets of many people worldwide. Now, more than 125 years later, it has yet to be totally controlled, mainly because of the bewildering array of new biotypes known as physiologic or pathologic races that appear with each succeeding introduction of race-specific, blight-resistant potato cultivars (Vanderplank 1963, 1984, 1988; Malcolmson 1976). The production of new races of *P. infestans* is discussed in Chapters 3 and 6. Volume 7 of *Advances in Plant Pathology* is devoted entirely to *P. infestans* (see Fry and Spielman 1991). Fry et al. (1993) relate new information on how the A2 mating type migrated to different areas of the world, most likely from its native home in central Mexico. Gillis (1993) presents an excellent popular rendition of how the A2 mating type migrated from Mexico to nearly every potato-producing nation of the world. Until the 1980s, only the A1 mating type was present outside Mexico. The book *Phytophthora infestans 150* contains numerous papers on *P. infestans* given at an international meeting on the 150th anniversary of the description of *P. infestans* (Dowley et. al. 1995).

Until the early 1980s, A1 was the predominant mating type detected in all regions of the world except central Mexico, where both A1 and A2 mating types coexisted at a 50:50 ratio (Gallegly and Galindo 1958; Gallegly and Niederhauser 1959). Since the report of Hohl and Islen (1984), A2 isolates are now found in nearly all potato-growing regions of the world, but usually at ratios less than 50:50 (Fry et al. 1992, 1993; Spielman et al. 1991; see also Chapter 3 for more citations). However, Sujkowski et al. (1994) reported an approximate 50:50 ratio of A1 to A2 mating types in Poland.

On the basis of isozymic studies of more than 200 isolates of *P. infestans* collected in 20 countries, S. B. Goodwin et al (1994) hypothesized that a single clonal line of *P. infestans* migrated in about 1842 or 1843 from Mexico first to the northeastern United States and from there in about 1845 to Europe, where late blight of potato became an international problem. Subsequent migrations of *P. infestans* to Africa and Asia were likely from Europe because most of the seed potatoes of the world emanated from there. Variants of the single clonal line of *P. infestans* most likely were the result of mutations or mitotic recombination. Fry and Goodwin (1995) summarize the migration of the A2 mating type and the new population of *P. infestans* from Mexico to the northeast United States and to Europe, East Asia, and South America. See also S. B. Goodwin et. al. (1994) for a description of the panglobal distribution of a single clonal line of *P. infestans* from Mexico to the United States and other parts of the world.

It is not understood why the A2 type did not spread

throughout the world with the movement of potato as a food crop during the 1800s. It was not until the early 1980s that A2 mating types were reported in several European countries, Egypt, the Middle East, Asia, and South America (Spielman 1991). Spielman et al. (1991) and Fry et al. (1993) summarize and map the migration of the new genetic types of *P. infestans* throughout the world. Use of molecular isozyme techniques (described in Chapter 3 in the section Genetics and Cytology of *Phytophthora,* in an excerpt from Spielman [1991], and in Chapter 4 under Molecular Technology for Differentiation of Species) allowed the determination of the change in the genetic makeup of *P. infestans.* The detection of particularly unique isozyme characteristics in isolates collected after 1980 that did not occur in isolates from previous years supports the assumption that a change in the genetic makeup of *P. infestans* is in progress. The available evidence indicates that the newer genotypes are pathologically more fit than populations made up of the older isolates (discussed by Spielman [1991] and Spielman et al. [1991]). These and the studies summarized by Fry et al. (1993) and Fry and Goodwin (1995) confirm that *P. infestans* is highly variable and because of its ability to become adapted to different hosts and different situations remains a major threat to world food production.

Recent information on migration of the A2 type into Great Britain, other areas of Europe, the United States, and British Columbia (K. L. Deahl in Gillis [1993]) raises the question of whether natural mating of the A1 and A2 types could lead to even greater genetic variability than existed when only the A1 mating type was present. Although the extensive monitoring of isozyme groups representing "old" and "new" populations appearing in Europe during the 1980s has not thus far indicated that recombinant types have arisen, the possibility exists. Fry et al. (1993) state that some evidence indicates that the high frequency of unique genetic biotypes in Poland resembles the diversity that exists in central Mexico where A1 and A2 types coexist (W. Sujkowsky, *unpublished*). Sujkowski et al. (1994) present evidence that the introduction of the A2 mating type into Europe has influenced the increase in virulent biotypes (see the section in Chapter 3 Genetics and Cytology of *Phytophthora*).

Characteristics of *Phytophthora infestans*

P. infestans is classified in group IV (Stamps et al. 1990). See Tables 4.2 and 4.3 for tabular keys and Appendix 4.9 for a dichotomous key (Ho 1992). Morphology is shown in Figures 33.1 and 4.12G. See Fry et al. (1993) and Hooker (1981) for photomicrographs of spore structures.

Sporangia

Sporangia are ovoid, ellipsoid to limoniform, tapering at the base, caducous (pedicel <3 μm), and semipapillate. Average size of sporangia ranges from 36 × 22 μm (Tucker 1931) to 29 × 19 μm (Waterhouse 1963). These dimensions are similar to those of de Bary (1876), Rosenbaum (1917), Haskell (1921), K. O. Müller (1928), and Leonian and Greer (1929). Sporangiophores are compound sympodial (Figure 4.5B) with a small characteristic swelling just below the sporangium (Figures 4.5C and 4.6).

Hyphal Swellings and Chlamydospores

Neither hyphal swellings nor chlamydospores have been reported, except in a paper from Russia by Patrikeyeva (1979), who noted chlamydospores with a two-layer wall after incubation for 4 to 9 months on oat-pea agar at 9 to 10°C.

Sex Organs

P. infestans is heterothallic. Antheridia are amphigynous; oogonia are 31 to 50 μm in diameter (average 38 μm); oospores formed in plant leaves are aplerotic, 24 to 35 μm in diameter (average 30 μm); in artificial culture they measure 24 to 56 μm in diameter. Until the early 1980s, A1 was the only mating type found in most of the world, but in central Mexico both A1 and A2 isolates coexisted at a 50:50 ratio (Niederhauser 1991). Inoculation of a leaf with an A2 isolate of *P. drechsleri* and *P. infestans* (A1) induced oospore production (Skidmore et al. 1984). Whether or not unique biotypes could arise from unrelated species is unknown. Most likely, formation of oospores results from stimulation by hormonelike substances emitted by the opposite mating type (see Chapter 3).

Growth Temperatures

The minimum temperature for growth is 4°C, optimum 20°C, and maximum 26°C.

Some Diseases Caused by *Phytophthora infestans*

Late blight of potato and tomato are the most important diseases caused by *P. infestans.* Both hosts are members of the plant family Solanaceae. Other hosts and their distribution are listed in Table 33.1.

Late Blight of *Solanum tuberosum* L. (Potato)

Late blight of potato is described in detail in the *Compendium of Potato Diseases* (Hooker 1981), by Rich (1983) in a text on potato diseases, and in a publication on potato health management (Rowe 1993). Shorter descriptions are given by Stamps (1985h) and Holliday (1980). Late blight with an emphasis on the disease in Mexico is discussed by Niederhauser and Cobb (1959) and Niederhauser et al. (1954).

Infected foliage first becomes yellow and then water soaked and eventually turns black (Plate 33.1). The leaf symptoms, which may appear on the foliage any time during the development of the plant, consist of purple black or brownish black lesions (Plate 33.2) and usually appear first at the tip or margins of the leaf. Later, lesions may occur anywhere on the leaf, petiole, or even the stem. Symptoms often resemble those caused by frost. If cool, moist conditions prevail, whitish masses of sporangia (Figure 33.1) appear on the underside of the leaf. A pungent odor usually becomes prevalent in potato fields before the more obvious symptoms of late blight are apparent.

The tubers become affected later in the season. In the early stages, slightly brown or purple blotches appear on the skin. In damp soils the disease progresses rapidly, and the tuber decays either before or after harvest. Tuber infection is followed by secondary invasion by bacteria and fungi. This type of rot is known as "wet rot." When soil is dry, the brown discoloration extends only to a depth of about 1 cm into the tuber; however, in storage, infection progresses throughout the tuber.

Disease Development. Sporangia are produced rapidly on infected leaves at temperatures near 21°C when relative humidity is near 100%. The deciduous sporangia are readily splashed by water or spread by wind. Disease development is favored by cool (16 to 21°C), cloudy, moist weather, during which new sporangia are continually being formed. Slightly warmer weather favors the infection process. If the weather clears and the relative humidity stays low, the progress of the disease is checked, and the characteristic white mycelium and

Table 33.1. Distribution of hosts of *Phytophthora infestans*

Host	Common name	Disease	Geographical distribution
Acer sp.	Maple	Leaf blight	Soviet Union (Pshedetskaya 1968)[a]
Anthocercis viscosa R.	Anthoceris	Leaf blight	Germany (Kuhn 1859)
Aster thomsonii Clarke	Aster	Leaf blight	India (Raj et al. 1976)
Atropa belladonna L.	Deadly nightshade	Leaf blight	United States (Peterson 1942)[a]
Bupleurum maddeni Clarke	Thoroughwax	Leaf blight	India (Raj et al. 1976)
Capsicum annuum L.	Red pepper	Leaf blight	United States (Cox 1948)
Datura metel L.	Datura	Late blight	Israel (Sztejnberg and Wahl 1966)[a]
Datura meteloides DC	Datura	Flower and fruit blight	United States (Vartanian and Endo 1985b)[a]
Datura stramonium L.	Jimsonweed; jamestown weed; thorn apple	Leaf blight	United Kingdom (Hirst and Moore 1957); India (Anon. 1962b)[a]; Israel (Sztejnberg and Wahl 1966); United States (Vartanian and Endo 1985b)[a]
Erigeron multicaulis DC	Fleabane	Leaf blight	India (Raj et al. 1976)
Galinsoga parviflora		Leaf blight	India (Raj et al. 1976)
Geranium nepalense Sweet	Cranebill	Leaf blight	India (Raj et al. 1976)
Hyoscyamus aureus L.		Late blight	Israel (Sztejnberg and Wahl 1966)[a]
Hyoscyamus niger L.	Henbane; stinking nightshade	Leaf blight	England (W. G. Smith 1884); Poland (Garbowski 1913); Germany (Vowinckel, 1926)[a]; United States (Reddick 1928)[a]
Ipomoea hederacea Jacq.	Morning-glory	Leaf blight	India (Raj et al. 1976)
Ipomoea purpurea (L.) Roth	Common morning-glory	Leaf blight	India (Raj et al. 1976)
Lycium chinense Mill.	Chinese matrimony vine	Leaf blight	United States (Vartanian and Endo 1985b)[a]
Lycium halimifolium Mill.	Matrimony vine	Leaf blight	Germany (Vowinckel 1926)[a]; United States (Reddick 1928)[a]; England (Moore 1945)
Lycium obovatum L.		Leaf blight	Germany (Vowinckel 1926)[a]
Lycium turcomanicum Turcz.	Box thorn; matrimony vine	Leaf blight	Germany (Vowinckel 1926)[a]; United Kingdom (Hirst and Moore 1975)[a]
Lycopersicon esculentum Mill.	Tomato	Late blight; damping-off	France (Tulasne 1854); Germany (Kuhn 1859); United Kingdom (W. E. Smith 1881); United States (Thaxter 1989b); Russia (Speschnew 1896); India (Butler 1903); Worldwide distribution by 1930
Mandragora officinarum L.	Mandrake	Late blight	Israel (Sztejnberg and Wahl 1966)[a]
Mirabilis jalapa L.	Four o-clock	Leaf blight	Mexico (Servin 1953; Romero and Fourton 1962; Perches and Galindo 1967)
Nicandra physalodes (L.) Gaertn.	Apple of Peru	Leaf blight	United States (Peterson 1947)[a]
Nicotiana acuminata (R.C. Grah.) Hook	Nicotiana	Leaf blight	United States (Vartanian and Endo 1985b)[a]
Nicotiana clevelandii Gray	Nicotiana	Leaf blight	United States (Vartanian and Endo 1985b)[a]
Nolana humifusa (Gouan) Johnst.	Nolana	Leaf blight	Peru (Turkensteen 1978)
Petunia Juss.	Petunia	Seedling blight	England (W. G. Smith 1884); Australia (Samuel 1932); United Kingdom (Hirst and Moore 1957)
Petunia hybrida Hort.	Petunia	Leaf blight	Sweden (Lagerheim 1891); United States (Peterson 1947, Cox 1948)[a]; United Kingdom (Hirst and Moore 1957); India (Anon. 1962b)[a]; Mauritius (Anon. 1965); Israel (Sztejnberg and Wahl 1969)[a]
Physalis alkekengi L.	Winter cherry; Chinese lantern plant	Leaf blight	Germany (Vowinckel 1926)[a]; United States (Reddick 1928)[a]
Physalis angulata L.	Ground cherry	Leaf blight	United States (Peterson 1947)
Physalis ixocarpa Brot.	Tomatillo	Leaf blight	Mexico (Gandara 1909)
Polygonum alatum Buch-Ham	Smartweed, knotweed	Leaf blight	India (Raj et al. 1976)
Rumex acetosa Linn.	Garden sorrel	Leaf blight	India (Raj et al. 1976)
Salpichroa origanifolia (Lam). Baill.	Cock's eggs	Leaf necrosis	Japan (Hori 1964)
Salpiglossis sp.		Leaf blight	United States (Peterson 1947)[a]
Salpiglossis sinuata Ruiz and Pav.	Painted tongue	Leaf blight	India (Dastur 1913)
Schizanthus sp.	Butterfly flower	Leaf blight	India (Anon. 1962b)
Schizanthus grahamii Gill.	Butterfly flower	Stem, leaf, and bud blight	Germany (de Bary 1876); United States (Reddick 1928)
Schizanthus pinnatus Ruiz and Pav.	Butterfly flower	Leaf blight	United States (Vartanian and Endo 1985b)[a]
Solanum andigenum Juz. and Bukasov.		Leaf blight	Peru (Niederhauser 1953)
Solanum antipoviczii		Blight	Mexico (Reddick 1932)
Solanum atropurpureum Schr.		Leaf, stem, and fruit infection	Sweden (Hammarlund 1933)
Solanum aviculare Forst. (*Solanum laciniatum* Ait.)	Kangaroo apple	Leaf and tuber blight	Germany (Kühn 1859); Australia (Brittlebank 1920); United States (Berg 1926); France (Marchal and Foex 1932); Sweden (Hammarlund 1933); New Zealand (Driver 1957); Czechoslovakia (Brejcha et al. 1959)
Solanum boreale Gray		Leaf blight	Mexico (Niederhauser and Mills 1953)
Solanum brachycarpum		Leaf blight	Mexico (Toxopeus 1960)
Solanum bulbocastaneum Ait.		Leaf blight	Germany (Kuhn 1859); Australia (Brittlebank 1920); France (Marchal and Foex 1932); Sweden (Hammarlund 1933)
Solanum caldasii Humb. and Bonpl.		Leaf blight	France (Marachal and Foex 1932)

(*continued on next page*)

Table 33.1. *(continued)*

Host	Common name	Disease	Geographical distribution
Solanum capripense Kunth.		Leaf blight	Ecuador (Lagerheim 1891)
Solanum cardiophyllum Lindl.		Leaf blight	United Kingdom (Smith 1884); Mexico (Niederhauser and Mills 1953)
Solanum commersonii Dun.		Leaf blight	France (Labergerie 1904); England (Jones 1905); United States (Stuart 1906)
Solanum crispum Ruiz and Pav.		Leaf blight	England (Beaumont and Staniland 1937); India (Butler and Jones 1949)
Solanum demissum Lindl.	Wild potato	Leaf blight	United Kingdom (Lindley 1848); Mexico (Gandera 1909)
Solanum dulcamara L.	Bittersweet, nightshade	Leaf blight	Germany (Corda 1847); Mexico (Gandara 1909); United Kingdom (Smith 1913; Hirst and Stedman 1960); Russia (Bondartzeva-Monteverde 1926); United States (Reddick 1928; Cox 1948)
Solanum edinense Berthault.		Leaf blight	United Kingdom (Salaman 1910); Germany (Broili 1921)
Solanum ehrenbergii		Leaf blight	Mexico (Fernández and Galindo 1989)
Solanum etuberosum Lindl.		Leaf blight	Netherlands (Bergsma 1845); France (Tulasne 1854); United States (Stuart 1906); United Kingdom (Salaman 1910)
Solanum fendleri Gray.		Leaf blight	United States (Reddick 1928)[a]
Solanum goniocalix Juz. and Bukasov		Leaf blight	Peru (Bazan de Segura and Carrera 1953)
Solanum humboldti Dun.		Leaf blight	Germany (Röder 1935)
Solanum incanum L.		Leaf blight	Kenya (Nattrass and Ryan 1951); Israel (Sztejnberg and Wahl 1966)[a]
Solanum indicum L.		Leaf blight	New Zealand (Driver 1957)
Solanum iopetalum (Bitter) Hawkes		Leaf blight	Mexico (Niederhauser and Mills 1953)
Solanum jamesii Torrey		Leaf blight	United States (Reddick 1928)[a]; Mexico (Nierderhauser and Mills 1953)
Solanum laciniatum Ait. (*S. ariculare* G. Forst.)	Kangaroo apple	Leaf blight	New Zealand (Driver 1957)
Solanum maglia Molin.	Darwin potato	Leaf blight	Germany (de Bary 1861); United Kingdom (Jones 1905); United States (Stuart 1906)
Solanum marginatum L.		Leaf, stem, and fruit infection	Sweden (Hammarlund 1933)
Solanum medians Bitt.		Leaf blight	Peru (Ochoa 1955)
Solanum melongena L.	Eggplant	Fruit rot and calyx blight	Germany (Corda 1847); United States (Haskell 1921); France (Simonet 1925); Russia (Bondartzeva-Monteverde 1926); Santo Domingo (Ciferri 1927b); Rhodesia (Bates 1959)
Solanum muricatum Ait.	Sweet pepino; Peruvian cucumber; pear melon	Leaf blight	Ecuador (Lagerheim 1891); Peru (Revilla 1963); Colombia (Guzman-Naranjo 1966)
Solanum nigrum L.	Black nightshade; poison-berry	Leaf necrosis	Germany (DeBary 1863); United States (Peterson 1947); United Kingdom (Hirst and Stedman 1960); Japan (Hori 1964)
Solanum panduraeforme Drege		Leaf blight	Kenya (Nattrass and Ryan 1951)
Solanum pinnatisectum Bitter.		Leaf blight	Mexico (Niederhauser and Mills 1953)
Solanum pyracanthum Jacq.		Leaf, stem, and fruit infection	Sweden (Hammarlund 1933); Israel (Sztejnberg and Wahl 1966)[a]
Solanum racemigerum Zodda.		Leaf blight	Germany (Röder 1935)
Solanum rostratum Dunal	Buffalobur	Leaf blight	United States (Peterson 1947)[a]
Solanum sambucinum Rusby.		Leaf blight	Mexico (Niederhauser and Mills 1953)
Solanum sarachioides Sendt.		Leaf and stem blight	United States (Gardner and Yarwood 1942; Vartanian and Endo 1985b)
Solanum senecioides Domb. ex Dun.		Leaf blight	Peru (Turkensteen 1978)
Solanum simile Muell.		Leaf and stem blight	New Zealand (Driver 1957)
Solanum sisymbriifolium		Leaf blight	United States (Vartanian and Endo 1985b)[a]
Solanum stoloniferum Schlect.		Leaf blight	France (Tulasne 1954); United States (Stuart 1906); Mexico (Niederhauser and Mills 1953)
Solanum tomatillo Phil.	Tomatillo	Leaf and stem blight	Germany (Röder 1935)
Solanum tuberiferum Dun.		Leaf blight	Peru (Turkensteen 1978)
Solanum tuberosum L.	Potato	Late blight	Europe (Eriksson 1917, 1918 , noted occurrence from 1830 to 1842); United States (G. Smith 1913, noted occurrence in 1842); Taiwan (Sawada 1919); world-wide by 1930 (Tucker 1933)
Solanum utile Klotzsch		Leaf blight	Germany (Munter 1849)
Solanum verrucosum Schl.		Leaf blight	France (Tulasne 1854); Mexico (Reddick 1932)
Sonchus oleraceus Linn.	Compositae	Leaf blight	India (Raj et al. 1976)
Tilia sp.	Lime tree	Leaf blight	Soviet Union (Pshedetskaya 1968)[a]
Withania somnifera (L.) Dun.		Late blight	Israel (Sztejnberg and Wahl 1966)

[a] Artificially inoculated host.

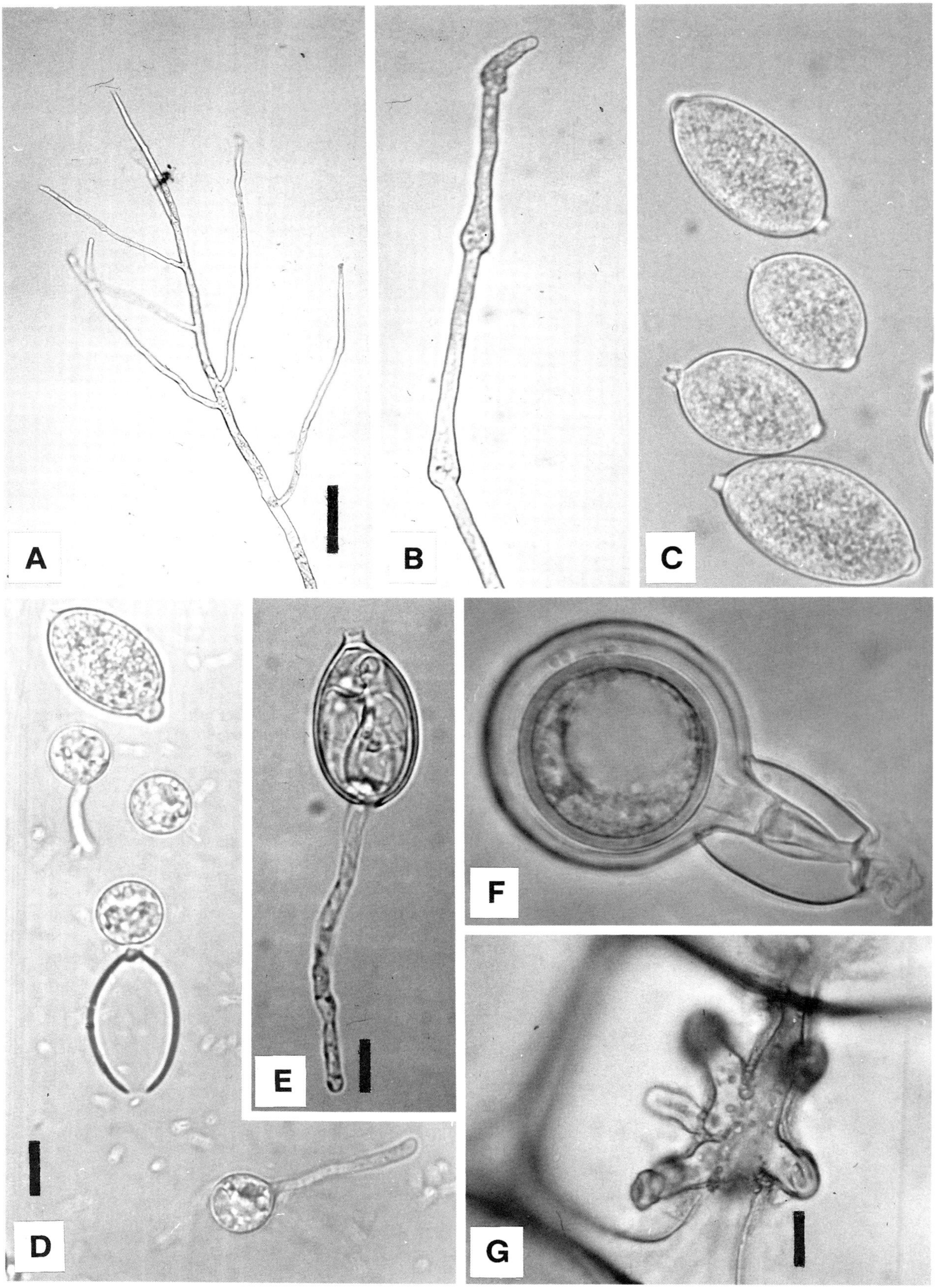

Figure 33.1. *Phytophthora infestans.* **A,** Sporangiophore; **B,** sporangiophore branch showing swellings at successive sites of sporangium formation; **C,** sporangia germinating by zoospores (**D**) and germ tube (**E**); **F,** oospore with antheridium; **G,** haustoria within tuber cells. Bar in A = 50 μm; bars in D, E, and G = 10 μm. (Reproduced by permission of the American Phytopathological Society, courtesy Hooker 1981)

sporangia are not produced on the undersides of the leaves. With the return of cool, moist weather, the fungus again becomes active and symptoms reappear. Long periods of cold weather are unfavorable for disease development, however. In warmer regions, the development of late blight depends on how long the temperature remains below 21°C. An extensive discussion of the environmental conditions favoring late blight of potato worldwide is given by Cox and Large (1960). The role of cultural factors on the development of late blight is discussed at more length in Chapter 5.

The mechanisms by which *P. infestans* persists in soil and becomes the primary source of inoculum have long been controversial (Andrivon 1995; Gregory 1983). According to many reports, inoculum survives in infected tubers and in areas where both A1 and A2 mating types exist (e.g., in Mexico). Oospores are considered to be a source of both inoculum and pathologic variability (see Andrivon [1995] for an intensive review of this subject). Andrivon (1995) states that despite 150 years of research, there are more questions than answers about the soil stages in the life cycle of *P. infestans*. The worldwide migration of a new and diverse group of genotypes and the ability to utilize a wider range of genetic markers (S. B. Goodwin et. al. 1995) should allow researchers to resolve many questions about the role of oospores in the soil phase of the life cycle of *P. infestans*. The reports of Chang and Ko (1991) and Drenth et. al. (1995) indicate that spores can form, survive, and germinate.

P. infestans is disseminated by airborne sporangia (distant spread) and from point sources by water splash of sporangia from infected leaves. Thus, epidemics can result from the dispersal of inoculum from distant fields or from infected seed pieces (Hirst and Steadman 1960).

The following conditions usually precede late blight epidemics: 1) night temperatures below the dew point for at least 4 h; 2) low night temperature that does not drop below 10°C; 3) the mean period of cloudiness not less than 0.8 of a day; and 4) rainfall of at least 0.1 mm on the following day.

Tubers, especially those near the soil surface, may be infected by sporangia or zoospores that wash downward from infected leaves during rain storms. Infection appears to be limited to periods when soil temperature is 18°C or below (Sato 1979).

Forecasting of late blight probability is complex. Hartill et al. (1990) determined the times required for production of lesions on leaves, for production of sporangia during periods of leaf wetness, and for development of new lesions over a range of 5 to 24°C. Equations were developed by use of the data to relate time to sporulate with temperature. A 2-h break in leaf wetness at any time during the first 3 h of incubation after inoculation reduced the numbers of lesions formed. Michaelides (1985) reported a simulation model of the production of disease by *P. infestans* on potato. His discussion contains a basic literature review and a mathematical model.

Control. The application of fungicides prior to infection in the field has been one of the main approaches to control (Fry 1977b; Fry et al. 1979; Rich 1983; Schwinn and Margot 1991). Chemical control has been greatly facilitated by the advent of late blight forecasting (Bourke 1970; Krause et al. 1975; Krause and Massier 1975; MacKenzie 1981; Fry and Fohner 1985; Fry and Doster 1991), which has resulted in more timely and more economical use of fungicides (see Chapter 7 for further discussion of chemical control and disease forecasting).

Until the late 1970s, fungicides commonly used for control of late blight included Bordeaux mixture, which is somewhat toxic to foliage, other copper-based compounds that are less toxic, pthalimides, and the dithiocarbamates (Rich 1983). The dosage rates and application times varied with the weather conditions (Schwinn 1983; Heitefuss 1989). Since then, improved control has been facilitated by use of the systemic phenylamide fungicides such as metalaxyl, which are usually applied with a protective fungicide such as mancozeb to help suppress the development and increase in the population of phenylamide-resistant biotypes of *P. infestans* (Young et al. 1979; Schwinn 1983; Schwinn and Morton 1990). See Chapter 7 for a full discussion of control of *P. infestans* with fungicides and in particular Table 7.6 for a listing of references on control with phenylamide compounds such as metalaxyl and furalaxyl. Fosetyl-Al, although effective against several other *Phytophthora*-caused diseases, has not been found to be effective for control of late blight (Chapter 7). Integration of polygenic host resistance and use of fungicides, as discussed by Fry (1975, 1977a, b, 1983) and Fry et al. (1983), can be efficacious.

Although metalaxyl and other phenylamide fungicides are the most highly effective of any that have been developed for control of late blight (see Chapter 7), the development of resistance within populations of *P. infestans* has become a limiting factor in the use of this fungicide. In the Netherlands, 11% of the isolates from community gardens where metalaxyl had not been used were resistant, but 45% of the isolates were resistant in commercial fields where metalaxyl had been used (Fry et al. 1991). Development of resistance to metalaxyl has been alleviated to some extent by combining a broad-spectrum fungicide, such as mancozeb, with metalaxyl and by limiting the number of times fields are sprayed with metalaxyl. Use of metalaxyl on seed fields is not allowed in some countries because of the danger that metalaxyl-resistant biotypes might be disseminated with seed tubers. See Chapter 7 (Table 7.7 for references on resistance to metalaxyl) for a more complete discussion of resistance and Chapter 3 for a discussion of the genetics of metalaxyl resistance showing that resistance is associated with a single gene. Principles governing the use of phenylamide fungicides as described by Delp (1980, 1984, 1988) and Delp and Dekker (1985) are reviewed in Chapter 7.

The development and wide dissemination of the "new" *P. infestans* biotypes (e.g., US7 A2 and other isolates) (S. B. Goodwin et al. 1995) to many areas of the world during the past decade is one of the most serious threats to the control of late blight since the disease was first noted during the 1840s. Many of these new biotypes, which Fry and his colleagues (e.g., Fry and Goodwin 1995; see other citations in Chapters 3 and 7) have detected by allozyme and DNA fingerprinting technology, are of the previously unknown A2 mating type, are more aggressive, and (a more serious problem) are resistant to metalaxyl, the most effective late blight fungicide developed to date. These new biotypes are now present on both potatoes and tomatoes in California (S. B. Goodwin et al. 1995) and in other western states (Deahl et al. 1991, 1995) where late blight had not previously been considered a serious threat to production. Serious attention will have to be applied by regulatory and experiment station agencies to overcome this problem. Some fungicides, to which resistance in *P. infestans* has not developed, are discussed in Chapter 7. Gisi (1991) presents research data that indicate that application of two or three fungicides elicits a synergistic response to control late blight in Europe.

Since sporangia of *P. infestans* increase and spread rapidly and over long distances (Hirst 1953; Hirst and Moore 1957; Hirst and Steadman 1953), attention to the problem of control should be given not only by individual growers but by regulatory agencies in large geographical regions.

Breeding for race-specific resistance to late blight was once considered an efficacious approach to control but has since proved to be of only limited use because many pathologic (physiologic) races of *P. infestans* have the ability to attack new cultivars with single-gene resistance. The early breakthrough involving incorporation of single dominant genes for resistance to the various races of *P. infestans* into *Solanum tuberosum* from other solanaceous species such as *S. demissum* was initially highly successful, but new pathogenic races of the fungus developed in the field that rendered the new cultivar susceptible. As the population of the new race increases and becomes dominant in the following years, the new cultivar is no longer resistant (see Chapters 3 and 7 for a more detailed discussion of race-specific and horizontal resistance).

General (horizontal) resistance, which is more stable but does not confer as high a level of resistance as single-gene resistance, has been identified in many cultivars (Niederhauser 1961, 1962, 1991, 1993; Niederhauser and Mills 1953) and after the failure of single-gene-type vertical resistance, has become the best approach to genetic control (Umaerus 1983; Dowley et al. 1991). As explained by Vanderplank (1963) and MacKenzie et al. (1983), general resistance, also referred to as rate-limiting, multigenic, field, or horizontal resistance (Chapter 6), reduces the rate of the epidemic and is equally effective against all races. General resistance is more stable because usually it is conditioned by more than one gene and there is less probability that mutations in populations of *P. infestans* will simultaneously occur in nature in more that one gene at a time. Since plants with general resistance are not immune, less selection pressure is placed on the population of *P. infestans* than by cultivars with a single gene for immunity. For instance, when the plant has immunity, the only isolates that will survive are the virulent mutants, because *P. infestans* does not sporulate on race-specific resistant plants. In most regions of the world, breeding programs now emphasize the development of general resistance instead of race-specific resistance. Methods of assessing general resistance in breeding lines have been described by Main and Gallegly (1964) and Malcolmson (1976). For detailed discussions of race-specific and multigenic or general resistance in relation to late blight of potato, see Gallegly and Niederhauser (1959), Vanderplank (1968), Day (1973), Umaerus (1969, 1970), and Umaerus et al. (1983) and Chapter 6, where more citations are given and the subject is discussed further. International cooperative ventures for the control of late blight of potato are discussed by Niederhauser (1993), a pioneer in the research effort to control late blight in Mexico and other regions around the world.

Many cultivars with general resistance (for example, Sebago) are available commercially in different regions of the world (O'Brien and Rich 1976; Rich 1983). Since some cultivars are better adapted to certain localities, such information should be obtained from local advisors before selecting resistant cultivars. Plants with general resistance require less fungicide than susceptible cultivars (Fry 1975, 1977a, 1982; Fry et al. 1983). After the crop matures, vines should be killed by an acceptable chemical method to prevent infection of tubers by inoculum in leaves and stems (Rich 1983).

Late Blight of *Lycopersicon esculentum* Mill. (Tomato)

Late blight of tomato and its control is described by J. B. Jones et al. (1991).

Leaf lesions are water soaked with definite margins and a gray green color. The infected tissue becomes brown and necrotic, and the leaf dies. Stems and petioles are also affected. Eventually, the plant will be killed if the disease is not controlled. The foliage symptoms are similar to those on potato. On tomato fruits, gray green, water-soaked spots enlarge rapidly. The lesions then turn dark brown, and although the diseased areas become wrinkled, the rotted tissue remains firm. The unaffected parts of the fruit ripen to a normal red color. Occasionally, the lesion may have rings in a zonate pattern that superficially resemble those of buckeye rot caused by *P. parasitica;* however, in fruit rot caused by *P. infestans,* the rings are much closer together. Seeds from infected fruit often give rise to infected tomato seedlings (Vartanian and Endo 1985a).

Disease Development. Disease development is similar to that of late blight on potato. Since rainfall during summer months is rare in California, late blight was seldom a problem before 1979. In certain areas, however, dew formation on staked tomatoes is sufficient to allow late blight to become limiting to production (Vartanian and Endo 1985a, b). Late blight became a problem after 1979 probably because of a series of relatively mild winters in which frost did not kill overwintering volunteer solanaceous host plants and because of intensive tomato-cropping practices (three crops per year in San Diego County). Late blight was found in the field throughout the year on volunteer tomato plants and on plants grown in home gardens. The prevalent tomato races were T0 and T1. Only the A1 mating type occurred in southern California. Race T1 grew and sporulated more rapidly than race T0 (Vartanian and Endo 1985b). In 1993, metalaxyl-resistant A2 isolates of *P. infestans* were found in California (M. D. Coffey, *personal communication*).

Control. Bordeaux mixture used as a foliar spray before infection is effective but is toxic to the foliage of tomato, so other, more insoluble copper-based compounds, pthalimide, dithiocarbamate, or phenylamide fungicides are more often used. The fungicides are applied on the basis of late blight forecasts similar to those employed for potato. Y. Cohen et al. (1979) found that the systemic phenylamide fungicide metalaxyl, used as a single soil drench, controlled late blight of potted tomato plants and gave protection from blight for at least 6 weeks. Many subsequent citations listed in Chapter 7 indicate that the phenylamide fungicides (e.g., metalaxyl) are effective for control of late blight.

Although metalaxyl is highly effective, resistance to metalaxyl within populations of *P. infestans* on potato has become a problem in Europe and now occurs widely in California (M. D. Coffey, *personal communication*). The addition of the broader spectrum fungicide mancozeb to metalaxyl or other phenylamide fungicides may at least partially alleviate this buildup of biotypes with resistance to *P. infestans.* This subject is discussed further in Chapters 3 and 7.

J. B. Jones et al. (1991) note that the disease-forecasting systems, such as the Hyre system, which predicts disease onset on the basis of temperature and rainfall, the Wallin system, which predicts disease on the basis of temperature and humidity, and BLITECAST, which integrates the two systems in a computer program, help to determine when application of a fungicide will be most effectual. Use of forecasting systems is discussed in Chapter 7.

Although cultivars with resistance to the known tomato races of *P. infestans* are commercially available, host resistance is not an important element of control in the United States (J. B. Jones et al. 1991). The early breeding work was complicated by the presence of potato and tomato races of *P. infestans* and the possible interrelationships between the two (Berg 1926). It is now known that some potato races behave as different tomato races when inoculated on tomato plants. Races T0 and T1 are recognized to be most prevalent in the

United States; T1 is the more aggressive. Race characteristics of a given isolate should be considered separately on the two hosts, since tomato genes for resistance to *P. infestans* differ from those known in the potato (Wilson and Gallegly 1955; Gallegly and Niederhauser 1959).

Some new biotypes of *P. infestans* can be aggressive on both tomato and potato. Legard et al. (1995) state, "Our results were consistent with the hypothesis that increased aggressiveness on tomato (*Lycopersicon* spp.) has evolved in isolates already pathogenic to *Solanum* spp." Their report also shows that some isolates from potato are relatively nonaggressive to tomato.

CHAPTER 34

Phytophthora inflata

Caroselli and Tucker (1949)

Phytophthora inflata was first described by Caroselli and Tucker (1949) as the causal agent of pit canker of elm trees. The disease had been reported since the early 1900s, but the causal agent remained undetermined until Caroselli and Tucker conclusively demonstrated that *P. inflata* was the cause. Pit or elm trunk canker has been reported only on *Ulmus americana* L. (American elm) and *U. fulva* Mich. (slippery elm) in the northern United States and eastern Canada. G. Hall et al. (1992) reported *P. inflata* from rotted roots of *Sambucus tenuifolium* and *Syringa vulgaris* in England.

Characteristics of *Phytophthora inflata*

P. inflata is classified in group III (Stamps et al. 1990). Morphology is shown in Figure 34.1. See Tables 4.2 and 4.3 for tabular keys.

Sporangia

Sporangia are semipapillate, broad, limoniform to elongate with occasional lateral attachments, 20 to 67 µm long × 15 to 32 µm wide, average 38 × 23 µm (length-breadth ratio 1.3:1), and persistent on the stalk and form on unbranched sporangiophores (Caroselli and Tucker 1949). G. Hall et al. (1992) found sporangia to be highly variable in size with an average length-breadth ratio of 1.12:1. Sporangia germinate by external proliferation.

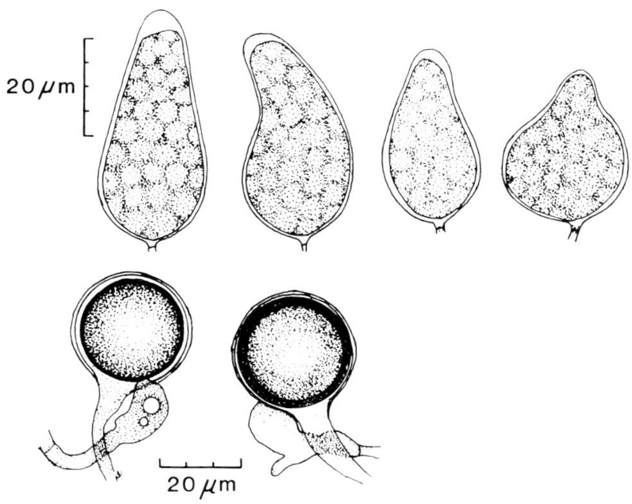

Figure 34.1. Morphology of *Phytophthora inflata.* Lower row, oogonia with paragynous, often inflated and contorted antheridia. Upper row, ovate to obpyriform, sometimes irregular in shape, semipapillate sporangia. Inflated antheridia are the main diagnostic feature. (Drawing by A. Vaziri after Caroselli and Tucker 1949)

Hyphal Swellings

Thin-walled, hyaline, intercalary hyphal swellings form in aqueous cultures.

Chlamydospores

Chlamydospores are not formed.

Sex Organs

P. inflata is homothallic; oogonia are 30 to 43 µm in diameter, average 34 µm; antheridia are paragynous, often contorted or inflated and twisted around the oogonial stalk or irregularly lobed or branched (up to 15 × 50 µm); oospores are 26 to 39.3 µm in diameter, average 31.3 µm, with a thick wall (3 to 4 µm) and are aplerotic. The inflated antheridia are highly characteristic (G. Hall et al. 1992).

Growth Temperatures

Optimum temperature for growth is 25 to 30°C, and the maximum is <35°C. Information on the minimum temperature is not available.

Distinguishing Characteristics

The large, contorted, inflated, variously lobed or branched paragynous antheridia differentiate *P. inflata* from other *Phytophthora* species. The inflated antheridia are most readily observed in young cultures. Although Caroselli and Tucker (1949) considered *P. megasperma* and *P. erythroseptica* to be similar to *P. inflata,* the semipapillate nature of the sporangia places *P. inflata* in group III of Stamps et al. (1990), whereas the other species produce nonpapillate sporangia, thus placing them in group VI. The inflated antheridia differentiate *P. inflata* from certain other species in group III, such as *P. citricola,* which also produces paragynous antheridia and semipapillate sporangia.

Disease Caused by *Phytophthora inflata*

Pit Canker of *Ulmus americana* L. (American Elm)

A general decline in growth, yellow or pale green foliage, and an occasional dieback of twigs and small branches are typical symptoms. The foliar symptoms follow the formation of concentric or zonate cankers and callus tissue on the trunk or scaffold branches. Each new formation of callus tissue is reinvaded by the fungus, resulting in the formation of a new ring of callus tissue; this repetitive process results in the development of the concentric cankers that are characteristic of the disease. Sometimes cankers are masked by the thick bark of the trunk. Fissures develop near the margins of cankers, from which a red to brown fluid exudes. In a severely infected tree, the depression formed by the cankers and the sloughing of the bark near the diseased areas lead to the formation of a characteristic rough, twisted, and irregularly shaped trunk (Caroselli and Tucker 1949).

Disease Development. Prolonged rainy conditions during periods of high temperatures (25 to 30°C) enhance the progress of the disease.

Control. Good drainage around elm trees should be provided and low-lying areas or heavy soils for new plantings should be avoided. If registered for use, systemic fungicides, metalaxyl, and/or fosetyl-Al should be efficacious.

CHAPTER 35

Phytophthora insolita

Ann and Ko (1980)

Phytophthora insolita was isolated from soil in a citrus orchard by baiting soil slurry samples with citrus leaves (method of Grimm and Alexander [1973]) by Ann and Ko (1980). Since the isolate produced unusual (Latin *insolita*) oogonia devoid of antheridia, it was described as a new species. No natural hosts have been reported; however, it was pathogenic to inoculated apple and cucumber fruits but not to sweet orange or cucumber fruits (Ko and Ann 1985). It was isolated infrequently from soil in Taiwan but was not pathogenic to sweet orange (Ann 1984).

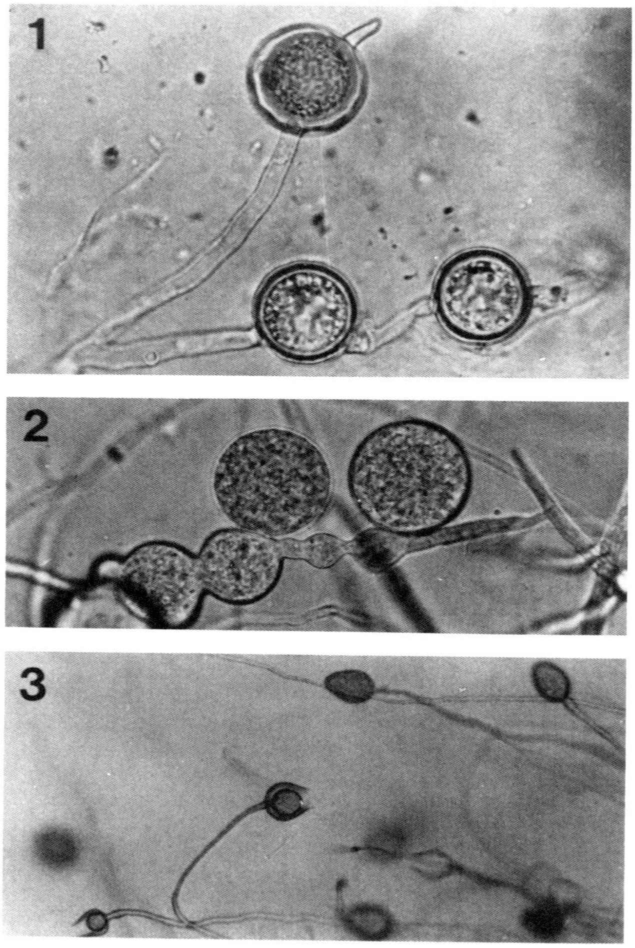

Figure 35.1. Morphology of *Phytophthora insolita.* **1,** Sex organs. Globose oogonia containing oospores that nearly fill the oogonia (plerotic). Antheridia are not formed. **2,** Globose chlamydospores and hyphal swellings. **3,** Nonpapillate sporangia. (Reproduced by permission from Mycologia 72:1180–1185, P. J. Ann and W. H. Ko, copyright 1980, The New York Botanical Garden, Bronx, New York)

Characteristics of *Phytophthora insolita*

P. insolita is classified in group V (Stamps et al. 1990). Photomicrographs of sporangia, chlamydospores, and oogonia are shown in Figure 35.1 (Ann and Ko 1980). See Figure 35.2 for a detailed drawing, Tables 4.2 and 4.3 for tabular keys, and Appendix 4.9 for a dichotomous key (Ho 1992).

Sporangia

Sporangia, which measure 38 to 68 × 29 to 39 μm, are produced readily when blocks of young mycelium on V8 juice agar are transferred to water and incubated at 25°C under light. Sporangia are terminal, noncaducous, ovoid, and nonpapillate. New sporangia proliferate both externally and internally from older sporangia after zoospores have been released.

Chlamydospores

Thin-walled, globose chlamydospores are formed terminally and measure 22 to 48 μm in diameter.

Hyphal Swellings

Irregularly shaped hyphal swellings of varying sizes are formed on V8 juice agar.

Growth Temperatures

The minimum temperature for growth is >9°C, the optimum is 32°C, and the maximum is 38°C. No growth occurs at 40°C. *P. insolita* is not sensitive to hymexazol (50 μg/ml) (Ann and Ko 1980).

Sex Organs

P. insolita is homothallic. Oospores are produced abundantly on 20% V8 juice agar neutralized with 0.2 g of calcium

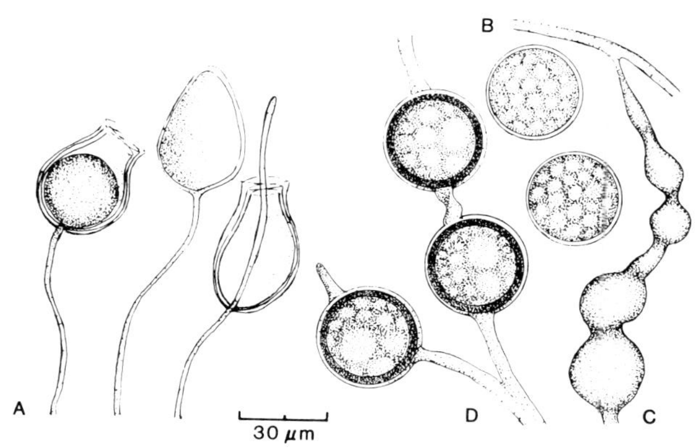

Figure 35.2. Morphology of *Phytophthora insolita.* **A,** Nonpapillate, ovoid sporangia: an empty sporangium producing secondary sporangia (internal proliferation), an ungerminated sporangium, and a sporangium giving rise to a germ tube emanating from within an empty sporangium (external proliferation). **B,** Globose chlamydospores with thin walls. **C,** Hyphal swellings. **D,** Globose, thick-walled oogonia without antheridia. (Drawing by A. Vaziri)

carbonate. Spherical oogonia (ranging in diameter from 29 to 36 μm), which form terminally, have smooth to slightly undulating walls. The oogonial stalks are variable in length and are branched sympodially from a place immediately beneath the older oogonium. Antheridia have not been seen. Thick-walled (approximately 2.4 μm) oospores (27 to 31 μm in diameter, average 30 μm) are globose and nearly fill the oogonium. Oospore formation was induced when *P. insolita* was paired with a *P. parasitica* A2 mating type on V8 juice agar but not when paired with an A1 mating type. On the basis of hormone production (Ko 1980), *P. insolita* was considered to belong to type 9 or 11 of group II, which forms oospores by self-induction (Ann and Ko 1980).

CHAPTER 36

Phytophthora iranica

Ershad (1971)

No synonyms are known for *Phytophthora iranica,* which was first isolated and described in Varamin, Iran, by Ershad (1971) from roots of eggplant, *Solanum melongena* L. When artificially inoculated, tomato fruits, sugar beet roots, and potato tubers were susceptible, but eggplant, cucumber, sweet orange, and apple fruits were not. *P. iranica* caused a soft rot of potato tubers, which turned pink when exposed to air. In a report by Belisario et al. (1993), a stem and root rot of myrtle (*Myrtus communis* L.) was attributed to *P. iranica,* but this isolate is now proposed to be a new species, *P. italica* (Chapter 37; Cacciola et al. 1994, 1996).

Characteristics of *Phytophthora iranica*

P. iranica is classified in group I (Stamps et al. 1990). Morphology is shown in Figure 36.1 for an isolate from Iran. See Tables 4.2 and 4.3 for tabular keys.

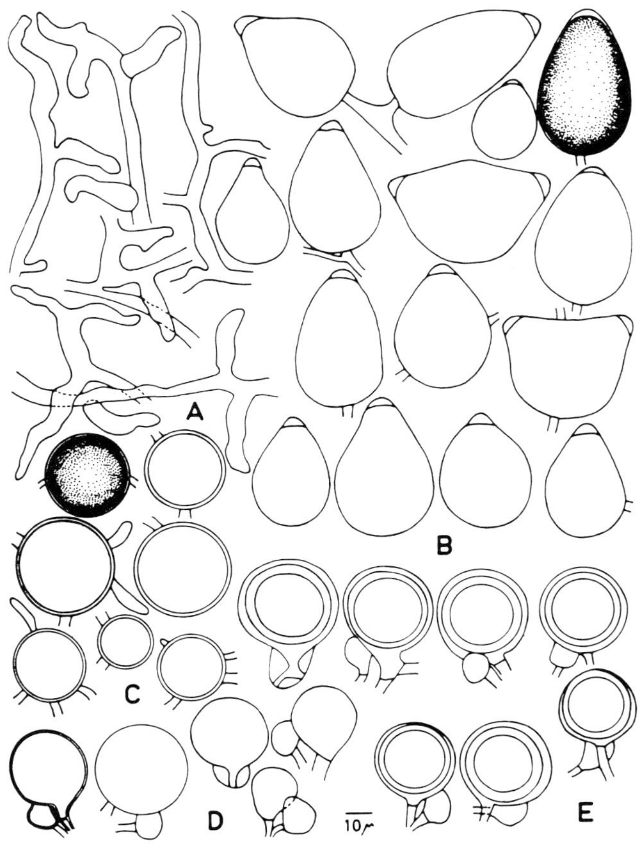

Figure 36.1. Morphology of *Phytophthora iranica.* **A,** Mycelium; **B,** papillate sporangia; **C,** chlamydospores; **D,** oogonia; and **E,** oospores. (Reproduced by permission of Blackwell Wissenschafts Verlag, Berlin, Germany, courtesy Ershad 1971)

Sporangia

Sporangia are papillate (sometimes with two papillae per sporangium); ovoid, obpyriform, ellipsoid to subspherical, 30 to 72 µm long × 22 to 51 µm wide (average 47.9 × 36.8 µm) (length-breadth ratio 1.3:1); and persistent on the stalk. Sporangiophores are short and sympodially branched (Ershad 1971).

Chlamydospores

Chlamydospores are rare; 17 to 41 µm in diameter, average 28.7 µm; mostly intercalary; and rarely terminal.

Hyphal Swellings

Hyphal swellings are not formed.

Sex Organs

Oogonia are formed in single culture (homothallic) and are subspherical and 21 to 45 µm in diameter, average 34 µm. Antheridia are mostly paragynous, but occasionally an amphigynous antheridium is found; oospores are aplerotic, 15 to 37 µm in diameter, average 29.3 µm; the oospore wall is 1 to 5 µm thick, average 3 µm (Ershad 1971).

Growth Temperatures

The minimum temperature for growth is 10°C, optimum 27.5°C, and maximum 35°C (Ershad 1971). Similar values were reported by Belisario et al. (1993).

Distinguishing Characteristics

P. iranica differs from *P. cactorum* by its the production of noncaducuous sporangia, sporangia with more than one papilla, and thicker oospore walls.

When tested by electrophoresis on slabs of polyacrylamide gel, *P. iranica* (now *P. italica*) produced protein patterns clearly distinguishable from those of *P. citrophthora, P. cactorum,* and *P. nicotianae* (Belisario et al. 1993).

Diseases Caused by *Phytophthora iranica*

Root Rot of *Solanum melongena* L. (Eggplant)

P. iranica was isolated from eggplant but was not pathogenic to eggplant by experimental inoculation.

By artificial inoculation, *P. iranica* causes a soft rot of potato tubers. After inoculation, cut surfaces of the tuber turn pink and with prolonged exposure to air, turn brown. These symptoms are also characteristic of pink rot of potatoes caused by *P. erythroseptica* (Chapter 25). *P. iranica* was weakly or moderately pathogenic on sugar beets and tomatoes, but cucumbers, oranges, and apples were not susceptible (Ershad 1971).

Root Rot of Myrtle (*Myrtus communis* L.)

An isolate that was considered to be *P. iranica* (Belisario et al. 1993) is now proposed as the new species *P. italica* (Cacciola et al. 1994, 1996; Chapter 37). It was isolated from roots of myrtle trees grown in pots in a greenhouse in Italy. Tests showed it to be weakly pathogenic to myrtle stems and roots.

No control methods have been described, but presumably they would be similar to those used against other root-rotting species of *Phytophthora.*

CHAPTER 37

Phytophthora italica

Cacciola, Magnano di San Lio, and Belisario (1996)

Phytophthora italica was originally identified as *P. iranica*. It was isolated from and proved to be pathogenic to roots of myrtle (*Myrtus communis* L.) in Italy (Belisario et al. 1993). Cacciola et al. (1996) compared the myrtle isolate with an authentic *P. iranica* isolate from the Centraalbureau Voor Schimmelcultures (Baarn, the Netherlands). The myrtle isolate could be distinguished from *P. iranica* as well as from *P. citricola* and *P. pseudotsugae* by morphological characteristics as well as by polyacrylamide gel electrophoresis of mycelial proteins and esterase.

Characteristics of *Phytophthora italica*

P. italica is classified in group I (Tables 4.2 and 4.3). See also Figure 37.1. The description of the species by Cacciola et al. (1996) follows.

Phytophthora italica sp. nov. Cacciola, Magnano di San Lio & Belisario

Colonies on agar media slow-growing with defined margins; colonies on potato dextrose agar (Oxoid) with a slight stellate pattern and on V8 juice agar (Campbell) uniform, white to yellowish-brown, with aerial mycelium woolly and dense-like felt; colonies on cornmeal agar (Oxoid) white, appressed and slightly radiate, with sparse aerial mycelium when mature. Hyphae hyaline, 3-6 µm in diameter, poorly branched, becoming septate with age. Sporangiophores simple, sympodial. Sporangia terminal, occasionally intercalary, germinating either directly or by zoospores, distinctly papillate, frequently bipapillate, sometimes with three papillae, persistent, varying in shape and dimensions. Sporangia produced on V8 juice agar ovoid, obpyriform, subsphaerical or broadly ellipsoid, 14-56 µm (modal value 39 µm) long by 11-38 µm (modal value 29 µm) wide; length-to-breadth ratio 1.1:1-1.5:1 (modal value 1.3:1). Chlamydospores rarely produced. Gametangia produced abundantly in single culture on V8 juice agar and cornmeal agar. Oogonia, subspherical, often

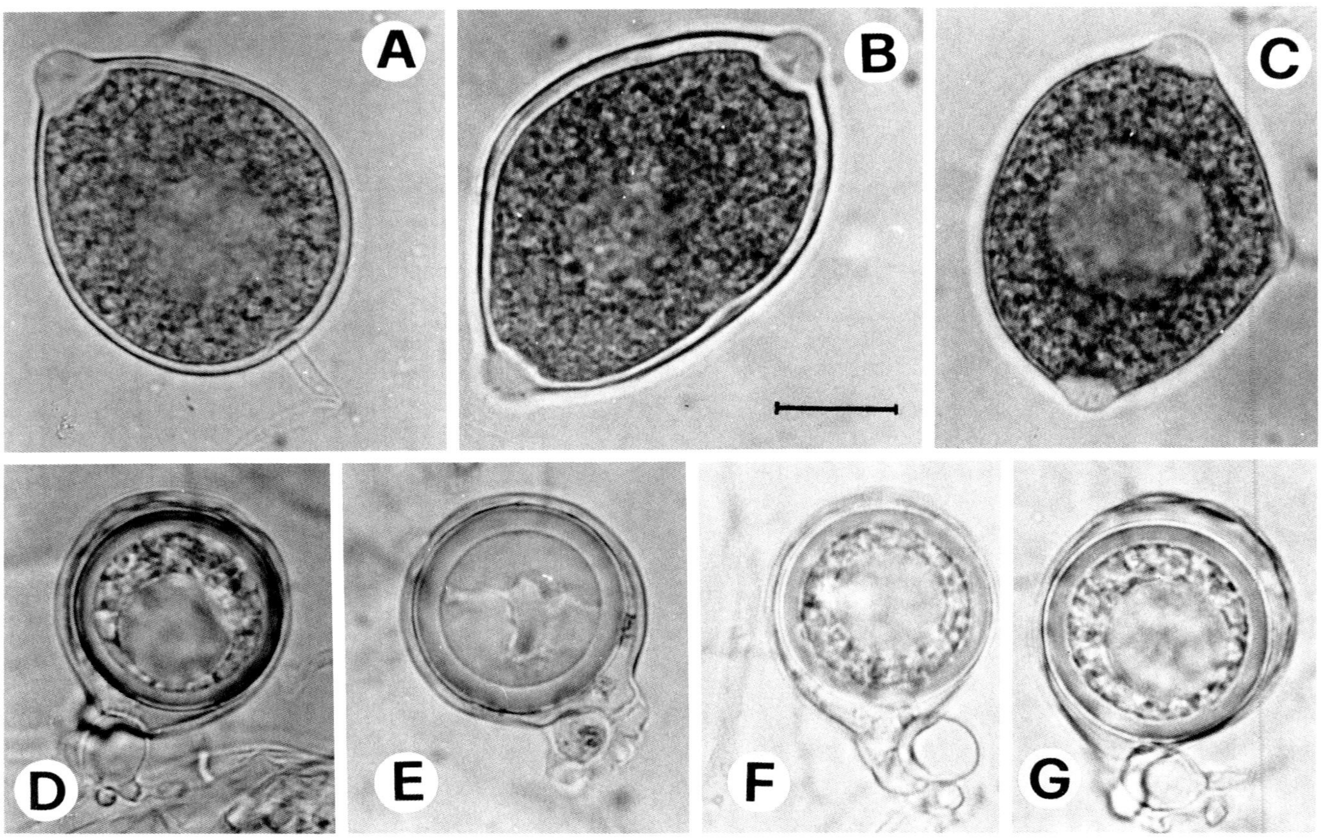

Figure 37.1. Morphology of *Phytophthora italica,* previously reported to be *P. iranica* (Cacciola et al. 1994). **A, B, and C,** Distinctly papillate, ovoid sporangia from myrtle plants in Italy stained with cotton blue in lactophenol. In **A,** note the inconspicuous basal plug at the insertion of the sporangiophore. **D, E, F, and G,** Unstained sex organs showing paragynous antheridia on globose oogonia containing aplerotic oospores. Bar = 10 µm. (Reproduced by permission of the American Phytopathological Society, St. Paul, Minnesota, courtesy Belisario et al. 1993)

slightly flattened or pyriform, with a short stalk and smooth walls, varying in size; diameter of oogonia produced on V8 juice agar 15-29 µm at the poles (mean 22.0 ± 3.38 µm; modal value 24 µm). Antheridia single, unicellular, paragynous, mostly subspherical (mean diameter 7.1 ± 1.60 µm), sometimes ovoid, 6-12 µm long by 4-10 µm wide. Oospores single, markedly aplerotic, 11-23 µm in diameter (mean 16.8 ± 2.96 µm; modal value 19 µm). Growth occurs between 10 and 35°C, with optimum at 26°C.

Habitat: The type is represented by a culture isolated from rotted roots of myrtle (*Myrtus communis* L.) in Sardinia, Italy, and is preserved as pure culture on PDA in the collection of the Institute of Plant Pathology, University of Catania, Italy.

Etymology: the etymology, *P. italica* which means 'from Italy', refers to the country where this species was first reported and underlines that this species is distinct from *P. iranica* as had originally been classified.

Distinguishing Characteristics

The oogonia of *P. italica* were similar in shape and size to those of *P. iranica.* The antheridia of *P. italica* were always paragynous, whereas antheridia of *P. iranica* were either amphigynous or paragynous (most frequent) and on cylindrical amphigynous antheridia, fingerlike appendages often formed.

The average diameter of antheridia was consistently smaller for *P. italica* than for *P. cactorum, P. citricola, P. iranica,* and *P. pseudotsugae.* Hyphae produced by *P. italica* were straight to irregularly branched, but hyphae of *P. iranica* were sinuate with ends that typically coiled. *P. italica* produced distinctly different protein and esterase patterns when compared with *P. cactorum, P. citricola, P. iranica,* and *P. pseudotsugae.*

CHAPTER 38

Phytophthora japonica

Waterhouse (1974a)

Synonyms include *Pythiomorpha oryzae* Ito and Nagai (1931) and *Phytophthora oryzae* (Ito and Nagai) Waterhouse (1958). Waterhouse (1958) proposed the name *P. oryzae* for *Pythiomorpha oryzae* but was unaware that Hara (1939), cited by Waterhouse (1974a), had transferred *Sclerospora oryzae* Brizi to the genus *Phytophthora* as *P. oryzae* (Brizi) Hara. Since this name had priority over *P. oryzae* (Ito and Nagai 1931), Waterhouse (1974a) proposed the new name *P. japonica* Waterhouse to replace *P. oryzae* (Ito and Nagai) Waterhouse. Katsura (1971) presented data for *P. oryzae,* which we assume to be *P. japonica.*

P. japonica was described as a pathogen of rice seeds and seedlings in Japan, but no other hosts are known. De Cock (1986) tentatively identified an isolate from Lake Grevelingen in the Netherlands as *P. japonica.*

Characteristics of *Phytophthora japonica*

P. japonica is classified in group VI (Stamps et al. 1990). Morphology is shown in Figure 38.1. See Tables 4.2 and 4.3 for tabular keys.

Sporangia

Sporangia are nonpapillate and persistent on the stalk, proliferate internally, and are ovoid and ellipsoid to elongated. The size, 41 to 84 µm long × 26 to 48 µm wide, average 60 × 36 µm, is considered to be unusually large for *Phytophthora* (Waterhouse 1958). Length-breadth ratio is 1.9:1. Sporangiophores are sympodial in water (Katsura 1971).

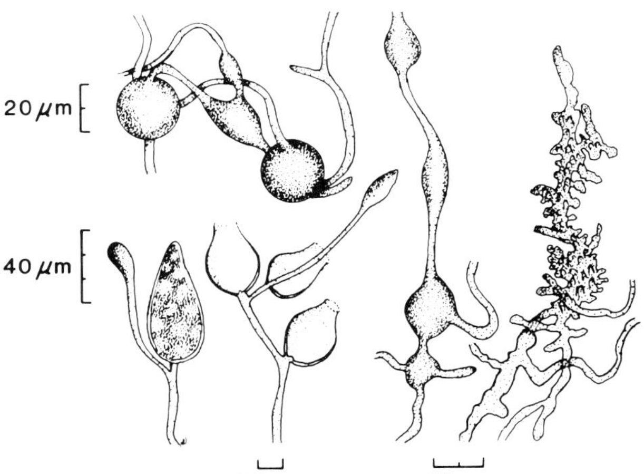

Figure 38.1. Morphology of *Phytophthora japonica.* Lower row, nonpapillate, ellipsoid sporangia formed sympodially on sporangiophores; irregular, knoblike hyphal swellings. Upper row, globular hyphal swellings. (Drawing by A. Vaziri after Ito and Nagai 1931; see also Waterhouse 1970a, Plate 12)

Hyphal Swellings

Hyphal swellings are globular or somewhat irregular and often in clumps (Katsura 1971).

Chlamydospores

Chlamydospores are not produced.

Sex Organs

Oospores are not seen in single culture but formed when paired with an isolate of *P. drechsleri* (IMI 40499). This pairing resulted in oospore production so abundant that the medium turned yellow (Waterhouse 1958). In paired cultures, some oogonia were hyaline or pale yellow and up to 45 µm in diameter (similar to those of *P. drechsleri*), and others were amber colored and up to 70 µm in diameter. Both amphigynous and paragynous antheridia were formed (Waterhouse 1958). Since oospores are not formed except when paired with another species, they are not considered to be useful for taxonomic purposes and are not cited by Stamps et al. (1990).

Growth Temperatures

The minimum temperature for growth is 9 to 11°C, the optimum is 26 to 28°C, and the maximum is 35°C (Katsura 1971).

Distinguishing Characteristics

The swollen, knoblike appearance of the mycelium is a distinguishing feature (Ito and Nagai 1931).

Disease Caused by *Phytophthora japonica*

Seedling Damping-off of *Oryza sativa* L. (Rice)

P. japonica is one of 12 water molds that cause damping-off of rice seedlings (Ou 1972). A white, cottony growth forms on the surfaces of seeds in the nursery. Mycelium can often be seen growing out from breaks in the seed coat that result from threshing or germination. Seed germination is effectively inhibited in cases of severe infestation. Damping-off results when seedlings are infected after the plumules have grown to approximately 1 cm in length.

Disease Development. The development of the disease is influenced by the weather and the vigor of the seedlings. Infection occurs readily at 18 to 35°C but not at 10 to 12°C (Ito and Nagai 1931). The disease progresses rapidly after the growth of the seedlings has been retarded by cool weather.

Mechanically damaged seeds are particularly prone to infection.

Control. Seed treatment with captan, thiram, or difolatan controls the disease (Ou 1972).

CHAPTER 39

Phytophthora katsurae

Ko and Chang (1979)

Phytophthora castaneae Katsura and Uchida (Katsura 1976), a synonym, was first isolated and described from diseased chestnuts (*Castanea crenata* Siebold and Zuch.) in Japan. It has since been isolated by Ko and Chang (1979) from forest soils in Taiwan. The name *P. castaneae* was changed to *P. katsurae* by Ko and Chang (1979) because the former name was found to be an illegitimate homonym of *P. castaneae* (Mangis) Clements and Shear (1931), which was rejected by Waterhouse (1963). A *Phytophthora* species causing fruit and heart rot of coconut was reported to resemble *P. katsurae* (Uchida et al. 1992).

A summary of *P. katsurae* and the diseases caused (Stamps 1985g) indicates that both chestnut and coconut are hosts and that the fungus is distributed in Japan, Taiwan, Hawaii, Ivory Coast, Australia, and Papua New Guinea. *P. katsurae* was isolated and proved to be pathogenic on cocoa pods in the Ivory Coast. The size of lesions caused was much smaller (2 cm^2) than those caused by *P. palmivora* (27 cm^2). This was a first record for cocoa from Ivory Coast (Liyanage and Wheeler 1989). Fruit rot of coconut in Hawaii usually starts at the stem and effects fruit of all ages (Ooka 1984).

Characteristics of *Phytophthora katsurae*

P. katsurae is classified in group II (Stamps et al. 1990). See Tables 4.2 and 4.3 for tabular keys and Appendix 4.9 for a dichotomous key (Ho 1992). Morphology is shown in Figures 39.1 and 39.2.

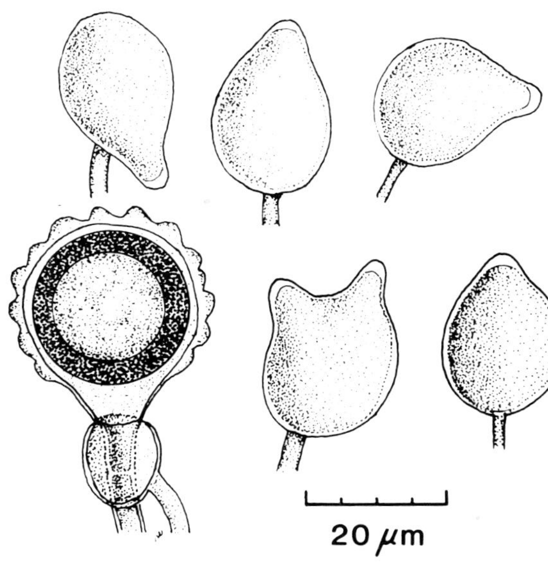

Figure 39.1. Morphology of *Phytophthora katsurae.* Upper row, papillate sporangia. Lower row, oogonium with verrucose (bullate) wall and amphigynous antheridium; and papillate sporangia, one of which is bipapillate. (Drawing by A. Vaziri)

Sporangia

Sporangia are mostly terminal but occasionally are laterally attached and sometimes formed intercalarily. They are limoniform, ovoid, obpyriform to obturbinate, and papillate, occasionally with two papillae per sporangium. Some sporangia are asymmetric (Figures 39.1 and 39.2); they are 10 to 42.5 μm long × 10 to 37.5 μm wide (average 27.5 × 22.5 μm) on the stalk (Katsura 1976). Ko and Chang (1979) recorded sizes of sporangia as 25.1 to 47.4 × 22.3 to 30.7 μm. Length-breadth ratio is <2.0. Sporangiophores are unbranched.

Hyphal Swellings

Hyphal swellings are not formed.

Chlamydospores

Chlamydospores are spherical and 12.0 to 19.2 μm in diameter.

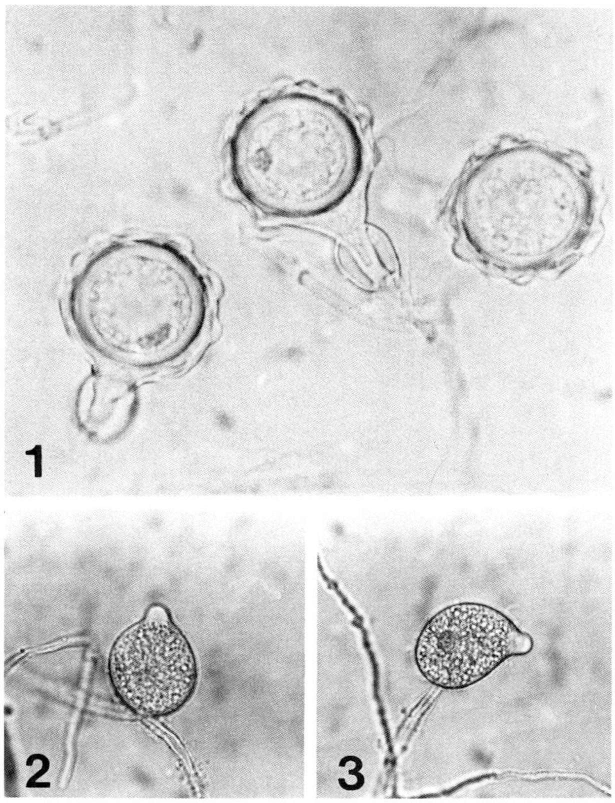

Figure 39.2. 1, Bullate oogonia of *Phytophthora katsurae* with amphigynous antheridia. **2 and 3,** Papillate sporangia. (Reproduced by permission from Mycologia 71:840–844, Ko and Chang, copyright 1979, The New York Botanical Garden, Bronx, New York)

Sex Organs

P. katsurae is homothallic. Oospores form abundantly in 10% V8 juice agar (Ko and Chang 1979). Antheridia are amphigynous and spherical to ovoid; oogonia have warty protuberances on the surface and a distinct funnel-shaped base (Figures 39.1 and 39.2); oogonia are 19 to 31 μm in diameter (average 27 × 25 μm); oospores are globose, 15 to 27.5 μm in diameter (average 20 μm). Up to 95% of 30-day-old oospores germinated when incubated under blue and green light (H. S. Chang and Shu 1988).

Growth Temperatures

The minimum temperature for growth is 9°C, optimum 26 to 28°C, and maximum 32°C.

Distinguishing Characteristics

P. katsurae differs from *P. cambivora,* which is also a pathogen of chestnut trees (see Chapter 13), in that it is homothallic and the oogonia and oospores are smaller. Sporangia of *P. katsurae* are papillate, whereas those of *P. cambivora* are nonpapillate. It differs from *P. heveae* by the presence of protuberances on the oogonial wall and the funnel-shaped oogonial stalk (Katsura 1976; also noted by Oudemans and Coffey [1991b]). See Chapter 4, Figure 4.28, for a photomicrograph comparing oogonia of *P. heveae* and *P. katsurae* (Oudemans and Coffey 1991b). Oudemans and Coffey (1991b) found by isozyme analysis that *P. heveae* and *P. katsurae* are closely related.

Disease Caused by *Phytophthora katsurae*

Trunk Rot of *Castanea crenata* Sieb. et Zucc. (Chestnut)

Lesions normally develop on the trunks of mature chestnut trees about 10 to 100 cm above the ground. The bark around the lesions gradually softens, and many fissures develop around the infected area. The fissures are usually covered with a black exudate, and a characteristic fermentation odor can be detected. The bark deteriorates rapidly. Diseased trees are normally killed within 4 years of the initial infection.

Disease Development. The disease develops rapidly after periods of prolonged rainfall. In Japan, this is usually during the month of June (Uchida 1967).

Control. Good drainage around trees should be provided. New plantings should not be made in areas subject to flooding or on heavy, poorly drained soils.

CHAPTER 40

Phytophthora lateralis

Tucker and Milbrath (1942)

No synonyms are known for *Phytophthora lateralis.* The fungus was first described by Tucker and Milbrath (1942) from diseased Lawson cypress, also known as Port Orford cedar (*Chamaecyparis lawsoniana* (Murr.) Parl.), in the northwestern United States, including northern California (Kliejunas and Adams 1981). It is a serious threat to cedar forests in the Northwest (Zobel et al. 1985) and in British Columbia, Canada (Atkinson 1965). The pathogen, with few exceptions, is limited to hosts in the family Cupressaceae (Table 40.1). In 1991, *P. lateralis* was found to cause root rot of a Pacific yew tree, *Taxus brevifolia* Nutt. (DeNitto and Kliejunas 1991). *P. lateralis* was first detected in the United States near Seattle in 1923 and was thought to have been imported from France (Roth et al. 1972).

Characteristics of *Phytophthora lateralis*

P. lateralis is classified in group V (Stamps et al. 1990). See Tables 4.2 and 4.3 for tabular keys. A description of the morphology of *P. lateralis* is given by G. Hall (1991c). Morphology is shown in Figures 40.1 and 40.2.

Sporangia

Sporangia are nonpapillate and borne sympodially on simple sporangiophores; they are ovate, obovate, obpyriform, and often elongate or distorted in shape; 20 to 60 µm long × 12 to 20 µm wide (average 26 × 15 µm); in water, elongated necks occur; sporangia are persistent on the stalk (Tucker and Milbrath 1942). The length-breadth ratio varies from 1.6 to 1.9:1 (average 1.71:1). Sporangia formed most abundantly in weak agar media such as V8 juice (10 ml/liter), cornmeal (25 g/liter), or whole milk (20 ml/liter), especially when these media were made up with Hoagland's mineral solution instead of distilled water (Trione 1974). Sporangium formation was markedly increased by incubation of cultures under combined Blacklight Blue and Cool White fluorescent lamps (680 μWcm^{-2}) continuously or for 12-h periods.

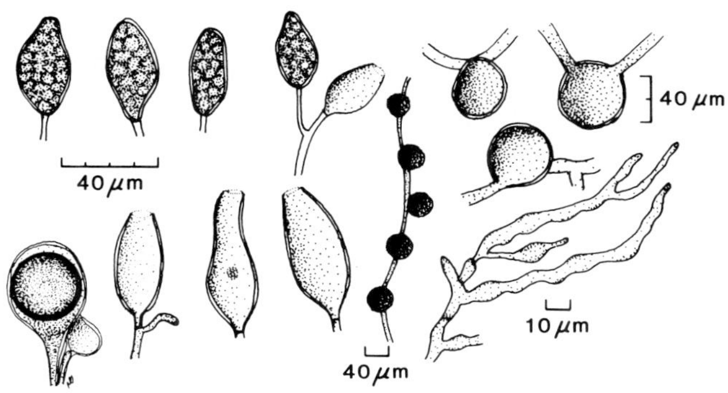

Figure 40.1. Morphology of *Phytophthora lateralis.* Lower row, globose oogonium with paragynous antheridium; sporangia after evacuation of zoospores; chlamydospores forming laterally on mycelium. Mycelium is usually smooth but can be gnarled. Upper row, ovoid sporangia with varying shapes; sporangia forming sympodially; chlamydospores forming laterally from hyphal swellings. A photograph of sporangia by Trione (1974) indicated that shapes were more uniform than those shown by Tucker and Milbrath (Figure 40.2). Most resembled the sporangia drawn above on the sympodial sporangiophores. (Drawing by A. Vaziri after Tucker and Milbrath 1942)

Table 40.1. Distribution of hosts of *Phytophthora lateralis*

Host	Common name	Disease	Geographical distribution
Actinidia chinensis Planch.	Kiwi fruit	Root and crown rot	New Zealand (Robertson 1982b)
Chamaecyparis lawsoniana (Murr.) Parl.	Lawson cypress, Port Orford cedar	Root and crown rot	United States (Tucker and Milbrath 1942); Canada (Anon. 1953a; Atkinson 1965)
C. lawsoniana var. *alumi*	Lawson cypress, Port Orford cedar	Root and crown rot	United States (Torgeson et al. 1954)
C. lawsoniana var. *cerula*	Lawson cypress, Port Orford cedar	Root and crown rot	United States (Torgeson et al. 1954)
C. lawsoniana var. *elwoodi*	Lawson cypress, Port Orford cedar	Root and crown rot	United States (Torgeson et al. 1954)
C. lawsoniana var. *erecta-aurea*	Lawson cypress, Port Orford cedar	Root and crown rot	United States (Torgeson et al. 1954)
C. lawsoniana var. *erecta-glauca*	Lawson cypress, Port Orford cedar	Root and crown rot	United States (Torgeson et al. 1954)
C. lawsoniana var. *erecta-veridis*	Lawson cypress, Port Orford cedar	Root and crown rot	United States (Torgeson et al. 1954)
C. lawsoniana var. *glauca*	Lawson cypress, Port Orford cedar	Root and crown rot	United States (Torgeson et al. 1954)
C. lawsoniana var. *lutea*	Lawson cypress, Port Orford cedar	Root and crown rot	United States (Torgeson et al. 1954)
C. lawsoniana var. *minima*	Lawson cypress, Port Orford cedar	Root and crown rot	United States (Torgeson et al. 1954)
C. lawsoniana var. *nestoides*	Lawson cypress, Port Orford cedar	Root and crown rot	United States (Torgeson et al. 1954)
C. lawsoniana var. *stewarti*	Lawson cypress, Port Orford cedar	Root and crown rot	United States (Torgeson et al. 1954)
C. lawsoniana var. *wisseli*	Lawson cypress, Port Orford cedar	Root and crown rot	United States (Torgeson et al. 1954)
C. obtusa Sieb. and Zucc. var. *crippsii*	Hinoki cypress	Root and crown rot	United States (Tucker and Milbrath 1942)
C. obtusa Sieb. and Zucc. var. *gracilis* Nash	Hinoki cypress	Root and crown rot	United States (Tucker and Milbrath 1942)
Taxus brevifolia Nutt.	Pacific yew	Crown rot	United States (DeNitto and Kliejunas 1991)

Hyphal Swellings

Hyphal swellings do not form, but the early stages of chlamydospore formation resemble hyphal swellings.

Chlamydospores

Chlamydospores are abundant in many agar media at temperatures between 15 and 25°C; they are terminal or intercalary, 20 to 77 μm in diameter (average 40 μm), and cinnamon brown in color when produced in cedar foliage agar (Trione 1974). Chlamydospore production was suppressed by light (Englander and Roth 1980).

Sex Organs

P. lateralis is homothallic. Antheridia are paragynous; oogonia are smooth, spherical, and terminal, 33 to 50 μm in diameter; oospores are plerotic with walls 6 μm thick, 28 to 46 μm in diameter (average 40 μm), and deeply pigmented; the color depends on the agar medium used, e.g., cinnamon colored in cedar foliage agar and amber in alfalfa agar medium. Oospores were formed most abundantly on agar of Port Orford cedar foliage (filtrate from 200 g of foliage steamed at 95°C for 1 h plus 15 g of glucose per liter of water) (Trione 1974).

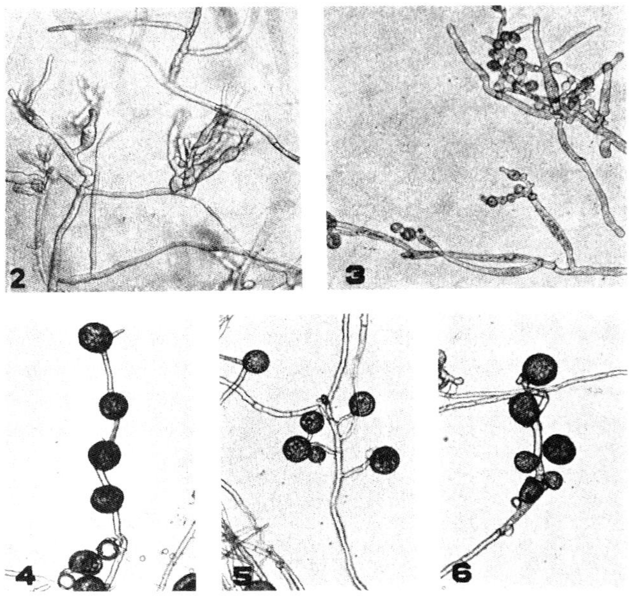

Figure 40.2. Photomicrographs of unstained water mounts of living material of *Phytophthora lateralis.* **2,** Young mycelium (82×) showing large, tortuous type of branching; **3,** mycelium (82×) from an older culture showing gnarled branching; **4–6,** characteristic development of chlamydospores (82×); **7–14,** characteristic development of sporangia (164×); **7,** evacuated sporangium and sporangium with contents; **8–10,** characteristic types of sporangia; **11–13,** germination of sporangia to produce new mycelium; **14,** an evacuated sporangium and the development of a second sporangium. (Reprinted from Figures 2–14 by permission from Mycologia 34:94–103, C. M. Tucker and J. A. Milbrath, copyright 1942, The New York Botanical Garden, Bronx, New York)

Growth Temperatures

The minimum temperature for growth is 3°C, the optimum is 20°C, and the maximum is <26°C.

Distinguishing Characteristics

The formation of the chlamydospores, which are lateral on the mycelium, differs markedly from the clustered chlamydospores of other nonpapillate species such as *P. cambivora, P. cinnamomi, P. cryptogea, P. drechsleri,* and *P. richardiae.* The chlamydospores of *P. lateralis* are often sessile on the hyphae or appear to be supported by two hyphae (Figure 40.1).

P. cinnamomi can be distinguished from *P. lateralis* by the presence of a characteristic grapelike cluster of hyphal swellings and chlamydospores and the production of amyphigynous antheridia in dual cultures (heterothallic). *P. lateralis* produces paragynous antheridia in single culture (homothallic). *P. cinnamomi* grows at a higher maximum temperature (>30°C) than *P. lateralis* (<26°C).

Disease Caused by *Phytophthora lateralis*

Root Rot of *Chamaecyparis lawsoniana* (Murr.) Parl. (Cedar)

Port Orford cedar is not only an important timber tree but is also used widely in ornamental plantings (Trione 1959). The disease was first reported in 1923 in nursery stock imported from France (Roth et al. 1972), but the causal agent was not discovered until 1942 when Tucker and Milbrath demonstrated that root rot was caused by an undescribed *Phytophthora* species, which they named *P. lateralis.* This disease was so severe in some localities that nurserymen were forced to abandon the planting of Port Orford cedar trees. It has also devastated whole areas of well-established cedar plantations. Except for one report on Pacific yew (DeNitto and Kliejunas 1991), Port Orford cedar is the only natural host.

The light color of the foliage is the first observable aboveground symptom after infection of the roots. The foliage wilts slightly on warm days and gradually changes in color from yellow to bronze (Plate 40.1) and finally to light brown or tan and becomes crisp and dry.

Infected roots appear water soaked and are often a deep cinnamon brown color. Spread of the disease to the main trunk and girdling of the crown (Plate 40.2) account for the color changes in the foliage. If the outer bark is removed from the infected area, a clear zone can be seen between the infected and healthy tissues. A black, resinous line can also be seen in the cambium. These symptoms readily distinguish this disease from the common problem of transplant failure and from infections caused by *P. cinnamomi.* In the latter, the margin between healthy and diseased tissues is less contrasted in both color and sharpness. Young trees are killed within a few weeks, while older trees survive for 2 to 4 years before death (Trione 1959). In addition to root and collar rot, aerial infection of cedar trees often occurs in moist, humid weather (Trione and Roth 1957).

Although *P. lateralis* is relatively sensitive to hymexazole, an antibiotic generally more toxic to *Pythium* species than to *Phytophthora* (Chapter 2, Table 2.2), isolation from a suspension of soil was improved by adding hymexazole (25 µg/ml) before floating Port Orford cedar branchlet baits for 6 days. The branchlet baits were then transferred to a selective cornmeal agar containing 20 µg each of vancomycin and streptomycin per milliliter (Hansen et al. 1979; Hamm and Hansen 1984). See Chapter 2 for more extensive coverage of isolation techniques.

Ostrofsky et al. (1977) isolated chlamydospores of *P. lateralis* from infested sites in the forest by wet sieving soil samples and floating Port Orford cedar branchlets on water suspensions of the soil material on the sieve. *P. lateralis* chlamydospores were largely associated with organic matter in soil. Some chlamydospores survived for 16 weeks in frozen organic matter and in dry soil.

Disease Development. *P. lateralis* is most active on cedar in areas where conditions are cool, but where temperatures are moderately high and conditions are moist, root rot of cedar is usually caused by *P. cinnamomi,* particularly in the southern United States and Europe. The optimum temperature for sporangia production of *P. lateralis* is between 15 and 20°C. The disease is therefore greatly enhanced by heavy rains during late winter and early spring, which favor the buildup of inoculum in soil. Infection can occur at 12 to 25°C, but with the end of the rainy season and the subsequent rise in temperatures, the disease is suppressed. The pathogen is inactive during the summer but appears to survive the long dry season as chlamydospores and oospores in plant tissues (Trione 1959).

Control. The incidence of disease can be minimized by planting disease-free propagating stock in areas where cedars have not been grown previously, preventing unnecessary movement of people, animals, and equipment from diseased to healthy areas of a plantation (Plates 40.3 and 40.4), and destroying both diseased and adjacent healthy cedar trees in affected areas of the forest. Resistance in cedar trees can be determined by inoculating rooted cuttings with zoospores and by inoculating branch wounds (Hansen et al. 1989).

In British Columbia, Canada, *Chaemaecyparis formosensis* (*C. obtusa* var. *formosensis* Matsum.), *C. nootkatensis* (D. Don) Spach., Alaska cedar, *C. pisifera* (Siebold and Zucc.) Endl., *C. taiwanensis, C. thyoides* (L.) B. S. P., Atlantic white cedar, and Leyland cypress were resistant to *P. lateralis;* however, scions of *C. lawsoniana* grafted to *C. formosensis, C. nootkatensis,* or *C. pisifera* were incompatible (Hunt and O'Reilly 1984).

An extensive publication on the ecology, pathology, and management of Port Orford cedar forests discusses the management of forests in the northwestern United States that are threatened by root rot caused by *P. lateralis* (Zobel et al. 1985).

CHAPTER 41

Phytophthora lepironiae

Sawada 1919

Synonyms of *Phytophthora lepironiae* include *Nozemia lepironiae* Sawada (1919a, b) and *Kawakamia lepironiae* (Sawada) Sawada (1942). *P. lepironiae* was first isolated from *Lepironia mucronata* Rich. by Sawada (1919), who originally named it *Nozemia lepironiae* on the basis of the formation of paragynous antheridia. Sawada (1942) then transferred the species to the genus *Kawakamia* on the basis of the pedicel-like appendage to the sporangium. He finally transferred it to the genus *Phytophthora.*

Characteristics of *Phytophthora lepironiae*

P. lepironiae is classified in group III (Stamps et al. 1990). Morphology is illustrated in Figure 41.1. See Tables 4.2 and 4.3 for tabular keys and Appendix 4.9 for a dichotomous key (Ho 1992; Ho et al. 1995).

Sporangia

Sporangia are ovoid, broadly ellipsoidal, or spherical; semipapillate; caducous (pedicel 5 to 20 µm in length); and 40 to 64 µm long × 36 to 52 µm wide. Length-breadth ratio is <1.6. They are borne on short, unbranched sporangiophores.

Hyphal Swellings and Chlamydospores

P. lepironiae does not produce hyphal swellings or chlamydospores.

Sex Organs

P. lepironiae is homothallic; antheridia are paragynous; oogonia are 40 to 52 µm in diameter; and oospores are aplerotic and 30 to 38 µm in diameter.

Growth Temperatures

Growth temperatures are not recorded.

Distinguishing Characteristics

P. lepironiae is known only on its host, *Lepironia mucronata,* and has thus far not been isolated or grown in axenic culture.

Disease Caused by *Phytophthora lepironiae*

Stem Blight of *Lepironia mucronata*

P. lepironiae causes dark brown spots on the stems of *Lepironia* in Japan and Korea. It has not been reported on any other host.

Disease Development. Precise conditions for disease development are not known.

Control. Control measures have not been described.

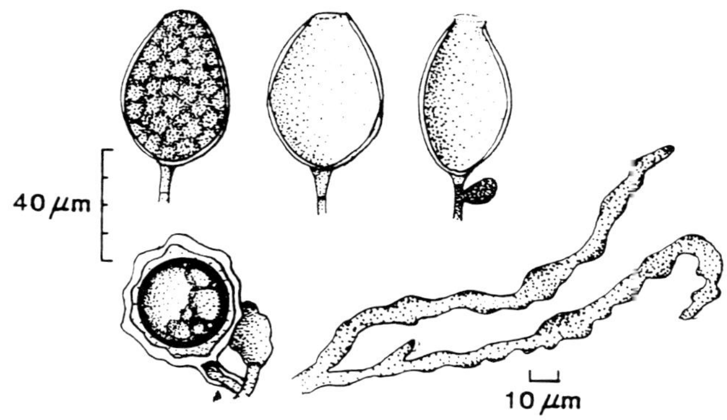

Figure 41.1. Morphology of *Phytophthora lepironiae.* Lower row, oogonium with a rough, irregular outer wall and a paragynous antheridium; and irregularly shaped mycelium. Upper row, broadly ellipsoid, nonpapillate sporangium; empty sporangium; and initiation of a sympodially placed sporangiophore. (Drawing by A. Vaziri)

CHAPTER 42

Phytophthora macrochlamydospora

Irwin (1991)

Phytophthora macrochlamydospora was isolated in 1974 from soybean, *Glycine max* L. cv. Hampton, affected by a stem and root rot. The fungus was pathogenic when tested back on soybean seedlings. Since sporangia could not be produced, the culture was stored until 1990 when sporangia were induced by transfer of blocks of mycelial cultures on V8 juice agar to sterile distilled water under diffuse light at 24°C. After change of the water 4 days later, sporangia formed. In addition, chlamydospores and hyphal swellings formed, but sex organs were not found (Irwin 1991).

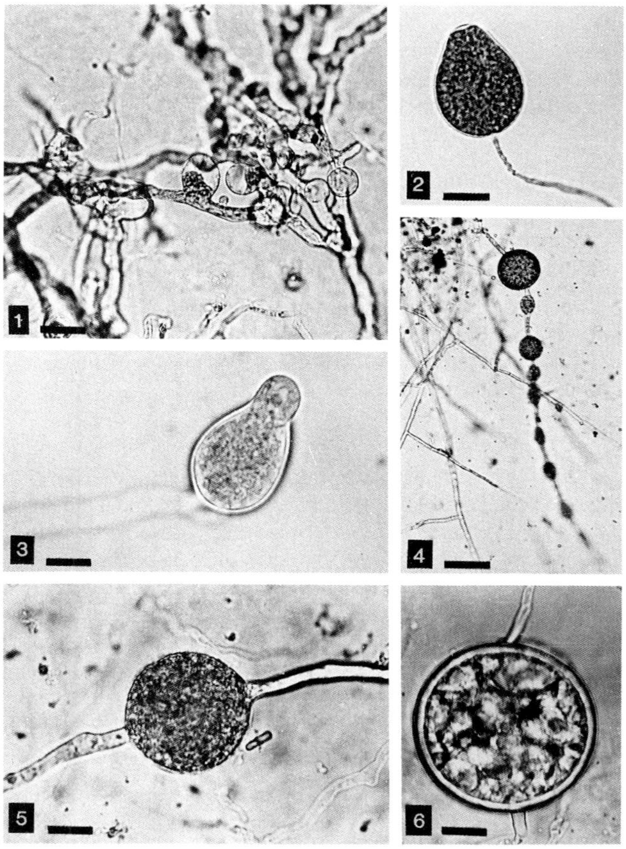

Figure 42.1. Morphology of *Phytophthora macrochlamydospora*. **1,** Coralloid hyphal swellings. **2,** Typical semipapillate sporangium. **3,** Atypical sporangium showing pronounced beak (only rarely observed). **4,** Catenulate hyphal swellings. **5,** Intercalary immature chlamydospore. **6,** Intercalary mature chlamydospore. All bars = 8.5 µm. (Reproduced by permission from Mycologia 83:517–518, copyright 1991, The New York Botanical Garden, Bronx, New York, courtesy J. Irwin 1991)

Characteristics of *Phytophthora macrochlamydospora*

P. macrochlamydospora would be classified in group III or IV (Stamps et al. 1990) because of its production of semipapillate sporangia; however, the absence of a sexual stage indicates that this is an arbitrary placement. Morphology is shown in Figure 42.1. See Tables 4.2 and 4.3 for tabular keys.

Sporangia

Sporangia are formed terminally on long (up to 500 µm) sporangiophores that proliferate either internally or sympodially. Sporangia are semipapillate, ellipsoid to obpyriform, and measure 20 to 50 × 18 to 36 µm (Figures 42.1.2 and 42.1.3). Zoospores that measure 9 to 13 × 8 to 10 µm are released from the sporangium.

Chlamydospores

Large, spherical chlamydospores that form on solid or liquid V8 juice media are terminal or intercalary (Figure 42.1.5), 24 to 84 µm in diameter (average 55 µm) with walls 2.5 to 4.0 µm thick (Figure 42.1.6). The large size of the chlamydospore is the most distinctive feature.

Hyphal Swellings

Thin-walled hyphal swellings are spheroidal to ellipsoid (Figure 42.1.1) and are often formed in chains (Figure 42.1.4).

Growth Temperatures

The minimum temperature for growth is 6°C, optimum 23 to 30°C, and maximum 34°C.

Distinguishing Characteristics

P. macrochlamydospora differs from *P. colocasiae,* which produces semipapillate sporangia (group IV), by its production of much larger chlamydospores. *P. colocasiae* also produces oogonia and oospores, but *P. macrochlamydospora* does not.

CHAPTER 43

Phytophthora meadii

McRae (1918)

Phytophthora meadii was first described by McRae (1918) from diseased rubber trees, *Hevea brasiliensis* Muel. Arg., in India. *P. meadii* has a limited natural host range (Table 43.1) and is probably best recognized as the cause of stripe canker, black thread, pod rot, shoot dieback, or secondary leaf fall disease of rubber trees. Dantanarayana et al. (1984) concluded that *P. meadii* was the major pathogen on rubber in Sri Lanka. The koleroga or fruit rot disease of arecanut is caused by *P. meadii* in Karnataka, India (Sastry and Hegde 1987a). Gadd (1924) reported *P. faberi* to be the causal agent of disease on rubber in Sri Lanka. Although Tucker (1931) merged *P. faberi* and *P. meadii* under *P. palmivora,* Waterhouse (1963) distinguished *P. meadii* from *P. faberi,* a synonym of *P. palmivora.*

Characteristics of *Phytophthora meadii*

P. meadii is classified in group II (Stamps et al. 1990). A summary of characteristics is given by Waterhouse (1974b), Stamps (1985d), and Holliday (1980). Morphology is shown in Figure 43.1. See Tables 4.2 and 4.3 for tabular keys, Appendix 4.9 for a dichotomous key (Ho 1992), and Tables 51A.4 and 51A.5 for morphological comparison with *P. palmivora* and related species. Sansome et al. (1991) present a hypothesis that *P. meadii* might be a species hybrid with *P. palmivora* as one parent because self-fertile isolates of *P. meadii* were polyploid but A1 and A2 types were diploid; however, Oudemans and Coffey (1991b) indicated by isozyme data that *P. palmivora* would be an unlikely parent because of the large genetic difference separating these species.

Sporangia

Sporangia are terminal or lateral; papillate, occasionally with two papillae; caducous (pedicel is 10 to 20 µm in length; Figure 4.15); ellipsoid or elongated, obpyriform, occasionally spherical, often distorted into lobed or hourglass shapes; on fruit they are 33 to 67 µm long × 14 to 28 µm wide (average 48 × 21 µm) and in water 20 to 44 µm long × 16 to 29 µm wide (average 32 × 23 µm). The length-breadth ratio varies from 1.3 to 2.0 (Waterhouse 1974a). Oudemans and Coffey (1991b) report a range of 1.4 to 1.7. Sporangia are produced sympodially on branched sporangiophores that are 10 to 20 µm long (Figure 43.1).

Hyphal Swellings

Hyphal swellings are not produced.

Chlamydospores

Chlamydospores are rare and range from 16 to 30 µm in diameter with the majority under 30 µm. They are usually produced at temperatures above 30°C after periods of incubation >18 days (Peries and Fernando 1966).

Sex Organs

P. meadii is predominantly homothallic (Waterhouse 1974b) with some evidence of heterothallism (Peries and Dantanarayana 1965). A1 and A2 types have been described in India (Rajalakshmy et al. 1985). A1 isolates of *P. meadii* paired with A2 isolates of *P. palmivora* produced oospores (Sastry and

Table 43.1. Distribution of hosts of *Phytophthora meadii*

Host	Common name	Disease	Geographical distribution
Allium cepa L.	Onion	Root rot	Hawaii in United States (Sideris 1929)[a]
Ananas comosus (L.) Merrill (*A. sativus* Schult.)	Pineapple	Top rot	Hawaii in United States (Sideris 1929)
Areca catechu Linn.	Arecanut	Koleroga; fruit rot	India (Sastry and Hedge 1987)
Clarkia elegans Doug.	Godetia	Seedling blight	India (Dastur 1916)[a]
Dianthus caryophyllus L.	Carnation; deillet	Wilt and stem rot	Mauritius (Shepherd 1937)
Elettaria cardamomum	Cardamom	Blight	India (Sastry and Hegde 1987a, b; 1989)
Euonymus japonicus Thunb.	Spindle tree	Blight	China (Zheng and Lu 1989)
Gilia spp.		Seedling blight	India (Dastur 1916)[a]
Hevea brasiliensis Muel. Arg.	Rubber	Black thread; stripe canker; bark and pod rot; secondary leaf fall	Indonesia (Rutgers and Arens 1914); Sri Lanka (Petch 1914); Burma (Dastur 1916); India (McRae 1918); Nigeria (Laycock 1922); Congo (Strong and Shattuck 1930); Malaysia (Chee et al. 1967); Thailand (Chee and Greenwood 1968). Also reported from Cambodia, Vietnam, Liberia, Ghana, Nigeria, Cameroon, Brazil, Peru, Nicaragua, Costa Rica, and Venezuela (Pillai 1982)
Manihot glaziovii Muel. Arg.	Ceara rubber tree	Leaf blight	India (McRae 1918)[a]
Piper betle L.	Betel nut	Foot rot	Malaysia (Kheng 1982)
Ricinus communis L.	Castor bean; castor oil plant	Seedling blight	India (McRae 1918)[a]
Salpiglossis spp.	Solanaceae	Seedling blight	India (Dastur 1916)[a]
Solanum melongena L.	Eggplant; brinjal; aubergine	Blight	India (Thankamma 1983)
Theobroma cacao	Cocoa	Black pod	India (Sastry and Hegde 1987b)

[a] Artificially inoculated host.

Hegde 1987c). Sex organs developed only at 15 to 20°C (Peries and Fernando 1966). Antheridia are rounded and amphigynous (12 × 13 up to 10 × 16 µm in diameter); oogonia are spherical to pyriform with smooth or wrinkled walls; those produced in agar culture measure 21 to 49 µm in diameter (average 33.4 µm); on fruit, oogonia measure 20 to 48 µm in diameter (average 25.1 µm); oospores are markedly aplerotic; those produced in agar media measure 16 to 32.8 µm in diameter (average 24 µm); on fruit, oospores measure 16 to 32 µm in diameter (average 21 µm), and the oospore wall is 4 µm thick.

Growth Temperatures

The minimum temperature for growth is >5°C, the optimum is 25 to 30°C, and the maximum is >33°C.

Distinguishing Characteristics

Dantanarayana et al. (1984) distinguished *P. meadii* from *P. palmivora* by production of spherical to ovoid, caducous sporan\gia on medium-length pedicels (*P. palmivora* sporangia have short pedicels; Figure 4.15), rare production of chlamydospores, absence of pigment production on agar media, formation of the sympodium on which sporangia formed, and formation of aplerotic oospores (Table 43.2). Prior to Waterhouse (1963, 1974a) and Dantanarayana et al. (1984), many isolations from rubber in Sri Lanka and southern India were erroneously ascribed to *P. palmivora* because *P. meadii* was not considered by some to be distinct (Holliday 1980).

Oudemans and Coffey (1991b) reported that sporangia of *P. meadii* (six isolates) were slightly larger (average 38.5 × 24.6 µm) than those of *P. botryosa* (four isolates; average 29.8 × 17.9 µm); however, isozyme data of Oudemans and Coffey (1991b) indicate that *P. meadii* and *P. botryosa* have a high level of similarity but that *P. meadii* is genetically distinct from *P. palmivora*. They suggest that *P. meadii* and *P. botryosa* probably represent two geographically separate populations of a single species, since *P. meadii* has been reported mainly from India and Sri Lanka and *P. botryosa* from Thailand, Malaysia, and Vietnam. See Figure 4.15, in which the sporangial morphology of these three species is compared. Tables 51A.4 and 51A.5 give morphological data for *P. meadii* and several other similar species.

Some Diseases Caused by *Phytophthora meadii*

Stripe Canker or Black Thread, Secondary Leaf Fall, and Green Pod Decay of *Hevea brasiliensis* Muel. Arg. (Para Rubber)

Stripe canker is caused by *P. palmivora* and *P. botryosa* as well as by *P. meadii*. Description of symptoms of stripe canker is given under *P. palmivora* (Chapter 51). Secondary leaf fall and green pod decay were first recognized in Burma at the turn of the century. Dastur (1916) was the first to associate a *Phytophthora* species with the disease, and McRae (1918) in India first described the pathogen as *P. meadii*.

P. palmivora, *P. botryosa*, *P. heveae*, and *P. parasitica* also cause diseases of rubber (Thankamma 1983; Pillai 1982).

P. meadii can be isolated from fruits, leaves, and bark. Symptoms normally occur on the fruits as dull, ash gray spots approximately 3 weeks after the start of the monsoon season. These spots frequently develop at the stem end of the fruit and gradually spread across the entire surface and eventually coalesce. In humid weather, a whitish fuzz of mycelium and sporangia often covers the infected portions. Drops of latex, which eventually turn black, also exude from these spots. The fruits turn dark, become soft, and remain attached to the trees for a considerable time. If the disease is severe, every fruit on a tree may become infected. During the fruit rot phase, infected leaves shed, thus causing trees to become almost or entirely denuded. Individual leaves become yellow or red, similar to the normal color of leaves shed prior to dormancy. Some affected leaves fall while still green. Dark brown or black lesions with white spots of coagulated latex are usually present in the center of each fallen leaf. Leaf fall is followed by dieback of the branches. The shoots that arise from these branches during the following season wilt early (Pillai 1982).

Late in the monsoon season after infection of the bark near the tapping cut, the tissue becomes soft, and the rot extends upward until suppressed by dry weather. The splitting of the bark and its renewed growth each year give this phase of the disease the name "stripe canker." In Brazil, *P. meadii* attacks the green shoots rather than the leaves. Young leaves wither but remain attached to the dead or dying shoot tip, although the leaves themselves are seldom infected. The disease is rapidly reactivated from old lesions after a few rainy days (Chee and Wastie 1980).

Disease Development. The disease develops rapidly about 2 to 3 weeks after the monsoon season has begun. Sporangia develop abundantly on diseased tissues, and new infections occur after rain drops splash sporangia and zoospores onto healthy tissue. During the hot, dry season, *P. meadii* becomes inactive, and the progress of the disease is arrested.

In arecanut, *P. meadii* survives in India for the entire dry season (January to June) in infected branches and tree tops and also in infected nuts lying on irrigated soil (Sastry and Hegde 1989b). Chlamydospores in infected rubber petioles were viable 22 weeks after being buried in soil (Liyanage and Wheeler 1991).

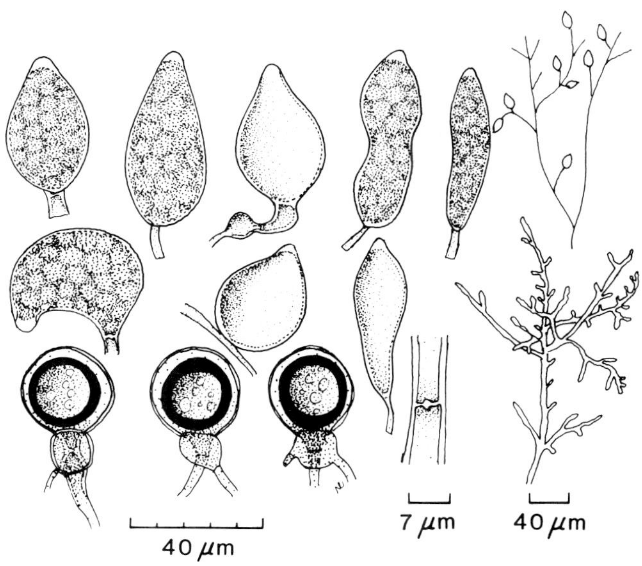

Figure 43.1. Morphology of *Phytophthora meadii.* Lower row, amphigynous antheridia on globose to obpyriform oogonia (oospores are distinctly aplerotic); old hyphal section with a septum; mycelial pattern. Center row, curved, papillate, caducous sporangium; sporangium sessile on the mycelium; sporangium with an elongated, ellipsoid, abnormal shape. Upper row, obpyriform to ellipsoid sporangia; sporangium with swollen sporangiophore; elongated, hourglass-shaped sporangia; sporangia produced on a simple sympodium (does not conform to size bars). (Drawing by A. Vaziri)

Table 43.2. Diagnostic characteristics of *Phytophthora meadii* and *P. palmivora*[a]

Characteristic	*P. meadii*	*P. palmivora*
Sporangial morphology	Near spherical to ellipsoid prominent papilla, base usually rounded, sometimes slightly tapering; mostly asymmetrical	Mostly ovoid to ellipsoid, prominent papilla, rounded base; often regular
Sporangial pedicel type on solid LBA[b]	Narrow, medium length, and not occluded; range 6.6–41.5 (mean 12.3 ± 0.12) µm	Broad, short, and occluded; length range 2.5–3.3 (mean 3.2 ± 0.1) µm
Chlamydospores[c]	Very rare; range 19.9–38.1 (mean 30.5 ± 1.7) µm, wall thin, 1.6 µm; hyaline to golden yellow	Formed promptly and abundantly; initally hyaline to golden brown at maturity, sometimes almost chocolate brown; range 24.9–48.1 (mean 40.0 ± 1.3) µm; wall thick, 3.3–3.9 µm
Pigmentation on LBA	Absent	Present
Sporangiophores	Irregular in water to sometimes simple sympodial and well defined on LBA; often with swellings at branching points, thickness ranging from 1.6 to 2.5 µm	Regularly simple sympodial and well defined with distinct swellings at the nodes where sporangia arise; thickness ranging from 1.6 to 3.2 µm
Oospore morphology and size[c]	Often aplerotic; range 18.3–33.2 (mean 22.0 ± 0.0 µm)	Mostly plerotic; range 19.9–28.2 µm (mean 22.8 ± 0.1 µm)
Chromosomes[d]	$n = 8–9$	$n = 9–12$
Compatibility type	Predominantly A1 (A2 uncommon)	Rubber A1 (?), cocoa A2 (?), coconut A2 (?)
Distribution on rubber in Sri Lanka	Common	Rare

[a] Reproduced by permission of the British Mycological Society, courtesy Dantanarayana et al. (1984).
[b] Lima bean agar.
[c] Significantly larger in *P. palmivora.*
[d] Estimated by E. Sansome (*personal communication*).

In Malaysia, prediction of the leaf fall disease of rubber caused by *P. botryosa* is based on a continuous 4-day rainy period of 2.5 mm or more, relative humidity greater than 90% for a minimum of 14 h, and the decrease of maximum temperatures to below 32.2°C (Lim 1982). In India, when rainfall reaches 360 mm in any month between April and August, leaf fall disease caused by *P. meadii* becomes severe (Agnihothrudu 1975). Since *P. meadii* and *P. botryosa* are considered to be closely related (Oudemans and Coffey 1991b), this information may be applicable to both *P. botryosa* and *P. meadii.*

Control. In southern India, secondary leaf fall and pod decay greatly reduce the yields if trees are not treated with fungicides (Pillai 1982). Control is difficult because of the many sources of primary inoculum and the presence of several physiologic races of the pathogen. Ideally, destruction of diseased branches and infected fruit will reduce the inoculum level (McRae 1919); however, these methods did not prove to be economical and effective for control of disease (Pillai 1982). A single premonsoon application of a copper-based fungicide such as Bordeaux mixture has been effective but must be used with a large volume of water. Copper oxychloride in an oil emulsion applied from an aircraft or mist blower at low volume has proved effective in some areas of India (Pillai 1977, 1982). Resistant clones budded on high-yielding susceptible clones has experimentally provided a measure of control in southern India (Pillai 1982). Clones with a high degree of resistance are listed by Chee and Wastie (1980).

Top Rot of *Ananas comosus* (Pineapple)

Top rot first appears as a brown spot on green fruit that gradually extends to cover the whole fruit. Diseased fruits have a dry and lighter-colored pulp than normal surrounded by a brown or discolored margin. Leaf bases and the stem may also become infected soon after young plants are set out in the field.

P. cinnamomi (Chapter 15), *P. palmivora* (Chapter 51), *P. nicotianae* (Chapter 50), *P. citrophthora* (Chapter 17), and *P. drechsleri* (Chapter 23) are also pathogens of pineapple. Of these, the disease known as heart rot, caused by *P. cinnamomi,* is perhaps the best recognized wherever pineapples are grown.

Disease Development. The disease is normally limited to areas of high rainfall and low temperatures (22 to 26°C), e.g., Hawaii. Poor soil drainage greatly enhances disease development.

Control. See Heart Rot of Pineapple (Chapter 15A).

CHAPTER 44

Phytophthora medicaginis

E. M. Hansen and Maxwell (1991)

Synonyms of *Phytophthora medicaginis* are *P. cryptogea* (Erwin 1954), *P. megasperma* (Erwin 1965), *P. megasperma* f. sp. *medicaginis* (Kuan and Erwin 1980a), and *P. sojae* f. sp. *medicaginis* (Faris et al. 1989).

P. medicaginis was first found to be the causal agent of a root rot of alfalfa (*Medicago sativa* L.) in California (Erwin 1954), but the disease had likely existed long before. The pathogen was first identified as *P. cryptogea* because of its production of hyphal swellings; nonpapillate, ellipsoid to ovoid sporangia; and amphigynous antheridia. In 1965 (Erwin), the fungus was reclassified as *P. megasperma* because of the discovery that paragynous as well as amphigynous antheridia were produced. Morphologically, the fungus could have been classified as *P. megasperma* var. *sojae* (Waterhouse 1963) on the basis of a small oogonium diameter of about 31 µm. Kuan and Erwin (1980a) questioned the separation of *P. megasperma* var. *sojae* from *P. megasperma* var. *megasperma* on the basis of production of oogonia larger than 40 µm in diameter (Waterhouse 1963) because oogonium diameters of isolates from alfalfa, soybean, cherry, and other hosts ranged from 30 to 58 µm in a continuum with no sharp distinction between those previously called *P. megasperma* var. *sojae* and those called *P. megasperma* var. *megasperma*. Kuan and Erwin (1980a) proposed calling the narrow-host-range isolates from alfalfa *P. megasperma* f. sp. *medicaginis* and those from soybean *P. megasperma* f. sp. *glycinea*. Faris et al. (1989) relegated the soybean pathogen from *P. megasperma* f. sp. *sojae* to *P. sojae* f. sp. *glycines* and the alfalfa pathogen, *P. megasperma* f. sp. *medicaginis*, to *P. sojae* f. sp. *medicaginis*. Several studies (Förster et al. 1987, 1989; Förster and Coffey 1991, 1992) based on restriction fragment length polymorphisms (RFLP) of DNA and isozyme technology (Nygaard et al. 1989) showed that *P. megasperma* f. sp. *sojae* and *P. megasperma* f. sp. *medicaginis* are distinctly different from each other and from other *P. megasperma* isolates. Also, both pathogens could be isolated from the same soil samples (Nygaard et al. 1989), and the protoplasts of the two pathogens could not be forced to fuse (Kuhn 1991). *P. sojae* is classed in mtDNA group I and *P. medicaginis* is in group H of Förster and Coffey (1993). These data conclusively indicate that *P. megasperma* f. sp. *medicaginis* and *P. megasperma* f. sp. *sojae* are not genetically related to each other or to *P. megasperma* and hence should not be classified as formae speciales of *P. megasperma*. Because of these differences, E. M. Hansen and Maxwell (1991) described the alfalfa pathogen as the new species, *P. medicaginis*, and the soybean pathogen was relegated back to *P. sojae* as originally described by Kaufmann and Gerdemann (1958) (Chapter 59). The history of these revisions is discussed in Chapter 4, in Chapter 46 on *P. megasperma*, and in Chapter 59 on *P. sojae*.

P. medicaginis (mtDNA group H of Förster and Coffey [1993]) should be distinguished from the large-oogonia isolates of *P. megasperma* that have been found on alfalfa in Canada (Faris et al. 1983) and Oregon (AL-2) (E. M. Hansen and Hamm 1983; mtDNA group B of Förster and Coffey [1993]). These *P. megasperma* isolates are mild pathogens that attack lateral roots but not the taproots of alfalfa (Faris et al. 1983, 1989; E. M. Hansen and Hamm 1983; Faris 1988). The *P. megasperma* isolates from alfalfa produce larger oogonia (43–60 µm in diameter) than *P. medicaginis* and grow more rapidly. Numerous studies using protein patterns, isozyme, and DNA RFLP data indicate that the two taxa are genetically different (Förster and Coffey 1993; see also the discussion in Chapter 46 on *P. megasperma*). Barr (1980) reported a large-oogonia isolate from alfalfa that was heterothallic. These isolates occur in mtDNA group C (Förster and Coffey 1993).

A *Phytophthora* isolate, reported as *P. megasperma* cv. HTI (high-temperature isolate) obtained from alfalfa in the inland desert valleys of southern California, grows optimally at 27 to 33°C with a maximum at 39°C (Ribeiro et al. 1978). This isolate has been found by DNA RFLP technology, however, to be totally unrelated to *P. medicaginis, P. sojae,* or other *P. megasperma* varieties or biotypes (Förster et al. 1989; Förster and Coffey 1993) (isolate EP 238 in mtDNA group with unique DNA patterns). Since this biotype does not produce oospores and is genetically unrelated to *P. medicaginis* or *P. megasperma,* it should be relegated back to an unidentified *Phytophthora* species.

P. medicaginis causes root rot of alfalfa, an important disease in nearly all regions of the world where alfalfa is grown (Table 44.2). Although alfalfa is the main host, it has also been found to cause root rot of sesbania (Erwin and Kennedy 1957), chickpea (*Cicer arietinum* L.) (Erwin 1965; Bhelwa 1962; Vock et al. 1980; Irwin and Dale 1982), and sainfoin (*Onobrychis vicifolia* Scop.) (Gray and Wofford 1987) (Table 44.2). Phytophthora root rot has become a major problem in Australia on chickpea (Irwin and Dale 1982) but has not been detected in the field on this crop in the United States. Wilcox and Mircetich (1987) showed that cultivar Mahaleb cherry trees were moderately susceptible to *P. medicaginis* in wet soil. They did not support the concept of the forma specialis to distinguish host specificity.

The optimal temperature (25°C) for some large-oogonia isolates is similar to that of *P. medicaginis* on V8 juice agar (Faris 1988); however, E. M. Hansen and Hamm (1983) report the optimal temperature to be 22.5°C.

Characteristics of *Phytophthora medicaginis*

P. medicaginis is classified in group V (Stamps et al. 1990) because of its production of nonpapillate sporangia and both paragynous and amphigynous antheridia. Spore morphology is shown in Figures 44.1–44.3. See Tables 4.2 and 4.3 for tabular keys. Photomicrographs of hyphal swellings are shown in Figure 4.4B, paragynous and amphigynous antheridia on globose oogonia in Figure 4.26A and B, a germinating oospore in Figure 44.3, and an ovoid, nonpapillate sporangium in Figure 44.2. Dimensions of sporangia, oogonia, antheridia, and oospores and cardinal temperatures for growth from different references are shown in Table 44.1. The cultural pattern of growth on potato-dextrose agar is shown in Figure 4.2.

Sporangia

Sporangia in aqueous solutions are ovoid to ellipsoid and nonpapillate (Figures 4.12L and 44.1) and vary in size from 40 to 60 µm in length and 24 to 30 µm in width. A new sporangium may proliferate from within another sporangium. In some isolates, an inflated hyphal element forms at the base of the sporangium (Erwin 1965).

Chlamydospores and Hyphal Swellings

Hyphal swellings are frequently seen in aqueous cultures on infected plant tissue or vegetable juice agar plugs incubated in water (Erwin 1954, 1965) (Figure 4.4B). Chlamydospores have been described in cultures that were incubated in soil and have been found among hyphal swellings. They were thin-walled and globular and varied from 6 to 16 µm in diameter; they also occurred singly, in chains, or in clusters on infected roots of alfalfa plants. Chlamydospores, hyphal swellings, and some portions of mycelium survived up to 7 months in soil at 5 to 30°C (Basu 1980). Z. A. El-Hamalawi (*unpublished*) noted chlamydospores while conducting a study of oospore germination (El-Hamalawi and Erwin 1986a), but they have not been readily producible in soil or media cultures (D. C. Erwin, *unpublished*).

Sex Organs

P. medicaginis is homothallic. Single-zoospore isolates produce oospores on V8 juice agar or other suitable media and in infected roots of alfalfa. Oogonia average 31 to 38 µm in diameter, and oospores average from 23 to 34 µm in diameter (Table 44.1). Oospore walls are about 2 to 4 µm thick. Oospores persist in a viable state from 4.5 months to 3.5 years (Table 5.3) (Pratt and Mitchell 1975; Stack and Millar 1985c).

Both amphigynous and paragynous antheridia are formed by single-zoospore cultures (Figure 4.26B and C). In some isolates, up to 91% of the antheridia are paragynous; in others, only 7% are paragynous (Erwin 1965). Percentages of paragy-

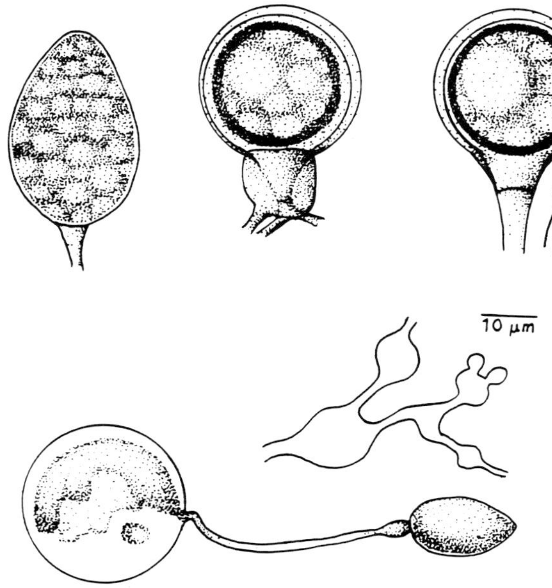

Figure 44.1. Morphology of *Phytophthora medicaginis.* Nonpapillate sporangia, hyphal swellings, and paragynous and amphigynous antheridia on globose oogonia. (Drawing by A. Vaziri)

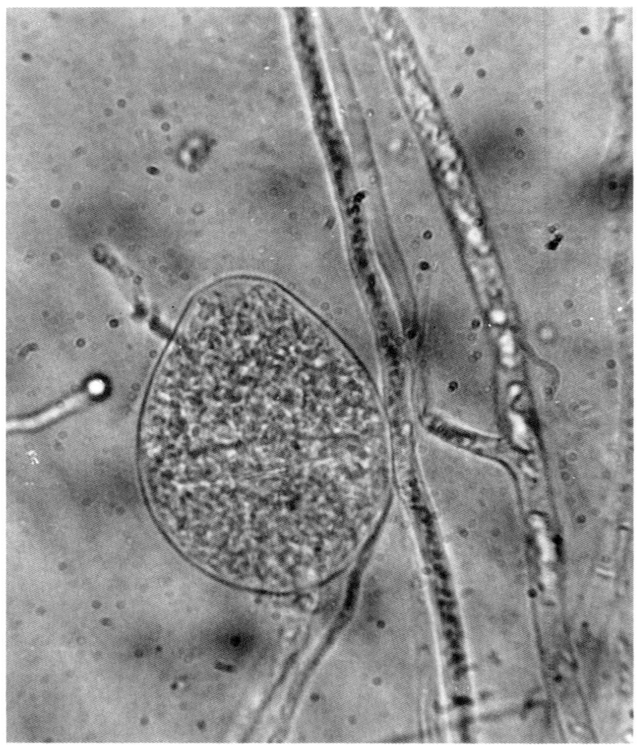

Figure 44.2. Nonpapillate, ovoid sporangium of *Phytophthora medicaginis.* (Reproduced by permission of the American Phytopathological Society, St. Paul, Minnesota, from Erwin 1954)

Table 44.1. Characteristics of *Phytophthora medicaginis* from alfalfa cited in different references

Sporangia length × breadth (µm)	Oogonia diameter (µm)	Oospore diameter (µm)	Antheridia[a]	Min : opt : max temperatures (°C)	Reference
27–63 × 23–38 (av. 31 × 39)	23–59 (av. 31)	19–37 (av. 25)	9 : 91	8 : 25 : 33	Erwin (1954)
25–58 × 17–37 (av. 37 × 30)	26.2–39.4 (av. 31.5)	17.0–29.6 (av. 24.3)	50 : 50	NS[b]	Leath and Baylor (1975)
55 × 37	Av. 30	Av. 23	8 : 24	5.7 : 22–28 : 32–35	Matsumoto and Aragaki (1978)
54.3 × 39.5	Av. 30.8	Av. 26.4	50 : 50	NS	Romero (1978)
50–60 × 24–30	Av. 37–39	Av. 28–34	NS	5 : 26 : 35	E. M. Hansen and Maxwell (1991)
. . .	. . .	. . .	. . .	5 : 25 : 35	Faris (1988)

[a] Percentage amphigynous to percentage paragynous.
[b] Not stated.

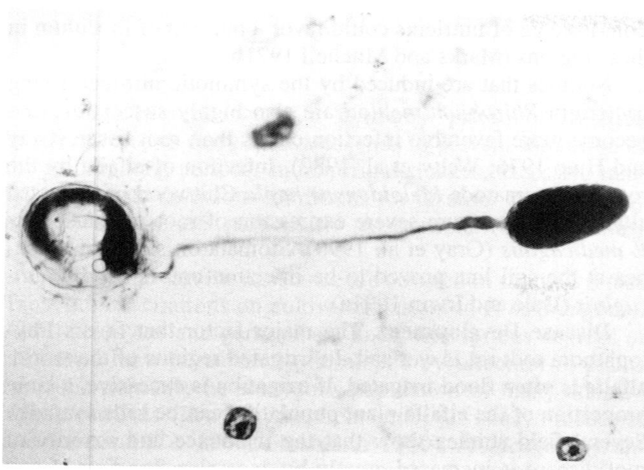

Figure 44.3. Germinating oospore of *Phytophthora medicaginis* producing an ellipsoid, nonpapillate sporangium. Spores are stained with trypan blue in 0.85% sodium chloride. (Reproduced by permission of the American Phytopathological Society, St. Paul, Minnesota, from Erwin 1966)

nous antheridia in other studies are variable (12 to 100%) among several isolates (Hamm and Hansen 1982b) (Table 44.1). Oospores germinate to form sporangia (Figure 44.3) (Salvatorre et al. 1973; Stack and Millar 1985a–c; El-Hamalawi and Erwin 1986a, b).

Growth Temperatures

Growth is relatively slow (4 to 5 mm per day) on agar media and is optimal at 25 to 26°C. Little growth occurs at 5 or >33°C (Erwin 1954, 1966; E. M. Hansen and Maxwell 1991). Growth at 30°C is only slightly less than that at 24 to 27°C (Table 44.1).

Growth Characteristics in Media

Radial growth is somewhat petallate on potato-dextrose agar (Figure 4.2), but on V8 juice agar, cornmeal agar, or synthetic agar, there is little pattern (Erwin 1966; Kuan and Erwin 1980). Some single-zoospore isolates that produce appressed mycelium form a much higher number of oospores than isolates that produce fluffy mycelium. This characteristic was variable in progeny derived from single oospores (Erwin and Li 1986). *P. medicaginis* is sensitive to hymexazole, which is used in some selective media to suppress *Pythium* spp. (see Chapter 2).

Distinguishing Characteristics

The morphological differences between *P. sojae* and *P. medicaginis* are not great; however, those with experience are able to differentiate the two species. Nygaard et al. (1989) showed that *P. sojae* and *P. medicaginis* can be isolated from soil samples by baiting with either soybean or alfalfa plant material. Mycelial growth of *P. sojae* tends to be more appressed than that of *P. medicaginis* (E. M. Hansen et al. 1986). Several reports show that *P. sojae* does not grow in potato-dextrose agar (see Chapter 57; Erwin et al. 1968; Kuan and Erwin 1980). The lower optimum temperature for growth of *P. sojae* (20°C) has been cited as different from 25°C reported for *P. medicaginis* (E. M. Hansen and Maxwell 1991), but this criterion might not be dependable because 25°C has also been cited as optimum for *P. sojae* (see Chapter 59). Hohl (1983) reported that *P. sojae* isolates grew 1.9 to 4.3 times more rapidly on a nitrate-containing synthetic agar medium than on an ammonium-containing medium. Growth rate of *P. medicaginis* was similar on both nitrate and ammonium agar media. Luo et al. (1988) reported that growth of *P. medicaginis* on a medium containing malachite green (125 μg/ml) was completely inhibited but was 60% of normal for *P. sojae.*

Oogonium diameter-sporangium length ratios were shown to differentiate the two taxa statistically, but examination of the data shown by E. M. Hansen and Hamm (1983) indicated the differences may be so small that they would not be noted by casual observation. Oogonia are produced in V8 juice or carrot agar more abundantly by isolates of *P. megasperma* than by isolates of *P. sojae.* The sporangia of *P. medicaginis* tend to be uniformly ellipsoid (egg shaped), but sporangia of *P. sojae* tend to vary from ellipsoid to obpyriform (shown in Figure 59.1). However, Figure 59.2 shows only ellipsoid sporangia. A basal appendage is formed on some sporangia of *P. medicaginis* (Erwin 1965) but not on those of *P. sojae.*

Isozyme analysis (Nygaard et al. 1989), protein patterns (Irwin and Dale 1982; Faris et al. 1986; E. M. Hansen et al. 1986), and DNA RFLP patterns (Förster et al. 1989; Förster and Coffey 1993) differentiate *P. medicaginis* from *P. sojae, P. trifolii,* and other biotypes classified as *P. megasperma.* See Chapter 46 on *P. megasperma* for a discussion of Förster and Coffey (1993).

Part of the description of *P. medicaginis* by E. M. Hansen and Maxwell (1991) is reproduced below (reproduced by permission from Mycologia 83:376–381, E. M. Hansen and D. P. Maxwell, copyright 1991, The New York Botanical Gardens, Bronx, New York):

> Homothallic, producing oospores in single strain culture, with paragynous or amphigynous antheridia. Oogonia 37–39 μm diam, occasionally 32–43 μm, oospores 28–34 μm diam, with oospore walls 2–4 μm thick. Sporangia formed in liquid culture, ovoid to ellipsoid or pear-shaped, without papillae, 50–60 μm long by 24–30 μm wide. Slow growing on agar media, 4–5 mm/day at optimum 25–26C. Little or no growth at 5C and 35C. Colony dense, patternless, with appressed margin. Pathogenic on alfalfa,, *Medicago sativa* L., and less commonly on chickpea. Named for the principal host. Type material deposited in the Oregon State University herbarium (OSC). Cultures are on file at The American Type Culture Collection.
>
> The species exhibits aggressive pathogenicity on alfalfa and chickpea but causes limited symptoms on other hosts. It resembles *P. sojae* in oogonium diam (average 37 μm), but sporangia are shorter, averaging 53 μm. *Phytophthora medicaginis* has a greater tendency to form amphigynous antheridia than other species in the *P. megasperma* complex, with some isolates being >50% amphigynous (Erwin, 1954, 1965; Hamm and Hansen, 1982b). Growth on agar media is similar to but readily distinguished from *P. sojae* and *P. trifolii.* Growth is slow (<5.0 mm/day) at 26C, patternless and compact. The colony margin is relatively appressed. Optimum growth occurs at 26C, with little or no growth at 5C or 35C. Faris (1988) obtained similar results on V8 and pea agar.

Diseases Caused by *Phytophthora medicaginis*

Phytophthora root rot occurs on alfalfa, chickpea, and sainfoin and is experimentally pathogenic to sesbania (Table 44.2). The disease is described by Stuteville and Erwin (1990) and Leath et al. (1988).

P. medicaginis causes damping-off of seedlings similar to that caused by *Pythium* species and also causes root rot on well-established plants (Erwin 1954; Schmitthenner 1964; Welty and Busbice 1976). The first indication of root rot is the appearance of a yellow and reddish brown discoloration of the lower leaves and wilting of the foliage. Lesions on the taproots *sperma* was differentiated from *P. syringae* by growth at 28°C (Tompkins et al. 1936b).

E. M. Hansen and Maxwell (1991) report the optimum temperature for growth of Douglas fir (DF1) isolates, 27.1°C; DF2, 21.6°C; large-oogonia isolates from alfalfa (A1F2), 22.5°C; rose, 25°C; apple, 22.5°C and 20°C; grape, 20°C; juniper, 22.5°C; cherry, 25°C; pear, 25°C; poplar, 20°C; *Brassica* spp., 20°C; and the type culture from hollyhock (Drechsler 1931), 20°C.

Some Diseases Caused by *Phytophthora megasperma*

Root Rot of *Brassica* spp. (e.g., Cabbage, Cauliflower, Brussel Sprouts, Turnip, Kale, and Rutabaga)

Root rot of cauliflower, *Brassica oleracea* L. var. *botrytis,* was described by Tompkins et al. (1936b). Plants are susceptible to attack at all stages of growth. After infection of roots, the leaves turn progressively purple or reddish from the margins inward, wilt, and shed. As the disease progresses, the stems become soft and flaccid. The lower end of the taproot is blackened and eventually rots completely. The cortex sloughs off the taproot and lateral roots. Adventitious roots often develop above the diseased portions of the taproot. In Oregon, severe root rot occurs in fall-planted cauliflower but much less so in spring-planted crops (Hamm and Koepsell 1984). The isolates that caused disease were included in the broad-host-range (BHR) group (E. M. Hansen et al. 1986). Karakaya et al. (1995) reported that *P. megasperma* caused root rot of kale and tyfon, both winter forage crops in Wyoming.

A storage rot of swedes (rutabaga and turnip), *B. napa* L., occurred in the United Kingdom (Geeson et al. 1990). This disease occurred on swedes that had been dipped in a benomyl suspension to control *Botrytis cineria* and in a storage area maintained at 0.5 to 1°C and 92% relative humidity. The infected roots were brown to dark gray brown and relatively firm. A similar disease was reported in Canada (Jones 1952) and in Ireland (Downes and Loughnane 1969).

Disease Development. Disease on cauliflower develops most rapidly on poorly drained or ponded soils, particularly after rainy periods or irrigation, at moderate soil temperatures of 20 to 30°C.

Control. Irrigation should be managed to prevent waterlogging of the soil. Cabbage was less severely affected than cauliflower. In greenhouse tests, the cultivar Armando April was tolerant to *P. megasperma* (Hamm and Koepsell 1984).

Crown Rot of *Malus pumila* Mill. (Apple)

Typical symptoms of the crown rot include delayed bud break, stunted growth, orange discoloration of the bark, and reduced crop yield (Robertson and Dance 1971). Crown cankers resemble those caused by *P. cactorum* (Chapter 11) but usually are not as extensive. Cankers eventually girdle the trunk, and within these cankers the bark rots completely. Lesions of varying lengths develop on lateral roots.

Disease symptoms are similar on *Prunus* spp. except that gum exudes from cankers. Delayed bud break and dieback of branches commonly occur (C. I. Robertson, DSIR Auckland, N.Z., *personal communication*).

Disease Development. The symptoms develop rapidly after prolonged heavy rains.

Control. See Crown, Collar, and Root Rot of *Malus pumila* Mill. (Apple) and *Pyrus communis* (Pear) in Chapter 11 (*P. cactorum*).

CHAPTER 47

Phytophthora melonis

Katsura (1976)

No synonyms are presently known for *Phytophthora melonis,* which was isolated from melon. The status of *P. melonis* as a species is problematic because it is similar morphologically to *P. drechsleri* and considered to be conspecific by Ho et al. (1984c), Ho (1986b), and Ho and Jong (1991). In a study of mtDNA RFLP (restriction fragment length polymorphism) and isozyme patterns by Mills et al. (1991), however, four isolates from China and/or Japan that had been classified as *P. melonis* from cucurbit plants grouped with one *P. sinensis* and eight *P. drechsleri* isolates from cucurbit plants (group F). This group of isolates was distinct from isolates that had been previously identified as *P. drechsleri* (grew at 35°C; their group A) and from isolates previously identified as *P. cryptogea* (their group B). Since the groups were genetically distinct, Mills et al. (1991) recommended not merging *P. melonis* into *P. drechsleri* and, alternatively, that an amended description of *P. melonis* might be appropriate. Technically, the original description of semipapillate sporangia by Katsura (1976) is erroneous. Sporangia were later found to be nonpapillate by Ho et al. (1984b, c), Ho (1986b), and Mills et al. (1991). If maintained as a species, *P. melonis* should be redescribed. The studies by Mills et al. (1991) and Ho and colleagues are discussed at more length in Chapter 20 (*P. cryptogea*), Chapter 23 (*P. drechsleri*), and Chapter 58 (*P. sinensis*).

P. melonis was first isolated by Katsura (1968) from diseased cucumber (*Cucumis sativus* L.) plants in Japan. It also affects watermelon and pumpkin. Katsura (1971) reported that inoculation of legumes and apples with *P. melonis* induced symptoms. The pathogen appears to have a limited geographical distribution and has been reported only from Japan.

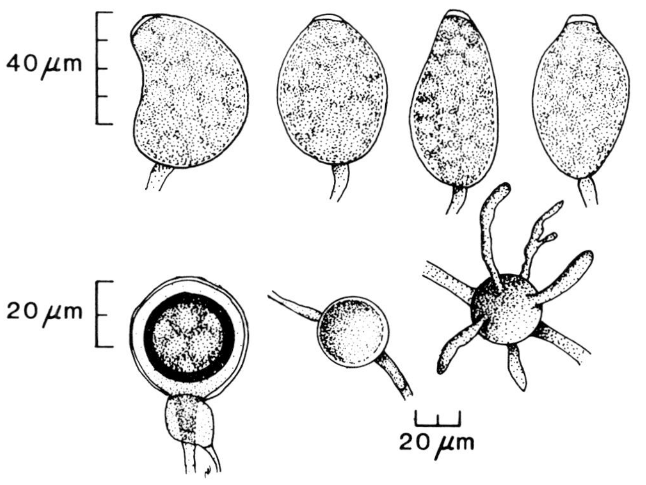

Figure 47.1. Morphology of *Phytophthora melonis.* Lower row, globose oogonium with an amphigynous antheridium; intercalary chlamydospore; and germinating chlamydospore. Upper row, sporangia of variable ovoid to ellipsoid shapes. Although drawn and interpreted by Katsura (1976) as semipapillate, Ho et al. (1984) reexamined the type isolate and found sporangia to be nonpapillate. (Drawing by A. Vaziri after Katsura 1976)

Characteristics of *Phytophthora melonis*

P. melonis is classified in group VI (Stamps et al. 1990). Morphology is shown in Figure 47.1. See Tables 4.2 and 4.3 for tabular keys.

Sporangia

Sporangia are ovoid, obpyriform, with rounded bases, sometimes with variable shapes; semipapillate; and 33 to 75 μm long × 23 to 46 μm wide (average 48.8 × 33.6 μm). The length-breadth ratio is 1.5:1. Although Katsura (1976) described sporangia as semipapillate, Ho et al. (1984b) and Ho (1986) restudied the type culture and found the sporangia to be nonpapillate and to produce new sporangia by proliferation. Sporangiophores are sympodial.

Hyphal Swellings

Hyphal swellings are not reported by Katsura (1976) but probably occur.

Chlamydospores

Chlamydospores are reported to be spherical, often intercalary, and 16.3 to 52.0 μm in diameter (average 31.2 μm) by Katsura (1976); however, Ho et al. (1984b, c) did not find chlamydospores.

Sex Organs

P. melonis was considered to be homothallic by Katsura (1976). Antheridia are amphigynous and often spherical; oogonia are 27.5 to 35.0 μm in diameter; oospores are 25 to 32.5 μm in diameter (average 29.3 μm); however, Ho et al. (1984b) found the type isolate to be heterothallic.

Growth Temperatures

The minimum temperature for growth is 9°C, the optimum is 28 to 32°C, and the maximum is 37°C.

Disease Caused by *Phytophthora melonis*

Late Blight or Foot Rot of *Cucumis sativus* L. (Cucumber)

Late blight of cucumber affects leaves, stems, and fruits. Soon after heavy rains, infection is initiated at the collar region, which turns greenish brown and becomes soft. The stem eventually shrinks, and the plant falls over. Lesions on leaves become darker green than the rest of the leaf and slowly progress toward the pedicel if moist conditions prevail. Infection on fruit is usually initiated at the tip, resulting in deeply depressed, dark green, water-soaked lesions. A soft rot of the fruit follows. Under moist conditions, fungal mycelium and powdery masses of sporangia form on the surface of the diseased fruit.

P. melonis is considered to be more pathogenic than *P. capsici* on cucurbits.

Disease Development. *P. melonis* is most active after heavy rains. Rain water accumulating around the stem of the plant results in invasion of the collar region. In Japan, the disease is most severe from the middle of May to the end of June.

Control. Cucumbers should be planted in well-drained areas. Grafting of cucumber plants onto Kurodane or Shintosa squash cultivars (Yoshio et al. 1976) has controlled the disease. This method is probably practical only in greenhouse culture.

CHAPTER 48

Phytophthora mexicana

Hotson and Hartge (1923)

No synonyms are known for *Phytophthora mexicana*, which was first isolated and described by Hotson and Hartge (1923) from diseased tomatoes in Mexico. Since then it has been reported only on *Dieffenbachia maculata* (Lodd.) G. Don (Kröber et al. 1983). Until the latter report, its geographical distribution was probably limited to Mexico. Kröber et al. (1983) suspected that the fungus was imported to a nursery from the Netherlands. The distribution and availability of isolates is limited, according to Mchau and Coffey (1994c), who stated that only one isolate was available for their study.

Characteristics of *Phytophthora mexicana*

P. mexicana is classified in group II (Stamps et al. 1990). Morphology is shown in Figure 48.1. See Tables 4.2 and 4.3 for tabular keys.

Sporangia

Sporangia are ovoid to fusiform, often with distorted shapes; occasionally there are two papillae on the same sporangium; lateral attachment of sporangium is possible; they are sometimes intercalary; sporangia are persistent on the hyphae and 46 to 77 × 16 to 33 μm (average 45.2 × 23.1 μm). Sporangia are usually formed on unbranched sporangiophores; however, formation of sporangia on simple sympodia is illustrated by Hotson and Hartge (1923).

Hyphal Swellings

Hyphal swellings are not produced.

Chlamydospores

Chlamydospores are rare; spherical, 28 to 44 μm in diameter; and mostly intercalary.

Sex Organs

P. mexicana was described as homothallic in the original description (Hotson and Hartge 1923) but later as heterothallic by Savage et al. (1968). Antheridia are amphigynous; oogonia are 30 to 37.5 μm in diameter (average 30.7 μm) and produced in abundance at temperatures between 8 and 10°C; oospores are 24 to 37.4 μm (average 25.4 μm).

Growth Temperatures

The minimum temperature for growth is 10°C, the optimum 27.5°C, and the maximum <35°C.

Distinguishing Characteristics

P. mexicana closely resembles *P. parasitica* but differs from it, according to Hotson and Hartge (1923), by its larger oospores and host specificity. Leonian and Geer (1929) proposed to combine *P. mexicana* with *P. erythroseptica*, a species that it resembles in many aspects. On the other hand, Tucker (1931) found that this species closely resembled *P. parasitica* in morphology and pathogenicity. Kröber et al. (1983) indicated that it resembled *P. palmivora*; however, sporangia of *P. palmivora* are caducous, but those of *P. mexicana* are not.

A report noted that only one culture (ATCC 46731) of *P. mexicana* was available for study. Rather surprisingly, *P. mexicana* isozymically grouped with *P. capsici* isolates in CAP1 even though *P. mexicana* produced sporangia that were more ovoid than those of *P. capsici* (length-breadth ratio was 1.3 compared with 1.3 to 2.0 for *P. capsici* CAP1 isolates) but also were noncaducous. *P. capsici* sporangia are all caducous with long pedicels (Oudemans and Coffey 1991c; Mchau and Coffey 1994c). In addition, no oospores were formed by the *P. mexicana* culture except when paired with P864 (A1 mating type) of *P. capsici* (Mchau and Coffey 1994c).

Disease Caused by *Phytophthora mexicana*

Fruit Rot of *Lycopersicon esculentum* Mill. (Tomato)

The symptoms caused by *P. mexicana* on tomato fruits are similar to those caused by *P. nicotianae* (*P. parasitica*) and are described in detail in the section on buckeye rot of tomato fruits (Chapter 50); however, fruits usually shrivel and decay without the appearance of the zonate markings typical of buckeye rot. A black wilt and damping-off of young plants also occur.

Disease Development. Similar to that of buckeye rot (Chapter 50).

Control. See section on buckeye rot (Chapter 50).

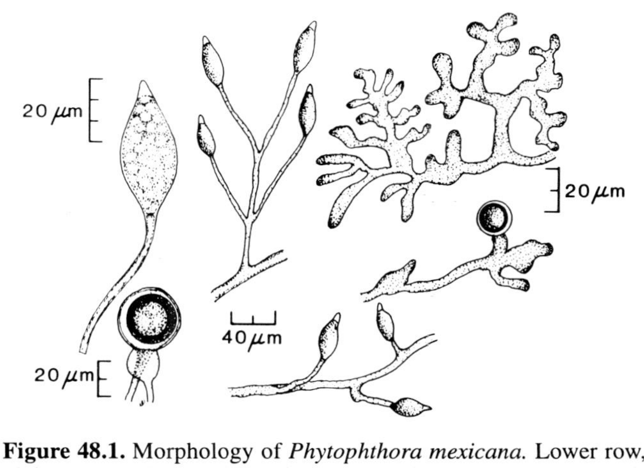

Figure 48.1. Morphology of *Phytophthora mexicana.* Lower row, globose oogonium with amphigynous antheridium; young, papillate sporangia formed on a sympodial sporangiophore; chlamydospore formed laterally on inflated hyphae. Upper row, fusiform, papillate sporangium; sporangia forming in a sympodium; inflated hyphae. Sporangia illustrated by Mchau and Coffey (1994c) were more ovoid. (Drawing by A. Vaziri after Hotson and Hartge 1923)

CHAPTER 49

Phytophthora mirabilis

Galindo and Hohl (1985)

Phytophthora mirabilis from *Mirabilis jalapa* L. was first classified as *P. infestans* by Servin (1958), although the fungus produced larger sporangia and was more sensitive to malachite green than *P. infestans*. *P. mirabilis* forms oospores when paired with opposite mating types of *P. infestans*. Romero and Fourton (1962) reported that leaf blight was caused by a specialized form of *P. infestans*. Fourton (1962, cited by Galindo and Hohl [1985]) isolated both A1 and A2 mating types in a natural population that was causing a leaf blight on *M. jalapa*, a nonagricultural plant, and showed that it was capable of causing blight on *M. jalapa* plants but not on potato. Since it was considered to be a variety or form of *P. infestans*, this taxa was not reported.

Characteristics of *Phytophthora mirabilis*

P. mirabilis is classified in group IV (Stamps et al. 1990). Sporangia are illustrated in Figure 49.1. See Tables 4.2 and 4.3 for tabular keys.

Sporangia

P. mirabilis isolates produce sporangia on infected leaves that are ellipsoid, caducous with short pedicels, and semipapillate. The average length of sporangia is 33.7 µm, range 26.4 to 38.5 µm, compared with a length of 28.8 µm, range 22.0 to 39.6 µm, for sporangia of *P. infestans*. Breadth data were not given by Galindo and Hohl (1985), but the length-breadth ratio was given as 1.9:1. Sporangiophores are compound sympodial with indeterminate branches.

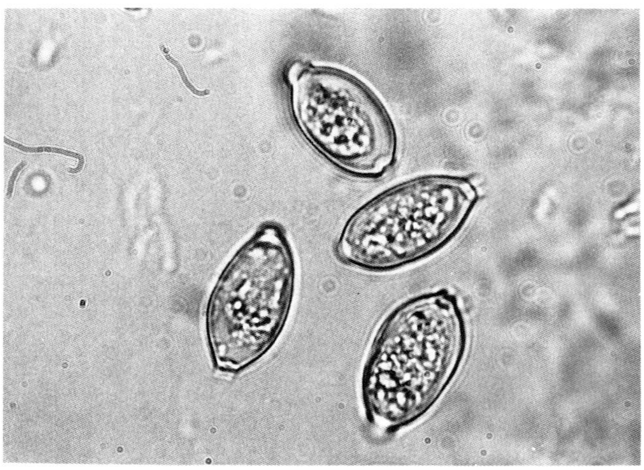

Figure 49.1. Caducous sporangia of *Phytophthora mirabilis* with short stalks. (Courtesy J. Galindo, Colegio de Postgraduados, Chapingo, Mexico)

Chlamydospores

Chlamydospores have not been reported.

Sex Organs

Paired A1 and A2 types (heterothallic) produce globose, smooth-walled oogonia, 26.4 µm in diameter, containing smooth-walled, colorless oospores, 21.3 µm in diameter. Antheridia are amphigynous and average 13.8 × 11.8 µm. A1 and A2 mating types occur in the field at a ratio of about 1:1.

Distinguishing Characteristics

All *P. mirabilis* isolates grow well on a synthetic medium containing nitrate as the nitrogen source, but *P. infestans* isolates do not (Hohl 1975, 1983). *P. mirabilis* is pathogenic on detached leaves of *M. jalapa* but not on leaves of potato or tomato. All except one of 12 *P. mirabilis* isolates showed esterase activity, but isolates of *P. infestans* did not. The *P. mirabilis* isolates tolerated the anti-*Pythium* fungicide hymexazole, but *P. infestans* did not (Galindo and Hohl 1985). See Chapter 2 for a discussion of hymexazole, a fungicide that is toxic to most *Pythium* species but allows most *Phytophthora* species to grow.

Sporangia of *P. mirabilis* are larger than those of *P. infestans* (length 34 versus 29 µm for *P. infestans*); oospores (diameter 22 versus 30 µm) and oogonia (diameter 30 versus 38 µm) are smaller than those of *P. infestans*. The sporangia of *P. mirabilis* are predominantly ellipsoid and thus have a greater length-breadth ratio (1.9:1) than *P. infestans* (1.7:1) (Galindo and Hohl 1985). The basal swelling at the point of the sporangium attachment is similar to that of *P. infestans* (see Chapter 4, Figure 4.6). *P. mirabilis* differs from other members of group IV, *P. colocasiae, P. hibernalis,* and *P. ilicis,* by the production of compound sympodial sporangiophores and typical basal swellings (similar in this respect to *P. phaseoli* and *P. infestans*). *P. mirabilis* is heterothallic, but *P. phaseoli* is homothallic (Galindo and Hohl 1985). Despite the differences between *P. infestans* and *P. mirabilis*, Moller et al. (1993) consider *P. mirabilis* to be a forma specialis of *P. infestans* on the basis of mtDNA RFLP similarities.

Disease Caused by *Phytophthora mirabilis*

The only known host of *P. mirabilis* is *M. jalapa,* which grows wild in the vicinity of Chapingo, Mexico, and on which leaves are blighted.

CHAPTER 50

Phytophthora nicotianae

Breda de Haan (1896)

= *Phytophthora parasitica*

Dastur (1913)

Synonyms include *Phytophthora melongenae* Sawada (1915b), *P. allii* Sawada (1915b), *P. terrestris* (as *P. terrestria*) Sherbakoff (1917), *Blephaspora terrestris* Sherbakoff Peyronel (1920), *P. parasitica* var. *rhei* Godfrey (1923), *P. jatrophae* Jensen (1923) (according to Waterhouse [1956], the original citation cannot be traced), *P. tabaci* Sawada (1927), *P. parasitica* var. *piperina* Dastur (1913), *P. formosana* Sawada (1942), *P. lycopersici* Sawada (1942), *P. ricini* Sawada (1942), and *P. parasitica* var. *sesami* Kale and Prasad (1957). Waterhouse (1963, 1974c) and Waterhouse and Waterston (1964d, e) describe the varieties *nicotianae* and *parasitica* of *P. nicotianae;* however, these are no longer accepted (G. Hall 1993).

Until 1963, the name *P. parasitica* Dastur (1913), the cause of a seedling blight of castor bean (*Ricinus communis* L.), had been accepted worldwide rather than *P. nicotianae* as the proper name for this species because the latter had been inadequately described by Breda de Haan (1896). Ashby (1928) noted that the drawing of paragynous antheridia by Breda de Haan (1896) (reproduced in Figure 50.1) was in error because the antheridia of *Phytophthora* isolates from tobacco and other hosts were then known to be amphigynous. (See Figures 50.2, which shows amphigynous antheridia and is drawn from Dastur's [1913] description of *P. parasitica,* and 50.3 for correct morphology.) This indicated that Breda de Haan's water culture of infected roots was contaminated with a *Pythium* species and was not a pure culture of *Phytophthora.* Ashby proposed that the name *P. nicotianae* be "eliminated" and that the name of all strains of this pathogen from tobacco and other hosts be merged under the properly described name *P. parasitica* Dastur (1913). Later, Rosenbaum (1917), Tucker (1931), Leonian (1934), and Meurs (1934) concurred. We agree that *P. parasitica* is the most appropriate name but are constrained by the International Code of Botanical Nomenclature as stated below.

Waterhouse (1963) replaced the name *P. parasitica* with the name *P. nicotianae* Breda de Haan (1896) because it had priority despite the error in the description by Breda de Haan. She also noted that the lack of a Latin description of *P. nicotianae,* as required now by the International Code of Botanical Nomenclature, was allowed in publications prior to 1908 (Stafleau et al. 1972; Voss 1983; also discussed by Ho and Jong [1989a]).

The change from the name *P. parasitica* to *P. nicotianae* (Waterhouse 1963) has not been universally accepted. Most publications in the United States continue to use the name *P. parasitica* (e.g., Tsao 1969b, 1971; Kuske and Benson 1983a, b; Blaker and MacDonald 1986; Blaker and Hewitt 1987; Timmer et al. 1988b, 1989; Matheron and Matejka 1989, 1990; Förster and Coffey 1990, 1991; Benson 1993, to name only a few). In France, several reports have been published under the name *P. parasitica* (e.g., Louvel 1975; Billard et al. 1988; Bonnet et al. 1989; Ricci et al. 1992). Indeed, the elicitin compound produced by *P. parasitica* that elicits the production of a resistance reaction in tobacco plants has been named "parasiticein" (Billard et al. 1988; Ricci et al. 1992; Kamoun et al. 1993b, 1994). A computer printout of Agricola abstracts (1970–1994) showed that there were 907 papers listing *P. parasitica* and 536 listing *P. nicotianae.*

Although Ho and Jong (1989a) also favored the widely used name *P. parasitica* over the name *P. nicotianae,* they concluded that the mandate of the current rules of the International Code of Botanical Nomenclature (Article 11; Voss et al. 1983) should be maintained "in spite of the ambiguity of the antheridial configuration of *P. nicotianae* (by Breda de Haan [1896])." To correct this error, they stated that a petition to preserve the species name *parasitica* could be made to the International Botanical Congress under Article 14. Ironically, Ho and Jong (1989a) stated that *P. nicotianae* could have been rejected because of the possible inclusion of two fungi (*Pythium* and *Phytophthora*) (Figure 50.1) in Breda de Haan's (1896) original description in accordance with Article 70 (Stafleau et al. 1972), but the article was deleted in the most recent code (Voss et al. 1983).

Thus, as discussed by Ho and Jong (1989a), if nomenclature of *Phytophthora* species is subject to the International Code of Botanical Nomenclature, the name *P. nicotianae* legalistically should prevail.

In our opinion, the name *P. parasitica* is preferable; however, such a change should not be mandated unilaterally by us but rather by a consensus of those in the field. We suggest that the question should be discussed by a workshop on *Phytophthora* and that the decision on the proper name be made on the basis of the rules of nomenclature for the kingdom Chromista.

Judging by the history of the past 30 some years, however, both names will probably continue to be used; for this reason we cite *P. parasitica* as equal to *P. nicotianae,* despite the dilemma this poses for the naming of the species of *Phytophthora* with the largest host range of any yet described.

G. Hall's (1993) redescription of *P. nicotianae,* which is based on a study of 31 morphological, physiological, and biochemical characters, concludes that the two varieties *nicotianae* and *parasitica* as described by Waterhouse (1963) should not be retained. This agrees with Ho and Jong (1989a) and several other studies to be discussed below.

The Waterhouse (1963, 1974b) separation of *P. nicotianae* var. *parasitica* from *P. nicotianae* var. *nicotianae* was based on small differences in morphology but not on pathogenicity to tobacco. Thus, *P. nicotianae* var. *nicotianae* is not necessarily a synonym of the tobacco epithet, *P. parasitica* var. *nicotianae sensu* Tucker (1931), although it is often misunderstood to be so (Ho and Jong 1989a).

P. nicotianae var. *nicotianae* was characterized (Waterhouse 1963, 1974b; Waterhouse and Waterston 1964d) by the production of spherical oogonia (20 to 35 µm in diameter, mostly 28 to 30) in single culture of some isolates or when paired with an isolate of an opposite mating type; spherical oospores not filling the oogonium, always more than 20 but mostly 24 to 30 µm in diameter; amphigynous antheridia in which the oogonial stalk is often eccentric, giving a superficial appearance of paragyny; hyphae up to 5 µm in diameter with frequent hyphal swellings, some of which are "spiky" because of radial growth of hyphae from them; abundant chlamydospores (20 to 40 µm in diameter) with walls about 1.5 µm thick; and spherical to obturbinate, noncaducous, papillate sporangia formed terminally on the sporangiophore and not in a sympodium.

P. nicotianae var. *parasitica* was characterized (Waterhouse 1963, 1974b; Waterhouse and Waterston 1964e) by production of spherical oogonia (16 to 31 µm, mostly 24 to 26 µm in diameter) either in single culture of some isolates or when paired with an opposite mating type; spherical oospores with an average diameter less than 20 µm (maximum 26 µm);

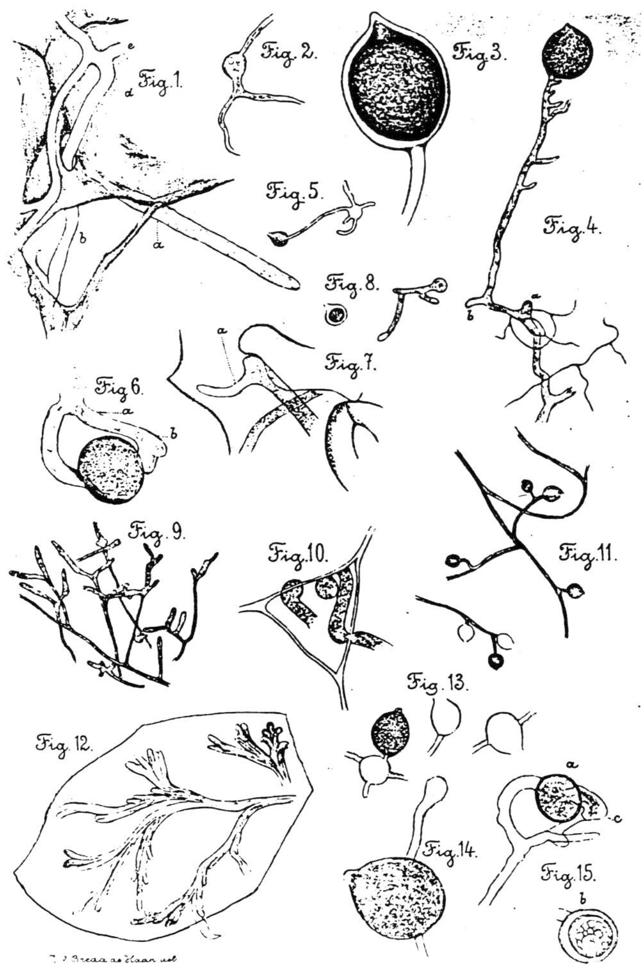

Figure 50.1. Morphology of *Phytophthora nicotianae* according to Breda de Haan (1896). **1,** Proliferation of the intercellular mycelium of *P. nicotianae.* (330×) **2,** Germinating oospore in sugar water. (236×) **3,** Sporangium of *P. nicotianae.* (330×) **4,** Sporangiophore and sporangium appear through a stomate. (137×) **5,** Germination of the zoospore and formation of a secondary sporangium. (275×) **6,** Oogonium and antheridium of *P. nicotianae.* (330×) (This is atypical for a *Phytophthora* species.) **7,** Penetration of the epidermis by a hyphal thread. (330×) **8,** Mature "egg" spore and germinating "egg" spore. (198×) **9,** Mycelium of *P. nicotianae* cultured in water. (65×) **10,** Penetration of the cell wall by mycelium. (330×) **11,** Mature and immature sporangia. (52×) **12,** Superficial mycelium with haustoria on a cell wall. (198×) **13,** Various forms of sporangia. (275×) **14,** Sporangium with proliferating mycelium. (330×) **15a,** Fertilization of the oospore and **b,** mature oospore. (330×) (From Breda de Haan 1896; legend translated from Dutch by P. Oudemans)

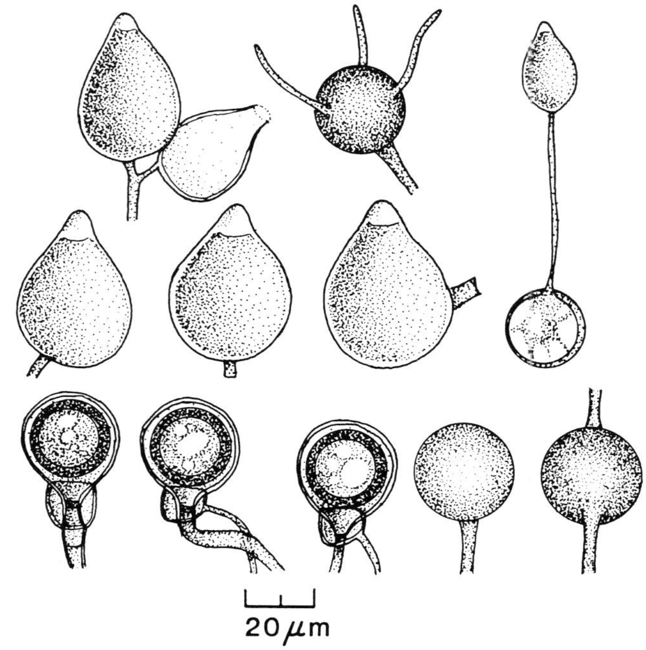

Figure 50.2. Morphology of *Phytophthora nicotianae* (= *P. parasitica*). Upper row, papillate, ovoid sporangia; germinating chlamydospore. Middle row, papillate, ovoid sporangia; germinating oospore. Lower row, oogonia with amphigynous antheridia containing oospores; terminal and intercalary chlamydospores. (Drawing by A. Vaziri)

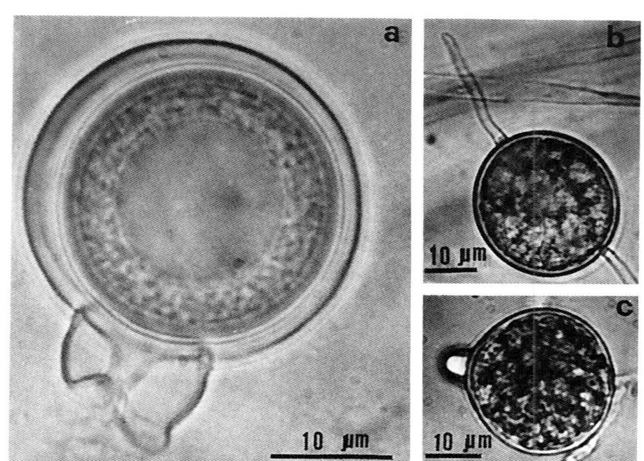

Figure 50.3. Morphology of *Phytophthora nicotianae* from prickly pear cactus. **a,** Oospore; **b,** intercalary chlamydospore; and **c,** papillate sporangium. (Reproduced by permission of the American Phytopathological Society, St. Paul, Minnesota, courtesy Cacciola and Magnano di San Lio 1988)

spherical or oval amphigynous antheridia; hyphae up to 9 µm in diameter without hyphal swellings; papillate sporangia broadly ovoid or ellipsoid to spherical formed on irregularly branched sporangiophores in a sympodium and spherical or occasionally ellipsoid sporangia that are caducous with short pedicels; and chlamydospores forming late (10 to 14 days) and usually about 22 to 30 µm in diameter with thicker walls (3 to 4 µm) than those of *P. nicotianae* var. *nicotianae*.

Despite the use of these two varietal epithets since 1963, several studies show that the differences (Waterhouse 1963) cannot be confirmed. Tsao and Sisemore (1983) showed that the morphological features in two isolates from tobacco, one isolate from tomato, and eight isolates from citrus did not differ in morphology. Ho and Jong (1989) compared 13 isolates from tobacco and 22 isolates from other hosts and concluded that var. *nicotianae* and var. *parasitica* could not be differentiated. Oudemans and Coffey (1991) and G. Hall (1993) reported that oogonium and oospore sizes of several isolates varied in a continuum with no valid separation associated with varieties and that sporangia produced by isolates of either variety were not caducous. The separation of the varieties *nicotianae* and *parasitica* by morphological criteria has also been questioned by others (Lucas 1975; Gallegly 1983b). Dimensions of spore structures for several *P. nicotianae* isolates (Table 50.1) are similar to those of tobacco isolates designated *P. parasitica* var. *nicotianae sensu* Tucker (1931).

Protein patterns of *P. nicotianae* isolates from tomato, carnation, and citrus were almost identical to those of *P. nicotianae* f. sp. *nicotianae* (their epithet) tobacco isolates from the United States (Erselius and de Vallavielle 1984). Several other biochemical comparisons of the tobacco isolates designated *P. parasitica* var. *nicotianae sensu* Tucker (1931) with isolates from other crops designated *P. parasitica* also showed that var. *nicotianae* and var. *parasitica* were indistinguishable. Serological analysis showed that isolates from tobacco and other crops were similar (Morton and Dukes 1967). Restriction fragment length polymorphism (RFLP)-DNA patterns (see Chapter 4) of eight isolates from tobacco (four from the United States, two from Cuba, and two from Australia), two isolates from carnation, and two from citrus were similar (Panabiéres et al. 1989). RFLP-DNA patterns of *P. parasitica* isolates from tobacco in the United States and South Africa and a wide range of other hosts (two of these isolates had been labeled *P. nicotianae* var. *nicotianae,* one from carnation in France and the others from soil in Taiwan) were similar (Figure 50.4) (Förster et al. 1990b; Förster and Coffey 1991).

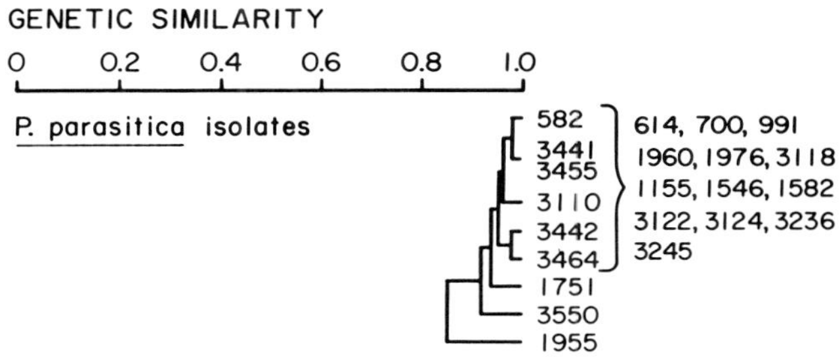

Figure 50.4. MtDNA RFLP phenogram of *Phytophthora parasitica* (*nicotianae*) isolates. Six tree topologies were generated for isolates of this species; the phenogram shown has a high cophenetic correlation (0.99878), thus indicating that the mtDNA diversity is very limited. The varieties *parasitica* and *nicotianae* are not separated. Isolate numbers P582, P3441, P3110, P3550, and P1955 had been identified as *P. nicotianae* var. *nicotianae.* P582, P3110, and P1955 were from tobacco. All others had been identified as *P. parasitica.* (Reproduced by permission of Academic Press, courtesy Förster et al. 1990b)

Table 50.1. Dimensions of various reproductive structures and cardinal growth temperatures of *Phytophthora nicotianae* or *P. parasitica* isolates from different hosts

Host	Sporangia (µm)	Length-breadth ratio	Chlamydospores diameter (µm)	Oogonia diameter (µm)	Oospores diameter (µm)	Min : opt : max temperatures (°C)	Reference
Castor bean	11–60 × 10–45 (av. 37 × 30)	NM[a]	20–60	15–27 (av. 23.8)	13–24 (av. 18.6)	NM	Dastur (1913)
Rhubarb	33–48 × 24–39 (av. 41 × 30.8)	1.3 : 1	18–45 (av. 33)	NM	21–27 (av. 24)	13 : 30 : 37	Godfrey (1923)
Tobacco	36.4 × 26.8	NM	av. 23.8	NM	NM	12 : 25–30 : 35	Tisdale and Kelley (1926)
Coconut palm	23–58 × 19–42 (av. 37.4 × 32.3)	1.4 : 1	25–57 (av. 36.4)	NM	12–24 (av. <20)	NM	Ashby (1928)
Walnut	30–82 × 19–49 (av. 55.9 × 36.6)	1.5 : 1	13–64 (av. 35.9)	27–52.5 (av. 42.2)	22.5–45 (av. 35.3)	NM	Cookson (1929)
Several	22–63 × 17–41 (av. 40 × 29)	NM	15–42 (av. 25.4)	21.5–30.4 (av. 26.4)	16.9–27.1 (av. 22.8)	10 : 30 : 35–37	Tucker (1931)
Custard apple	28–53 × 15–25 (av. 40.4 × 19.2)	NM	19–38.2 (av. 27.7)	NM	16–25 (av. 21)	NM	Rao et al. (1962b)
Tobacco	40.2 × 28.7	NM	av. 28.7	av. 28.7	av. 29.4	NM	Tucker (1931)
Loquat	20–57 × 18–49 (av. 37 × 30.5)	1.21 : 1	17–45 (av. 34.8)	NM	24–35 (av. 29)	5 : 27–31 : 35	Weltzien and Schwinn (1966)
Palm (*Washingtonia*)	33–49 × 13–41 (av. 33.2 × 25)	1.32 : 1	13–30 (av. 19.4)	16–30 (av. 25.6)	15–30 (av. 24.5)	7 : 27–32 : 37	Ershad (1971)
African violet	23–55 × 17–43 (av. 41 × 30.6)	1.33 : 1	14–40 (av. 28.5)	NM	NM	7 : 28–31 : 36	Kröber and Plate (1973)
Fig	27–48 × 16–27 (av. 36 × 229)	NM	15.9–26 (av. 20.7)	NM	NM	11 : 25–27 : 33	Rondon (1975)
Citrus	27–60 × 18.7–40 (av. 43.4 × 31.2)	1.4 : 1	NM	28.9 ± 2.0	22.9–32.2 (av. 25.9)	NM	Feichtenberger (1980)
Several	Variable in a continuum	1.1–1.3 : 1	25–40	23–28	21–25	5 : NM : 35–40	G. Hall (1993)

[a] Not mentioned by reference.

Table 50.2. Distribution of hosts of *Phytophthora nicotianae* (= *P. parasitica*)[a]

Host	Common name	Disease	Geographical distribution
Abelmoschus manihot (L.) Medic. (*Hibiscus manihot* L.)	Manihot	Leaf, stem, and root rot	Japan (Tasugi and Ikeda 1939); Fiji (K. M. Graham 1971)
Acacia mearnsii DeWild.	Black wattle	Black butt or gummosis	South Africa (Zeijlemaker 1968, 1971)
Agathis australis Salisb. Hort. ex Lindl.	Kauri pine	Root rot	New Zealand (Brien and Dingley 1959)
Agapanthus africanus L'Her	Lily of the Nile	Leaf and seedling blight	Japan (Asuyama 1935)
Agave sisalana Perrine	Sisal	Zebra disease	East Africa (Peregrine 1963); China (Ho et al. 1983b)
Aglaia domestica Pelleg.	Lang sat	Foot rot	Thailand (Kamjaipai and Chompoorat 1984)
Agonis flexuosa Schav.	Peppermint tree	Root rot	Australia (Hardy and Sivasithamparam 1988)
Ailanthus altissima (Mill.) Swingle	Tree-of-heaven	Root rot	Iran (Ershad 1971)
Allium cepa L.	Onion	Damping-off, root rot, and leaf spot	Formosa (Sawada 1915b); South Rhodesia (Hopkins 1951)
Allium fistulosum L.	Welsh onion	Leaf and flower stalk blight	Formosa (Sawada 1915b)
Aloe barbadensis Mill. (*A. vera* (L.) Webb and Berth.)	Unguentine cactus, medicinal aloe	Root, stem, and base rot	United States (Averre and Reynolds 1964)
Anacardium occidentale L.	Cashew	Shoot rot, leaf fall	India (Thankamma 1974)
Ananas comosus (L.) Merrill (*A. sativus* Schult.)	Pineapple	Heart, leaf, base, and stem rot	West Indies (Ashby 1920); Cuba (Bruner 1923); Hawaii (Sideris 1929); Argentina (Blanchard 1930); India (Chowdhury 1945); El Salvador (Crandall et al. 1951); Venezuela (Pontis Videla 1953); Taiwan (Chen 1955); Ivory Coast (Frossard 1967); Fiji (K. M. Graham 1971); Mexico (Garcia and Adam 1972); Cook Islands (Joseph and Purea 1973); Thailand (Suzui et al. 1979)
Annona squamosa L.	Custard apple or sitaphal	Fruit rot	India (V. G. Rao et al. 1962b, 1963)
Antirrhinum majus L.	Snapdragon, 'Guele de loup,' muflier	Damping-off	South Rhodesia (Hopkins 1931a); Mauritius (Shepherd 1933); Malagasy Republic (Bouriquet 1950); United States (Gill 1960); Japan (Katsura 1971); Iran (Ershad 1979); Germany (Kröber 1981a)
Anthurium scherzeranum Schott.	Pigtail anthurium, flamingo flower	Root rot, petiole dry rot	Germany (Schwinn 1961b); Hawaii (Uchida and Aragaki 1979)
Aphelandra squarrosa Nees cv. Dania	Zebra plant	Stem rot	United States (Chase 1981)
Aralia L.	Aralia	Collar rot	Greece (Kouyeas and Chitzanidis 1978)
Araucaria araucana (Mol.) C. Koch	Monkey-puzzle tree	Root rot, dieback	Iran (Ershad 1971)
Araucaria columnaris (Forst.) Hook	New Caladonia pine	Seedling rot	Peru (Bazan de Segura 1951)
Areca catechu L.	Areca palm	Foliage blight	United States (Nagata and Aragaki 1989)
Artocarpus integrifolia (= *A. heterophylus* Lam.)	Jack fruit	Fruit rot	India (V. G. Rao 1965, 1970)
Asmina triloba (L.) Dun.	Common pawpaw	Root rot, leaf fall	Malaysia (Anon. 1971a)
Atropa belladonna L.	Deadly nightshade	Root, crown, and leaf rot	United States (Middleton 1943b)
Avicennia marina (Fors. K.) Vierh.	Avicennia	Root rot	Australia (Weste et al. 1982)
Banksia dryandroides Baxt. and Sweet.	Banksia	Root rot	Australia (Hardy and Sivasithamparam 1988)
Banksia hookerana Meissn.	Banksia	Root rot	Australia (Hardy and Sivasithamparam 1988)
Banksia laricina Gardn.	Banksia	Root rot	Australia (Hardy and Sivasithamparam 1988)
Banksia occidentalis R. Br.	Banksia	Root rot	Australia (Hardy and Sivasithamparam 1988)
Banksia praemorsa Andrews	Banksia	Root rot	Australia (Hardy and Sivasithamparam 1988)
Banksia prionotes Lind.	Banksia	Root rot	Australia (Hardy and Sivasithamparam 1988)
Basella rubra L.	Malobar spinach	Leaf rot	India (Saksena and Mehrotra 1970)[b]
Bauera sissiflora F. Muell.		Root rot	Australia (Hardy and Sivasithamparam 1988)
Beaufortia eriocaphala W. V. Fitzg.		Root rot	Australia (Hardy and Sivasithamparam 1988)
Begonia L.	Mother-of-pearl	Leaf necrosis, stem rot	Argentina (Frezzi 1950); United Kingdom (Moore 1959)
Begonia L. spp.	Begonia	Collar rot	Malaysia (Liu 1977)
Bougainvillea spp.	Bougainvillea	Blight	United States (Alfieri 1970)
Brachychiton diversifolius R. Br.	Bottle tree	Root, stem necrosis	Argentina (Frezzi 1950)
Brassaia actinophylla Endl. (*Schefflera actinophylla* Harms.)	Australian umbrella tree	Leaf spot	United States (Wisler et al. 1978)
Brassica juncea (L.) Czerniak	Brown mustard	Root rot	Philippines (Clara 1928)[b]
Brassica rapa L.	Turnip	Black root rot	United States (Godfrey 1923)[b]
Bryophyllum pinnatum (Lam.) Kurz.	Kalanchoe 'Brujas'	Stem, leaf rot	Puerto Rico (Tucker 1927); Bermuda (Tucker 1933); India (V. G. Rao et al. 1962b)[b]
Buxus sempervirens L. var. *arborescens* and var. *suffruticosa*	Boxwood	Root rot, stem necrosis	United States (Andrus 1933)
Cactus species	Cactus	Brown leaf rot	Puerto Rico (Matz 1920); India (V. G. Rao et al. 1962b)[b]

(*continued on next page*)

Table 50.2. (*continued*)

Host	Common name	Disease	Geographical distribution
Cajanus cajan (L.) Millsp.	Pigeon pea	Stem canker	Hawaii (Merlich 1936)[b]; Puerto Rico (Kaiser and Melendez 1978)
Caladium colocasia (L.) Wight	Dasheen, taro	Corm rot	India (Dastur 1916)
Callistephus chinensis (L.) Nees.	China aster	Root rot, wilt	Argentina (Frezzi 1950); United Kingdom (Moore 1959)
Calluna vulgaris Hull.	Heather	Root rot	Australia (Hardy and Sivasithamparam 1988)
Calothamnus gracilis R. Br.	Netbrush, one-sided bottlebrush	Root rot	Australia (Hardy and Sivasithamparam 1988)
Capsicum annuum L.	Red or chili pepper	Foot rot, fruit drop, seedling blight	United States (Kendrick 1923); Puerto Rico (Nolla 1929)[b]; Japan (Tasugi and Ikeda 1939); Italy (Borzini 1956); Mauritius (Anon. 1962a); India (V. G. Rao 1970)
Caralluma diffusa Wight	Succulent herb	Root rot	India (Agharkar and Bannerjee 1932)
Carica papaya L.	Papaya, pawpaw	Stem girdling, fruit and root rot	Mauritius (Tucker 1933); Java (Thung 1938); Malaysia (A. Thompson 1940; Liu 1977); Hawaii (Parris 1942); East Africa (Wallace and Wallace 1949); Brazil (Campacci 1951); India (V. G. Rao et al. 1962b)[b]; Papua New Guinea (Trujillo 1971); Fiji (Firman 1975)
Carica sp.	Paw paw	Stem blight	United States (Tucker 1931)[b]
Carnegiea spp.		Root rot	Japan (Katsura 1971)
Carthamus tinctorius L.	Safflower, false saffron	Root rot	Argentina (Frezzi 1950); United States (Klisiewicz 1970)
Cassia obtussifolia L.		Root rot	Japan (Katsura 1971)
Catharanthus roseus (L.) G. Don (see also *Vinca rosea* L.)	Vinca	Foliage blight	United States (Gill et al. 1977)
Catharanthus roseus (L.) Don. (*Vinca rosea* L.)	Rose periwinkle, old maid	Leaf, stem, fruit, and flower blight	India (Dastur 1916); Java (Schwarz 1926); Argentina (Blanchard 1930); Venezuela (Malaguti 1951); Japan (Katsura 1971); New Calendonia (Firman 1975); United States (Gill et al. 1977; Keim 1977)
Cattleya sp.	Orchid	Black rot	Taiwan (Chen and Hsieh 1978); United States (Uchida and Aragaki 1991)
Cedrus deodara (D. Don) G. Don	Deodar cedar	Root rot	China (Ho et al. 1984d)
Ceiba pentandra (L.) Gaertn. (*Eriodendron anfractuosum* DC)	Silk-cotton or Kapok tree	Leaf necrosis	Java (Thung 1938)[b]; Italy (Lio and Davino 1977)
Celosia cristata L.	Cockscomb	Flower, leaf, and root rot	Japan (Katsura 1970); Malaysia (Liu 1977)
Celosia plumosa Hort. ex Burvenich	Woolflower	Root rot, stem necrosis	Argentina (Frezzi 1950)
Cereus spp.	Cactus	Root rot	Japan (Katsura 1971)
Chamaelaucium uncinatum 'Alba'		Root rot	Australia (Hardy and Sivasithamparam 1988)
Chamaelaucium uncinatum Schau.		Root rot	Australia (Hardy and Sivasithamparam 1988)
Chrysalidocarpus lutescens (Bory) H. Wendl.	Areca or golden-fruited palm	Leaf blight	United States (Nagata and Aragaki 1989)
Chrysalidocarpus lutescens H. Wendl.	Yellow palm, areca palm	Leaf blight	India (Dastur 1935)[b]
Cinchona pubescens Vahl.	Quinine tree	Wilt stem canker, seedling blight	Puerto Rico, Colombia, Peru, Guatemala (Crandall and Davis 1945; Darley and Flores 1951); French Guinea (Chevaugeon and Merny 1956)
Citrullus citrullus (L.) Karst.	Watermelon	Fruit rot	United States (Sherbakoff 1917)[b]
Citrus aurantiifolia (Christm.) Swingle	Lime	Fruit and foot rot, mal di gomma	Sierra Leone (Deighton 1933); United States (H. S. Fawcett 1936); Malaysia (Liu 1977)
Citrus aurantium L.	Sour, bitter, or Seville orange	Top wilt of seedlings, foot rot, gummosis	Trinidad (Baker 1934); United States (H. S. Fawcett 1915, 1936); Argentina (Frezzi 1950); Venezuela (Malaguti 1951)
Citrus bergamia	Bergamot		Italy (Magnano di San Lio and Davino 1977)
Citrus limon (L.) Burm.	Lemon	Foliar wilt, stem, fruit, root rot, mal di gomma	United States (H. S. Fawcett 1915, 1936); Paraguay (Fawcett and Bitancourt 1940); Argentina (Frezzi 1950); Venezuela (Malaguti 1951); Greece (Kouyeas and Chitzanidis 1978)
Citrus maxima (Burm.) Merrill.	Shaddock pommelo	Foot rot	United States (H. S. Fawcett 1936); Philippines (Lee 1925)
Citrus medica L.	Citron	Foot rot, collar canker, mal di gomma	United States (H. S. Fawcett 1915, 1936); Puerto Rico (Matz 1920); Corsica (Dufrenoy 1925); India (V. G. Rao et al. 1962b)
Citrus nobilis L.	Mandarin	Foot rot	Argentina (Frezzi 1950); Malaysia (Liu 1977)
Citrus paradisi Macfady	Grapefruit	Foliar wilt, stem and foot rot, mal di gomma	Trinidad (Baker 1934); Venezuela (Malaguti 1951); United States (H. S. Fawcett 1915, 1936; Timmer 1973)
Citrus reticulata Blanco.	Satsuma orange, tangerine	Leaf fall disease	Venezuela (Malaguti 1951); India (Kumbhare and Moghe 1976)

(*continued on next page*)

Table 50.2. *(continued)*

Host	Common name	Disease	Geographical distribution
Citrus sinensis (L.) Osbeck	Sweet orange	Foliar wilt, foot and fruit rot, mal di gomma, dieback	United States (H. S. Fawcett 1915, 1936; Sherbakoff 1917; Klotz et al. 1958); Philippines (Lee 1922); Portugal (Moniz de Maia 1925); Malaysia (Liu 1977); Java (Toxopeus 1930); Argentina (Blanchard 1930); Venezuela (Malaguti 1951); Papua New Guinea (Trujillo 1971); Taiwan (Ann 1984); Fiji (Firman 1975)
Citrus spp.	Citrus	Foot and root rot, brown rot of fruit, gummosis	Tunisia (Chabrolin 1932); El Salvador (Crandall 1948); Argentina (G. L. Fawcett 1916; Condado et al. 1949); Papua New Guinea (Dumbleton 1954); Australia (Simmonds (1955); West Indies (Burke 1956); Morocco (Moreau and Moreau 1960); Venezuela (Knorr et al. 1964); Guatemala (Sanchez 1965); India (Saksena and Mehrotra 1970)[b]; Malaysia (Anon. 1971a); Iran (Ershad 1971); Japan (Katsura 1971); Greece (Kouyeas and Chitzanidis 1978); Thailand (Suzui et al. 1979); India (Lele and Kapoor 1980); Portugal (Telhada 1990)
Clarkia elegans Doug.	Annual herbs	Seedling blight	India (Dastur 1913)[b]; Malagasy Republic (Bouriquet 1950)
Cocos nucifera L.	Coconut	Bud rot, leaf and stalk rot	Philippines (Reinking 1918); Jamaica (Ashby 1920)
Coleus barbatus	Coleus	Blight	India (Parakhia et al. 1986)
Colocasia esculenta (L.) Schott	Taro	Leaf rot	India (Dastur 1916); Fiji (Graham 1971)
Columnea gloriosa T. Sprague	Gesneriad	Root and collar rot	United States (Ploetz and Englehard 1980)
Commelina nudiflora L.	Creeping day flower		Sumatra (Palm 1923); Bulgaria (Atansoff and Kovacevski 1929)
Cordyline terminalis Kunth.	Dracena palm, ti	Leaf spot	Hawaii (Trujillo et al. 1975)
Cotyledon sp. L.	Cotyledon	Purple rot	South Africa (Wager 1931); Zimbabwe (Hopkins 1931b, 1932)
Crotalaria incana L.	Rattle box	Seedling blight	Hawaii (Mehrlich 1936)
Crowea saligna Andr.		Root rot	Australia (Hardy and Sivasithamparam 1988)
Cryptomeria japonica (L.)	Japanese cedar	Apical necrosis	Argentina (Frezzi 1950)
Cucumis melo var. *inodorus* Naud.	Honeydew melon	Fruit rot	United States (Drechsler 1929a); United Kingdom (Williams 1927; Moore 1959); Greece (Kouyeas and Chitzanidis 1978)
Cucumis sativus L.	Cucumber	Friut rot	India (V. G. Rao et al. 1962b)[b]
Cyphomandra betacea (Cav.) Sendtn.	Tree tomato	Root rot	Gold Coast (Dade and Wright 1931)
Daphne odora Thunb.	Winter daphne	Stem and root rot	United States (Tompkins and Tucker 1951)
Datura inoxia Mill.	Indian apple	Leaf blight	India (Sattar et al. 1982)
Daucua carota L.	Carrot	Soft rot of roots	United States (Godfrey 1923)[b]
Dedonea speciosa Meissn.		Root rot	Australia (Hardy and Sivasithamparam 1988)
Dendrobium spp. Schwarz	Orchid	Wilt, foliage blight	Mauritius (Wiehe 1948); United States (Uchida and Aragaki 1991)
Dianthus caryophyllus L.	Carnation	Root rot	Argentina (Frezzi 1950); Greece (Sarejanni 1952); United States (Hine and Aragaki 1962); France (Tramier and Mercier 1965); Iran (Ershad 1971); Italy (Garibaldi and Rapetti 1975); Netherlands (Rattink 1979, 1981)
Dianthus chinenis L.	Rainbow pink	Root rot	Argentina (Frezzi 1950)
Dianthus spp.	Carnation	Collar rot	Greece (Sarejanni 1952); United States (Hine and Aragaki 1962)
Dolichos lablab L.	Hyacinth, Indian or Egyptian bean	Storage rot	India (Rao 1965)
Durio zibethinus Murr.	Durian tree	Patch canker, slow decline	Malaysia (Navarathnam 1966; Liu 1977); Thailand (Suzui et al. 1979)
Echinocactus spp.	Echinocactus	Root rot	Japan (Katsura 1971)
Elettaria cardamomum (L.) Maton	Cardamom	Leaf blight, fruit and leaf rot, 'Azhukal'	India (Menon et al. 1972; Thankamma and Pillai 1973)
Epidendron sp.	Orchid	Root rot, foliage blight	United States (Uchida and Aragaki 1991)
Epiphyllum spp.	Epiphyllum	Root rot	Japan (Tanaka 1917)
Episcia cupreata (Hook.) Hanst.	Flame violet	Root and collar rot	United States (Ploetz and Engelhard 1980)
Eriobotrya japonica (Thunb.) Lindl.	Loquat tree, Japanese plum	Trunk rot	Corsica (Dufrenoy 1927); Italy (Cristinzio 1935); Lebanon (Weltzien and Schwinn 1966)
Eriodendron anfractuosum DC	Kapok tree	Leaf blight	Java (Thung 1938)[b]
Eucalyptus citridora Hook.	Lemon scented gum	Gummosis	Brazil (Arruda 1943)
Eucalyptus delagatensus R. T. Bak.	Alpine ash	Foliage blight	Italy (Belisario 1990)
Eucalyptus globulus Labill.	Blue gum	Foliage blight	Italy (Belisario 1990)
Eucalyptus regnans	Mountain ash, giant gum	Foliage blight	Italy (Belisario 1990)

(continued on next page)

Table 50.2. (*continued*)

Host	Common name	Disease	Geographical distribution
Eucalyptus viminalis Labill.	Manna gum	Apical necrosis	Argentina (Frezzi 1950)
Euphorbia pulcherrima Willd. ex Klotzsch.	Poinsettia	Flower, bract, and leaf blight, root rot	United States (Uchida and Aragaki 1979; Engelhard and Ploetz 1979)
Euryclis amboinsis		Leaf blight	India (Dastur 1935)[b]
Eutaxia oborata Gardn.		Root rot	Australia (Hardy and Sivasithamparam 1988)
Fagophrum esculentum Moench.	Buckwheat	Seedling blight	India (Dastur 1913)[b]
Ficus carica L.	Fig	Leaf, shoot, and fruit blight	India (V. G. Rao et al. 1962b)[b]; Venezuela (Rondon 1975)
Forsythia viridissima	Forsythia	Root and collar rot	Italy (Cacciola et al. 1994)
Fortunella spp. Swingle	Kumquat	Root rot	Japan (Katsura 1971)
Fragaria chiloensis ananassa L.	Strawberry	Root and crown rot	Argentina (Frezzi 1950); Australia (McCarty and Teakle 1959); Japan (Suzui et al. 1980)
Fremontodendron californicum (Torr.) Cov. (*Fremontia california* Torr.)	Flannel bush	Stem canker, crown and fruit rot	United States (Zentmyer et al. 1979)
Fuchia species L.	Lady's eardrops	Root rot	Argentina (Frezzi 1950); United States (Patil and Young 1960)
Gastrolobium calycinum Benth.		Root rot	Australia (Hardy and Sivasithamparam 1988)
Gerbera species L.	Daisy	Collar and root rot	Greece (Kouyeas and Chitzanidis 1978)
Gerbera jamesonii Bolus ex Hook.	Transvaal or barberton daisy	Root rot	Japan (Katsura 1971)
Gloxinia species L'Her	Canterbury bells	Stem and leaf rot	Austria (Vukovits 1956); Greece (Kouyeas and Chitzanidis 1978); United States (Ploetz and Engelhard 1980)
Godetia spp.	Godetia	Foot rot	Rhodesia (Hopkins 1935)
Gossypium barbadense L.	Sea Island cotton	Boll rot	West Indies and Puerto Rico (Tucker 1931)
Gossypium hirsutum L.	Upland cotton	Boll rot, seedling damping-off	United States (Tucker 1931)[b];India (V. G. Rao et al. 1962b)[b]; United States (Pinckard and Guidroz 1973); Thailand (Kaosiri and Siddhipongse 1985)
Gossypium species	Cotton	Seedling damping-off, boll rot	West Indies (Hopkins 1925)[c]; India (Mitra 1929a)
Grammatophyllum speciosum Blume	Orchid	Heart and leaf rot	Java (Schwarz 1927)
Grevillea biternala Meissn.		Root rot	Australia (Hardy and Sivasithamparam 1988)
Grevillea drummondii Meissn.		Root rot	Australia (Hardy and Sivasithamparam 1988)
Grevillea thelemanniana Endl.		Root rot	Australia (Hardy and Sivasithamparam 1988)
Gypsophila paniculata L.	Baby's breath	Decay of stems and crowns	United States (Engelhard 1974a, b)
Hakea francisiana F. Muell.		Root rot	Australia (Hardy and Sivasithamparam 1988)
Hakea laurina R. Br.	Sea uchin tree	Root rot	Australia (Hardy and Sivasithamparam 1988)
Hakea platysperma Hook.		Root rot	Australia (Hardy and Sivasithamparam 1988)
Hebe speciosa 'La Sequisante'		Root rot	Australia (Hardy and Sivasithamparam 1988)
Hebe speciosa Lock. and Al. 'Buxifolia'		Root rot	Australia (Hardy and Sivasithamparam 1988)
Hedera helix L.	Ivy	Leaf necrosis	Argentina (Frezzi 1950); Greece (Kouyeas and Chitzanidis 1978)
Helianthus annus L.	Sunflower	Root rot	Egypt (Tarabeih 1971)
Heliotropium arborescens L. (*H. peruvianum* L.)	Heliotrope, cherry pie	Necrotic leaf lesions	Argentina (Frezzi 1950)
Hevea brasiliensis Muell. Arg.	Rubber	Leaf fall	India (Edathil and George 1976)
Hibiscus cannabilis L.	Kenaf, Java jute	Collar rot and wilt	Java (Muller and Van Eek 1939); Cuba (Crandall et al. 1953); Puerto Rico (Alconero and Stone 1969)
Hibiscus esculentus L.	Okra	Root and fruit rot	India (V. G. Rao et al. 1962b); United States (Erwin 1964)
Hibiscus rosa-sinensis L.	Chinese hibiscus, China rose	Collar rot	West Australia (MacNish 1963)
Hibiscus sabdariffa L. var. *altissima*	Roselle	Collar, root, and basal stem rot, wilt	Philippines (Reinking 1918); Malaysia (Thomson 1929; Navaratnam 1970); Java (Muller and Van Eek 1939); Central African Republic (Follin et al. 1966); Puerto Rico (Alconero and Stone 1969); Ivory Coast (Boccas and Pellegrin 1976)
Hypocalyma angustifolium Endl.		Root rot	Australia (Hardy and Sivasithamparam 1988)
Hypocalyma tetrapterum Turcz.		Root rot	Australia (Hardy and Sivasithamparam 1988)
Impatiens balsamina L.	Touch-me-not, jewel weed	Flower blight	Argentina (Frezzi 1950)
Ipomoea batatas (L.) Lam.	Sweet potato	Spongy root rot	United States (Tisdale and Kelley 1926)
Iris sp.	Iris	Leaf blight	India (Dastur 1935)[b]
Jatropha curcas L.	Barbados or physic nut	Leaf blight, fruit rot	Malagasy Republic (Bouriquet 1933)[b]; Java, India (Thung 1938[b]; Saksena and Mehrotra 1970)[b]; Malaysia (Liu 1977)
Juglans regia L.	English walnut	Root and crown rot	Australia (Cookson 1929); United States (Matheron and Mircetich 1985c)

(*continued on next page*)

Table 50.2. (*continued*)

Host	Common name	Disease	Geographical distribution
Kalanchoe pinnata Pers. (*Bryophyllum calycinum* Salisb.)	Succulent herb	Leaf and stem black rot	Puerto Rico (Tucker 1928)
Lactuca sativa L.	Lettuce	Leaf blight	Philippines (Ocfemia 1925)[b]
Laeliocattyleya Lindl. sp.	Orchid	Root and foliage blight	United States (Uchida and Aragaki 1991)
Lavendula angustifolia Mill.	Lavender	Root rot	United States (Putnam 1991)
Lavendula species L.	Lavender	Collar rot	Greece (Pappas 1978)
Leea coccinea Planch.	West Indian holly	Blight	France (Vegh and Berre 1985)
Leucoena glauca Benth.	Acacia	Wilt	Mauritius (Wiehe 1948)
Leucophyllum frutescens Johnston	Texas ranger	Crown rot	United States (Wheeler and Boyle 1971)
Lilium elegans L.	Lily	Root rot	Japan (Asuyama 1935)
Lilium longiflorum var. *maximum* Nickols	Bermuda easter lily	Stump rot	West Indies (Whetzel 1922); United States (Ogilvie 1924); Japan (Tasugi and Kumazawa 1938); United Kingdom (Moore 1959)
Lilium regale Wilson	Regal lily		United States (Wilson 1936)
Lilium speciosum Thunb.	Showy lily	Stump rot	United States (Guterman 1930)
Luffa acutangula (L.) Roxb.	Angled loofah, Sing-kwa, rag gourd	Fruit rot	Mauritius (Shepherd 1933)
Lupinus albus L.	White lupine	Stem rot	United States (Sherbakoff 1917); Italy (Peyronel 1920)
Lupinus luteus L.	White lupine	Foot rot	Italy (Petri 1932)
Lycopersicon esculentum L.	Tomato	Buckeye, brown, and root rot, stem girdling, seedling damping-off, seedling, leaf, and shoot blight	United States (Sherbakoff 1917; Kendrick 1923); Cuba (Bruner 1918); Ceylon (Petch 1920); England (Bewley 1920); Netherlands (van Poeteren 1925); Ireland (Pethybridge 1925); Puerto Rico (Nolla 1928[b]); Argentina (Blanchard 1930); South Africa (Wager 1935; Ferreira et al. 1991); Canada (Conners 1938); China (Wei and Cheo 1944); Rhodesia (Hopkins 1945); Malaysia (McIntosh 1951; Liu 1977); Venezuela (Malguti 1951); India (Singh and Srivastava 1953); Tasmania (Wade 1954a); Fiji and Samoa (K. M. Graham 1971); Iran (Ershad 1971); Japan (Katsura 1971); Germany (Skadow 1985)
Macadamia integrifolia Maiden and Betche	Macadamia nut	Foliage and fruit blight	United States (Hunter et al. 1971)
Magnolia grandiflora L.	Magnolia	Stem and apical necrosis	Argentina (Frezzi 1950)
Malus pumila Mill.	Apple	Fruit rot, stem canker	United States (Tucker 1931[b]; Matheron et al. 1988); India (Padwick 1938); Iran (Ershad 1971)
Mamillaria spp.		Root rot	Japan (Katsura 1971)
Manihot esculenta Crantz.	Cassava, manioc, tapioca	Leaf blight	Java (Thung 1938)[b]
Martinya annua L.		Leaf rot	India (Saksena and Mehrotra 1970)[b]
Martinya diandra Glox.		Seedling blight	India (Dastur 1935)[b]
Martynia sp. L.		Fruit rot	United States (Tucker 1931)[b]
Meconopsis spp. Vig.	Asiatic poppy	Damping-off	United Kingdom (Moore 1959)
Medicago sativa L.	Alfalfa, lucerne	Root rot	Australia (Purss 1959)
Melalevea hypercifolia Sm.		Root rot	Australia (Hardy and Sivasithamparam 1988)
Melalevea laterifola Benth.		Root rot	Australia (Hardy and Sivasithamparam 1988)
Melia azedarach L.	Chinaberry, paradise tree	Stem and apical necrosis	Argentina (Frezzi 1950)
Mentha cardifolia Opiz.	Kitchen mint	Wilt, foliar blight	Thailand (Kaosiri 1984)
Monstera deliciosa Liebm. (*Philodendron pertusum* Kunth. and Bouch)	Cutleaf philodendron	Leaf spot	United States (Ridings and McRitchie 1974)[b]
Musa × *paradisiaca*	Plantain	Fruit rot	India (V. G. Rao et al. 1962b)[b]
Musa sapientum Kuntze.	Banana	Pseudostem rot	West Indies (Ashby 1925); South Africa (Thompson 1981)
Myoporum parvifolium R. Br.		Crown rot	United States (Gill and Esparza 1977)
Myrtus communis L.	Myrtle	Seedling root rot	Italy (Belisario et al. 1991)
Nemesia species		Root rot	United Kingdom (Moore 1959)
Nicotiana rustica L.	Wild tobacco	Root rot	Iran (Ershad 1971)
Nicotiana tabacum L.	Tobacco	Black shank disease, lanas-ziekte, bibitzickte, pataprieta, faitapia	Sumatra and Java (Breda de Haan 1893; Lodewijks 1909); Cuba (Horne 1909); Philippines (Reinking 1918); Ceylon (Petch 1922); United States (Tisdale 1922); Puerto Rico (Cook 1924); Japan (Ito and Tokunaga 1935)
Oenothera spp.	Evening primrose	Stem and leaf spot	India (Dastur 1913)[b]
Opuntia spp.	Prickly pear cactus	Root rot	Japan (Katsura 1971)
Opuntia ficus-indica Mill.	Prickly pear cactus	Foot rot	Italy (Cacciola and Magnano di San Lio 1988)

(*continued on next page*)

Table 50.2. *(continued)*

Host	Common name	Disease	Geographical distribution
Paaphiopedilum sp.	Orchid	Root rot, foliage blight	United States (Uchida and Aragaki 1991)
Passiflora edulis Sims.	Passion flower	Wilt, crown rot defoliation, fruit drop, damping-off	Australia (Simmonds 1959); South Africa (Van den Boom and Huller 1970); Sarawak (G. J. Turner 1974); India (Ullasa and Sohi 1975); Taiwan (Hsieh 1979)
Paulownia fortunei (Seem.) Hance		Foot rot, leaf blight	Taiwan (Hsieh 1979)
Paulownia kawakamii Ito		Foot rot, leaf blight	Taiwan (Hseih 1979)
Paulownia tomentosa Steud.		Foot rot, leaf blight	Taiwan (Hseih 1979)
Peperomia arifolia Miq.	Peperomia	Leaf, petiole, and stem rot	India (Dastur 1935)
Peperomia species Ruiz and Pav.	Peperomia	Crown rot	United States (Sirdhana et al. 1968)
Persea americana Mill.	Avocado	Collar rot	United States (Stevens and Piper 1941); French Polynesia (Reddy 1973)
Petroselinum crispum (Mill.) Nym.	Parsley	Crown rot	United States (Hine and Aragaki 1963)
Petroselinum sativum Hoffm.	Parsley, persil	Foot rot and wilt	Mauritius (Wiehe 1948)
Petunia × *hybrida* Hort. Vilm.-Andr.	Petunia	Crown rot	India (Dastur 1916); United States (Phillips and Baker 1962)
Phalaenopsis Blume	Orchid	Black rot	Taiwan (Chen and Hsieh 1978)
Phaseolus lunatus L.	Butter or civet bean	Pod blight	Argentina (Frezzi 1950)
Phaseolus vulgaris L.	Garden bean	Wet pod rot, wilt, damping-off of seedlings	Puerto Rico (Stevenson 1917); India (Venkatakrishnaiya 1950; V. G. Rao et al. 1962b)[b]
Philodendron oxycardium	Heart-leaf philodendron	Leaf spot	United States (Ridings and McRitchie 1974)[b]
Philodendron panduraeforme (HBK) Kunth.	Fiddle-leaf philodendron	Leaf spot	United States (Ridings and McRitchie 1974)[b]
Philodendron selloum L. Koch	Philodendron	Leaf spot	United States (Ridings and McRitchie 1974)[b]
Philodendron scardens C. Koch and H. Sello	Philodendron	Leaf spot	United States (Ridings and McRitchie 1974)[b]
Phlox species L.	Phlox	Damping-off	United Kingdom (Moore 1959)
Pilea imperialis L.	Moon valley pilea	Leaf, bud, and stem blight	United States (McCoy et al. 1976)
Pimilea ferruginea Labill.		Root rot	Australia (Hardy and Sivasithamparam 1988)
Pinanga Labill. sp.	Palm	Foliage blight	United States (Nagata and Aragaki 1989)[b]
Pinus halepensis Mill.	Apleppo or Jerusalem pine	Root rot	Argentina (Frezzi 1950)
Pinus radiata D. Don	Monterey pine	Root rot	United States (Munnecke and Bricker 1976)
Pinus taiwanensis Hayata	Pine	Damping-off	Malaysia (Liu 1977)
Piper betle L.	Betel nut, pan	Root rot, leaf blight, wilt	India (Dastur 1927; McRae 1932); Burma (Tucker 1933); Bangladesh (Mridha and Alamgir 1991)
Piper betle var. *densum*	Betel vine	Foot rot	Thailand (Silayoi et al. 1983)[b]
Piper colubrinum		Foot rot	Thailand (Silayoi et al. 1983)[b]
Piper chaba		Foot rot	Thailand (Silayoi et al. 1983)[b]
Piper longum L.	Pepper	Leaf rot	India (Saksena and Mehrotra 1970)[b]
Piper nigrum L.	Black pepper	Fruit, crown, foot, and root rot	India (Dastur 1935; Saxena 1977; Thankamma 1983); Malaysia (Liu 1977); China (Ho et al. 1983b); Thailand (Silayoi et al. 1983)[b]; Philippines (Tsao et al. 1993)
Piper sarmentosum		Foot rot	Thailand (Silayai et al. 1983)[b]
Pisonia brunoniana 'Variegata' Endl.	Bird-catcher tree, para-para	Basal stalk rot	Germany (Sautoff 1965)
Pistacia vera L.	Pistachio	Foot rot, gummosis	Greece (Anon. 1964; Kouyeas 1973)
Pisum sativum L.	Garden pea	Pod rot, seedling damping-off	Puerto Rico (Matz 1920); Brazil (Muller 1934); Mauritius (Wiehe 1948); India (V. G. Rao et al. 1962b)[b]; Papua New Guinea (Trujillo 1971)
Pollia sp.			Sumatra (Ashby 1929a)
Polyalthia longifolia L.		Leaf spot	India (Mitra 1929a)
Poncirus species Raf.	Trifoliate orange	Root rot	Japan (Katsura 1971)
Primula species L.	Primrose	Collar rot	Australia (Brittlebank and Fish 1927); United Kingdom (Moore 1959)
Pritchardia thurstonii F.J. Meull. and Drude	Palm	Foliage blight	United States (Nagata and Aragaki 1989)[b]
Protea spp. L.		Stem and root rot	United States (J. M. Uchida and M. Aragaki, *personal communication*)
Prunus amygdalus Batsch. (*Prunus dulcis* (Mill.) Webb)	Almond	Collar and trunk rot	Greece (Helena 1978)
Prunus armeniaca L.	Apricot	Collar and trunk rot	Greece (Kouyeas and Chitzanidis 1978)
Prunus persica (L.) Batsch.	Peach	Fruit rot	India (Padwick 1938)
Psidium guajava L.	Guava	Fruit rot	India (Mitra 1929a)
Ptychosperma macarthurii (H. Wendl.) Nicholas	Palm	Foliage blight	United States (Nagata and Aragaki 1989)[b]
Pyrus communis L.	Pear	Fruit rot	Greece (Kouyeas and Chitzanidis 1978)
Punica granatum L.	Pomegranate	Fruit rot, leaf blight	Argentina (Frezzi 1950); India (Mishra et al. 1980; Thankamma 1983)

(continued on next page)

Table 50.2. *(continued)*

Host	Common name	Disease	Geographical distribution
Rheum emodi Wall.	Rhubarb	Root rot	China (Ho et al. 1983b)
Rheum rhaponticum L.	Rhubarb, pie plant	Crown and root rot, wilt	United States (Godfrey 1923); South Africa (Doidge 1924); Zimbabwe (Hopkins 1931a); Australia (Anon. 1943)
Rhododendron L. sp.	Rhododendron	Dieback	United States (Benson and Jones 1980)
Rhynchostylis Blume	Orchid	Black rot	Taiwan (Chen and Hsieh 1978)
Richardia scabra L. (*Richardsonia scabra* (L.) St. -Hil.)	Mexican clover	Seedling blight	United States (Mehrlich 1936)[b]
Ricinus communis L.	Castor bean, castor oil plant	Seedling blight	India (Dastur 1913); Sumatra (Palm 1923); Soviet Union (Tropova 1928); Puerto Rico (Nolla 1929)[b]; Argentina (Blanchard 1930); United States (Mehrlich 1936)[b]; Java (Thung 1938)[b]; Taiwan (Sawada 1942); Australia (Anon. 1953d); China (Liang 1964); Japan (Katsura 1971); Egypt (Tarabeih 1971)
Robinia pseudoacacia L.	Black locust	Pre-emergence damping-off	United States (Lambert and Crandall 1936)
Rosmarinus officinalis L.	Rosemary	Collar rot	United States (Wheeler and Boyle 1971)
Saintpaulia ionantha Herm. Wendl.	African violet	Crown and root rot	Germany (Kröber and Plate 1973); United States (Raabe and Havlimann 1973; Chase 1981); Canada (Busch and Smith 1978)
Salpiglossis sinuata Ruiz. and Pav.	Painted tongue	Seedling blight	India (Dastur 1913)[b]
Salvia officinalis L.	Garden salvia, condiment sage	Root rot	United States (Middleton 1947); Japan (Katsura 1971)
Salvia splendens Sellow ex Roem and Schult.	Scarlet salvia	Flower blight	Argentina (Frezzi 1950); Japan (Katsura 1971)
Sauropus androgynus Merr.		Collar rot	Malaysia (Liu 1977)
Schinus molle L.	Pepper tree, pirul, mastic tree	Bud and flower necrosis	Argentina (Frezzi 1950)
Schizanthus retusus Hook.	Butterfly flower	Seedling blight	India (Dastur 1913)[b]
Secium edule (Jacq.) Schwartz.	Chayote	Fruit rot	Mauritius (Anon. 1963c)
Senecio species L.	Cineraria; groundsel	Foot rot	United Kingdom (Moore 1959)
Sequoidendron giganteum (Lindl.) Buchholz	Giant redwood	Foliar blight	United States (Sandlin and Ferrin 1993)[b]
Sesamum indicum L. (*S. orientale* L.)	Sesame	Leaf, stem, and root disease, seedling damping-off	India (Butler 1918; Kale and Prasad 1959); Puerto Rico (Cifferi 1930); Argentina (Frezzi 1950); Venezuela (Malaguti 1953); Iran (Ershad 1971)
Sesamum japonicum Nakai		Root rot	Japan (Katsura 1971)
Sida cordifolia		Flower blight	India (Dastur 1913)
Simmondsia chinensis (Link.) Schneid.	Jojoba, goat nut	Seedling and leaf blight	United States (Stanghellini 1978; Alcorn et al. 1989)
Sinningia speciosa (Lodd.) Hiern.	Gloxinia	Crown and root rot	United Kingdom (Anon. 1936); Canada (Busch and Smith 1978)
Solanum capsicastrum Link. ex Schauer	False Jerusalem cherry	Wilt	United Kingdom (Orchard 1935)

(continued on next page)

We do not advocate maintenance of the two varieties, and we list the distribution of hosts as they were reported for *P. parasitica* or for each variety of *P. nicotianae* (Table 50.2).

Tucker's (1931) designation of *P. parasitica* var. *nicotianae* for tobacco isolates was based only on specific pathogenicity of these isolates to tobacco. He agreed with Tisdale (1922) that *P. parasitica* isolates from several hosts were similar to those reported to cause black shank of tobacco (*P. nicotianae;* Breda de Haan 1896). Tucker's choice of *P. parasitica* var. *nicotianae* for tobacco isolates was unfortunate and in error because he moved the formerly used species name *nicotianae* to the lower rank of variety; also, host-specific pathogenicity should not have been used as a basis for a variety, because variety status should be based on morphological but not on pathological grounds (International Code of Botanical Nomenclature [Voss et al. 1983]). Although tobacco pathologists in the United States designate the black shank pathogen *P. parasitica* var. *nicotianae,* Lucas (1975) in a discussion of the nomenclature problem used the name *P. parasitica* f. *nicotianae* rather than var. *nicotianae.* If he meant f. sp. (forma specialis), this would be more correct than the term "var." Erselius and de Vallavielle (1984) used the name *P. parasitica* f. sp. *nicotianae* for tobacco isolates. Such a designation may be useful for communicative purposes; however, Ho and Jong (1989a) and Brasier (1983) express reservations.

Lucas (1975) states that tobacco is the only natural host plant of *P. parasitica* var. *nicotianae sensu* Tucker (1931) and in support of this concept states that pepper, tomato, eggplant, castor bean, and potato have all been grown without loss in fields in North Carolina in which the black shank pathogen had occurred on tobacco. Although artificial inoculation of many different hosts with tobacco isolates induced disease in a large number of hosts other than tobacco (such as apple, tomato, or eggplant fruits, cotton bolls, potato tubers, stems of

Table 50.2. *(continued)*

Host	Common name	Disease	Geographical distribution
Solanum melongena L.	Eggplant, brinjal, aubergine	Fruit, branch, and collar rot, seedling blight and damping-off	Formosa (Sawada 1915); Philippines (Ocfemia 1925); United States (Kendrick 1923; Tucker 1931[b]); West Indies (Ashby 1925); Puerto Rico (Nolla 1928, 1929[b]); Brazil (Muller 1936); India (Padwick 1938; V. G. Rao et al. 1962b[b]; Jain et al. 1982); Java (Reitsma and Sloof 1947); Mauritius (Wiehe 1948); Greece (Sarejanni 1952); Papua New Guinea (Trujillo 1971); Iran (Ershad 1971); China (Ho et al. 1983b)
Solanum pseudocapsicum L.	Jerusalem cherry	Root rot	China (Ho et al. 1984b)
Solanum tuberosum L.	Potato	Tuber rot, leaf and stem blight	India (Dastur 1913a); United States (Tucker 1931[b]; Mehrlich 1934); Japan (Tasugi and Ikeda 1939); Venezuela (Malaguti 1951); Iran (Ershad 1971)
Sollya heterophylla Lindl.		Root rot	Australia (Hardy and Sivasithamparam 1988)
Spathiphyllum Schott. sp.	Spathiphyllum	Leaf blight	United States (Uchida 1989)
Stephania sinica Diels.		Root rot	China (Ho et al. 1983b)
Striga asiatica Kuntze	Herbe teu	Collar rot	Mauritius (Wiehe 1948)
Syringa vulgaris L.	Lilac	Stem blight, leaf spots	Italy (Scurti 1962)
Tephrosia vogelii Hook.	Hoary pea	Wilt	Puerto Rico (Ruppel and Barnes 1965)
Theobroma cacao L.	Cacao	Pod rot	Malaysia (Liu 1977)
Thrinax sp. Swartz.	Palm	Foliage blight	United States (Nagata and Aragaki 1989)[b]
Trema ambionensis Blume			Sumatra (Palm 1923)
Trifolium vesiculosum Savi	Arrow leaf clover	Root rot	United States (Pratt 1981)
Vanda amonie	Orchid	Crown rot	Malaysia (Liu 1977)
Vanda sp.	Orchid	Root rot and foliage blight	United States (Uchida and Aragaki 1991)
Vanilla planifolia Andr.	Vanilla	Black disease blast, pod rot	Reunion (Maublanc and Barat 1928); Malagasy Republic (Bouriquet 1933); Puerto Rico (Cibes and Childers 1949)
Verbena hybrida Voss.	Garden verbena	Wilt	United Kingdom (P. H. Williams et al. 1937)
Vicia faba L.	Broad bean, English bean	Root rot	Japan (Katsura 1971)
Vigna unguiculata (L.) Walp. (*V. sinensis* (Torner) Savi.)	Cowpea, black-eyed pea	Root rot	Java (Schwarz 1931)
Vinca rosea L.	Periwinkle	Blight	India (Dastur 1916); Argentina (Blanchard 1930)
Viola species	Pansy	Root rot	United Kingdom (Moore 1959)
Vitis species	Grape	Seedling damping-off	India (Sridhar et al. 1975); South Africa (Marais 1980)
Washingtonia filifera (Lind.) Wendl.	Desert fan palm, petticoat palm	Basal rot	United States (Darley and Wilbur 1953); Iran (Ershad 1971)
Washingtonia spp.	Palms	Bud rot	Iran (Ershad 1971); United States (Wheeler and Boyle 1971)
Westringia rosemariniformis Smith		Root rot	Australia (Hardy and Sivasithamparam 1988)
Wickstroemia indica Mey	Herbe tourlerelle	Wilt	Mauritius (Wiehe 1941)
Zea mays	Corn	Root and stalk rot	United States (Ulstrup and Tucker 1948)
Zygocactus truncatus Schum.	Christmas cactus	Basal stem and root rot	United States (Alfieri and Miller 1971)

[a] The varieties *nicotianae* and *parasitica* of *P. nicotianae* are no longer recognized (G. Hall 1993). Note: Tucker (1933) lists many more references.
[b] Artificially inoculated hosts.
[c] Although Hopkins (1925) was the first to identify *P. parasitica* as the causal agent of boll rot of cotton in the West Indies, Robson (1912) was the first to mention a *Phytophthora* sp. in association with boll rot in the West Indies.

pawpaw [*Carica* spp.], and eggplant seedlings), the tobacco isolates were only slightly virulent on stems of castor bean, the host on which *P. parasitica* was originally described by Dastur (1913), cacao pods (*Theobroma cacao* L.), and tomato seedlings. The roots of these plants were resistant (Tucker 1931). Although G. Hall's (1993) report did not show conclusive evidence for specificity to tobacco via a laboratory seedling test, Lucas's (1975) field data may be more definitive.

Isolates morphologically identified as *P. parasitica* (*sensu* Dastur 1913) or *P. nicotianae* (*sensu* Waterhouse 1963) are pathogenic to a wide range of genera and plant families (Table 50.2). Since considerable evidence supports host preference by some isolates, any one isolate should not be considered to be pathogenic to all the plants listed in Table 50.2. For instance, an isolate of *P. parasitica* from okra was not pathogenic to citrus and vice versa (Erwin 1964). Isolates from sesame (*Sesamum orientale* L.) were specific to sesame (Kale and Prasad 1957; Gemawat and Prasad 1964). Isolates identified as *P. nicotianae* var. *nicotianae* (not from tobacco) from *Peperomia obtusifolia, Saintpaulia ionantha,* and a *Sinningia* hybrid had different host ranges, but none were host specific (Rattink 1981). *P. parasitica* isolates from citrus in France produced necrotic symptoms on citrus leaves, but isolates from carnation and tomato did not. Isolates from carnation were more virulent on roots of carnation than were other isolates, and isolates from tomato were more virulent on tomato than on other hosts. The degree of susceptibility among hosts was not clear, but the greatest degree of virulence was shown by isolates on their own hosts (Bonnet et al. 1978). Isolates from citrus (*Citrus* spp.) were more virulent on roots of rough lemon than isolates from petunia (*Petunia hybrida* Vilm.), tomato (*Lycopersicon esculentum* Mill.), walnut (*Juglans hindsii* (Jeps.) Jeps.), silk tree (*Albizia julibrissin* (Durazz.) Schneider), jojoba (*Simmondsia chinensis* (Link.) Schneider), hibiscus (*Hibiscus rosa-sinensis* L.), and peach (*Prunus persica* (L.) Batsch). Tomato, however, was highly susceptible to many isolates, including citrus isolates (Matheron and Matejka 1990). Thus, it cannot be presumed that any one isolate would attack all hosts listed in Table 50.2.

The redescription of *P. nicotianae* is reproduced by permission of the British Mycological Society, courtesy of G. Hall (1993):

Neotypus of *Phytophthora nicotianae* van Breda de Haan; from *Simmondsia chinensis,* California, United States of America, 1977, collector unknown. IMI 335174. Living cultures are preserved at IMI [International Mycological Institute, Egham, United Kingdom] and the University of California at Riverside (P1098).

Mycelium. Colonies dense or loose rosette, or with no pattern, and dome-shaped or low, spreading, aerial mycelium. Radial growth rate on V8 agar at 25°C (3)–6.0 ± 1.57–(9.5) mm/day. Radiating hyphal swellings with walls continuous with the somatic mycelium forming within 5–14 days in water in 75% of cultures. Growth or stasis at 5°C, good growth at 35°, *ca.* 50% of cultures showing stasis or growth at 40°. Sporangia. Produced abundantly in liquid culture in light or dark, very rarely and then very weakly on solid agar in light or dark; predominantly spherical, ovoid, obturbinate or ellipsoid with a prominent papilla, (14)–43 ± 8.8–(74) × 12–36 ± 7.0–(60) μm (n = 2720), length:breadth ratio (1.12:1)–1.20:1 ± 0.050–(1.33:1); spherical sporangia often with 2 (very occasionally 3) papillae. In water, sporangia usually borne singly or in very loose sympodia of 2–4 sporangia; seldom in clusters, often laterally attached to sporangiophore; (0)–2% ± 2.8–(11%) caducous with a short (5 μm) pedicel. Chlamydospores. Present in *ca.* 50% of cultures, spherical, (22)–33 ± 5.5–(52) μm diam. (n = 700), forming within 5–14 days, wall 3–4 μm thick with irregular subcellular inclusions, delimited from somatic mycelium by a septum. Oogonia. Spherical, (16)–26 ± 3.2–(36) μm diam. (n = 1720), wall 1–2 μm thick, homogeneous contents, delimited from somatic hypha by a septum. Antheridia. Always amphigynous, cylindrical, 10–12 × 9–10 μm, unicellular. Oospores. Spherical, pale yellow when mature, (12)–24 ± 3.2–(34) μm diam. (n = 1710); aplerotic index (61)–77% ± 8.0–(95)%; oospore wall 2 μm thick, wall index (35)–42% ± 2.9–(50)%; containing irregularly shaped subcellular inclusions.

Characteristics of *Phytophthora nicotianae* (= *P. parasitica*)

P. nicotianae is classified in group II (Stamps et al. 1990) and redescribed by G. Hall (1993). See Tables 4.2 and 4.3 for tabular keys, Appendix 4.9 for a dichotomous key (Ho 1992), Tables 51A.4 and 51A.5 for morphological comparison with *P. palmivora* and related species, and Waterhouse and Waterston (1964d, e) for Commonwealth Mycological Institute descriptions of *P. nicotianae.*

Sporangia

Shapes of sporangia vary from ellipsoid, ovoid, pyriform, obpyriform, to spherical with a prominent papilla (see Figure 4.12C); occasionally two papillae occur on a single sporangium. Sporangia are not caducous. Papillate sporangia are produced singly or in a loose sympodium on long stalks 100 to 595 μm in length (average 375 μm) in some isolates (Thomson and Hine 1972); sporangia are 11 to 60 μm long × 20 to 45 μm wide (average 40.18 × 28.53 μm) with a length-breadth ratio of 1.1 to 1.7 (average 1.34). G. Hall (1993) reports highly variable sizes for different isolates and a length-breadth ratio of 1.1 to 1.3. Other ratios include 1.3 (Godfrey 1923) and 1.4 (Ashby 1928; Waterhouse 1963; Feichtenberger 1980). Sporangiophores are irregularly or sympodially branched. Waterhouse (1974c) indicated that sporangia of the variety *parasitica* were caducous with short stalks (2.0 μm) but the sporangia of variety *nicotianae* were not; however, this was not confirmed by Al-Hediathy and Tsao (1979a, b) and Trichilo and Aragaki (1982). Sporangia from isolates labeled either variety were persistent on the stalk (noncaducous). G. Hall (1993) noted that some sporangia broke off sporangiophores; but there was no septum at the breakage point, and thus they concluded that sporangia are noncaducous.

A study of the morphology of an isolate from crepe ginger, *Costus speciosus* (J. König) Sm., by electron and light microscopy revealed reticulate ridges on the surfaces of sporangia that had not been previously described (Khan et al. 1988).

Hyphal Swellings

Hyphal swellings with thin walls were noted by G. Hall (1993).

Chlamydospores

Chlamydospores are usually abundant; however, G. Hall (1993) reported that only 50% of 81 isolates produced them abundantly. Chlamydospores are terminal or intercalary and 13 to 60 μm in diameter, average 28 μm. G. Hall's (1993) neotype averaged 33 μm in diameter. Dastur (1913) reported a range of 20 to 60 μm.

Sex Organs

Most isolates are heterothallic, but some isolates form oogonia and oospores in single culture when inoculum is transferred from old cultures (Brasier 1972; Tsao et al. 1980). Isolates from vinca, *Catharanthus roseus,* were homothallic (Schubert and Leahy 1989). Oospores were found in stelar tissues of infected Washington Navel orange roots in Tulare County, California (Lutz and Menge 1991). Antheridia are amphigynous and spherical or oval; oogonia are smooth and spherical and range from 15 to 64 μm in diameter (average 26.8 μm). G. Hall (1993) reports 26 ± 3.2 μm. Oospores are aplerotic; Dastur (1913) reports a range of 13 to 24 μm. Ranges of 13 to 35 μm in diameter with an average of 22.6 μm (see Table 50.1) are common. G. Hall (1993) reports a diameter of 24 ± 3.2 μm.

Growth Temperatures

The minimum temperature for growth varies from 5 to 7°C with different isolates; the optimum is 27 to 32°C; and maximum is 37°C. G. Hall (1993) reports that most isolates stopped growth at 5 and 35°C, but about half the cultures were killed at 40°C. A few isolates grew at 40°C.

Growth Characteristics

Colonies of some isolates on potato-dextrose agar (PDA) are typically arachnoid (weblike; Figure 4.2), but on V8 juice agar, cultures are more fluffy. On PDA, a branch of mycelium progresses approximately 0.5 to 1.0 mm and appears to proliferate while another branch progresses radially and the process is repeated. Since petallate growth patterns do not occur on PDA, *P. parasitica* can be rapidly differentiated from *P. citrophthora,* both of which are pathogens of citrus, because the latter species produces a petallate colony on PDA. On cornmeal agar, growth types were labeled rosette (72.9%), stellate (11.1%), lanose (2.4%), and indeterminate (intermediate between rosette and lanose, 13.5%) by G. Hall (1993).

Some Diseases Caused by *Phytophthora nicotianae* (= *P. parasitica*)

Different isolates of *P. nicotianae* cause root disease of numerous crops (Table 50.2). Some isolates are specifically pathogenic to a single host, for instance, *P. parasitica* var. *nicotianae* to tobacco (Apple 1957, 1962; G. B. Lucas 1965, 1975; Shew and Lucas 1991). In some cases, blight of foliar or fruit tissue occurs, as exemplified by brown rot of citrus species (Fawcett 1915, 1936; Klotz 1978). The pathogen has a long history of persistence in soil and infected tissue, which is probably the result of the abundant production of chlamydospores. Kröber (1980) reported survival for up to 4- and 6-year periods in soil (see Table 5.3 for other references). In infected pepper fruits, *P. nicotianae* was viable in infected seeds for more than a year (Bhardwaj et al. 1987).

Root and crown rots are probably the most common manifestations of disease. However, flowers, fruits, and leaves are also infected; for instance, *P. parasitica* isolates from citrus cause fibrous root rot, crown rot gummosis, and brown rot of fruit (Frasier 1942; Klotz 1943, 1973; Klotz et al. 1958; Broadbent 1977b). Diseases on citrus trees are similar to those caused by *P. citrophthora* except that *P. parasitica* is active at higher temperatures (see Chapter 17 on *P. citrophthora*).

Black Shank of *Nicotiana tabacum* L. (Tobacco)

The causal agent of black shank of tobacco has been known as *P. parasitica* var. *nicotianae* since Tucker (1931) gave it varietal status because of its host preference for tobacco. As explained above, the variety epithets *parasitica* and *nicotianae* should be discontinued (G. Hall 1993) and the epithet *P. parasitica* var. *nicotianae* is not valid. The term "f. sp. *nicotianae*" may be better for communication purposes to indicate host preference for tobacco. Lucas (1975) referred to the black shank pathogen as *P. parasitica* f. *nicotianae,* Herr and Sutton (1984) as *P. parasitica* f. sp. *nicotianae,* and Erselius and de-Vallavielle (1984) as *P. nicotianae* f. sp. *nicotianae.* Ho and Jong (1989) express reservations about using the forma specialis designation because of several exceptions to host specificity expressed in the literature (Tisdale and Kelley 1926; Ashby 1928; Nolla 1928; Bell and Haasis 1967; Munnecke and Bricker 1976; Jain et al. 1982). Tobacco is considered to be the only natural host (Lucas 1975; Shew and Lucas 1991).

The black shank disease has gradually spread throughout the tobacco-growing regions of the world and is now considered to be a serious problem nearly every place that tobacco is grown. Black shank was first described on the North American continent in Florida in 1922 (Tisdale 1922; Tisdale and Kelley 1926) and is present on all tobacco farms of the southeastern United States where flue-cured tobacco is grown. In most of the southeastern United States, black shank is considered a potential hazard. Lucas (1975) comprehensively described the disease and estimated that the loss in North Carolina alone was 1% of the total crop (about $7 million). Black shank can cause a complete loss of tobacco crops (Tisdale and Kelley 1926). Severe black shank losses have been exacerbated by association with high populations of the root-knot nematode in the soil (Sasser et al. 1955). Black shank occurs as far north as Connecticut (McIntyre and Hankin 1977). See Shew and Lucas (1991) for a complete description of the disease.

Disease Development. Plants at all stages of growth, whether in the seedbed or in the field, are susceptible to attack. Young seedlings in particular are highly susceptible to damping-off in the nursery; however, if cool weather prevails, the disease may develop so slowly that symptoms are missed. When infected seedlings are transplanted, root rot usually occurs soon after. Leaves of infected plants wilt in warm weather (Plate 50.1). In advanced stages, dry, dark brown or black lesions develop in the cortical tissue on the stem near the soil line. These lesions rapidly expand upward and may cover as much as half the length of the stem (Plate 50.2).

Leaves of infected plants nearing maturity suddenly wilt irreversibly (Plate 50.1), turn yellow, and droop. As the disease progresses, the entire root system and the base of the stem decay (Plate 50.2). When the stem of a diseased plant is cut longitudinally, the pith is characteristically separated into platelike disks (Plate 50.3). On some resistant varieties, disease occurs only on the roots with little visible damage aboveground other than stunting of growth.

Temperatures above 20°C, high humidity, and rainfall favor disease development. High temperatures are rarely limiting because *P. nicotianae* grows at temperatures up to 35 to 36°C. The severity of disease is enhanced by increase in soil moisture; 100% seedling loss occurred when soil was saturated for 24 h, but in nonsaturated soil, only 7% disease occurred. Symptoms can appear on young seedlings in the field within 48 h after initial infection (Shew 1983).

Spread of *P. parasitica* on tobacco is slower on resistant than on susceptible cultivars (Shew 1987; Ferrin and Mitchell 1986b). Disease buildup occurred most rapidly from small inoculum levels on the susceptible cultivar Hicks, but on a resistant cultivar, Speight G-28, disease increased only when inoculum density in the soil was high (Ferrin and Mitchell 1986a).

Control. According to Lucas (1975), transplants from infested beds are the main method by which the pathogen has spread. Nurseries should use known pathogen-free soil or soil treated with methyl bromide (see Chapter 7). Probably the reason that *P. nicotianae* has become so widespread is the contamination of transplant nurseries. Control by short crop rotation is relatively ineffective because the pathogen is known to survive in the soil for 5 years or longer in the absence of tobacco. Long-term rotations (more than 5 years) with nonsusceptible crops such as peanuts or cotton have reduced the population of the pathogen and that of the root-knot nematode (Lucas 1965).

Strict control of the movement of people, implements, and water between infested and noninfested areas is an important means of limiting the dissemination of the pathogen.

Physiologic specialization was reported by Apple (1962). Four pathologic races have been described, but race 0 is predominant. Race 0 (not pathogenic to *Nicotianae plumbaginifolia*) and race 1 (pathogenic to *N. plumbaginifolia*) occur in most tobacco-growing regions of the world, race 2 is present in South Africa (Prinsloo and Pauer 1973), and race 3 occurs in Connecticut in the United States. *N. nesophila* and cultivar 1071 of *N. tabaccum* are resistant to race 3 (McIntyre and Taylor 1978). Resistant cultivars with genes conditioning high resistance to some or all of the known physiological races of the pathogen are now commercially available. The resistance of the cultivars that are most effective against races 0 and 1 is quantitatively inherited and obtained from the cultivar Florida 301, originally developed from Cuban cigar cultivars. Cultivars developed from *N. longiflora* and *N. plumbaginifolia* have extremely high resistance to race 0 that is controlled by a single dominant gene; but race 1, to which these cultivars are susceptible, often increases in population when these cultivars are planted on land highly infested with race 0. Therefore, cultivars with Florida 301-type resistance, which is effective against both races 0 and 1, are the primary means of control (Apple 1962; Lucas 1975).

Metalaxyl and propamocarb applied as transplant water treatments significantly reduced losses from black shank (Reilly 1980). Prothiocarb as a row treatment induced over 50% control in a heavily infested field in which a susceptible cultivar was grown (Prinsloo et al. 1979). See Chapter 7 for examples of control with a phenlyamide fungicide, e.g., metalaxyl or the phosphonate fosetyl-Al.

Brown Rot, Foot Rot, Gummosis, and Root Rot of *Citrus* species

The diseases described in Chapter 17 under *P. citrophthora* are similar to those caused by *P. parasitica* except for the difference in adaptability of the two species to different temperature regimes. *P. citrophthora* is more active at moderate temperatures (<30°C) and *P. parasitica* at higher temperatures (<35°C). For instance, in California, where the climate is a Mediterranean type with high temperatures but no rainfall during the summer, *P. parasitica* is active during the summer but *P. citrophthora* is active during the fall, winter, and spring months.

Disease Development. Development of diseases caused by both *P. citrophthora* and *P. parasitica* is discussed in relation to inoculum density in soil in Chapter 5 (see Table 5.4 for references).

Control. Control measures for diseases caused by both *P. citrophthora* and *P. parasitica* are similar and are discussed in Chapter 17. Control by cultural means is covered in Chapter 5. Control of gummosis by use of resistant rootstocks is discussed in Chapter 6. Control by use of fungicides, including the effective modern systemic fungicides metalaxyl and fosetly-Al is covered in Chapter 7. Useful references include Broadbent (1978), Klotz (1978), Klotz and Calavan (1978), and Ricci et al. (1990).

Heart Rot of *Ananas comosus* (L.) Mill. (Pineapple)

The symptoms and control of this disease are similar to those caused by *P. cinnamomi.* See Chapter 15.

Foliar Blight of *Bougainvillea* Species

Disease Development. Small, irregular, dark green, water-soaked spots occur at the tips and margins of the newly formed leaves. The leaves turn flaccid, blacken, and curl inward 12 to 24 h after initial infection (Alfieri 1970). Older, mature leaves remain healthy. Floral bracts appear to be particularly susceptible and often become covered with a cottony growth of mycelium during periods of high humidity. The infection progresses into the petioles and, in some cases, invades the young stems, resulting in lesions with a linear shape. Infected stems eventually blacken, collapse, and droop.

Control. The cultivar Sanderiana is highly resistant, while After Glow is moderately resistant and Barbara Karst and Gold are very susceptible (Alfieri 1970). Daconil 2787 is an effective protectant fungicide (Alfieri 1970), and captan has shown some control. If metalaxyl or fosetyl-Al are registered for use, they should be effective.

Root Rot of *Buxus harlandi* Hanc. and *B. sempervirens* L. (Boxwood)

The disease was described by Andrus (1933) on *Buxus sempervirens.*

Disease Development. Root rot (Plate 50.4) occurs before top symptoms become apparent. The feeder roots are first infected followed by infection of the larger primary roots. The base of the main stem often becomes discolored up to 4 cm above the soil level. Foliage becomes off-green on one or more branches of the plant. Leaves then turn bronze or dull yellow and eventually drop. The disease develops rapidly in water-logged soils, both in the field and in the greenhouse. Temperatures above 20°C favor the disease.

Control. In container-grown plants, particular attention must be paid to planting in a well-drained soil mix, which increases the aeration of the soil. Excessively wet or poorly drained soil should be avoided. In the field, planting in land with impermeable soil layers or in areas with poorly drained soils should be avoided.

Root Rot of *Carthamus tinctorius* L. (Safflower)

Disease symptoms and control measures are discussed under *P. drechsleri.* See Chapter 23.

Blight of *Catharanthus roseus* (L.) Don. (Periwinkle)

Periwinkle is an important ground cover used in landscape plantings. The disease has been found frequently in California (Keim 1977; Gill et al. 1977), Florida (Schubert and Leahy 1989), and several other areas in the world (Table 50.2).

Disease Development. Infection begins in terminal shoots, which become necrotic and blighted; the flowers wilt, turn brown, and shrivel. The disease then progresses downward along the stems. In advanced stages, dark brown to black lesions or cankers encircle the stems, resulting in wilt and eventually death of the terminal branches. Roots of diseased plants often remain healthy.

Prolonged warm weather accompanied by high humidity, rain, or frequent irrigation are contributing factors to the development and severity of the disease.

Control. Clean nursery stock, free of infection, should be planted in well-drained soil, and excessive watering should be avoided (Schubert and Leahy 1988). Where registration permits, systemic fungicides such as metalaxyl or fosetyl-Al are efficacious. Although metalaxyl at the label rate was effective for control by soil drench, Ferrin and Rhode (1992) isolated several naturally resistant strains (effective concentration at the 50% level of metalaxyl [EC_{50}] values ranged from 0.255 to 3.080 mg a.i./ml). This EC value was 19 times the rate recommended by the registered label.

Collar Rot of *Dianthus caryophyllus* L. (Carnation)

Collar rot of carnation was described in Greece in 1952 (Sarejanni), in Hawaii in 1962 (Hine and Aragaki), and in France in 1965 (Tramier and Mercier). Carnations are grown widely for ornamental planting and for cut flowers.

Disease Development. Diseased plants wilt suddenly. The leaves become grayish and wrinkle. A soft rot accompanied by shrinkage of the tissues occurs at the crown (Plate 50.6). When plants are gently pulled, stems break readily at the crown, leaving the whole root system behind in the soil. Phytophthora collar rot can be distinguished from *Fusarium* and bacterial wilt diseases by the bleaching of the upper ends of healthy shoots and by the lack of discoloration of the vascular tissues of stems that is caused by the latter pathogens.

Collar rot is most common in newly planted cuttings and is

seldom seen on second-year plants. Rooted cuttings are most susceptible until approximately 3 months of age. Symptoms generally occur 1 to 4 weeks after initial infection, depending on the amount of inoculum in the soil and the temperature (above 20°C). Plant mortality approaches 100% with an initial inoculum concentration of less than one propagule per gram of dry soil (Ricci 1972). The disease is limited at temperatures below 20°C. The pathogen often enters through wounds in stems when the tips of young plants are pinched to stimulate lateral bud development.

Control. Fungicides such as maneb, captan, basic copper sulfate, or a phenylamide fungicide (e.g., metalaxyl) or phosphonate (e.g., fosetyl-Al) applied as a spray after pruning are effective. Soil fumigation of the propagation beds with methyl bromide before planting prevents carryover of inoculum.

Blight or Azhukal Disease of *Elettaria cardamomum* (L.) Maton (Cardamom)

Cardamom, an expensive crop, is often referred to as the "queen of spices." Though rarely attacked by other pests or pathogens, cardamom production is reduced in India by blight caused by *P. parasitica* (Nair and Menon 1982). The disease was first described by Menon et al. (1972) in Kerala, India. A fruit and leaf rot was reported by Thankamma and Pillai (1973).

Disease Development. Blight is indicated by dark green or water-soaked lesions on immature leaves. The lesions gradually enlarge and turn necrotic. Under wet and warm conditions, the whole leaf becomes infected, shrivels, or shreds but remains attached to the plant. The panicles and capsules become water soaked. When young, immature capsules are infected, they usually shed within 3 to 5 days. Infection of mature capsules results either in shriveling or reduction of the quality of the crop. The disease progresses under favorable conditions into the panicles. The infected areas of the panicle usually turn yellow and eventually shrivel.

The tissue on infected rhizomes becomes dark brown, soft, and watery. Shoots arising from the rhizome either are infected early or may survive for a season before succumbing to the disease. Infected roots are discolored at the tips. The disease then spreads toward the base of the root. In advanced stages, the epidermis and cortex may disintegrate, exposing the vascular tissues.

Blight is usually observed at the beginning of the monsoon season and becomes severe 2 to 3 months later. The pathogen survives in infected rhizomes for long periods. Affected rhizomes and seeds are sources of inoculum for spread of the pathogen (Nair and Menon 1982). *P. nicotianae* survives up to 32 weeks in dry soil.

P. palmivora from coconut was reported to cause disease on cardamom (Radha and Joseph 1974), but this finding could not be repeated by Nair and Menon (1982).

Control. Nambiar and Sarma (1974) report that blight can be effectively controlled by spraying twice with Bordeaux mixture or with copper oxychloride (0.2%). If registered, a systemic phenylamide (e.g., metalaxyl) or phosphonate (e.g., fosetyl-Al) compound should be effective.

Boll Rot and Damping-off of *Gossypium hirsutum* L. (Cotton)

Cotton bolls are susceptible to attack by *P. parasitica* (Tucker 1931; Rao et al. 1962b; Pinckard and Guidroz 1973), but the disease occurs infrequently in the United States. *P. cactorum* (Patil-Kulkarney and Aswathaiah 1962), *P. palmivora* (Liang 1964), and *P. capsici* are also reported to be causal agents of boll rot. *P. capsici* caused a severe epidemic in California after sprinkler irrigation with water from a farm reservoir (Garber et al. 1986).

Disease Development. The entire surface of the boll becomes blue black to black within a few days after infection. The surface becomes spongy, and the contents disintegrate into a watery decay. Bolls near maturity rot within 2 to 3 days. In damp weather, the white, cottony mycelium of the pathogen grows on the surface of the boll. No other part of the plant is affected.

Symptoms of Phytophthora boll rot often resemble those caused by *Rhizoctonia solani* but can readily be distinguished from it by color and texture. *Phytophthora* spp. cause a black, soft, watery rot, whereas *Rhizoctonia* causes a brown, dry rot.

Damping-off of cotton seedlings has also been reported. Irregular, water-soaked areas that occur on the cotyledons and first true leaves rapidly turn pale yellow or become colorless and are surrounded by yellow green halos. In advanced stages, the infection spreads to the petioles, resulting in dropping of leaves. Invasion of the terminal portion kills the seedlings. Both boll rot and damping-off develop most rapidly during periods of high humidity, after rainfall or sprinkler irrigation, and at temperatures above 25°C.

Control. Effective control methods have not been reported.

Crown Rot of *Gypsophila paniculata* L. (Baby's Breath)

The first observable symptoms are wilting of the leaves, decrease in the size of the stems, and/or the presence of chlorotic leaves. Later, a soft, watery decay develops in the crown tissue (Engelhard 1974a, b).

Disease Development. Moderate to high temperatures (25 to 30°C) and high humidity enhance disease development. Decay of the crown region develops in 2 to 3 days after initial infection. Losses of up to 30% of the crop have been recorded during the first month after transplanting (Engelhard 1974a, b).

Control. The fungistat 5-ethoxy-3-(tricholoromethyl)-1,2,4-thiadiazole (Ethazol 30W) gave good control under greenhouse conditions (Engelhard 1974a, b). If registered, metalaxyl and fosetyl-Al should be effective.

Foot and Stem Rot of *Hibiscus sabdariffa* L. var. *altissima* (Roselle)

Roselle is often used as a high-yielding substitute for jute in India and is particularly suited to areas that undergo both extreme drought and high rainfall; however, when soil moisture is excessive, roselle can be severely affected by Phytophthora foot and stem rot (Mukherjee 1959; Mukherjee and Basak 1973).

Disease Development. Two distinct sites of infection on the plant are known. Symptoms initially appear at or below the collar region, and water-soaked, brown lesions develop on the taproot and adventitious roots. In advanced stages, the entire root system becomes soft and moist. Severe root rot causes a general wilting of the plant. On stems, the pathogen causes necrotic lesions that gradually coalesce to encircle the stem. The infection then progresses both up and down the stem, eventually girdling the collar region. Roots of these diseased plants usually remain healthy.

In advanced stages of disease, the stem splits at several places and a gummy exudate oozes from the cracks. The fibers initially remain white and healthy but eventually become infected and turn brown; individual fibers separate, and some rot completely. Affected plants wilt and topple over in windy weather. Disease development is enhanced by temperatures above 20°C and frequent rainfall, which occurs during the monsoon season in India (Mukherjee 1967).

Control. Effective control measures have not been reported. Boccas and Pellegrin (1976) found that six cultivars of *Hibiscus sabdariffa* had horizontal (multigenic) resistance.

Buckeye Rot and Root Rot of *Lycopersicon esculentum* L. (Tomato)

Buckeye rot of tomato fruit was first described in the United States by Sherbakoff (1917). Basal stem rot was first described in South Africa in 1991 (Ferreira et al. 1991). It is now known to occur in almost all tomato-growing regions of the world. Root rot is prevalent in many areas of California, where tomatoes are grown under furrow irrigation (Ristaino and Duniway 1989) (Plate 50.5). These diseases can also be caused by *P. capsici* and *P. drechsleri* (J. B. Jones et al. 1991).

Disease Development. High soil moisture and warm (above 20°C) soil or air temperatures are required for rapid disease development. The disease occurs most frequently in waterlogged areas of the field. Incipient infections on early-picked green fruit, which are overlooked, result in fruit rot during transit.

Control. Tomato plants can be staked to prevent contact of fruit with soil; however, this will not prevent rain-splashed soil from reaching the fruit. Roots of plants grown on raised beds are less severely affected.

In some areas, Bordeaux mixture and other copper-based fungicides prevent infection on fruit. If registered for use, metalaxyl or fosetyl-Al should be effective. In Australia, phosphorous acid (10 g a.i./liter) reduced the incidence and severity of root rot of tomato seedlings in a greenhouse. Field applications by soil drench increased the yield of tomatoes and reduced the severity of root rot (Flett et al. 1990). See also Flett et al. (1991) for control of root rot with fosetyl-Al and metalaxyl. See Chapter 7 for discussion and references on fungicide use.

Root rot has become a serious problem, especially in fields in California where the amount of irrigation water often becomes excessive (Ristaino et al. 1988; Ristaino and Duniway 1989). Management of soil water to prevent excess soil saturation reduces the severity and incidence of root rot.

Resistant cultivars are commercially available. Methods of testing for resistance in cultivars are reported by Skadow (1985) in Germany. See also Blaker and Hewitt (1987a, b) and the discussion of resistance in Chapter 6. Seedling resistance was a good predictor of mature plant resistance identified in commercial lines and in *Lycopersicon esculentum* var. *cerasiforme* (Dun.) A. Gray (Blaker and Hewitt 1987b).

Crown Rot and Blight of *Macadamia integrifolia* Maiden and Betche (Macadamia Nut)

See Chapter 15 on *P. cinnamomi* for symptoms and disease development.

Bud Rot of Plants in the Family Orchidaceae (Orchids)

Bud rot, leaf blight, root rot, stem blight (Plate 50.7), and damping-off of seedlings are commonly found in orchid nurseries. The pathogen can destroy the bud in less than a week, after which the infection kills the plant. The publication of Uchida and Aragaki (1991) illustrates diseases of orchids in color and also describes control measures. Both *P. parasitica* and *P. palmivora* are commonly associated with the diseases. Orchids in the genera *Dendrobium, Cattylea, Epidendrum, Vanda, Paphiopedilum,* and *Laeliocatyleya* are affected (Uchida and Aragaki 1991). Uchida (1994) discusses diseases of orchids caused by *P. cactorum* and *P. palmivora.*

Disease Development. Bud rot develops rapidly during warm weather (above 20°C) when rainfall or irrigation by sprinklers is frequent. The cultivars in the *Vanda* and *Arachnis* groups are most susceptible.

Control. Infection can be prevented by drenching or spraying the plants with the systemic fungicides metalaxyl, fosetyl-Al, or ethazole (Uchida and Aragaki 1991).

Wilt and Crown Rot of *Passiflora edulis* Sims (Passion Fruit)

The first indication of infection is either the appearance of leaves that seem to be scalded or the presence of chlorotic leaves scattered randomly throughout the foliage. Tips of vines are often blackened. As infection progresses, veinal necrosis, wilting, and extensive defoliation occur. In advanced stages, branches die back and the plants collapse as a result of rotting of the crown. As the disease progresses, the bark and cortex slough off. The vascular tissues at this stage become completely necrotic.

When crowns are affected, the bark cracks and the stem swells at the ground level to a size three to five times the diameter of the normal stem. On green fruit, gray green, water-soaked spots often coalesce to cover the surface. Infected fruits drop prematurely. See Table 50.2 for references.

Disease Development. The disease develops rapidly in low-lying areas of the field that are subject to waterlogging. Infested soil beneath the vines is often splashed onto lower leaves during rains. Soilborne inoculum can be disseminated during wet, windy weather.

Control. Plantings should be restricted to soil with adequate drainage or to sites on sloping land. Significant reduction of the disease has been obtained in South Africa by utilizing yellow granadilla (*Passiflora edulis* forma *flavicarpa*) as a rootstock for purple granadilla (Brodrick et al. 1976b). The bud union should be at least 2 feet above the soil level to prevent infection of the scion by soil splash. Fungicides have not proved to be totally effective and serve mainly to reduce the rate of spread. If fungicides are used, the foliage should be pruned to allow better penetration.

Crown Rot of *Petunia hybrida* Hort. Vilm-Andr. (Petunia)

Annual petunia transplants to ornamental flower beds are used widely in landscapes. *P. parasitica* may cause up to 80% loss of the stand of petunia in ornamental flower beds.

Disease Development. On blighted or wilted plants in the field, water-soaked lesions or cankers develop in the crown region and gradually extend up the stems. When crowns become girdled (Plate 50.8), the plants die. Roots are often unaffected and remain healthy. The disease is widely prevalent in nursery stock, from which it spreads to the field. In greenhouses, the disease causes a black discoloration and dry rot of the crown and lateral branches followed by wilting and death of the plant (Phillips and Baker 1962). The disease is most severe where infested beds are cropped continuously with petunia and watered excessively. Plants with latent infections that are sold through retail outlets account for a high incidence of the disease in home gardens.

Control. Pathogen-free stock should be planted; petunia

should be rotated frequently with other bedding plants that are less susceptible. Dense or close plantings should be avoided. If registered for use, treating soil in pots or beds with metalaxyl as a soil drench or fosetyl-Al should suppress but is unlikely to eradicate the pathogen.

In Florida, soil drench and foliar spray with pyroxychlor was the most effective fungicidal control; however, pyroxychlor is not currently marketed (Knauss 1962). Captafol as a foliar spray was effective when applied after a soil drench with pyroxychlor. Copper hydroxide was effective but too phytotoxic to be useful. Captan, chlorothalonil, and zinc ion maneb complex (Dithane M45) provided moderate control. Zinc ion maneb complex may be phytotoxic.

Foot Rot and Leaf Rot of *Piper betle* L. (Pan)

Piper betle is an important crop grown throughout most of the Far East. It is utilized mainly for masticatory purposes. Foot and leaf rot diseases were first described by Dastur (1927) who later (1935) named the pathogen *P. parasitica* var. *piperina.* Although the name is still used (Mehrotra 1982), Waterhouse (1963) rejected the variety name since it was invalidly published without a Latin diagnosis and reduced it to a synonym of *P. nicotianae* var. *nicotianae.* An extensive review of the current status of disease on pan is given by Khare et al. (1988).

Disease Development. Both foot rot and leaf rot phases of the disease are known. The foot rot phase develops when the plant is subjected to waterlogged conditions, and the leaf rot phase develops during the rainy season when both temperatures and relative humidity are high.

Foot rot results in a sudden yellowing of otherwise healthy-looking vines followed by a wilting of the tender shoots and leaves. When such plants are uprooted, the roots appear dark brown or black. The disease usually begins in the feeder roots and gradually progresses through the main roots to the collar region of the plant. Foot rot can result in a 30 to 100% loss of the crop (Saksena 1977).

Under humid conditions, circular, brown, water-soaked spots rapidly enlarge to cover the entire leaf and eventually turn black. A white fuzz of mycelium develops on the underside of the leaf if wet conditions prevail. When wet periods are interspersed with warm, dry periods, lesions develop concentrically on the leaves, since the progress of the disease is arrested by dryness. The disease may also extend into the petioles and occasionally into the stem. Leaf rot may result in a 10 to 30% loss of the crop (Saksena 1977).

The most rapid development of the disease occurs at high humidity and temperatures above 20°C. Waterlogged soils and overhead shade provide ideal conditions for the development of foot rot.

Control. Irrigation water that is contaminated by debris infested with *P. parasitica* should not be used. Weeds should be removed from the vicinity of the planted area, since *P. parasitica* overwinters in several weed hosts (Saksena and Mehrota 1970).

Amending the soil with corn straw supplemented with ammonium nitrate or with the fungus *Trichoderma viride* was effective for experimental control of the disease (Mehrotra and Tiwari 1976).

Cuttings dipped in a streptomycin solution followed by sprays with a 1% Bordeaux mixture at 15-day intervals were protected from the disease at planting time (Saksena 1977). If registered for use, a phenylamide (e.g., metalaxyl) or phosphonate (fosetyl-Al) fungicide should be effective.

Foot Rot of *Pistacia vera* L. (Pistachio)

Disease Development. Young foliage becomes slightly chlorotic in the spring after foot rot infection. Leaves eventually turn yellow but do not fall from the tree. The growth of the tree is stunted followed by an unusually heavy set of fruit. Occasionally, infected trees may suddenly wilt without any previous indication of disease symptoms. As the disease progresses, a heavy concentration of resin oozes from the lower portion of the trunk and stains the soil around the base of the tree. Although the cambium in infected areas of the tree is stained black, the xylem does not become discolored. Trunks of infected trees eventually become girdled (Kouyeas 1973). *P. citrophthora* and *P. citricola* also cause foot rot of pistachio (Kouyeas 1973). Disease development is similar to that of foot rot of citrus caused by *P. citrophthora* (see Chapter 17).

Control. Control measures are similar to those recommended for gummosis of citrus (Chapter 17). *Pistacia vera* and the rootstock *P. terbinthius* are both susceptible (Kouyeas 1973). A resistant rootstock is not presently available.

Crown Rot of *Saintpaulia ionantha* Herm. Wendl. (African Violet)

Disease Development. *P. parasitica* infects the crown and roots of African violets. The usual aboveground symptom is a soft brown to black rot, beginning in the crown area, that progresses into petioles and leaf blades. Roots become rotted, and the plants become stunted and often wilt. If only the roots are infected, the plants become stunted; leaf and stem rot is usually followed by death of the plant. Disease develops rapidly at temperatures between 20 and 30°C and at high relative humidity (Kröber and Plate 1973).

Control. Propagating stock should be grown in noncontaminated or steam-sterilized soil, and excessive irrigation should be avoided. According to Strider (1979), the most resistant cultivars are New York and California of the Optimara series and Kathy of the Melodic series. The most susceptible cultivars are Suzanne and Adeline.

Blight of *Sesamum indicum* L. (Sesame)

Sesame is an important oilseed crop in many countries. Blight was first recorded in India by Butler (1918) and subsequently by Kale and Prasad (1957) and Gemewat and Prasad (1964). It is now known to occur wherever sesame is cultivated (Table 50.2).

Disease Development. The appearance of water-soaked spots on the leaves is the first indication of blight. These spots gradually coalesce until the leaf is entirely destroyed. Brown to black lesions occur near the base of the stem close to the soil. Pods on the affected branch are malformed, and seeds shrivel and turn brown. Infected plants are eventually killed. Losses as high as 66% of the crop occur in some areas.

In humid weather, the cottony mycelium of the fungus grows on the pods. The pathogen is believed to be host specific and has been referred to as *P. parasitica* var. *sesami* by Kale and Prasad (1957), but this variety is not recognized (Stamps et al. 1990). If host specificity is confirmed, the fungus could be designated a forma specialis rather than a variety.

The fungus has been known to survive in the soil in the form of chlamydospores, even in the absence of crop residues. Seedborne inoculum in infected fruit can initiate the disease.

Control. Experimentally, spraying plants with zineb or Bordeaux mixture decreased the percentage of diseased plants (Gemewat and Prasad 1964). If registered for use, metalaxyl and/or fosetyl-Al should be efficacious.

CHAPTER 51A

Phytophthora palmivora (Butler) Butler (1919) var. *palmivora*

Synonyms of *Phytophthora palmivora* include *P. omnivora* de Bary (1881), *Pythium palmivorum* Butler (1907), *P. faberi* Maublanc (1909), *P. theobromae* Coleman (1910), *Kawakamia carica* Hara (1915), *P. fici* Hori (1915), *P. carica* Hara (1916), *P. palmivora* var. *piperis* Muller (1936), and *P. palmivora* var. *theobromae* (Coleman) Orellana (1959). The undescribed species *P. agaves* (Gandora 1909) on *Agaves americana* L. may be a synonym of *P. palmivora*. Waterhouse (1963) lists it as *nomen nudum*.

The black pod disease of cocoa, *Theobroma cacao* L., was attributed to *P. omnivora* (Massee 1899a, b), and later a destructive disease of palms was reported to be caused by *Pythium palmivorum* (Butler 1907). The fungus was later renamed *P. palmivora* (Butler) Butler (1919). *P. palmivora* is a ubiquitous pathogen with a wide host range (Table 51A.1) (see also Chee 1966 for a host range). *P. palmivora* probably originated in Central or South America and was transported on infected plants throughout the world by man (Zentmyer 1988).

The book on *Phytophthora* diseases of cocoa by Gregory et al. (1974) contains many useful chapters such as that by Waterhouse (1974b), in which the characteristics of *P. palmivora* are compared with those of similar species. The book on tropical diseases by Holliday (1980) includes a discussion of several diseases caused by *P. palmivora*. Stamps et al. (1985a) present a brief description of *P. palmivora*. The volume on the epidemiology of *Phytophthora* on cocoa in Nigeria (Gregory and Maddison 1981) contains the results of several years of research by a number of workers on the causal agents, spread, and dissemination of *Phytophthora* pathogens on cocoa. The relationships of *P. palmivora*, *P. megakarya*, and *P. capsici* (*P. palmivora* MF4) to black pod disease of cocoa are discussed by Brasier et al. (1981). Sreenivasan (1982) reviewed the status of the *Phytophthora* species that attack cocoa. Some diseases of fruit trees in the tropics caused by *P. palmivora* are summarized by Zentmyer and Mitchell (1985). Foliar and root diseases of orchids that are caused by both *P. palmivora* and *P. parasitica* in Hawaii are beautifully illustrated in the bulletin by Uchida and Aragaki (1991), and symptoms and control are discussed.

Because of the considerable morphological and pathological variation within *P. palmivora*, several strains of *P. palmivora* have been described. The history of strains, varieties, and biotypes of *P. palmivora* is reviewed in detail by Zentmyer (1974) and Brasier et al. (1981). (See Table 51A.2.) Babacauh (1972) described seven strains from crops grown in western and central Africa and described *P. palmivora* var. *heterocystica* (Babacauh 1983a) (Chapter 51B).

In the early literature, Gadd (1924) and Ashby (1929) separated isolates attributed to *P. palmivora* into the "cocoa" group and the "rubber" group. Oospores were observed only when isolates from the cocoa and rubber groups were grown together. More recent studies, however, show that the designation of the cocoa group as the A1 and the rubber group as the A2 mating type is not consistent, since both A1 and A2 mating types are found within both cocoa and rubber isolates (Savage et al. 1968; Zentmyer and Mitchell 1970; Zentmyer et al. 1973; Zentmyer 1974). In the 1970s, it became evident that other *Phytophthora* species as well as *P. palmivora* were causal agents of black pod of cocoa. The history of this research as described by Brasier et al. (1981) is covered below.

Table 51A.2 summarizes the characteristics of different isolates of *P. palmivora* from several hosts that have been described by several investigators. The black pod disease of cocoa is caused not only by *P. palmivora* but also by isolates previously considered to be atypical *P. palmivora* and now reclassified as *P. megakarya* or *P. capsici*. Waterhouse (1974b) lists two morphological forms, MF1 (the so-called "typical" form of *P. palmivora*) and MF2, on the basis of different shapes of sporangia and oospores. Later work by Brasier et al. (1981) showed that MF2 could not be accepted as a valid taxon and is considered to be a variant of *P. palmivora*. At an international workshop, isolates were subdivided into MF1, MF3, and MF4 groups (M. J. Griffin 1977; Brasier and Griffin 1979), mainly on the basis of the length of the sporangial stalk (Zentmyer et al. 1977; Idosu and Zentmyer 1978; Kaosiri et al. 1978), sporangial and oogonial morphology (Brasier et al. 1981), and the size and number of chromosomes (Sansome et al. 1975). The small (S) size of chromosomes in the oogonia and oospores of MF1 separated it from MF3, which had large (L) chromosomes. The MF1 form was considered to be typical of *P. palmivora*, and the MF3 form was described as the new species *P. megakarya* (Chapter 45) on the basis of larger nuclei and five to six large chromosomes per nucleus compared with nine to 12 small chromosomes for *P. palmivora*. Also, pedicels on the caducous sporangia of *P. megakarya* were 10 to 30 µm in length compared with the short pedicels (about 5 µm) on sporangia of *P. palmivora* (Brasier and Griffin 1979; Brasier et al. 1981) (Table 51A.3). The MF4 type was considered to resemble *P. capsici* (Zentmyer et al. 1977; Idosu and Zentmyer 1978; Kaosiri et al. 1978). It has since been placed in the redescribed species *P. capsici* (Tsao and Alizadeh 1987; Tsao 1991; Mchau and Coffey 1995). The MF designations that served to identify these groups while research was in progress should not be used any longer. They are listed in Tables 51A.3–51A.5 for historical purposes. See Brasier and Griffin (1979) for photomicrographs and definitive data that differentiate *P. palmivora*, *P. megakarya*, and *P. capsici*.

Phytophthora palmivora as a Mycoherbicide

P. palmivora has been found to be an important biocontrol agent for the weed called strangle vine or milkweed, *Morrenia odorata*. This isolate, first identified by Burnett et al. (1974) in Florida as *P. citrophthora*, was later corrected to *P. palmivora* (Feichtenberger et al. 1984). Chlamydospores are formulated in a commercial preparation known as a mycoherbicide under the trade name DeVine (Ridings et al. 1976; Ridings 1986; Raghavan-Charudattan and Charudattan 1987; Auld 1990).

The weed is highly susceptible to the *P. palmivora* isolate and is efficiently controlled by application of chlamydospore inoculum. Kenney (1986) describes the development of chlamydospores in the makeup of a stable mycoherbicide formulation. During the years 1978–1980, 95 to 100% control of milkweed vine was achieved. TeBeest and Templeton (1985) reviewed the history and progress in weed control by DeVine. As would be expected, successful use of DeVine depends on the environmental factors that are necessary for disease development, including frequent rainfall and/or high humidity and optimal temperatures.

Characteristics of *Phytophthora palmivora*

Since the morphological and physiological characteristics of *P. palmivora* vary considerably, precise descriptions of sporangia and oospores of all forms and strains cannot be made under one heading. A range in the variation is to be expected for isolates from different hosts (Table 51A.2). See Brasier and Griffin (1979) for excellent photomicrographs of sporangia and oospores.

P. palmivora is classified in group II (Stamps et al. 1990) and described in detail by Waterhouse (1974b) and Stamps (1985a). The dimensions of sporangia, chlamydospores, sex organs, and growth temperatures for several isolates from different hosts are summarized in Table 51A.2. Characteristics of *P. palmivora, P. megakarya,* and *P. capsici* are compared in Table 51A.3. See Tables 4.2 and 4.3 for tabular keys and Appendix 4.9 for a dichotomous key. Morphology is shown in Figure 51A.1. Brasier and Griffin (1979) illustrate the wide variation in shapes of sporangia. Dimensions, shapes, and caducity of sporangia from several different reports on *P. palmivora, P. megakarya, P. capsici, P. arecae, P. meadii, P. botryosa, P. heveae,* and *P. nicotianae* (*P. parasitica*) are given in Table 51A.4. Table 51A.5 compares the chlamydospore and sex organ characteristics for these species.

Sporangia

Sporangia are variable in shape, depending on the isolate, mostly elliptical to ovoid, prominently papillate, and form on sympodial sporangiophores (Figures 51A.1, 4.9C, 4.12E, and 4.16A). As many as 20 sporangia can occur on one sympodium. Sporangia are caducous (deciduous) and after they fall off the sporangiophore are characterized by short (5-μm) pedicels. Sporangia are variable in size (Table 51A.2) but average 40 to 60 μm in length and 25 to 35 μm in breadth with a length-breadth ratio of 1.4 to 2.0 (Holliday 1980). Isolates from cocoa had a length-breadth ratio of 1.67 to 1.93 (Brasier and Griffin 1979). A recent study of a wide range of isolates from several hosts showed that sporangia were ellipsoid to spherical (predominantly ellipsoid with a broad base) with average lengths of 45.3 μm and breadths of 29.8 μm. The length-breadth ratio averaged 1.5 (range 1.3 to 1.8) (Mchau and Coffey 1994a; Table 51A.4). About 10 to 40 zoospores are discharged from sporangia incubated in water (Holliday 1980). Sporangia are stimulated by light (Brasier 1969a). According to Waterhouse (1974b), sporangia are more elongated in light.

Chlamydospores

Globose to subglobose chlamydospores are terminal or intercalary on mycelium of most isolates of *P. palmivora* (Figure 4.24C and D); however, six of 93 isolates from widely distributed areas did not form chlamydospores on V8 juice agar (Mchau and Coffey 1994a). Chlamydospore diameters measure 32 to 42 μm (Holliday 1980), average 33 μm (Waterhouse 1974b), 36 μm (Ashby 1929a), and 36.2 ± 9.6 μm (Mchau and Coffey 1994a) (Table 51A.2).

Sex Organs

P. palmivora forms oogonia and oospores when A1 and A2 mating strains are paired (Ashby 1929; Brasier and Griffin 1979). Oogonia are spherical, and antheridia are amphigynous (Waterhouse 1974a). The mean oogonial diameter ranges from 22.3 to 34.8 μm (Brasier and Griffin 1979). Oospores average 22.8 ± 0.1 μm (Dantanarayana et al. 1984) (Table 51A.5). Oospores are loose in the oogonia (aplerotic). The distribution of A1 and A2 mating types is reported by Zentmyer et al. (1973) and Zentmyer (1974).

Babacauh (1983) made crosses of a *P. palmivora* A2 isolate with a "near *P. citrophthora*" A1 isolate, both from cocoa, and noted that amphigyny may result from the ability of the same hyphal tip to differentiate both an antheridium and an oogonium.

Growth Temperatures

The minimum temperature for growth is 11°C, the optimum is 27.5 to 30°C, and the maximum is near 35°C (Waterhouse 1974b). See Table 51A.2 for more data.

Distinguishing Characteristics

P. palmivora is distinguished from other heterothallic species with conspicuous papillate sporangia by production of chlamydospores and elongated, ellipsoidal to ovoid deciduous sporangia with short pedicels (Ho 1990). Tables 51A.3–51A.5 summarize some of the distinguishing characteristics. Ho agrees with Waterhouse (1963) that *P. carica,* which has been reported on fig (*Ficus carica* L.) in Taiwan, is a synonym of *P. palmivora.* The review of Dantanarayana et al. (1984) on the taxonomy of *Phytophthora* species isolated from rubber in Sri Lanka indicates that *P. meadii,* the principal pathogen on rubber, is distinguished from *P. palmivora* by its nearly spherical sporangia compared with the ovoid sporangia of *P. palmivora.* Pedicels of *P. meadii* sporangia are medium in length (mean of 12.3 μm), while pedicels of *P. palmivora* are short (mean of

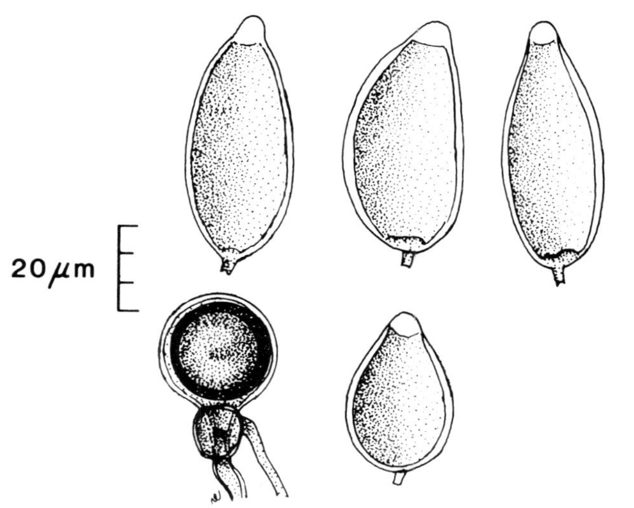

Figure 51A.1. Morphology of *Phytophthora palmivora* (MF1 or S type). Lower row, globose oogonium with an amphigynous antheridium and an ovoid sporangium with a short pedicel. Upper row, variations in shapes of sporangia. (Drawing by A. Vaziri)

Table 51A.1. Distribution of hosts reported for *Phytophthora palmivora*[a]

Host	Common name	Disease	Geographical distribution
Abelmoschus esculentus (L.) Moench	Okra; gumbo	Flower and fruit rot	India (Ballakrishnan 1947a)
Achras sapota L.	Sapodilla; chiku	Fruit rot	India (V. G. Rao et al. 1962a)
Agave sisalana Perrine	Sisal	Zebra disease	E. Africa (Peregrine 1963, 1964, 1969)
Agave wightii Dr. and Pr.	Agave; century plant	Leaf blight	India (Ramakrishnan and Seethalakshmi 1956b)
Aleurites fordii Hemsl.	Tung-oil tree; China wood-oil tree	Leaf spot	India (Venkatarayan 1932)
Allium tuberosum Roxb. ex Sprenge	Chives	Blight	China (Ho et al. 1983b)
Anacardium occidentale L.	Cashew nut	Seedling damping-off	India (Kumararaj and Bhide 1962)
Ananas comosus (L.) Merr.	Pineapple	Heart rot; root rot	Hawaii (Mehrlich 1932, 1934); Ivory Coast (Ravisé 1966); Thailand (Suzui et al. 1979)
Annona glabra L.	Pond or monkey apple	Seedling blight	Philippines (Reinking 1923)[b]
Annona muricata L.	Sour sop; prickly custard apple	Seedling and fruit rot	Philippines (Reinking 1923; Tsao et al. 1993); Malaysia (Chee 1969b)
Annona squamosa L.	Custard apple	Purple blotch; fruit rot	Australia (Purss 1953)
Antirrhinum majus L.	Snapdragon	Root rot	Puerto Rico (Tucker 1929); Iran (Ershad 1979)
Aquilegia vulgaris L.	Columbine	Root rot	Mauritius (Wiehe 1948)
Archontophoenix alexandrae (F. J. Meull.) H. Wendl. and Drude	Alexandra palm	Seedling blight	Philippines (Reinking 1923)
Areca catechu L.	Betel nut; pinang	Fruit rot	India (Butler 1907); Sri Lanka (Petch 1924)
Areca lutescens L.	Areca palm	Bud rot	India (Ramakrishnan and Seethalakshmi 1956b)
Arecastrum romanzofficianum (Cham.) Becc.	Queen palm; plumy-coconut	Bud rot	United States (Burger 1924)
Artocarpus altilis (Parkins) Fosb.	Breadfruit	Fruit rot	Western Samoa (Gerlach and Salevao 1984); India (Ramakrishnan 1960)
Artocarpus camansi	Camansi	Crown and root rot	Philippines (Tsao et al. 1993)
Artocarpus communis Forst.	Breadfruit	Leaf fall; fruit rot; seedling blight	Sri Lanka (Petch 1907); South Pacific (D. B. Reddy 1973; Trujillo 1960)
Artocarpus heterophyllus Lam. (*A. integer* Thunb.)	Jackfruit	Crown and fruit rot	Sri Lanka (Bertus 1946); India (Ramakrishnan 1960); Philippines (Tsao et al. 1993)
Azadirachta indica	Neem	Crown and fruit rot	Philippines (Tsao et al. 1993)
Baccaurea motteyana Muell. Arg.	Rambai	Fruit rot	Malaysia (Chee 1969b)
Borassus flabellifer L.	Palmyra palm	Bud rot	India (Butler 1906)
Borassus flabelliformis	Godavari	Bud rot	India (McRae 1934)
Bougainvillea spectabilis Willd.	Bougainvillea	Leaf and flower blight	Southern Rhodesia (Ashby 1929a); India (Ramakrishnan and Seethalakshmi 1956b)
Brachychiton diversifolius Hort.	Bottle tree	Root and stem necrosis	Argentina (Frezzi 1950)
Brachychiton populneus (Schott and Endler.) R. B. (Hortus Third lists *B. diversifolius* as a synonym)	Bottle tree, kurrajong	Root rot	Argentina (Frezzi 1950); United States (Gill and Zentmyer 1979; Gill 1982)
Bryophyllum calycinum Salisb.	Airplant	Leaf blight	India (Ramakrishnan and Seethalakshmi 1956b)
Capsicum annuum L.	Chili pepper	Fruit rot	West Africa (P. D. Turner 1960b)
Capsicum annuum L. var. *grossum* Sendt.	Bell or green pepper; pimento	Fruit rot	India (Ramakrishnan and Seethalakshmi 1956b)
Carica papaya L.	Papaya	Soft fruit rot; stem canker; foot rot	Philippines (Reinking 1918); Ceylon (Gadd 1924); Santo Domingo (Ciferri 1927b, 1930); Malaysia (Thompson 1939); India (Balakrishnan 1947b); Mauritius (Wiehe 1948); Indonesia (Reitsma and Sloof 1947); Australia (Teakle 1957, Anon. 1958a); United States (Hawaii), Costa Rica, Mexico (Hine et al. 1965); Brazil (Zentmyer et al. 1973); United States (Kliejunas and Ko 1974; Ko 1980)
Carthamus tinctorius L.	Safflower; false saffron	Leaf blight	Venezuela (Malaguti 1950)
Cattelya sp.	Orchid	Heart and leaf rot	Java (Schwarz 1927); Hawaii (Hine 1952)[b]; Sri Lanka (Zentmyer et al. 1973); Taiwan (Chen and Hsieh 1978)
Ceiba casearia Medic.	Silk-cotton tree	Bark canker; fruit rot	Java (Hartley 1924)
Cereus pitijaga DC.	Cactus	Leaf blight	India (Coleman 1910)
Chamaedorea elegans Mart.	Chamaedorea palm	But rot and leaf blight	Italy (Magnano di San Lio et al. 1987a)
Chrysalidocarpus lutescens Wendl.	Areca; yellow or butterfly palm	Seedling blight; bud rot	Philippines (Roldan and Luczon 1953); India (Ramakrishnan and Seethalakshmi 1956b)
Cinchona calisaya Wedd.	Calisya tree	Seedling blight	Philippines (Celino 1934)
Cinchona ledgeriana Moens	Cinchona	Seedling blight	India (Kheswalla 1935)
Cinchona spp.	Cinchona	Top blight of seedlings; girdle canker	Philippines (Celino 1934); India (Kheswalla 1935); Congo (Stoffels 1945); Bolivia, Costa Rica and Guatemala (Crandall and Davis 1945, 1950; Crandall 1954)
Citrullus vulgaris Schrad.	Watermelon	Fruit rot	Malaysia (Chee 1969b)
Citrus aurantifolia (Christm.) Swing.	Lime	Seedling blight	Philippines (Ocfemia and Roldan 1927)
Citrus aurantium L.	Sour orange, mosambi	Seedling blight; crown rot and root rot	Philippines (Ocfemia and Roldan 1927); Trinidad (R. E. D. Baker 1934); India (Kamat 1958); United States (Zitco et al. 1991)

(*continued on next page*)

Table 51A.1. (*continued*)

Host	Common name	Disease	Geographical distribution
Citrus grandis	Pomelo	Foot rot	Thailand (Kamjaipai and Intavamolsri 1983)
Citrus limon (L.) Burm.	Lemon	Seedling blight	Philippines (Ocfemia and Roldan 1927)
Citrus maxima Merrill	Pommelo	Seedling blight	Philippines (Ocfemia and Roldan 1927)
Citrus medica L.	Citron	Seedling blight; stem rot	Philippines (Ocfemia and Roldan 1927); United States (Zitco et al. 1991)
Citrus nobilis Lour.	Mandarin orange	Fruit and leaf rot	India (Thomas et al. 1947)[c]
Citrus paradisii Macf.	Grapefruit	Brown fruit rot; seedling blight	Puerto Rico (Tucker 1927); Trinidad (R. E. D. Baker 1934)
Citrus reticulata Blanco	Tangerine	Seedling blight	United States (Puerto Rico) (Anon. 1960); Sarawak (Holliday and Mowat 1963)
Citrus sinensis (L.) Osbeck.	Sweet orange	Shoot dieback; leaf-fall; fruit rot; root rot	Tanganyika (Wallace 1934); India (Uppal and Kamat 1936; Deva Rajan and Aiyappa 1945); United States (Zitco et al. 1991)
Citrus spp.	Citrus	Seedling blight; foot rot; gummosis	Philippines (Reinking 1921); India (Uppal 1933); Argentina (Frezzi and Macola 1943); Ivory Coast (Moreau and Moreau 1954); Thailand (Suzui et al. 1979); India (Lele and Kapoor 1980)
Citrus suhuiensis	Orange	Fruit rot	Malaysia (Chee 1969b)
Clianthus dampieri A. Cunn.			Zimbabwe (Chee 1969b)
Cocos campestris (probably *Butia capitata* (Mart.) Bacc.)	Jelly palm	Bud rot	Morocco (Malencon and Deleceuse 1937; Alfieri et al. 1984)
Cocos nucifera L.	Coconut palm	Bud rot; leaf drop; calzy-end rot; "leaf-bitten disease"	Jamaica (W. Fawcett 1889); Philippines (Reinking 1919b); India (Butler 1906; McRae 1922); Fiji and Tonga (H. W. Simmonds 1925); Puerto Rico (Tucker 1931); Portuguese East Africa, United States, Malabar, W. Indies (Ocfemia 1934); Papua New Guinea (Dwyer 1937); Brazil (Chaves Batista 1946); Mauritius (Wiehe 1948); Diego Garcia (Orian 1959); Indonesia (Bennett et al. 1986)
Cocos plumosa Hook		Bud rot	United States (Burger 1924)
Colocasia antiquorum Schott.	Elephants' ear; taro; coco-yam	Leaf blight	India (Narasimhan 1930)
Cyphomandran betacea Sendt.	Tree tomato	Patch canker	India (Balakrishnan 1947b)
Dendrobium maccarthia Thwaites	Orchid	Wilt	Sri Lanka (Gadd 1924); United States (Hine 1962)
Dianthus caryophullus L.	Carnation	Shoot wilt	Italy (Andreucci 1959); Malawi (Chee 1969b)
Dieffenbachia picta (Lodd.) Schott.	Dieffenbachia; dumb cane; mother-in-law's tongue	Root and stem rot	United States (Tompkins and Tucker 1947)
Durio lowianus Scoort. ex King	Wild durian	Root rot; gummosis	Thailand (Kamjaipai et al. 1984)
Durio zibethinus Murr.	Durian tree	Trunk canker; patch canker; decline	Malaysia (Thompson 1934; Navaratnam 1966); Thailand (Suzui et al. 1979); Philippines (Tsao et al. 1993)
Dypsis madagascariensis Nichols.	Palm	Bud rot	Philippines (Reinking 1923)[b]
Elaeis guineensis Jacq.	Oil palm; macaw fat	Spear dieback	Papua New Guinea (Dwyer 1953)
Epidendrum sp.	Orchid	Leaf and stem blight	United States (Hine 1962)
Erythrina sp.	Legume; dadap	Bark canker	Java (Hartley 1924)
Eugenia aquea	Water apple; rose apple	Fruit rot	Malaysia (Chee 1969b)
Eugenia aromatica Baill.	Clove	Blight	Malaysia (Ashby 1922)
Euonymus japonicus (Thunb.)	Spindle tree	Blight	China (Zheng and Lu 1989)
Euphorbia peplus L.	Spurge	Root rot	Mauritius (Wiehe 1948)
Euphoria longana Lam.	Longan	Crown and root rot	Philippines (Tsao et al. 1993)
Ficus capensis Thunb.	Cape fig	Fruit rot	West Africa (P. D. Turner 1960b)[b]
Ficus carica L.	Fig	White powdery rot of fruits	Japan (Hori 1915; Katsura 1969); Taiwan (Sawada 1919); Australia (Anon. 1950); United States (El-Gholl and Alfieri 1984)
Ficus religiosa L.	Peepul tree; Banyan tree	Leaf spot	India (Singh 1959)
Garcinea mangostana	Mangosteen	Crown and root rot	Philippines (Tsao et al. 1993)
Gerbera jamesonii Bolus	Transvaal daisy	Blossum blight	India (V. G. Rao and Ullasa 1971)
Gmelina arborea Roxb.	Malay bushbeech	Fruit rot	West Africa (P. D. Turner 1960b)[b]
Gossypium barbadense L.	Sea island cotton	Boll rot	West Indies (Harland 1917)
Gossypium hirsutum L.	Cotton	Seedling blight; damping-off	Spain (Melero-Vara and Jiménez-Diaz 1990)
Gossypium sp.	Cotton	Boll rot	West Indies (Ashby 1922)
Grammatophyllum scriptum Blume	Orchid	Leaf blight; heart rot	Java (Schwarz 1927)
Hedera helix L.	English ivy	Root rot and stem blight	United States (Keim et al. 1976b); Hawaii (Uchida and Aragaki 1978)
Herrania sp. Goudot		Fruit rot	West Africa (P. D. Turner 1960b)[b]
Hevea brasiliensis Muel. Arg.	Para rubber	Black stripe; black thread; patch canker; leaf fall and fruit rot	Burma (Dastur 1903); Sri Lanka (Petch 1903); Cameroon (Peters 1912); Uganda (Small 1915); Malaysia (Belgrave and Norris 1917); Indonesia (Van Hall 1923); Philippines (Reinking 1923)[b]; Brazil (Weir 1926); Fiji (Campbell 1927);

(*continued on next page*)

Table 51A.1. *(continued)*

Host	Common name	Disease	Geographical distribution
Hevea brasiliensis Muel. Arg. *(continued)*			Indonesia (Schooneveldt 1939); Congo (Staner 1941); Indochina (Borget 1949); Liberia (Darley and Silverborg 1952); Papua New Guinea (Mann 1953); Costa Rica (J. B. Carpenter 1954); India (Ramakrishnan and Radhakrishna 1963;Pillai 1980); Thailand (Tsao et al. 1976); China (Ho et al. 1984a); several areas of the world (Liyanage and Wheeler 1989)
Hibiscus esculentus L.	Gombo; ladies' finger; okra	Fruit, seedling and collar rot	India (Balakrishnan 1947a)
Hibiscus rosa-sinensis L.	Hibiscus; show flower	Leaf, stem, and flower rot	United States (James and Plakidas 1952); India (Ramakrishnan and Seethalakshmi 1956b)
Hibiscus sabdariffa L.	Roselle	Seedling blight; stem gummosis	Philippines (Reinking 1921); Malaysia (Thompson 1929); Congo (Babacauh 1972)
Hibiscus schizopetalus L.	Hibiscus	Leaf and stem blight	United States (James and Plakidas 1952)
Hivistona rotundifolia Mart.		Bud rot	Philippines (Reinking 1923)
Howea belmonreana Becc. (*Kentia belmonreana* Muell.)	Belmore sentry palm; curly palm	Bud rot	France (Labrousse 1931)
Howea forsterana (Moore and Muell.) Becc. (*Kentia*)	Sentry palm; thatch-leaf palm	Bud rot	Italy (Verneau 1953)
Impatiens balsamina L.	Garden balsam	Leaf rot	India (Arya 1959)[b]
Jatropha curas Kent	Physic nut	Fruit, leaf, and stem rot	India (Ramakrishnan and Seethalakshmi 1956b)[b]
Juglans spp.	Walnut	Root and collar rot	Italy (Curzi 1933)
Lablab niger Med. K. (= *L. purpureus* (L.) Sweet)	Hyacinth bean; bonavista bean	Fruit rot	Malaysia (Chee 1969b)
Lansium domesticum	Lanzones, langstat	Crown and root rot	Philippines (Tsao et al. 1993)
Leucaena leucocephala L.		Leaf blight	Solomon Islands (Jackson and Newhock 1978)[b]
Livistona rotundifolia Mart.	Palm	Seedling blight	Philippines (Reinking 1923)
Lycopersicon esculentum L.	Tomato	Fruit rot	India (Coleman 1910); Bermuda (Waterston 1938); Ivory Coast (Ravisé 1966); Brazil (Zentmyer et al. 1973)
Macadamia integrifolia Maiden and Betche.	Macadamia nut	Leaf blight	United States (Hunter et al. 1971; Kunimoto et al. 1976, reclassified to *P. capsici*)
Malus pumila Mill.	Apple	Fruit rot	Philippines (Reinking 1923)
Mangifera foetida	Horse mango	Fruit rot	Malaysia (Chee 1969b)
Mangifera indica L.	Mango	Canker; gummosis; seedling blight; black rot	Philippines (Tsao et al. 1994a); West Africa (P. D. Turner 1960b[b]); Ivory Coast (Lourd and Keuli 1975); Thailand (Kueprokone et al. 1986); United States (Matheron and Matejka 1988c)
Manihot glaziovii Müll. Arg.	Ceara rubber	Seedling blight	India (Ramakrishnan and Seethalakshmi 1956b)
Mimusops elengi L.	Spanish cherry; medlar	Fruit rot	West Africa (Bunting 1924)
Mimusops hexandra	Palla wood	Seedling blight	India (Ramakrishnan and Seethalakshmi 1956b)
Morrenia odorata Lindl. (erronoeously reported as *P. citrophthora* (Burnett et al. 1974])	Milkweed vine	Stem and root rot	United States (Feichtenberger et al. 1984b)
Musa textilis Née.	Manila hemp, abaca	Heart rot of pseudostems	Philippines (Hernandez 1919)
Myristica fragrans Houtt.	Nutmeg	Dieback of seedlings; trunk canker	Java (Rutgers 1917); West Indies (Nowell 1917)
Nicotiana tabacum L.	Tobacco	Stem and seedling blight	India (Thomas et al. 1947)[c]; Italy (Marcelli 1959); Nigeria (Chee 1969b)
Normanbya merrillie F. Muell	Palm	Seedling blight	Philippines (Reinking 1923)[b]
Odontadenia grandiflora Schum. (*O. speciosa* Benth.)	(Apocynaceae)	Shoot dieback; leaf, flower, and bud rot	Sri Lanka (Gadd 1924); United States (Tompkins and Tucker 1947b)
Ormosia calavensis Azaolaex Blco.		Leaf blight	Solomon Islands (Jackson and Newhook 1978)
Paphiopedilum sp.	Orchid	Leaf and stem blight	United States (Hine 1962)[b]
Parkia speciosa Hort ex Hassk.	Petai; mimosa	Fruit rot	Malaysia (Chee 1969b)
Paulownia Siebold and Zucc. *taiwaniana*	Princess tree	Root rot	Taiwan (Hsieh 1979)
Pelargonium L'Her - *Zonale* (Hybrid)	Geranium	Root rot	Iran (Ershad 1971)
Peperomia spp.	Peperomia	Leaf, stem, and root rot	United States (Ark and Dewolfe 1951)
Persea americana Mill.	Avocado; alligator pear	Seedling blight	Philippines (Reinking 1921; Tsao et al. 1993); United States (Conover 1948); West Africa (P. D. Turner 1960)[b]
Petunia × *hybrida* Hort. Vilm.-Andr.	Petunia	Root rot	Iran (Ershad 1971)
Phalaenopsis amabilis (L.) Blume	Moth orchid	Root rot; black rot	Indonesia (Schwarz 1937)
Phaseolus vulgaris L.	Garden bean		Philippines (Youngberg 1926)
Phoenix dactylifera L.	Date palm	Seedling blight	Philippines (Reinking 1923)
Pinanga insignis	Palm	Seedling blight	Philippines (Reinking 1923)[b]
Piper betle L.	Betel vine	Stem canker; foot rot; leaf blight	India (McRae 1934; Thomas et al. 1947)[c]; West Africa (P. D. Turner 1960); Indonesia (Tsao et al. 1985) (causal agent may be *P. capsici sensu* Tsao 1991 formerly considered to be atypical *P. palmivora* MF4)

(continued on next page)

Table 51A.1. *(continued)*

Host	Common name	Disease	Geographical distribution
Piper nigrum L.	Black pepper	Quick wilt; leaf and collar rot	Indonesia (H. R. A. Muller 1936); Sarawak (Newman 1941); Indochina (Barat 1952); Puerto Rico (Gregory et al. 1960); India (Raj and Jose 1966); Brazil (Holliday 1965); Jamaica (Leather 1967); Ghana (Popisil et al. 1972); Pacific Islands (Firman 1975); Thailand (Tsao and Tummakate 1977); China (Ho et al. 1983b) (isolates would probably now be identified as *P. capsici sensu* Tsao 1991 = *P. palmivora* MF4)
Pisum sativum L.	Garden pea	Seedling blight	Philippines (Reinking 1923)[b]
Poncirus trifoliata (L.) Raf. × *Citrus paradisi* MacFady	Citrumelo	Root rot	United States (Zitco et al. 1991)
Pseudospondias microcarpa Engl.	Anacardiaceae	Fruit rot	West Africa (P. D. Turner 1960b)[b]
Psidium guajava L.	Guava	Seedling blight	Philippines (Reinking 1921); West Africa (P. A. Turner 1960b)[b]
Psophocarpus tetragonolobus (L.) DC.	Winged bean; goa bean	Fruit rot	Malaysia (Chee 1969b)[b]
Ptychosperma macarthuri (H. Wendl.) Nichols	Palm	Seedling blight	Philippines (Reinking 1923)[b]
Ptyocarpus indicus Willd.		Leaf blight	Solomon Islands (Jackson and Newhook 1978)[b]
Punica granatum L.	Pomegranate	Brown rot of fruit	India (More et al. 1989)
Quercus spp.	Oak	Root rot	England (Day 1939)[b]
Quisqualis indica L.	Rangoon creeper	Blight	India (Arya 1959)
Rhopalostylis baueri Wendl. and Drude	Palm	Bud rot	France (Labrousse 1931)
Rhynchostylis sp. Blume	Orchid	Black rot	Taiwan (Chen and Hsieh 1978)
Ricinus communis L.	Castor bean	Seedling and leaf blight	India (McRae 1933; Ramakrishnan and Seethalakshmi 1956b)[b]; United States (Uchida and Aragaki 1988)
Sabal causiarum (Cook.) Becc.	Puerto Rican hat palm	Bud rot	Puerto Rico (Tucker 1927)
Sandoricum koetjape	Santol	Seedling blight	Philippines (Reinking 1923)
Santalum album L.	Sandalwood	Leaf blight	India (Narasimhan 1931)
Schizanthus wisetonensis Hort.	Butterfly flower		India (Coleman 1910)
Solanum melongena L.	Eggplant; brinjal	Fruit and root rot	Philippines (Ocfemia 1925); Java (Reitsma and Slooff 1947); Ivory Coast (Ravisé 1966); Puerto Rico (Zentmyer et al. 1973)
Solanum tuberosum L.	Potato	Tuber rot	India (Ramakrishnan and Seethalakshmi 1956b)[b]
Spondias lutea L. (*S. mombin* L.)	Hog plum; jobo	Seedling blight	Philippines (Reinking 1919a)[b]; India (Thomas et al. 1947)[c]
Spondias mangifera	Hog plum	Fruit rot	India (Thomas et al. 1947)[c]
Syagrus campestris (Mart.) Wendl. (*Cocos campestris* Mart.)	Palm	Root rot	Morocco (Malencon and Delecluse 1937)
Syringa vulgaris L.	Common lilac	Collar and trunk rot	Greece (Kouyeas and Chitzanidis 1978)
Tamarindus indica	Tamarind	Crown and root rot	Philippines (Tsao et al. 1993)
Tecoma smithii Wats.	Yellow bells; trumpet bush	Root rot	Sri Lanka (H. F. Davidson 1934)
Theobroma cacao L.	Cocoa (cacao)	Pod rot; trunk and branch canker; cherelle wilt; seedling wilt; black pod; pourriture brune; podridao parda; podredumbre negra; chupon wilt	Sri Lanka (Willis and Green 1897); West Indies (Massee 1899b); West Africa (Büsse 1905); Mexico (Martinez 1912); Java (Rutgers 1912); Philippines (Ghofulpo 1915); Uganda (Small 1915); Fiji (Knowles 1916); Brazil (d'Utra 1917); New Guinea (Bryce 1924; Dwyer 1937, 1953); Congo (Bellefroid 1928); Guatemala (Palm 1932); Cameroon (Voelcker 1940); India (Thomas et al. 1947; K. Ramakrishnan and Thankappan 1965); Costa Rica (Newhall 1948); Sierra Leone (Deighton 1949); Malaysia (Buckley and Allen 1951; Liu 1977); Columbia (Manotas 1953); Indonesia (Hubert 1957); Caroline Islands (Zaiger and Zentmyer 1965); Cambodia and Vietnam (Ravisé 1966); Sarawak (Brasier 1969b); most of South America (Zentmyer et al. 1973); New Zealand (Newhook and Jackson 1978); Thailand (Chomensilpe et al. 1983); Mexico (Lozano et al. 1984)
Theobroma pentagona Bern.	Alligator cacao	Brown rot of fruit	Java (Rutgers 1912)
Trichosanthes anquina	Snake gourd	Fruit rot	Malaysia (Chee 1969b)
Vanda caerulea Griff.	Orchid	Leaf rot	Java (Schwarz 1927); United States (Hine 1962)[b]

(continued on next page)

Table 51A.1. *(continued)*

Host	Common name	Disease	Geographical distribution
Vanda limbata Blume	Orchid	Leaf and heart rot	Java (Schwarz 1927)
Vanilla planifolia Andr.	Vanilla	Leaf, pod, stem, and root rot	Reunion (Kopp 1930); French Polynesia (Tsao and Mu 1987; Mu and Tsao 1987)
Vinca rosea L.	Periwinkle	Blight	Java (Schwartz 1926)
Washingtonia filifera H. Wendl.	Washington palm	Bud rot	United States (Anon. 1941)
Yucca gloriosa L.	Spanish dagger; palm lily; Roman candle	Top rot	Mauritius (Orian 1953)
Zinnia elegans Jacq.	Common zinnia	Leaf blight	India (Arya 1959)
Zizyphus jujuba Mill.	Jujube	Root rot	Taiwan (G. Hall 1989b)
Zizyphus mauritiana Lam.	Indian jubube	Fruit rot	Taiwan (Lin 1984)

[a] Because the *P. palmivora* complex has been reevaluated, many of the citations on black pod of cocoa before 1979 may refer to *P. megakarya* (Brasier and Griffin 1979; Brasier et al. 1981) or *P. capsici* (Tsao 1991; Mchau and Coffey 1995).
[b] Artificially inoculated host.
[c] Thomas et al. (1947) listed *P. palmivora* isolates obtained from the indicated hosts but did not give pathogenicity information.

Table 51A.2. Characteristics of *Phytophthora palmivora* as described by various investigators

Host	Sporangia (μm)	Length-breadth ratio	Chlamydospores diameter (μm)	Oogonia diameter (μm)	Oospores diameter (μm)	Min : opt : max temperatures (°C)	Reference
Palm	38–72 × 33–42 (av. 50 × 35)	1.43 : 1	25–45 (av. 40)	NO[a]	NO	NO	Butler (1907)
Cacao	34–63 × 19–49 (av. 48.5 × 32.3)	1.47 : 1	23–55 (av. 38.9)	NO	NO	NO	Rosenbaum (1917)
Coconut	19–83 × 13–45 (av. 52 × 31)	1.68 : 1	19–61 (av. 41.6)	NO	NO	NO	Reinking (1923)
Rubber	31–62 × 22–39 (av. 39.6 × 24)	1.65 : 1	22–48 (av. 35.5)	NO	17–29	NO	Gadd (1924)
Citrus	23–65 × 15–39 (av. 44.7 × 29.5)	1.45 : 1	19–49 (av. 31)	NO	NO	NO	Ocfemia and Roldan (1927)
Cacao	36–62 × 24–36 (av. 49 × 30.6)	1.6 : 1	21–49 (av. 38)	NO	19–29 (av. 24.1)	NO	Ashby (1929a, b)
Coconut	36–66 × 24–27 (av. 53.5 × 31)	1.73 : 1	19–49 (av. 36.1)	NO	27–41 (av. 31.5)	NO	Ashby (1929a, b)
Cacao	30–78 × 18–42 (av. 49.7 × 30.3)	1.66 : 1	21–45 (av. 35.9)	NO	NO (av. 23)	NO	A. Thompson (1929)
Rubber	36–75 × 21–36 (av. 50.9 × 26.9)	1.87 : 1	27–34	NO	NO (av. 24.1)	NO	A. Thompson (1929)
Coconut	26–88 × 18–41 (av. 52.7 × 31)	1.7 : 1	19–49 (av. 34.8)	NO	NO	NO	Tucker (1931)
Bougainvillea	28–56 × 19–45 (av. 40 × 31)	NO	25–40 (av. 34)	24–31 (av. 26)	16–25 (av. 22)	NO	Ramakrishnan and Seethalakshmi (1956a)
Hibiscus	22–102 × 19–53 (av. 62 × 37)	NO	31–71 (av. 50)	23–31 (av. 28)	16–23 (av. NO)	NO	Ramakrishnan and Seethalakshmi (1956a)
Black pepper[b]	28–76 × 20–40 (av. 43 × 30)	1.5 : 1	12–40 (av. 25)	21–31 (av. 25.5)	17–30 (av. 22.7)	10 : 25–29 : 35	Holliday and Mowat (1963)
Petunia	27–62 × 19–37 (av. 45.1 × 23.9)	1.88 : 1	14–43 (av. 28.9)	23–37 (av. 28.6)	18–31 (av. 23.4)	12 : 25–27 : 35	Ershad (1971)
Cacao 'S' type (MF1)	27–65 × 17–34 (av. 43.8 × 19.6)	1.82 : 1	NO	25–34 (av. 29)	19.6–34 (av. 24.7)	10 : 28–30 : 34	Brasier and Griffin (1979)
Wide range of hosts	31.0–56.4 × 20.7–36.7	1.2 : 1–1.8 : 1	36.2 ± 9.6	NO	NO	5 : 24–30 : 35	Mchau and Coffey (1994a)

[a] Not observed by reference cited.
[b] Probably MF4 = *P. capsici*.

3.2 μm). Chlamydospores are rare on *P. meadii* but abundant on *P. palmivora*. Both species are heterothallic. The morphology of sporangia and patterns of cultures for these species are illustrated by Dantanarayana et al. (1984). Tucker (1931) considered *P. arecae* and *P. meadii* to be synonymous with *P. palmivora*. Oudemans and Coffey (1991b, c) found that *P. arecae* and *P. palmivora* are closely related genetically (isozyme data) and that they are likely to be synonymous.

The previous finding that *P. arecae* isolates clustered closely with *P. palmivora* (Oudemans and Coffey 1991b, c) was supported by Mchau and Coffey (1994a), who found that the six isolates, previously identified as *P. arecae*, clustered with *P. palmivora* (ET7 and ET8). Since the isolates of both species were similar both morphologically and in isozyme patterns, they concluded that the two species are conspecific and that isolates previously designated *P. arecae* should be included in *P. palmivora*.

Mchau and Coffey (1994a) suggested that southwest Asia may be a region from which *P. palmivora* originated, since durian and coconut are indigenous to this area of the world.

The isozyme study of *P. palmivora* isolates from cacao by Oudemans and Coffey (1991b, c) showed that the isolates they studied were genetically very similar; however, a limited number of isolates from coconut and one from durian were genetically diverse from the *P. palmivora* isolates from cacao. A subsequent study by Mchau and Coffey (1994a) included 93

Table 51A.3. Summary of important diagnostic characteristics of *Phytophthora palmivora, P. megakarya,* and MF4 (*P. capsici sensu* Tsao 1991) from cacao[a,b]

Characteristic	*P. palmivora* (S-type or MF1)	*P. megakarya* (L-type or MF3)	*P. capsici* (MF4)
Chromosomes	n = 9–12 small	n = 5–6 large	n = 9–12 small
Colony morphology	Aerial mycelium sparse except in center; stellate, striate pattern; sharply defined edge	Fairly uniform, deep, cotton-woollike aerial mycelium over entire colony; faint lobed/floral pattern; diffuse edge	Uniform dense aerial mycelium over entire colony; petalloid pattern; diffuse edge
Sporangium morphology	Near spherical to ovate-elongate; prominent papillae; rounded base	Near spherical to ovate-elongate; prominent papillae; rounded base	Elongate and boat-shaped; less prominent papillae; base tapered toward pedicel
Sporangium pedicel type	Broad, short, and occluded; length mostly 2–5 μm	Narrow, medium length, and not occluded; mostly 10–30 μm	Narrow, long, and not occluded; length 20–150 μm, sometimes up to 250 μm
Length-breadth ratio of sporangia	1.2–1.8 (up to 2.2 on pods)	1.2–1.6	1.6–2.0 (–2.4)
Compatibility type	Predominantly A2 (A1 uncommon)	Predominantly A1 (A2 uncommon)	A1 and A2
Oogonial morphology	Spherical, meeting stalk fairly abruptly; stalk narrow, tubular, or broadening toward oogonium	Pyriform, tapering downward to a funnel-shaped base	Spherical to slighly oval, meeting stalk fairly abruptly; stalk narrow, tubular, or broadening toward oogonium
Staining of oogonial wall with aceto-orcein	Nil	Stains darkly	Stains darkly (see Brasier and Medeiros 1978)
Response to *Trichoderma* by A2 isolates	Negative	Negative	Positive
Cardinal temperatures (°C) (min., opt., max.)	10–11, 28–30, 34	10–11, 24–26, 29–30	8, 28–30, 33
Distribution on cacao	Worldwide	West Africa	Central and South America, West Indies

[a] Based on tests carried out on carrot agar. Reproduced by permission of Cambridge University Press, Cambridge, United Kingdom; courtesy C. M. Brasier and M. J. Griffin (1979).
[b] *P. palmivora* MF4 has been redescribed as *P. capsici* (Tsao and Alizadeh 1987; Tsao 1991; Mchau and Coffey 1995).

Table 51A.4. Comparison of sporangium characteristics of the morphological forms of *Phytophthora palmivora* with some species that closely resemble *P. palmivora*

Species or form	Sporangiophores[a]	Sporangia shape[b]	Sporangia size (μm)[c]	L : B[d]	Caducous	Pedicel length (μm)	Reference[e]
P. palmivora sensu Butler (MF1)	Sympodium	E, O, broad base	35–60 × 20–40 (55.5 × 33)	1.7–1.9	+	2–5	Brasier and Griffin (1979); Stamps (1985a)
P. megakarya (MF3)	Loose sympodium	OBP, O, E	20–60 × 13–41 (36 × 26)	1.2–1.6	+	10–30	Brasier and Griffin (1979); Stamps (1985b)
P. palmivora (MF4) (= *P. capsici*)	Compound sympodium; umbellate	E, boat-shaped	34–69 × 19.30 (43 × 22)	1.6–2.0	+	10–150	Brasier and Griffin (1979)
P. capsici sensu Tsao 1991 (MF4) (= *P. capsici*)	Compound sympodium; umbellate		37–50 × 19–30[f] (42 × 25)	1.14–2.19	+	43–148	Tsao (1991)
P. capsici sensu Tucker (1931)	Compound sympodium	E, OBP, OBT, S, variable	16–71 × 12–57 (35 × 23.6)	7.3–2.0 (1.7)	+	10–200	Tucker (1931); Stamps (1985f)
P. arecae[f]	Irregular sympodium	E, OBT, tapered base	30–71 × 20–48 (45 × 33)	1.1–1.6 (1.35)	+	1.6	Coleman (1910); Stamps (1985c)
P. meadii	Sympodium	E, OBP, S, distorted shapes	25–72 × 14–40 (48 × 24)	1.3–2.0	+	10–20	McRae (1918); Waterhouse (1974b)
P. botryosa	Sympodium with nodal swellings	O, in clumps	28–31 × 15–16 (28 × 15)	1.6–2.1 (1.8)	+	5–20	Chee (1969); Stamps (1985e)
P. heveae	Irregular sympodium; occasional swellings	E, OBP, broad base	27–66 × 20–49 (44 × 29.6)	1.1–2.9 (1.45)	+	<10	Thompson (1929); Waterhouse (1963); Stamps (1978d)
P. nicotianae (*P. parasitica*)	Irregular sympodium	E, O, OBP, S	11–60 × 10–45 (40 × 28.5)	1.1–1.7 (1.34)	–		Waterhouse (1963)
P. colocasiae	Irregular sympodium	E, O	40–70 × 17–28 (59 × 23)	(1.6)	+	3.5–10	Butler and Kulkarni (1913)
P. arecae[g]		E, O	45.4 ± 4.0 × 29.5 ± 2.8	1.6	+	3.4 ± 0.3	Mchau and Coffey (1994a)
P. palmivora[g]		E, O	45.3 ± 5.0 × 29.8 ± 3.6	1.5	+	3.5 ± 1.0	Mchau and Coffey (1994a); Stamps (1985a)

[a] See Figure 4.5 (Chapter 4) for illustrations of sporangiophores of the different forms and species.
[b] E, ellipsoidal; O, ovoid; OBP, obpyriform; OBT, obturbinate; S, spherical (Figures 4.9 and 4.10; Chapter 4).
[c] Averages are in parentheses.
[d] Length-breadth ratio.
[e] Second author listed has summarized pertinent data on that species.
[f] Carrot agar disk in water under light (Tsao 1991).
[g] *P. arecae* and *P. palmivora* are considered to be conspecific (Mchau and Coffey 1994a). See Chapter 8.

isolates of *P. palmivora* from cacao and many other hosts, as well as six isolates described as *P. arecae* from arecanut. These isolates were compared morphologically and by isozyme patterns. The isozyme patterns consisted of 18 electrophoretic types (ETs). The greatest amount of genetic diversity was found among the isolates from coconut (eight ETs), durian (five ETs), and other hosts.

Mchau and Coffey (1994a) found that chlamydospores were formed abundantly by some isolates considered to be *P. arecae* but not by others; likewise, some isolates of *P. palmivora* formed chlamydospores but others did not. Diameters of chlamydospores were similar. Chlamydospores of both types were similar, as indicated much earlier by Tucker (1931), who treated *P. arecae* and *P. palmivora* as conspecific. Although sporangia size and shape and production of irregular sporangiophores by *P. arecae* have been used to differentiate the two species (Waterhouse 1963), Mchau and Coffey (1994a) did not note any significant difference in the shapes (predominantly ellipsoid or ovoid with a rounded base) or sizes (31.0 to 56.4 µm length and 20.7 to 36.7 µm breadth). The length-breadth ratios ranged from 1.2:1 to 1.8:1 (Tables 51A.4 and 51A.5). *P. meadii* was distinctly separated from *P. palmivora* but showed a close relationship to *P. botryosa.*

P. palmivora and *P. megakarya* can be differentiated by growth under white light. *P. palmivora* is stimulated, but *P. megakarya* is inhibited by white light (Blanc Industrie U40RS Claude fluorescent bulbs). These differences were even more marked by exposure to green light (General Electric F40 G-green fluorescent bulbs emitting light at a wavelength of 525 nm). The change in growth pattern was the result not only of the effect of light on the medium (clear V8 juice agar) but also of an effect on the isolates. The reaction to light was suggested as a simple diagnostic means for separation of *P. palmivora* from *P. megakarya* (Blaha 1983).

Turner (1969) and Waterhouse (1974b) considered the *Phytophthora* isolated from *Piper nigrum* (black pepper) to be an atypical strain of *P. palmivora.* Further research by Tsao and Tummakate (1977) indicated that the atypical strain from black pepper is similar to that designated MF4 by Brasier and Griffin (1979). These isolates are now included in the redescription of *P. capsici* (Tsao and Alizadeh 1987; Tsao 1991; Mchau and Coffey 1995; see Chapter 14 on *P. capsici*). The MF4 group was placed in *P. capsici* because of the presence of long, caducous sporangiophores and the similarity of sporangial shapes (Tsao and Alizadeh 1987; Tsao 1991).

In the literature before 1974, many isolates from cocoa that were designated *P. palmivora* are now known to be *P. megakarya* or *P. capsici.* The principal characteristics of *P. palmivora, P. megakarya,* and *P. capsici* are summarized in Tables 51A.3–51A.5.

Since the discovery previous to 1981 that isolates now considered to be *P. megakarya* and *P. capsici* were misidentified as *P. palmivora* (Brasier et al. 1981), many citations from Southeast Asia that attribute the cause of a disease such as foot rot of *Piper niger* or *P. betle* to *P. palmivora* are probably wrong. In Nigeria, an important cocoa-growing area, *P. megakarya* is much more prevalent as the cause of severe black pod rot than *P. palmivora* (Gregory and Maddison 1981).

Epidemiology and Disease Development

Development of individual diseases is covered below; however, in principle, the development and epidemiology of all of the variants and different biotypes of *P. palmivora* share many characteristics.

The most important developmental factor is the ability of *P. palmivora* to form sporangia and abundant zoospores on diseased tissue when free water is present. Because the sporangia are caducous, they are readily dislodged and disseminated by rain splash and wind-driven rain or by other agents,

Table 51A.5. Comparison of chlamydospores and sex organs of the morphological forms of *Phytophthora palmivora* with species that closely resemble *P. palmivora*

Species or form	Chlamydospores diameter (µm)	Mating types	Oogonia diameter[a] (µm)	Oospores diameter[a] (µm)	Min : opt : max temperatures (°C)	References[b]
P. palmivora sensu Butler (MF1)	30–45, abundant	A2; A1 rare	21–40 (29.5)	16–30 (23.0)	11 : 28–30 : <35	Brasier and Griffin (1979); Stamps (1985a)
P. megakarya (MF3)	20–40	A1; A2 rare	19–37 (26.8)	18–34 (23–28)	10 : 24–26 : 30	Brasier and Griffin (1979); Stamps (1985b)
P. palmivora (MF4) (= *P. capsici*)	Rare	A1 and A2	(26.5)	(24.2)	8 : 28–30 : 33	Brasier and Griffin (1979); Stamps (1985a)
P. capsici sensu Tsao 1991 (MF4) (*P. palmivora* atypical type)	Rare in some; abundant in others; 28–29	A1 and A2	27–43 (33.0)	22–37 (28.0)	6–9 : 27–30 : 33–39	Tsao (1991)
P. capsici sensu Tucker (1931)	Unknown	A1, A2, and H[c]	27–46 (28.7)	28–35 (24.4)	11 : 28 : >35	Tucker (1931); Stamps (1985f)
P. arecae	24–60, rare	A1 and A2	23–44 (30.0)	23–36 (28–30)	10 : 27–30 : 35	Coleman (1910); Stamps (1985c)
P. meadii	16–36	A1, A2, and H	21–49 (33.4)	15–23 (24.0)	75 : 25–30 : <33	McRae (1918); Waterhouse (1974b)
P. botryosa	14–30, rare	A1 and A2	(23–30) (25.0)	21–24 (22.3)	9 : 26 : 32	Chee (1969); Stamps (1985e)
P. heveae	Unknown	H	17–32 (25.9)	15–26.8 (21.5)	11 : 25 : 32–32.5	Thompson (1929); Stamps (1978d)
P. nicotianae (*P. parasitica*)	13–60, abundant	A1, A2, and H	15–64 (26.8)	13–35 (22.6)	5 : 27–32 : >37	Tucker (1931); Waterhouse (1963)
P. colocasiae	17–38, abundant	A1 and A2	20–35 (29.0)	18–30 (23.0)	>10 : 27–30 : >35	Butler and Kulkarni (1913)

[a] Average is in parentheses.
[b] Second author listed has summarized pertinent data on that species.
[c] Homothallic.

such as clippers, knives, rodents, or ants (Taylor and Adedoyin 1981; Taylor and Griffin 1981) that carry inoculum from diseased to healthy plant tissue.

Because of their ability to produce sporangia rapidly, *P. palmivora* and similar species have a short regeneration time in tropical areas where frequent rainfall prevails and temperatures range from 20 to 30°C. Gregory (1983) considered black pod of cocoa to be a typical multicyclic disease; this concept, based on his philosophy, is discussed below.

In the terminology of Vanderplank (1963), the growth of a typical multicycle epidemic can be expressed by the continuous compound interest equation

$$x = x_0 e^{rt},$$

where x is the proportion of plants or units of the plant affected by the disease; x_0 is the total amount of initial capital, analogous to the amount of diseased tissue or inoculum at the start of the epidemic; e is the natural logarithm; r is the rate of increase, also called the apparent infection rate; and t is time in days, weeks, or months. Vanderplank (1963) used this equation as a model to explain the control of multicyclic diseases. If sanitary practices, such as burning infected cocoa pods, are used, x_0 is reduced but under field conditions is probably never reduced to zero. If the plant has vertical resistance (near immunity), the x_0 is reduced to near zero because the pathogen does not sporulate. If chemicals that eradicate the inoculum, such as methyl bromide, are available, x_0 can be reduced to zero and no disease should occur; however, in the real world, this seldom if ever happens. Cocoa pods with black pod lesions can be removed, but what percentage of them can be removed economically? Is *P. palmivora* still lurking in flower cushions or in plant debris on the orchard floor?

Although vertical resistance (near immunity and complete suppression of sporangia formation on tissue) reduces x_0 to zero, vertical resistance in cocoa trees to *P. palmivora* is difficult to obtain. Even if it is obtained, the history of development of new pathological or physiological races, as occurs in *P. infestans* on potato with R-genes for vertical resistance, indicates that vertical resistance may not be dependable. What is left in the equation is r, the rate factor. The rate of disease progress can be reduced by applying preventative fungicides before the inoculum reaches susceptible tissue. Copper fungicides such as the old Bordeaux mixture, which still has the greatest tenacity under rainfall, and the fixed copper fungicides are somewhat effective and, where the cost is not too great, have a measurable level of effectiveness because the population of the pathogen is reduced even though it is not eradicated. The new systemic phenylamide fungicides such as metalaxyl (see Chapter 7) have a greater capacity to penetrate plant tissue than the older protectant fungicides. This feature favors greater persistence than that achieved by fungicides that are not systemic. The main problems with the newer systemic fungicides are the expense, the possibility for development of resistant biotypes of *P. palmivora,* and, in tropical areas, wash-off by rainfall. Fosetyl-Al is systemic and moves both upward and downward in plants. If registered for use, it should be effective. See Chapter 7 for a discussion of the use of fungicides.

Another factor that decreases the rate (r) of the epidemic is horizontal resistance or, as termed by others, field resistance or rate-limiting resistance (MacKenzie et al. 1983). There are many examples of this type of resistance in the numerous tests that have been conducted over the years; however, on perennial tree crops, implementation of horizontal resistance, which must also satisfy demands for certain important horticultural qualities and adaptation to different geographical areas in the world, is difficult. Despite the difficulty, attaining horizontal resistance should be a goal and, if obtained, should be used additively with other means of control such as fungicides and cultural control practices. See Chapter 6 for a discussion of these concepts. Cultural practices involve correction of the adverse effects of shading, close spacing, and excessive fertilization, all of which either favor the pathogen or increase the susceptibility of the host plant. Gregory's chapters (Gregory and Maddison 1981) base control strategies on the above theoretical approach. Although there are exceptions, these concepts provide a working model for development of control methods.

Diseases Caused by *Phytophthora palmivora*

Black Pod, Stem Canker, and Chupon Wilt of *Theobroma cacao* L. (Cocoa)

P. palmivora causes three types of disease on cocoa trees, black pod, stem canker, and chupon wilt (chupons are small suckers at the base of the tree). These diseases, the most important of which is black pod, are estimated to be responsible for a 20 to 30% loss of the world's cocoa crop. In some areas, up to 90 to 95% of the crop is rendered unusable. Black pod is now known to be endemic in all cocoa-growing countries (Table 51A.1). The early history of the disease has been documented by Thorold (1974). The comprehensive book *Phytophthora Diseases of Cocoa* (Gregory et al. 1974) contains chapters (based on a worldwide conference) on subjects such as taxonomy, epidemiology, and control. A subsequent book (Gregory and Maddison 1981) covers the results of several years of team research based in Nigeria on black pod caused by *P. megakarya,* which previous to Brasier and Griffin (1979) had been considered to be part of the *P. palmivora* complex. Firman (1974, 1975, 1978a, b) discusses the black pod and canker diseases of cocoa in Fiji and the South Pacific. Fulton (1989) aptly summarizes the cocoa disease trilogy, black pod, monilia pod rot, and witches'-broom, in the Americas.

Several other species of *Phytophthora* cause black pod on cocoa. *P. megakarya* occurs in Nigeria and Cameroon (principal pathogen) and is also found in Ghana, Fernando Po, and Gabon. In Brazil, *P. capsici* is the main causal agent and also occurs commonly in Mexico, Guatemala, El Salvador, Venezuela, and Cameroon. *P. citrophthora* as well as *P. palmivora* causes black pod in Brazil; *P. megasperma* has been reported in Venezuela (reviewed by Zentmyer 1987). In Mexico, *P. capsici, P. citrophthora,* and *P. heveae* were isolated from cocoa trees (Lozano and Romero 1984).

Black Pod

Black pod was probably known as early as 1833 in Sri Lanka (formerly Ceylon) (Thorold 1974). Pods are susceptible at all stages of development. Pods or cherelles (immature pods) may be infected at any place on the surface, but infection is most often initiated at the tip or stem end. The first symptom is a brown to black spot on the pod, which spreads rapidly in all directions and eventually covers the whole pod (Plate 51A.1), hence the name black pod. When humid conditions prevail, a white bloom consisting of mycelium and sporangia forms on the surface of diseased pods. In advanced stages, *P. palmivora* invades the internal tissues and causes discoloration and shriveling of the cocoa beans. Diseased pods eventually become black and mummify. A report from Trinidad indicates that black pod causes a distinct, characteristic seaweed-like odor (Sreenivasan and Quesnel 1975).

Disease Development. The inoculum that initiates black pod can emanate from soil and infected roots, stems, and leaves and can be spread on harvesting or pruning equipment or by ants or rodents. Persistent lesions on stems can allow *P. palmivora* to grow downward into the peduncle of a pod (discussed by Jackson and Newhook [1978]). When free water is present, sporangia and zoospores form and infect the pods. The caducous sporangia may become airborne and spread to other sites. Contact of diseased pods with healthy ones will also allow the pathogen to spread. Windborne rain is important; however, the range of dispersal is rarely beyond 4 m (Hicks 1975). Ants, *Anoplolepsis longipes* (Jerdon), a ground, non-tent-building species, were shown in laboratory tests to transport inoculum, but in the field they did not increase the incidence of black pod infection; however, the tent-building species, *Technomyrmex albipes* (F. Sm.), transported inoculum and increased the number of fruits with black pod symptoms in Papua New Guinea (McGregor and Moxon 1985). McGregor and Moxon (1985) review reports, e.g., Dade (1927), on the role of ants in the epidemiology of black pod. The book edited by Gregory and Maddison (1981) discusses the role of ants in the epidemiology of the black pod disease in Nigeria caused principally by *P. megakarya.* Ants (*Pheidole megacephala* (Fabr.)) spread *P. palmivora* propagules vertically and horizontally in the Ivory Coast (Babacauh 1982). In the Solomon Islands, rats (*Rattus rattus* L. and *R. exulans*) disseminate *P. palmivora* inoculum (Jackson and Newhook 1978).

Narrow spacing between trees, heavy foliage canopy, and cool, damp weather favor the rapid spread of the disease. In Ghana, an increase in spacing from 8 × 8 feet to 12 × 12 feet reduced pod rot from 12.4 to 5.2% (P. D. Turner 1965b).

Control. The account by Fulton (1989) succinctly addresses the problems associated with control of black pod. The black pod disease is difficult to control because *P. palmivora* persists in the soil and debris more than 3 years after black pod-diseased trees have been removed. P. D. Turner (1965a) reported that *P. palmivora* is a strong competitor for cocoa tissue in soil and survives in soil more than 2 years. Ramarao and Umabala (1982) showed that *P. palmivora* survived for 60 days in soil. Chlamydospores germinated well in natural soil but not in autoclaved soil. The multicycle nature of *P. palmivora* allows rapid increases in inoculum. Sanitary measures such as removal of diseased pods frequently reduces the final number of black pod-infected fruit (Holiday's 1980 review). A guideline from the South Pacific Commission in New Caledonia (Firman 1978b) recommends that infected green pods be removed frequently because they are a source of inoculum and because they will rot completely within 2 weeks. Ripe pods with a few black spots can be harvested, however, because the cocoa beans will still be useful. Despite the general recommendation to remove infected pods, a study in the Solomon Islands showed that heavy roguing did not significantly reduce loss caused by black pod (Jackson and Newhook 1978), probably because the inoculum (detected by baiting) also exists on the bark, healthy canopy leaves, and even on leaves of a shade tree, *Leucaena leucocephala* L. Sanitary measures such as disinfesting harvesting and pruning tools before use and removing and burning infected pods have some effect, but under high rainfall conditions, it is difficult to suppress the inoculum sufficiently to attain control.

Pruning of the canopy followed by spraying with Bordeaux mixture or other copper-based compounds at 2-week intervals has been somewhat effective, but in some areas, such as Nigeria (Hislop and Park 1982), the degree of control was not good. Bordeaux mixture was better than other formulations, probably because it withstands weathering. Fungicide application by spraying is particularly necessary prior to rainfall and during periods of high relative humidity. Since *P. palmivora* rarely spreads beyond 4 m, control is more economical if individually diseased trees are sprayed rather than entire areas (Hicks 1975).

McGregor (1984a) showed in small-scale evaluations on cocoa pods in Papua New Guinea that several fungicides (including copper and metalaxyl) were equally effective against artificial inoculation with zoospores except when rain fell after spraying. Cuprous oxide rapidly leached away, but metalaxyl was rapidly absorbed by pods and remained effective in prevention of black pod infection. McGregor proposed that metalaxyl should be used during critical wet periods and cuprous oxide for the remainder of the year. Persad (1987) in Trinidad recommended metalaxyl together with a copper fungicide in four applications during the crop season. Fosetyl-Al was ineffective as a foliar and pod spray. In Papua New Guinea, fosetyl-Al was highly effective when applied by trunk injection (Holderness 1992), and in Australia, potassium phosphonate controlled black pod, canker, and seedling blight (R. D. Anderson and Guest 1990; R. D. Anderson et al. 1989).

In Papua New Guinea, seedling blight in nurseries is effectively controlled by a seed treatment and a foliar spray with metalaxyl (10 g a.i./1,000 ml) (McGregor 1984b).

Cultivars that are resistant to black pod are listed by Soria (1974) and Lawrence (1978); however, variability in virulence of *P. palmivora* isolates to different cultivars, such as that reported in Jamaica by Fagan (1988), can reduce the effectiveness of host resistance. See Table 6.1 in Chapter 6.

Stem Canker

Stem canker was first thought to be caused by a pathogen different from *P. palmivora,* which causes black pod; however, Rorer (1910) proved that *P. palmivora* was capable of producing both diseases. The history, symptoms, and control measures for stem canker were described in detail by Firman (1974). Stem canker is often difficult to detect in the early stages because external symptoms are not always visible on the outer bark. Defoliation is usually the first visible symptom and is caused by coalescence of the cankers that girdle the stem or fan branches. In early stages of canker development, the bark can be peeled to reveal wood of a watery gray color, often with reddish streaks that intensify in color after exposure to air. In advanced stages of the disease, a reddish fluid that exudes from cankers dries into a rusty deposit. The wood turns brown, and blackish streaks can often be seen. Sometimes the disease may remain localized in the bark, especially if conditions become unfavorable for disease development (dry weather). In some instances, infected trees flower more abundantly than healthy trees. This is probably because of the girdling effect that kills the phloem tissue.

In Brazil, a characteristic protuberance or "button bulge" has been observed on the surface of diseased trunks. Thumb pressure on these protuberances yields a reddish brown exudate. Some trees also display vertical fissures that often extend to more than 40 cm in length (Pereira and Pizzigatti 1980).

A sudden death of trees occurs in Papua New Guinea, where *P. palmivora* and the bark beetle (*Eccoptopterus spinosus* (Oliver)) are associated. The sudden death syndrome is initiated by *P. palmivora*-induced cankers on trunks and main branches. Although the role of the bark beetle is not proved experimentally, observations indicate that beetles are attracted to *Phytophthora* cankers (Prior 1986).

Disease Development. Sporangia may be spread from infected pods to stems during rain storms or by physical contact; however, the pathogen does not usually penetrate healthy bark directly. *P. palmivora* often grows from a stem canker through the peduncle to a pod and causes a lesion (illustrated in Figure 1 of Jackson and Newhook [1978]).

Control. Excision of stem cankers was effective in the Caroline Islands; however, roguing of diseased pods was not sufficient for control of black pod in the Solomon Islands (Jackson and Newhook 1978). Spraying with a fungicide was necessary to prevent the spread of the fungus not only to pods but also into newly formed flower cushions and branches (Zaiger and Zentmyer 1965). Application of metalaxyl controlled black pod in Trinidad (Persad 1987) and bark canker in Papua New Guinea (Prior and Smith 1982). Although it would be expected that fosetyl-Al should also be effective, it was found in Trinidad to be ineffective as a foliar spray both in the laboratory and the field (Persad 1987). Fosetyl-Al was highly effective when applied by trunk infection (Holderness 1992). See Table 7.8 for references on phenylamide fungicides (e.g., metalaxyl) and Table 7.12 for references on phosphonates (e.g., fosetyl-Al).

The Criollo cultivar is the most susceptible, Forastero hybrids are intermediate in susceptibility, and Trunitanio has shown some resistance to canker. Although reported to be susceptible in some localities, the cultivar Amelonado has shown resistance in Fiji and the Solomon Islands (Firman 1978).

Chupon Wilt

The chupon is a sucker that emerges at the base of the cocoa tree and because of its soft tissue is susceptible to infection by *P. palmivora*. Usually, the axil of a young leaf is infected, and then an angular, necrotic lesion forms. The young shoots are then rapidly girdled, resulting in wilting of the shoot. Diseased chupons are also an important source of inoculum (Jackson and Newhook 1978).

Purple Blotch and Fruit Rot of *Annona squamosa* L. (Sugar or Sweetsop Apple)

P. palmivora causes water-soaked lesions on green fruit. Fruits of all ages are susceptible; when affected, the fruits drop and often mummify rapidly. Excessive dropping of small fruits and the presence of discolored fruit on the tree are generally indicative of purple blotch. As the disease progresses, lesions on fruits gradually turn purple brown to black and eventually extend over the entire surface of the fruit. Internally, affected fruits become brown, and in severe cases, the seeds shrivel (Purss 1953). See reviews by Cook (1975) and Zentmyer and Mitchell (1985/1986). Purple blotch has also been reported to be caused by *P. parasitica* (V. G. Rao et al. 1962).

Disease Development. Rain favors infection of fruit near the ground because of splashing of inoculum. The disease is most prevalent after heavy rainfall and is usually restricted to the fruit on lower branches, which usually drop about 4 days after the appearance of purple lesions.

Control. Branches close to the ground should be pruned. Mulching around trees reduces the intensity of rain splash by which caducous sporangia are carried up to fruits. Use of fungicides, if cost effective and registered, should be useful.

Fruit Rot of *Artocarpus communis* (= *A. altilis* (Parkins) Fosb.) (Breadfruit)

Water-soaked lesions with light brown centers appear 48 to 72 h after infection, enlarge rapidly, and completely rot the fruit. The fruit then mummifies on the tree (Trujillo 1960).

Disease Development. Lesions enlarge most rapidly at temperatures of 27 to 29°C. Diurnal fluctuations in temperature that cause changes in the growth rate of the fungus result in characteristic light and dark regions in some lesions.

Control. Effective control measures have not been developed. Infection of rough-skinned cultivars appears to be independent of wounds, but smooth-skinned cultivars become infected only through wounds (see summary by Zentmyer and Mitchell [1985/1986]).

Root and Fruit Rot of *Carica papaya* L. (Papaya)

P. palmivora, the causal agent of diseases of papaya in Hawaii, was at one time misidentified as *P. parasitica* (Hine et al. 1965; see Tokunaga and Bartnicki-Garcia 1971). Plants are susceptible to root rot at all ages (Plate 51A.2), but seedlings are particularly susceptible (Plate 51A.3). Leaves of affected plants wilt, turn yellow, and hang limply; in advanced stages, all of the leaves drop, leaving the trunk bare (Plate 51A.2). The lateral roots and taproots are entirely destroyed, and a foul odor often emanates from diseased trees. Trees with rotted roots are susceptible to drought and fall during high winds.

P. palmivora causes a trunk rot when plants are wounded. The infection spreads through the upper portion of the trunk, thus causing the leaf- and fruit-bearing branches to collapse.

On ripening fruit, the rot initially appears as small, circular, water-soaked lesions 5 to 10 mm in diameter. These rapidly expand into white, circular or elliptic lesions 70 mm or more in diameter (Plate 51A.4). On damp, humid days, a whitish mass of sporangia and mycelia is visible on the surface of the fruit. Ripe fruits are more susceptible to infection than green fruits because of the reduction in the proteolytic enzyme papain with ripening (Hine et al. 1965).

P. cinnamomi has also been implicated as a cause of papaya root rot in Peru (Bazan de Segura 1951).

Disease Development. Excessive soil moisture and ambient temperatures between 20 and 30°C result in rapid disease development. Epidemics of fruit rot usually occur during rainy periods because the caducous sporangia on diseased fruit are splashed onto healthy fruit.

Control. Papaya should be planted on land not previously planted to papaya or on land that has been rotated to nonsusceptible crops. Seedbeds in nurseries should be steamed or fumigated prior to planting. Good soil drainage is essential for control of damping-off.

Fungicides such as metalaxyl, Difolatan 4-F, Dithane 4S, or Kocide 10 prevent infection when applied weekly or biweekly a few weeks before the rainy season. Additional applications are required after heavy rains. Weekly sprays with chlorotholonil and zinc-maneb fungicides reduce the incidence of trunk rot and fruit rot (Alvarez and Nelson 1982). If registered for use, metalaxyl or fosetyl-Al should be effective.

Fruit rots can be controlled by hot water treatment (48°C for 20 min), provided the infections are not older than 24 h. Two-day-old infections were not controlled by hot water treatment (Aragaki et al. 1981). Alvarez and Nelson (1982) found that chlorothalonil (3.4 kg. a.i./ha) sprayed weekly was effective in protecting trees before trunk cankers developed. After cankers formed, trees continued to die during wet weather, regardless of the fungicide applied.

Seedling roots of papaya are susceptible to *P. palmivora* only during the first 2 to 3 months of growth. Therefore, the seedling disease phase on papaya can be controlled by planting seeds in pathogen-free virgin soil placed in the planting hole (Ko 1971, 1982a). This method is advantageous in Hawaii, where the porous lava soils make fumigation with methyl bromide impractical.

The incidence of trunk rot can be reduced by avoiding damage to trunks during cultivation. *Carica cauliflora* and hybrids with *C. monoica* have resistance to root rot (Zentmyer and Mitchell 1985/1986).

Bud Rot and Premature Nut Fall of *Cocos nucifera* L. (Coconut) and Other Palms

P. palmivora was first described by Butler (1906) as causing bud rot on coconut palm in India; however, bud rot had been noted as early as 1834 in the West Indies. In Jamaica, where bud rot was described by Ashby (1920), *P. palmivora* causes both bud rot of coconuts and black pod of cocoa. The status of the bud rot disease of coconut was reviewed by Radha and Joseph (1982).

In Hawaii, the cause of bud rot of coconut was tentatively ascribed to *P. katsurae* (Uchida et al. 1992). *P. palmivora* was isolated from 10% of the crowns of coconut trees and *P. katsurae* from 12% by using carrot-potato-dextrose agar (Steer and Coates-Beckford 1990). The actual incidence of isolation of the two fungi might have been even higher if a selective medium had been used (Chapter 2). Chase and Broschat (1991) describe root rot, bud rot, and leaf blights caused by *P. palmivora, P. nicotianae,* and *P. katsurae* on *Cocos nucifera* and several ornamental palms such as *Chaemaedora elegans* and *Chrysalidocarpus lutescens.* They also discuss control measures. Bud rot appears to be endemic with coconut culture and is one of the most destructive diseases affecting palm trees (Sharples 1926; Ocfemia 1934). See also Cook (1975) and Zentmyer and Mitchell (1985/1986). Coffey et al. (1991) published a comprehensive review on the epidemiology and control of bud rot and premature nut fall in Indonesia.

Trees from 15 to 40 years of age are most susceptible. The first observable symptom is the yellow brown color of the emerging young heart leaf. When infected, this leaf soon wilts and dies. By the time these symptoms can be seen, the growing point of the coconut has already rotted. The heart leaf can be easily removed by a gentle pull soon after the first signs of chlorosis appear. Bud rot caused by *P. palmivora* spreads from the heart leaf outward and finally invades the entire central whorl of leaves and spathes (Ashby 1920). Older leaves often remain green and fall after maturity; however, since no new leaves are formed, the trunk soon becomes bare. Invasion by secondary organisms such as bacteria cause an offensive odor (Ocfemia 1934). Radha and Joseph (1982) found that *Xanthomonas, Pseudomonas,* and *Erwinia* species are associated with bud rot and tend to follow the drier rot caused by *P. palmivora.* They found that success of isolation of *P. palmivora* was always greatest when the sample was taken from the drier tissue.

Rows of brown spots across the leaf pinnae denote incipient infection in some cultivars of palm (e.g., Palmyra). A dark red to brown line marks the limits of the advance of the fungus in the woody tissue. The affected portions of the trunk turn soft. In some palms, infection of the growing point results in a watery rot that is often mistaken for bacterial infection.

P. palmivora can also cause disease on young, immature nuts, which then fail to develop and fall prematurely. More lesions developed on artificially inoculated 4-month-old nuts than on 10-month-old nuts. Infection seldom occurs without wounding (Bennett et al. 1986).

Disease Development. D. C. Harris et al. (1984) showed that *P. palmivora* is soilborne and persists in infected roots. They noted that plants imported to Great Britain from Jamaica were frequently contaminated with *P. palmivora.* They conjectured that soilborne inoculum in roots might be a source of outbreaks of the bud rot disease. Severity of bud rot is greatest at temperatures of 16 to 20°C accompanied by high relative humidity. Whole plantations have been destroyed when conditions have remained favorable. Primary lesions are visible within 72 h after *P. palmivora* penetrates the tissue, and secondary sporangia form on the primary lesions within the next 48 h. An infection cycle can be completed in approximately 1 week if moist conditions prevail. Oospores develop within 4 weeks in diseased tissue and also in culture when contaminated with *Thielaviopsis* (Joseph and Radha 1975; Radha and Joseph 1975, 1982).

Control. As a sanitary measure, diseased trees should be cut down and destroyed by bud scorching with flame throwers to reduce the supply of the primary inoculum. Coffey et al. (1991) describe an integrated control plan for Indonesia that combines sanitation and use of protectant fungicidal sprays. Prophylactic sprays with Bordeaux mixture may be beneficial if applied before infection occurs. Metalaxyl should be efficacious if registered for use and economically feasible. Root drench with fosetyl-Al and phosphonic acid is effective but may be too expensive for small growers (Coffey et al. 1991). Spacing trees at sufficient distances from each other allows more rapid drying of the soil. Palmyra palm (*Borassus flabellifer*) is more susceptible than coconut palm to infection but less susceptible than *Areca* palms.

Patch Canker of *Durio zibethinus* Murr. (Durian)

The durian tree is grown mainly for its fruit. In some areas, the cultivation of the durian tree is seriously threatened by patch canker that occurs on the trunk of the tree (reported first by Thompson in 1934 and confirmed by Navaratnam in 1966). Exudation of dark red fluid around the collar region is usually the first indication of infection. Cankers grow and eventually girdle the tree. The infected cortical tissue, normally pink or light brown, changes to red, dark brown, or black. The lesion is bordered by a well-defined wavy margin. Leaves fall and branches die back even before the crown of the tree is completely girdled. Some infected trees may persist for a year or more, but eventually all the leaves drop, and the tree dies (see Zentmyer and Mitchell [1985/1986] and Holliday [1980] for a summary of information on this disease).

Disease Development. Infection results mainly from zoospores that are produced during periods of high rainfall. Affected trees may survive for long periods in dry weather because the pathogen becomes inactive. Patch canker is more prevalent in plantations in which trees have been planted close together and the foliar canopy is dense.

Control. Excision of diseased bark tissue on the trunk followed by treatment of the exposed areas with a suitable fungicide is effective if the disease is detected early. Wide spacing of trees and weed prevention around the bases of trees helps to allow drying of the trunks and foliage.

Some clones of durian are less susceptible than others (Navaratnam 1966).

Diseases of *Hevea brasiliensis* (Para Rubber)

Several diseases of rubber are caused by various *Phytophthora* species. Sporangial characteristics of these species are shown in Table 51.4. *P. botryosa, P. heveae, P. meadii, P. palmivora,* and *P. parasitica* have been described as pathogens of the rubber tree; however, *P. palmivora* and *P. meadii* are the most frequently isolated and described as the causal agents of black stripe (Plate 51A.5), patch canker, green pod rot (Plate 51A.6), green twig blight, and abnormal leaf fall. Opinions among countries differ about which *Phytophthora* species are responsible. For instance, in India abnormal leaf fall and pod rot are attributed to *P. meadii,* while in Malaysia these diseases are considered to be caused by *P. palmivora* and *P. botryosa.* A summary of rubber diseases is given by Holliday (1980).

Liyanage and Wheeler (1989) examined isolates collected from infected rubber trees from several areas of the world (Brazil, China, India, Indonesia, the Ivory Coast, Malaysia, Sri Lanka, and Thailand) and distinguished *P. palmivora* (MF1 type), *P. meadii, P. botryosa, P. citricola, P. citrophthora* (first report from the Ivory Coast of Africa and Indonesia), and *P. palmivora* MF4 (*P. capsici sensu* Tsao 1991; Mchau and Coffey 1995). To differentiate species, Liyanage and Wheeler (1989) placed disks from a sporangium-forming culture in a cotton blue-lactophenol-water mixture and shook the mixture to determine the length-breadth ratios and pedicel lengths of sporangia.

P. palmivora was noted to occur in China, Indonesia, and Malaysia on rubber. Reports by Chee (1969b, 1970, 1972, 1973a, b, 1976), Chee and Greenwood (1968), Chee et al. (1967), and Chee and Wastie (1970, 1980) cover the impact of *P. palmivora* and other species on rubber diseases in Malaysia.

Black Stripe or Black Thread

Black stripe is the most severe of the *Phytophthora* diseases of para rubber. Superficially, the symptoms are often confused with tapping injury. Initially, sunken, discolored areas occur on the bark just above the tapping cut (Plate 51A.5). In advanced stages, vertical fissures in the bark also occur. When the bark is scraped away, the characteristic vertical black line, from which the disease gets its name, extends beyond the tapping cut. Splits and cracks often occur in the bark.

Black thread has often been attributed to *P. meadii,* although in most cases the symptoms caused by either *P. palmivora* or *P. meadii* on rubber trees are indistinguishable.

Disease Development. Since moist air currents disseminate the caducous sporangia, the disease spreads rapidly during drizzling rain and wind storms. Infested tapping knives are also an important means of dissemination.

Control. Tapping knives should be disinfested by dipping them in a suitable fungicide before each tree is tapped. Infected trees should not be tapped until the weather is dry. Since *P. palmivora* penetrates deeply into the wood, protective fungicides are not usually effective. Some reports indicate that the disease can be suppressed during the tapping period by the application of captafol. In Malaysia, metalaxyl (Ridomil 25 WP) gave improved control over captafol as a protectant and therapeutic fungicide against black stripe (Tan 1983).

A summary of 5 years of field trials indicated that Sandofan (10% oxadixyl and 56% mancozeb), Fruivit (10% oxadixyl and 56% propineb), and Caltan (45% folpet and 6% ofurace) were effective protectants and curatives against the tapping panel disease. Fentin oxide and Ridomil Combi (10% metalaxyl and 40% folpet), when applied twice weekly, were as effective as Difolatan (captafol). Formulation with the sticker Methocel F4M (3%) reduced the necessity to spray twice weekly to once. Experimentally, phosphorous acid (active component of fosetyl-Al, see Chapter 6) conferred good control (A. M. Tan 1990). If registered for use, fosetyl-Al sprays should be effective control measures.

Patch Canker, Lump Canker, or Foot Disease

Diseases on the untapped bark occur mainly after injuries to the bark. Bark from which the cork has been removed is particularly susceptible. The infection begins under the bark and is not visible until the outer bark is peeled back. The cortex often, but not always, has a distinctive claret color. The disease can spread up and down the trunk for a considerable distance before external symptoms are visible. The cambium under healthy bark can be killed, resulting in a coagulation of the latex. During the dry season, when the disease is normally arrested, the regeneration of callus tissue over the coagulated latex results in a characteristic lumpy canker (Chee 1968).

Disease Development. *P. palmivora* enters through wounds in the trunk and spreads rapidly during cool (16 to 20°C), moist weather.

Control. Practices that wound trees should be avoided, but if trees are wounded, they should be treated soon after with a suitable fungicide. Surgical excision of the diseased tissue prior to treatment with a fungicide is effective if the disease is detected at an early stage.

Secondary Leaf Fall and Green Pod Decay

Secondary leaf fall and green pod decay of *Hevea* are described under *P. meadii* (Chapter 43).

Foot Rot of *Piper nigrum* L. (Black Pepper)

As explained in the introduction to this chapter, the causal agent of foot rot of black pepper was attributed in the past to an atypical *P. palmivora,* later designated MF4, but since then many of the *Phytophthora* isolates causing foot rot of black pepper have been determined to be *P. capsici* (redescribed by Tsao and Alizadeh [1988]; Tsao 1991; Mchau and Coffey 1995). See Chapter 14 for a discussion of this disease and Tables 51A.4 and 51A.5 for a comparison of the morphological characteristics of *P. palmivora* and several other species: *P. megakarya, P. capsici, P. arecae, P. meadii, P. botryosa, P. hevea,* and *P. nicotianae* (= *P. parasitica*).

Das and Cheeran (1985) reported that all isolates collected from black pepper in the Kerala region of India were identified as *P. palmivora* MF1, MF2, or MF3 but that no MF4 (now considered to be *P. capsici*) type isolates were found.

CHAPTER 51B

Phytophthora palmivora var. *heterocystica*

Babacauh 1983

Seven strains of *Phytophthora palmivora* from crops grown in western and central Africa were delineated by different degrees of virulence on different hosts (Babacauh 1972). A variant of *P. palmivora* isolated from *Theobroma cacao* in the Ivory Coast that produces a nonpapillate, conidial type of sporangia was described as a new variety, *P. palmivora* var. *heterocystica* (Babacauh 1983a). Some of these isolates produce chlamydospores but not sporangia, and some produce conidium-like sporangia without papillae. Babacauh attributes this change in morphology to an evolutionary trend toward less dependence on free water, as exists in *Peronospora,* a downy mildew. Symptoms produced on cocoa trees are similar to those caused by *P. palmivora* var. *palmivora.* There is some doubt about the status of *P. palmivora* var. *heterocystica* because cultures are not available and no subsequent citations have appeared in the literature.

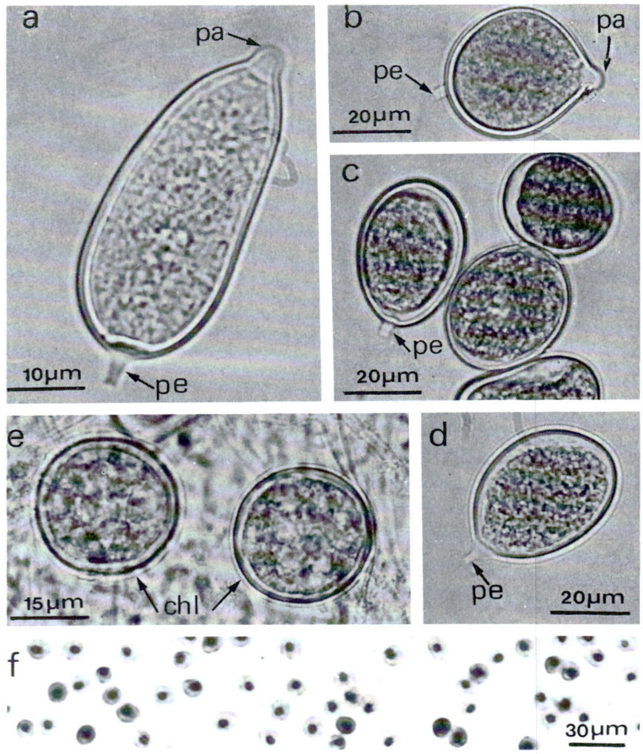

Figure 51B.1. Morphology of *Phytophthora palmivora* and *P. palmivora* var. *heterocystica.* **a and b,** Papillate (pa) sporangia with short pedicels (pe) typical of *P. palmivora.* **c and d,** *P. palmivora* var. *heterocystica.* Conidial, caducous sporangia are similar to the typical sporangia of *P. palmivora* but are without papillae. Walls of conidial sporangia are not as thick as those of chlamydospores (chl) illustrated in **e**. The exclusive presence of a mixture of conidial and typical papillate sporangia characterizes *P. palmivora* var. *heterocystica.* **e,** Chlamydospores. **f,** Mononucleate zoospores of *P. palmivora* stained with Giemsa. (Reproduced by permission of Bulletin de la Societe Botanique de France, courtesy Babacauh 1983a)

Characteristics of *Phytophthora palmivora* var. *heterocystica*

P. palmivora var. *heterocystica* is included in group II in the Stamps et al. (1990) key. See Tables 4.2 and 4.3 for tabular keys. Morphology is illustrated in Figure 51B.1. Sex organs were not described by Babacauh (1983a).

Sporangia (Sporocysts)

Sporocysts are without papillae and are subglobose with a length-breadth ratio of 1.2 to 1.5 and deciduous with pedicels 5 µm in length. Some isolates produced both typical papillate sporangia as well as sporocysts.

Chlamydospores

Chlamydospores are globose and similar to those described for *P. palmivora.* Some isolates form only chlamydospores and not sporangia.

Sex Organs

Oogonia and antheridia do not occur (Stamps et al. 1990).

CHAPTER 52

Phytophthora phaseoli

Thaxter (1889)

There are no synonyms for *Phytophthora phaseoli,* which was first described by Thaxter (1889a) and amended by Clinton (1906) as causing a destructive disease of lima beans in Connecticut, the United States. Since then, *P. phaseoli* has been reported in the Philippines, the former Soviet Union, Puerto Rico, Italy, and the Congo (now Zaire) (Table 52.1). Distribution is shown in CMI Map 201.

Characteristics of *Phytophthora phaseoli*

P. phaseoli is classified in group IV (Stamps et al. 1990). Morphology is shown in Figure 52.1. See Tables 4.2 and 4.3 for tabular keys.

Sporangia

Sporangia are semipapillate, oval, or elliptical with a truncate base. The size varies with the substrate on which they are produced (Hyre and Cox 1953). Thaxter (1889) reports dimensions of 35 to 50 × 20 to 24 μm (average 28 × 19 μm). The length-breadth ratio averages 1.4:1. Sporangia are nonproliferating and caducous with short (5 to 20 μm) pedicels. Sporangiophores are sympodially branched and slightly swollen at the base.

Hyphal Swellings and Chlamydospores

Neither hyphal swellings nor chlamydospores are produced.

Sex Organs

P. phaseoli is homothallic and produces oospores abundantly on lima bean or host tissue. Antheridia are amphigynous and shaped like a flattened sphere; oogonia are subspherical, measuring 16.2 to 35.6 μm in diameter (average 23.7 μm); oospores measure 14.4 to 26.6 μm in diameter (average 19.2 to 22.5 μm) (Hyre and Cox 1953).

Growth Temperatures

The minimum temperature for growth is 5°C, optimum 15 to 20°C, and maximum 25 to 30°C. Incubation at 30°C for 2 weeks was lethal. Data on the effect of temperature on sporangia and zoospore germination (Hyre and Cox 1953) supported the prediction that disease forecasting could be based on expected temperature and rainfall parameters (Hyre 1957). *P. phaseoli* can be maintained for long periods on sterile lima bean seeds in water (Goth and Wester 1963).

Disease Caused by *Phytophthora phaseoli*

Downy Mildew of *Phaseolus lunatus* L. (Lima Bean)

After its first appearance in the United States, the downy mildew disease caused by *P. phaseoli* spread rapidly and posed a serious threat to the lima bean industry in the production

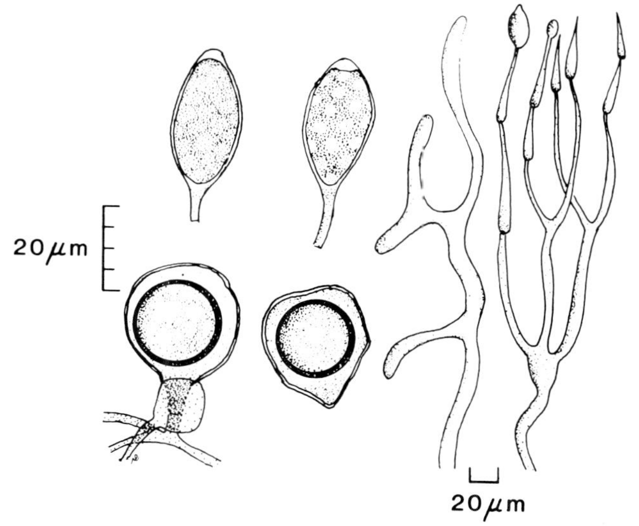

Figure 52.1. Morphology of *Phytophthora phaseoli.* Upper row, ovoid to elliptical sporangia with short pedicels. Lower row, subspherical, globose oogonia with amphigynous antheridia. Right, mycelium and sympodially branched sporangiophores. (Drawing by A. Vaziri after Thaxter 1889a)

Table 52.1. Distribution of hosts of *Phytophthora phaseoli*

Host	Common name	Disease	Geographical distribution
Brassica juncea (L.) Czerniak	Brown mustard	Seedling blight	Philippines (Clara 1928)[a]
Hevea brasiliensis (H.B.K.) Muell. Arg.	Paral rubber	Seedling blight	Philippines (Clara 1928)[a]
Lycopersicon esculentum L.	Tomato	Seedling blight	Philippines (Clara 1928)[a]
Phaseolus lunatus L.	Lima bean, butter bean	Downy mildew	United States (Thaxter 1889a); Russia (Speschnew 1897); Puerto Rico (Henricksen 1907); Philippines (Clara 1928); Congo (Hendrikx 1939); Italy (Casarini et al. 1960)
Raphanus sativus L.	Radish	Seedling blight	Philippines (Clara 1928)[a]
Sandoricum indicum Cav.	Santol	Seedling blight	Philippines (Clara 1928)[a]
Solanum melongena L.	Eggplant	Seedling blight	Philippines (Clara 1928)[a]
Vigna unguiculata subsp. *sesquipedalis* (L.) Verdc.	Asparagus bean	Seedling blight	Philippines (Clara 1928)[a]

[a] Artificially inoculated host.

areas along the North Atlantic seaboard. Entire fields in New Jersey and Delaware were devastated (Wester 1968); however, the disease is of minor importance in warmer and drier areas of the country.

The disease is most destructive on the pods. Shoots, leaves, and petioles are also occasionally infected. Early lesions on the leaves are irregularly shaped and purplish. A white felt of mycelium and sporangia gradually progresses to cover the whole surface of the pod. Reddish bands often surround the infected areas. Infected pods eventually shrivel and are worthless commercially.

Disease Development. Precise conditions for disease development have been defined and implemented into a downy mildew forecasting program in many areas of the northeastern United States. Optimal infection occurs when the 5-day moving mean temperature ending on that day is less than 26.1°C, the minimum temperature is 7.2°C or above, and the 10-day total rainfall ending on that day is 3.05 cm or more. The initial appearance of symptoms can be expected about 8 days later (Hyre 1957). Downy mildew can also be predicted to occur when the dew point remains at 20.5°C or above for 12 consecutive hours and the weather forecast indicates that it will last at least another 48 h. Occasionally in arid regions, which were previously thought to be free of the disease, humidity becomes sufficiently high for infection to occur (Scarpa and Raniere 1964).

Sporangia are airborne and can also be disseminated by insects. Sturgis (1898) illustrated in detail the mechanisms involved in the dissemination of sporangia of *P. phaseoli* by bees that visit flowers. The pathogen can also be transferred by slugs that feed on young lima bean leaves and cotyledons (Wester et al. 1964). The pathogen overwinters in the soil as oospores or as mycelium in infected lima bean pods. Oospores, which had formed in pods and seeds, overwintered and were capable of causing disease; however, sporangia and mycelium did not survive in infected tissue (Wester et al. 1966).

Control. Four physiologic races of the pathogen, designated A, B, C, and D, are known. Resistance to races A, B, and C has been incorporated into many commercially available cultivars. Wester (1968) reported cultivars resistant to race B, and Wester et al. (1972) reported multiple race resistance to downy mildew. In Costa Rica, the cultivars Brunca and Porillo Sintetico showed the lowest disease index of eight cultivars tested in the field (Vargas et al. 1985).

Protective fungicides such as Bordeaux mixture and other copper-based fungicides have been somewhat effective if applied before disease occurs. Maneb applied at the time of a downy mildew forecast followed by a second application 7 to 12 days later is an effective control measure (Crossan et al. 1957). Streptomycin controlled the disease in greenhouse tests (Zaumeyer and Wester 1956). Metalaxyl or fosetyl-Al should be effective if registered for use.

CHAPTER 53

Phytophthora porri

Foister (1931)

No synonyms are known for *Phytophthora porri*, which was originally isolated from diseased leek plants (*Allium porrum* L.) in Scotland by Foister (1931). It was first reported from North America on stored carrots (Stelfox and Henry 1978; Ho 1983a). Although *P. porri* has been reported on several other plant species (Table 53.1) such as carnation (Kouyeas 1977) (Plate 53.1), it is recognized to be a pathogen of plants primarily in Amaryllidaceae. Distribution is shown in CMI Map 204. The recent report by de Cock et al. (1992) alludes to the heterogeneity in morphology, pathogenicity, and DNA restriction fragment length polymorphism (RFLP) patterns of isolates from different hosts.

Characteristics of *Phytophthora porri*

P. porri is classified in group III (Stamps et al. 1990). Morphology is illustrated in Figure 53.1. See Tables 4.2 and 4.3 for tabular keys. A description of *P. porri* is given by Stamps (1978e) and Holliday (1980). Some isolates are host specific. Isolates from *Campanula* were not pathogenic on leek and vice versa (Legge 1951).

P. porri as it is known currently, is not a homogeneous species. Within isolates from leek, cabbage, and carrot from different geographical areas, all of which were considered to be *P. porri,* de Cock et al. (1992) found variations in temperature-growth relations; sporangial production; oogonium, antheridium, and oospore production; pathogenicity to leek (*A. porrum*) and cabbage (*Brassica oleracea* L.); and RFLP of mitochondrial DNA (mtDNA). RFLP patterns were significantly different in isolates from *Allium* and *Brassica.* The RFLP patterns of the *Allium* isolates differed from the *Brassica* and *Daucus* isolates (Figure 53.2), all of which correlated well with differences in growth at temperatures higher than 24 to 27°C. Isolates from *A. porrum* produced oogonia consistently, but those from *B. oleracea* produced oogonia shortly after isolation and only occasionally afterward. Isolates from *B. oleracea* produced sporangia readily on agar media, but isolates from *A. porrum* produced sporangia only when transferred to hemp seed-water culture. De Cock et al. (1992) suggested that the isolates from *Brassica* and *Daucus carota* (carrot) might represent a different, as yet unnamed, species or possibly a forma specialis because they are host specific. Thus, some isolates recorded from carrot and cabbage in Table 53.1 may eventually be reclassified.

Mycelium

Coiling of mycelium (Figure 4.1B, Chapter 4, and Figure A of Stamps [1978e]) is typical; hyphae measure up to 8 μm in diameter but are not uniform.

Table 53.1. Distribution of hosts of *Phytophthora porri*

Host	Common name	Disease	Geographical distribution
Allium ascalonicum L.	Shallot	Shanking	United Kingdom (Hickman 1943)
Allium bakeri Regel.	Scallion	Leaf blight	Japan (Katsura et al. 1969)
Allium cepa L.	Onion	Leaf blight	Netherlands (Tichelaar and Van Kesteren 1967); United Kingdom (Griffin and Jones 1977)
Allium cepa L.	Onion	Shanking	United Kingdom (Hickman 1943); Japan (Katsura et al. 1969)[a]; South Africa (Maltitz and von Broembsen 1984)
Allium fistulosum L.	Welsh onion	Leaf blight	Japan (Katsura et al. 1969)[a]
Allium fistulosum var. *caespitosum* Mak.	Japanese bunching onion	Leaf blight	Japan (Katsura 1971)
Allium geyeri S. Wats		Leaf blight	Japan (Yokoyama 1976)
Allium nipponicum Franch. et Sav.		Leaf blight	Japan (Katsura et al. 1969)[a]
Allium porrum L.	Leek	White tip disease, leaf dieback	United Kingdom (Foister 1931); Netherlands (Van Bakel 1964); Northern Ireland (Taylor 1965); Norway (Semb 1969); Ireland (Anon. 1974)
Allium sativum L.	Garlic	Leaf blight	Japan (Katsura et al. 1969)[a]
Allium tuberosum Rottl. ex Spreng.	Chinese chive, oriental garlic	Leaf blight	Japan (Katsura et al. 1969)[a]
Allium wakegi Araki		Leaf blight	Japan (Yokoyama 1976)
Brassica oleracea L. var. *capitata* L.	White cabbage	Storage rot	Germany (Semb 1971); United Kingdom (Geeson 1976)
Campanula persicifolia L.	Willow bellflower	Crown rot	United Kingdom (Legge 1951)
Daucus carota L. subsp. *sativus*	Carrot	Root rot	Germany (Semb 1971); Canada (Stelfox and Henry 1978; Ho 1983a; Thomson et al. 1986)
Dianthus caryophyllus L.	Carnation	Collar rot	Greece (Kouyeas 1977)
Gladiolus L.	Sword lily	Collar and bulb rot	Greece (Kouyeas 1977)
Hyacinthus orientalis L.	Hyacinth	Leaf blight	Japan (Katsura 1971)
Tulipa gesneriana L.	Tulip	Blight and rot	Japan (Katsura 1970)

[a] Artificially inoculated host.

Sporangia

Sporangia are obpyriform, ovoid, or ellipsoid, occasionally with distorted shapes; semipapillate; persistent on the stalk; nonproliferating; and borne successively. Length-breadth ratio is 1.29 to 1.46. For dimensions of sporangia from different hosts, see Table 53.2. Sporangiophores are undifferentiated, often with intercalary swellings. Sporangia and zoospores can be induced by incubating fast-growing mycelium with soil extract at 15°C for 2 days followed by replacement of soil extract with cold (3°C) distilled water (Smilde and Van Ness 1992).

Hyphal Swellings

Hyphal swellings that are round, ellipsoid, or angular and occurring singly or in chains form abundantly in culture.

Chlamydospores

Chlamydospores form in culture media after prolonged incubation in water (8 weeks) and measure 20.8 to 35.3 µm in diameter, average 30.0 µm.

Sex Organs

P. porri is homothallic. Globose oogonia and oospores form abundantly in culture. Both amphigynous and paragynous antheridia occur. Oogonia average about 34 µm in diameter. Oospores are spherical and aplerotic and average about 30 µm in diameter. For dimensions of oogonia and oospores from different isolates, see Table 53.2. Wheat and barley straw was a favorable substrate for production of oospores of isolates from carrot (Thomson et al. 1986). DeCock (1992) noted that most oogonia were abortive. Healthy oospores were rare.

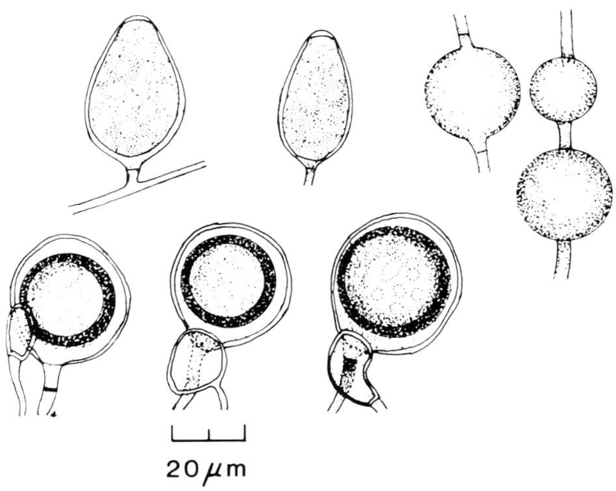

Figure 53.1. Morphology of *Phytophthora porri.* Lower row, globose oogonia with either paragynous (left) or amphigynous (right) antheridia. Oospore wall is thick (3 to 4.5 µm). Upper row, semipapillate sporangia; globose chlamydospores. (Drawing by A. Vaziri)

Growth Temperatures

Although Foister (1931) originally stated that the maximum temperature for growth of *P. porri* was <35°C, most investigators (de Cock et al. 1992) now report a maximum temperature of 27°C or less (also reported for a carnation isolate by Kouyeas [1977]). Leonian (1934) also found this to be the case with Foister's original isolate. Examples of growth at minimum, optimum, and maximum temperatures (°C) include

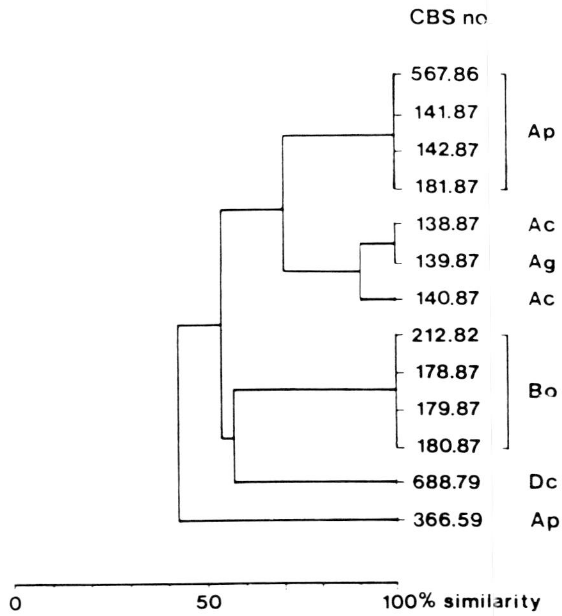

Figure 53.2. Dendogram from group-average cluster analysis based on similarity coefficient of restriction patterns of mtDNA of 13 isolates of *Phytophthora porri* from different hosts. Isolates are indicated by their CBS accession numbers and their respective hosts. Ap = *Allium porrum,* Ac = *A. cepa,* Ag = *A. grayi,* Bo = *Brassica oleracea,* and Dc = *Daucus carota.* Isolate 366.59 Ap was considered to be closer to *P. nicotianae* that to *P. porri* on the basis of mtDNA RFLP. (Reproduced by permission of Netherlands Journal of Plant Pathology, courtesy de Cock et al. 1992)

Table 53.2. Dimensions of reproductive structures of *Phytophthora porri* from various hosts

Host	Sporangia length and breadth (µm)	Oogonia diameter (µm)	Oospores diameter (µm)	Reference
Leek	35–82 × 23–52 (av. 46 × 37)	29–46 (av. 38.5)	22–39 (av. 32.5)	Foister (1931)
Campanula	15–60 × 12–35 (av. 41 × 30)	30–42 (av. 36.0)	24–32 (av. 31.0)	Legge (1951)
Scallion	25–58 × 17–39 (av. 36 × 25.2)	18.9–43.8 (av. 34.1)	18.0–36.7 (av. 29.3)	Katsura et al. (1969)
Tulip	17–57 × 10–45 (av. 31.9 × 23.4)	17.5–42.5 (av. 34.0)	17.5–42.5 (av. 29.3)	Katsura (1970)
Tulip	17–47 × 10–37 (av. 28.3 × 20)	22.5–42.5 (av. 34.3)	22.5–35.0 (av. 29.5)	Katsura (1970)
Carnation	av. 46 × 37	av. 40–46	(av. 37.0)	Kouyeas (1977)
Salad onion	38.6 ± 8.6 × 27.5 ± 6.2	38.2 ± 3.8)	33.7 ± 5.3	Griffin and Jones (1977)

<5:25:35 (*Allium porrum,* Foister 1931); <8:20:27 (*A. porrum* from Foister, Leonian 1934); no data:20:27 (*Campanula persicofolia,* Legge 1951); 3:21:27 (*A. porrum,* de Cock et al. 1992); 3:18:27 (*A. grayi,* de Cock et al. 1992); 3:21:24 and 3:21:27 (*B. oleracea*) and 3:24:30 (*Daucus carota*) (de Cock et al. 1992). RFLP patterns of an isolate (366.59) from *A. porrum* with a maximum growth at 36°C had a greater similarity to *P. nicotianae* than to *P. porri* (Figure 53.2). De Cock et al. (1992) concluded that this isolate could not be identified as *P. porri.*

Disease Caused by *Phytophthora porri*

White Tip of *Allium porrum* L. (Leek)

White tip of leek was known in the United Kingdom long before its formal description by Foister (1931). The first observable symptom of the disease is a yellowing of the leaf tips that extends for several centimeters down the leaf blade. These leaves die rapidly and turn white, hence the term "white tip." Leaves are often distorted and twisted. If infection occurs early in their development, the plants become stunted. Infection in older plants results in a rapid wilt; severe infection in mature plants often results in breakage of the stem at the soil line.

A similar disease of onions and shallots known as "shanking" has been reported from England. In this case infection is spread through roots and bulbs. The roots are completely destroyed, the scales of the bulbs become soft and water soaked, and the leaves turn yellow and shrivel.

P. porri also causes a storage rot of carrots (Ho 1983a) and cabbage (Kiewnick 1985) and a stem rot of lettuce (Sitepu and Wallace 1984). Root rot of *Campanula* is also caused by *P. porri,* but isolates do not infect leek and vice versa (Legge 1951). Recent work by de Cock et al. (1992) indicates that some of these isolates may have been misidentified.

Disease Development. The disease develops rapidly in damp, humid weather at temperatures of 18 to 20°C. As much as 50% of the crop can be destroyed. The pathogen survives in the soil between crops as oospores formed in plant tissue. Yokoyama (1976) discusses epidemiology and forecasting of disease based on temperature and rainfall. Smilde and Van Ness (1992) reported that up to 80% infection of leek plants occurred 30 to 60 days after inoculation in the field with either zoospores or diseased leaf tissue. Oospore inoculum was less successful (12%).

Control. Protective sprays with captafol, Bordeaux mixture, or the dithiocarbamates have provided effective control of the disease. The newer systemic fungicides metalaxyl and fosetyl-Al should be effective if registered for use.

CHAPTER 54

Phytophthora primulae

Tomlinson (1952)

No synonyms are presently known for *Phytophthora primulae,* which was first isolated and described by Tomlinson (1952) from diseased polyanthus (*Primula polyantha* Mill.) roots in England. *P. primulae* has a narrow host range and has thus far been detected only from *Primula* spp. in England, Denmark, and New Zealand (Stamps 1985i). Norin (1987) tentatively identified an isolate causing root disease of primula as *P. primulae.* Pathogenicity to apple and tomato fruits has been demonstrated but only by artificial inoculation (Tomlinson 1952). The disease on polyanthus was named brown core by Tomlinson.

Characteristics of *Phytophthora primulae*

P. primulae (Tomlinson 1952) is classified in group III (Stamps et al. 1990). Morphology is shown in Figures 54.1 and 54.2. See Tables 4.2 and 4.3 for tabular keys. A description of *P. primulae* is given by Stamps (1985i).

Sporangia

Sporangia are variable in shape, ovoid (Figures 54.1 and 54.2), bluntly ellipsoid, limoniform, or elongated with several constrictions; semipapillate; persistent on the stalk; and 33 to 165 µm long × 23 to 52 µm wide (average 77 × 44 µm). Sporangia with constrictions measure 100 to 351 µm long × 6 to 36 µm wide (Figure 54.1). Sporangia do not form new sporangia by internal or external proliferation. Sporangiophores are unbranched and undifferentiated. Occasionally, sporangia are formed on a monochasial sympodium (Tomlinson 1952). The length-breadth ratio is about 2.1:1.

Hyphal Swellings and Chlamydospores

Hyphal swellings and chlamydospores do not occur.

Sex Organs

P. primulae is homothallic. Antheridia are mostly paragynous; however, occasionally an amphigynous antheridium has been found (Stamps 1985i). Oogonia are abundant in diseased tissue or culture and are terminal or lateral and smooth and measure 23 to 43 µm in diameter (average 37 µm); oospores measure 17 to 33 µm in diameter (average 30 µm).

Growth Temperatures

The minimum temperature for growth is 1°C, the optimum is 15 to 20°C, and the maximum is <27°C.

Distinguishing Characteristics

P. primulae often forms coiled hyphae on solid agar media similar to those of *P. porri* (Figure 4.1B, Chapter 4). Sporangia do not form new sporangia by external or internal proliferation. *P. fragariae* gives rise to new sporangia by proliferation;

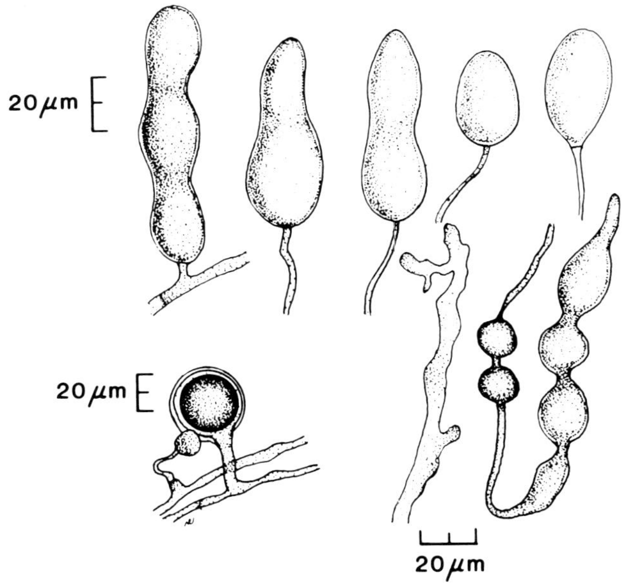

Figure 54.1. Morphology of *Phytophthora primulae.* Lower row, globose oogonium with a paragynous antheridium; swollen hyphae. Upper row, sporangia, some of which are constricted; others are oval to ellipsoid. (Drawing by A. Vaziri from Tomlinson 1952)

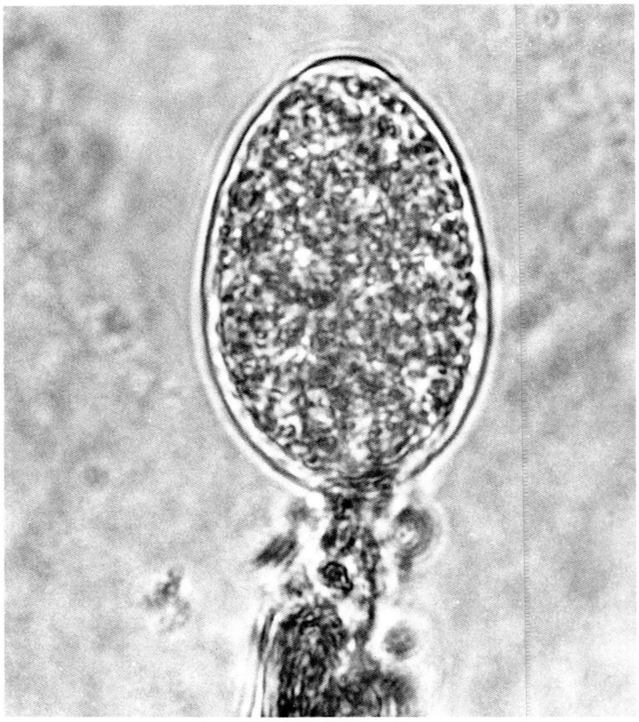

Figure 54.2. Oval sporangium of *Phytophthora primulae.* (Courtesy M. J. Griffin, ADAS, Cambridge, United Kingdom)

P. porri has both paragynous and amphigynous antheridia and produces new sporangia by sympodial branching (Tomlinson 1952).

Disease Caused by *Phytophthora primulae*

Brown Core Root Rot of *Primula polyantha* Mill. (Polyanthus)

The first symptom is a wilting of the outer leaves in late spring. Wilting gradually progresses until all the leaves from the outside to the center of the rosette are affected. Stunted plants are easily distinguished from surrounding healthy plants. When diseased plants are pulled, they are devoid of lateral roots, giving a "rat-tail" effect. When the stele is slit longitudinally, a typical brown discoloration in the xylem tissue is clearly visible, thus the name "brown core." This tissue contains large numbers of oospores (illustrated by Tomlinson [1952]). The symptoms are similar to those associated with the red core disease of strawberry (Chapter 26A), except that in this instance the stele is brown rather than red (Tomlinson 1952).

Disease Development. The pathogen enters through the lateral roots and eventually infects the stele. The disease progresses rapidly in cool (18 to 22°C), damp weather.

Control. Since *P. primulae* can remain viable in the soil as oospores for several years, polyanthus should not be planted in soil where the disease has occurred previously. Infested soil can be treated by fumigation with methyl bromide or by steam.

CHAPTER 55

Phytophthora pseudotsugae

Hamm and Hansen (1983)

Phytophthora pseudotsugae was first isolated along with several other species of *Phytophthora* from rotted roots of Douglas fir (*Pseudotsuga menziesii* (Mirb.) Franco) in forest tree nurseries but was identified only as *Phytophthora* species 1 (R. G. Pratt et al. 1976).

It was noted in a subsequent study to be an important component of the complex of *Phytophthora* species that cause root rot of Douglas fir (E. M. Hansen et al. 1979). *P. pseudotsugae* was described as a new species by Hamm and Hansen (1983).

Characteristics of *Phytophthora pseudotsugae*

P. pseudotsugae is classified in group I (Stamps et al. 1990) and briefly described in G. Hall (1989a). Morphology is shown in Figure 55.1. See Tables 4.2 and 4.3 for tabular keys. Distribution is in the northwestern United States (Washington and Oregon) (CMI Distribution Maps of Plant Diseases 600).

Sporangia

Sporangia are spherical to ovoid, sometimes obturbinate or ampulliform, and distinctly papillate. Sporangia are seldom formed on solid media but are formed when colonies grown on pea broth are washed and incubated in distilled water or soil extract at 10 or 20°C. Sporangia formed at 10°C within 1 to 2 days and averaged 49 × 39 μm (range 42 to 51 × 32 to 43 μm) but at 20°C were smaller and averaged 39 × 32 μm (range 34 to 45 × 28 to 35 μm). The length-breadth ratios, average 1.23:1 (range 1.18 to 1.27:1), were similar at 10 and 20°C. Sporangia are persistent on simple, long (29 to 632 μm) sporangiophores. Intercalary sporangia are formed by some isolates but not by others.

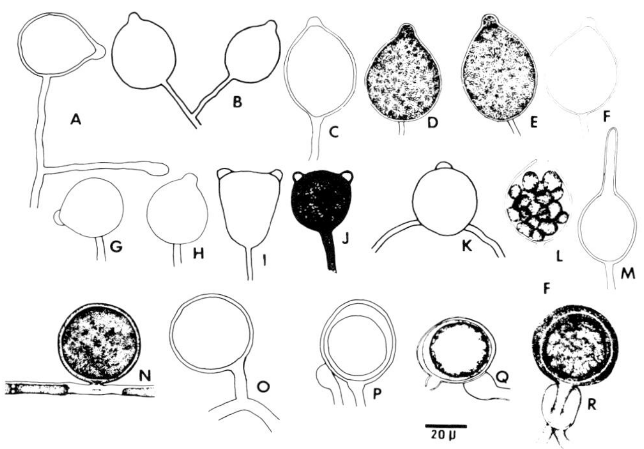

Figure 55.1. *Phytophthora pseudotsugae.* **A–H,** Ovoid, papillate sporangia. **I and J,** Abnormally shaped sporangia. **K,** An intercalary sporangium. **L and M,** Germination of sporangia by production of zoospores. **N,** Oogonium formed adjacent to the mycelium. **O,** Oogonium on a stalk. **P and Q,** Oogonia containing oospores that, although aplerotic, nearly fill the oogonia; antheridia are paragynous. **R,** Oogonium containing an oospore; an amphigynous antheridium is attached. (Reproduced by permission of the Canadian Journal of Botany, courtesy Hamm and Hansen 1983)

Chlamydospores and Hyphal Swellings

Hyphal swellings occur on some isolates (Hamm and Hansen 1983). Irregular hyphal swellings are illustrated by G. Hall (1989a).

Sex Organs

P. pseudotsugae produces sex organs in single culture (homothallic) in cornmeal agar, lima bean agar, pea broth, and V8 juice agar. Average diameter of oogonia on V8 juice agar and lima bean agar is 35 μm and on cornmeal agar and pea broth 31 μm. On V8 juice agar, the diameter ranges from 24 to 45 μm, with isolate means ranging from 32 to 40 μm. Oogonia are spherical and smooth walled (about 1.2 μm thick) and borne terminally on stalks varying from 20 to 120 μm in diameter, widening to 5.8 μm at the point of oogonial attachment.

Antheridia are hyaline and usually spherical or club shaped. Dimensions range from 6 to 21 × 10 to 24 μm (average 11 × 15 μm). Antheridia are predominantly paragynous and attached to the oogonia at a point near the oogonial stalk; occasionally antheridia are amphigynous (Figure 55.1).

Oospores are pigmented, spherical, smooth walled, and aplerotic but nearly fill the oogonium. Diameters of oospores ranged from 27 to 33 μm, averaging 24 μm, on V8 juice agar. Hamm and Hansen (1983) noted that most oospores were reported to be abortive.

Growth Temperatures

Growth occurs at temperatures ranging from 2 to 30°C with an optimum at 20 to 25°C.

Distinguishing Characteristics

P. pseudotsugae lacks any uniquely diagnostic characteristic but can be readily separated from *P. cactorum*. *P. pseudotsugae* produces long and loosely branched sporangiophores unlike those produced by *P. cactorum* or *P. iranica*. Both *P. cactorum* and *P. iranica* produce chlamydospores, but *P. pseudotsugae* does not. *P. cactorum* produces caducous sporangia, but *P. pseudotsugae* does not (Hamm and Hansen 1983). *P. pseudotsugae* differs from *P. iranica* by production of smaller sporangia and from *P. clandestina* by the absence of subterminal antheridia and markedly aplerotic oospores (G. Hall 1989a). Protein patterns were similar for several isolates of *P. pseudotsugae* and differed distinctly from those of *P. cactorum* (also classified in group I).

Disease Caused by *Phytophthora pseudotsugae*

Root Rot of *Pseudotsuga menziesii* (Douglas fir)

In a study of root diseases of Douglas fir trees in forest nurseries in Oregon and Washington, isolates of an unidentified species were labeled *Phytophthora* species 1 (R. G. Pratt et al. 1976) but are now considered to be *P. pseudotsugae.* Tests of pathogenicity of *P. pseudotsugae* on 3-year-old seedlings transplanted to steamed soil artificially infested with cornmeal-sand inoculum followed by daily irrigation of the soil to capacity for 8 weeks resulted in a disease rating of less than 10% root discoloration, a rating much lower than that for *P. cinnamomi* (51 to 75%). After inoculation by stem puncture with *P. pseudotsugae,* the length of lesions progressed about 0.95 mm per day compared with 3.4 mm per day for *P. cactorum.*

Disease Development. Rotting of lateral roots of Douglas fir trees occurs most readily in heavy nursery soils that are saturated for long periods. Symptoms are not usually evident on foliage during the first year, but roots may become infected, causing them to turn reddish brown. During the second year, infected plants become stunted and roots become dark brown (Cooley et al. 1985).

Control. Control methods are similar to those described for other *Phytophthora* pathogens of forest trees. Infection usually occurs in nurseries prior to transplanting to forests. According to Cooley et al. (1985), improvement of vertical and horizontal soil drainage is an important control measure. In some nurseries, subsurface tiles are installed for drainage systems and irrigation is managed to avoid standing water or long-lasting saturation of the soil. See Chapter 5 for a discussion of control of diseases in nurseries.

If transplanted Douglas fir seedlings are only lightly infected, survival rates in the forest are relatively good, although *Phytophthora* spp. may remain viable and be recoverable for up to 2 years (E. M. Hansen et al. 1980; Cooley et al. 1985).

In a general discussion of control of root rot caused by several *Phytophthora* species, including *P. pseudotsugae,* Cooley et al. (1985) indicated that western red cedar is resistant; the hardwoods, pine, spruce, larch, and incense cedar are tolerant; and western and mountain hemlock, Douglas fir, and true fir are highly susceptible. *P. pseudotsugae* is pathogenic only on Douglas fir, however.

Nursery soil can be fumigated with methyl bromide-chloropicrin, metham-sodium (Vapam), or dazomet (Mylone). These treatments reduce the population of *Phytophthora* species but do not guarantee that the nursery will remain free of *Phytophthora* or *Pythium* spp. because the degree of fungal kill might be incomplete or the land could become reinfested by introduction of infected plants, by mechanically transferred inoculum in drain water, or by machinery (Cooley et al. 1985). See also Chapter 6 on control by use of fungicides.

As indicated in Chapter 7, both metalaxyl and fosetyl-Al are effective systemic fungicides that are useful for control of *Phytophthora* root diseases by application as soil drenches; fosetyl-Al is also effective as a foliar application. According to Cooley et al. (1985), only metalaxyl (Subdue) has been shown to be practical in numerous tests in the field. Application of metalaxyl as a granule or as a liquid soil drench effectively suppresses the progress of root rot in infected seedlings; however, metalaxyl does not eradicate it from infected roots (Hamm et al. 1984). Metalaxyl does not affect the development of mycorrhizae in Frasier fir plants (Kenerley et al. 1984a).

According to Cooley et al. (1985), one well-timed treatment with metalaxyl per year controls the progress of root rot. Treatment during the spring or fall when roots are most active is the most effective. They warn against too frequent treatment with metalaxyl because of the danger that metalaxyl-resistant isolates might become dominant. Also, repeated treatment of soil with metalaxyl can lead to the buildup of a microflora that biodegrades metalaxyl (Bailey and Coffey 1985, 1986).

CHAPTER 56

Phytophthora quininea

Crandall (1947)

No synonyms are presently known for *Phytophthora quininea,* which was first isolated from diseased roots of *Cinchona* trees in Peru by Crandall in 1947. *P. quininea* appears to be host specific and is limited in geographical distribution to Peru, Bolivia, and Puerto Rico.

Characteristics of *Phytophthora quininea*

P. quininea is classified in group V (Stamps et al. 1990). See Tables 4.2 and 4.3 for tabular keys and Figure 56.1 for morphology.

Sporangia

Sporangia are ovate to obpyriform, widest at the base, nonpapillate, and persistent on the stalk and proliferate through the previously evacuated sporangium; they measure 45.9 to 67.5 µm in length × 25.5 to 37.5 µm in width (average 53.8 × 31.1 µm). Sporangiophores are simple and undifferentiated.

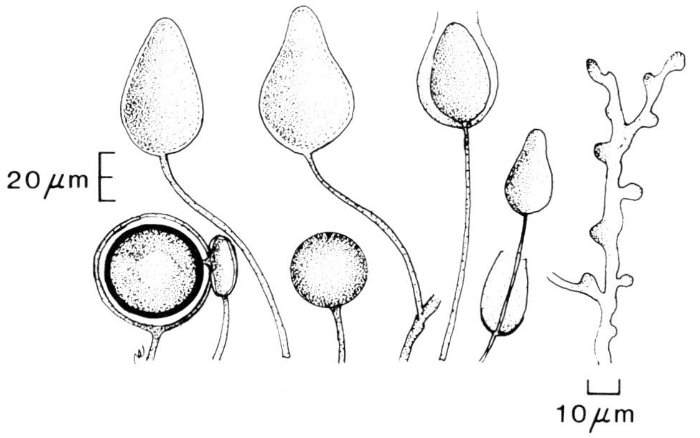

Figure 56.1. Morphology of *Phytophthora quininea.* Left to right, globose oogonium with a paragynous antheridium; nonpapillate, ovoid sporangium; globose chlamydospore; nonpapillate, obpyriform sporangium; internal and external proliferation of sporangia; and swollen, gnarled mycelium. (Drawing by A. Vaziri)

Hyphal Swellings

Hyphal swellings are irregular in shape and size.

Chlamydospores

Chlamydospores are terminal or intercalary, are produced abundantly on cornmeal and oatmeal agars, and measure 45 to 90 µm in diameter (Crandall 1947).

Sex Organs

P. quininea is homothallic. Antheridia are paragynous and irregular in shape and size; oogonia are subspherical to spherical and smooth and measure 67.5 to 82.5 µm in diameter; oospores are markedly aplerotic and range from 48.7 to 63.7 µm in diameter.

Growth Temperatures

The minimum temperature for growth is 15°C, the optimum is 25°C, and the maximum is 30°C.

Disease Caused by *Phytophthora quininea*

Root and Collar Rot of *Cinchona officinalis* L. and *C. pubescens* Vahl (Cinchona)

Disease Development. Plants of all ages, from seedlings to mature trees, are susceptible. Leaves of trees with root rot turn pale green and gradually take on a reddish tinge before wilting, and all but the terminal leaves eventually fall. The collar region of the trunk usually becomes swollen above the infected area. Wedge-shaped, reddish brown regions are frequently visible on the stem. Infected plants may remain alive for several months, often producing new roots from the collar region.

Control. Since the disease can be spread in infected nursery stock, it is essential that transplants be grown in nurseries that are disease free. If infested, the soil should be adequately fumigated or steamed before planting. Trees should be planted in soil with good vertical and horizontal drainage.

CHAPTER 57

Phytophthora richardiae

Buisman (1927)

Phytophthora richardiae was first isolated and described from diseased calla lilies by Buisman (1927) in the Netherlands. Ashby (1929b) named it *P. cryptogea* var. *richardiae* (Buism.), but it has been recognized as *P. richardiae* for many years (Tucker 1931; Waterhouse 1963; Stamps et al. 1990). It is pathogenic only on plants in the genus *Zantedeschia*. *Z. aethiopica* (L.) Spreng. at one time was designated *Richardia aethiopica* Hort., hence the name *P. richardiae* (Tucker 1933). See IMI Distribution Map 171.

Characteristics of *Phytophthora richardiae*

P. richardiae is classified in group VI (Stamps et al. 1990). G. Hall (1991d) gives descriptions of the disease and the pathogen. See Tables 4.2 and 4.3 for tabular keys. Morphology is shown in Figure 57.1.

Sporangia

Sporangia are nonpapillate, variable in shape, ovoid, obpyriform, obturbinate, often with more than one apex, and occasionally have a lateral attachment. They are persistent on the stalk and proliferate internally. Sizes vary from 40 to 60 µm long × 20 to 40 µm wide, average 52 × 33 µm (Buisman 1927) or 41.7 × 26.2 µm (Ashby 1929b). Sporangiophores are unbranched.

Hyphal Swellings

Hyphal swellings are large, spherical, rounded, or angular, often with radiating hyphae, and in chains (catanulate).

Chlamydospores

Chlamydospores are not produced.

Sex Organs

P. richardiae is homothallic. Antheridia are large, often >20 µm wide, and most often amphigynous with characteristic protuberances but occasionally paragynous. Oogonia measure 34 to 38 µm in diameter, average 32 µm (Buisman 1927); 29 to 48 µm, average 39 µm (Ashby 1929b); average 28.9 µm (Tucker 1931). Oospores form readily in host tissue and average 29.0 µm in diameter (Buisman 1927); 19 to 41 µm, average 30.7 µm (Ashby 1929b); average 25.4 µm (Tucker 1931). Oospore walls are very thick (about 5 µm) (G. Hall 1991d).

Growth Temperatures

The minimum temperature for growth is 10°C, optimum 24 to 25°C, and maximum 30 to 32°C. Tucker (1931) reports minimal growth at 30°C, optimal at 20°C, and no growth at 30°C. He notes that *P. richardiae* resembles *P. cryptogea,* which grows at 10°C, optimally at 25°C, and has some growth at 30°C.

Distinguishing Characteristics

P. richardiae is morphologically similar to *P. vignae* and *P. erythroseptica* and is separated mainly by the size, shape, and orientation of the antheridia (G. Hall 1991d).

Disease Caused by *Phytophthora richardiae*

The disease on *Z. aethiopica* had appeared in England by 1927 and in America by 1930 (see Tucker [1933] for citations). In the United States, it occurs in California, Florida, Illinois, Indiana, Massachusetts, New Jersey, New York, Ohio, Pennsylvania, and Washington (Anonymous 1979).

In addition to calla lily, *P. richardiae* has been associated with root diseases on *Asparagus* sp., carrot (*Daucus carota*), and tomato (*Lycopersicon esculentum*). It is distributed in Asia, the Philippines, Australasia and Oceana, Australia, Europe, Belgium, Czechoslovakia, the Netherlands, and the United Kingdom (IMI Distribution Map 171; G. Hall 1991d).

Root Rot of *Zantedeschia aethiopica* (L.) Spreng. (Calla Lily)

Phytophthora root rot is a destructive disease of the calla lily. The disease first becomes noticeable when a yellow discoloration on the outer leaves slowly spreads over the whole leaf, followed by wilting and necrosis. Later, inner leaves become infected. Blossoms become either small or malformed with a brown discoloration at the distal end of the spathe.

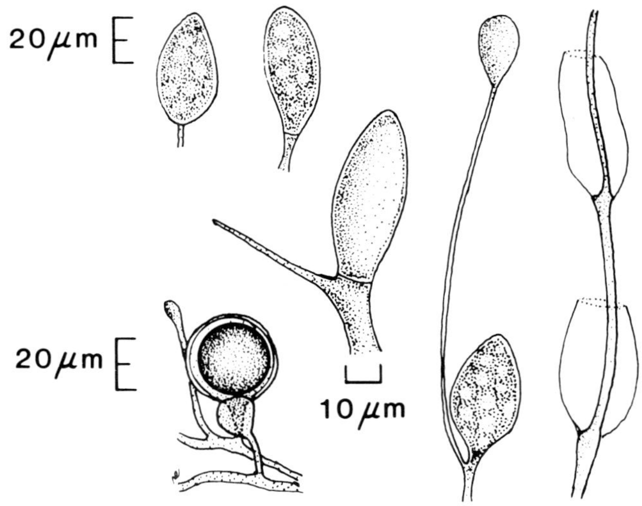

Figure 57.1. Morphology of *Phytophthora richardiae.* Lower row, globose oogonium with an amphigynous antheridium (often up to 20 µm wide). Oospore (not shown) has a thick (~5-µm) wall. Sporangium with a subterminal protuberance. Sporangia forming sympodially and proliferating externally. Upper row, ovoid to ellipsoid, nonpapillate sporangia. (Drawing by A. Vaziri from Buisman 1927 in Waterhouse 1970a)

When diseased plants are pulled, decayed roots readily break off. Some roots may have a slight red discoloration. Partially infected roots have a translucent, water-soaked appearance. Corms are rarely affected but in advanced stages may be partly destroyed. The texture of the rotted corms remains fairly dry and spongy. Diseased plants do not immediately collapse but linger in a decadent condition. Subsequently formed leaves and flowers have progressively more severe symptoms until the plants eventually die.

Disease Development. The fungus penetrates the corm at the point of attachment of the roots and remains viable in such lesions from one season to the next. Oospores form in diseased tissue and, although their role in not known, very likely serve to perpetuate the pathogen in soil.

Control. Dipping corms in water at 50°C for 1 h has been effective in suppression of the pathogen in propagative tissue. Soaking corms in a 2% mixture of formaldehyde for 1 h or mercuric chloride (28 g in 250 liters of water) is also effective (cited in G. Hall 1991d). These treatments may not currently be registered for use because of their toxicity to humans. Metalaxyl or related phenylamide compounds or fosetyl-Al should be effective (see Chapter 7).

CHAPTER 58

Phytophthora sinensis

Yu and Zhuang (1982)

Phytophthora sinensis was isolated from stems, vines, leaves, and fruit of cucumber plants, *Cucumis sativus* L., in China and described by Yu and Zhuang (1982). Because it is morphologically and physiologically (growth at 35°C) very similar to *P. drechsleri,* Ho (1986b) reported that *P. sinensis* is synonymous with both *P. drechsleri* and *P. melonis,* but *P. drechsleri* has nomenclatural priority. Until the evaluation of isozyme and mitochondrial DNA restriction fragment length polymorphism (mtDNA RFLP) patterns of these species and *P. cryptogea* by Mills et al. (1991), merging *P. melonis* and *P. sinensis* under *P. drechsleri* seemed to be the logical solution; however, Mills et al. (1991) found that isozyme and mtDNA RFLP patterns of cucumber isolates identified as *P. sinensis, P. melonis,* and certain but not all isolates of *P. drechsleri* were similar. They concurred with Ho (1986b) that all of the isolates from cucumber were similar morphologically, grew at 35°C, and were highly pathogenic to cucumber; however, this group differed markedly from a mtDNA RFLP group A that contained isolates considered to be authentic *P. drechsleri.* They suggested that the cucumber isolates not be merged with *P. drechsleri* because they differed genetically. For this reason, this species is retained until the discrepancy is resolved.

Meanwhile, there is no clear consensus about whether the cucumber isolates should be classified as *P. melonis* (Katsura 1968, 1975), which would have priority over *P. sinensis.* A technical difficulty is that Katsura (1968) described *P. melonis* as producing semipapillate sporangia, but both Ho (1986b) and Mills et al. (1991) agree that sporangia are indeed nonpapillate. Mills et al. (1991) state, "Alternatively an amended description of *P. melonis* Katsura (1968) may be appropriate."

Description of *Phytophthora sinensis*

P. sinensis is classified in group VI (Stamps et al. 1990). The morphology of *P. sinensis* is similar to that of *P. drechsleri* (Chapter 23, Figure 23.1). See Tables 4.2 and 4.3 for tabular keys. The following is summarized from Yu and Zhuang (1982).

Sporangia

Sporangia are induced by incubation of mycelial blocks of agar media (not specified) in Petri's solution (see Appendix 3.34 in Chapter 3). Sporangia are ovoid to elongate, persistent, and nonpapillate and proliferate from the empty sporangia.

Sporangia measure 24 to 80 × 20 to 40 µm, average 51.7 ± 13.1 × 31.1 ± 5.5 µm. The length-breadth ratio is approximately 1.7 (range 1.2 to 2.2). The exit pore is 9 to 18 µm in width. Sporangia form on undifferentiated, simple or branched sporangiophores.

Hyphal Swellings

Hyphal swellings are globose to subglobose and 10 to 30 µm in diameter.

Chlamydospores

Chlamydospores were not observed by Yu and Zhuang (1982) or by Ho (1986b).

Sex Organs

According to Yu and Zhuang (1982), *P. sinensis* isolates were homothallic; however, Ho (1986b) stated that they acted as heterothallic isolates because oospores were formed when paired with A1 or A2 isolates of *P. melonis, P. drechsleri,* or *P. cinnamomi.* Yu and Zhuang (1982) reported that oogonia produced on oatmeal agar were spherical or subspherical, tapering at the base, with a smooth wall. Sizes ranged from 19 to 31 µm in diameter, average 27.4 ± 2.8 µm. Antheridia are amphigynous, single, hyaline, and subglobose, short cylindrical, or variable in shape. Antheridia are mostly one celled, but occasionally a two-cell antheridium is seen. Antheridia measure 7 to 37 × 9 to 25 µm, average 19.3 × 18.5 µm. Oospores are spherical to subspherical and plerotic in the oogonium. Oospores measure 16.5 to 28.0 µm in diameter, average 23.9 ± 2.8 µm. Ho et al. (1984b) reexamined the type culture of *P. sinensis* but stated that oogonia were abortive.

Growth Temperatures

The minimum temperature for growth is 9°C, optimum 29 to 32°C, and maximum 36.5°C.

Distinguishing Characteristics

Yu and Zhuang (1982) noted the similarity of *P. sinensis* to *P. melonis* and *P. drechsleri.* They distinguish *P. sinensis* from *P. melonis* on the following evidence: Katsura's (1968) report that *P. sinensis* produces nonpapillate sporangia versus semipapillate for *P. melonis;* proliferation of sporangia (up to 17 times); larger antheridia averaging 19.3 × 18.5 µm versus 15 × 20 µm for *P. melonis;* production of an occasional two-celled antheridium by *P. sinensis;* and absence of chlamydospores in *P. sinensis.* Ho (1986b) reexamined the type culture of *P. sinensis,* authentic cultures of cucumber isolates of *P. melonis* from Japan, and a *P. drechsleri* isolate from Iran (Ershad and Mostowfipoor 1969). He found that all isolates were predominantly heterothallic but formed a few oospores in single culture, that proliferation of nonpapillate sporangia occurred, that antheridia were occasionally two celled, and that chlamydospores were not formed by any culture when mycelial disks of V8 agar were incubated in sterile distilled water at 15°C for 5 weeks. Ho (1986b) concluded that *P. sinensis* should be included in *P. drechsleri.*

We conclude that there is little rationale for maintaining *P. sinensis* as a species. The mtDNA RFLP and isozyme pattern data of Mills et al. (1991), however, indicate that several biotypes within the *P. drechsleri-P. cryptogea* complex differ markedly from the *P. melonis, P. sinensis, P. drechsleri* group from melons. Until a more definitive evaluation and redescription of *P. sinensis* or *P. melonis* are made, the cucumber isolates should be classified as *P. drechsleri.*

CHAPTER 59

Phytophthora sojae

Kaufmann and Gerdemann (1958)

Synonyms include *Phytophthora megasperma* var. *sojae* Hildebrand (1959), *P. megasperma* f. sp. *glycinea* Kuan and Erwin (1980a), and *P. sojae* f. sp. *glycines* Faris et al. (1989).

The causal agent of root and stem rot of soybean, *Glycine max* (L.) Merrill, occurring in the Midwest region of the United States, was first identified as *P. cactorum* (Herr 1957) but was later redescribed as the new species *P. sojae* (Kaufmann and Gerdemann 1958). The disease had been noted previously (Suhovecky 1955; Suhovecky and Schmitthenner 1955). The host range is relatively narrow; soybean is the only major host (Table 59.1).

Klein (1959) discussed the etiology of the disease but apparently did not accept the name *P. sojae.* Hildebrand (1959) rejected the species name *P. sojae* and placed the soybean isolates in *P. megasperma* var. *sojae* because the size of oogonia was similar to that of *P. megasperma* in the revised description by Tompkins et al. (1936). Tompkins et al. (1936) revised the Tucker (1931) key to include isolates with oogonia less than 40 µm in diameter in *P. megasperma* (see Chapter 46 on *P. megasperma* and Chapter 4 for a historical account). Hildebrand (1959) created the new variety *P. megasperma* var. *sojae* to accommodate the soybean pathogen mainly on the basis of its specific pathogenicity to soybean; however, since pathogenicity is not a valid criterion for a botanical variety, *P. megasperma* var. *sojae* was established by Waterhouse (1963) to be characterized not by host specificity to soybean but by oogonia <45 µm in diameter compared with *P. megasperma* var. *megasperma,* which was characterized by oogonia >45 µm in diameter. Thus, the revised *P. megasperma* var. *sojae* included not only soybean isolates but also those from many other crops such as alfalfa and cauliflower, while the "large-spored" *P. megasperma* var. *megasperma* included isolates from Douglas fir and other crops (E. M. Hansen and Hamm 1983; E. M. Hansen 1991). Waterhouse's choice of nomenclature for the small-oogonia *P. megasperma* var. *sojae* was confusing because "var. *sojae*" no longer referred specifically to the soybean pathogen as Hildebrand had earlier implied (Erwin 1983). Because oogonial diameters of isolates of *P. megasperma* from

Table 59.1. Distribution of hosts of *Phytophthora sojae*

Host	Common name	Disease	Geographical distribution
Geranium carolinianum L.	Cranesbill	Stem rot	Morgan 1964[a]
Glycine max (L.) Merrill	Soybean	Root and stem rot	United States (Suhovecky 1955; Suhovecky and Schmitthenner 1955; Kaufmann and Gerdemann 1958); Canada (Hildebrand 1959); Australia (Pegg et al. 1980; Ryley et al. 1989)
Lupinus adsurgens	Lupine	Seedling wilt	United States (J. P. Jones 1969)[a]
L. albicaulis Dougl. ex Hook	Sickle-keeled lupine	Seedling wilt	United States (J. P. Jones 1969)[a]
L. albifrons Benth. ex Lindl.	Lupine	Seedling wilt	United States (J. P. Jones 1969)[a]
L. albus L.	White lupine	Seedling wilt	United States (J. P. Jones and Johnson 1969)
L. angustifolius L.	Blue lupine	Seedling wilt	United States (J. P. Jones and Johnson 1969)
L. argenteus Pursh.	Lupine	Seedling wilt	United States (J. P. Jones 1969)[a]
L. bicolor Lindl.	Bicolor lupine	Seedling wilt	United States (J. P. Jones and Johnson 1969)[a]
L. concinnus J. Agardh.	Lupine	Seedling wilt	United States (J. P. Jones 1969)[a]
L. cruickshanksii (= *L. mutabilis* Sweet)	Lupine	Seedling wilt	United States (J. P. Jones 1969)[a]
L. densiflorus Benth.	Lupine	Seedling wilt	United States (J. P. Jones and Johnson 1969)[a]
L. elegans	Lupine	Seedling wilt	United States (J. P. Jones 1969)[a]
L. hartwegii Lindl.	Lupine	Seedling wilt	United States (J. P. Jones 1969)[a]
L. hirsutissimus Benth.	Lupine	Seedling wilt	United States (J. P. Jones 1969)[a]
L. hirsutus L.	Blue lupine	Seedling wilt	United States (J. P. Jones 1969)[a]
L. humicola	Lupine	Seedling wilt	United States (J. P. Jones 1969)[a]
L. latifolius Lindl. ex J. Agardh.	Lupine	Seedling wilt	United States (J. P. Jones 1969)[a]
L. luteolus	Lupine	Seedling wilt	United States (J. P. Jones 1969)[a]
L. luteus L.	Yellow lupine	Seedling wilt	United States (J. P. Jones and Johnson 1969)
L. mutabilis Sweet	Tarwi	Seedling wilt	United States (J. P. Jones 1969)[a]
L. nootkatensis Donn ex Sims	Nootka lupine	Seedling wilt	United States (J. P. Jones 1969)[a]
L. pachylobus	Lupine	Seedling wilt	United States (J. P. Jones 1969)[a]
L. perennis L.	Wild lupine	Seedling wilt	United States (J. P. Jones 1969)[a]
L. pilosus J. A. Murray	Lupine	Seedling blight	United States (J. P. Jones 1969)[a]
L. polyphyllus Lindl.	Washington lupine	Seedling blight	United States (J. P. Jones 1969)[a]
L. rivularis Douglas ex Lindl.	River bank lupine	Seedling blight	United States (J. P. Jones 1969)[a]
L. rothmaleri	Lupine	Seedling blight	United States (J. P. Jones 1969)[a]
L. sericeus Pursh.	Lupine	Seedling blight	United States (J. P. Jones 1969)[a]
L. succulentus Dougl. ex K. Koch	Lupine	Seedling wilt	United States (J. P. Jones and Johnson 1969)[a]
Phaseolus lunatus L.	Lima bean	Root and stalk rot	Canada (Hildebrand 1959)[a]
P. vulgaris L.	String bean	Collar rot	Canada (Hildebrand 1959)[a]

[a] Pathogenicity shown by artificial inoculation.

alfalfa, soybean, cherry, and other hosts varied from 30 to 58 μm in a continuum with no discrete division among sizes of various isolates, Kuan and Erwin (1980a) created the formae speciales *glycinea* for soybean isolates and *medicaginis* for alfalfa isolates. Pathogenicity readily differentiated the alfalfa and soybean isolates, and they were considered to be somewhat similar morphologically. DNA RFLP patterns show, however, that both formae speciales as well as certain isolates of *P. megasperma* from other crops are markedly different (Förster et al. 1987–1989; see Chapter 4 for further discussion and Figures 4.29 and 4.30 for DNA RFLP patterns that differentiate the alfalfa and soybean isolates). Therefore, these formae speciales are unrelated. Faris et al. (1989) reinstated the name *P. sojae* and designated the soybean isolates *P. sojae* f. sp. *glycines* because protein patterns and some morphological and physiological differences distinguished the alfalfa and soybean pathogens. Hansen and Maxwell (1991) also accepted the species name *P. sojae* of Kaufmann and Gerdemann (1958) for the host-specific soybean pathogen, and they created the new species *P. medicaginis* for the alfalfa pathogen and *P. trifolii* for the subterranean clover pathogen formerly designated *P. megasperma* f. sp. *trifolii* by R. G. Pratt (1981). They designated other isolates *P. megasperma* as described by Drechsler (1931). The redescription of *P. sojae* is reproduced by permission from Mycologia 83:376–381, E. M. Hansen and D. P Maxwell, copyright 1991, The New York Botanical Garden, Bronx, New York.

This species (*Phytophthora sojae*) exhibits aggressive, race-specific pathogenicity to soybean, but causes few or no symptoms on other hosts. Oogonia average 37–39 μm in diam and are similar in size to those from some isolates from numerous other hosts. Sporangia average 58–61 μm long. Oogonium diam and sporangial length are distinctive when considered together. Hyphal growth rate is slow (<5 mm/day) at 20 C, which is a lower temperature for maximum growth than for other groups of the complex. Kaufmann and Gerdemann (1958) reported no growth at 5 and 35 C; E. M. Hansen and Hamm (1983) report slight growth (<1 mm/day). Colony appearance is distinctive on carrot agar (E. M. Hansen *et al.*, 1986), with dense cottony, patternless growth to the colony margin. Hohl (1983) distinguished *P. sojae* (as *P. megasperma* f. sp. *glycinea*) from other *P. megasperma* isolates by its growth on nitrate nitrogen.

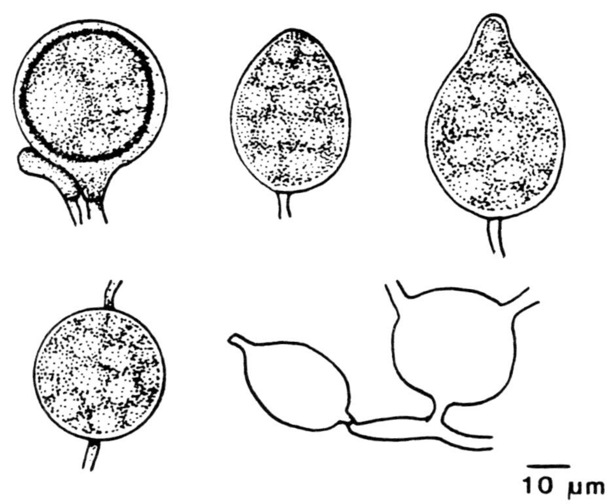

Figure 59.1. Morphology of *Phytophthora sojae.* Upper row, globose oogonium with paragynous antheridium; ovoid and obpyriform sporangia. Lower row, chlamydospore (note: Hildebrand 1959 is the only citation on chlamydospore formation); empty remnants of hyphal swellings. (Drawing by A. Vaziri from Hildebrand 1959)

Phytophthora sojae isolates are readily identified by total protein pattern with polyacrylamide gel electrophoresis (SDS PAGE); the SOY protein group of E. M. Hansen *et al.*, 1986), isoenzyme patterns (Nygaard *et al.*, 1989), and RFLPs (Förster *et al.*, 1989). They have a chromosome number (n = 12–15) similar to other small-spored species of the *P. megasperma* complex.

Characteristics of *Phytophthora sojae*

P. sojae is classified in group V (Stamps et al. 1990). Morphology is shown in Figures 59.1 and 59.2. See Tables 4.2 and 4.3 for tabular keys.

Sporangia

Sporangia are nonpapillate and ovoid, ellipsoid, and sometimes obpyriform. Sporangia range in size from 23.3 to 88.8 μm × 16.6 to 51.8 μm, average 58 × 38.3 (Kaufmann and Gerdemann 1958) (Figures 59.1 and 59.2E–J). Sporangiophores are simple and undifferentiated. Sporangia are formed both sympodially and by internal and external proliferation (Hildebrand 1959).

Hyphal Swellings

Intercalary hyphal swellings, which are spherical to irregular in shape, occur especially in aqueous culture.

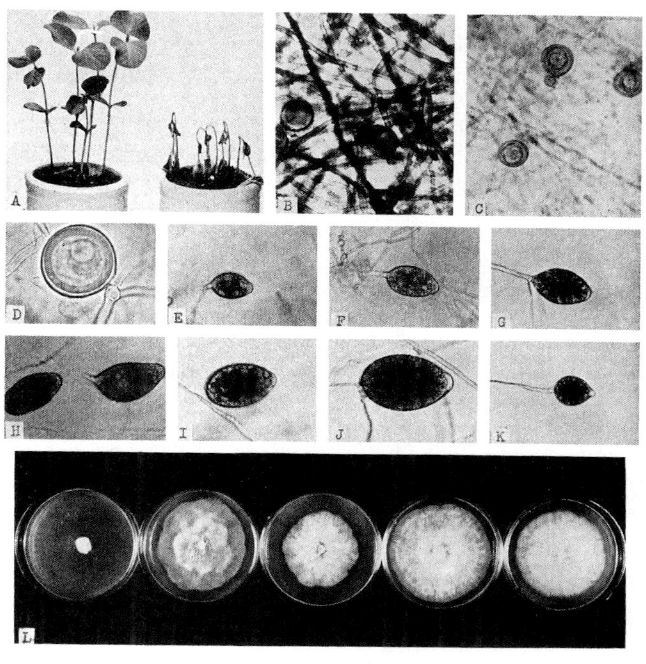

Figure 59.2. A, Root and stem rot of soybean caused by *Phytophthora sojae;* left, noninoculated plants; right, inoculated plants. **B,** Mycelium of *P. sojae* stained after treatment with chloroiodide of zinc. **C,** Mycelium of *P. cactorum* not stained after treatment with chloroiodide of zinc. **D,** Globose oogonium containing an oospore that fills the oogonium. A paragynous antheridium can be faintly seen on the lower hemisphere of the oogonium. **E–J,** Ovoid, nonpapillate sporangia of *P. sojae.* **K,** Papillate sporangium of *P. cactorum.* **L,** Left to right, lack of growth of *P. sojae* after 7 days on potato-dextrose agar compared with growth of *P. megasperma* and isolates of *P. cactorum* from sweet clover, lemon, and lilac. (Reproduced by permission of the American Phytopathological Society, St. Paul, Minnesota, courtesy Kaufmann and Gerdemann 1958)

Chlamydospores

Chlamydospores are not mentioned in the original description by Kaufmann and Gerdemann (1958), but Hildebrand (1959) showed a photomicrograph of a chlamydospore (Figure 59.1) and stated that chlamydospores conformed to the definition of Blackwell (1949). They were similar in size to oogonia. Schmitthenner in Sinclair and Backman (1989) stated that chlamydospores occurred, but he did not give dimensions.

Sex Organs

P. sojae is homothallic. Oospores are formed in infected root tissue of susceptible and tolerant cultivars (Slusher and Sinclair 1973). Oogonia are spherical to subspherical and vary from 29.4 to 45.7, average 36.3 µm (Hildebrand 1959); 19.2 to 38.3, average 31 µm (Kaufmann and Gerdemann 1958); and 32 to 38 µm (Faris et al. 1989). Antheridia are mostly paragynous, but some are amphigynous (Hildebrand 1959; Faris et al. 1989). The diameters of oospores range from 19.2 to 38.3 µm (average 31.4 µm) (Kaufmann and Gerdemann 1958). Oospores nearly fill the oogonium (plerotic) (Figure 59.2D). Oospores germinate at relatively high percentages (Erwin and McCormick 1971; Schechter and Gray 1987; Bhat et al. 1990, 1992, 1993; Bhat 1991; Bhat and Schmitthenner 1992, 1993a, b). Figure 3.6 shows the progress of germinating oospores at different stages (Erwin and McCormick 1971). Figure 3.8 shows that after two nuclei fuse and the inner wall of the oospore erodes, the oospore swells and germinates (Jiang et al. 1989).

Growth Temperatures

Reports of optimum temperatures for growth include 25°C (Hildebrand 1959; Faris et al. 1989) and 20°C (Kaufmann and Gerdemann 1958; Hansen and Maxwell 1991). No growth occurs at 5 and 35°C (Kaufmann and Gerdemann 1958), but slight growth was reported by Hansen and Maxwell (1991). According to Schmitthenner in Sinclair and Backman (1989), 25–28°C is optimum and 32–35°C is maximum for growth.

Growth Characteristics

P. sojae does not grow on potato-dextrose agar (Figure 59.2L) (Kaufmann and Gerdemann 1958; Hildebrand 1959; Kuan and Erwin 1980a); however, addition of sitosterol (30 µg/ml) or dilution of potato-dextrose agar to 1:10 increases growth to normal (Erwin et al. 1968).

Distinguishing Characteristics

Although *P. sojae, P. medicaginis,* and *P. trifolii* have been classified within the species *P. megasperma* as formae speciales and have been considered to be relatively similar, several features readily distinguish *P. sojae* from *P. medicaginis.* Because *P. trifolii* has been noted only in one area of Mississippi, its characteristics are less known (see Chapter 62). Table 59.2 is a compilation of physiological and morphological features that distinguish *P. sojae* from *P. medicaginis* (see also Nygaard et al. [1989] and Chapter 44 on *P. medicaginis*).

Diseases Caused by *Phytophthora sojae*

Soybean is the only main crop on which *P. sojae* causes severe outbreaks of root and stem rot (Table 59.1). The disease is widely distributed in soybean-growing regions of Canada (Hildebrand 1959) and the United States (Schmitthenner 1985). In recent years, the disease has become an agricultural problem in Queensland, Australia (Pegg et al. 1980), and in eastern Australia (Ryley et al. 1989, 1992). How the pathogen arrived there has not been determined. Although southeastern Asia is considered to be the area from which soybean emanated, we have not seen any reports of root and stem rot caused by *P. sojae* from that region.

Other legumes and a large number of nonleguminous hosts were not susceptible when inoculated by stem puncture (Kaufmann and Gerdemann 1958; Hildebrand 1959). Slight disease was noted in some legumes inoculated by stem puncture, but none of these crops are seriously affected.

Although *P. sojae* is considered to be specific to soybean, it has been isolated from diseased lupine plants that had been grown in potted Sharkey clay soil collected in a field near Stoneville, Mississippi, in which root rot on soybean had occurred (J. P. Jones and Johnson 1969). The blue lupine (*Lupinus angustifolius* L.), yellow lupine (*L. luteus* L.), and white lupine (*L. albus* L.) were susceptible to inoculation by artificial soil infestation and by stem puncture. Because several lupine species are susceptible (Table 59.1), J. P. Jones and Johnson (1969) postulated that *P. sojae* might have existed on naturally occurring lupine plants prior to the introduction of soybeans to North America.

Root and Stem Rot of *Glycine max* (L.) Merrill (Soybean)

Symptoms are described by Hildebrand (1959), Sinclair and Backman (1989), and McGee (1993). Ward (1990) reviewed the literature on pathogenicity with major emphasis on the nature of race-specific resistance. A review on control by host resistance by Athow (1987) is discussed in Chapter 6.

Soybean plants of any age, from seedling to maturity (Plates 59.1 and 59.2), are susceptible to *P. sojae;* however, susceptibility declines somewhat with age. Preemergence damping-off, as shown in Plate 59.2, occurs frequently. At the postemergence stage, seedlings may be killed quickly, espe-

Table 59.2. Features that differentiate *Phytophthora sojae* from *P. medicaginis* [a]

Characteristic	*P. sojae*	*P. medicaginis*
Mycelial growth on V8 juice agar	Somewhat appressed radiate (Luo et al. 1988)	Tends to be fluffy uniform (no pattern)
Mycelial growth on potato-dextrose agar	Very little growth (Figure 59.2L)	Profuse growth (Kuan and Erwin 1980a)
Growth on malachite green	60%	0% (Luo et al. 1988)
Growth rate on V8 juice medium	Slow; 7.8 mm/day	Fast; 10 mm/day (Luo et al. 1988)
Sporangia shape	Ellipsoid to obpyriform; less symmetrical (Figure 59.1)	Symmetrical; egg shaped (Figure 4.12L)
Basal appendage to sporangia	Not seen	Occurs frequently (Erwin 1965)
Oogonia and oospores in culture media	Abundant	Tend to be less abundant in wild type except in selected isolates (Erwin and Li 1986)
Antheridia	Mostly paragynous	Mostly amphigynous but variable with isolate (Erwin 1965)

[a] Compiled in part with the assistance of unpublished observations, courtesy A. F. Schmitthenner. See also Nygaard et al. (1989).

cially after heavy rainfall. On older plants, wilting of the leaves occurs first (Plate 59.3); chlorotic areas develop between the veins and along margins of the leaves. A brown, girdling rot often extends up the stem as high as 10 nodes above the soil (Plate 59.4). Occasionally, lesions may be confined to the stem with no visible root symptoms. Once infected, lateral and branch roots turn brown, and the cortex and vascular tissues become discolored. When the soil is wet for prolonged periods, even tolerant cultivars may become yellow and stunted. Although symptoms are expressed in susceptible and to some extent in tolerant cultivars, those with *Rps* genes for race-specific resistance do not develop symptoms. Oospores normally form in both susceptible and tolerant cultivars (Figure 59.3) (Slusher and Sinclair 1973), but in resistant cultivars oospores form sparsely.

Disease Development. The primary inoculum comes from oospores that form in roots and stems of both tolerant and susceptible cultivars (Figure 59.3) (Slusher and Sinclair 1973; Kittle and Gray 1979b, c; Schmitthenner 1985). Oospores are capable of persisting in infected tissue or soil. *P. sojae* can be recovered from soil that has been frozen (freezing kills mycelium and sporangia but not oospores) by baiting premoistened soil with soybean leaf disks (Canaday and Schmitthenner 1982; Schmitthenner 1985; Schmitthenner et al. 1994; Appendix 2.8). Moistening soil before flooding and baiting with soybean leaf disks apparently allows oospores to germinate, and flooding enhances the production of sporangia and zoospores. Although this has been proved only in the laboratory, a similar process most likely occurs in the field. See Chapter 2 for methods of isolation of *Phytophthora* by baiting and Chapter 3 for a discussion of oospore germination.

P. sojae was isolated from immature seeds collected from pods that were produced on an infected stem (Klein 1959). It has been isolated from mature seeds but only after a 2-month cold-wet treatment. Oospores had been noted in seed coats and were probably stimulated to germinate by the cold-wet treatment. *P. sojae* was also isolated from soil mixed in seed lots (Klein 1959). This finding could explain dissemination of *P. sojae* over wide distances. There is little information on this aspect of distribution over long distances, however.

Although rotation to nonsusceptible crops such as wheat should help to reduce the inoculum, cropping with wheat for as many as 4 years did not eliminate *P. sojae* (Schmitthenner 1985). This result further suggests that *P. sojae* can persist as oospores in soil.

Schmitthenner (1985) postulates that oospores may have an internal timing mechanism that prevents synchronous germination and prolongs dormancy for long periods. Several studies indicate that germination of laboratory-produced oospores is often erratic and can be induced only by applying certain methods, for example, short periods of increased temperature (Long et al. 1975; Long and Keen 1977), incubation under cool conditions prior to germination (Erwin and McCormick 1971), use of different substrates such as soybean root exudate or soil extract (Jimenez and Lockwood 1982), and other methods (see Chapter 3 for a more detailed discussion of oospore germination). More information is needed on the mechanism of oospore persistence and germination in field-developed oospores. The reports of Bhat (1991), Bhat et al. (1992, 1993b), Bhat and Schmitthenner (1992), and Schmitthenner (1994) show that the oospore is important not only for persistence but also as a source of variability.

Phytophthora root and stem rot develops rapidly at temperatures between 25 and 30°C, particularly in areas with poorly drained, heavy soils and after heavy rainfall. *P. sojae* infects lateral roots and progresses into the taproots and hypocotyls in susceptible cultivars, but in resistant cultivars the fungus is limited to the cortex and stele of lateral roots (Beagle-Ristaino 1983b). Sporangia and zoospores form on infected roots, and zoospores are transported in free water. Losses up to 50% or more occur, depending on the susceptibility of the cultivar and the environmental conditions. The incidence and severity of the disease is reduced during dry weather and at low soil moisture.

Control. According to Athow (1987), cultural practices, systemic fungicides, and resistant cultivars are effectively used for control. Improvement of soil tilth and drainage tend to reduce the severity of Phytophthora root rot; however, the agricultural changes in soybean culture that emphasize larger equipment, earlier planting, reduced tillage, increased use of herbicides, and less rotation enhance disease development (Abney et al. 1979; Kittle and Gray 1979a; D. R. Duncan and Paxton 1981; Tachibana and Van Deist 1983).

Management of soil water, where possible, is important, since heavy soils in which free water is held for long periods are most conducive to root rot. The incidence of root rot increases with the increase in bulk density (weight per volume) of soil. In a glasshouse study in which soil from a field in Ontario, Canada, was used, the incidence of Phytophthora root rot increased from 2.5% at a bulk density of 1.20 g/cc to 45% at a bulk density of 1.49 g/cc (Fulton et al. 1961). Bulk density is increased in soil that has been packed by machinery (Kittle and Gray 1979a; L. E. Gray and Pope 1986). Heavy soils and especially soils on which ponding occurs should be avoided.

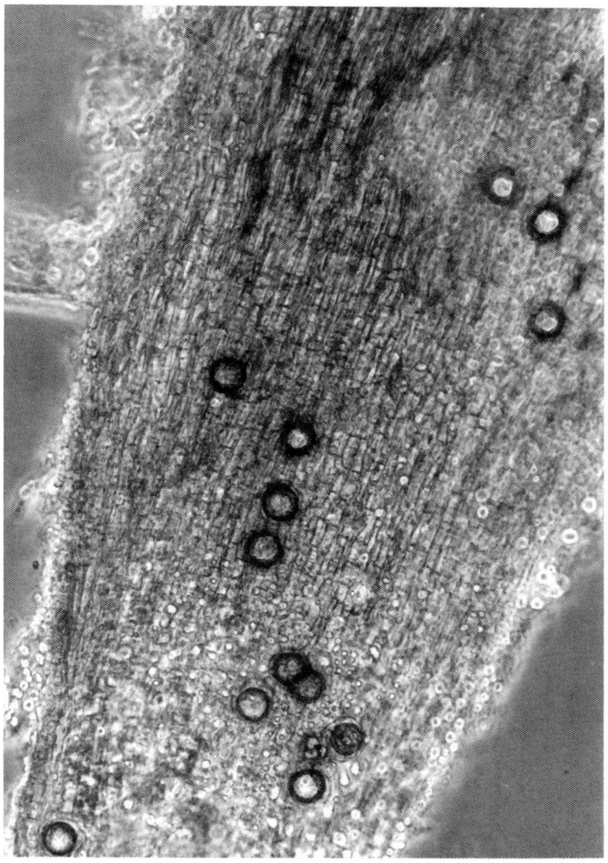

Figure 59.3. Oospores formed in a root tip of a soybean plant infected by *Phytophthora sojae.* (Courtesy R. L. Slusher and J. B. Sinclair, Department of Plant Pathology, University of Illinois, Urbana; see also Slusher and Sinclair 1973)

Control by use of race-specific cultivars has been the most important method and the most actively researched by universities, the United States Department of Agriculture, and private companies in the soybean seed industry.

Cultivars with race-specific resistance are not affected by *P. sojae* until a new pathologic (physiologic) race develops in high enough population to cause disease. Fortunately, plant breeders and plant pathologists have been able to select a wide array of resistance genes (*Rps*) that have been incorporated into agronomically adapted soybean cultivars. Pathologic (physiologic) races of *P. sojae* were first described by Morgan and Hartwig (1965). Control by use of cultivars with single-gene (*Rps*)-controlled resistance is reviewed by Athow (1987), who also discusses the development of new races that infect new soybean cultivars selected for resistance. See Chapter 6, Control by Host Resistance, and Table 6.1 for a listing of references on host resistance.

According to Athow (1987), after the report of races 1 and 2 by Morgan and Hartwig (1965), race 3 was reported in Ohio by Schmitthenner (1972), race 4 from Kansas by Schwenk and Sim (1974), races 5 and 6 from Ontario, Canada, by Haas and Buzzell (1976), races 7, 8, and 9 by Laviolette and Athow from Indiana (1977), races 10 to 16 from Mississippi by Keeling (1980), races 17 to 20 from Mississippi and Arkansas by Keeling (1982a), races 21 and 22 from Indiana by Laviolette and Athow (1983), and race 23 from Nebraska by D. M. White et al. (1983). Since Athow's (1987) report, two additional races have been identified. As shown in Table 6.3, Ward (1990) lists 25 races. Layton et al. (1986) described race 26, and Wagner and Wilkinson (1992a) described race 27. A note in *Phytopathology News* (December 1993, page 171) states that 37 races of *P. sojae* have been characterized. Schmitthenner et al. (1994) summarized race data over a 10-year period in Ohio and concluded that several new races, isolated from the field, were capable of defeating all the *Rps* genes or combinations that are now available in the soybean industry.

Race-specific resistance is associated with the rapidity of production of the host chemical glyceollin (phytoalexin) in response to inoculation. The role of phytoalexins is discussed in Chapter 6. See Table 6.4. The proposed sequence of biochemical events that occur in resistant soybean plants is shown in Table 6.5 (Keen and Yoshikawa 1983a).

Given that new, previously undescribed races of *P. sojae* were detected by isolation from field soils (Hobe and Schmitthenner 1981; Schmitthenner et al. 1994), the potential for the appearance of new races is extremely great. Rutherford et al. (1985) detected new races in successive generations of isolates from single zoospores.

Schmitthenner et al. (1994) suggest that the continual development of new races could be a result of segregation of isolates from a natural heterokaryotic culture (Long et al. 1975; Long and Keen 1977; Layton and Kuhn 1990) or from hybrid oospores (Bhat and Schmitthenner 1993; Förster et al. 1994).

Because all *Phytophthora* species are diploid or even polyploid (see Chapter 3), it would be expected that the genome of *P. sojae* would have a potential for variability; however, it has been shown only recently that oospores may be a source of variation. Several reports by Bhat and his colleagues at Ohio State University (Bhat et al. 1990; Bhat 1991; and see below) show conclusively that genetic crosses between *P. sojae* isolates with different race characteristics can result in hybrids, even though *P. sojae* is homothallic and mating types have not been described (Bhat and Schmitthenner 1993b). Bhat and Schmitthenner (1993a) found that the race characteristic of single-oospore progeny of a race 1 isolate was single gene controlled and segregated 3:1, virulent to avirulent. Isolates of races 3 and 4, however, did not yield any avirulent isolates. Thus, isolates can be homozygous (breed true) or heterozygous and/or heterokaryotic for race characteristics. Reports on the genetics of *P. sojae* are reviewed further in Chapters 3 and 6.

Mutant lines with resistance to metalaxyl and para *p*-fluorophenylalanine were developed and shown to be capable of cross fertilizing in mixed cultures to produce hybrid progeny (6 to 76%) (Bhat and Schmitthenner 1993a, b). These studies, among others, give the best explanation to date of why new races may develop rapidly in the field. See the discussions in Chapter 3 and Chapter 6.

Use of moderate rate-reducing resistance (tolerance) (Schmitthenner and Walker 1979) in addition to race-specific resistance has become a successful control measure that is used widely in all soybean-growing regions (Schmitthenner 1985; Schmitthenner and Van Doren 1985; Athow 1987; Ward 1990; see Table 6.1 for more references). Specific recommendations for selection of resistant and/or tolerant cultivars for field use should be obtained from local sources of information. As explained more fully in Chapter 6, tolerance (Schmitthenner and Walker 1979), also called rate-limiting resistance (Tooley and Grau 1982, 1984a, b; Wagner et al. 1993), is usually controlled by more than one gene, and thus there is less likelihood that races will evolve that defeat tolerance. Both types of resistance are utilized in the production of commercial cultivars of soybeans.

Because tolerance is controlled by multiple genes, the probability that new races will become adapted to multiple genes in the soybean is much lower than the probability that a *P. sojae* race will become adapted to a single gene for race-specific resistance in soybean.

The generally accepted strategy for breeding involves the selection of single-gene resistance to races that exist in a particular locality and, in conjunction, selection for tolerance (Buzzell and Anderson 1982; Schmitthenner 1985; Athow 1987). Methods used for selection of race-specific resistance and tolerance in soybean cultivars are discussed in Chapter 6.

The systemic fungicide metalaxyl (Ciba-Geigy Corporation) is effective as a seed treatment (Apron) and an in-furrow row treatment (Ridomil). Seed treatment significantly reduces root rot and damping-off in greenhouse tests (Papavizas et al. 1979) and is used frequently in the field. Although furrow treatment with metalaxyl is effective (T. R. Anderson and Buzzell 1982; Schmitthenner 1985; Schmitthenner and Van Doren 1985; Sinclair and Backman 1989), it may not be cost effective.

In an integrated control study in Ohio involving resistance, tolerance, row treatment with metalaxyl, and soil drainage, metalaxyl at 1.12 kg/ha was the most effective measure that increased yield; however, at this dosage, metalaxyl was not cost effective. Tolerance of cultivars was the next most effective factor; however, when cultural conditions were highly conducive for Phytophthora root rot (e.g., land was not tilled, tillage was reduced, and the seedbed was flat), tolerance was not as effective as race-specific resistance (Schmitthenner and Van Doren 1985).

CHAPTER 60

Phytophthora syringae

(Klebahn) Klebahn (1909)

Synonyms of *Phytophthora syringae* include *Ovularia syringae* Berkeley (1881), *Phloeophthora syringae* Klebahn (1905), *Nozemia syringae* (Klebahn) Pethybridge (1913), *P. cactorum* subvar. *syringae* (Klebahn) Sarejanni (1936), *Pythiomorpha fischeriana* Höhnk (1936), and *P. fischeriana* (Höhnk) Sparrow (1960). *P. syringae* was first isolated and described by Klebahn (1905) from diseased lilac trees, *Syringa vulgaris* L., in Germany. In addition to lilac, *P. syringae* has a fairly wide host range that includes 31 plants in 19 genera of flowering plants (Table 60.1). It is best known, however, as a pathogen of plants in the family Roseacae. Its distribution is mainly in temperate climates (CMI Map 174, ed. 3, 1977).

P. syringae and the diseases caused are reviewed by Waterhouse and Waterston (1964b) and D. C. Harris (in I. M. Smith et al. 1988; D. C. Harris 1991b). Fruit rot of apple is reviewed by Upstone (1978). Ho and Jong (1993) compared *P. syringae* and *P. hibernalis* and showed that these species are not conspecific as stated by Tucker (1931).

Characteristics of *Phytophthora syringae*

P. syringae is classified in group III (Stamps et al. 1990). A summary is given by Waterhouse and Waterston (1964b). See Tables 4.2 and 4.3 for tabular keys. Morphology is shown in Figure 60.1.

Sporangia

Sporangia are broadly ovoid or obpyriform, semipapillate, and persistent and form in succession from a single sporangium. Dimensions are given in Table 60.2. The length-breadth ratio varies from 1.32:1 to 1.85:1. Sporangiophores form in a close monochasial sympodium with intercalary swellings.

Hyphal Swellings

Hyphal swellings are rounded or angular and often in chains (catenulate) and are occasionally delimited by septa. They often germinate by production of slender hyphae.

Chlamydospores

Chlamydospores are not produced according to Waterhouse and Waterston (1964b); however, Stamps et al. (1990) report chlamydospores 25 µm in diameter. Ho and Jong (1993) did not report chlamydospores.

Sex Organs

P. syringae is homothallic. Oogonia form abundantly in culture and host tissues; antheridia are predominately paragynous and occasionally amphigynous; and oospores fill the oogonium (plerotic). Dimensions of oogonia and oospores from different reports are given in Table 60.2. Oogonium diameter averages 33 µm, and oospore diameter averages about 30 µm. Germination of oospores in vitro was described by D. C. Harris and Cole (1982).

Growth Temperatures

The minimum temperature for growth is <5°C, the optimum is 15 to 20°C, and the maximum is 23 to 25°C.

Distinguishing Characteristics

Hyphae of old cultures are often knotty, vesicular, and coiled. The coiling of hyphae is similar to that observed in *P. porri* and *P. primulae*. *P. syringae* was erroneously considered by Tucker (1931) to be conspecific with *P. hibernalis*. These two species have often been confused. *P. syringae* produces noncaducous, ellipsoid sporangia and predominantly paragynous antheridia. *P. hibernalis* produces caducous, ovoid, obpyriform to elongated sporangia (larger than those of *P. syringae*) characterized by long pedicels (23–73 µm) and both paragynous and amphigynous antheridia in varying proportions (Ho and Jong 1993). Ho and Jong (1993) review the examples in the published literature where misidentification has led to confusion, e.g., Nadel-Schiffman (1947), who was corrected by Kouyeas and Chitzanidis (1968). Ho and Jong (1993) state that an N37 isolate from California labeled *P. hibernalis* actually produced amphigynous antheridia, indicating that it was probably *P. syringae*. Other citations in which *P. hibernalis* was identified as *P. syringae* include Wager (1942b) and Frezzi (1950). *P. syringae* differs from *P. porri* by its lower maximum temperature for growth and lack of chlamydospores. See Ho and Jong (1993) for a comparative study of *P. syringae* and *P. hibernalis*.

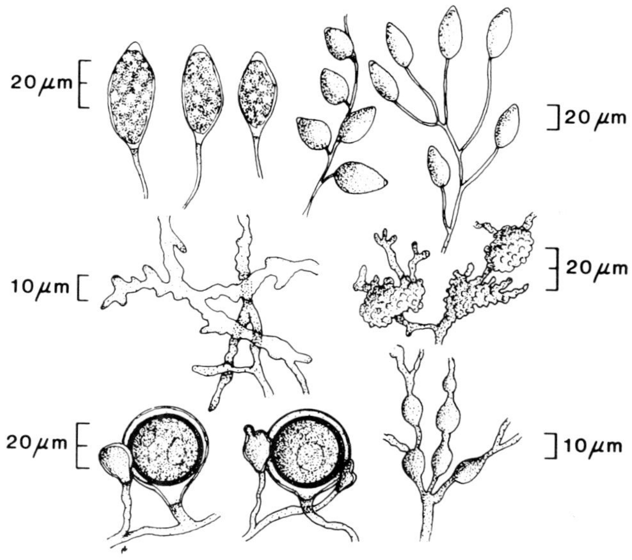

Figure 60.1. Morphology of *Phytophthora syringae*. Lower row, globose oogonia and paragynous antheridia (sometimes two antheridia per oogonium); inflated, catenulate hyphal swellings. Center row, irregular, torulose mycelium. Upper row, ovoid, nondeciduous sporangia formed sympodially on sporangiophores. (Drawing by A. Vaziri)

Some Diseases Caused by *Phytophthora syringae*

Collar and Fruit Rot of *Malus pumila* Mill. (Apple)

P. syringae was described in Switzerland by Osterwalder (1906) and by Lafferty and Pethybridge (1922) in Ireland as the causal agent of a rot of apple fruits. In southeast England, *P. syringae* was found to be the cause of an important and widespread bark rot of mature apple trees (Sewell and Wilson 1963) and later found to cause a high incidence of bark rotting in young trees during the "heeled in" phase between lifting in the nursery and planting in the orchard (Sewell and Wilson 1964). According to D. C. Harris (1991b), this disease is now rare in England. The occurrence of *P. syringae* in stool-beds was reported by Tidball and Linderman (1990) in Oregon. The flesh of infected fruit turns dark brown, but the rotted tissue remains firm. Fruit rot occurs normally during late summer and autumn while fruit is on the tree.

Collar rot occurs on young and mature bearing trees during

Table 60.1. Distribution of hosts of *Phytophthora syringae*

Host	Common name	Disease	Geographical distribution
Aesculus hippocastanum L.	Horse chestnut	Twig blight	Germany (Klebahn 1909)[a]
Alnus glutinosa (L.) Gaertn.	Black alder	Twig blight	Germany (Klebahn 1909)[a]
Castanea crenata Siebold and Zucc.	Japanese chestnut	?	United Kingdom (W. R. Day 1939)[a]
Castanea sativa Mill.	Sweet or Eurasian chestnut	Root rot	United Kingdom (W. R. Day 1938)
Cereus martianus Zucc.	Cactus	Wilt	Germany (Himmelbaur 1911)[a]
Cereus tetracanthus Lab.	Cactus	Wilt	Germany (Himmelbaur 1911)[a]
Chionanthus virginicus L.	Old-man's beard	Wilt	Germany (Himmelbaur 1911)[a]
Citrus sp.	Citrus	Gummosis, root rot	Portugal (Telhada 1990)
Citrus spp.	Citrus	Fruit rot	United States (H. S. Fawcett 1936); Argentina (Frezzi 1950); Italy (Verneau 1954); Morocco (Van der Weyen 1966)
Citrus aurantium L.	Sour orange	Fruit rot	Greece (Kouyeas and Chitzanidis 1978)
Citrus limon (L.) Burm.	Lemon	Fruit rot	Greece (Kouyeas and Chitzanidis 1978)
Citrus sinensis (L.) Osbeck.	Sweet orange	Fruit rot	Greece (Kouyeas and Chitzanidis 1978)
Corylus avellana L.	Hazelnut	Twig blight	Germany (Klebahn 1909)[a]
Crataegus oxyacantha L.	Hawthorn	Twig blight	Germany (Klebahn 1909)[a]
Echinopsis eyriesii Zucc.	Sea-urchin cactus	Wilt	Germany (Himmelbaur 1911)[a]
Erica hiemalis Nicols	White winter heather	Blight	United Kingdom (Bewley and Oyler 1935)
Erica nivalis Andr.	Heather	Blight	United Kingdom (Bewley and Oyler 1935)
Fagus L.	Beech	Root rot	United Kingdom (W. R. Day 1939)[a]
Fagus sylvatica L.	European beech	Seedling blight	Germany (Klebahn 1909); United Kingdom (W. R. Day 1938)
Foeniculum vulgare Mill.	Fennel	Leaf blight	Italy (Noviello and Snyder 1962)
Forsythia viridissima Lindl.	Golden bells	Twig blight	Germany (Klebahn 1909)[a]
Jasminum nudiflorum Lindl.	Jasmine	Twig blight	Germany (Klebahn 1909)[a]
Malus sp.	Crabapple	Stem canker of nursery stock	United States (Young and Milbrath 1959)
Malus pumila Mill. and *Malus sylvestris* Mill.	Apple	Stem canker, fruit rot	Switzerland (Osterwalder 1906; Bolay 1977); Ireland (Lafferty and Pethybridge 1922; Kavanagh et al. 1969); United Kingdom (Ogilvie 1931); Northern Ireland (Colhoun 1938); United States (Young and Milbrath 1959); New Zealand (Dingley 1969); Canada (Ross and Gowrley 1969). See D. C. Harris (1991) for more citations.
Pinus radiata D. Don	Monterey pine	Root rot	New Zealand (Newhook 1959)
Prunus dulcis (Mill.) D. Webb (*P. amygdalus* Batsch.)	Almond	Pruning wound canker	United States (Bostock and Doster 1985)
Prunus armeniaca L.	Apricot	Apoplexy, wilt, stem canker	New Zealand (H. C. Smith 1956); United States (Young and Milbrath; Doster and Bostock 1988c[a]); Greece (Kouyeas and Chitzanidis 1968)
Prunus avium (L.) L.	Sweet cherry	Twig blight	Germany (Klebahn 1909)[a]
Prunus cereasus L.	Sour cherry	Root and crown rot	United States (Bielenin and Jones 1988)
Prunus mahaleb L.	Mahaleb cherry	Apoplexy, wilt	Greece (Kouyeas and Chitzanidis 1968)
Prunus persica (L.) Batsch.	Peach	Apoplexy, wilt, stem canker, collar rot	New Zealand (H. C. Smith 1956); United States (Young and Milbrath 1959; Doster and Bostock 1988c[a]); Greece (Kouyeas and Chitzanidis 1968); Italy (D'Ercole and Flori 1978)
Prunus sp.	Plum	Apoplexy, wilt	Greece (Kouyeas and Chitzanidis 1968)
Pyrus communis L.	Pear	Fruit rot	Ireland (Lafferty and Pethybridge 1922); United Kingdom (Ogilivie 1931)
Quercus sp.	Oak	Twig blight	Germany (Klebahn 1909)[a]
Syringa vulgaris L.	Common lilac	Downy mildew, leaf spot, shoot dieback	Scotland (Berkeley 1881); Germany (Klebahn 1905); France (Arnaud 1918a, b); Italy (Peglion and Sacchetti 1927); Belgium (Marchal and Verplancke 1926); most of Europe by 1930; United States (Hedges 1929); Soviet Union (Jermalaviciute 1958); Canada (Conners 1967)
Tilia sp.	Lime tree	Twig blight	Germany (Klebahn 1909)[a]

[a] Artificially inoculated host.

late autumn. The trunk of the tree is often girdled from the collar to several inches above the ground. Exudation of gum from affected bark is a common symptom. Buds of young trees fail to open the following spring, and mature trees put forth weak, small shoots, which do not develop further. In nursery stock, dark, slightly sunken cankers occur on stems, and in severe cases the stem is girdled.

Symptoms of fruit rot and collar rot are similar to those caused by *P. cactorum* (see Chapter 11). D. C. Harris's (1991b) article compares and contrasts the diseases caused by both *P. syringae* and *P. cactorum.* Upstone (1978) describes symptoms on apple fruit, development of fruit rot, and control. Sewell et al. (1974) compared the activity of *P. syringae, P. cactorum,* and *P. citricola* in soil on collar rot of apple. A greater number of apple cultivars are susceptible to *P. syringae* than to *P. cactorum* (Sewell and Wilson 1963).

Disease Development. Since *P. syringae* grows most rapidly at temperatures of 15 to 20°C with a maximum of 23°C, it is most active during the cooler months of the year (September to November in the northern hemisphere). It is thus of year-round importance in temperate regions of the world. In the United Kingdom, the activity of *P. syringae* is greatest during September to October (Sewell and Wilson 1964, 1973). According to D. C. Harris (1991b), *P. syringae* is more important in northern France, where conditions are relatively cooler, than in southern France, where *P. cactorum* is the dominant causal agent of fruit rot. Infection of fruit is caused by rain splash of contaminated soil just before or during the harvest (Upstone and Gunn 1978). Infection of fruit in the orchard leads to further rotting, even during cold storage (Upstone 1978). The incidence of disease on fruits increases rapidly from the contact of healthy fruit by diseased fruit.

Oospores that develop in fallen leaves may survive for more than 15 years and when germinated produce sporangia, which constitute an important primary source of inoculum in the soil (D. C. Harris 1979a, 1985a, c, 1991b). Sporangia and zoospores are disseminated by water and in irrigation canals. The formation of bark or root canker is exacerbated by wet soil, a trait that is typical of all *Phytophthora*-caused diseases (D. C. Harris 1991b).

Control. As stated in Chapter 11 on *P. cactorum,* planting disease-free trees on noninfested land is the first line of control, because *P. syringae* is not universally distributed. When fruit is first placed in storage, it should be dried as rapidly as possible since fruit rot occurs only when the fruit remains moist for 3 days or longer at 3.3°C (Edney 1978). Apple fruit stored in a mixture of carbon dioxide and oxygen had a lower incidence of fruit rot than that stored in air (Edney and Chambers 1981a).

The recent review and discussion of Phytophthora collar rot and fruit rot diseases of apple trees by D. C. Harris (1991b) covers the control of *P. syringae* and *P. cactorum,* both of which are important causal agents in Great Britain. D. C. Harris states that protective fungicides such as Bordeaux mixture and other copper-based fungicides have long been recommended for use, but there is little critical evidence that they are effective for prevention of crown or bark rots or as treatments of wounds created by pruning. Since copper does not move systemically into plant tissue, treatment of infected tissue is not effective below the treated surface. Captan applied to trunks of trees is an effective protectant but does not have eradicatory action (Edney 1978).

The recently introduced systemic phenylamide fungicides (e.g., metalaxyl and furalaxyl) and phosphonate (e.g., fosetyl-Al or phosphorous acid) have improved the quality of chemical control immensely (see Chapter 7 for a discussion and listing of references on systemic fungicides, Table 7.6 for phenylamides, and Table 7.11 for phosphonates). A single application of furalaxyl completely suppressed *P. syringae* in naturally infested soil for 13 weeks to a depth of 10 cm (D. C. Harris 1979b).

In Great Britain, a mixture of metalaxyl (for control of *Phytophthora*) and carbendazim (for control of fruit rot caused by *Penicillium expansum*) is used commercially as a postharvest fruit-dip treatment. Fruit rot caused by *P. syringae* was experimentally controlled by treatment of soil under bearing trees with metalaxyl or furalaxyl (D. C. Harris 1979b; Edney and Chambers 1981b). Postharvest dip with metalaxyl or furalaxyl effectively controls fruit rot caused by *P. syringae* (D. C. Harris 1991b).

Field treatments with site-specific fungicides to control fruit rots are not considered to be ecologically sound because of the danger that resistant mutants of the target pathogen could inadvertently be selected (D. C. Harris 1991b). Thus, if a metalaxyl-resistant population were to be propagated in packing houses, the efficacy of control by phenylamide fungicides would be lost. This aspect is discussed at more length in Chapter 7 with reference to control of late blight of potato with phenylamide fungicides.

Leaf Spot and Shoot Dieback of *Syringa vulgaris* L. (Lilac)

P. syringae was described first on lilacs by Berkeley in 1881. By the time *P. syringae* was recognized as the causal agent of leaf spot and shoot dieback, the disease had spread through most of Europe. It was first reported in the United States by Hedges in 1929.

Chester (1932) comprehensively reviewed the history and early literature on leaf blight of lilac and compared the morphology and physiology of *P. syringae, P. cactorum,* and a newly described variety, *P. cactorum* var. *applanata* (now known as *P. citricola*) (Waterhouse 1963), all of which cause blight of lilac. *P. syringae* was pathogenic mainly on lilac tissue during the winter dormant season but not during the summer. *P. cactorum* and *P. cactorum* var. *applanata* (*P. citricola*) were more active during the spring and summer months when temperatures were about 25°C.

Small, water-soaked lesions on leaves are the first symptoms that occur during winter or early spring. The lesions gradually increase in size during periods of frequent rainfall

Table 60.2. Dimensions of reproductive structures of *Phytophthora syringae* from various hosts

Host	Average sporangium size (μm)	Average oogonium diameter (μm)	Average oospore diameter (μm)	Reference
Lilac	39.9 × 25.3	. . .	29.5	Rosenbaum (1917)
Apple	39.0 × 26.5	26.4	24.4	Lafferty and Pethybridge (1922)
Lilac	24.4 × 19.7	28.4	25.0	Tucker (1931)
Fennel	32.0 × 48.9	33.5	30.3	Noviello and Snyder (1962)
Fennel	31.6 × 51.3	37.8	32.8	Noviello and Snyder (1962)

and eventually become irregular in shape. The brown patches are surrounded by lighter colored margins. Entire leaves may be invaded. The fungus subsequently grows down the petioles into the cortex of the branches. Dark brown or black cankers form in the bark. The vascular system is affected only in advanced stages of the disease. Occasionally the infection may be limited to a single bud (Chester 1932).

Disease Development. Leaf blight and shoot dieback caused by *P. syringae* are favored by cool weather (18 to 23°C) and have become epidemic in Holland during the winter months. The disease is usually most prevalent in plants that are stored and closely packed in cool chambers for the winter, preparatory to forcing in the early spring.

The rate of spread of the disease in the field is increased by rainfall and high humidity. A prolonged, rainy fall delays hardening of the new stems and correspondingly extends the period of greatest susceptibility. When lilacs are heeled in so that buds are in close proximity to the soil, they are particularly susceptible to infection.

Oospores form abundantly in diseased tissues; sporangia often occur on the surfaces of affected tissues during moist weather.

Control. Control by development of resistant cultivars is complicated because several physiological races are known. Sanitary measures include development of disease-free plants before planting in noninfested land. On existing bushes, all infected plant parts should be removed or burned. Close heeling in of stored lilac seedlings should be eliminated. When planting stock is stored, it should be left uncovered and exposed to the air to reduce the relative humidity. In outdoor nurseries, the plants should be spaced so that diseased plants can be removed and destroyed before infection spreads to healthy trees. Infected suckers should be removed.

Plants can be sprayed with a copper-based fungicide or with sulfur to prevent infection; however, these fungicides must be used with caution to prevent phytotoxicity (Chester 1932). Spraying trees with a phenylamide fungicide such as metalaxyl or a phosphonate such as fosetyl-Al should be efficacious if registered for use. See Chapter 7.

Pruning Wound Canker of *Prunus dulcis* (Mill.) D. Webb (Almond)

In California, which has a Mediterranean type of climate with rain occurring only during the winter months, *P. syringae* causes pruning wound cankers during the winter months, but during summer months the lesions become inactive and *P. syringae* cannot be recovered (Bostock and Doster 1985). Pruning wound cankers are not as damaging to the tree as collar or crown rot caused by *P. cactorum* (see Chapter 11), probably because the progress of the disease is arrested by high summer temperatures. In New Zealand, *P. syringae* causes stem and branch cankers on young peach and apricot trees (H. C. Smith 1953).

Disease Development. *P. syringae* can be isolated during the winter but not during the summer from the margins of lesions surrounding pruning wounds by plating tissue on a medium containing cornmeal and PVP (pimaricin, vancomycin, and pentachloronitrobenzine; 5–10, 300, and 25 mg/liter, respectively). In two orchards, 23.4 and 10.5% of pruning wounds yielded *P. syringae.* Pruning wound cankers can also be produced by *Ceratocystis fimbriata* Ell. and Halst.; however, this fungus is most active under warm conditions, whereas *P. syringae* is active only under cool conditions. *P. syringae* actively caused gumming cankers at temperatures ranging from 2 to 20°C. *C. fimbriata* primarily infects wounds made during mechanical harvest of almonds. How *P. syringae,* which does not produce caducous sporangia, is disseminated to wound sites high in the scaffolds of trees is still a moot question. The possibility exists that inoculum comes from previously infected cankers (Doster and Bostock 1988). Although *P. cactorum, P. citricola,* and *P. megasperma* also cause collar rot on almond trees, none was capable of causing pruning cankers at 2°C (Bostock and Doster 1985; Doster and Bostock 1988d).

P. syringae produces oospores in fallen almond leaves, and since oospores can be recovered by isolation on PVP-cornmeal agar, it can be assumed that they are a source of inoculum (Doster and Bostock 1988d).

P. syringae did not infect unwounded tissue. The number of secondary infections increased linearly with the increase in diameter of the pruning wound (Doster and Bostock 1988d). Pruning wounds were susceptible to inoculation up to 6 weeks after pruning, but susceptibility of pruning wounds to infection subsequently decreased with time after pruning up to 6 weeks (Doster and Bostock 1988b). This increase in resistance varied with temperature. When stem wounds were aged for 2 weeks at 25°C before inoculation, only 38% of the wounds developed cankers, but when aged at 6°C, cankers developed in all of the inoculated wounds. In the field, inoculation of most 2-week-old wounds and of many 4-week-old wounds resulted in cankers, but almost all of the 6-week-old wounds were immune. This result was related to the slower formation of lignin, suberin tissue, and ligninthioglycollic acid (LTGA) at lower temperatures (Doster and Bostock 1988c).

Control. All almond cultivars grown in California were susceptible when tested in the field, but the expansion rate of cankers on stems was slower on the cultivar Ripon. Nonpareil was the only cultivar tested that developed cankers when 3-week-old wounds were inoculated with *P. syringae,* thus indicating that it might be the most susceptible. When freshly wounded stems of apricot and peach trees were inoculated, cankers were produced, but on plum and French plum trees, the cankers were small (Doster and Bostock 1988b).

In 1-day-old wounds, the concentration of LTGA was 71% higher than in unwounded tissue. Since wounds inoculated with *P. syringae, P. hibernalis,* and *P. cactorum* (all pathogens on almond tissue) yielded substantially more LTGA 3 days after inoculation than uninoculated wounds, LTGA was implicated as part of the resistance mechanism. However, in wounds inoculated with *P. infestans,* a nonpathogen, the concentration of LTGA was not increased (Doster and Bostock 1988e). Low temperatures (5–9°C) reduced the resistance of almond trees to infection by *P. syringae.* This was associated with a reduction of lignification and suberization in bark tissue (Doster and Bostock 1988f).

Chemical control of cankers produced in pruning wounds appears to be effective. Fosetyl-Al (3 and 30 mg/liter) applied to pruning wounds completely prevented formation of pruning wound cankers. When Kocide 101 (77WP; cupric oxide; 740 g a.i. per liter of boiled linseed oil) was used to treat wounds 1 day before inoculation, 20% of the inoculations resulted in cankers compared with 73.5% in the untreated, inoculated trees. However, copper treatment frequently induced one or more signs of phytotoxicity: formation of clear gum, xylem discoloration, excessive bark dieback, or abnormal wound-induced lignification (Doster and Bostock 1988c). If registered for use, fosetyl-Al or metalaxyl should be an effective control measure.

CHAPTER 61

Phytophthora tentaculata

Kröber and Marwitz 1993

Phytophthora tentaculata was first isolated from roots and stalks of the greenhouse-grown ornamental plants *Chrysanthemum frutescens* hybrid, *C. leucanthemum* L., *Delphinium ajacis* L., and *Verbena* hybrid and described by Kröber and Marwitz (1993). Koch's postulates on pathogenicity were not reported.

Description of *Phytophthora tentaculata*

P. tentaculata is classified in group I (Stamps et al. 1991) because of its production of papillate sporangia and paragynous antheridia. Morphology is shown in Figure 61.1. See Tables 4.2 and 4.3 for tabular keys.

Mycelium

On carrot agar, mycelium is arachnoid (like a spider web). Hyphal swellings are relatively small and occur where mycelium branches.

Chlamydospores

Chlamydospores are intercalary or terminal, spherical, thin-walled, and 10 to 45 µm in diameter (average 26.6 µm).

Sporangia

Sporangia are papillate; some are bipapillate. They are spherical or ovoid to obpyriform; some are distorted. A few are caducous with a short pedicel. They are 10 to 81 × 13 to 52 µm (average 35.7 × 27.4 µm); the length-breadth ratio is 1.30:1.

Sex Organs

P. tentaculata is homothallic. Oogonia form abundantly in culture and are 20 to 49 µm in diameter (average 34.0 µm). One to two, rarely three, paragynous antheridia (16.3 × 11.9 µm) are attached to the oogonium; a few are amphigynous. Antheridia are diclinous (the oogonium and antheridium emanate from different hyphae). Antheridia stalks are long and sometimes branched. A few stalks completely encircle the oogonium (Figure 61.1d). Some stalks have tooth-shaped projections. Oospores are spherical, hyaline, aplerotic, and 14 to 38 µm in diameter (average 28.1 µm).

Growth Temperatures

The minimum temperature for growth is 7°C, optimum 15 to 25°C, and maximum 32°C. The growth rate is 3 to 5 mm per day at optimum temperature.

Distinguishing Characteristics

In group I, *P. tentaculata* differs from *P. cactorum* by production of hyphal swellings, the different shape of sporangia, and by its low number of caducous sporangia. *P. cactorum* sporangia are nearly all caducous. *P. tentaculata* has larger oogonia and oospores than *P. cactorum* and also differs from *P. clandestina* in that it produces chlamydospores whereas *P. clandestina* does not. *P. tentaculata* also differs from *P. pseudotsugae*, which does not produce chlamydospores.

Disease Development and Control

No information is available on the development and control of disease, but presumably control measures would be similar to those of other species of *Phytophthora*.

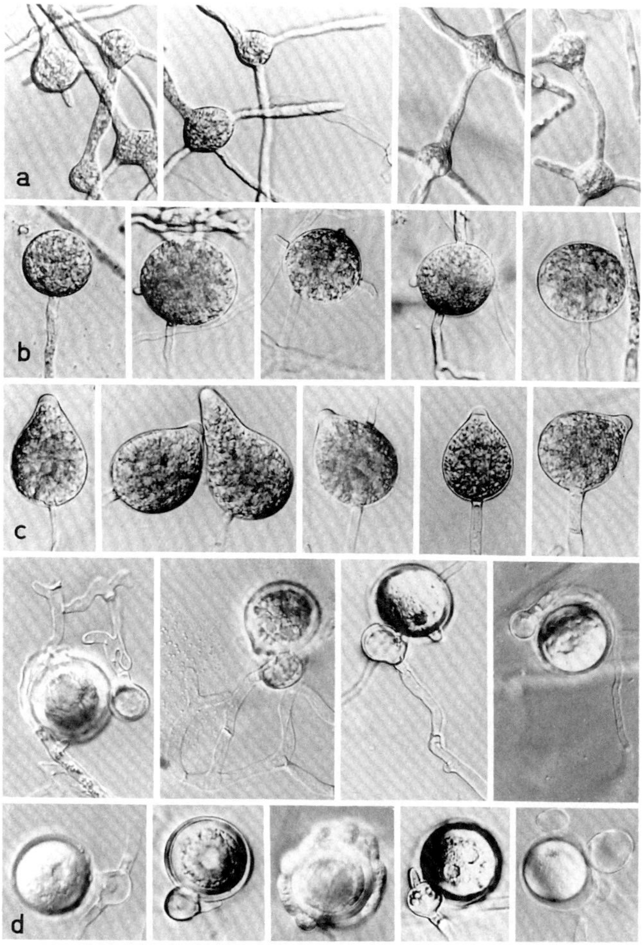

Figure 61.1. Morphology of *Phytophthora tentaculata* sp. nov. **a,** Hyphal swellings; **b,** chlamydospores; **c,** zoosporangia; **d,** oogonia, antheridia, and oospores. (347×) (Reproduced by permission of Verlag Eugen Ulmer GmbH & Co., Stuttgart, Germany, courtesy Kröber and Marwitz 1993)

CHAPTER 62

Phytophthora trifolii

Hansen and Maxwell (1991)

No synonyms are presently known for *Phytophthora trifolii,* which was first proposed by R. G. Pratt (1981) as *P. megasperma* f. sp. *trifolii* to distinguish isolates of *P. megasperma* that are pathogenic to clover (*Trifolium* spp.) from what was then known as *P. megasperma* f. sp. *medicaginis* on alfalfa (*Medicago sativa*) and *P. megasperma* f. sp. *glycinea* on soybean (*Glycine max*). See Chapter 46 (*P. megasperma*) for a general discussion.

P. trifolii is limited in distribution to an area in Mississippi where it causes severe loss of stands of arrowleaf clover, *T. vesiculosum* Savi, an annual plant used in pastures (Table 62.1). R. G. Pratt (1981) reported that this pathogen was different from *P. sojae* (*P. megasperma* f. sp. *glycinea*) and *P. medicaginis* (*P. megasperma* f. sp. *medicaginis*) and caused a severe stand-depleting root disease on arrowleaf clover but was not pathogenic to alfalfa or soybean. *P. erythroseptica,* which was virulent on crimson clover (*T. incarnatum* L.) but mild on arrowleaf clover, and *P. parasitica,* which was weakly virulent on all clovers tested, were also isolated. Since 1981, there have not been any other reports of root disease caused by *P. trifolii,* but according to R. G. Pratt (*unpublished*), outbreaks of root disease still occur in Mississippi. E. M. Hansen and Hamm (1983) reported the morphology, pathogenicity, and sensitivity to metalaxyl of isolates of *P. trifolii* obtained from R. G. Pratt (1981).

Description of *Phytophthora trifolii*

P. trifolii is classified in group V (Stamps et al. 1990). No photomicrographs are available in the literature. See Tables 4.2 and 4.3 for tabular keys. The English diagnosis of *P. trifolii* follows (reproduced by permission from Mycologia 83:376–381, E. M. Hansen and Maxwell, copyright 1991, The New York Botanical Garden, Bronx, New York):

Homothallic, producing oospores in single strain culture, with predominantly paragynous antheridia. Oogonia 43 μm (37–46 μm) diam, oospores 33–37 μm diam. Sporangia formed in liquid culture, ovoid to ellipsoid, nonpapillate, 44–50 μm long by 30–35 μm wide. Slow growing on agar media, about 3–4 mm/day at 25C, no growth at 35C. Colony dense, tufted, patternless. Pathogenic on arrowleaf clover, *Trifolium vesiculosum* Savi, and other *Trifolium* species. Named for the principal host. Holotype materials are deposited at the Oregon State University herbarium. Cultures are on file at the American Type Culture Collection.

On carrot agar, oogonia of *P. trifolii* average 43 μm diam. Sporangia formed on colonies grown in pea broth are smaller (44 μm in length) than other species of the *P. megasperma* complex. The combination of relatively short sporangia and larger oogonia is distinctive. Growth rate is slow (3–4 mm/day at 25C), and optimal between 20 and 25C. No growth occurred at 30C. Colonies exhibit dense, patternless growth on carrot agar, often with a tufted appearance. The fungus is sensitive to metalaxyl on cornmeal agar, with no growth at 1 mg/liter (E. M. Hansen and Hamm, 1983).

Isolates of *P. trifolii* have a unique protein pattern with SDS PAGE (polyacrilamide gel electrophoresis) and isozyme analysis (Nygaard *et al.,* 1989). Thus, they are readily distinguished from other isolates of the *P. megasperma* complex. There are 11–15 metaphase chromosomes in oogonia, which is similar in number to *P. sojae* and *P. medicaginis* (Hansen *et al.,* 1986).

Table 62.1. Distribution of hosts of *Phytophthora trifolii* causing root rot in the United States[a]

Host	Common name
Trifolium agrarium L.	Hop clover
T. incarnatum L.	Crimson clover
T. repens L.	White clover
T. subterraneum L.	Subterranean clover
T. vesiculosum Savi	Arrowleaf clover

[a] Source: R. G. Pratt (1981).

Sporangia

Sporangia are ovoid to slightly obpyriform (R. G. Pratt 1981) and average 49.3 μm in length; the length-breadth ratio is 1.4:1 (E. M. Hansen and Hamm 1983).

Hyphal Swellings

Hyphal swellings have not been described.

Chlamydospores

Chlamydospores have not been described.

Sex Organs

Oogonia form in single culture (homothallic), are globose, and measure 21 to 51 μm in diameter (R. G. Pratt 1981). E. M. Hansen and Hamm (1983) report an average of 44.9 μm. A majority of the antheridia are paragynous, but some are amphigynous.

Growth Temperatures

Growth is optimal at 22.5°C, and no growth occurs at 5, 35 (E. M. Hansen and Hamm 1983), or 30°C (E. M. Hansen and Maxwell 1991). Growth at 25°C is relatively slow (0.5 cm per day on V8 juice agar) (R. G. Pratt 1981).

Distinguishing Characteristics

P. trifolii grows optimally at a lower temperature (22.5°C) and has larger oogonia (44.9 μm in diameter) and shorter sporangia (43.9 μm) than *P. medicaginis* and *P. sojae.* There are few gross differences in morphology among the three species, however. Growth of all three species is inhibited by metalaxyl (1 μg/ml), but only 44% inhibition is reported for DF (isolates of Douglas fir) and ALF2 (large-oogonia isolates of *P. megasperma* from alfalfa) (E. M. Hansen and Hamm 1983).

Disease Caused by *Phytophthora trifolii*

Root Rot of *Trifolium vesiculosum* (Arrowleaf Clover)

The rotted roots of diseased plants are red brown in color. Lateral roots are also destroyed, and stems become red after which extensive defoliation occurs. Remaining leaves are chlorotic or necrotic. When roots are split longitudinally, a distinct margin between healthy and diseased tissues can be observed. The disease frequently extends up the taproot into the crown of the plant. Plants are eventually killed. Subterranean clover (*T. subterraneum* L.), white clover (*T. repens* L.), and crimson clover (*T. incarnatum* L.) are less susceptible than arrowleaf clover (R. G. Pratt 1981).

Disease Development. The disease is most prevalent in low, poorly drained areas of the field.

Control. Control measures have not been recommended except for management of soil water to allow for better drainage without ponding.

The cultivars Alsike and Berseem and cultivars of red clover are highly resistant or immune to Phytophthora root rot (R. G. Pratt 1981).

CHAPTER 63

Phytophthora undulata

H. E. Petersen (1909) M. W. Dick comb. nov. (1989)

Dick (1989) transferred *Pythium undulatum* to *Phytophthora undulata* and lists *Pythium undulatum* H. E. Petersen (1909) as the basionym and *Pythiomorpha undulata* (H. E. Petersen) Apinis (1929) (the genus *Pythiomorpha* was invalidated by Blackwell et al. [1941]) and *Pythium undulatum* H. E. Petersen var. *litorale* Hohnk (1953) as synonyms. We quote from Petersen (1910) because he repeats the description of *Pythium undulatum* sp. nov. in English:

> It is difficult to trace the intramatrical mycelium; the extramatrical consists of long (often several millimetres), unbranched, more or less undulating hyphae. Terminal sporangia, seldom lateral. They open at the apex, sometimes with a papilla. The zoospores are similar to those in *Pythium.* When the zoospores have escaped sometimes new sporangia are formed within the old ones; sometimes the hyphae may penetrate and not form any sporangia until later on. The protoplasm is refractive; the membranes of the hyphae and sporangia more or less brownish coloured. This species seems to live especially on decaying leaves and stems of *Nuphar* and *Nymphaea.* Occasionally I found it on old fruits of Iris and on branches of trees. Several localities in North Sealand and Jutland.

P. undulata has been reported from soil in Alaska (Hamm et al. 1988) and several other areas of the world (see Plaats-Niterink [1981] and Dick [1989] for citations) but has not been proven to be a pathogen of any plant. This is somewhat of an exception for species in the genus *Phytophthora.* Dick (1989) decided to transfer *Pythium undulatum* to *Phytophthora* because it did not show ribosomal DNA bands in DNA preparations that were characteristic of *Pythium* species. All assignable *Pythium* species produced three DNA bands, but *Pythium undulatum* produced only two (Belkhiri and Dick 1988). Van der Plaats-Niterink (1981) reviewed the literature on *Pythium undulatum* and noted that it had been transferred to *Pythiomorpha* because zoospores were differentiated in the sporangium and not in the vesicle that emerges from the sporangium (see Chapter 4 for a comparison of *Pythium* and *Phytophthora*). Blackwell et al. (1941) showed that *Pythiomorpha* was an invalid genus with features already incorporated in *Phytophthora.* Drechsler's (1946) drawing of elliptical sporangia of *Pythium undulatum* strongly resembled those of *Phytophthora.* Dick (1989) noted that sporangial production by the isolates previously called *Pythium undulatum* was exceptional and that the diameter of the encysted zoospore was at the upper range compared with other species of *Pythium. P. undulata* is also resistant to hymexazole, which is toxic to many *Pythium* species (see Table 2.2 in Chapter 2).

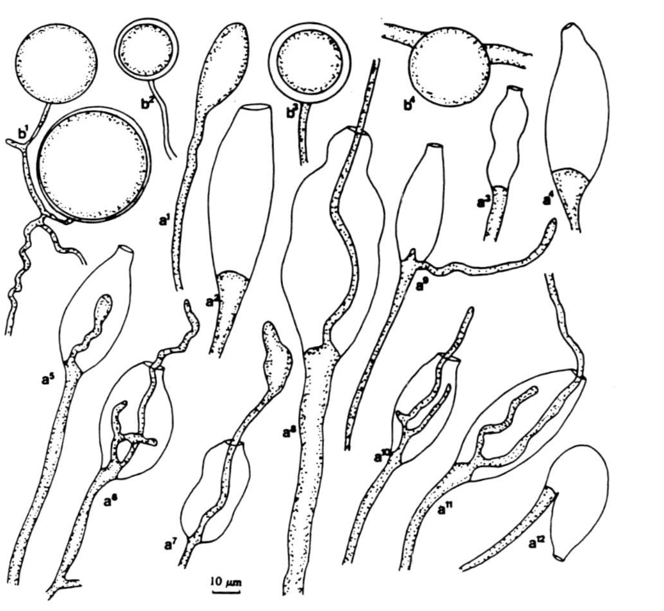

Figure 63.1. Morphology of *Phytophthora undulata* (*Pythium undulatum,* CBS 157.69). $\mathbf{a}^{1-12}$, Sporangia, some with proliferations. $\mathbf{b}^{1-4}$, Chlamydospores. (Reproduced by permission, courtesy Van der Plaats-Niterink 1981)

Characteristics of *Phytophthora undulata*

This species is classified in group VI in the key of Stamps et al. (1990). See Figure 63.1 for morphology and and Tables 4.2 and 4.3 for tabular keys.

Hyphae

Undulating hyphae are formed (Petersen 1910).

Sporangia

Sporangia are long and elliptical, proliferate internally, and are sometimes papillate. Size averages 77.5 × 33 μm. Zoospores are often formed in the sporangium (Petersen 1910; Van der Plaats-Niterink 1981).

Chlamydospores

Chlamydospores are dark yellow, thick walled (2 to 5 μm), subglobose, both intercalary and terminal, and average 36 μm in diameter (Petersen 1910; Van der Plaats-Niterink 1981). Drechsler (1946) compared chlamydospores of *Pythium undulatum* with oospores of *Pythium helicoides* and considered them to be parthenospores because of the similarity in wall structure to that of oospores.

Sex Organs

No oogonia, antheridia, or oospores have been seen.

Growth Temperatures

The minimum temperature for growth on cornmeal agar is 5°C, the optimum is 20 to 25°C, and the maximum is 20 to 25°C. Daily growth rate on potato-carrot agar is 20 mm at 25°C.

CHAPTER 64

Phytophthora verrucosa

Alcock and Foister (1940) (see Foister 1940)

There are no synonyms for *Phytophthora verrucosa,* which was first isolated and described by Foister in 1940 from diseased tomato, *Lycopersicon esculentum* Mill., in Scotland, United Kingdom. *P. verrucosa* has a limited host range and geographical distribution since it is reported only from the United Kingdom. Besides tomato, *P. verrucosa* has been reported as the causal agent of root rot on *Meconopsis* spp. (Foister 1940) and of seedling wilt of *Dahlia* (Byford 1958).

Characteristics of *Phytophthora verrucosa*

P. verrucosa is classified in group V (Stamps et al. 1990). See Tables 4.2 and 4.3 for tabular keys. Morphology is shown in Figure 64.1. *P. verrucosa* is not culturable (Stamps et al. 1990).

Sporangia

Sporangia are obpyriform or oval and are borne terminally or laterally. They are broadly rounded without pedicels (noncaducous), are nonpapillate, and proliferate internally. They are 31 to 56 µm long × 24 to 36 µm wide (average 41.5 × 29.5 µm) and are borne on undifferentiated sporangiophores.

Hyphal Swellings and Chlamydospores

Hyphal swellings and chlamydospores are not produced.

Sex Organs

Oospores form abundantly in host tissue (homothallic). Both amphigynous and paragynous antheridia are produced. The oogonial wall is characteristically verrucose, occasionally smooth, measuring 23 to 47 µm in diameter (average 37 µm); oospores are markedly aplerotic, measuring 17 to 31 µm (average 24 µm).

Growth Temperatures

Growth temperatures have not been recorded.

Disease Caused by *Phytophthora verrucosa*

Toe Rot of *Lycopersicon esculentum* Mill. (Tomato)

Wilting is the first observable symptom of toe rot during the day; however, the plants subsequently revive during the night. The upper sides of the leaves and the stem become dark green to dark blue, and the undersides of the leaves often turn purple. Leaves develop slowly and are usually pointed, and the stems become woody. The roots become brown and hard and in advanced cases of disease development are often killed. Affected roots are characteristically pointed at the tip.

Disease Development. *P. verrucosa* is soilborne and persists as oospores that form in host tissue. The oospores are the primary source of inoculum. The disease spreads rapidly by contact of healthy with diseased plants when conditions are cool (18 to 23°C) and humidity is high.

Control. Because of the rapid spread of the disease by contact, it is best to discard the entire box of seedlings if any one plant in it is diagnosed to have toe rot. If soil is infested, it should be fumigated with methyl bromide or steamed to eliminate inoculum before plants are propagated in a nursery. The phenylamide or phosphonate fungicides should be effective if registered for use.

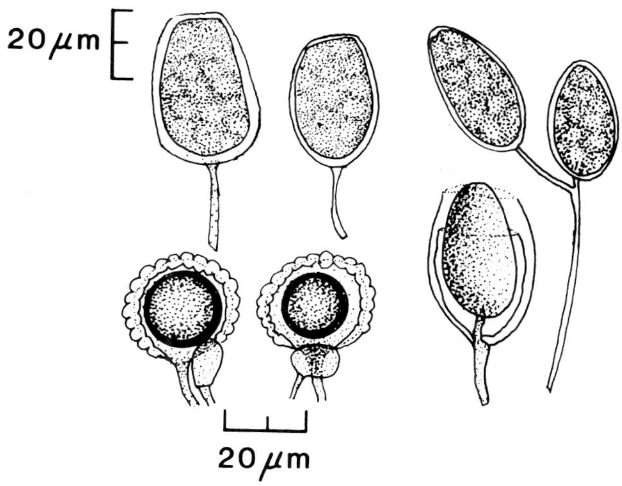

Figure 64.1. Morphology of *Phytophthora verrucosa.* Lower row, oogonia with verrucose walls and paragynous and amphigynous antheridia; sporangium proliferating within an empty sporangiophore. Upper row, nonpapillate sporangia; sporangia forming on a simple sympodium. (Drawing by A. Vaziri)

CHAPTER 65

Phytophthora vignae

Purss (1957)

There are no synonyms for *Phytophthora vignae,* which was first isolated from cowpeas with root and stem rot in Queensland, Australia, and described by Purss in 1957. *P. vignae* is pathogenic under natural conditions only to cowpea, *Vigna unguiculata* (L.) Walp. (= *V. sinensis* (L.) Endl. ex Hassk), and adzuki bean, *V. angularis* (Willd.) Ohwi & Ohashi, referred to in some references as *Phaseolus angularis* ((Willd.) W. Wight). *P. vignae* has been reported to be a pathogen of apple fruit by artificial inoculation. Isolates from adzuki bean differ pathogenically from those of cowpea. *P. vignae* f. sp. *adzukicola* was reported by Tsuchiya et al. (1986) to be specific to adzuki bean. See also Kitazawa et al. (1978).

Cowpea is also a host of *P. cactorum,* which causes the red stem canker disease (Weimer 1949). In California, *P. drechsleri* was found to be a cause of root rot in heavily irrigated fields (Erwin et al. 1991).

Widespread outbreaks of disease caused by *P. vignae* appear to be limited geographically to Australia (Purss 1957, 1958) on cowpea and to Japan (Kitazawa et al. 1978, 1979) on adzuki bean (*P. angularis*). Adzuki bean is cited in *Hortus Third* (1976) as *Vigna angularis* (Willd.) Ohwi & Ohashi and by Tsuchiya et al. (1986) as *Phaseolus radiatus* var. *aurea* Prain (this epithet is listed in *Hortus Third* [1976] as mung bean but may be botanically incorrect). Sivakadadcham and Fernando (1991) detected *P. vignae* in riverbank soil used to grow cowpea plants in a greenhouse. Fernando and Linderman (1993) reported isolation of *P. vignae* in several areas of Sri Lanka by baiting soil with California Blackeye plants and tea leaves. The disease was seen in only one area of Sri Lanka. In the Guangzhou area of China, *P. vignae* was isolated from and proved to be pathogenic to yard-long bean, *V. sesquipedalis* (L.) Fruw (J. K. Hwang and Qi 1984).

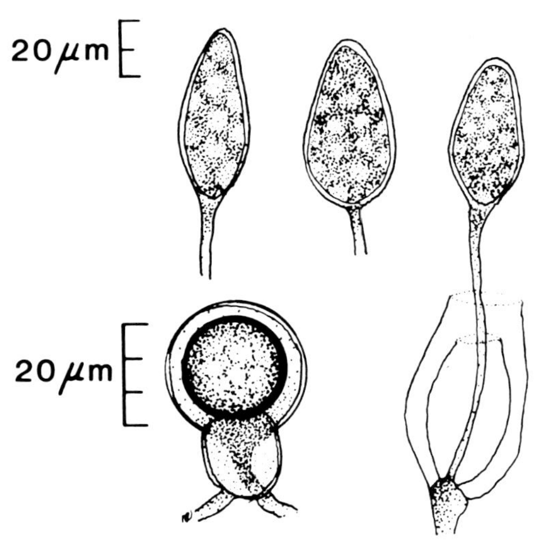

Figure 65.1. Morphology of *Phytophthora vignae.* Lower row, globose oogonium containing a spherical, aplerotic oospore; sporangium germinating by internal and external proliferation. Upper row, ovoid and ellipsoid sporangia. (Drawing by A. Vaziri from Purss 1957)

Host range studies by Purss (1958) indicated that *Cajanus cajan* (L.) Millsp. (pigeon pea), *Calopogonium mucunoides* Desv. (calopo), *Glycine jaranica* L., *Glycine max* (L.) Merr. (soybean), *Medicago sativa* L. (lucerne or alfalfa), *Mucana* spp. (velvet bean), *Phaseolus lathyroides* L. (phasey bean), *Phaseolus mungo* L. (mung bean), *Phaseolus vulgaris* L. (bean), *Pisum arvense* L. (field pea), *Pueraria phaseoloides* (Roxb.) Benth. (puero), and *Vicia sativa* L. (common vetch) were not susceptible to stem inoculation with *P. vignae* from cowpea. Kitazawa (1978) showed that the adzuki bean isolate was not pathogenic to *Phaseolus vulgaris* L., *P. coccineus* L., *P. lunatus* L., *P. radiatus* L. (bean), *Vigna sinensis* Endl. (cowpea), *Glycine max* (soybean), *Pisum sativum* L. (pea), *Vicia faba* L. (broad bean), *Dochlios lablab* L., *Canavalia gladiata* (Jacq.) D.C. (sword bean), *Medicago sativa* L. (alfalfa), *Capsicum annuum* L. (pepper), *Solanum melongena* L. (eggplant), or *Lycopersicon esculentum* Mill. (tomato).

Characteristics of *Phytophthora vignae*

P. vignae is classified in group VI (Stamps et al. 1990). See Tables 4.2 and 4.3 for tabular keys. The mycological description by Purss (1957) is summarized here. Morphology is shown in Figure 65.1.

Sporangia

Sporangia are variable in shape, ovoid to obpyriform, tapering toward the base; nonpapillate; persistent on the stalk; and proliferate both externally and internally. They are 24 to 72 μm long × 15 to 54 μm wide (average 48 × 27 μm) with a length-breadth ratio of 1.3:1 to 2.0:1 (average 1.8:1). Sporangia form in water or soil extract but do not form on solid culture media. They form on undifferentiated sporangiophores in aqueous culture, but on diseased tissues they form on monochasial or compound sympodia.

Hyphal Swellings

Hyphal swellings form abundantly and are large, spherical or angular, intercalary, and rarely terminal.

Chlamydospores

Chlamydospores are sparse in culture, terminal or intercalary, and measure 12 to 21 μm in diameter (average 17 μm).

Sex Organs

P. vignae is homothallic; oospores form abundantly in host tissues and in axenic culture; antheridia are variable in shape from spherical to ovate (average 16 × 15 μm) and amphigynous; oogonia are smooth walled and measure 24 to 46 μm in diameter (average 32 μm); oospores are markedly aplerotic (typically loose and eccentric within the oogonium) and measure 18 to 32 μm in diameter (average 26 μm); wall thickness is 3 μm.

Oospores of *P. vignae* were produced on V8 juice agar amended with sitosterol (30 mg/liter) at 20°C and isolated free of mycelial debris by flotation and sedimentation. Oospores germinated readily on water agar to form nonpapillate sporangia. Germination was optimal at 25°C. Oospore germination was as high as 70 to 90% at 20 to 30 days of age; but at 120 to 210 days of age, germination was reduced to less than 10% (Ishiguro and Ui 1981a). Oospore germination was reduced markedly by an increase in the osmotic potential of water agar with KCl or NaCl to –0.1 bar. An increase to –3.0 bars totally inhibited germination.

Growth Temperatures

The minimum temperature for growth is 10 to 12°C, optimum 28 to 30°C, and maximum 32°C, with no growth at 34.5°C.

Diseases Caused by *Phytophthora vignae*

Stem Rot of *Vigna unguiculata* (Cowpea)

Outbreaks of the disease occur first as isolated patches of yellow or wilted plants that spread rapidly during moist weather and optimal temperature (19 to 28°C) (Plate 65.1). In areas where the disease has occurred previously, infected plants are scattered throughout the field. Plants of all ages, from seedling to maturity, are susceptible. Individual plants turn yellow and if the weather stays dry, wilt and die. During moist weather, infected plants live for a few weeks before succumbing. Characteristic brown lesions, beginning at ground level and extending upward, are sunken and gradually progress until the stem is entirely girdled (Figure 65.2 and Plate 65.2). During moist weather, a white "fuzz" of mycelium can be seen on the margins of lesions. Internal discoloration of the stems also occurs, and the root system is often completely destroyed. Brown spots may also develop on leaves (Purss 1957).

Disease Development. *P. vignae* persists in diseased tissue, probably as oospores, and develops rapidly between 19 and 28°C. As temperatures drop during the fall, the incidence of the disease lessens. Growing successive crops of susceptible cultivars results in an increase in the incidence of disease. Waterlogged conditions in the soil enhance rapid disease development, however, when the inoculum level is high (previous crop was diseased), stem rot occurs even in areas that are not waterlogged (Purss 1957).

Control. Where possible, well-drained land should be selected. Rotation as long as possible between cowpea crops may be useful.

The most effective control of root and stem rot is by use of resistant cultivars. Soon after the discovery of *P. vignae* as the cause of a new root rot of cowpea, Purss (1958) screened a large number of cultivars and categorized them into three groups. The first included the cultivars Poona and Reeves, which were susceptible. The second included the cultivars Malabar and Giant as well as others, which were susceptible when tested as potted plants but showed variable reaction in the field. When these cultivars were inoculated by placing mycelium in a potato-dextrose agar block on the lower stems, they showed good field resistance, but resistance tended to break down at high inoculum levels or in soils that were frequently wet. This type of resistance is also known as general resistance or tolerance (see Chapter 6). The third group included the cultivars Blackeye 5 (not California Blackeye 5), Havana, Santiago, and Chinese Red, which appear to be highly resistant both in the field and as potted plants. Fernando and Linderman (1993) reported that several cultivars of cowpea were resistant to some isolates but not to others. This suggested that more than one race might exist in Sri Lanka despite the fact that the disease has not become endemic.

The cultivar Caloona, with resistance to race 1, was obtained in progeny from crossing Poona with Blackeye 5 and was released in 1963 (Purss 1963; Plate 65.3). This cultivar controlled stem rot; but several years after the release of Caloona, four new pathogenic races were detected. The cultivar Chinese Red was resistant to all races, Blackeye 5 was resistant to races 1 and 2, and Bechuana White was resistant to races 1, 2, and 3. In 1975, Purss released the cultivar Red Caloona, into which the gene for resistance from Chinese Red had been incorporated by crossing with the cultivar Caloona. Further proof that the inheritance of resistance to races of *P. vignae* is conditioned by a single dominant gene is given by Ralton et al. (1988) and Bateman et al. (1989).

As discussed in Chapter 6, single-gene resistance in the soybean-*P. sojae* system is conditioned by production of a phytoalexin (host response to a pathogen), glyceollin. The mechanism of host resistance to *P. vignae* appears to be simi-

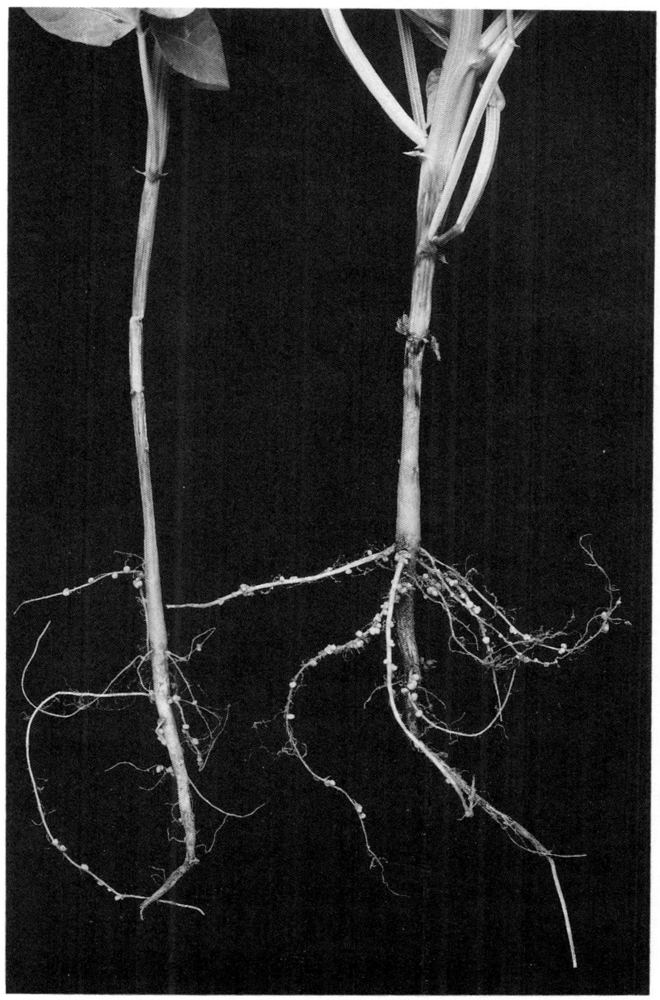

Figure 65.2. Stem rot of cowpea caused by *Phytophthora vignae*. (Reproduced by permission, Queensland Department of Agriculture, Queensland, Australia, courtesy G. S. Purss)

lar. The cultivar Caloona is near isogenic (closely related) to the susceptible cultivar Poona; it differs only in its resistance to race 2. The single-gene resistance is associated with a rapid increase in the production of a phytoalexin called kievetone (Partridge and Keen 1976). An enzyme, phenylalamide-ammonia-lyase (PAL), a precursor necessary for the rapid production of kievetone in stem tissue, is produced in much greater quantity in Caloona than in the susceptible Poona cultivar. In response to race 3 of *P. vignae,* which is pathogenic to Caloona, only low levels of PAL are formed. Experimental treatment of cut hypocotyls with an inhibitor of the PAL enzyme, L-2-aminooxy-3-phenylpropionic acid hydrobromide (L-AOPP), changed the resistance of the Caloona cultivar to race 2 to susceptibility. This result gave added proof that PAL is the active component conditioning production of kievetone and resistance to race 2 (Ralton et al. 1988).

Stem Rot of *Vigna angularis* (Adzuki Bean) and *Vigna sesquipedalis* (Yard-Long Bean)

Stem rot of adzuki bean is similar to the disease on cowpea. Reddish brown streaks appear on affected stems, and after the stem is girdled, the plant wilts and dies (Kitazawa et al. 1978). As stated above, the cowpea and adzuki bean biotypes are different pathogenically but similar in morphology and protein and isozyme patterns.

Disease Development. The conditions that favor disease development are similar to those that favor disease on cowpea.

Control. Although selection of well-drained land is important, this is not always a practical solution in high-rainfall areas. In pot trials, the systemic fungicide fosetyl-Al controlled stem rot on yard-long bean when applied after inoculation (Hwang and Qi 1984).

Host resistance has been obtained and appears to be the most efficacious control measure. See Chapter 6, Control by Host Resistance, and Table 6.1.

The discovery of *P. vignae* in Japan on adzuki bean resulted in the demonstration that isolates from cowpea and adzuki bean were host specific to some extent. Since the isolates from cowpea and adzuki bean differed in pathogenicity (Kitazawa et al. 1978, 1979; Tsuchiya et al. 1986) but were shown to be similar on the basis of isozyme and protein patterns, Tsuchiya et al. (1986) designated the adzuki bean isolates *P. vignae* f. sp. *adzukicola* and the cowpea isolates *P. vignae* f. sp. *vignae.* Within the forma specialis *adzukicola,* three pathologic races were demonstrated by their differential pathogenicity to a series of cultivars. Race 1 is unable to attack the cultivars Notoshozu and Kotobukishozu but is virulent on all other cultivars; race 2 is virulent on all cultivars except Notoshozu; and race 3 is pathogenic to the above cultivars as well as Wasetairyu 1, Chagarawase, Hayateshozu, and Takarashozu.

CHAPTER 66

Sclerophthora macrospora

(Saccardo) Thirumalachar, Shaw, and Narashimhan 1953

= *Phytophthora macrospora*

Ito and Tanaka (1940)

Synonyms include *Sclerospora kriegeriana* Magnus (1895; cited by Ito and Tanaka [1940]), *Sclerospora macrospora* Saccardo (1890), *Kawakamia macrospora* (Saccardo) Hara (1930), *Sclerospora oryzae* Brizi (1919), *Phytophthora oryzae* (Brizi) Hara (1930), *Nozemia macrospora* (Saccardo) Tasugi (1931) (cited by Ito and Tanaka [1940]; but according to Waterhouse [1970a], this combination was not included), and *P. macrospora* Ito and Tanaka (1940).

The fungus was first described by Saccardo (1890) from *Alopecurus* sp. plants in Australia with downy mildew symptoms. Tasugi (1931; cited in Ito and Tanaka [1940]) later transferred the fungus to the genus *Nozemia,* since the antheridia were found to be paragynous. Ito and Tanaka (1940) concluded, however, that the fungus belonged to the genus *Phytophthora* because it produced typical sporangia and zoospores and erected the binomial *P. macrospora* (Sacc.) Ito and Tanaka. Thirumalachar et al. (1953) disagreed with this classification because of the failure of Ito and Tanaka (1940) to recognize the thick oogonial wall, which Thirumalachar et al. stated was confluent with the oospore wall as indicated by McDonough (1946; quoted by Thirumalachar et al. [1953]). They concluded that the oogonial stage was similar to that of *Sclerospora* but that the sporangial stage (from germinating oospores) was similar to that of *Phytophthora.* For this reason, they created the new genus *Sclerophthora.*

Thirumalachar (1969), in defense of the name *Sclerophthora macrospora,* stated that the sporangiophore of *Sclerophthora* is determinate as is that of *Sclerospora* but that the "conidiophores are micronemous" (not differentiated from the vegetative mycelium). However, the sporangiophores of *Phytophthora* are also micronemous and indefinite in growth. Extensive branching may take place with sporangia borne on the branches.

Waterhouse (1963) claimed that the oogonial and oospore walls could be separated by pressing on the coverslip and thus were not confluent. She maintained that the thickness of the oogonial wall is dependent on the substrate in which oospores are produced. Ho (1990) examined Sawada's preserved specimens in Taiwan and also concluded that the walls were not fused and that the oospore was plerotic (filled the oogonium).

Although *Phytophthora macrospora* was listed as the valid name by Waterhouse (1963) and Newhook et al. (1978) (group III), the key of Stamps et al. (1990) accepted the taxon *Sclerophthora macrospora* and omitted *P. macrospora.* We are listing this pathogen as *S. macrospora,* but we are not convinced, on the basis of Waterhouse (1963) and Ho (1990), that it is not a *Phytophthora* species, even though the name *S. macrospora* is now widely used (Kenneth 1981).

S. macrospora is recognized as a serious pathogen of several economically important members of the family Gramineae (Table 66.1). Diseases of more than 43 genera in Gramineae are described by Ou (1972). See also Kenneth (1981), who reviewed downy mildew diseases on maize, rice, wheat, oats, barley, and about 40 other species, including *Eleusine, Echinochloa, Eragrostis, Iseilema, Micanthus, Paspalum, Pennisetum, Saccharum, Setaria,* and *Sorghum.* Kenneth (1981) cited other species of *Sclerophthora.*

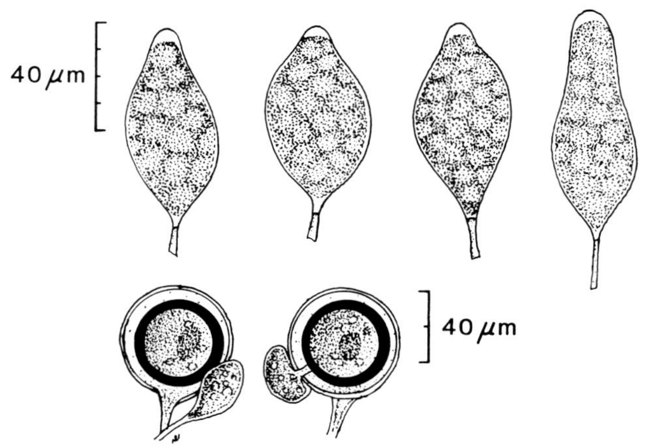

Figure 66.1. Morphology of *Sclerophthora macrospora.* Lower row, globose oogonia with paragynous antheridia. The drawing of the oospore in the oogonium is similar to that of Ito and Tanaka (1940), which did not show that the oospore wall and the oogonium wall were confluent as stated by Thirumalachar et al. (1953). Upper row, semipapillate, caducous sporangia with shapes varying from ellipsoid to obpyriform. (Drawing by A. Vaziri)

Characteristics of *Sclerophthora macrospora*

The name *Sclerophthora macrospora* is accepted by Kenneth (1981) and I. M. Smith et al. (1988) as a cause of downy mildew of several gramineous crops. Morphology is shown in Figure 66.1.

S. macrospora is an obligate parasite and cannot be readily cultured; however, Kenneth (1981) cites a report by Tokura (1975) in which *S. macrospora* was grown in axenic culture on a semisynthetic medium. If *S. macrospora* is culturable, it would be an unusual character for a downy mildew. Except for the culture reported by Tokura, no other isolates have been reported to be culturable.

Sporangia

Sporangia are caducous (pedicel is 14 μm in length), semipapillate, nonproliferating, obpyriform, ellipsoid, or spherical, with a tapered base and occasionally with elongated necks. In water they are 65 to 112 μm long × 32 to 55 μm wide (average

87 × 44 µm) and on wheat leaves 57.5 to 97 µm long × 30 to 65 µm wide (average 74.6 × 46.8 µm) and are probably the largest of all the downy mildews (Kenneth 1981). They form in nature mainly between midnight and dawn, probably because temperature and moisture conditions are favorable for formation during that time. Sporangiophores are formed in a sympodium (Thirumalachar et al. 1953).

Hyphal Swellings and Chlamydospores

Neither hyphal swellings nor chlamydospores are produced.

Sex Organs

S. macrospora is homothallic; antheridia are paragynous; oogonia are globose to oblong, form abundantly in diseased tissues, and measure 55 to 100 µm long × 50 to 95 µm wide (average 69 × 65 µm); oospores are plerotic and measure 42.5 to 72.5 µm long × 42.5 to 70 µm wide (average 60.25 × 56.67 µm). The wall thickness varies from 3.8 to 10 µm (Ito and Tanaka 1940; Waterhouse 1970a). The confluence of the oogonial wall with the oospore wall is a diagnostic character according to Thirumalachar et al. (1953). Oospores of *S. macrospora* are much larger than those of most other mildew-causing fungi, e.g., *Sclerospora graminicola.* Oospores have been noted to germinate in large numbers (Kenneth 1981).

Growth Temperatures

Ito and Tanaka (1940) report abundant production of sporangia at 11–24°C. *S. macrospora* is not culturable.

Diseases Caused by *Sclerophthora macrospora*

The malformation and excessive tillering of the plant is characteristic of many diseases caused by *S. macrospora,* e.g., crazy top of corn, yellow dwarf of wheat, and downy mildew

Table 66.1. Distribution of hosts of *Sclerophthora macrospora*

Host	Common name	Disease	Geographical distribution
Agropyron repens (L.) Beauv.	Quack grass	Downy mildew	United States (Miles and Epps 1942)
Agropyron tsukusiense (Monda) Ohwi.	Wheat grass	Downy mildew	Japan (Katsura et al. 1954)
Agrostis alba L.	Redtop grass	Downy mildew	United States (Miles and Epps 1942)
Agrostis clavata Trin. var. *nukabo* Ohwi.	Bentgrass	Downy mildew	Japan (Katsura et al. 1954)
Alopecurus L. sp.	Foxtail	Downy mildew	Australia (Saccardo 1890)
Alopecurus aequalis Sobol.	Short hawn	Downy mildew	Japan (Katsura et al. 1954)
Alopecurus agrestris Sieber ex Kunth	Foxtail	Downy mildew	United States (Miles and Epps 1942)
Avena sativa L.	Oats	Downy mildew; yellow dwarf	Italy (Peglion 1901); France (Arnaud 1915); Australia (Noble 1926); United States (Miles and Epps 1942)
Beckmonoria syzigachne (Steud.) Fernald.		Downy mildew	Japan (Katsura et al. 1954)
Bromus commutatis Schrad.		Downy mildew	United States (Miles and Epps 1942)
Cyperus articulatus L.		Downy mildew	United States (Miles and Epps 1942)
Eleucine coracana Gaertn.	Finger millet; raji	Downy mildew	India (Thirumalachar et al. 1953)
Elytrophorus spicatus Camus		Downy mildew	India (Thirumalachar et al. 1953)
Eragrostis major Host.	Love grass	Downy mildew	United States (Miles and Epps 1942)
Glyceria acutifolia Torr.	Manna grass	Downy mildew	Japan (Katsura et al. 1954)
Glyceria maritima Wahlb.		Downy mildew	United States (Miles and Epps 1942)
Holcus lunatus L.	Velvet grass	Downy mildew	United States (Miles and Epps 1942)
Holcus mollis L.		Downy mildew	United States (Miles and Epps 1942)
Hordeum vulgare L. (*H. sativum* Pers.)	Barley	Downy mildew; yellow dwarf	Italy (Peglion 1930); United States (Mackie 1930)
Lollium perenne L.	Perennial ryegrass	Downy mildew	United States (Miles and Epps 1942)
Lollium temulentum L.	Ryegrass	Downy mildew	United States (Miles and Epps 1942)
Miscanthus sinensis Anderss.	Eulalia; perennial grass	Downy mildew	Japan (Katsura et al. 1954)
Oryza sativa L.	Rice	Downy mildew; yellow dwarf; sheath blight	Italy and Japan (Brizi 1919); Formosa (Sawada 1927); Australia (Noble 1937)
Panicum antidotale Retz.	Giant or blue panic grass	Downy mildew	United States (Miles and Epps 1942)
Paspalm thunbergii Kunth.	Paspalm	Donwy mildew	Japan (Katsura et al. 1954)
Phalaris arundinaceae L.	Reed canary grass	Downy mildew	Germany (Magnus 1895); Japan (Katsura et al. 1954)
Phalaris canariensis L.	Canary grass	Downy mildew	United States (Miles and Epps 1942)
Phalaris communis Trin.	Common reed	Downy mildew	Japan (Katsura et al. 1954)
Poa acroleuca Steud.	Blue grass	Downy mildew	Japan (Katsura et al. 1954)
Poa annua L.	Annual blue grass	Downy mildew	Japan (Katsura et al. 1954)
Poa nipponica Koidz.	Blue grass	Downy mildew	Japan (Katsura et al. 1954)
Saccolepsis interrupta Stapf.		Downy mildew	India (Thirumalachar et al. 1953)
Secale cereale L.	Rye	Downy mildew	Bulgaria (Atansoff and Kovachevsky 1929); United States (Mackie 1930); Australia (Noble 1937); Japan (Katsura et al. 1954)
Setaria viridis Beauv.		Downy mildew	Germany and United States (Magnus 1895)
Stenotaphrum secundatum (Walt) O. Kuntze	St. Augustine grass	Downy mildew	United States (Toler et al. 1983)
Syntherisma sanguinale (*Panicum*) Dulac.	Panic grass	Downy mildew	Japan (Yamada 1911)
Triticum vulgare Vill.	Wheat	Downy mildew; yellow dwarf	France (Arnaud 1915); United States (Weston 1921); Australia (Carne 1924; Noble 1926); Japan (Tanaka 1940)
Zea mays L.	Corn; maize	Crazy top	Italy (Cugini and Traverso 1902); United States (Ullstrup 1952); Japan (Katsura et al. 1954)
Zizania latifolia Turcz.	Wild rice	Downy mildew	Japan (Katsura et al. 1954)

of barley. Often, the entire head is converted into bushy, leaf-like structures, giving a witches'-broom effect. Oospores form abundantly in diseased leaves but seldom in roots.

Downy Mildew or Yellow Dwarf of *Oryza sativa* (Rice)

The downy mildew disease of rice is described by Ou (1972). Infection first occurs in young rice plants, either in the nursery or in paddy fields, and becomes systemic in the plant. Infected plants are slightly dwarfed and often bear leaves that are oriented at almost right angles to the stems. The leaves are chlorotic and often narrower than normal, healthy leaves. Diseased leaves may also be thick or leathery in texture. After the plants are transplanted, the disease progresses into the buds. There is often an excessive amount of tillering, and yellowish white spots may occasionally appear on the leaves. Diseased plants do not normally form spikelets. If heading occurs, excessive proliferation may occur, and the rachis sometimes bends back on itself, causing the head to form a circle or compact knot. If the spikelet is formed, it becomes severely malformed, curled, or distorted.

Disease Development. The fungus overwinters in diseased leaves as oospores. Rice seedlings in the nursery are infected in the spring by inoculum produced on wild grasses. The disease develops rapidly during the cool seasons of the year when temperatures are between 18 and 20°C (Katsura 1965).

Control. The disease is usually severe enough to require special control measures (Ou 1972). Infected seedlings regain health if temperatures increase to 35°C for 20 days or if the night temperature remains at 30°C or higher for 30 days. Fungicidal treatment of seeds has helped to prevent infection of young seedlings (Ou 1972). Metalaxyl should be effective if registered for use. Metalaxyl applied to rice foliage on young plants reduced the incidence of disease and increased yield 133% (Kang and Lee 1987).

Diseases of Other Gramineous Crops

Kenneth (1981) discusses downy mildew diseases caused by *S. macrospora* on a wide range of gramineous hosts and traces the evolution of the taxonomy of this species from *P. macrospora* to *S. macrospora.* Plants are infected by means of zoospores splashed by rain or other means to susceptible tissue. Young tissue is more susceptible than old. Waterlogged soil has been associated with occurrence of the disease. Sporangia formation occurs during the period from midnight to dawn after dew formation. On most hosts, infection is systemic, and symptoms vary with the host infected. On maize plants, excessive numbers of tillers form, and leaves become narrow and chlorotic. The tassel proliferates abnormally and becomes a mass of leafy structures known as crazy top. Wheat plants also tiller excessively; the stems become twisted, and the awns on the head are malformed. In some cases inoculum can become seedborne (Bains and Shooty 1985). See I. M. Smith et al. (1988) for a general discussion of the disease in Europe. They note that the disease seldom becomes of economic importance in Europe except in fields that become waterlogged. The sporadic nature of the disease has made selection for resistance difficult.

CHAPTER 67

Hosts That Are Susceptible to More Than One Species of *Phytophthora*

This chapter consists of a single table (Table 67.1) that lists many plant hosts known to be affected by more than one *Phytophthora* species. The literature cited consists of either an original publication on the disease or a key publication that presents general information (many older citations are listed in Tucker [1933]). Information about different *Phytophthora* species that attack a single host may be of critical importance for diagnostic purposes, especially because some cultivars are resistant only to one particular species of *Phytophthora.* For instance, according to Mircetich and Browne (1987) and Browne and Mircetich (1993), resistance of apple rootstocks to *P. cactorum* does not ensure that they will be resistant to other species. For more information, the reader should consult the chapter in which an individual species is discussed.

Table 67.1. Hosts that are susceptible to more than one *Phytophthora* species

Host	*Phytophthora* spp.	Diseases caused	Early occurrence and/or significant information on species and host[a]	Reference
Actinidia deliciosa (Chevalier) Liang et Ferguson (kiwi fruit)	*cactorum* *citrophthora* *cryptogea* *drechsleri* *megasperma* *citricola*	Root and crown rot	United States	Conn et al. (1991)
	cinnamomi	Root and crown rot	New Zealand	G. I. Robertson (1982b)
	citricola, cryptogea, gonapodyides, Phytophthora species (non-sporulating), *megasperma* (not pathogenic)	Root and crown rot	New Zealand	Stewart and McCarrison (1991)
Aesculus hippocastanum L. (horse chestnut)	*cactorum*	Bleeding canker	United States	Caroselli (1953)
	cambivora	Root rot	United Kingdom	Brasier and Strouts (1976)
	citricola	Bark canker	United Kingdom	Brasier and Strouts (1976)
	megasperma	Root rot	United Kingdom	Brasier and Strouts (1976)
	syringae	Twig blight	Germany	Klebahn (1909[b])
Agave sisalana Perrine (sisal)	*arecae, palmivora, parasitica* var. *nicotianae*	Zebra disease	East Africa	Peregrine (1969)
Ananas comosus (L.) Merr. (pineapple)	*drechsleri, citrophthora*	Heart rot	United States	Sideris (1932)
	meadii	Top rot	United States	Sideris (1929)
	palmivora	Heart rot	United States	Mehrlich (1934, 1936)
	parasitica	Heart rot	West Indies	Ashby (1920)
	cinnamomi	Heart rot	United States	Hine et al. (1964)
	cinnamomi	Heart and top rot	United States	Mehrlich (1934)
Antirrhinum majus L. (snapdragon)	*cactorum*	Stem girdling, root rot	India	Sundararaman and Ramakrishnan (1928)
	cinnamomi	Root rot	United Kingdom	Oyler and Bewley (1937)
	cryptogea	Stem rot	Australia	Brittlebank and Fish (1927)
	palmivora	Root rot	Puerto Rico	Tucker (1933)
	parasitica	Damping-off	Zimbabwe	Hopkins (1931a)
Asparagus offinalis L. (asparagus)	*richardiae, megasperma*	Root rot	Australia	Cited by Falloon and Grogan (1988)
	cactorum	Storage rot	France	
	megasperma var. *sojae, cryptogea*	Root rot	United States	Falloon et al. (1983); Falloon and Grogan (1988)
Brassica spp. (cabbage, cauliflower, broccoli, etc.)	*cactorum*	Leaf blight	United States	Young (1926[b])
	cryptogea	Root rot	United Kingdom	Pethybridge et al. (1934)
	erythroseptica	Root rot	Ireland	Whelan and Loughnane (1969)
	megasperma	Root rot, damping-off	United States	Tompkins et al. (1936b); Hamm and Koepsell (1984)
	parasitica	Black root rot	United States	Godfrey (1923)
	porri	Storage rot	United Kingdom	Geeson (1976)

(*continued on next page*)

Table 67.1. (*continued*)

Host	*Phytophthora* spp.	Diseases caused	Early occurrence and/or significant information on species and host[a]	Reference
Callistephus chinensis L. (China aster)	*cactorum*	Blight	Czechoslovakia	Nicklova-Navratilova (1949)
	cryptogea	Blackleg, foot rot, damping-off	Ireland	Pethybridge and Lafferty (1919)
Capsicum annuum L. (pepper)	*cactorum*	Seedling rot	Italy	Trotter (1924)
	cactorum	Fruit rot	El Salvador	Crandall et al. (1951)
	capsici	Fruit and root rot	United States	Leonian (1922)
	capsici	Seedling damping-off	Argentina	Godoy (1940)
	citrophthora	Root rot, stem necrosis	Argentina	Frezzi (1950)
	palmivora	Fruit rot	India	Ramakrishnan and Seethalalakshmi (1965b)
	parasitica	Fruit rot	United States	Kendrick (1923)
Carica papaya L. (papaya, pawpaw, melon tree)	*cinnamomi*	Root rot, leaf fall	Peru	Bazan de Segura (1951)
	palmivora	Fruit rot, stem canker	Philippines	Reinking (1918)
	parasitica	Stem girdling, fruit rot	United States	Parris (1942)
	cactorum	Root rot	United States	Banihashemi and Mitchell (1975)
Carthamus tinctorius L. (safflower)	*cryptogea*	Root and stem rot	United States	Duniway (1976); Klisiewicz (1977)
	drechsleri	Root and stem rot	United States	Erwin (1950, 1952)
	palmivora	Leaf blight	Venezuela	Malaguti (1950)
	parasitica	Root rot	Argentina	Frezzi (1950)
		Root rot	United States	Klisiewicz (1977)
Catharanthus roseus L. (periwinkle)	*citrophthora*	Root rot, stem necrosis	Argentina	Frezzi (1950)
	parasitica	Leaf, stem, flower, and fruit blight	India	Dastur (1916)
Cereus spp. (cactus)	*cactorum*	Root rot	Germany	DeBary 1881)
	parasitica	Root rot	Japan	Katsura (197[illegible])
	syringae	Wilt	Germany	Himmelbaur (1911[b])
Chamaecyparis lawsoniana (A. Murr.) Parl. (Port Orford cedar)	*citricola*	Wilt	Netherlands	Rattink and Van Steekelenburg (1973)
	hibernalis	Trunk canker	United Kingdom	Young and Strouts (1976)
	lateralis	Root rot	United States	Tucker and Milbrath (1942)
	eriugena	Blight	Ireland	Clancy and Kavanaugh (1977)
Chrysanthemum cinerarifolium (Trev.) Vis. (pyrethrum)	*cactorum*	Stem canker	Peru	Crandall and Dieguez (1953)
	cambivora	Wilt	India	Ramakrishnan and Soumini (1948)
	cinnamomi	Root rot	Russia	Novatelynova (1974)
	cryptogea	Blight	Zimbabwe	Whiteside (1966)
	drechsleri	Root rot	Argentina	Frezzi (1950)
Cicer arietinum L. (chickpea, garbanzo)	*citrophthora*	Root rot, stem necrosis	Argentina	Frezzi (1950)
	medicaginis	Root rot	Australia	Irwin and Dale (1982)
	drechsleri	Root rot	Iran	Ershad (1971)
	megasperma	Root rot	Argentina	Frezzi (1950)
Cinchona spp. (cinchona)	*cinnamomi*	Root rot, stripe canker	Malaysia	Thompson (1940)
	palmivora	Top blight	Philippines	Celino (1934)
	parasitica	Stem canker, seedling blight	South America	Crandall and Davis (1945)
Citrullus spp. (melon)	*cactorum*	Fruit rot, wilt	United States	Brown and Evans (1933)
	capsici	Fruit rot, wilt	Japan	Katsura and Tokura (1954)
	capsici	Crown rot	France	Clerjeau (1973)
	cryptogea	Root and crown rot	Egypt	Satour and El-Shinnawy (1973)
	drechsleri	Blight	Japan	Katsura (1953)
	palmivora	Fruit rot	Malaysia	Chee (1969)
	parasitica	Fruit rot	United States	Sherbakoff (1917[b])
Citrus spp. (citrus) (see also Table 17.1)	*boehmeriae*	Brown rot of fruit	Argentina	Frezzi (1941, 1950)
	cactorum	Stem canker, fruit rot	United States	Fawcett (1920)
	capsici	Root rot	United States	Wiant and Tucker (1940[b])
	cinnamomi	Root rot	Brazil	H. S. Fawcett and Bitancourt (1940)
	citricola	Brown rot of fruit	Taiwan	Sawada (1927)
	citrophthora	Brown rot of fruit	France	Boccas and Laville (1978)
		Brown rot of fruit, root rot, gummosis	United States	R. E. Smith and Smith (1906)
	drechsleri	Root rot	United States	Fawcett (1936)
	hibernalis	Brown rot of fruit	Australia	Carne (1925)

(*continued on next page*)

Table 67.1. *(continued)*

Host	*Phytophthora* spp.	Diseases caused	Early occurrence and/or significant information on species and host[a]	Reference
Citrus spp. (*continued*)	*megasperma*	Root rot and stem canker	United States	H. S. Fawcett (1933)
	palmivora	Seedling blight	Philippines	Reinking (1921)
	parasitica	Gummosis, root rot, brown rot of fruit	United States	H. S. Fawcett (1915)
	syringae	Fruit rot	Argentina	Frezzi (1950)
	All of above	Gummosis, root rot, brown rot	World	Klotz (1973, 1978); Ricci et al. (1990) (reviews)
Cocos nucifera L. (coconut)	*arecae*	Nut-fall, mahali disease	India	Sundararaman and Ramakrishnan (1924)
	palmivora	Bud rot, leaf drop	United States	W. Fawcett 1889)
		Bud rot	Philippines	Reinking (1918); Tucker (1931, 1933)
		Bud rot	Jamaica	Ashby (1920)
		Bud rot	India	Butler (1906)
	palmivora and *katsurae*	Bud rot	Jamaica	Steer and Coates-Beckford (1990)
	katsurae	Bud rot	United States	Uchida et al. (1992)
Colocasia antiquorum Schott. (elephant's ear)	*arecae*	Leaf spot	India	Narasimhan (1927)
	colocasia	Corm rot, leaf blight	India	Butler and Kulkarni (1913)
	palmivora	Leaf blight	India	Thomas et al. (1947)
	parasitica	Leaf rot	Fiji	Graham (1971)
Cucumis melo var. *inodorus* Naud. (honeydew melon)	*cactorum*	Fruit rot	United States	Wiant (1939)
	capsici	Fruit and root rot	United States	Tompkins and Tucker (1937a, 1941b)
	citrophthora	Fruit rot	United States	Drechsler (1929a)
	cryptogea	Root and crown rot	Egypt	Satour and El-Shinnawy (1973)
	drechsleri	Fruit and stem rot	Iran	Ershad (1971)
	parasitica	Fruit rot	United States	Drechsler (1929a)
Cucumis sativus L. (cucumber)	*cactorum*	Fruit rot	United States	Brown and Evans (1933[b])
	capsici	Fruit rot	United States	Kreutzer et al. (1940)
	cryptogea	Root and crown rot	United Kingdom	Bewley (1930)
	drechsleri	Fruit rot	Iran	Ershad (1971)
	parasitica	Fruit rot	India	Rao et al. (1962[b])
Dianthus caryophyllus L. (carnation)	*cactorum*	Stem blight	Switzerland	Muller-Thurgau and Osterwalder (1924)
	citrophthora	Collar rot	Greece	Anon. (1965a)
	cryptogea	Root rot	United States	Middleton et al. (1944)
	meadii	Wilt and stem rot	Mauritius	Shepherd (1937)
	megasperma	Crown rot	New Zealand	Dingley (1969)
	palmivora	Wilt	Italy	Andreucci (1959)
	parasitica	Root rot	Argentina	Frezzi (1950)
	porri	Collar rot	Greece	Kouyeas (1977)
Eucalyptus L'Her (Australian gum tree)	*cactorum*	Root rot	United States	C. O. Smith (1937[b])
	cinnamomi	Root rot	Australia	Podger et al. (1965); see Shea et al. (1983), Shearer et al. (1988), Shearer (1990), Weste (1974, 1979, 1984), Weste and Law (1973) for reviews
	citrophthora	Stem canker	United States	C. O. Smith (1937[b])
	parasitica	Gummosis	Brazil	Arruda (1943)
	cactorum	Root rot	Australia	Shearer et al. (1988); Shearer (1990) (reviews)
	cambivora	Root rot	Australia	
	citricola	Root rot	Australia	
	cryptogea (A1 and A2)	Root rot	Australia	
	drechsleri	Root rot	Australia	
	megasperma	Root rot	Australia	
	nicotianae	Root rot	Australia	
Fragaria × *ananassa* (strawberry)	*cactorum*	Leather rot of fruit, stem and root rot	Switzerland	Osterwalder (1912); Madden et al. (1991)
			Iran	Ershad (1971)
	fragariae	Red stele	Scotland	Alcock (1929)
	parasitica	Root and crown rot	Argentina	Frezzi (1950)
Gossypium spp. (cotton)	*boehmeriae*	Boll blight	China	Liang (1964)
	capsici	Boll rot	United States	Garber et al. (1986)

(*continued on next page*)

Table 67.1. (*continued*)

Host	*Phytophthora* spp.	Diseases caused	Early occurrence and/or significant information on species and host[a]	Reference
Gossypium spp. (*continued*)	*palmivora*	Boll rot	West Indies	Ashby (1922)
		Seedling blight	Spain	Melaro-Vara and Jimenez Díaz (1990)
	parasitica	Boll rot, seedling blight	West Indies	Hopkins (1925)
		Boll rot	United States	Pinckard and Guidroz (1973)
Hevea brasiliensis Muell. (para rubber)	*arecae*	Fruit rot	India	Narasimhan (1927)
	botryosa	Leaf fall disease	Malaysia	Chee (1969)
	heveae	Black stripe	Malaysia	Thompson (1929)
	meadii	Stripe canker	Indonesia	Rutgers (1912)
	palmivora	Black stripe, leaf fall	Sri Lanka	Petch (1914)
	parasitica	Patch canker	Malaysia	Thompson (1929)
	citrophthora	Black stripe	Indonesia; Ivory Coast	Liyanage and Wheeler (1989a)
	botryosa, heveae, meadii, palmivora	Several diseases	World	Holliday (1980), Radziah and Hashim (1990) (reviews)
Hibiscus spp.	*cactorum*	Foot rot	United States	James and Plakidas (1951)
	cinnamomi	Root rot	Australia	Anon. (1955c)
	citricola	Root rot, wilt	New Zealand	Brien (1938)
	palmivora	Leaf, stem, and flower rot	United States	James and Plakidas (1952)
	parasitica	Collar rot	Philippines	Reinking (1918)
	parasitica var. *nicotianae*	Root and collar rot	Malaysia	Navaratnam (1970)
Juglans L. (walnut)	*cactorum*	Black root	United States	Barrett and Smith (1926)
	cinnamomi	Root rot	United States	Crandall (1936)
	citricola	Seedling root rot	United States	Green and Pratt (1970)
	citrophthora	Stem canker	United States	C. O. Smith and Barrett (1931); Mircetich and Matheron (1983)
	megasperma	Root and crown rot	United States	Mircetich and Matheron (1983)
	citricola	Root and crown rot	Australia	Cookson (1929)
	cryptogea			
	cinnamomi			
	cactorum			
	parasitica			
	parasitica	Root and crown rot	United States	Matheron and Mircetich (1985b)
	cryptogea			
Lilium spp. (lily)	*cactorum*	Foot rot	United States	Drechsler (1931)
	cinnamomi	Root rot	Australia	Anon. (1948)
	parasitica	Stump rot	West Indies	Whetzel (1922)
Lupinus spp. (lupin)	*cambivora*	Blight	France	Allain (1935)
	cinnamomi	Root rot	W. Australia	Titze and Palzer (1969)
	cryptogea	Wilt	Australia	Brittlebank and Fish (1927)
Lycopersicon esculentum L. (tomato)	*arecae*	Stem dieback, foot rot	Netherlands	Verhoeff and Weber (1966)
	capsici	Fruit, crown, and root rot	United States	Kreutzer et al. (1940)
	cinnamomi	Root rot	Argentina	Frezzi (1950)
	citricola	Seedling wilt	Italy	Salerno and Calabretta (1960)
	citrophthora	Fruit rot	Argentina	Frezzi (1950)
	cryptogea	Fruit rot, seedling damping-off	Ireland	Pethybridge and Lafferty (1919)
	drechsleri	Fruit rot	United States	Tompkins and Tucker (1941a)
	infestans	Late blight	France	Tulasne (1854)
			World	Ingram and Williams (1991) (review)
	palmivora	Fruit rot	India	Ramakrishnan and Soumini (1947)
	parasitica	Buckeye rot, seedling damping-off	United States	Sherbakoff (1917)
	mexicana	Root rot	Mexico	Hotson and Hartge (1923)
Malus pumila Mill. (apple)	*cactorum*	Fruit, collar, and root rot	Switzerland	Osterwalder (1906)
		Root and crown rot	Canada	Utkhede (1986) (review)
		Root and crown rot	United States	Baines (1935, 1939); Jeffers and Aldwinckle (1988)
	cambivora	Root rot, trunk canker	Canada	McIntosh (1966)
		Root and crown rot	United States	Jeffers and Aldwinckle (1988)
	cinnamomi	Fruit rot	United Kingdom	Oyler and Bewley (1937[b])
	citricola	Root rot	Canada	McIntosh (1966)

(*continued on next page*)

Table 67.1. (*continued*)

Host	*Phytophthora* spp.	Diseases caused	Early occurrence and/or significant information on species and host[a]	Reference
Malus pumila Mill. (*continued*)	*cryptogea*	Root rot	United States	Browne and Mircetich (1988, 1993); Mircetich and Brown (1987, review of several species)
	drechsleri	Root rot, trunk canker	United States	Mircetich et al. (1974)
	gonapodyides	Fruit rot	Denmark	Petersen (1910)
	megasperma	Root rot, trunk canker	United States	Mircetich et al. (1974); Jeffers et al. (1982)
		Root rot, trunk canker	New Zealand	Robertson and Dance (1971)
	palmivora	Fruit rot	Philippines	Reinking (1923)
	parasitica	Fruit rot, stem canker	India	Padwick (1938)
	syringae	Stem canker, fruit rot	Ireland	Lafferty and Pethybridge (1922)
			World	Harris (1991) (review)
Medicago sativa L. (alfalfa)	*citricola*	Root rot	South Africa	Thompson (1988)
	cryptogea[c]	Root rot	United States	Erwin (1954)
	drechsleri	Root rot	Australia	Cother (1975[b])
	drechsleri	Root rot	South Africa	Thompson (1987, 1988)
	medicaginis[c]	Root rot	United States	E. M. Hansen and Maxwell (1991)
	megasperma[c]	Root rot	United States	Erwin (1965)
	megasperma f. sp. *medicaginis*[c]	Root rot	United States	Kuan and Erwin (1980)
	megasperma f. sp. *medicaginis*[c]	Root rot	South Africa	Thompson (1987, 1988)
	megasperma (ALF)	Root rot	United States	Hamm and Hansen (1981, 1982b)
	megasperma (ALF)	Root rot	Canada	Faris (1985)
	megasperma (AL2)	Root rot	United States	E. M. Hansen and Hamm (1983)
	megasperma (2i and 2ii types)	Root rot	Canada	Faris (1985)
	parasitica	Root rot	Australia	Purss (1959)
Melilotus L. spp. (clover)	*cactorum*	Root rot	United States	F. R. Jones (1939)
	megasperma f. sp. *trifolii*	Root rot	United States	R. G. Pratt (1981)
Nicotiana L. spp. (tobacco)	*cinnamomi*	Root rot	Russia	Novatelynova (1974)
	fragariae	Seedling rot	Canada	Anon. (1955b)
	nicotianae (*parasitica*)	Black shank	Java	Breda de Haan (1893, 1896)
		Black shank	United States	Tisdale (1922); Lucas (1975), Shew and Lucas (1991) (reviews)
Olea europea L. (olive)	*cactorum*	Canker	United States	C. O. Smith (1937[b])
	cinnamomi	Root rot	Australia	Anon. (1949)
	citricola	Collar and trunk canker	Greece	Kouyeas (1978)
	megasperma	Collar and trunk canker	Greece	Kouyeas and Chitzanidis (1968)
Persea americana Mill. (avocado)	*cactorum*	Bark canker	United States	Barrett (1917)
	cinnamomi	Root rot	Puerto Rico	Tucker (1928)
			United States	Wager (1942a)
			World	Zentmyer (1980) (review)
	citricola	Trunk canker	United States	Zentmyer et al. (1974)
	citrophthora	Bark canker	United States	H. S. Fawcett (1916)
	heveae	Trunk canker	Guatemala	Zentmyer et al. (1978)
	palmivora	Seedling blight	Philippines	Reinking (1921)
	parasitica	Collar rot	United States	Stevens and Piper (1941)
Phaseolus spp. (bean)	*capsici*	Root rot	Argentina	Frezzi (1950)
	cinnamomi	Root rot	Argentina	Frezzi (1950)
	cryptogea	Stem and root rot	United States	Flowers et al. (1973)
	drechsleri	Seedling blight	Japan	Katsura (1958)
	palmivora	Blight	Philippines	Youngberg (1926)
	parasitica	Wet pod rot, damping-off	Puerto Rico	Stevenson (1917)
Pieris japonica (Thunb.) Don. (Japanese andromeda)	*cinnamomi*	Root rot	New Zealand	Dingley (1969)
	citricola, citrophthora	Root rot, shoot blight, stem dieback	United States	Gerlach et al. (1974, 1976a, b)
	parasitica	Shoot blight	United States	Gerlach et al. (1974)
Pinus spp. (pine)	*boehmeriae*	Root rot	Australia	Anon. (1962c)
	cactorum	Seedling root rot	United Kingdom	Collinge (1912)
	cinnamomi	Root rot	United States	Crandall (1936)
		Root rot	New Zealand	Newhook (1959, 1982)
	citricola	Root rot	United States	Sandlin et al. (1992)
	cryptogea	Decline	Australia	Bumbieris (1976)
	citrophthora	Preemergence seedling disease	United States	Rathbun-Gravatt (1931)
		Root rot	Argentina	Frezzi (1950)

(*continued on next page*)

Table 67.1. (*continued*)

Host	*Phytophthora* spp.	Diseases caused	Early occurrence and/or significant information on species and host[a]	Reference
Pinus spp. (*continued*)	*drechsleri*	Seedling disease	Australia	Heather et al. (1977)
	heveae	Root rot	Australia	Oxenham and Winks (1963)
	parasitica	Root rot	United States	Munnecke and Bricker (1976)
Piper betle L. (pan, betel nut)	*colocasiae*	Wilt	Malaysia	Thompson (1928)
	palmivora	Stem canker	India	Thomas et al. (1947)
	parasitica	Root rot	India	McRae (1932)
	capsici	Root rot	Taiwan	Ho (1990)
Pistacia vera L. (pistachio tree)	*citricola*	Wilt	Greece	Kouyeas (1973)
	nicotianae	Leaf necrosis, root rot	United States	French (1987)
	capsici, cryptogea, parasitica	Trunk and branch canker	United States	MacDonald et al. (1992)
Pisum spp. (pea)	*cactorum*	Root rot	United States	Clinton (1920)
	cambivora	Seedling blight	Italy	Arnaudi (1928[b])
	erythroseptica	Soft rot	United States	Bywater and Hickman (1959)
Primula spp. (primula)	*cactorum*	Stem canker	United States	Williams et al. (1937)
	citricola	Root rot	United Kingdom	Dennis and Foister (1942)
	parasitica	Collar rot	Australia	Brittlebank and Fish (1927)
	primula	Brown core	United Kingdom	Tomlinson (1952)
Prunus armeniaca L. (apricot)	*cactorum*	Bark canker, fruit rot	United States	E. H. Smith (1915)
	cambivora	Root and trunk rot	United States	Mircetich et al. (1974)
	cinnamomi	Root rot	Australia (NSW)	Anon. (1949)
	citrophthora	Stem canker	United States	R. E. Smith and Smith (1925)
	drechsleri	Root rot, trunk canker	United States	Mircetich et al. (1974)
	megasperma	Root rot, trunk canker	Australia	Anon. (1958b)
	parasitica	Root rot, trunk canker	Greece	Kouyeas (1978)
	syringae	Wilt, stem canker	New Zealand	H. C. Smith (1956)
Prunus avium (L.) L. (sweet cherry)	*cactorum*	Crown rot, trunk canker	Spain	Unamuno (1911)
	cambivora *drechsleri* *megasperma*	Root rot, trunk canker	United States	Mircetich et al. (1974); Mircetich and Matheron (1976); Wilcox and Mircetich (1985)
	citrophthora	Stem canker	United States	R. E. Smith and Smith (1925)
	syringae	Twig blight	Germany	Klebahn (1909[b])
Prunus domestica L. (plum)	*cactorum*	Stem canker	United States	R. E. Smith and Smith (1925)
	cambivora	Root rot, trunk canker	United States	Mircetich et al. (1974)
	citrophthora	Collar rot	Greece	Kouyeas and Chitzanidis (1978)
	drechsleri	Root rot, trunk canker	United States	Mircetich et al. (1974)
	megasperma	Root rot, trunk canker	Greece	Kouyeas and Chitzanidis (1968)
Prunus dulcis (Mill.) D. A. Webb (almond)	*cactorum*	Bark canker	United States	E. H. Smith (1915)
	citricola	Collar and trunk rot	Greece	Kouyeas (1973)
	citricola	Canker	Australia	Wicks and Lee (1985)
	citrophthora	Root rot	Greece	Anon. (1965)
	cryptogea	Collar and trunk canker	United States	Middleton et al. (1944)
	drechsleri	Collar and trunk canker	Greece	Kouyeas (1978)
	megasperma	Collar and trunk canker	Greece	Kouyeas and Chitzanidis (1968)
	parasitica	Collar and trunk rot	Greece	Helena (1978)
Prunus persica (L.) Batsch. (peach)	*cactorum*	Bark canker, seedling blight	United States	R. E. Smith and Smith (1925)
	cambivora	Root rot, trunk canker	United States	Mircetich et al. (1974)
	cinnamomi	Root rot	Australia (NSW)	Anon. (1948)
	citricola	Root and crown rot	Zimbabwe	Rothwell (1975)
		Root and crown rot	United States	Jeffers and Aldwinckle (1987, 1988)
	citrophthora	Trunk canker	United States	E. H. Smith (1915)
	drechsleri	Root rot, trunk canker	United States	Mircetich et al. (1974)
	megasperma	Root rot, wilt	Italy	Govi and Tassinari (1954)
	parasitica	Fruit rot	India	Padwick (1938)
	syringae	Wilt, stem canker	New Zealand	H. C. Smith (1956)
Pyrus communis L. (pear)	*cactorum*	Crown rot	United States	D. H. Rose and Lindgren (1925)
		Crown rot	Germany	Braun (1954); Braun and Schwinn (1963)
		Root and collar rot	Canada	McIntosh (1960, 1965)
		Root rot	United States	Cameron (1962)
	cinnamomi	Root rot	United States	Cameron (1962)

(*continued on next page*)

Table 67.1. (*continued*)

Host	*Phytophthora* spp.	Diseases caused	Early occurrence and/or significant information on species and host[a]	Reference
Rhododendron spp.	*cactorum*	Stem and seedling rot	United States	White (1930)
	cambivora	Blight	United States	White (1937)
	cinnamomi	Root rot	United States	White (1930)
	citricola	Bud, leaf, and twig blight	Germany	Kröber (1959)
	citrophthora	Dieback	United States	Hoitink and Schmitthenner (1974a)
	cryptogea	Root rot	United States	White and Hamilton (1935)
	gonapodyides	Root rot	United States	Hoitink and Schmitthenner (1974a)
	heveae	Dieback	United States	Benson and Jones (1980)
	parasitica	Dieback	United States	Benson and Jones (1980)
Rubus idaeus (red raspberry)	*fragariae* var. *rubi*	Root rot	Scotland, United States	Duncan et al. (1991); Wilcox et al. (1993)
	citricola, citrophthora	Root rot	Chile	Latorre and Muñoz (1993)
	citrophthora, various species	Root rot	United States, Scotland	Duncan et al. (1991) (review)
	cryptogea	Root rot	Australia	Washington (1988)
Senecio cruentus (Masson) DC. (cineraria, groundsel)	*cinnamomi*	Root rot	United Kingdom	Munro (1945)
	cryptogea	Foot rot, wilt	Australia	Brittlebank and Fish (1927)
	drechsleri	Rot, collar rot	Argentina	Carranza (1964)
	erythroseptica	Wilt	New Zealand	R. Lucas (1977)
	megasperma	Root rot	United States	Tompkins et al. (1936b)
	parasitica	Root rot	United Kingdom	Moore (1959)
Solanum melongena L. var. *esculentum* Ness (eggplant)	*capsici*	Fruit rot	Argentina	Frezzi (1950)
	cryptogea	Collar and stem rot	Greece	Chitzanidis and Kouyeas (1970)
	drechsleri	Seedling blight	United States	Tucker (1931[b])
	infestans	Fruit rot, calyx blight	Germany	Corda 1847)
	palmivora	Fruit rot	Java	Reitsma and Slooff (1947)
	parasitica	Fruit rot, seedling blight	Taiwan	Sawada (1915)
	parasitica var. *nicotianae*	Fruit rot	Taiwan	Sawada (1915)
Solanum tuberosum L. (potato)	*arecae*	Pink tuber rot	United States	Rosenbaum (1914[b])
	cactorum	Pink tuber rot	South Africa	Wijers (1937)
	cinnamomi	Pink tuber rot	United States	Tucker (1933)
	drechsleri	Pink tuber rot	United States	Drechsler (1929b)
	erythroseptica	Pink tuber rot	Ireland	Pethybridge (1913)
	infestans	Late blight	Europe	de Bary 1876)
	iranica	Tuber rot	Iran	Ershad (1971[b])
	megasperma	Pink tuber rot	Northern Ireland	Cairns and Muskett (1933)
	palmivora	Tuber rot	India	Ramskrishnan and Seethalakshmi (1956b[b])
	parasitica	Tuber rot, leaf and stem blight	United States	Tucker (1931[b])
	parasitica var. *nicotianae*	Tuber rot	United States	Tucker (1931[b])
Spinacia oleracea L. (spinach)	*cryptogea*	Root rot, seedling damping-off	Germany	Kröber and Beckman (1973)
	erythroseptica	Seedling rot	Canada	W. Jones (1954[b])
	fragariae	Seedling rot	Canada	Anon. (1955a[b])
	megasperma	Root rot	United States	Shanor (1944)
Syringa vulgaris L. (lilac)	*cactorum*	Leaf and twig blight	United States	R. P. White (1930)
	citricola	Shoot blight	United States	Chester (1932)
	palmivora	Collar and stem rot	Greece	Kouyeas and Chitzanidis (1978)
	parasitica	Stem blight, leaf spots	Italy	Scurti (1962)
	syringae	Downy mildew, leaf spot, shoot dieback	Scotland	Berkeley (1881)
Theobroma cacao L. (cocoa)	*citrophthora*	Pod and seedling rot, cankers	Brazil	Kellam and Zentmyer (1981, 1986b)
	heveae	Pod rot	Malaysia	P. D. Turner (1968)
	megakarya	Pod and seedling rot	West Africa	Brasier and Griffin (1979)
	megasperma	Pod rot	Venezuela	Anon. (1965)
	palmivora	Black pod	Sri Lanka	Willis and Green 1897)
	capsici	Black pod	Brazil and other Latin American countries, Cameroun	Griffin (1977); Tsao and Alizadeh (1988)

(*continued on next page*)

Table 67.1. *(continued)*

Host	*Phytophthora* spp.	Diseases caused	Early occurrence and/or significant information on species and host[a]	Reference
Tulipa spp. (tulip)	*cactorum*	Blossom blight, stem canker	United States	Stevens and Plunkett (1925)
	cryptogea	Root and bulb rot, blossom blight	United Kingdom	Alcock (1929)
	porri	Blight, root rot	Japan	Katsura (1970)
Vanda spp. (orchid)	*cactorum*	Stem canker	Philippines	Schwarz (1927)
	palmivora	Leaf and stem rot	Java	Schwarz (1927)
Vicia spp. (vetch)	*cactorum*	Black rot	Canada	Conners (1955)
	cinnamomi	Root rot	United Kingdom	Williams et al. (1937)
Vigna unguiculata (L.) Walp. (cowpea)	*cactorum*	Red stem canker	United States	Weimer (1949)
	drechsleri	Root rot	United States	Erwin et al. (1991)
	parasitica	Root rot	Java	Schwarz cited as unpublished in Tucker (1931)
	vignae	Stem rot	Australia	Purss (1957)
Vitis spp. (grape)	*cactorum, cinnamomi, cryptogea*	Root rot	South Africa	Marais (1980)
	parasitica	Seedling damping-off, root rot	India	Sridhar et al. (1975); Marais (1979)
Zantesdeschia aethiopica Spreng. (calla lily)	*cryptogea*	Root rot	United Kingdom	Pethybridge et al. (1934)
	erythroseptica	Rhizome rot, leaf and petiole blight	United States	Tompkins and Tucker (1947)
	richardiae	Root rot	Netherlands	Buisman (1927)

[a] For geographical distribution of each disease, see tables under each *Phytophthora* species (except *P. cinnamomi*) in the text.
[b] Artificially inoculated host.
[c] *P. cryptogea* on *Medicago sativa* was reclassified to *P. megasperma* (Erwin 1965), to *P. megasperma* f. sp. *medicaginis* (Kuan and Erwin 1980), and to *P. medicaginis* (Hansen and Maxwell 1991).

Literature Cited

Abney, T. S., Scott, D. H., and Bergeson, G. B. 1979. Carbofuram increases *Phytophthora* damage in soybean (Abstr.). Proc. 9th Int. Congr. Plant Prot. Proc. Amer. Phytopathol. Soc. Abstr. 561.

Abo-el-Dahab, M. K., and El-Goorani, M. A. 1969. The pink rot disease of potato tubers in U.A.R. (Egypt). J. Phytopathol. U.A.R. 1:53–58.

Adams, G. C. Jr., and Bielenin, A. 1988. First report of *Phytophthora cactorum* and *P. citricola* causing crown rot of fir species in Michigan (Abstr.). Plant Dis. 72:79.

Afek, U., and Sztejnberg, A. 1986. A citrus phytoalexin, 6,7-dimethoxy-courmarin, as a defense mechanism against *Phytophthora citrophthora,* and influence of Aliette® and phosphorous acid on its production. Phytoparasitica 14:246.

Afek, U., and Sztejnberg, A. 1988. Accumulation of scoparone, a phytoalexin associated with resistance of citrus to *Phytophthora citrophthora.* Phytopathology 78:1678–1682.

Afek, U., and Sztejnberg, A. 1989. Effects of fosetyl-Al and phosphorous acid on scoparone, a phytoalexin associated with resistance of citrus to *Phytophthora citrophthora.* Phytopathology 79:736–739.

Afek, U., and Sztejnberg, A. 1993. Temperature and gamma irradiation effects on scoparone, a citrus phytoalexin conferring resistance to *Phytophthora citrophthora.* Phytopathology 83:753–758.

Afek, U., Sztejnberg, A., and Carmely, S. 1986. 6,7-Dimethoxy-coumarin, a citrus phytoalexin conferring resistance against Phytophthora gummosis. Phytochemistry 25:1855–1856.

Afek, U., Sztejnberg A., and Solel, Z. 1990. A rapid method for evaluating citrus seedlings for resistance to foot rot caused by *Phytophthora citrophthora.* Plant Dis. 74:66–68.

Aggarwal, A. and Mehrotra, R. S. 1987. Control of Phytophthora leaf blight of taro (*Colocasia esculenta*) by fungicides and rogueing. Phytoparasitica 15:299–305.

Aggarwal, A., Narula, K. L., Gurinderjit, K., Mehrotra, R. S., and Kaur, G. 1990. *Phytophthora colocasiae* Racib.—Its taxonomy, physiology, pathology and control. Pages 105–134 in: Perspectives in Mycological Research. Vol. II. S. K. Hasija and K. S. Bilgrami. Internat. Biosci. Ser. Vol. XV. Today and Tomorrow Printers, New Delhi, India.

Agharkar, S. P., and Bannerjee, S. N. 1932. The occurrence of *Phytophthora parasitica* Dast. on *Boucerosia diffusa.* Page 298 in: Proc. 19th Indian Sci. Congr.

Agnihothrudu, V. 1975. Abnormal leaf fall of rubber. Pages 223–230 in: Advances in Mycology and Plant Pathology. S. P. Raychaudhuri, A. Varma, K. S. Bhargava, and B. S. Mehrotra, eds. Harsh Kumar at Sagar Printers, New Delhi. 285 pp.

Agostini, J. P., Timmer, L. W., Castle, W. S., and Mitchell, D. J. 1991. Effect of citrus rootstocks on soil populations of *Phytophthora parasitica.* Plant Dis. 75:296–300.

Agrawal, S. C. 1989. Effect of landscape on the severity of stem blight of pigeon pea caused by *Phytophthora drechsleri* f. sp. *cajani.* Ind. J. Pulse. Res. 2:179–180.

Agrawal, S. C., and Khare, M. N. 1987. Comparative epidemic growth of stem blight on four genotypes of pigeon pea. Ind. J. Mycol. Plant Pathol. 17:79–80.

Aguirre, R. J., Hine, R. B., and Schonhorst, M. H. 1983. Distribution of Phytophthora root rot of alfalfa in central Mexico and development of disease resistance in Mexican cultivars of alfalfa. Plant Dis. 67:91–94.

Ainsworth, G. C. 1968. Plant Pathologist's Pocket Book. Commonw. Mycol. Inst., Kew, Surrey, U.K. 267 pp.

Al-Hedaithy, S. S. A., and Tsao, P. H. 1979a. The effects of culture media and sporulation methods on caducity and pedicel length of sporangia in selected species of *Phytophthora.* Mycologia 71:392–401.

Al-Hedaithy, S. S. A., and Tsao, P. H. 1979b. Sporangium pedicel length in *Phytophthora* species and the consideration of its uniformity in determining sporangium caducity. Trans. Br. Mycol. Soc. 72:1–13.

Al-Heeti, M. B., and Sinclair, J. B. 1988. Antagonism between *Gliocladium roseum, Trichoderma harzianum,* or *Trichothecium roseum* and *Phytophthora megasperma* f. sp. *glycinea.* Mycopathologia 103:135–140.

Al-Kherb, S. M., Fininsa, C., Shattock, R. C., and Shaw, D. S. 1995. The inheritance of virulence of *Phytophthora infestans* to potato. Plant Pathol. 44:552–562.

Alavi, A., and Strange, R. N. 1979. A baiting technique for isolating *Phytophthora drechsleri,* causal agent of crown rot of *Cucumis* species in Iran. Plant Dis. Rep. 63:1084–1086.

Albert, G., Curtze, J., and Drandarevski, C. A. 1988. Dimethomorph (CME 151), a novel curative fungicide. Proc. Br. Crop Prot. Conf. Pests Dis. 1:17–24.

Albuquerque, F. C. 1968. *Piper colubrinum* link porta-enxêrto para *Piper nigrum* L. resistente as enfermedades causadas por *Phytophthora palmivora* Butl. e *Fusarium solani* f. sp. *piperi* (*Piper colubrinum,* a grafting rootstock for *Piper nigrum* resistant to diseases caused by *Phytophthora palmivora* and *Fusarium solani* f. sp. *piperi*). Pesqui. Agropecu. Bras. 3:141–145. (In Portuguese). (Rev. Plant Pathol. 1970, 49:2955.)

Albuquerque, F. C., Duarte, M. de L. R., Manco, G. R., and Martins e Silva, H. 1974. Requeima das folhas da castanheira do para (*Bertholletia excelsa*) causado por *Phytophthora heveae* (Leaf blight of Brazil nut (*Bertholletia excelsa*) caused by *Phytophthora heveae*). Pesqui. Agropecu. Bras. Agron. 9:101–105. (In Portuguese). (Rev. Plant Pathol. 1965, 55:5369.)

Alcock, N. L. 1926. A preliminary note on a *Phytophthora* on *Atropa belladonna.* Pharm. J. 116(3254):232.

Alcock, N. L. 1929. A root disease of the strawberry. Gard. Chron. 86:14–15.

Alcock, N. L., Howells, D. V., and Foister, C. E. 1930. Strawberry disease in Lanarkshire. Scott. J. Agric. 10:242–252.

Alconero, R., and Stone, E. 1969. *Phytophthora parasitica nicotianae* in roselle and kenaf in Puerto Rico. Plant Dis. Rep. 53:702–705.

Alconero, R., Albuquerque, F., Almeyda, N., and Santiago, A. G. 1972. Phytophthora foot rot of black pepper in Brazil and Puerto Rico. Phytopathology 62:144–148.

Alcorn, S. M., Orum, T. V., and Matheron, M. E. 1989. Phytophthora leaf blight of jojoba (*Simmondsia chinensis*) in Arizona. Plant Dis. 73:444.

Aleksic, Z., Aleksic, D., and Miladinovic, Z. 1975. Bolesti paprike i mogucnosti njihovog suzbijanja gajenjem otpornih sorti (Diseases of pepper and the possibilities of their control by cultivating resistant varieties). Agron. Glas. 37:73–82. (In Slavic). (Rev. Plant Pathol. 1976, 55:3831.)

Alexander, M. 1977. Microbial transformations of phosphorus. Pages 347–348 in: Introduction to Soil Microbiology. John Wiley & Sons, New York.

Alexander, M., Daggett, P. M., Gherna, R., Jong, S., Simione, F., and Hatt, H. 1980. American Type Culture Collection Methods. I. Laboratory Manual on Freezing and Freeze Drying. American Type Culture Collection, Rockville, MD.

Alexopoulos, C. J., and Mims, C. W. 1979. Introductory Mycology. John Wiley & Sons, New York. 632 pp.

Alfaro Moreno, A., and Vegh, I. 1971. La "tristeza" o "seca" del pimiento producida por *Phytophthora capsici* Leonian (Tristeza or seca of *Capsicum* caused by *Phytophthora capsici* Leonian). Ann. Inst. Nac. Investnes Agrar. Ser. Prot. Veg. 1:9–42. (In Spanish). (Rev. Plant Pathol. 1972, 51:3685.)

Alfieri, S. A., Jr. 1970. Bougainvillea blight, a new disease caused by *Phytophthora parasitica*. Phytopathology 60:1806–1808.

Alfieri, S. A., and Miller, J. W. 1971. Basal stem and root rot of Christmas cactus caused by *Phytophthora parasitica*. Phytopathology 61:804–806.

Ali-Shtayeh, M. S., Len, L. H. C., and Dick, M. W. 1986. An improved method and medium for quantitative estimates of populations of *Pythium* species from soil. Trans. Br. Mycol. Soc. 86:39–47.

Ali-Shtayeh, M. S., MacDonald, J. D., and Kabashima, J. 1991. A method for using commercial ELISA tests to detect zoospores of *Phytophthora* and *Pythium* species in irrigation water. Plant Dis. 75:305–311.

Alizadeh, A., and Tsao, P. H. 1985a. Effect of light on sporangium formation, morphology, ontogeny, and caducity of *Phytophthora capsici* and *P. palmivora* MF4 isolates from black pepper and other hosts. Trans. Br. Mycol. Soc. 85:47–69.

Alizadeh, A., and Tsao, P. H. 1985b. Chlamydospore formation in *Phytophthora palmivora* MF4. Trans. Br. Mycol. Soc. 85:71–79.

Allain, A. 1935. Contribution a l'étude du *Phytophthora cambivora*. Morphologie, cytologie, et action pathogene du parasite (A contribution to the study of *Phytophthora cambivora*. The morphology, cytology, and pathogenic action of the parasite). Typographie Firmin-Didot et Cie., Paris. 127 pp. (In French). (Rev. Appl. Mycol. 1936, 15:408).

Allen, D. J., and Nandra, S. S. 1975. Effects of pH and calcium concentration on the sporulation of *Phytophthora* isolates from agave. Plant Dis. Rep. 59:555–558.

Allen, E. H., and Thomas, C. A. 1971. Time course of safynol accumulation in resistant and susceptible safflower infected with *Phytophthora drechsleri*. Physiol. Plant Pathol. 1:235–240.

Allen, R. N., and Harvey, J. D. 1974. Negative chemotaxis of zoospores of *Phytophthora cinnamomi*. J. Gen. Microbiol. 84: 28–38.

Allen, R. N., and Newhook, F. J. 1973. Chemotaxis of zoospores of *Phytophthora cinnamomi* to ethanol in capillaries of soil pore dimensions. Trans. Br. Mycol. Soc. 61:287–302.

Allen, R. N., and Newhook, F. J. 1974. Suppression by ethanol of spontaneous turning activity in zoospores of *Phytophthora cinnamomi*. Trans. Br. Mycol. Soc. 63:383–385.

Allen, R. N., Pegg, K. G., Forsberg, L. I., and Firth, D. J. 1980. Fungicidal control of pineapple and avocado diseases caused by *Phytophthora cinnamomi*. Aust. J. Exp. Agric. Anim. Husb. 20:119–124.

Allison, J. L. 1952. Black shank of tobacco. F.A.O. Plant Prot. Bull. 1:44.

Alt, D. 1980. Histological studies on the bark of apple trees after infection with *Phytophthora cactorum* (Leb. et Cohn) Schroet. Angew. Bot. 54:393–401.

Altman, F. P. 1976. Tetrazolium salts and formazans. Prog. Histochem. Cytochem. 9:1–56.

Alva, A. K., Lanyon, L. E., and Leath, K. T. 1985. Excess soil water and Phytophthora root rot stresses of Phytophthora root rot sensitivity and resistant alfalfa cultivars. Agron. J. 77:437–442.

Alva, A. K., Lanyon, L. E., and Leath, K. T. 1986. Production of alfalfa in Pennsylvania soils of differing wetness. Agron. J. 78:469–473.

Alvarez, A. M., and Nelson, M. G. 1982. Control of *Phytophthora palmivora* in papaya orchards with weekly sprays of chlorothalonil. Plant Dis. 66:37–39.

Alvarez, G. L. A. 1962. Phytophthora damping-off of "brujas" (*Bryophyllum* spp.). J. Agric. Univ. P.R. 46:363.

Amin, K. S., Baldev, B., and Williams, F. J. 1976. Differentiation of Phytophthora stem blight from Fusarium wilt of pigeon pea by field symptoms. F.A.O. Plant Prot. Bull. 24:123–124.

Amin, K. S., Baldev, B. and Williams, F. J. 1978. *Phytophthora cajani*, a new species causing stem blight on *Cajanus cajan*. Mycologia 70:171–176.

Amouzou-Alladaye, E., Dunez, J., and Clerjeau, M. 1988. Immunoenzymatic detection of *Phytophthora fragariae* in infected strawberry plants. Phytopathology 78:1022–1026.

Anandaraj, M. 1985. Effect of adhesives of the retention of Bordeaux mixture on the arecanut bunches. Ann. Trop. Res. 7:62–65.

Anandaraj, M., and Sarma, Y. R. 1990. A simple baiting technique to detect and isolate *Phytophthora capsici* (*P. palmivora* MF4) from soil. Mycol. Res. 94:1003–1004.

Anastasiou, C. J. and Churchland, L. M. 1969. Fungi on decaying leaves in marine habitats. Can. J. Bot. 47:251–257.

Andersen, R. A., Barr, D. J. S., Lynn, D. H., Melkonian, M., Moestrup, Ø., and Sleigh, M. A. 1991. Terminology and nomenclature of the cytoskeletal elements associated with the flagellar/ciliary apparatus in protists. Protoplasma 164:1–8.

Anderson, E. J. 1951a. Effect of nutrient salt concentrations on infection of pineapple roots by *Phytophthora cinnamomi* (Abstr.). Phytopathology 41:2.

Anderson, E. J. 1951b. A simple method for detecting the presence of *Phytophthora cinnamomi* Rands in soil. Phytopathology 41:187–189.

Anderson, H. W. 1917. Notes on apple diseases in Illinois. Trans. Ill. Hortic. Soc. 51:413–420.

Anderson, H. W. 1940. Red stele root rot of the strawberry. Trans. Ill. Hortic. Soc. 74:383–393.

Anderson, R. D., and Guest, D. I. 1990. The control of black pod, canker and seedling blight of cocoa, caused by *Phytophthora palmivora* with potassium phosphonate. Australas. Plant Pathol. 19:127–129.

Anderson, R. D., Middleton, R. M., and Guest, D. I. 1989. Development of a bioassay to test the effect of phosphorous acid on black pod of cocoa. Mycol. Res. 93:110–112.

Anderson, T. R., and Buzzell, R. I. 1982. Efficacy of metalaxyl in controlling Phytophthora root and stalk rot of soybean cultivars differing in field tolerance. Plant Dis. 66:1144–1145.

Anderson, T. R., and Buzzell, R. I. 1988. Defeated Rps1 gene in soybean. Can. J. Plant Sci. 68:225–227.

Anderson, T. R., and Buzzell, R. I. 1992. Inheritance and linkage of the *Rps*7 gene for resistance to Phytophthora rot of soybean. Plant Dis. 76:958–959.

Andreucci, E. 1959. Un avvizzimento dei germogli del Garofano prodotto da *Phytophthora palmivora* Butl. (A wilt of carnation shoots produced by *Phytophthora palmivora* Butl.). Riv. Ortoflorofruttic. Ital. Vol. 43. 11 pp. (In Italian). (Rev. Appl. Mycol. 1959, 38:696–697).

Andrilenas, P. A. 1974. Farmers use of pesticides in 1971—Quantities. Econ. Res. Ser. U.S. Dep. Agric. Econ. Rep. 2T2.

Andrivon, D. 1995. Biology, ecology, and epidemiology of the potato late blight pathogen *Phytophthora infestans* in soil. Phytopathology 85:1053–1056.

Andrus, C. F. 1933. Fungus flora accompanying decline of boxwood. Plant Dis. Rep. 17:169–170.

Ann, P. J. 1984. Species, mating types and pathogenicity of *Phytophthora* distributed in citrus orchards in Taiwan. Trans. Br. Mycol. Soc. 82:631–634.

Ann, P. J. 1992. *Phytophthora* diseases of ornamental plants in Araceae in Taiwan. Plant Pathol. Bull. 1:79–89.

Ann, P. J., and Ko, W. H. 1980. *Phytophthora insolita*, a new species from Taiwan. Mycologia 72:1180–1185.

Ann, P. J., and Ko, W. H. 1985. Variants of *Phytophthora cinnamomi* extend the known limits of the species. Mycologia 77:946–950.

Ann, P. J., and Ko, W. H. 1987. Phenotypic behavior of progeny from the cross between chloramphenicol-resistant A1 and streptomycin-resistant A2 mating types of *Phytophthora parasitica*. (Abstr.). Phytopathology 77:1711.

Ann, P. J., and Ko, W. H. 1988a. Induction of oospore germination of *Phytophthora parasitica*. Phytopathology 78:335–338.

Ann, P. J., and Ko, W. H. 1988b. Hormonal heterothallism in *Phytophthora parasitica:* A novel model of sexual reproduction? J. Gen. Microbiol. 134:2985–2992.

Ann, P. J., and Ko, W. H. 1989. Effect of chloroneb and ethazol on mating type of *Phytophthora parasitica* and *P. cinnamomi*. Bot. Bull. Acad. Sin. Taiwan 30:207–210.

Ann, P. J., and Ko, W. H. 1990a. Survey of enzyme activity on solid media in *Phytophthora*. Can. J. Bot. 68:139–143.

Ann, P. J., and Ko, W. H. 1990b. Growth rate and colony morphology of progenies of zoospores and selfed oospores of *Phytophthora parasitica*. Mycologia 82:693–697.

Ann, P. J., and Ko, W. 1994. An asexual variant of *Phytophthora insolita*. Can. J. Microbiol. 40:810-815.

Ann, P. J., Kao, C. W., and Ko, W. H. 1986. Mating type distribution of *Phytophthora colocasiae* in Taiwan. Mycopathologica 93:193–194.

Anonymous. 1932. Tenth Annu. Rep. Minist. Agric. North. Irel. 1930–31. pp. 38–41. (Rev. Appl. Mycol. 1933, 12:265–266.)

Anonymous. 1934. England and Wales: New and interesting phytopathological records for the year 1933. Int. Bull. Plant Prot. 8:51. (Rev. Appl. Mycol. 1934, 13:493.)

Anonymous. 1936. Minist. Agric. Fish., England and Wales. New and interesting phytopathological records for the year 1935. Int. Bull. Plant Prot. 10:49–50.

Anonymous. 1938. Plant Pathology. Rep. Hawaii Agric. Exp. Stn. 1937. Pages 34–45.

Anonymous. 1940. Annu. Rep. of the director of plant industry Commonwealth of the Philippines. 221 pp. (Rev. Appl. Mycol. 1942, 21:66.)

Anonymous. 1941. Plant Pathology. Fifty-first Annu. Rep. for the year ending June 30, 1940. Ariz. Agric. Exp. Stn. Annu. Rep.: 90–102.

Anonymous. 1943. Plant Diseases. Notes contributed by the Biological Branch. Agric. Gaz. N.S.W. 54:223–226. (Rev. Appl. Mycol. 1943, 22:422).

Anonymous. 1948. Plant Diseases. Notes contributed by the Biological Branch. Agric. Gaz. N.S.W. 59:32–36, 46. (Rev. Appl. Mycol. 1948, 27:275.)

Anonymous. 1949a. Plant Diseases. Notes contributed by the Biological Branch. Agric. Gaz. N.S.W. 60:595–600. (Rev. Appl. Mycol. 1950, 29:496–497.)

Anonymous. 1949b. Twenty-third Annu. Rep. Dep. Sci. Ind. Res., New Zealand. 88 pp. (Rev. Appl. Mycol. 1951, 30:19–21.)

Anonymous. 1950. Plant Diseases. Agric. Gaz. N.S.W. 61:465–467. (Rev. Appl Mycol. 1951, 30:216.)

Anonymous. 1953a. Annu. Rep. For. Insect, Dis. Surv. Can. Dep. Agric. 1952. 154 pp. (Rev. Appl. Mycol. 1953, 32:518–521.)

Anonymous. 1953b. Orders issued by the Royal Danish Ministry of Agric. Copenhagen. 53 pp.

Anonymous. 1953c. Diseases of *Antirrhinum*. Tasman. J. Agric. 24: 366–369. (Rev. Appl. Mycol. 1955, 34:649.)

Anonymous. 1953d. Plant disease survey for the twelve months ending 30th June, 1953. Twenty-third Annu. Rep. N.S.W. Dep. Agric. 38 pp. (Rev. Appl. Mycol. 1955, 34:283–285.)

Anonymous. 1954. Twenty-seventh Annu. Rep., Dep. Sci. Ind. Res., New Zealand, 1953. 107 pp. (Rev. Appl. Mycol. 1955, 34:212–214.)

Anonymous. 1955a. Plant Pathol. Div. Res. Exp. Rec. Minist. Agric. North. Irel. 5:129–151. (Rev. Appl. Mycol. 1958, 37:3–4.)

Anonymous. 1955b. Rep. Minist. Agric. for Canada for the year ended March 31, 1955. 166 pp. (Rev. Appl. Mycol. 1956, 35:156–160.)

Anonymous. 1955c. Plant disease survey for the twelve months ending 30th June, 1954. Twenty-fourth Annu. Rep. N.S.W. Dep. Agric. Biol. Br. Div. Sci. Serv. Sydney, Australia. 37 pp. (Rev. Appl. Mycol. 1956, 35:2–4.)

Anonymous. 1957. New plant diseases. Agric. Gaz. N.S.W. 68:204.

Anonymous. 1958a. Diseases of the papaw. Agric. Gaz. N.S.W. 69: 534–538. (Rev. Appl. Mycol. 1959, 38:269.)

Anonymous. 1958b. Rep. Dep. Agric. N.S.W., for the year ended 30th June, 1957. 110 pp. (Rev. Appl. Mycol. 1959, 38:120–121.)

Anonymous. 1960. Index of Plant Diseases in the United States. U.S. Dep. Agric. Handbook. 165. 531 pp. (Rev. Appl. Mycol. 1961, 40:511–512.)

Anonymous. 1961. Quarterly Rep. Plant Prot. Comm. S.E. Asia and the Pacific. F.A.O.

Anonymous. 1962a. Rep. Dep. Agric., Mauritius, 1961. Pages 45–50. (Rev. Appl. Mycol. 1963, 42:516.)

Anonymous. 1962b. Cent. Potato Res. Inst., Simla (India). Agric. Res., New Delhi. 2:160–165. (Rev. Appl. Mycol. 1963, 42:698.)

Anonymous. 1962c. Rep. Victorian Plant Res. Inst. No. 3, July 1962–December 1965; No. 4, January 1966–1967. 99 pp. (Rev. Plant Pathol. 1972, 51:2127.)

Anonymous. 1963a. List of important diseases of economic plants in the Philippines including ornamentals. Proc. 4th Meeting Plant Prot. Comm. S.E. Asia.

Anonymous. 1963b. Resistencia a la enfermedad de la 'tinta' de algunas variedades de castaños gallegos (Resistance of some Galician varieties of chestnut to 'ink' disease). Ann. Inst. For. Invest. Exp. 35:189–197. (In Spanish.) (Rev. Appl. Mycol. 1966, 45: 1214.)

Anonymous. 1963c. Plant Pathology. Mauritius Dep. Agric. Rep. 1962:43–47.

Anonymous. 1964. Plant Pathology. Rep. Agric. Dep. Univ. Coll., Dublin. 1962–1963:45–46.

Anonymous. 1965a. Rep. Dep. Agric. Mauritius, 1963. Pages 59–63. (Rev. Appl. Mycol. 1965, 44:3232.)

Anonymous. 1965b. Pineapple root rot. Agric. Rev. 4:241. (In Chinese).

Anonymous. 1971a. Annu. Rep. Dep. Agric., Sabah (Malaysia), for the year 1969. 172 pp. (Rev. Plant Pathol. 1973, 52:3152.)

Anonymous. 1971b. Diseases of cultivated plants observed in Greece during 1962–1969. Ann. Inst. Phytopathol. Benaki, N.S. 10:70–118.

Anonymous. 1972. Records of new introduction, discovery or dissemination. Estimates of losses caused by insects and diseases. Q. Rep. F.A.O. Caribbean Plant Prot. Comm. 5:1–5.

Anonymous. 1974. Res. Rep. 1974. Hortic. Dublin. 121 pp.

Anonymous. 1977. New records of diseases and fungi. Rep. Bot. Dep. Victoria Univ. Wellington No. 10. 20 pp.

Anonymous. 1978a. Res. Rep. 1977. Horticulture, Dublin, Irish Republic. 118 pp. (Rev. Plant Pathol. 1978, 57:5260.)

Anonymous. 1978b. Annu. Rep. Inst. Bot. July 1977–June 1978. Taipei, Taiwan. Acad. Sin. 34 pp. (Rev. Plant Pathol. 1979, 58:22.)

Ansani, C. V. 1981. Sobrevivencia de *Phytophthora capsici* Leonian, no solo (Survival of *Phytophthora capsci* Leonian in soil). Thesis, Univ. F. Viçosa, Viçosa, M.G., Brasil. 36 pp. (In Portuguese).

Ansani, C. V., and Matsuoka, K. 1983a. Sobrevivencia de *Phytophthora capsici* no solo (Survival of *Phytophthora capsici* in soil). Fitopatol. Bras. 8:269–276. (In Portuguese). (Rev. Plant Pathol. 1984, 63:4205.)

Ansani, C. V., and Matsuoka, K. 1983b. Infectividade e viabilidade de oosporos de *Phytophthora capsici* no solo (Infectivity and viability of oospores of *Phytophthora capsici* in soil). Fitopatol. Bras. 8:137–146. (In Portuguese). (Rev. Plant Pathol. 1984, 63:2641.)

Ansani, C. V., and Matsuoka, K. 1983c. Efeito de densidade de zoosporos e idade de mudas de pimentão (*Capsicum annuum*) na infectividade de *Phytophthora capsici* (Effects of zoospore density and age of sweet pepper [*Capsicum annuum*] plants on the infectivity of *Phytophthora capsici*). Fitopatol. Bras. 8:263–268. (In Portuguese). (Rev. Plant Pathol. 1984, 63:4204.)

Ansani, C. V., and Matsuoka, K. 1987. Producao de oosporos de *Phytophthora capsici* em meio líquido (Oospore production by *Phytophthora capsici* in liquid medium). Fitopatol. Bras. 12:360–364. (In Portuguese). (Rev. Plant Pathol. 1988, 67:5357.)

Apinis, A. E. 1929. Untersuchungen über die in Lettland gefundended Saprolegniaceen, nebst Bemerkungen über andere Wasserpilze (Investigation about Saprolegniaceae that are found in clay soil in addition to an observation about other water molds). Acta Hortic. Botanici Universitatis Latviensis 4:201–246.

Apple, J. L. 1957. Pathogenic, cultural and physiological variation within *Phytophthora parasitica* var. *nicotianae*. Phytopathology 47:733–739.

Apple, J. L. 1961. The development of black shank in tobacco as influenced by host nutrition. Phytopathology 51:386–389.

Apple, J. L. 1962. Physiological specialization with *Phytophthora parasitica* var. *nicotianae*. Phytopathology 52:351–354.

Apple, J. L. 1963. Persistence of *Phytophthora parasitica* var. *nicotianae* in soil. Plant Dis. Rep. 47:632–634.

Apple, J. L. 1964. Changes in the black shank disease reaction of susceptible tobacco (*Nicotiana tabacum*) induced by gamma irradiation. Phytopathology 54:23–28.

Aragaki, M. and Uchida, J. Y. 1980. Foliar stage of Phytophthora blight of macadamia. Plant Dis. 64:483–484.

Aragaki, M., and Uchida, J. Y. 1992. Recent findings in *Phytophthora capsici* (*P. tropicalis* complex) (Abstr.). Phytopathology 82: 1164.

Aragaki, M., Kimoto, W. S., and Uchida, J. Y. 1981. Limitations of hot water treatment in the control of Phytophthora fruit rot of papaya. Plant Dis. 65:744–745.

Arentz, F., and Simpson, J. A. 1986. Distribution of *Phytophthora cinnamomi* in Papua, New Guinea and notes on its origins. Trans. Br. Mycol. Soc. 87:289–295.

Ark, P. A., and DeWolfe, T. A. 1951. Phytophthora rot of *Peperomia*. Plant Dis. Rep. 35:46–47.

Ark, P. A., and Dickey, R. S. 1950. A modification of the van Tiegham cell for purification of contaminated fungus cultures. Phytopathology 40:389–390.

Arnaud, G. 1915. Le mildiou des cereales (*Sclerospora macrospora* Sacc.) en France (Mildew of cereals [*Sclerospora macrospora* Sacc.] in France). Bull. Soc. Pathol. Veg. Fr. 2:105:111. (In French).

Arnaud, G. 1918a. Le mildiou des lilas et la maladie des cotylédons d'erable (The mildew of lilac and the disease of maple). Bull. Soc. Pathol. Veg. Fr. 5:58–60. (In French). (Cited in Tucker 1933).

Arnaud, G. 1918b. Maladies nouvelles ou peu connues en France, Ser. II. Le mildiou (*Phytophthora syringae* Kleb) et maladies diverses du lilas (New or little known diseases in France, Ser. II. The mildew [*Phytophthora syringae* Kleb.] and different diseases of lilac). Ann. Serv. Epiphyties 6:214–227. (In French).

Arnaud, G., and Arnaud, M. 1930. La pourriture des fraises et les Phytophthorées (Strawberry rot and species of *Phytophthora*). Rev. Pathol. Vég. Entomol. Agric. 17:349–351. (In French). (Rev. Appl. Mycol. 1931, 10:42.) (Cited in Tucker 1933).

Arnaud, G., and Barthelet, J. 1929. Le mildiou des poires en France (*Phytophthora* sp.) (Mildew of pears in France [*Phytophthora* sp.]. Rev. Pathol. Vég. Entomol. Agric. 16:303–308. (In French). (Rev. Appl. Mycol. 1930, 9:392–393.) (Cited in Tucker 1933).

Arnaudi, C. 1928. Nouve esperienze sulla vaccinazione delle piante (New experiments on the vaccination of plants). Riv. Patol. Veg. 18:161–168. (In Italian). (Cited in Tucker 1933).

Arruda, S. C. 1943. Observacoes sobre algumas doencas do eucalipto no estado de S. Paulo (Observations on some eucalyptus diseases in the state of Sao Paulo). Biologico 9:140–144. (In Portuguese). (Rev. Appl. Mycol. 1943, 22:505.)

Arya, H. C. 1959. A note on the pathology of an isolate of *Phytophthora palmivora* (Butler) Butler on *Bougainvillea spectabilis*. Sci. Cult. 25:156–158.

Ashby, S. F. 1920. Notes on two diseases of the coconut palm in Jamaica caused by fungi of the genus *Phytophthora*. West Indian Bull. 18:61–73. (Cited in Tucker 1933).

Ashby, S. F. 1922. Oospores in cultures of *Phytophthora faberi*. Kew Bull. Misc. Inf. 9:257–262. (Cited in Tucker 1933).

Ashby, S. F. 1925. Researches on Panama disease. Proc. 9th West Indian Agric. Conf. 1924:51–53. (Cited in Tucker 1933).

Ashby, S. F. 1928. The oospores of *Phytophthora nicotianae* Br. de Haan, with notes on the taxonomy of *P. parasitica* Dastur. Trans. Br. Mycol. Soc. 13:86–95. (Cited in Tucker 1933).

Ashby, S. F. 1929a. Strains and taxonomy of *Phytophthora palmivora* Butler (*P. faberi* Maubl.). Trans Br. Mycol. Soc. 14:18–38. (Cited in Tucker 1933).

Ashby, S. F. 1929b. Further note on the production of sexual organs in paired cultures of species and strains of *Phytophthora*. Trans. Br. Mycol. Soc. 14:254–260. (Cited in Tucker 1933).

Ashby, S. F. 1929c. The production of sexual organs in pure cultures of *Phytophthora cinnamomi* Rands. and *Blepharospora cambivora* Petri. Trans. Br. Mycol. Soc. 14:260–263. (Cited in Tucker 1933).

Aslyng, H. C. 1963. Soil physics terminology. Int. Soc. Soil Sci. Bull. 23:1–4.

Asuyama, H. 1935. New diseases and pathogens reported in the year of 1934 on our cultivated plants in Japan. Ann. Phytopathol. Soc. Jpn. 4:191–197. (In Japanese).

Atanasoff, D., and Kovacevski, I. C. 1929. Nova parazitri gubiza Bulgaria (Parasitic fungi new for Bulgaria). Izv. Bulg. Bot. Druzh. 3:45–52. (In Bulgarian). (Rev. Appl. Mycol. 1930, 9:203.)

Athow, K. L. 1987. Fungal diseases. Pages 687–727 in: Soybeans: Improvement, Production and Uses. J. R. Wilcox, H. R. Boerma, E. J. Kamprath and L. E. Schrader, eds. Agron. Monogr. No. 16. ASA-CSSA-SSSA, Madison, Wisconsin. 888 pp.

Athow, K. L., and Laviolette, F. A. 1982. Rps_6, a major gene for resistance to *Phytophthora megasperma* f. sp. *glycinea* in soybean. Phytopathology 72:1564–1567.

Athow, K. L., and Laviolette, F. A. 1983. Breeding soybeans for race specific *Phytophthora* resistance. Proc. Thirteenth Soybean Res. Conf. 12–21.

Athow, K. L., Laviolette, F. A., and Wilcox, J. R. 1979. Genetics of resistance to physiologic races of *Phytophthora megasperma* var. *sojae* in the soybean cultivar Tracy. Phytopathology 69:641–642.

Athow, K. L., Laviolette, F. A., Mueller, E. H., and Wilcox, J. R. 1980. A new major gene for resistance to *Phytophthora megasperma* var. *sojae* in soybean. Phytopathology 70:977–980.

Atkinson, R. G. 1965. *Phytophthora* species inciting root rot of *Chamaecyparis lawsoniana* and other ornamentals in coastal British Columbia. Can. J. Bot. 43:1471–1475.

Auld, B. A. 1990. Biocontrol of weeds using insects and fungi. Plant Prot. Q. 5:60–61.

Aveling, T. A. S., and Rijkenberg, F. H. J. 1989. Behaviour of *Phytophthora cinnamomi* zoospores on roots of four avocado cultivars. J. Phytopathol. 125:157–164.

Aveling, T. A. S., and Rijkenberg, F. H. J. 1991. A quantitative study of *Phytophthora cinnamomi* zoospore encystment and germination on the roots of four avocado cultivars. Phytophylactica 23:229–231.

Averna-Saccá, R. 1912. Informe de la division de patologia vegetal (Information from the division of plant pathology). Bol. Agrr. Sao Paulo, Brasil. 13(Ser. 3):208–247. (In Portuguese). (Cited in Tucker 1933).

Averre, C. W. and Reynolds, J. E. 1964. Phytophthora root and stem rot of Aloe. Proc. Fla. State Hortic. Soc. 77:438–440.

Awan, A. B., and Struchtemeyer, R. A. 1957. The effect of fertilization on the susceptibility of potatoes to late blight. Am. Potato J. 34:315–319.

Ayers, A. R., Ebel, J., Finelli, F., Berger, N., and Albersheim, P. 1976a. Host-pathogen interactions. IX. Quantitative assays of elicitor activity and characterization of the elicitor present in the extracellular medium of cultures of *Phytophthora megasperma* var. *sojae*. Plant Physiol. 57:751–759.

Ayers, A. R., Ebel, J., Valent, B., and Albersheim, P. 1976b. Host-pathogen interactions. X. Fractionation and biological activity of an elicitor isolated from the mycelial walls of *Phytophthora megasperma* var. *sojae*. Plant Physiol. 57:760–765.

Ayers, A. R., Valent, B., Ebel, J., and Albersheim, P. 1976c. Host-pathogen interactions. XI. Composition and structure of wall-released elicitor fractions. Plant Physiol. 57:766.

Ayers, W. A. 1971. Induction of sporangia in *Phytophthora cinnamomi* by a substance from bacteria and soil. Can. J. Microbiol. 17:1517–1523.

Ayers, W. A., and Lumsden, R. D. 1975. Factors affecting production and germination of oospores of three *Pythium* species. Phytopathology 65:1094–1100.

Ayers, W. A., and Zentmyer, G. A. 1971. Effect of soil solution and two soil pseudomonads on sporangium production by *Phytophthora cinnamomi*. Phytopathology 61:1188–1193.

Babacauh, K. D. 1972. Interactions entre le *Phytophthora palmivora* (Butl.) Butl. et les plantes supérieures. II. Variabilité du pouvoir pathogène (Interaction between *Phytophthora palmivora* (Butl.) Butl. and higher plants. II. Variability of the pathogenic capacity). Cah. ORSTOM Ser. Biol. 16:17–29. (In French).

Babacauh, K. D. 1982. Rôle des communités d'insectes et de l'eau dans la dissémination de *Phytophthora palmivora* (Butl.) Butl. emend. Bras. et Griff. dans les cacaoyères de la Côte d'Ivoire (Role of communities of insects and water on the dissemination of *Phytophthora palmivora* (Butl.) Butl. emend Bras. and Griff. on cacao plantations in the Ivory Coast). Café Cacao Thé (Paris) 26: 31–36. (In French).

Babacauh, K. D. 1983a. Structure des populations de *Phytophthora palmivora* (Butl.) Butl. emend. Bras. et Griff. parasite du cacaoyer (*Theobroma cacao* L.) (Population structure of *Phytophthora palmivora* from cacao [*Theobroma cacao* L.]) Bull. Soc. Bot. Fr. Lett. Bot. 130:15–25. (In French).

Babacauh, K. D. 1983b. Autre modalite de l'amphigynie in *Phytophthora* de Bary (Another mode of amphigyny in *Phytophthora* de Bary). Bull. Soc. Bot. Lett. Bot. 130:131–136. (In French).

Bahnweg, G. 1978. Physiology of Peronosporales (*Phytophthora* spp. and *Pythium* sp.) associated with the decay of leaf litter in marine environment. Veroeff. Inst. Meeresforsch. Bremerhaven 17:33–44.

Bailey, A. M., and Coffey, M. D. 1984. A sensitive bioassay for quantification of metalaxyl in soils. Phytopathology 74:667–669.

Bailey, A. M., and Coffey, M. D. 1985. Biodegradation of metalaxyl in avocado soils. Phytopathology 75:135–137.

Bailey, A. M., and Coffey, M. D. 1986. Characterization of microorganisms involved in accelerated biodegradation of metalaxyl and metalochlor in soils. Can. J. Microbiol. 32:562–569.

Bailey, A. M., Mena, G. L. and Herrera-Estrella, L. 1991. Genetic transformation of the plant pathogens *Phytophthora capsici* and *Phytophthora parasitica.* Nucleic Acids Res. 19:4273–4278.

Bain, H. F., and Demaree, J. B. 1938. Isolation of the fungus causing the red stele or red core disease of strawberries. Science N.S. 88: 151–152.

Bain, H. F., and Demaree, J. B. 1945. Red stele root disease of the strawberry caused by *Phytophthora fragariae.* J. Agric. Res. 70: 11–30.

Baines, R. C. 1935. Phytophthora trunk canker of apple. Phytopathology 25:5.

Baines, R. C. 1939. Phytophthora trunk canker or collar rot of apple trees. J. Agric. Res. 59:159–184.

Bains, S. S., and Shooty, J. S. 1985. Seed transmission of *Sclerophthora macrospora* in wheat. Seed Res. 13:154–156.

Baker, J. J. 1976. Tech. Bull. Ministry Agric. Fish. Fd. No. 25. H.M.S.O., London. (Cited in Griffin and Jones, 1977.)

Baker, K. F. 1978. Biological control of *Phytophthora cinnamomi.* Pages 72–79 in: International Plant Propagators Soc. Comb. Proc. Vol. 28 for 1978.

Baker, K. F., ed. 1957. The U.C. System for producing healthy container-grown plants. Calif. Agric. Exp. Stn. Man. 23. 332 pp.

Baker, K. F., and Cook, R. J. 1974. Biological Control of Plant Pathogens. W. H. Freeman and Co., San Francisco. 433 pp.

Baker, K. F., and Roistacher, C. N. 1957. Heat treatment of soil. Pages 123–137 in: The U.C. System for Producing Healthy Container-Grown Plants, K. F. Baker, ed. Calif. Agric. Exp. Stn. Man. 23. 332 pp.

Baker, K. F., Roistacher, C. N., and Chandler, P. A. 1957. Treatment of nursery containers. Pages 210–216 in: The U.C. System for Producing Healthy Container-Grown Plants. K. F. Baker, ed. Calif. Agric. Exp. Stn. Ext. Serv. Man. 23. 332 pp.

Baker, R. 1968. Mechanisms of biological control of soil-borne pathogens. Annu. Rev. Phytopatol. 6:263–294.

Baker, R. 1987. Mycoparasitism: Ecology and physiology. Can. J. Plant Pathol. 9:370–379.

Baker, R. 1990. An overview of current and future strategies and models for biological control. Pages 375–388 in: Biological Control of Soilborne Plant Pathogens. D. Hornby, ed. CAB International, London. 479 pp.

Baker, R., and Dunn, P. E., eds. 1990. New Directions in Biological Control. Alan R. Liss, New York. 837 pp.

Baker, R. E. D. 1934. Gummosis of citrus in Trinidad. I. Marsh grapefruit on sour orangestock. Trop. Agric. (Trinidad) 11:236–239.

Baker, R. E. D. 1935. Citrus fruit-rots in Trinidad. Trop. Agric. (Trinidad) 12:145–152.

Baker, R. E. D. 1936. Notes on Trinidad fungi. I. *Phytophthora.* Trop. Agric. Trin. 13:330–332.

Balakrishnan, M. S. 1947a. *Phytophthora palmivora* Bulter causing a seedling blight of *Hibiscus esculentus* L. Proc. Indian Acad. Sci. Sect. B 26:142–146.

Balakrishnan, M. S. 1947b. *Phytophthora palmivora* Butler on *Cyphomandra betacea* Lendt. and *Carica papaya* Linn. Curr. Sci. 16:146–147.

Banihashemi, Z. 1970. A new technique for isolation of *Phytophthora* and *Pythium* species from soil. Plant Dis. Rep. 54:261–262.

Banihashemi, Z., and Mitchell, J. E. 1975. Use of safflower seedlings for the detection and isolation of *Phytophthora cactorum* from soil and its application to population studies. Phytopathology 65:1424–1430.

Banihashemi, Z., and Mitchell, J. E. 1976. Factors affecting oospore germination in *Phytophthora cactorum,* the incitant of apple collar rot. Phytopathology 66:443–448.

Bar-Joseph, M., Roistacher, C. N., Garnsey, S. M., and Gumpf, D. J. 1981. A review on tristeza, an ongoing threat to citriculture. Proc. Int. Soc. Citric. 1:419–423.

Barash, I., Klisiewicz, J. M., and Kosuge, T. 1965. Utilization of carbon compounds by zoospores of *Phytophthora drechsleri* and their effect on motility and germination. Phytopathology 55:1257–1261.

Barat, H. 1952. Étude sur le dépérissement des Poivrières en Indochine (A study on the decline of the pepper plantings in Indo-China). Arch. Rech. Agron. Cambodge, Laos, Vietnam No. 13, 92 pp. (In French). (Rev. Appl. Mycol. 1953, 32:509.)

Barbercheck, M. E., and Von Broembsen, S. L. 1986. Effects of soil solarization on plant-parasitic nematodes and *Phytophthora cinnamomi* in South Africa. Plant Dis. 70:945–950.

Barbetti, M. J. 1989. Response of *Medicago* cultivars to fungal root pathogens associated with *Trifolium subterraneum.* Plant Prot. Q. 4:75–77.

Barchietto, T., Saindrenan, P., and Bompeix, G. 1988. Uptake and utilization of phosphonate ions by *Phytophthora citrophthora* and *Nectria haematococca* in relation to their selective toxicity. Pestic. Sci. 22:159–167.

Barchietto, T., Saindrenan, P., and Bompeix, G. 1989. Characterization of phosphonate uptake in two *Phytophthora* spp. and its inhibition by phosphate. Arch. Microbiol. 151:54–58.

Barksdale, T. H., Papavizas, G. C., and Johnston, S. A. 1984. Resistance to foliar blight and crown rot of pepper caused by *Phytophthora capsici.* Plant Dis. 68:506–509.

Barnes, B., and Melville, R. 1932. Notes on British aquatic fungi. Trans. Br. Mycol. Soc. 17:82–96.

Barooah, S. R. 1955. Annual report of the grow more food campaign of the Department of Agriculture. Assam. 1953–1954. Part II. 155 pp. (Rev. Appl. Mycol. 1956, 35:422–423.)

Barr, D. J. S. 1980. Heterothallic-like reaction in the large-oospore form of *Phytophthora megasperma.* Can J. Plant Pathol. 2:116–118.

Barr, D. J. S. 1981. The phylogenetic and taxonomic implications of flagellar rootlet morphology among zoosporic fungi. BioSystems 14:359–370.

Barr, D. J. S. 1982a. Eumycota. Page 178 in: Synopsis and Classification of Living Organisms. Vol. I. S. P. Parker, ed. McGraw-Hill Book Co., New York. 1166 pp.

Barr, D. J. S. 1982b. Mastigomycotina. Page 178 in: Synopsis and Classification of Living Organisms. Vol. I. S. P. Parker, ed. McGraw-Hill Book Co., New York. 1166 pp.

Barr, D. J. S. 1983. The zoosporic grouping of plant pathogens—Entity or non-entity. Pages 43–83 in: Zoosporic Plant Pathogens, A Modern Perspective. S. T. Buczaki, ed. Academic Press. 352 pp.

Barr, D. J. S. 1992. Evolution and kingdoms of organisms from the perspective of a mycologist. Mycologia 84:1–11.

Barr, D. J. S., and Allan, P. M. E. 1985. A comparison of the flagellar apparatus in *Phytophthora, Saprolegnia, Thraustochytrium* and *Rhizidiomyces.* Can. J. Bot. 63:138–154.

Barr, D. J. S., and Désaulniers, N. L. 1989. The flagellar apparatus in zoospores of *Phytophthora sojae* f. sp. *glycinea* and *P. megasperma.* Can. J. Bot. 67:1916–1926.

Barr, D. J. S., and Désaulniers, N. L. 1990. The flagellar apparatus in *Phytophthora infestans* zoospores. Can J. Bot. 68:2112–2118.

Barr, D. J. S., and Désaulniers, N. L. 1992. The flagellar apparatus in zoospores of *Phytophthora, Pythium,* and *Halophytophthora.* Can. J. Bot. 70:2163–2169.

Barrett, J. T. 1917. *Pythiacystis* related to *Phytophthora.* Phytopathology 7:150–151.

Barrett, J. T. 1928. *Phytophthora* in relation to crown rot of walnut (Abstr.). Phytopathology 18:948–949.

Barrett, J. T., and H. S. Fawcett. 1922. Withertip, tear-stain and control of brown rot. Calif. Citrogr. 7:232–233, 254.

Barrett, J. T., and Smith, C. O. 1926. Walnut tree crown rot; its development and suggested treatment. Diamond Walnut News 8:4–5,7.

Barritt, B. H., Crandall, P. C., and Bristow, P. R. 1979. Breeding for root rot resistance in red raspberry. J. Am. Soc. Hortic. Sci. 104: 92–94.

Barritt, B. H., Covey, R. P., and Dilley, M. A. 1990. In vitro testing of the reaction of apple rootstocks to *Phytophthora cactorum.* Fruit Var. J. 44:23–25.

Barta, A. L., and Schmitthenner, A. F. 1986. Interaction between flooding stress and Phytophthora root rot among alfalfa cultivars. Plant Dis. 70:310–313.

Barthelet, J. 1934. Sur une maladie des rhododendrons (On a disease of rhododendrons). Rev. Pathol. Veg. 21:31–35. (In French). (Rev. Appl. Mycol. 1935, 14:314.)

Bartnicki-Garcia, S. 1966. Chemistry of hyphal walls of *Phytophthora.* J. Gen. Microbiol. 42:57–69.

Bartnicki-Garcia, S. 1969. Cell wall differentiation in the Phycomycetes. Phytopathology 59:1065–1071.

Bartnicki-Garcia, S. 1970. Cell wall composition and other biochemical markers in fungus phylogeny. Pages 81–102 in: Phytochemical Phylogeny. J. B. Harborne, ed. Academic Press, London. 335 pp.

Bartnicki-Garcia, S. 1973. Cell wall genesis in a natural protoplast: The zoospore of *Phytophthora palmivora.* Pages 77–91 in: Yeast, Mould, and Plant Protoplasts. J. R. Villanueva, I. García-Acha, S. Gascón, and F. Uruburu, eds. Academic Press, London. 381 pp.

Bartnicki-Garcia, S. 1990. Role of vesicles in apical growth and a new mathematical model of hyphal morphogenesis. Pages 211–232 in: Tip Growth in Fungal Cells. I. B. Heath, ed. Academic Press, San Diego. 351 pp.

Bartnicki-Garcia, S., and Hemmes, D. E. 1976. Some aspects of the form and function of oomycete spores. Pages 593–641 in: The Fungal Spore: Form and Function. D. J. Weber and W. M. Hess, eds. John Wiley & Sons, New York. 895 pp.

Bartnicki-Garcia, S., and Lippmann, E. 1966. Liberation of protoplasts from the mycelium of *Phytophthora.* J. Gen. Microbiol. 42: 411–416.

Bartnicki-Garcia, S., and Wang, M. C. 1983. Biochemical aspects of morphogenesis in *Phytophthora.* Pages 121–137 in: *Phytophthora:* Its Biology, Taxonomy, Ecology, and Pathology. D. C. Erwin, S. Bartnicki-Garcia, and P. H. Tsao, eds. American Phytopathological Society, St. Paul, Minn. 392 pp.

Bartual, R., Marsal, J. I., Carbonell, E. A., Tello, J. C., and Campos, T. 1991. Genética de la resistencia a *Phytophthora capsici* Léon. en pimiento (Genetics of resistance in pepper to *Phytophthora capsici* Leon.). Bol. San. Veg. Plagas 17:3–124. (In Spanish).

Bashan, B., Kadish, D., Levy, Y., and Cohen, Y. 1989. Infectivity to potato, sporangial germination, and respiration of isolates of *Phytophthora infestans* from metalaxyl-sensitive and metalaxyl-resistant populations. Phytopathology 79:832–836.

Bashan, B., Levy, Y., and Cohen, Y. 1990. Variation in sensitivity of *Phytophthora infestans* to fosetyl-Al. Plant Pathol. 39:134–140.

Bashan, B., Levy, Y., and Cohen, Y. 1991. Synergistic interactions between Phenylamide fungicides and respiration inhibitors in controlling phenylamide-resistant populations of *Phytophthora infestans* on potato tuber disks. Phytoparasitica 1991, 19:1.

Bastiaansen, M. G., Pieroh, E. A., and Aelbers, E. 1974. Prothiocarb, a new fungicide to control *Phytophthora fragariae* in strawberries and *Pythium ultimum* in flower bulbs. Meded. Fac. Landbouwwet. Rijksuniv. Genet. 39:1019–1025.

Basu, P. K. 1980. Production of chlamydospores of *Phytophthora megasperma* and their possible role in primary infection and survival in soil. Can. J. Plant Pathol. 2:70–75.

Basu, P. K. 1988. Relationship between Phytophthora root rot severity index and the percentage of resistant alfalfa plants. Can. Plant Dis. Surv. 68:23–26.

Bateman, K. S., Hinch, J. M., Ralton, J. E., Clarke, A. E., McKenzie, J. A., Imrie, B. C., and Howlett, B. J. 1989. Inheritance of resistance of cowpea to *Phytophthora vignae* in whole plants, cuttings, and stem callus cultures. Austr. J. Bot. 37:511–517.

Bates, G. R. 1959. Branch of Botany, Plant Pathology, and Seed Testing. Rep. Minist. Agric. Rhod. Nyasald 1957–1958:26–33. (Rev. Appl. Mycol. 39:210).

Bates, G. R. 1961. Branch of Botany, Plant Pathology, and Seed testing. Rep. Minist. Agr. Rhod. Nyasald 1959–1960:50–57.

Baudys, E. 1930. Hniti ovoce ve Skladkach (Fruit rots in storage). Cesky Odbor. Zemedelske Rady Moravske. Brno. Leafl. 26. 2 pp. (In Bohemian). (Rev. Appl. Mycol. 1931, 10:253.)

Baudys, E. 1935. Nejdulezitejsi choroby a skudci Jirin (The more important diseases and pests of *Dahlia*). Publ. Phytopathol. Sec. Reg. Agric. Exp. Stn. Brno. 149. 7 pp. (In Bohemian). (Rev. Appl. Mycol. 1936, 15:369.)

Baumert, D., and Buschhaus, H. 1982. Cyprofuram, a new fungicide for the control of Phycomycetes. Meded. Fac. Landbouwwet. Rijksuniv. Genet. 47:979–983.

Bazán de Segura, C. 1951. Plant diseases new to Peru. Plant Dis. Rep. 35:465–466.

Bazán de Segura, C. 1958. La marchitez o 'wilt' del Aji en el Peru (*Capsicum* wilt in Peru). Inf. Estac. Agric. La Molina 105. 19 pp. (In Spanish). (Rev. Appl. Mycol. 1959, 38:52.)

Bazán de Segura, C. 1970. Ataque de *Phytophthora palmivora* a macadamia en Costa Rica (Attack of *P. palmivora* on *Macadamia* in Costa Rica). Turrialba 20:375–376. (In Spanish). (Rev. Appl. Mycol. 1972, 51:2901.)

Bazán de Segura, C., and Carrera, J. M. L. 1953. Experimento comparativo de fungicidas para el control del 'hielo' de la papaien Valle de Carabayllo (Comparative experiment with fungicides for the control of potato 'blight' in the Valley of Carabayllo). Inf. Estac. Exp. Agric., Lima, Peru. 81. 16 pp. (Rev. Appl. Mycol 1953, 33:622.)

Beach, B. G. W., Chalandon, A., Gallinelli, G., and Horriere, D. 1979. The control of various *Phytophthora* diseases in tropical crops with aluminum tris (ethyl phosphonate). Pages 319–329 in: Proc. Br. Crop Prot. Conf. Pests Dis. 19–22 Nov. 1979. Brighton, England. 891 pp.

Beach, W. S. 1922. The crown rot of rhubarb caused by *Phytophthora cactorum.* Penn. Agric. Exp. Stn. Bull. 174. 28 pp.

Beagle-Ristaino, J. E., and Rissler, J. F. 1983a. Effect of *Rhizobium japonicum* nodulation on severity of Phytophthora root rot of soybean. Plant Dis. 67:651–654.

Beagle-Ristaino, J. E., and Rissler, J. F. 1983b. Histopathology of susceptible and resistant soybean roots inoculated with zoospores of *Phytophthora megasperma* f. sp. *glycinea.* Phytopathology 73: 590–595.

Beakes, G. W. 1987. Oomycete phylogeny: Ultrastructural perspectives. Pages 405–421 in: Evolutionary Biology of the Fungi. A. D. M. Rayner, C. M. Brasier and D. Moore, eds. Cambridge Univ. Press, Cambridge, U.K. 465 pp.

Beakes, G. W., and Bartnicki-Garcia, S. 1989. Ultrastructure of mature oogonium-oospore wall complexes in *Phytophthora megasperma*—A comparison of *in vivo* and *in vitro* dissolution of the oospore wall. Mycol. Res. 93:321–334.

Beakes, G. W., El-Hamalawi, Z. A., and Erwin, D. C. 1986. Ultrastructure of mature oospores of *Phytophthora megasperma* f. sp. *medicaginis:* Preparation, protocols and effects of MTT vital staining and permanganate pre-treatment. Trans. Br. Mycol. Soc. 86:195–206.

Beaumont, A. 1950. Diseases of China asters. Gard. Chron. Ser. 3. 128:140–141. (Rev. Appl. Mycol. 1951, 30:42–43.)

Beaumont, A., and Hodson, W. E. H. 1931. Seventh Annu. Rep. Seale-Hayne Agric. Coll., Newton Abbot Devon, for the year ending September 1930. 36 pp. (Rev. Appl. Mycol. 1931, 10:434–435.)

Beaumont, A., and Staniland, L. N. 1937. 13th Annu. Rep. Dep. Plant Pathol., Seale-Hayne Agric. Coll., Devon, for the year ending September 1936. 35 pp. (Rev. Appl. Mycol. 1937, 16:514–515.)

Beaupré, C. M.-S., and Cheo, P. C. 1983. Phytophthora root rot disease and other pests of guayule culture. Pages 477–482 in: Proc. Third Int. Guayule Conf. E. C. Gregg, J. L. Tipton and H. T. Huong, eds. Calif., USA.

Belgrave, W. N. C., and Norris, F. de la M. 1917. Notes on bark cankers and their treatment. Fed. Malay States Agric. Bull. 6:2–10. (Cited in Tucker 1933).

Belisario, A. 1990. A new *Phytophthora* disease of the aerial parts of *Eucalyptus* species. EPPO Bull. 20:129–132.

Belisario, A., Magnano di San Lio, G., and Pane, A. 1991. *Phytophthora nicotianae* van B. de Haan agente di un Marciume delle radici e del colletto del Mirto (*Phytophthora nicotianae* van B. de Haan the causal organism of a root and collar rot of myrtle). Micol. Ital. 20:108–114. (In Italian).

Belisario, A., Magnano di San Lio, G., and Cacciola, S. O. 1993.

Phytophthora iranica, a new root pathogen of myrtle from Italy. Plant Dis. 77:1050–1055.

Belkhiri, A., and Dick, M. W. 1988. Comparative studies on the DNA of *Pythium* species and some possibly related taxa. J. Gen. Microbiol. 134:2673–2683.

Bell, D. K., and Haasis, F. A. 1967. Etiology and epiphytology of root rot, stem necrosis, and foliage blight of boxwood caused by *Phytophthora parasitica* Dastur. N. Carolina Agr. Exp. Sta. Bull. 177:1–47.

Bell, F. H., and Alandia, B. S. 1957. Diseases of temperate climate crops in Bolivia. Plant Dis. Rep. 41:646–649.

Bellefroid, V. 1928. Note sur la culture du cacao dans les terres rouges de Lukolela (Note on the culture of cacao on the red soil of Lukolela). Bull. Agric. Congo Belge 19:3–58. (In French).

Benhamou, N., and Côte, F. 1992. Ultrastructure and cytochemistry of pectin and cellulose degradation in tobacco roots infected by *Phytophthora parasitica* var. *nicotianae.* Phytopathology 82:468–478.

Bennett, C. P. A., Roboth, O., Sitepu, G., and Lolong, A. 1986. Pathogenicity of *Phytophthora palmivora* (Butler) Butler causing premature nutfall disease of coconut (*Cocos nucifera* L.). Indones. J. Crop Sci. 2:59–70.

Benson, D. M. 1978. Thermal inactivation of *Phytophthora cinnamomi* for control of Fraser fir root rot. Phytopathology 68:1373–1376.

Benson, D. M. 1979. Efficacy and in vitro activity of two systemic acylalanines and ethazole for control of *Phytophthora cinnamomi* root rot of azalea. Phytopathology 69:174–178.

Benson, D. M. 1980. Chemical control of rhododendron dieback caused by *Phytophthora heveae.* Plant Dis. 64:684–686.

Benson, D. M. 1982. Cold inactivation of *Phytophthora cinnamomi.* Phytopathology 72:560–563.

Benson, D. M. 1984. Influence of pine bark, matric potential, and pH on sporangium production by *Phytophthora cinnamomi.* Phytopathology 74:1359–1363.

Benson, D. M. 1985. Fungicides for control of Phytophthora root rot of azalea in landscape beds. Plant Dis. 69:697–699.

Benson, D. M. 1990. Landscape survival of fungicide-treated azaleas inoculated with *Phytophthora cinnamomi.* Plant Dis. 74:635–637.

Benson, D. M. 1991. Detection of *Phytophthora cinnamomi* in azalea with commercial serological assay kits. Plant Dis. 75:478–482.

Benson, D. M. 1993. Suppression of *Phytophthora parasitica* on *Catharanthus roseus* with aluminum. Phytopathology 83:1303–1308.

Benson, D. M., and Blazich, F. A. 1989. Control of Phytophthora root rot of *Rhododendron chapmanii.* J. Environ. Hort. 7:73–75.

Benson, D. M., and Campbell, C. L. 1985. Spatial pattern of Phytophthora root rot of azalea in container-grown nursery stock. Plant Dis. 69:1049–1054.

Benson, D. M., and Cochran, F. D. 1980. Resistance of evergreen hybrid azaleas to root rot caused by *Phytophthora cinnamomi.* Plant Dis. 64:214–215.

Benson, D. M., and Jones, R. K. 1980. Etiology of rhododendron dieback caused by four species of *Phytophthora.* Plant Dis. 64:687–691.

Benson, D. M., Jones, R. K., and Barker, K. R. 1982. Disease loss assessment for azalea, rhododendron, and Japanese holly in North Carolina nurseries. Plant Dis. 66:125–128.

Benson, D. M., Shew, H. D. and Jones, R. K. 1982. Effects of raised and ground separated level beds and pine bark on survival of azalea and population dynamics of *Phytophthora cinnamomi.* Can. J. Plant Pathol. 4:278–280.

Berg, A. 1926. Tomato late blight and its relation to late blight of potato. West Va. Agric. Exp. Stn. Bull. 205. 31 pp.

Berg, L. A. 1966. Influence of light and other factors on oospore germination in species of *Phytophthora.* Ph.D. thesis, West Va. Univ., Morgantown. 107 pp.

Berg, L. A., and Gallegly, M. E. 1966. Effect of light on oospore germination in species of *Phytophthora* (Abstr.). Phytopathology 56:583.

Bergamaschi, P., Borsari, T., Garaveglia, C., and Mirenna, L. 1981. Methyl N-phenyl-acetyl-N-2,6xylyl-dl-alaninate (M98834), a new systemic fungicide controlling downy mildew and other diseases caused by Peronosporales. Proc. 11th. Br. Crop Prot. Conf. Pages 11–18.

Berger, R. D. 1981. Comparison of the Gompertz and logistic equations to describe soil-borne pathogens. Phytopathology 71:716–719.

Berggren, B., Widmark, A. K., and Umaerus, V. 1988. The expression of general resistance to late blight (*Phytophthora infestans*) in potato leaves. Potato Res. 31:611–616.

Bergquist, R. R. 1974. Effect of fungicide rate, spray interval, timing of spray application and precipitation in relation to control of Phytophthora leaf blight of taro. Ann. Bot. 38:213–221.

Bergsma, 1845. De aardappel-epidemie in Nederland in den jare 1845, Utrecht (The potato epidemic in the Netherlands in the year 1845). (In Dutch). (Cited in Tucker 1933, 184:80.)

Berkeley, M. J. 1881. Lilac fungus. Gard. Chron. II. 16:665. (Cited in Tucker 1933).

Bernard, R. L., and Cremeens, C. R. 1981. An allele at the *rps* locus from the variety 'Kingwa.' Soybean Gen. Newsl. 8:40–42.

Bernard, R. L., Smith, P. E., Kaufmann, M. J., and Schmitthenner, A. F. 1957. Inheritance of resistance to Phytophthora root and stem rot in soybean. Agron. J. 49:391.

Bernhardt, E. A., and Grogan, R. G. 1982. Effect of soil matric potential on the formation and indirect germination of sporangia of *Phytophthora parasitica, P. capsici,* and *P. cryptogea.* Phytopathology 72:507–511.

Bertrand, A., Ducret, J., Debourge, J. C., Horriére, D. 1977. Étude des proprietés d'une nouvelle famille de fongicides: Les monethylphosphites metalliques. Charactestiques physichochimiques et propriétes biologiques (Study of the properties of a new family of fungicides: The metallic monoethyl phosphites. Physiochemical characteristics and biological properties). Phytiatr. Phytopharm. 26:3–18. (In French).

Bertus, L. S. 1946. Plant Pathology. Ceylon Admin. Rep. Dir. Agric. Ceylon, 1945. D9–D11. (Rev. Appl. Mycol. 1947, 26:530–531.)

Bewley, W. 1922. Tomato diseases. J. R. Hortic. Soc. 57:169–174.

Bewley, W. F. 1920. "Damping off" and "foot rot" of tomato seedlings. Ann. Appl. Biol. 7:156–172. (Cited in Tucker 1933).

Bewley, W. F. 1927. Mycological Investigations. Annu. Rep. Exp. Res. Stn., Cheshunt. Herts, England. 12:26.

Bewley, W. F. 1930. Mycological investigations. Exp. Res. Stn. Cheshunt. Herts, England. Annu. Rep. 1929, 15:7–37.

Bewley, W. F., and Buddin, W. 1921. On the fungus flora of glasshouse water supplies in relation to plant disease. Ann. Appl. Biol. 8:10–19.

Bewley, W. F., and Oyler, E. 1935. A disease of cultivated Heaths. Rep. Exp. Res. Stn. Cheshunt. Herts, England. 1934. Pages 55–60. (Rev. Appl. Mycol. 1935, 14:637.)

Bhardwaj, S. S., Sharma, S. L., and Tyagi, S. N. S. 1987. Perpetuation of *Phytophthora nicotianae* var. *nicotianae* in bell pepper seeds. Ind. J. Plant Pathol. 5:94.

Bhat, R. G. 1991. Genetics of virulence in *Phytophthora megasperma* f. sp. *glycinea.* Ph.D. dissertation, Dept. of Plant Pathology, Ohio State Univerity, Columbus. 199 pp.

Bhat, R. G., and Schmitthenner, A. F. 1992. Virulence evaluation of single-oospore cultures of various physiologic races of *Phytophthora sojae* (Abstr.). Phytopathology 82:1082.

Bhat, R. G., and Schmitthenner, A. F. 1993a. Selection and characterization of inhibitor-resistant mutants of *Phytophthora sojae.* Exp. Mycol. 17:109–121.

Bhat, R. G., and Schmitthenner, A. F. 1993b. Genetic crosses between physiologic races of *Phytophthora sojae.* Exp. Mycol. 17: 122–129.

Bhat, R. G., Schmitthenner, A. F., and McBlain, B. A. 1990. Genetic analysis of virulence in *Phytophthora megasperma* f. sp. *glycinea* (Abstr.). Phytopathology 80:968.

Bhat, R. G., Olah, A. F., and Schmitthenner, A. F. 1992. Characterization of universally avirulent strains of *Phytophthora sojae.* Can. J. Bot. 70:1175–1185.

Bhat, R. G., McBlain, B. A., and Schmitthenner, A. F. 1993a. The inheritance of resistance to metalaxyl and to fluorophenylalanine in matings of homothallic *Phytophthora sojae.* Mycol. Res. 97:865–870.

Bhat, R. G., McBlain, B. A., and Schmitthenner, A. F. 1993b. Development of pure lines of *Phytophthora sojae* races. Phytopathology 83:473–477.

Bhelwa, P. W. 1962. Seed decay, seedling blight, and root rot of *Cicer arietinum* caused by *Phytophthora cryptogea.* Diss. Abstr. 23:389–390. (Rev. Appl. Mycol. 1963, 42:287–288.)

Bielenin, A. 1977. Collar rot of apple trees: II. Seasonal fluctuations of susceptibility of apple trees to *Phytophthora cactorum* (Leb. & Cohn) Schroet. Fruit Sci. Rep. 4:27–39.

Bielenin, A., and Borecki, Z. 1970. Zgnilizna pierscieniouva podstawy pniadrzew owocowych powodowana przez grzyb *Phytophthora cactorum* (Leb. et Cohn) Schroeter (Collar rot of the trunk of fruit trees caused by *Phytophthora cactorum*). Acta Agrobot. 23:353–366. (In Polish).

Bielenin, A., and Jones, A. L. 1988a. Prevalence and pathogenicity of *Phytophthora* spp. from sour cherry trees in Michigan. Plant Dis. 72:473–476.

Bielenin, A., and Jones, A. L. 1988b. Efficacy of sprays of fosetyl-Al and drenches of metalaxyl for the control of Phytophthora root and crown rot of cherry. Plant Dis. 72:477–480.

Bielenin, A., Jeffers, S. N., Wilcox, W. F., and Jones, A. L. 1988. Separation by protein electrophoresis of six species of *Phytophthora* associated with deciduous fruit crops. Phytopathology 78: 1402–1408.

Billard, V., Bruneteau, M., Bonnet, Ph., Ricci, P., Pernollet, J. C., Huet, S. C., Vergne, A., Richard, G., and Michel, G. 1988. Chromographic purification and characterization of elicitors of necrosis on tobacco produced by incompatible *Phytophthora* species. J. Chromatogr. 44:87–94.

Bimpong, C. E. 1975. Changes in metabolic reserves and enzyme activities during zoospore motility and cyst germination in *Phytophthora palmivora.* Can. J. Bot. 53:1411–1416.

Bimpong, C. E., and Clerk, G. C. 1970. Motility and chemotaxis in zoospores of *Phytophthora palmivora* (Butl.) Butl. Ann. Bot. 34: 617–624.

Bimpong, C. E., and Hickman, C. J. 1975. Ultrastructural and cytochemical studies of zoospores, cysts and germinating cysts of *Phytophthora palmivora.* Can. J. Bot. 53:1310–1327.

Bingham, F. T., Zentmyer, G. A., and Martin, J. P. 1958. Host nutrition in relation to Phytophthora root rot of avocado seedlings. Phytopathology 48:144–148.

Bischt, V. S., and Nene, Y. L. 1988. A selective medium for *Phytophthora drechsleri* f. sp. *cajani* causing pigeon pea blight. Int. Pigeon Pea Newsl. 1988 No. 8:12–13.

Bjor, T., and Mulelid, K. 1991. Differential resistance to tuber late blight in potato cultivars without R-genes. Potato Res. 34:3–8.

Black, W. 1952. A genetical basis for the classification of strains of *Phytophthora infestans.* Proc. Roy. Soc. Edinb. B 65:36–51.

Black, W., Mastenbroek, C., Mills, W. R., and Peterson, L. C. 1953. A proposal for an international nomenclature of races of *Phytophthora infestans* and of genes controlling immunity in *Solanum demissum* derivatives. Euphytica 2:73–179.

Blackwell, E. 1943. The life history of *Phytophthora cactorum* (Leb. and Cohn) Schroet. Trans. Br. Mycol. Soc. 26:71–89.

Blackwell, E. 1949. Terminology in *Phytophthora.* Commonw. Mycol. Inst. Mycol. Pap. No. 30. 24 pp.

Blackwell, E. M., and Waterhouse, G. M. 1931. Spores and spore germination in the genus *Phytophthora.* Trans. Br. Mycol. Soc. 15: 294–310.

Blackwell, E. M., Waterhouse, G. M., and Thompson, M. V. 1941. The invalidity of the genus *Pythiomorpha.* Trans. Br. Mycol. Soc. 25:148–165.

Blaha, G. 1983. Effet de la lumière sur *Phytophthora palmivora* et *Phytophthora megakarya,* agents de la pourriture brune des cabosses du cacaoyer (Effect of light on *Phytophthora palmivora* and *Phytophthora megakarya,* agents causing brown rot on cocoa pods). Café Cacao Thé (Paris) 27:91–112. (In French).

Blaha, G. 1990. Profils isoenzymatiques et RFLP pour l'identification des *Phytophthora* du cacoyer (Use of isoenzyme patterns and RFLP to identify the *Phytophthora* spp. on cacao). EPPO (Eur. Mediterr. Plant Prot. Organ.) Bull. 20:59–65. (In French).

Blaha, G., and Lotodé, R. 1982. Étude comparée du pouvoir pathogène sur semenceaux de *Theobroma cacao* des espèces de *Phytophthora* parasites du cacaoyer. Analyse des éléments du pouvoir pathogène de *Phytophthora palmivora* et de *P. megakarya* (Comparative study on the pathogenicity of cocoa tree pest species of *Phytophthora* to seedlings of *Theobroma cacao.* Analysis of the factors involved in the pathogenicity of *Phytophthora palmivora* and *P. megakarya*). Café Cacao Thé (Paris) 26:237–272. (In French).

Blaker, N. S., and Hewitt, J. D. 1987a. A comparison of resistance to *Phytophthora parasitica* in tomato. Phytopathology 77:1113–1116.

Blaker, N. S., and Hewitt, J. D. 1987b. Comparison of seedling and mature plant resistance to *Phytophthora parasitica* in tomato. Hort. Sci. 22:103–105.

Blaker, N. S., and MacDonald, J. D. 1981. Predisposing effects of soil moisture extremes on the susceptibility of rhododendron to Phytophthora root and crown rot. Phytopathology 71:831–834.

Blaker, N. S., and MacDonald, J. D. 1986. The role of salinity in the development of Phytophthora root rot of citrus. Phytopathology 76:970–975.

Blakesley, R. W., and Boezi, J. A. 1977. A new staining technique for proteins in polyacrylamide gels using Coomassie brilliant blue 8250. Anal. Biochem. 82:580–582.

Blanchard, E. E. 1930. Principales insectos y enfermedades que perjudican los cultivos citricos en la Republica Argentina (Principal insects and diseases which damage the citrus cultivations in the Argentine Republic). Minist. Agric. Nac. (Buenos Aires) Secc. Prop. Inform. Circ. 815. 114 pp. (In Spanish). (Rev. Appl. Mycol. 1931, 10:24). (Cited in Tucker 1933).

Bloch, C. B., De Wit., P. J. G. M., and Kuć, J. 1984. Elicitation of phytoalexins by arachidonic and eicosapentaenoic acids: A host survey. Physiol. Plant Pathol. 25:199–208.

Blodgett, E. C. 1945. Water rot of potatoes. Plant Dis. Rep. 29:124–126.

Blondell, L. 1978. Botanical classification of species of the genus *Citrus.* Fruits D'Outre-Mer (Paris) 33:695–720.

Boccas, B. 1972. Contribution à l'étude du cycle chez les *Phytophthora* (Contribution to the study of the life cycle of *Phytophthora*). C. R. Acad. Sci. (Paris). 275D:663–666. (In French).

Boccas, B. 1976. La reproduction sexuée chez les *Phytophthora,* heterothalliques. Ses voies et quelques-unes de ses conséquences génétiques (The sexual reproduction in the heterothallic *Phytophthora,* its course and some of its genetic consequences) Cah. ORSTOM Ser. Biol. 11:129–133. (In French).

Boccas, B. R. 1981. Interspecific crosses between closely related heterothallic *Phytophthora* species. Phytopathology 71:60–65.

Boccas, B., and Laville, E. 1978. Les maladies a *Phytophthora* des agrumes (*Phytophthora* diseases of fruits). Inst. Rech. Fruits Agrumes (IRFA). 162 pp. (In French).

Boccas, B., and Pellegrin, F. 1976. Evaluation de la résistance de quelques varieties de roselle au *Phytophthora parasitica* Dast. (Evaluation of the resistance of some varieties of roselle to *Phytophthora parasitica* Dast.). Coton Fibres Trop. 31:231–234. (In French). (Rev. Plant Pathol. 1977, 56:5070.)

Boccas, B., and Zentmyer, G. A. 1976. Genetical studies with interspecific crosses between *Phytophthora cinnamomi* and *Phytophthora parasitica.* Phytopathology 66:477–484.

Boesewinkel, H. J. 1974. *Phytophthora* on asparagus in New Zealand. Plant Dis. Rep. 58:525–529.

Boesewinkel, H. J. 1976. Storage of fungal cultures in water. Trans. Br. Mycol. Soc. 66:183–185.

Boesewinkel, H. J. 1977. New plant disease records in New Zealand: Records in the period 1969–76. N.Z. J. Agric. Res. 20:583–589.

Bohm, W. 1979. Methods of Studying Root Systems. Springer-Verlag, New York. 188 pp.

Bolay, A. 1962. Le Mildiou de l'abricotier (Mildew on apricot). Agric. Rom. 1:35–37. (In French). (Rev. Appl. Mycol. 1962, 41:608.)

Bolay, A. 1977. *Phytophthora syringae,* agent d'une grave pourriture des pommes en conservation (*Phytophthora syringae,* causal agent of a serious rot of stored apples). Rev. Suisse Vitic. Arboric. Hortic. 9:161–169. (In French). (Rev. Plant Pathol. 1978, 57:664.)

Bolkan, H. A. 1985. A technique to evaluate tomatoes for resistance to Phytophthora root rot in the greenhouse. Plant Dis. 69:708–709.

Bollen, G. J., and Fuchs, A. 1970. On the specificity of the *in vitro* and *in vivo* antifungal activity of benomyl. Neth. J. Plant Pathol. 76:299–312.

Bompeix, G., and Saindrenan., P. 1984. *In vitro* antifungal activity of fosetyl Al and phosphorous acid on *Phytophthora* species. Fruits 39:777–786.

Bompeix, G., Ravisé, A., Raynal, G., Fettouche, F., and Durand, M. C. 1980. Modalités del'obtention des necroses bloquantes sur feuilles detacheés de tomate par l'action du tris-O-ethyl phosphonate d'aluminium (phosethyl d'aluminium), hypothéses sur son mode d'action in vivo (Blocking necrosis formation in tomato leaves infected by *Phytophthora* spp. after treatment by aluminum tris-O-ethyl phosphonate *in vivo* [phosethyl of aluminum], hypotheses on the mode of action *in vivo*). Ann. Phytopathol. 12:337–351. (In French).

Bompeix, G., Fettouche, F., and Saindrenan, P. 1981. Mode d'action du phosethyl Al (Mode of action of fosetyl Al). Phytiar. Phytopharm. 30:257–72. (In French).

Bonavia, E. 1888. Oranges and Lemons of India and Ceylon. W. H. Allen Co., London. 334 pp.

Bondartzeva-Monteverde, V. N. 1926. *Phytophthora infestans* (Mont.) DeBy on tomatoes. Reprinted from Morbi Plantarum, Leningrad, 15. 27 pp. (In Russian). (Rev. Appl. Mycol. 1926, 5:703–704.) (Cited in Tucker 1933).

Bonde, R., and Schultz, E. S. 1943. Potato refuse piles as a factor in the dissemination of late blight. Maine Agric. Exp. Stn. Bull. 416: 230–246.

Bonde, R., and Schultz, E. S. 1944. Potato refuse piles and late blight epidemics. Maine Agric. Exp. Stn. Bull. 426:233–234.

Bonhoff, A., Rieth B., Golecki, J., and Griseback, H. 1987. Race cultivar-specific differences in callose deposition in soybean roots following infection with *Phytophthora megasperma* f. sp. *glycinea.* Planta 172:101–105.

Bonnet, P. 1988. Purification de divers filtrats de culture de *Phytophthora* et activités biologiques sur le tabac des différentes fractions (Purification of different filtrates of cultures of *Phytophthora* and biological activity of different fractions on tobacco). Agronomie 8:347–350. (In French).

Bonnet, P., Maia, N., Tello-Marquina, J., and Venard, P. 1978. Pouvoir pathogène du *Phytophthora parasitica* (Dastur): Facteurs de variabilité et notion de spécialisation parasitaire (Pathogenic capacity of *Phytophthora parasitica* [Dastur]: Factors on variability and concept of parasitic specialization). Ann. Phytopathol. 10:15–29. (In French).

Bonnet, P., Ricci, P, and Mercier, S. 1980. Le dépérissement du *Gerbera* causé par *Phytophthora cryptogea* Pethyb. et Laff. Diagnostic à partir du materiel végétal par isolement et determination rapide (The decline of *Gerbera* caused by *Phytophthora cryptogea* Pethyb. et Laff. Diagnosis from vegetative material by isolation and rapid determination). Ann. Phytopathol. 12:109–118. (In French).

Bootsma, A. 1979. Potato late blight forecasting in Prince Edward Island in 1978. Can. Plant Dis. Surv. 59:63–66.

Borget, M. 1949. Nouvelles recherches sur les champignons parasites de l'*Hevea* en Indochine (Further studies on the fungi parasitic on *Hevea* in Indo-China). Rev. Appl. Bot. 29:317–318. (In French). (Rev. Appl. Mycol. 1950, 29:229.)

Borner, H., Schatz, G., and Grisebach, H. 1983. Influence of the systemic fungicide metalaxyl on glyceollin accumulation of soybean infected with *Phytophthora megasperma* f. sp. *glycinea.* Physiol. Plant Pathol. 23:145–152.

Borzini, G. 1956. Note fitopatologiche per l'anno 1955 (Phytopathological notes for the year 1955). Boll. Lab. Sper. Fitopatol. Torino, N.S. 19:3–61. (In French). (Rev. Appl. Mycol. 1957, 36:680–681.)

Bose, S. K., and Metha, P. R. 1951. Records of disease occurrence. Plant Prot. Bull. (India) 3:46–48.

Bostock, R. M. 1989. Metabolism of lipids containing arachidonic and eicosapentaenoic acids in race-specific interactions between *Phytophthora infestans* and potato. Phytopathology 79:898–902.

Bostock, R. M., and Doster, M. A. 1985. Association of *Phytophthora syringae* with pruning wound cankers of almond trees. Plant Dis. 69:568–571.

Bostock, R. M., Kuć, J. A., and Laine, R. A. 1981. Eicosapentaenoic and arachidonic acids from *Phytophthora infestans* elicit fungitoxic sesquiterpenes in the potato. Science, Rep. Ser. 212:67–69.

Bostock, R. M., Laine, R. A., and Kuć, J. A. 1982. Factors affecting the elicitation of sesquiterpenoid phytoalexin accumulation by eicosapentaenoic and arachidonic acids in potato. Plant Physiol. 70:1417–1424.

Botha, T., Wehner, F. C., and Kotzé, J. M. 1990. Evaluation of new and existing techniques for *in vitro* screening of tolerance to *Phytophthora cinnamomi* avocado rootstocks. Phytophylactica 22:335–338.

Bouchibi, N., van Bruggen, A. H. C., and MacDonald, J. D. 1990. Effect of ion concentration and sodium:calcium ratio of a nutrient solution on Phytophthora root rot of tomato and zoospore motility and viability of *Phytophthora parasitica.* Phytopathology 80:1323–1329.

Boughton, T. J., Malajczuk, N., and Robinson, A. D. 1978. Suppression of the infection of jarrah roots by *Phytophthora cinnamomi* with application of calcium carbonate. Aust. J. Bot. 26:611–615.

Boukema, I. W. 1983. Inheritance of resistance to foot and root rot caused by *Phytophthora nicotianae* v. Breda de Haan var. *nicotianae* in tomato (*Lycopersicon* Mill.). Euphytica 32:103–109.

Bouriquet, G. 1933. Les maladies du vanillier a Madagascar (Diseases of vanilla in Madagascar). Ann. Cryptogam. Exot. 6:59–78. (In French). (Rev. Appl. Mycol. 1934, 13:57.)

Bouriquet, G. 1950. Les maladies des plantes et la végétation a Madagascar (The diseases of plants and the vegetation in Madagascar). Rev. Bot. Appl. 31:341–442. (In French). (Rev. Appl. Mycol. 1951, 30:630.)

Bouriquet, G., Dadant, R., Bassino, J. P., and Lasoa, L. 1959. Un chancre de l'ecorce del l'*Albizzia stipulata* a Madagascar (A bark canker of *A. stipulata* in Madagascar). Agron. Trop. (Paris) 14:711–720. (In French). (Rev. Appl. Mycol. 1960, 39:442.)

Bourke, A. 1991. Potato blight in Europe in 1845: The scientific controversy. Pages 12–24 in: *Phytophthora.* J. A. Lucas, R. C. Shuttock, D. S. Shaw, and L. R. Cooke, eds. Cambridge Univ. Press, Cambridge, U.K. 447 pp.

Bourke, P. M. A. 1970. Use of weather information in the prediction of plant disease epiphytotics. Annu. Rev. Phytopathol. 8:345–370.

Bowen, G. D. 1980. Misconceptions, concepts and approaches in rhizosphere biology. Pages 283–304 in: Contemporary Microbial Ecology. D. C. Ellwood, J. N. Hedger, M. J. Latham, J. M. Lynch, and J. H. Slater, eds. Academic Press, London. 438 pp.

Bower, L. A., and Coffey, M. D. 1985. Development of laboratory tolerance to phosphorous acid, fosetyl-Al, and metalaxyl in *Phytophthora capsici.* Can. J. Plant Pathol. 7:1–6.

Bowers, J. H., and Mitchell, D. J. 1990. Effect of soil-water matric potential and periodic flooding on mortality of pepper caused by *Phytophthora capsici.* Phytopathology 80:1447–1450.

Bowers, J. H., and Mitchell, D. J. 1991. Relationship between inoculum level of *Phytophthora capsici* and mortality of pepper. Phytopathology 81:178–184.

Bowers, J. H., Papavizas, G. C., and Johnston, S. A. 1990. Effect of soil temperature and soil-water matric potential on the survival of *Phytophthora capsici* in natural soil. Plant Dis. 74:771–777.

Bowers, J. H., Sonaday, R. M., and Mitchell, D. J. 1990. Path coefficient analysis of the effect of rainfall variables on the epidemiology of Phytophthora blight of pepper caused by *Phytophthora capsici.* Phytopathology 80:1439–1446.

Boyd, A. E. W. 1980. Development of potato blight (*Phytophthora infestans*) after planting infected seed tubers. Ann. Appl. Biol. 95: 301–309.

Brasier, C. M. 1969a. The effect of light and temperature on reproduction in vitro in two tropical species of *Phytophthora.* Trans. Br. Mycol. Soc. 52:105–113.

Brasier, C. M. 1969b. Formation of oospores in vivo by *Phytophthora palmivora.* Trans. Br. Mycol. Soc. 52:273–279.

Brasier, C. M. 1971a. Induction of sexual reproduction in single A^2 isolates of *Phytophthora* species by *Trichoderma viride.* Nature New Biol. 231:283.

Brasier, C. M. 1971b. Report of Forest Research for the year ended March 1971. H. M. Stationery Office, London. 77:84. (Rev. Plant Pathol. 1972, 51:2871.)

Brasier, C. M. 1972. Observations on the sexual mechanism in *Phytophthora palmivora* and related species. Trans. Br. Mycol. Soc. 58:237–251.

Brasier, C. M. 1975a. Stimulation of sex organ formation in *Phytoph-*

damage in the Fallbrook area. Calif. Avocado Soc. Yearb. 44:110–113.

Burrell, R. G., Clayton, C. W., Gallegly, M. E., and Lilly, V. G. 1966. Factors affecting the antigenicity of the mycelium of three species of *Phytophthora*. Phytopathology 56:422–426.

Busch, L. V., and Smith, E. A. 1978. Control of root and crown rot of African violet and of *Gloxinia* caused by *Phytophthora nicotianae* var. *nicotianae*. Can. Plant Dis. Surv. 58:73–74.

Büsse, W. 1905. Reisebericht über die pflanzenpathologischen Expedition des Kolonial-wirtschaftlichen Komitees nach West Afrika (Field trip report on plant pathological expedition by the colonial research committee to West Africa). Tropenpflanzer 9:25–37. (In German).

Butcher, T. B., Stukely, M. J. C., and Chester, G. W. 1984. Genetic variation in resistance of *Pinus radiata* to *Phytophthora cinnamomi*. For. Ecol. Manage. 8:197–220.

Butler, E. J. 1919. Report of the Imperial Mycologist, 1918–1919. Sci. Rep. Res. Inst. Pusa 1918–1919. p. 82.

Butler, E. J. 1903. Potato diseases in India. Agric. Ledger 4:87–124. (Cited in Tucker 1933).

Butler, E. J. 1906. Some diseases of palms. Agric. J. (India) 1:299–310. (Cited in Tucker 1933).

Butler, E. J. 1907. An account of the genus *Pythium* and some Chytridiaceae. Mem. Dep. Agric. India, Bot. Ser. 1:82–84.

Butler, E. J. 1910. The bud-rot of palms in India. Mem. Dep. Agric. India, Bot. Ser. 3:221–280.

Butler, E. J. 1918. Fungi and disease in plants, 330 Thacker Spink Co., Calcutta. (Cited in Kale and Prasad 1957).

Butler, E. J., and Bisby, G. R. 1931. The fungi of India. Imp. Counc. Agric. Res. Sci. Monogr. 1. 237 pp.

Butler, E. J., and Jones, S. G. 1949. Plant Pathology. Macmillan & Co. 979 pp.

Butler, E. J., and Kulkarni, G. S. 1913. Colocasia blight caused by *Phytophthora colocasiae*. Rac. Bot. Mem. Dep. Agric. India 5:233–261. (Cited in Tucker 1933).

Buzzell, R. I., and Anderson, T. R. 1982. Plant loss response of soybean cultivars to *Phytophthora megasperma* f. sp. *glycinea* under field conditions. Plant Dis. 66:1146–1148.

Byford, W. J. 1958. *Phytophthora verrucosa* on dahlia (Abstr.). Plant Pathol. 7:38.

Byrd, D. W., Jr., Barker, K. R., Ferris, H., Nusbaum, C. J., Griffin, W. E., Small, R. H., and Stone, C. A. 1976. Two semi-automatic elutriators for extracting nematodes and certain fungi from soil. J. Nematol. 8:206–212.

Byrt, P. N., and Grant, B. R. 1979. Some conditions governing zoospore production in axenic cultures of *Phytophthora cinnamomi* Rands. Aust. J. Bot. 27:103–115.

Byrt, P. N., and Holland, A. A. 1978. Infection of axenic *Eucalyptus* seedlings with *Phytophthora cinnamomi* zoospores. Aust. J. Bot. 26:169–176.

Byrt, P. N., Irving, H. R., and Grant, B. R. 1982a. The effect of cations on zoospores of the fungus *Phytophthora cinnamomi*. J. Gen. Microbiol. 128:1189–1198.

Byrt, P. N., Irving, H. R., and Grant, B. R. 1982b. The effect of organic compounds on the encystment, viability and germination of zoospores of *Phytophthora cinnamomi*. J. Gen. Microbiol. 128: 2343–2351.

Bywater, J., and Hickman, C. J. 1959. A new variety of *Phytophthora erythroseptica*, which causes a soft rot of pea roots. Trans. Br. Mycol. Soc. 42:513–524.

Cacciola, S. O., and Magnano di san Lio, G. 1988. Foot rot of prickly pear cactus caused by *Phytophthora nicotianae*. Plant Dis. 72:793–796.

Cacciola, S.O., Magnano di san Lio, Greco, G., and Pane, A. 1990. Electrophoretic study of three related *Phytophthora* species. EPPO (Eur. Mediterr. Plant Prot. Organ.) Bull. 20:47–58.

Cacciola, S. O., Belisario, A., Pane, A., and Magnano di San Lio, G. 1994. Forsythia: A new host of *Phytophthora nicotianae* in Italy. Plant Dis. 78:525–528.

Cacciola, S. O., Magnano di san Lio, G., and Belisario, A. 1994. A new species of *Phytophthora, P. italica*. (Abstr.) Page 13. Proc. 5th Internat. Mycol. Congr., Vancouver, B.C., Canada.

Cacciola, S. O., Magnano di san Lio, G., and Belisario, A. 1996. *Phytophthora italica* sp. nov. on myrtle. Phytopathol. Mediterr. 35: in press.

Cafe Filho, A. C., and Duniway, J. M. 1995. Dispersal of *Phytophthora capsici* and *P. parasitica* in furrow-irrigated rows of bell pepper, tomato and squash. Plant Pathol. 44:1025–1032.

Cahill, D. M., and Hardham, A. R. 1994. Exploitation of zoospore taxis in the development of a novel dipstick immunoassay for the specific detection of *Phytophthora cinnamomi*. Phytopathology 84:193–200.

Cahill, D. M., and Ward, E. W. B. 1989. Effects of metalaxyl on elicitor activity, stimulation of glyceollin production and growth of sensitive and tolerant isolates of *Phytophthora megasperma* f. sp. *glycinea*. Physiol. Mol. Plant Pathol. 35:97–112.

Cahill, D., Weste, G., and Grant, B. 1985. Leakage from seedling roots inoculated with *Phytophthora cinnamomi*. Phytopathol. Z. 114:348–364.

Cahill, D. M., Weste, G. M., and Grant, B. R. 1986. Changes in cytokinin concentrations in xylem exudate following infection of *Eucalyptus marginata* Donn ex Sm. with *Phytophthora cinnamomi* Rands. Plant Physiol. 81:1103–1109.

Cahill, D. M., Legge, N., Grant, B. R., and Weste, G. M. 1989. Cellular and histological changes induced by *Phytophthora cinnamomi* in a group of plant species ranging from fully susceptible to fully resistant. Phytopathology 79:417–424.

Cahill, D. M., Bennett, I. J., and McComb, J. A. 1992. Resistance of micropropagated *Eucalyptus marginata* to *Phytophthora cinnamomi*. Plant Dis. 76:630–632.

Cairns, H., and Muskett, A. E. 1933. *Phytophthora megasperma* causing pink rot of the potato. Nature 131:277.

Caligari, P. D. S., MacKay, G. R., Sterwart, H. E., and Wastie, R. L. 1985. Confirmatory evidence for the efficacy of a seedling progeny test for resistance to potato foliage blight (*Phytophthora infestans* [Mont.] de Bary). Potato Res. 28:439–442.

Calvert, O. H., Williams, L. F., and Whitehead, M. D. 1960. Frozen-lima-bean agar for culture and storage of *Phytophthora sojae*. Phytopathology 50:136–137.

Cameron, H. R. 1962. Susceptibility of pear roots to *Phytophthora*. Phytopathology 52:1295–1297.

Cameron, J. N. and Carlile, M. J. 1977. Negative geotaxis of zoospores of the fungus *Phytophthora*. J. Gen. Microbiol. 98:599–602.

Campacci, C. A. 1951. Ocorrencia de *Phytophthora parasitica* Dast., em Mamoeiro (Occurrence of *Phytophthora parasitica* Dast. on papaw). Biológico 17:142–143. (In Portuguese). (Rev. Appl. Mycol. 1952, 31:339.)

Campbell, A. I. 1969. The effect of some apple viruses on the susceptibility of two clonal rootstocks to collar rot caused by *Phytophthora cactorum*. J. Hortic. Sci. 44:69–73.

Campbell, A. M., Duncan, J. M., and Malcolmson, J. F. 1985. Production of oospores in vitro by self fertilization in single isolates of A1 mating types of *Phytophthora infestans*. Trans. Br. Mycol. Soc. 84:533–535.

Campbell, A. M., Moon, R. P., Duncan, J. M., Gurr, S. J., and Kinghorn, J. R. 1989. Protoplast formation and regeneration from sporangia and encysted zoospores of *Phytophthora infestans*. Physiol. Mol. Plant Pathol. 34:299–307.

Campbell, C. L. 1986. Interpretation and uses of disease progress curves for root diseases. Pages 38–54 in: Plant Disease Epidemiology: Population Dynamics and Management. Vol. I. K. J. Leonard and W. E. Fry, eds. Macmillan, New York. 372 pp.

Campbell, C. L., and Powell, N. T. 1980. Progression of disease induced by soil-borne pathogens: Tobacco black shank. Prot. Ecol. 2:177–182.

Campbell, C. L., Jacobi, W. R., Powell, N. T., and Main, C. E. 1984. Analysis of disease progression and the randomness of occurrence of infected plants during tobacco black shank epidemics. Phytopathology. 74:230–235.

Campbell, J. G. C. 1925. Report by the mycologist for the year 1924. Fiji Dep. Agric. Annu. Rep. 1924:13–14. (Rev. Appl. Mycol. 1926, 5:215.) (Cited in Tucker 1933).

Campbell, J. G. C. 1927. Report by the mycologist, 1926. Fiji Dep. Agric. Annu. Rep. 1926:3–4.

Campbell, S. J., and Hamm, P. B. 1989. Susceptibility of Pacific

Northwest conifers to Phytophthora root rot. Tree Planters Notes 40:15–18.

Campbell, W. A. 1949. A method of isolating *Phytophthora cinnamomi* directly from soil. Plant Dis. Rep. 33:134–135.

Campbell, W. A., and Copeland, O. L., Jr. 1954. Littleleaf disease of shortleaf and loblolly pine. U.S. Dep. Agric. Circ. 940. 41 pp.

Campbell, W. A., and Gallegly, M. E. 1965. *Phytophthora heveae* from eastern Tennessee and western North Carolina. Plant Dis. Rep. 49:233–234.

Campbell, W. A., and Presley, J. T. 1946. Diseases of cultivated guayule and their control. U.S. Dep. Agric. Circ. No. 749. 42 pp.

Campêlo, A. M. F. L., and Luz, E. D. M. N. 1981. Etiologia da podridão parda do cacaneiro nos estados da Bahiae e Espírito Santo, Brazil (Etiology of black pod of cocoa in the states of Bahia and Espirito Santo, Brazil). Fitopatol. Bras. 6:313–321. (Rev. Plant Pathol. 61:4013.)

Can, H. C., and Baxter, L. W. 1977. The status of our knowledge of root rot of camellia. Am. Camellia Soc. Yearb. 1977. 173–178.

Canaday, C. H., and Schmitthenner, A. F. 1979. The effect of nitrogen on Phytophthora root rot of soybean (Abstr.). Phytopathology 69:539.

Canaday, C. H., and Schmitthenner, A. F. 1982. Isolating *Phytophthora megasperma* f. sp. *glycinea* from soil with a baiting method that minimizes *Pythium* contamination. Soil Biol. Biochem. 14: 67–68.

Canova, A., and D'Ercole, N. 1972. Midollo rosso delle radici di fragola da *Phytophthora fragariae* Hickman, in Italia (Strawberry red stele caused by *Phytophthora fragariae* Hickman, in Italy). Inf. Fitopatologic. 22:9–11. (In Italian).

Cardoso, J. G. A. 1934. Mozambique: Diseases and pests of citrus in the district of Lourenco Marques. Int. Bull. Plant Prot. 8:126.

Carefoot, G. L., and Sprott, E. R. 1967. Famine on the Wind. Man's Battle Against Plant Disease. Rand McNally. 231 pp.

Carlile, M. J. 1983. Motility, Taxis, and Tropism in *Phytophthora.* Pages 95–107 in: *Phytophthora:* Its Biology, Taxonomy, Ecology, and Pathology. D. C. Erwin, S. Bartnicki-Garcia, and P. H. Tsao, eds. American Phytopathological Society, St. Paul, Minn. 392 pp.

Carlile, M. J., and Tew, P. M. 1988. Chemotropism of germ-tubes of *Phytophthora citricola.* Trans. Br. Mycol. Soc. 90:644–646.

Carne, W. M. 1924. Downy mildew (*Sclerospora macrospora*), an unusual disease of wheat. J. Agric. West. Aust. (2nd ser.) 1:315–316.

Carne, W. M. 1925. A brown rot of citrus in Australia (*Phytophthora hibernalis* n. sp.). J. R. Soc. West. Aust. 12:13–41. (Cited in Tucker 1933).

Carnes, D. W. 1976. The sensitivity of *Phytophthora palmivora* to chloramphenicol. Ph.D. dissertation, Univ. of Calif., Riverside. 75 pp.

Caroselli, N. E. 1953. Bleeding canker disease of hardwoods. Sci. Tree Top. 2:1–6.

Caroselli, N. E., and Howard, F. L. 1939. Bleeding canker of maples. Bull. Bartlett Tree Res. Lab. 3:44–48. (Rev. Appl. Mycol. 1940, 19:570–571.)

Caroselli, N. E., and Tucker, C. M. 1949. Pit canker of elm. Phytopathology 39:481–488.

Carpenter, C. W. 1920. Report of the plant pathologist. Hawaii Agric. Exp. Stn. Rep. 1919:49–54. (Cited in Tucker 1933).

Carpenter, J. B. 1954. An epidemic of Phytophthora leaf fall of *Hevea* rubber trees in Costa Rica. Phytopathology 44:597–601.

Carpenter, J. B., and Furr, J. R. 1962. Evaluation of tolerance to root rot caused by *Phytophthora parasitica* in seedlings of *Citrus* and related genera. Phytopathology 52:1277–1285.

Carranza, J. M. 1964. Podredumbre del cuello y raices de la cineraria (*Senecio cruentus*) producida por *Phytophthora drechsleri* (Collar and root rots of Cineraria [*S. cruentus*] produced by *Phytophthora drechsleri*). Rev. Fac. Agron. Univ. Nac. La Plata 39:1–9. (In Spanish). (Rev. Appl. Mycol. 1964, 44:461.)

Carter, D. A., Archer, S. A., Buck, K. W., Shaw, D. S., and Shattock, R. C. 1990. Restriction fragment length polymorphisms of mitochondrial DNA of *Phytophthora infestans.* Mycol. Res. 94:1123–1128.

Carter, D. A., Archer, S. A., Buck, K. W., Shaw, D. S., and Shattock, R. C. 1991. DNA polymorphisms in *Phytophthora infestans:* The UK experience. Pages 272–295 in: *Phytophthora.* J. A. Lucas, R. C. Shattock, D. S. Shaw, and L. R. Cooke, eds. Cambrige Univ. Press, Cambridge, U.K. 447 pp.

Carter, J. C. 1953. Phytophthora canker of Russian olive (Abstr.). Phytopathology 43:468.

Caruso, F. L., and Wilcox, W. F. 1990. *Phytophthora cinnamomi* as a cause of root rot and die back of cranberry in Massachusetts. Plant Dis. 74:664–667.

Carvalho, A. M. S. 1982. Transmissao e sobrevivencia de *Phytophthora capsici* em sementes de pimentao (*Capsicum annuum*) (Transmission and survival of *Phytophthora capsici* in sweet pepper seed [*Capsicum annuum*]). Thesis, Univ. Viçosa, Viçosa, M.G., Brasil. 23 pp. (In Portuguese).

Casarini, B., Quaglia, A., and Silvestri, G. 1960. La 'Peronospora' del 'Fagiolo di lima' (A *Peronospora* of lima bean). Phytopathol. Mediterr. 1:47–48. (In Italian).

Caspary, R. 1852. Verhandlungen des Verein zur Beförderung des Gartenbaues in den Koniglich Preussischen Staaten (Proceedings of the Association for Horticulture in the Royal Prussian States). p. 327. (In German). (Cited in Tucker 1931).

Caspary, R. 1855. Über einige Hyphomyceten mit zwei-und dreierlei Fruchten (On a hyphomycete with two and three kinds of fruits). Ber. Verhandl. Königl. Preufs. Akad. Wiss. Berlin 1855:308–333. (In German). (Cited in Tucker 1931).

Cassells A. C., Deadman, M. L., Brown, C. A., and Griffin, E. 1991. Field resistance to late blight (*Phytophthora infestans* (Mont.) De Bary) in potato (*Solanum tuberosum* L.) somaclones associated with instability and pleiotropic effects. Euphytica 56:75–80.

Castellani, E. 1939. Considerazioni fitopatologiche sull' Africa Orientale Italiana (Phytopathological notes on Italian East Africa). Agricoltura Colon. 33:486–492. (In Italian). (Rev. Appl. Mycol. 1940, 19:8.)

Castro, J., and Zentmyer, G. A. 1969. Mortality of germinated F2 oospores from crosses of F1 single oospore cultures from *Phytophthora infestans* (Abstr.). Phytopathology 59:10.

Castronovo, N. 1955. Produccion de zoosporos binucleados por *Phytophthora infestans* (Production of binucleate zoospores in *Phytophthora infestans*). Rev. Argent. Agron. 22:177–182. (In Spanish).

Caten, C. E. 1974. Intra-racial variation in *Phytophthora infestans* and adaptation to field resistance for potato blight. Ann. Appl. Biol. 77:259–270.

Caten, C. E., and Jinks, J. L. 1968. Spontaneous variability of single isolates of *Phytophthora infestans.* I. Cultural variation. Can. J. Bot. 46:329–348.

Cavalier-Smith, T. 1986. The kingdom Chromista: Origin and systematics. Pages 309–347 in: Progress in Phycological Research. Vol. 4. I. Round and D. J. Chapman, eds. Biopress, Bristol, England. 481 pp.

Cavalier-Smith, T. 1987. The origin of fungi and pseudofungi. Pages 334–359 in: Evolutionary Biology of Fungi. A. D. M. Rayner, C. M. Brasier, and D. Moore, eds. Cambrige Univ. Press, Cambridge, U.K. 465 pp.

Cavalier-Smith, T. 1988. Eukaryotic cell evolution. Pages 203–223 in: Proc. XIV Int. Bot. Congr. W. Greuter and B. Zimmer, eds. Koeltz, Königstein/Taunus. 479 pp.

Cavalier-Smith, T. 1989. The kingdom Chromista. Pages 381–407 in: The Chromophyte Algae: Problems and Perspectives. J. C. Green, B. S. C. Leadbeater, and W. L. Diver, eds. Syst. Assoc. Spec. Vol. 38. Clarendon Press, Oxford. 429 pp.

Cejp, K. 1961. Dve nebezpecne a skodlive Fytoftory v eskoslovensku (Two dangerous and harmful phytophthoras in Czechoslovakia). Ceska. Mykol. 15:246–252. (In Czech). (Rev. Appl. Mycol. 1962, 41:203.)

Cejp, K., and Jechova, V. 1962. Subtropical species of the genus *Phytophthora* de Bary imported to us with fruit from the south: *Phytophthora citrophthora* (Sm. & Sm.) Leon. and *P. citricola* Sawada. Ceska. Mykol. 16:198–202. (Rev. Appl. Mycol. 1963, 42:197.)

Celino, M. S. 1934. Blight of Cinchona seedlings. Philipp. Agric. 23:111–127.

Chabrolin, C. 1932. Contribution a l'étude des maladies des arbres fruitiers in Tunisie (A contribution to the study of fruit tree diseases in Tunis). Ann. Ser. Bot. Agron. Tunisie. 9:177–200. (In French).

Chamberlain, D. W. 1972. Heat-induced susceptibility to nonpathogens and cross-protection against *Phytophthora megasperma* var. *sojae* in soybean. Phytopathology 62:645–646.

Chamberlain, D. W., and Gerdemann, J. W. 1966. Heat-induced susceptibility of soybeans to *Phytophthora megasperma* var. *sojae, Phytophthora cactorum,* and *Helminthosporium sativum.* Phytopathology 56:70–73.

Chamberlain, D. W., and Paxton, J. D. 1968. Protection of soybean plants by phytoalexin. Phytopathology 58:1349–1350.

Chambers, S. C. 1958. Pink rot of potatoes. J. Agric. West Aust. (3rd ser.) 7:405–406.

Chang, H. S. 1988. *Phytophthora* species associated with strawberry fruit rot in Taiwan. Bot. Bull. Acad. Sin. 29:61–67.

Chang, H. S., and Shu, I. M. 1988. Oospore germination of homothallic *Phytophthora* species and the identity of *Phytophthora heveae* from Taiwan. Bot. Bull. Acad. Sin. Taiwan 29:15–21. (Rev. Plant Pathol. 1990, 69:11)

Chang, T. T., and Ko, W. H. 1990. Resistance to fungicides and antibiotics in *Phytophthora parasitica:* Genetic nature and use in hybrid determination. Phytopathology 80:1414–1421.

Chang, T. T., and Ko, W. H. 1991. Factors affecting the germination of oospores of *Phytophthora infestans.* J. Phytopathol. 133:29–35.

Chang-Ho, Y., and Hickman, C. J. 1970. Some factors involved in the accumulation of Phycomycete zoospores on plant roots. Pages 103–108 in: Root Diseases and Soil-Borne Pathogens. T. A. Toussoun, R. V. Bega, and P. E. Nelson, eds. Univ. Calif. Press, Berkeley. 252 pp.

Channon, A. G. 1972. Annu. Rep. West of Scotland Agric. Coll. for the year ending 30th September, 1972. Glasgow. 130 pp. (Rev. Plant Pathol. 1974, 53:1.)

Chase, A. R. 1981. Phytophthora stem rot of *Aphelandra squarrosa.* Plant Dis. 65:921–922.

Chase, A. R. 1987. Compendium of Ornamental Foliage Plant Diseases. American Phytopathological Society, St. Paul, Minnesota. 114 pp.

Chase, A. R., and Broschat, T. K. 1991. Diseases and Disorders of Ornamental Palms. American Phytopathological Society, St. Paul, Minnesota. 56 pp.

Chaube, H. S., and Singh, U. S. 1991. Plant Disease Management, Principles and Practice. CRC Press, Boca Raton, Florida. 319 pp.

Chaudhary, R. G., Mathura, R., and Rai, M. 1988. A note on the varietal screening of Taro to Phytophthora blight. Haryana J. Hort. Sci. 17:278–279.

Chauhan, V. B., and Singh, U. P. 1987. A naturally occurring resistant forma specialis of *Phytophthora drechsleri* to metalaxyl. J. Phytopathol. 120:93–96.

Chauhan, V. B., and Singh, U. P. 1988. In vitro effect of some fungicides on mycelial growth formation and germination of sporangia and zoospores of *Phytophthora drechsleri* f. sp. *cajani.* Pesticides 22:23–26.

Chauhan, V. B., and Singh, U. P. 1991. Effect of weed infestation on the severity of Phytophthora blight of pigeon pea. Plant Dis. 75: 1230–1232.

Chaurasia, S. C., and Thind, T. S. 1975. Activity of some antibiotics against *Phytophthora parasitica* var. *piperina,* incitant of foot-rot and leaf-rot of pan. Hind. Antibiot. Bull. 18:28–30.

Chaves Batista, A. 1946. Principais doencas das plantas no nordeste (Principal plant diseases in the northeast). Bol. Agric., Pernambuco 13:195–252. (In Portugese). (Rev. Appl. Mycol. 1947, 26: 287).

Chee, K. H. 1968. Patch canker of *Hevea brasiliensis* caused by *Phytophthora palmivora* in Malaysia. Plant Dis. Rep. 52:132–133.

Chee, K. H. 1969a. Variability of *Phytophthora* species from *Hevea brasiliensis.* Trans. Br. Mycol. Soc. 52:425–436.

Chee, K. H. 1969b. Hosts of *Phytophthora palmivora.* Rev. Appl. Mycol. 48:337–344.

Chee, K. H. 1970. *Phytophthora heveae* and *Pythium vexans* of *Hevea.* J. Rubber Res. Inst. Malays. 23:13–14.

Chee, K. H. 1972. Recent investigations in the control of black stripe. Pages 155–162 in: Proc. Rubber Res. Inst. Plant. Conf. Malays.

Chee, K. H. 1973a. Production, germination, and survival of chlamydospores of *Phytophthora palmivora* from *Hevea brasiliensis.* Trans. Br. Mycol. Soc. 61:21–26.

Chee, K. H. 1973b. Phenotypic differences among single-oospore cultures of *Phytophthora palmivora* and *P. botryosa* from *Hevea brasiliensis.* Mycopathol. Mycol. Appl. 50:275–292.

Chee, K. H. 1976. Micro-organisms associated with rubber (*Hevea brasiliensis* Muell. Arg.) Agr. Ser. Rep. Rubber Res. Inst. Malays. No. 4. 78 pp.

Chee, K. H., and Foong, K. M. 1968. Use of cacao pod for recovering *Phytophthora* species pathogenic to *Hevea brasiliensis.* Plant Dis. Rep. 52:5.

Chee, K. H., and Greenwood, J. M. F. 1968. Phytophthora leaf fall and pod rot of *Hevea brasiliensis* in Thailand. F.A.O. Plant Prot. Bull. 16:1–5.

Chee, K. H., and Newhook, F. J. 1965a. Improved methods for use in studies on *Phytophthora cinnamoni* Rands and other *Phytophthora* species. N.Z. J. Agric. Res.:88–95.

Chee, K. H., and Newhook, F. J. 1965b. Variability in *Phytophthora cinnamomi* Rands. N.Z.J. Agric. Res. 8:96–103.

Chee, K. H., and Newhook, F. J. 1965c. Nutritional studies with *Phytophthora cinnamomi* Rands. N.Z. J. Agric. Res. 8:523–529.

Chee, K. H., and Newhook, F. J. 1966. Relationship of microorganisms to sporulation of *Phytophthora cinnamomi* Rands. N.Z. J. Agric. Res. 9:32–43.

Chee, K. H., and Turner, N. A. 1965. A steroid in pea (*Pisum sativum* L.) influencing growth and sporulation of *Phytophthora cinnamomi* Rands. N.Z. J. Agric. Res. 8:104–108.

Chee, K. H., and Wastie, R. L. 1970. Black pod disease of cacao. Planter, Kuala Lumpur 46:294–297.

Chee, K. H., and Wastie, R. L. 1980. The status and future prospects of rubber diseases in tropical America. Rev. Plant Pathol. 59:541–548.

Chee, K. H., Lim, T. M., and Wastie, R. L. 1967. An outbreak of Phytophthora leaf fall and pod rot on *Hevea brasiliensis* in Malaysia. Plant Dis. Rep. 51:443–446.

Chee, K. H., Zentmyer, G. A., Foong, K. M., and Klure, L. J. 1976. Mating types of *Phytophthora palmivora* in Malaysia. Plant Dis. Rep. 60:866–867.

Chen, D.-W. 1955. Studies on heart rot of pineapple. J. Agric. For. 9:114–136.

Chen, D.-W., and Zentmyer, G. A. 1970. Production of sporangia by *Phytophthora cinnamomi* in axenic culture. Mycologia 62:397–402.

Chen, H. F. 1985. Common diseases on Ramie and their control. China's Fibre Crops No. 3:25–27. (In Chinese). (Rev. Plant Pathol. 1988, 67:2420).

Chen, J. S., and Hsieh, S. P. Y. 1978. Phytophthora black rot of *Phalaenopsis* in Taiwan. Plant Prot. Bull. 20:161–170.

Chen, W., Hoitink, H. A. J., and Madden, I. V. 1988. Microbial activity and biomass in container media for predicting suppressiveness to damping off and root rot caused by *Pythium ultimum.* Phytopathology 78:1447–1450.

Chester, K. S. 1930. The *Phytophthora* disease of the calla in America. J. Arnold Arbor. Harv. Univ. 11:169–171. (Cited in Tucker 1933).

Chester, K. S. 1932. A comparative study of the three *Phytophthora* diseases of lilac and of their pathogens. J. Arnold Arbor. Harv. Univ. 13:232–269.

Chesters, C. G. C. 1969. Morphology as a taxonomic criterion. Pages 517–542 in: The Fungi, an Advanced Treatise. Vol. III. G. C. Ainsworth and A. S. Sussman, eds. Academic Press, New York. 738 pp.

Chevaugeon, J., and Merny, G. 1956. Maladies des arbres à quinquina en Guinée française (Diseases of Chinchona trees in French Guinea). J. Agric. Trop. Bot. Appl. 3:605–626. (In French). (Rev. Appl. Mycol. 1959, 38:94.)

Chi, C. C., and Sabo, F. E. 1978. Chemotaxis of zoospores of *Phytophthora megasperma* to primary roots of alfalfa seedlings. Can. J. Bot. 56:795–800.

Child, J. J., Dé Fago, G., and Haskins, R. H. 1969. The effect of cholesterol and polyene antibiotics on the permeability of the protoplasmic membrane of *Pythium* PRL 2142. Can. J. Microbiol. 15:599–603.

Chitzanidis, A., and Argyri, I. 1990. Spread of Phytophthora rot on stored citrus and its prevention by fungicide dips. Bull. OEPP/ EPPO Bull. 20:163–168.

Chitzanidis, A., and Kouyeas, H. 1970. Notes on Greek species of *Phytophthora.* II. Ann. Inst. Phytopathol. Benaki, N.S. 9:267–274.

Chitzanidis, A., and Lascaris, D. 1981. Research on Phytophthora rots of oranges in Greece. Proc. Int. Soc. Citric. 2:796–799.

Cho, C. W., and Fuller, M. S. 1989a. Ultrastructural studies of encystment and germination in *Phytophthora palmivora.* Mycologia 81:539–548.

Cho, C. W., and Fuller, M. S. 1989b. Observations of the water expulsion vacuole of *Phytophthora palmivora.* Protoplasma 149:47–56.

Cho, J. J. 1981. Phytophthora root rot of *Banksia:* Host range and chemical control. Plant Dis. 65:830–833.

Chomenansilpe, S., Doungpunya, A., and Nilprapai, C. 1983. Black pod disease of cocoa. J. Thai Phytopathol. Soc. 3:222–224.

Chou, L. G., and Schmitthenner, A. F. 1974. Effect of *Rhizobium japonicum* and *Endogone mosseae* on soybean root rot caused by *Phythium ultimum* and *Phytophthora megasperma* var. *sojae.* Plant Dis. Rep. 58:221–225.

Chowdhury, S. 1945. Heart or stem-rot of pineapple. Indian J. Agric. Sci. 15:139–140.

Christensen, M., and Raper, K. B. 1978. Synoptic key to *Aspergillus nidulans* group species and related *Emericella* species. Trans. Br. Mycol. Soc. 71:177–191.

Christensen, T. 1990. Plants, animals, algae and fungi, four non-taxonomic group designations. Taxon 39:93–94.

Christie, T. 1956. Black root-rot, *Phytophthora cactorum* (Leb. & Cohn) Schroet., of the hop plant. N.Z. J. Sci. Technol. 38:277–284.

Chung, Y. R., Kim, H. J., Ohh, S. H., and Park, K. J. 1984. Comparison of rhizosphere environments in soils suppressive and conducive to ginseng root rot. Korean J. Plant Prot. 23:142–146.

Cibes, H. R., and Childers, N. F. 1949. Report of the Federal Experiment Station in Puerto Rico 1948. 27 pp. (Rev. Appl. Mycol. 1949, 28:377.)

Ciferri, R. 1927a. Notae mycologicae et phytopathologicae (Mycological and phytopathological notes). Riv. Patol. Veg. 17:209–294. (In French). (Rev. Appl. Mycol. 1928, 7:229–230.) (Cited in Tucker 1933).

Ciferri, R. 1927b. Informe de fitopatologia principales enfermedades de las plantas cultivadas; observadas en el curso del año 1926 (Phytopathological report of the main diseases of cultivated plants observed during 1926). Inf. Annu. Estac. Nac. Agron. Moca 1926: 36–44. (In Spanish). (Cited in Tucker 1933).

Ciferri, R. 1930. Phytopathological survey of Santo Domingo, 1925–1929. J. Dep. Agric. Puerto Rico 14:5–44. (Cited in Tucker 1933).

Ciferri, R. 1954. Schedae Mycologicae XII–XXXIV (Mycological specimens XII–XXXIV). Sydowia 8:245–270. (In Italian).

Çinar, A., and Bicici, M. 1977. Control of *Phytophthora capsici* Leonian on red peppers. J. Turk. Phytopathol. 6:119–124.

Çinar, A., Göksedef, M., and Bicici, M. 1990. Control of *Phytophthora citrophthora* by fentin acetate in laboratory experiments and by soil drench application in a lemon orchard. EPPO (Eur. Mediterr. Plant Prot. Organ.) Bull. 20:149–152.

Çinar, A., Tuzcu, O., and Göksedef, M. O. 1976. Resistance study of the citrus rootstocks to *Phytophthora citrophthora* (Smith & Smith) Leonian. I: A research on the resistance of 14 different rootstocks to *Phytophthroa citrophthora.* (Smith & Smith) Leonian. J. Turk. Phytopathol. 5:49–59.

Clancy, K. J. 1979. Studies on a disease of seedlings of *Chamaecyparis lawsoniana.* Ph.D. dissertation, Nat. Univ. Ireland, Dublin.

Clancy, K. J., and Kavanagh, J. A. 1977. The isolation and pathogenicity of a new *Phytophthora* species. Univ. Coll. Dublin Fac. Gen. Agric. Res. Rep. 1976–1977:27–28.

Clara, F. M. 1928. A *Phytophthora* disease of Santol seedlings. Philipp. J. Sci. 35:411–425.

Clare, B. G. 1963. Starch-gel electrophoresis of proteins as an aid in identifying fungi. Nature 200:803–804.

Clare, B. G., and Zentmyer, G. A. 1966. Starch gel electrophoresis of proteins from species of *Phytophthora.* Phytopathology 56:1334–1335.

Clark, J. R., Moore, J. N., and Draper, A. D. 1986. Inheritance of resistance to Phytophthora root rot in highbush blueberry. J. Am. Soc. Hort. Sci. 111:106–109.

Clark, M. F. 1981. Immunosorbent assays in plant pathology. Annu. Rev. Phytopathol. 19:83–106.

Clark, M. F., and Adams, A. N. 1977. Characterisitics of the microplate method of enzyme-linked immunosorbent assay for the detection of plant viruses. J. Gen. Virol. 34:475–483.

Clarke, D. D. 1966. Factors affecting the development of single zoospore colonies of *Phytophthora infestans.* Trans. Br. Mycol. Soc. 49:177–184.

Clarke, D. D. 1986. Tolerance of parasites and disease in plants and its significance in host-parasite interactions. Pages 161–196 in: Advances in Plant Pathology. Vol. 5. D. S. Ingram and P. H. Williams, eds. 270 pp.

Clarke, R. G., and Greenhalgh, F. C. 1986. Screening for alternate hosts of *Phytophthora clandestina.* Australas. Plant Pathol. 15:61–63.

Clements, F. E., and Shear, C. L. 1931. The Genera of Fungi. H. W. Wilson Co., New York. 496 pp.

Clerjeau, M. 1973. A new collar rot disease of melon caused by *Phytophthora nicotianae* Br. de H. var. *parasitica* (Dastur) Waterh. and *Phytophthora capsici* Leon. C. R. Seances Acad. Agr. Fr. 59: 54–58.

Clerjeau, M., and Beyries, A. 1977. Étude comparee de l'action preventive et du pouvoir systemique de quelques fongicides nouveaux (phosphites-prothiocarbe-pyroxychlore) sur poivron vis-a-vis de *Phytophthora capsici* (Leon) (Comparative study of the preventive action and systemic capacity of some new fungicides [phosphites, prothiocarb, pyroxychlor] against *Phytophthora capsici* (Leon.)]. Stn. Pathol. Veg. INRA, Montfavet. Phytiatr. Phytopharm. 26:73–83.

Clinton, G. P. 1906. Downy mildew, *Phytophthora phaseoli* Thaxt., of lima beans. Rep. Conn. Agric. Exp. Stn. 1905. pp. 278–303.

Clinton, G. P. 1911. Oospores of potato blight. *Phytophthora infestans.* Conn. Agric. Exp. Stn. Bienn. Rep. 1909–1910 part 10: 753–754.

Clinton, G. P. 1920. New or unusual plant injuries and diseases found in Connecticut, 1916–1919. Conn. Agric. Exp. Stn. Bull. 222:397–482. (Cited in Tucker 1933).

Cockayne, A. H., and Cunningham, G. H. 1921. Lemon brown rot and its control. N.Z. J. Agric. 22:271–274. (Cited in Tucker 1933).

Coffey, M., and Cohen, Y. 1984. Crown and collar rot of avocado: A need for more research. Calif. Avocado Soc. Yearb. 68:69–74.

Coffey, M. D. 1987. Phytophthora root rot of avocado. An integrated approach to control in California. Plant Dis. 71:1046–1052.

Coffey, M. D. 1991. Strategies for integrated control of soilborne *Phytophthora* species. Pages 441–432 in: *Phytophthora.* J. A. Lucas, R. C. Shattock, D. S. Shaw, and L. R. Cooke, eds. Cambridge Univ. Press, Cambridge, U.K. 447 pp.

Coffey, M. D. 1992. Phytophthora root rot of avocado. Pages 423–444 in: Plant Diseases of International Importance. Volume III. Diseases of Fruit Crops. J. Kumar, H. S. Chaube, U. S. Singh, and A. N. Mukhopadhyay, eds. Prentice Hall, Englewood Cliffs, N.J. 456 pp.

Coffey, M. D., and Bower, L. A. 1984a. In vitro variability among isolates of six *Phytophthora* species in response to metalaxyl. Phytopathology 74:502–506.

Coffey, M. D., and Bower, L. A. 1984b. In vitro variability among isolates of eight *Phytophthora* species in response to phosphorus acid. Phytopathology 74:738–742.

Coffey, M. D., and Gees, R. 1991. The cytology of development. Pages 31–51 in: Advances in Plant Pathology. Vol. VII. D. S. Ingram and P. H. Williams, eds. Academic Press, London. 284 pp.

Coffey, M. D., and Joseph, M. C. 1985. Effects of phosphorous acid and fosetyl-A1 on the life cycle of *Phytophthora cinnamomi* and *P. citricola.* Phytopathology 75:1042–1046.

Coffey, M. D., and Ouimette, D. G. 1989. Phosphonates: Antifungal. compounds against Oomycetes. Pages 109–129 in: Nitrogen, Phosphorus and Sulphur Utilization by Fungi. Symposium of the British Mycological Soc. L. Boddy, R. Marchant, and D. J. Reed, eds. Cambridge Univ. Press, Cambridge, U.K. 361 pp.

Coffey, M. D., and Wilson, U. E. 1983a. Histology and cytology of infection and disease caused by *Phytophthora.* Pages 289–301 in:

Phytophthora: Its Biology, Taxonomy, Ecology, and Pathology. D. C. Erwin, S. Bartnicki-Garcia, and P. H. Tsao, eds. American Phytopathological Society, St. Paul, Minn. 392 pp.

Coffey, M. D., and Wilson, U. E. 1983b. An ultrastructural study of the late-blight fungus *Phytophthora infestans* and its interaction with the foliage of two potato cultivars possessing different levels of general (field) resistance. Can. J. Bot. 61:2669–2685.

Coffey, M. D., and Young, L. H. 1984. Responses to metalaxyl of sensitive and resistant isolates of *Phytophthora infestans.* Phytopathology 74:615–620.

Coffey, M. D., Klure, L. J., and Bower, L. A. 1984a. Variability in sensitivity to metalaxyl of isolates of *Phytophthora cinnamomi* and *Phytophthora citricola.* Phytopathology 74:417–422.

Coffey, M. D., Ohr, H. D., Campbell, S. D., and Guillemet, F. B. 1984b. Chemical control of *Phytophthora cinnamomi* on avocado rootstocks. Plant Dis. 68:956–958.

Coffey, M. D., Sitepu, D., Kharie, S., Mangindaan, H., Waroka, J. S., and Motulo, H. G. J. 1992. Integrated control of Phytophthora bud rot and premature nutfall of coconut in Indonesia. Proc. III. Int. Conf. Plant Prot. Tropics. Mar. 20–23, 1990. Vol. V. 132–135.

Cohen, E. 1981a. Metalaxyl for postharvest control of brown rot of citrus fruit. Plant Dis. 65:672–675.

Cohen, E. 1981b. Postharvest control of *Phytophthora citrophthora* with metalaxyl and its relation to other fungi pathogenic to citrus fruit. Proc. Int. Soc. Citric. Vol. 2. Pages 793–796.

Cohen, E. 1982. Prevention of spread and contact infection of brown rot disease in citrus fruit by metalaxyl postharvest treatment. Phytopathol. Z. 103:120–125.

Cohen, E., and Schiffmann-Nadel, M. 1978a. Prevention of brown rot spread by contact infection in packed citrus fruits. Plant Dis. Rep. 62:386–388.

Cohen, E., and Schiffmann-Nadel, M. 1978b. Effect of different chemical compounds on *Phytophthora citrophthora* in vitro and in citrus fruit. Plant Dis. Rep. 62:388–389.

Cohen, S. D. 1984. Detection of mycelium and oospores of *Phytophthora megasperma* forma specialis *glycinea* by vital stains in soil. Mycologia 76:34–39.

Cohen, Y. 1979. A new systemic fungicide against the downy mildew of cucumber. Phytopathology 69:433–435.

Cohen, Y. 1986. Fungicide mixtures controlling late blight in potatoes. Meded. Fac. Landouwwet. Rijksuniv. Gent 51:715–718.

Cohen, Y., and Coffey, M. D. 1984. Pathogenicity of *Phytophthora citricola* from avocado to *Persea indica* (Abstr.). Phytopathology 74:846.

Cohen, Y., and Coffey, M. D. 1986. Systemic fungicides and the control of Oomycetes. Annu. Rev. Phytopathol. 24:311–338.

Cohen, Y., and Perl, M. 1973. Stage specificity in streptomycin action against some plant pathogenic Peronosporales. Phytopathology 63:1172–1180.

Cohen, Y., and Reuveni, M. 1983. Occurrence of metalaxyl-resistant isolates of *Phytophthora infestans* in potato fields in Israel. Phytopathology 73:925–927.

Cohen, Y., and Samoucha, Y. 1984. Cross-resistance to four systemic fungicides in metalaxyl-resistant strains of *Phytophthora infestans* and *Pseudoperonospora cubensis.* Plant Dis. 68:137–139.

Cohen, Y., and Samoucha, Y. 1990. Competition between oxadixyl-sensitive and -resistant field isolates of *Phytophthora infestans* on fungicide-treated potato crops. Crop Prot. 9:15–20.

Cohen, Y., Eyal, H., and Sadon, T. 1975. Light-induced inhibition of sporangial formation of *Phytophthora infestans* on potato leaves. Can. J. Bot. 53:2680–2686.

Cohen, Y., Reuveni, M., and Eyal, H. 1979. The systemic antifungal activity of Ridomil against *Phytophthora infestans* on tomato plants. Phytopathology 69:645–649.

Cohen, Y., Baider, A., and Cohen, B.-H. 1995. Dimethomorph activity against oomycete fungal plant pathogens. Phytopathology 85:1500–1506.

Coleman, L. C. 1910. Diseases of the areca palm. I. Koleroga or root disease. Ann. Mycol. 8:591–626.

Colhoun, J. 1938. Fungi causing rots of apple fruits in storage in Northern Ireland. Ann. Appl. Biol. 25:88–99.

Collinge, W. E. 1912. Plant diseases due to fungi. Rep. Econ. Biol. 2:41–49. (Cited in Tucker 1933).

Collins, D. J., Pacumbaba, R. P., and Sapra, V. T. 1985. Races of *Phytophthora megasperma* f. sp. *glycinea* (Phytophthora rot) affecting soybeans in north Alabama. Phytopathol. Z. 113:43–46.

Collins, G. B., Legg, P. D., Litton, C. C., and Kasperbauer, M. J. 1971. Inheritance of resistance to black shank in *Nicotiana tabacum* L. Can. J. Genet. Cytol. 13:422–428.

Colon, L. T., and Budding, D. J. 1988. Resistance to late blight (*Phytophthora infestans*) in ten wild *Solanum* species. Euphytica:77–86.

Colwell, J. D. 1958. Observations on the pedology and fertility of some krasnozems in northern New South Wales. J. Soil Sci. 9: 46–57.

Commonwealth Mycological Institute. Distribution Maps of Plant Diseases. Maps are released periodically by The Commonwealth Mycological Institute (International Mycological Institute, Bakeham Lane, Egham, Surrey, United Kingdom). Maps are cited in the text by number.

Conn, K. E., Gubler, W. D., Mircetich, S. M., and Hasey, J. K. 1991. Pathogenicity and relative virulence of nine *Phytophthora* spp. from kiwifruit. Phytopathology 81:974–979.

Conners, I. L. 1938. Seventeenth annual report of the Can. Plant Dis. Surv. 11. 87 pp.

Conners, I. L. 1939. Eighteenth annual report of the Can. Plant Dis. Surv. 1938. 12. 112 pp.

Conners, I. L. 1955. Thirty-fourth Annual Report of the Can. Plant Dis. Surv. 1954. 20. 140 pp.

Conners, I. L. 1967. An annotated index of plant diseases in Canada and fungi recorded on plants in Alaska, Canada, and Greenland. Can. Dep. Agric. Res. Branch Publ. 1251. 381 pp.

Conners, I. L., and Saville, D. B. O. 1946. Twenty-fifth Annu. Rep. Can. Plant Dis. Surv., 1945. 126 pp. (Rev. Appl. Mycol. 1947, 26:7–9.)

Conners, I. L., and Saville, D. B. O. 1948. Twenty-seventh Annu. Rep. Can. Plant Dis. Sur., 1947. 118 pp. (Rev. Appl. Mycol. 1948, 27:553–555.)

Conover, R. A. 1948. A seedling blight of avocado caused by *Phytophthora palmivora.* Phytopathology 38:1032–1034.

Constantinescu, O. 1988. An instrument and procedure for single-spore isolation. Trans. Br. Mycol. Soc. 91:700–702.

Converse, R. H. 1967. Physiologic races of *Phytophthora fragariae* on strawberry in California, Oregon, and Washington. Phytopathology 57:173–177.

Converse, R. H. 1970. Occurrence of *Phytphthora fragariae* race A-10 in Caliornia. Plant Dis. Rep. 54:969–971.

Converse, R. H., and Moore, J. N. 1966. Susceptibility of certain *Potentilla* and *Geum* species to infection by various races of *Phytophthora fragariae.* Phytopathology 56:637–639.

Converse, R. H., and Schwartze, C. D. 1968. A root rot of red raspberry caused by *Phytophthora erythroseptica.* Phytopathology 58:56–59.

Converse, R. H., and Scott, D. H. 1962. Physiologic specialization in *Phytophthora fragariae.* Phytopathology 52:802–807.

Converse, R. H., and Shiroishi, K. K. 1962. Oospore production by single zoospore isolates of *Phytophthora fragariae* in culture. Phytopathology 52:807–809.

Converse, R. H., Adams, A. N., Barbara, D. J., Clark, M. F., Casper, R., Hepp, R. F., Martin, R. R., Morris, T. J., Spiegel, S., and Yoshikawa, N. 1988. Laboratory detection of viruses and mycoplasma-like organisms in strawberry. Plant Dis. 72:744–749.

Cook, A. A. 1975. Diseases of Tropical and Subtropical Fruits and Nuts. Hafner Press, New York. 317 pp.

Cook, M. T. 1913. The Diseases of Tropical Plants. Macmillan & Co., London. 317 pp.

Cook, M. T. 1924. Annual Report of the division of plant pathology and botany for the fiscal year of 1923–24. Annu. Rep. Insular Exp. Stn. Puerto Rico 24:79–87.

Cook, M. T. 1939. Enfermedades de las plantas económicas de las Antilles (Diseases of economic crops in the Antilles). Monogr. Univ. Puerto Rico B, 4. 530 pp. (Cited in Waterhouse 1957).

Cook, R. J., and Baker, K. F. 1983. The nature and practice of biological control of plant pathogens. American Phytopathological Society, St. Paul, Minn. 539 pp.

Cook, R. J., and Duniway, J. M. 1981. Water relations in the life-cycles of soilborne plant pathogens. Pages 119–139 in: Water Potential Relations in Soil Microbiology. J. F. Parr, W. R. Gardner, and L. F. Elliot, eds. Soil Sci. Soc. Am. Special Publ. 9. 151 pp.

Cook, R. J., and Papendick, R. I. 1971. Effect of soil water on microbial growth, antagonism, and nutrient availability in relation to soil-borne fungal diseases of plants. Pages 81–88 in: Root Diseases and Soil-Borne Pathogens. T. A. Tousson, R. V. Began, and P. E. Nelson, eds. Univ. Calif. Press, Berkeley. 252 pp.

Cook, R. J., and Papendick, R. I. 1972. Influence of water potential of soils and plants on root disease. Annu. Rev. Phytopathol. 10:349–374.

Cooke, L. R. 1991. Current problems in chemical control of late blight: The Northern Ireland experience. Pages 337–348 in: *Phytophthora.* J. A. Lucas, R. C. Shattock, D. S. Shaw, and L. R. Cooke, eds. Cambridge Univ. Press, Cambridge, U.K. 447 pp.

Cooke, L. R., Clifford, D. R., and Holgate, M. E. 1982. Control of potato late blight (*Phytophthora infestans*) with systemic fungicides. Pestic. Sci. 12:678–684.

Cookson, I. C. 1929. An account of a crown rot of English walnut trees in Victoria. Proc. R. Soc. Victoria 42 (N.S.):5–26.

Cooley, S. J, Hamm, P. B., and Hansen, E. M. 1985. Management guide to Phytophthora root rot in bareroot conifer nurseries of the Pacific Northwest. U.S. Dep. Agric. For. Serv. Publ. Pacific Northwest Region. 13 pp.

Cooper, D. C., and Porter, C. L. 1928. Phytophthora blight of peony. Phytopathology 18:881–899.

Corda, A. 1847. Beiträge zur Kunde der Kartoffel in Beziehung auf ihre Organisation und Krankheiten (Contribution to potato science in connection with its organization and disease). Hlubecks Oekonom. Neuigk. Verhandl. 59:465–471. (Cited in Tucker 1933).

Cordeiro, Z. J. M. 1981. Histopathologia de *Capsicum annuum* L. resistente e suscetivel a *Phytophthora capsici* L. (Histopathology of *Capsicum annuum* L. resistant and susceptible to *Phytophthora capsici* L.). Univ. Viçosa, Viçosa, M.G., Brasil. 57 pp. (In Portuguese).

Corliss, J. O. 1984. The kingdom Protista and its 45 phyla. BioSys 17:87–126.

Cormack, M. W. 1940. *Phytophthora cactorum* as a cause of root rot in sweetclover. Phytopathology 30:700–701.

Cother, E. J. 1975. *Phytophthora drechsleri:* Pathogenicity testing and determination of effective host range. Aust. J. Bot. 23:87–94.

Cother, E. J., and Griffin, D. M. 1973a. The role of alternative hosts in survival of *Phytophthora drechsleri.* Aust. J. Biol. Sci. 26:1109–1113.

Cother, E. J., and Griffin, D. M. 1973b. Formation of chlamydospores by *Phytophthora drechsleri.* Trans. Br. Mycol. Soc. 61:379–383.

Cother, E. J., and Griffin, D. M. 1974. Chlamydospore germination in *Phytophthora drechsleri.* Trans. Br. Mycol. Soc. 63:273–279.

Cotton, A. D. 1921. Report of the occurence of insect and fungus pests on plants in England and Wales for the year 1919. Minist. Agric. Fish. Misc. Publ. 33:26–68. (Cited in Tucker 1933).

Cotton, A. D. 1922. Fungus diseases of crops 1920–21. Minist. Agric. Fish. Misc. Publ. 38. 104 pp. (Cited in Tucker 1933).

Covey, R. P., Jr., and Harris, D. C. 1990. Phytophthora fruit rot. Pages 30–31 in: Compendium of Apple and Pear Diseases. A. L. Jones and H. S. Aldwinckle, eds. American Phytopathological Society, St. Paul, Minn. 100 pp.

Cox, A. E., and Large, E. C. 1960. Potato blight epidemics throughout the world. U.S. Dep. Agric. Handb. 174. 230 pp.

Cox, C. E. 1948. Natural infection of *Solanum dulcamara* by *Phytophthora infestans* (Abstr.). Phytopathology 38:575.

Cox, P. G., and Kasimani, C. 1988. Control of taro leaf blight using metalaxyl. Trop. Pest Manage. 34:81–84.

Coyier, D. L. 1981. Control of *Phytophthora ilicis* on English holly with sprays or soil drenches (Abstr.). Phytopathology 71:868.

Coyier, D. L., and Roane, M. K., eds. 1986. Compendium of Rhododendron and Azalea Diseases. American Phytopathological Society, St. Paul, Minnesota. 77 pp.

Crandall, B. S. 1936. Root disease of some conifers and hardwoods caused by *Phytophthora cambivora* (*P. cinnamomi*). Plant Dis. Rep. 20:202–204.

Crandall, B. S. 1947. A new *Phytophthora* causing root and collar rot of *Cinchona* in Peru. Mycologia 39:218–223.

Crandall, B. S. 1948. Gomosis de los Citrus (Citrus gummosis). El café de El Salvador 210:881–883. (In Spanish). (Rev. Appl. Mycol. 1965, 30:35.)

Crandall, B. S. 1950. Cinchona root and collar rot in Peru and Bolivia. U.S. Dep. Agric. Circ. 855. 16 pp.

Crandall, B. S. 1954. The diseases of the cinchona tree. F.A.O. Plant Prot. Bull. 3:33–37.

Crandall, B. S., and Davis, W. C. 1945. Phytophthora wilt and stem canker of cinchona. Phytopathology 35:138–140.

Crandall, B. S., and Dieguez, C. J. 1948. Phytophthora stem canker of sesame in Peru. Phytopathology 38:753–755.

Crandall, B. S., and Dieguez, C. J. 1953. Pyrethrum collar rot in Peru. Plant Dis. Rep. 37:482.

Crandall, B. S., and Gravatt, G. F. 1967a. The distribution of *Phytophthora cinnamomi.* Part I. Host distribution. Ceiba 13:43–53.

Crandall, B. S., and Gravatt, G. F. 1967b. The distribution of *Phytophthora cinnamomi.* Part II. Geographical distribution. Ceiba 13: 57–78.

Crandall, B. S., and Hartley, C. 1938. *Phytophthora cactorum* associated with seedling diseases in forest nurseries. Phytopathology 28:358–360.

Crandall, B. S., Gravatt, G. F., and Ryan, M. M. 1945. Root disease of *Castanea* species and some coniferous and broadleaf nursery stocks, caused by *Phytophthora cinnamomi.* Phytopathology 35: 162–180.

Crandall, B. S., Abrego, L., and Patiño, B. 1951. A check list of the diseases of economic plants of El Salvador, Central America. Plant Dis. Rep. 35:545–554.

Crandall, B. S., Parrado Alvarez, J. L., and Roque Pérez, R. 1953. Las enfermedades del kenaf y su control (Diseases of kenaf and their control). Circ. Agric. Minist. Agric. Cuba 4. 30 pp. (In Spanish). (Rev. Appl. Mycol. 1955, 34:786–787.)

Creager, D. B. 1937. Phytophthora crown rot of dogwood. J. Arnold Arbor. Harv. Univ. 18:344–348.

Creamer, J. R., and Bostock, R. M. 1986. Characterization and biological activity of *Phytophthora infestans* phospholipids in the hypersensitive response of potato tuber. Physiol. Mol. Plant Pathol. 28:215–225.

Cristinzio, M. 1935. Alcune malattie crittogamiche del Nespolo del Giappone ed in particolare la 'tricchiolatura' (Some fungal diseases of loquat and, in particular, scab). Ric. Ossvz. Divulg. Fitopatol. Compania Mezzogiorno (Portici). 4:25–50. (In Italian). (Rev. Appl. Mycol. 1935, 14:777–778.)

Cristinzio, M., and Verneau, R. 1954. L'eziologia del 'mal nero' del Noce in Campania (The etiology of 'black disease' of walnut in Campania). Ric. Fitopol. Campan. 12:3–34. (In Italian). (Rev. Appl. Mycol. 1956, 35:561.)

Critopoulos, P. D. 1955. Foot rot of tomato incited by *Phytophthora capsici.* Bull. Torrey Bot. Club 82:168–182.

Crooks, K. M. 1937. Studies on Australian aquatic Phycomycetes. Proc. R. Soc. Victoria 49 (N.S.):206–232.

Crosier, W. 1933. Culture of *Phytophthora infestans.* Phytopathology 23:713–720.

Crosier, W. 1934. Studies in the biology of *Phytophthora infestans* (Mont.) De Bary. Cornell Univ. Agric. Exp. Stn. Mem. 155. 40 pp.

Crossan, D. F., Lloyd, P. J., Hyre, R. A., and Heuberger, J. W. 1957. Control of downy mildew of lima bean. Plant Dis. Rep. 41:156–159.

Cruickshank, I. A. M. 1963. Phytoalexins. Annu. Rev. Phytopathol. 1:351–374.

Cruz, B. P. B., and Da Silveira, A. P. 1965. A requeima-do-pimentao vem do solo (Black rot of pepper comes from the soil). Coopercotia 22:37. (In Italian).

Csinos, A., and Hendrix, J. W. 1977. Nonparasitic stunting of tobacco plants by *Phytophthora cryptogea.* Can. J. Bot. 55:26–29.

Cugini, G., and Traverso, G. B. 1902. La *Sclerospora macrospora* Sacc. parassita della *Zea mays* (*Sclerospora macrospora* Sacc., parasitic on *Zea mays*). Linn. Staz. Sper. Agr. Ital. 35:46. (In Italian).

Cummins, J. N., Aldwinckle, H. S., and Byers, R. E. 1983. 'Novole' apple. Hort. Sci. 18:772–774.

Cunliffe, C., Lonsdale, D., and Epton, H. A. S. 1977. Transmission of *Phytophthora erythroseptica* on stored potatoes. Trans. Br. Mycol. Soc. 69:27–30.

Desjardins, P. R., Zentmyer, G. A., and Reynolds, D. A. 1969. Electron microscopic observations of the flagellar hairs of *Phytophthora palmivora* zoospores. Can. J. Bot. 47:1077–1079.

Desjardins, P. R., Zentmyer, G. A., and Reynolds, D. A. 1970. On the binucleate condition of the quadriflagellated zoospores of *Phytophthora palmivora.* Mycologia 62:421–427.

Desjardins, P. R., Wang, M. C., and Bartnicki-Garcia, S. 1973a. Electron microscopy of zoospores and cysts of *Phytophthora palmivora:* Morphology and surface texture. Arch. Mikrobiol. 88:61–70.

Desjardins, P. R., Zentmyer, G. A., Chen, D.-W., DeWolfe, T. A., Klotz, L. J., and Reynolds, D. A. 1973b. Flagellar hairs on zoospores of *Phytophthora* species: Tip hairs on the whiplash flagellum. Experientia 29:240–241.

Desmazieres, J. B. H. 1845. Sur la maladie des pommes de terre (On a disease of potatoes). Cryptogam. Fr. 1:492 (1846). (In French).

Despreaux, D., Cambrony, D., Clement, D., Nyasse, S., and Partiot, M. A. D. 1987. Etude de la pourriture brune des cabosses su cacaoyer au Cameroun: Definition de nouvelles methods de lutte (Study of black pod in Cameroon: Description of new control methods). Proc. Tenth Int. Cocoa Res. Conf. Santo Domingo, Dominican Republic, 17–23 May 1987. 924 pp.

Deutschmann, F. 1954. Eine wurzelfäule an Erdbeeren, hervorgerufen durch *Phytophthora cactorum* (Leb & Cohn) Schroet (A root rot of strawberries caused by *Phytophthora cactorum* [Leb. & Cohn.] Schroet.). Nachrichtenbl. Dtsch. Pflanzenschutzdienst (Braunschweig) 6:7–9. (Cited in Nienhaus 1960).

Deva Rajan, M. R., and Aiyappa, K. M. 1945. Leaf fall and fruit-rot disease of oranges (*Citrus aurantium*). Indian Farming 5:512–513.

Dhingra, O. D., and Sinclair, J. B. 1985. Detection and estimation of inoculum. Pages 70–73 in: Basic Plant Pathology Methods. CRC Press, Boca Raton, Fla. 355 pp.

Dick, M. W. 1969. Morphology and taxonomy of the Oomycetes, with special reference to Saprolegniaceae, Leptomitaceae, and Pythiaceae. I. Sexual reproduction. New Phytol. 68:751–775.

Dick, M. W. 1972. Morphology and taxonomy of the Oomycetes, with special reference to Saprolegniaceae, Leptomitaceae, and Pythiaceae. II. Cytogenetic systems. New Phytol. 71:1151–1159.

Dick, M. W. 1982. Oomycetes. Pages 179–180 in: Synopsis and Classification of Living Organisms. Vol. I. S. B. Parker, ed. McGraw-Hill Book Co., New York. 1166 pp.

Dick, M. W. 1987. Sexual reproduction: Nuclear cycles and life histories with particular reference to lower eukaryotes. Biol. J. Linn. Soc. 30:181–192.

Dick, M. W. 1989. *Phytophthora undulata* comb. nov. Mycotaxon 35:449–453.

Dick, M. W. 1990a. Phylum Oomycota. Pages 661–685 in: Handbook of Protoctista. L. Margulis, J. O. Corliss, M. Melkonian, D. J. Chapman, eds. Jones and Bartlett Pub. Boston. 914 pp.

Dick, M. W. 1990b. Keys to *Pythium.* Published by the author. Reading, U.K. 64 pp.

Dick, M. W. 1992. Patterns of phenology in populations of zoosporic fungi. Chapter 20, pages 355–382 in: The Fungal Community. Its Organization and Role in the Ecosystem. G. C. Carroll and D. T. Wicklow, eds. Marcell Dekker, New York. 976 pp.

Dick, M. W. 1995a. Sexual reproduction in the Perononosporomycetes (chromistan fungi). Can. J. Bot. 73(Suppl. 1):5712–5724.

Dick, M. W. 1995b. The Straminipilous Fungi. A New Classification for the Biflagellate Fungi and Their Uniflagellate Relatives with Particular Reference to Lagenidiaceous Fungi. C.A.B. Internat. Mycol. Pap. No. 168.

Dick, M. W., and Win-Tin. 1973. The development of cytological theory in the Oomycetes. Biol. Rev. 48:133–158.

Diener, U. L. 1955. Sporulation in pure culture by *Stemphylium solani.* Phytopathology 45:141–145.

Dietrich, S. M. C. 1976. Presence of polyphosphate of low molecular weight in Zygomycetes. J. Bacteriol. 127:1408–1413.

Dimitroff, T. 1926. Fungi injurious to our forests and silviculture (Translated title). Bull. Soc. Bot. Bulgarie 1:53–66. (Rev. Appl. Mycol. 1926, 5:525). (In Bulgarian). (Cited in Tucker 1933).

Dimock, A. W., and Baker, K. F. 1944. Hot-water treatment for control of Phytophthora root rot of calla. Phytopathology 34:979–981.

Dingley, J. M. 1969. Records of plant diseases in New Zealand. N.Z. Dep. Sci. Ind. Res. Bull. 192. 298 pp.

Dixon, R. A., Dey, P. M., and Lamb, C. J. 1983. Phytoalexins: Enzymology and molecular biology. Adv. Enzymol. 55:1–136.

Do Amaral, J. F. 1952. Requeima-do-pimentao (Chili blight). Biologico 18:160–161. (In Spanish).

Doepel, R. F. 1966. *Phytophthora* species on citrus in Western Australia. Plant Dis. Rep. 50:494–496.

Doidge, E. M. 1924. A preliminary check list of plant diseases occurring in South Africa. Bot. Surv. S. Afr. Mem. 6. 56 pp. (Cited in Tucker 1933).

Doidge, E. M. 1925. Brown rot in citrus fruits (*Pythiacystis citrophthora* R. and E. Smith). J. Dep. Agric. S. Afr. 10:499–503. (Cited in Tucker 1933).

Doke, N., and Tomiyama, K. 1977. Effect of high molecular substances released from zoospores of *Phytophthora infestans* on hypersensitive response of potato tubers. Phytopathol. Z. 90:236–242.

Doke, N., and Tomiyama, K. 1980a. Effect of hyphal wall components from *Phytophthora infestans* on protoplasts of potato tuber tissues. Physiol. Plant Pathol. 16:169–176.

Doke, N., and Tomiyama, K. 1980b. Suppression of the hypersensitive response of potato tuber protoplasts to hyphal wall components by water soluble glucans isolated from *Phytophthora infestans.* Physiol. Plant Pathol. 16:177–186.

Doke, N., Nakae, Y., and Tomiyama, K. 1976. Effect of blasticidin S on the production of rishitin in potato tuber tissue infected by an incompatible race of *Phytophthora infestans.* Phytopathol. Z. 87: z337–344.

Doke, N., Garas, N. A., and Kuć, J. 1979. Partial characterization and aspects of the mode of action of a hypersensitivity-inhibiting factor (HIF) isolated from *Phytophthora infestans.* Physiol. Plant Pathol. 15:127–140.

Doke, N., Garas, N. A., and Kuć, J. 1980. Effect of host hypersensitivity of suppressors released during the germination of *Phytophthora infestans* cytospores. Phytopathology 70:35–39.

Doke, N., Ramirez, A. V., and Tomiyama, K. 1987. Systematic induction of resistance in potato plants against *Phytophthora infestans* by local treatment with hyphal wall components of the fungus. J. Phytopathol. 119:232–239.

Dolan, T. E., and Coffey, M. D. 1984. Biocontrol of *Phytophthora cinnamomi* on *Persea indica* and *P. americana* by prior inoculation with *Phytophthora parasitica* (Abstr.). Phytopathology 74:807.

Dolan, T. E., and Coffey, M. D. 1986. Laboratory screening technique for assessing resistance of four avocado rootstocks to *Phytophthora cinnamomi.* Plant Dis. 70:115–118.

Dolan, T. E., and Coffey, M. D. 1988. Correlative *in vitro* and *in vivo* behavior of mutant strains of *Phytophthora palmivora* expressing different resistances to phosphorus acid and fosetyl-Na. Phytopathology 78:974–978.

Dolan, T. E., Cohen, Y., and Coffey, M. D. 1986. Protection of *Persea* species against *Phytophthora cinnamomi* and *P. citricola* by prior inoculation with a citrus isolate of *P. parasitica.* Phytopathology 76:194–198.

Dolgova, A. V., and D'yakov, Y. T. 1984. Genetic evidence of the diploid nature of nuclei in the mycelium of the phytopathogenic fungus *Phytophthora infestans.* Doklady Akademii Nauk SSSR 276:1494–1496. (Translated from Russian).

Doster, M. A., and Bostock, R. M. 1988a. The effect of temperature and type of medium on oospore production by *Phytophthora syringae.* Mycologia 80:77–81.

Doster, M. A., and Bostock, R. M. 1988b. Susceptibility of almond cultivars and stone fruit species to pruning wound cankers caused by *Phytophthora syringae.* Plant Dis. 72:490–492.

Doster, M. A., and Bostock, R. M. 1988c. Chemical protection of almond pruning wounds from infection by *Phytophthora syringae.* Plant Dis. 72:492–494.

Doster, M. A., and Bostock, R. M. 1988d. Incidence, Distribution, and Development of pruning wound cankers caused by *Phytophthora syringae* in almond orchards in California. Phytopathology 78:468–472.

Doster, M. A., and Bostock, R. M. 1988e. Quantification of lignin formation in almond bark in response to wounding and infection by *Phytophthora* species. Phytopathology 78:473–477.

Doster, M. A., and Bostock, R. M. 1988f. Effects of low temperature on resistance of almond trees to *Phytophthora* pruning wound cankers in relation to lignin and suberin formation in wounded bark tissue. Phytopathology 78:478–483.

Doster, M. A., and Fry, W. E. 1991. Evaluation by computer simulation of strategies to time metalaxyl applications for improved control of potato late blight. Crop Prot. 10:209–214.

Doster, M. A., Milgroom, M. G., and Fry, W. E. 1990. Quantification of factors influencing potato late blight suppression and selection for metalaxyl resistance in *Phytophthora infestans:* A simulation approach. Phytopathology 80:1190–1198.

Douchet, J. P., Absi, M., Hay, S. J. B., Muntan, L., and Villani, A. 1977. European results with DPX 3217, a new fungicide for the control of grape downy mildew and potato late blight. Proc. Br. Crop Prot. Conf. Pests Dis. 1977, 2:535–540. Croydon, England. Br. Crop Prot. Counc. Pub.

Dowley, L. J., and O'Sullivan, E. 1981. Metalaxyl-resistant strains of *Phytophthora infestans* (Mont.) de Bary in Ireland. Potato Res. 24: 417–421.

Dowley, L. J., and O'Sullivan, E. 1991. Changes in the distribution of phenylamide-resistant strains of *Phytophthora infestans* (Mont.) de Bary in Ireland 1985–89. Potato Res. 34:67–69.

Dowley, L. J., O'Sullivan, E., and Kehoe, H. W. 1991. Development and evaluation of blight resistant potato cultivars. Pages 373–382 in: *Phytophthora.* J. A. Lucas, R. C. Shattock, D. S. Shaw, and L. R. Cooke, eds. Cambridge Univ. Press, Cambridge, U.K. 447 pp.

Dowley, L. J., Bannon, E., Cooke, L. R., Keane, T., and O'Sullivan, E., eds. 1995. *Phytophthora infestans* 150. Boole Press, Dublin, Ireland. 392 pp.

Downes, M. J., and Loughnane, J. B. 1969. New or uncommon plant diseases and pests *Phytophthora megasperma* Drechsl. on broccoli and swede in the Republic of Ireland. Plant Pathol. 18:48.

Dowson, W. J. 1934. *Phytophthora megasperma* Drechsler in Tasmania. Trans. Br. Mycol. Soc. 19:89–90.

Drapeau, R., Fortin, J. A., and Gagnon, C. 1973. Antifungal activity of *Rhizobium.* Can. J. Bot. 51:681–682.

Drechsler, C. 1926. Foot-rot of *Lilium candidum* and *Lilium pyrenaicum* caused by *Phytophthora cactorum.* Phytopathology 16:51–53.

Drechsler, C. 1929a. A fruit rot of honey dew melons due to a species of *Phytophthora* (Abstr.). Phytopathology 19:85.

Drechsler, C. 1929b. A diplanetic species of *Phytophthora* causing pink rot of potato tubers (Abstr.). Phytopathology 19:92.

Drechsler, C. 1930. Repetitional diplanetism in the genus *Phytophthora.* U.S. Dep. Agric. J. Agric. Res. 40:557–573.

Drechsler, C. 1931. A crown rot of hollyhocks caused by *Phytophthora megasperma* n. sp. J. Wash. Acad. Sci. 21:513–526.

Drechsler, C. 1938. Two hyphomycetes parasitic on oospores of root-rotting Oomycetes. Phytopathology 28:81–103.

Drechsler, C. 1946. Several species of *Pythium* peculiar in their sexual development. Phytopathology 36:781–864.

Drenth, A., Janssen, E. M., and Govers, F. 1995. Formation and survival of oospores of *Phytophthora infestans* under natural conditions. Plant Pathol. 44:86–94.

Drilias, M. J., Kuntz, J. E., and Worf, G. L. 1982. Collar rot and basal canker of sugar maple. J. Arboric. 8:29–33.

Drilias, M. J., Kuntz, J. E., and Worf, G. L. 1984. Sugar maple cotyledons used to detect *Phytophthora citricola* and *P. cactorum* in soils (Abstr.). Phytopathology 74:845.

Driver, C. M. 1957. Infection of native *Solanum* species by the potato blight fungus. Nature (London) 180:1367–1368.

Dufrénoy, J. 1925. Maladies du cedratier et du citronnier en Corse (Diseases of the citron and of the lemon in Corsica). Bull. Off. Reg. Agric. Midi. 26 pp. (Rev. Appl. Mycol. 1926, 5:549.)

Dufrénoy, J. 1927. Un *Phytophthora* parasite du néflier du Japon (A *Phytophthora* parasite on the loquat). Rev. Pathol. Veg. Entomol. Agric. Fr. 14:201–206 (In French). (Cited in Tucker 1933).

Dukes, P. D. 1969. The influence of some nonhost crops on the incidence of black shank of flue-cured tobacco (Abstr.). Phytopathology 59:113.

Dukes, P. D., and Apple, J. L. 1961. Chemotaxis of zoospores of *Phytophthora parasitica* var. *nicotianae* by plant roots and certain chemical solutions. Phytopathology 51:195–197.

Dumbleton, L. J. 1954. A list of plant diseases recorded in South Pacific territories. Tech. Pap. S. Pacific Comm. 78 pp. (Rev. Appl. Mycol., 1956, 35:423.

Duncan, D. R., and Paxton, J. D. 1981. Trifluralin enhancement of Phytophthora root rot of soybean. Plant Dis. 65:435–436.

Duncan, J., Fordyce, W., Harper, P. C., and Rankin, P. A. 1986. Eliminating red core (*Phytophthora fragariae*) from Scottish certified stock strawberries. Res. Dev. Agric. 3:43–46.

Duncan, J. M. 1975a. Germination of oospores of *Phytophthora fragariae.* Trans. Br. Mycol. Soc. 65:338–341.

Duncan, J. M. 1975b. Red core of strawberry. Annu. Rep. Scott. Hortic. Res. Inst. 1974. Page 55.

Duncan, J. M. 1976. The use of bait plants to detect *Phytophthora fragariae* in soil. Trans. Br. Mycol. Soc. 66:85–89.

Duncan, J. M. 1977. Germination in vitro of *Phytophthora fragariae* oospores from infected root tissue. Trans Br. Mycol. Soc. 69:391–395.

Duncan, J. M. 1979. *In situ* bait plants as a method for detecting red core in strawberry fields. Ann. Appl. Biol. 92:307–312.

Duncan, J. M. 1980a. A technique for detecting red stele (*Phytophthora fragariae*) infection of strawberry stocks before planting. Plant Dis. 64:1023–1025.

Duncan, J. M. 1980b. Persistence of mycelium of *Phytophthora fragariae* in soil. Trans. Br. Mycol. Soc. 75:383–387.

Duncan, J. M. 1985a. Effect of temperature and other factors on *in vitro* germination of *Phytophthora fragariae* oospores. Trans. Br. Mycol. Soc. 85:455–462.

Duncan, J. M. 1985b. Effect of fungicides on survival, infectivity and germination of *Phytophthora fragarie* oospores. Trans. Br. Mycol. Soc. 85:585–593.

Duncan, J. M. 1988. A colour reaction associated with formation of oospores by *Phytophthora* spp. Trans. Br. Mycol. Soc. 90:336–337.

Duncan, J. M., and Cowan, J. B. 1980. Effect of temperature and soil moisture content on persistence of infectivity of *Phytophthora fragariae* in naturally infested field soil. Trans. Br. Mycol. Soc. 75:133–139.

Duncan, J. M., and Kennedy, D. M. 1989. The effect of waterlogging on Phytophthora root rot of red raspberry. Plant Pathol. 38:161–168.

Duncan, J. M., Kennedy, D. M., and Seemüller, E. 1987. Identities and pathogenicities of *Phytophthora* spp. causing root rot of red raspberry. Plant Pathol. 36:276–289.

Duncan, J. M., Kennedy, D. M., and Scott, P. H. 1991. Relationships between non-papillate soilborne species of *Phytophthora:* Root rot of raspberry. Pages 129–147 in: *Phytophthora.* J. A. Lucas, R. C. Shattock, D. S. Shaw, and L. R. Cooke, eds. Cambridge Univ. Press, Cambridge, U.K. 447 pp.

Duniway, J. M. 1975a. Formation of sporangia by *Phytophthora drechsleri* in soil at high matric potentials. Can. J. Bot. 53:1270–1275.

Duniway, J. M. 1975b. Limiting influence of low water potential on the formation of sporangia by *Phytophthora drechsleri* in soil. Phytopathology 65:1089–1093.

Duniway, J. M. 1976. Movement of zoospores of *Phytophthora cryptogea* in soils of various textures and matric potentials. Phytopathology 66:877–882.

Duniway, J. M. 1977a. Changes in resistance to water transport in safflower during the development of Phytophthora root rot. Phytopathology 67:331–337.

Duniway, J. M. 1977b. Predisposing effect of water stress on the severity of Phytophthora root rot in safflower. Phytopathology 67: 884–889.

Duniway, J. M. 1979. Water relations of water molds. Annu. Rev. Phytopathol. 67:431–460.

Duniway, J. M. 1983. Role of physical factors in the development of *Phytophthora* diseases. Pages 175–187 in: *Phytophthora:* Its Biology, Taxonomy, Ecology, and Pathology. D. C. Erwin, S. Bartnicki-Garcia, and P. H. Tsao, eds. American Phytopathological Society, St. Paul, Minn. 392 pp.

Duniway, J. M., and Gordon, T. R. 1986. Water relations and pathogen activity in soil. In: Symposium on Water, Fungi, and Plants held at the 3rd British Mycological Society General Meeting, Lancaster, England, Bull. Br. Mycol. Soc. 19(Suppl.1):11.

Dunstan, R. H., and Grant, B. R. 1990. The presence of phosphonate sensitive macromolecules on the surface of *Phytophthora* spp. (Abstr.). Proc. IV Int. Mycol. Congr., Regensburg, Germany. Page 181.

Dunstan, R. H., Smillie, R. H., and Grant, B. R. 1990a. The effects of subtoxic levels of phosphate on the metabolism and potential virulence factors of *Phytophthora palmivora.* Physiol. Mol. Plant Pathol. 36:205–220.

Dunstan, R. H., Iser, J., and Grant, B. R. 1990b. Suppression of the wound response in potato leaves by surface components from the mycelium of *Phytophthora infestans* (Abstr.). Proc. IV Int. Mycol. Congr., Regensburg, Germany. Page 180.

Durand, M.-C., and Sallé, G. 1981. Effet du tris-o-éthyl phosphonate d'aluminium sur le couple *Lycopersicum esculentum* Mill. *Phytophthora capsici* Leon. Étude cytologique et cytochimique (Effect of tris-o-ethyl phosphonate of aluminum on *Phytophthora capsici* Leon. on tomato. A cytologic and cytochemical study). Agronomie 1:723–731. (In French).

Durrell, L. W. 1968. Hyphal invasion by *Trichoderma viride.* Mycopathol. Mycol. Appl. 35:138–144.

Dutt, B. L. 1982. Physiologic specialization in *Phytophthora* species. Pages 139–147 in: *Phytophthora* Diseases of Tropical Cultivated Plants. K. K. K. Nambiar, ed. Cent. Plant. Crops Res. Inst., Kasaragod, Kerala, India. 284 pp.

Dutta, P. K., Hedge, R. K., and Anahosur, K. H. 1984. An assured *in vitro* technique for the germination of oospores of *Phytophthora palmivora.* Curr. Sci. (India) 53:655–656.

Dwyer, R. E. P. 1937. The diseases of coconuts (*Cocos nucifera*) in New Guinea. New Guinea Agric. Gaz. 3:28–93.

Dwyer, R. E. P. 1953. Diseases of coconuts (*Cocos nucifera*) in Papua and New Guinea. Papua New Guinea Agric. Gaz. 8:24–40.

Easton, G. D., and Nagle, M. E. 1985. Timing and root absorption affecting efficiency of metalaxyl in controlling *Phytophthora infestans* on potato in northwestern Washington State. Plant Dis. 69: 499–500.

Ebel, J. 1986. Phytoalexin synthesis: The biochemical analysis of the induction process. Annu. Rev. Phytopathol. 24:235–264.

Ebel, J., and Grisbach, H. 1988. Defense strategies of soybean against the fungus *Phytophthora megasperma* f. sp. *glycinea:* A molecular analysis. Trends Biochem. Sci. 13:23–27.

Ebel, J., Stäb, M. R., and Schmidt, W. E. 1985. Induction of enzymes of phytoalexin synthesis in soybean cells by fungal elicitor. Pages 247–254 in: Primary and Secondary Metabolism of Plant Cell Cultures. Neumann et al., eds. Springer-Verlag, Berlin. 377 pp.

Eckert, J. W., and Brown, G. E. 1986. Postharvest citrus diseases and their control. Pages 315–360 in: Fresh Citrus Fruits. W. F. Wardowski, S. Nagy, and W. Grierson, eds. AVI, Westport, Conn. 571 pp.

Eckert, J. W., and Tsao, P. H. 1960. A preliminary report on the use of pimaricin in the isolation of *Phytophthora* spp. from root tissues. Plant Dis. Rep. 44:660–661.

Eckert, J. W., and Tsao, P. H. 1962. A selective antibiotic medium for isolation of *Phytophthora* and *Pythium* from plant roots. Phytopathology 52:771–777.

Edathil, T. T., and George, M. K. 1976. *Phytophthora nicotianae* var. *parasitica* (Dastur) Waterhouse on *Hevea brasiliensis* in South India. Rubber Board Bull. 13:3–4.

Edgington, L. V., Khew, K. L., and Barron, G. L. 1971. Fungitoxic spectrum of benzimidazole compounds. Phytopathology 61:42–44.

Edmiston, R. J. 1975. A selection of species resistant to *Phytophthora cinnamomi.* For. Focus. 14:15.

Edney, K. L. 1978. The infection of apples by *Phytophthora syringae.* Ann. Appl. Biol. 88:31–36.

Edney, K. L., and Chambers, D. A. 1981a. Post-harvest treatments for the control of *Phytophthora syringae* storage rot of apples. Ann. Appl. Biol. 97:237–241.

Edney, K. L., and Chambers, D. A. 1981b. The use of metalaxyl to control *Phytophthora syringae* rot of apple fruits. Plant Pathol. 30: 167–170.

Egan, A. R., Murray, A., and Mullins, S. 1995. Past history and future prospects for fungicides for the control of *Phytophthora infestans* on potatoes. Pages 160–170 in: *Phytophthora infestans* 150. L. J. Dowley, E. Bannon, L. R. Cooke, T. Keane, and E. O'Sullivan, eds. Boole Press, Dublin, Ireland.

El-Din, G., Ahmed, K. G. M., Fadl, F. A. M., and Khatagi, Y. S. 1981. Physiological studies and host range of *Phytophthora cactorum*—The causal organism of leather rot of fruits and collar rot of strawberry, *Fragaria* sp., in Egypt. Air Shams Univ. Fac. Agric. Res. Bull. O(1465):1–18.

El-Gholl, N. E., and Alfieri, S. A., Jr. 1984. Fruit rot of fig caused by *Phytophthora palmivora.* Proc. Fla. State. Hortic. Soc. 97:327–328.

El-Goorani, M. A., Farag, S. A., and Shehata, M. R. A. 1976. The effect of *Bacillus subtilis* and *Penicillium patulum* on in vitro growth and pathogenicity of *Rhizoctonia solani* and *Phytophthora cryptogea.* Phytopathol. Z. 85:345–352.

El-Hamalawi, Z. A., and Erwin, D. C. 1986a. Physical, enzymic, and chemical factors affecting viability and germination of oospores of *Phytophthora megasperma* f. sp. *medicaginis.* Phytopathology 76: 503–507.

El-Hamalawi, Z. A., and Erwin, D. C. 1986b. Components in alfalfa root extract and root exudate that increase oospore germination of *Phytophthora megasperma* f. sp. *medicaginis.* Phytopathology 76: 508–513.

El-Hamalawi, Z. A., and Menge, J. A. 1994a. Avocado trunk canker disease caused by *Phytophthora citricola:* Investigation of factors affecting infection and disease development. Plant Dis. 78:260–264.

El-Hamalawi, Z. A., and Menge, J. A. 1994b. Effect of wound age and fungicide treatment of wounds on susceptibility of avocado stems to infection by *Phytophthora citricola.* Plant Dis. 78:700–704.

El-Hamalawi, Z. A., and Menge, J. A. 1995. Seasonal fluctuations in the extent of colonization of avocado plants by the stem canker pathogen *Phytophthora citricola.* J. Am. Soc. Hortic. Sci. 120:157–162.

El-Helaly, A. F., Assawah, M. W., Elarosi, H. M., and Wasfy, E. E. H. 1968. Fruit rots of vegetable marrow in Egypt (United Arab Republic). Phytopathol. Mediterr. 7:107–109.

El-Shehedi, A. A., Satour, M. M., Abduo, Y. A., and Hassan, M. M. 1976. Studies on Phytophthora root and crown rot of sesame. Egypt. J. Phytopathol. 8:1–8.

Elenkov, E. 1975. Nyakoi izmene niya pri bolestite po zelenchukovite kulturioranzheriite (Some changes in diseases of vegetable crops in the glass house). Rastit. Zash. 23:5–7. (Rev. Plant Pathol. 1976, 55:1160.)

Elenkov, E., and Bakharieva, V. 1975. Mildew, *Phytophthora capsici,* a dangerous disease of pepper in Bulgaria. Rastit. Zash. 23:31–35.

Elfving, D. C., Kaufmann, M. R., and Hall, A. E. 1972. Interpreting leaf water potential measurements with a model of the soil-plant-atmosphere continuum. Physiol. Plant. 27:161–168.

Elgin, J. H. Jr., Welty, R. E., and Gilchrist, D. B. 1988. Breeding for disease and nematode resistance. Pages 827–858 in: Alfalfa and Alfalfa Improvement. A. A.Hanson, D. K. Barnes, and R. R. Hill, eds. Agron. Monogr. No. 29. ASA-CSSA-SSSA, Madison, Wis. 1084 pp.

Elliott, C. G. 1972. Calcium chloride and growth and reproduction of *Phytophthora cactorum.* Trans. Br. Mycol. Soc. 58:169–172.

Elliott, C. G. 1977. Sterols in fungi: Their functions in growth and reproduction. Adv. Microbiol. Physiol. 15:121–173.

Elliott, C. G. 1983. Physiology of sexual reproduction in *Phytophthora.* Pages 71–80 in: *Phytophthora:* Its Biology, Taxonomy, Ecology, and Pathology. D. C. Erwin, S. Bartnicki-Garcia, and P. H. Tsao, eds. American Phytopathological Society, St. Paul, Minn. 392 pp.

Elliott, C. G. 1986. Inhibition of reproduction of *Phytophthora* by the calmodulin-interacting compounds trifluoperazine and ophiobolin. A. J. Gen. Microbiol. 132:2781–2785.

Elliott, C. G. 1988. Stages in oosporogenesis of *Phytophthora* sensitive to inhibitors of calmodulin and phosphodiesterase. Trans. Br. Mycol. Soc. 90:187–192.

Elliott, C. G. 1989. Some aspects of nitrogen nutrition and reproduction in *Phytophthora.* Mycol. Res. 92:34–44.

Elliott, C. G., and MacIntyre, D. 1973. Genetical evidence on the life-history of *Phytophthora.* Trans. Br. Mycol. Soc. 60:311–316.

Elliott, C. G., Hendrie, M. R., Knights, B. A., and Parker, W. 1964. A steroid growth factor requirement in a fungus. Nature (London) 203:427–428.

Elliott, C. K., and Maxwell, D. P. 1984. Evaluation of isozymes for the identification of isolates of *Phytophthora megasperma* (Abstr.). Phytopathology 74:866.

Ellis, M. A., and Grove, G. G. 1983. Leather rot in Ohio strawberries. Plant Dis. 67:549.

Ellis, M. A., Ferree, D. C., and Madden, L. V. 1986. Evaluation of metalaxyl and captafol soil drenches, composted hardwood bark soil amendments, and graft union placement on control of apple collar rot. Plant Dis. 70:24–26.

Ellis, M. A., Converse, R. H., Williams, R. N., and Williamson, B. 1991. Compendium of Raspberry and Blackberry Diseases and Insects. American Phytopathological Society, St. Paul, Minnesota. 122 pp.

Ellstrand, N. C. 1984. Multiple paternity within the fruits of wild radish, *Raphinus sativus*. Am. Nat. 123:819–828.

Engel, G. 1977. Kragenfaüleanfalligkeit verschiedener neuer unterlagen (Collar rot susceptibility of different new rootstocks) (Abstr.). Plant Breed. 48:681.

Engelhard, A. W. 1974a. Crown rot and wilt of baby's breath (*Gypsophila paniculata* L.) caused by the soil fungus *Phytophthora parasitica* Dast. Proc. Fla. State Hortic. Soc. 86:428–431.

Engelhard, A. W. 1974b. A serious new crown rot and wilt of baby's breath (*Gypsophila paniculata*) incited by *Phytophthora parasitica*. Plant Dis. Rep. 58:669–672.

Engelhard, A. W., and Ploetz, R. C. 1979. Phytophthora crown and stem rot, an important new disease of poinsettia (*Euphorbia pulcherrima*). Proc. Fla. State Hortic. Soc. 92:348–350.

Englander, L., and Roth, L. F. 1980. Interaction of light and sterol on sporangium and chlamydospore production by *Phytophthora lateralis*. Phytopathology 70:650–654.

Englander, L., and Turbitt, W. 1979. Increased chlamydospore production by *Phytophthora cinnamomi* using sterols and near-ultraviolet light. Phytopathology 69:813–817.

Englander, L., Merlino, J. A., and McGuire, J. J. 1980. Efficacy of two new systemic fungicides and ethazole for control of Phytophthora root rot of Rhododendron, and spread of *Phytophthora cinnamomi* in propagation benches. Phytopathology 70:1175–1179.

Engvall, E., and Perlmann, P. 1971. Enzyme-linked immunosorbent assay (ELISA). Quantitative assay of immunoglobulin G. Immunochemistry 8:871–874.

Ericksson, B., and Laane, M. M. 1982. Examinations of nuclear behaviour during sexual cycle in the diplontic oomycete *Phytophthora palmivora* (Butl.) Butl. Hereditas 97:73–86.

Eriksson, J. 1917. Développement primaire du mildou (*Phytophthora infestans*) au cours de la végétation de la pomme de terre (Primary development of mildew [*Phytophthora infestans*] in the progress of growth of potato). Rev. Gen. Bot. 29:257–260; 305–320; 333–349; 376–380; 1918, Rev. Gén. Bot. 30:16–30. (Cited by Tucker 1933).

Érsek, T. 1975. The sensitivity of *Phytophthora infestans* to several antibiotics. Z. Pflanzenkr. Pflanzenschutz 82:614–617.

Érsek, T. 1982. A burgonyavész kórokazója: A *Phytophthora infestans* (The causative agent of potato blight: *Phytophthora infestans*). J. Hungarian Solanum Soc., Budapest 2:19–47 (In Hungarian).

Érsek, T., Hölker, U., and Höfer, M. 1991. Non-lethal immobilization of zoospores of *Phytophthora infestans* by Li^+. Mycol. Res. 95: 970–972.

Érsek, T., Schoelz, J. E., and English, J. T. 1993. Fusion and transformation of zoospores of *Phytophthora* spp. by PEG/Li^+ treatment. (Abstr.) 6th Int. Congr. Plant Pathology, 1993. Montreal, Quebec. p. 279.

Érsek, T., Schoelz, J. E., and English, J. T. 1994a. PCR amplification of species-specific DNA sequences can distinguish among *Phytophthora* species. Appl. Environ. Microbiol. 60:2616–2621.

Érsek, T., Schoelz, J. E., and English, J. T. 1994b. Characterization of selected drug-resistant mutants of *Phytophthora capsici, P. parasitica,* and *P. citrophthora.* Acta Phytopathol. et Entomol. Hung. 29:215–229.

Érsek, T., Galal, A. A., and Kevei, F. 1994c. Mucidin, an antibiotic highly effective on *Phytophthora infestans.* Acta Phytopathol. et Entomol. Hung. 29:239–247.

Érsek, T., English, J. T., and Schoelz, J. E. 1995. Creation of species hybrids of *Phytophthora* with modified host ranges by zoospore fusion. Phytopathology 85:1343–1347.

Erselius, L. J., and Boccas, B. 1978. Gametangial meiosis in cocoa black pod *Phytophthora* isolates. Ann. Phytopathol. 10:287–294.

Erselius, L. J., and de Vallavieille, C. 1984. Variation in protein profiles of *Phytophthora:* Comparison of six species. Trans. Br. Mycol. Soc. 83:463–472.

Erselius, L. J., and Shaw, D. S. 1982. Protein and enzyme differences between *Phytophthora palmivora* and *Phytophthora megakarya:* Evidence for self-fertilization in pairings of the two species. Trans. Br. Mycol. Soc. 78:227–238.

Ershad, D. 1971. Beitrag zur Kenntnis der *Phytophthora* Arten in Iran und ihrer Phytopathologischen Bedeutung (Contribution to the knowledge of *Phytophthora* species in Iran and their phytopathogenic importance). Mitt. Biol. Bundesanst. Land u. Forstw. (Berl.-Dahlem) 140. 84 pp. (Rev. Plant Pathol. 1972, 51:198)

Ershad, D., and Mostowfipoor, P. 1969. The die-back or root rot of cucurbits in Iran by *Phytophthora drechsleri* Tucker. Iran. J. Plant Pathol. 5:38–45.

Erwin, D. C. 1950. Phytophthora root rot of safflower in Nebraska caused by *Phytophthora drechsleri.* Plant Dis. Rep. 34:306.

Erwin, D. C. 1952. Phytophthora root rot of safflower. Phytopathology 42:32–35.

Erwin, D. C. 1954. Root rot of alfalfa caused by *Phytophthora cryptogea.* Phytopathology 44:700–704.

Erwin, D. C. 1964. A strain of *Phytophthora parasitica* from okra and its sexual compatibility with isolates from citrus. Phytopathology 54:114–115.

Erwin, D. C. 1965. Reclassification of the causal agent of root rot of alfalfa from *Phytophthora cryptogea* to *P. megasperma.* Phytopathology 55:1139–1143.

Erwin, D. C. 1966. Varietal reaction of alfalfa to *Phytophthora megasperma* and variation in virulence of the causal fungus. Phytopathology 56:653–657.

Erwin, D. C. 1968. The effect of calcium on mycelial growth of *Phytophthora megasperma* and *P. cinnamomi.* Mycologia 60:112–116.

Erwin, D. C. 1983. Variability within and among species of *Phytophthora.* Pages 149–165 in: *Phytophthora:* Its Biology, Taxonomy, Ecology, and Pathology. D. C. Erwin, S. Bartnicki-Garcia, and P. H. Tsao, eds. American Phytopathological Society, St. Paul, Minn. 392 pp.

Erwin, D. C., and Katznelson, H. 1961a. Studies on the nutrition of *Phytophthora cryptogea.* Can. J. Microbiol. 7:15–25.

Erwin, D. C., and Katznelson, H. 1961b. Suppression and stimulation of mycelial growth of *Phytophthora cryptogea* by certain thiamine-requiring and thiamine-synthesizing bacteria. Can. J. Microbiol. 7:945–950.

Erwin, D. C., and Kennedy, B. W. 1957. Studies on Phytophthora root rot of alfalfa (Abstr.). Phytopathology 47:520.

Erwin, D. C., and Kraft, J. M. 1967. Dynamic life processes of seven plant pathogenic fungi shown by cinephotomicrography (Abstr.). Phytopathology 57:98.

Erwin, D. C., and Li, X. 1986. Variability in cultural type, oospore formation, and virulence on alfalfa of single zoospore and oospore isolates of *Phytophthora megasperma* f. sp. *medicaginis* (PMM) (Abstr.). Phytopathology 76:1102.

Erwin, D. C., and McCormick, W. H. 1971. Germination of oospores produced by *Phytophthora megasperma* var. *sojae.* Mycologia 63: 972–977.

Erwin, D. C., Kennedy, B. W., and Lehman, W. F. 1959. Xylem necrosis and root rot of alfalfa associated with excessive irrigation and high temperatures. Phytopathology 49:572–578.

Erwin, D. C., Zentmyer, G. A., Galindo, J., and Niederhauser, J. S. 1963. Variation in the genus *Phytophthora.* Annu. Rev. Phytopathol. 1:375–396.

Erwin, D. C., Partridge, J. E., and McCormick, W. H. 1968. Mycelial growth and production of oospores as affected by β-sitosterol, and the germination of oospores of *Phytophthora megasperma* var. *sojae* (Abstr.). Phytopathology 58:1049.

Erwin, D. C., Baumer, J. S., Jr., Khan, R. A., Lehman, W. F., and Vaziri, A. 1982. Registration of A77-10B alfalfa germplasm (Reg. No. GP118). Crop Sci. 22:1267.

Erwin, D. C., Bartnicki-Garcia, S., and Tsao, P. H., eds. 1983. *Phy-*

tophthora: Its Biology, Taxonomy, Ecology, and Pathology. American Phytopathological Society, St. Paul, Minn. 392 pp.

Erwin, D. C., Khan, R. A., Ver Linden, C., Frate, C., and Munier, D. 1991. Blackeye bean root rot diseases identified. Calif. Agric. 45(4):27–29.

Evans, H. C. 1973. Invertebrate vectors of *Phytophthora palmivora,* causing black pod disease of cocoa in Ghana. Ann. Appl. Biol. 75:331–345.

Evans, H. C., and Prior, C. 1987. Cocoa pod diseases: Causal agents and control. Outlook Agric. 16:35–41.

Evans, S. G. 1979. Susceptibility of plants to fungal pathogens when grown by the nutrient film technique (NFT). Plant Pathol. 28:45–48.

Evola Maltese, C., Conigliaro, G., and Shaw, D. S. 1995. The development of sporangia of *Phytophthora infestans.* Mycol Res. 99: 1175–1181.

Eye, L. L., Sneh, B., and Lockwood, J. L. 1976. Isolation and identification of *Phytophthora megasperma* var. *sojae* from soil using soybean seedlings as baits (Abstr.). Proc. Am. Phytopathol. Soc. 3:239.

Eye, L. L., Sneh, B., and Lockwood, J. L. 1978a. Factors affecting zoospore production by *Phytophthora megasperma* var. *sojae.* Phytopathology 68:1766–1768.

Eye, L. L., Sneh, B., and Lockwood, J. L. 1978b. Inoculation of soybean seedlings with zoospores of *Phytophthora megasperma* var. *sojae* for pathogenicity and race determination. Phytopathology 68:1769–1773.

Fagan, H. J. 1988. Species and mating types of *Phytophthora* on cocoa in Jamaica and their effect on selected cocoa varieties. Trans. Br. Mycol. Soc. 91:389–401.

Falcon, M. F., Fox, R. L., and Trujillo, E. E. 1984. Interactions of soil pH, nutrients and moisture on Phytophthora root rot of avocado. Plant Soil 81:165–176.

Falloon, P. G., and Fraser, H. A. 1991. Control of establishment failures in asparagus (*Asparagus officinalis* L.) caused by Phytophthora rot. N.Z. J. Crop Hortic. Sci. 19:47–52.

Falloon, P. G., and Grogan, R. G. 1988. Isolation, distribution, pathogenicity and identification of *Phytophthora* spp. on asparagus in California. Plant Dis. 72:495–497.

Falloon, P. G., Falloon, L. M., Mullen, R. J., Benson, B. L., and Grogan, R. G. 1983. Effect of Phytophthora spear rot on asparagus yield. Calif. Agric. 37(7–8):16–17.

Falloon, P. G., Mullen, R. J., Benson, B. L., and Grogan, R. G. 1985. Control of Phytophthora rot with metalaxyl in established asparagus. Plant Dis. 69:921–923.

Falloon, P. G., Greathead, A. S., Mullen, R. J., Benson, B. L., and Grogan, R. G. 1991. Individual and combined effects of flooding, Phytophthora rot, and metalaxyl on asparagus establishment. Plant Dis. 75:514–518.

Farih, A., and Vanderweyen, A. 1983. Occurrence of *Phytophthora cactorum* on strawberry in Morocco. Phytophthora Newsl. 11:14.

Farih, A., Menge, J. A., Tsao, P. H., and Ohr, H. D. 1981a. Metalaxyl and efosite aluminum for control of *Phytophthora gummosis* and root rot on citrus. Plant Dis. 65:654–657.

Farih, A., Tsao, P. H, and Menge, J. A. 1981b. Fungitoxic activity of efosite aluminum on growth sporulation, and germination of *Phytophthora parasitica* and *P. citrophthora.* Phytopathology 71:934–936.

Faris, M. A. 1985. Variability and interaction between alfalfa cultivars and isolates of *Phytophthora megasperma.* Phytopathology 75:390–394.

Faris, M. A. 1988. Variability in growth of *Phytophthora megasperma* f. sp. *medicaginis* in relation to temperature. Can. J. Plant Pathol. 10:99–104.

Faris, M. A., and Sabo, F. E. 1981. Effect of *Phytophthora megasperma* on yield and survival of resistant and susceptible alfalfa cultivars. Can. J. Plant Sci. 61:955–960.

Faris, M. A., and Sabo, F. E., and Barr, D. J. S. 1983. Studies on *Phytophthora megasperma* isolates with different levels of pathogenicity on alfalfa cultivars. Can. J. Plant Pathol. 5:29–33.

Faris, M. A., Sabo, F. E., and Cloutier, Y. 1986. Intraspecific variation in gel electrophoresis patterns of soluble mycelial proteins of *Phytophthora megasperma* isolated from alfalfa. Can. J. Bot. 64: 262–265.

Faris, M. A., Sabo, F. E., Barr, D. J. S., and Lin, C. S. 1989. The systematics of *Phytophthora sojae* and *P. megasperma.* Can. J. Bot. 67:1442–1447.

Farr, D. F., Bills, G. F., Charmuris, G. P., Rossman, A. Y. 1989. Fungi on Plants and Plant Products in the United States. American Phytopathological Society, St. Paul, Minnesota. 1252 pp.

Farrell, G. M., Preece, T. F., and Wren, M. J. 1969. Effects of infection by *Phytophthora infestans* (Mont.) de Bary on the stomata of potato leaves. Ann. Appl. Biol. 63:265–275.

Fatemi, J. 1974. A simple technique for isolation of *Phytophthora* and *Pythium* species from soil. Phytopathol. Mediterr. 13:120–121.

Fatemi, J. 1980. The role of *Phytophthora* species in almond decline in Iran. Phytopathol. Z. 99:97–100.

Favaloro, M., and Somma, V. 1970. Ricerche sul marciume del colletto e radicale degli agrumi. Specie di *Phytophthora* presenti negli agrumeti della Sicilia occidentale (Research on collar and root rot of citrus. Species of *Phytophthora* present on citrus near western Sicily). Phytopathol. Mediterr. 9:54–56. (In Italian).

Fawcett, G. L. 1916. Algunos organísmos de la gomosis y de la podredumbre del pie del género *Citrus* (Some causal agents of gummosis and foot rot in the genus *Citrus*). Rev. Ind. Agric. Tucumán 7:272. (In Spanish). (Cited in Tucker 1933).

Fawcett, H. S. 1915. Citrus diseases of Florida and Cuba compared with those of California. Calif. Agric. Exp. Stn. Bull. 262:153–210.

Fawcett, H. S. 1916. A *Pythiacystis* on avocado trees. Phytopathology 5:66–67. (Cited in Tucker 1933).

Fawcett, H. S. 1920. *Pythiacystis* and *Phytophthora.* Phytopathology 10:397–399.

Fawcett, H. S. 1923. Gummosis of citrus. J. Agric. Res. 24:191–236.

Fawcett, H. S. 1925. Bark diseases of citrus trees in California. Calif. Agric. Exp. Stn. Bull. 395. 59 pp.

Fawcett, H. S. 1930. Brown rot of citrus in Mediterranean countries identical with that here. Calif. Citrogr. 16:81. (Cited in Tucker 1933).

Fawcett, H. S. 1933. New locations for *Phytophthora citrophthora* and *P. hibernalis* on citrus. Phytopathology 23:667–669.

Fawcett, H. S. 1936. Citrus Diseases and Their Control. McGraw-Hill Book Co., New York. 656 pp.

Fawcett, H. S., and Bitancourt, A. A. 1940. Occurrence, pathogenicity, and temperature relations of *Phytophthora* species on citrus in Brazil and other South American countries. Arq. Inst. Biol. Sao Paulo, Brasil, 11:107–118.

Fawcett, H. S., and Klotz, L. J. 1934. A procedure for inducing the production of the sporangial and swarm stages in certain species of *Phytophthora.* Phytopathology 24:693–694.

Fawcett, W. 1889. Report on the Cayman Islands. Bot. Dep. Jamaica, Bull. 11. 10 pp.

Fehr, W. R., Caviness, C. E., Burmood, D. T., and Pennington, J. S. 1971. Stage of development descriptions for soybeans, *Glycine max* (L.) Merrill. Crop Sci. 11:929–931.

Feichtenberger, E. 1990. Control of Phytophthora gummosis of citrus with systemic fungicidess in Brazil. EPPO (Eur. Mediterr. Plant Prot. Organ.) Bull. 20:139–148.

Feichtenberger, E., Rossetti, V., Salibe, A. A., and Oliveira, D. A. 1984a. Reaction of pera sweet orange trees preimmunized against tristeza to inoculations with *Phytophthora citrophthora.* Pages 49–52 in: Proc. Ninth Conf. Int. Org. Citrus Virologists. S. M. Garnsey, L. W. Timmer, and J. A. Dodds, eds.

Feichtenberger, E., Zentmyer, G. A., and Menge, J. A. 1984b. Identity of *Phytophthora* isolated from milkweed vine. Phytopathology 74:50–55.

Feld, S. J., Menge, J. A., and Pehrson, J. E. 1979. Brown rot of citrus: A review of the disease. Citrograph 64:101–106.

Feld, S. J., Menge, J. A., and Stolzy, L. H. 1990. Influence of drip and furrow irrigation on Phytophthora root rot of citrus under field and greenhouse conditions. Plant Dis. 74:21–27.

Felix, E. L. 1962. Culture media for sporangial production in *Phytophthora fragariae* (Abstr.). Phytopathology 52:9.

Fell, J. W., and Master, I. M. 1975. Phycomycetes (*Phytophthora* spp. nov. and *Pythium* sp. nov.) associated with degrading mangrove (*Rhizophora mangle*) leaves. Can. J. Bot. 53:2908–2922.

Fenn, M. E., and Coffey, M. D. 1984. Studies on the *in vitro* and *in vivo* antifungal activity of fosetyl-Al and phosphorus acid. Phytopathology 74:606–611.

Fenn, M. E., and Coffey, M. D. 1985. Further evidence for the direct mode of action of fosetyl-Al and phosphorous acid. Phytopathology 75:1064–1068.

Fernández, P. S., and Galindo, A. J. 1989. Caracterización de la resistencia en diversos clones de papita güera (*Solanum cardiophyllum* Lindl. *Solanum ehrenbergii* (Bitt.) Rydb. al ataque de *Phytophthora infestans* (Mont.) de Bary (Resistance characterization in different clones of papita güera [*Solanum cardiophyllum* Lindl. and *Solanum ehrenbergii* (Bitt.) Rydb.] against *Phytophthora infestans* [Mont.] de Bary). Agrocienca 77:157–161. (In Spanish).

Fernando, W. G. D., and Linderman, R. G. 1993. Occurrence, distribution, and pathogenicity of the cowpea root and stem rot pathogen, *Phytophthora vignae*, in soils of Sri Lanka. Plant Dis. 77:1158–1164.

Ferreira, J. F., Naude, S. P., and Thompson, A. H. 1991. First report of *Phytophthora nicotianae* var. *nicotianae* on tomatoes in South Africa. Phytophylactica 23:233–234.

Ferrin, D. M., and Kabashima, J. N. 1991. In vitro insensitivity to metalaxyl of isolates of *Phytophthora citricola* and *P. parasitica* from ornamental hosts in southern California. Plant Dis. 75:1041–1044.

Ferrin, D. M., and Mitchell, D. J. 1986a. Influence of initial density and distribution of inoculum on the epidemiology of tobacco black shank. Phytopathology 76:1153–1158.

Ferrin, D. M., and Mitchell, D. J. 1986b. Influence of soil water status on the epidemiology of tobacco black shank. Phytopathology 76:1213–1217.

Ferrin, D. M., and Rohde, R. G. 1992a. Population dynamics of *Phytophthora parasitica*, the cause of root and crown rot of *Catharanthus roseus* in relation to fungicide use. Plant Dis. 76:60–63.

Ferrin, D. M., and Rohde, R. G. 1992b. In vivo expression of resistance to metalaxyl by a nursery isolate of *Phytophthora parasitica* from *Catharanthus roseus*. Plant Dis. 76:82–84.

Ferrin, D. M., and Wadsworth, M. L. 1992. Effect of metalaxyl on sporulation and growth of metalaxyl-resistant and metalaxyl-sensitive isolates of *Phytophthora parasitica* in vitro. Plant Dis. 76:492–495.

Fevre, M. 1979. Glucanases, glucan synthases and wall growth in *Saprolegnia monoica*. Pages 225–263 in: Fungal Walls and Hyphal Growth. J. H Burnett and A. P. J. Trinci, eds. Cambridge Univ. Press, London. 418 pp.

Figueiredo, M. M., and Albuquerque, F. C. 1970. Podridao mole das raizes da mandioca (*Manihot esculenta*) (Soft rot of cassava roots [*Manihot esculenta*]). Bol. Tec. Inst. Pesqui. Agropecu. Bras. 5: 389–393. (In Portuguese).

Filmer, C. L. R., MacDonald, J. D., Paul, J. L., and Leiser, A. T. 1986. Influence of air-filled porosity of container media on Phytophthora root rot of toyon. HortScience 21:1010–1011.

Filonow, A. B., and Lockwood, J. L. 1985. Evaluation of several actinomycetes and the fungus *Hypochytrium catenoides* as biocontrol agents for Phytophthora root rot of soybean. Plant Dis. 69: 1033–1036.

Fincham, J. R. S. 1989. Transformation in fungi. Microbiol. Rev. 53: 148–170.

Finlay, A. R., and McCracken, A. R. 1991. Microbial suppression of *Phytophthora cinnamomi*. Pages 383–398 in: *Phytophthora*. J. A. Lucas, R. C. Shattock, D. S. Shaw, and L. R. Cooke, eds. Cambridge Univ. Press, Cambridge, U.K. 447 pp.

Firman, I. D. 1974. Cocoa canker. Pages 131–139 in: *Phytophthora* Disease of Cocoa. P. H. Gregory, ed. Longman, London. 348 pp.

Firman, I. D. 1975. *Phytophthora* and *Pythium* species and the diseases caused by them in the area of the South Pacific Commission. Fiji Agric. J. 37:1–8.

Firman, I. D. 1978a. Natural incidence of canker on amelonado and hybrid cocoa in Fiji and results from artificial inoculation with *Phytophthora palmivora*. Trop. Agric. 55:269–272.

Firman, I. D. 1978b. Black pod and canker of cocoa. Advisory Leafl., South Pac. Comm. No. 7. 4 pp. (Rev. Plant Pathol. 1979, 58:1688.)

Fisher, D. J., and Hayes, A. L. 1982. Mode of action of the systemic fungicides furalaxyl, metalaxyl and ofurace. Pestic. Sci. 13:330–339.

Fisher, D. J., and Hayes, A. L. 1984. Studies of mechanisms of metalaxyl fungitoxicity and resistance to metalaxyl. Crop Prot. 3:177–185.

Fisher, D. J., and Hayes, A. L. 1985. A comparison of the biochemical and physiological effects of the systemic fungicide cyprofuram with those of the related compounds metalaxyl and metolachlor. Crop Prot. 4:501–510.

Fleetwood Walker, P. M. 1954. Observations on the physiology of *Phytophthora fragariae* (Abstr.). Trans. Br. Mycol. Soc. 38:169.

Fletcher, W. A. 1957. Citrus varieties and rootstocks for New Zealand. Orchardist N.Z. 30. 33 pp.

Fletcher, W. A. 1975. Growing Tamarillos. N.Z. Minist. Agric. Fish. Bull. 306. 28 pp.

Flett, S. P., Ashcroft, W., Lim, T., and Jerie, P. 1990. Evaluation of phosphonic (phosphorous) acid for control of Phytophthora root rot of tomatoes. Australas. Plant Pathol. 19:131–132.

Flett, S. P., Ashcroft., W. J., Jerie, P. H., and Taylor, P. A. 1991. Control of Phytophthora root rot in processing tomatoes by metalaxyl and fosetyl-Al. Aust. J. Exp. Agric. 31:279–283.

Flor, H. H. 1956. The complementary genic systems in flax and flax rusts. Adv. Genet. 8:29–54.

Flowers, R. A., and Hendrix, H. W. 1969. Gallic acid in a procedure for isolation of *Phytophthora parasitica* var. *nicotianae* and *Pythium* spp. from soil. Phytopathology 59:725–731.

Flowers, R. A., Erwin, D. C., and Hendrix, J. W. 1973. Isolation of *Phytophthora cryptogea* from bean in Kentucky. Plant Dis. Rep. 57:90–191.

Fohner, G. R., Fry, W. E., and White, G. B. 1984. Computer simulation raises questions about timing protectant fungicide application frequency according to a potato late blight forecast. Phytopathology 74:1145–1147.

Foister, C. E. 1930. Blossom blight and shanking of tulips. Gard. Chron. 87:171–172.

Foister, C. E. 1931. The white tip disease of leeks and its causal fungus, *Phytophthora porri* n. sp. Trans. Proc. Bot. Soc. Edinb., Scotland, U.K. 30 (Part 4):257–281.

Foister, C. E. 1940. Descriptions of new fungi causing economic diseases in Scotland. Trans. Bot. Soc. Edinb. 33 (Part 1):65–68.

Foister, C. E. 1961. The economic plant diseases of Scotland: A survey and check list covering the years 1924–1957. Dep. Agric. Fish. Scotland Tech. Bull. 1. 209 pp.

Follin, J. C. 1971. L'utilisation du benomyl pour l'isolement selectif des Pythiacees (Utilization of benomyl for the selective isolation of Pythiaceae). Coton Fibres Trop. 26:467. (In French).

Follin, J. C., Mildner, P., and Cauquil, J. 1966. Le chancre de la Roselle [*Hibiscus sabdariffa*] en Republique Centrafricaine. Premieres observations (Roselle [*Hibiscus sabdariffa*] canker in the Central African Republic. Preliminary observations). Coton Fibres Trop. 21:189–194. (In French).

Forlot, P., Harranger, J., Payen, J., and Schwinn, F. J. 1966. *Phytophthora cryptogea* Pethyb. & Laff. nouvel agent de la pourriture des endives (*Cichorium intybus* L. var. *foliosum* Hegi) en cours de forçage (*Phytophthora cryptogea* Pethyb. & Laff. (A new causal agent of heart rot of foliage of endives [*Cichorium intybus* L. var. *foliosum* Hegi]). Phytopathol. Z. 56:1–18. (In French).

Forsberg, L. I. 1985. *Phytophthora* species on ornamental plants in Queensland. Australas. Plant Pathol. 14:18–20.

Förster, H., and Coffey, M. D. 1990. Mating behavior of *Phytophthora parasitica:* Evidence for sexual recombination in oospores using DNA restriction fragment length polymorphisms as genetic markers. Exp. Mycol. 14:351–359.

Förster, H., and Coffey, M. D. 1991. Approaches to the taxonomy of *Phytophthora* using polymorphisms in mitochondrial and nuclear DNA. Pages 164–183 in: *Phytophthora*. J. A. Lucas, R. C. Shattock, D. S. Shaw, and L. R. Cooke, eds. Cambrige Univ. Press, Cambridge, U.K. 447 pp.

Förster, H., and Coffey, M. D. 1992. Molecular characterization of *Phytophthora* isolates with non-papillate sporangia causing root rot of raspberry using mtDNA restriction fragment length polymorphisms. Mycol. Res. 96:571–577.

Förster, H., and Coffey, M. D. 1993. Molecular taxonomy of *Phytophthora megasperma* based on mitochondrial and nuclear DNA polymorphisms. Mycol. Res. 97:1101–1112.

Förster, H., Ribeiro, O. K., and Erwin, D. C. 1983. Factors affecting oospore germination of *Phytophthora megasperma* f. sp. *medicaginis*. Phytopathology 73:442–448.

Förster, H., Kinscherf, T. G., Leong, S. A., and Maxwell, D. P. 1987. Molecular analysis of the mitochondrial genome of *Phytophthora*. Curr. Genet. 12:215–218.

Förster, H., Kinscherf, T. G., Leong, S. A., and Maxwell, D. P. 1988. Estimation of relatedness between *Phytophthora* species by analysis of mitochondrial DNA. Mycologia 80:466–478.

Förster, H., Kinscherf, T. G., Leong, S. A., and Maxwell, D. P. 1989. Restriction fragment length polymorphisms of the mitochondrial DNA of *Phytophthora megasperma* isolated from soybean, alfalfa, and fruit trees. Can. J. Bot. 67:529–537.

Förster, H., and Coffey, M. D., Elwood, H., and Sogin, M. L. 1990a. Sequence analysis of the small subunit ribosomal RNAs of three zoosporic fungi and implications for fungal evolution. Mycologia 82:306–312.

Förster, H., Oudemans, P., and Coffey, M. D. 1990b. Mitochondrial and nuclear DNA diversity within six species of *Phytophthora*. Exp. Mycol. 14:18–31.

Förster, H., Tyler, B. M., and Coffey, M. D. 1994. *Phytophthora sojae* races have arisen by clonal evolution and by rare outcrosses. Mol. Plant-Microbe Interact. 7:780–791.

Förster, H., Learn, G., and Coffey, M. D. 1995. Towards a better understanding of the evolutionary history of species of the genus *Phytophthora* using isozymes, DNA RFLPs and ribosomal DNA spacer sequences. Pages 42–54 in: *Phytophthora infestans* 150. L. J. Dowley, E. Bannon, L. R. Cooke, T. Keane, and E. O'Sullivan, eds. Boole Press, Dublin, Ireland.

Fortes J. F., and Pecknold, P. C. 1981. Evaluation of media for sporangial production and radial growth of *Phytophthora cactorum*. Pesqui. Agropecu. Bras. 16:213–218.

Fothergill, P. G., and Child, J. H. 1964. Comparative studies of the mineral nutrition of three species of *Phytophthora*. J. Gen. Microbiol. 36:67–78.

Fox, R. T. V. 1990. Rapid methods for diagnosis of soil-borne plant pathogens. Soil Use Manage. 6:179–184.

Fraedrich, S. W., and Tainter, F. H. 1989. Effect of dissolved oxygen concentration on the relative susceptibility of shortleaf and loblolly pine root tips to *Phytophthora cinnamomi*. Phytopathology 79:1114–1118.

Fraedrich, S. W., Tainter, F. H., and Miller, A. E. 1989. Zoospore inoculum density of *Phytophthora cinnamomi* and the infection of lateral root tips of shortleaf and loblolly pine. Phytopathology 79: 1109–1113.

Fraser, L. 1942. Phytophthora root rot of citrus. J. Aust. Inst. Agric. Sci. 8:101–105.

French, A. M. 1987. California plant disease host index. Part I. Fruits and nuts. Calif. Dep. Food Agric. 39 pp.

Frezzi, M. J. 1940. La *Phytophthora citrophthora* causante de la podredumbre del pie del naranjo y la gomosis del tronco del limonero en Corrientes (*Phytophthora citrophthora* agent of foot rot of orange and bark gummosis of lemon in Corrientes). Rev. Argent. Agron. 7:165. (In Spanish).

Frezzi, M. J. 1941. *Phytophthora boehmeriae,* causante de la podredumbre morena de los frutos citricos, en la Republica Argentina (*Phytophthora boehmeriae,* the agent of brown rot of citrus fruits in the Argentine Republic). Rev. Argent. Agron. 8:200–205. (In Spanish). (Rev. Appl. Mycol. 1944, 23:296.)

Frezzi, M. J. 1942. Podredumbre morena de los frutos citricos y parasitos que la producen en Corrientes, Argentina (Brown rot of citrus fruits and the parasites that produce it in Corrientes, Argentina). Rev. Argent. Agron 9:216–220. (In Spanish). (Rev. Appl. Mycol. 1944, 23:294.)

Frezzi, M. J. 1950. Las especies de *Phytophthora* en la Argentina (The species of *Phytophthora* in Argentina). Rev. Invest. Agric. Buenos Aires 4:47–133. (In Spanish).

Frezzi, M. J. 1956. Especies de *Pythium* fitopatogenas identificadas en la Republica Argentina (The phytopathogenic species of *Pythium* identified in the Argentine Republic). Rev. Invest. Agric. 10: 113–241. (In Spanish).

Frezzi, M. J. 1957. *Phytophthora cryptogea*, causante de la muerte de '*Populus simonii*' en Mendoza (Argentina) (*Phytophthora cryptogea,* cause of death of '*Populus simonii*' in Mendoza [Argentina]). Rev. Argent. Agron. 24:136–143. (In Spanish).

Frezzi, M. J., and Macola, T. 1943. *Phytophthora palmivora,* causante de la 'podredumbre morena' de los frutos citricos en Cordoba (Argentina). (*Phytophthora palmivora* the agent of 'brown rot' of citrus fruits in Cordoba, [Argentina]). Rev. Argent. Agron. 3:227–230. (In Spanish). (Rev. Appl. Mycol 1944, 23:15.)

Friend, J. 1991. Host-pathogen interactions: Current questions. Pages 46–49 in: *Phytophthora*. J. A. Lucas, R. C. Shattock, D. S. Shaw, and L. R. Cooke, eds. Cambridge Univ. Press, Cambridge, U.K. 447 pp.

Fröhlich, G., and Rodewald, W. 1970. Pests and Diseases of Tropical Crops and Their Control. Pergamon Press, New York. 371 pp.

Frosheiser, F. I. 1980. Conquering Phytophthora root rot with resistant alfalfa cultivars. Plant Dis. 64:909–912.

Frosheiser, F. I., and Barnes, D. K. 1973. Field and greenhouse selection for Phytophthora root rot resistance in alfalfa. Crop Sci. 13:735–738.

Frosheiser, F. I., and Barnes, D. K. 1984. Phytophthora root rot resistance. Page 25 in: Standard tests to characterize pest resistance in alfalfa cultivars. J. H. Elgin, Jr., and D. J. Barnes, eds. U.S. Dep. Agric. Misc. Publ. 1434. 38 pp.

Frossard, P. 1967. Lutte contre la pourriture du coeur des plants d'ananas en Cote d'Ivoire (Control of heart rot in pineapple plants in the Ivory Coast). Fruits 22:535–542. (In French).

Frossard, P. 1978. Lutte contre la pourriture du coeur a *Phytophthora* de l'ananas. Fongicides classiques et fongicides nouveaux (Control of heart rot of pineapple. Classical fungicides and new fungicides). Fruits 33:183–191. (In French).

Frossard, P., and Bourdeaut, J. 1974. Lutte contre le deperissement des avocatiers en Cote d'Ivoire (essai Dexon-IFAC-Azaguie) (Control of avocado decline in the Ivory Coast [Test of Dexon-IFAC-Azaguie]). Fruits 29:525–529. (In French).

Fry, W. E. 1975. Integrated effects of polygenic resistance and a protective fungicide on development of potato late blight. Phytopathology 65:908–911.

Fry, W. E. 1977a. Integrated control of potato late blight: Effects of polygenic resistance and techniques of timing fungicide applications. Phytopathology 67:415–420.

Fry, W. E. 1977b. Management with chemicals. Pages 213–238 in: Plant Disease. Vol. I. How Disease Is Managed. J. G. Horsfall and E. B. Cowling, eds. Academic Press, New York. 465 pp.

Fry, W. E. 1982. Principles of Disease Management. Academic Press, New York. 378 pp.

Fry, W. E., and Doster, M. A. 1991. Potato late blight: Forecasts and disease suppression. Pages 326–336 in: *Phytophthora*. J. A. Lucas, R. C. Shattock, D. S. Shaw, and L. R. Cooke, eds. Cambridge Univ. Press, Cambridge, U.K. 447 pp.

Fry, W. E., and Fohner, G. R. 1985. Construction of predictive models: I. Forecasting disease development. Pages 161–178 in: Adv. Plant Pathol. Vol. 3. C. A. Gilligan, ed. Academic Press, London. 255 pp.

Fry, W. E., and Goodwin, S. B. 1995. Recent migrations of *Phytophthora infestans*. Pages 89–95 in: *Phytophthora infestans* 150. L. J. Dowley, E. Bannon, L. R. Cooke, T. Keane, and E. O'Sullivan, eds. Boole Press, Dublin, Ireland.

Fry, W. E., Bruck, R. I., and Mundt, C. C. 1979. Retardation of potato late blight epidemics by fungicidess with eradicant and protectant properties. Plant Dis. Rep. 63:970–974.

Fry, W. E., Apple, A. E., and Bruhn, J. A. 1983. Evaluation of potato late blight forecasts modified to incorporate host resistance and fungicide weathering. Phytopathology 73:1054–1059.

Fry, W. E., Drenth, A., Spielman, L. J., Mantel, B. C., Davidse, L. C., and Goodwin, S. B. 1991. Population genetic structure of *Phytophthora infestans* in the Netherlands. Phytopathology 81:1330–1336.

Fry, W. E., Goodwin, S. B., Matuszak, J. M., Spielman, L. J., and Drenth, A. 1992. Population genetics and intercontinental migrations of *Phytophthora infestans*. Annu. Rev. Phytopathol. 30:107–129.

Fry, W. E., Goodwin, S. B., Dyer, A. T., Matuszak, J. M., Drenth, A., Tooley, P. W., Sujkowski, L. S., Koh, Y. J., Cohen, B. A., Spielman, L. J., Deahl, K. L., Inglis, D. A., and Sandlan, K. P. 1993. Historical and recent migrations of *Phytophthora infestans:* Chronology, pathways, and implications. Plant Dis. 77:653–661.

Fry, W. L., and Spielman, L. J. 1991. Population biology. Pages 171–192 in: *Phytophthora infestans,* the Cause of Late Blight of Potato. Advances in Plant Pathology. Vol. 7. D. S. Ingram and P. H. Williams, eds. Academic Press, New York. 273 pp.

Fucikovsky, L., and Luna, I. 1987. Avocado diseases and their control in Mexico. Yearb. S. Afr. Avocado Growers' Assoc. 10:119–121.

Fuller, M. S., and Gisi, U. 1985. Comparative studies of the *in vitro* activity of the fungicides oxadixyl and metalaxyl. Mycologia 77: 424–432.

Fuller, M. S., and Poyton, R. O. 1964. A new technique for the isolation of aquatic fungi. Bioscience 14(9):45–46.

Fulton, J. M., Mortimore, C. G., and Hildebrand, A. A. 1961. Note on the relation of soil bulk density to the incidence of Phytophthora root and stalk rot of soybeans. Can. J. Soil Sci. 41:247.

Fulton, J. P., and Fulton, N. D. 1951. Phytophthora root rot of tomatoes. Phytopathology 41:99–101.

Fulton, R. H. 1989. The cacao disease trilogy: Black pod, Monilia pod rot, and witches' broom. Plant Dis. 73:601–603.

Gabor, B. K., and Coffey, M. D. 1990. Quantitative analysis of the resistance to *Phytophthora cinnamomi* in five avocado rootstocks under greenhouse conditions. Plant Dis. 74:882–885.

Gabor, B. K., and Coffey, M. D. 1991. Comparison of rapid methods for evaluating resistance to *Phytophthora cinnamomi* in avocado rootstocks. Plant Dis. 75:118–120.

Gabor, B. K., Guillemet, F. B., and Coffey, M. D. 1990. Comparison of field resistance to *Phytophthora cinnamomi* in twelve avocado rootstocks. HortScience 25:1655–1656.

Gabor, B. K., O'Gara, E. T., Philip, B. A., Horan, D. P., and Hardham, A. R. 1993. Specificities of monoclone antibodies to *Phytophthora cinnamomi* in two rapid diagnostic assays. Plant Dis. 77:1189–1197.

Gabriel, C. J., and Cook, R. J. 1990. Biological control—The need for a new scientific framework. BioScience 40:204–207.

Gadd, C. H. 1924. *Phytophthora faberi* Maubl. Ann. R. Bot. Gard. Peradeniya (Ceylon) 9:47–89.

Gadgil, P. D. 1974a. Report of Forest Research Institute for 1973. N.Z. For. Serv., Wellington. 1974. 93 pp.

Gadgil, P. D. 1974b. *Phytophthora heveae,* a pathogen of Kauri. N.Z. J. For. Sci. 4:59–63.

Gahan, P. B., and Kalinda, M. 1968. The use of tetrazolium salts in the histochemical demonstration of succinic dehydrogenase activity in plant tissues. Histochemie 14:81–88.

Galindo, A. J., and Gallegly, M. E. 1960. The nature of sexuality in *Phytophthora infestans.* Phytopathology 50:123–128.

Galindo, A. J., and Hohl, H. R. 1985. *Phytophthora mirabilis,* a new species of *Phytophthora.* Sydowia Ann. Mycol. Ser. II. 38:87–96.

Galindo, A. J., and Zentmyer, G. A. 1964. Mating types in *Phytophthora cinnamomi.* Phytopathology 54:238–239.

Galindo, A. J., and Zentmyer, G. A. 1967a. Genetics and cytology of *Phytophthora.* Nature 214:1356.

Galindo, A. J., and Zentmyer, G. A. 1967b. Genetical and cytological studies of *Phytophthora* strains pathogenic to pepper plants. Phytopathology 57:1300–1304.

Gallegly, M. E. 1968. Genetics of pathogenicity of *Phytophthora infestans.* Annu. Rev. Phytopathol. 6:375–396.

Gallegly, M. E. 1970. Genetics of *Phytophthora.* Phytopathology 60:1135–1141.

Gallegly, M. E. 1983a. New criteria for classifying *Phytophthora* and critique of existing approaches. Pages 167–172 in: *Phytophthora:* Its Biology, Taxonomy, Ecology, and Pathology. D. C. Erwin, S. Bartnicki-Garcia, and P. H. Tsao, eds. American Phytopathological Society, St. Paul, Minn. 392 pp.

Gallegly, M. E. 1983b. Summary of the open discussion session on taxonomy of *Phytophthora.* Pages 173–174 in: *Phytophthora:* Its Biology, Taxonomy, Ecology, and Pathology. D. C. Erwin, S. Bartnicki-Garcia, and P. H. Tsao, eds. American Phytopathological Society, St. Paul, Minn. 392 pp.

Gallegly, M. E., and Galindo. A. J. 1958. Mating types and oospores of *Phytophthora infestans* in nature in Mexico. Phytopathology 48: 274–277.

Gallegly, M. E., and Niederhauser, J. S. 1959. Genetic controls of host-parasite interactions in the Phytophthora late blight disease. Pages 168–182 in: Plant Pathology: Problems and Progress 1908–1958. C. S. Holton, G. W. Fischer, R. W. Fulton, H. Hart, and S. E. A. McCallan, eds. Univ. Wis. Press, Madison. 588 pp.

Gandara, G. 1909. Nota acerca de las enfermedades fungosas del maguey (Notes on fungal diseases of maguey). Mem. 1908–1909 y Rev. Nac. Cient. "Antonio Alzate." 25:293–305. (In Spanish). (Cited in Tucker 1933).

Garas, N. A., Doke, N., and Kuć, J. 1979. Suppression of the hypersensitive reaction in potato tubers by mycelial components from *Phytophthora infestans.* Physiol. Plant Pathol. 15:117–126.

Garber, R. H., DeVay, J. E., Hake, K., and Jeffers, D. 1986. A new boll rot disease in California caused by *Phytophthora capsici.* Page 27 in: Proc 46th Annu. Cotton Disease Council, Beltwide Cotton Res. Conf. National Cotton Council, Memphis, TN. 543 pp.

Garbowski, L. 1913. Keimungsversuche mit Konidien von *Phytophthora infestans* de Bary (Germination research on conidia of *Phytophthora infestans* de Bary). Centralbl. Bakt. (II Abt.) 36:500–508. (In German). (Cited in Tucker 1933).

García, E. R., and Adam, A. V. 1972. Major diseases of pineapple in Oaxaca, Mexico and their control. F.A.O. Plant Prot. Bull. 20:79–87.

Gardner, M. W., and Yarwood, C. E. 1942. *Phytophthora infestans* on the weed *Solanum sarachoides.* Plant Dis. Rep. 26:501–502.

Gardner, W. R., and Ehlig, C. F. 1965. Physical aspects of the internal water relations of plant tissues. Plant Physiol. 40:705–710.

Garibaldi, A., and Gullino, G. 1974. Preliminary observations on a new disease of *Anemone coronaria* L. caused by *Phytophthora cactorum* (Leb. et Cohn.) Schroet. Pages 951–956 in: Twenty-sixth Int. Symp. on Phytopharmacie and Phytocyatrie, 7 May 1974. Part 2. (Rev. Plant Pathol. 1975, 54:4726.)

Garibaldi, A., and Rapetti, S. 1975. Osservazioni sul marciume pedale del garofano (Observation on the foot rot of carnation). Colture Prolette 4:17–20. (In Italian).

Gaulliard, J. M., and Pelossier, R. 1983. Efficacite de phosethyl Al en trempage des argumes (fruits) contre *Phytophthora parasitica* agent de la pourriture burne et contre *Penicillium digitatum* (Effectiveness of fosetyl-Al by soaking citrus fruits against *Phytophthora parasitica*, an agent of brown rot, and against *Penicillium digitatum* [green mold]). Fruits 38:693–697, 735–738. (In French, English summary).

Gees, R., and Coffey, M. D. 1989. Evaluation of a strain of *Myrothecium roridum* as a potential biocontrol agent against *Phytophthora cinnamomi.* Phytopathology 79:1079–1084.

Geeson, J. D. 1976. Storage rot of white cabbage caused by *Phytophthora porri.* Plant Pathol. 25:115–116.

Geeson, J. D., Browne, K. M., and McKeown, B. 1990. Storage rot of swede caused by *Phytophthora* sp. Plant Pathol. 39:629–631.

Geissler, A. E., and Katekar, G. F. 1983. Effect of fungicides on stages of the life cycle of *Phytophthora cinnamomi.* Pestic. Sci. 14:501–507.

Gemawat, P. D., and Prasad, N. 1964. Further studies on Phytophthora blight of *Sesamum.* Indian Phytopathol. 17:273–283.

George, S. W., and Milholland, R. D. 1986a. Inoculation and evaluation of strawberry plants with *Phytophthora fragariae.* Plant Dis. 70:371–375.

George, S. W., and Milholland, R. D. 1986b. Growth of *Phytophthora fragariae* on various clarified natural media and selected antibiotics. Plant Dis. 70:1100–1104.

Georgescu, C. C., Ene, M., Petrescu, M., Stefanescu, M., and Miron, V. 1957. Bolile si daunatorii padurilor. Biologie si combatere (Diseases and pests of trees. Biology and control). Agrosilvica de Stat. Bucharest, Romania. 638 pp. (In Romanian). (Rev. Appl. Mycol. 1959, 38:424.)

Gerdemann, J. W. 1950. The effect of fertilizer treatment on sweetclover root rot. Plant Dis. Rep. 34:261–262.

Gerlach, W. W. P., Hoitink, H. A. J., and Ellett, C. W. 1974. Shoot blight and stem dieback of *Pieris japonica* caused by *Phytophthora citricola, P. citrophthora* and *Botryosphaeria dothidea.* Phytopathology 64:1368–1370.

Gerlach, W. W. P., Hoitink, H. A. J., and Schmitthenner, A. F. 1976a. *Phytophthora citrophthora* on *Pieris japonica:* Infection, sporulation, and dissemination. Phytopathology 66:302–308.

Gerlach, W. W. P., Hoitink, H. A. J., and Schmitthenner, A. F. 1976b. Survival and host range of *Phytophthora citrophthora* in Ohio nurseries. Phytopathology 66:309–311.

Gerrettson-Cornell, L. 1974. A comparative test of isolation of *Phytophthora cinnamomi* Rands between the lupin baiting and a newly devised apple trap. Phyton 32:35–36.

Gerrettson-Cornell, L. 1975. Notes on the formation of sporangia in *Phytophthora drechsleri* Tucker. Aust. Plant Pathol. Soc. Newsl. 4:32–33.

Gerrettson-Cornell, L. 1976a. *Phytophthora heveae* and *P. boehmeriae* in a *Eucalyptus pilularis* plantation in New South Wales. Aust. Plant Pathol. Soc. Newsl. 5:8–9.

Gerrettson-Cornell, L. 1976b. Effect of different methods of water demineralization on the recovery of *Phytophthora cinnamomi* Rands by lupin baiting. Phyton 34:85–88.

Gerrettson-Cornell, L. 1977. Preliminary results of a study of the sexuality in *Phytophthora cambivora* (Petri) Buis. Phyton 35:153–157.

Gerrettson-Cornell, L. 1978. Notes on the morphology of an isolate of *Phytophthora cambivora* (Petri) Buis. from Australia. Phyton 36:85–89.

Gerrettson-Cornell, L. 1979. Notes on the morphology of five isolates of *Phytophthora cryptogea* Pethyb. and Laff. from Australia. Phyton 37:105–118.

Gerrettson-Cornell, L. 1980. Notes on the morphology of some isolates of *Phytophthora* from Australia. Cryptogam. Mycol. 1:139–154.

Gerrettson-Cornell, L. 1983. A compendium of the morphology of *Phytophthora cinnamomi* Rands from Australia. Acta Bot. Hung. 29:91–105.

Gerrettson-Cornell, L. 1985. A working key to the species of *Phytophthora* de Bary. Acta Bot. Hung. 31:89–97.

Gerrettson-Cornell, L. 1989. A compendium and classification of the species of the genus *Phytophthora* de Bary by the canons of the traditional taxonomy. For. Comm. N.S.W. Tech. Pap. 45. 103 pp.

Gerrettson-Cornell, L., and Lind, P. 1981. Observations on the sporangial shape in *Phytophthora*. Phyton 40:31–35.

Gerrettson-Cornell, L., and Simpson, J. 1984. Three new marine *Phytophthora* species from New South Wales. Mycotaxon 19:453–470.

Gerrettson-Cornell, L., and Townsend, S. R. 1974. A study of a lupin baiting technique for isolation of *Phytophthora cinnamomi* Rands under different lighting conditions. Phyton 32:39–43.

Gerrettson-Cornell, L., Humphreys, F. R., and Townsend, S. R. 1976. Results of a preliminary investigation on the use of *Pinus radiata* bark against *Phytophthora cinnamomi* Rands. Phyton. 34:3–6.

Ghofulpo, T. G. 1915. Cacao and its local diseases. Philipp. Agric. For. 4:162–172. (Cited in Tucker 1933).

Ghosh, S. K., Sitansu, P., and Pan, S. 1989. A comprehensive account of the fungal diseases of *Colocasia esculenta* (L.) Schott. Indian J. Mycol. Res. 27:107–119.

Gil Ortega, R., Palazón Español, C., and Cuartero Zueco, J. 1990. Genetics of resistance to *Phytophthora capsici* in the Mexican pepper 'Line 29.' EPPO (Cur. Mediterr. Plant Prot. Organ.) Bull. 20:117–122.

Gil Ortega, R., Palazón Español, C., and Cuartero Zueco, J. 1991. Genetics of resistance to *Phytophthora capsici* in the pepper line 'SMC-334.' Plant Breed. 107:50–55.

Gilbert, G. S., Handelsman, J., and Parke, J. L. 1990. Role of ammonia and calcium in lysis of zoospores of *Phytophthora cactorum* by *Bacillus cereus* strain UW85. Exp. Mycol. 14:1–8.

Gill, D. L. 1960. A stem and branch rot of snapdragon. Plant Dis. Rep. 44:946–947.

Gill, H. S. 1979. New diseases caused by *Phytophthora* species. Phytophthora Newsl. 7:39–40.

Gill, H. S. 1980. Phytophthora root rot of jojoba seedlings. Calif. Plant Pathol. 4:114–115.

Gill, H. S. 1982. *Phytophthora palmivora* on bottle-tree. Calif. Plant Pest Dis. Rep. 1:15.

Gill, H. S., and Esparza, J. F. 1977. A *Phytophthora* disease of *Myoporum parvifolium* in California. Phytophthora Newsl. 5:36.

Gill, H. S., and Powell, D. 1968. The use of polyacrylamide gel disc electrophoresis in delimiting three species of *Phytophthora*. Phytopathol. Z. 63:23–29.

Gill, H. S., and Zentmyer, G. A. 1978. Identification of *Phytophthora* species by disc electrophoresis. Phytopathology 68:163–167.

Gill, H. S., and Zentmyer, G. A. 1979. Detection of *Phytophthora palmivora* on bottle tree (*Brachychiton populneum*) in California. Phytophthora Newsl. 7:34.

Gill, H. S., and Zentmyer, G. A. 1981. *Phytophthora cryptogea* associated with mortality of jojoba seedlings. Phytophthora Newsl. No. 9:23.

Gill, H. S., Zentmyer, G. A., Ribeiro, O. K., and Klure, L. J. 1976. A *Phytophthora* disease of African daisies (*Osteospermum* spp.) in California. Plant Dis. Rep. 60:647–649.

Gill, H. S., Ribeiro, O. K., and Zentmyer, G. A. 1977. Phytophthora blight of periwinkles in the Coachella Valley of California. Plant Dis. Rep. 61:560–561.

Gillis, A. M. 1993. The magnificent devastator gets around. BioScience 43:368–371.

Gilpatrick, J. D. 1969a. Role of ammonia in the control of avocado root rot with alfalfa meal soil amendment. Phytopathology 59:973–978.

Gilpatrick, J. D. 1969b. Effect of soil amendments upon inoculum survival and function in Phytophthora root rot of avocado. Phytopathology 59:979–985.

Gisi, U. 1975a. Eine neue Methode zur Quantitativen Direktbeobachtung der Sporangien von *Phytophthora cactorum* (Leb. et Cohn) Schroet. im boden (A new method of quantitating direct germination of spores of *Phytophthora cactorum* [Leb. & Cohn.] Schroet. in soil). Z. Pflanzenkr. Pflanzenschutz 82:30–47. (In German).

Gisi, U. 1975b. Untersuchungen über die bod enphase von *Phytophthora cactorum* (Leb. et Cohn) Schroet. mit Fluoreszenzoptischer Direktbeobachtung (Investigations on the soil phase of *Phytophthora cactorum* [Leb. & Cohn.] Schroet. with fluorescent or direct observation. Z. Pflanzenkr. Pflanzenschutz 82:355–377. (In German).

Gisi, U. 1983. Biophysical aspects of the development of *Phytophthora*. Pages 109–119 in: *Phytophthora:* Its Biology, Taxonomy, Ecology, and Pathology. D. C. Erwin, S. Bartnicki-Garcia, and P. H. Tsao, eds. American Phytopathological Society, St. Paul, Minn. 392 pp.

Gisi, U. 1991. Synergism between fungicides for control of *Phytophthora*. Pages 361–372 in: *Phytophthora*. J. A. Lucas, R. C. Shattock, D. S. Shaw, and L. R. Cooke, eds. Cambridge Univ. Press, Cambridge, U.K. 447 pp.

Gisi, U., and Meyer, D. 1973. Ökologische Untersuchungen an *Phytophthora cactorum* (Leb. et Cohn) Schroet. im Boden mit direkten Beobachtungsmethoden (Research on direct observation of *Phytophthora cactorum* [Leb. et Schroet.] in soil). Phytopathol. Z. 76: 276–279.

Gisi, U., and Schwinn, F. J. 1976. Die Eignung von Vitalfarbstoffen und optischen Aufhellern zur fluorenszenzmikroskopischen Beobactung von *Phytophthora cactorum* (Leb. et Cohn) Schroet. in vitro und im Boden (The suitability of vital stains and optical brighteners to fluorescent microscopical observation of *Phytophthora cactorum* [Leb. et Cohn] Schroet. in vitro and in soil). Microsc. Acta 77:402–419.

Gisi, U., and Zentmyer, G. A. 1980. Mechanism of zoospore release in *Phytophthora* and *Pythium*. Exp. Mycol. 4:362–377.

Gisi, U., Oertli, J. J., and Schwinn, F. J. 1977. Wasser- und Salzbeziehungen der sporangien von *Phytophthora cactorum* (Leb. et Cohn) Schroet. in vitro (Water and salt relations of sporangia of *Phytophthora cactorum* in vitro). Phytopathol. Z. 89:261–284. (In German. English summary).

Gisi, U., Hemmes, D. E., and Zentmyer, G. A. 1979a. Origin and significance of the discharge vesicle in *Phytophthora*. Exp. Mycol. 3:321–339.

Gisi, U., Schwinn, F. J., and Oertli, J. J. 1979b. Dynamics of indirect germination in *Phytophthora cactorum* sporangia. Trans. Br. Mycol. Soc. 72:437–446.

Gisi, U., Zentmyer, G. A., and Klure, L. J. 1980. Production of sporangia by *Phytophthora cinnamomi* and *P. palmivora* in soils at different matric potentials. Phytopathology 70:301–306.

Gisi, U., Harr, J., Sandmeier, R., and Wiedmer, H. 1983. A new systemic oxazolidinone fungicide (San 731) against diseases caused by Peronosporales. Meded. Fac. Landbouwwet. Rijksuniv. Gent 48:541–549.

Gisi, U., Binder, H., and Rimbach, E. 1985. Synergistic interactions of fungicides with different modes of action. Trans. Br. Mycol. Soc. 85:299–306.

Gisi, U., Binder, H., and Rimbach. E. 1986. Synergistic effects in mixtures of fungicides with different modes of action. Trans. Br. Mycol. Soc. 85:299–306.

Gjaerum, H. B., Tjamos, E. C., and Viranyi, F. 1988. Oomycetes. Pages 199–239 in: European Handbook of Plant Diseases. I. M. Smith, J. Dunez, D. H. Phillips, R. A. Lelliott, and S. A. Archer, eds. Blackwell Scientific Publications, Oxford, U,K. 583 pp.

Glynne, M. D. 1935. Incidence of take-all on wheat and barley on experimental plots at Woburn. Ann. Appl. Biol. 22:225–235.

Glynne, M. D. 1965. Crop sequence in relation to soil-borne pathogens. Pages 423–435 in: Ecology of Soil-Borne Plant Pathogens. K. F. Baker and W. C. Snyder, eds. Univ. Calif. Press, Berkeley. 571 pp.

Godfrey, G. H. 1923. A Phytophthora foot rot of rhubarb. J. Agric. Res. 23:1–26.

Godoy, E. F. 1940. El "mildew" o "tizón" del pimiento producido por la *Phytophthora capsici* en la República Argentina (The 'mildew' or 'blight' of chili produced by *Phytophthora capsici* in the Argentine Republic). Rev. Fac. Agron. Univ. Nac. La Plata (Tercera epoca) 24:235–280. (In Spanish). (Cited in Frezzi 1950).

Gollifer, D. E., Jackson, G. V. H., and Newhook, F. J. 1980. Survival of inoculum of the leaf blight fungus *Phytophthora colocasiae* infecting taro, *Colocasia esculenta* in the Solomon Islands. Ann. Appl. Biol. 94:379–390.

Gomez, E. T. 1925. Leaf blight of Gabi. Philipp. Agric. 14:429–440.

Gomez-Moreno, M. L. 1942. Araceas de Fernando Poo (Araceae of Fernando Poo). Ann. Agric. Terr. Esp. Golfo Guinea: 7–37. (Rev. Appl. Mycol. 1945, 24:175.)

Gondell, M. A. 1946. La susceptibilidad de diferentes especies y variedades citricas a la *Phytophthora citrophthora* (Sm. y Sm.) Leon., *P. parasitica* Dastur, y *P. megasperma* Leon. en la zona de Concordia (Entre Rios) (The susceptibility of different citrus species and varieties to *Phytophthora citrophthora* [Sm. & Sm.] Leon., *P. parasitica* Dastur and *P. megasperma* Leon. in the region of Concordia [Entre Rios]). Publ. Misc. Minist. Agric. B. Aires, Ser. A 19. 24 pp. (In Spanish). (Rev. Appl. Mycol. 1947, 26:102).

Goodall, G. E. 1955. Avocado root disease in Santa Barbara County. Calif. Avocado Soc. Yearb. 39:149–153.

Goodall, G. E., Ohr, H. D., and Zentmyer, G. A. 1987. Mounding benefits replanting avocado root rot orchards. Calif. Avocado Soc. Yearb. 71:147–151.

Goode, P. M. 1956. Infection of strawberry roots by zoospores of *Phytophthora fragarie*. Trans. Br. Mycol. Soc. 39:367–377.

Goodell, J. J., De Angelis, P. L., and Ayers, A. R. 1985. Immunochemical identification of antigens involved in plant/pathogen interactions. Pages 447–457 in: Cellular and Molecular Biology of Plant Stress. Alan R. Liss, New York. 494 pp.

Gooding, G. V., and Lucas, G. B. 1959a. Factors influencing sporangial formation and zoospore activity in *Phytophthora parasitica* var. *nicotianae*. Phytopathology 49:277–281.

Gooding, G. V., and Lucas, G. B. 1959b. Effect of inoculum level on the severity of tobacco black shank. Phytopathology 49:274–276.

Goodwin, P. H., Kirkpatrick, B. C., and Duniway, J. M. 1989. Cloned DNA probes for identification of *Phytophthora parasitica*. Phytopathology 79:716–721.

Goodwin, P. H.. English, J. T., Neher, D. A., Duniway, J. M., and Kirkpatrick, B. C. 1990a. Detection of *Phytophthora parasitica* from soil and host tissue with a species-specific DNA probe. Phytopathology 80:277–281.

Goodwin, P. H., Kirkpatrick, B. C., and Duniway, J. M. 1990b. Identification of *Phytophthora citrophthora* with cloned DNA probes. Appl. Environ. Microbiol. 56:669–674.

Goodwin, S. B. 1991. DNA polymorphism in *Phytophthora infestans:* The Cornell experience. Pages 256–271 in: *Phytophthora.* J. A. Lucas, R. C. Shattock, D. S. Shaw, and L. R. Cooke, eds. Cambridge Univ. Press, Cambridge, U.K. 447 pp.

Goodwin, S. B., Spielman, L. J., Matuszak, J. M., Bergeron, S. N., and Fry, W. E. 1992. Clonal diversity and genetic differentiation of *Phytophthora infestans* populations in Northern and Central Mexico. Phytopathology 82:955–961.

Goodwin, S. B., Cohen, B. A., Deahl, K. L., and Fry, W. E. 1994a. Migration from northern Mexico as the probable cause of recent genetic changes in populations of *Phytophthora infestans* in the United States and Canada. Phytopathology 84:553–558.

Goodwin, S. B., Cohen, B. A., and Fry, W. E. 1994b. Panglobal distribution of a single clonal lineage of the Irish potato famine fungus. Proc. Natl. Acad. Sci. 91:11591–11595.

Goodwin, S. B., Sujkowski, L. S., Dyer, A. T., Fry, B. A., and Fry, W. E. 1995. Direct detection of gene flow and probable sexual reproduction of *Phytophthora infestans* in northern North America. Phytopathology 85:473–479.

Gorborova Ye, V., Bagirova, S. F., Dolgova, A. V. and Dyakov Yu, T. 1989. Vegetative incompatibility of *Phytophthora infestans* (Mont.) de Bary. Rep. Acad. Sci. USSR 304:1245–1248. (In Russian).

Goth, R. W. 1981. An efficient technique for prolonged storage of *Phytophthora infestans.* Am. Potato J. 58:257–260.

Goth, R. W., and Wester, R. E. 1963. Culture of *Phytophthora phaseoli* on living and sterilized media. Phytopathology 53:233–234.

Gottlieb, D., Knaus, R. J., and Wood, S. G. 1978. Differences in the sterol synthesizing pathways of sterol-producing and non-sterol producing fungi. Phytopathology 68:1168–1169.

Götz, E. 1985. Pathotypendefinition bei *Phytophthora infestans* (Mont.) de Bary mittels in-vitro-Testsortiment (Definition of pathotypes of *Phytophthora infestans* (Mont.) de Bary by an in vitro test collection). Arch. Phytopathol. Pflanzenschutz 21:265–271. (In German, English summary).

Götz, E. 1988. Nachweis der Oosporenbildung bei *Phytophthora infestans* (Mont.) de Bary. (Detection of oospore formation by *Phytophthora infestans* [Mont.] de Bary. Importance to potato research). Kartoffelforschung Aktuell. Pages 85–92.

Götz, E. 1991. Untersuchungen zum Auftreten des A_2-Paarungstyps bei *Phytophthora infestans* (Mont.) de Bary in Ostdeutschland (Research on the occurrence of A2 mating types by *Phytophthora infestans* [Mont.] de Bary in East Germany). Potato Res. 34:233–237.

Govi, G., and Tassinari, M. M. 1954. *Phytophthora megasperma* Drech. agente di marciume radicale del pesco (*Phytophthora megasperma* Drech., the cause of root rot of peach). Ann. Sper. Agrar. N.S. 8:1187–1196. (In Italian). (Rev. Appl. Mycol. 1956, 35:200.)

Govorova, G. F. 1983. Sources of resistance in strawberry breeding. Mikol. Fitopatol. 17:213–218.

Graberg, M. 1979. Rödröta pa jordgubbar funnen i Sverige (Red core of strawberries found in Sweden). Vaxtskyddsnotiser 43:113–115. (In Swedish). (Rev. Plant Pathol. 1980, 59:5844.)

Grabski, C., and Gisi, U. 1987. Quantification of synergistic interactions of fungicides against *Plasmopora* and *Phytophthora.* Crop Protect. 6:64–71.

Graham, J. H. 1990. Evaluation of tolerance of citrus rootstocks to Phytophthora root rot in chlamydospore-infested soil. Plant Dis. 74:743–746.

Graham, J. H., and Egel, D. S. 1988. Phytophthora root rot development on mycorrhizal and phosphorous-fertilized nonmycorrhizal sweet orange seedlings. Plant Dis. 72:611–614.

Graham, K. M. 1971. Plant Diseases of Fiji. H.M. Stationery Off., London. 256 pp.

Gramshaw, D., Langdon, P.W., Lowe, K. F., and Lloyd, D. L. 1985. Performance of North American and Australian lucernes in the Queensland subtropics. 1. Field reaction to important root and crown diseases. Aust. J. Exp. Agric. 25:76–81.

Grant, B. R. 1984. *Phytophthora cinnamomi* a plant pathogen offering possibilities for the biochemist. Trends Biol. Sci. 9:83–85.

Grant, B. R., and Byrt, P. N. 1984. Root temperature effects on the growth of *Phytophthora cinnamomi* in the roots of *Eucalyptus marginata* and *E. calophylla.* Phytopathology 74:179–184.

Grant, B. R., Griffith, J. M., Irving, H. R., and Radda, M. 1984. An improved method for the synchronous production of zoospores from *Phytophthora palmivora.* Exp. Mycol. 8:382–385.

Grant, B. R., Irving, H. R., and Radda, M. 1985. The effect of pectin and related compounds on encystment and germination of *Phytophthora palmivora* zoospores. J. Gen. Microbiol. 131:669–676.

Grant, B. R., Griffith, J. M., and Irving, H. R. 1986. A model to explain ion-induced differentiation in zoospores of *Phytophthora palmivora*. Exp. Mycol. 10:89–98.

Grant, V. 1971. Organismic Evolution. Freeman and Co., San Francisco. 418 pp.

Gray, F. A., and Hine, R. B. 1976. Development of Phytophthora root rot of alfalfa in the field and the association of *Rhizobium* nodules with early root infections. Phytopathology 66:1413–1417.

Gray, F. A., and Wofford, D. S. 1987. Phytophthora seed rot and seedling blight of sainfoin (*Onobrychis viciifolia*). Plant Dis. 71: 74–78.

Gray, F. A., Hine, R. B., Schonhorst, M. H., and Naik, J. D. 1973. A screening technique useful in selecting for resistance in alfalfa to *Phytophthora megasperma*. Phytopathology 63:1185–1188.

Gray, F. A., Bohl, W. H., and Wofford, D. S. 1988. Relationships of increasing levels of resistance of Phytophthora root rot in alfalfa to stand longevity and yield. Plant Dis. 72:1064–1067.

Gray, F. A., Griffin, G. D., Johnson, D. A., Eckert, J. W., and Kazimir, J. E. 1990. Interrelationships between *Meloidogyne hapla* and *Phytophthora megasperma* f. sp. *medicaginis* in seedling damping-off and root infection of alfalfa. Phytopathology 80:228–232.

Gray, L. E., and Pope, R. A. 1986. Influence of soil compaction on soybean stand, yield, and Phytophthora root rot incidence. Agron. J. 78:189–191.

Grech, N. M., and Rijkenberg, F. H. J. 1992. Injection of electrolytically generated chlorine into citrus micro-irrigation systems for the control of certain waterborne root pathogens. Plant Dis. 76: 457–461.

Green, R. J., Jr., and Pratt, R. G. 1970. Root rot of black walnut seedlings caused by *Phytophthora citricola*. Plant Dis. 54:583–585.

Greenhalgh, F. C. 1978. Evaluation of techniques for quantitative detection of *Phytophthora cinnamomi*. Soil Biol. Biochem. 10:257–259.

Greenhalgh, F. C., and Clarke, R. C. 1985. Use of fungicides to study significance and etiology of root rot of subterranean clover in dryland pastures of Victoria. Pages 234–236 in: Ecology and Management of Soilborne Plant Pathogens. C. A. Parker, A. D. Rovira, K. J. Moore, P. T. W. Wong, and J. F. Kollmorgen, eds. American Phytopathological Society, St. Paul, Minnesota.

Greenhalgh, F. C., and Taylor, P. A. 1985. *Phytophthora clandestina*, cause of severe taproot rot of subterranean clover in Victoria, Australia. Plant Dis. 69:1002–1004.

Greenhalgh, F. C., Merriman, P. R., and Keane, P. J. 1988. Relative importance of root rots of subterranean clover caused by *Aphanomyces euteiches* and *Phytophthora clandestina*. Plant Pathol. 37: 344–350.

Gregg, M. 1957. Germination of oospores of *Phytophthora erythroseptica*. Nature (London) 180:150.

Gregory, L. E., Almeyda, N., and Theis, T. 1960. The black pepper research programme in Puerto Rico. Proc. Am. Soc. Hortic. Sci. 4: 64–65.

Gregory, P. H. 1966. The fungus spore: What it is and what it does. Pages 1–13 in: The Fungus Spore. M. F. Madelin, ed. Butterworths, London. 337 pp.

Gregory, P. H., ed. 1974. *Phytophthora* Disease of Cocoa. Longman, London. 336 pp.

Gregory, P. H. 1983. Some major epidemics caused by *Phytophthora*. Pages 271–278 in: *Phytophthora:* Its Biology, Taxonomy, Ecology, and Pathology. D. C. Erwin, S. Bartnicki-Garcia, and P. H. Tsao, eds. American Phytopathological Society, St. Paul, Minn. 392 pp.

Gregory, P. H., and Maddison, A. C., eds. 1981. Epidemiology of *Phytophthora* on cocoa in Nigeria. Commonw. Mycol. Inst. Kew. Phytopathol. Pap. No. 25. 188 pp.

Gregory, P. H., Griffin, M. J., and Maddison, M. R. 1984. Cocoa black pod: A reinterpretation. Cocoa Growers' Bull. 35:5–22.

Griffin, D. M. 1963. Soil moisture and the ecology of soil fungi. Biol. Rev. 38:141–166.

Griffin, D. M. 1969. Soil water in the ecology of fungi. Annu. Rev. Phytopathol. 7:289–310.

Griffin, D. M. 1972. Ecology of Soil Fungi. Syracuse Univ. Press, Syracuse, New York. 193 pp.

Griffin, D. M. 1978. Effect of soil moisture on survival and spread of pathogens. Pages 175–197 in: Water Deficits and Plant Growth. Vol. 5. T. T. Kozlowski, ed. Academic Press, New York. 323 pp.

Griffin, M. J. 1977. Cocoa *Phytophthora* Workshop. Rothamsted Exp. Stn. England, 24–26 May, 1976. PANS 23:107–110.

Griffin, M. J., and Jones, O. W. 1977. *Phytophthora porri* on autumn-sown salad onions. Plant Pathol. 26:149–150.

Griffin, M. J., Idowu, O. L., Maddison, A. C., Taylor, B., and Word, M. R. 1981. Sources of infection. Pages 75–95 in: Epidemiology of *Phytophthora* on Cocoa in Nigeria. P. H. Gregory and A. C. Maddison, eds. Phytopathol. Pap. 25 Commonw. Mycol. Inst. Kew, Surrey, U.K.

Griffith, G. W., Snell, R., and Shaw, D. S. 1995. Late blight (*Phytophthora infestans*) on tomato in the tropics. The Mycologist 9:87–89.

Griffith, J. M., Davis, A. J., and Grant, B. R. 1992. Target sites of fungicides to control Oomycetes. Pages 69–100 in: Target Sites of Fungicide Action. W. Köller, ed. CRC Press, Boca Raton, Fla. 328 pp.

Griffith, J. M., Coffey, M. D., and Grant, B. R. 1993. Phosphonate inhibition as a function of phosphate concentration in isolates of *Phytophthora palmivora*. J. Gen. Microbiol. 139:2109–2116.

Grimm, G. R., and Alexander, A. F. 1973. Citrus leaf pieces as traps for *Phytophthora parasitica* from soil slurries. Phytopathology 63: 540–541.

Grimm, G. R., and Hutchison, D. J. 1973. A procedure for evaluating resistance of citrus seedlings to *Phytophthora parasitica*. Plant Dis. Rep. 57:669–671.

Grimm, G. R., and Whidden, R. 1966. Preventing infection from the foot rot fungus. Proc. Fla. State Hortic. Soc. 79:73–75.

Grisham, M. P., Taber, R. A., and Barnes, L. W. 1983. Phytophthora rot of potatoes in Texas caused by *Phytophthora parasitica* and *P. cryptogea*. Plant Dis. 67:1258–1261.

Grohmann, U. J., and Hoffmann, G. M. 1989. Blattinfektion und Gewebebesiedelung durch *Phytophthora infestans* an Kartoffeln unter Einfluß steigender Dosierungen von Metalaxyl (Leaf infection and tissue colonization in potato plants by *Phytophthora infestans* influenced by increasing dosages of metalaxyl). Z. Pflanzenkr. Pflanzenschutz 96:585–603.

Gross, D. 1979. Phytoalexine der Kartoffel (The phytoalexins of the potato). Biochem. Physiol. Pflanz. 174:327–344.

Grove, G. G., and Boal, R. J. 1991. Influence of temperature and wetness duration on infection of immature apple and pear fruit by *Phytophthora cactorum*. Phytopathology 81:1465–1471.

Grove, G. G., Madden, L. V., Ellis, M. A., and Schmitthenner, A. F. 1985a. Influence of temperature and wetness duration on infection of immature strawberry fruit by *Phytophthora cactorum*. Phytopathology 75:165–169.

Grove, G. G., Ellis, M. A., Madden, L. V., and Schmitthenner, A. F. 1985b. Overwinter survival of *Phytophthora cactorum* in infected strawberry fruit. Plant Dis. 69:514–515.

Grove, G. G., Ellis, M. A., and Madden, L. V. 1985c. Leather rot of strawberry: A summary of research in Ohio. Adv. Strawberry Prod. 4 (Spring): 4–9.

Grove, S. N., and Bracker, C. E. 1970. Protoplasmic organization of hyphal tips among fungi: Vesicles and Spitzenkörper. J. Bacteriol. 104:989–1009.

Guest, D. I. 1984a. The influence of cultural factors on the direct antifungal activities of fosetyl-Al, propamocarb, metalaxyl, SN 75196 and Dowco 444. Phytopathol. Z. 111:155–164.

Guest, D. I. 1984b. Modification of defence responses in tobacco and capsicum following treatment with fosetyl-Al [aluminium tris (O-ethyl phosphonate)]. Physiol. Plant Pathol. 25:125–134.

Guest, D. I., and Grant, B. 1991. The complex action of phosphonates as antifungal agents. Biol. Rev. 66:159–187.

Guest, D. I., Upton, J. C. R., and Rowan, K. S. 1989. Fosetyl-Al alters the respiratory response in *Phytophthora nicotianae* var. *nicotianae*-infected tobacco. Physiol. Mol. Plant Pathol. 34:257–265.

Guiscafre-Arrillaga, J. R. 1931. Gomosis de la toronja en la Horienda Eugenia (Grapefruit gummosis in the Eugenia Horienda). (P.R.) Rev. Obras Publ. 8:124–125. (In Spanish).

Guiscafre-Arrillaga, J. R. 1932. The brown rot fungus in Puerto Rico. J. Dep. Agric. P.R. 16:193–202

Gujer, H. U. 1991. Integrated control of potato late blight (*Phytoph-

thora infestans) in Switzerland: Concept and first results. EPPO (Eur. Mediterr. Plant Prot. Organ.) Bull. 21:61–66.

Gullino, G. 1975. Observazioni sulle principali malattie fungine delle *Kalanchoë* in Italia (Observations on the principal fungus diseases of *Kalanchoe* in Italy). Rev. Econ. Agrar. (Italy) 6:14–19. (In Italian).

Gunnell, P. S., and Webster, R. K. 1988. Crown and root rot of cultivated wild rice in California caused by *Phytophthora erythroseptica.* Plant Dis. 72:909–910.

Günther, E., and Grümmer, G. 1958. Untersuchungen über die fruchtfäulen der tomate (Investigations on fruit rot of tomato). Gartenbauwissenschaft 23:130–159. (In German). (Rev. Appl. Mycol. 1959, 38:280.)

Gupta, J. P., Erwin, D. C., Eckert, J. W., and Zaki, A. I. 1985. Translocation of metalaxyl in soybean plants and control of stem rot caused by *Phytophthora megasperma* f. sp. *glycinea.* Phytopathology 75:865–869.

Gupta, V. K., and Utkhede, R. S. 1986. Factors affecting the production of antifungal compounds by *Enterobacter aerogenes* and *Bacillus subtilis,* antagonists of *Phytophthora cactorum.* J. Phytopathol. 117:9–16.

Gupta, V. K., and Utkhede, R. S. 1987. Nutritional requirement for production of antifungal substance by *Enterobacter aerogenes* and *Bacillus subtilis* antagonists of *Phytophthora cactorum.* J. Phytopathol. 120:143–153.

Gürtler, H. 1984. Physiological races of *Phytophthora infestans* in Denmark and low temperature storage of isolates. Potato Res. 27: 25–31.

Gurunath Rao, V. 1970. Contributions to the genus *Phytophthora* de Bary from India. Mycopathol. Mycol. Appl. 42:248–258.

Güssow, H. T. 1920. Phytopathological notes. Phytopathology 10:50.

Guterman, C. E. F. 1930. Final summary of the work on diseases of lilies for the lily disease investigation fellowship. Hortic. Soc. N.Y. Yearb. 1930. Pages 51–102.

Gutter, Y. 1983. Supplementary antimold activity of fosetyl-Al, a new brown rot fungicide for citrus fruits. Phytopathol. Z. 107:301–308.

Guy, S. O., Grau, C. R., and Oplinger, E. S. 1989. Effect of temperature and soybean cultivar on metalaxyl efficacy against *Phytophthora megasperma* f. sp. *glycinea.* Plant Dis. 73:236–239.

Guzmán-N., J. 1964. Nature of partial resistance of certain clones of three *Solanum* species to *Phytophthora infestans.* Phytopathology 54:1398–1404.

Guzman-Naranjo, J. 1966. El pepino dulce un nuevo hospedante del *Phytophthora infestans* (Mont.) de Bary (Sweet cucumber a new host of *Phytophthora infestans*). Fitopatol. Chile 1:10–14. (In Spanish).

Haas, J. H. 1964. Isolation of *Phytophthora megasperma* var. *sojae* in soil dilution plates (Abstr.). Phytopathology 54:894.

Haas, J. H., and Buzzell, R. I. 1976. New race 5 and 6 of *Phytophthora megasperma* var. *sojae* and differential reactions of soybean cutivars for race 1 to 6. Phytopathology 66:1361–1362.

Haasis, F. A., and Nelson, R. R. 1963. Studies on the biological relationship of species of *Phytophthora* as measured by oospore formation in intra- and interspecific crosses. Plant Dis. Rep. 47:705–709.

Hächler, H., and Hohl, H. R. 1984. Temporal and spatial distribution patterns of collar and papillae wall appositions in resistant and susceptible tuber tissue of *Solanum tuberosum* infected by *Phytophthora infestans.* Physiol. Plant Pathol. 24:107–118.

Hahn, M. G., Bonhoff, A., and Griseback, H. 1985. Quantitative localization of the phytoalexin glyceollin I in relation to fungal hyphae in soybean roots infected with *Phytophthora megasperma* f. sp. *glycinea.* Plant Physiol. 77:591–601.

Hall, A. M. 1959. The culture of *Phytophthora infestans* in artificial media. Trans. Br. Mycol. Soc. 42:15–26.

Hall, G. 1989a. *Phytophthora pseudotsugae.* Mycopathologia 106: 203–204.

Hall, G. 1989b. Unusual or interesting records of plant pathogenic Oomycetes. Plant Pathol. 38:604–611.

Hall, G. 1991a. *Phytophthora clandestina.* Mycopathologia 115:223–224.

Hall, G. 1991b. *Phytophthora ilicis.* Mycopathologia 115:225–226.

Hall, G. 1991c. *Phytophthora lateralis.* Mycopathologia 115:227–229.

Hall, G. F. 1991d. *Phytophthora richardiae.* Mycopathologia 115: 231–232.

Hall, G. 1993. An integrated approach to the analysis of variation in *Phytophthora nicotianae* and a redescription of the species. Mycol. Res. 97:559–574.

Hall, G., Dobson, S., and Nicholls, C. 1992. First record of *Phytophthora inflata* in the United Kingdom. Plant Pathol. 41:95–97.

Hall, R. 1969. Molecular approaches to taxonomy of fungi. Bot. Rev. (Lancaster) 35:285–304.

Hall, R., Zentmyer, G. A., and Erwin, D. C. 1969. Approach to taxonomy of *Phytophthora* through acrylamide gel-electrophoresis of proteins. Phytopathology 59:770–774.

Hallett, I. C., and Dick, M. W. 1981. Seasonal and diurnal fluctuations of oomycete propagule numbers in the free water of a freshwater lake. J. Ecol. 69:671–692.

Halsall, D. M. 1976a. Specificity of cytoplasmic and cell-wall antigens from four species of *Phytophthora.* J. Gen. Microbiol. 94: 149–158.

Halsall, D. M. 1976b. Zoospore chemotaxis in Australian isolates of *Phytophthora* species. Can. J. Microbiol. 22:409–422.

Halsall, D. M. 1977. Effects of certain cations on the formation and infectivity of *Phytophthora* zoospores. 2. Effects of copper, boron, cobalt, manganese, molybdenum, and zinc ions. Can. J. Microbiol. 23:1002–1010.

Halsall, D. M. 1978. A comparison of *Phytophthora cinnamomi* infection in *Eucalyptus sieberi,* a susceptible species, and *Eucalyptus maculata,* a field resistant species. Aust. J. Bot. 26:643–655.

Halsall, D. M. 1982a. A forest soil suppressive to *Phytophthora cinnamomi* and conducive to *Phytophthora cryptogea.* I. Survival, germination and infectivity of mycelium and chlamydospores. Aust. J. Bot. 30:11–25.

Halsall, D. M. 1982b. A forest soil suppressive to *Phytophthora cinnamomi* and conducive to *Phytophthora cryptogea.* II. Suppression of sporulation. Aust. J. Bot. 30:27–37.

Halsall, D. M., and Forrester, R. I. 1977. Effects of certain cations on the formation and infectivity of *Phytophthora* zoospores. 1. Effects of calcium, magnesium, potassium and iron ions. Can. J. Microbiol. 23:994–1001.

Hamm, P. B., and Hansen, E. M. 1981. Host specificity of *Phytophthora megasperma* from Douglas fir, soybean, and alfalfa. Phytopathology 71:65–68.

Hamm, P. B., and Hansen, E. M. 1982a. Pathogenicity of *Phytophthora* species to Pacific Northwest conifers. Eur. J. For. Pathol. 12:167–174.

Hamm, P. B, and Hansen, E. M. 1982b. Single-spore isolate variation: The effect on varietal designation in *Phytophthora megasperma.* Can. J. Bot. 60:2931–2938.

Hamm, P. B., and Hansen, E. M. 1983. *Phytophthora pseudotsugae* a new species causing root rot of Douglas-fir. Can. J. Bot. 61:2626–2631.

Hamm, P. B., and Hansen, E. M. 1984. Improved method for isolating *Phytophthora lateralis* from soil. Plant Dis. 68:517–519.

Hamm, P. B., and Hansen, E. M. 1987. Identification of *Phytophthora* spp. known to attack conifers in the Pacific Northwest. Northwest Sci. 61:103–109.

Hamm, P. B., and Koepsell, P. A. 1984. Phytophthora root rot of cabbage and cauliflower in Oregon. Plant Dis. 68:533–535.

Hamm, P. B., Cooley, S. J., and Hansen, E. M. 1984. Response of *Phytophthora* spp. to metalaxyl in forest tree nurseries in the Pacific Northwest. Plant Dis. 68:671–673.

Hamm, P. B., Hansen, E. M., and Sutherland, J. R. 1985. *Phytophthora cryptogea* and *Phytophthora drechsleri* associated with root-rotted Douglas fir seedlings. Plant Dis. 69:361.

Hamm, P. B., Hansen, E. M., Hennon, R. M., and Shaw, C. G. 1988. *Pythium* species from forest and muskeg areas of southeast Alaska. Trans. Br. Mycol. Soc. 91:385–388.

Hammarlund, C. 1933. Beiträge zur Kenntnis der Mikromycetenflora der Provinz Skane (Schonen) (Research on the germination of the micromyceteflora of the province Skane [Schonen]). Ark. Bot. 25:1–126. (In German). (Cited in Tucker 1933).

Handelsman, J., Raffel, S., Mester, E. H., Wunderlich, L., and Grau,

C. R. 1990. Biological control of damping-off of alfalfa seedlings with *Bacillus cereus* UW85. Appl. Environ. Microbiol. 56:713–718.

Handelsman, J., Nesmith, W. C., and Raffel, S. J. 1991. Microassay for biological and chemical control of infection of tobacco by *Phytophthora parasitica* var. *nicotianae*. Curr. Microbiol. 22:317–319.

Handreck, K., and Black, N. 1984. Growing media for ornamental plants and turf. N.S.W. Univ. Press, Sydney, Australia.

Hansen, E. M. 1985. Forest pathogens on N.W. North America and their potential for damage in Britain. For. Comm. For. Rec. 129. (U.K.) 14 pp.

Hansen, E. M. 1987. Speciation in *Phytophthora:* Evidence from the *Phytophthora megasperma* complex. Pages 325–337 in: Evolutionary Biology of the Fungi. A. D. M. Rayner, C. M. Brasier, and D. Moore, eds. Cambridge Univ. Press, Cambridge, U.K. 465 pp.

Hansen, E. M. 1991. Variation in the species of the *Phytophthora megasperma* complex. Pages 148–163 in: *Phytophthora.* J. A. Lucas, R. C. Shattock, D. S. Shaw, and L. R. Cooke, eds. Cambridge Univ. Press, Cambrige, U.K. 447 pp.

Hansen, E. M., and Hamm, P. B. 1983. Morphological differentiation of host-specialized groups of *Phytophthora megasperma.* Phytopathology 73:129–134.

Hansen, E. M., and Maxwell, D. P. 1991. Species of the *Phytophthora megasperma* complex. Mycologia 83:376–381.

Hansen, E. M., Hamm, P. B., Julis, A. J., and Roth, L. F. 1979. Isolation, incidence and management of *Phytophthora* in forest tree nurseries in the Pacific Northwest. Plant Dis. Rep. 63:607–611.

Hansen, E. M., Roth, L. F., Hamm, P. B., and Julis, A. J. 1980. Survival, spread, and pathogenicity of *Phytophthora* spp. on Douglas fir seedlings planted on forest sites. Phytopathology 70:422–425.

Hansen, E. M., Brasier, C. M., Shaw, D. S., and Hamm, P. B. 1986. The taxonomic structure of *Phytophthora megasperma:* Evidence for emerging biological species groups. Trans. Br. Mycol. Soc. 87: 557–573.

Hansen, E. M., Hamm, P. B., Shaw, C. G., III, and Hennon, P. E. 1988. *Phytophthora drechsleri* in remote areas of southeast Alaska. Trans. Br. Mycol. Soc. 91:379–384.

Hansen, E. M., Hamm, P. B., and Roth, L. F. 1989. Testing Port-Orford-cedar for resistance to *Phytophthora.* Plant Dis. 73:791–794.

Hansen, H. N., and Smith, R. E. 1932. The mechanism of variation in imperfect fungi. *Botrytis cinerea.* Phytopathology 22:953–964.

Hansen, H. N., and Thomas, H. E. 1945. Diseases of *Fremontia.* Madroño 8:39–42.

Hansen, J. D. (Chairman of Committee on Standardization of Common Names for Plant Diseases) 1985. Common names for plant diseases. Plant Dis. 69:649–676.

Hara, K. 1918. *Kawakamia carica.* J. Hortic. Soc. Jpn. 30(4):20–22. (In Japanese). (Cited in Tanaka 1920).

Hara, K. 1930. Rice Diseases. Tokyo. (In Japanese). (Cited in Ou 1972).

Hardham, A. R. 1987. Microtubules and the flagellar apparatus in zoospores and cysts of the fungus *Phytophthora cinnamomi.* Protoplasma 137:109–124.

Hardham, A. R., Suzaki, E., and Perkin, J. L. 1985. The detection of monoclonal antibodies specific for surface components on zoospores and cysts of *Phytophthora cinnamomi.* Exp. Mycol. 9:264–268.

Hardham, A. R., Suzaki, E., and Perkin, J. L. 1986. Monclonal antibodies to isolate species, and genus specific components on the surface of zoospores and cysts of the fungus *Phytophthora cinnamomi.* Can. J. Bot. 64:311–321.

Hardham, A. R., Gubler, F., and Duniec, J. 1991a. Ultrastructural and immunological studies of zoospores of *Phytophthora.* Pages 50–69 in: *Phytophthora.* J. A. Lucas, R. C. Shattock, D. S. Shaw, and L. R. Cooke, eds. Cambridge Univ. Press, Cambridge, U.K. 446 pp.

Hardham, A. R., Gubler, F., and Duniec, J., and Elliott, J. 1991b. A review of methods for the production and use of monoclonal antibodies to study zoosporic plant pathogens. J. Microsc. 162:305–318.

Harding, P. R., and Savage, D. C. 1964. Investigation of possible correlation of hot-water washing with excessive storage decay in coastal California lemon packing houses. Plant Dis. Rep. 48:808–810.

Hardy, G. E. St. J., and Sivasithamparam, K. 1988. *Phytophthora* spp. associated with container-grown plants in nurseries in Western Australia. Plant Dis. 72:435–437.

Hardy, G. E. St. J., and Sivasithamparam, K. 1989. Microbiological, chemical, and physical changes during decomposing of a eucalyptus (*Eucalyptus calophylla* and *E. diversicolor* bark mix). Biol. Fertil. Soils 8:260–270.

Hardy, G. E. St. J., and Sivasithamparam, K. 1991a. Effects of sterile and non-sterile leachates extracted from composted eucalyptus bark and pine-bark container media on *Phytophthora* spp. Soil Biol. Biochem. 23:25–30.

Hardy, G. E. St. J., and Sivasithamparam, K. 1991b. How container media and matric potential affect the production of sporangia, oospores and chlamydospores by three *Phytophthora* species. Soil Biol. Biochem. 23:31–39.

Hardy, G. E. St. J., and Sivasithamparam, K. 1991c. Suppression of Phytophthora root rot by a composted eucalyptus bark mix. Aust. J. Bot. 39:153–159.

Hardy, G. E. St. J., and Sivasithamparam, K. 1991d. Sporangial responses do not reflect microbial suppression of *Phytophthora drechsleri* in composted eucalyptus bark mix. Soil Biol. Biochem. 23:757–765.

Hargreaves, A. J., and Duncan, J. M. 1978. Detection of *Phytophthora* species in field soils by simple baiting procedures. Soil Biol. Biochem. 10:343–345.

Harkness, R. W., and Reynolds, J. E. 1964. Effect of nitrogen and potassium nutrition on the Phytophthora leaf spot of *Philodendron.* Proc. Fla. State Hortic. Soc. 77:475–481.

Harland, S. C. 1917. Relation of cryptogamic diseases to cotton yield. West Indies Bull. 16:197–199.

Harnish, W. N. 1965. Effect of light on production of oospores and sporangia in species of *Phytophthora.* Mycologia 57:85–90.

Harrigan, E. K. S., Mc Rae, C. F., and Heritage, A. D. 1982. Sources of resistance to *Phytophthora cryptogea* and *Alternaria carthami* for safflower breeding in Australia. J. Aust. Inst. Agric. Sci. 48: 158–160.

Harris, D. C. 1979a. The occurrence of *Phytophthora syringae* in fallen apple leaves. Ann. Appl. Biol. 91:309–312.

Harris, D. C. 1979b. The suppression of *Phytophthora syringae* in orchard soil by furalaxyl as a means of controlling fruit rot of apple and pear. Ann. Appl. Biol. 91:331–336.

Harris, D. C. 1985a. The colonisation of fallen apple leaves by *Phytophthora syringae* in relation to inoculum levels in orchard soil. Ann. Appl. Biol. 107:179–188.

Harris, D. C. 1985b. *Microdochium fusarioides* sp. nov. from oospores of *Phytophthora syringae.* Trans. Br. Mycol. Soc. 84:358–361.

Harris, D. C. 1985c. Survival of *Phytophthora syringae* oospores in and on apple orchard soil. Trans. Br. Mycol. Soc. 85:153–155.

Harris, D. C. 1986. Methods for preparing, estimating and diluting suspensions of *Phytophthora cactorum* zoospores. Trans. Br. Mycol. Soc. 86:482–486.

Harris, D. C. 1987. Orchard trials on the control of apple collar rot. J. Hortic. Sci. 62:441–447.

Harris, D. C. 1991a. A comparison of dazomet, chloropicrin, and methyl bromide as soil disinfectants. J. Hortic. Sci. 66:51–58.

Harris, D. C. 1991b. The *Phytophthora* diseases of apple. J. Hortic. Sci. 66:513–544.

Harris, D. C., and Bielenin, A. 1986. Evaluation of selective media and bait methods for estimating *Phytophthora cactorum* in apple orchard soils. Plant Pathol. 35:565–574.

Harris, D. C., and Cole, D. M. 1982. Germination of *Phytophthora syringae* oospores. Trans. Br. Mycol. Soc. 79:527–530.

Harris, D. C., and Stickels, J. E. 1981. Crown rot (*Phytophthora cactorum*) in glasshouse-grown strawberries at East Malling Res. Stn. Plant Pathol. 30:205–212.

Harris, D. C., and Tobutt, K. R. 1986. Factors influencing the mortality of apple seedlings inoculated with zoospores of *Phytophthora cactorum* (L. & C.) Schroet. J. Hortic. Sci. 61:457–464.

Harris, D. C., Cardon, J. A., Justin, S. H. F. W., and Passey, A. J. 1984. *Phytophthora palmivora* on cultured roots of coconut. Trans. Br. Mycol. Soc. 82:249–255.

Harris, J. A., Kassaby, F. Y., and Smith, I. W. 1985. Variations in mortality in families of *Eucalyptus regnans* caused by *Phytophthora cinnamomi,* up to 5 years after planting. Aust. Fos. Res. 15:57–65.

Harris, M. R. 1934. A *Phytophthora disease* of snapdragons. Phytopathology 24:412–417.

Harrison, J. G., Barker, H., Lowe, R., and Rees, E. A. 1990. Estimation of amounts of *Phytophthora infestans* mycelium in leaf tissue by enzyme-linked immunosorbent assay. Plant Pathol. 39:274–277.

Harrison, J. G., Lowe, R., and Duncan, J. M. 1991. Use of ELISA for assessing major gene resistance of potato leaves to *Phytophthora infestans.* Plant Pathol. 40:431–435.

Hart, R. M., Keast, D., and Smythe, G. 1986. Metabolic stability of *Phytophthora cinnamomi* Rands in Western Australia. Can. J. Microbiol. 32:820–822.

Hartig, R. 1876. Die Buchencotyledonen-Krankheit (Cotyledon disease of beech). Ztschr. Forst. Jadgw. 8:117–123. (In German).

Hartig, R. 1880. Der Buchenkeimlingspilz, *Phytophthora* (*Peronospora fagi* M.) (The beech seedling fungus, *Phytophthora* [*Peronospora fagi* M.]). Untersuch. Forstb. Inst. München. 1:33–57. (In German).

Hartig, R. 1882. *Phytophthora omnivora* (*Phytophthora fagi* und *Peronospora sempervivi*). Lehrb. Baumkrankheiten, Berl.:42–46. (In German). (Cited in Tucker 1933).

Hartig, R. 1883. Beschädigung der Nadelholzsaatbeete durch *Phytophthora omnivora* (*Fagi*) (Damage in conifer seed beds by *Phytophthora omnivora* [*Fagi*]). Forstwiss. Centralbl. 27:593–596. (In German). (Cited in Nienhaus 1960.)

Hartill, W. F. T., Young, K., Allan, D. J., and Henshall, W. R. 1990. Effects of temperature and leaf wetness on the potato late blight. N.Z. J. Crop Hortic. Sci. 18:181–184.

Hartley, C. 1921. Damping-off in forest nurseries. U.S. Dep. Agric. Bull. 934. 99 pp. (Cited in Tucker 1933).

Hartley, C. 1924. De ziekten van de cacao (A disease of cocoa). Dep. Landb. Inst. Plantenziekt en. Nijv. en Handel, Meded. 19. 16 pp. (In Dutch). (Cited in Tucker 1933).

Haskell, R. J. 1921. *Phytophthora infestans* on eggplant in the United States. Phytopathology 11:504–505. (Cited in Tucker 1933).

Haskins, R. H., and Gardner, N. R. 1978. Effects of *Trichoderma* on sexual reproduction of some species of *Pythium* and *Phytophthora.* Can. J. Bot. 56:1651–1654.

Haskins, R. H., Tulloch, A. P., and Mircetich, R. G. 1964. Steroids and the stimulation of sexual reproduction of a species of *Pythium.* Can. J. Microbiol. 10:187–195.

Havey, M. J., and Grau, C. R. 1985. Decline of established alfalfa in soils naturally infested with *Phytophthora megasperma* f. sp. *medicaginis* and level of correlation by seedling assay. Plant Dis. 69:221–224.

Havey, M. J., and Maxwell, D. P. 1987. Inheritance of Phytophthora root rot resistance in two diploid alfalfa species. Crop Sci. 27:225–228.

Havey, M. J., and Maxwell, D. P. 1988. Transfer of disease resistance from diploid to tetraploid alfalfa by unreduced female gametes. Plant Dis. 72:603–604.

Havey, M. J., Maxwell, D. P., and Irwin, J. A. G. 1987. Independent inheritance of genes conditioning resistance to Phytophthora root rot from diploid and tetraploid alfalfa. Crop Sci. 27:873–879.

Hawksworth, D. L. 1991. The fungal dimension of biodiversity: Magnitude, significance, and conservation. Mycol. Res. 95:641–655.

Hawksworth, D. L., Sutton, B. C., and Ainsworth, G. C. 1983. Ainsworth and Bisby's Dictionary of the Fungi. Commonw. Mycol. Inst. Kew, Surrey. 445 pp.

Hawthorne, H. T. 1988. Fungi causing storage rots of *Cucurbita* spp. N.Z. J. Exp. Agric. 16:151–157.

Heaney, S. P., Alger, J. C., and Roques, J. F. 1988a. A novel fungicide for use against diseases on vines. Proc. Br. Crop Prot. Conf. Pests Dis. 1:245.

Heaney, S. P., Shepard, M. C., Crowley, P. J., and Shearing, S. J. 1988b. A novel benzamide fungicide. Proc. Br. Crop Prot. Conf. Pests Dis. 2:551.

Heather, W. A., and Pratt, B. H. 1975. Association of *Phytophthora drechsleri* Tucker with death of *Pinus radiata* in southern New South Wales. Aust. J. Bot. 23:285–288.

Heather, W. A., Pratt, B. H., and Chin, T. Y. 1977. Pre- and post-emergence damping-off of seedlings of *Pinus* species by *Phytophthora cinnamomi* and *P. drechsleri.* Aust. J. Bot. 25:385–393.

Heddergott, H. 1953. Plant Protection Conf. Fed. Biol. Inst. Brunswick, Westphalia (1952). Mitt. Biol. ZentAnst. Berl. 75:74–79. (Rev. Appl. Mycol. 1954, 33:13).

Hedges, F. 1929. Phytophthora blight of lilac. Plant Dis. Rep. 13:27.

Heisey, R. F., and Murphy, R. P. 1985. Phenotypic recurrent selection for resistance to Phytophthora root rot in two diploid alfalfa populations. Crop Sci. 25:693–694.

Heitefuss, R. 1989. Crop and Plant Protection: The Practical Foundations. Halsted Press, New York. 261 pp.

Helena, K. 1978. New diseases caused by *Phytophthora* spp. Phytophthora Newsl. 6:72–73.

Helgeson, J. P., Kemp, J. D., Haberlach, G. T., and Maxwell, D. P. 1972. A tissue culture system for studying disease resistance: The black shank disease in tobacco callus cultures. Phytopathology 62: 1439–1443.

Helgeson, J. P., Haberlach, G. T., and Upper, C. D. 1976. A dominant gene conferring disease resistance to tobacco plants is expressed in tissue cultures. Phytopathology 66:91–96.

Helgeson, J. P., Hunt, G. J., Haberlach, G. T., and Austin, S. 1986. Somatic hybrids between *Solanum brevidens* and *Solanum tuberosum:* Expression of a late blight resistance gene and potato leaf roll resistance. Plant Cell Rep. 3:212–214.

Helton, A. W., and Dilbeck, R. 1987. Phytophthora collar-rot of orchard trees. Current Info. Series #752. Univ. Idaho Co-op. Extn. Service. 4 pp.

Helton, A. W., Dilbeck, R., and Thostenson, A. A. 1984. Phytophthora collar rot in Idaho orchards (Abstr.). Phytopathology 74:1138.

Hemmes, D. E. 1983. Cytology of *Phytophthora.* Pages 9–40 in: *Phytophthora:* Its Biology, Taxonomy, Ecology, and Pathology. D. C. Erwin, S. Bartnicki-Garcia, and P. H. Tsao, eds. American Phytopathological Society, St. Paul, Minn. 392 pp.

Hemmes, D. E., and Bartnicki-Garcia, S. 1975. Electron microscopy of gametangial interaction and oospore development in *Phytophthora capsici.* Arch. Microbiol. 103:91–112.

Hemmes, D. E., and Hohl, H. R. 1969. Ultrastructural changes in directly germinating sporangia of *Phytophthora parasitica.* Am. J. Bot. 56:300–313.

Hemmes, D. E., and Hohl, H. R. 1971. Ultrastructural aspects of encystation and cyst germination in *Phytophthora parasitica.* J. Cell. Sci. 9:175–191.

Hemmes, D. E., and Hohl, H. R. 1972. Flagellum degeneration in the fungus *Phytophthora palmivora* (formerly *Phytophthora parasitica*). J. Gen. Microbiol. 73:345–351.

Hemmes, D. E., and Hohl, H. R. 1975. Ultrastructural changes and a second mode of flagellar degeneration during aging of *Phytophthora palmivora* sporangia. J. Cell Sci. 19:563–577.

Hemmes, D. E., and Lerma, A. P., Jr. 1985. The ultrastructure of developing and germinating chlamydospores of *Phytophthora palmivora.* Mycologia 77:743–755.

Hemmes, D. E., and Ribeiro, O. K. 1977. Electron microscopy of early gametangial interaction in *Phytophthora megasperma* var. *sojae.* Can. J. Bot. 55:436–447.

Hemmes, D. E., and Wong, L. D. S. 1975. Utrastructure of chlamydospores of *Phytophthora cinnamomi* during development and germination. Can. J. Bot. 53:2945–2957.

Hendrikx, F. L. 1939. Observations phytopathologiques a la station de Mulungu en 1938 (Phytopathological observations at the station of Mulungu in 1938). Publ. Inst. Nat. Etude Agron. Congo Belge. Pages 117–128 (1939). (In French). (Rev. Appl. Mycol. 1940, 19:329–330).

Hendrix, F. F., Jr., and Campbell, W. A. 1970. *Phytophthora* and *Pythium* species from old growth forest sites in Kentucky, North Carolina, and Tennessee (Abstr.). Phytopathology 60:1295.

Hendrix, F. F., Jr., and Campbell, W. A. 1973. Pythiums as plant pathogens. Annu. Rev. Phytopathol. 11:77–98.

Hendrix, F. F., Jr., and Kuhlman, E. G. 1965. Factors affecting direct recovery of *Phytophthora cinnamomi* from soil. Phytopathology 55:1183–1187.

Hendrix, F. F., Jr., and Papa K. E. 1974. Taxonomy and genetics of *Pythium.* Proc. Am. Phytopathol. Soc. 1:200–207.

Hendrix, J. W. 1964. Sterol induction of reproduction and stimulation of growth of *Pythium* and *Phytophthora.* Science 144:1028–1029.
Hendrix, J. W. 1965. Influence of sterols on growth and reproduction of *Pythium* and *Phytophthora* spp. Phytopathology. 55:790–797.
Hendrix, J. W. 1967. Light-cholesterol relationships in morphogenesis of *Phytophthora palmivora* and *P. capsici* sporangia. Mycologia 59:1107–1111.
Hendrix, J. W. 1970. Sterols in growth and reproduction of fungi. Annu. Rev. Phytopathol. 8:111–130.
Hendrix, J. W. 1974. Physiology and biochemistry of growth and reproduction in *Pythium.* Proc. Am. Phytopathol. Soc. 1:207–210.
Hendrix, J. W. 1976. Inhibition of growth of tobacco plants by *Phytophthora cryptogea.* Tob. Sci. 20:152–153. Also Tob. Int. 178(20): 50–51.
Hendrix, J. W., and Apple, J. L. 1964. Fats and fatty acid derivatives as growth stimulants and carbon sources for *Phytophthora parasitica* var. *nicotianae.* Phytopathology 54:987–994.
Hendrix, J. W., and Guttman, S. M. 1970. Sterol or calcium requirement by *Phytophthora parasitica* var. *nicotianae* for growth on nitrate nitrogen. Mycologia 62:195–198.
Hendrix, J. W., and Kuhlman, E. G. 1965. Factors affecting direct recovery of *Phytophthora cinnamomi* from soil. Phytopathology 55:1183–1187.
Hendrix, J. W., Norman, C., and Apple, J. L. 1966. Chemical and physical factors influencing growth of *Phytophthora parasitica* var. *nicotianae* on vegetable oils. Physiol. Plant. 19:159–166.
Hendrix, J. W., Guttman, S. M., and Wightman, D. L. 1969. Cation and sterol effects on growth of *Phytophthora parasitica* var. *nicotianae.* Phytopathology 59:1620–1624.
Henfling, J. W. D. M., Bostock, R., and Kuć, J. 1980. Effect of abscisic acid on rishitin and lubimin accumulation and resistance to *Phytophthora infestans* and *Cladosporium cucumerinum* in potato tuber tissue slices. Phytopathology 70:1074–1078.
Henniger, H. 1959. Versuche zur Kultur verscheidener Rassen von *Phytophthora infestans* (Mont.) de Bary auf Künstlichen Nahrboden (Research on the culture of different races of *Phytophthora infestans* [Mont.] de Bary on artificial culture medium). Phytopathol. Z. 34:285–306. (In German).
Henricksen, H. C. 1907. El cultivo de legumbres en Puerto Rico (Vegetable growing in Puerto Rico). P.R. Agric. Exp. Stn. Bull. 7. 58 pp.
Henry, A. W., and Stelfox, D. 1966. *Phytophthora citricola* Sawada, in relation to shoot blight of lilacs and crown rot of elders in Alberta. Can. Plant Dis. Surv. 46:146.
Henry, A. W., and Stelfox, D. 1968. Comparative behavior of the oospores and oogonia of *Phytophthora citricola* on an artificial medium. Can. J. Bot. 46:1419–1421.
Hepting, G. H., Buchanan, T. S., and Jackson, L. W. R. 1945. Little leaf disease of pine. U.S. Dep. Agric. Circ. 716. 15 pp.
Heritage, A. D., and Harrigan, E. K. S. 1984. Environmental factors influencing safflower screening for resistance to *Phytophthora cryptogea.* Plant Dis. 68:767–769.
Herlihy, M. 1970. Contrasting effects of nitrogen and phosphorus on potato tuber blight. Plant Pathol. 19:65–68.
Herlihy, M., and Carroll, P. J. 1969. Effects of N, P and K and their interactions on yield, tuber blight and quality of potatoes. J. Sci. Food Agric. 20:513–517.
Herman, R. P., and Herman, C. A. 1985. Prostaglandins or prostaglandin-like substances are implicated in normal growth and development in Oomycetes. Prostaglandins 29:819–830.
Hernandez, A. 1919. Plant diseases. Philipp. Bur. Agric. Annu. Rep. 19:50–54. (Cited in Tucker 1933).
Herr, L. J. 1957. Factors affecting a root rot of soybeans incited by *Phytophthora cactorum* (Abstr.). Phytopathology 47:15–16.
Hickman, C. J. 1940. The red core root disease of the strawberry caused by *Phytophthora fragariae* n. sp. J. Pomol. Hortic. Sci. 18: 89–118.
Hickman, C. J. 1943. Shanking: A new disease of onion and shallot. Gard. Chron. 114:140.
Hickman, C. J. 1958. *Phytophthora*—Plant destroyer. Trans. Br. Mycol. Soc. 41:1–13.
Hickman, C. J. 1962. Physiologic races of *Phytophthora fragariae.* Ann. Appl. Biol. 50:95–103.
Hickman, C. J. 1970. Biology of *Phytophthora* zoospores. Phytopathology 60:1128–1135.
Hickman, C. J., and English, M. P. 1951a. Factors influencing the development of red core in strawberries. Trans. Br. Mycol. Soc. 34:223–236.
Hickman, C. J., and English, M. P. 1951b. The susceptibility of strawberry varieties to red core. Trans. Br. Mycol. Soc. 34:356–359.
Hickman, C. J., and Ho, H. H. 1966. Behaviour of zoospores in plant-pathogenic Phycomcetes. Annu. Rev. Phytopathol. 4:195–220.
Hicks, P. G. 1975. *Phytophthora palmivora* pod rot of cacao in Papua, New Guinea; investigations 1962–1971. Papua New Guinea Agric. J. 26:10–16.
Hide, G. A., and Cayley, G. R. 1991. Control of late blight (*Phytophthora infestans* [Mont.] de Bary) with fungicide applied by hydraulic and electrostatic sprayers. Potato Res. 34:183–186.
Hildebrand, A. A. 1959. A root and stalk rot of soybeans caused by *Phytophthora megasperma* Drechsler var. *sojae* var. nov. Can. J. Bot. 37:927–957.
Hildebrand, E. M. 1950. Techniques for the isolation of single micro-organisms. II. Bot. Rev. 16:181–207.
Hiley, W. E. 1919. The Fungal Diseases of the Common Larch. Clarendon Press. Oxford, England. 204 pp.
Hilty, J. W., and Schmitthenner, A. F. 1962. Pathogenic and cultural variability of single zoospore isolates of *Phytophthora megasperma* var. *sojae.* Phytopathology 52:859–862.
Himmelbaur, W. 1911. Zur Kenntnis der Phytophthoreen (Information on the phytophthoras) Jahrb. Hamburg Wiss. Anst. 28:39–61. (In German). (Cited in Tucker 1933).
Hinch, J. M., Wetherbee, R., Mallet, S. E., and Clarke, A. E. 1985. Response of *Zea mays* roots to infection with *Phytophthora cinnamomi.* 1. The epidermal layer. Protoplasma 126:178–188.
Hinch, J., and Weste, G. 1979. Behaviour of *Phytophthora cinnamomi* zoospores on roots of Australian forest species. Aust. J. Bot. 27:679–691.
Hine, R. B. 1961. Trunk canker of Macadamia in Hawaii caused by *Phytophthora cinnamomi* Rands. Plant Dis. Rep. 45:868.
Hine, R. B. 1962. Pathogenicity of *Phytophthora palmivora* in the Orchidaceae. Plant Dis. Rep. 46:643–645.
Hine, R. B., and Aragaki, M. 1962. A new disease of carnations in Hawaii caused by *Phytophthora parasitica.* Hawaii Farm Sci. 11:6.
Hine, R. B., and Aragaki, M. 1963. Influence of soil temperature on a crown rot disease of parsley caused by *Phytophthora parasitica.* Phytopathology 53:1113–1114.
Hine, R. B., Alaban, C., and Klemmer, H. 1964. Influence of soil temperatures on root and heart rot of pineapple caused by *Phytophthora cinnamomi* and *Phytophthora parasitica.* Phytopathology 54:1287–1289.
Hine, R. B., Aragaki, M., and Tokunaga, J. 1965. Enzymatic inactivation of the papaya blight fungus, *Phytophthora parasitica,* by papain and other proteolytic enzymes. Phytopathology 55:1223–1226.
Hine, R. B., Holtzmann, O. V., Raabe, R. D. 1965. Diseases of papaya in Hawaii. Hawaii Agric. Exp. Sta. Bull. 136:26.
Hine, R. B., Gray, F. A., Schonhorst, M. H., and Sanders, J. S. 1975. Resistance to Phytophthora root rot in selected lines of non-dormant alfalfa. Phytopathology 65:840–844.
Hirst, J. M. 1953. Changes in atmospheric spore content: Diurnal periodicity and the effects of weather. Trans. Br. Mycol. Soc. 36: 375–393.
Hirst, J. M., and Moore, W. C. 1957. *Phytophthora infestans* on *Petunia* and *Datura.* Plant Pathol. 6:76.
Hirst, J. M., and Stedman, O. J. 1960. The epidemiology of *Phytophthora infestans.* II. The source of inoculum. Ann. Appl. Biol. 48:489–517.
Hislop, E. C., and Park, P. O. 1962a. Studies on the chemical control of *Phytophthora palmivora* (Butl.) Butl. on *Theobroma cacao* L. in Nigeria. I. Laboratory bio-assay of fungicides on detached pods. Ann. Appl. Biol. 50:57–65.
Hislop, E. C., and Park, P. O. 1962b. Studies on the chemical control of *Phytophthora palmivora* (Butl.) Butl. on *Theobroma cacao* L. in Nigeria. II. Persistence of fungicides on pods. Ann. Appl. Biol. 50: 67–76.
Hislop, E. C., and Park, P. O. 1962c. Studies on the chemical control

of *Phytophthora palmivora* (Butl.) on *Theobroma cacao* L. in Nigeria. III. Field Trials. Ann. Appl. Biol. 50:77–88.

Ho, H. H. 1969a. Effects of root substances on the growth and sporulation of *Phytophthora megasperma* var. *sojae*. J. Elisha Mitchell Sci. Soc. 85:97–100.

Ho, H. H. 1969b. Notes on the behavior of *Phytophthora megasperma* var. *sojae* in soil. Mycologia 61:835–838.

Ho, H. H. 1978. Hyphal branching systems in *Phytophthora* and other Phycomycetes. Mycopathologia 64:83–86.

Ho, H. H. 1979a. The nature of amphigyny in *Phytophthora*. Mycologia 71:1057–1064.

Ho, H. H. 1979b. Electron microscope studies of oogonial wall and elemental composition of oogonia in *Phytophthora*. Mycopathologia 68:17–21.

Ho, H. H. 1981. Synoptic keys to the species of *Phytophthora*. Mycologia 73:705–714.

Ho, H. H. 1982. Affinity groups among plant pathogenic species of *Phytophthora* in culture. Mycopathologia 79:141–146.

Ho, H. H. 1983a. *Phytophthora porri* from stored carrots in Alberta. Mycologia 75:747–751.

Ho, H. H. 1983b. The nature of paragyny in *Phytophthora*. Mycopathologia 83:119–123.

Ho, H. H. 1986a. Notes on the heterothallic behavior of *Phytophthora megasperma* from alfalfa. Mycologia 78:306–309.

Ho, H. H. 1986b. *Phytophthora melonis* and *P. sinensis* synonymous with *P. drechsleri*. Mycologia. 78:907–912.

Ho, H. H. 1987. Effect of hymexazol on growth and reproduction of low temperature *Phytophthora* species. Mycopathologia 98:17–20.

Ho., H. H. 1990. Taiwan *Phytophthora*. Bot. Bull. Acad. Sin. 31:89–106.

Ho, H. H. 1992. Keys to the species of *Phytophthora* in Taiwan. Plant Pathol. Bull. (Taiwan) 1:104–109.

Ho, H. H., and Chang, H. S. 1992. A re-evaluation of *Phytophthora* species described by K. Sawada in Taiwan. Mycotaxon 43:297–316.

Ho, H. H., and Hickman, C. J. 1967a. Asexual reproduction and behavior of zoospores of *Phytophthora megasperma* var. *sojae*. Can. J. Bot. 45:1963–1981.

Ho, H. H., and Hickman, C. J. 1967b. Factors governing zoospore responses of *Phytophthora megasperma* var. *sojae* to plant roots. Can. J. Bot. 45:1983–1994.

Ho, H. H., and Jong, S. C. 1986. A comparison between *Phytophthora cryptogea* and *P. drechsleri*. Mycotaxon 27:289–319.

Ho, H. H., and Jong, S. C. 1987. Critical problems in the taxonomy of *Phytophthora* species in culture. Mycotaxon 29:207–232.

Ho, H. H., and Jong, S. C. 1988. *Phytophthora fragariae*. Mycotaxon 31:305–322.

Ho, H. H., and Jong, S. C. 1989a. *Phytophthora nicotianae* (*P. parasitica*). Mycotaxon 35:243–276.

Ho, H. H., and Jong, S. C. 1989b. *Phytophthora erythroseptica*. Mycotaxon 36:73–90.

Ho, H. H., and Jong, S. C. 1990. *Halophytophthora* gen. nov., a new member of the family Pythiaceae. Mycotaxon 36:377–382.

Ho, H. H., and Jong, S. C. 1991. Species concepts of *Phytophthora cryptogea* and *P. drechsleri*. Mycotaxon 40:35–39.

Ho, H. H., and Jong, S. C. 1993. *Phytophthora hibernalis* and *P. syringae*. Mycotaxon 47:439–460.

Ho, H. H., and Zentmyer, G. A. 1977a. Morphology of *Phytophthora cinnamomi*. Mycologia 69:701–713.

Ho, H. H., and Zentmyer, G. A. 1977b. Infection of avocado and other species of *Persea* by *Phytophthora cinnamomi*. Phytopathology 67:1085–1089.

Ho, H. H., Zachariah, K., and Hickman, C. J. 1968a. The ultrastructure of zoospores of *Phytophthora megasperma* var. *sojae*. Can. J. Bot. 46:37–41.

Ho, H. H., Hickman, C. J., and Teleford, R. W. 1968b. The morphology of zoospores of *Phytophthora megasperma* var. *sojae* and other Phycomycetes. Can. J. Bot. 46:88–89.

Ho, H. H., Zentmyer, G. A., and Erwin, D. C. 1977. Morphology of sex organs of *Phytophthora cambivora*. Mycologia 69:641–646.

Ho, H. H. Zhuang, W. Y., Liang, Z. R., and Yu, Y. N. 1983a. *Phytophthora cinnamomi* on black locust (*Robinia pseudoacacia*) in Jiangsu province of China. Mycologia 75:881–886.

Ho, H. H., Yu, Y. N, Zhuang, W. Y., and Liang, Z. R. 1983b. Mating types of heterothallic species of *Phytophthora* in China I. Acta Mycol. Sin. 2:187–191.

Ho, H. H., Liang, Z. R., Zhuang, W. Y., and Yu, Y. N. 1984a. *Phytophthora* spp. from rubber tree plantations in Yunnan province of China. Mycopathologia 86:121–124.

Ho, H. H., Lu, J. Y., and Gong, L. Y. 1984b. Mating types of heterothallic species of *Phytophthora* in China. II. Acta Mycol. Sin. 3: 29–32.

Ho, H. H., Lu, J. Y., and Gong, L. Y. 1984c. *Phytophthora drechsleri* causing blight of *Cumumis* species in China. Mycologia 76:115–121.

Ho, H. H., Lu, J. Y., and Gong, L. Y. 1984d. *Phytophthora* spp. associated with the decline of *Cedrus deodara* in Nanjing, China. Mycopathologia 85:59–64.

Ho, H. H., Yu, Y. N., Zhuang, W. Y., Liang, Z. R., Lu, J. Y., and Gong, L. Y. 1984e. *Phytophthora* spp. in China (Abstr.). Phytopathology 74:883.

Ho, H. H., Ann, F. J., and Chang, H. S. 1995. The Genus *Phytophthora* in Taiwan. Institute of Botany, Academia Sinica, Monograph Series 15. 86 pp.

Hoagland, D. R., and Arnon, D. I. 1950. The water culture method for growing plants without soil. Calif. Agric. Exp. Stn. Circ. No. 347. 32 pp.

Hobe, M. A. 1981. Pathogenic variability of *Phytopathora megasperma* f. sp. *glycinea* isolated from northwest Ohio soils. M.S. thesis, Ohio State Univ., Columbus. 32 pp.

Hodgson, W. A. 1962. Studies on the nature of partial resistance to *Phytophthora infestans*. Am. Potato J. 39:8–13.

Hodgson, W. A., and Grainger, P. N. 1964. Culture of *Phytophthora infestans* on artificial media prepared from rye seeds. Can. J. Plant Sci. 44:583.

Hoerner, G. R. 1949. Hop diseases in the United States. Brew. Dig. 24:45–51. (Rev. Appl. Mycol. 28:542–593).

Hohl, H. R. 1975. Levels of nutritional complexity in *Phytophthora:* Lipids, nitrogen sources and growth factors. Phytopathol. Z. 84: 18–33.

Hohl, H. R. 1983. Nutrition of *Phytophthora*. Pages 41–54 in: *Phytophthora:* Its Biology, Taxonomy, Ecology, and Pathology. D. C. Erwin, S. Bartnicki-Garcia, and P. H. Tsao, eds. American Phytopathological Society, St. Paul, Minn. 392 pp.

Hohl, H. R. 1988. Geographical distribution of the A^2 mating type of *Phytophthora infestans*. Phytophthora Newsl. 14:27–28.

Hohl, H. R. 1991a. Nutrition. Pages 53–58 in: *Phytophthora infestans* the Cause of Late Blight of Potato. Adv. Plant Pathol. Vol. 7. D. S. Ingram and P. H. Williams, eds. Academic Press, London. 265 pp.

Hohl, H. R. 1991b. Surface-related host-pathogen interactions in *Phytophthora*. Pages 70–89 in: *Phytophthora*. J. A. Lucas, R. C. Shattock, D. S. Shaw, and L. R. Cooke, eds. Cambridge Univ. Press, Cambridge, U.K. 447 pp.

Hohl, H. R., and Iselin, K. 1984. Strains of *Phytophthora infestans* from Switzerland with A^2 mating type behaviour. Trans. Br. Mycol. Soc. 83:529–530.

Hohl, H. R., and Iselin, K. 1987. Liquid nitrogen preservation of zoosporic fungi. Pages 143–145 in: Zoosporic Fungi in Teaching and Research. M. S. Fuller and A. Jaworski, eds. Southeastern Pub. Co., Athens, Ga. 302 pp.

Hohl, H. R., and Stössel, P. 1976. Host-parasite interfaces in a resistant and a susceptible cultivar of *Solanum tuberosum* inoculated with *Phytophthora infestans:* Tuber tissue. Can J. Bot. 54:900–912.

Hohl, H. R., and Suter, E. 1976. Host-parasite interfaces in a resistant and a susceptible cultivar of *Solanum tuberosum* inoculated with *Phytophthora infestans:* Leaf tissue. Can. J. Bot. 54:1956–1970.

Höhnk, W. 1936. On three pythiaceous Oomycetes. Beih. Bot. Centralbl. 55(Abt. A):89–99.

Hohrein, B. A., Bean, G. A., and Graham, J. H. 1983. Greenhouse technique to evaluate alfalfa resistance to *Phytophthora megasperma* f. sp. *medicaginis*. Plant Dis. 67:1332–1333.

Hoitink, H. A. J. 1980. Composted bark, a lightweight growth medium with fungicidal properties. Plant Dis. 64:142–147.

Hoitink, H. A. J., and Fahy, P. C. 1986. Basis for the control of soil-borne plant pathogens with composts. Annu. Rev. Phytopathol. 24: 93–114.

Hoitink, H. A. J., and Powell, C. C. 1989. Control of *Phytophthora* diseases of woody ornamentals. Buckeye Nurseryman (June). Pages 3–7.

Hoitink, H. A. J., and Powell, C. C., Jr. 1990. Fighting *Phytophthora:* A guide to combating Phytophthora root rot and dieback of ericaceous crops. Am. Nurserymen 171:69–73.

Hoitink, H. A. J., and Schmitthenner, A. F. 1969. Rhododendron wilt caused by *Phytophthora citricola.* Phytopathology 59:708–709.

Hoitink, H. A. J., and Schmitthenner, A. F. 1970. *Phytophthora* species involved in rhododendron wilt and sources of resistance (Abstr.). Phytopathology 60:1296.

Hoitink, H. A. J., and Schmitthenner, A. F. 1974a. Relative prevalence and virulence of *Phytophthora* species involved in rhododendron root rot. Phytopathology 64:1371–1374.

Hoitink, H. A. J., and Schmitthenner, A. F. 1974b. Resistance of *Rhododendron* species and hybrids to Phytophthora root rot. Plant Dis. Rep. 58:650–653.

Hoitink, H. A. J., Daft, G., and Gerlach, W. W. P. 1975a. Phytophthora shoot blight and stem dieback of azalea and *Pieris* and its control. Plant Dis. Rep. 59:235–237.

Hoitink, H. A. J., Schmitthenner, A. F., and Herr, L. J. 1975b. Composted bark for control of root rot of ornamentals. Ohio Rep. 60:25–26.

Hoitink, H. A. J., Herr, L. J., and Schmitthenner, A. F. 1976. Survival of some plant pathogens during composting of hardwood tree bark. Phytopathology 66:1369–1372.

Hoitink, H. A. J., VanDoren, D. M., Jr., Schmitthenner, A. F. 1977. Suppression of *Phytophthora cinnamomi* in a composted hardwood bark potting medium. Phytopathology 67:561–565.

Hoitink, H. A. J., Watson, M. E., and Faber, W. R. 1986. Effect of nitrogen concentration in juvenile foliage of rhododendron on *Phytophthora* dieback severity. Plant Dis. 70:292–294.

Holdaway, B. F., and Tsao, P. H. 1971. Survival of *Phytophthora parasitica* in soils (Abstr.). Phytopathology 61:1321.

Holdaway, B. F., and Tsao, P. H. 1972. Longevity of chlamydospores of *Phytophthora parasitica* in water. Plant Dis. Rep. 56:971–973.

Holderness, M. 1992. Comparison of metalaxyl, cuprous oxide sprays, and potassium phosphonate as sprays and trunk injections for control of *Phytophthora palmivora* pod rot and canker of cocoa. Crop Prot. 11:141–147.

Hölker, U., Érsek, T., and Hofer, M. 1993. Changes in ion fluxes and the energy demand during zoospore development in *Phytophthora infestans* zoospores. Folia Microbiol. 38:193–200.

Holliday, P. 1965. A wilt of *Piper nigrum* in Brazil. Commonw. Phytopathol. News 5:4.

Holliday, P. 1980. Fungus Diseases of Tropical Crops. Cambridge Univ. Press, Cambridge, U.K. 607 pp.

Holliday, P. 1989. A Dictionary of Plant Pathology. Cambridge Univ. Press, Cambridge, U.K. 369 pp.

Holliday, P., and Mowat, W. P. 1963. Foot rot of *Piper nigrum* L. (*Phytophthora palmivora*). Phytopathol. Pap. 5. Commonw. Mycol. Inst. Kew, Surrey, England. 62 pp.

Hollomon, D. W. 1965. A medium for the direct isolation of *Phytophthora infestans.* Plant Pathol. 14:34–35.

Holmes, S. J. I., and Channon, A. G. 1984. Studies on metalaxyl-resistant *Phytophthora infestans* in potato crops in southwest Scotland. Plant Pathol. 33:347–54.

Holub, E. B., and Grau, C. R. 1990a. Ability of *Aphanomyces euteiches* to cause disease of seedling alfalfa compared with *Phytophthora megasperma* f. sp. *medicaginis.* Phytopathology 80:331–335.

Holub, E. B., and Grau, C. R. 1990b. Productivity and survival of alfalfa ramets in soil naturally infested with *Aphanomyces euteiches* and *Phytophthora megasperma.* Can. J. Plant Pathol. 12:83–91.

Honour, R. C., and Tsao, P. H. 1974. Production of oospores by *Phytophthora parasitica* in liquid medium. Mycologia 66:1030–1038.

Hooker, A. L. 1967. The genetics and expression of resistance in plants to rusts of the genus *Puccina.* Annu. Rev. Phytopathol. 5: 163–182.

Hooker, W. J., ed. 1981. Compendium of Potato Diseases. American Phytopathological Society, St. Paul, Minn. 141 pp.

Hooley, P., and Shaw, D. S. 1985. Inheritance of sensitivity to heavy metals in *Phytophthora drechsleri.* Trans. Br. Mycol. Soc. 85:677–681.

Hooley, P., Fyfe, A. M., Eviola-Maltese, C., Shaw, D. S. 1982. Duplication cycle in nuclei of germinating zoospores of *Phytophthora drechsleri* as revealed by DAPI staining. Trans. Br. Mycol. Soc. 79:563–566.

Hooley, P., Shaw, D. S., and Fyfe, A. M. 1985. A comparison of the DNA contents of nuclei of *Phytophthora drechsleri* and *Aspergillus nidulans* using the fluorochrome DAPI. Phytophthora Newsl. 13:15–16.

Hopkins, J. C. 1925. Notes on the soft rot of cotton bolls in the West Indies caused by *Phytophthora.* Ann. Bot. 39:267–280.

Hopkins, J. C. F. 1931a. Plant pathology in Southern Rhodesia during the year 1930. Rhod. Agric. J. 28:384–389.

Hopkins, J. C. F. 1931b. A list of plant diseases occurring in Southern Rhodesia. Rhod. Agric. J. 28:663–669.

Hopkins, J. C. F. 1932. A list of plant diseases occurring in Southern Rhodesia. Suppl. 2. Minist. Agric. Lands. South. Rhodesia Bull. 856. 8 pp. (Cited in Tucker 1933).

Hopkins, J. C. F. 1935. Annual report of the branch of Plant Pathology for the year ending 31st December, 1934. Rhod. Agric. J. 32:397–405.

Hopkins, J. C. F. 1940. Annual report of the plant pathologist for the year ending 31st December, 1939. Rhod. Agric. J. 37:411–423.

Hopkins, J. C. F. 1945. Diseases of fruits, flowers, and vegetables in Southern Rhodesia. 9. Diseases of tomatoes. Rhod. Agric. J. 42: 81–87.

Hopkins, J. C. F. 1951. Summary of annual report for the branch of botany and plant pathology for the year ended 30th September, 1950. Rhod. Agric. J. 48:454–459.

Hord, M. J., and Ristaino, J. B. 1991. Effects of physical and chemical factors on the germination of oospores of *Phytophthora capsici* in vitro. Phytopathology 81:1541–1546.

Hord, M. J., and Ristaino, J. B. 1992. Effect of the matric component of soil water potential on infection of pepper seedlings in soil infested with oospores of *Phytophthora capsici.* Phytopathology 82:792–798.

Hori, M. A. 1964. Studies on the disease cycle of potato late blight caused by *Phytophthora infestans* (Mont.) de Bary. Bull. Agric. Chem. Insp. Stn. (Tokyo). 69 pp. (In Japanese).

Hori, S. 1907. A disease of the Japanese ginseng caused by *Phytophthora cactorum.* (Con. et Leb.) Schrot. Bull. Imp. Cent. Agric. Exp. Stn. Jpn. 1:152–162. (In Japanese). (Cited in Wilson 1914 and Tucker 1933).

Hori, S. 1915. *Phytophthora fici.* J. Plant Prot. (Tokyo) 2(11):930–932. (In Japanese). (Cited in Tanaka 1920).

Horne, W. T. 1909. Informe del departamento de patologia vegetal (Report of the Plant Pathology Department). Estac. Central Agron. (Cuba) Inf. 2:77–98. (In Spanish).

Horst, R. K. 1979. Westcott's Plant Disease Handbook. 4th ed. Van Nostrand Reinhold, New York. 803 pp.

Hotson, J. W., and Hartge, L. 1923. A disease of tomato caused by *Phytophthora mexicana,* sp. nov. Phytopathology 13:520–531. (Cited in Tucker 1933).

Houck, L. G. 1962. Factors influencing development and control of *P. fragariae* Hick., the cause of red stele disease of strawberries. Ph.D. thesis, Oregon State Univ., Corvallis. 162 pp.

Hough, A., Mulder, N. J., and La Grange, J. M. 1979. Heat treatment for the control of *Phytophthora gummosis* in citrus. Plant Dis. Rep. 63:40–43.

Howard, A. 1901. Fungoid diseases of cacao in the West Indies. West Indian Bull. 2:190–211. (Cited in Tucker 1933).

Howard, F. L. 1941. The bleeding canker disease of hardwoods and possibilities of control. Proc. 8th W. Shade Tree Conf. 8. 10 pp. (Rev. Appl. Mycol. 1942, 21:394).

Hsaio, T. C. 1973. Plant responses to water stress. Annu. Rev. Plant Physiol. 24:519–570.

Hsieh, H. J. 1979. New diseases caused by *Phytophthora* species. Phytophthora Newsl. 7:39–40.

Hsu, S. C., and Lockwood, J. L. 1984. Biological control of Phytophthora root rot of soybean by *Hypochytrium catenoides* in greenhouse tests. Phytopathol. Z. 109:139–146.

Huang, T. H., and Chen, D. W. 1974. Survival and behavior of *Phytophthora* species Plant Prot. Bull. (Taiwan) 16:171. (Abstract in Chinese). (Cited in Ho 1992).

Huang, T. H., Chen, D. W., and Leu, L. S. 1976. Phytophthora fruit and root rot of papaya in Taiwan. Plant Prot. Bull. (Taiwan) 18: 293–308.

Huber, D. M., Fortes, J. F., Pecknold, P. C., and Harmon, M. C. 1982. Aminopeptidase profiles of four species of *Phytophthora*. Appl. Environ. Microbiol. 43:597–602.

Hubert, F. P. 1957. Diseases of some export crops in Indonesia. Plant Dis. Rep. 41:55–64.

Hudspeth, M. E. S., Shumard, D. S., Bradford, C. J. R., and Grossman, L. J. 1983. Organization of *Achyla* mtDNA: A population with two orientations and a large inverted repeat containing both rRNA genes. Proc. Natl. Acad. Sci. USA 80:142–146.

Hughes, I. K. 1963. Troublesome tobacco diseases on southern border. Queensl. Agric. J. 89:326–330.

Huguenin, B. 1974. Influence des conditions de culture sur la formation et la germination in vitro des chlamydospores du *Phytophthora palmivora* (Influence of cultural conditions on the formation and germination in vitro of chlamydospores of *Phytophthora palmivora*). Ann. Phytopathol. 6:425–440. (In French).

Huguenin, B., and Boccas, B. 1971. Rôle de quelques facteurs dans la formation et la germination des oospores chez le *Phytophthora palmivora* Butl. (Influence of some factors on the formation and germination of oospores in strains of *Phytophthora palmivora* Butl.) Ann. Phytopathol. 3:353–371. (In French).

Huguenin, B., and Jaques, R. 1973. Etablissement du spectre d'action de la lumière sur la sporogenèse et l'oogonèse de souches du *Phytophthora palmivora* Butl. (Establishment of the action spectra of light on sporangia and oogonia in cultures of *Phytophthora palmivora* Butl.). C. R. Hebd. Seances Acad. Sci. Ser. D 276:725–728. (In French).

Humphrey, H. B., and Wood, J. I. 1935. Diseases of plants in the United States in 1933. Plant Dis. Rep. Suppl. 86. 107 pp.

Humphrey, N. 1951. Potato Blight. Leafl. Agric. For. Dep. St. Helena 6. 4 pp. (Rev. Appl. Mycol. 1958, 37:303).

Hunger, R. M., Hamm, P. B., Horner, C. E., and Hansen, E. M. 1982. Tolerance of *Phytophthora megasperma* isolates to metalaxyl. Plant Dis. 66:645–649.

Hunt, R. S., and O'Reilly, H. J. 1984. Evaluation of control of Lawson cypress root rot with resistant root stocks. Can. J. Plant Pathol. 6:172–174.

Hunter, J. E., and Kunimoto, R. K. 1972. Development of a field bioassay and evaluation of fungicides for control of Phytophthora blight of macadamia racemes. Plant Dis. Rep. 56:689–691.

Hunter, J. E., and Kunimoto, R. K. 1974. Dispersal of *Phytophthora palmivora* sporangia by wind-blown rain. Phytopathology 64:202–206.

Hunter, J. E., Kunimoto, R. K., and Rohrbach, K. G. 1971. Phytophthora blight, a new disease of macadamia. Phytopathology 61: 1130–1133.

Hwang, B. K., and Sung, N. K. 1989. Effect of metalaxyl on capsidiol production in stems of pepper plants infected with *Phytophthora capsici*. Plant Dis. 73:748–751.

Hwang, B. K., Ebrahim-Nesbat, F., Wolf-Dieter, I., and Heitefuss, R. 1990. An ultrastructural study of the effect of metalaxyl on *Phytophthora capsici* infected stems of *Capsicum annuum*. Pestic. Sci. 29:151–162.

Hwang, B. K., de Cock, A. W. A. M., Bahnweg, G., Prell, H. H., and Heitefuss, R. 1991. Restriction fragment length polymorphisms of mitochondrial DNA among *Phytophthora capsici* isolates from pepper (*Capsicum annuum*). Syst. Appl. Microbiol. 14:111–116.

Hwang, H., Hsu, T. Y., and Lin, C. K. 1963. A favourable synthetic liquid medium for *Phytophthora infestans*. Acta Phytopathol. Sin. 6:225–228.

Hwang, H., Hsu, T. Y., and Lin, C. K. 1965. Further studies on synthetic media for *Phytophthora infestans*. Acta Microbiol. Sin. 11: 463–469. (In Chinese, English summary).

Hwang, J. K., and Qi, P. K. 1984. A study on *Phytophthora* of yardlong bean. Acta Phytopathol. Sin. 14:193–199. (In Chinese).

Hwang, S. C., and Ko, W. H. 1978. Biology of chlamydospores, sporangia, and zoospores of *Phytophthora cinnamomi* in soil. Phytopathology 68:726–731.

Hwang, S. C., Ko, W. H., and Aragaki, M. 1975. A simplified method for sporangial production by *Phytophthora cinnamomi*. Mycologia 67:1233–1234.

Hylin, J. W., Hylin, V., and Aragaki, M. 1970. A chemically defined medium for the study of growth, sporulation and presumptive sporogenic substances in *Alternaria tomato*. Mycologia 62:200–202.

Hyre, R. A. 1954. Progress in forecasting late blight of potato and tomato. Plant Dis. Rep. 38:245–253.

Hyre, R. A. 1957. Forecasting downy mildew of lima bean. Plant Dis. Rep. 41:7–9.

Hyre, R. A., and Cox, R. S. 1953. Factors affecting viability and growth of *Phytophthora phaseoli*. Phytopathology 43:419–425.

Ideta, I. 1903. Sinnoho New Agric. News 57:31–32. (Cited in G. Waterhouse 1956).

Idosu, G. O., and Zentmyer, G. A. 1978. *Phytophthora palmivora:* A comparative study of "typical" and "atypical" isolates from cacao (*Theobroma cacao*). Mycologia 70:1101–1112.

Ilieva, E. 1979. *Phytophthora crytogea* Pethyb. et Laff. nov. patogen po domatite, pipera i Krastavitsite v B'lgoniya (*Phytophthora cryptogea* Pethyb. et Laff., a new pathogen of tomatoes, pepper, and cucumber in Bulgaria). Gradinar. Lozar. Nauka 16:58–64. (In Bulgarian). (Rev. Plant Pathol. 1981, 60:4327).

Ilieva, E., and Koleva, N. 1984. *Phytophthora citricola* Sawada—A soil pathogen of strawberry. Gradinar. Lozar. Nauka 21(7):42–48. Plant Prot. Inst., Kostinbrod, Bulgaria. (In Bulgarian). (Rev. Plant Pathol. 64:1661).

Ingham, J. L., Keen, N. T., and Hymowitz, T. 1977. A new isoflavone from fungus inoculated stems of *Glycine wightii*. Phytochemistry 16:1943–1946.

Ingram, D. S, and Williams, P. H., eds. 1991. *Phytophthora infestans* the Cause of Late Blight of Potato. Adv. Plant Pathol. Vol. 7. Academic Press, New York. 273 pp.

International Mycological Institute. Distribution Maps of Plant Diseases (see Commonwealth Mycological Institute).

Ioannou, N., and Grogan, R. G. 1977. The influence of soil matric potential on the production of sporangia by *Phytophthora parasitica* in relation to its isolation from soil by baiting techniques (Abstr.). Proc. Am. Phytopathol. Soc. 4:173.

Ioannou, N., and Grogan, R. G. 1984a. Control of Phytophthora root rot of processing tomato with ethazol and metalaxyl. Plant Dis. 68:429–435.

Ioannou, N., and Grogan, R. G. 1984b. Water requirements for sporangium formation by *Phytophthora parasitica* in relation to bioassay in soil. Plant Dis. 68:1043–1048.

Ioannou, N., and Grogan, R. G. 1985. Formation and germination of chlamydospores of *Phytophthora parasitica* under various oxygen and carbon dioxide tensions. Plant Dis. 69:400–403.

Ippolito, A., DeCicco, V., Vanadia, S., Maurantonio, V., and Salerno, M. 1990. Inoculum density and citrus root infection of *Phytophthora* spp. as related to some chemical and physical soil factors. Pages 99–101 in: 8th Congr. Mediterr. Phytopathol. Union.

Irving, H. R., and Grant, B. R. 1984a. The effects of pectin and plant root surface carbohydrates on encystment and development of *Phytophthora cinnamomi* zoospores. J. Gen. Microbiol. 130:1015–1018.

Irving, H. R., and Grant, B. R. 1984b. The effect of calcium on zoospore differentiation in *Phytophthora cinnamomi*. J. Gen. Microbiol. 130:1569–1576.

Irwin, J. A. G. 1974. Reaction of lucerne cultivars to *Phytophthora megasperma,* the cause of a root rot in Queensland. Aust. J. Exp. Agric. Anim. Husb. 14:561–565.

Irwin, J. A. G. 1976. Observations on the mode of infection of lucerne roots by *Phytophthora megasperma*. Aust. J. Bot. 24:447–451.

Irwin, J. A. G. 1977. Factors contributing to poor lucerne persistence in southern Queensland. Aust. J. Exp. Agric. Anim. Husb. 17:998–1002.

Irwin, J. A. G. 1991. *Phytophthora macrochlamydospora* (Pythiaceae), a new species from Australia. Mycologia 83:517–519.

Irwin, J. A. G., and Dale, J. L. 1982. Relationships between *Phytophthora megasperma* isolates from chick pea, lucerne and soybean. Aust. J. Bot. 30:199–210.

Irwin, J. A. G., and Langdon, P. W. 1982. A laboratory procedure for determining relative levels of field resistance in soybeans to *Phytophthora megasperma* f. sp. *glycinea.* Aust. J. Agric. Res. 33:33–39.

Irwin, J. A. G., and Maxwell, D. P. 1980. Resistance of *Medicago* species accessions to *Phytophthora megasperma* f. sp. *medicaginis.* Plant Dis. 64:396–397.

Irwin, J. A. G., Miller, S. A., and Maxwell, D. P. 1979. Alfalfa seedling resistance to *Phytophthora megasperma.* Phytopathology 69:1051–1055.

Irwin, J. A. G., Maxwell, D. P., and Bingham, E. T. 1981a. Inheritance of resistance to *Phytophthora megasperma* in diploid alfalfa. Crop Sci. 21:271–276.

Irwin, J. A. G., Maxwell, D. P., and Bingham, E. T. 1981b. Inheritance of resistnce to *Phytophthora megasperma* in tetraploid alfalfa. Crop Sci. 21:277–283.

Irwin, J. A. G., Cahill, D. M., and Drenth, A. 1995. *Phytophthora* in Australia. Aust. J. Agric. Res. 46:1311–1337.

Ishiguro, K., and Ui, T. 1981a. Factors influencing the germination of oospore of *Phytophthora vignae* Purss. Ann. Phytopathol. Soc. Jpn. 47:213–217.

Ishiguro, K., and Ui, T. 1981b. The effect of osmotic potential on the germination of oospore of *Phytophthora vignae* Purss. Ann. Phytopathol. Soc. Jpn. 47:218–221.

Ito, S., and Nagai, M. 1931. On the root disease of the seeds and seedlings of rice plant caused by some aquatic fungi. J. Fac. Agric. Hokkaido Univ. (Sapporo) 32:45–69. (In Japanese).

Ito, S., and Tanaka, I. 1940. *Phytophthora macrospora* (Sacc.). Ann. Phytopathol. Soc. Jpn. 10:127–128. (In Japanese).

Ito, S., and Tokunaga, Y. 1935. Notae mycologicae Asiae Orientalis I. (Mycological notes from Eastern Asia). Trans. Sapporo Nat. Hist. Soc. 14:11–33. (Rev. Appl. Mycol. 1936, 15:57–58).

Ivanova, D. G., Gughova, N. V., Merzlyak, M. N., and Rassadina, G. V. 1991a. Effect of *Phytophthora infestans* infection on superoxide dismutase dependent cytochrome c–reducing activities of leaves as related to resistance of potato plants to late blight. Plant Sci. 78: 151–156.

Ivanova, D. G., Gughova, N. V., Merzlyak, M. N., Rassadina, G. V., and Gusev, M. V.1991b. Oxygen radical generation in different types of potato–*Phytophthora infestans* interactions. Sov. Plant Physiol. 38:281–289.

Iwan, J., Hertel, B., and Radzuweit, J. 1977. Transformations of prothiocarb in plant and soil. Neth. J. Plant Pathol. 83(Suppl. 1):277–284.

Jacks, H. 1951. Soil disinfection. XI. Control of foot rot (*Phytophthora cryptogea* Pethyb. & Laff.) of tomatoes. N.Z. J. Sci. Technol. Sec. A 33:71–75.

Jackson, G. V. H. 1977. Taro leaf blight. Advisory Leafl. No. 3. South Pacific Commission (New Caledonia). 4 pp.

Jackson, G. V. H., and Newhook, F. J. 1978. Sources of *Phytophthora palmivora* inoculum in Solomon Island cocoa plantations. Trans. Br. Mycol. Soc. 71:239–249.

Jacobson, L. 1951. Maintenance of iron supply in nutrient solutions by a single addition of ferric potassium ethylenediamine tetraacetate. Plant Physiol. 26:411–413.

Jahnke, K. D., Leipoldt, G., and Prell, H. H. 1987. Studies on the preparation and viability of *Phytophthora parasitica* spheroplasts. Trans. Br. Mycol. Soc. 89:213–220.

Jain, M. P., Bhatnagar, M. K., and Jain, K. L. 1980. Nutritional and pathological studies on fruit rot of brinjal caused by *Phytophthora nicotianae* B. de Haan var *nicotianae.* Proc. Workshop on *Phytophthora* Diseases of Tropical Cultivated Plants. K. K. N. Nambiar, ed. Pages 109–117 in: Cent. Plant. Crops Res. Inst., Kasaragod, Kerala, India. 284 pp.

Jain, M. P., Bhatnagar, M. K., and Jain, K. L. 1982. Fruit rot of brinjal in India caused by *Phytophthora nicotianae* var. *nicotianae.* Indian J. Mycol. Plant Pathol. 12:253–254.

Jamart, G., Kamoen, O., and van Voerenbergh, J. 1985. The control of red core caused by *Phytophthora fragariae* on strawberries in N.F.T. Meded. Fac. Landbouwwet. Rijksuniv. Gent 50:1087–1096.

James, M. O., and Plakidas, A. G. 1952. Two new *Phytophthora* diseases of ornamental hibiscus. Phytopathology 42:144–146.

James, R. V., and Fry, W. E. 1982. Potato late blight (Film). Dept. of Plant Pathology, Cornell Univ., Ithaca, New York.

Jancarik, V. 1961. Damping-off of seedlings in forest nurseries and its control in Czechoslovakia. Za. Sots. Sel.-Khoz. Nauk. 10:87–112. (Rev. Appl. Mycol. 1962, 41:413).

Jang, J. C., and Tainter, F. H. 1990a. Cellular responses of pine callus to infection by *Phytophthora cinnamomi.* Phytopathology 80:1347–1352.

Jang, J. C., and Tainter, F. H. 1990b. Hyphal growth of *Phytophthora cinnamomi* on pine callus tissue. Plant Cell Rep. 8:714–744.

Jarvis, W. R. 1992. Managing Diseases in Greenhouse Crops. American Phytopathological Society, St. Paul, Minn. 325 pp.

Jeffers, S. N. 1992. Preplant root treatments to reduce the incidence of *Phytophthora* species on dormant apple rootstocks. Plant Dis. 76:12–19.

Jeffers, S. N., and Aldwinckle, H. S. 1984. *Phytophthora* species associated with nursery-grown apple root stocks and trees (Abstr.). Phytopathology 74:845–846.

Jeffers, S. N., and Aldwinckle, H. S. 1987. Enhancing detection of *Phytophthora cactorum* in naturally infested soil. Phytopathology 77:1475–1482.

Jeffers, S. N., and Aldwinckle, H. S. 1988. Phytophthora crown rot of apple trees: Sources of *Phytophthora cactorum* and *P. cambivora* as primary inoculum. Phytopathology 78:328–335.

Jeffers, S. N., Aldwinckle, H. S., Burr, T. J., and Arneson, P. A. 1981. Excised twig assay for the study of apple tree crown rot pathogens in vitro. Plant Dis. 65:823–825.

Jeffers, S. N., Aldwinckle, H. S., Burr, T. J., and Arneson, P. A. 1982. *Phytophthora* and *Pythium* species associated with crown rot in New York apple orchards. Phytopathology 72:533–538.

Jeffers, S. N., and Martin, S. B. 1986. Comparison of two media selective for *Phytophthora* and *Pythium* species. Plant Dis. 70:1038–1043.

Jeffers, S. N., and Wilcox, W. F. 1990. Phytophthora crown, collar, and root rots. Pages 43–45 in: Compendium of Apple and Pear Diseases. A. L. Jones and H. S. Aldwinckle, eds. American Phytopathological Society, St. Paul, Minn. 100 pp.

Jeffrey, S. I. B., Jinks, J. L, and Grindle, M. 1962. Intraracial variation in *Phytophthora infestans* and field resistance to potato blight. Genetica 32:323–338.

Jenkins, S. F., Jr. 1962. Preliminary studies estimating the disease potential of *Phytophthora parasitica* var. *nicotianae* in infested tobacco soils. Plant Dis. Rep. 46:825–826.

Jermalavičiūtė, S. 1958. Fungi of the order Peronosporales of the Lithuanian S.S.R. Botanical studies II. Papers on the myco- and lichenflora of the Baltic. Pages 101–107. Inst. Zool. Bot. Akad. Nauk, Estonian S.S.R. (Rev. Appl. Mycol. 1963, 42:590).

Jeun, Y. C., and Hwang, B. K. 1991. Carbohydrate, amino acid, phenolic and mineral contents of pepper plants in relation to age-related resistance to *Phytophthora capsici.* J. Phytopathol. 131:40–52.

Jiang, J. 1991. *Phytophthora* oospore germination *in vitro* and *in situ* and beta-1,3 glucanase activity in oospores and mycelium of *Phytophthora cactorum.* Ph.D. dissertation, Univ. Calif., Riverside. 189 pp.

Jiang, J., and Erwin, D. C. 1990. Morphology, plasmolysis, and tetrazolium bromide stain as criteria for determining viability of *Phytophthora* oospores. Mycologia 82:107–113.

Jiang, J., and Erwin, D. C. 1993. The effect of nutrients on germination of cold-treated oospores of *Phytophthora cactorum in vitro.* Mycol. Res. 97:293–298.

Jiang, J., Stephenson, L. W., Erwin, D. C., and Leary, J. V. 1989. Nuclear changes in *Phytophthora* during oospore maturation and germination. Mycol. Res. 92:463–469.

Jiang, J., Keen, N. T., and Erwin, D. C. 1991. β-1,3-Glucanase activity in oospores and mycelium of *Phytophthora cactorum* (Abstr.). Phytopathology 81:1225.

Jiang, Y. 1990. Alterations in the host-parasite interface between *Phytophthora infestans* and tomato leaf cells after treatment with fosetyl-Al. Plant Sci. 71:267–276.

Jiang, Y., and Grossmann, F. 1991. Cellular damage to *Phytophthora infestans* in tomato leaves treated with oxadixyl: An ultrastructural investigation. J. Pytopathol. 132:116–128.

Jimenez, B., and Lockwood, J. L. 1980. Laboratory method for assessing field tolerance of soybean seedlings to *Phytophthora megasperma* var. *sojae*. Plant Dis. 64:775–778.

Jimenez, B., and Lockwood, J. L. 1982. Germination of oospores of *Phytophthora megasperma* f. sp. *glycinea* in the presence of soil. Phytopathology 72:662–666.

Johnson E. M., and Valleau, W. D. 1954. Disappearance of black shank from Kentucky soils. Ky. Agric. Exp. Stn. Bull. 615. 11 pp.

Johnson, H. W., and Keeling, B. L. 1969. Pathogenicity of *Phytophthora megasperma* isolated from subterranean clover roots. Phytopathology 59:1279–1283.

Johnson, H. W., and Morgan, F. L. 1965. Phytophthora root and crown rot of alfalfa in the Yazoo-Mississippi delta. Plant Dis. Rep. 49:753–755.

Johnson, L. B., and Klisiewicz, J. M. 1969. Environmental effects on safflower reaction to *Phytophthora drechsleri*. Phytopathology 59: 469–472.

Johnson, L. F., and Curl, E. A. 1972. Methods for Research on the Ecology of Soil-Borne Plant Pathogens. Burgess Publ., Minneapolis, Minn. 247 pp.

Johnson, L. F., Qian, P., and Ferriss, R. S. 1990. Soil matric potential effects on changes in wall morphology, germination and lysis of oospores of *Pythium ultimum*. Phytopathology 80:1357–1361.

Johnston, A. 1959. Plant disease investigation in the Federation of Malaya. F.A.O. Plant Prot. Bull. 8:1–4.

Johnston, A. 1960. A preliminary plant disease survey in the British Solomon Islands Protectorate. F.A.O. Rome, Italy. (Mimeographed). 36 pp. (Rev. Appl. Mycol. 1960, 39:536.)

Johnston, A. 1964. Quarterly report for October–December 1963 of the Plant Protection Committee for the South East Asia and Pacific region. F.A.O. Publ. Bangkok, Thailand. 20 pp.

Jones, A. L., and Aldwinckle, H. S., eds. 1990. Compendium of Apple and Pear Diseases. American Phytopathological Society, St. Paul, Minn. 100 pp.

Jones, D. R., Graham, W. G., and Ward, E. W. B. 1975a. Ultrastructural changes in pepper cells in an incompatible interaction with *Phytophthora infestans*. Phytopathology 65:1274–1285.

Jones, D. R., Unwin, C. H., and Ward, E. W. B. 1975b. The significance of capsidiol induction in pepper fruit during an incompatible interaction with *Phytophthora infestans*. Phytopathology 65:1286–1288.

Jones, F. R. 1939. Evidence of resistance in sweetclover to a Phytophthora root rot. Phytopathology 29:909–911.

Jones, J. B., Jones, J. P., Stall, R. E., and Zitter, T. A., eds. 1991. Compendium of Tomato Diseases. American Phytopathological Society, St. Paul, Minn. 100 pp.

Jones, J. P. 1969. Reaction of *Lupinus* species to *Phytophthora megasperma* var. *sojae*. Plant Dis. Rep. 53:907–909.

Jones, J. P., and Johnson, H. W. 1969. Lupine, a new host for *Phytophthora megasperma* var. *sojae*. Phytopathology 59:504–507.

Jones, L. R. 1905. Disease resistance of potatoes. U.S. Dep. Agric. Bur. Plant Ind. Bull. 87. 39 pp. (Cited in Tucker 1933).

Jones, L. R., Giddings, N. J., and Lutman, B. F. 1912. Investigations of the potato fungus, *Phytophthora infestans*. Vt. Agric. Exp. Stn. Bull. 168. 100 pp.

Jones, O. W., and Griffin, M. J. 1974. Root and stem rot (*Phytophthora* spp.) of *Hebe* 'Pagei.' Plant Pathol. 23:47.

Jones, W. 1945. Pink rot disease of potatoes in British Columbia. Sci. Agric. 25:596–600.

Jones, W. 1952. An unreported Phytophthora rot of Swede turnips (Abstr.). Phytopathology 42:515.

Jones, W. 1954. Pink rot of potato tubers on Vancouver Island. Can. J. Agric. Sci. 34:504–506.

Jones, W. 1955. Foot rot of wallflower caused by *Phytophthora megasperma*. Can. J. Bot. 33:531–533.

Jong, S. C., and Atkins, W. B. 1985. Conservation, collection, and distribution of cultures. Pages 153–194 in: Fungi Pathogenic for Humans and Animals. Vol. B, Part 2. D. H. Howard, ed. Marcell Dekker, New York.

Jorgensen, C. A. 1934. Bogens Kimbladskinnel og dens Bekaempelse. Dan. Skovforen (The cotyledonary leaf fungus of the beech and its control). Tidsskr. 4:123–127. (In Danish). (Cited in Nienhaus 1960). (Rev. Appl. Mycol. 1934, 13:605.)

Jorgensen, H. A., Kristensen, H. R., Begtrup, J., and Dahl, M. H. 1964. Plantesydomme i Danmark 1963 (Plant diseases in Denmark in 1963). Tidsskr. Planteavl. 68:369–429. (In Danish). (Rev. Appl. Mycol. 1964, 44:596a).

Joseph, M. C., and Coffey, M. D. 1984. Development of laboratory resistance to metalaxyl in *Phytophthora citricola*. Phytopathology 74:1411–1414.

Joseph, T., and Radha, K. 1975. Role of *Phytophthora palmivora* in bud rot of coconut. Plant Dis. Rep. 59:1014–1017.

Juárez, P. C., Teliz, O. D., and Garcia, E. R. 1989. Antagonismo microbiano sobre la dinamica poblacional y supervivencia de *Phytophthora cinnamomi* Rands, e influencia en la productividad de arboles de aguacate (*Persea americana* Mill) (Influence of microbial anatagonism on the population dynamics and survival of *Phytophthora cinnamomi* Rands, and on the productivity of avocado trees [*Persea americana* Mill.]). Agrociencia 77:43–55. (In Spanish).

Juárez-Palacios, C., Felix-Gastelum, R., Wakeman, R. J., Paplomatas, E. J., and De Vay, J. E. 1991. Thermal sensitivity of three species of *Phytophthora* and the effect of soil solarization on their survival. Plant Dis. 75:1160–1164.

Judelson, H. S. 1993. Intermolecular ligation mediates efficient cotransformation in *Phytophthora infestans*. Mol. Gen. Genet. 239: 241–250.

Judelson, H. S., Coffey, M. D., Arredondo, F. R., and Tyler, B. M. 1993a. Transformation of the oomycete pathogen, *Phytophthora megasperma* f. sp. *glycinea* occurs by DNA integration into single or multiple chromosomes. Curr. Genet. 23:211–218.

Judelson, H. S., Dudler, R., Pieterse, C. M. J., Unkles, S. E., and Michelmore, R. W. 1993b. Expression and antisense inhibition of transgenes in *Phytophthora infestans* is modulated by choice of promoter and position effects. Gene 138:63–69.

Judelson, H. S., and Michelmore, R. W. 1989. Structure and expression of a gene encoding *Hsp70* from the oomycete fungus *Bremia lactucae*. Gene 79:207–217.

Judelson, H. S., and Michelmore, R. W. 1990. Highly abundant and stage specific mRNAs in the obligate pathogen *Bremia lactucae*. Mol. Plant Microbe Interact. 3:225–232.

Judelson, H. S., and Michelmore, R. W. 1991. Transient expression of foreign genes in the oomycete *Phytophthora infestans* using *Bremia lactucae* regulatory sequences. Curr. Gen. 19:453–459.

Judelson, H. S., Tyler, B. M., and Michelmore, R. W. 1991. Transformation of the oomycete pathogen, *Phytophthora infestans*. Mol. Plant Microbe Interact. 4:602–607.

Judelson, H. S., Tyler, B. M., and Michelmore, R. W. 1992. Regulatory sequences for expressing genes in oomycetes. Mol. Gen. Genet. 234:138–146.

Julis, A. J., Clayton, C. N., and Sutton, T. B. 1978. Detection and distribution of *Phytophthora cactorum* and *P. cambivora* on apple rootstocks. Plant Dis. Rep. 62:516–520.

Kadir, S., and Umaerus, V. 1987. *Phytophthora infestans* A^2 compatibility type recorded in Sweden (Abstr.). 10th Triannual Conf. Eur. Assoc. Potato Res., Aalborg, Denmark. 223 pp.

Kadish, D., and Cohen, Y. 1988a. Competition between metalaxyl-sensitive and metalaxyl-resistant isolates of *Phytophthora infestans* in the absence of metalaxyl. Plant Pathol. 37:558–564.

Kadish, D., and Cohen, Y. 1988b. Fitness of *Phytophthora infestans* from metalaxyl-sensitive and resistant populations. Phytopathology 78:912–915.

Kadish, D., and Cohen, Y. 1988c. Estimation of metalaxyl resistance in *Phytophthora infestans*. Phytopathology 78:915–919.

Kadish, D., and Cohen, Y. 1989. Population dynamics of metalaxyl-sensitive and metalaxyl-resistant isolates of *Phytophthora infestans* in untreated crops of potato. Plant Pathol. 38:271–276.

Kadish, D., and Cohen, Y. 1992. Overseasoning of metalaxyl-sensitive and metalaxyl-resistant isolates of *Phytophthora infestans* in potato tubers. Phytopathology 82:887–889.

Kadish, D., Grinberger, M., and Cohen, Y. 1990. Fitness of metalaxyl-sensitive and metalaxyl-resistant isolates of *Phytophthora infestans* on susceptible and resistant potato cultivars. Phytopathology 80:200–205.

Kadooka, J. Y., and Ko, W. H. 1973. Production of chlamydospores by *Phytophthora palmivora* in culture media. Phytopathology 63: 559–562.

Kaewruang, W., Sivasithamparam, K., and Hardy, G. E. 1988. *Phytophthora cryptogea,* an additional pathogen of gerbera in western Australia. Aust. Plant Pathol. 17:67–68.

Kaewruang, W., Sivasithamparam, K., and Hardy, G. E. 1989. Use of soil solarization to control root rots of gerberas (*Gerbera jamesonii*). Biol. Fertil. Soils 8:38–47.

Kaiser, W. J., and Meléndez, P. L. 1978. A Phytophthora stem canker disease of pigeon pea in Puerto Rico. Plant Dis. Rep. 62:240–242.

Kale, G. B., and Prasad, N. 1957. Phytophthora blight of *Sesamum.* Indian Phytopathol. 10:38–47.

Kalmowitz, K., Whitwell, T., Zehr, E., and Toler, J. 1991. Pesticides and weeds influence *Phytophthora cinnamomi* presence and growth in container-grown azaleas. HortScience 26:1428.

Kalshoven, L. G. E. 1981. Pests of Crops in Indonesia. Revised and translated by P. A. van der Laan, assisted by G. H. L. Rothchild. Ichtiar Baru, Jakarta, Indonesia. 701 pp.

Kamat, M. N. 1958. Handbook of Tropical Crop Diseases. Prakash Publ. House, Poona, India. 84 pp.

Kamjaipai, W., and Chompoorat, C. 1984. Foot rot of *Aglaia domestica.* J. Thai Phytopathol. Soc. 4:202–203.

Kamjaipai, W., and Intavimolsri, S. 1983. Foot rot of pomelo. J. Thai Phytopathol. Soc. 3:201–203.

Kamjaipai, W., Leelasethakul, K., and Silayoi, I. 1984. Stem rot of *Durio lowianus.* J. Thai Phytopathol. Soc. 4:191–193.

Kamjaipai, W., and Ui, T. 1978. Mating types of *Phytophthora capsici* Leonian, the causal fungus of pumpkin rot in Hokkaido. Ann. Phytopathol. Soc. Jpn. 44:440–446.

Kamoun, S., Young, C. B., Gelascock, B., and Tyler, B. M. 1993a. Extracellular protein elicitors from *Phytophthora* host-specificity and induction of resistance to bacterial and fungal phytopathogens. Mol. Plant Microbe Interact. 6:15–25.

Kamoun, S., Klucher, K. M., Coffey, M. D., and Tyler, B. M. 1993b. A gene encoding a host-specific elicitor protein of *Phytophthora parasitica.* Mol. Plant-Microbe Interact. 6:573–581.

Kamoun, S., Young, M., Förster, H., Coffey, M. D., and Tyler, B. M. 1994. Potential role of elicitins in the interaction between *Phytophthora* and tobacco. Appl. Environ. Microbiol. 60:1593–1598.

Kang, S. W., and Lee, K. H. 1987. Efficacy of metalaxyl for control of rice downy mildew caused by *Sclerophthora macrospora.* Korean J. Plant Pathol. 3:17–19.

Kanlong, S., and Hendrix, J. W. 1977. Lack of relationship between ability to kill plants and ability to inhibit plant growth among *Phytophthora* species. Can. J. Bot. 55:17–22.

Kannaiyan, J., and Nene, Y. L. 1984. Efficacy of metalaxyl for control of Phytophthora blight of pigeon pea. Indian Phytopathol. 37:506–510.

Kannaiyan, J., Ribeiro, O. K., Erwin, D. C., and Nene, Y. L. 1980. Phytophthora blight of pigeon pea in India. Mycologia 72:169–181.

Kannaiyan, J., Nene, Y. L., Raju, T. N., and Sheila, V. K. 1981. Screening for resistance to Phytophthora blight of pigeon pea. Plant Dis. 65:61–62.

Kannwischer, M. E., and Mitchell, D. J. 1978. The influence of a fungicide on the epidemiolgy of black shank of tobacco. Phytopathology 68:1760–1765.

Kanniwischer, M. E., and Mitchell, D. J. 1981. Relationships of numbers of spores of *Phytophthora parasitica* var. *nicotianae* to infection and mortality of tobacco. Phytopathology 71:69–73.

Kanouse, B. B. 1925. Physiology and morphology of *Pythiomorpha gonapodioides.* Bot. Gaz. 79:196–206.

Kao, C. W. 1978. Root rot of *Gerbera jamesonii* caused by *Phytophthora cryptogea.* Plant Prot. Bull. (Taiwan). 20(4):(1977 Abstract, p. 11, in Chinese). (Cited in Ho 1992).

Kao, C. W., and Leu, L. S. 1977. Three unreported species of *Phytophthora* in Taiwan. Plant Prot. Bull. (Taiwan). 19:302–303. (Abstract in Chinese). (Cited in Ho 1992).

Kao, C. W., and Leu, L. S. 1979. Strawberry fruit rot caused by *Phytophthora cactorum* and *P. citrophthora.* Plant Prot. Bull. (Taiwan) 21:239–243.

Kaosiri, T. 1983. *Gossypium hirsutum,* a new host of *Phytophthora cinnamomi.* J. Thai Phytopathol. Soc. 207–210.

Kaosiri, T. 1984. Phytophthora blight of kitchen mint. J. Thai Phytopathol. Soc. 4:204–208.

Kaosiri, T., and Siddhipongse, S. 1985. *Phytophthora* boll rot of cotton in Thailand. J. Thai Phytopathol. 5:95–101.

Kaosiri, T., and Zentmyer, G. A. 1980. Protein, esterase, and peroxidase patterns in the *Phytophthora palmivora* complex from cacao. Mycologia 72:988–1000.

Kaosiri, T., Zentmyer, G. A., and Erwin, D. C. 1978. Stalk length as a taxonomic criterion for *Phytophthora palmivora* isolates from cacao. Can. J. Bot. 56:1730–1738.

Kaosiri, T., Zentmyer, G. A., and Erwin, D. C. 1980. Oospore morphology and germination in the *Phytophthora palmivora* complex from cacao. Mycologia 72:888–907.

Kaplan, D. T., Keen, N. T., and Thomason, I. J. 1980. Studies on the mode of action of glyceollin in soybean incompatibility to the root knot nematode, *Meloidogyne incognita.* Physiol. Plant Pathol. 16: 319–325.

Karakaya, A., Gray, F. A., and Koch, D. W. 1995. Characterization of two *Phytophthora* spp. from kale and tyfon. Mycotaxon 56:483–490.

Kassaby, F. Y. 1985. Solar-heating soil for control of damping-off diseases. Soil Biol. Biochem. 17:429–434.

Kassaby, F. Y., Fagg, P. C., and Marks, G. C. 1977. Survival in, and colonization of mountain forest by *Phytophthora cinnamomi* Rands in eastern Victoria. Aust. For. Res. 7:173–183.

Katan, J. 1981. Solar heating (solarization) of soil for control of soilborne pests. Annu. Rev. Phytopathol. 19:211–236.

Katan, J., and De Vay, J. E., eds. 1991. Soil Solarization. CRC Press, Boca Raton, Fla. 256 pp.

Katan, J., Greenberger, A., Alan, H., and Grinstein, A. 1976. Solar heating by polyethylene mulching for the control of diseases caused by soil-borne pathogens. Phytopathology 66:683–688.

Kato, S., and Coe, R., New, L., and Dick, M. W. 1990. Sensitivities of various Oomycetes to hymexazol and metalaxyl. J. Gen. Microbiol. 136:2127–2134.

Katsura, K. 1958. A Phytophthora rot of watermelon caused by *Phytophthora drechsleri.* Sci. Rep. Saikyo Univ. Agric. 10:77–85.

Katsura, K. 1965. Downy mildew of rice plant—Infection and disease development. Ann. Phytopathol. Soc. Jpn. 31:186–192. (In Japanese).

Katsura, K. 1968. *Phytophthora melonis* n. sp. of cucumber. Ann. Phytopathol. Soc. Jpn. 34:167.

Katsura, K. 1969. Two species of *Phytophthora* on fig. Sci. Rep. Kyoto Prefect. Univ. 21:24–31. (In Japanese).

Katsura, K. 1970. Studies on the blight and rot of tulip caused by *Phytophthora porri* Foister. I. On the disease and its causal fungus. Sci. Rep. Kyoto Prefect. Univ. Agric. 22:9–16 (In Japanese).

Katsura, K. 1971. *Phytophthora* Diseases of Plants, Theory and Practice. Seibundo Shkinkosha, Tokyo. 128 pp. (In Japanese).

Katsura, K. 1976. Two new species of *Phytophthora* causing damping-off of cucumber and trunk rot of chestnut. Trans. Mycol. Soc. Jpn. 17:238–242. (In Japanese).

Katsura, K., and Maeda, S. 1971. Chlamydospore formation of *Phytophthora porri* Foister. Sci. Rep. Kyoto Prefect. Univ. Agric. 23:34–38. (In Japanese).

Katsura, K., and Miyata, Y. 1971. Swimming behavior of *Phytophthora capsici* zoospores. Pages 107–128 in: Morphological and Biochemical Events in Plant-Parasite Interaction. S. Akai and S. Ouchi, eds. Phytopathol. Soc. Jpn., Tokyo. 415 pp.

Katsura, K., and Tokura, R. 1954. Studies on *Phytophthora* disease of economic plants (VII). A brown rot of watermelon caused by *Phytophthora capsici* Leonian. Sci. Rep. Saikyo Univ. Agric. 6:38–48. (In Japanese).

Katsura, K., and Tokura, R. 1955. Studies on *Phytophthora* disease of economic plants (VIII), a brown rot of eggplants caused by *Phytophthora capsici* Leonian. Jubilee Publ. in Commemoration of 60th Birthdays of Prof. Tochinai and Prof. Fukushi. Pages 167–172. (In Japanese).

Katsura, K., and Yamamoto, K. 1969. Two species of *Phytophthora* on fig (*Ficus carica* L.). Sci. Rep. Kyoto Univ. Agric. 21:24–31. (In Japanese).

Katsura, K., Tokura, R., and Furuya, T. 1954. Studies on the downy mildew of rice plant. II. Wild grasses as the source of primary infection to rice plant in nursery beds. Sci. Rep. Saikyo Agric. Univ. 6:49–66. (In Japanese).

Katsura, K., Masago, H., and Miyata, Y. 1966. Movements of zoospores of *Phytophthora capsici.* I. Electrotaxis in some organic solutions. Ann. Phytopathol. Soc. Jpn. 32:215–220. (In Japanese).

Katsura, K., Isaka, M., and Miyagoshi, M. 1969. *Phytophthora porri* Foister, the causal fungus of the leaf blight and bulb rot of scallion, *Allium bakeri* Regel. Ann. Phytopathol. Soc. Jpn. 35:55–61. (In Japanese).

Katznelson, H., Rouatt, J. W., and Payne, T. M. B. 1955. The liberation of amino acids and reducing compounds by plant roots. Plant Soil 7:35–48.

Kaufmann, M. J., and Gerdemann, J. W. 1958. Root and stem rot of soybean caused by *Phytophthora sojae* n. sp. Phytopathology 48: 201–208.

Kavanagh, J. A., O'Malley, M., and Simmonds, A. 1969. Some observations on the rotting of apple fruits by *Phytophthora syringae* Kleb. Ir. J. Agric. Res. 8:439–441.

Kaven, G. 1934. Krankheiten und Schädlinge an Rhododendron (Plant disease and damage on rhododendron). Die Kranke Pflanze 11:123–126. (In German).

Kawakami, T., and Suzuki, R. 1908. List of fungi on cultivated plants in Formosa. Part I. Bull. Div. Plant Pathol. Agric. Exp. Sta. Gov. Formosa 1:1–64. (In Japanese).

Keast, D., and Walsh, L. G. 1979. Passage and survival of chlamydospores of *Phytophthora cinnamomi* Rands, the causal agent of forest dieback disease, through the gastrointestinal tracts of termites and wild birds. Appl. Environ. Microbiol. 37:661–664.

Keeling, B. L. 1976. A comparison of methods used to test soybeans for resistance to *Phytophthora megasperma* var. *sojae.* Plant Dis. Rep. 60:800–802.

Keeling, B. L. 1980. Research on Phytophthora root and stem rot: Isolation, testing proceedures, and seven new physiologic races. Pages 367–370 in: Proc. World Soybean Conf. II. F. T. Corbin, ed. Westview Press, Boulder, Colo. 897 pp.

Keeling, B. L. 1982. Four new physiologic races of *Phytophthora megasperma* f. sp. *glycinea.* Plant Dis. 66:334–335.

Keeling, B. L. 1984. A new physiologic race of *Phytophthora megasperma* f. sp. *glycinea.* Plant Dis. 66:626–627.

Keen, N. T. 1971. Hydroxyphaseollin production by soybeans resistant and susceptible to *Phytophthora megasperma* var. *sojae.* Physiol. Plant Pathol. 1:265–275.

Keen, N. T. 1975. Specific elicitors of plant phytoalexin production: Determinants of race specificity in pathogens? Science 187: 74–75.

Keen, N. T. 1982. Specific recognition in gene-for-gene host-parasite systems. Adv. Plant Pathol. 1:35–82.

Keen, N. T. 1990. Phytoalexins and their elicitors. Pages 114–129 in: Microbes and Microbial Products as Herbicides. R. E. Hoagland, ed. Am. Chem. Soc. Symp. Ser. 439. Washington, D.C. 341 pp.

Keen, N. T., and Bruegger, B. 1977. Phytoalexins and chemicals that elicit their production in plants. Pages 1–26 in: Host Plant Resistance to Pests. P. A. Hedin, ed. Am. Chem. Soc., Washington, D.C. 286 pp.

Keen, N. T., and Legrand, M. 1980. Surface glycoproteins: Evidence that they may function as the race specific phytoalexin elicitors of *Phytophthora megasperma* f. sp. *glycinea.* Physiol. Plant Pathol. 17:175–192.

Keen, N. T., and Yoshikawa, M. 1983a. Physiology of disease and the nature of resistance to *Phytophthora.* Pages 279–287 in: *Phytophthora:* Its Biology, Taxonomy, Ecology, and Pathology. D. C. Erwin, S. Bartnicki-Garcia, and P. H. Tsao, eds. American Phytopathological Society, St. Paul, Minn. 392 pp.

Keen, N. T., and Yoshikawa, M. 1983b. β-1,3-Endoglucanase from soybean releases elicitor-active carbohydrates from fungus cell walls. Plant Physiol. 71:460–465.

Keen, N. T., Wang, M. C., Bartnicki-Garcia, S., and Zentmyer, G. A. 1975. Phytotoxicity of mycolaminarins—β-1,3-Glucans from *Phytophthora* spp. Physiol. Plant Pathol. 7:91–97.

Keen, N. T., Yoshikawa, M., and Wang, M. C. 1983. Phytoalexin elicitor activity of carbohydrates from *Phytophthora megasperma* f. sp. *glycinea* and other sources. Plant Physiol. 71:466–471.

Keen, N. T., Lyne, R. L., and Hymowitz, T. 1986. Phytoalexin production as a chemosystematic parameter within the genus *Glycine.* Biochem. Syst. Ecol. 14:481–486.

Keen, N. T., and Yoshikawa, M. 1990. The expression of resistance in soya beans to *Phytophthora megasperma* f. sp. *glycinea.* Pages 329–344 in: Biological Control of Soil-Borne Plant Pathogens. D. Hornby, R. J. Cook, Y. Henis, W. H. Ko, A. D. Rovira, B. Schippers, and P. R. Scott, eds. CAB International, Oxford, U.K. 479 pp.

Keenan, P. J., Ellis, J. S., Rathmell, W. G., and Friend, J. 1986. Carbohydrate and lipid-containing elicitors from *Phytophthora infestans.* Do they have a common mechanism of action? Pages 185–189 in: Biology and Molecular Biology of Plant-Pathogen Interactions. J. Bailey, ed. Springer Verlag, Berlin. 415 pp. (Cited in Friend 1991).

Keim, R. 1977. Foliage blight of periwinkle in Southern California. Plant Dis. Rep. 61:182–184.

Keim, R., Mock, T., and Guenther, H. 1976a. Phytophthora crown rot of toyon. Plant Dis. Rep. 60:88–90.

Keim, R., Zentmyer, G. A., and Klure, L. J. 1976b. *Phytophthora palmivora* on ivy in California and its control with pyroxychlor. Plant Dis. Rep. 60:632–633.

Keim, R., Klure, L. J., and Zentmyer, G. A. 1979. Collar rot of Washington palms in containers. Plant Dis. Rep. 63:718–720.

Kellam, M. K., and Coffey, M. D. 1985. Quantitative comparison of the resistance to Phytophthora root rot in three avocado rootstocks. Phytopathology 75:230–234.

Kellam, M. K., and Zentmyer, G. A. 1981. Isolation of *Phytophthora citrophthora* from cocoa in Brazil (Abstr.). Phytopathology 71:230.

Kellam, M. K., and Zentmyer, G. A. 1986a. Comparisons of single-oospore isolates of *Phytophthora* species from naturally infected cocoa pods in Brazil. Mycologia 78:351–358.

Kellam, M. K., and Zentmyer, G. A. 1986b. Morphological, physiological, ecological, and pathological comparisons of *Phytophthora* species isolated from *Theobroma cacao.* Phytopathology 76:159–164.

Kelley, W. D. 1976. Evaluation of *Trichoderma harzianum*-impregnated clay granules as a biocontrol for *Phytophthora cinnamomi* causing damping-off of pine seedlings. Phytopathology 66:1023–1027.

Kelley, W. D. 1977. Interactions of *Phytophthora cinnamomi* and *Trichoderma* spp. in relation to propagule production in soil cultures at 26°C. Can. J. Microbiol. 23:288–294.

Kelley, W. D., and Rodriguez-Kabana, R. 1976. Competition between *Phytophthora cinnamomi* and *Trichoderma* spp. in autoclaved soil. Can. J. Microbiol. 22:1120–1127.

Kendrick, J. B. 1923. Phytophthora rot of tomato, eggplant, and pepper. Proc. Indiana Acad. Sci. 38:299–306. (Cited in Tucker 1933).

Kenerley, C. M., and Bruck, R. I. 1983. Overwintering and survival of *Phytophthora cinnamomi* in Fraser fir and cover cropped nursery beds in North Carolina. Phytopathology 73:1643–1647.

Kenerley, C. M., Bruck, R. I., and Grand, L. F. 1984a. Effects of metalaxyl on growth and ectomycorrhizae of Fraser fir seedlings. Plant Dis. 68:32–35.

Kenerley, C. M., Papke, K., and Bruck, R. I. 1984b. Effect of flooding on development of Phytophthora root rot in Fraser fir seedlings. Phytopathology 74:401–404.

Kennedy, B. W., and Erwin, D. C. 1961. Some factors influencing sporangium formation of a *Phytophthora* species isolated from lucerne in certain salt solutions. Trans. Br. Mycol. Soc. 44:291–297.

Kennedy, D. M., and Duncan, J. M. 1993. Occurrence of races in *Phytophthora fragariae* var. *rubi* on raspberry. Acta Hortic. 352:555–562.

Kennedy, D. M., and Duncan, J. M. 1995. A papillate *Phytophthora* species with specificity to *Rubus.* Mycol. Res. 99:57–68.

Kennedy, R., and Pegg, G. F. 1990. *Phytophthora cryptogea* root rot of tomato in rockwool nutrient culture. II. Effect of root zone temperature on infection, sporulation and symptom development. Ann. Appl. Biol. 117:537–551.

Kenneth, R. G. 1981. Downy mildew of graminaceous crops. Pages 367–394 in: The Downy Mildews. D. M. Spencer, ed. Academic Press, London. 636 pp.

Kenneth, R. 1982. *Phlyctochytrium* sp. as a potential biological control agent against oomycetous fungi in soil. Phytoparasitica 10: 120.

Kenney, D. S. 1986. De Vine—The way it was developed—An industrialist's view. Weed Sci. 34(suppl. 1):15–16.

Kerkenaar, A. 1981. On the antifungal mode of action of metalaxyl, an inhibitor of nucleic acid synthesis in *Pythium splendens*. Pestic. Biochem. Physiol. 16:1–13.

Kerkenaar, A., and Kaars Sijpesteijn, A. 1977. On the mode of action of prothiocarb. Neth. J. Plant Pathol. 83(suppl. 1):145–152.

Kerr, A. 1957. 'Red core' disease of strawberries. J. Dep. Agric. South Aust. 60:354–356.

Kerr, A. 1963. The root rot Fusarium wilt complex of peas. Aust. J. Biol. Sci. 16:55–69.

Kerwin, J. L., and Duddles, N. D. 1989. Reassessment of the role of phospholipids in sexual reproduction by sterol-auxotrophic fungi. J. Bacteriol. 171:3831–3839.

Kerwin, J. L., Simmons, C. A., and Washino, R. K. 1986. Eicosanoid regulation of oosporogenesis by *Lagenidium giganteum*. Prostaglandins Leukotrienes Med. 23:173–178.

Kerwin, J. L., and Washino, R. K. 1986. Oosporogenesis by *Lagenidium giganteum:* Induction and maturation are regulated by calcium and calmodulin. Can. J. Microbiol. 32:663–672.

Keyworth, W. G. 1943. A *Phytophthora* disease of the hop in Great Britain. Gard. Chron. (Ser. 3) 113:238.

Khan, A. J., and Ravise, A. 1985. Stimulation of defense reations in citrus by fosetyl-Al and fungal elicitors against *Phytophthora* spp. Pages 281–281 in: Proc. 1985 Common Coll. Int. Fungicides Crop Prot. Bordeaux, France.

Khan, A. J., Vernenghi, A., and Ravise, A. 1986. Incidence of fosetyl-Al and elicitors on the defense reactions of citrus attacked by *Phytophthora* spp. Fruits 41:587–595.

Khan, A. J., Weber, D. J., and Husain, A. 1988. Morphology of *Phytophthora nicotianae* variety *nicotianae*. Indian J. Plant Pathol. 6:1–7.

Khare, M. N., Bhale, M. S., Tiwari, D. P., and Naya, K. 1988. Current status of Phytophthora foot and leaf rot of betelvine. Pages 164–177 in: Perspectives in Plant Pathology and Mycology. V. P. Agnihotre, A. K. Sarbhoy, and Dinesh Kumar, eds. Malhotra Pub. House, New Delhi, India.

Kheng, K. T. 1978. Pepper foot rot. Tech. Bull. No. 2. Dep. Agric. Sarawak, Malaysia. 21 pp.

Kheng, K. T., and Ling, K. K. 1980. Evaluation of chemicals for the control of *Phytophthora* from *Piper nigrum*. Malays. Agric. J. 52: 263–272.

Kheng-Hoy, C., and Turner, N. A. 1965. A steroid factor in pea (*Pisum sativum* L.) influencing growth and sporulation of *Phytophthora cinnamomi* Rands. N.Z. J. Agric. Res. 8:104–108.

Kheswalla, K. F. 1935. Seedling blight of *Cinchona ledgeriana* Moens caused by *Phytophthora palmivora* Butl. in Darjeeling district. Indian J. Agric. Sci. 5:485–495.

Khew, K. L., and Zentmyer, G. A. 1973. Chemotactic response of zoospores of five species of *Phytophthora*. Phytopathology 63: 1511–1517.

Khew, K. L., and Zentmyer, G. A. 1974. Electrophoretic response of zoospores of seven species of *Phytophthora*. Phytopathology 64: 500–507.

Kiewnick, L. 1985. *Phytophthora porri* Foister als Ursuche einer Lagerfaule bei China Kohl (*Phytophthora porri* as the cause of a storage rot in Chinese cabbage). Nachrichtenbl. Dtsch. Pflanzenschutz. 37:136–137. (In German).

Kiewnick, L., Schickedanz, F., Kröber, H. 1978. Eine neue *Phytophthora*-Wurzel-und-Stengelfäule an *Elatior begonia* (A new Phytophthora root and stem rot disease of *Elatior begonia*). Nachrichtenbl. Dtsch. Pflanzenschutzdienstes 30:42. (Rev. Plant Pathol. 1978, 57:4992).

Kilen, T. C. 1986. A favorable linkage combination in the soybean. J. Hered. 77:275–277.

Kilen, T. C., and Barrentine, W. L. 1983. Linkage relationships in soybean between genes controlling reactions to Phytophthora root rot and metribuzin. Crop Sci. 23:894–896.

Kilen, T. C., and Keeling, B. L. 1977. Simultaneous screening of soybeans against three races of *Phytophthora megasperma* var. *sojae*. Crop Sci. 17:185–186.

Kilen, T. C., and Keeling, B. L. 1990. Gene frequency changes in soybean bulk populations exposed to Phytophthora rot. Crop Sci. 30:575–578.

Kilen, T. C., Hartwig, E. E., and Keeling, B. L. 1974. Inheritance of a second major gene for resistance to Phytophthora rot in soybeans. Crop Sci. 14:260–262.

Kim, B. S., Lee, E. K., and Chung, B. K. 1975. An investigation on the brown rot of eggplant caused by *Phytophthora capsici* Leonian. Korean J. Plant Prot. 14:77–79 (In Korean).

Kim, E. S., and Hwang, B. K. 1992. Virulence to Korean pepper cultivars of isolates of *Phytophthora capsici* from different geographic areas. Plant Dis. 76:486–489.

Kim, Y. J., Hwang, B. K., and Park, K. W. 1989. Expression of age-related resistance in pepper plants infected with *Phytophthora capsici*. Plant Dis. 73:737–747.

Kinal, J., Shearer, B. L., and Fairman, R. G. 1993. Dispersal of *Phytophthora cinnamomi* through lateritic soil by laterally flowing subsurface water. Plant Dis. 77:1085–1090.

Kincaid, R. R., Martin, F. G., Gammon, N., Jr., Breland, H. L., and Pritchett, W. L. 1970. Multiple regression of tobacco black shank, root knot, and coarse root indexes on soil pH, potassium, calcium, and magnesium. Phytopathology 60:1513–1516.

King, G. D. 1980. Ecology of sporangium production by *Phytophthora cinnamomi*. Ph.D. thesis, University of California, Riverside. 142 pp.

Kinghorn, J. R., Moon, R. P., Unkles, S. E., and Duncan, J. M. 1991. Gene structure and expression in *Phytophthora infestans* and the development of gene-mediated transformation. Pages 295–311 in: *Phytophthora*. J. A. Lucas, R. C. Shattock, D. S. Shaw, and L. R. Cooke, eds. Cambridge Univ. Press, Cambridge, U.K. 447 pp.

Kitazawa, K., Tsuchiya, S., Kodama, F., Kamjaipai, W., Ogoshi, A., and Yanagita, K. 1978. Phytophthora stem rot of adzuki-bean (*Phaseolus angularis*) caused by *Phytophthora vignae* Purss. Ann. Phytopathol. Soc. Jpn. 44:528–531.

Kitazawa, K., Suzui, T., and Yanagita, K. 1979. Pathogenicity of *Phytophthora vignae* Purss to adzuki-bean and cowpea. Ann. Phytopathol. Soc. Jpn. 45:406–408.

Kittle, D. R., and Gray, L. E. 1979a. The influence of soil temperature, moisture, porosity, and bulk density on the pathogenicity of *Phytophthora megasperma* var. *sojae*. Plant Dis. Rep. 63:231–234.

Kittle, D. R., and Gray, L. E. 1979b. Production and pathogenicity of oospores of *Phytophthora megasperma* var. *sojae*. Plant Dis. Rep. 63:628–631.

Kittle, D. R., and Gray, L. E. 1979c. Storage and use of *Phytophthora megasperma* var. *sojae* oospores as inoculum. Phytopathology 69: 821–823.

Klarman, W. L., and Gerdemann, J. W. 1963a. Induced susceptibility in soybean plants genetically resistant to *Phytophthora sojae*. Phytopathology 53:863–864.

Klarman, W. L., and Gerdemann, J. W. 1963b. Resistance of soybeans to three *Phytophthora* species due to the protection of a phytoalexin. Phytopathology 53:1317–1320.

Klarman, W. L., and Gerdemann, J. W. 1974. Histopathology of resistant and susceptible soybean hypocotyls inoculated with *Phytophthora megasperma* var. *sojae*. Phyotpathology 64:971–975.

Klebahn, H. 1905. Eine neue Pilzkrankheit der *Syringae* (A new fungal disease of *Syringae*). Zbl. Bakt. 15, II:335–336. (In German). (Cited in Tucker 1933).

Klebahn, H. 1909. Die neue Zweig-und Knospenkrankheit (A new twig and bud disease). Krankheiten Flieders. Berlin. Pages 18–75. (In German). (Cited in Tucker 1933).

Klebs, G. 1899. Zur Physiologie der Fortpflanzung einiger Pilzer. II. *Saprolegnia mixta* de Bary (On the physiology of reproduction of several fungi. II. *Saprolegnia mixta* de Bary). Jahrb. Wiss. Bot. 33:513–593. (In German). (Cited in C. G. Elliott 1989).

Klebs, G. 1900. Zur Physiologie der Fortpflanzung einiger Pilzer. III. Allgemeine Betrachtungen (On the physiology of reproduction of several fungi. III. General observations). Jahrb. Wiss. Bot. 35:80–203. (In German). (Cited in C. G. Elliott 1989).

Klein, H. H. 1959. Etiology of the *Phytophthora* disease of soybeans. Phytopathology 49:380–383.

Kleinschmidt, G. D., and Gerdemann, J. W. 1972. Stunting of citrus seedlings in fumigated nursery soils related to the absence of endomycorrhizae. Phytopathology 62:1447–1453.

Klemmer, H. W., and Nakano, R. Y. 1962. Techniques in isolation of pythiaceous fungi from soil and diseased pineapple tissue. Phytopathology 52:955–956.

Klemmer, H. W., and Nakano, R. Y. 1964. Distribution and pathogenicity of *Phytophthora* and *Pythium* in pineapple soils of Hawaii. Plant Dis. Rep. 48:848–852.

Kliejunas, J., and Adams, D. H. 1981. Phytophthora root rot of Port-Orford-cedar in California. Plant Dis. 65:446–447.

Kliejunas, J. T., and Ko, W. H. 1974. Effect of motility of *Phytophthora palmivora* zoospores on disease severity in papaya seedlings and substrate colonization in soil. Phytopathology 64:426–428.

Kleijunas, J. T., and Ko, W. H. 1976. Dispersal of *Phytophthora cinnamomi* on the island of Hawaii. Phytopathology 66:457–460.

Kliejunas, J. T., and Nagata, J. T. 1979. *Phytophthora cinnamomi* in Hawaiian forest soils: Seasonal variations in population levels. Phytopathology 69:1268–1272.

Kliejunas, J. T., and Nagata, J. T. 1980. *Phytophthora cinnamomi* in Hawaiian forest soils: Technique for enumeration and types of propagules recovered. Soil Biol. Biochem. 12:89–91.

Klimczak, L. J., and Prell, H. H. 1984. Isolation and characterization of mitochondrial DNA of the oomycetous fungus *Phytophthora infestans*. Curr. Genet. 8:323–326.

Klisiewicz, J. M. 1970. Factors affecting production and germination of oospores of *Phytophthora drechsleri*. Phytopathology 60:1738–1742.

Klisiewicz, J. M. 1977. Identity and relative virulence of some heterothallic *Phytophthora* species associated with root and stem rot of safflower. Phytopathology 67:1174–1177.

Klisiewicz, J. M., and Beard, B. H. 1976. Diseases of sunflower in California. Plant Dis. Rep. 60:298–301.

Klopping, H. L., and Delp, C. J. 1980. 2-Cyano-*N*-[(ethylamino) carbonyl]-2-(methoxyimino) acetamide, a new fungicide. J. Agric. Food Chem. 28:467–468.

Klotz, L. J. 1940. Studies on *Phytophthora citrophthora* (Abstr.). Phytopathology 30:14.

Klotz, L. J. 1973. Color Handbook of Citrus Diseases. Univ. Calif. Press, Berkeley. 122 pp.

Klotz, L. J. 1978. Fungal, bacterial, and nonparasitic diseases and injuries originating in the seedbed, nursery, and orchard. Pages 1–66 in: The Citrus Industry. Vol. IV. Crop Protection. W. Reuther, E. C. Calavan, and G. E. Carman, eds. Rev. ed. Univ. of Calif. Div. of Agric. Sci. 362 pp.

Klotz, L. J., and Calavan, E. C. 1969. Gum diseases of citrus in California. Calif. Agric. Exp. Stn. Ext. Serv. Circ. 396. 26 pp.

Klotz, L. J., and Calavan, E. C. 1978. Review of gum diseases of citrus in California. Calif. Div. Agric. Sci. Univ. Calif. Agric. Sci. Leafl. 2593. 27 pp.

Klotz, L. J., and DeWolfe, T. A. 1958. Techniques for isolating *Phytophthora* spp. which attack citrus. Plant Dis. Rep. 42:675–676.

Klotz, L. J., and DeWolfe, T. A. 1960. The production and use of zoospore suspensions of *Phytophthora* spp. for investigations on diseases of citrus. Plant Dis. Rep. 44:572–573.

Klotz, L. J., and DeWolfe, T. A. 1961a. Limitations of the hot water immersion treatment for the control of Phytophthora brown rot of lemons. Plant Dis. Rep. 45:264–267.

Klotz, L. J., and DeWolfe, T. A. 1961b. Brown rot contact infection of citrus fruits prior to hot water treatment. Plant Dis. Rep. 45:268–271.

Klotz, L. J., and Fawcett, H. S. 1939. Isolation of *Phytophthora* spp. Phytopathology 29:290–291.

Klotz, L. J., DeWolfe, T. A., and Wong P. P. 1958. Decay of fibrous roots of citrus. Phytopathology 48:616–622.

Klotz, L. J., Wong, P. P., and DeWolfe, T. A. 1959. Survey of irrigation water for the presence of *Phytophthora* spp. pathogenic to citrus. Plant Dis. Rep. 43:830–832.

Klotz, L. J., DeWolfe, T. A., Roistacher, C. N., Nauer, E. M., and Carpenter, J. B. 1960. Heat treatments to destroy fungi in infected seeds and seedlings of citrus. Plant Dis. Rep. 44:858–861.

Klotz, L. J., DeWolfe, T. A. and MacRill, J. R. 1961. Controlling brown rot of citrus fruit. Calif. Citrogr. 47:52–55.

Klotz, L. J., Bitters, W. P., and DeWolfe, T. A. 1965a. Citrus rootstocks resistant to Phytophthora root rot. Calif. Agr. 19(2):10–12.

Klotz, L. J., DeWolfe, T. A., and MacRill, J. R. 1965b. Control of brown rot of citrus in orchard and packing houses. Calif. Citrogr. 50:456, 474–476.

Klotz, L. J., Richards, S. J, and DeWolfe, T. A. 1967. Irrigation effects on root rot of young citrus trees. Calif. Citrogr. 52:91.

Klotz, L. J., Bitters, W. P., DeWolfe, T. A., and Garber, M. J. 1968. Some factors in resistance of citrus to *Phytophthora* spp. Plant Dis. Rep. 52:952–955.

Knauss, J. F. 1975. Control of basal stem and root rot of Christmas cactus caused by *Pythium aphanidermatum* and *Phytophthora parasitica*. Proc. Fla. State Hortic. Soc. 88:567–571.

Knorr, L. C., Malaguti, G., Serpa, D., and Leal, F. 1964. World citrus problems. F.A.O. Plant Prot. Bull. 12:121–128.

Knowles, C. H. 1916. Bud rot of the coconut palm. Dep. Agric. Fiji Pamphlet 20. 2 pp. (Cited in Tucker 1933).

Knowles, P. F., Miller, M. D., Henderson, D. W., Foy, C. L., Carlson, E. C., Klisiewicz, J. M., Goss, J. R., Jones, L. G., and Edwards, R. T. 1965. Safflower. Calif. Agric. Exp. Stn. Ext. Serv. Circ. 532. 50 pp.

Ko, W. H. 1971. Biological control of seedling root rot of papaya caused by *Phytophthora palmivora*. Phytopathology 61:780–782.

Ko, W. H. 1978. Heterothallic *Phytophthora:* Evidence for hormonal regulation of sexual reproduction. J. Gen. Microbiol. 107:15–18.

Ko, W. H. 1979. Mating type distribution of *Phytophthora colocasiae* on the island of Hawaii. Mycologia 71:434–437.

Ko, W. H. 1980. Hormonal regulation of sexual reproduction in *Phytophthora*. J. Gen. Microbiol. 116:459–463.

Ko, W. H. 1981. Reversible change of mating type in *Phytophthora parasitica*. J. Gen. Microbiol. 125:451–454.

Ko, W. H. 1982a. Biological control of Phytophthora root rot of payaya with virgin soil. Plant Dis. 66:446–448.

Ko, W. H. 1982b. Fruit rot of guava caused by *Phytophthora citricola*. Plant Dis. 66:854–855.

Ko, W. H. 1983. Isolation and partial characterization of a hormones produced by *Phytophthora parasitica*. J. Gen. Microbiol. 129: 1397–1401.

Ko, W. H. 1985. Stimulation of sexual reproduction of *Phytophthora cactorum* by phospholipids. J. Gen. Microbiol. 131:2591–2594.

Ko, W. H. 1988. Hormonal heterothallism and homothallism in *Phytophthora*. Annu. Rev. Phytopathol. 26:57–73.

Ko, W. H., and Ann, P. J. 1985. *Phytophthora humicola,* a new species from soil of a citrus orchard in Taiwan. Mycologia 77:631–636.

Ko, W. H., and Arakawa, C. K. 1980. Oospore germination of *Phytophthora katsurae*. Trans. Mycol. Soc. Jpn. 21:215–219.

Ko, W. H., and Chan, M. J. 1974. Infection and colonization potential of sporangia, zoospores, and chlamydospores of *Phytophthora palmivora* in soil. Phytopathology 64:1307–1309.

Ko, W. H., and Chang, H. S. 1979. *Phytophthora katsurae,* a new name for *P. castanae*. Mycologia 71:840–844.

Ko, W. H., and Ho, W. C. 1983. Reassessment of the apparent sterol requirement for sexual reproduction in *Phytophthora*. Ann. Phytopathol. Soc. Jpn. 49:316–321.

Ko, W. H., and Kunimoto, R. K. 1981. Hormone production and reception among different isolates of *Phytophthora parasitica* and *P. palmivora*. Mycologia 73:440–444.

Ko, W. H., Chase, L. L., and Kunimoto, R. K. 1973. A microsyringe method for determining concentration of fungal propagules. Phytopathology 63:1206–1207.

Ko, W. H., Kunimoto, R. K., and Nishijima, W. T. 1982. Fruit rot of guava caused by *Phytophthora citricola*. Plant Dis. 66:854–855.

Ko, W. H., Lee, C. J., and Su, H. J. 1986. Chemical regulation of mating type in *Phytophthora parasitica*. Mycologia 78:134–136.

Koch, J. 1971. Rhododendron blight caused by *Phytophthora cambivora*. Friesia (Copenhagen). 9:418–425.

Kodama, T., and Yoshioka, A. 1974. Phytophthora rot of cucumber and its control in Nara Prefecture. Proc. Kansai Plant Prot. Soc. 16:1–6.

Koike, S. T., Ouimette, D. G., and Coffey, M. D. 1987. First report of avocado fruit rot caused by *Phytophthora citricola* in California. Plant Dis. 71:1045.

Kopp, A. 1930. Les maladies des plantes a la Reunion (Maladies of plants in Reunion). Rev. Bot. Appl. Agric. Trop. 10:281–287. (In French). (Cited in Tucker 1931).

Kotthoff, P. 1937. Neue Topflanzkranheiten (New pot plant diseases). Kranke Pfl. 14:28–30.

Kouyeas, H. 1952. The foot rot of pistachio tree (*Pistacia vera* L.). Ann. Inst. Phytopathol. Benaki, N.S. 6:81–87.

Kouyeas, H. 1973. Pathogenicity of *Phytophthora* species to pistachio tree. Annls. Inst. Phytopathol. Benaki, N.S. 10:333–341.

Kouyeas, H. 1977. Two new hosts of *Phytophthora porri* Foister. Ann. Inst. Phytopathol. Benaki, N.S. 11:357–360.

Kouyeas, H., and Chitzanidis, A. 1968. Notes on Greek species of *Phytophthora*. Ann. Inst. Phytopathol. Benaki, N.S. 8:175–192.

Kouyeas, H., and Chitzanidis, A. 1978. Host list of *Phytophthora* spp. identified in Greece. Phytophthora Newsl. 6:53–54.

Kozlowska, C., Brennejzen, B., and Benben, K. 1961. Stan zagrozenia lasów polskich wr. anto przez wazniejsze choroby pochodzenia grzybowego (The menace to Polish forests of the more serious fungus diseases in 1960). Pr. Inst. Badaw. Lesn. 226:47–56. (In Polish). (Rev. Appl. Mycol. 1962, 41:551–552.)

Kramer, C. L., and Pady, S. M. 1966. A new 24-hour spore sampler. Phytopathology 56:517–520.

Krause, R. A., and Massie, L. B. 1975. Predictive systems: Modern approaches to disease control. Annu. Rev. Phytopathol. 13:31–47.

Krause, R. A., Massie, L. B., and Hyre, R. A. 1975. Blitecast: A computerized forecast of potato late blight. Plant Dis. Rep. 59:95–98.

Kreutzer, W. A., and Bryant, L. R. 1946. Certain aspects of the epiphytology and control of tomato fruit rot caused by *Phytophthora capsici* Leonian. Phytopathology 36:329–339.

Kreutzer, W. A., Bodine, E. W., and Durrell, L. W. 1940. Cucurbit diseases and rot of tomato fruit caused by *Phytophthora capsici.* Phytopathology 30:972–976.

Krikun, J., and Frank, Z. R. 1982. Metham sodium applied by sprinkler irrigation to control pod rot and Verticillium wilt of peanut. Plant Dis. 66:128–130.

Kröber, H. 1959. *Phytophthora cactorum* (Leb. et Cohn) Schroet. var. *applanata* Chest. als Erreger einer Zweigkrankheit an rhododendron (*Phytophthora cactorum* [Leb. et Cohn] Schroet var. *applanta* Chest., agent of a twig disease of rhododendron). Phytopathol. Z. 36:381–393. (In German). (Rev. Appl. Mycol. 1960, 39:318).

Kröber, H. 1970. Annual report for 1969. Fed. Biol. Inst. Agric. For. at Berlin and Brunswick. 144 pp. (In German). (Rev. Plant Pathol. 1971, 50:2077e).

Kröber, H. 1971. Phytophthora-welke an *Gypsophila paniculata* L. Erreger *Phytophthora cryptogea* Pethybr. et. Laff. (A crown rot of *Gypsophila paniculata* L. caused by *Phytophthora cryptogea* Pethyb. and Laff). Nachrichtenbl. Dtsch. Pflanzenschutzdienst. (Braunschweig) 23:74–77. (In German).

Kröber, H. 1979. Anfälligkeit von apfelsorten gegenüber *Phytophthora cactorum* und ihre Abhängigkeit von verschiedenen Faktoren (Susceptibility of apple cultivars to *Phytophthora cactorum* as influenced by different factors). Z. Pflanzenkr. Pflanzenschutz 86:1–11. (In German).

Kröber, H. 1980. Uberdauerung einiger *Phytophthora*—Arten im Boden (Survival of some *Phytophthora* species in soil). Z. Pflanzenkr. Pflanzenschutz 87:227–235. (In German).

Kröber, H. 1981. Vergleichende untersuchungen vom grundtyp *Phytophthora cryptogea* und *P. drechsleri* abweichender isolate (Comparative studies of isolates different from the common type of *Phytophthora cryptogea* and of *P. drechsleri.* Phytopathol. Z. 102:219–231. (In German).

Kröber, H. 1985. Erfahrungen mit *Phytophthora* de Bary und *Pythium* Pringsheim (Experiences with *Phytophthora* de Bary and *Pythium* Pringsheim). Mitt. Biol. Bundesanst. Land. Forstwirtsch. Berl.-Dahlem 225. 175 pp. (In German).

Kröber, H., and Beckmann, E. O. 1973. *Phytophthora cryptogea* Pethybr. et Laff. als Erreger einer wurzelfäule an Spinat und deren abhängigkeit vom bodenzustand und dem witterungsverlauf (Root rot of spinach caused by *Phytophthora cryptogea* Pethyb. & Laff. and its dependence upon the nature of the soil and the meteorological conditions). Phytopathol. Z. 78:160–169. (In German).

Kröber, H., and Marwitz, R. 1993. *Phytophthora tentaculata* sp. nov. und *Phytophthora cinnamomi* var. *parvispora* var. nov., zwei neue Pilz von Zierpflanzen in Deutschland (*Phytophthora tentaculata* sp. nov. and *Phytophthora cinnamomi* var. *parvispora* var. nov., two new fungi from ornamental plants in Germany). J. Plant Dis. Prot. 100:250–258.

Kröber, H., and Plate, H. P. 1973. *Phytophthora*-fäule an Saintpaulien (Erreger: *Phytophthora nicotianae* var. *parasitica* (Dast.) Waterh.) (*Phytophthora*-rot of saintpaulia caused by *Phytophthora nicotianae* var. *parasitica* [Dast.] Waterh.). Phytopathol. Z. 76:348–355. (In German).

Kröber, H., Kiewnick, L., and Richter, J. 1983. Phytophthora-Stammbasisfäule an *Dieffenbachia maculata* (Erreger: *P. mexicana* Hots. et Hart.) (Stem base rot of *Dieffenbachia maculata* [Cause: *P. mexicana* Hots. et Hart]). Phytopathol. Z. 107:244–249. (In German).

Kuan, T. L., and Erwin, D. C. 1980a. Formae speciales differentiation of *Phytophthora megasperma* isolates from soybean and alfalfa. Phytopathology 70:333–338.

Kuan, T. L., and Erwin, D. C. 1980b. Predisposition effect of water saturation of soil on Phytophthora root rot of alfalfa. Phytopathology 70:981–986.

Kuan, T. L., and Erwin, D. C. 1982. Effect of soil matric potential on Phytophthora root rot of alfalfa. Phytopathology 72:543–548.

Kuć, J. 1972. Phytoalexins. Annu. Rev. Phytopathol. 10:207–232.

Kuć, J. 1982. Phytoalexins from the Solanaceae. Pages 81–105 in: Phytoalexins. J. A. Bailey and J. W. Mansfield, eds. Blackie, London and Glasgow. 334 pp.

Kuć, J., and Preisig, C. 1984. Fungal regulation of disease resistance mechanisms in plants. Mycologia 76:767–784.

Kuć, J., Currier, W. W., and Shih, M. J. 1976. Terpenoid phytoalexins. Pages 225–237 in: Biochemical Aspects of Plant-Parasite Relationships. J. Friend and D. R. Threlfall, eds. Academic Press, London.

Kuć, J., Tjamos, E., and Bostock, R. 1984. Metabolic regulation of terpenoid accumulation and disease resistance in potato. Pages 103–126 in: Isopentenoids in Plants: Biochemistry and Function. W. D. Nes, G. Fuller, and L. S. Tsai, eds. Marcel Dekker, New York.

Kueh, T. K. 1976. Plant diseases in Sarawak, recorded during 1973 and 1974. Tech. Document, F.A.O. Plant Prot. Committee for S.E. Asia & Pacific Region 1976. 104. 9 pp. (Rev. Plant Pathol. 1977, 56:2927).

Kueh, T. K., and Khew, K. L. 1980. A screening technique useful in selecting for resistance in black pepper to *Phytophthora palmivora.* Malays. Agric. J. 52(4):37–45.

Kueh, T. K., and Khew, K. L. 1982. Survival of *Phytophthora palmivora* in soil and after passing through alimentary canals of snails. Plant Dis. 66:897–899.

Kueprakone, U., Saengkong, S., Pienpuck, K., and Choobumroong, W. 1986. *Phytophthora palmivora* (Butl.) Butl., the causal organism of black rot of mango seedlings. Thai Agric. Res. J. 4:67–73.

Kuhlman, E. G. 1964. Survival and pathogenicity of *Phytophthora cinnamomi* in several western Oregon Soils. For. Sci. 10:151–158.

Kuhn, D. N. 1991. Parasexuality in *Phytophthora*? Pages 242–255 in: *Phytophthora.* J. A. Lucas, R. C. Shattock, D. S. Shaw, and L. R. Cooke, eds. Cambridge Univ. Press, Cambridge, U,K. 447 pp.

Kühn, J. 1859. Die Krankheiten de Kulturgewächse, ihre Ursachen und ihre Verhütung (The diseases of cultivated plants, their causes, and their prevention). Gustav Bosselmann, Berl. 312 pp. (Cited in Tucker 1933; Schwinn and Margot 1991).

Kuhn, P. J., Albert, G., Wyness, L. E., Lee, S. A., and Lindsley, M. C. 1989a. Studies on the antifungal activity in vitro of dimethomorph, a novel fungicide active against downy mildews and *Phytophthora* (Abstr.). 9th International Symposium on Systemic Fungicides and Antifungal Compounds. Reinhardsbrunn, GDR. (Cited in Kühn et al. 1991).

Kuhn, P. J., Westwell, J. M., Albert, G., Lee, S. A., and Wyness, L. E. 1989b. Effects of dimethomorph on the growth and development of *Phytophthora infestans* in vitro and in vivo (Abstr.) International Symposium on *Phytophthora.* Dublin, Eire. (Cited in Kühn et al. 1991).

Kuhn, P. J., Pitt, D., Lee, S. A., Wakley, G., and Sheppard, A. N. 1991. Effects of dimethomorph on the morphology and ultrastructure of *Phytophthora.* Mycol. Res. 95:333–340.

Kulibaba, Y. F. 1972. Bolezni gerbery v zakrytom grunte (Diseases of gerbera under glass). Bull. Glavnogo Botanicheskogo Sada U.S.S.R. 83:115–119. (In Russian). (Rev. Plant Pathol. 1973, 52: 761).

Kulish, V. B. and D'yakov, Y. T. 1979. Isolation of diploid strains from mixed cultures of two physiological races of *Phytophthora infestans* [Mont.] de Bary. (Translated title). Dokl. Akad. Nauk. S.S.S.R. 244:689–691. (In Russian). (Rev. Plant Pathol. 1980, 59: 881.)

Kumararaj, K., and Bhide, V. P. 1962. 'Damping-off' of cashew nut *Anacardium occidentale* L.) seedlings caused by *Phytophthora palmivora* Butler in Maharashtra State. Curr. Sci. 31:23.

Kumari, P. S., Hegde, R. K., Basarkar, P. W., Srinivasan, C. N., Srikant-Kulkarni, Kulkarni, S. 1988. Role of enzymes in the gut region of the garden snail (*Cryptozona semirugata* Beck.) on the germination of oospores of *Phytophthora* spp. Indian J. Plant Pathol. 6:49–51.

Kumbhare, G B., and Moghe, P. G. 1976. Leaf fall disease of Nagpur oranges caused by *Phytophthora nicotianae* var. *parasitica* Waterhouse. Curr. Sci. 45:561–562.

Kunimoto, R. K., Aragaki, M., Hunter, J. E., and Ko, W. H. 1976. *Phytophthora capsici,* corrected name for the cause of Phytophthora blight of macadamia racemes. Phytopathology 66:546–548.

Kuske, C. R., and Benson, D. M. 1983a. Survival and splash dispersal of *Phytophthora parasitica,* causing dieback of rhododendron. Phytopathology 73:1188–1191.

Kuske, C. R., and Benson, D. M. 1983b. Overwintering and survival of *Phytophthora parasitica,* causing dieback of rhododendron. Phytopathology 73:1192–1196.

Kuske, C. R., Benson, D. M., and Jones, R. K. 1983. A gravel container base for control of *Phytophthora* dieback in rhododendron nurseries. Plant Dis. 67:1112–1113.

Labergerie, M. 1904. Le *Solanum* commersoni et ses variations en 1904 (Commercial *Solanum* and its variations in 1904). Soc. Nat. d'Agr. 64:811–825. (In French). (Cited in Tucker 1933).

Labrousse, F. 1931. La pourriture du coeur des palmiers en France (*Phytophthora palmivora* Butl.) (Heart rot of palms in France [*Phytophthora palmivora* Butl.]). Rev. Pathol. Veg. Entomol. Agric. Fr. 18:13–24. (In French). (Cited in Tucker 1933).

Lacey, J. 1965. The infectivity of soils containing *Phytophthora infestans.* Ann. Appl. Biol. 56:363–380.

Laemmli, U. K. 1970. Cleavage of structural proteins during the assembly of the head of bacteriophage T4. Nature 227:680–685.

Lafferty, H. A., and Pethybridge, G. H. 1922. On a *Phytophthora* parasitic on apples which has both amphigynous and paragynous antheridia; and on allied species which show the same phenomenon. Sci. Proc. R. Dublin Soc. 17:29–43.

Lagerheim, C. 1891. Las enfermedad de los pepinos, su causa y su curación (Diseases of cucumbers, causes and cure). Rev. Ecuat. 2, No. 24. (Abstr. in Z. Pflanzenkrankh. 2:161, 1892). (In Spanish). (Cited in Tucker 1933).

Lamb, R. C., and Wills, W. H. 1983. Root rot of Japanese umbrella pine, *Sciadopitys verticillata.* Plant Dis. 67:698.

Lamb, R. C., Aldwinckle, H. S., and Terry, P. E. 1985. 'Freedom' a disease resistant apple. HortScience 20:774–775.

Lambert, E. B., and Crandall, B. S. 1936. A seedling wilt of black locust caused by *Phytophthora parasitica.* J. Agric. Res. 53:467–476.

Lamboy, J. S., and Paxton, J. D. 1992. Metalaxyl sensitivity selection within *Phytophthora megasperma* f. sp. *glycinea.* Plant Dis. 76: 932–936.

Landaluze, P. U. 1941. Report on the work performed at the Coruña station of Agricultural Phytopathology during the year 1941. Publ. Estac. Fitopatol. Agric. Coruña 20. 48 pp. (Rev. Appl. Mycol. 1943, 22:52–53).

Langbein, H., and Pehl, P. 1962. Einfluss der Stickstoffversorgung auf den *Phytophthora*-befall der Kartoffeln (Influence of nitrogen concentration on *Phytophthora* infection of potato). Meded. Landbouwhogesch. Opzoekingsstn. Staat Gent 27:1186–1198. (In German).

Lange, L., and Olson, L. W. 1983. The fungal zoospore: Its structure and biological significance. Pages 2–41 in: Zoosporic Plant Pathogens: A Modern Perspective. S. T. Buczaki, ed. Academic Press, London. 352 pp.

Lapwood, D. H. 1961a. Potato haulm resistance to *Phytophthora infestans.* II. Lesion production and sporulation. Ann. Appl. Biol. 49:316–330.

Lapwood, D. H. 1961b. Potato haulm resistance to *Phytophthora infestans.* III. Lesion distrubution and leaf destruction. Ann. Appl. Biol. 49:704–716.

Lapwood, D. H. 1982. Field studies on late blight (*Phytophthora infestans*) of potato. Pages 90–93 in: Proc. Workshop on *Phytophthora* Diseases of Tropical Cultivated Plants. K. K. K. Nambiar, ed. Cent. Plant. Crops Res. Inst., Kasaragod, Kerala, India. 284 pp.

Large, E. C. 1940. The Advance of the Fungi. Jonathan Cape, London. 488 pp.

Larkin, R. P., Ristaino, J. B., and Campbell, C. L. 1995. Detection and quantification of *Phytophthora capsici* in soil. Phytopathology 85:1057–1063.

Larsson, M., and Gerhardson, B. 1990. Isolates of *Phytophthora cryptogea* pathogenic to wheat and some other crop plants. J. Phytopathol. 129:303–315.

Latin, R. X., Mackenzie, D. R., and Cole, H., Jr. 1981. The influence of host and pathogen genotypes on the apparent infection rates of potato late blight epidemics. Phytopathology 71:82–85.

Latorre, B. A., Alvarez, C., and Ribeiro, O. K. 1991. Phytophthora root rot of kiwifruit in Chile. Plant Dis. 75:949–952.

Laville, E. 1979. Utilisation d'un nouveau fongicide systemique: l'Aliette, dans la lutte contre la gommose a *Phytophthora* des agrumes (Use of a new systemic fungicide, Aliette, in the control of citrus gummosis due to *Phytophthora*). Fruits 34:35–41. (In French).

Laviola, C. 1968. Studies on the genetics of *Phytophthora infestans.* Ph.D. thesis, West Virginia University, Morgantown. 72 pp.

Laviola, C., and Gallegly, M. E. 1983. Genetic recombination and mode of inheritance of pathogenic characters by *Phytophthora infestans* through sexual reproduction. Pages 339–345 in: Durable Resistance in Crops. F. Lamberti, J. M. Waller, and N. A. Van der Graaff, eds. Plenum, New York. 454 pp.

Laviola, C., Gallegly, M. E., and Young, R. J. 1978. Genetics of *Phytophthora infestans.* I. Single-zoospore cultures from germsporangia and their significance in genetical studies. Phytopathol. Mediterr. 17:39–44.

Laviolette, F. A., and Athow, K. L. 1977. Three new physiologic races of *Phytophthora megasperma* var. *sojae.* Phytopathology 67: 267–268.

Laviolette, F. A., and Athow, K. L. 1981. Physiologic races of *Phytophthora megasperma* f. sp. *glycinea* in Indiana, 1973–1979. Plant Dis. 65:884–885.

Laviolette, F. A., and Athow, K. L. 1983. Two new physiologic races of *Phytophthora megasperma* f. sp. *glycinea.* Plant Dis. 67:497–498.

Laviolette, F. A., Athow, K. L., Mueller, E. H., and Wilcox, J. R. 1979. Inheritance of resistance in soybeans to physiologic races 5, 6, 7, 8, and 9 of *Phytophthora megasperma* var. *sojae.* Phytopathology 69:270–271.

Law, T. F., and Milholland, R. D. 1991. Production of sporangia and oospores of *Phytophthora fragariae* in roots of strawberry plants. Plant Dis. 75:475–478.

Law, T. F., and Milholland, R. D. 1992. Susceptibility of strawberry genotypes to infection and colonization by races of *Phytophthora fragariae* and the growth responses of inoculated genotypes. Plant Dis. 76(4):335–339.

Lawrence, J. S. 1978. Screening of cacao cultivars for resistance to *Phytophthora palmivora* in the collection at CATIE, Costa Rica. Rev. Theobroma (Brasil) 8(4):125–131.

Laycock, T. 1922. Mycology. Agric. Dep. South. Provs. Nigeria, Annu. Rep. 1920–21. Pages 23–25. (Cited in Tucker 1933).

Layton, A. C., and Kuhn, D. N. 1988a. Heterokaryon formation by protoplast fusion of drug-resistant mutants in *Phytophthora megasperma* f. sp. *glycinea.* Exp. Mycol. 12:180–194.

Layton, A. C., and Kuhn, D. N. 1988b. The virulence of interracial heterokaryons of *Phytophthora megasperma* f. sp. *glycinea.* Phytopathology 78:961–966.

Layton, A. C., and Kuhn, D. N. 1990. *In planta* formation of heterokaryons of *Phytophthora megasperma* f. sp. *glycinea.* Phytopathology 80:602–606.

Layton, A. C., Athow, K. L., and Laviolette, F. A. 1984. Inheritance of resistance to *Phytophthora megasperma* f. sp. *glycinea* in a soybean plant introduction. Plant Dis. 68:1080–1083.

Layton, A. C., Athow, K. L., and Laviolette, F. A. 1986. New physiologic race of *Phytophthora megasperma* f. sp. *glycinea.* Plant Dis. 70:500–501.

Lazarovits, G., Unwin, C. H., and Ward, E. W. B. 1980. Rapid assay of systemic fungicides against Phytophthora rot of soybeans. Plant Dis. 64:163–165.

Leach, S. S., and Rich, A. E. 1969. The possible role of parasexuality and cytoplasmic variation in race differentiation in *Phytophthora infestans.* Phytopathology 59:1360–1365.

Leal, J. A., and Gomez-Miranda, B. 1965. The effect of light and darkness on the germination of the oospores of certain species of *Phytophthora* on some synthetic media. Trans. Br. Mycol. Soc. 48:491–494.

Leal, J. A., Friend, J., and Holliday, P. 1964. A factor controlling sexual reproduction in *Phytophthora.* Nature 203:545–546.

Leal, J. A., Gallegly, M. E., and Lilly, V. G. 1967a. The relation of the carbon-nitrogen ratio in the basal medium to sexual reproduction in species of *Phytophthora.* Mycologia 59:953–964.

Leal, J. A., Lilly, V. G., and Gallegly, M. E. 1967b. The utilization of L-asparagine by two species of *Phytophthora.* Proc. W. Va. Acad. Sci. 39:95–101.

Leal, J. A., Lilly, V. G., and Gallegly, M. E. 1970. Some effects of ammonia on species of *Phytophthora.* Mycologia 62:1041–1056.

Leal, J. A., Gallegly, M. E., and Lilly, V. G. 1971. The value of 21 amino acids as nitrogen sources for *Phytophthora cactorum* and *P. heveae.* Can. J. Microbiol. 17:1319–1325.

Leary, J. V., Roheim, J. R., and Zentmyer, G. A. 1974. Ribosome content of various spore forms of *Phytophthora* spp. Phytopathology 64:404–409.

Leary, J. V., Zentmyer, G. A., Klure, L. J., Pond, E. C., and Grantham, G. L. 1982. Variability in growth of *Phytophthora cinnamomi* isolates in response to antibiotics. Phytopathology 72:750–754.

Leath, K. T., and Baylor, J. E. 1975. Phytophthora root rot of alfalfa in Pennsylvania. Plant Dis. Rep. 59:972–974.

Leath, K. T., Erwin, D. C., and Griffin, G. D. 1988. Diseases and nematodes. Pages 621–670 in: Alfalfa and Alfalfa Improvement. A. A. Hanson, D. K. Barnes, and R. R. Hill, Jr., eds. Agron. Monogr. 29. ASA-CSSA-SSSA, Madison, Wis. 1084 pp.

Leather, R. I. 1967. The occurrence of a Phytophthora root and leaf disease of black pepper in Jamaica. F.A.O. Plant Prot. Bull. 15: 15–16.

Lebert, H., and Cohn, F. 1870. Über die Fäule der Cactusstämme (On the rot of cactus stems). Beitr. Biol. Pflanz. 1:51–57. (In German). (Cited in Tucker 1933).

Lederer, W., and Seemüller, E. 1992. Untersuchungen zur Prädisposition der Erdbeere für die Rhizomefäule (*Phytophthora cactorum*) (Research on the predisposition of strawberry to root rot [*Phytophthora cactorum*]). J. Plant Dis. Prot. 99:225–233. (In German).

Lee, B. S. 1973. Sexuality of *Phytophthora palmivora* on *Piper nigrum* in Malaysia. MARDI Res. Bull. 1:29–32.

Lee, B. S., and Varghese, G. 1974. Studies on the genus *Phytophthora* in Malaysia. I. Isolation techniques, comparative morphology and physiology and reaction to antibiotics. Malays. Agric. Res. 3:13–21. (Rev. Plant Pathol. 54:790)

Lee, B. S., and Zentmyer, G. A. 1982. Influence of calcium nitrate and ammonium sulfate on Phytophthora root rot of *Persea indica.* Phytopathology 72:1558–1564.

Lee, H. A. 1922. Observations on previously unreported or noteworthy diseases in the Philippines. Philipp. Agric. Rev. 14:422–434. (Cited in Tucker 1933).

Lee, H. A. 1925. The comparative resistance to foot rot of various citrus species as root stocks. Philipp. J. Sci. 27:243–254. (Cited in Tucker 1933).

Lee, S. B., and Taylor, J. W. 1992. Phylogeny of five fungus-like protoctistan *Phytophthora* species, inferred from the internal transcribed spacers of ribosomal DNA. Mol. Biol. Evol. 9:636–653.

Lee, S. B, White, T. J., and Taylor, J. W. 1993. Detection of *Phytophthora* species by oligonucleotide hybridization to amplified ribosomal DNA spacers. Phytopathology 83:177–181.

Leeper, G. W. 1967. Introduction to Soil Science. Melbourne Univ. Press, Parkville, Victoria. 253 pp.

Legard, D. E., Lee, T. Y., and Fry, W. E. 1995. Pathogenic specialization in *Phytophthora infestans:* Aggressiveness on tomato. Phytopathology 85:1356–1361.

Legge, B. J. 1951. A Phytophthora crown rot of campanula. Trans. Br. Mycol. Soc. 34:293–303.

Le Grand-Pernot, F. 1986a. Quelques réflexions sur les sources de variations d'isolates A1 de *Phytophthora infestans* (Mont.) de Bary (Some reflections on sources of variability in A1 isolates of *Phytophthora infestans* [Mont.] de Bary). Agronomie 6:321–324. (In French).

Le Grand-Pernot, F. 1986b. Les isolats de *Phytophthora infestans* (Mont.) de Bary sont-ils naturellement hétérozygotes? (Are isolates of *Phytophthora infestans* [Mont.] de Bary heterozygous in nature?). Agronomie 6:877–884. (In French).

Le Grand-Pernot, F. 1988. Hétérozygotie des isolats de *Phytophthora infestans;* conséquences in dans l'apparition des races physiologiques (Natural heterozygosity of *Phytophthora infestans* [Mont.] de Bary and its consequences for the appearance of physiologic races). Agronomie 8:163–168. (In French).

Le Grand-Pernot, F., and Pellegrin, F. 1976. Nature du cycle végétatif des *Phytophthora parasitica* Dastur et *palmivora* (Butler) Butler (Nature of the vegetative cycle of *Phytophthora parasitica* Dastur and *palmivora* [Butler] Butler). Ann. Phytopathol. 8:379–388. (In French).

Lehman, W. F., Erwin, D. C., and Stanford, E. H. 1967. Root rot tolerance in new alfalfa strains now available to plant breeders. Calif. Agric. 21(4):6.

Lehman, W. F., Richards, S. J., Erwin, D. C., and Marsh, A. W. 1968. Effect of irrigation in treatments on alfalfa (*Medicago sativa* L.) production, persistence, and soil salinity in southern California. Hilgardia 39:277–295.

Lele, V. C., and Kapoor, J. N. 1982. Occurrence of *Phytophthora* on citrus in India (Abstr.). Pages 273–274 in: Proc. *Phytophthora* Diseases of Tropical Cultivated Plants. K. K. N. Nambiar, ed. Cent. Plant. Crops Res. Inst., Kasaragod, Kerala, India. 284 pp.

Leonian, L. H. 1922. Stem and fruit blight of peppers caused by *Phytophthora capsici* sp. nov. Phytopathology 12:401–408.

Leonian, L. H. 1925. Physiological studies on the genus *Phytophthora.* Am. J. Bot. 12:444–498.

Leonian, L. H. 1934. Identification of *Phytophthora* species. W. Va. Univ. Agric. Exp. Stn. Bull. 262. 36 pp.

Leonian, L. H., and Geer, H. L. 1929. Comparative value of the size of *Phytophthora* sporangia obtained under standard conditions. J. Agric. Res. 39:293–311.

Leonian, L. H., and Lilly, V. G. 1937. Partial purification of a vitamin-like substance which stimulates sexual reproduction in certain fungi. Am. J. Bot. 24:700–702.

Leonian, L. H., and Lilly, V. G. 1938. Studies on the nutrition of fungi. I. Thiamin, its constituents, and the source of nitrogen. Phytopathology 28:531–548.

Leonian, L. H., and Lilly, V. G. 1940. Studies on the nutrition of fungi. IV. Factors affecting the growth of some thiamin-requiring fungi. Am. J. Bot. 27:18–26.

Le Roux, H. F., Wehner, F. C., Kotzé, J. M., and Grech, N. M. 1991. Combining fosetyl-Al trunk injection or metalaxyl soil drenching with soil application of aldicarb for control of citrus decline. Plant Dis. 75:1233–1236.

Leu, L. S., and Kao, C. W. 1981. Pepperblight induced by *Phytophthora capsici.* Plant Prot. Bull. Taiwan 23:59–66. (In Chinese).

Leu, L. S., and Lee, C. C. 1976. Phytophthora blight of *Passiflora edulis* Sims. Plant Prot. Bull. (Taiwan) 18:286–292.

Levy, Y., Benderly, M., Cohen, Y., Gisi, U., and Bassand, D. 1986. The joint action of fungicides in mixtures: Comparison of two methods for synergy calculations. EPPO Bull. 16:651–657.

Levy, Y., Cohen, Y., and Benderly, M. 1991. Disease development and buildup of resistance to oxadixyl in potato crops inoculated with *Phytophthora infestans* as affected by oxadixyl and oxadixyl mixtures: Experimental and simulation studies. J. Phytopathol. 132:219–229.

Liang, P. Y. 1964. Identification of *Phytophthora* species causing cotton boll blight and castor bean blight in North China. Acta

Phytopathol. Sin. 7:11–20. (In Chinese). (Rev. Appl. Mycol. 1965, 44:1029).

Liew, E. C. Y, Maclean, D. J., Manners, J. M., Dawson, D., and Irwin, J. A. G. 1991. Use of restriction fragment length polymorphisms to study genetic relationships between Australian and Japanese isolates of *Phytophthora vignae.* Aust. J. Bot. 39:335–346.

Lilly, V. G. 1966. The effects of sterols and light on the production and germination of *Phytophthora* spores. Pages 259–271 in: The Fungus Spore. M. F. Madelin, ed. Colston Papers No. 18. Proc. 18th Symp. Colston Res. Soc. Univ. Bristol, Bristol, U.K. Butterworths, London. 337 pp.

Lim, T. M. 1982. A forecasting system for use in the chemical control of Phytophthora leaf fall on plantation rubber in Malaysia (Abstr.). Page 272 in: Proc. Workshop on *Phytophthora* diseases of tropical cultivated plants. K. K. N. Nambiar ed. Cent. Plant. Crops Res. Inst., Kasaragod, Kerala, India. 284 pp.

Lin, C. C. 1984. Phytophthora fruit rot of Indian jujube (*Zizyphus mauritiana* L.). Plant Prot. Bull. Taiwan 24:427–429.

Lin, C. K., and Liang, P. Y., 1965. Studies on nitrogen, calcium, and organic acid requirements with reference to pH relations in the nutrition of some species of *Phytophthora.* Acta Microbiol. Sin. 11:470–479. (In Chinese).

Lin, C. K., Hwang, H., Liang, P. V., and Hsu, T. Y. 1982. Nutritional studies on various species of *Phytophthora,* with special reference to *P. infestans.* Pages 313–322 in: The Collected Works of Dr. Ch'wan-Kwang Lin. Hsintu. Publ. Co., New York. 336 pp.

Linde, A. R., Stanghellini, M. E., and Matheron, M. E. 1990. Root rot of hydroponically grown lettuce caused by *Phytophthora cryptogea.* Plant Dis. 74:1037.

Linderman, R. G. 1985. Microbial interactions in the mycorrhizosphere. Pages 117–120 in: Proc. 6th North American Conf. Mycorrhizae. Bend, Oregon.

Linderman, R. G. 1987. Response of shade tree seedlings, VA mycorrhizal fungi, and *Pythium* to soil fumigation with vapam and methyl bromide. Page 30 in: Proc. VII N. Am. Conf. Mycorrhizae. D. M. Sylvia, L. L. Hung, and J. H. Graham, eds. Inst. Food Agric. Sci., Univ. Florida, Gainesville. 364 pp.

Linderman, R. G. 1989. Organic amendments and soil-borne diseases. Can. J. Plant Pathol. 11:180–183.

Linderman, R. G., and Ribeiro, O. K. 1991. Chemical and biological control of *Phytophthora* species in woody plants. Pages 399–410 in: *Phytophthora.* J. A. Lucas, R. C. Shattock, D. S. Shaw, and L. R. Cooke, eds. Cambridge Univ. Press, Cambridge, U.K. 447 pp.

Linderman, R. G., and Zeitoun, F. 1977. *Phytophthora cinnamomi* causing root rot and wilt of nursery-grown native western azalea and salal. Plant Dis. Rep. 61:1045–1048.

Linderman, R. G., Moore, L. W., Baker, K. F., and Cooksey, D. A. 1983. Strategies for detecting and characterizing systems for biological control of soilborne plant pathogens. Plant Dis. 67:1058–1064.

Lindfors, T., and Holmberg, C. H. 1941. Växtsjukdomar i Sverige 1933–1937. Skrifter utgivna av Statens Växtskydds-Anstalt 4, 1939–1940 (Plant diseases in Sweden). Medd. Växtskydds Anst. (Stockholm) 33:131. (In Swedish). (Cited in Nienhaus 1960).

Lindley, J. 1848. Notes on the wild potato. J. Hortic. Soc. (London) 3:65–72. (Cited in Tucker 1933).

Lindquist, J. C. 1938. Muerte de manzanas ocasionada por *Phytophthora cactorum* (Death of apples caused by *Phytophthora cactorum*). Rev. Argent. Agron. 21:195–199. (In Spanish).

Lippman, E., Erwin, D. C., and Bartnicki-Garcia, S. 1974. Isolation and chemical composition of oospore-oogonium walls of *Phytophthora megasperma* var. *sojae.* J. Gen. Microbiol. 80:131–141.

Liu, P. S. W. 1977. Diseases caused by *Phytophthora* and *Pythium* in Sabah, Malaysia. Tech. Bull. 3. 48 pp.

Liyanage, N. I. S. 1989. *Phytophthora citricola* on rubber in Sri Lanka. Plant Pathol. 38:438–439.

Liyanage, N. I. S., and Wheeler, B. E. J. 1989a. Comparative morphology of *Phytophthora* species on rubber. Plant Pathol. 38:592–597.

Liyanage, N. I. S., and Wheeler, B. E. J. 1989b. *Phytophthora katsurae* from cocoa. Plant Pathol. 38:627–629.

Liyange, N. I. S., and Wheeler, B. E. J. 1991. Survival of *Phytophthora meadii* in Sri Lankan soils. Plant Pathol. 40:436–444.

Lockwood, J. L., and Cohen, S. D. 1978. Race determination of *Phytophthora megasperma* var. *sojae* using differential soybean varieties inoculated with zoospores or incubated on flooded soil samples. Plant Dis. Rep. 62:1019–1021.

Lodewijks, J. A. 1909. Onderzoekingen over de *Phytophthora* der Vorstenlandsche tabak (Investigation about the *Phytophthora* of tobacco in Vorstenlanden). Meded. Alg. Proefstn. Salatiga, Java. Ser. 2. 23. 5 pp. (In Dutch). (Cited in Tucker 1931).

Loest, F. C. 1950. Orchard practices: Relation to collar rot of citrus. Farming S. Afr. 25:331–333.

Long, K., Fraser, L., Bacon, P., and Broadbent, P. 1977. The Benton citrange: A promising *Phytophthora* resistant rootstock for citrus trees. Proc. Int. Soc. Citric. 2:541–544.

Long, M., and Keen, N. T. 1977. Evidence for heterokaryosis in *Phytophthora megasperma* var. *sojae.* Phytopathology 67:670–674.

Long, M., Keen, N. T., Ribeiro, O. K., Leary, J. V., Erwin, D. C., and Zentmyer, G. A. 1975. *Phytophthora megasperma* var. *sojae*: Development of wild-type strains for genetic research. Phytopathology 65:592–597.

Lonsdale, D., Cunliffe, C., and Epton, H. A. S. 1980. Possible routes of entry of *Phytophthora erythroseptica* Pethyb. and its growth within potato plants. Phytopathol. Z. 97:109–117.

Lopatecki, L. E., and Newton, W. 1956. The nutrition of *Phytophthora.* Can. J. Bot. 34:751–757.

Lourd, M., and Keuli, S. D. 1975. Note sur un chancre a *Phytophthora* du manguier en Cote d'Ivoire (Note on a canker of mango caused by *Phytophthora* in the Ivory Coast). Fruits 30:541–544. (In French).

Louvel, D. 1975. Etude des relations entre l'ananas et le *Phytophthora parasitica:* Mise au point d'une technique d'inoculation, localisation des niveaux de la résistance (Study on the relationship of pineapple and *Phytophthora parasitica:* Development of an inoculation technique, localization of the level of resistance). Fruits 30:669–680. (In French).

Lowe, K. F., Gramshaw, D., Bowdler, T. M., and Lydke, D. H. 1985. Performance of North American and Australian lucernes in the Queensland subtropics. 2. Yield and plant survival in irrigated stands. Aust. J. Exp. Agric. 25:82–90.

Lowings, P. H., and Acha, I. G. 1959. Some factors affecting growth of *Phytophthora infestans* (Mont.) de Bary. I. *P. infestans* on living potato leaves. Trans. Br. Mycol. Soc. 42:491–501.

Lozano-T., Z. E., and Romero-C., S. 1984. Estudio taxonomico de aislamientos de *Phytophthora* patogenos de cacao (Taxonomic studies of *Phytophthora* isolates pathogenic to cacao). Agrociencia 56:175–182. (In Spanish).

Lu, N. S. J., Barnes, D. K., and Frosheiser, F. I. 1973. Inheritance of Phytophthora root rot resistance in alfalfa. Crop Sci. 9:714–717.

Lucas, G. B. 1965. Diseases of Tobacco. Scarecrow Press, New York. 778 pp.

Lucas, G. B. 1975. Black shank. Pages 115–141 in: Diseases of Tobacco. Biological Consulting Associates, Raleigh, N.C. 621 pp.

Lucas, J. A., Bower, L. A., and Coffey, M. D. 1990a. Fungicide resistance in soil-borne *Phytophthora* species. E.P.P.O. (Eur. Mediterr. Plant Prot. Organ.) Bull. 20:199–206.

Lucas, J. A., Greer, G., Oudemans, P. V., and Coffey, M. D. 1990b. Fungicide sensitivity in somatic hybrids of *Phytophthora capsici* obtained by protoplast fusion. Physiol. Mol. Plant Pathol. 36:175–187.

Lucas, J. A., Shattock, R. C., Shaw, D. S., and Cooke, L. R., eds. 1991. *Phytophthora.* Cambridge Univ. Press, Cambridge, U,K. 447 pp.

Lucas, R. 1977. New plant disease record—*Phytophthora erythroseptica* Pethy. wilt of cineraria (*Senecio cruentus* D.C.) cultivars. N.Z. J. Agric. Res. 20:253–254.

Lueschen, W. E., Barnes, D. K., Rabas, D. L., Frosheiser, F. I., and Smith, D. M. 1976. Field performance of alfalfa cultivars resistant and susceptible to Phytophthora root rot. Agron. J. 68:281–285.

Lukens, R. J., Cham, D. C. K., and Etter, G. 1978. Ortho 20615, a new systemic for the control of plant diseases caused by Oomycetes. Phytopathology News 12:142.

Luo, L., Ho, H. H., and Jong, S. C. 1988. Study on the physiological characteristics of the genus *Phytophthora.* Mycotaxon 32:199–217.

Lutz, A. L. 1987. Propagule survival and population fluctuations of

Phytophthora parasitica and control of root rot in irrigated citrus groves. Ph.D. dissertation, Univ. of Calif., Riverside. 187 pp.

Lutz, A., and Menge, J. A. 1986a. Breaking winter dormancy of *Phytophthora parasitica* progagules using heat shock. Mycologia 78:148–150.

Lutz, A., and Menge, J. A. 1986b. Citrus root health II: Phytophthora root rot. Citrograph 72:173–179.

Lutz, A., and Menge, J. A. 1991. Population fluctuations and the numbers and types of propagules of *Phytophthora parasitica* that occur in irrigated citrus groves. Plant Dis. 75:173–179.

Lutz, A., Menge, J. A., and O'Connell, N. 1986. Hardpans, claypans, and other mechanical impedences. Citrograph 71:57–61.

Lutz, A., Menge J. A., and Bender, G. 1989. Phytophthora root rot in citrus: Can it be controlled by manipulation of irrigation practices? Calif. Grower 13 (5):8–10.

Lutz, A., Menge, J. A., and Ferrin, D. M. 1991. Increased germination of propagules of *Phytophthora parasitica* by heating citrus soils sampled during winter. Phytopathology 81:865–872.

Lyne, R. L., Mulheirn, L. J., and Leworthy, D. P. 1976. New pterocarpinoid phytoalexins of soybean. Chem. Commun. J. Chem. Soc. Sect. D. 13:497–498.

Lyr, H., ed. 1995. Modern Selective Fungicides—Properties, Applications, Mechanisms of Action. 2nd ed. Gustav Fischer Verlag, Jena, Germany. 595 pp.

Maas, J. L. 1972. Growth and reproduction in culture of ten *Phytophthora fragariae* races. Mycopathol. Mycol. Appl. 48:323–334.

Maas, J. L., ed. 1984. Compendium of Strawberry Diseases. American Phytopathological Society, St. Paul, Minn. 159 pp.

MacDonald, J. D. 1982. Effect of salinity stress on the development of Phytophthora root rot of chrysanthemum. Phytopathology 72: 214–219.

MacDonald, J. D. 1984. Salinity effects on the susceptibility of chrysanthemum roots to *Phytophthora cryptogea.* Phytopathology 74:621–624.

MacDonald, J. D. 1991. Heat stress enhances Phytophthora root rot severity in container-grown chrysanthemums. J. Am. Soc. Hortic. Sci. 116:36–41.

MacDonald, J. D., and Duniway, J. M. 1978a. Influence of the matric and osmotic components of water potential on zoospore discharge in *Phytophthora.* Phytopathology 68:751–757.

MacDonald, J. D., and Duniway, J. M. 1978b. Temperature and water stress effects on sporangium viability and zoospore discharge in *Phytophthora cryptogea* and *P. megasperma.* Phytopathology 68: 1449–1455.

MacDonald, J. D., and Duniway, J. M. 1978c. Influence of soil texture and temperature on the motility of *Phytophthora cryptogea* and *P. megasperma* zoospores. Phytopathology 68:1627–1630.

MacDonald, J. D., and Duniway, J. M. 1979. Use of fluorescent antibodies to study the survival of *Phytophthora megasperma* and *P. cinnamomi* zoospores in soil. Phytopathology 69:436–441.

MacDonald, J. D., Swiecki, T. J., Blaker, N. S., and Shapiro, J. D. 1984. Effects of salinity stress on the development of Phytophthora root rots. Calif. Agric. 38(10):23–24.

MacDonald, J. D., Stites, J., and Kabashima, J. 1990. Comparison of serological and culture plate methods for detecting species of *Phytophthora, Pythium,* and *Rhizoctonia* in ornamental plants. Plant Dis. 74:655–659.

MacDonald, J. D., Banihashemi, Z., Mircetich, S. M., Browne, G., and Bolkan, L. 1992. Trunk and branch canker of pistachio caused by *Phytophthora* spp. (Abstr.). Phytopathology 82:1084.

MacHardy, W. E. 1979. A simplified, non-computerized program for forecasting potato late blight. Plant Dis. Rep. 63:21–25.

MacIntyre, D., and Elliot, C. G. 1974. Selection for growth-rate during asexual and sexual propagation in *Phytophthora cactorum.* Genet. Res. Camb. 24:295–309.

Mackay, A., Weste, G., and Sharpe, K. 1985. Survival of *Phytophthora cinnamomi* in buried eucalyptus roots. Phytopathol. Z. 114:214–223.

MacKenzie, D. R. 1981. Scheduling fungicide applications for potato late blight with blitecast. Plant Dis. 65:394–399.

MacKenzie, D. R., Elliott, V. J., Kidney, B. A., King, E. D., Royer, M. H., and Theberge, R. L. 1983. Application of modern approaches to the study of the epidemiology of diseases caused by *Phytophthora.* Pages 303–313 in: *Phytophthora:* Its Biology, Taxonomy, Ecology, and Pathology. D. C. Erwin, S. Bartnicki-Garcia, and P. H. Tsao, eds. American Phytopathological Society, St. Paul, Minn. 392 pp.

Mackie, W. W. 1930. *Sclerospora macrospora* in barley. Phytopathology 20:107.

MacNish, G. C. 1963. Hibiscus collar rot. J. Agric. West Aust. 4:519–523.

Madden, L. V., and Ellis, M. A. 1990. Effect of ground cover on splash dispersal of *Phytophthora cactorum* from strawberry fruits. J. Phytopathol. 129:170–174.

Madden, L. V., Ellis, M. A., Grove, G. G., Reynolds, K. M., and Wilson, L. L. 1991. Epidemiology and control of leather rot of strawberries. Plant Dis. 75:439–446.

Magnano di San Lio, G., and Davino, M. 1977. An isolate of *Phytophthora nicotianae* var. *nicotianae* pathogenic to citrus plants. Riv. Patol. Veg. Ser. IV. 13:85–93.

Magnano di San Lio, G., and Perrotta, G. 1983. Applicazioni del metado 'MPN' nella determinazione quantitativa delle popolazioni di specie di *Phytophthora* patogene per gli Agrumi (Quantitative determinations of citrus *Phytophthora* populations by the MPN method). Phytopathol. Mediterr. 22:181–187. (In Italian, English summary).

Magnano di San Lio, G., and Tuttobene, R. 1985. Un marciume bruno dei frutti di feijoa (A brown rot of feijoa fruit). Inf. Fitopatol. 35:43–47. (In Italian).

Magnano di San Lio, G., Pennisi, A. M., and Perrotta, G. 1984. Effect of some fungicides on population of citrus phytophthoras in soil. Proc. Int. Soc. Citric. 1:415–419.

Magnano di San Lio G., Cacciola, S. O., and Perrota, G. 1986a. Note critiche su alcuni isolati di *Phytophthora palmivora* rinvenuti in Italia (Isolates of *Phytophthora palmivora* from Italy). Phytopathol. Mediterr. 25:92–100. (In Italian).

Magnano di San Lio, G., Pennisi, A. M., and Tuttobene, R. 1986b. Phytophthora blight of ornamental citrus: Chemical control of soilborne inoculum. Pages 471–478 in: Proc. of the Experts Meeting. 1985. Integrated Pest Control in Citrus-Groves. R. Cavalloro and E. Di Martino, eds. A. A. Balkema, Rotterdam. 616 pp.

Magnano di San Lio, G., Tuttobene, R., and Pennisi, A. M. 1986c. Phytophthora blight—A destructive disease of ornamental citrus. Pages 279–286 in: Proc. of the Experts Meeting. 1985. Integrated Pest Control in Citrus-Groves. R. Cavalloro and E. Di Martino, eds. A. A. Balkema, Rotterdam. 616 pp.

Magnano di San Lio, G., Cacciola, S. O., Pennisi, A. M., and Tuttobene, R. 1987a. Applicazioni di elettroforesi nello studio di specie di *Phytophthora* patogene per gli agrumi (Electrophoresis as a tool for studying *Phytophthora* species pathogenous to citrus). Pages 661–667 in: Il Recente contributo della Ricerca allo Sviluppo dell'Agrumicoltura Italiana. C. Delfino, ed. Sassari, Italy. (In Italian). 786 pp.

Magnano di San Lio, G., Reforgiato, G., and Russo, F. 1987b. Inoculo di *Phytophthora* spp. in campi sperimentali di portinnesti degli agrumi (Inoculum of *Phytophthora* spp. in experimental plots of citrus root stocks). Pages 699–675 in: Il Recente Contributo della Ricerca allo Sviluppo dell'Agrumicoltura Italiana. C. Delfino, ed. Sasari, Italy. (In Italian). 786 pp.

Magnano di San Lio, G., Tirro, A., Cacciola, S. O., and Rapisarda, A. M. 1987c. Il marciume pedale della gerbera in Sicilia (Foot-rot of gerbera in Sicily). Dif. Piante 10:227–236. (In Italian).

Magnano di San Lio, G., Tuttobene, R., and Cacciola, S. O. 1987d. *Phytophthora palmivora:* Una nuova avversita della *Chamaedorea* (*Phytophthora palmivora:* A new pathogen on *Chamaedorea*). Inf. Fitopatol. 6:55–58. (In Italian).

Magnano di San Lio, G., Perrotta, G., Cacciola, S. O., and Tuttobene, R. 1988. Factors affecting soil populations of *Phytophthora* in citrus orchards. Pages 767–774 in: Proc. Sixth Int. Citrus Congr. R. Goren and K. Mendel, eds. Balaban Publishers, Philadelphia. 1851 pp.

Magnano di San Lio, G., Messina, F., Greco, G., and Perrotta, G. 1990. Effect of irrigation on the dynamics of *Phytophthora citrophthora* populations in soil of citrus orchards. EPPO Bull. 20: 83–89.

Magnano di San Lio, G., Belisario, A., and Cacciola, S. O. 1992. Il

Lentisco un nuovo ospite di *Phytophthora cryptogea* Pethyb. et Laff. (Lentisk, a new host of *Phytophthora cryptogea* Pethyb. and Laff.). Micol. Ital. 21:13–14. (In Italian).
Magnus, P. 1895. Über das Mycelium und der Parasitismus einer neuen *Sclerospora* (On the mycelium and the parasitism of a new *Sclerospora*). Art. Bot. Centralbl. 64. 111. (In German).
Maia, N., Venard, P., and Lavrut, F. 1976. Etude des divisions mitotiques et méiotiques du cycle de *Phytophthora capsici* (Mitotic and meiotic division in the cycle of *Phytophthora capsici*). Ann. Phytopathol. 8:141–146.
Main, C. E., and Gallegly, M. E. 1964. The disease cycle in relation to multigenic resistance of potato to late blight. Am. Potato J. 41:387–400.
Maklakova, G. F. 1958. *Phytophthora* on strawberries. Zashch. Rast. (Mosc.) 6:54. (Rev. Appl. Mycol. 1959, 38:324).
Malaguti, G. 1950. Phytophthora blight of safflower. Phytopathology 40:1154–1156.
Malaguti, G. 1951. Epifitias de 'tizon' en plantas citricas, causadas por *Phytophthora parasitica* ('Tizon' epidemic in citrus plants caused by *Phytophthora parasitica*). Agron. Trop. (Maracay) 1: 213–226. (In Spanish).
Malaguti, G. 1953. Una podredumbre del tallo del Ajonjoli (*Sesamum indicum* L.) causada por *Phytophthora* (Stem rot of sesame [*Sesamum indicum* L.] caused by *Phytophthora*). Agron. Trop. (Maracay) 3:201–204. (In Spanish).
Malaguti, G., and Pontis-Videla, R. E. 1950a. El tizon o 'mildew' del pimiento en Venezuela ('Blight' or 'mildew' of pimento pepper in Venezuela). Agric. Venez. 14:4–5. (In Spanish).
Malaguti, G., and Pontis-Videla, R. E. 1950b. *Phytophthora capsici* in Venezuela. Riv. Agric. Subtrop. Trop. 44:4–12. (In Italian). (Rev. Appl. Mycol. 1951, 30:212).
Malajczuk, N. 1979a. Biological suppression of *Phytophthora cinnamomi* in eucalyptus and avocados in Australia. Pages 635–652 in: Soil-Borne Plant Pathogens. B. Schippers and W. Gams, eds. Acadmic Press, London. 686 pp.
Malajczuk, N. 1979b. The microflora of unsuberized roots of *Eucalyptus calophylla* R. Br. and *Eucalyptus marginata* Donn. ex Sm. seedlings grown in soils suppressive and conducive to *Phytophthora cinnamomi* Rands. II. Mycorrhizal roots and associated microflora. Aust. J. Bot. 27:255–272.
Malajczuk, N. 1983. Microbial antagonism to *Phytophthora*. Pages 197–218 in: *Phytophthora:* Its Biology, Taxonomy, Ecology, and Pathology. D. C. Erwin, S. Bartnicki-Garcia, and P. H. Tsao, eds. American Phytopathological Society, St. Paul, Minn. 392 pp.
Malajczuk, N. 1985. Perspective on ectomycorrhizal-pathogen interaction. Pages 123–124 in: Proc. 6th North American Conf. Mycorrhizae. 471 pp.
Malajczuk, N. 1988. Interaction between *Phytophthora cinnamomi* zoospores and micro-organisms on non-mycorrhizal and ectomycorrhizal roots of *Eucalyptus marginata*. Trans. Br. Mycol. Soc. 375–382.
Malajczuk, N., and McComb, A. J. 1977. Root exudates from *Eucalyptus calophylla* R. Br. and *Eucalyptus marginata* Donn. ex Sm. seedlings and their effect on *Phytophthora cinnamomi* Rands. Aust. J. Bot. 25:501–514.
Malajczuk, N., and Theodorou, C. 1979. Influence of water potential on growth and cultural characteristics of *Phytophthora cinnamomi*. Trans. Br. Mycol. Soc. 72:15–18.
Malajczuk, N., McComb. A. J., and Parker, C. A. 1975. An immunofluorescence technique for detecting *Phytophthora cinnamomi* Rands. Aust. J. Bot. 23:289–309.
Malajczuk, N., McComb, A. J., and Parker, C. A. 1977. Infection by *Phytophthora cinnamomi* Rands on roots of *Eucalyptus calophylla* R. Br. and *Eucalyptus marginata* Donn. ex Sm. Aust. J. Bot. 25: 483–500.
Malajczuk, N., Bowen, G. D., and Greenhalgh, F. C. 1978. A combined fluorescent antibody and soil sieving technique to count chlamydospores of *Phytophthora cinnamomi* in soil. Soil Biol. Biochem. 10:437–438.
Malcolmson, J. F. 1969a. Races of *Phytophthora infestans* occurring in Great Britain. Trans. Br. Mycol. Soc. 53:417–423.
Malcolmson, J. F. 1969b. Factors involved in resistance to blight (*Phytophthora infestans* [Mont.] de Bary) in potatoes and assessment of resistance using detached leaves. Ann. Appl. Biol. 64: 461–468.
Malcolmson, J. F. 1976. Assessment of field resistance to blight (*Phytophthora infestans*) in potatoes. Trans. Br. Mycol. Soc. 67: 321–325.
Malcolmson, J. F. 1979. Isolation and identification of races of *Phytophthora infestans* from breeders' assessment plots. Trans. Br. Mycol. Soc. 73:155–188.
Malcolmson, J. F. 1985. *Phytophthora infestans* A2 compatibility type recorded in Great Britain. Trans. Br. Mycol. Soc. 85:531.
Malcolmson, J. F., and Black, W. 1966. New R genes in *Solanum demissum* Lindl. and their complementary races of *Phytophthora infestans* (Mont.) de Bary. Euphytica 15:199–203.
Malencon, G., and Delecluse, R. 1937. Champignons pathogenes observes au Maroc (Fungal pathogens observed in Morocco). Bull. Soc. Sci. Nat. Phys. Maroc. 17:132–144. (In French).
Maloney, K. E., Wilcox, W. F., and Sanford, J. C. 1993. Raised beds and metalaxyl for controlling Phytophthora root rot of raspberry. HortScience 28:1106–1108.
Maloy, O. C. 1972. Plant Diseases. Page 6 in: Extension Multilith 3540. Cooperative Extn. Serv. College of Agric. Wash. State Univ., Pullman. 48 pp.
Maloy, O. C., and Alexander, M. 1958. The most probable number method for estimating populations of plant pathogenic organisms in the soil. Phytopathology 48:126–128.
Mancini, G., Moretti, F., Capello, P. F., and de Punzio, F. 1977. Esperienza preliminare di lotta contro *Phytophthora capsici* (Leonian) (Preliminary control trials against *Phytophthora capsici* [Leonian]). Oss. Mal. Piante, Turin, Italy. 593–599. (In Italian). (Rev. Plant Pathol. 1977, 56:5353.)
Maniatis, T., Fritsch, E. F., and Sambrook, J. 1982. Molecular Cloning: A Laboratory Manual. Cold Spring Harbor Laboratory, Cold Spring Harbor, NY.
Manis, W. E. 1954. Phytophthora leaf fall and dieback. Plant Dis. Rep. Suppl. 225:49–52.
Mann, C. E. T. 1953. Investigation of the rubber industry in Papua and New Guinea. Papua and New Guinea Agric. Gaz. 8:40–56.
Manning, W. J., and Crossan, D. F. 1966. Effects of a particular soil bacterium on sporangial production in *Phytophtora cinnamomi* in liquid culture. Phytopathology 56:235–237.
Manotas, A. L. E. 1953. El contenido en azucares y ácidos orgánicos de la mazorca de cacao y su probable relacion con ataques de enfermedades (The sugar and organic acid content of cacao pods and its probable relation to disease attacks). Acta Agron. (Palmira) 3:177–188. (In Spanish).
Mansoori, B., and Banihashemi, Z. 1982. Evaluating cucurbit seedling resistance to *Phytophthora drechsleri*. Plant Dis. 66:373–376.
Marais, L. J., and Kotzé, J. M. 1976. Ectomycorrhizae of *Pinus patula* as biological deterrents to *Phytophthora cinnamomi*. S. Afr. For. J. 99:35–39.
Marais, P. G. 1979. Fungi associated with root rot in vineyards in the Western Cape, mostly *Phytophthora* and *Pythium*. Phytophylactica 11:65–68.
Marcelli, E. 1959. Un marciume del piede del Tabacco in semenzaio e in campo causato da *Phytophthora* sp. (A foot rot of tobacco in the seed bed and in the field caused by *Phytophthora* sp.). Tabacco (Roma) 63:235–257.
Marchal, E. 1908. Sur une maladie nouvelle du poirier (On a new disease of pear). Bull. Bot. Roy. Soc. Belg. 45:343–344. (In French). (Cited in Tucker 1933).
Marchal, E., and Verplancke, G. 1926. Champignons parasites nouveaux pour la flore Belge observés de 1919 á 1925 (Parasitic fungi new to the Belgian flora observed from 1919 to 1925). Bull. Roy. Soc. Bot. Belg. 59:19–25. (In French). (Rev. Appl. Mycol. 1927, 6:258.) (Cited in Tucker 1933).
Marchal, P., and Foex, E. 1932. Rapport phytopathologique pour l'annee 1931 (Phytopathological report for the year 1931). Ann. Epiphyt. (Paris) 18:1–53. (In French).
Marchionatto, J. B. 1938. Argentine Republic researches on the pepper diseases at Salta and Jujuy. Int. Bull. Plant Prot. 12:169–170.
Margot, P. 1983. Control of seed-borne diseases with metalaxyl. Seed Sci. Technol. 11:921–933.
Margot, P., Eckhardt, W., and Dahmen, H. 1988. CGA 80000: A new

phenylamide fungicide against soil-borne Peronosporales. Proc. Int. Br. Crop Prot. Conf. Pests Dis. 1988. Pages 527–534.

Margulis, L., and Schwartz, K. V. 1988. Five Kingdoms: An Illustrated Guide to the Phyla of Life on Earth. Freeman and Co. New York. 338 pp.

Margulis, L., Corliss, J. O., Melkonian, M., and Chapman, D. J., eds. 1990. Handbook of Protoctista. Jones and Bartlett Pub. Boston, Mass. 914 pp.

Marks, G. C. 1978. Review of the host-parasite relations of *Eucalyptus* spp. infected by *Phytophthora cinnamomi* Rands. Pages 28–42 in: *Phytophthora* and Forest Management in Australia. K. M. Old, ed. Rep. CSIROC Div. Forest Res. Canberra, Australia. 18–20 October 1978. 114 pp.

Marks, G. C., and Kassaby, F. Y. 1974. Detection of *Phytophthora cinnamomi* in soils. Aust. For. 36:198–203.

Marks, G. C., and Mitchell, J. E. 1970. Detection, isolation, and pathogenicity of *Phytophthora megasperma* from soils and stimulation of inoculum levels. Phytopathology 60:1687–1690.

Marks, G. C., and Mitchell, J. E. 1971a. Factors involved with the reaction of alfalfa to root rot caused by *Phytophthora megasperma*. Phytopathology 61:510–514.

Marks, G. C., and Mitchell, J. E. 1971b. Penetration and infection of alfalfa roots by *Phytophthora megasperma* and the pathological anatomy of infected roots. Can. J. Bot. 49:63–67.

Marks, G. C., and Smith, I. W. 1981. Factors influencing suppression of root rot of *Eucalyptus sieberi* and *E. obliqua* caused by *Phytophthora cinnamomi*. Aust. J. Bot. 29:483–495.

Marks, G. C., and Smith, I. W. 1990. Control of experimental *Phytophthora cinnamomi* stem infections of *Rhododendron, Leucadendron,* and *Eucalyptus* by dimethomorph, fosetyl-Al and metalaxyl. Aust. J. Exp. Agric. 30:139–143.

Marks, G. C., and Kassaby, F. Y., and Reynolds, S. T. 1972. Die-dack in mixed hardwood forests of eastern Victoria. A preliminary report. Aust. J. Bot. 20:141–154.

Marks, G. C., Kassaby, F. Y., and Fagg, P. C. 1973. Dieback tolerance in eucalypt species in relation to fertilization and soil populations of *Phytophthora cinnamomi*. Aust. J. Bot. 21:53–65.

Marks, G. C., Smith, I. W., and Kassaby, F. Y. 1981. Trunk infection of *Eucalyptus* species by *Phytophthora cinnamomi* Rands. Aust. Forest Res. 11:257–267.

Maronek, D. M., and Hendrix, J. W. 1978. Resistance to race O of *Phytophthora parasitica* var. *nicotianae* in tissue cultures of a tobacco breeding line with black shank resistance derived from *Nicotiana longiflora*. Phytopathology 68:233–234.

Marsh, A. W. 1978. Questions and answers about tensiometers. Div. Agric. Sci. Univ. Calif. Leafl. 2264. 12 pp.

Marshall, R. C., and Cole, A. L. J. 1987. Infection of *Camellia* species by *Phytophthora cinnamomi*. N.Z. Camellia Bull. 15(2):6–11 (See also Am. Camellia Soc. Yearb. 1988:104–108).

Marshall, R. P., and Waterman, A. M. 1948. Common diseases of important shade trees. U.S. Dep. Agric. Farmer's Bull. 1987. 53 pp.

Martin, G. H. 1927. Diseases of forest and shade trees, ornamental and miscellaneous plants in the United States in 1926. U.S. Dep. Agric. Plant Dis. Rep. Suppl. 37:349–446.

Martin, J. P., Baines, R. C., and Page, A. L. 1963. Observations on the occasional temporary growth inhibition of citrus seedlings following heat or fumigation treatment of soil. Soil Sci. 95:175–185.

Martinez, L. 1912. Diseases of cacao and means for their control. Bol. Dir. Gen. Agr. (Mexico) Rev. Agr. 2,6:520–532. (In Spanish). (Cited in Tucker 1933).

Martius, C. F. P. von. 1842. Die Kartoffel-epidemie der letzten Jahre oder die Stockfaule und Räude der Kartoffeln (The potato epidemic of the past year or potato stem rot and scab). Munich. (In German). (Cited in L. R. Jones et al. 1912).

Marudarajan, D. 1941. Observations on the production of sexual organs in paired cultures of *Phytophthora* species of the *Palmivora* group. Proc. Indian Acad. Sci. Sect. B. 14:384–389.

Marx, D. H. 1969. The influence of ectotrophic mycorrhizal fungi on the resistance of pine roots to pathogenic infections. I. Antagonism of mycorrhizal fungi to root pathogenic fungi and soil bacteria. Phytopathology 59:153–163.

Marx, D. H. 1970. The influence of ectotrophic mycorrhizzal fungi on the resistance of pine roots to pathogenic infections. V. Resistance of mycorrhizae to infection by vegetative mycelium of *Phytophthora cinnamomi*. Phytopathology 60:1472–1473.

Marx, D. H. 1972. Ectomycorrhizae as biological deterrents to pathogenic root infections. Annu. Rev. Phytopathol. 10:429–454.

Marx, D. H. 1973a. Mycorrhizae and feeder root diseases. Pages 351–382 in: Ectomycorrhizae, Their Ecology and Physiology. G. C. Marks and T. T. Kozlowski, eds., Academic Press, New York. 444 pp.

Marx, D. H. 1973b. Growth of ectomycorrhizal and nonmycorrhizal shortleaf pine seedlings in soil with *Phytophthora cinnamomi*. Phytopathology 63:18–23.

Marx, D. H., and Bryan, W. C. 1969. *Scleroderma bovista,* an ectotrophic mycorrhizal fungus of pecan. Phytopathology 59:1128–1132.

Marx, D. H., and Haasis, F. A. 1965. The induction of aseptic sporangial formation in *Phytophthora cinnamomi* by metabolic diffusates of soil microorganisms. Nature 206:673–674.

Masago, H., Yoshikawa, M., Fukada, M., and Nakanishi, N. 1977. Selective inhibition of *Pythium* spp. on a medium for direct isolation of *Phytophthora* spp. from soils and plants. Phytopathology 67:425–428.

Masago, H., Yoshikawa, M., Matsumoto, T., Katsumoto, M., Ogata, H., and Kishibe, Y. 1989. Differentiation and identification of *Phytophthora* species by two-dimensional electrophoresis. Ann. Phytopathol. Soc. Jpn. 55:336–343.

Massee, G. 1899. Cacao diseases in Trinidad. Kew Bull. Misc. Inf. 1899. Pages 1–6. (Cited in Tucker 1933).

Matheron, M. E., and Matejka, J. C. 1988a. In vitro activity of sodium tetrathiocarbonate on sporulation and growth of six *Phytophthora* species. Phytopathology 78:1234–1237.

Matheron, M. E., and Matejka, J. C. 1988b. Persistence of systemic activity for fungicides applied to citrus trunks to control Phytophthora gummosis. Plant Dis. 72:170–174.

Matheron, M. E., and Matejka, J. C. 1988c. Root and crown rot on nursery-grown mango trees caused by *Phytophthora palmivora*. Plant Dis. 74:530.

Matheron, M. E., and Matejka, J. C. 1989. Temporal changes in susceptibility of citrus phloem tissue to colonization by *Phytophthora citrophthora* and *P. parasitica*. Plant Dis. 73:408–411.

Matheron, M. E., and Matejka, J. C. 1990. Differential virulence of *Phytophthora parasitica* recovered from citrus and other plants to rough lemon and tomato. Plant Dis. 74:138–140.

Matheron, M. E., and Matejka, J. C. 1991. Effect of sodium tetrathiocarbonate, metalaxyl, and fosetyl-Al on development and control of Phytophthora root rot of citrus. Plant Dis. 75:264–268.

Matheron, M. E., and Mircetich, S. M. 1985a. Seasonal variation in susceptibility of *Juglans hindsii* and Paradox rootstocks of English walnut trees to *Phytophthora citricola*. Phytopathology 75:970–972.

Matheron, M. E., and Mircetich, S. M. 1985b. Influence of flooding duration on development of Phytophthora root and crown rot of *Juglans hindsii* and Paradox walnut rootstocks. Phytopathology 75:973–976.

Matheron, M. E., and Mircetich, S. M. 1985c. Pathogenicity and relative virulence of *Phytophthora* spp. from walnut and other plants to rootstocks of English walnut trees. Phytopathology 75:977–981.

Matheron, M. E., and Mircetich, S. M. 1985d. Relative resistance of different rootstocks of English walnut to six *Phytophthora* spp. that cause root and crown rot in orchard trees. Plant Dis. 69:1039–1041.

Matheron, M. E., and Mircetich, S. M. 1985e. Control of Phytophthora root and crown rot and trunk canker in walnut with metalaxyl and fosetyl-Al. Plant Dis. 69:1042–1043.

Matheron, M. E., Young, D. J., and Matejka, J. C. 1988. Phytophthora root and crown rot of apple trees in Arizona. Plant Dis. 72:481–484.

Matkin, O. A., and Chandler, P. A. 1957. The U.C.-type soil mixes. Pages 68–85 in: The UC System for Producing Healthy Container-Grown Plants. K. F. Baker, ed. Calif. Agric. Exp. Stn. Man. 23. 332 pp.

Matsumoto, N., and Araki, T. 1978. Alfalfa root rot caused by *Phy-*

tophthora megasperma Drechsler in Japan. Ann. Phytopathol. Soc. Jpn. 44:214–217.

Matsumoto, N., and Sato, T. 1979. *Phytophthora cryptogea* Pethyb. & Laff. found in alfalfa-field soil. Ann. Phytopathol. Soc. Jpn. 45:362–368.

Matsuoka, K., and Ansani, C. V. 1981. Ausencia de efeito estimulante a formacao de oosporos por *Phytophthora capsici* em decoctos de raizes de pimentao (*Capsicum annuum*) (Absence of stimulatory effect of root extracts from pepper [*Capsicum annuum*] on oospore formation by *Phytophthora capsici*). Fitopatol. Bras. 6:425–431. (In Portuguese).

Matsuoka, K. and Ansani, C. V. 1984. Formação de oosporos de *Phytophthora capsici* nos tecidos de pimentão (*Capsicum annuum*) (Formation of oospores by *Phytophthora capsici* in the roots and stems of sweet pepper [*Capsicum annuum*]). Fitopatol. Bras. 9:27–36. (In Portuguese).

Matsuoka, K., Casali, V. W. D., and Saraiva, T. R. C. B. 1984. Fontes de resistência a *Phytophthora capsici* em *Capsicum annuum* (Disease resistance in *Capsicum annuum* sources to *Phytophthora capsici*). Fitopatol. Bras. 9:193–201. (In Portuguese).

Matthews, E. M., Kroontje, W., and Henderson, R. G. 1960. Effect of length of rotation on losses from black shank in flue-cured tobacco varieties. Tobacco N.Y. 151:224–226.

Matz, J. 1920 Investigations of root disease of sugar cane. J. Dep. Agric. Puerto Rico. 4(1):28–40.

Maublanc, A. 1909. Les maladies des plantes cultivees dans les pays chauds. Maladies du cacaoyer (Diseases of plants in warm countries. Diseases of cocoa). L'Agric Prat. Pays Chauds 9:314–324. (In French).

Maublanc, A., and Barat, H. 1928. Une maladie nouvelle de la vanille (A new disease of vanilla). Agron. Colon. 17:77–82. (In French). (Rev. Appl. Mycol. 1928, 7:599–600).

Mazzani, B., Nava, C., Martinez, A., Layrisse, A., Rivas, N., and Malaguti, G. 1975. Incorporacion de la resistencia a *Phytophthora* y *Macrophomina* a la variedad de ajonjoli 'Aceitera' (Incorporation of resistance to *Phytophthora* and *Macrophomina* in the sesame variety 'Aceitera'). Agron. Trop. 25:11–21. (In Spanish).

McAlpine, D. 1910. Some points of practical importance in connection with the life history of *Phytophthora infestans* (Mont.) de Bary. Ann. Mycol. 8:156–166. (Cited in Tucker 1933).

McBlain, B. A., Zimmerly, M. M., Schmitthenner, A. F., and Hacker, J. K. 1991a. Tolerance to Phytophthora rot in soybean: I. Studies of the cross 'Ripley' × 'Harper.' Crop Sci. 31:1405–1411.

McBlain, B. A., Hacker, J. K., Zimmerly, M. M., and Schmitthenner, A. F. 1991b. Tolerance to Phytophthora rot in soybean: II. Evaluation of three tolerance screening methods. Crop Sci. 31:1412–1417.

McCain, A. H., and Scharpf, R. F. 1986. Phytophthora shoot blight and canker disease of *Abies* spp. Plant Dis. 70:1036–1037.

McCain, A. H., Holtzmann, O. V., and Trujillo, E. E. 1967. Concentration of *Phytophthora cinnamomi* chlamydospores by soil sieving. Phytopathology 57:1134–1135.

McCarter, S. M. 1965. Effect of certain soil environmental factors and host nutrition on black shank disease development in tobacco. Diss. Abstr. 26:6955.

McCarthy, G. J. P., and Teakle, D. S. 1959. A crown rot of strawberry caused by *Phytophthora parasitica*. Queensl. J. Agric. Sci. 16: 73–77.

McComb, J. A., Bennett, I. J., Crane, C., and Stukely, M. J. C. 1991. Selection and propagation of jarrah dieback resistance. Combined Proc. Int. Plant Progagators Soc. 40:86–90.

McCoy, J. 1991. The world food prize for UA adjunct professor. Ariz. Alumnus 68:11–13.

McCoy, R. E., Mitchell, D. J., and Subirats, F. J. 1976. A new disease of *Pilea* caused by *Phytophthora parasitica*. Plant Dis. Rep. 60: 680–681.

McCracken, A. R. 1985. The isolation of *Phytophthora* species from apple orchard soils. Rec. Agric. Res. 33:31–36.

McCredie, T. A., Kingsley, W. D., and Sivasithamparam, K. 1985. Variability in the resistance of *Banksia* L. f. species to *Phytophthora cinnamomi* Rands. Aust. J. Bot. 33:629–637.

McGee, D. C. 1993. Soybean Diseases: A Reference Source for Seed Technologists. American Phytopathological Society, St. Paul, Minnesota. 151 pp.

McGregor, A. J. 1984a. Comparison of cuprous oxide and metalaxyl with mixtures of these fungicides for the control of *Phytophthora* pod rot of cocoa. Plant Pathol. 33:81–87.

McGregor, A. J. 1984b. Control of seedling blight of cocoa. Papua New Guinea J. Agric. For. Fish. 33:1–2,39–50.

McGregor, A. J., and Moxon, J. E. 1985. Potential for biological control of tent building species of ants associated with *Phytophthora palmivora* pod rot of cocoa in Papua, New Guinea. Ann. Appl. Biol. 107:271–277.

Mchau, G. R. A., and Coffey, M. D. 1994a. Isozyme diversity in *Phytophthora palmivora:* Evidence for a southeast Asia centre of origin. Mycol. Res. 98:1035–1043.

Mchau, G. R. A., and Coffey, M. D. 1994b. An integrated study of morphological and isozyme patterns found within a worldwide collection of *Phytophthora citrophthora* and a redescription of the species. Mycol. Res. 98:1269–1299.

Mchau, G. R. A., and Coffey, M. D. 1995. Evidence for the existence of two distinct subpopulations in *Phytophthora capsici* and a redescription of the species. Mycol. Res. 99:89–102.

McIntosh, A. E. S. 1951. Annu. Rep. Dep. Agric. Malaya, 1949. 87 pp. (Rev. Appl. Mycol. 1951, 30:508–509).

McIntosh, D. L. 1953. A trunk and crown rot of sweet cherry in British Columbia. Phytopathology 43:402–403.

McIntosh, D. L. 1959. Collar rot of pear trees in British Columbia. Phytopathology 49:795–797.

McIntosh, D. L. 1960. The infection of pear rootlets by *Phytophthora cactorum*. Plant Dis. Rep. 44:262–264.

McIntosh, D. L. 1964. *Phytophthora* spp. in soils of the Okanagan and Similkameen valleys of British Columbia. Can. J. Bot. 42:1411–1415.

McIntosh, D. L. 1965. The susceptibility of rootlets of several pear species and varieties to infection by *Phytophthora cactorum*. Proc. Am. Soc. Hortic. Sci. 87:113–116.

McIntosh, D. L. 1966. The occurrence of *Phytophthora* spp. in irrigation systems in British Columbia. Can. J. Bot. 44:1591–1596.

McIntosh, D. L. 1968. Resistance to *Phytophthora cactorum* among open-pollinated seedlings of apple varieties and species. Proc. Am. Soc. Hortic. Sci. 93:71–76.

McIntosh, D. L. 1972. Effects of soil water suction, soil temperature, carbon and nitrogen amendments, and host rootlets on survival in soil of zoospores of *Phytophthora cactorum*. Can. J. Bot. 50:269–272.

McIntosh, D. L. 1977. *Phytophthora cactorum* propagule density levels in orchard soil. Plant Dis. Rep. 61:528–532.

McIntyre, J. L., and Hankin, L. 1977. Lack of kestose (fructosylsucrose) production by race 3 of *Phytophthora parasitica* var. *nicotianae* differentiates it from races 0 and 1. Mycologia 69:756–760.

McIntyre, J. L., and Lukens, R. J. 1977. Screening chemicals to control *Phytophthora parasitica* var. *nicotianae*. Plant Dis. Rep. 61: 366–370.

McIntyre, J. L., and Taylor, G. S. 1976. Screening tobacco seedlings for resistance to *Phytophthora parasitica* var. *nicotianae*. Phytopathology 66:70–73.

McIntyre, J. L., and Taylor, G. S. 1978. Race 3 of *Phytophthora parasitica* var. *nicotianae*. Phytopathology 68:35–38.

McIntyre, J. L., and Walton, G. S. 1981. Control of strawberry red stele caused by *Phytophthora fragariae*. Plant Dis. 65:835–836.

McKee, R. K. 1962. Identification of R-genes in *Solanum stoloniferum*. Euphytica 11:42–46.

McKee, R. K. 1964. Observations on infection by *Phytophthora infestans*. Trans. Br. Mycol. Soc. 47:365–374.

McKeen, W. E. 1955. Interaction product of glycine and dextrose toxic to *Phytophthora fragariae*. Science 123:509.

McKeen, W. E. 1958. Red stele root disease of the loganberry and strawberry caused by *Phytophthora fragariae*. Phytopathology 48: 129–132.

McKenzie, D. 1985. The influence of metalaxyl (Ridomil®) on Phytophthora root and collar rot of citrus in South Africa. Citrus J. (August):16–19.

McLennan, E. I. 1936. Notes on the organisms causing brown rot of citrus fruit in Victoria (*Phytophthora citrophthora* (Sm & Sm.) Leon. and *P. hibernalis* Carne.) Proc. R. Soc. Victoria 48:96–102.

McMorris, T. C. 1978. Sex hormones on the aquatic fungus *Achlya*. Lipids 13:716–722.

McNabb, S. A., and Klassen, G. R. 1988. Uniformity of mitochondrial DNA complexity in Oomycetes and the evolution of the inverted repeat. Exp. Mycol. 12:233–242.

McRae, W. 1918. *Phytophthora meadii* n. sp. on *Hevea brasiliensis*. Mem. Dep. Agric. India Bot. Ser. 9:219–273. (Cited in Tucker 1933).

McRae, W. 1932. Scientific reports of the Imper. Inst. Agric. Res. Pusa, 1930–31. Pages 73–86. (Rev. Appl. Mycol. 1932, 11:424–426.) (Cited in Tucker 1933).

McRae, W. 1933. Report of the imperial mycologist. Sci. Rep. Imper. Inst. Agric. Res., Pusa. 1931–32. Pages 122–140. (Rev. Appl. Mycol. 1934, 13:11–13.)

McRae, W. 1934. Foot-rot diseases of *Piper betle* L. in Bengal. Indian J. Agric. Sci. 4:585–617.

Mehrlich, F. P. 1932. *Pseudopythium phytophthoron,* a synonym of *Phytophthora cinnamomi*. Mycologia 24:453–454. (Cited in Tucker 1933).

Mehrlich, F. P. 1934. Control of *Phytophthora* heart rot of pineapple plants. Phytopathology 24:173–196.

Mehrlich, F. P. 1935. Nonsterile soil leachate stimulating to zoosporangia production by *Phytophthora* sp. Phytopathology 25:432–434.

Mehrlich, F. P. 1936. Pathogenicity and variation in *Phytophthora* species causing heart rot of pineapple plants. Phytopathology 26: 23–43.

Mehrotra, R. S. 1970. Techniques for demonstrating accumulation of zoospores of *Phytophthora* species on roots in soil. Can. J. Bot. 48:879–882.

Mehrotra, R. S. 1972. Behavior of zoospores of *Phytophthora megasperma* var. *sojae* and *P. drechsleri* in soil. Can. J. Bot. 50:2125–2130.

Mehrotra, R. S. 1982. *Piper betle* wilt caused by *Phytophthora parasitica* var. *piperina* Dastur. Pages 50–62 in: *Phytophthora* Diseases of Tropical Cultivated Plants. K. K. N. Nambiar, ed. Cent. Plant. Crops Res. Inst., Kasaragod, Kerala, India. 283 pp.

Mehrotra, R. S., and Prasad, G. 1982. Saprophytic survival and mating types of *Phytophthora nicotianae* var. *parasitica* causing fruit rot of guava (Abstr.). Page 273 in: Proc. Workshop on *Phytophthora* Diseases of Tropical Cultivated Plants. K. K. N. Nambiar, ed. Cent. Plant. Crops Res. Inst., Kasaragod, Kerala, India. 284 pp.

Mehrotra, R. S., and Tiwari, D. P. 1967. Studies on the root rot and leaf rot of *Piper betle*. I. Behaviour of *Phytophthora parasitica* var. *piperina* in soil. Indian Phytopathol. 20:161–167.

Mehrotra, R. S., and Tiwari, D. P. 1976. Organic amendments and control of foot rot of *Piper betle* caused by *Phytophthora parasitica* var. *piperina*. Ann. Microbiol. 127 A:415–421.

Melero-Vara, J. M., and Jimenez-Díaz, R. M. 1990. Etiology, incidence, and distribution of cotton seedling damping-off in southern Spain. Plant Dis. 74:597–600.

Melo, M. B. de. 1987. Sobrevivencia e localizacao de *Phytophthora capsici* Leonian em Sementes de Abobora (*Cucurbita maxima* Drech.) (Survival and histopathology of squash seed [*Cucurbita maxima* Drech.], infected by *Phytophthora capsici*). Thesis, Univ. Viçosa, Viçosa, M.G., Brasil. 46 pp. (In Portuguese).

Mendiola, N., and Espino, R. B. 1916. Some Phycomycetous diseases of cultivated plants in the Philippines. Philipp. Agric. For. 5:67–72. (Cited in Tucker 1933).

Menge, J. A. 1982. Effect of soil fumigants and fungicides on vesicular-orbuscular fungi. Phytopathology 72:1125–1132.

Menge, J. A. 1983. Utilization of vesicular-arbuscular mycorrhizal fungi in agriculture. Can. J. Bot. 61:1015–1024.

Menge, J. A. 1986. Use of systemic fungicides on citrus. Citrograph 71:245–250, 252.

Menge, J. A., Lembright, H., and Johnson, E. L. V. 1977. Utilization of mycorrhizal fungi in citrus nurseries. Proc. Int. Soc. Citric. 1: 129–132.

Menge, J. A., Labanauskas, C. K. Johnson, E. L. V., and Platt, R. G. 1978a. Partial substitution of mycorrhizal fungi for phosphorous fertilization in the greenhouse culture of citrus. Soil Sci. Soc. Am. J. 42:926–930.

Menge, J. A., Munnecke, D. E., Johnson, E. L. V., and Carnes, D. W. 1978b. Dosage response of the vesicular-arbuscular mycorrhizal fungi *Glomus fasciculatus* and *G. constrictus* to methyl bromide. Phytopathology 68:1368–1372.

Menge, J. A., Johnson, E. L. V., Pond, E., Ferrin, D., Liu, H., Lutz, A., Strother, M., Bartnicki, D., Afek, U., and Sjoerdsma, J. 1988. Distribution and frequency of *Phytophthora parasitica* and *P. citrophthora* associated with root rot of citrus in California (Abstr.). Phytopathology 78:1576.

Menge, J. A., Pehrson, J., Dodds, A., Strawn, A., Morse, J., Meyer, J., Arpaia, M. L., Pond, E., Hane, D., Embleton, T., Takele, E., Atkin, D., Coggins, C., Van Gundy, S., and Adams, C. 1990. Integrated crop management increases citrus growth and yields. Calif. Agric. 44:(5)11–12.

Menon, M. R., Sajoo, B. V., Ramakrishnan, C. K., and Rema Devi, L. 1972. A new *Phytophthora* disease of cardamom (*Elettaria cardamomum* [L.] Maton.). Curr. Sci. 41:231.

Menon, M. R., Sajoo, B. V., Ramakrishnan, C. K., and Rema Devi, L. 1973. Control of *Phytophthora* disease of cardamom. Agric. Res. J. Kerala 11:93–94.

Menyonga, J. M., and Tsao, P. H. 1966. Production of zoospore suspensions of *Phytophthora parasitica*. Phytopathology 56:359–360.

Menzies, J. D. 1963. The direct assay of plant pathogen populations in soil. Annu. Rev. Phytopathol. 1:127–142.

Mercado Mejia, J. A., and Bustamente, E. 1993. Evaluacion de cultivares criollos de chile dulce (*Capsicum annuum*) a *Phytophthora capsici* (Evaluation of resistance of cultivars of sweet pepper [*Capsicum annuum*] to *Phytophthora capsici*). Manejo Integrado de Plagas (Costa Rica) 27:5–10.

Merrill, E. D. 1929. Plant Pathology. Annu. Rep. Calif. Agric. Exp. Stn. for the year 1927–28. Pages 40–44.

Merwin, I. A., Wilcox, W. F., and Stiles, W. C. 1992. Influence of orchard ground cover management on the development of Phytophthora crown and root rots of apple. Plant Dis. 76:199–205.

Merz, W. G., Burrell, R. G., and Gallegly, M. E. 1969. A serological comparison of six heterothallic species of *Phytophthora*. Phytopathology 59:367–370.

Mes, M. G. 1934. A wilt of snapdragon (*Antirrhinum majus*) in South Africa. S. Afr. J. Sci. 31:281–287.

Messina, F., Cacciola, S. O., and Magnano di San Lio, G. 1989. *Phytophthora cactorum* (Leb. et Cohn.) Schroet., agente della necrosi del colletto e del rizoma della fragola in Sicilia (*Phytophthora cactorum* [Leb. et Cohn.] Schroet., the agent causing crown and root necrosis of strawberry in Sicily). Dif. Piante 12:175–180.

Meurs, A. 1934. Parasitic stemburn of Deli tobacco. Phytopathol. Z. 7:169–185.

Meyer, D. 1975. Zur Keimung der Oosporen von *Phytophthora cactorum* (Leb. et Cohn) Schroet. (Studies on the germination of the oospores of *Phytophthora cactorum* [Leb. et Cohn] Schroet.). Z. Pflanzenkr. Pflanzenschutz 82:168–172. (In German).

Meyer, D., and Schönbeck, F. 1975. Untersuchungen über die Entwicklung von *Phytophthora cactorum* (Leb. et Cohn) Schroet. im Boden (Investigations on the development of *Phytophthora cactorum* [Leb et. Cohn] Schroet. in soil). Z. Pflanzenkr. Pflanzenschutz 82:337–354. (In German).

Meyer, R., Parish, R. W., and Hohl, H. R. 1976. Hyphal tip growth in *Phytophthora*. Arch. Microbiol. 110:215–224.

Micales, J. A., Bonde, M. R., and Peterson, G. L. 1986. The use of isozyme analysis in fungal taxonomy and genetics. Mycotaxon 27:405–449.

Michaelides, S. C. 1985. A simulation model of the fungus *Phytophthora infestans* (Mont.) de Bary. Ecol. Model. 28:121–137.

Middleton, J. T. 1943a. The taxonomy, host range, and geographic distribution of the genus *Pythium*. Mem. Torrey Bot. Club 20(1). 171 pp.

Middleton, J. T. 1943b. Phytophthora rot of belladonna. Bull. Torrey Bot. Club 70:244–251.

Middleton, J. T. 1947. Root rot of condiment sage (Abstr.). Phytopathology 37:363.

Middleton, J. T., and Tucker, C. M. 1939. A disease of gloxinia caused by *Phytophthora cryptogea* (Abstr.). Phytopathology 29: 17–18.

Middleton, J. T., Tucker, C. M., and Tompkins. C. M. 1944. A disease of gloxinia caused by *Phytophthora cryptogea.* J. Agric. Res. 68:405–413.

Milat, M. L., Ducruet, J. M., Ricci, R., Marty, F., and Blein, J. P. 1991a. Physiological and structural changes in tobacco leaves treated with cryptogein, a proteinaceous elicitor from *Phytophthora cryptogea.* Phytopathology 81:1364–1368.

Milat, M. L., Ricci, P., Bonnet, P., and Blein, J. P. 1991b. Capsidiol and ethylene production by tobacco cells in response to cryptogein, an elicitor from *Phytophthora cryptogea.* Phytochemistry 30:2171–2173.

Milburn, M., and Gravatt. G. F. 1932. Preliminary note on a Phytophthora root disease of chestnut. Phytopathology 22:977–978. (Cited in Tucker 1933).

Miles, L. E., and Epps, J. M. 1942. The downy mildew disease of oats, caused by *Sclerospora macrospora.* Phytopathology 32:867–878.

Milgroom, M. G., and Fry, W. E. 1987. A metalaxyl bioassay for large numbers of small foliar samples. Phytopathology 77:237–240.

Milgroom, M. G., and Fry, W. E. 1988a. A model for the effects of metalaxyl on potato late blight epidemics. Phytopathology 78:559–565.

Milgroom, M. G., and Fry, W. E. 1988b. A simulation analysis of the epidemiological principles for fungicide resistance management in pathogen populations. Phytopathology 78:565–570.

Milgroom, M. G., McCulloch, C. E., and Fry, W. E. 1988. Distribution and temporal dynamics of metalaxyl in potato foliage. Phytopathology 78:555–559.

Milholland, R. D., Cline, W. O., and Daykin, M. E. 1989. Criteria for identifying pathogenic races of *Phytophthora fragariae* on selected strawberry genotypes. Phytopathology 79:535–538.

Millardet, A. 1885. The discovery of Bordeaux mixture. I. Treatment of mildew and rot. II. Treatment of mildew with copper sulfate and lime mixture. III. Concerning the history of the treatment of mildew with copper sulfate. Translated from French by F. J. Schneiderhan. Phytopathological Classics No. 3. American Phytopathological Society, St. Paul, Minnesota. 25 pp.

Miller, A. L., Shand, E., and Gow, N. A. R. 1988. Ion currents associated with root tips emerging laterals and induced wound sites in *Nicotiana tabacum:* Spatial relationship proposed between resulting electrical fields and phytophthoran zoospore infection. Plant Cell Environ. 11:21–25.

Miller, H. N., and Noegel, K. A. 1967. Phytophthora crown rot of petunia in Florida. Proc. Fla. State Hortic. Soc. 80:449–451.

Miller, P. A. 1940. Notes on diseases of ornamental plants in Southern California. Plant Dis. Rep. 24:219–222.

Miller, P. M. 1955. V-8 juice agar as a general purpose medium for fungi and bacteria. Phytopathology 45:461–462.

Miller, R. W. R. 1945. Annu. Rep. Dep. Agric. Tanganyika Territory. 1944. 9 pp. (Rev. Appl. Mycol. 1946, 25:100.)

Miller, S. A., and Maxwell, D. P. 1983. Evaluation of disease resistance. Chapter 31. In: Handbook of Plant Cell Culture. Vol. I. Techniques for Propagation and Breeding. D. A. Evans, W. R. Sharp, and P. V. Armirato, eds. Macmillan, New York. 970 pp.

Miller, S. A., and Maxwell, D. P. 1984a. Light microscope observations of susceptible, host resistant, and nonhost resistant interactions of alfalfa with *Phytophthora megasperma.* Can. J. Bot. 62: 109–116.

Miller, S. A., and Maxwell, D. P. 1984b. Ultrastructure of susceptible, host-resistant, and nonhost resistant interactions of alfalfa with *Phytophthora megasperma.* Can. J. Bot. 62:117–128.

Miller, S. A., and Martin, R. R. 1988. Molecular diagnosis of plant disease. Annu. Rev. Phytopathol. 26:409–432.

Miller, S. A., Korjagin, V., Miller, S. M., Petersen, F. P., Klopmeyer, M., Lankow, R. F., and Grothaus, G. D. 1989. Detection and quantitation of *Phytophthora megasperma* f. sp. *glycinea* in field soil by immunoassay (Abstr.). Phytopathology 79:1139.

Miller, S. M., Petersen, F. P., Miller, S. A., Rittenburg, J. H., Wood, S. C., and Grothaus, G. D. 1989. Development of a direct immunoassay to detect *Phytophthora megasperma* f. sp. *glycinea* in soil (Abstr.). Phytopathology 79:1139.

Mills, S. D., Förster, H., and Coffey, M. D. 1991. Taxonomic structure of *Phytophthora cryptogea* and *P. drechsleri* based on isozyme and mitochondrial DNA analysis. Mycol. Res. 95:31–48.

Milne, D. L., and Chamberlain, J. 1971. Experimental control of avocado Phytophthora root rot in South Africa. S. Afr. Citrus J. 449:15–16.

Milne, D. L., Kuhne, F. A., Brodrick, H. T., Logie, J. M., de Villiers, E. A., and Wood, R. 1975. Yellow grandilla outshines purple grandilla in yield and disease resistance. Citrus and Subtrop. Fruit J. 502:11, 12, 24. (In Hortic. Abstr. 46:8027.)

Minogue, K. P., and Fry, W. E. 1981. Effect of temperature, relative humidity, and rehydration rate on germination of dried sporangia of *Phytophthora infestans.* Phytopathology 71:1181–1184.

Mircetich, S. M. 1981. Phytophthora root rot of dediduous fruit trees in California. Proc. British Columbia Fruit Grower's Assoc. Ann. Hortic. Forum 13:1–14.

Mircetich, S. M. 1982. Phytophthora root rot and crown rot of apricot trees. Acta Hortic. 21:272–276.

Mircetich, S. M., and Browne, G. T. 1987. Phytophthora root and crown rot of deciduous fruit trees: Progress and problems in etiology, epidemiology and control. Pages 64–95 in: Proc. of the Summerland Res. Stn. Commemorative Symposium: Challenges and Opportunities in Fruit Production, Protection and Utilization Research. N. E. Looney, ed. Summerland, B.C., Canada. 111 pp.

Mircetich, S. M., and Keil, H. L. 1970. *Phytophthora cinnamomi* root rot and stem canker of peach trees. Phytopathology 60:1376–1382.

Mircetich, S. M., and Matheron, M. E. 1976. Phytophthora root and crown rot of cherry trees. Phytopathology 66:549–558.

Mircetich, S. M., and Matheron, M. E. 1981. Differential resistance of various cherry rootstocks to *Phytophthora* species (Abstr.). Phytopathology 71:243.

Mircetich, S. M., and Matheron, M. E. 1983. Phytophthora root and crown rot of walnut trees. Phytopathology 73:1481–1488.

Mircetich, S. M., and Zentmyer, G. A. 1966. Production of oospores and chlamydospores of *Phytophthora cinnamomi* in roots and soil. Phytopathology 56:1076–1078.

Mircetich, S. M., and Zentmyer, G. A. 1969. Effect of carbon and nitrogen compounds on germination of chlamydospores of *Phytophthora cinnamomi* in soil. Phytopathology 59:1732–1735.

Mircetich, S. M., and Zentmyer, G. A. 1970. Germination of chlamydospores of *Phytophthora.* Pages 112–115 in: Root Diseases and Soil-Borne Pathogens. T. A. Toussoun, R. V. Bega, and P. E. Nelson, eds. Univ. Calif. Press, Berkeley. 252 pp.

Mircetich, S. M., Zentmyer, G. A., and Kendrick, J. B., Jr. 1968. Physiology of germination of chlamydospores of *Phytophthora cinnamomi.* Phytopathology 58:666–671.

Mircetich, S. M., Matheron, M. E., and Schreader, W. R. 1974. Sweet cherry root rot and trunk canker caused by *Phytophthora* spp. (Abstr.). Proc. Am. Phytopathol. Soc. 1:58.

Mircetich, S. M., Matheron, M. E., and Tyler, R. H. 1976. Apple root rot and crown rot caused by *Phytophthora cambivora* and *Phytophthora* sp. (Abstr.). Proc. Am. Phytopathol. Soc. 3:114.

Mircetich, S. M., Campbell, R. N., and Matheron, M. E. 1977. Phytophthora trunk canker of coast live oak and cork oak trees in California. Plant Dis. Rep. 61:66–70.

Mircetich, S. M., Browne, G. T., Krueger, W., and Schreader, W. 1985. *Phytophthora* spp. isolated from surface water-irrigation sources in California (Abstr.). Phytopathology 75:1346–1347.

Mishra, A., Shivpuri, A., and Mahla, R. N. 1982. A new fruit rot of pomegranate caused by *Phytophthora nicotianae* var. *nicotianae* A2 and its control (Abstr.). Proc. Workshop on *Phytophthora* Diseases of Tropical Cultivated Plants. K. K. K. Nambiar, ed. Cent. Plant. Crops Res. Inst., Kasaragod, Kerala, India. 284 pp.

Mitchell, D. J. 1978. Relationships of inoculum levels of several soilborne species of *Phytophthora* and *Pythium* to infection of several hosts. Phytopathology 68:1754–1759.

Mitchell, D. J., and Kannwischer-Mitchell, E. 1983. Relationship of inoculum density of *Phytophthora* species to disease incidence in various hosts. Pages 259–269 in: *Phytophthora:* Its Biology, Taxonomy, Ecology, and Pathology. D. C. Erwin, S. Bartnicki-Garcia, and P. H. Tsao, eds. American Phytopathological Society, St. Paul, Minn. 392 pp.

Mitchell, D. J., and Zentmyer, G. A. 1971a. Effects of oxygen and carbon dioxide tensions on growth of several species of *Phytophthora.* Phytopathology 61:787–791.

Mitchell, D. J., and Zentmyer, G. A. 1971b. Effects of oxygen and carbon dioxide tensions on sporangium and oospore formation by *Phytophthora* spp. Phytopathology 61:807–812.

Mitchell, D. J., Kannwischer, M. E., and Moore, E. S. 1978a. Relationship of numbers of zoospores of *Phytophthora cryptogea* to infection and mortality of watercress. Phytopathology 68:1446–1448.

Mitchell, D. J., Strandberg, J. O., and Kannwischer, M. E. 1978b. Root and stem rot of watercress (*Nasturtium officinale*) caused by *Phytophthora cryptogea*. Plant Dis. Rep. 62:599–600.

Mitra, M. 1929a. *Phytophthora parasitica* Dast. causing 'damping-off' disease of cotton seedlings and 'fruit-rot' of Guava in India. Trans. Br. Mycol. Soc. 14:249–254. (Cited in Tucker 1933).

Mitra, M. 1929b. Some diseases of crops in the Andaman Islands, Agric. Res. Inst. Pusa, Bull. 195. 14 pp. (Cited in Tucker 1933).

Mohan, S. B. 1988. Evaluation of antisera raised against *Phytophthora fragariae* for detecting the red core disease of strawberries by enzyme-linked immunosorbent assay (ELISA). Plant Pathol. 47:206–216.

Mohan, S. B. 1989a. Cross-reactivity of antiserum raised against *Phytophthora fragariae* with other *Phytophthora* species and its evaluation as a genus-detecting antiserum. Plant Pathol. 38:352–363.

Mohan, S. B. 1989b. Analysis of cross-reactive antigens of *Phytophthora fragariae* and strawberry and their relation to resistance and susceptibility. J. Phytopathol. 127:316–330.

Moller, E. M., DeCock, A. W. A. M., and Prell, H. H. 1993. Mitochondrial and nuclear DNA restriction enzyme analysis of the closely related *Phytophthora* species *P. infestans, P. mirabilis,* and *P. phaseoli*. J. Phytopathol. 139:309–321.

Molot, P. M., and Beyries, A. 1976. Contamination in vitro de la fraise et de la tomate immature par diverses especes de *Phytophthora* (Artificial inoculation of unripe strawberry and tomato fruits with several species of *Phytophthora*). Ann. Phytopathol. 8:91–94. (In French).

Moniz da Maia, R. 1925. Um novo Oomycete parasita de árvores do género *Citrus*? (A new Oomycete parasite on trees of genus *Citrus*?). Rev. Agron. (Lisb.) Serv. 41. 11 pp. (In Portuguese). (Cited in Tucker 1933).

Montagne, J. F. C. 1845. Note sur la maladie qui ravage les pommes de terre et caracteres du *Botrytis infestans* (Note on the disease that ravages potatoes and is characterized as *Botrytis infestans*). Mem. Inst. Fr. 609:98–101, 312–313. (In French).

Montgomerie, I. G. 1961. Mycological investigations. Ann. Rep. Scottish Hort. Res. Inst. 8:69–70.

Montgomerie, I. G. 1967. Pathogenicity of British isolates of *Phytophthora fragariae* and their relationship with American and Canadian races. Trans. Br. Mycol. Soc. 50:57–67.

Montgomerie, I. G. 1977. Red core disease of strawberry. Hortic. Rev. No. 5. Scottish Hortic. Res. Inst., Invergowrie, Dundee, U.K. 47 pp.

Montgomerie, I. G., and Kennedy, D. M. 1975. Races of *Phytophthora fragariae* from Essex, Hampshire and Kent. Plant Pathol. 24: 191–193.

Montgomerie, I. G., and Kennedy, D. M. 1977. A new systemic fungicide to control red core of strawberry. Plant Pathol. 26:139–143.

Montgomerie, I. G., and Kennedy, D. M. 1980. The pathogenicity of *Phytophthora* species to red raspberry. Acta Hortic. 112:167–176.

Montgomerie, I. G., and Kennedy, D. M. 1983. An improved method of isolating *Phytophthora fragariae*. Trans. Br. Mycol. Soc. 80: 178–183.

Moore, J. N., Scott, D. H., and Converse, R. H. 1964. Pathogenicity of *Phytophthora fragariae* to certain *Potentilla* species. Phytopathology 54:173–176.

Moore, L. D., and Orcutt, D. M. 1982. Free sterol and total lipids in stems of susceptible and resistant tobacco cultivars colonized by *Phytophthora parasitica* var. *nicotianae*. Phytopathology 72:1048–1051.

Moore, W. C. 1945. New and interesting plant diseases. Trans. Br. Mycol. Soc. 28:3–4.

Moore, W. C. 1948. Report on fungus, bacterial and other diseases of crops in England and Wales for the years 1943–1946. Minist. Agric. Fish. Bull. U.K. 139:73–74, 82–83.

Moore, W. C. 1959. British Parasitic Fungi. Cambridge Univ. Press, London. 430 pp.

Moots, C. K., Nickell, C. D., Gray, L. E., and Lim, S. M. 1983. Reaction of soybean cultivars to 14 races of *Phytophthora megasperma* f. sp. *glycinea*. Plant Dis. 67:764–767.

Moots, C. K., Nickell, C. D., and Gray, L. E. 1988. Effects of soil compaction on the incidence of *Phytophthora megasperma* f. sp. *glycinea* in soybean. Plant Dis. 72:896–900.

More, W. D., Bangar, G., and Khetmalas, M. B. 1989. A new fruit rot of pomegranate in Maharashtra. J. Maharashtra Univ. 14:386.

Moreau, C., and Moreau, M. 1954. Le deperissement des Agrumes en Cote d'Ivoire (The wilting of citrus in the Ivory Coast). Rev. Mycol. 19, Suppl. Colon. 2:55–57. (Rev. Appl. Mycol. 1956, 35:447).

Moreau, C., and Moreau, M. 1957. Un cas de fletrissement des cineraires (A case of cineraria wilt). Bull. Soc. Mycol. Fr. 73:94–96. (In French).

Moreau, C., and Moreau, M. 1960. Pourritures brunes des Agrumes en entrepots liees a l'etat sanitaire des plantations (Brown rot of citrus in storage related to the sanitary condition of the plantations). Fruits 15:478–480.

Morgan, F. L. 1964. Infection of cranesbill by the soybean *Phytophthora*. Plant Dis. Rep. 48:140–141.

Morgan, F. L., and Hartwig, E. E. 1965. Physiologic specialization in *Phytophthora megasperma* var. *sojae*. Phytopathology 55:1277–1279.

Morgan, F. L., and Johnson, H. W. 1965. Phytophthora root and crown rot of vetch. Plant Dis. Rep. 49:84–88.

Morita, H. 1965. Red stele root disease of strawberry caused by *Phytophthora fragariae*. I. Identification of the causal fungus. Ann. Phytopathol. Soc. Jpn. 30:239–245. (In Japanese, English summary).

Morris, B. M., and Gow, N. A. R. 1993. Mechanisms of electrotaxis of zoospores of phytopathogenic fungi. Phytopathology 83:877–882.

Morris, P. F., and Ward, E. W. B. 1992. Chemoattraction of zoospores of the soybean pathogen, *Phytophthora sojae*, by isoflavones. Physiol. Mol. Plant Pathol. 40:17–22.

Morris, P. F., Savard, M. E., and Ward, E. W. B. 1991. Identification and accumlation of isoflavonoids and isoflavone glucosides in soybean leaves and hypocotyls in resistance responses to *Phytophthora megasperma* f. sp. *glycinea*. Physiol. Mol. Plant Pathol. 39:229–244.

Morrow, J. J., and Martin, J. B. 1977. Effective measurement of chlorine residual. Effluent Water Treat. J. 17:238–242.

Mortimer, A. M., and Shaw, D. S. 1975. Cytofluorimetric evidence for meiosis in gametangial nuclei of *Phytophthora drechsleri*. Genet. Res. Camb. 25:201–205.

Mortimer, A. M., Shaw, D. S., and Sansome, E. R. 1977. Genetical studies of secondary homothallism in *Phytophthora drechsleri*. Arch. Microbiol. 111:255–259.

Morton, D. J., and Dukes, P. D. 1967. Serological differentiation of *Pythium aphanidermatum* from *Phytophthora parasitica* var. *nicotianae* and *Phytophthora parasitica*. Nature 213:923.

Morton, H. V., Kern, C. L., and Taylor, T. D. 1982. Control of Pythium damping-off and Phytophthora root rot of soybeans with metalaxyl (Abstr.). Phytopathology 72:971.

Morwood, R. B. 1956. Notes on plant diseases listed for Fiji. Agric. J. (Fiji) 27:83–86.

Mridha, M. A. U., and Alamgir, S. M. 1991. Occurrence of foot and leaf rot disease of betelvine (*Piper betle* L.) caused by *Phytophthora parasitica* Dastur at Chittagong, Bangladesh. Bangladesh J. Bot. 20:91–92.

Mu, L., and Tsao, P. H. 1987. Identities of *Phytophthora* isolates causing vanilla blight and root rot in French Polynesia (Abstr.). Phytopathology 77:1704.

Muchovej, J. J., Maffia, L. A., and Muchovej, R. M. C. 1980. Effect of exchangeable soil aluminum and alkaline calcium salts on the pathogenicity and growth of *Phytophthora capsici* from green pepper. Phytopathology 70:1212–1214.

Mueller, S. C., and Fick, G. 1987. Response of susceptible and resistant alfalfa cultivars to Phytophthora root rot in the absence of measurable flooding damage. Agron. J. 79:201–204.

Mukherjee, N. 1959. Root and stem rot of 'Roselle' caused by a strain of *Phytophthora parasitica* Dast. Proc. 46th Indian Sci. Congr. III:501.

Mukherjee, N. 1974. Antifungal activities of some oils and detergents. I. Actions on spore germination, growth and sporulation of some plant pathogenic fungi. Z. Pflanzenkr. Pflanzenschutz 81: 468–471.

Mukherjee, N. 1976. Antifungal activities of some oils and detergents. II. Effect on germ tube growth, sclerotial germination and growth of sclerotial plant pathogens. Z. Pflanzenkr. Pflanzenschutz 83:305–308.

Mukherjee, N., and Basak, M. 1973. Chemotherapeutic control of foot and stem rot of roselle (*Hibiscus sabdariffa* var. *altissima*) caused by *Phytophthora parasitica* var. *sabdariffae*. Indian Phytopathol. 26:576–577.

Mukherjee, N., and Roy, A. B. 1962. Microbial influence on the formation of oospores in culture by *Phytophthora parasitica* var. *sabdariffae*. Phytopathology 52:583–584.

Muller, A. S. 1934. Brazil: Lista preliminar das doencas de plantas de estado de Minas Gerais (Brazil: Preliminary list of diseases of plants in the state of Minas Gerais). Int. Bull. Plant Prot. 8:193–198. (In Portuguese).

Muller, A. S. 1936. Brazil: Novos relatos de doencas de plantas no estado de Minas Gerais (Brazil: Some new records of plant diseases in the state of Minas Gerais). Int. Bull. Plant Prot. 10:98–99. (In Portuguese).

Muller, H. R. A. 1936. Het *Phytophthora*-voetrot van peper (*Piper nigrum* L.) in Nederlandsche-Indie (The Phytophthora foot rot of black pepper [*Piper nigrum* L.] in the Netherlandish Indies). Meded. Inst. Plantenziekten., Batavia 88. 73 pp. (In Dutch). (Rev. Appl. Mycol. 1937, 16:559).

Muller, H. R. A., and Van Eek, T. 1939. Aanteekeningen over eenige ziekten van Roselle en Java Jute op Java (Notes on some diseases of Roselle and Java jute in Java). Meded. Alg. Proefst. Landb. Batavia 32. 21 pp. (In Dutch). (Rev. Appl. Mycol. 1940, 19:219–221).

Müller, K. O. 1928. Untersuchungen über die Kartoffelkrautfaule und die Biologie ihres Erregers (Investigations on potato blight and the biology of its causal organism). Arb. Biol. Reichsanst. Land. Forstwirtsch. (Berlin) 16:197–211 (In German).

Müller, K. O. 1953. The nature of resistance of the potato plant to blight *Phytophthora infestans*. Nat. Inst. Agric. Bot. J. 6:346–360.

Müller, K. O., and Borger, H. 1940. Experimentelle Untersuchungen über die *Phytophthora*-Resistenz der Kartoffel (Experimental investigations of the *Phytophthora* resistance of potato). Arb. Biol. Reinchsanst. Land. Forstwirtsch. (Berlin) 23:189–231 (In German).

Müller-Thurgau, H., and Osterwalder, A. 1924. Bericht der Schweizerischen Versuchsanstalt für Obst-, Wein- und Gartenbau in Wädenswil für die Jahre 1921–1923. III. Pflanzenphysiologische und Pflanzenpathologische Abteilung (Report of the Swiss research institute for fruit horticulture, viticulture, and garden horticulture for the years 1921–1923. III. Plant physiology and plant pathology). Landw. Jahrb. Schweiz 38:573–581. (In German). (Cited in Tucker 1933).

Müller-Thurgau, H., Osterwalder, A., and Jegen, G. 1922. Bericht der Schweizerischen Versuchsanstalt für Obst-, Wein-, und Gartenbau im Wädenswil für die Jahre 1917–1920 (Report of the Swiss research institute for fruit horticulture, viticulture, and garden horticulture in Wadenswil for the years 1917–1920). Landw. Jahrb. Schweiz. 36:774–784. (In German). (In Nienhaus 1960; Tucker 1933).

Munn, C. B., and Drysdale, R. B. 1975. Kievitone production and phenylalanine ammonia-lyase activity in cowpea. Phytochemistry 14:1303–1307.

Munnecke, D. E. 1967. Fungicides in the soil environment. Pages 509–559 in: Fungicides. Vol. 1. D. C. Torgeson, ed. Academic Press, New York. 697 pp.

Munnecke, D. E. 1972. Factors affecting the efficacy of fungicides in soil. Annu. Rev. Phytopathol. 10:375–397.

Munnecke, D. E. 1984. Establishment of micro-organisms in fumigated avocado soil to attempt to prevent reinvasion of the soils by *Phytophthora cinnamomi*. Trans. Br. Mycol. Soc. 83:287–294.

Munnecke, D. E., and Bricker, J. L. 1976. Phytophthora root rot of *Pinus radiata* in Christmas tree plantings. Plant Dis. Rep. 60:928–932.

Munnecke, D. E., and Ferguson, J. 1953. Methyl bromide for nursery soil fumigation. Phytopathology 43:375–377.

Munnecke, D. E., Domsch, K. H., and Eckert, J. W. 1962. Fungicidal activity of air passed through columns of soil treated with fungicides. Phytopathology 52:1298–1306.

Munnecke, D. E., Bricker, J. L., and Kolbezen, M. J. 1974. Dosage response of *Phytophthora cinnamomi* to methyl bromide. Phytopathology 64:1007–1009.

Munnecke, D. E., Bricker, J. L., and Kolbezen, M. J. 1978. Comparative toxicity of gaseous methyl bromide to ten soilborne phytopathogenic fungi. Phytopathology 68:1210–1216.

Munro, M. C. D. 1945. A root rot of *Cineraria,* and a study of the species of *Phytophthora* concerned. Trans. Br. Mycol. Soc. 28: 115–126.

Munter, J. 1849. Beitrag zu dem *Solanum utile* Klotzsch (Contribution to *Solanum utile* Klotzsch). Allg. Gartenzeit. 7:356. (In German). (In Tucker 1933).

Murch, R. S., and Paxton, J. D. 1979. Rhizosphere anaerobiosis and glyceollin accumulation in soybean. Phytopathol. Z. 96:91–94.

Murch, R. S., and Paxton, J. D. 1980. Rhizosphere salinity and phytoalexin accumulation in soybean. Plant Soil 54:163–167.

Murphy, P. A. 1927. The production of resting spores of *Phytophthora infestans* on potato tubers. Sci. Proc. R. Dublin Soc. 18:407–412.

Murray, D. I. L. 1987. Rhizosphere microorganisms from the jarrah forest of Western Australia and their effects on vegetative growth and sporulation in *Phytophthora cinnamomi* Rands. Aust. J. Bot. 35:567–580.

Mussell, H. 1980. Tolerance to disease. Pages 39–52 in: Plant Disease. Vol. V. How Plants Defend Themselves. J. G. Horsfall and E. B. Cowling, eds. Academic Press, New York. 534 pp.

Mussell, H., and Fay, F. 1973. Aseptic zoospore production by *Phytophthora fragariae*. Phytopathology 63:1081–1082.

Mussell, H. W., and Staples, R. C. 1971. Phytoalexin-like compounds apparently involved in strawberry resistance to *Phytophthora fragariae*. Phytopathology 61:515–517.

Musumeci, M. R., and Olvieira, A. R. 1975. Accumulation of phenols and phytoalexins in citrus tissues inoculated with *P. citrophthora* (Sm. & Sm.) Leonian. Summa Phytopathol. 1:275–282.

Musumeci, M. R., Feichtenberger, E., Ruegg, E. F., and Campacci, C. A. 1982. Absorcao e translocacao sistemica de metalaxyl-^{14}C por plantuas de *Citrus sinensis* (L.) (Osbeck) apos aplicacoes em diferentes solos (Uptake and systemic translocation of metalaxyl-^{14}C in seedlings of *Citrus sinensis* [L.] [Osbeck] from different soils). Fitopatol. Bras. 7:393–400. (In Portuguese).

Myers, D. F., and Subramanya, R. 1984. Chemical control of Phytophthora crown rot of pepper in Florida. Phytopathology 74:855.

Nadel-Schiffmann, M. 1947. *Phytophthora hibernalis*. Palest. J. Bot. Rehovot Ser. 6:148–157.

Nagata, N. M., and Aragaki, M. 1989. Etiology and control of Phytophthora leaf blight of golden-fruited palm. Plant Dis. 73:661–663.

Nagai, Y., Takeuchi, T., and Watanabe, T. 1978. A stem blight of rose caused by *Phytophthora megasperma*. Phytopathology 68: 684–688.

Naidu, R., Ramakrishnan, K., and Govindu, H. C. 1982. Reaction of citrus and related genera to *Phytophthora nicotianae* var. *parasitica*. Pages 199–205 in: Workshop on *Phytophthora* Diseases of Tropical Cultivated Plants. K. K. N. Nambiar, ed. Cent. Plant. Crops Res. Inst., Kasaragod, Kerala, India. 284 pp.

Nair, R. R., and Menon, M. R. 1982. 'Azhukal' disease of cardamom. Pages 24–33 in: Proc. Workshop on *Phytophthora* Diseases of Tropical Cultivated Plants. K. K. N. Nambiar, ed. Cent. Plant. Crops Res. Inst., Kasaragod, Kerala, India. 284 pp.

Nakanishi, T., and Sisler, H. D. 1983. Mode of action of hymexazol in *Pythium aphanidermatum*. J. Pestic. Sci. 8:173–181.

Nakata, K., and Takimoto, S. 1922. Studies on ginseng diseases in Korea. Bull. Agric. Exp. Stn. Chosen 5:1–18. (In Korean). (Rev. Appl. Mycol. 1923, 2:502).

Nakov, B. 1976. Fitoftorozata—Nova bolest po yagodata u nas (Phytophthorosis, a new disease of strawberry in Bulgaria). Rastitelna Zashchita 24:30–32. (In Bulgarian). (Rev. Plant Pathol. 1977, 56: 304).

Nambiar, E. P., Nair, T. J., and Money, N. S. 1965. Preliminary studies on the incidence of wilt disease of pepper and its relationship to the nitrogen and base status of the soil. Indian J. Agric. Sci. 35:276–281.

Nambiar, K. K. N., ed. 1982. Proceedings of the Workshop on *Phytophthora* Diseases of Tropical Cultivated Plants. 19–23 Sept. 1980. Cent. Plant. Crops Res. Inst., Kasaragod, Kerala, India. 284 pp.

Nambiar, K. K. N., and Sarma, Y. R. 1974. Chemical control of capsule rot of cardamom. J. Plant. Crops 2:30–31.

Nambiar, K. K. N., and Sarma, Y. R. 1977. Wilt diseases of black pepper. J. Plant. Crops 5:92–103.

Nambiar, K. K. N., and Sarma, Y. R. 1982. Some aspects of epidemiology of foot rot of black pepper. Pages 225–232 in: Workshop on *Phytophthora* Diseases of Cultivated Plants. K. K. N. Nambiar, ed. Cent. Plant. Crops Res. Inst., Kasaragod, Kerala, India. 284 pp.

Narasimhan, M. J. 1926. Stamping out the koleroga of Areca. Mysore Agric. Calendar 1926. Pages 25, 28. (Rev. Appl. Mycol. 1926, 5:361).

Narasimhan, M. J. 1927. Wild plants affected by koleroga. Mysore Agric. Calendar 1927. Pages 36–37. (Rev. Appl. Mycol. 1927, 6: 289). (Cited in Tucker 1933).

Narasimhan, M. J. 1930. Studies in the genus *Phytophthora* in Mysore. I. Heterothallic strains of *Phytophthora*. Phytopathology 20: 201–214. (Cited in Tucker 1933).

Narasimhan, M. J. 1931. Sexuality of the koleraga fungus *Phytophthora arecae* (Cole.) Pethy. J. Mysore Agric. Exp. Union 12:4–7.

Narasimhan, M. J. 1932. Report of work done in the mycological section during 1930–1931. Admin. Rep. Mysore Agric. Dep. for 1930–31:24–27.

Narasimham, V., and Ramakrishnan, K. 1969. Studies on wilt disease of betelvine. Madras Agric. J. 56:68–73.

Narasimhan, V., Venkata Rao, A., Subramanian, K. S., and Vidhyasekaran, P. 1976. Fungicidal control of betelvine wilt. Pesticides 10:34–35.

National Academy of Sciences. 1987. Report of the Research Briefing Panel on Biological Control in Managed Ecosystems. R. C. Cook, Chair. National Academy Press, Washington, D.C. 12 pp.

Nattrass, R. M., and Ryan, M. 1951. New hosts of *Phytophthora infestans* in Kenya. Nature (London) 168:85–86.

Navaratnam, S. J. 1966. Patch canker of the durian tree. Malay. Agric. J. 45:291–294.

Navaratnam, S. J. 1970. A root and collar rot of ornamental hisbiscus in West Malaysia. Planter, Kuala Lumpur 46:220–221.

Neema, C., Bompeix, G., and Mourichon, X. 1988. Induction of *Phytophthora citrophthora* mutants resistant to fosetyl-Al and phosphorus acid and study of their pathogenicity towards citrus. Fruits 43:499–505.

Neher, D., and Duniway, J. M. 1991. Relationship between amount of *Phytophthora parasitica* added to field soil and the development of root rot in processing tomatoes. Phytopathology 81:1124–1129.

Neher, D., and Duniway, J. M. 1992. Dispersal of *Phytophthora parasitica* in tomato fields by furrow irrigation. Plant Dis. 582–586.

Nei, M., and Tajima, F. 1983. Maximum likelihood estimation of the number of nucleotide substitutions from restriction sites data. Genetics 105:207–217.

Nemestothy, G. S., and Guest, D. I. 1990. Phytoalexin accumulation, phenylalanine ammonia lyase activity and ethylene biosynthesis in fosetyl-Al treated resistant and susceptible tobacco cultivars infected with *Phytophthora nicotianae* var. *nicotianae*. Physiol. Mol. Plant Pathol. 37:207–219.

Nes, W. D. 1987. Biosynthesis and requirement for sterols in the growth and reproduction of Oomycetes. Pages 304–328 in: Ecology and Metabolism of Plant Lipids. G. Fuller and W. D. Nes, eds. ACS Symp. Ser. 325. Am. Chem. Soc., Washington, D.C. 374 pp.

Nes, W. D. 1988. Phytophthorols—Novel lipids produced by *Phytophthora cactorum*. Lipids 23:9–16.

Nes, W. D. 1990. Control of sterol biosynthesis and its importance to developmental regulation and evolution. Recent Adv. Phytochem. 24:283–327.

Nes, W. D., and Nes, W. R. 1983. Dissociation of oogonial formation from oospore production in *Phytophthora cactorum* by sterols with varying side chain lengths. Experientia 39:276–278.

Nes, W. D., and Stafford, A. E. 1983. Evidence for metabolic and functional discrimination of sterols by *Phytophthora cactorum*. Proc. Nat. Acad. Sci. 80:3227–3231.

Nes, W. D., and Stafford, A. E. 1984. Side-chain structural requirements for sterol-induced regulation of *Phytophthora cactorum* physiology. Lipids 19:544–549.

Nes, W. D., Patterson, G. W., and Bean, G. A. 1979. The effect of steroids and their solubilizing agents on mycelial growth of *Phytophthora cactorum*. Lipids 14:458–462.

Nes, W. D., Patterson, G. W., and Bean, G. A. 1980. Effect of steric and nuclear changes in steroids and triterpenoids on sexual reproduction in *Phytophthora cactorum*. Plant Physiol. 66:1008–1011.

Nes, W. D., Saunders, G. A., and Heftmann, E. 1982a. Role of steroids and triterpenoids in the growth and reproduction of *Phytophthora cactorum*. Lipids 17:178–183.

Nes, W. D., Hanners, P. K., Bean, G. A., and Patterson, G. W. 1982b. Inhibition of growth and sitosterol-induced sexual reproduction in *Phytophthora cactorum* by steroidal alkaloids. Phytopathology 72: 447–450.

Nes, W. D., Le, P. H., Berg, L. R., Patterson, G. W., and Kerwin, J. L. 1986. A comparison of cycloartenol and lanosterol biosynthesis and metalbolism by the Oomycetes. Experientia 42:556–558.

Nes, W. R. 1974. Role of sterols in membranes. Lipids 9:596–612.

Nes, W. R. 1977. The biochemistry of plant sterols. Adv. Lipid Res. 15:233–324.

Nes, W. R., and Nes, W. D. 1980. Lipids in Evolution. Plenum Press, New York. 224 pp.

Nesbitt, H. J., Malajaczuk, N., and Glenn, A. R. 1985. Bacterial colonization of *Phytophthora cinnamomi* Rands. Pages 371–375 in: Microbial Ecology. M. W. Loutit and J. A. R. Miles, eds. Springer-Verlag, Berlin. 452 pp.

Newhall, A. G. 1948. Research at Turrialba on cacao diseases. Cacao Inf. Bull. 1. 7 pp.

Newhook, F. J. 1958. Phytophthora root rot of *Pinus radiata*. Commonw. Phytopathol. News. 4:9–10.

Newhook, F. J. 1959. The association of *Phytophthora* spp. with mortality of *Pinus radiata* and other conifers. I. Symptoms and epidemiology in shelter belts. N.Z. J. Agric. Res. 2:808–843.

Newhook, F. J. 1960. Some comments on the relationship of climate to *Phytophthora* infection of *Pinus radiata*. N.Z. J. For. 8:261–265.

Newhook, F. J. 1970. *Phytophthora cinnamomi* in New Zealand. Pages 173–176 in: Root Diseases and Soilborne Pathogens. T. A. Toussoun, R. V. Bega, and P. E. Nelson, eds. Univ. Calif. Press, Berkeley. 252 pp.

Newhook, F. J. 1978. *Phytophthora cinnamomi* in native forests of Australia and New Zealand: Indigenous or introduced? Pages 353–359 in: Microbial Ecology. M. W. Loutit and J. A. R. Miles, eds. Springer-Verlag, Heidelberg, Germany. 450 pp.

Newhook, F. J. 1982. Epidemiology in the genus *Phytophthora*. Pages 83–89 in: Proc. Workshop on *Phytophthora* Diseases of Tropical Cultivated Plants. K. K. N. Nambiar, ed. Cent. Plant. Crops Res. Inst., Kasarogod, Kerala, India. 284 pp.

Newhook, F. J. 1988. *Phytophthora* prevention and cure. Growing Today (Dec/Jan):31–33.

Newhook, F. J., and Jackson, G. V. H. 1977. *Phytophthora palmivora* in cocoa plantation soils in the Solomon Islands. Trans. Br. Mycol. Soc. 69:31–68.

Newhook, F. J., and Podger, F. D. 1972. The role of *Phytophthora cinnamomi* in Australian and New Zealand forests. Annu. Rev. Phytopathol. 10:299–326.

Newhook, F. J., Waterhouse, G. M., and Stamps, D. J. 1978. Tabular key to the species of *Phytophthora* de Bary. Mycol. Pap. 143. Commonw. Mycol. Inst. Kew, Surrey, U.K. 20 pp.

Newhook, F. J., Young, B. R., Allen, S. D., and Allen, R. N. 1981. Zoospore motility of *Phytophthora cinnamomi* in particulate substrates. Phytopathol. Z. 101:202–209.

Newhouse, J. R., Tooley, P. W., Smith, O. P., and Fishel, R. A. 1992. Characterization of double-stranded RNA in isolates of *Phytophthora infestans* from Mexico, the Netherlands, and Peru. Phytopathology 82:164–169.

Newman, C. L. 1941. Ann. Rep. Dep. Agric. Sarawak for 1940. (Cited in Sarma and Nambiar 1982).

Newton, W. 1956. The nonutilization of d-alanine and growth stimulation by a-alanine in the nutrition of *Phytophthora parasitica.* Can. J. Bot. 34:759–761.

Ngo, H. C., and Baxter, L. W. 1977. The status of our knowledge of root rot of Camellias caused by *Phytophthora cinnamomi* in 1977. Am. Camellia Soc. Yearb. 1977:172–187.

Nguyen, M. V., Rode, M. W., and Payne, D. W. 1985. Inoculation of seedling cuttings for screening the soybean against races of *Phytophthora megasperma* f. sp. *glycinea.* Crop Sci. 25:121–123.

Nicholls, D. D., and Tucker, B. M. 1956. Pedology and chemistry of the balsaltic soils of the Lismore district, New South Wales. Page 153 in: C.S.I.R.O. Soil Publ. 7. CSIRO, Melbourne, Australia.

Nicklová-Navrátilová, H. 1949. Tri ctvrti roku zahradnicke prae ocima fytopathologa (Three quarters of a century of gardening practice through the eyes of a phytopathologist). Ochr. Rostl. 22: 238–256. (In Czechoslovakian). (Rev. Appl. Mycol. 1950, 29:619).

Niederhauser, J. S. 1954. Observations on late blight and other potato diseases in Peru. Plant Dis. Rep. 38:81–82.

Niederhauser, J. S. 1956. The blight, the blighter, and the blighted. Trans. N.Y. Acad. Sci. 19:55–63.

Niederhauser, J. S. 1961. Genetic studies on *Phytophthora infestans* and *Solanum* species in relation to late blight resistance in the potato. Recent Adv. Bot. 1:491–497.

Niederhauser, J. S. 1962. Evaluation of multigenic "field resistance" of the potato to *Phytophthora infestans* in 10 years of trials at Toluca, Mexico (Abstr.). Phytopathology 52:746.

Niederhauser, J. S. 1991. *Phytophthora infestans:* The Mexican connection. Pages 25–45 in: *Phytophthora.* J. A. Lucas, R. C. Shattock, D. S. Shaw, and L. R. Cooke, eds. Cambridge Univ. Press, Cambridge, U.K. 447 pp.

Niederhauser, J. S. 1993. International cooperation in potato research and development. Annu. Rev. Phytopathol. 31:1–21.

Niederhauser, J. S., and Cobb, W. C. 1959. The late blight of potatoes. Sci. Am. 200:100–112.

Niederhauser, J. S., and Mills, W. R. 1953. Resistance of *Solanum* species to *Phytophthora infestans* in Mexico. Phytopathology 43: 456–457.

Niederhauser, J. S., Cervantes, J., and Servín, L. 1954. Late blight in Mexico and its implications. Phytopathology 44:406–408.

Nienhaus, F. 1959. Rhabarberfäule durch *Phytophthora cactorum* (Rhubarb rot caused by *Phytophthora cactorum*). Nachrichtenbl. Dtsch. Pflanzenschutzdienst (Berl.) N. F. 11:58–59. (In German). (Rev. Appl. Mycol. 1960, 39:142).

Nienhaus, F. 1960. Das Wirtsspektrum von *Phytophthora cactorum* (Leb. et Cohn) Schroet. (The host spectrum of *Phytophthora cactorum* [Leb. et Cohn] Schroet.). Phytopathol. Z. 38:33–68. (In German).

Niere, J. O., Griffith, J. M., and Grant, B. R. 1990. NMR studies on the effect of phosphite on *Phytophthora palmivora.* Gen. Microbiol. 136:147–156.

Nirwan, R. S., and Upadhyaya, J. 1972. Phytophthora blight of cowpea new to India. Indian Phytopathol. 25:162–163.

Nisikado, Y., Hirata, K., and Kimura, K. 1941. On a Phytophthora rot of fig. Ber. Ohara Inst. Landwirstsch. Biol. 8:427–442.

Noble, R. J. 1926. Downy mildew of wheat (*Sclerospora macrospora* Sacc.) Agric. Gaz. N.S.W. 37:204–208.

Noble, R. J. 1929. Australia: Plant diseases observed in New South Wales. Int. Bull. Plant Prot. 3:6–7.

Noble, R. J. 1937. The occurrences of plant diseases in New South Wales. Dep. Agric. N.S.W. Sci. Bull. 57. 42 pp.

Nolla, J. A. B. 1928. The black shank of tobacco in Puerto Rico. J. Dep. Agric. P.R. 12:185–215.

Nolla, J. A. B. 1929. The black-shank of tobacco in Puerto Rico. Phytopathology 19:93–94.

Noon, J. P., and Hickman, C. J. 1974. Oospore production by a single isolate of *Phytophthora capsici* in the presence of chloroneb. Can. J. Bot. 52:1591–1595.

Norin, I. 1987. Plant protection year 1986—Horticulture. Vaxtskyddsnotiser 51:38–40. (In Swedish).

Noronha, E. de A. 1959. Lista de doencas das plantas onservadas em Angola durante os anos de 1949–1956 (List of plant diseases observed in Angola during the years 1949–1956). Agron. Angol. 11:25–28. (In Portuguese).

Norton, D. C., and Rosberg, D. W. 1954. Watermelon fruit rot caused by *Phytophthora parasitica* in Texas. Plant Dis. Rep. 38:854.

Noske, G. L., and Shearer, B. L. 1985. Quaternary ammonium compounds were more effective than a phenolic compound or sodium hypochlorite in inhibiting growth of *Phytophthora cinnamomi* Rands. Aust. Plant. Pathol. 14:37–40.

Nourisseau, J. G., and Baudry, A. 1987. Un *Phytophthora* cause de dépérissement du framboisier en France (A *Phytophthora* cause of a decline of raspberry in France). Phytoma No. 384:39–41. (In French).

Noviello, C., and Snyder, W. C. 1962. A *Phytophthora* disease of fennel. Phytopathol. Z. 46:139–163.

Noviello, C., Cristinzio, G., and Aloj, B. 1977. Una grave malattia della zucca in Campania (A serious disease of squash in Campania). Ann. Fac. Sci Agrar. Univ. Stud. Napoli Portici 11:11–22. (In Italian). (Rev. Plant Pathol. 1979, 58:3536).

Novotel'nova, N. S. 1974. The genus *Phytophthora.* Sov. Acad. Sci. Bot. Inst., Leningrad. 206 pp. (In Russian).

Nowell, W. 1917. Diseases of Crop Plants in the Lesser Antilles. West India Comm., London. 274 pp. (Cited in Tucker 1933).

Nygaard, S. L., Elliot, C. K., Cannon, S. J., and Maxwell, D. P. 1989. Isozyme variability among isolates of *Phytophthora megasperma.* Phytopathology 79:773–780.

Obbink, J. G., and Alexander, D. M. 1976. Calcium deficiency and *Phytophthora cinnamomi* symptoms in young avocado seedlings. J. Aust. Inst. Agric. Sci. 42:131–134.

O'Brien, M. J., and Rich, A. E. 1976. Potato Diseases. Agric. Handbook 474. U.S. Dep. Agric. 79 pp.

Ocaña, G. 1967. Relation of sterols to the differential sensitivity of spores and mycelia of *Phytophthora* to polyene antibiotics. Ph.D. dissertation, Univ. of California, Riverside. 94 pp.

Ocaña, G., and Tsao, P. H. 1966. A selective agar medium for the direct isolation and enumeration of *Phytophthora* in soil (Abstr.). Phytopathology 56:893.

Ocfemia, G. O. 1925. The *Phytophthora* disease of eggplant in the Philippine Islands. Philipp. Agric. 14:317–328.

Ocfemia, G. O. 1934. Bud rot of coconut. Philipp. Agric. 23:4–10.

Ocfemia, G. O., and Roldan, E. F. 1927. Phytophthora blight of citrus. Am. J. Bot. 14:1–15.

Ochoa, C. 1955. Species of *Solanum* (*Tuberarium*) of South America. Present taxonomic status and species used in plant breeding with special reference to Peru. Phytopathology 45:247–250.

Ochoa, C. 1981. *Solanum irosinum,* new Peruvian tuber bearing species resistant to *Phytophthora infestans.* Am. Potato J. 58:131–133.

Odermatt, M., Röthlisberger, A., Werner, C., and Hohl, H. 1988. Interactions between agarose-embedded plant protoplasts and germ tubes of *Phytophthora.* Physiol. Mol. Plant Pathol. 33:209–220.

Odigie, E. E., and Ikotun, T. 1982. In vitro and in vivo inhibition of growth of *Phytophthora palmivora* (Butl.) Butl. by antagonistic microorganisms. Fitopatol. Bras. 7:157–167.

Oertel, B., and Jelke, E. 1986. Formation of multinucleate zoospores in *Phytophthora infestans* (Mont.) de Bary (*Oomycetes, Phytiaceae*). Protoplasma 135:173–179.

Ogilvie, L. 1924. Control of plant pests and diseases. Bermud. Dep. Agric. Bull. 3:2–3. (Cited in Tucker 1933).

Ogilvie, L. 1931. A fruit rot of apples and pears due to a variety of *Phytophthora syringae.* Annu. Rep. Agric. Hortic. Res. Stn. Univ. of Bristol, U.K. for 1930. pp. 147–150. (Rev. Appl. Mycol. 1932, 11:111–112). (Cited in Tucker 1933).

Ohr, H. W., Murphy, M. K., and Bender, G. 1986. Control of Phytophthora root rot in container-grown citrus. Calif. Agric. 40:18–19.

Okaisabor, E. K. 1971a. The mechanism of initiation of *Phytophthora* pod rot epiphytotics. Pages 398–404 in: Proc. 3rd Int. Cocoa Res. Conf., Accra, Ghana. 1969.

Okaisabor, E. K. 1971b. Studies on some aspects of the soil phase of *Phytophthora palmivora* (Butl.) Butl. Pages 484–489 in: Proc. 3rd Int. Cocoa Res. Conf., Accra, Ghana. 1969.

Olah, A. F., and Schmitthener, A. F. 1985. A growth chamber test for [illegible] tolerance in soybean seedlings. Phytopathology 75:546–548.

Olah, A. F., Schmitthenner, A. F., and Walker, A. K. 1985. Glyceollin accumulation in soybean lines tolerant to *Phytophthora megasperma* f. sp. *glycinea.* Phytopathology 75:542–546.

Old, K. M., ed. 1979. *Phytophthora* and Forest Management in Australia. C.S.I.R.O. Melbourne, Australia. 114 pp.

Old, K. M., Moran, G. F., and Bell, J. S. 1984. Isozyme variability among isolates of *Phytophthora cinnamomi* from Australia and Papua, New Guinea. Can. J. Bot. 62:2016–2022.

Old, K. M., Dudzinsky, M. J., and Bell, J. K. 1988. Isozyme variability in field populations of *Phytophthora cinnamomi* in Australia. Aust. J. Bot. 36:355–360.

Oliveros, B., Lozano, J. C., and Booth, R. H. 1974. A Phytophthora root rot of cassava in Colombia. Plant Dis. Rep. 58:703–705.

Olofsson, B. 1991. Chemical late blight control in Sweden. E.P.P.O. (Eur. Mediterr. Plant Prot. Organ.) Bull. 21:57–60.

Olunloyo, O. A., and Adeniji, M. O. 1976. The influence of NPK fertilizers on the root and stem rot disease of roselle caused by *Phytophthora parasitica* var. *nicotianae*. Niger. J. Plant Prot. 2:84–87.

O'Malley, D. M., Wheeler, N. E., and Guries, R. P. 1980. A manual for starch gel electrophoresis. Univ. Wis., Madison, Dep. For. Staff Pap. Ser. 11. 15 pp.

O'Neill, T. M., and Griffin, G. W. 1987. Effect of fosetyl aluminum and captafol on red core disease and fruit rot of established strawberries. Plant Pathol. 36:258–263.

Ooka, J. J., Uchida, J. Y., and Yamamoto, B. T. 1984. *Phytophthora katsurae* fruit rot of coconut in Hawaii (Abstr.). Phytopathology 74:816.

Orchard, O. B. 1935. A disease of winter cherry. Rep. Exp. Res. Stn., Cheshunt, 1934. Pages 54–55. (Rev. Appl. Mycol. 1935, 14:636–637).

Orellana, R. G. 1954. Contribution to the study of survival, dissemination and control of *Phytophthora* of cacao. Proc. Inter-Am. Cacao Conf. 5th Turrialba, Costa Rica. Sec. Fitopatol. 1:1–10.

Orellana, R. G. 1956. Cacao diseases in Haiti. F.A.O. Plant Prot. Bull. 4:148–151.

Orellana, R. G. 1959. Variation in *Phytophthora palmivora* isolated from cacao and rubber. Phytopathology 49:210–213.

Oren, Y., and Solel, Z. 1978. Control of brown rot of citrus fruit by application of fungicides via sprinkler irrigation systems. Phytoparasitica 6:65–70.

Orian, A. J. E. 1959. Report on a visit to Diego Garcia. Rev. Agric. Sucr. 'Ile Maurice 38:127–143. (Rev. Appl. Mycol. 1960, 39:32–33).

Orian, G. 1951. Rep. Dep. Agric. Mauritius, 1949. Pages 66–72, 1950; pages 80–85, 1951. Plant Pathology Division. (Rev. Appl. Mycol. 1952, 31:537–538).

Orian, G. 1953. Botanical Division. Rep. Agric. Mauritius, 1952. Pages 37–40.

Orlikowski, L. 1978. Occurrence of *Phytophthora cryptogea* Pethyb. et Laff. in gerbera (*Gerbera jamesonii* Bolus) Bull. Pol. Acad. Sci. Biol. 26:495–498.

Orlikowski, L. 1980/1981. Studies on the biological control of *Phytophthora cryptogea* Pethybr. et Laff. The mycoflora associated with gerbera production in Polish greenhouses and effects of its main components on the development of the pathogen. Prot. Ecol. 2:285–286.

Orlikowski, L. B., and Schmidle, A. 1985. Zur biologischen Bekämpfung von *Phytophthora cactorum* mit *Trichoderma viride* (On the biologic control of *Phytophthora cactorum* with *Trichoderma viride*). Nachrichtenbl. Dtsch. Pflanzenschutzdienst 37:78–79. (In German).

Orlikowski, L. B., and Wojdyla, A. 1988. Action of Sandofan M toward *Phytophthora cryptogea*. Acta Hortic. 221:213–216.

Orlikowski, L. B., Skrzypczak, C., and Wojdyla, A. 1984. Occurrence, biology, pathogenicity, and control of *Phytophthora cryptogea* on cineraria. Pr. Inst. Sadow. Poland. Ser. B. 9:79–85.

Orlikowski, L. B., Leoni-Ebelling, M., and Schmidle, A. 1986. Über die wirkung von Metalaxyl und Phosethyl-Aluminium auf *Phytophthora cactorum* an Apfelbaumen (Efficacy of metalaxyl and phosethyl-aluminium in the control of *Phytophthora cactorum* on apple trees). Z. Pflanzenkr. Pflanzenschutz 93:202–209. (In German).

Ornstein, L. 1964. Disc electrophoresis I. Background and theory. Ann. N.Y. Acad. Sci. 121:321–349.

Osburn, R. M., Milner, J. L., Oplinger, J. L., Smith, R. S., and Handelsman, J. 1995. Effect of *Bacillus cereus* UW85 on the yield of soybean at two field sites in Wisconsin. Plant Dis. 79:551–556.

Osnitzkaya, E. A. 1949. *Phytophthora* of tomatoes in protected soil. Orchard Kitchen Garden 2:62–64. (Rev. Appl. Mycol. 1949, 28: 648).

Osterwalder, A. 1906. Die *Phytophthora*-Fäule beim Kernobst (Phytophthora rot of pome fruit). Centralbl. Bakt. II. 15:435–440. (In German).

Osterwalder, A. 1909. Unbekannte Krankheiten an Kulturpflanzen und deren Ursachen 2. Die *Phytophthora*-Krankheit bei *Calceolaria rugosa* (Hort.) verursacht durch *Phytophthora omnivora* de Bary (Unknown diseases of cultivated plants and their causes. 2. The *Phytophthora* disease of *Calceolaria rugosa* [Hort.] caused by *Phytophthora omnivora* de Bary). Centralbl. Bakt. II. 25:265–267. (Cited in Tucker 1933).

Osterwalder, A. 1912. Das Absterben von Veredlungen verursacht durch *Phytophthora omnivora* de Bary (The die-off of graftings caused by *Phytophthora omnivora* de Bary). Landw. Jb. Schweiz 26:321–322. (In German). (Cited in Tucker 1933).

Ostrofsky, W. D., Pratt, R. G., and Roth, L. F. 1977. Detection of *Phytophthora lateralis* in soil organic matter and factors that affect its survival. Phytopathology 67:79–84.

O'Sullivan, E., and Dowley, L. J. 1983. Physiological specialisation in strains of *Phytophthora infestans* sensitive and resistant to metalaxyl. Ir. J. Agric. Res. 22:105–107.

O'Sullivan, E., and Dowley, L. J. 1991. A note of the occurrence of the A2 mating type and self-fertile isolates of *Phytophthora infestans* in the Republic of Ireland. Ir. J. Agric. Res. 30:67–69.

Ou, S. H. 1972. Rice Diseases. Commonw. Mycol. Inst., Kew, Surrey, U.K. 368 pp.

Oudemans, P., and Coffey, M. D. 1991a. Isozyme comparison within and among worldwide sources of three morphologically distinct species of *Phytophthora*. Mycol. Res. 95:19–30.

Oudemans, P., and M. D. Coffey. 1991b. A revised systematics of twelve papillate *Phytophthora* species based on isozyme analysis. Mycol. Res. 95:1025–1046.

Oudemans, P., and Coffey, M. D. 1991c. Relationships between *Phytophthora* species: Evidence from isozyme analysis. Pages 184–203 in: *Phytophthora*. J. A. Lucas, R. C. Shattock, D. S. Shaw, and L. R. Cooke, eds. Cambridge Univ. Press, Cambridge, U.K. 447 pp.

Oudemans, P., Förster, H., and Coffey, M. D. 1994. Evidence for distinct isozyme subgroups within *Phytophthora citricola* and close relationships with *P. capsici* and *P. citrophthora*. Mycol. Res. 98:189–199.

Ouimette, D. G., and Coffey, M. D. 1988. Quantitative analysis of organic phosphonates, phosphonate, and other inorganic anions in plants and soil by using high-performance ion chromatography. Phytopathology 78:1150–1155.

Ouimette, D. G., and Coffey, M. D. 1989a. Comparative antifungal activity of four phosphonate compounds against isolates of nine *Phytophthora* species. Phytopathology 79:761–767.

Ouimette, D. G., and Coffey, M. D. 1989b. Phosphonate levels in avocado (*Persea americana*) seedlings and soil following treatment with fosetyl-Al or potassium phosphonate. Plant Dis. 73:212–215.

Ouimette, D. G., and Coffey, M. D. 1990. Symplastic entry and phloem translocation of phosphonate. Pestic. Biochem. Physiol. 38:18–25.

Ownley, B. H., and Benson, D. M. 1991. Relationship of matric water potential and air-filled porosity of container media to development of Phytophthora root rot of rhododendron. Phytopathology 81:936–941.

Ownley, B. H., Benson, D. M., and Bilderback, T. E. 1990. Physical properties of container media and relaton to severity of Phytophthora root rot of rhododendron. J. Am. Soc. Hortic. Sci. 115:564–570.

Oxenham, B. L. 1963. Rep. Dep. Agric., Queensland, Australia, for 1962–63, Plant Pathology Section. (Rev. Appl. Mycol. 1964, 43: 1228).

Oxenham, B. L., and Stone, Y. L. 1969. Citrus brown rot in Queensland. Queensl. J. Agric. Anim. Sci. 26:174–180.

Oxenham, B. L., and Winks, B. L. 1963. Phytophthora root rot of *Pinus* in Queensland. Queensl. J. Agric. Sci. 20:355–366.

Oyler, E., and Bewley, W. F. 1937. A disease of cultivated heaths caused by *Phytophthora cinnamomi* Rands. Ann. Appl. Biol. 24: 1–16.

Padwick, G. W. 1938. India: New plant diseases recorded in 1937. Int. Bull. Plant Prot. 12:122–123.

Pady, S. M., Kramer, C. L., Pathak, V. K. 1960. Suppression of fungi by light on media containing rose bengal. Mycologia 52:347–350.

Pag, H. 1960. Untersuchungen über das 'Gerbera-Sterben' (Investigations on the 'Death of Gerbera'). Nachrichtenbl. Dtsch. Pflanzenschutzdienst. Stuttgart 12:74–77. (In German). (Rev. Appl. Mycol. 1961, 40:50).

Paharia, K. D., and Mathur, P. N. 1961. New host plant of Colocasia blight (*Phytophthora colocasiae* Racib.). Curr. Sci. 30:354.

Paktitis, S., Grant, B., and Lawrie, A. 1986. Surface changes in *Phytophthora palmivora* zoospores following induced differentiation. Protoplasma 135:119–129.

Pal, M., and Grewal, J. S. 1975. Utilization of different nitrogen sources by *Phytophthora drechsleri* var. *cajani.* Indian Phytopathol. 28:499–501.

Pal, M., and Grewal, J. S. 1976. Effect of NPK fertilizers on Phytophthora blight of pigeon-pea. Indian J. Agric. Sci. 46:32–35.

Pal, M., Grewal, J. S., and Sarbhoy, A. K. 1970. A new stem rot of arhar caused by *Phytophthora.* Indian Phytopathol. 23:583–587.

Palazón, C., Gil, R., Palazón, I., and Delgado, I. 1990. Efficacité de quelques fongicides dan la lutte contre *Phytophthora capsici* sur poivron (Efficacy of several fungicides against *Phytophthora capsici* on pepper). E.P.P.O. (Eur. Mediterr. Plant Prot. Organ.) Bull. 20:177–184 (In French).

Palloix, A., Daubeze, A. M., and Pochard, E. 1988a. Time sequences of root infection and resistance expression in an artificial inoculation method of pepper with *Phytophthora capsici.* J. Phytopathol. 123:12–24.

Palloix, A., Daubeze, A. M., and Pochard, E. 1988b. Phytophthora root rot of pepper: Influence of host genotype and pathogen strain on the inoculum density-disease severity relationships. J. Phytopathol. 123:25–35.

Palloix, A., Daubeze, A. M., Phaly, T., and Pochard, E. 1990. Breeding transgressive lines of pepper for resistance to *Phytophthora capsici* in a recurrent selection system. Euphytica 51:141–150.

Palm, B. T. 1923. Verslag van het Deli Proefstation over 1 Juli 1922–30 Juni 1923 (Report of the Deli Experiment station for the period 1st July, 1922. to 30th June, 1923). Meded. Deli Proefstat. Medan-Sumatra Ser. 2, 29. 41 pp. (In Dutch). (Rev. Appl. Mycol. 1924, 3:107).

Palm, B. T. 1932. Pflanzenkrankheiten aus Guatemala (Plant disease in Guatemala). Z. Pflanzenkr. Pflanzenschutz 42:11–17. (In German). (Cited in Tucker 1933).

Pan, J., Ho, H. H., and Jong, S. C. 1994. A scanning electron microscopy study of *Phytophthora* and *Halophytophthora* species. Mycotaxon 51:257–279.

Panabiéres, F., Marais, A., Trentin, F., Bonnet, P., and Ricci, P. 1989. Repetitive DNA polymorphism analysis as a tool for identifying *Phytophthora* species. Phytopathology 79:1105–1109.

Papaioannoy, A. J. 1954. Notes Phytopathologiques. I. Un hote nouveau du *Phytophthora cactorum* (L. et C.) Schröter. (Phytopathologic notes. I. A new host of *Phytophthora cactorum* [L. et C.] Schröter.) Ann. Inst. Phytopathol. Benaki, N.S. 8:96–102. (In French).

Papavizas, G. C. 1985. *Trichoderma* and *Gliocladium:* Biology, ecology, and potential for biocontrol. Annu. Rev. Phytopathol. 23: 23–54.

Papavizas, G. C., and Bowers, J. H. 1981. Comparative fungitoxicity of captafol and metalaxyl to *Phytophthora capsici.* Phytopathology 71:123–128.

Papavizas, G. C., and Lewis, J. A. 1976. Acetone infusion of pyroxychlor into soybean seed for the control of *Phytophthora megasperma* var. *sojae.* Plant Dis. Rep. 601:484–488.

Papavizas, G. C., Schwenk, F. W., Locke, J. C., and Lewis, J. A. 1979. Systemic fungicides for controlling Phytophthora root rot and damping-off of soybean. Plant Dis. Rep. 63:708–712.

Papavizas, G. C., Bowers, J.H., and Johnston, S. A. 1981. Selective isolation of *Phytophthora capsici* from soils. Phytopathology 71: 129–133.

Pape, H. 1933. Krankheiten und Schädlinge der Begonien. Seiten 149–186 in: Die Begonien, ihre Beschreibung, Kultur, Züchtung und Geschichte (Diseases and pests of begonias. Pages 149–186 in: The Begonias, their Description, Culture, Cultivation and History). Ulmer-Verl., Stuttgart. (In German). (Cited in Nienhaus 1960). (Rev. Appl. Mycol. 1937, 16:692–693).

Paplomatas, E. J., Elena, K., and Lascaris, D. 1995. First report of *Phytophthora boehmeriae* causing boll rot of cotton. Plant Dis. 79:860.

Pappas, A. 1978. New diseases caused by *Phytophthora* spp. Phytophthora Newsl. 6:72–73.

Pappas, A. C. 1985. Metalaxyl resistance in *Phytophthora infestans* on greenhouse tomatoes in Greece. Plant Pathol. 34:293–296.

Parakhia, A. M., Savalia, R. L., and Vaishnav, M. U. 1986. A new disease of coleus. Indian Phytopathol. 39:316.

Parham, B. E. V. 1942. Plant Protection notes from the pathological laboratory. III. Agric. J. (Fiji) 13:27–28.

Parham, B. E. V. 1947. Economic botany notes. III. Disease of taro. Agric. J. (Fiji) 18:77–80.

Park, M. 1930. Report of the Mycological Division, Ceylon Dep. Agric. Tech. Rep. for the year 1929. Pages 1–6.

Park, M. 1939. Report on the work of the division of plant pathology. Admin. Rep. Dir. Agric. Ceylon, 1937. Pages 42–48. (Rev. Appl. Mycol. 1939, 18:655).

Parker, J. E., Knogge, W., and Scheel, D. 1991. Molecular aspects of host-pathogen interactions in *Phytophthora.* Pages 90–103 in: *Phytophthora.* J. A. Lucas, R. C. Shattock, D. S. Shaw, and L. R. Cooke, eds. Cambridge Univ. Press, Cambridge, U.K. 447 pp.

Parker, S. P. 1982. Synopsis and Classification of Living Organisms. McGraw-Hill Book Co., New York. 1166 pp.

Parlevliet, J. E. 1979. Components of resistance that reduce the rate of epidemic development. Annu. Rev. Phytopathol. 17:203–222.

Parmentier, G. 1970. L'incorporation de charbon actif aux milieux de culture de *Phytophthora infestans* (The incorporation of activated carbon in the culture medium of *Phytophthora infestans*). Parasitica 26:31–40. (In French).

Parris, G. K. 1941. Diseases of taro in Hawaii and their control with notes on field production. Hawaii Agric. Exp. Stn., Circ. No. 18. 29 pp.

Parris, G. K. 1942. *Phytophthora parasitica* on papaya (*Carica papaya*) in Hawaii. Phytopathology 32:314–320.

Partridge, J. E., and Erwin, D. C. 1969. Preparation of mycelium-free suspensions of oospores of *Phytophthora megasperma* var. *sojae* (Abstr.). Phytopathology 59:14.

Partridge, J. E., and Keen, N. T. 1976. Association of the phytoalexin kievitone with single-gene resistance of cowpeas to *Phytophthora vignae.* Phytopathology 66:426–429.

Passavalli, L. P. 1937. Difesa preventiva e curativa della Fragola da malattie e insetti novici (Different preventative and curative defenses of the strawberry against diseases and pests). Boll. Soc. Tosc. Octic. 62:55–60. (In Italian).

Patil, S. S., and Young, R. A. 1960. The influence of temperature on development of *Phytophthora parasitica* root rot of *Fuchsia.* Phytopathology 50:386–388.

Patil-Kulkarni, B. G., and Aswathaiah, B. 1962. Phytophthora boll rot on cotton (*Gossypium barbadense* L.) in Mysore. Sci. Cult. 28: 233–234. (Rev. Appl. Mycol. 1963, 42:24).

Patrikeyeva, M. V. 1979. Obrazovanie khlamidospor u *Phytophthora infestans* (Mont.) de Bary (The formation of chlamydospores in *Phytophthora infestans* [Mont.] de Bary). Mikol. Fitopatol. 13:333–337. (In Russian).

Paxton, J. D. 1976. A new technique for isolating *Phytophthora megasperma* var. *sojae* from field soil (Abstr.). Proc. Am. Phytopathol. Soc. 3:238.

Paxton, J. D. 1983. Phytophthora root and stem rot of soybean: A case study. Pages 19–30 in: Biochemical Plant Pathology. J. A. Callow, ed. John Wiley & Sons, New York. 484 pp.

Paxton, J. D. 1988. Phytoalexins in plant-parasite interactions. Pages 537–549 in: Experimental and Conceptual Plant Pathology. Vol. 3. W. H. Hess, R. S. Singh, U. S. Singh, and D. J. Weber, eds. Gordon and Breach Sci. Pub. 599 pp.

Paxton, J. D., and Chamberlain, D. W. 1967. Acquired local resistance of soybean plants to *Phytophthora* spp. Phytopathology 57:352–353.

Paxton, J. D., and Chamberlain, D. W. 1969. Phytoalexin production and disease resistance in soybeans as affected by age. Phytopathology 59:775–777.

Payette, P. A., and Perrault, C. 1944. Action de la thiamine sur le *Phytophthora infestans* (Mont.) de Bary (Effect of thiamine on *Phytophthora infestans* [Mont.] de Bary. Can. J. Res. Sect. C 22:127–132. (In French).

Pearce, R. B., and Ride, J. P. 1980. Specificity of induction of the lignification response in wounded wheat leaves. Physiol. Plant Pathol. 16:197–204.

Pegg, G. F., and Jordan, M. M. 1990. *Phytophthora cryptogea* root rot of tomato in rock wool nutrient culture. I. Analysis of root infection. Ann. Appl. Biol. 117:525–535.

Pegg, K. G. 1973. Phytophthora blight in passion fruit. Queensl. Agric. J. 99:655–656.

Pegg, K. G. 1977a. Soil application of elemental sulphur as a control of *Phytophthora cinnamomi* root and heart rot of pineapple. Aust. J. Exp. Agric. Anim. Husb. 17:859–865.

Pegg, K. G. 1977b. Biological control of *Phytophthora cinnamomi* root rot of avocado and pineapple in Queensland. Pages 7–12 in: Seminar Papers. Annu. Conf. Aust. Nurserymen's Assoc., Hobart. 38 pp.

Pegg, K. G. 1978. Disease-free avocado nursery trees. Queensl. Agric. J. 104:134–136.

Pegg, K. G., and Alcorn, J. L. 1982. *Phytophthora operculata* sp. nov., a new marine fungus. Mycotaxon 16:99–102.

Pegg, K. G., and Whiley, A. W. 1987. *Phytophthora* control in Australia. S. Afr. Avocado Growers' Assoc. Yearrb. 10:94–96.

Pegg, K. G., Kochman, J. K., and Vock, N. T. 1980. Root and stem rot of soybean caused by *Phytophthora megasperma* var. *sojae* in Queensland. Aust. Plant Pathol. 9:15.

Pegg, K. G., Whiley, A. W., Saranah, J. B., and Glass, R. J. 1985. Control of Phytophthora root rot of avocado with phosphorus acid. Australas. J. Plant Pathol. 14:25–29.

Pegg, K. G., Whiley, A. W., and Hargreaves, P. A. 1990. Phosphonic acid (phosphorous) acid treatments control *Phytophthora* diseases in avocado and pineapple. Aust. Plant Pathol. 19:122–124.

Peglion, V. 1901. La *Peronospora* del frumento. Nuove ricerche (A *Peronospora* of wheat. New research). Stat. Sper. Agric. Ital. 34: 506–532. (In Italian).

Peglion, V. 1930. La formazione del conidi e la germinazione delle oospore della '*Sclerospora macrospora*' Sacc. Boll. della R. Staz. Patol. Veg. Roma, N.S. 10:153–164.

Peglion, V., and Sacchetti, M. 1927. Intorno alla *Peronospora* del'illa. (*Phytophthora syringae* Klebahn) (On the downy mildew of lilac [*Phytophthora syringae* Klebahn]). Rendic. Accad. Lincei. Ser. 6, 5:696–698. (In Italian). (Rev. Appl. Mycol. 1927, 6:668). (Cited in Tucker 1933).

Pepin, H. S. 1967. Susceptibility of members of the Rosaceae to races of *Phytophthora fragariae.* Phytopathology 57:782–784.

Pepin, H. S., and Daubeny, H. A. 1964. The relationships of English and American races of *Phytophthora fragariae.* Phytopathology 54:241.

Perches, S., and Galindo, A. J. 1967. Supervivencia del *Phytophthora infestans* (Mont.) de Bary, causante del tizón tardío de la papa y jitomate (Survival of *Phytophthora infestans* [Mont.] de Bary, causing late blight of potato and tomato). Agrociencia 1:92–98. (In Italian).

Peregrine, W. T. H. 1963. Zebra disease studies. Rep. Tanganyika Sisal Grow. Assoc. 1962–1963:48–49.

Peregrine, W. T. H. 1964. Annual Report, Sisal Research Station, Mlingano, 1964–65. Rep. Tanganyika Sisal Grow. Assoc. Part III:35–59.

Peregrine, W. T. H. 1969. Investigations on chemical control of zebra disease in *Agave* hybrid no. 11648. Ann. Appl. Biol. 63:45–51.

Peregrine, W. T. H., and Wienk, J. F. 1964. Annual report of the Sisal Research Station, Mlingano, 1963–1964. Rep. Tanganyika Sisal Grow. Assoc. 1963–1964:31–61. (Rev. Appl. Mycol. 1965, 44:723).

Pereira, J. S., and Pizzigatti, R. 1980. Symptomatology of Phytophthora canker of cacao following river flooding. Rev. Theobroma 10:203–211.

Pergola, G., and Garibaldi, A. 1977. Control of collar rot of carnation caused by *Phytophthora nicotianae* var. *parasitica.* Inst. Sper. Floricoltura, San Remo, Imperia, Italy. (Rev. Plant Pathol. 1978, 57:2971).

Peries, O. S., and Fernando, T. M. 1966. Studies on the biology of *Phytophthora meadii.* Trans. Br. Mycol. Soc. 49:311–325.

Peries, O. S., and Dantanarayana, D. M. 1965. Compatibility and variation in *Phytophthora* cultures isolated from *Hevea brasiliensis* in Ceylon. Trans. Br. Mycol. Soc. 48:631–637.

Persad, C. 1987. Studies on the control of black pod disease of cocoa in Trinidad. 10th Int. Cocoa Res. Conf. Santo Domingo, Dominican Repbulic, 17–23 May, 1987. Pages 437–440.

Pesti, M., and Ferenczy, L. 1979. Formation and regeneration of protoplasts from *Phytophthora infestans.* Acta Phytopathol. Acad. Sci. Hung. 14:1–4.

Petch, T. 1907. Report of the government mycologist. Ceylon Admin. Rep. 1906:C5. (Cited in Tucker 1933).

Petch, T. 1909. New diseases of rubber. Trop. Agric. 33:377–381.

Petch, T. 1914. Diseases of Hevea in Ceylon Proc. Int. Congr. Trop. Agric. 3:172. (Cited in Tucker 1933).

Petch, T. 1918. Fungous diseases of food crops in Ceylon. Trop. Agric. 50:159–163. (Cited in Tucker 1933).

Petch, T. 1920. Fungous diseases. Ceylon Agric. Soc. Yearb., 1919–1920:114–120. (Cited in Tucker 1933).

Petch, T. 1922. Additions to Ceylon Fungi II. Ann. Roy. Bot. Gard Peradeniya 7:279–322. (Cited in Tucker 1933).

Petch, T. 1924. Report of the Division of Botany and Mycology. Annu. Rep. Ceylon Dep. Agric. 1923. Pages D18–D19. (Cited in Tucker 1933).

Peters, L. 1912. Über eine Fruchtfäule von *Hevea brasiliensis* in Kamerun (On a fruit rot of *Hevea brasiliensis* in Cameroon). Mitt. Kais Biol. Anst. 12:18–22 (In German). (Cited in Tucker 1933).

Petersen, H. E. 1909. Studier over Ferskvands-Phycomyceter. Botanisk Tidskrift 29:345–440. (In Danish)

Petersen, H. E. 1910. An account of Danish freshwater-Phycomycetes with biological and systematical remarks. Ann. Mycol. 8:494–560.

Peterson, L. C. 1942. Studies on the late blight of potatoes and its causal organism, *Phytophthora infestans* (Mont.) de Bary (Abstr. of Ph.D. thesis, Cornell Univ. Pages 380–382.) (Rev. Appl. Mycol. 1945, 24:285).

Peterson, L. C. 1947. The overwintering of *Phytophthora infestans* (Mont.) de Bary under Long Island conditions. Am. Potato J. 24: 188–197.

Peterson, P. D., Jr., Campbell, C. L., and Griffith, C. S. 1992. James E. Teschemacher and the cause and management of potato blight in the United States. Plant Dis. 76:754–756.

Pethybridge, G. H. 1913. On the rotting of potato tubers by a new species of *Phytophthora* having a method of sexual reproduction hitherto undescribed. Sci. Proc. Roy. Dublin Soc. 13:529–565. (Cited in Tucker 1933).

Pethybridge, G. H. 1914. Further observations on *Phytophthora erythroseptica* Pethyb. and on the disease produced by it in the potato plant. Sci. Proc. Roy. Dublin Soc. 14:179–193.

Pethybridge, G. H. 1925. Plant Diseases Division. Annu. Gen. Rep. Ireland Dep. Agric. Tech. Instr. 23:62–63. (Cited in Tucker 1933).

Pethybridge, G. H., and Lafferty, H. A. 1919. A disease of tomato and other plants caused by a new species of *Phytophthora.* Sci. Proc. Roy. Dublin Soc. 15:487–503. (Cited in Tucker 1933).

Pethybridge, G. H., Moore, W. C., and Smith, A. 1934. Report on fungus, bacterial, and other disease crops in England and Wales for the years 1928–32. Min. Agric. Fish. Bull. 79. 117 pp.

Petri, L. 1917. Ricerche sulla morfologia e biologia della *Blepharospora cambivora, parasitica* del Castagno (Research on the morphology and biology of *Blepharospora cambivora, parasitica* from Chestnut). Atti. R. Accad. Lincei, Rend. Cl. Sci. Fis. Mat. Nat. Ser. 5. 26:297–299. (Cited in Tucker 1933).

Petri, L. 1925. L'agente del marciume radicale degli agrumi (The agent of root rot of citrus). Ann. R. Inst. Sup. For. Ser. II. 7 pp. (In Italian).

Petri, L. 1932. Rassegna dei casi fitapatologici osservati nel 1931 (Review of phytopathological records observed in 1931). Boll. Stn. Patol. Veg. Roma 12:1–64. (In Italian). (Rev. Appl. Mycol. 1932, 11:765–766). (Cited in Tucker 1933).

Petri, L. 1935. Rassegna dei casi fitopatologici osservati nel 1934

(Review of phytopathological records observed in 1934). Boll. Stn. Patol. Veg. Roma 15:1–95. (In Italian). (Rev. Appl. Mycol. 1935, 14:679–681).

Petri, L. 1941. Rassegna dei casi fitopatologici osservati nel 1940 (Review of phytopathological records noted in 1940). Boll. Stn. Patol. Veg. Roma 21:1–56. (In Italian). (Rev. Appl. Mycol. 1946, 25:492).

Pettit, T. R., and Pegg, G. F. 1991. The quantitative estimation of *Phytophthora cactorum* in infected strawberry tissue. Mycol. Res. 95:233–238.

Peyronel, B. 1920. Un interessante parasita del Lupino non ancora segnalato in Italia, *Blepharospora terrestris* (Sherb.) Peyr. Atti R. Acad. Lincei, Ser. 5. Rend. Cl. Sci. Fis. Mat. Nat. 29. I:194–197. (In Italian). (Cited in Tucker 1933).

Pfender, W. F. 1982. Monocyclic and polycyclic root diseases distinguishing between the nature of the disease cycle and the shape of the disease progress curve. Phytopathology 72:31–32.

Pfender, W. F., Hine, R. B., and Stanghellini, M. E. 1977. Production of sporangia and release of zoospores by *Phytophthora megasperma* in soil. Phytopathology 67:657–663.

Phillips, D. J., and Baker, R. 1962. Phytophthora crown rot of petunia. Plant Dis. Rep. 46:506–508.

Phillips, D., Grant, B. R., and Weste, G. 1987. Histological changes in the roots of an avocado cultivar, Duke 7, infected with *Phytophthora cinnamomi.* Phytopathology 77:691–698.

Phillips, D., Weste, G., and Hinch, J. M. 1991. Resistance to *Phytophthora cinnamomi* in callus tissue derived from three avocado cultivars. Can. J. Bot. 69:2026–2032.

Piccone, M. F., Whiley, A. W., and Pegg, K. 1987. Curing root rot in avocados. Aust. Hortic. (May). Pages 5–15.

Pieper, M., and Smith, M. A. L. 1988. A model whole plant microculture selection system for Kentucky bluegrass. Crop Sci. 28:611–614.

Pieroh, E. A., Krass, W., and Hemmen, C. 1978. Propamocarb, ein neues Fungizid zur Abwehr von Oomyceten im Zierpflanzen- und Gemüsebau (Propamocarb, a new fungicide for protection in ornamental plants and garden vegetables from Oomycetes). Meded. Fac. Landbouwwet. Rijksuniv. Gent 43:933–942. (In German).

Pietkiewicz, J. B. 1978. Stability of *Phytophthora infestans* (Mont.) de Bary races in culture. Prot. Ecol. 1:55–62.

Pillai, P. N. R. 1982. Abnormal leaf fall of rubber caused by *Phytophthora* spp. Pages 17–23 in: Proc. Workshop on *Phytophthora* Diseases of Cultivated Plants. K. K. N. Nambiar, ed. Cent. Plant. Crops. Res. Inst., Kasaragod, Kerala, India. 284 pp.

Pillai, P. N. R., and Radha, K. 1977. Aerial spraying against abnormal leaf fall disease of rubber in India. Planters' Bull. 148:10–14.

Pinching, H. C. 1923. Rubber in Burma. Bull. Rubber Grow. Assoc. 10:647–655.

Pinckard, J. A., and Guidroz, G. F. 1973. A boll rot of cotton caused by *Phytophthora parasitica.* Phytopathology 63:896–899.

Pinkas, Y., Kariv, A., and Katan, J. 1984. Soil solarization for the control of *Phytophthora cinnamomi:* Thermal and biological effects. (Abstr.). Phytopathology 74:796.

Pinto, De T. A. 1966. *Phytophthora cactorum* (L. et C.) Schroet., Nuevo patogeno del almendro en Chile (*Phytophthora cactorum* [L. et. C.] Schroet., new pathogen of almond in Chile). Fitopatol. Chile 1:24–26. (In Spanish).

Pirone, P. P. 1937. Marigold wilt (Abstr.). Phytopathology 27:138.

Pittis, J. E., and Colhoun, J. 1984. Isolation and identification of Pythiaceous fungi from irrigation water and their pathogenicity to *Antirrhinum*, tomato and *Chamaecyparis lawsoniana.* Phytopathol. Z. 110:301–318.

Pittis, J. E., and Shattock, R. C. 1994. Viability, germination and infection potential of oospores of *Phytophthora infestans.* Plant Pathol. 43:387–396.

Pivot, V., Bruneteau, M., Mas, P., Molot, P. M., and Michel, G. 1991. Isolement, identification et activité biologique de phospholipids isolés du mycélium de *Phytophthora capsici* (Isolation, identification and biological activity of phospholipids from *Phytophthora capsici* mycelium). C. R. Acad. Sci. Paris 313:259–264.

Plaats-Niterink, A. J. van der. 1981. Monograph of the genus *Pythium.* Centralbureau voor Schimmelcultures, Baarn, Studies in Mycology 21:1–242.

Plakidas, A. G. 1964. Strawberry Diseases. Louisiana State Univ. Press, Baton Rouge. 195 pp.

Platt, H. W. 1983. Effects of metalaxyl, mancozeb, and chlorothalonil on blight yield, and tuber rot of potato. Can. J. Plant Pathol. 5:38–42.

Platt, H. W. 1985a. Failure of metalaxyl to eradicate late blight on potatoes (cv. Green Mountain). Am. Potato J. 62:39–46.

Platt, H. W. 1985b. Controlling potato late blight with systemic-protectant fungicide combinations of metalaxyl and mancozeb. Am. Potato J. 62:499–510.

Platt, H. W., and McRae, K. B. 1990. Assessment of field responses of potato cultivars and breeder seedlings to potato late blight epidemics. Am. Potato J. 67:427–441.

Ploetz, R. C., and Engelhard, A. W. 1980. Crown and root rot of gloxinia and other gesneriads caused by *Phytophthora parasitica.* Plant Dis. 64:487–490.

Ploetz, R. C., and Mitchell, D. J. 1989. Root rot of bamboo palm caused by *Phytophthora arecae.* Plant Dis. 73:266–269.

Ploetz, R. C., and Parrado, J. L. 1988. Quantitation and detection of *Phytophthora cinnamomi* in avocado production areas of South Florida. Plant Dis. 72:981–984.

Plourde, D. F., and Green, R. J., Jr. 1982. Effect of monochromatic light on germination of oospores and formation of sporangia of *Phytophthora citricola.* Phytopathology 72:58–61.

Podger, F. D. 1972. *Phytophthora cinnamomi,* a cause of lethal disease in indigenous plant communities in Western Australia. Phytopathology 62:972–981.

Podger, F. D., and Newhook, F. J. 1971. *Phytophthora cinnamomi* in indigenous plant communities in New Zealand. N.Z. J. Bot. 9:625–638.

Podger, F. D., Doepel, R. F., and Zentmyer, G. A. 1965. Association of *Phytophthora cinnamomi* with a disease of *Eucalyptus marginata* forest in Western Australia. Plant Dis. Rep. 49:943–947.

Poedinok, N. L., Dolgova, A. V., and D'yakov, Y. T. 1982. Parasexual process in phytopathogenic fungus *Phytophthora infestans* (Mont.) de Bary. Genetika 18:1423–1428. (Translated from Russian).

Polach, F. J., and Webster, R. K. 1972. Identification of strains and inheritance of pathogenicity in *Phytophthora capsici.* Phytopathology 62:20–26.

Ponchet, J., Ricci, P., Andreoli, C., and Augé, G. 1972. Méthodes sélectives d'isolment du *Phytophthora nicotianae* f. sp. *parasitica* (Dastur) Waterh. á partir du sol (Selective methods of isolation for *Phytophthora nicotianae* f. sp. *parasitica* [Dastur] Waterh. from the soil). Ann. Phytopathol. 4:97–108. (In French).

Ponchet, J., Andreoli, C., and Augé, G. 1975. Essai de caractérisation du pouvoir pathogene chez quelques especes du genre *Phytophthora* (An attempt to characterize the pathogenicity of some *Phytophthora* species). Ann. Phytopathol. 7:105–114. (In French).

Pond, E., Menge, J. A., and Ohr, H. D. 1984. The effect of metalaxyl and efosite-al applied through the drip irrigation system on *Phytophthora parasitica* in the soil and on the yield of navel oranges (Abstr.). Phytopathology 74:854.

Pontecorvo, G. 1956. The parasexual cycle in fungi. Annu. Rev. Microbiol. 10:393–400.

Pontis, R. E. 1941. El "mal de la tinta" del nogal en la República Argentina ("Ink disease" of walnut in the Argentine Republic). Rev. Argent. Agron. 8:317–325. (In Spanish).

Pontis, R. E. 1945. *Phytophthora capsici* en frutos de zapallito de ronco (*Phytophthora capsici* on squash fruits). Rev. Argent. Agron. 12:17–21. (In Spanish).

Pontis Videla, R. E. 1953. Outbreaks and new records. Venezuela. F.A.O. Plant Prot. Bull. 1:91–92, 140.

Pontis, R. E., and Feldman, J. M. 1959. Una podredumbre del cuello y las raices de *Callistephus chinensis* en la Argentina, causada por *Phytophthora cryptogea* (Collar and root rot of *Callistephus chinensis* in Argentina caused by *Phytophthora cryptogea*). Rev. Argent. Agron. 26:17–22. (In Spanish).

Pontis, R. E., and Rodriguez-Landaeta, A. 1953. Una podredumbre de los frutos de la berenjena (*Solanum melongena* L.) en Venezuela causada por *Phytophthora capsici* (Fruit rot of eggplant [*Solanum melongena* L.] in Venezuela caused by *Phytophthora capsici* Leonian). Agron. Trop. (Maracay) 3:117–119. (In Spanish).

Popisil, F., Karikari, S. K., and Boamah-Mensah, E. 1972. Black pepper and its potential as a spice in Ghana. The Ghana Farmer 16:70–81.

Powell, N. T. 1979. Internal synergisms among organisms including disease. Pages 113–133 in: Plant Disease. Vol. IV. J. G. Horsfall and E. B. Cowling, eds. Academic Press, New York.

Powell, N. T., and Nusbaum, C. J. 1960. The black shank-root-knot complex in flue-cured tobacco. Phytopathology 50:899–906.

Pratt, B. H. 1971. Isolation of Basidomycetes from Australian eucalyptus forest and assessment of their antagonism to *Phytophthora cinnamomi.* Trans. Br. Mycol. Soc. 56:243–250.

Pratt, B. H., and Heather, W. A. 1972. Method for rapid differentiation of *Phytophthora cinnamomi* from other *Phytophthora* species isolated from soil by lupin baiting. Trans. Br. Mycol. Soc. 59:87–96.

Pratt, B. H., and Heather, W. A. 1973a. The origin and distribution of *Phytophthora cinnamomi* Rands in Australian native plant communities and the significance of its association with particular plant diseases. Aust. J. Biol. Sci. 26:559–573.

Pratt, B. H., and Heather, W. A. 1973b. Recovery of potentially pathogenic *Phytophthora* and *Pythium* spp. from native vegetation in Australia. Aust. J. Biol. Sci. 26:575–582.

Pratt, B. H., Sedgley, J. H., Heather, W. A., and Shepherd, C. J. 1972a. Oospore production in *Phytophthora cinnamomi* in the presence of *Trichoderma koningii.* Aust. J. Biol. Sci. 25:861–863.

Pratt, B. H., Heather, W. A., and Shepherd, C. J. 1972b. Transcontinental occurrence of A_1 and A_2 strains of *Phytophthora cinnamomi* in Australia. Aust. J. Biol. Sci. 25:1099–1100.

Pratt, B. H., Heather, W. A., and Shepherd, C. J. 1973. Recovery of *Phytophthora cinnamomi* from native vegetation in a remote area of New South Wales. Trans. Br. Mycol. Soc. 60:197–204.

Pratt, R. G. 1981. Morphology, pathogenicity, and host range of *Phytophthora megasperma, P. erythroseptica,* and *P. parasitica* from arrowleaf clover. Phytopathology 71:276–282.

Pratt, R. G., and Mitchell, J. E. 1973. Conditions affecting the detection of *Phytophthora megasperma* in soils of Wisconsin alfalfa fields. Phytopathology 63:1374–1379.

Pratt, R. G., and Mitchell, J. E. 1975. The survival and activity of *Phytophthora megasperma* in naturally infested soils. Phytopathology 65:1267–1272.

Pratt, R. G., and Mitchell, J. E. 1976. Interrelationships of seedling age, inoculum, soil moisture level, temperature, and host and pathogen genotype in Phytophthora root rot of alfalfa. Phytopathology 66:81–85.

Pratt, R. G., Mitchell, J. E., and Willis, D. A. 1975. Resistance and susceptibility to *Phytophthora megasperma* expressed in alfalfa cotyledons. Phytopathology 65:365–369.

Pratt, R. G., Roth, L. F., Hansen, E. M., and Ostrofsky, W. D. 1976. Identity and pathogenicity of species of *Phytophthora* causing root rot of Douglas fir in the Pacific Northwest. Phytopathology 66: 710–714.

Preisig, C. L., and Kuć, J. 1987. Phytoalexins, elicitors, enhancers, suppressors, and other considerations in the regulation of R-gene resistance to *Phytophthora infestans* in potato. Pages 203–221 in: Molecular Determinants of Plant Diseases. S. Nishimura, C. P. Vance, and N. Doke, eds. Japan Sci. Press, Tokyo, and Springer-Verlag, Berlin. 293 pp.

Prell, H. H., Karlovsky, P., and Bahnweg, G. 1991. Towards transformation in *Phytophthora nicotianae.* Pages 312–325 in: *Phytophthora.* J. A. Lucas, R. C. Shattock, D. S. Shaw, and L. R. Cooke, eds. Cambridge Univ. Press, Cambridge, U.K. 447 pp.

Price, T. V., and Fox, P. 1986. Studies on the behaviour of furalaxyl on Pythiaceous fungi and cucumbers in recirculating hydroponic systems. Aust. J. Agric. Res. 37:65–77.

Prillieux, E. 1895. Maladies des plantes agricoles et des arbres fruitiers et forestiers causees par des parasites vegetaux (Diseases of agricultural plants and fruit and forest trees caused by plant parasites). Vol. I. Paris. (Cited in Hartley 1921). (In French).

Prinsloo, G. C., and Pauer, G. D. C. 1973. Die identifikasie van rasse van *Phytophthora nicotianae* (B. de Haan) var. *nicotianae* Wat in Suid Afrika voirkom (The identification of a race of *Phytophthora nicotianae* [B. de Haan] var. *nicotianae* found in South Africa). Phytophylactica 6:217–220. (In Dutch).

Prinsloo, G. C., de Villiers, D. A., and Norton, A. 1979. Field trials with prothiocarb (SN 41703) for the control of black shank (*Phytophthora nicotianae* var. *nicotianae*) of tobacco. Phytophylactica 11:149–150.

Prior, C. 1986. Sudden death of cocoa in Papua New Guinea associated with *Phytophthora palmivora* cankers invaded by bark beetles. Ann. Appl. Biol. 109:535–543.

Pristou, R., and Gallegly, M. E. 1954. Leaf penetration by *Phytophthora infestans.* Phytopathology 44:81–86.

Psarros, E. 1978. New diseases caused by *Phytophthora* spp. Phytophthora Newsl. 6:72–73.

Psarros, E. E., and Tzavella-Khomari, K. 1982. Attack of a hazelnut by a *Phytophthora.* Proc. Greek Sci. Soc. Hortic. 1:35–36.

Pscheidt, J. W., Burket, J. Z., Fischer, S. L., and Hamm, P. B. 1992. Sensitivity and clinical use of *Phytophthora*-specific immunoassay kits. Plant Dis. 76:928–932.

Pshedetskaya, L. I. 1968. Infection of maple and lime leaves by different races of *Phytophthora infestans* (Mont.) de Bary. Mikol Fitopatol. 2:329–330 (In Russian).

Pullman, G. S., De Vay, J. E., Garber, R. H., and Weinhold, A. R. 1979. Control of soil borne pathogens by plastic tarping of soil. Pages 439–446 in: Soil-Borne Plant Pathogens. B. Schippers and W. Gams, eds. Academic Press, London. 686 pp.

Pulli, S. K., and Tesar, M. B. 1975. Phytophthora root rot in seedling-year alfalfa as affected by management practices inducing stress. Crop Sci. 15:861–864.

Punja, Z. K., and Ormrod, D. J. 1979. New or noteworthy plant diseases in coastal British Columbia 1975 to 1977. Can. Plant Dis. Surv. 59:22–32.

Purss, G. S. 1953. The fruit rots of the custard apple. Queensl. J. Agric. Sci. 10:247–265.

Purss, G. S. 1957. Stem rot: A disease of cowpeas caused by an undescribed species of *Phytophthora.* Queensl. J. Agric. Sci. 14:125–154.

Purss, G. S. 1958. Studies on varietal resistance to stem rot (*Phytophthora vignae* Purss) in the cowpea. Queensl. J. Agric. Sci. 15:1–14.

Purss, G. S. 1959. Root rot of lucerne. Queensl. Agric. J. 85:767–770.

Purss, G. S. 1963. Coloona—Stem rot resistant cowpea. Queensl. Agric. J. 89:756–759.

Purss, G. S. 1972. Pathogenic specialization in *Phytophthora vignae.* Aust. J. Agric. Res. 23:453–456.

Purss, G. S. 1975. A new stem-rot resistant cowpea cultivar. Queensl. Agric. J. 101:551.

Putnam, M. 1991. Root rot of lavender caused by *Phytophthora nicotianae.* Plant Pathol. 40:480–482.

Putnam, M. L., and Mitchell, J. E. 1984. The use of metalaxyl to control Phytophthora root rot of ginseng (*Panax quinquefolius*) (Abstr.). Phytopathology 74:854–855.

Puttemans, A. 1937. Reivindicacao visando a deno minacao scientifica da doenea da batateira *Phytophthora infestans* (Mont.) de Bary (A plea for the revision of the scientific name of potato blight *Phytophthora infestans* [(Mont.] de Bary). Rodriguesia 2:341–350. (In Portuguese). (Rev. Appl. Mycol. 1938, 17:60–61).

Quillec, G., Renard, J. L., and Ghesquiere, H. 1984. Le *Phytophthora heveae* du cocotier: Son role dans la pourriture du coeur et dans la chute des noix (*Phytophthora heveae* of coconut: Role in bud rot and nutfall). Oleagineux 39:477–485. (In French).

Raabe, R. D., Conners, L., and Martinez, A. P. 1981. Check list of Plant Diseases in Hawaii. Hawaii Inst. Agric. Human Res. Inf. Text Series No. 22. 313 pp.

Raciborski, M. 1900. Parasitische Algen und Pilze, Java's (Java's parasitic algae and fungi). I. Batavia. Page 9. (Cited in Waterhouse 1970a under *P. colocasiae*).

Rader, W. E. 1952. Diseases of stored carrots in New York State. N.Y. (Cornell) Agric. Exp. Stn. Bull. 889. 64 pp.

Radha, K., and Joseph, T. 1974. Investigation on the bud rot disease (*Phytophthora palmivora*) of coconut. PL-480 Final Rep. Cent. Plant. Crops Res. Inst., Kayangulam, Kerala, India. 30 pp.

Radha, K., and Joseph, T. 1982. Bud rot of coconut. Pages 11–16 in: Proc. Workshop on *Phytophthora* Diseases of Tropical Cultivated Plants 19–23 September 1980. K. K. N. Nambiar, ed. Cent. Plant. Crops Res. Inst., Kasaragod, Kerala, India. 284 pp.

Radziah, N. Z., and Hashim, I. 1990. Major leaf diseases of rubber and their management. Plant. Bull. Rubber Res. Inst. Malays. 204:67–79.

Raghaven-Charudattan, and Charudattan, R. 1987. DeVine TM leads the way. Sprays of living fungi can help Florida in its $360 million war with weeds. Univ. Florida. Research 87. Biocontrol. Pages 38–39. (Rev. Plant Pathol. 1987, 66:2168.)

Raghavendra Rao, N. N. 1985. Relative efficacy of a single pre-monsoon spray of some fungicides for the control of brown fruit-rot of citrus. Indian J. Agric. Sci. 55:189–192.

Rahman, M. A., and Uddin, K. S. 1985. Rubber diseases in Thailand and their bearing on planting stocks imported to Thailand. Bano-Biggyan-Patrika 14:5–11.

Rahimian, M. K., and Banihashemi, Z. 1979. A method for obtaining zoospores of *Pythium aphanidermatum* and their use in determining cucurbit seedling resistance to damping-off. Plant Dis. Rep. 63:658–661.

Rahimian, M. K., and Mitchell, J. E. 1988. Detection and quantification of *Phytophthora cactorum* in naturally infested soils. Phytopathology 78:949–952.

Raj, J. S., and Jose, P. C. 1966. A Phytophthora wilt of pepper, *Piper nigrum.* Sci. Cult. 32:90–92.

Raj, S. A., Padmanabhan, C., and Narasimhan, V. 1973. Efficacy of certain fungicides in controlling the wilt disease of betelvine (*Piper betle* L.). Madras Agric. J. 60:632–633.

Raj, S. A., Bhattacharyya, S. K., and Sharma, S. R. 1976. New hosts of *Phytophthora infestans* in nature. Indian Phytopathol. 29:342–343.

Rajalakshmy, V. K., Joseph, A., and Arthassery, S. 1985. Occurrence of two mating groups in *Phytophthora meadii* causing abnormal leaf fall disease of rubber in South India. Trans. Br. Mycol. Soc. 85:723–725.

Ralton, J. E., Howlett, B. J., Clarke, A. E., Irwin, J. A. G., and Imrie, B. 1988. Interaction of cowpea with *Phytophthora vignae:* Inheritance of resistance and production of phenylaline ammonia-lyase as a resistance response. Physiol. Mol. Plant Pathol. 32:89–103.

Ram, R. D. 1983. Relative resistance in wild species of pome and stone fruits to collar rot of apple caused by *Phytophthora cactorum* (Lebert & Cohn) Schroet. Prog. Hortic. 15:283–285.

Ramachandran, N., and Sarma, Y. R. 1985. Efficacy of three systemic fungicides in controlling *Phytophthora* infections of black pepper. Indian Phytopathol. 38:160–162.

Ramakrishnan, K., and Thankappan, M. 1965. First report of black pod disease of cocoa in India. S. Indian Hortic. 13:33–34.

Ramakrishnan, T. S. 1960. Notes on some fungi from South India. VIII. Proc. Indian Acad. Sci. Sect. B 51:164–168.

Ramakrishnan, T. S., and Radhakrishna Pillai, P. N. 1963. Black stripe, black thread, or bark rot. Rubber Bd. Bull. 6:110–112.

Ramakrishnan, T. S., and Seethalakshmi, V. 1956a. Studies in the genus *Phytophthora.* III. Homothallic strains of *Phytophthora* on *Areca catechu.* Proc. Indian Acad. Sci. Sect. B 43:308–313.

Ramakrishnan, T. S., and Seethalakshmi, V. 1956b. Studies in the genus *Phytophthora.* IV. New hosts for *Phytophthora palmivora* from South India. Proc. Indian Acad. Sci. Sect. B 44:79–84.

Ramakrishnan, T. S., and Soumini, C. K. 1947. Fruit rot of tomatoes caused by *Phytophthora palmivora* Butl. Proc. Indian Acad. Sci. Sect. B 25:39–42.

Ramakrishnan, T. S., and Soumini, C. K. 1948. Wilt disease of *Pyrethrum.* Indian Phytopathol. 1:27–33.

Ramarao, P., and Umabala, K. 1982. Survival of *Phytophthora palmivora* in soil. Indian Phytopathol. 35:65–67.

Ramirez, B. N., and Mitchell, D. J. 1975. Relationship of density of chlamydospores and zoospores of *Phytophthora palmivora* in soil to infection of papaya. Phytopathology 65:780–785.

Ramsfjell, T. 1951. Soppsjukdomar pa Eplefrukt (Fungal diseases of apple fruit). Frukt. og Boer 4:79–94. (In Dutch). (Cited in Nienhaus 1960).

Rana, K. S., and Gupta, V. K. 1984. Effect of fungicides on the viability of *Phytophthora cactorum* propagules in the soil. Phytopathol. Z. 110:245–250.

Rands, R. D. 1922. Streepkanker van Kaneel, veroorzaakt door *Phytophthora cinnamomi* n. sp. (Stripe canker of cinnamon caused by *Phytophthora cinnamomi* n. sp.). Meded. Inst. Plantenziekten 54. 41 pp. (In Dutch).

Rangaswami, G. 1962. Pythiaceous Fungi. Indian Council Agric. Res. New Delhi. 276 pp.

Rao, N. N. G. 1985. Relative efficacy of a single pre-monsoon spray of some fungicides for the control of brown fruit rot of citrus. Indian J. Agric. Sci. 55:189–192.

Rao, V. G. 1965. Some new records of market and storage diseases of vegetables in Bombay, Maharashtra. Mycopathol. Mycol. Appl. 27:49–59.

Rao, V. G. 1970. Contributions to the genus *Phytophthora* de Bary from India. Mycopathol. Mycol. Appl. 42:241–258.

Rao, V. G., and Ullasa, B. A. 1971. A new *Phytophthora* blossom blight of *Gerbera* from India. Indian Phytopathol. 24:386–387.

Rao, V. G., Desai, M. K., and Kulkarni, N. B. 1962a. A new Phytophthora fruit rot of *Achras sapota* from India. Plant Dis. Rep. 46:381–382.

Rao, V. G., Desai, M. K., and Kulkarni, N. B. 1962b. A new Phytophthora fruit rot of *Annona squamosa* from India. Plant Dis. Rep. 46:874–876.

Rao, V. G., Desai, M. K., and Kulkarni, N. B. 1963. A new Phytophthora fruit rot of *Anona squamosa* L. Sci. Cult. 29:199–200.

Rathbun-Gravatt, A. 1925. Direct inoculation of coniferous stems with damping-off fungi. J. Agric. Res. 30:327–339.

Rathbun-Gravatt, A. 1931. Germination loss of coniferous seeds due to parasites. J. Agric. Res. 42:71–92.

Rattink, H. 1979. Phytophthora wilt of carnation in the Netherlands. Neth. J. Plant Pathol. 85:83–84.

Rattink, H. 1981. Characteristics and pathogenicity of six *Phytophthora* isolates from pot plants. Neth. J. Plant Pathol. 87:83–90.

Rattink, H., and Van Steekelenburg, N. A. M. 1973. The occurrence of *Phytophthora citricola* Sawada sensu Waterhouse (syn. *P. cactorum* [Leb. et Cohn] Schroet. var. *applanata* Chest.) in ornamentals in the Netherlands. Neth. J. Plant Pathol. 79:277–278.

Ravisé, A. 1966. Observations sur la reproduction sexueé de souches du *Phytophthora palmivora* (Butl.) Butl. parasite de cultures tropicales (Observation on the sexual reproduction of isolates of *Phytophthora palmivora* [Butl.] Butl., parasite of tropical crops). Cah. ORSTOM Ser. Biol. 2:91–101. (In French).

Ravisé, A., and Boccas, B. 1969. Premiere liste annotee des Pythiacees parasites des plantes cultivees au Congo (First lists of Pythiaceous parasites of plants cultivated in the Congo). Cah. Maboke 7:41–69. (In French).

Rawlins, S. L. 1971. Some new methods for measuring the components of water potential. Soil Sci. 112:8–16.

Rawn, C. D., and Van Etten, J. L. 1978. A shake culture method for Pythiaceae applicable to rapid, small-scale assay of vegetative physiology. Phytopathology 68:1384–1388.

Reddick, D. 1928. Blight-resistant potatoes. Phytopathology 18:483–502.

Reddick, D. 1932. Some diseases of wild potatoes in Mexico. Phytopathology 22:609–612.

Reddick, D. 1943. Development of blight-immune varieties. Am. Potato J. 20:118–126.

Reddy, D. B., ed. 1970. Breadfruit diseases in the Trust Territory of Pacific Islands. Inf. Letter F.A.O. Plant Prot. Comm. S.E. Asia and Pac. Reg. Tech. Doc. 79. 8 pp. (Rev. Plant Pathol. 1973, 52:2697).

Reddy, D. B., ed. 1973. Preliminary list of pests and diseases of principal crops in French Polynesia. F.A.O. Plant Prot. Comm. S.E. Asia and Pac. Reg. Tech. Doc. 89. 5 pp.

Reddy, M. K., and Anandaraj, M. 1982. Koleroga of areca nut. Pages 71–79 in: Proc. Workshop *Phytophthora* Diseases of Tropical Cultivated Plants. K. K. N. Nambiar, ed. Cent. Plant. Crops Res. Inst., Kasaragod, India. 284 pp.

Reddy, M. K., Saraswathy, N., and Chandramohanan, R. 1978. Diseases of arecanut in India—A review and further considerations. J. Plant. Crops 6:28–43.

Redondo, E. J., and Galindo, J. A. 1989. Obtensión de plantas diferenciales para determinar razas pathogénicas del Hongo *Phytophthora capsici* (Obtaining differential plants to determine pathogenic races of *Phytophthora capsici*) (Abstr.). Fitopatologica 24:20–21.

Redondo, E., Rodriguez, R., Ortega, M., Larqué, A., and Engleman, E. M. 1989. Mecanismo de infeccion y patologica de las plantas de Chile susceptibles y resistentes al hongo *Phytophthora capsici*

Leo. (Mechanism of infection and pathology of plants of Chile [pepper] susceptible and resistant to the fungus *Phytophthora capsici* Leo.). Agrociencia 77:123–137. (In Spanish).

Reeves, R. J. 1975. Behaviour of *Phytophthora cinnamomi* Rands in different soils and water regimes. Soil Biol. Biochem. 7:19–24.

Reeves, R. J., and Jackson, R. M. 1972. Induction of *Phytophthora cinnamomi* oospores in soil by *Trichoderma viride.* Trans. Br. Mycol. Soc. 59:156–159.

Reeves, R. J., and Jackson, R. M. 1974. Stimulation of sexual reproduction in *Phytophthora* by damage. J. Gen. Microbiol. 84:303–310.

Reichle, R. E. 1969a. Retraction of flagella by *Phytophthora parasitica* var. *nicotianae* zoospores. Arch. Mikrobiol. 66:340–347.

Reichle, R. E. 1969b. Fine structure of *Phytophthora parasitica* zoospores. Mycologia 61:30–51.

Reilly, C. C., Byron, G. A., Hendrix, F. F., and Hotchkiss, M. W. 1989. First report of *Phytophthora cactorum* causing a shuck and kernel rot of pecan fruit (Abstr.). Plant Dis. 73:701.

Reilly, J. J. 1980. Chemical control of black shank of tobacco. Plant Dis. 64:274–277.

Reiner, R. 1982. Antibiotics, an Introduction. Georg Thieme Verlag, Stuttgart and New York. 172 pp.

Reinking, O. A. 1918. Philippine economic-plant diseases. Philipp. J. Sci. Sect. A 13:165–216; 217–274. (Cited in Tucker 1933).

Reinking, O. A. 1919a. Diseases of economic plants in Southern China. Philipp. Agric. 8:109–134. (Cited in Tucker 1933).

Reinking, O. A. 1919b. *Phytophthora faberi* Maubl., the cause of coconut bud rot in the Philippines. Philipp. J. Sci. 14:131–150. (Cited in Tucker 1933).

Reinking, O. A. 1921. Citrus diseases in the Philippines, Southern China, IndoChina, and Siam. Philipp. Agric. 9:121–179. (Cited in Tucker 1933).

Reinking, O. A. 1923. Comparative study of *Phytophthora faberi* on coconut and cacao in the Philippine Islands. J. Agric. Res. 25:267–284. (Cited in Tucker 1933).

Reitsma, J., and Sloof, W. C. 1947. A disease of eggplant fruits caused by *Phytophthora parasitica* Dastur and *P. palmivora* Butler. Chron. Nat. 103:60–63.

Revilla, V. A. 1963. Nueva enfermedad del pepino en el Perú (New disease of cucumber in Peru). Agronomia. (La Molina) 30:26–29. (In Spanish).

Reynolds, K. M., Bulger, M. A., Madden. L. V., and Ellis, M. A. 1987. New methods using simulated rain to study the splash dispersal of plant pathogens. Phytopathology 77:921–926.

Reynolds, K. M., Madden, L. V., and Ellis, M. A. 1988. Effect of weather variables on strawberry leather rot epidemics. Phytopathology 78:822–827.

Reynolds, K. M., Madden, L. V., Reichard, D. L., and Ellis, M. A. 1989. Splash dispersal of *Phytophthora cactorum* from infected strawberry fruit by simulated canopy drip. Phytopathology 79:425–432.

Rhodes, L. H., and Myers, D. K. 1989. Effect of seed treatment with metalaxyl or pyroxyfur on damping-off of alfalfa caused by *Phytophthora megasperma* f. sp. *medicaginis.* Crop Prot. 8:369–372.

Ribeiro, O. K. 1978. A Source Book of the Genus *Phytophthora.* Cramer, Vaduz, Liechtenstein. 417 pp.

Ribeiro, O. K. 1983. Physiology of asexual sporulation and spore germination in *Phytophthora.* Pages 55–70 in: *Phytophthora:* Its Biology, Taxonomy, Ecology, and Pathology. D. C. Erwin, S. Bartnicki-Garcia, and P. H. Tsao, eds. American Phytopathological Society, St. Paul, Minn. 392 pp.

Ribeiro, O. K., and Linderman, R. G. 1991. Chemical and biological control of *Phytophthora* species in woody plants. Pages 399–410 in: *Phytophthora.* J. A. Lucas, R. C. Shattock, D. S. Shaw, and L. R. Cooke, eds. Cambridge Univ. Press, Cambridge, U.K. 447 pp.

Ribeiro, O. K., Erwin, D. C., and Zentmyer, G. A. 1975a. An improved synthetic medium for oospore production and germination of several *Phytophthora* species. Mycologia 67:1012–1019.

Ribeiro, O. K., Zentmyer, G. A., and Erwin, D. C. 1975b. Comparative effects of monochromatic radiation on the germination of oospores of three *Phytophthora* spp. Phytopathology 65:904–907.

Ribeiro, O. K., Zentmyer, G. A., and Erwin, D. C. 1976a. The influence of qualitative and quantitative radiation on reproduction and spore germination of four *Phytophthora* species. Mycologia 68:1162–1173.

Ribeiro, O. K., Zentmyer, G. A., and Erwin, D. C. 1976b. Cytochromes in the mycelium and oospores of *Phytophthora capsici.* Phytopathology 66:172–174.

Ribeiro, O. K., Erwin, D. C., and Khan, R. A. 1978. A new high-temperature *Phytophthora* pathogenic to roots of alfalfa. Phytopathology 68:155–161.

Ricci, P. 1972. Moyens d'etude de l'inoculum du *Phytophthora nicotianae* f. sp. *parasitica* (Dastur) Waterh., parasite de l'oeillet dans le sol (*Phytophthora nicotianae* f. sp. *parasitica* [Dastur] Waterh. of carnation: Quantitative estimation of inoculum in the soil). Ann. Phytopathol. 4:257–276. (In French).

Ricci, P., Bonnet, P., Ponchet, M., Molot, P. M., and Bruneteau, M. 1988. Recherche et intérêt d'eliciteurs fongiques dans les interactions de trois especes végétales avec des champignons phytopathogenes (Search for and importance of fungal elicitors in the interaction of three plant species with plant pathogenic fungi of the genus *Phytophthora*). Cryptogam. Mycol. 9:191–200. (In French).

Ricci, P., Pope-de-Vallavielle, C., Panabieres, F., Marais, A., and Augé, G. 1990. Caracteres comparés des especes de *Phytophthora* pathogenes des agrumes (Characterization of the *Phytophthora* spp. attacking citrus). EPPO (Eur. Mediterr. Plant Prot. Organ.) Bull. 20:19–28. (In French).

Ricci, P., Trentin, F., Bonnet, P., Venard, P., Mouton-Perronnet, F., and Bruneteau, M. 1992. Differential production of parasiticein, an elicitor of necrosis and resistance in tobacco, by isolates of *Phytophthora parasitica.* Plant Pathol. 41:298–307.

Rich, A. E. 1983. Potato Diseases. Academic Press, New York. 238 pp.

Rich, S. 1978. External morphology of zoospores of *Phytophthora parasitica* var. *nicotianae* race 3. Mycologia 70:679–684.

Richards, S. J., and Hagen, R. M. 1958. Soil moisture tensiometer: How it works, how to use it. Calif. Agric. Exp. Stn. Ext. Serv. Leafl. 100. 8 pp.

Richardson, L. T. 1976. Variability in sensitivity to the fungicide Pyroxychlor within the genera *Phytophthora* and *Pythium.* Can. J. Plant Sci. 56:365–369.

Ridings, W. H. 1986. Biological control of stranglevine in citrus—A researcher's view. Weed Sci. 34 (Suppl. 1):31–32.

Ridings, W. H., and McRitchie, J. J. 1974. Phytophthora leaf spot of *Philodendron oxycardium* and related species. Proc. Fla. State Hortic. Soc. 87:442–447.

Ridings, W. H., Gallegly, M. E., and Lilly, V. G. 1969. Thiamine requirements helpful in distinguishing isolates of *Pythium* from those of *Phytophthora.* Phytopathology 59:737–742.

Ridings, W. H., Mitchell, D. J., Schoulties, C. L., and Elcholl, N. E. 1976. Biological control of milkweed vine for Florida citrus groves with a pathotype of *Phytophthora citrophthora.* Pages 224–240 in: Proc. IV Int. Symp. Biol. Control of Weeds. T. E. Freeman, ed. Center for Environmental Programs, Inst. Food Agric. Serv., Univ. Florida, Gainesville. 298 pp.

Rimfeldt, K. R. 1979. Root rot of *Chamaecyparis* Gartneryrket 69:368–370.

Ristaino, J. B. 1990. Intraspecific variation among isolates of *Phytophthora capsici* from pepper and cucurbit fields in North Carolina. Phytopathology 80:1253–1259.

Ristaino, J. B. 1991. Influence of rainfall, drip irrigation, and inoculum density on the development of Phytophthora root and crown rot epidemics and yield in bell pepper. Phytopathology 81:922–929.

Ristaino, J. B., and Duniway, J. M. 1989. Effect of preinoculation and post inoculation water stress on the severity of Phytophthora root rot in processing tomatoes. Plant Dis. 73:349–352.

Ristaino, J. B., Duniway, J. M., and Marois, J. J. 1988. Influence of frequency and duration of furrow irrigation on the development of Phytophthora root rot and yield in processing tomatoes. Phytopathology 78:1701–1706.

Ristaino, J. B., Duniway, J. M., and Marois, J. J. 1989. Phytophthora root rot and irrigation schedule influence growth and phenology of processing tomatoes. J. Am. Soc. Hortic. Sci. 114:556–561.

Ristaino, J. B., Larkin, R. P., and Campbell, C. L. 1993. Spatial and temporal dynamics of *Phytophthora* epidemics in commercial bell pepper fields. Phytopathology 83:1312–1320.

Rivera-Peña, A. 1990a. Wild tuber-bearing species of *Solanum* and incidence of *Phytophthora infestans* (Mont.) de Bary on the western slopes of the volcano Nevado de Toluca. II. Distribution of *Phytophthora infestans.* Potato Res. 33:341–347.

Rivera-Peña, A. 1990b. Wild tuber-bearing species of *Solanum* and incidence of *Phytophthora infestans* (Mont.) de Bary on the western slopes of the volcano Nevado de Toluca. III. Physiological races of *Phytophthora infestans.* Potato Res. 33:349–355.

Rivera-Peña, A., and Molina-Galan, J. 1989. Wild tuber-bearing species of *Solanum* and incidence of *Phytophthora infestans* (Mont.) de Bary on the western slopes of the volcano Nevado de Toluca. I. *Solanum* species. Potato Res. 32:181–195.

Robbins, W. J. 1938. Thiamin and growth of species of *Phytophthora.* Bull. Torrey Bot. Club 65:267–276.

Robbins, W. J., and Kavanagh, V. 1942. Vitamin deficiencies of the filamentous fungi. Bot. Rev. 8:411–471.

Robertson, A. G., and Ogilvie, L. 1953. Root rot of brassicas caused by *Phytophthora megasperma.* Plant Pathol. 2:15.

Robertson, G. I. 1975. A paper disc technique for the recovery of *Pythium* spp. from soil or water. N.Z. J. Agric. Res. 18:409–410.

Robertson, G. I. 1977. *Phytophthora cinnamomi* root rot of avocado. Orchardist, N.Z. 50:54–55.

Robertson, G. I. 1982a. New plant disease records in New Zealand: Additional host records of fungal pathogens. N.Z. J. Agric. Res. 25:109–111.

Robertson, G. I. 1982b. Kiwifruit can tolerate *Phytophthora,* but not 'wet feet.' Orchardist, N.Z. 55:148–151.

Robertson, G. I., and Dance, H. M. 1971. The association of *Phytophthora megasperma* with crown rot of apple trees. N.Z. J. Agric. Res. 14:509–551.

Robertson, H. E., and Boyer, P. D. 1956. The biological inactivity of glucose 6-phosphite, inorganic phosphites and other phosphites. Arch. Biochem. Biophys. 62:380–395.

Robertson, N. F. 1953. Pepper disease in Sarawak. Commonw. Phytopathol. News. 1:20.

Robinson, J. 1957. Outdoor carnations and Gerberas in New Zealand. Reed, Wellington. 72 pp.

Robinson, R. A. 1969. Disease resistance terminology. Rev. Appl. Mycol. 48:593–606.

Robinson, R. A. 1976. Plant Pathosystems. Springer-Verlag, Berlin, Germany. 184 pp.

Robinson, W. 1915. "Black neck" or wilt disease of asters. Ann. Appl. Biol. 2:125–137. (Cited in Tucker 1933).

Robson, W. 1912. Report of the Botanic Station, Montserrat. Imp. Dep. Agric. West Indies (1911–1912):17.

Röder, K. 1935. Untersuchungen über die *Phytophthora* Krankheit (*Phytophthora infestans*) der tomate (Studies on the *Phytophthora* disease [*Phytophthora infestans*] of the tomato with particular reference to biologic specialization in the causal organism). Phytopathol. Z. 8:589–614. (In German).

Rodriguez-Kabana, R., Backman, P. A., and Curl, E. A. 1977. Control of seed and soilborne plant diseases. Pages 117–161 in: Antifungal Compounds. Vol. 1, Discovery, Development, and Uses. M. R. Siegel and H. D. Sisler, eds. Marcel Dekker, New York. 600 pp.

Rogers, V. E., Irwin, J. A. G., and Stovold, G. 1978. The development of lucerne with resistance to root rot in poorly aerated soils. Aust. J. Exp. Agric. Anim. Husb. 18:434–441.

Rohrbach, K. G., and Schenck, S. 1985. Control of pineapple heart rot, caused by *Phytophthora parasitica* and *P. cinnamomi,* with metalaxyl, fosetyl-Al, and phosphrous acid. Plant Dis. 69:320–23.

Roiger, D. J., and Jeffers, S. N. 1991. Evaluation of *Trichoderma* spp. for biological control of Phytophthora crown rot and root rot of apple seedlings. Phytopathology 81:910–917.

Roiger, D. J., and Jeffers, S. N., and Caldwell, R. W. 1991. Occurrence of *Trichoderma* spp. in apple orchard and woodland soils. Soil Biol. Biochem. 23:353–359.

Roistacher, C. N. 1981. A blueprint for disaster, part one: The history of seedling yellow disease. Citrograph 67(1):4–5,24.

Roldan, E. F., and Luczon, C. V. 1953. Phytophthora blight of ornamental Palmera palm seedlings (*Chrysalidocarpus lutescens* Wendl.). Philipp. J. For. 9:43–53.

Romero, C. S. 1978. Tres especies de *Phytophthora* de Bary de nuevo registro en Mexico (Three *Phytophthora* de Bary species of new record in Mexico). Agrociencia 34:45–55. (In Spanish).

Romero, S., and Erwin, D. C. 1969. Variation in pathogenicity among single-oospore cultures of *Phytophthora infestans.* Phytopathology 59:1310–1317.

Romero, S., and Fourton, H. 1962. Leaf blight of *Mirabilis jalapa* in Mexico caused by a specialized form of *Phytophthora infestans* (Abstr.). Phytopathology 53:25.

Rondon, G. A. J. 1975. A new fungus disease of fig (*Ficus carica* L.) in Venezuela. Agron. Trop. 25:487–494.

Rooke, D. M., and Shattock, R. C. 1983. Effect of chloramphenicol and streptomycin on developmental stages of *Phytophthora infestans.* J. Gen. Microbiol. 129:3401–3410.

Roos, H., and Buchenauer, H. 1985. Resistance of *Plasmopara viticola* to cyprofuram and metalaxyl. Pages 331–335 in: Fungicides for Crop Protection. BCPC Monograph 31. I. A. Smith, ed. Croydon BCPC Publ. 504 pp.

Roosje, G. S. 1962. *Phytophthora syringae,* geisoleerld nit bast van appel en peer (*Phytophthora syringae* isolated from apple and pear bark). Tijdschr. Plantenziekten 68:246–247. (In Dutch).

Roper, J. A. 1966. Mechanisms of inheritance. III. The parasexual cycle. Pages 589–617 in: The Fungi: An Advanced Treatise. Vol. II. G. C. Ainsworth and S. S. Alfred, eds. Academic Press, New York. 805 pp.

Rorer, J. B. 1910. Pod-rot, canker and chupon wilt of cacao caused by *Phytophthora* sp. Trinidad and Tobago Dep. Agric. Bull. 9:38.

Rose, D. H. 1924. Leather rot of strawberries. J. Agric. Res. 28:357–376. (Cited in Tucker 1933).

Rose, D. H., and Lindegren, C. C. 1925. Phytophthora rot of pears and apples. J. Agric. Res. 30:463–468. (Cited in Tucker 1933).

Rose, J. L., Irwin, J. A. G., Ryley, M. J., Langdon, P. W., and Jenner, L. B. 1982. Reaction of soybean cultivars to races of *Phytophthora megasperma* f. sp. *glycinea* present in Queensland. Aust. J. Agric. Res. 33:763–771.

Rose, S. L., Li, C. Y., and Hutchins, A. S. 1980. A streptomycete antagonist to *Phellinus weirii, Fomes annosus,* and *Phytophthora cinnamomi.* Can. J. Microbiol. 26:583–587.

Rosenbaum, J. 1914. *Phytophthora arecae* (Colem.), Pethyb., causing a rot of potato tubers (Abstr.). Phytopathology 4:387.

Rosenbaum, J. 1915. *Phytophthora* disease of ginseng. N.Y. Cornell Agric. Exp. Stn. Bull. 363:65–106. (Cited in Tucker 1933).

Rosenbaum, J. 1917. Studies of the genus *Phytophthora.* J. Agric. Res. 8:233–276.

Ross, R. G., and Gourley, C. O. 1969. *Phytophthora syringae* fruit rot of apples in Nova Scotia. Can. Plant Dis. Surv. 49:33–35.

Rossetti, V., Feichtenberger, E., Salibe, A. A., and Oliveira, D. A. 1984. Reaction of exocortis and rumple diseased lemon trees to *Phytophthora citrophthora* inoculations. Pages 180–183 in: Proc. Ninth Conf. Int. Org. Virol. S. M. Garnsey, L. W. Timmer, and J. A. Dodds, eds.

Rotem, J., and Cohen, Y. 1974. Epidemiological patterns of *Phytophthora infestans* under semi-arid conditions. Phytopathology 64:711–714.

Rotem, J., and Palti, J. 1969. Irrigation and plant diseases. Annu. Rev. Phytopathol. 7:267–288.

Rotem, J., and Sari, A. 1983. Fertilization and age-conditioned predisposition of potatoes to sporulation of and infection by *Phytophthora infestans.* Z. Pflanzenkr. Pflanzenschutz 90:83–88.

Rotem, J., Cohen, Y., and Bashi, E. 1978. Host and environmental influences on sporulation in vivo. Annu. Rev. Phytopathol. 16:83–101.

Roth, E. R., Toole, E. R., and Hepting, G. H. 1948. Nutrition aspects of the littleleaf disease of pine. J. For. 46:578–587.

Roth, L. F., Bynum, H. H., and Nelson, E. E. 1972. Phytophthora root rot of Port-Orford-cedar. U.S. Dep. Agric. For. Pest Leafl. 131. 7 pp.

Rothwell, A. 1975. A revised list of plant diseases in Rhodesia—Additions 1966–72. Kirkia 10:295–307.

Rowe, R. C. 1982. Translocation of metalaxyl and RE26745 in potato and comparison of foliar and soil application for control of *Phytophthora infestans.* Plant Dis. 66:989–993.

Rowe, R. C. 1993. Potato Health Management. American Phytopathological Society, St. Paul, Minnesota. 193 pp.

Rowe, R. C., and Schmitthenner, A. F. 1977. Potato pink rot in Ohio caused by *Phytophthora erythroseptica* and *P. cryptogea*. Plant Dis. Rep. 61:807–810.

Royle, D. J., and Hickman, C. J. 1964. Observations on *Phytophthora cinnamomi*. Can. J. Bot. 42:311–318.

Ruben, D. M., Frank, Z. R., and Chet, I. 1980. Factors affecting behavior and developmental synchrony of germinating oospores of *Pythium aphanidermatum*. Phytopathology 70:54–59.

Rullich, G., and Schöber, B. 1988. Auftreten und Verbreitung des A2-Paarungstyps von *Phytophthora infestans* (Mont.) de Bary in der Bundesrepublik Deutschland (Occurrence and distribution of A2-mating types of *Phytophthora infestans* [Mont.] de Bary in the republic of Germany). Kartoffelbau 39:244–246. (In German).

Ruppel, E. G., and Almeyda, N. 1965. Susceptibility of native *Piper* species to the collar-rot pathogen of black pepper in Puerto Rico. Plant Dis. Rep. 49:550–551.

Ruppel, E. G., and Barnes, D. K. 1965. Phytophthora wilt of *Tephrosia vogelii* in Puerto Rico. Plant Dis. Rep. 49:676–679.

Rutgers, A. A. L. 1912. Onderzoekingen over den cacaokanker (Investigations of cacao canker). Meded. Plantenziekten 1:1–31. (In Dutch). (Cited in Tucker 1933).

Rutgers, A. A. L. 1917. Hevea-Kanker III (Hevea canker III). Meded. Lab. Plantenziekten, Buitenz. Org. 28. 49 pp. (In Dutch). (Cited in Tucker 1933).

Rutherford, F. S., and Ward, E. W. B. 1985a. Estimation of relative DNA content in nuclei of races of *Phytophthora megasperma* f. sp. *glycinea* by quantitative fluorescence microscopy. Can. J. Genet. Cytol. 27:614–616.

Rutherford, F. S., and Ward, E. W. B. 1985b. Evidence for genetic control of oospore abortion in *Phytophthora megasperma* f. sp. *glycinea*. Can. J. Bot. 63:1671–1673.

Rutherford, F. S., Ward, E. W. B., and Buzzell, R. I. 1985. Variation in virulence in successive single-zoospore propagations of *Phytophthora megasperma* f. sp. *glycinea*. Phytopathology 75:371–374.

Ryley, M. J., and Obst, N. R. 1992. Race-specific resistance in soybean cv. Davis to *Phytophthora megasperma* f. sp. *glycinea*. Plant Dis. 76:665–668.

Ryley, M. J., Mosetter, H. F., and Rose, J. L. 1989. Yield losses of soybeans due to *Phytophthora megasperma* f. sp. *glycinea*. Aust. J. Agric. Res. 40:1161–1169.

Sabatini, V., Feichtenberger, E., Pimentel, C. V. P., and Feitosa, M. I. 1987. Etiologia da podridao parda do cacaueiro na regiao do litoral paulista (Etiology of black pod disease in the seacoast of the state of São Paulo, Brasil). (In Portuguese). Summa Phytopathol. 13:21.

Saccardo, P. A. 1888. Sylloge Fungorum 7:237–238.

Saccardo, P. A. 1890. Fungi aliquot Australienses. Hedwigia 29:155.

Saimmaime, I., Coulomb, C., and Coulomb, P. J. 1991. Trans-cinnamate 4-hydroxylase activity in host-parasite interaction: *Capsicum annuum-Phytophthora capsici*. Plant Physiol. Biochem. 29:481–487.

Saindrenan, P., and Bompeix, G. 1985. Rôle des phytoalexines dans la réponse de *Vigna unguiculata* traitue par le phosestyl-Al, à l'infection par *Phytophthora cryptogea* (Role of phytoalexins in the response of *Vigna unguiculata* treated with phosetyl-Al on infection by *Phytophthora cryptogea*). C. R. Acad. Sci. Ser. 3. 10:411–414.

Saindrenan, P., Darakis, G., Bompeix, G. 1985. Determination of ethyl phosphite, phosphite, and phosphate in plant tissues by anion-exchange high performance liquid chromatography and gas chromatography. J. Chromatogr. 347:267–73.

Saindrenan, P., Barchietto, T., and Bompeix, G. 1988a. Modification of the phosphite induced resistance response in leaves of cowpea infected with *Phytophthora cryptogea* by and aminooxyacetate. Plant Sci. 58:245–252.

Saindrenan, P., Barchietto, T., Avelino, J., and Bompeix, G. 1988b. Effects of phosphite on phytoalexin accumulation in leaves of cowpea infected with *Phytophthora cryptogea*. Physiol. Mol. Plant Pathol. 32:425–435.

Saindrenan, P., Barchietto, T., and Bompeix, G. 1990. Effects of phosphonate on the elicitor activity of culture filtrates of *Phytophthora cryptogea* in *Vigna unguiculata*. Plant Sci. 67:245–251.

Sakai, R. 1961. Studies on the nutritional physiology of *Phytophthora infestans* (Mont.) de Bary. Hokkaido Nat. Agric. Exp. Stn. Rep. 57:1–58.

Sakata, K., and Rickards, R. W. 1980. Synthesis of homothallin II. Pages 165–172 in: 23rd Symp. Chemistry of Natural Products. Nagoya, Japan.

Saksena, S. B. 1977. *Phytophthora parasitica* the scourge of pan (*Piper betle*) L. Indian Phytopathol. 30:1–16.

Saksena, S. B., and Mehrotra, R. S. 1970. Studies on foot-rot and leaf-rot of *Piper betle* L. Host range and role of cuttings in the spread of the disease. J. Indian Bot. Soc. 49:24–29.

Salaman, R. N. 1910. The inheritance of colour and other characters in the potato. J. Genet. 1:7–46. (Cited in Tucker 1933).

Sale, P. R. 1985. Kiwifruit Culture. Gov. Printing Off. Wellington, New Zealand. 96 pp.

Salerno, M., and Calabretta, C. 1960. Grave infezione da *Phytophthora citricola* Sawada (*P. cactorum* var. *applanata* Chester) su piantine di Pomodoro in Sicilia (Severe infection by *Phytophthora citricola* Sawada [*P. cactorum* var. *applanata* Chester] of tomato seedlings in Sicily). Atti. 1st Bot. Univ. Pavia, Italy. Ser. 5. 18: 222–233. (Rev. Appl. Mycol. 1961, 40:433).

Salih, H. 1974. A *Phytophthora* species new for Turkey determined in citrus orchards in Adana. J. Turk. Phytopathol. 3:113–115.

Salisbury, P. J. 1954. A study of the Phytophthora root rots of Lawson cypress in British Columbia. Can. Dep. Agric. For. Biol. Div. Sci. Serv. Victoria, British Columbia. 17 pp.

Salmon, E. S., and Ware, W. M. 1927. Root rot of *Richardia*. Gard. Chron. 81:234–235. (Cited in Tucker 1933).

Salvatore, M. A., Gray, F. A., and Hine, R. B. 1973. Enzymatically induced germination of oospores of *Phytophthora megasperma*. Phytopathology 63:1083–1084.

Sambrook, J., Fritsch, T., and Maniatis, T. 1989. Molecular Cloning: A Laboratory Manual. 2nd ed. Cold Spring Harbor Laboratory, Cold Spring Harbor, NY.

Samoucha, Y., and Cohen, Y. 1988. Synergistic interactions of cymoxanil mixtures in the control of metalaxyl-resistant *Phytophthora infestans* of potato. Phytopathology 78:636–640.

Samoucha, Y., and Cohen, Y. 1989. Field control of potato late blight by synergistic fungicidal mixtures. Plant Dis. 73:751–753.

Samoucha, Y., Hugelschofer, U., and Gisi, U. 1987. Effects of disease intensity and application type on efficacy and synergy of fungicide mixtures against *Phytophthora infestans*. J. Phytopathol. 120:44–52.

Samraj, J., and Jose, P. C. 1966. A Phytophthora wilt of pepper. Sci. Cult. 32:90–92.

Samuel, G. 1932. Summary of plant disease records in south Australia for the two years ending June 30, 1932. J. Dep. Agric. S. Aust. 36:300–301. (Cited in Tucker 1933).

San Antonio, J. P., and Blount, V. 1973. Use of liquid nitrogen to preserve downy mildew (*Phytophthora phaseoli*) inoculum. Plant Dis. Rep. 57:724.

Sanchez, A. 1965. Enfermedades mas communes de los citricos en Guatemala (The most common citrus diseases in Guatemala). Bol. Fec. Min. Agric. Guatemala 16:1–28. (In Spanish).

Sanders, P. L., Coffey, M. D., Greer, G. D., and Soika, M. D. 1990. Laboratory-induced resistance to fosetyl-Al in a metalaxyl-resistant field isolate of *Pythium aphanidermatum*. Plant Dis. 74:690–692.

Sandler, H. A., Timmer, L. W., Graham, J. H., and Zitko, S. E. 1989. Effect of fungicide applications on populations of *Phytophthora parasitica* and on feeder root densities and fruit yields of citrus trees. Plant Dis. 73:902–906.

Sandlin, C. M., and Ferrin, D. M. 1993. Foliar blight and root rot of container-grown giant redwood caused by *Phytophthora citrophthora*. Plant Dis. 77:591–594.

Sandlin, C. M., Wadsworth, M. L., Ferrin, D. M., and Sanchez, L. 1992. First report of root rot of Monterey pine in California caused by *Phytophthora citricola*. Plant Dis. 76:643.

Sansavini, S., and Rosati, P. 1986. Genetic resistance of fruit crops to parasitic diseases. Riv. Fruttic. Ortofloric. 48:37–54.

Sansome, E. 1961. Meiosis in the oogonium and antheridium of *Pythium debaryanum* Hesse. Nature 191:827–828.

Sansome, E. 1963. Meiosis in *Pythium debaryanum* Hesse and its significance in the life-history of the biflagellatae. Trans. Br. Mycol. Soc. 46:63–72.

Sansome, E. 1965. Meiosis in diploid and polyploid sex organs of *Phytophthora* and *Achlya*. Cytologia 30:103–117.

Sansome, E. 1970. Selfing as a possible cause of disturbed ratios in *Phytophthora* crosses. Trans. Br. Mycol. Soc. 54:101–107.

Sansome, E. 1976. Gametangial meiosis in *Phytophthora capsici*. Can. J. Bot. 54:1535–1545.

Sansome, E. 1977. Polyploidy and induced gametangial formation in British isolates of *Phytophthora infestans*. J. Gen. Microbiol. 99: 311–316.

Sansome, E. 1980. Reciprocal translocation heterozygosity in heterothallic species of *Phytophthora* and its significance. Trans. Br. Mycol. Soc. 74:175–185.

Sansome, E. 1985. Cytological studies on *Phytophthora nicotianae* var. *parasitica* in relation to mating type. Trans. Br. Mycol. Soc. 84:87–93.

Sansome, E. R. 1987. Fungal chromosomes as observed with the light microscope. Pages 97–113 in: Evolutionary Biology of the Fungi. A. D. M. Rayner, C. M. Brasier, and D. Moore, eds. Cambridge Univ. Press, Cambridge, U.K. 465 pp.

Sansome, E., and Brasier, C. M. 1974. Polyploidy associated with varietal differentiation in the *megasperma* complex of *Phytophthora*. Trans. Br. Mycol. Soc. 63:461–467.

Sansome, E., Brasier, C. M., and Griffin, M. J. 1975. Chromosome size differences in *Phytophthora palmivora,* a pathogen of cocoa. Nature (London) 255:704–705.

Sansome, E., Brasier, C. M., and Sansome, F. W. 1979. Further cytological studies on the 'L' and 'S' types of *Phytophthora* from cocoa. Trans. Br. Mycol. Soc. 73:293–302.

Sansome, E., Brasier, C. M., and Hamm, P. B. 1991. *Phytophthora meadii* may be a species hybrid. Mycol. Res. 95:273–277.

Saraiva, T. R. C. B. 1982. Heranca da resistencia a murcha do pimentao (*Capsicum annuum* L.), causada por *Phytophthora capsici* Leonian (Heredity of resistance to wilt of pepper (*Capsicum annuum* L.] caused by *Phytophthora capsici* Leonian). 40 pp. Thesis, Univ. Viçosa, Viçosa, M.G., Brasil. (In Portuguese).

Sarejanni, J. A. 1935. Notes phytopathologiques (Phytopathological notes). Ann. Phytopathol. Benaki, Greece 1:67–76. (In French).

Sarejanni, J. A. 1936. La pourriture du Collet des Solanées cultivees et la classification du genre *Phytophthora* (A collar rot of cultivated *Solanum* and the classification of the genus *Phytophthora*). Ann. Phytopathol. Benaki, Greece 2:35–52. (In French).

Sarejanni, J. A. 1952a. Le *Phytophthora* des aubergines en Grece (A *Phytophthora* on eggplant in Greece). Ann. Phytopathol. Benaki, Greece 6:14–18. (In French).

Sarejanni, J. A. 1952b. Un *Phytophthora* du collet de oeillets (A *Phytophthora* on the crown of carnations). Ann. Phytopathol. Benaki, Greece 6:32–36. (In French).

Sarejanni, J. A., and Demetriades, S. D. 1951. Catalogue commente No. 2 des champignons et bacteries rencontres sur les plantes cultivees en Grece (Annotated list No. 2 of the fungi and bacteria occurring on cultivated plants in Greece). Ann. Phytopathol. Benaki, Greece 5:5–11. (In French).

Sarma, Y. R., Ramachandran, N., and Anandaraj, M. 1991. Black pepper diseases in India. Proc. Int. Pepper Community Workshop. Control of Black Pepper Diseases 1988. Y. R. Sarma and T. Premkumar, eds. Nat. Res. Cent. Spices, Calicut, Kerala, India. 213 pp.

Sasser, J. N., Lucas, G. B., and Powers, H. R., Jr. 1955. The relationship of root-knot nematodes to black-shank resistance in tobacco. Phytopathology 45:459–461.

Sastry, M. N. L., and Hegde, R. K. 1987a. *Phytophthora* associated with arecanut (*Areca catechu* Linn.) in Uttara Kannada. Karnataka. Curr. Sci. 56:367–368.

Sastry, M. N. L., and Hegde, R. K. 1987b. Pathogenic variation in *Phytophthora* species affecting plantation crops. Indian Phytopathol. 40:365–369.

Sastry, M. N. L., and Hegde, R. K. 1987c. Distribution of mating types and their role in perpetuation of *Phytophthora palmivora* and *P. meadii*. Indian Phytopathol. 40:370–373.

Sastry, M. N. L., and Hegde, R. K. 1988. Control of fruit rot or koleroga disease of arecanut (*Areca catechu* L.). Trop. Agric. 65:150–152.

Sastry, M. N. L., and Hegde, R. K. 1989a. Variability of *Phytophthora* species obtained from plantations of Karnataka. Indian Phytopathol. 42:421–425.

Sastry, M. N. L., and Hegde, R. K. 1989b. Survival of *Phytophthora meadii* causing fruit rot of arecanut. Indian Phytopathol. 42:573–574.

Satchuthananthevale, V. 1963. Complementary strains of *Phytophthora palmivora* from Ceylon rubber. Phytopathology 53:729.

Sato, N. 1979. Effect of soil temperature on the field infection of potato tubers by *Phytophthora infestans*. Phytopathology 69:989–993.

Sato, N., and Tomiyama, K. 1976. Relationship between inhibition of intracellular hyphal growth of *Phytophthora infestans* and rishitin concentration in infected potato cells. Ann. Phytopathol. Soc. Jpn. 43:598–600.

Sato, N., Kitazawa, K., and Tomiyama, K. 1971. The role of rishitin in localizing the invading hyphae of *Phytophthora infestans* in infection sites at the cut surfaces of potato tubers. Physiol Plant Pathol. 1:289–295.

Satour, M. M., and Butler, E. E. 1967. A root and crown rot of tomato caused by *Phytophthora capsici* and *P. parasitica*. Phytopathology 57:510–515.

Satour, M. M., and Butler, E. E. 1968. Comparative morphological and physiological studies of the progenies from intraspecific matings of *Phytophthora capsici*. Phytopathology 58:183–192.

Satour, M. M., and El-Shinnawy, S. A. 1973. A new root and crown rot of cucurbits caused by *Phytophthora cryptogea*. Egypt. J. Phytopathol. 5:65–71.

Sattar, A., Khan, A. J., Alam, M., and Hussein, A. 1982. Phytophthora leaf blight disease of *Datura inoxia*. Indian Phytopathol. 35:162.

Satyprasad, K., and Ramaro, P. 1980. A simple technique for isolating *Phytophthora palmivora* from soil. Curr. Sci. 49:360–361.

Sauthoff, W. 1964. Untersuchungen über eine für Europa neue Pilzkrankheit der Azaleen (Research on a new fungal disease of azalea for Europe). Biol. Bundensanst. Land-Fortwirtsch. 1963. A96. (In German).

Sauthoff, W. 1965. Untersuchungen über eine Stengelgrund fäule an *Pisonia* (Research on a stalk rot of *Pisonia*). Biol. Bundesanst. Land-Forstwirtsch. (Berlin and Braunschw.) 1964:A988. (In German).

Sauve, R. J., and Mitchell, D. J. 1977. An evaluation of methods for obtaining mycelium-free oospores of *Pythium aphanidermatum* and *P. myriotylum*. Can. J. Microbiol. 23:643–648.

Savage, E. J., Clayton, C. W., Hunter, J. H., Brenneman, J. A., Laviola, C., and Gallegly, M. E. 1968. Homothallism, heterothallism, and interspecific hybridization in the genus *Phytophthora*. Phytopathology 58:1004–1021.

Sawada, K. 1911. *Phytophthora* disease of taro. Spec. Bull. Agric. Exp. Sta. Gov. Formosa 2:75–84. (In Japanese). (Cited in Ho 1992).

Sawada, K. 1915a. Materials of Formosan fungi (5). Trans. Nat. Hist. Soc. Formosa 20:14–19. (In Japanese). (Cited in Ho 1992).

Sawada, K. 1915b. Two new species of the genus *Phytophthora* causing diseases of onion and eggplant. Spec. Bull. Agric. Exp. Stn. Govt. Formosa 11:1–139. (In Japanese). (Cited in Ho 1992).

Sawada, K. 1919a. Descriptive catalogue of the Formosan fungi. I. Agric. Gov. Res. Inst. Formosa Spec. Bull. 19:1–695. (In Japanese). (Cited in Tucker 1933; Ho 1992).

Sawada, K. 1919b. Preliminary report on *Phytophthora* diseases of Cyperaceae. Formosan Agric. Mag. 146:8–14. (In Japanese). (Cited in Ho 1992).

Sawada, K. 1927a. Descriptive catalogue of the Formosan fungi. III. Rep. Dep. Agric. Gov. Res. Inst. Formosa Bull. 27:1–62. (In Japanese). (Rev. Appl. Mycol. 1928, 7:273). (Cited in Tucker 1933).

Sawada, K. 1927b. Descriptive catalogue of the Formosan fungi (IV). Rep. Agric. Gov. Res. Inst. Formosa 27:1–62. (In Japanese). (Cited in Ho 1992).

Sawada, K. 1931. Descriptive catalogue of the Formosan fungi (IV). List of fungi found in Formosa. Rep. Dep. Agric. Gov. Res. Inst. Formosa 51:1–131. (In Japanese).

Sawada, K. 1941. Materials of the Formosan fungi (41). Trans. Nat. Hist. Soc. Formosa 31:261–267. (In Japanese). (Cited in Ho 1992).

Sawada, K. 1942a. On the species of the genus *Kawakamia*. Formosan Agric. Rev. 38:351–355. (In Japanese). (Cited in Waterhouse 1970a).

Sawada, K. 1942b. *Phytophthora* species on tobacco. Formosa Agric. Rev. 38:7–25. (In Japanese). (Cited in Ho 1992).

Saxena, S. B. 1977. *Phytophthora parasitica* the scourge of pan (*Piper betle* L.). Indian Phytopathol. 30:1–16.

Say, J. S. 1963. Gerberas for pleasure and profit. N.Z. J. Agric. 106:343–351.

Scarpa, M. J., and Raniere, L. C. 1964. The use of consecutive hourly dewpoints in forecasting downy mildew of lima bean. Plant Dis. Rep. 48:77–81.

Schaffnit, E., and Volk, A. 1927. Über den Einfluss der Ernahrung auf die Empfanglichkeit der Pflanzen fur Parasiten (1. Teil) (On the effect of nutrition on susceptibility of plants against parasites [Part I]). Forsch. Geb. Pflanzenkr. Immun. Pflanzenreich 3:1–45. (In German).

Schechter, S. E., and Gray, L. E. 1987. Oospore germination in *Phytophthora megasperma* f. sp. *glycinea*. Can. J. Bot. 65:1465–1467.

Schenk, A. 1875. Neue *Peronospora: P. sempervivi* (New *Peronospora: P. sempervivi*). Bot. Zeit. 33:690–693. (In German). (Cited in Nienhaus 1960; Tucker 1933).

Schenk, P. J. 1926. Parasitair omvallen van tulpen (Parasitic wilt of tulips). Floralia 47:88–89. (In Dutch). (Cited in Tucker 1933).

Schick, R. 1932. Über das Verhalten von *Solanum demissum, Solanum tuberosum* und ihren Bastarden gegenüber verschiedenen Herkünften von *Phytophthora infestans* (Behavior of *Solanum demissum, Solanum tuberosum* and their hybrids to different isolates of *Phytophthora infestans*). Züchter 4:233–237.

Schiffmann-Nadel, M., and Cohen, E. 1968. Method for production of sporangia of *Phytophthora citrophthora*. Phytopathology 58:550.

Schlösser, E., and Gottlieb, D. 1968. The effect of sterols on the metabolism of *Pythium* species. Arch. Microbiol. 61:246–253.

Schlösser, E., Shaw, P. D., and Gottlieb, D. 1969. Sterols in species of *Pythium*. Arch. Mikrobiol. 66:147–153.

Schlub, R. L. 1983. Epidemiology of *Phytophthora capsici* on bell pepper. J. Agric. Sci. Camb. 100:7–11.

Schmidle, A. 1961. Schäden an Stachelbeertrieben durch *Phytophthora cactorum* (Damage to gooseberry shoots by *Phytophthora cactorum*). Nachrichtenbl. Dtsch. Pflanzenschutzdienstes. (Stuttg.) 13:90–91. (Rev. Appl. Mycol. 1962, 41:49).

Schmitthenner, A. F. 1964. Prevalence and virulence of *Phytophthora, Aphanomyces, Pythium, Rhizoctonia,* and *Fusarium* isolated from diseased alfalfa seedlings. Phytopathology 54:1012–1018.

Schmitthenner, A. F. 1970. Significance of populations of *Pythium* and *Phytophthora* in soil. Pages 25–27 in: Root Diseases and Soil-Borne Pathogens. T. A. Toussoun, R. V. Bega, and P. E. Nelson, eds. Univ. Calif. Press, Berkeley. 252 pp.

Schmitthenner, A. F. 1972. Evidence for a new race of *Phytophthora megasperma* var. *sojae* pathogenic to soybean. Plant Dis. Rep. 56:536–539.

Schmitthenner, A. F. 1973. Isolation and identification methods for *Phytophthora* and *Pythium*. Pages 94–110 in: Proc. First Woody Ornamental Dis. Workshop Jan. 24–25, 1973. Univ. Mo. 128 pp.

Schmitthenner, A. F. 1985. Problems and progress in control of Phytophthora root rot of soybean. Plant Dis. 69:362–368.

Schmitthenner, A. F., and Canaday, C. H. 1983. Role of chemical factors in development of *Phytophthora* diseases. Pages 189–196 in: *Phytophthora:* Its Biology, Ecology, Taxonomy, and Pathology. D. C. Erwin, S. Bartnicki-Garcia, and P. H. Tsao, eds. American Phytopathological Society, St. Paul, Minn. 392 pp.

Schmitthenner, A. F., and Hilty, J. W. 1962. A modified dilution technique for obtaining single-spore isolates of fungi from contaminated material. Phytopathology 52:582–583.

Schmitthenner, A. F., and Van Doren, D. M., Jr. 1985. Integrated control of root rot of soybean caused by *Phytophthora megasperma* f. sp. *glycinea*. Pages 263–266 in: Ecology and Management of Soilborne Plant Pathogens. C. A. Parker, A. D. Rovira, K. J. Moore, P. T. W. Wong, and J. F. Kollmorgen, eds. American Phytopathological Society, St. Paul, Minn. 358 pp.

Schmitthenner, A. F., and Walker, A. K. 1979. Tolerance versus resistance for control of Phytophthora root rot of soybeans. Pages 35–44 in: Ninth Soybean Seed Res. Conf. Chicago, Illinois.

Schmitthenner, A. F., Hobe, M., and Bhat, R. G. 1994. *Phytophthora sojae* races in Ohio over a 10-year interval. Plant Dis. 78:269–276.

Schöber, B., and Rullich, G. 1986. Oosporenbildung von *Phytophthora infestans* (Mont.) de Bary (Oospore formation of *Phytophthora infestans* [Mont.] de Bary). Potato Res. 29:395–398.

Schoenweiss, D. F. 1975. Predisposition, stress, and plant disease. Annu. Rev. Phytopathol. 13:193–211.

Scholten, G. 1970. Wilt disease in *Gerbera*. Neth. J Plant Pathol. 76:212–218.

Schönbeck, F. 1967. Untersuchungen über Bluteninfektionen. V. Untersuchungen au Tulpen (Investigation on flower infection. V. Research on tulips). Phytopathol. Z. 59:205–224.

Schots, A., Dewey, M., and Oliver, R. 1994. Modern Assays for Plant Pathogenic Fungi. C.A.B.I. Books. Univ. Arizona Press, Tucson. 270 pp.

Schoulties, C. L., Baker, K. F., and Sabersky-Lehmann. 1980. Factors influencing zoospore production by *Phytophthora cinnamomi* in axenic culture. Can. J. Bot. 58:2117–2122.

Schreier, O. 1954. Das auftreten wichtiger schadensursachen an kulturpflanzen in österreich im Jahre 1954 (The occurrence of important causes of injury to cultivated plants in Austria in the year 1954). Pflanzenschutzberichte 14:23–33. (In German). (Rev. Appl. Mycol. 1955, 34:210).

Schrödter, H., and Ullrich, J. 1966. Weitere Untersuchungen zur Biometeorologie und Epidemiologie von *Phytophthora infestans* (Mont.) de By. Ein neues Conzept zur Lösung des Problems der epidemiologischen Prognoses (Further studies on the biometeorology and epidemiology of *Phytophthora infestans* [Mont.] de By. A new concept for solving the problem of forecasting epidemics). Phytopathol. Z. 56:265–278. (In German).

Schröeter, J. 1889. Gattung *Phytophthora* de Bary (The genus *Phytophthora* de Bary). Kryptogamenflora Schlesien 3:235–236. (In German).

Schubert, T. S., and Leahy, R. M. 1989. Phytophthora blight of *Catharanthus roseus*. Fla. Dep. Agric. Consum. Serv., Plant Pathol. Circ. No. 321:174–175.

Schutte, G. C., Bezuidenhout, J. J., and Kotzé, J. M. 1991. Timing of application of phosphonate fungicides using different application methods as determined by means of gas-liquid-chromatography for Phytophthora root rot control of citrus. Phytophylactica 23:69–71.

Schwarz, M. B. 1926. Instervingsverschijnselen bij *Lochnera rosea* veroorzaakt door *Phytophthora* spec. (Symptoms of die-back in *Lochnera rosea* caused by *Phytophthora* spec.). Indische Culturen (Tasmania) 11:288–290. (In Dutch). (Cited in Tucker 1933).

Schwarz, M. B. 1927. Het hart en bladrot van orchideën, veroorzaakt door *Phytophthora omnivora* De By (The heart and leaf rot of orchids caused by *Phytophthora omnivora* De By). H. van Ingen, Soerabaja. 7 pp. (In Dutch). (Rev. Appl. Mycol 1928, 7:100). (Cited in Tucker 1933).

Schwenk, F. W., and Sim, T. 1974. Race 4 of *Phytophthora megasperma* var. *sojae* from soybean proposed. Plant Dis. Rep. 58:352–354.

Schwenk, F. W., Ciaschini, C. A., Nickell, C. D., and Trombold, D. G. 1979. Inoculation of soybean plants by injection with zoospores of *Phytophthora megasperma* var. *sojae*. Phytopathology 69:1233–1234.

Schwinn, F. J. 1959. Untersuchungen zur Systematik der Gattung *Phytophthora* de Bary (Research on the systematics of the genus *Phytophthora* de Bary). Arch. Microbiol. 33:223–252. (In German).

Schwinn, F. J. 1960. *Phytophthora cactorum* (Leb. et Cohn) Schroet. als Erreger einer Stengelfäule an *Pelargonium* Arten (*Phytophthora cactorum* [Leb. et Cohn] as the agent of stem rot of *Pelargonium* spp.). Nachrichtenbl. Dtsch. Pflanzenschutzdienstes (Stuttg.) 12: 168–170. (In German). (Rev. Appl. Mycol. 1961, 40:365).

Schwinn, F. J. 1961a. Der Nachweis von *Phytophthora cactorum* (Leb. et Cohn) Schroet. im Boden (The detection of *Phytophthora cactorum* [Leb. et Cohn] Schroet. in soil). Mitt. Biol. Bundesanst. Land-Forstwirtsch. (Berl.-Dahlem) 104:42–44. (In German).

Schwinn, F. J. 1961b. Über eine durch *Phytophthora parasitica* Dastur verursachte fäule an anthurien (On a rot of *Anthurium* caused by *Phytophthora parasitica* Dastur). Nachrichtenbl. Dtsch. Pflanzenschutzdienstes (Stuttg.) 13:129–132. (In German). (Rev. Appl. Mycol. 1962, 41:307–308).

Schwinn, F. J. 1979. Control of Phycomycetes: A changing scene. Proc. Br. Crop Prot. Conf. 10th 3:791–802.

Schwinn, F. J. 1983. New developments in chemical control of *Phytophthora.* Pages 327–334 in: *Phytophthora:* Its Biology, Taxonomy, Ecology, and Pathology. D. C. Erwin, S. Bartnicki-Garcia, and P. H. Tsao, eds. American Phytopathological Society, St. Paul, Minn. 392 pp.

Schwinn, F. J., and Margot, P. 1991. Control with chemicals. Pages 225–265 in: *Phytophthora infestans,* Cause of Late Blight of Potato. Adv. Plant Pathol. Vol. 7. D. S. Ingram and P. H. Williams, eds. Academic Press. 273 pp.

Schwinn, F. J., and Morton, H. V. 1990. Antiresistance strategies. Pages 170–183 in: Managing Resistance to Agrochemicals. M. B. Green, H. M. LeBaron, and W. K. Moberg, eds. Am. Chem. Soc. Symp. Ser. 421. Washington, D.C. 496 pp.

Schwinn F. J., and Staub, T. 1983. Biological properties of metalaxyl. Pages 123–133 in: Systemic Fungicides and Antifungal Compounds. H. Lyr and C. Polter, eds. Akademie Verlag, Berlin. 451 pp.

Schwinn, F. J., and Staub, T. 1987. Phenylamides and other fungicides against Oomycetes. Pages 259–273 in: Modern Selective Fungicides, Properties, Applications and Mechanisms of Action. L. Horst, ed. Longman Scientific and Technical, England. 383 pp.

Schwinn, F. J., and Staub, T. 1995. Oomycetes fungicides. Chapter 16. Phenylamides and other fungicides against *Oomycetes.* Pages 323–346 in: Modern Selective Fungicides—Properties, Applications, Mechanisms of Action. 2nd ed. H. Lyr, ed. Gustav Fischer Verlag, Jena, Germany.

Schwinn, F. J., Staub, T., and Urech, P. A. 1977a. A new type of fungicide against diseases caused by Oomycetes. Meded. Fac. Landbouwwet. Rijksuniv. Gent 42:1181–1188.

Schwinn, F. J., Staub, T., and Urech, P. A. 1977b. Die Bekampfung falscher Mehltau-Krankheiten mit einem Wirkstoff aus der Gruppe der Acylalanine (The control of false-mildew with an effective material from the acylalanine group). Mitt. Biol. Bundesanst. Land-Forstwirtsch. Berl.-Dahlem. 178:145–146. (In German).

Schwinn, F. J., and Urech, P. A. 1986. Progress in the chemical control of diseases caused by Oomycetes. Pages 89–106 in: Fungicide Chemistry. Advances and Practical Applications. M. B. Green and D. A. Spilker, eds. Am. Chem. Soc. Symp. Ser. 304. Washington, D.C. 173 pp.

Scott, D. H., Jeffers, W. F., Darrow, G. M., and Ink, D. P. 1950. Occurrence of strains of the strawberry red stele fungus, *Phytophthora fragariae* Hickman, as shown by differential variety response. Phytopathology 40:194–198.

Scott, D. H., Draper, A. D., and Maas, J. L. 1976. Mass screening of young strawberry seedlings for resistance to *Phytophthora fragariae* Hickman. Hortic. Sci. 11:257–258.

Scott, D. H., Draper, A. D., and Galletta, G. J. 1984. Breeding strawberries for resistance for red stele resistance. Plant Breed. Rev. 2:195–214.

Scurti, J. C. 1962. Contributo alla conoscenza delle Phytophthorae del Lilla (A contribution to the knowledge of *Phytophthora* spp. on lilac). Boll. Lab. Sper. Fitopatol. Torino N.S. 25:43–49.

Seemüller, E., and Schmidle, A. 1979. Einfluss der Herkunft von *Phytophthora cactorum* Isolaten auf ihre Virulenz and Apfelrinde, Erdbeerhizomen und Erdbeerfrüchten (Influence of the origin of *Phytophthora cactorum* isolates on their virulence on apple bark, strawberry rhizomes and strawberry fruits). Phytopathol. Z. 94: 218–225. (In German, English summary).

Seemüller, E., and Sun, C. 1989. Auftreten von Metalaxyl-Resistenz bei *Phytophthora fragariae* (Occurrence of metalaxyl resistance in *Phytophthora fragariae*). Nach. Dtsch. Pflanzenschutzdienstes 41:71–73 (In German).

Seemüller, E., Duncan, J. M., Kennedy, D. M., and Reidl, M. 1986. *Phytophthora* sp. als Ursache einer Wurzerfaul an Himbeere (*Phytophthora* sp. as a cause of root rot of raspberry). Nachrichtenbl. Dtsch. Pflanzenschutzdienstes 38:17–21. (In German, English summary).

Seethalakshmi, K. V. 1953. Blight of *Cyperus rotundus* L. and *C. bulbosus* Vahl. Indian Phytopathol. 6:57–62.

Seethalakshmi, V., and Ramakrishnan, T. S. 1953. *Phytophthora cyperi-bulbosi,* sp. nov. on *Cyperus bulbosus* Vahl. Curr. Sci. 22: 149–150.

Sehgal, S. P., and Prasad, N. 1972. Note on testing of different varieties, strains, and species of *Sesamum* for resistance to Phytophthora blight. Indian J. Agric. Sci. 42:122.

Semb, L. 1969. White tip disease of leeks in Norway. Gartneryrket 59:1042–1044.

Semb, L. 1971. A rot of stored cabbage caused by a *Phytophthora* sp. Acta Hortic. 20:32–35.

Serra, S., Franceschini, A., and Foddai, A. 1988a. Valutazione di alcuni substrati selettivi per la determinazione quantitativa di *Phytophthora crytogea* Peth. et Laff. nel Terreno (Evaluation of some selective media for the quantitative determination of *Phytophthora cryptogea*). Phytopathol. Mediterr. 27:23–27. (In Italian, English summary).

Serra, S., Franceschini, A., and Foddai, A. 1988b. Valutazione dell'efficacia del metodo di "diluizione in piastra" (SDP) nella determinazione quantitativa delle popolazioni di *Phytophthora cryptogea* Peth. et Laff. nel Terreno (Efficiency of the soil dilution plate (SDP) method for assessing the inoculum density (ID) of *Phytophthora cryptogea* Peth. et Laff. in soil). Phytopathol. Mediterr. 27:173–177. (In Italian, English summary).

Serres, J. M., and Carraro, G. A. 1976. DPX-3217, a new fungicide for the control of grape downy mildew, potato blight and other Peronosporales. Meded. Fac. Landbouwwet. Rijksuniv. Gent 42: 645–650.

Servin, L. 1958. Especie de *Phytophthora* atacando a *Mirabilis jalapa* (Species of *Phytophthora* that attack *Mirabilis jalapa*). Memoria del Congreso Nacional de Entomologia y Fitopatologia, Mexico. Pages 491–500.

Sewell, G. W. F., and Wilson, J. F. 1959. Resistance trials of some apple rootstock varieties to *Phytophthora cactorum* (L. & C.) Schroet. J. Hortic. Sci. 34:51–59.

Sewell, G. W. F., and Wilson, J. F. 1963. Branch, stem, and collar rot of apple caused by *Phytophthora* species of the '*cactorum* group.' Nature 200:1229.

Sewell, G. W. F., and Wilson, J. F. 1964. Death of maiden apple trees caused by *Phytophthora syringae* Kleb. and a comparison of the pathogen with *P. cactorum.* (L & C). Schroet. Ann. Appl. Biol. 53: 275–280.

Sewell, G. W. F., and Wilson, J. F. 1973. Phytophthora collar rot: Seasonal effects on infection and disease development. Ann. Appl. Biol. 74:149–158.

Sewell, G. W. F., Wilson, J. F., and Dakwa, J. T. 1974. Seasonal variations in the activity in soil of *Phytophthora cactorum, P. syringae,* and *P. citricola* in relation to collar rot disease of apple. Ann. Appl. Biol. 76:179–186.

Shah, A. A., and Hasan, S. 1991. First report of Phytophthora root and crown rot of apple trees caused by *Phytophthora cactorum* in Pakistan. Plant Dis. 75:537.

Shahzad, S., Coe, R., and Dick, M. W. 1992. Biometry of oospores and oogonia of *Pythium* (Oomycetes): The independent taxonomic value of calculated ratios. Bot. J. Linn. Soc. 108:143–165.

Shanor, L. 1944. *Phytophthora megasperma* root rot of spinach in North Carolina. Plant Dis. Rep. 28:351.

Sharangapani, S. G. 1930 Appendix I. Annual report of the economic botanist to the Govt. of Bengal for the year 1929–30: Annu. Rep. Dep. Agric. Bengal. Pages 37–46. (Rev. Appl. Mycol. 1931, 10: 233). (Cited in Tucker 1933).

Sharma, N. 1990. Aseptic germination of oospores of *Phytophthora palmivora.* Can. J. Bot. 68:2548–2552.

Sharp, J. K., McNeil, M., and Albersheim, P. 1984. The primary structures of one elicitor-active and seven elicitor-inactive hexa (β-D-glucopyranosyl)-D-glucitols isolated from the mycelial walls of *Phytophthora megasperma* f. sp. *glycinea.* J. Biol. Chem. 259: 11321–11336.

Sharp, R. E., and Davies, W. J. 1979. Solute regulation and growth by roots and shoots of water-stressed maize plants. Planta 147:43–49.

Sharples, A. 1926. Diseases of coconut palms. Malay. Agric. J. 14: 65–73.

Shattock, R. C. 1988. Studies on the inheritance of resistance to metalaxyl in *Phytophthora infestans.* Plant Pathol. 37:4–11.

Shattock, R. C., and Shaw, D. S. 1975. Mutants of *Phytophthora infestans* resistant to, and dependent upon, antibiotics. Trans. Br. Mycol. Soc. 64:29–41.

Shattock, R. C., Janssen, B. D., Whitbread, R., and Shaw, D. S. 1977. An interpretation of the frequencies of host specific phenotypes of *Phytophthora infestans* in North Wales. Ann. Appl. Biol. 86:249–260.

Shattock, R. C., Tooley, P. W., and Fry, W. E. 1986a. Genetics of *Phytophthora infestans:* Determination of recombination, segregation, and selfing by isozyme analysis. Phytopathology 76:410–413.

Shattock, R. C., Tooley, P. W., and Fry, W. E. 1986b. Genetics of *Phytophthora infestans:* Characterization of single-oospore cultures from A1 isolates induced to self by intraspecific stimulation. Phytopathology 76:407–410.

Shattock, R. C., Tooley, P. W., Sweigard, J., and Fry, W. E. 1987. Genetic studies of *Phytophthora infestans.* Pages 175–185 in: Genetics and Plant Pathogenesis. P. R. Day and G. J. Jellis, eds. Blackwell Sci. Pub., Oxford. 353 pp.

Shattock, R. C., Shaw, D. S., Fyfe, A. M., Dunn, J. R., Loney, K. H., and Shattock, J. A. 1990. Phenotypes of *Phytophthora infestans* collected in England and Wales from 1985 to 1988: Mating type, response to metalaxyl and isoenzyme analysis. Plant Pathol. 39: 242–248.

Shaw, C. R., and Prasad, R. 1970. Starch gel electrophoresis of enzymes—A compilation of recipes. Biochem. Genet. 4:297–320.

Shaw, D. S. 1967. A method of obtaining single-oospore cultures of *Phytophthora cactorum* using live water snails. Phytopathology 57:454.

Shaw, D. S. 1983. The cytogenetics and genetics of *Phytophthora.* Pages 81–94 in: *Phytophthora:* Its Biology, Taxonomy, Ecology, and Pathology. D. C. Erwin, S. Bartnicki-Garcia, and P. H. Tsao, eds. American Phytopathological Society, St. Paul, Minn. 392 pp.

Shaw, D. S. 1987. The breeding system of *Phytophthora infestans:* The role of the A2 mating type. Pages 161–174 in: Genetics and Plant Pathogenesis. P. R. Day and G. J. Jellis, eds. Blackwell Sci. Pub., Oxford. 352 pp.

Shaw, D. S. 1988. The *Phytophthora* species. Pages 27–51 in: Advances in Plant Pathology. Vol. 6. D. S. Ingram and P. H. Williams, eds. Academic Press, London. 566 pp.

Shaw, D. S. 1990. The comparative genetics of *Phytophthora.* EPPO (Eur. Mediterr. Plant Prot. Organ.) Bull. 20:67–76.

Shaw, D. S. 1991. Genetics. Pages 131–170 in: *Phytophthora infestans,* the Cause of Late Blight of Potato. Advances in Plant Pathology. Vol. 7. D. S. Ingram and P. H. Williams, eds. Academic Press, London. 273 pp.

Shaw, D. S., and Elliot, C. G. 1968. Streptomycin resistance and morphological variation in *Phytophthora cactorum.* J. Gen. Microbiol. 51:75–84.

Shaw, D. S., and Khaki, I. A. 1971. Genetical evidence for diploidy in *Phytophthora.* Genet. Res. 17:165–167.

Shaw, D. S., and Shattock, R. C. 1991. Genetics of *Phytophthora infestans:* The Mendelian approach. Pages 218–230 in: *Phytophthora.* J. A. Lucas, R. C. Shattock, D. S. Shaw, and L. R. Cooke, eds. Cambridge Univ. Press, Cambridge, U.K. 447 pp.

Shaw, D. S., Fyfe, A. M., Hibberd, P. G., and Abdel-Sattar, M. A. 1985. Occurrence of the rare A_2 mating type of *Phytophthora infestans* on imported Egyptian potatoes and the production of sexual progeny with A1 mating types from the UK. Plant Pathol. 34:552–556.

Shazad, S., Coe, R., and Dick, M. W. 1992. Biometry of oospores and oogonia of *Pythium* (Oomycetes): The independent taxonomic value of calculated ratios. Bot. J. Linn. Soc. 108:143–165.

Shea, S. R. 1975a. Environmental factors of the northern jarrah forest in relation to pathogenicity and survival of *Phytophthora cinnamomi.* For. Dep. Bull. 85. Perth, West. Aust. 83 pp.

Shea, S. R. 1975b. Focus on jarrah dieback—A threat to W. Australia's unique jarrah forests. For. Focus No. 14:3–15.

Shea, S. R. 1977. Environmental factors of the northern jarrah forest in relation to pathogenicity and survival of *Phytophthora cinnamomi.* For. Dept. West. Aust. Res. Stn. Bull. 85. 83 pp.

Shea, S. 1988. Controlling dieback—Research and management to curb a major forest threat. Chapter 17. Pages 163–177 in: Case Studies in Environmental Hope. P. Newman, S. Neville, and L. Duxbury, eds. Environ. Prot. Authority of West. Aust. State Conservation Strategy. 185 pp.

Shea, S. R., and Broadbent, P. 1983. Developments in cultural and biological control of *Phytophthora* diseases. Pages 335–350 in: *Phytophthora:* Its Biology, Taxonomy, Ecology, and Pathology. D. C. Erwin, S. Bartnicki-Garcia, and P. H. Tsao, eds. American Phytopathological Society, St. Paul, MN. 392 pp.

Shea, S. R., and Malajczuk, N. 1977. Potential for control of eucalypti dieback in Western Australia. Pages 13–19 in: Aust. Nurs. Assoc. Ann. Conf. Seminar Papers. 38 pp.

Shea, S. R., Gillen, K. J., and Kitt, R. J. 1978. Variation in sporangial production of *Phytophthora cinnamomi* Rands on jarrah (*E. marginata* Sm.) forest sites with different understorey compositions. Aust. For. Res. 8:219–226.

Shea, S. R., Gillen, K. J., and Leppard, W. I. 1980. Seasonal variation in population levels of *Phytophthora cinnamomi* Rands in soil in diseased, freely-drained *Eucalyptus marginata* Sm. sites in the northern jarrah forest of south-western Australia. Prot. Ecol. 2: 135–156.

Shea, S. R., Shearer, B., and Tippett, J. 1982. Recovery of *Phytophthora cinnamomi* Rands from vertical roots of Jarrah (*Eucalyptus marginata* Sm.). Aust. Plant Pathol. 11:25–28.

Shea, S. R., Shearer, B. L., Tippett, J. T., and Deegan, P. M. 1983. Distribution, reproduction, and movement of *Phytophthora cinnamomi* on sites highly conducive to jarrah dieback in south Western Australia. Plant Dis. 67:970–973.

Shea, S. R., Shearer, B. L., Tippett, J. T., and Deegan, P. M. 1984. A new perspective on Jarrah dieback. For. Focus 31:3–11.

Shearer, B. 1990. Dieback of native plant communities caused by *Phytophthora* species—A major factor affecting land use in South-Western Australia. Land and Water Res. News. No. 5:15–26.

Shearer, B., Wills, R., and Stakely, M. 1991. Wildflower killers. Landscope 7:28–34.

Shearer, B. L., and Bailey, R. 1989. The fight against jarrah dieback. Landscope 5:38–44.

Shearer, B. L., and Shea, S. R. 1987. Variation in seasonal population fluctuations of *Phytophthora cinnamomi* within and between infected *Eucalyptus marginata* sites of southwestern Australia. For. Ecol. Manage. 21:209–230.

Shearer, B. L., Shea, S. R., and Fairman, R. G. 1981. Infection of the stem and large roots of *Eucalyptus marginata* by *Phytophthora cinnamomi.* Australas. Plant Pathol. 10:2–3.

Shearer, B. L., Michaelson, B. J., and Somerford, P. J. 1988. Effects of isolate and time of inoculation on invasion of secondary phloem of *Eucalyptus* spp. and *Banksia grandis* by *Phytophthora* spp. Plant Dis. 72:121–126.

Sheila, V. K., Nene, Y. L., and Kannaiyan, J. 1983. A simple culture medium for *Phytophthora drechsleri* f. sp. *cajani.* Indian Phytopathol. 36:152–154.

Shen, C. Y., and Tsao, P. H. 1983. Sensitivity of *Phytophthora infestans* to hymexazol in selective media. Trans. Br. Mycol. Soc. 80:567–570.

Shen, C. Y., and Bower, L. A., Erwin, D. C., and Tsao, P. H. 1983. Formation of sex organs in the A_1 mating type of *Phytophthora infestans* induced chemically by A_2 isolates of other species of *Phytophthora.* Can. J. Bot. 61:1462–1466.

Shepard, J. F., Bidney, D., and Shahin, E. 1980. Potato protoplasts in crop Improvement. Science 208:17–23.

Shepherd, C. J. 1976. Pigment production from tyrosine by Australian isolates of *Phytophthora* species. Aust. J. Bot. 24:607–617.

Shepherd, C. J. 1978. Mating behaviour of Australian isolates of *Phytophthora* species. I. Inter- and intraspecific mating. Aust. J. Bot. 26:123–138.

Shepherd, E. F. S. 1933. Annu. Rep. Agric. Mauritius 1932:32–38. (Rev. Appl. Mycol. 1934, 13:494–495).

Shepherd, E. F S. 1937. A revised list of plant diseases occurring in Mauritius. Bull. Dep. Agric. Mauritius (Sci. Ser.) 23. 14 pp. (Rev. Appl. Mycol. 1938, 17:295).

Sherbakoff, C. D. 1917. Buckeye rot of tomato fruit. Phytopathology 7:119–129.

Shew, H. D. 1983. Effects of soil matric potential on infection of tobacco by *Phytophthora parasitica* var. *nicotianae.* Phytopathology 73:1160–1163.

Shew, H. D. 1985. Response of *Phytophthora parasitica* var. *nicotianae* to metalaxyl exposure. Plant Dis. 69:559–562.

Shew, H. D. 1987. Effect of host resistance on spread of *Phytoph-*

thora parasitica var. *nicotianae* and subsequent development of tobacco black shank under field conditions. Phytopathology 77: 1090–1093.

Shew, H. D., and Benson, D. M. 1981. Fraser fir root rot induced by *Phytophthora citricola.* Plant Dis. 65:688–689.

Shew, H. D., and Benson, D. M. 1982. Qualitative and quantitative soil assays for *Phytophthora cinnamomi.* Phytopathology 72:1029–1032.

Shew, H. D., and Lucas, G. B., eds. 1991. Compendium of Tobacco Diseases. American Phytopathological Society, St. Paul, Minn. 90 pp.

Shew, H. D., Benson, D. M., and Grand, L. F. 1979a. A comparison of baits for isolating *Phytophthora cinnamomi* from soil (Abstr.). Phytopathology 69:532.

Shew, H. D., Benson, D. M., and Grand, L. F. 1979b. A quantitative soil assay for *Phytophthora cinnamomi* (Abstr.). Phytopathology 69:1045.

Shieszko, S. F., Carpenter, J. B., Lowe, E. P., and Jakob, J. G. 1947. Improved methods for the cultivation and storage of *Phytophthora infestans.* Phytopathology 37:635–649.

Shimony, C., and Friend, J. 1975. Ultrastructure of the interaction between *Phytophthora infestans* and leaves of two cultivars of potato (*Solanum tuberosum* L.) Orion and Majestic. New Phytol. 74:59–65.

Shimony, C., and Friend, J. 1976. Ultrastructure of the interaction between *Phytophthora infestans* and tuber slices of resistant and susceptible cultivars of potato (*Solanum tuberosum* L.) Orion and Majestic. Isr. J. Bot. 25:174–183.

Shimony, C., and Friend, J. 1977. The ultrastructure of the interaction between *Phytophthora infestans* (Mont.) de Bary and tuber discs of potato (*Solanum tuberosum* L.) cv. King Edward. Physiol. Plant Pathol. 11:243–249.

Shipton, P. J. 1977. Monoculture and soilborne plant pathogens. Annu. Rev. Phytopathol. 15:387–407.

Shishkina, A. K. 1969. New causal agents of seedling diseases of *Cedrus deodara* (Loud.) *Phytophthora cactorum* and *Pythium artotrogus.* In Proc. Inst. Plant Prot., Georgian S.S.R. 21:95–100. 357 pp. (Rev. Plant Pathol. 1971, 50:2083).

Shtienberg, D., and Fry, W. E. 1990. Field and computer simulation evaluation of spray-scheduling methods for control of early and late blight of potato. Phytopathology 80:772–777.

Shtienberg, D., Doster, M. A., Pelletier, J. R., and Fry, W. E. 1989. Use of simulation models to develop a low-risk strategy to suppress early and late blight in potato foliage. Phytopathology 79: 590–595.

Shumard-Hudspeth, D. S., and Hudspeth, M. E. S. 1990. Genic rearrangements in *Phytophthora* mitochondrial DNA. Curr. Genet. 17:413–415.

Siciliano, M. J., and Shaw, C. R. 1976. Separation and visualization of enzymes on gels. Pages 185–209 in: Chromatographic and Electrophoretic Techniques. Vol. 2. I. Smith, ed. Heinemann Medical Books, London.

Sidebottom, J. R., and Shew, H. D. 1985a. Effects of soil texture and matric potential on sporangium production by *Phytophthora parasitica* var. *nicotianae.* Phytopathology 75:1435–1438.

Sidebottom, J. R., and Shew, H. D. 1985b. Effect of soil type and soil matric potential on infection of tobacco by *Phytophthora parasitica* var. *nicotianae.* Phytopathology 75:1439–1443.

Sideris, C. P. 1929. Stem rot of pineapple plants (Abstr.). Phytopathology 19:1146. (Cited in Tucker 1933).

Sideris, C. P. 1932. Taxonomic studies in the family Pythiaceae. II. *Pythium.* Mycologia 24:14–61. (Cited in Tucker 1933).

Sideris, C. P., and Paxton, G. E. 1930. Heart rot of pineapple plants. Phytopathologoy 20:951–958. (Cited in Tucker 1933).

Silayoi, I., Krengpiem, P., Boonthai, S., and Foongkiatpaiboon, A. 1983. Root rot disease of betel vine. J. Thai Phytopathol. Soc. 3:10–13.

Simmonds, H. W. 1925. Pests and diseases of the coconut palm in the islands of the southern Pacific. Fiji Dep. Agric. Bull. 16. 31 pp.

Simmonds, J. H. 1955. Report of the Plant Pathology Section for 1954–55, Dep. Agric. Queensl. Pages 51–52.

Simmonds, J. H. 1959. Report of the Plant Pathology Section for 1958–59, Dep. Agric. Queensl. Pages 49–50.

Simonet, M. 1925. Note sur le *Phytophthora infestans* (Note on *Phytophthora infestans*). J. Soc. Nat. Hortic. Fr. Ser. 4. 26:272–274.

Sims, J. J., Keen, N. T., and Honwad, V. K. 1972. Hydroxyphaseollin, an induced antifungal compound from soybeans. Phytochemistry 11:827–828.

Sinclair, J. B., and Backman, P. A., eds. 1989. Compendium of Soybean Diseases, 3rd ed. American Phytopathological Society, St. Paul, Minn. 116 pp.

Sing, V. O., and Bartnicki-Garcia, S. 1975a. Adhesion of *Phytophthora palmivora* zoospores: Detection and ultrastructural visualization of concanavalin A-receptor sites appearing during encystment. J. Cell Sci. 19:11–20.

Sing, V. O., and Bartnicki-Garcia, S. 1975b. Adhesion of *Phytophthora palmivora* zoospores: Electron microscopy of cell attachment and cyst wall fibril formation. J. Cell Sci. 18:123–132.

Singh, B., and Srivastava, H. C. 1953. Damping-off of tomato seedlings. J. Indian Bot. Soc. 32:1–16.

Singh, N. B., and Lambert, J. W. 1985. Effect of gene Rps_1 for resistance to Phytophthora rot on yield and other characteristics of soybean. Crop Sci. 25:494–496.

Singh, R. S. 1959. The occurrence of *Phytophthora palmivora* on leaves of *Ficus religiosa.* Sci. Cult. 25:382–384.

Singh, U. B. 1943. Some diseases of fruits and fruit trees in Kumaon—I. I.C.A.R. Misc. Bull. 51:16. (In V. G. Rao 1970).

Singh, U. B. 1946. Pythium collar rot of field pea at Cawnpore, United Provinces. Curr. Sci. 15:195–196.

Singh, U. P., and Chauhan, V. B. 1988. Effect of temperature on germination of zoospores of *Phytophthora drechsleri* f. sp. *cajani.* Indian Phytopathol. 41:80–85.

Singh, U. P., and Chauhan, V. B. 1992. Phytophthora blight of pigeon pea. Pages 375–387 in: Plant Diseases of International Importance: Diseases of Cereals and Pulses. U. S. Singh, A. N. Mukhopadhyay, J. Kumar, and H. S. Chaube, eds. Vol. I. Prentice Hall, Englewood Cliffs, N.J. 488 pp.

Singleton, L. L., Mihail, J. D., and Rush, L. M. 1992. Methods for Research on Soilborne Phytopathogenic Fungi. American Phytopathological Society. St. Paul, Minnesota. 266 pp.

Siqueira, C. B., and Reifscheider, J. B. 1981. Bell pepper (*Capsicum annuum* 'Cascadura ikeda') damping-off; *Phytophthora capsici.* Fungicides and Nematicide Tests (American Phytopathological Society) 36:65.

Siradhana, B. S., Ellet, C. W., and Schmitthenner. A. F. 1968. Crown rot of *Peperomia.* Plant Dis. Rep. 52:244.

Siradhana, B. S., Schmitthenner, A. F., and Ellet, C. W. 1969. Formation of phytoalexin in *Peperomia* in relation to resistance to *Phytophthora nicotianae* var. *parasitica.* Phytopathology 59:405–410.

Sitepu, D., and Wallace, H. R. 1984. Crop disease studies. Lettuce diseases. Bienn. Rep. Univ. Adelaide, South Australia. Waite Agric. Res. Inst. 1982–1983. 145 pp.

Sivakadadcham, B., and Fernando, W. G. D. 1991. Root and stem rot on greenhouse-grown cowpea caused by *Phytophthora vignae* in Sri Lanka. Plant Dis. 75:215.

Sivasithamparam, K. 1981. Some effects of extracts from tree barks and sawdust on *Phytophthora cinnamomi* Rands. Aust. Plant Pathol. 10:18–20.

Sivasithamparam, K., Smith, L. D. J., and Goss, O. M. 1981. Effect of potting media containing fresh sawdust and composted tree-barks on *Phytophthora cinnamomi* Rands. Aust. Plant Pathol. 10:20–21.

Sjoerdsma, J. D., Menge, J. A., and Johnson, E. V. L. 1988. Ecology of *Phytophthora citrophthora* root rot of citrus (Abstr.). Phytopathology 78:1519.

Skadow, K. 1985. Eine Methode der Resistenzprüfung von Tomaten gegen *Phytophthora nicotianae* v. Breda de Haan var. *nicotianae* (A method of testing tomato for resistance against *Phytophthora nicotianae* v. Breda de Haan var. *nicotianae*). Arch. Phytopathol. Pflanzenschutz, Berlin 21:143–150.

Skidmore, D. I., Shattock, R. C., and Shaw, D. S. 1984. Oospores in cultures of *Phytophthora infestans* resulting from selfing induced by the presence of *P. drechsleri* isolated from blighted potato foliage. Plant Pathol. 33:173–183.

Skipp, R. A., and Cristensen, M. J. 1982. Invasion of white clover roots by fungi and other soil microorganisms. III. The capacity of fungi isolated from white clover roots to invade seedling root tissue. N.Z. J. Agric. Res. 25:97–101.

Skotland, C. B. 1955. A *Phytophthora* damping-off disease of soybean. Plant Dis. Rep. 39:682–683.

Skoudridakis, M. T., and Bourbos, V. A. 1990. Comparative effectiveness of metalaxyl and phosethyl-Al to *Phytophthora cambivora* (Petri) Buism. Meded. Fac. Landbouwwet. Rijksuniv. Gent 55:963–966.

Slatyer, R. O. 1957. The influence of progressive increase in total soil moisture stress, on transpiration, growth, and internal water relationships of plants. Aust. J. Biol. Sci. 10:320–336.

Slatyer, R. O. 1967. Plant Water Relationships. Academic Press, New York. 366 pp.

Slavík, B., ed. 1974. Methods of Studying Plant Water Relations. Springer-Verlag, New York. 449 pp.

Sleigh, M. A. 1981. Flagellar beat patterns and their possible evolution. BioSystems 14:423–431.

Slusher, R. L., and Sinclair, J. B. 1973. Development of *Phytophthora megasperma* var. *sojae* in soybean roots. Phytopathology 63:1168–1171.

Slusher, R. L., Haas, D. L., Carothers, Z. B., and Sinclair, J. B. 1974. Ultrastructure at the host-parasite interface of *Phytophthora megasperma* var. *sojae* in soybean rootlets. Phytopathology 64:834–840.

Small, W. 1915. Report by the botanist. Annu. Rep. Dep. Agric. Uganda, 1914. Pages 59–62. (Cited in Tucker 1933).

Smilde, W. D., and Van Ness, M. 1992. Field inoculation of leek with *Phytophthora porri.* Neth. J. Plant Pathol. 98:157–159.

Smillie, R. H., Grant, B. R., and Guest, D. 1989. The mode of action of phosphite: Evidence for both direct and indirect modes of action on three *Phytophthora* spp. in plants. Phytopathology 79:921–926.

Smillie, R. H., Dunstan, R. H., Grant, B. R., Griffith, J. M., Iser, J., and Niere, J. O. 1990. The mode of action of the antifungal agent phosphite. Bull. O.E.P.P./E.P.P.O. 20:185–192.

Smith, C. O. 1937. Inoculation of some economic plants with *Phytophthora cactorum* and *P. citrophthora.* Phytopathology 27:1106–1109.

Smith, C. O., and Barrett, J. T. 1930. Some inoculations of *Juglans* with *Phytophthora*-like fungi (Abstr.). Phytopathology 20:849–850. (Cited in Tucker 1933).

Smith, C. O., and Barrett, J. T. 1931. Crown rot of *Juglans* in California. J. Agric. Res. 43:885–904.

Smith, D. 1982. Liquid nitrogen storage of fungi. Trans. Br. Mycol. Soc. 79:415–421.

Smith, D. 1983. Cryoprotectants and the cryopreservation of fungi. Trans. Br. Mycol. Soc. 80:360–363.

Smith, D., and Onions, A. H. S. 1983. A comparison of some preservation techniques for fungi. Trans. Br. Mycol. Soc. 81:535–540.

Smith, D., Coulson, G. E., and Morris, G. J. 1986. A comparative study of the morphology and viability of hyphae of *Penicillium expansum* and *Phytophthora nicotianae* during freezing and thawing. J. Gen. Microbiol. 132:2013–2021.

Smith, E. H. 1915. *Pythiacystis* infection of deciduous nursery stock. Phytopathology 5:317–322. (Cited in Tucker 1933).

Smith, G. 1913. Infection experiments with the potato 'blight' fungus. J. Southeast Agric. Coll. Wye, Kent, U.K. 22:494–496. (Cited in Tucker 1933).

Smith, G. S., Hutchinson, D. J., and Henderson, C. T. 1991. Comparative use of soil infested with chlamydospores to screen for relative susceptibility to Phytophthora foot rot in citrus cultivars. Plant Dis. 75:402–405.

Smith, H. C. 1950. Collar rot of apples and gooseberries. Orchardist N.Z. 23:11–14.

Smith, H. C. 1953. Collar rot of apple trees in Essex. Plant Pathol. 2:85–86.

Smith, H. C 1956. Collar rot of apricots, peaches and cherries. Orchardist N.Z. 29:22–23.

Smith, I. M., Dunez, J., Phillips, D. H., Lelliott, R. A., and Arther, S. A. 1988. European Handbook of Plant Diseases. Blackwell Sci. Publ., London. 583 pp.

Smith, I. W., and Marks, G. C. 1986. Effect of moisture stress in *Eucalyptus sieberi* on growth of lesions caused by *Phytophthora cinnamomi.* Aust. For. Res. 16:273–279.

Smith, L. D. 1981. Chemical control of Phytophthora root and crown rot of petunia (Abstr.). Phytopathology 71:256.

Smith, M. A. L., Spomer, L. A., Meyer, M. J., and McClelland, M. T. 1989. Non-invasive image analysis evaulation of growth during plant micropropagation. Plant Cell Tissue Organ Cult. 19:91–102.

Smith, M. A. L., Spomer, L. A., and McClelland, M. T 1990. Direct analysis of root zone data in a microculture system. Plant Cell Tissue Organ Cult. 23:21–26.

Smith, M. A. L., Wagner, R. E., Anderson, J. S., and Spomer, L. A. 1991. In vitro asay for evaluation of Phytophthora rot on soybean taproots. Crop Sci. 31:1364–1366.

Smith, P. G., Kimble, K. A., Grogan, R. G., and Millet, A. H. 1967. Inheritance of resistance in peppers to Phytophthora root rot. Phytopathology 57:377–379.

Smith, P. M. 1979. A study of the effects of fungitoxic compounds on *Phytophthora cinnamomi* in water. Ann. Appl. Biol. 93:149–157.

Smith, P. M. 1980. An assessment of fungicides for the control of wilt and die-back caused by *Phytophthora cinnamomi* in container-grown *Chamaecyparis lawsoniana* cv. Ellwoodii. Ann. Appl. Biol. 94:225–234.

Smith, R. E., and Butler, O. 1908. Gum disease of citrus trees in California. Calif. Agric. Exp. Stn. Bull. 200:235–270

Smith, R. E., and Smith, E. H. 1906. A new fungus of economic importance. Bot. Gaz. 42:215–221. (Cited in Tucker 1933).

Smith, R. E., and Smith, E. H. 1925. Further studies on Pythiaceous infection of deciduous fruit trees in California. Phytopathology 15:389–404. (Cited in Tucker 1933).

Smith, R. E., Smith, C. O., and Ramsey, H. J. 1912. Walnut culture in California. Walnut blight. Calif. Agric. Exp. Stn. Bull. 231:119–398.

Smith, V. L., Wilcox, W. F., and Harman, G. E. 1990. Potential for biological control of Phytophthora root and crown rots of apple by *Trichoderma* and *Gliocladium* spp. Phytopathology 80:880–885.

Smith, W. G. 1881. Diseases of tomatoes. Gard. Chron. 16:346. (Cited in Tucker 1933).

Smith, W. G. 1884. Diseases of Field and Garden Crops. Macmillan, London. 343 pp. (Cited in Tucker 1933).

Smolak, J. 1925. Zpráva o cinnosti stanice pro choroby rostlin prist vyšši škole ovocnicovinarške a zahradnické na Melnicezarak 1923 a 1924 (Report on the activity of the Phytopathological Station at the High School of Pomiculture, Viticulture and Horticulture at Melnik, for the years 1923–1924). Ochr. Rostl. 5:41–46. (In Polish). (Rev. Appl. Mycol. 1925, 5:146). (Cited in Tucker 1933).

Smoot, J. J., Gough, F. J., Lamey, H. A., Eichenmuller, J. J., and Gallegly, M. E. 1958. Production and germination of oospores of *Phytophthora infestans.* Phytopathology 48:165–171.

Snapp, S. S., Shennan, C., and van Bruggen, A. H. C. 1991. Effects of salinity on severity of infection by *Phytophthora parasitica* Dast., ion concentrations, and growth of tomato, *Lycopersicon esculentum* Mill. New Phytol. 119:275–284.

Sneh, B. 1972a. An agar medium for the isolation and macroscopic recognition of *Phytophthora* spp. from soil on dilution plates. Can. J. Microbiol. 18:1389–1392.

Sneh, B. 1972b. Use of cellulase and hemicellulase for the separation of *Phytophthora cactorum* oospores from mycelium mats. Can. J. Bot. 50:2685–2686.

Sneh, B. 1974. Recovery of sporangia of *Phytophthora cactorum* from soil. Can. J. Bot. 52:1777–1778.

Sneh, B., and Katz, D. A. 1988. Behaviour of *Phytophthora citrophthora* and *P. nicotianae* var. in soil, and differences in their tolerance to antimicrobial components of selective media used for isolation of *Phytophthora* spp. J. Phytopathol. 122:208–221.

Sneh, B., and McIntosh, D. L. 1974. Studies on the behavior and survival of *Phytophthora cactorum* in soil. Can. J. Bot. 52:795–802.

Sneh, B., Humble, S. J., and Lockwood, J. L. 1977. Parasitism of oospores of *Phytophthora megasperma* var. *sojae, P. cactorum, Pythium* sp., and *Aphanomyces euteiches* in soil by oomycetes, chytridiomycetes, hyphomycetes, actinomycetes, and bacteria. Phytopathology 67:622–628.

Sneh, B., Eye, L. L., and Lockwood, J. L. 1981. Factors affecting germination of oospores of *Phytophthora megasperma* var. *sojae.* Phytopathol. Z. 101:314–322.

Snieszko, S. F., Carpenter, J. B., Lowe, E. P., and Jakob, J. G. 1947. Improved methods for the cultivation and storage of *Phytophthora infestans.* Phytopathology 37:635–649.

Solel, Z., and Pinkas, Y. 1984. A modified selective medium for detecting *Phytophthora cinnamomi* on avocado roots. Phytopathology 74:506–508.

Soltis, D. E., Haufler, C. H., Darrow, D. C., and Gastony, G. J. 1983. Starch gel electrophoresis of ferns: A compilation of grinding buffers, gel and electrode buffers and staining schedules. Am. Fern J. 73:9–27.

Sonoda, R. M., and Ogawa, J. M. 1968. Phytophthora crown and root rot of hops in California. Plant Dis. Rep. 52:76–78.

Sorauer, P. 1928. Handbuch der Pflanzenkrankheiten (Handbook of Plant Diseases). 5th ed. Verlang. Paul Parey, Berlin. (In German).

Soria, V. J. 1974. Sources of resistance to *Phytophthora palmivora*. Pages 197–202 in: *Phytophthora* Disease of Cocoa. P. H. Gregory, ed. Longman, London. 348 pp.

Soulie, M.C., Lameta, A. A., and Bompeix, G. 1989. Phosphite-induced resistance of tomato leaflets to *Phytophthora capsici* and changes in fatty acids of host and fungal origin. Plant Physiol. Biochem. 27:523–527.

Sozzi, D., and Staub, T. 1987. Accuracy of methods to monitor sensitivity of *Phytophthora infestans* to phenylamide fungicides. Plant Dis. 71:422–425.

Sparrow, F. K., Jr. 1933. Inoperculate chytridiaceous organisms collected in the vicinity of Ithaca, N.Y. with notes on other aquatic fungi. Mycologia 25:513–535.

Sparrow, F. K., Jr. 1960. Aquatic Phycomycetes. Univ. Mich. Press, Ann Arbor. 1187 pp.

Sparrow, F. K. 1973. Mastigomycotina (zoosporic fungi). Chapter 4. Pages 61–73 in: The Fungi: An Advanced Treatise. Vol. IVb. G. C. Ainsworth, F. K. Sparrow, and A. C. Sussman, eds. Academic Press, New York. 504 pp.

Sparrow, F. K. 1976. The present status of classification in biflagellate fungi. Pages 213–222 in: Recent Advances in Aquatic Mycology. E. B. Gareth Jones, ed. John Wiley & Sons, New York. 749 pp.

Spencer, S., and Benson, D. M. 1981. Root rot of *Aucuba japonica* caused by *Phytophthora cinnamomi* and *P. citricola* and suppressed with bark media. Plant Dis. 65:918–921.

Spencer, S., and Benson, D. M. 1982. Pine bark, hardwood bark compost and peat amendment effects on development of *Phytophthora* spp. and lupine root rot. Phytopathology 72:346–351.

Speschnew, N. N. 1897. Les parasites vegetaux de la Cakhetie (The parasites of vegetables in Cakhetie). Arb. Bot. Gart. Tiflis 2. 70 pp. (In French). (Abstr. in Z. Pflanzenkr. 11:113–115). (Cited in Tucker 1933).

Spielman, L. J. 1991. Isozymes and the population genetics of *Phytophthora infestans*. Pages 231–241 in: *Phytophthora*. J. A. Lucas, R. C. Shattock, D. S. Shaw, and L. R. Cooke, eds. Cambridge Univ. Press, Cambridge, U.K. 447 pp.

Spielman, L. J., McMaster, B. J., and Fry, W. E. 1989. Dominance and recessiveness at loci for virulence against potato and tomato in *Phytophthora infestans*. Theor. Appl. Genet. 77:832–838.

Spielman, L. J., Swiegard, J. A., Shattock, R. C., and Fry, W. E. 1990. The genetics of *Phytophthora infestans:* Segregation of allozyme markers in F2 and backcross progeny and the inheritance of virulence against potato resistance genes R2 and R4 in F1 progeny. Exp. Mycol. 14:57–69.

Spielman, L. J., Drenth, A., Davidse, L. C., Sujkowski, L. J., Gu, W., Tooley, P. W., and Fry, W. E. 1991. A second world-wide migration and population displacement of *Phytophthora infestans*? Plant Pathol. 40:422–430.

Spielman, L. J., McMaster, B. J., and Fry, W. E. 1992. Relationships among measurements of fitness and disease severity in *Phytophthora infestans*. Plant Pathol. 41:317–324.

Spring, D. E., Ellis, M. A., Spotts, R. A., Hoitink, H. A. J., and Schmitthenner, A. F. 1980. Suppression of the apple collar rot pathogen in composted hardwood bark. Phytopathology 70:1209–1212.

Springer, J. K., and Johnston, S. A. 1982. Black polyethylene mulch and Phytophthora blight of pepper. Plant Dis. 66:281.

Sreenivasan, T. N. 1982. Present status of cocoa *Phytophthora*. Pages 63–70 in: *Phytophthora* Diseases of Tropical Cultivated Plants. K. K. N. Nambiar, ed. Cent. Plant. Crops Res. Inst., Kasaragod, Kerala, India. 284 pp.

Sreenivasan, T. N., and Quesnel, V. C. 1977. Field differentiation of black pod from other cacao pod diseases. Trop. Agric. 54:371–372.

Sridhar, T. S., Ullasa, B. A., and Sohi, H. S. 1975. Occurrence of a new disease on grape seedlings caused by *Phytophthora nicotianae* var. *parasitica* (Dastur) Waterhouse from India. Curr. Sci. 44:406.

Stack, J. P., and Millar, R. L. 1985a. Isolation and characterization of a metalaxyl-insensitive isolate of *Phytophthora megasperma* f. sp. *medicaginis*. Phytopathology 75:1387–1392.

Stack, J. P., and Millar, R. L. 1985b. Competitive colonization of organic matter in soil by *Phytophthora megasperma* f. sp. *medicaginis*. Phytopathology 75:1393–1398.

Stack, J. P., and Millar, R. L. 1985c. Relative survival potential of propagules of *Phytophthora megasperma* f. sp. *medicaginis*. Phytopathology 75:1398–1404.

Stafleu, F. A., Bonner, C. E. B., McVaugh, R., Merkle, R. D., Rollins, R. C., Ross, R., Schopt, J. M., Schulze, G. M., and DeVilmorin, R. 1972. International Code of Nomenclature. Vol. 82. XI. Int. Bot. Congr., August 1969. A. Oosthoek's Uitgeversmaatschappij N.V., Utrecht, The Netherlands. 426 pp.

Stahl, M. 1965. Wurzelhalsfäule bei Stiefmütterchen (Viola × Wittrockiana Gams) (Collar rot on pansy [Viola × Wittrockiana Gams]). Z. Pflanzenkr. Pflanzenschutz 72:5–29. (In German).

Stammler, G., Seemüller, E., and Duncan, J. M. 1993. Analysis of RFLPs in nuclear and mitochondrial DNA and the taxonomy of *Phytophthora fragariae*. Mycol. Res. 97:151–156.

Stamps, D. J. 1978a. *Phytophthora boehmeriae*. Commonw. Mycol. Inst. Descriptions of Pathogenic Fungi and Bacteria No. 591. 2 pp.

Stamps, D. J. 1978b. *Phytophthora cryptogea*. Commonw. Mycol. Inst. Descriptions of Pathogenic Fungi and Bacteria No. 592. 2 pp.

Stamps, D. J. 1978c. *Phytophthora erythroseptica*. Commonw. Mycol. Inst. Descriptions of Pathogenic Fungi and Bacteria No. 593. 2 pp.

Stamps, D. J. 1978d. *Phytophthora hevea*. Commonw. Mycol. Inst. Descriptions of Pathogenic Fungi and Bacteria No. 594. 2 pp.

Stamps, D. J. 1978e. *Phytophthora porri*. Commonw. Mycol. Inst. Descriptions of Pathogenic Fungi and Bacteria No. 595. 2 pp.

Stamps, D. J. 1985a. *Phytophthora palmivora*. Commonw. Mycol. Inst. Descriptions of Pathogenic Fungi and Bacteria No. 831. 2 pp.

Stamps, D. J. 1985b. *Phytophthora megakarya*. Commonw. Mycol. Inst. Descriptions of Pathogenic Fungi and Bacteria No. 832. 2 pp.

Stamps, D. J. 1985c. *Phytophthora arecae*. Commonw. Mycol. Inst. Descriptions of Pathogenic Fungi and Bacteria No. 833. 2 pp.

Stamps, D. J. 1985d. *Phytophthora meadii*. Commonw. Mycol. Inst. Descriptions of Pathogenic Fungi and Bacteria No. 834. 2 pp.

Stamps, D. J. 1985e. *Phytophthora botryosa*. Commonw. Mycol. Inst. Descriptions of Pathogenic Fungi and Bacteria No. 835. 2 pp.

Stamps, D. J. 1985f. *Phytophthora capsici*. Commonw. Mycol. Inst. Descriptions of Pathogenic Fungi and Bacteria No. 836. 2 pp.

Stamps, D. J. 1985g. *Phytophthora katsurae*. Commonw. Mycol. Inst. Descriptions of Pathogenic Fungi and Bacteria No. 837. 2 pp.

Stamps, D. J. 1985h. *Phytophthora infestans*. Commonw. Mycol. Inst. Descriptions of Pathogenic Fungi and Bacteria No. 838. 2 pp.

Stamps, D. J. 1985i. *Phytophthora primulae*. Commonw. Mycol. Inst. Descriptions of Pathogenic Fungi and Bacteria No. 839. 2 pp.

Stamps, D. J. 1985j. *Phytophthora drechsleri*. Commonw. Mycol. Inst. Descriptions of Pathogenic Fungi and Bacteria No. 840. 2 pp.

Stamps, D. J., Waterhouse, G. M., Newhook, F. J., and Hall, G. S. 1990. Revised tabular key to the species of *Phytophthora*. Commonw. Agric. Bur. Int. Mycol, Inst. Mycol. Pap. 162. 28 pp.

Standish, E. D., MacDonald, J. D., and Humphrey, W. A. 1982. Phytophthora root and crown rot of junipers in California. Plant Dis. 66:925–928.

Staner, P. 1929. Maladies du citrus au Congo Belge (Citrus diseases in the Belgian Congo). Bull. Agric. Congo. Belge. 20:364–373. (In French). (Cited in Tucker 1933).

Stanghellini, M. E. 1974. Spore germination, growth and survival of *Pythium* in soil. Proc. Am. Phytopathol. Soc. 1:211–214.

Stanghellini, M. E. 1978. *Phytophthora parasitica* root rot of jojoba. Phytophthora Newsl. 6:47.

Stanghellini, M. E., and Kronland, W. C. 1982. Root rot of chicory caused by *Phymatotrichum omnivorum* and *Phytophthora cryptogea*. Plant Dis. 66:262–263.

Staples, R. R. 1958. Rep. Minst. Agr. Rhod.-Nyasaland 1956–1956: 7–86.

Stapleton. J. J., and DeVay, J. E. 1982. Effect of soil solarization on populations of selected soilborne microorganisms and growth of deciduous fruit tree seedlings. Phytopathology 72:323–326.

Stapleton, J. J., and DeVay, J. E. 1986. Soil solarizaration: A non-chemical approach for management of plant pathogens and pests. Crop Prot. 5:190–198.

Staub, T. 1991. Fungicide resistance: Practical experience with anti-resistance strategies and the role of integrated use. Annu. Rev. Phytopathol. 29:421–442.

Staub, T., and Hubele, A. 1981. Recent advances in the chemical control of Oomycetes. Pages 389–422 in: Chemie der Pflanzenschutz und Schaedlingsbekämpfungsmittel. Vol. 6. R. Wegler, ed. Springer-Verlag, Heidelberg.

Staub, T., and Stozzi, D. 1984. Fungicide resistance: A continuing challenge. Plant Dis. 68:1026–1031.

Staub, T., Dahmen, H., Urech, P., and Schwinn, F. J. 1979. Failure to select for in vivo resistance in *Phytophthora infestans* to acylalanine fungicides. Plant Dis. Rep. 63:385–389.

Staub, T. H., and Young, T. R. 1980. Fungitoxicity of metalaxyl against *Phytophthora parasitica* var. *nicotianae.* Phytopathology 70:797–801.

Staub, T. H., Dahmen, H., and Schwinn, F. J. 1978. Biological characterization of uptake and translocation of fungicidal acylalanine in grape and tomato plants. J. Plant Dis. Prot. 85:162–168.

Staub, T. H., Dahmen, H., and Schwinn, F. J. 1980. Effects of ridomil on the development of *Plasmopara viticola* and *Phytophthora infestans* on their host plants. Z. Pflanzenkr. Pflanzenschutz 87:83–91.

Stavely, J. R. 1979. Disease resistance. Pages 87–110 in: *Nicotiana:* Procedures for Experimental Use. U.S. Dep. Agric. Tech. Bull. 1586.

Steer, J., and Coates-Beckford, P. L. 1990. Role of *Phytophthora katsurae, P. palmivora, Thielaviopsis paradoxa* and *Enterobacter* sp. in budrot disease of coconuts in Jamaica. Oleagineaux 45:539–545.

Steib, R. J., and Chilton, S. J. P. 1951. The Phytophthora rot of sugarcane seed pieces in Louisiana. Int. Soc. Sugar Cane Technol. Proc. 7:524–528.

Steinberg, R. A. 1948. Essentiality of calcium in the nutrition of fungi. Science 107:423.

Steinhausen, W. R. 1979. On the importance of plant pathogenic organisms assessed by literature studies. Z. Pflanzenkr. Pflanzenschutz 86:86–92.

Stelfox, D., and Henry, A. W. 1978. Occurrence of the rubbery brown rot of stored carrots in Alberta. Can. Plant Dis. Surv. 58:87–91.

Stephenson, L. W., and Erwin, D. C. 1972. Encirclement of the oogonial stalk by amphigynous antheridium in *Phytophthora capsici.* Can J. Bot. 50:2539–2441.

Stephenson, L. W., Erwin, D. C., and Leary, J. V. 1974a. Hyphal anastomosis in *Phytophthora capsici.* Phytopathology 64:149–150.

Stephenson, L. W., Erwin, D. C., and Leary, J. V. 1974b. Meiotic configurations in the oospore of *Phytophthora capsici.* Can. J. Bot. 52:2141–2143.

Sterne, R. E., Zentmyer, G. A., and Kaufman, M. R. 1977a. The effect of matric and osmotic potential of soil on Phytophthora root disease of *Persea indica.* Phytopathology 67:1491–1494.

Sterne, R. E., Zentmyer, G. A., and Kaufmann, M. R. 1977b. The influence of matric potential, soil texture, and soil amendment on root disease caused by *Phytophthora cinnamomi.* Phytopathology 67:1495–1500.

Sterne, R. E., Kaufman, M. R., and Zentmyer, G. A. 1978. Effect of Phytophthora root rot on water relations of avocado: Interpretation with a water transport model. Phytopathology 68:595–602.

Stevens, F. L., and Plunkett, O. A. 1925. Tulip blossom blight. Illinois Agric. Exp. Stn. Bull. 265:299–307. (Cited in Tucker 1933).

Stevens, H. E., and Piper, R. B. 1941. Avocado diseases in Florida. U.S. Dep. Agric. Circ. 582. 46 pp.

Stevenson, J. A. 1917. Report of the department of pathology and botany. Annu. Rep. Inst. Exp. Stn. (Puerto Rico) 1916:37–98. (Cited in Tucker 1933).

Stevenson, J. A. 1926. Foreign Plant Diseases. A Manual of Economic Plant Diseases Which Are New or Widely Distributed in the United States. U.S. Dep. Agric. Washington, D.C. 198 pp. (Cited in Tucker 1933).

Stewart, A., and McCarrison, A. M. 1991. Pathogenicity and relative virulence of seven *Phytophthora* species on kiwifruit. N.Z. J. Crop Hortic. Sci. 19:73–76.

Stewart, H. E. 1990. Effect of plant age and inoculum concentration on expression of major gene resistance to *Phytophthora infestans* in detached potato leaflets. Mycol. Res. 94:823–826.

Stewart, H. E., and McClamont, D. C. 1982. The effect of metalaxyl and mancozeb on the susceptibility of potato tubers to late blight. Ann. Appl. Biol. 100. Suppl. 54–55.

Stewart, H. E., and Flavelle, P. H., McCalmont, D. C., and Wastie, R. L. 1983. Correlation between glasshouse and field tests for resistance to foliage blight caused by *Phytophthora infestans.* Potato Res. 26:41–48.

Stirling, A. M., and Irwin, J. A. G. 1986. Etiology of a newly described root rot of guar (*Cyamopsis tetragonoloba*) in Australia caused by *Phytophthora cryptogea.* Plant Pathol. 35:527–534.

Stirrup, H. H. 1939. Sugar beet diseases. Ann. Appl. Biol. 26:402–404.

Stoffels, E. H. J. 1945. Le Quinquina (*Chinchona*). Publ. Inst. Nat. Etude Agron. Congo Belge, Ser. Tech. 24. 57 pp. (In Spanish). (Rev. Appl. Mycol. 1946, 25:45).

Stössel, A., Unwin, C. H., and Ward, E. W. B. 1972. Postinfectional inhibitors from plants. I. Capsidiol, an antifungal compound from *Capsicum frutescens.* Phytopathol. Z. 74:141–152.

Stössel, P. 1983. In vitro effects of glyceollin on *Phytophthora megasperma* f. sp. *glycinea.* Phytopathology 73:1603–1607.

Stössel, P. 1988. Pathogenesis and host-parasite specificity in *Phytophthora.* Pages 139–456 in: Experimental and Conceptual Plant Pathology. Vol. 2. W. M. Hess, R. S. Singh, C. S. Singh, and D. J. Weber, eds. Gordon and Breach Sci. Pub. 599 pp.

Stössel, P., and Hohl, H. R. 1980. Sterols in *Phytophthora infestans* and their role in the parasitic interactions with *Solanum tuberosum.* Ber. Schweiz. Bot. Ges. 90:118–128.

Stössel, P., Lazarovits, G., and Ward, E. W. B. 1981. Electron microscope study of race-specific and age-related resistant and susceptible reactions of soybeans to *Phytophthora megasperma* var. *sojae.* Phytopathology 71:617–623.

Stovold, G. 1973. *Phytophthora drechsleri* Tucker and *Pythium* spp. as pathogens of safflower in New South Wales. Aust. J. Exp. Agric. Anim. Husb. 13:455–459.

Stovold, G. E., and Carratt, D. E. 1978. Occurrence of *Phytophthora megasperma* in alfalfa growing soils of New South Wales. Plant Dis. Rep. 662:742–744.

Strider, D. L. 1978. Reaction of African violet cultivars to *Phytophthora nicotianae* var. *parasitica.* Plant Dis. Rep. 62:112–114.

Strider, D. L. 1979. Evaluation of Melodie and Optimara African violets for resistance to Phytophthora rot. Plant Dis. Rep. 63:382–384.

Strömberg, A., and Brishammar, S. 1991. Induction of systemic resistance in potato (*Solanum tuberosum* L.) plants to late blight by local treatment with *Phytophthora infestans* (Mont.) de Bary, *Phytophthora cryptogea* Pehtyb. & Laff., or dipotassium phosphate. Potato Res. 34:219–225.

Strong, R. P., and Shattuck, G. C. 1930. Harvard African Expedition 1926–1927, Contr. Dep. Trop. Med. Inst. Trop. Biol. Med. 5:389–410. (In Rev. Appl. Mycol. 1931, 10:165).

Stuart, W. 1906. Disease resistance of potatoes. Vt. Agric. Exp. Stn. Bull. 122:107–136. (Cited in Tucker 1933).

Stuntz, D. E., and Seliskar, C. E. 1943. A stem canker of dogwood and madrona. Mycologia 35:207–221.

Sturgis, W. C. 1898. On some aspects of vegetable pathology and the conditions which influence the dissemination of plant diseases. Bot. Gaz. 25:187–194.

Stuteville, D. L., and Erwin, D. C., eds. 1990. Compendium of Alfalfa Diseases. American Phytopathological Society, St. Paul, Minn. 104 pp.

Su, M. T. 1933. Report of the mycologist Burma, Mandalay for the year ended March 1933. 2 pp. (Rev. Appl. Mycol. 1934, 13:78).

Suhovecky, A. J. 1955. Phytophthora root rot of soybeans. Ph.D. dissertation, Ohio State Univ., Columbus. 83 pp.

Suhovecky, A. J., and Schmitthenner, A. F. 1955. Soybeans affected by early root rot. Ohio Farm Home Res. 40:85–86.

Sujkowski, L. S. 1990. Effect of some factors on detectability of virulence genes in *Phytophthora infestans* (Mont.) de Bary. J. Phytopathol. 129:9–18.

Sujkowski, L. S., Goodwin, S. B., Dyer, A. T., and Fry, W. E. 1994. Increased genotypic diversity via migration and possible occurrence of sexual reproduction of *Phytophthora infestans* in Poland. Phytopathology 84:201–207.

Sujkowski, L. S., Fry, B. A., Power, R. J., Goodwin, S. B., Peever, T. L., Hamlen, R. A., and Fry, W. E. 1995. Sensitivities of Mexican isolates of *Phytophthora infestans* to chlorothalonil, cymoxanil, and metalaxyl. Plant Dis. 79:1117–1120.

Sulistyowati, L., and Keane, P. J. 1992. Effect of soil salinity and water content on stem rot caused by *Phytophthora citrophthora* and accumulation of phytoalexin in citrus root stalks. Phytopathology 82:771–777.

Sulistyowati, L., Keane, P. J., and Anderson, J. W. 1990. Accumulation of the phytoalexin, 6,7-dimethoxycoumarin, in roots and stems of citrus seedlings following inoculation with *Phytophthora citrophthora*. Physiol. Mol. Plant Pathol. 37:451–461.

Sundararaman, S., and Ramakrishnan, T. S. 1924. The "mahali" disease of coconuts in Malabar. Mem. Dep. Agric. India Bot. Ser. 13:87–97.

Sundararaman, S., and Ramakrishnan, T. S. 1928. Foot-rot and wilt of Antirrhinums. Mem. Dep. Agric. India, Bot. Ser. 16:83–100.

Sussman, A. S. 1966. Dormancy and spore germination. Pages 733–764 in: The Fungi. Vol. II. C. G. Ainsworth and A. S. Sussman, eds. Academic Press, New York. 805 pp.

Sutherland, C. F., Newhook, F. J., and Levy, J. 1959. The association of *Phytophthora* spp. with mortality of *Pinus radiata* and other conifers. II. Influence of soil drainage on disease. N.Z. J. Agric. Res. 2:844–858.

Sutherland, E. D., and Cohen, S. D. 1983. Evaluation of tetrazolium bromide as a vital stain for fungal oospores. Phytopathology 73: 1532–1535.

Sutherland, E. D., and Lockwood, J. L. 1984. Hyperparasitism of oospores of some Peronosporales by *Actinoplanes missouriensis* and *Humicola fuscoatra* and other Actinomycetes and fungi. Can. J. Plant Pathol. 6:139–145.

Sutherland, E. D., Baker, K. K., and Lockwood, J. L. 1984. Ultrastructure of *Phytophthora megasperma* f. sp. *glycinea* oospores parasitized by *Actinoplanes missouriensis* and *Humicola fuscoatra*. Trans. Br. Mycol. Soc. 82:726–729.

Suzui, T. 1984. Ecology of *Phytophthora* diseases in vegetable crops in Japan. Pages 137–144 in: Soilborne Crop Diseases in Asia. Food and Technology Center for Asian and Pacific Region (FFTC) Book Series No. 26. Taipei.

Suzui, T., and Hoshino, Y. 1979. Collar rot of apple caused by *Phytophthora cambivora* (Petri) Buism. Ann. Phytopathol. Soc. Jpn. 45:344–352.

Suzui, T., Kueprakone, U., and Kamphangridthrong, T. 1979. *Phytophthora* spp. isolated from some economic plants in Thailand. Tech. Bull. Trop. Agric. Res. Cent. 12:32–41. (Rev. Plant Pathol. 1980, 59:2090).

Suzui, T., Makino, Y., and Ogoshi, A. 1980. Phytophthora rot of strawberry caused by *Phytophthora nicotianae* var. *parasitica*. Ann. Phytopathol. Soc. Jpn. 46:169–178.

Swiecki, T. J., and MacDonald, J. D. 1988. Histology of chrysanthemum roots exposed to salinity stress and *Phytophthora cryptogea*. Can. J. Bot. 66:280–288.

Swiecki, T. J., and MacDonald, J. D. 1991. Soil salinity enhances Phytophthora root rot of tomato but hinders asexual reproduction by *Phytophthora parasitica*. J. Am. Soc. Hortic. Sci. 116:471–477.

Swiezynski, K. M. 1990. Resistance to *Phytophthora infestans* in potato cultivars and its relation to maturity. Gen. Polonika 31:99–106.

Swiezynski, K. M., Sieczka, M. T., Sujkowski, L. S., Zarzycka, H., and Zimonoch-Guzowska. 1991. Resistance to *Phytophthora infestans* in potato genotypes originating from wild species. Plant Breed. 107:28–38.

Sydow, H., and Butler, E. J. 1906. Fungi Indiae Orientalis, Pars I. (Fungi of oriental India, Part I.). Ann. Mycol. 4:424–452. (In German).

Sztejnberg, A., and Wahl, I. 1969. Studies on the host range of the potato late blight fungus and its oversummering on wild and ornamental Solanaceae plants. Proc. 1st Congr. Mediterr. Phytopathol. Union, Bari-Naples 1:452–460 (1966).

Tachibana, H. D., and Van Diest, A. 1983. Association of Phytophthora root rot of soybean with conservation tillage. Phytopathology 73:844.

Taher, M. A. 1976. Studies on the efficacy of some new fungicides in controlling gummosis diseases of lemon caused by *Phytophthora citrophthora* L. Bangladesh Hortic. 4:50–53.

Tainter, F. H., Weste, G. M., and Mistretta, P. A. 1987. Annotated bibliography of little leaf and tree decline diseases caused by *Phytophthora cinnamomi* Rands. Dept. Forestry Clemson Univ. Tech. Pap. No. 18. 97 pp.

Takahi, Y., Nakanishi, T., and Kamimura, S. 1974a. Characteristics of hymexazol as a soil fungicide. Ann. Phytopathol. Soc. Jpn. 40:362–367.

Takahi, Y., Nakanishi, T., Tomita, K., and Kamimura, S. 1974b. Effects of 3-hydroxyisoxazoles as soil fungicides in relation to their chemical structure. Ann. Phytopathol. Soc. Jpn. 40:354–361.

Takimoto, S. 1931. Gummosis or brown rot disease of the kumquat. Stud. Citrologica. Tanaka Citrus Exp. Stn. 5:55–60.

Tan, A. M. 1983. A new fungicide for the control of black stripe. Plant. Bull. Rubber Res. Inst. Malays. 174:13–16.

Tan, A. M. 1990. New recommendations for the control of black stripe. Plant. Bull. Rubber Res. Inst. Malays. 202:14–17.

Tan, G. Y., and Tan, W. K. 1990. Additive inheritance of resistance to pod rot caused by *Phytophthora palmivora* in cocoa. Theor. Appl. Genet. 80:258–264.

Tanaka, I. 1940. *Phytophthora macrospora* (Sacc.) S. Ito et I. Tanaka on wheat plant. Ann. Phytopathol. Soc. Jpn. 10:127–138. (In Japanese, English summary).

Tanaka, T. 1917. New Japanese fungi. Notes and translations. II. Mycologia 9:249–251. (Cited in Tucker 1933).

Tanaka, T. 1920. New Japanese fungi. Notes and translations. Mycologia 12:25–32. (Cited in Tucker 1933).

Tantius, P. H., Fyfe, A. M., Shaw, D. S., and Shattock, R. C. 1986. Occurrence of the A2 mating type and self-fertile isolates of *Phytophthora infestans* in England and Wales. Plant Pathol. 35:578–581.

Tarabeih, A. M. 1971. Studies on certain root rots of *Melianthus annuus* L. and *Ricinus communis* L. in Egypt. Ph.D. thesis, Alexandria Univ., Alexandria, Egypt.

Tariq, V. N. 1990. Factors influencing zoosporangia formation by *Phytophthora fragariae* in vitro. Mycol. Res. 94:205–210.

Tasugi, H., and Ikeda, Y. 1939. Phytophthora blight of *Hibiscus manihot*. Ann. Phytopathol. Soc. Jpn. 9:69–85. (In Japanese, English summary).

Tasugi, H., and Kumazawa, M. 1931. Phytophthora root rot of lily. J. Imp. Agric. Exp. Stn. 13:207–223. (In Japanese).

Tasugi, H., and Kumazawa, M. 1932. Phytophthora blight of peony. Studies on Japanese Peronosporales, I. J. Imp. Agric. Exp. Stn. Nisigahara, Tokyo 2:75–96. (In Japanese). (Rev. Appl. Mycol. 1933, 12:176).

Tay, F. C. S., Nandapalan, K., and Davison, E. M. 1983. Growth and zoospore germination of *Phytophthora* spp. on $P_{10}VP$ agar with hymexazol. Phytopathology 73:234–240.

Taylor, B., and Adedoyin, S. F. 1981. A general view of ant biology in relation to cocoa. Chapter 9. Epidemiology of Phytophthora Pod Rot of Cocoa in Nigeria. P. H. Gregory and A. C. Maddison, eds. Commonw. Mycol. Inst. Phytopathol. Pap. 25:132–144.

Taylor, B., and Griffin, M. J. 1981. The role and relative importance of different ant species in the dissemination of black pod disease of cocoa. Chapter 8. Epidemiology of Phytophthora Pod Rot of Cocoa in Nigeria. P. H. Gregory and A. C. Maddison, eds. Commonw. Mycol. Inst. Phytopathol. Pap. 25:114–131.

Taylor, G. S., and Waggoner, P. E. 1979. Resistance to black shank in the field predicted by a test of tobacco seedlings. Phytopathology 69:1132–1134.

Taylor, J. C. 1965. White tip of leeks (*Phytophthora porri* Foister). Plant Pathol. 14:189.

Taylor, P. A. 1977. *Phytophthora* spp. in irrigation water in the Goulburn Valley, Victoria. Austr. Phytopathol. Newsl. 6:41–42. (Rev. Plant Pathol. 1978, 57:2174).

Taylor, P. A. 1984. An unusual *Phytophthora* associated with root rot of subterranean clover. Plant Dis. 68:450.

Taylor, P. A., and Greenhalgh, F. C. 1987a. Effect of temperature and watering regimes on Phytophthora and Aphanomyces root rots of subterranean clover (Abstr.). Page 78 in: Proc. VI Australas. Plant Pathol. Soc. Conf., Adelaide.

Taylor, P. A., and Greenhalgh, F. C. 1987b. Significance, causes and control of root rots of subterranean clover. Pages 249–251 in: Temperate Pastures; Their Production, Use and Management. Aust. Wool Corp. Tech. Publ. Ser. J. L. Wheeler, C. J. Pearson, and G. E. Robards, eds. Commonw. Sci. Ind. 609 pp.

Taylor, P. A., and Washington, W. S. 1984. Curative treatments for *Phytophthora cactorum* in peach trees using metalaxyl and phosetyl-Al. Australas. Plant Pathol. 13:31–33.

Taylor, P. A., Pascoe, I. G., and Greenhalgh, F. C. 1985a. *Phytophthora clandestina* sp. nov. in roots of subterranean clover. Mycotaxon 22:77–85.

Taylor, P. A., Barbetti, M. J., and Wong, D. H. 1985b. Occurrence of *Phytophthora clandestina* in Western Australia. Plant Prot. Q. 1:57–58.

Taylor, P. A., Clarke, R. G., Kelly, K., and Smiley, R. 1985c. Root rot of irrigated subterranean clover in northern Victoria: Significance and prospects for control. Pages 271–273 in: Ecology and Management of Soilborne Plant Pathogens. C. A. Parker, A. D. Rovira, K. J. Moore, P. T. W. Wong, and J. F. Kollmorgen, eds. American Phytopathological Society, St. Paul, Minn. 358 pp.

Teakle, D. S. 1957. Papaw root rot caused by *Phytophthora palmivora* Butl. Queensl. J. Agric. Sci. 14:81–91.

Tedford, E. C., Miller, T.L., and Nielsen, M. T. 1990. A detached-leaf technique for detecting resistance to *Phytophthora parasitica* var. *nicotianae* in tobacco. Plant Dis. 74:313–316.

Telhada, J. A. 1990. Les maladies a *Phytophthora* des agrumes au Portugal: Caracteristiques des especes isolees (*Phytophthora* diseases of citrus in Portugal: Characteristics of the isolates obtained). E.P.P.O. (Eur. Mediterr. Plant Prot. Organ.) Bull. 20:77–82.

Teng, S. C. 1932. Some fungi from Canton. Contrib. Biol. Lab. Sci. Soc. China Bot. Ser. 8:121–128. (In Chinese). (Rev. Appl. Mycol. 1933, 12:724–725).

Thakur, R. N., Pandotra, V. R., and Sastry, K. S. M. 1976. Stem burn—A new disease of *Hibiscus sabdariffa* from India. Indian Phytopathol. 29:322–323.

Thankamma, L. 1974. *Phytophthora nicotianae* var. *nicotianae* on *Anacardium occidentale* in South India. Plant Dis. Rep. 58:767–768.

Thankamma, L. 1983. *Phytophthora* species on eight indigenous host species in South India and their pathogenicity on rubber. Indian Phytopathol. 36:17–23.

Thankamma, L., and Pillai, P. N. R. 1973. Fruit rot and leaf rot diseases of cardamom in India. F.A.O. Plant Prot. Bull. 21:83–84.

Thankappan, M. 1985. Leaf blight of taro—A review. J. Root Crops 11:1–8.

Thaxter, R. 1889a. A new American *Phytophthora*. Bot. Gaz. 14:273–274. (Cited in Tucker 1933).

Thaxter, R. 1889b. Report of the mycologist. Conn. State Agric. Exp. Stn. Annu. Rep. 1889:127–177.

Therrien, C. D., Ritch, D. L., Davidse, L. C., Jespers, Ad B. K., and Spielman, L. J. 1989. Nuclear DNA content, mating type and metalaxyl sensitivity of eighty-three isolates of *Phytophthora infestans* from the Netherlands. Mycol. Res. 92:140–146.

Thirumalachar, M. J. 1969. Morphological basis for the characterization and separation of the genera *Phytophthora, Sclerophthora,* and *Sclerospora* (Abstr.). Indian Phytopathol. 17:155.

Thirumalachar, M. J., Shaw, C. G., and Narasimhan, M. J. 1953. The sporangial phase of the downy mildew on *Eleusine coracana* with a discussion of the identity of *Sclerospora macrospora* Sacc. Bull. Torrey Bot. Club. 80:299–307.

Thomas, C. A. 1951. The occurrence and pathogenicity of *Phytophthora* species which cause root rot of safflower. Plant Dis. Rep. 35:454–455.

Thomas, C. A. 1976. Resistance of VFR 1 safflower to Phytophthora root rot and its inheritance. Plant Dis. Rep. 60:123–125.

Thomas, C. A., and Blount, V. L. 1976. Race D of *Phytophthora phaseoli*. Plant Dis. Rep. 60:308.

Thomas, C. A., and Hill, R. F. 1977. Reaction of cotyledons of safflower cultivars to *Phytophthora drechsleri:* Effect of temperature and inheritance. Phytopathology 67:698–699.

Thomas, C. A., and Klisiewicz, J. M. 1963. Selective pathogenesis within *Phytophthora drechsleri*. Phytopathology 53:368.

Thomas, C. A., and Zimmer, D. E. 1970. Resistance of Biggs safflower to Phytophthora root rot and its inheritance. Phytopathology 60:63–64.

Thomas, C. A., Rubis, D. D., and Black, D. S. 1960. Development of safflower varieties resistant to Phytophthora root rot. Phytopathology 50:129–130.

Thomas, K. M., Ramakrishnan, T. S., Soumini, C. K., and Balakrishnan, M. S. 1947. Studies in the genus *Phytophthora*. I. Oospore formation and taxonomy of *Phytophthora palmivora* Butl. Proc. Indian Acad. Sci. Sect. B. 26:147–163.

Thomison, P. R., Thomas, C. A., Kenworthy, W. J., and McIntosh, M. S. 1988. Evidence of pathogen specificity in tolerance of soybean cultivars to Phytophthora rot. Crop Sci. 28:714–715.

Thomison, P. R., Thomas, C. A., and Kenworthy, W. J. 1991. Tolerant and root-resistant soybean cultivars: Reactions to Phytophthora rot in inoculum-layer tests. Crop Sci. 31:73–75.

Thompson, A. 1928. A preliminary note on *Phytophthora* spp. found in Malaya. Malay. Agric. J. 16:40–47. (Cited in Tucker 1933).

Thompson, A. 1929. *Phytophthora* species in Malaya. Malay. Agric. J. 17:53–100. (Cited in Tucker 1933).

Thompson, A. 1934. A disease of the durian tree. Malay. Agric. J. 22:369–371.

Thompson, A. 1939. Notes on plant diseases in 1937–38. Malay. Agric. J. 27:86–98.

Thompson, A. 1940. Notes on plant diseases in 1939. Malay. Agric. J. 28:400–407.

Thompson, A. H. 1981. *Phytophthora* finger rot, a post harvest disease of banana in South Africa. Phytophylactica 13:161–163.

Thompson, A. H. 1987. Phytophthora root rot of lucerne in the Transvaal, South Africa. Phytophylactica 19:319–322.

Thompson, A. H. 1988. Phytophthora root rot of lucerne in the Cape Province, South Africa. Phytophylactica 20:181–182.

Thompson, A. H., and Phillips, A. J. L. 1988. Root rot of cabbage caused by *Phytophthora drechsleri*. Plant Pathol. 37:297–299.

Thomson, L. W., Henry, A. W., and Zalik, S. 1986. Sterol content of mature cereal straw: Possible role in oospore formation in *Phytophthora*. Phytochemistry 25:2505–2508.

Thomson, S. V., and Allen, R. M. 1974. Occurrence of *Phytophthora* species and other potential plant pathogens in recycled irrigation water. Plant Dis. Rep. 58:945–949.

Thomson, S. V., and Allen, R. M. 1976. Mechanisms of survival of zoospores of *Phytophthora parasitica* in irrigation water. Phytopathology 66:1198–1202.

Thomson, S. V., and Hine, R. B. 1972. Atypical sporangium-like structures of *Phytophthora parasitica*. Mycologia 64:457–460.

Thorn, W. A., and Zentmyer, G. A. 1954. Hosts of *Phytophthora cinnamomi* Rands. Plant Dis. Rep. 38:47–52.

Thorold, C. A. 1955. Observations on black-pod disease (*Phytophthora palmivora*) of cacao in Nigeria. Trans. Br. Mycol. Soc. 38: 435–452.

Thorold, C. A. 1959. Methods of controlling black-pod disease (caused by *Phytophthora palmivora*) of *Theobroma cacao* in Nigeria. Ann. Appl. Biol. 47:708–715.

Thorold, C. A. 1974. History up to 1930 of black pod and other cocoa diseases caused by *Phytophthora palmivora*. Pages 13–21 in: *Phytophthora* Disease of Cocoa. P. H. Gregory, ed. Longman, London. 348 pp.

Thung, T. H. 1938. De epidemiologie van de *Phytophthora parasitica* var. *nicotianae* op de Vorstenlandsche Tabaksonderneming (The epidemiology of *Phytophthora parasitica* var. *nicotianae* on the Vorstenland tobacco plantations). Meded. Proefstn. Vorstenl.

Tab. 86. 55 pp. (In Dutch). (Rev. Appl. Mycol. 1939, 18:419–420).

Thurston, H. D. 1971. Relationship of general resistance: Late blight of potato. Phytopathology 61:620–626.

Thurston, H. D. 1990. Plant disease management by practices of traditional farmers. Plant Dis. 74:96–102.

Thurston, H. D., and Schultz, O. 1981. Late blight. Pages 40–41 in: Compendium of Potato Diseases. W. J. Hooker, ed. American Phytopathological Society, St. Paul, Minn. 125 pp.

Thurston, H. D., Plaisted, R. L., Brodie, B. B., Jones, E. D., Loria, R., Halseth, D., and Sieczka, J. B. 1985. Elba: A late maturing, blight-resistant potato variety. Am. Potato J. 62:653–656.

Thurston, H. W., Jr., and Orton, C. R. 1921. A *Phytophthora* parasitic on peony. Science 54:170–171.

Thyagarajan, P., Rao, A. V., Varadarajan, S., and Sundararajan, R. 1972. Studies on betelvine wilt disease—Influence of nitrogen and phosphorus in the control of betelvine wilt disease. Madras Agric. J. 59:187–189.

Tichelaar, G. M., and Van Kesteren, H. A. 1967. Aantasting van Ui door *Phytophthora porri* (Attack of onions by *Phytophthora porri*). Neth. J. Plant Pathol. 73:103–104. (In Dutch).

Tidball, C. J., and Linderman, R. G. 1990. Phytophthora root and stem rot of apple rootstocks from stool beds. Plant Dis. 74:141–146.

Tilford, P. E. 1932. Calla lily root rot and its control. Ohio Agric. Exp. Stn. Bi-mon. Bull. 157:138–140.

Timmer, L. W. 1973. Characteristics of *Phytophthora* isolates from Texas citrus orchards. J. Rio Grande Valley Hortic. Soc. 27:44–48.

Timmer, L. W. 1977. Preventive and curative trunk treatments for control of Phytophthora foot rot of citrus. Phytopathology 67:1149–1154.

Timmer, L. W. 1979. Preventive and systemic activity of experimental fungicides against *Phytophthora parasitica* on citrus. Plant Dis. Rep. 63:324–327.

Timmer, L. W., and Castle, W. S. 1985. Effectiveness of metalaxyl and fosetyl-Al against *Phytophthora parasitica* on sweet orange. Plant Dis. 69:741–743.

Timmer, L. W., and Fucik, J. E. 1975. The effect of rainfall, drainage, tree spacing, and fungicide application on the incidence of citrus brown rot. Phytopathology 65:241–242.

Timmer, L. W., and Leyden, R. F. 1976. Effect of irrigation and soil management practices on the incidence of Phytophthora foot rot of Citrus. J. Rio Grande Val. Hortic. Soc. 30:19–25.

Timmer, L. W., and Menge, J. A. 1988. *Phytophthora*-induced diseases. Pages 22–24 in: Compendium of Citrus Diseases. J. O. Whiteside, S. M. Garnsey, and L. W. Timmer, eds. American Phytopathological Society, St. Paul, Minn. 105 pp.

Timmer, L. W., Castro, J., Erwin, D. C., Belser, W. L., and Zentmyer, G. A. 1970a. Genetic evidence for zygotic meiosis in *Phytophthora capsici*. Am. J. Bot. 57:1211–1218.

Timmer, L. W., Erwin, D. C., and McCormick, W. H. 1970b. Glucose inhibition and inheritance of pigment-producing ability in *Phytophthora*. Mycologia 62:967–977.

Timmer, L. W., Sandler, H. A., Graham, J. H., and Zitko, S. E. 1988a. Sampling citrus orchards in Florida to estimate populations of *Phytophthora parasitica*. Phytopathology 78:940–944.

Timmer, L. W., Graham, J. H., Sandler, H. A., and Zitko, S. E. 1988b. Populations of *Phytophthora parasitica* in bearing citrus orchards in Florida and response to fungicide applications. Citrus Ind. 69(11):40–44,54.

Timmer, L. W., Zitko, S. E., Sandler, H. A., and Graham, J. H. 1989. Seasonal and spatial analysis of populations of *Phytophthora parasitica* in citrus orchards in Florida. Plant Dis. 73:810–813.

Timmer, L. W., Menge, J. A., Zitko, S. E., Pond, E., Miller, S. A., and Johnson, E. L. V. 1993. Comparison of ELISA techniques and standard isolation methods for *Phytophthora* detection in citrus orchards in Florida and California. Plant Dis. 77:791–796.

Tinline, R. D., and MacNeill, B. H. 1969. Parasexuality in plant pathogenic fungi. Annu. Rev. Phytopathol. 7:147–170.

Tippett, J. 1978. Interactions between *Phytophthora cinnamomi* and plant roots. PhD. thesis, Monash Univ., Clayton, Victoria, Australia. 217 pp.

Tippett, J., and Malajczuk, N. 1979. Interaction of *Phytophthora cinnamomi* and a resistant host, *Acacia pulchella*. Phytopathology 69:764–772.

Tippett, J. T., and Hill, T. C. 1983. The relationship between bark moisture and invasion of *Eucalyptus marginata* by *Phytophthora cinnamomi*. Australas. Plant Pathol. 12:40–41.

Tippett, J. T., Hill, T. C., and Shearer, B. L. 1985. Resistance of *Eucalyptus* spp. to invasion by *Phytophthora cinnamomi*. Aust. J. Bot. 33:409–418.

Tippett, J. T., Crombie, D. S., and Hill, T. C. 1987. Effect of phloem water relations on the growth of *Phytophthora cinnamomi* in *Eucalyptus marginata*. Phytopathology 77:246–250.

Tisdale, W. B. 1922. Tobacco diseases in Gadsden County in 1922. Fla. Agric. Exp. Stn. Bull. 166:77–118.

Tisdale, W. B. 1931. Development of strains of cigar-wrapper tobacco resistant to black shank (*Phytophthora nictotianae* Breda de Haan). Fla. Agric. Exp. Stn. Bull. 226. 45 pp.

Tisdale, W. B., and Kelley, J. G. 1926. A *Phytophthora* disease of tobacco. Fla. Agric. Exp. Stn. Bull. 179:159–219.

Titze, J. F., and Palzer, C. R. 1969. Host list of *Phytophthora cinnamomi* Rands with special reference to Western Australia. Tech. Note 1. West. Reg. Stn. For. Res. Inst. Kelmscott, W. Australia. 58 pp.

Titze, J. F., and Palzer, C. R. 1970. Host list of *Phytophthora cinnamomi* Rands with special reference to Western Australia. Tech. Note 1. Addenda 1970. West. Reg. Stn. For. Res. Inst. Kelmscott, W. Australia. 7 pp.

Tokunaga, J., and Bartnicki-Garcia, S. 1971. Cyst wall formation and endogenous carbohydrate utilization during synchronous encystment of *Phytophthora palmivora* zoospores. Arch. Mikrobiol. 79: 283–292.

Toler, R. W., Bruton, B. D., and Grisham, M. P. 1983. Evaluation of St. Augustine grass accessions and cultivars for resistance to *Sclerophthora macrospora*. Plant Dis. 67:1008–1010.

Tollenaar, D. 1958. *Phytophthora palmivora* of cocoa and its control. Neth. J. Agric. Sci. 6:24–38.

Tomiyama, K. 1973. Effect of temperature on the process of the hypersensitive death of a potato plant cell infected by an incompatible race of *Phytophthora infestans*. Ann. Phytopathol. Soc. Jpn. 39:73–78.

Tomiyama, K., Sakuma, T., Ishizaka, N., Sato, N., Katsui, N., Takasugi, M., and Masamune, T. 1968. A new antifungal substance isolated from resistant potato tuber infected by pathogens. Phytopathology 58:115–116.

Tomlinson, J. A. 1952. Brown core root rot of *Primula* caused by *Phytophthora primulae* n. sp. Trans. Br. Mycol. Soc. 35:221–235.

Tompkins, C. M., and Tucker, C. M. 1937a. Phytophthora rot of honeydew melon. J. Agric. Res. 54:933–944.

Tompkins, C. M., and Tucker, C. M. 1937b. Foot rot of China-aster, annual stock, and Transvaal daisy caused by *Phytophthora cryptogea*. J. Agric. Res. 55:563–574.

Tompkins, C. M., and Tucker, C. M. 1941a. Buckeye rot of tomato in California. J. Agric. Res. 62:467–474.

Tompkins, C. M., and Tucker, C. M. 1941b. Root rot of pepper and pumpkin caused by *Phytophthora capsici*. J. Agric. Res. 63:417–426.

Tompkins, C. M., and Tucker, C. M. 1947a. Leaf blight of pink calla caused by *Phytophthora erythroseptica*. Phytopathology 37:382–389.

Tompkins, C. M., and Tucker, C. M. 1947b. Stem rot of *Dieffenbachia picta* caused by *Phytophthora palmivora* and its control. Phytopathology 37:868–874.

Tompkins, C. M., and Tucker, C. M. 1950. Rhizome rot of white calla caused by *Phytophthora erythroseptica*. Phytopathology 40:712–714.

Tompkins, C. M., and Tucker, C. M. 1951. Stem and root rot of *Daphne odora* caused by *Phytophthora parasitica*. Phytopathology 41:654–655.

Tompkins, C. M., Tucker, C. M., and Clarke, A. E. 1935. Root rot of aster caused by *Phytophthora cryptogea* (Abstr.). Phytopathology 25:895.

Tompkins, C. M., Richards, B. L., Tucker, C. M., and Gardner, M. W. 1936a. Phytophthora rot of sugar beet. J. Agric. Res. 52:205–216.

Tompkins, C. M., Tucker, C. M., and Gardner, M. W. 1936b. Phytophthora root rot of cauliflower. J. Agric. Res. 53:685–692.

Tooley, P. W. 1988. Use of uncontrolled freezing for liquid nitrogen storage of *Phytophthora* species. Plant Dis. 72:680–682.

Tooley, P. W. 1990. Variation in resistance to *Phytophthora infestans* among 21 *Solanum verrucosum* plant introductions. Am. Potato J. 67:491–498.

Tooley, P. W., and Fry, W. E. 1985. Field assessment of fitness of isolates of *Phytophthora infestans*. Phytopathology 75:982–988.

Tooley, P. W., and Grau, C. R. 1982. Identification and quantitative characterization of rate-reducing resistance to *Phytophthora megasperma* f. sp. *glycinea* in soybean seedlings. Phytopathology 72: 727–733.

Tooley, P. W., and Grau, C. R. 1984a. Field characterization of rate-reducing resistance to *Phytophthora megasperma* f. sp. *glycinea* in soybean. Phytopathology 74:1201–1208.

Tooley, P. W., and Grau, C. R. 1984b. The relationship between rate-reducing resistance to *Phytophthora megasperma* f. sp. *glycinea* and yield of soybean. Phytopathology 74:1209–1216.

Tooley, P. W., and Therrien. 1987. Cytophotometric determination of the nuclear DNA content of 23 Mexican and 18 non-Mexican isolates of *Phytophthora infestans*. Exp. Mycol. 11:19–26.

Tooley, P. W., and Therrien, C. D. 1991. Variation in ploidy in *Phytophthora infestans*. Pages 204–217 in: *Phytophthora*. J. A. Lucas, R. C. Shattock, D. S. Shaw, and L. R. Cooke, eds. Cambridge Univ. Press, Cambridge, U.K. 447 pp.

Tooley, P. W., Grau, C. R., and Stough, M. C. 1982. Races of *Phytophthora megasperma* f. sp. *glycinea* in Wisconsin. Plant Dis. 66:472–475.

Tooley, P. W., Bergstrom, G. C., and Wright, M. J. 1984. Widely virulent isolates of *Phytophthora megasperma* f. sp. *glycinea* causing root and stem rot of soybeans in New York. Plant Dis. 68:726–727.

Tooley, P. W., Fry, W. E., and Villareal-González, M. J. 1985. Isozyme characterization of sexual and asexual *Phytophthora infestans* populations. J. Hered. 76:431–435.

Tooley, P. W., Sweigard, J. A., and Fry, W. E. 1986. Fitness and virulence of *Phytophthora infestans* isolates from sexual and asexual populations. Phytopathology 76:1209–1212.

Tooley, P. W., Therrien, C. D., and Ritch, D. L. 1989. Mating type, race composition, nuclear DNA content, and isozyme analysis of Peruvian isolates of *Phytophthora infestans*. Phytopathology 79: 478–481.

Torgeson, D. C. 1954. Root rot of Lawson cypress and other ornamentals caused by *Phytophthora cinnamomi*. Contrib. Boyce Thompson Inst. 17:359–373.

Torgeson, D. C., Young, R. A., and Milbrath, J. A. 1954. Phytophthora root rot diseases of Lawson cypress and other ornamentals. Oregon Agric. Exp. Stn. Bull. 573. 18 pp.

Torres, H., French, E. R., and Nielsen, L. W. 1970. Potato diseases in Peru, 1965–1968. Plant Dis. Rep. 54:315–318.

Torres, H., Martin, C., and Henfling, J. 1985. Chemical control of pink rot of potato (*Phytophthora erythroseptica* Pethyb.). Am. Potato J. 62:355–361.

Toxopeus, H. J. 1930. De gomziekte in Djeroek manis (*Citrus sinensis* Osbeck.) en hare bestrijding (Gummosis of Djeroek manis [*Citrus sinensis* Osbeck.] and its control). Korte Meded Inst. Plantenziekten 13. 6 pp. (In Dutch). (Rev. Appl. Mycol. 1930, 9:646–647).

Toxopeus, H. J. 1956. Reflections on the origin of new physiologic races in *Phytophthora infestans* and the breeding for resistance in potatoes. Euphytica 5:221–237.

Toxopeus, H. J. 1960. Studies on the resistance of tuber-bearing *Solanaceae* from Mexico to *Phytophthora infestans*. Euphytica 9: 39–56.

Toxopeus, H. 1974. Breeding for black pod resistance in *Theobroma cacao* L. Pages 203–210 in: *Phytophthora* Diseases of Cocoa. P. H. Gregory, ed. Longman, London. 348 pp.

Tramier, R., and Mercier, S. 1965. *Phytophthora nicotianae* f. sp. *parasitica* (Dastur) Waterh. nouveau parasite de l'Oeillet en France (*Phytophthora nicotianae* f. sp. *parasitica* [Dastur] Waterh., a new parasite of carnation in France). Phytopathol. Mediterr. 41:176–177. (In French).

Trébuchet, G. 1950. La selection sanitaire du fraisier (The selection of sanitation for the strawberry). Bull. Soc. Nat. Hortic. Fr. Ser. 7:119–126 and 141–148. (In French). (Rev. Appl. Mycol. 1951, 30:174).

Trichilo, P. J., and Aragaki, M. 1982. Sporangial caducity and pedicel length of *Phytophthora nicotianae* variety *parasitica*. Mycologia 74:927–931.

Trione, E. J. 1959. The pathology of *Phytophthora lateralis* on native *Chamaecyparis lawsoniana*. Phytopathology 49:306–310.

Trione, E. J. 1974. Sporulation and germination of *Phytophthora lateralis*. Phytopathology 64:1531–1533.

Trione, E. J. 1988. The use of monoclonal antibodies in plant pathology. Pages 33–43 in: Experimental and Conceptual Plant Pathology. Vol. 1. W. M. Ness, R. S. Singh, U. S. Singh, and D. J. Weber, eds. Gordon and Breach Science Publishers, New York. 136 pp.

Trione, E. J., and Roth, L. F. 1957. Aerial infection of *Chamaecyparis* by *Phytophthora lateralis*. Plant Dis. Rep. 41:211–215.

Troedson, R. J., Ryley, M. J., Byth, D. E., and Irwin, J. A. G. 1991. Effect of Phytophthora root and stem rot on the response of field-grown soybean to saturated soil culture. Aust. J. Agric. Res. 42: 791–799.

Tropova, A. T. 1928. Fungous diseases of newly introduced crops and endeavours to find means for their control (Translated title). Proc. Pan-Soviet Congr. Bot., Leningrad, 1928. Page 188. (In Russian). (Rev. Appl. Mycol. 1930, 9:489). (Cited in Tucker 1933).

Trotter, A. 1924. "Cancrena pedale" del peperone e melanzana nella campania (*Capsicum annuum* e *Solanum melongena*) (Foot rot of chili and eggplant [*Capsicum annuum* and *Solanum melongena*] in the Campania). Riv. Patol. Veg. 14:125–130. (In Italian). (Cited in Tucker 1933).

Troutman, J. L., and Wills, W. H. 1964. Electrotaxis of *Phytophthora parasitica* zoospores and its possible role in infection of tobacco by the fungus. Phytopathology 54:225–228.

Trujillo, E. 1960. A Phytophthora fruit rot of breadfruit. Phytopathology 60:1542.

Trujillo, E. E. 1965. The effects of humidity and temperature on Phytophthora blight of Taro. Phytopathology 55:183–188.

Trujillo, E. E. 1971. A list of diseases of economic plants in the Trust Territory of the Pacific Islands. TTPI Dep. Resources and Development, Div. Agric., Saipan. 23 pp.

Trujillo, E. E., and Aragaki, M. 1964. Taro blight and its control. Hawaii Farm Ser. 13:3–5.

Trujillo, E. E., and Marcley, M. 1967. Effect of soil temperature and moisture on survival of *Phytophthora parasitica* and *Pythium aphanidermatum* (Abstr.). Phytopathology 57:9.

Trujillo, E. E., Alvarez, A. M., and Swindale, D. N. 1975. Phytophthora leaf spot of Ti. Plant Dis. Rep. 59:452–453.

Tsao, P. H. 1960. A serial dilution end-point method for estimating disease potentials of citrus phytophthoras in soil. Phytopathology 50:717–724.

Tsao, P. H. 1969a. Vital fluorescent labeling as applied to soil ecological studies of *Phytophthora parasitica* (Abstr.). Phytopathology 59:16.

Tsao, P. H. 1969b. Studies on the saprophytic behavior of *Phytophthora parasitica* in soil. Proc. 1st Int. Citrus Symp. 3:1221–1230.

Tsao, P. H. 1970a. Applications of the vital fluorescent labeling technique with brighteners to studies of saprophytic behavior of *Phytophthora* in soil. Soil Biol. Biochem. 2:247–256.

Tsao, P. H. 1970b. Selective media for isolation of pathogenic fungi. Annu. Rev. Phytopathol. 8:157–186.

Tsao, P. H. 1971. Chlamydospore formation in sporangium-free liquid cultures of *Phytophthora parasitica*. Phytopathology 61:1412–1413.

Tsao, P. H. 1976. Recovery of *Phytophthora* species from old, badly decayed, infected tissues of *Hevea brasiliensis*. Trans. Br. Mycol. Soc. 66:557–558.

Tsao, P. H. 1977a. Prospects of biological control of citrus root disease fungi. Proc. Int. Soc. Citric. 3:857–863.

Tsao, P. H. 1977b. Importance of sporangium caducity, pedicel length, and ontogeny in *Phytophthora* speciation (Abstr.). Page 678 in: Proc. 2nd Int. Mycol. Congr., Tampa, Florida. 27 August–3 September, 1977.

Tsao, P. H. 1983. Factors affecting isolation and quantitation of *Phy-*

tophthora from soil. Pages 219–236 in: *Phytophthora:* Its Biology, Taxonomy, Ecology, and Pathology. D. C. Erwin, S. Bartnicki-Garcia, and P. H. Tsao, eds. American Phytopathological Society, St. Paul, Minn. 392 pp.

Tsao, P. H. 1990. Why many Phytophthora root rots and crown rots of tree and horticultural crops remain undetected. EPPO (Eur. Mediterr. Plant Prot. Organ.) Bull. 20:11–17.

Tsao, P. H. 1991. The identities, nomenclature and taxonomy of *Phytophthora* isolates from black pepper. Pages 185–211 in: Diseases of Black Pepper. Proc. Int. Pepper Comm. Workshop on Black Pepper Diseases. Goa, India. Y. R. Sarma and T. Premkumor, eds. 209 pp.

Tsao, P. H., and Alizadeh, A. 1988. Recent advances in the taxonomy and nomenclature of the so-called "*Phytophthora palmivora*" MF4 occurring on cocoa and other tropical crops. Pages 441–445 in: 10th Int. Cocoa Res. Conf. Proc. Santo Domingo. 17–23 May, 1987.

Tsao, P. H., and Bricker, J. L. 1968. Germination of chlamydospores of *Phytophthora parasitica* in soil (Abstr.). Phytopathology 58:1070.

Tsao, P. H., and Garber, M. J. 1960. Methods of soil infestation, watering, and assessing the degree of root infection for greenhouse *in situ* ecological studies with citrus phytophthoras. Plant Dis. Rep. 44:710–715.

Tsao, P. H., and Guy, S. O. 1977. Inhibition of *Mortierella* and *Pythium* in a *Phytophthora* isolation medium containing hymexazol. Phytopathology 67:796–801.

Tsao, P. H., and Menyonga, J. M. 1966. Response of *Phytophthora* spp. and soil microflora to antibiotics in the pimaricin-vancomycin medium (Abstr.). Phytopathology 56:152.

Tsao, P. H., and Mu, L. 1987. Involvement of *Phytophthora* in vanilla root rot in French Polynesia (Abstr.). Phytopathology 77:1704.

Tsao, P. H., and Ocana, G. 1969. Selective isolation of species of *Phytophthora* from natural soils on an improved antibiotic medium. Nature (London) 223:636–638.

Tsao, P. H., and Oster, J. J. 1981. Relation of ammonia and nitrous acid to suppression of *Phytophthora* in soils amended with nitrogenous organic substances. Phytopathology 71:53–59.

Tsao, P. H., and Sisemore, D. J. 1978. Morphological variability in *Phytophthora parasitica* (*P. nicotianae*) isolates from citrus, tomato, and tobacco (Abstr.). Phytopathol. News 12:213.

Tsao, P. H., and Tummakate, A. 1977. The identity of a *Phytophthora* species from black pepper in Thailand. Mycologia 69:631–637.

Tsao, P. H., and Zentmyer, G. A. 1979. Suppression of *Phytophthora cinnamomi* and *P. parasitica* in urea-amended soils. Pages 191–199 in: Soil-Borne Plant Pathogens. B. Schippers and W. Gams, eds. Academic Press, London. 686 pp.

Tsao, P. H., Chew-Chin, N., and Syamananda, R. 1975. Occurrence of *Phytophthora palmivora* on *Hevea rubber* in Thailand. Plant Dis. Rep. 59:955–958.

Tsao, P. H., Tummakate, A., and Bhavakul, K. 1976. Recovery of *Phytophthora* species from old, badly decayed, infected tissues of *Hevea brasiliensis.* Trans. Br. Mycol. Soc. 66:557–558.

Tsao, P. H., Ugale, R., Hobbs, S., and Farih, A. 1980. Control of homothallic oospore formation in *Phytophthora parasitica* by culture manipulations. Trans. Br. Mycol. Soc. 75:153–156.

Tsao, P. H., Kasim, R., and Mustika, I. 1985. Morphology and identity of black pepper *Phytophthora* isolates in Indonesia. F.A.O. Plant Prot. Bull. 33:61–66.

Tsao, P. H., Luzaran, P. B., de los Santos, A. B., Portales, L. A., Gochangco, A. M., and Gruber, L. C. 1994a. Phytophthora crown and root rot of mango detected in Philippine nurseries. Plant Dis. 78:100.

Tsao, P. H., Gruber, L. C., Portales, L. A., Gochangco, A. M., Luzaran, P. B., de los Santos, A. B., and Pag, H. 1994b. Some new records of Phytophthora crown and root rots in the Philippines and in world literature (Abstr.). Phytopathology 84:871.

Tsuchiya, S., Yanagawa, M., and Ogoshi, A. 1986. Formae speciales differentiation of *Phytophthora vignae* isolates from cowpea and adzuki bean. Ann. Phytopathol. Soc. Jpn. 52:577–584.

Tu, J. C. 1978. Protection of soybean from severe Phytophthora root rot by *Rhizobium.* Physiol. Plant Pathol. 12:233–240.

Tu, J. C. 1979. Evidence of differential tolerance among some root rot fungi to rhizobial parasitism in vitro. Physiol. Plant Pathol. 14:171–177.

Tu, J. C. 1980. Incidence of root rot and overwintering of alfalfa as influenced by rhizobia. Phytopathol. Z. 97:97–108.

Tucker, C. M. 1927. *Sabal causiarum* (Cook) Beccari: A new host of the coconut bud-rot fungus. J. Agric. Res. 34:879–888.

Tucker, C. M. 1928. Report of the plant pathologist. Puerto Rico Agric. Exp. Stn. 1927:25–27.

Tucker, C. M. 1929. Report of the plant pathologist. Puerto Rico Agric. Exp. Stn. Rep. 1928:29–35.

Tucker, C. M. 1931. Taxonomy of the genus *Phytophthora* de Bary. Univ. Mo. Agric. Exp. Stn. Res. Bull. 153. 207 pp.

Tucker, C. M. 1933. Distribution of the genus *Phytophthora.* Univ. Mo. Agric. Exp. Stn. Res. Bull. 184. 80 pp.

Tucker, C. M. 1936. Work of the Agricultural Experiment Station. Report of the Director for the year 1935. Mo. Agric. Exp. Stn. Bull. 370. 100 pp.

Tucker, C. M., and Milbrath, J. A. 1942. Root rot of *Chamaecyparis* caused by a species of *Phytophthora.* Mycologia 34:94–103.

Tudosescu-Banescu, V. 1966. Bolile cactuslor si combaterea lor (Diseases of cactus and their control). Natura (Buchar.) Ser. Biol. 6:76–78. (In Romanian).

Tuite, J. 1969. Plant Pathological Methods—Fungi and Bacteria. Burgess Publishing, Minneapolis, Minn. 239 pp.

Tulasne, L. R. 1854. Note sur le champignons endophytes, tels que celui de la pomme de terre (Note on fungal endophytes such as that of the potato). Compt. Rendu Acad. Sci. Paris. 38:1101–1104. (In French).

Tullio, V., and Talame, M. 1977. Prove di lotta contro i marciumi del finocchio (Control trials against fennel rots). Oss. Mal. Piante, Pescara, Italy. Pages 631–635. (In Italian). (Rev. Plant Pathol. 1977, 56:5356).

Turelli, M., Coulomb, C., Coulomb, P. J., Roggero, J. P., and Bounias, M. 1984. Effects of capsidiol on the lipid and protein content of isolated membranes of *Phytophthora capsici.* Physiol. Plant Pathol. 24:211–221.

Turelli, M., Coulomb, C., Mutaftschiev, S., and Coulomb, P. J. 1986. Etude cytologique du mode d'action du capsidiol sur les hyphes de *Phytophthora capsici* (A cytological study of the mode of action of capsidiol in the hyphae of *Phytophthora capsici*). Can. J. Bot. 64:701–709. (In French).

Turkensteen, L. J. 1978. *Phytophthora infestans:* Three new hosts and a specialized form causing a foliar blight of *Solanum muricatum* in Peru. Plant Dis. Rep. 62:829.

Turner, G. J. 1969. *Phytophthora palmivora* from *Piper betle* in Sarawak. Trans. Br. Mycol. Soc. 52:411–418.

Turner, G. J. 1971. Resistance in *Piper* species and other plants to infection by *Phytophthora palmivora* from *Piper nigrum.* Trans. Br. Mycol. Soc. 57:61–66.

Turner, G. J. 1973. Pathogenic variation in isolates of *Phytophthora palmivora* from *Piper nigrum.* Trans. Br. Mycol. Soc. 60:583–585.

Turner, G. J. 1974. Phytophthora wilt and crown rot of *Passiflora edulis.* Trans. Br. Mycol. Soc. 62:59–63.

Turner, P. D. 1960a. Saprophytic activity of *Phytophthora palmivora.* Annu. Rep. West Afr. Cocoa Res. Inst. 1958–1959:25–26.

Turner, P. D. 1960b. Strains of *Phytopthora palmivora* (Butl.) Butl. from *Theobroma cacao* L. I. Isolates from West Africa. Trans. Br. Mycol. Soc. 43:665–672.

Turner, P. D. 1965a. Behavior of *Phytophthora palmivora* in soil. Plant Dis. Rep. 49:135–137.

Turner, P. D. 1965b. Effect of spacing and pruning on Phytophthora pod rot of cacao. Plant Dis. Rep. 49:257.

Turner, P. D. 1968. Pod rot of cocoa in Malaysia caused by *Phytophthora heveae.* F.A.O. Plant Prot. Bull. 16:32–33.

Turner, P. D. 1974. Allied pathogens of other crops. Pages 287–300 in: *Phytophthora* Diseases of Cocoa. P. H. Gregory, ed. Longman, London. 348 pp.

Tuset, J. J., and Portilla, M. T. 1990. Control of Phytophthora brown rot of citrus fruits. EPPO (Eur. Mediterr. Plant Prot. Organ.) Bull. 20:153–161.

Tuset-Barrachina, J. J. 1977. Contribución al conocimiento del género *Phytophthora* De Bary en España (Contribution to the knowledge of the genus *Phytophthora* De Bary in Spain). Ann. Inst. Nac. Invest. Agrar. Prot. Veg. 7:11–106. (In Spanish). (Rev. Plant Pathol. 1979, 58:564).

Tyler-Whittle, M. 1970. The Plant Hunters. Chilton Book Co., Philadelphia. 281 pp.

Tzamos, E. 1978. New diseases caused by *Phytophthora* spp. Phytophthora Newsl. 6:72–73.

Uchida, J. Y. 1989. Foliar blight of *Spathiphyllum* caused by *Phytophthora*. HITAHR Brief. No. 084. 3 pp.

Uchida, J. Y. 1994. Diseases of orchids in Hawaii. Plant Dis. 78:220–224.

Uchida, J. Y., and Aragaki, M. 1978. Leaf and stem blight of ivy caused by *Phytophthora palmivora*. Plant Dis. Rep. 62:699–701.

Uchida, J. Y., and Aragaki, M. 1979. New diseases caused by *Phytophthora* species. Phytophthora Newsl. 7:39–40.

Uchida, J. Y., and Aragaki, M. 1980. Chemical stimulation of oospore formation in *Phytophthora capsici*. Mycologia 72:1103–1108.

Uchida, J. Y., and Aragaki, M. 1985. Occurrence of chlamydospores in *Phytophthora capsici*. Mycologia 77:832–835.

Uchida, J. Y., and Aragaki, M. 1988. Seedling blight of castor bean in Hawaii caused by *Phytophthora palmivora*. Plant Dis. 72:994.

Uchida, J. Y., and Aragaki, M. 1989. Comparison of pepper isolates of *Phytophthora capsici* from New Mexico to other solanaceous and non-solanaceous isolates (Abstr.). Phytopathology 79:1212.

Uchida, J. Y., and Aragaki, M. 1991. *Phytophthora* diseases of orchids in Hawaii. Res. Ext. Ser. 129. HITAHR, University of Hawaii. 7 pp.

Uchida, J. Y., Aragaki, M., Ooka, J., and Nagata, N. M. 1992. Phytophthora fruit and heart rots of coconut in Hawaii. Plant Dis. 76: 925–927.

Uchida, K. 1967. *Phytophthora* disease of chestnut. Plant Prot. 21: 383–387.

Ulstrup, A. J. 1952. Observation on crazy top of corn. Phytopathology 42:675–680.

Ulstrup, A. J., and Tucker, C. M. 1948. The occurrence of *Phytophthora parasitica* on corn. Phytopathology 38:1029–1031.

Umaerus, V. 1969. Studies on field resistance to *Phytophthora infestans*. IV. Influence of mineral nutrition. Z. Pflanzenzuecht. 62:357–369.

Umaerus, V. 1970. Studies on field resistance to *Phytophthora infestans*. V. Mechanisms of resistance and applications to potato breeding. Z. Pflanzenzuecht. 63:1–23.

Umaerus, V., and Lihnell, D. 1976. A laboratory method for measuring the degree of attack by *Phytophthora infestans*. Potato Res. 19:91–107.

Umaerus, V., Umaerus, M., Erjefalt, L., and Nilsson, B. A. 1983. Control of *Phytophthora* by host resistance: Problems and progress. Pages 315–326 in: *Phytophthora:* Its Biology, Taxonomy, Ecology, and Pathology. D. C. Erwin, S. Bartnicki-Garcia, and P. H. Tsao, eds. American Phytopathological Society, St. Paul, Minn. 392 pp.

Unamuno, L. 1911. Los estragos de la *Phytophthora cactorum* en las peras y ciruelas. España y America (Damage of *Phytophthora cactorum* on pears and plums. Spain and America). Z. Pflanzenkr. 21:379. (In Spanish). (Cited in Tucker 1933).

Unger, F. 1847. Beitrag zur Kenntnis der in der Kartoffelkrankheit vorkommenden Pilze und der Ursache ihres Entstehens (Contributions to the knowledge of the fungi occurring in the potato disease and the cause of their emergence). Bot. Zeitschr. 5:314.

Uppal, B. N. 1933. India: Plant diseases in the Bombay presidency. Int. Bull. Plant Prot. 7:187.

Uppal, B. N., and Desai, M. K. 1939. *Koleroga* disease of arecanut. Curr. Sci. 8:122–124.

Uppal, B. N., and Kamat, M. N. 1936. Gummosis of citrus in Bombay. Indian J. Agric. Sci. 6:803–822.

Upstone, M. E. 1974. The link between Phytophthora rot in stored apples and rain splash. Grower 82:248.

Upstone, M. E. 1976. Field trials of fungicides to control red core of strawberries. Ann. Appl. Biol. 84:103–106.

Upstone, M. E. 1978. *Phytophthora syringae* fruit rot of apples. Plant Pathol. 27:24–30.

Upstone, M. E., and Finney, J. 1966. Tomato rootstocks and the control of Phytophthora root rot. Plant Pathol. 15:15–19.

Upstone, M. E., and Gunn, E. 1978. Rainfall and the occurrence of *Phytophthora syringae* fruit rot of apples in Kent 1973–1975. Plant Pathol. 27:30–35.

Uraz, E. 1950. Kestane murekkep hastaligi *Phytophthora cambivora* (Chestnut ink disease caused by *Phytophthora cambivora*). Mahsul Hekimi. Plant Prot. J. Turkey 3:116. (In Turkish).

Urech, P. A., Schwinn, F. J., and Staub, T. 1977. CGA 48988, a novel fungicide for the control of late blight, downy mildews and related soil-borne diseases. Proc. 9th Br. Crop Prot. Conf. 2:623–631.

Utkhede, R. S. 1982. Effects of six herbicides on the growth of *Phytophthora cactorum* and a bacterial antagonist. Pestic. Sci. 13: 693–695.

Utkhede, R. S. 1983. Inhibition of *Phytophthora cactorum* by bacterial isolates and effects of chemical fungicides on their growth and antagonism. Z. Pflanzenkr. Pflanzenschutz 90:140–145.

Utkhede, R. S. 1984a. Enhancement of crown rot of apple trees with nitrogen fertilizers and composts (Abstr.). Phytopathology 74:814.

Utkhede, R. S. 1984b. Effect of nitrogen fertilizers and wood composts on the incidence of apple crown rot in British Columbia. Can J. Plant Pathol. 6:329–332.

Utkhede, R. S. 1984c. Effects of fungicides on apple crown rot caused by *Phytophthora cactorum*. Pestic. Sci. 15:241–246.

Utkhede, R. S. 1984d. Antagonism of isolates of *Bacillus subtilis* to *Phytophthora cactorum*. Can. J. Bot. 62:1032–1035.

Utkhede, R. S. 1984e. Effect of bacterial antagonist on *Phytophthora cactorum* and apple crown rot. Phytopathol. Z. 109:169–175.

Utkhede, R. S. 1986. Biology and control of apple crown rot caused by *Phytophthora cactorum:* A review. Phytoprotection 67:1–13.

Utkhede, R. S. 1987a. Control of crown rot (*Phytophthora cactorum*) of apple trees with the systemic fungicides metalaxyl and fosetyl-aluminium. Pestic. Sci. 19:289–295.

Utkhede, R. S. 1987b. Chemical and biological control of crown and root rot of apple caused by *Phytophthora cactorum*. Can. J. Plant Pathol. 9:295–300.

Utkhede, R. S., and Gaunce, A. P. 1983. Inhibition of *Phytophthora cactorum* by a bacterial antagonist. Can. J. Bot. 61:3343–3348.

Utkhede, R. S., and Gupta, V. K. 1988. In vitro selection of strains of *Phytophthora cactorum* resistant to metalaxyl. J. Phytopathol. 122: 35–44.

Utkhede, R. S., and Quamme, H. A. 1988. Use of the excised shoot assay to evaluate resistance to *Phytophthora cactorum* of apple rootstock cultivars. Can. J. Plant Sci. 68:851–857.

Utkhede, R. S., and Smith, E. M. 1991a. Effects of fosetyl-Al, metalaxyl, and *Enterobacter aerogenes* on crown and root rot of apple trees caused by *Phytophthora cactorum* in British Columbia. Plant Dis. 75:406–409.

Utkhede, R. S., and Smith, E. M. 1991b. *Phytophthora* and *Pythium* species associated with root rot of young apple trees and their control. Soil Biol. Biochem. 23:1059–1063.

Vaartaja, O. 1955. *Phytophthora cactorum* in Saskatchewan. Can. Dep. Agric. Div. For. Biol. Bi-Mon. Progr. Rep. 11:2.

Vaartaja, O., Pitblado, R. E., Buzzell, R. I., and Crawford, L. G. 1979. Chemical and biological control of Phytophthora root and stalk rot of soybean. Can. J. Plant Sci. 59:307–311.

Valdebenito-Sanhuenza, R. M. 1987. Uso de formaldeído e *Trichoderma* para prevenir a recolonização de solo por *Phytophthora* em pomares do macieras (Use of formaldehyde and *Trichoderma* to prevent recolonization of soil by *Phytophthora* in apple orchards). Page 55 in: Anais da II. Reduniao Sobre Controle Biologico de Doencas de Plantas. Campinas Funação Cargill. (In Portuguese).

de Vallavieille, C., and Erselius, L. J. 1984. Variation in protein profiles of *Phytophthora:* Survey of a composite population of three species on citrus. Trans. Br. Mycol. Soc. 83:473–479.

van Bakel, J. M. M. 1964. Papiervlekkenziekte in Prei (*Phytophthora porri* F.) [White tip disease of leek (*Phytophthora porri*)]. Meded. Dir. Tuinb. 27:198–200. (In Dutch). (Rev. Appl. Mycol. 1964, 43: 2464).

van den Boom, T., and Huller, I. M. 1970. Phytophthora stem rot of passion fruit, *Passiflora edulis*, in South Africa. Phytophylactica 2:71–74.

Vanderplank, J. E. 1963. Plant Diseases: Epidemics and Control. Academic Press, New York. 349 pp.

Vanderplank, J. E. 1971. Stability of resistance to *Phytophthora infestans* in cultivars without R genes. Potato Res. 14:263–270.

Vanderplank, J. E. 1975. Principles of Plant Infection. Academic Press, New York. 216 pp.

Vanderplank, J. E. 1984. Disease Resistance in Plants. Academic Press, New York. 194 pp.

Vanderplank, J. E. 1986. Specific susceptibility and specific feeding in gene-for-gene systems. Pages 200–223 in: Advances in Plant Pathology. Vol. 5. D. S. Ingram and P. H. Williams, eds. 270 pp.

Vanderplank, J. E. 1988. Resistance to disease and maximum potential yield. Pages 593–599 in: Experimental and Conceptual Plant Pathology. Vol. 3. W. M. Hess, R. S. Singh, V. S. Singh, and D. J. Weber, eds. Gordon and Breach Sci. Pub. 599 pp.

Vanderplank, J. E. 1991. The two gene-for-gene hypotheses and a test to distinguish them. Plant Pathol. 40:1–3.

van der Plaats-Niterink, A. J. 1981. Monograph of the genus *Pythium*. Studies in Mycology, 21. Centraalbureau voor schimmel cultures. Baarn, Netherlands. 242 pp.

van der Scheer, H. A. T. 1971. Isolation of *Phytophthora cactorum* from soil in orchards and strawberry fields and differences in pathogenicity to apple. Neth. J. Plant Pathol. 77:65–72.

van der Walle, R. 1946. Observations et recherches effectuees a la station de phytopathologie de l'État pendant l'annee 1945 (Observations and research carried out at the State Phytopathological Station during the year 1945). Bull. Inst. Agron. Gembloux 15: 130–138. (In French). (Rev. Appl. Mycol. 1949, 28:51).

van de Weg, W. E., Wassenaar, L. M., and van de Lindeloof, C. P. J. 1989. Inheritance of resistance to *Phytophthora fragariae* Hickman in strawberry. Euphytica 42:25–30.

van der Weyen, A. 1966. *Phytophthora syringae* sur Agrumes au Maroc (*Phytophthora syringae* on citrus fruits in Morocco). Acad. Agric. Fr. (Paris) Compt. Rend. Hebd. Seances 52:687–689. (In French).

van der Zaag, D. E. 1956. Overwintering en epidemiologie van *Phytophthora infestans,* Tevens enige nieuwe bestrijdingsmogelijkheden (Overwintering and epidemiology of *Phytophthora infestans,* and some new possibilities of control). Tijdschr. Plantenziekten 62:89–156. (In Dutch).

van der Zwet, T. 1991. World food prize awarded to Niederhauser. Phytopathol. News 25:18–19.

van der Zwet, T., and Forbes, I. L. 1961. *Phytophthora megasperma,* the principal cause of seed-piece rot of sugarcane in Louisiana. Phytopathology 51:634–640.

van der Zwet, T., Forbes, I. L., and Steib, R. J. 1960. Studies on the Phytophthora rot of sugarcane seed pieces in Louisiana. Plant Dis. Rep. 44:519–523.

van Hall, C. J. J. 1923. Ziekten en plagen der cultuurgewassen in Nederlandsch-Indie in 1922 (Diseases and pests of economic plants in the Dutch East Indies in 1922). Meded. Inst. Plantenziekten 58. 42 pp. (In Dutch). (Rev. Appl. Mycol. 1924, 3:189–190).

van Hall, C. J. J. 1924. Ziekten en plagen der cultuurgewassws in Nederlandsch-Indie in 1923 (Diseases and pests of economic crops in the Dutch East Indies in 1923). Meded. Inst. Plantenziekten 64. 47 pp. (In Dutch). (Rev. Appl. Mycol. 1925, 4:79–80).

van Hook, J. M. 1906. A disease of ginseng due to *Phytophthora* sp. Crops. (N.S.) N.S. 5:94.

van Poeteren, N. 1922. Verslag over de werkzaamheden van den Plantenziekten kundigen Dienst in de jaren 1920 en 1921 (Report of the work of the phytopathological service during the years 1920 and 1921). Versl. Meded. Plantenziektentkund. Dienst. Wageningen 27. 90 pp. (In Dutch). (Cited in Tucker 1933).

van Poeteren, N. 1925. Verslag over de werkzaamheden van den Plantenziektenkundigen Dienst in het jaar 1924 (Report of the activities of the phytopathological service in the year 1924). Versl. Meded. Plantenzeiktenkund. Dienst. Wageningen. 41. 62 pp. (In Dutch). (Cited in Tucker 1933).

van Poeteren, N. 1926. Verslag over de werkzaamheden van den Plantenziekten kundigen Dienst in het jaar 1925. Versl. Meded. Plantenziektenkund. Dienst. Wageningen 44:23. (In Dutch). (Cited in Tucker 1933).

Vardi, A., Epstein, E., and Breiman, A. 1986. Is the *Phytophthora citrophthora* culture filtrate a reliable tool for the in vitro selection of resistant *Citrus* variants? Theor. Appl. Genet. 72:569–574.

Vargas, R., Fernandez, A., Mora, B., and Araya, R. 1985. Assessment of French bean cultivars for the incidence of *Phytophthora phaseoli* Thaxt. Phytopathology 75:1179–1180.

Varney, E. H. 1961. Vermiculite media for growing fungi. Plant Dis. Rep. 45:393.

Varns, J. L., Currier, W. W., and Kuć, J. 1971a. Specificity of rishitin and phytuberin accumulation by potato. Phytopathology 61:968–971.

Varns, J. L., Kuć, J., and Williams, E. B. 1971b. Terpenoid accumulation as a biochemical response of the potato tuber to *Phytophthora infestans*. Phytopathology 61:174–177.

Vartanian, V. G., and Endo, R. M. 1985a. Survival of *Phytophthora infestans* in seeds extracted from infected tomato fruits. Phytopathology 75:375–378.

Vartanian, V. G., and Endo, R. M. 1985b. Overwintering hosts, compatability types, and races of *Phytophthora infestans* on tomato in southern California. Plant Dis. 69:516–519.

Vaughan, E. K., Roberts, A. N., and Mellenthin, W. M. 1954. The influence of Douglas fir sawdust and certain fertilizer elements on the incidence of red stele disease of strawberry. Phytopathology 44:601–603.

Vaziri, A., Keen, N. T., and Erwin, D. C. 1981. Correlation of medicarpin production with resistance to *Phytophthora megasperma* f. sp. *medicaginis* in alfalfa seedlings. Phytopathology 71:1235–1238.

Vaziri, A., Keen, N. T., and Erwin, D. C. 1983. Stimulation of oospore production in *Phytophthora megasperma* f. sp. *medicaginis* by medicarpin. Phytopathology 73:730–734.

Vegh, I., and Berre, A., Le. 1985. Quelques maladies nouvelles a *Phytophthora* des arbustes d'ornement en France (Some new *Phytophthora* diseases of ornamental shrubs in France). Rev. Hortic. 258:37–39. (In French).

Vegh, I., and Bourgeois, M. 1975. Observations preliminaires sur l'etiologie du deperissement des coniferes d'ornement dans les pepinieres francaises:role du *Phytophthora cinnamomi* Rands (Preliminary observations on the etiology of wilt of ornamental conifers in French nurseries: Role of *Phytophthora cinnamomi* Rands). Pepinier Hortic. Maraich 153:39–49. (In French).

Vegh, I., Leroux, P., LeBerre, A., and Lanen C. 1985. Détection sur *Chamaecyparis lawsoniana* 'Ellwoodii' d'une souche de *Phytophthora cinnamomi* Rands résistante au phoséthyl-Al (Detection on *Chamaecyparis lawsoniana* 'Ellwoodii' an isolate of *Phytophthora cinnamomi* resistant to phosethyl-Al). Pépiniériste, Horticulteurs, Maraïchers Rev. Hortic. 262:19–21. (In French).

Venkata Rao, M. K. 1930. Annual report of work done in the mycological section during the year 1928–1929. Annu. Rep. Agric. Dep. Mysore for the year 1928–1929. Part I:18–21.

Venkatarayan, S. V. 1932. *Phytophthora arecae,* parasitic on *Areca* tops, and a strain of *P. palmivora* Butl. (*P. faberi* Maubl.) on a new host, *Aleurites fordi.* Phytopathology 22:217–227. (Cited in Tucker 1933).

Venkatakrishnaiya, N. S. 1950. *Phytophthora parasitica* on French bean. *Phaseolus vulgaris* Linn. Indian J. Agric. Sci. 20:391–394.

Verhoeff, K., and Weber, L. 1966. Foot rot of tomatoes caused by two *Phytophthora* species. Neth. J. Plant Pathol. 72:317–318.

Verneau, R. 1953. Marciume della corona fogliare di *Howea forsteriana* (Bud rot of *Howea forsteriana*). Ann. Sper. Agrar. (N.S.) 7:525–538. (In Italian). (Rev. Appl. Mycol. 1955, 34:369).

Verneau, R. 1954. Le *Phytophthora* parasite degli Agrumi nella Campania (The *Phytophthora* species parasitic on citrus in the Campania). Ann. Sper. Agrar. (N.S.) 8:133–162. (In Italian). (Rev. Appl. Mycol. 1955, 34:717).

Vincelli, P. C., Wilcox, W. F., Beaupré, C. M. S. 1990. First report of *Phytophthora cryptogea* causing root rot of sugar beet in Wyoming. Plant Dis. 74:614.

Visarathanonth, N. 1976. Phytophthora foot rot of roselle. Kasetsart J. 10:5–13. (Rev. Plant Pathol. 1979, 58:2260).

Vlasveld, W. P. N. 1954. De ziektebestrijding in 1954 (Disease control in 1954). Fruitteelt 43:1056–1057. (In Dutch).

Vlasveld, W. P. N. 1969. Twee nieuwe ziekten bij aardbeien (Two new fungi on aged bean). Fruitteelt 58:1018–1019. (In Dutch).

Vlk, W. 1939. Über die Geisselstructur der Saprolegniaceanseen schwärmer (On the flagellar structure of Saprolegniaceae zoospores). Arch. Protistenkunde 92:157–160. (In German).

Vock, N. T., Langdon, P. W., and Pegg, K. G. 1980. Root rot of chickpea caused by *Phytophthora megasperma* var. *sojae* in Queensland. Aust. Plant Pathol. 9:117.

Voelcker, O. J. 1940. A review of cacao selection in the Cameroons. Trop. Agric. (Trinidad) 17:223–225.

Voglino, P. 1916. Eumycetes. Ann. R. Accad. Agric. Torino 59:253. (Cited in Tucker 1933).

Voller, A., Bidwell, D. E., and Bartlett, A. 1976. Enzyme immunoassays in diagnostic medicine. Theory and practice. Bull. World Health Organization 53:55–65.

Voller, A., Bidwell, D. E., and Bartlett, A. 1979. The Enzyme Linked Immunosorbent Assay (ELISA). Dynatech Lab., Virginia. 125 pp.

Von Broembsen, S. 1984a. Occurrence of *Phytophthora cinnamomi* on indigenous and exotic hosts in South Africa, with special reference to the South-Western Cape Province. Phytophylactica 16: 221–225.

Von Broembsen, S. 1984b. Distribution of *Phytophthora cinnamomi* in rivers of the South-Western Cape Province. Phytophylactica 16: 227–229.

Von Broembsen, S. L., and Brits, G. J. 1985. Phytophthora root rot of commercially cultivated proteas in South Africa. Plant Dis. 69: 211–213.

von Malitz, P. M., and Von Broembsen, S. L. 1984. *Phytophthora porri* in South Africa. Plant Dis. 68:732.

von Malitz, P. M., and Von Broembsen, S. 1985. Lemon shoot tip blight caused by *Phytophthora citricola* in propagation tunnels. Phytophylactica 17:47–48.

Vorob'eva, Y. V., and Gridnev, V. V. 1974. Interspecific hybridization in *Phytophthora*. Mikol. Fitopatol. 8:184. (Translation from Russian).

Vorob'eva, Y. V., and Gridnev, V. V. 1981. Genetics of *Phytophthora* fungi. I. Crossability of various heterothallic species of *Phytophthora* fungi. Genetika 17:992–995. (In Russian).

Vorob'eva, Y. V., and Gridnev, V. V. 1983. Genetics of *Phytophthora* species. II. Interspecific hybridization of *Phytophthora* species and its role in the formative process in nature. Genetika 19:1786–1789. (Translated from Russian).

Voros, J. 1965. Streptomycin sensitivity of Oomycetes due to the increased absorption of streptomycin by their mycelia. Phytopathol. Z. 54:249–257.

Voss, E. G. (ed.). 1983. International Code of Botanical Nomenclature. Regnum Veg. 111:1–426.

Vo-Thi-Hai, Bompeix, G., and Ravise, A. 1979. Role du tris-O-ethyl phosphonate d'aluminium dans la stimulation des reactions de defense des tissues de tomate contre le *Phytophthora capsici* (Role of tris-O-ethyl phosphonate of aluminum on the stimulation of defence reactions in the tissues of tomato against *Phytophthora capsici*). C. R. Acad. Sci. Ser. D. 288:1171–1174. (In French).

Vujicić, R., and Colhoun, J. 1966. Asexual reproduction in *Phytophthora erythroseptica*. Trans. Br. Mycol. Soc. 49:245–254.

Vukovits, G. 1956. Eine gefährliche Gloxinienkrankheit (A dangerous disease of *Gloxinia*). Pflanzenarzt 9:105–106. (In German). (Rev. Appl. Mycol. 1957, 36:102–103).

Waard, P. W. F., de. 1980. Problem areas and prospects of production of pepper (*Piper nigrum* L.). An overview. Dep. Agric. Res. Bull. 308:1–29. R. Trop. Inst. Amsterdam.

Wade, G. C. 1954a. Tomato diseases. Part I. Soil-borne diseases. Tasman. J. Agric. 25:26–31.

Wade, G. C. 1954b. Potato diseases in Tasmania. Tasman. J. Agric. 25:240–252.

Wagener, W. W., and Cave, M. S. 1944. Phytophthora canker of madrone in California. Plant Dis. Rep. 28:328.

Wager, V. A. 1931. Diseases of plants in South Africa due to members of the Pythiaceae. S. Afr. Dep. Agric. Sci. Bull. 105. 43 pp. (Rev. Appl. Mycol. 1932, 11:330–331).

Wager, V. A. 1935. Brown rot of tomato fruits due to *Phytophthora parasitica* Dast. S. Afr. J. Sci. 32:235–237.

Wager, V. A. 1940. Descriptions of South African Pythiaceae with records of their occurrence. Bothalia 4:3–35.

Wager, V. A. 1942a. *Phytophthora cinnamomi* and wet soil in relation to the dying-back of avocado trees. Hilgardia 14:519–532.

Wager, V. A. 1942b. Pythiaceous fungi on citrus. Hilgardia 14:535–548.

Wagner, R. E., and Wilkinson, H. T. 1992a. A new physiological race of *Phytophthora sojae* on soybean. Plant Dis. 76:212.

Wagner, R. E., and Wilkinson, H. T. 1992b. An aeroponics system for investigating disease development on soybean taproots infected with *Phytophthora sojae*. Plant Dis. 76:610–614.

Wagner, R. E., and Wilkinson, H. T. 1993. Spheroplast and protoplast formation from cysts of *Phytophthora sojae* treated with digestive enzymes. Mycologia 85:131–136.

Wagner, R. E., Carmen, S. G., and Wilkinson, H. T. 1993. Evaluation and modeling of rate-reducing resistance of soybean seedlings to *Phytophthora sojae*. Phytopathology 83:187–192.

Walerych, H., Wojciechowski, J., and Borys, M. W. 1970. Resistance of tomato leaves to *Phytophthora infestans* de By as affected by N-NO_3, N-NH_4NO_3 or N-$(NH_2)_2CO$. Phytopathol. Z. 68:244–257.

Walker, A. K., and Schmitthenner, A. F. 1982. Developing soybean varieties tolerant to Phytophthora rot. Pages 67–78 in: Proc. 12th Soybean Seed Res. Conf., Chicago, Ill. 7–8 December. Am. Seed Trade Assoc., Washington D.C.

Walker, G. E. 1988. Phytophthora root rot of container-grown citrus as affected by foliar sprays and soil drenches of phosphorous and acetyl salicylic acids. Plant Soil 107:107–112.

Walker, J., and McLeod. 1969. New records of plant diseases in New South Wales. Agric. Gaz. N.S.W. 80:237–242.

Wallace, G. B. 1934. Report of the Mycologist, 1933. Ann. Rep. Dep. Agric. Tanganyika Territory 1933:76–78. (Rev. Appl. Mycol. 1934, 13:746).

Wallace, G. B., and Wallace, M. M. 1949. A list of plant diseases of economic importance in Tanganyika territory. Commonw. Mycol. Inst. Pap. 26. 21 pp.

Wallin, J. R. 1962. Summary of recent progress in predicting late blight epidemics in United States and Canada. Am. Potato J. 39: 306–312.

Walton, G. S., and Rich, S. 1974. Serious and unusual plant diseases in Connecticut in 1973. Plant Dis. Rep. 58:428–429.

Wang, C. C., and Chien, H. S. 1975. Studies on the infection and control of *Phytophthora palmivora* of papaw fruit. Taiwan Agric. Q. 11:109–116.

Wang, J. S., and Lin, M. S. 1979. Study on biological characteristics of zoospores of *Phytophthora fragariae* Hickm. var. *oryzobladis*. Wang and Lin. Acta Phytopathol. Sin. 9:55–60.

Wang, J. S., and Lu, J. Y. 1978. Phytophthora leaf blight of rice seedling, a new disease of rice. Acta Microbiol. Sin. 18:95–101. (In Chinese with English summary).

Wang, M. C., and Bartnicki-Garcia, S. 1973. Novel phosphoglucans from the cytoplasm of *Phytophthora palmivora* and their selective occurrence in certain life cycle stages. J. Biol. Chem. 248:4112–4118.

Wang, M. C., and Bartnicki-Garcia, S. 1974. Mycolaminarans: Storage (1→3)-β-D-glucans from the cytoplasm of the fungus *Phytophthora palmivora*. Carbohydr. Res. 37:331–338.

Wang, P., and Lu, J. Y. 1988. Homothallic and heterothallic behavior of *Phytophthora drechsleri* Tucker. Acta Mycol. Sin. 7:102–111. (In Chinese). (Rev. Plant Pathol. 1989, 68:703).

Ward, E. W. B. 1976. Capsidiol production in pepper leaves in incompatible interactions with fungi. Phytopathology 66:175–176.

Ward, E. W. B. 1990. The interaction of soya beans with *Phytophthora megasperma* f. sp. *glycinea:* Pathogenicity. Pages 311–327, Chapter 21 in: Biological Control of Soil-Borne Pathogens. D. Hornby, R. J. Cook, Y. Henis, W. H. Ko, A. D. Rovira, B. Schippers, and P. R. Scott, eds. C.A.B. International, Wallingford, Oxon, U.K. 479 pp.

Ward, E. W. B., and Buzzell, R. I. 1983. Influence of light, temperature and wounding on the expression of soybean genes for resistance to *Phytophthora megasperma* f. sp. *glycinea*. Physiol. Plant Pathol. 23:401–409.

Ward, E. W. B., and Lazarovits, G. 1982. Temperature-induced changes in specificity in the interaction of soybeans with *Phytophthora megasperma* f. sp. *glycinea*. Phytopathology 72:826–830.

Ward, E. W. B., and Stössel, A. 1974. Isolation of the phytoalexin, capsidiol, from pepper leaves and stems. 66th Ann. Meet. Am. Phytopathol. Soc. 11–15.

Ward, E. W. B., and Stössel, A. 1976. On the question of 'elicitors' in incompatible interactions between plants and fungal pathogens. Phytopathology 66:940–941.

Ward, E. W. B., Lazarovits, G., Stössels, P., Barrie, S. D., and Unwin, C. H. 1980. Glyceollin production associated with control of Phytophthora rot of soybeans by the systemic fungicide, metalaxyl. Phytopathology 70:738–740.

Ward, E. W. B., Stössel, P., and Lazarovits, G. 1981. Similarities between age-related and race-specific resistance of soybean hypocotyls to *Phytophthora megasperma* var. *sojae*. Phytopathology 71:504–508.

Ward, E. W. B., Cahill, D. M., and Bhattacharyya, M. K. 1989. Early cytological differences between compatible and incompatible interactions of soybeans with *Phytophthora megasperma* f. sp. *glycinea*. Physiol. Mol. Plant Pathol. 34:267–283.

Ward, M. R., and Griffin, M. J. 1981. Soil phase of cocoa *Phytophthora*. Chapter 4. Pages 50–61 in: Epidemiology of Cocoa in Nigeria. P. H. Gregory and A. C. Maddison, eds. Commonw. Mycol. Inst. Phytopathol. Pap. No. 25. 188 pp.

Wardlaw, C. W. 1926. Lankarshire strawberry disease. A report for the use of growers. Bot. Dep. Univ. Glasgow. Pages 1–38.

Wardlaw, T. J., and Palzer, C. 1985. Stem diseases in nursery seedlings caused by *Phytophthora cactorum, P. citricola* and *Pythium anandrum*. Australas. Plant Pathol. 14:57–59.

Washington, W. S. 1988. *Phytophthora cryptogea* as a cause of root rot of raspberry in Australia; resistance of raspberry cultivars and control by fungicides. Plant Pathol. 37:225–230.

Wastie, R. L. 1991. Breeding for resistance. Pages 193–224 in: *Phytophthora infestans,* the Cause of Late Blight of Potato. Adv. Plant Pathol. Vol. 7. D. S. Ingram and P. H. Williams, eds. Academic Press, London. 273 pp.

Waterhouse, G. M. 1931. The production of conidia in the genus *Phytophthora*. Trans. Br. Mycol. Soc. 15:311–321.

Waterhouse, G. M. 1940. A chytrid allied to *Pleolpidium inflatum* Butler. Trans. Br. Mycol. Soc. 24:7–19.

Waterhouse, G. M. 1956. The genus *Phytophthora*. Commonw. Mycol. Inst. Misc. Publ. No. 12. 120 pp.

Waterhouse, G. M. 1957. *Phytophthora citricola* Sawada (syn. *P. cactorum* var. *applanata* Chester). Trans. Br. Mycol. Soc. 40:349–357.

Waterhouse, G. M. 1958. The invalidity of the genus *Pythiomorpha*. II. Trans. Br. Mycol. Soc. 41:196–202.

Waterhouse, G. M. 1962. Presidential address: The zoospore. Trans. Br. Mycol. Soc. 45:1–20.

Waterhouse, G. M. 1963. Key to the species of *Phytophthora* de Bary. Mycol. Pap. 92. 22 pp. Commonw. Mycol. Inst. Kew, U.K.

Waterhouse, G. M. 1970a. The genus *Phytophthora* de Bary. Mycol. Pap. 122:1–59. Commonw. Mycol. Inst. Kew, Surrey, England. 104 pp.

Waterhouse, G. M. 1970b. Taxonomy in *Phytophthora*. Phytopathology 60:1141–1143.

Waterhouse, G. M. 1973. Chapter 10. Peronosporales. Pages 165–183 in: The Fungi: An Advanced Treatise. Vol. IVa. G. C. Ainsworth, F. K. Sparrow, and A. S. Sussman, eds. Academic Press, New York. 504 pp.

Waterhouse, G. M. 1974a. *Phytophthora japonica,* a new name for *Pythiomorpha oryzae*. Trans. Br. Mycol. Soc. 63:419–420.

Waterhouse, G. M. 1974b. *Phytophthora palmivora* and some related species. Pages 51–70 in: *Phytophthora* Disease of Cocoa. P. H. Gregory, ed. Longman, London. 336 pp.

Waterhouse, G. M. 1974c. Other *Phytophthora* species recorded on cacao. Pages 71–79 in: *Phytophthora* Disease of Cocoa. P. H. Gregory, ed. Longman, London. 348 pp.

Waterhouse, G. M., and Blackwell, E. M. 1954. Key to the species of *Phytophthora* recorded in the British Isles. Commonw. Mycol. Inst. Pap. 57. 9 pp.

Waterhouse, G. M., and Stamps, D. J. 1969. Isolation of *Phytophthora* and *Pythium*. Pages 99–102 in: Isolation Methods for Microbiologists. D.A. Shapton and G. W. Gould, eds. Academic Press, London. 178 pp.

Waterhouse, G. M., and Waterston, J. M. 1964a. *Phytophthora hibernalis*. Commonw. Mycol. Inst. Descriptions of Pathogenic Fungi and Bacteria. No. 31. 2 pp.

Waterhouse, G. M., and Waterston, J. M. 1964b. *Phytophthora syringae*. Commonw. Mycol. Inst. Descriptions of Pathogenic Fungi and Bacteria. No. 32. 2 pp.

Waterhouse, G. M., and Waterston, J. M. 1964c. *Phytophthora citrophthora*. Commonw. Mycol. Inst. Descriptions of Pathogenic Fungi and Bacteria. No. 33. 2 pp.

Waterhouse, G. M., and Waterston, J. M. 1964d. *Phytophthora nicotianae* var. *nicotianae*. Commonw. Mycol. Inst. Descriptions of Pathogenic Fungi and Bacteria. No. 34. 2 pp.

Waterhouse, G. M., and Waterston, J. M. 1964e. *Phytophthora nicotianae* var. *parasitica*. Commonw. Mycol. Inst. Descriptions of Pathogenic Fungi and Bacteria. No. 35. 2 pp.

Waterhouse, G. M., and Waterston, J. M. 1966a. *Phytophthora cactorum*. Commonw. Mycol. Inst. Descriptions of Pathogenic Fungi and Bacteria. No. 111. 2 pp.

Waterhouse, G. M., and Waterston, J. M. 1966b. *Phytophthora cambivora*. Commonw. Mycol. Inst. Descriptions of Pathogenic Fungi and Bacteria. No. 112. 2 pp.

Waterhouse, G. M., and Waterston, J. M. 1966c. *Phytophthora cinnamomi*. Commonw. Mycol. Inst. Descriptions of Pathogenic Fungi and Bacteria. No. 113. 2 pp.

Waterhouse, G. M., and Waterston, J. M. 1966d. *Phytophthora citricola*. Commonw. Mycol. Inst. Descriptions of Pathogenic Fungi and Bacteria. No. 114. 2 pp.

Waterhouse, G. M., and Waterston, J. M. 1966e. *Phytophthora megasperma*. Commonw. Mycol. Inst. Descriptions of Pathogenic Fungi and Bacteria. No. 115. 2 pp.

Waterhouse, G. M., Newhook, F. J., and Stamps, D. J. 1983. Present criteria for classification of *Phytophthora*. Pages 139–147 in: *Phytophthora:* Its Biology, Taxonomy, Ecology, and Pathology. D. C. Erwin, S. Bartnicki-Garcia, and P. H. Tsao, eds. American Phytopathological Society, St. Paul, Minn. 392 pp.

Waters, R. 1916. Dying of young fruit trees. N.Z. J. Agric. 12:112–121.

Wearing, A. H., and Birch, C. J. 1988. Root rot of pigeon pea. Int. Pigeonpea Newsl. 8:13.

Weber, A. 1922. Tomatsygdomme (Tomato diseases). Copenhagen, N.C. Rom. 66 pp. (In Danish). (Rev. Appl. Mycol. 1923, 2:245).

Wegman, J. 1966. Annu. Rep. 1965. Res. Stn. for floriculture, Aalsmeer, Netherlands. 134 pp. (Rev. Appl. Mycol. 1966, 45:2858).

Wei, C. T., and Cheo, P. C. 1944. Diseases of tomato in the vicinity of Chengtu. Chin. J. Agric. Sci. 4:288–291. (In Chinese). (Rev. Appl. Mycol. 1946, 25:367–368).

Weicht, T. R., and MacDonald, J. D. 1992. Effect of Phytophthora root rot on Na^+ uptake and accumulation by safflower. Phytopathology 82:520–526.

Weimer, J. L. 1949. Red stem canker of cowpea, caused by *Phytophthora cactorum*. J. Agric. Res. 78:65–75.

Weindlmayr, J. 1965. Untersuchungen über Einflusz gesteigerter Stickstoff-, Kali- und Phosphorgaben auf die *Phytophthora* Anfalligkeit von Kartoffelpflanzen in Nahrlosungskultur (Research on the influence of nitrogen, potassium, and phosphorous on the susceptibility of potato plants to *Phytophthora* in nutrient culture). Bodenkultur 16:144–168. (In German).

Weir, J. R. 1926. A pathological survey of the para rubber tree (*Hevea brasiliensis*) in the Amazon Valley. U.S. Dep. Agric. Bull. 1380. 129 pp. (Cited in Tucker 1933).

Weiss, F. 1930. A root rot of the white calla new to the United States. Flor. Exch. 73:11, 13. (Cited in Tucker 1933).

Weiss, F. 1943. Rhododendron dieback and canker. Plant Dis. Rep. 27:254.

Welsh, M. F. 1942. Studies of crown rot of apple trees. Can. J. Res. Sect. C. 20:457–490.

Welty, R. E., Barker, K. R., and Lindsey, D. L. 1980. Effects of *Meloidogyne hapla* and *M. incognita* on Phytophthora root rot of alfalfa. Plant Dis. 64:1097–1099.

Weltzien, H. C., and Schwinn, F. J. 1966. Phytophthora trunk rot on loquat trees, *Eriobotrya japonica* in Lebanon. Phytopathol. Z. 56:331–339.

Wernham, C. C., and Miller, H. J. 1948. Longevity of fungus cultures under mineral oil. Phytopathology 38:931–934.

Werres, S., and Casper, R. 1987. Nachweis von *Phytophthora fragariae* Hickman in Wurzeln der Erdbeerkultursorte "Tenira" mit Hilfe des enzyme-linked immunosorbent assay (ELISA) (Detection of *Phytophthora fragariae* in roots of the strawberry variety "Tenira" by enzyme-linked immunosorbent assay [ELISA]). J. Phytopathol. 118:367–369. (In German).

Wescott, C. 1950. Plant Disease Handbook. D. Van Nostrand, New York. 746 pp.

Weste, G. 1974. *Phytophthora cinnamomi:* The cause of severe disease in certain native communities in Victoria. Aust. J. Bot. 22:1–8.

Weste, G. 1975. Pathogenicity of *Phytophthora cinnamomi* towards *Nothofagus cunninghamii.* Aust. J. Bot. 23:277–283.

Weste, G. 1979. The plant communities—susceptible—vulnerable—resistant. Pages 7–13 in: *Phytophthora* and Forest Management in Australia. K. M. Old, ed. Rep. Conf. CSIRO Div. Forest Res. Canberra, Australia. 114 pp.

Weste, G. 1983a. *Phytophthora cinnamomi:* The dynamics of chlamydospore formation and survival. Phytopathol. Z. 106:163–176.

Weste, G. 1983b. Population dynamics and survival of *Phytophthora.* Pages 237–257 in: *Phytophthora:* Its Biolology, Taxonomy, Ecology, and Pathology. D. C. Erwin, S. Bartnicki-Garcia, and P. H. Tsao, eds. American Phytopathological Society, St. Paul, Minn. 392 pp.

Weste, G. 1984. Damage and loss caused by *Phytophthora* species in forest crops. Pages 273–284 in: Plant Diseases: Infection, Damage and Loss. R. K. S. Woods and G. J. Jellis, eds. 327 pp.

Weste, G. 1994. Impact of *Phytophthora* species on native vegetation of Australia and Papua, New Guinea. Austr. Plant Pathol. 23:190–209.

Weste, G., and Law, C. 1973. The invasion of native forest by *Phytophthora cinnamomi.* III. Threat to the National Park, Wilson's Promontory, Victoria. Aust. J. Bot. 21:31–50.

Weste, G., and Marks, G. C. 1974. The distribution of *Phytophthora cinnamomi* in Victoria. Trans. Br. Mycol. Soc. 63:559–572.

Weste, G., and Marks, G. C. 1987. The biology of *Phytophthora cinnamomi* in Australian forests. Annu. Rev. Phytopathol. 25:207–229.

Weste, G., and Ruppin, P. 1975. Factors affecting the population density of *Phytophthora cinnamomi* in native forests in the Brisbane Ranges, Victoria. Aust. J. Bot. 23:77–85.

Weste, G., and Ruppin, P. 1977. *Phytophthora cinnamomi:* Population densities. Aust. J. Bot. 25:461–475.

Weste, G. M., and Taylor, P. 1971. The invasion of native forest by *Phytophthora cinnamomi.* I. Brisbane Ranges, Victoria. Aust. J. Bot. 19:281–294.

Weste, G., and Vithanage, K. 1977. Microbial populations of three forest soils: Seasonal variations and changes associated with *Phytophthora cinnamomi.* Aust. J. Bot. 25:377–383.

Weste, G., and Vithanage, K. 1978a. Effect of *Phytophthora cinnamomi* on microbial populations associated with the roots of forest flora. Aust. J. Bot. 26:153–167. Accessory host list for *Phytophthora cinnamomi* (mimeographed) available from the author.

Weste, G., and Vithanage, K. 1978b. Seasonal variation in numbers of the chlamydospores in Victorian forest soils infected with *Phytophthora cinnamomi.* Aust. J. Bot. 26:657–662.

Weste, G., and Vithanage, K. 1979. Survival of chlamydospores of *Phytophthora cinnamomi* in several non-sterile, host-free forest soils and gravels at different soil water potentials. Aust. J. Bot. 27:1–9.

Weste, G., Cooke, D., and Taylor, P. 1973. The invasion of native forest by *Phytophthora cinnamomi.* II. Post-infection vegetation patterns, regeneration, decline in inoculum, and attempted control. Aust. J. Bot. 21:13–29.

Weste, G., Cahill, D., and Stamps, J. 1982. Mangrove dieback in North Queensland, Australia. Trans. Br. Mycol. Soc. 79:165–167.

Wester, R. E. 1968. New lima beans resistant to "B" strain of *Phytophthora phaseoli.* Plant Dis. Rep. 52:563–564.

Wester, R. E., Goth, R. W., and Webb, R. E. 1964. Transmission of downy mildew of lima beans by slugs (Abstr.). Phytopathology 54:749.

Wester, R. E., Goth, R. W., and Drechsler, C. 1966. Overwintering of *Phytophthora phaseoli.* Phytopathology 56:95–97.

Wester, R. E., Fisher, V. J., and Blount, V. L. 1972. Multiple resistance in lima beans to downy mildew (*Phytophthora phaseoli*). Plant Dis. Rep. 56:65–66.

Westerdijk, J., and Luijk, A. V. 1920. *Phytophthora erythroseptica* Peth. als parasit von *Atropa belladona* (*Phytophthora erythroseptica* parasite of *Atropa belladona*). Meded. Phytopathol. Lab. "Willie Commelin Scholten" 4:31–32. (In Dutch).

Weston, W. H., Jr. 1918. Report on plant diseases in Guam. Guam Agric. Exp. Stn. Rep. 1917:45–62.

Weston, W. H., Jr. 1921. The occurrence of wheat downy mildew in the United States. U.S. Dep. Agric. Dep. Circ. 186. 6 pp.

Weststeijn, G. 1973. *Phytophthora nicotianae* var. *nicotianae* on tomatoes. Neth. J. Plant Pathol. 79. Suppl. 1. 86 pp.

Wheeler, J. E., and Boyle, A. M. 1971. Indentification of four *Phythophthora* isolates previously unreported from Arizona. Phytopathology 61:1293–1296.

Whelan, J., and Loughnane, J. B. 1969. Non-solanaceous hosts of *Phytophthora erythroseptica.* Sci. Proc. R. Dublin Soc. Ser. B. 2: 171–177.

Whetzel, H. H. 1922. Report of the Pathologist. Rep. Bd. Dep. Agric., Bermuda 1921:30–64.

Whiley, A. W., Pegg, K. G., Saranah, J. B., and Forsberg, L. I. 1986. The control of Phytophthora root rot of avocado with fungicides and the effect of this disease on water relations, yield and ring neck. Aust. J. Exp. Agric. 26:249–253.

Whisson, S. C., Maclean, D. J., Manners, J. M., and Irwin, J. A. G. 1992. Genetic relationships among Australian and North American isolates of *Phytophthora megasperma* f. sp. *glycinea* assessed by multicopy DNA probes. Phytopathology 82:863–868.

Whisson, S. C., Drenth, A., MacLean, D. J., and Irwin, J. A. G. 1994. Evidence for outcrossing in *Phytophthora sojae* and linkage of a DNA marker to two avirulence genes. Curr. Genet. 27:77–82.

White, D. M., Partridge, J. E., and Williams, J. H. 1983. Races of *Phytophthora megasperma* f. sp. *glycinea* on soybeans in eastern Nebraska. Plant Dis. 67:1281–1282.

White, R. P. 1930. Two *Phytophthora* diseases of rhododendron (Abstr.). Phytopathology 20:131.

White, R. P. 1937. Rhododendron wilt and root rot. N.J. Agric. Exp. Stn. Bull. 615. 32 pp. (Rev. Appl. Mycol. 1939, 18:3[illegible]).

White, R. P., and Hamilton, C. C. 1935. Diseases and insect pests of Rhododendron and Azalea. N.Z. Agric. Exp. Stn. Circ. 350. 23 pp.

Whiteside, J. O. 1966. A revised list of plant diseases in Rhodesia. Kirkia 5:87–196.

Whiteside, J. O. 1970. Factors contributing to the restricted occurrence of citrus brown rot in Florida. Plant Dis. Rep. 54:608–612.

Whiteside, J. O. 1971. Some factors affecting the occurrence and development of foot rot on citrus trees. Phytopathology 61:1233–1238.

Whiteside, J. O., and Oswalt, T. W. 1973. An unusual brown rot outbreak in a Florida citrus grove following sprinkler irrigation with *Phytophthora*-infested water. Plant Dis. Rep. 57:391–393.

Whiteside, J. O., Garnsey, S. M., and Timmer, L. W., eds. 1988. Compendium of Citrus Diseases. American Phytopathological Society, St. Paul, Minn. 105 pp.

Whitney, E. D., and Duffus, J. E., eds. 1986. Compendium of Beet Diseases and Insects. American Phytopathological Society, St. Paul, Minn. 107 pp.

Whittaker, R. H. 1969. New concepts of kingdoms of organisms. Science 163:150–160.

Whittaker, S. L., Shattock, R. C., and Shaw, D. S. 1991a. Variation in DNA content of nuclei of *Phytophthora infestans* as measured by a microfluorimetric method using the fluorochrome DAPI. Mycol. Res. 95:602–610.

Whittaker, S. L., Shattock, R. C., and Shaw, D. S. 1991b. Inheritance of DNA contents in sexual progenies of *Phytophthora infestans.* Mycol. Res. 95:1094–1100.

Whittaker, S. L., Shattock, R. C., and Shaw, D. S. 1992. The duplication cycle and DAPI-DNA contents in nuclei of germinating zoospore cysts of *Phytophthora infestans.* Mycol. Res. 96:355–358.

Wiant, J. S. 1939. Species of *Phytophthora* responsible for market decay of Western Honey Dew melons and cantaloups. Plant Dis. Rep. 23:322.

Wiant, J. S., and Tucker, C. M. 1940. A rot of winter queen watermelons caused by *Phytophthora capsici.* J. Agric. Res. 60:73–88.

Wicks, T. J. 1988a. Effect of solarisation on the control of *Phytophthora cambivora* in almond and cherry. Aust. J. Exp. Agric. 28:539–545.

Wicks, T. J. 1988b. Effect of metalaxyl on the control of Phytophthora crown rot of almonds. Aust. J. Exp. Agric. 28:547–552.

Wicks, T. J., and Hall, B. 1988. Preliminary evaluation of phosphorous acid, fosetyl-Al and metalaxyl for controlling *Phytophthora cambivora* on almond and cherry. Crop Prot. 7:314–318.

Wicks, T. J., and Hall, B. 1990. Control of Phytophthora canker with phosphonate in artificially inoculated almond and cherry trees. Aust. J. Exp. Agric. 30:313–320.

Wicks, T. J., and Lee, T. C. 1985. Effects of flooding, rootstocks and fungicides on Phytophthora crown rot of almonds. Aust. J. Exp. 25:705–710.

Wicks, T. J., and Lee, T. C. 1986. Phytophthora crown rot in almond trees. Aust. J. Agric. Res. 37:277–287.

Wicks, T. J., and Volle, D. 1976. Phytophthora wilt of chestnuts in South Australia. Plant Dis. Rep. 60:700–702.

Wiebe, H. H., Campbell, G. S., Gardner, W. H., Rawlins, S. L., Cary, J. W., and Brown, R. W. 1971. Measurement of plant and soil water status. Utah Agric. Exp. Stn. Bull. No. 484. 71 pp.

Wiehe, P. O. 1941. Report of the Department of Agriculture. Mauritius, Division of Plant Pathology, 1939. Pages 10–14. (Rev. Appl. Mycol. 1941, 20:291).

Wiehe, P. O. 1948. The plant diseases and fungi recorded from Mauritius. Mycol. Pap. 24. 39 pp. Commonw. Mycol. Inst. Kew, U.K.

Wijers, E. E. 1937. Phytophthora wilt in carnation plants. S. Afr. J. Sci. 34:194–213.

Wilcox, J. R. 1976. Breeding for root rot resistance. Pages 485–490 in: Proc. World Soybean Res. Conf. L. D. Hill, ed.

Wilcox, W. F. 1989. Identity, virulence, and isolation frequency of seven *Phytophthora* spp. causing root rot of raspberry in New York. Phytopathology 79:93–101.

Wilcox, W. F., and Ellis, M. A. 1989. Phytophthora root and crown rots of peach trees in the eastern Great Lakes region. Plant Dis. 73:794–798.

Wilcox, W. F, and Mircetich, S. M. 1985a. Pathogenicity and relative virulence of seven *Phytophthora* spp. on Mahaleb and Mazzard cherry. Phytopathology 75:221–226.

Wilcox, W. F., and Mircetich, S. M. 1985b. Influence of soil water matric potential on the development of Phytophthora root and crown rots of Mahaleb cherry. Phytopathology 75:648–653.

Wilcox, W. F., and Mircetich, S. M. 1985c. Effects of flooding duration on the development of Phytophthora root and crown rots of cherry. Phytopathology 75:1451–1455.

Wilcox, W. F., and Mircetich, S. M. 1987. Lack of host specificity among isolates of *Phytophthora megasperma.* Phytopathology 77: 1132–1137.

Wilcox, W. F., Scott, P. H., Hamm, P. B., Kennedy, D. M., Duncan, J. M., Brasier, C. M., and Hansen, E. M. 1993. Identity of *Phytophthora* species attacking raspberry in Europe and North America. Mycol. Res. 97:817–831.

Wilde, P. 1961. Ein beitrag zur kenntnis der variabilitat von *Phytophthora infestans* (Mont.) de Bary (A contribution to knowledge of variability of *Phytophthora infestans* [Mont.] de Bary). Arch. Mikrobiol. 40:163–195. (In German).

Wilhelm, S., and Paulus, A. O. 1980. How soil fumigation benefits the California strawberry industry. Plant Dis. 64:264–270.

Wilhelm, S., and Sagen, J. E. 1974. A History of the Strawberry. Univ. of California, Div. Agric. Sci., Berkeley. 298 pp.

Wilhelm, S., Storkan, R. C., and Wilhelm, J. M. 1974. Preplant soil fumigation with methylbromide-chloropicrin mixtures for control of soil-borne diseases of strawberries—A summary of fifteen years of development. Agric. Environ. 1:227–236.

Wilkinson, H. T., and Millar, R. L. 1982. Effects of soil temperature and moisture on activity of *Phytophthora megasperma* f. sp. *medicaginis* and alfalfa root rot in the field. Phytopathology 72:790–793.

Wilkinson, H, T., Miller, R. D., and Millar, R. L. 1981. Infiltration of fungal and bacterial propagules into soil. Soil Sci. Soc. Am. J. 45:1034–1039.

Williams, B. L., and Pascoe, L. G. 1994. *Phytophthora megasperma,* a pathogen of *Trifolium repens* in irrigated pastures of northern Victoria, Australia. Australas. Plant Pathol. 23:88–96.

Williams, D. J., Beach, B. G. W., Horriere, D., and Marechal, G. 1977. LS 74-783, a new systemic fungicide with activity against Phycomycete diseases. Proc. Br. Crop Prot. Conf., 9th. 2:565–573.

Williams, F. J., Amin, K. S., and Baldev, B. 1975. Phytophthora stem blight of *Cajanus cajan.* Phytopathology 65:1029–1030.

Williams, P. H. 1927. "Damping-off" of tomatoes and cucumbers. Ann. Rep. Exp. Stn. Chestnut, Herts 1926. Pages 26–32.

Williams, P. H., Orchard, O. B., White, H. L., Oyler, E., Ainsworth, G. C., and Read, W. H. 1938. Plant diseases. Rep. Exp. Res. Stn. Cheshunt, 1937:58–59. (Rev. Appl. Mycol. 1937, 16:726–728).

Willis, J. C., and Green, E. E. 1897. The cacao canker. Trop. Agric. (Colombo) 17:272–273.

Willis, W. G. 1983. New developments in cereal and soybean seed treatment fungicides. Plant Dis. 67:257–258.

Wills, W. H. 1954a. The utilization of carbon and nitrogen compounds by *Phytophthora parasitica* Dastur var. *nicotianae* (Breda de Haan) Tucker. J. Elisha Mitchell Sci. Soc. 70:231–269.

Wills, W. H. 1954b. Sporangium formation of *Phytophthora parasitica* Dastur var. *nicotianae* (Breda de Haan) Tucker. J. Elisha Mitchell Sci. Soc. 70:235–243.

Wills, W. H. 1971. Expression of black shank resistance of tobacco measured in inoculated whole plants and excised leaf tissue. Tob. Sci. 15:47–50.

Wills, W. H., and Crews, J. W. 1964. Expression of black shank resistance in leaves of flue-cured tobacco. Phytopathology 54:1356–1358.

Wilson, G. W. 1907. Studies in North American Peronosporales. II. Phytophthoreae and Rhysotheceae. Bull. Torrey Bot. Club 34:387–416.

Wilson, G. W. 1914. Studies in North American Peronosporales. V. A review of the genus *Phytophthora.* Mycologia 6:54–83.

Wilson, A. R., and Boyd, A. E. W. 1945. Potato wastage in clamps. J. Minist. Agric. 51:507–512. (Rev. Appl. Mycol. 1945, 24:163–164).

Wilson, J. B., and Gallegly, M. E. 1955. The interrelationship of potato and tomato races of *Phytophthora infestans.* Phytopathology 45:473–476.

Wilson, J. D. 1936. Vegetable diseases in Ohio. Plant Dis. Rep. 20: 257–258.

Wilson, V. E., and Coffey, M. D. 1980. Cytological evaluation of general resistance to *Phytophthora infestans* in potato foliage. Ann. Bot. (London) 45:81–90.

Wisler, G. C., Ridings, W. H., and Cox, R. S. 1978. Phytophthora leafspot of *Brassaia actinophylla.* Proc. Fla. State Hortic. Soc. 91:240–242.

Wolf, F. A. 1960. Notes on tobacco diseases in Colombia, South America. Plant Dis. Rep. 44:62–63.

Wong, D. H., Sivasithamparam, K., and Barbetti, M. J. 1984. Effects of soil temperature and moisture on the pathogenicity of fungi associated with root rot of subterranean clover. Aust. J. Agric. Res. 35:675–684.

Wong, D. H., Sivasithamparam, K., and Barbetti, M. J. 1985. Fungi associated with root rot of subterranean clover in Western Australia. Aust. J. Exp. Agric. 25:574–579.

Wong, D. H., Sivasithamparam, K., and Barbetti, M. J. 1986a. Influence of environmental factors on the growth and survival of *Phytophthora clandestina.* Can. J. Microbiol. 32:553–556.

Wong, D. H., Sivasithamparam, K., and Barbetti, M. J. 1986b. Soil-behaviour of *Phytophthora clandestina.* J. Phytopathol. 116:67–73.

Wong, D. H., Sivasithamparam, K., and Barbetti, M. J. 1986c. Influence of soil temperature, moisture and other fungal root pathogens on pathogenicity of *Phytophthora clandestina* to subterranean clover. Trans. Br. Mycol. Soc. 86:479–482.

Wong, D. H., Barbetti, M. J., and Sivasithamparam, K. 1986d. Behaviour of *Phytophthora clandestina* propagules at a field site in Western Australia. Aust. J. Soil Res. 24:485–491.

Wood, S. G., and Gottlieb, D. 1978. Evidence from mycelial studies for differences in sterol biosynthetic pathways of *Rhizoctonia solani* and *Phytophthora cinnamomi*. Biochem. J. 170:343–354.

Woodward, J. R., Keane, P J., and Stone, B. A. 1980. β-Glucans and β-glucan hydrolases in plant pathogenesis with special reference to wilt-inducing toxins from *Phytophthora* species. Pages 113–142 in: Fungal Polysaccharides. P. Sandford and K. Matsuda, eds. Am. Chem. Soc. 284 pp.

Wormald, H. 1919. A Phytophthora rot of pears and apples. Ann. App. Biol. 6:89–100. (Cited in Tucker 1933).

Wormald, H. 1936. Notes on plant diseases in 1935. Rept. E. Malling Res. Stn. 1935:142–145.

Worthing C. R., and Walker, S. B., eds. 1987. The Pesticide Manual. A World Compendium XIX. Br. Crop Prot. Counc., Thornton Heath. 1081 pp.

Wycoff, K. L., and Ayers, A. R. 1990. Monoclonal antibodies to surface and extracellular antigens of a fungal plant pathogen, *Phytophthora megasperma* f. sp. *glycinea*, recognize specific carbohydrate epitopes. Physiol. Mol. Plant Pathol. 37:55–79.

Wycoff, K. L., and Ayers, A. R. 1991. Localization of carbohydrate antigens in the walls of *Phytophthora megasperma* f. sp. *glycinea* by monoclonal antibodies. Physiol. Mol. Plant Pathol. 39:95–109.

Wycoff, K. L., Jellison, J., and Ayers, A. R. 1987. Monoclonal antibodies to glycoprotein antigens of a fungal plant pathogen, *Phytophthora megasperma* f. sp. *glycinea*. Plant Physiol. 85:508–515.

Wynn, A. R., and Epton, H. A. S. 1979. Parasitism of oospores of *Phytophthora erythroseptica* in soil. Trans Br. Mycol. Soc. 73:255–259.

Wynn, W. K. 1968. Development of controlled conditions for the study of red stele disease of strawberries. Contrib. Boyce Thompson Inst. 24:95–102.

Xu, J., Lu, J., and Fang, Z. 1990. Studies on *Phytophthora* species from black locust (*Robinia pseudoacacia* L.) in China. J. Nanjing Agric. Univ. 13:17–23.

Yamada, G. 1911. *Sclerospora* Krankheit der Reispflanzen (*Sclerophthora* disease of rice plants). Spec. Bull. Imp. Coll. Agric. For. Morioka. 10 pp. (In German).

Yamamoto, M. 1982. Recent approach to studies of potato late blight, with special reference to the nature of resistance of potatoes to the invasion of *Phytophthora infestans*. Shokubutsu Byogai Kenkyu (Forsch. Gebiet Pflanzenkrankh.) 9:1–37.

Yang, S. S., Sung, N. K., Choi, D. I., and Kim, C. H. 1989. Pathogenic variation of *Phytophthora capsici* on red-pepper in Korea. Korean J. Plant Pathol. 5:370–376.

Yarwood, C. E. 1975. Temperature coefficients in plant pathology. Phytopathology 65:1198–1201.

Yokoyama, S. 1976. Studies on the leaf blight of onion (Tamanegi Shiroiro Ekibyo) caused by *Phytophthora porri* Foister. Bull. Fukuoka Agric. Exp. Stn., Chikushino, Jpn. 22. 55 pp. (In Japanese).

Yokoyama, S., and Yoshida, K. 1937. Onion *Phytophthora* disease. Kyushu Assoc. Plant Prot. Proc. 13:35–36. (In Japanese). (Rev. Appl. Mycol. 56:4799).

Yoshikawa, M. 1983. Macromolecules, recognition, and the triggering of resistance. Pages 267–298 in: Biochemical Plant Pathology. J. A. Callow, ed. John Wiley & Sons, Chichester. 484 pp.

Yoshikawa, M., and Masago, H. 1977. Effect of cyclic AMP in catabolite-repressed zoosporagial formation in *Phytophthora capsici*. Can. J. Bot. 55:840–843.

Yoshikawa, M., Nishida, A., and Masago, H. 1977. Synchronized zoosporangial formation in liquid culture of *Phytophthora capsici*. Can. J. Bot. 55:844–847.

Yoshikawa, M., Yamauchi, K., and Masago, H. 1978a. Glyceollin: Its role in restricting fungal growth in resistant soybean hypocotyls infected with *Phytophthora megasperma* var. *sojae*. Physiol. Plant Pathol. 12:73–82.

Yoshikawa, M., Yamauchi, K., and Masago, H. 1978b. De novo messenger RNA and prtein synthesis are required for phytoalexin-mediated disease resistance in soybean hypocotyls. Plant Physiol. 61:314–317.

Yoshikawa, M., Keen, N. T., and Wang, M. C. 1983. A receptor on soybean membranes for a fungal elicitor of phytoalexin accumulation. Plant Physiol. 73:497–506.

Yoshikawa, M., Masago, H., Onoe, T., and Matsuda, K. 1987a. Mode of biochemical action of phytoalexins. Pages 253–269 in: Molecular Determinants of Plant Diseases. S. Nishimura, C. P. Vance, and N. Doke, eds. Springer-Verlag, Berlin.

Yoshikawa, M., Onoe, T., Masago, H., and Sagawa, H. 1987b. Ultrastructural *in situ* evidence for the role of glyceollin in the expression of soybean resistance to fungal invasion. Ann. Phytopathol. Soc. Jpn. 53:227–241.

Yoshimura, M. A., Uchida, J. Y., and Aragaki, M. 1985. Etiology and control of poinsettia blight caused by *Phytophthora nicotianae* and *P. drechsleri*. Plant Dis. 69:511–513.

Yoshio, M., Terada, T., Umekage, D., and Masago, H. 1976. An acute Phytophthora wilt of cucumber in Sagano district of Kyoto. II. Field experiments for control of this disease by Utsugi's grafting-cutting method. Proc. Kansai Plant Prot. Soc. 18:33–40.

Young, C. W. T., and Strouts, R. G. 1975. Report on forest research for the year ended March 1975. Rep. For. Comm. U.K. H. M. Stationery Office. (Rev. Plant Pathol. 1976, 56:3751.)

Young, C. W. T., and Strouts, R. G. 1976. Report on Forest. Res. 1976. For. Comm. U.K. H.M. Stationery Office. 101 pp. (Rev. Plant Pathol. 1977, 56:4702).

Young, D. H. 1991. Effects of zarilamide on microtubules and nuclear division in *Phytophthora capsici* and tobacco suspension-cultured cells. Pestic. Biochem. Physiol. 40:149–161.

Young, P. A. 1926. Facultative parasitism and host ranges of fungi. Am. J. Bot. 13:502–520.

Young, R. A., and Milbrath, J. A. 1959. A stem canker disease of fruit tree nurserystock caused by *Phytophthora syringae* (Abstr.). Phytopathology 49:114–115.

Young, R. J., Bhatia, S. K., and Pencis, S. I. 1979. Control of potato late blight by the systemic fungicide N-(2,6-dimethylphenyl)-N-(methoxy acetyl)-alanine methyl ester (CGA 48988) (Abstr.). Phytopathology 69:538.

Youngberg, S. 1926. Plant diseases. Philipp. Bur. Agric. Annu. Rep. 1925:62–72.

Youssef, A., and Bergamini, A. 1989. L'Actinidia sa culture (The culture of actinidia). La Maison Rustique, Paris, France. 63 pp (In French).

Yu, J. Y., Chang, H. S., and Ko, W. H. 1981. Factors affecting the induction of sexual reproduction in *Phytophthora parasitica* by *P. colocasiae*. J. Gen. Microbiol. 123:249–252.

Yu, Y. N., and Zhuang, W. 1982. *Phytophthora sinensis*, a new species causing blight on *Cucumis sativus*. Mycotaxon 14:181–188.

Yu, Y. N., Li, J. L., and Yang, X. F. 1986. Studies on *Phytophthora* species from rubber tree (*Hevea brasiliensis*) in China. Acta Mycol. Sin. 5:193–207. (In Chinese, English summary).

Yuen, G. Y., Schroth, M. N., Weinhold, A. R., and Hancock, J. G. 1991. Effects of soil fumigation with methyl bromide and chloropicrin on root health and yield of strawberry. Plant Dis. 75:416–420.

Zaiger, D., and Zentmyer, G. A. 1965. Phytophthora canker of cacao in the Caroline Islands. Plant Dis. Rep. 49:565–567.

Zak, B. 1961. Aeration and other soil factors affecting southern pines as related to littleleaf disease. U.S. Dep. Agric. For. Serv. Tech. Bull. No. 1248. 30 pp.

Zak, B. 1971. Characterization and classification of mycorrhizae of Douglas fir. II. *Pseudotsuga menziesii* + *Rhizopogon vinicolor*. Can. J. Bot. 49:1079–1084.

Zaki, A. I., and Zentmyer, G. A. 1976. An approach to synchronous mycelial growth for sporangium production of *Phytophthora cinnamomi*. Phytophthora Newsl. 4:31–32.

Zaki, A. I., Zentmyer, G. A., and LeBaron, H. M. 1981. Systemic translocation of ^{14}C-labeled metalaxyl in tomato, avocado, and *Persea indica*. Phytopathology 71:509–514.

Zaki, A. I., Zentmyer, G. A., Sims, J. J., and Keen, N. T. 1983. Stimulation of sexual reproduction in the A2 mating type of *Phytophthora cinnamomi* by oleic acid and lipids from avocado roots. Phytopathology 73:199–203.

Zan, K. 1962. Activity of *Phytophthora infestans* in soil in relation to tuber infection. Trans. Br. Mycol. Soc. 45:205–221.

Zanetti, A., Beauvais, F., Huet, J. C., and Pernollet, J. C. 1992. Movement of elicitins, necrosis-inducing proteins secreted by *Phytophthora* sp. in tobacco. Planta 187:163–170.

Zaumeyer, W. J., and Wester, R. E. 1956. Control of downy mildew of lima beans with streptomycin. Plant Dis. Rep. 40:776–780.

Zaviezo, T., Latorre, B. A., and Torres, R. 1993. Effectiveness of three phenylamide fungicides against *Phytophthora cryptogea* isolated from kiwi and their mobility in soil. Plant Dis. 77:1239–1243.

Zeijlemaker, F. C. J. 1968. The gummosis of black wattle, a complex of diseases. Rep. Wattle Res. Inst. for 1967–68:40–43.

Zeijlemaker, F. C. J. 1971. Black-butt of black wattle caused by *Phytophthora nicotianae* var. *parasitica*. Phytopathology 61:144–145.

Zentmyer, G. A. 1952. A substance stimulating sexual reproduction in *Phytophthora cinnamomi* (Abstr.). Phytopathology 42:24.

Zentmyer, G. A. 1961. Chemotaxis of zoospores for root exudates. Science 133:1595–1596.

Zentmyer, G. A. 1963. Biological control of Phytophthora root rot of avocado with alfalfa meal. Phytopathology 53:1383–1387.

Zentmyer, G. A. 1965. Bacterial stimulation of sporangium production in *Phytophthora cinnamomi*. Science 150:1178–1179.

Zentmyer, G. A. 1966. Role of amino acids in chemotaxis of zoospores of three species of *Phytophthora* (Abstr.). Phytopathology 56:907.

Zentmyer, G. A. 1970. Tactic responses of zoospores of *Phytophthora*. Pages 109–111 in: Root Diseases and Soil-Borne Pathogens. T. A. Toussoun, R. V. Bega, and P. E. Nelson, eds. Univ. Calif. Press, Berkeley. 252 pp.

Zentmyer, G. A. 1972. Avocado root rot. Calif. Avocado Soc. Yearb. 55:29–36.

Zentmyer, G. A. 1973. Control of Phytophthora root rot of avocado with p-dimethylaminobenzenediazo sodium sulfonate (Dexon). Phytopathology 73:267–272.

Zentmyer, G. A. 1979. Stimulation of sexual reproduction in the A2 mating type of *Phytophthora cinnamomi* by a substance in avocado roots. Phytopathology 69:1129–1131.

Zentmyer, G. A. 1980. *Phytophthora cinnamomi* and the Diseases It Causes. Monogr. No. 10. American Phytopathological Society, St. Paul, Minn. 96 pp.

Zentmyer, G. A. 1983. The world of *Phytophthora*. Pages 1–8 in: *Phytophthora*: Its Biology, Taxonomy, Ecology, and Pathology. D. C. Erwin, S. Bartnicki-Garcia, and P. H. Tsao, eds. American Phytopathological Society, St. Paul, Minn. 392 pp.

Zentmyer, G. A. 1984. Avocado diseases. Trop. Pest Mgmt. 30(4): 388–400.

Zentmyer, G. A. 1987. Taxonomic relationships and distribution of species of *Phytophthora* causing black pod of cacao. Proc. Tenth Int. Cocoa Res. Conf., Santo Domingo, Dominican Republic. 17–23 May, 1987. Pages 391–395. Cocoa Prod. Alliance, London. 925 pp.

Zentmyer, G. A. 1988. Origin and distribution of four species of *Phytophthora*. Trans. Br. Mycol. Soc. 91:367–378.

Zentmyer, G. A., and Bingham, F. T. 1956. The influence of nitrite on the development of Phytophthora root rot of avocado. Phytopathology 46:121–124.

Zentmyer, G. A., and Erwin, D. C. 1970. Development and reproduction of *Phytophthora*. Phytopathology 60:1120–1127.

Zentmyer, G. A., and Marshall, L. A. 1959. Factors affecting sporangial production by *Phytophthora cinnamomi* (Abstr.). Phytopathology 49:556.

Zentmyer, G. A., and Mircetich, S. M. 1965. Testing for resistance of avocado to *Phytophthora* in nutrient solution. Phytopathology 55: 487–489.

Zentmyer, G. A., and Mircetich, S. M. 1966. Saprophytism and persistence in soil by *Phytophthora cinnamomi*. Phytopathology 56: 710–712.

Zentmyer, G. A., and Mitchell, D. J. 1985/86. *Phytophthora* diseases of fruit trees in the tropics. Rev. Trop. Plant Pathol. 2:287–309.

Zentmyer, G. A., and Munnecke, D. E. 1952. Phytophthora root rot of nursery stock. Plant Dis. Rep. 36:211–212.

Zentmyer, G. A., and Ohr, H. D. 1978. Avocado root rot. Univ. Calif. Div. Agric. Sci., Berkeley. Leafl. 2440. 15 pp.

Zentmyer, G. A., and Ribeiro, O. K. 1977. The effect of visible and near-visible radiation on sporangium production by *Phytophthora cinnamomi*. Phytopathology 67:91–95.

Zentmyer, G. A., and Richards, S. J. 1952. Pathogenicity of *Phytophthora cinnamomi* to avocado trees and the effect of irrigation on disease development. Phytopathology 42:35–37.

Zentmyer, G. A., and Schieber, E. 1987. The search for resistance to Phytophthora root rot in Latin America. S. Afr. Avocado Growers' Assoc. Yearrb. 10:109–110.

Zentmyer, G. A., and Storey, W. B. 1961. Phytophthora canker of macadamia trees. Calif. Macadamia Soc. Yearb. 7:65–67.

Zentmyer, G. A., and Thompson, C. R. 1967. The effect of saponins from alfalfa on *Phytophthora cinnamomi* in relation to control of root rot of avocado. Phytopathology 57:1278–1279.

Zentmyer, G. A., and Thorn, W. A. 1967. Hosts of *P. cinnamomi*. Calif. Avocado Soc. Yearb. 51:177–186.

Zentmyer, G. A., Gilpatrick, J. D., and Thorn, W. A. 1960. Methods of isolating *Phytophthora cinnamomi* from soil and from host tissue (Abstr.). Phytopathology 50:87.

Zentmyer, G. A., Mitchell, D. J., Jefferson, L., Roheim, J., and Carnes, D. 1973. Distribution of mating types of *Phytophthora palmivora*. Phytopathology 63:663–667.

Zentmyer, G. A., Jefferson, L., Hickman, C. J., and Yung, C. H. 1974. Studies of *Phytophthora citricola*, isolated from *Persea americana*. Mycologia 66:830–845.

Zentmyer, G. A., Kaosiri, T., and Idosu, G. 1977. Taxonomic variants in the *Phytophthora palmivora* complex. Trans. Br. Mycol. Soc. 69:329–332.

Zentmyer, G. A., Klure, L. J., and Pond, E. C. 1978. A new canker disease of avocado caused by *Phytophthora heveae*. Plant Dis. Rep. 62:918–922.

Zentmyer, G. A., Klure, L. J., and Pond, E. C. 1979a. The influence of temperature and nutrition on formation of sexual structures by *Phytophthora cinnamomi*. Mycologia 71:55–65.

Zentmyer, G. A., Pond, E. C., and Klure, L. J. 1979b. A new *Phytophthora* disease of ornamental flannel bush (*Fremontia californica*). Plant Dis. Rep. 63:721–724.

Zhang, K. M., Ho, H. H., and Lu, J. V. 1995. The occurrence of *Phytophthora heveae* in mainland China. Mycopathol. 129:87–90.

Zhang, Q., Griffith, J. M., and Grant, B. R. 1992. Role of phosphatidic acid during differentiation of *Phytophthora palmivora* zoospores. J. Gen. Microbiol. 138:451–459.

Zhuang, W. Y. 1981. Taxonomic studies on the genus *Phytophthora* of China. Master thesis, Inst. Microbiol. Acad. Sin., Beijing. 96 pp. (In Chinese).

Zheng, X. B., and Lu, J. Y. 1989. Studies on *Phytophthora* species in Fujian, Zhejiang, Jiangsu provinces and Shanghai, China. Acta Mycol. Sin. 8:161–168. (In Chinese). (Rev. Plant Pathol. 1991, 70:247)

Ziegler, E., and Pontzen, R. 1982. Specific inhibition of glucan-elicited glyceollin accumulation in soybeans by an extracellular mannan-glycoprotein of *Phytophthora megasperma* f. sp. *glycinea*. Physiol. Plant Pathol. 20:321-331.

Zimmer, D. E., and Urie, A. L. 1967. Influence of irrigation and soil infestation with strains of *Phytophthora drechsleri* on root rot resistance of safflower. Phytopathology 57:1056–1059.

Zimmermann, A. 1901. Uber einige an tropischen Kulturpflanzen beobachtete Pilze. I. (About some fungi observed on tropical cultivated plants. I.). Centralbl. Bakt. 2:101–106, 139–147.

Zinkernagel, V., and Krober, H. 1978. *Phytophthora cinnamomi* Rands als erregereiner triebfäule an *Gaultheria procumbens* L. (*Phytophthora cinnamomi* Rands as pathogen of stem rot on *Gaultheria procumbens* L.). Nachrichtenbl. Dtsch. Pflanzenschutzdienst 30:181–183. (Rev. Plant Pathol. 1979, 58:2897).

Zitko, S. E., Timmer, L. W., and Sandler, H. A. 1991. Isolation of *Phytophthora palmivora* pathogenic to citrus in Florida. Plant Dis. 75:532–535.

Zobel, D. B., Roth, L. F., and Hawk, G. M. 1985. Ecology, pathology, and management of Port-Orford-cedar (*Chamaecyparis lawsoniana*). Gen. Tech. Rep. Pac. Northwest For. Range Exp. Stn. U.S.D.A. For. Serv. No. PNW-184. 161 pp.

Zook, D. M., Erwin, D. C., and Stolzy, L. H. 1986. Anatomical, morphological, and physiological responses of alfalfa to flooding. Plant Soil 96:293–296.

Glossary

C—centigrade or Celsius (°C = [°F – 32] × 5/9)
cm—centimeter (0.39 in.; 10 mm = 1 cm)
F—Fahrenheit
g—gram
h—hour
in.—inch (2.54 cm)
kg—kilogram (1 kg = 1,000 g = 2.205 pounds)
km—kilometer (1 km = 1,000 m = 0.6214 mile)
L—liter (1 L = 1.057 quarts liquid)
m—meter (1 m = 39.37 inches)
meq—milliequivalent
mg—milligram (1 mg = 0.001 g)
min—minute
ml—milliliter (1 ml = 0.001 L)
mm—millimeter (1 mm = 0.001 m)
µg—microgram (one millionth of a gram)
µm—micrometer (one millionth of a meter)
ppm—parts per million
s—second

activation—stage of oospore germination during which the endospore wall becomes eroded and the ooplast swells
allozymes—isozymes whose variant forms are coded at the same locus; assayed by starch-gel electrophoresis (see Chapter 3 and Spielman 1991)
amphigynous—pertaining to the sexual stage in which the oogonial stalk is surrounded by the antheridium
antheridium—the male gametangium, a multinucleate, swollen hyphal tip affixed firmly to the basal side of the oogonium
aplerotic—pertaining to a mature oospore that does not fill the oogonium; when the oospore fills the oogonium it is plerotic. This characteristic has been used more frequently to characterize *Pythium* spp. than *Phytophthora* spp.
axoneme—main core of a flagellum in which two central tubules are surrounded by nine double microtubules (see Chapter 4)

caducous or **deciduous**—pertaining to sporangia that become dislodged readily and separate from the sporangiophore (e.g., *Phytophthora palmivora*) in contrast to being persistent (e.g., *P. cinnamomi*)
chemotaxis—movement of zoospores toward a chemical stimulus
chlamydospore—spore, usually globose but occasionally ovoid, that is delimited from the mycelium by a septum; may be terminal or intercalary with a thickened inner wall; survives for long periods in soil
coenocytic—pertaining to multinucleate mycelium in which the nuclei are not separated by cell walls
conidium—nonmotile spore that is usually caducous (deciduous). Occasionally the caducous sporangia of *Phytophthora infestans* are called conidia because they are capable of being spread by splashing water to adjacent plants and may infect a leaf before zoospores form; however, the term "sporangium" is more appropriate and most frequently used.

deciduous—*see* caducous
DNA—deoxyribonucleic acid; a constituent of chromosomes in nuclei that acts as a blueprint for the synthesis of proteins and contains the hereditary information for the development of the organism

electrophoresis—chromatography process by which particles (especially proteins and enzymes) migrate at different rates through a starch or acrylamide gel in an electric field. The gel is stained to mark the location of the protein and then photographed.
electrotaxis—movement of zoospores toward weak electrical fields on plant roots
epidemiology—study of disease that spreads widely and affects a large population of individuals
epiphytology—study of the character, ecology, and outbreaks of plant diseases; since the principles are similar to those of epidemiology, the latter term is more frequently used (Vanderplank 1963)

flagellum—whiplike appendage on the zoospore. The zoospore has a long whiplash flagellum with a smooth surface and a shorter tinsel flagellum on which mastigonemes (hairlike structures) occur.

gene-for-gene—pertaining to genetic relationships in which a host plant (e.g., soybean) possesses a race-specific resistance gene and the pathogen (e.g., *Phytophthora sojae*) possesses a corresponding gene for virulence

haustorium—branch of the mycelium within an infected plant that does not break the host cell but becomes appressed to it and through which nutrients are absorbed
heterokaryon—cell containing genetically unlike nuclei
heterokont—pertaining to zoospore fungi with different types of flagella; e.g., in *Phytophthora,* the zoospore has both a whiplash and a tinsel flagellum (Barr 1983)
heterothallic—pertaining to sexual reproduction in which conjugation is possible only through interaction of different thalli (Hawksworth et al. 1983). In *Phytophthora,* in most cases, oospores form only when the A1 and A2 mating types are paired; however, several reports indicate that oospores form in single A2 mating types following certain stimuli (see Chapter 3).
homothallic—pertaining to sexual reproduction that can take place without the interaction of two different thalli (Hawksworth et al. 1983)
hyphal swelling—globular swelling that forms at a branch point or within the mycelium. It is by definition (Blackwell 1949) not separated from the coenocytic mycelium by a septum. When separated by a septum and having a thickened inner wall, it is termed a chlamydospore.

isokont—pertaining to zoosporic microorganisms with two flagella of equal length (Barr 1983)

isozymes—variant forms of an enzyme that share the same catalytic function; assayed by use of starch gel electrophoresis (see Chapter 3 and Spielman 1991); *see also* allozymes

mastigoneme—short, hairlike, lateral protuberance on tinsel flagella (Barr 1983)

matric potential—component of water potential; the force that holds water in soil resulting from the absorption or capillary effects characteristic of various pore sizes in soil (see Duniway 1983 and Appendix 5.1)

mycelium—tubular strands that make up the body of the fungal microorganism. In *Phytophthora,* mycelium is nonseptate; however, plugs (often called false septa) can be seen in old mycelium.

oogonium—the female gametangium in which the oospore forms after fertilization by the antheridium

oospore—zygote or thick-walled spore that forms within the oogonium after fertilization; may be long-lived

paragynous—pertaining to the sexual stage in which the antheridium is attached to the side of the oogonium

pellucid body or **pellucid spot**—translucent, nucleus-containing body, two or more of which can be seen inside and near the periphery of the mature oospore

phloem—plant tissue in which carbohydrates are transported

plerotic—pertaining to an oospore that fills the oogonium; one that does not is aplerotic

proliferation—formation of a sporangium within an empty sporangium that has previously emitted zoospores (internal or nested) or after the sporangiophore has emerged from the empty sporangium (external)

RFLP—restriction fragment length polymorphism. After DNA is purified and treated with certain enzymes that react at different specific regions on the DNA molecule, the lengths of the resulting fragments are categorized by use of starch or acrylamide-gel electrophoresis. The relative differences in fragment lengths are characteristic of genetic groups (see Chapter 4).

RNA—ribonucleic acid; a molecule similar to DNA that functions in the decoding of genetic information as it is used in the cell

sporangiophore—stalk on which the sporangium is formed; may be unbranched or branched to form a simple sympodium or a compound sympodium

sporangium—sac within which zoospores form, especially when water is cooled to about 10°C below ambient temperature. In solid substrates, sporangia usually germinate by germ tubes. See Chapter 4 for illustrations of various shapes of sporangia.

xylem—plant tissue containing vessels and tracheids through which water is transported

zoospore—spore that forms within the sporangium and exits through the terminal pore, has a tinsel and a whiplash flagellum, and is capable of swimming for several hours (see Chapters 3 and 4 for illustrations and discussion)

Index